Continuous and Discrete-Time Signals and Systems

Athanasios I. Margaris

Continuous and Discrete-Time Signals and Systems

Volume 1

Springer

Athanasios I. Margaris
Trikala, Thessaly, Greece

ISBN 978-3-031-81208-8 ISBN 978-3-031-81209-5 (eBook)
https://doi.org/10.1007/978-3-031-81209-5

This Springer imprint is published by the registered company Springer Nature Switzerland AG
The registered company address is: Gewerbestrasse 11, 6330 Cham, Switzerland

In loving memory of my parents, Ioannis and Agoris

Preface

Digital signal processing (DSP) is a foundational discipline in modern engineering, powering advancements in telecommunications, audio and video processing, biomedical engineering, and numerous other fields. Despite the abundance of DSP literature, there remains a need for a comprehensive, well-structured resource that seamlessly blends theoretical rigor with practical application, accessible to students, educators, and professionals alike. This two-volume work aims to address this need, offering a systematic, authoritative exploration of DSP that serves as both a pedagogical textbook and a technical reference. By combining mathematical precision with real-world examples, it equips readers with the tools to tackle complex signal processing challenges while fostering a deep understanding of the field.

The book is meticulously organized to provide a logical progression from fundamental concepts to advanced techniques, catering to readers at various levels of expertise. Volume 1 establishes the core principles of continuous- and discrete-time signals and systems, while Volume 2 focuses on the design and implementation of analog and digital filters, concluding with an introduction to statistical signal processing. Each chapter is designed to be self-contained yet interconnected, supporting both linear study and selective reference. Extensive problem sets at the end of each chapter reinforce theoretical concepts through practical exercises, making the text ideal for classroom use, self-directed learning, and professional development.

Structure of Volume 1

Volume 1 lays the groundwork for DSP by introducing the essential principles of signal and system analysis. **Continuous-Time Signals** (Chap. 1) covers foundational concepts such as signal energy and power, periodicity, even and odd signals, and elementary functions like exponentials, sinusoids, and generalized functions (e.g., unit step, ramp, and impulse). It introduces key operations including convolution, correlation, and vector space representations, providing a robust basis for signal manipulation. **Continuous-Time Systems** (Chap. 2) builds on this foundation, exploring system classifications (e.g., linear vs. nonlinear, causal vs. noncausal, time-invariant vs. time-varying), stability, and responses (impulse, step, and frequency). It also addresses system interconnections, schematic representations, and state-space descriptions, emphasizing practical design considerations.

Fourier Analysis in the Continuous-Time Domain (Chap. 3) introduces frequency-domain analysis through Fourier series and transforms. It covers the exponential and trigonometric forms of Fourier series, convergence conditions, the Gibbs phenomenon, and their application to linear time-invariant (LTI) systems. The continuous-time Fourier transform is explored in depth, including its properties, convergence, and applications to periodic signals and frequency selection filters. **The Laplace Transform** (Chap. 4) provides a powerful tool for system analysis, detailing regions of convergence, poles and zeros, properties, and inverse transforms. It emphasizes solving differential equations for first- and second-order systems, bridging time- and frequency-domain techniques.

System Analysis in the Frequency Domain (Chap. 5) integrates these tools to analyze transfer functions, stability (e.g., Routh-Hurwitz criterion), and frequency responses. It includes practical methods like Bode diagrams, signal-flow graphs, and state-space solutions, equipping readers with design-oriented skills. The focus then shifts to discrete-time systems with **Discrete-Time Signals** (Chap. 6), which parallels Chap. 1 by addressing sampling, energy, periodicity, and operations like convolution and correlation. **Discrete-Time Systems** (Chap. 7) explores system classifications, LTI system responses, and state-space representations, with a focus on stability and recursive systems.

The Z Transform (Chap. 8) introduces a key tool for discrete-time system analysis, covering poles, zeros, convergence, and inverse transforms. **Systems in the Z Domain: Digital Processing Techniques** (Chap. 9) delves into digital signal processing, including transfer functions, difference equations, and multirate processing techniques like downsampling and oversampling. Volume 1 concludes with **Fourier Analysis in the Discrete-Time Domain** (Chap. 10), addressing discrete-time Fourier transforms (DTFT), discrete Fourier transforms (DFT), and their applications in frequency analysis and short-time signal processing.

Structure of Volume 2

Volume 2 builds on Volume 1's foundations, focusing on practical design and advanced topics. **Design of Analog Filters** (Chap. 11) provides a thorough exploration of filter types (e.g., low-pass, high-pass, band-pass), characteristics, and design methodologies, including Butterworth, Chebyshev, elliptic, and Bessel-Thomson approximations. It covers frequency transformations, circuit theory, and the Paley-Wiener theorem, ensuring a practical perspective on filter implementation. **Design of Digital Filters** (Chap. 12) contrasts analog and digital approaches, addressing finite and infinite impulse response (FIR and IIR) filters, quantization effects, and advanced design techniques such as the Parks-McClellan algorithm, Kaiser window method, and bilinear transformation. It emphasizes implementation through signal-flow graphs, hardware considerations, and error analysis (e.g., quantization and overflow).

Elements of Statistical Signal Processing (Chap. 13) introduces stochastic processes, power spectral density, and system responses to random inputs. It covers key concepts like the Wiener-Khinchin theorem, white noise, and quantization error analysis, providing a foundation for advanced signal processing applications. The book concludes with Appendix A, a comprehensive reference consolidating fundamental mathematical concepts, including complex numbers, difference equations, Chebyshev polynomials, elliptic functions, and probability theory. This appendix supports readers by providing quick access to essential mathematical tools used throughout the text.

Objectives and Audience

This book is designed for a diverse audience, including undergraduate and graduate students in electrical and computer engineering, practicing engineers, and researchers seeking a definitive DSP reference. Its balanced approach—combining theoretical depth with practical examples—makes it suitable for introductory courses, advanced study, and professional applications. The inclusion of contemporary topics, such as multirate processing, statistical signal analysis, and advanced filter design, ensures alignment with current industry and research trends.

By offering a unified treatment of continuous and discrete-time systems, analog and digital filters, and statistical processing within a cohesive pedagogical framework, this work fills a critical gap in DSP literature. Our goal is to inspire readers to explore the dynamic and evolving field of digital signal processing, fostering both intellectual curiosity and technical proficiency. We hope this book will serve as a trusted companion for learning, teaching, and applying DSP in real-world contexts.

Trikala, Thessaly, Greece
April 2025

Athanasios I. Margaris

Contents

Volume 1

1 Continuous-Time Signals

A classic example of the application of continuous-time signals is the radar systems used for the detection and tracking of moving and stationary targets with the help of electromagnetic waves

1.1 Introduction to Continuous-Time Signals

One of the concepts that hold a central position in signal and system theory is the concept of the signal. Generally speaking, a *signal* is defined as any quantity that varies in function of time, or more generally, in function of any other physical quantity. Typical examples of such signals are *the current flowing through an alternating voltage circuit* or *the position of a body moving in space under the influence of some force*. From the mathematical side of the subject, a signal is nothing more than a function of one or more independent variables that convey information about the nature or behavior of the phenomenon it describes. For example, an *image* can be described as a function of two variables $f(x, y)$, the value of which at each point is the measure of the *intensity* of the image at the position of the pixel that is described by the coordinates (x, y).

However, in the following description we limit ourselves only to *one-dimensional signals* that are described by a function of a single variable that also represents the *time*. In addition, in this chapter we assume that time is *continuous* in the sense that the function describing the signal *is defined for any time*. In order to understand the concept of continuity with respect to time, let us consider Fig. 1.1, which depicts a typical curve of a function, and let us remember the way in which we find the value of the function $f(t)$ for some time $t = t_1$. We know well that, in order to do this, we must bring a line perpendicular

A. I. Margaris, *Continuous and Discrete-Time Signals and Systems*,
https://doi.org/10.1007/978-3-031-81209-5_1

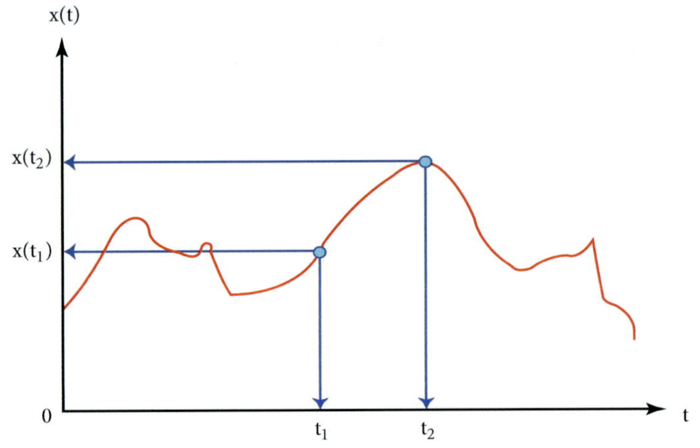

Fig. 1.1 Typical example of a continuous-time signal $x(t)$. Between any two points on the time axis t_1 and t_2, no matter how close they are, there are infinite intermediate points t_i for each of which the signal $x(t)$ has some value $x_i = x(t_i)$

to the horizontal axis at the position $t = t_1$, to determine the point of intersection of the curve of the function with this straight line, and finally, to calculate the ordinate of this point by drawing from this point of intersection the perpendicular to the vertical axis. Let us now consider a time interval (t_1, t_2) with the time instances t_1 and t_2 corresponding to two points on the horizontal axis. We know that the number of points located between these two points is infinite and that for each of these infinite points we can determine the value of the function $f(t)$ by applying the process we described above. If we now subdivide the interval (t_1, t_2) into two equal parts, the number of points that belong to each of the two equal intervals that emerged in the above way will obviously be infinite, something that will always be true regardless of how small the segment of the horizontal axis we consider is. In the limited case in which we have subdivided the interval (t_1, t_2) into infinite segments, each of which has infinitesimal size, the number of points that will be located within each segment will continue to be infinite and for each of these we will be able to find the value of the function $f(t)$ by applying the above process. Based on all the above, it now becomes apparent what we mean by saying that time is continuous and that a continuous function $f(t)$ is defined at every time instance.

The theory of *continuous* and *discrete*[1] signals and systems has experienced great growth in recent years and constitutes the basis for the design and implementation of a plethora of applications that are used both in our daily life and in specialized experiments and simulations. As examples of such applications, we mention indicatively:

- The *radar* that records the presence and position of an object by measuring the time interval required for a signal to approach the object and return back reflected by it.
- The *sonar* that operates in a similar way and allows the detection of objects located under the surface of the sea.
- All kinds of *telecommunication systems* that rely on the transmission of analog and digital signals that have been shaped and encoded in the appropriate way in each case.
- The *GPS* type systems that allow the detection of the position of moving objects.

As for the nature and characteristics of such a theory, they consist of three distinct and independent components: modeling, analysis, and design.

The *modeling* deals with the development of an abstract conceptual description of the evolution of a signal over time, in order to determine the characteristics of the physical process that created it. This is usually accomplished by recording the content of the signal, which is then processed using advanced hardware and software arrangements.

The *analysis* involves extracting useful information from the content of the signal that is related to the physical process that created it and describes it.

Finally, the *design* is the reverse of modeling and analysis and involves the composition of a model for the physical process that created the signal. This process is particularly useful, as it allows the design and implementation of systems and arrangements that are in every case suitable for the efficient exploitation of the signal in the appropriate way in each case.

[1] As we will see in a later chapter, signals and systems of discrete-time are not defined for every time instance, but only for a set distinguished and different from each other in time.

The following sections present fundamental concepts related to the field of continuous-time signals, such as their energy and power, their symmetry properties, the elementary signals that constitute the building blocks of all kinds of continuous-time signals, as well as the basic operations that can be applied to them.

1.2 Energy and Power of Continuous-Time Signals

The *total energy* E of a continuous-time signal $x(t)$ for the time interval $[t_1, t_2]$ is defined as

$$E = \int_{t_1}^{t_2} |x(t)|^2 dt \ , \tag{1.1}$$

whereas the *average power* of it-that is, the energy per unit of time-is defined as the ratio of the signal's energy to the duration of the aforementioned time interval, or in mathematical form[2]

$$P = \langle |x(t)|^2 \rangle = \frac{1}{t_2 - t_1} \int_{t_1}^{t_2} |x(t)|^2 dt. \tag{1.2}$$

In the above equations, the quantity $|x(t)|$ expresses the *magnitude* of the signal $x(t)$ and covers the case of both *real* and *complex* signals. If the signal is a complex number, it follows that

$$|x(t)| = \sqrt{\Re\{x(t)\}^2 + \Im\{x(t)\}^2},$$

where $\Re\{x(t)\}$ and $\Im\{x(t)\}$ are the real and the imaginary parts of the complex quantity $x(t)$. On the other hand, if the signal contains real values, it will obviously be $|x(t)| = x(t)$ and the above relations take the simple form

$$E = \int_{t_1}^{t_2} x^2(t)\, dt \quad \text{and} \quad P = \frac{1}{t_2 - t_1} \int_{t_1}^{t_2} x^2(t)\, dt.$$

In cases where we want to calculate the values of the above quantities *for the entire infinite range of values* $(-\infty, +\infty)$ the above relations take the form

$$E_\infty = \lim_{T\to\infty} \int_{-T}^{+T} |x(t)|^2 dt = \int_{-\infty}^{+\infty} |x(t)|^2 dt \quad \text{and} \quad P_\infty = \lim_{T\to\infty} \frac{1}{2T} \int_{-T}^{+T} |x(t)|^2 dt = \lim_{T\to\infty} \frac{E_\infty}{2T}.$$

Note that the above equations do not allow the determination of the *dimensions* of energy and power, as these depend on the nature of the signal. For example, if the signal $x(t)$ expresses a time-varying voltage $v(t)$, the units of energy and power will be Volts^2 · second and Volts^2 respectively, whereas, if it expresses some current $\iota(t)$, these units will be Amperes^2 · second and Amperes^2. The concepts of *energy* and *power*, as formulated, express *the magnitude of the signal* or alternatively *its energy capability* (for example, they could describe the energy consumed on a resistance of one Ohm, if the signal $x(t)$ expresses the voltage applied to the ends of the resistance or the current that flows through it[3]) and they do not have the usual meaning

[2] The energy and power of the signal exist as quantities, only when the above integral converges toward some finite value. For this to happen, the amplitude of the signal must tend to zero at the limit $|t| \to \infty$. In the opposite case, the above integral does not converge and the energy of the signal becomes infinite.

[3] The defining equations of energy for the signals $x(t) = v(t)$ (voltage) and $x(t) = \iota(t)$ (current intensity) are

$$E_v = \int_{t_1}^{t_2} \frac{v^2(t)}{R}\, dt = \frac{1}{R} \int_{t_1}^{t_2} v^2(t)\, dt \quad \text{and} \quad E_\iota = \int_{t_1}^{t_2} \iota^2(t) R\, dt = R \int_{t_1}^{t_2} \iota^2(t)\, dt,$$

whereas the power defining equations are similar.

of energy as we know it from physics. The reason why energy is defined in the way we present it has to do with the fact that since the signal is characterized not only by an amplitude but also by a certain duration, the duration of the signal should be used in the definition of the signal's magnitude (since, as we mentioned, energy can be considered as such a magnitude) and since the signal is defined in the field of continuous-time, that is, it is defined at every time instance, it is reasonable to use an integral, the limits of which are related to the duration of the signal. The choice of the second power of the signal-e.g., $x^2(t)$ -in the definition of the signal's energy has its origin in physics (observe, for example, how the dynamic energy of a spring $E = (1/2)kx^2$, the kinetic energy $E = (1/2)mv^2$, and the energy consumed in a resistance $E = i^2R$, are characterized by the same type of dependence), whereas in the above definition it is not taken into account at all what exactly the considered signal represents.

Using the concepts of energy and power, we can classify continuous-time signals into three categories:

- Signals with finite total energy for which the property $E_\infty < 0$ holds and which are known as *energy signals*. From the definition of average power, the average power of these signals is equal to zero:

$$P_\infty = \lim_{T\to\infty} \frac{E_\infty}{2T} = 0. \tag{1.3}$$

- Signals of finite average power for which $P_\infty < 0$ are known as *power signals*. If the value of the average power is positive, the total energy will be *infinite*. This conclusion is expected, since the average power of the signal expresses its *average energy per unit of time*. Therefore, if the average energy of the signal per unit of time is positive, its integral over the entire range of values $(-\infty, +\infty)$ will have an infinite value.
- Signals for which both the total energy and the average power have an *infinite* value.

Problem 1.2.1 Determine the energy of the signal

$$x(t) = \begin{cases} e^{-\xi t} & (\xi > 0) \text{ for } t \geq 0 \\ 0 & \text{for } t < 0. \end{cases}$$

Solution From the defining equation of the energy of the signal, it immediately follows that

$$E = \int_{-\infty}^{+\infty} |x(t)|^2 dt = \int_0^\infty e^{-2\xi t} dt = -\frac{1}{2\xi} e^{-2\xi t}\Big|_0^\infty = \frac{1}{2\xi}.$$

We observe that the energy of the signal is finite and therefore the signal is an energy signal.

Taking into account that a finite energy signal has zero power whereas a finite power signal has infinite energy, it is obvious that a signal cannot simultaneously be *both* an energy signal *and* a power signal, but must necessarily belong to one of these two signal categories. We also mention that all signals used in practice have *finite energy* and for this reason they are *energy signals*. Conversely, power signals exist only theoretically, since they have infinite duration and therefore it is impossible to construct them.

Problem 1.2.2 Determine the energy of the signal

$$x(t) = \begin{cases} e^{-\alpha t} \sin(\beta t) & \text{for } t \geq 0 \\ 0 & \text{for } t < 0. \end{cases}$$

Solution From the signal defining equation we find that

$$|x(t)|^2 = [e^{-\alpha t} \sin(\beta t)]^2 = e^{-2\alpha t} \sin^2(\beta t) = \frac{1}{2} e^{-2\alpha t} \Big[1 - \cos(2\beta t)\Big],$$

where to derive the last expression we used the identity $\cos(2t) = 1 - 2\sin^2 t$. Replacing this relationship in the energy defining equation and taking into account that the signal is defined only in the area $t \geq 0$, from the energy defining equation of the signal we will have

$$E = \int_{-\infty}^{+\infty} |x(t)|^2 dt = \frac{1}{2}\int_0^{\infty} e^{-2\alpha t}\Big[1 - \cos(2\beta t)\Big]\, dt = \frac{1}{2}\int_0^{\infty} e^{-2\alpha t} dt - \frac{1}{2}\int_0^{\infty} e^{-2\alpha t}\cos(2\beta t)\, dt.$$

Using integral calculation techniques, it turns out that[4]

$$\int_0^{\infty} e^{-2\alpha t} dt = -\frac{1}{2\alpha} e^{-2\alpha t}\Big|_0^{\infty} = \frac{1}{2\alpha} \quad \text{and} \quad \int_0^{\infty} e^{-2\alpha t}\cos(2\beta t) dt = \frac{\alpha}{2(\alpha^2 + \beta^2)}$$

and by substituting the above expressions into the energy defining equation we find

$$E = \frac{1}{4\alpha} - \frac{1}{4}\frac{\alpha}{\alpha^2 + \beta^2} = \frac{1}{4}\left(\frac{1}{\alpha} - \frac{\alpha}{\alpha^2 + \beta^2}\right) = \frac{\beta^2}{4\alpha(\alpha^2 + \beta^2)}.$$

From the last relation we observe that the nature of the signal depends on the values of the parameters α and β. If $\alpha = 0$, the signal becomes purely sinusoidal and is characterized by infinite energy, whereas for any other real value of these parameters the energy of the signal is finite and therefore the signal in question is an *energy signal.*

Problem 1.2.3 Determine the energy of the signal $x(t) = e^{-|t|}$.

Solution From the defining equation of the signal $x(t)$ we observe that

$$|x(t)|^2 = e^{-2|t|} = \begin{cases} e^{+2t} & \text{for } t < 0 \\ e^{-2t} & \text{for } t > 0 \end{cases} \quad \text{and therefore it is}$$

$$E = \int_{-\infty}^{+\infty} |x(t)|^2 dt = \int_{-\infty}^{0} e^{2t} dt + \int_0^{\infty} e^{-2t} dt = \frac{1}{2}e^{2t}\Big|_{-\infty}^{0} - \frac{1}{2}e^{-2t}\Big|_0^{\infty} = \frac{1}{2} + \frac{1}{2} = 1 \quad .$$

We observe that the energy of the signal is *finite* and for this reason, the signal is an *energy signal.*

---

[4] The calculation of the first integral is trivial, as can easily be seen, whereas for the second integral, applying the method of integration by parts successively, we have

$$I = \int_0^{\infty} e^{-2\alpha t}\cos(2\beta t)dt = \int_0^{\infty} \frac{d}{dt}\left(-\frac{1}{2\alpha}e^{-2\alpha t}\right)\cos(2\beta t)dt = -\frac{1}{2\alpha}e^{-2\alpha t}\cos(2\beta t)\Big|_0^{\infty} - \int_0^{\infty}\left(-\frac{1}{2\alpha}e^{-2\alpha t}\right)\frac{d}{dt}\cos(2\beta t)dt$$

$$= -\frac{1}{2\alpha}(0-1) + \frac{1}{2\alpha}\int_0^{\infty} e^{-2\alpha t}[-\sin(2\beta t)](2\beta)dt = \frac{1}{2\alpha} - \frac{\beta}{\alpha}\int_0^{\infty} e^{-2\alpha t}\sin(2\beta t)dt$$

$$= \frac{1}{2\alpha} - \frac{\beta}{\alpha}\int_0^{\infty}\frac{d}{dt}\left(-\frac{1}{2\alpha}e^{-2\alpha t}\right)\sin(\beta t)dt = \frac{1}{2\alpha} - \frac{\beta}{\alpha}\left\{-\frac{1}{2\alpha}e^{-2\alpha t}\sin(2\beta t)\Big|_0^{\infty} - \int_0^{\infty}\left(-\frac{1}{2\alpha}e^{-2\alpha t}\right)\frac{d}{dt}\sin(2\beta t)dt\right\}$$

$$= \frac{1}{2\alpha} - \frac{\beta}{\alpha}\left\{-\frac{1}{2\alpha}(0-0) + \frac{1}{2\alpha}\int_0^{\infty} e^{-2\alpha t}\cos(2\beta t)dt\right\} = \frac{1}{2\alpha} - \frac{\beta^2}{\alpha^2}\int_0^{\infty} e^{-2\alpha t}\cos(2\beta t)dt = \frac{1}{2\alpha} - \frac{\beta^2}{\alpha^2}I$$

Moving the second term of the second member to the first member and doing the operations we will have

$$I + \frac{\beta^2}{\alpha^2} = \frac{1}{2\alpha} \Longrightarrow I\left(1 + \frac{\beta^2}{\alpha^2}\right) = \frac{1}{2\alpha} \Longrightarrow I\frac{\alpha^2 + \beta^2}{\alpha^2} = \frac{1}{2\alpha} \Longrightarrow I = \frac{1}{2\alpha}\frac{\alpha^2}{\alpha^2 + \beta^2} = \frac{\alpha}{2(\alpha^2 + \beta^2)}.$$

Problem 1.2.4 Determine the power of the signal

$$x(t) = A\cos(\omega_1 t + \vartheta_1) + B\cos(\omega_2 t + \vartheta_2), \quad (\omega_1 \neq \omega_2)$$

Solution From the power defining equation of the signal we will get

$$P_\infty = \lim_{T\to\infty}\int_{-T}^{+T} x^2(t)\,dt = \lim_{T\to\infty}\frac{1}{2T}\int_{-T}^{+T}\Big[A\cos(\omega_1 t + \vartheta_1) + B\cos(\omega_2 t + \vartheta_2)\Big]^2 dt =$$

$$= \lim_{T\to\infty}\frac{A^2}{2T}\int_{-T}^{+T}\cos^2(\omega_1 t + \vartheta_1)\,dt + \lim_{T\to\infty}\frac{B^2}{2T}\int_{-T}^{+T}\cos^2(\omega_2 t + \vartheta_2)\,dt +$$

$$+ \lim_{T\to\infty}\frac{AB}{T}\int_{-T}^{+T}\cos(\omega_1 t + \vartheta_1)\cos(\omega_2 t + \vartheta_2)\,dt.$$

Using the trigonometric identity $\cos(2x) = 2\cos^2 x - 1$ for $x = \omega_1 t + \vartheta_1$ we will get

$$\cos^2(\omega_1 t + \vartheta_1) = \frac{1}{2}\Big[1 + \cos\Big(2(\omega_1 t + \vartheta_1)\Big)\Big]$$

and the first integral is written in the form

$$\lim_{T\to\infty}\frac{A^2}{4T}\int_{-T}^{+T}\Big[1 + \cos[2(\omega_1 t + \vartheta_1)]\Big]dt = \lim_{T\to\infty}\frac{A^2}{4T}\Big\{\int_{-T}^{+T}dt + \int_{-T}^{+T}\cos[2(\omega_1 t + \vartheta_1)]\,dt\Big\}.$$

It is not difficult to see that the second term of the last equation

$$\lim_{T\to\infty}\frac{A^2}{4T}\int_{-T}^{+T}\cos[2(\omega_1 t + \vartheta_1)]\,dt$$

has a value equal to zero. This happens because the value of the above integral expresses the area of the surface defined by the horizontal axis and the graphical representation of the cosine, which is finite and therefore if it is multiplied by the quantity $A^2/4T$ at the limit $T \to \infty$, it will give a value equal to zero. Therefore, we have

$$\lim_{T\to\infty}\frac{A^2}{4T}\int_{-T}^{+T}\Big[1 + \cos[2(\omega_1 t + \vartheta_1)]\Big]dt = \lim_{T\to\infty}\frac{A^2}{4T}\int_{-T}^{+T}dt = \lim_{T\to\infty}\frac{A^2}{4T}(2T) = \frac{A^2}{2}.$$

In a completely analogous way, it is proven that

$$\lim_{T\to\infty}\frac{B^2}{2T}\int_{-T}^{+T}\cos^2(\omega_2 t + \vartheta_2)\,dt = \frac{B^2}{2}.$$

As for the value of the third integral, it is equal to zero, a property that is proven if we use the trigonometric identity[5]

$$\cos(\omega_1 t + \vartheta_1)\cos(\omega_2 t + \vartheta_2) = \cos[(\omega_1 + \omega_2)t + (\vartheta_1 + \vartheta_2)] + \cos[(\omega_1 - \omega_2)t + (\vartheta_1 - \vartheta_2)]$$

and we rely on the argumentation we presented previously.[6]

[5] In this case, the above integral is written as

$$\lim_{T\to\infty}\frac{AB}{T}\int_{-T}^{+T}\cos(\omega_1 t+\vartheta_1)\cos(\omega_2 t+\vartheta_2)dt = \lim_{T\to\infty}\frac{AB}{T}\int_{-T}^{+T}\cos[(\omega_1+\omega_2)t + (\vartheta_1+\vartheta_2)]dt + \lim_{T\to\infty}\frac{AB}{T}\int_{-T}^{+T}\cos[(\omega_1-\omega_2)t + (\vartheta_1-\vartheta_2)]dt.$$

[6] Let it be noted that this is true, only when $\omega_1 \neq \omega_2$, since in the opposite case it is proven that the third term tends to the value $2AB\cos(\theta_1 - \theta_2)$ in the limit $T \to \infty$.

Combining the above results, the power of the signal in question is formulated as

$$P_\infty = \lim_{T\to\infty} \int_{-T}^{+T} x^2(t)dt = \frac{A^2}{2} + \frac{B^2}{2}$$

Note the result we arrived at is very important. Indeed, if we extend the above reasoning for the case of a continuous-time signal that can be formulated as a linear combination of infinite terms

$$x(t) = \sum_{n=1}^{\infty} C_n \cos(\omega_n t + \vartheta_n) \quad \text{it is proven in the same way that} \quad P_\infty = \frac{1}{2}\sum_{n=1}^{\infty} C_n^2$$

(in the above relation we assume that all frequencies ω_n are different from each other; see the Subnote 6). In the case where the signal also has a constant term C_0, that is, it is written as

$$x(t) = C_0 + \sum_{n=1}^{\infty} C_n \cos(\omega_n t + \vartheta_n) \quad \text{its power is equal to} \quad P_\infty = C_0^2 + \frac{1}{2}\sum_{n=1}^{\infty} C_n^2$$

1.3 Periodic Signals

A continuous-time signal $x(t)$ is called *periodic*, if there is a positive number T such that the relation holds

$$x(t) = x(t+T) \tag{1.4}$$

for every time instance t. The power of this property means that the signal remains unchanged under a process of time shift with a value equal to T, or equivalently, that after the passage of T time units, the signal repeats in exactly the same way. The magnitude T that appears in the above equation is known as the *period* of the signal.

It is not difficult to understand that since the signal repeats exactly after T time units[7] it will be characterized by the property

$$x(t) = x(t+T) = x[(t+T)+T] = x\{[(t+T)+T]+T\} = \cdots = x\Big(t + \underbrace{T+T+T+\cdots+T}_{m \text{ times}}\Big)$$

or in equivalent notation

$$x(t) = x(t+T) = x(x+2T) = \cdots = x(t+mT),$$

where $m = 1, 2, \ldots$. Therefore, the signal is characterized by many periods of the form $T' = mT$ $(m = 1, 2, \ldots)$. Of all these periods, the one that corresponds to the smallest value of m for which the above relationship holds, is (usually) symbolized by T_0 and is known as the *fundamental period* of the signal.

The number of repetitions of the signal per unit of time is usually symbolized by the letter f and is called *frequency*.[8] The unit of measurement of this quantity is called Hertz (Hz) and has dimensions of inverse time (i.e., Hz = sec^{-1}). It is not difficult to see that the period and the frequency of a periodic signal are connected by a relationship of inverse proportionality, whereas their product is equal to the unit. In order for this to be understood, let us consider Fig. 1.2, which depicts a simple sinusoidal signal with a frequency of 5Hz. This frequency value means that according to its definition, within 1 s the signal

[7] This property of the periodic signal forces us to assume that a periodic signal always starts at time $t = -\infty$ and lasts forever, constantly repeating itself. Indeed, if we accept that the signal started at a specific time, for example, at time $t = 0$, then the signal $x(t+T)$ would have to start at time $t = -T$ and would not be the same as the signal $x(t)$.

[8] Closely related to the frequency f is the angular frequency ω, which is defined as $\omega = 2\pi/T_0 = 2\pi f_0$, where T_0 and f_0 are the fundamental period and the frequency of the signal.

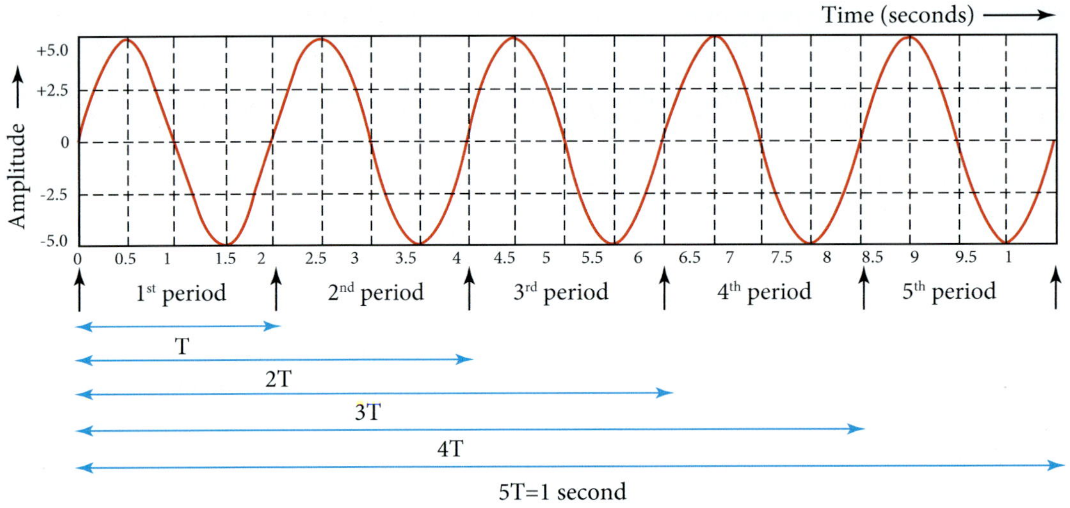

Fig. 1.2 The correlation between the period and the frequency of a periodic signal

repeats five times. If T is the period of the signal, that is, the duration of each repetition, then the total duration of the five repetitions will be equal to $T + T + T + T + T = 5T$. According to the definition of frequency, These repetitions should take place within a time span of 1 s, and therefore $5T = 1$ (second). In the general case of a frequency f signal in the above time span, f repetitions will take place. Thus, $fT = 1$, from which it follows that $f = 1/T$ or equivalently that $T = 1/f$. Therefore, these two quantities are connected by an *inverse proportionality* relationship.

If a signal is not periodic, it is called an *aperiodic* continuous-time signal. There are also cases in which the fundamental period of the signal cannot be defined. Considering, for example, a constant value signal of the form $x(t) = A$ $(\forall t \in R)$, where A is a constant, it is obvious that the value of T for which the relation $x(t) = x(t + T)$ holds can be any, so that the smallest positive value for which the above relation is satisfied and which, according to what we mentioned, would constitute the fundamental period of the signal, is not defined.

Problem 1.3.1 Prove that the continuous-time signal $x(t) = e^{j\omega_0 t}$ is periodic with a period value of $T_0 = 2\pi/\omega_0$.

Solution For the signal $x(t)$ to be periodic, there must be a positive number T such that $x(t+T) = x(t)$. Using the defining equation of the complex exponential signal the above relation takes the form

$$e^{i\omega_0(t+T)} = e^{j\omega_0 t} \text{ or equivalently } e^{j\omega_0 t} e^{j\omega_0 T} = e^{j\omega_0 t}$$

From the last relation, it follows that for the signal $x(t)$ to be periodic, it must be $e^{j\omega_0 T} = 1$. We distinguish, therefore, two cases:

- $\omega_0 = 0$: in this case, the complex exponential signal degenerates into the constant value signal $x(t) = 1$ and therefore is periodic with any period value.
- $\omega_0 \neq 0$: in this case, the complex exponential signal is periodic for those values of the period T that satisfy the relation[9]

$$\omega_0 T = 2\pi m \text{ from which it follows that } T = m\frac{2\pi}{\omega_0} \text{ for } m = 1, 2, \ldots$$

Therefore, the fundamental period of the signal is the smallest of the above values, which corresponds to the value $m = 1$ and is $T_0 = 2\pi/\omega_0$.

[9] From Euler's formula $e^{ix} = \cos x + i \sin x$, it follows that for $x = 2\pi m$, $e^{i(2\pi m)} = \cos(2\pi m) + i \sin(2\pi m) = 1$ since it is known that $\cos(2\pi m) = 1$ and $\sin(2\pi m) = 0$.

Problem 1.3.2 Prove that the cosine signal $x(t) = A\cos(\omega_0 t + \varphi)$ is periodic with a fundamental period of $T_0 = 2\pi/\omega_0$.

Solution For the signal $x(t)$ to be periodic, there must be a positive number T such that $x(t+T) = x(t)$. Using the defining equation of the aforementioned signal, the above relation is written as

$$x(t+T) = A\cos(\omega_0(t+T)+\varphi) = A\cos(\omega_0 t + \omega_0 T + \varphi)$$

From trigonometry, however, it is known that the cosine function is a periodic function with period $T = 2\pi m$. Therefore, for $x(t+T) = x(t)$ or equivalently $A\cos(\omega_0 t + \omega_0 T + \varphi) = A\cos(\omega_0 t + \varphi)$, it must be $\omega_0 T = 2\pi m$. Therefore, $T = m(\omega_0/2\pi)$ and thus the fundamental period of the signal is equal to $T = \omega_0/2\pi$, as in the previous case.

Using the exact same argument, we can prove that the sinusoidal signal

$$x(t) = A\sin(\omega_0 t + \varphi)$$

is also periodic with a fundamental period $T_0 = 2\pi/\omega_0$.

Problem 1.3.3 Considering two continuous periodic signals $x(t)$ and $y(t)$ with fundamental periods T_1 and T_2 respectively, examine under what conditions the sum of the two signals $z(t) = x(t) + y(t)$ is a periodic signal and determine its fundamental period.

Solution Since the signals $x(t)$ and $y(t)$ are periodic with fundamental periods T_1 and T_2, they will satisfy the known relations

$$x(t) = x(t + mT_1), \quad m = 1, 2, 3, \ldots \quad \text{and} \quad y(t) = y(t + nT_2), \quad n = 1, 2, 3, \ldots.$$

Considering that the sum of the two signals, $z(t) = x(t) + y(t)$, is a periodic signal with a fundamental period T, it must satisfy the condition $z(t+T) = z(t)$, which can be written in the form

$$z(t+T) = x(t+mT_1) + y(t+nT_2).$$

But the above relation is satisfied only when it is

$$T = mT_1 = nT_2 \quad \text{or equivalently} \quad \frac{T_1}{T_2} = \frac{n}{m}$$

From the last equation, it follows that the signal $z(t) = x(t) + y(t)$ is a periodic signal, when the ratio of the values of the fundamental periods of the two signals can be expressed as the quotient of two integers and therefore constitutes a rational number. In this case, the fundamental period of the signal $z(t)$ is the least common multiple of the numbers m and n, whereas if the two numbers are prime, it is calculated from the relation $T = mT_1 = nT_2$. On the other hand, if the ratio T_1/T_2 is an irrational number, then the above common multiple does not exist, that is, the signal is not periodic.

Problem 1.3.4 Examine whether the sum of two periodic signals $x(t)$ and $y(t)$ with period values $T_1 = 15$ ms and $T_2 = 20$ ms is periodic and if it is, calculate its period.

Solution The ratio of the values of the fundamental periods of the two signals

$$\frac{T_1}{T_2} = \frac{15}{20} = \frac{3}{4}$$

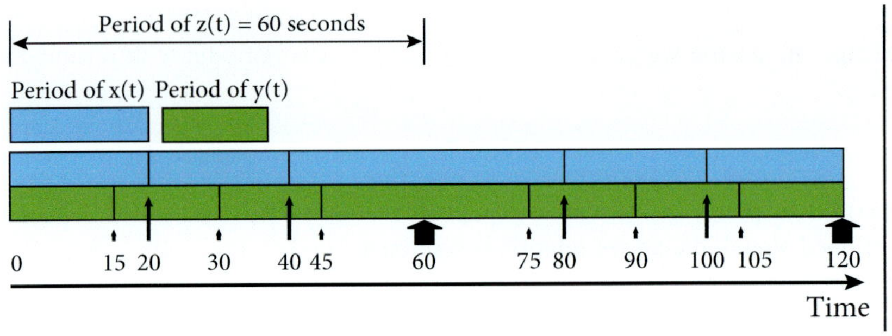

Fig. 1.3 The period of a signal resulting from the sum of two periodic signals with fundamental periods of 15 and 20 ms, is equal to 60 ms

is a rational number-since it is the quotient of two integers-and therefore the sum of the two signals is a periodic signal. This periodic signal repeats each time the periods of the two signals coincide and its fundamental period is determined by the least common multiple of the two periods, since at this point in time the aforementioned coincidence occurs for the first time. This situation is presented diagrammatically in Fig. 1.3. From this figure, we observe that the signal $z(t) = x(t) + y(t)$ repeats periodically at 60, 120 etc. msec and therefore its fundamental period is equal to 60 ms.

Problem 1.3.5 Find the fundamental period of the signal $x(t) = \cos(15\pi t) + 3\sin(8\pi t)$.

Solution If T is the period of the signal $x(t)$, then the condition of periodicity $x(t) = x(t+T)$ will apply where the signal $x(t+T)$ is defined as

$$x(t+T) = \cos[15\pi(t+T)] + 3\sin[8\pi(t+T)] = \cos(15\pi t + 15\pi T) + 3\sin(8\pi t + 8\pi T)$$

Taking into account that the periods of the two sinusoidal signals are of the form $T_1 = 2\pi m$ and $T_2 = 2\pi n$, the condition of periodicity is satisfied if it is

$$15\pi T = 2\pi m \quad \text{and} \quad 8\pi T = 2\pi n,$$

from which it follows that[10]

$$T = \frac{2m}{15} = \frac{n}{4} \quad \text{or equivalently} \quad \frac{n}{m} = \frac{8}{15}.$$

But the two numbers above do not have a common divisor; therefore, $n = 8$ and $m = 15$, from which it follows that $T = n/2 = 4$.

Problem 1.3.6 Prove that the integral of any periodic signal $x(t)$ of period T over a range of values $[\alpha, \beta]$ remains unchanged if the integration limits are shifted by a distance T, or in mathematical writing, that

$$\int_{\alpha}^{\beta} x(t)\, dt = \int_{\alpha+T}^{\beta+T} x(t)\, dt$$

for every pair of real values (α, β).

[10] In the general case of two periodic signals $x(t)$ and $y(t)$ with period values T_1 and T_2 the sum of the two signals $z(t) = x(t) + y(t)$ is a periodic signal with period T, if the relation $T = mT_1 = nT_2$ holds, from which it follows that $T_1/T_2 = n/m$. Therefore, the property of periodicity for the signal $z(t)$ is ensured when the ratio of the periods of the two signals is a rational number-that is, it can be expressed as the quotient of two integers-whereas in the opposite case the signal $z(t)$ is not periodic.

Solution Proceeding with the change of variable $\xi = t + T$ the periodicity relation $x(t) = x(t+T)$ is formulated as $x(\xi - t) = x(\xi)$, whereas differentiating we get $dt = d\xi$. As for the limits of integration for $t_1 = \alpha$ will be $\xi_1 = \alpha + T$ and for $t_2 = \beta$ they will be $\xi_2 = \beta + T$. Therefore, the above integral is written as

$$\int_{\alpha}^{\beta} x(t)\,dt = \int_{\alpha+T}^{\beta+T} x(\xi - T)d\,\xi = \int_{\alpha+T}^{\beta+T} x(\xi)\,d\xi = \int_{\alpha+T}^{\beta+T} x(t)\,dt.$$

To derive the last relation, we simply proceeded to change the integration variable from ξ to t. ■

Note In the special case where $\beta = 0$, the above relation is written as

$$\int_{\alpha}^{0} x(t)\,dt = \int_{\alpha+T}^{T} x(t)\,dt$$

and integrating over the interval $[0, T]$ we get

$$\int_{0}^{T} x(t)\,dt = \int_{0}^{\alpha+T} x(t)\,dt + \int_{\alpha+T}^{T} x(t)\,dt = \int_{0}^{\alpha+T} x(t)\,dt + \int_{\alpha}^{0} x(t)\,dt = \int_{\alpha}^{\alpha+T} x(t)\,dt,$$

an equation that describes another useful property of periodic signals according to which *the area enclosed by the graphical representation of a periodic signal $x(t)$ is the same for all time intervals, the length of which is equal to the period T of the signal.*

Problem 1.3.7 Check if the following continuous-time signals are periodic and if so, determine their period.

$$x_1(t) = \sin\left(t + \frac{\pi}{10}\right) \quad \text{and} \quad x_2(t) = 5\cos\left(\frac{4\pi t}{15}\right)$$

Solution The signal $x_1(t)$ is of the form $x(t) = \cos(\omega_0 t + \varphi)$ with parameter values $\omega_0 = 1$ and $\phi = \pi/10$, that is, it is periodic with a period value $T = 2\pi/\omega_0 = 2\pi$. As for the signal $x_2(t)$, it is also of the same form with a circular frequency $\omega_0 = 4\pi/15$ and therefore it is periodic with a period value $T = 2\pi/\omega_0 = 15/2$.

Problem 1.3.8 Check if the continuous-time signal

$$x(t) = \cos\left(\frac{\pi t}{4}\right) + \cos\left(\frac{3\pi t}{7}\right)$$

is periodic and if so, determine its period.

Solution Following the methodology of the previous exercise, we find that the signal $\cos(\pi t/2)$ is of the form $\cos(\omega_0 t)$ with a circular frequency $\omega_0 = \pi/2$ and therefore it is periodic with a period value $T_1 = 2\pi/\omega_0 = 4$ whereas similarly the signal $\cos(2\pi t/7)$ is characterized by a frequency value $\omega_0 = 2\pi/7$ and therefore it is also periodic with a period value $T_2 = 2\pi/\omega_0 = 7$. We observe that the ratio of the periods $T_1/T_2 = 4/7$ **is a rational number and therefore the sum of the two signals $x(t)$ will also be periodic with a period value $T = 4T_1 = 3T_2 = 24$ (the value 24 is the least common multiple of the numbers** 6 **and** 8**).**

Problem 1.3.9 Check if the continuous-time signals

$$x_1(t) = \sin t + \cos(\sqrt{3}t) \quad \text{and} \quad x_2(t) = 4\sin^2 t$$

are periodic and if so, specify their period.

Solution The signal $\sin t$ is obviously periodic with a period of $T_1 = 2\pi$ whereas the signal $\cos(\sqrt{2}t)$ is of the form $\cos(\omega_0 t)$ with a circular frequency $\omega_0 = \sqrt{2}$ and a period value of $T = 2\pi/\omega_0 = \pi\sqrt{2}$. However, in this case the ratio $T_1/T_2 = \sqrt{2}$ is not a rational number and therefore the signal $x_1(t)$ is not periodic.

On the other hand, using the known trigonometric identity $\cos(2t) = 1 - 2\sin^2 t$ the signal $x_2(t)$ is written in the form

$$x_2(t) = 2\sin^2 t = 1 - \cos(2t)$$

and consists of a sum of two signals, the first of which, being a constant value signal—since its value for every t is equal to 1—is characterized by any period, whereas the second has a circular frequency $\omega_0 = 2$ and a period $T = 2\pi/\omega_0 = \pi$. Therefore, the signal $x_2(t)$ is periodic with a period equal to π.

1.4 Even and Odd Signals

A second division of continuous-time signals is based on the symmetry that characterizes them under the transformation of *time reversal*, which is implemented if in the defining equation of the signal we replace the symbol t with the symbol $-t$. If after performing this transformation it is found that the relation $x(-t) = x(t)$ holds, the signal is called *even*, whereas if it is found that $x(-t) = -x(t)$, the signal is characterized as an *odd* continuous-time signal. If a signal is odd, then its value at the time $t = 0$ is equal to zero. Indeed, from the defining equation of the continuous-time signal for this time we will get $x(0) = -x(0)$, from which it finally turns out that $x(0) = 0$.

If we construct the graph of an even and an odd signal, we will find that *the graph of the even signal is symmetrical with respect to the vertical axis*, whereas *the graph of the odd signal is symmetrical to the origin of the coordinate system*, a property that characterizes any even or odd function respectively, as is well known from mathematics.

Although there are signals that are not characterized by either even or odd symmetry, however, it is proven that any continuous-time signal $x(t)$ can be expressed as the sum of an even component $\mathcal{E}\text{v}\{x(t)\}$ and an odd component $\mathcal{O}\text{d}\{x(t)\}$. Indeed, if we write the signal $x(t)$ in the form

$$x(t) = \frac{1}{2}[x(t) + x(-t)] + \frac{1}{2}[x(t) - x(-t)] = \mathcal{E}\text{v}(t) + \mathcal{O}\text{d}(t) \tag{1.5}$$

where the signals $\mathcal{E}\text{v}(t)$ and $\mathcal{O}\text{d}(t)$ are defined as

$$\mathcal{E}\text{v}(t) = \mathcal{E}\text{v}\{x(t)\} = \frac{1}{2}[x(t) + x(-t)] \quad \text{and} \quad \mathcal{O}\text{d}(t) = \mathcal{O}\text{d}\{x(t)\} = \frac{1}{2}[x(t) - x(-t)]$$

It is not difficult to ascertain that the signal $\mathcal{E}\text{v}(t)$ is characterized by even symmetry, whereas the signal $\mathcal{O}\text{d}(t)$ is characterized by odd symmetry. Indeed, if we replace t with $-t$ in the defining equations of these two signals, we will have

$$\mathcal{E}\text{v}(-t) = \frac{1}{2}[x(-t) + x[-(-t)]] = \frac{1}{2}[x(t) + x(-t)] = \mathcal{E}\text{v}(t) \quad \text{and}$$

$$\mathcal{O}\text{d}(-t) = \frac{1}{2}[x(-t) - x[-(-t)]] = -\frac{1}{2}[x(t) - x(-t)] = -\mathcal{O}\text{d}(t),$$

meaning that the signals $\mathcal{E}\mathrm{v}(t)$ and $\mathcal{O}\mathrm{d}(t)$ are characterized by even and odd symmetry respectively. Combining all the above, we can conclude that *every continuous-time signal $x(t)$ can be written as the sum of an even and an odd component.*

It is proven that *the product of two even or two odd signals is an even signal, whereas the product of an even and an odd signal is an odd signal.* In order to prove the truth of this assertion, let us consider a continuous-time signal $z(t)$ of the form $z(t) = x(t)y(t)$, where $x(t)$ and $y(t)$ are any two continuous-time signals. If the signals $x(t)$ and $y(t)$ are even, they will satisfy the symmetry property $x(-t) = x(t)$ and $y(-t) = y(t)$ and therefore the signal $z(-t)$ will be calculated as

$$z(-t) = x(-t)y(-t) = x(t)y(t) = z(t)$$

meaning that $z(t)$ is an even signal. On the other hand, if the signals $x(t)$ and $y(t)$ are odd signals, they will satisfy the relations $x(-t) = -x(t)$ and $y(-t) = -y(t)$ and the signal $z(-t)$ will have the form

$$z(-t) = x(-t)y(-t) = [-x(t)][-y(t)] = x(t)y(t) = z(t)$$

so in this case too, $z(t)$ is again an even signal. Finally, if the signal $x(t)$ is even and the signal $y(t)$ is odd, $x(-t) = x(t)$ and $y(-t) = -y(t)$, so the value $z(-t)$ will be calculated as

$$z(-t) = x(-t)y(-t) = x(t)[-y(t)] = -x(t)y(t) = -z(t)$$

meaning that $z(t)$ is an odd signal. This result obviously also applies in the case where the even one of the two signals is the $y(t)$, whereas the odd one of the two signals is the $x(t)$.

Problem 1.4.1 Determine the even and odd components of the continuous-time signal $x(t) = e^{j\omega t}$.

Solution From the defining equations of the components $\mathcal{E}\mathrm{v}(t)$ and $\mathcal{O}\mathrm{d}(t)$ and for the continuous-time signal $x(t) = e^{i\omega t}$ we have

$$\mathcal{E}\mathrm{v}(t) = \frac{x(t) + x(-t)}{2} = \frac{e^{j\omega t} + e^{-j\omega t}}{2} = \cos(\omega t) \quad \text{and} \quad \mathcal{O}\mathrm{d}(t) = \frac{x(t) - x(-t)}{2} = \frac{e^{j\omega t} - e^{-j\omega t}}{2} = j\sin(\omega t),$$

where for the derivation of the final relations we used the Euler formulas

$$e^{+j\omega t} = \cos(\omega t) + j\sin(\omega t) \quad \text{and} \quad e^{-j\omega t} = \cos(\omega t) - j\sin(\omega t)$$

Problem 1.4.2 Considering any continuous-time signal that is completed in the interval $[-\xi, +\xi]$ $(\forall \varepsilon \in R)$, prove that

$$I = \int_{-\xi}^{+\xi} x(t)\,dt = \begin{cases} 2\displaystyle\int_{0}^{\xi} x(t)\,dt & \text{if the signal } x(t) \text{ is even} \\ 0 & \text{if the signal } x(t) \text{ is odd} \end{cases}$$

(*Hint*: formulate the integral in the interval $[-\xi, +\xi]$ as the sum of two integrals defined in the intervals $[-\xi, 0]$ and $[0, +\xi]$ and use the property of symmetry of even and odd signals—consult also Sect. 3.2.1 on page 190.)

Solution The integral I of the continuous-time signal $x(t)$ can be written in the form

$$I = \int_{-\varepsilon}^{+\varepsilon} x(t)dt = \int_{-\varepsilon}^{0} x(t)dt + \int_{0}^{+\varepsilon} x(t)dt$$

Proceeding with a change of integration variable $t = -\xi$ in the first integral of the left-hand side, the integration limits $t_1 = -\varepsilon$ and $t_2 = 0$ will take the values $\xi_1 = \epsilon$ and $xi_2 = 0$ respectively, whereas differentiating the above relation we will get $dt = -d\xi$. Therefore,

$$I = -\int_{\varepsilon}^{0} x(-\xi)d\xi + \int_{0}^{+\varepsilon} x(t)dt = \int_{0}^{+\varepsilon} x(t)dt + \int_{0}^{+\varepsilon} x(-t)dt,$$

where to derive the last relation we made the simple substitution $\xi = t$ and reversed the limits and the sign in the integral of $x(-\xi)$. If the signal $x(t)$ is even, $x(-t) = x(t)$ and the above relation will take the form

$$I = \int_{0}^{+\varepsilon} x(t)dt + \int_{0}^{+\varepsilon} x(-t)dt = \int_{0}^{+\varepsilon} x(t)dt + \int_{0}^{+\varepsilon} x(t)dt = 2\int_{0}^{+\varepsilon} x(t)dt.$$

On the other hand, if the signal $x(t)$ is odd, $x(-t) = -x(t)$ and the above relation will become

$$I = \int_{0}^{+\varepsilon} x(t)dt + \int_{0}^{+\varepsilon} x(-t)dt = \int_{0}^{+\varepsilon} x(t)dt - \int_{0}^{+\varepsilon} x(t)dt = 0.$$

This property of even and odd continuous-time signals can be demonstrated graphically taking into account that: (a) the graphical representation of an even signal is symmetric with respect to the vertical axis while the graphical representation of an odd signal is symmetric with respect to the origin of the coordinate system and (b) the value of the integral of a function over a specific range of values is equal to the area of the region defined by the curve of the function, the horizontal axis, and the vertical lines on the horizontal axis at the points corresponding to the limits of integration. It is obvious that in even signals the vertical axis separates this area into two equal areas and therefore the total area of the above region is equal to twice the area of the region related to the positive values of t, whereas in odd signals These areas have equal and opposite areas and therefore their sum is zero. This situation is diagrammatically presented in the left and right images of Fig. 1.4.

The above relations are very useful and will be used in a subsequent chapter to simplify the defining equations of the coefficients of the Fourier expansion of even and odd functions.

Problem 1.4.3 Considering any continuous-time signal $x(t)$ to prove that

$$\int_{-\infty}^{+\infty} x^2(t)\, dt = \int_{-\infty}^{+\infty} x_e^2(t)\, dt + \int_{-\infty}^{+\infty} x_o^2(t)\, dt,$$

where $x_e(t)$ and $x_o(t)$ are the even and odd components of it (*Hint*: formulate the signal $x(t)$ as a sum of the form $x_e(t) + x_o(t)$ and use the symmetry property for the product of the two signals).

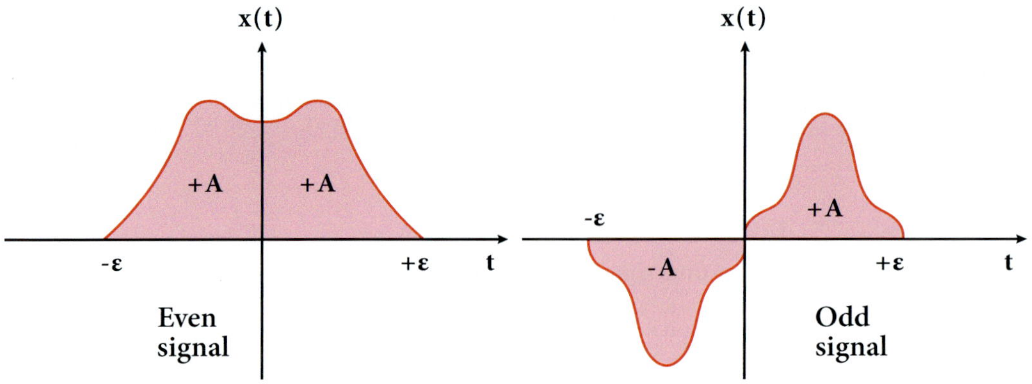

Fig. 1.4 Calculation of the integral of even and odd signals for the value range $[-\varepsilon, +\varepsilon]$ (Adapted from Oppenheim, Willsky, Nawab, Fig. 1.17)

Solution In the previous section we proved that any continuous-time signal $x(t)$ can be written in the form $x(t) = x_e(t) + x_o(t)$, where $x_e(t)$ and $x_o(t)$ are its even and odd components. Squaring it we will get

$$x^2(t) = [x_e(t) + x_o(t)]^2 = x_e^2(t) + x_o^2(t) + 2x_e(t)x_o(t)$$

and by taking the integral of the two terms over the interval $(-\infty, +\infty)$ we end up with a relation of the form

$$\int_{-\infty}^{\infty} x^2(t)dt = \int_{-\infty}^{\infty} x_e^2(t)dt + \int_{-\infty}^{\infty} x_o^2(t)dt + 2\int_{-\infty}^{\infty} x_e(t)x_o(t)dt$$

However, since the signals $x_e(t)$ and $x_o(t)$ are characterized by even and odd symmetry respectively, their product $\xi(t) = x_e(t)x_o(t)$ as we have previously proven, will be an odd signal. Also, in the previous exercise we showed that the integral of an odd signal $\xi(t)$ of the form

$$I = \int_{-\varepsilon}^{+\varepsilon} \xi(t)dt, \quad \forall \xi \in R$$

has a value equal to zero. Combining all of the above, it is not difficult to realize that

$$\int_{-\infty}^{\infty} x_e(t)x_o(t)dt = \lim_{\varepsilon \to \infty} \int_{-\varepsilon}^{+\varepsilon} \xi(t)dt = 0$$

and therefore we will get

$$\int_{-\infty}^{+\infty} x^2(t)dt = \int_{-\infty}^{+\infty} x_e^2(t)dt + \int_{-\infty}^{+\infty} x_o^2(t)dt,$$

which is the desired result. If we consider that the integral of the square of the signal over the infinite range of values $(-\infty, +\infty)$ is nothing but its total energy, the above relation leads us to the very useful and interesting conclusion that the energy of any continuous-time signal equals the sum of the energies of its even and odd components.

Problem 1.4.4 Considering the continuous-time signals depicted in Fig. 1.5, determine their even and odd components.

Solution To calculate the even and odd components of the above signals, we will rely on the definitions of these two components; however, in this case, since we do not have the defining equations of these signals but only their graphical representation, we will necessarily work in a diagrammatic way. In order to determine the signal $x(-t)$ resulting from the operation of time reversal, we should draw the mirror image of the above representation with respect to the vertical axis (which is equivalent to replacing t with $-t$), thus constructing the time-reversed signal depicted in Fig. 1.6. If we add together the original signal $x(t)$ and the time-reversed signal $x(-t)$, the resulting signal will describe twice the even component $x_e(t)$, whereas subtracting the above signals term by term, we will construct twice the odd component $x_o(t)$. Therefore, to arrive at these two components, we should divide the ordinates of all points of the graphical representations of the signals $2x_e(t)$ and $2x_o(t)$ by two, thus ending up with the signals depicted in the lower part of Fig. 1.6. From the graphical representations of these signals, we confirm the type of symmetry that characterizes them, since the even signal is symmetrical with respect to

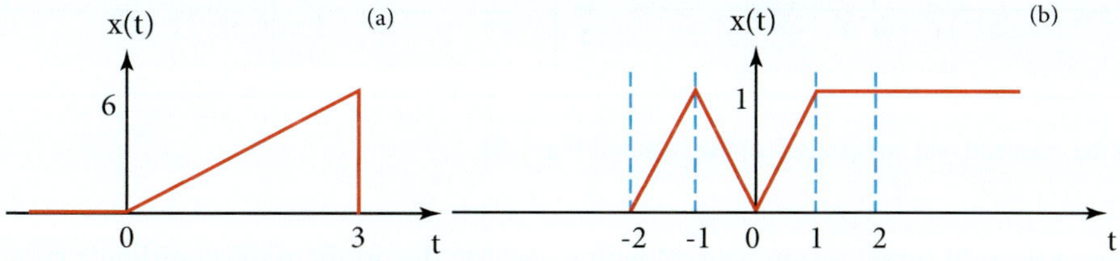

Fig. 1.5 The continuous-time signals of Exercise 1.4.4

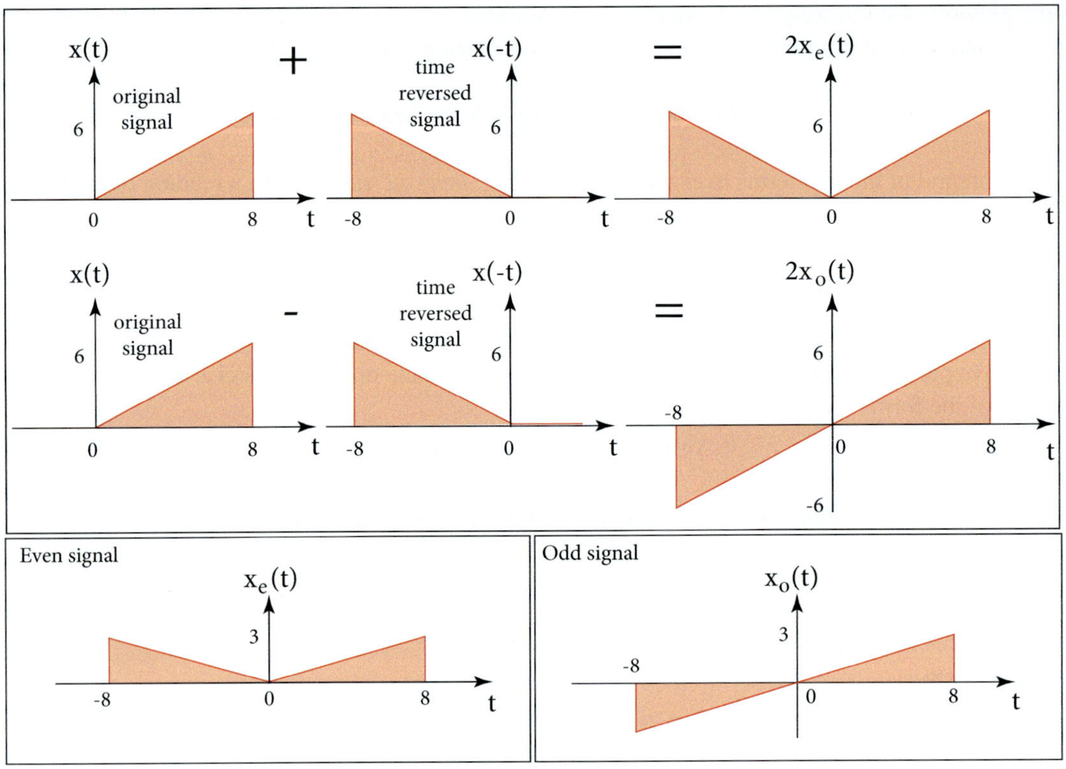

Fig. 1.6 Calculation of the even and odd components of the signal of Fig. 1.5a

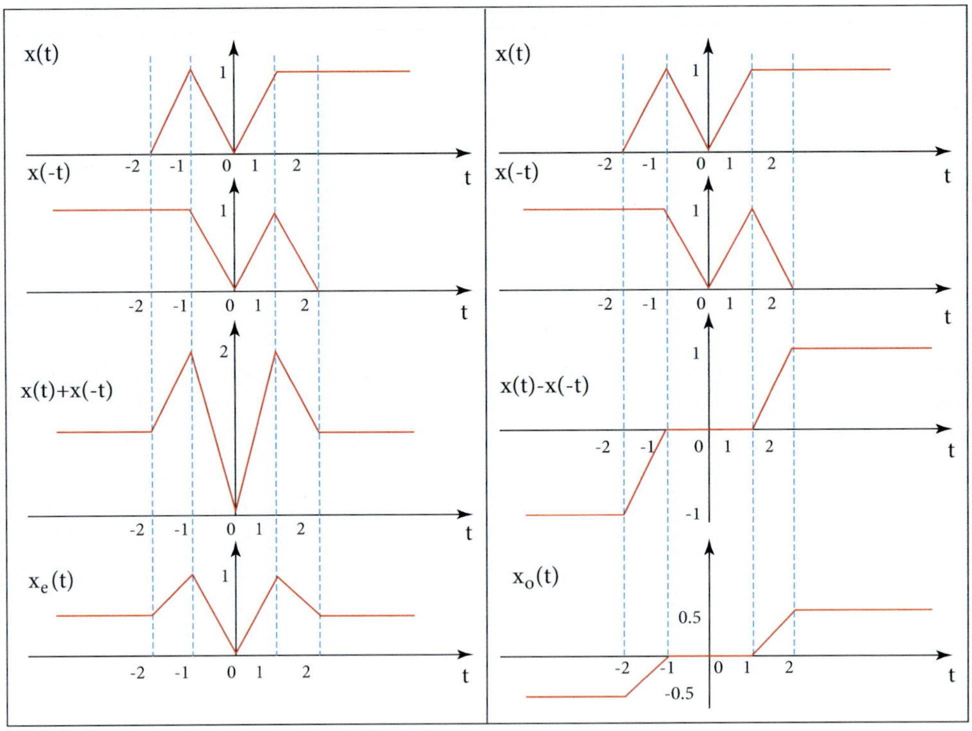

Fig. 1.7 Calculation of the even and odd components of the signal of Fig. 1.5b

the vertical axis, whereas the odd signal is symmetrical with respect to the origin of the coordinate system. The same process applied to the second signal leads to the signals depicted in Fig. 1.7.

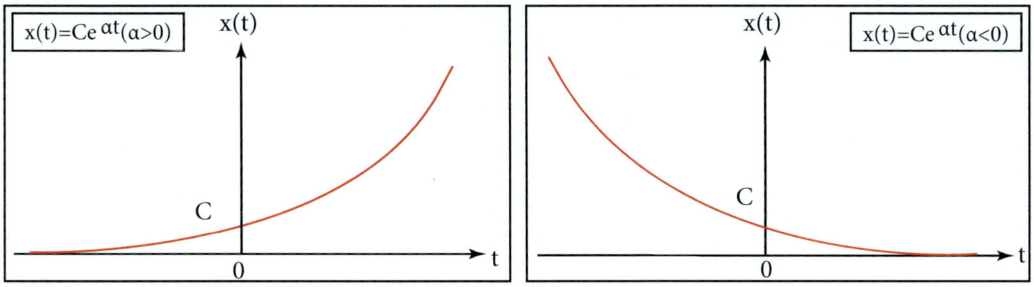

Fig. 1.8 Typical examples of real exponential signals characterized by an exponential increase and an exponential decrease in relation to time

1.5 Elementary Exponential and Trigonometric Signals

Although a continuous-time signal can have any form, in practice, many of these signals can be described with the help of certain special or fundamental signals, which will be presented in detail in this and the next sections. These fundamental signals are the elementary *exponential* and *trigonometric* signals of the sine and cosine that will be presented in this section, as well as the *step function* $u(t)$ and the *function* $\delta(t)$ of Dirac (or *impulse function*), which will be presented immediately after.

1.5.1 Exponential Signals

The most general form of an *exponential continuous-time signal* is described by the equation

$$x(t) = Ce^{\alpha t} \tag{1.6}$$

in which both the amplitude C and the exponent parameter α are complex numbers. Therefore, depending on the nature and the value of these parameters, there are many categories of continuous exponential signals, which in general terms are the following:

Real Exponential Signals

In these signals the parameters C and α are *real numbers* and therefore our signal is a real continuous-time signal. Signals of this type are characterized by two distinct types of behavior: if the value of α is positive, the amplitude of the signal increases exponentially in relation to time, whereas conversely, if the value of α is negative, the amplitude of the signal decreases exponentially over time. Typical examples of phenomena characterized by this kind of exponential increase are the *chain reactions* in nuclear reactors and complex *chemical reactions*, whereas on the other hand, phenomena such as *radioactive decay* and the process of *discharging* a capacitor, are characterized by an exponential decrease. These two different types of temporal evolution are presented in Fig. 1.8. It is obvious that, if the value of α is equal to zero, the amplitude of the signal acquires the constant value $x(t) = C$ for all time instances.

Imaginary Exponential Signals

In the special case where the parameter α is an imaginary number of the form $\alpha = j\omega_0$, the exponential signal is written as

$$x(t) = Ce^{j\omega_0 t} \tag{1.7}$$

and as we have proven in the section of periodic signals, this signal is always periodic with a fundamental period value $T = 2\pi/\omega_0$. The above relationship holds for positive values of frequency ω, whereas when this magnitude is negative, the periodicity condition is written as $|\omega_0|T = 2\pi m$ and the fundamental period corresponding to the value $m = 1$ is $T = 2\pi/|\omega_0|$. From the above analysis, it is concluded that the imaginary exponential signals $x_1(t) = Ce^{j\omega_0 t}$ and $x_2(t) = Ce^{-j\omega_0 t}$ are characterized by the same fundamental period value.

A very important property of imaginary exponential signals is that they constitute an *orthogonal basis of functions*. This property is the theoretical basis of the Fourier development of continuous-time signals and is demonstrated in the corresponding section.

Periodic complex exponential signals are characteristic examples of signals characterized by infinite total energy and finite average power. It is not difficult to find that the total energy of such a signal calculated for a time period is

$$E_P = \int_0^T |x(t)|^2 dt = \int_0^T |e^{j\omega_0 t}|^2 dt = \int_0^T dt = T$$

(since it is known that $|e^{j\omega_0 t}| = 1$), whereas the average power is equal to $P_T = E_T/T = 1$. However, since there are an infinite number of periods, as the variable t moves from the value $-\infty$ to the value $+\infty$, the total energy of the signal integrated over the infinite range of values $(-\infty, +\infty)$ will have an *infinite value*. On the other hand, each of all these periods will look exactly the same as the rest—since the signal is periodic—and therefore the complex exponential signal will have finite average power with a value equal to

$$P_\infty = \lim_{T\to\infty} \frac{1}{2T} \int_{-T}^{+T} |e^{j\omega_0 t}|^2 dt = \lim_{T\to\infty} \frac{1}{2T} \int_{-T}^{+T} dt = \lim_{T\to\infty} \frac{2T}{2T} = 1.$$

Complex Exponential Signals

In the general case where both the amplitude of the signal C and the exponent parameter α are complex numbers, the resulting signal can be described with the help of the real and imaginary signals that we studied in the previous sections. Indeed, if we express the width of the signal in polar form $C = |C|e^{j\vartheta}$ (where $|C|$ is the magnitude of C and ϑ is its phase) and the parameter α in Cartesian form $\alpha = r + j\omega_0$ (where r is the real and ω_0 is the imaginary part) the exponential signal $x(t)$ is formulated as

$$x(t) = Ce^{\alpha t} = |C|e^{j\vartheta}e^{r+j\omega_0 t} = |C|e^{j\vartheta}e^{rt}e^{j\omega_0 t} = (|C|e^{rt})e^{j(\omega_0 t+\vartheta)} = (|C|e^{rt})\Big[\cos(\omega_0 t + \vartheta) + j\sin(\omega_0 t + \vartheta)\Big], \quad (1.8)$$

where for the derivation of the last relation we used the Euler formula.

In the case where $r = 0$ we will have $e^{rt} = 1$ and the signal will take the form

$$x(t) = |C|\cos(\omega_0 t + \vartheta) + j|C|\sin(\omega_0 t + \vartheta)$$

from which it follows that both the real and the imaginary parts of the signal are *sinusoidal* signals (unless, of course, $\omega_0 = 0$—essentially $\alpha = 0$—then we will have $x(t) = |C|\cos\theta + j|C|\sin\theta = |C|e^{j\theta} = C$ and the signal degenerates into a constant complex value signal). On the other hand, it is not difficult to see that for values $r > 0$, the real and the imaginary parts of the signal are sinusoidal signals multiplied by an increasing exponential signal and therefore their amplitude varies as a function of time, whereas for values $r < 0$ these sinusoidal signals are multiplied by a decreasing exponential factor, and therefore their width decreases as a function of time. These two types of temporal variation are presented in Fig. 1.9. In this figure, the dashed curves represent the graphical representations of the functions $\pm|C|e^{rt}$ and express the *envelopes* of the widths of the two signals, as they evolve over time.

1.5.2 Sinusoidal Signals

The elementary signals of the sine and the cosine are connected with the imaginary exponential signals that we presented in a previous section through the known type of Euler

$$Ae^{j\omega_0 t} = A\cos(\omega_0 t) + jA\sin(\omega_0 t).$$

If we combine this equation with the also known relation

$$Ae^{-j\omega_0 t} = A\cos(\omega_0 t) - jA\sin(\omega_0 t)$$

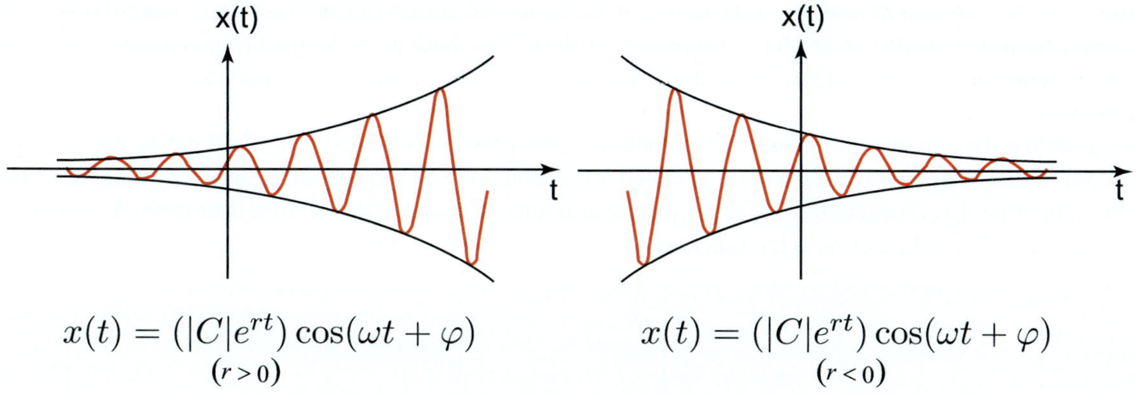

Fig. 1.9 Typical examples of complex exponential signals whose widths increase and decrease as a function of time

these two periodic signals arise in the following way:

- By adding the two relations term by term, we will get

$$A(e^{j\omega_0 t} + e^{-j\omega_0 t}) = 2A\cos(\omega_0 t) \quad \text{that is} \quad A\cos(\omega_0 t) = \frac{A}{2}\left(e^{j\omega_0 t} + e^{-j\omega_0 t}\right).$$

- By subtracting the two relations term by term, we will get

$$A(e^{j\omega_0 t} - e^{-j\omega_0 t}) = 2Aj\sin(\omega_0 t) \quad \text{that is} \quad A\sin(\omega_0 t) = \frac{A}{2j}\left(e^{j\omega_0 t} - e^{-j\omega_0 t}\right).$$

Alternatively, we can add to the above equations and some phase φ and express the trigonometric functions in the most general form

$$A\cos(\omega_0 t + \varphi) = \frac{A}{2}\left[e^{j(\omega_0 t+\varphi)} + e^{-j(\omega_0 t+\varphi)}\right] = \frac{A}{2}e^{j\varphi}e^{j\omega_0 t} + \frac{A}{2}e^{-j\varphi}e^{-j\omega_0 t}$$

$$A\sin(\omega_0 t + \varphi) = \frac{A}{2j}\left[e^{j(\omega_0 t-\varphi)} + e^{-j(\omega_0 t+\varphi)}\right] = \frac{A}{2j}e^{i\varphi}e^{j\omega_0 t} - \frac{A}{2j}e^{-j\varphi}e^{-j\omega_0 t}$$

whereas, of course, we can also use the expressions

$$A\cos(\omega_0 t + \varphi) = \Re\left\{Ae^{j(\omega_0 t+\varphi)}\right\} \quad \text{and} \quad A\sin(\omega_0 t + \varphi) = \Im\left\{Ae^{j(\omega_0 t+\varphi)}\right\},$$

where $\Re\{z\}$ and $\Im\{z\}$ are the real and imaginary parts of the complex quantity z.

In the above equations, the phase ϕ constitutes *a measure of the relative position in time* for a period of time: if the sine is characterized by some phase ϕ, then its value at time $t = 0$ is not equal to zero but equal to $A\cos(\phi)$, whereas the situation is similar with the signal of cosine. As we proved in the section on periodic signals, these two signals are periodic with a period value of $T_0 = 2\pi/\omega_0$ with the cyclic frequency ω_0, which is known as fundamental frequency, and can alternatively be written as $\omega_0 = 2\pi f_0$. In the last relation, the magnitude f_0 expresses the number of repetitions per unit of time and is related to the period of the signal with the known relation $f_0 = 1/T_0$.

1.6 Generalized Functions

Apart from the above elementary exponential and trigonometric signals, there are many other such signals that are basic tools for describing any continuous-time signal, such as the step function $u(t)$ and the impulse function (or Dirac's δ function) $\delta(t)$. Functions of this type do not exist as functions in the usual sense of the term—something that is particularly true for the function $\delta(t)$—but are generalizations of the classical functions and are known as *generalized functions* or *distributions*. Because even a basic knowledge of the basic properties of these distributions is deemed necessary for understanding the behavior of the functions $u(t)$ and $\delta(t)$, in the following pages we present a summary of the basic theory that characterizes these distributions.

In order to define the concept of a generalized function or distribution, we must first of all define the concept of a *test function*. A real function of a real variable $\varphi(t)$ is called a test function when it admits an arbitrary number of derivatives, while at the same time it is characterized by a zero value at points that are outside a finite and defined range of values. A typical example of such a test function is the function

$$\varphi(t,\alpha) = \begin{cases} \exp\left(-\dfrac{\alpha^2}{\alpha^2 - t^2}\right) & |t| < \alpha \\ 0 & |t| > \alpha \end{cases} . \tag{1.9}$$

In this function the symbol t is the independent variable, whereas α is a parameter of the function.

Considering a test function $\varphi(t)$, which is defined in the above way, it is characterized by the following properties:

- If any arbitrary real function of a real variable $f(t)$ admits an arbitrary number of derivatives, then as a test function we can use the function $\psi(t) = f(t)\varphi(t)$.
- If additionally the function $f(t)$ has a zero value at points that are outside a finite range of values, then

$$\psi(t) = \int_{-\infty}^{+\infty} f(\tau)\varphi(t-\tau)\,d\tau, \qquad -\infty < t < +\infty$$

  can also be used as a test function.
- A sequence of test functions of the form $\{\varphi_n\}$ $(1 \leq n < \infty)$ tends to zero, if all the functions φ_n are identically zero outside of a certain range of values that is independent of the value of n, whereas additionally, each function φ_n, as well as all its derivatives, tend uniformly to the zero value.
- All the test functions φ_n are members of a linear vector space D. Therefore, if we consider any two functions $\varphi_1, \varphi_2 \in D$, then the functions $\varphi_2 + \varphi_2$ and $\alpha\phi_1$ will also belong to D for any value of the parameter α.

Having defined the concept of the test function $\varphi(t)$ we can define the generalized function or distribution $g(t)$ as the *correlation* of an arbitrary test function $\varphi(t)$ with some number $N_g[\varphi(t)]$. Considering, for example, any one function $f(t)$, this will be a distribution if it satisfies a relation of the form

$$\int_{-\infty}^{+\infty} f(t)\varphi(t)\,dt = N_f[\varphi(t)], \tag{1.10}$$

where $N_f[\varphi(t)]$ is some number, for each test function $\varphi(t)$ that belongs to the vector space D.

Generalized functions are characterized by a set of very useful properties, the defining equations of which are listed below.

- Linearity—Homogeneity

$$\int_{-\infty}^{+\infty} g(t)\left[\alpha_1\varphi_1(t) + \alpha_2\varphi_2(t)\right] dt = \alpha_1 \int_{-\infty}^{+\infty} g(t)\varphi_1(t)\,dt + \alpha_2 \int_{-\infty}^{+\infty} g(t)\varphi_2(t)\,dt$$

- Sum

$$\int_{-\infty}^{+\infty} \left[g_1(t) + g_2(t)\right]\varphi(t)\,dt = \int_{-\infty}^{+\infty} g_1(t)\varphi(t)\,dt + \int_{-\infty}^{+\infty} g_2(t)\varphi(t)\,dt$$

- Shift with respect to the variable t

$$\int_{-\infty}^{+\infty} g(t-t_0)\varphi(t)\,dt = \int_{-\infty}^{+\infty} g(t)\varphi(t+t_0)\,dt$$

- Scaling (multiplication of the argument by α)

$$\int_{-\infty}^{+\infty} g(\alpha t)\varphi(t)\,dt = \frac{1}{|\alpha|}\int_{-\infty}^{+\infty} g(t)\varphi\left(\frac{t}{\alpha}\right)dt$$

- Symmetry (provided that the function $\varphi(t)$ is an even or odd function)

$$\int_{-\infty}^{+\infty} g(t)\varphi(t)\,dt = 0$$

- Differentiation of a generalized function. It is proven that

$$\int_{-\infty}^{+\infty} \frac{dg(t)}{dt}\varphi(t)\,dt = -\int_{-\infty}^{+\infty} g(t)\frac{d\varphi(t)}{dt}\,dt$$

as far as the first derivative is concerned, whereas for the general case of the derivative of order n

$$\int_{-\infty}^{+\infty} \frac{d^n g(t)}{dt^n}\,\varphi(t)\,dt = (-1)^n\int_{-\infty}^{+\infty} g(t)\frac{d^n\varphi(t)}{dt^n}\,dt$$

- Product of a generalized function with a regular function

$$\int_{-\infty}^{+\infty}\Big[g(t)f(t)\Big]\varphi(t)dt = \int_{-\infty}^{+\infty} g(t)\Big[f(t)\varphi(t)\Big]dt$$

The above relation holds under the condition that the function $f(t)\varphi(t)$ is a member of the set of test functions.

- Convolution (an operation that will be studied in a subsequent section)

$$\int_{-\infty}^{+\infty}\left[\int_{-\infty}^{+\infty} g_1(\tau)g_2(t-\tau)d\tau\right]\varphi(t)\,dt = \int_{-\infty}^{+\infty} g_1(\tau)\left[\int_{-\infty}^{+\infty} g_2(t-\tau)\varphi(t)\right]d\tau$$

- Convergence of a sequence of generalized functions: a sequence of generalized functions $\{g_n(t)\}$ $(1 \le 1 < \infty)$ is aforementioned to converge to a distribution $g(t)$, if the relation holds

$$\lim_{n\to\infty}\int_{-\infty}^{+\infty} g_n(t)\varphi(t)\,dt = \int_{-\infty}^{+\infty} g(t)\varphi(t)\,dt$$

for all functions $\varphi(t)$ that belong to the set of test functions.

The proof of these properties can be found in the literature and is left as an exercise to the reader.

1.6.1 The Unit Step Function

The *unit step function* or *step function* $u(t)$ is defined by a mathematical relation of the form

$$u(t) = \begin{cases} 1 \text{ for } t > 0 \\ 0 \text{ for } t < 0 \end{cases} \tag{1.11}$$

and its graphical representation is shown in Fig. 1.10a. From the above equation, we observe that at the position $t = 0$, the step function presents a discontinuity and its value is not defined. In any case, we can use a variation of the step function known as the *Heaviside function* and is defined as[11]

$$H(t) = \begin{cases} 1 & \text{for } t > 0 \\ 0.5 & \text{for } t = 0 \\ 0 & \text{for } t < 0, \end{cases} \tag{1.12}$$

thus allowing the determination of the function's behavior at the position $t = 0$. However, it should be noted that in many books the value of the function $u(t)$ at the time $t = 0$ is considered equal to 1, an approach that we will adopt in several of the exercises of this chapter.

The unit step function is a very useful tool because it allows the definition of the so-called *causal signals*. A signal is called causal when it is defined only for times $t > 0$. This, of course, is expected, as the signals studied and used in practice are created by natural systems and arrangements and obviously in later time instances from the moment of initiation of the operation of the arrangement, which is considered to be the time instance $t = 0$. If a signal is not causal, it is called a non*causal signal*. Noncausal signals are often used as idealized mathematical models for the study of causal signals.

In order to demonstrate the concept of causal signals, let us consider the signal $x(t) = e^{-\alpha t}$ defined in the infinite range of values $-\infty < t < +\infty$. In this case, the causal version of the signal is $y(t) = e^{-\alpha t}u(t)$ with the signal $y(t)$ having nonzero values in the range of values $0 < t < \infty$. Generally speaking, the causal version $y(t)$ of a noncausal signal $x(t)$ can be defined with the help of the step function from an equation of the form

$$y(t) = x(t)u(t) = \begin{cases} x(t) & \text{for } t > 0 \\ 0 & \text{for } t < 0 \end{cases} \tag{1.13}$$

The two most important properties that characterize the unit step function $u(t)$ are the act of scaling defined as

$$Au(t) = \begin{cases} A & \text{for } t > 0 \\ 0 & \text{for } t < 0 \end{cases}$$

and the act of time shifting by t_0 described by an equation of the form

$$u(t - t_0) = \begin{cases} 1 & \text{for } t > t_0 \\ 0 & \text{for } t < t_0. \end{cases}$$

If we proceed with the change of variable $t \to -t$, we immediately find that

$$u(-t) = \begin{cases} 1 & \text{for } t < 0 \\ 0 & \text{for } t > 0 \end{cases}$$

and by adding the functions $u(t)$ and $u(-t)$, we find that

---

[11] The Heaviside function can also be described with the help of the function

$$\upsilon(t, \xi) = \begin{cases} 1 & \text{for } t > \xi \\ 0.5[(t/\xi) + 1] & \text{for } -\xi \le t \le \xi \\ 0 & \text{for } t < -\xi \end{cases}$$

as $H(t) = \lim\limits_{\xi \to 0} \upsilon(t, \xi)$, or from the function $\upsilon(t, n) = (1 + e^{-nt})^{-1}$ $H(t) = \lim\limits_{n \to \infty} \upsilon(t, n)$.

$$u(t) - u(-t) = \begin{cases} +1 \text{ for } t > 0 \\ 0 \text{ for } t = 0 \\ -1 \text{ for } t < 0 \end{cases} = \text{sgn(t)}, \tag{1.14}$$

where sgn(t) is the *sign function*, which takes a numerical value as an argument and returns one of the values +1, −1, and 0, depending on whether this value is positive, negative, or zero respectively. ■

1.6.2 The Unit Ramp Function

The *unit ramp function* or *climbing function* is defined by the equation

$$r(t) = tu(t) = \begin{cases} t \text{ for } t > 0 \\ 0 \text{ for } t < 0 \end{cases} \tag{1.15}$$

and is the integral of the *unit step function*. Indeed, differentiating the above relation, we will have

$$\frac{dr(t)}{dt} = \frac{d}{dt}[tu(t)] = u(t) = \begin{cases} 1 \text{ for } t > 0 \\ 0 \text{ for } t < 0, \end{cases}$$

that is, the functions $r(t)$ and $u(t)$ are connected by a relation of the form

$$r(t) = \int_{-\infty}^{t} u(\tau)\, d\tau. \tag{1.16}$$

The graphical representation of the ramp function is shown in Fig. 1.10b and is also characterized by the properties of scaling and time shifting by t_0, which are given by the relations

$$kr(t) = \begin{cases} kt \text{ for } t > 0 \\ 0 \ \text{ for } t < 0 \end{cases} \quad \text{and} \quad r(t - t_0) = \begin{cases} t \text{ for } t > t_0 \\ 0 \text{ for } t < t_0. \end{cases}$$

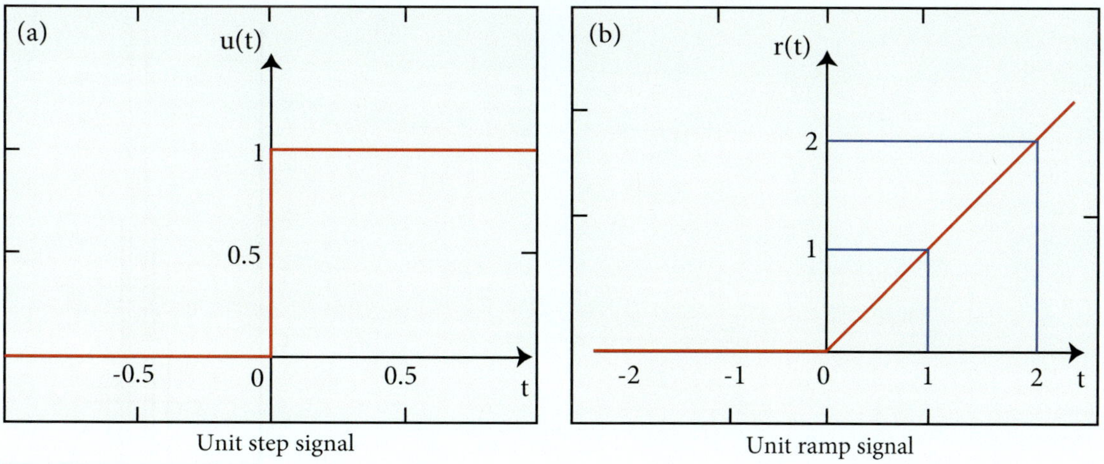

Fig. 1.10 The unit step function and the unit ramp function

1.6.3 The Unit Impulse Function

The *unit impulse function* or *Dirac's function* $\delta(t)$ was invented by the English physicist Paul A.M. Dirac in 1930 as a mathematical tool for his work on quantum mechanics and is defined as

$$\delta(t) = \begin{cases} \infty \text{ for } t = 0 \\ 0 \quad \text{for } t \neq 0. \end{cases} \tag{1.17}$$

The behavior of this type can be simulated by a *narrow square pulse of great height*, such that the area enclosed by it is *equal to one*. The width of this pulse is equal to ε, whereas its height is equal to $1/\varepsilon$, so that in the limit $\varepsilon \to 0$ the width of the pulse tends to zero, but its height tends to infinity, with the area, however, enclosed by it being in every case constant and equal to one. Therefore, the function $\delta(t)$ is equal to zero for $t \neq 0$ and tends to infinity at the position $t = 0$. This situation is presented in Fig. 1.11. In this figure, image (a) depicts the symbolism used for the function $\delta(t)$, image (b) shows the pulse that describes it, whereas image (c) shows the way in which this pulse changes as the range ε tends to zero. The above description in mathematical language is formulated as

$$\delta(t) = \lim_{\varepsilon \to 0} \left[\frac{1}{\varepsilon} \Pi \left(\frac{t}{\varepsilon} \right) \right], \tag{1.18}$$

where the square pulse $\Pi(t)$ is defined by the relation

$$\Pi(t) = \begin{cases} 1 \ |t| > 1/2 \\ 0 \text{ anywhere else} \end{cases}$$

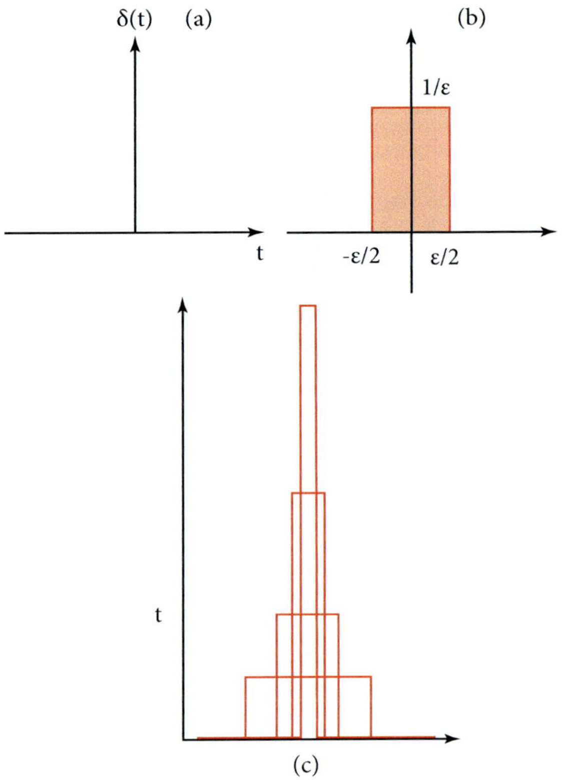

Fig. 1.11 Representation of a unit impulse function by a square pulse

Note that although a *square pulse* was used in the above description, the function δ can be described using a pulse of *any shape*—e.g., *triangular pulse*—since the important characteristic of this function is not its shape but the fact that as the range of the curve that describes it tends to zero, the area enclosed by it remains constantly stable and equal to one. We also mention that the above definition is not strict from the formal mathematical perspective and it is a fact that it has led to many difficulties. It is obvious that there is no usual function characterized by such properties and in reality, the function $\delta(t)$ *is not a function with the usual sense of the term*. Indeed, although a typical function is fully determined by its values at every time instance t, the function $\delta(t)$ has zero value for every $t \neq 0$, whereas at the position $t = 0$ (which is the only one of interest) the function is undefined. However, all these difficulties are lifted if we treat the function $\delta(t)$ not as a usual function but as an *operator acting on a function*, or alternatively, as a *generalized function* or *distribution*. Recalling that a distribution is a special type of function that acts on other functions, we can describe the function $\delta(t)$, without specifying its values at every time instance—an approach that leads to the previous dead ends—but describing the result of its effect on the usual typical functions.

Adopting this mathematical way of thinking, we can consider the function $\delta(t)$ as a *generalized function* or *distribution*, which associates each function $\varphi(t)$, smooth at the position $t = 0$, with its value at this time instance—which obviously will be equal to $\varphi(0)$—or in mathematical writing

$$\int_{-\infty}^{+\infty} \delta(t)\, \varphi(t)\, dt = \varphi(0) \tag{1.19}$$

The above definition can be generalized to describe the *time-shifted function* $\delta(t-t_0)$, which is defined through the relation[12]

$$\int_{-\infty}^{+\infty} \delta(t-t_0)\, \varphi(t)\, dt = \varphi(t_0). \tag{1.20}$$

In the last equation—which leads to our original definition of the function $\delta(t)$ for the value $t_0 = 0$—the generalized function $\delta(t-t_0)$ correlates the function $\varphi(t)$ with its value at the time t_0, which of course is equal to $\varphi(t_0)$. If we consider the function $\varphi(t) = 1$ and set $t_0 = 0$, the above relation takes the form

$$\int_{-\infty}^{+\infty} \delta(t)\, dt = 1. \tag{1.21}$$

But the above integral calculated in the range of values $(-\infty, +\infty)$ is nothing but the *area* enclosed by the curve of the function $\delta(t)$, the value of which is equal to one, in full agreement with what we mentioned earlier.

This definition of the function $\delta(t)$ is particularly useful, as among other things it allows the determination of the derivative of the unit step function at the position $t = 0$, which under normal conditions cannot be calculated (since the function $u(t)$ at this position presents discontinuity). However, under the prism of generalized functions, this derivative exists and is indeed equal to the unit impulse function $\delta(t)$.

Indeed, considering any function $\varphi(t)$ that meets the properties we presented in the section on generalized functions and using the derivative rule for these functions, where we use the function $u(t)$ as a distribution, we will have

$$\int_{-\infty}^{+\infty} \frac{du(t)}{dt}\, \varphi(t)\, dt = u(t)\, \varphi(t)\Big|_{-\infty}^{+\infty} - \int_{-\infty}^{+\infty} u(t) \frac{d\varphi(t)}{dt} dt = \varphi(\infty) - 0 - \int_0^{\infty} \frac{d\varphi(t)}{dt} dt = \varphi(\infty) - \varphi(t)\Big|_0^{\infty} = \varphi(0)$$

[12] The above equation defining the function $\delta(t)$ in three dimensions is generalized as

$$\iiint_R \varphi(\boldsymbol{s})\delta(\boldsymbol{s}-\boldsymbol{r})d\boldsymbol{s} = \varphi(\boldsymbol{r}),$$

where $\varphi(\boldsymbol{r}) = \varphi(x, y, z)$, $d\boldsymbol{r} = dxdydz$, whereas the integration space R contains the point $\boldsymbol{s} = \boldsymbol{r}$.

If we compare the relations

$$\int_{-\infty}^{+\infty} \frac{du(t)}{dt}\, \varphi(t)\, dt = \int_{-\infty}^{+\infty} u'(t)\, \varphi(t)\, dt = \varphi(0) \quad \text{and} \quad \int_{-\infty}^{+\infty} \delta(t)\, \varphi(t)\, dt = \varphi(0)$$

it immediately follows that $u'(t) = \delta(t)$ and by performing the integration we get

$$u(t) = \int_{-\infty}^{t} \delta(\tau)\, d\tau.$$

Therefore, *the unit impulse function is the derivative of the unit step function* $u(t)$*, which in turn is the integral of the impulse function* $\delta(t)$. This is the way in which these two functions are related to each other.
The impulse function $\delta(t)$ is characterized by several interesting properties, some of which are proven in the solved exercises that follow.

Problem 1.6.1 Prove that $x(t)\, \delta(t - t_0) = x(t_0)\, \delta(t - t_0)$ for every function $x(t)$ continuous at the position $t = t_0$.

Solution Let us consider a function $f(t) = x(t)\varphi(t)$ with the functions $x(t)$ and $\varphi(t)$ being continuous at the position $t = t_0$—as is $f(t)$ itself. Replacing this function in the equation defining the shifted function $\delta(t)$

$$\int_{-\infty}^{+\infty} f(t)\, \delta(t - t_0)\, dt = f(t_0) \quad \text{we get} \quad \int_{-\infty}^{+\infty} \Big[x(t)\, \varphi(t)\Big]\delta(t - t_0)dt = x(t_0)\, \varphi(t_0).$$

However, it is

$$\varphi(t_0) = \int_{-\infty}^{+\infty} \delta(t - t_0)\, \varphi(t_0)\, dt$$

and the above relation is written

$$\int_{-\infty}^{+\infty} \Big[x(t)\, \varphi(t)\Big]\, \delta(t - t_0)\, dt = \int_{-\infty}^{+\infty} \varphi(t)\Big[x(t)\, \delta(t - t_0)\Big]dt = x(t_0) \int_{-\infty}^{+\infty} \delta(t - t_0)\, \varphi(t_0)\, dt = \int_{-\infty}^{+\infty} \varphi(t)\Big[x(t_0)\delta(t - t_0)\Big]dt$$

Therefore, we have

$$\int_{-\infty}^{+\infty} \varphi(t)\, \Big[x(t)\, \delta(t - t_0)\Big]\, dt = \int_{-\infty}^{+\infty} \varphi(t)\, \Big[x(t_0)\, \delta(t - t_0)\Big]\, dt,$$

from which it follows that[13] $x(t)\, \delta(t - t_0) = x(t_0)\, \delta(t - t_0)$.

Note 1 if in the above relation we set $t_0 = 0$ it immediately follows that

$$x(t)\, \delta(t) = x(0)\, \delta(t) \tag{1.22}$$

equation, which constitutes another interesting property of the function $\delta(t)$.

Note 2 in the special case of the function $x(t) = t$ it is obvious that $x(0) = 0$ and the last relation gives $t\delta(t) = 0$. A generalization of the above property is the

[13] This conclusion, as well as all those based on the comparison of two integrals and the equation of the corresponding quantities, is a consequence of the *property of equivalence of distributions* according to which two generalized functions $g_1(t)$ and $g_2(t)$ are equal to each other if and only if for every suitably defined test function $\varphi(t)$ the relation holds

$$\int_{-\infty}^{+\infty} \varphi(t) g_1(t) dt = \int_{-\infty}^{+\infty} \varphi(t) g_2(t) dt.$$

$$t^n \frac{d^m \delta(t)}{dt^m} = \begin{cases} (-1)^n n! \delta(t) & m = n \\ (-1)^n \dfrac{m!}{(m-n)!} \dfrac{d^{(m-n)} \delta(t)}{dt^{m-n}} & m > n \\ 0 & m < n \end{cases}$$

Problem 1.6.2 Prove the property of the impulse function $\delta(\alpha t) = \frac{1}{|\alpha|}\delta(t)$.

Solution Proceeding with the change of variables $\xi = \alpha t$, by differentiating we get $d\xi = \alpha dt$, so $t = \xi/\alpha$ and $dt = d\xi/\alpha$. We distinguish two cases:

- $\alpha > 0$: in this case we will have

$$\int_{-\infty}^{+\infty} \varphi(t)\,\delta(\alpha t)\,dt = \frac{1}{\alpha}\int_{-\infty}^{+\infty} \varphi\left(\frac{\xi}{\alpha}\right)\delta(\xi)\,d\xi = \frac{1}{\alpha}\phi\left(\frac{\xi}{\alpha}\right)\bigg|_{\xi=0} = \frac{\varphi(0)}{\alpha} = \frac{\varphi(0)}{|\alpha|}.$$

- $\alpha < 0$: in this case, the limits of the integral are reversed; therefore, we have

$$\int_{-\infty}^{+\infty} \varphi(t)\,\delta(\alpha t)\,dt = \frac{1}{\alpha}\int_{+\infty}^{-\infty} \varphi\left(\frac{\xi}{\alpha}\right)\delta(\xi)d\,\xi = -\frac{1}{\alpha}\int_{-\infty}^{+\infty} \varphi\left(\frac{\xi}{\alpha}\right)\delta(\xi)\,d\xi = -\frac{1}{\alpha}\phi\left(\frac{\xi}{\alpha}\right)\bigg|_{\xi=0} = -\frac{\varphi(0)}{\alpha} = \frac{\varphi(0)}{|\alpha|}.$$

Combining the above two results and using the equation

$$\varphi(0) = \int_{-\infty}^{+\infty} \varphi(t)\delta(t)dt \qquad \text{we get}$$

$$\int_{-\infty}^{+\infty} \varphi(t)\,\delta(\alpha t)\,dt = \frac{1}{|\alpha|}\int_{-\infty}^{+\infty} \varphi(t)\,\delta(t)\,dt = \int_{-\infty}^{+\infty} \varphi(t)\left[\frac{1}{|\alpha|}\delta(t)\right]dt,$$

from which, with the help of the equivalence property, we arrive at the desired relation.

Note If we set $\alpha = -1$ in the above relation, we will have $\delta(-t) = \delta(t)$, that is, the function $\delta(t)$ is characterized by even symmetry. Proceeding with the change of variable $t \to t - t_0$, this relation generalizes as $\delta(t - t_0) = \delta(t_0 - t)$.

Problem 1.6.3 It is proven that if $f(t)$ is a generalized function, then the derivative of order n, which is $f^{(n)}(t) = d^n f(t)/dt^n$, is defined by the equation

$$\int_{-\infty}^{+\infty} \varphi(t) f^{(n)}(t)dt = (-1)^n \int_{-\infty}^{+\infty} \varphi^{(n)}(t) f(t)dt = (-1)^n \frac{d^{(n)} f(t)}{dt^n}\bigg|_{t=0}.$$

Based on the above equation, prove that

$$\text{(a)} \quad \int_{-\infty}^{+\infty} \varphi(t)\frac{d\delta(t)}{dt}dt = -\frac{d\varphi(t)}{dt}\bigg|_{t=0} \quad \text{and} \quad \text{(b)} \quad \frac{d\delta(t)}{dt} = -\frac{1}{t}\delta(t).$$

Solution (a) If in the equation of the derivative of order n of the generalized function $f(t) = \delta(t)$ we use the value $n = 1$ it immediately follows that

$$\int_{-\infty}^{+\infty} \varphi(t) \frac{d\delta(t)}{dt} dt = -\int_{-\infty}^{+\infty} \frac{d\varphi(t)}{dt} \delta(t) dt = -\left. \frac{d\varphi(t)}{dt} \right|_{t=0},$$

where to derive the result we used the defining equation of the function $\delta(t)$. ■.

Note Using the defining equation of the shifted function $\delta(t - t_0)$ it is proven that

$$\int_{-\infty}^{+\infty} \varphi(t) \frac{d\delta(t - t_0)}{dt} dt = -\left. \frac{d\varphi(t)}{dt} \right|_{t=t_0}.$$

(b) Using the equation we proved previously for the function $\varphi(t) = tf(t)$ we have

$$\int_{-\infty}^{+\infty} \left[tf(t) \right] \frac{d\delta(t)}{dt} dt = \int_{-\infty}^{+\infty} f(t) \left[t \frac{d\delta(t)}{dt} \right] dt = -\frac{d}{dt} \left[tf(t) \right]_{t=0} = -\left[f(t) + t \frac{df(t)}{dt} \right]_{t=0} = -f(0).$$

However, from the definition of the function $\delta(t)$, we have

$$f(0) = \int_{-\infty}^{+\infty} f(t)\delta(t)dt$$

so we will take

$$\int_{-\infty}^{+\infty} f(t) \left[t \frac{d\delta(t)}{dt} \right] dt = -\int_{-\infty}^{+\infty} f(t)\delta(t)dt = \int_{-\infty}^{+\infty} f(t) \left[-\delta(t) \right] dt$$

and applying the property of equivalence we will have $t\delta'(t) = -\delta(t)$ from which the desired result follows.

Problem 1.6.4 An alternative way of defining the function $\delta(t)$ in which the limits of the integral that define it are finite has the form

$$\int_{\alpha}^{\beta} x(t)\delta(t)dt = \begin{cases} x(0) & \text{for } \alpha < 0 < \beta \\ 0 & \text{for } \alpha < \beta < 0 \text{ or } 0 < \alpha < \beta \\ \text{undefined for } \alpha = 0 & \text{or} \quad \beta = 0 \end{cases}$$

Based on the above equation, calculate the integrals

$$I_1 = \int_{-1}^{+1} (3t^2 + 1)\delta(t)dt \quad \text{and} \quad I_2 = \int_{1}^{2} (3t^2 + 1)\delta(t)dt.$$

Solution

(a) The integral I_1 falls into the first of the three cases of the above equation, with the function $x(t)$ defined as $x(t) = 3t^2 + 1$ whereas the limits of the integral have the values $\alpha = -1$ and $\beta = +1$. Therefore, we have

$$I_1 = \int_{-1}^{+1} (3t^2 + 1)\delta(t)dt = (3t^2 + 1)\Big|_{t=0} = 1$$

(b) The integral I_2 falls into the second of the three cases of the above equation with the function $x(t)$ being the same one we used previously, whereas the integration limits have the values $\alpha = 1$ and $\beta = 2$. Therefore, the value of this integral is equal to zero.

Problem 1.6.5 Using the defining equation of the time-shifted impulse function, calculate the values of the integrals

$$I_1 = \int_{-\infty}^{+\infty} \left[t^2 + \cos(\pi t) \right] \delta(t-1)dt \quad \text{and} \quad I_2 = \int_{-\infty}^{+\infty} e^{-t}\delta(2t-2)dt$$

Solution From the defining equation of the time-shifted impulse function $\delta(t - t_0)$

$$\int_{-\infty}^{+\infty} x(t)\delta(t-t_0)dt = x(t_0)$$

we will have: (a) For the integral I_1 is $x(t) = t^2 + \cos(\pi t)$ and $t_0 = 1$ and therefore we have

$$I_1 = \int_{-\infty}^{+\infty} \left[t^2 + \cos(\pi t) \right] \delta(t-1)dt = \left[t^2 + \cos(\pi t) \right]_{t=1} = 1 + \cos(\pi) = 1 - 1 = 0.$$

(b) The integral I_2 is transformed as

$$I_2 = \int_{-\infty}^{+\infty} e^{-t}\delta(2t-2)dt = \int_{-\infty}^{+\infty} e^{-t}\delta[2(t-1)]dt = \frac{1}{2}\int_{-\infty}^{+\infty} e^{-t}\delta(t-1)dt,$$

where to derive the last expression we used the scaling property $\delta(\alpha t) = \delta(t)/|\alpha|$ for the value $\alpha = 2$. Then, using the equation of the time-shifted function $\delta(t - t_0)$ with $x(t) = e^{-t}$ and $t_0 = 1$ we get

$$I_2 = \frac{1}{2}\int_{-\infty}^{+\infty} e^{-t}\delta(t-1)dt = \left. \frac{e^{-t}}{2} \right|_{t=1} = \frac{1}{2e},$$

which is the desired result. ■

We conclude the presentation of the function $\delta(t)$ with the observation that this function can be represented in many alternative ways, some of which are

$$\delta(t) = \lim_{\varepsilon \to 0} \frac{e^{-t^2/\varepsilon}}{\sqrt{\varepsilon\pi}}, \qquad \delta(t) = \lim_{\omega \to \infty} \frac{\sin(\omega t)}{\pi t}, \qquad \delta(t) = \lim_{\varepsilon \to 0} \left(\frac{1}{\pi} \frac{\varepsilon}{t^2 + \varepsilon^2} \right), \qquad \delta(t) = \frac{1}{2\pi}\int_{-\infty}^{+\infty} \cos(\omega t)d\omega.$$

The proof of these relations is beyond the scope of this book and can be found in the literature. We simply mention that in all these cases the function $\delta(t)$ is considered the limit of sequences of functions of the form $\delta(t) = \lim_{n\to\infty} D_n(t)$ with the functions $D_n(t)$ having the appropriate mathematical form in each case.

1.7 Other Continuous-Time Signal Representation Functions

In addition to the basic exponential and trigonometric functions, as well as the generalized functions we presented in the previous sections, there are some other useful mathematical functions that are used for the representation and description of continuous-time signals. Two significant functions that belong to this category are the *square pulse* $\Pi_\alpha(t)$ and the *triangular pulse* $\Lambda_\alpha(t)$, which are defined by the relations

$$\Pi_\alpha(t) = \begin{cases} 1 \text{ for } |t| < \alpha \\ 0 \text{ for } |t| > \alpha \end{cases} \quad \text{and} \quad \Lambda_\alpha(t) = \begin{cases} 1 - \dfrac{|t|}{\alpha} & \text{for } |t| < \alpha \\ 0 & \text{for } |t| > \alpha \end{cases}$$

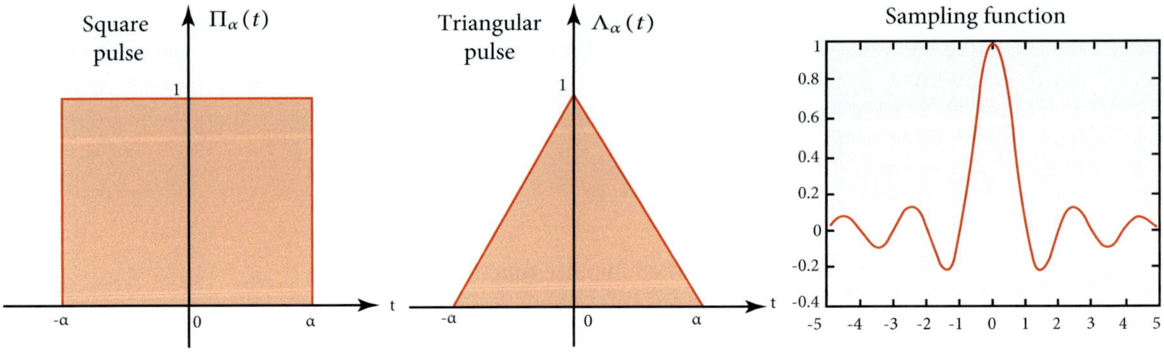

Fig. 1.12 The square and the triangular pulse and the sampling function

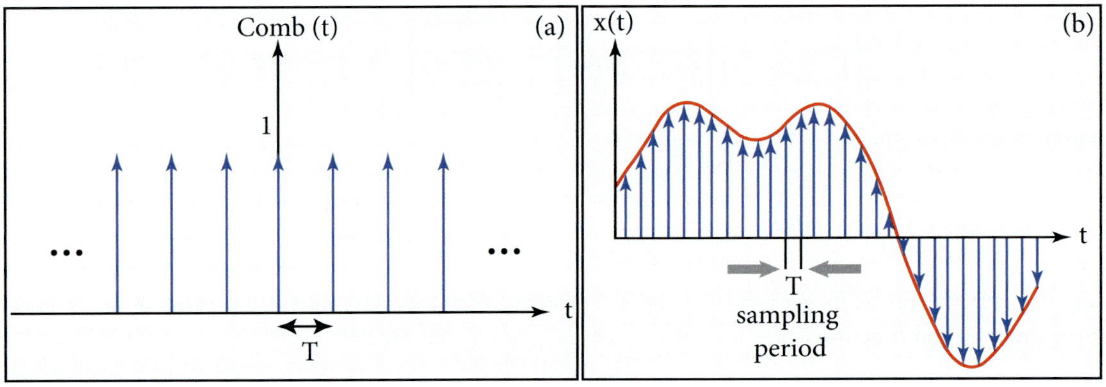

Fig. 1.13 The comb function and its use for the sampling of a continuous-time signal

respectively. These two functions are related to each other through the operation of *convolution*—this correlation will be presented in the homonymous section—and are diagrammatically represented in Fig. 1.12a and b respectively.

A third interesting function depicted in Fig. 1.12c is the sampling function

$$\operatorname{sinc}(t) = \begin{cases} \dfrac{\sin t}{t} & \text{for } t \neq 0 \\ 1 & \text{for } t = 0. \end{cases}$$

This function is very important in signal theory and systems, particularly in the *Fourier analysis* and will be used several times in the future.

Closely related to the sampling of continuous-time signals is the *comb function*, which is defined as a *sequence of temporally equidistant unit impulse responses* of the form

$$\operatorname{comb}(t) = \sum_{n=-\infty}^{+\infty} \delta(t - t_n) = \sum_{n=-\infty}^{+\infty} \delta(t - nT), \tag{1.23}$$

where $t_n = nT$ $(n = -\infty, \ldots, -1, 0, +1, \ldots, +\infty)$ a set of *discrete-time instants*, which should increase at a rate at least equal to $\sqrt{|n|}$. The comb function is graphically represented in Fig. 1.13a from which it appears that this function is periodic with period T. It is proven that the above series *converges* so that the function can be defined. Using this function, we can proceed to a *sampling* process of any continuous-time signal $x(t)$ with a sampling period T equal to the period of the comb function. Indeed, using the known properties of the function $\delta(t)$, it is not difficult to find that the signal $y(t)$ resulting from the sampling of the signal $x(t)$ with a sampling period T is given by the relation

$$y(t) = \sum_{n=-\infty}^{+\infty} x(nT)\delta(t - nT) = x(t) \sum_{n=-\infty}^{+\infty} \delta(x - nT).$$

Taking into account the definition of the temporally shifted unit in the context of the impulse function, we understand that the functions $\delta(t - nT)$ allow the recording of the signal values $x(t)$ at the time instances nT (where $n = -\infty, \ldots, -1, 0, +1, \ldots, +\infty$), thus leading to *its sampling*—this situation is presented in Fig. 1.13b. We will return to this topic later.

1.8 Elementary Transforms of Continuous-Time Signals

As already mentioned, a continuous-time signal is described by a real function of a real variable of the form $x = x(t)$ (although the use of complex signals is also very common). Using this mathematical representation we can describe certain transformations that may be applied to these signals and that in general are the following:

- *Time scaling*: this transformation leads to the creation of the signal $y(t) = x(\alpha t)$ and consists in multiplying the independent variable t by the value of a positive real constant α. The original signal $x(t)$ under the influence of this transformation slows down or equivalently extends over time, if it is $\alpha < 1$; on the contrary, if it is $\alpha > 1$, the signal $x(t)$ accelerates or equivalently compresses over time.
- *Time reversal*: this transformation leads to the creation of the signal $y(t) = x(-t)$, which is reversed in time compared with the original signal. This transformation is equivalent to the transformation of time scaling, if the value of the constant α is equal to -1.
- *Time shift*: this transformation consists in shifting the signal by a time interval of duration T, which can either be *time delay*, if $T < 0$, or *time advance*, if $T > 0$.

It is interesting to mention at this point that the above elementary transforms of the time variable t can be combined with each other, thus allowing the implementation of more complex processes. The most general of these processes is based on the use of all three of the above processes and leads to the creation of the signal $x(\alpha t - b)$, which can be implemented in two ways:

1 Time shift of the signal $x(t)$ by a time interval of duration b, so as to create the signal $x(t - b)$, which then undergoes the process of time scaling—which it is suggested to replace $t \to \alpha t$—so as to construct the signal $x(\alpha t - b)$.
2 Temporal scaling of the signal $x(t)$, so as to create the signal $x(\alpha t)$ and temporal displacement of the scaled signal by b/α, so as to create the signal $x(\alpha[t - (b/\alpha)]) = x(\alpha t - b)$.

Although the above processes are based only on the transformations of temporal scaling and displacement, if the value of α is negative, the temporal scaling includes the transformation of time reversal.

The use of the above transformations is presented diagrammatically in Fig. 1.14.

1.9 Representation with Elementary Functions

Using the elementary signals presented in the previous sections as the basic building blocks, we can express any continuous-time signal as a function of these elementary signals. This way of representation depends, of course, *on the type of function used* and details on how this is done are presented in the following pages. Considering that the representation of signals with the help of series and integrals of *trigonometric* and *exponential* functions falls under the theory of *Fourier analysis*, which is extensively covered in a subsequent chapter, the description that follows is limited only to the representation of signals with the help of the *unit step function* and the *unit impulse function*.

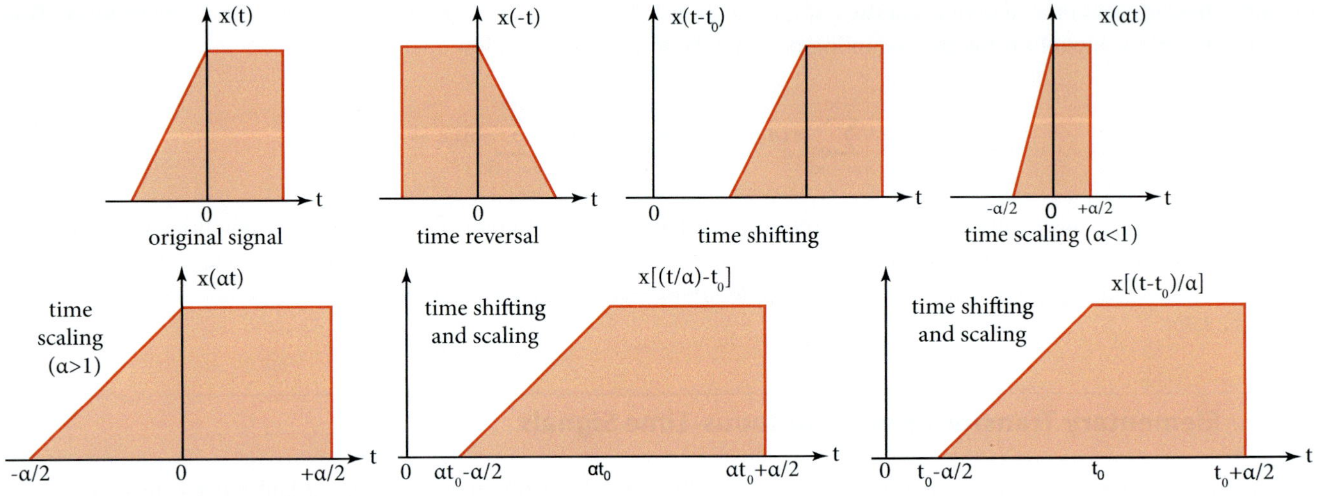

Fig. 1.14 Demonstration of the elementary transforms of continuous-time signals (adapted from Boulet, Sect. 1.3)

1.9.1 Representation with the Help of the Step Function $u(t)$

Depending on the form and characteristics of the continuous-time signal we consider, it can be written as a *sum of finite or infinite terms of time-shifted step functions* of the form $u(t - t_0)$ with the value of the constant t_0 depending on the value ranges in which the signal is defined. Thus, a relatively simple signal of the form

$$x(t) = \begin{cases} 1 \text{ for } 2 \leq t \leq 4 \\ 0 \text{ anywhere else} \end{cases}$$

is described by the simple sum $x(t) = u(t-2) - u(t-4)$, whereas a more complex signal, such as

$$x(t) = \begin{cases} \kappa \text{ for } t \leq 1 \\ \ell \text{ for } 1 < t \leq 3 \\ \mu \text{ for } 3 < t \leq 6 \\ \nu \text{ for } t > 6 \end{cases}$$

is described by a linear combination of time-shifted step functions of the form

$$x(t) = \kappa u(1-t) + \ell\Big[u(3-t) - u(1-t)\Big] + \mu\Big[u(6-t) - u(3-t)\Big] + \nu u(6-t).$$

In the general case of an arbitrary continuous-time signal $x(t)$, the form of which can be anything, it can be described by *dividing it into small segments corresponding to time intervals of duration T and approximating each of these segments with the help of the appropriate time-shifted step function* $u(t)$. This situation is diagrammatically presented in Fig. 1.15. It is not difficult to understand that the smaller the value of the time duration T, the smaller the approximation error of the real signal $x(t)$ from its approximate equivalent $\tilde{x}(t)$. In mathematical description, this way of representing the signal is described by the linear combination of an infinite number of difference terms *time-shifted step functions* with the coefficients of this linear combination constituting samples of the considered signal $x(t)$, so that we can write that

$$\tilde{x}(t) = \sum_{n=-\infty}^{+\infty} x(nT)\Big[u(t-nT) - u[t-(n+1)T]\Big]. \tag{1.24}$$

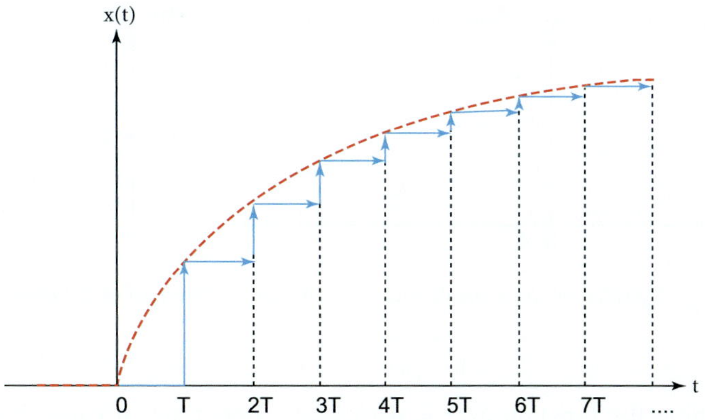

Fig. 1.15 Approximation of a continuous-time signal as the overlap of infinite time-shifted step functions (Shmaliy, Continuous-Time Signals, Fig. 1.15c)

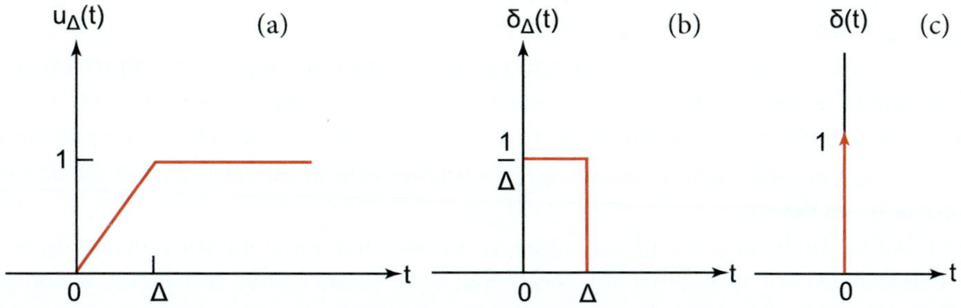

Fig. 1.16 (**a**) Approximation of the function $u(t)$ by the function $u_\Delta(t)$, (**b**) approximation of the function $\delta(t)$ by the function $\delta_\Delta(t)$, and (**c**) the actual function $\delta(t)$ and the way it is symbolized (Oppenheim, Willsky, Nawab, Figures 1.33 and 1.34)

1.9.2 Representation Using the Impulse Function $\delta(t)$

The description of the way in which we can represent any continuous-time signal with the help of the *impulse function* $\delta(t)$ requires us to deal with this function in a different way than the one we have defined. As already mentioned, the function $\delta(t)$ is the derivative of the step function $u(t)$, with the difficulty of defining this derivative, owing to the discontinuity that characterizes the step function at the position $t = 0$ having been removed, considering the impulse function as a *generalized function* or *distribution*. Alternatively, the relation $\delta(t) = u'(t)$ can be treated in a different way, using not the real step function, which falls *instantaneously* from the value 0 to the value 1, but an approximation of it—which we symbolize with $u_\Delta(t)$—in which this transition takes place over a time interval of duration Δ [this approximation is shown in Fig. 1.16a].[14] In this case, the derivative $\delta_\Delta(t) = u'_\Delta(t)$ will have the form presented in Fig. 1.16b[15] and is mathematically described by an equation of the form

$$\delta_\Delta(t) = \begin{cases} 1/\Delta & \text{for } 0 \le t < \Delta \\ 0 & \text{anywhere else.} \end{cases}$$

It is not difficult to understand that, following this approach, the *step function* $u(t)$ and the *impulse function* $\delta(t)$ will be approximated as the limits

$$u(t) = \lim_{\Delta \to 0} u_\Delta(t) \quad \text{and} \quad \delta(t) = \lim_{\Delta \to 0} \delta_\Delta(t)$$

and therefore they constitute *idealizations* of the functions $u_\Delta(t)$ and $\delta_\Delta(t)$ respectively. The function $\delta(t)$ is represented by our now familiar symbol, which is reproduced in Fig. 1.16c.

[14] The description and the diagrams used can be found in Section 1.4.2 of the book by Oppenheim, Willsky and Nawab.

[15] The straight line segment that connects the points (0, 0) and $(\Delta, 1)$ in image (a) has a straight line equation $y = x/\Delta$, whereas after setting $x = \Delta$ the function is constant and equal to one. Therefore, for the interval $[0, \Delta]$ the derivative of $u_\Delta(t)$ will be equal to the slope $1/\Delta$ whereas subsequently it will obviously be zero, thus leading to the graphical representation of image (b).

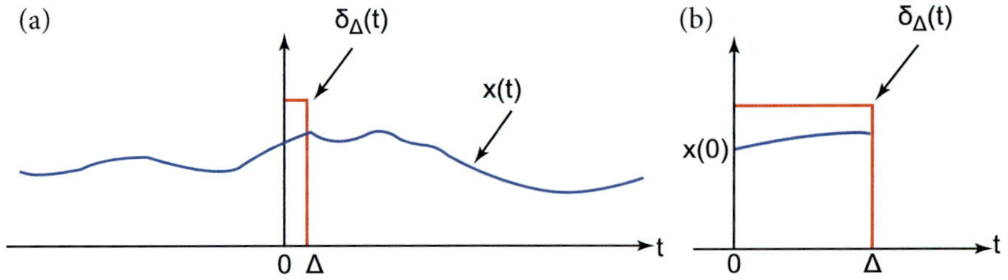

Fig. 1.17 (**a**) The signals $x(t)$ and $\delta_\Delta(t)$ plotted on the same diagram and (**b**) the nonzero part of the product $y(t)$ in magnification (Oppenheim, Willsky, Nawab, Figure 1.39)

The above way of approaching the function $\delta(t)$ is particularly important, because it gives us the ability to describe any continuous-time signal $x(t)$ *as a linear combination of time-shifted functions* $\delta_\Delta(t)$. Indeed, if we consider the product $y(t) = x(t)\delta_\Delta(t)$ of the signal $x(t)$ with the function $\delta_\Delta(t)$ and take into account the equation defining this function, it is obvious that the signal $y(t)$ will be zero everywhere, except for the range of values $0 \leq t < \Delta$ [this situation is presented in Fig. 1.17a, whereas Fig. 1.17b depicts in magnification the nonzero part of the signal $y(t)$]. If we assume that the range Δ of this interval is very small, so that the value of the signal $y(t)$ within this interval is approximately constant and equal to the value $y(0)$ of the signal at the position $t = 0$, then we can approximately write that $x(t)\delta_\Delta(t) \approx x(0)\delta_\Delta(t)$. This equation for the limit $\Delta \to 0$ leads to our known property $x(t)\delta(t) = x(0)\delta(t)$, which for the general case is formulated as $x(t)\delta(t - t_0) = x(t_0)\delta(t - t_0)$; we have also encountered this relationship in the past, during the description of the impulse function in the homonymous section.

Taking into account that the multiplication of the signal $x(t)$ with some time-shifted function $\delta_\Delta(t)$ cuts off a part of the signal (in the sense we described earlier) and that any continuous-time signal can be considered to consist of the combination of many such elementary signals, we can formulate the claim that any arbitrary continuous-time signal $x(t)$ can be described by an infinite sum of such products of the form

$$\tilde{x}(t) = \sum_{n=-\infty}^{+\infty} x(n\Delta)\delta_\Delta(t - n\Delta)\Delta \tag{1.25}$$

in the way presented in Fig. 1.18. In this figure, four elementary parts of the signal $x(t)$ are depicted, which are defined in the intervals $(-2\Delta, -\Delta)$, $(-\Delta, 0)$, $(0, \Delta)$ and $(\Delta, 2\Delta)$, whereas the generalization of this image, so as to determine all the elementary parts of the signal $x(t)$, is obvious.

In the limiting case where $\Delta \to 0$, the approximate signal $\tilde{x}(t)$ approaches the actual signal $x(t)$, which, therefore, can be described by an equation of the form

$$x(t) = \lim_{\Delta \to 0} \sum_{n=-\infty}^{+\infty} x(n\Delta)\delta_\Delta(t - n\Delta)\Delta, \tag{1.26}$$

with the above sum being converted into an integral. Indeed, if for simplicity we set $\tau = n\Delta$, the m-th elementary part of the signal $x(t)$ results from the multiplication of the signal $x(t)$ with the pulse $\delta_\Delta(t - \tau)$, which is defined in the range $t_1 < m\Delta < t_2$, where $t_1 = t - \Delta$ and $t_2 = t$. The width of this pulse is equal to $\Delta t = t_2 - t_1 = \Delta$ and, since the height of it according to the analysis we presented earlier is equal to $X(m\Delta)/\Delta$, the area of the surface enclosed by it will be equal to $\Delta t[X(m\Delta)/\Delta] = X(m\Delta)$.[16] As the width Δ tends to zero, the value of this area tends to the value of the surface defined by the graphical representation of the product $x(\tau)\delta_\Delta(t - \tau)$ and the straight lines $t = t_1$ and $t = t_2$. In addition, for this value of time t, the only nonzero term among the infinite terms of the above sum is the one corresponding to the value $n = m$ and therefore the right-hand side of the above equation is equal to $x(m\Delta)$.

Combining all of the above, we conclude that in the limit $\Delta \to 0$ the signal $x(t)$ will be equal to the area of the surface defined by the graphical representation of the function $x(\tau)\delta(t)$, which in turn is equal to the integral of this function with limits from $-\infty$ to $+\infty$. The above conclusion in mathematical writing is formulated as[17]

[16] Following the argumentation we presented earlier, we assume that the width Δ is small enough, so that the value of the signal $x(t)$ in the mth segment of it is approximately constant and equal to $x(m\Delta)$.

[17] In reality, this relation is nothing but the defining equation of the time-shifted function $\delta(t - \tau)$.

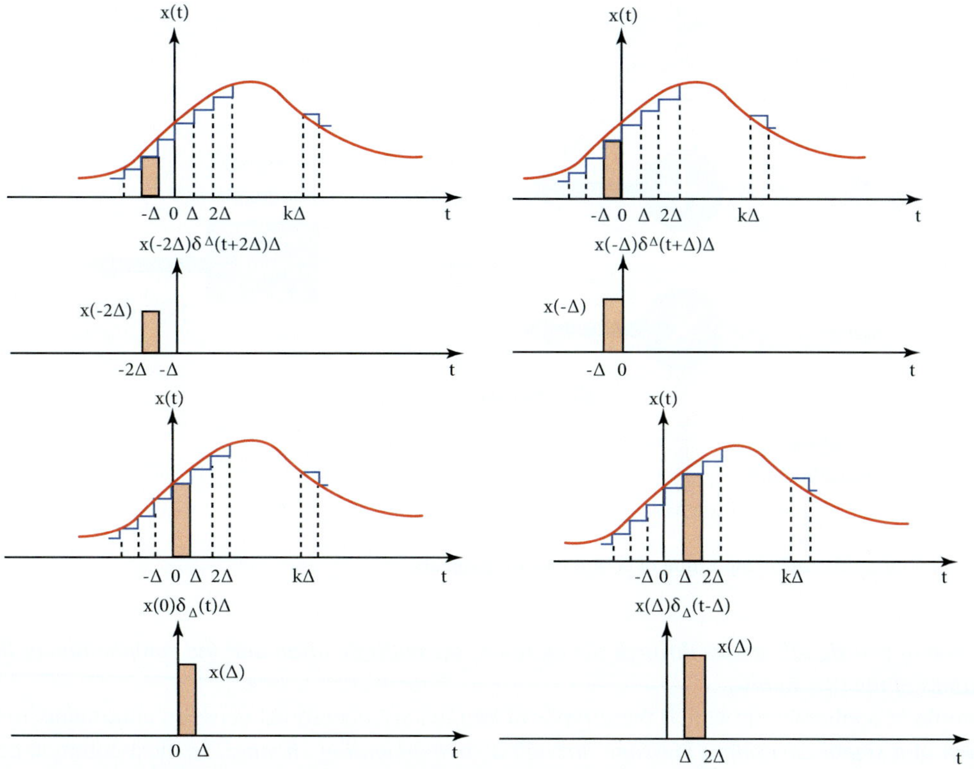

Fig. 1.18 Approximation of an arbitrary signal $x(t)$ by a linear combination of time-shifted functions $\delta_\Delta(t)$ (adapted from Oppenheim, Willsky, Nawab, Figure 2.12)

$$x(t) = \int_{-\infty}^{+\infty} x(\tau)\delta(t-\tau)d\tau \tag{1.27}$$

and represents the way in which any continuous-time signal $x(t)$ is described with the help of the impulse function $\delta(t-\tau)$. This equation, which is one of the fundamental equations in the field of continuous-time signals, with the operation between the signals $x(t)$ and $\delta(t)$ defined by it, is known as *convolution*; this operation will be studied in detail in a subsequent section.

1.10 Fundamental Operations of Continuous-Time Signals

Although in the previous sections the basic continuous-time signals were mainly presented, it is possible to construct more complex or composite signals from the overlap of signals characterized by less complexity. There are several fundamental operations that can be performed between simpler signals to construct more complex signals, which are presented in detail in the following sections.

1.10.1 Addition, Multiplication, and Composition of Signals

These fundamental operations allow the construction of complex signals from the addition and multiplication of simpler signals, while additionally the signals can be combined with each other using the known rules of *function composition*. The basic operations of addition, multiplication, and composition are represented with the help of special symbols, which are presented in Fig. 1.19. From this figure, it is not difficult to understand that *the signal resulting from the sum, the product*

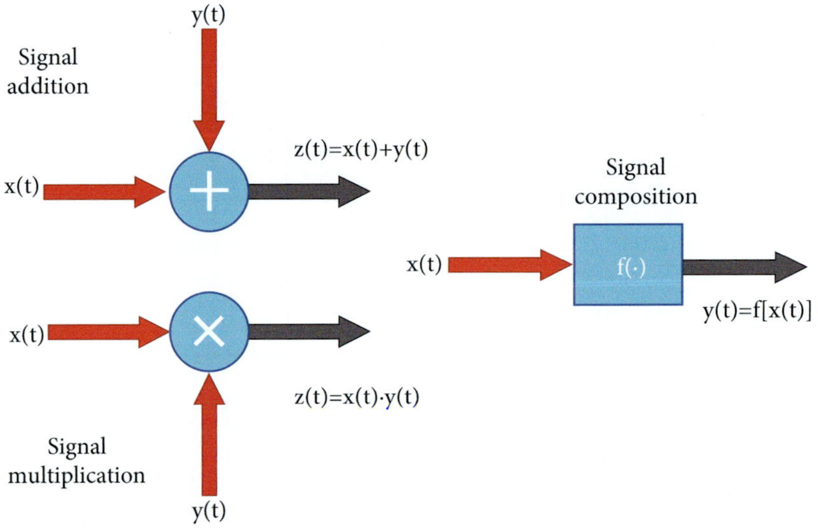

Fig. 1.19 Addition, multiplication, and composition of continuous-time signals

and the composition of two signals arises through the addition, the multiplication and the composition of the corresponding instantaneous values of the two signals.

A classic example of application in which these types of elementary operations between continuous-time signals appear is *the transmission of a signal carrying a message through a communication channel*. In most common cases, the signal *is not transmitted as is* but instead is combined with another signal that flows through the channel. This signal is known as the *carrier signal*, consists of a sinusoidal waveform of the form $x_C(t) = A_C(\cos\omega t + \varphi)$, and is modulated in such a way so as to carry the content of the message sent from the sender to the recipient. Depending on the characteristic of the signal that is modulated for the transmission of information, we have three types of modulation: the *amplitude modulation*, the *frequency modulation* and the *phase modulation*. In each case, the recipient must know the type of modulation used, so as to perform the reverse process of *demodulation* and reconstruct the original signal. In this whole description, we should also add the *noise* that is inevitably introduced by the channel and that for the transmission of an analog signal is considered to be white noise with normal distribution and variance that depends on the power of the transmitted signal. More details on this issue can be found in the bibliography.

1.10.2 Convolution of Continuous-Time Signals

The convolution of two continuous-time signals $x(t)$ and $y(t)$ is defined by the integral[18]

$$z(t) \doteq x(t) * y(t) = \int_{-\infty}^{+\infty} x(\tau) \cdot y(t-\tau)d\tau, \tag{1.28}$$

which is known as the *convolution integral*. This operation between signals is extremely important in the field of signal processing, because it allows the calculation of the output of a *linear and time-invariant continuous-time system* (although as we will see later, the situation is completely analogous in the field of *discrete-time*). We will return to this topic later.

In the special case where the signals $x(t)$ and $y(t)$ are both *periodic* with a common period T, the integral of the convolution, as defined by the above relation, *does not converge*—that is, it *diverges*—and therefore the operation

[18] In this definition, the limits of the integral are extended to the entire range of values $(-\infty, +\infty)$, because the two signals are assumed to have infinite duration. Conversely, if the two signals have finite durations T_1 and T_2, then, as we will see in a subsequent section, the convolution of the two signals also has a finite duration, which, moreover, is equal to $T_1 + T_2$.

$z(t) = x(t) * y(t)$ cannot be defined (from the above definition it is obvious that if the two signals are periodic, then $x(t) * y(t) \to \infty$). In this case, a variation of the convolution operation called *periodic convolution* is defined as

$$z(t) = x(t) \otimes y(t) = \int_0^T x(t)y(t-\tau)d\tau, \tag{1.29}$$

where T is the *common period* of the two signals (let us note, however, that the above integral can be calculated over any period, for example, from $-T/2$ to $+T/2$). It is not difficult to prove that the convolution $x(t) \otimes y(t)$ is periodic with period T. Indeed, since the signal $y(t)$ is periodic with this period value, $y(t-\tau+T) = y(t-\tau)$ and from the above equation we will have

$$z(t+T) = \int_0^T x(\tau)y(t-\tau+T)d\tau = \int_0^T x(\tau)y(t-\tau)d\tau = z(t)$$

that is, the function $z(t)$ is also periodic with period T. Let us note, however, that this type of convolution is valid, *only when both signals are periodic*, whereas the general definition we mentioned at the beginning of this section is used for aperiodic signals.

Problem 1.10.1 Calculate the convolution between the continuous-time signals $x(t) = e^{-\alpha t}u(t)$ and $y(t) = e^{-\beta t}u(t)$ for the cases (1) $\alpha \neq \beta$ and (2) $\alpha = \beta$ and then for the cases (3) $\beta = 0$ (for $\alpha \neq 0$) and (4) $\alpha = \beta = 0$.

Solution

(1) The convolution of the two signals $x(t)$ and $y(t)$ will be given by the relation

$$z(t) = x(t) * y(t) = \int_{-\infty}^{+\infty} x(\tau)y(t-\tau) = \int_{-\infty}^{+\infty} e^{-\alpha\tau}u(\tau)e^{-\beta(t-\tau)}u(t-\tau)d\tau,$$

where the step functions $u(\tau)$ and $u(t-\tau)$ are defined as

$$u(\tau) = \begin{cases} 1 \text{ for } \tau > 0 \\ 0 \text{ for } \tau < 0 \end{cases} \quad \text{and} \quad u(t-\tau) = \begin{cases} 1 \text{ for } t-\tau > 0 \text{(i.e. for } \tau < t) \\ 0 \text{ for } t-\tau < 0 \text{(i.e. for } \tau > t) \end{cases}$$

We observe that although in the general case the above integral should be calculated in the infinite interval $(-\infty, +\infty)$, *the existence of the two step functions limits these limits*, since these functions are both different from zero and equal to one only in the interval $0 < \tau < t$ and zero anywhere else. By direct calculation of the integral—which is calculated in the above range of values—we find that

$$z(t) = \int_0^t e^{-\alpha\tau}e^{-\beta(t-\tau)}d\tau = e^{-\beta t}\int_0^t e^{(\beta-\alpha)\tau}d\tau = e^{-\beta t}\left[\frac{e^{(\beta-\alpha)\tau}}{\beta-\alpha}\right]_0^t = \frac{e^{-\beta t}}{\beta-\alpha}\left[e^{(\beta-\alpha)t} - 1\right] = \frac{e^{-\alpha t} - e^{-\beta t}}{\beta-\alpha}$$

The above relation holds for the range of values $t > 0$ and since for $t < 0$ it is $z(t) = 0$, we can combine the above cases into a simple equation, writing

$$z(t) = x(t) * y(t) = \begin{cases} \dfrac{e^{-\alpha t} - e^{-\beta t}}{\beta-\alpha} & \text{for } t > 0 \\ 0 & \text{for } t < 0 \end{cases} = \frac{e^{-\alpha t} - e^{-\beta t}}{\beta-\alpha}u(t).$$

(2) If we set $\alpha = \beta$ in the above relation, the convolution $z(t)$ takes the indeterminate form $0/0$. Applying, therefore, the rule of L' Hospital's we find that

$$z(t) = \lim_{\alpha\to\beta}\left\{\frac{\frac{d}{d\alpha}\left(e^{-\alpha t} - e^{-\beta t}\right)}{\frac{d}{d\alpha}(\beta - \alpha)}u(t)\right\} = -\lim_{\alpha\to\beta}\left\{-te^{-\alpha t}u(t)\right\} = te^{-\beta t}u(t).$$

(3) Using the result of question (1) for the special case $\beta = 0$, we find that

$$z(t) = \left[\frac{e^{-\alpha t} - e^{-\beta t}}{\beta - \alpha}u(t)\right]_{\beta=0} = \frac{e^{-\alpha t} - 1}{-\alpha}u(t) = \frac{1}{\alpha}\left(1 - e^{-\alpha t}\right)u(t).$$

(4) Using the result of question (2) for the special case $\alpha = \beta = 0$, we find that

$$z(t) = \left[te^{-\beta t}u(t)\right]_{\beta=0} = tu(t) = r(t),$$

where $r(t) = ru(t)$ is the ramp or climbing function that we have studied in a previous section. Taking into account that for $\alpha = \beta = 0$, we will have $x(t) = y(t) = u(t)$, leading to the important conclusion that *the convolution of the step function $u(t)$ with itself leads to the ramp function $r(t)$.*

Problem 1.10.2 Calculate the convolution between the signals $x(t) = te^{\alpha t}u(t)$ and $y(t) = e^{\alpha t}u(t)$.

Solution From the defining equation of convolution we have

$$z(t) = x(t) * y(t) = \int_{-\infty}^{+\infty} x(\tau)y(t-\tau)d\tau = \int_{-\infty}^{+\infty} \tau e^{\alpha\tau}u(\tau)e^{\alpha(t-\tau)}u(t-\tau)d\tau$$

Following the argumentation of the previous exercise we observe that the step functions $u(\tau)$ and $u(t-\tau)$ have a value equal to one in the interval $0 < \tau < t$, and therefore, the convolution of the two signals will be nonzero only within this interval. Calculating the above integral for this range of values we find that

$$z(t) = \int_0^t \tau e^{\alpha\tau}e^{\alpha(t-\tau)}d\tau = e^{\alpha t}\int_0^t \tau d\tau = e^{\alpha t}\left[\frac{t^2}{2}\right]_0^t = \frac{1}{2}t^2e^{\alpha t}$$

or equivalently $z(t) = (1/2)t^2e^{\alpha t}u(t)$, where in the last equation we have incorporated the case $t < 0$ for which it is $z(t) < 0$.

Problem 1.10.3 Calculate the convolution between the signals $x(t) = te^{\alpha t}u(t)$ and $y(t) = e^{\beta t}u(t)$.

Solution Using the convolution defining equation and limiting ourselves to the integration interval $0 < \tau < t$, we have

$$z(t) = x(t) * y(t) = \int_{-\infty}^{+\infty} x(\tau)y(t-\tau)d\tau = \int_{-\infty}^{+\infty} \tau e^{\alpha\tau}u(\tau)e^{\beta(t-\tau)}u(t-\tau)d\tau = \int_0^t \tau e^{\alpha\tau}e^{\beta(t-\tau)}d\tau = e^{\beta t}\int_0^t \tau e^{(\alpha-\beta)\tau}d\tau$$

Initially, let us calculate the indefinite integral that appears in the last relation. Setting for simplicity $\lambda = \alpha - \beta$ and applying the method of integration by parts, we get

$$\int \tau e^{\lambda\tau}d\tau = \int \tau d\left(\frac{e^{\lambda\tau}}{\lambda}\right) = \frac{\tau e^{\lambda\tau}}{\lambda} - \frac{1}{\lambda}\int e^{\lambda\tau}d\tau = \frac{\tau e^{\lambda\tau}}{\lambda} - \frac{e^{\lambda\tau}}{\lambda^2} = \frac{\lambda\tau e^{\lambda\tau} - e^{\lambda\tau}}{\lambda^2} = \frac{(\lambda\tau - 1)e^{\lambda\tau}}{\lambda^2}$$

Table 1.1 Convolution of known continuous-time signals

$x(t)$	$y(t)$	$x(t)\,^{\mathrm{a}}\,y(t)$
$e^{\alpha t}u(t)$	$e^{\beta t}u(t)$	$\dfrac{e^{\alpha t}-e^{\beta t}}{\alpha-b}u(t), \quad (\alpha\neq\beta)$
$te^{\alpha t}u(t)$	$e^{\alpha t}u(t)$	$\dfrac{1}{2}t^2e^{\alpha t}u(t)$
$t^N u(t)$	$e^{\alpha t}u(t)$	$\dfrac{N!\,e^{\alpha t}}{\alpha^{N+1}}u(t)-\sum_{k=0}^{N}\dfrac{N!\,t^{N-k}}{\alpha^{k+1}(N-k)!}u(t)$
$te^{\alpha t}u(t)$	$e^{\beta t}u(t)$	$\dfrac{e^{\beta t}-e^{\alpha t}+(\alpha-\beta)te^{\alpha t}}{(\alpha-\beta)^2}u(t)$
$t^M e^{\alpha t}u(t)$	$t^N e^{\alpha t}u(t)$	$\dfrac{M!N!}{(N+M+1)!}t^{M+N+1}e^{\alpha t}u(t)$
$t^M e^{\alpha t}u(t)$	$t^N e^{\beta t}u(t)$	$\sum_{k=0}^{M}\dfrac{(-1)^k M!(N+k)!t^{M-k}e^{\alpha t}}{k!(M-k)!(\alpha-\beta)^{N+k+1}}u(t)+$
		$\sum_{k=0}^{N}\dfrac{(-1)^k N!(M+k)!t^{N-k}e^{\beta t}}{k!(N-k)!(\beta-\alpha)^{M+k+1}}u(t)$
$e^{-\alpha t}\cos(\beta t+\vartheta)u(t)$	$e^{\lambda t}u(t)$	$\dfrac{\cos(\vartheta-\phi)e^{\lambda t}-e^{-\alpha t}\cos(\beta \mathrm{t}+\vartheta-\phi)}{\sqrt{(\alpha+\lambda)^2+\beta^2}}u(t)^{\mathrm{a}}$
$e^{\alpha t}u(t)$	$e^{\beta t}u(-t)$	$\dfrac{e^{\alpha t}u(t)+e^{\beta t}u(-t)}{\beta-\alpha}, \quad \Re(\beta)>\Re(\alpha)$
$e^{\alpha t}u(-t)$	$e^{\beta t}u(-t)$	$\dfrac{e^{\alpha t}-e^{\beta t}}{\beta-\alpha}u(-t)$

[a] The angle ϕ is defined as $\phi=\tan^{-1}\left(-\dfrac{\beta}{\alpha+\lambda}\right)$

Therefore, we have

$$\int_0^t \tau e^{(\alpha-\beta)\tau}d\tau=\left[\frac{\{(\alpha-\beta)\tau-1\}e^{(\alpha-\beta)\tau}}{(\alpha-\beta)^2}\right]_0^t=\frac{(\alpha-\beta)te^{(\alpha-\beta)t}-e^{(\alpha-\beta)t}+1}{(\alpha-\beta)^2}$$

and therefore the convolution of the two signals will have the form

$$z(t)=\frac{e^{\beta t}}{(\alpha-\beta)^2}\left[(\alpha-\beta)te^{(\alpha-\beta)t}-e^{(\alpha-\beta)t}+1\right]=\frac{(\alpha-\beta)te^{\alpha t}-e^{\alpha t}+e^{\beta t}}{(\alpha-\beta)^2}$$

(for $t\geq 0$) or equivalently

$$z(t)=\frac{(\alpha-\beta)te^{\alpha t}-e^{\alpha t}+e^{\beta t}}{(\alpha-\beta)^2}u(t)$$

to include the negative time instances where $z(t)=0$. ■

The above exercises present in detail the process of calculating the convolution of elementary signals, whereas the expressions of convolution between more complex continuous-time signals that are used very often in practice are included in Table 1.1.

Properties of Convolution

The operation of convolution between two continuous-time signals $x(t)$ and $y(t)$ is characterized by several interesting properties, the most important of which are the following:

- **Homogeneity:** Considering two continuous-time signals $x(t)$ and $y(t)$, it is proven that

$$\left[\alpha x(t)\right]*y(t)=\alpha\left[x(t)*y(t)\right]=x(t)*\left[\alpha y(t)\right], \tag{1.30}$$

where α is any constant. ***Proof***: From the defining equation of convolution it immediately follows that

$$\left[\alpha x(t)\right] * y(t) = \int_{-\infty}^{+\infty} \alpha x(\tau)y(t-\tau)d\tau =$$

$$= \alpha\left\{\int_{-\infty}^{+\infty} x(\tau)y(t-\tau)d\tau\right\} = \alpha\left[x(t) * y(t)\right] = \int_{-\infty}^{+\infty} x(\tau)\left[\alpha y(t-\tau)\right]d\tau = x(t) * \left[\alpha y(t)\right]$$

- **Commutative property:** Considering two continuous-time signals $x(t)$ and $y(t)$, it is proven that

$$x(t) * y(t) = y(t) * x(t) \tag{1.31}$$

Proof If in the convolution defining equation we proceed to the substitution $\xi = t - \tau$, then $\tau = t - \xi$, and differentiating we get $d\tau = -d\xi$. As for the integration limits for $\tau = -\infty$, $\xi = +\infty$, whereas for $\tau = +\infty$, $\xi = -\infty$. Therefore,

$$x(t) * y(t) = \int_{-\infty}^{\infty} x(t)y(t-\tau) = -\int_{+\infty}^{-\infty} x(t-\xi)y(\xi)d\xi = \int_{-\infty}^{+\infty} x(t-\tau)y(\tau)d\tau = y(t) * x(t),$$

where for the derivation of the last relation we replaced the integration variable from ξ to τ.

- **Associative property:** Considering three continuous-time signals $x(t)$, $y(t)$, and $z(t)$, it is proven that

$$[x(t) * y(t)] * z(t) = x(t) * [y(t) * z(t)] \tag{1.32}$$

Proof Using the convolution defining equation, the first member of the above equation is written

$$[x(t) * y(t)] * z(t) = \left[\int_{-\infty}^{+\infty} x(\tau)y(t-\tau)\, d\tau\right] * z(t) = \int_{-\infty}^{+\infty}\left\{\int_{-\infty}^{+\infty} x(\tau)y(\xi-\tau)\, d\tau\right\}z(t-\xi)\, d\xi$$

Proceeding to the change of the integration variable from ξ to $\lambda = \xi - \tau$, $\xi = \lambda + \tau$ and differentiating we get $d\xi = d\lambda$. Therefore, changing the order of integration the above relation is written as

$$[x(t) * y(t)] * z(t) = \int_{-\infty}^{+\infty} x(\tau)\left\{\int_{-\infty}^{+\infty} y(\lambda)z(t-\lambda-\tau)\, d\lambda\right\} d\tau = \int_{-\infty}^{+\infty} x(\tau)\left\{\int_{-\infty}^{+\infty} y(\lambda)z\left[(t-\tau)-\lambda\right] d\lambda\right\} d\tau$$

If we now proceed simultaneously with the substitutions $\lambda \to t$ and $t' \to t - \tau$ it is not difficult to realize that the inner integral is actually

$$\int_{-\infty}^{+\infty} y(t)z(t'-t)dt = \left[y(t) * z(t)\right]_{t=t'} = [y * z](t-t')$$

and the above relation will take the form

$$[x(t) * y(t)] * z(t) = \int_{-\infty}^{+\infty} x(\tau)[y * z](t-\tau)d\tau = x(t) * [y(t) * z(t)].$$

- **Distributive property:** Considering three continuous-time signals $x(t)$, $y(t)$ and $z(t)$, it is proven that

$$x(t) * [y(t) + z(t)] = x(t) * y(t) + x(t) * z(t) \tag{1.33}$$

Proof From the defining equation of the convolution we have

$$x(t) * [y(t) + z(t)] = \int_{-\infty}^{+\infty} x(\tau)\left[y(t-\tau) + z(t-\tau)\right] d\tau =$$

$$= \int_{-\infty}^{+\infty} x(\tau)y(t-\tau)\, d\tau + \int_{-\infty}^{+\infty} x(\tau)z(t-\tau)\, d\tau = x(t) * y(t) + x(t) * z(t)$$

- **Linearity:** Let signals $x_1(t)$, $x_2(t)$, $z_1(t)$, $z_2(t)$, and $y(t)$ be such that

$$z_1(t) = x_1(t) * y(t) \quad \text{and} \quad z_2(t) = x_2(t) * y(t).$$

In this case, if we consider the signal $x(t) = \alpha x_1(t) + \beta x_2(t)$ it is proven that

$$z(t) = x(t) * y(t) = \alpha z_1(t) + \beta z_2(t), \tag{1.34}$$

where α and β are any real numbers.
Proof From the defining equation of the signal $z(t)$ and using the property of homogeneity and the distributive property we have

$$\begin{aligned} z(t) &= x(t) * y(t) = \left[\alpha x_1(t) + \beta x_2(t)\right] * y(t) = \left[\alpha x_1(t)\right] * y(t) + \left[\beta x_2(t)\right] * y(t) \\ &= \alpha\left[x_1(t) * y(t)\right] + \beta\left[x_2(t) * y(t)\right] = \alpha z_1(t) + \beta z_2(t) \end{aligned}$$

- **Surface area:** Considering three continuous-time signals $x(t)$, $y(t)$, and $z(t)$ such that $z(t) = x(t) * y(t)$, the area defined by the curve of the function $z(t)$ is equal to the product of the two areas defined by the curves of the functions $x(t)$ and $y(t)$.
Proof The area defined by the curve of the function $z(t)$ is given, it is well known by the integral

$$E_z = \int_{-\infty}^{+\infty} z(t)dt,$$

with the function $z(t)$ defined as

$$z(t) = x(t) * y(t) = \int_{-\infty}^{+\infty} x(\tau)y(t-\tau)d\tau.$$

Replacing the above expression in the equation defining the area E_z, we get

$$\begin{aligned} E_z &= \int_{-\infty}^{+\infty} z(t)dt = \int_{-\infty}^{+\infty} \left(\int_{-\infty}^{+\infty} x(\tau)y(t-\tau)d\tau\right) dt = \int_{-\infty}^{+\infty}\int_{-\infty}^{+\infty} x(\tau)y(t-\tau)d\tau dt \\ &= \left(\int_{-\infty}^{+\infty} x(\tau)d\tau\right)\left(\int_{-\infty}^{+\infty} y(t-\tau)dt\right) = E_x E_y \end{aligned}$$

with the areas E_x and E_y defined in a similar way by the relationships

$$E_x = \int_{-\infty}^{+\infty} x(\tau)d\tau \quad \text{and} \quad E_y = \int_{-\infty}^{+\infty} y(t-\tau)dt = \int_{-\infty}^{+\infty} y(\xi)d\xi$$

(the variable change $\xi = t - \tau$ in the second integral obviously makes no difference).
- **Range:** The duration of the convolution $z(t) = x(t) * y(t)$ of two continuous-time signals $x(t)$ and $y(t)$ characterized by finite durations T_1 and T_2 respectively, is equal to $T_1 + T_2$. This property is represented diagrammatically in Fig. 1.20 and its proof will be given qualitatively in the section on the graphical computation of convolution.
- **Scaling:** Considering three continuous-time signals $x(t)$, $y(t)$ and $z(t) = x(t) * y(t)$, it is proven that

$$x\left(\frac{t}{\alpha}\right) * y\left(\frac{t}{\alpha}\right) = |\alpha| z\left(\frac{t}{\alpha}\right) \tag{1.35}$$

Proof From the convolution defining equation we have

$$x\left(\frac{t}{\alpha}\right) * y\left(\frac{t}{\alpha}\right) = \int_{-\infty}^{+\infty} x\left(\frac{t}{\alpha}\right) y\left(\frac{t-\tau}{\alpha}\right) d\tau = \int_{-\infty}^{+\infty} x\left(\frac{t}{\alpha}\right) y\left(\frac{t}{\alpha} - \frac{\tau}{\alpha}\right) d\tau.$$

Proceeding with the variable change $\tau = r\alpha$, $r = \tau/\alpha$ and $dr = d\tau/\alpha$. We distinguish the following cases:

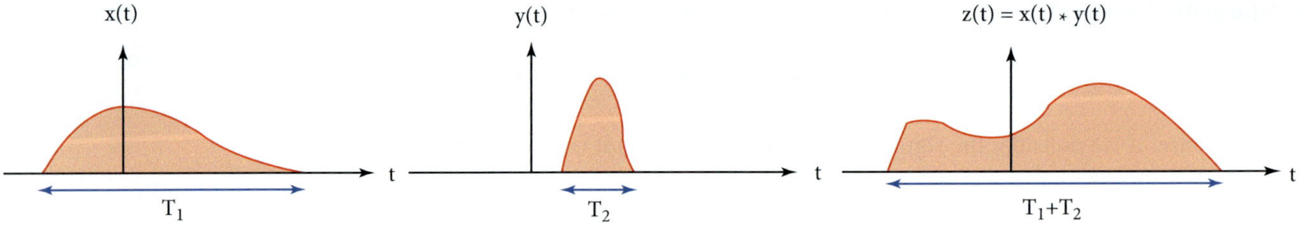

Fig. 1.20 Graphical demonstration of the range property of the convolution of signals of finite duration

(a) $\alpha > 0$: in this case

$$x\left(\frac{t}{\alpha}\right) * y\left(\frac{t}{\alpha}\right) = \alpha \int_{-\infty}^{+\infty} x\left(\frac{t}{\alpha}\right) y\left(\frac{t}{\alpha} - r\right) dr = \alpha z\left(\frac{t}{\alpha}\right) = |\alpha| z\left(\frac{t}{\alpha}\right)$$

(b) $\alpha < 0$: In this case the integration limits are reversed and we get

$$x\left(\frac{t}{\alpha}\right) * y\left(\frac{t}{\alpha}\right) = \alpha \int_{+\infty}^{-\infty} x\left(\frac{t}{\alpha}\right) y\left(\frac{t}{\alpha} - r\right) dr = -\alpha \int_{-\infty}^{+\infty} x\left(\frac{t}{\alpha}\right) y\left(\frac{t}{\alpha} - r\right) dr = -\alpha z\left(\frac{t}{\alpha}\right) = |\alpha| z\left(\frac{t}{\alpha}\right)$$

Combining the above results, we arrive at the desired property.

- **Differentiation:** Considering three continuous-time signals $x(t)$, $y(t)$ and $z(t) = x(t) * y(t)$, it is proven that

$$\frac{dz(t)}{dt} = \frac{dx(t)}{dt} * y(t) = x(t) * \frac{dy(t)}{dt} \tag{1.36}$$

Proof By differentiating the defining equation of convolution

$$z(t) = x(t) * y(t) = \int_{-\infty}^{+\infty} x(\tau) y(t - \tau) d\tau$$

it immediately follows that

$$\frac{dz(t)}{dt} = \frac{d}{dt}\left\{\int_{-\infty}^{+\infty} x(\tau) y(t - \tau) d\tau\right\} = \int_{-\infty}^{+\infty} x(\tau) \left.\frac{dy(t)}{dt}\right|_{t=t-\tau} d\tau = x(t) * \frac{dy(t)}{dt} = \frac{dx(t)}{dt} * y(t),$$

where for the derivation of the last relation we used the commutative property of convolution.

- **Neutral element:** The impulse function $\delta(t)$ constitutes the neutral element of the convolution operation in the signal space. This means that for any continuous-time signal $x(t)$ the property holds

$$x(t) * \delta(t) = \delta(t) * x(t) = x(t). \tag{1.37}$$

Proof This property is directly proven from the defining equation of the time-shifted function $\delta(t - \tau)$ according to which for any continuous function $x(t)$ it holds that

$$\int_{-\infty}^{+\infty} x(\tau) \delta(t - \tau) d\tau = x(\tau)\Big|_{\tau=t} = x(t).$$

But the above integral is nothing but the defining equation of the convolution $x(t) * \delta(t)$ from which the desired result follows. In a completely analogous way it is proven that

$$x(t) * \delta(t - t_0) = \delta(t - t_0) * x(t) = x(t - t_0),$$

a property that is particularly useful in modulation processes.

- **Time invariance:** Considering three signals $x(t)$, $y(t)$ and $z(t) = x(t) * y(t)$ it is proven that

$$z(t - t_0) = x(t - t_0) * y(t) = \int_{-\infty}^{+\infty} x(\tau - t_0)y(t - \tau)d\tau \tag{1.38}$$

Proof From the defining equation of the convolution for $t \to t - t_0$ we will get

$$z(t - t_0) = \int_{-\infty}^{+\infty} x(\tau)y(t - t_0 - \tau)d\tau.$$

Proceeding with the variable change $\xi = \tau + t_0$, $d\xi = d\tau$ and $\tau = \xi - t_0$ and by substitution we have

$$z(t - t_0) = \int_{-\infty}^{+\infty} x(\tau)y\Big[t - (t_0 + \tau)\Big]d\tau = \int_{-\infty}^{+\infty} x(\xi - t_0)y(t - \xi)d\xi = \int_{-\infty}^{+\infty} x(\tau - t_0)y(t - \tau)d\tau = x(t - t_0) * y(t)$$

where for the derivation of the last relation we proceeded to the substitution $\xi \longrightarrow \tau$. ■

Observation: Using this property in combination with the commutative property of convolution $x(t) * y(t) = y(t) * x(t)$, it follows in exactly the same way that

$$z(t - t_0) = x(t) * y(t - t_0).$$

We also mention that the property of time invariance in a more general formulation is described by the equation

$$x(t - t_1) * y(t - t_2) = z(t - t_1 - t_2),$$

which, of course, holds under the condition that $z(t) = x(t) * y(t)$. In order to prove the above property, we consider the equation

$$x(t - t_1) * y(t - t_2) = \int_{-\infty}^{+\infty} x(\tau - t_1)y(t - \tau - t_2)d\tau$$

and we make the substitution $\lambda = \tau - t_1$. Then, $\tau = \lambda + t_1$ and $d\lambda = d\tau$ and finally we arrive at the desired property

$$x(t - t_1) * y(t - t_2) = \int_{-\infty}^{+\infty} x(\lambda)y(t - \lambda - t_1 - t_2)\,d\lambda = \int_{-\infty}^{+\infty} x(\lambda)y[(t - t_1 - t_2) - \lambda]\,d\lambda = z(t - t_1 1 - t_2).$$

Problem 1.10.4 Using the continuous-time signals $x(t) = \sin(\omega t)\,u(t)$ and $y(t) = e^{-\alpha t}\,u(t)$ verify the commutative property of convolution.

Solution To solve the exercise we must prove that

$$\Big[\sin \omega t\; u(t)\Big] * \Big[e^{-\alpha t}\; u(t)\Big] = \Big[e^{-\alpha t}\; u(t)\Big] * \Big[\sin \omega t\; u(t)\Big]$$

From the defining equation of convolution and integrating over the interval $0 < \tau < t$ in which the step functions $u(t)$ and $u(t - \tau)$ are both different than zero, the first term is written as

$$z(t) = x(t) * y(t) = \int_{-\infty}^{+\infty} x(\tau)y(t - \tau)d\tau = \int_{-\infty}^{+\infty} \sin(\omega\tau)u(\tau)e^{-\alpha(t-\tau)}u(t - \tau)d\tau$$

$$= \int_0^t \sin(\omega\tau)e^{-\alpha(t-\tau)}d\tau = e^{-\alpha t}\int_0^t e^{\alpha\tau}\sin(\omega\tau)d\tau$$

The integral of the last relation is computed using the well-known method of integration by parts, applied successively to get the result[19]

$$\int_0^t e^{\alpha\tau}\sin(\omega\tau)d\tau = \frac{e^{\alpha t}[\alpha\sin(\omega t) - \omega\cos(\omega t)] + \omega}{\alpha^2+\omega^2}.$$

Substituting the above relation into the convolution defining equation, we find

$$z(t) = e^{-\alpha t}\left\{\frac{e^{\alpha t}[\alpha\sin(\omega t) - \omega\cos(\omega t)] + \omega}{\alpha^2+\omega^2}\right\} = \frac{\omega e^{-\alpha t} + \alpha\sin(\omega t) - \omega\cos(\omega t)}{\alpha^2+\omega^2}$$

On the other hand, for the second member of the commutative property we have

$$z(t) = y(t) * x(t) = \int_{-\infty}^{+\infty} y(\tau)x(t-\tau)d\tau = \int_{-\infty}^{+\infty} e^{-\alpha\tau}u(\tau)\sin[\omega(t-\tau)]u(t-\tau)d\tau = \int_0^t e^{-\alpha\tau}\sin[\omega(t-\tau)]d\tau.$$

Using the trigonometric identity $\sin(\alpha-\beta) = \sin\alpha\cos\beta - \cos\alpha\sin\beta$ the function $\sin[\omega(t-\tau)]$ of the last integral develops as

$$\sin[\omega(t-\tau)] = \sin(\omega t - \omega\tau) = \sin(\omega t)\cos(\omega\tau) - \cos(\omega t)\sin(\omega\tau)$$

and the above relationship takes the form

$$z(t) = \int_0^t e^{-\alpha\tau}\Big[\sin(\omega t)\cos(\omega\tau) - \cos(\omega t)\sin(\omega\tau)\Big]d\tau = \sin(\omega t)\int_0^t e^{-\alpha\tau}\cos(\omega\tau)d\tau - \cos(\omega t)\int_0^t e^{-\alpha\tau}\sin(\omega\tau)d\tau.$$

The first of the two integrals is calculated, as before, by the method of integration by parts, which leads to a relationship of the form[20]

$$\int_0^t e^{-\alpha\tau}\cos(\omega\tau)d\tau = \frac{e^{-\alpha t}[\omega\sin(\omega t) - \alpha\cos(\omega t)] + \alpha}{\alpha^2+\omega^2}.$$

---

[19] In this case, the application of the method gives us

$$\begin{aligned} I &= \int_0^t e^{\alpha\tau}\sin(\omega\tau)d\tau = \int_0^t \frac{d}{d\tau}\left(\frac{e^{\alpha\tau}}{\alpha}\right)\sin(\omega\tau)d\tau = \frac{e^{\alpha\tau}}{\alpha}\sin(\omega\tau)\Big|_0^t - \frac{\omega}{\alpha}\int_0^t e^{\alpha\tau}\cos(\omega\tau)d\tau \\ &= \frac{e^{\alpha t}}{\alpha}\sin(\omega t) - \frac{\omega}{\alpha}\int_0^t \frac{d}{d\tau}\left(\frac{e^{\alpha\tau}}{\alpha}\right)\cos(\omega\tau)d\tau = \frac{e^{\alpha t}}{\alpha}\sin(\omega t) - \frac{\omega}{\alpha}\left(\frac{e^{\alpha\tau}}{\alpha}\cos(\omega t)\Big|_0^t + \frac{\omega}{\alpha}\int_0^t e^{\alpha t}\sin(\omega\tau)d\tau\right) \\ &= \frac{e^{\alpha\tau}}{\alpha}\sin(\omega t) - \frac{\omega}{\alpha^2}\left(e^{\alpha t}\cos(\omega t) - 1\right) - \frac{\omega^2}{\alpha^2}\int_0^t e^{\alpha\tau}\sin(\omega\tau)d\tau = \frac{e^{\alpha\tau}}{\alpha}\sin(\omega t) - \frac{\omega}{\alpha^2}\left(e^{\alpha t}\cos(\omega t) - 1\right) - \frac{\omega^2}{\alpha^2}I \end{aligned}$$

Transferring the last expression of the right member to the left member, it is easily derived that

$$\left(1+\frac{\omega^2}{\alpha^2}\right)I = \frac{\alpha^2+\omega^2}{\alpha^2}I = \frac{\alpha e^{\alpha t}\sin(\omega t) - \omega e^{\alpha t}\cos(\omega t) + \omega}{\alpha^2} \quad \text{and finally} \quad I = \frac{e^{\alpha t}[\alpha\sin(\omega t) - \omega\cos(\omega t)] + \omega}{\alpha^2+\omega^2}.$$

[20] Applying the same process as before, this time we have

$$\begin{aligned} I &= \int_0^t e^{-\alpha\tau}\cos(\omega\tau)d\tau = \int_0^\tau d\left(-\frac{e^{-\alpha\tau}}{\alpha}\right)\cos(\omega\tau) = -\frac{e^{-\alpha\tau}}{\alpha}\cos(\omega\tau)\Big|_0^t - \frac{\omega}{\alpha}\int_0^t e^{-\alpha\tau}\sin(\omega\tau)d\tau \\ &= \frac{1}{\alpha} - \frac{e^{-\alpha t}}{\alpha}\cos(\omega t) - \frac{\omega}{\alpha}\int_0^t d\left(-\frac{e^{-\alpha\tau}}{\alpha}\right)\sin(\omega\tau) = \frac{1}{\alpha} - \frac{e^{-\alpha t}}{\alpha}\cos(\omega t) - \frac{\omega}{\alpha}\left(-\frac{e^{-\alpha\tau}}{\alpha}\sin(\omega\tau)\Big|_0^t + \frac{\omega}{\alpha}\int_0^t e^{-\alpha\tau}\cos(\omega\tau)d\tau\right) \\ &= \frac{1}{\alpha} - \frac{e^{-\alpha t}}{\alpha}\cos(\omega t) + \frac{\omega}{\alpha^2}e^{-\alpha t}\sin(\omega t) - \frac{\omega^2}{\alpha^2}\int_0^t e^{-\alpha\tau}\cos(\omega\tau)d\tau = \frac{1}{\alpha} - \frac{e^{-\alpha t}}{\alpha}\cos(\omega t) + \frac{\omega}{\alpha^2}e^{-\alpha t}\sin(\omega t) - \frac{\omega^2}{\alpha^2}I \end{aligned}$$

Transferring the last expression of the right-hand side to the left-hand side, it easily follows that

$$\left(1+\frac{\omega^2}{\alpha^2}\right)I = \frac{\alpha^2+\omega^2}{\alpha^2}I = \frac{\alpha - \alpha e^{-\alpha t}\cos(\omega t) + \omega e^{-\alpha t}\sin(\omega t)}{\alpha^2} \quad \text{and finally} \quad I = \frac{e^{-\alpha t}[\omega\sin(\omega t) - \alpha\cos(\omega t)] + \alpha}{\alpha^2+\omega^2}.$$

On the other hand, the second integral can arise from what we calculated in the first term, if we make the substitution $\alpha \to -\alpha$. Therefore,

$$\int_0^t e^{-\alpha\tau}\sin(\omega\tau)d\tau = \left\{\frac{e^{\alpha t}[\alpha\sin(\omega t) - \omega\cos(\omega t)] + \omega}{\alpha^2+\omega^2}\right\}_{\alpha\to-\alpha} = \frac{\omega - e^{-\alpha t}[\alpha\sin(\omega t)+\omega\cos(\omega t)]}{\alpha^2+\omega^2}.$$

Replacing these two integrals in the expression of the convolution of the second term, we find that

$$\begin{aligned} z(t) &= \sin(\omega t)\frac{e^{-\alpha t}[\omega\sin(\omega t) - \alpha\cos(\omega t)] + \alpha}{\alpha^2+\omega^2} - \cos(\omega t)\frac{\omega - e^{-\alpha t}[\alpha\sin(\omega t)+\omega\cos(\omega t)]}{\alpha^2+\omega^2} \\ &= \frac{\omega e^{-\alpha t}\sin^2\omega t - \alpha e^{-\alpha t}\sin(\omega t)\cos(\omega t) + \alpha\sin(\omega t)}{\alpha^2+\omega^2} - \frac{\omega\cos(\omega t) + \alpha e^{-\alpha t}\sin(\omega t)\cos(\omega t) + \omega e^{-\alpha t}\cos^2(\omega t)}{\alpha^2+\omega^2} \\ &= \frac{\omega e^{-\alpha t}[\cos^2(\omega t)+\sin^2\omega t] + \alpha\sin(\omega t) - \omega\cos(\omega t)}{\alpha^2+\omega^2} = \frac{\omega e^{-\alpha t} + \alpha\sin(\omega t) - \omega\cos(\omega t)}{\alpha^2+\omega^2}, \end{aligned}$$

an expression that is identical to the one we calculated for the first term. In this way, we have practically proven the validity of the commutative property of convolution.

Problem 1.10.5 Calculate the convolution between the signals $x(t) = \sin t u(t)$ and $y(t) = u(t)$.

Solution This exercise is a special case of the previous one for the values $\alpha = 0$ and $\omega = 1$. Using the result we arrived at for the aforementioned values of the parameters we will get

$$z(t) = x(t) * y(t) = \left\{\frac{\omega e^{-\alpha t} + \alpha\sin(\omega t) - \omega\cos(\omega t)}{\alpha^2+\omega^2}\right\}_{\alpha=0,\omega=1} = 1 - \cos t.$$

Problem 1.10.6 Calculate the convolution between any continuous-time signal $x(t)$ and the unit step function $u(t)$ and comment on the result.

Solution Using the defining equation of the convolution we have

$$z(t) = x(t) * u(t) = \int_{-\infty}^{+\infty} x(\tau)u(t-\tau)d\tau.$$

Relying on the definition of the unit step function

$$u(t) = \begin{cases} 1 \text{ for } t > 0 \\ 0 \text{ for } t < 0 \end{cases} \quad \text{and} \quad u(t-\tau) = \begin{cases} 1 \text{ for } t-\tau > 0 \text{ or } \tau < t \\ 0 \text{ for } t-\tau < 0 \text{ or } \tau > t \end{cases}$$

and by substituting in our initial equation, we will get

$$z(t) = \int_{-\infty}^{t} x(\tau)d\tau.$$

Therefore, the convolution between a continuous-time signal $x(t)$ and the unit step function $u(t)$ results in *the completion of the signal over all previous time instances up to the current one*. The system that performs this process is known as *integrator*.

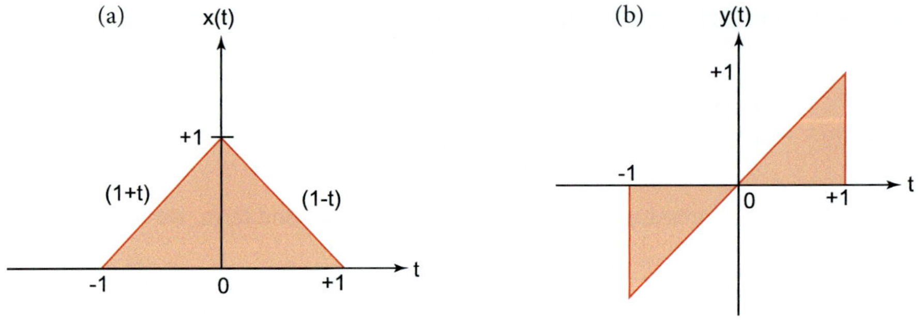

Fig. 1.21 The continuous-time signals $x(t)$ and $y(t)$ of Exercise 1.10.7

Graphical Computation of Convolution

Although the computation of the convolution $z(t)$ of two continuous-time signals $x(t)$ and $y(t)$ can be performed directly from the integral that defines it, as we have seen in previous examples, in practice this computation can be performed in a diagrammatic way and using the basic operations of shifting and time reversal that we presented in a previous section. Generally speaking, the process that we apply in this approach consists of the following steps:

- We choose one of the two signals, say, $y(\tau)$, and apply the operation of time reversal on it, so that we get the signal $y(-\tau)$.
- We apply the operation of time shifting by a time interval t on the time-reversed signal, so that we get the signal $y(t-\tau)$.
- We multiply the time-reversed and simultaneously shifted signal $y(t-\tau)$ with the other signal $x(\tau)$, so that we get the product $x(\tau)y(t-\tau)$.
- We integrate the product $x(\tau)y(t-\tau)$ over the interval $(-\infty, +\infty)$ (or generally over the appropriate interval of values in each case), so that we get the convolution of the two signals

$$z(t) = x(t) * y(t) = \int_{-\infty}^{+\infty} x(\tau)y(t-\tau) = \int_{-\infty}^{+\infty} y(\tau)x(t-\tau)d\tau$$

with the last integral of the above equation resulting from the power of the commutative property of convolution. This graphical method of calculating the convolution is demonstrated in the next two examples in order for the reader to fully understand the way it works and is used in practice.

Problem 1.10.7 Calculate the convolution between the signals $x(t)$ and $y(t)$ of Fig. 1.21.

Solution The graphical calculation of the convolution will be carried out by applying the process we previously described and which is depicted in Fig. 1.22. In this figure, image (a) depicts the original signal $y(\tau)$, image (b) shows the time-reversed signal $y(-\tau)$, which is a mirror image with respect to the y axis compared with our original signal, whereas image (c) shows the time-reversed and shifted signal $y(t-\tau)$, which is shifted with respect to the reversed signal $y(-\tau)$ by a time interval of duration t. On the other hand, image (d) presents the product of the signals $x(\tau)y(t-\tau)$ for some values of the parameter t that in our example are related to the interval $[-2, +2]$. In order to construct the product of the two signals, we need to draw them one on top of the other and then multiply between them their corresponding instantaneous values. It is obvious that in areas of the t axis in which the two signals do not overlap, the product $x(\tau)y(t-\tau)$ and therefore the convolution $x(t) * y(t)$ will be equal to zero—since in these areas $x(\tau) = 0$ or $y(t-\tau) = 0$.

Based on the contents of Fig. 1.22, we distinguish the following cases:

- $t < -2$: In this range of values, the signals $x(\tau)$ and $y(t-\tau)$ do not overlap and therefore both their product and their convolution $z(t)$ will have a value equal to zero.

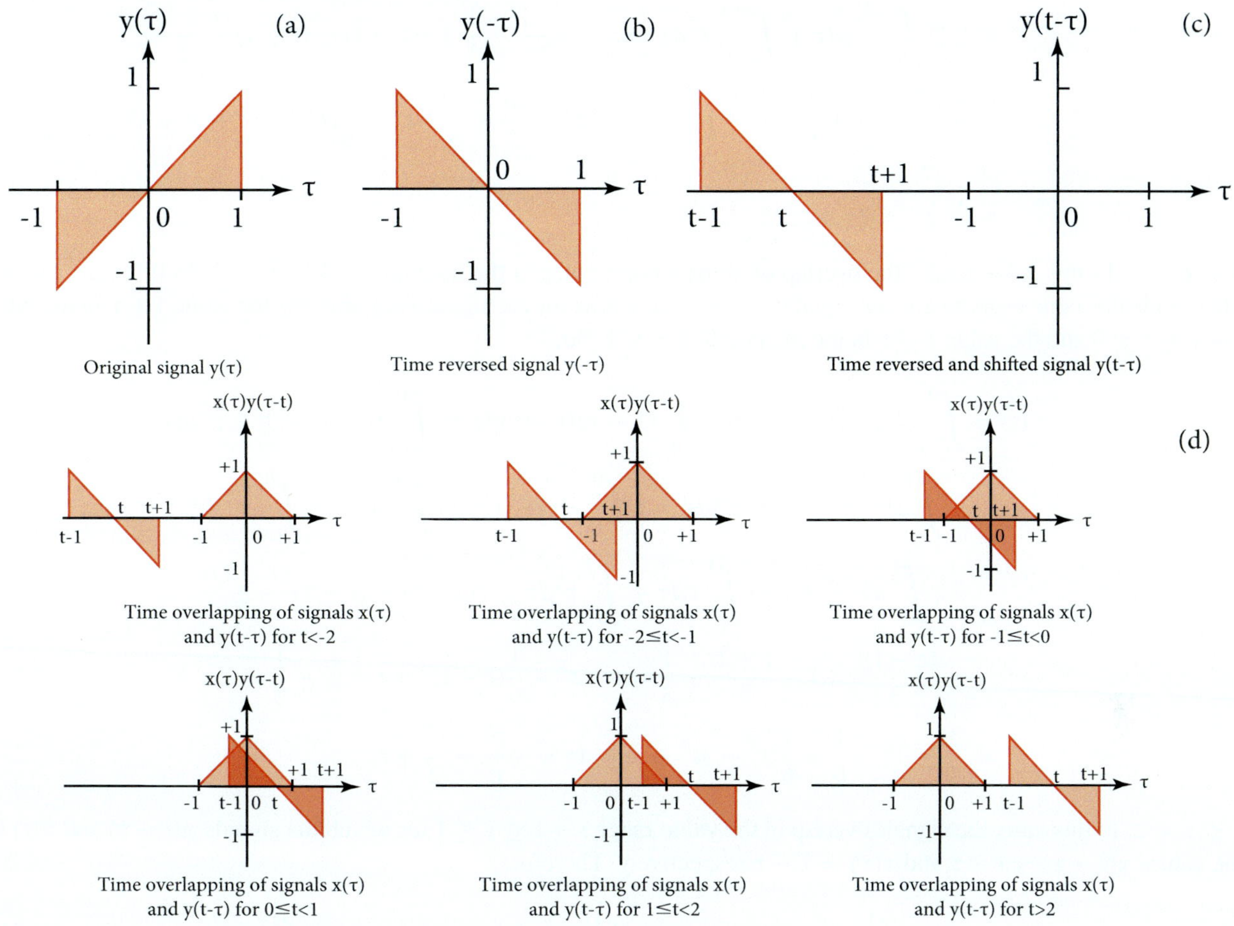

Fig. 1.22 Graphical calculation of the convolution of the signals $x(t)$ and $y(t)$ of Exercise 1.10.7

- $-2 \le t \le -1$: In this case, the signals overlap in the range of values $-1 \le \tau \le t+1$. For this range of values, it is $x(\tau) = 1 + \tau$ and $y(t-\tau) = t - \tau$ and therefore, we have

$$z(t) = \int_{-1}^{t+1} (1+t)(t-\tau)d\tau = \int_{-1}^{t+1} (t - \tau + t\tau - \tau^2)d\tau = t\int_{-1}^{t+1} d\tau$$
$$+ (t-1)\int_{-1}^{t+1} \tau d\tau - \int_{-1}^{t+1} \tau^2 d\tau = t\Big[\tau\Big]_{-1}^{\tau+1} + (t-1)\Big[\frac{\tau^2}{2}\Big]_{-1}^{t+1} - \Big[\frac{\tau^3}{3}\Big]_{-1}^{t+1}$$
$$= t(t+2) + \frac{1}{2}(t-1)(t^2+2t) - \frac{1}{3}(t^3+3t^2+3t+2) = \frac{t^3}{6} + \frac{t^2}{2} - \frac{2}{3}$$

- $-1 < t < 0$: Here the signals overlap in the value range $-1 \le \tau \le t+1$. For this value range, the signal $y(\tau)$ has the same value everywhere, which is equal to $t - \tau$. However, the signal $x(\tau)$ is not the same everywhere. More specifically, for $-1 \le \tau \le 0$ it is $x(\tau) = 1 + \tau$, whereas for $0 \le \tau \le t+1$ it is $x(\tau) = 1 - \tau$. So,

$$z(t) = \int_{-1}^{0} (1+\tau)(t-\tau)d\tau + \int_{0}^{t+1} (1-\tau)(t-\tau)d\tau = \int_{0}^{-1} (t - \tau + t\tau - \tau^2)d\tau$$
$$+ \int_{0}^{t+1} (t - \tau - t\tau + \tau^2)d\tau = t\int_{-1}^{0} d\tau + (t-1)\int_{-1}^{0} \tau d\tau - \int_{-1}^{0} \tau^2 d\tau + t\int_{0}^{t+1} d\tau$$

$$- (t+1)\int_{0}^{t+1} \tau d\tau + \int_{0}^{t+1} \tau^2 d\tau = t - \frac{t-1}{2} - \frac{1}{3} + t(t+1) - (t+1)\frac{(t+1)^2}{2}$$
$$+ \frac{(t+1)^3}{3} = \frac{t}{2} + \frac{1}{6} + t^2 + t - \frac{1}{6}(t+1)^3 = \frac{3t}{2} + \frac{1}{6} + t^2 - \frac{1}{6}(t^3 + 3t^2 + 3t + 1) =$$
$$= -\frac{t^3}{6} + \frac{t^2}{2} + t$$

- $0 < t < 1$: In this value range, the overlap of signals takes place in the interval $t - 1 \leq \tau \leq 1$. In this case, the signal $y(t - \tau)$ is the same everywhere and equal to $t - \tau$, whereas as for the signal $x(\tau)$, this has the value $1 + \tau$ in the interval $t - 1 \leq \tau \leq 0$ and the value $1 - \tau$ in the interval $0 \leq \tau \leq 1$. So,

$$z(t) = \int_{t-1}^{0} (1+\tau)(t-\tau)d\tau + \int_{0}^{1} (1-\tau)(t-\tau)d\tau = \int_{t-1}^{0} (t - \tau + t\tau - \tau^2)d\tau$$
$$+ \int_{0}^{1} (t - \tau - t\tau + \tau^2)d\tau = t\int_{t-1}^{0} d\tau + (t-1)\int_{t-1}^{0} \tau d\tau - \int_{t-1}^{0} \tau^2 d\tau$$
$$+ t\int_{0}^{1} d\tau - (t+1)\int_{0}^{1} \tau d\tau + \int_{0}^{1} \tau^2 d\tau = t(1-t) - (t-1)\frac{(t-1)^2}{2}$$
$$+ \frac{(t-1)^3}{3} + t - \frac{t+1}{2} + \frac{1}{3} = t - t^2 - \frac{1}{6}(t-1)^3 + t - \frac{t}{2} - \frac{1}{6} =$$
$$= \frac{3t}{2} - t^2 - \frac{1}{6} - \frac{1}{6}(t^3 - 3t^2 + 3t - 1) = -\frac{t^3}{6} - \frac{t^2}{2} + t.$$

- $1 \leq \tau < 2$: In this case, the signals overlap in the value range $t - 1 \leq \tau \leq 1$ for which the signals $y(t - \tau)$ and $x(t)$ have the values $y(t - \tau) = t - \tau$ and $x(\tau) = 1 - \tau$ respectively. Therefore,

$$z(t) = \int_{t-1}^{1} (1-\tau)(t-\tau)d\tau = \int_{t-1}^{1} (t - \tau - t\tau + \tau^2)d\tau = t\int_{t-1}^{1} d\tau$$
$$- (t+1)\int_{t-1}^{1} \tau d\tau + \int_{t-1}^{1} \tau^2 d\tau = t(2-t) - \frac{t+1}{2}(2t - t^2)$$
$$+ \frac{1}{3}(-t^3 + 3t^2 - 3t + 2) = 2t - t^2 - \frac{1}{2}(2t^2 - t^3 + 2t - t^2)$$
$$+ \frac{1}{3}(2 - t^3 + 3t^2 - 3t) = \frac{t^3}{6} - \frac{t^2}{2} + \frac{2}{3}$$

- $t > 2$: In this value range the signals $x(\tau)$ and $y(t-\tau)$ do not overlap and therefore both their product and their convolution $z(t)$ will have a value equal to zero.

Combining all the above results, we conclude that the convolution of the signals $x(t)$ and $y(t)$ will be given by the relation

$$z(t) = \begin{cases} +(t^3/6) + (t^2/2) - (2/3) & -2 \leq t \leq -1 \\ -(t^3/6) + (t^2/2) + t & -1 \leq t < 0 \\ -(t^3/6) - (t^2/2) + t & 0 \leq t < 1 \\ +(t^3/6) - (t^2/2) + (2/3) & 1 \leq t < 2 \\ 0 & \text{anywhere else} \end{cases}$$

The convolution of the signals $x(t)$ and $y(t)$ is presented diagrammatically in Fig. 1.23.

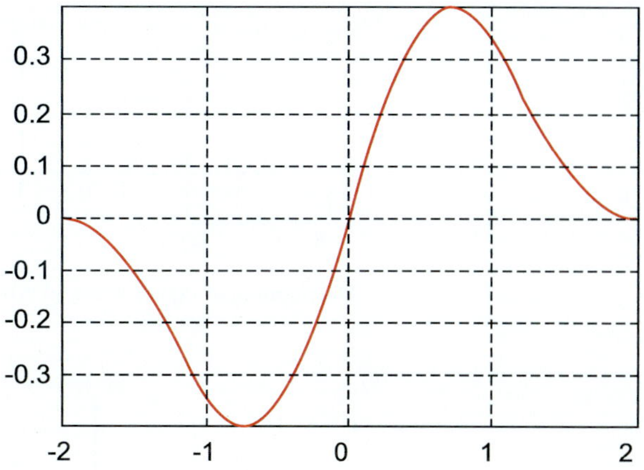

Fig. 1.23 The convolution of Exercise 1.10.7

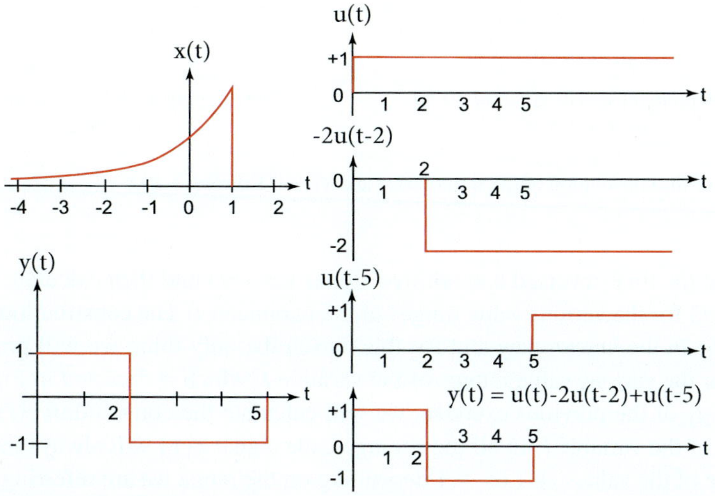

Fig. 1.24 Images (**a**)—top left—and (**b**)—bottom left—show the continuous-time signals $x(t)$ and $y(t)$ of Exercise 1.10.8 whose convolution is requested. Image (**c**)—top and bottom right—shows the process of constructing the signal $y(t)$ as the sum of time-shifted step functions

Problem 1.10.8 Calculate graphically the convolution between the signals $x(t)$ and $y(t)$, which are defined by the relations

$$x(t) = e^{2t}u(t) \quad \text{and} \quad y(t) = u(t) - 2u(t-2) + u(t-5)$$

Solution The continuous-time signals $x(t)$ and $y(t)$, whose convolution is requested are presented diagrammatically in Fig. 1.24. The same figure also depicts the process of constructing the signal $y(t)$ as the sum of the elementary signals $u(t)$, $-2u(t-2)$ and $u(t-5)$, which are obviously defined as

$$u(t) = \begin{cases} 1 \text{ for } t > 0 \\ 0 \text{ for } t < 0 \end{cases} \qquad -2u(t-2) = \begin{cases} -2 \text{ for } t > 2 \\ 0 \text{ for } t < 2 \end{cases} \qquad u(t-5) = \begin{cases} 1 \text{ for } t > 5 \\ 0 \text{ for } t < 5 \end{cases}$$

In order to calculate the convolution of the two signals

$$z(t) = \int_{-\infty}^{+\infty} x(\tau)y(t-\tau)d\tau = \int_{-\infty}^{+\infty} \Big[u(t-\tau) - 2u(t-2-\tau) + u(t-5-\tau)\Big] e^{2\tau}u(1-\tau)d\tau$$

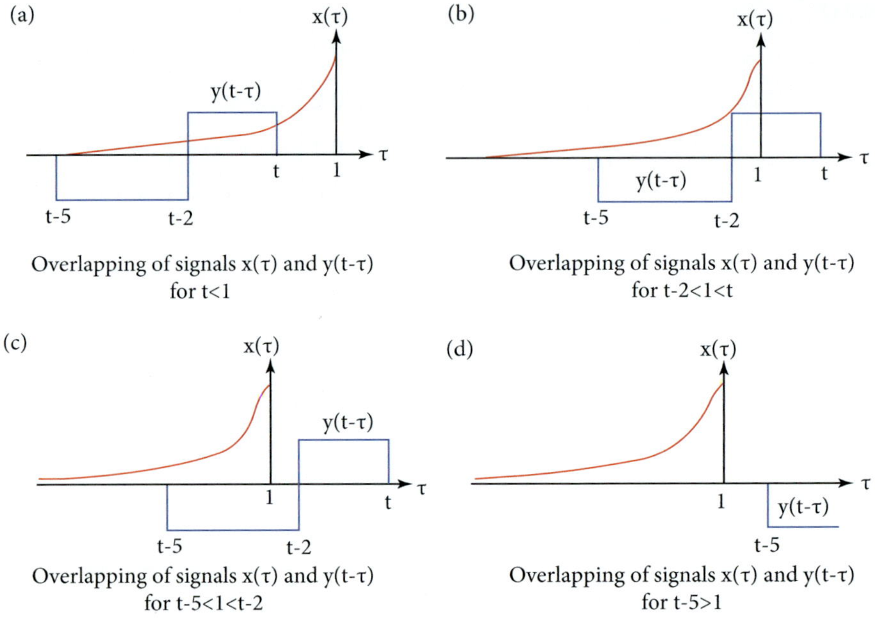

Fig. 1.25 Graphical calculation of the convolution of the signals $x(t)$ and $y(t)$ of Exercise 1.10.8

we must, as before, construct the time-reversed and -shifted signal $y(t-\tau)$ and then calculate the integral of the product of this signal with the signal $x(\tau)$ for the various value ranges of the parameter t. The construction of the signal $y(t-\tau)$ is now considered trivial and is done in the known way and for this reason the only thing we will present diagrammatically is the overlap of the two signals for the various value ranges of the variable t, which is depicted in Fig. 1.25.

Following the methodology of the previous exercise, we will calculate the convolution $z(t)$ of the signals $x(t)$ and $y(t)$ for the various value ranges of the variable t. In all these ranges, the signal $x(\tau)$ will always have the value $e^{2\tau}$ whereas the signal $y(t-\tau)$ will have one of the values $+1$ or -1 depending on the range we are referring to. In detail, we will have:

- $t < 1$: In this case, the signals $x(\tau)$ and $y(t-\tau)$ overlap in the value range $t-5 \le \tau < t$. For $t-5 \le \tau \le t-2$, $y(t-\tau) = -1$ whereas for the range $t-2 \le \tau < t$, $y(t-\tau) = +1$ and therefore we will have

$$\begin{aligned} z(t) &= -\int_{t-5}^{t-2} e^{2\tau} d\tau + \int_{t-2}^{t} e^{2\tau} d\tau = \frac{e^{2\tau}}{2}\bigg|_{t-2}^{t} - \frac{1}{2}e^{2\tau}\bigg|_{t-5}^{t-2} \\ &= \frac{1}{2}\Big[e^{2t} - e^{2(t-2)} - e^{2(t-2)} + e^{2(t-5)}\Big] = \frac{1}{2}\Big(e^{2t} - e^{2t}e^{-4} - e^{2t}e^{-4} + e^{2t}e^{-10}\Big) \\ &= \frac{1}{2}e^{2t}(1 - 2e^{-4} + e^{-10}) \end{aligned}$$

- $t-2 < 1 < t$ (or equivalently $1 < t < 3$): In this case, the signals $x(\tau)$ and $y(t-\tau)$ overlap in the value range $t-5 \le \tau < 1$. For $t-5 \le \tau < t-2$ it is $y(t-\tau) = -1$ whereas for $t-2 \le \tau < 1$ it is $y(t-\tau) = +1$. Therefore, we will have

$$\begin{aligned} z(t) &= -\int_{t-5}^{t-2} e^{2\tau} d\tau + \int_{t-2}^{1} e^{2\tau} d\tau = \frac{e^{2\tau}}{2}\bigg|_{t-2}^{1} - \frac{e^{2\tau}}{2}\bigg|_{t-5}^{t-2} \\ &= \frac{1}{2}\Big[e^{2} - e^{2(t-2) - e^{2(t-2)}} + e^{2(t-5)}\Big] = \frac{1}{2}\Big(e^{2} - e^{2t}e^{-4} - e^{2t}e^{-4} + e^{2t}e^{-10}\Big) \\ &= \frac{e^2}{2} + \frac{1}{2}(e^{-10} - 2e^{-4})e^{2t}. \end{aligned}$$

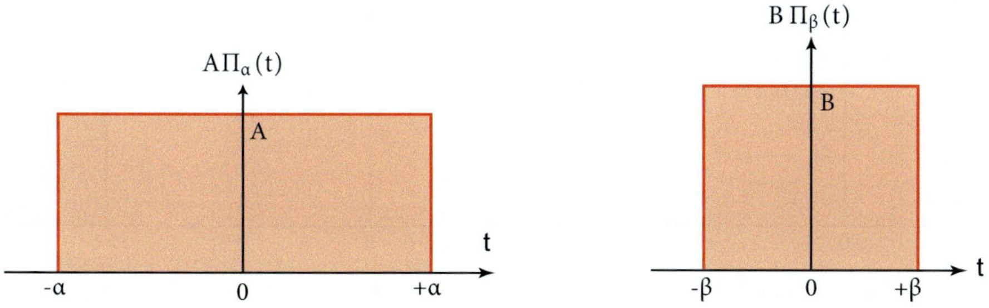

Fig. 1.26 The square pulses $A\Pi_\alpha(t)$ and $B\Pi_\beta(t)$ of Exercise 1.10.9

- $t-5<1<t-2$ (or equivalently $3<t<6$): In this case, the signals $x(\tau)$ and $y(t-\tau)$ overlap in the value range $t-5\le\tau<1$. In this value range, it is $y(t-\tau)=-1$ and therefore we will have

$$z(t)=-\int_{t-5}^{1}e^{2\tau}d\tau=\frac{1}{2}\Big[e^{2(t-5)}-e^{2}\Big].$$

- $t-5>1$ (or equivalently $t>6$): In this case, the two signals do not overlap and therefore their convolution $z(t)$ will be equal to zero.

Combining the above results, we conclude that the convolution of the signals $x(t)$ and $y(t)$ will be given by the relation

$$z(t)=\begin{cases}(e^{2t}/2)(1-2e^{-4}+e^{-10}) & \text{for } t\le 1\\ (e^{2}/2)+(e^{2t}/2)(e^{-10}-2e^{-4}) & \text{for } 1<t<3\\ [e^{2(t-5)}-e^{2}]/2 & \text{for } 3<t<6\\ 0 & \text{for } t>6.\end{cases}$$

The graphical representation of the function $z(t)$ can be drawn using the appropriate software application and is left as an exercise to the reader.

Problem 1.10.9 Calculate graphically the convolution between the rectangular pulses

$$x(t)=A\Pi_\alpha(t)=\begin{cases}A & |t|<\alpha\\ 0 & |t|>\alpha\end{cases}\quad\text{and}\quad y(t)=B\Pi_\beta(t)=\begin{cases}B & |t|<\beta\\ 0 & |t|>\beta,\end{cases}$$

which are depicted in Fig. 1.26 (for the solution of the exercise consider the case $\alpha>\beta$).

Solution Following the methodology of the previous exercise, we calculate the convolution $z(t)$ of the rectangular pulses $x(t)$ and $y(t)$ for the various value ranges of the variable. t. In this case, there are five scenarios that need to be examined, which are presented diagrammatically in Fig. 1.27. In all cases, the two signals will always have the values $x(t)=A$ and $y(t-\tau)=B$ and the only thing that changes are the limits of integration that we will use for each value range. In a detailed description, the graphical method of calculating the convolution is as follows:

- Case 1: $t+\beta\le-\alpha$ or equivalently $t\le-(\alpha+\beta)$ [image (a) of the figure]. In this case, the two signals do not overlap and therefore the convolution $z(t)$ will be equal to zero.
- Case 2: $-(\alpha+\beta)<t\le-(\alpha-\beta)$ [figure (b) of the figure]. In this case, the two signals overlap for the value range $-\alpha\le\tau\le t+\beta$ and therefore will be

$$z(t)=\int_{-\alpha}^{t+\beta}ABd\tau=AB\int_{-\alpha}^{t+\beta}d\tau=AB(t+\alpha+\beta)$$

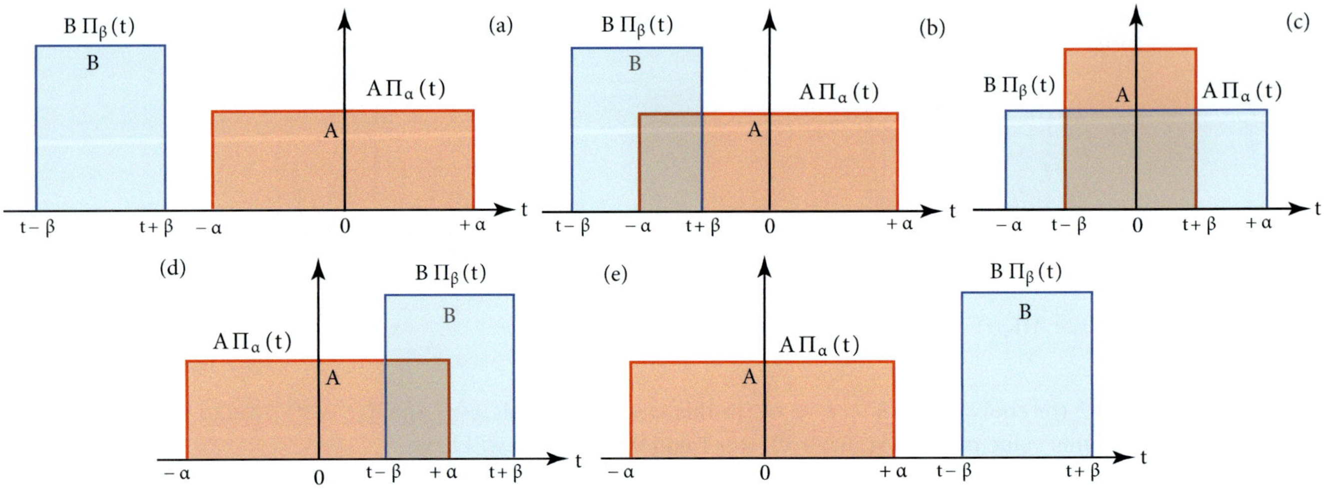

Fig. 1.27 The five cases of interest for calculating the convolution of the rectangular pulses $A\Pi_\alpha(t)$ and $B\Pi_\beta(t)$ of Exercise 1.10.9

- Case 3: $-(\alpha-\beta) < t \le (\alpha-\beta)$ [figure (c) of the figure]. In this case, the two signals overlap for the value range $-\beta \le \tau \le +\beta$ and therefore we will have

$$z(t)=\int_{-\beta}^{+\beta} ABd\tau = AB\int_{-\beta}^{+\beta} d\tau = 2\beta AB.$$

- Case 4: $(\alpha-\beta) < \tau \le (\alpha+\beta)$ [image (d) of the diagram]. In this case, the square pulses overlap for the value range $t-\beta \le \tau < \alpha$ and therefore will be

$$z(t)=\int_{t-\beta}^{\alpha} ABd\tau = AB\int_{t-\beta}^{\alpha} d\tau = AB(-t+\alpha+\beta).$$

- Case 5: $t > (\alpha+\beta)$ [image (e) of the diagram]. In this case, the two signals do not overlap and therefore the convolution $z(t)$ will be equal to zero.

Combining the above results, we conclude that the convolution of the rectangular pulses $x(t) = A\Pi_\alpha(t)$ and $y(t) = B\Pi_\beta(t)$ is given by the relation

$$z(t)=\begin{cases} 0 & \text{for } t \le -(\alpha+\beta) \\ AB(t+\alpha+\beta) & \text{for } -(\alpha+\beta) \le t < -(\alpha-\beta) \\ 2\beta AB & \text{for } -(\alpha-\beta) \le t < (\alpha-\beta) \\ AB(-t+\alpha+\beta) & \text{for } (\alpha-\beta) \le t < (\alpha+\beta) \\ 0 & \text{for } t > (\alpha+\beta). \end{cases}$$

The graphical representation of the function $z(t)$ is shown in Fig. 1.28a from which it is clear that the convolution of the rectangular pulses $x(t) = A\Pi_\alpha(t)$ and $y(t) = B\Pi_\beta(t)$ is a *trapezoid* pulse whose large base is defined by the points $[-(\alpha+\beta), 0]$ and $[(\alpha+\beta), 0]$, the small base is defined by the points $[-(\alpha-\beta), 2\beta AB]$ and $[(\alpha-\beta), 2\beta AB]$, whereas its height is equal to $2\beta AB$.

In the special case where the relations $\alpha = \beta$ and $A = B$ hold, the length of the trapezoid's smaller base becomes zero and the trapezoid degenerates into a *triangle* defined by the points $A(-2\alpha, 0)$, $B(2\alpha, 0)$ and $\Gamma(0, 2\alpha A^2)$. This triangle is

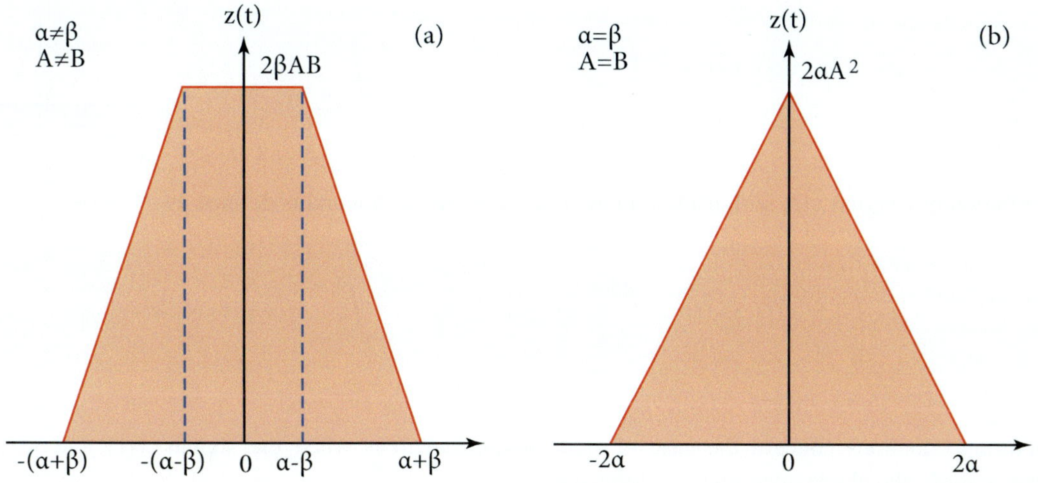

Fig. 1.28 The convolution of the rectangular pulses $A\Pi_\alpha(t)$ and $B\Pi_\beta(t)$ of Exercise 1.10.9, for the cases ($\alpha \neq \beta$, $A \neq B$) and ($\alpha = \beta$, $A = B$)

depicted in Fig. 1.28b, where the height of the triangle is 20 times smaller than its actual value for illustrative purposes. We therefore conclude that *the convolution of two identical rectangular pulses is a triangular pulse*, which is described by the above characteristics. As for the equation describing this pulse, it arises from the above general result, if we proceed with the substitutions $\beta = \alpha$ and $B = A$—in this case, the equation in question is reduced to its simplest form

$$z(t) = \begin{cases} 0 & \text{for } t \le -2\alpha \\ A^2(2\alpha + t) & \text{for } -2\alpha < t < 0 \\ A^2(2\alpha - t) & \text{for } 0 < t < 2\alpha \\ 0 & \text{for } t \ge 2\alpha \end{cases} = \begin{cases} A^2(2\alpha - |t|) & \text{for } |t| < 2\alpha \\ 0 & \text{for } |t| \ge 2\alpha, \end{cases}$$

which is recognized as the defining equation of a triangular pulse.

Note In all the exercises of graphical computation of the convolution, the method applied worked as follows: one of the two signals remained constant, whereas the other moved along the axis of the year from left to right and for the various regions values of the variable t for which the two signals overlapped, their convolution $z(t)$ was calculated. This situation was particularly evident in the last exercise, where the moving square pulse began to move from the left; starting from a position characterized by a complete absence of overlap and bypassing the other pulse, it finally stopped in a position on the right for which again the two pulses did not overlap. It is obvious that, if the signals whose convolution is requested have finite time durations T_1 and T_2, then the time required for the above movement and during which the two signals overlap, so that their convolution is different from zero, is equal to $T_1 + T_2$. Conversely, for time instances $t > (T_1 + T_2)$ the two signals no longer overlap and therefore their convolution is equal to zero. This observation explains in a qualitative way the property of the range of the convolution according to which *the time duration of the convolution of two finite signals $x(t)$ and $y(t)$ with durations T_1 and T_2 is equal to $T_1 + T_2$* and an example of its application is the convolution of the two identical rectangular pulses of duration 2α, which is a triangular pulse of duration 4α.

1.10.3 Cross-Correlation and Autocorrelation of Continuous-Time Signals

Another significant operation that can be applied between two continuous-time signals $x(t)$ and $y(t)$ is the *correlation* or *cross-correlation*, which is defined by an equation of the form

$$\varphi_{xy}(t) \doteq x(t) \diamond y(t) = \int_{-\infty}^{+\infty} x(\tau)y(\tau - t)d\tau = \int_{-\infty}^{+\infty} x(t + \tau)y(\tau)d\tau \tag{1.39}$$

The cross-correlation of a signal $x(t)$ with itself is known as *auto-correlation* and is defined as

$$\varphi_{xx}(t) \doteq x(t) \diamond x(t) = \int_{-\infty}^{+\infty} x(\tau)x(\tau - t)d\tau = \int_{-\infty}^{+\infty} x(t + \tau)x(\tau)d\tau \tag{1.40}$$

The above defining equations concern the case of *real signals*, whereas when the signals $x(t)$ and $y(t)$ are *complex continuous-time signals*, the above quantities are defined as

$$\varphi_{xy}(t) \doteq x(t) \diamond y^*(t) = \int_{-\infty}^{+\infty} x(\tau)y^*(\tau - t)d\tau \tag{1.41}$$

$$\varphi_{xx}(t) \doteq x(t) \diamond x^*(t) = \int_{-\infty}^{+\infty} x(\tau)x^*(\tau - t)d\tau \tag{1.42}$$

with the symbol $(*)$ referring to the operation of *complex conjugation*.

It is not difficult to ascertain that in the special case for which $t = 0$, *the auto-correlation function coincides with the energy of the signal* E_∞. Indeed, if we set $t = 0$ in the general defining equation of auto-correlation for a complex signal, it immediately follows that

$$\varphi_{xx}(0) = \int_{-\infty}^{+\infty} x(\tau)x^*(\tau)d\tau = \int_{-\infty}^{+\infty} |x(\tau)|^2 d\tau = E_\infty. \tag{1.43}$$

For this reason, the autocorrelation function—which is also the one mainly used in practice—is often known as the *energy autocorrelation function*. Of course, this definition is valid only for *energy signals* whose energy E_∞ is finite. Conversely, for *power signals* characterized by infinite energy and finite power, the autocorrelation function—which in this case is called the *power autocorrelation function*—is defined by the equation

$$R_{xx}(t) = \lim_{T\to\infty} \frac{1}{2T} \int_{-T}^{+T} x(\tau)x^*(\tau - t)d\tau \tag{1.44}$$

and which for $t = 0$ coincides with the average power of the signal, since it is proven that

$$R_{xx}(0) = \lim_{T\to\infty} \frac{1}{2T} \int_{-T}^{+T} x(\tau)x^*(\tau)d\tau = \lim_{T\to\infty} \frac{1}{2T} \int_{-T}^{+T} |x(\tau)|^2 d\tau = P_x.$$

In the special case of *periodic* signals, which are power signals, the power autocorrelation function arises from the above integral *calculated for a time period of one period*, that is,

$$R_{xx}(t) = \frac{1}{T} \int_{-T/2}^{+T/2} x(\tau)x^*(\tau - t)d\tau, \tag{1.45}$$

where $T = 2\pi/\omega_0$ is the period of the signal.

The above differentiation of autocorrelation for energy signals and the power signals naturally extend to cross-correlation. Thus, the relationships given at the beginning of this section refer to the cross-correlation of energy signals, whereas the corresponding expression for the power signals takes the form

$$R_{xy}(t) = \frac{1}{T}\int_{-T/2}^{+T/2} x(\tau)y^*(\tau - t)d\tau \tag{1.46}$$

Problem 1.10.10 Calculate the cross-correlation of the signals $x(t) = A\cos(\omega t)$ and $y(t) = B\sin(\omega t)$.

Solution The signals $x(t)$ and $y(t)$ are periodic signals and therefore are power signals. What is essentially required is the *power cross-correlation*, which is calculated as

$$R_{xy}(t) = \frac{1}{T}\int_{-T/2}^{+T/2} x(\tau)y^*(\tau - t)d\tau = \frac{AB}{T}\int_{-T/2}^{+T/2} \cos(\omega\tau)\sin\Big[\omega(\tau - t)\Big]d\tau$$

Using the known trigonometric identity $2\cos\alpha\sin\beta = \sin(\alpha+\beta) - \sin(\alpha-\beta)$, the expression to be integrated is formulated as

$$\cos(\omega\tau)\sin[\omega(\tau - t)]d\tau = \frac{1}{2}\Big[\sin\Big(\omega(2\tau - t)\Big) - \sin(\omega t)\Big]$$

and the cross-correlation of the two signals takes the form

$$R_{xy}(t) = \frac{AB}{2T}\left\{\int_{-T/2}^{+T/2} \sin[\omega(2\tau - t)]d\tau - \int_{-T/2}^{T/2} \sin(\omega t)d\tau\right\}.$$

The first of the two integrals is equal to zero, as it is easy to prove[21] and therefore

$$R_{xy}(t) = -\frac{AB}{2T}\int_{-T/2}^{T/2} \sin(\omega t)d\tau = -\frac{AB}{2T}\sin(\omega t)\int_{-T/2}^{T/2} d\tau = -\frac{AB}{T}\sin(\omega t),$$

which is the desired result. We observe that for $t = 0$ it is $R_{xy}(0) = 0$, which is expected, since the functions of sine and cosine are orthogonal to each other. ■

It is interesting to note that the calculation of the cross-correlation of two continuous-time signals $x(t)$ and $y(t)$ can be carried out with the help of a *diagrammatic method*, similar to the one we used for the calculation of the convolution. This fact is expected, since these two operations are characterized by similar mathematical form. Following the argumentation we presented then, we can formulate the conclusion that the cross-correlation function is defined—that is, it has nonzero values—*only in those areas of the time axis where the two signals overlap*.[22] A typical example of this situation is depicted in the Fig. 1.29, in which the time shift of the signal $y(t)$ is depicted in relation to the signal $x(t)$ for various values of the time shift τ, through the application of a process similar to the one we encountered during the graphical calculation of the convolution. Observing Fig. 1.29, we find that, when the variable t is greater than or equal to zero, the two signals overlap in the interval $t \le \tau \le \xi$, whereas for the negative values of time t the overlap interval is $0 \le \tau \le \xi - |t|$. Therefore, in

[21] This result is based on the observation that $\omega T = 2\pi$ and it is proven as follows:

$$\int_{-T/2}^{+T/2} \sin[\omega(2\tau - t)]d\tau = -\frac{\cos[\omega(t - 2\tau)]}{2\omega}\Bigg|_{-T/2}^{+T/2} = -\frac{1}{2\omega}\Big[\cos[\omega(t - T)] - \cos[\omega(t + T)]\Big] = 0.$$

[22] In the case where the cross-correlation function of two signals $x(t)$ and $y(t)$ is equal to zero for all values of the parameter t, the two signals are characterized as *fully uncorrelated*.

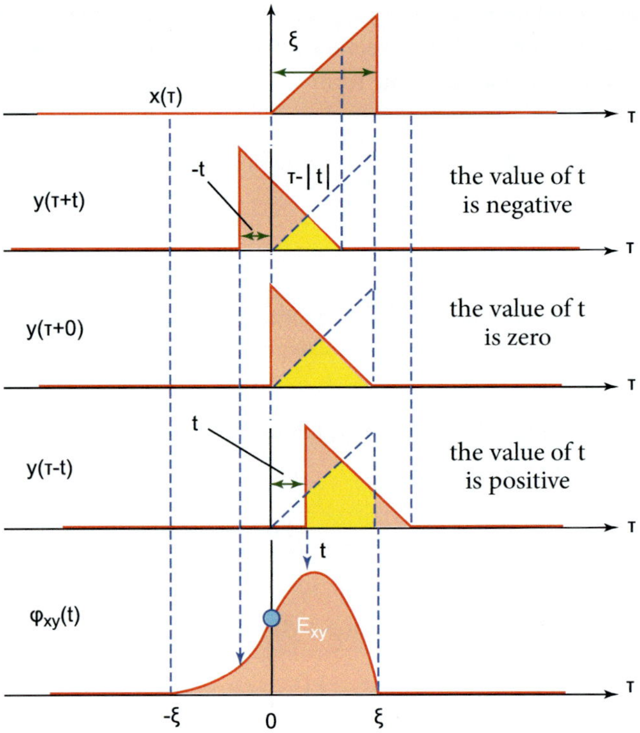

Fig. 1.29 The cross-correlation of the signals $x(t)$ and $y(t)$ has nonzero values only in those areas of the time axis in which the two signals overlap (Source: Shmaliy, *continuous-time Signals*, Fig. 4.29)

calculating the cross-correlation function for these two cases, the limits of the integral will be determined by the above value intervals, as shown in the next exercise.

Problem 1.10.11 Calculate the cross-correlation between the continuous-time signals

$$x(t) = Au(t) - Au(t-\xi) \quad \text{and} \quad y(t) = \begin{cases} At/\xi \text{ for } 0 \leq t \leq \xi \\ 0 \quad \text{elsewhere} \end{cases}$$

Solution From the defining equations of the two signals, we find that the signal $x(t)$ is a square pulse of duration ξ and height A, whereas the signal $y(t)$ is a ramp function characterized by the same properties. Working with a diagrammatic method similar to the one we used for the calculation of the convolution, we come to the conclusion that for the case $t \geq 0$ the cross-correlation of the two signals is calculated as

$$\varphi_{xy}(t) = \int_{-\infty}^{+\infty} x(\tau)y(\tau - t)d\tau = \int_t^{\xi} A \cdot \frac{A}{\xi}(\tau - t)d\tau = \frac{A^2}{\xi}\left\{\int_t^{\xi} \tau d\tau - t\int_t^{\xi} d\tau\right\} =$$
$$= \frac{A^2}{\xi}\left\{\frac{\xi^2 - t^2}{2} - t(\xi - t)\right\} = \frac{A^2}{\xi}(\xi - t)\left[\frac{\xi + t}{2} - t\right] = \frac{A^2}{2\xi}(\xi - t)^2$$

whereas for the case $t < 0$ we will have

$$\varphi_{xy}(t) = \int_{-\infty}^{+\infty} x(\tau)y(\tau - t)d\tau = \int_0^{\xi-|t|} A \cdot \frac{A}{\xi}(\tau - |t|)d\tau = \frac{A^2}{\xi}\left\{\int_0^{\xi-|t|} \tau d\tau - |t|\int_0^{\xi-|t|} d\tau\right\}$$
$$= \frac{A^2}{\xi}\left\{\frac{(\xi - |t|)^2}{2} - |t|(\xi - |t|)\right\} = \frac{A^2}{\xi}(\xi - |t|)\left\{\frac{(\xi - |t|)}{2} - |t|\right\} = \frac{A^2}{2\xi}(\xi - |t|)(\xi - 3|t|)$$

It is proven that the cross-correlation function $\varphi_{xy}(t)$ is characterized by the property $\varphi_{xy}(t) \neq \varphi_{yx}(t)$ and more specifically that $\varphi_{xy}(t) = \varphi_{yx}(-t)$. Therefore, the operation of cross-correlation in the general case *is not characterized by the commutative property*. This conclusion arises directly from the defining equation of cross-correlation, as it is found that

$$\varphi_{xy}(t) = \int_{-\infty}^{+\infty} x(t+\tau)y(\tau)d\tau = \int_{-\infty}^{+\infty} y(-t+\tau)x(\tau)d\tau = \varphi_{yx}(-t)$$

If the two signals coincide, then the above relationship is reduced to the simpler form $\varphi_{xx}(t) = \varphi_{xx}(-t)$ from which follows the conclusion that the autocorrelation function is characterized by *even symmetry*.

The act of cross-correlation is considered particularly important, because it allows the quantitative determination of the degree of *similarity* that exists between two signals. In order for this to be perceived, let us consider two signals $x(t)$ and $y(t)$, which differ by a time shift on the horizontal axis. In this case, the calculation of the cross-correlation allows the determination of the degree of shift of one signal in relation to the other, so that the two signals coincide with each other as much as possible. This is something that is done by sliding one signal along the horizontal axis and calculating the value of the product of the two signals for each possible value of the time shift τ and for each time instance t. The moment in time at which the two signals coincide, their degree of similarity becomes the *maximum possible* and the cross-correlation function takes its maximum value. This is something that is graphically explained by saying that, when the two signals coincide, the positive areas defined by their graphical representations align to the maximum possible degree, thus leading to the maximization of the integral value, which naturally is equal to the area defined by the graphical representation of the product of the two signals (note that this is something that also applies to the *negative surfaces*, since the product of two negative numbers is a positive quantity).

A characteristic example of the application of the cross-correlation function is the operating principle of *radar*, which is used for the detection of the presence and position of an object anywhere in space. In a radar, an electromagnetic signal is emitted toward the target, which after being reflected from it, returns to its source with the duration of transmission from and to the target being a function of the distance of the target from the source. Because in the ideal case the incident and the reflected signal are exactly the same, with the reflected signal simply being a time-delayed version of the incident signal, the function that is actually used is not the cross-correlation but the autocorrelation.

Properties of the Autocorrelation Function

The energy autocorrelation function is characterized by a multitude of useful properties,[23] which in general terms are the following:

- The autocorrelation function $\varphi_{xx}(t)$ is characterized by even symmetry, that is $\varphi_{xx}(-t) = \varphi_{xx}(t)$. This property was proven in the previous pages.
- The maximum value of the autocorrelation function $\varphi_{xx}(t)$ corresponds to the argument value $t = 0$, or in mathematical notation $|\varphi_{xx}(t)| \leq \varphi_{xx}(0)$, with the value $\varphi_{xx}(0)$ corresponding, as we have seen previously, to the energy of the signal $x(t)$. This property is a consequence of the *Cauchy-Schwartz inequality*.
  Proof Let us consider a continuous-time signal of the form $z(\tau) = \alpha x(\tau) + \beta y(\tau - t)$, where α and β are any two real constants. In this case, the energy of the signal $z(\tau)$ is calculated as

$$E_z = \int_{-\infty}^{+\infty} |z(\tau)|^2 d\tau = \int_{-\infty}^{+\infty} \left[\alpha x(\tau) + \beta y(\tau - t)\right]^2 d\tau$$

$$= \alpha^2 \int_{-\infty}^{+\infty} x^2(\tau)d\tau + 2\alpha\beta \int_{-\infty}^{+\infty} x(\tau)y(\tau - t)d\tau + \beta^2 \int_{-\infty}^{+\infty} y^2(\tau - t)d\tau.$$

[23] Similar properties obviously characterize the power autocorrelation function.

However,

$$\int_{-\infty}^{+\infty} x^2(\tau)d\tau = E_x = \varphi_{xx}(0), \quad \int_{-\infty}^{+\infty} x^y(\tau - t)d\tau = E_y = \varphi_{yy}(0), \quad \int_{-\infty}^{+\infty} x(\tau)y(\tau - t)d\tau = \varphi_{xy}(t)$$

and the above equation becomes $E_z = \alpha^2 E_x + 2\alpha\beta R_{xy}(t) + \beta^2 E_y \geq 0$ or dividing by the quantity β^2 $(\beta \neq 0)$

$$\varphi_{xx}(0)\left(\frac{\alpha}{\beta}\right)^2 + 2\varphi_{xy}(t)\left(\frac{\alpha}{\beta}\right) + \varphi_{yy}(0) \geq 0.$$

If we set $\alpha/\beta = \xi$, we end up with the quadratic inequality

$$\varphi_{xx}(0)\xi^2 + 2\varphi_{xy}(t)\xi + \varphi_{yy}(0) \geq 0,$$

which, as is well known, is valid only when the discriminant is less than or equal to zero. Therefore, we have

$$\Delta = 4\varphi_{xy}^2(t) - 4\varphi_{xx}(0)\varphi_{yy}(0) \leq 0 \quad \text{or equivalently} \quad \varphi_{xy}^2(t) - \varphi_{xx}(0)\varphi_{yy}(0) \leq 0,$$

and finally

$$\varphi_{xy}(t) \leq \sqrt{\varphi_{xx}(0)}\sqrt{\varphi_{yy}(0)}.$$

The above equation is valid for the general case of the cross-correlation of any two continuous-time signals $x(t)$ and $y(t)$. If the two signals coincide, it immediately follows that $|\varphi_{xx}(t)| \leq |\varphi_{xx}(t)| \equiv E_x$, which is the desired result. According to this result, *the autocorrelation function takes its maximum value, which is equal to the energy of the signal for the case where* $t = 0$. This conclusion is, of course, expected, since for $t = 0$ the signals $x(\tau)$ and $x(\tau - t)$ coincide completely.■

- The autocorrelation function does not contain information about the phase of the signal and is independent of the choice of the origin of the time axis corresponding to the value $\tau = 0$.
- If the signal $x(t)$ is periodic with period T, then the autocorrelation function is also periodic and is characterized by the same period value T.
  Proof The autocorrelation of the signal $x(t)$ at the time $t + T$ will be given by the relation

$$\varphi_{xx}(t + T) = \int_{-\infty}^{+\infty} x(\tau)x(\tau - t - T)d\tau.$$

However, if the signal is periodic with period T, it will satisfy the relation $x(t) = x(t - T)$ and therefore we have $x(\tau - t - T) = x(\tau - t)$. Using this property in the previous equation, we will get

$$\varphi_{xx}(t + T) = \int_{-\infty}^{+\infty} x(\tau)x(\tau - t)d\tau = \varphi_{xx}(t),$$

that is, the autocorrelation $\varphi_{xx}(t)$ is also a periodic signal with period T.

- If the signal $x(t)$ has zero mean and is aperiodic, then it is proven that $\lim_{t\to\infty} \varphi_{xx}(t) = 0$.
- The autocorrelation of the sum of two completely uncorrelated signals—that is, signals whose cross-correlation is equal to zero for all values of its argument—is equal to the sum of the autocorrelations of the two signals separately.
  Proof Let us consider two continuous-time signals $x(t)$ and $y(t)$ and the signal $z(t) = x(t) + y(t)$ defined as the sum of the two signals. In this case, the autocorrelation of the signal $z(t)$ will be given by the relation

$$\begin{aligned}
\varphi_{zz}(t) &= \int_{-\infty}^{+\infty} z(\tau)z(\tau - t)d\tau = \int_{-\infty}^{+\infty} \Big[x(\tau) + y(\tau)\Big]\Big[x(\tau - t) + y(\tau - t)\Big]d\tau \\
&= \int_{-\infty}^{+\infty} \Big[x(\tau)x(\tau - t) + x(\tau)y(\tau - t) + y(\tau)x(\tau - t) + y(\tau)y(\tau - t)\Big]d\tau \\
&= \int_{-\infty}^{+\infty} x(\tau)x(\tau - t) + \int_{-\infty}^{+\infty} x(\tau)y(\tau - t)d\tau + \int_{-\infty}^{+\infty} y(\tau)x(\tau - t)d\tau + \int_{-\infty}^{+\infty} y(\tau)y(\tau - t)d\tau.
\end{aligned}$$

However, if the signals $x(t)$ and $y(t)$ are completely uncorrelated, we have

$$\varphi_{xy}(t) = \int_{-\infty}^{+\infty} x(\tau)y(\tau - t)d\tau = 0 \quad \text{and} \quad \varphi_{yx}(t) = \int_{-\infty}^{+\infty} y(\tau)x(\tau - t)d\tau = 0$$

and the above relation will take the form

$$\varphi_{zz}(t) = \int_{-\infty}^{+\infty} x(\tau)x(\tau - t) + \int_{-\infty}^{+\infty} y(\tau)y(\tau - t)d\tau = \varphi_{xx}(t) + \varphi_{yy}(t) \quad ,where$$

$$\varphi_{xx}(t) = \int_{-\infty}^{+\infty} x(\tau)x(\tau - t)d\tau \quad \text{and} \quad \varphi_{yy}(t) = \int_{-\infty}^{+\infty} y(\tau)y(\tau - t)d\tau$$

the autocorrelations of the signals $x(t)$ and $y(t)$ respectively.

1.11 Vector Spaces of Signal Representation

Taking into account that a continuous-time signal is nothing more than a mathematical function $f(x)$, we can proceed to the study of these signals using known tools from the field of functional analysis, such as linear vector spaces (see the appropriate section of the Appendix). Specifically, considering a finite set of N real continuous-time signals $x_1(t), x_2(t), \ldots, x_N(t)$, which share some common properties, we can consider that these belong to some suitably defined signal space $\mathscr{R}$, which is an example of a functional vector space and is characterized by all the properties we listed in the relevant section. For example, (a) the sum $z = x(t) + y(t)$ of two signals $x(t), y(t) \in \mathscr{R}$ also belongs to the space $\mathscr{R}$, (b) for every signal $x(t) \in \mathscr{R}$, the signal $y(t) = \alpha x(t)$, where α, any real number, also belongs to the space $\mathscr{R}$ and (γ) the space $\mathscr{R}$ is characterized by the existence of a special zero signal $\emptyset(t)$ such that for every $x(t) \in \mathscr{R}$ the relation $x(t) + \emptyset(t) = x(t)$ holds. In the above description, the signals are real functions of a real variable, whereas if they are complex quantities, then according to what we mentioned, the vector space to which they belong is a complex vector space.

Considering any linear signal space $\mathscr{R}$, we can choose a set of linearly independent basis vectors $\{x_1(t), x_2(t), x_3(t), \ldots\}$ the dimension of which can be finite or infinite and represent each vector/signal $x(t)$ of this space as a linear combination of the form $x(t) = \sum_i \alpha_i x_i(t)$ with the coefficients α_i of the linear combination being the projections of the signal $x(t)$ onto the elements of the vector basis of the space. In practice, the entire interval $(-\infty, \infty)$ is chosen as the domain of the functions $x(t)$, whereas in the general case this vector signal space is considered to be a Hilbert vector space of infinite dimensions, that is, a complete space equipped with a function of a vector norm that is induced by a suitably defined inner product. Specifically, the functions $\|x(t)\|_1$, $\|x(t)\|_2$ and $\|x(t)\|_\infty$ are used as the norm function, which arise from the general norm function.

$$\|\boldsymbol{x}\|_p = \sqrt[p]{\sum_{i=1}^{n} |x_i|^p} = \left(\sqrt{\sum_{i=1}^{n} |x_i|^p}\right)^{1/p}$$

for the values $p = 1$, $p = 2$ and $p = \infty$, but also the function

$$\|x(t)\|_{\text{rms}} = \sqrt{\lim_{T\to\infty} \frac{1}{T} \int_{-T/2}^{+T/2} x^2(t)dt} \tag{1.47}$$

(for the properties that these norm functions fulfill, see the relevant section of the appendix). The above definitions apply to scalar signals, whereas for vector signals which will also concern us in this book, the corresponding expressions are

$$\|\boldsymbol{x}(t)\|_1 = \int_{-\infty}^{+\infty} \left(\sum_i |x_i(t)|\right) dt = \int_{-\infty}^{+\infty} \|x(t)\|_i dt = \sum_i \|x(t)\|_i, \qquad \|\boldsymbol{x}(t)\|_2 = \sqrt{\int_{-\infty}^{+\infty} \boldsymbol{x}^T(t)\boldsymbol{x}(t)d\tau}$$

$$\|\boldsymbol{x}(t)\|_\infty = \sup_t \max_i |x_i(t)| = \sup_t \|x(t)\|_\infty, \qquad \|\boldsymbol{x}(t)\|_{\infty,p} = \sup_t \|\boldsymbol{x}(t)\|_p$$

$$\|\boldsymbol{x}(t)\|\text{rms} = \sqrt{\lim_{T\to\infty} \frac{1}{T} \int_{-T/2}^{+T/2} \boldsymbol{x}^T(t)\boldsymbol{x}(t)dt}$$

As for the inner product that implies the above norm functions, this, for the general case of two continuous complex energy signals $x(t)$ and $y(t)$ is defined as

$$< x, y >= \int_{-\infty}^{+\infty} x(t)y^*(t)dt \tag{1.48}$$

and is characterized by all the properties of the inner product we presented in the appendix section. In the above equation, the two signals are assumed to have *infinite duration*, whereas if these are defined in a finite time interval $\alpha \le t \le \beta$, the limits of the integrals are formed accordingly. If the inner product of two signals is equal to zero, then as known, the two signals are characterized as *orthogonal* to each other.

1.12 Problems

1. Determine the energy of the signal

$$x(t) = \begin{cases} e^{-2t} \sin 5t & \text{for } t \ge 0 \\ 0 & \text{for } t < 0. \end{cases}$$

2. Determine the energy of the signal

$$x(t) = \begin{cases} 0.5[\cos(2\pi f_c t) + 1] & \text{for } -(1/2f_c) \le t \le (1/2f_c) \\ 0 & \text{anywhere else,} \end{cases}$$

which is widely used in telecommunications system applications.

3. Determine the energy and power of the following signals:

$$x_1(t) = -5 + 10\cos\left(10t + \frac{\pi}{3}\right) \quad \text{and} \quad x_2(t) = 10\cos\left(140t + \frac{\pi}{16}\right) - 4\sin\left(150t + \frac{2\pi}{5}\right).$$

4. Determine the energy and power of the following signals:

$$x_1(t) = [1 + 4\sin(5t)]\cos(2t), \quad x_2(t) = 5\cos(7t)\cos(10t), \quad x_3(t) = \sin(5t)\cos(40t), \quad x_4(t) = e^{i\alpha t}\cos(2\omega t).$$

5. Determine the power of the signals $x(t) = A\cos(2\pi f_0 t + \vartheta)$ and $x(t) = (A/2)\exp(j2\pi f_0 t)$ $(A \in \mathbb{C})$.
6. Decide whether the following continuous-time signals are energy signals or power signals, or neither of the two: (a) $x(t) = 5\cos(\pi t)\text{rect}(t)$ (where rect(t) is the rectangular pulse), (b) $x(t) = 2\sin(5\pi t) + 3\cos(\pi t)$, and (c) $x(t) = e^{2t}$.
7. Determine the fundamental period of the signal $x(t) = 2\cos(12t + 1) - 4\sin(4t - 5)$.
8. Given the signals $x(t) = \sin(t)$ and $y(t) = \cos(\pi t)$, determine the fundamental periods, as well as the period of the sum $x(t) + y(t)$, if you determine that this is a periodic signal.
9. Determine if the signal $x(t) = 5\cos(t) + 3\sin(2t)$ is periodic and if so, calculate its period.
10. Determine the even and odd components of the continuous-time signal defined as $x(t) = \cos t \sin^3 t + \sin t \cos^3 t$.
11. Determine the even and odd components of the continuous-time signals

$$x(t) = \sin(\omega t), \quad x(t) = \cos(\omega t), \quad x(t) = \sin(\omega t + \vartheta), \quad x(t) = \cos(\omega t + \vartheta)$$

12. Considering a continuous-time signal $x(t)$ with an even component $x_e(t)$ prove that

$$\int_{-\infty}^{+\infty} x(t)\, dt = \int_{-\infty}^{+\infty} x_e(t)\, dt.$$

13. Considering the continuous-time signal defined by the equation

$$y(t) = \frac{x(t) + x^*(-t)}{2},$$

where $x^*(t)$ is the complex conjugate signal of $x(t)$, prove that the real part of $y(t)$ is characterized by even symmetry, whereas the imaginary part of $y(t)$ is characterized by odd symmetry.

14. Considering the continuous-time signal defined by the equation

$$y(t) = \frac{x(t) - x^*(-t)}{2},$$

where $x^*(t)$ is the complex conjugate signal of $x(t)$, prove that the real part of $y(t)$ is characterized by odd symmetry, whereas the imaginary part of $y(t)$ is characterized by even symmetry.

15. Prove that the signal $x(t) = t^2 \cos(t^2 + 1)$ is an even signal, whereas the signal $x(t) = \sin(2t^4)/t$ is an odd signal.
16. Considering the aperiodic signal $x(t) = \sin(\pi t)$ defined in the value range $[0, \infty)$, determine whether the odd component of $\mathcal{O}d\{x(t)\}$ constitutes a periodic signal.
17. Considering the aperiodic signal $x(t) = \cos(\pi t)$ defined in the value range $[0, \infty)$, determine whether the even component of $\mathcal{E}v\{x(t)\}$ constitutes a periodic signal.
18. Determine the even and odd components of the signal

$$x(t) = \begin{cases} -2t & t < 0 \\ t & t > 0. \end{cases}$$

19. For each of the following signals, determine whether it is even or odd, or alternatively if it is not characterized by either of these two types of symmetry.

$$x(t) = \tan^{-1} t, \quad x(t) = t^4 \sin^3 t \cos t, \quad x(t) = \cos(7\pi t), \quad x(t) = e^{\xi t}, \quad x(t) = |t|,$$

$$x(t) = \sin(3t), \quad x(t) = |\sin(2t)|, \quad x(t) = t^2 \cos t, \quad x(t) = cost \sin^2 t, \quad x(t) = t \sin t$$

20. Prove that the step function $u(t)$ and the sign function sgn(t) are related as

$$u(t) = \frac{1}{2} + \frac{1}{2}\text{sgn(t)}.$$

21. Determine the even and odd components of the step function $u(t)$.
22. Using the methodology of Exercise 1.6.2 (refer to page 27), prove that

$$\delta\left(\frac{t - t_0}{\alpha}\right) = |\alpha|\delta(t - t_0) \quad \text{and} \quad \delta(\alpha t - t_0) = \frac{1}{|\alpha|}\delta\left(t - \frac{t_0}{\alpha}\right).$$

23. Based on the properties of the function $\delta(t)$, prove that for every constant A it holds

$$\int_{-\infty}^{+\infty} A\delta(t)dt = \int_{-\infty}^{+\infty} A\delta(t - t_0)dt = A.$$

24. Prove that $\sin(t)\delta(t) = 0$ and $\cos(t)\delta(t - \pi) = -\delta(t - \pi)$.
25. Prove that

$$\varphi(t)\frac{d\delta(t)}{dt} = -\left.\frac{d\varphi(t)}{dt}\right|_{t=0}\delta(t) + \varphi(0)\frac{d\delta(t)}{dt}.$$

26. Prove that

$$\int_{-\infty}^{+\infty} \frac{d\delta(t)}{dt} = 0 \quad \text{and} \quad \frac{d^{(n)}\delta(-t)}{dt^n} = (-1)^n \frac{d^{(n)}\delta(t)}{dt^n}.$$

27. Prove that

$$\varphi(t)\frac{d^{(n)}\delta(t)}{dt^n} = \sum_{k=0}^{n}(-1)^k \frac{n!}{k!(n-k)!}\frac{d^k\varphi(0)}{dt^k}\frac{d^{(n-k)\delta(t)}}{dt^{n-k}}.$$

28. Prove that

$$\delta(t^2-\alpha^2)=\frac{1}{2\alpha}\Big[\delta(t+\alpha)+\delta(t-\alpha)\Big] \quad\text{and}\quad \delta(t^2-1)=\frac{1}{2}\Big[\delta(t+1)+\delta(t-1)\Big].$$

29. Using the defining equation of the time-shifted impulse function, prove that

$$\int_{-\infty}^{+\infty}\delta(t)e^{-i\omega t}dt=1 \quad\text{and}\quad \int_{-\infty}^{+\infty}\delta(t-2)\cos\left(\frac{\pi t}{4}\right)dt=0.$$

30. Using the appropriate properties of the function $\delta(t)$, calculate the values of the integrals

$$I_1=\int_{-\infty}^{+\infty}\delta(2t-3)\sin(\pi t)dt \quad\text{and}\quad I_2=\int_{-\infty}^{+\infty}x(2-t)\delta(3-t)dt.$$

31. Using the appropriate properties of the function $\delta(t)$, prove that

$$\int_{-\alpha}^{+\alpha}\int_{-\infty}^{+\infty}\delta(t)e^{-i\omega(t+\vartheta)}dtd\omega=2\alpha\frac{\sin(\alpha\vartheta)}{\alpha\vartheta}.$$

32. Using the appropriate properties of the function $\delta(t)$, prove that

$$\int_{-\alpha}^{+\alpha}\int_{-\infty}^{+\infty}\delta\left(\frac{t}{\vartheta}\right)e^{-i\omega(t+\vartheta)}dtdw=2\alpha|\vartheta|\frac{\sin(\alpha\vartheta)}{\alpha\vartheta}.$$

33. Using the defining equation of the shifted function $\delta(t-t_0)$, simplify the expressions

$$x(t)=e^{-t}\cos(10t)\delta(t) \quad\text{and}\quad x(t)=\sin(2\pi t)\sum_{k=0}^{\infty}\delta(t-k).$$

34. Using the defining equation of the shifted function $\delta(t-t_0)$, simplify the expressions

$$x(t)=\left(\frac{\sin t}{t^2+2}\right)\delta(t),\quad x(t)=\frac{\delta(t+3)}{it+2},\quad x(t)=\frac{\delta(t)\sin(kt)}{t}.$$

35. Prove that (a) $(t^4+4)\delta(t)=4\delta(t)$, (b) $\cos^2(t^2-\pi)\delta(t)=-\delta(t)$, and (c) $e^{-3t}\delta(t-4)=e^{-12}\delta(t-4)$.
36. Show that the function

$$x(t)=\frac{1}{\varepsilon\sqrt{2\pi}}\exp\left(-\frac{t^2}{2c^2}\right)$$

in the limit $\varepsilon\to 0$, is a δ function.
37. Show that the function $x(t)=\alpha e^{-\alpha t}u(t)$ in the limit $\alpha\to+\infty$ is a δ function.
38. Calculate the energy E_∞ of the signal

$$y(t)=\int_{-\infty}^{t}x(\tau)d\tau,$$

where the signal $x(t)$ is defined as $x(t)=\delta(t+2)-\delta(t-2)$.
39. Prove that $\delta'(-t)=-\delta'(t)$, where $\delta'(t)=d\delta(t)/dt$ is the first-order derivative of the function $\delta(t)$.
40. Prove that $\delta(x-\alpha)\delta(x-\beta)=0$ (this relation holds only for the case $\alpha\neq\beta$ whereas if it is $\alpha=\beta$, this product is undefined).
41. Prove that the sampling function is characterized by the property

$$\int_{-\infty}^{+\infty}\sin c(x)dx=\int_{-\infty}^{+\infty}\frac{\sin x}{x}dx=\pi.$$

42. Prove the scaling property for the comb function, which is described by an equation of the form

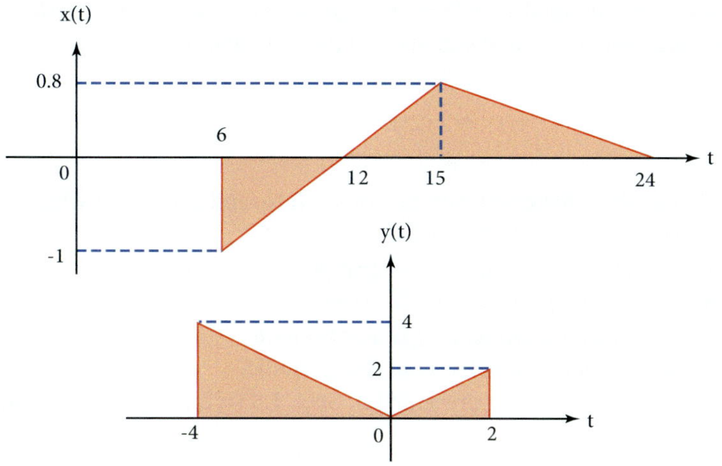

Fig. 1.30 The continuous-time signals $x(t)$ and $y(t)$ of Problem 43

$$\sum_{n=-\infty}^{+\infty} \delta(t-nT) = |\alpha| \cdot \sum_{n=-\infty}^{+\infty} \delta\Big[\alpha(t-nT)\Big].$$

(*Hint*: This property arises directly from the corresponding property of the unit impulse function).

43. Considering the continuous-time signals $x(t)$ and $y(t)$ depicted in Fig. 1.30, draw the signals (a) $x(-t)$, (b) $x(t+8)$, (c) $x(t-8)$, (d) $x(5t)$, (e) $x(4t-3)$, (f) $x(t/2)$, (g) $x(t/4)$, (h) $y(t-5)$, (i) $y(t/2)$, (j) $y(-3t)$, (k) $y(3-4t)$, (l) $y(2-t)$.
44. Draw the signal $x(t) = \exp(j2t)$.
45. Considering the triangular pulse of duration $2T$ and amplitude A defined as

$$x(t) = A\text{tri}\left(\frac{t}{T}\right)$$

show that the first derivative of this signal is given by the expression

$$\frac{dx(t)}{dt} = \frac{A}{T} \times \text{rect}\left(\frac{t+0.5T}{T}\right) - \frac{A}{T} \times \text{rect}\left(\frac{t-0.5T}{T}\right),$$

where rect is the rectangular pulse, whereas the second derivative has the form

$$\frac{d^x(t)}{dt^2} = \frac{A}{T}\delta(t+T) + \frac{A}{T}\delta(t-T) - \frac{2A}{T}\delta(t).$$

46. Considering the continuous-time signal defined as

$$x(t) = \begin{cases} M & |t| \leq 3 \\ 0 & \text{anywhereelse} \end{cases}$$

draw the signals $x(t)$, $x(t+2)$, $x(t-2)$ as well as their sum.

47. Considering the continuous-time signal defined as

$$x(t) = \begin{cases} 2t & 0 \leq t < 2 \\ 4-3t & 2 \leq t < 5 \\ 0 & \text{anywhereelse} \end{cases}$$

design the signals $x(t)$, $x(3t)$, $x(t/3)$, and $x(-t)$.

48. Design the continuous-time signal defined as $x(t) = (t - 1)u(t - 1) - (t - 3)u(t - 3)$.
49. Let there be a continuous-time signal $x(t)$ and signals $y(t)$ and $z(t)$ defined as

$$y(t) = x(2t) \quad \text{and} \quad z(t) = x\left(\frac{t}{2}\right).$$

For each of the following sentences, check whether it is true or false and in the case where the sentence is true, determine the relationship that correlates the periods of the above signals.
(a) If the signal $x(t)$ is periodic, then the signal $y(t)$ is also periodic.
(b) If the signal $y(t)$ is periodic, then the signal $x(t)$ is also periodic.
(c) If the signal $x(t)$ is periodic, then the signal $z(t)$ is also periodic.
(d) If the signal $z(t)$ is periodic, then the signal $x(t)$ is also periodic.
50. Based on the theory of elementary transforms presented in Sect. 1.8 (refer to page 31), design the following signals:

$$x(t) = 2[u(t) - u(t - 1)]$$

$$x(t) = u(t) + u(t - 1) - 4u(t - 2) + u(t - 4)$$

$$x(t) = u(t + 1) - u(t - 1)$$

$$x(t) = 2u(2t) - u(t - 1)$$

$$x(t) = u[\sin(\omega_0 t)]$$

$$x(t) = u(t - 2.1) + u(2t - 3)$$

51. Considering two signals $x(t)$ and $y(t)$ with energies $E\{x(t)\}$ and $E\{y(t)\}$ respectively and using the defining equation of the energy of a continuous-time signal, prove that:
The scaling of the signal's amplitude by the factor A leads to the multiplication of the signal's energy by the quantity A^2, that is, if it is $z(t) = Ax(t)$, then we will have $E\{z(t)\} = A^2 E\{x(t)\}$.
The transformation of time shift does not change the energy of the signal and therefore

$$E\{x(t)\} = E\{x(t \pm T)\}.$$

If the signals $x(t)$ and $y(t)$ are not characterized by time overlap, the energy of the sum of the two signals is equal to the sum of the energies of the two signals, that is,

$$E\{x(t) + y(t)\} = E\{x(t)\} + E\{y(t)\}.$$

The transformation of time scaling by a constant α leads to the multiplication of the signal's energy by the quantity $1/|\alpha|$ or in mathematical notation

$$E\{x(\alpha t)\} = \frac{1}{|\alpha|} E\{x(t)\}.$$

The physical significance of this property is that the compression of a signal in time (which corresponds to values $\alpha > 1$) leads to a reduction of the signal's energy, whereas the extension of a signal in time (which corresponds to values $\alpha < 1$) leads to an increase in the energy of the signal.
The transformation of time reversal does not change the energy of the signal, that is

$$E\{x(t)\} = E\{x(-t)\}.$$

52. Prove that the square pulse $\Pi_\alpha(t)$ is related to the step function $u(t)$ as $\Pi_\alpha(t) = u(t + \alpha) - u(t - \alpha)$
53. Express the triangular pulse $\Lambda_\alpha(t)$ with the help of the step function $u(t)$.
54. Prove that the transformation of the time scaling of a sinusoidal signal by a factor α does not change the amplitude and phase of the signal, but leads to a change in its frequency, which increases or decreases depending on the value of the parameter α.

55. Let there be a continuous-time signal $x(t)$ with a value $x(t) = 0$ for $t < 3$. For each of the signals (a) $x(1-t)$, (b) $x(1-t) + x(2-t)$, (c) $x(3t)$, and (d) $x(t/3)$, determine the range of values of t for which their value is guaranteed to be zero.
56. Describe the following continuous-time signals using the unit step function

$$x(t) = \begin{cases} +1 \text{ for } 1 \leq t < 4 \\ -2 \text{ for } 4 \leq t < 9 \\ -1 \text{ for } t \geq 9 \end{cases} \quad \text{and} \quad x(t) = \begin{cases} -5 \text{ for } -3 \leq t < 0 \\ -1 \text{ for } 0 \leq t < 4 \\ +7 \text{ for } 4 \leq t < 8 \\ -3 \text{ for } t \geq 8. \end{cases}$$

57. Calculate the convolution between the discrete-time signals

$$\text{(a) } x(t) = e^{-t}u(t) \text{ and } y(t) = e^{-2t}u(t)$$
$$\text{(b) } x(t) = e^{3t}u(t) \text{ and } y(t) = e^{-3t}u(t)$$
$$\text{(c) } x(t) = e^{-2t}u(t) \text{ and } y(t) = e^{t}u(t)$$

58. Prove that the convolution between an even and an odd signal is an odd signal, whereas the convolution between two even or two odd signals is an even signal.
59. Following the methodology of Exercise 1.10.4 (refer to page 43) prove the commutative property of convolution, using continuous-time signals.
$$x(t) = \cos(\omega t)\, u(t) \quad \text{and} \quad y(t) = e^{-\alpha t}\, u(t)$$
60. Calculate the convolution between $x(t) = \cos t u(t)$ and $y(t) = u(t)$.
61. Calculate the convolution between the signals $x(t) = \sin(\beta t)\, u(t)$ and $y(t) = \cos(\alpha t)\, u(t)$ for the cases $\alpha = \beta$ and $\alpha \neq \beta$ and verify the result through the commutative property.
62. Using the continuous-time signals $x(t) = e^{\alpha t}\, u(t)$, $y(t) = e^{\beta t}\, u(t)$ and $z(t) = e^{\gamma t}\, u(t)$ verify the associative and distributive properties of convolution.
63. Prove that the operation of convolution between any continuous-time signal $x(t)$ and the function comb(t) leads to the creation of a periodic signal.
64. Prove that the operation of convolution between two time-shifted impulse functions $\delta(t-\alpha)$ and $\delta(t-\beta)$ satisfies the property $\delta(t-\alpha) * \delta(t-\beta) = \delta(t-\alpha-\beta)$.
65. Prove that
$$\left\{\sum_{n=-N}^{N} \delta(t-nT)\right\} * \left\{\sum_{n=-N}^{N} \delta(t-nT)\right\} = \sum_{n=-2N}^{2N} (2N+1-|n|)\delta(t-nT).$$
66. Using the appropriate software application, design the signals $x(t) = e^{-3t}u(t)$ and $y(t) = e^{-t}u(t)$ and then calculate their convolution using the graphical method.
67. Graphically calculate the convolution between the signals $x(t)$ and $y(t)$, which are defined as

$$x(t) = e^{3t}u(t) \quad \text{and} \quad y(t) = u(t) - 2u(t-3) + u(t-6).$$

68. Repeat the process for the signals $x(t) = u(t)$ and $y(t) = e^{-\alpha t}u(t)$ $(\alpha > 0)$.
69. Calculate the convolution between the signals $x(t) = e^{-\alpha t}u(t)$ and $y(t) = e^{\alpha t}u(-t)$ $(\alpha > 0)$, using the integral calculation method and the graphical method, in order to verify their equivalence.
70. For each pair of signals $x(t)$ and $y(t)$ whose convolution $x(t) * y(t)$ was graphically calculated in Exercises 1.10.7 (refer to page 46) and 1.10.8 (refer to page 49), calculate the expression $y(t) * x(t)$ using the graphical method again in order to verify the validity of the commutative property of the convolution.
71. Using the graphical method of calculating the convolution, prove that

$$\left\{u(t)\right\} * \left\{\cos(\pi t)u(t)\right\} = \frac{\sin(\pi t)}{\pi}$$

72. Determine the power autocorrelation function of the periodic signals

$$x(t) = 7\sin\left(\omega t + \frac{\pi}{6}\right) \quad \text{and} \quad x(t) = 2\sin\left(\omega t - \frac{\pi}{10}\right)$$

73. Determine the power autocorrelation function of periodic signals

$$x(t) = 2\cos(\omega t) + 3\sin\left(\omega t + \frac{\pi}{14}\right) \quad \text{and} \quad x(t) = \sin\left(\omega t - \frac{\pi}{5}\right) - \cos(\omega t).$$

74. Calculate the energy autocorrelation function of a symmetric triangular pulse of amplitude A and duration τ.
75. Prove that the cross-correlation of continuous-time signals

$$x(t) = Au(t) - Au(t-\tau) \quad \text{and} \quad y(t) = Au(t) - 2Au(t - \frac{\tau}{2}) + Au(t-\tau)$$

is given by the relation

$$\varphi_{xy}(t) = \begin{cases} -\dfrac{A^2\tau}{2}\left(1 - \dfrac{2}{\tau}\left|t + \dfrac{\tau}{2}\right|\right) & \text{for } -\tau \le t < 0 \\ +\dfrac{A^2\tau}{2}\left(1 - \dfrac{2}{\tau}\left|t - \dfrac{\tau}{2}\right|\right) & \text{for } -\tau \le t < 0 \\ 0 & \text{anywhere else.} \end{cases}$$

76. Following the methodology of Exercises 1.10.10 (refer to page 55) and 1.10.11 (refer to page 56), determine the autocorrelation of a square pulse of duration τ and height A.
77. Prove that the autocorrelation of the signal $x(t) = \sin(\omega t + \vartheta)$ is $R_{xx}(t) = \cos(\omega t)/2$.
78. Prove that the cross-correlation of two signals $x(\tau)$ and $y(\tau)$, which are connected by the relation $y(\tau) = \alpha x(\tau - T)$, presents its maximum value at the position $t = T$.
79. Using the defining equations for the measure of a continuous-time signal to calculate the L_2 norm of continuous-time signals

$$x(t) = \begin{cases} \alpha t \text{ for } 0 < t < \xi \\ 0 \quad \text{elsewhere} \end{cases} \quad \text{and} \quad y(t) = \begin{cases} A\cos(\omega t + \varphi) \text{ for } 0 < t < \xi \\ 0 \qquad \text{elsewhere.} \end{cases}$$

80. Using the defining equations for the norm of a continuous-time signal to calculate the L_2 norm of the continuous-time signal

$$x(t) = \begin{cases} A\cos(\omega t + \varphi) \text{ for } 0 < t < \xi \\ 0 \qquad \text{elsewhere.} \end{cases}$$

2 Continuous-Time Systems

A typical example of a continuous-time system is the sound intensity attenuation systems used in various sound systems

2.1 Introduction to Continuous-Time Systems

The second fundamental concept related to the field of continuous-time signals and systems is the concept of the *system*. A system can be defined using two continuous-time signals $x(t)$ and $y(t)$ as an entity (implemented in *hardware* or *software*) that transforms the signal $x(t)$ (considered *the input signal*) into the signal $y(t)$ (considered the *output signal*). The concept of the system is very broad and is used to describe a plethora of entities of all kinds, such as a *physical process*, a *device*, or even a *set* of many devices and arrangements that are interconnected. A typical example of a simplified continuous-time system is a *resistor* R, which, according to Ohm's law, correlates the intensity $i(t)$ that flows through it with the *voltage* $v(t)$ that is applied at its ends by the known relationship $v(t) = Ri(t)$. In this case, the system's input is the current $x(t) = i(t)$, while the system's output is the voltage $y(t) = v(t)$.

According to the foregoing definition of the system, the output produced by it should qualitatively be different from its input, Could we, for example, consider that a system's input represents a physical process that is created independently of the system, while its output is created by the process that implements the operation of the system, when we transmit some input to it? Depending on the nature of the process implemented by the system, it can be classified into many and different categories. For example, a system can transform energy from one form to another (such as, for example, a system that converts *chemical* energy into *mechanical* energy) or carry out some process, such as, for example, an *arm of a robot*. However, the concept of a system, at least as will be discussed in this chapter, is one according to which the system proceeds to process an input signal, *enhancing* or *removing* some of its features. Such systems are known as *filters*. A typical example

A. I. Margaris, *Continuous and Discrete-Time Signals and Systems*,
https://doi.org/10.1007/978-3-031-81209-5_2

of such a system is a *digital image processing system* that enhances certain features of the image, such as, for example, its *brightness*, while removing others, such as *noise*. Other systems are used to extract useful features and properties of input signals, such as, for example, *voice processing and recognition systems*.

In the following sections, we will proceed to a detailed study of so-called *continuous-time systems* and *continuous-time systems*, in which both the system's input $x(t)$ and its output $y(t)$ are continuous-time signals and therefore are characterized by all those properties of continuous-time signals discussed in the previous chapter. From the mathematical side of the subject, a system is supposed to implement a transform $\mathcal{T}$, which is applied to the input signal $x(t)$, so as to create the output signal $y(t)$, a property that in mathematical notation is formulated as

$$y(t) = \mathcal{T}\{x(t)\} \tag{2.1}$$

and is diagrammatically represented in Fig. 2.1.

In Fig. 2.1, the direction of the arrows is particularly important because it informs us that the transmission of some input to the system leads to the creation of a corresponding output. In an equivalent formulation, we can assert that the output signal is the *result* of the appearance in the system's input of some input signal. In this sense, the input and output signals of a continuous-time system are related through a *cause–effect* relationship.

The theory of continuous-time systems can be considered a set of four different thematic fields each of which deals with a specific category of problems related to this cognitive area. These thematic fields, in general terms, are as follows:

- *system analysis*, which analyzes and interprets the properties of a continuous-time system;
- *system design*, which makes it possible to determine the structure and operation of a system, so that it meets defined specifications;
- *system synthesis*, which allows for the analysis of system structure using special diagrams; and
- *system implementation*, which defines the rules for constructing real systems, so that the existing requirements in each case are covered.

A typical example of a continuous-time system is presented in Fig. 2.2. This figure depicts a *resistor–capacitor circuit (*RC circuit*)*, which receives a sinusoidal alternating voltage $x(t)$ and produces as a response at its output another voltage $y(t)$. The mathematical equation that correlates these two voltages arises from the laws of Kirchoff and is given by the relation

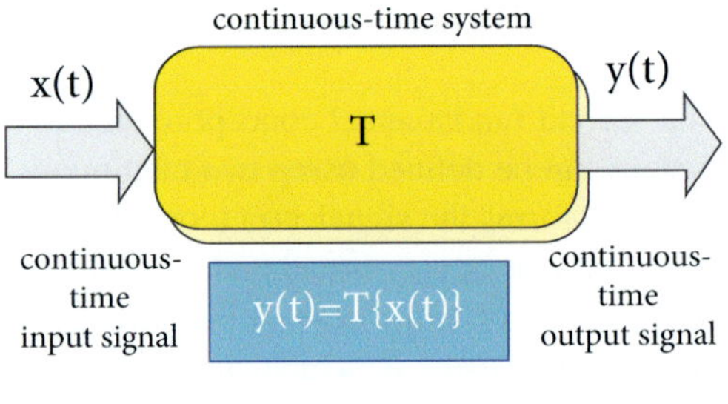

Fig. 2.1 Typical schematic diagram of a general continuous-time system

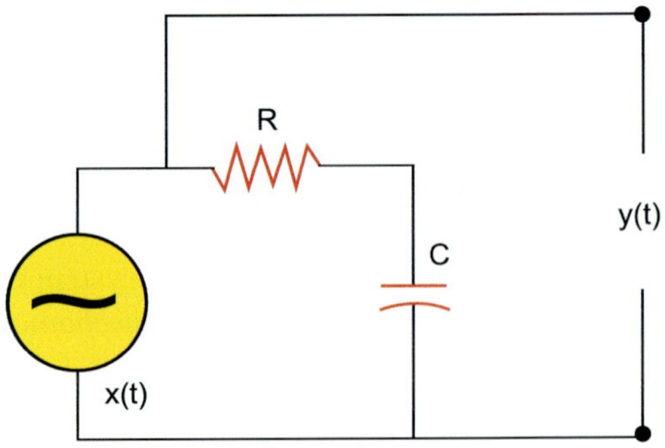

Fig. 2.2 Typical example of a continuous-time system

$$y(t) = v_c(0) + Rx(t) + \frac{1}{C}\int_0^t x(\tau)d\tau \tag{2.2}$$

(where $v_c(0)$ is the voltage across the capacitor at time $t = 0$) and quantitatively describes the transformation undergone by the system's input signal, so as to give as its output the corresponding output signal. Other examples of continuous-time systems that are used very often in practice are *wired or wireless telecommunication systems* that allow the transmission of a signal from a sender to a receiver through a communication channel characterized by the presence of noise using the appropriate coding and modulation techniques, *radar systems* that allow the measurement of the distance between objects moving in space, and *control systems* that are used to control the operation of specific devices and arrangements. A detailed description of the operation of all these systems can be found in the bibliography.

2.2 Types and Categories of Continuous-Time Systems

As already mentioned, a system is defined as a mathematical model that applies to each input signal $x(t)$ a *transformation* (or *operator*) $\mathcal{T}$, so as to create the output signal $y(t) = \mathcal{T}\{x(t)\}$. This definition, of course, is purely theoretical; in practice, the structure and the way a system operates are described in more practical ways, such as, for example, a *differential equation* that determines the correlation between its input and output, a *transfer function* that describes the same information but in different representation fields, or a mathematical description of the *state* of the system in a suitably defined multidimensional space. All these different ways of describing a continuous-time system are equivalent to each other and allow the complete determination of the state of the system, as well as its characterization in many different ways. The following pages describe in great detail both the set of characteristics that describe a system and the various categories in which it can be placed.

2.2.1 Causal and Stochastic Systems

A continuous-time system can be characterized as *causal* or *stochastic*, depending on the nature of the transformation it applies to its input. If this transformation is described exactly using *smooth mathematical functions*, while the application of the system to the same input always gives the same output, then the system is aforementioned to be *causal*. Conversely, if this transformation is described using *statistical terms* and is governed by *probabilistic mechanisms* in such a way that all recorded outputs of the system for the same input differ among themselves characterized by some variance in their values, then the system is aforementioned to be *stochastic* or *random*. The basic characteristic of a stochastic system is that *knowledge of the state of the system at any time is not sufficient to predict its output at later times*. Note that all examples of systems that will be analyzed and used in this and subsequent chapters refer only to causal systems.

A typical example of a causal system is an *electronic circuit*, which, if used in exactly the same way and with the same parameter values, will always create the same output. On the other hand, an example of a stochastic system would be a *communication channel* that introduces some noise into the signal it carries. Although the signal entering the channel is fully deterministic, the noise added to it constitutes a *stochastic process*,[1] resulting in the output produced by the system being characterized by elements of randomness.

2.2.2 Single- and Multiple-Input and Output Systems

Although in our description so far we have silently assumed that a system receives exactly one input and produces exactly one output, in practice a system can be designed so as to accept an *arbitrary number of inputs* and produce an *arbitrary number of outputs*. This particular characteristic, namely, the number of inputs and outputs, allows us to classify systems into four different categories, which in general terms are the following:

- *SISO (single-input single-output) systems* are systems that at each time instance t receive a single input $x(t)$ and produce a single output $y(t)$.

[1] A stochastic process is a sequence of random variables characterized by a combined probability density; see the relevant section.

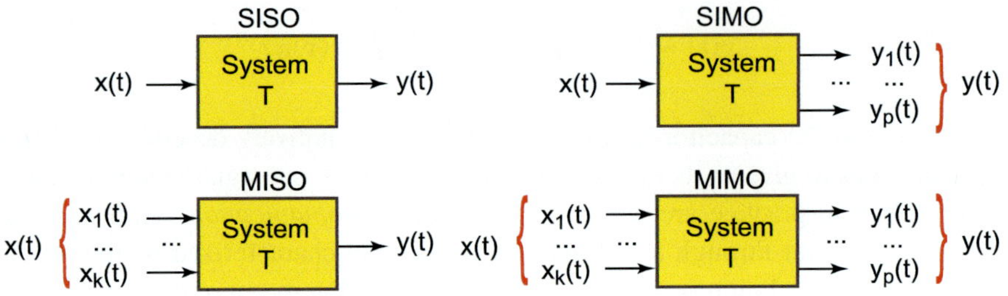

Fig. 2.3 Categories of systems based on number of inputs and outputs

- *SIMO (single-input multiple-output) systems* are systems that at each time instance t receive a single input $x(t)$ and produce p output signals that form the vector signal $\mathbf{y}(t) = \{y_1(t), y_2(t), \ldots, y_p(t)\}$.
- *MISO (multiple-input single-output) systems* are systems that at each time instance t receive k input signals defining a vector signal of the form $\boldsymbol{x}(t) = \{x_1(t), x_2(t), \ldots, x_k(t)\}$ and produce a single output $y(t)$.
- *MIMO (multiple-input multiple-output) Systems* are systems that at each time instance t receive k input signals of the form

$$\{x_1(t), x_2(t), \ldots, x_k(t)\}$$

and produce p output signals of the form

$$\{y_1(t), y_2(t), \ldots, y_p(t)\}$$

These four categories of systems are presented diagrammatically in Fig. 2.3.

2.2.3 Static and Dynamic Systems

A continuous-time system is aforementioned to be a *static* (or *memoryless*) system if, at every time instance t, its output $y(t)$ depends only on its input at time instance t. If, on the contrary, the current output of the system depends not only on the current input but also on *previous inputs*, that is, inputs that were transmitted to the system at earlier times, then the system is aforementioned to be *dynamic* (or a *system with memory*). If the number of these inputs is *finite*, the system is aforementioned to be a *finite memory system*, while in the opposite case the system is known as an *infinite memory system*. Examples of static systems are those described by the simple equations

$$y(t) = \mathcal{T}\{x(t)\} = \sin[x(t)] \quad \text{and} \quad y(t) = \mathcal{T}\{x(t)\} = x^2(t)$$

while examples of dynamic systems are the systems

$$y(t) = \mathcal{T}\{x(t)\} = \sin[x(2t)] + x(t-10) \quad \text{and} \quad y(t) = \mathcal{T}\{x(t)\} = 5x(t-5) + x^5(t-1)$$

The concept of memory in a continuous dynamic system refers to the existence of a mechanism that can maintain or store information related to the inputs of the system at times different from the current one. In some physical systems, this memory is associated with a process of storing energy. Considering, for example, a capacitor of capacity C, we know that the voltage $v(t)$ at the ends of the capacitor at any time t is given by the relation

$$v(t) = \frac{1}{C}\int_{-\infty}^{t} i(\tau)d\tau \tag{2.3}$$

where $i(\tau)$ is the *intensity of the electric current* flowing through the circuit. In this case, the capacitor proceeds to actual energy storage by accumulating electric charge, which, as is well known, is defined as the integral of the electric current. Note that this definition of memory can be extended to cover those cases in which the current output of a system depends not only on *previous* but also on *subsequent* values of its inputs and outputs.

Problem 2.2.1 For each of the following systems, examine whether it is static or dynamic:

$$y(t) = \sin(t)\cos(t), \quad y(t) = \int_{-\infty}^{t} x(\tau)d\tau, \quad y(t) = \frac{x(t)}{1 + x(t-1)}, \quad y(t) = tx(t)$$

Solution The first and fourth of these systems are static because their output at each time t depends only on the system's input at the same time. Conversely, the second and third systems are dynamic because the current output of the system depends on previous inputs. More specifically in the third system, the output $y(t)$ depends on the system's input at time $t-1$, while the output of the second system depends on all previous inputs, which are infinite in number.

Note As we have already mentioned, dynamic systems are also known as *systems with memory* because the calculation of their output requires a knowledge of previous values, which must therefore be "remembered" by the system. If the number of these values is *finite*, as is the case with the third system, then the system is aforementioned to be a *finite memory system*, while if this number is infinite, as is the case with the second system, the system is known as an *infinite memory system*.

2.2.4 Causal and Noncausal Systems

A system is aforementioned to be *causal* when its output $y(t)$ at any time t depends only on the inputs $x(t-\tau)$ transmitted to the system at *earlier* times—and therefore correspond to values $\tau \geq 0$—and not on future inputs—which correspond to the parameter values $\tau < 0$. If, on the contrary, the current output of the system depends on inputs of the form $x(t+\tau)$ for values $\tau > 0$, then the system is aforementioned to be *noncausal*, and the calculation of its output requires knowledge of the system's future input values.

In an equivalent description, we can characterize a system as causal when it obeys the *cause–effect* relationship. Taking into account that the output of a system is the result of transmitting some input to it (which therefore can be treated as the cause that led to the creation of the output), we can describe a causal system as a system whose output cannot appear before some input is applied to it. It is obvious that all systems implemented in practice are *causal*. However, it should be noted that the existence of noncausal systems is possible in cases where the independent variable is not time, as happens, for example, in *image processing applications*. Another special category of noncausal systems are those used in applications such as, for example, *voice processing*, in which all of a system's input has been recorded in advance and is stored in some storage arrangement. In this case, since we know *the entire contents of the system*, we can, for each sample of the input signal that corresponds to a certain time, retrieve and use sample values recorded at later times but already stored and known in advance.

Problem 2.2.2 For each of the following systems, examine whether it is causal or noncausal:

$$y(t) = \frac{2x(t)}{1 + x(t) + x(t-2)}, \quad y(t) = tx(t), \quad y(t) = x^2(t+1), \quad y(t) = e^{\alpha t}u(t)$$

Solution The first, second, and fourth systems are causal because the current output of the system depends only on the current input and previous inputs. Conversely, the third system is not causal because the system's output at time instance t depends on the later input $x(t+1)$. A noncausal system is not physically implementable because it violates the cause–effect relationship.

Note The characterization of the last system as causal is due to the existence of the unit step function $u(t)$, which has a nonzero value only in the range of values $t \geq 0$.

2.2.5 Linear and Nonlinear Systems

A continuous-time system is aforementioned to be *linear* when the transform $\mathcal{T}$ applied to the input signal $x(t)$ to produce the output signal $y(t)$ obeys the principle of *homogeneity* (or *scaling*), as well as the *principle of superposition*. Conversely, if this does not happen, the system is aforementioned to be *nonlinear*. According to the principle of homogeneity, if a system's response to the signal $x(t)$ is the signal $y(t)$, then the system's response to the signal $\alpha x(t)$, where α is an arbitrary constant coefficient, will be the signal $\alpha y(t)$, or, in mathematical notation,

$$\mathcal{T}\{\alpha x(t)\} = \alpha \mathcal{T}\{x(t)\} \tag{2.4}$$

On the other hand, the principle of superposition, which in a sense includes the property of homogeneity,[2] is formulated as follows: *A system's response to an input signal, defined as the linear combination of a number of elementary inputs, is equal to the linear combination of the corresponding elementary outputs.* In mathematical notation, considering N elementary continuous input signals $x_1(t), x_2(t), \ldots, x_N(t)$ with corresponding outputs of the form $y_1(t) = \mathcal{T}\{x_1(t)\}, y_2(t) = \mathcal{T}\{x_2(t)\}, \ldots, y_N(t) = \mathcal{T}\{x_N(t)\}$, the system's response to an input signal $x(t)$ defined as

$$x(t) = \alpha_1 x_1(t) + \alpha_2 x_2(t) + \cdots + \alpha_N x_N(t) = \sum_{i=1}^{N} \alpha_i x_i(t)$$

where α_i $(i = 1, 2, \ldots, N)$ are constant coefficients, will be the signal

$$y(t) = \alpha_1 y_1(t) + \alpha_2 y_2(t) + \cdots + \alpha_N y_N(t) = \sum_{i=1}^{N} \alpha_i y_i(t) = \sum_{i=1}^{N} \alpha_i \mathcal{T}\left\{x_i(t)\right\}$$

If we take into account that the signal $y(t)$ is defined as

$$y(t) = T\{x(t)\} = \mathcal{T}\left\{\sum_{i=1}^{N} \alpha_i x_i(t)\right\}$$

we can, by combining the last two equations, formulate the principle of parallelism in the form[3]

---

[2] In reality, the principle of superposition, in addition to the property of *homogeneity*, also includes the property of *additivity*, according to which if a system's response to the inputs $x_1(t)$ and $x_2(t)$ are the outputs $y_1(t)$ and $y_2(t)$, then, if the sum $x_1(t) + x_2(t)$ is transmitted to the system, then the output that will result from the system will be the sum $y_1(t) + y_2(t)$.

[3] This equation refers to a SISO-type system, which accepts a single input and produces a simple output, while in the general case of a MIMO system, whose input and output are vectors of suitable dimensions, the following relationship will apply:

$$\mathcal{T}\left\{\sum_{i=1}^{N} \alpha_i \boldsymbol{x}_i(t)\right\} = \sum_{i=1}^{N} \alpha_i \mathcal{T}\left\{\boldsymbol{x}_i(t)\right\}.$$

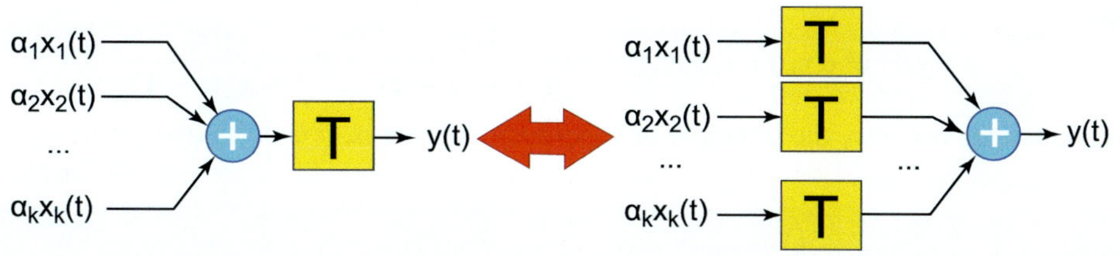

Fig. 2.4 Diagrammatic demonstration of principle of superposition

$$\mathcal{T}\left\{\sum_{i=1}^{N}\alpha_i x_i(t)\right\} = \sum_{i=1}^{N}\alpha_i \mathcal{T}\left\{x_i(t)\right\} \tag{2.5}$$

The principle of superposition is demonstrated diagrammatically in Fig. 2.4.

Problem 2.2.3 Considering the continuous-time systems described by the equations

$$\text{(a) } y(t) = Rx(t) \qquad \text{(b) } y(t) = x^2(t) \qquad \text{(c) } y(t) = \frac{x(t)}{1+x(t-1)}$$

examine whether these are linear or nonlinear.

Solution To determine whether each of the preceding systems is linear or not, we must examine whether it obeys *the principle of superposition*. According to this principle, if a system's response to the inputs $x_1(t)$ and $x_2(t)$ are the outputs $y_1(t) = \mathcal{T}\{x_1(t)\}$ and $y_2(t) = \mathcal{T}\{x_2(t)\}$ and the system is linear, then the system's response to the input $x(t) = \alpha x_1(t) + \beta x_2(t)$ should be $y(t) = \alpha y_1(t) + \beta y_2(t)$. In detail, then, we have:

(a) Considering two inputs $x_1(t)$ and $x_2(t)$, the response of the system to these is the outputs $y_1(t) = Rx_1(t)$ and $y_2(t) = Rx_2(t)$. If we now transmit the linear combination $x(t) = \alpha x_1(t) + \beta x_2(t)$ to the system, then the output of the system will have the form

$$y(t) = Rx(t) = R[\alpha x_1(t) + \beta x_2(t)] = \alpha[Rx_1(t)] + \beta[Rx_2(t)] = \alpha y_1(t) + \beta y_2(t)$$

We observe that the system obeys the principle of superposition, so it is *linear*.

(b) Considering two inputs $x_1(t)$ and $x_2(t)$, the response of the system to these will lead to the outputs $y_1(t) = x_1^2(t)$ and $y_2(t) = x_2^2(t)$. On the other hand, the response of the system to the linear combination $x(t) = \alpha x_1(t) + \beta x_2(t)$ is

$$y(t)=x^2(t)=[\alpha x_1(t)+\beta x_2(t)]^2=\alpha^2 x_1^2(t)+2\alpha\beta x_1(t)x_2(t)+\beta^2 x_2^2(t)=\alpha^2 y_1(t)+2\alpha\beta x_1(t)x_2(t)+\beta^2 y_2(t)\neq\alpha y_1(t)+\beta y_2(t)$$

We observe, therefore, that our system does not obey the principle of superposition, that is, *it is not linear*.

(c) Considering two inputs $x_1(t)$ and $x_2(t)$, the response of the system to these will lead to the outputs

$$y_1(t) = \frac{x_1(t)}{1+x_1(t-1)} \quad \text{and} \quad y_2(t) = \frac{x_2(t)}{1+x_2(t-1)}$$

On the other hand, the response of the system to the combined input $x(t) = \alpha x_1(t) + \beta x_2(t)$ will lead to an output of the form

$$y(t) = \frac{x(t)}{1+x(t-1)} = \frac{\alpha x_1(t) + \beta x_2(t)}{1+\alpha x_1(t-1)+\beta x_2(t-1)}$$

We observe that

$$y(t) \neq \alpha y_1(t) + \beta y_2(t) = \alpha \frac{x_1(t)}{1 + x_1(t-1)} + \beta \frac{x_2(t)}{1 + x_2(t-1)}$$

Therefore, the system does not obey the principle of superposition, that is, *it is not linear*.

Problem 2.2.4 Examine whether the continuous-time system $\frac{dy(t)}{dt} + 6y(t) = x(t)$ is linear or nonlinear.

Solution In this case, we do not know the analytical form of the system's output $y(t)$, only the *differential equation* that correlates the system's input with its output. To prove that our system is linear, we will consider two inputs, $x_1(t)$ and $x_2(t)$, that correspond to outputs $y_1(t)$ and $y_2(t)$ and, we will check whether the combined input $x(t) = \alpha x_1(t) + \beta x_2(t)$ correlates with the corresponding combined output $y(t) = \alpha y_1(t) + \beta y_2(t)$ through the same differential equation. If we find that the differential equation is satisfied by the functions $x(t)$ and $y(t)$, then it is linear, and so the system it describes will also be a linear continuous-time system.

Let us consider, then, two inputs $x_1(t)$ and $x_2(t)$ with corresponding outputs $y_1(t)$ and $y_2(t)$. In this case, the pairs of functions $[x_1(t), y_1(t)]$ and $[x_2(t), y_2(t)]$ will satisfy the foregoing differential equation, so that we can write

$$\frac{dy_1(t)}{dt} + 6y_1(t) = x_1(t) \quad \text{and} \quad \frac{dy_2(t)}{dt} + 6y_2(t) = x_2(t)$$

If we multiply the first equation by the value of a constant α and the second equation by the value of another constant β, we will get

$$\alpha \frac{dy_1(t)}{dt} + 6\alpha y_1(t) = \alpha x_1(t) \quad \text{or, equivalently,} \quad \frac{d[\alpha y_1(t)]}{dt} + 6\alpha y_1(t) = \alpha x_1(t)$$
$$\beta \frac{dy_2(t)}{dt} + 6\beta y_2(t) = \beta x_2(t) \quad \text{or equivalently} \quad \frac{d[\beta y_2(t)]}{dt} + 6\beta y_2(t) = \beta x_2(t)$$

and adding the last equations term by term, it turns out that

$$\frac{d}{dt}\Big[\alpha y_1(t) + \beta y_2(t)\Big] + 6\Big[\alpha y_1(t) + \beta y_2(t)\Big] = \alpha x_1(t) + \beta x_2(t)$$

We observe, therefore, that the differential equation of the system is satisfied by the combined input $x(t) = \alpha x_1(t) + \beta x_2(t)$ and the corresponding combined output $y(t) = \alpha y_1(t) + \beta y_2(t)$. It is therefore a linear differential equation, meaning the system described by it will be linear. ■

Continuous-time linear systems are usually described by *complete ordinary linear differential equations* of the form

$$\alpha_N \frac{d^N y(t)}{dt^N} + \alpha_{N-1} \frac{d^{N-1} y(t)}{dt^{N-1}} + \cdots + \alpha_0 y(t) = \beta_M \frac{d^M x(t)}{dt^M} + \cdots + \beta_1 \frac{dx(t)}{dt} + \beta_0 x(t)$$

with the values of the coefficients α_i and β_i being constant or varying as a function of time. Therefore, to determine the response $y(t)$ of the system, when a signal $x(t)$ is transmitted to its input, *we must replace the function $x(t)$ and all its derivatives in the preceding differential equation and determine its solution, which will have the form $y = y(t)$*. According to the fundamental theory of differential equations, this solution (which also represents the output of the system for time instances $t \geq 0$) can be considered the result of two different and independent causes: (a) the *initial conditions* (often known as *the state of the system*) at time $t = 0$ and (b) *the input of the system $x(t)$* for time instances $t \geq 0$. Taking into account that the system is linear, and therefore will be characterized by the power of the principle of superposition, we can state that the response $y(t)$ of such a system for time instances $t \geq 0$ will consist of the sum of two terms: a term known as a *response to* *zero input*, which is associated with the existence of initial conditions corresponding to the time $t = 0$ and arises by setting $x(t) = 0$ for every $t \geq 0$, and a term referred to as a *response to zero state*, which arises from the input $x(t)$ for $t \geq 0$ and for which all initial conditions are set to a zero value. Based on all the foregoing considerations, we can formulate the following conclusion:

The total response of a continuous-time linear system is equal to the sum of the response to the zero state and the response to the zero input.

The foregoing property of linear systems, which allows us to separate a system's response due to the existence of initial conditions from that due to the transmission of some input to the system, is known as *decomposition*. It should also be noted that the property of linearity extends to these two distinct types of response, each of which must obey separately and independently from the other, to the principle of superposition. A typical example of a system output characterized by the aforementioned behavior is the one that describes the output of the RC circuit that we presented earlier and that is based on the foregoing description can be formulated as

$$\underbrace{y(t)}_{\text{total response}} = \underbrace{v_c(0)}_{\text{response to the zero-input}} + \underbrace{Rx(t) + \frac{1}{C}\int_0^t x(\tau)d\tau}_{\text{response to the zero-state}} \tag{2.5}$$

2.2.6 Time-Varying and Time-Invariant Systems

A continuous-time system is called *time invariant* when the transform $\mathcal{T}$ applied to the input signal $x(t)$ to create the output signal $y(t)$ *does not change as a function of time*. This means that if at time t the signals $x(t)$ and $y(t)$ are correlated through the relationship

$$y(t) = \mathcal{T}\{x(t)\} \tag{2.6}$$

then at time $t \pm \tau$—for any value of the time shift τ—we should have

$$y(t \pm \tau) = \mathcal{T}\{x(t \pm \tau)\} \tag{2.7}$$

If the system does not satisfy the preceding property, then it is aforementioned to be *time varying*.

A typical example of a time-invariant physical system is a RC circuit in which the values of resistance R and capacitance C do not change as a function of time. In this case, the correlation that exists between the input and the output of the system will always be the same and independent of time. If, on the contrary, the values of the foregoing quantities are functions of time of the form $R = R(t)$ and $C = C(t)$, then the output of the system will be a function of the time t for which we calculate it, and the system is *time varying*.

Problem 2.2.5 For each of the systems described by the equations

$$\text{(a)}\, y(t) = \sin[x(t)], \quad \text{(b)}\, y(t) = x(t-6), \quad \text{(c)}\, y(t) = x^2(t)$$

examine whether it is time-varying or time-invariant.

Solution To examine the character of the system in terms of the characteristic of time invariance, we must check whether or not it satisfies the property $y(t-\tau) = \mathcal{T}[x(t-\tau)]$ (the situation is completely analogous if, instead of $t-\tau$, we use $t+\tau$), which is done as follows.

Initially we make the substitution $t \to t - \tau$ to calculate the expression $y_\tau(t) = y(t - \tau)$ and then determine the value of the expression $\mathcal{T}[x(t - \tau)]$. Taking into account that the transform $\mathcal{T}$ is applied on the delayed version of the signal $x(t - \tau)$ we should in the equation of the system definition make the substitution $x(t) \to x(t - \tau)$. If these two processes lead to the same result, then the system is time-invariant, while in the opposite case it is time-varying. In detail, then, we have:

(a) Making the substitution $t \to t - \tau$, we find that $y(t - \tau) = \sin[x(t - \tau)]$. On the other hand, the substitution $x(t) \to x(t - \tau)$ leads to the result $\mathcal{T}\{x(t - \tau)\} = \sin[x(t - \tau)]$. We observe that $y(t - \tau) = \mathcal{T}[x(t - \tau)]$, that is, the system is *time-invariant*.
(b) In this case, the substitution $t \to t - \tau$ leads us to the result $y(t - \tau) = x(t - 6 - \tau)$, while on the other hand, setting $x(t) \to x(t - \tau)$, we will get $\mathcal{T}\{x(t - \tau)\} = x(t - 6 - \tau)$. We observe that the two results coincide, meaning the system is *time-invariant*.
(c) In a completely analogous way, for $t \to t - \tau$ we have $y(t - \tau) = x^2(t - \tau)$, while the substitution $x(t) \to x(t - \tau)$ leads us to the result $\mathcal{T}\{x(t - \tau)\} = x^2(t - \tau)$, so the system is *time-invariant*.

Problem 2.2.6 For each of the systems described by the equations

$$\text{(a) } y(t) = x(t)\cos(t), \qquad \text{(b) } y(t) = \frac{dx(t)}{dt}, \qquad \text{(c) } y(t) = \int_{-\infty}^{t} x(\tau)d\tau$$

examine whether they are time-varying or time-invariant.

Solution To examine the character of the system with respect to the characteristic of time invariance, we must check whether or not it satisfies the property $y(t - \tau) = \mathcal{T}[x(t - \tau)]$ (the situation is completely analogous if, instead of $t - \tau$, we use $t + \tau$), which is done as follows.

Initially we make the substitution $t \to t - \tau$ to calculate the expression $y_\tau(t) = y(t - \tau)$ and then determine the value of the expression $\mathcal{T}[x(t - \tau)]$. Taking into account that the transform $\mathcal{T}$ is applied on the delayed version of the signal $x(t - \tau)$, in the system defining equation we should make the substitution $x(t) \to x(t - \tau)$. If these two processes lead to the same result, the system is time-invariant, while in the opposite case it is time-varying. In detail, then, we have:

(a) Proceeding with the substitution $t \to t - \tau$ results in $y(t - \tau) = x(t - \tau)\cos(t - \tau)$. However, the substitution $x(t) \to x(t - \tau)$ leads to a result of the form $\mathcal{T}\{x(t - \tau)\} = x(t - \tau)\cos(t)$. We observe that the two results do not coincide, that is, the system is *time-varying*.
(b) In this case, the substitution $t \to t - \tau$ leads to the result

$$y(t - \tau) = \frac{dx(t - \tau)}{d(t - \tau)} \quad \text{while for} \quad x(t) \to x(t - \tau) \quad \text{we have} \quad \mathcal{T}\{x(t - \tau)\} = \frac{d}{dt}\Big[x(t - \tau)\Big]$$

Proceeding with the substitution $t - \tau = \xi$, we will have $\xi = \xi(t) = t - \tau$ and $\xi'(t) = 1$, and therefore

$$\mathcal{T}\{x(t - \tau)\} = \frac{d}{dt}\Big[x(t - \tau)\Big] = \frac{dx(\xi)}{dt} = \frac{dx(\xi)}{d\xi}\frac{d\xi}{dt} = \frac{dx(t - \tau)}{d(t - \tau)}$$

We observe that the two results coincide, that is, the system is *time-invariant*.
(c) Following the methodology of the previous exercises, we set $t \to t - \vartheta$ and find that

$$y(t - \vartheta) = \int_{-\infty}^{t-\vartheta} x(\tau)d\tau$$

while the substitution $x(\tau) \to x(\tau - \vartheta)$ leads to a result of the form

$$\mathcal{T}\{x(\tau - \vartheta)\} = \int_{-\infty}^{t} x(\tau - \vartheta)d\tau$$

Proceeding with the variable change $\tau - \vartheta = \xi$, we will have $d\tau = d(\xi + \vartheta) = d\xi$. As for the limits of integration, for $\tau \to -\infty$ we will obviously also have $\xi \to -\infty$, while for $\tau = t$ we will have $\xi + \vartheta = t$ or, equivalently, $\xi = t - \vartheta$, and the preceding relation will take the form

$$\mathcal{T}\{x(\tau - \vartheta)\} = \int_{-\infty}^{t} x(\tau - \vartheta) d\tau = \int_{-\infty}^{t-\vartheta} x(\xi) d\xi$$

We observe that the two results coincide, that is, the system is *time-invariant*. ■

The preceding categories of continuous-time systems concern the description of systems using the characteristics of *linearity*, *causality*, and *time invariance*. Of course, apart from the preceding characteristics, systems can be described and categorized by other characteristics, such as *controllability*, *observability*, and *stability*. Of the preceding characteristics, controllability and observability are related to the description of systems in state space and will be studied in the homonymous section, while stability, due to its particular importance as far as the mode of operation of continuous-time systems, will be studied separately from the other characteristics.

Problem 2.2.7 Prove that the response of a time-invariant continuous-time system to an input signal that is periodic with period T is also a periodic signal that, moreover, is characterized by the same period value.

Solution Since the system is time-invariant and produces the output $y(t) = \mathcal{T}\{x(t)\}$, it must satisfy the condition of time invariance $y(t - T) = \mathcal{T}\{x(t - T)\}$. However, if the input signal $x(t)$ is periodic with period T, it will satisfy the property $x(t) = x(t - T)$, and the last relation becomes

$$y(t - T) = \mathcal{T}\{x(t - T)\} = \mathcal{T}\{x(t)\} = y(t)$$

Therefore, the output $y(t)$ of the system will be periodic with a period equal to that of the input signal.

2.3 Stability of Continuous-Time Systems

In almost all cases and regardless of the type and characteristics of the system used in each case, the *stability* of the system constitutes a strongly desirable characteristic and is considered a necessary prerequisite for the smooth operation of the system, so that the requirements for which it has been constructed are met. Speaking generally, a system is called *stable* when the application of small disturbances to it does cause it to deviate from its equilibrium state.

From physics, it is known that there are three types of equilibrium—stable, unstable, and indifferent equilibrium—and obviously a stable system should be characterized by the so-called stable equilibrium. There are two different ways in which we can characterize a system in terms of its stability. The first way is related to the so-called *asymptotic stability*, which involves an examination of the internal structure of the system, while the second relates to the *bounded-input bounded-output (BIBO) stability*, which involves an examination of the way in which the system's output behaves when we change slightly the input we transmit to it.

Considering a continuous linear system described by a general differential equation of the form

$$\alpha_N \frac{d^N y(t)}{dt^N} + \cdots + \alpha_1 \frac{dy(t)}{dt} + \alpha_0 y(t) = \beta_M \frac{d^M x(t)}{dt^M} + \cdots + \beta_1 \frac{dx(t)}{dt} + \beta_0 y(t)$$

or, in summary notation,

$$\sum_{k=0}^{N} \alpha_k \frac{d^k y(t)}{dt^k} = \sum_{k=0}^{M} \beta_k \frac{d^k x(t)}{dt^k} \tag{2.8}$$

we can characterize it in terms of its stability by studying its internal structure, that is, by determining the positions of the roots λ_i $(i = 1, 2, \ldots, N)$ of the *characteristic polynomial* of the system on the complex plane. Without being able to give a detailed description of the prevailing theory, we simply mention that the system is considered *asymptotically stable* when all the roots of the characteristic polynomial (some of which may possibly be repeated, leading to multiplicity $p > 1$) have negative real parts, that is, when the relation $\Re\{\lambda_i\} < 0$ holds (for $i = 1, 2, \ldots, N$). Conversely, when even a single root has a positive real part, then the system is aforementioned to be *asymptotically unstable*. The reason why the system is characterized in the aforementioned way is that each root of the characteristic polynomial λ_i corresponds to a mode of system oscillation of the form $y_i(t) = e^{\lambda_i t}$ $(i = 1, 2, \ldots, N)$, which in the limit $t \to \infty$ tends to infinity if $\Re\{\lambda_i\} > 0$ and to zero if $\Re\{\lambda_i\} < 0$. In the case where the characteristic polynomial of the system has roots with multiplicity $p > 1$, some of which are on the imaginary axis, it is proven that the system is again asymptotically unstable, while finally, the system is aforementioned to be *marginally stable* if and only if there are roots on the imaginary axis with multiplicity $p = 1$. In a subsequent section we will prove mathematically some of these properties that we now describe only qualitatively.

On the other hand, the second way of studying the stability of a system allows for characterizing the system as *BIBO stable*, when the transmission to it of a *blocked input* $x(t)$ leads to the creation by the system of a *blocked output* $y(t)$. In other words, a system is considered to be characterized by BIBO stability if for every input–output pair $[x(t), y(t)]$ there are finite real constants α and β such that

$$\text{for every } x(t) \text{ such that } |x(t)| \le \alpha \quad \Longleftrightarrow \quad |y(t)| \equiv |\mathcal{T}\{x(t)\}| \le \beta$$

If there is even one system input for which the system output is not blocked, then the system is aforementioned to be *BIBO unstable*.

From these two different ways of characterizing the stability of a system, the first consists in studying the internal characteristics of the system, *relying only on the nonzero initial conditions that describe it and without transmitting any input to it*, while, in contrast, by the criterion of BIBO stability *we completely ignore the internal structure of the system and any initial conditions that describe it*—and which are considered zero—*and we study the system only in terms of the way it behaves when we transmit a blocked input to it*. For this reason, the first method of characterization is associated with so-called *zero-input stability*, while the second is associated with so-called *zero-state stability*.. In the following pages, as well as in all the remaining chapters, we will focus solely on BIBO stability and not deal at all with the second type of stability.

To characterize a continuous-time system as stable or unstable according to BIBO, we can either determine a bounded input for which the system's output is not bounded, meaning the system is unstable, or ensure that for every bounded input the system produces a bounded output, meaning it is stable. This way of characterizing the system in terms of its stability is presented through the exercises that follow.

Problem 2.3.1 Determine the asymptotic stability of the continuous-time systems

$$\text{(a)} \quad \frac{d^3y(t)}{dt^3} + 5\frac{d^2y(t)}{dt^2} + 12\frac{dy(t)}{dt} + 8y(t) = \frac{dx(t)}{dt} - 3x(t);$$

$$\text{(b)} \quad \frac{d^3y(t)}{dt^3} + 2\frac{d^2y(t)}{dt^2} + 4\frac{dy(t)}{dt} + 8y(t) = \frac{d^2x(t)}{dt^2} + \frac{dx(t)}{dt} + x(t).$$

Solution To characterize the asymptotic stability of the foregoing systems, also known as zero-input stability, we must set the system's input, as well as all its derivatives, to zero, which will allow us to determine the roots of the characteristic polynomial of the corresponding homogeneous equation and check the sign of their real part. Thus, we will have the following results:

(a) The homogeneous equation of the system will have the form

$$\frac{d^3y(t)}{dt^3} + 5\frac{d^2y(t)}{dt^2} + 12\frac{dy(t)}{dt} + 8y(t) = 0$$

Considering an exponential solution of the form $y(t) = e^{\lambda t}$ and differentiating three times, we find that $y'(t) = \lambda e^{\lambda t}$, $y''(t) = \lambda^2 e^{\lambda t}$, and $y'''(t) = e^{\lambda^3 t}$. If we substitute these functions into the differential equation and eliminate the exponential function $e^{\lambda t}$, which is always different from zero, we end up with the characteristic polynomial of the equation

$$\lambda^3 + 5\lambda^2 + 12\lambda + 8 = 0 \quad \text{which is also written} \quad (\lambda + 1)(\lambda^2 + 4\lambda + 8) = 0$$

From the last relation it follows that $\lambda + 1 = 0$ or, equivalently, $\lambda_1 = -1$ or $\lambda^2 + 4\lambda + 8 = 0$, with roots $\lambda_2 = -2 + 2i$ and $\lambda_3 = -2 - 2i$. We observe that the real parts of all roots are negative, that is, the system is *asymptotically stable*.

(b) Following the same process, we find that the characteristic polynomial of the homogeneous equation of the system

$$\frac{d^3y(t)}{dt^3}+2\frac{d^2y(t)}{dt^2}+4\frac{dy(t)}{dt}+8y(t)=0 \quad \text{has the form} \quad \lambda^3+2\lambda^2+4\lambda+8=0 \quad \text{or equivalently} \quad (\lambda+2)(\lambda^2+4) = 0$$

The roots of this polynomial are $\lambda_1 = -2$, $\lambda_2 = -2i$ and $\lambda_3 = +2i$. We observe that one of the roots is negative, while the other two are characterized by a multiplicity value of $p = 1$ and are located on the imaginary axis of the complex plane. Therefore, the system in question is *marginally stable*. ■

If we know the type of asymptotic stability of a system, we can determine its BIBO stability, since it has been proven that *an asymptotically stable system is BIBO stable, while a marginally stable or unstable system is BIBO unstable*. Therefore, of the two systems we considered earlier, the first is stable and the second is BIBO unstable. However, it should be noted that the converse of the foregoing proposition is not necessarily true, and therefore *knowledge of a system's BIBO stability generally does not suffice for the characterization of a system in terms of its asymptotic stability*.

In cases where we do not know the differential equation of a system, which would allow us to determine its asymptotic stability by applying the method described previously, and all we have is the relationship that connects the input with the output, we can characterize it in terms of its BIBO stability, checking whether the system meets the criterion imposed by this. The next exercise demonstrates the application of this approach.

Problem 2.3.2 For each of the systems described by the equations

$$y(t) = tx(t) \quad \text{and} \quad y(t) = e^{x(t)}$$

examine whether it is stable or unstable according to BIBO.

Solution

(a) As already mentioned, if there is even one bounded input $x(t)$ for which the output $y(t)$ is not bounded, the system is aforementioned to be unstable. In this case, we find that if we transmit the bounded input $x(t) = 1$ to the system, the system's output will be $y(t) = t$, which is not bounded since it tends to infinity as $t \to \infty$. Therefore, the system is *unstable*.

(b) In this case, it is quite difficult to determine a specific bounded input that leads to the creation of an unbounded output, and for this reason we will try to characterize the system by examining *whether every bounded input leads to a bounded output*. Considering any bounded input of the form $|x(t)| \leq \alpha$ or, equivalently, $-\alpha \leq x(t) \leq +\alpha$, we find that the system's output will be in every case bounded, since from the equation that describes the system it immediately follows that

$$e^{-\alpha} \leq |y(t)| \leq e^{+\alpha}$$

Therefore, this system is *BIBO stable*.

Problem 2.3.3 For each of the systems described by the equations

$$y(t) = \sin[x(t)], \qquad y(t) = x^2(t), \qquad y(t) = x(t)\cos(\omega t)$$

examine whether it is stable or unstable according to BIBO.

Solution

(a) As is well known from trigonometry, the sine function is bounded since it is known to be $|\sin(t)| \leq 1$ for every $t \in R$. This means that whatever the system's input, its output will always be characterized by the property $|y(t)| \leq 1$, that is, the system is *BIBO stable*.
(b) If we transmit a bounded input of the form $|x(t)| \leq \alpha$ to the system, then the system's output will also be bounded since it obviously satisfies the relation $|y(t)| \leq \alpha^2$. Therefore, the system is *BIBO stable*.
(c) From the system's defining equation and taking into account that $|\cos(\omega t)| \leq 1$, we will get

$$|y(t)| = |x(t)\cos(\omega t)| = |x(t)||\cos(\omega t)| \leq |x(t)|$$

But this means that if we transmit a bounded input $|x(t)| \leq \alpha$ to the system, then the output will also be bounded since it will hold that $|y(t)| \leq |x(t)| \leq |\alpha|$, that is, the system is *BIBO stable*.

Problem 2.3.4 For each of the systems described by the equations

$$y(t) = x(t-2) + x(2-t), \quad y(t) = \int_{-\infty}^{2t} x(\tau)d\tau, \quad y(t) = \frac{dx(t)}{dt}$$

examine whether they are stable or unstable according to BIBO.

Solution

(a) Transmitting the bounded input $|x(t)| \leq \alpha$, we observe that

$$|y(t)| = |x(t-2) + x(2-t)| \leq |x(t-2)| + |x(2-t)| = \alpha + \alpha = 2\alpha$$

Therefore, since the system's output is bounded, the system is *BIBO stable*.
(b) If we transmit the bounded input $|x(t)| \leq \alpha$, the output will be such that

$$|y(t)| \leq \int_{-\infty}^{2t} \alpha d\tau = \alpha \int_{-\infty}^{2t} d\tau \equiv \infty$$

We observe that the system's output is not bounded; hence, the system is *BIBO unstable*.
(c) In this case, we can characterize the system as unstable *if we find even one bounded input for which the system's output is not bounded*. If the step function $u(t)$, which is obviously bounded, is transmitted to the system, then the system's output will be equal to the derivative of the step function, which is known to be the impulse function $\delta(t)$. However, we know that this function has a value equal to zero everywhere, except for the position $t = 0$, where it *becomes infinite*. Therefore, since we have determined some input for which the output is not bounded, even for a specific time instance, the system should be characterized as *unstable according to BIBO*.

2.4 Impulse, Step, and Frequency Response

A very useful tool often used in the quantitative characterization of the behavior of continuous-time linear systems is the knowledge of the system's response to elementary continuous-time signals presented in the previous chapter. Limiting ourselves to continuous-time signals of the *unit step function*, the *unit impulse function*, and the *elementary complex exponential signal* that are used in almost all cases, we can give the following definitions.

Definition 2.4.1 (Impulse Response) The *impulse response* $h(t)$ of a continuous-time linear system is called the output produced by the system when the unit impulse function $\delta(t)$ is transmitted to its input, or, in mathematical notation,

$$h(t) = \mathcal{T}\{\delta(t)\}. \tag{2.9}$$

Definition 2.4.2 (Step Response) The *step response* $s(t)$ of a continuous-time linear system is called the output produced by the system when the unit step function $u(t)$ is transmitted to its input, or, in mathematical notation,

$$s(t) = \mathcal{T}\{u(t)\}. \tag{2.10}$$

Definition 2.4.3 (Frequency Response) The *frequency response* $\mathcal{H}(j\omega)$ of a continuous-time linear system is called the ratio of the output signal produced by the system when a basic complex exponential signal with amplitude equal to A and arbitrary frequency ω is transmitted to this input signal, or, in mathematical notation,[4]

$$\mathcal{H}(j\omega) = \frac{\mathcal{T}\{Ae^{j\omega t}\}}{Ae^{j\omega t}} \tag{2.11}$$

In almost all cases, the frequency response is a complex function, and if it is expressed in its polar form $\mathcal{H}(j\omega) = |\mathcal{H}(j\omega)| \exp[\measuredangle\mathcal{H}(j\omega)]$, then its magnitude $|\mathcal{H}(j\omega)|$ is known as an amplitude response, while its phase $\measuredangle\mathcal{H}(j\omega)$ is known as a phase response.

The foregoing definitions are valid *only for LTI systems* in which the transform $\mathcal{T}$ applied by the system to the input signal *does not change over time*. If, on the contrary, this transformation changes over time as a function of the form $\mathcal{T} = \mathcal{T}(t)$, then the foregoing responses are defined as follows:

- Impulse response: $h(t, \tau) = \mathcal{T}(t)\{\delta(t - \tau)\}$
- Step response: $s(t, \tau) = \mathcal{T}(t)\{u(t - \tau)\}$
- Frequency response: $\mathcal{H}(j\omega, t) = \dfrac{\mathcal{T}(t)\{e^{j\omega t}\}}{e^{j\omega t}} = |\mathcal{H}(j\omega, t)| \exp[\measuredangle\mathcal{H}(j\omega, t)]$

We therefore observe that in the case of linear time-varying systems, the amplitude and phase response functions depend on both the frequency ω and time t.

Problem 2.4.1 Find the impulse, step, and frequency responses of the RC circuit depicted in Fig. 2.5.

Solution From Kirchoff's law it is known that the input voltage $x(t)$ equals the voltage drop across the resistance $i(t)R$ plus the voltage across the capacitor $y(t)$, that is, $x(t) = i(t)R + y(t)$. However,

$$i(t) = \frac{dq(t)}{dt} = C\frac{dy(t)}{dt}$$

(since, as is well known $i(t) = q'(t)$, with $q = CV$), and by substitution we have

[4] In the theory of continuous-time signals, the frequency response of a linear time-invariant (LTI) system is symbolized not by $\mathcal{H}(\omega)$ but by $\mathcal{H}(j\omega)$, for reasons that will be explained in a subsequent chapter.

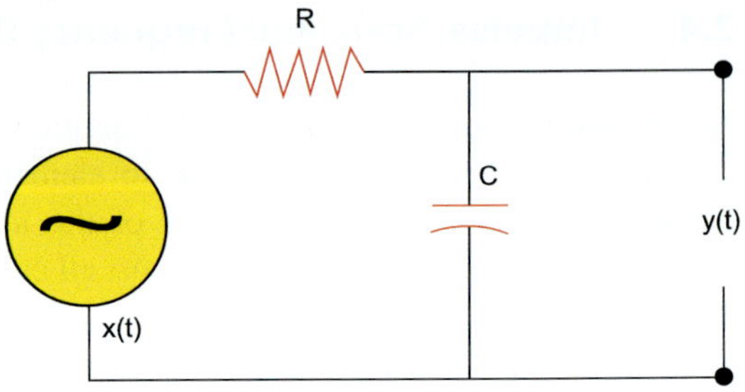

Fig. 2.5 Circuit from Exercise 2.4.1

$$x(t) = RC\frac{dy(t)}{dt} + y(t)$$

or, equivalently,

$$\frac{dy(t)}{dt} + \frac{1}{RC}y(t) = \frac{1}{RC}x(t)$$

which is also the differential equation that describes the circuit. This differential equation is a first-order linear equation with constant coefficients, and its general solution is proven[5] to be

$$y(t) = y(0)\exp\left(-\frac{t}{RC}\right) + \exp\left(-\frac{t}{RC}\right)\frac{1}{RC}\int_{0^-}^{t}\exp\left(\frac{\tau}{RC}\right)x(\tau)d\tau$$

If for simplicity we assume that for $t = 0$ it holds that $y(0) = 0$, then the preceding relation will be written

$$y(t) = \exp\left(-\frac{t}{RC}\right)\frac{1}{RC}\int_{0^-}^{t}\exp\left(\frac{\tau}{RC}\right)x(\tau)d\tau$$

To determine the required responses, we need to replace in the foregoing equation the input $x(t)$ with (a) the *impulse function* $\delta(t)$, (b) the *step function* $u(t)$, and (c) some *harmonic function*—let us assume $x(t) = \sin(\omega t)$. The foregoing process for each of the above cases is carried out in the following way:

(a) Impulse response: Proceeding with the replacement $x(t) = \delta(t)$, it will hold that $y(t) = h(t)$, where

$$h(t) = \exp\left(-\frac{t}{RC}\right)\frac{1}{RC}\int_{0^-}^{t}\exp\left(\frac{\tau}{RC}\right)\delta(\tau)d\tau$$

Using the defining equation of the impulse function for the specific integration limits and for the function $\varphi(\tau) = \exp(\tau/RC)$, it will hold that

$$\int_{0^-}^{t}\varphi(\tau)\delta(\tau)d\tau = \int_{0^-}^{t}\exp(\tau/RC)\delta(\tau)d\tau = \varphi(0) = \exp(\tau/RC)|_{\tau=0} = 1$$

and therefore the impulse response of the circuit will be given by the relation

$$h(t) = \begin{cases} \dfrac{e^{-t/RC}}{RC} & \text{for } t \geq 0 \\ 0 & \text{for } t < 0 \end{cases} = \frac{1}{RC}e^{-t/RC}u(t)$$

[5] This equation has the form $y'(t) + g(t)y(t) = h(t)$, where $g(t) = 1/RC$ and $h(t) = x(t)/RC$, and according to the *Lagrange method*, its solution has the form

$$y(t) = \left[C - \int h(t)\exp\left(\int g(t)\,dt\right)dt\right]\exp\left(-\int g(t)dt\right)$$

with the integration limits and the constant C appropriate in each case calculated with the help of the initial conditions.

where we took into account that for $t < 0$ it holds that $h(t) = 0$.

(b) Step response: Proceeding with the replacement $x(t) = u(t)$, we have the case that $y(t) = s(t)$, where

$$s(t) = \exp\left(-\frac{t}{RC}\right)\frac{1}{RC}\int_{0^-}^{t} \exp\left(\frac{\tau}{RC}\right)u(\tau)d\tau$$

However, $u(t) = 1$ for $t \geq 0$ and the preceding integral is calculated as

$$\int_{-0}^{t} \exp\left(\frac{\tau}{RC}\right) u(\tau)d\tau = \int_{0}^{t} \exp\left(\frac{\tau}{RC}\right) d\tau = RC\left[\exp\left(\frac{\tau}{RC}\right)\right]_0^t = RC\left(e^{t/RC} - 1\right)$$

Therefore, the step response will have the form

$$s(t) = \exp\left(-\frac{t}{RC}\right)\frac{1}{RC}\left\{RC\left[\exp\left(\frac{t}{RC}\right) - 1\right]\right\} = 1 - \exp\left(-\frac{t}{RC}\right)$$

or, in equivalent form,

$$s(t) = \begin{cases} 1 - e^{-t/RC} & \text{for } t \geq 0 \\ 0 & \text{for } t < 0 \end{cases} = \left[1 - e^{-t/RC}\right]u(t)$$

where we took into account that for $t < 0$ it holds that $s(t) = 0$.

(c) **Frequency response**: Proceeding with the substitution $x(t) = \sin(\omega t)$, we will have $y(t) \equiv H(\omega)$, where

$$\mathcal{H}(j\omega) = \exp\left(-\frac{t}{RC}\right)\frac{1}{RC}\int_{0^-}^{t} \exp\left(\frac{\tau}{RC}\right) \sin(\omega\tau)d\tau$$

In a previous exercise we used the definite integral

$$\int_0^t e^{\alpha\tau} \sin(\omega\tau)d\tau = \frac{e^{\alpha t}[\alpha \sin(\omega t) - \omega\cos(\omega t)] + \omega}{\alpha^2 + \omega^2}$$

In this case, $\alpha = 1/RC$, and by substitution we get

$$\mathcal{H}(j\omega) = \frac{\alpha^2 \sin(\omega t) + \alpha\omega[e^{-\alpha t} - \cos(\omega t)]}{\alpha^2 + \omega^2} = \frac{\sin(\omega t) + \omega RC\left[e^{-t/RC} - \cos(\omega t)\right]}{\omega^2 R^2 C^2 + 1}$$

a function that also expresses the frequency response of the system. As an exercise, the reader is encouraged to repeat the preceding calculations and calculate the impulse, step, and frequency responses for the circuit of Fig. 2.6.

The preceding responses of continuous-time systems to elementary signals are particularly important, mainly as far as the special category of *linear* systems is concerned, which are characterized by the power of the *principle of superposition*. According to this principle, *the response of a system to a linear combination of some elementary inputs is the linear combination of the corresponding elementary outputs*. If we take into account that any continuous-time signal can be expressed with the help of the elementary signals of the step function, the impulse function, and the simple trigonometric functions in the way presented in the corresponding sections, we can formulate the conclusion that *the response of a system to any signal can be expressed using the step response, the impulse response, and the frequency response functions*.

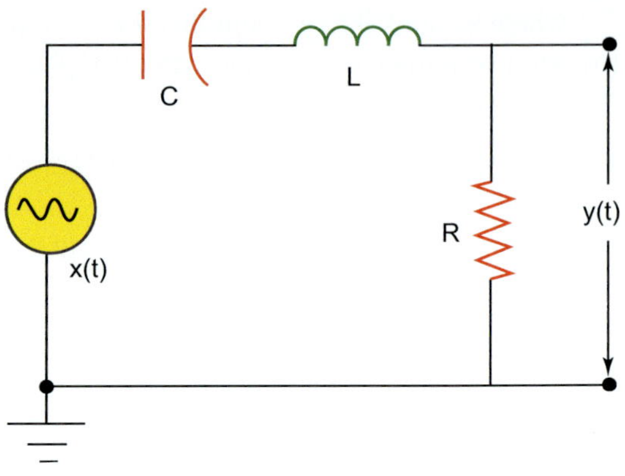

Fig. 2.6 A typical RLC circuit

To demonstrate this idea, let us calculate the response of a linear system to a signal expressed using the impulse function $\delta(t)$. As we proved in the previous section, each continuous-time signal $x(t)$ can be approximated by a linear combination of time-shifted impulse functions of the form

$$\tilde{x}(t) = \sum_{n=-\infty}^{+\infty} x(n\Delta)\delta_\Delta(t - n\Delta)\Delta \tag{2.12}$$

If this continuous-time signal passes through a time-varying linear system—the case of time-invariant systems will be studied later—the output of the system will be

$$\tilde{y}(t) = \mathcal{T}(t)\{\tilde{x}(t)\} = \mathcal{T}(t)\left\{\sum_{n=-\infty}^{+\infty} x(n\Delta)\delta_\Delta(t - n\Delta)\Delta\right\} \tag{2.13}$$

Using the *principle of superposition* and the definition of *impulse response* of the time-varying system that we previously mentioned, it turns out that

$$\tilde{y}(t) = \sum_{n=-\infty}^{+\infty} x(n\Delta)\Big[\mathcal{T}(t)\{\delta_\Delta(t - n\Delta)\}\Big]\Delta = \sum_{n=-\infty}^{+\infty} x(n\Delta)\tilde{h}(t, n\Delta)\Delta \tag{2.14}$$

and taking the limit of the sum for $\Delta \to 0$, we end up with the relation

$$y(t) = \lim_{\Delta\to 0} \sum_{n=-\infty}^{+\infty} x(n\Delta)\tilde{h}(t, n\Delta)\Delta = \int_{-\infty}^{+\infty} x(\tau)h(t, \tau)d\tau \tag{2.15}$$

an equation that expresses the response $y(t)$ of a linear time-varying continuous-time system, as a function of the impulse response $h(t, \tau)$. Similar arguments can be formulated for those cases where the input signal is expressed using the step function, as well as the elementary trigonometric functions.

The foregoing description refers to the simple case of linear systems of the SISO type, which accept a simple input $x(t)$ and produce a simple output $y(t)$, although the generalization for the case of MIMO systems that accept multiple inputs and produce multiple outputs is obvious. Considering a MIMO system with p inputs and q outputs that accepts an input $\boldsymbol{x}(t) = [x_1(t)\ x_2(t)\ \ldots\ x_p(t)]^T$ and produces an output $\boldsymbol{y}(t) = [y_1(t)\ y_2(t)\ \ldots\ y_q(t)]^T$, the last relation is generalized in the expression[6]

[6] This description assumes that the system is in a *state of rest* and that its response depends exclusively and only on the input we transmit to it (hence, in reality, it is the *zero-state response*). If this is not the case, then the output of the system does not depend unambiguously on the input it receives, since it will be different for different initial conditions. This is something that obviously applies to SISO-type systems as well.

$$\boldsymbol{y}(t) = \mathcal{T}(t)\{\boldsymbol{x}(t)\} = \int_{-\infty}^{+\infty} \boldsymbol{x}(\tau)\boldsymbol{h}(t,\tau)d\tau \tag{2.16}$$

where the parameter $\boldsymbol{h}(t,\tau)$ is a matrix of dimensions $p \times q$ defined as

$$\boldsymbol{h}(t,\tau) = \begin{bmatrix} h_{11}(t,\tau) & h_{12}(t,\tau) & \dots & h_{1p}(t,\tau) \\ h_{21}(t,\tau) & h_{22}(t,\tau) & \dots & h_{2p}(t,\tau) \\ \vdots & \vdots & & \vdots \\ h_{q1}(t,\tau) & h_{q2}(t,\tau) & \dots & q_{1p}(t,\tau) \end{bmatrix} \tag{2.17}$$

and is called the *impulse response matrix*. In this matrix, the element h_{ij} $(i = 1, 2, \dots, p,\ \ j = 1, 2, \dots q)$ expresses the impulse response between the jth input and the ith output, that is, *the ith output of the system at time t, when the jth input transmitted to it at time* τ *is the impulse function* $\delta(\tau)$*, while all other inputs are zero*. More details about the impulse response matrix and the general behavior of linear MIMO systems can be found in the literature.[7]

2.5 Linear Time-Invariant Systems

An interesting and particularly useful category of continuous-time systems are those characterized *both by the property of linearity and by the property of time invariance*. In other words, this means that these systems, on the one hand, will satisfy the principle of superposition, while, on the other hand, the transform $\mathcal{T}$ they apply to each input signal does not change as a function of time. In such a case, the calculation of the system's output is simplified significantly. Indeed, as we saw in detail in the previous section, the zero-state response of a linear system to an arbitrary input signal $x(t)$ is given by the relation

$$y(t) = \int_{-\infty}^{+\infty} x(\tau)h(t,\tau)d\tau$$

with the impulse response $h(t,\tau)$ defined as $h(t,\tau) = \mathcal{T}(t)\{\delta(t-\tau)\}$. However, if we assume that the system is both simultaneous and time-invariant, then it is obvious that

$$h(t,\tau) = \mathcal{T}(t)\{\delta(t-\tau)\} \equiv \mathcal{T}\{\delta(t-\tau)\} = h(t-\tau)$$

and the system output takes the form

$$y(t) = \int_{-\infty}^{+\infty} x(\tau)h(t-\tau)d\tau \tag{2.18}$$

But the integral of the right-hand side of the last equation simply expresses the *convolution* of the continuous-time signals $x(t)$ and $h(t)$, which was considered in detail in the previous chapter. We arrive, therefore, at the very useful and interesting conclusion that *the response* $y(t)$ *of a continuous-time LTI system to the arbitrary input signal* $x(t)$ *is equal to the convolution between the input signal* $x(t)$ *and the system's impulse response* $h(t)$, or, in mathematical notation, $y(t) = h(t) * x(t)$.

[7] See, for example, the book by Chi-Tsong Chen, *Linear System Theory and Design*, Holt, Rinehart and Winston, 1970.

Based on the foregoing conclusion, we can claim that *the impulse response of a continuous-time LTI system fully describes the system*. Indeed, whatever the system's input, its output will always result from the convolution between the input signal and the system's impulse response. Therefore, for every input of the system we can always calculate its output, something that is also demonstrated in the following exercises.

Problem 2.5.1 Let us consider a continuous-time LTI system whose impulse response is given by the relation $h(t) = e^{-\alpha t}u(t)$. Determine the output of this system when the input signal $x(t) = e^{-\beta t}u(t)$ is transmitted to it. Examine the cases $\alpha \neq \beta$ and $\alpha = \beta$.

Solution In a previous exercise (Exercise 1.10.1, page 37), we proved that the convolution between continuous-time signals $e^{-\alpha t}u(t)$ and $e^{-\beta t}u(t)$ is given by the relation

$$\left\{e^{-\alpha t}u(t)\right\} * \left\{e^{-\beta t}u(t)\right\} = \begin{cases} \dfrac{e^{-\alpha t} - e^{-\beta t}}{\beta - \alpha} & \text{for } \alpha \neq \beta \\ te^{-\alpha t}u(t) & \text{for } \alpha = \beta \end{cases}$$

If, therefore, we consider the first of the two signals as the *impulse response* of the system and the second as the *input signal* that is transmitted to it, it is obvious that the system's output for $\alpha \neq \beta$ and $\alpha = \beta$ will be given by the corresponding expressions of the right-hand side of the preceding equation. Generalizing this conclusion, we can formulate the claim that *if all pairs of signals whose convolution we calculated in the exercises in the homonymous section are considered as the impulse response and the input of a system (or vice versa, as the input and the impulse response, due to the power of the commutative property of the convolution), then the signal resulting from the convolution of these two signals can be considered as the output of the involved system.* ■

Note In almost all cases the systems we consider are causal, which means that their output is calculated only for time instances $t \geq 0$. Therefore, if we want to be more meticulous in our claims, we should write the output of the foregoing system for $\alpha \neq \beta$ not as

$$y(t) = \frac{e^{-\alpha t} - e^{-\beta t}}{\beta - \alpha} \quad \text{but as} \quad y(t) = \frac{e^{-\alpha t} - e^{-\beta t}}{\beta - \alpha}u(t)$$

where multiplication by the unit step function $u(t)$ indicates precisely that the system output $y(t)$ is nonzero only for time instances $t \geq 0$. The same, of course, applies to the case $\alpha = \beta$. This manner of writing will be adopted without comment in the rest of this section.

Problem 2.5.2 Let us consider a continuous-time LTI system whose impulse response is given by the relation

$$h(t) = \begin{cases} \alpha e^{-\beta t} & \text{for } t \geq 0 \\ 0 & \text{for } t < 0 \end{cases} = \alpha e^{-\beta t}u(t), \quad b > 0$$

Determine the output $y(t)$ when a square pulse of amplitude A and duration ξ is transmitted to its input, defined as

$$x(t) = \begin{cases} A \text{ for } 0 \leq t \leq \xi \\ 0 \text{ anywhere else} \end{cases} = Au(t) - Au(t - \xi)$$

Solution Using the convolution defining equation, we can write the system output in the form

$$y(t) = x(t) * h(t) = \int_{-\infty}^{+\infty} x(\tau)h(t-\tau)d\tau = \int_{-\infty}^{+\infty} \left[Au(\tau) - Au(\tau - \xi)\right]\left\{\alpha e^{-\beta(t-\tau)}u(t-\tau)\right\}d\tau$$

or, expanding the expression,

$$y(t) = \alpha A e^{-\beta t}\left\{\int_{-\infty}^{+\infty} e^{\beta\tau}u(\tau)u(t-\tau)d\tau - \int_{-\infty}^{+\infty} e^{\beta\tau}u(\tau-\xi)u(t-\tau)d\tau\right\}$$

The unit step functions $u(\tau)$ and $u(t-\tau)$ have a nonzero value only in the interval $0 \le \tau \le t$, and therefore the first integral is calculated as

$$\int_{-\infty}^{+\infty} e^{\beta\tau}u(\tau)u(t-\tau)d\tau = \int_0^t e^{\beta\tau}d\tau = \frac{1}{\beta}\left[e^{\beta\tau}\right]_0^t = \frac{e^{\beta t}-1}{\beta}$$

To calculate the second integral, we proceed with the change variable $\tau - \xi = \lambda$. If we calculate the new integration limits and express all the quantities that appear in this function of the new variable λ, it is proven that[8]

$$\int_{-\infty}^{+\infty} e^{\beta\tau}u(\tau-\xi)u(t-\tau)d\tau = \frac{e^{\beta t}-e^{\beta\xi}}{\beta}$$

Before we proceed to the substitution of the expressions we calculated for the preceding integrals in the defining equation of the output $y(t)$ so that we can end up calculating it, we need to remark on the use of the step function, which makes the system output a causal signal. As we have already mentioned, if when calculating the output of a LTI system we end up with a relation of the form $y(t) = f(t)$, we usually write it as $y(t) = f(t)u(t)$ so as to emphasize the fact that it is defined only for time instances $t \ge 0$. In this case, the input signal $x(t) = Au(t) - Au(t-\xi)$ can be considered to arise from the overlap of two elementary inputs and, more specifically, of $x_1(t) = Au(t)$ and $x_2(t) = -Au(t-\xi)$. Therefore, due to the linear nature of the system, its output will arise from the overlap of the corresponding elementary outputs $y_1(t)$ and $y_2(t)$, which, using the expressions of the integrals we calculated earlier, have the form

$$y_1(t) = +\alpha A e^{-\beta t}\frac{e^{\beta t}-1}{\beta} = +\frac{\alpha A}{\beta}\left(1-e^{-\beta t}\right) \quad \text{and} \quad y_2(t) = -\alpha A e^{-\beta t}\frac{e^{\beta t}-e^{\beta\xi}}{\beta} = -\frac{\alpha A}{\beta}\left[1-e^{-\beta(t-\xi)}\right]$$

However, as we can see, the input $x_2(t)$ is temporally shifted with respect to the input $x_1(t)$ by ξ time units—since $x_1(t)$ is multiplied by $u(t)$, while $x_2(t)$ is multiplied by $u(t-\xi)$. Therefore, the output $y_2(t)$ should be shifted by ξ time units in relation to the output $y_1(t)$. Thus, to make these outputs causal signals, we will make the substitutions $y_1(t) \to y_1(t)u(t)$ and $y_2(t) \to y_2(t)u(t-\xi)$, and therefore the total output of the system will have the form

$$y(t) = \frac{\alpha A}{\beta}\left\{\left(1-e^{-\beta t}\right)u(t) - \left[1-e^{-\beta(t-\xi)}\right]u(t-\xi)\right\}$$

which is also the desired result.

---

[8] Indeed, in this case it will hold that $\tau = \lambda + \xi$, and differentiating we get $d\tau = d(\lambda+\xi) = d\lambda$. The exponential expression becomes $e^{\beta\tau} = e^{\beta(\lambda+\xi)} = e^{\beta\lambda}e^{\beta\xi}$, while the two step functions will take the form $u(\tau-\xi) = u(\lambda)$ and $u(t-\tau) = u(t-\lambda-\xi)$. Making all these substitutions, the second integral is written

$$\int_{-\infty}^{+\infty} e^{\beta\tau}u(\tau-\xi)u(t-\tau)d\tau = e^{\beta\xi}\int_{-\infty}^{+\infty} e^{\beta\lambda}u(\lambda)u(t-\lambda-\xi)d\lambda$$

Observing that the step functions $u(\lambda)$ and $u(t-\lambda-\xi)$ are defined as

$$u(\lambda) = \begin{cases} 1 \text{ for } \lambda \ge 0 \\ 0 \text{ for } \lambda < 0 \end{cases} \quad \text{and} \quad u(t-\lambda-\xi) = \begin{cases} 1 \text{ for } \lambda \le t-\xi \\ 0 \text{ for } \lambda > t-\xi \end{cases}$$

the last integral has a nonzero value in the range of values $0 \le \lambda \le t-\xi$, which is calculated as

$$\int_{-\infty}^{+\infty} e^{\beta\lambda}u(\lambda)u(t-\lambda-\xi)d\lambda = \int_0^{t-\xi} e^{\beta\lambda}d\lambda = \frac{1}{\beta}\left[e^{\beta\lambda}\right]_0^{t-\xi} = \frac{e^{\beta(t-\xi)}-1}{\beta}$$

Consequently, the second integral in the convolution equation will have a value equal to

$$\int_{-\infty}^{+\infty} e^{\beta\tau}u(\tau-\xi)u(t-\tau)d\tau = e^{\beta\xi}\cdot\frac{e^{\beta(t-\xi)}-1}{\beta} = e^{\beta\xi}\cdot\frac{e^{\beta t}\cdot e^{\beta\xi}-1}{\beta} = \frac{e^{\beta t}-e^{\beta\xi}}{\beta}.$$

2.5.1 Consequences of Properties of Convolution

As was already mentioned, the convolution $x(t) * y(t)$ of two continuous-time signals $x(t)$ and $y(t)$ is characterized by the commutative, associative, and distributive properties, which are formulated as follows:

- Commutative property: $x(t) * y(t) = y(t) * x(t)$.
- Associative property: $[x(t) * y(t)] * z(t) = x(t) * [y(t) * z(t)]$.
- Distributive property: $x(t) * [y(t) + z(t)] = x(t) * y(t) + x(t) * z(t)$.

From a mathematical point of view, the foregoing properties are indeed interesting, but they offer nothing special, nor do they find any practical application. However, if we apply them to a LTI system whose output $y(t)$ results from the convolution of the system's impulse response $h(t)$ with its input $x(t)$, then leads to some interesting and useful conclusions, and more specifically:

- According to the commutative property, which is formulated as

$$y(t) = x(t) * h(t) = h(t) * x(t) \tag{2.19}$$

  the interchange of the roles of the signals $h(t)$ and $x(t)$ does not change the output $y(t)$ of the system, which remains the same. We can therefore formulate the conclusion that *a LTI system with impulse response $h(t)$, which receives the input signal $x(t)$, is fully equivalent to a system with impulse response $x(t)$, which receives the input signal $h(t)$.*
- Let us consider two systems with impulse responses $h_1(t)$ and $h_2(t)$, which are connected in series, in the sense that *the output of the first system is used as input of the second system.* If we symbolize by $x(t)$ the input of the first system, then the output of this system will be equal to $w(t) = x(t) * h_1(t)$, while the output of the second system—which receives as input the signal $w(t)$—will be $y(t) = h_2(t) * w(t) = w(t) * h_2(t) = [x_1(t) * h_1(t)] * h_2(t)$. In this case, the power of the associative property

$$y(t) = [x_1(t) * h_1(t)] * h_2(t) = x(t) * [h_1(t) * h_2(t)] = [h_1(t) * h_2(t)] * x(t) \tag{2.20}$$

  leads us to the conclusion that *two LTI systems with impulse responses $h_1(t)$ and $h_2(t)$ that are connected in series are equivalent to a simple system whose impulse response $h(t)$ arises from the convolution of the impulse responses of the two systems, or, in mathematical notation, $h(t) = h_1(t) * h_2(t)$.*
- Finally, considering two LTI systems with impulse responses $h_1(t)$ and $h_2(t)$, the transmission of the common input $x(t)$ to them will lead to the creation of the corresponding outputs, which obviously will be $y_1(t) = x(t) * h_1(t)$ and $y_2(t) = x(t) * h_2(t)$. If these two systems are connected in parallel, then *the output of the composite system that will result will obviously be equal to the sum of the two outputs*, or, in mathematical notation, $y(t) = y_1(t) + y_2(t) = x(t) * h_1(t) + x(t) * h_2(t)$. In this case, the power of the distributive property

$$y(t) = x(t) * h_1(t) + x(t) * h_2(t) = x(t) * [h_1(t) + h_2(t)] \tag{2.21}$$

  leads us to the conclusion that *two LTI systems with impulse responses $h_1(t)$ and $h_2(t)$ that are connected in parallel are equivalent to a simple system whose impulse response arises from the sum of the impulse responses of the two systems, or, in mathematical notation, $h(t) = h_1(t) + h_2(t)$.*

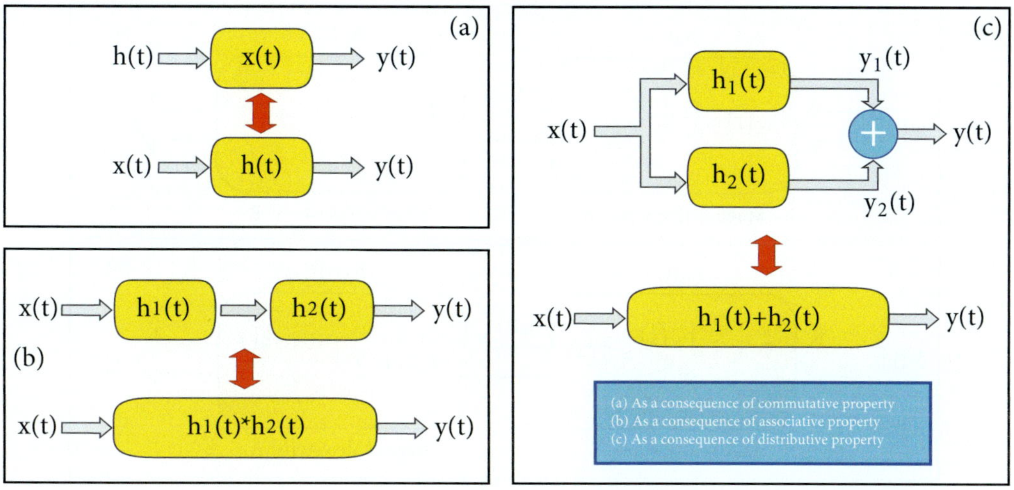

Fig. 2.7 The consequences of the three properties of the operation of convolution in continuous-time LTI systems

The foregoing consequences of these three properties of the operation of convolution are presented in Fig. 2.7.

2.5.2 The Impulse Response of LTI Systems

As was mentioned earlier, in a LTI system, the output signal $y(t)$ can be calculated from the *convolution* between the system's impulse response $h(t)$ and its input signal $x(t)$. Therefore, calculating the impulse response of such a system is a very important process, since we can then calculate the corresponding output signal for each input signal in the way mentioned earlier. There are two different ways to calculate the impulse response, the first relies *on the use of the differential equation that describes the system* and the second *on the differentiation of the function of the system's step response*. These two ways of calculating the impulse response are presented in detail in the following pages.

Calculating the Impulse Response from the System's Differential Equation

Let us consider a LTI system described by a linear differential equation of order N with constant coefficients of the form

$$\sum_{k=0}^{N} \alpha_k \frac{d^k y(t)}{dt^k} = \sum_{k=0}^{M} \beta_k \frac{d^k x(t)}{dt^k} \tag{2.22}$$

If we take into account the properties of *linearity* and *commutativity* that describe such a system, we can formulate the assertion that *this system consists of two LTI subsystems, A and B, of which subsystem A implements the right-hand side of the preceding equation, while subsystem B implements the left-hand side*. In other words, the first subsystem takes the signal $x(t)$ *as input and returns as its output the signal* $w(t)$ *which is defined as the linear combination of the signal* $x(t)$, as well as all its derivatives up to order M, so that we can write

$$w(t) = \mathcal{T}_1\{x(t)\} = \sum_{k=0}^{M} \beta_k \frac{d^k x(t)}{dt^k} \tag{2.23}$$

(where $\mathcal{T}_1$ is the transformation implemented by subsystem A). On the other hand, subsystem B, which is connected in series with subsystem A, takes the signal $w(t)$ as input and produces as its output the signal $y(t)$, which is connected with the input $w(t)$, with the relation

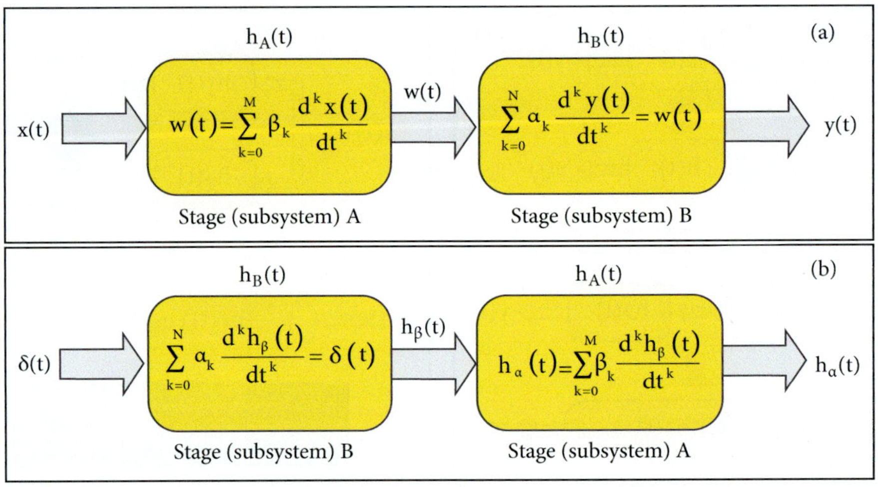

Fig. 2.8 A continuous-time LTI system can be described as the composition of two subsystems, each of which implements a member of the differential equation that describes it (Boulet, Figures 3.4 and 3.5)

$$\sum_{k=0}^{N} \alpha_k \frac{d^k y(t)}{dt^k} = w(t) \tag{2.24}$$

This situation is presented diagrammatically in Fig. 2.8α.

Taking into account that the two subsystems A and B are characterized by the *commutative property* regarding their positions,[9] it is obvious that we can alternate their order without changing anything in the operation of the system. If we then use the new system that emerged after the alternation to transmit as input to subsystem B the impulse function $\delta(t)$, so as to receive as the output of subsystem A the impulse response $h(t)$, then, according to the analysis presented previously, the functions $\delta(t)$ and $h(t)$ will be defined as

$$\delta(t) = \sum_{k=0}^{N} \alpha_k \frac{d^k h_\beta(t)}{dt^k} \quad \text{and} \quad h(t) \equiv h_\alpha(t) = \sum_{k=0}^{M} \beta_k \frac{d^k h_\beta(t)}{dt^k}$$

[see Fig. 2.8b]. This fact allows us to calculate the *impulse response* of the system through the differential equation that describes it, applying the following procedure:

1. Implementing the operation of the first subsystem of Fig. 2.8b, we replace the right-hand side of the differential equation with the impulse function $\delta(t)$. In this case, the output of this subsystem will be the impulse response $h_\beta(t)$, with the two functions being connected by the relation

$$\sum_{k=0}^{N} \alpha_k \frac{d^k h_\beta(t)}{dt^k} = \delta(t) \tag{2.25}$$

A necessary prerequisite for making this substitution is that the maximum orders of the derivatives of the functions $y(t)$ and $x(t)$ satisfy the condition $N \geq M$ and that at the time $t = 0$ the system is in a state of rest, so that we can write that

[9] In other words, this means that *if the composite system is implemented by first connecting subsystem A and then subsystem B, then this is equivalent to the system that would result if we first placed B and then A*. The equivalence of these two systems arises using the associative in combination with the commutative property of convolution and is presented in a subsequent section.

$$y(0^-) = \frac{dy(0^-)}{dt} = \frac{d^2y(0^-)}{dt^2} = \cdots = \frac{d^{(N-1)}y(0^-)}{dt^{(N-1)}} = 0 \tag{2.26}$$

2. The preceding differential equation must be solved with respect to the function $h_\beta(t)$, so that it can be used subsequently for the description of the second subsystem. In order for this process to take place, we must first observe that *the impulse function $\delta(t)$ appearing in the last equation must be generated only from the derivative $h_\beta^{(N)}(t)$, since, if it were generated from derivatives of lower order, then derivatives of the impulse function would appear in the equation, something that obviously does not happen.*[10] However, this means that the function $h_\beta(t)$ as well as all its derivatives $h'_\beta(t), h''_\beta(t), \ldots, h_\beta^{(N-1)}(t)$ up to order $N-1$ will be smooth (or, in the worst case, will have finite discontinuities), that is, their integral for an infinitesimally small integration interval will tend to zero, so that we can write

$$\int_{0^-}^{0^+} \frac{d^k h_\beta(t)}{dt^k} dt = \frac{d^{(k-1)} h_\beta(0^+)}{dt^{k-1}} = 0, \quad (k = 1, 2, \ldots, N-1) \tag{2.27}$$

Integrating, therefore, our equation in the interval from $t = 0^-$ to $t = 0^+$, we will have[11]

$$\int_{0^-}^{0^+} \sum_{k=0}^{N} \alpha_k \frac{d^k h_\beta(t)}{dt^k} dt = \sum_{k=0}^{N} \alpha_k \int_{0^-}^{0^+} \frac{d^k h_\beta(t)}{dt^k} dt = \alpha_N \int_{0^-}^{0^+} \frac{d^N h_\beta(t)}{dt^N} dt = \alpha_N \frac{d^{(N-1)} h_\beta(0^+)}{dt^{N-1}} = \int_{0^-}^{0^+} \delta(t) dt = 1$$

from which, of course, it immediately follows that

$$\frac{d^{(N-1)} h_\beta(0^+)}{dt^{N-1}} = \frac{1}{\alpha_N} \tag{2.28}$$

Therefore, the N initial conditions for the function $h_\beta(t)$ will be

$$h_\beta(0+) = \frac{dh_\beta(0^+)}{dt} = \cdots = \frac{d^{(N-2)} h_\beta(0^+)}{dt^{N-2}} = 0 \quad \text{and} \quad \frac{d^{(N-1)} h_\beta(0^+)}{dt^{N-1}} = \frac{1}{\alpha_N}$$

3. Having determined the initial conditions of the problem, we are now in a position to solve the homogeneous differential equation

$$\sum_{k=0}^{N} \alpha_k \frac{d^k h_\beta(t)}{dt^k} = 0 \tag{2.29}$$

for the time $t = 0^+$ and for the preceding initial conditions (the equation to be solved becomes homogeneous because the function $\delta(t)$ appearing on the right side of the full differential equation is equal to zero for the time we consider). Considering the well-known complex exponential solution of the form $h_\beta(t) = Ae^{st}$, where $A, s \in C$ and $t > 0$, deriving up to N times, we get $h'_\beta(t) = Ase^{st}$, $h''_\beta(t) = As^2e^{st}, \ldots, h_\beta^{(N)}(t) = As^N e^{\beta t}$, and substituting into the differential equation we will get

$$\alpha_0(Ae^{st}) + \alpha_1(Ase^{st}) + \alpha_2(As^2e^{st}) + \cdots + \alpha_N(As^N e^{st}) = 0$$

[10] If, for example, the function $\delta(t)$ were created by the derivative $h_\beta^{(N-2)}(t)$, then the derivative $h_\beta^{(N-1)}(t)$ would create the function $\delta'(t)$, while the function $h_\beta^{(N)}(t)$ would create the function $\delta''(t)$, but these do not appear in the equation we consider.

[11] In a real physical system, such as an *electrical circuit*, the time at which some input is transmitted to the system is treated as the starting point for time measurement. In this case, the conditions that exist immediately before the application of this input are aforementioned to correspond to the time $t = 0^-$, while those that apply immediately after this application are aforementioned to correspond to the time $t = 0^+$. It should be noted that in practice, we are mainly interested in the initial conditions at time $t = 0^-$, especially if the input to be applied to the system is characterized by discontinuities at a point in time $t = 0$ as is, for example, the impulse function $\delta(t)$.

or, equivalently,

$$Ae^{st}\left(\alpha_0 + \alpha_1 s + \alpha_2 s^2 + \cdots + \alpha_N s^N\right) = 0$$

If we assume that the proportionality constant A is different from zero—something that obviously also happens with the exponential expression $e^{st} = 0$—we will have

$$\alpha_0 + \alpha_1 s + \alpha_2 s^2 + \cdots + \alpha_N s^N = \sum_{k=0}^{N} \alpha_k s^k = 0 \tag{2.30}$$

an equation that is also the characteristic polynomial of the differential equation under consideration. This polynomial has at most N distinct complex roots of the form $\{s_k\}_{k=1}^{N}$, and assuming for simplicity that things are indeed like this—that is, there are no repeated roots, but N distinct roots[12]—there will be N discrete functions of the form $h_\beta^k(t) = e^{s_k t}$ $(k = 1, 2, \ldots, N)$, which satisfy the homogeneous differential equation. Taking into account that this equation is a *linear differential equation with constant coefficients*, we can formulate the conclusion that its general solution—which is the desired function $h_\beta(t)$—can be written as a linear combination of these elementary complex exponential functions, a conclusion that in mathematical notation is formulated as

$$h_\beta(t) = \sum_{k=0}^{N} A_k h_\beta^k(t) = \sum_{k=0}^{N} A_k e^{s_k t} \tag{2.31}$$

with the N complex coefficients $\{A_k\}_{k=1}^{N}$ being able to be calculated from the N initial conditions of the problem determined previously. These conditions, if applied to the function $h_\beta(t)$ just determined, lead to the system of N equations with N unknowns

$$h_\beta(0+) = \sum_{k=1}^{N} A_k e^{s_k t}\Big|_{t=0^+} = \sum_{k=1}^{N} A_k$$

$$\frac{dh_\beta(0^+)}{dt} = \sum_{k=1}^{N} A_k s_k e^{s_k t}\Big|_{t=0^+} = \sum_{k=1}^{N} A_k s_k$$

$$\frac{d^2 h_\beta(0^+)}{dt^2} = \sum_{k=1}^{N} A_k s_k^2 e^{s_k t}\Big|_{t=0^+} = \sum_{k=1}^{N} A_k s_k^2$$

....

$$\frac{d^{(N-1)} h_\beta(0^+)}{dt^{N-1}} = \sum_{k=1}^{N} A_k s_k^{N-1} e^{s_k t}\Big|_{t=0^+} = \sum_{k=1}^{N} A_k s_k^{N-1} = \frac{1}{\alpha_N}$$

---

[12] In the case where some of the roots of the characteristic polynomial have multiplicity $p > 1$, the form of the solution we will get is slightly different. If, for example, some root s of the characteristic polynomial has multiplicity $p = 2$, then from this will arise the solutions e^{st} and te^{st}, while in the general case of a root with multiplicity value m, the solutions that will arise from this will be $e^{st}, te^{st}, t^2e^{st}, \ldots, t^{m-1}e^{st}$. Finally, in the case where the characteristic polynomial has *complex roots* and *real coefficients*, these will always appear in pairs. Therefore, if the quantity $\alpha + j\beta$ is a root of the characteristic polynomial, then the same should happen with the quantity $\alpha - j\beta$. In this case and for linear systems whose response $y(t)$ is a real signal, it is proven that this function is given by the relation $y(t) = ce^{\alpha t}\cos(\beta t + \vartheta)$, with the values of the parameters c and ϑ being defined in the appropriate way.

which in matrix form is expressed as

$$\begin{bmatrix} 0 \\ 0 \\ 0 \\ \vdots \\ 1/\alpha_N \end{bmatrix} = \begin{bmatrix} 1 & 1 & \dots & 1 \\ s_1 & s_2 & \dots & s_N \\ s_1^2 & s_2^2 & \dots & s_N^2 \\ \dots & \dots & \dots & \dots \\ s_1^{N-1} & s_2^{N-1} & \dots & s_N^{N-1} \end{bmatrix} \begin{bmatrix} A_1 \\ A_2 \\ A_3 \\ \vdots \\ A_N \end{bmatrix}$$

In the preceding equation, the $N \times N$ dimension matrix defined as

$$V = \begin{bmatrix} 1 & 1 & \dots & 1 \\ s_1 & s_2 & \dots & s_N \\ s_1^2 & s_2^2 & \dots & s_N^2 \\ \dots & \dots & \dots & \dots \\ s_1^{N-1} & s_2^{N-1} & \dots & s_N^{N-1} \end{bmatrix} \tag{2.32}$$

is known as the *Vandermonde matrix*.[13] It is proven that this matrix is always invertible, so that the calculation of the coefficients A_k and, by extension, the function $h_\beta(t)$ corresponding to this group of initial conditions is possible.

4. In the last step of the process, all we have to do is use the function $h_\beta(t)$ that we determined earlier to calculate the impulse response of the system $h(t) \equiv h_\alpha(t)$ from an equation of the form

$$h(t) = \sum_{k=0}^{M} \beta_k \frac{d^k h_\beta(t)}{dt^k} \tag{2.33}$$

The following exemplary solved exercise demonstrates the application of this method.

Problem 2.5.3 Find the impulse response of a system described by the differential equation

$$\frac{dy(t)}{dt} + \alpha y(t) = \frac{dx(t)}{dt} - 2x(t), \quad \alpha > 0$$

(see Boulet, Exercise 3.1).

Solution The preceding equation is a partial case of the general differential equation with parameter values $M = N = 1$, $\alpha_0 = \alpha$, $\alpha_1 = 1$, $\beta_0 = -2$, and $\beta_1 = 1$. Applying the process we described earlier, we set the right side of the preceding equation to the function $\delta(t)$ so that it takes the form

$$\frac{dh_\beta(t)}{dt} + \alpha h_\beta(t) = \delta(t)$$

and we integrate over the interval $(0^-, 0^+)$; in this case, we will end up with the initial condition

$$\left.\frac{d^{(N-1)} h_\beta(t)}{dt^{N-1}}\right|_{N=1} = h_\beta(0^+) = \left.\frac{1}{\alpha_N}\right|_{N=1} = \frac{1}{\alpha_1} = 1$$

[13] See, for example, the book by Roget A. Horn & Charles R. Johnson, *Topics in Matrix Analysis*, Cambridge University Press, 1991.

and then we will solve the corresponding homogeneous equation, which has the form

$$\frac{dh_\beta(t)}{dt} + \alpha h_\beta(t) = 0$$

so that the aforementioned condition is satisfied. Considering the well-known complex exponential solution of the form $h_\beta(t) = Ae^{st}$, where $A, s \in C$, and working in the interval $t \geq 0$, it is proven that the characteristic polynomial of the differential equation is given by the relation $s + \alpha = 0$, and it has a unique root with value $s = -\alpha$, that is, the general solution of the homogeneous differential equation will be $h_\beta(t) = Ae^{-\alpha t}$. Using the initial condition $h_\beta(0^+) = 1$, it immediately follows that $A = 1$, and therefore the function $h_\beta(t)$ will have the form $h_\beta(t) = e^{-\alpha t}$ for $t \geq 0$, or, in shorthand, $h_\beta(t) = e^{-\alpha t}u(t)$.

Substituting the function $h_\beta(t)$ into the defining equation of the impulse response, it turns out that

$$h(t) = \sum_{k=0}^{M} \beta_k \frac{d^k h_\beta(t)}{dt^k}\bigg|_{M=1} = \frac{dh_\beta(t)}{dt} - 2h_\beta(t) = \frac{d}{dt}\left\{e^{-\alpha t}u(t)\right\} - 2e^{-\alpha t}u(t) =$$

$$= -\alpha e^{-\alpha t}u(t) + e^{-\alpha t}\frac{du(t)}{dt} - 2e^{-\alpha t}u(t) = -(\alpha+2)e^{-\alpha t}u(t) + e^{-\alpha t}\delta(t)$$

—since $u'(t) = \delta(t)$—which is the desired result. Let us note that in practice this result is almost always simplified; indeed, using the property of the function $\delta(t)$ according to which the relation $x(t)\delta(t) = x(0)\delta(t)$ holds for the function $x(t) = e^{-\alpha t}$, we will have

$$x(t)\delta(t) = e^{-\alpha t}\delta(t) = x(0)\delta(t) = \left[e^{-\alpha t}\right]_{t=0}\delta(t) = \delta(t)$$

from which it finally follows that $h(t) = -(\alpha+2)e^{-\alpha t}u(t) + \delta(t)$.

Problem 2.5.4 Determine the impulse response of the system described by the equation

$$\frac{d^2y(t)}{dt^2} + 3\frac{dy(t)}{dt} + 2y(t) = x(t)$$

(see Boulet, Example 3.6, but also Oppenheim, Willsky, Nawab, Exercise 2.56δ).

Solution The foregoing differential equation arises from the general differential equation for parameter values $N = 2$, $M = 0$, and coefficients $\alpha_0 = 2$, $\alpha_1 = 3$, $\alpha_2 = 1$, and $\beta_0 = 1$. Following the methodology described earlier, we solve the equation

$$\frac{d^2h_\beta(t)}{dt^2} + 3\frac{dh_\beta(t)}{dt} + 2h_\beta(t) = \delta(t)$$

which for the range of values $t > 0$ is reduced to the homogeneous differential equation

$$\frac{d^2h_\beta(t)}{dt^2} + 3\frac{dh_\beta(t)}{dt} + 2h_\beta(t) = 0$$

which must be solved with initial conditions

$$h_\beta(0^+) = 0 \quad \text{and} \quad \frac{d^{(N-1)}h_\beta(t)}{dt^{N-1}}\bigg|_{N=2} = \frac{dh_\beta(0^+)}{dt} = \frac{1}{\alpha_N}\bigg|_{N=2} = \frac{1}{\alpha_2} = 1$$

Considering a complex exponential solution of the form $h_\beta(t) = Ae^{st}$ and differentiating twice with respect to time, we get $h'_\beta(t) = Ase^{st}$ and $h''_\beta(t) = As^2e^{st}$. Substituting these expressions into the homogeneous equation, it turns out that

$$As^2e^{st} + 3Ase^{st} + 2Ae^{st} = 0 \quad \text{or, equivalently,} \quad Ae^{st}(s^2 + 3s + 2) = 0 \quad \text{and finally}$$

$$s^2 + 3s + 2 = 0$$

with the last expression also being the characteristic polynomial of the homogeneous differential equation. The foregoing polynomial is written $(s + 2)(s + 1) = 0$ and has two roots, $s_1 = -2$ and $s_2 = -1$. Therefore, the general solution of the homogeneous differential equation is given in the form

$$h_\beta(t) = \sum_{k=1}^{2} A_k e^{s_k t} = A_1 e^{-2t} + A_2 e^{-t}, \quad (t \geq 0)$$

The constant coefficients A_1 and A_2 appearing in the foregoing relation can be calculated with the help of the initial conditions, which in this case are written

$$h_\beta(0^+) = A_1 e^{-2t} + A_2 e^{-t}\Big|_{t=0^+} = A_1 + A_2 = 0$$

$$\frac{dh_\beta(0^+)}{dt} = -2A_1 e^{-2t} - A_2 e^{-t} = \Big|_{t=0^+} = -2A_1 - A_2 = 1$$

The preceding system of two equations with unknowns coefficients A_1 and A_2 is easily solved, leading to the values $A_1 = -1$ and $A_2 = 1$; therefore, the function $h_\beta(t)$ will have the form

$$h_\beta(t) = e^{-t} - e^{-2t} \quad (\text{for } t \geq 0) = \left(e^{-t} - e^{-2t}\right)u(t)$$

and substituting it into the defining equation of the impulse response $h(t)$, it turns out that

$$h(t) = \sum_{k=0}^{M} \beta_k \frac{d^k h_\beta(t)}{dt^k}\Bigg|_{M=0} = h_\beta(t) = \left(e^{-t} - e^{-2t}\right)u(t)$$

which is also the desired result.

Correlation of Values of Parameters M and N

During the solution of the previous exercise, we found that the impulse response of the system

$$\frac{d^2y(t)}{dt^2} + 3\frac{dy(t)}{dt} + 2y(t) = x(t)$$

was calculated as a solution of the homogeneous differential equation that has the form $h(t) = (e^{-t} - e^{-2t})u(t)$. Repeating the same process for the system

$$\frac{dy(t)}{dt} + \alpha y(t) = \frac{dx(t)}{dt} - 2x(t)$$

(the solution of this exercise is left to the reader), it turns out that $h(t) = -(\alpha + 2)e^{-\alpha t}u(t) + \delta(t)$. We find, therefore, that in the second case, the impulse response $h(t)$ of the system also contains an impulse function $\delta(t)$. The reason for the appearance of this additional term is the validity of the condition $N = M$, where N and M are the maximum orders of the derivatives of the functions $y(t)$ and $x(t)$, respectively. Generally speaking, the correlation that exists between the values of the parameters M and N determines the character of the impulse response of the system. We distinguish three cases:

- $N > M$: In this case, the impulse response contains *only the solution* $y_h(t)$ *of the homogeneous differential equation of the system* and therefore will have the form $h(t) = \alpha y_h(t)u(t)$.

- $N = M$: In this case, the impulse response contains, *besides the solution of the homogeneous differential equation of the system, an impulse function term*—but not its derivatives—so that we can write that $h(t) = \alpha y_h(t)u(t) + \beta\delta(t)$.
- $N < M$: In this case, the impulse response, besides from the solution of the homogeneous differential equation of the system and the term of the impulse function $\delta(t)$, will also contain *all the derivatives of the impulse function with step value up to and including* $M - N$, so that we can write that

$$h(t) = \alpha y_h(t)u(t) + \beta\delta(t) + c_1\frac{d\delta(t)}{dt} + c_2\frac{d^2\delta(t)}{dt^2} + \cdots + \frac{d^{(M-N)}\delta(t)}{dt^{M-N}} = \alpha y_h(t)u(t) + \beta\delta(t) + \sum_{k=1}^{M-N} c_k\frac{d^k\delta(t)}{dt^k}$$

To understand this behavior of the impulse response, let us return to the theory we described earlier. As we have already mentioned, the impulse response of the system is calculated from the relation

$$h(t) = \sum_{k=0}^{M} \beta_k \frac{d^k h_\beta(t)}{dt^k}$$

with the function $h_\beta(t)$ resulting from the solution of the homogeneous version of the differential equation

$$\sum_{k=0}^{N} \alpha_k \frac{d^k h_\beta(t)}{dt^k} = \delta(t)$$

As we pointed out during the foundation of our theory, the existence in the preceding relation only of the impulse function $\delta(t)$ and not of some derivatives leads us to the conclusion that *this function must be created by the highest-order derivative of the impulse response, which is* $h^{(N)}h_\beta(t)$. Thus, if $M = N$, then the same reasoning allows us to conclude that the function $\delta(t)$ must also appear in the defining equation of the function $h(t)$—since for $M = N$ the derivative $h_\beta^{(M)}(t)$ will again be the highest-order derivative that will create the function $\delta(t)$. We are led, therefore, to the conclusion that the impulse response will contain the term of the function $\delta(t)$, so that we can write that $h(t) = \alpha y_h(t)u(t) + \beta\delta(t)$.

Finally, in the case $M > N$, we can generalize the foregoing assertion and formulate the conclusion that since the function $h_\beta^{(N)}(t)$ will create the function $\delta(t)$, the function $h_\beta^{(N+1)}(t)$ will create the function $\delta'(t)$, the function $h_\beta^{(N+2)}(t)$ will create the function $\delta''(t)$, and the function $h_\beta^{(N-M)}(t)$ will create the function $\delta^{(M-N)}(t)$. Therefore, in addition to the solution of the homogeneous equation, the impulse response $h(t)$ for the case $M > N$ will also contain a linear combination of the derivatives $\delta^{(i)}(t)$ $(i = 0, 1, 2, \ldots, M - N)$, so that we can write that

$$h(t) = \alpha y_h(t)u(t) + \beta\delta(t) + \sum_{k=1}^{M-N} c_k\frac{d^k\delta(t)}{dt^k} \tag{2.34}$$

A typical example of the case $M > N$ is presented in the next exercise. However, in these exercises, the derivatives of the impulse function, which have not been studied so far, are used. We will discuss these functions briefly and describe them so that readers can fully familiarize themselves with their use. ■

Derivatives and integrals of impulse function: Let us consider a LTI system that accepts an arbitrary signal $x(t)$ at its input and produces the first derivative of the signal with respect to time $x'(t)$ as its output, so that we can write that[14]

$$y(t) = h(t) * x(t) = \frac{dx(t)}{dt}$$

where $h(t)$ is the *impulse response* of the system. If we transmit the impulse function $\delta(t)$ as the input of this system, it is obvious that

$$h(t) * \delta(t) = \frac{d\delta(t)}{dt}$$

[14] This system is known as a *first-order differentiator*.

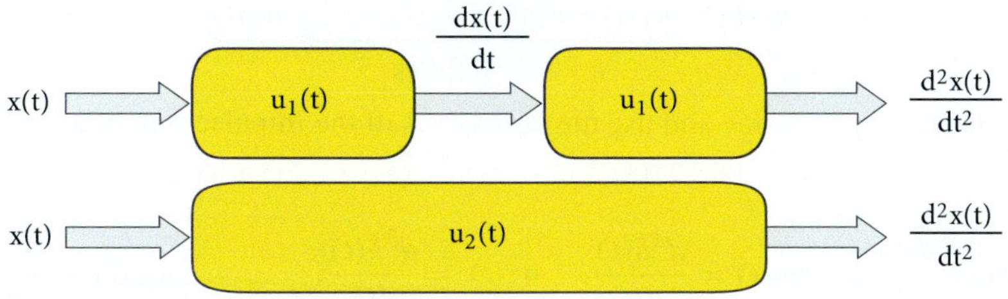

Fig. 2.9 A double differentiator is equivalent to two simple differentiators connected in series, and according to the associative property of convolution will be $u_2(t) = u_1(t) * u_1(t)$

and comparing the preceding equation with the known expression $x(t) * \delta(t) = x(t)$, it immediately follows that

$$h(t) = \frac{d\delta(t)}{dt} = u_1(t) \tag{2.35}$$

Therefore, *the impulse response of the simple first-order differentiator is the first derivative with respect to time of the impulse function* $\delta(t)$, which is symbolized by $u_1(t)$ and in the international literature is known as a *unit doublet*. Using this definition, we can reformulate the input/output equation of the system, writing it in the form

$$\frac{dx(t)}{dt} = x(t) * u_1(t)$$

Based on the same arguments and considering a second LTI system that for every input signal $x(t)$ returns its second time derivative $x''(t)$ (which is known as a *second-order differentiator*), we can write

$$y(t) = h(t) * x(t) = \frac{d^2x(t)}{dt^2}$$

Therefore, for the input signal $x(t) = \delta(t)$ we will have $h(t) = \delta''(t) = u_2(t)$, and the input/output relationship of the system is reformulated as

$$\frac{d^2x(t)}{dt^2} = x(t) * u_2(t)$$

If we differentiate the defining equation of the function $u_1(t)$, we will get

$$\frac{d^2x(t)}{dt^2} = \frac{d}{dt}\left[\frac{dx(t)}{dt}\right] = \frac{d}{dt}\left\{x(t) * u_1(t)\right\} = x(t) * u_1(t) * u_1(t)$$

and comparing the last two relations, it immediately follows that

$$u_2(t) = u_1(t) * u_1(t) \tag{2.36}$$

From the foregoing relation, it is concluded that *the impulse response of the double differentiation system arises from the convolution of the impulse responses* $u_1(t)$ *of the simple differentiators*. This conclusion is, of course, expected: The second derivative of the function $x(t)$ arises as a result of two successive differentiations, which can be performed by two simple differentiators that are connected in series (Fig. 2.9); according to the physical significance of the *associative property* of convolution, *the impulse response of the double differentiator will be equal to the convolution of the impulse responses of the two simple differentiators*.

Generalizing the foregoing description, we can formulate the conclusion that the impulse response $u_k(t)$ $(k > 0)$ of a LTI system that receives a continuous-time signal $x(t)$ at its input and produces as its output the k-order derivative of the signal, that is, the function $x^{(k)}(t)$, can be calculated from the successive convolution of k number of functions $u_1(t)$, or, in mathematical notation,

$$u_k(t) = \underbrace{u_1(t) * u_1(t) * u_1(t) * \cdots * u_1(t)}_{\text{k times}} \tag{2.37}$$

In this way, then, we have the ability to define and use the derivatives of the impulse function $\delta(t)$ up to some order value N through the functions

$$u_1(t) = \frac{d\delta(t)}{dt}, \quad u_2(t) = \frac{d^2\delta(t)}{dt^2}, \quad u_3(t) = \frac{d^3\delta(t)}{dt^3}, \quad \ldots \quad , u_N(t) = \frac{d^N\delta(t)}{dt^N}$$

Although the theory we have developed so far allows us to study the characteristics of the system's impulse response for the case $M > N$, let us mention for completeness that, apart from the derivatives of the impulse function $\delta(t)$, we can also define the *integrals* of this function. However, to do this, we will no longer consider a *differentiator*, but, on the contrary, an *integrator*, that is, a system that accepts at its input a continuous-time signal $x(t)$ and produces as its output the integral of this signal in the interval $(-\infty, t]$. This means that if the impulse response of this system is some function $h(t)$, then the input/output relationship for this system will be described by an equation of the form

$$y(t) = \mathcal{T}\{x(t)\} = h(t) * x(t) = x(t) * h(t) = \int_{-\infty}^{t} x(\tau)d\tau$$

Taking into account that the convolution of any signal $x(t)$ with the step function $u(t)$ is calculated as

$$y(t) = x(t) * u(t) = \int_{-\infty}^{+\infty} x(\tau)u(t-\tau)d\tau = \int_{-\infty}^{t} x(\tau)d\tau$$

comparison of the last relations leads us to the conclusion that the step function $u(t)$ can be considered the impulse response of this integration system, so that we can write

$$x(t) * u(t) = \int_{-\infty}^{t} x(\tau)d\tau$$

a relation that constitutes the defining equation of the function $u(t)$. But the impulse response is simply the output of the system, when at its input the impulse function $\delta(t)$ is applied and therefore

$$u(t) = \int_{-\infty}^{t} \delta(t)d\tau$$

a property that was proven in a previous section dealing with the impulse function as a *generalized function* or *distribution*.

Let us now extend this reasoning, to describe a system of a *double integrator* that arises from *two simple integrators connected in series*. If we symbolize the impulse response of this system as $u_{-2}(t)$[15] and use the argumentation connected with the associative property of convolution, we can write that

$$u_{-2}(t) = u(t) * u_{-1}(t) = \int_{-\infty}^{+\infty} u_{-1}(\tau)u(t-\tau)d\tau = \int_{-\infty}^{t} u_{-1}(\tau)d\tau = tu(t) \tag{2.38}$$

since the function $u_{-1}(t) \equiv u(t)$ is equal to one for $t \geq 0$ and zero for $t < 0$. But the function $tu(t)$ is simply the *ramp function* $r(t)$, which, therefore, is equal to *the integral of the step function*, a conclusion at which we had also arrived in the section on these functions.

[15] From the perspective of this symbolism, the step function $u(t)$ can be written as $u_{-1}(t)$, while the impulse function $\delta(t)$ can be written as $u_0(t)$. In this way we can use the notation $u_r(t)$ $(r = \ldots, -2, -1, 0, +1, +2, \ldots)$ to describe both the *derivatives* and the *integrals* of the impulse function, with the negative values of the index referring to the integrals and the positive values of the index describing the derivatives, while the function $\delta(t)$ itself corresponds to the value $r = 0$. Using this symbolism, the relationship between the impulse and step functions can be written in the form

$$u(t) \equiv u_{-1}(t) = u(t) * u_0(t) = \int_{-\infty}^{+\infty} \delta(\tau)u(t-\tau)d\tau = \int_{-\infty}^{t} \delta(\tau)d\tau$$

a relationship that in the general case is formulated as $u_{-m} = u(t) * u_{-(m-1)}(t)$, $(m = 1, 2, 3, \ldots)$.

Combining the preceding equations, we can write that

$$x(t) * u_{-2}(t) = \left(\int_{-\infty}^{t} x(\tau)d\tau\right) * u(t) = \int_{-\infty}^{t}\left(\int_{-\infty}^{\xi} x(\tau)d\tau\right) d\xi$$

an expression that can be considered the defining equation of the function $u_{-2}(t)$. Finally, generalizing the preceding relationship for the general case of an integrator of order k, we can write that

$$u_{-k}(t) = u(t) * u_{-(k-1)}(t) = \underbrace{u(t) * u(t) * u(t) * \cdots * u(t)}_{\text{k times}} = \int_{-\infty}^{t} u_{-(k-1)}(\tau)d\tau$$

($k = 0, 1, 2, 3, \ldots$), an equation that can be written in the form[16]

$$u_{-k}(t) = \frac{t^{k-1}}{(k-1)!}u(t) \tag{2.39}$$

The proof of the last relation can be made by applying the *inductive method*. Initially, we observe that the relation holds for $k = 1$, since in this case it immediately follows that $u_{-1}(t) = u(t)$, a property that is obviously true. Then we accept that the relation holds for some value $k = m > 0$, that is, we accept that

$$u_{-m}(t) = \int_{-\infty}^{t} u_{-(m-1)}(\tau)d\tau$$

and, finally, we check whether this relation also holds for $k = m + 1$, so we must prove that

$$u_{-(m+1)}(t) = \frac{t^m}{m!}u(t)$$

But this is easily derived, since, if we make the substitution $k = m + 1$ in the preceding relation, we will get

$$u_{-(m+1)}(t) = u(t) * u_{-m}(t) = \int_{-\infty}^{+\infty} u_{-m}(\tau)u(t-\tau)d\tau = \int_{-\infty}^{t} u_{-m}(\tau)d\tau = \int_{0}^{t} \frac{\tau^{m-1}}{(m-1)!}d\tau = \left.\frac{\tau^m}{m(m-1)!}\right|_0^t = \frac{t^m}{m!}u(t)$$

We observe, therefore, that this property holds *for every integer number k*, and therefore it is true.

Problem 2.5.5 Determine the impulse response of the system

$$\frac{d^2y(t)}{dt^2} + 5\frac{dy(t)}{dt} + 6y(t) = \frac{d^3x(t)}{dt^3} + 2\frac{d^2x(t)}{dt^2} + 4\frac{dx(t)}{dt} + 3x(t)$$

(see Oppenheim, Willsky, Nawab, Exercise 2.56st).

[16] This equation can arise immediately from the direct calculation of the functions $u_{-k}(t)$ in the same way that was done earlier for the function $u_{-2}(t)$. We will have, for example,

$$u_{-3}(t) = u(t) * u_{-2}(t) = \int_0^t u_{-2}(\tau)d\tau = \int_0^t \tau u(\tau)d\tau = \frac{t^2}{2}u(t) = \frac{t^{3-1}}{(3-1)!}$$

$$u_{-4}(t) = u(t) * u_{-3}(t) = \int_0^t u_{-3}(t)d\tau = \frac{1}{2}\int_0^t \tau^2 u(t)d\tau = \frac{t^3}{6}u(t) = \frac{t^{4-1}}{(4-1)!}u(t), \quad \text{etc.}$$

Solution The foregoing equation arises from the general differential equation of the system for parameter values $N = 2$, $M = 3$ and coefficients $\alpha_0 = 6$, $\alpha_1 = 5$, $\alpha_2 = 1$, $\beta_0 = 3$, $\beta_1 = 4$, $\beta_2 = 2$, and $\beta_3 = 1$. We observe that in this case, the relation $M > N$ holds, and for this reason the impulse response of the system is expected to contain the first derivative of the impulse function, which is $u_1(t)$.

Applying the methodology we previously presented, we set the right side of the equation to the function $\delta(t)$, writing it in the form

$$\frac{d^2h_\beta(t)}{dt^2} + 5\frac{dh_\beta(t)}{dt} + 6h_\beta(t) = \delta(t)$$

and then we solve the homogeneous equation that will arise for $t > 0$, with initial conditions $h_\beta(0^+) = 1$ and $h'_\beta(0^+) = 1$. Considering as usual complex exponential solutions of the form $h_\beta(t) = Ae^{st}$ $(A, s \in C, t > 0)$, we end up with the characteristic polynomial

$$s^2 + 5s + 6 = 0 \Rightarrow (s+2)(s+3) = 0 \quad \text{with roots} \quad \begin{cases} s_1 = -2 \\ s_2 = -3 \end{cases}$$

Therefore, the function $h_\beta(t)$ can be expressed as a linear combination of the form

$$h_\beta(t) = A_1 e^{-2t} + A_2 e^{-3t}$$

Deriving the foregoing relation, we end up with the equation

$$\frac{dh_\beta(t)}{dt} = -2A_1 e^{-2t} - 3A_2 e^{-3t}$$

and applying the initial conditions $h_\beta(0^+) = 1$ and $h'_\beta(0^+) = 1$, we end up with the system of equations

$$h_\beta(0^+) = A_1 e^{-2t} + A_2 e^{-3t}\Big|_{t=0^+} = A_1 + A_2 = 0$$

$$\frac{dh_\beta(0^+)}{dt} = -2A_1 e^{-2t} - 3A_2 e^{-3t}\Big|_{t=0^+} = -2A_1 - 3A_2 = 1$$

whose solution is proven to be $A_1 = 1$ and $A_2 = -1$. Therefore, the function $h_\beta(t)$ that satisfies the aforementioned initial conditions is

$$h_\beta(t) = e^{-2t} - e^{-3t} \quad (\text{for } t \geq 0) = \left(e^{-2t} - e^{-3t}\right) u(t)$$

Therefore, to determine the impulsive response of the system $h(t)$, we must replace the function $h_\beta(t)$ in the differential equation

$$h(t) = \frac{d^3h_\beta(t)}{dt^3} + 2\frac{d^2h_\beta(t)}{dt^2} + 4\frac{dh_\beta(t)}{dt} + 3h_\beta(t)$$

Deriving three times with respect to time and simplifying, as before, the expressions that arise, using the property $x(t)\delta(t) = x(0)\delta(t)$, where the function $x(t)$ will be either e^{-2t} or e^{-3t} (in both cases it is the case that $x(0) = 1$), successively we have

$$\frac{dh_\beta(t)}{dt} = \left(3e^{3t} - 2e^{-2t}\right)u(t) + \left(e^{-2t} - e^{-3t}\right)\delta(t) = \left(3e^{3t} - 2e^{-2t}\right)u(t)$$

$$\frac{d^2h_\beta(t)}{dt^2} = \left(4e^{-2t} - 9e^{-3t}\right)u(t) + \left(3e^{-3t} - 2e^{-2t}\right)\delta(t) = \left(4e^{-2t} - 9e^{-3t}\right)u(t) + \delta(t)$$

$$\frac{d^3h_\beta(t)}{dt^3} = \left(27e^{-3t} - 8e^{-2t}\right)u(t) + \left(4e^{-2t} - 9e^{-3t}\right)\delta(t) + u_1(t) = \left(27e^{-3t} - 8e^{-2t}\right)u(t) - 5\delta(t) + u_1(t)$$

Replacing these expressions in the defining equation of the impulse response $h(t)$, we find

$$h(t) = \left(27e^{-3t} - 8e^{-2t}\right)u(t) - 5\delta(t) + u_1(t) + 2\left\{\left(4e^{-2t} - 9e^{-3t}\right)u(t) + \delta(t)\right\}$$
$$+ 4\left\{\left(3e^{3t} - 2e^{-2t}\right)u(t)\right\} + 3\left(e^{-2t} - e^{-3t}\right)u(t)$$

which is the desired result. We observe that indeed the impulse response contains the first derivative of the impulse function $u_1(t)$, in full agreement with prevailing theory.

Calculation of Impulse Response from Derivative of Step Response

The second way to calculate the impulse response of a LTI system is based on the derivative of the step response, as these two responses are related to each other in the way described by the following theorem.

Theorem 2.5.1 *The step response $s(t)$ of a causal continuous-time LTI system is equal to the integral of the impulse response $h(t)$, which in turn constitutes the derivative of the step response $s(t)$. The foregoing property in mathematical notation is expressed as*

$$s(t) = \int_0^t h(\tau)d\tau \quad \text{or equivalently} \quad h(t) = \frac{ds(t)}{dt}.$$

Proof From basic theory it is known that *the step response is the output of a system, when the unit step function $u(t)$ is transmitted to its input*. If the system is LTI, then the step response will be equal to the *convolution* between the impulse response of the system $h(t)$ and the input signal $u(t)$, or, in mathematical notation,

$$s(t) = h(t) * u(t) = \int_{-\infty}^{+\infty} h(\tau)u(t-\tau)d\tau = \int_{-\infty}^{t} h(\tau)d\tau$$

with the last expression resulting from the fact that the function $u(t-\tau)$ has a value equal to one for the range of values $\tau \leq t$, while it is equal to zero anywhere else. Therefore, *the step response is equal to the integral of the impulse response*, calculated for the interval $(-\infty, t]$. For the special case of causal systems whose impulse response $h(\tau)$ is equal to zero for $\tau < 0$, the lower limit in the preceding integral corresponds to the value $t = 0$, from which the desired result is derived. Finally, it is obvious that since the step response is the integral of the impulse response, the impulse response will in turn be the derivative of the step response, so we can write that $h(t) = ds(t)/dt$.[17]

This correlation that exists between these two fundamental responses is very important, because it allows us to calculate the step response of a LTI system, when we know its impulse response, and vice versa. Considering, for example, the RC circuit of Fig. 2.5 (page 82), whose step response is equal to $s(t) = (1 - e^{-t/RC})u(t)$, we can directly calculate the impulse response as the derivative of the step response; it will therefore be the case that

$$h(t) = \frac{ds(t)}{dt} = \frac{d}{dt}\left\{\left(1 - e^{-t/RC}\right)u(t)\right\} = \left\{\frac{d}{dt}\left(1 - e^{-t/RC}\right)\right\}u(t) + \left(1 - e^{-t/RC}\right)\frac{du(t)}{dt} =$$
$$= \frac{1}{RC}e^{-t/RC}u(t) + \left(1 - e^{-t/RC}\right)\delta(t) = \frac{1}{RC}e^{-t/RC}u(t) + \delta(t) - e^{-t/RC}\delta(t)$$

[17] The experienced reader will immediately recognize that this correlation is due to the fact that since both the transform $\mathcal{T}$ of the system and the derivative operator are linear, it is possible to change their order—we will therefore have

$$h(t) = \mathcal{T}\{\delta(t)\} = \mathcal{T}\left\{\frac{du(t)}{dt}\right\} = \frac{d}{dt}\left\{\mathcal{T}\{u(t)\}\right\} = \frac{ds(t)}{dt} = s'(t).$$

where for the derivation of the last relation we used the fact that $\delta(t) = u'(t)$. If we now take into account that the function $\delta(t)$ is defined only for the time $t = 0$, while it has zero value for $t \neq 0$, we realize that the last two terms are nonzero only for $t = 0$, and therefore at this time we will have[18]

$$\left[\delta(t) - e^{-t/RC}\delta(t)\right]_{t=0} = \delta(0) - e^{-0/RC}\delta(0) = \delta(0) - \delta(0) = 0$$

Therefore, the last two terms cancel each other out and the impulse response is written

$$h(t) = \frac{1}{RC}e^{-t/RC}u(t)$$

a result that is the same as the one we got from the equation defining the impulse response.

On the other hand, if we only know the impulse response, we can determine from it the step response, calculating its integral over the interval $(-\infty, t]$. In this case, using the previous result it immediately follows that

$$s(t) = \int_{-\infty}^{t} h(\tau)d\tau = \frac{1}{RC}\int_{-\infty}^{t} e^{-\tau/RC}u(\tau)d\tau = \begin{cases} \dfrac{1}{RC}\displaystyle\int_{0}^{t} e^{-\tau/RC}d\tau & \text{for } \tau \geq 0 \\ 0 & \text{for } \tau < 0 \end{cases} = \begin{cases} 1 - e^{-t/RC} & \text{for } \tau \geq 0 \\ 0 & \text{for } \tau < 0 \end{cases}$$

or, equivalently, $s(t) = [1 - e^{-t/RC}]u(t)$, which is the same result as that obtained from the equation defining the step response.

Note The foregoing property can be proved in a different way, using the property of the differentiation that characterizes the operation of convolution and according to which, if the signals $x(t)$, $y(t)$, and $z(t)$ are related to the operation of convolution as $z(t) = x(t) * y(t)$, then it will hold that

$$z'(t) = x'(t) * y(t) = x(t) * y'(t)$$

Taking into account the definition of impulse response and considering input–output pairs of the form $[x(t), y(t)] = [\delta(t), h(t)]$ and $[x(t), y(t)] = [u(t), s(t)]$, these will be related to the operation of convolution as

$$h(t) = h(t) * \delta(t) \quad \text{and} \quad s(t) = h(t) * u(t)$$

Differentiating the last relation and taking into account that the impulse function is the derivative of the step function, that is, that $\delta(t) = u'(t)$, we will get $s'(t) = h(t) * u'(t) = h(t) * \delta(t)$ which, compared with the $h(t) = h(t) * \delta(t)$, leads us directly to the conclusion that $h(t) = s'(t)$.

2.5.3 Frequency Response of LTI Systems

As already mentioned, the frequency response of a continuous-time system is defined as the output of the system, when we transmit to its input the elementary *complex exponential signal* of frequency ω_0 of the form $x(t) = Ae^{j\omega_0 t}$. In the special case where the system we consider is LTI, then its output $y(t)$ will result from the convolution between the impulse response of the system $h(t)$ and the input signal $x(t)$. We will thus have

$$y(t) = h(t) * x(t) = \int_{-\infty}^{+\infty} h(\tau)x(t-\tau)d\tau = A\int_{-\infty}^{+\infty} h(\tau)e^{j\omega_0(t-\tau)}d\tau = Ae^{j\omega_0 t}\int_{-\infty}^{+\infty} h(\tau)e^{-j\omega_0\tau}d\tau$$

[18] We can reach the same conclusion using the property $x(t)\delta(t) = x(0)\delta(t)$, where $x(t) = e^{-t/RC}$.

Defining the function

$$\mathcal{H}(j\omega) = \int_{-\infty}^{+\infty} h(\tau)e^{-j\omega\tau}d\tau \tag{2.40}$$

the foregoing relation will take the form

$$y(t) = \mathcal{T}\{x(t)\} = Ae^{j\omega_0 t}\mathcal{H}(j\omega_0) = \mathcal{H}(j\omega_0)x(t) \tag{2.41}$$

But the last relation is *the eigenvalue equation of the transform $\mathcal{T}$ applied by the system to the input signal*. We can therefore make the claim that *the elementary complex exponential signals of the form $x(t) = Ae^{j\omega_0 t}$ are characteristic functions or eigenfunctions of the system, with the corresponding eigenvalues being the values of the function $\mathcal{H}(j\omega)$, calculated at the position of the frequency ω_0 of the input signal*. The function $\mathcal{H}(j\omega)$ is known as a *frequency response function* of the system, and, as we will see in the next chapter, it is simply the *Fourier transform* of the impulse response $h(t)$.

To determine the effect of a system on the exponential input signal, we must take into account that the function $\mathcal{H}(j\omega)$ is a complex function of the real frequency ω and therefore can be given in its exponential form

$$\mathcal{H}(j\omega) = |\mathcal{H}(j\omega)|e^{j\measuredangle\mathcal{H}(j\omega)} \tag{2.42}$$

where $|\mathcal{H}(j\omega)|$ is the magnitude of the function $\mathcal{H}(j\omega)$ and $\measuredangle\mathcal{H}(j\omega)$ is its phase. Therefore, by substituting the function $\mathcal{H}(j\omega)$ expressed in its exponential form in the system's eigenvalue equation, we will get

$$y(t) = Ae^{j\omega_0 t}\left\{|\mathcal{H}(j\omega_0)|e^{j\measuredangle\mathcal{H}(j\omega)}\right\} = \left(A|\mathcal{H}(j\omega_0)|\right)\exp\left\{j\left(\omega_0 t + \measuredangle\mathcal{H}(j\omega)\right)\right\} \tag{2.43}$$

We therefore arrive at the very important result that *the effect of a continuous-time LTI system on a fundamental complex exponential signal of frequency ω is the multiplication of the signal's amplitude by the magnitude of the function $\mathcal{H}(j\omega)$ and the phase shift of the signal by the phase of the function $\mathcal{H}(j\omega)$, while in contrast, its frequency ω does not change*. This means that the output signal $y(t)$ will be characterized by the same frequency as the input signal $x(t)$.

From the equation $y(t) = \mathcal{H}(j\omega)x(t)$ it immediately follows that

$$\mathcal{H}(j\omega) = \frac{y(t)}{x(t)} = \frac{\mathcal{T}\{x(t)\}}{x(t)} = \frac{\mathcal{T}\{Ae^{j\omega t}\}}{Ae^{j\omega t}} \tag{2.44}$$

thereby arriving at the definition of the frequency response that we had given in a previous section. In continuous-time signal theory, the function $|\mathcal{H}(j\omega)|$ is known as the *amplitude response*, while the function $\measuredangle\mathcal{H}(j\omega)$ is known as the *phase response* of the system.

2.5.4 Eigenvalues and Eigenfunctions: Transfer Function

Noting that the elementary complex exponential function $x(t) = Ae^{j\omega t}$ is a special case of the more general complex exponential signal of the form $x(t) = Ae^{st}$, where $s = \alpha + j\omega$ (the two signals coincide for the case $\alpha = 0$), we can, by applying the same process, generalize the results of the previous section. Therefore, if in a LTI system with impulse response $h(t)$ we transmit the exponential complex signal $x(t) = Ae^{st}$ ($s \in C$), the resulting output from this will be

$$y(t) = h(t) * x(t) = \int_{-\infty}^{+\infty} h(\tau)x(t-\tau)d\tau = A\int_{-\infty}^{+\infty} h(\tau)e^{s(t-\tau)}d\tau = Ae^{st}\int_{-\infty}^{+\infty} h(\tau)e^{-s\tau}d\tau = Ae^{st}\mathcal{H}(s) = \mathcal{H}(s)x(t)$$

with the function $\mathcal{H}(s)$ (which is known as a *transfer function*) being defined as

$$\mathcal{H}(s) = \int_{-\infty}^{+\infty} h(\tau)e^{-s\tau}d\tau \tag{2.45}$$

(in a later chapter we will find that the preceding equation constitutes the defining equation of the *Laplace transform*). It is obvious that the equation $y(t) = \mathcal{H}(s)x(t)$ is valid only for those values of the complex quantity s for which the integral of the function $\mathcal{H}(s)$ *converges* to some finite value. The set of values of s for which the foregoing integral converges is known as the *region of convergence*. We will return to this topic later.

In this general case, the eigenvalue equation of the system takes the form

$$y(t) = \mathcal{T}\{x(t)\} = \mathcal{H}(s)x(t) \tag{2.46}$$

and therefore *the complex exponential signal* $x(t) = Ae^{st}$ *constitutes an eigenfunction of the system with the eigenvalue being the corresponding value of the transfer function* $\mathcal{H}(s)$. It is not difficult to prove that this function for the case of a linear system described by the general linear differential equation

$$\sum_{k=0}^{N} \alpha_k \frac{d^k y(t)}{dt^k} = \sum_{k=0}^{M} \beta_k \frac{d^k x(t)}{dt^k}$$

is a rational function, that is, a quotient of two polynomial functions of s. Taking into account that $x(t) = e^{st}$ and $y(t) = \mathcal{H}(s)e^{st}$ (for simplicity we set $A = 1$) and differentiating the functions $x(t)$ and $y(t)$, M and N times, respectively, we will have

$$\begin{array}{ll} x'(t) = se^{st} & y'(t) = s\mathcal{H}(s)e^{st} \\ x''(t) = s^2e^{st} & y''(t) = s^2\mathcal{H}(s)e^{st} \\ x'''(t) = s^3e^{st} & y'''(t) = s^3\mathcal{H}(s)e^{st} \\ \cdots\cdots\cdots & \cdots\cdots\cdots \\ x^{(M)}(t) = s^M e^{st} & y^{(N)}(t) =! s^N\mathcal{H}(s)e^{st} \end{array}$$

Replacing the foregoing expressions in the differential equation of the system, it will take the form

$$\alpha_0\mathcal{H}(s)e^{st} + \alpha_1 s\mathcal{H}(s)e^{st} + \alpha_2 s^2\mathcal{H}(s)e^{st} + \cdots + \alpha_{N-1}s^{N-1}\mathcal{H}(s)e^{st} + \alpha_N s^N\mathcal{H}(s)e^{st} =$$
$$= \beta_0 e^{st} + \beta_1 s e^{st} + \beta_2 s^2 e^{st} + \cdots + \beta_{N-1}s^{N-1}e^{st} + \beta_M s^M e^{st}$$

If we eliminate the exponential quantity e^{st} from both sides and solve for $\mathcal{H}(s)$, we find

$$\mathcal{H}(s) = \frac{\beta_0 + \beta_1 s + \beta_2 s^2 + \cdots + \beta_{N-1}s^{N-1} + \beta_M s^M}{\alpha_0 + \alpha_1 s + \alpha_2 s^2 + \cdots + \alpha_{N-1}s^{N-1} + \alpha_N s^N} = \left\{\sum_{k=0}^{M} \beta_k s^k\right\} \Big/ \left\{\sum_{k=0}^{N} \alpha_k s^k\right\} = \frac{P(s)}{Q(s)} \tag{2.47}$$

where obviously the polynomials $P(s)$ and $Q(s)$ are defined as

$$P(s) = \sum_{k=0}^{M} \beta_k s^k \quad \text{and} \quad Q(s) = \sum_{k=0}^{N} \alpha_k s^k$$

If we factorize the polynomials $P(s)$ and $Q(s)$ and determine their roots, then the roots of the polynomial $P(s)$ that zero the transfer function are called *zeros* of the function, while the roots of the polynomial $Q(s)$ that cause the transfer function to diverge are known as *poles* of the function. The positions of the zeros and poles of the transfer function on the complex plane are of exceptional importance for the *stability* and more generally the behavior of the system and will be analyzed in great detail in subsequent chapters.

Problem 2.5.6 Prove that the transfer function for an ideal differentiator is $\mathcal{H}(s) = s$, while for an ideal integrator it is $\mathcal{H}(s) = 1/s$.

Solution The differential equation that describes the operation of the ideal differentiator is $y(t) = dx(t)/dt$ and arises from the general differential equation for parameter values $N = 0$ and $M = 1$ and coefficients $\alpha_0 = 1$, $\beta_0 = 0$, and $\beta_1 = 1$:

$$\mathcal{H}(s) = \sum_{k=0}^{1} \beta_k s^k = \beta_1 s = s$$

Similarly, the differential equation that describes the ideal integrator is $dy(t)/dt = x(t)$ and arises from the general differential equation for parameter values $N = 1$ and $M = 0$ and coefficients $\alpha_0 = 0$, $\alpha_1 = 1$, and $\beta_0 = 1$:

$$\mathcal{H}(s) = 1 \Big/ \left\{ \sum_{k=0}^{1} \alpha_k s^k \right\} = \frac{1}{\alpha_1 s} = \frac{1}{s}$$

The physical significance of the relation $y(t) = \mathcal{H}(s)x(t) = A\mathcal{H}(s)e^{st}$ is that *if an exponential signal of the form* $x(t) = Ae^{st}$ *passes through a continuous-time LTI system, then the output signal that will result will differ from the input signal only in the value of its amplitude*, which will be equal to the amplitude of the input signal multiplied by the magnitude of the function $\mathcal{H}(s)$. Considering, then, three input signals of the form

$$x_1(t) = A_1 e^{s_1 t}, \quad x_2(t) = A_2 e^{s_2 t}, \quad x_3(t) = A_3 e^{s_3 t}$$

the corresponding responses of the system to these will be

$$y_1(t) = A_1 \mathcal{H}(s_1) e^{s_1 t}, \quad x_2(t) = A_2 \mathcal{H}(s_2) e^{s_2 t}, \quad x_3(t) = A_3 \mathcal{H}(s_3) e^{s_3 t}$$

Taking into account that the system is linear and therefore will be characterized by the power of the principle of superposition, it is obvious that if we consider an input signal $x(t)$ that arises from a linear combination of the three preceding signals of the form

$$x(t) = C_1 x_1(t) + C_2 x_2(t) + C_3 x_3(t) = C_1(A_1 e^{s_1 t}) + C_2(A_2 e^{s_2 t}) + C_3(A_3 e^{s_3 t})$$

then the response of the system to this signal will arise from the linear combination of the corresponding elementary outputs, so that we can write that

$$y(t) = C_1 y_1(t) + C_2 y_2(t) + C_3 y_3(t) = C_1[A_1 \mathcal{H}(s_1) e^{s_1 t}] + C_2[A_2 \mathcal{H}(s_2) e^{s_2 t}] + C_3[A_3 \mathcal{H}(s_3) e^{s_3 t}]$$

a property that for the case of k elementary signals is generalized as

$$y(t) = \mathcal{T}\{x(t)\} = \mathcal{T}\Big\{\sum_k C_k(A_k e^{s_k t})\Big\} = \sum_k C_k \mathcal{H}(s_k)(A_k e^{s_k t}) \tag{2.48}$$

In this way we can find the response of a LTI system to any input signal if we analyze it in its simple components and use the principle of superposition in the preceding way.

Problem 2.5.7 Verify that the complex exponential signal $x(t) = e^{st}$ constitutes an eigenfunction of the system described by the equation

$$y(t) = \int_{-\infty}^{t} e^{-(t-\tau)} x(\tau) d\tau$$

and determine its eigenvalue.

Solution If in the equation defining the output of the system we replace the input signal $x(t) = e^{st}$, we write

$$y(t) = \int_{-\infty}^{t} e^{-(t-\tau)} e^{s\tau} d\tau = e^{-t} \int_{-\infty}^{t} e^{(s+1)\tau} d\tau = \frac{e^{-t}}{s+1} e^{(s+1)\tau} \Big|_{-\infty}^{t} = \frac{e^{-t} e^{st} e^{t}}{s+1} = \frac{1}{s+1} e^{st}$$

with the preceding integral converging for values of s for which we have $\Re\{s\} > -1$.[19]

We observe, therefore, that the input signal $x(t)$ satisfies the eigenvalue equation $y(t) = \mathcal{H}(s)e^{st}$ and therefore constitutes an eigenfunction of the system with eigenvalue $\mathcal{H}(s) = 1/(s+1)$.

Problem 2.5.8 Prove that the continuous-time signals $x_1(t) = \cos(\omega_0 t)$ and $x_2(t) = \sin(\omega_0 t)$ constitute eigenfunctions of a LTI system characterized by even and real impulse response $h(t)$ and that they correspond, moreover, to the same eigenvalue.

Solution In a previous section we proved that the elementary complex exponential signal $x(t) = e^{j\omega_0 t}$ constitutes the eigenfunction of every LTI system, with the eigenvalue being the transfer function $\mathcal{H}(j\omega)$ calculated at the position of the input signal frequency, a property that is formulated as

$$y(t) = \mathcal{H}(j\omega_0)x(t) = \left(\int_{-\infty}^{+\infty} h(\tau) e^{-j\omega_0 d\tau}\right) e^{j\omega t}$$

Using the Euler formula $e^{-j\omega_0 t} = \cos(\omega_0 t) - j\sin(\omega_0 t)$ the function $\mathcal{H}(\omega_0)$ takes the form

$$\mathcal{H}(j\omega_0) = \int_{-\infty}^{+\infty} h(\tau) e^{-j\omega_0 \tau} d\tau = \int_{-\infty}^{+\infty} h(\tau)[\cos(\omega_0 t) - j\sin(\omega_0 \tau)] d\tau = \int_{-\infty}^{+\infty} h(\tau)\cos(\omega_0 \tau) d\tau - j \int_{-\infty}^{+\infty} h(\tau)\sin(\omega_0 \tau) d\tau$$

Taking into account that the impulse response of the system is even, while the trigonometric functions of the cosine and the sine are characterized by even and odd symmetry, respectively, the two integrals of the foregoing equation are calculated as (see also Sect. 3.2.1 on page 190)

$$I_1 = \int_{-\infty}^{+\infty} h(\tau)\cos(\omega_0 \tau) d\tau = \int_{-\infty}^{0} h(\tau)\cos(\omega_0 \tau) d\tau + \int_{0}^{+\infty} h(\tau)\cos(\omega_0 \tau) d\tau = \int_{\infty}^{0} h(-\tau)\cos(-\omega_0 \tau) d(-\tau)$$
$$+ \int_{0}^{+\infty} h(\tau)\cos(\omega_0 \tau) d\tau = \int_{0}^{+\infty} h(\tau)\cos(\omega_0 \tau) d\tau + \int_{0}^{+\infty} h(\tau)\cos(\omega_0 \tau) d\tau = 2\int_{0}^{+\infty} h(\tau)\cos(\omega_0 \tau) d\tau$$

[19] The convergence conditions of integrals of this form are presented in the chapter on the *Laplace transform*.

$$I_2 = \int_{-\infty}^{+\infty} h(\tau)\sin(\omega_0\tau)d\tau = \int_{-\infty}^{0} h(\tau)\sin(\omega_0\tau)d\tau + \int_{0}^{+\infty} h(\tau)\sin(\omega_0\tau)d\tau$$

$$= \int_{\infty}^{0} h(-\tau)\sin(-\omega_0\tau)d(-\tau) + \int_{0}^{+\infty} h(\tau)\sin(\omega_0\tau)d\tau = \int_{0}^{+\infty} h(\tau)\sin(\omega_0\tau)d\tau - \int_{0}^{+\infty} h(\tau)\sin(\omega_0\tau)d\tau = 0$$

and therefore the function $\mathcal{H}(j\omega_0)$ will take the form

$$\mathcal{H}(j\omega_0) = 2\int_{0}^{+\infty} h(\tau)\cos(\omega_0\tau)d\tau$$

Let us now assume that in the considered LTI system we transmit the complex exponential signal $x(t) = e^{-j\omega_0 t}$. In this case, following the methodology we presented in previous sections, we can prove that this signal also constitutes an *eigenfunction* of the system with the eigenvalue given by the function

$$\mathcal{H}(-j\omega_0) = \int_{-\infty}^{+\infty} h(\tau)e^{j\omega_0\tau}d\tau = \left(\int_{-\infty}^{+\infty} h(\tau)e^{-j\omega_0\tau}\right)^* = \mathcal{H}^*(j\omega_0)$$

$$= 2\int_{0}^{+\infty} h(\tau)\cos(-\omega_0\tau)d\tau = 2\int_{0}^{+\infty} h(\tau)\cos(\omega_0\tau)d\tau = \mathcal{H}(j\omega_0)$$

We observe, therefore, that the input signals $x_1(t) = e^{j\omega_0 t}$ and $x_2(t) = e^{-j\omega_0 t}$ are eigenfunctions of the system and correspond to the same eigenvalue $\mathcal{H}(j\omega_0)$, so that we can write

$$y_1(t) = \mathcal{T}\{x_1(t)\} = \mathcal{T}\{e^{j\omega_0 t}\} = \mathcal{H}(j\omega_0)e^{j\omega_0 t} \quad \text{and} \quad y_2(t) = \mathcal{T}\{x_2(t)\} = \mathcal{T}\{e^{-j\omega_0 t}\} = \mathcal{H}(j\omega_0)e^{-j\omega_0 t}$$

Taking into account now that our system is *linear* and therefore will be characterized by the power of the *principle of superposition*, we will have correspondingly

$$\mathcal{T}\{\cos(\omega_0 t)\} = \mathcal{T}\left\{\frac{e^{j\omega_0 t} + e^{-j\omega_0 t}}{2}\right\} = \frac{1}{2}\mathcal{T}\{e^{j\omega_0 t}\} + \frac{1}{2}\mathcal{T}\{e^{-j\omega_0 t}\} =$$

$$= \frac{1}{2}\mathcal{H}(j\omega_0)e^{j\omega_0 t} + \frac{1}{2}\mathcal{H}(j\omega_0)e^{-j\omega_0 t} = \mathcal{H}(j\omega_0)\frac{e^{j\omega_0 t} + e^{-j\omega_0 t}}{2} = \mathcal{H}(j\omega_0)\cos(\omega_0 t)$$

$$\mathcal{T}\{\sin(\omega_0 t)\} = \mathcal{T}\left\{\frac{e^{j\omega_0 t} - e^{-j\omega_0 t}}{2i}\right\} = \frac{1}{2j}\mathcal{T}\{e^{j\omega_0 t}\} - \frac{1}{2j}\mathcal{T}\{e^{-j\omega_0 t}\} =$$

$$= \frac{1}{2j}\mathcal{H}(j\omega_0)e^{j\omega_0 t} - \frac{1}{2j}\mathcal{H}(j\omega_0)e^{-j\omega_0 t} = \mathcal{H}(j\omega_0)\frac{e^{j\omega_0 t} - e^{-j\omega_0 t}}{2j} = \mathcal{H}(j\omega_0)\sin(\omega_0 t)$$

Therefore, the elementary trigonometric signals $\cos(\omega_0 t)$ and $\sin(\omega_0 t)$ constitute eigenfunctions of each LTI system and, moreover, with the same eigenvalue that is equal to the value of the function $\mathcal{H}(\omega_0)$.

2.5.5 Stability of LTI Systems

As already mentioned, a continuous-time system is considered *BIBO stable* when for every bounded input $x(t)$ that is transmitted to the system the output $y(t)$ produced by it is also bounded. Considering a bounded input signal $x(t)$, such that $|x(t)| < M < \infty$ for every time instance t, and taking into account that in a LTI system the output $y(t)$ is given by the integral of the convolution

$$y(t) = h(t) * x(t) = \int_{-\infty}^{+\infty} h(\tau)x(t-\tau)d\tau \tag{2.49}$$

the determination of the system's stability criterion consists in calculating the *measure of the output signal* $y(t)$, which obviously will satisfy the condition

$$|y(t)| = \left| \int_{-\infty}^{+\infty} h(\tau)x(t-\tau)d\tau \right| \leq \int_{-\infty}^{+\infty} |h(\tau)||x(t-\tau)|d\tau \leq M \int_{-\infty}^{+\infty} |h(\tau)|d\tau$$

where to derive the last relation we used the fact that the input $x(t)$ is *bounded*. Therefore, for the system's output to be bounded, it must be

$$\int_{-\infty}^{+\infty} |h(\tau)|d\tau < \infty \tag{2.50}$$

a condition that in turn ensures the system's stability according to BIBO. Consequently, *a system is BIBO stable when its impulse response* $h(t)$ *is absolutely integrable*.

The foregoing condition, besides being sufficient, is also necessary. To prove this, let us assume that, although the impulse response is not absolutely integrable [i.e., the integral of its measure over the interval $(-\infty, +\infty)$ diverges], nevertheless for every bounded input the system produces a bounded output, and let us try to find a signal that refutes this claim. Indeed, considering the bounded input signal

$$x(t) = \begin{cases} +1 \text{ for } h(t_0 - t) > 0 \\ 0 \text{ for } h(t_0 - t) = 0 \\ -1 \text{ for } h(t_0 - t) < 0 \end{cases}$$

the system's output at time $t = t_0$ will have the form

$$|y(t_0)| = \left| \int_{-\infty}^{+\infty} h(t_0 - \tau)d\tau \right| = \int_{-\infty}^{+\infty} |h(\xi)|d\xi \to \infty$$

where in the last expression we proceeded to the change of variable $\xi = t_0 - \tau$. We conclude, therefore, that if the impulse response is not absolutely integrable, there is at least one bounded input signal that leads to an unbounded output signal, and therefore the system is not BIBO stable. Therefore, *the condition of absolute integrability of the system's impulse response, besides being sufficient, is also necessary*.

In the case where we know the differential equation of the system that, as we have already mentioned, in the general case has the form

$$\sum_{k=0}^{N} \alpha_k \frac{d^k y(t)}{dt^k} = \sum_{k=0}^{M} \beta_k \frac{d^k x(k)}{dt^k} \tag{2.51}$$

BIBO stability can be examined as a function of the roots $\{s_k\}_{k=1}^{N}$ of the characteristic polynomial

$$\alpha_N s^N + \alpha_{N-1} s^{N-1} + \cdots + \alpha_2 s^2 + \alpha_1 s + \alpha = 0 \tag{2.52}$$

These roots appear in the exponents of the exponential solutions of the homogeneous equation (see the analysis carried out on the preceding pages) and therefore allow the characterization of the system in terms of its internal stability.

As was already mentioned, the impulse response of a system whose differential equation we know is given by the relation

$$h(t) = \sum_{k=0}^{M} \beta_k \frac{d^k h_\beta(t)}{dt^k} = \sum_{k=0}^{M} \beta_k \frac{d^k}{dt^k} \Bigg\{ \underbrace{\sum_{m=1}^{N} \alpha_m e^{s_m t} u(t)}_{h_\beta(t)} \Bigg\} \tag{2.53}$$

where for simplicity we assume that the N roots of the characteristic polynomial are all distinct from each other. Considering the case $N \geq M$ for which the impulse response will have at worst an anomaly at the position $t = 0$ (whose integral in the interval $(-\infty, 0]$ is finite), the stability of the system is fully determined by the exponential quantities of the foregoing relation. Indeed, replacing the impulse response from the preceding expression in the condition of absolute integrability, we will successively obtain[20]

$$\int_{0^+}^{+\infty} |h(t)|dt = \int_{0^+}^{+\infty} \left| \sum_{k=0}^{M} \beta_k \frac{d^k}{dt^k} \left\{ \sum_{m=1}^{N} \alpha_m e^{s_m t} \right\} \right| dt \leq \int_{0^+}^{+\infty} \sum_{k=0}^{M} |\beta_k| \left| \frac{d^k}{dt^k} \left\{ \sum_{m=1}^{N} \alpha_m e^{s_m t} \right\} \right| dt =$$

$$= \sum_{k=0}^{M} |\beta_k| \int_{0^+}^{+\infty} \left| \frac{d^k}{dt^k} \left\{ \sum_{m=1}^{N} \alpha_m e^{s_m t} \right\} \right| dt = \sum_{k=0}^{M} |\beta_k| \int_{0^+}^{+\infty} \left| \sum_{m=1}^{N} \alpha_m (s_m)^k e^{s_m t} \right| dt \leq$$

$$= \sum_{k=0}^{M} |\beta_k| \int_{0^+}^{+\infty} \sum_{m=1}^{N} |\alpha_m (s_m)^k e^{s_m t}| dt \leq \sum_{k=0}^{M} |\beta_k| \int_{0^+}^{+\infty} \sum_{m=1}^{N} |\alpha_m (s_m)^k| e^{\Re\{s_m\}t} dt$$

From the last expression we observe that the upper limit of the foregoing quantity becomes infinite only if the integral of the function $\exp(\Re\{s_m\}t)$ becomes infinite for one or more values of m in the interval $1 \leq m \leq M$. But this happens only when the real parts of the roots s_m of the characteristic polynomial are positive. We conclude, therefore, that a continuous-time LTI system[21] is stable when all the roots of the characteristic polynomial of the system's differential equation have negative real parts, something we pointed out in the section on the stability of linear systems. The preceding condition, besides being sufficient, is proven to be necessary in the sense that if even one root of the characteristic polynomial has a positive real part, then the continuous-time system it is associated with is unstable.

To prove the necessity of the preceding condition, let us consider the LTI system described by the differential equation

$$\frac{dy(t)}{dt} - y(t) = \frac{dx(t)}{dt} - x(t)$$

In this case, the system is characterized by the condition $M = N$, and therefore we expect its impulse response to contain the impulse function $\delta(t)$. Indeed, by setting the second term of the preceding equation to the impulse function $\delta(t)$ and solving the homogeneous equation

$$\frac{dh_\beta(t)}{dt} - h_\beta(t) = 0$$

for time instances $t > 0$, we end up with a characteristic polynomial of the form $s - 1 = 0$ with the simple root $s = 1$, which obviously has a positive real part. Therefore, $h_\beta(t) = Ae^{st}u(t) = e^t u(t)$, since applying the initial condition, which is $h_\beta(0^+) = 1$, results in $A = 1$. Substituting the preceding function into the defining equation of the impulse response, it turns out that

$$h(t) = \frac{dh_\beta(t)}{dt} - h_\beta(t) = \frac{d}{dt}\left\{e^t u(t)\right\} - e^t u(t) = e^t u(t) + e^t \delta(t) - e^t u(t) = e^t \delta(t) = \delta(t)$$

where for the derivation of the last relation we used the known property $x(t)\delta(t) = x(0)\delta(t)$, with $x(t) = e^t$.

[20] The last equation in the following derivation results from the fact that the measure of the complex exponential quantity $e^{\lambda t}$, where λ is a complex number of the form $\Re\{\lambda\} + i\Im\{\lambda\}$ is calculated as

$$|e^{\lambda t}| = |e^{(\Re\{\lambda\}+i\Im\{\lambda\})t}| = |e^{\Re\{\lambda\}t} e^{i\Im\{\lambda\}t}| = |e^{\Re\{\lambda\}t}||e^{i\Im\{\lambda\}t}| = |e^{\Re\{\lambda\}t}|$$

since, as is well known,

$$|e^{i\Im\{\lambda\}t}| = |\cos[\Im\{\lambda\}t] + i \sin[\Im\{\lambda\}t]| = \sqrt{cos^2[\Im\{\lambda\}t] + \sin^2[\Im\{\lambda\}t]} = 1.$$

[21] Note that this is something that applies to all linear systems and not only to LTI systems since the proof is based on the differential equation that describes these systems and that in all cases is linear with constant coefficients.

We observe, therefore, that $h(t) = \delta(t)$,[22] and therefore the impulse response $h(t)$ is absolutely integrable (since, as is well known, the integral of the impulse function $\delta(t)$ over the interval $(-\infty, +\infty)$ is equal to one and, therefore, finite). Consequently, according to the criterion of *absolute integrability*, this system is stable. In reality, of course, this is not the case, because even if the input signal is identically zero for every time instance, any nonzero initial conditions—even if they have infinitesimal value—will excite the exponential term e^t, which diverges in the limit $t \to \infty$, and therefore the system's output is not bounded and the system is unstable. We will revisit this issue in a later chapter, where we will investigate the stability of continuous-time systems in terms of the positions of the *poles* and the *zero values* of the *transfer function* on the complex plane.

Note The preceding discussion applies only for the case $M \leq N$, a condition that is sufficient (but not necessary) for the stability of a system according to BIBO. In contrast, if $M > N$, then the system is *always* unstable according to BIBO, and for this reason such systems are not used in practice. This property of continuous-time systems will be discussed in a subsequent chapter using the system's transfer function.
In the following exercises, the method of characterizing the stability of a system according to the criterion of absolute integrability of its impulse response is demonstrated.

Problem 2.5.9 For each LTI system whose impulse response has the form

$$\text{(a) } h(t) = e^{-4t}u(t-2), \quad \text{(b) } h(t) = e^{-6t}u(3-t), \quad \text{(c) } h(t) = e^{-2t}u(t+50)$$

examine whether it is stable or unstable according to BIBO. (see Oppenheim, Willsky, Nawab, Exercise 2.29).

Solution

(a) The step function $u(t-2)$ is defined as

$$u(t-2) = \begin{cases} 1 \text{ for } t-2 \geq 0 \Rightarrow t \geq 2 \\ 0 \text{ for } t-2 < 0 \Rightarrow t < 2 \end{cases}$$

Therefore, we will have

$$\int_{-\infty}^{+\infty} |h(t)|dt = \int_{2}^{+\infty} e^{-4t}dt = -\frac{1}{4}e^{-4t}\Big|_{2}^{+\infty} = \frac{e^{-8}}{4} < \infty$$

We observe that the impulse response of the system is absolutely integrable, that is, the system is *BIBO stable*.

(b) The step function $u(3-t)$ is defined as

$$u(3-t) = \begin{cases} 1 \text{ for } 3-t \geq 0 \Rightarrow t \leq 3 \\ 0 \text{ for } 3-t < 0 \Rightarrow t > 3 \end{cases}$$

Therefore, we will have

$$\int_{-\infty}^{+\infty} |h(t)|dt = \int_{-\infty}^{3} e^{-6t}dt = -\frac{1}{6}e^{-6t}\Big|_{-\infty}^{3} = \infty$$

We observe that the system's impulse response is not absolutely integrable, and therefore the system is *BIBO unstable*.

(c) The step function $u(t+50)$ is defined as

$$u(t+50) = \begin{cases} 1 \text{ for } t+50 \geq 0 \Rightarrow t \geq -50 \\ 0 \text{ for } t+50 < 0 \Rightarrow t < -50 \end{cases}$$

[22] In the general case, we will have $y(t) = x(t)$, and therefore the equation in question describes an identical system.

Therefore, we will have

$$\int_{-\infty}^{+\infty} |h(t)|dt = \int_{-50}^{+\infty} e^{-2t}dt = -\frac{1}{2}e^{-2t}\Big|_{-50}^{+\infty} = -\frac{1}{2}\left(-e^{-100}\right) = \frac{e^{100}}{2} < \infty$$

We observe that the system's impulse response is absolutely integrable, meaning the system is *BIBO stable*.

Problem 2.5.10 Examine the BIBO stability of the LTI systems

$$\text{(a) } h(t) = e^{2t}u(-1-t), \quad \text{(b) } h(t) = e^{-6|t|}, \quad \text{(c) } h(t) = te^{-t}u(t)$$

(see Oppenheim, Willsky, Nawab, Exercise 2.29).

Solution

(a) The step function $u(-1-t)$ is defined as

$$u(-1-t) = \begin{cases} 1 \text{ for } -1-t \geq 0 \Rightarrow t \leq -1 \\ 0 \text{ for } -1-t < 0 \Rightarrow t > -1 \end{cases}$$

Therefore, we will have

$$\int_{-\infty}^{+\infty} |h(t)|dt = \int_{-\infty}^{-1} e^{2t}dt = \frac{1}{2}e^{2t}\Big|_{-\infty}^{-1} = \frac{1}{2}e^{-2} < \infty$$

The impulse response is absolutely integrable, meaning the system is *BIBO stable*.

(b) From the defining equation of absolute value

$$|t| = \begin{cases} +t \text{ for } t \geq 0 \\ -t \text{ for } t < 0 \end{cases} \quad \text{it follows that} \quad h(t) = \begin{cases} e^{-6t} \text{ for } t \geq 0 \\ e^{+6t} \text{ for } t < 0 \end{cases}$$

Therefore, we will have

$$\int_{-\infty}^{+\infty} |h(t)|dt = \int_{-\infty}^{0} e^{6t}dt + \int_{0}^{+\infty} e^{-6t}dt = \frac{1}{6}e^{6t}\Big|_{-\infty}^{0} - \frac{1}{6}e^{-6t}\Big|_{0}^{+\infty} = \frac{2}{6} = \frac{1}{3} < \infty$$

The impulse response is absolutely integrable, meaning the system is *BIBO stable*.

(c) Applying the method of integration by parts, it turns out that

$$\int te^{-t}dt = -\int (-e^{-t})tdt = -\int d(e^{-t})tdt = -\left(te^{-t} - \int e^{-t}dt\right) = -te^{-t} + e^{-t} = -(t+1)e^{-t}$$

and therefore

$$\int_{-\infty}^{+\infty} |h(t)|dt = \int_{0}^{+\infty} te^{-t}dt = -(t+1)e^{-t}\Big|_{0}^{\infty} = 1 < \infty$$

The system's impulse response is absolutely integrable, meaning the system is *BIBO stable*.

Problem 2.5.11 Considering a causal continuous-time system that arises from the series connection of two systems with impulse responses $h_1(t) = e^{-2t}u(t)$ and $h_2(t) = 2e^{-t}u(t)$, check whether it is BIBO stable or not.

Solution Since the two systems are connected in series, the impulse response of the composite system will result from the convolution of the impulse responses $h_1(t)$ and $h_2(t)$ of the two systems. We will therefore have

$$h(t) = h_1(t) * h_2(t) = \int_{-\infty}^{+\infty} h_1(\tau)h_2(t-\tau)d\tau = 2\int_{-\infty}^{+\infty} e^{-2\tau}u(\tau)e^{-(t-\tau)}u(t-\tau)d\tau =$$

$$= 2e^{-t}\int_{-\infty}^{+\infty} e^{-\tau}u(\tau)u(t-\tau)d\tau = 2(e^{-t} - e^{-2t})u(t)$$

Applying, therefore, the criterion of absolute integrability, it turns out that

$$\int_{-\infty}^{+\infty} |h(t)|dt = 2\int_{0}^{+\infty} (e^{-t} - e^{-2t})dt = 2\left(1 - \frac{1}{2}\right) = 1 < \infty$$

We observe that the impulse response of the system is *absolutely integrable*, and therefore the system is *BIBO stable*.

Problem 2.5.12 Prove that a LTI system, whose impulse response is a periodic signal, is unstable.

Solution This conclusion arises directly from the characterization of the system's stability according to the criterion of absolute integrability of the impulse response $h(t)$. Indeed, if the impulse response is a nonzero periodic signal defined in the interval $(-\infty, +\infty)$, then the system described by this is unstable, since it immediately follows that

$$\int_{-\infty}^{+\infty} |h(t)dt| \to \infty$$

2.6 Interconnection of Continuous-Time Systems

The whole theory that has been described in the preceding pages refers to individual continuous-time systems that accept as their input some signal $x(t)$, apply some transformation to it, and produce as their output a second signal $y(t)$. It is obvious that these individual systems can be interconnected in many different ways so as to form complex systems that perform more complex functions.

In almost all cases, a continuous-time system can be constructed from the interconnection of a multitude of elementary continuous-time subsystems that apply to the signal or signals they receive as input all the elementary signal operations described in previous sections, such as *addition*, *subtraction*, and *multiplication* of signals, and *scaling* of the amplitude of the signals, as well as the *time delay* that it may undergo. Other subsystems apply more complex forms of processing to the input signal, such as, for example, the *differentiation* and the *integration*. The most important of these elementary subsystems that are used as basic building blocks for the construction of more complex systems are presented in Fig. 2.10.

In a more detailed description, elementary continuous-time subsystems operate as follows:

- *Amplitude scaling subsystem*: This subsystem receives an input signal $x(t)$ and produces as its output the signal $y(t) = \mathcal{T}\{x(t)\} = \alpha x(t)$, where α is a constant coefficient known as *gain*. Considering that the output $y(t)$ depends only on the current input $x(t)$, it is obvious that this subsystem is *static*, or, in other words, a system *without memory*. The electrical equivalent of the amplitude scaling subsystem is a resistance R that receives as input a current $i(t)$ and produces as its output the voltage $v(t) = Ri(t)$.
- *Addition, subtraction, and multiplication of signals subsystems*: These subsystems—which are also *static*—receive as their input two continuous-time signals, $x_1(t)$ and $x_2(t)$, and return respectively *their sum* $x_1(t) + x_2(t)$, *their difference* $x_1(t) - x_2(t)$, and *their product* $x_1(t) \cdot x_2(t)$.
- The *ideal integrator*, which accepts the input signal $x(t)$ and produces the output signal

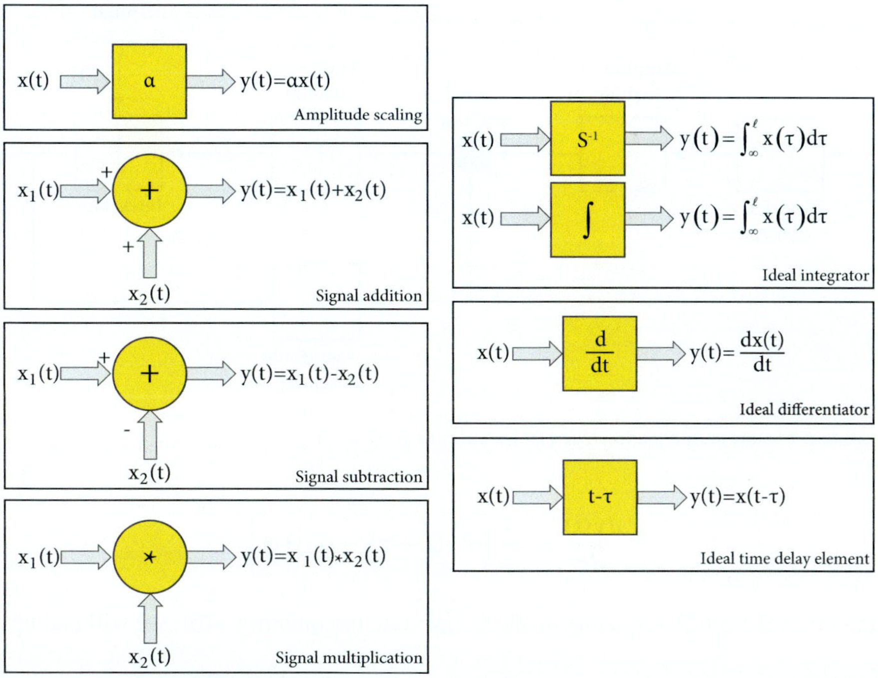

Fig. 2.10 The elementary continuous-time subsystems that are used as the fundamental building blocks for the construction of more complex systems

$$y(t) = \mathcal{T}\{x(t)\} = x(t) * u(t) = \int_{-\infty}^{t} x(\tau)d\tau \tag{2.54}$$

Therefore, the ideal integrator is a system of *infinite memory*. For the subsystem of the ideal integrator, the use of two different symbols has prevailed, of which the first uses the icon of the integral and the second the label S^{-1}. The electrical equivalent of the ideal integrator is a *capacitor* of capacity C, which correlates the electric current $i(t)$ with the voltage $v(t)$ by the known relation

$$v(t) = \frac{1}{C}\int_{-\infty}^{t} i(\tau)d\tau \tag{2.55}$$

- The ideal differentiator that accepts as input a signal $x(t)$ and produces as output its derivative $y(t) = x'(t)$ is, therefore, a *static* continuous subsystem. Its electrical equivalent is an inductor L that correlates the current $i(t)$ that flows through it with the voltage at its ends with a relation of the form $v(t) = L[di(t)/dt]$.
- *Ideal time delay element*: This subsystem accepts at its input the signal $x(t)$ and creates its time-delayed version $y(t) = x(t - \tau)$; therefore, it is a *dynamic* system.

The circuit that results from the interconnection of the fundamental subsystems, so that we can construct complex systems with a higher degree of complexity, is known as a *schematic diagram*.

To understand the way in which the fundamental building blocks are used to describe continuous-time systems, let us consider a system described by a common linear first-order differential equation with constant coefficients of the form

$$\alpha_1 \frac{dy(t)}{dt} + \alpha_0 y(t) = \beta x(t - \tau)$$

which accepts at its input the time-delayed signal $x(t - \tau)$ and produces as its output the signal $y(t)$. If we reformulate the preceding differential equation in the form

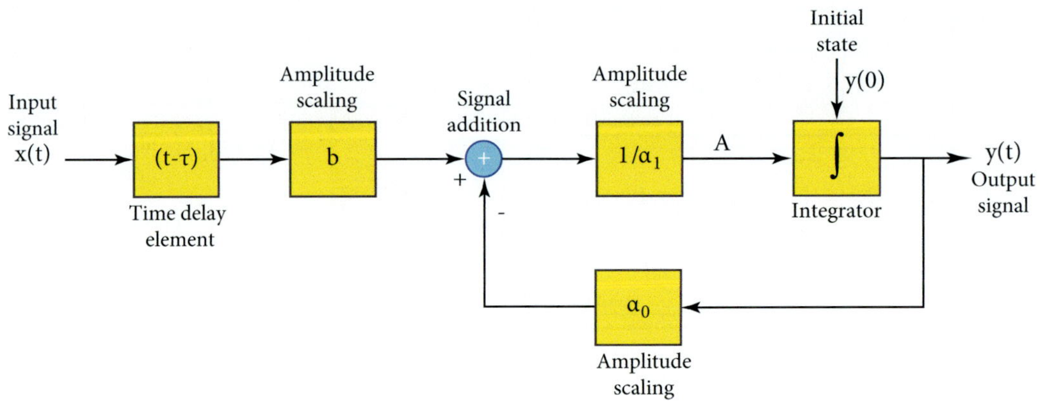

Fig. 2.11 Schematic diagram for solution of equation $\alpha_1 y'(t) + \alpha_0 y(t) = \beta x(t - \tau)$

$$\frac{dy(t)}{dt} = \frac{1}{\alpha_1}\Big[\beta x(t-\tau) - \alpha_0 y(t)\Big]$$

and integrate it over the interval from 0 to t, using as the initial state the quantity $y(0)$, we will end up with the expression

$$y(t) = y(0) + \int_0^t \underbrace{\frac{1}{\alpha_1}\Big[\beta x(\xi-\tau) - \alpha_0 y(\xi)\Big]}_{y'(\xi)} d\xi$$

which is described by the schematic diagram in Fig. 2.11.

Observing the foregoing schematic diagram, it is not difficult to understand how it works. Initially, the input signal $x(t)$ passes through a time-delay element, so as to create the time-delayed input $x(t - \tau)$, which in turn passes through an amplitude scaling element, so as to create the expression $\beta x(t - \tau)$. When the signal reaches the signal addition element, it will be added to it but with a negative sign (i.e., it will essentially be subtracted), a signal that rises from the lower branch of the circuit. But what is this signal? To determine it, we observe that this signal must be exactly the same as the one that exits the circuit, since the lower branch of the circuit starts just before the exit, and there is no other element to modify the signal. Therefore, the signal that moves on the other branch will be the output $y(t)$,[23] which, passing through the scaling element, will be multiplied by the constant α_0, and then the signal that will emerge—$\alpha_0 y(t)$—will meet the signal $\beta x(t - \tau)$ and will be subtracted from it, creating the signal $\beta x(t - \tau) - \alpha_0 y(t)$ [we observe, therefore, the way in which the expression gradually forms that is included in the expression for $y(t)$]. Subsequently, the output of the adder will pass through a second amplitude scaling element, which will multiply its input by the constant $1/\alpha_1$, thereby creating the expression found inside the integral of the last equation.

From the foregoing analysis, it becomes apparent that the signal flowing through the circuit at position A is simply the derivative of the system's output, that is, the signal $y'(t)$. All we have to do next is pass this signal through an integrator, so that as its output we get the signal $y(t)$, which ultimately issues from the system. This is also the way in which a schematic diagram describes the structure of a continuous-time system.

Note It is interesting to point out at this point that because the functions of differentiators and integrators are the inverse of each other, any differential equation can be represented both using differentiators and integrators. For example, the second-order differential equation

$$2\frac{d^2y(t)}{dt^2} + 5\frac{dy(t)}{dt} + 4y(t) = x(t)$$

can be represented by the two alternative equivalent ways presented in Fig. 2.12, with image (a) referring to the use of differentiators and image (b) referring to the use of integrators. However, in practice, integrators are much easier to implement compared to differentiators, and for this reason in almost all cases the approach of image (b) is preferred. For this reason, in all the diagrams that follow, the simulation of continuous-time systems is based on the use of integrators.

[23] continuous-time systems whose schematic diagrams include closed loops with their output feeding back their input are known as *recursive systems* and will be studied in detail in a subsequent section.

Fig. 2.12 Representation of a differential equation using differentials and integrators

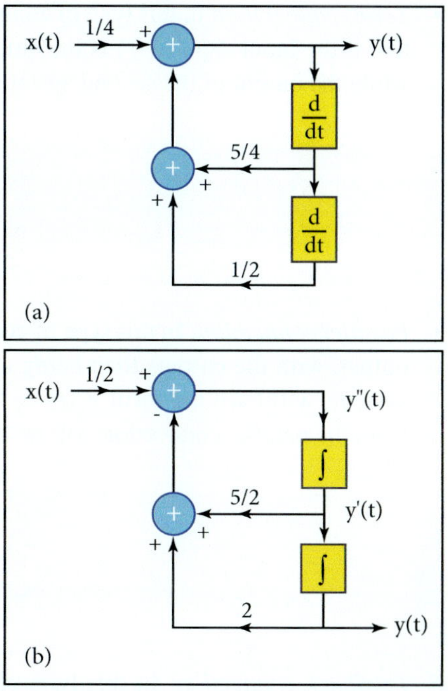

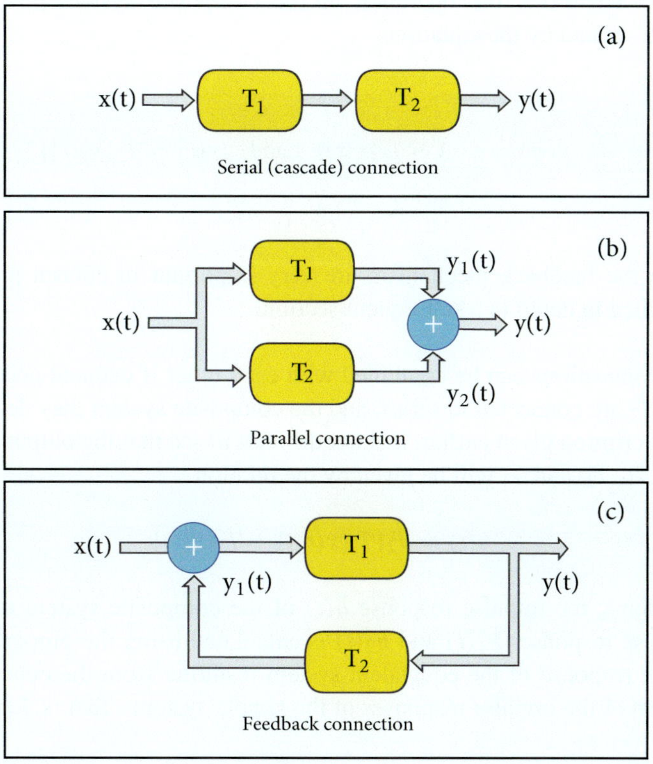

Fig. 2.13 The three different ways of interconnecting continuous-time systems: (**a**) series connection, (**b**) parallel connection, and (**c**) feedback connection

Using the preceding fundamental building blocks, we can construct any continuous-time system that implements the appropriate transform $\mathcal{T}$ in each case. However, we can connect larger systems together so as to construct more complex structures. Considering, for example, two systems of the form $y(t) = \mathcal{T}_1\{x(t)\}$ and $y(t) = \mathcal{T}_2\{x(t)\}$, these can be connected in the three different ways presented in Fig. 2.13 and in general terms are described as follows:

- *Series connection*: In this type of connection, the output of the first system becomes input to the second system. This means that if the input signal $x(t)$ is transmitted to the first system, then the output of the first system will be $y_1(t) = \mathcal{T}_1\{x(t)\}$, while the output of the second system will result from the application of both transformations, that is,

$$y_2(t) = \mathcal{T}_2\{y_1(t)\} = \mathcal{T}_2\{\mathcal{T}_1\{x(t)\}\} \tag{2.56}$$

- *Parallel connection*: In this type of connection, the two systems receive the common input $x(t)$, and each produces its own output, with the outputs then being added to give the final output of the system. This means that the outputs of the two systems will have the form $y_1(t) = \mathcal{T}_1\{x(t)\}$ and $y_2(t) = \mathcal{T}_2\{x(t)\}$, while the output of the equivalent system resulting from the parallel connection will be

$$y(t) = y_1(t) + y_2(t) = \mathcal{T}_1\{x(t)\} + \mathcal{T}_2\{x(t)\} \tag{2.57}$$

- *Feedback connection*: In this type of connection, the output of the first system feeds back its input, after first passing through the second system. Therefore, the operation of the composite system resulting from the aforementioned connection method will be described by the equations

$$y_1(t) = x(t) + \mathcal{T}_1\{x(t)\} \quad \text{and} \quad y(t) = \mathcal{T}_1\{y_1(t)\}$$

  The systems resulting from the feedback mechanism are very important in control processes, and due to their great importance, they will be studied in detail in a subsequent section.

Note that the aforementioned connections can be combined with each other if deemed necessary, as shown in Fig. 2.14. In this figure, the systems $\mathcal{T}_1$ and $\mathcal{T}_2$ are connected in series, and the composite system they define is connected in parallel with the system $\mathcal{T}_3$. Based on the description given earlier, it is not difficult to see that the output of the system resulting from the interconnection of the systems $\mathcal{T}_1$, $\mathcal{T}_2$, and $\mathcal{T}_3$ will be given by the relation

$$y(t) = \mathcal{T}_2\{\mathcal{T}_1\{x(t)\}\} + \mathcal{T}_3\{x(t)\}$$

In the special case of LTI systems, the impulse response $h(t)$ of the composite system resulting from the connection of two simple systems with impulse responses $h_1(t)$ and $h_2(t)$ is calculated using the properties of convolution. As we have already mentioned, the impulse response of the equivalent system resulting from the connection of the simple systems in series is equal to the convolution of the impulse responses of the simple systems, that is, $h(t) = h_1(t) * h_2(t)$, while for the

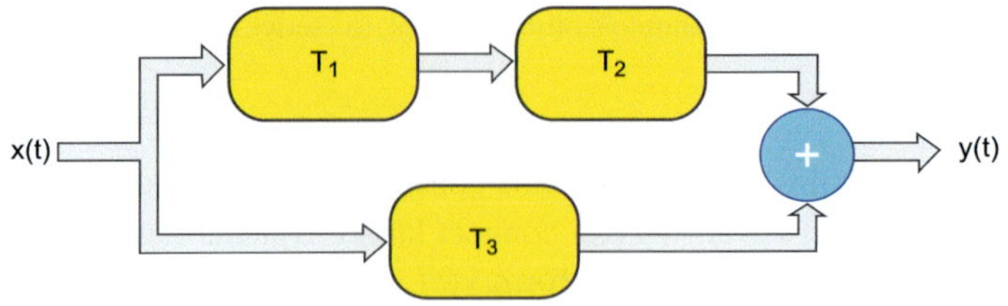

Fig. 2.14 Combined use of parallel and series connections of continuous-time systems

parallel connection the impulse response $h(t)$ is equal to the sum of the impulse responses $h_1(t)$ and $h_2(t)$, so we can write $h(t) = h_1(t) + h_2(t)$. In addition, the series interconnection of two LTI systems is characterized by another property: *If we change the positions of the two systems, that is, we place the second first and then the first, the composite system that will emerge will be equivalent to the original.*

The aforementioned property is demonstrated by the use of the associative and commutative properties of convolution. Considering an input signal $x(t)$ that first passes through the system $h_1(t)$ and then through the system $h_2(t)$, the power of the associative property tells us that

$$y(t) = [x(t) * h_1(t)] * h_2(t) = x(t) * [h_1(t) * h_2(t)]$$

However, since convolution is characterized by the commutative property, we will have $h_1(t) * h_2(t) = h_2(t) * h_1(t)$ and the foregoing relation is written

$$y(t) = x(t) * [h_2(t) * h_1(t)] = [x(t) * h_2(t)] * h_1(t)$$

where at the end we used the associative property again. But the last relation is simply the output of the composite system that results from the series connection of the systems $h_2(t)$ and $h_1(t)$, that is, *the alternation of the order in which we connect the systems does not change the output of the composite system, which is always the same.*

2.7 Representation with Schematic Diagrams

The schematic diagrams presented in the preceding pages constitute a valuable tool for describing continuous-time systems, as they allow the diagrammatic representation of the solution $y(t)$ of the general differential equation of a LTI system that has the form

$$\sum_{n=0}^{N} \alpha_n \frac{d^n y(t)}{dt^n} = \sum_{m=0}^{M} \beta_m \frac{d^m x(t)}{dt^m} \tag{2.58}$$

using such elements as *differentiators* and *integrators*, as well as *adders* and *multipliers* of signals. Next, we will present two different ways of representing a system that rely on the use of integrators and adders, the first of which is known as the *noncanonical form* or *direct form I realization*, while the second is known as the *canonical* form or *direct form II realization*. A representation is called canonical *when the number of integrators used is the same as the order of the system, so that there are no redundant integrators.*

To describe these two alternative forms of implementation, let us formulate the differential equation of the system in the form

$$\sum_{n=0}^{N} \alpha^n D^n y(t) = \sum_{m=0}^{M} \beta_m D^m x(t) \tag{2.59}$$

where the symbol $D^n \triangleq d^n/dt^n$ describes *the operator of multiple differentiation*. Based on the last equation, we can define the two alternative forms of system representation as follows:

2.7.1 Noncanonical Form

To determine the nonnormal form, we can assume without loss of generality that $N = M$. Defining the operator of multiple integration from the relation

$$D^{-N} \triangleq \underbrace{\int_{-\infty}^{t} \int_{-\infty}^{t} \cdots\cdots \int_{-\infty}^{t}}_{\text{N times}} \tag{2.60}$$

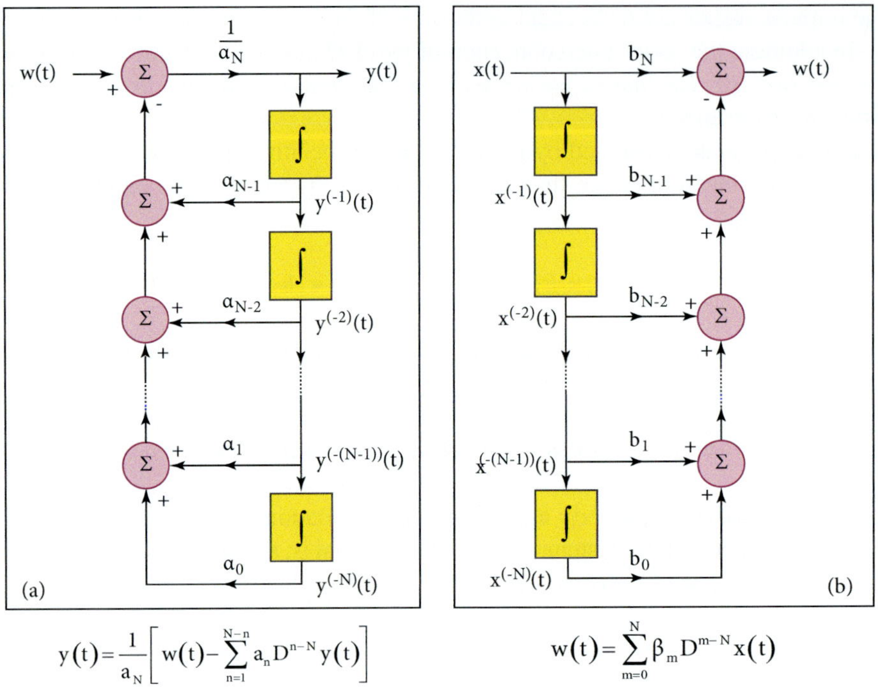

Fig. 2.15 Schematic diagrams for functions $y(t)$ and $w(t)$ of the nonstandard form

and multiplying it by both sides of the foregoing equation (essentially integrating both sides of the equation N times), it takes the form

$$\sum_{n=0}^{N} \alpha_n D^{n-N} y(t) = \sum_{m=0}^{N} \beta_m D^{m-N} x(t)$$

If we replace the right side of the foregoing equation with the function

$$w(t) = \sum_{m=0}^{N} \beta_m D^{m-N} x(t) \tag{2.61}$$

and expand the sum of the left side for the value $n = N$, the foregoing equation becomes

$$\alpha_N y(t) + \sum_{n=0}^{N-1} \alpha_n D^{n-N} y(t) = w(t)$$

or, equivalently,

$$y(t) = \frac{1}{\alpha_N}\left[w(t) - \sum_{n=1}^{N-1} \alpha_n D^{n-N} y(t)\right] \tag{2.62}$$

Using the rules for describing differential equations from schematic diagrams, it is not difficult to see that the last equation is described by the schematic diagram of Fig. 2.15a, while in a completely similar way the function $w(t)$ defined by the foregoing equation is described by the schematic diagram of Fig. 2.15b. The interpretation of the way in which these diagrams represent the differential equations that define the preceding functions should not be difficult to understand and is left as an exercise to the reader.

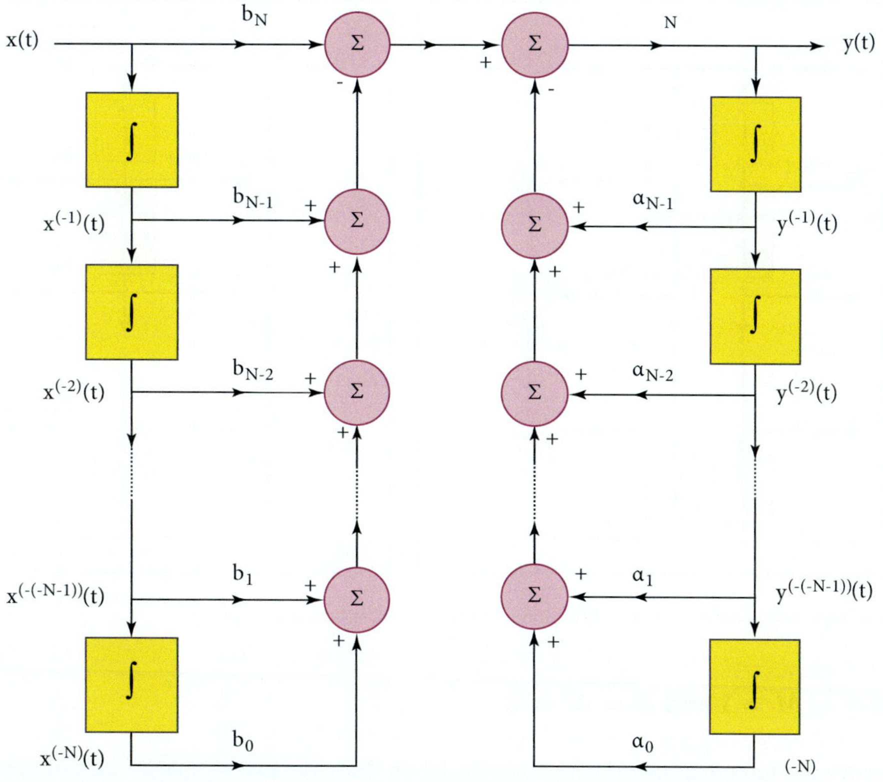

Fig. 2.16 Nonstandard form of representation of differential system equation from schematic diagram

By carrying out the preceding mathematical analysis, we managed to describe our system as a composite system consisting of two simpler subsystems connected in series. From these subsystems—let us call them A and B—the first one receives as input the signal $x(t)$ and produces as output the signal $w(t)$, while the second one receives as input the signal $w(t)$ and produces as output the signal $y(t)$. It is, therefore, obvious that the schematic diagram of our original system will result from the *series connection* of the schematic diagrams of the two subsystems, with the result of this connection being depicted in Fig. 2.16. This composite diagram that arises in the aforementioned way is known as the *nonstandard form* of the representation of the differential equation of a system using schematic diagrams, as it is characterized by the existence of $2N$ integrators for the representation of a differential equation of order N.

The foregoing description was based, for simplicity's sake, on the assumption that the relation $M = N$ holds, while if $M < N$, then the schematic diagram will also contain other branches that will be related to the coefficients β_m for the values $M < m \leq N$.

2.7.2 Normal Form

As was already mentioned in a previous section, *the series connection of two LTI systems A and B is characterized by the property of commutativity, regarding the positions of the two systems*. In other words, this means that the order in which we connect the two systems to construct a more complex system does not matter, and therefore, if instead of the connection $A \rightarrow B$ we make the connection $B \rightarrow A$, the result we get will be the same. This property of commutativity allows us to change the order in which the two schematic diagrams are connected and, thus, construct the diagram depicted in Fig. 2.17α.

From this figure we observe that the two integrators that exist in each line, on the one hand, are continuous, one after the other, on the other hand, they act on the same signal. We can therefore simplify this diagram if in each line *we replace the pair of elementary integrators* that exist there *with a simple integrator*, thereby constructing the diagram depicted in Fig. 2.17b. This schematic diagram is known as the *normal form of the representation of the system's differential equation*.

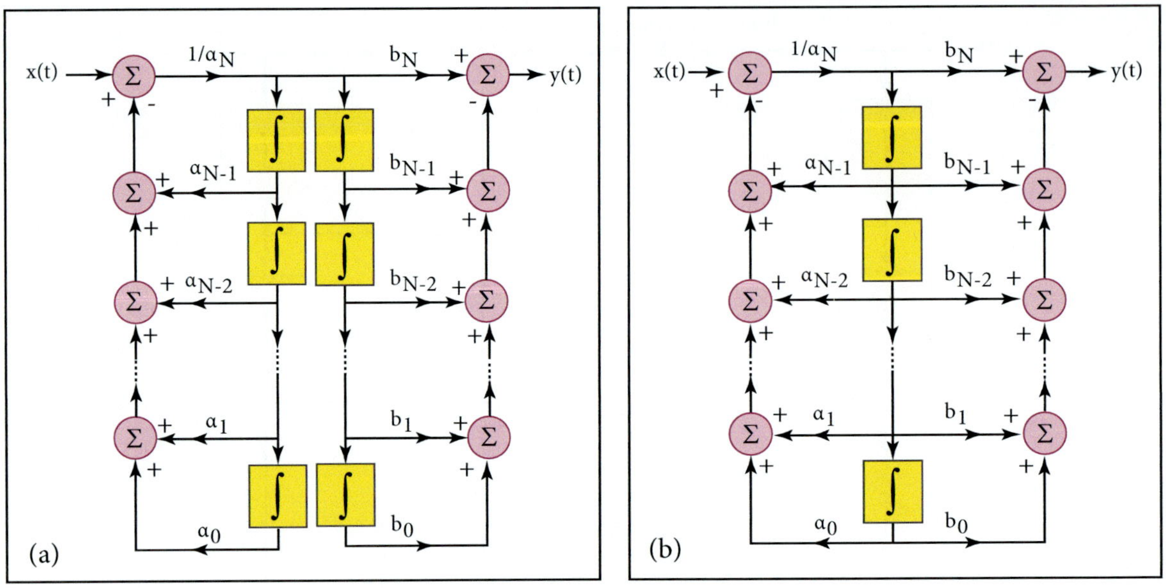

Fig. 2.17 Normal form of representation of system's differential equation from schematic diagram

2.7.3 The Cases $N = M = 1$ and $N = M = 2$

To understand in practice the foregoing theoretical analysis, let us study the two simplest cases of continuous-time systems that correspond to parameter values $N = M = 1$ and $N = M = 2$. Starting with the case $M = N = 1$, which is also the simplest, the general differential equation of the system is reduced to the simplest form

$$\alpha_1 \frac{dy(t)}{dt} + \alpha_0 y(t) = \beta_0 x(t) + \beta_1 \frac{dx(t)}{dt} \tag{2.63}$$

Integrating the preceding equation $N = 1$ times, it will take the form

$$\alpha_1 y(t) + \alpha_0 \int_{-\infty}^{t} y(\tau)d\tau = \beta_0 \int_{-\infty}^{t} x(\tau)d\tau + \beta_1 x(t)$$

Following the previously presented methodology, we set

$$w(t) = \beta_0 \int_{-\infty}^{t} x(\tau)d\tau + \beta_1 x(t) \quad \text{so we will have} \quad y(t) = \frac{1}{\alpha_1}\left[w(t) - \alpha_0 \int_{-\infty}^{t} y(\tau)d\tau\right]$$

The schematic diagrams of the subsystems described by the equations defining the signals $w(t)$ and $y(t)$ are presented in Fig. 2.18a and b, while the nonstandard and standard forms of the overall system that were constructed in the way we presented earlier are shown in Fig. 2.18c and d, respectively.

Taking into account that the output of the first system $w(t)$ becomes input to the second subsystem, we can find the input/output correlation of the overall system if in the second equation we make the substitution $x(t) = w(t)$, resulting in the equation

$$y(t) = \frac{1}{\alpha_1}\left[\beta_0 \int_{-\infty}^{t} x(\tau)d\tau + \beta_1 x(t) - \alpha_0 \int_{-\infty}^{t} y(\tau)d\tau\right] \tag{2.64}$$

On the other hand, for the case $M = N = 2$, the differential equation of the system takes the form

$$\alpha_2 \frac{d^2y(t)}{dt^2} + \alpha_1 \frac{dy(t)}{dt} + \alpha_0 y(t) = \beta_0 x(t) + \beta_1 \frac{dx(t)}{dt} + \beta_2 \frac{d^2x(t)}{dt^2} \tag{2.65}$$

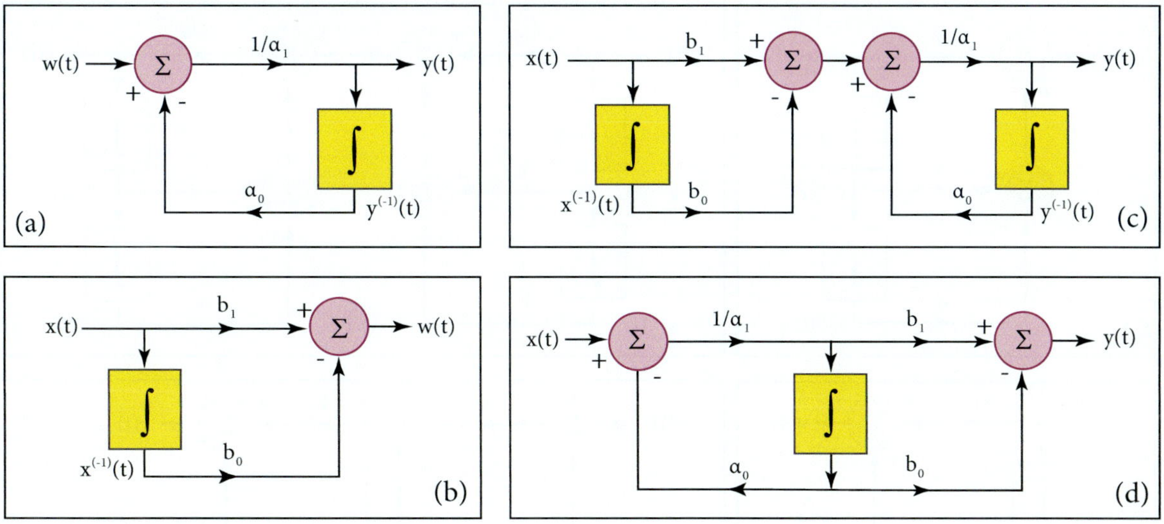

Fig. 2.18 Schematic diagrams for case $N = M = 1$

Applying the methodology presented earlier, we integrate $N = 2$ times. Integrating for the first time we get

$$\alpha_2 \frac{dy(t)}{dt} + \alpha_1 y(t) + \alpha_0 \int_{-\infty}^{\tau} y(\xi) d\xi = \beta_0 \int_{-\infty}^{\tau} x(\xi) d\xi + \beta_1 x(t) + \beta_2 \frac{dx(t)}{d\tau}$$

while the integration of the foregoing equation for the second time leads to the result

$$\alpha_2 y(t) + \alpha_1 \int_{-\infty}^{t} y(\tau)\, d\tau + \alpha_0 \int_{-\infty}^{t} \left(\int_{-\infty}^{\tau} y(\xi)\, d\xi \right) d\tau = \beta_0 \int_{-\infty}^{t} \left(\int_{-\infty}^{\tau} x(\xi) d\xi \right) d\tau + \beta_1 \int_{-\infty}^{t} x(\tau)\, d\tau + \beta_2 x(t)$$

In full analogy with the previous case, we set

$$w(t) = \beta_0 \int_{-\infty}^{t} \left(\int_{-\infty}^{\tau} x(\xi)\, d\xi \right) d\tau + \beta_1 \int_{-\infty}^{t} x(\tau)\, d\tau + \beta_2 x(t)$$

so the system output $y(t)$ will take the form

$$y(t) = \frac{1}{\alpha_2} \left[w(t) - \alpha_1 \int_{-\infty}^{t} y(\tau)\, d\tau - \alpha_0 \int_{-\infty}^{t} \left(\int_{-\infty}^{\tau} y(\xi)\, d\xi \right) d\tau \right]$$

The schematic diagrams of the elementary subsystems that define the signals $w(t)$ and $y(t)$ are presented in Fig. 2.19a and b, while the nonstandard and standard forms of the overall system are presented in Fig. 2.19c and d, respectively.

Combining the equations for the signals $w(t)$ and $y(t)$, we can determine the correlation that exists between the signals $x(t)$ and $y(t)$, leading to the equation

$$y(t) = \frac{1}{\alpha_2} \left[\beta_0 \int_{-\infty}^{t} \left(\int_{-\infty}^{\tau} x(\xi)\, d\xi \right) d\tau + \beta_1 \int_{-\infty}^{t} x(\tau)\, d\tau + \beta_2 x(t) - \alpha_1 \int_{-\infty}^{t} y(\tau)\, d\tau - \alpha_0 \int_{-\infty}^{t} \left(\int_{-\infty}^{\tau} y(\xi)\, d\xi \right) d\tau \right]$$

To understand how the foregoing equations relate to the corresponding schematic diagrams, the symbols used in the diagrams are defined as

$$y^{(-1)}(t) = \int_{-\infty}^{t} y(\tau) d\tau \quad \text{and} \quad y^{(-2)}(t) = \int_{-\infty}^{t} \left(\int_{-\infty}^{\tau} y(\xi) d\xi \right) d\tau$$

while similar symbols apply to the functions $x^{(-1)}$ and $x^{(-2)}$.

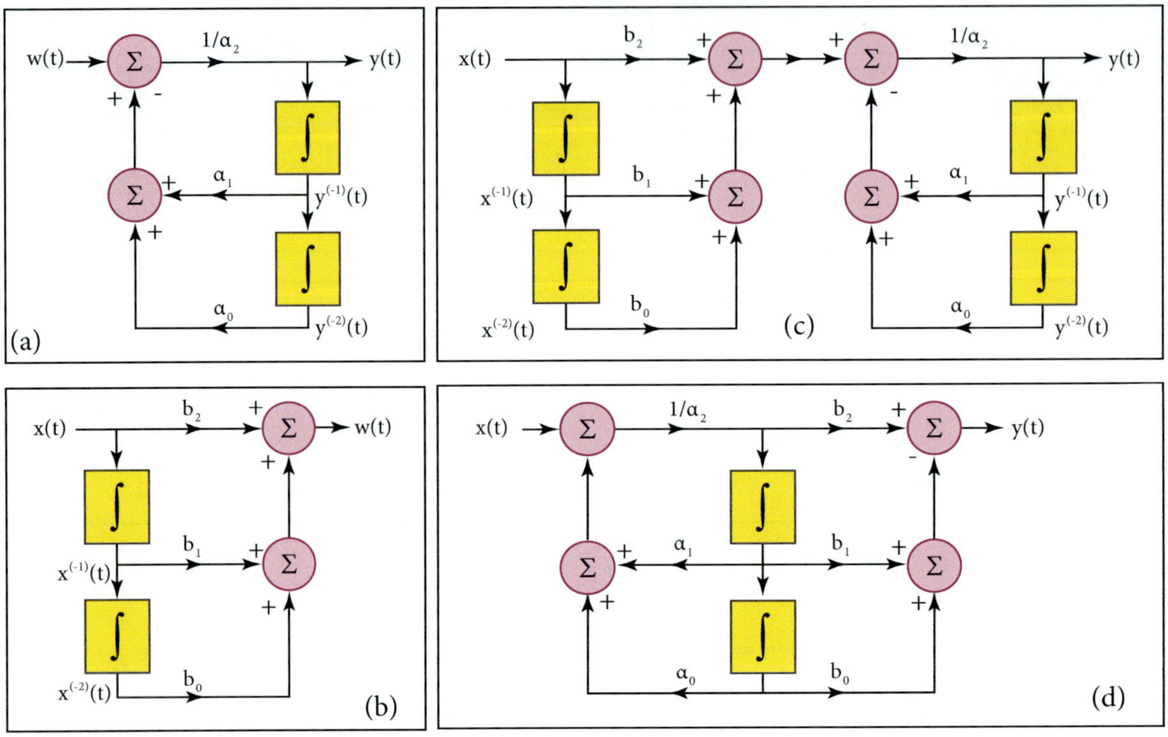

Fig. 2.19 Schematic diagrams for case $N = M = 2$

2.8 Reversibility and Inverse Systems

A continuous-time system that implements some transform $\mathcal{T}$ is aforementioned to be reversible *if the transmission of discrete inputs always leads to the creation of corresponding discrete outputs*, so that by observing its output we can reproduce its input (in other words, this means that it is not possible for two different input signals to lead to the same output and, therefore, for two input signals, $x_1(t)$ and $x_2(t)$, such that $x_1(t) \neq x_2(t)$, it must also be the case that $y_1(t) \neq y_2(t)$, where $y_1(t) = \mathcal{T}\{x_1(t)\}$ and $y_2(t) = \mathcal{T}\{x_2(t)\}$ are the corresponding output signals).[24] A typical example of such a system is the one described by the equation $y(t) = (1/2)x(t)$, whose inverse is the system $w(t) = 2x(t)$. On the other hand, an example of a noninvertible system is a system described by the equation $y(t) = x^2(t)$, since in this case the inputs $x_1(t) = x(t)$ and $x_2(t) = -x(t)$ lead to the same output signal $y(t)$. It is obvious that *the inverse of the inverse of a system is the system itself.*

Taking into account that the inverse of a system will implement the inverse transform $\mathcal{T}^{-1}$, it is not difficult to realize that *if a system is connected in series with its inverse, then the equivalent system that will result will be the identity system that copies the input to its output* and therefore is described by a relation of the form $y(t) = x(t)$. Indeed, if we take into account that in the series connection the transformations of the two systems are applied successively on the input signal, the output of the system will have the form

$$y(t) = \mathcal{T}^{-1}\{\mathcal{T}\{x(t)\}\} = x(t) \tag{2.66}$$

which indeed describes the identity system. This situation is presented in Fig. 2.20.

It is proven that *if a continuous-time LTI system is reversible, then the inverse system will also be LTI* and therefore will have its own impulse response $g(t)$. To prove the truth of this claim, let us consider a system LTI that is connected

[24] If we consider the system as a function that maps input signals (from its domain) to output signals (in its range), then the above requirement about distinct inputs producing distinct outputs is equivalent to requiring the function to be unambiguous, that is, both one-to-one (injective) and onto (surjective), so that its inverse exists.

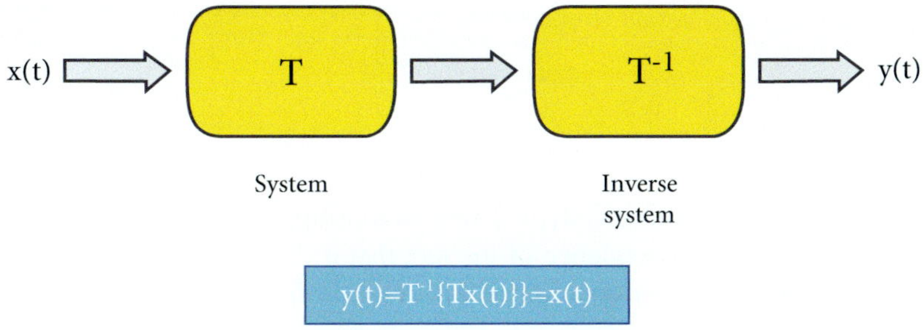

Fig. 2.20 The series connection of a continuous-time system with its inverse leads to the creation of the identity system

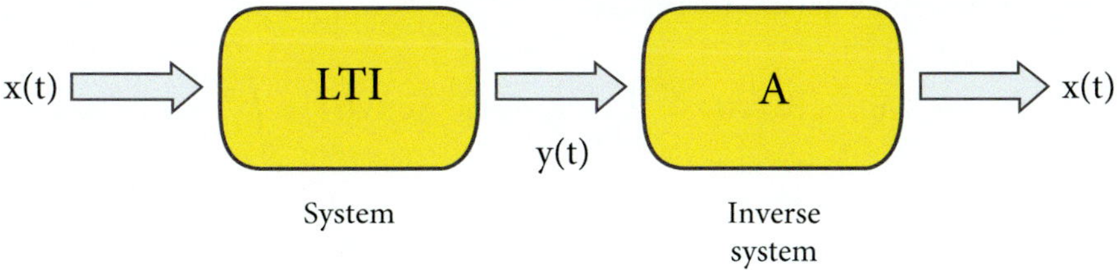

Fig. 2.21 The inverse of a continuous-time LTI system is also a LTI system

in series with a system A that we know is the inverse of the first system, and therefore, their series connection will lead to the identity system, as shown in Fig. 2.21. Since the first system is linear, it will be characterized by the principle of superposition. Therefore, considering two elementary inputs $x_1(t)$ and $x_2(t)$ with corresponding outputs $y_1(t)$ and $y_2(t)$, we see that the response of the system to the linear combination $x(t) = \alpha x_1(t) + \beta x_2(t)$ will be the linear combination $y(t) = \alpha y_1(t) + \beta y_2(t)$, which will be transmitted as input to system A. But this system, being the inverse of the first system, will output the signal $x(t)$; therefore, this system will be a linear system, since the response to the linear combination of the elementary inputs $\alpha y_1(t) + \beta y_2(t)$ gives the linear combination of the corresponding elementary outputs $\alpha x_1(t) + \beta x_2(t)$. With completely similar arguments we discover that system A must also be time-invariant. Indeed, if in the first system we transmit the time-delayed input $x(t - \tau)$, then because this is time-invariant, it will give as output the time-delayed output $y(t - \tau)$, which will be transmitted as input to system B. On the other hand, the output of this system will coincide with the output of the identity system resulting from the series connection of the two systems and therefore will be equal to $x(t - \tau)$. Therefore, since the response of system A to the signal $y(t - \tau)$ is the signal $x(t - \tau)$, it is obvious that this system fulfills the property of time invariance, and since we previously proved that it is also linear, we can characterize it as a continuous-time LTI system.

If we now proceed to the interconnection of these two systems in series, then the impulse response $\xi(t)$ of the equivalent system that will arise will be equal to the convolution of the impulse responses of these systems, so we can write that $\xi(t) = h(t) * g(t)$. However, as mentioned previously, this equivalent system is the identity system that copies the input to its output and therefore it will hold that

$$y(t) = \xi(t) * x(t) = x(t) * [h(t) * g(t)] = x(t)$$

If we compare the last equation with the known identity $x(t) = x(t) * \delta(t)$, it immediately follows that

$$h(t) * g(t) = \delta(t) \tag{2.67}$$

and therefore, *the convolution of the impulse response of a system with the impulse response of the inverse system is equal to the impulse function* $\delta(t)$. This also constitutes the fundamental relationship that characterizes the impulse responses of these two systems.

The process of calculating the input of a reversible system when its output is known is referred to as *deconvolution* and is described by an equation of the form

$$x(t) = y(t) * g(t) \tag{2.68}$$

where $g(t)$ is the impulse response of the *inverse* system. This relationship, which in reality also constitutes the correlation of input/output of the inverse system, is a consequence of the fact that if the inverse system is connected in series with the corresponding direct system, then the equivalent system that will result is the *identity* system.

Problem 2.8.1 For each of the following systems, examine whether it is reversible or not, and for those that are characterized as reversible, determine the inverse system:

$$\text{(a) } y(t) = x(t-6), \quad \text{(b) } y(t) = \int_{-\infty}^{t} x(\tau)d\tau, \quad \text{(c) } y(t) = \int_{-\infty}^{t} e^{-(t-\tau)}x(\tau)d\tau.$$

Solution

(a) The system is reversible, with the inverse system described by the equation $y(t) = x(t+6)$, which is obvious given that the inverse transform of the time delay of the signal by four time units is the time advance of the signal by four time units.

(b) Taking into account that the inverse operation of integration is differentiation, it is not difficult to realize that *the inverse system of the integrator is the differentiator* described by an equation of the form $y(t) = dx(t)/dt$.

(c) The system is reversible. The inverse system is described by the relation $y(t) = x(t) + dx(t)/dt$.

For each given LTI system, the determination of the inverse system—if it exists—is relatively easily done through the use of the *Laplace transform*, and for this reason the systematic study of these systems will be presented in detail in the corresponding chapter.

2.9 Description of Systems in State Space

In the preceding sections we presented the basic theory that characterizes continuous-time systems, studying them from different perspectives and using for their analysis several tools and techniques, which include the following:

- *The system's response to basic continuous-time signals*, such as step and impulse functions, as well as simple exponential and trigonometric functions.
- *The input/output correlation* that describes the way a system's output is calculated when its input is known. This correlation can be determined either using a differential equation of order N whose solution gives the output of the system or by directly determining this output and the way it depends on the input $x(t)$ through the use of some function of the form $y(t) = f[x(t)]$.
- *The schematic representation of the system's structure* using appropriate schematic diagrams that describe the way in which the basic subsystems are connected and interact with each other, so as to describe the operation of the system dictated by the corresponding differential equation.
- *The convolution integral* provided that the continuous-time system we consider is LTI.
- *The system's transfer function*, which is defined as the ratio of the system's output to its input, when we transmit to it the basic exponential signal $x(t) = e^{st}$. This kind of description mainly concerns LTI systems.

The main disadvantage of the aforementioned methods of studying a system is that they describe the system *through the correlations that exist between its input and output, without delving into its internals and studying its internal structure*. This weakness often leads us to erroneous conclusions about the characteristics of the system, such as, for example, characterizing a system as *stable*, while in reality it is *unstable*. To prevent the emergence of all these problems, we can study a system using the concept of *state space* and resorting to the use of a set of N variables, which are known as *state variables*. These variables are a function of time, so we can describe them using notation of the form $q_1(t), q_2(t), \ldots, q_N(t)$.

What exactly is the role of these variables? To answer this question, we must remember that the response of a continuous-time system at any time t depends on all the inputs that have been transmitted to the system for the entire previous time period extending from $-\infty$ to time t. For example, if a system is LTI, then its output will be calculated according to the integral known from the convolution

$$y(t) = \int_{-\infty}^{t} x(\tau)h(t-\tau)d\tau \tag{2.69}$$

If we assume that the system inputs are not known for the entire previous infinite time period, but only for times $t > t_0$, we can still calculate the system output for times $t > t_0$, *provided that the system at time* $t = t_0$ *is in a state of rest*. Indeed, if we reformulate the foregoing integral in the form

$$y(t) = \int_{-\infty}^{t_0} x(\tau)h(t-\tau)d\tau + \int_{t_0}^{t} x(\tau)h(t-\tau)d\tau = y(t, t_0) + \int_{t_0}^{t} x(\tau)h(t-\tau)d\tau \tag{2.70}$$

(with the function $y(t, t_0)$ symbolizing the first of the two integrals), it is obvious that, only when $y(t, t_0) = 0$—which means that at time t_0 the system is in a *state of rest*—is it possible to calculate the system output $y(t)$ for times $t > t_0$, as in the opposite case the calculation of the output requires knowledge of the function $y(t, t_0)$. The set of initial conditions of the system at time $t = t_0$ is known as the *initial state* of the system.

Based on the foregoing description, we can define the state variables as the *minimum amount of information* (i.e., the minimum number of variables) *that we need to know about the system, so that their values at any time* t_0 *are sufficient to determine the behavior of the system at any time* $t \geq t_0$, provided that for these times the input $x(t)$ of the system is known. In other words, this means that *the output of the system at any time is fully determined by knowledge of the values of the state variables of the system* (which are collectively referred to as the state or state vector of the system), as well as the inputs of the system that we transmit to it.

Note that the choice of the set of variables to be used to determine the state of a system is not unique; on the contrary, there are many sets of such variables that are equally suitable for use in such a description. Usually, natural quantities of the system that are easy to measure are chosen as variables, such as, for example, positions and velocities or, alternatively, voltages and currents. Regardless of the process of their selection, the state variables must meet two basic prerequisites: (a) *their minimum number must be equal to the order of the differential equation that describes the system under study* and (b) *they must be linearly independent of each other*. Typical examples of differential physical systems that can be described in state space are as follows:

Example 2.9.1 Consider a vehicle moving on a plane at speed υ under the influence of force F and the presence of friction f proportional to the speed (Fig. 2.22). From physics it is known that the motion equation of the vehicle is

$$x(t) = \upsilon t + \frac{1}{2}\gamma t^2 + x_0$$

where $\gamma = d\upsilon/dt$ is the acceleration of the vehicle and x_0 its initial position. Knowing the initial position and speed of the vehicle, as well as the value of the force F, we can calculate its final position since the speed results from the differentiation of the position vector, the acceleration is calculated as

$$\text{Acceleration} = \frac{\text{Force}}{\text{Mass}} - \text{Friction}$$

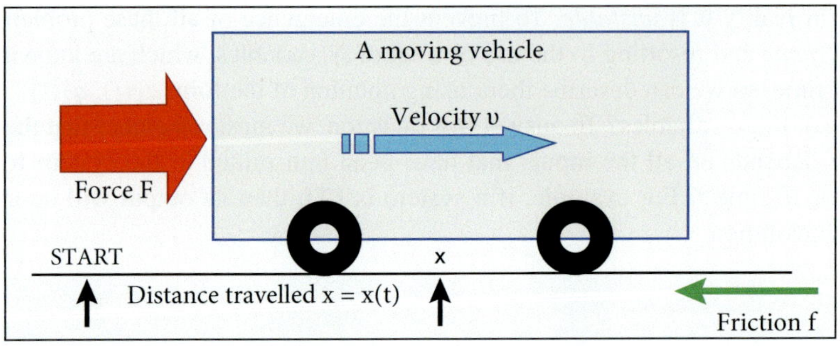

Fig. 2.22 Vehicle moving on a plane under the influence of force F

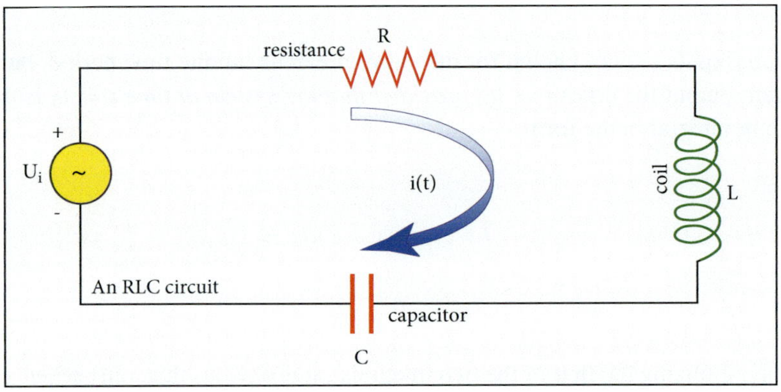

Fig. 2.23 RLC circuit traversed by a current of intensity $i(t)$

while the friction is proportional to the speed. Therefore, at any time we can fully describe the state of the vehicle, knowing its position x and its speed υ. Using, therefore, the state variables $q_1(t) = x(t)$ and $q_2(t) = \upsilon(t)$, the foregoing equation is transformed into the system of differential equations

$$\frac{dq_1(t)}{dt} = q_2(t) \quad \text{and} \quad \frac{dq_2(t)}{dt} = \frac{F}{m} - fq_2(t)$$

or, in matrix form,

$$\begin{bmatrix} dq_1(t)/dt \\ dq_2(t)/dt \end{bmatrix} = \begin{bmatrix} 0 & 1 \\ 0 & -f \end{bmatrix} \begin{bmatrix} q_1(t) \\ q_2(t) \end{bmatrix} + \begin{bmatrix} 0 \\ 1/m \end{bmatrix} F$$

Example 2.9.2 Consider the RLC circuit of Fig. 2.23, which is traversed by a current of intensity $i(t)$. To calculate the value of this current for each time instance, we need to know the input voltage $\upsilon_i(t)$, the initial remaining charge of the capacitor q_0, and the initial current of the coil i_0. In this case, the application of circuit theory leads to the conclusion that the electric current of the circuit will be given by the equation

$$i(t) = \frac{\exp(-t/LC)}{4RLC^2 - 1} \left\{ \frac{V}{L} \Big[(4RLC^2 - 1)\cos(Mt) - 2LCM\sin(Mt) \Big] \right.$$
$$\left. + i_0 \Big[(4RLC^2 - 1)\cos(Mt) - 2LCM\sin(Mt) \Big] - 4RLC^2 \sin(Mt) q_0 \right\}$$

Fig. 2.24 Body of mass m performing vertical oscillation with damping coefficient β

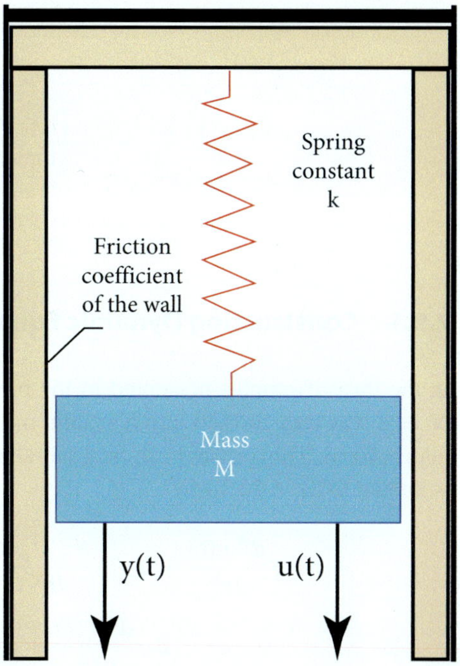

where $M = \sqrt{4RLC^2 - 1}/2LC$. Choosing as *state variables* the current $i(t)$ and the electric charge $q(t)$, the application of *Kirchhoff*'s law leads us to the equation

$$V = Ri(t) + L\frac{di(t)}{dt} + \frac{1}{C}\int i(t)dt$$

which, taking into account that $q_1(t) = q(t) = \int i(t)dt$, $q_2(t) = i(t)$, leads to $\frac{dq_1(t)}{dt} = q_2(t)$, which can be formulated in matrix form as

$$\begin{bmatrix} dq_1(t)/dt \\ dq_2(t)/dt \end{bmatrix} = \begin{bmatrix} 0 & 1 \\ -R/L & -1/LC \end{bmatrix} \begin{bmatrix} q_1(t) \\ q_2(t) \end{bmatrix} + \begin{bmatrix} 0 \\ 1/L \end{bmatrix} V$$

Example 2.9.3 Let there be a mass M that performs vertical oscillation suspended on a spring of constant k and subjected to the friction of the wall β (Fig. 2.24). From physics, it is known that the equation of motion is

$$m\frac{d^2x(t)}{dt^2} + \beta\frac{dx(t)}{dt} + kx(t) = F(t)$$

where $F(t)$ is the total force applied to the body and $x(t)$ is the vertical displacement of the mass from its equilibrium position. Taking into account that the differential equation of the system is of second order, it is obvious that for its solution and the finding of the function $x(t)$ for every $t > t_0$, we need the initial conditions $x(t_0)$ and $\dot{x}(t_0)$, and therefore we must resort to the use of two state variables. We consider $q_1(t) = x(t)$ and $q_2(t) = x'(t)$ as such variables. In this case, the differential equation of the system is reduced to a system of two first-order differential equations of the form

$$\frac{dq_1(t)}{dt} = \frac{dx(t)}{dt} = q_2(t) \quad \text{and} \quad \frac{dq_2(t)}{dt} = \frac{d^2x(t)}{dt^2} = -\frac{k}{m}q_1(t) - \frac{\beta}{m}q_2(t) + \frac{F(t)}{m}$$

or, in matrix form,

$$\begin{bmatrix} dq_1(t)/dt \\ dq_2(t)/dt \end{bmatrix} = \begin{bmatrix} 0 & 1 \\ -k/m & \beta/m \end{bmatrix} \begin{bmatrix} q_1(t) \\ q_2(t) \end{bmatrix} + \begin{bmatrix} 0 \\ 1/m \end{bmatrix} F$$

2.9.1 Constructing Dynamic Equations

In the three examples presented in the preceding pages, the differential equation of a system characterized by order value $N = 2$ was transformed into a system of two coupled first-order differential equations, which in turn were reformulated in matrix form. This process can be generalized for the general case of a SISO system of order N, whose differential equation is known to have the form

$$\alpha_N \frac{d^N y(t)}{dt^N} + \alpha_{N-1} \frac{d^{N-1} y(t)}{dt^{N-1}} + \cdots + \alpha_2 \frac{d^2 y(t)}{dt^2} + \alpha_1 \frac{dy(t)}{dt} + \alpha_0 y(t) =$$
$$\beta_N \frac{d^N x(t)}{dt^N} + \beta_{N-1} \frac{d^{N-1} x(t)}{dt^{N-1}} + \cdots + \beta_2 \frac{d^2 x(t)}{dt^2} + \beta_1 \frac{dx(t)}{dt} + \beta_0 x(t)$$

In the most common cases, the construction of a system's state equations is based on its diagrammatic representation both in its *normal* and in its *nonnormal* forms. In this approach, as state variables of the system $q_1(t), q_2(t), \ldots, q_N(t)$ we define *the outputs of the N integrators* included in this diagram, a process that is depicted diagrammatically in Fig. 2.25 (this scheme refers to the normal form of a system, which is usually used in practice, while the corresponding process for the nonnormal form is briefly presented below). Studying this scheme and taking into account that the output of an integrator is the integral of its input, it is not difficult to discover that the state variables of the system will be related to each other by a system of equations of the form

$$\begin{aligned} q_1(t) &= \int q_2(t)dt \Rightarrow \frac{dq_1(t)}{dt} = q_2(t) \\ q_2(t) &= \int q_3(t)dt \Rightarrow \frac{dq_2(t)}{dt} = q_3(t) \\ q_3(t) &= \int q_4(t)dt \Rightarrow \frac{dq_3(t)}{dt} = q_4(t) \\ &\cdots\cdots\cdots\cdots\cdots\cdots\cdots\cdots \\ q_{N-1}(t) &= \int q_N(t)dt \Rightarrow \frac{dq_{N-1}(t)}{dt} = q_N(t) \end{aligned} \tag{2.71}$$

In the preceding equations, we should also add the equation for the time derivative of the state variable $q_N(t)$, which from the foregoing scheme is found to be given by the relation

$$\frac{dq_N(t)}{dt} = -\frac{\alpha_0}{\alpha_N} q_1(t) - \frac{\alpha_1}{\alpha_N} q_2(t) - \cdots - \frac{\alpha_{N-2}}{\alpha_N} q_{N-1}(t) - \frac{\alpha_{N-1}}{\alpha_N} q_N(t) + \frac{1}{\alpha_N} x(t)$$

We end up, therefore, with the system of N coupled first-order differential equations

$$\frac{dq_1(t)}{dt} = q_2(t), \quad \frac{dq_2(t)}{dt} = q_3(t), \quad \frac{dq_3(t)}{dt} = q_4(t), \quad \cdots\cdots\cdots, \quad \frac{dq_{N-1}(t)}{dt} = q_N(t)$$
$$\frac{dq_N(t)}{dt} = -\frac{\alpha_0}{\alpha_N} q_1(t) - \frac{\alpha_1}{\alpha_N} q_2(t) - \cdots - \frac{\alpha_{N-2}}{\alpha_N} q_{N-1}(t) - \frac{\alpha_{N-1}}{\alpha_N} q_N(t) + \frac{1}{\alpha_N} x(t)$$

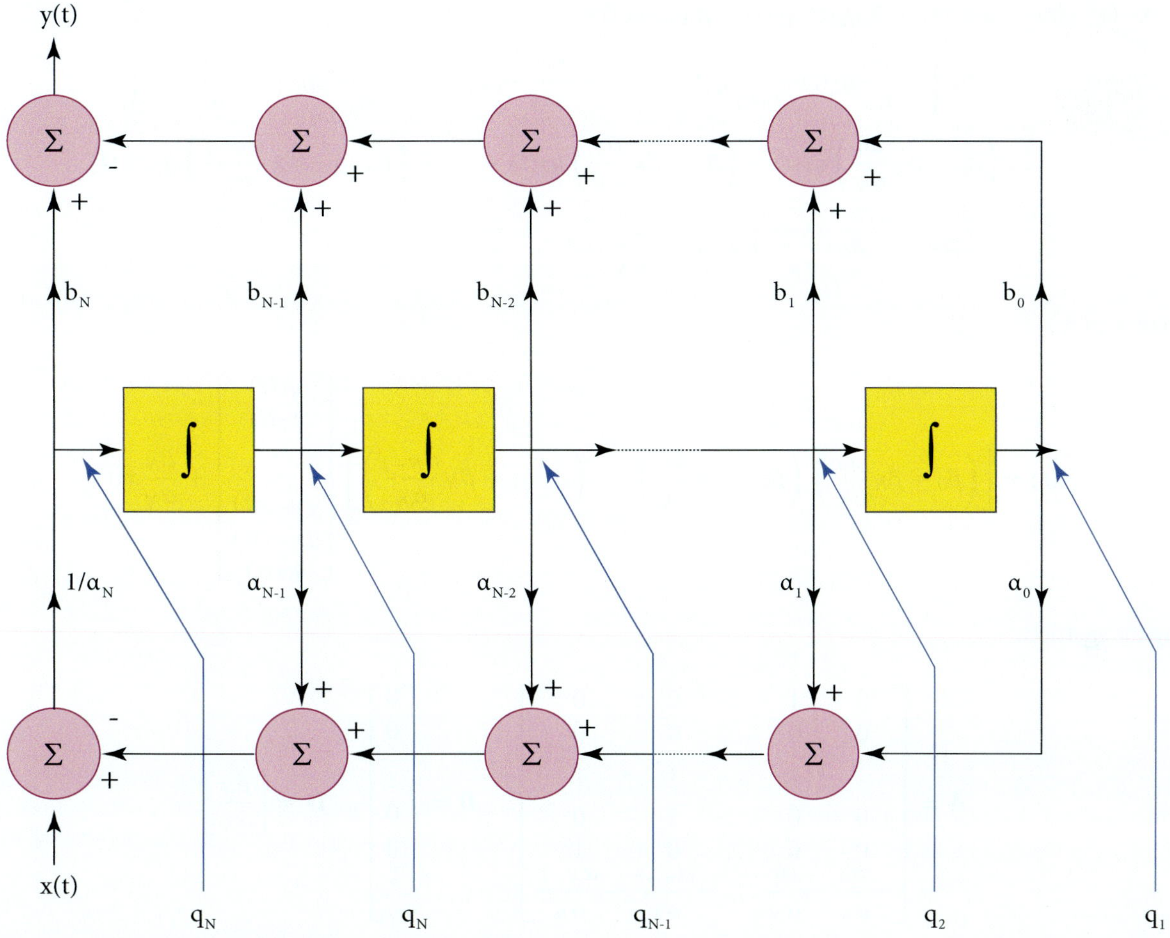

Fig. 2.25 Definition of system's state variables from its normal form

which in matrix form is expressed as

$$\begin{bmatrix} \dot{q}_1(t) \\ \dot{q}_2(t) \\ \vdots \\ \dot{q}_{N-2}(t) \\ \dot{q}_{N-1}(t) \\ \dot{q}_N(t) \end{bmatrix} = \begin{bmatrix} 0 & 1 & \dots & 0 & 0 \\ 0 & 0 & \dots & 0 & 0 \\ \vdots & \vdots & \ddots & \vdots & \vdots \\ 0 & 0 & \dots & 1 & 0 \\ 0 & 0 & \dots & 0 & 1 \\ -\dfrac{\alpha_0}{\alpha_N} & -\dfrac{\alpha_1}{\alpha_N} & \dots & -\dfrac{\alpha_{N-2}}{\alpha_N} & -\dfrac{\alpha_{N-1}}{\alpha_N} \end{bmatrix} \begin{bmatrix} q_1(t) \\ q_2(t) \\ \vdots \\ q_{N-2}(t) \\ q_{N-1}(t) \\ q_N(t) \end{bmatrix} + \begin{bmatrix} 0 \\ 0 \\ \vdots \\ 0 \\ 0 \\ \dfrac{1}{\alpha_N} \end{bmatrix} x(t)$$

and its solution leads to the determination of the state variables of the system.

On the other hand, the system's output $y(t)$ as a function of its state variables is given by the relation

$$y(t) = \beta_0 q_1(t) + \beta_1 q_2(t) + \dots + \beta_{N-1} q_N(t) + \beta_N \frac{dq_N(t)}{dt}$$

as can be seen by studying the highest horizontal branch of the adders of the schematic diagram of Fig. 2.25. If in the last relation we replace the derivative $\dot{q}_N(t)$ from the defining equation, we will get the equation

$$
\begin{aligned}
y(t) &= \beta_0 q_1(t) + \beta_1 q_2(t) + \cdots + \beta_{N-1} q_N(t) \\
&\quad + \beta_N \left\{ -\frac{\alpha_0}{\alpha_N} q_1(t) - \frac{\alpha_1}{\alpha_N} q_2(t) - \cdots - \frac{\alpha_{N-2}}{\alpha_N} q_{N-1}(t) - \frac{\alpha_{N-1}}{\alpha_N} q_N(t) + \frac{1}{\alpha_N} x(t) \right\} \\
&= \left(\beta_0 - \beta_N \frac{\alpha_0}{\alpha_N}\right) q_1(t) + \left(\beta_1 - \beta_N \frac{\alpha_1}{\alpha_N}\right) q_2(t) + \cdots + \left(\beta_{N-2} - \beta_N \frac{\alpha_{N-2}}{\alpha_N}\right) q_{N-1}(t) \\
&\quad + \left(\beta_{N-1} - \beta_N \frac{\alpha_{N-1}}{\alpha_N}\right) q_N(t) + \frac{\beta_N}{\alpha_N} x(t)
\end{aligned}
$$

or, in matrix form,

$$
y(t) = \left[\left(\beta_0 - \beta_N \frac{\alpha_0}{\alpha_N}\right) \left(\beta_1 - \beta_N \frac{\alpha_1}{\alpha_N}\right) \cdots \left(\beta_{N-1} - \beta_N \frac{\alpha_{N-1}}{\alpha_N}\right) \right] \begin{bmatrix} q_1(t) \\ q_2(t) \\ \vdots \\ q_{N-2}(t) \\ q_{N-1}(t) \\ q_N(t) \end{bmatrix} + \frac{\beta_N}{\alpha_N} x(t)
$$

Defining the matrices

$$
\boldsymbol{A} = \begin{bmatrix} 0 & 1 & \cdots & 0 & 0 \\ 0 & 0 & \cdots & 0 & 0 \\ \vdots & \vdots & \ddots & \vdots & \vdots \\ 0 & 0 & \cdots & 1 & 0 \\ 0 & 0 & \cdots & 0 & 1 \\ -\frac{\alpha_0}{\alpha_N} & -\frac{\alpha_1}{\alpha_N} & \cdots & -\frac{\alpha_{N-2}}{\alpha_N} & -\frac{\alpha_{N-1}}{\alpha_N} \end{bmatrix}, \quad \boldsymbol{B} = \begin{bmatrix} 0 \\ 0 \\ \vdots \\ 0 \\ 0 \\ \frac{1}{\alpha_N} \end{bmatrix}, \quad \boldsymbol{D} = \left[\frac{\beta_N}{\alpha_N} \right]
$$

and $\boldsymbol{C} = \left[\left(\beta_0 - \beta_N \frac{\alpha_0}{\alpha_N}\right) \left(\beta_1 - \beta_N \frac{\alpha_1}{\alpha_N}\right) \cdots \left(\beta_{N-1} - \beta_N \frac{\alpha_{N-1}}{\alpha_N}\right) \right]$

The two matrix equations we arrived at above take the form[25]

$$
\dot{\boldsymbol{q}}(t) = \boldsymbol{A}\boldsymbol{q}(t) + \boldsymbol{B}x(t) \quad \text{and} \quad y(t) = \boldsymbol{C}\boldsymbol{q}(t) + \boldsymbol{D}x(t) \tag{2.72}
$$

where $\dot{\boldsymbol{q}}(t) = [\dot{q}_1(t), \dot{q}_2(t), \ldots, \dot{q}_N(t)]^T$ is the time derivative of the state vector $\boldsymbol{q}(t)$, while the functions $x(t)$ and $y(t)$ obviously refer to the input and output of the system. The first of the two preceding equations is known as the *state equation* of the system, while the second is called the *output equation*, while the matrices that appear in these are characterized by the following properties:

- Matrix $\boldsymbol{A}$ has dimensions $N \times N$ and is called the *system matrix*.
- Matrix $\boldsymbol{B}$ has dimensions $N \times 1$ and is called the *input matrix*.
- Matrix $\boldsymbol{C}$ has dimensions $1 \times N$ and is called the *measurement matrix*.
- Matrix $\boldsymbol{D}$ has dimensions 1×1 and is called the *output matrix*.

[25] These equations concern continuous-time LTI systems, while in the general case of arbitrary time-varying systems they take the form

$$
\dot{\boldsymbol{q}}(t) = f[\boldsymbol{q}(t), x(t), t] \quad \text{and} \quad y(t) = g[\boldsymbol{q}(t), x(t), t]
$$

with the functions $f(\ldots)$ and $g(\ldots)$ not necessarily being linear.

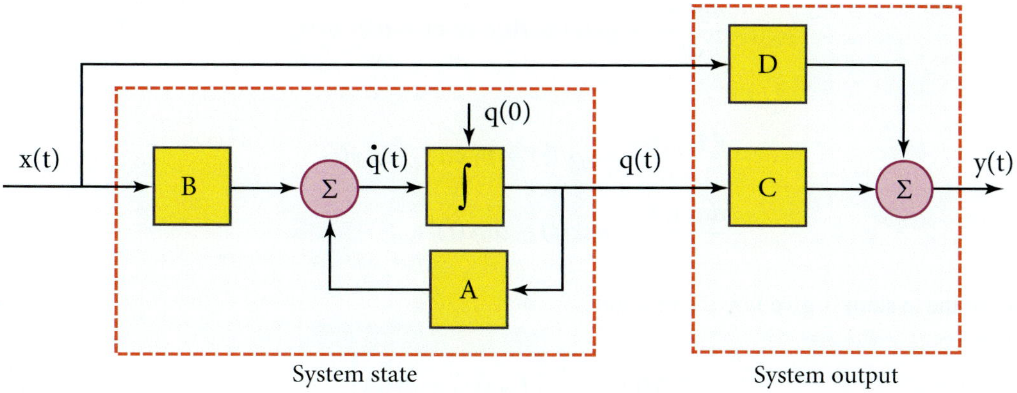

Fig. 2.26 Schematic diagram of SISO system in state space for the case $M = N$ (Source: Shmaliy, *Continuous-time Systems*, Fig. 4.22). In the same book, corresponding schematic diagrams can be found for the case of SISO systems with $M < N$ as well as for the case of a closed-loop LTI system)

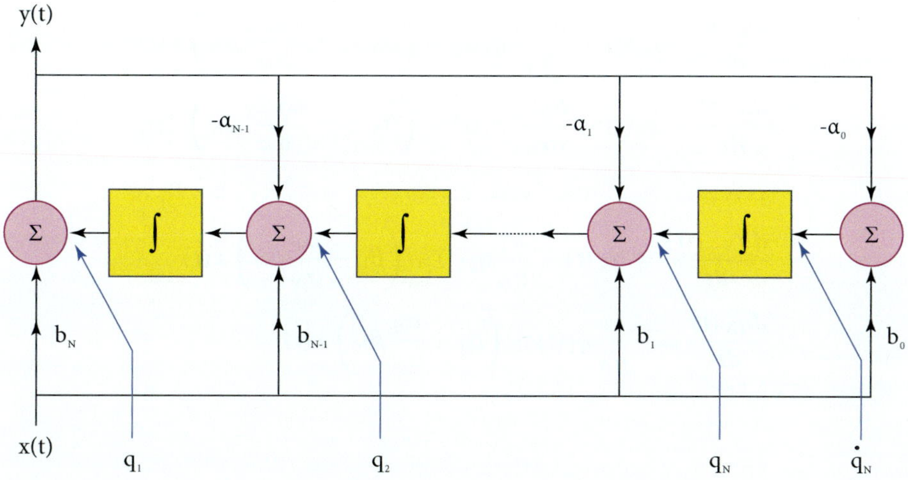

Fig. 2.27 Definition of system state variables from nonstandard form

A schematic diagram that describes the structure of a SISO system in the state space is presented in Fig. 2.26. Note that the analysis we did is, for simplicity's sake, in the case of $M = N$, while if it is $M < N$ or $\beta_N = 0$, the matrix $\boldsymbol{D}$ transforms into a zero matrix and the description of the system is significantly simplified.

2.9.2 Dynamic Equations and Nonnormal Form

Using the methodology presented in previous pages, we can construct the dynamic equations of the system in the state space based on the *nonnormal form* of its representation with schematic diagrams. In full analogy with the previous case, we again choose as state variables *the outputs of the N integrators*, as represented in Fig. 2.27. In this figure, the schematic diagram of the nonstandard form is presented in an equivalent condensed form, so the way in which the equations were constructed becomes immediately understandable.

Examination of the structure of the foregoing diagram makes it easy to ascertain how the state variables of the system are related to each other through the following equations:

$$\frac{dq_1(t)}{dt} = q_2(t) + \beta_{N-1}x(t) - \alpha_{N-1}y(t)$$

$$\frac{dq_2(t)}{dt} = q_3(t) + \beta_{N-2}x(t) - \alpha_{N-2}y(t)$$

$$\frac{dq_3(t)}{dt} = q_4(t) + \beta_{N-3}x(t) - \alpha_{N-3}y(t) \tag{2.73}$$

$$\cdots\cdots\cdots\cdots\cdots\cdots\cdots\cdots$$

$$\frac{dq_{N-1}(t)}{dt} = q_N(t) + \beta_1 x(t) - \alpha_1 y(t)$$

$$\frac{dq_N(t)}{dt} = \beta_0 x(t) - \alpha_0 y(t)$$

while the output of the system is given by the relation

$$y(t) = \frac{1}{\alpha_N}\Big[\beta_N x(t) + q_1(t)\Big] \tag{2.74}$$

If we eliminate the function $y(t)$ from the preceding system of equations using the last relation, it is not difficult to prove that it will take the form

$$\frac{dq_1(t)}{dt} = q_2(t) - \frac{\alpha_{N-1}}{\alpha_N}q_1(t) + \left(\beta_{N-1} - \frac{\alpha_{N-1}}{\alpha_N}\beta_N\right)x(t)$$

$$\frac{dq_2(t)}{dt} = q_3(t) - \frac{\alpha_{N-2}}{\alpha_N}q_1(t) + \left(\beta_{N-2} - \frac{\alpha_{N-2}}{\alpha_N}\beta_N\right)x(t)$$

$$\cdots\cdots\cdots\cdots\cdots\cdots\cdots\cdots\cdots\cdots\cdots\cdots$$

$$\frac{dq_{N-1}(t)}{dt} = q_N(t) - \frac{\alpha_1}{\alpha_N}q_1(t) + \left(\beta_1 - \frac{\alpha_1}{\alpha_N}\beta_N\right)x(t)$$

$$\frac{dq_N(t)}{dt} = -\frac{\alpha_0}{\alpha_N}q_1(t) + \left(\beta_0 - \frac{\alpha_0}{\alpha_N}\beta_N\right)x(t)$$

or, using matrices,

$$\dot{\boldsymbol{q}}(t) = \boldsymbol{A}\boldsymbol{q}(t) + \boldsymbol{B}x(t) \tag{2.75}$$

where $\dot{\boldsymbol{q}}(t) = [\dot{q}_1(t), \dot{q}_2(t), \ldots, \dot{q}_N(t)]^T$ is a derivative with respect to time of the state vector $\boldsymbol{q}(t)$, while the matrices $\boldsymbol{A}$ and $\boldsymbol{B}$ are defined as

$$\boldsymbol{A} = \begin{bmatrix} -\frac{\alpha_{N-1}}{\alpha_N} & 1\,0 \ldots 0 \\ -\frac{\alpha_{N-2}}{\alpha_N} & 0\,1 \ldots 0 \\ \vdots & \vdots\,\vdots \ddots \vdots \\ -\frac{\alpha_1}{\alpha_N} & 0\,0 \ldots 1 \\ -\frac{\alpha_0}{\alpha_N} & 0\,0 \ldots 0 \end{bmatrix} \quad \text{and} \quad \boldsymbol{B} = \begin{bmatrix} \beta_{N-1} - \frac{\alpha_{N-1}}{\alpha_N}\beta_N \\ \beta_{N-2} - \frac{\alpha_{N-2}}{\alpha_N}\beta_N \\ \cdots\cdots\cdots \\ \beta_1 - \frac{\alpha_1}{\alpha_N}\beta_N \\ \beta_0 - \frac{\alpha_0}{\alpha_N}\beta_N \end{bmatrix}$$

On the other hand, the defining equation of the system's output $y(t)$ can also be formulated in matrix form:

$$y(t) = \boldsymbol{C}\boldsymbol{q}(t) + \boldsymbol{D}x(t) \tag{2.76}$$

where the matrices $\boldsymbol{C}$ and $\boldsymbol{D}$ are defined as

$$\boldsymbol{C} = \left[\frac{1}{\alpha_N} \; 0 \; 0 \cdots 0 \right] \quad \text{and} \quad \boldsymbol{D} = \left[\frac{\beta_N}{\alpha_N} \right]$$

Thus, we again end up with the dynamic equations of the system, with the matrices $\boldsymbol{A}$, $\boldsymbol{B}$, $\boldsymbol{C}$, and $\boldsymbol{D}$ being defined in the way previously described. The schematic diagram that depicts the structure of the system in the state space is presented, as noted, in Fig. 2.26.

2.9.3 The Case of MIMO Systems

Although the preceding analysis concerned SISO systems that receive one input and produce one output, the generalization of the description to MIMO systems is obvious. Therefore, the state equations of a MIMO system that consists of K inputs and L outputs define a system of N differential equations of the form

$$\begin{aligned}
\dot{q}_1(t) &= g_1[q_1(t), q_2(t), \ldots, q_N(t), x_1(t), x_2(t), \ldots, x_K(t)] \\
\dot{q}_2(t) &= g_2[q_1(t), q_2(t), \ldots, q_N(t), x_1(t), x_2(t), \ldots, x_K(t)] \\
&\cdots\cdots\cdots\cdots\cdots\cdots\cdots\cdots\cdots\cdots \\
\dot{q}_N(t) &= g_N[q_1(t), q_2(t), \ldots, q_N(t), x_1(t), x_2(t), \ldots, x_K(t)]
\end{aligned} \tag{2.77}$$

in which the vector $\boldsymbol{q}(t) = [q_1(t), q_2(t), \ldots, q_N(t)]^T$ is, as before, the state vector of the system, while the simple input $x(t)$ is replaced by the vector of the K inputs $\boldsymbol{x}(t) = [x_1(t), x_2(t), \ldots, x_K(t)]^T$. If the system under consideration is linear, then the functions $g_i(\boldsymbol{q}, \boldsymbol{x})$ $(i = 1, 2, \ldots, N)$ will be linear functions of their variables and the foregoing equations can be expressed in their simplest form:

$$\begin{aligned}
\dot{q}_1(t) &= \alpha_{11}q_1(t) + \alpha_{12}q_2(t) + \cdots + \alpha_{1N}q_N(t) + \beta_{11}x_1(t) + \beta_{12}x_2(t) + \cdots + \beta_{1K}x_K(t) \\
\dot{q}_2(t) &= \alpha_{21}q_1(t) + \alpha_{22}q_2(t) + \cdots + \alpha_{2N}q_N(t) + \beta_{21}x_1(t) + \beta_{22}x_2(t) + \cdots + \beta_{2K}x_K(t) \\
&\cdots\cdots\cdots\cdots\cdots\cdots\cdots\cdots\cdots\cdots\cdots\cdots \\
\dot{q}_N(t) &= \alpha_{N1}q_1(t) + \alpha_{N2}q_2(t) + \cdots + \alpha_{NN}q_N(t) + \beta_{N1}x_1(t) + \cdots + \beta_{NK}x_K(t)
\end{aligned}$$

In a completely analogous way, the L outputs of the system that form the output vector that has the form $\boldsymbol{y}(t) = [y_1(t), y_2(t), \ldots, y_L(t)]^T$ will satisfy the relations

$$\begin{aligned}
y_1(t) &= \gamma_{11}q_1(t) + \gamma_{12}q_2(t) + \cdots + \gamma_{1N}q_N(t) + \delta_{11}x_1(t) + \delta_{12}x_2(t) + \cdots + \delta_{1K}x_K(t) \\
y_2(t) &= \gamma_{21}q_1(t) + \gamma_{22}q_2(t) + \cdots + \gamma_{2N}q_N(t) + \delta_{21}x_1(t) + \delta_{22}x_2(t) + \cdots + \delta_{2K}x_K(t) \\
&\cdots\cdots\cdots\cdots\cdots\cdots\cdots\cdots\cdots\cdots\cdots\cdots \\
y_L(t) &= \gamma_{L1}q_1(t) + \gamma_{L2}q_2(t) + \cdots + \gamma_{LN}q_N(t) + \delta_{L1}x_1(t) + \cdots + \delta_{LK}x_K(t)
\end{aligned}$$

The preceding equations that constitute *the dynamic equations of the system in the case of MIMO systems*, can, as before, be formulated in a more concise way in tabular form:

$$\begin{bmatrix} \dot{q}_1(t) \\ \dot{q}_2(t) \\ \vdots \\ \dot{q}_N(t) \end{bmatrix} = \begin{bmatrix} \alpha_{11} & \alpha_{12} & \ldots & \alpha_{1N} \\ \alpha_{21} & \alpha_{22} & \ldots & \alpha_{2N} \\ \vdots & \vdots & \ddots & \vdots \\ \alpha_{N1} & \alpha_{N2} & \ldots & \alpha_{NN} \end{bmatrix} \begin{bmatrix} q_1(t) \\ q_2(t) \\ \vdots \\ q_N(t) \end{bmatrix} + \begin{bmatrix} \beta_{11} & \beta_{12} & \ldots & \beta_{1K} \\ \beta_{21} & \beta_{22} & \ldots & \beta_{2K} \\ \vdots & \vdots & \ddots & \vdots \\ \beta_{N1} & \beta_{N2} & \ldots & \beta_{NK} \end{bmatrix} \begin{bmatrix} x_1(t) \\ x_2(t) \\ \vdots \\ x_K(t) \end{bmatrix}$$

$$\begin{bmatrix} y_1(t) \\ y_2(t) \\ \vdots \\ y_L(t) \end{bmatrix} = \begin{bmatrix} \gamma_{11} & \gamma_{12} & \dots & \gamma_{1N} \\ \gamma_{21} & \gamma_{22} & \dots & \gamma_{2N} \\ \vdots & \vdots & \ddots & \vdots \\ \gamma_{L1} & \gamma_{L2} & \dots & \gamma_{LN} \end{bmatrix} \begin{bmatrix} q_1(t) \\ q_2(t) \\ \vdots \\ q_N(t) \end{bmatrix} + \begin{bmatrix} \delta_{11} & \delta_{12} & \dots & \delta_{1K} \\ \delta_{21} & \delta_{22} & \dots & \delta_{2K} \\ \vdots & \vdots & \ddots & \vdots \\ \delta_{L1} & \delta_{L2} & \dots & \delta_{LK} \end{bmatrix} \begin{bmatrix} x_1(t) \\ x_2(t) \\ \vdots \\ x_K(t) \end{bmatrix}$$

or in an even more concise form:

$$\begin{aligned} \dot{\boldsymbol{q}}(t) &= \boldsymbol{A}\boldsymbol{q}(t) + \boldsymbol{B}\boldsymbol{x}(t) \\ \boldsymbol{y}(t) &= \boldsymbol{C}\boldsymbol{q}(t) + \boldsymbol{D}\boldsymbol{x}(t) \end{aligned} \tag{2.78}$$

where the expression $\dot{\boldsymbol{q}}(t) = [\dot{q}_1(t), \dot{q}_2(t), \dots, \dot{q}_N(t)]$ represents the time derivative of the state vector $\boldsymbol{q}(t)$, while matrices $\boldsymbol{A}$, $\boldsymbol{B}$, $\boldsymbol{C}$, and $\boldsymbol{D}$ in this case are defined as follows:

$$\boldsymbol{A} = \begin{bmatrix} \alpha_{11} & \alpha_{12} & \dots & \alpha_{1N} \\ \alpha_{21} & \alpha_{22} & \dots & \alpha_{2N} \\ \vdots & \vdots & \ddots & \vdots \\ \alpha_{N1} & \alpha_{N2} & \dots & \alpha_{NN} \end{bmatrix} \quad \boldsymbol{B} = \begin{bmatrix} \beta_{11} & \beta_{12} & \dots & \beta_{1K} \\ \beta_{21} & \beta_{22} & \dots & \beta_{2K} \\ \vdots & \vdots & \ddots & \vdots \\ \beta_{N1} & \beta_{N2} & \dots & \beta_{NK} \end{bmatrix}$$

$$\boldsymbol{C} = \begin{bmatrix} \gamma_{11} & \gamma_{12} & \dots & \gamma_{1N} \\ \gamma_{21} & \gamma_{22} & \dots & \gamma_{2N} \\ \vdots & \vdots & \ddots & \vdots \\ \gamma_{L1} & \gamma_{L2} & \dots & \gamma_{LN} \end{bmatrix} \quad \boldsymbol{D} = \begin{bmatrix} \delta_{11} & \delta_{12} & \dots & \delta_{1K} \\ \delta_{21} & \delta_{22} & \dots & \delta_{2K} \\ \vdots & \vdots & \ddots & \vdots \\ \delta_{L1} & \delta_{L2} & \dots & \delta_{LK} \end{bmatrix}$$

and are characterized by the following properties:

- Matrix $\boldsymbol{A}$ has dimensions $N \times N$ and is called *system matrix.*
- Matrix $\boldsymbol{B}$ has dimensions $N \times K$ and is called *input matrix.*
- Matrix $\boldsymbol{C}$ has dimensions $L \times N$ and is called *measurement matrix.*
- Matrix $\boldsymbol{D}$ has dimensions $L \times K$ and is called *output matrix.*

Problem 2.9.1 Determine the dynamic equations of a system described by the equation

$$\frac{d^3y(t)}{dt^3} - 5\frac{d^2y(t)}{dt^2} + 2\frac{dy(t)}{dt} + 4y(t) = 3\frac{dx(t)}{dt} - 4x(t)$$

based on both its normal and nonnormal forms.

Solution The differential equation of the system arises from the general differential equation of linear systems with order values $N = 3$ and $M = 1$ and parameter values $\alpha_0 = 4$, $\alpha_1 = 2$, $\alpha_2 = -5$, $\alpha_3 = 1$ and $\beta_0 = -4$, $\beta_1 = 3$ and $\beta_2 = \beta_3 = 0$. Therefore, using the defining equations of matrices $\boldsymbol{A}$, $\boldsymbol{B}$, $\boldsymbol{C}$, and $\boldsymbol{D}$, which are associated with the *normal form* of the system, we will have

$$\boldsymbol{A} = \begin{bmatrix} 0 & 1 & 0 \\ 0 & 0 & 1 \\ -\dfrac{\alpha_0}{\alpha_3} & -\dfrac{\alpha_1}{\alpha_3} & -\dfrac{\alpha_2}{\alpha_3} \end{bmatrix} = \begin{bmatrix} 0 & 1 & 0 \\ 0 & 0 & 1 \\ -4 & -2 & 5 \end{bmatrix}, \quad \boldsymbol{B} = \begin{bmatrix} 0 \\ 0 \\ \dfrac{1}{\alpha_3} \end{bmatrix} = \begin{bmatrix} 0 \\ 0 \\ 1 \end{bmatrix}$$

$$\boldsymbol{C} = \left[\left(\beta_0 - \beta_3 \frac{\alpha_0}{\alpha_3} \right) \left(\beta_1 - \beta_3 \frac{\alpha_1}{\alpha_3} \right) \left(\beta_2 - \beta_3 \frac{\alpha_2}{\alpha_3} \right) \right] = [-4 \; 3 \; 0], \quad \text{and} \quad \boldsymbol{D} = [0]$$

This will produce

$$\begin{bmatrix}\dot{q}_1(t)\\ \dot{q}_2(t)\\ \dot{q}_3(t)\end{bmatrix}=\begin{bmatrix}0 & 1 & 0\\ 0 & 0 & 1\\ -4 & -2 & 5\end{bmatrix}\begin{bmatrix}q_1(t)\\ q_2(t)\\ q_3(t)\end{bmatrix}+\begin{bmatrix}0\\ 0\\ 1\end{bmatrix}x(t)=\begin{bmatrix}q_2(t)\\ q_3(t)\\ -4q_1(t)-2q_2(t)+5q_3(t)+x(t)\end{bmatrix}$$

from which it naturally follows that

$$\frac{dq_1(t)}{dt}=q_2(t),\quad \frac{dq_2(t)}{dt}=q_3(t),\quad \frac{dq_3(t)}{dt}=-4q_1(t)-2q_2(t)+5q_3(t)+x(t)$$

while the output of the system is calculated as

$$y(t)=\begin{bmatrix}-4 & 3 & 0\end{bmatrix}\begin{bmatrix}q_1(t)\\ q_2(t)\\ q_3(t)\end{bmatrix}+[0]x(t)=-4q_1(t)+3q_2(t)$$

On the other hand, the defining equations of the matrices for the *nonnormal form* lead to the expressions

$$\boldsymbol{A}=\begin{bmatrix}-\dfrac{\alpha_2}{\alpha_3} & 1 & 0\\ -\dfrac{\alpha_1}{\alpha_3} & 0 & 1\\ -\dfrac{\alpha_0}{\alpha_3} & 0 & 0\end{bmatrix}=\begin{bmatrix}5 & 1 & 0\\ -2 & 0 & 1\\ -4 & 0 & 0\end{bmatrix},\quad \boldsymbol{B}=\begin{bmatrix}\beta_2-\dfrac{\alpha_2}{\alpha_3}\beta_3\\ \beta_1-\dfrac{\alpha_1}{\alpha_3}\beta_3\\ \beta_0-\dfrac{\alpha_0}{\alpha_3}\beta_3\end{bmatrix}=\begin{bmatrix}0\\ 3\\ -4\end{bmatrix},\quad \boldsymbol{C}=\begin{bmatrix}\dfrac{1}{\alpha_3} & 0 & 0\end{bmatrix}=\begin{bmatrix}1 & 0 & 0\end{bmatrix},\quad \boldsymbol{D}=\begin{bmatrix}0\end{bmatrix}$$

This will lead to

$$\begin{bmatrix}\dot{q}_1(t)\\ \dot{q}_2(t)\\ \dot{q}_3(t)\end{bmatrix}=\begin{bmatrix}5 & 1 & 0\\ -2 & 0 & 1\\ -4 & 0 & 0\end{bmatrix}\begin{bmatrix}q_1(t)\\ q_2(t)\\ q_3(t)\end{bmatrix}+\begin{bmatrix}0\\ 3\\ -4\end{bmatrix}x(t)=\begin{bmatrix}5q_1(t)+q_2(t)\\ -2q_1(t)+q_3(t)+3x(t)\\ -4q_1(t)-4x(t)\end{bmatrix}$$

from which it naturally follows that

$$\frac{dq_1(t)}{dt}=5q_1(t)+q_2(t),\quad \frac{dq_2(t)}{dt}=-2q_1(t)+q_3(t)+3x(t),\quad \frac{dq_3(t)}{dt}=-4q_1(t)-4x(t)$$

while the output of the system is calculated as

$$y(t)=\begin{bmatrix}1 & 0 & 0\end{bmatrix}\begin{bmatrix}q_1(t)\\ q_2(t)\\ q_3(t)\end{bmatrix}+\begin{bmatrix}0\end{bmatrix}(t)=q_1(t)$$

2.9.4 Solution of the State Equation

Having constructed the state equation of the system which, as is well known, has the form

$$\frac{d\boldsymbol{q}(t)}{dt}=\boldsymbol{Aq}(t)+\boldsymbol{Bx}(t) \tag{2.79}$$

let us now describe the method for solving it. However, before we describe this process, let us provide a brief theoretical introduction to *polynomial* and *exponential* matrix functions, as these are necessary tools for our study. Let us consider, therefore, the set of square matrices of order N that are defined in the set $R^{N\times N}$. In this case, we define the *polynomial function* $g(\boldsymbol{M})$ of degree n of the square matrix $\boldsymbol{M}$ as a mapping $g: R^{N\times N}\to R^{N\times N}$, such that

$$g(\boldsymbol{M})=\alpha_0\boldsymbol{M}^n+\alpha_1\boldsymbol{M}^{n-1}+\alpha_2\boldsymbol{M}^{n-2}+\cdots+\alpha_{n-1}\boldsymbol{M}+\alpha_n\boldsymbol{I} \tag{2.80}$$

where the power i of the matrix $\boldsymbol{M}$ $(0 \leq i \leq n)$ is defined as

$$\boldsymbol{M}^i = \underbrace{\boldsymbol{M} \cdot \boldsymbol{M} \cdot \boldsymbol{M} \cdot \dots \cdot \boldsymbol{M}}_{i \text{ times}}$$

while the symbol $\boldsymbol{I}$ describes the unit matrix of order N. In an entirely similar way, we can define the *power series* of the square matrix $\boldsymbol{M}$ from a relation of the form

$$g(\boldsymbol{M}) = \sum_{i=0}^{\infty} \alpha_i \boldsymbol{M}^i \tag{2.81}$$

which, of course, exists as long as the foregoing series of infinite terms converges to a finite value.

Closely connected to the power series function $g(\boldsymbol{M})$ is the *exponential function* $e^{\boldsymbol{M}t}$, which in reality and in full analogy with the usual exponential function $e^{\alpha t}$ $(\alpha \in R)$ is defined as an infinite power series of terms of the form

$$\mathcal{J}(\boldsymbol{M}) = e^{\boldsymbol{M}t} = \sum_{k=0}^{\infty} \frac{(\boldsymbol{M}t)^k}{k!} = \boldsymbol{I} + \boldsymbol{M}t + \frac{(\boldsymbol{M}t)^2}{2!} + \frac{(\boldsymbol{M}t)^3}{3!} + \ldots + \frac{(\boldsymbol{M}t)^n}{n!} + \ldots \tag{2.82}$$

from which it follows that $e^{\boldsymbol{0}} = \boldsymbol{I}$ where $\boldsymbol{I}$ is the *unit matrix of order N*. The exponential function of the square matrix $\boldsymbol{M}$ is characterized by the same properties as the usual exponential function $e^{\alpha t}$ and therefore will be

$$e^{\boldsymbol{A}(t_1+t_2)} = e^{\boldsymbol{A}t_1} e^{\boldsymbol{A}t_2} \quad \text{and} \quad e^{(\boldsymbol{A}+\boldsymbol{B})t} = e^{\boldsymbol{A}t} e^{\boldsymbol{B}t}$$

If in the first of the preceding equations we set $t_2 = -t_1$, we will have

$$e^{\boldsymbol{A}0} = e^{\boldsymbol{0}} = \boldsymbol{I} = e^{\boldsymbol{A}t_1} e^{-\boldsymbol{A}t_1} \quad \text{and therefore} \quad (e^{\boldsymbol{A}t_1})^{-1} = e^{-\boldsymbol{A}t_1}$$

while in a completely analogous way, if in the second equation we set $\boldsymbol{B} = -\boldsymbol{A}$, it will take the form

$$(e^{\boldsymbol{A}t})(e^{-\boldsymbol{A}t}) = (e^{-\boldsymbol{A}t})(e^{\boldsymbol{A}t}) = e^{\boldsymbol{0}} = \boldsymbol{I}$$

The time derivative of the exponential function of a matrix is calculated as

$$\frac{d}{dt}(e^{\boldsymbol{M}t}) = \frac{d}{dt}\left(\sum_{k=0}^{\infty} \frac{(\boldsymbol{M}t)^k}{k!}\right) = \frac{d}{dt}\left(\boldsymbol{I} + \sum_{k=1}^{\infty} \frac{(\boldsymbol{M}t)^k}{k!}\right) = \sum_{k=1}^{\infty} \frac{kt^{k-1}\boldsymbol{M}^k}{k!} = \sum_{k=1}^{\infty} \frac{t^{k-1}\boldsymbol{M}^k}{(k-1)!} = \boldsymbol{M} \sum_{k=1}^{\infty} \frac{(\boldsymbol{M}t)^{k-1}}{(k-1)!}$$

Defining the dummy index $\ell = k - 1$, for $k = 1$, will yield $\ell = 0$, and the preceding relation will finally take the form

$$\frac{d}{dt}(e^{\boldsymbol{M}t}) = \boldsymbol{M} \sum_{\ell=0}^{\infty} \frac{(\boldsymbol{M}t)^\ell}{\ell!} = \boldsymbol{M} e^{\boldsymbol{M}t}$$

a property that is entirely analogous to the corresponding property of the derivative of the ordinary exponential function. As for the integral of this function, it is given by a relation of the form

$$\int \boldsymbol{\Phi}(t)dt = \int e^{\boldsymbol{A}t} dt = \boldsymbol{A}^{-1}\boldsymbol{\Phi}(t) = \boldsymbol{\Phi}(t)\boldsymbol{A}^{-1}$$

It is obvious that if matrix $\boldsymbol{A}$ is a *diagonal matrix* of the form $\boldsymbol{A} = \{\alpha_{ii}\} (i = 1, 2, \ldots, n)$, then the exponential function is also a diagonal matrix defined as $e^{\boldsymbol{A}t} = \{\varphi_{ii}(t)\} = \{\exp(\alpha_{ii}t)\}$ $(i = 1, 2, \ldots, n)$. Finally, if matrix $\boldsymbol{A}$ has eigenvalues $\lambda_1, \lambda_2, \ldots, \lambda_n$ with corresponding multiplicities $m_1, m_2, \ldots, m_n$, then it is proven that there are matrices F_{ki}, such that

$$\boldsymbol{\Phi}(t) = \sum_{k=1}^{n} \sum_{i=0}^{m_k-1} t^i e^{\lambda_k t} F_{ki}$$

We now return to the state equation to be solved. Let us multiply both sides of it by the quantity e^{-At}. We will then get

$$e^{-At}\frac{d\boldsymbol{q}(t)}{dt} = e^{-At}\boldsymbol{A}\boldsymbol{q}(t) + e^{-At}\boldsymbol{B}\boldsymbol{x}(t) \quad \text{or equivalently} \quad -e^{-At}\boldsymbol{A}\boldsymbol{q}(t) + e^{-At}\frac{d\boldsymbol{q}(t)}{dt} = e^{-At}\boldsymbol{B}\boldsymbol{x}(t) \tag{2.83}$$

Taking into account that the time derivative of the product of two matrices $\boldsymbol{A}$ and $\boldsymbol{B}$ satisfies the property

$$\frac{d}{dt}(\boldsymbol{A}\boldsymbol{B}) = \frac{d\boldsymbol{A}}{dt}\boldsymbol{B} + \boldsymbol{A}\frac{d\boldsymbol{B}}{dt}$$

it is not difficult to realize that the left-hand side of the preceding equation can take the form

$$-e^{-At}\boldsymbol{A}\boldsymbol{q}(t) + e^{-At}\frac{d\boldsymbol{q}(t)}{dt} = \frac{d}{dt}(e^{-At})\boldsymbol{q}(t) + e^{-At}\frac{d\boldsymbol{q}(t)}{dt} = \frac{d}{dt}\left[e^{-At}\boldsymbol{q}(t)\right]$$

Therefore, we will have

$$\frac{d}{dt}\left[e^{-At}\boldsymbol{q}(t)\right] = e^{-At}\boldsymbol{B}\boldsymbol{x}(t) \tag{2.84}$$

and by integrating we get

$$e^{-At}\boldsymbol{q}(t)\Big|_{t_0}^{t} = e^{-At}\boldsymbol{q}(t) - e^{-At_0}\boldsymbol{q}(t_0) = \int_{t_0}^{t} e^{-A\tau}\boldsymbol{B}\boldsymbol{x}(\tau)d\tau$$

If we transfer the quantity $e^{-At_0}\boldsymbol{q}(t_0)$ to the right side of the equation and multiply both sides by the quantity e^{At}, it will take the form

$$\boldsymbol{q}(t) = e^{A(t-t_0)}\boldsymbol{q}(t_0) + \int_{t_0}^{t} e^{A(t-\tau)}\boldsymbol{B}\boldsymbol{x}(\tau)d\tau = e^{A(t-t_0)}\boldsymbol{q}(t_0) + e^{At} * \boldsymbol{B}\boldsymbol{x}(t) \tag{2.85}$$

—where the symbol $*$ refers to the *convolution* of matrices[26]—which is also the solution of the system's state equation. Of the two terms appearing on the right side of the last relation, the first is known as the *zero-input component*—since it is independent of the input $\boldsymbol{x}(t)$—while the second term is known as the zero state component, since it is independent of the system's initial condition vector $\boldsymbol{q}(t_0)$. In the simple case where we assume that $t_0 = 0$, the last relation takes its simplest form:

$$\boldsymbol{q}(t) = e^{At}\boldsymbol{q}(0) + \int_{0}^{t} e^{A(t-\tau)}\boldsymbol{B}\boldsymbol{x}(\tau)d\tau = e^{At}\boldsymbol{q}(0) + e^{At} * \boldsymbol{B}\boldsymbol{x}(t) \tag{2.86}$$

The preceding equations hold for the general case of MIMO systems, while for the case of SISO systems, their form remains the same, of course, replacing the vector function $\boldsymbol{x}(t)$ with the scalar function $x(t)$.

In the special case where the system input $\boldsymbol{x}(t)$ is equal to zero, the preceding relationship will take the form

[26] The *convolution of matrices* is defined in a similar way to their product, with the difference that *the operation of multiplication between the individual elements has been replaced by the operation of convolution*. In other words, the convolution of two matrices $\boldsymbol{A}$ and $\boldsymbol{B}$ of dimensions $M \times N$ and $N \times L$ respectively is a matrix $\boldsymbol{C}$ of dimensions $M \times L$ whose elements are given by the relation

$$c_{ij} = \sum_{k=1}^{N} \alpha_{ik} * \beta_{kj}, \quad (1 \le i \le M, 1 \le j \le L)$$

Taking into account that in the integral of the convolution of the matrices the integration interval is $[0, t]$, it is silently implied that all elements of matrix e^{At} have been multiplied by the step function $u(t - t_0)$.

$$q(t) = e^{A(t-t_0)} q(t_0) = \Phi(t, t_0) q(t_0) \tag{2.87}$$

with the matrix $\Phi(t, t_0) = e^{A(t-t_0)}$ being known as the *state transition matrix*, as it determines the way in which the state vector $q(t)$ changes from time t_0 to time t. Replacing the quantity $\Phi(t, t_0)$ in the solution of the state equation, we write

$$q(t) = \Phi(t, t_0) q(t_0) + \int_{t_0}^{t} \Phi(t, \tau) B x(\tau) d\tau \tag{2.88}$$

and using the foregoing relationship in the output equation of the system, we will get

$$y(t) = C q(t) + D x(t) = C \Phi(t, t_0) q(t_0) + C \int_{t_0}^{t} \Phi(t, \tau) B x(\tau) d\tau + D x(t) \tag{2.89}$$

an equation that allows us to calculate the output $y(t)$ of the system for the given input $x(t)$ that we transmit to it.

2.9.5 Calculation of Exponential Function e^{At}

The calculation of the *exponential function* e^{At}, which is required for finding the solution of the state equation $q(t)$, is generally difficult to perform directly from the defining equation of the exponential matrix function that we presented earlier, and in the most common cases it relies on the use of the *eigenvalues* of matrix A. Recall from basic theory that for every square matrix A of order N a vector x of N elements is called an *eigenvector* of matrix A, with the corresponding *eigenvalue* λ, if these two quantities satisfy the equation

$$A x = \lambda x$$

which can also be written as

$$(A - \lambda I) x = \begin{bmatrix} \alpha_{11} - \lambda & \alpha_{12} & \alpha_{13} & \dots & \alpha_{1N} \\ \alpha_{21} & \alpha_{22}-\lambda & \alpha_{23} & \dots & \alpha_{2N} \\ \alpha_{31} & \alpha_{32} & \alpha_{33} - \lambda & \dots & \alpha_{3N} \\ \vdots & \vdots & \vdots & \ddots & \vdots \\ \alpha_{N1} & \alpha_{N2} & \alpha_{N3} & \dots & \alpha_{NN} - \lambda \end{bmatrix} \begin{bmatrix} x_1 \\ x_2 \\ x_3 \\ \vdots \\ x_N \end{bmatrix} = 0$$

This algebraic system of equations has a solution if and only if the following relation holds:

$$\det(A - \lambda I) = \begin{vmatrix} \alpha_{11} - \lambda & \alpha_{12} & \alpha_{13} & \dots & \alpha_{1N} \\ \alpha_{21} & \alpha_{22}-\lambda & \alpha_{23} & \dots & \alpha_{2N} \\ \alpha_{31} & \alpha_{32} & \alpha_{33} - \lambda & \dots & \alpha_{3N} \\ \vdots & \vdots & \vdots & \ddots & \vdots \\ \alpha_{N1} & \alpha_{N2} & \alpha_{N3} & \dots & \alpha_{NN} - \lambda \end{vmatrix} = 0$$

This polynomial equation can take the form

$$Q(\lambda) = \lambda^N + \alpha_{N-1}\lambda^{N-1} + \cdots + \alpha_1\lambda + \alpha_0 = 0$$

and its solution leads to the calculation of the N eigenvalues of matrix $\boldsymbol{A}$. Note that, according to the *Cayley–Hamilton theorem*, the foregoing characteristic equation is satisfied *also by matrix* $\boldsymbol{A}$ *itself*, so that we can write

$$\boldsymbol{Q}(\boldsymbol{A}) = \boldsymbol{A}^N + \alpha_{N-1}\boldsymbol{A}^{N-1} + \cdots + \alpha_1\boldsymbol{A} + \alpha_0\boldsymbol{I} = 0 \tag{2.90}$$

Let us now consider any eigenvalue λ_i $(i = 1, 2, \ldots, N)$, and let us assume any *polynomial function* of it

$$f(\lambda_i) = \beta_0 + \beta_1\lambda_i + \beta_2\lambda_i^2 + \cdots = \sum_{j=1}^{\infty} \beta_j\lambda_i^j$$

Taking into account that the eigenvalue λ_i is a *root of the characteristic equation* of the matrix $\boldsymbol{A}$, it is obvious that it will satisfy the relation

$$\lambda_i^N = -\alpha_{N-1}\lambda_i^{N-1} - \alpha_{N-2}\lambda_i^{N-2} - \cdots - \alpha_1\lambda_i - \alpha_0$$

If we now start to continuously multiply the preceding relation by the eigenvalue λ_i, then in each iteration cycle the first expression of the right-hand side will always contain the quantity λ_i^N. In this case, it is not difficult to realize that if we continuously replace this expression from the preceding relation (thereby creating a recursive process), then at any given time *the greatest power of the eigenvalue* λ_i *that will appear in it will always be the* $N-1$. In this way, then, we can express the entire power series of infinite terms

$$\sum_{i=0}^{\infty} \beta_i\lambda^i = \beta_0 + \beta_1\lambda_i + \beta_2\lambda_i^2 + \ldots$$

depending only on the quantities $1, \lambda_i, \lambda_i^2, \ldots, \lambda_i^{N-1}$, so we can write that

$$f(\lambda_i) = \beta_0 + \beta_1\lambda_i + \beta_2\lambda_i^2 + \cdots + \beta_{N-1}\lambda_i^{N-1} = \sum_{j=0}^{N-1} \beta_i\lambda_i^j \tag{2.91}$$

If we record the foregoing relationship for each of the N eigenvalues of the matrix $\boldsymbol{A}$, we will end up with the system of equations

$$\begin{bmatrix} f(\lambda_1) \\ f(\lambda_2) \\ f(\lambda_3) \\ \vdots \\ f(\lambda_N) \end{bmatrix} = \begin{bmatrix} 1 & \lambda_1 & \lambda_1^2 & \ldots & \lambda_1^{N-1} \\ 1 & \lambda_2 & \lambda_2^2 & \ldots & \lambda_2^{N-1} \\ 1 & \lambda_3 & \lambda_3^2 & \ldots & \lambda_3^{N-1} \\ \vdots & \vdots & \vdots & \ddots & \vdots \\ 1 & \lambda_N & \lambda_N^2 & \ldots & \lambda_N^{N-1} \end{bmatrix} \begin{bmatrix} \beta_0 \\ \beta_1 \\ \beta_2 \\ \vdots \\ \beta_{N-1} \end{bmatrix} \tag{2.92}$$

Based on the Cayley–Hamilton theorem, we can state the claim that the foregoing property can be extended for any function $f(\boldsymbol{A})$ of a square matrix $\boldsymbol{A}$ of order N, which can be formulated as

$$f(\boldsymbol{A}) = \sum_{i=0}^{N-1} \beta_i\boldsymbol{A}^i = \beta_0\boldsymbol{I} + \beta_1\boldsymbol{A} + \cdots + \beta_{N-1}\boldsymbol{A}^{N-1}$$

If we take into account that the exponential function $e^{\boldsymbol{A}t}$ is described by a similar power series of infinite terms, the foregoing property allows us to calculate the function in question[27] from a relation of the form

[27] The solution of the state equation and the calculation of the exponential function $e^{\boldsymbol{A}t}$ can also be calculated using the *Laplace transform*, as we will see in the homonymous chapter.

$$e^{At} = \sum_{k=0}^{\infty} \frac{(At)^k}{k!} = \sum_{i=0}^{N-1} \beta_i A^i$$

with the coefficients β_i $(i = 0, 1, 2, \ldots, N-1)$ being calculated from the equations

$$\begin{bmatrix} \beta_0 \\ \beta_1 \\ \beta_2 \\ \vdots \\ \beta_{N-1} \end{bmatrix} = \begin{bmatrix} 1 & \lambda_1 & \lambda_1^2 & \ldots & \lambda_1^{N-1} \\ 1 & \lambda_2 & \lambda_2^2 & \ldots & \lambda_2^{N-1} \\ 1 & \lambda_3 & \lambda_3^2 & \ldots & \lambda_3^{N-1} \\ \vdots & \vdots & \vdots & \ddots & \vdots \\ 1 & \lambda_N & \lambda_N^2 & \ldots & \lambda_N^{N-1} \end{bmatrix}^{-1} \begin{bmatrix} f(\lambda_1) \\ f(\lambda_2) \\ f(\lambda_3) \\ \vdots \\ f(\lambda_N) \end{bmatrix} \tag{2.93}$$

for the functions $f(\lambda_i) = e^{\lambda_i t}$, and this is what usually happens in practice. The preceding relation holds when *the eigenvalues of matrix* A *are all distinct from each other*, that is, they have multiplicity equal to one.

On the other hand, if the eigenvalues λ_i are not distinct from each other, the process followed is this: Suppose that some eigenvalue λ_k has multiplicity equal to m, while all the other eigenvalues are distinct from each other. In this case, we determine the defining equation of the polynomial $f(\lambda_k)$ for this particular eigenvalue and we differentiate m times with respect to λ. If we then make the substitution $\lambda = \lambda_k$, we end up with a group of m equations involving the eigenvalue λ_k of the form

$$\left.\frac{d^{n-1}}{d\lambda^{n-1}} f(\lambda)\right|_{\lambda=\lambda_k} = \left.\frac{d^{n-1}}{d\lambda^{n-1}} \left(\sum_{j=1}^{N-1} \beta_j \lambda^j \right)\right|_{\lambda=\lambda_k} \quad (n = 1, 2, \ldots, m) \tag{2.94}$$

which, in combination with the previous equations, allow us to determine the coefficients β_i.

The second way of calculating the function e^{At} is based again on the use of the *eigenvalues* of matrix A, which are now used for its *diagonalization*. Considering a square matrix A of order N with *distinct* eigenvalues $\lambda_1, \lambda_2, \ldots, \lambda_N$ and corresponding eigenvectors $x_1, x_2, \ldots, x_N$, let us define the matrix

$$P = \begin{bmatrix} x_1 & x_2 & \cdots & x_N \end{bmatrix} = \begin{bmatrix} x_{11} & x_{21} & x_{31} & \ldots & x_{N1} \\ x_{12} & x_{22} & x_{32} & \ldots & x_{N2} \\ x_{13} & x_{23} & x_{33} & \ldots & x_{N3} \\ \vdots & \vdots & \vdots & \ddots & \vdots \\ x_{1N} & x_{2N} & x_{3N} & \ldots & x_{NN} \end{bmatrix} \tag{2.95}$$

In this case, the product of matrices AP will have the form

$$AP = A[x_1 \, x_2 \, \ldots \, x_N] = [Ax_1 \, Ax_2 \, \ldots \, Ax_N]$$

However, since the vector x_i is an eigenvector of matrix A with the corresponding eigenvalue λ_i, we will have $Ax_i = \lambda_i x_i$ $(i = 1, 2, \ldots, N)$ and the preceding relationship will take the form

$$AP = [\lambda_1 x_1 \, \lambda_2 x_2 \, \ldots \, \lambda_N x_N] = [x_1 \, x_2 \, \ldots \, x_N] \begin{bmatrix} \lambda_1 & 0 & 0 & \ldots & 0 \\ 0 & \lambda_2 & 0 & \ldots & 0 \\ \vdots & \vdots & \vdots & \ddots & \vdots \\ 0 & \lambda_2 & 0 & \ldots & 0 \\ 0 & 0 & 0 & \ldots & \lambda_N \end{bmatrix} = P\Lambda$$

In the foregoing equation, the matrix Λ, which is defined as

$$\boldsymbol{\Lambda} = \begin{bmatrix} \lambda_1 & 0 & 0 \dots & 0 \\ 0 & \lambda_2 & 0 \dots & 0 \\ \vdots & \vdots & \vdots \ddots & \vdots \\ 0 & 0 & 0 \dots & \lambda_N \end{bmatrix} \tag{2.96}$$

is called the *eigenvalue matrix* and is a *diagonal matrix* whose elements constitute the *eigenvalues* of the matrix $\boldsymbol{A}$. Therefore, $\boldsymbol{AP} = \boldsymbol{P\Lambda}$, from which it immediately follows that $\boldsymbol{\Lambda} = \boldsymbol{P}^{-1}\boldsymbol{AP}$, while matrix $\boldsymbol{A}$ can be written in the form

$$\boldsymbol{A} = \boldsymbol{P\Lambda P}^{-1} \tag{2.97}$$

In the foregoing equations, matrix $\boldsymbol{P}$ is known as the *diagonalization matrix*, since its columns constitute the *eigenvectors* of matrix $\boldsymbol{A}$, which are known to be *linearly independent* of each other, it is obvious that its determinant is different from zero, so its inverse exists.

Now, to calculate the exponential function $e^{\boldsymbol{A}t}$, we must first observe that the use of the equation $\boldsymbol{A} = \boldsymbol{P\Lambda P}^{-1}$ allows us to write the expression $\boldsymbol{A}^k$ in the form

$$\boldsymbol{A}^k = (\boldsymbol{P\Lambda P}^{-1})^k = \underbrace{(\boldsymbol{P\Lambda P}^{-1})\dots(\boldsymbol{P\Lambda P}^{-1})}_{k \text{ times}} = \boldsymbol{P\Lambda}(\boldsymbol{P}^{-1}\boldsymbol{P})\boldsymbol{\Lambda}(\boldsymbol{P}^{-1}\boldsymbol{P}) dots \boldsymbol{\Lambda P}^{-1} =$$
$$= \boldsymbol{P}(\boldsymbol{\Lambda I})(\boldsymbol{\Lambda I})\dots(\boldsymbol{\Lambda I})\boldsymbol{\Lambda P}^{-1} = \boldsymbol{P}\,\underbrace{\boldsymbol{\Lambda\Lambda\Lambda}\dots\boldsymbol{\Lambda}}_{k \text{ times}}\,\boldsymbol{P}^{-1} = \boldsymbol{P\Lambda}^k\boldsymbol{P}^{-1}$$

and the defining equation of the exponential function will take the form

$$e^{\boldsymbol{A}t} = \sum_{k=0}^{\infty} \frac{t^k \boldsymbol{A}^k}{k!} = \sum_{k=0}^{\infty} \frac{t^k}{k!}(\boldsymbol{P\Lambda}^k\boldsymbol{P}^{-1}) = \boldsymbol{P}\left(\sum_{k=0}^{\infty} \frac{t^k \boldsymbol{\Lambda}^k}{k!}\right)\boldsymbol{P}^{-1}$$

However, since matrix $\boldsymbol{\Lambda}$ is diagonal, it is obvious that

$$\boldsymbol{\Lambda}^k = \underbrace{\begin{bmatrix} \lambda_1 & 0 & 0 \dots & 0 \\ 0 & \lambda_2 & 0 \dots & 0 \\ \vdots & \vdots & \vdots \ddots & \vdots \\ 0 & 0 & 0 \dots & \lambda_N \end{bmatrix} \cdots \begin{bmatrix} \lambda_1 & 0 & 0 \dots & 0 \\ 0 & \lambda_2 & 0 \dots & 0 \\ \vdots & \vdots & \vdots \ddots & \vdots \\ 0 & 0 & 0 \dots & \lambda_N \end{bmatrix}}_{k \text{ times}} = \begin{bmatrix} \lambda_1^k & 0 & 0 \dots & 0 \\ 0 & \lambda_2^k & 0 \dots & 0 \\ \vdots & \vdots & \vdots \ddots & \vdots \\ 0 & 0 & 0 \dots & \lambda_N^k \end{bmatrix}$$

Therefore, we will have

$$\sum_{k=0}^{\infty} \frac{t^k \boldsymbol{\Lambda}^k}{k!} = \begin{bmatrix} \sum_{k=0}^{\infty} \frac{t^k \lambda_1^k}{k!} & 0 & 0 \dots & 0 \\ 0 & \sum_{k=0}^{\infty} \frac{t^k \lambda_2^k}{k!} & 0 \dots & 0 \\ \vdots & \vdots & \vdots \ddots & \vdots \\ 0 & 0 & 0 \dots \sum_{k=0}^{\infty} \frac{t^k \lambda_N^k}{k!} \end{bmatrix} \quad \text{or} \quad \sum_{k=0}^{\infty} \frac{t^k \boldsymbol{\Lambda}^k}{k!} = \begin{bmatrix} e^{\lambda_1 t} & 0 & 0 \dots & 0 \\ 0 & e^{\lambda_2 t} & 0 \dots & 0 \\ \vdots & \vdots & \vdots \ddots & \vdots \\ 0 & 0 & 0 \dots & e^{\lambda_N t} \end{bmatrix}$$

since from the analysis it is known that

$$\sum_{k=0}^{\infty} \frac{t^k \lambda_i^k}{k!} = e^{\lambda_i t}, \quad (i = 1, 2, \ldots, N)$$

Replacing the foregoing relation in the defining equation of the function e^{At}, it will take the form

$$e^{At} = P\left(\sum_{k=0}^{\infty} \frac{t^k \Lambda^k}{k!}\right) P^{-1} = P \begin{bmatrix} e^{\lambda_1 t} & 0 & 0 \ldots & 0 \\ 0 & e^{\lambda_2 t} & 0 \ldots & 0 \\ \vdots & \vdots & \vdots \ddots & \vdots \\ 0 & 0 & 0 \ldots & e^{\lambda_N t} \end{bmatrix} P^{-1} \tag{2.98}$$

an equation that constitutes an alternative way of calculating the foregoing function.

The exponential functions $f(\lambda_i) = e^{\lambda_i t}$ $(i = 1, 2, 3, \ldots, N)$ that appear in both methods of calculating the exponential function e^{At} are known as system *eigenfunctions*, with the eigenvalues $\lambda_1, \lambda_2, \ldots, \lambda_N$ constituting the corresponding *eigenfrequencies*. These eigenfrequencies constitute an inherent characteristic of the system, and their role is particularly important for its characterization, since for some arbitrary initial condition $q(0) \neq x_i$ $(i = 1, 2, 3, \ldots, N)$ there occurs an excitation of all these frequencies, and therefore the solution $q(t)$ of the homogeneous state equation $\dot{q}(t) = Aq(t)$ is the result of their *superposition*. To prove the truth of this claim, let us consider again matrices P and P^{-1}, which diagonalize matrix A (a process that presupposes that the matrix in question has *distinct* eigenvalues), and let us symbolize the rows of the matrix P^{-1} as yi^T $(i = 1, 2, 3, \ldots, N)$, so we can write that

$$P = \begin{bmatrix} x_1 \; x_2 \ldots x_N \end{bmatrix} \quad \text{and} \quad P^{-1} = \begin{bmatrix} y_1^T \\ y_2^T \\ \vdots \\ y_N^T \end{bmatrix}$$

Taking into account that the product of matrices P and P^{-1} is the unit matrix I of order N, we have

$$y_i^T x_j = \begin{cases} 1 \text{ for } i = j \\ 0 \text{ for } i \neq j \end{cases} \tag{2.99}$$

and therefore the solution of the homogeneous state equation will take the form

$$q(t) = e^{At} q(0) = P e^{\Lambda t} P^{-1} q(0) = \begin{bmatrix} x_1 \; x_2 \ldots x_N \end{bmatrix} e^{\Lambda t} \begin{bmatrix} y_1^T \\ y_2^T \\ \vdots \\ y_N^T \end{bmatrix} q(0)$$

or, equivalently,

$$q(t) = \sum_{k=1}^{N} x_k e^{\lambda_k t} [y_k^T q(0)] \tag{2.100}$$

From the last relation it follows that if the initial value of the state vector $q(0)$ coincides with one of the eigenvectors x_i $(i = 1, 2, 3, \ldots, N)$, then out of the N terms of the foregoing sum, the only one that will survive will be the one corresponding to the value $k = i$. Because the relation $y_i^T p_i = 1$ holds, we will have $q(t) = e^{\lambda_i t} x_i$, and therefore the state vector $q(t)$ will have the same direction as the eigenvector x_i in the N-dimensional space describing the system, while its magnitude will vary according to the exponential function $e^{\lambda_i t}$. In the opposite case, as we have already mentioned, all N terms of the foregoing sum will survive, leading to the excitation of all the eigenfrequencies of the system. As for the coefficients $\boldsymbol{y}_k^T \boldsymbol{q}(0)$ $(k = 1, 2, 3, \ldots, N)$, these actually describe the initial state vector $\boldsymbol{q}(0)$ as a linear combination of the system's eigenvectors, since, if we assume that it has the form

$$\boldsymbol{q}(0) = \sum_{i=1}^{N} \alpha_i \boldsymbol{x}_i = \alpha_1 \boldsymbol{x}_1 + \cdots + \alpha_N \boldsymbol{x}_N = [\boldsymbol{x}_1\ \boldsymbol{x}_2\ \ldots\ \boldsymbol{x}_N] \begin{bmatrix} \alpha_1 \\ \alpha_2 \\ \vdots \\ \alpha_N \end{bmatrix} = \boldsymbol{P} \begin{bmatrix} \alpha_1 \\ \alpha_2 \\ \vdots \\ \alpha_N \end{bmatrix}$$

it will obviously be the case that

$$\begin{bmatrix} \alpha_1 \\ \alpha_2 \\ \vdots \\ \alpha_N \end{bmatrix} = \boldsymbol{P}^{-1} \boldsymbol{q}(0) = \begin{bmatrix} \boldsymbol{y}_1^T \boldsymbol{q}(0) \\ \boldsymbol{y}_2^T \boldsymbol{q}(0) \\ \vdots \\ \boldsymbol{y}_N^T \boldsymbol{q}(0) \end{bmatrix}$$

from which it follows that

$$\boldsymbol{q}(0) = \sum_{k=1}^{N} [\boldsymbol{y}_k^T \boldsymbol{q}(0)] \boldsymbol{x}_k = [\boldsymbol{y}_1^T \boldsymbol{q}(0)] \boldsymbol{x}_1 + [\boldsymbol{y}_2^T \boldsymbol{q}(0)] \boldsymbol{x}_2 + \cdots + [\boldsymbol{y}_N^T \boldsymbol{q}(0)] \boldsymbol{x}_N \tag{2.101}$$

The third and final method of calculating matrix $e^{\boldsymbol{A}t}$ that we will mention here again refers to the case of discrete eigenvalues of matrix $\boldsymbol{A}$ and consists in finding N matrices $\boldsymbol{E}_i$ $(i = 1, 2, 3 \ldots, N)$, such that matrix $\boldsymbol{A}$ can be written in the form

$$\boldsymbol{A} = \sum_{i=1}^{N} \lambda_i \boldsymbol{E}_i = \lambda_1 \boldsymbol{E}_1 + \lambda_2 \boldsymbol{E}_2 + \cdots + \lambda_N \boldsymbol{E}_N \tag{2.102}$$

where λ_i $(i = 1, 2, \ldots, N)$ are the eigenvalues of matrix $\boldsymbol{A}$.[28] In this case, it is proven that the exponential function $e^{\boldsymbol{A}t}$ can be calculated from the relation

$$e^{\boldsymbol{A}t} = \sum_{i=1}^{N} e^{\lambda_i t} \boldsymbol{E}_i = e^{\lambda_1 t} \boldsymbol{E}_1 + e^{\lambda_2 t} \boldsymbol{E}_2 + \cdots + e^{\lambda_N t} \boldsymbol{E}_N \tag{2.103}$$

More methods for calculating the exponential function $e^{\boldsymbol{A}t}$ can be found in the literature.[29]

[28] This process is known as the *spectral analysis* of matrix $\boldsymbol{A}$, with the set of eigenvalues $\{\lambda_i\}$ $(i = 1, 2, \ldots, N)$ known as the *spectrum* of matrix $\boldsymbol{A}$. Matrices $\boldsymbol{E}_i$ are proven capable of being calculated from the relation

$$\boldsymbol{E}_i = \left\{ \prod_{\substack{k=1 \\ k \neq i}}^{M} (\boldsymbol{A} - \lambda_i \boldsymbol{I}) \right\} \bigg/ \left\{ \prod_{\substack{k=1 \\ k \neq i}}^{M} (\lambda_i - \lambda_k) \right\}, \quad (i = 1, 2, 3, \ldots, N)$$

and are characterized by a number of useful properties, the most important of which are

$$\sum_{i=1}^{N} \boldsymbol{E}_i = \boldsymbol{I}, \quad \boldsymbol{E}_j \boldsymbol{E}_j = 0\ (i \neq j), \quad \boldsymbol{E}_i^2 = \boldsymbol{E}_i, \quad \boldsymbol{A}\boldsymbol{E}_i = \boldsymbol{E}_i \boldsymbol{A} = \lambda_i \boldsymbol{E}_i.$$

[29] See, for example, the work by Moler C. and C. Van Loan, "Nineteen dubious ways to compute the exponential of a matrix," *SIAM Review*, 20(4), pp. 801–836, 1978.

2.9.6 The Impulse Response in State Space

Using the defining equation of the state vector in the form of matrix convolution, we can construct an equation that will describe the *impulse response* of a system in state space. Assuming for simplicity that $t_0 = 0$, the state vector $\boldsymbol{q}(t)$ will take the form

$$\boldsymbol{q}(t) = e^{\boldsymbol{A}t}\boldsymbol{q}(0) + e^{\boldsymbol{A}t} * \boldsymbol{B}\boldsymbol{x}(t) = e^{\boldsymbol{A}t}\boldsymbol{q}(0) + e^{\boldsymbol{A}t}\boldsymbol{B} * \boldsymbol{x}(t)$$

with the last equation resulting from the fact that the elements of matrix $\boldsymbol{B}$ are constant. Replacing the foregoing relation in the output equation of the system, it is written as

$$\boldsymbol{y}(t) = \boldsymbol{C}\Big[\boldsymbol{\Phi}(t)\boldsymbol{q}(0) + \boldsymbol{\Phi}(t)\boldsymbol{B} * \boldsymbol{x}(t)\Big] + \boldsymbol{D}\boldsymbol{x}(t) \tag{2.104}$$

where for simplicity we used the notation $\boldsymbol{\Phi}(t, 0) \equiv \boldsymbol{\Phi}(t) = e^{\boldsymbol{A}t}$.

Let us now consider a diagonal matrix $\boldsymbol{\delta}(t)$ of order K in which all the diagonal elements contain the impulse function $\delta(t)$. Taking into account the definition of the convolution operation between matrices and using the known relation $x(t) * \delta(t) = x(t)$, we find that

$$\boldsymbol{\delta}(t) * \boldsymbol{x}(t) = \begin{bmatrix} \delta(t) & 0 & 0 & \dots & 0 \\ 0 & \delta(t) & 0 & \dots & 0 \\ 0 & 0 & \delta(t) & \dots & 0 \\ \vdots & \vdots & \vdots & \ddots & \vdots \\ 0 & 0 & 0 & \dots & \delta(t) \end{bmatrix} \begin{bmatrix} x_1(t) \\ x_2(t) \\ x_3(t) \\ \vdots \\ x_K(t) \end{bmatrix} = \begin{bmatrix} \delta(t) * x_1(t) \\ \delta(t) * x_2(t) \\ \delta(t) * x_3(t) \\ \vdots \\ \delta(t) * x_K(t) \end{bmatrix} = \begin{bmatrix} x_1(t) \\ x_2(t) \\ x_3(t) \\ \vdots \\ x_K(t) \end{bmatrix} = \boldsymbol{x}(t)$$

and the preceding relation will take the form

$$\boldsymbol{y}(t) = \boldsymbol{C}\Big[\boldsymbol{\Phi}(t)\boldsymbol{q}(0) + \boldsymbol{\Phi}(t)\boldsymbol{B} * \boldsymbol{x}(t)\Big] + \boldsymbol{D}\boldsymbol{\delta}(t) * \boldsymbol{x}(t) = \boldsymbol{C}\boldsymbol{\Phi}(t)\boldsymbol{q}(0) + \Big[\boldsymbol{C}\boldsymbol{\Phi}(t)\boldsymbol{B} + \boldsymbol{D}\boldsymbol{\delta}(t)\Big] * \boldsymbol{x}(t) \tag{2.105}$$

To determine the zero-state response of the system, we must set $\boldsymbol{q}(0) = 0$ in the foregoing relation. We will therefore have

$$\boldsymbol{y}(t) = \Big[\boldsymbol{C}\boldsymbol{\Phi}(t)\boldsymbol{B} + \boldsymbol{D}\boldsymbol{\delta}(t)\Big] * \boldsymbol{x}(t) = \boldsymbol{h}(t) * \boldsymbol{x}(t)$$

We have arrived, therefore, once again at our familiar relation expressed in its vector form (since it refers to a MIMO system) with the quantity

$$\boldsymbol{h}(t) = \boldsymbol{C}\boldsymbol{\Phi}(t)\boldsymbol{B} + \boldsymbol{D}\boldsymbol{\delta}(t) \tag{2.106}$$

known as the *impulse response matrix*. Taking into account that the considered system consists of K inputs and L outputs, it is obvious that the matrix $\boldsymbol{h}(t)$ will have dimensions $L \times K$, while the corresponding equation for SISO systems results in the dimensions of the matrices being equal to unity.

2.9.7 The State Transition Matrix

As already mentioned, the solution of the state equation has the form

$$\boldsymbol{q}(t) = \boldsymbol{\Phi}(t, t_0)\boldsymbol{q}(t_0) + \int_{t_0}^{t} \boldsymbol{\Phi}(t, \tau)\boldsymbol{B}\boldsymbol{x}(\tau)d\tau \tag{2.107}$$

with the matrix $\boldsymbol{\Phi}(t, t_0) = e^{\boldsymbol{A}(t-t_0)}$ being known as the state transition matrix, in the sense that it allows the determination of the temporal evolution of the state vector if its initial value $\boldsymbol{q}(0)$ is known. In reality, the foregoing definition applies only to time-invariant systems, while for time-varying systems this matrix is defined as $\boldsymbol{\Phi}(t, \tau) = \boldsymbol{U}(t)\boldsymbol{U}^{-1}(\tau)$, where the matrix $\boldsymbol{U}(t)$ is a solution of the equation $\dot{\boldsymbol{U}}(t) = \boldsymbol{A}(t)\boldsymbol{U}(t)$.

The state transition matrix is characterized by a multitude of useful properties, the most important of which are the following:

- $\boldsymbol{\Phi}^{-1}(t, \tau) = \boldsymbol{\Phi}(\tau, t)$
- $\boldsymbol{\Phi}^{-1}(t, \tau)\boldsymbol{\Phi}(t, \tau) = \boldsymbol{I}$
- $\dot{\boldsymbol{\Phi}}(t, \tau) = \boldsymbol{A}(t)\boldsymbol{\Phi}(t, \tau)$
- $\boldsymbol{\Phi}(\tau, \tau) = \boldsymbol{I}$
- $\boldsymbol{\Phi}(t_2, t_1)\boldsymbol{\Phi}(t_1, t_0) = \boldsymbol{\Phi}(t_2, t_0)$

In the simple case of the exponential function $\boldsymbol{\Phi}(t) = e^{\boldsymbol{A}(t-t_0)}$, the proof of the first properties is done in the manner presented in the preceding section, while the proof of the last property is also trivial, since it is found that

$$\boldsymbol{\Phi}(t_2 - t_0) = e^{\boldsymbol{A}(t_2-t_0)} = e^{\boldsymbol{A}(t_2-t_1+t_1-t_0)} = e^{\boldsymbol{A}(t_2-t_1)}e^{\boldsymbol{A}(t_1-t_0)} = \boldsymbol{\Phi}(t_2 - t_1)\boldsymbol{\Phi}(t_1 - t_0)$$

This property is very important because it allows us to study the temporal evolution of the system *by dividing it into successive time periods and studying them independently of each other.*

Problem 2.9.2 Check whether the following table is a valid state transition table:

$$\boldsymbol{\Phi}(t, \tau) = \begin{bmatrix} e^{\tau-t} & e^{\tau^2-t^2} - 1 \\ 0 & e^{\tau-t} \end{bmatrix}$$

Solution Since matrix $\boldsymbol{\Phi}$ is a valid state transition matrix, it must satisfy the property

$$\boldsymbol{\Phi}(t_2, t_1)\boldsymbol{\Phi}(t_1, t_0) = \boldsymbol{\Phi}(t_2, t_0)$$

However, we find that

$$\boldsymbol{\Phi}(t_2, t_1)\boldsymbol{\Phi}(t_1, t_0) = \begin{bmatrix} e^{t_2-t_1} & e^{t_2^2-t_1^2} - 1 \\ 0 & e^{t_2-t_1} \end{bmatrix} \begin{bmatrix} e^{t_1-t_0} & e^{t_1^2-t_0^2} - 1 \\ 0 & e^{t_1-t_0} \end{bmatrix} =$$

$$= \begin{bmatrix} e^{t_2-t_0} & e^{t_1^2-t_0^2+t_2-t_1} + e^{t_2^2-t_1^2+t_1-t_0} - e^{t_2-t_1} - e^{t_1-t_0} \\ 0 & e^{t_2-t_0} \end{bmatrix} \neq \boldsymbol{\Phi}(t_2 - t_0)$$

and consequently, this matrix does not constitute a valid state transition matrix.

2.9.8 Similarity Transformations

An interesting property of systems in state space is that their description can be carried out using more than one set of *state variables* that are related to each other through *linear correlations*. Considering, for example, two state vectors $\boldsymbol{q}(t) = [q_1(t), q_2(t), \ldots, q_N(t)]^T$ and $\boldsymbol{p}(t) = [p_1(t), p_2(t), \ldots, p_N(t)]^T$, which describe the same system in state space, there is a matrix $\boldsymbol{W}$ such that

$$\begin{bmatrix} p_1(t) \\ p_2(t) \\ \vdots \\ p_N(t) \end{bmatrix} = \begin{bmatrix} w_{11} & w_{12} & \dots & w_{1N} \\ w_{21} & w_{22} & \dots & w_{2N} \\ \vdots & \vdots & \ddots & \vdots \\ w_{N1} & w_{N2} & \dots & w_{NN} \end{bmatrix} \begin{bmatrix} q_1(t) \\ q_2(t) \\ \vdots \\ q_N(t) \end{bmatrix} \quad (2.108)$$

or, in summary notation, $\boldsymbol{P}(t) = \boldsymbol{W}\boldsymbol{q}(t)$. If the N linear equations resulting from the foregoing matrix equation are all independent of each other, then matrix $\boldsymbol{W}$ has *nonzero determinant* and the *inverse matrix* $\boldsymbol{W}^{-1}$ exists, so we can write that $\boldsymbol{q}(t) = \boldsymbol{W}^{-1}\boldsymbol{p}(t)$. Differentiating the last relation, we will get $\dot{\boldsymbol{q}}(t) = \boldsymbol{W}^{-1}\dot{\boldsymbol{p}}(t)$, and when we substitute this into the system's state equation, it will take the form

$$\boldsymbol{W}^{-1}\dot{\boldsymbol{p}}(t) = \boldsymbol{A}\boldsymbol{W}^{-1}\boldsymbol{p}(t) + \boldsymbol{B}\boldsymbol{x}(t) \quad \text{or, equivalently,} \quad \boldsymbol{W}\boldsymbol{W}^{-1}\dot{\boldsymbol{p}}(t) = \boldsymbol{W}\boldsymbol{A}\boldsymbol{W}^{-1}\boldsymbol{p}(t) + \boldsymbol{W}\boldsymbol{B}\boldsymbol{x}(t)$$

and finally

$$\dot{\boldsymbol{p}}(t) = \tilde{\boldsymbol{A}}\boldsymbol{p}(t) + \tilde{\boldsymbol{B}}\boldsymbol{x}(t) \quad (2.109)$$

—where $\widetilde{\boldsymbol{A}} = \boldsymbol{W}\boldsymbol{A}\boldsymbol{W}^{-1}$ and $\widetilde{\boldsymbol{B}} = \boldsymbol{W}\boldsymbol{B}$—which constitutes the state equation of the system expressed in terms of the new state vector $\boldsymbol{p}(t)$.

In a completely analogous way, the system's output equation is written

$$\boldsymbol{y}(t) = \boldsymbol{C}\boldsymbol{W}^{-1}\boldsymbol{p}(t) + \boldsymbol{D}\boldsymbol{x}(t) = \widetilde{\boldsymbol{C}}\boldsymbol{p}(t) + \boldsymbol{D}\boldsymbol{x}(t) \quad (2.110)$$

where $\widetilde{\boldsymbol{C}} = \boldsymbol{C}\boldsymbol{W}^{-1}$. We observe, therefore, that the dynamic equations of the system *maintain the same form as the original*, with the new matrices appearing in them given by the preceding relations. The transformations of the state vectors in the foregoing way are known as *similarity transformations* and allow for the transition of the system to the state space from one type of representation to another, without cause even the slightest change in the nature and characteristics of the system.

It is interesting to note at this point that the eigenvalues of matrix $\boldsymbol{A}$ are always the same, regardless of the choice of the state vector that will be used to describe the system in state space, and therefore will coincide with the eigenvalues of the matrix $\widetilde{\boldsymbol{A}} = \boldsymbol{W}\boldsymbol{A}\boldsymbol{W}^{-1}$. Taking into account that *the eigenvalues constitute the roots of the characteristic polynomial of matrix* $\boldsymbol{A}$, it is enough to show that the characteristic equations of matrices $\boldsymbol{A}$ and $\widetilde{\boldsymbol{A}}$ are the same. But this is something that is proven directly, since we find that

$$\boldsymbol{W}(\boldsymbol{A} - \lambda\boldsymbol{I})\boldsymbol{W}^{-1} = \boldsymbol{W}\boldsymbol{A}\boldsymbol{W}^{-1} - \boldsymbol{W}\lambda\boldsymbol{I}\boldsymbol{W}^{-1} = \boldsymbol{W}\boldsymbol{A}\boldsymbol{W}^{-1} - \lambda\boldsymbol{W}\boldsymbol{I}\boldsymbol{W}^{-1} = \widetilde{\boldsymbol{A}} - \lambda\boldsymbol{I}$$

If we take the determinant of the foregoing relation and take into account that the determinants of matrices $\boldsymbol{W}$ and $\boldsymbol{W}^{-1}$ are each other's inverse, we will get

$$|\boldsymbol{W}||\boldsymbol{A} - \lambda\boldsymbol{I}||\boldsymbol{W}^{-1}| = |\boldsymbol{A} - \lambda\boldsymbol{I}| = |\widetilde{\boldsymbol{A}} - \lambda\boldsymbol{I}|$$

Therefore, matrices $\boldsymbol{A}$ and $\widetilde{\boldsymbol{A}}$ will have the same eigenvalues, which will be calculated as the roots of the equation

$$|\boldsymbol{A} - \lambda\boldsymbol{I}| = |\widetilde{\boldsymbol{A}} - \lambda\boldsymbol{I}| = 0$$

that is, these eigenvalues will not depend on the choice of the state vector that will be used to describe the system. In the next chapter, we will show how these similarity transformations, in addition to the eigenvalues of matrix $\boldsymbol{A}$, leave the transfer function of the system unchanged.

Problem 2.9.3 Solve the state equation $\dot{\boldsymbol{q}}(t) = \boldsymbol{A}\boldsymbol{q}(t) + \boldsymbol{B}x(t)$ for parameter values

$$\boldsymbol{A} = \begin{bmatrix} 0 & 2 \\ -1 & -3 \end{bmatrix}, \quad \boldsymbol{B} = \begin{bmatrix} 0 \\ 1 \end{bmatrix}, \quad \boldsymbol{q}(0) = \begin{bmatrix} 2 \\ 1 \end{bmatrix}, \quad x(t) = 0$$

(to solve this exercise with the Laplace transform method, see Exercise 5.11.1 on page 466).

Solution The solution of the state equation has the form

$$\boldsymbol{q}(t) = e^{\boldsymbol{A}t}\boldsymbol{q}(0) + \int_0^t e^{\boldsymbol{A}(t-\tau)}\boldsymbol{B}x(\tau)d\tau = e^{\boldsymbol{A}t}\boldsymbol{q}(0)$$

since in this case $x(t) = 0$. Therefore, the solution of the exercise essentially consists in finding the exponential function $e^{\boldsymbol{A}t}$.

To determine this function, we will use both methods presented in the preceding pages, so as to familiarize ourselves with their use. In the first step of the process, we need to calculate the eigenvalues of matrix $\boldsymbol{A}$ as the roots of the equation

$$\det(\boldsymbol{A} - \lambda\boldsymbol{I}) = \begin{vmatrix} -\lambda & 2 \\ -1 & -3-\lambda \end{vmatrix} = \lambda^2 + 3\lambda + 2 = (\lambda + 1)(\lambda + 2) = 0$$

and therefore, matrix $\boldsymbol{A}$ will have the distinct eigenvalues $\lambda_1 = -1$ and $\lambda_2 = -2$.

Starting with the first method, the coefficients β_i $(i = 0, 1)$ will be determined by the equation

$$\begin{bmatrix} \beta_0 \\ \beta_1 \end{bmatrix} = \begin{bmatrix} 1 & \lambda_1 \\ 1 & \lambda_2 \end{bmatrix}^{-1} \begin{bmatrix} e^{\lambda_1 t} \\ e^{\lambda_2 t} \end{bmatrix} = \begin{bmatrix} 1 & -1 \\ 1 & -2 \end{bmatrix}^{-1} \begin{bmatrix} e^{-t} \\ e^{-2t} \end{bmatrix}$$

The calculation of the preceding inverse matrix of dimensions 2×2 leads us to the result[30]

$$\begin{bmatrix} 1 & -1 \\ 1 & -2 \end{bmatrix}^{-1} = \frac{1}{\begin{vmatrix} 1 & -1 \\ 1 & -2 \end{vmatrix}} \begin{bmatrix} -2 & 1 \\ -1 & 1 \end{bmatrix} = -\begin{bmatrix} -2 & 1 \\ -1 & 1 \end{bmatrix} = \begin{bmatrix} 2 & -1 \\ 1 & -1 \end{bmatrix}$$

It will therefore hold that

$$\begin{bmatrix} \beta_0 \\ \beta_1 \end{bmatrix} = \begin{bmatrix} 2 & -1 \\ 1 & -1 \end{bmatrix} \begin{bmatrix} e^{-t} \\ e^{-2t} \end{bmatrix} = \begin{bmatrix} 2e^{-t} - e^{-2t} \\ e^{-t} - e^{-2t} \end{bmatrix}$$

from which it follows that

$$\beta_0 = 2e^{-t} - e^{-2t} \quad \text{and} \quad \beta_1 = e^{-t} - e^{-2t}$$

Substituting the values of the parameters β_1 and β_2 into the equation of the exponential function, we will get

$$e^{\boldsymbol{A}t} = \sum_{i=0}^{1} \beta_i \boldsymbol{A}^i = \beta_0\boldsymbol{I} + \beta_1\boldsymbol{A} = \left(2e^{-t} - e^{-2t}\right)\begin{bmatrix} 1 & 0 \\ 0 & 1 \end{bmatrix} + \left(e^{-t} - e^{-2t}\right)\begin{bmatrix} 0 & 2 \\ -1 & -3 \end{bmatrix}$$

---

[30] From matrix algebra, it is known that the inverse of a 2×2 matrix of the form

$$\boldsymbol{M} = \begin{bmatrix} \alpha_{11} & \alpha_{12} \\ \alpha_{21} & \alpha_{22} \end{bmatrix} \quad \text{is the matrix} \quad \boldsymbol{M}^{-1} = \frac{1}{\det(\boldsymbol{M})} \begin{bmatrix} \alpha_{22} & -\alpha_{12} \\ -\alpha_{21} & \alpha_{11} \end{bmatrix}$$

where $\det(\boldsymbol{M})$ is the determinant of matrix $\boldsymbol{M}$, which naturally must be different from zero.

$$= \begin{bmatrix} 2e^{-t} - e^{-2t} & 0 \\ 0 & 2e^{-t} - e^{-2t} \end{bmatrix} + \begin{bmatrix} 0 & 2e^{-t} - 2e^{-2t} \\ -e^{-t} + e^{-2t} & -3e^{-t} + 3e^{-2t} \end{bmatrix} = \begin{bmatrix} 2e^{-t} - e^{-2t} & 2e^{-t} - 2e^{-2t} \\ -e^{-t} + e^{-2t} & -e^{-t} + 2e^{-2t} \end{bmatrix}$$

To verify the correctness of the foregoing result, let us calculate the function e^{At}, also using the second method. First, we observe that since the eigenvalues of matrix $\boldsymbol{A}$ are distinct from each other, the matrix is diagonalizable, so the method can be applied. Let $\boldsymbol{x}_1$ and $\boldsymbol{x}_2$ be the eigenvectors corresponding to the eigenvalues $\lambda_1 = -1$ and $\lambda_2 = -2$. Then it will hold that

$$\boldsymbol{A}\boldsymbol{x}_1 = \lambda_1 \boldsymbol{x}_1 \Rightarrow \begin{bmatrix} 0 & 2 \\ -1 & -3 \end{bmatrix} \begin{bmatrix} x_{11} \\ x_{12} \end{bmatrix} = - \begin{bmatrix} x_{11} \\ x_{12} \end{bmatrix} \Rightarrow \begin{bmatrix} 2x_{12} \\ -x_{11} - 3x_{12} \end{bmatrix} = - \begin{bmatrix} x_{11} \\ x_{12} \end{bmatrix}$$

Equating the corresponding quantities, we end up with the equation $x_{11} = -2x_{12}$; for $x_{12} = 1$ we have $x_{11} = -2$, that is, the eigenvector $\boldsymbol{x}_1$ is $\boldsymbol{x}_1 = [\,-2 \;\; 1\,]^T$.

In a completely analogous way, for the second eigenvector we will have

$$\boldsymbol{A}\boldsymbol{x}_2 = \lambda_2 \boldsymbol{x}_2 \Rightarrow \begin{bmatrix} 0 & 2 \\ -1 & -3 \end{bmatrix} \begin{bmatrix} x_{21} \\ x_{22} \end{bmatrix} = -2 \begin{bmatrix} x_{21} \\ x_{22} \end{bmatrix} \Rightarrow \begin{bmatrix} 2x_{22} \\ -x_{21} - 3x_{22} \end{bmatrix} = -2 \begin{bmatrix} x_{11} \\ x_{12} \end{bmatrix}$$

In this case, we find that $x_{22} = -x_{21}$. For $x_{21} = 1$, we will have $x_{22} = -1$, so the eigenvector $\boldsymbol{x}_2$ is $\boldsymbol{x}_2 = \begin{bmatrix} 1 & -1 \end{bmatrix}^T$.

Having determined the two eigenvectors, we can construct matrix $\boldsymbol{P}$. We will therefore have

$$\boldsymbol{P} = [\boldsymbol{x}_1 \;\; \boldsymbol{x}_2] = \begin{bmatrix} x_{11} & x_{21} \\ x_{12} & x_{22} \end{bmatrix} = \begin{bmatrix} -2 & 1 \\ 1 & -1 \end{bmatrix} \quad \text{and therefore} \quad \boldsymbol{P}^{-1} = \begin{bmatrix} -1 & -1 \\ -1 & -1 \end{bmatrix}$$

as can be proven. Using the preceding values of matrices $\boldsymbol{P}$ and $\boldsymbol{P}^{-1}$, matrix e^{At} is calculated as

$$e^{At} = \boldsymbol{P} \begin{bmatrix} e^{\lambda_1 t} & 0 \\ 0 & e^{\lambda_2 t} \end{bmatrix} \boldsymbol{P}^{-1} = \begin{bmatrix} -2 & 1 \\ 1 & -1 \end{bmatrix} \begin{bmatrix} e^{-t} & 0 \\ 0 & e^{-2t} \end{bmatrix} \begin{bmatrix} -1 & -1 \\ -1 & -1 \end{bmatrix}$$

$$= \begin{bmatrix} -2e^{-t} & e^{-2t} \\ e^{-t} & -e^{-2t} \end{bmatrix} \begin{bmatrix} -1 & -1 \\ -1 & -1 \end{bmatrix} = \begin{bmatrix} 2e^{-t} - e^{-2t} & 2e^{-t} - 2e^{-2t} \\ -e^{-t} + e^{-2t} & -e^{-t} + 2e^{-2t} \end{bmatrix}$$

which is the same result as the one obtained previously. Finally, by substituting matrix e^{At} into the solution of the system $\boldsymbol{q}(t) = e^{At}\boldsymbol{q}(0)$, we will get

$$\begin{bmatrix} q_1(t) \\ q_2(t) \end{bmatrix} = \begin{bmatrix} 2e^{-t} - e^{-2t} & 2e^{-t} - 2e^{-2t} \\ -e^{-t} + e^{-2t} & -e^{-t} + 2e^{-2t} \end{bmatrix} \begin{bmatrix} 2 \\ 1 \end{bmatrix} = \begin{bmatrix} 4e^{-t} - 2e^{-2t} + 2e^{-t} - 2e^{-2t} \\ -2e^{-t} + 2e^{-2t} - e^{-t} + 2e^{-2t} \end{bmatrix} = \begin{bmatrix} 6e^{-t} - 4e^{-2t} \\ -3e^{-t} + 4e^{-2t} \end{bmatrix}$$

or, equivalently, $q_1(t) = 6e^{-t}u(t) - 4e^{-2t}u(t)$ and $q_2(t) = -3e^{-t}u(t) + 4e^{-2t}u(t)$, which is also the desired result.

Problem 2.9.4 Solve the state equation $\dot{\boldsymbol{q}}(t) = \boldsymbol{A}\boldsymbol{q}(t) + \boldsymbol{B}x(t)$ for parameter values

$$\boldsymbol{A} = \begin{bmatrix} -5 & -6 \\ 1 & 0 \end{bmatrix}, \quad \boldsymbol{B} = \begin{bmatrix} 1 \\ 0 \end{bmatrix}, \quad \boldsymbol{q}(0) = \begin{bmatrix} 5 \\ 4 \end{bmatrix}, \quad x(t) = \sin(100t)$$

Solution The solution to the state equation has the form

$$\boldsymbol{q}(t) = e^{At}\boldsymbol{q}(0) + \int_0^t e^{A(t-\tau)} \boldsymbol{B} x(\tau) d\tau$$

for the values of $\boldsymbol{A}$, $\boldsymbol{B}$, $\boldsymbol{q}(0)$, and $x(t)$, which are given in the statement. Following the methodology of the previous exercise, we calculate the eigenvalues of matrix $\boldsymbol{A}$. They will therefore be

$$\det(\boldsymbol{A} - \lambda\boldsymbol{I}) = \begin{vmatrix} -5-\lambda & -6 \\ 1 & -\lambda \end{vmatrix} = \lambda^2 + 5\lambda + 6 = (\lambda+2)(\lambda+3) = 0$$

and therefore matrix $\boldsymbol{A}$ will have the distinct eigenvalues $\lambda_1 = -2$ and $\lambda_2 = -3$. Using these values, the coefficients β_0 and β_1 are calculated as

$$\beta_0 = 3e^{-2t} - 2e^{-3t} \quad \text{and} \quad \beta_1 = e^{-2t} - e^{-3t}$$

leading finally to the result

$$e^{\boldsymbol{A}t} = \sum_{i=0}^{1} \beta_i \boldsymbol{A}^i = \beta_0 \boldsymbol{I} + \beta_1 \boldsymbol{A} = \begin{bmatrix} -2e^{-2t} + 3e^{-3t} & -6e^{-2t} + 6e^{-3t} \\ e^{-2t} - e^{-3t} & 3e^{-2t} - 2e^{-3t} \end{bmatrix}$$

(reader are encouraged to perform all the intermediate steps and verify the result by applying the second method so as to familiarize themselves with how it is applied). We will therefore have

$$e^{\boldsymbol{A}t}\boldsymbol{q}(0) = \begin{bmatrix} -2e^{-2t} + 3e^{-3t} & -6e^{-2t} + 6e^{-3t} \\ e^{-2t} - e^{-3t} & 3e^{-2t} - 2e^{-3t} \end{bmatrix} \begin{bmatrix} 5 \\ 4 \end{bmatrix} = \begin{bmatrix} -34e^{-2t} + 39e^{-3t} & 17e^{-2t} - 13e^{-3t} \end{bmatrix}$$

as can be proven.

On the other hand, the product of the matrices within the integral is calculated as

$$e^{\boldsymbol{A}(t-\tau)}\boldsymbol{B} = \begin{bmatrix} -2e^{-2(t-\tau)} + 3e^{-3(t-\tau)} & -6e^{-2(t-\tau)} + 6e^{-3(t-\tau)} \\ e^{-2(t-\tau)} - e^{-3(t-\tau)} & 3e^{-2(t-\tau)} - 2e^{-3(t-\tau)} \end{bmatrix} \begin{bmatrix} 1 \\ 0 \end{bmatrix} = \begin{bmatrix} -2e^{-2(t-\tau)} + 3e^{-3(t-\tau)} \\ e^{-2(t-\tau)} - e^{-3(t-\tau)} \end{bmatrix}$$

and therefore we will have

$$\int_0^t e^{\boldsymbol{A}(t-\tau)}\boldsymbol{B}x(\tau)d\tau = \int_0^t \begin{bmatrix} -2e^{-2(t-\tau)} + 3e^{-3(t-\tau)} \\ e^{-2(t-\tau)} - e^{-3(t-\tau)} \end{bmatrix} \sin(100\tau)d\tau = \begin{bmatrix} \int_0^t \left(-2e^{-2(t-\tau)} + 3e^{-3(t-\tau)} \right) \sin(100\tau)d\tau \\ \int_0^t \left(e^{-2(t-\tau)} - e^{-3(t-\tau)} \right) \sin(100\tau)d\tau \end{bmatrix}$$

with the integrals of the last expression being calculated as[31]

$$\begin{aligned} I_1 &= -2e^{-2t} \int_0^t e^{2\tau} \sin(100\tau)d\tau + 3e^{-3t} \int_0^t e^{3\tau} \sin(100\tau)d\tau \\ &= \frac{-4\sin(100t) + 200\cos(100t) - 200e^{-2t}}{10004} + \frac{9\sin(100t) - 300\cos(100t) + 300e^{-3t}}{10009} \\ I_2 &= -e^{-2t} \int_0^t e^{2\tau} \sin(100\tau)d\tau - e^{-3t} \int_0^t e^{3\tau} \sin(100\tau)d\tau \\ &= \frac{2\sin(100t) - 100\cos(100t) + 100e^{-2t}}{10004} - \frac{3\sin(100t) - 100\cos(100t) + 100e^{-3t}}{10009} \end{aligned}$$

By substituting the foregoing expressions into the defining equation of the vector $\boldsymbol{q}(t)$ and equating the corresponding members, we finally determine the state variables $q_1(t)$ and $q_2(t)$.

---

[31] For the calculation of the integrals the equation

$$\int_0^t e^{\alpha t} \sin(\beta t)dt = \frac{e^{\alpha t}[\alpha \sin(\beta t) - \beta \cos \beta t] + \beta}{\alpha^2 + \beta^2}$$

was used, which was proven in a previous exercise of this chapter.

Problem 2.9.5 Solve the state equation $\dot{\boldsymbol{q}}(t) = \boldsymbol{A}\boldsymbol{q}(t) + \boldsymbol{B}x(t)$ for parameter values

$$\boldsymbol{A} = \begin{bmatrix} -2 & 0 \\ 1 & -1 \end{bmatrix}, \quad \boldsymbol{B} = \begin{bmatrix} 1 \\ 0 \end{bmatrix}, \quad \boldsymbol{q}(0) = \begin{bmatrix} 0 \\ -1 \end{bmatrix}, \quad x(t) = u(t).$$

(to solve this exercise with the Laplace transform method, see Exercise 5.11.2 on page 468).

Solution In this case, matrix $\boldsymbol{A}$ has eigenvalues $\lambda_1 = -1$ and $\lambda_2 = -1$, with corresponding eigenvectors $\boldsymbol{x}_1 = [\,0\ \ 1\,]^T$ and $\boldsymbol{x}_2 = [\,1\ \ -1\,]^T$. Using either of the two methods for calculating the exponential function of a matrix, we find that[32]

$$e^{\boldsymbol{A}t} = \begin{bmatrix} e^{-2t} & 0 \\ e^{-t} - e^{-2t} & e^{-t} \end{bmatrix} \quad \text{and therefore} \quad e^{\boldsymbol{A}t}\boldsymbol{q}(0) = \begin{bmatrix} 0 \\ -e^{-t} \end{bmatrix}$$

On the other hand, the product $e^{\boldsymbol{A}(t-\tau)}\boldsymbol{B}$ is calculated as

$$e^{\boldsymbol{A}(t-\tau)}\boldsymbol{B} = \begin{bmatrix} e^{-2(t-\tau)} & 0 \\ e^{-(t-\tau)} - e^{-2(t-\tau)} & e^{-t} \end{bmatrix} \begin{bmatrix} 1 \\ 0 \end{bmatrix} = \begin{bmatrix} e^{-2(t-\tau)} \\ e^{-(t-\tau)} - e^{-2(t-\tau)} \end{bmatrix}$$

and therefore

$$\int_0^t e^{\boldsymbol{A}(t-\tau)}\boldsymbol{B}\,x(\tau)\,d\tau = \int_0^t \begin{bmatrix} e^{-2(t-\tau)} \\ e^{-(t-\tau)} - e^{-2(t-\tau)} \end{bmatrix} u(\tau)\,d\tau = \begin{bmatrix} \displaystyle\int_0^t e^{-2(t-\tau)}u(\tau)\,d\tau \\ \displaystyle\int_0^t \left(e^{-(t-\tau)} - e^{-2(t-\tau)}\right) u(\tau)\,d\tau \end{bmatrix} = \begin{bmatrix} I_1 \\ I_2 \end{bmatrix}$$

with the two integrals being calculated as

$$I_1 = \int_0^t e^{-2(t-\tau)}d\tau = e^{-2t}\int_0^t e^{2\tau}d\tau = \frac{e^{-2t}}{2}(e^{2t}-1) = \frac{1}{2}(1-e^{-2t})$$

$$I_2 = \int_0^t e^{-(t-\tau)}d\tau - \int_0^t e^{-2(t-\tau)}d\tau = e^{-t}\int_0^t e^{\tau}d\tau - e^{-2t}\int_0^t e^{2\tau}d\tau$$

$$= e^{-t}(e^t - 1) - \frac{e^{-2t}}{2}(e^{2t}-1) = 1 - e^{-t} - \frac{1}{2}(1-e^{-2t})$$

Therefore,

$$\int_0^t e^{\boldsymbol{A}(t-\tau)}\boldsymbol{B}\,x(\tau)\,d\tau = \begin{bmatrix} \frac{1}{2}\left(1-e^{-2t}\right) \\ 1 - e^{-t} - \frac{1}{2}\left(1-e^{-2t}\right) \end{bmatrix}$$

Substituting into the solution of the state equation, it turns out that

$$\boldsymbol{q}(t) = e^{\boldsymbol{A}t}\boldsymbol{q}(0) + \int_0^t e^{\boldsymbol{A}(t-\tau)}\boldsymbol{B}x(\tau)d\tau = \begin{bmatrix} 0 \\ -e^{-t} \end{bmatrix} + \begin{bmatrix} \frac{1}{2}\left(1-e^{-2t}\right) \\ 1 - e^{-t} - \frac{1}{2}\left(1-e^{-2t}\right) \end{bmatrix} = \begin{bmatrix} \frac{1}{2}\left(1-e^{-2t}\right) \\ 1 - 2e^{-t} - \frac{1}{2}\left(1-e^{-2t}\right) \end{bmatrix}$$

from which we arrive at the desired result:

$$q_1(t) = \frac{1}{2}\left(1-e^{-2t}\right)u(t) \quad \text{and} \quad q_2(t) = u(t) - 2e^{-t}\,u(t) - \frac{1}{2}\left(1-e^{-2t}\right)u(t)$$

[32] The values of the coefficients β_0 and β_1 that are related to the first method are equal to $\beta_0 = 2e^{-t} - e^{-2t}$ and $\beta_1 = e^{-t} - e^{-2t}$.

Problem 2.9.6 Solve the state equation $\dot{\boldsymbol{q}}(t) = \boldsymbol{A}\boldsymbol{q}(t) + \boldsymbol{B}\boldsymbol{x}(t)$ for parameter values

$$\boldsymbol{A} = \begin{bmatrix} -1 & 1 \\ 0 & -2 \end{bmatrix}, \quad \boldsymbol{B} = \begin{bmatrix} 1 & 1 \\ 0 & 1 \end{bmatrix}, \quad \boldsymbol{q}(0) = \begin{bmatrix} 1 \\ 2 \end{bmatrix}, \quad \boldsymbol{x}(t) = \begin{bmatrix} u(t) \\ \delta(t) \end{bmatrix}$$

(To solve this exercise with the Laplace transform method, see Exercise 5.11.3 on page 469.)

Solution First, we observe that the system in question is characterized by the existence of two inputs and therefore will be either a MISO or MIMO type of system. Applying the process that we previously described, we find that matrix $\boldsymbol{A}$ has eigenvalues $\lambda_1 = -1$ and $\lambda_2 = -2$ with corresponding eigenvectors $\boldsymbol{x}_1 = [\,1 \;\; 0\,]^T$ and $\boldsymbol{x}_2 = [\,1 \;\; -1\,]^T$ (in this case, the values of the coefficients β_0 and β_2 associated with the first method of calculating the exponential function turn out to be equal to $\beta_0 = 2e^{-t} - e^{-2t}$ and $\beta_1 = e^{-t} - e^{-2t}$). The expressions $e^{\boldsymbol{A}t}$ and $e^{\boldsymbol{A}t}\boldsymbol{q}(0)$ turn out to be respectively equal to

$$e^{\boldsymbol{A}t} = \begin{bmatrix} e^{-t} & e^{-t} - e^{-2t} \\ 0 & e^{-2t} \end{bmatrix} \quad \text{and} \quad e^{\boldsymbol{A}t}\boldsymbol{q}(0) = \begin{bmatrix} 3e^{-t} - 2e^{-2t} \\ 2e^{-2t} \end{bmatrix}$$

while in a similar way it turns out that

$$e^{\boldsymbol{A}(t-\tau)}\boldsymbol{B} = \begin{bmatrix} e^{-(t-\tau)} & e^{-(t-\tau)} - e^{-2(t-\tau)} \\ 0 & e^{-2(t-\tau)} \end{bmatrix} \begin{bmatrix} 1 & 1 \\ 0 & 1 \end{bmatrix} = \begin{bmatrix} e^{-(t-\tau)} & 2e^{-(t-\tau)} - e^{-2(t-\tau)} \\ 0 & e^{-2(t-\tau)} \end{bmatrix}$$

Therefore, the integral that appears in the solution of the state equation is calculated as

$$I = \int_0^t e^{\boldsymbol{A}(t-\tau)}\boldsymbol{B}\boldsymbol{x}(\tau)d\tau = \int_0^t \begin{bmatrix} e^{-(t-\tau)} & 2e^{-(t-\tau)} - e^{-2(t-\tau)} \\ 0 & e^{-2(t-\tau)} \end{bmatrix} \begin{bmatrix} u(\tau) \\ \delta(\tau) \end{bmatrix} d\tau$$

$$= \int_0^t \begin{bmatrix} e^{-(t-\tau)}u(\tau) + \left(2e^{-(t-\tau)} - e^{-2(t-\tau)}\right)\delta(\tau) \\ e^{-2(t-\tau)}\delta(\tau) \end{bmatrix} d\tau = \begin{bmatrix} \int_0^t \left\{ e^{-(t-\tau)}u(\tau) + \left(2e^{-(t-\tau)} - e^{-2(t-\tau)}\right)\delta(\tau) \right\} d\tau \\ \int_0^t e^{-2(t-\tau)}\delta(\tau)d\tau \end{bmatrix}$$

The first of the two integrals is calculated as[33]

$$I_1 = e^{-t}\int_0^t e^{\tau}u(\tau)d\tau + 2\int_0^t e^{-(t-\tau)}\delta(\tau)d\tau - \int_0^t e^{-2(t-\tau)}\delta(\tau)d\tau = e^{-t}(e^t - 1) + 2e^{-t} - e^{-2t} = 1 + e^{-t} - e^{-2t}$$

while for the other integral we have

$$I_2 = \int_0^t e^{-2(t-\tau)}\delta(\tau)d\tau = e^{-2t}$$

Substituting the preceding quantities into the solution of the system's state equation, we will get

$$\boldsymbol{q}(t) = e^{\boldsymbol{A}t}\boldsymbol{q}(0) + \int_0^t e^{\boldsymbol{A}(t-\tau)}\boldsymbol{B}\boldsymbol{x}(\tau)d\tau = \begin{bmatrix} 3e^{-t} - 2e^{-2t} \\ 2e^{-2t} \end{bmatrix} + \begin{bmatrix} 1 + e^{-t} - e^{-2t} \\ e^{-2t} \end{bmatrix}$$

[33] Setting $t - \tau = \xi$ will result in $d\xi = -d\tau$, and the new integration interval will be $[t, 0]$. Therefore, using the property of displacement and the even symmetry that characterizes the impulse function, we will get

$$\int_0^t e^{-(t-\tau)}\delta(\tau)d\tau = -\int_t^0 e^{-\xi}\delta(t-\xi)d\xi = \int_0^t e^{-\xi}\delta(\xi - t)d\xi = e^{-t}$$

and the other integral is calculated in a similar way.

and finally

$$\boldsymbol{q}(t) = \begin{bmatrix} 4e^{-t} - 3e^{-2t} + 1 \\ 3e^{-2t} \end{bmatrix}$$

Replacing the preceding quantities in the solution of the system's state equation, we will get

$$\boldsymbol{q}(t) = e^{\boldsymbol{A}t}\boldsymbol{q}(0) + \int_0^t e^{\boldsymbol{A}(t-\tau)}\, \boldsymbol{B}\, \boldsymbol{x}(\tau)\, d\tau = \begin{bmatrix} 4e^{-t} - 3e^{-2t} + 1 \\ 3e^{-2t} \end{bmatrix}$$

or, equivalently, $q_1(t) = 4e^{-t} - 3e^{-2t} + 1$ and $q_2(t) = 3e^{-2t}$, which is the desired result.

Problem 2.9.7 Determine the output and the impulse response of a system described by the matrices

$$\boldsymbol{A} = \begin{bmatrix} 0 & 2 \\ -1 & -3 \end{bmatrix}, \quad \boldsymbol{B} = \begin{bmatrix} 0 \\ 1 \end{bmatrix}, \quad \boldsymbol{C} = \begin{bmatrix} 2 & 3 \end{bmatrix}, \quad \boldsymbol{D} = [-1], \quad \boldsymbol{q}(0) = \begin{bmatrix} 2 \\ 1 \end{bmatrix}$$

and characterized by the output value $x(t) = 0$.

Solution In Exercise 2.9.3 (page 147) we proved that the state vector of this system is

$$\boldsymbol{q} = \begin{bmatrix} 6e^{-t} - 4e^{-2t} \\ -3e^{-t} + 4e^{-2t} \end{bmatrix}$$

while the state transition matrix is

$$\boldsymbol{\Phi}(t) = e^{\boldsymbol{A}t} = \begin{bmatrix} 2e^{-t} - e^{-2t} & 2e^{-t} - 2e^{-2t} \\ -e^{-t} + e^{-2t} & -e^{-t} + 2e^{-2t} \end{bmatrix}$$

Therefore, the output of the system is calculated as

$$y(t) = \boldsymbol{C}\boldsymbol{q}(t) + \boldsymbol{D}x(t) = \begin{bmatrix} 2 & 3 \end{bmatrix} \begin{bmatrix} 6e^{-t} - 4e^{-2t} \\ -3e^{-t} + 4e^{-2t} \end{bmatrix} = 3e^{-t} + 4e^{-2t}$$

while its impulse response will have the form

$$h(t) = \boldsymbol{C}\boldsymbol{\Phi}(t)\boldsymbol{B} + \boldsymbol{D}\delta(t) = \begin{bmatrix} 2 & 3 \end{bmatrix} \begin{bmatrix} 2e^{-t} - e^{-2t} & 2e^{-t} - 2e^{-2t} \\ -e^{-t} + e^{-2t} & -e^{-t} + 2e^{-2t} \end{bmatrix} \begin{bmatrix} 0 \\ 1 \end{bmatrix} - \delta(t)$$

$$= \begin{bmatrix} e^{-t} + e^{-2t} & e^{-t} + 2e^{-2t} \end{bmatrix} \begin{bmatrix} 0 \\ 1 \end{bmatrix} - \delta(t) = e^{-t} + 2e^{-2t} - \delta(t)$$

as can be proven.

Problem 2.9.8 Determine the output and the impulse response of a system described by the matrices

$$\boldsymbol{A} = \begin{bmatrix} -1 & 1 \\ 0 & -2 \end{bmatrix}, \quad \boldsymbol{B} = \begin{bmatrix} 1 & 1 \\ 0 & 1 \end{bmatrix}, \quad \boldsymbol{C} = \begin{bmatrix} 5 & -6 \\ 2 & 4 \end{bmatrix}, \quad \boldsymbol{D} = \begin{bmatrix} 1 & -1 \\ 0 & 1 \end{bmatrix}$$

while its input $\boldsymbol{x}(t)$ and the initial value of the state vector $\boldsymbol{q}(0)$ have the form

$$\boldsymbol{x}(t) = \begin{bmatrix} u(t) \\ \delta(t) \end{bmatrix} \quad \text{and} \quad \boldsymbol{q}(0) = \begin{bmatrix} 1 \\ 2 \end{bmatrix}$$

(to solve this exercise with the Laplace transform method, see Exercise 5.11.5 on page 472).

Solution This system is of the MIMO type with two inputs and two outputs. In Exercise 2.9.4 (page 148) we proved that the state vector of this system is in the form

$$\boldsymbol{q} = \begin{bmatrix} 4e^{-t} - 3e^{-2t} + 1 \\ 3e^{-2t} \end{bmatrix}$$

while the state transition matrix is

$$\boldsymbol{\Phi}(t) = e^{\boldsymbol{A}t} = \begin{bmatrix} e^{-t} & e^{-t} - e^{-2t} \\ 0 & e^{-2t} \end{bmatrix}$$

Therefore, the system output is calculated as

$$\begin{aligned} y(t) = \boldsymbol{C}\boldsymbol{q}(t) + \boldsymbol{D}\boldsymbol{x}(t) &= \begin{bmatrix} 5 & -6 \\ 2 & 4 \end{bmatrix} \begin{bmatrix} 4e^{-t} - 3e^{-2t} + 1 \\ 3e^{-2t} \end{bmatrix} + \begin{bmatrix} 1 & -1 \\ 0 & 1 \end{bmatrix} \begin{bmatrix} u(t) \\ \delta(t) \end{bmatrix} \\ &= \begin{bmatrix} 20e^{-t} - 15e^{-2t} + 5 - 18e^{-2t} \\ 8e^{-t} - 6e^{-2t} + 2 + 12e^{-2t} \end{bmatrix} + \begin{bmatrix} u(t) - \delta(t) \\ u(t) \end{bmatrix} = \begin{bmatrix} 20e^{-t} - 33e^{-2t} + u(t) - \delta(t) + 5 \\ 8e^{-t} + 6e^{-2t} + \delta(t) + 2 \end{bmatrix} \end{aligned}$$

while the impulse response matrix is calculated as

$$\begin{aligned} \boldsymbol{h}(t) = \boldsymbol{C}\boldsymbol{\Phi}(t)\boldsymbol{B} + \boldsymbol{D}\delta(t) &= \begin{bmatrix} 5 & -6 \\ 2 & 4 \end{bmatrix} \begin{bmatrix} e^{-t} & e^{-t} - e^{-2t} \\ 0 & e^{-2t} \end{bmatrix} \begin{bmatrix} 1 & 1 \\ 0 & 1 \end{bmatrix} + \begin{bmatrix} 1 & -1 \\ 0 & 1 \end{bmatrix} \begin{bmatrix} \delta(t) & 0 \\ 0 & \delta(t) \end{bmatrix} \\ &= \begin{bmatrix} 5e^{-t} & 5e^{-t} - 11e^{-2t} \\ 2e^{-t} & 2e^{-t} + 2e^{-2t} \end{bmatrix} \begin{bmatrix} 1 & 1 \\ 0 & 1 \end{bmatrix} + \begin{bmatrix} \delta(t) & 0 \\ 0 & \delta(t) \end{bmatrix} = \begin{bmatrix} 5e^{-t} & 10e^{-t} - 11e^{-2t} \\ 2e^{-t} & 4e^{-t} + 2e^{-2t} \end{bmatrix} + \begin{bmatrix} \delta(t) & 0 \\ 0 & \delta(t) \end{bmatrix} \end{aligned}$$

and finally

$$\begin{bmatrix} 5e^{-t} + \delta(t) & 10e^{-t} - 11e^{-2t} - \delta(t) \\ 2e^{-t} & 4e^{-t} + 2e^{-2t} + \delta(t) \end{bmatrix}$$

2.9.9 Controllability and Observability

A continuous-time system is aforementioned to be *controllable* if *all its state variables $q_1(t), q_2(t), \ldots, q_N(t)$ can be controlled with the help of its inputs, so that we have the ability to direct the system to a specific state by transmitting the appropriate input to it.* If this cannot be achieved for some or even all of its state variables, the system is aforementioned to be *partially controllable* or *uncontrollable*, respectively.

Closely related to the concept of controllability is that of *observability*, which is a measure of the degree of characterization of the state of a system when its output is known. A continuous-time system is aforementioned to be *observable* when *for every possible sequence of state vectors and input we can determine its current state within a finite time period, knowing only its output.* In the opposite case, the system is aforementioned to be unobservable. The concepts of controllability and observability were first introduced by *Rudolph Emil Kalman* in 1960.

For the special case of continuous-time LTI systems, controllability and observability are defined in the following way:

Controllability: A LTI system described by the equations

$$\begin{aligned} \dot{\boldsymbol{q}}(t) &= \boldsymbol{A}\boldsymbol{q}(t) + \boldsymbol{B}\boldsymbol{x}(t) \\ \boldsymbol{y}(t) &= \boldsymbol{C}\boldsymbol{q}(t) + \boldsymbol{D}\boldsymbol{x}(t) \end{aligned} \tag{2.111}$$

is aforementioned to be *fully controllable* within a finite time interval $[t_0, t_1]$ if for every initial state $\boldsymbol{q}_0 = \boldsymbol{q}(t_0)$ and for every state $\boldsymbol{q}_1$ there is an input $\boldsymbol{x}(t)$ $(t \in [t_0, t_1])$ that causes the system to transition to the state $\boldsymbol{q}(t_1) = \boldsymbol{q}_1$.

To establish the mathematical theory of *controllability*, let us start with the defining equation of the state $\boldsymbol{q}(t_1)$, which, being a solution of the state equation constructed in a previous section, will be given by the relation

$$\boldsymbol{q}(t_1) = e^{\boldsymbol{A}(t_1-t_0)}\boldsymbol{q}(t_0) + \int_{t_0}^{t_1} e^{\boldsymbol{A}(t_1-\tau)}\,\boldsymbol{B}\,\boldsymbol{x}(\tau)\,d\tau$$

But the foregoing relation can also be given in the form

$$\boldsymbol{0} = \left(e^{\boldsymbol{A}(t_1-t_0)}\boldsymbol{q}(t_0) - \boldsymbol{q}(t_1)\right) + \int_{t_0}^{t_1} e^{\boldsymbol{A}(t_1-\tau)}\,\boldsymbol{B}\,\boldsymbol{x}(\tau)\,d\tau$$

(where obviously we have transferred the quantity $\boldsymbol{q}(t_1)$ to the left side) or, alternatively, as

$$\boldsymbol{0} = e^{\boldsymbol{A}(t_1-t_0)}\,\boldsymbol{p}(t_0) + \int_{t_0}^{t_1} e^{\boldsymbol{A}(t_1-\tau)}\boldsymbol{B}\,\boldsymbol{x}(\tau)\,d\tau \tag{2.112}$$

with the quantity $\boldsymbol{p}(t_0)$ being defined by the equation

$$\boldsymbol{p}(t_0) = \boldsymbol{I}\boldsymbol{q}(t_0) - e^{\boldsymbol{A}(t_0-t_1)}\boldsymbol{q}(t_1) \tag{2.113}$$

as can be proven. But what is the consequence of transforming the solution of the state equation in this way? Simply that *the transition of the system from the state* $\boldsymbol{q}_0$ *at time* t_0 *to the state* $\boldsymbol{q}_1$ *at time* t_1 *has become equivalent to the transition of the system from the state* $\boldsymbol{p}_0$ *at time* t_0*to the state* $\boldsymbol{0}$[34] *at time* t_1 [the placement of the zero state on the left and not on the right side of the last equation was made expressly to emphasize the preceding equivalence by comparing the last relation with that which defines the state $\boldsymbol{q}(t_1)$].

To simplify things even further, let us proceed to the definition of the magnitude $\widetilde{\boldsymbol{P}}(t_1, t_0) = e^{\boldsymbol{A}(t_1-t_0)}\boldsymbol{p}(t_0)$, so the last equation will take the form

$$\widetilde{\boldsymbol{P}}(t_0, t_1) = \int_{t_0}^{t_1} e^{\boldsymbol{A}(t_1-\tau)}\,\boldsymbol{B}\,\boldsymbol{x}(\tau)\,d\tau \tag{2.114}$$

However, according to the Cayley-Hamilton theorem, the exponential function of the matrix $e^{\boldsymbol{A}(t_1-\tau)}$ is formulated as

$$e^{\boldsymbol{A}(t_1-\tau)} = \sum_{i=0}^{N-1} \beta_i(t_1-\tau)\boldsymbol{A}^i \tag{2.115}$$

and by substituting in the last equation we will get

$$\widetilde{\boldsymbol{P}}(t_0, t_1) = \int_{t_0}^{t_1}\left(\sum_{i=0}^{N-1}\beta_i(t_1-\tau)\boldsymbol{A}^i\right)\boldsymbol{B}\,\boldsymbol{x}(\tau)\,d\tau = \sum_{i=0}^{N-1}\boldsymbol{A}^i\boldsymbol{B}\left(\int_{t_0}^{t_1}\beta_i(t_1-\tau)\,\boldsymbol{x}(\tau)\,d\tau\right)$$

The preceding equation can also be written

$$\widetilde{\boldsymbol{P}}(t_0, t_1) = \left[\boldsymbol{B}\ \boldsymbol{A}\boldsymbol{B}\ \boldsymbol{A}^2\boldsymbol{B}\ \ldots\ \boldsymbol{A}^{N-1}\boldsymbol{B}\right]\begin{bmatrix}\int_{t_0}^{t_1}\beta_0\,(t_1-\tau)\,\boldsymbol{x}(\tau)\,d\tau\\ \int_{t_0}^{t_1}\beta_1\,(t_1-\tau)\,\boldsymbol{x}(\tau)\,d\tau\\ \vdots\\ \int_{t_0}^{t_1}\beta_{N-1}(t_1-\tau)\boldsymbol{x}(\tau)\,d\tau\end{bmatrix}$$

[34] From this perspective, the controllability of a system can be defined as *the ability we have starting from any state* $\boldsymbol{q}$ *of the state space to return to its beginning, that is, to the state* $\boldsymbol{0}$.

or, equivalently,

$$\widetilde{\boldsymbol{P}}(t_0, t_1) = \boldsymbol{S_c} \cdot \boldsymbol{\Gamma} \tag{2.116}$$

where

$$\boldsymbol{\Gamma} = \left[\gamma_1\ \gamma_2 \ldots \gamma_{N-1}\right]^T \quad with \quad \gamma_i = \int_{t_0}^{t_1} \beta_i(t_1 - \tau)\, \boldsymbol{x}(\tau)\, d\tau, \quad (i = 1, 2, \ldots, N-1)$$

In the last equation, the quantity $\widetilde{\boldsymbol{P}}(t_0, t_1)$ is a vector of dimensions $N \times 1$, the matrix

$$\boldsymbol{S_c} = \left[\boldsymbol{B}\ \boldsymbol{AB} \ldots \boldsymbol{A}^{N-1}\boldsymbol{B}\right] \tag{2.117}$$

which is known as a *controllability matrix*, has dimensions $N \times NK$—the description refers to a MIMO system with K inputs and N state variables—while the last table with dimensions $NK \times 1$ [since each of the N rows contains the input vector $\boldsymbol{x}(t)$, which has K elements] is a function of the system's input $\vec{x}(t)$. According to a fundamental theorem of control theory, *the system described by the last equation is fully controllable if the controllability matrix* $\boldsymbol{S_c}$ *has a rank value equal to N*, or, in mathematical notation,

$$\text{rank}\ \boldsymbol{S_c} = N \tag{2.118}$$

—recall that the rank of a matrix expresses the number of its linearly independent rows[35]—so that the inverse matrix $\boldsymbol{S_c}^{-1}$ exists. The proof of this theorem is quite extensive and relies on the use of many lemmas and theorems from the theory of algebraic spaces, so it is impossible to present here.[36] However, a qualitative interpretation of this theorem can be given if we observe that the state $\widetilde{\boldsymbol{P}}(t_0, t_1) \in R^N$ has been formulated as a *linear combination of the columns of the controllability matrix* $\boldsymbol{S_c}$. Assuming that the state equation of the system is controllable, the foregoing property must hold for every state $\widetilde{\boldsymbol{P}}(t_0, t_1) \in R^N$, that is, the image of the matrix of controllability must be the whole space R^N. However, according to linear algebra, this can happen only when the rank of the matrix $\boldsymbol{S_c}$ is equal to N.

Observability: A continuous-time LTI system described by the dynamic equations $\dot{\boldsymbol{q}}(t) = \boldsymbol{Aq}(t) + \boldsymbol{Bx}(t)$ and $\boldsymbol{y}(t) = \boldsymbol{Cq}(t) + \boldsymbol{Dx}(t)$ is aforementioned to be *fully observable* within a finite time interval $[t_0, t_1]$ if for every time t_0 it is possible to determine an initial state $\boldsymbol{q}(t_0)$ knowing only the system's input $\boldsymbol{x}(t)$ and the corresponding output $\boldsymbol{y}(t)$ for the given time interval.

It is proven that a MIMO system with L outputs and N state variables is observable when the matrix

$$\boldsymbol{S_o} = \begin{bmatrix} \boldsymbol{C} \\ \boldsymbol{CA} \\ \boldsymbol{CA}^2 \\ \vdots \\ \boldsymbol{CA}^{N-1} \end{bmatrix} \tag{2.119}$$

---

35 In reality, for a matrix $\boldsymbol{A}$ two rank values are defined: the *row rank* R_r, which expresses the number of linearly independent rows, and the *column rank* R_c, which expresses the number of linearly independent columns. If matrix $\boldsymbol{A}$ has dimensions $m \times n$, then it is obvious that $R_r \leq m$ and $R_c \leq n$. These two rank values can be equated if we transform the matrix into the so-called *reduced row echelon form*, so it will be characterized by the following properties: (a) all nonzero rows of the matrix (i.e., those containing at least one nonzero element) are above its zero rows and (b) the first nonzero element from the left of each row is further inside the matrix than the corresponding element of the row that is located right above it. The conversion of the table to reduced row echelon form is done using elementary row transformations, which do not change the rank of the original matrix $\boldsymbol{A}$. However, when a matrix is in this special form, its rank is equal to the number of its nonzero rows.

36 For a detailed presentation of the proof see the book by Polderman and Willems *Introduction to Mathematical Systems Theory: A Behavioral Approach*, Texts in Applied Mathematics, Volume 26, Springer Verlag, 2009, ISBN 978-0-387-35763-8.

of dimensions $NL \times N$, known as an observability matrix, has a rank value equal to N, that is,

$$\text{rank } \boldsymbol{S_o} = N \tag{2.120}$$

so that the inverse matrix $\boldsymbol{S_o}^{-1}$ exists. For the proof of this theorem, we must take into account that the initial value of the state vector $\boldsymbol{q}(t_0)$ depends both on the output $\boldsymbol{y}(t)$ and its derivatives $\boldsymbol{y}^{(i)}(t)$ $(i = 1, 2, \ldots, N-1)$. By differentiating the output equation $N-1$ times and using the state equation successively, we will get

$$\boldsymbol{y}(t) = \boldsymbol{Cq}(t) + \boldsymbol{Dx}(t)$$

$$\dot{\boldsymbol{y}}(t) = \boldsymbol{C}\dot{\boldsymbol{q}}(t) + \boldsymbol{D}\dot{\boldsymbol{x}}(t) = \boldsymbol{CAq}(t) + \boldsymbol{CBx}(t) + \boldsymbol{D}\dot{\boldsymbol{x}}(t)$$

$$\ddot{\boldsymbol{y}}(t) = \boldsymbol{CA}\dot{\boldsymbol{q}}(t) + \boldsymbol{CB}\dot{\boldsymbol{x}}(t) + \boldsymbol{D}\ddot{\boldsymbol{x}}(t) = \boldsymbol{CA}^2\boldsymbol{q}(t) + \boldsymbol{CABx}(t) + \boldsymbol{CB}\dot{\boldsymbol{x}}(t) + \boldsymbol{D}\ddot{\boldsymbol{x}}(t)$$

$$\cdots\cdots\cdots\cdots\cdots\cdots\cdots\cdots\cdots\cdots\cdots\cdots\cdots\cdots$$

$$\boldsymbol{y}^{(N-1)}(t) = \boldsymbol{CA}^{N-1}\boldsymbol{q}(t) + \boldsymbol{CA}^{N-1}\boldsymbol{bx}(t) + \boldsymbol{CA}^{N-3}\dot{\boldsymbol{x}}(t) + \ldots + \boldsymbol{CABx}^{(N-3)}(t) + \boldsymbol{CBx}^{(N-2)}(t) + \boldsymbol{Dx}^{(N-1)}(t)$$

or in tabular form

$$\begin{bmatrix} \boldsymbol{y}(t) \\ \dot{\boldsymbol{y}}(t) \\ \ddot{\boldsymbol{y}}(t) \\ \vdots \\ \boldsymbol{y}^{(N-1)}(t) \end{bmatrix} = \begin{bmatrix} \boldsymbol{C} \\ \boldsymbol{CA} \\ \boldsymbol{CA}^2 \\ \vdots \\ \boldsymbol{CA}^{N-1} \end{bmatrix} \boldsymbol{q}(t) + \begin{bmatrix} \boldsymbol{D} & \boldsymbol{0} & \ldots & \boldsymbol{0} & \boldsymbol{0} \\ \boldsymbol{CB} & \boldsymbol{D} & \ldots & \boldsymbol{0} & \boldsymbol{0} \\ \boldsymbol{CAB} & \boldsymbol{CB} & \ldots & \boldsymbol{0} & \boldsymbol{0} \\ \vdots & \vdots & \vdots & \ddots & \vdots \\ \boldsymbol{CA}^{N-2}\boldsymbol{B} & \boldsymbol{CA}^{N-3}\boldsymbol{B} & \ldots & \boldsymbol{CB} & \boldsymbol{D} \end{bmatrix} \begin{bmatrix} \boldsymbol{x}(t) \\ \dot{\boldsymbol{x}}(t) \\ \ddot{\boldsymbol{x}}(t) \\ \vdots \\ \boldsymbol{x}^{(N-1)}(t) \end{bmatrix}$$

From the last relation it is obvious that the calculation of the vector $\boldsymbol{q}(t)$ as a function of the inputs and outputs of the system, as well as their derivatives, requires the property rank $\boldsymbol{S_o} = N$ for the matrix $\boldsymbol{S_o}$, so that it can be inverted.

The foregoing criteria that allow the characterization of a system as controllable and/or observable concern the general case in which system matrix $\boldsymbol{A}$ is characterized by *arbitrary* eigenvalues and corresponding eigenvectors. However, there are other ways of characterizing a system with respect to these properties for more specialized cases, such as a system whose matrix $\boldsymbol{A}$ can be diagonalized. Let us consider, therefore, a system with dynamic equations

$$\dot{\boldsymbol{q}}(t) = \boldsymbol{Aq}(t) + \boldsymbol{Bx}(t) \quad \text{and} \quad \boldsymbol{y}(t) = \boldsymbol{Cq}(t) + \boldsymbol{Dx}(t)$$

and suppose there is a *state vector* $\boldsymbol{p}(t)$ and *transformation matrix* $\boldsymbol{U}$ such that $\boldsymbol{q}(t) = \boldsymbol{U}\,\boldsymbol{p}(t)$, which transform the dynamic equations of the system into the form

$$\dot{\boldsymbol{p}}(t) = \boldsymbol{\Lambda p}(t) + \widetilde{\boldsymbol{B}}\boldsymbol{x}(t) \quad \text{and} \quad \boldsymbol{y}(t) = \widetilde{\boldsymbol{C}}\,\boldsymbol{p}(t) + \boldsymbol{Dx}(t)$$

where the matrix $\boldsymbol{\Lambda} = \boldsymbol{U}^{-1}\boldsymbol{AU}$ is a *diagonal matrix*, while matrices $\widetilde{\boldsymbol{C}}$ and $\widetilde{\boldsymbol{B}}$ are defined as $\widetilde{\boldsymbol{C}} = \boldsymbol{CU} = \{\tilde{c}_{ij}\}$ and $\widetilde{\boldsymbol{B}} = \boldsymbol{U}^{-1}\boldsymbol{B} = \{\tilde{\beta}_{ij}\}$ (for a description of the process applied see Sect. 2.9.8 on the *equivalence of state vectors*). Assuming that the diagonal matrix $\boldsymbol{\Lambda}$ has *distinct* eigenvalues $\lambda_1, \lambda_2, \ldots, \lambda_N$—in the opposite case, other procedures are applied that, due to a lack of space, cannot be presented here—the new state equation $\dot{\boldsymbol{p}}(t) = \boldsymbol{\Lambda p}(t) + \widetilde{\boldsymbol{B}}\boldsymbol{x}(t)$ is reduced to a system of first-order differential equations of the form

$$\frac{dp_i(t)}{dt} = \lambda_i p_i(t) + \tilde{\beta}_{i1}x_1(t) + \tilde{\beta}_{i2}x_2(t) + \cdots + \tilde{\beta}_{iK}x_K(t) = \lambda_i p_i(t) + \sum_{j=1}^{K} \tilde{\beta}_{ij}x_j(t)$$

$(i = 1, 2, 3, \ldots, N)$, where K is the number of inputs of the considered MIMO system.

Let us now see what will happen if we assume that the ξ row of matrix $\widetilde{\boldsymbol{B}}$ has all its elements equal to zero. In this case, the corresponding differential equation will take the form $\dot{p}_\xi(t) = \lambda_\xi p_\xi(t)$. But this means that *the state variable* $p_\xi(t)$ *is no longer related to any of the system's inputs, either directly or indirectly*—through some of the remaining $N-1$ state variables with which it is also uncorrelated—*and therefore it is uncontrollable*. On the other hand, if even one element of the ξ rows of matrix $\widetilde{\boldsymbol{B}}$ has a value different from zero, then the variable $p_\xi(t)$ will be related to at least one input and therefore will be controllable. We are led, therefore, to the conclusion that *a continuous-time LTI system whose matrix* $\boldsymbol{A}$ *is diagonalizable is fully controllable if and only if matrix* $\widetilde{\boldsymbol{B}}$ *does not have rows of zero elements*.

Using completely analogous reasoning, we can formulate the criterion of *observability* using the output equation of a system that can be analyzed into a system of L equations of the form

$$y_i(t) = \tilde{c}_{i1}p_1(t) + \tilde{c}_{i2}p_2(t) + \cdots + \tilde{c}_{iN}p_N(t) + \sum_{k=1}^{L} d_{ik}x_K(t), \quad (i = 1, 2, 3, \ldots, L)$$

where L is the number of outputs of the MIMO system and $\boldsymbol{D} = \{d_{ij}\}$. In this case, we observe that if any of the coefficients $\tilde{c}_{im}$ is equal to zero, then the state variable $p_m(t)$ will not appear in the defining equation of the output $y_i(t)$, and because this variable, due to the diagonal nature of the matrix $\boldsymbol{\Lambda}$, does not depend on the remaining $N-1$ state variables, *it will never be observed in the output number i of the system* ($i = 1, 2, 3, \ldots, L$), *either directly or indirectly*. Generalizing this conclusion, we can formulate the claim that if all the elements of column number m of matrix $\widetilde{C}$ are equal to zero, then the variable $p_m(t)$ will never appear in any of the L outputs of the system, and therefore it is unobservable. Conversely, if even one of the elements of this column has a value different from zero, then the variable $p_m(t)$ will be observed in at least one output, and therefore it is observable. We are led, therefore, to the conclusion that *a continuous-time LTI system whose matrix* $\boldsymbol{A}$ *allows for diagonalization is fully observable if and only if matrix* $\widetilde{\boldsymbol{C}}$ *does not have columns of zero elements*.

Problem 2.9.9 Examine whether or not a LTI system described by the state equation

$$\dot{\boldsymbol{q}}(t) = \begin{bmatrix} 0 & 1 & 0 & 0 \\ 0 & 0 & -1 & 0 \\ 0 & 0 & 0 & 1 \\ 0 & 0 & 5 & 0 \end{bmatrix} \boldsymbol{q}(t) + \begin{bmatrix} 0 \\ 1 \\ 0 \\ -2 \end{bmatrix} \boldsymbol{x}(t)$$

is fully controllable (Chen C-T, *Linear System Theory and Design*, Example 6.2).

Solution The state equation of the system has the general form $\dot{\boldsymbol{q}}(t) = \boldsymbol{A}\boldsymbol{q}(t) + \boldsymbol{B}\boldsymbol{x}(t)$ with parameter values

$$\boldsymbol{A} = \begin{bmatrix} 0 & 1 & 0 & 0 \\ 0 & 0 & -1 & 0 \\ 0 & 0 & 0 & 1 \\ 0 & 0 & 5 & 0 \end{bmatrix} \quad \text{and} \quad \boldsymbol{B} = \begin{bmatrix} 0 \\ 1 \\ 0 \\ -2 \end{bmatrix}$$

for the system matrices and dimension value $N = 4$. To determine whether the system is fully controllable, we need to construct the *controllability matrix*, which in this case has the form

$$\boldsymbol{S_c} = \begin{bmatrix} \boldsymbol{B} & \boldsymbol{AB} & \boldsymbol{A}^2\boldsymbol{B} & \boldsymbol{A}^3\boldsymbol{B} \end{bmatrix}$$

and determine its rank value. If the rank of matrix R is found to be equal to 4, it means that the system is fully controllable, which will not be the case otherwise.

To construct the controllability matrix $\boldsymbol{S_c}$, we need to calculate the products of all the matrices appearing in it. Therefore, we have

$$AB = \begin{bmatrix} 0 & 1 & 0 & 0 \\ 0 & 0 & -1 & 0 \\ 0 & 0 & 0 & 1 \\ 0 & 0 & 5 & 0 \end{bmatrix} \begin{bmatrix} 0 \\ 1 \\ 0 \\ -2 \end{bmatrix} = \begin{bmatrix} 1 \\ 0 \\ -2 \\ 0 \end{bmatrix}$$

$$A^2 = \begin{bmatrix} 0 & 1 & 0 & 0 \\ 0 & 0 & -1 & 0 \\ 0 & 0 & 0 & 1 \\ 0 & 0 & 5 & 0 \end{bmatrix} \begin{bmatrix} 0 & 1 & 0 & 0 \\ 0 & 0 & -1 & 0 \\ 0 & 0 & 0 & 1 \\ 0 & 0 & 5 & 0 \end{bmatrix} = \begin{bmatrix} 0 & 0 & -1 & 0 \\ 0 & 0 & 0 & -1 \\ 0 & 0 & 5 & 0 \\ 0 & 0 & 0 & 5 \end{bmatrix}$$

$$A^3 = \begin{bmatrix} 0 & 0 & -1 & 0 \\ 0 & 0 & 0 & -1 \\ 0 & 0 & 5 & 0 \\ 0 & 0 & 0 & 5 \end{bmatrix} \begin{bmatrix} 0 & 1 & 0 & 0 \\ 0 & 0 & -1 & 0 \\ 0 & 0 & 0 & 1 \\ 0 & 0 & 5 & 0 \end{bmatrix} = \begin{bmatrix} 0 & 0 & 0 & -1 \\ 0 & 0 & -5 & 0 \\ 0 & 0 & 0 & 5 \\ 0 & 0 & 25 & 0 \end{bmatrix}$$

and using the last two expressions, we find that

$$A^2B = \begin{bmatrix} 0 & 0 & -1 & 0 \\ 0 & 0 & 0 & -1 \\ 0 & 0 & 5 & 0 \\ 0 & 0 & 0 & 5 \end{bmatrix} \begin{bmatrix} 0 \\ 1 \\ 0 \\ -2 \end{bmatrix} = \begin{bmatrix} 0 \\ 2 \\ 0 \\ -10 \end{bmatrix} \quad \text{and} \quad A^3B = \begin{bmatrix} 0 & 0 & 0 & -1 \\ 0 & 0 & -5 & 0 \\ 0 & 0 & 0 & 5 \\ 0 & 0 & 25 & 0 \end{bmatrix} \begin{bmatrix} 0 \\ 1 \\ 0 \\ -2 \end{bmatrix} = \begin{bmatrix} 2 \\ 0 \\ -10 \\ 0 \end{bmatrix}$$

Replacing the preceding expressions in the controllability matrix, it immediately follows that

$$S_c = \left[B \ AB \ A^2B \ A^3B\right] = \begin{bmatrix} 0 & 1 & 0 & 2 \\ 1 & 0 & 2 & 0 \\ 0 & -2 & 0 & -10 \\ -2 & 0 & -10 & 0 \end{bmatrix}$$

It is proven (as we will see later) that the rank of this matrix is equal to the dimension of matrix A—that is $R = N = 4$—and therefore, the system is *fully controllable*. Linear algebra is known to provide the easiest way to calculate the rank of a matrix, which is through the use of the *Gauss elimination method*, which is commonly used to solve algebraic systems. In the first stage of its application, this method transforms the matrix of the algebraic system to be solved into reduced row echelon form, so all we have to do to determine the rank of the matrix is to count the number of nonzero rows. Alternatively, the rank of matrix A can be calculated using the appropriate software application functions, such as the function rank(A) of the programming environment of MATLAB.

Let us see how the foregoing process is carried out. Initially, matrix A can be considered the coefficient matrix of a homogeneous linear algebraic system, the number of unknowns of which is equal to N, so we can write that

$$\begin{bmatrix} 0 & 1 & 0 & 2 \\ 1 & 0 & 2 & 0 \\ 0 & -2 & 0 & -10 \\ -2 & 0 & -10 & 0 \end{bmatrix} \equiv \left\rangle \begin{cases} 0x + 1y + 0z + 2k = 0 \\ 1x + 0y + 2z + 0k = 0 \\ 0x - 2y + 0z - 10k = 0 \\ -2x + 0y - 10z + 0k = 0 \end{cases} \right. \equiv \left\rangle \begin{matrix} y + 2k = 0 \\ x + 2z = 0 \\ -2y - 10k = 0 \\ -2x - 10z = 0 \end{matrix} \right.$$

To apply the Gauss elimination method, we need to rearrange the preceding equations—we know from algebra that rearranging the rows of a matrix leaves it unchanged—so that the first one necessarily contains x, the second necessarily contains y, the third necessarily contains z, and the fourth necessarily contains k. Then, applying the appropriate transformations in each case, we need to eliminate x from the second, third, and fourth equations, eliminate y from the third and fourth equations, and finally eliminate z from the fourth equation, so as to transform matrix A into reduced row echelon form. Thus, we have successively:

$$\left.\begin{aligned} x + 2z &= 0 \\ y + 2k &= 0 \\ -2x - 10z &= 0 \\ -2y - 10k &= 0 \end{aligned}\right\} \equiv \quad \left.\begin{aligned} x + 2z &= 0 \\ y + 2k &= 0 \\ -6z &= 0 \\ -2y - 10k &= 0 \end{aligned}\right\} \equiv \quad \begin{aligned} x + 2z &= 0 \\ y + 2k &= 0 \\ -6z &= 0 \\ -6k &= 0 \end{aligned}$$

(a) (b) (c)

In the system of equations (a) we proceeded with the initial rearrangement of the equations in the way mentioned earlier, while in system (b) we eliminated x from the third equation (since it does not appear in the second and fourth equations anyway). To do this, we multiplied the first equation by 2 and added it to the third. Finally, in system (c), we eliminated y from the last equation by multiplying the second by 2 and adding it to the fourth.

The final system we ended up with can be given in full form as

$$\begin{aligned} 1x + 0y + 2z + 0k &= 0 \\ 0x + 1y + 0z + 2k &= 0 \\ 0x + 0y - 6z + 0k &= 0 \\ 0x + 0y + 0z - 6k &= 0 \end{aligned} \quad \text{and corresponds to the matrix} \quad \boldsymbol{A} = \begin{bmatrix} 1 & 0 & 2 & 0 \\ 0 & 1 & 0 & 2 \\ 0 & 0 & -6 & 0 \\ 0 & 0 & 0 & -6 \end{bmatrix}$$

which is in reduced row echelon form, as can be verified. We observe that in this matrix, all rows contain at least one nonzero element, that is, the rank of the matrix has a value $R = 4$. This is also how the ranks of the controllability and observability matrices should be calculated in all the exercises that follow.

Problem 2.9.10 Examine whether a LTI system described by the following dynamic equations is fully controllable and observable:

$$\dot{\boldsymbol{q}}(t) \begin{bmatrix} 0 & 1 & -1 \\ -1 & -1 & 1 \\ -1 & -1 & 0 \end{bmatrix} \boldsymbol{q}(t) + \begin{bmatrix} 0 \\ 1 \\ 0 \end{bmatrix} \boldsymbol{x}(t) \quad \text{and} \quad \boldsymbol{y}(t) = \begin{bmatrix} 1 & 0 & 1 \end{bmatrix} \boldsymbol{q}(t)$$

Solution Observing that the dimension of matrix $\boldsymbol{A}$ is $N = 3$, the controllability matrix will have the form

$$\boldsymbol{S_c} = \begin{bmatrix} \boldsymbol{B} & \boldsymbol{AB} & \boldsymbol{A}^2\boldsymbol{B} \end{bmatrix}$$

with matrices $\boldsymbol{A}$ and $\boldsymbol{B}$ defined as

$$\boldsymbol{A} = \begin{bmatrix} 0 & 1 & -1 \\ -1 & -1 & 1 \\ -1 & -1 & 0 \end{bmatrix} \quad \text{and} \quad \boldsymbol{B} = \begin{bmatrix} 0 \\ 1 \\ 0 \end{bmatrix}$$

Calculating as before the products of the matrices that appear in the controllability matrix, we find

$$\boldsymbol{AB} = \begin{bmatrix} 1 \\ -1 \\ -1 \end{bmatrix}, \quad \boldsymbol{A}^2 = \begin{bmatrix} 0 & 0 & 1 \\ 0 & -1 & 0 \\ 1 & 0 & 0 \end{bmatrix}, \quad \boldsymbol{A}^2\boldsymbol{B} = \begin{bmatrix} 0 \\ -1 \\ 0 \end{bmatrix}$$

and by substitution in the controllability matrix we find that it has the form

$$\boldsymbol{S_c} = \begin{bmatrix} \boldsymbol{B} & \boldsymbol{AB} & \boldsymbol{A}^2\boldsymbol{B} \end{bmatrix} = \begin{bmatrix} 0 & 1 & 0 \\ 1 & -1 & -1 \\ 0 & -1 & 0 \end{bmatrix}$$

Applying the aforementioned method, we find that the rank of matrix $\boldsymbol{S_c}$ is $R = 2 < N$, that is, the system *is not controllable*.

On the other hand, to characterize the system in terms of its observability, we should calculate the rank of the observability matrix. In this case, it turns out that

$$\boldsymbol{C} = \begin{bmatrix}1 \ 0 \ 1\end{bmatrix}, \quad \boldsymbol{CA} = \begin{bmatrix}-1 \ 0 \ -1\end{bmatrix}, \quad \boldsymbol{CA}^2 = \begin{bmatrix}1 \ 0 \ 1\end{bmatrix}$$

and therefore the observability matrix will have the form

$$\boldsymbol{S_o} = \begin{bmatrix}\boldsymbol{C} \\ \boldsymbol{CA} \\ \boldsymbol{CA}^2\end{bmatrix} = \begin{bmatrix}1 & 0 & 1 \\ -1 & 0 & -1 \\ 1 & 0 & 1\end{bmatrix}$$

It is proven that the rank of matrix $\boldsymbol{S_o}$ is $R = 1 < N$, meaning the system is *not observable*.

Problem 2.9.11 Examine whether a LTI system described by the following dynamic equations is fully controllable and observable:

$$\dot{\boldsymbol{q}}(t) = \begin{bmatrix}1 & 0 & -2 \\ 0 & 1 & -2 \\ 0 & 0 & 2\end{bmatrix}\boldsymbol{q}(t) + \begin{bmatrix}0 & 1 \\ 0 & 1 \\ 1 & 0\end{bmatrix}\boldsymbol{x}(t) \quad \text{and} \quad \boldsymbol{y}(t) = \begin{bmatrix}1 \ 0 \ 1\end{bmatrix}\boldsymbol{q}(t)$$

Solution The dimension of matrix $\boldsymbol{A}$ is $N = 3$. Following the methodology of the previous exercise, we find

$$\boldsymbol{AB} = \begin{bmatrix}-2 & 1 \\ -2 & 1 \\ 2 & 0\end{bmatrix}, \quad \boldsymbol{A}^2 = \begin{bmatrix}1 & 0 & -6 \\ 0 & 1 & -6 \\ 0 & 0 & 4\end{bmatrix}, \quad \boldsymbol{A}^2\boldsymbol{B} = \begin{bmatrix}-6 & 1 \\ -6 & 1 \\ 4 & 0\end{bmatrix}$$

Therefore, the controllability matrix is calculated as

$$\boldsymbol{S_c} = \begin{bmatrix}\boldsymbol{B} \ \boldsymbol{AB} \ \boldsymbol{A}^2\boldsymbol{B}\end{bmatrix} = \begin{bmatrix}0 & 1 & \vdots & -2 & 1 & \vdots & -6 & 1 \\ 0 & 1 & \vdots & -2 & 1 & \vdots & -6 & 1 \\ 1 & 0 & \vdots & 2 & 0 & \vdots & 4 & 0\end{bmatrix}$$

and has dimensions 3×6. It is proven that the rank of this table is equal to $R = 2 < N$, meaning the system *is not controllable*.

On the other hand, to characterize the system in terms of its observability, we must, as before, calculate *the rank of the observability matrix*. In this case, it turns out that

$$\boldsymbol{C} = \begin{bmatrix}1 \ 0 \ 1\end{bmatrix}, \quad \boldsymbol{CA} = \begin{bmatrix}1 \ 0 \ 0\end{bmatrix}, \quad \boldsymbol{CA}^2 = \begin{bmatrix}1 \ 0 \ -2\end{bmatrix}$$

and therefore the observability matrix will have the form

$$\boldsymbol{S_o} = \begin{bmatrix}\boldsymbol{C} \\ \boldsymbol{CA} \\ \boldsymbol{CA}^2\end{bmatrix} = \begin{bmatrix}1 & 0 & 1 \\ 1 & 0 & 0 \\ 1 & 0 & -2\end{bmatrix}$$

It is proven that the rank of matrix $\boldsymbol{S_o}$ is $R = 2 < N$, meaning the system *is not observable*.

Duality of Controllability and Observability

Let us consider a MIMO system with K inputs and L outputs described by the pair of dynamic equations

$$\dot{\boldsymbol{q}}(t) = \boldsymbol{Aq}(t) + \boldsymbol{Bx}(t) \quad \text{and} \quad \boldsymbol{y}(t) = \boldsymbol{Cq}(t) + \boldsymbol{Dx}(t)$$

in which, as is well known, matrices $\boldsymbol{A}$, $\boldsymbol{B}$, $\boldsymbol{C}$, and $\boldsymbol{D}$ have dimensions $N \times N$, $N \times K$, $L \times N$, and $L \times K$, respectively, where N is the number of the system's state variables, while the input and output vectors $\boldsymbol{x}(t)$ and $\boldsymbol{y}(t)$ contain respectively K and L elements. Based on this system, we can define a *dual* system as one described by the equations

$$\dot{\boldsymbol{p}}(t) = \boldsymbol{A}^T \boldsymbol{p}(t) + \boldsymbol{C}^T \boldsymbol{v}(t) \quad \text{and} \quad \boldsymbol{w}(t) = \boldsymbol{B}^T \boldsymbol{p}(t) + \boldsymbol{D}^T \boldsymbol{v}(t)$$

in which the state vector $\boldsymbol{p}(t)$ contains N elements, the input and output vectors $\boldsymbol{v}(t)$ and $\boldsymbol{w}(t)$ contain L and K elements respectively, while the symbol $\boldsymbol{M}^T$ refers to the transpose of the matrix $\boldsymbol{M}$. If we symbolize these two systems as (1) and (2), it is not difficult to ascertain that the controllability and observability matrices for system (1) have the form

$$\boldsymbol{S_c}^{(1)} = \begin{bmatrix} \boldsymbol{B} & \boldsymbol{AB} & \boldsymbol{A}^2\boldsymbol{B} & \ldots & \boldsymbol{A}^{N-1}\boldsymbol{B} \end{bmatrix} \quad \text{and} \quad \boldsymbol{S_o}^{(1)} = \begin{bmatrix} \boldsymbol{C} \\ \boldsymbol{CA} \\ \boldsymbol{CA}^2 \\ \vdots \\ \boldsymbol{CA}^{N-1} \end{bmatrix}$$

while the corresponding matrices for system (2) are defined as

$$\boldsymbol{S_c}^{(2)} = \begin{bmatrix} \boldsymbol{C} & \boldsymbol{A}^T\boldsymbol{C}^T & (\boldsymbol{A}^T)^2\boldsymbol{C}^T & \ldots & (\boldsymbol{A}^T)^{N-1}\boldsymbol{C}^T \end{bmatrix} = \begin{bmatrix} \boldsymbol{C} \\ \boldsymbol{CA} \\ \boldsymbol{CA}^2 \\ \vdots \\ \boldsymbol{CA}^{N-1} \end{bmatrix}^T \tag{2.121}$$

$$\boldsymbol{S_o}^{(2)} = \begin{bmatrix} \boldsymbol{B}^T \\ \boldsymbol{B}^T\boldsymbol{A}^T \\ \boldsymbol{B}^T(\boldsymbol{A}^T)^2 \\ \vdots \\ \boldsymbol{B}^T(\boldsymbol{A}^T)^{N-1} \end{bmatrix} = \begin{bmatrix} \boldsymbol{B} & \boldsymbol{AB} & \boldsymbol{A}^2\boldsymbol{B} & \ldots & \boldsymbol{A}^{N-1}\boldsymbol{B} \end{bmatrix}^T \tag{2.122}$$

It is not difficult to see that the controllability and observability matrices of systems (1) and (2) are related through the relations

$$\boldsymbol{S_c}^{(2)} = \left(\boldsymbol{S_o}^{(1)}\right)^T \quad \text{and} \quad \boldsymbol{S_o}^{(2)} = \left(\boldsymbol{S_c}^{(1)}\right)^T$$

that is, *between the concepts of controllability and observability, there is a kind of duality*. In other words, this means *that the system described by matrix* $\boldsymbol{M}$ *is controllable if and only if the system described by the inverse matrix* $\boldsymbol{M}^T$ *is observable*. Formulating the dynamic equations of the two systems using matrix symbolism of the form

$$\begin{bmatrix} \dot{\boldsymbol{q}}(t) \\ \boldsymbol{y}(t) \end{bmatrix} = \begin{bmatrix} \boldsymbol{A} & \boldsymbol{B} \\ \boldsymbol{C} & \boldsymbol{D} \end{bmatrix} \begin{bmatrix} \boldsymbol{q}(t) \\ \boldsymbol{x}(t) \end{bmatrix} \quad \text{and} \quad \begin{bmatrix} \dot{\boldsymbol{p}}(t) \\ \boldsymbol{w}(t) \end{bmatrix} = \begin{bmatrix} \boldsymbol{A}^T & \boldsymbol{C}^T \\ \boldsymbol{B}^T & \boldsymbol{D}^T \end{bmatrix} \begin{bmatrix} \boldsymbol{p}(t) \\ \boldsymbol{v}(t) \end{bmatrix} = \begin{bmatrix} \boldsymbol{A} & \boldsymbol{B} \\ \boldsymbol{C} & \boldsymbol{D} \end{bmatrix}^T \begin{bmatrix} \boldsymbol{p}(t) \\ \boldsymbol{v}(t) \end{bmatrix}$$

the matrices $\boldsymbol{M}$ and $\boldsymbol{M}^T$ used in the last sentence are defined as

$$\boldsymbol{M} = \begin{bmatrix} \boldsymbol{A} & \boldsymbol{B} \\ \boldsymbol{C} & \boldsymbol{D} \end{bmatrix} \quad \text{and} \quad \boldsymbol{M}^T = \begin{bmatrix} \boldsymbol{A}^T & \boldsymbol{C}^T \\ \boldsymbol{B}^T & \boldsymbol{D}^T \end{bmatrix} = \begin{bmatrix} \boldsymbol{A} & \boldsymbol{B} \\ \boldsymbol{C} & \boldsymbol{D} \end{bmatrix}^T$$

Canonical Forms for Controllability

Considering a *controllable* LTI system with a state equation of the form $\boldsymbol{q}(t) = \boldsymbol{A}\boldsymbol{q}(t) + \boldsymbol{B}\boldsymbol{x}(t)$, it is proven that it can be represented in two different ways, the first of which refers to the so-called *controllability normal form*, while the second is known as the *controller's normal form*. In the following analysis, we will assume for simplicity that the system accepts a simple input $x(t)$, although the extension of the theory to the case of multiple input systems is done in a completely analogous way.

Theorem 2.9.1 *Every controllable continuous-time LTI system with a number of inputs $K = 1$ which is described by the state equation*

$$\boldsymbol{q}(t) = \boldsymbol{A}\boldsymbol{q}(t) + \boldsymbol{B}x(t)$$

can always be transformed into controllable form (for simplicity we assume that $\alpha_N = 1$)

$$\dot{\boldsymbol{q}}_c(t) = \widetilde{\boldsymbol{A}}\boldsymbol{q}_c(t) + \widetilde{\boldsymbol{B}}x(t) = \begin{bmatrix} 0\,0\,0 \ldots & -\alpha_0 \\ 1\,0\,0 \ldots & -\alpha_1 \\ 0\,1\,0 \ldots & -\alpha_2 \\ \vdots\;\vdots\;\vdots\;\ddots & \vdots \\ 0\,0\,0 \ldots & -\alpha_{N-1} \end{bmatrix} \boldsymbol{q}_c(t) + \begin{bmatrix} 1 \\ 0 \\ 0 \\ \vdots \\ 0 \end{bmatrix} x(t)$$

where α_i $(i = 1, 2, \ldots, N-1)$ are the coefficients of the characteristic polynomial

$$det(\lambda \boldsymbol{I} - \boldsymbol{A}) = \lambda^N + \alpha_{N-1}\lambda^{N-1} + \cdots + \alpha_1\lambda + \alpha_0$$

while the similarity transformation that causes the transition of the system in the preceding normal form is the controllability matrix

$$\boldsymbol{T} \equiv S_c = \left[\boldsymbol{B}\ \boldsymbol{A}\boldsymbol{B} \ldots \boldsymbol{A}^{N-1}\boldsymbol{B}\right] \tag{2.123}$$

Proof Since the state vectors $\boldsymbol{q}(t)$ and $\boldsymbol{q}_c(t)$ are related through the similarity transform $\boldsymbol{T}$, they must satisfy the relation $\boldsymbol{q}(t) = \boldsymbol{T}\boldsymbol{q}_c(t)$, while matrices $\boldsymbol{A}$ and $\boldsymbol{B}$ should be related to the matrices of the new representation $\widetilde{\boldsymbol{A}}$ and $\widetilde{\boldsymbol{B}}$ through the relations $\widetilde{\boldsymbol{A}} = \boldsymbol{T}^{-1}\boldsymbol{A}\boldsymbol{T}$ and $\widetilde{\boldsymbol{B}} = \boldsymbol{T}^{-1}\boldsymbol{B}$, which can also be written as

$$\boldsymbol{T}\widetilde{\boldsymbol{A}} = \boldsymbol{A}\boldsymbol{T} \quad \text{and} \quad \boldsymbol{T}\widetilde{\boldsymbol{B}} = \boldsymbol{B}$$

It is sufficient, then, to prove the validity of the foregoing relations.

Starting from the first relation and substituting into it the expressions of matrices $\boldsymbol{T}$ and $\widetilde{\boldsymbol{A}}$, we will get

$$\left[\boldsymbol{B}\ \boldsymbol{A}\boldsymbol{B} \ldots \boldsymbol{A}^{N-1}\boldsymbol{B}\right] \begin{bmatrix} 0\,0\,0 \ldots & -\alpha_0 \\ 1\,0\,0 \ldots & -\alpha_1 \\ 0\,1\,0 \ldots & -\alpha_2 \\ \vdots\;\vdots\;\vdots\;\ddots & \vdots \\ 0\,0\,0 \ldots & -\alpha_{N-1} \end{bmatrix} = \boldsymbol{A}\left[\boldsymbol{B}\ \boldsymbol{A}\boldsymbol{B} \ldots \boldsymbol{A}^{N-1}\boldsymbol{B}\right]$$

The product of the matrices of the left-hand side of the foregoing relation is a matrix with elements

$$\left[\boldsymbol{A}\boldsymbol{B}\ \boldsymbol{A}^2\boldsymbol{B} \ldots (-\alpha_0\boldsymbol{B} - \alpha_1\boldsymbol{A}\boldsymbol{B} - \alpha_2\boldsymbol{A}^2\boldsymbol{B} - \cdots - \alpha_{N-1}\boldsymbol{A}^{N-1}\boldsymbol{B})\right]$$

while the product of the matrices of the right member is calculated as

$$\left[\boldsymbol{A}\boldsymbol{B}\ \boldsymbol{A}^2\boldsymbol{B}\ \boldsymbol{A}^3\boldsymbol{B} \ldots \boldsymbol{A}^N\boldsymbol{B}\right]$$

We observe that the two products of matrices have identical corresponding elements, except for those in their last column. Therefore, it is enough to show that

$$\boldsymbol{A}^N\boldsymbol{B} = -\alpha_0\boldsymbol{B} - \alpha_1\boldsymbol{A}\boldsymbol{B} - \alpha_2\boldsymbol{A}^2\boldsymbol{B} - \cdots - \alpha_{N-1}\boldsymbol{A}^{N-1}\boldsymbol{B}$$

But this is something that is easily proven, since from the *Cayley–Hamilton theorem* we know that

$$\boldsymbol{A}^N = -\alpha_0\boldsymbol{I} - \alpha_1\boldsymbol{A} - \alpha_2\boldsymbol{A}^2 - \cdots - \alpha_{N-1}\boldsymbol{A}^{N-1}$$

and therefore we will have

$$\boldsymbol{A}^N\boldsymbol{B} = -\alpha_0\boldsymbol{B} - \alpha_1\boldsymbol{A}\boldsymbol{B} - \alpha_2\boldsymbol{A}^2\boldsymbol{B} - \cdots - \alpha_{N-1}\boldsymbol{A}^{N-1}\boldsymbol{B}$$

that is, the equality $\boldsymbol{T}\widetilde{\boldsymbol{A}} = \boldsymbol{A}\boldsymbol{T}$ is a valid equality. In a similar way, it turns out that

$$\boldsymbol{T}\widetilde{\boldsymbol{B}} = \begin{bmatrix}\boldsymbol{B} & \boldsymbol{A}\boldsymbol{B} & \dots & \boldsymbol{A}^{N-1}\boldsymbol{B}\end{bmatrix}\begin{bmatrix}1\\0\\0\\\vdots\\0\end{bmatrix} = \boldsymbol{B}$$

and therefore the theorem is true. ■

The foregoing theorem refers to the *normal form of controllability*, while the normal form of the controller is defined by the next theorem.

Theorem 2.9.2 *Every controllable continuous-time LTI system with a number of inputs* $K = 1$ *that is described by the state equation*

$$\boldsymbol{q}(t) = \boldsymbol{A}\boldsymbol{q}(t) + \boldsymbol{B}x(t)$$

can always be transformed into the so-called controller form, which is defined as

$$\dot{\boldsymbol{q}}_c(t) = \widetilde{\boldsymbol{A}}\boldsymbol{q}_c(t) + \widetilde{\boldsymbol{B}}x(t) = \begin{bmatrix}0 & 1 & 0 & \dots & 0\\0 & 0 & 1 & \dots & 0\\0 & 0 & 0 & \dots & 0\\\vdots & \vdots & \vdots & \ddots & \vdots\\-\alpha_0 & -\alpha_1 & -\alpha_2 & \dots & -\alpha_{N-1}\end{bmatrix}\boldsymbol{q}_c(t) + \begin{bmatrix}0\\0\\0\\\vdots\\1\end{bmatrix}x(t)$$

where α_i $(i = 1, 2, \dots, N-1)$ *are the coefficients of the characteristic polynomial*

$$det(\lambda\boldsymbol{I} - \boldsymbol{A}) = \lambda^N + \alpha_{N-1}\lambda^{N-1} + \cdots + \alpha_1\lambda + \alpha_0$$

As for the similarity transform $\boldsymbol{T}$ *that causes the transition of the system to the normal form of the controller, it is defined by the relation*

$$\boldsymbol{T} = \begin{bmatrix}\boldsymbol{r}_N\\\boldsymbol{r}_N\boldsymbol{A}\\\boldsymbol{r}_N\boldsymbol{A}^2\\\vdots\\\boldsymbol{r}_N\boldsymbol{A}^{N-1}\end{bmatrix}^{-1} \quad \text{or, equivalently,} \quad \boldsymbol{T}^{-1} = \begin{bmatrix}\boldsymbol{r}_N\\\boldsymbol{r}_N\boldsymbol{A}\\\boldsymbol{r}_N\boldsymbol{A}^2\\\vdots\\\boldsymbol{r}_N\boldsymbol{A}^{N-1}\end{bmatrix} \tag{2.124}$$

where $\boldsymbol{r}_N$ *is the last—that is, the* N*th—row of the inverse of the controllability matrix* $\boldsymbol{S_c}$*; in other words, it will hold that* $\boldsymbol{r}_N\boldsymbol{S_c} = \begin{bmatrix}0\ 0\ 0 \dots 1\end{bmatrix}$.

Proof In full analogy with the previous case, we will again have $\boldsymbol{q}(t) = \boldsymbol{T}\boldsymbol{q}_c(t)$, while matrices $\boldsymbol{A}$ and $\boldsymbol{B}$ will be related to the matrices of the new representation $\widetilde{\boldsymbol{A}}$ and $\widetilde{\boldsymbol{B}}$ through the relations

$$\widetilde{\boldsymbol{A}} = \boldsymbol{T}^{-1}\boldsymbol{A}\boldsymbol{T} \quad \text{and} \quad \widetilde{\boldsymbol{B}} = \boldsymbol{T}^{-1}\boldsymbol{B}$$

Assuming that the transform $\boldsymbol{T}$ has the form $\boldsymbol{T} = \begin{bmatrix}\boldsymbol{c}_1\ \boldsymbol{c}_2 \ldots \boldsymbol{c}_N\end{bmatrix}$, where $\boldsymbol{c}_i$ $(i = 1, 2, \ldots, N)$ are column vectors with N elements each, the first of the foregoing relations is written

$$\widetilde{\boldsymbol{A}} = \boldsymbol{T}^{-1}\boldsymbol{A}\boldsymbol{T} = \begin{bmatrix} \boldsymbol{r}_N \\ \boldsymbol{r}_N\boldsymbol{A} \\ \boldsymbol{r}_N\boldsymbol{A}^2 \\ \vdots \\ \boldsymbol{r}_N\boldsymbol{A}^{N-1} \end{bmatrix} \boldsymbol{A}\ \begin{bmatrix}\boldsymbol{c}_1\ \boldsymbol{c}_2 \ldots \boldsymbol{c}_N\end{bmatrix}$$

or, equivalently,

$$\widetilde{\boldsymbol{A}} = \begin{bmatrix} \boldsymbol{r}_N\boldsymbol{A}\boldsymbol{c}_1 & \boldsymbol{r}_N\boldsymbol{A}\boldsymbol{c}_2 & \boldsymbol{r}_N\boldsymbol{A}\boldsymbol{c}_3 & \ldots & \boldsymbol{r}_N\boldsymbol{A}\boldsymbol{c}_N \\ \boldsymbol{r}_N\boldsymbol{A}^2\boldsymbol{c}_1 & \boldsymbol{r}_N\boldsymbol{A}^2\boldsymbol{c}_2 & \boldsymbol{r}_N\boldsymbol{A}^2\boldsymbol{c}_3 & \ldots & \boldsymbol{r}_N\boldsymbol{A}^2\boldsymbol{c}_N \\ \boldsymbol{r}_N\boldsymbol{A}^3\boldsymbol{c}_1 & \boldsymbol{r}_N\boldsymbol{A}^3\boldsymbol{c}_2 & \boldsymbol{r}_N\boldsymbol{A}^3\boldsymbol{c}_3 & \ldots & \boldsymbol{r}_N\boldsymbol{A}^3\boldsymbol{c}_N \\ \vdots & \vdots & \vdots & \ddots & \vdots \\ \boldsymbol{r}_N\boldsymbol{A}^N\boldsymbol{c}_1 & \boldsymbol{r}_N\boldsymbol{A}^N\boldsymbol{c}_2 & \boldsymbol{r}_N\boldsymbol{A}^N\boldsymbol{c}_3 & \ldots & \boldsymbol{r}_N\boldsymbol{A}^N\boldsymbol{c}_N \end{bmatrix}$$

However, we know that the matrix $\boldsymbol{A}^N$ appearing in the foregoing relation satisfies the *Cayley–Hamilton theorem*

$$\boldsymbol{A}^N = -\alpha_0\boldsymbol{I} - \alpha_1\boldsymbol{A} - \alpha_2\boldsymbol{A}^2 - \cdots - \alpha_{N-1}\boldsymbol{A}^{N-1}$$

Multiplying, first, the last relation from the left by $\boldsymbol{r}_N$ and then from the right by $\boldsymbol{c}_k$, we will get successively

$$\boldsymbol{r}_N\boldsymbol{A}^N = -\alpha_0\boldsymbol{r}_N - \alpha_1\boldsymbol{r}_N\boldsymbol{A} - \alpha_2\boldsymbol{r}_N\boldsymbol{A}^2 - \cdots - \alpha_{N-1}\boldsymbol{r}_N\boldsymbol{A}^{N-1} \qquad \text{and}$$

$$\boldsymbol{r}_N\boldsymbol{A}^N\boldsymbol{c}_k = -\alpha_0\boldsymbol{r}_N\boldsymbol{c}_k - \alpha_1\boldsymbol{r}_N\boldsymbol{A}\boldsymbol{c}_k - \alpha_2\boldsymbol{r}_N\boldsymbol{A}^2\boldsymbol{c}_k - \cdots - \alpha_{N-1}\boldsymbol{r}_N\boldsymbol{A}^{N-1}\boldsymbol{c}_k$$

Now taking into account that matrices $\boldsymbol{T}$ and $\boldsymbol{T}^{-1}$ are related through the relation $\boldsymbol{T}\boldsymbol{T}^{-1} = \boldsymbol{T}^{-1}\boldsymbol{T} = \boldsymbol{I}$, it follows that

$$\boldsymbol{T}^{-1}\boldsymbol{T} = \begin{bmatrix} \boldsymbol{r}_N \\ \boldsymbol{r}_N\boldsymbol{A} \\ \boldsymbol{r}_N\boldsymbol{A}^2 \\ \vdots \\ \boldsymbol{r}_N\boldsymbol{A}^{N-1} \end{bmatrix} \begin{bmatrix}\boldsymbol{c}_1\ \boldsymbol{c}_2 \ldots \boldsymbol{c}_N\end{bmatrix} = \begin{bmatrix} 1\ 0\ 0 \ldots 0 \\ 0\ 1\ 0 \ldots 0 \\ 0\ 0\ 1 \ldots 0 \\ \vdots\ \vdots\ \vdots\ \ddots\ \vdots \\ 0\ 0\ 0 \ldots 1 \end{bmatrix}$$

from which it immediately follows that

$$\boldsymbol{r}_N\boldsymbol{A}^{j-1}\boldsymbol{c}_k = \delta_{kj} \quad (0 \leq j, k \leq N) \quad \text{where} \quad \delta_{kj} = \begin{cases} 1 \text{ for } j = k \\ 0 \text{ for } j \neq k \end{cases}$$

the *Kronecker's* δ. Based on the last relation, we find that

$$\boldsymbol{r}_N\boldsymbol{A}^N\boldsymbol{c}_k = -\alpha_{k-1}$$

In the next step of the process, we use the relations

$$\boldsymbol{r}_N\boldsymbol{A}^{j-1}\boldsymbol{c}_k = \delta_{kj} \quad (0 \leq j, k \leq N) \quad \text{and} \quad \boldsymbol{r}_N\boldsymbol{A}^N\boldsymbol{c}_k = -\alpha_{k-1}$$

to calculate the values of all the elements of matrix $\widetilde{A}$; the result will have the form

$$r_N A c_2 = r_N A^2 c_3 = r_N A^3 c_4 = \cdots = r_N A^{N-1} c_N = 1$$

$$r_N A^N c_1 = -\alpha_0,\ r_N A^N c_2 = -\alpha_1,\ r_N A^N c_3 = -\alpha_2,\ \ldots\ r_N A^N c_N = -\alpha_{N-1}$$

while all the other elements have values equal to zero. It will therefore hold that

$$\widetilde{A} = \begin{bmatrix} 0 & 1 & 0 & \ldots & 0 \\ 0 & 0 & 1 & \ldots & 0 \\ 0 & 0 & 0 & \ldots & 0 \\ \vdots & \vdots & \vdots & \ddots & \vdots \\ -\alpha_0 & -\alpha_1 & -\alpha_2 & \ldots & -\alpha_{N-1} \end{bmatrix}$$

that is, matrix $\widetilde{A}$ will have the form suggested by the foregoing theorem.

In a completely analogous way for matrix B we have

$$\widetilde{B} = T^{-1}B = \begin{bmatrix} r_N \\ r_N A \\ r_N A^2 \\ \vdots \\ r_N A^{N-1} \end{bmatrix} B = \begin{bmatrix} r_N B \\ r_N AB \\ r_N A^2 B \\ \vdots \\ r_N A^{N-1} B \end{bmatrix} = \begin{bmatrix} 0 \\ 0 \\ 0 \\ \vdots \\ 1 \end{bmatrix}$$

To derive the last relation, we took into account the definition of the vector r_N as the Nth row of the inverse of the controllability matrix S_C, which is mathematically formulated as

$$r_N \left[B\ AB\ A^2B \ldots A^{N-1}B \right] = \left[0\ 0\ 0 \ldots 1 \right]$$

or, equivalently, $r_N A^{k-1} B = \delta_{kN} \quad (1 \le k \le N)$, from which the desired result follows. ■

Problem 2.9.12 Transform into controllable form and controller form a LTI system described by the state equation

$$\dot{q}(t) = \begin{bmatrix} 0 & 1 & 0 & 0 \\ 0 & 0 & -1 & 0 \\ 0 & 0 & 0 & 1 \\ 0 & 0 & 5 & 0 \end{bmatrix} q(t) + \begin{bmatrix} 0 \\ 1 \\ 0 \\ -2 \end{bmatrix} x(t).$$

Solution In a previous exercise (Exercise 2.9.9 on page 157), we proved that the system in question was controllable, since the controllability matrix that was calculated as

$$S_c = \left[B\ AB\ A^2B\ A^3B \right] = \begin{bmatrix} 0 & 1 & 0 & 2 \\ 1 & 0 & 2 & 0 \\ 0 & -2 & 0 & -10 \\ -2 & 0 & -10 & 0 \end{bmatrix}$$

has rank $R = 4 = N$, and therefore the transformation into the two preceding forms is possible, with the transformation matrix being the matrix S_c itself. Applying the theory presented earlier, the new matrix $\widetilde{A}$ for the representation of the system in controllability form will be given by the relation $\widetilde{\boldsymbol{A}} = \boldsymbol{T}^{-1}\boldsymbol{A}\boldsymbol{T} = \boldsymbol{S_c}^{-1}\boldsymbol{A}\boldsymbol{S_c}$, with the inverse of the controllability matrix being calculated as[37]

$$\boldsymbol{S_c}^{-1} = \begin{bmatrix} 0 & 5/3 & 0 & 1/3 \\ 5/3 & 0 & 1/3 & 0 \\ 0 & -1/3 & 0 & -5/3 \\ -1/3 & 0 & -5/3 & 0 \end{bmatrix}$$

Substituting the preceding matrices into the defining equation of matrix $\widetilde{\boldsymbol{A}}'$, it turns out that

$$\widetilde{\boldsymbol{A}} = \begin{bmatrix} 0 & 5/3 & 0 & 1/3 \\ 5/3 & 0 & 1/3 & 0 \\ 0 & -1/3 & 0 & -5/3 \\ -1/3 & 0 & -5/3 & 0 \end{bmatrix} \begin{bmatrix} 0 & 1 & 0 & 0 \\ 0 & 0 & -1 & 0 \\ 0 & 0 & 0 & 1 \\ 0 & 0 & 5 & 0 \end{bmatrix} \begin{bmatrix} 0 & 1 & 0 & 2 \\ 1 & 0 & 2 & 0 \\ 0 & -2 & 0 & -10 \\ -2 & 0 & -10 & 0 \end{bmatrix}$$

and performing the operations it turns out that

$$\widetilde{\boldsymbol{A}} = \boldsymbol{S_c}^{-1}\boldsymbol{A}\boldsymbol{S_c} = \begin{bmatrix} 0 & 0 & 0 & 0 \\ 1 & 0 & 0 & 0 \\ 0 & 1 & 0 & 5 \\ 0 & 0 & 1 & 0 \end{bmatrix}$$

If we compare the last expression with the form of matrix $\widetilde{\boldsymbol{A}}$ as predicted by theory, we are led to the conclusion that the coefficients of the characteristic polynomial of the matrix will have the values $\alpha_0 = \alpha_1 = \alpha_3 = 0$ and $\alpha_2 = -5$. To verify whether this result is correct, let us calculate the characteristic polynomial of matrix $\boldsymbol{A}$ using the known method. It is not hard to see that this polynomial takes the form

$$f(\lambda) = \det(\boldsymbol{A} - \lambda \boldsymbol{I}) = \begin{vmatrix} -\lambda & 1 & 0 & 0 \\ 0 & -\lambda & -1 & 0 \\ 0 & 0 & -\lambda & 1 \\ 0 & 0 & 5 & -\lambda \end{vmatrix} = \lambda^4 - 5\lambda^2$$

from which it follows that the result we obtained is correct.

On the other hand, matrix $\widetilde{\boldsymbol{B}}$ of the new representation is calculated as

$$\widetilde{\boldsymbol{B}} = \boldsymbol{S_c}^{-1}\boldsymbol{B} \begin{bmatrix} 0 & 5/3 & 0 & 1/3 \\ 5/3 & 0 & 1/3 & 0 \\ 0 & -1/3 & 0 & -5/3 \\ -1/3 & 0 & -5/3 & 0 \end{bmatrix} \begin{bmatrix} 0 \\ 1 \\ 0 \\ -2 \end{bmatrix} = \begin{bmatrix} 1 \\ 0 \\ 0 \\ 0 \end{bmatrix}$$

which, of course, is expected. Therefore, the considered system expressed in *controllable form* is described by the state equation

$$\dot{\boldsymbol{q}}_c(t) = \widetilde{\boldsymbol{A}}\boldsymbol{q}_c(t) + \widetilde{\boldsymbol{B}}x(t) = \begin{bmatrix} 0 & 0 & 0 & 0 \\ 1 & 0 & 0 & 0 \\ 0 & 1 & 0 & 5 \\ 0 & 0 & 1 & 0 \end{bmatrix} \boldsymbol{q}_c(t) + \begin{bmatrix} 1 \\ 0 \\ 0 \\ 0 \end{bmatrix} x(t)$$

with the vectors $\boldsymbol{q}(t)$ and $\boldsymbol{q}_c(t)$ being related through the transformation

$$\boldsymbol{q}(t) = \boldsymbol{S_c}\boldsymbol{q}_c(t) = \begin{bmatrix} 0 & 1 & 0 & 2 \\ 1 & 0 & 2 & 0 \\ 0 & -2 & 0 & -10 \\ -2 & 0 & -10 & 0 \end{bmatrix} \boldsymbol{q}_c(t)$$

---

[37] Although there is an analytical formula for calculating the inverse of a matrix 4×4, it is too large to be included here and to be used to solve the exercise, making it practically inapplicable (however, it can be found in several books). For this reason, for the calculation of the inverse matrices, we used the function inv($\boldsymbol{A}$) in MATLAB.

On the other hand, the representation of the system *in the form of the controller* will take place using transform $\boldsymbol{T}$, which will be calculated according to its defining equation. Taking into account that the last row of the inverse of the controllability matrix is

$$\boldsymbol{r}_4 = \begin{bmatrix} -1/3 \; 0 \; -5/3 \; 0 \end{bmatrix}$$

and that for matrix $\boldsymbol{A}$ it will hold that—as it turns out—

$$\boldsymbol{A}^2 = \begin{bmatrix} 0 & 0 & -1 & 0 \\ 0 & 0 & 0 & -1 \\ 0 & 0 & 5 & 0 \\ 0 & 0 & 0 & 5 \end{bmatrix} \quad \text{and} \quad \boldsymbol{A}^3 = \begin{bmatrix} 0 & 0 & 0 & -1 \\ 0 & 0 & -5 & 0 \\ 0 & 0 & 0 & 5 \\ 0 & 0 & 25 & 0 \end{bmatrix}$$

it is not difficult to ascertain that

$$\boldsymbol{r}_4\boldsymbol{A} = \begin{bmatrix} 0 \; -1/3 \; 0 \; -5/3 \end{bmatrix}, \quad \boldsymbol{r}_4\boldsymbol{A}^2 = \begin{bmatrix} 0 \; 0 \; -1/2 \; 0 \end{bmatrix}, \quad \boldsymbol{r}_4\boldsymbol{A}^3 = \begin{bmatrix} 0 \; 0 \; 0 \; -1/2 \end{bmatrix}$$

Therefore, matrices $\boldsymbol{T}^{-1}$ and $\boldsymbol{T}$ will have the form

$$\boldsymbol{T}^{-1} = \begin{bmatrix} \boldsymbol{r}_4 \\ \boldsymbol{r}_4\boldsymbol{A} \\ \boldsymbol{r}_4\boldsymbol{A}^2 \\ \boldsymbol{r}_4\boldsymbol{A}^3 \end{bmatrix} = \begin{bmatrix} -1/3 & 0 & -5/3 & 0 \\ 0 & -1/3 & 0 & -5/3 \\ 0 & 0 & -1/2 & 0 \\ 0 & 0 & 0 & -1/2 \end{bmatrix} \quad \text{and} \quad \boldsymbol{T} = \begin{bmatrix} \boldsymbol{r}_4 \\ \boldsymbol{r}_4\boldsymbol{A} \\ \boldsymbol{r}_4\boldsymbol{A}^2 \\ \boldsymbol{r}_4\boldsymbol{A}^3 \end{bmatrix}^{-1} = \begin{bmatrix} -3 & 0 & 1 & 0 \\ 0 & -3 & 0 & 1 \\ 0 & 0 & -2 & 0 \\ 0 & 0 & 0 & -2 \end{bmatrix}$$

and by substituting in the equations of the transformation matrices we finally arrive at the relations

$$\widetilde{\boldsymbol{A}} = \begin{bmatrix} 0 & 1 & 0 & 0 \\ 0 & 0 & 1 & 0 \\ 0 & 0 & 0 & 1 \\ 0 & 0 & 5 & 0 \end{bmatrix} \quad \text{and} \quad \widetilde{\boldsymbol{B}} = \begin{bmatrix} 0 \\ 0 \\ 0 \\ 1 \end{bmatrix}$$

in full agreement with prevailing theory. We are led, therefore, to the conclusion that the representation of the system *in the form of the controller* will be described by the state equation

$$\dot{\boldsymbol{q}}_c(t) = \widetilde{\boldsymbol{A}}\boldsymbol{q}_c(t) + \widetilde{\boldsymbol{B}}x(t) = \begin{bmatrix} 0 & 1 & 0 & 0 \\ 0 & 0 & 1 & 0 \\ 0 & 0 & 0 & 1 \\ 0 & 0 & 5 & 0 \end{bmatrix} \boldsymbol{q}_c(t) + \begin{bmatrix} 0 \\ 0 \\ 0 \\ 1 \end{bmatrix} x(t)$$

with the vectors $\boldsymbol{q}(t)$ and $\boldsymbol{q}_c(t)$ being related through the transformation

$$\boldsymbol{q}(t) = \boldsymbol{T}\boldsymbol{q}_c(t) = \begin{bmatrix} -3 & 0 & 1 & 0 \\ 0 & -3 & 0 & 1 \\ 0 & 0 & -2 & 0 \\ 0 & 0 & 0 & -2 \end{bmatrix} \boldsymbol{q}_c(t)$$

Normal Forms for Observability

Considering an *observable* continuous-time LTI system described by dynamic equations of the form[38]

$$\dot{\boldsymbol{q}}(t) = \boldsymbol{A}\boldsymbol{q}(t) \quad \text{and} \quad \boldsymbol{y}(t) = \boldsymbol{C}\boldsymbol{q}(t)$$

[38] From what we have mentioned so far, it is obvious that *the characteristic of controllability concerns the inputs and the state vector of the system, but it has nothing to do with its outputs*, while the characteristic of observability *is not related in any way to the inputs of the system but is something that concerns the state vector and its outputs*. For this reason, and to simplify our description, in our study of controllability, we only dealt with the system's state equation, while for the same reason in this section we will simplify the system's dynamic equations by setting the output $\boldsymbol{x}(t)$ to zero.

where $\boldsymbol{q}(t)$ is the state vector of the system with length N elements, and $\boldsymbol{y}(t)$ is the output vector of the system with length $\boldsymbol{L}$ elements, it is proven that this can always be transformed into two different forms, the first of which is known as the *observable form*, while the second is known as *the observer's form*. In the following discussion, we will consider the simple case of a system with only one output, although the generalization to multiple output systems is done in a similar way.

Theorem 2.9.3 *Every observable continuous-time LTI system with a number of outputs $L = 1$, which is described by the dynamic equations*

$$\dot{\boldsymbol{q}}(t) = \boldsymbol{A}\boldsymbol{q}(t) \quad \textit{and} \quad y(t) = \boldsymbol{C}\boldsymbol{q}(t)$$

can always be transformed into the observable form

$$\dot{\boldsymbol{q}}_o(t) = \widetilde{\boldsymbol{A}}\boldsymbol{q}_o(t) = \begin{bmatrix} 0 & 1 & 0 & \dots & 0 \\ 0 & 0 & 1 & \dots & 0 \\ 0 & 0 & 0 & \dots & 0 \\ \vdots & \vdots & \vdots & \ddots & \vdots \\ -\alpha_0 & -\alpha_1 & -\alpha_2 & \dots & -\alpha_{N-1} \end{bmatrix} \boldsymbol{q}_o(t)$$

$$y(t) = \widetilde{\boldsymbol{C}}\boldsymbol{q}_o(t) = \begin{bmatrix} 1\,0\,0\dots0 \end{bmatrix} \boldsymbol{q}_o(t)$$

where α_i $(i = 1, 2, \dots, N-1)$ are the coefficients of the characteristic polynomial $(\alpha_N = 1)$

$$det(\lambda \boldsymbol{I} - \boldsymbol{A}) = \lambda^N + \alpha_{N-1}\lambda^{N-1} + \dots + \alpha_1\lambda + \alpha_0$$

while the similarity transformation that causes the transition of the system to the observable form is the inverse of the observability matrix

$$\boldsymbol{T} \equiv \boldsymbol{S_o}^{-1} = \begin{bmatrix} \boldsymbol{C} \\ \boldsymbol{CA} \\ \boldsymbol{CA}^2 \\ \vdots \\ \boldsymbol{CA}^{N-1} \end{bmatrix}^{-1} \tag{2.125}$$

Proof Based on the *duality* of controllability–observability, let us consider the controllable form of a dual—with respect to the considered—system, which, as is well known, is defined as $\dot{\boldsymbol{p}}(t) = \boldsymbol{A}^T\boldsymbol{p}(t) + \boldsymbol{C}^T\upsilon(t)$. As we proved in the previous section, the similarity transform $\boldsymbol{W}$ that allows the representation of this dual system in its *controllable* form, formulated as $\dot{p}_c(t) = \widetilde{\boldsymbol{A}}'\boldsymbol{p}_c(t) + \widetilde{\boldsymbol{B}}'\upsilon(t)$, where matrices $\widetilde{\boldsymbol{A}}'$ and $\widetilde{\boldsymbol{B}}'$ are defined as

$$\widetilde{\boldsymbol{A}}' = \boldsymbol{W}^{-1}(\boldsymbol{A}^T)\boldsymbol{W} = \begin{bmatrix} 0\,0\,0\dots & -\alpha_0 \\ 1\,0\,0\dots & -\alpha_1 \\ 0\,1\,0\dots & -\alpha_2 \\ \vdots\,\vdots\,\vdots\,\ddots & \vdots \\ 0\,0\,0\dots & -\alpha_{N-1} \end{bmatrix}, \quad \widetilde{\boldsymbol{B}}' = \boldsymbol{W}^{-1}(\boldsymbol{C}^T) = \begin{bmatrix} 1 \\ 0 \\ 0 \\ \vdots \\ 0 \end{bmatrix}$$

is $\boldsymbol{W} = \boldsymbol{S_c} = (\boldsymbol{S_o})^T$. Now, taking the inverses of the preceding matrices, so as to return to our original system, we get

$$\widetilde{\boldsymbol{A}} = \widetilde{\boldsymbol{A}}'^T = \boldsymbol{W}^T\boldsymbol{A}(\boldsymbol{W}^{-1})^T = \begin{bmatrix} 0 & 1 & 0 & \dots & 0 \\ 0 & 0 & 1 & \dots & 0 \\ \vdots & \vdots & \vdots & \ddots & \vdots \\ -\alpha_0 & -\alpha_1 & -\alpha_2 & \dots & -\alpha_{N-1} \end{bmatrix} \quad \text{and}$$

$$\widetilde{\boldsymbol{C}} = \widetilde{\boldsymbol{B}}'^T = \boldsymbol{C}(\boldsymbol{W}^{-1})^T = \begin{bmatrix} 1\,0\,0\dots0 \end{bmatrix}$$

which is simply the *observable normal form*. If we assume that the transformation that transfers the system to this form is $\boldsymbol{T}$, then from the last relation it immediately follows that

$$\boldsymbol{T} = (\boldsymbol{W}^{-1})^T = [(\boldsymbol{S_o}^T)^{-1}]^T = \boldsymbol{S_o}^{-1} = \begin{bmatrix} \boldsymbol{C} \\ \boldsymbol{CA} \\ \boldsymbol{CA}^2 \\ \vdots \\ \boldsymbol{CA}^{N-1} \end{bmatrix}^{-1}$$

a conclusion that completes the proof of the theorem. ■

The preceding theorem refers to the observable form of the system, while the second form in which it can be found is defined as follows.

Theorem 2.9.4 *Every observable continuous-time LTI system with a number of outputs $L = 1$, which is described by the dynamic equations*

$$\dot{\boldsymbol{q}}(t) = \boldsymbol{A}\boldsymbol{q}(t) \quad \textit{and} \quad y(t) = \boldsymbol{C}\boldsymbol{q}(t)$$

can always be transformed into the so-called observer form, which is formulated as

$$\dot{\boldsymbol{q}}_o(t) = \widetilde{\boldsymbol{A}}\boldsymbol{q}_o(t) = \begin{bmatrix} 0\,0\,0 \dots & -\alpha_0 \\ 1\,0\,0 \dots & -\alpha_1 \\ 0\,1\,0 \dots & -\alpha_2 \\ \vdots\,\vdots\,\vdots\,\ddots & \vdots \\ 0\,0\,0 \dots & -\alpha_{N-1} \end{bmatrix} \boldsymbol{q}_o(t)$$

$$y(t) = \widetilde{\boldsymbol{C}}\boldsymbol{q}_o(t) = \begin{bmatrix} 0\,0\,0 \dots 1 \end{bmatrix} \boldsymbol{q}_o(t)$$

where α_i $(i = 1, 2, \dots, N-1)$ are the coefficients of the characteristic polynomial

$$det(\lambda \boldsymbol{I} - \boldsymbol{A}) = \lambda^N + \alpha_{N-1}\lambda^{N-1} + \dots + \alpha_1\lambda + \alpha_0$$

As for the similarity transformation that causes the system's transition to the observer form, it is defined as

$$\boldsymbol{T} = \begin{bmatrix} \boldsymbol{c}_N \; \boldsymbol{A}\boldsymbol{c}_N \; \boldsymbol{A}^2\boldsymbol{c}_N \dots \boldsymbol{A}^{N-1}\boldsymbol{c}_N \end{bmatrix} \tag{2.126}$$

where $\boldsymbol{c}_N$ is the last—that is, the Nth—column of the observability matrix $\boldsymbol{S_o}$, which therefore satisfies the condition

$$\boldsymbol{S_o}\boldsymbol{c}_N = \begin{bmatrix} 0 \\ 0 \\ 0 \\ \vdots \\ 1 \end{bmatrix}$$

Proof Relying once again on the *duality* that characterizes the concepts of controllability and observability, let us consider the controller form of the dual—with respect to the considered—system described by a state equation of the form

$$\dot{\boldsymbol{p}}(t) = \boldsymbol{A}^T\boldsymbol{p}(t) + \boldsymbol{C}^T\upsilon(t)$$

In this case, the transform $\boldsymbol{W}$ that transforms the representation of the dual system into the form of the controller has, as we have proven, the form

$$\boldsymbol{W} = \begin{bmatrix} \boldsymbol{r}_N \\ \boldsymbol{r}_N \boldsymbol{A}^T \\ \boldsymbol{r}_N (\boldsymbol{A}^T)^2 \\ \vdots \\ \boldsymbol{r}_N (\boldsymbol{A}^T)^{N-1} \end{bmatrix}$$

where $\boldsymbol{r}_N$ is the row number N of the inverse of the controllability matrix $\boldsymbol{S_c} = (\boldsymbol{S_o})^T$, that is

$$\boldsymbol{r}_N \boldsymbol{S_c} = \boldsymbol{r}_N (\boldsymbol{S_o})^T = [0\ 0\ 0\ \ldots\ 1]$$

We also know that a dual system transformed *in the form of the controller* is described by the equation

$$\dot{\boldsymbol{p}}_c(t) = \widetilde{\boldsymbol{A}}' \boldsymbol{p}_c(t) + \widetilde{\boldsymbol{B}}' \upsilon(t)$$

with matrices $\widetilde{\boldsymbol{A}}'$ and $\widetilde{\boldsymbol{B}}'$ being defined as

$$\widetilde{\boldsymbol{A}}' = \boldsymbol{W}^{-1} (\boldsymbol{A}^T) \boldsymbol{W} = \begin{bmatrix} 0 & 1 & 0 & \ldots & 0 \\ 0 & 0 & 1 & \ldots & 0 \\ 0 & 0 & 0 & \ldots & 0 \\ \vdots & \vdots & \vdots & \ddots & \vdots \\ -\alpha_0 & -\alpha_1 & -\alpha_2 & \ldots & -\alpha_{N-1} \end{bmatrix} \quad \text{and} \quad \widetilde{\boldsymbol{B}}' = \boldsymbol{W}^{-1} (\boldsymbol{C})^T = \begin{bmatrix} 0 \\ 0 \\ 0 \\ \vdots \\ 1 \end{bmatrix}$$

If we now take the inverses of the foregoing matrices, so as to return to our original system, we will end up with the matrices

$$\widetilde{\boldsymbol{A}} = \widetilde{\boldsymbol{A}}'^T = \boldsymbol{W}^T \boldsymbol{A} (\boldsymbol{W}^{-1})^T = \begin{bmatrix} 0\,0\,0 \ldots & -\alpha_0 \\ 1\,0\,0 \ldots & -\alpha_1 \\ 0\,1\,0 \ldots & -\alpha_2 \\ \vdots\,\vdots\,\vdots\,\ddots & \vdots \\ 0\,0\,0 \ldots & -\alpha_{N-1} \end{bmatrix} \quad \text{and}$$

$$\widetilde{\boldsymbol{C}} = \widetilde{\boldsymbol{B}}'^T = \boldsymbol{C} (\boldsymbol{W}^{-1})^T = [0\,0\,0 \ldots 1]$$

which are simply those that define *the form of the observer*. Assuming that the transformation that leads the system to this form is $\boldsymbol{T}$, it is not difficult to discover that this is defined as

$$\boldsymbol{T} = (\boldsymbol{W}^{-1})^T = \left[\left(\begin{bmatrix} \boldsymbol{r}_N \\ \boldsymbol{r}_N \boldsymbol{A}^T \\ \boldsymbol{r}_N (\boldsymbol{A}^T)^2 \\ \vdots \\ \boldsymbol{r}_N (\boldsymbol{A}^T)^{N-1} \end{bmatrix}^{-1} \right)^{-1} \right]^T = \left[\boldsymbol{r}_N^T\ \boldsymbol{A} \boldsymbol{r}_N^T\ \boldsymbol{A}^2 \boldsymbol{r}_N^T \ldots \boldsymbol{A}^{N-1} \boldsymbol{r}_N^T \right]$$

But the vector $\boldsymbol{r}_N$ expresses the row number N of the inverse of the controllability matrix $\boldsymbol{S_c} = (\boldsymbol{S_o})^T$, that is, the magnitude $\boldsymbol{r}_N^T$ expresses the column number N $\boldsymbol{c}_N$ of the inverse observability matrix. We will therefore have

$$\boldsymbol{T} = \left[\boldsymbol{c}_N\ \boldsymbol{A} \boldsymbol{c}_N\ \boldsymbol{A}^2 \boldsymbol{c}_N \ldots \boldsymbol{A}^{N-1} \boldsymbol{c}_N \right]$$

a conclusion that completes the proof of the theorem. ■

Problem 2.9.13 Consider a LTI system described by the dynamic equations

$$\dot{\boldsymbol{q}}(t) = \begin{bmatrix} 0 & 0 & 1 & 0 \\ 0 & 0 & 0 & 1 \\ -1 & 1 & 0 & 0 \\ 1 & -1 & 0 & 0 \end{bmatrix} \boldsymbol{q}(t) + \begin{bmatrix} 0 \\ 0 \\ -1 \\ 1 \end{bmatrix} x(t), \quad y(t) = \begin{bmatrix} 1 & 0 & 0 & 0 \end{bmatrix} \boldsymbol{q}(t)$$

Examine whether it is observable or not, and, if it is, transform it into the observable form and the normal form of the observer.

Solution The rank of matrix $\boldsymbol{A}$ of the system is $N = 4$. From the elements of the problem it emerges that

$$\boldsymbol{A}^2 = \begin{bmatrix} -1 & 1 & 0 & 0 \\ 1 & -1 & 0 & 0 \\ 0 & 0 & -1 & 1 \\ 0 & 0 & 1 & -1 \end{bmatrix} \quad \text{and} \quad \boldsymbol{A}^3 = \begin{bmatrix} 0 & 0 & -1 & 1 \\ 0 & 0 & 1 & -1 \\ 2 & -2 & 0 & 0 \\ -2 & 2 & 0 & 0 \end{bmatrix}$$

from which we then arrive at the values

$$\boldsymbol{CA} = \begin{bmatrix} 0 & 0 & 1 & 0 \end{bmatrix}, \quad \boldsymbol{CA}^2 = \begin{bmatrix} -1 & 1 & 0 & 0 \end{bmatrix}, \quad \boldsymbol{CA}^3 = \begin{bmatrix} 0 & 0 & -1 & 1 \end{bmatrix}$$

Therefore, the observability matrix of the system will have the form

$$\boldsymbol{S_o} = \begin{bmatrix} \boldsymbol{C} \\ \boldsymbol{CA} \\ \boldsymbol{CA}^2 \\ \boldsymbol{CA}^3 \end{bmatrix} = \begin{bmatrix} 1 & 0 & 0 & 0 \\ 0 & 0 & 1 & 0 \\ -1 & 1 & 0 & 0 \\ 0 & 0 & -1 & 1 \end{bmatrix}$$

It is proven that the rank of this matrix has a value $R = 4 = N$. Therefore, the considered system is observable, and it is possible to transform it into the two normal forms with respect to observability presented in the preceding pages.

Starting from the *observable form*, the transform $\boldsymbol{T}$ that allows the representation of the system in this form is the inverse of the observability matrix, which is calculated as

$$\boldsymbol{T} = \boldsymbol{S_o}^{-1} \begin{bmatrix} 1 & 0 & 0 & 0 \\ 1 & 0 & 1 & 0 \\ 0 & 1 & 0 & 0 \\ 0 & 1 & 0 & 1 \end{bmatrix}$$

and therefore the new matrix $\widetilde{\boldsymbol{A}}$ of the representation of the system in this form will have—according to theory—a value equal to

$$\widetilde{\boldsymbol{A}} = \boldsymbol{T}^{-1}\boldsymbol{AT} = \boldsymbol{S_o}\boldsymbol{A}\boldsymbol{S_o}^{-1} = \begin{bmatrix} 0 & 1 & 0 & 0 \\ 0 & 0 & 1 & 0 \\ 0 & 0 & 0 & 1 \\ 0 & 0 & -2 & 0 \end{bmatrix}$$

If we compare the preceding form with that predicted by theory, we come to the conclusion that the coefficients of the characteristic polynomial of matrix $\boldsymbol{A}$ should be $\alpha_0 = \alpha_1 = \alpha_3 = 0$ and $\alpha_2 = 2$. This result is correct, since, if we construct the aforementioned polynomial, it turns out that

$$f(\lambda) = \det(\boldsymbol{A} - \lambda \boldsymbol{I}) = \begin{vmatrix} -\lambda & 0 & 1 & 0 \\ 0 & -\lambda & 0 & 1 \\ -1 & 1 & -\lambda & 0 \\ 1 & -1 & -\lambda & 0 \end{vmatrix} = -\lambda^4 + 2\lambda^2$$

from which emerge the values of the coefficients we calculated previously.

On the other hand, matrix $\widetilde{\boldsymbol{C}}$ of the new representation will have a value equal to

$$\widetilde{\boldsymbol{C}} = \boldsymbol{C}\boldsymbol{S_o}^{-1} = \begin{bmatrix} 1 & 0 & 0 & 0 \end{bmatrix}$$

and therefore the observable form of this system is described by the dynamic equations

$$\dot{\boldsymbol{q}}_o(t) = \begin{bmatrix} 0 & 1 & 0 & 0 \\ 0 & 0 & 1 & 0 \\ 0 & 0 & 0 & 1 \\ 0 & 0 & -2 & 0 \end{bmatrix} \boldsymbol{q}_o(t) \quad \text{and} \quad \boldsymbol{y}(t) = \begin{bmatrix} 1 & 0 & 0 & 0 \end{bmatrix} \boldsymbol{q}_o(t)$$

(for reasons of simplicity we have omitted the input $x(t)$, which is not related to observability), with the state vectors $\boldsymbol{q}(t)$ and $\boldsymbol{q}_o(t)$ being related through the transform $\boldsymbol{T} \equiv \boldsymbol{S_o}^{-1}$.

On the other hand, for the transformation of the system into the form of the observer we need the fourth column of the inverse of the observability matrix $\boldsymbol{c_4}$, as well as the quantities $\boldsymbol{A}\boldsymbol{c}_4$, $\boldsymbol{A}^2\boldsymbol{c}_4$, and $\boldsymbol{A}^3\boldsymbol{c}_4$. These expressions are calculated as

$$\boldsymbol{c}_4 \begin{bmatrix} 0 \\ 0 \\ 0 \\ 1 \end{bmatrix}, \quad \boldsymbol{A}\boldsymbol{c}_4 = \begin{bmatrix} 0 \\ 1 \\ 0 \\ 0 \end{bmatrix}, \quad \boldsymbol{A}^2\boldsymbol{c}_4 = \begin{bmatrix} 0 \\ 0 \\ 1 \\ -1 \end{bmatrix}, \quad \boldsymbol{A}^3\boldsymbol{c}_4 = \begin{bmatrix} 1 \\ -1 \\ 0 \\ 0 \end{bmatrix}$$

and therefore, the transform $\boldsymbol{T}$ that converts the representation of the system into the form of the observer is

$$\boldsymbol{T} = [\, \boldsymbol{c}_N \;\; \boldsymbol{A}\boldsymbol{c}_N \;\; \boldsymbol{A}^2\boldsymbol{c}_N \;\; \boldsymbol{A}^3\boldsymbol{c}_N \,] = \begin{bmatrix} 0 & 0 & 0 & 1 \\ 0 & 1 & 0 & -1 \\ 0 & 0 & 1 & 0 \\ 1 & 0 & -1 & 0 \end{bmatrix}$$

the inverse of which is the matrix

$$\boldsymbol{T}^{-1} \begin{bmatrix} 0 & 0 & 0 & 1 \\ 0 & 1 & 0 & -1 \\ 0 & 0 & 1 & 0 \\ 1 & 0 & -1 & 0 \end{bmatrix}^{-1} = \begin{bmatrix} 0 & 0 & 1 & 1 \\ 1 & 1 & 0 & 0 \\ 0 & 0 & 1 & 0 \\ 1 & 0 & 0 & 0 \end{bmatrix}$$

as can be demonstrated by the use of the function inv($\boldsymbol{T}$) in MATLAB. Therefore, the matrices that allow the representation of the system in the form of the observer will be

$$\widetilde{\boldsymbol{A}} = \boldsymbol{T}^{-1}\boldsymbol{A}\boldsymbol{T} = \begin{bmatrix} 0 & 0 & 0 & 0 \\ 1 & 0 & 0 & 0 \\ 0 & 1 & 0 & -2 \\ 0 & 0 & 1 & 0 \end{bmatrix} \quad \text{and} \quad \widetilde{\boldsymbol{C}} = \boldsymbol{C}\boldsymbol{T} = \begin{bmatrix} 0 & 0 & 0 & 1 \end{bmatrix}$$

that is, the dynamic equations of the system in this form will be

$$\dot{\boldsymbol{q}}_o(t) = \begin{bmatrix} 0 & 0 & 0 & 0 \\ 1 & 0 & 0 & 0 \\ 0 & 1 & 0 & -2 \\ 0 & 0 & 1 & 0 \end{bmatrix} \boldsymbol{q}_o(t) \quad \text{and} \quad \boldsymbol{y}(t) = \begin{bmatrix} 1 & 0 & 0 & 0 \end{bmatrix} \boldsymbol{q}_o(t)$$

which is also the desired result. ■

Based on the proofs of the foregoing theorems and referring to the initial sections where we constructed the dynamic equations for the case of the representations *direct form I* and *direct form II*, it is not difficult to see that the equations that result from direct form I refer to the form of the observer and are known as the observability scheme, while those related to direct form II refer to the form of the controller and are known as the controllability scheme. It is obvious that *the controllability scheme is by definition controllable, while the observability scheme is by definition observable*. Controllability schemes play a decisive role in the theory of systems of control, as they allow for the design and representation of controllable systems, while observability diagrams are particularly useful in processes of *simulation* of linear dynamic systems, because they allow for the integration and use of the initial conditions that are appropriate in each case.

More details on the transformations that allow for the transition from one form of representation to the other, as well as on the generalization of the theory in the case of nonlinear and time-varying systems, can be found in the literature.

2.9.10 Stability in State Space

In a previous section, we examined the *stability* of continuous-time systems using criteria that concerned the *impulse response* and the *roots of the characteristic polynomial* that is related to the differential equations that describe them. However, the study of stability based only on input/output characteristics does not always lead to safe conclusions, with the sole exception of systems that are *simultaneously controllable and observable*. This characteristic of the *simultaneous controllability and observability* of a system will be studied using a *transfer function* that describes the system, and for this reason the discussion is postponed to the chapter of the *Laplace transform*. In this section, we will talk simply about the characterization of a continuous-time system in terms of its stability, using its properties in state space.

The study of the stability of systems in state space is based on the theory formulated by *Lyapunov* according to which *the stability of a system with a state equation of the form* $\dot{\boldsymbol{q}}(t) = \boldsymbol{A}\boldsymbol{q}(t) + \boldsymbol{B}\boldsymbol{x}(t)$ *can be fully determined by studying the homogeneous state equation* $\dot{\boldsymbol{q}}(t) = \boldsymbol{A}\boldsymbol{q}(t)$ in the sense that the system matrix $\boldsymbol{A}$ contains all the necessary information regarding the characterization of the system's stability. In most cases, we are interested in the so-called asymptotic stability, which is defined in the following way.

A continuous and time-invariant system with a homogeneous state equation $\dot{\boldsymbol{q}}(t) = \boldsymbol{A}\boldsymbol{q}(t)$ is called asymptotically stable if and only if the solution of this equation with initial condition $\boldsymbol{q}(t_0) = \boldsymbol{q}_0$ tends to zero for every initial condition $\boldsymbol{q}_0$, that is, when

$$\lim_{t\to\infty} \boldsymbol{q}(t) = 0, \quad \forall \boldsymbol{q}(t_0). \tag{2.127}$$

The most important of the stability criteria that have been formulated over time is the one formulated by Lyapunov according to which a continuous-time system described by the aforementioned characteristics is considered asymptotically stable when all the roots of the characteristic polynomial of the matrix $\boldsymbol{A}$

$$f(\lambda) = \alpha_N\lambda^N + \alpha_{N-1}\lambda^{N-1} + \cdots + \alpha_2\lambda^2 + \alpha_1\lambda + \alpha_0 = 0$$

—in other words, its eigenvalues—have *negative real parts*. The proof of this criterion arises from the observation that since the solution of the homogeneous state equation $\dot{\boldsymbol{q}}(t) = \boldsymbol{A}\boldsymbol{q}(t)$ has the form $\boldsymbol{q}(t) = e^{\boldsymbol{A}t}\boldsymbol{q}(t_0)$, the requirement for the property $\lim \boldsymbol{q}(t) = 0$ to hold at the limit $t \to \infty$ and for every initial condition $\boldsymbol{q}(t_0)$ is reformulated as

$$\lim_{t\to\infty} e^{\boldsymbol{A}t} = \boldsymbol{O}$$

where $\boldsymbol{O}$ is the zero matrix of appropriate dimensions. However, we know that the exponential matrix $e^{\boldsymbol{A}t}$ depends on the eigenvalues of the characteristic polynomial of matrix $\boldsymbol{A}$, since, as we proved in a previous section, it is given by the relation

$$e^{\boldsymbol{A}t} = \boldsymbol{P} \begin{bmatrix} e^{\lambda_1 t} & 0 & 0 \dots & 0 \\ 0 & e^{\lambda_2 t} & 0 \dots & 0 \\ \vdots & \vdots & \vdots \ddots & \vdots \\ 0 & 0 & 0 \dots & e^{\lambda_N t} \end{bmatrix} \boldsymbol{P}^{-1}$$

where $\boldsymbol{P}$ and $\boldsymbol{P}^{-1}$ are the matrices that diagonalize the system's matrix $\boldsymbol{A}$. From the foregoing relation, it becomes apparent that in order for the exponential function $e^{\boldsymbol{A}t}$ to tend to the zero matrix at the limit $t \to \infty$, all the eigenvalues of matrix $\boldsymbol{A}$ must have negative real parts,[39] or I,n mathematical notation, $\Re\{\lambda_i\} < 0\ (1 \leq i \leq N)$. The last conclusion refers to the case of *asymptotic stability*, in contrast, if $\Re\{\lambda_i\} \leq 0\ (1 \leq i \leq N)$, then the system is *marginally stable*. Finally, if there is even one eigenvalue of matrix $\boldsymbol{A}$ with a positive real part, then the system is aforementioned to be *unstable*.

Problem 2.9.14 Considering a continuous-time LTI system described by the matrix

$$\boldsymbol{A}_1 = \begin{bmatrix} -1 & 2 & 1 \\ 1 & -1 & 1 \\ 0 & 0 & 1 \end{bmatrix} \quad \boldsymbol{A}_2 = \begin{bmatrix} 1 & 2 & 1 \\ 0 & -1 & 1 \\ 1 & 2 & -1 \end{bmatrix} \quad \boldsymbol{A}_3 = \begin{bmatrix} 1 & 2 & 3 \\ 0 & 1 & 2 \\ 0 & 0 & 1 \end{bmatrix}$$

examine whether it is stable or unstable (Shmaliy, *Continuous-Time Systems*, Problem 4.28).

Solution The solution of the exercise consists in determining the eigenvalues of the matrix and characterizing its stability based on the sign of its real part. For this specific matrix, its characteristic polynomial is calculated as

$$f(\lambda) = \det(\boldsymbol{A}_1 - \lambda \boldsymbol{I}) = \begin{vmatrix} -1-\lambda & 2 & 1 \\ 1 & -1-\lambda & 1 \\ 0 & 0 & 1-\lambda \end{vmatrix} = (\lambda - 1)(\lambda^2 + 2\lambda - 1)$$

and has three distinct roots with values

$$\lambda_1 = 1, \quad \lambda_2 = -1 + \sqrt{2} = 0.414213, \quad \lambda_3 = -1 - \sqrt{2} = -2.414213$$

We observe that there are eigenvalues with positive real parts, that is, the system is *unstable*.

For matrix $\boldsymbol{A}_2$, its characteristic polynomial is calculated as

$$f(\lambda) = \det(\boldsymbol{A}_2 - \lambda \boldsymbol{I}) = \begin{vmatrix} 1-\lambda & 2 & 1 \\ 0 & -1-\lambda & 1 \\ 1 & 2 & -1-\lambda \end{vmatrix} = -\lambda^3 - \lambda^2 + 4\lambda + 2$$

and it has three distinct roots with values

$$\lambda_1 = 1.8136, \qquad \lambda_2 = -0.4707, \qquad \lambda_3 = -1 - \sqrt{2} = -2.3429$$

We observe that there are eigenvalues with positive real parts, that is, the system is *unstable*.

For matrix $\boldsymbol{A}_3$, its characteristic polynomial is calculated as

[39] A matrix characterized by this property is known as a Hurwitz matrix.

$$f(\lambda) = \det(\boldsymbol{A}_3 - \lambda \boldsymbol{I}) = \begin{vmatrix} 1-\lambda & 2 & 3 \\ 0 & 1-\lambda & 2 \\ 0 & 0 & 1-\lambda \end{vmatrix} = (1-\lambda)^3$$

and it has three distinct roots with values

$$\lambda_1 = \lambda_2 = \lambda_3 = 1$$

We observe that there are eigenvalues with positive real parts, that is, the system is *unstable*.

2.10 Problems

1. For each of the following systems, determine the category it belongs to (i.e., whether it is SISO, SIMO, MISO, or MIMO). In the following equations, the parameters $\boldsymbol{A}$, $\boldsymbol{B}$, $\boldsymbol{C}$, and $\boldsymbol{D}$ are coefficient matrices with the appropriate dimensions in each case, while elements referring to vectors are displayed in bold:

   (a) $\dfrac{d\boldsymbol{q}(t)}{dt} = \boldsymbol{A}\boldsymbol{q}(t) + \boldsymbol{B}x(t)$ where $\boldsymbol{y}(t) = \boldsymbol{C}\boldsymbol{q}(t) + \boldsymbol{D}x(t)$, (b) $\alpha x(t) = \boldsymbol{B}\boldsymbol{y}(t)$

   (c) $y(t) = \boldsymbol{A}\boldsymbol{x}(t) + \beta\dfrac{d^2y(t)}{dt^2} + \dfrac{d}{dt}\left\{c(t)x(t)\right\}$, (d) $\dfrac{d\boldsymbol{q}(t)}{dt} = \boldsymbol{A}\boldsymbol{q}(t) + \boldsymbol{B}x(t)$ where $\boldsymbol{y}(t) = \boldsymbol{C}\boldsymbol{q}(t)$

2. Determine whether each of the following systems is static or dynamic:

$$y(t) = 2tx(2t) \quad \text{and} \quad y(t) = t^2\frac{dx(t)}{dt}$$

3. Determine whether each of the continuous-time systems described by the following equations is static or dynamic, time-varying or time-invariant, linear or nonlinear, and causal or noncausal:

$$y(t) = x(t-3) + x(3-t), \quad y(t) = \cos(5t)x(t), \quad y(t) = \int_{-\infty}^{5t} x(\tau)d\tau$$

4. Determine whether each of the continuous-time systems described by the following equations is static or dynamic, time-varying or time-invariant, linear or nonlinear, and causal or noncausal:

$$y(t) = \begin{cases} 0 & \text{for } t < 0 \\ x(t) + x(t-3) & \text{for } t \geq 0 \end{cases}, \quad y(t) = x\left(\frac{t}{5}\right), \quad y(t) = \frac{dx(t)}{dt}$$

5. Consider the continuous-time systems described by the differential equations

$$\text{(a) } y(t) = \alpha x(t) + e^{\alpha t}\int_{-\infty}^{t+\vartheta} x(\tau)d\tau + e\frac{dx(t)}{dt}, \quad (\vartheta \geq 0), \quad \text{(b) } y(t) = \alpha\frac{dy(t)}{dt} + \beta\frac{d^2y(t)}{dt^2} + \gamma\frac{d^3y(t)}{dt^3} + \delta x(t)$$

   Determine whether each system is static or dynamic, time-varying or time-invariant, linear or nonlinear, and causal or noncausal:

6. Determine whether each of the following systems is linear or nonlinear:

$$\frac{dy(t)}{dt} + 5y(t) = x^2(t) \quad \frac{dy(t)}{dt} + y^2(t) = x(t) \quad \frac{dy(t)}{dt} + 2y(t) = x(t)\frac{dx(t)}{dt}$$

7. Consider a continuous-time LTI system whose impulse response is given by the relation

$$h(t) = \begin{cases} \alpha e^{-\beta t} & \text{for } t \geq 0 \\ 0 & \text{for } t < 0 \end{cases} = \alpha e^{-\beta t}u(t), \quad b > 0$$

Determine the output $y(t)$ when a square pulse of amplitude A and duration ξ is used at its input, defined as

$$x(t) = \begin{cases} A \text{ for } 0 \leq t \leq \xi \\ 0 \text{ elsewhere} \end{cases} = Au(t) - Au(t-\xi)$$

8. Determine whether the continuous-time systems described by the following differential equations are linear or nonlinear:

$$y(t)\frac{dy(t)}{dt} + 5y(t) = 2x(t) \quad \text{and} \quad \frac{dy(t)}{dt} + t^2 y(t) = (2t+3)x(t)$$

9. Determine whether each of the following systems is time-varying or time-invariant:

$$y(t) = x(t-5), \quad y(t) = x(-2t), \quad y(t) = \left[\frac{dx(t)}{dt}\right]^3$$

10. Determine whether each of the systems described by the input/output relations

$$y(t) = \frac{1}{5}x(5-t) + x(t), \qquad y(t) = \frac{1}{x(t+2)}, \qquad y(t) = 3x(t+2) - 2\cos[x(t)]$$

is LTI.

11. Determine whether each of the systems described by the input/output relations

$$y(t) = t^2 x(t), \qquad y(t) = x\cos[x(t)], \qquad y(t) = x(t)\sin(t)$$

is linear, time-invariant, causal, stable, and dynamic.

12. Consider the continuous-time systems described by the differential equations

$$y(t) = tx(t)u(t) \quad \text{and} \quad y(t) = \frac{d}{dt}\left\{x(t-2)\right\}$$

Determine whether each of these systems is static or dynamic, time-varying or time-invariant, linear or nonlinear, and causal or noncausal.

13. Determine whether each of the systems described by the following input/output relations is static or dynamic, time-varying or time-invariant, linear or nonlinear, and causal or noncausal:

$$y(t) = t\frac{dx(t)}{dt}, \qquad y(t) = x(t) + \frac{dx(t)}{dt}, \qquad \frac{dy(t)}{dt} + 6y(t) = 4x(t)$$

14. Determine the asymptotic stability of a continuous-time systems described by the differential equations

$$\frac{d^2y(t)}{dt^2}+8\frac{dy(t)}{dt}+12y(t)=\frac{dx(t)}{dt}-x(t), \quad \frac{d^4y(t)}{dt^4}+2\frac{d^2y(t)}{dt^2}=x(t), \quad \frac{d^3y(t)}{dt^3}-5\frac{d^2y(t)}{dt^2}-\frac{dy(t)}{dt}+5y(t)=3\frac{dx(t)}{dt}+x(t)$$

15. For each of the systems described by the equations

$$y(t) = \frac{1}{2}\int_{-\infty}^{+\infty} x(\tau)\Big[\delta(t-\tau) - \delta(t+\tau)\Big] \quad \text{and} \quad y(t) = \frac{d}{dt}(x-1)$$

determine whether it is stable or unstable according to BIBO.

16. Calculate the output $y(t)$ of a LTI system with impulse response

$$h(t) = u(t+1) - u(t-1)$$

when the continuous-time signal $x(t) = u(t) + u(t-1)$ is transmitted to its input.

17. What is the output of a continuous-time LTI system when not the signal $x(t)$ but the signal $x'(t)$ is transmitted to its input?
18. Consider the pairs of signals $[h(t), x(t)]$, which express the impulse response and the input signal of a continuous-time LTI system respectively:

$$\text{(a)}\quad h(t) = u(t)\ \&\ x(t) = te^{-3t}u(t), \quad \text{(b)}\quad h(t) = u(t)\ \&\ x(t) = t^2e^{-2t}u(t), \quad \text{(c)}\quad h(t) = e^{-t}u(t)\ \&\ x(t) = u(t)$$

Calculate the output of the system $y(t)$ for all possible combinations of signals $x(t)$ and $h(t)$.
19. Let us consider a continuous-time LTI system that upon receiving the signal $x(t) = \exp(j2\pi f_0 t)$ produces the output $y(t) = 2\exp(j2\pi 2 f_0 t)$, while upon receiving the signal $x(t) = \exp(-j2\pi f_0 t)$ produces the output $y(t) = 2\exp(-j2\pi 2 f_0 t)$. What is the output of the system when the signal $x(t) = 2\cos(2\pi f_0 t)$ is transmitted to its input?
20. A continuous-time LTI system is given with an impulse response of the form $h(t) = e^{-\alpha t}u(t)$ $(\alpha > 0)$. Calculate the output $y(t)$ of the system when the signals (a) $x(t) = \cos(\omega t)$, (b) $x(t) = \cos(\omega t)u(t)$, (c) $x(t) = \sin(\omega t)$, and (d) $y(t) = \sin(\omega t)u(t)$ are transmitted to its input.
21. Show that the impulse response of a system described by the differential equation

$$\frac{dy(t)}{dt} + 2y(t) = 3\frac{dx(t)}{dt} + x(t)$$

is $h(t) = -5e^{-2t}u(t) + 3\delta(t)$.
22. Prove that the impulse response of the system

$$\frac{d^2y(t)}{dt^2} + 3\frac{dy(t)}{dt} + y(t) = \frac{dx(t)}{dt}$$

is given by the relation $h(t) = (1 - t)e^{-t}u(t)$.
23. Determine the impulse response of a system described by the equation $y'(t) + \alpha y(t) = x(t)$, where α is a constant.
24. Considering the system of Problem 23 to determine its step response if its impulse response is (a) known and (b) unknown. In the second case, after calculating the step response, determine the impulse response of the system.
25. Considering again the system of Problem 23, determine the output of the system when the input signal is $x(t) = Ae^{-\beta t}u[t]$ and considering an initial condition of the form $y(0) = M$.
26. If the step response of a system has the form $s(t) = \cos(\omega_0 t)u(t)$, what is its impulse response?
27. Determine the impulse response of the systems

$$\text{(a)}\quad \frac{d^2y(t)}{dt^2} + 5\frac{dy(t)}{dt} + 6y(t) = \frac{d^2x(t)}{dt^2} + 7\frac{dx(t)}{dt} + 11x(t)$$

$$;(b)\quad \frac{d^2y(t)}{dt^2} + 6\frac{dy(t)}{dt} + 9y(t) = 2\frac{dx(t)}{dt} + 9x(t)$$

$$;(c)\quad \frac{d^2y(t)}{dt^2} + \frac{dy(t)}{dt} + y(t) = x(t)$$

28. Considering a system with impulse response

$$h(t) = 4e^{-3t}\cos(4t)u(t)$$

determine the zero-state response for the input signals $x_1(t) = u(t)$, $x_2(t) = e^{-t}u(t)$, and $x_3(t) = e^{-3t}u(t)$.
29. Prove that the function $u_1(t)$ satisfies the properties

$$\int_{-\infty}^{+\infty} u_1(t)dt = 0 \quad \text{and} \quad \int_{-\infty}^{+\infty} x(\tau - t)u_1(t)d\tau + \frac{dx(-t)}{dt} = 0$$

30. Show that for any two positive integers k and r it holds that $u_k(t) * u_r(t) = u_{k+r}(t)$, and investigate the consequences of it. If we set $k = -1$ and $r = +1$ in this relationship, it follows that $u_{-1} * u_1(t) = \delta(t)$, which is, of course, expected, since the operation of the integrator is the inverse of the operation of the differentiator.
31. Considering two LTI systems with impulse responses $u_1(t)$ and $h(t)$ connected in series—this connection is equivalent to a system that for each input signal $x(t)$ generates the derivative of the output signal $y'(t)$—prove that

$$y(t) = \left(\int_{-\infty}^{t} x(\tau)d\tau\right) * \frac{dh(t)}{dt} = \int_{-\infty}^{t}\left[\frac{dx(\tau)}{d\tau} * h(t)\right]d\tau = \frac{dx(t)}{dt} * \left(\int_{-\infty}^{t} h(\tau)d\tau\right)$$

32. Determine which of the systems described by the input/output equations

$$y(t) = 2x(t-3), \qquad y(t) = \cos[x(t)], \qquad y(t) = \int_{-\infty}^{5\tau} x(\tau)d\tau$$

are reversible, and for those that are proven to be, determine the input/output relationship of the inverse system.
33. Considering two systems with impulse responses $h_1(t) = e^{-2t}u(t)$ and $h_2(t) = e^{-2t}u(t)$ that are connected in series, determine the impulse response of the composite system and decide whether it is LTI.
34. Considering two systems with impulse responses $h_1(t) = e^{-2t}u(t)$ and $h_2(t) = e^{-2t}u(t)$ that are connected in parallel, determine the impulse response of the composite system and decide whether it is LTI.
35. Prove that

$$x(t) = \int_{-\infty}^{+\infty} \frac{dx(\tau)}{d\tau}u(t-\tau)d\tau \quad \text{and} \quad y(t) = \int_{-\infty}^{+\infty} \frac{dx(\tau)}{d\tau}s(t-\tau)d\tau$$

where $u(t)$ and $s(t)$ are the step function and the step response of the system.
36. Demonstrate that the response of a LTI system to elementary functions

$$x_1(t) = A\cos(\omega_0 t + \varphi) \quad \text{and} \quad x_2(t) = A\sin(\omega_0 t + \varphi)$$

is given by the relations

$$y_1(t) = (A|\mathcal{H}(j\omega_0)|)\cos[\omega_0 t + \varphi + \measuredangle\mathcal{H}(j\omega_0)] \quad \text{and} \quad y_2(t) = (A|\mathcal{H}(j\omega_0)|)\sin[\omega_0 t + \varphi + \measuredangle\mathcal{H}(j\omega_0)]$$

respectively, with the function $\mathcal{H}(j\omega)$ defined as

$$\mathcal{H}(j\omega) = |\mathcal{H}(j\omega)|e^{j\measuredangle\mathcal{H}(j\omega)} = \int_{-\infty}^{+\infty} h(t)e^{-j\omega t}dt$$

where $h(t)$ is the impulse response of the system. This conclusion is very important because it reveals to us that a fundamental harmonic signal undergoes from a continuous-time LTI system the same effect that a fundamental exponential complex signal undergoes, that is, multiplication of its amplitude by the quantity $|\mathcal{H}(j\omega_0)|$ and a phase shift by the quantity $\measuredangle\mathcal{H}(j\omega_0)$.
37. Determine the impulse response of a system described by the equation

$$\frac{d^2y(t)}{dt^2} + \frac{dy(t)}{dt} + 9y(t) = \frac{d^4x(t)}{dt^4} + \frac{d^3x(t)}{dt^3} - 5x(t)$$

38. Let there be two LTI systems with impulse responses $h_1(t) = e^{-\alpha t}u(t)$ and $h_2(t) = e^{-\beta t}u(t)$. Calculate the response to the input signal $x(t) = u(t) - u(t-1) + \delta(t+1)$ of the composite system that will result from the connection of the foregoing systems (a) in series and (b) in parallel and for the cases $\alpha \neq \beta$ and $\alpha = \beta$.
39. For each of the systems defined in the previous exercises, determine the step response function $s(t)$ (a) from the differential equation of the system, making the substitution $x(t) = u(t)$, and (b) by integrating the impulse response $h(t)$. In this way, you will verify in practice that both these methods lead to the same result.

40. Prove that the step and impulse response of the system

$$\frac{d^2y(t)}{dt^2} + 3\frac{dy(t)}{dt} + 2y(t) = x(t)$$

are given by the relations

$$s(t) = \left(\frac{1}{2}e^{-2t} + e^{-t} + \frac{1}{2}\right)u(t) \quad \text{and} \quad h(t) = \left(e^{-t} - e^{-2t}\right)u(t)$$

respectively, in two different ways: (a) using as input signal the step function $x(t) = u(t)$ so as to determine the step response from the derivative of which you will calculate the impulse response and (b) using as input signal the impulse function $x(t) = \delta(t)$ so as to determine the impulse response, from whose completion you will subsequently calculate the step response. To familiarize yourself with the process, repeat it for all systems whose differential equations are given in the previous exercises.

41. Considering a system with step response

$$s(t) = e^{-t}u(t) - e^{-2t}u(t)$$

determine its response to the input signal

$$x(t) = \delta(\pi - t) - \cos(\sqrt{3})u(t)$$

42. Determine the roots of the characteristic polynomial of a system with impulse response of the form $h(t) = u(t)$, and characterize it in terms of its stability.

43. Examine the stability according to BIBO of the LTI systems

$$(a)h(t) = u(t), \quad (b)h(t) = u(t+6) - u(t-6), \quad (c)h(t) = \sum_{i=0}^{\infty}(0.25)^i\delta(2t - i)$$

44. Examine the stability according to BIBO of the LTI systems

$$(a)h(t) = e^{-5t}u(t-3), \quad (b)h(t) = e^{-8t}u(7-t), \quad (c)h(t) = e^{-3t}u(t+5)$$

45. Calculate the impulse response of the composite system for each of the pairs of simple systems with impulse responses

$$\begin{array}{lll} (a) & h_1(t) = e^{-4t}u(t-3) & h_2(t) = e^{-6t}u(3-t) \\ (b) & h_1(t) = e^{-2t}u(t+80) & h_2(t) = e^{2t}u(-1-t) \\ (c) & h_1(t) = e^{-6|t|} & h_2(t) = te^{-t}u(t) \end{array}$$

which are connected (a) in series and (b) in parallel.

46. Determine whether each of the following systems is reversible or not, and for those that are reversible, determine the inverse system:

$$h_1(t) = \delta(t+6), \quad h_2(t) = u(t), \quad h_3(t) = \delta(t-4), \quad h_4(t) = e^{-t}u(t)$$

47. Let there be a LTI system A that is connected in series with its inverse system B in the way depicted in Fig. 2.21.

- What is the impulse response of system B when the impulse response of system A is (a) $h(t) = u(t)$ and (b) $h(t) = e^{-t}u(t)$?
- What is the impulse response of system B when the input transmitted to it is the signal $w(t) = 10x(t-3)$, where $x(t)$ is the input signal of system A?

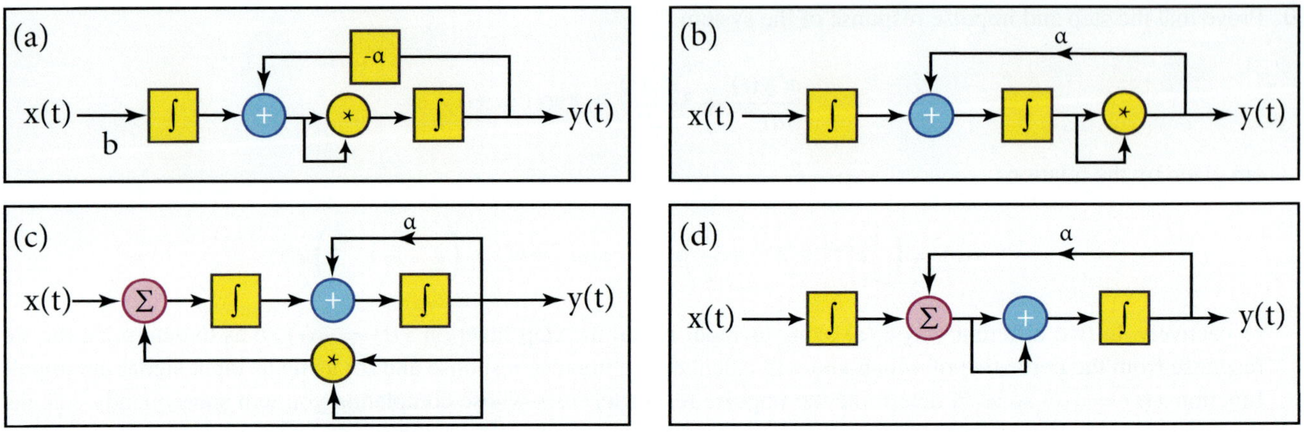

Fig. 2.28 Examples of schematic diagrams describing differential equations

48. Construct a schematic diagram for each of the differential equations

$$\frac{d^2y(t)}{dt^2} - 4\frac{dy(t)}{dt} + 5y(t) = -2\frac{dx(t)}{dt} + 3x(t)$$

$$2x(t) + \frac{d^2y(t)}{dt^2} = y(t)\frac{dy(t)}{dt}$$

$$\frac{d^2y(t)}{dt^2} + 2\frac{dy(t)}{dt} - y(t) = \frac{d^2x(t)}{dt^2} - 2\frac{dx(t)}{dt} + 2x(t)$$

49. The existence of a signal multiplier in the schematic diagrams of Fig. 2.28 means that the systems described by images (a), (b), and (c) are nonlinear. Considering that this element does not exist so that the systems described by the diagrams become linear, construct a differential equation that describes them and then determine their impulse, step, and frequency responses.
50. Determine the differential equation described by each of the schematic diagrams of Fig. 2.28..
51. For each of the LTI systems described by the differential equations

$$\text{(a)}\quad 2y(t) + \int_{-\infty}^{t} x(\tau)d\tau - 2\frac{dy(t)}{dt} = \int_{-\infty}^{t}\left(\int_{-\infty}^{\tau} y(\xi)d\xi\right)d\tau - 2y(t)$$

$$\text{(b)}\quad 3\frac{d^3y(t)}{dt^3} + \frac{dx(t)}{dt} + 2\frac{d^2x(t)}{dt^2} - 4\frac{dy(t)}{dt} + y(t) = 2x(t)$$

$$\text{(c)}\quad 2\frac{d^2y(t)}{dt^2} + x(t) + 2y(t) = 2\frac{d^3y(t)}{dt^3} + 4\frac{dx(t)}{dx}$$

construct schematic diagrams that represent their normal and nonnormal form.

52. For each of the LTI systems described by the differential equations

$$\text{(a)}\quad 4\left(\frac{d^2y(t)}{dt^2} - x(t)\right) = 3\left(\frac{dy(t)}{dt} - \frac{d^2x(t)}{dt^2}\right)$$

$$\text{(b)}\quad 3\int_{-\infty}^{t} y(\tau)d\tau + 2\frac{d^2y(t)}{dt^2} - 4\frac{dy(t)}{dt} + y(t) = 3\frac{d^2x(t)}{dt^2} - x(t);$$

$$\text{(c)}\quad 2\frac{d^2x(t)}{dt^2} + 4x(t) + y(t) = 2\frac{d^3y(t)}{dt^3} + 4\frac{dx(t)}{dt}$$

construct schematic diagrams representing its normal and nonnormal form.

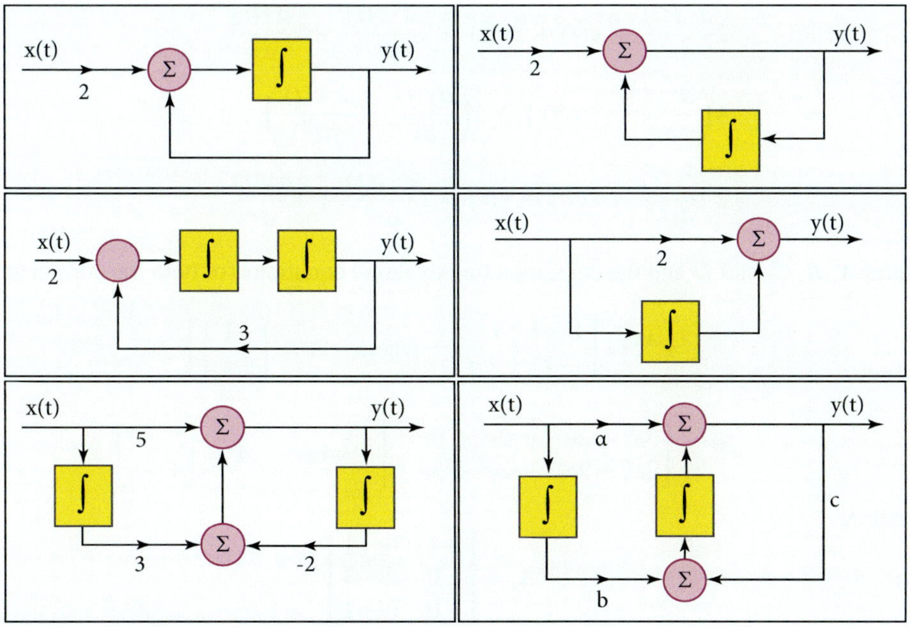

Fig. 2.29 Schematic diagrams of continuous-time systems of Exercise 54

53. Construct schematic diagrams to describe the following continuous-time systems:

$$\frac{dy(t)}{dt} + \alpha y(t) = \beta x(t), \quad \alpha\frac{d^2y(t)}{dt^2} + \beta\frac{dy(t)}{dt} + \gamma y(t) = \frac{dx(t)}{dt}$$

$$\frac{d^2y(t)}{dt^2} - y(t) = 2x(t), \quad \frac{dy(t)}{dt} + y(t) = \frac{dx(t)}{dt}$$

54. Construct a differential equation that characterizes each of the systems described by the diagrams of nonnormal form in Fig. 2.29.
55. Construct dynamic equations and a schematic diagram for both the normal and nonnormal forms of a systemfor the case in which $M < N$ or $\beta_N = 0$, which is also the one usually encountered in practice.
56. Consider a continuous-time system described by the differential equation

$$2\frac{d^3y(t)}{dt^3} + 4\frac{d^2y(t)}{dt^2} - 2\frac{dy(t)}{dt} + 6y(t) = 3\frac{d^3x(t)}{dt^3} + \frac{d^2x(t)}{dt^2} + 6x(t)$$

Prove that matrices $\boldsymbol{A}$, $\boldsymbol{B}$, $\boldsymbol{C}$, and $\boldsymbol{D}$ for the nonnormal form are given by the relations

$$\boldsymbol{A} = \begin{bmatrix} -2 & 1 & 0 \\ 1 & 0 & 1 \\ -3 & 0 & 0 \end{bmatrix}, \quad \boldsymbol{B} = \begin{bmatrix} -5/2 \\ +3/2 \\ -3/2 \end{bmatrix}, \quad \boldsymbol{C} = \begin{bmatrix} 1 & 0 & 0 \end{bmatrix} \quad \text{and} \quad \boldsymbol{D} = \begin{bmatrix} \frac{3}{2} \end{bmatrix}$$

while the corresponding expressions for the normal form are

$$\boldsymbol{A} = \begin{bmatrix} 0 & 1 & 0 \\ 0 & 0 & 1 \\ -3 & 1 & -2 \end{bmatrix}, \quad \boldsymbol{B} = \begin{bmatrix} 0 \\ 0 \\ 1 \end{bmatrix}, \quad \boldsymbol{C} = \begin{bmatrix} -\frac{3}{2} & +\frac{3}{2} & -\frac{5}{2} \end{bmatrix} \quad \text{and} \quad \boldsymbol{D} = \begin{bmatrix} \frac{3}{2} \end{bmatrix}$$

57. For each of the continuous-time LTI systems described by the differential equations

$$3\frac{d^3y(t)}{dt^3} + \frac{dx(t)}{dt} + 2\frac{d^2x(t)}{dt^2} - 4\frac{dy(t)}{dt} + y(t) = 2x(t)$$

$$2\frac{d^2y(t)}{dt^2} + x(t) + 2y(t) = 2\frac{d^3y(t)}{dt^3} + 4\frac{dx(t)}{dx}$$

$$4\left(\frac{d^2y(t)}{dt^2} - x(t)\right) = 3\left(\frac{dy(t)}{dt} - \frac{d^2x(t)}{dt^2}\right)$$

$$2\frac{d^2x(t)}{dt^2} + 4x(t) + y(t) = 2\frac{d^3y(t)}{dt^3} + 4\frac{dx(t)}{dt}$$

determine matrices $\boldsymbol{A}$, $\boldsymbol{B}$, $\boldsymbol{C}$, and $\boldsymbol{D}$ and the corresponding dynamic equations for both the normal and nonnormal forms.

58. Prove that

$$e^{At} = \begin{bmatrix} 1-t & t \\ -t & 1+t \end{bmatrix} \quad \text{where} \quad \boldsymbol{A} = \begin{bmatrix} -1 & 1 \\ -1 & 1 \end{bmatrix}$$

Prove that

$$e^{At} = \begin{bmatrix} 3e^t - 2e^{-t} & -3e^t + 3e^{-t} \\ 2e^t - 2e^{-t} & -2e^t + 3e^{-t} \end{bmatrix} \quad \text{where} \quad \boldsymbol{A} = \begin{bmatrix} 5 & -6 \\ 4 & -5 \end{bmatrix}$$

59. Prove that the matrix

$$\boldsymbol{A} = \begin{bmatrix} -1 & 0 & 1 \\ 0 & -2 & -3 \\ 0 & 0 & -1 \end{bmatrix}$$

has eigenvalues $\lambda_1 = \lambda_2 = -1$ and $\lambda_3 = -2$ that lead to coefficient values[40]

$$\beta_0 = 2te^{-t} + e^{-2t}, \quad \beta_1 = -2e^{-t} + 3te^{-t} + 2e^{-2t}, \quad \beta_2 = -e^{-t} + te^{-t} + e^{-2t}$$

and that the function e^{At} has the form

$$e^{At} = \beta_0 \boldsymbol{I} + \beta_1 \boldsymbol{A} + \beta_2 \boldsymbol{A}^2 = \begin{bmatrix} e^{-t} & 0 & te^{-t} \\ 0 & e^{-2t} & -3e^{-t} + 3e^{-2t} \\ 0 & 0 & e^{-t} \end{bmatrix}$$

(note that the eigenvalue λ_1 has multiplicity $m = 2$, and therefore you will need to use the variation of the first calculation method, which refers to repeated eigenvalues).

60. Prove that the matrix

$$\boldsymbol{A} = \begin{bmatrix} 0 & -1 & 1 \\ -2 & 1 & 2 \\ -1 & 1 & 2 \end{bmatrix}$$

has eigenvalues $\lambda_1 = 3$, $\lambda_2 = 1$ and $\lambda_3 = -1$ with corresponding eigenvectors $\boldsymbol{x}_1 = \begin{bmatrix} 0 & 1 & 1 \end{bmatrix}^T$, $\boldsymbol{x}_2 = \begin{bmatrix} 1 & 0 & 1 \end{bmatrix}^T$, and $\boldsymbol{x}_3 = \begin{bmatrix} 1 & 1 & 0 \end{bmatrix}^T$. Then calculate matrices $\boldsymbol{P}$ and $\boldsymbol{P}^{-1}$ and the value of the exponential function e^{At}.

61. Prove that the solution of the state equation $\dot{\boldsymbol{q}}(t) = \boldsymbol{A}\boldsymbol{q}(t) + \boldsymbol{B}x(t)$ for parameter values

$$\boldsymbol{A} = \begin{bmatrix} 0 & 1 \\ -2 & -3 \end{bmatrix}, \quad \boldsymbol{B} = \begin{bmatrix} 0 \\ 1 \end{bmatrix}, \quad \boldsymbol{q}(0) = \begin{bmatrix} 0 \\ 0 \end{bmatrix}, \quad x(t) = u(t)$$

[40] To solve the exercise, it is necessary to calculate the inverse of a 3 × 3 matrix. From linear algebra, it is known that the inverse of a matrix $\boldsymbol{M}$ of dimensions 3 × 3 of the form

$$\boldsymbol{M} = \begin{bmatrix} \alpha_{11} & \alpha_{12} & \alpha_{13} \\ \alpha_{21} & \alpha_{22} & \alpha_{23} \\ \alpha_{31} & \alpha_{32} & \alpha_{33} \end{bmatrix}$$

is calculated from the equation

$$\boldsymbol{M}^{-1} = \frac{1}{\det(\boldsymbol{M})} \begin{bmatrix} \alpha_{33}\alpha_{22} - \alpha_{32}\alpha_{23} & -(\alpha_{33}\alpha_{12} - \alpha_{32}\alpha_{13}) & \alpha_{23}\alpha_{12} - \alpha_{22}\alpha_{13} \\ -(\alpha_{33}\alpha_{21} - \alpha_{31}\alpha_{23}) & \alpha_{33}\alpha_{11} - \alpha_{31}\alpha_{13} & -(\alpha_{23}\alpha_{11} - \alpha_{21}\alpha_{13}) \\ \alpha_{32}\alpha_{21} - \alpha_{31}\alpha_{22} & -(\alpha_{32}\alpha_{11} - \alpha_{31}\alpha_{12}) & \alpha_{22}\alpha_{11} - \alpha_{21}\alpha_{12} \end{bmatrix}$$

where $\det(\boldsymbol{M}) = \alpha_{11}(\alpha_{33}\alpha_{22} - \alpha_{32}\alpha_{23}) - \alpha_{21}(\alpha_{33}\alpha_{12} - \alpha_{32}\alpha_{13}) + \alpha_{31}(\alpha_{23}\alpha_{12} - \alpha_{22}\alpha_{13})$ is the determinant of matrix $\boldsymbol{M}$.

is given by the relation

$$q(t) = \begin{bmatrix} \frac{1}{2} - e^{-t} + \frac{1}{2}e^{-2t} \\ e^{-t} - e^{-2t} \end{bmatrix}$$

62. Determine whether the LTI system described by the following state equation is fully controllable or not:

$$\dot{q}(t) = \begin{bmatrix} -1 & 0 & 0 \\ 0 & 2 & -1 \\ 0 & -1 & 2 \end{bmatrix} q(t) + \begin{bmatrix} 0 \\ 1 \\ 1 \end{bmatrix} x(t)$$

63. Characterize the following LTI systems in terms of their controllability:

$$A = \begin{bmatrix} -4 & 0 \\ 0 & -5 \end{bmatrix}, \quad B = \begin{bmatrix} 1 \\ 1 \end{bmatrix} \quad \text{and} \quad A = \begin{bmatrix} 2 & 0 \\ -1 & 1 \end{bmatrix}, \quad B = \begin{bmatrix} 1 \\ -1 \end{bmatrix}$$

64. Characterize the LTI system described by the following dynamic equations in terms of its controllability and observability:

$$\dot{q}(t) = \begin{bmatrix} -6 & 1 & 0 \\ -11 & 0 & 1 \\ -6 & 0 & 0 \end{bmatrix} q(t) + \begin{bmatrix} 0 \\ 1 \\ 2 \end{bmatrix} x(t) \quad \text{and} \quad y(t) = \begin{bmatrix} 1 & 0 & 0 \end{bmatrix} q(t)$$

65. Characterize the following LTI system in terms of its controllability and observability:

$$\dot{q}(t) = \begin{bmatrix} -1 & 2 & 1 \\ 0 & -1 & 1 \\ 0 & 0 & 1 \end{bmatrix} q(t) + \begin{bmatrix} 1 \\ 0 \\ 1 \end{bmatrix} x(t) \quad \text{and} \quad y(t) = \begin{bmatrix} 1 & 0 & 0 \end{bmatrix} q(t)$$

66. Characterize the following LTI system in terms of its observability:

$$\dot{q}(t) = \begin{bmatrix} 0 & 0 & 0 & 0 & 0 \\ 1 & 0 & 0 & 0 & 0 \\ 0 & 0 & 0 & 0 & 0 \\ 0 & 0 & 1 & 0 & 0 \\ 0 & 0 & 0 & 1 & 0 \end{bmatrix} q(t) \quad \text{and} \quad y(t) = \begin{bmatrix} 0 & 1 & 0 & 0 & 0 \\ 1 & 0 & 0 & 0 & 1 \end{bmatrix}$$

67. Characterize the following LTI system in terms of its controllability and observability:

$$\dot{q}(t) = \begin{bmatrix} 0 & 1 & 0 \\ 0 & 0 & 1 \\ -2 & -4 & -3 \end{bmatrix} q(t) + \begin{bmatrix} 1 & 0 \\ 0 & 1 \\ -1 & 1 \end{bmatrix} x(t) \quad \text{and} \quad y(t) = \begin{bmatrix} 0 & 1 & -1 \\ 1 & 2 & 1 \end{bmatrix} q(t)$$

68. Characterize the LTI system described by the following dynamic equations in terms of its controllability and observability:

$$\dot{q}(t) = \begin{bmatrix} 0 & 0 & 1 & 0 \\ 0 & 0 & 0 & 1 \\ 0 & -1 & 0 & 0 \\ 0 & 21.6 & 0 & 0 \end{bmatrix} q(t) + \begin{bmatrix} 0 \\ 0 \\ 1 \\ -2 \end{bmatrix} x(t) \quad \text{and} \quad y(t) = \begin{bmatrix} 1 & 0 & 0 & 0 \\ 0 & 1 & 0 & 0 \end{bmatrix} q(t)$$

69. Characterize in terms of its controllability and observability the LTI system

$$\dot{q}(t) = \begin{bmatrix} -1 & 1 & 0 & 0 \\ 0 & -1 & 0 & 0 \\ 0 & 0 & -2 & 1 \\ 0 & 0 & 0 & -2 \end{bmatrix} q(t) + \begin{bmatrix} 1 \\ 0 \\ 0 \\ 1 \end{bmatrix} x(t) \quad \text{and} \quad y(t) = \begin{bmatrix} 1 & 1 & 1 & 0 \\ 0 & 1 & 0 & 0 \end{bmatrix} q(t)$$

70. Examine in terms of its controllability and observability the LTI system described by the dynamic equations

$$\dot{\boldsymbol{q}}(t) = \begin{bmatrix} -3 & -2 & -0.75 & 0 & 0 \\ -3 & 0 & 0 & 0 & 0 \\ 0 & 2 & 0 & 0 & 0 \\ 0 & 0 & 1 & 0 & 0 \\ 0 & 0 & 0 & 2 & 0 \end{bmatrix} \boldsymbol{q}(t) + \begin{bmatrix} 1 \\ 0 \\ 0 \\ 0 \\ 0 \end{bmatrix} x(t) \qquad \text{and} \qquad y(t) = \begin{bmatrix} 0 & 0 & 0 & 0 & 2.75 \end{bmatrix} \boldsymbol{q}(t)$$

and if you find it to be controllable and observable, transform it into the corresponding normal forms presented in the preceding pages.

3 Fourier Analysis in the Continuous-Time Domain

Jean Baptiste Joseph Fourier (1768–1830)

3.1 Introduction

As was already mentioned in the previous chapter, the ability to represent any input signal $x(t)$ by a linear combination of infinitely time-shifted impulse functions allows the calculation of the output of a linear system since the linear combinations of the system's responses to each impulse function separately (this observation is a consequence of the power of the *principle of superposition*). In addition, if the system is also time-invariant, then, as we have proven, its output is defined as the *convolution of the system's impulse response and the input signal*, or, in mathematical notation,

A. I. Margaris, *Continuous and Discrete-Time Signals and Systems*,
https://doi.org/10.1007/978-3-031-81209-5_3

$$y(t) = h(t) * x(t) = \int_{-\infty}^{+\infty} h(t)x(t-\tau)d\tau \tag{3.1}$$

We observe, therefore, that the impulse response provides a complete description of the system since the knowledge of it allows the calculation of the system's output for each input signal $x(t)$.

In this chapter, we will present an alternative way of representing a continuous-time signal, *as a series of simple trigonometric functions of specific frequency*, which is known as *Fourier series or expansion* . The foregoing representation refers to *periodic* signals, while, in contrast, *aperiodic* signals are described using a function known as a *Fourier transform*. Note that this description applies to both continuous and discrete-time signals and systems, and for this reason *the Fourier transform can be formulated in both its continuous and discrete forms*.

Historically, the fundamental concepts of *Fourier analysis*—as this way of describing continuous and discrete-time signals is typically called—were first used by the ancient Babylonians, who are rumored to have used linear combinations of simple trigonometric functions to predict astronomical phenomena. On the other hand, limiting ourselves to the modern era, the concepts of Fourier analysis first appeared in the works of *Euler* in 1748 in connection with the study of the motion of the *vibrating string*. These ideas were supported and extended by *Bernoulli* in 1753. However, they did not receive wide acceptance as they were formulated descriptively and not mathematically; on the contrary, in 1759 they met the strong reaction of *Lagrange*, who considered that the use of trigonometric functions in this way was very limited. However, about half a century later, this methodology was reformulated with a strict formalism by the French mathematician *Jean Baptiste Joseph Fourier*—from whom it took its name—who formulated his conclusions studying the *problem of heat transmission and diffusion*. Fourier's theory was completed in 1807, and its main conclusion was that *the distribution of heat on a body could be represented by a series of harmonically correlated sinusoidal and cosinusoidal functions*. Furthermore, he asserted that every continuous periodic function can be represented by such a series, provided it satisfies the conditions established by Dirichlet in 1829 for such a representation to be feasible. However, it should be noted that significant contributions to the latter issue were also made by *Poisson* and *Cauchy*. After formulating his theory, Fourier went a step further and proved that *every aperiodic continuous function could be represented by a multitude of weighted integrals of non-harmonically correlated sinusoidal functions*, thereby introducing the eponymous transform.

Today it is commonly accepted that Fourier's theory has exercised a profound influence on a wide range of directions in physics and mathematics, for example, the theory of *integration*, the methodology of developing an arbitrary function in a series of *operator eigenfunctions*, and the study of the motion of planets and the evolution of Earth's climate. It should also be noted that, although Fourier's theory was concerned exclusively with continuous-time signals, it did not take long for it to be extended to the field of discrete-time, definitively influencing techniques related to the field of numerical analysis. The last significant milestone of this long and exciting journey is the year 1960, when *Cooley* and *Tukey* invented a fast method of calculating the *discrete Fourier transform* known as the *fast Fourier transform (FFT)*, which can be implemented on computers, allowing the use of Fourier techniques in problems that had previously been considered practically unsolvable due to their high computational cost.

3.2 Fourier Expansion

From the theory of vector spaces it is known that the expansion of any vector in terms of the vectors of an orthonormal basis of a Hilbert functional space of infinite dimensions is known as a *Fourier expansion*, in honor of the great French physicist and mathematician Jean Baptiste Joseph Fourier (1768–1830), and the coefficients of this expansion are known as *Fourier coefficients*. Considering that for a vector space infinite orthonormal bases can be defined, it is not difficult to understand that the form of the coefficients and the Fourier expansion of a function will depend on the form of the functions that constitute the orthonormal basis used in each case. In this section, we will construct *exponential* and *trigonometric* bases in the space $L^2([-\pi, \pi])$, which is a special case of the Hilbert vector space $L^2([\alpha, \beta])$ of infinite dimensions defined for the interval $[\alpha, \beta] = [-\pi, \pi]$ (this interval was chosen because trigonometric and exponential functions are periodic functions with period $T = 2\pi$, which are defined in this interval).

Let us start, then, with the sequence of exponential functions $x_n(t) = e^{jnt}$ $(n \in \mathbb{Z})$—or, equivalently, $-\infty < n < \infty$—and let us show that this constitutes an *orthogonal set* of functions of the Hilbert space $L^2([-\pi, \pi])$, determining the *inner product* $\langle x_n(t), x_m(t)\rangle$, which, according to the defining equation, is calculated as

$$\langle x_n(t), x_m(t)\rangle = \int_{-\pi}^{\pi} x_n(t)x_m^*(t)dt = \int_{-\pi}^{\pi} e^{jnt}e^{-jmt}dt = \int_{-\pi}^{\pi} e^{j(n-m)t}dt$$

$$= \frac{1}{j(n-m)}e^{j(n-m)t}\bigg|_{-\pi}^{\pi} = \frac{e^{j(n-m)\pi} - e^{-j(n-m)\pi}}{j(n-m)} = \frac{2\sin[(n-m)\pi]}{n-m}$$

In the special case where $n \neq m$, that is, $k = n - m \neq 0$, this will be $\sin(k\pi) = 0$, and therefore we will have $\langle x_n(t), x_m(t)\rangle = 0$. On the other hand, if $n = m$, then the foregoing relation takes the indeterminate form $0/0$, and the application of L'Hospital's rule will give us

$$\langle x_n(t), x_m(t)\rangle = \lim_{n\to m} \frac{2\sin[(n-m)\pi]}{n-m} = \lim_{n\to m} \frac{\frac{d}{dm}\{2\sin[(n-m)\pi]\}}{\frac{d}{dm}\{n-m\}} = \lim_{n\to m}\left\{2\pi\cos[(n-m)\pi]\right\} = 2\pi$$

Combining the foregoing results, we arrive at the expression

$$\langle x_n(t), x_m(t)\rangle = 2\pi\delta_{mn}$$

(where δ_{mn} is the Kronecker delta), and therefore the functions $x_n(t) = e^{jnt}$ $(n \in \mathbb{Z})$ indeed constitute an *orthogonal* set of the Hilbert function space $L^2([-\pi, \pi])$. As for the corresponding *orthonormal set*, it is not difficult to see that it takes the form

$$e_n(t) = \frac{1}{\sqrt{2\pi}}x_n(t) = \frac{1}{\sqrt{2\pi}}e^{jnt} \quad (n \in \mathbb{Z})$$

since, using the preceding result and resorting to the properties of the inner product, we will have

$$\langle e_n(t), e_m(t)\rangle = \left\langle \frac{1}{\sqrt{2\pi}}x_n(t), \frac{1}{\sqrt{2\pi}}x_m(t)\right\rangle = \frac{1}{2\pi}\langle x_n(t), x_m(t)\rangle = \frac{1}{2\pi}(2\pi)\,\delta_{nm} = \delta_{nm}$$

It is proven that the foregoing orthonormal set is *complete* and therefore constitutes a basis of the vector space $L^2([-\pi, \pi])$, which also has infinite elements. Therefore, every function $x(t) \in L^2([-\pi, \pi])$ can be expressed as a *linear combination of the basis vectors*. Therefore, we can write that

$$x(t) = \sum_{n=infty}^{\infty} \langle x(t), e_n(t)\rangle\, e_n(t) = \frac{1}{\sqrt{2\pi}} \sum_{n=-\infty}^{\infty} c_n e^{jnt}$$

The preceding equation is known as the *exponential Fourier series* of the function $x(t)$, with Fourier coefficients of the form

$$c_n = \langle x(t), e_n(t)\rangle = \left\langle x(t), \frac{1}{\sqrt{2\pi}}e^{jnt}\right\rangle = \frac{1}{\sqrt{2\pi}}\langle x(t), e^{jnt}\rangle = \frac{1}{\sqrt{2\pi}}\int_{-\pi}^{\pi} x(t)\,e^{-jnt}dt$$

In order for the preceding series of infinite terms to be a convergent series, it must satisfy a set of conditions that are known as Dirichlet conditions. One of these conditions requires that the function $x(t)$ be a periodic function, and therefore, the exponential and the trigonometric equivalent to it, the Fourier expansion, are defined only for periodic continuous-time signals with a period value of $L = 2T$. We will return to this issue later.

Using the orthonormal basis $e_n(t)$, we can construct a second basis for the space $L^2([-\pi, \pi])$ that relies on the use of the *trigonometric functions of sine and cosine*. Indeed, using Euler's formulas and the defining equation of the basis $e_n(t)$, we easily find that

$$\cos(nt) = \frac{e^{jnt} + e^{-jnt}}{2} = \frac{\sqrt{2\pi}e_n(t) + \sqrt{2\pi}e_{-n}(t)}{2} = \sqrt{\frac{\pi}{2}}e_n(t) + \sqrt{\frac{\pi}{2}}e_{-n}(t) \quad \text{and}$$

$$\sin(nt) = \frac{e^{jnt} - e^{-jnt}}{2j} = \frac{\sqrt{2\pi}e_n(t) - \sqrt{2\pi}e_{-n}(t)}{2j} = -j\sqrt{\frac{\pi}{2}}e_n(t) + j\sqrt{\frac{\pi}{2}}\sqrt{2}e_{-n}(t)$$

We therefore observe that the functions $c_n(t) = \cos(nt)$ and $s_n(t) = \sin(nt)$ can be expressed as linear combinations of the basis functions $e_n(t)$ and $e_{-n}(t)$ and, consequently, *they can also be used as a basis*,[1] something that can, of course, be verified without resorting to the use of exponential functions. Initially, we observe that for values $n \neq m$, the preceding functions are orthogonal to each other, as it is easily observed that

$$\langle c_n(t), c_m(t)\rangle = \int_{-\pi}^{\pi} \cos(nt)\cos(mt)dt = \frac{1}{2}\int_{-\pi}^{\pi} \cos[(n-m)t]dt + \frac{1}{2}\int_{-\pi}^{\pi} \cos[(n+m)t]dt = 0$$

$$\langle s_n(t), s_m(t)\rangle = \int_{-\pi}^{\pi} \sin(nt)\sin(mt)dt = \frac{1}{2}\int_{-\pi}^{\pi} \cos[(n-m)t]dt - \frac{1}{2}\int_{-\pi}^{\pi} \cos[(n+m)t]dt = 0$$

$$\langle s_n(t), c_m(t)\rangle = \int_{-\pi}^{\pi} \sin(nt)\cos(mt)dt = \frac{1}{2}\int_{-\pi}^{\pi} \sin[(n-m)t]dt + \frac{1}{2}\int_{-\pi}^{\pi} \sin[(n+m)t]dt = 0$$

Note that for each value of n there are two functions, $c_n(t)$ and $s_n(t)$, except for the value $n = 0$, for which there is only the function $c_0(t) = \cos(0) = 1$, since the function $s_n(0)$ has a value equal to zero. This case is usually treated separately. It is obvious that the *orthogonality condition* still holds now, since, as can easily be verified, it is $\langle 1, \cos(nt)\rangle = \langle 1, \sin(nt)\rangle = 0$. On the other hand, these functions *do not constitute an orthonormal set* since their measures is not equal to unity and, more specifically,

$$\|c_0(t)\| = \sqrt{\langle c_0(t), c_0(t)\rangle} = \sqrt{\langle 1, 1\rangle} = \int_{-1}^{1} dt = \sqrt{2\pi}$$

$$\|c_n(t)\| = \sqrt{\langle c_n(t), c_n(t)\rangle} = \sqrt{\int_{-\pi}^{\pi} \cos^2(nt)dt} = \sqrt{\pi}$$

$$\|s_n(t)\| = \sqrt{\langle s_n(t), s_n(t)\rangle} = \sqrt{\int_{-\pi}^{\pi} \sin^2(nt)dt} = \sqrt{\pi}$$

To construct the corresponding orthonormal set, *we must divide the preceding functions by their measures*, and therefore the required orthonormal set of functions is

$$\tilde{c}_0(t) = \frac{1}{\sqrt{2\pi}} \qquad \tilde{c}_n(t) = \frac{c_n(t)}{\sqrt{\pi}} = \frac{1}{\sqrt{\pi}}\cos(nt) \qquad \tilde{s}_n(t) = \frac{s_n(t)}{\sqrt{\pi}} = \frac{1}{\sqrt{\pi}}\sin(nt)$$

It has been proven that the preceding set of functions is *complete*,[2] and therefore the functions $\tilde{c}_0(t)$, $\tilde{c}_n(t)$, and $\tilde{s}_n(t)$ $(n \geq 1)$ constitute *a basis* of the space $L^2([-\pi, \pi])$. Thus, using the theory of the previous sections, we can develop the random function $x(t) \in L^2([-\pi, \pi])$ into

$$x(t) = \langle x(t), \tilde{c}_0(t)\rangle\tilde{c}_0(t) + \sum_{n=1}^{\infty}\langle x(t), \tilde{c}_n(t)\rangle\tilde{c}_n(t) + \sum_{n=1}^{\infty}\langle x(t), \tilde{s}_n(t)\rangle\tilde{s}_n(t) = \alpha_0\tilde{c}_0(t) + \sum_{n=1}^{\infty}\alpha_n\tilde{c}_n(t) + \sum_{n=1}^{\infty}\beta_n\tilde{s}_n(t)$$

with the Fourier coefficients being calculated as

$$\alpha_0 = \langle x(t), \tilde{c}_0(t)\rangle = \int_{-\pi}^{\pi} x(t)\frac{1}{\sqrt{2\pi}}dt = \frac{1}{\sqrt{2\pi}}\int_{-\pi}^{\pi} x(t)dt$$

[1] In many linear algebra books, this basis (as well as other infinite-dimensional bases) is known as the *Hilbert basis* in order to distinguish it from a basis with a finite number of elements, known as the Hamel basis. Note, however, that there is generally no agreement on this issue, as in some books Hamel bases are also treated as *sets of orthogonal vectors with an infinite number of elements.*

[2] For the proof of this property, see Athanasios I. Margaris, *Linear Algebra*, Tziola Publications, ISBN 978-960-418-471-2, 2016, pages 262–264.

$$\alpha_n = \langle x(t), \tilde{c}_n(t)\rangle = \int_{-\pi}^{\pi} x(t)\frac{1}{\sqrt{\pi}}\cos(nt)dt = \frac{1}{\sqrt{\pi}}\int_{-\pi}^{\pi} x(t)\cos(nt)dt$$

$$\beta_n = \langle x(t), \tilde{s}_n(t)\rangle = \int_{-\pi}^{\pi} x(t)\frac{1}{\sqrt{\pi}}\sin(nt)dt = \frac{1}{\sqrt{\pi}}\int_{-\pi}^{\pi} x(t)\sin(nx)dt$$

We will therefore have

$$x(t) = \left(\frac{1}{\sqrt{2\pi}}\int_{-\pi}^{\pi} x(t)dt\right)\frac{1}{\sqrt{2\pi}} + \sum_{n=1}^{\infty}\left(\frac{1}{\sqrt{\pi}}\int_{-\pi}^{\pi} x(t)\cos(nt)dt\right)\frac{\cos(nt)}{\sqrt{\pi}} + \sum_{n=1}^{\infty}\left(\frac{1}{\sqrt{\pi}}\int_{-\pi}^{\pi} x(t)\sin(nt)dt\right)\frac{\sin(nt)}{\sqrt{\pi}}$$

$$= \frac{1}{2\pi}\int_{-\pi}^{\pi} x(t)dt + \sum_{n=1}^{\infty}\left(\frac{1}{\pi}\int_{-\pi}^{\pi} x(t)\cos(nt)dt\right)\cos(nt) + \sum_{n=1}^{\infty}\left(\frac{1}{\pi}\int_{-\pi}^{\pi} x(t)\sin(nt)dt\right)\sin(nt)$$

or, equivalently,

$$x(t) = \frac{A_0}{2} + \sum_{n=1}^{\infty} A_n\cos(nt) + \sum_{n=1}^{\infty} B_n\sin(nt) = \frac{A_0}{2} + \sum_{n=1}^{\infty}(A_n\cos(nt) + B_n\sin(nt))$$

where

$$A_0 = \frac{1}{\pi}\int_{-\pi}^{\pi} x(t)dt, \qquad A_n = \frac{1}{\pi}\int_{-\pi}^{\pi} x(t)\cos(nt)dt, \qquad B_n = \frac{1}{\pi}\int_{-\pi}^{\pi} x(t)\sin(nt)dt$$

which is the expression of the trigonometric Fourier expansion that is usually encountered in practice. It is not difficult to understand how the *exponential and the trigonometric Fourier expansion are completely equivalent* and the choice of which one will be used has to do with which of the two is more convenient in terms of performing mathematical operations. For example, when the function $x(t)$ we use is of an *exponential nature*, then obviously the use of the *exponential expansion* is recommended, while, if it is characterized, for example, by some *symmetry* (even or odd), it is beneficial to use the *trigonometric expansion* since in this case it is proven that the coefficients B_n and A_n, respectively, have a value equal to zero, which significantly simplifies the process.

It is interesting to note that the foregoing equations refer to the special case in which the considered $x(t)$ has a period $L = 2\pi$, while for the general case of a continuous-time signal with any period $L = 2T$, the orthonormal exponential base $e_n(t)$ has the form

$$e_n(t) = \frac{1}{\sqrt{2T}}\exp\left(j\frac{\pi nt}{T}\right) = \frac{1}{\sqrt{L}}\exp\left(jn\frac{2\pi}{L}t\right) = \frac{1}{\sqrt{L}}\exp(in\omega_0 t) \quad n \in \mathbb{Z}$$

($\omega_0 = 2\pi/L$), while the inner product that is used for the proof of the property of *orthogonality* is

$$\langle e_n(t), e_m(t)\rangle = \int_{-T}^{T} e_n(t)\, e_m^*(t)dt \equiv \int_{T_0}^{T_0+L} e_n(t)\, e_m^*(t)dt$$

where T_0 is any point in time. On the other hand, the orthonormal basis for the corresponding *trigonometric Fourier expansion* includes the functions

$$\tilde{c}_0(t) = \frac{1}{\sqrt{2T}}, \qquad \tilde{c}_n(t) = \sqrt{\frac{1}{T}}\cos\left(\frac{\pi nt}{T}\right) = \sqrt{\frac{2}{L}}\cos(n\omega_0 t), \qquad \tilde{s}_n(t) = \sqrt{\frac{1}{T}}\sin\left(\frac{\pi nt}{T}\right) = \sqrt{\frac{2}{L}}\sin(n\omega_0 t)$$

($n = 1, 2, 3, \ldots,\ \omega_0 = 2\pi/L$), with the corresponding coefficients of the expansion defined as

$$A_0 = \frac{1}{T}\int_{-T}^{T} x(t)dt = \frac{2}{L}\int_{-L/2}^{L/2} x(t)dt$$

Table 3.1 Basic properties of Fourier series

Property	Signal $x(t)$	Fourier coefficients
Homogeneity	$\alpha x(t)$	αx_n
Addition	$x(t)+y(t)$	x_n+y_n
Linearity	$\alpha x(t)+\beta y(t)$	$\alpha x_n+\beta y_n$
Multiplication	$x(t)y(t)$	$\sum_{m=-\infty}^{\infty} x_k y_{n-k}$
Convolution	$\frac{1}{T}\int_0^T x(t-\tau)y(\tau)d\tau$	$x_n y_n$
Time shifting	$x(t-\tau)$	$\exp\left(\frac{-2j\pi n\tau}{T}\right)x_n$
Frequency shifting	$\left(\frac{2j\pi mt}{T}\right)x(t)$	x_{n-m}
Time reversal	$x(-t)$	x_{-n}
Complex conjugation	$x^*(t)$	x^*_{-n}
Time scaling	$x(\alpha t)$	x_n
Differentiation	$x'(t)$	$jn\omega_0 x_n$
Integration	$\int_{-\infty}^{t} x(\tau)d\tau$	$\frac{x_n}{jn\omega_0}$

$$A_n=\frac{1}{T}\int_{-T}^{T}x(t)\cos\left(\frac{\pi nt}{T}\right)dx=\frac{2}{L}\int_{-L/2}^{L/2}x(t)\cos(n\omega_0 t)dt$$

$$B_n=\frac{1}{T}\int_{-T}^{T}x(t)\sin\left(\frac{\pi nx}{T}\right)dt=\frac{2}{L}\int_{-L/2}^{L/2}x(t)\sin(n\omega_0 t)dt$$

and the expansion itself has the form

$$x(t)=\frac{A_0}{2}+\sum_{n=1}^{\infty}\left(A_n\cos\frac{n\pi t}{T}+B_n\sin\frac{n\pi t}{T}\right)=\frac{A_0}{2}+\sum_{n=1}^{\infty}\left(A_n\cos(n\omega_0 t)+B_n\sin(n\omega_0 t)\right) \tag{3.2}$$

If the period $L=2\pi$ is used in the preceding equations, that is, essentially we set $T=\pi$, we will have $\omega_0=1$, and the foregoing equations are reduced to those we constructed previously, as can be easily determined. If in the defining equations of the Fourier coefficients A_n and B_n we proceed to the replacement $n\to -n$, it turns out that $A_{-n}=A_n$ and $B_{-n}=B_n$.

The most important of the properties associated with the Fourier development of one or more continuous-time signals are presented in Table 3.1, while their detailed proof is presented in a subsequent section.

3.2.1 Fourier Series of Even and Odd Functions

If the considered periodic function $x(t)$ is an *even* or *odd* function, then the use of the existing symmetry makes it possible to simplify the defining equation of the Fourier series, which will contain *only terms of sines or cosines*. Considering, for example, an even periodic signal $x(t)$ for which, as is well known, the relation $x(-t)=x(t)$ holds, the coefficients B_n of the Fourier series are shown to have a value equal to zero:

$$B_n=\frac{1}{\pi}\int_{-\pi}^{\pi}x(t)\sin(nt)\,dt=\frac{1}{\pi}\int_{-\pi}^{0}x(t)\sin(nt)\,dt+\frac{1}{\pi}\int_{0}^{\pi}x(t)\sin(nt)\,dt=$$

$$=\frac{1}{\pi}\int_{\pi}^{0}x(-t)\sin(-nt)\,d(-t)+\frac{1}{\pi}\int_{0}^{\pi}x(t)\sin(nt)\,dt=-\frac{1}{\pi}\int_{0}^{\pi}x(t)\sin(nt)\,dt+\frac{1}{\pi}\int_{0}^{\pi}x(t)\sin(nt)\,dt=0$$

From the preceding relationship it follows that the *Fourier expansion of an even function will only contain cosine terms* since the coefficients of the cosines are equal to zero for every value of the index n. Therefore, we have

$$x(t) = \frac{A_0}{2} + \sum_{n=1}^{\infty} A_n \cos(nt), \quad \text{where} \quad A_n = \frac{2}{\pi}\int_0^{\pi} x(t)\cos(nt)\,dt$$

as can be proven. Indeed, following the same calculation methodology that we used for the coefficients B_n, we will obtain

$$A_n = \frac{1}{\pi}\int_{-\pi}^{\pi} x(t)\cos(nt)\,dt = \frac{1}{\pi}\int_{-\pi}^{0} x(t)\cos(nt)\,dt + \frac{1}{\pi}\int_{0}^{\pi} x(t)\cos(nt)\,dt = \frac{1}{\pi}\int_{\pi}^{0} x(-t)\cos(-nt)\,d(-t) +$$

$$+\frac{1}{\pi}\int_{0}^{\pi} x(t)\sin(nt)\,dt = \frac{1}{\pi}\int_{0}^{\pi} x(t)\cos(nt)\,dt + \frac{1}{\pi}\int_{0}^{\pi} x(t)\cos(nt)\,dt = \frac{2}{\pi}\int_{0}^{\pi} x(t)\cos(nt)\,dt$$

In the case where the considered periodic signal is characterized by odd symmetry—which means that it satisfies the condition $x(-t) = -x(t)$—the Fourier expansion that describes it will only contain sine terms since then we will have $A_n = 0\ (\forall n \in N)$. Thus, we have

$$x(t) = \sum_{n=1}^{\infty} B_n \sin(nt) \quad \text{where} \quad B_n = \frac{2}{\pi}\int_0^{\pi} x(t)\sin(nt)\,dt$$

The proof of the preceding relationships is done in exactly the same way and is left as an exercise to the reader.

3.2.2 Exponential Form of Fourier Series

Although the exponential Fourier series was calculated directly from the defining equation of this series, it can also be constructed from the trigonometric Fourier series

$$x(t) = \frac{A_0}{2} + \sum_{n=1}^{\infty}\left(A_n \cos\frac{n\pi t}{T} + B_n \sin\frac{n\pi t}{T}\right)$$

an approach, that verifies the fact that these two series are completely equivalent to each other. Let us present this construction process, which starts by presenting the expression that appears in the sum of the preceding equation as

$$A_n \cos\frac{n\pi t}{T} + B_n \sin\frac{n\pi t}{T} = \sqrt{A_n^2 + B_n^2}\left[\frac{A_n}{\sqrt{A_n^2 + B_n^2}}\cos\frac{n\pi t}{T} + \frac{B_n}{\sqrt{A_n^2 + B_n^2}}\sin\frac{n\pi t}{T}\right]$$

If we now define the quantities

$$\cos\varphi_n = \frac{A_n}{\sqrt{A_n^2 + B_n^2}} \quad \text{and} \quad \sin\varphi_n = \frac{B_n}{\sqrt{A_n^2 + B_n^2}}$$

and use the known trigonometric identity

$$\cos(\alpha - \beta) = \cos\alpha\cos\beta + \sin\alpha\sin\beta$$

then the last relation takes the form

$$\sqrt{A_n^2+B_n^2}\left[\frac{A_n}{\sqrt{A_n^2+B_n^2}}\cos\frac{n\pi t}{T} + \frac{B_n}{\sqrt{A_n^2+B_n^2}}\sin\frac{n\pi t}{T}\right] = C_n\left[\cos\varphi_n\cos\frac{n\pi t}{T} + \sin\varphi_n\sin\frac{n\pi t}{T}\right] = C_n\cos\left(\frac{n\pi t}{T} - \varphi_n\right)$$

where

$$C_n = \sqrt{A_n^2 + B_n^2} \qquad \text{and} \qquad \tan\varphi_n = \frac{\cos\varphi_n}{\sin\varphi_n} = \frac{B_n}{A_n}$$

[from the last relation it follows that $\varphi_n = \arctan(B_n/A_n)$]. Replacing these expressions in the trigonometric form of the Fourier expansion and using the known identity $\cos t = (e^{jt} + e^{-jt})/2$, we obtain (for simplicity we consider the case $T = \pi$)

$$x(t) = \frac{A_0}{2} + \sum_{n=1}^{\infty} C_n \cos(nt - \varphi_n) = \frac{A_0}{2} + \sum_{n=1}^{\infty} \frac{C_n}{2}\left[e^{-j\varphi_n} e^{jnt} + e^{j\varphi_n} e^{-jnt}\right]$$

or, equivalently,

$$x(t) = \frac{A_0}{2} + \sum_{n=1}^{\infty} \frac{C_n}{2} e^{-j\varphi_n} e^{jnt} + \sum_{n=1}^{\infty} \frac{C_n}{2} e^{j\varphi_n} e^{-jnt}$$

$$= \frac{A_0}{2} + \sum_{n=1}^{\infty} \frac{C_n}{2} e^{-j\varphi_n} e^{jnt} + \sum_{n=-1}^{-\infty} \frac{C_n}{2} e^{-j\varphi_n} e^{jnt} = \frac{A_0}{2} + \left[\sum_{n=1}^{\infty} + \sum_{n=-1}^{-\infty}\right] \frac{C_n}{2} e^{-j\varphi_n} e^{jnt}$$

(since, as mentioned earlier, we will have that $A_{-n} = A_n$ and $B_{-n} = -B_n$ and, therefore, $\varphi_{-n} = -\varphi_n$). In addition, because for $n = 0$ it follows that $B_0 = 0$, we have that $\varphi_0 = 0$ and $C_0 = A_0$, and therefore the last relation is written

$$x(t) = \sum_{n=-\infty}^{\infty} \frac{C_n}{2} e^{-j\varphi_n} e^{jnt} = \sum_{n=-\infty}^{\infty} \alpha_n e^{jnt} \tag{3.3}$$

where $\alpha_n = (C_n/2)e^{-j\varphi_n}$. This equation is known as the *exponential form of the Fourier expansion*. The complex coefficients α_n are associated with the coefficients A_n and B_n of the trigonometric expansion through the relation

$$\alpha_n = \frac{1}{2} C_n e^{-j\varphi_n} = \frac{1}{2}\sqrt{A_n^2 + B_n^2}[\cos(\varphi_n) - j\sin(\varphi_n)] = \frac{1}{2}\sqrt{A_n^2 + B_n^2}\left[\frac{A_n}{\sqrt{A_n^2 + B_n^2}} - j\frac{B_n}{\sqrt{A_n^2 + B_n^2}}\right] = \frac{1}{2}(A_n - jB_n)$$

from which it follows that $\alpha_{-n} = (1/2)(A_n + jB_n) = \alpha_n^*$. From the two foregoing relations we have $A_n = \alpha_n + \alpha_{-n}$ and $B_n = j(\alpha_n - \alpha_{-n})$, as well as $C_n = \alpha_n + |\alpha_{-n}|$.

Substituting the expressions of A_n and B_n into the last equation, we obtain

$$\alpha_n = \frac{1}{2}\left[\frac{1}{\pi}\int_{-\pi}^{\pi} x(t)\cos(nt)\,dt - j\frac{1}{\pi}\int_{-\pi}^{\pi} x(t)\sin(nt)\,dt\right]$$

$$= \frac{1}{2\pi}\int_{-\pi}^{\pi} x(t)[\cos(nt) - j\sin(nt)]\,dt = \frac{1}{2\pi}\int_{-\pi}^{\pi} x(t)e^{-jnt}\,dt = \frac{1}{\sqrt{2\pi}}\frac{1}{\sqrt{2\pi}}\int_{-\pi}^{\pi} x(t)e^{-jnt}\,dt = \frac{1}{\sqrt{2\pi}} c_n$$

(see the expression for the coefficients c_n in our initial construction of the exponential expansion), and substituting the last expression into the relation for the signal $x(t)$ results in

$$x(t) = \sum_{n=-\infty}^{\infty} \alpha_n e^{jnt} = \frac{1}{\sqrt{2\pi}} \sum_{n=-\infty}^{\infty} c_n e^{int}$$

which is the same result as the one we obtained from the defining equation of the Fourier expansion. The foregoing relation concerns, as mentioned earlier, the case $T = \pi$, while the equation for the general case is constructed in the same way and is left as an exercise to the reader. Note that this process can also be carried out in the reverse direction, that is, starting with the exponential Fourier expansion and ending up in the equivalent trigonometric expansion.

Problem 3.2.1 Prove that the set of *exponential functions* of the form $x_n(t) = e^{jn\omega_0 t}$ (where $n = 0, \pm 1, \pm 2, \ldots$) constitutes an *orthogonal basis* over any time duration $T_0 = 2\pi/\omega_0$, and then, using the *orthogonality condition* that you determine, derive the defining equation of the complex Fourier coefficients.

Solution Considering any two exponential functions $x_m(t) = e^{jm\omega_0 t}$ and $x_k(t) = e^{jk\omega_0 t}$, their inner product[3] will be given according to the defining equation from the relation

$$< x_m(t), x_k(t) >= \int_{t_0}^{t_0+T_0} x_m(t)x_k^*(t)\,dt = \int_{t_0}^{t_0+T_0} e^{jm\omega_0 t}(e^{jk\omega_0 t})^* dt = \int_{t_0}^{t_0+T_0} e^{j(m-k)\omega_0 t}\,dt$$

where we set $w(t) = 1$ for every $t \in (t_0, t_0 + T_0)$. The last integral is calculated as

$$\int_{t_0}^{t_0+T_0} e^{j(m-k)\omega_0 t} dt = \frac{1}{j(m-k)\omega_0}\left[e^{j(m-k)\omega_0 t}\right]_{t_0}^{t_0+T_0} = \frac{1}{j(m-k)\omega_0}\left[e^{j(m-k)\omega_0(t_0+T_0)} - e^{j(m-k)\omega_0 t_0}\right]$$

and since $\omega_0 = 2\pi/T_0$, the last relation is written

$$\int_{t_0}^{t_0+T_0} e^{j(m-k)\omega_0 t} dt = \frac{T_0}{2\pi j(m-k)}\left[e^{j(m-k)(2\pi+\omega_0 t_0)} - e^{j(m-k)\omega_0 t_0}\right] = \frac{T_0 e^{j(m-k)\omega_0 t_0}}{2\pi j(m-k)}\left[e^{2\pi j(m-k)} - 1\right]$$

From the last relation, it is not difficult to see that for $m \neq k$, the value of the foregoing integral is equal to zero since, as is well known, it is hold that $e^{2\pi j(m-k)} = 1$. On the other hand, for $m = k$ the foregoing integral takes the indeterminate form $0/0$. Applying the rule of L'Hospital, we obtain

$$\lim_{m\to k} \frac{T_0 e^{j(m-k)\omega_0 t_0}}{2\pi j(m-k)}\left[e^{2\pi j(m-k)} - 1\right] = \lim_{m\to k} \frac{\dfrac{d}{dm}\{T_0 e^{j(m-k)\omega_0 t_0}[e^{2\pi j(m-k)} - 1]\}}{\dfrac{d}{dm}\{2\pi j(m-k)\}}$$

$$= T_0 \lim_{m\to k} \frac{j\omega_0 t_0 e^{j(m-k)\omega_0 t_0}[e^{2\pi j(m-k)} - 1] + 2\pi j e^{j(m-k)\omega_0 t_0} e^{2\pi j(m-k)}}{2\pi j} = T_0$$

Combining the foregoing results, we can conclude that the inner product of any two imaginary exponential functions has the form

$$< x_m(t), x_k(t) >= \int_{t_0}^{t_0+T_0} x_m(t)x_k^*(t)dt = \int_{t_0}^{t_0+T_0} e^{j(m-k)\omega_0 t} dt = \begin{cases} 0 & \text{for } m \neq k \\ T_0 & \text{for } m = k \end{cases}$$

that is, the set of these functions constitutes an orthogonal basis. For this basis to be not only orthogonal but also orthonormal, it must hold that $T_0 = 1$, and therefore $\omega_0 = 2\pi$. We end up, then, with the very useful conclusion that the *exponential functions of the form* $x_n(t) = e^{j2\pi nt}$ *constitute an orthonormal basis of functions* and, according to the basic theory, every continuous-time signal can be written as a linear combination of these functions. We will return to this topic later.

Let us now see how we can use the orthogonality condition of the imaginary exponential functions that we proved earlier to arrive at the defining equation of the complex Fourier coefficients. Multiplying the equation

[3] This topic is discussed in Sect. 1.5.1, page 17.

$$x(t) = \sum_{n=-\infty}^{+\infty} \alpha_n e^{jn\omega_0 t}$$

by the quantity $e^{-jm\omega_0 t}$ and integrating from t_0 to $t_0 + T_0$, we obtain

$$\int_{t_0}^{t_0+T_0} x(t)e^{-jm\omega_0 t}dt = \int_{t_0}^{t_0+T_0} \left(\sum_{n=-\infty}^{+\infty} \alpha_n e^{jn\omega_0 t} \right) e^{-jm\omega_0 t}dt = \sum_{n=-\infty}^{+\infty} \alpha_n \int_{t_0}^{t_0+T_0} e^{j(n-m)\omega_0 t}dt$$

Now, using the *orthogonality condition* of the imaginary exponential functions, we can state the claim that of the infinite terms of the last sum, the only nonzero term is the one that corresponds to the value of the index $n = m$ for which the integral of the sum has a value equal to T_0. Therefore,

$$\int_{t_0}^{t_0+T_0} x(t)\, e^{-jm\omega_0 t}dt = \alpha_m\, T_0 \qquad \text{or equivalently} \qquad \alpha_n = \frac{1}{T_0}\int_{t_0}^{t_0+T_0} x(t)\, e^{-j\omega_0 nt}dt$$

where in the last relation we changed the dummy index from m to n.

3.2.3 Linear Spectra

Let us consider the simple trigonometric function $x(t) = A\sin(\omega t + \varphi)$, which is characterized as a harmonic function with amplitude $|A|$, angular frequency ω, and initial phase φ. This function is periodic with period $L = 2\pi/\omega$, since it is easily proven that for every value of t,

$$x(t+L) = A\sin[(\omega(t+L)+\varphi)] = A\sin\left[\omega\left(t+\frac{2\pi}{\omega}\right)+\varphi\right] = A\sin[(\omega t+\varphi)+2\pi] = A\sin(\omega t+\varphi)$$

with the terms amplitude, frequency, and phase having their origin in classical mechanics and, more specifically, in simple harmonic motion. Indeed, referring to a differential equations book, it is not difficult to find that the function $x(t) = A\sin(\omega t + \varphi)$ is a solution of the second-order linear differential equation with constant coefficients $mx''(t) + kx(t) = 0$ or, equivalently, $x''(t) + \omega^2 x(t) = 0$, where $\omega = \sqrt{k/m}$, which describes the motion of a body of mass m moving along the x-axis under the influence of a restoring force of the form $F(x) = -kx$.

Resorting to the well-known trigonometric identity of the sum of sines, the foregoing relation is written

$$x(t) = A\sin(\omega t + \varphi) = A(\cos\omega t\sin\varphi + \sin\omega t\cos\varphi) = \alpha\cos(\omega t) + \beta\sin(\omega t)$$

where $\alpha = A\sin\varphi$ and $\beta = A\cos\varphi$. Taking into account that $\omega = 2\pi/L$ and setting $L = 2T$, we will have $\omega = 2\pi/2T = \pi/T$, and the foregoing relation is written

$$x(t) = \alpha\cos\frac{\pi t}{T} + \beta\sin\frac{\pi t}{T}$$

Based on all the preceding points, let us define the signal

$$x_n(t) = \alpha_n\cos\frac{n\pi t}{T} + \beta_n\sin\frac{n\pi t}{T}, \qquad n = 1, 2, 3, \ldots$$

which, according to our earlier discussion, constitutes a harmonic with frequency $\omega_n = n\pi/T = n\omega$ $(n = 1, 2, 3, \ldots)$, which is an integer multiple of the frequency ω and period $L_n = 2\pi/\omega_n = 2T/n = L/n$ $(n = 1, 2, 3, \ldots)$. From the last relation it follows that $L = nL_n = 2T$, and consequently the period $L = 2T$ constitutes the common period of all the harmonics $x_n(t)$ $(n = 1, 2, 3, \ldots)$, since, as is well known, every integer multiple of the fundamental period of a periodic signal also constitutes the same period of the signal. Observing now that the Fourier expansion is simply the sum of an infinite number of such harmonics that share the same period $L = 2T$, it is obvious that both this and the signal $x(t)$ represented by this Fourier series, provided that this series actually converges to this function, will themselves also constitute periodic functions with period $L = 2T$. We therefore find that the trigonometric Fourier series is defined only for periodic signals, although it should be noted that there are periodic signals for which this series is not defined; however, this expansion is defined for a very wide range of functions and, in reality, is defined for all functions of practical interest. This series allows for the formulation of a periodic signal with period $L = 2T$ as a linear combination of an infinite number of harmonics whose frequencies ω_n, of which $(n = 1, 2, 3, \dots)$, constitute an integer multiple of a fundamental frequency $\omega = \pi/T$.

In cases where we have calculated the coefficients of the exponential Fourier expansion of a continuous periodic signal $x(t)$, we have the ability to construct special diagrams, known as *spectra*, and depict the coefficients of the Fourier expansion at the positions of the frequencies corresponding to each of the aforementioned harmonics. Taking into account that the Fourier coefficients are complex quantities, we can formulate them in the form $\alpha_n = |\alpha_n|e^{j\varphi_n}$, where $|\alpha_n|$ and φ_n are the amplitude and phase of the complex quantity α_n. In this case, the graphical representation of the variation of the function $|\alpha_n|$ for the various frequency values $\omega_n = n\omega_0$ $(n = \pm 1, \pm 2, \dots)$ is known as the *amplitude spectrum*, while the corresponding diagram for the phase φ_n is known as the *phase spectrum*. Taking into account that the parameter n takes integer values, it is obvious that the amplitude and phase spectra will not be continuous curves but will consist of a set of vertical lines whose intersection with the horizontal axis corresponds to the positions $\omega_n = n\omega$ $(n = \pm 1, \pm 2, \dots)$. For this reason, these diagrams are known as *line spectra*.

In a more rigorous mathematical description, the Fourier expansion of the signal $x(t)$ can, as we have already proven, be formulated in the form

$$x(t) = \frac{A_0}{2} + \sum_{n=1}^{\infty} C_n \cos(nt - \varphi_n) = \sum_{n=-\infty}^{+\infty} \alpha_n e^{jnt} \tag{3.4}$$

where $C_n = \sqrt{A_n^2 + B_n^2}$ and $\varphi_n = \tan^{-1}(B_n/A_n)$, while the exponential Fourier coefficients are defined as $\alpha_n = (C_n/2)e^{-j\varphi_n}$. Therefore, $|\alpha_n| = C_n/2$, meaning the amplitude spectrum of the signal $x(t)$ is simply the diagrammatic representation of the quantity $C_n/2$ at the positions of the discrete frequencies $\omega_n = n\omega = 2\pi n/T$, where T is the period of the signal. Note that in the foregoing description, the index n can take negative values. This means that the *amplitude spectrum can be extended to the region of negative frequencies*. Using the property $|\alpha_{-n}| = |\alpha_n^*| = |\alpha_n|$, it is not difficult to see that the height of the lines at the positions of the negative frequencies is the same as that of the lines associated with the corresponding positive frequencies.

Using the symmetry of the Fourier coefficients for the various cases, we can derive useful conclusions about the form of the amplitude and phase spectra. For example, in the case where the signal is real, it will hold that $\alpha_{-n} = \alpha_n^*$, from which it follows that $|\alpha_{-n}| = |\alpha_n|$ and $\varphi_{-n} = -\varphi_n$. From the last relations it follows that the *amplitude spectrum of a real signal is an even function of the frequency* ω, *while the corresponding phase spectrum is an odd function of* ω.

Problem 3.2.2 Starting from the exponential representation of the Fourier series of a continuous-time signal $x(t)$, derive the equation that describes its trigonometric representation.

Solution To prove the desired result, we formulate the exponential representation of the Fourier series in the form

$$x(t) = \sum_{n=-\infty}^{\infty} \alpha_n e^{in\omega_0 t} = \alpha_0 + \sum_{n=-\infty}^{-1} \alpha_n e^{in\omega_0 t} + \sum_{n=1}^{\infty} \alpha_n e^{in\omega_0 t}$$

and then, changing the index of the first sum from n to $-n$, we get

$$x(t) = \sum_{n=-\infty}^{\infty} \alpha_n e^{in\omega_0 t} = \alpha_0 + \sum_{n=1}^{\infty} \alpha_{-n} e^{-in\omega_0 t} + \sum_{n=1}^{\infty} \alpha_n e^{in\omega_0 t} = \alpha_0 + \sum_{n=1}^{\infty} \left(\alpha_n e^{in\omega_0 t} + \alpha_{-n} e^{-in\omega_0 t} \right)$$

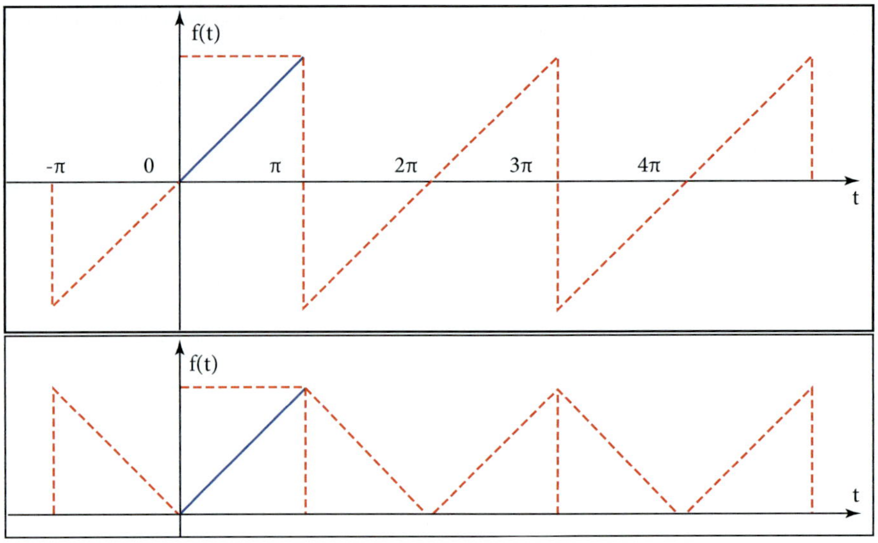

Fig. 3.1 Extension of signal $x(t) = t$ as an even and odd function in the interval $-\pi < t < \pi$

In the next step of the process, we use the Euler formulas

$$e^{+in\omega_0 t} = \cos(n\omega_0 t) + i\,\sin(\omega_0 t) \quad \text{and} \quad e^{-in\omega_0 t} = \cos(n\omega_0 t) - i\,\sin(\omega_0 t)$$

and the previous relation is written

$$x(t) = \alpha_0 + \sum_{n=1}^{\infty}\left[\alpha_n \cos(n\omega_0 t) + i\alpha_n \sin(n\omega_0 t) + \alpha_{-n}\cos(n\omega_0 t) - i\alpha_{-n}\sin(n\omega_0 t)\right] =$$

$$= \alpha_0 + \sum_{n=1}^{\infty}\left[(\alpha_n + \alpha_{-n})\cos(n\omega_0 t) + i(\alpha_n - \alpha_{-n})\sin(n\omega_0 t)\right]$$

Finally, setting $\alpha_0 = A_0/2$, $\alpha_n + \alpha_{-n} = A_n$, and $i(\alpha_n - \alpha_{-n}) = B_n$, we obtain

$$x(x) = \frac{A_0}{2} + \sum_{n=1}^{\infty}\Big(A_n \cos(n\omega_0 t) + B_n \sin(n\omega_0 t)\Big)$$

which is the desired trigonometric representation of the Fourier series of the signal $x(t)$.

Problem 3.2.3 Find the Fourier series of the identity function $x(t) = t$ defined in the interval $0 < t < \pi$, in both its trigonometric and exponential forms (Retzeperis, Section 2.3.5, Example 1).

Solution The solution to the problem requires extending the definition of the signal $x(t)$ in the interval $-\pi < t < \pi$. Given that we have no information about the characteristics of this signal in the interval $-\pi < t < 0$, we have the freedom to define these characteristics ourselves and consider the function that describes the signal as *even* or as *odd*. The graphical representation of the signal function for these two cases is presented in Fig. 3.1a, b.

Assuming that the signal function $x(t)$ is extended as an *even* function in the interval $[-\pi, \pi]$, the Fourier series will contain *only cosine terms* since, as is well known, $B_n = 0$; on the other hand, the coefficients A_0 and A_n will be calculated according to the theory from the relations

$$A_0 = \frac{2}{\pi}\int_0^{\pi} t\,dt = \frac{2}{\pi}\left[\frac{t^2}{2}\right]_0^{\pi} = \frac{2}{\pi}\frac{\pi^2}{2} = \pi \qquad \text{and}$$

$$A_n = \frac{2}{\pi}\int_0^{\pi} t\cos(nt)\,dt = \frac{2}{\pi}\left[\frac{\cos(nt)}{n^2}\right]_0^{\pi} = \begin{cases} 0 & n = 2k \\ -\dfrac{4}{\pi n^2} & n = 2k+1 \end{cases}$$

Therefore, the Fourier expansion of the function $x(t) = t$, considered as an *even* function, will have the form

$$x(t) = t = \frac{\pi}{2} - \frac{4}{\pi}\sum_{k=0}^{\infty}\frac{\cos[(2k+1)t]}{(2k+1)^2}$$

On the other hand, if the signal $x(t)$ is extended to the interval $[-\pi, \pi]$ as an *odd* function, then the Fourier expansion will contain *only sinusoidal terms* since, as is well known, $A_n = 0$, while the coefficients B_n are calculated as

$$B_n = \frac{2}{\pi}\int_0^{\pi} t\,\sin(nt)\,dt = \frac{2}{\pi}\left[-\frac{t\cos(nt)}{n}\right]_0^{\pi} + \frac{1}{n}\int_0^{\pi}\cos(nt)\,dt = (-1)^{n+1}\frac{2}{n}$$

Therefore, the Fourier expansion of the signal $x(t) = t$, considered as an *odd* function, is defined as

$$x(t) = 2\sum_{n=1}^{n}(-1)^{n+1}\frac{\sin(nt)}{n}$$

The foregoing expressions refer to the *trigonometric form* of the Fourier expansion, while in order to determine *its exponential form*, we must calculate the complex coefficients from their defining equation. Using the method of integration by parts, it turns out that

$$\alpha_n = \frac{1}{2\pi}\int_{-\pi}^{\pi} te^{-jnt}dt = -\frac{1}{2jn\pi}\left\{\left[te^{-jnt}\right]_{-\pi}^{\pi} - \int_{-\pi}^{\pi} e^{-jnt}dt\right\} = \frac{j}{2n\pi}\left[te^{-jnt} + \frac{1}{jn}e^{-jnt}\right]_{-\pi}^{\pi} =$$

$$= \frac{j}{2n\pi}\left[\pi(e^{-jn\pi} + e^{jn\pi}) + \frac{1}{jn}(e^{-jn\pi} - e^{jn\pi})\right] = \frac{j}{2n}2\cos(n\pi) = j\frac{(-1)^n}{n}$$

We can arrive at the same result using the relation $\alpha_n = (A_n - jB_n)/2$ along with the expressions of the coefficients A_n and B_n for the case of extending the function as odd in the interval $[-\pi, \pi]$. On the other hand, for the value $n = 0$, it will hold that $\alpha_0 = A_0/2 = 0$, that is, the *exponential* form of the Fourier expansion of the signal $x(t) = t$ will be given by the equation

$$x(t) = j\sum_{n=-\infty}^{\infty}\frac{(-1)^n}{n}e^{jnt} \quad (n \neq 0)$$

Problem 3.2.4 Find the Fourier expansion of the triangular signal of Fig. 3.2 defined as

$$x(t) = \begin{cases} E + \dfrac{E}{\pi}\omega t \text{ for } -\pi < \omega t < 0 \\ E - \dfrac{E}{\pi}\omega t \text{ for } 0 < \omega t < \pi \end{cases}$$

(Retzeperis, Section 3.1.2)

Solution The signal function $x(t)$ is *even* in its domain, and therefore $B_n = 0$. To calculate the coefficients A_0 and A_n we define the variable $x = \omega t$. Then we will have

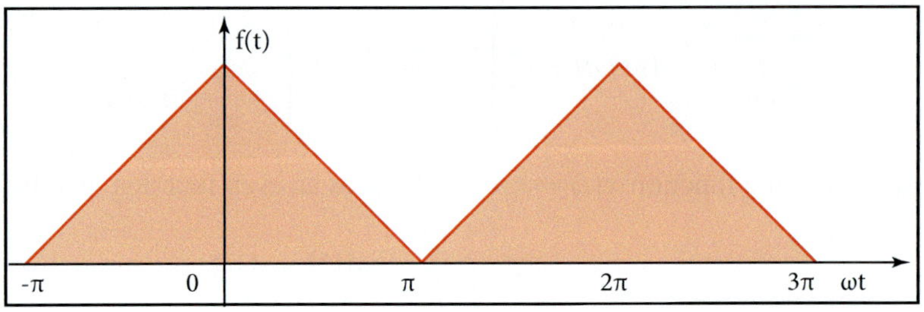

Fig. 3.2 The triangular signal of Exercise 3.2.4

$$A_0 = \frac{2}{\pi}\int_0^{\pi}\left(E - \frac{E}{\pi}x\right)dx = \frac{2E}{\pi}\left[x - \frac{x^2}{2\pi}\right]_0^{\pi} = E \quad \text{and}$$

$$\begin{aligned}A_n &= \frac{2}{\pi}\int_0^{\pi}\left(E - \frac{E}{\pi}x\right)\cos(nx)dx = \frac{2E}{\pi}\left[\int_0^{\pi}\cos(nx)dx - \frac{1}{\pi}\int_0^{\pi}x\cos(nx)dx\right]\\ &= -\frac{2E}{n\pi^2}\left[\left[x\sin(nx)\right]_0^{\pi} - \int_0^{\pi}\sin(nx)dx\right] = -\frac{2E}{n\pi^2}\left[x\sin(nx) + \frac{\cos(nx)}{n}\right]_0^{\pi} = -\frac{2E}{n\pi^2}\left[\frac{\cos(n\pi)}{n} - \frac{1}{n}\right] =\\ &= \frac{2E}{n^2\pi^2}[1-\cos(n\pi)] = \begin{cases} 0 & n = 2k \\ \dfrac{4E}{n^2\pi^2} & n = 2k+1 \end{cases}\end{aligned}$$

where for the calculation of the result we applied the method of integration by parts. Therefore, the *trigonometric form* of the Fourier expansion of the signal $x(t)$ will be given by the equation

$$x(t) = \frac{E}{2} + \frac{4E}{\pi^2}\sum_{k=0}^{\infty}\frac{\cos[(2k+1)\omega t]}{(2k+1)^2}$$

while the *exponential form* of this expansion is proven to have the form

$$x(t) = \frac{E}{2} + \frac{2E}{\pi^2}\sum_{n=-\infty}^{\infty}\frac{1}{n^2}e^{jn\omega t} \quad (n \neq 0)$$

(the proof of the last relation is left as an exercise to the reader).

3.2.4 Convergence Conditions of Fourier Series

From what was mentioned in previous sections, it is not difficult to realize that the *Fourier series* of a continuous-time signal $x(t)$ has practical value, *only when at every time instance t it converges to a finite value*, which of course is simply the value of the signal $x(t)$. Although Fourier's claim that every periodic signal can be represented by the homonymous series is not absolutely correct, it does apply to a very large category of periodic signals, so that it finds practical application in almost all cases of interest. Of course, nothing prevents us from calculating the Fourier coefficients from their defining equation, regardless of whether the signal we are examining has a Fourier series. However, if some basic conditions are not met, some of these coefficients may be estimated as having an infinite value, while in other cases, even if all the coefficients are estimated as finite, their substitution into the defining equation of the Fourier series may not lead to the desired result. Based on the foregoing discussion, it becomes obvious that the convergence conditions of the Fourier series of any signal $x(t)$ is an issue that requires special investigation.

A particular category of periodic signals for which their Fourier series exists in every case are those whose energy as calculated over a period of time is finite, a property that is mathematically formulated as

$$\int_T |x(t)|^2 dt < \infty \tag{3.5}$$

Without being able to proceed to an analytical mathematical proof of this assertion, we can formulate the proposition that, according to theory, the power of the foregoing property ensures that the value of the Fourier coefficients, as they are calculated from the integral

$$\alpha_n = \frac{1}{T}\int_T x(t)e^{jn\omega_0 t}dt = \frac{1}{T}\int_T x(t)e^{jn(2\pi/T)t}dt$$

(where T is the period of the signal and $\omega_0 = 2\pi/T$ its cyclic frequency), will be finite. If we symbolize by

$$x_N(t) = \sum_{n=-N}^{+N} \alpha_n e^{jn\omega_0 t} \tag{3.6}$$

the approximation of the signal $x(t)$, by the $2N+1$ terms of the Fourier series, the power of the property on the finiteness of the energy that we formulated previously ensures that the energy

$$E_N = \int_T |e_N(t)|^2 dt \tag{3.7}$$

of the approximation error

$$e_N(t) = x(t) - x_N(t) = x(t) - \sum_{n=-N}^{+N} \alpha_n e^{jn\omega_0 t} \tag{3.8}$$

will tend to zero for the limit $N \to \infty$, a property that is proven to hold for several cases of discontinuous-time signals. However, let us note that the foregoing property does not ensure that the values of the signal $x(t)$ and its representation according to Fourier will coincide for every time instance t, only that the energy characterizing the approximation error will have a value equal to zero.

An alternative set of conditions whose fulfillment ensures that for every time instance t the value of the signal $x(t)$ will coincide with that to which its Fourier expansion converges are the *Dirichlet conditions*, which are satisfied by essentially all functions that are of practical interest, a fact that allows us to calculate the Fourier expansion of any signal $x(t)$ without examining whether it converges to a finite value, since such a thing is taken for granted. Note that the Dirichlet conditions ensure the existence of the Fourier expansion even at points where the signal presents discontinuities; in this case, the convergence value of the series results from the average of the values of the signal on both sides of the discontinuity, which, of course, must be finite. The last property is expressed in mathematical form as follows:

$$\frac{A_0}{2} + \sum_{n=1}^{\infty}\left(A_n \cos\frac{n\pi t}{T} + B_n \sin\frac{n\pi t}{T}\right) = \frac{1}{2}\Big[x(t_0+) + x(t_0-)\Big]$$

where $t = t_0$ is a point of discontinuity of the function $x(t)$, while the quantities $x(t_0+)$ and $x(t_0-)$ are defined as

$$x(t_0+) = \lim_{\substack{t\to t_0\\ t>t_0}} x(t) \qquad \text{and} \qquad x(t_0-) = \lim_{\substack{t\to t_0\\ t<t_0}} x(t)$$

A continuous-time signal $x(t)$ satisfies the *Dirichlet conditions* if it is characterized by the following properties:

- The function $x(t)$ is *absolutely integrable* over any time interval of a period's duration, that is, it satisfies the condition

$$\int_T |f(t)|dt < \infty$$

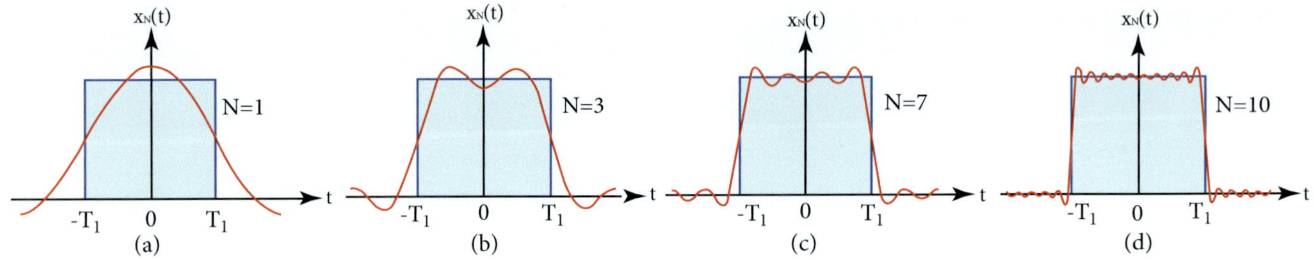

Fig. 3.3 The Gibbs phenomenon

It is not difficult to see that the power of this condition ensures that the coefficients of the Fourier expansion will be finite since they will be

$$|\alpha_n| \leq \frac{1}{T}\int_T |x(t)\,e^{-jn\omega_0 t}|dt = \frac{1}{T}\int_T |x(t)|dt < \infty$$

- Over any finite time interval, the signal $x(t)$ should be characterized by *bounded variation*, that is, it should have a finite number of maxima and minima over a period's duration.
- Over any finite time interval, the signal $x(t)$ should have a *finite number of discontinuity points with finite value*.

As already mentioned, the preceding conditions are essentially satisfied by all continuous-time signals of practical interest, while, on the other hand, signals that do not meet these conditions are considered problematic and find no practical application.

3.2.5 The Gibbs Phenomenon

The existence of discontinuities in the graphical representation of a continuous-time signal $x(t)$ causes the appearance of an undesirable phenomenon known as the *Gibbs phenomenon* (in honor of mathematician *Josiah Gibbs*,[4] who formulated its interpretation in 1899) and significantly affects the process of designing *digital filters*, as we will see in the next chapter. In Fig. 3.3, the approximation of the function of a rectangular signal $x(t)$ of period $T = 4T_1$ (which is characterized by symmetry with respect to the vertical axis) from its Fourier series $x_N(t)$ for an increasingly large number of terms N is presented. It is not difficult to see that this function satisfies the *Dirichlet conditions* and that for every time instance t the property $\lim x_N(t) = x(t)$ will hold for the limit $N \to \infty$. However, as observed in the figure, the existence of discontinuities at positions $t = -T_1$ and $t = T_1$ results in the appearance of *oscillations* whose amplitude increases as we approach the points of discontinuity. As for the maximum amplitude of these oscillations, it has been proven that it remains constant and does not decrease with increases in the number of terms N, as one might expect. More specifically, if we assume that the width of the signal is equal to one unit, then the maximum width of the fluctuations is proven to exceed this value by a percentage whose value will be calculated in the analysis that follows.

To explain this phenomenon, we must take into account that at every time instance t, the Fourier expansion converges to the value of the signal $x(t)$, except for points of discontinuity, at which it converges *to the average of the function values on both sides of the discontinuity* (this situation is diagrammatically presented in Fig. 3.3, from where it can be seen that at positions $t = -T_1$ and $t = T_1$ the curve of the Fourier expansion intersects with the straight line segments defined by the preceding equations, exactly in the middle). However, as the variable t tends toward the positions of discontinuity, the number of terms N of the expansion must continuously increase, so that the approximation error of the signal $x(t)$ from its Fourier expansion is smaller than some limit. But this, in turn, means that the number of oscillations will also increase, with the maximum width of the fluctuation remaining constant. The main consequence of this phenomenon, known as the *Gibbs phenomenon*, is that if we consider only the N before the terms of a function's expansion according to Fourier, the remaining

[4] In reality, the Gibbs phenomenon was first observed by Wilbraham in 1848, but the work he compiled was not taken seriously by the scientific community, and was studied again by Michelson in 1899, who observed it in practice with the help of experimental arrangements. The explanation of the phenomenon was given by Gibbs a year later, while in 1906 Bôcher presented a detailed mathematical analysis of the phenomenon and established its name.

expansion will show this variance, and the value of N should be chosen in such a way that the energy associated with it is as small as possible.

In a more rigorous mathematical description, let us consider the function of a rectangular signal

$$x(t) = \begin{cases} 1 \text{ for } 0 < t < T \\ 0 \text{ for } t > T \end{cases} \tag{3.9}$$

which has a magnitude equal to one for the time period $(0, T)$ and equal to zero for the rest of the time. Applying the theory associated with Fourier analysis, it is proven that the existence of the boundary conditions $x(0) = x(T) = 0$ causes the appearance of only sinusoidal terms in the signal's expansion $x(t)$, which can therefore be written in the form

$$x(t) = \sum_{n=1}^{\infty} B_n \sin\left(\frac{n\pi t}{T}\right) = 1 \quad \text{for every } t \in (0, T)$$

To calculate the coefficients B_n, we will use the well-known *orthogonality relations* of trigonometric functions. By multiplying the foregoing equation by the quantity $\sin(m\pi t/T)$ and integrating from 0 to T, we obtain

$$\int_0^T dt \sin\left(\frac{m\pi t}{T}\right) \sum_{n=1}^{\infty} B_n \sin\left(\frac{n\pi t}{T}\right) = \int_0^T dt \sin\left(\frac{m\pi t}{T}\right)$$

By interchanging the operations of summation and integration and taking into account the orthogonality condition of the sine functions, the left-hand side of the foregoing equation is written

$$\sum_{n=1}^{\infty} B_n \int_0^T \sin\left(\frac{m\pi t}{T}\right) \sin\left(\frac{n\pi t}{T}\right) dt = \sum_{n=1}^{\infty} B_n \cdot \begin{cases} 0 & \text{for } n \neq m \\ T/2 & \text{for } n = m \end{cases} = B_m \frac{T}{2}$$

On the other hand, the right-hand side of the preceding equation is calculated as

$$\int_0^T dt \sin\left(\frac{m\pi t}{T}\right) = -\frac{T}{m\pi}\left[\cos\left(\frac{m\pi t}{T}\right)\right]_0^T = \frac{T[1-\cos(m\pi)]}{m\pi} = \frac{T[1-(-1)^m]}{m\pi} = \frac{2T}{(2k+1)\pi}$$

since the foregoing equation is different from zero only for odd values of m, as can be verified. Equating the results that emerged from the calculation of the two sides of our initial equation, we obtain

$$B_{2k+1}\frac{T}{2} = \frac{2T}{(2k+1)\pi} \qquad \text{and finally} \qquad B_{2k+1} = \frac{4}{(2k+1)\pi}$$

Therefore, the Fourier expansion of the continuous-time signal $x(t)$ will be given by the equation

$$x(t) = \frac{4}{\pi} \sum_{k=0}^{\infty} \frac{1}{2k+1} \sin\left[\frac{(2k+1)\pi t}{T}\right] = 1 \tag{3.10}$$

This expansion for small values of the parameter N is presented together with the signal $x(t)$ in Fig. 3.4.

Examination of Fig. 3.4 reveals that the *width of the approximation of the signal $x(t)$ from its equivalent Fourier representation is greater than one*, something that normally should not happen. Additionally, the study of the problem has shown that, *even if we increase the number of terms of the expansion, the range and energy of the approximation error are reduced, something that does not apply to the maximum width*. To calculate the value of the maximum width—which

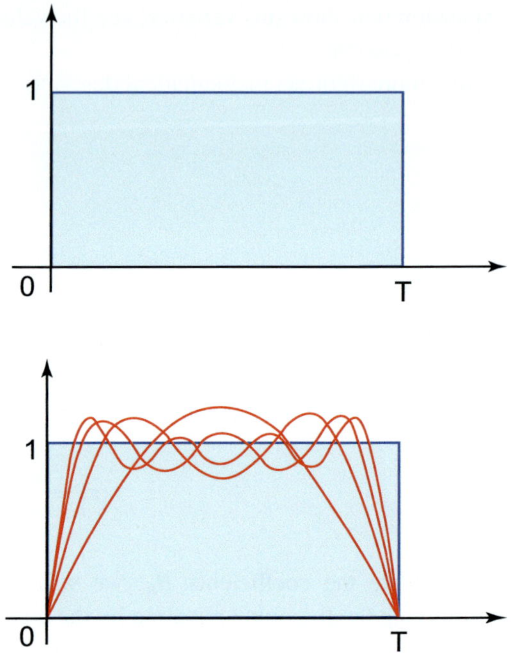

Fig. 3.4 Demonstration of Gibbs phenomenon for continuous-time signal $x(t) = 1$, where $t \in (0, T)$

constitutes the first of the extremes of the preceding function—we must determine the value of t for which the first derivative of the signal $x(t)$ is equal to zero. This process will be applied on the approximation of the signal $x(t)$ from the first N terms of the expansion, which will be given by the relation

$$x_N(t) = \frac{4}{\pi} \sum_{k=0}^{N} \frac{1}{2k+1} \sin\left[\frac{(2k+1)\pi t}{T}\right] = 1 \tag{3.11}$$

and then we will calculate the preceding value at the limit $N \to \infty$.

Therefore, calculating the derivative of the foregoing relation and setting the mathematical expression that arises to the zero value, we are led to the equation

$$\begin{aligned} x'_N(t) &= \frac{4}{\pi} \sum_{k=0}^{N} \frac{1}{2k+1} \frac{(2k+1)\pi}{T} \cos\left[\frac{(2k+1)\pi t}{T}\right] \\ &= \frac{4}{T} \sum_{k=0}^{N} \cos\left[\frac{(2k+1)\pi t}{T}\right] = 0 \end{aligned}$$

and taking into account that

$$\cos\left[\frac{(2k+1)\pi t}{T}\right] = \Re\left\{ \exp\left[\frac{j(2k+1)\pi t}{T}\right]\right\}$$

We are led to the conclusion that the determination of the value for which the function $x_N(t)$ presents the first extremum can be calculated indirectly by calculating the sum of the exponential quantities and taking the real part of the quantity that results.

If we set for simplicity reasons $\xi = e^{j\pi t/T}$ and use the appropriate relationship from the mathematical typology (see Table 6.2 on page 500), we are led to the equation

$$\sum_{k=0}^{\infty} \exp\left[\frac{j(2k+1)\pi t}{T}\right] = \sum_{k=0}^{N} \xi^{2k+1} = \xi \sum_{k=0}^{N} \xi^{2k} = \xi \frac{1-\xi^{2N+2}}{1-\xi^2} = \xi^{N+1} \frac{\xi^{N+1} - \xi^{-(N+1)}}{\xi - \xi^{-1}}$$

and, returning to our original variables, we can write that

$$\sum_{k=0}^{N} \exp\left[\frac{j(2k+1)\pi t}{T}\right] = e^{j\pi t(N+1)/T}\left[\frac{e^{j\pi t(N+1)/T} - e^{-j\pi t(N+1)/T}}{e^{j\pi t/T} - e^{-j\pi t/T}}\right] = \frac{\sin\left[(N+1)\dfrac{\pi t}{T}\right]}{\sin\left(\dfrac{\pi t}{T}\right)} \, exp\left[\frac{j\pi(N+1)t}{T}\right]$$

Thus, we have

$$\sum_{k=0}^{N} \cos\left[\frac{(2k+1)\pi t}{T}\right] = \Re\left\{\frac{\sin\left[(N+1)\dfrac{\pi t}{T}\right]}{\sin\left(\dfrac{\pi t}{T}\right)} \, exp\left[\frac{j\pi(N+1)t}{T}\right]\right\}$$

$$= \frac{\sin\left[(N+1)\dfrac{\pi t}{T}\right]}{\sin\left(\dfrac{\pi t}{T}\right)} \cos\left[\frac{(N+1)\pi t}{T}\right] = \frac{\sin\left[2(N+1)\dfrac{\pi t}{T}\right]}{2\sin\left(\dfrac{\pi t}{T}\right)}$$

For the derivation of the last relationship we used the known identity $\sin(2\alpha) = 2\sin(\alpha)\cos(\alpha)$ and replacing in the equation of the function $f'_N(t)$, we finally obtain

$$x'_N(t) = \frac{2}{T}\left\{\frac{\sin[2(N+1)\pi t/T]}{2\sin(\pi t/T)}\right\} = 0 \tag{3.12}$$

From the last relation it is not difficult to see that the values at which the function $x'_N(t)$ presents extremes are those that satisfy the condition

$$\frac{2(N+1)\pi t_m}{T} = \pm m\pi \qquad \text{from which it follows that} \qquad t_m = \pm\frac{mT}{2(N+1)}$$

The value of m that leads to the value of t for which the function presents the first maximum is obviously $m = 1$, and the corresponding value of t is $t_1 = T/[2(N+1)]$. Replacing the preceding value of t in the equation of the function $x_N(t)$, we find that the value of the first maximum is

$$f_N\left[\frac{T}{2(N+1)}\right] = \frac{4}{\pi}\sum_{k=0}^{N}\frac{1}{2k+1}\sin\left[\frac{(2k+1)\pi T}{2T(N+1)}\right] = \frac{4}{\pi}\sum_{k=0}^{N}\frac{1}{2k+1}\sin\left[\frac{(2k+1)\pi}{2(N+1)}\right]$$

To complete the description of the problem, let us see what happens in the limit $N \to \infty$. It is not difficult to see that as N tends to infinity, the number of terms in the preceding sum also tends to infinity, since the index k varies in the interval $[0, N]$, while the argument of the sine varies in the interval $(\varepsilon, \pi - \varepsilon)$, where $\varepsilon \to 0$. Setting $t_k = k/N$ results in $\Delta t_k = 1/N$, and from the foregoing relation we obtain

$$\lim_{N\to\infty}\left[\frac{4}{\pi}\sum_{k=0}^{N}\frac{\sin(\pi t_k)}{2Nt_k}\right] = \frac{2}{\pi}\lim_{N\to\infty}\left[\frac{2}{\pi}\sum_{k=0}^{\infty}\frac{\sin(\pi t_k)}{t_k}\Delta t_k\right] = \frac{2}{\pi}\int_0^1\frac{\sin(\pi t)}{t}dt$$

Proceeding with a change of the integration variable from t to $x = \pi t$, the foregoing relation will take the form

$$\underset{(N\to\infty)}{f_N\left[\frac{T}{2(N+1)}\right]} = \frac{2}{\pi}\int_0^1\frac{\sin(\pi t)}{t}dt = \frac{2}{\pi}\int_0^{\pi}\frac{\sin x}{x}dx = \frac{2}{\pi}Si(\pi) = 1.17898... \tag{3.13}$$

where the function $Si(x)$ is defined by the equation

$$Si(x) = \int_0^x \frac{\sin t}{t} dt \qquad \text{while, as proven,} \qquad Si(\pi) = \int_0^\pi \frac{\sin t}{t} dt = 1.852...$$

and is known as the *sine integral*. The preceding result also constitutes the quantitative interpretation of the *Gibbs phenomenon*, according to which the existence of discontinuities in the function $x(t)$ causes the appearance of *oscillations*, the maximum amplitude of which exceeds the maximum value of the signal, by a percentage of the order of 18%. The foregoing result is generalized for every smooth and piecewise continuous function $x(t)$ since, as Gibbs proved, the maximum height of this overshoot is equal to $0.09[x(t_0+) - x(t_0-)]$, where t_0 is a point of discontinuity of the function $x(t)$. More details about the Gibbs phenomenon can be found in the literature.

3.2.6 Properties of Continuous-Time Fourier Expansion

For reasons of simplicity and understanding of the text, the coefficients α_n of the exponential expansion of the signal $x(t)$ are symbolized as x_n, while the corresponding coefficients of the exponential expansion of the signal $y(t)$ are symbolized as y_n. Using this symbolism, the Fourier series of the continuous-time signals $x(t)$ and $y(t)$ are given by the equations

$$x(t) = \sum_{n=-\infty}^{\infty} x_n \exp\left(j\frac{2\pi nt}{T}\right) \qquad \text{and} \qquad y(t) = \sum_{n=-\infty}^{\infty} y_n \exp\left(j\frac{2\pi nt}{T}\right)$$

where T is the period of these two signals. The presentation of the foregoing properties follows next.

- **Homogeneity**: If the coefficients of the Fourier expansion of the signal $x(t)$ are the x_n, then the coefficients of the Fourier expansion of the signal $y(t) = \alpha x(t)$ are $y_n = \alpha x_n$.
  Proof From the defining equation of the Fourier coefficients of the signals $x(t)$ and $y(t)$ we get

$$x_n = \frac{1}{T}\int_T x(t)\, e^{-jn\omega_0 t} dt \quad \text{and} \quad y_n = \frac{1}{T}\int_T y(t)\, e^{-jn\omega_0 t} dt$$

  If the function $y(t)$ is of the form $y(t) = \alpha x(t)$, then the last relation leads to the desired result:

$$y_n = \frac{1}{T}\int_T \alpha x(t)\, e^{-jn\omega_0 t} dt = \alpha \frac{1}{T}\int_T x(t)\, e^{-jn\omega_0 t} dt = \alpha x_n$$

- **Addition**: If the coefficients of the Fourier expansion of the signals $x(t)$ and $y(t)$ are the x_n and y_n, then the coefficients of the Fourier expansion of the signal $z(t) = x(t) + y(t)$ will be $z_n = x_n + y_n$.
  Proof From the defining equation of the Fourier coefficients we will have

$$z_n = \frac{1}{T}\int_T z(t)\, e^{-jn\omega_0 t} dt = \frac{1}{T}\int_T \Big[x(t) + y(t)\Big]\, e^{-jn\omega_0 t} dt = \frac{1}{T}\int_T x(t)\, e^{-jn\omega_0 t} dt + \frac{1}{T}\int_T y(t)\, e^{-jn\omega_0 t} dt = x_n + y_n$$

- **Linearity**: If the coefficients of the Fourier expansion of the signals $x(t)$ and $y(t)$ are the x_n and y_n, then the coefficients of the Fourier expansion of the signal $z(t) = \alpha x(t) + \beta y(t)$ will be $z_n = \alpha x_n + \beta y_n$.
  Proof The foregoing property arises directly by combining the properties of *homogeneity* and *addition*. Alternatively, following the methodology of the previous proofs, it turns out that

$$\begin{aligned} z_n &= \frac{1}{T}\int_T z(t)\, e^{-jn\omega_0 t} dt = \frac{1}{T}\int_T \Big[\alpha x(t) + \beta y(t)\Big]\, e^{-jn\omega_0 t} dt \\ &= \alpha \frac{1}{T}\int_T x(t)\, e^{-jn\omega_0 t} dt + \beta \frac{1}{T}\int_T y(t)\, e^{-jn\omega_0 t} dt = \alpha x_n + \beta y_n \end{aligned}$$

- **Multiplication**: If the coefficients of the Fourier expansion of the signals $x(t)$ and $y(t)$ are the x_n and y_n, then the coefficients of the Fourier expansion of the signal $z(t) = x(t)y(t)$ will be given by the equation of discrete convolution:

$$z_n = \sum_{k=-\infty}^{+\infty} x_k y_{n-k}$$

Proof Taking into account the exponential Fourier representation of the signals $x(t)$ and $y(t)$

$$x(t) = \sum_{k=-\infty}^{+\infty} x_k e^{jk\omega_0 t} \quad \text{and} \quad y(t) = \sum_{m=-\infty}^{+\infty} y_m e^{jm\omega_0 t}$$

their product will be written

$$z(t) = x(t)y(t) = \left(\sum_{k=-\infty}^{+\infty} x_k e^{jk\omega_0 t}\right)\left(\sum_{m=-\infty}^{+\infty} y_m e^{jm\omega_0 t}\right)$$
$$= \sum_{k=-\infty}^{\infty}\sum_{m=-\infty}^{\infty} x_k e^{jk\omega_0 t} y_m e^{jm\omega_0 t} = \sum_{k=-\infty}^{\infty}\sum_{m=-\infty}^{\infty} x_k y_m e^{j\omega_0 (m+k)t}$$

Defining the dummy index $n = m + k$ will result in $m = n - k$, and the last relation will take the form

$$z(t) = \sum_{k=-\infty}^{\infty}\sum_{n=-\infty}^{\infty} x_k y_{n-k} e^{jn\omega_0 t} = \sum_{n=-\infty}^{+\infty}\left(\sum_{k=-\infty}^{+\infty} x_k y_{n-k}\right) e^{-jn\omega_0 t}$$

If we compare the last relation with the exponential representation, according to Fourier, of the function $z(t)$, which is given by the equation

$$z(t) = \sum_{n=-\infty}^{+\infty} z_n e^{jn\omega_0 t}$$

then the desired result is directly derived.

- **Convolution**: If the coefficients of the Fourier expansion of the signals $x(t)$ and $y(t)$ are the x_n and y_n, then the coefficients of the Fourier expansion of the function

$$z(t) = \frac{1}{T}\int_T x(t-\tau)y(\tau)d\tau \quad \text{will be the} \quad z_n = x_n y_n$$

Proof From the equation of the definition of the coefficients z_n we have

$$z_n = \frac{1}{T}\int_T z(t)e^{-jn\omega_0 t}dt = \frac{1}{T}\int_T \left\{\frac{1}{T}\int_T x(t-\tau)y(\tau)d\tau\right\} e^{-jn\omega_0 t}dt = \frac{1}{T}\int_T y(\tau)\left\{\frac{1}{T}\int_T x(t-\tau)\, e^{-jn\omega_0 t}dt\right\}d\tau$$

In the next step of the process, we set $\xi = t - \tau$ (from which it follows that $t = \xi + \tau$ and $dt = d\xi$), and the foregoing integral is written

$$z_n = \frac{1}{T}\int_T y(\tau)\left\{\frac{1}{T}\int_T x(\xi)e^{-jn\omega_0(\xi+\tau)}d\xi\right\}d\tau = \left(\frac{1}{T}\int_T x(\xi)\, e^{-jn\omega_0\xi}d\xi\right)\left(\frac{1}{T}\int_T y(\tau)\, e^{-jn\omega_0\tau}d\tau\right) = x_n y_n$$

(as follows from the defining equations of the coefficients x_n and y_n), which is also the desired result.

- **Time shift**: If the coefficients of the Fourier expansion of the signal $x(t)$ are the x_n, then the coefficients of the Fourier expansion of the signal $y(t) = x(t-\tau)$ will be $y_n = e^{-jn\omega_0\tau}x_n$.

Proof From the defining equation of the coefficients y_n we have

$$y_n = \frac{1}{T}\int_T y(t)e^{-jn\omega_0 t}dt = \frac{1}{T}\int_T x(t-\tau)e^{-jn\omega_0 t}dt$$

Setting $\xi = t - \tau$, we have $t = \xi + \tau$ and $dt = d\xi$. Moreover, it is not difficult to see that if the variable t varies over a period, the same will apply to the variable ξ; more specifically, it is proven that if t varies in the interval $[0, T]$, then ξ will vary in the interval $[-\tau, T-\tau]$. Making the foregoing substitutions, the last relation is written

$$y_n = \frac{1}{T}\int_T x(\xi)e^{-jn\omega_0(\xi+\tau)}d\xi = e^{-jn\omega_0\tau}\left\{\frac{1}{T}\int_T x(\xi)e^{-jn\omega_0\xi}d\xi\right\} = e^{-jn\omega_0\tau}x_n$$

as derived from the defining equation of the coefficients x_n, which is also the desired result.

- **Frequency shift**: If the coefficients of the Fourier expansion of the signal $x(t)$ are the x_n, then the coefficients of the Fourier expansion of the signal $y(t) = e^{j\omega_0 mt}x(t)$ will be $y_n = x_{n-m}$.
Proof From the defining equation of the coefficients y_n we have

$$y_n = \frac{1}{T}\int_T y(t)\,e^{-jn\omega_0 t}dt = \frac{1}{T}\int_T e^{j\omega_0 mt}x(t)\,e^{-in\omega_0 t}dt = \frac{1}{T}\int_T x(t)\,e^{-j\omega_0(n-m)t}dt = x_{n-m}$$

- **Time reversal**: If the coefficients of the Fourier expansion of the signal $x(t)$ are the x_n, then the coefficients of the Fourier expansion of the signal $y(t) = x(-t)$ will be given by the relation $y_n = x_{-n}$.
Proof From the defining equation of the exponential Fourier series of signals $x(t)$ and $y(t)$ we have

$$x(t) = \sum_{k=-\infty}^{+\infty} x_n\,e^{jn\omega_0 t} \quad \text{and} \quad y(t) = \sum_{k=-\infty}^{+\infty} y_n\,e^{jn\omega_0 t}$$

From the first relation it follows that

$$x(-t) = \sum_{k=-\infty}^{+\infty} x_n\,e^{-jn\omega_0 t} \quad \text{or, equivalently,} \quad x(-t) = \sum_{k=-\infty}^{+\infty} x_{-n}\,e^{jn\omega_0 t}$$

with the second relation resulting from the first, proceeding to the replacement $n \to -n$. However, $y(t) = x(-t)$, and equating the defining equations of these two functions, we obtain

$$\sum_{k=-\infty}^{+\infty} y_n\,e^{jn\omega_0 t} = \sum_{k=-\infty}^{+\infty} x_{-n}\,e^{jn\omega_0 t}$$

from which it directly follows that $y_n = x_{-n}$.

- **Complex conjugation**: If the coefficients of the Fourier expansion of the signal $x(t)$ are the x_n, then the coefficients of the Fourier expansion of the signal $y(t) = x^*(t)$ will be given by the relation $y_n = x^*_{-n}$, where the symbol $*$ refers to the operation of complex conjugation.
Proof Applying the operation of complex conjugation to the equation of the exponential representation, according to Fourier, of the signal $x(t)$, we obtain

$$x^*(t) = \sum_{n=-\infty}^{+\infty} x^*_n e^{-jn\omega_0 t} \quad \text{or, equivalently,} \quad x^*(t) = \sum_{n=-\infty}^{+\infty} x^*_{-n}\,e^{jn\omega_0 t}$$

where in the last equation we changed the silent index from n to $-n$. However, $y(t) = x^*(t)$, and therefore

$$y(t)=\sum_{k=-\infty}^{+\infty} y_n e^{jn\omega_0 t}=\sum_{n=-\infty}^{+\infty} x^*_{-n} e^{jn\omega_0 t}$$

from which it immediately follows that $y_n = x^*_{-n}$. We observe, then, that *if we apply the operation of complex conjugation to a continuous-time signal $x(t)$, then the Fourier coefficients undergo both the operation of complex conjugation and the operation of time reversal.*

In the special case where the signal $x(t)$ is real and therefore satisfies the condition $x(t) = x^*(t)$, then the application of the foregoing process leads directly to the relationship $x_{-n} = x_n^*$, that is, the Fourier coefficients are symmetrically conjugate quantities. This characteristic results in the appearance of certain symmetry properties, which in general are as follows:

$$\Re(x_n)=\Re(x_{-n}),\qquad \Im(x_n)=-\Im(x_{-n}),\qquad |x_n|=|x_{-n}|,\qquad \measuredangle x_n=\measuredangle x_{-n}$$

On the other hand, if the signal $x(t)$ is an imaginary value signal, it is similarly proven that $\Re(x_{-n}) = -\Re(x_n)$ and $\Im(x_{-n}) = \Im(x_n)$ or, in condensed form, $x_{-n} = -x_n^*$, while for the magnitude and phase of the Fourier coefficients, equations similar to those written above arise. Finally, if the values of the signal $x(t)$ are complex, then the coefficients of the Fourier expansion x_n can, under certain conditions, be both *real* and *imaginary* numbers, and more specifically:

- If the complex values of the signal $x(t)$ are characterized by an even symmetry with respect to their real part and by an odd symmetry with respect to their imaginary part, the coefficients of the Fourier expansion will be real numbers and therefore will satisfy the property $\Im(x_n) = 0$.
- If the complex values of the signal $x(t)$ are characterized by an odd symmetry with respect to their real part and by an even symmetry due to their symmetry with respect to their imaginary part, the Fourier coefficients will be imaginary numbers and therefore will satisfy the property $\Re(x_n) = 0$.

In the preceding equations, the quantities $\Re(s)$, $\Im(s)$, and $\measuredangle s$ refer respectively to the real part, the imaginary part, and the phase of the complex quantity s.

- **Time scaling**: If the coefficients of the Fourier expansion of the signal $x(t)$ are the x_n, then the coefficients of the Fourier expansion of the signal $y(t) = x(\alpha t)$ will be given by the relation $y_n = x_n$, that is, they will remain unchanged. It should be noted that in contrast to the other properties where we have silently assumed that the signals $x(t)$ and $y(t)$ have the same period T, in this case the signal $y(t)$ has a period $T' = T/\alpha$. Generally speaking, if the signal $x(t)$ has a period T and a cyclic frequency $\omega_0 = 2\pi/T$, then the signal $x(\alpha t)$ will have a period $T' = T/\alpha$ and a cyclic frequency $\omega_0' = \alpha\omega_0$.
Proof From the defining equation of the Fourier coefficients of the signal $y(t) = x(\alpha t)$ we get

$$y_n=\frac{1}{T'}\int_{T'} y(t)\, e^{-jn\omega_0' t}dt=\frac{\alpha}{T}\int_{T'} x(\alpha t)\, e^{-jn\alpha\omega_0 t}dt$$

In the next step of the process, we set $\xi = \alpha t$. Then $dt = d\xi/\alpha$, and the foregoing relation is written

$$y_n=\frac{\alpha}{T}\int_{T'} x(\xi)\, e^{-jn\xi t}\frac{d\xi}{\alpha}=\frac{1}{T}\int_{T} x(\xi)\, e^{-jn\xi t}dt=x_n$$

which is the desired result. It should be noted, however, that even though the act of temporal scaling does not change the value of the Fourier series coefficients, the Fourier representation of the signal $x(\alpha t)$

$$x(\alpha t)=\sum_{k=-\infty}^{\infty} x_n\, e^{jn(\alpha\omega_0)t}$$

will change since the cyclic frequency from ω_0 will become $\omega' = \alpha\omega_0$.

- **Differentiation**: If the coefficients of the Fourier series of the signal $x(t)$ are the x_n, then the coefficients of the Fourier series of the signal $y(t) = x'(t)$ will be given by the relation $y_n = jn\omega_0 x_n$.
Proof Differentiating the defining equation of the Fourier representation of the signal $x(t)$ we obtain

$$y(t)=\frac{dx(t)}{dt}=\frac{d}{dt}\left\{\sum_{n=-\infty}^{+\infty} x_n e^{jn\omega_0 t}\right\}=\sum_{n=-\infty}^{+\infty}(jn\omega_0 x_n)\, e^{jn\omega_0 t}=\sum_{n=-\infty}^{+\infty} y_n\, e^{jn\omega_0 t}$$

from which it follows that $y_n = jn\omega_0 x_n$, which is the desired result.

- **Integration**: If the coefficients of the Fourier series of the signal $x(t)$ are the x_n, then the coefficients of the Fourier series of the signal

$$y(t) = \int_{-\infty}^{t} x(\tau)d\tau \qquad \text{are given by the relation} \qquad y_n = x_n/jn\omega_0$$

Proof Integrating the defining equation of the Fourier representation of the signal $x(t)$, we obtain

$$y(t) = \int_{-\infty}^{t} x(\tau)\, d\tau = \int_{-\infty}^{t} \left(\sum_{n=-\infty}^{+\infty} x_n e^{jn\omega_0 \tau} \right) d\tau = \sum_{n=-\infty}^{\infty} x_n \int_{-\infty}^{t} e^{jn\omega_0 \tau} d\tau$$

$$= \sum_{n=-\infty}^{+\infty} x_n \left[\frac{e^{jn\omega_0 \tau}}{jn\omega_0} \right]_{-\infty}^{t} = \sum_{n=-\infty}^{+\infty} \left(\frac{x_n}{jn\omega_0} e^{jn\omega_0 t} \right) = \sum_{n=-\infty}^{+\infty} y_n e^{jn\omega_0 t}$$

from which it follows that $y_n = x_n/jn\omega_0$, which is also the desired result.

- **Parseval's equation**: From basic theory it is known that the average power of a continuous-time signal described by the function $x(t)$ is given by the relation

$$\langle x^2 \rangle = \lim_{T\to\infty} \frac{1}{2T} \int_{-T}^{+T} |x(t)|^2 dt \tag{3.14}$$

If we replace in the foregoing relation the continuous-time signal $x(t)$ from its exponential representation according to Fourier, it will take the form

$$\langle x \rangle = \lim_{T\to\infty} \frac{1}{2T} \int_{-T}^{+T} \left(\sum_{n=-\infty}^{+\infty} x_n e^{jn\omega_0 t} \right)^2 dt = \lim_{T\to\infty} \frac{1}{2T} \int_{-T}^{+T} dt \left(\sum_{n=-\infty}^{+\infty} x_n e^{jn\omega_0 t} \right) \left(\sum_{m=-\infty}^{+\infty} x_m e^{jm\omega_0 t} \right) =$$

$$= \sum_{n=-\infty}^{+\infty} \sum_{m=-\infty}^{+\infty} x_n x_m \left(\lim_{T\to\infty} \frac{1}{2T} \int_{-T}^{+T} e^{j(n+m)\omega_0 t} \right) dt$$

However, as we proved in a previous section, *imaginary exponential signals constitute an orthogonal set of functions*. This means that the integral of the last relation will be equal to zero for all values of the index m, except for the value $m = -n$, for which it will have a value equal to $2T$. Therefore, the foregoing relation will take the form

$$\langle x \rangle = \sum_{n=-\infty}^{+\infty} x_n x_{-n} = \sum_{n=-\infty}^{+\infty} x_n x_n^* = \sum_{n=-\infty}^{+\infty} |x_n|^2$$

where we silently assumed that since the signal was characterized by the existence of power, it will be *real* and therefore we will have $x_{-n} = x_n^*$ (as we have already proven). Reformulating the foregoing relationship for a time period of one period, it is formulated as

$$\frac{1}{T} \int_{T} |x(t)|^2 dt = \sum_{n=-\infty}^{+\infty} |x_n|^2 \tag{3.15}$$

The preceding equation is known as *Parseval's theorem*. The physical meaning of this relationship is that *the power of the signal for a time period equal to one period is equal to the sum of the measures of the coefficients of the Fourier expansion of the signal* $x(t)$.

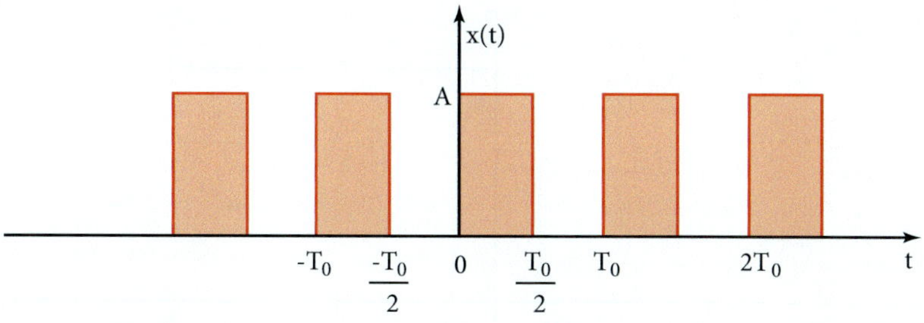

Fig. 3.5 The rectangular signal of Exercise 3.2.5

Problem 3.2.5 Find the Fourier expansion of rectangular signal $x(t)$ of period T_0 presented in Fig. 3.5 in both its exponential and trigonometric forms (Hsu, *Signals and Systems*, Exercise 5.5).

Solution From Fig. 3.5 we find that the signal $x(t)$ is described by the equation

$$x(t) = \begin{cases} A \text{ for } 0 < t < T_0/2 \\ 0 \text{ for } T_0/2 < t < T_0 \end{cases}$$

and from the defining equation of the exponential coefficients Fourier we obtain

$$\alpha_n = \frac{1}{T_0}\int_0^{T_0} x(t)e^{-jn\omega_0 t}\,dt = \frac{1}{T_0}\int_0^{T_0/2} Ae^{-jn\omega_0 t}\,dt = -\frac{A}{jn\omega_0 T_0}e^{-jn\omega_0 t}\Big|_0^{T/2} = \frac{A}{jn\omega_0 T_0}(e^{-jn\omega_0 T_0/2} - 1)$$

However, $\omega_0 = 2\pi/T_0$, and the foregoing relation will take the form

$$\alpha_n = \frac{A}{2\pi in}(1 - e^{-i\pi n}) = \frac{A}{2\pi in}[1 - (-1)^n] = \frac{A}{\pi i(2k+1)} = \frac{-iA}{(2k+1)\pi}$$

since the foregoing expression is different from zero only for odd values of n of the form $n = 2k + 1$.

On the other hand, for $n = 0$ we have

$$\alpha_0 = \frac{1}{T_0}\int_0^{T_0} x(t)\,dt = \frac{1}{T_0}\int_0^{T_0/2} Adt = \frac{A}{2}$$

and therefore the exponential form of the Fourier expansion of the signal $x(t)$ will be described by the equation

$$x(t) = \frac{A}{2} + \frac{A}{i\pi}\sum_{k=-\infty}^{+\infty}\frac{1}{2k+1}e^{i(2k+1)\omega_0 t}$$

To determine the coefficients of the trigonometric Fourier expansion, we will use the known relations $A_0 = \alpha_0/2$, $A_n = \alpha_n + \alpha_{-n} = \Re(\alpha_n)$, and $B_n = i(\alpha_n - \alpha_{-n}) = -2\Im(\alpha_n)$, from which it follows that $A_0 = A$, $A_{2k} = B_{2k} = 0$ $(k \neq 0)$, $A_{2k+1} = 2\Re(\alpha_{2m+1}) = 0$, and $B_{2k+1} = -2\Im(\alpha_{2m+1}) = 2A/[(2k+1)\pi]$. Therefore, the *trigonometric Fourier expansion* of the foregoing signal will have the form

$$x(t) = \frac{A}{2} + \frac{2A}{\pi}\sum_{k=0}^{\infty}\frac{\sin[(2k+1)\omega_0 t]}{2k+1}$$

which is also the desired result.

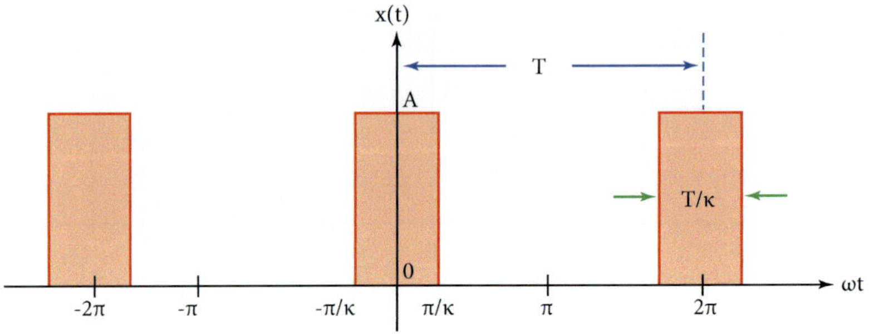

Fig. 3.6 The rectangular signal from Exercise 3.2.6

Problem 3.2.6 Design the linear amplitude spectrum of the signal from Fig. 3.6.

Solution Studying Fig. 3.6, we easily ascertain that the duration of each square pulse is equal to T/k, and therefore the period of the signal—which is equal to 2π, leading to a fundamental frequency value $\omega = 2\pi/T = 1$—is k times greater than the duration of each pulse. As for the square signal $x(t)$, it is described by the equation

$$x(t) = A \quad \text{for} \ -\frac{\pi}{k} \leq t \leq \frac{\pi}{k}$$

that is, the coefficients of the exponential Fourier expansion will be calculated as

$$\alpha_n = \frac{1}{T}\int_T x(t)\, e^{-jnt} dt = \frac{1}{2\pi}\int_{-\pi}^{+\pi} Ae^{-jnt} dt = \frac{A}{2\pi}\int_{-\pi/k}^{+\pi/k} e^{-jnt} dt$$

For $n = 0$ we have

$$\alpha_0 = \frac{A}{2\pi}\int_{-\pi/k}^{+\pi/k} dt = \frac{A}{2\pi}\left(\frac{\pi}{k} + \frac{\pi}{k}\right) = \frac{A}{2\pi}\frac{2\pi}{k} = \frac{A}{k}$$

while for $n \neq 0$ the Fourier coefficients are calculated as

$$\alpha_n = \frac{A}{2\pi}\int_{-\pi/k}^{+\pi/k} e^{-jnt} dt = \frac{A}{-2j\pi n}\left[e^{-j\pi n}\right]_{-\pi/k}^{+\pi/k} = \frac{A(e^{j\pi n/k}) - e^{-j\pi n/k}}{2j\pi n} =$$

$$= \frac{A}{\pi n}\frac{e^{j\pi n/k} - e^{-j\pi n/k}}{2i} = \frac{A}{\pi n} sin\left(\frac{\pi n}{k}\right) = \frac{A}{k}\frac{sin(\pi n/k)}{\pi n/k} = \frac{A}{k}\text{sinc}\left(\frac{\pi n}{k}\right)$$

with the function sinc(x) defined by the relation $\text{sinc}(x) = \sin(x)/x$.

A graphical representation of the amplitude of the Fourier coefficients for the various frequency values and for the values $k = 2$ and $k = 5$ is given in Fig. 3.7. We observe that the Fourier coefficients are *real numbers*, that is, they are characterized by zero phase value, and therefore the phase diagram does not exist.

Problem 3.2.7 Find the exponential and trigonometric forms of the Fourier series of the function

$$x(t) = \sum_{k=-\infty}^{+\infty} \delta\,(t - kT)$$

(Hsu, *Signals and Systems*, Exercise 5.8).

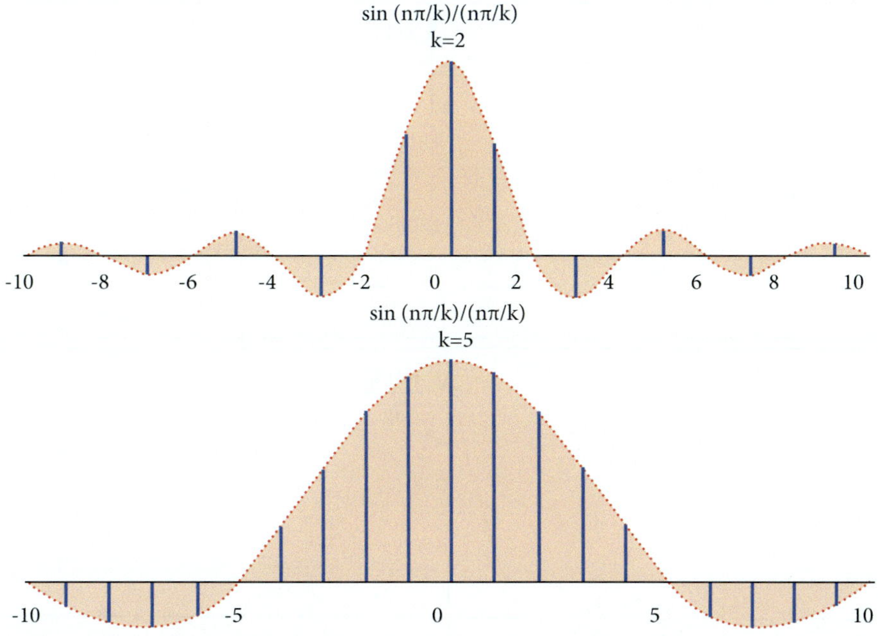

Fig. 3.7 The graphical representation of the amplitude of the Fourier coefficients of the square signal of Exercise 3.2.6 for the values $k = 2$ and $k = 5$

Solution The *exponential form* of the Fourier series of the foregoing signal will be given by the relation

$$x(t) = \sum_{n=-\infty}^{+\infty} \alpha_n \, e^{jn\omega t} = \frac{1}{T} \sum_{n=-\infty}^{+\infty} e^{jn\omega t}$$

since the coefficients α_n are calculated as

$$\alpha_n = \frac{1}{T} \int_{-T/2}^{+T/2} \delta(t) \, e^{-jn\omega t} dt = \frac{1}{T}$$

To determine the *trigonometric form* of the expansion, we must take into account that the function $\delta(t)$ is an *even function* and therefore the coefficients $B_n = 0$ will be equal to zero. As for the coefficients A_n, these are calculated as

$$A_n = \frac{2}{T} \int_{-T/2}^{+T/2} \delta(t) \, \cos(k\omega t) \, dt = \frac{2}{T}$$

Therefore, we will have

$$x(t) = \frac{1}{T} + \frac{2}{T} \sum_{n=1}^{\infty} \cos(n\omega t)$$

which is also the desired result.

Problem 3.2.8 Determine the Fourier series and the energy density spectrum of the square pulse sequence of Fig. 3.8α (Proakis and Manolakis, Example 4.1.1).

Solution From the figure we see that the signal $x(t)$ is periodic with fundamental period T and satisfies the *Dirichlet conditions*. Therefore, it can be developed into a Fourier series of the form

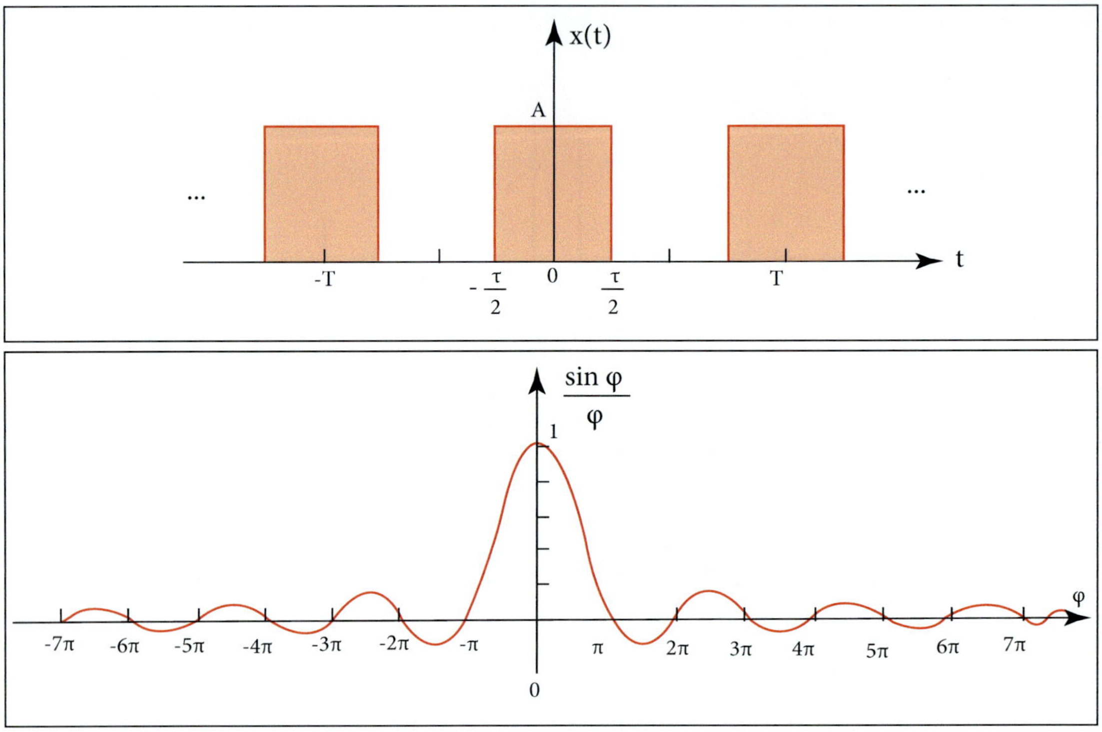

Fig. 3.8 Example of a square pulse and its corresponding Fourier series (see Exercise 3.2.8) (Proakis and Manolakis, Figures 4.3 and 4.4)

$$x(t) = \sum_{n=-\infty}^{\infty} \alpha_n e^{jn\omega t}, \qquad \text{where} \qquad \alpha_n = \frac{1}{T}\int_T x(t)e^{-jn\omega t}$$

Since the signal $x(t)$ is an *even function of time*, we choose the integration interval to be $[-T/2, +T/2]$. Therefore,

$$\alpha_0 = \frac{1}{T}\int_{-T/2}^{+T/2} x(t)dt = \frac{A}{T}\int_{-\tau/2}^{+\tau/2} dt = \frac{A\tau}{T}$$

for $n = 0$ (the coefficient α_0 expresses the average value of the signal), while for $n \neq 0$ the Fourier coefficients are calculated as

$$\alpha_n = \frac{1}{T_p}\int_{T_p} x(t)e^{-in\omega_0 t} = \frac{A}{T_p}\int_{-\tau/2}^{+\tau/2} e^{-in\omega_0 t} = \frac{A}{T_p}\left[\frac{e^{-in\omega_0 t}}{-in\omega_0}\right]_{-\tau/2}^{+\tau/2} =$$

$$= \frac{2A}{T_p n\omega_0}\frac{e^{in\omega_0\tau/2} - e^{-in\omega_0\tau/2}}{2i} = \frac{2A}{T_p n\omega_0}\sin\left(\frac{n\omega_0\tau}{2}\right) = \frac{A\tau}{T_p}\frac{\sin(n\omega_0\tau/2)}{n\omega_0\tau/2} = \frac{A\tau}{T_p}\frac{\sin\varphi}{\varphi}$$

(for values $n = \pm 1, \pm 2, \ldots$), where $\varphi = n\omega\tau/2$. The graphical representation of the function $\sin\varphi/\varphi$ is shown in Fig. 3.8β for the interval $-\infty < \varphi < +\infty$. We observe that this function tends to zero at the limits $\varphi \to \pm\infty$, while it has a maximum value equal to one for $\varphi = 0$ and zero values at the positions $\varphi = m\pi$ $(m = \pm 1, \pm 2, \ldots)$.

It is obvious that since the signal $x(t)$ is an *even* function, the Fourier coefficients are *real numbers*. These coefficients are depicted diagrammatically in Fig. 3.9α for constant T and variable τ and in Fig. 3.9β for variable T and constant τ. Finally, as far as the power density spectrum is concerned, this will obviously be given by the relation

$$|\alpha_n|^2 = \begin{cases} \left(\dfrac{A\tau}{T}\right)^2 & n = 0 \\ \left(\dfrac{A\tau}{T}\right)^2\left[\dfrac{\sin(n\omega\tau/2)}{n\omega\tau/2}\right]^2 & n > 0 \end{cases}$$

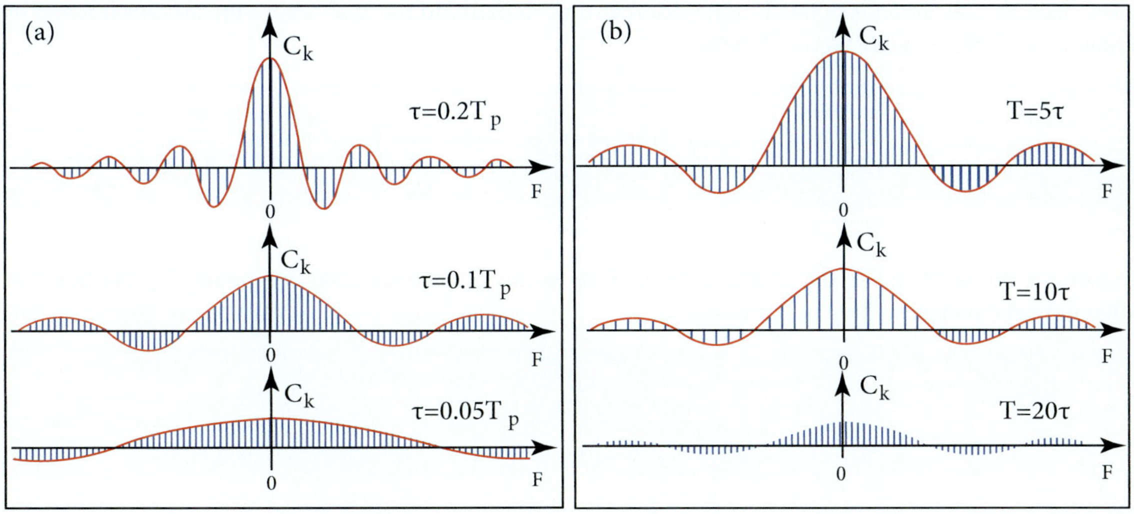

Fig. 3.9 Fourier coefficients of square pulse from Fig. 3.8α for constant T and variable τ [image (**a**), Proakis and Manolakis, Figure 4.6] and for variable T and constant τ [image (**b**), Proakis and Manolakis, Figure 4.6]

3.2.7 Fourier Expansion and LTI Systems

As already mentioned, every periodic signal $x(t)$ with period T that satisfies the Dirichlet conditions can be expressed as a Fourier expansion of the form

$$x(t) = \sum_{n=-\infty}^{+\infty} \alpha_n e^{jn(2\pi/T)t} dt = \sum_{n=-\infty}^{+\infty} \left(\frac{1}{T} \int_T x(t) e^{-jn(2\pi/T)t} dt \right) e^{jn(2\pi/T)t}$$

that is, as a *linear combination of elementary imaginary exponential signals*, which, as has been proven, constitute an *orthogonal basis* of functions for any time period of duration T. It is obvious that if the signal $x(t)$ passes through a linear and time-invariant continuous-time system, then, according to the principle of superposition, *the effect of the system on the input signal will result from the linear combination of the effects of the system on each elementary component separately.*

In the previous chapter, we proved that every elementary exponential signal of the form $x(t) = e^{st}$ (with the complex variable s defined as $s = \sigma + j\omega$) constitutes an *eigenfunction* of a linear time-invariant (LTI) continuous-time system whose eigenvalue is the *transfer function*

$$\mathcal{H}(s) = \int_{-\infty}^{+\infty} h(\tau)\, e^{-st} dt \qquad \text{or equivalently} \qquad \mathcal{H}(\sigma + j\omega) = \int_{-\infty}^{+\infty} h(\tau)\, e^{-(\sigma+j\omega)t} dt$$

which for the special case $\sigma = 0$ transforms into the *frequency response function*[5]

$$\mathcal{H}(j\omega) = \int_{-\infty}^{+\infty} h(t)\, e^{-j\omega t} dt \tag{3.16}$$

According to the analysis presented in the chapter on continuous-time systems, the effect of a LTI system on an input signal $x(t)$ is the multiplication of the signal's amplitude by the function $\mathcal{H}(s)$, a behavior that, as we will see in a subsequent

[5] This is precisely why the frequency response function is not symbolized by $\mathcal{H}(\omega)$ but by $\mathcal{H}(j\omega)$.

chapter, also characterizes systems defined in the discrete-time domain. In the case where the aforementioned input signal can be expressed as a linear combination of the form

$$x(t) = \sum_n \alpha_n \, e^{s_n t} \tag{3.17}$$

Then the system's effect on this is the multiplication of the width of each separate component by the function $\mathcal{H}(s_n)$, a property that in mathematical writing is expressed as

$$y(t) = \sum_n \alpha_n \mathcal{H}(s_n) \, e^{s_n t} \tag{3.18}$$

Therefore, *the system's output will also be expressed as a linear combination of elementary complex exponential functions*.

The preceding behavior of continuous-time systems is demonstrated in the following exercises.

Problem 3.2.9 Let us consider a LTI system with frequency response

$$\mathcal{H}(j\omega) = \int_{-\infty}^{+\infty} h(t) \, e^{-j\omega t} \, dt = \frac{\sin(4\omega)}{\omega}$$

If the input to the system is the periodic signal

$$x(t) = \begin{cases} +1 \text{ for } 0 \leq t < 4 \\ -1 \text{ for } 4 \leq t < 8 \end{cases}$$

with a period of $T = 8$, determine the system's output $y(t)$ (Oppenheim, Willsky, Nawab, Problem 3.13).

Solution Determining the system's output for the given input signal requires formulating the signal in the form of a *exponential Fourier expansion*, so that we can then apply the *principle of superposition* in the way described earlier. From the defining equation of the signal $x(t)$, it is not difficult to see that it is a *real* signal characterized by *odd symmetry*, and therefore, according to the symmetry properties of exponential Fourier coefficients, these will be imaginary and also characterized by *odd symmetry*. Specifically, for the value $n = 0$ we will have $\alpha_n = 0$, while for $n \neq 0$ it has been proven that

$$\alpha_n = \frac{1}{T} \int_T x(t) e^{-jn(2\pi/T)t} dt = \frac{1}{8} \int_0^8 x(t) e^{-jn(2\pi/8)t} dt = \frac{1}{8} \int_0^4 e^{-jn(\pi/4)t} dt - \frac{1}{8} \int_4^8 e^{-jn(\pi/4)t} dt$$
$$= \frac{1}{8} \left[\frac{e^{-jn(\pi/4)t}}{-jn(\pi/4)} \right]_0^4 - \frac{1}{8} \left[\frac{e^{-jn(\pi/4)t}}{-jn(\pi/4)} \right]_4^8 = -\frac{e^{-jn\pi} - 1}{2in\pi} + \frac{e^{-2jn\pi} - e^{-jn\pi}}{2jn\pi} = \frac{1 - e^{-jn\pi}}{jn\pi}$$

and finally

$$\alpha_n = \begin{cases} 0 & \text{for } n = \pm 2m \\ \dfrac{2}{jn\pi} & \text{for } n = \pm(2m+1) \end{cases}$$

that is, the exponential Fourier expansion of the signal $x(t)$ will have the form

$$x(t) = 2 \sum_{m=-\infty}^{+\infty} \frac{1}{j(2m+1)\pi} \exp\left[\frac{j(2m+1)\pi t}{4} \right]$$

Therefore, according to the theory presented earlier, the output of the system for the given input signal will be calculated from the equation

$$y(t) = 2 \sum_{m=-\infty}^{+\infty} \frac{\mathcal{H}[(j(2m+1)\pi/4]}{j(2m+1)\pi} \exp\left[\frac{j(2m+1)\pi t}{4}\right] = 0$$

since from the defining equation of the system's frequency response it immediately follows that

$$\mathcal{H}\left[j(2m+1)\pi/4\right] = 4\,\frac{\sin[(2m+1)\pi]}{(2m+1)\pi} = 0$$

Problem 3.2.10 Considering a LTI system described by the differential equation

$$\frac{dy(t)}{dt} + 4y(t) = x(t)$$

determine the coefficients of the Fourier series expansion of the system's output $y(t)$ when the signals (a) $x(t) = \cos(2\pi t)$ and (b) $x(t) = \sin(4\pi t) + \cos[6\pi t + (\pi/4)]$ are transmitted to its input (Oppenheim, Willsky, Nawab, Problem 3.33).

Solution To solve the problem, we must first determine the *frequency response function of the system*, $\mathcal{H}(j\omega)$. Assuming that we transmit to our system as input the *elementary imaginary exponential function* $x(t) = e^{j\omega t}$, the output of the system will have—as we have proven—the form $y(t) = \mathcal{H}(j\omega)x(t)$. Differentiating the foregoing equation, we obtain $y'(t) = j\omega\mathcal{H}(j\omega)e^{j\omega t}$, and following substitution into the differential equation, it will take the form

$$j\omega\mathcal{H}(j\omega)\,e^{j\omega t} + 4\mathcal{H}(j\omega)\,e^{j\omega t} = e^{j\omega t}$$

from which it finally follows that

$$\mathcal{H}(j\omega) = \frac{1}{j\omega + 4}$$

On the other hand, it is known that the *Fourier series expansion* of the system's output $y(t)$ has the form

$$y(t) = \sum_n \alpha_n \mathcal{H}(jn\omega_0)\,e^{jn\omega_0 t}$$

where α_n are the coefficients of the Fourier expansion of the input signal whose fundamental frequency is equal to ω_0. Therefore, the coefficients of the Fourier expansion of the system's output will result from the relation $\beta_n = \alpha_n\mathcal{H}(jn\omega_0)$, that is, their calculation requires the determination of both the coefficients α_n and the function $\mathcal{H}(jn\omega_0)$ for the various values of the parameter n. The process of calculating these quantities for the input signals that have been given is presented in what follows.

(a) The input signal $x(t) = \cos(2\pi t)$ is periodic with fundamental frequency $\omega_0 = 2\pi$ and period value $T = 2\pi/\omega = 1$. Therefore, the coefficients α_n of its Fourier expansion will be calculated as

$$\alpha_n = \frac{1}{T}\int_T x(t)\,e^{-jn\omega_0 t}dt = \int_0^1 \cos(2\pi t)\,e^{-jn2\pi t}dt$$
$$= \int_0^1 \frac{e^{2j\pi t} + e^{-2j\pi t}}{2}\,e^{-jn2\pi t}dt = \frac{1}{2}\int_0^1 e^{2\pi j(1-n)t}dt + \frac{1}{2}\int_0^1 e^{-2\pi j(1+n)t}dt$$

From the last equation and using the *orthogonality condition* of the exponential imaginary functions, it is not difficult to find that the only nonzero Fourier coefficients are those corresponding to the values $n = 1$ and $n = -1$, with their values being equal to $\alpha_1 = \alpha_{-1} = 1/2$. Therefore, the values of the corresponding coefficients of the Fourier expansion of the system's output will be calculated as

$$\beta_1 = \alpha_1 \mathcal{H}(j\omega_0) = \alpha_1 \mathcal{H}(j2\pi) = \frac{1}{2}\left\{\frac{1}{4+2\pi j}\right\} = \frac{1}{8+4\pi j} \quad \text{and}$$

$$\beta_{-1} = \alpha_{-1} \mathcal{H}(-j\omega_0) = \alpha_{-1} \mathcal{H}(-2\pi j) = \frac{1}{2}\left\{\frac{1}{4-2\pi j}\right\} = \frac{1}{8-4\pi j}$$

(b) The input signal $x(t) = \sin(4\pi t) + \cos[6\pi t + (\pi/4)]$ is also periodic with fundamental frequency $\omega_0 = 2\pi$ and period $T = 1$. Therefore, the coefficients of the Fourier expansion will be calculated as

$$\alpha_n = \frac{1}{T}\int_T x(t)e^{-jn\omega_0 t}dt = \int_0^1 \sin(4\pi t)e^{-jn\omega_0 t}dt + \int_0^1 \cos\left(6\pi t + \frac{\pi}{4}\right)e^{-jn\omega_0 t}dt =$$

$$= \frac{1}{2j}\int_0^1 e^{j4\pi t}e^{-jn2\pi t}dt - \frac{1}{2j}\int_0^1 e^{-j4\pi t}e^{-jn2\pi it}dt + \frac{1}{2}\int_0^1 e^{j6\pi t}e^{j\pi/4}e^{-jn2\pi t}dt +$$

$$+ \frac{1}{2}\int_0^1 e^{-j6\pi t}e^{-j\pi/4}e^{-jn2\pi t}dt = \frac{1}{2j}\int_0^1 e^{2\pi j(2-n)t}dt - \frac{1}{2j}\int_0^1 e^{-2\pi j(2+n)t}dt +$$

$$+ \frac{1}{2}e^{j\pi/4}\int_0^1 e^{2\pi j(3-n)t}dt + \frac{1}{2}e^{-j\pi/4}\int_0^1 e^{-2\pi j(3+n)t}dt$$

From the last equation and again using the orthogonality condition of complex exponential functions, it is concluded that the only nonzero coefficients of the Fourier expansion of the input signal $x(t)$ are $\alpha_2 = 1/2j$, $\alpha_{-2} = -1/2j$, $\alpha_3 = e^{j\pi/4}/2$, and $\alpha_{-3} = e^{-j\pi/4}/2$, and therefore the corresponding coefficients of the output of the system $y(t)$ are calculated as

$$\beta_2 = \alpha_2 \mathcal{H}(2j\omega_0) = \alpha_2 \mathcal{H}(4\pi j) = \frac{1}{2j(4+4\pi j)} \beta_{-2} = \alpha_{-2}\mathcal{H}(-2j\omega_0) = \alpha_{-2}\mathcal{H}(-4\pi j) = -\frac{1}{2j(4-4\pi j)}$$

$$\beta_3 = \alpha_3 \mathcal{H}(3j\omega_0) = \alpha_3 \mathcal{H}(3\pi j) = \frac{e^{j\pi/4}}{2(4+6\pi j)} \beta_{-3} = \alpha_{-3}\mathcal{H}(-3j\omega_0) = \alpha_{-3}\mathcal{H}(-6\pi j) = \frac{e^{-j\pi/4}}{2(4-6\pi j)}$$

Problem 3.2.11 Considering a LTI system with impulse response $h(t) = e^{-4|t|}$, determine the coefficients of the Fourier expansion of the system's output $y(t)$ for the input signals

$$x(t) = \sum_{n=-\infty}^{+\infty} \delta(t-n) \quad \text{and} \quad x(t) = \sum_{n=-\infty}^{+\infty} (-1)^n \delta(t-n).$$

[Oppenheim, Willsky, Nawab, Problem 3.34, questions (a) and (b)].

Solution Using the defining equation of the absolute value function, we can write the impulse response of the system in the form

$$h(t) = e^{-4|t|} = \begin{cases} e^{-4t} & \text{for } t \geq 0 \\ e^{+4t} & \text{for } t < 0 \end{cases}$$

and therefore the frequency response function of the system will be calculated as

$$H(\omega) = \int_{-\infty}^{+\infty} e^{-4|t|}e^{-j\omega t}dt = \int_{-\infty}^{0} e^{4t}e^{-j\omega t}dt + \int_0^{\infty} e^{-4t}e^{-j\omega t}dt =$$

$$= \int_{-\infty}^{0} e^{(4-j\omega)t}dt + \int_0^{\infty} e^{-4+j\omega}dt = \left[\frac{e^{(4-j\omega)}}{4-j\omega}\right]_{-\infty}^{0} - \left[\frac{e^{-(4+j\omega)}}{4+j\omega}\right]_0^{\infty} = \frac{1}{4-j\omega} + \frac{1}{4+j\omega}$$

Therefore, according to the methodology we used in the previous exercise, the coefficients of the Fourier expansion of the system's output will be calculated from the relation $\beta_k = \alpha_k \mathcal{H}(jk\omega_0)$, where α_k are the Fourier coefficients of the input signal and $\mathcal{H}(jk\omega_0)$ is the frequency response of the system calculated for the fundamental frequency ω_0 of the input signal. In detail, therefore, we will have the following results:

(a) The input signal is periodic with period $T = 1$. To understand this property of the signal, let us develop the sum that appears in the defining equation by writing it as

$$x(t) = \sum_{n=-\infty}^{+\infty} \delta(t-n) = \cdots + \delta(t-2) + \delta(t-1) + \delta(t) + \delta(t+1) + \delta(t+2) + \ldots$$

and let us observe that its graphical representation is a set of semi-straight lines of infinite height that belong to the upper half-plane and repeat at regular time intervals [this signal is known as a *train of impulse functions*, it has a period T and is described by the equation $x(t) = \sum_n (t - nT)$ (where $n = \ldots, -2, -1, 0, +1, +2, \ldots$)]. Therefore, the fundamental frequency of the signal is $\omega_0 = 2\pi/T = 2\pi$, and the coefficients of the Fourier expansion of the foregoing signal will be calculated as

$$\alpha_k = \frac{1}{T}\int_T x(t)e^{-jk\omega_0 t}dt = \int_{-1/2}^{+1/2} \sum_{n=-\infty}^{+\infty} \delta(t-n)\, e^{-2\pi jkt}dt$$

However, of all the infinitely time-shifted functions $\delta(t-n)$ that appear in the foregoing relation, the only one that is included within the preceding integration interval is the function $\delta(t)$ corresponding to the value $n = 0$ (that's why the interval $[-1/2, +1/2]$ was chosen as the integration interval), and the foregoing relation takes the form

$$\alpha_k = \int_{-1/2}^{+1/2} \delta(t)\, e^{-2\pi jkt}dt = e^{-2\pi jkt}\Big|_{t=0} = e^0 = 1$$

where we used the property of the function $\delta(t)$

$$\int_\alpha^\beta \delta(t)\, \varphi(t)\, dt = \varphi(0)$$

for the values $(\alpha = -1/2) < 0 < (\beta = 1/2)$ and $\varphi(t) = e^{-2\pi jkt}$. Therefore, $\alpha_k = 1$ for all values of the parameter k, and consequently, the coefficients β_k will be given by the relation

$$\beta_k = \alpha_k \mathcal{H}(2\pi jk) = \frac{1}{4 + 2\pi jk} + \frac{1}{4 - 2\pi jk}$$

(b) Developing the input signal in the form

$$x(t) = \sum_{n=-\infty}^{+\infty} (-1)^n \delta(t-n) = \cdots + \delta(t-2) - \delta(t-1) + \delta(t) - \delta(t+1) + \delta(t+2) + \ldots$$

and using the arguments we formulated previously, it is not difficult to find that the input signal is periodic with period $T = 2$, and therefore its fundamental frequency is $\omega = 2\pi/T = \pi$. From the defining equation of the Fourier coefficients we have

$$\alpha_k = \frac{1}{T}\int_T x(t)\, e^{-jk\omega_0 t}dt = \frac{1}{2}\int_{-0.5}^{+1.5} \sum_{n=-\infty}^{+\infty} (-1)^n \delta(t-n)\, e^{-j\pi kt}dt$$

where once again we chose the limits of integration so that the functions $\delta(t-n)$ are not calculated on the limits of the integration interval. In this case, from all the infinitely many time-shifted functions δ of the foregoing sum, the only ones included in the preceding interval are the functions $\delta(t)$ and $\delta(t+1)$ that correspond to the values $n = 0$ and $n = -1$, that is, the foregoing equation is written

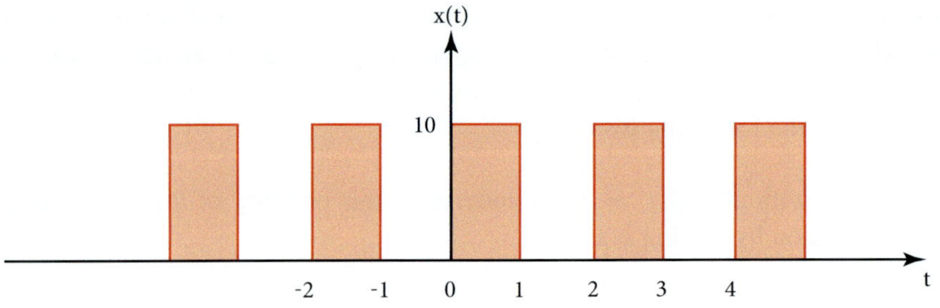

Fig. 3.10 The continuous-time signal $x(t)$ from Exercise 3.2.12

$$\alpha_k = \frac{1}{2}\int_{-0.5}^{+1.5} [\delta(t) - \delta(t+1)]\, e^{-j\pi kt} dt = \frac{1}{2}\int_{-0.5}^{+1.5} \delta(t) e^{-j\pi kt} dt - \frac{1}{2}\int_{-0.5}^{+1.5} \delta(t+1) e^{-j\pi kt} dt$$

Using again the property of the function $\delta(t)$ that we used in the previous question, the first integral is calculated as

$$\int_{-0.5}^{+1.5} \delta(t)\, e^{-j\pi kt} dt = e^{-j\pi kt}\Big|_{t=0} = e^0 = 1$$

while similarly to the second integral we will have

$$\int_{-0.5}^{+1.5} \delta(t+1)\, e^{-j\pi kt} dt = \int_{+0.5}^{+2.5} \delta(\xi)\, e^{-j\pi k(\xi-1)} d\xi = e^{jk\pi}\int_{+0.5}^{+2.5} \delta(\xi) e^{-jk\pi\xi} d\xi = e^{jk\pi} e^{-i\pi k\xi}\Big|_{\xi=0} = e^{jk\pi} = (-1)^k$$

where for the derivation of the final result we proceeded to the substitution $t \longrightarrow \xi = t + 1$. Replacing the foregoing expressions into the equation defining the Fourier coefficients, we will obtain

$$\alpha_k = \frac{1}{2} - \frac{1}{2}(-1)^k = 0.5 - 0.5\,(-1)^k = \begin{cases} 0 \text{ for } k = 2m \\ 1 \text{ for } k = 2m+1 \end{cases}$$

and therefore the coefficients of the Fourier expansion of the system output $y(t)$ are calculated as

$$\beta_k = \alpha_k \mathcal{H}(jk\pi) = \begin{cases} 0 & \text{for } k = 2m \\ \dfrac{1}{4+jk\pi} + \dfrac{1}{4-jk\pi} & \text{for } k = 2m+1 \end{cases}$$

Problem 3.2.12 Considering a LTI system described by the differential equation

$$\frac{dy(t)}{dt} + 2y(t) = x(t)$$

determine the Fourier expansion of the system's output $y(t)$ when the signal $x(t)$, depicted in Fig. 3.10 (Hsu, *Signals and Systems*, Exercises 5.4.5 and 5.4.6), is transmitted to its input.

Solution In the first step of the process, we should determine the coefficients of the exponential Fourier expansion of the continuous-time signal $x(t)$. From the defining equation it is not difficult to see that this signal is described by the equation

$$x(t) = \begin{cases} 10 \text{ for } 0 \le t \le 1 \\ 0 \ \text{ for } 1 \le t \le 2 \end{cases}$$

and therefore it is characterized by a period value $T = 2$ and a fundamental frequency $\omega_0 = 2\pi/T = \pi$. From the defining equation of the exponential Fourier coefficients, we will have, then,

$$\alpha_n = \frac{1}{T}\int_T x(t)e^{-jn\omega_0 t}dt = \frac{10}{2}\int_0^1 e^{-jn\pi t}dt = -\frac{5}{jn\pi}\left[e^{-jn\pi t}\right]_0^1 = \frac{5(1-e^{-jn\pi})}{jn\pi} = \frac{5[1-(-1)^n]}{in\pi} = \frac{10}{j\pi(2m+1)}$$

since the last relation is calculated at a value different from zero only for odd values of the index n that have the form $n = 2m+1$.

On the other hand, for $n = 0$ it is proven that $\alpha_0 = 5$, and therefore the Fourier expansion of the input signal will have the form

$$x(t) = 5 + \frac{10}{i\pi}\sum_{m=-\infty}^{+\infty}\frac{1}{2m+1}e^{i(2m+1)\pi t}$$

In the next step of the process, we need to determine the *frequency response function* of the system. Following the methodology we used in a previous exercise, let us assume that we transmit to the system the *elementary exponential function* $x(t) = e^{i\omega_0 t}$, so, according to the theory the response of the system, will have the form $y(t) = H(\omega_0)e^{i\omega_0 t}$. Differentiating the preceding equation we obtain $y'(t) = i\omega_0 H(\omega_0)e^{i\omega_0 t}$, and substituting in the differential equation of the system we finally obtain

$$H(\omega_0) = \frac{1}{i\omega_0 + 2}$$

Based on the foregoing results, we can determine the coefficients of the Fourier expansion of the output signal from the known relation $\beta_n = \alpha_n H(in\omega_0)$, where

$$H(in\omega_0) = H(in\pi) = \frac{1}{in\pi + 2} = \frac{1}{i(2m+1)\pi + 2}$$

since the coefficients α_n are defined only for odd values of the index n. Thus, we will have

$$\beta_n = \alpha_n H(in\pi) = \frac{10}{i\pi(2m+1)}\frac{1}{[i(2m+1)\pi + 2]}$$

For $m = 0$ it is proven that $\beta_0 = \alpha_0 H(0) = 5/2$, and therefore, finally, the Fourier expansion of the system's output will take the form

$$y(t) = \frac{5}{2} + \frac{10}{i\pi}\sum_{m=-\infty}^{+\infty}\frac{e^{i(2m+1)\pi t}}{(2m+1)[(2m+1)i\pi + 2]}$$

which is also the desired result.

3.3 Continuous-Time Fourier Transform

As already mentioned, every periodic signal of continuous-time with period $T = T_0$ can be expressed in the form of the *Fourier exponential series*

$$x(t) = \sum_{n=-\infty}^{+\infty}\alpha_n\, e^{jn\omega_0 t} = \sum_{n=-\infty}^{+\infty}\alpha_n\, e^{jn(2\pi F_0)t} \qquad (3.19)$$

with the *fundamental frequency* F_0 being equal to $1/T_0$ and the *Fourier coefficients* determining the shape of the waveform $x(t)$, which results from the superposition of the elementary exponential functions $e^{jn\omega_0 t}$.

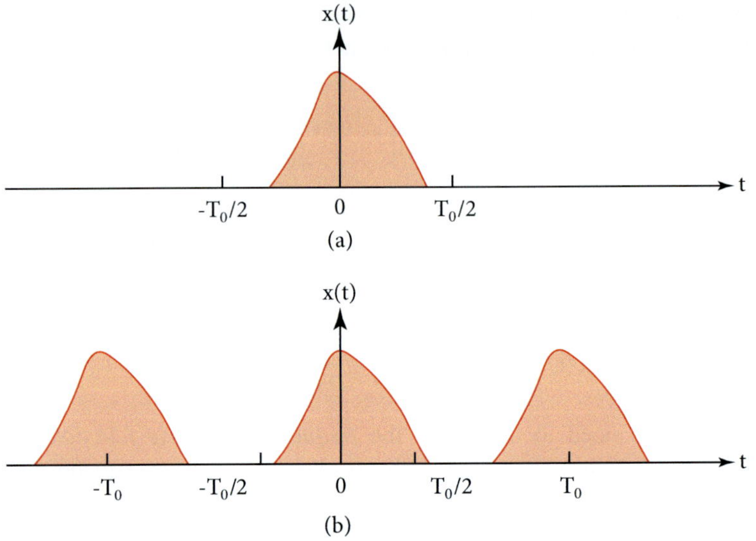

Fig. 3.11 Creation of periodic signal $x_p(t)$ from repetitions of aperiodic signal $x(t)$

If we plot the power of the signal, which, as is well known, is given by the relation

$$P_x = \frac{1}{T_0}\int_{T_0} |x(t)|^2 dt = \sum_{n=-\infty}^{+\infty} |\alpha_n|^2 \tag{3.20}$$

for the various harmonic components, we will obtain the *power spectrum*, which graphically presents the way in which the power of the signal is distributed to the various frequencies. It is obvious that, because the power of the signal is defined only at the discrete values of frequencies $\{0, \pm F_0, \pm 2F_0, \dots\}$, the power spectrum will consist of a set of lines parallel to the vertical axis, with the height of the line number k—which corresponds to the frequency $F_k = kF_0$—being equal to the value of the quantity $|\alpha_k|^2$. For this reason, the power spectrum of a periodic signal is also known as the *linear spectrum*. It is not difficult to determine that the distance between two successive lines of the power spectrum is equal to the *fundamental frequency* F_0, that is, it varies inversely proportional to the period of the signal T_0.

Let us now assume that the period of the signal begins to increase infinitely, a process that results in a *reduction of the distance between two successive spectral lines*. If the period takes an infinite value, this distance will become equal to zero, the spectrum of the signal will change from *discrete* to *continuous*, while the signal itself will change from *periodic* to *aperiodic*, that is, it will be characterized by a *complete absence of periodicity*.

Assuming that the aperiodic signal $x(t)$ is symmetric with respect to the *vertical axis* and that it has a finite duration $T < T_0$, we can construct from it a periodic signal $x_p(t)$ with successive iterations of the aperiodic signal every T_0 time units; this situation is shown diagrammatically in Fig. 3.11. It is obvious that the periodic signal $x_p(t)$ approximates the aperiodic signal $x(t)$ in the limit $T_0 \to \infty$, a property that is mathematically formulated as

$$x(t) = \lim_{T_0 \to \infty} x_p(t) \tag{3.21}$$

From the last property it follows that the *spectrum of the aperiodic signal* $x(t)$ *will result from the spectrum of the periodic signal* $x_p(t)$ *in the limit* $T_0 \to \infty$.

If we develop the periodic signal $x_p(t)$ in an *exponential Fourier series*, we will be led to the equation

$$x_p(t) = \sum_{n=-\infty}^{+\infty} \alpha_n e^{jn\omega_0 t}$$

with the Fourier coefficients being calculated from the relation

$$\alpha_n = \frac{1}{T_0}\int_{-T_0/2}^{+T_0/2} x_p(t)\, e^{-jn\omega_0 t}\, dt = \frac{1}{T_0}\int_{-T_0/2}^{+T_0/2} x(t)\, e^{-jn\omega_0 t}\, dt = \frac{1}{T_0}\int_{-\infty}^{+\infty} x(t)\, e^{-jn\omega_0 t}\, dt$$

In the foregoing relation, the second sum results from the first considering that in the interval $-T_0/2 \le t \le T_0/2$ it holds that $x_p(t) = x(t)$, while the third sum results from the second based on the fact that for $|t| > T_0/2$, we have $x(t) = 0$. Now defining the function

$$\mathcal{X}(j\omega) = \int_{-\infty}^{+\infty} x(t)\, e^{-j\omega t}\, dt \tag{3.22}$$

which is known as the *exponential Fourier transform* of the signal $x(t)$, the foregoing relation will take the form $\alpha_n = (1/T_0)\mathcal{X}(n\omega_0)$, from which it follows that

$$T_0\alpha_n = \mathcal{X}(n\omega_0) = \mathcal{X}\left(\frac{2\pi n}{T_0}\right)$$

From the last relation it follows that the Fourier coefficients α_n (multiplied by the period T_0) can be treated as samples of the function $\mathcal{X}(j\omega)$ recorded at integer multiples of the circular frequency $\omega_0 = 2\pi F_0 = 2\pi/T_0$.

Substituting the last relation into the defining equation of the periodic signal $x_p(t)$, it becomes

$$x_p(t) = \frac{1}{T_0}\sum_{n=-\infty}^{+\infty} \mathcal{X}\left(\frac{2\pi n}{T_0}\right) e^{jn\omega_0 t} = \frac{1}{2\pi}\sum_{n=-\infty}^{+\infty} \mathcal{X}(n\Delta\omega)\, e^{jn(\Delta\omega)t}\,\Delta\omega$$

with the second sum resulting from the first by setting $\Delta f = 1/T_0$ or, equivalently, $\Delta\omega = 2\pi/T_0$. To obtain the aperiodic signal $x(t)$ now, we need to calculate the foregoing relation in the limit $T_0 \to \infty$ or, equivalently, in the limit $\Delta\omega \to 0$. Therefore, we have

$$x(t) = \lim_{T_0\to\infty} x_p(t) = \frac{1}{2\pi}\lim_{\Delta\omega\to 0}\sum_{n=-\infty}^{+\infty} \mathcal{X}(n\Delta\omega)\, e^{jn(\Delta\omega t)}\,\Delta\omega = \frac{1}{2\pi}\int_{-\infty}^{+\infty} \mathcal{X}(j\omega)\, e^{j\omega t}\, d\omega$$

a relation known as the *inverse Fourier transform* of the signal $x(t)$. The relations

$$\mathcal{X}(j\omega) \equiv \mathcal{F}\{x(t)\} = \int_{-\infty}^{+\infty} x(t)\, e^{-j\omega t}\, dt \quad \text{and} \tag{3.23}$$

$$x(t) \equiv \mathcal{F}^{-1}\{X(\omega)\} = \frac{1}{2\pi}\int_{-\infty}^{+\infty} \mathcal{X}(j\omega)\, e^{j\omega t}\, d\omega \tag{3.24}$$

constitute a pair of equations referred to as a *direct-inverse transform*, where the first equation is known as an *analysis equation* and the second equation as a *synthesis equation*. An alternative way of formulating these two equations would be

$$\mathcal{X}(j\omega) = \frac{1}{\sqrt{2\pi}}\int_{-\infty}^{+\infty} x(t)\, e^{-j\omega t}\, dt \qquad \text{and} \qquad x(t) = \frac{1}{\sqrt{2\pi}}\int_{-\infty}^{+\infty} \mathcal{X}(j\omega)\, e^{j\omega t}\, d\omega$$

where we simply changed the value of the multiplicative coefficients in the two integrals.[6] In the foregoing equations, the *cyclic frequency* ω is used as an independent variable, ordinary frequency will result in $df = d\omega/2\pi$, and the relations of the direct-inverse transform will take the form

$$\mathcal{X}(f) = \int_{-\infty}^{+\infty} x(t)\, e^{-j2\pi f t} dt \qquad \text{and} \qquad x(t) = \int_{-\infty}^{+\infty} \mathcal{X}(f)\, e^{j2\pi f t t} df$$

3.3.1 Integral and Trigonometric Fourier Transforms

If we combine the equations of a direct-inverse Fourier transform and, more specifically, if we replace the function $\mathcal{X}(j\omega)$ that appears in the synthesis equation from its defining equation, the continuous-time signal $x(t)$ will be formulated as

$$x(t) = \frac{1}{2\pi}\int_{-\infty}^{+\infty} e^{j\omega t}\left\{\int_{-\infty}^{+\infty} x(\xi)\, e^{-j\omega\xi} d\xi\right\} d\omega = \frac{1}{2\pi}\int_{-\infty}^{+\infty}\left\{x(\xi)\, e^{-j\omega(\xi - t)} d\xi\right\} d\omega$$

or, equivalently,

$$\begin{aligned} x(t) &= \frac{1}{2\pi}\int_{-\infty}^{+\infty}\left\{x(\xi)\Big[\cos[\omega(\xi - t)] - j\sin[\omega(\xi - t)]\Big] d\xi\right\} d\omega = \\ &= \frac{1}{2\pi}\int_{-\infty}^{+\infty}\left\{\int_{-\infty}^{+\infty} x(\xi)\cos[\omega(\xi - t)]\right\} d\omega - \frac{j}{2\pi}\int_{-\infty}^{+\infty}\left\{\int_{-\infty}^{+\infty} x(\xi)\sin[\omega(\xi - t)]\right\} d\omega = \\ &= \frac{1}{2\pi}\int_{-\infty}^{+\infty}\left\{\int_{-\infty}^{+\infty} x(\xi)\cos[\omega(\xi - t)]\right\} d\omega \end{aligned}$$

where we used Euler's formula

$$e^{-j\omega(\xi - t)} = \cos[\omega(\xi - t)] - j\sin[\omega(\xi - t)]$$

and we silently assumed that the signal $x(t)$ was a real-valued signal, an assumption that led to the removal of the second term of the foregoing equation.

It is not difficult to see that the last integral of the foregoing relation is an *even function of the frequency* ω and therefore can be written in the form

$$x(t) = \frac{1}{2\pi}\int_{-\infty}^{+\infty}\left\{\int_{-\infty}^{+\infty} x(\xi)\cos[\omega(\xi - t)]\right\} d\omega = \frac{1}{\pi}\int_{0}^{\infty}\left\{\int_{-\infty}^{+\infty} x(\xi)\cos[\omega(\xi - t)]\right\} d\omega$$

Using the well-known trigonometric identity

$$\cos[\omega(\xi - t)] = \cos(\omega\xi - \omega t) = \cos(\omega\xi)\cos(\omega t) + \sin(\omega\xi)\sin(\omega t)$$

we finally get

$$x(t) = \frac{1}{\pi}\int_{0}^{\infty}\left\{\int_{-\infty}^{+\infty} x(\xi)\Big[\cos(\omega\xi)\cos(\omega t) + \sin(\omega\xi)\sin(\omega t)\Big] d\xi\right\} d\omega =$$

[6] Generally speaking, the equations of a direct-inverse Fourier transform can be defined for any two multiplicative coefficients α and β through the relations

$$X(\omega) = \sqrt{\frac{|\beta|}{(2\pi)^{1-\alpha}}}\int_{-\infty}^{+\infty} x(t)\, e^{j\beta\omega t} dt \qquad \text{and} \qquad x(t) = \sqrt{\frac{|\beta|}{(2\pi)^{1+\alpha}}}\int_{-\infty}^{+\infty} X(\omega)\, e^{-j\beta\omega t} d\omega$$

so that the product $\alpha \cdot \beta$ is always equal to $1/2\pi$.

$$= \int_0^{\infty} \left\{ \frac{1}{\pi} \int_{-\infty}^{+\infty} x(\xi) \cos(\omega \xi) d\xi \right\} \cos(\omega t) d\omega + \int_0^{\infty} \left\{ \frac{1}{\pi} \int_{-\infty}^{+\infty} x(\xi) \sin(\omega \xi) d\xi \right\} \sin(\omega t) d\omega$$

or, equivalently,

$$x(t) = \int_0^{\infty} \Big[\mathcal{A}(\omega) \cos(\omega t) + \mathcal{B}(\omega) \sin(\omega t) \Big] d\omega \tag{3.25}$$

with the functions $\mathcal{A}(\omega)$ and $\mathcal{B}(\omega)$ defined by the relations

$$\mathcal{A}(\omega) = \frac{1}{\pi} \int_{-\infty}^{+\infty} x(\xi) \cos(\omega \xi)\, d\xi \qquad \mathcal{B}(\omega) = \frac{1}{\pi} \int_{-\infty}^{+\infty} x(\xi) \sin(\omega \xi)\, d\xi \tag{3.26}$$

This way of writing the signal $x(t)$ is known as the *Fourier integral*,[7] while the functions $\mathcal{A}(\omega)$ and $\mathcal{B}(\omega)$ are known as the *cosine* and the *sine* Fourier transform, respectively, and can be considered as *trigonometric form* of the Fourier transform.

Following the methodology we used in the theory of the Fourier series, we define

$$\cos[\varphi(\omega)] = \frac{\mathcal{A}(\omega)}{\sqrt{[\mathcal{A}(\omega)]^2 + [\mathcal{B}(\omega)]^2}} \qquad \text{and} \qquad \sin[\varphi(\omega)] = \frac{\mathcal{B}(\omega)}{\sqrt{[\mathcal{A}(\omega)]^2 + [\mathcal{B}(\omega)]^2}}$$

In this case, the Fourier integral can be written in the form

$$x(t) = \int_0^{\infty} \sqrt{[\mathcal{A}(\omega)]^2 + [\mathcal{B}(\omega)]^2} \left\{ \frac{\mathcal{A}(\omega) \cos(\omega t)}{\sqrt{[\mathcal{A}(\omega)]^2 + [\mathcal{B}(\omega)]^2}} + \frac{\mathcal{B}(\omega) \sin(\omega t)}{\sqrt{[\mathcal{A}(\omega)]^2 + [\mathcal{B}(\omega)]^2}} \right\} d\omega =$$

$$= \int_0^{\infty} \mathcal{C}(\omega) \Big[\cos(\omega t) \cos[\varphi(\omega)] + \sin(\omega t) \sin[\varphi(\omega)] \Big] d\omega = \int_0^{\infty} \mathcal{C}(\omega) \cos[\omega t - \varphi(\omega)]\, d\omega$$

This equation constitutes an alternative way of formulating the *inverse Fourier transform.* The functions $\mathcal{C}(\omega)$ and $\varphi(\omega)$ appearing in the foregoing relation are defined by the equations

$$\mathcal{C}(\omega) = \sqrt{[\mathcal{A}(\omega)]^2 + [\mathcal{B}(\omega)]^2} \qquad \text{and} \qquad \varphi(\omega) = \arctan\left[\frac{\mathcal{B}(\omega)}{\mathcal{A}(\omega)}\right]$$

From the foregoing relations, the close analogy between the Fourier series and the Fourier transform becomes apparent: The sum symbol appearing in the series equation has been replaced by the symbol of the integral, while the constant Fourier coefficients A_n and B_n have been replaced by the Fourier transforms $\mathcal{A}(\omega)$ and $\mathcal{B}(\omega)$, which are a function of the cyclic frequency ω and, therefore, determine the continuous frequency spectrum of the function $x(t)$.

If the signal $x(t)$ is an even or odd function, the Fourier transforms are given by the simplified expressions

$$\mathcal{A}(\omega) = \int_0^{\infty} x(t) \sin(\omega t)\, dt, \quad \mathcal{B}(\omega) = 0 \quad \text{(for even symmetry)} \quad \text{and}$$

$$\mathcal{A}(\omega) = 0, \quad \mathcal{B}(\omega) = \int_0^{\infty} x(t) \cos(\omega t)\, dt \quad \text{(for odd symmetry)}$$

[7] In the international literature, the *Fourier integral* is also referred to as a *Fourier series expansion* due to the great similarity that exists between its defining equation and the corresponding equation of the Fourier series.

while the signal $x(t)$ for the even and odd cases respectively will take the forms

$$x(t) = \int_0^{\infty} \mathcal{A}(\omega)\cos(\omega t)d\omega \quad \text{and} \quad x(t) = \int_0^{\infty} \mathcal{B}(\omega)\sin(\omega t)d\omega$$

The *exponential Fourier transform* $\mathcal{X}(j\omega)$ of the function $x(t)$ is related to the *sine* and the *cosine* transforms through the relation

$$\mathcal{X}(j\omega) = \int_{-\infty}^{\infty} x(t)\, e^{-j\omega t} dt = \int_{-\infty}^{\infty} x(t)\Big[\cos(\omega t) - j\sin(\omega t)\Big] = \int_{-\infty}^{\infty} x(t)\cos(\omega t)\, dt - j\int_{-\infty}^{\infty} x(t)\sin(\omega t)\, dt = \mathcal{A}(\omega) - j\mathcal{B}(\omega)$$

the proof of which obviously relied on the use of the known Euler equation. Taking into account that the trigonometric Fourier transforms satisfy the symmetry equations $\mathcal{A}(-\omega) = \mathcal{A}(\omega)$ and $\mathcal{B}(-\omega) = -\mathcal{B}(\omega)$—as can be derived from their defining equations—it is obvious that

$$\mathcal{X}(-\omega) = \mathcal{A}(-\omega) - j\mathcal{B}(-\omega) = \mathcal{A}(\omega) + j\mathcal{B}(\omega) = \mathcal{X}^*(\omega)$$

where the symbol $(*)$ refers to the complex conjugate. This is one of the many properties of the Fourier transform that are presented collectively in Table 3.2, while their analytical proof is presented in Sect. 3.3.5 (refer to page 235).

3.3.2 Magnitude and Phase of Fourier Transform

Taking into account that the Fourier transform is in general a complex function of the real frequency ω, it is obvious that like any other complex function, it can be expressed in its exponential form

Table 3.2 Basic properties of Fourier transform

Property	Signal $x(t)$	Fourier transform
Homogeneity	$\alpha x\ (t)$	$\alpha X\ (j\omega)$
Linearity	$x(t) + y(t)$	$X\ (j\omega) + Y\ (j\omega)$
Addition	$\alpha x\ (t) + \beta y\ (t)$	$\alpha\, X\ (j\omega) + \beta\ Y\ (j\omega)$
Differentiation	$\dfrac{d^n x(t)}{dt^n}$	$(j\omega)^n X\ (j\omega)$
Integration	$\int_{-\infty}^{t} x(\tau)d\tau$	$\dfrac{X\ (j\omega)}{i\omega} + \pi X\ (0)\ \delta(\omega)$
Frequency shifting	$\exp(j\omega t)\, x(t)$	$X\ (\omega - \omega_0)$
Cosine modulation	$x\ (t)\cos\ (\omega_0 t)$	$\frac{1}{2}\,[X\,(\omega + \omega_0) + X\ (\omega - \omega_0)]$
Sine modulation	$x(t)\ \sin\ (\omega_0 t)$	$\frac{1}{2}\,[X\ (\omega + \omega_0) - X\ (\omega - \omega_0)]$
Time shifting	$x(t - \tau)$	$\exp(-j\omega\tau)X(\omega)$
Convolution	$x(t) * y(t)$	$X(j\omega)Y(j\omega)$
Multiplication	$x(t)y(t)$	$\frac{1}{2\pi}\ [X(j\omega) * Y(j\omega)]$
Time scaling	$x(\alpha t)$	$\frac{1}{\lvert\alpha\rvert} X\left(\frac{\omega}{\alpha}\right)$
Duality	X (t)	$2\pi x(-\omega)$
Complex conjugation	$x^*(t)$	$X^*(-\omega)$
Real part	$\Re\{x(t)\}$	$\frac{1}{2}\big[X\ (j\omega) + X^*(-\omega)\big]$
Imaginary part	$\Im\{x(t)\}$	$\frac{1}{2j}\ \big[X(\omega) - X^*(-\omega)\big]$

$$\mathcal{X}(j\omega) = |\mathcal{X}(j\omega)|e^{j\measuredangle \mathcal{X}(j\omega)} \tag{3.27}$$

with the functions $|\mathcal{X}(j\omega)|$ and $\measuredangle \mathcal{X}(j\omega)$ expressing the magnitude and phase of the transform $\mathcal{X}(\omega)$, respectively. Of these two functions, the magnitude of the Fourier transform is related to the signal's energy (see the Bessel equation in the transform properties section). More specifically, the magnitude

$$dE = \frac{1}{2\pi}|\mathcal{X}(j\omega)|^2 d\omega \tag{3.28}$$

expresses the *amount of signal energy corresponding to an infinitesimal frequency band* $d\omega$, while to calculate the energy of the signal corresponding to a frequency band $[\omega_1, \omega_2]$, we must calculate the *integral* of the foregoing equation with limits of integration from ω_1 to ω_2. Therefore, the magnitude of the transform $\mathcal{X}(j\omega)$ describes the *frequency content of the signal* and gives us information regarding the relative magnitudes of the complex exponential signals from which the continuous-time signal $x(t)$ we are dealing with is composed.

On the other hand, the phase of the transform $\measuredangle \mathcal{X}(j\omega)$ does not affect in any way the amplitudes of the individual components of the signal but, on the contrary, gives us information regarding the relative phases of these components. This function has a very significant impact on the characteristics of the signal in the time domain, and, as can be proven, two signals whose Fourier transforms have exactly the same magnitude, but different phases, are completely different from each other.[8] The functions of the measure and the phase of the Fourier transform fully identify the effect of a LTI continuous-time system on the input signal that is transmitted to it; the analytical study of these characteristics is included in the section on the frequency analysis of signals and systems that will be presented later in this chapter.

3.3.3 Convergence Conditions of Fourier Transform

The *convergence conditions* of the Fourier transform are similar to those that characterize the convergence of the Fourier series. Generally speaking, a necessary prerequisite for the existence of the Fourier transform of a continuous-time signal $x(t)$ is the satisfaction of the three *Dirichlet conditions*, which, as mentioned previously, are as follows:

- The signal $x(t)$ has a finite number of discontinuity points of finite magnitude.
- The signal $x(t)$ has a finite number of maximum and minimum values within any finite time interval.
- The signal $x(t)$ is *absolutely integrable*, that is, it satisfies the condition

$$\int_{-\infty}^{+\infty} |x(t)|dt < \infty \tag{3.29}$$

Note that the proof of the validity of the last condition is relatively easy, since from the defining equation of the Fourier transform we obtain

$$|X(\omega)| = \left|\int_{-\infty}^{+\infty} x(t)e^{-j\omega t}dt\right| \leq \int_{-\infty}^{+\infty} |x(t)||e^{-j\omega t}|dt = \int_{-\infty}^{+\infty} |x(t)|dt < \infty$$

and therefore if the aforementioned condition is valid, then the measure of the Fourier transform is *finite*.

[8] An excellent example of the way in which the phase of the Fourier transform affects the appearance of an image can be found in the book by Gonzalez & Woods: *Digital Image Processing* (see Example 4.14 on page 270 of the third edition).

An alternative, but not very strong, condition for the existence of the aforementioned transform requires the signal to be *square integrable*, that is, to satisfy the property

$$\int_{-\infty}^{+\infty} |x(t)|^2 dt < \infty \tag{3.30}$$

which in turn ensures that the signal will have *finite energy*; indeed, from the defining equation of the signal's energy it immediately follows that

$$E_x = \int_{-\infty}^{+\infty} |x(t)|^2 dt < \infty$$

If the foregoing condition is met, the existence of the Fourier transform is ensured, in the sense that if the transform of the signal $x(t)$ converges to the value

$$y(t) = \frac{1}{2\pi} \int_{-\infty}^{+\infty} X(\omega) e^{-j\omega t} dt \tag{3.31}$$

then the *approximation error* $e(t) = x(t) - y(t)$ will be characterized by *zero energy*, hence, resulting in

$$E_e = \int_{-\infty}^{+\infty} |e(t)|^2 dt = \int_{-\infty}^{+\infty} |x(t) - y(t)|^2 dt = 0 \tag{3.32}$$

The foregoing condition of finite energy is characterized as *less powerful* compared to the *Dirichlet conditions* in the sense that it is *sufficient but not necessary*. In other words, this means that an absolutely integrable signal is characterized by the existence of finite energy, but on the other hand, *a signal of finite energy is not always absolutely integrable*. A typical example of this case is the function

$$x(t) = \frac{\sin(2\pi t)}{\pi t} = 2\frac{\sin(2\pi t)}{2\pi t} = 2\mathrm{sinc}(2\pi t)$$

which, although it has finite energy, is not absolutely integrable, since at the position $t = 0$ the integral of the condition of absolute integrability tends to infinity. It should also be noted that, although the third Dirichlet condition does not apply to this function, its transform according to Fourier exists and is given by the relation

$$\mathcal{X}(j\omega) = \begin{cases} 1 \text{ for } |f| \le 1 \\ 0 \text{ for } |f| > 1 \end{cases} \tag{3.33}$$

where $f = \omega/2\pi$. Generally speaking, *almost all signals of finite energy have a Fourier transform*, so that we do not check whether the conditions for its existence are met, since these apply in almost all cases.

In addition, it is interesting to note that, although the derivation of the direct-inverse Fourier transform relationship was based on the use of a finite-duration signal, in practice the foregoing equations hold for a very large category of signals whose duration can be infinite, provided the equation of the Fourier transform of these signals is characterized by the existence of *impulse functions*. More details on this subject can be found in the bibliography.

3.3.4 Examples of Fourier Transform Calculations

After presenting the basic theory related to the Fourier transform and the prerequisites for its existence, examples of solved exercises follow that will help readers familiarize themselves with the calculation process of this transform. The function $\mathcal{X}(j\omega)$ that will result as the outcome of this process is often known as the *spectrum of the signal* $x(t)$, as it graphically presents *the distribution of a signal's energy at various frequencies*. The design of this spectrum in an $\{\omega, X(\omega)\}$ axis system can be carried out with various applications, such as *Mathematica* or *MATLAB*, and is left as an exercise to the reader.

Problem 3.3.1 Find the Fourier transform of the function

$$x(t) = \delta(t+1) + \delta(t-1)$$

(Opeenheim, Willsky, Nawab, Problem 4.2α).

Solution From the defining equation of the Fourier transform we have

$$\mathcal{X}(j\omega) = \int_{-\infty}^{+\infty} x(t)\, e^{-j\omega t} dt = \int_{-\infty}^{+\infty} \delta(t+1)\, e^{-j\omega t} dt + \int_{-\infty}^{+\infty} \delta(t-1)\, e^{-j\omega t} dt$$

Using the property of the function $\delta(t)$

$$\int_{-\infty}^{+\infty} \delta(t-t_0)\, \varphi(t)\, dt = \varphi(t_0)$$

for the function $\varphi(t) = e^{-i\omega t}$ we obtain

$$\mathcal{X}(j\omega) = e^{-j\omega t}\Big|_{t=-1} + e^{-j\omega t}\Big|_{t=1} = e^{j\omega} + e^{-j\omega} = 2\cos(\omega)$$

which is the desired result.

Problem 3.3.2 Find the Fourier transform of the function $x(t) = Ae^{-\alpha t}u(t)$ (Shmaliy, *Continuous-Time Signals*, Section 2.4.1).

Solution From the defining equation of the Fourier transform we have

$$\mathcal{X}(j\omega) = \int_{-\infty}^{+\infty} x(t)e^{-j\omega t} dt = A\int_{-\infty}^{+\infty} e^{-\alpha t}e^{-j\omega t}u(t)dt = A\int_{0}^{\infty} e^{-(\alpha+j\omega)t}dt = A\left[\frac{e^{-(\alpha+j\omega)t}}{\alpha+j\omega}\right]_{\infty}^{0} = \frac{A}{\alpha+j\omega}$$

Problem 3.3.3 Find the Fourier transform of the function $x(t) = e^{-\alpha t}\cos(\omega_0 t)u(t)$.

Solution From the defining equation of the Fourier transform and replacing the cosine function from the Euler formula, it turns out that

$$\mathcal{X}(j\omega) = \int_{-\infty}^{+\infty} x(t)e^{-j\omega t} dt = \int_{-\infty}^{+\infty} e^{-\alpha t}\cos(\omega_0 t)e^{-j\omega t}u(t)dt = \int_{0}^{\infty} e^{-(\alpha+j\omega t)}\frac{e^{j\omega_0 t} - e^{-j\omega_0 t}}{2}dt$$

$$= \frac{1}{2}\int_0^{\infty} e^{-(\alpha+j\omega-j\omega_0)t}dt + \frac{1}{2}\int_0^{\infty} e^{-(\alpha+j\omega+j\omega_0)t}dt = \frac{1}{2}\left[\frac{e^{-(\alpha+j\omega-j\omega_0)t}}{\alpha+j\omega-j\omega_0}\right]_{\infty}^{0} + \frac{1}{2}\left[\frac{e^{-(\alpha+j\omega+j\omega_0)t}}{\alpha+j\omega+j\omega_0}\right]_{\infty}^{0}$$

and finally

$$\frac{1}{2}\left[\frac{1}{\alpha+j(\omega-\omega_0)} + \frac{1}{\alpha+j(\omega+\omega_0)}\right]$$

Problem 3.3.4 Find the Fourier transform of the function

$$x(t) \equiv \Pi_\alpha(t) = \begin{cases} 1 \text{ for } |t| \le \alpha \\ 0 \text{ for } |t| > \alpha \end{cases}$$

which describes a square signal with amplitude $h = 1$ and duration $T = 2\alpha$ (e.g., Hsu, *Signals and Systems*, Exercise 5.19).

Solution From the defining equation of the Fourier transform we have

$$\mathcal{X}(j\omega) = \int_{-\infty}^{+\infty} x(t)e^{-j\omega t}dt = \int_{-\alpha}^{+\alpha} e^{-j\omega t}dt = \left[-\frac{e^{-j\omega t}}{j\omega}\right]_{-\alpha}^{+\alpha} = \frac{e^{j\omega\alpha} - e^{-j\omega\alpha}}{j\omega}$$

$$= \frac{2}{\omega}\frac{e^{j\omega\alpha} - e^{-j\omega\alpha}}{2j} = \frac{2}{\omega}\sin(\omega\alpha) = 2\alpha\frac{\sin(\omega\alpha)}{\omega\alpha} = 2\alpha\,\mathrm{sinc}(\omega\alpha)$$

Problem 3.3.5 Find the Fourier transform of the signal $x(t) = e^{-\alpha|t|}$ $(\alpha > 0)$.

Solution Using the definition of the absolute value function

$$|t| = \begin{cases} +t \text{ for } t \ge 0 \\ -t \text{ for } t < 0 \end{cases}$$

we can write the signal $x(t)$ in the form

$$x(t) = e^{-\alpha|t|} = e^{-\alpha t}u(t) + e^{\alpha t}u(-t)$$

and from the defining equation of the Fourier transform we get

$$\mathcal{X}(j\omega) = \int_{-\infty}^{+\infty} x(t)e^{-j\omega t}dt = \int_{-\infty}^{+\infty} e^{-\alpha t}e^{-j\omega t}u(t)dt + \int_{-\infty}^{+\infty} e^{\alpha t}e^{-j\omega t}u(-t)dt =$$

$$= \int_0^{\infty} e^{-(\alpha+i\omega)t}dt + \int_{-\infty}^{0} e^{(\alpha-j\omega)t}dt = \left[\frac{e^{-(\alpha+j\omega)t}}{\alpha+j\omega}\right]_0^{\infty} + \left[\frac{e^{(\alpha-j\omega)t}}{\alpha-j\omega}\right]_{-\infty}^{0} = \frac{1}{\alpha+i\omega} + \frac{1}{\alpha-j\omega} = \frac{2\alpha}{\alpha^2+\omega^2}$$

as can be proven.

Problem 3.3.6 Find the Fourier transform of the function $\delta(t)$.

Solution From the defining equation of the Fourier transform we have

$$\mathcal{F}\{\delta(t)\} = \int_{-\infty}^{+\infty} \delta(t)\, e^{-j\omega t}\, dt = \int_{-\infty}^{+\infty} \delta(t-0)\, e^{-j\omega t}\, dt = e^{-j\omega t}\Big|_{t=0} = 1$$

where for the function $\varphi(t) = e^{-j\omega t}$ we used the property of the impulse function

$$\int_{-\infty}^{+\infty} \delta(t-t_0)\, \varphi(t)\, dt = \varphi(t_0)$$

Problem 3.3.7 Find the Fourier transform and the amplitude and phase spectra of the function

$$x(t) \equiv \Lambda_\alpha(t) = \begin{cases} 1 - \dfrac{|t|}{\alpha} & \text{for } |t| \le \alpha \\ 0 & \text{anywhere else} \end{cases}$$

which expresses a triangular continuous-time signal with amplitude $h = 1$ and duration $T = 2\alpha$ (Proakis and Manolakis, Exercise for solution 4.3α).

Solution From the defining equation of the Fourier transform we have

$$\begin{aligned} \mathcal{X}(j\omega) &= \int_{-\infty}^{+\infty} x(t)\, e^{-j\omega t} dt = \int_{-\alpha}^{+\alpha} \left(1 - \frac{|t|}{\alpha}\right) e^{-j\omega t} dt = \int_{-\alpha}^{0} \left(1 + \frac{t}{\alpha}\right) e^{-j\omega t} dt \\ &+ \int_{0}^{\alpha} \left(1 - \frac{t}{\alpha}\right) e^{-j\omega t} dt = \int_{0}^{\alpha} \left(1 - \frac{t}{\alpha}\right) e^{-j\omega t} dt - \int_{0}^{-\alpha} \left(1 + \frac{t}{\alpha}\right) e^{-j\omega t} dt \end{aligned}$$

If in the second integral we proceed to the substitution $t \to -t$, the function $\mathcal{X}(j\omega)$ will take the form

$$\mathcal{X}(j\omega) = \int_{0}^{\alpha} \left(1 - \frac{t}{\alpha}\right) e^{-j\omega t} dt + \int_{0}^{\alpha} \left(1 - \frac{t}{\alpha}\right) e^{j\omega t} dt = \int_{0}^{\alpha} \left(1 - \frac{t}{\alpha}\right)\left[e^{j\omega t} + e^{-j\omega t}\right] dt = 2\int_{0}^{\alpha} \left(1 - \frac{t}{\alpha}\right) \cos(\omega t) dt$$

For the value $\omega = 0$ it immediately follows that

$$\mathcal{X}(j\omega) = 2\int_{0}^{\alpha} \left(1 - \frac{t}{\alpha}\right) dt = 2\left[t - \frac{t^2}{2\alpha}\right]_0^{\alpha} = 2\left(\alpha - \frac{\alpha^2}{2\alpha}\right) = \alpha$$

while for $\omega \neq 0$, integrating by parts, we end up with the expression

$$\mathcal{X}(j\omega) = \frac{2}{\omega}\int_{0}^{\alpha} \left(1 - \frac{t}{\alpha}\right) d\Big[\sin(\omega t)\Big] = \left[\frac{2}{\omega}\left(1 - \frac{t}{\alpha}\right)\sin(\omega t)\right]_0^{\alpha} - \frac{2}{\omega}\int_{0}^{\alpha} \sin(\omega t) d\left(1 - \frac{t}{\alpha}\right)$$

It is not difficult to see that the value of the first expression is equal to zero, and as for the second expression, taking into account that

$$d\left(1 - \frac{t}{\alpha}\right) = -\frac{1}{\alpha}$$

the foregoing equation is written

$$\mathcal{X}(j\omega) = \frac{2}{\omega\alpha}\int_{0}^{\alpha} \sin(\omega t) dt = -\frac{2}{\alpha\omega^2}\Big[\cos(\omega t)\Big]_0^{\alpha} = \frac{2}{\alpha\omega^2}[1 - \cos(\alpha\omega)] = \frac{4}{\alpha\omega^2}\sin^2\left(\frac{\alpha\omega}{2}\right)$$

where for the derivation of the last expression we used the known trigonometric identity $1 - \cos(\omega\alpha) = 2\sin^2(\omega\alpha/2)$. From the last equation we observe that the Fourier transform is a real number and therefore the amplitude and phase spectra will be given by the equations

$$|\mathcal{X}(j\omega)| = \frac{4}{\alpha\omega^2}\sin^2\left(\frac{\alpha\omega}{2}\right) \quad \text{and} \quad \measuredangle\mathcal{X}(j\omega) = 0$$

which is also the desired result.

Problem 3.3.8 Determine the amplitude and phase spectra of the signal

$$x(t) = \begin{cases} Ae^{-\alpha t} & \text{for } t \geq 0 \\ 0 & \text{for } t < 0 \end{cases} \quad \text{with} \quad \alpha > 0.$$

Solution The signal $x(t)$ can be written in the form $x(t) = Ae^{-\alpha t}u(t)$ and as we showed in Exercise 3.3.2 (refer to page 227), it has a Fourier transform given by the relation

$$\mathcal{X}(j\omega) = \frac{A}{\alpha + j\omega}$$

To calculate the amplitude and phase of the complex function $\mathcal{X}(j\omega)$, we need to express it in its Cartesian form $\mathcal{X}(j\omega) = \Re\{\mathcal{X}(j\omega)\} + j\Im\{\mathcal{X}(j\omega)\}$. Thus, it holds that

$$\mathcal{X}(j\omega) = \frac{A}{\alpha + j\omega} = \frac{A}{\alpha + j\omega}\frac{\alpha - j\omega}{\alpha - j\omega} = \frac{A\alpha - jA\omega}{\alpha^2 + \omega^2} = \frac{A\alpha}{\alpha^2 + \omega^2} - j\frac{A\omega}{\alpha^2 + \omega^2} = \Re\{\mathcal{X}(j\omega)\} + j\Im\{\mathcal{X}(j\omega)\}$$

and therefore the real and imaginary parts of this function have the form

$$\Re\{\mathcal{X}(j\omega)\} = \frac{A\alpha}{\alpha^2 + \omega^2} \quad \text{and} \quad \Im\{\mathcal{X}(j\omega)\} = -\frac{A\omega}{\alpha^2 + \omega^2}$$

Consequently, the amplitude and phase of this function will be

$$|\mathcal{X}(j\omega)| = \sqrt{\Re\{\mathcal{X}(j\omega)\}^2 + \Im\{\mathcal{X}(j\omega)\}^2} = \sqrt{\frac{A^2\alpha^2 + A^2\omega^2}{[\alpha^2 + \omega^2]^2}} = \sqrt{\frac{A^2[\alpha^2 + \omega^2]}{[\alpha^2 + \omega^2]^2}} = \sqrt{\frac{A^2}{\alpha^2 + \omega^2}} = \frac{A}{\sqrt{\alpha^2 + \omega^2}}$$

$$\measuredangle\mathcal{X}(j\omega) = \tan^{-1}\left(\frac{\Im\{\mathcal{X}(j\omega)\}}{\Re\{\mathcal{X}(j\omega)\}}\right) = \tan^{-1}\left(-\frac{\omega}{\alpha}\right)$$

Problem 3.3.9 Determine the Fourier transform of the function

$$x(t) = e^{-3|t|}\sin(2t)$$

Solution Using the defining equation of the absolute value function, we can formulate the signal $x(t)$ in the form

$$x(t) = e^{-3t}\sin(2t)u(t) + e^{3t}\sin(2t)u(-t)$$

and from the defining equation of the Fourier transform we obtain

$$\mathcal{X}(j\omega) = \int_{-\infty}^{+\infty} x(t)^{-j\omega t} dt = \int_{-\infty}^{+\infty} e^{-3t} \sin(2t)\, e^{-j\omega t} u(t) dt$$

$$+ \int_{-\infty}^{+\infty} e^{3t} \sin(2t)\, e^{-j\omega t} u(-t) dt = \int_{0}^{\infty} e^{-(3+j\omega)t} \sin(2t) dt + \int_{-\infty}^{0} e^{(3-j\omega)t} \sin(2t) dt$$

The first integral is calculated as

$$I_1 = \int_0^\infty e^{-(3+j\omega)t} \sin(2t) dt = \int_0^\infty e^{-(3+j\omega)t} \frac{e^{j2t} - e^{-j2t}}{2i} dt = \frac{1}{2j} \int_0^\infty e^{-(3+j\omega-2j)t} dt - \frac{1}{2j} \int_0^\infty e^{-(3+j\omega+2j)t} dt$$

$$= \frac{1}{2j} \left[\frac{e^{-(3+j\omega-2j)}}{3 + j\omega - 2j} \right]_\infty^0 - \frac{1}{2j} \left[\frac{e^{-(3+j\omega+2j)}}{3 + j\omega + 2j} \right]_\infty^0 = \frac{1}{2j} \left(\frac{1}{3 + j\omega - 2j} \right) - \frac{1}{2j} \left(\frac{1}{3 + j\omega + 2j} \right)$$

and finally

$$I_1 = \frac{1/2j}{3 + j(\omega - 2)} - \frac{1/2j}{3 + j(\omega + 2)}$$

In a completely analogous way, for the second integral we have

$$I_2 = \int_{-\infty}^0 e^{(3-j\omega)t} \sin(2t) dt = \int_{-\infty}^0 e^{(3-j\omega)t} \frac{e^{j2t} - e^{-j2t}}{2j} dt = \frac{1}{2j} \int_{-\infty}^0 e^{(3-j\omega+2j)t} dt - \frac{1/2j^0}{\int_{-\infty}} e^{(3-j\omega-2j)t} dt$$

$$= \frac{1}{2j} \left[\frac{e^{(3-j\omega+2j)}}{3 - j\omega + 2j} \right]_{-\infty}^0 - \frac{1}{2j} \left[\frac{e^{(3-j\omega-2j)}}{3 - j\omega - 2j} \right]_{-\infty}^0 = \frac{1}{2j} \left(\frac{1}{3 - j\omega + 2i} \right) - \frac{1}{2j} \left(\frac{1}{3 - j\omega - 2i} \right)$$

and finally

$$I_2 = \frac{1/2j}{3 - j(\omega - 2)} - \frac{1/2j}{3 - j(\omega + 2)}$$

By substituting the foregoing expressions into the defining equation of the Fourier transform we get

$$\mathcal{X}(j\omega) = \left[\frac{1/2j}{3 + j(\omega - 2)} + \frac{1/2j}{3 - j(\omega - 2)} \right] - \left[\frac{1/2j}{3 + j(\omega + 2)} + \frac{1/2j}{3 - j(\omega + 2)} \right] = \frac{3j}{9 + (\omega + 2)^2} - \frac{3j}{9 + (\omega - 2)^2}$$

which is the desired result.

Problem 3.3.10 Determine the Fourier transform of the signal $x(t) = te^{-2t} \sin(4t)u(t)$ (Oppenheim, Willsky and Nawab, Exercise for solution 4.21e).

Solution To solve the exercise, it is necessary to calculate the value of the integral

$$I = \int_0^\infty te^{\alpha t} dt$$

Applying the method of integration by parts, we find that

$$\int e^{\alpha t} dt = \int d\left(\frac{e^{\alpha t}}{\alpha} \right) t = \frac{t}{\alpha} e^{\alpha t} - \frac{1}{\alpha} \int e^{\alpha t} dt = \frac{t}{\alpha} e^{\alpha t} - \frac{1}{\alpha^2} e^{\alpha t} = \frac{(\alpha t - 1) e^{\alpha t}}{\alpha^2}$$

In order for the preceding integral to converge at the limit $t \to \infty$, the value of the parameter α must be negative. Setting $\alpha = -\beta$ ($\beta > 0$), the preceding integral takes the form

$$I = \int te^{-\beta t}dt = -\left(\frac{\beta t + 1}{\beta^2}\right)e^{-\beta t}$$

from which it follows that

$$\int_0^{\infty} te^{-\beta t}dt = -\left[\left(\frac{\beta t + 1}{\beta^2}\right)e^{-\beta t}\right]_0^{\infty} = \frac{1}{\beta^2}$$

Returning to the problem and expressing the sine function using exponential functions through the Euler equations, the defining equation of the Fourier transform of the signal $x(t)$ allows us to write that

$$\mathcal{X}(j\omega) = \int_{-\infty}^{+\infty} x(t)e^{-j\omega t}dt = \int_{-\infty}^{+\infty} te^{-2t}\sin(4t)e^{-j\omega t}u(t)dt = \int_0^{\infty} te^{-(2+j\omega)t}\sin(4t)dt$$

$$= \int_0^{\infty} te^{(2+j\omega)t}\frac{e^{j4t} - e^{-j4t}}{2j}dt = \frac{1}{2j}\int_0^{\infty} te^{-(2+j\omega-4j)t}dt - \frac{1}{2j}\int_0^{\infty} te^{-(2+j\omega+4j)t}dt$$

from which we finally have

$$\mathcal{X}(j\omega) = \frac{1}{2j}\left[\frac{1}{[2 + j(\omega - 4)]^2} - \frac{1}{[2 + j(\omega + 4)]^2}\right]$$

where the final result was derived using the value of the auxiliary integral that we previously calculated for the values $\beta = 2 + j\omega - 4j$ and $\beta = 2 + j\omega + 4j$.

Problem 3.3.11 Determine the Fourier transform of the function

$$x(t) = \begin{cases} 1 - t^2 & \text{for } 0 < t < 1 \\ 0 & \text{elsewhere} \end{cases}$$

(Oppenheim, Willsky and Nawab, Exercise for solution 4.21θ).

Solution From the defining equation of the Fourier transform we obtain

$$\mathcal{X}(j\omega) = \int_{-\infty}^{+\infty} x(t)\, e^{-j\omega t}dt = \int_0^1 (1 - t^2)\, e^{-j\omega t}dt = \int_0^1 e^{-j\omega t}dt - \int_0^1 t^2e^{-j\omega t}dt$$

The process of calculating the first integral is trivial and leads to the result

$$\int_0^1 e^{-j\omega t}dt = -\frac{1}{j\omega}e^{-j\omega t}\Big|_0^1 = \frac{1 - e^{-j\omega}}{j\omega}$$

On the other hand, the second integral is calculated by the method of integration by parts and using the definite integral I of the previous exercise (but for the interval [0,1]) for the value $\beta = j\omega$. The result of this calculation has the form

$$\int t^2e^{-j\omega t}dt = \int d\left(\frac{e^{-j\omega t}}{-j\omega}\right)t^2 = \frac{e^{-j\omega t}}{-j\omega}t^2 - \int \frac{e^{-j\omega t}}{-j\omega}d(t^2) = -\frac{t^2}{j\omega}e^{-j\omega t} + \frac{2}{j\omega}\int te^{-j\omega t}dt$$

However, the last integral is calculated as

$$\int te^{-j\omega t}dt = \left(\frac{1 + j\omega t}{\omega^2}\right)e^{-j\omega t}$$

where we used the indefinite integral I of the previous exercise for parameter value $\beta = j\omega$, and finally we obtain

$$\int t^2 e^{-j\omega t} dt = -\frac{t^2}{j\omega} e^{-j\omega t} + \frac{2}{j\omega} \frac{1 + j\omega t}{\omega^2} e^{-j\omega t} = \frac{e^{-j\omega t}}{j\omega} \left(\frac{2 + 2j\omega t - t^2\omega^2}{\omega^2} \right)$$

from which it follows that

$$\int_0^1 t^2 e^{-j\omega t} dt = \frac{e^{-j\omega t}}{j\omega} \left(\frac{2 + 2j\omega t - t^2\omega^2}{\omega^2} \right) \Bigg|_0^1 = \frac{e^{-j\omega}}{j\omega} \cdot \frac{2 + 2j\omega - \omega^2}{\omega^2} - \frac{1}{j\omega} \cdot \frac{2}{\omega^2} = \frac{e^{-j\omega}(2 + 2j\omega - \omega^2) - 2}{j\omega^3}$$

Substituting the foregoing expressions into the defining equation of the Fourier transform, we are led to the expression

$$\mathcal{X}(j\omega) = \frac{1 - e^{-j\omega}}{j\omega} - \frac{e^{-j\omega}(2 + 2j\omega - \omega^2) - 2}{j\omega^3} = \frac{\omega^2 - 2e^{-j\omega} - 2j\omega e^{-j\omega} + 2}{j\omega^3} = \frac{1}{j\omega} - \frac{2e^{-j\omega}}{\omega^2} - \frac{2e^{-j\omega} - 2}{i\omega^3}$$

which is the desired result.

Problem 3.3.12 Determine the Fourier transform of the normal distribution

$$x(t) = \frac{1}{\sqrt{2\pi\sigma^2}} \exp\left(-\frac{t^2}{2\sigma^2} \right).$$

Solution From the defining equation of the Fourier transform we obtain

$$\mathcal{X}(j\omega) = \int_{-\infty}^{+\infty} x(t) e^{-j\omega t} dt = \frac{1}{\sqrt{2\pi\sigma^2}} \int_{-\infty}^{+\infty} exp\left(-\frac{t^2}{2\sigma^2} \right) e^{-j\omega t} dt = \frac{1}{\sqrt{2\pi\sigma^2}} \int_{-\infty}^{+\infty} exp\left[-\left(\frac{t^2}{2\sigma^2} + j\omega t \right) \right] dt$$

To determine the value of the foregoing integral, we must make the expression of the exponential function a perfect square, which is easily achieved by adding and subtracting the quantity $(j\omega\sigma/\sqrt{2})^2$. We then have that

$$\frac{t^2}{2\sigma^2} + j\omega t = \frac{t^2}{2\sigma^2} + j\omega t + \left(\frac{j\omega\sigma}{\sqrt{2}} \right)^2 - \left(\frac{j\omega\sigma}{\sqrt{2}} \right)^2 = \left(\frac{t}{\sigma\sqrt{2}} + \frac{j\omega\sigma}{\sqrt{2}} \right)^2 - \left(\frac{j\omega\sigma}{\sqrt{2}} \right)^2 = \left(\frac{t}{\sigma\sqrt{2}} + \frac{j\omega\sigma}{\sqrt{2}} \right)^2 + \frac{\omega^2\sigma^2}{2}$$

and the foregoing relation will take the form

$$\mathcal{X}(j\omega) = \frac{1}{\sqrt{2\pi\sigma^2}} \exp\left(-\frac{\omega^2\sigma^2}{2} \right) \int_{-\infty}^{+\infty} \exp\left[-\left(\frac{t}{\sigma\sqrt{2}} + \frac{j\omega\sigma}{\sqrt{2}} \right)^2 \right] dt$$
$$= \frac{1}{\sqrt{2\pi\sigma^2}} \exp\left(-\frac{\omega^2\sigma^2}{2} \right) \int_{-\infty}^{+\infty} \exp\left[-\frac{1}{2} \left(\frac{t}{\sigma} + j\omega\sigma \right)^2 \right] dt$$

or, equivalently,

$$\mathcal{X}(j\omega) = \frac{1}{\sqrt{2\pi\sigma^2}} \exp\left(-\frac{\omega^2\sigma^2}{2} \right) \int_{-\infty + j\omega\sigma^2}^{+\infty + j\omega\sigma^2} \exp\left(-\frac{\xi^2}{2\sigma^2} \right) d\xi$$

where we changed the integration variable $\xi = t + j\omega\sigma^2$. Taking into account now that the function $\mathcal{X}(j\omega)$ is of a complex nature, the foregoing integration operation should be performed on the complex plane, using *complex integration techniques* and the *Cauchy's integral theorem*, which allows for the calculation of contour integrals of complex functions of the form $\oint_C f(z)dz$ (provided the function $f(z)$ and its partial derivatives are analytic within a simple connected region that contains the closed path C). Based on the foregoing theorem, we can write the preceding integral in the general form

$$\int_{-x+jy}^{x+jy} e^{-\xi^2/(2\sigma^2)} d\xi = -\int_{x+jy}^{x} e^{-\xi^2/(2\sigma^2)} d\xi - \int_{x}^{-x} e^{-\xi^2/(2\sigma^2)} d\xi - \int_{-x}^{-x+jy} e^{-\xi^2/(2\sigma^2)} d\xi =$$
$$-\int_{-x}^{+x} e^{-\xi^2/(2\sigma^2)} d\xi - \int_{x+jy}^{x} e^{-\xi^2/(2\sigma^2)} d\xi - \int_{-x}^{-x+jy} e^{-\xi^2/(2\sigma^2)} d\xi$$

where $y = \omega\sigma^2$. It is obvious that the contributions of the vertical paths to the value of the integral become zero (in reality, they cancel each other out, since they have opposite values), and if we calculate the preceding integral in the limit $x \to \infty$, it will take the simple form

$$\int_{-\infty+j\omega\sigma^2}^{+\infty+j\omega\sigma^2} \exp\left(-\frac{\xi^2}{2\sigma^2}\right) d\xi = \lim_{x\to\infty} \int_{-x}^{+x} e^{-\xi^2/(2\sigma^2)} d\xi = \int_{-\infty}^{+\infty} \exp\left(-\frac{\xi^2}{2\sigma^2}\right) d\xi = \sqrt{2\pi\sigma^2}$$

since the last integral is simply the *Gaussian integral*[9] for the value $\alpha = 1/2\sigma^2$, that is,

$$\int_{-\infty}^{+\infty} e^{-\alpha\xi^2} d\xi = \sqrt{\frac{\pi}{\alpha}}$$

Replacing the value of the integral in the Fourier transform equation $\mathcal{X}(j\omega)$, it will take the form

$$\mathcal{X}(j\omega) = \frac{1}{\sqrt{2\pi\sigma^2}} \exp\left(-\frac{\omega^2\sigma^2}{2}\right) \int_{-\infty+i\omega\sigma^2}^{+\infty+i\omega\sigma^2} \exp\left(-\frac{\xi^2}{2\sigma^2}\right) d\xi = \frac{1}{\sqrt{2\pi\sigma^2}} \exp\left(-\frac{\omega^2\sigma^2}{2}\right)\sqrt{2\pi\sigma^2} = \exp\left(-\frac{\omega^2\sigma^2}{2}\right)$$

which is also the desired result. This result is very interesting, as it leads us to the conclusion that the *Fourier transform of a normal distribution is also a normal distribution*. This is obviously not the case in general, as any signal $x(t)$ generally has a different shape from its Fourier transform.

Taking into account that the normal distribution extends to the infinite time interval $(-\infty, +\infty)$, it follows that the signal $x(t)$ that describes it *is not causal*. However, we can define a pulse duration τ as the *range of the interval defined by the time instances $t_1 = -\tau/2$ and $t_2 = +\tau/2$ at the positions at which the height of the normal distribution pulse has become equal to $1/\beta$ of its maximum value*—which is equal to $A = 1/\sqrt{2\pi\sigma^2}$—for some value of the parameter β ($\beta > 1$). For simplicity, we define the parameter $\alpha = \sqrt{1/2\sigma^2}$ [this means that the normal distribution is formulated as $x(t) = A\exp(-\alpha^2 t^2)$], and this parameter value can be calculated if, in the equation defining the normal distribution, we set $t = \tau/2$ and $x(t) = A/\beta$. Then we have

[9] A simple way to calculate the *Gaussian integral* is based on integration in two-dimensional space $\mathbb{R}^2$ and the use of polar coordinates. Setting $r^2 = x^2 + y^2$, we calculate the integral

$$\int_{R^2} e^{-\alpha r^2} dxdy = \int_{-\infty}^{+\infty}\int_{-\infty}^{+\infty} e^{-\alpha(x^2+y^2)} dxdy = \left(\int_{-\infty}^{+\infty} e^{-\alpha x^2} dx\right)\left(\int_{-\infty}^{+\infty} e^{-\alpha y^2} dy\right) = \left(\int_{-\infty}^{+\infty} e^{-\alpha x^2} dx\right)^2$$

The same integral in polar coordinates $x = r\cos\vartheta$, $y = r\sin\vartheta$ is calculated as

$$\int_{R^2} e^{-\alpha r^2} dxdy = \int_0^{2\pi}\int_0^{\infty} re^{-\alpha r^2} dr d\vartheta = \int_0^{2\pi} d\vartheta \int_0^{\infty} re^{-\alpha r^2} dr = 2\pi \int_0^{\infty} re^{-\alpha r^2} dr$$

Proceeding with the change of variable $s = -r^2$ will lead to $ds = -2rdr$, and the foregoing integral is written in the form

$$\int_{R^2} e^{-\alpha r^2} dxdy = 2\pi \int_{-\infty}^{0} \frac{e^{\alpha s}}{2} ds = \frac{\pi}{\alpha}\left[e^{\alpha s}\right]_{-\infty}^{0} = \frac{\pi}{\alpha}(e^0 - e^{-\infty}) = \frac{\pi}{\alpha}$$

Comparing the two results we obtain

$$\left(\int_{-\infty}^{+\infty} e^{-\alpha x^2} dx\right)^2 = \frac{\pi}{\alpha} \quad \text{from where finally we get} \quad \int_{-\infty}^{+\infty} e^{-\alpha x^2} dx = \sqrt{\frac{\pi}{\alpha}}.$$

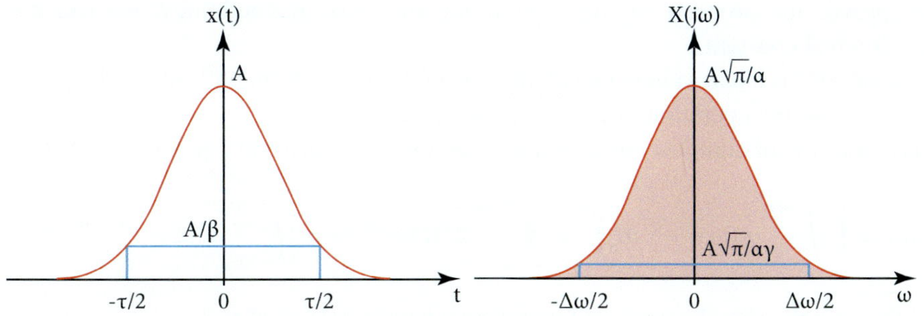

Fig. 3.12 The normal distribution $x(t)$ and its spectrum $\mathcal{X}(j\omega)$ (Shmaliy, *Continuous-Time Signals*, Fig. 2.26)

$$A\exp\left[-\left(\frac{\alpha\tau}{2}\right)^2\right]=\frac{A}{\beta}$$

from which it follows that

$$\alpha=2\frac{\sqrt{\ln\beta}}{\tau}$$

In a completely analogous way, we can define similar limits in the Fourier transform of the function $x(t)$, whose maximum height is obviously equal to $A\sqrt{\pi}/\alpha=1$. Considering a parameter $\gamma>1$, we can define the *spectrum width* $\Delta\omega$ as the *interval defined by the frequencies* $\omega_1=-\Delta\omega/2$ *and* $\omega_2=+\Delta\omega/2$ *at the positions whose pulse height has become equal to the* $1/\gamma$ *of its maximum value*. If in the equation defining the function $X(\omega)$ we set $\omega=\Delta\omega/2$ and $X(\omega)=1/\gamma$, we obtain

$$\exp\left(-\frac{\Delta\omega}{16\alpha^2}\right)=\frac{1}{\gamma}$$

from which it follows that

$$\frac{\Delta\omega}{2}=2\alpha\sqrt{\ln\gamma}$$

The curves of the distributions $x(t)$ and $\mathcal{X}(j\omega)$, along with the time duration τ and the bandwidth $\Delta\omega$, are presented diagrammatically in Fig. 3.12. Let us also note that for positive values of the parameter α the function $X(\omega)$ is real, which means that the *spectrum of the distribution coincides with the amplitude spectrum, while the phase spectrum has a value equal to zero for all frequencies*.

3.3.5 Properties of the Fourier Transform

The Fourier transform is a *generalized linear integral transform* characterized by a set of useful properties. These properties significantly simplify the process of calculating the spectrum of continuous-time signals in relation to the trivial method of calculation (examples of which we presented in the previous section), and their analytical proof is presented below. In the description that follows, the symbolism $x(t) \leftrightarrows \mathcal{X}(j\omega)$ or $x(t) \leftrightarrows \mathcal{F}\{x(t)\}$ implies that the Fourier transform of the function $x(t)$ is the function $\mathcal{X}(j\omega)\equiv\mathcal{F}\{x(t)\}$, and vice versa.

- **Linearity**: If the Fourier transforms of the continuous-time signals $x(t)$ and $y(t)$ are the functions $\mathcal{X}(j\omega)$ and $\mathcal{Y}(j\omega)$, then the Fourier transform of the signal $z(t)=\alpha x(t)+\beta y(t)$ will be the function $\mathcal{Z}(j\omega)=\alpha\mathcal{X}(j\omega)+\beta\mathcal{Y}(j\omega)$.
  Proof From the defining equation of the Fourier transform we have

$$\mathcal{Z}(j\omega)=\int_{-\infty}^{+\infty}z(t)\,e^{-j\omega t}dt=\int_{-\infty}^{+\infty}\Big[\alpha x(t)+\beta y(t)\Big]\,e^{-j\omega t}dt$$
$$=\alpha\int_{-\infty}^{+\infty}x(t)\,e^{-j\omega t}dt+\beta\int_{-\infty}^{+\infty}y(t)\,e^{-j\omega t}dt=\alpha\mathcal{X}(j\omega)+\beta\mathcal{Y}(j\omega)$$

This equation also proves the property of linearity. Using the same methodology, we can prove that $\mathcal{F}\{\gamma x(t)\} = \gamma \mathcal{F}\{f(t)\}$, where γ is a real constant.

- **Symmetry**: If the signal $x(t)$ is a real-valued signal, then its Fourier transform will satisfy the property $\mathcal{X}^*(\omega) = \mathcal{X}(-\omega)$, where the symbol $*$ refers to the operation of complex conjugation.

Proof Applying the operation of complex conjugation to the Fourier transform equation, we get

$$\mathcal{X}^*(\omega) = \left(\int_{-\infty}^{+\infty} x(t)\, e^{-j\omega t} dt \right)^* = \int_{-\infty}^{+\infty} x^*(t)\, e^{j\omega t} dt = \int_{-\infty}^{+\infty} x(t)\, e^{-j(-\omega)t} dt = \mathcal{X}(-\omega)$$

where we used the fact that the signal is real, and therefore we have $x(t) = x^*(t)$.

Expressing the complex function $\mathcal{X}(j\omega)$ in the form

$$\mathcal{X}(j\omega) = \Re\{\mathcal{X}(j\omega)\} + j\Im\{\mathcal{X}(j\omega)\}$$

from the foregoing relation it immediately follows that

$$\Re\{\mathcal{X}(j\omega)\} = \Re\{\mathcal{X}(-\omega)\} \qquad \text{and} \qquad \Im\{\mathcal{X}(j\omega)\} = -\Im\{\mathcal{X}(-\omega)\}$$

that is, the *real part of the Fourier transform of a real signal is an even function of the frequency* ω, *while its imaginary part is an odd function of the frequency* ω. On the other hand, if the Fourier transform is expressed in its polar form

$$\mathcal{X}(j\omega) = |\mathcal{X}(j\omega)| e^{j\measuredangle \mathcal{X}(j\omega)}$$

then the property $\mathcal{X}^*(\omega) = \mathcal{X}(-\omega)$ leads directly to the relations

$$|\mathcal{X}(j\omega)| = |\mathcal{X}(-\omega)| \qquad \text{and} \qquad \measuredangle \mathcal{X}(j\omega) = -\measuredangle \mathcal{X}(-\omega)$$

from which it follows that the *magnitude of the Fourier transform of a real signal is an even function of frequency, while conversely the phase of* $\measuredangle \mathcal{X}(j\omega)$ *is an odd function of frequency*. These symmetry properties are particularly useful because they allow us to calculate the magnitude and phase of the Fourier transform *only for the region of positive frequencies* and then, using these properties, to extend them to the region of negative frequencies.

In the special case of an even real signal $x(t)$, its Fourier transform will also be *a real and even function of frequency* ω. To prove the truth of this assertion, we start from the defining equation of the function $\mathcal{X}(-\omega)$

$$\mathcal{X}(-\omega) = \int_{-\infty}^{+\infty} x(t)\, e^{-j(-\omega)t} dt = \int_{-\infty}^{+\infty} x(t)\, e^{j\omega t} dt$$

and proceed to the change of variable $t \longrightarrow -t$. In this case, the foregoing equation is written

$$\mathcal{X}(-\omega) = \int_{-\infty}^{+\infty} x(-t)\, e^{-j\omega t} dt = \int_{-\infty}^{+\infty} x(t)\, e^{-j\omega t} dt = \mathcal{X}(j\omega)$$

where we used the assumption that the signal $x(t)$ is even and therefore will satisfy the property $x(-t) = x(t)$. If we combine the last relation with the property $\mathcal{X}^*(\omega) = \mathcal{X}(-\omega)$, which we proved to be valid for *real signals*, it immediately follows that $\mathcal{X}^*(\omega) = \mathcal{X}(j\omega)$, that is, the function $\mathcal{X}(j\omega)$, apart from being *even*, is also *real*. Using similar arguments, we can prove that the *Fourier transform of a signal with real values characterized by odd symmetry is imaginary and characterized by odd symmetry*, a process that is left to the reader as an exercise.

- **Even and odd part**: If a continuous-time signal $x(t)$ has a Fourier transform $\mathcal{X}(j\omega) = \mathcal{F}\{x(t)\}$, then the Fourier transform of the even component of the signal will be equal to the real part of the function $\mathcal{X}(j\omega)$, while the Fourier transform of the odd component of the signal will be equal to the imaginary part of the function $\mathcal{X}(j\omega)$, or, in mathematical notation,

$$\mathcal{F}\{\mathcal{E}\text{v}(t)\} = \Re\{\mathcal{X}(j\omega)\} \qquad \text{and} \qquad \mathcal{F}\{\mathcal{O}\text{d}(t)\} = j\Im\{\mathcal{X}(j\omega)\}$$

Proof Starting from the defining equation of the even component, its Fourier transform will have the form

$$\mathcal{F}\{\mathcal{E}\mathrm{v}(t)\} = \int_{-\infty}^{+\infty} \mathcal{E}\mathrm{v}(t) e^{-j\omega t} et = \frac{1}{2}\int_{-\infty}^{+\infty} x(t) e^{-j\omega t} dt + \frac{1}{2}\int_{-\infty}^{+\infty} x(-t) e^{-j\omega t} dt$$

But the first integral is the Fourier transform $\mathcal{X}(j\omega)$ of the signal $x(t)$, while the second integral is expressed as

$$\int_{-\infty}^{+\infty} x(-t) e^{-j\omega t} dt = \int_{-\infty}^{+\infty} x(\xi) e^{j\omega\xi} (-d\xi) = \int_{-\infty}^{+\infty} x(\xi) e^{j\omega\xi} d\xi$$

$$= \int_{-\infty}^{+\infty} x^*(\xi) e^{j\omega\xi} d\xi = \left(\int_{-\infty}^{+\infty} x(\xi) e^{j\omega\xi} d\xi \right)^* = \mathcal{X}^*(\omega)$$

where we proceeded with the variable change $t \longrightarrow -\xi$ and took into account that the signal $x(t)$ was real, so it equaled its complex conjugate $x^*(t)$. Combining the foregoing results, it immediately follows that

$$\mathcal{F}\{\mathcal{E}\mathrm{v}(t)\} = \frac{\mathcal{X}(j\omega) + \mathcal{X}^*(j\omega)}{2} = \Re\{\mathcal{X}(j\omega)\}$$

while in a similar way we find

$$\mathcal{F}\{\mathcal{O}\mathrm{d}(t)\} = \frac{\mathcal{X}(j\omega) - \mathcal{X}^*(j\omega)}{2} = j\Re\{\mathcal{X}(j\omega)\}$$

- **Time shift**: If the Fourier transform of the signal $x(t)$ is the function $\mathcal{X}(j\omega)$, then the Fourier transform of the signal defined as $y(t) = x(t - t_0)$ will be $\mathcal{Y}(j\omega) = e^{-j\omega t_0}\mathcal{X}(j\omega)$.
Proof The Fourier transform of the signal $y(t)$ is given by the relation

$$\mathcal{Y}(j\omega) = \int_{-\infty}^{+\infty} y(t)\, e^{-j\omega t} dt = \int_{-\infty}^{+\infty} x(t - t_0)\, e^{-j\omega t} dt$$

Proceeding with a change of the integration variable from t to $\xi = t - t_0$, we have $t = \xi + t_0$ and $dt = d\xi$, and the foregoing relationship will take the form

$$\mathcal{Y}(j\omega) = \int_{-\infty}^{+\infty} x(\xi)\, e^{-j\omega(\xi + t_0)} d\xi = e^{-j\omega t_0} \int_{-\infty}^{+\infty} x(\xi)\, e^{-j\omega\xi} d\xi = e^{-j\omega t_0}\mathcal{X}(j\omega)$$

which is also the desired result. Applying the same process, we can prove that the Fourier transform of the signal $y(x) = x(t + t_0)$ is the function $\mathcal{Y}(j\omega) = e^{j\omega t_0}\mathcal{X}(j\omega)$. The basic consequence of the foregoing property is that the *act of time shifting a signal $x(t)$ does not change the magnitude of its Fourier transform*. Indeed, if we write that transform in its polar form

$$\mathcal{F}\{x(t)\} \equiv \mathcal{X}(j\omega) = |\mathcal{X}(j\omega)| e^{j\measuredangle\mathcal{X}(j\omega)}$$

similarly we will have

$$\mathcal{F}\{x(t - t_0)\} = e^{-j\omega t_0}\mathcal{X}(j\omega) = e^{-j\omega t_0}|\mathcal{X}(j\omega)| e^{j\measuredangle\mathcal{X}(j\omega)} = |\mathcal{X}(j\omega)| e^{-j\omega[\measuredangle\mathcal{X}(j\omega) - t_0]}$$

From the last relationship it follows that the *act of time shifting simply introduces a phase shift in the Fourier transform of the signal $x(t)$, while the magnitude of that transform remains unchanged.*

- **Differentiation**: If the Fourier transform of the signal $x(t)$ is the function $\mathcal{X}(j\omega)$, then the Fourier transform of the signal $y(t) = x'(t)$ will be $\mathcal{Y}(j\omega) = j\omega\mathcal{X}(j\omega)$.
Proof From the defining equation of the Fourier transform $\mathcal{Y}(j\omega)$ we obtain

$$\mathcal{Y}(j\omega) = \int_{-\infty}^{+\infty} y(t)\, e^{-j\omega t} dt = \int_{-\infty}^{+\infty} \frac{dx(t)}{dt}\, e^{-j\omega t} dt = \int_{-\infty}^{+\infty} e^{-j\omega t} d\{x(t)\}$$

Integrating by parts, the foregoing equation will take the form

$$\mathcal{Y}(j\omega) = x(t)e^{-j\omega t}\Big|_{-\infty}^{+\infty} - \int_{-\infty}^{+\infty} x(t)\, d\{e^{-j\omega t}\} = j\omega \int_{-\infty}^{+\infty} x(t)\, e^{-j\omega t} dt = j\omega\mathcal{X}(j\omega)$$

which is the desired result. To derive the last equation, we silently assumed that the value of the signal $x(t)$ at the limits $t \to \pm\infty$ was equal to zero and that this signal was described by an analytic function, so that its derivative exists.

Applying this property repeatedly, it is proven that for the general case we have

$$\mathcal{F}\{x^{(n)}(t)\} = \mathcal{F}\left\{\frac{d^n x(t)}{dt^n}\right\} = (j\omega)^n \mathcal{X}(j\omega)$$

a property that is particularly useful in solving linear differential equations with constant coefficients and calculating the output of linear and time-invariant continuous-time systems.

- **Time scaling**: If the Fourier transform of the signal $x(t)$ is the function $\mathcal{X}(j\omega)$, then the Fourier transform of the signal defined as $y(t) = x(\alpha t)$ will be

$$\mathcal{Y}(j\omega) = \frac{1}{|\alpha|}\mathcal{X}\left(\frac{\omega}{\alpha}\right)$$

Proof From the defining equation of the Fourier transform $\mathcal{Y}(j\omega)$ we obtain

$$\mathcal{Y}(j\omega) = \int_{-\infty}^{+\infty} y(t)\, e^{-j\omega t} dt = \int_{-\infty}^{+\infty} x(\alpha t)\, e^{-j\omega t} dt$$

Changing the integration variable from t to $\xi = \alpha t$, we have $t = \xi/\alpha$ and $dt = d\xi/\alpha$, and therefore

$$\mathcal{Y}(j\omega) = \frac{1}{\alpha}\int_{-\infty}^{+\infty} x(\xi)\, e^{-j\xi(\omega/\alpha)} d\xi = \frac{1}{\alpha}\mathcal{X}\left(\frac{\omega}{\alpha}\right)$$

The foregoing relation holds only for positive values of α, while applying the same process we finally arrive at the general relation

$$\mathcal{Y}(j\omega) \equiv \mathcal{F}\{x(\alpha t)\} = \begin{cases} +\dfrac{1}{\alpha}\mathcal{X}\left(\dfrac{\omega}{\alpha}\right) & \text{for } \alpha > 0 \\ \\ -\dfrac{1}{\alpha}\mathcal{X}\left(\dfrac{\omega}{\alpha}\right) & \text{for } \alpha < 0 \end{cases} = \frac{1}{|\alpha|}\mathcal{X}\left(\frac{\omega}{\alpha}\right)$$

which holds for both positive and negative values of the parameter α.

From the foregoing property it follows that for values $\alpha > 1$ the signal $x(t)$ expands in time and the function $\mathcal{X}(j\omega)$ compresses in the frequency domain, while for values $\alpha < 1$ the opposite is true. It is also interesting to note that for the value $\alpha = -1$ the foregoing relation takes the form

$$\mathcal{F}\{x(-t)\} = \mathcal{X}(-\omega)$$

from which it follows that the *inversion of a signal in the time domain results in the inversion of its spectrum in the frequency domain.*

- **Duality**: If the Fourier transform of the signal $x(t)$ is the function $\mathcal{X}(j\omega)$, then the Fourier transform of the function $\mathcal{X}(t)$ will be the function $y(\omega) = 2\pi x(-\omega)$. Using for the formulation of this property the simple frequency $f = \omega/2\pi$, this is written

$$\mathcal{F}\{x(t)\} = \mathcal{X}(f) \Leftrightarrow \mathcal{F}\{\mathcal{X}(t)\} = x(-f)$$

Proof If in the equation defining the inverse Fourier transform

$$x(t) \equiv \mathcal{F}^{-1}\{\mathcal{X}(j\omega)\} = \frac{1}{2\pi}\int_{-\infty}^{+\infty} \mathcal{X}(j\omega)\, e^{j\omega t} d\omega$$

we proceed to change the integration variable from t to $-t$, it will take the form

$$x(-t) = \frac{1}{2\pi}\int_{-\infty}^{+\infty} \mathcal{X}(j\omega)\, e^{-j\omega t} d\omega \qquad \text{and finally} \qquad 2\pi x(-\omega) = \int_{-\infty}^{+\infty} \mathcal{X}(t)\, e^{-j\omega t} dt = \mathcal{F}\{\mathcal{X}(t)\}$$

where to derive the last relation we switched the positions of the variables ω and t. If we use the simple frequency f in place of the circular frequency ω, the foregoing relation will take the form

$$x(-f) = \int_{-\infty}^{+\infty} \mathcal{X}(t)\, e^{-j2\pi f t} dt = \mathcal{F}\{\mathcal{X}(t)\}$$

If the signal $x(t)$ is characterized by even symmetry, it immediately follows that

$$\mathcal{F}\{x(t)\} = \mathcal{X}(f) \Leftrightarrow \mathcal{F}\{\mathcal{X}(t)\} = x(f)$$

This property of the Fourier transform is particularly important, as it allows us to determine or even predict many of its properties, based on the fact that *if the continuous-time signal $x(t)$ has specific characteristics in the frequency domain, then, due to this symmetry, which characterizes the equations of the direct and inverse transform, the same characteristics will be associated with the function $\mathcal{X}(j\omega)$ in the time domain.* To understand this property, we recall that a simple point in the frequency domain with the frequency value ω_0 corresponds to a certain sinusoidal signal in the time domain, whose frequency is equal to ω_0. Based on the property of the duality of the Fourier transform, we can conclude that a simple point in the time domain—that is, a simple time instance t_0—corresponds to a sinusoidal signal in the frequency domain, which can be proven from the equations of the direct and inverse transform. Another example of this dual nature of the transform relates to the Fourier transform of a *periodic function*, which, as we will see in the next section, takes the form of a train of weighted impulse functions that are equally spaced in time. Relying on the property of duality, we can formulate the assertion that the *Fourier transform of a train of weighted and temporally equidistant impulse functions will be a function that is periodic with respect to frequency* whose power can be proven by the application of the trivial method of calculating that transform.

An example of the application of the property of duality concerns the process of calculating the Fourier transform of the signal $x(t) = \text{sinc}(\alpha t)$, which, if based on the use of the defining equation

$$\mathcal{X}(j\omega) = \int_{-\infty}^{+\infty} x(t)\, e^{-j\omega t} dt = \frac{1}{\alpha}\int_{-\infty}^{+\infty} \frac{\sin(\alpha t)}{t}\, e^{-j\omega t} dt$$

may become quite complex. However, if we take into account that the Fourier transform of the rectangular window has the form of the function sinc, a property that is mathematically formulated as

$$x(t) = \begin{cases} 1 \text{ for } |t| < T \\ 0 \text{ for } |t| > T \end{cases} \quad \overset{\mathcal{F}}{\longleftrightarrow} \quad \mathcal{X}(j\omega) = \frac{2\sin(\omega T)}{\omega} = 2T\,\text{sinc}(\omega T)$$

then, invoking the property of duality, we can claim that the spectrum of the function sinc will have the form of a rectangular window, or, in mathematical terms,

$$x(t) = \frac{\sin(\alpha t)}{\pi t} = \frac{\alpha}{\pi}\text{sinc}(\alpha t/\pi) \quad \overset{\mathcal{F}}{\longleftrightarrow} \quad \mathcal{X}(j\omega) = \begin{cases} 1 \text{ for } |\omega| < \alpha \\ 0 \text{ for } |\omega| > \alpha \end{cases}$$

a proposition that is proven to be true. The foregoing example highlights the great usefulness of the property of duality in calculation processes of *signal spectra*, and many of the exercises that will be solved later rely on the use of this property.

- **Frequency shift**: If the Fourier transform of the signal $x(t)$ is the function $\mathcal{X}(j\omega)$, then the Fourier transform of the signal $y(t) = e^{j\omega_0 t}x(t)$ will be $\mathcal{Y}(j\omega) = \mathcal{X}(\omega - \omega_0)$.

  Proof From the equation defining the function $\mathcal{Y}(j\omega)$ we have

$$\mathcal{Y}(j\omega) = \int_{-\infty}^{+\infty} y(t)\, e^{-j\omega t} dt = \int_{-\infty}^{+\infty} e^{j\omega_0 t}x(t)\, e^{-j\omega t} dt = \int_{-\infty}^{+\infty} x(t)\, e^{-j(\omega-\omega_0)t} dt = \mathcal{X}(\omega - \omega_0)$$

which is the desired result. In a completely similar way, it turns out that the Fourier transform of the signal $y(t) = e^{-j\omega_0 t}x(t)$ is $\mathcal{Y}(j\omega) = \mathcal{X}(\omega + \omega_0)$.

- **Time reversal**: If the Fourier transform of the signal $x(t)$ is the function $\mathcal{X}(j\omega)$, then the Fourier transform of the signal defined as $y(t) = x(-t)$ will have the form $\mathcal{Y}(j\omega) = \mathcal{X}(-\omega)$.
  Proof From the equation defining the function $\mathcal{Y}(j\omega)$ we have

$$\mathcal{Y}(j\omega) = \int_{-\infty}^{+\infty} y(t)\, e^{-j\omega t}\, dt = \int_{-\infty}^{+\infty} x(-t)\, e^{-j\omega t}\, dt$$

  Changing the integration variable from t to $-t$, the foregoing equation will take the form

$$\mathcal{Y}(j\omega) = -\int_{+\infty}^{-\infty} x(t)\, e^{j\omega t}\, dt = \int_{-\infty}^{+\infty} x(t)\, e^{-j(-\omega)t}\, dt = \mathcal{X}(-\omega)$$

  which is the desired result. We reached the same result by applying the property of *time scaling* for the scaling parameter $\alpha = -1$.

- **Multiplication**: If the Fourier transforms of the signals $x(t)$ and $y(t)$ are the functions $\mathcal{X}(j\omega)$ and $\mathcal{Y}(j\omega)$ respectively, then the Fourier transform of the signal $z(t) = x(t)y(t)$ will be

$$\mathcal{Z}(j\omega) \equiv \mathcal{F}\{z(t)\} = \frac{1}{2\pi}\Big[\mathcal{X}(j\omega) * \mathcal{Y}(j\omega)\Big]$$

  where the symbol $*$ refers to the operation of convolution.
  Proof From the defining equation of the Fourier transform $\mathcal{Z}(j\omega)$ we have

$$\mathcal{Z}(j\omega) = \int_{-\infty}^{+\infty} z(t)\, e^{-j\omega t}\, dt = \int_{-\infty}^{+\infty} x(t)\, y(t)\, e^{-j\omega t}\, dt$$

  Replacing the signal $y(t)$ from the equation that describes it as the inverse Fourier transform of the function $\mathcal{Y}(j\omega)$, which is

$$y(t) = \frac{1}{2\pi}\int_{-\infty}^{+\infty} \mathcal{Y}(\xi)\, e^{jt\xi}\, d\xi$$

  the foregoing relation will take the form

$$\begin{aligned}\mathcal{Z}(j\omega) &= \int_{-\infty}^{+\infty} x(t)\left\{\frac{1}{2\pi}\int_{-\infty}^{+\infty} \mathcal{Y}(\xi)\, e^{jt\xi}\, d\xi\right\} e^{-j\omega t}\, dt \\ &= \frac{1}{2\pi}\int_{-\infty}^{+\infty} \mathcal{Y}(\xi)\left\{\int_{-\infty}^{+\infty} x(t)\, e^{-j(\omega-\xi)t}\, dt\right\} d\xi = \frac{1}{2\pi}\int_{-\infty}^{+\infty} \mathcal{Y}(\xi)\mathcal{X}(\omega-\xi)\, d\xi\end{aligned}$$

  and finally

$$\mathcal{Z}(j\omega) = \frac{1}{2\pi}\Big[\mathcal{X}(j\omega) * \mathcal{Y}(j\omega)\Big]$$

- **Convolution**: If the Fourier transforms of the signals $x(t)$ and $y(t)$ are the functions $\mathcal{X}(j\omega)$ and $\mathcal{Y}(j\omega)$, then the Fourier transform of the convolution $z(t) = x(t) * y(t)$ of the two signals will be

$$\mathcal{Z}(j\omega) \equiv \mathcal{F}\{z(t)\} = \mathcal{X}(j\omega)\mathcal{Y}(j\omega)$$

  Proof From the defining equation of the Fourier transform $\mathcal{Z}(j\omega)$ we have

$$\mathcal{Z}(j\omega) = \int_{-\infty}^{+\infty} z(t)\, e^{-j\omega t}\, dt = \int_{-\infty}^{+\infty} \Big(x(t) * y(t)\Big)\, e^{-j\omega t}\, dt = \int_{-\infty}^{+\infty}\left\{\int_{-\infty}^{+\infty} x(\tau)\, y(t-\tau)\, d\tau\right\} e^{-j\omega t}\, dt$$

Proceeding to the substitution $t \to \xi = t - \tau$ we have $t = \tau + \xi$ and $dt = d\xi$, which results in

$$\mathcal{Z}(j\omega) = \int_{-\infty}^{+\infty} \left\{ \int_{-\infty}^{+\infty} x(\tau)\, y(\xi)\, e^{-j\omega(\tau+\xi)} d\tau \right\} d\xi = \left(\int_{-\infty}^{+\infty} x(\tau)\, e^{-j\omega\tau} d\tau \right) \left(\int_{-\infty}^{+\infty} y(\xi)\, e^{-j\omega\xi} d\xi \right)$$

and finally

$$\mathcal{Z}(j\omega) = \mathcal{X}(j\omega)\mathcal{Y}(j\omega)$$

which is also the desired result. From the last two properties it follows that the *Fourier transform of the product of two signals is proportional to the convolution of their transforms, while the Fourier transform of the convolution of two signals equals the product of their transforms*. This symmetry characterizing the operations of the product and convolution is yet another manifestation of the duality property of the Fourier transform.

An alternative way of proving the convolution theorem is based on the property of elementary exponential functions to constitute the *eigenfunctions* of a *LTI* continuous-time system with eigenvalues that are those of the *frequency response* function of the system calculated at the positions of the input signal frequencies (this property was analytically proven in the previous chapter). Starting from the defining equation of the inverse Fourier transform and expressing the integral appearing in it as a limit of a sum of the form

$$x(t) = \frac{1}{2\pi} \int_{-\infty}^{+\infty} \mathcal{X}(j\omega)\, e^{j\omega t} d\omega = \frac{1}{2\pi} \lim_{\omega_0 \to 0} \sum_{n=-\infty}^{+\infty} \mathcal{X}(n\omega_0)\, e^{jn\omega_0 t}\, \omega_0$$

the elementary response of the system to the simple exponential complex sequence $x(t) = e^{jn\omega_0 t}$ will be the function $y(t) = \mathcal{H}(jn\omega_0)\, e^{jn\omega_0 t}$, where

$$\mathcal{H}(jn\omega_0) = \int_{-\infty}^{+\infty} h(t) e^{-jn\omega_0 t}$$

Generalizing the foregoing description and using the *principle of superposition*, we can formulate the claim that the *system's response to the signal $x(t)$, which is defined as the superposition of the exponential complex signals, will result from the superposition of the foregoing elementary responses*, or, in mathematical terms,

$$y(t) = \frac{1}{2\pi} \lim_{\omega_0 \to 0} \sum_{n=-\infty}^{+\infty} \mathcal{X}(n\omega_0)\, \mathcal{H}(jn\omega_0)\, e^{jn\omega_0}\, \omega_0 = \frac{1}{2\pi} \int_{-\infty}^{+\infty} \mathcal{X}(j\omega)\, \mathcal{H}(j\omega)\, e^{j\omega t} d\omega$$

If we now take into account that the output $y(t)$—which is defined as the convolution of the signals $x(t)$ and $h(t)$—is the inverse Fourier transform of the function $\mathcal{Y}(j\omega)$, we can express the signal $y(t)$ in the form

$$y(t) = \frac{1}{2\pi} \int_{-\infty}^{+\infty} \mathcal{Y}(j\omega)\, e^{j\omega t} d\omega$$

which, if compared with the previous equation [which is an alternative way of writing the signal $y(t)$], leads directly to the relationship $\mathcal{Y}(j\omega) = \mathcal{H}(j\omega)\mathcal{X}(j\omega)$, which is the desired result. This alternative way of proving the convolution theorem is particularly important, and its physical meaning is that the frequency response function of the system $\mathcal{H}(j\omega)$, which is defined as the Fourier transform of its impulse response $h(t)$, is simply the *change in the complex amplitude that an elementary exponential complex signal undergoes when it passes through a LTI continuous-time system.*

- **Differentiation in the frequency domain**: If the Fourier transform of the signal $x(t)$ is the function $\mathcal{X}(j\omega)$, then the Fourier transform of the signal $y(t) = tx(t)$ is $\mathcal{Y}(j\omega) = i\mathcal{X}'(j\omega)$.
  Proof From the defining equation of the function $\mathcal{Y}(j\omega)$ we have

$$\mathcal{Y}(j\omega) = \int_{-\infty}^{+\infty} y(t)\, e^{-j\omega t} dt = \int_{-\infty}^{+\infty} tx(t)\, e^{-j\omega t} dt = -\frac{1}{j} \int_{-\infty}^{+\infty} x(t)\, \frac{d(e^{-j\omega t})}{d\omega}\, dt$$

$$= j\frac{d}{d\omega} \left\{ \int_{-\infty}^{+\infty} x(t)\, e^{-j\omega t} dt \right\} = j\frac{d\mathcal{X}(j\omega)}{d\omega}$$

which is the desired result. By repeatedly applying the foregoing property, it is proven that in the general case the following relation holds:

$$\mathcal{F}\{t^n x(t)\} = j^n F^{(n)}(\omega) = j^n \frac{d^n \mathcal{X}(j\omega)}{d\omega^n}$$

- **Modulation**: The property of *modulation* is simply an alternative name for the property of the product in the sense that the *multiplication of two signals results in the modulation of the amplitude of one signal in a way that is determined by the content of the other signal*. In the international literature, equations related to the act of modulation have the form

$$\mathcal{F}\{x(t)\cos(\omega_0 t)\} = \frac{1}{2}\Big[\mathcal{X}(\omega+\omega_0) + \mathcal{X}(\omega-\omega_0)\Big] \quad \text{and} \quad \mathcal{F}\{x(t)\sin(\omega_0 t)\} = \frac{j}{2}\Big[\mathcal{X}(\omega+\omega_0) - \mathcal{X}(\omega-\omega_0)\Big]$$

and they result from combining the properties of *linearity* and *frequency domain shift*. Indeed, replacing the cosine function in the first equation with Euler's formula, it will take the form

$$x(t)\cos(\omega_0 t) = x(t)\,\frac{e^{j\omega_0 t} + e^{-j\omega_0 t}}{2} = \frac{1}{2}\Big[x(t)\,e^{j\omega_0 t}\Big] + \frac{1}{2}\Big[x(t)\,e^{-j\omega_0 t}\Big]$$

Taking the Fourier transform of the two terms and using the property of *linearity*, we are led to the equation

$$\mathcal{F}\{x(t)\cos(\omega_0 t)\} = \frac{1}{2}\mathcal{F}\Big[x(t)\,e^{j\omega_0 t}\Big] + \frac{1}{2}\mathcal{F}\Big[x(t)\,e^{-j\omega_0 t}\Big]$$

However, from the application of the property of *frequency domain shift* we obtain

$$\mathcal{F}\{e^{j\omega_0 t}x(t)\} = \mathcal{X}(\omega-\omega_0) \quad \text{and} \quad \mathcal{F}\{e^{-j\omega_0 t}x(t)\} = \mathcal{X}(\omega+\omega_0)$$

and replacing the foregoing expressions in the last representation, we end up with the first modulation equation. In a completely similar way, replacing the sine function with Euler's formula $\sin(\omega_0 t) = (e^{j\omega_0 t} - e^{-j\omega_0 t})/2j$ we get

$$\mathcal{F}\{x(t)\sin(\omega_0 t)\} = \frac{j}{2}\mathcal{F}\Big[x(t)\,e^{-j\omega_0 t}\Big] - \frac{j}{2}\mathcal{F}\Big[x(t)\,e^{j\omega_0 t}\Big]$$

and using the property of frequency shift, we prove the second equation.

- **Integration**: If the Fourier transform of the signal $x(t)$ is the function $\mathcal{X}(j\omega)$, then the Fourier transform of the signal

$$y(t) = \int_{-\infty}^{t} x(\tau)\,d\tau \quad \text{will be the function} \quad \mathcal{Y}(j\omega) = \frac{1}{j\omega}\mathcal{X}(j\omega) + \pi\mathcal{X}(0)\delta(\omega)$$

Proof From the analytical form of the signal $y(t)$ it follows that this is simply the convolution between the signal $x(t)$ and the step function $u(t)$, since, as we saw in a previous chapter,

$$y(t) = x(t) * u(t) = \int_{-\infty}^{+\infty} x(\tau)\,u(t-\tau)\,d\tau = \int_{-\infty}^{t} x(\tau)\,d\tau$$

Applying, therefore, the theorem of convolution, it immediately follows that

$$\mathcal{Y}(j\omega) = \mathcal{X}(j\omega)\mathcal{U}(j\omega)$$

where $\mathcal{U}(j\omega)$ is the Fourier transform of the step function $u(t)$. Therefore, the proof of the property of integration requires the determination of the function $\mathcal{U}(j\omega)$.

But how will the calculation of the Fourier transform of the step function $u(t)$ take place? In reality, this process is reduced to determining the Fourier transform of the *sign function* upon which the step function $u(t)$ functionally depends. If we compare the definitions of the two functions

$$u(t) = \begin{cases} 1 \text{ for } t \geq 0 \\ 0 \text{ for } t < 0 \end{cases} \quad \text{and} \quad sgn(t) = \begin{cases} +1 \text{ for } t > 0 \\ -1 \text{ for } t < 0 \end{cases}$$

it is not difficult to see that they are related by an equation of the form[10]

$$u(t) = \frac{1}{2}\Big[1 + sgn(t)\Big]$$

and by applying the property of *linearity* of the Fourier transform, we obtain

$$\mathcal{U}(j\omega) \equiv \mathcal{F}\{u(t)\} = \frac{1}{2}\mathcal{F}\{1\} + \frac{1}{2}\mathcal{F}\{sgn(t)\}$$

Of the two transforms that appear in the preceding equation, the first one is calculated as

$$\mathcal{F}\{1\} = \int_{-\infty}^{+\infty} e^{-j\omega t}\, dt = 2\pi\delta(-\omega) = 2\pi\delta(\omega)$$

where we used the definition of the *impulse function* $\delta(t)$

$$\delta(t) = \frac{1}{2\pi}\int_{-\infty}^{+\infty} e^{j\omega t}\, dt$$

and the fact that the function $\delta(t)$ is characterized by *even symmetry*. On the other hand, to calculate the Fourier transform of the sign function, let us consider the function

$$\varphi_\alpha(t) = \begin{cases} +e^{-\alpha t} \text{ for } t > 0 \\ -e^{+\alpha t} \text{ for } t < 0 \end{cases} \quad (\alpha > 0)$$

which approximates the sign function in the limit $\alpha \to 0$. The Fourier transform of the function $\varphi_\alpha(t)$ is calculated as

$$\mathcal{F}\{\varphi_\alpha(t)\} = \int_{-\infty}^{\infty} \varphi_\alpha(t)\, e^{-j\omega t}\, dt = \int_{0}^{\infty} e^{-\alpha t} e^{-j\omega t}\, dt - \int_{-\infty}^{0} e^{\alpha t} e^{-j\omega t}\, dt = -\frac{2j\omega}{\alpha^2 + \omega^2}$$

and therefore the spectrum of the sign function will result from the spectrum of the function $\varphi_\alpha(t)$ in the limit $\alpha \to 0$, from which we obtain

$$\mathcal{F}\{sgn(t)\} = \lim_{\alpha \to 0} \mathcal{F}\{\varphi_\alpha(t)\} = -\lim_{\alpha \to 0} \frac{2j\omega}{\alpha^2 + \omega^2} = -\frac{2j}{\omega} = \frac{2}{j\omega}$$

Replacing the values of the functions $\mathcal{F}\{1\}$ and $\mathcal{F}\{sgn(t)\}$ in the defining equation of the function $\mathcal{U}(j\omega)$, it is calculated as

$$\mathcal{U}(j\omega) \equiv \mathcal{F}\{u(t)\} = \frac{1}{2}\mathcal{F}\{1\} + \frac{1}{2}\mathcal{F}\{sgn(t)\} = \frac{1}{2}\left(2\pi\delta(\omega) + \frac{2}{j\omega}\right) = \pi\delta(\omega) + \frac{1}{j\omega}$$

and determines the *spectrum of the step function*, while if the same function is replaced in the defining equation of our initial function $\mathcal{Y}(j\omega)$, it will take the form

$$\mathcal{Y}(j\omega) \equiv \mathcal{F}\left\{\int_{-\infty}^{t} f(\tau)\, d\tau\right\} = \mathcal{X}(j\omega)\mathcal{U}(j\omega) = \frac{\mathcal{X}(j\omega)}{j\omega} + \pi\mathcal{X}(0)\delta(\omega)$$

which is also the desired result. The last equation was derived using the property of the impulse function, which is formulated as $x(t)\delta(t) = x(0)\delta(t)$.

[10] This correlation between the functions $u(t)$ and $sgn(t)$ can be seen if we plot the graphs of the two functions. In this case, it is not difficult to see that the *graph of the step function arises from the graph of the sign function by raising the curve by one unit and dividing by 2.*

- **Parseval's identity**: If the Fourier transforms of the signals $x(t)$ and $y(t)$ are the functions $\mathcal{X}(j\omega)$ and $\mathcal{Y}(j\omega)$, then it is proven that

$$\int_{-\infty}^{+\infty} x(t)y^*(t)\,dt = \frac{1}{2\pi}\int_{-\infty}^{+\infty} \mathcal{X}(j\omega)\mathcal{Y}^*(\omega)\,d\omega$$

Proof Expressing the signals $x(t)$ and $y(t)$ as the inverse Fourier transforms of the functions $\mathcal{X}(j\omega)$ and $\mathcal{Y}(j\omega)$, we get

$$x(t) = \frac{1}{2\pi}\int_{-\infty}^{+\infty} \mathcal{X}(j\omega)\,e^{j\omega t}d\omega \qquad \text{and} \qquad y(t) = \frac{1}{2\pi}\int_{-\infty}^{+\infty} \mathcal{Y}(\xi)\,e^{j\xi t}d\xi$$

and the integral on the left-hand side of the equation to be proven takes the form

$$\int_{-\infty}^{+\infty} x(t)y^*(t)\,dt = \int_{-\infty}^{+\infty}\left(\frac{1}{2\pi}\int_{-\infty}^{+\infty}\mathcal{X}(j\omega)e^{j\omega t}d\omega\right)\left(\frac{1}{2\pi}\int_{-\infty}^{+\infty}\mathcal{Y}(\xi)e^{j\xi t}d\xi\right)^* dt =$$
$$= \int_{-\infty}^{+\infty}\left(\frac{1}{2\pi}\int_{-\infty}^{+\infty}\mathcal{X}(j\omega)e^{j\omega t}d\omega\right)\left(\frac{1}{2\pi}\int_{-\infty}^{+\infty}\mathcal{Y}^*(\xi)e^{-j\xi t}d\xi\right) dt =$$
$$= \frac{1}{(2\pi)^2}\int_{-\infty}^{+\infty}\int_{-\infty}^{+\infty}\int_{-\infty}^{+\infty}\mathcal{X}(j\omega)\,\mathcal{Y}^*(\xi)\,e^{j(\omega-\xi)t}d\omega d\xi dt$$

Using the definition of the function $\delta(t)$ in the form

$$\delta(\omega-\xi) = \frac{1}{2\pi}\int_{-\infty}^{+\infty} e^{-j(\omega-\xi)t}dt$$

the foregoing relation is formulated as

$$\int_{-\infty}^{+\infty} x(t)\,y^*(t)\,dt = \frac{1}{2\pi}\int_{-\infty}^{+\infty}\int_{-\infty}^{+\infty}\mathcal{X}(j\omega)\,\mathcal{Y}^*(\xi)\,\delta(\omega-\xi)\,d\omega d\xi = \frac{1}{2\pi}\int_{-\infty}^{+\infty}\mathcal{X}(j\omega)\,\mathcal{Y}^*(\omega)\,d\omega$$

which is also the final result. To derive the last relation, the following property was used:

$$\int_{-\infty}^{+\infty}\mathcal{Y}^*(\xi)\,\delta(\omega-\xi)\,d\xi = \mathcal{Y}^*(\omega)$$

- **Bessel's equation**: If the Fourier transform of the continuous-time signal $x(t)$ is the function $\mathcal{X}(j\omega)$, then it is proven that

$$\int_{-\infty}^{+\infty}|x(t)|^2dt = \frac{1}{2\pi}\int_{-\infty}^{+\infty}|\mathcal{X}(j\omega)|^2d\omega$$

Proof *Bessel's equation* arises directly from *Parseval's theorem* if we set $x(t) = y(t)$. We will then have $\mathcal{X}(j\omega) = \mathcal{Y}(j\omega)$, and Parseval's theorem takes the simple form

$$\int_{-\infty}^{+\infty} x(t)\,x^*(t)\,dt = \frac{1}{2\pi}\int_{-\infty}^{+\infty}\mathcal{X}(j\omega)\,\mathcal{X}^*(j\omega)\,d\omega \qquad \text{or equivalently} \qquad \int_{-\infty}^{+\infty}|x(t)|^2dt = \frac{1}{2\pi}\int_{-\infty}^{+\infty}|\mathcal{X}(j\omega)|^2d\omega$$

which is also the desired result. Alternatively, Bessel's equation Bessel can be proved as follows:

$$\int_{-\infty}^{+\infty} |x(t)|^2 dt = \int_{-\infty}^{+\infty} x(t)\, x^*(t)\, dt = \int_{-\infty}^{+\infty} x(t) \left(\frac{1}{2\pi} \int_{-\infty}^{+\infty} \mathcal{X}^*(j\omega) e^{-j\omega t} d\omega \right) dt$$
$$= \frac{1}{2\pi} \int_{-\infty}^{+\infty} \mathcal{X}^*(j\omega) \left(\int_{-\infty}^{+\infty} x(t)\, e^{-j\omega t} dt \right) d\omega = \frac{1}{2\pi} \int_{-\infty}^{+\infty} \mathcal{X}^*(j\omega)\, \mathcal{X}(j\omega)\, d\omega$$

and finally

$$\int_{-\infty}^{+\infty} |x(t)|^2 dt = \frac{1}{2\pi} \int_{-\infty}^{+\infty} |\mathcal{X}(j\omega)|^2 d\omega$$

a method that is almost similar to the previous one, but relatively simpler, since the final result emerges directly and without the use of the function $\delta(t)$. Taking into account that the quantity of the left-hand side of the Bessel equation—which in the international literature is also referred to as *Parseval's theorem* since it is simply a special case of it—expresses the total energy of the signal, the physical meaning of this equation is the *principle of energy conservation* during the transition from the time domain to the frequency domain. According to *Bessel's theorem*, to calculate the total energy of the signal, we can either calculate the *energy per unit of time* $|x(t)|^2$ and integrate for all time instants or calculate the *energy per unit of frequency* $|\mathcal{X}(j\omega)|^2/2\pi$ and integrate for all frequencies. In the foregoing description, the quantity $|\mathcal{X}(j\omega)|^2$ is known as the *energy density spectrum* and, as we will prove later, is simply the *Fourier transform of the autocorrelation of the signal* $x(t)$.

It is important to mention at this point that the preceding equation of Bessel applies only to *aperiodic* and not to *periodic* signals, which, as is well known, have *infinite energy*. For this special category of signals, the corresponding Parseval relationship has, as we have proven, the form

$$\frac{1}{T} \int_T |x(t)|^2 dt = \sum_{n=-\infty}^{+\infty} |\alpha_n|^2$$

where T is the period of the signal and α_n is the coefficient of the *exponential Fourier expansion* of the periodic signal associated with the harmonic component numbered n. The physical significance of Parseval's equation for periodic signals is that the *average energy of the signal for one period equals the energy associated with the coefficients of the Fourier expansion*, with the quantity $|\alpha_n|^2$ expressing the contribution of the harmonic component numbered n to the total energy of the signal.

- **Correlation**: If the Fourier transforms of the signals $x(t)$ and $y(t)$ are the functions $\mathcal{X}(j\omega)$ and $\mathcal{Y}(j\omega)$, then the Fourier transform of the correlation function

$$z(t) = x(t) \star y(t) = \int_{-\infty}^{+\infty} x^*(\tau)\, y(t+\tau)\, d\tau$$

will be $\mathcal{Z}(j\omega) = \mathcal{X}^*(j\omega)\, \mathcal{Y}(j\omega)$. This property is known as the *Wiener–Khintchine* theorem.

Proof From the defining equation of the function $\mathcal{Z}(j\omega)$ we have

$$\mathcal{Z}(j\omega) = \int_{-\infty}^{+\infty} z(t)\, e^{-j\omega t} dt = \int_{-\infty}^{+\infty} \left(\int_{-\infty}^{+\infty} x^*(\tau)\, y(t+\tau)\, d\tau \right) e^{-j\omega t} dt = \int_{-\infty}^{+\infty} \int_{-\infty}^{+\infty} x^*(\tau)\, y(t+\tau)\, e^{-j\omega t} d\tau dt$$

Changing the integration variable from t to $\xi = t + \tau$, we have $t = \xi - \tau$ and $dt = d\xi$, and the foregoing relation leads to the desired result as

$$\mathcal{Z}(j\omega) = \int_{-\infty}^{+\infty} \int_{-\infty}^{+\infty} x^*(\tau)\, y(\xi)\, e^{-j\omega(\xi-\tau)} d\xi d\tau = \left(\int_{-\infty}^{+\infty} x^*(\tau)\, e^{j\omega\tau} d\tau \right) \left(\int_{-\infty}^{+\infty} y^*(\xi)\, e^{-j\omega\xi}\, d\xi \right) =$$
$$= \left(\int_{-\infty}^{+\infty} x(\tau)\, e^{-j\omega\tau} d\tau \right)^* \left(\int_{-\infty}^{+\infty} y^*(\xi)\, e^{-j\omega\xi}\, d\xi \right) = \mathcal{X}^*(j\omega)\, \mathcal{Y}(j\omega)$$

- **Autocorrelation**: If the Fourier transform of the signal $x(t)$ is the function $\mathcal{X}(j\omega)$, then the Fourier transform of the autocorrelation function

$$y(t) = x(t) \star x(t) = \int_{-\infty}^{+\infty} x^*(\tau)\, x(t+\tau)\, d\tau$$

will be the energy density function $\mathcal{Y}(j\omega) = |\mathcal{X}(j\omega)|^2$.
Proof If in the equation of the *Wiener–Khintchine* we make the substitution $x(t) = y(t)$, it immediately follows that $\mathcal{Y}(j\omega) = \mathcal{X}^*(\omega)\mathcal{X}(j\omega) = |\mathcal{X}(j\omega)|^2$, which is also the desired result. Alternatively, we can arrive at the same result with the help of the *Parseval identity*, which allows for the formulation of the *autocorrelation function* in the form

$$y(t) = \int_{-\infty}^{+\infty} x^*(\tau)\, x(t+\tau)\, d\tau = \frac{1}{2\pi}\int_{-\infty}^{+\infty} \mathcal{F}\{x(\tau+t)\}\, \mathcal{F}^*\{x(t)\}\, d\omega$$

However, $\mathcal{F}^*\{x(t)\} = \mathcal{X}^*(j\omega)$ and $\mathcal{F}\{x(\tau+t)\} = e^{j\omega t}\mathcal{X}(j\omega)$, and the autocorrelation function becomes

$$y(t) = \frac{1}{2\pi}\int_{-\infty}^{+\infty} \mathcal{X}(j\omega)\, \mathcal{X}^*(j\omega)\, e^{j\omega t}\, d\omega = \frac{1}{2\pi}\int_{-\infty}^{+\infty} |\mathcal{X}(j\omega)|^2 e^{j\omega t} d\omega$$

But the last relation is recognized as an inverse transform equation. Therefore, the *autocorrelation function is the inverse Fourier transform of the energy density function* from which the desired result arises.

- **The areas of the curves** $x(t)$ **and** $\mathcal{X}(j\omega)$: If the Fourier transform of the signal $x(t)$ is the function $\mathcal{X}(j\omega)$, then:

1. The area of the curve of the signal $x(t)$ equals the value of the transform $\mathcal{X}(j\omega)$ for the frequency value $\omega = 0$, or, in mathematical notation,

$$\mathcal{X}(0) = \int_{-\infty}^{+\infty} x(t)\, dt$$

2. The area of the curve of the transform $\mathcal{X}(j\omega)$ equals the value of the signal $x(t)$ for the time point $t = 0$ multiplied by the quantity $1/2\pi$, or, in mathematical notation,

$$x(0) = \frac{1}{2\pi}\int_{-\infty}^{+\infty} \mathcal{X}(j\omega)\, d\omega$$

The foregoing relations are easily proven if in the defining equations of the direct and inverse Fourier transform we set $\omega = 0$ and $t = 0$, respectively.

Examples of Applying the Properties of the Transform

Using the properties of the Fourier transform, the process of calculating the spectrum of continuous-time signals is significantly simplified compared to the conventional calculation method that relies on the use of its defining equation. This claim is proven by solving a set of exemplary solved exercises that are presented in the following pages.

Problem 3.3.13 Find the Fourier transform of the signal $x(t) = e^{j\alpha t}$.

Solution From the defining equation of the Fourier transform we find that

$$\mathcal{X}(j\omega) = \int_{-\infty}^{+\infty} x(t)\, e^{-j\omega t} dt = \int_{-\infty}^{+\infty} e^{j\alpha t} e^{-j\omega t} dt = \int_{-\infty}^{+\infty} e^{j(\alpha-\omega)t} dt = 2\pi\,\delta(\omega - \alpha)$$

We can reach the same result using the property of *frequency shift*, which is described by the equation $\mathcal{F}\{e^{j\omega_0 t} x(t)\} = \mathcal{X}(\omega - \omega_0)$ and the result $\mathcal{F}\{1\} = 2\pi\,\delta(\omega)$, which we proved during the description of the *integration* property. Defining the function $y(t) = 1 \cdot x(t) = e^{j\alpha t} \cdot 1$ we obtain

$$\mathcal{Y}(j\omega) = \mathcal{F}\{e^{j\alpha t} \cdot 1\} = \mathcal{F}\{1\}_{\omega \to \omega - \alpha} = 2\pi\,\delta(\omega - \alpha)$$

thereby concluding the result that emerged from the trivial calculation method. Using the same process, we can prove that $\mathcal{F}\{e^{-j\alpha t}\} = 2\pi\,\delta(\omega + \alpha)$.

Problem 3.3.14 Find the Fourier transform of the harmonic signal $x(t) = \cos(\alpha t)$.

Solution From the first equation of *modulation*

$$\mathcal{F}\{x(t)\cos(\alpha t)\} = \frac{1}{2}\Big[\mathcal{X}(\omega + \omega_0) + \mathcal{X}(\omega - \omega_0)\Big]$$

—for $x(t) = 1$ and $\omega_0 = \alpha$, and using also the property of *frequency shift*, we find

$$\mathcal{F}\{\cos(\alpha t)\} = \frac{1}{2}\Big[\mathcal{F}\{1\}_{\omega \to \omega + \alpha} + \mathcal{F}\{1\}_{\omega \to \omega - \alpha}\Big] = \pi\Big[\delta(\omega + \alpha) + \delta(\omega - \alpha)\Big]$$

Problem 3.3.15 Find the Fourier transform of the harmonic signal $x(t) = \sin(\alpha t)$.

Solution From the second equation of *modulation*

$$\mathcal{F}\{x(t)\sin(\alpha t)\} = \frac{j}{2}\Big[\mathcal{X}(\omega + \omega_0) - \mathcal{X}(\omega - \omega_0)\Big]$$

for $x(t) = 1$ and $\omega_0 = \alpha$ and using the property of *frequency shift*, we have

$$\mathcal{F}\{\sin(\alpha t)\} = \frac{j}{2}\Big[\mathcal{F}\{1\}_{\omega \to \omega + \alpha} - \mathcal{F}\{1\}_{\omega \to \omega - \alpha}\Big] = j\pi\Big[\delta(\omega + \alpha) - \delta(\omega - \alpha)\Big]$$

Problem 3.3.16 Find the Fourier transform of the signal $x(t) = t^n$.

Solution The property of *differentiation in the frequency domain*

$$\mathcal{F}\{t^n x(t)\} = j^n \frac{d^n \mathcal{X}(j\omega)}{d\omega^n} = j^n \mathcal{X}^{(n)}(\omega)$$

and the result $\mathcal{F}\{1\} = 2\pi\delta(\omega)$ immediately imply that

$$\mathcal{F}\{t^n\} = \mathcal{F}\{t^n \cdot 1\} = j^n \frac{d^n \mathcal{F}\{1\}}{d\omega^2} = 2\pi j^n \frac{d^n \delta(\omega)}{d\omega^n}$$

Problem 3.3.17 Find the Fourier transform of the signal $x(t) = 1/t$.

Solution In the proof of the property of *integration* we showed that the Fourier transform of the *sign function* has the form $F(\omega) = 2/j\omega$. Therefore, using the *property of duality* of the transform, which is formulated as

$$\mathcal{F}\{x(t)\} = \mathcal{X}(j\omega) \Leftrightarrow \mathcal{F}\{\mathcal{X}(t)\} = 2\pi x(-\omega)$$

it immediately follows that

$$\frac{2}{j}\mathcal{F}\left\{\frac{1}{t}\right\} = 2\pi \cdot \text{sgn}(-\omega) \quad \text{and therefore} \quad \mathcal{F}\left\{\frac{1}{t}\right\} = -j\pi \cdot \text{sgn}(\omega)$$

Generalizing the foregoing property, it is proven that

$$\mathcal{F}\left\{\frac{1}{t^n}\right\} = -j\pi \cdot \frac{(-j\omega)^{n-1}}{(n-1)!} \text{sgn}(\omega)$$

Problem 3.3.18 Find the Fourier transform of the triangular signal

$$x(t) \equiv \Lambda_\alpha(t) = \begin{cases} 1 - \frac{|t|}{\alpha} & \text{for } |t| \le \alpha \\ 0 & \text{anywhere else} \end{cases}$$

using (a) the property of convolution and (b) the properties of differentiation and time shift. To solve the exercise, consider that the triangular signal results from the convolution of two square pulses, or, in mathematical notation,

$$\Lambda_\alpha(t) = \frac{1}{\alpha}\left[\Pi_{\alpha/2}(t) * \Pi_{\alpha/2}(t)\right] \quad \text{where} \quad \Pi_{\alpha/2}(t) = \begin{cases} 1 \text{ for } |t| < \frac{\alpha}{2} \\ 0 \text{ anywhere else} \end{cases}$$

and that the second derivative with respect to time of the triangular signal is given by the equation

$$\frac{d^2\Lambda_\alpha(t)}{dt^2} = \frac{1}{\tau}\left[\delta(t+\alpha) - 2\delta(t) + \delta(t-\alpha)\right]$$

Solution (a) In Exercise 3.3.4 (refer to page 228), we showed that the Fourier transform of the rectangular window $\Pi_\alpha(t)$ has the form $\mathcal{X}(j\omega) = 2\sin(\omega\alpha)/\omega$—so it holds that $\mathcal{F}\{\Pi_{\alpha/2}(t)\} = 2\sin(\omega\alpha/2)/\omega$. Therefore, applying the *convolution theorem* we obtain

$$\mathcal{F}\{\Lambda_\alpha(t)\} = \frac{1}{\alpha}\mathcal{F}\{\Pi_{\alpha/2}(t) * \Pi_{\alpha/2}(t)\} = \frac{1}{\alpha}\mathcal{F}\{\Pi_{\alpha/2}(t)\}\mathcal{F}\{\Pi_{\alpha/2}(t)\}$$

$$= \frac{1}{\alpha}\left[\frac{2}{\omega}\sin\left(\frac{\omega\alpha}{2}\right)\right]\left[\frac{2}{\omega}\sin\left(\frac{\omega\alpha}{2}\right)\right] = \frac{4}{\alpha\omega^2}\sin^2\left(\frac{\omega\alpha}{2}\right)$$

(b) Taking the Fourier transform of the two terms of the differential equation satisfied by the second derivative of the triangular signal and using the property of *linearity* we get

$$\mathcal{F}\left\{\frac{d^2\Lambda_\alpha(t)}{dt^2}\right\} = \frac{1}{\alpha}\Big[\mathcal{F}\{\delta(t+\alpha)\} - 2\mathcal{F}\{\delta(t)\} + \mathcal{F}\{\delta(t-\alpha)\}\Big]$$

Using the result $\mathcal{F}\{\delta(t)\} = 1$ and the property of *time shift* we obtain

$$\mathcal{F}\{\delta(t+\alpha)\} = e^{i\omega\alpha}\mathcal{F}\{\delta(t)\} = e^{i\omega\alpha}$$

$$\mathcal{F}\{\delta(t-\alpha)\} = e^{-i\omega\alpha}\mathcal{F}\{\delta(t)\} = e^{-i\omega\alpha}$$

On the other hand, applying the property of *differentiation*, we find that

$$\mathcal{F}\left\{\frac{d^2\Lambda_\alpha(t)}{dt^2}\right\} = (i\omega)^2\mathcal{F}\{\Lambda_\alpha\}(t) = -\omega^2\mathcal{F}\{\Lambda_\alpha\}(t)$$

Replacing the foregoing expressions in the last equation, it will take the form

$$-\omega^2\mathcal{F}\{\Lambda_\alpha(t)\} = \frac{1}{\alpha}(e^{i\omega\alpha} - 2 + e^{-i\omega\alpha}) = \frac{2}{\alpha}[\cos(\omega\alpha - 1)] = -\frac{4}{\alpha}\sin^2\left(\frac{\omega\alpha}{2}\right)$$

from which it finally follows that

$$\mathcal{F}\{\Lambda_\alpha(t)\} = \frac{4}{\alpha\omega^2}\sin^2\left(\frac{\omega\alpha}{2}\right)$$

which is the same result as the one we obtained previously as well as with the application of the defining equation of the Fourier transform (see Problem 3.3.7 in page 231).

Problem 3.3.19 Determine the spectra of the continuous-time signals

$$x_1(t) = \Pi_1\left(\frac{t}{2}\right)e^{i20\pi t} \qquad \text{and} \qquad x_2(t) = \Pi_1\left(\frac{t}{2}\right)\cos(20\pi t)$$

Solution The signal function $\Pi_1(t/2)$ is defined as

$$\Pi_1\left(\frac{t}{2}\right) = \begin{cases} 1 \text{ for } \left|\frac{t}{2}\right| < 1 \\ 0 \text{ anywhere else} \end{cases} = \begin{cases} 1 \text{ for } |t| < 2 \\ 0 \text{ anywhere else} \end{cases} = \Pi_2(\tau)$$

and therefore its spectrum will have the form

$$\mathcal{F}\left\{\Pi_1\left(\frac{t}{2}\right)\right\} = \mathcal{F}\{\Pi_2(t)\} = \frac{2}{\omega}\sin(\omega\alpha)\Big|_{\alpha=2} = \frac{2}{\omega}\sin(2\omega)$$

Based on this result and using the properties of the Fourier transform, the spectra of the functions $x_1(t)$ and $x_2(t)$ are calculated in the following way:

(a) From the property of *frequency shift* $\mathcal{F}\{e^{i\omega_0 t}x(t)\} = \mathcal{X}(\omega - \omega_0)$ for the value $\omega_0 = 20\pi$ we get

$$\mathcal{X}_1(\omega) = \mathcal{L}\{x_1(t)\} = \mathcal{L}\left\{\Pi_1\left(\frac{t}{2}\right)e^{i20\pi t}\right\} = \mathcal{F}\left\{\Pi_1\left(\frac{t}{2}\right)\right\}_{\omega\to\omega-20\pi}$$

$$= \frac{2}{\omega}\sin(2\omega)\bigg|_{\omega\to\omega-20\pi} = \frac{2}{\omega - 20\pi}\sin[2(\omega - 20\pi)]$$

which is the desired result.

(b) From the first equation of the property of *modulation* for the signal $x(t) = \Pi_1(t/2)$ and the quantity $\omega_0 = 20\pi$ we get

$$\mathcal{X}_2(\omega) = \mathcal{L}\{x_2(t)\} = \frac{1}{2}\left[\mathcal{F}\left\{\Pi_1\left(\frac{t}{2}\right)\right\}_{\omega\to\omega+20\pi} + \mathcal{F}\left\{\Pi_1\left(\frac{t}{2}\right)\right\}_{\omega\to\omega-20\pi}\right] =$$

$$= \frac{1}{2}\left[\frac{2}{\omega}\sin(2\omega)\bigg|_{\omega\to\omega+20\pi} + \frac{2}{\omega}\sin(2\omega)\bigg|_{\omega\to\omega-20\pi}\right] = \frac{\sin[2(\omega + 20\pi)]}{\omega + 20\pi} + \frac{\sin[2(\omega - 20\pi)]}{\omega - 20\pi}$$

which is the desired result.

Problem 3.3.20 To verify the result of question (b) of Exercise 3.3.19 using the property of *multiplication* of the Fourier transform.

Solution The Fourier transforms of the signals $\Pi_1(t/2)$ and $\cos(20\pi t)$ are given as we proved in previous exercises from the relations

$$\mathcal{F}\left\{\Pi_1\left(\frac{t}{2}\right)\right\} = \frac{2}{\omega}\sin(2\omega) \quad \text{and} \quad \mathcal{F}\{\cos(20\pi t)\} = \pi\left[\delta(\omega + 20\pi) + \delta(\omega - 20\pi)\right]$$

Therefore, applying the *multiplication theorem* we obtain

$$\mathcal{X}_2(\omega) = \mathcal{F}\left\{\Pi_1\left(\frac{t}{2}\right)\cos(20\pi t)\right\} = \frac{1}{2\pi}\mathcal{F}\left\{\Pi_1\left(\frac{t}{2}\right)\right\} * \mathcal{F}\{\cos(20\pi t)\}$$

$$= \frac{1}{2\pi}\left(\frac{2}{\omega}\sin(2\omega)\right) * \left\{\pi\left[\delta(\omega + 20\pi) + \delta(\omega - 20\pi)\right]\right\} = \frac{\sin(2\omega)}{\omega} * \delta(\omega + 20\pi) + \frac{\sin(2\omega)}{\omega} * \delta(\omega - 20\pi)$$

The last expressions are calculated using the properties of the function $\delta(t)$ as

$$\frac{\sin(2\omega)}{\omega} * \delta(\omega + 20\pi) = \frac{\sin[2(\omega + 20\pi)]}{\omega + 20\pi} \quad \text{and} \quad \frac{\sin(2\omega)}{\omega} * \delta(\omega - 20\pi) = \frac{\sin[2(\omega - 20\pi)]}{\omega - 20\pi}$$

from which the desired result is derived.

3.3.6 Fourier Transform of Periodic Signals

Although the Fourier transform has been defined only for *aperiodic* signals, using the *exponential Fourier series* of a periodic signal, we can define the Fourier transform for this special category of signals, thereby creating a unified way of representing signals in the frequency domain regardless of whether they are periodic or not. Note that the Fourier transforms of periodic signals are typically not defined, since they do not meet the condition of *absolute integrability* whose power is a necessary prerequisite for the convergence of the Fourier transform. However, this problem is addressed by generalizing the Fourier transform, so that the use of impulse functions in the frequency field is permissible. In a mathematical description, let us consider a periodic signal of period T that, as is well known, can be represented by an *exponential Fourier series* of the form

$$x(t) = \sum_{n=-\infty}^{+\infty} \alpha_n \, e^{jn\omega_0 t}, \qquad \omega_0 = \frac{2\pi}{T}$$

Substituting the preceding signal into the defining equation of the Fourier transform, we obtain

$$\mathcal{F}\{x(t)\} = \int_{-\infty}^{+\infty} x(t)\, e^{-j\omega t} dt = \int_{-\infty}^{+\infty} \left(\sum_{n=-\infty}^{+\infty} \alpha_n \, e^{jn\omega_0 t} \right) e^{-j\omega t} dt$$

$$= \sum_{n=-\infty}^{+\infty} \alpha_n \left(\int_{-\infty}^{+\infty} e^{jn\omega_0 t} e^{-j\omega t} dt \right) = \sum_{n=-\infty}^{+\infty} \alpha_n \mathcal{F}\{e^{jn\omega_0 t}\}$$

where

$$\mathcal{F}\{e^{jn\omega_0 t}\} = \int_{-\infty}^{+\infty} e^{jn\omega_0 t} e^{-j\omega t} dt = \int_{-\infty}^{+\infty} e^{j(n\omega_0 - \omega)t} dt = \left[\frac{e^{j(n\omega_0-\omega)t}}{j(n\omega_0 - \omega)} \right]_{-\infty}^{+\infty}$$

is the Fourier transform of the function $e^{jn\omega_0 t}$, which obviously does not converge. To bypass this problem, we can use the result $\mathcal{F}\{1\} = \delta(\omega)$ and the property of *frequency shift* of the transform, so the preceding expression is calculated as

$$\mathcal{F}\{e^{jn\omega_0 t}\} = \mathcal{F}\{e^{jn\omega_0 t} \cdot 1\} = \mathcal{F}\{1\}_{\omega \to \omega - n\omega_0} = \delta(\omega) \Big|_{\omega \to \omega - n\omega_0} = 2\pi \delta(\omega - n\omega_0)$$

ultimately leading to a result of the form

$$\mathcal{F}\{x(t)\} = \sum_{n=-\infty}^{+\infty} (2\pi\alpha_n)\delta(\omega - n\omega_0) \tag{3.34}$$

The foregoing equation expresses the *Fourier transform of the periodic signal* $x(t)$ and, according to this, *the Fourier transform of a periodic signal with Fourier coefficients* α_n *is simply a train of infinite in number impulse functions, each of which is associated with some of the infinite frequencies* $n\omega_0$ *and has a multiplicative coefficient with a value of* $2\pi\alpha_n$.

We already saw this form of the Fourier transform in the exercise calculating the transform of the signal $x(t) = \cos(\omega t)$ (see Exercise 3.3.14 on page 247 for $\omega = \alpha$), which is a periodic function with period $T = 2\pi/\omega$. From the Euler formula

$$\cos(\omega t) = \frac{e^{j\omega t} + e^{-j\omega t}}{2} = \frac{1}{2} e^{j\omega t} + \frac{1}{2} e^{-j\omega t}$$

it immediately follows that the only nonzero Fourier coefficients of this function correspond to the values $n = 1$ and $n = -1$ and are given by the relations $\alpha_1 = \alpha_{-1} = 1/2$. From the foregoing equation, therefore,

$$\mathcal{F}\{x(t)\} = 2\pi\alpha_1\delta(\omega + \omega_0) + 2\pi\alpha_{-1}\delta(\omega - \omega_0) = \pi \Big[\delta(\omega + \omega_0) + \delta(\omega - \omega_0) \Big]$$

which is the result we had calculated using the first equation related to the property of *modulation*. Similarly, and using the Euler equation for the sine function, it is proven that

$$\mathcal{F}\{\sin(\omega t)\} = j\pi \Big[\delta(\omega + \omega_0) - \delta(\omega - \omega_0) \Big]$$

which is the result we had arrived at, using the second equation of the property of *modulation* (see Exercise 3.3.15 on page 247 for $\omega = \alpha$). Therefore, the Fourier transform of the signals $\cos(\omega t)$ and $\sin(\omega t)$ emerge as special cases of the general defining equation of the *Fourier transform of periodic functions* and their relatively simple form is due to the fact that both have only two nonzero coefficients in their exponential expansion according to Fourier.

Problem 3.3.21 Find the Fourier transform of the impulse train functions

$$x(t) = \sum_{n=-\infty}^{+\infty} \delta(t - nT)$$

Solution The continuous-time signal $x(t)$ is a periodic function with period T, and its coefficients of the *exponential expansion* according to Fourier are given by the relation

$$\alpha_n = \frac{1}{T} \int_{-T/2}^{+T/2} x(t)\, e^{-jn\omega_0 t} dt = \frac{1}{T} \int_{-T/2}^{+T/2} \sum_{n=-\infty}^{+\infty} \delta(t - nT)\, e^{-jn\omega_0 t} dt$$

However, as we saw in a previous exercise (see Exercise 3.2.11 on page 216), from all the infinite time-shifted impulse functions of the foregoing sum, the only one that belongs to the preceding integration interval is the one corresponding to the index value $n = 0$. This results in

$$\alpha_n = \frac{1}{T} \int_{-T/2}^{+T/2} \delta(t)\, e^{-jn\omega_0 t} dt = \frac{1}{T} \Big[e^{-jn\omega_0 t} \Big]_{t=0} = \frac{1}{T}$$

where we used the property of the function $\delta(t)$

$$\int_{\alpha}^{\beta} \delta(t)\, \varphi(t)\, dt = \varphi(0)$$

for values $\alpha = -\beta = T/2$ and for the function $\varphi(t) = e^{-jn\omega_0 t}$. We are therefore led to the conclusion that all the Fourier coefficients are equal to each other with a value of $1/T$, and from the defining equation of the Fourier transform of periodic signals we obtain

$$\mathcal{X}(j\omega) = 2\pi \sum_{n=-\infty}^{+\infty} \alpha_n \delta(\omega - n\omega_0) = \frac{2\pi}{T} \sum_{n=-\infty}^{+\infty} \delta(\omega - n\omega_0) = \frac{2\pi}{T} \sum_{n=-\infty}^{+\infty} \delta\left(\omega - \frac{2\pi n}{T}\right)$$

From the last relation it follows that the *Fourier transform of a periodic train of impulse functions with a period of T in the time domain is a periodic train of impulse functions with a period of $2\pi/T$ in the frequency domain*, which is another indication of the symmetry that exists between the time and frequency domains. In addition, it is not difficult to see that these two periods are inversely proportional, and therefore the more distant the impulse functions are in the time domain, the denser are the corresponding impulse responses in the frequency domain.

Problem 3.3.22 Given the continuous-time signal

$$x(t) = \sum_{n=-\infty}^{+\infty} \frac{\sin(n\pi/4)}{n\pi/4}\, \delta\left(t - \frac{n\pi}{4}\right)$$

(a) Determine the signal $y(t)$ so that

(continued)

$$x(t) = \left(\frac{\sin t}{\pi t}\right) g(t)$$

(b) Using the property of the multiplication of the Fourier transform, prove that the function $\mathcal{X}(j\omega) \equiv \mathcal{F}\{x(t)\}$ is periodic and determine its value (Opeenheim, Willsky, Nawab, Exercise for solution 4.16).

Solution (a) Setting $\varphi(t) = \sin t/t$ and $t_0 = n\pi/4$ and using the known property

$$\varphi(t)\delta(t - t_0) = \varphi(t_0)\delta(t - t_0)$$

the expression within the sum is transformed as

$$\varphi(t)\,\delta(t - t_0) = \frac{\sin(n\pi/4)}{n\pi/4}\,\delta\left(t - \frac{n\pi}{4}\right) = \frac{\sin t}{t}\,\delta\left(t - \frac{n\pi}{4}\right) = \varphi(t)\,\delta(t - t_0)$$

Replacing the foregoing expression in the equation defining the signal, it will take the form

$$\sum_{n=-\infty}^{+\infty} \frac{\sin(n\pi/4)}{n\pi/4}\,\delta\left(t - \frac{n\pi}{4}\right) = \sum_{n=-\infty}^{+\infty} \frac{\sin t}{t}\,\delta\left(t - \frac{n\pi}{4}\right) = \left(\frac{\sin t}{\pi t}\right)\sum_{n=-\infty}^{+\infty} \pi\,\delta\left(t - \frac{n\pi}{4}\right)$$

from which it immediately follows that

$$y(t) = \sum_{n=-\infty}^{+\infty} \pi\,\delta\left(t - \frac{n\pi}{4}\right) = \pi\sum_{n=-\infty}^{+\infty} \delta\left(t - \frac{n\pi}{4}\right)$$

From the foregoing relationship we conclude that the signal $y(t)$ is periodic with period $T = \pi/4$.

(b) According to the result of the previous exercise, the Fourier transform of the signal $y(t)$ will be given by the relationship

$$\mathcal{Y}(j\omega) = \pi\left\{\frac{2\pi}{T}\sum_{n=-\infty}^{+\infty} \delta\left(\omega - \frac{2\pi n}{T}\right)\right\} = \pi\left\{2\pi\frac{4}{\pi}\sum_{n=-\infty}^{+\infty} \delta\left(\omega - 2\pi n\frac{4}{\pi}\right)\right\} = 8\pi\sum_{n=-\infty}^{+\infty} \delta\left(\omega - 8n\right)$$

from which it follows that the function $\mathcal{Y}(j\omega)$ is periodic with period $T = 8$. Applying now the multiplication theorem

$$\left(\frac{\sin t}{\pi t}\right) y(t) \overset{\mathcal{F}}{\longleftrightarrow} \frac{1}{2\pi}\left\{\mathcal{F}\left\{\frac{\sin t}{\pi t}\right\} * \mathcal{F}\{g(t)\}\right\}$$

the transform of the signal $x(t)$ is calculated as

$$\mathcal{X}(j\omega) = \frac{1}{2\pi}\left\{\mathcal{F}\left\{\frac{\sin t}{\pi t}\right\} * \left(8\pi\sum_{n=-\infty}^{+\infty} \delta\left(\omega - 8n\right)\right)\right\} = 4\left\{S(\omega) * \sum_{n=-\infty}^{+\infty} \delta\left(\omega - 8n\right)\right\}$$

where for simplicity we set $S(\omega) = \mathcal{F}\{\sin t/\pi t\}$. Using now the property

$$S(\omega) * \delta(\omega - \omega_0) = S(\omega_0)$$

the foregoing equation will take the form

$$\mathcal{X}(j\omega) = 4\sum_{n=-\infty}^{+\infty} S(\omega - 8n) = 4\sum_{n=-\infty}^{+\infty} \mathcal{F}\left\{\frac{\sin t}{\pi t}\right\}\Bigg|_{\omega\to\omega-8n}$$

which is the desired result. According to the preceding equation, the function $\mathcal{X}(j\omega)$ results from the repetition of the quantity $4S(\omega)$ every 8 rad/s and therefore is periodic.

To calculate the value of the function $\mathcal{X}(j\omega)$ for one time period, we will use the known result

$$S(\omega) = \mathcal{F}\left\{\frac{\sin t}{\pi t}\right\} = \begin{cases} 1 & |\omega| \le 1 \\ 0 & \text{anywhere else} \end{cases}$$

from which it follows that

$$\mathcal{X}(j\omega) = \mathcal{F}\{x(t)\} = \begin{cases} 4 & |\omega| \le 1 \\ 0 & 1 < |\omega| \le 4 \end{cases}$$

Problem 3.3.23 Determine the Fourier transform of the autocorrelation function $R_x(t)$ of a periodic signal $x(t)$.

Solution From basic theory it is known that the autocorrelation function

$$R_x(t) = \frac{1}{T}\int_T x^*(\xi)x(\xi + t)d\xi$$

of a periodic signal with period T is also a periodic function and is characterized by the same period value. Therefore, it can be expressed as a Fourier exponential series of the form

$$R_x(t) = \sum_{n=-\infty}^{+\infty} \rho_n\, e^{jn\omega_0 t}, \quad \omega_0 = \frac{2\pi}{T}$$

while its Fourier transform—being a periodic function—will have the form of a train of infinite and temporally equidistant impulse functions of the form

$$\mathcal{R}(\omega) = \sum_{n=-\infty}^{+\infty} (2\pi\rho_n)\, \delta\left(\omega - n\omega_0\right) = \sum_{n=-\infty}^{+\infty} (2\pi\rho_n)\, \delta\left(\omega - \frac{2\pi n}{T}\right)$$

If we assume that the exponential Fourier series of the function $x(t)$ has the form

$$x(t) = \sum_{n=-\infty}^{+\infty} \alpha_n\, e^{jn\omega_0 t}$$

then we can express the coefficients ρ_n as a function of the coefficients α_n by replacing in the equation of their definition the equation of the definition of the function $R_x(t)$. This will then yield

$$\alpha_n = \frac{1}{T}\int_T R_x(t)\, e^{-jn\omega_0 t} dt = \frac{1}{T}\int_T \left\{\frac{1}{T}\int_T x^*(\xi)\, x(\xi + t)\, d\xi\right\} e^{-jn\omega_0 t} dt = \frac{1}{T}\int_T x^*\xi \left\{\int_T x(\xi + t)\, e^{-jn\omega_0 t} dt\right\} d\xi$$

Setting $\tau = t + \xi$, we obtain $d\tau = dt$, and the foregoing relation will take the form

$$\begin{aligned} \alpha_n &= \int_T \frac{1}{T} x^*(\xi) \left\{\int_T x(\tau)\, e^{-jn\omega_0(\tau-\xi)}\right\} d\xi = \left(\frac{1}{T}\int_T x^*(\xi)\, e^{jn\omega_0 \xi} d\xi\right)\left(\frac{1}{T}\int_T x(t)\, e^{-jn\omega_0 \tau} d\tau\right) = \\ &= \left(\frac{1}{T}\int_T x(t)\, e^{-jn\omega_0 \tau} d\tau\right)^* \left(\frac{1}{T}\int_T x(t)\, e^{-jn\omega_0 \tau} d\tau\right) = \alpha_n^* \alpha_n = |\alpha_n|^2 \end{aligned}$$

from which it ultimately follows that

$$\mathcal{R}(\omega) = 2\pi \sum_{n=-\infty}^{+\infty} |\alpha_n|^2 \delta\left(\omega - \frac{2\pi n}{T}\right)$$

Problem 3.3.24 Determine the Fourier transform of the signal

$$x(t) = \sin\left(2\pi t + \frac{\pi}{4}\right)$$

Solution Using Euler's formula for the sine function, the signal $x(t)$ is written in the form

$$x(t) = \sin\left(2\pi t + \frac{\pi}{4}\right) = \frac{e^{i[2\pi t+(\pi/4)]} - e^{-i[2\pi t+(\pi/4)]}}{2i} = \frac{1}{2i}e^{i\pi/4}e^{i2\pi t} - \frac{1}{2i}e^{-i\pi/4}e^{-i2\pi t}$$

From the preceding relation it follows that the only nonzero coefficients of the exponential Fourier series correspond to values $n = 1$ and $n = -1$ and are $\alpha_1 = e^{i\pi/4}/2i$ and $\alpha_{-1} = -e^{-i\pi/4}/2i$. Therefore,

$$\mathcal{X}(j\omega) = 2\pi\alpha_1\delta(\omega-2\pi) + 2\pi\alpha_{-1}\delta(\omega+2\pi) = i\pi e^{-i\pi/4}\delta(\omega+2\pi) - i\pi e^{i\pi/4}\delta(\omega-2\pi)$$

which is the desired result.

Problem 3.3.25 Determine the Fourier transform of the signal $x(t) = 1 + \cos[6\pi t + (\pi/8)]$.

Solution From the property of *linearity* of the Fourier transform we have

$$\mathcal{F}\{x(t)\} = \mathcal{F}\{1\} + \mathcal{F}\left\{\cos\left(6\pi t + \frac{\pi}{8}\right)\right\}$$

However, $\mathcal{F}\{1\} = \delta(\omega)$, and using the Euler formula, we obtain

$$\cos\left(6\pi t + \frac{\pi}{8}\right) = \frac{e^{j[6\pi t+(\pi/8)]} + e^{-j[6\pi t+(\pi/8)]}}{2} = \frac{1}{2}e^{j\pi/8}e^{j6\pi t} + \frac{1}{2}e^{-j\pi/8}e^{-j6\pi t}$$

From the last relation it follows that the only nonzero Fourier coefficients of the function $\cos[6\pi t + (\pi/8)]$ are $\alpha_1 = e^{j\pi/8}/2$ and $\alpha_{-1} = e^{-j\pi/8}/2$, and, following the methodology of the previous exercise, it turns out that

$$\mathcal{F}\left\{\cos\left(6\pi t + \frac{\pi}{8}\right)\right\} = \pi e^{j\pi/8}\delta(\omega - 6\pi) + \pi e^{-j\pi/8}\delta(\omega + 6\pi)$$

and therefore the Fourier transform of the signal $x(t)$ will have the form

$$\mathcal{X}(j\omega) = \delta(\omega) + \pi e^{j\pi/8}\delta(\omega - 6\pi) + \pi e^{-j\pi/8}\delta(\omega + 6\pi)$$

3.3.7 Bandwidth Theorem

In the calculation of the Fourier transform of the normal distribution, we defined the concepts of *active duration* $\Delta t = t_2 - t_1$ and *active bandwidth* $\Delta\omega = \omega_2 - \omega_1$ of the functions $x(t)$ and $\mathcal{X}(j\omega) \equiv \mathcal{F}\{x(t)\}$ as the *regions of the time and frequency axes outside of which the functions* $x(t)$ *and* $X(\omega)$ *have such small values as to be considered negligible*. According to the *bandwidth theorem*, there is a global constant γ such that the active duration and the active bandwidth of any signal $x(t)$ satisfy the equation

$$\Delta t \cdot \Delta\omega \geq \gamma \tag{3.35}$$

The physical significance of this theorem is that these two quantities related to the signal $x(t)$ and its Fourier transform $\mathcal{X}(j\omega)$ *cannot simultaneously take an arbitrarily small value since their product* must be at least equal to the value of the constant γ. Because the foregoing relation is formulated in a way similar to that of *Heisenberg's indeterminacy principle* in *quantum mechanics*, the bandwidth theorem is often characterized as the *principle of Fourier analysis indeterminacy*.

The bandwidth theorem is formulated in many different forms depending on the category of functions it relates to; and in the following analysis we will limit ourselves to *continuous-time signals* whose energy E is *finite*. To prove this theorem, we must first determine the defining equation of the *active duration* Δt of a continuous-time signal $x(t)$, a process that will be carried out in the following way. Assuming that the Fourier transform of the signal $x(t)$ is the function $\mathcal{X}(j\omega) = |\mathcal{X}(j\omega)|\exp[j\measuredangle\mathcal{X}(j\omega)]$, where $|\mathcal{X}(j\omega)|$ and $\measuredangle\mathcal{X}(j\omega)$ are the amplitude and phase of the complex function $\mathcal{X}(j\omega)$, respectively, the application of the property of differentiation in the frequency domain $tx(t) \leftrightarrow i\mathcal{X}'(j\omega)$ leads to an equation of the form

$$\mathcal{Y}(j\omega) = \mathcal{F}\{y(t)\} = \mathcal{F}\{-jtx(t)\} = \frac{d\mathcal{X}(j\omega)}{d\omega} = \left[\frac{d|\mathcal{X}(j\omega)|}{d\omega} + j|\mathcal{X}(j\omega)|\frac{d\measuredangle\mathcal{X}(j\omega)}{d\omega}\right]e^{j\measuredangle\mathcal{X}(j\omega)}$$

with the continuous-time signal $y(t)$ defined as $y(t) = -jtx(t)$. Using now the Parseval equation

$$\int_{-\infty}^{+\infty}|y(t)|^2dt = \frac{1}{2\pi}\int_{-\infty}^{+\infty}|\mathcal{Y}(j\omega)|^2d\omega$$

where $|y(t)|^2 = t^2|x(t)|^2$ and

$$|\mathcal{Y}(j\omega)|^2 = \left|\frac{d\mathcal{X}(j\omega)}{d\omega}\right|^2 = \left|\frac{d|\mathcal{X}(j\omega)|}{d\omega} + j|\mathcal{X}(j\omega)|\frac{d\measuredangle\mathcal{X}(j\omega)}{d\omega}\right|^2|e^{j\measuredangle\mathcal{X}(j\omega)}|^2 = \left[\frac{d|\mathcal{X}(j\omega)|}{d\omega}\right]^2 + [|\mathcal{X}(j\omega)|]^2\left[\frac{d\measuredangle\mathcal{X}(j\omega)}{d\omega}\right]^2$$

this will take the form

$$(\Delta t)^2 = \int_{-\infty}^{+\infty}t^2|x(t)|^2dt = \frac{1}{2\pi}\int_{-\infty}^{+\infty}\left\{\left[\frac{d|\mathcal{X}(j\omega)|}{d\omega}\right]^2 + [|\mathcal{X}(j\omega)|]^2\left[\frac{d\measuredangle\mathcal{X}(j\omega)}{d\omega}\right]^2\right\}d\omega \tag{3.36}$$

which is also the defining equation of the *active duration* Δt of the signal $x(t)$. In a completely analogous way, the active bandwidth of the Fourier transform $\mathcal{X}(j\omega)$ is defined by the equation

$$(\Delta\omega)^2 = \int_{-\infty}^{+\infty}\omega^2|\mathcal{X}(j\omega)|^2d\omega \tag{3.37}$$

According to the bandwidth theorem, as it is formulated for signals of finite energy, if the signal $x(t)$ is characterized by the property $\lim_{t\to\pm\infty}[\sqrt{t}x(t)] = 0$, then it will hold that

$$(\Delta t)(\Delta\omega) \geq \sqrt{\pi/2} \tag{3.38}$$

with equality corresponding to signals that follow a Gaussian-type distribution, that is, they are described by the equation $x(t) = \sqrt{\alpha/\pi}\exp(-\alpha t^2)$.

Proof From the Schwarz inequality

$$\left| \int_{\alpha}^{\beta} f(t)\, g(t)\, dt \right|^2 \le \left(\int_{\alpha}^{\beta} |f(t)|^2 dt \right) \left(\int_{\alpha}^{\beta} |g(t)|^2 dt \right)$$

and for any two functions $f(t)$ and $g(t)$ such that $g(t) - xf(t) \neq 0$ we get

$$\int_{\alpha}^{\beta} \left[xf(t) - g(t) \right]^2 dt = x^2 \int_{\alpha}^{\beta} [f(t)]^2 dt - 2x \int_{\alpha}^{\beta} f(t)\, g(t) dt + \int_{\alpha}^{\beta} [g(t)]^2 dt \ge 0$$

If we use as integration limits the values $\alpha = -\infty$ and $\beta = +\infty$ and as functions $f(t)$ and $g(t)$ we use $f(t) = tx(t)$ and $g(t) = x'(t)$, the foregoing relation will take the form

$$\left| \int_{-\infty}^{+\infty} tx(t) \frac{dx(t)}{dt} dt \right|^2 \le \left(\int_{-\infty}^{+\infty} t^2 |x(t)|^2 dt \right) \left(\int_{-\infty}^{+\infty} \left| \frac{dx(t)}{dt} \right|^2 dt \right)$$

Integrating the left-hand side of the foregoing equation by parts and considering without loss of generality that the *total energy of the signal is equal to one*, or, in mathematical notation,

$$\int_{-\infty}^{+\infty} |x(t)|^2 dt = \frac{1}{2\pi} \int_{-\infty}^{+\infty} |\mathcal{X}(j\omega)|^2 \, d\omega = 1$$

this will take the form

$$\int_{-\infty}^{+\infty} tx(t) \frac{dx(t)}{dt} dt = t \frac{x^2(t)}{2} \bigg|_{-\infty}^{+\infty} - \frac{1}{2} \int_{-\infty}^{+\infty} |x(t)|^2 dt = -\frac{1}{2}$$

On the other hand, using the property $\mathcal{F}\{x'(t)\} = j\omega\mathcal{X}(j\omega)$ and the theorem of Parseval, we have

$$\int_{-\infty}^{+\infty} \left| \frac{dx(t)}{dt} \right|^2 dt = \frac{1}{2\pi} \int_{-\infty}^{+\infty} |\omega\mathcal{X}(j\omega)|^2 d\omega$$

Substituting the preceding relations into our final expression, it will take the form

$$\frac{1}{4} \le \left(\int_{-\infty}^{+\infty} t^2 |x(t)|^2 dt \right) \frac{1}{2\pi} \int_{-\infty}^{+\infty} \omega^2 |\mathcal{X}(j\omega)|^2 d\omega$$

from which it immediately follows that $(\Delta t)(\Delta\omega) \ge \sqrt{\pi/2}$, where we used the expressions of Δt and $\Delta\omega$, which we proved previously. ■

The preceding theorem can be generalized by defining the quantities Δt and $\Delta\omega$ from the relations

$$(\Delta t)^2 = \frac{\displaystyle\int_{-\infty}^{+\infty} (t - \bar{t})^2 |x(t)|^2 dt}{\displaystyle\int_{-\infty}^{+\infty} |x(t)|^2 dt} \quad \text{and} \quad (\Delta\omega)^2 = \frac{\displaystyle\int_{-\infty}^{+\infty} (\omega - \overline{\omega})^2 |\mathcal{X}(\omega)|^2 d\omega}{\displaystyle\int_{-\infty}^{+\infty} |\mathcal{X}(j\omega)|^2 d\omega}$$

where $\bar{t}$ and $\overline{\omega}$ are the *average values* of the parameters t and ω, which are defined as

$$\bar{t} = \frac{\displaystyle\int_{-\infty}^{+\infty} t|x(t)|^2 dt}{\displaystyle\int_{-\infty}^{+\infty} |x(t)|^2 dt} \quad \text{and} \quad \overline{\omega} = \frac{\displaystyle\int_{-\infty}^{+\infty} \omega|\mathcal{X}(j\omega)|^2 d\omega}{\displaystyle\int_{-\infty}^{+\infty} |\mathcal{X}(j\omega)|^2 d\omega}$$

If we consider that the expressions in the denominators of the foregoing relations are simply the energy of the signal, which is defined as

$$E = \int_{-\infty}^{+\infty} |x(t)|^2 dt = \frac{1}{2\pi} \int_{-\infty}^{+\infty} |\mathcal{X}(j\omega)|^2 d\omega$$

the foregoing relations take the simple form

$$\Delta t = \sqrt{\frac{1}{E} \int_{-\infty}^{+\infty} (t-\bar{t})^2 |x(t)|^2 dt} \quad \text{and} \quad \Delta\omega = \sqrt{\frac{1}{2\pi E} \int_{-\infty}^{+\infty} (\omega-\bar{\omega})^2 |\mathcal{X}(j\omega)|^2 d\omega}$$

Following a process similar to the one we applied previously and considering that the function $x(t)$ satisfies the property

$$\lim_{t\to\pm\infty} \Big[t|x(t)| \Big]^2 = 0 \quad \text{it is proven that} \quad (\Delta t)(\Delta\omega) \geq \frac{1}{2}$$

a relation that constitutes an *alternative formulation of the bandwidth theorem*. In full analogy with the previous case, the equality symbol in the last relation refers to signals that follow a normal distribution.

The preceding formulation of the bandwidth theorem refers to *continuous square-integrable signals of finite energy*, while for other categories of signals the theorem is formulated in a different way. Considering, for example, the case of *absolutely integrable signals*, it has been proven that they satisfy the property $(\Delta\omega)(\Delta t) \geq 2\pi$. To prove the truth of this claim, let us consider an absolutely integrable signal $x(t)$ with an absolutely integrable Fourier transform $\mathcal{X}(j\omega)$. For this special category of signals, the *active duration* Δt and the *active bandwidth* $\Delta\omega$ are defined through the equations

$$\int_{-\infty}^{+\infty} |x(t)| dt = |x(\bar{t})| dt \quad \text{and} \quad \int_{-\infty}^{+\infty} |\mathcal{X}(j\omega)| d\omega = |\mathcal{X}(\bar{\omega})| d\omega$$

where the quantities $\bar{t}$ and $\bar{\omega}$ are any two constant values of the variables t and ω for which $x(\bar{t}) \neq 0$ and $X(\bar{\omega}) \neq 0$. Following the methodology of the previous analysis, let us define the parameters $\bar{t}$ and $\bar{\omega}$ as the centroids of the functions $x(t)$ and $\mathcal{X}(j\omega)$, which are given by the relations

$$\bar{t} = \frac{\int_{-\infty}^{+\infty} t|x(t)| dt}{\int_{-\infty}^{+\infty} |x(t)| dt} \quad \text{and} \quad \bar{\omega} = \frac{\int_{-\infty}^{+\infty} \omega|\mathcal{X}(j\omega)| d\omega}{\int_{-\infty}^{+\infty} |\mathcal{X}(j\omega)| d\omega}$$

The proof of the bandwidth theorem is relatively simple and is based on the assumption that since the signals $x(t)$ and $\mathcal{X}(j\omega)$ are absolutely integrable, they will satisfy the relations

$$|x(\bar{t})| \leq \frac{1}{2\pi} \int_{-\infty}^{+\infty} |\mathcal{X}(j\omega)| d\omega = \frac{1}{2\pi} |\mathcal{X}(\bar{\omega})| \Delta\omega \quad \text{and} \quad |\mathcal{X}(j\omega)| \leq \int_{-\infty}^{+\infty} |x(t)| dt = |x(\bar{t})| t$$

Solving the preceding equations for Δt and $\Delta\omega$ and multiplying term by term, we get

$$(\Delta t)(\Delta\omega) \geq \frac{|\mathcal{X}(\bar{\omega})|}{|x(\bar{t})|} \frac{2\pi |x(\bar{t})|}{|\mathcal{X}(\bar{\omega})|} = 2\pi \tag{3.39}$$

which is the desired result. The foregoing inequality refers to the general case of values $\bar{t} \neq 0$ and $\bar{\omega} \neq 0$, while for the values $\bar{t} = \bar{\omega} = 0$ it is proven in the same way that the quantities Δt and $\Delta\omega$ satisfy the equation $(\Delta\omega)(\Delta t) = 2\pi$.

3.4 Inverse Fourier Transform

As was already mentioned in a previous section, the *inverse Fourier transform* of the function $\mathcal{X}(j\omega)$ is given by the equation

$$x(t) = \frac{1}{2\pi}\int_{-\infty}^{+\infty} \mathcal{X}(j\omega)e^{j\omega t}\,d\omega \tag{3.40}$$

Considering a given function $\mathcal{X}(j\omega)$ and using the foregoing equation along with the properties of the Fourier transform that we presented in the preceding pages, we can directly calculate the function $x(t)$, a process that is applied in the exercises that follow.

Problem 3.4.1 Determine the inverse Fourier transform of the function

$$\mathcal{X}(j\omega) = \frac{2\sin[3(\omega - 2\pi)]}{\omega - 2\pi}.$$

(Oppenheim, Willsky, Nawab, Exercise for solution 4.22α)

Solution Setting $\omega_0 = 2\pi$ and $\mathcal{Y}(j\omega) = 2\sin(3\omega)/\omega$, we observe that the function $\mathcal{X}(j\omega)$ has the form

$$\mathcal{X}(j\omega) = \mathcal{Y}(j\omega)|_{\omega\to\omega-\omega_0}$$

and according to the property of *frequency shift*, if $x(t)$ and $y(t)$ are the inverse transforms of the functions $\mathcal{X}(j\omega)$ and $\mathcal{Y}(j\omega)$, they will be related as $x(t) = e^{j\omega_0 t}y(t) = e^{j2\pi t}y(t)$. Observing the function $\mathcal{Y}(j\omega)$, we find that it has the form $\mathcal{Y}(j\omega) = 2\sin(\omega\alpha)/\alpha$ for the value $\alpha = 3$, which, as we have proven, constitutes the Fourier transform of the function

$$x(t) \equiv \Pi_\alpha(t) = \begin{cases} 1 \text{ for } |t| \le \alpha \\ 0 \text{ for } |t| > \alpha \end{cases}$$

Combining the preceding results, the inverse Fourier transform of the function $\mathcal{X}(j\omega)$ is

$$x(t) = \begin{cases} e^{j2\pi t} & \text{for } |t| \le 3 \\ 0 & \text{for } |t| > 3 \end{cases}$$

Problem 3.4.2 Find the inverse Fourier transform of the function $\mathcal{X}(j\omega) = \cos[4\omega + (\pi/3)]$ (Oppenheim, Willsky, Nawab, Exercise for solution 4.22β).

Solution Developing the function $\mathcal{X}(j\omega)$ using Euler's formula, we obtain

$$X(\omega) = \cos\left(4\omega + \frac{\pi}{3}\right) = \frac{e^{j[4\omega+(\pi/3)]} + e^{-j[4\omega+(\pi/3)]}}{2} = \frac{1}{2}e^{j4\omega}e^{j\pi/3} + \frac{1}{2}e^{-j4\omega}e^{-j\pi/3}$$

and from the defining equation of the inverse Fourier transform it immediately follows that

$$x(t) = \frac{1}{2\pi}\int_{-\infty}^{+\infty} \mathcal{X}(j\omega)e^{j\omega t}\,d\omega = \frac{e^{j\pi/3}}{4\pi}\int_{-\infty}^{+\infty} e^{j\omega(t+4)}dt + \frac{e^{-j\pi/3}}{4\pi}\int_{-\infty}^{+\infty} e^{j\omega(t-4)}dt = \frac{e^{j\pi/3}}{4\pi}\delta(t+4) + \frac{e^{-j\pi/3}}{4\pi}\delta(t-4)$$

where we used the definition of the impulse function

$$\delta(t) = \frac{1}{2\pi}\int_{-\infty}^{+\infty} e^{j\omega t}\, d\omega$$

Problem 3.4.3 Determine the inverse Fourier transform of the function

$$\mathcal{X}(j\omega) = 2[\delta(\omega-1) - \delta(\omega+1)] + 3[\delta(\omega-2\pi) + \delta(\omega+2\pi)].$$

(Oppenheim, Willsky, Nawab, Exercise for solution 4.22γ)

Solution To solve this specific problem, we do not need to use the defining equation of the inverse Fourier transform, since we know that the Fourier transform of the signal $x(t) = e^{j\alpha t}$ has the form $\mathcal{X}(j\omega) \equiv \mathcal{F}\{e^{j\alpha t}\} = 2\pi\delta(\omega-\alpha)$, and therefore

$$\mathcal{F}^{-1}\{\delta(\omega-\alpha)\} = \frac{1}{2\pi}e^{j\alpha t}$$

Thus, using the *principle of superposition* that also characterizes the inverse Fourier transform (since it is a linear transformation), we have

$$x(t) = \mathcal{F}^{-1}\{\mathcal{X}(j\omega)\} = 2\Big[\mathcal{F}^{-1}\{\delta(\omega-1)\} - \mathcal{F}^{-1}\{\delta(\omega+1)\}\Big] + 3\Big[\mathcal{F}^{-1}\{\delta(\omega-2\pi)\} - \mathcal{F}^{-1}\{\delta(\omega+2\pi)\}\Big] =$$

$$= \frac{3}{2\pi}(e^{j2\pi t} - e^{-j2\pi t}) + \frac{2}{2\pi}(e^{jt} - e^{-jt}) = \frac{6\cos(2\pi t)}{2\pi} + \frac{4j\sin(t)}{2\pi} = \frac{3}{\pi}\cos(2\pi t) + \frac{2j}{\pi}\sin(t)$$

Problem 3.4.4 Determine the inverse Fourier transform of the function

$$\mathcal{X}(j\omega) = \begin{cases} A \text{ for } |\omega| \le T \\ 0 \text{ for } |\omega| > T. \end{cases}$$

Solution From the defining equation of the inverse Fourier transform we have

$$x(t) = \frac{1}{2\pi}\int_{-\infty}^{+\infty} \mathcal{X}(j\omega)e^{i\omega t} d\omega = \frac{1}{2\pi}\int_{-T}^{+T} Ae^{i\omega t} d\omega = \frac{A}{2\pi}\left[\frac{e^{i\omega t}}{it}\right]_{-T}^{+T}$$

$$= \frac{A}{\pi t}\frac{e^{iTt} - e^{-iTt}}{2i} = \frac{A}{\pi t}\sin(Tt) = \frac{AT}{\pi}\frac{\sin(Tt)}{Tt} = \frac{AT}{\pi}\operatorname{sinc}(Tt)$$

as can be proven, which is the desired result.

Problem 3.4.5 Determine the inverse Fourier transform of the function

$$\mathcal{X}(j\omega) = 20\,\frac{\sin(5\omega)}{5\omega}e^{-3i\omega}$$

Solution Setting $\mathcal{Y}(j\omega) = 20\sin(5\omega)/5\omega$ and using the property of time shift, we find that the function $x(t)$ can be formulated in the form $x(t) = \mathcal{F}^{-1}\{\mathcal{Y}(j\omega)\}_{t\to t-3}$. Taking into account that the function

$$\mathcal{Y}(j\omega) = 2\,\frac{\sin(\omega\alpha)}{\omega} = 2\alpha\,\frac{\sin(\omega\alpha)}{\omega\alpha}$$

is the Fourier transform of the square signal $\Pi_\alpha(t)$, as we proved in a previous exercise, it is not difficult to realize that the inverse Fourier transform of the function

$$\mathcal{Y}(j\omega) = 20\frac{\sin(5\omega)}{5\omega} = 2\left[10\frac{\sin(5\omega)}{\omega}\right] = 2\left[2\cdot 5\frac{\sin(5\omega)}{\omega}\right]$$

will be the function

$$y(t) \equiv 2\Pi_5(t) = \begin{cases} 2 \text{ for } |t| \le 5 \\ 0 \text{ for } |t| > 5 \end{cases}$$

Combining the foregoing results, it turns out that the inverse Fourier transform of the function $\mathcal{X}(j\omega)$ is the continuous-time signal

$$x(t) \equiv 2\Pi_5(t-3) = \begin{cases} 2 \text{ for } |t-3| \le 5 \\ 0 \text{ for } |t-3| > 5 \end{cases}$$

Problem 3.4.6 Determine the inverse Fourier transform of the function

$$\mathcal{X}(j\omega) = \frac{\cos\omega}{\omega^2 + 8\omega + 20}.$$

Solution Using the Euler formula and noticing that the expression of the denominator is formulated as $\omega^2 + 8\omega + 20 = (\omega+4)^2 + 4$, the preceding function can be written in the form

$$\mathcal{X}(j\omega) = \frac{e^{j\omega} + e^{-j\omega}}{2[(\omega+4)^2+4]} = \frac{e^{j\omega}}{2}\left\{\frac{1}{(\omega+4)^2+4}\right\} + \frac{e^{-j\omega}}{2}\left\{\frac{1}{(\omega+4)^2+4}\right\}$$

In the first step of the process, let us calculate the inverse Fourier transform of the first fraction of the preceding equation. Defining the function

$$\mathcal{Y}(j\omega) = \frac{1}{\omega^2+4} = \frac{1}{4}\left\{\frac{4}{\omega^2+4}\right\} = \frac{1}{4}\left\{\frac{2\cdot 2}{\omega^2+2^2}\right\}$$

we find that this results from the general equation

$$\mathcal{Y}(j\omega) = \frac{2\alpha}{\omega^2+\alpha^2} = \mathcal{F}^{-1}\left\{e^{-\alpha|t|}\right\}$$

for parameter value $\alpha = 2$ (this result comes from the direct application of the defining equation of the direct transform and the definition of the absolute value function; its derivation is left as an exercise to the reader). Based on the last equation, the inverse transform of the function $\mathcal{Y}(j\omega)$ is calculated as

$$y(t) = \frac{1}{4}\mathcal{F}^{-1}\{\mathcal{Y}(j\omega)\} = \frac{1}{4}\mathcal{F}^{-1}\left\{\frac{2\alpha}{\omega^2+\alpha^2}\right\}_{\alpha=2} = \frac{1}{4}e^{-2|t|}$$

Now, defining the function

$$\mathcal{Z}(j\omega) = \frac{1}{2}\left\{\frac{1}{(\omega+4)^2+4}\right\} = \frac{1}{2}\mathcal{Y}(\omega+4)$$

and using the property of *frequency shift* $z(t) = e^{j\omega_0 t}y(t) \Leftrightarrow \mathcal{Z}(j\omega) = \mathcal{Y}(\omega - \omega_0)$ for value $\omega_0 = -4$, it immediately follows that

$$z(t) = \frac{1}{2}e^{-j4t}y(t) = \frac{1}{8}e^{-j4t}e^{-2|t|}$$

In the last step of the process, we will use the property of the *time shift* of the transform

$$x_1(t) = z(t - t_0) \Leftrightarrow \mathcal{X}_1(\omega) = e^{-j\omega t_0}\mathcal{Z}(j\omega)$$

for the value $t_0 = 1$, and we obtain the result

$$x_1(t) = \mathcal{F}^{-1}\left\{\frac{e^{j\omega}}{2}\left[\frac{1}{(\omega+4)^2+4}\right]\right\} = \frac{1}{8}e^{-4j(t+1)}e^{-2|t+1|}$$

Using the same process, it is proven that the inverse Fourier transform of the second fraction of the function $\mathcal{X}(j\omega)$ will have the form

$$x_2(t) = \mathcal{F}^{-1}\left\{\frac{e^{-j\omega}}{2}\left[\frac{1}{(\omega+4)^2+4}\right]\right\} = \frac{1}{8}e^{-4j(t-1)}e^{-2|t-1|}$$

Therefore, adding these two expressions together, we arrive at the final result:

$$x(t) = \mathcal{F}^{-1}\{X(\omega)\} = \frac{1}{8}\left(e^{-4j(t+1)}e^{-2|t+1|} + e^{-4j(t-1)}e^{-2|t-1|}\right)$$

Problem 3.4.7 Determine the inverse Fourier transform of the function

$$\mathcal{X}(j\omega) = \frac{2}{1+2i(\omega-1)}$$

Solution Setting $\mathcal{Y}(j\omega) = 2/(1+2i\omega)$ and using the property of *frequency shift* for $\omega_0 = 1$, we find that the inverse Fourier transform of the function $\mathcal{X}(j\omega)$ will have the form $x(t) = e^{it}\mathcal{F}^{-1}\{\mathcal{Y}(j\omega)\}$. However,

$$\mathcal{Y}(j\omega) = \frac{2}{1+2i\omega} = \frac{1}{(1/2)+i\omega}$$

and using the result $\mathcal{F}\{e^{-\alpha t}u(t)\} = (\alpha + i\omega)^{-1}$ for the parameter value $\alpha = 1/2$, it follows that $y(t) = e^{-(1/2)t}u(t)$ or, equivalently,

$$x(t) = e^{it}y(t) = e^{it}e^{-(1/2)t}u(t)$$

Problem 3.4.8 Determine the inverse Fourier transform of the function

$$\mathcal{X}(j\omega) = \frac{\omega^2+21}{\omega^2+9}$$

Solution Although the solution to the exercise can be carried out using the method of partial fraction expansion, the signal $x(t)$ can be found very easily by writing the preceding function in the form

$$\mathcal{X}(j\omega) = \frac{\omega^2+21}{\omega^2+9} = 1 + \frac{12}{\omega^2+9}$$

hence, from the inverse transform relation we will have

$$x(t) = \mathcal{F}^{-1}\{\mathcal{X}(j\omega)\} = \mathcal{F}^{-1}\{1\} + 2\mathcal{F}^{-1}\left\{\frac{6}{\omega^2+9}\right\}$$

However, $\mathcal{F}^{-1}\{1\} = \delta(\omega)$, while the expression of the second term is written

$$\mathcal{Y}(j\omega) = \frac{2\cdot 3}{\omega^2+3^2} = \left.\frac{2\alpha}{\omega^2+\alpha^2}\right|_{\alpha=3} = \left.\mathcal{F}\{e^{-\alpha|t|}\}\right|_{\alpha=3}$$

as we proved in an earlier exercise. Therefore, $y(t) = e^{-3|t|}$, and finally $x(t) = \delta(t) + 2e^{-3|t|}$, which is the desired result.

Problem 3.4.9 Determine the inverse Fourier transform of the function

$$\mathcal{X}(j\omega) = \frac{4+10j\omega}{-\omega^2+6j\omega+8}.$$

Solution To solve the problem, we factorize the polynomial of the denominator as

$$-\omega^2+6j\omega+8 = (2+j\omega)(4+j\omega)$$

Using the method of partial fraction expansion, we obtain

$$\frac{4+10j\omega}{-\omega^2+6j\omega+8} = \frac{A}{2+j\omega} + \frac{B}{4+j\omega} = \frac{(A+B)j\omega+(4A+2B)}{(2+j\omega)(4+j\omega)}$$

Equating the coefficients of the corresponding powers of ω in the first and last fractions, we obtain the system of equations $A+B=10$ and $4A+2B=4$ with solution $(A, B) = (-8, 18)$. Thus,

$$\mathcal{X}(j\omega) = -\frac{8}{2+j\omega} + \frac{18}{4+j\omega}$$

and finally

$$x(t) = \mathcal{F}^{-1}\{\mathcal{X}(j\omega)\} = -8\mathcal{F}^{-1}\left\{\frac{1}{2+j\omega}\right\} + 18\mathcal{F}^{-1}\left\{\frac{1}{4+j\omega}\right\} = (18e^{-4t} - 8e^{-2t})u(t)$$

as easily emerges, referring to tables of elementary transforms.

3.4.1 The Method of Complex Integration

An alternative method of calculating the integral that appears in the defining equation of the inverse Fourier transform is based on the use of *residue calculus* and the method of *Cauchy's complex integration*.[11] According to this method, if the Fourier transform is a rational function of frequency ω and has N *poles*[12] at positions $\omega_1, \omega_2, \ldots, \omega_N$ on the complex plane, then it is proven that the integral of the function $\mathcal{X}(j\omega)\, e^{j\omega t}$ along a closed path C that contains the poles of the preceding function is given by the relation

[11] See the book by G.Arfken: Mathematical Methods for Physicists, Academic Press, ISBN 0-12-059820-5, 1985.

[12] As we will see in the section on the *transfer function*, the poles of a rational complex function are those values of its argument that zero the denominator of the function and therefore lead to its infinity.

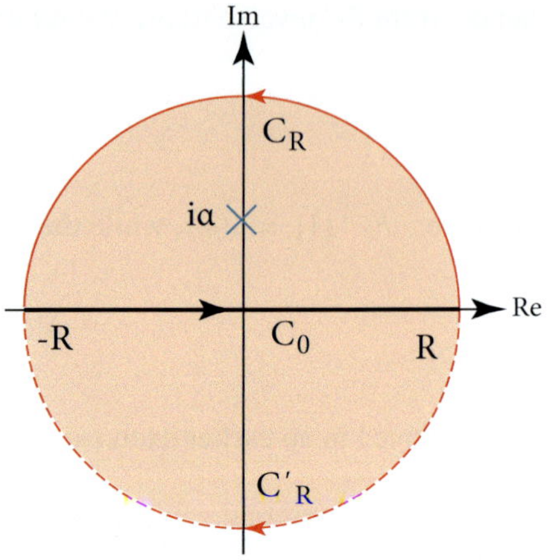

Fig. 3.13 The closed path C of Exercise 3.4.10

$$\oint_C \mathcal{X}(j\omega)\, e^{j\omega t} d\omega = 2\pi j \sum_{i=1}^{N} \text{Res}\left[\mathcal{X}(j\omega)e^{j\omega t}\right]_{\omega=\omega_i} \tag{3.41}$$

In the most common cases, the closed path of the contour integral on the left-hand side consists of a *semicircular path of radius* R and a *section of the horizontal axis defined by the values of the interval* $[-R, +R]$. If we calculate the integral at the limit $R \to \infty$, the integral along the semicircle takes a value equal to zero, as predicted by the *Jordan lemma*,[13] while the remaining integral is the one that will ultimately yield the desired function $x(t)$. For more information on the process of complex integration, the reader is referred to an introductory book on complex analysis. The following exercise demonstrates the application of this method, which is also used for the calculation of other inverse transforms, such as the inverse transforms of Laplace and $\mathcal{Z}$ that are studied in subsequent chapters.

Problem 3.4.10 Find the inverse Fourier transform of the function $\mathcal{X}(j\omega) = 1/(\alpha + j\omega)$.

Solution The function $\mathcal{X}(j\omega)$ is a rational function and has a simple pole at the position $\omega = j\alpha$, as can be verified. If we calculate the contour integral of the function $\mathcal{X}(j\omega)e^{j\omega t}$ along the path C depicted in Fig. 3.13 that consists of the upper semicircle C_R and the straight line C_0 and contains the pole $\omega = j\alpha$, the application of the residue theorem will give us

$$\oint_C \mathcal{X}(j\omega)\, e^{j\omega t} d\omega = 2\pi j \;\text{Res}\left[\mathcal{X}(j\omega)e^{j\omega t}\right]_{\omega=j\alpha} =$$

$$= 2\pi j \lim_{\omega\to j\alpha}\left[(\omega - j\alpha)\frac{e^{j\omega t}}{i\,(\omega - j\alpha)}\right] = 2\pi \lim_{\omega=j\alpha} e^{j\omega t} = 2\pi e^{-\alpha t}$$

On the other hand, the integral along the path C', which consists of the lower semicircle C'_R and the straight line C_0, is equal to zero because there are no poles of the function $\mathcal{X}(j\omega)e^{j\omega t}$ enclosed by it.

[13] According to Jordan's lemma, the value of the integral

$$I = \int_{-\infty}^{+\infty} f(t)\, e^{j\alpha t}\, dt$$

along a semicircle C_R of radius $R \to \infty$ located in the upper complex half-plane is equal to zero if the function $f(t)$ satisfies the property $\lim_{R\to\infty} |f(Re^{i\vartheta})| = 0$. The foregoing relation holds for positive values of the parameter α. As for the negative values of α, Jordan's lemma applies provided the semicircle C_R is located in the lower complex half-plane.

For values $t > 0$ the integral is calculated along the semicircular path C_R, and according to *Jordan's lemma*, we will have

$$\lim_{R\to\infty} \int_{C_R} \mathcal{X}(j\omega)\, e^{j\omega t} d\omega = 0$$

since the function $\mathcal{X}(j\omega)$ tends to zero at the limit $\omega \to \infty$. Therefore, we will have

$$\begin{aligned} x(t) &= \frac{1}{2\pi} \int_{-\infty}^{+\infty} \mathcal{X}(j\omega)\, e^{j\omega t} d\omega = \frac{1}{2\pi} \lim_{R\to\infty} \int_{C_0} \mathcal{X}(j\omega)\, e^{j\omega t} d\omega \\ &= \frac{1}{2\pi} \lim_{R\to\infty} \left(\int_{C} \mathcal{X}(j\omega)\, e^{j\omega t} d\omega - \int_{C_R} \mathcal{X}(j\omega)\, e^{j\omega t} d\omega \right) = e^{-\alpha t} \end{aligned}$$

On the other hand, for $t < 0$ the integral will be calculated along the path C' and, as mentioned previously, its value will be equal to *zero* because the function $\mathcal{X}(j\omega)e^{j\omega t}$ *does not have poles* within this path. Following the analysis we applied previously, we will get

$$x(t) = \frac{1}{2\pi} \int_{-\infty}^{+\infty} \mathcal{X}(j\omega) e^{j\omega t} d\omega = \frac{1}{2\pi} \lim_{R\to\infty} \int_{C_0} \mathcal{X}(j\omega) e^{j\omega t} d\omega = \frac{1}{2\pi} \lim_{R\to\infty} \left(\int_{C'} \mathcal{X}(j\omega) e^{j\omega t} d\omega - \int_{C'_R} \mathcal{X}(j\omega) e^{j\omega t} d\omega \right) = 0$$

Combining the foregoing results, we will obtain[14]

$$x(t) = \frac{1}{2\pi} \int_{-\infty}^{+\infty} \mathcal{X}(j\omega)\, e^{j\omega t} d\omega = \begin{cases} e^{-\alpha t} & \text{for } t > 0 \\ 0 & \text{for } t < 0 \end{cases}$$

3.4.2 The Method of Partial Fraction Expansion

The theoretical background required for the application of the *method of partial fraction expansion* can be found in any mathematics book, while extensive examples of this method's use are also included in the corresponding sections of the chapters dealing with the study of Laplace and $\mathcal{Z}$ transforms, whose inverse transforms are calculated in exactly the same way. This method is based on the property of the *linearity* of the *direct* and *inverse* Fourier transforms according to which the Fourier transform of the continuous-time signal

$$x(t) = \sum_{i=1}^{N} \alpha_i x_i(t) = \alpha_1 x_1(t) + \alpha_2 x_2(t) + \cdots + \alpha_N x_N(t) \tag{3.42}$$

is the function

$$\mathcal{X}(j\omega) = \sum_{i=1}^{N} \alpha_i \mathcal{X}_i(\omega) = \alpha_1 \mathcal{X}_1(\omega) + \alpha_2 \mathcal{X}_2(\omega) + \cdots + \alpha_N \mathcal{X}_N(\omega) \tag{3.43}$$

where $\mathcal{X}_i(\omega) = \mathcal{F}\{x_i(t)\}$ $(i = 1, 2, \ldots, N)$. Reversing the formulation of the foregoing proposition, we can formulate the claim that if the function $\mathcal{X}(j\omega)$ whose inverse transform is sought can be formulated in the form of a linear combination of simpler Fourier transforms $\mathcal{X}_i(\omega)$, then the continuous-time signal $x(t)$ will result from the equation

[14] For more exercises on calculating inverse transforms, via the application of the method of complex integration, the reader is referred to the chapters dealing with the presentation of the Laplace and $\mathcal{Z}$ transforms, in which this method is extensively used for the calculation of the inverse of those transforms.

Table 3.3 Examples of elementary Fourier transforms

Signal $x(t)$	Fourier transform $X(j\omega)$
$\sum_{n=-\infty}^{+\infty} \alpha_n e^{jn\omega_0 t}$	$2\pi \sum_{n=-\infty}^{+\infty} \alpha_n \delta(\omega - n\omega_0)$
$e^{j\omega_0 t}$	$2\pi\delta(\omega - \omega_0)$
$\cos(\omega_0 t)$	$\pi[\delta(\omega - \omega_0) + \delta(\omega + \omega_0)]$
$\sin(\omega_0 t)$	$j\pi[\delta(\omega + \omega_0) - \delta(\omega - \omega_0)]$
$x(t) = 1$	$2\pi\delta(\omega)$
$x(t) = x(t+T) = \begin{cases} 1 & \lvert t\rvert < \alpha \\ 0 & \alpha < \lvert t\rvert < \frac{T}{2} \end{cases}$	$\sum_{n=-\infty}^{+\infty} \frac{2\sin n\omega_0\alpha}{n} \delta(\omega - n\omega_0)$
$\sum_{n=-\infty}^{+\infty} \delta(t - nT)$	$\frac{2\pi}{T} \sum_{k=-\infty}^{+\infty} \delta\left(\omega - \frac{2\pi k}{T}\right)$
$\begin{cases} 1 & \lvert t\rvert < \alpha \\ 0 & \lvert t\rvert > \alpha \end{cases}$	$\frac{2\sin(\omega\alpha)}{\omega}$
$\frac{\sin(Wt)}{\pi t}$	$\begin{cases} 1 & \lvert\omega\rvert < W \\ 0 & \lvert\omega\rvert > W \end{cases}$
$\delta(t)$	1
$u(t)$	$\frac{1}{j\omega} + \pi\delta(\omega)$
$\delta(t - t_0)$	$e^{-j\omega t_0}$
$e^{-\alpha t}u(t),\ \Re(a) > 0$	$\frac{1}{\alpha + j\omega}$
$te^{-\alpha t}u(t),\ \Re(\alpha) > 0$	$\frac{1}{(\alpha + j\omega)^2}$
$\frac{t^{n-1}}{(n-1)!}e^{-\alpha t}u(t),\ \Re(\alpha) > 0$	$\frac{1}{(\alpha + j\omega)^n}$

$$x(t) = \alpha_1 \mathcal{F}^{-1}\{\mathcal{X}_1(\omega)\} + \alpha_2 \mathcal{F}^{-1}\{\mathcal{X}_2(\omega)\} + \cdots + \alpha_N \mathcal{F}^{-1}\{\mathcal{X}_N(\omega)\} \tag{3.44}$$

where the elementary inverse transforms $\mathcal{F}^{-1}\{\mathcal{X}_i(\omega)\}$ $(i = 1, 2, \ldots, N)$ can be found from tables of elementary transforms, such as, for example, Table 3.3. The process of calculating the inverse Fourier transform by applying the method of partial fraction expansion is presented in detail in the example solved exercises that follow.

Problem 3.4.11 Determine the inverse Fourier transform of the function

$$\mathcal{X}(j\omega) = \frac{\sin(2\omega)}{\omega^2 + 8\omega + 20}.$$

Solution Using Euler's formula for the sine function and expanding the expression of the denominator as a product of the form $\omega^2 + 8\omega + 20 = (\omega - 2)(\omega - 10)$, the function $\mathcal{X}(j\omega)$ is written

$$\mathcal{X}(j\omega) = \frac{e^{2j\omega} - e^{-2j\omega}}{2j(\omega - 10)(\omega - 2)} = \frac{e^{2j\omega}}{2j(\omega - 10)(\omega - 2)} - \frac{e^{-2j\omega}}{2j(\omega - 10)(\omega - 2)}$$

and then, applying the method of partial fraction expansion, , we obtain

$$\frac{1}{(\omega-10)(\omega-2)}=\frac{A}{\omega-10}+\frac{B}{\omega-2}=\frac{(A+B)\omega-(2A+10B)}{(\omega-10)(\omega-2)}$$

If we equate the coefficients of the corresponding powers of ω in the first and last fractions, we get the system of equations $A+B=0$ and $2A+10B=-1$ with the solution $(A,B)=(1/8,-1/8)$, as can be proved, which allows us to define the function

$$\mathcal{Y}(j\omega)=\frac{1}{2j(\omega-10)(\omega-2)}=\frac{1}{32}\left[\frac{2}{j(\omega-10)}-\frac{2}{j(\omega-2)}\right]$$

In the next step of the process, referring to tables of elementary transforms (e.g., Table 3.3; refer to page 266) we find that $\mathcal{F}\{sgn(t)\}=2/j\omega$, where $sgn(t)$ is the *sign function*—this property was proved in a previous section. Thus, using the property of the *frequency shift* $z(t)=e^{j\omega_0 t}y(t)\Leftrightarrow\mathcal{Z}(j\omega)=\mathcal{Y}(\omega-\omega_0)$ for the values $\omega_0=10$ and $\omega_0=2$, we obtain

$$z_1(t)=\mathcal{F}^{-1}\left\{\frac{2}{j(\omega-10)}\right\}=e^{10jt}\mathcal{F}^{-1}\left\{\frac{2}{j\omega}\right\}=e^{10jt}\mathrm{sgn}(t)\qquad\text{and}$$

$$z_2(t)=\mathcal{F}^{-1}\left\{\frac{2}{j(\omega-2)}\right\}=e^{2jt}\mathcal{F}^{-1}\left\{\frac{2}{j\omega}\right\}=e^{2jt}\mathrm{sgn}(t)$$

respectively, from which follows

$$z(t)=z_1(t)+z_2(t)=\frac{1}{32}\left(e^{10jt}-e^{2jt}\right)\mathrm{sgn}(t)$$

In the last step of the process, using the property of *time shift*, formulated as $x_1(t)=z(t-t_0)\Leftrightarrow\mathcal{X}_1(\omega)=e^{-j\omega t_0}\mathcal{Z}(j\omega)$, we obtain

$$x_1(t)=\mathcal{F}^{-1}\left\{\frac{e^{2j\omega}}{2j(\omega-10)(\omega-2)}\right\}=\frac{1}{32}\left(e^{10jt}-e^{2jt}\right)\mathrm{sgn}(t)\Big|_{t\to t+2}=\frac{1}{32}\left(e^{10j(t+2)}-e^{2j(t+2)}\right)\mathrm{sgn}(t+2)$$

$$x_2(t)=\mathcal{F}^{-1}\left\{\frac{e^{-2j\omega}}{2j(\omega-10)(\omega-2)}\right\}=\frac{1}{32}\left(e^{10jt}-e^{2jt}\right)\mathrm{sgn}(t)\Big|_{t\to t-2}=\frac{1}{32}\left(e^{10j(t-2)}-e^{2j(t-2)}\right)\mathrm{sgn}(t-2)$$

and adding the foregoing expressions, we arrive at the desired result:

$$x(t)=\frac{1}{32}\left[\left(e^{10j(t+2)}-e^{2j(t+2)}\right)\mathrm{sgn}(t+2)-\left(e^{10j(t-2)}-e^{2j(t-2)}\right)\mathrm{sgn}(t-2)\right]$$

Problem 3.4.12 Determine the inverse Fourier transform of the function

$$\mathcal{X}(j\omega)=\frac{1}{(1+2j\omega)(1+j\omega)}.$$

Solution To solve the problem, develop the function $\mathcal{X}(j\omega)$ in a series of partial fractions, writing it in the form

$$\frac{1}{(1+2j\omega)(1+j\omega)}=\frac{A}{1+2j\omega}+\frac{B}{1+j\omega}=\frac{(A+2B)j\omega+(A+B)}{(1+2j\omega)(1+i\omega)}$$

Equating the coefficients of the corresponding powers of ω in the first and last fractions, we obtain the system of equations $(A+2B)j=0$ and $A+B=1$ with the solution $(A,B)=(2,-1)$, as can be easily proven. Therefore,

$$\mathcal{X}(j\omega)=\frac{2}{1+2j\omega}-\frac{1}{1+j\omega}=\frac{1}{(1/2)+j\omega}-\frac{1}{1+j\omega}$$

and therefore, the continuous-time signal $x(t)$ is calculated as

$$x(t) = \mathcal{F}^{-1}\{\mathcal{X}(j\omega)\} = \mathcal{F}^{-1}\left\{\frac{1}{(1/2) + j\omega}\right\} - \mathcal{F}^{-1}\left\{\frac{1}{1 + j\omega}\right\}$$

Referring to tables of elementary transforms, we find that

$$\mathcal{F}^{-1}\left\{\frac{1}{(1/2) + j\omega}\right\} = e^{-t/2}u(t) \quad \text{and} \quad \mathcal{F}^{-1}\left\{\frac{1}{1 + j\omega}\right\} = e^{-t}u(t)$$

from which it finally follows that $x(t) = (e^{-t/2} - e^{-t})u(t)$, which is the desired result.

Problem 3.4.13 Using the property of convolution and the techniques for calculating the inverse Fourier transform, determine the convolution between the discrete-time signals

$$x(t) = te^{-2t}u(t) \text{ and } h(t) = e^{-4t}u(t)$$

$$x(t) = te^{-2t}u(t) \text{ and } h(t) = te^{-4t}u(t)$$

$$x(t) = e^{-t}u(t) \text{ and } h(t) = e^{t}u(-t)$$

(Oppenheim, Willsky, Nawab, Exercise for solution 4.26).

Solution According to the *convolution theorem* of the Fourier transform, if the signals $y(t)$, $x(t)$, and $h(t)$ are related through an equation of the form $y(t) = h(t) * x(t)$, where the symbol $*$ refers to the operation of convolution, and then the corresponding Fourier transforms will satisfy the relation $\mathcal{Y}(j\omega) = \mathcal{H}(j\omega)\mathcal{X}(j\omega)$. Therefore, an alternative way of calculating the convolution $y(t)$ apart from the direct use of the defining equation is through the relation $y(t) = \mathcal{F}^{-1}\{\mathcal{Y}(j\omega)\} = \mathcal{F}^{-1}\{\mathcal{H}(j\omega)\mathcal{X}(j\omega)\}$, which obviously requires the prior calculation of the Fourier transforms of the functions $h(t)$ and $x(t)$.

Referring to tables of elementary transforms, we find that

$$\mathcal{L}\{e^{-\alpha t}u(t)\}_{\Re(\alpha)>0} = \frac{1}{\alpha + j\omega} \quad \text{and} \quad \mathcal{L}\{te^{-\alpha t}u(t)\}_{\Re(\alpha)>0} = \frac{1}{(\alpha + j\omega)^2}$$

from which it follows that

$$\mathcal{F}\{te^{-2t}u(t)\} = \frac{1}{(2 + j\omega)^2} \quad \mathcal{F}\{e^{-4t}u(t)\} = \frac{1}{4 + j\omega} \quad \mathcal{F}\{te^{-4t}u(t)\} = \frac{1}{(4 + j\omega)^2} \quad \mathcal{F}\{e^{-t}u(t)\} = \frac{1}{1 + j\omega}$$

while from the last relation and the property of *time reversal* $x(t) \leftrightarrow \mathcal{X}(j\omega) \Leftrightarrow x(-t) \leftrightarrow \mathcal{X}(-\omega)$ we will have $\mathcal{F}\{e^{t}u(-t)\} = 1/(1 - j\omega)$. Using the foregoing relations, the convolution $y(t)$ for the three given cases is calculated as follows:

(a) The Fourier transforms of the signals $x(t)$ and $h(t)$ are calculated as

$$\mathcal{X}(j\omega) = \frac{1}{(2 + j\omega)^2} \quad \text{and} \quad \mathcal{H}(j\omega) = \frac{1}{4 + j\omega}$$

and from the convolution theorem we obtain the equation

$$\mathcal{Y}(j\omega) = \mathcal{X}(j\omega)\mathcal{H}(j\omega) = \frac{1}{(2 + j\omega)^2(4 + j\omega)}$$

Applying the method of partial fraction expansion, it turns out that

$$\mathcal{Y}(j\omega) = \frac{1}{(2+j\omega)^2(4+j\omega)} = \frac{A}{4+j\omega} + \frac{B}{2+j\omega} + \frac{C}{(2+j\omega)^2}$$
$$= \frac{-(A+B)\omega^2 + (4A+6B+C)j\omega + (4A+8B+4C)}{(2+j\omega)^2(4+j\omega)}$$

as easily follows from simple operations. Equating the coefficients of the corresponding powers of ω in the first and last fractions, we obtain the system of equations

$$A + B = 0$$
$$4A + 6B + C = 0$$
$$4A + 8B + 4C = 1$$

with the solution $(A, B, C) = (1/4, -1/4, 1/2)$. Therefore,

$$\mathcal{Y}(j\omega) = \frac{1}{4}\left\{\frac{1}{4+j\omega}\right\} - \frac{1}{4}\left\{\frac{1}{2+j\omega}\right\} + \frac{1}{2}\left\{\frac{1}{(2+j\omega)^2}\right\}$$

and referring to tables of inverse transforms, we arrive at the desired result:

$$y(t) = \frac{1}{4}\mathcal{F}^{-1}\left\{\frac{1}{4+j\omega}\right\} - \frac{1}{4}\mathcal{F}^{-1}\left\{\frac{1}{2+j\omega}\right\} + \frac{1}{2}\mathcal{F}^{-1}\left\{\frac{1}{(2+j\omega)^2}\right\} = \left(\frac{1}{4}e^{-4t} - \frac{1}{4}e^{-2t} + \frac{1}{2}te^{-2t}\right)u(t)$$

(b) The Fourier transforms of the signals $x(t)$ and $h(t)$ are calculated as

$$\mathcal{X}(j\omega) = \frac{1}{(2+j\omega)^2} \quad \text{and} \quad \mathcal{H}(j\omega) = \frac{1}{(4+j\omega)^2}$$

and from the convolution theorem we get the equation

$$\mathcal{Y}(j\omega) = \mathcal{X}(j\omega)\mathcal{H}(j\omega) = \frac{1}{(2+j\omega)^2(4+j\omega)^2}$$

Applying the method of partial fraction expansion, we get

$$\mathcal{Y}(j\omega) = \frac{1}{(2+j\omega)^2(4+j\omega)^2} = \frac{A}{2+j\omega} + \frac{B}{(2+j\omega)^2} + \frac{C}{4+j\omega} + \frac{D}{(4+j\omega)^2}$$
$$\frac{-j\omega^3(A+C) - (10A+B+8C+D)\omega^2}{(2+j\omega)^2(4+j\omega)^2} + \frac{(32A+8B+20C+4D)i\omega + (32A+16B+16C+4D)}{(2+j\omega)^2(4+j\omega)^2}$$

as emerges after several operations, and equating the coefficients of the corresponding powers of ω in the first and last fractions, we get the following system of four equations:

$$A + C = 0$$
$$10A + B + 8C + D = 0$$
$$32A + 8B + 20C + 4D = 0$$
$$32A + 16B + 16C + 4D = 1$$

The solution of the system has the form $(A, B, C, D) = (-1/4, 1/4, 1/4, 1/4)$, and therefore the function $\mathcal{Y}(j\omega)$ is written

$$\mathcal{Y}(j\omega) = -\frac{1}{4}\left\{\frac{1}{2+j\omega}\right\} + \frac{1}{4}\left\{\frac{1}{(2+j\omega)^2}\right\} + \frac{1}{4}\left\{\frac{1}{4+j\omega}\right\} + \frac{1}{4}\left\{\frac{1}{(4+j\omega)^2}\right\}$$

Taking the inverse transform of the foregoing relation and referring to tables of elementary transforms, it turns out that

$$\begin{aligned} x(t) &= -\frac{1}{4}\mathcal{F}^{-1}\left\{\frac{1}{2+j\omega}\right\} + \frac{1}{4}\mathcal{F}^{-1}\left\{\frac{1}{(2+j\omega)^2}\right\} + \frac{1}{4}\mathcal{F}^{-1}\left\{\frac{1}{4+j\omega}\right\} + \frac{1}{4}\mathcal{F}^{-1}\left\{\frac{1}{(4+j\omega)^2}\right\} \\ &= \frac{1}{4}\left(te^{-2t} - e^{-2t} + te^{-4t} + e^{-4t}\right) u(t) \end{aligned}$$

(c) The Fourier transforms of the signals $x(t)$ and $h(t)$ are calculated as

$$\mathcal{X}(j\omega) = \frac{1}{1+j\omega} \quad \text{and} \quad \mathcal{H}(j\omega) = \frac{1}{1-j\omega}$$

and from the convolution theorem we get the equation

$$\mathcal{Y}(j\omega) = \mathcal{X}(j\omega)\mathcal{H}(j\omega) = \left\{\frac{1}{1+j\omega}\right\}\left\{\frac{1}{1-j\omega}\right\} = \frac{1}{2}\frac{2\cdot 1}{\omega^2+1^2}$$

Therefore, the function $y(t)$ will have the form $y(t) = (1/2)e^{-|t|}$, since the function $2\mathcal{Y}(j\omega)$ results from the Fourier transform of the function $x(t) = e^{-\alpha|t|}$ for a parameter value $\alpha = 1$.

3.5 Description of Systems in Frequency Domain

Following the analytical description of the theoretical background related to the Fourier series and the Fourier transform, let us move on to the presentation of some study techniques and representation of signals and systems in the frequency domain. These techniques are mainly based on the *convolution theorem* and the *differentiation property* of the Fourier transform, which are formulated as

$$y(t) = h(t) * x(t) \Longleftrightarrow \mathcal{Y}(j\omega) = \mathcal{H}(j\omega)\mathcal{X}(j\omega) \quad \text{and} \tag{3.45}$$

$$\mathcal{F}\{x^{(n)}(t)\} = \mathcal{F}\left\{\frac{d^n x(t)}{dt^n}\right\} = (j\omega)^n \mathcal{X}(j\omega) \tag{3.46}$$

and they are used in the following way:

- If we know the impulse response $h(t)$ of a LTI system, then the output $y(t)$ of the system for a given input signal $x(t)$ will be calculated from the *convolution* of the signals $h(t)$ and $y(t)$, or, in mathematical notation, $y(t) = h(t) * x(t)$. Alternatively, we can use the *convolution theorem* to calculate the Fourier transform of the output signal as the product $\mathcal{Y}(j\omega) = \mathcal{H}(j\omega)\mathcal{X}(j\omega)$ and then the system output as the inverse transform $y(t) = \mathcal{F}^{-1}\{\mathcal{Y}\}(\omega)$.
- If for a LTI system we know the output $y(t)$ that corresponds to a given input signal $x(t)$, then, by calculating the *Fourier transforms* of the signals $\mathcal{X}(j\omega)$ and $\mathcal{Y}(j\omega)$, we can determine the *frequency response* of the system from the relation $\mathcal{H}(j\omega) = \mathcal{Y}(j\omega)/\mathcal{X}(j\omega)$, which obviously constitutes a different formulation of the convolution theorem. Taking into account that the *frequency response of the system constitutes the Fourier transform of the impulse response* $h(t)$, we can calculate the latter signal from the *inverse transform* $h(t) = \mathcal{F}^{-1}\{\mathcal{H}(j\omega)\}$.
- If we know the *frequency response* of the system, we can, by formulating it in the known form $\mathcal{H}(j\omega) = \mathcal{Y}(j\omega)/\mathcal{X}(j\omega)$, determine the correlation of the Fourier transforms of the signals $x(t)$ and $y(t)$ and by calculating the inverse Fourier transform of the expression that will emerge, we can construct the *differential equation* that describes the system.
- If we know the *differential equation* of the system, we can calculate the Fourier transform of the two sides, and then, using the *property of differentiation*, we can construct an algebraic equation from which we can calculate the *frequency response* $\mathcal{H}(j\omega)$, as well as the *impulse response* as $h(t) = \mathcal{F}^{-1}\{\mathcal{H}(j\omega)\}$.

Examples of the application of the aforementioned techniques are presented in the following pages. However, before we proceed to the study of these examples, it would be good to summarize the theory that describes the general behavior of LTI systems in the frequency domain, and immediately after this we will define the systems *linear* and *nonlinear phase*.

3.5.1 Amplitude-Phase Representation of Fourier Transform and Convolution Theorem

According to our discussion in the previous chapter, the output $y(t)$ of a *LTI system* with impulse response $h(t)$, when the input signal $x(t)$ is transmitted, is given by the convolution of the last two signals, or, in mathematical notation,

$$y(t) = h(t) * x(t) \tag{3.47}$$

In this case, according to the convolution theorem, the Fourier transforms $\mathcal{Y}(j\omega)$, $\mathcal{H}(j\omega)$, and $\mathcal{X}(j\omega)$ of the signals $y(t)$, $h(t)$, and $x(t)$, respectively, will be related by a simple product of the form

$$\mathcal{Y}(j\omega) = \mathcal{H}(j\omega)\mathcal{X}(j\omega) \tag{3.48}$$

with the function $\mathcal{H}(j\omega)$, which is called the *frequency response of the system* and defined as the Fourier transform of the impulse response $h(t)$, so that we can write

$$\mathcal{H}(j\omega) = \int_{-\infty}^{+\infty} h(t)\, e^{-j\omega t} dt \tag{3.49}$$

According to the convolution theorem, the *spectrum of the output signal of a LTI system results from the product of the spectrum $\mathcal{H}(j\omega)$ of the impulse response $h(t)$ and the spectrum of the input signal $\mathcal{X}(j\omega)$*. Expressing the preceding transforms in their exponential form, they are written

$$\mathcal{Y}(j\omega) = |\mathcal{Y}(j\omega)|e^{j\measuredangle\mathcal{Y}(j\omega)}, \qquad \mathcal{H}(j\omega) = |\mathcal{H}(j\omega)|e^{j\measuredangle\mathcal{H}(j\omega)}, \qquad \mathcal{X}(j\omega) = |\mathcal{X}(j\omega)|e^{j\measuredangle\mathcal{X}(j\omega)}$$

and replacing the foregoing expressions in the equation of the *convolution theorem*, we obtain

$$\mathcal{Y}(j\omega) = |\mathcal{Y}(j\omega)|e^{j\measuredangle\mathcal{Y}(j\omega)} = \left[|\mathcal{H}(j\omega)|e^{j\measuredangle\mathcal{H}(j\omega)}\right]\left[|\mathcal{X}(j\omega)|e^{j\measuredangle\mathcal{X}(j\omega)}\right] = \Big(|\mathcal{H}(j\omega)||\mathcal{X}(j\omega)|\Big)\, e^{j[\measuredangle\mathcal{H}(j\omega)+\measuredangle\mathcal{X}(j\omega)]}$$

from which it immediately follows that

$$|\mathcal{Y}(j\omega)| = |\mathcal{H}(j\omega)||\mathcal{X}(j\omega)| \quad \measuredangle\mathcal{Y}(j\omega) = \measuredangle\mathcal{H}(j\omega) + \measuredangle\mathcal{X}(j\omega) \tag{3.50}$$

From the foregoing analysis, it is concluded that the effect of a continuous-time LTI system on a signal input is the *multiplication of the magnitude* of the Fourier transform of the signal with the magnitude of the frequency response of the system and the *phase shift* of the Fourier transform of the signal by the phase of the frequency response of the system. Taking into account the significant impact that the phase of the signal transform generally has on its characteristics, when this is expressed in the time domain, we can surmise that even if the magnitude of the system's frequency response is constant and likewise for all frequencies, the foregoing phase shift can affect significantly the characteristics of the input signal $x(t)$. In the technical terminology used in the field of signals and systems, the magnitude of a system's frequency response $|\mathcal{H}(j\omega)|$ is known as the *gain*, while the graphical representations of the functions $|\mathcal{H}(j\omega)|$ and $\measuredangle\mathcal{H}(j\omega)$ with respect to the frequency ω of the input signal are called the *amplitude response* and *phase response* and graphically depict the way in which the considered LTI system changes the amplitudes and phases of the individual components of the input signal.

Having defined the phase $\measuredangle\mathcal{H}(j\omega)$ of the frequency response, we can define two very important quantities that characterize the behavior of the system and, more specifically, the phase delay defined as $\tau_p(\omega) = -(1/\omega)\measuredangle\mathcal{H}(j\omega)$ and the group delay defined as $\tau_g(\omega) = -\frac{d}{d\omega}\{\measuredangle\mathcal{H}(j\omega)\}$. These two quantities allow the characterization of the system's behavior and are discussed in the chapters on analog and digital filters.

Problem 3.5.1 Calculate and plot the amplitude and phase response of a LTI system whose impulse response is given by the relation $h(t) = u(t-2) - u(t+1)$.

Solution The frequency response of the system will emerge as known from the Fourier transform of the impulse response. It will therefore be

$$\mathcal{H}(j\omega) = \int_{-\infty}^{+\infty} h(t)\, e^{-j\omega t}dt = \int_{-\infty}^{+\infty} u(t-2)\, e^{-j\omega t}dt - \int_{-\infty}^{+\infty} u(t+1)\, e^{-j\omega t}dt$$

and since the step functions $u(t-2)$ and $u(t+1)$ are defined as

$$u(t-2) = \begin{cases} 1 & t > 2 \\ 0 & t < 2 \end{cases} \quad \text{and} \quad u(t+1) = \begin{cases} 1 & t > -1 \\ 0 & t < -1 \end{cases}$$

we have

$$\mathcal{H}(j\omega) = \int_{2}^{\infty} e^{-j\omega t}dt - \int_{-1}^{\infty} e^{-j\omega t}dt = -e^{-j\omega/2}\frac{2\sin(3\omega/2)}{\omega} = -3e^{-j\omega}\,\mathrm{sinc}\left(\frac{3\omega}{2\pi}\right)$$

from which it immediately follows that

$$|\mathcal{H}(j\omega)| = 3\left|\mathrm{sinc}\left(\frac{3\omega}{2\pi}\right)\right| \quad \text{and} \quad \measuredangle\mathcal{H}(j\omega) = -\frac{\omega}{2} = -\pi f$$

The graphical representations of the amplitude and phase responses of the system are shown in Fig. 3.14.

Problem 3.5.2 Calculate and plot the amplitude and phase response of a LTI system whose output, when the input signal $x(t) = (e^{-t} + e^{-3t})u(t)$ is transmitted, takes the form $y(t) = (2e^{-t} - 2e^{-4t})u(t)$.

Solution To solve the problem, we need, as before, the *frequency response* $\mathcal{H}(j\omega)$ which this time we need to calculate. From the well-known convolution theorem $y(t) = h(t) * x(t) \Longleftrightarrow \mathcal{Y}(j\omega) = \mathcal{H}(\omega)\mathcal{X}(j\omega)$ it immediately follows that $\mathcal{H}(j\omega) = \mathcal{Y}(j\omega)/\mathcal{X}(j\omega)$, and therefore we must first calculate the transforms $\mathcal{X}(j\omega)$ and $\mathcal{Y}(j\omega)$. Specifically, we will have:

$$\mathcal{X}(j\omega) = \int_{-\infty}^{+\infty} x(t)\, e^{-j\omega t}dt = \int_{0}^{\infty} e^{-j(1+j\omega)t}dt + \int_{0}^{\infty} e^{-j(3+j\omega)t}dt = \frac{4+2j\omega}{(1+j\omega)(3+j\omega)}$$

$$\mathcal{Y}(j\omega) = \int_{-\infty}^{+\infty} y(t)\, e^{-j\omega t}dt = 2\int_{0}^{\infty} e^{-j(1+j\omega)t}dt - 2\int_{0}^{\infty} e^{-j(4+j\omega)t}dt = \frac{6}{(1+j\omega)(4+j\omega)}$$

as is easily proven. Therefore, the impulse response of the system is calculated as

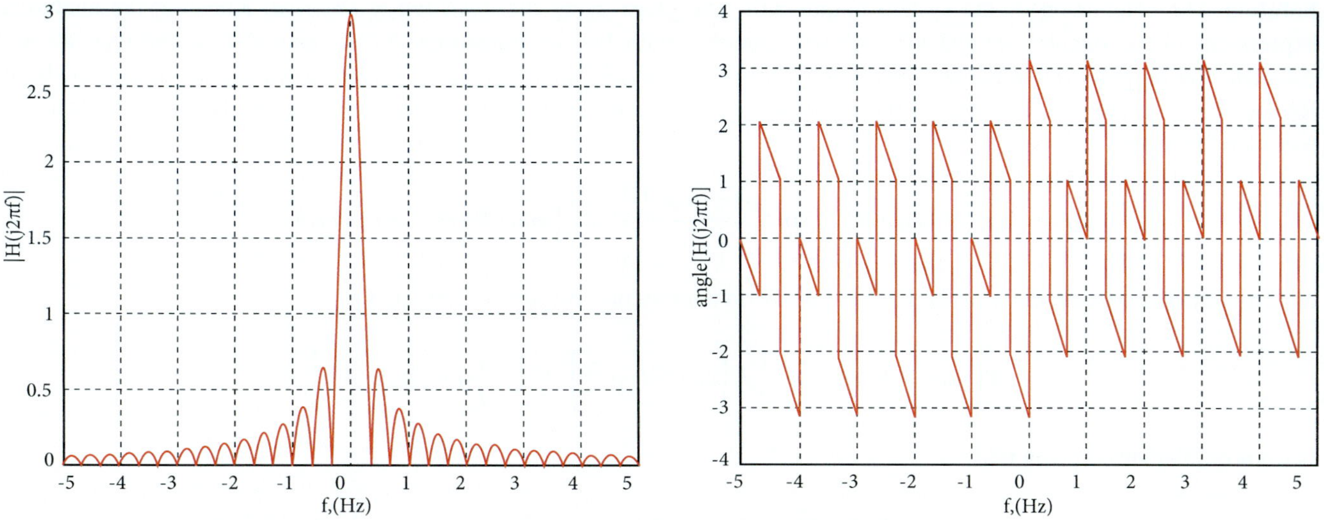

Fig. 3.14 Amplitude and phase response functions of system from Exercise 3.5.1

$$\mathcal{H}(j\omega)=\frac{\mathcal{Y}(j\omega)}{\mathcal{X}(j\omega)}=\frac{6}{(1+j\omega)(4+j\omega)}\times\frac{(1+j\omega)(3+j\omega)}{2(2+j\omega)}=\frac{9+3j\omega}{8+6j\omega-\omega^2}$$

If we expand the function $\mathcal{H}(j\omega)$ in a series of partial fractions, it turns out that

$$\mathcal{H}(j\omega)=\frac{9(\omega^2+8)}{\omega^4+20\omega^2+64}-j\omega\frac{3(\omega^2+10)}{\omega^4+20\omega^2+64}$$

The foregoing equation is expressed in Cartesian form, while for its formulation in exponential form and the determination of amplitude and phase responses we must use the known relations

$$|\mathcal{H}(j\omega)|=\sqrt{\Re\{\mathcal{H}(j\omega)\}^2+\Im\{\mathcal{I}(\omega)^2\}} \qquad \text{and} \qquad \measuredangle\mathcal{H}(j\omega)=-\tan^{-1}\left\{\frac{\Im\{\mathcal{H}(j\omega)\}}{\Re\{\mathcal{H}(j\omega)\}}\right\}$$

The calculation of these functions and the construction of the graphs that describe them is left as an exercise to the reader. For the construction of the graphs, any software application that offers design capabilities can be used, such as, for example, the applications *Mathematica* and *MATLAB*.

Note: As was already mentioned, once we calculate the frequency response of the system $\mathcal{H}(j\omega)$, we can then calculate its impulse response from the inverse transform $h(t)=\mathcal{F}^{-1}\{\mathcal{H}(j\omega)\}$. Developing the function $\mathcal{H}(j\omega)$ in a partial fraction expansion, it turns out that

$$\mathcal{H}(j\omega)=\frac{3}{2}\left\{\frac{1}{4+j\omega}\right\}+\frac{3}{2}\left\{\frac{1}{2+j\omega}\right\}$$

and from the linearity property of the inverse transform we get

$$h(t)=\mathcal{F}^{-1}\{\mathcal{H}(j\omega)\}=\frac{3}{2}\mathcal{F}^{-1}\left\{\frac{1}{4+j\omega}\right\}+\frac{3}{2}\mathcal{F}^{-1}\left\{\frac{1}{2+j\omega}\right\}=\frac{3}{2}\left(e^{-4t}+e^{-2t}\right)u(t)$$

where to derive the last relation we used the relation $\mathcal{F}\{e^{-\alpha t}u(t)\}=1/(\alpha+j\omega)$ for $\alpha=4$ and $\alpha=2$.

Problem 3.5.3 Calculate the output of a LTI system with impulse response $h(t)=\text{sinc}(4t-2)$ when the continuous input signal $x(t)=\text{sinc}(2t)\cos(2t)$ is transmitted to its input.

Solution The calculation of the system output will take place using the *convolution theorem*. Knowing the analytical expressions of the signals $x(t)$ and $h(t)$, we will calculate their Fourier transforms $\mathcal{X}(j\omega)$ and $\mathcal{H}(j\omega)$, and then we will calculate the output $y(t)$ from the inverse transform $y(t) = \mathcal{F}^{-1}\{\mathcal{Y}(j\omega)\} = \mathcal{F}^{-1}\{\mathcal{H}(j\omega)\mathcal{X}(\omega)\}$. In detail, we have: Replacing in the defining equation of the input signal $x(t)$ the function of the cosine from Euler's formula, the equation will take the form

$$x(t) = \text{sinc}(2t)\,\cos(2t) = \text{sinc}\,\frac{e^{j2t} + e^{-j2t}}{2} = \frac{1}{2}\Big[\text{sinc}(2t)\,e^{j2t} + \text{sinc}(2t)\,e^{-j2t}\Big]$$

and from the linearity property of the Fourier transform the function $\mathcal{X}(j\omega)$ is calculated as

$$\mathcal{X}(j\omega) = \mathcal{F}\{x(t)\} = \frac{1}{2}\mathcal{F}\Big\{\text{sinc}(2t)\,e^{j2t}\Big\} + \frac{1}{2}\mathcal{F}\Big\{\text{sinc}(2t)\,e^{-j2t}\Big\}$$

If we now take into account that

$$\mathcal{F}\{\text{sinc}(\alpha t)\} = \mathcal{F}\left\{\frac{\sin(\alpha\pi t)}{\alpha\pi t}\right\} = \frac{1}{\alpha}\mathcal{F}\left\{\frac{\sin(\alpha\pi t)}{\pi t}\right\} = \frac{1}{\alpha}\Pi_{\alpha\pi}(\omega) = \frac{1}{\alpha}\begin{cases} 1 \; if|\omega| \le \pi \\ 0 \; if|\omega| > \pi \end{cases}$$

and we use the *property of frequency shift* of the Fourier transform, which is formulated as

$$\mathcal{F}\{x(t)\,e^{j\omega_0 t}\} = \mathcal{X}(\omega - \omega_0)$$

(for the values $\alpha = \omega_0 = 2$), it turns out that

$$\mathcal{F}\Big\{\text{sinc}(2t)\,e^{j2t}\Big\} = \frac{1}{2}\Pi_{2\pi}(\omega - 2) = \frac{1}{2}\begin{cases} 1 \; if|\omega - 2| \le 2\pi \\ 0 \; if|\omega - 2| > 2\pi \end{cases}$$

$$\mathcal{F}\Big\{\text{sinc}(2t)\,e^{-j2t}\Big\} = \frac{1}{2}\Pi_{2\pi}(\omega + 2) = \frac{1}{2}\begin{cases} 1 \; if|\omega + 2| \le 2\pi \\ 0 \; if|\omega + 2| > 2\pi \end{cases}$$

Replacing these expressions in the equation of the transform $\mathcal{X}(j\omega)$, it turns out to be equal to

$$\mathcal{X}(j\omega) = \frac{1}{2}\mathcal{F}\Big\{\text{sinc}(2t)\,e^{j2t}\Big\} + \frac{1}{2}\mathcal{F}\Big\{\text{sinc}(2t)\,e^{-j2t}\Big\} = \frac{1}{4}\Big\{\Pi_{2\pi}(\omega - 2) + \Pi_{2\pi}(\omega + 2)\Big\}$$

Note: We can reach the same result using the property of multiplication of the Fourier transform, which in this case is formulated as

$$\mathcal{F}\Big\{\text{sinc}(2t)\cdot\cos(2t)\Big\} = \frac{1}{2\pi}\Big\{\mathcal{F}\{\text{sinc}(2t)\} * \mathcal{F}\{\cos(2t)\}\Big\}$$

and let us calculate the convolution between the transforms of the two functions, which have the form

$$\mathcal{F}\{\text{sinc}(2t)\} = \frac{1}{2}\Pi_{2\pi}(\omega) \qquad \text{and} \qquad \mathcal{F}\{\cos(2t)\} = \pi\Big[\delta(\omega + 2) + \delta(\omega - 2)\Big]$$

On the other hand, the impulse response of the considered system has the form

$$h(t) = \text{sinc}(4t - 2) = \text{sinc}\left[4\left(t - \frac{1}{2}\right)\right] = \text{sinc}(4t)_{t \to t - \frac{1}{2}}$$

and using the *property of time shift* of the Fourier transform, which is formulated as $\mathcal{F}\{x(t - t_0)\} = e^{-j\omega t_0}\mathcal{X}(j\omega)$ for the value $t_0 = 1/2$, it immediately follows that

$$\mathcal{H}(j\omega) = \mathcal{F}\left\{\text{sinc}\left[4\left(t - \frac{1}{2}\right)\right]\right\} = e^{-j\omega/2}\text{sinc}(4t) = \frac{1}{4}e^{-j\omega/2}\Pi_{4\pi}(\omega)$$

The graphical representations of the transforms $\mathcal{X}(j\omega)$ and $\mathcal{H}(j\omega)$ lead to the equations

$$\mathcal{X}(j\omega) = \begin{cases} 0 & \omega \leq -2 - 2\pi \\ 1/4 & -2 - 2\pi \leq \omega < 2 - 2\pi \\ 1/2 & 2 - 2\pi \leq \omega < -2 + 2\pi \\ 1/4 & -2 + 2\pi \leq \omega < 2 + 2\pi \\ 0 & \omega \geq 2 + 2\pi \end{cases}$$

$$\mathcal{H}(j\omega) = \begin{cases} 1/4 & -4\pi \leq \omega \leq +4\pi \\ 0 & \text{anywhereelse} \end{cases}$$

Therefore, the Fourier transform of the output will be defined for the range of values $-2 - 2\pi \leq \omega \leq 2 + 2\pi$ since outside this the transform $\mathcal{X}(j\omega)$ is equal to zero. For this range of values we will have

$$\mathcal{Y}(\omega) = \mathcal{H}(j\omega)\mathcal{X}(j\omega) = \frac{1}{4}e^{-j\omega/2}X(\omega)$$

If we take the inverse transform and use the *property of time shift*, it turns out that

$$y(t) = \mathcal{F}^{-1}\{\mathcal{Y}(j\omega)\} = \frac{1}{4}\mathcal{F}\{e^{-j\omega/2}X(\omega)\} = \frac{1}{4}x(t)_{t\to t-\frac{1}{2}} = \frac{1}{4}\text{sinc}(2t-1)\cos(2t-1)$$

which is also the desired result.

Problem 3.5.4 Calculate the output of a LTI system with impulse response

$$h(t) = e^{-2|t-1|}$$

when the input signal is transmitted

$$x(t) = \cos\left(2\pi t + \frac{\pi}{2}\right)$$

Solution Following the previous methodology, let us calculate the Fourier transforms of the signals $h(t)$ and $x(t)$. To calculate the frequency response $\mathcal{H}(j\omega)$, we will use the equation

$$\mathcal{F}\{e^{-\alpha|t|}\} = \frac{2\alpha}{\alpha^2 + \omega^2}$$

which we proved in a previous exercise for the value $\alpha = 2$, in combination with the *property of time shift* of the transform, which is formulated as $\mathcal{F}\{x(t-t_0)\} = e^{-i\omega t_0}\mathcal{X}(j\omega)$ for the value $t_0 = 1$. It is not difficult to ascertain that the implementation of this process will lead to a frequency response of the form

$$\mathcal{H}(j\omega) = \mathcal{F}\{h(t)\} = \mathcal{F}\{e^{-2|t-1|}\} = \frac{4e^{-i\omega}}{\omega^2 + 4}$$

In a completely analogous way, formulating the input signal $x(t)$ in the form

$$x(t) = \cos\left(2\pi t + \frac{\pi}{2}\right) = \cos\left[2\pi\left(t + \frac{1}{4}\right)\right]$$

and using the result

$$\mathcal{F}\{\cos(\alpha t)\} = \pi\Big[\delta(\omega + \alpha) + \delta(\omega - \alpha)\Big]$$

for the value $\alpha = 2\pi$, along with the *property of time shift* $\mathcal{F}\{x(t + t_0)\} = e^{i\omega t_0}\mathcal{X}(j\omega)$ for the value $t_0 = 1/4$, it follows that

$$\mathcal{X}(j\omega) = \mathcal{F}\{x(t)\} = \mathcal{F}\left\{\cos\left[2\pi\left(t + \frac{1}{4}\right)\right]\right\} = \pi e^{i\omega/4}\Big[\delta(\omega + 2\pi) + \delta(\omega - 2\pi)\Big]$$

Substituting the foregoing expressions into the convolution theorem we will get

$$\mathcal{Y}(j\omega) = \mathcal{H}(j\omega)\mathcal{X}(j\omega) = \frac{4\pi e^{-3i\omega/4}}{\omega^2 + 4}\Big[\delta(\omega + 2\pi) + \delta(\omega - 2\pi)\Big]$$

and therefore the desired output of the system $y(t)$ will result from the inverse transform

$$y(t) = \mathcal{F}^{-1}\{\mathcal{Y}(j\omega)\} = \frac{1}{2\pi}\int_{-\infty}^{+\infty}\mathcal{Y}(j\omega)\, e^{i\omega t}\, d\omega = 2\int_{-\infty}^{+\infty}\frac{e^{i\omega(t-3/4)}}{\omega^2 + 4}\Big[\delta(\omega + 2\pi) + \delta(\omega - 2\pi)\Big]\, d\omega$$

Using the *property of the selection of the impulse function*, which is formulated as

$$\int_{-\infty}^{+\infty} f(\omega)\,\delta(\omega - \omega_0)\, d\omega = f(\omega_0)$$

for the value $\omega_0 = 2\pi$ and for the function

$$f(\omega) = \frac{1}{\omega^2 + 4}\exp\left[i\omega\left(t - \frac{3}{4}\right)\right]$$

it immediately follows that

$$\int_{-\infty}^{+\infty}\frac{e^{i\omega(t-3/4)}}{\omega^2 + 4}\,\delta(\omega + 2\pi)\, d\omega = \left.\frac{e^{i\omega(t-3/4)}}{\omega^2 + 4}\right|_{\omega=-2\pi} = \frac{e^{-i2\pi(t-3/4)}}{4\pi^2 + 4} \quad \text{and}$$

$$\int_{-\infty}^{+\infty}\frac{e^{i\omega(t-3/4)}}{\omega^2 + 4}\,\delta(\omega - 2\pi)\, d\omega = \left.\frac{e^{i\omega(t-3/4)}}{\omega^2 + 4}\right|_{\omega=2\pi} = \frac{e^{i2\pi(t-3/4)}}{4\pi^2 + 4}$$

and by substituting into the defining equation of the signal $y(t)$ we will get

$$y(t) = \frac{\frac{1}{2}\, e^{i2\pi(t-3/4)} + e^{i2\pi(t-3/4)}}{\pi^2 + 1} = \frac{\cos\left[2\pi\left(t - \frac{3}{4}\right)\right]}{\pi^2 + 1} = \frac{x(t-1)}{\pi^2 + 1}$$

Problem 3.5.5 Construct the differential equation of a LTI system that, when it receives the signal $x(t) = (e^{-t} + e^{-3t})u(t)$ at its input, produces the signal $y(t) = (2e^{-2t} - 2e^{-4t})u(t)$ as its output.

Solution In a previous exercise we proved that the *frequency response* of a system was

$$\mathcal{H}(j\omega) = \frac{\mathcal{Y}(j\omega)}{\mathcal{X}(j\omega)} = \frac{9+3j\omega}{8+6j\omega+(j\omega)^2}$$

which, as it turns out, can also be expressed in the form

$$(j\omega)^2\mathcal{Y}(j\omega) + 6(j\omega)\mathcal{Y}(j\omega) + 8\mathcal{Y}(j\omega) = 3(j\omega)\mathcal{X}(j\omega) + 9\mathcal{X}(j\omega)$$

If we take the inverse transform of the foregoing equation, we will end up with the expression

$$\mathcal{F}^{-1}\{(j\omega)^2\mathcal{Y}(j\omega)\} + 6\mathcal{F}^{-1}\{(j\omega)\mathcal{Y}(j\omega)\} + 8\mathcal{F}^{-1}\{\mathcal{Y}(j\omega)\} = 3\mathcal{F}^{-1}\{(j\omega)\mathcal{X}(j\omega)\} + 9\mathcal{F}^{-1}\{\mathcal{X}(j\omega)\}$$

However, $\mathcal{F}^{-1}\{\mathcal{Y}(j\omega)\} = y(t)$ and $\mathcal{F}^{-1}\{\mathcal{X}(j\omega)\} = x(t)$, while from the *property of differentiation* of the Fourier transform we will get

$$\mathcal{F}^{-1}\{(j\omega)^2\mathcal{Y}(j\omega)\} = \frac{d^2y(t)}{dt^2}, \quad \mathcal{F}^{-1}\{(j\omega)\mathcal{Y}(j\omega)\} = \frac{dy(t)}{dt}, \quad \mathcal{F}^{-1}\{(j\omega)\mathcal{X}(j\omega)\} = \frac{dx(t)}{dt}$$

and substituting these expressions into the foregoing relation, we will end up with the equation

$$\frac{d^2y(t)}{dt^2} + 6\frac{dy(t)}{dt} + 8y(t) = 3\frac{dx(t)}{dt} + 9x(t)$$

which is the differential equation that describes the operation of the considered system.

Problem 3.5.6 Consider a LTI system described by the equation

$$\frac{d^2y(t)}{dt^2} + 8\frac{dy(t)}{dt} + 7y(t) = \frac{dx(t)}{dt} + 4x(t).$$

Determine the frequency and impulse response of the system and then its output when the continuous-time signal $x(t) = e^{-4t}u(t)$ is transmitted to its input.

Solution Taking the Fourier transform of the two terms of the differential equation, it turns out that

$$\mathcal{F}\left\{\frac{d^2y(t)}{dt^2}\right\} + 8\mathcal{F}\left\{\frac{dy(t)}{dt}\right\} + 7\mathcal{F}\{y(t)\} = \mathcal{F}\left\{\frac{dx(t)}{dt}\right\} + 4\mathcal{F}\{x(t)\}$$

However, $\mathcal{F}\{x(t)\} = \mathcal{X}(j\omega)$ and $\mathcal{F}\{y(t)\} = \mathcal{Y}(j\omega)$, while from the *property of differentiation* of the Fourier transform we will get

$$\mathcal{F}\left\{\frac{d^2y(t)}{dt^2}\right\} = (j\omega)^2\mathcal{Y}(j\omega), \quad \mathcal{F}\left\{\frac{dy(t)}{dt}\right\} = (j\omega)\mathcal{Y}(j\omega), \quad \mathcal{F}\left\{\frac{dx(t)}{dt}\right\} = (j\omega)\mathcal{X}(j\omega)$$

Substituting these expressions into the preceding equation, it turns out that

$$\mathcal{Y}(j\omega)\Big[(j\omega)^2 + 8j\omega + 7\Big] = \mathcal{X}(j\omega)(4+j\omega)$$

and therefore the *frequency response* of the system is calculated as

$$\mathcal{H}(j\omega) = \frac{\mathcal{Y}(j\omega)}{\mathcal{X}(j\omega)} = \frac{4+j\omega}{7+8j\omega-\omega^2} = \frac{4+j\omega}{(7+j\omega)(1+j\omega)}$$

Having calculated the *frequency response* of the system, we can determine its *impulse response* as the *inverse Fourier transform* of the function $\mathcal{H}(j\omega)$. Developing this function into a sum of partial fractions, it will take the form

$$\mathcal{H}(j\omega) = \frac{4+j\omega}{(7+j\omega)(1+j\omega)} = \frac{1}{2}\left\{\frac{1}{7+j\omega}\right\} + \frac{1}{2}\left\{\frac{1}{1+j\omega}\right\}$$

and referring to tables of elementary transforms, we finally arrive at the equation

$$h(t) = \mathcal{F}^{-1}\{\mathcal{H}(j\omega)\} = \frac{1}{2}\mathcal{F}^{-1}\left\{\frac{1}{7+j\omega}\right\} + \frac{1}{2}\mathcal{F}^{-1}\left\{\frac{1}{1+j\omega}\right\} = \frac{1}{2}\left(e^{-7t} + e^{-t}\right)u(t)$$

Finally, to calculate the output of the system corresponding to the input $x(t) = e^{-4t}u(t)$, we will use the convolution theorem. Referring to tables of elementary transform, we will have

$$\mathcal{X}(j\omega) = \mathcal{F}\{x(t)\} = \mathcal{F}\{e^{-4t}u(t)\} = \frac{1}{4+j\omega}$$

and therefore the Fourier transform of the system output will have the form

$$\mathcal{Y}(j\omega) = \mathcal{H}(j\omega)\mathcal{X}(j\omega) = \frac{4+j\omega}{(7+j\omega)(1+j\omega)} \times \frac{1}{4+j\omega} = \frac{1}{(7+j\omega)(1+j\omega)}$$

Developing the transform $\mathcal{Y}(j\omega)$ into a sum of partial fractions and taking the inverse transform of the foregoing relation, the system response $y(t)$ to the input signal $x(t) = e^{-4t}u(t)$ is calculated to be equal to

$$y(t) = \mathcal{F}^{-1}\{\mathcal{Y}(j\omega)\} = \frac{1}{6}\mathcal{F}^{-1}\left\{\frac{1}{1+j\omega}\right\} - \frac{1}{6}\mathcal{F}^{-1}\left\{\frac{1}{7+j\omega}\right\} = \frac{1}{6}\left(e^{-t} - e^{-7t}\right)u(t)$$

3.5.2 Linear and Nonlinear Phase Systems

One of the simplest cases of continuous-time LTI systems are those systems that are described by a simple frequency response of the form[15]

$$\mathcal{H}(j\omega) = |\mathcal{H}(j\omega)|e^{j\measuredangle\mathcal{H}(j\omega)} = e^{-j\omega\tau} \tag{3.51}$$

from which it immediately follows that $|\mathcal{H}(j\omega)| = 1$ and $\measuredangle\mathcal{H}(j\omega) = -\omega\tau$. It is not difficult to realize that these systems are characterized by *a gain value equal to one* and a phase that is *a linear function of the frequency* ω; these systems are known as *linear phase systems*. According to our discussion in the preceding sections, if a signal passes through such a system input with a Fourier transform of the form

$$\mathcal{X}(j\omega) = |\mathcal{X}(j\omega)|e^{j\measuredangle\mathcal{X}(j\omega)} \tag{3.52}$$

[15] Although for simplicity we consider the case of *unit gain*, the situation is exactly the same if we consider a frequency response of the form $\mathcal{H}(j\omega) = Ge^{-i\omega\tau}$ which leads to a gain value equal to G. In signal transmission processes in telecommunications systems, the signal is considered to be transmitted *without distortion* when its amplitude is multiplied by a constant and/or its phase is shifted by a constant amount independent of frequency.

then the Fourier transform of the output signal will be calculated as

$$\mathcal{Y}(j\omega) = \mathcal{H}(j\omega)\mathcal{X}(j\omega) = \Big(|\mathcal{H}(j\omega)||\mathcal{X}(j\omega)|\Big)\, e^{j[\measuredangle\mathcal{H}(j\omega)+\measuredangle\mathcal{X}(j\omega)]} = |\mathcal{X}(j\omega)|e^{j[\measuredangle\mathcal{X}(\omega)-\omega\tau]} =$$

$$= e^{-j\omega\tau}\Big\{|\mathcal{X}(j\omega)|e^{j\measuredangle\mathcal{X}(j\omega)}\Big\} = e^{-j\omega\tau}\mathcal{X}(j\omega)$$

Taking the inverse Fourier transform of the preceding relationship and using the property of *time shift*, we find that

$$y(t) = x(t-\tau) \tag{3.53}$$

that is, *the effect of a LTI system with unit gain and phase that is a linear function of frequency* ω *on an input signal is the time delay of the signal by* τ *time units* or, more generally, by the slope of the linear phase $\measuredangle\mathcal{H}(j\omega) = -\omega\tau$. In the special case where this slope is an integer, this behavior characterizes also *discrete-time systems*, with the situation however becoming more complex when we use noninteger slope values.

On the other hand, if the phase of the LTI system under consideration *is not linear* but has *nonlinear characteristics*, the fundamental complex exponential components of the input signal (each of which is characterized by its own frequency) will be shifted in a way that will change their relative phases, and therefore, if combined, to construct the input signal, the result we will get will be different from the original; therefore, the signal *is distorted*. Considering, for example, an input signal characterized by *limited bandwidth* (see Sect. 3.3.7 on page 255), we know that its Fourier transform will have zero or at least very small values for frequencies that are located outside a narrow frequency band centered at a specific frequency with value, let us say, ω_0. Although the phase of the system is nonlinear, if we assume that the range of the aforementioned frequency band is very small, then we can approximate the nonlinear phase of the system with a *linear approximation* of the form

$$\measuredangle\mathcal{H}(j\omega) \approx -\varphi - \omega\tau \tag{3.54}$$

and calculate the Fourier transform of the system's output from the approximate relation

$$\mathcal{Y}(j\omega) \approx \mathcal{X}(j\omega)|\mathcal{H}(j\omega)|e^{-j\varphi}e^{-j\omega\tau} \tag{3.55}$$

From the last relation it follows that the conclusion that the effect of the system on the Fourier transform of the input signal is the *modulation of its measure* according to the measure of the function of the frequency response and multiplication (a) by a *constant complex exponential factor* of the form $e^{-j\varphi}$ and (b) by a *linear phase term* of the form $e^{-j\omega\tau}$, which will cause a time delay equal to τ time units. This time delay is known as *group delay* for the value of the center of the frequency band ω_0 and is defined as the *negative slope of the phase of the system, calculated at the frequency of interest*, or, in mathematical notation,

$$\tau(\omega) = -\frac{d}{d\omega}\Big\{\measuredangle\mathcal{H}(j\omega)\Big\} \tag{3.56}$$

If the value of the function $\tau(\omega)$ is constant and independent of frequency, then all components of the signal will undergo the same time delay, while in the opposite case the different frequency components will undergo different delays, resulting in a signal that will emerge from their synthesis as different from the original. We conclude, therefore, that if we want to construct a system that reproduces the original signal as its output (such as, for example, a signal transmission telecommunication system), it is not enough to use only a constant gain value; we must simultaneously impose the presence of linear phase response.

Taking into account that the frequency response $\mathcal{H}(j\omega)$ has the form $\mathcal{H}(j\omega) = |\mathcal{H}(j\omega)|e^{j\measuredangle\mathcal{H}(j\omega)}$ we will have

$$\begin{aligned}\ln\mathcal{H}(j\omega) &= \ln\left(|\mathcal{H}(j\omega)|e^{j\measuredangle\mathcal{H}(j\omega)}\right) = \ln|\mathcal{H}(j\omega)| + j\measuredangle\mathcal{H}(j\omega)\\ &= \frac{1}{2}\ln|\mathcal{H}(j\omega)|^2 + j\measuredangle\mathcal{H}(j\omega) = \frac{1}{2}\ln\left[\mathcal{H}(j\omega)\mathcal{H}(-j\omega)\right] + j\measuredangle\mathcal{H}(j\omega)\end{aligned}$$

Therefore, the phase response will be equal to the imaginary part of the logarithm of the frequency response, or, alternatively, according to the foregoing relationship, it will have the form

$$\measuredangle\mathcal{H}(j\omega) = \frac{1}{j}\ln\mathcal{H}(j\omega) - \frac{1}{2j}\ln\mathcal{H}(j\omega) - \frac{1}{2j}\ln\mathcal{H}(-j\omega) = \frac{1}{2j}\ln\mathcal{H}(j\omega) - \frac{1}{2j}\ln\mathcal{H}(-j\omega) = -\frac{j}{2}\ln\frac{\mathcal{H}(j\omega)}{\mathcal{H}(-j\omega)}$$

Proceeding with the substitution $s = j\omega$, we define the generalized phase as

$$\measuredangle\mathcal{H}(s) = -\frac{1}{2}\ln\frac{\mathcal{H}(s)}{\mathcal{H}(-s)}$$

After calculating the phase response, we can calculate the group delay from the relationship

$$\begin{aligned}\tau(\omega) &= -\frac{d}{d\omega}\left\{\Im\{\ln\mathcal{H}(j\omega)\}\right\} = -\Im\left\{\frac{d}{d\omega}\ln\mathcal{H}(j\omega)\right\} = -\Im\left\{j\frac{d}{d(j\omega)}\ln\mathcal{H}(j\omega)\right\} =\\ &= -\Re\left[\frac{d}{ds}\{\ln\mathcal{H}(s)\}\right]_{s=j\omega} = -\mathcal{E}\text{v}\left\{\left[\frac{1}{\mathcal{H}(s)}\frac{d\mathcal{H}(s)}{ds}\right]\right\}_{s=j\omega}\end{aligned}$$

with the prefix $\mathcal{E}\text{v}$ describing the even component of the expression that follows and that contains the transfer function of the system $\mathcal{H}(s) = \mathcal{H}(j\omega)_{\omega=s/j}$. We are led, therefore, to the conclusion, that in linear phase systems, the group delay is an even function of frequency and therefore can be calculated from the equation

$$\tau(\omega) = -\frac{1}{2}\left[\frac{\mathcal{H}'(s)}{\mathcal{H}'(s)} + \frac{\mathcal{H}'(-s)}{\mathcal{H}'(-s)}\right]_{s=j\omega}$$

Taking into account that in almost all cases the transfer function $\mathcal{H}(s)$ is a rational function of the form $N(s)/D(s)$, we will have

$$\frac{d\ln\mathcal{H}(s)}{ds} = \frac{d\ln N(s)}{ds} - \frac{d\ln D(s)}{ds} = \frac{1}{N(s)}\frac{dN(s)}{ds} - \frac{1}{D(s)}\frac{dD(s)}{ds}$$

and therefore the group delay will be calculated from the relation

$$\tau(\omega) = \left[-\frac{d\measuredangle\mathcal{H}(s)}{ds}\right]_{s=j\omega} = -\frac{1}{2}\left[\frac{1}{N(s)}\frac{dN(s)}{ds} + \frac{1}{N(-s)}\frac{dN(-s)}{ds} - \frac{1}{Q(s)}\frac{dQ(s)}{ds} - \frac{1}{Q(-s)}\frac{dQ(-s)}{ds}\right]$$

Problem 3.5.7 Calculate the magnitude, phase, and group delay for a LTI system whose frequency response has the form

$$\mathcal{H}(j\omega) = \frac{1 - j\omega}{1 + j\omega}$$

Solution From the defining equation of the system's frequency response it immediately follows that

$$|\mathcal{H}(j\omega)| = \left|\frac{1-j\omega}{1+j\omega}\right| = \frac{|1-j\omega|}{|1+j\omega|} = \frac{\sqrt{1+\omega^2}}{\sqrt{1+\omega^2}} = 1$$

$$\measuredangle\mathcal{H}(j\omega) = \measuredangle\{1-j\omega\} - \measuredangle\{1+j\omega\} = \tan^{-1}(-\omega) - \tan^{-1}(\omega) = -2tan^{-1}(\omega)$$

and therefore the group delay is calculated as

$$\tau(\omega) = -\frac{d}{d\omega}\left\{\measuredangle\mathcal{H}(j\omega)\right\} = 2\frac{d}{d\omega}\left[\tan^{-1}(\omega)\right] = \frac{2}{1+\omega^2}$$

Problem 3.5.8 Calculate the magnitude, phase, and group delay for a LTI system whose frequency response has the form

$$\mathcal{H}(j\omega) = e^{-j\omega}\frac{2-e^{j\omega}}{2-e^{-j\omega}}$$

Solution Taking into account that

$$2-e^{j\omega} = [2-\cos(\omega)] - j\sin(\omega) \qquad \text{and} \qquad 2-e^{-j\omega} = [2-\cos(\omega)] + j\sin(\omega)$$

the magnitudes of the complex quantities of the numerator and the denominator are calculated as

$$|2-e^{j\omega}| = |2-e^{-j\omega}| = \sqrt{[2-\cos(\omega)]^2+\sin^2(\omega)} \quad \text{and}$$

$$\measuredangle\{2-e^{j\omega}\} = \tan^{-1}\left\{-\frac{\sin(\omega)}{2-\cos(\omega)}\right\}, \quad \measuredangle\{2-e^{-j\omega}\} = \tan^{-1}\left\{\frac{\sin(\omega)}{2-\cos(\omega)}\right\}$$

while it additionally holds that $|e^{-j\omega}| = 1$ and $\measuredangle\{e^{-j\omega}\} = -\omega$. Therefore, we will have

$$|\mathcal{H}(j\omega)| = |e^{-j\omega}|\frac{|2-e^{j\omega}|}{|2-e^{-j\omega}|} = 1 \quad \text{and}$$

$$\measuredangle\mathcal{H}(j\omega) = \measuredangle\{e^{-j\omega}\} + \measuredangle\{2-e^{j\omega}\} - \measuredangle\{2-e^{-j\omega}\} = -\omega + \tan^{-1}\left\{-\frac{\sin(\omega)}{2-\cos(\omega)}\right\} -$$

$$-\tan^{-1}\left\{\frac{\sin(\omega)}{2-\cos(\omega)}\right\} = -\omega - 2\tan^{-1}\left\{\frac{\sin(\omega)}{2-\cos(\omega)}\right\}$$

and therefore the group delay will be calculated from the derivative

$$\tau(\omega) = -\frac{d}{d\omega}\left\{\measuredangle\mathcal{H}(j\omega)\right\} = -\frac{d}{d\omega}\left[-\omega - 2\tan^{-1}\left\{\frac{\sin(\omega)}{2-\cos(\omega)}\right\}\right] =$$

$$= -1 + 2\frac{d}{d\omega}\tan^{-1}\left\{\frac{\sin(\omega)}{2-\cos(\omega)}\right\}\frac{d}{d\omega}\left\{\frac{\sin(\omega)}{2-\cos(\omega)}\right\} = \frac{3}{5-4\cos(\omega)}$$

as it turns out after operations, which is also the desired result. For the derivation of the last relation, we used the rules of differentiation of composite functions and the analytical expression for the derivative of the inverse tangent function, which, as is well known, has the form $[\tan^{-1}(\omega)]' = 1/(1+\omega^2)$.

3.6 Frequency Selection Filters

According to the theory of Fourier analysis, a *periodic* signal $x(t)$ can be expressed as a sum of infinite harmonic functions with frequency values $0, \omega_0, 2\omega_0, \ldots, n\omega_0, \ldots$ (i.e., integer multiples of the *fundamental frequency* ω_0), whose amplitudes depend on the coefficients of the Fourier series and define the so-called *line spectrum*. On the other hand, if the signal is *aperiodic*, it can again be represented as the superposition of infinite frequency components, with the spectrum in this case being continuous and determined through calculation of the Fourier transform $\mathcal{X}(j\omega)$. We also mentioned that in practice, the frequency content of a signal does not include the infinite number of frequency components that result from the foregoing analysis, but only those that belong to the *bandwidth of the signal*, that is, the frequency range, outside of which the functions $x(t)$ and $\mathcal{X}(\omega)$ have such small value as to be considered negligible.

In some signal processing applications, we are only interested in certain frequency ranges, and we ignore all others. We know, for example, that the human ear perceives sound signals with frequencies in a range of 16 Hz–16 KHz, while it is extremely sensitive in the frequency range of 500 Hz–5 KHz. For this reason, when we deal with the problem of *processing* and *recognition* of the human voice by a computer, we usually apply a preprocessing stage to the vocal signal during which we only keep those frequency components with frequency values in a range of 300 Hz–3400 Hz and remove all other frequencies. This process is carried out by a system implemented in hardware or software, which is known as a *filter*.

Generally speaking, a filter is simply a *LTI system of continuous or discrete-time*, which performs one of the two fundamental functions, the first of which includes *the alteration of the relative widths of the frequency components of the input signal*, and the second involves the *complete elimination*–or at least significant *attenuation*—of some of these. Considering that the process of altering the relative widths of the components of the signal leads to a *change in the shape of the spectrum that describes it*, we can characterize the filters that perform this function as *frequency shaping filters*. On the other hand, the filters that eliminate some frequencies of the input signal while simultaneously allowing the passage of some others are known as *frequency selection filters*. It is obvious that the determination of the mode of operation of a filter requires the design of the *frequency response function* in the appropriate way, so that the existing requirements in each case are met.

Although frequency shaping filters find many applications, mainly in audio and image processing systems, in the following section we will focus exclusively on the *frequency selection filters*, which allow the passage of only certain frequency regions of the input signal, cutting off all the rest. Depending on their mode of operation regarding the frequency ranges that they promote at their output or cutoff, frequency selection filters can be grouped into four categories as follows:

- *Low-pass filters*, which allow the passage of frequencies with a low value while cutting off higher frequencies.
- *High-pass filters*, which allow the passage of frequencies with a high value while cutting off lower frequencies.
- *Band-pass filters*, which allow the passage of all frequencies that belong to a specific band while cutting off all other frequencies.
- *Band-stop filters*, which cut off the frequencies that belong to a specific frequency band while allowing the passage of all other frequencies.

Examples of applying *low-pass*, *high-pass*, and *band-pass* filters are presented in Fig. 3.15. In this figure, the input signal consists of three simple harmonic components with frequencies of 50, 110, and 210 Hz. We observe that at the output of these filters, the *low*, *high*, and *mid* frequencies, appear successively, a result that also demonstrates their usage method.

In the terminology of *analog* and *digital* filters, the frequencies that pass through the filter constitute the so-called *passband*, while those that are cut off by it define the *stopband* or *cutoff band*. Closely connected to the operation of the filters is also the *cutoff frequency* ω_c, which determines the area or areas of operation of the filter used in each case and constitutes the boundary between the cutoff band and the passband. For example, a low-pass filter allows the passage of all frequencies with values $|\omega| \leq \omega_c$, while it cuts off the frequencies $|\omega| > \omega_c$, while a high-pass filter allows the passage of frequencies with the value $|\omega| \geq \omega_c$ and cuts off signal components with the frequency $|\omega| < \omega_c$. Finally, regarding band-pass and band-stop filters, these are described using two cutoff frequencies, ω_{c_1} and ω_{c_2}, with band-pass filters allowing the passage of components with frequencies $\omega_{c_1} \leq |\omega| \leq \omega_{c_2}$ and cutting off all the rest and band-stop filters cutting off the frequencies that belong to the aforementioned range of values and allowing the passage of those that have a value $|\omega| \leq \omega_{c_1}$ or $|\omega| \geq \omega_{c_2}$.

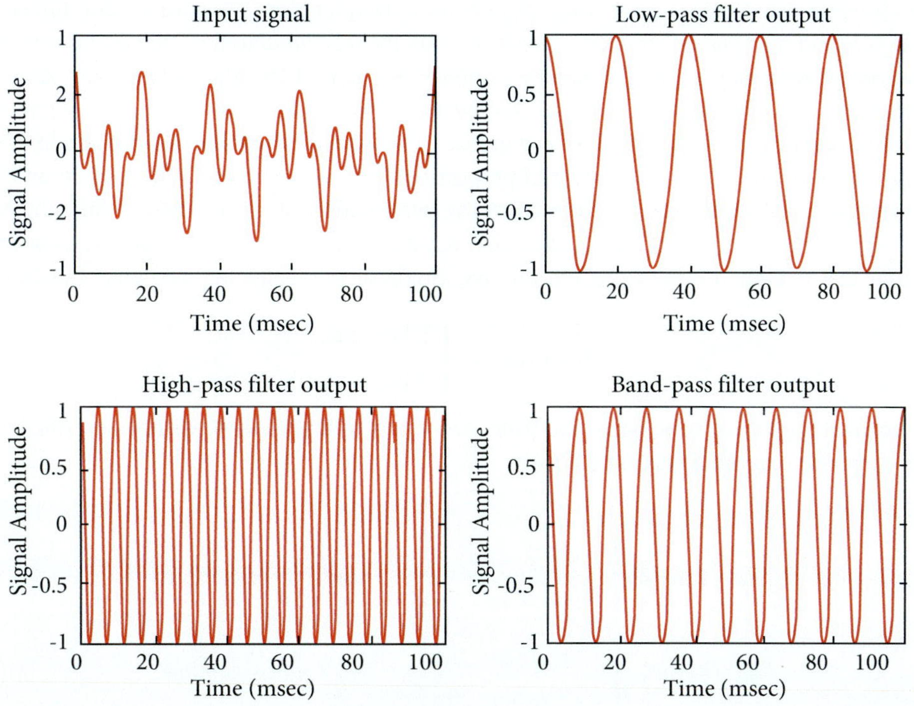

Fig. 3.15 Examples of the use of low-pass, high-pass, and band-pass filters

3.6.1 Ideal Filters

According to convolution theorem expressed in the amplitude-phase representation, the Fourier transform of the output of a LTI system has the form

$$\mathcal{Y}(j\omega) = \mathcal{H}(j\omega)\mathcal{X}(j\omega) = |\mathcal{H}(j\omega)||\mathcal{X}(j\omega)|e^{j[\measuredangle\mathcal{H}(j\omega)+\measuredangle\mathcal{X}(j\omega)]} \tag{3.57}$$

while the same output of the system can be calculated taking the inverse transform of the preceding relation. It is not difficult to understand that the elimination of a component of the signal $x_{\omega_0}(t)$ with frequency ω_0 can be easily achieved if we design the frequency response function in such a way that $\mathcal{H}(j\omega_0) = 0$; in this case, we will have $\mathcal{Y}(j\omega_0) = 0$ and therefore $y_{\omega_0}(t) = 0$, that is, the frequency component $x_{\omega_0}(t)$ is eliminated during the passage of the signal through the filter. On the other hand, if we want this frequency to pass through the filter, without it undergoing the *slightest distortion*, we must impose the requirement $\mathcal{H}(j\omega_0) = 1$; in this case, we will have $\mathcal{Y}(j\omega_0) = \mathcal{X}(j\omega_0)$, and from the inverse transform it immediately follows that $y_{\omega_0}(t) = x_{\omega_0}(t)$, that is, the output signal of the system will be an exact copy of the input signal.

Generalizing this conclusion, we can formulate the assertion that the frequency response of a low-pass filter with cutoff frequency ω_c should have the form

$$\mathcal{H}_{\ell p}(j\omega) = \begin{cases} 1 & |\omega| \leq \omega_c \\ 0 & |\omega| > \omega_c \end{cases} \tag{3.58}$$

so as to cut off the components of the signal with frequencies $\omega > \omega_c$, letting all the remaining frequencies pass. Expressing the frequency response in its exponential form, $\mathcal{H}_{\ell p}(j\omega) = |\mathcal{H}_{\ell p}(j\omega)|e^{j\measuredangle\mathcal{H}_{\ell p}(j\omega)}$, it is obvious that for $\mathcal{H}_{\ell p}(j\omega) = 1$, it must hold that $|\mathcal{H}_{\ell p}(j\omega)| = 1$ and $\measuredangle\mathcal{H}_{\ell p}(j\omega) = 0$. Low-pass filters described by the preceding function are known as *ideal low-pass filters*, as they allow the passage of frequency components with frequencies *lower than the cutoff frequency* without the slightest distortion while completely eliminating all remaining frequencies. We observe that filters of this category are characterized by *unit gain* and *zero phase*, while the transition from the *passband* to the *stopband* is instantaneous. Therefore, at the position of the cutoff frequency ω_c, the frequency response function of the filter will present *discontinuity*.

A second category of ideal filters are those characterized by a *unit gain* and phase value that vary *linearly in relation to the frequency ω of the input signal*; that is, the frequency response of these filters will have a magnitude of $|\mathcal{H}_{\ell p}(j\omega)| = 1$ and a phase of $\measuredangle\mathcal{H}_{\ell p}(j\omega) = -\alpha\omega$, where α is a constant of proportionality. As we noted in the section on linear phase systems, the effect of these systems on an input signal is limited to the introduction of a time delay, while the content of the signal remains unchanged. Therefore, in a sense, this case is also related to a distortion-free transmission process. Considering an ideal low-pass filter that belongs to this category, the magnitude of the frequency response function that describes it is formulated as

$$|\mathcal{H}_{\ell p}(j\omega)| = \begin{cases} 1 & |\omega| \leq \omega_c \\ 0 & |\omega| > \omega_c \end{cases} \tag{3.59}$$

On the other hand, the ideal high-pass, band-pass, and band-stop filters of linear phase will be described by similar equations, with their frequency response magnitude taking the form

$$|\mathcal{H}_{hp}(j\omega)| = \begin{cases} 1 & |\omega| \geq \omega_c \\ 0 & |\omega| < \omega_c \end{cases} \qquad \text{for high-pass filters}$$

$$|\mathcal{H}_{bp}(j\omega)| = \begin{cases} 1 & \omega_{c_1} \leq |\omega| \leq \omega_{c_2} \\ 0 & \text{anywhere else} \end{cases} \qquad \text{for band-pass filters}$$

$$|\mathcal{H}_{bs}(j\omega)| = \begin{cases} 0 & \omega_{c_1} \leq |\omega| \leq \omega_{c_2} \\ 1 & \text{anywhere else} \end{cases} \qquad \text{for band-stop filters}$$

The foregoing expressions concern the *measure of the frequency response* of the four fundamental types of ideal filters, while in all cases the phase of the filters is a linear function of the frequency and is given by a relation of the form

$$\measuredangle\mathcal{H}(j\omega) = \begin{cases} -\alpha\omega & \text{for the passband} \\ 0 & \text{for the stopband} \end{cases} \tag{3.60}$$

where α is a proportionality constant. From the preceding definitions it follows that the frequency responses of the ideal low-pass and the ideal high-pass filter are related to each other through the expression

$$\mathcal{H}_{hp}(j\omega) = 1 - \mathcal{H}_{\ell p}(j\omega) \tag{3.61}$$

while in a completely analogous way for the frequency responses of the ideal band-pass and the ideal band-stop filter we can write that

$$\mathcal{H}_{bs}(j\omega) = 1 - \mathcal{H}_{bp}(j\omega) \tag{3.62}$$

The graphical representation of the measure of the frequency response for these four categories of filter are presented in Fig. 3.16, while the corresponding phase diagrams are presented in Fig. 3.17.

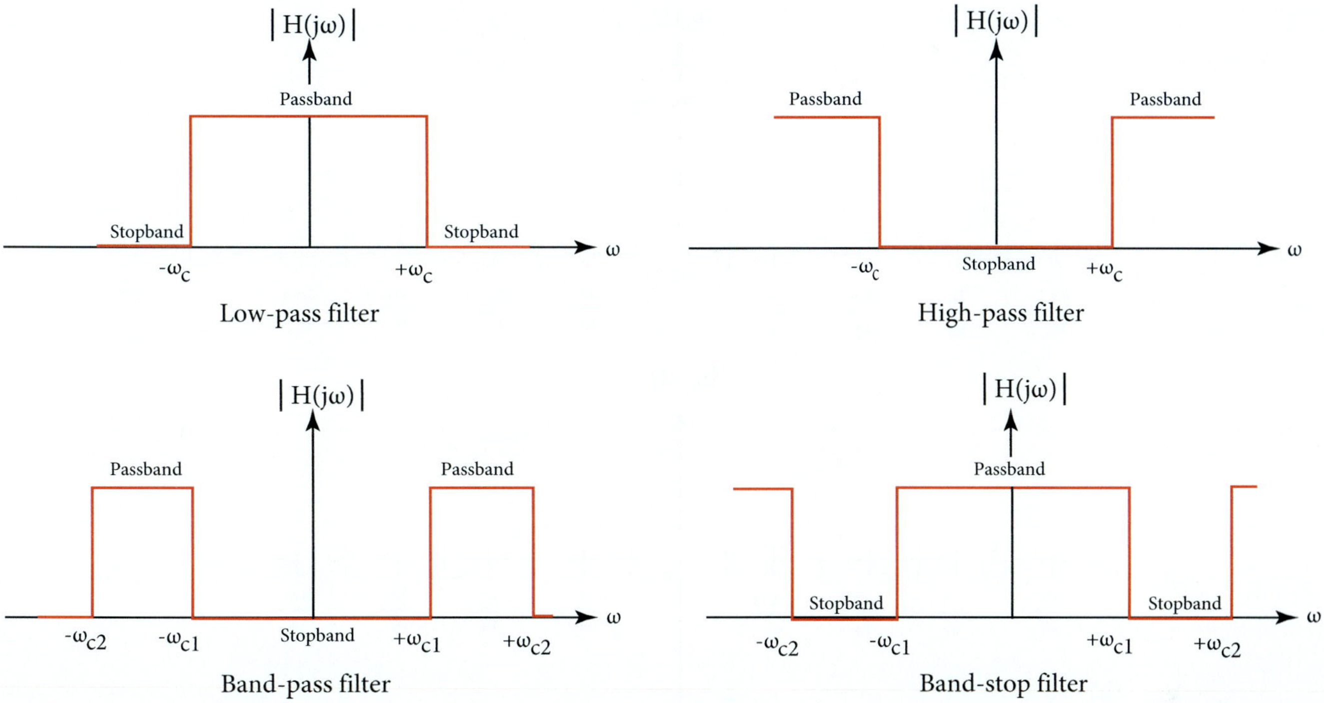

Fig. 3.16 Change *of magnitude* of function *frequency response* of the four different categories of ideal filter as function of frequency ω of input signal

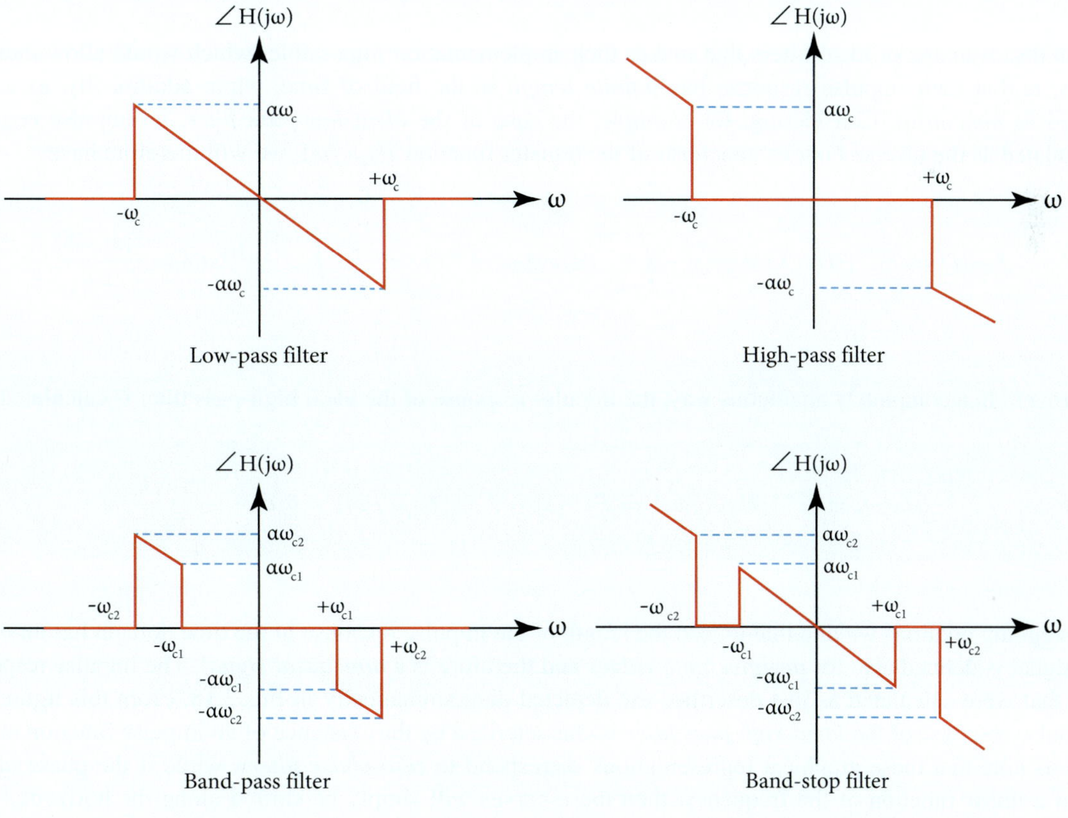

Fig. 3.17 Change *of phase* of function *frequency response* of the four different categories of ideal filter as function of frequency ω of input signal

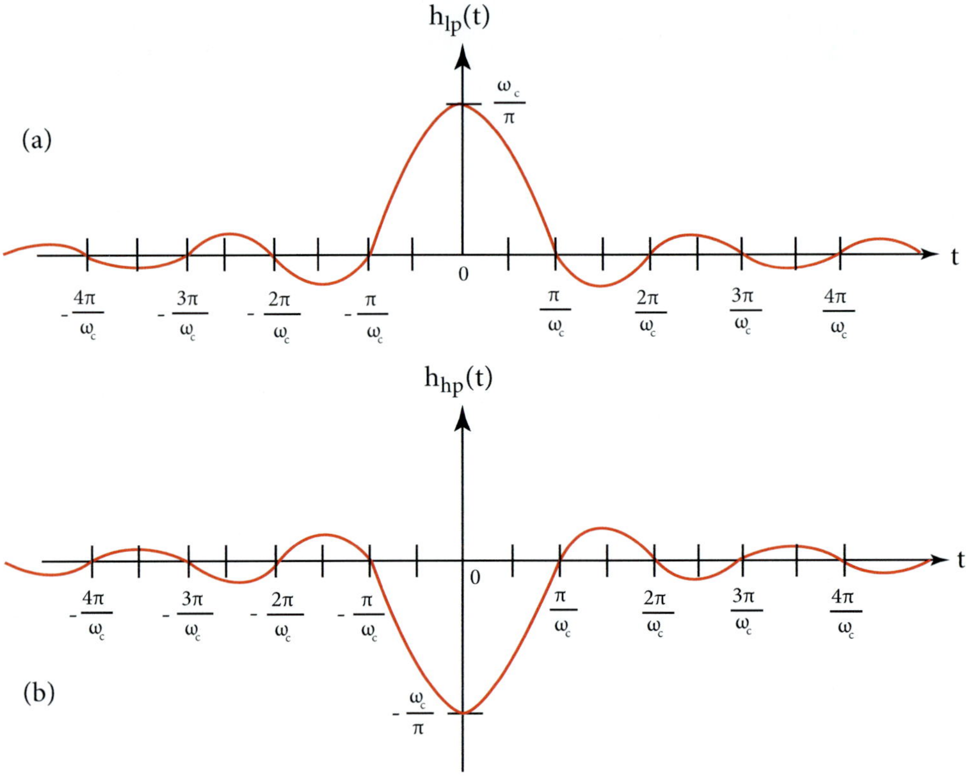

Fig. 3.18 Impulse response of (**a**) *ideal low-pass filter* and (**b**) *ideal high-pass filter*

The main disadvantage of ideal filters that makes their implementation impossible, which would allow their use in real applications, is that their impulse response has *infinite length* in the field of time, while additionally, as a signal, it is characterized as *noncausal*. Considering, for example, the case of the *ideal low-pass filter*, its impulse response $h_{\ell p}(t)$ will be calculated as the inverse Fourier transform of the transfer function $\mathcal{H}_{\ell p}(j\omega)$; we will therefore have

$$h_{\ell p}(t) = \mathcal{F}^{-1}\{\mathcal{H}_{\ell p}(j\omega)\} = \frac{1}{2\pi}\int_{-\infty}^{+\infty} \mathcal{H}_{\ell p}(j\omega)\, e^{j\omega t} d\omega = \frac{1}{2\pi}\int_{-\omega_c}^{+\omega_c} e^{j\omega t} d\omega = \frac{\sin(\omega_c t)}{\pi t}$$

as can be proven. In a completely analogous way, the impulse response of the ideal high-pass filter is calculated as

$$h_{hp}(\omega) = \mathcal{F}^{-1}\{\mathcal{H}_{hp}(j\omega)\} = \mathcal{F}^{-1}\{1\} - \mathcal{F}^{-1}\{\mathcal{H}_{\ell p}(j\omega)\} = \delta(t) - \frac{\sin(\omega_c t)}{\pi t}$$

From the foregoing relations we find that indeed the length of the impulse response in the time domain has an *infinite value*, while this signal is defined also for *negative time values* and therefore is a *noncausal signal*. The impulse responses $h_{\ell p}(t)$ and $h_{hp}(t)$ that were calculated as just described are depicted diagrammatically in Fig. 3.18. From this figure we observe that the impulse response of the *ideal high-pass filter* is characterized by the existence of an impulse function at the position $t = 0$. Let us note that these graphical representations correspond to *zero-phase filters*, while if the phase of the filter is not zero but a linear function of the frequency, then these curves will simply be shifted along the horizontal axis, in full agreement with our discussion in previous sections. Indeed, if we assume that the frequency response of the low-pass filter is given by the relation

$$\mathcal{H}_{\ell p}(\omega) = \begin{cases} e^{-j\omega\alpha} & |\omega| \leq \omega_c \\ 0 & |\omega| > \omega_c \end{cases} \quad (3.63)$$

so as to be characterized by *unit magnitude* and by a *linear phase proportional to frequency* of the form $\measuredangle\mathcal{H}(j\omega) = -\alpha\omega$ for frequency values $-\omega_c \leq \omega \leq \omega_c$, the application of the foregoing analysis will lead us to an impulse response of the form

$$h(t) = \frac{\sin[\omega_c(t-\alpha)]}{\pi(t-\alpha)} = \frac{\omega_c}{\pi}\mathrm{sinc}(t-\alpha) \quad (3.64)$$

a conclusion that confirms our claim and can of course be extended to describe the behavior of the other types of ideal filters.

3.6.2 Real Filters

As was already mentioned, the *ideal filters* are not physically implementable, since they are not causal, while their impulse response expressed in the time domain has *infinite length*. Let us note, however, that even if the construction of these filters was possible, certain characteristics make them unsuitable for their use in practical cases. Let us assume, for example, that we want to *separate* two signals—a process that is a typical example of the use of filters—whose spectra are characterized by some degree of overlap. It is obvious that if an ideal low-pass filter with a cutoff frequency located at the upper limit of the spectrum of one signal will result in the complete elimination of certain frequency components of the other signal (due to the existing overlap of the spectra), which will therefore be distorted under the influence of the filter.

A second disadvantage of ideal filters is that the frequency response function presents *discontinuities* at the positions of the cutoff frequencies, and therefore their step response will have a value near the limit of discontinuity greater than one, according to the *Gibbs phenomenon* (Sect. 3.2.5, page 200). We find, therefore, that ideal frequency selection filters are only *theoretical entities*, while in practice we construct low-pass, high-pass, band-pass, and band-stop filters that approximate as much as possible the operation of the corresponding ideal filter. It is obvious that the more accurate the approximation of the ideal filter by its realistic equivalent, the greater the degree of complexity and the cost of implementing the arrangement (we informally mention at this point that analog filters are constructed using *resistors*, *capacitors*, and *operational amplifiers*, while digital filters are constructed using *memory registers*, *multipliers*, and *adders*).

The main differences between real filters and their ideal counterparts, regarding their frequency response width, are that the value of this width *is not constant within the passbands*, while the transition from the passbands to the stopbands *is not abrupt but happens gradually*. More specifically, the following apply:

- The magnitude of the frequency response function $|\mathcal{H}(j\omega)|$ does not have a constant value along the filter's passbands but shows *fluctuations*. If the width of these fluctuations is equal to δ_p, then the value of the function's width will fluctuate between $1 \pm \delta_p$; in filter terminology, this width is known as a *passband ripple*. In a completely analogous way, the value of the function's magnitude $|\mathcal{H}(j\omega)|$ in the stopbands is usually not constant and equal to zero, but it also shows fluctuations whose width is equal to δ_s, and this width is known as a *stopband ripple*.
- The transition from a passband to a stopband is not abrupt, and therefore in no case is there a discontinuity at the cutoff frequency ω_c, as happens in ideal filters. On the contrary, the drop of the magnitude $|\mathcal{H}(j\omega)|$ is quite smooth, and the actual nullification of the value occurs at a different frequency ω_d, with the frequency band $\omega_d - \omega_c$ being called the *transition band*. The frequencies ω_c and ω_d are known as the *end of the passband* and the *end of the cutoff band*, respectively (this description, of course, depends on the type of filter being used), while the frequency range defined by these is known as the *width of the transition band*. As for the range of the passband frequency area, this is known as the *bandwidth of the filter*.

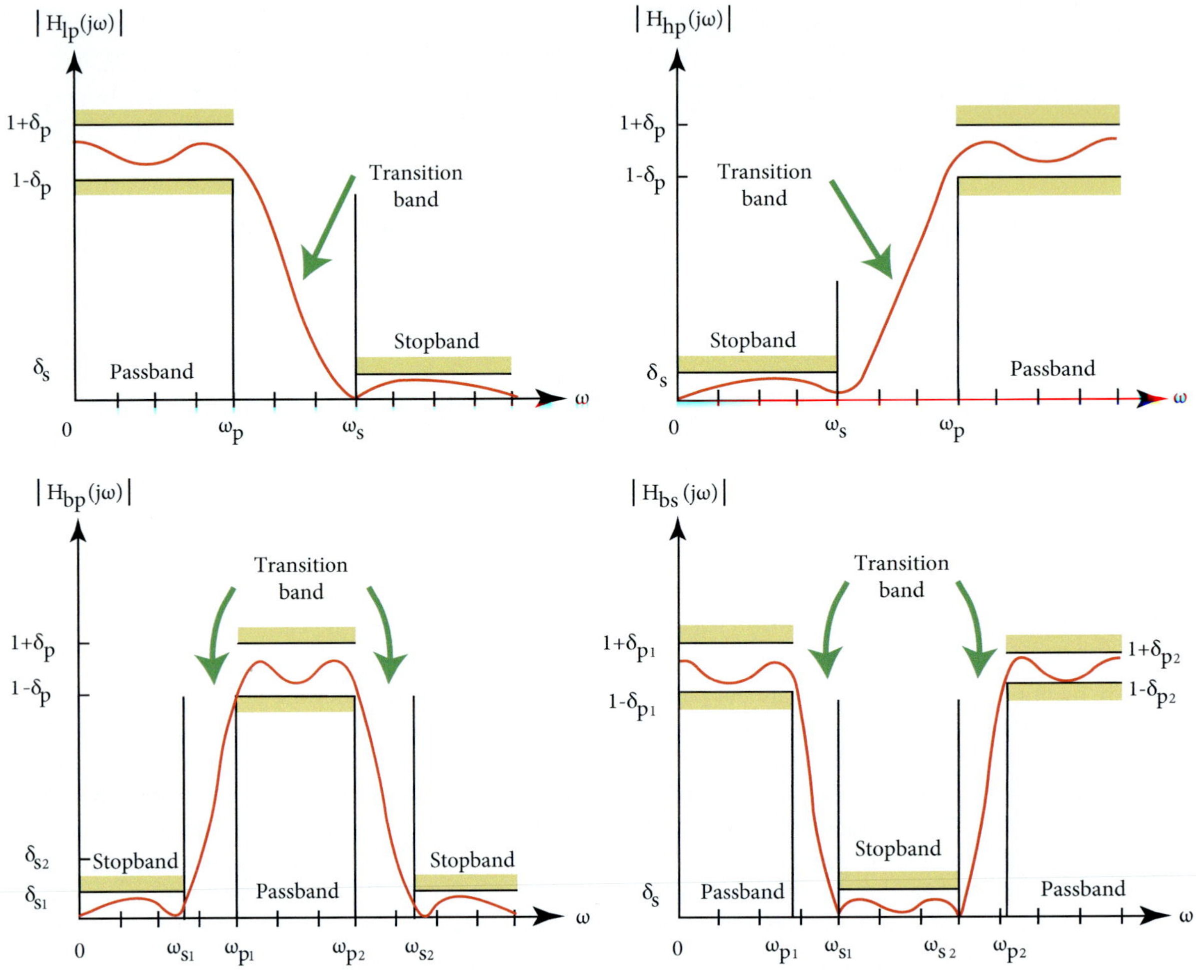

Fig. 3.19 Change in magnitude of frequency response function of the four different categories of real filters as function of frequency ω of input signal

The change in the magnitude of the frequency response function for the case of a realistic low-pass, high-pass, band-pass, and band-stop filter is presented in Fig. 3.19.

In the case of real filters, the cutoff frequency ω_c is usually located within the transition band and is defined as the value of the frequency at the position where the amplitude of the frequency response has been reduced to 0.7071 of its maximum value; this reduction expressed in logarithmic scale corresponds to a value equal to -3 dB. Therefore, for the case of low-pass filters and referring to Fig. 3.19, the value of the cutoff frequency will belong to the range of values $\omega_p \leq \omega_c \leq \omega_s$. As for the signal phase, it should also be determined in some cases and is usually chosen so as to be *linear*—or at least *almost linear*—a function of frequency (for those frequency values that belong to the pass bands of the filter) for reasons explained earlier.

First-Order Resistor–Capacitor Type Real Filters

A simple example of an electrical circuit that can be used as both a *low-pass* and *high-pass* filter is a *resistor–capacitor (RC)* circuit consisting of a *resistor*, R, and a *capacitor*, C, connected in series and contains a voltage source $V_s(t)$ that is treated as the input signal of the circuit $x(t)$. Assuming that the potential differences at the ends of the resistor and the capacitor are $V_R(t)$ and $V_c(t)$, it is proven that this circuit operates as a *low-pass* filter, if as output signal $y(t)$ we consider the voltage $V_c(t)$, and as *high-pass filter*, if as output signal we consider the voltage $V_R(t)$. Indeed, in the first case, the signals $x(t) = V_s(t)$ and $y(t) = V_c(t)$ are related through the equation

$$RC\frac{dV_c(t)}{dt} + V_c(t) = V_s(t) \tag{3.65}$$

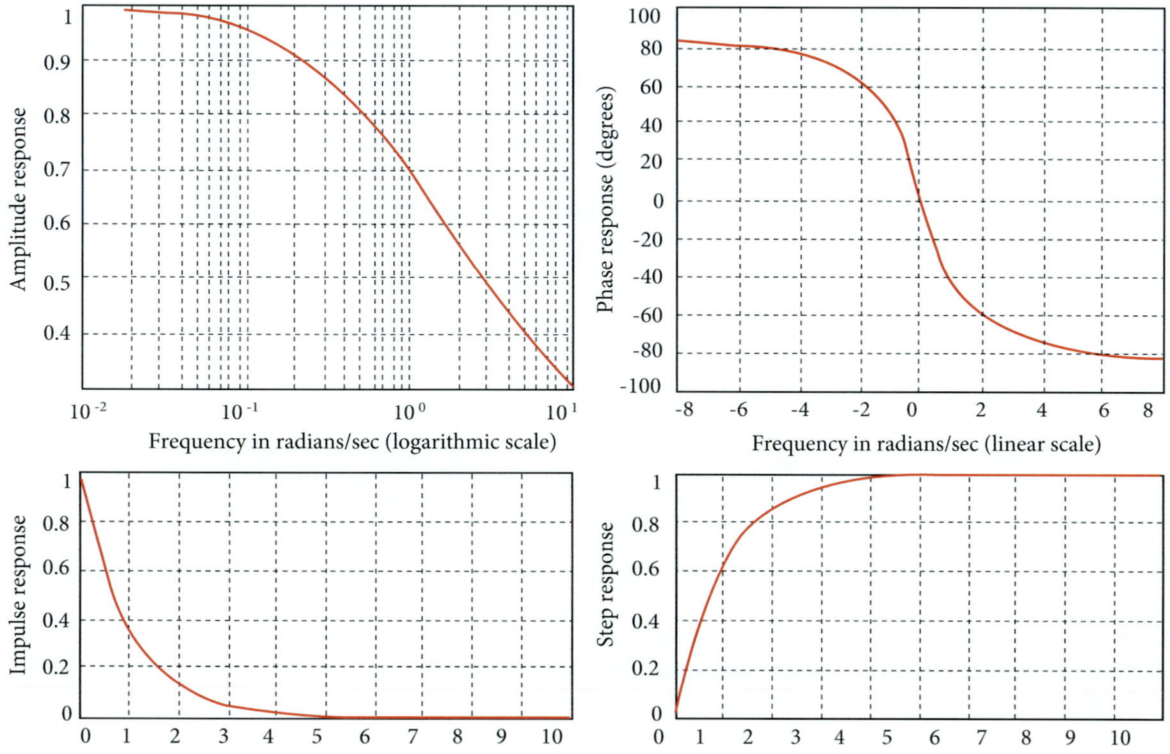

Fig. 3.20 Change in magnitude and phase of low-pass RC filter as function of (**a** and **b**) the frequency ω of the input signal and (**c** and **d**) the time change of the impulse and step response of the filter forparameter value $\tau = RC = 1$

which, *under zero initial conditions*, describes a LTI system whose frequency response $\mathcal{H}(j\omega)$ correlates the input with the output of the system as

$$V_c(t) = \mathcal{H}(j\omega)\, e^{j\omega t}$$

(in the foregoing relation we assumed that the input function of the system was $V_s(t) = e^{j\omega t}$). Replacing the foregoing expressions in the differential equation of the system it turns out that

$$\mathcal{H}(j\omega) = \frac{1}{1 + j\omega RC} = \frac{1}{\sqrt{1+\omega^2 R^2 C^2}} \exp\Big[- j\tan^{-1}(\omega RC)\Big] \tag{3.66}$$

from which we can calculate the impulse and step response from the known relations

$$h(t) = \mathcal{F}^{-1}\{\mathcal{H}(j\omega)\} = \mathcal{F}^{-1}\left\{\frac{1}{1 + j\omega RC}\right\} = \frac{1}{RC} e^{-t/RC} u(t)$$

$$s(t) = \int_{-\infty}^{t} h(\tau)\, d\tau = \frac{1}{RC}\int_{-\infty}^{t} e^{-\tau/RC} u(\tau)\, d\tau = \frac{1}{RC}\int_{0}^{t} e^{-\tau/RC} d\tau = (1 - e^{-t/RC})\, u(t)$$

The graphical representation of the magnitude and phase of the frequency response $\mathcal{H}(j\omega)$ is presented in Fig. 3.20a and b, while the time variation of the impulse and step responses are presented in Fig. 3.20c and d. Initially, we observe that the magnitude of the frequency response is equal to one for frequency value $\omega = 0$, has a value equal to $1/\sqrt{2} \approx 0.707$ for frequency value $\omega = 1/RC$, and tends to zero in the limit $\omega \to \infty$; therefore, this circuit indeed represents a *simple realistic low-pass filter*. However, we identify a problem if we want to allow the passage of very low frequency values, we must use a large value for the parameter $\tau = RC$, which in turn will significantly increase the time interval required for the step response to approach its final value. Conversely, if we want a faster step response, we must increase the value of the parameter τ, something that will result in the passage through the filter of components with a higher frequency. In other words, *improving the system's behavior in the time domain causes the deterioration of the corresponding behavior in the frequency domain*, and during the design of the appropriate filters in each case, we must choose the appropriate parameter values, so that we end up with a satisfactory solution.

On the other hand, if we consider the voltage $V_R(t)$ across the resistor as the system output, the differential equation that correlates the input and output signals $x(t) = V_s(t)$ and $y(t) = V_R(r)$ will have the form

$$RC\frac{dV_R(t)}{dt} + V_R(t) = RC\frac{dV_S(t)}{dt} \tag{3.67}$$

with the frequency response of the system given by the equation

$$\mathcal{H}(j\omega) = \frac{j\omega RC}{1 + j\omega RC} = \frac{\omega RC}{\sqrt{1 + \omega^2 R^2 C^2}} \exp\left[j \tan^{-1}\left(\frac{1}{\omega RC}\right)\right] \tag{3.68}$$

as can be proven by considering the input signal voltage $V_s(t) = e^{j\omega t}$.

If the frequency response function is expressed in the form

$$\mathcal{H}(j\omega) = \frac{j\omega RC}{1 + j\omega RC} = 1 - \frac{1}{j\omega RC}$$

then the impulse and step response of the circuit in question will be calculated as

$$h(t) = \mathcal{F}^{-1}\{\mathcal{H}(j\omega)\} = \mathcal{F}^{-1}\{1\} - \mathcal{F}^{-1}\left\{\frac{1}{j\omega RC}\right\} = \delta(t) - \frac{1}{RC}e^{-t/RC}u(t)$$

$$s(t) = \int_{-\infty}^{t} h(\tau)\, d\tau = \int_{-\infty}^{t} \delta(\tau)\, d\tau - \frac{1}{RC}\int_{-\infty}^{t} e^{-\tau/RC}u(\tau)\, d\tau = e^{-t/RC}u(t)$$

The graphical representation of the magnitude and phase of the frequency response $\mathcal{H}(j\omega)$ is presented in Fig. 3.21a and β, while the temporal variation of the impulse and step response are presented in Fig. 3.21γ and δ. Note that, while for the value $\omega = 1/RC$ it holds that $|\mathcal{H}(j\omega)| = 1/\sqrt{2} \approx 0.707$, things are different for the areas of low and high frequencies: At the limit $\omega \to 0$, it holds that $|\mathcal{H}(j\omega)| \approx 0$ while at the limit $\omega \to \infty$ it is $|\mathcal{H}(j\omega)| \approx 1$. Therefore, the system favors the passage of components with high frequencies and behaves as a *high-pass realistic filter*. The filters we presented in this section are known as *first-order continuous-time filters*, because the differential equation that describes their operation is a first-order linear differential equation with constant coefficients.

Problem 3.6.1 Let us consider a continuous low-pass filter whose frequency response has a magnitude

$$|\mathcal{H}(j\omega)| = \begin{cases} 1 & |\omega| \le 150\pi \\ 0 & \text{anywhere else} \end{cases}$$

Determine the impulse response of the filter $h(t)$ considering that it constitutes a real-value signal and that the group delay of the filter has the constant value $\tau(\omega) = 12$.

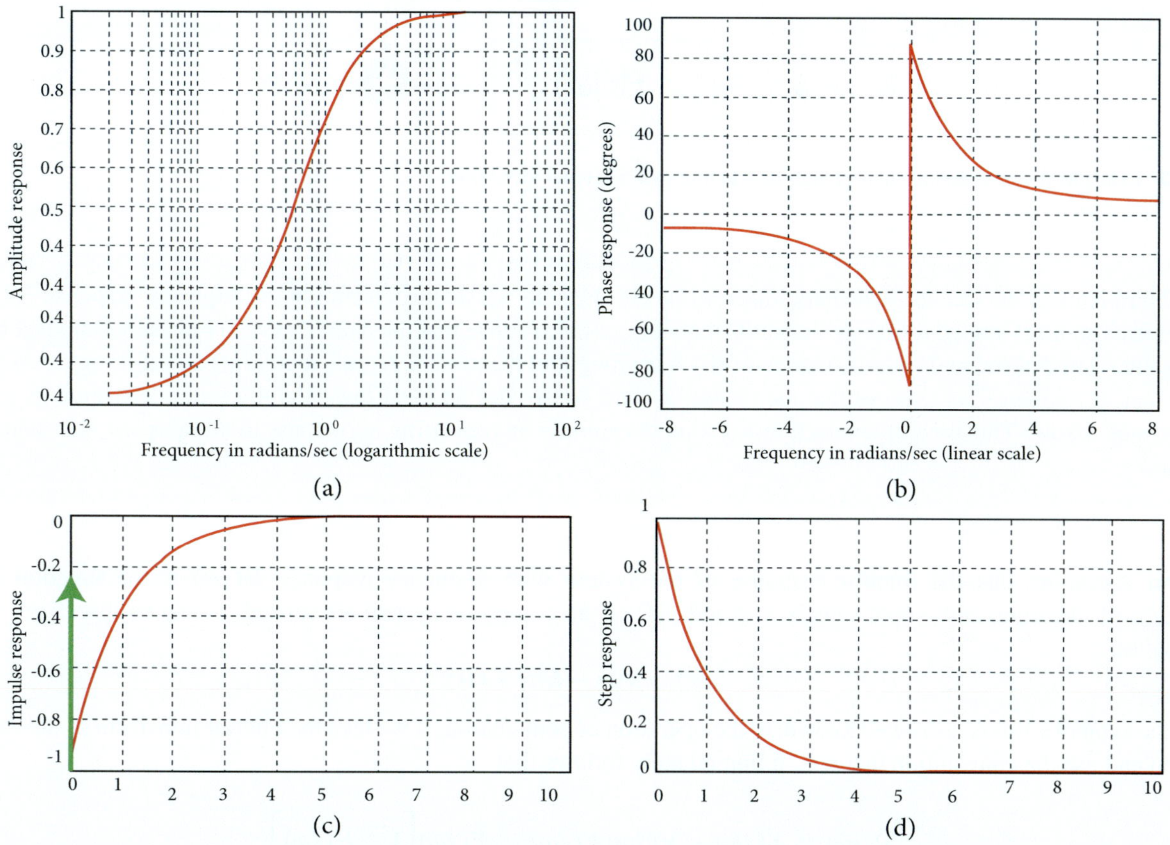

Fig. 3.21 Variation of magnitude and phase of high-pass RC filter as function of (**a** and **b**) frequency ω of input signal and (**c** and **d**) temporal variation of impulse and step response of filter for parameter value $\tau = RC = 1$

Solution From basic theory, it is known that as long as the *group delay* has a constant value regardless of frequency, the *phase* of the filter's frequency response—of which the group delay is the *negative first derivative*—will be a *linear function of frequency* of the form

$$\measuredangle\mathcal{H}(j\omega) = -\alpha\omega + \beta$$

However, considering that the impulse response $h(t)$ is a signal of *real values* and relying on the symmetry properties of the Fourier transform, we can conclude that *the phase of the system's frequency response will be characterized by odd symmetry*; therefore, $\measuredangle\mathcal{H}(-\omega) = -\measuredangle\mathcal{H}(j\omega)$, from which it immediately follows that $\beta = 0$. Therefore, the phase of the system's frequency response will have the form $\measuredangle\mathcal{H}(j\omega) = -\alpha\omega = -\tau(\omega)\omega = -12\omega$, where we used the definition of group delay and replaced its value.

On the other hand, the inverse transform of the function $|\mathcal{H}(j\omega)|$ is calculated as

$$g(t) = \mathcal{F}^{-1}\{|\mathcal{H}(j\omega)|\} = \frac{\sin(150\pi t)}{\pi t}$$

(as we proved in a previous exercise for the general case of the rectangular window function). Therefore, according to the *property of time shift* of the Fourier transform, the impulse response of the system will have the form

$$h(t) = \mathcal{F}^{-1}\{\mathcal{H}(j\omega)\} = \mathcal{F}^{-1}\{|\mathcal{H}(j\omega)|e^{-i12t}\} = g(t-12) = \frac{\sin[150\pi(t-12)]}{\pi(t-12)}$$

which is also the desired result. Note that this proof presupposes the characterization of the impulse response as a *real-valued signal*; in the opposite case, the constant β will not be equal to zero since the phase of the frequency response will no longer be characterized by odd symmetry and we will need to obtain more information in order to uniquely determine its value.

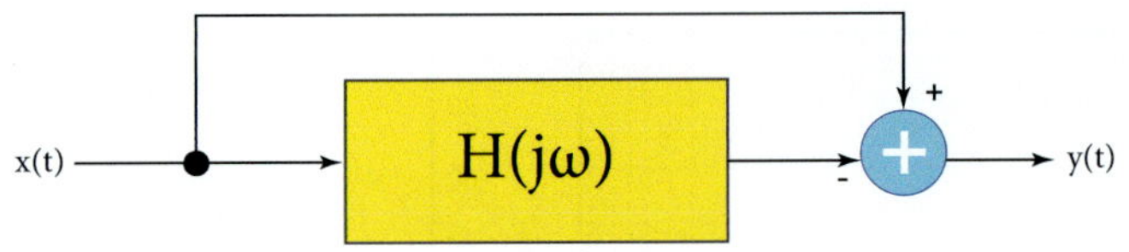

Fig. 3.22 Example of conversion from ideal high-pass to low-pass filter and vice versa

Problem 3.6.2 Consider the schematic diagram in Fig. 3.22, in which the system, with frequency response $\mathcal{H}(j\omega)$, represents an ideal low-pass filter with cutoff frequency $\omega_{\ell p}$. (a) Prove that the overall system represents an ideal high-pass filter and determine its cutoff frequency. (b) Assuming that the system $\mathcal{H}(j\omega)$ represents an ideal high-pass filter with cutoff frequency ω_{hp}, prove that the overall system constitutes an ideal low-pass filter and determine its cutoff frequency. (Note: This type of connection is a simple example of converting a low-pass to a high-pass filter and vice versa.)

Solution Assuming that the impulse response of the system with frequency response $\mathcal{H}(j\omega)$ is the function $h(t) = \mathcal{F}^{-1}\{\mathcal{H}(j\omega)\}$, the input and output signals $x(t)$ and $y(t)$ of the composite system are related, as can be discerned, through the equation

$$y(t) = x(t) - h(t) * x(t)$$

where the symbol $*$ refers, as is well known, to the operation of convolution. If we take the Fourier transform of the foregoing equation and use the convolution theorem, it immediately follows that

$$\mathcal{Y}(j\omega) = \mathcal{X}(j\omega) - \mathcal{H}(j\omega)\mathcal{X}(j\omega) = \mathcal{X}(j\omega)\Big[1 - \mathcal{H}(j\omega)\Big]$$

and therefore the frequency response function of the composite system will have the form

$$\mathcal{H}'(j\omega) = \frac{\mathcal{Y}(j\omega)}{\mathcal{X}(j\omega)} = 1 - \mathcal{H}(j\omega)$$

Taking into account that the frequency responses of an ideal low-pass and an ideal high-pass filter with a common cutoff frequency ω_c are related through the relationship $\mathcal{H}_{hp}(\omega) = 1 - \mathcal{H}_{\ell p}(\omega)$, we can conclude that if the function $\mathcal{H}(j\omega)$ represents an ideal low-pass filter with some cutoff frequency $\omega_{\ell p}$, then the function $\mathcal{H}'(j\omega) = 1 - \mathcal{H}(j\omega)$ will represent an ideal high-pass filter with cutoff frequency $\omega_{hp} = \omega_{\ell p}$ and impulse response of the form

$$h_{hp}(t) = \mathcal{F}^{-1}\{\mathcal{H}'(j\omega)\} = \mathcal{F}^{-1}\Big[1 - \mathcal{H}(j\omega)\Big] = \delta(t) - h(t) = \delta(t) - \frac{\sin(\omega_{\ell p}t)}{\pi t}$$

where for the extraction of the last relation we used the analytical expression for the impulse response of the ideal low-pass filter that we proved in previous pages (Sect. 3.6.1, page 283). On the other hand, if the function $\mathcal{H}(j\omega)$ represents an ideal high-pass filter with some cutoff frequency ω_{hp}, then we can claim that the function $\mathcal{H}'(j\omega) = 1 - \mathcal{H}(j\omega)$ will represent an ideal low-pass filter with cutoff frequency $\omega_{\ell p} = \omega_{hp}$ and impulse response of the form

$$h_{\ell p}(t) = \mathcal{F}^{-1}\{\mathcal{H}'(j\omega)\} = \mathcal{F}^{-1}\Big[1 - \mathcal{H}(j\omega)\Big] = \delta(t) - h(t) = \frac{\sin(\omega_{\ell p}t)}{\pi t}$$

These conversion processes of an ideal low-pass to an ideal high-pass filter and vice versa are presented in a diagrammatic way in Fig. 3.23a and b.

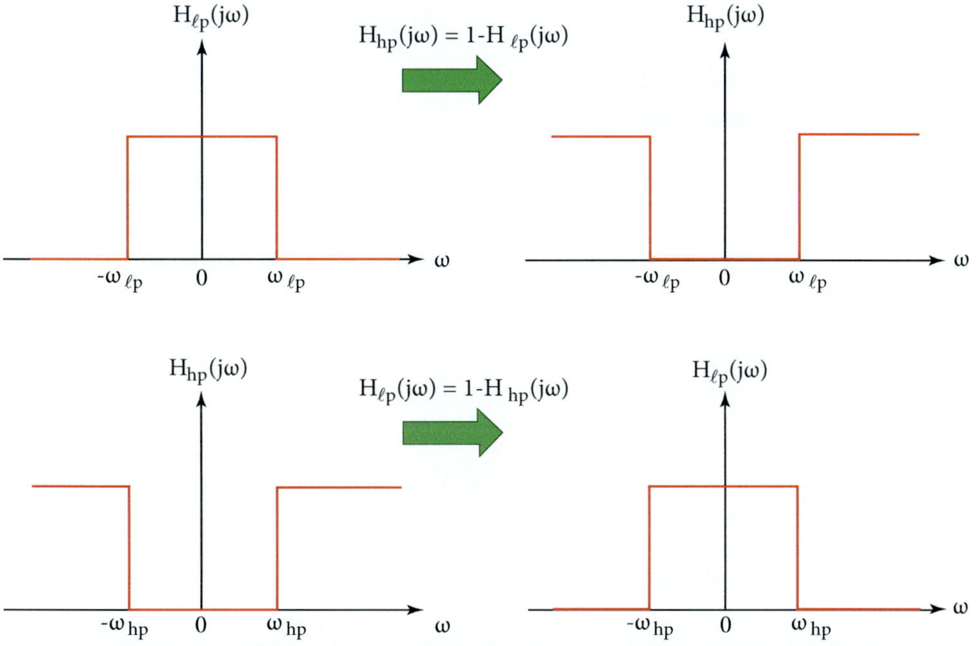

Fig. 3.23 The process of (**a**) converting the transfer function of an ideal low-pass filter to the transfer function of an ideal high-pass filter and (**b**) vice versa

Problem 3.6.3 Repeat Exercise 3.6.2, assuming that the schematic diagram of Fig. 3.22 refers to a realistic (nonideal) low-pass filter $\mathcal{H}(j\omega)$ characterized by the existence of a transition zone for the range of values $\omega_{\ell p1} \leq \omega \leq \omega_{\ell p2}$ and ripple width for the pass and cutoff regions with value δ_1 and δ_2, respectively. What is the effect of the phase of the frequency response on the character of the filter described by the aforementioned figure, assuming that it refers only to the magnitude of the frequency response?

Solution On the assumption that the system of Fig. 3.22 refers to a realistic filter, the magnitude of its frequency response will not abruptly go to zero at the position of the cutoff frequency $\omega = \omega_{\ell p}$ but will gradually decrease within a transition zone defined by the frequencies $\omega_{\ell p1} \leq \omega \leq \omega_{\ell p2}$. In addition, its value within the pass and cutoff zones will not be constant and equal to 1 and 0, respectively, but will be such that the following relations hold:

$$1 - \delta_1 < \mathcal{H}(j\omega) < 1 + \delta_1 \text{ for values } 0 \leq \omega \leq \omega_{\ell p1}$$
$$-\delta_2 < \mathcal{H}(j\omega) < \delta_2 \text{ for values } \omega > \omega_{\ell p2}$$

Applying the preceding analysis, the frequency responses $\mathcal{H}(j\omega)$ and $\mathcal{H}'(j\omega)$ of the original low-pass filter and the high-pass filter resulting from the aforementioned circuit are connected by the relation

$$\mathcal{H}'(j\omega) = 1 - \mathcal{H}(j\omega)$$

These functions are graphically represented in Fig. 3.24.

Let us now assume that the left image of Fig. 3.24 only describes the magnitude of the system's frequency response; let us symbolize this function with $\mathcal{G}(j\omega)$. In this case, there will be some unknown phase function $\measuredangle\mathcal{H}(j\omega)$, so that we can formulate the frequency response of the low-pass filter in the form $\mathcal{H}(j\omega) = \mathcal{G}(j\omega)e^{j\measuredangle\mathcal{H}(j\omega)}$. It is obvious that this phase function does not change the character of the filter, which will continue to be low-pass, since we will have

$$|\mathcal{H}(j\omega)| = |\mathcal{G}(j\omega)| \cdot |e^{j\measuredangle\mathcal{H}(j\omega)}| = |\mathcal{G}(j\omega)|$$

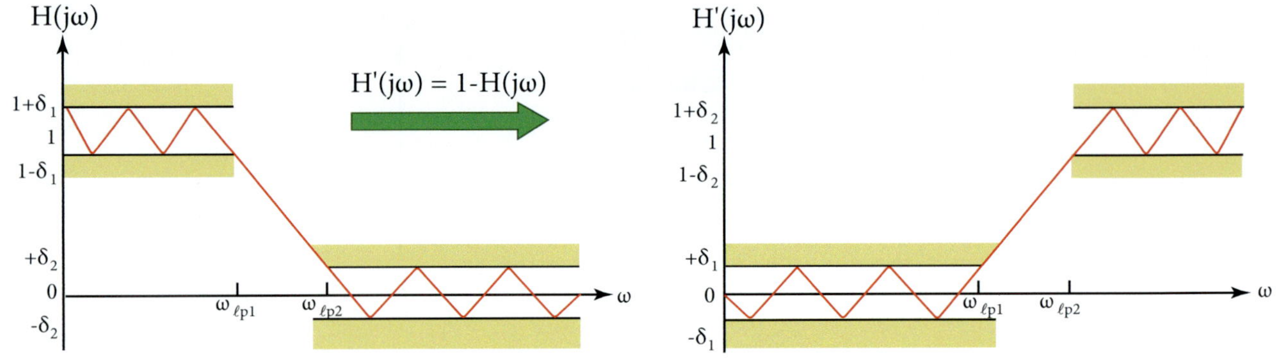

Fig. 3.24 Frequency response functions $\mathcal{H}(j\omega)$ and $\mathcal{H}'(j\omega)$ from Exercise 3.6.3

However, for the case of the composite system described by Fig. 3.22, things are different, and the resulting system does not necessarily refer to a high-pass filter. Indeed, in this case, the function of the composite system will have the form

$$\mathcal{H}'(j\omega) = 1 - \mathcal{H}(j\omega) = 1 - \mathcal{G}(j\omega)\, e^{j\measuredangle\mathcal{H}(j\omega)}$$

where it is not at all certain that it refers to a high-pass filter. In reality, if we take into account that

$$1 - |\mathcal{G}(j\omega)| \le |1 - \mathcal{G}(j\omega)\, e^{j\measuredangle\mathcal{H}(j\omega)}| \le 1 + |\mathcal{G}(j\omega)|$$

where obviously $|1 - \mathcal{G}(j\omega)e^{j\measuredangle\mathcal{H}(j\omega)}| = |\mathcal{H}'(j\omega)|$, it immediately follows that

$$1 - |\mathcal{G}(j\omega)| \le |\mathcal{H}'(j\omega)| \le 1 + |\mathcal{G}(j\omega)|$$

The foregoing relationship determines the range of values of the frequency response of the composite system, which in no case refers to a high-pass filter.

3.7 Problems

1. Determine the coefficients of the exponential complex Fourier expansion for each of the following continuous-time signals:
$$x(t) = \cos(\omega_0 t), \quad x(t) = \sin(\omega_0 t), \quad x(t) = \cos(2t) + \sin(9t), \quad x(t) = \sin^3 t$$
2. Prove that (a) if the signal $x(t)$ is real and even, then the coefficients of the Fourier expansion are real and characterized by even symmetry, that is, they satisfy the property $x_n = x_n^* = x_{-n}$, and (b) if the signal $x(t)$ is real and odd, then the coefficients of the Fourier expansion are imaginary and characterized by odd symmetry, that is, they satisfy the property $x_n = -x_{-n} = -(x_n^*)$.
3. Using the trigonometric Fourier expansion of the signal $x(t)$, prove that
$$\frac{1}{T}\int_{-T}^{+T} |x(t)|^2 dt = \frac{A_0^2}{2} + \sum_{n=1}^{\infty}(A_n^2 + B_n^2) \tag{3.69}$$
an equation that constitutes an alternative formulation of the *Parseval identity*.
4. Prove that for every positive integer $M = 1, 2, 3, \ldots$ the relation
$$\frac{A_0^2}{2} + \sum_{n=1}^{M}(A_n^2 + B_n^2) \le \frac{1}{T}\int_{-T}^{+T} |x(t)|^2 dt \tag{3.70}$$

holds, where A_n and B_n are the coefficients of the trigonometric Fourier expansion of the piecewise continuous-time signal $x(t)$, which is considered periodic with period T.

5. Prove that the Fourier expansion of the signal

$$x(t) = \begin{cases} 0 \text{ for } -5 < t < 0 \\ 3 \text{ for } 0 < t < 5 \end{cases}$$

with period $T = 10$ has the form

$$x(t) = \frac{3}{2} + \sum_{n=1}^{\infty} \frac{3[1 - \cos(n\pi)]}{n\pi} \sin\left(\frac{n\pi t}{5}\right)$$

6. Prove that the Fourier expansion of the signal $x(t) = t^2$ for the interval $0 < t < 2\pi$ with period $t = 2\pi$ is given by the relation

$$x(t) = \frac{4\pi^2}{3} + \sum_{n=1}^{\infty} \left[\frac{4}{n^2} \cos(nt) - \frac{4\pi}{n} \sin(nt)\right]$$

7. Prove that the Fourier expansion of the signal $x(t) = \sin t$ $(0 < t < \pi)$ in a cosine series has the form

$$x(t) = \frac{2}{\pi}\left\{1 - \sum_{n=2}^{\infty} \left[\frac{1 + \cos(n\pi)}{n^2 - 1}\right] \cos(nt)\right\}$$

8. Determine the exponential Fourier coefficients of the periodic signal

$$x(t) = \begin{cases} +1.5 \text{ for } 0 \le t \le 1 \\ -1.5 \text{ for } 1 \le t \le 2 \end{cases}$$

with fundamental frequency $\omega_0 = \pi$.

9. The exponential Fourier coefficients are given as follows:

$$\alpha_n = \begin{cases} 0 & \text{for } n = 0 \\ i^n \dfrac{\sin(n\pi/4)}{n\pi} & \text{for } n \neq 0 \end{cases} \quad \text{and} \quad \alpha_n = (-1)^n \frac{\sin(n\pi/8)}{2\pi n}, \quad \alpha_0 = \frac{1}{16}$$

Determine the periodic signal $x(t)$ that results for each case, with period value $T = 4$.

10. Let $x(t)$ be a real periodic signal whose Fourier series has the form

$$x(t) = \alpha_0 + 2\sum_{n=1}^{\infty} \left[B_n \cos(n\omega_0 t) - C_n \sin(n\omega_0 t)\right]$$

where $\alpha_n = B_n + iC_n$. Determine the coefficients α_n and β_n of the Fourier series of the *even* and *odd* components of the signal $x(t)$, which are defined by the relations

$$\text{Even}\{x(t)\} = \sum_{k=-\infty}^{+\infty} \alpha_n e^{in\omega_0 t} \quad \text{and} \quad \text{Odd}\{x(t)\} = \sum_{k=-\infty}^{+\infty} \beta_n e^{in\omega_0 t}$$

11. Prove that the Fourier series of the waveforms depicted in Fig. 3.25a,b are given respectively by the relations

$$x(t) = \frac{4A}{\pi} \sum_{n \text{ odd}} (-1)^{(n-1)/2} \frac{\cos(n\omega t)}{n} \quad \text{and} \quad x(t) = \frac{2A}{\pi} \sum_{n=1}^{\infty} (-1)^{n-1} \frac{\sin(n\omega t)}{n}$$

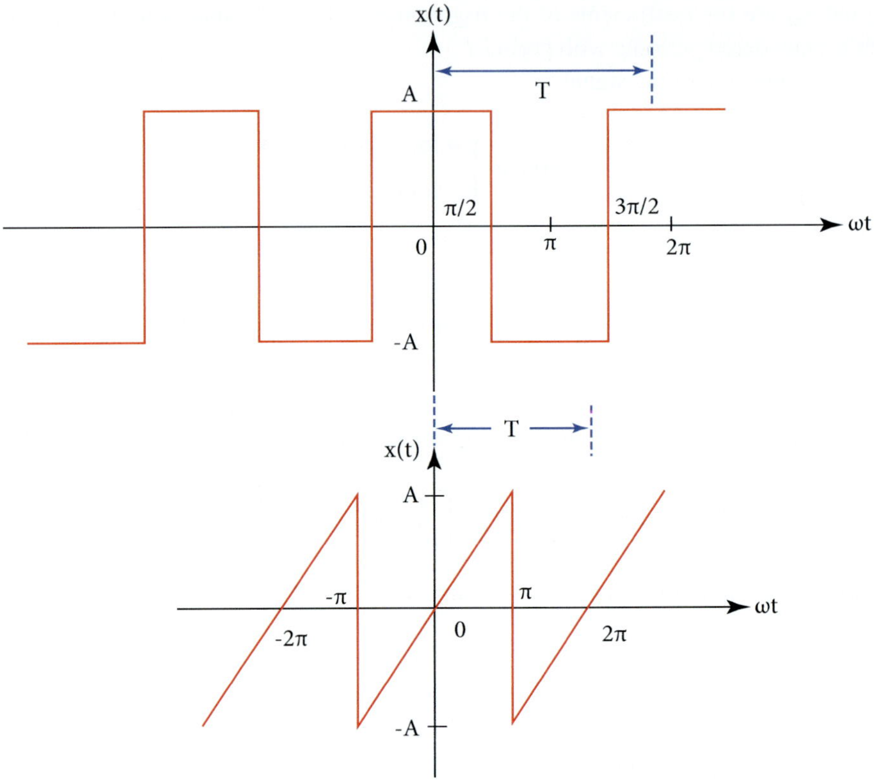

Fig. 3.25 Waveforms of Problem 11

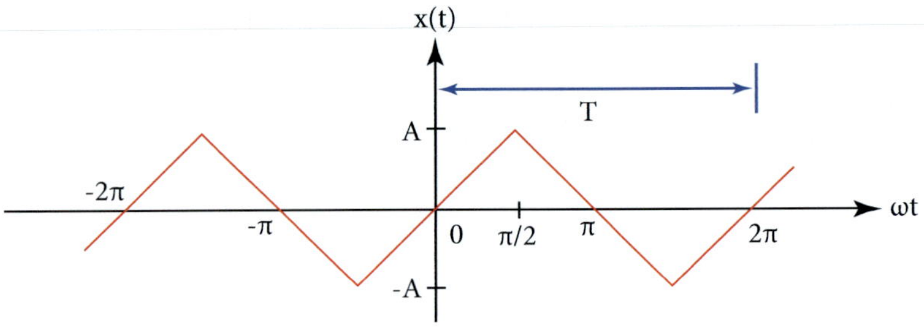

Fig. 3.26 Triangular waveform of Problem 12

12. Prove that the Fourier expansion of the triangular waveform of Fig. 3.26 is given by the equation

$$f(t) = \frac{8A}{\pi^2} \sum_{n \text{ odd}} (-1)^{(n-1)/2} \frac{\sin(n\omega t)}{n^2}$$

13. Prove that the Fourier expansion of the exponential waveform $x(t) = e^{-2t}$ $(-1 \leq t < 1)$ with period $T = 2$ depicted in Fig. 3.27 is given by the equation

$$x(t) = \sum_{n=-\infty}^{+\infty} \frac{(-1)^n (e^2 - e^{-2})}{2(2 + in\pi)} e^{in\pi t}$$

14. Prove that the Fourier expansion of the triangular waveform depicted in Fig. 3.28 is given by the equation

$$x(t) = \frac{8A}{\pi^2} \sum_{n=1}^{\infty} \frac{(-1)^{n+1}}{4n^2 - 1} \cos\left[\frac{(2n-1)t}{T}\right]$$

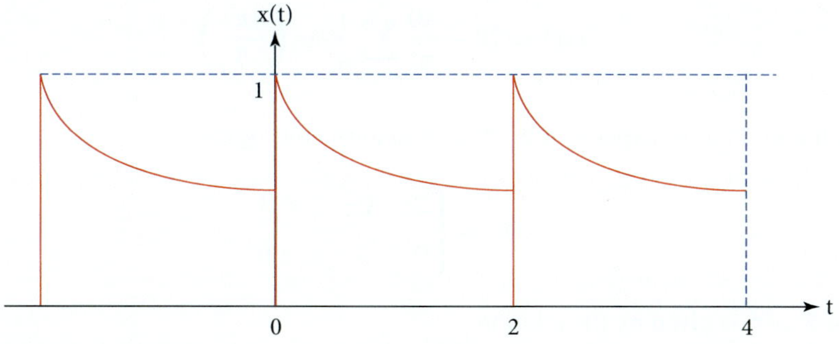

Fig. 3.27 Exponential waveform of Problem 13

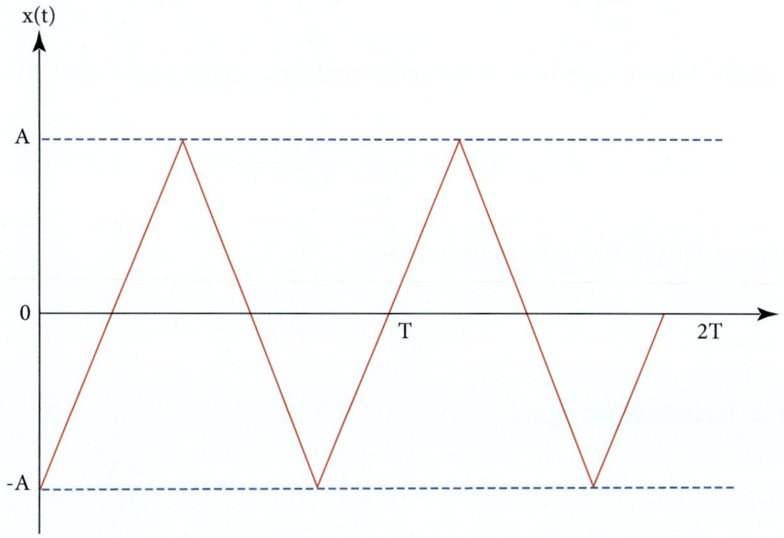

Fig. 3.28 Triangular waveform of Problem 14

15. Prove that the trigonometric Fourier expansion of the continuous-time signal

$$x(t) = \begin{cases} +8 & 0 < t < 2 \\ -8 & 2 < t < 4 \end{cases}$$

with a period value of $T = 4$ is given by the relation

$$x(t) = \frac{16}{\pi} \sum_{n=1}^{\infty} \frac{1 - \cos(n\pi)}{n} \sin\left(\frac{n\pi t}{2}\right)$$

16. Prove that the trigonometric expansion Fourier of the continuous-time signal

$$x(t) = \begin{cases} -t & -4 \le t \le 0 \\ +t & 0 \le t \le 4 \end{cases}$$

with a period value of $T = 8$ is given by the relation $x(t) = 2 - \frac{8}{\pi^2} \sum_{n=1}^{\infty} \frac{1 - \cos(n\pi)}{n^2} \cos\left(\frac{n\pi t}{4}\right)$.

17. Prove that the trigonometric Fourier expansion of the continuous-time signal $x(t) = 4t$ (for the value range $0 < t < 10$) with period value $T = 10$ has the form

$$x(t) = 20 - \frac{40}{\pi} \sum_{n=1}^{\infty} \frac{1}{n} \sin\left(\frac{n\pi t}{5}\right)$$

18. Prove that the trigonometric Fourier expansion of the continuous-time signal

$$x(t) = \begin{cases} 2t & 0 \le t < 3 \\ 0 & -3 < t < 0 \end{cases}$$

with a period value of $T = 6$ is given by the relation

$$x(t) = \frac{3}{2} + \sum_{n=1}^{\infty} \left\{ \frac{6[\cos(n\pi) - 1]}{n^2\pi^2} \cos\left(\frac{n\pi t}{3}\right) - \frac{6\cos(n\pi)}{n\pi} \sin\left(\frac{n\pi t}{3}\right) \right\}$$

19. Prove that the trigonometric Fourier expansion of the continuous-time signal $x(t) = \cos t \ (0 < t < \pi)$ has the form

$$x(t) = \frac{8}{\pi} \sum_{n=1}^{\infty} \frac{n \sin(2nt)}{4n^2 - 1}$$

20. Using Parseval's theorem, calculate the value of the integral

$$I = \int_0^{2\pi} \sin^{10}(t) dt$$

21. Calculate the Fourier coefficients of the signal

$$x(t) = 3\cos\left(2\pi f_0 t - \frac{\pi}{8}\right) + \sin^2\left(2\pi f_0 t + \frac{\pi}{3}\right)$$

22. If the complex signal $x(t)$ has Fourier coefficients of the form $\alpha_n = (-4 + j)/(\pi n)$, determine the Fourier coefficients of the complex conjugate signal $y(t) = x^*(t)$.
23. Decide whether the signal $x(t)$ with Fourier coefficients $\alpha_n = -\pi^2/(2jn^2)$ is a real or complex signal.
24. Determine and plot the amplitude and phase spectra of the signals

$$x(t) = \begin{cases} \cos(t) \text{ for } & -\frac{\pi}{2} \le t \le \frac{\pi}{2} \\ 0 & \text{elsewhere} \end{cases} \quad \text{and} \quad y(t) = \begin{cases} |t| \text{ for } & -1 \le t \le 1 \\ 0 & \text{elsewhere} \end{cases}$$

25. Considering the continuous-time signal given by the equation

$$x(t) = \frac{\sin(2t) + \sin(3t)}{2\sin(t)}$$

determine its period, calculate its Fourier series, and plot the amplitude and phase spectra.
26. Determine and plot the amplitude and phase spectra of the signal

$$x(t) = 4\cos\left(600\pi t + \frac{\pi}{3}\right) - 2\cos\left(1500\pi t - \frac{\pi}{8}\right)$$

27. Calculate the period and determine the amplitude and phase spectra of the signal

$$x(t) = 10 - 4\sin\left(300\pi t - \frac{\pi}{6}\right) + 6\cos(500\pi t)$$

28. What is the relationship between the spectra of the signals $x(t)$ and $y(t) = 2x(t)\cos(2\pi f_0 t)$?

29. Calculate the period of the signal $x(t) = \cos^2(5\pi t)\sin(16\pi t)$.
30. Verify the following direct-inverse Fourier transform relations

$$\mathcal{F}\{e^{-\alpha t^2+\beta t}\} = \sqrt{\frac{\pi}{\alpha}}\exp\left(-\frac{\omega^2}{4\alpha} - i\frac{\beta\omega}{2\alpha} + \frac{\beta^2}{4\alpha}\right)$$

$$\mathcal{F}\{sech(\alpha t)\} = \frac{\pi}{\alpha}\text{sech}\left(\frac{\pi\omega}{2\alpha}\right)$$

$$\mathcal{F}\{\sin(\alpha t^2)\} = -\sqrt{\frac{\pi}{\alpha}}\sin\left(\frac{\omega^2 - \alpha\pi}{4\alpha}\right)$$

$$\mathcal{F}\{\cos(\alpha t^2)\} = +\sqrt{\frac{\pi}{\alpha}}\cos\left(\frac{\omega^2 - \alpha\pi}{4\alpha}\right)$$

$$\mathcal{F}\{e^{-|\alpha|t}\} = \frac{2|\alpha|}{\alpha^2 + \omega^2}$$

$$\mathcal{F}\left\{\cos\left(\frac{\pi t}{2\alpha}\right)\right\} \quad (\text{for } -\alpha \le t \le +a) = \frac{4\alpha\pi\cos(\alpha\omega)}{\pi^2 - 4\alpha^2\omega^2}$$

$$\mathcal{F}\left\{\frac{1}{\alpha^2 + t^2}\right\} = \frac{\pi}{\alpha}e^{-\alpha|\omega|}$$

$$\mathcal{F}\{e^{-\alpha t}\sin(\omega_0 t)u(t)\} = \frac{\omega_0}{(\alpha + i\omega^2) + \omega_0^2}$$

31. Find the Fourier transform of the signal $x(t) = AT\text{sinc}(Tt)$.
32. Find the Fourier transform of the signal

$$x(t) = 2A\frac{\sin(\pi t)\sin(\pi t/2)}{\pi^2 t^2}$$

33. Find the Fourier transform of the periodic signal

$$x(t) = \sum_{n=1}^{3}\frac{A}{\pi n^2}\cos(800\pi nt)$$

34. Find the Fourier transform of the signals

$$x(t) = 2te^{-|t|} \qquad x(t) = \begin{cases}\sin(20\pi t) & \text{for } 0 \le t \le 2\\ 0 & \text{for } t < 0, t > 2\end{cases} \qquad x(t) = e^{-1.5|t|}\cos(3\pi t)$$

35. Find the Fourier transform of the signals

$$x(t) = \begin{cases}1 + \cos(5\pi t/T) & \text{for } |t| \le T/2\\ 0 & \text{for } |t| > T/2\end{cases} \quad \text{and} \quad x(t) = \begin{cases}e^{2t} & \text{for } t \le 0\\ 0 & \text{for } t > 0\end{cases}$$

36. Let us consider a periodic train of impulse functions of the form

$$x(t) = \sum_{k=-\infty}^{+\infty}\alpha_k\,\delta\,(t - k\Delta t)$$

where Δt is the distance between the graphs of two successive impulse functions. In order for this function to be periodic with period T, that is, to satisfy the condition $x(t + T) = x(t)$, there must be an integer number N such that the coefficients α_k satisfy the periodicity condition $\alpha_k = \alpha_{k+N}$ $(\forall k)$. In this case, the period T of the function $x(t)$, the value of the parameter N, and the time distance Δt are connected by a relation of the form $T = N\Delta t$.

Since the function $x(t)$ is periodic, then its Fourier transform will have the form of a train of functions infinitely shifted in the frequency domain $\delta(t)$, that is, it will be given by the relation

$$F(\omega) = \sum_{n=-\infty}^{+\infty} X_n \delta(\omega - n\Delta\omega)$$

where X_n are the coefficients of the Fourier expansion of the function $x(t)$ given by the equation

$$x(t) = \frac{1}{2\pi} \sum_{n=-\infty}^{+\infty} X_n \, e^{in\Delta\omega t}$$

Based on the foregoing description, prove that the coefficients X_n and α_k are connected with the direct-inverse transform relations

$$X_n = \frac{2\pi}{N\Delta t} \sum_{k=0}^{N-1} \alpha_k \, e^{-i(2\pi/N)nk} \quad \text{and} \quad \alpha_k = \frac{1}{N\Delta\omega} \sum_{n=0}^{N-1} X_n \, e^{i(2\pi/N)kn}$$

37. Determine the inverse Fourier transform of the function

$$\mathcal{X}(j\omega) = \frac{\beta - \alpha}{(\alpha + j\omega)(\beta + j\omega)}, \quad \beta \neq \alpha$$

38. Determine the inverse Fourier transform of the function

$$\mathcal{X}(j\omega) = \frac{1}{2 - \omega^2 + j3\omega}$$

39. Determine the inverse Fourier transform of the functions

$$\mathcal{X}_1(j\omega) = \sqrt{\frac{2}{\pi}} \frac{\sin(\omega t)}{\omega}, \quad \mathcal{X}_2(j\omega) = \sqrt{\frac{2}{\pi}} \frac{1}{\omega^2 + \alpha^2}, \quad \mathcal{X}_3(j\omega) = \frac{1}{\omega^4 + 1}, \quad \mathcal{X}_4(j\omega) = \sqrt{\frac{2}{\pi}} \frac{1}{(\omega^2 + \alpha^2)^2}$$

40. Prove that the inverse Fourier transform of the function

$$\mathcal{X}(j\omega) = \frac{6(3 + 2j\omega)}{(1 + j\omega)(4 + j\omega)(2 + j\omega)} \quad \text{is the signal} \quad x(t) = (2e^{-t} + 3e^{-2t} - 5e^{-4t})u(t)$$

41. Prove that the inverse Fourier transform of the function

$$\mathcal{X}(j\omega) = \pi\delta(\omega) + \frac{2(1 + j\omega)}{(1 + j\omega)^2 + 16} \quad \text{is the signal} \quad x(t) = (1 + 2e^{-t}\cos(4t))u(t)$$

42. Using the proof techniques of the properties of the Fourier transform of the function $f(t)$, prove the corresponding properties of the transform for the function $f(\mathbf{t}) = f(t_1, t_2, \ldots, t_n)$.
43. Calculate the output of a LTI system with an impulse response of the form $h(t) = e^{-2|t-1|}$ when the input signal is transmitted to its input

$$x(t) = \cos\left(2\pi t + \frac{\pi}{2}\right)$$

44. If for the signal $x(t)$ and its Fourier transform $\mathcal{X}(j\omega)$ we know that (a) the signal $x(t)$ is real and has only positive values, (b) it holds that $\mathcal{F}^{-1}\{(2+j\omega)\mathcal{X}(j\omega)\} = Ae^{-3t}u(t)$ where A is a constant value, and (c) it holds that $\int_{-\infty}^{\infty} |\mathcal{X}(j\omega)|^2 d\omega = (4/15)$, determine a closed expression for the signal $x(t)$.
45. The following pairs of signals $[h(t), x(t)]$ are given, which express the impulse response and the input signal of a continuous-time LTI system, respectively:

(a) $h(t) = u(t), \; x(t) = te^{-3t}u(t)$, (b) $h(t) = u(t), \; x(t) = t^2e^{-2t}u(t)$, (c) $h(t) = e^{-t}u(t), \; x(t) = u(t)$

(a) Using the convolution theorem, calculate the output of the system $y(t)$ for each pair of input signal and impulse response. (b) For each system with impulse response $h(t)$, calculate the frequency response function $\mathcal{H}(j\omega)$ and the amplitude and phase responses that arise from it.

46. For each of the systems described by the differential equations

$$\frac{d^2y(t)}{dt^2}+8\frac{dy(t)}{dt}+12y(t)=\frac{dx(t)}{dt}-x(t)$$
$$\frac{d^4y(t)}{dt^4}+2\frac{d^2y(t)}{dt^2}=x(t)$$
$$\frac{d^3y(t)}{dt^3}-5\frac{d^2y(t)}{dt^2}-\frac{dy(t)}{dt}+5y(t)=3\frac{dx(t)}{dt}+x(t)$$

(a) Determine its frequency and impulse response and (b) determine and plot the amplitude and phase response as a function of the frequency ω of the input signal.

47. Calculate the magnitude, phase, and group delay of the LTI systems with frequency response

$$\mathcal{H}_1(\omega)=\frac{1}{\left(1-\frac{1}{4}e^{-j\omega}\right)\left(1-\frac{3}{4}e^{-j\omega}\right)}\quad\text{and}\quad\mathcal{H}_2(\omega)=\frac{1}{\left(1-\frac{1}{4}e^{-j\omega}\right)\left(1+\frac{3}{4}e^{-j\omega}\right)}$$

48. Considering two LTI systems with frequency responses

$$\mathcal{H}_1(\omega)=\frac{1+\frac{1}{2}e^{-j\omega}}{\left(1+\frac{1}{4}e^{-j\omega}\right)}\quad\text{and}\quad\mathcal{H}_2(\omega)=\frac{\frac{1}{2}+e^{-j\omega}}{\left(1+\frac{1}{4}e^{-j\omega}\right)}$$

prove that $|\mathcal{H}_1(\omega)|=|\mathcal{H}_2(\omega)|$ and that for $\omega>0$ it holds that $\tau_2(\omega)>\tau_1(\omega)$, where $\tau_1(\omega)$ and $\tau_2(\omega)$ are the group delays of the two systems.

49. Starting from the defining equations of the frequency response of the ideal band-pass and band-stop filters, calculate the corresponding impulse response.

50. Reformulate the defining equation of the impulse response of the ideal, high-pass, band-pass, and band-stop filters, so that you can describe filters whose phase is a linear function of the frequency of the form $\measuredangle\mathcal{H}(j\omega)=-\alpha\omega$ for the appropriate frequency ranges.

51. The construction of a RC band-pass filter can be achieved by connecting in series a RC circuit that operates as a low-pass filter with a second RC circuit that operates as a high-pass filter. If R_1 and C_1 are the values of the resistance and the capacitor of the low-pass filter, while R_2 and C_2 are the values of the resistance and the capacitor of the high-pass filter, design the RC circuit of the band-pass filter and prove that its frequency response is given by the relation

$$\mathcal{H}(j\omega)=\left(\frac{1}{1+j\omega R_1C_1}\right)\left(\frac{i\omega R_2C_2}{1+j\omega R_2C_2}\right)$$

52. Determine the impulse and step response of the RC band-pass filter from Exercise 51, and draw their time variation along with the amplitude and phase responses of the frequency response function $\mathcal{H}(j\omega)$.

53. Considering a system with a frequency response function of the form

$$\mathcal{H}(j\omega)=\begin{cases}e^{-j\pi 2} & \omega>0\\ e^{j\pi 2} & \omega<0\end{cases}$$

which causes a phase shift by an angle $\varphi=-90^\circ$, determine (a) its impulse response $h(t)$, (b) the output $y(t)$ of the system for an arbitrary input signal $x(t)$, and (c) the output $y(t)$ of the system when the input signal $x(t)=\cos(\omega_0 t)$ is transmitted.

54. Considering a causal continuous-time LTI system with a transfer function of the form $\mathcal{H}(j\omega) = \mathcal{A}(\omega) + j\mathcal{B}(\omega)$, it should be shown that if its impulse response does not contain impulse functions at the beginning $t = 0$, then the functions $\mathcal{A}(\omega)$ and $\mathcal{B}(\omega)$ are related to each other as

$$\mathcal{A}(\omega) = \frac{1}{\pi}\int_{-\infty}^{+\infty} \frac{\mathcal{B}(\lambda)}{\omega - \lambda} d\lambda \qquad \text{and} \mathcal{B}(\omega) = -\frac{1}{\pi}\int_{-\infty}^{+\infty} \frac{\mathcal{A}(\lambda)}{\omega - \lambda} d\lambda$$

(see also Sect. 11.8 on the Hilbert transform on page 979).

55. Find the output of a low-pass filter with a frequency response of the form

$$\mathcal{H}(j\omega) = \begin{cases} e^{-i\omega} & -2 < \omega < 2 \\ 0 & \text{anywhere else} \end{cases}$$

when the continuous-time signal

$$x(t) = \sum_{k=1}^{\infty} \frac{1}{k} \cos\left(\frac{k\pi}{2} + 30^\circ\right)$$

56. Calculate the cutoff frequency of an ideal low-pass filter that receives a continuous input signal at its input of the form $x(t) = e^{-4t}u(t)$ allows only 50% of its energy to pass through.

57. Prove that the response of a low-pass RC filter with parameter value $\tau = RC = 1$ to the input signal

$$x(t) = \cos(0.5t) + \cos(10t)$$

has the form

$$y(t) = 0.995\cos(0.5t - 0.46) + 0.01\cos(10t - 1.47)$$

and comment on the result.

4 The Laplace Transform

Pierre-Simon, marquis de Laplace (1749–1827)

4.1 Introduction

A very useful transform that finds application in problems of studying *linear and time-invariant continuous-time systems*, *control systems based on the feedback mechanism*, as well as *electrical circuits*, is the *Laplace transform*, which was invented by the French physicist-mathematician *Pierre-Simon Laplace* (1749–1827) and allows the representation of a continuous-time signal $x(t)$ in the complex plane. Considering the general case of a signal $x(t)$ that meets certain conditions—which, let us note are sufficient, but not necessary—its transform according to Laplace is defined as

$$\mathcal{X}(s) = \mathcal{L}\{x(t)\} \triangleq \int_{-\infty}^{+\infty} x(t)\, e^{-st} dt. \tag{4.1}$$

A. I. Margaris, *Continuous and Discrete-Time Signals and Systems*,
https://doi.org/10.1007/978-3-031-81209-5_4

In the above equation, the variable s is of *complex nature*, it has the form $s = \sigma + j\omega$—where σ and ω are its real and imaginary parts—and is known as *complex frequency*. The characterization is justified by the fact that if we replace the variable s in the above relation with its equivalent notation $\sigma + j\omega$, it will take the form

$$\mathcal{X}(s) = \mathcal{X}(\sigma + j\omega) = \int_{-\infty}^{+\infty} x(t)\, e^{-(\sigma+j\omega)t}\, dt = \int_{-\infty}^{+\infty} \left\{ x(t)\, e^{-\sigma t} \right\} e^{-j\omega t}\, dt, \tag{4.2}$$

which is nothing more than the *Fourier transform* of the signal $y(t) = x(t)\, e^{-\sigma t}$. The presence of the exponential term $e^{-\sigma t}$ in the defining equation of the Laplace transform allows the convergence of the above integral, even in cases where the corresponding Fourier transform *does not exist*, a fact that allows the existence of signals that have a Laplace transforms, but not a Fourier transform. It is obvious that in the case where the relation $\sigma = 0$ holds, the two transformations coincide.

Generally speaking, the Fourier and Laplace transforms are special cases of a general category of transformations known as *generalized linear integral transformations*. These transformations are understood to be mathematical operators that act on an input signal $x(t)$ and produce an output signal $y(\xi)$, which is defined as

$$y(\xi) = T[x(t)] = \int_{\xi_1}^{\xi_2} k(\xi, t)\, x(t)\, dt. \tag{4.3}$$

In the above equation, the function $k(\xi, t)$ is called the *transformation kernel*. If the function of the kernel has an inverse function $k^{-1}(t, \xi)$, then there is also the *inverse integral transform*, which is defined as

$$x(t) = T^{-1}[y(\xi)] = \int_{t_1}^{t_2} k^{-1}(t, \xi)\, y(\xi)\, d\xi. \tag{4.4}$$

Let us note that in practice there are also other *generalized linear integral transformations* that correspond to specific values of the limits (ξ_1, ξ_2) and (t_1, t_2) and the kernel function $k(\xi, t)$, and find application in various problems of physics and mathematics. The most important of these transformations are presented in Table 4.1.

Table 4.1 Generalized linear integral transformations

Name	$k(\xi, t)$	ξ_1	ξ_2	$k^{-1}(t, \xi)$	t_1	t_2
Fourier	$\frac{e^{-j\xi t}}{\sqrt{2\pi}}$	$-\infty$	$+\infty$	$\frac{e^{j\xi t}}{\sqrt{2\pi}}$	$-\infty$	$+\infty$
Mellin	$t^{\xi-1}$	0	$+\infty$	$\frac{t^{-\xi}}{2\pi j}$	$c - j\infty$	$c + j\infty$
Laplace	$e^{-\xi t}$	$-\infty$	$+\infty$	$\frac{e^{\xi t}}{2\pi j}$	$c - j\infty$	$c + j\infty$
Hilbert	$\frac{1}{\pi(\xi - t)}$	$-\infty$	$+\infty$	$\frac{1}{\pi(\xi - t)}$	$-\infty$	$+\infty$
Hartley	$\frac{\cos(\xi t) + \sin(\xi t)}{2\pi}$	$-\infty$	$+\infty$	$\frac{\cos(\xi t) + \sin(\xi t)}{2\pi}$	$-\infty$	$+\infty$
Identity	$\delta(\xi - t)$	$t_1 < \xi$	$t_2 > \xi$	$\delta(t - \xi)$	$\xi_1 < t$	$\xi_2 > t$
Hankel	$t J_v(\xi t)$	0	$+\infty$	$\xi J_v(\xi t)$	0	$+\infty$
Abel	$\frac{2t}{\sqrt{t^2 - \xi^2}}$	ξ	$+\infty$	$-\frac{1}{\pi\sqrt{\xi^2 - t^2}} \frac{d}{d\xi}$	t	$+\infty$

A special form of the Laplace transform is the one that arises if we use the value $\alpha = 0$ as the lower limit in the defining equation and not the value $\alpha = -\infty$ that we used previously. In this case, the transform

$$\mathcal{X}(s) \equiv \mathcal{L}\{x(t)\} = \int_{0^-}^{\infty} x(t)\, e^{-st} dt = \lim_{\varepsilon \to 0^+} \int_{-\varepsilon}^{\infty} x(t)\, e^{-st} dt \tag{4.5}$$

is called a *one-sided unilateral (or one-sided) transform,* so as to distinguish it from the one we defined initially and which is known as a *bilateral (or two-sided) Laplace transform.*

From the defining equation of the unilateral Laplace transform, it is not difficult to see that although we can treat the lower limit both as a *right limit* ($\alpha \to 0^+$) and as a *left limit* ($\alpha \to 0^-$), in practice, the *second* approach is followed. In order to understand why this happens, we must take into account that the unilateral Laplace transform, being a generalized linear integral transform, accepts as input a signal $x(t)$ and produces as output another signal $\mathcal{L}\{x(t)\}$. Assuming now that the input signal of the transform is the function $\delta(0)$—that is, the *impulse function* $\delta(t)$ at time $t = 0$—it is obvious that, if we want to include in the problem this excitation and any initial conditions associated with it, we must set the lower limit of integration to the value 0^-, so that the time $t = 0$ is taken into account. Conversely, if such excitations do not exist, then we can without any problem use as the lower limit of the integral the value 0^+, since obviously the value of the integral in the interval $(0^-, 0^+)$ is equal to zero.

It is obvious that two continuous-time signals that differ only in the range of values $t < 0$, whereas they are identical for $t > 0$, *have different bilateral but the same unilateral transform.* On the other hand, a causal signal $x(t)$—with values $x(t) = 0$ for $t < 0$—has a unilateral Laplace transform, which *coincides* with the corresponding bilateral transform, in terms of both their functional representation and their properties, many of which are the same for both transforms; it should be noted, however, that there are certain properties in which the two transforms differ. For this reason, the proof of the properties of the Laplace transform will be carried out using the defining equation of the unilateral transform that leads to easier proofs, whereas wherever there is differentiation between the two transforms this will be explicitly mentioned.

The *unilateral Laplace transform* is a very valuable tool for analyzing and calculating the response of a causal system to a causal input signal, when the system is described by *linear differential equations with constant coefficients,* which are characterized by the existence of *nonzero initial conditions.* Detailed examples of using this transform in problems of this type will be given in the following sections.

4.2 Convergence Area and Existence Theorem

As already mentioned, during the calculation processes of the Laplace transform, in addition to the analytical form of the function $\mathcal{X}(s)$, we should also determine the region of convergence of the transform, that is, *the area of the complex level for which the above integral exists, converging toward some finite value.* Let it be noted that the integrals that appear in the defining equations of the Laplace transform are *generalized integrals* that are defined on nonfinite intervals and are calculated as *limits of integrals over finite intervals of values.* A typical case of a generalized integral is the integral

$$\int_{\alpha}^{\infty} x(t)\, dt = \lim_{\xi \to \infty} \int_{\alpha}^{\xi} x(t)\, dt, \tag{4.6}$$

where ξ is a positive real number. If the integral over the interval $[\alpha, \xi]$ exists for every $\xi > \alpha$ and if the limit of the integral exists for the case $\xi \to \infty$, then the generalized integral converges to this limit value, whereas in the opposite case, the integral does not converge, that is, it essentially does not exist.

In a completely analogous way, we can define the generalized integral for the interval $(-\infty, \beta]$ as

$$\int_{-\infty}^{\beta} x(t)\, dt = \lim_{\xi \to \infty} \int_{\xi}^{\beta} x(t)\, dt \tag{4.7}$$

and the generalized integral for the interval $(-\infty, +\infty)$ as

$$\int_{-\infty}^{+\infty} x(t)\,dt = \lim_{\xi\to-\infty}\int_{\xi}^{\alpha} x(t)\,dt + \lim_{\xi\to+\infty}\int_{\alpha}^{\xi} x(t)\,dt \tag{4.8}$$

It turns out that, for the Laplace transform of a continuous-time signal $x(t)$ to exist, this signal must in principle be a *sectionally continuous function* over some interval of values $[\alpha, \beta]$, i.e., there must be points $\alpha = t_0 < t_1 < t_2 < t_n < t_n = \beta$ with the following properties: (a) the signal $x(t)$ is continuous in every open subspace $t_{i-1} < t < t_i$ $(i = 1, 2, \ldots, n)$, i.e., there are functions $x_i(t)$ $(i = 1, 2, \ldots, n)$ continuous in the interval $(-\infty, +\infty)$ such that $\forall t \notin \{t_0, t_1, \ldots, t_n\}$ is

$$x(t) = \begin{cases} x_1(t) & t_0 < t < t_1 \\ x_2(t) & t_1 < t < t_2 \\ \ldots & \ldots\ldots\ldots \\ x_n(t) & t_{n-1} < t < t_n \end{cases} \tag{4.9}$$

and (b) the limit of the signal $x(t)$ should be *finite* for the cases $t \to t_{i-1}^{+}$ and $t \to t_i^{-}$. If the signal $x(t)$ is a piecewise continuous function in the interval $\alpha \le t \le \beta$ for every $t \ge \alpha$, then the property of *piecewise continuity* exists for every $\beta > \alpha$, a fact that, as proven, ensures the existence of the integral $\int_{\alpha}^{\beta} x(t)\,dt$ for every $\beta > \alpha$. However, this property is not sufficient to ensure the existence of the Laplace transform, as the integral appearing in it is generalized. The convergence conditions of this integral are described by the following theorem.

Theorem 4.2.1 *The generalized integral*

$$I = \int_{\alpha}^{\infty} x(t)\,dt$$

converges to a finite value, when the following conditions are met simultaneously:

- *The signal function $x(t)$ is piecewise continuous for every $t \ge \alpha$.*
- *There is a function $y(t)$, such that $|x(t)| \le y(t)$ for every $t \ge M$, where M is a positive constant.*
- *The generalized integral*

$$J = \int_{M}^{\infty} y(t)\,dt$$

exists and converges to a finite value.

Conversely, the integral I does not exist, that is, it diverges, if for every $t \ge M$ the property $x(t) \ge y(t) \ge 0$ holds, and the integral J also diverges.

From the above analysis, it is obvious that a significant role in determining the behavior of a generalized integral regarding its convergence is played by *the appropriate choice of the function $y(t)$*. Two interesting and useful forms of functions used in problems of this type are $y(t) \sim t^{-p}$ and $y(t) \sim e^{\alpha t}$. Closely connected with the functions of the last category is the definition of *exponential order functions*, which is formulated as follows.

Definition 4.2.1 (Exponential Order Function) *A signal function $x(t)$ is called* of exponential order α *at the limit $t \to \infty$, if real constants $M \ge 0$ and $K > 0$ exist, as well as a real parameter α, such that $|x(t)| \le Ke^{\alpha t}$ for values $t \ge M$. Let us* *note that in practice, to characterize a function as of exponential order* α, *it is sufficient to prove that the function* $x(t)/e^{\alpha t}$ *is bounded for large values of the parameter* t, *a property that in mathematical writing is expressed as*

$$\lim_{t\to\infty}\left\{x(t)\,e^{-\alpha t}\right\} = 0.$$

Based on the theoretical concepts we presented in the previous pages, we can formulate and prove the next theorem regarding the existence of the Laplace transform.

Theorem 4.2.2 *Let us consider a continuous-time signal* $x(t)$ *that meets the following conditions:*

- *It is piecewise continuous in the interval* $0 \le t \le A$ *for every value of the positive parameter* A.
- *It is of exponential order* α, *that is, there are real constants* $M \ge 0$ *and* $K > 0$ *and a positive real parameter* α *such that* $|x(t)| \le K e^{\alpha t}$ *for every* $t \ge M$.

In this case, the Laplace transform exists, denoted as $\mathcal{L}\{x(t)\} = \mathcal{X}(s)$ *with a region of convergence* $\sigma > \alpha$, *where the symbol* σ *expresses the real part of the complex variable* s.

Proof from the defining equation of the Laplace transform it follows that

$$\mathcal{X}(s) = \int_0^\infty x(t)e^{-st}dt = \int_0^M x(t)e^{-st}dt + \int_M^\infty x(t)e^{-st}dt.$$

From the two integrals of the last equation, the first one exists because of the property of *piecewise continuity* satisfied by the function $x(t)$[1] and therefore the existence of the Laplace transform depends on the convergence of the second integral. Using the basic theorems of integral calculus we will have

$$\left|\int_M^\infty x(t)e^{-st}dt\right| \le \int_M^\infty |x(t)\,e^{-st}|dt \le \int_M^\infty |x(t)|\cdot|e^{-st}|dt.$$

However, $s = \sigma + j\omega$ and therefore

$$|e^{-st}| = |e^{-(\sigma+j\omega)t}| = |e^{-\sigma t}|\cdot|e^{-j\omega t}| = e^{-\sigma t}$$

since, as is well known, $|e^{-j\omega t}| = 1$. On the other hand, relying on the property of the signal function $x(t)$ to be a function of *exponential order* α, we can write that $|x(t)| \le K e^{\alpha t}$. Replacing the above expressions in our initial equation we will get

$$\left|\int_M^\infty x(t)\,e^{-st}dt\right| \le \int_M^\infty K\,e^{\alpha t}e^{-\sigma t}dt \equiv K\int_M^\infty e^{-(\sigma-\alpha)t}dt,$$

where $\sigma = \Re(s)$ is the real part of the complex variable s.

From the above analysis it is obvious that in order for the second integral of our initial equation to converge so that the *Laplace transform* of the signal $x(t)$ exists, the quantity of the exponent in the above equation must be *positive*, so that the integral tends to zero for the limit $t \to \infty$. But this means that it must be $\sigma \equiv \Re(s) > \alpha$. Therefore, we conclude that if a continuous-time signal $x(t)$ is simultaneously *piecewise continuous* and *of exponential order*, it has a Laplace transform that converges for all complex quantities s for which the real part satisfies the property $\sigma > \alpha$. The last equation constitutes the defining equation of the region of convergence of this transform.

[1] In the above equation, the function to be integrated is not $x(t)$ but $x(t)\,e^{-st}$. However, the requirement for the function $x(t)$ to tend to zero for the limit $t \to \infty$, is obviously also satisfied by the function $x(t)\,e^{-st}$, so that the property of *piecewise continuity* can be used for the proof of the existence of the Laplace transform.

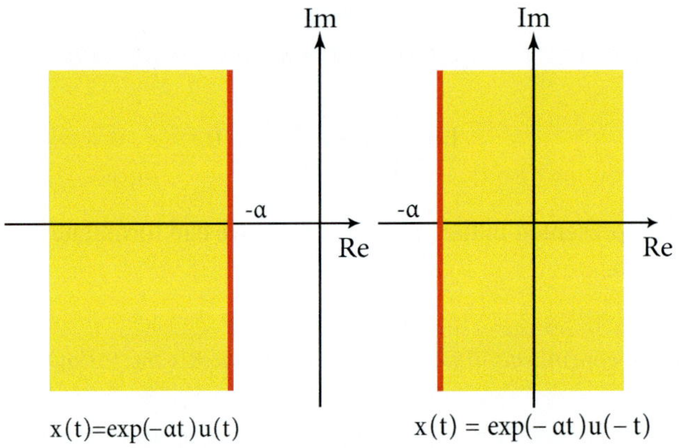

Fig. 4.1 The Region of Convergence (ROC) of the Laplace transform for the signals $x(t) = e^{-\alpha t}u(t)$ and $y(t) = -e^{-\alpha t}u(-t)$

Note In order to understand the significance of determining the region of convergence of the Laplace transform, let us calculate this transform for the signals $x(t) = e^{-\alpha t}u(t)$ and $y(t) = -e^{-\alpha t}u(-t)$, using the defining equation. For the first signal, therefore, we will have

$$\mathcal{X}(s) = \int_{-\infty}^{+\infty} x(t)\, e^{-st} dt = \int_{0}^{\infty} e^{-(\alpha+s)t} dt = \frac{1}{\alpha + s} = \frac{1}{(\alpha + \sigma) + j\omega}.$$

In order for the above integral to converge on the value in question, it must be $\alpha + \sigma > 0$, and therefore $\sigma \equiv \Re(s) > -\alpha$, since as is well known, the relation holds:

$$|e^{-(\alpha+s)t}| = |e^{-(\alpha+\sigma)t}| \cdot |e^{-j\omega t}| = |e^{-(\alpha+\sigma)t}|.$$

In a completely analogous way, for the second signal we will have

$$\mathcal{Y}(s) = \int_{-\infty}^{+\infty} y(t)\, e^{-st} dt = \int_{-\infty}^{0} -e^{-(\alpha+s)t} dt = \frac{1}{\alpha + s} = \frac{1}{(\alpha + \sigma) + j\omega}.$$

In this case, the existence of the negative sign in the exponent in combination with the limit $t \to -\infty$ imposes as a convergence criterion of the transform the condition $\alpha + \sigma < 0$ and therefore $\sigma \equiv \Re(s) < -\alpha$.

From the above analysis we observe that although the two signals are characterized by the same Laplace transform, the region of convergence is different. Therefore, when determining the Laplace transform of a signal $x(t)$, *in addition to the analytical form of the transform* $\mathcal{X}(s)$, *we must also determine the region of the complex plane for which it converges*. In cases where there is more than one region of convergence, we must also determine the one that interests us, because depending on the region that we choose, the characteristics and behavior of the system will be different. This feature will be presented in a subsequent section. The region of convergence of the Laplace transforms for the continuous-time signals $x(t)$ and $y(t)$ is presented diagrammatically in Fig. 4.1.

4.2.1 Properties of the Region of Convergence

The region of convergence of the Laplace transform is characterized by a set of useful properties, which, in general terms, are the following:

- The region of convergence of the Laplace transform consists of zones that are parallel to the imaginary axis of the complex plane.
  The power of this property is demonstrated by relying on the observation that the Laplace transform of a signal $x(t)$ converges only for those complex variables $s = \sigma + j\omega$ for which the function $x(t)\, e^{-\sigma t}$ is *absolutely integrable*, that is,

it satisfies the condition

$$\int_{-\infty}^{+\infty} |x(t)|e^{-\sigma t}dt < \infty.$$

But the above condition depends *only on the real part σ of the complex quantity s* and therefore the equation defining the region of convergence will have the general form $\sigma > \xi$ (where ξ is a real number) from which the desired result follows.

- The region of convergence of the Laplace transform cannot contain poles.
As we will see in the next section, an *explicit Laplace transform* can be expressed as a quotient of two polynomial functions for which the roots of the numerator are called *zero values*, and the roots of the denominator are called *poles*. It is obvious that at the poles the transform $\mathcal{L}\{x(t)\}$ takes *infinite value*—that is, it does not converge—and therefore *a pole cannot belong to the region of convergence*.
- If the continuous-time signal $\boldsymbol{x(t)}$ has *finite duration* and is *absolutely integrable*, then the region of convergence of the transform $\mathcal{L}\{x(t)\}$ is *the entire complex plane*.
Assuming that the signal $x(t)$ is defined in the interval $[T_1, T_2]$, the condition of *absolute integrability* is mathematically formulated as

$$\int_{T_1}^{T_2} |x(t)|dt < \infty.$$

On the other hand, according to the basic theory, the region of convergence of the Laplace transform is defined as the set of complex variables of the form $s = \sigma + j\omega$ for which the function $x(t)\ e^{-\sigma t}$ is *absolutely integrable* in its domain of definition, that is, it satisfies the condition

$$\int_{T_1}^{T_2} |x(t)|e^{-\sigma t}dt < \infty.$$

In order to prove the desired property, we must show that the last relation holds *for all values of the parameter σ*.

It is obvious that the above relationship holds for $\sigma = 0$: indeed, in this case, the previous equation reduces to the defining equation of *absolute integrability*, which is satisfied by the continuous-time signal $x(t)$, since it has been assumed to be absolutely integrable. On the other hand, for $\sigma > 0$ the maximum value of the function $e^{-\sigma t}$ in the interval $[T_1, T_2]$—in which the signal $x(t)$ is different from zero—is $e^{-\sigma T_1}$, so that we can write

$$\int_{T_1}^{T_2} |x(t)|e^{-\sigma t}dt < e^{-\sigma T_1} \int_{T_1}^{T_2} |x(t)|dt.$$

However, since the second term of the above equation is bounded—owing to the condition of *absolute integrability* satisfied by the signal $x(t)$—the same will apply to the first term, that is, the region of the complex plane for which the relation $\sigma \equiv \Re(s) > 0$ holds will belong to the region of convergence of the transform $\mathcal{L}\{x(t)\}$. With similar arguments, for negative values of the parameter σ we will have

$$\int_{T_1}^{T_2} |x(t)|e^{-\sigma t}dt < e^{-\sigma T_2} \int_{T_1}^{T_2} |x(t)|dt,$$

and thus the function $x(t)\ e^{-\sigma t}$ will also be *absolutely integrable* for the region of the complex plane for which $\sigma \equiv \Re(s) < 0$ holds. Therefore, ultimately, the region of convergence of the Laplace transform of the continuous-time signal $x(t)$ is *the entire complex plane*.
- If the continuous-time signal $x(t)$ is *right-sided*—that is, there is a time T_1 such that $x(t) = 0$ for $t < T_1 < \infty$—and the line $\Re(s) = \sigma_0$ (parallel to the imaginary axis) belongs to the region of convergence of the transform $\mathcal{L}\{x(t)\}$, then all the points s of the complex plane for which the relation $\Re(s) > \sigma_0$ holds will belong to the region of convergence of the transform.
In order to prove the validity of the above property, we use the condition of *absolute integrability* of the continuous-time signal $x(t)$, which, for the case of a *right-sided signal*, is formulated as

$$\int_{T_1}^{\infty} |x(t)|e^{-\sigma_0 t}dt < \infty,$$

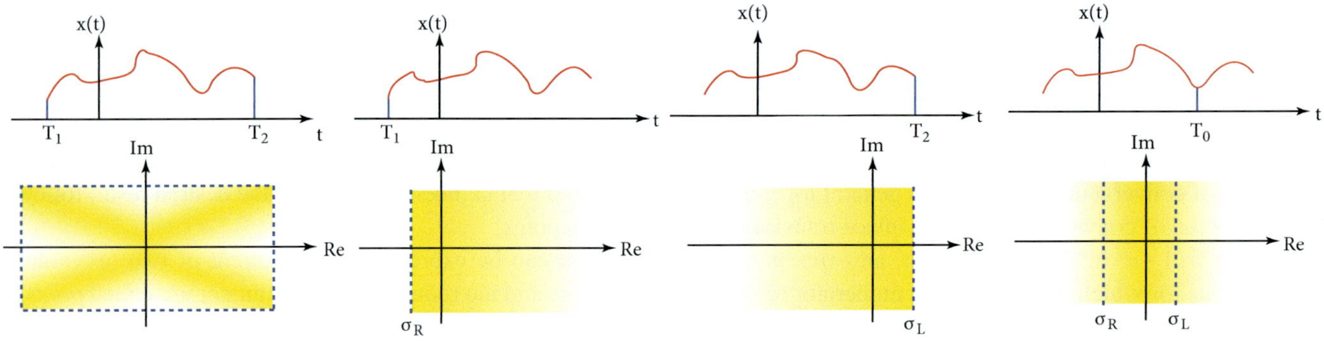

Fig. 4.2 The region of convergence of the Laplace transform for the four cases of functions that have such a transformation

where $\sigma_0 \equiv \Re(s)$ is a value for which the Laplace transform converges. If we now consider a value $\sigma_1 > \sigma_0$, we can conclude that the function $x(t)\, e^{-\sigma_1 t}$ will also be absolutely integrable, since the quantity $e^{-\sigma_1 t}$ will tend to zero faster than $e^{-\sigma_0 t}$ as $t \to \infty$. Indeed, for every $\sigma_1 > \sigma_0$ we find that

$$\int_{T_1}^{\infty} |x(t)|e^{-\sigma_1 t} dt = \int_{T_1}^{\infty} |x(t)|e^{-\sigma_0 t} e^{-(\sigma_1 - \sigma_0)t} dt \le e^{-(\sigma_1 - \sigma_0)T_1} \int_{T_1}^{\infty} |x(t)|e^{-\sigma_0 t} dt.$$

Taking into account that the time T_1 is finite, it is obvious that the right side of the above equation is bounded, that is, the function $|x(t)|e^{-\sigma_1 t}$ is *absolutely integrable*, a conclusion that also verifies the aforementioned property. Because for every point s that belongs to the region of convergence of the transform, all points located to the right of s—that is, they have a larger real part—also belong to the region of convergence; this region is often also called the *right half-plane*.

- If the continuous-time signal $x(t)$ is a *left-sided signal*—that is, there is a time moment $\boldsymbol{T_2}$ such that $x(t) = 0$ for $t > T_2 > -\infty$—and the line $\Re(s) = \sigma_0$ (parallel to the imaginary axis) belongs to the region of convergence of the transform $\mathcal{L}\{x(t)\}$, then all points s of the complex plane for which the relation $\Re(s) < \sigma_0$ holds will belong to the region of convergence of the transform.
The proof of this property is made, using similar arguments to those we used previously, and is left as an exercise to the reader. Because for every point s that belongs to the region of convergence of the transform, all points located to the left of s—that is, they have a smaller real part—also belong to the region of convergence, this region is often also called *left half-plane*.
- If the continuous-time signal $x(t)$ has *infinite duration* and is defined both for $t \geq 0$ and for $t < 0$, whereas in addition the line $\sigma_0 = \Re(s)$ belongs to the region of convergence of the transform $\mathcal{L}\{x(t)\}$, then this region is a zone of the complex that contains the straight line $\sigma_0 = \Re(s)$.
In the case of a signal with infinite duration, the determination of the region of convergence is based on the selection of an arbitrary time moment T_0 and the separation of the signal $x(t)$ into a *right-sided signal* $x_R(t)$ and a *left-sided signal* $x_L(t)$. In this case, the region of convergence of the transform $\mathcal{L}\{x(t)\}$ is defined as *the set of points s of the complex plane for which both the transform $\mathcal{L}\{x_R(t)\}$ and the transform $\mathcal{L}\{x_L(t)\}$ converge*. According to the properties of the region of convergence, there should be values σ_R and σ_L such that the regions of convergence of the transforms $\mathcal{L}\{x_R(t)\}$ and $\mathcal{L}\{x_L(t)\}$ are described by the equations $\Re(s) > \sigma_R$ and $\Re(s) < \sigma_L$ respectively. Therefore, the region of convergence of the transform $\mathcal{L}\{x(t)\}$ will result from the overlap of the two above regions (provided, of course, the relation $\sigma_R < \sigma_L$ holds, so that this overlap exists). In the opposite case, the two regions do not overlap and the Laplace transform $\mathcal{L}\{x(t)\}$ is not defined, even if the transforms $\mathcal{L}\{x_R(t)\}$ and $\mathcal{L}\{x_L(t)\}$ exist.
Speaking generally, a continuous-time signal $x(t)$ either will not have a Laplace transform at all, or if it does, it will fall into one of the four previous cases and more specifically: *(a) it will have finite duration and the region of convergence of the transform Laplace will be the entire complex plane (except perhaps for the points $s = 0$ and $s = \infty$), (b) it will be a right-sided function and the region of convergence will have the form $\sigma > \sigma_R$, (c) it will be a left-sided function and the region of convergence will have the form $\sigma < \sigma_L$, and (d) it will have infinite duration and the region of convergence will have the form $\sigma_R < \sigma < \sigma_L$*. The regions of convergence of the Laplace transform for these four cases of functions are presented in the images (a), (b), (c), and (d) respectively of Fig. 4.2.

- If the Laplace transform $\mathcal{L}\{x(t)\}$ of the signal $x(t)$ is a *rational function*, the region of convergence will either be limited by poles or will extend to infinity, if there are no poles. In any case, if there are poles, they cannot be located within the region of convergence.
- If the Laplace transform $\mathcal{L}\{x(t)\}$ of the signal $x(t)$ is a rational function, then the following apply: (a) if the signal $x(t)$ is a *right-sided signal*, then the region of convergence will be described by the equation $\sigma > \sigma_{max}$, where the quantity σ_{max} refers to the pole with the largest real part and (b) if the signal $x(t)$ is a *left-sided signal*, then the region of convergence will be described by the equation $\sigma < \sigma_{min}$, where the quantity σ_{min} refers to the pole with the smallest real part.

4.3 Examples of Laplace Transforms

In this section, we present some exemplary solved exercises of calculating the Laplace transform, in order for the reader to familiarize themselves with this process, as well as with the way of determining the region of convergence for each case.

Problem 4.3.1 *Determine the Laplace transform of the function $\delta(t)$.*

Solution From the defining equation of the Laplace transform we will have

$$\mathcal{L}\{\delta(t)\} = \int_{-\infty}^{+\infty} \delta(t)\, e^{-st}\, dt = e^{-st}\Big|_{t=0} = 1$$

The above relation arises through the application of *the property of selection* of the function $\delta(t)$ for the function $x(t) = e^{-st}$, which is described by the equation

$$\int_{-\infty}^{+\infty} \delta(t)\, x(t)\, dt = x(0).$$

Problem 4.3.2 *Determine the Laplace transform of the continuous-time signal $x(t) = e^{t} u(t)$.*

Solution From the defining equation of the Laplace transform we will have

$$\mathcal{X}(s) = \int_{-\infty}^{+\infty} x(t)\, e^{-st}\, dt = \int_{-\infty}^{+\infty} e^{t} u(t)\, e^{-st}\, dt = \int_{0}^{\infty} e^{(1-s)t}\, dt.$$

In order for the convergence property $\lim_{t\to\infty}[e^{(1-s)t}] = 0$ to hold, it must be $\sigma \equiv \Re(s) > 1$, since as we know we will have $|e^{(1-s)t}| = |e^{(1-\sigma)t}| \cdot |e^{-j\omega t}| = e^{(1-\sigma)t}$. For this range of values of the complex variable s, the above integral is calculated as

$$\mathcal{X}(s) = \int_{0}^{\infty} e^{(1-s)t}\, dt = \left[\frac{e^{(1-s)t}}{1-s}\right]_{0}^{\infty} = 1.$$

Problem 4.3.3 *Determine the Laplace transform of the continuous-time signal $x(t) = \cos(\omega t)u(t)$.*

Solution Using the well-known Euler relation $\cos(\omega t) = (e^{j\omega t} + e^{-j\omega t})/2$ and the definition of the function $u(t)$, the Laplace transform of the signal $x(t)$ is calculated as

$$\mathcal{X}(s)) = \int_{-\infty}^{+\infty} f(t)e^{-st}dt = \int_{-\infty}^{+\infty} \cos(\omega t)u(t)e^{-st}dt = \frac{1}{2}\int_0^{\infty} e^{j\omega t}e^{-st}dt$$

$$+\frac{1}{2}\int_0^{\infty} e^{-j\omega t}e^{-st}dt = \frac{1}{2}\int_0^{\infty} e^{(j\omega-s)t}dt + \frac{1}{2}\int_0^{\infty} e^{-(j\omega+s)t}dt = \frac{1}{2}\left[\frac{e^{(j\omega-s)t}}{j\omega-s}\right]_0^{\infty} + \frac{1}{2}\left[-\frac{e^{-(j\omega+s)t}}{j\omega+s}\right]_0^{\infty}$$

The region of convergence of the above transform will arise from the requirement to be

$$\lim_{t\to\infty}[e^{(j\omega-s)t}] = \lim_{t\to\infty}[e^{(j\omega-\sigma-j\omega)t}] = \lim_{t\to\infty} e^{-\sigma t} = 0$$

$$\lim_{t\to\infty}[e^{-\sigma t}e^{-2i\omega}] = e^{-2i\omega}\lim_{t\to\infty}[e^{-\sigma t}] = 0.$$

The above properties obviously apply for positive values of the parameter σ, that is, the region of convergence of the Laplace transform is determined by the equation $\sigma > 0$. Therefore, we have

$$\mathcal{X}(s) = -\frac{1}{2}\left[-\frac{1}{j\omega+s} - \frac{1}{j\omega-s}\right] = \frac{s}{s^2+\omega^2}.$$

Problem 4.3.4 *Determine the Laplace transform of the continuous-time signal* $x(t) = \sin(\omega t)u(t)$.

Solution In full analogy with the previous case, from the Euler relation $\sin(\omega t) = (e^{i\omega t} - e^{-i\omega t})/2i$ and the defining equations of the function $u(t)$ and the Laplace transform we will have

$$X(s) = \int_{-\infty}^{+\infty} f(t)e^{-st}dt = \int_{-\infty}^{+\infty} \sin(\omega t)u(t)e^{-st}dt = \frac{1}{2j}\int_0^{\infty} e^{j\omega t}e^{-st}dt - \frac{1}{2j}\int_0^{\infty} e^{-j\omega t}e^{-st}dt$$

$$= \frac{1}{2j}\int_0^{\infty} e^{(j\omega-s)t}dt - \frac{1}{2j}\int_0^{\infty} e^{-(j\omega+s)t}dt = \frac{1}{2j}\left[\frac{e^{(j\omega-s)t}}{j\omega-s}\right]_0^{\infty} - \frac{1}{2j}\left[-\frac{e^{-(j\omega+s)t}}{j\omega+s}\right]_0^{\infty}$$

.The region of convergence of the above transform will result from the requirement to be

$$\lim_{t\to\infty}[e^{(i\omega-s)t}] = \lim_{t\to\infty}[e^{(i\omega-\sigma-i\omega)t}] = \lim_{t\to\infty} e^{-\sigma t} = 0$$

$$\lim_{t\to\infty}[e^{-(i\omega+s)t}] = \lim_{t\to\infty}[e^{(-i\omega-\sigma-i\omega)t}] = \lim_{t\to\infty}[e^{-\sigma t}e^{-2i\omega}] = e^{-2i\omega}\lim_{t\to\infty}[e^{-\sigma t}] = 0.$$

The above properties obviously apply for positive values of the parameter σ, that is, the region of convergence of the Laplace transform is determined by the equation $\sigma > 0$. Therefore, we have

$$\mathcal{X}(s) = -\frac{1}{2i}\left[\frac{1}{i\omega-s} + \frac{1}{i\omega+s}\right] = \frac{\omega}{\omega^2+s^2}.$$

Problem 4.3.5 *Determine the Laplace transform of the continuous-time signal* $x(t) = tu(t)$

Solution From the defining equation of the transformation Laplace we will have

$$\mathcal{X}(s) = \int_{-\infty}^{+\infty} x(t)\,e^{-st}dt = \int_{-\infty}^{+\infty} tu(t)\,e^{-st}dt = \int_0^{\infty} te^{-st}dt.$$

Integrating by parts leads us to the equation

$$\int te^{-st}dt = \int td\Big(-\frac{e^{-st}}{s}\Big) = -\frac{t}{s}e^{-st} + \frac{1}{s}\int e^{-st}dt = -\frac{t}{s}e^{-st} - \frac{1}{s^2}e^{-st}$$

and therefore, we have

$$\mathcal{X}(s) = \int_0^{\infty} te^{-st}dt = -\frac{1}{s}\Big[te^{-st}\Big]_0^{\infty} - \frac{1}{s^2}\Big[e^{-st}\Big]_0^{\infty} = \frac{1}{s^2}$$

with region of convergence $\sigma > 0$ as evidenced by the form of the second term of the above equation.

Problem 4.3.6 *Determine the Laplace transform of the continuous-time signal* $x(t) = e^{-\alpha t}\cos(\omega t)u(t)$.

Solution From the defining equation of the Laplace transform we will have

$$\mathcal{X}(s) = \int_{-\infty}^{\infty} x(t)\, e^{-st}dt = \int_{-\infty}^{\infty} e^{-\alpha t}\cos(\omega t)\, u(t)\, e^{-st}dt = \int_0^{+\infty} e^{(s+\alpha)t}\cos(\omega t)\, dt$$

Using the Euler formula $\cos(\omega t) = (e^{j\omega t} + e^{-j\omega t})/2$, the above equation is written as

$$\mathcal{X}(s) = \frac{1}{2}\int_0^{\infty} e^{-(s+\alpha)t}e^{j\omega t}dt + \frac{1}{2}\int_0^{\infty} e^{-(s+\alpha)t}e^{-j\omega t}dt = \frac{1}{2}\int_0^{\infty} e^{(j\omega-s-\alpha)t}dt$$

$$= \frac{1}{2}\int_0^{\infty} e^{-(j\omega+s+\alpha)t}dt = \frac{1}{2}\Big[\frac{e^{(j\omega-s-\alpha)t}}{j\omega-s-\alpha}\Big]_0^{\infty} + \frac{1}{2}\Big[-\frac{e^{-(j\omega+s+\alpha)t}}{j\omega+s+\alpha}\Big]_0^{\infty}$$

It is obvious that, for the transform $X(s)$ to converge, it must be

$$\lim_{t\to\infty}\Big[e^{(j\omega-s-\alpha)t}\Big] = \lim_{t\to\infty}\Big[e^{-(j\omega+s+\alpha)t}\Big] = 0$$

For the first limit to be zero, the relation $j\omega - s - \alpha = j\omega - \sigma - j\omega - \alpha = -\sigma - \alpha < 0$ must hold, from which it follows that $\sigma > -\alpha$. The same condition applies to the second limit, that is, the region of convergence of the transform $X(s)$ is described by the equation $\sigma > -\alpha$. For this range of s values, the Laplace transform is calculated as

$$\mathcal{X}(s) = \frac{1}{2}\Big[\frac{1}{j\omega+(s+\alpha)} - \frac{1}{j\omega-(s+\alpha)}\Big] = \frac{s+\alpha}{\omega^2+(s+\alpha)^2}$$

Problem 4.3.7 *Determine the Laplace transform of the continuous-time signal* $x(t) = 2\sqrt{(t/\pi)}u(t)$ *(Poularikas,* The Transforms and Applications Handbook, *Example 5.2.2).*

Solution From the defining equation of the Laplace transform we will have

$$\mathcal{X}(s) = \int_{-\infty}^{+\infty} x(t)\, e^{-st}dt = \int_{-\infty}^{+\infty} 2\sqrt{\frac{t}{\pi}}u(t)\, e^{-st}dt = \frac{2}{\sqrt{\pi}}\int_0^{\infty}\sqrt{t}e^{-st}dt.$$

To calculate the above integral we set $\xi = \sqrt{t}$ and by differentiating, it turns out that $dt = 2\xi d\xi$. Therefore, the Laplace transform is formulated as

$$\mathcal{X}(s) = \frac{2}{\sqrt{\pi}} \int_0^\infty \xi e^{-s\xi^2} (2\xi d\xi) = \frac{4}{\sqrt{\pi}} \int_0^\infty \xi^2 e^{-s\xi^2} d\xi$$

From indefinite integral tables we find that

$$\int \xi^2 e^{-s\xi^2} d\xi = \frac{\sqrt{\pi}}{4s^{3/2}} \operatorname{erf}(s\sqrt{\xi}) - \frac{\xi e^{-s\xi^2}}{2s},$$

where $\operatorname{erf}(x)$ is the error function, which is defined as

$$\operatorname{erf}(x) = \frac{2}{\sqrt{\pi}} \int_0^x e^{-u^2} du, \quad \operatorname{erf}(0) = 0, \quad \operatorname{erf}(\infty) = 1$$

Therefore, we have

$$\int_0^\infty \xi^2 e^{-s\xi^2} d\xi = \left[\frac{\sqrt{\pi}}{4s^{3/2}} \operatorname{erf}(s\sqrt{\xi}) - \frac{\xi e^{-s\xi^2}}{2s} \right]_0^\infty = \frac{\sqrt{\pi}}{4s^{3/2}}$$

and by substituting in the defining equation of the transform, we arrive at the final result:

$$\mathcal{X}(s) = \frac{4}{\sqrt{\pi}} \frac{\sqrt{\pi}}{4s^{3/2}} = \frac{1}{s^{3/2}}.$$

Problem 4.3.8 *Considering the continuous-time signal*

$$x(t) = e^{-5t} u(t-1)$$

determine the Laplace transform $\mathcal{X}(s)$ and its region of convergence. Then determine the values of the parameters A and t_0 so that the Laplace transform $\mathcal{Y}(s)$ of the signal

$$y(t) = Ae^{-5t} u(-t - t_0)$$

has the same algebraic form as the transform $\mathcal{X}(s)$. Determine the region of convergence of the transform $\mathcal{Y}(s)$ (Oppenheim, Willsky, Nawab, Exercise 9.2).

Solution From basic theory, it is known that the function $u(t-1)$ is defined as

$$u(t-1) = \begin{cases} 1 \text{ for } t > 1 \\ 0 \text{ for } t < 1 \end{cases}$$

and therefore the Laplace transform of the signal $x(t)$ will be given by the relation

$$\mathcal{X}(s) = \int_{-\infty}^{+\infty} x(t)\, e^{-st} dt = \int_{-\infty}^{+\infty} e^{-5t} u(t-1)\, e^{-st} dt = \int_1^\infty e^{-(s+5)t} dt.$$

The region of convergence of the above transform will result from the requirement to be

$$|\mathcal{X}(s)| = \int_1^\infty |e^{-(s+5)t}| dt = \int_1^\infty |e^{-(\sigma+5)t}| dt < \infty$$

since $|e^{-i\omega t}| = 1$. The above property is obviously satisfied when $\sigma + 5 > 0$, i.e., $\sigma \equiv \Re(s) > -5$, which is also the equation of the region of convergence of the Laplace transform. For this range of values of s, the above integral is calculated as

$$\mathcal{X}(s) = \int_1^{\infty} e^{-(s+5)t}\,dt = \left[-\frac{e^{-(s+5)t}}{s+5} \right]_1^{\infty} = \frac{e^{-(s+5)}}{s+5}.$$

In a completely analogous way, the function $u(-t-t_0)$ is defined as

$$u(-t-t_0) = \begin{cases} 1 \text{ for } t < -t_0 \\ 0 \text{ for } t > -t_0 \end{cases}$$

and therefore the Laplace transform of the signal $y(t)$ will be calculated as

$$\mathcal{Y}(s) = \int_{-\infty}^{+\infty} y(t)\,e^{-st}\,dt = \int_{-\infty}^{+\infty} Ae^{-5t}u(-t-t_0)\,e^{-st}\,dt = A\int_{-\infty}^{-t_0} e^{-(s+5)t}\,dt.$$

In this case, the existence of the limit $t \to \infty$ in combination with the negative sign of the exponent leads us to the conclusion that the convergence condition of the integral will be $\sigma + 5 < 0$ from where it follows that $\sigma \equiv \Re(s) < -5$. This equation determines the region of convergence of the transform $\mathcal{Y}(s)$. For values of the variable s that belong to the above region, the above integral is calculated as

$$\mathcal{Y}(s) = A\int_{-\infty}^{-t_0} e^{-(s+5)t}\,dt = A\left[-\frac{e^{-(s+5)t}}{s+5} \right]_{-\infty}^{-t_0} = -A\frac{e^{-(s+5)t_0}}{s+5}$$

If we compare the transforms $\mathcal{X}(s)$ and $\mathcal{Y}(s)$, we find that for them to have the same form the constants A and t_0 must have the values $A = -1$ and $t_0 = 1$.

Problem 4.3.9 *Determine the Laplace transform of the continuous-time signal* $x(t) = t^n$.

Solution From the defining equation of the Laplace transform we will have

$$\mathcal{L}\{t^n\} = \int_0^{\infty} t^n e^{-st}\,dt = \int_0^{\infty} t^n d\left(-\frac{e^{-st}}{s}\right) = -t^n \frac{e^{-st}}{s}\Bigg|_0^{\infty} + \frac{1}{s}\int_0^{\infty} e^{-st}d(t^n) = \frac{n}{s}\int_0^{\infty} t^{n-1}e^{-st}\,dt = \frac{n}{s}\mathcal{L}\{t^{n-1}\},$$

where integration by parts took place, whereas the zero value was set as the lower limit. As we have already mentioned, this variation of the Laplace transform is known as the unilateral Laplace transform and is the one usually used in practice, as it refers to causal signals and systems whose behavior and content refer to positive time moments.

Applying the above relation recursively, we get

$$\begin{aligned} \mathcal{L}\{t^n\} &= \frac{n}{s}\mathcal{L}\{t^{n-1}\} \\ \mathcal{L}\{t^{n-1}\} &= \frac{n-1}{s}\mathcal{L}\{t^{n-2}\} \\ \mathcal{L}\{t^{n-2}\} &= \frac{n-2}{s}\mathcal{L}\{t^{n-3}\} \\ &\cdots\cdots\cdots\cdots \\ \mathcal{L}\{t\} &= \frac{1}{s}\mathcal{L}\{t^0\} = \frac{1}{s}\mathcal{L}\{1\} = \frac{1}{s} \end{aligned}$$

Table 4.2 Examples of elementary Laplace transforms

Continuous-time signal $x(t)$	Laplace transform	Region of convergence
$\delta(t)$	1	$\Re(s) > -\infty$
$u(t)$	$1/s$	$\Re(s) > 0$
$-u(-t)$	$1/s$	$\Re(s) < 0$
$tu(t)$	$1/s^2$	$\Re(s) > 0$
$e^{-\alpha t}u(t)$	$\dfrac{1}{s+\alpha}$	$\Re(s) > -\Re(\alpha)$
$e^{+\alpha t}u(t)$	$\dfrac{1}{s-\alpha}$	$\Re(s) < \Re(\alpha)$
$te^{-\alpha t}u(-t)$	$\dfrac{1}{(s+\alpha)^2}$	$\Re(s) < \Re(\alpha)$
$\cos(\omega t)u(t)$	$\dfrac{s}{s^2+\omega^2}$	$\Re(s) > 0$
$\sin(\omega t)u(t)$	$\dfrac{\omega}{s^2+\omega^2}$	$\Re(s) > 0$
$e^{-\alpha t}\cos(\omega t)u(t)$	$\dfrac{s+\alpha}{(s+\alpha)^2+\omega^2}$	$\Re(s) > -\Re(\alpha)$
$e^{-\alpha t}\sin(\omega t)u(t)$	$\dfrac{\omega}{(s+\alpha)^2+\omega^2}$	$\Re(s) > -\Re(\alpha)$
$t^n u(t)$	$\dfrac{n!}{s^{n+1}}$	$\Re(s) > 0$

since it is proven that

$$\mathcal{L}\{1\} = \int_0^\infty e^{-st}dt = \left[-\frac{e^{-st}}{s}\right] = \frac{1}{s}$$

with convergence area $\sigma > 0$. By recursively replacing in the first of the above relations the remaining relations, we get

$$\mathcal{L}\{t^n\} = \frac{n}{s}\frac{n-1}{s}\frac{n-2}{s}\cdots\frac{1}{s}\frac{1}{s} = \frac{n!}{s^{n+1}}.$$

Table 4.2 contains the Laplace transforms of indicative continuous-time signals $x(t)$ along with the convergence area for each of them.

4.4 Poles and Zero Values of the Laplace Transform

From the exercises of calculating Laplace transforms that were presented on the previous pages, it is not difficult to realize that in the general case of such transformations, they appear in the form of *rational functions*, that is, *quotients of polynomial functions* of the form

$$\mathcal{L}\{x(t)\} = \frac{N(s)}{D(s)} = \frac{\beta_0+\beta_1 s^{-1}+\beta_2 s^{-2}+\cdots+\beta_M s^{-M}}{\alpha_0+\alpha_1 s^{-1}+\alpha_2 s^{-2}+\cdots+\alpha_N s^{-N}} = \frac{\beta_0}{\alpha_0}\frac{(s-z_1)(s-z_2)\ldots(s-z_M)}{(s-p_1)(s-p_2)\ldots(s-p_N)}$$

and finally

$$\mathcal{L}\{x(t)\} = \frac{\beta_0}{\alpha_0}\frac{\prod_{k=1}^{M}(s-z_k)}{\prod_{k=1}^{N}(s-p_k)}s^{M-N}, \tag{4.10}$$

where for the derivation of the last relation we have proceeded to factorization of the numerator and the denominator. In the above equation, the parameters α_i and β_j ($i = 1, 2, \ldots, N$ and $j = 1, 2, \ldots, M$) are considered *real constants*, whereas obviously the degrees of the polynomials N and M are *positive integer numbers*. According to the basic theory of complex functions, the roots of the numerator $z_1, z_2, \ldots, z_M$ are known as *zero values* of the function $\mathcal{L}\{x(t)\}$, whereas the roots of the denominator $p_1, p_2, \ldots, p_N$ are known as *poles* of the aforementioned function. Therefore, according to the above relation, the Laplace transform $\mathcal{L}\{x(t)\}$ has *N poles and M zero values*,[2] whereas it also has $|M - N|$ zero values (for $N > M$) or poles (for $N < M$). The representation of the poles and the zero values of the function $\mathcal{L}\{f(t)\}$ is known as *pole-zero plot* of the Laplace transform.

The process of determining the poles and zero values of the Laplace transform and the design of the corresponding diagram are presented in detail in the following examples.

Problem 4.4.1 *Determine the Laplace transform of the signals*

$$x_1(t) = e^{-2t}u(t) + e^{-3t}u(t), \qquad x_2(t) = e^{-3t}u(t) + e^{2t}u(-t), \qquad x_3(t) = e^{2t}u(t) + e^{-3t}u(-t)$$

and design the convergence area and the pole-zero diagram.

Solution From Table 4.2 we find that

$$\mathcal{L}\{e^{-\alpha t}u(t)\} = \frac{1}{s+\alpha}$$

with convergence area $\Re(s) > -\Re(\alpha)$ and

$$\mathcal{L}\{e^{+\alpha t}u(t)\} = \frac{1}{s-\alpha}$$

with convergence area $\Re(s) < \Re(\alpha)$. Therefore, for the cases of the above signals, the following apply:

(a) The above relations for the values $\alpha = -2$ and $\alpha = -3$ are written as

$$\mathcal{L}\{e^{-2t}u(t)\} = \frac{1}{s+2} \quad \begin{matrix}\text{with convergence}\\ \text{area}\end{matrix} \quad \Re(s) > -2 \quad \text{and} \quad \mathcal{L}\{e^{-3t}u(t)\} = \frac{1}{s+3} \quad \begin{matrix}\text{with convergence}\\ \text{area}\end{matrix} \quad \Re(s) > -3$$

and using the property of *linearity* of the Laplace transform—the properties of this transformation will be presented in the next section—we are led to the result

$$\mathcal{L}\{x_1(t)\} = \mathcal{L}\{e^{-2t}u(t)\} + \mathcal{L}\{e^{-3t}u(t)\} = \frac{1}{s+2} + \frac{1}{s+3} = \frac{2s+5}{(s+2)(s+3)}.$$

The region of convergence of the transform $\mathcal{L}\{x_1(t)\}$ arises *from the overlap of the regions of convergence of the two transforms* and is described by the equation $\Re(s) > -2$. Furthermore, from the form of the Laplace transform, we find that the function $\mathcal{L}\{x_1(t)\}$ has a zero value at the position $s = -5/2$ and two poles at the positions $s = -2$ and $s = -3$. The pole-zero diagram of the Laplace transform of the function $x_1(t)$ is presented diagrammatically in image (a) of Fig. 4.3.

(b) Using the relations of Table 4.2 for the values $\alpha = -3$ and $\alpha = 2$ respectively, we get

$$\mathcal{L}\{e^{-3t}u(t)\} = \frac{1}{s+3} \quad \begin{matrix}\text{with region of}\\ \text{convergence}\end{matrix} \quad \Re(s) > -3 \quad \text{and} \quad \mathcal{L}\{e^{2t}u(-t)\} = -\frac{1}{s-2} \quad \begin{matrix}\text{with region of}\\ \text{convergence}\end{matrix} \quad \Re(s) < 2$$

[2] A rational complex function is called *regular* if it is $\beta_M \neq 0$ and $M < N$, that is *when the number of zero values of the function is less than the number of its poles*. If the function is not regular, then it can always be written as *the sum of a polynomial and a regular function*.

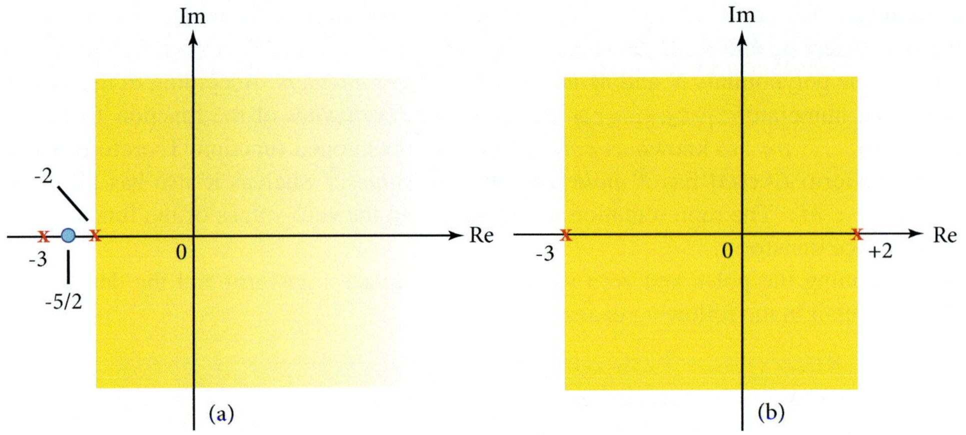

Fig. 4.3 The region of convergence of the Laplace transform for the functions of Exercise 4.4.1

and therefore the Laplace transform of the signal $x_2(t)$ is calculated as

$$\mathcal{L}\{x_2(t)\} = \mathcal{L}\{e^{-3t}u(t)\} + \mathcal{L}\{e^{2t}u(-t)\} = \frac{1}{s+3} - \frac{1}{s-2} = \frac{-5}{(s+2)(s+3)}$$

The region of convergence of the transform $\mathcal{L}\{x_2(t)\}$ arises *from the overlap of the regions of convergence of the two transforms* and is described by the equation $-3 < \Re(s) < 2$. Furthermore, from the form of the Laplace transform, we find that the function $\mathcal{L}\{x_2(t)\}$ does not have zero values, but it has two poles at positions $s = -2$ and $s = -3$. The pole-zero diagram of the Laplace transform of the signal $x_2(t)$ is presented diagrammatically in image (b) of Fig. 4.3.

(c) Finally, using the relations of Table 4.2 for the values $\alpha = 2$ and $\alpha = -3$ respectively, we will get

$$\mathcal{L}\{e^{2t}u(t)\} = \frac{1}{s-2} \quad \text{with region of convergence} \quad \Re(s) > 2 \quad \text{and} \quad \mathcal{L}\{e^{-3t}u(-t)\} = -\frac{1}{s+3} \quad \text{with region of convergence} \quad \Re(s) < -3.$$

We observe that the regions of convergence of the elementary transforms do not overlap and therefore the Laplace transform of the signal $x_3(t)$ does not exist.

Problem 4.4.2 *Determine the Laplace transform of the continuous-time signal $x(t) = e^{-\alpha|t|}$ and draw the pole-zero diagram for the cases $\alpha > 0$ and $\alpha < 0$.*

Solution From the property of the absolute value we will have

$$|t| = \begin{cases} +t & t > 0 \\ -t & t < 0 \end{cases}$$

and therefore the continuous-time signal $x(t)$ can be written in the form

$$x(t) = e^{-\alpha t}u(t) + e^{(-\alpha)(-t)}u(-t) = e^{-\alpha t}u(t) + e^{\alpha t}u(-t)$$

with a Laplace transform

$$\mathcal{L}\{x(t)\} = \mathcal{L}\{e^{-\alpha t}u(t)\} + \mathcal{L}\{e^{\alpha t}u(t)\} = \frac{1}{s+\alpha} - \frac{1}{s-\alpha} = -\frac{2\alpha}{s^2-\alpha^2}$$

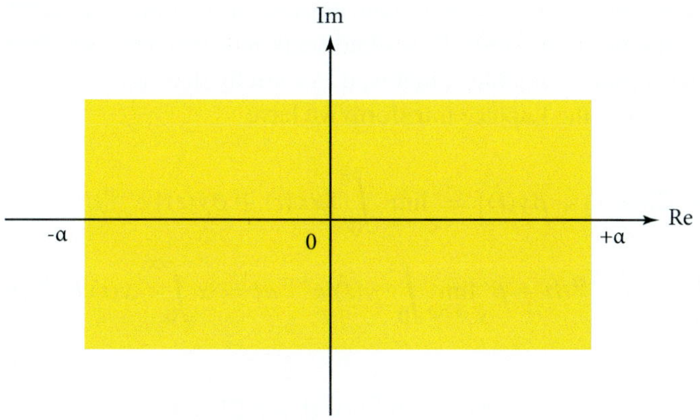

Fig. 4.4 The region of convergence of the Laplace transform of the signal $x(t) = e^{-\alpha|t|}$ for values $\alpha > 0$

as derived from the contents of Table 4.2. It is not difficult to see that for values $\alpha > 0$ the regions of convergence of the transforms $\mathcal{L}\{e^{-\alpha t}u(t)\}$ and $\mathcal{L}\{e^{\alpha t}u(t)\}$ are described by the equations $\Re(s) > -\alpha$ and $\Re(s) < \alpha$, that is, the region of convergence of the transform $\mathcal{L}\{x(t)\}$ will be *the overlap of the two above regions*, which is described by the equation $-\alpha < \Re(s) < \alpha$. In addition, from the mathematical form of the function $\mathcal{L}\{x(t)\}$ we find that it does not have zero values, whereas it does has two poles at the positions $s = -\alpha$ and $s = +\alpha$. The region of convergence and the pole-zero diagram of the transform $\mathcal{L}\{x(t)\}$ for the case $\alpha > 0$ are presented in Fig. 4.4.

On the other hand, for values $\alpha < 0$ the regions of convergence of the transforms $\mathcal{L}\{e^{-\alpha t}u(t)\}$ and $\mathcal{L}\{e^{\alpha t}u(t)\}$ are described by the equations $\Re(s) > \alpha$ and $\Re(s) < -\alpha$ and since in this case there is no overlap, the Laplace transform of the signal $x(t)$ does not exist.

It is important to note at this point that in the general case of a rational Laplace transform that has more than one pole, it is possible to have more than one region of convergence on the complex plane. Considering, for example, the function

$$\mathcal{X}(s) = \mathcal{L}\{x(t)\} = \frac{1}{(s^2 + 3s + 3)(s - 3)},$$

this has three simple poles at the positions $s_1 = 3$, $s_2 = -1.5 - j\sqrt{3}$, and $s_3 = -1.5 + j\sqrt{3}$. This function will therefore have three regions of convergence, the $\Re(s) < -1.5$, $1.5 < \Re(s) < 3$, and $\Re(s) > 3$ and in describing the system we must determine which of these three regions we will use.

4.5 Properties of the Laplace Transform

In full analogy with the Fourier transform that we presented in the previous chapter, the Laplace transform has a multitude of useful properties that significantly simplify the calculation process, while at the same time contributing to a deeper understanding of the correlations that exist between the field of *continuous-time* t and the field of *complex frequency* s. The detailed presentation of all these properties, along with their analytical proof, constitute the subject of the following pages.

- **Linearity:** If the Laplace transforms of the continuous-time signals $x(t)$ and $y(t)$ are the functions $\mathcal{X}(s) = \mathcal{L}\{x(t)\}$ and $\mathcal{Y}(s) = \mathcal{L}\{y(t)\}$ with regions of convergence $\sigma > \alpha_x$ and $\sigma > \alpha_y$ respectively, with $\alpha_y > \alpha_x$, then the Laplace transform of the function $z(t) = \alpha x(t) + \beta y(t)$ is the

$$\mathcal{Z}(s) = \mathcal{L}\{z(t)\} = \mathcal{L}\{\alpha x(t) + \beta y(t)\} = \alpha\mathcal{L}\{x(t)\} + \beta\mathcal{L}\{y(t)\} = \alpha\mathcal{X}(s) + \beta\mathcal{Y}(s),$$

where α and β are real numbers. As for the region of convergence of the transform $\mathcal{Z}(t)$, it is at least the intersection of the regions of convergence of the two transforms. If this intersection is the empty set, then the Laplace transform does not converge for any region of the complex variable s, that is, it essentially does not exist.
Proof From the defining equation of the Laplace transform we have

$$\mathcal{Z}(s) = \mathcal{L}\{z(t)\} = \mathcal{L}\{\alpha x(t) + \beta y(t)\} = \lim_{\xi\to\infty}\int_0^{\xi} [\alpha x(t) + \beta y(t)]\, e^{-st} dt =$$

$$= \alpha \lim_{\xi\to\infty}\int_0^{\xi} x(t)\, e^{-st} dt + \beta \lim_{\xi\to\infty}\int_0^{\xi} y(t)\, e^{-st} dt = \alpha \int_0^{\infty} x(t)\, e^{-st} dt + \beta \int_0^{\infty} y(t)\, e^{-st} dt$$

and finally

$$\mathcal{Z}(s) = \alpha\mathcal{L}\{x(t)\} + \beta\mathcal{L}\{y(t)\}$$

equation that proves the property of *linearity*. The above relationship for the general case of N continuous-time signals $x_1(t), x_2(t), \ldots, x_N(t)$ is formulated as

$$\mathcal{L}\left\{\sum_{i=1}^{N} c_i x_i(t)\right\} = \sum_{i=1}^{N} c_i \mathcal{L}\{x_i(t)\}$$

Closely related to the property of linearity is the property of homogeneity $\mathcal{L}\{\gamma x(t)\} = \gamma\mathcal{L}\{x(t)\}$, where γ is a real constant.

- **Time shift:** If the Laplace transform of the continuous-time signal $x(t)$ is the function $\mathcal{X}(s)$ with a region of convergence some subset R of the complex plane, then the Laplace transform of the signal $y(t) = x(t - t_0)$ will be the function $\mathcal{Y}(s) = e^{-st_0}\mathcal{X}(s)$, which, moreover, will be characterized by the same region of convergence.
Proof From the defining equation of the Laplace transform we have

$$\mathcal{Y}(s) = \mathcal{L}\{y(t)\} = \mathcal{L}\{x(t - t_0)\} = \lim_{\xi\to\infty}\int_0^{\xi} x(t - t_0)\, e^{-st} dt = \int_0^{\infty} x(t - t_0)\, e^{-st} dt$$

Proceeding with the variable change $t - t_0 = \xi, t = t_0 + \xi$ and $dt = d\xi$, whereas for $t = 0$ $\xi = -t_0$. Therefore, the above integral is written as

$$\mathcal{Y}(s) = \mathcal{L}\{x(t - t_0)\} = \int_{-t_0}^{\infty} x(\xi)\, e^{-s(\xi+t_0)} d\xi = e^{-st_0}\int_{-t_0}^{\infty} x(\xi)\, e^{-s\xi} d\xi$$

Considering that the signal $x(t)$ is causal, $x(t) = 0$ for $-t_0 \leq t \leq 0$ and the above relationship is written

$$\mathcal{Y}(s) = \mathcal{L}\{x(t - t_0)\} = e^{-st_0}\int_0^{\infty} x(\xi)\, e^{-s\xi} d\xi = e^{-st_0}\mathcal{L}\{x(t)\} = e^{-st_0}\mathcal{X}(s).$$

Following the same methodology, we can prove that

$$\mathcal{L}\{x(t + t_0)\} = e^{st_0}\int_0^{\infty} x(\xi)\, e^{-s\xi} d\xi = e^{st_0}\mathcal{L}\{x(t)\} = e^{st_0}\mathcal{X}(s).$$

- **Displacement in the complex plane:** If the Laplace transform of the signal $x(t)$ is the function $\mathcal{X}(s)$, then the Laplace transform of the signal $y(t) = e^{s_0 t}x(t)$ will be $\mathcal{Y}(s) = \mathcal{X}(s - s_0)$.
Proof from the defining equation of the Laplace transform we will have

$$\mathcal{Y}(s) = \mathcal{L}\{e^{s_0 t}x(t)\} = \lim_{\xi\to\infty}\int_0^{\xi} e^{s_0 t}x(t)\, e^{-st} dt = \int_0^{\infty} x(t)\, e^{-(s-s_0)t} dt = \mathcal{X}(s - s_0),$$

which is also the desired result. It is proven that the region of convergence R' of the transform $\mathcal{X}(s - s_0)$ arises from the region of convergence R of the transform $\mathcal{X}(s)$, displaced by the quantity $\sigma_0 \equiv \Re(s_0)$, and therefore for every $s \in R$ $s + \sigma_0 \in R'$. In addition, if the function $\mathcal{X}(s)$ has a pole or zero value at the position $s = \alpha$, then the function $\mathcal{X}(s - s_0)$ will have a pole or zero value at the position $s - s_0 = \alpha$, that is, at the position $s = s_0 + \alpha$.

Note An interesting application of the above property arises for the case $s_0 = j\omega$, for which the property of *displacement* is formulated as

$$\mathcal{L}\{e^{j\omega_0 t}x(t)\} = \mathcal{X}(s - j\omega_0)$$

In this case, the region of convergence of the transform $\mathcal{X}(s - j\omega_0)$ results from shifting the region R *parallel to the imaginary axis of the complex plane*. This means that if the function $\mathcal{X}(s)$ has a pole or zero value at the position $s = \alpha$, then the function $\mathcal{X}(s - j\omega_0)$ will have a pole or zero value at the position $s = \alpha + j\omega$.

- **Time scaling:** If the Laplace transform of the signal $x(t)$ is the function $\mathcal{X}(s)$, then the Laplace transform of the signal $y(t) = x(\alpha t)$ will be the function $\mathcal{Y}(s) = (1/|\alpha|)\mathcal{X}(s/\alpha)$.
Proof From the defining equation of the Laplace transform we will have

$$\mathcal{Y}(s) = \mathcal{L}\{y(t)\} = \mathcal{L}\{x(\alpha t)\} = \lim_{\xi\to\infty}\int_0^{\xi} x(\alpha t)\,e^{-st}dt = \int_0^{\infty} x(\alpha t)\,e^{-st}dt.$$

Proceeding with the variable change $\xi = \alpha t$, $t = \xi/\alpha$ and $dt = d\xi/\alpha$ and the above integral is written

$$\mathcal{Y}(s) = \int_0^{\infty} x(\xi)\,e^{-(\xi/\alpha)s}\frac{d\xi}{\alpha} = \frac{1}{\alpha}\int_0^{\infty} e^{-(\xi/\alpha)s}d\xi = \frac{1}{\alpha}\mathcal{X}\left(\frac{s}{\alpha}\right).$$

If the region of convergence of the transform $\mathcal{X}(s)$ is the R, then the region of convergence of the transform $\mathcal{L}\{x(\alpha t)\}$ will be the $R' = \alpha R$ and therefore for every $s \in R$ $\alpha s \in R'$. It is obvious that for values of the parameter α in the interval $0 < \alpha < 1$ the region R will be *compressed* by a factor equal to α, whereas for values $\alpha > 1$ the aforementioned region will *expand* by the value of the same factor.

The above analysis applies to values of α such that $\alpha > 0$. For negative values of α $-\alpha > 0$ and therefore the above relationship will take the form $\mathcal{L}\{x(\alpha t)\} = (-1/\alpha)\mathcal{X}(s/\alpha)$. Combining the above relationships for positive and negative values of α, we will have

$$\mathcal{Y}(s) = \mathcal{L}\{x(\alpha t)\} = \frac{1}{|\alpha|}\mathcal{X}\left(\frac{s}{\alpha}\right).$$

Note It is interesting to mention at this point that for negative values of the parameter α the region of convergence apart from *scaling* also undergoes *reversal*. Specifically, for the value $\alpha = -1$ the above relationship will take the form $\mathcal{L}\{x(-t)\} = \mathcal{X}(-s)$, with the region of convergence $R' = -R$.

- **Complex conjugation:** If the Laplace transform of the continuous-time signal $x(t)$ is the function $\mathcal{X}(s)$, then the Laplace transform of the signal $y(t) = x^*(t)$ will be the function $\mathcal{Y}(s) = \mathcal{X}^*(s^*)$.
Proof From the defining equation of the Laplace transform we will have

$$\mathcal{X}^*(s^*) = \left[\int_0^{\infty} x(t)\,e^{-s^*t}dt\right]^* = \int_0^{\infty} x^*(t)\,e^{-st}dt = \mathcal{L}\{x^*(t)\} = \mathcal{Y}(s),$$

which is also the desired result. The region of convergence of the transform $\mathcal{X}^*(s^*)$ is the same as that of the transform $\mathcal{X}(s)$. Therefore, if the signal $x(t)$ is a real-valued signal and the transform $\mathcal{X}(s)$ has a pole or zero value at the position $s = s_0$, then the aforementioned transform also has a pole or zero value respectively at the position $s = s_0^*$.

- **Differentiation:** Assuming that the signal $x(t)$ is (a) *piecewise continuous function with piecewise continuous derivative* $x'(t)$ in every interval $0 \leq t \leq T$ and (b) *of exponential order* at the limit $t \to \infty$, then for the region $\Re(s) > \alpha$ the Laplace transform $\mathcal{L}\{x'(t)\}$ of the derivative $x'(t)$ exists and is defined as

$$\mathcal{L}\left\{\frac{dx(t)}{dt}\right\} = s\mathcal{L}\{x(t)\} - x(0^+).$$

Proof from the defining equation of the Laplace transform we have

$$\mathcal{L}\left\{\frac{dx(t)}{dt}\right\} = \lim_{T\to\infty}\int_0^T \frac{dx(t)}{dt}\, e^{-st}dt$$

If there are values $0 = t_0 < t_1 < \cdots < t_{n-1} < t_n = T$, such that the function $x'(t)\, e^{-st}$ is continuous in the intervals $[t_{i-1}, t_i]$ $(i = 0, 1, \ldots, n)$, then the above integral can be written as a sum of n integrals of the form

$$\int_0^T \frac{dx(t)}{dt}\, e^{-st}dt = \int_0^{t_1} \frac{dx(t)}{dt}\, e^{-st}dt + \int_{t_1}^{t_2} \frac{dx(t)}{dt}\, e^{-st}dt + \cdots + \int_{t_{n-1}}^{T} \frac{dx(t)}{dt}\, e^{-st}dt.$$

Calculating each of the above integrals by the method of integration by parts, we arrive at the result

$$\int_{t_{i-1}}^{t_i} \frac{dx(t)}{dt}\, e^{-st}dt = \int_{t_{i-1}}^{t_i} e^{-st}dx(t) = x(t)\, e^{-st}\Big|_{t_{i-1}}^{t_i} - \int_{t_{i-1}}^{t_i} x(t)\, d(e^{-st}) = x(t)\, e^{-st}\Big|_{t_{i-1}}^{t_i} + s\int_{t_{i-1}}^{t_i} x(t)\, e^{-st}dt$$

—for the values $(i = 1, 2, \ldots, n)$—and therefore the above integral is written as

$$\begin{aligned} I = {} & x(t)\, e^{-st}\Big|_0^{t_1} + s\int_0^{t_1} x(t)\, e^{-st}dt + x(t)\, e^{-st}\Big|_{t_1}^{t_2} + s\int_{t_1}^{t_2} x(t)\, e^{-st}dt + \cdots + x(t)\, e^{-st}\Big|_{t_{n-1}}^{T} \\ & + s\int_{t_{n-1}}^{T} x(t)\, e^{-st}dt = \Big[x(t_1)\, e^{-st_1} - x(0^+)\Big] + \Big[x(t_2)\, e^{-st_2} - x(t_1)\, e^{-st_1}\Big] + \ldots \\ & + \Big[x(t_{n-1})\, e^{-st_{n-1}} - x(t_{n-2})\, e^{-st_{n-2}}\Big] + \Big[x(T)\, e^{-sT} - x(t_{n-1})\, e^{-st_{n-1}}\Big] \\ & + s\left\{\int_0^{t_1} x(t)\, e^{-st}dt + \int_{t_1}^{t_2} x(t)\, e^{-st}dt + \cdots + \int_{t_{n-1}}^{T} x(t)\, e^{-st}dt\right\} \end{aligned}$$

or equivalently

$$I = \int_0^T \frac{x(t)}{dt}\, e^{-st}dt = x(T)\, e^{-sT} - x(0^+) + s\int_0^T x(t)\, e^{-st}dt.$$

Substituting the above expression into our original equation, we get

$$\mathcal{L}\left\{\frac{dx(t)}{dt}\right\} = \lim_{T\to\infty}\int_0^T \frac{dx(t)}{dt}\, e^{-st}dt = \lim_{T\to\infty}\left\{x(T)\, e^{-sT}\right\} - x(0^+) + s\lim_{T\to\infty}\int_0^T x(t)\, e^{-st}dt = s\,\mathcal{L}\{x(t)\} - x(0^+),$$

where we used the exponential order property of the signal $x(t)$, which leads to the nullification of the quantity $\lim[e^{-sT}x(t)]$ at the limit $T \to \infty$.

An immediate consequence of the above property is the following theorem.

Theorem 4.5.1 *Let us assume that the signal $x(t)$ is a* piecewise continuous function *and has a* continuous derivative $x^{(n-1)}(t)$ *of order $n-1$, as well as a* piecewise continuous derivative $x^{(n)}(t)$ *of order n in every finite interval $0 \leq t \leq T$. Let us also assume that the signal $x(t)$, as well as all its derivatives $x^{(i)}(t)$ of order $(i = 1, 2, \ldots, n-1)$, are of* exponential order α *at the limit $t \to \infty$. If the above conditions are met, then there is the Laplace transform of the function $x^{(n)}(t)$ with a region of convergence $\Re(s) > \alpha$, which is given by the relation*

$$\mathcal{L}\{x^{(n)}(t)\} = s^n \mathcal{L}\{x(t)\} - s^{n-1}x(0^+) - s^{n-2}x'(0^+) - \cdots - x^{(n-1)}(0^+).$$

Proof By generalizing the equation

$$\mathcal{L}\{x^{(1)}(t)\} = s\mathcal{L}\{x^{(0)}(t)\} - x^{(0)}(0^+)$$

that we proved in the previous section for the derivative of order i of the signal $x(t)$, we will get

$$\mathcal{L}\{x^{(i)}(t)\} = s\mathcal{L}\{x^{(i-1)}(t)\} - x^{(i-1)}(0^+).$$

By repeatedly using the above relation for the values $i = 1, 2, \ldots, n$, we get

$$\begin{aligned}
\mathcal{L}\{x^{(1)}(t)\} &= s\mathcal{L}\{x^{(0)}(t)\} - x^{(0)}(0^+) \\
\mathcal{L}\{x^{(2)}(t)\} &= s\mathcal{L}\{x^{(1)}(t)\} - x^{(1)}(0^+) = s^2\mathcal{L}\{x^{(0)}(t)\} - sx^{(0)}(0^+) - x^{(1)}(0^+) \\
\mathcal{L}\{x^{(3)}(t)\} &= s\mathcal{L}\{x^{(2)}(t)\} - x^{(2)}(0^+) = s^3\mathcal{L}\{x^{(0)}(t)\} - s^2x^{(0)}(0^+) - x^{(1)}(0^+) - x^{(2)}(0^+)
\end{aligned}$$

and finally

$$\mathcal{L}\{x^{(n)}(t)\} = s^n \mathcal{L}\{x(t)\} - s^{n-1}x(0^+) - s^{n-2}x'(0^+) - \cdots - x^{(n-1)}(0^+),$$

which is the desired result.

Note It is important to mention at this point that the property of differentiation is formulated in a different way, if instead of the *unilateral transform (or one-sided)* we use the *bilateral (or two-sided)* Laplace transform. However, since the proof of the property requires the use of the *inverse transform*, this will be given in the section presenting the theory related to the aforementioned transform.

- **Differentiation in the space of s**: if the signal $x(t)$ is a *piecewise continuous function* and has a Laplace transform, then the *derivative* of the aforementioned transform with respect to the variable s is

$$\frac{d\mathcal{X}(s)}{ds} = \frac{d\mathcal{L}\{x(t)\}}{ds} = \mathcal{L}\{-tx(t)\}.$$

Proof Differentiating with respect to s the equation defining the Laplace transform, we get

$$\frac{d\mathcal{X}(s)}{ds} = \frac{d\mathcal{L}\{x(t)\}}{ds} = \frac{d}{ds}\int_0^\infty x(t)\, e^{-st}dt = \int_0^\infty x(t)\frac{d(e^{-st})}{ds}dt = \int_0^\infty (-t)x(t)\, e^{-st}dt = \mathcal{L}\{-tx(t)\}.$$

This relationship can be generalized for higher-order derivatives. By differentiating for the second time, we get

$$\frac{d^2\mathcal{L}\{x(t)\}}{ds^2} = \int_0^\infty (-t)x(t)\frac{d(e^{-st})}{ds}dt = \int_0^\infty t^2x(t)\, e^{-st}dt = \mathcal{L}\{t^2x(t)\},$$

whereas in the general case the relationship will hold

$$\frac{d^n \mathcal{L}\{x(t)\}}{ds^n} = \mathcal{L}\{(-t)^n x(t)\} \qquad (n = 1, 2, 3, \ldots).$$

- **Integration:** If the signal $x(t)$ is a *piecewise continuous function* and has a Laplace transform, then the Laplace transform of the signal

$$y(t) = \int_0^t x(\xi)\, d\xi$$

exists and is given by the relationship

$$\mathcal{Y}(s) = \mathcal{L}\left\{\int_0^t x(\xi)\, d\xi\right\} = \frac{\mathcal{L}\{x(t)\}}{s} + \frac{1}{s}x^{(-1)}(0^+) = \frac{\mathcal{X}(s)}{s} + \frac{1}{s}x^{(-1)}(0^+),$$

where the negative order derivative $x^{(-1)}(t)$ refers to the process of *integration*.
Proof From the defining equation of the Laplace transform we will have

$$\mathcal{L}\left\{\int_0^t x(\xi)\, d\xi\right\} = \int_0^\infty \left[\int_0^t x(\xi)\, d\xi\right] e^{-st} dt = \int_0^\infty \left[\int_0^t x(\xi)\, d\xi\right] d\left(-\frac{e^{-st}}{s}\right)$$
$$= -\frac{e^{-st}}{s}\int_{-\infty}^t x(\xi)\, d\xi\,\Bigg|_0^\infty + \frac{1}{s}\int_0^\infty x(t)\, e^{-st} dt = \frac{\mathcal{L}\{x(t)\}}{s} + \frac{1}{s}\int_{-\infty}^0 x(\xi)\, d\xi$$

and finally

$$\mathcal{L}\left\{\int_0^t x(\xi)\, d\xi\right\} = \frac{\mathcal{L}\{x(t)\}}{s} + \frac{1}{s}x^{(-1)}(0^+),$$

where we proceeded to integrate by parts and used the symbolism $x^{(-1)}(0^+)/s$ to describe the initial value of the integral of the signal $x(t)$ at the limit $x \to 0^+$.
Closely related to the above integration process is the following theorem:

Theorem 4.5.2 *If the continuous-time signal $x(t)$ has a Laplace transform and there is a limit of the function $x(t)/t$ for $t \to 0^+$, then*

$$\mathcal{L}\left\{\frac{x(t)}{t}\right\} = \int_0^\infty \mathcal{L}\{x(t)\} ds = \int_0^\infty \mathcal{X}(s)\, ds.$$

Proof Assuming that the function $\mathcal{X}(s) \equiv \mathcal{L}\{x(t)\}$ is *piecewise continuous* in every finite interval and *of exponential order* α, the integral

$$\mathcal{X}(s) = \int_0^\infty e^{-st} x(t)\, dt$$

will converge uniformly with respect to s and therefore for values of s such that $\Re(s) > \alpha$ and $\beta > \alpha$, it follows that

$$\int_s^\beta \mathcal{X}(s)\, ds = \int_s^\beta \left[\int_0^\infty e^{-st}\, x(t)\, dt\right] ds = \int_0^\infty x(t) \left[\int_s^\beta e^{-st} ds\right] dt = \int_0^\infty \frac{x(t)}{t}\left(e^{-st} - e^{-\beta t}\right) dt.$$

If we assume that the limit of the function $x(t)/s$ exists for $t \to \infty$, then the function in question is *piecewise continuous and of exponential order*, meaning that the last integral converges uniformly with respect to β. Specifically, at the limit $\beta \to \infty$ the above integral reduces to the form

$$\lim_{\beta\to\infty} \int_0^\infty \frac{x(t)}{t}\left(e^{-st} - e^{-\beta t}\right) dt = \int_0^\infty \frac{x(t)}{t}\, e^{-st} dt = \mathcal{L}\left\{\frac{x(t)}{t}\right\},$$

from which the desired result follows.

- **Convolution:** Let us consider two continuous-time signals $x(t)$ and $y(t)$ with Laplace transforms $\mathcal{X}(s) \equiv \mathcal{L}\{x(t)\}$ and $\mathcal{Y}(s) \equiv \mathcal{L}\{y(t)\}$. In this case, the transform of the convolution of the two signals, $z(t) = x(t) * y(t)$, will be given by the relation

$$\mathcal{Z}(s) = \mathcal{L}\{z(t)\} = \mathcal{L}\{x(t) * y(t)\} = \mathcal{L}\{x(t)\}\mathcal{L}\{y(t)\} = \mathcal{X}(s)\mathcal{Y}(s).$$

The above property is also known as the *convolution theorem*.

Proof From the basic theory it is known that the convolution of the signals $x(t)$ and $y(t)$ is defined as

$$x(t) * y(t) = \int_0^\infty x(t-\tau)\, y(\tau)\, d\tau = \int_0^\infty x(\tau)\, y(t-\tau)\, d\tau$$

and therefore using the defining equation of the Laplace transform, we get

$$\mathcal{L}\{x(t) * y(t)\} = \mathcal{L}\left\{\int_0^\infty x(\tau)\, y(t-\tau)\, d\tau\right\} = \int_0^\infty \left[\int_0^\infty x(\tau)\, y(t-\tau)\, d\tau\right] e^{-st} dt = \int_0^\infty y(\tau)\, d\tau \int_0^\infty x(t-\tau)\, e^{-st} dt.$$

Defining the variable $t - \tau = \xi$, $t = \tau + \xi$ and $dt = d\xi$, and the above relation is written

$$\mathcal{L}\{x(t) * y(t)\} = \int_0^\infty y(\tau)\, d\tau \int_{-\tau}^\infty x(\xi)\, e^{-s(\tau+\xi)}\, d\xi = \left(\int_0^\infty y(\tau)\, e^{-s\tau} d\tau\right)\left(\int_0^\infty x(\xi)\, e^{-s\xi}\, d\xi\right) = \mathcal{L}\{x(t)\}\mathcal{L}\{y(t)\},$$

which is also the desired result. To derive the last relation we silently assumed that the signal $x(t)$ is *causal*, so that $x(\xi) = 0$ for $-\tau \le \xi \le 0$.

A direct consequence of the convolution property is the following theorem:

Theorem 4.5.3 *The product of the Laplace transforms of three piecewise continuous-time signals $x(t)$, $y(t)$ and $z(t)$ corresponds to the Laplace transform of the convolution of the three signals, that is*

$$\mathcal{L}\{x(t) * y(t) * z(t)\} = \mathcal{L}\{x(t)\}\mathcal{L}\{y(t)\}\mathcal{L}\{z(t)\}.$$

Proof From the application of the convolution theorem $\mathcal{L}\{y(t) * z(t)\} = \mathcal{L}\{y(t)\}\mathcal{L}\{z(t)\}$ it immediately follows that

$$y(t) * z(t) = \mathcal{L}^{-1}\{\mathcal{L}\{y(t)\}\mathcal{L}\{z(t)\}\}$$

and therefore we will have

$$\begin{aligned}\mathcal{L}\{x(t) * y(t) * z(t)\} &= \mathcal{L}\left\{x(t) * \mathcal{L}^{-1}\left\{\mathcal{L}\{y(t)\}\mathcal{L}\{z(t)\}\right\}\right\} \\ &= \mathcal{L}\{x(t)\}\mathcal{L}\left\{\mathcal{L}^{-1}\left\{\mathcal{L}\{y(t)\}\mathcal{L}\{z(t)\}\right\}\right\} = \mathcal{L}\{x(t)\}\mathcal{L}\{y(t)\}\mathcal{L}\{z(t)\}\end{aligned}$$

- **Periodicity:** If the signal $x(t)$ is a *piecewise continuous function* and *of exponential order*, whereas at the same time it is also *periodic* with period T, then its transform according to Laplace satisfies the property

$$\mathcal{L}\{x(t)\} = \frac{1}{1-e^{-sT}} \int_0^T x(t)\, e^{-st} dt.$$

Proof By subdividing the interval $[0, \infty)$ into infinite intervals of time duration T each, the defining equation of the Laplace transform of the continuous-time signal $x(t)$ takes the form

$$\mathcal{L}\{x(t)\} = \int_0^{\infty} x(t)\, e^{-st} dt = \int_0^{T} x(t)\, e^{-st}\, dt + \int_T^{2T} x(t)\, e^{-st} dt + \int_{2T}^{3T} x(t)\, e^{-st} dt + \cdots = \sum_{n=0}^{\infty} \int_{nT}^{nT+T} x(t)\, e^{-st} dt.$$

In the next step we proceed to change the variable from t to $t = \xi + nT$, so by differentiating we get $dt = d\xi$. For $t = nT$ $\xi = 0$, whereas for $t = nT + T$ $\xi = T$ and the above integral becomes

$$\mathcal{L}\{x(t)\} = \sum_{n=0}^{\infty} \int_0^T x(\xi + nT)\, e^{-(\xi+nT)s} d\xi = \sum_{n=0}^{\infty} e^{-nTs} \int_0^T x(\xi + nT)\, e^{-\xi s} d\xi.$$

However, since the signal $x(\xi)$ is a periodic function with a period value equal to T, $x(\xi + nT) = x(\xi)$ and the above equation will take the form

$$\mathcal{L}\{x(t)\} = \left(\sum_{n=0}^{\infty} e^{-nTs}\right)\left(\int_0^T x(\xi)\, e^{-\xi s} d\xi\right) = \frac{1}{1-e^{-sT}} \int_0^T x(t)\, e^{-ts} dt,$$

where we returned to our original variable and used the identity

$$\sum_{n=0}^{\infty} e^{-nTs} = \sum_{n=0}^{\infty} \left(e^{-sT}\right)^n = \frac{1}{1-e^{-Ts}}.$$

The most important properties of the Laplace transform are included in Table 4.3.

Table 4.3 The properties of the Laplace transform

Property	Continuous-time signal $x(t)$	Laplace transform	Region of convergence
Linearity	$\alpha_1 x_1(t) + \alpha_2 x_2(t)$	$\alpha_1 X_1(s) + \alpha_2 X_2(s)$	$R_1 \cap R_2$
Time shift	$x(t - t_0)$	$e^{-st_0} X(s)$	R
Frequency shift	$e^{s_0 t} x(t)$	$X(s - s_0)$	$R + \sigma_0$
Time scaling	$x(\alpha t)$	$\frac{1}{\lvert\alpha\rvert} X\left(\frac{s}{\alpha}\right)$	αR
Time reversal	$x(-t)$	$X(-s)$	– R
Conjugation	$x^*(t)$	$X^*(s^*)$	R
Time derivative	$x'(t)$	$sX(s) - x(0^+)$	R
Frequency derivative	$-tx(t)$	$X'(s)$	R
Convolution	$x_1(t) * x_2(t)$	$X_1(s)X_2(s)$	$R_1 \cap R_2$
Notation	$L\{x(t)\} = X(s)$, $L\{x_1(t)\} = X_1(s)$, $L\{x_2(t)\} = X_2(s)$ with regions of convergence R, R_1, and R_2 respectively.		

Using the properties of the Laplace transform, we can significantly simplify the calculation process of transformations of this kind, as shown by the following exercises.

Problem 4.5.1 *Determine the Laplace transform of the continuous-time signal* $x(t) = t^2 e^{5t}$

Solution From the property of *differentiation in the s domain*

$$\frac{d^n \mathcal{L}\{x(t)\}}{ds^n} = \mathcal{L}\{(-t^n)x(t)\}$$

for the value $n = 2$ we get

$$\mathcal{X}(s) = \mathcal{L}\{t^2 e^{5t}\} = \frac{d^2}{ds^2}\left\{\mathcal{L}\{e^{5t}\}\right\}$$

with the Laplace transform of the expression e^{5t} being calculated as

$$\mathcal{L}\{e^{5t}\} = \int_0^{\infty} e^{5t} e^{-5s} dt = \int_0^{\infty} e^{-(s-5)t} dt = -\frac{e^{-(s-5)t}}{s-5}\bigg|_0^{\infty} = \frac{1}{s-5}.$$

Differentiating twice we get

$$\frac{d}{ds}\left\{\mathcal{L}\{e^{5t}\}\right\} = -\frac{1}{(s-5)^2} \quad \text{and} \quad \frac{d^2}{ds^2}\left\{\mathcal{L}\{e^{5t}\}\right\} = \frac{2}{(s-5)^3}$$

from which it follows that $\mathcal{X}(s) = \mathcal{L}\{t^2 e^{5t}\} = 2/(s-5)^3$.

Problem 4.5.2 *Prove that the Laplace transform of an even or odd signal is an even or odd function respectively.*

Solution The Laplace transform of the continuous-time signal $x(-t)$ is given by the known relation

$$\mathcal{L}\{x(-t)\} = \int_{-\infty}^{+\infty} x(-t)\, e^{-st} dt.$$

Changing the variable from t to $-t$, the limits of the above integral are reversed, whereas the relation $d(-t) = -dt$ will obviously hold. Therefore, we have

$$\mathcal{L}\{x(-t)\} = \int_{+\infty}^{-\infty} x(t)\, e^{st} d(-t) = \int_{-\infty}^{+\infty} x(t)\, e^{st} dt = \mathcal{X}(-s).$$

If the signal $x(t)$ is *even*, then $x(t) = x(-t)$ and therefore, $\mathcal{L}\{x(t)\} = \mathcal{L}\{x(-t)\}$ or equivalently $\mathcal{X}(s) = \mathcal{X}(-s)$, that is, the Laplace transform is an *even function*. Conversely, if the signal $x(t)$ is *odd*, then $x(t) = -x(-t)$ and from the above relation we will have $\mathcal{L}\{x(t)\} = -\mathcal{L}\{x(-t)\}$ or equivalently $\mathcal{X}(s) = -\mathcal{X}(s)$, that is, the Laplace transform is an *odd function*.

Problem 4.5.3 *Determine the Laplace transform of the continuous-time signal* $x(t) = t\cos(\omega t)$.

Solution From the property of *differentiation in the s domain* for $n = 1$ it follows that

$$\mathcal{L}\{-tx(t)\} = \frac{d\mathcal{L}\{x(t)\}}{ds} \quad \text{or equivalently} \quad \mathcal{L}\{tx(t)\} = -\frac{d\mathcal{L}\{x(t)\}}{ds},$$

and therefore, we have

$$\mathcal{L}\{t\cos(\omega t)\} = -\frac{d}{ds}\left\{\mathcal{L}\{\cos(\omega t)\}\right\}.$$

From tables of elementary transforms we find that $\mathcal{L}\{\cos(\omega t)\} = s/(s^2+\omega^2)$ and therefore, we have

$$\mathcal{L}\{t\cos(\omega t)\} = -\frac{d}{ds}\left(\frac{s}{s^2+\omega^2}\right) = \frac{s^2-\omega^2}{(s^2+\omega^2)^2}.$$

Problem 4.5.4 *Determine the Laplace transform of the continuous-time signal* $x(t) = t^2\cos(\omega t)$.

Solution Using the property of *differentiation in the s domain* twice, we are led to the result

$$\mathcal{L}\{t^2\cos(\omega t)\} = \mathcal{L}\{(-t)[(-t)\cos(\omega t)]\} = -\frac{d}{ds}\left\{\mathcal{L}\{-t\cos(\omega t)\}\right\}$$

$$= -\frac{d}{ds}\left[\frac{d}{ds}\mathcal{L}\{\cos(\omega t)\}\right] = -\frac{d}{ds}\left[\frac{d}{ds}\left(\frac{s}{s^2+\omega^2}\right)\right] = -\frac{d}{ds}\left[\frac{s^2-\omega^2}{(s^2+\omega^2)^2}\right] = \frac{2s^3-6s\omega^2}{(s^2+\omega^2)^3},$$

which is also the desired result. To solve the problem, we used the equation $\mathcal{L}\{\cos(\omega t)\} = s/(s^2+\omega^2)$, which was retrieved from tables of elementary transforms.

Problem 4.5.5 *Using the property of the periodicity of the Laplace transform, determine the transformation of the continuous-time signal* $x(t) = \cos(\omega t)$.

Solution The signal $x(t) = \cos(\omega t)$ is a *periodic trigonometric function* with a period of $T = 2\pi/\omega$. Using the property of *periodicity* and the Euler's formula for the cosine function, the integral of the function in the interval $[0, T] = [0, 2\pi/\omega]$ is calculated as

$$I = \int_0^{2\pi/\omega} \cos(\omega t)e^{-st}dt = \frac{1}{2}\int_0^{2\pi/\omega} e^{i\omega t}e^{-st}dt + \frac{1}{2}\int_0^{2\pi/\omega} e^{-i\omega t}e^{-st}dt = \frac{1}{2}\int_0^{2\pi/\omega} e^{(i\omega-s)t}dt + \frac{1}{2}\int_0^{2\pi/\omega} e^{-(i\omega+s)t}dt$$

$$= \frac{1}{2}\left[\frac{e^{(i\omega-s)t}}{i\omega-s}\right]_0^{2\pi/\omega} - \frac{1}{2}\left[\frac{e^{-(i\omega+s)t}}{i\omega+s}\right]_0^{2\pi/\omega} = \frac{1}{2(i\omega-s)}[e^{(i\omega-s)2\pi/\omega}-1] - \frac{1}{2(i\omega+s)}[e^{-(i\omega+s)2\pi/\omega}-1]$$

$$= \frac{1}{2}\left[\frac{1}{i\omega+s} - \frac{1}{i\omega-s}\right] + \frac{1}{2}\left[\frac{e^{-2\pi s/\omega}}{i\omega-s} - \frac{e^{-2\pi s/\omega}}{i\omega+s}\right] = \frac{s}{s^2+\omega^2} - \frac{se^{-2\pi s/\omega}}{s^2+\omega^2} = \frac{s(1-e^{-2\pi s/\omega})}{s^2+\omega^2}.$$

Replacing the above relation in the property of periodicity of the Laplace transform, we get

$$\mathcal{L}\{\cos(\omega t)\} = \frac{\int_0^{2\pi/\omega} \cos(\omega t)\, e^{-st}dt}{1-e^{-2\pi s/\omega}} = \frac{s\,(1-e^{-2\pi s/\omega})}{(s^2+\omega^2)(1-e^{-2\pi s/\omega})} = \frac{s}{s^2+\omega^2},$$

which is the same result as the one we got from the definition of the Laplace transform.

Problem 4.5.6 *Determine the Laplace transform of the continuous-time signal* $x(t) = |\sin(\omega t)|$.

Solution From the basic theory, it is known that the function $f(t) = \sin(\omega t)$ is periodic with period $T = 2\pi/\omega$. For this reason, the calculation of the Laplace transform will take place with the help of the property of *periodicity*, which in this case is formulated as

$$\mathcal{L}\{|\sin(\omega t)|\} = \frac{1}{1 - e^{-2\pi s/\omega}} \int_0^{2\pi/\omega} |\sin(\omega t)| e^{-st} dt.$$

Using the definition of absolute value for the sine function, it turns out that

$$|\sin(\omega t)| = \begin{cases} +\sin(\omega t) \; 0 \le t \le (\pi/\omega) \\ -\sin(\omega t) \; (\pi/\omega) \le t \le (2\pi/\omega) \end{cases}$$

and therefore the integral of the numerator will take the form

$$\int_0^{2\pi/\omega} |\sin(\omega t)| e^{-st} dt = \int_0^{\pi/\omega} \sin(\omega t)\, e^{-st} dt - \int_{\pi/\omega}^{2\pi/\omega} \sin(\omega t)\, e^{-st} dt.$$

From tables of elementary integrals we find that

$$\int \sin(\omega t)\, e^{-st} dt = -\frac{e^{-st}\, [\omega \cos(\omega t) + s \sin(\omega t)]}{\omega^2 + s^2}.$$

Therefore, the two integrals of the above equation are calculated as

$$\int_0^{\pi/\omega} \sin(\omega t)\, e^{-st} dt = \left[\frac{e^{-st}\, [\omega \cos(\omega t) + s \sin(\omega t)]}{\omega^2 + s^2} \right]_0^{\pi/\omega} = \frac{\omega(1 + e^{-s\pi/\omega})}{s^2 + \omega^2} \quad \text{and}$$

$$\int_{\pi/\omega}^{2\pi/\omega} \sin(\omega t)\, e^{-st} dt = \left[\frac{e^{-st}\, [\omega \cos(\omega t) + s \sin(\omega t)]}{\omega^2 + s^2} \right]_{\pi/\omega}^{2\pi/\omega} = -\frac{\omega(e^{-2\pi s/\omega} + e^{-\pi s/\omega})}{\omega^2 + s^2}$$

and substituting into our original equation, it takes the form

$$\int_0^{2\pi/\omega} \|\sin(\omega t)\| e^{-st} dt = \frac{\omega(1 + e^{-s\pi/\omega})}{s^2 + \omega^2} + \frac{\omega(e^{-2\pi s/\omega} + e^{-\pi s/\omega})}{\omega^2 + s^2} = \frac{\omega(1 + 2e^{-s\pi/\omega} + e^{-2s\pi/\omega})}{\omega^2 + s^2} = \frac{\omega(1 + e^{-s\pi/\omega})^2}{\omega^2 + s^2}.$$

On the other hand, the denominator in the defining equation of the Laplace transform is written as

$$1 - e^{-2s\pi/\omega} = 1 - (e^{-s\pi/\omega})^2 = (1 + e^{-s\pi/\omega})(1 - e^{-s\pi/\omega})$$

and therefore finally we get

$$\mathcal{L}\{|\sin(\omega t)|\} = \frac{\omega(1 + e^{-s\pi/\omega})(1 + e^{-s\pi/\omega})}{(\omega^2 + s^2)(1 + e^{-s\pi/\omega})(1 - e^{-s\pi/\omega})} = \frac{\omega(1 + e^{-s\pi/\omega})}{(\omega^2 + s^2)(1 - e^{-s\pi/\omega})} = \frac{\omega}{\omega^2 + s^2} \coth\left(\frac{s\pi}{2\omega}\right),$$

which is also the desired result. To derive the last relation, we used the known identity

$$\frac{1 + e^{-s\pi/\omega}}{1 - e^{-s\pi/\omega}} = \frac{e^{s\pi/\omega} + 1}{e^{s\pi/\omega} - 1} = \coth\left(\frac{s\pi}{2\omega}\right).$$

Problem 4.5.7 *Determine the Laplace transform of the continuous-time signal* $x(t) = e^{\alpha t}\cos(\beta t)$.

Solution From the property of complex shifting $\mathcal{L}\{e^{s_0 t}x(t)\} = \mathcal{X}(s - s_0)$ for $s_0 = \alpha$ and $x(t) = \cos(\beta t)$ we have

$$\mathcal{L}\{e^{\alpha t}\cos(\beta t)\} \leftrightarrows \mathcal{L}\{\cos(\beta t)\}_{s \to s-\alpha}$$

Referring to tables of elementary transforms, we find that $\mathcal{L}\{\cos(\beta t)\} = s/(s^2+\beta^2)$, and making the substitution $s \to s-\alpha$, we get

$$\mathcal{L}\left\{e^{\alpha t}\cos(\beta t)\right\} = \left.\frac{s}{s^2+\beta^2}\right|_{s \to s-\alpha} = \frac{s-\alpha}{(s-\alpha)^2+\beta^2}.$$

Problem 4.5.8 *Determine the Laplace transform of the continuous-time signal* $x(t) = te^{\alpha t}\cos(\beta t)$.

Solution Combining the properties of *shifting* and *differentiation in the complex plane*

$$\mathcal{L}\{e^{s_0 t}x(t)\} \leftrightarrows \mathcal{X}(s - s_0) \qquad \text{and} \qquad \mathcal{L}\{tx(t)\} = -\frac{d\mathcal{L}\{x(t)\}}{ds}$$

we get

$$\mathcal{L}\{te^{\alpha t}\cos(\beta t)\} = \mathcal{L}\{t\cos(\beta t)\}_{s \to s-\alpha} = \left.\left(-\frac{d}{ds}\mathcal{L}\{\cos(\beta t)\}\right)\right|_{s \to s-\alpha}$$

$$= -\left.\frac{d}{ds}\left(\frac{s}{s^2+\beta^2}\right)\right|_{s \to s-\alpha} = \left.\frac{s^2-\beta^2}{(s^2+\beta^2)^2}\right|_{s \to s-\alpha} = \frac{(s-\alpha)^2-\beta^2}{[(s-\alpha)^2+\beta^2]^2},$$

which is the desired result. To solve the above problem we used the equation $\mathcal{L}\{\cos(\beta t)\} = s/(s^2+\beta^2)$, which was retrieved from tables of elementary transforms.

Problem 4.5.9 *Determine the Laplace transform of the continuous-time signals*

$$x(t) = \cosh(\alpha t) \qquad \textit{and} \qquad y(t) = t\cosh(\alpha t).$$

Solution (a) From the theory it is known that the hyperbolic cosine is defined as $\cosh(\alpha t) = (e^{\alpha t} + e^{-\alpha t})/2$ and therefore from the property of *linearity* of the Laplace transform we have

$$\mathcal{L}\{\cosh(\alpha t)\} = \frac{1}{2}\Big(\mathcal{L}\{e^{\alpha t}\} + \mathcal{L}\{e^{-\alpha t}\}\Big) = \frac{1}{2}\left(\frac{1}{s-\alpha} + \frac{1}{s+\alpha}\right) = \frac{s}{s^2-\alpha^2},$$

where we used the elementary transforms $\mathcal{L}\{e^{\alpha t}\} = 1/(s-\alpha)$ and $\mathcal{L}\{e^{-\alpha t}\} = 1/(s+\alpha)$.
(b) From the previous result and the property of *differentiation in the complex plane* we have

$$\mathcal{L}\{t\cosh(\alpha t)\} = -\frac{d}{ds}\Big(\mathcal{L}\{\cosh(\alpha t)\}\Big) = -\frac{d}{ds}\left(\frac{s}{s^2-\alpha^2}\right) = \frac{s^2+\alpha^2}{(s^2-\alpha^2)^2}.$$

Problem 4.5.10 *Verify the validity of the differentiation property* $\mathcal{L}\{x'(t)\} = s\mathcal{L}\{x(t)\} - x(0^+)$ *using the signal* $x(t) = e^{-3t}\sin(2t)$.

Solution To solve the exercise, we need to determine the Laplace transform of both the signal $x(t)$ and its derivative $x'(t)$, which is calculated as

$$x'(t) = -3e^{-3t}\sin(2t) + 2e^{-3t}\cos(2t)$$

Referring to tables of elementary transforms, we find that

$$\mathcal{L}\{e^{-\alpha t}\sin(\omega t)\} = \frac{\omega}{(s+\alpha)^2+\omega^2} \quad \text{and} \quad \mathcal{L}\{e^{-\alpha t}\cos(\omega t)\} = \frac{s+\alpha}{(s+\alpha)^2+\omega^2},$$

while from the defining equation of the signal $x(t)$ we will have $x(0^+) = 0$. Therefore, the Laplace transform of the signal $x(t)$—that is, for the values $\alpha = 3$ and $\omega = 2$—is calculated as

$$\mathcal{L}\{e^{-3t}\sin(2t)\} = \frac{2}{(s+3)^2+2^2} = \frac{2}{s^2+6s+13},$$

from which it immediately follows that the right-hand side of the property of *differentiation* is given from the relationship

$$s\mathcal{L}\{x(t)\} - x(0^+) = s\mathcal{L}\{e^{-3t}\sin(2t)\} = \frac{2s}{s^2+6s+13}$$

On the other hand, the Laplace transform of the derivative $x'(t)$ will be given by the relationship

$$\mathcal{L}\{x'(t)\} = \mathcal{L}\{-3e^{-3t}\sin(2t) + 2e^{-3t}\cos(2t)\} = -3\mathcal{L}\{e^{-3t}\sin(2t)\} + 2\mathcal{L}\{e^{-3t}\cos(2t)\}.$$

However, it is

$$\mathcal{L}\{e^{-3t}\cos(2t)\} = \frac{s+3}{(s+3)^2+2^2} = \frac{s+3}{s^2-6s+13}$$

and using the expression for the transform $\mathcal{L}\{e^{-3t}\sin(2t)\}$ that we found earlier,

$$\mathcal{L}\{x'(t)\} = -3\frac{2}{s^2+6s+13} + 2\frac{s+3}{s^2+6s+13} = \frac{2s}{s^2+6s+13} = s\mathcal{L}\{x(t)\} - x(0^+),$$

a result that also verifies the validity of the property of *differentiation*.

Problem 4.5.11 *Calculate the Laplace transform of the continuous-time signal given by the equation* $y(t) = x(t-2) * g(3-t)$ *where* $x(t) = e^{-2t}u(t)$ *and* $g(t) = e^{-3t}u(t)$.

Solution From the property of *convolution* of the Laplace transform we have

$$y(t) = x(t-2) * g(3-t) \leftrightarrows \mathcal{L}\{y(t)\} = \mathcal{L}\{x(t-2)\}\mathcal{L}\{g(3-t)\}.$$

However, it is

$$\mathcal{L}\{x(t)\} = \mathcal{L}\{e^{-2t}u(t)\} = \frac{1}{s+2} \quad \text{with region of convergence } \Re(s) > -2$$

$$\mathcal{L}\{g(t)\} = \mathcal{L}\{e^{-3t}u(t)\} = \frac{1}{s+3} \quad \text{with region of convergence } \Re(s) > -3$$

(as derived from tables of elementary transforms), whereas from the application of the properties of *time scaling* and *time shifting* we get

$$\mathcal{L}\{x(t-2)\} = e^{-2s}\mathcal{X}(s) = \frac{e^{-2s}}{s+2} \quad \text{with region of convergence } \Re(s) > -2$$

$$\mathcal{L}\{g(3-t)\} = e^{-3t}\mathcal{G}(-s) = \frac{e^{-2s}}{3-s} \quad \text{with region of convergence } \Re(s) > -3,$$

where $\mathcal{X}(s) \equiv \mathcal{L}\{x(t)\}$ and $\mathcal{G}(s) \equiv \mathcal{L}\{g(t)\}$. Therefore, we have

$$\mathcal{L}\{y(t)\} = \mathcal{L}\{x(t-2)\}\mathcal{L}\{g(3-t)\} = \frac{e^{-2s}}{s+2}\frac{e^{-2s}}{3-s},$$

which is also the desired result.

4.6 Initial and Final Value Theorems

In some cases, it is required to find the value of the signal $x(t)$ for the limiting cases $t \to 0^+$ and $t \to \infty$. The finding of the above two values is based on the use of the theorems of initial and final values, which are formulated in the following way:

Theorem 4.6.1 *Let us consider a causal signal $x(t)$—which satisfies the property $x(t) = 0$ for $t < 0$—which does not contain impulsive functions or higher order anomalies at the time $t = 0$. In this case, if the functions $x(t)$ and $x'(t)$ have a Laplace transform, it is proven that*

$$\lim_{s\to\infty}\left[s\mathcal{L}\{x(t)\}\right] = \lim_{t\to 0^+} x(t) \equiv x(0^+).$$

This theorem is known as the initial value theorem.

Proof From the property of *differentiation* of the Laplace transform

$$\mathcal{L}\left\{\frac{dx(t)}{dt}\right\} = s\mathcal{L}\{x(t)\} - x(0^+)$$

we get (for the limit $s \to \infty$)

$$\lim_{s\to\infty}\mathcal{L}\left\{\frac{dx(t)}{dt}\right\} = \lim_{s\to\infty}\int_0^\infty \frac{dx(t)}{dt}e^{-st}dt = \lim_{s\to\infty}[s\mathcal{L}\{x(t)\} - x(0^+)].$$

From the above relation, it is not difficult to see that the integral of the Laplace transform tends to zero at the limit $s \to \infty$—owing to the existence of the quantity e^{-st}—and therefore, we have

$$\lim_{s\to\infty}\int_0^\infty \frac{dx(t)}{dt}e^{-st}dt = \lim_{s\to\infty}[s\mathcal{L}\{x(t)\} - x(0^+)] = 0.$$

Taking into account the fact that the function $x(0^+) \equiv \lim_{t\to 0^+} x(t)$ is independent of the variable s, we get

$$\lim_{s\to\infty}\left[s\mathcal{L}\{x(t)\}\right] = \lim_{t\to 0^+} x(t) \equiv x(0^+),$$

which is the desired result.

Theorem 4.6.2 *Let us consider a causal signal $x(t)$—which, as is well known, satisfies the property $x(t) = 0$ for $t < 0$—which has a finite limit value $t \to \infty$. In this case, if the functions $x(t)$ and $x'(t)$ have a Laplace transform, it is proven that*

$$\lim_{s \to 0}\left[s\mathcal{L}\{x(t)\}\right] = \lim_{t \to \infty} x(t) \equiv x(\infty).$$

This theorem is known as the final value theorem.

Proof Following the methodology of the previous theorem, we start from the equation of the *differentiation* theorem and take its limit for $s \to 0$. In this case, we have

$$\lim_{s \to 0} \mathcal{L}\left\{\frac{dx(t)}{dt}\right\} = \lim_{s \to 0} \int_0^\infty \frac{dx(t)}{dt} e^{-st} dt = \lim_{s \to 0} [s\mathcal{L}\{x(t)\} - x(0^+)].$$

Taking into account the fact that the variables s and t are independent of each other and that for $s \to 0$, $e^{-st} \to 1$, the integral of the above equation is written

$$\lim_{s \to 0} \int_0^\infty \frac{dx(t)}{dt} e^{-st} dt = \lim_{t \to \infty} x(t) - x(0^+).$$

Therefore, we have $\lim_{t \to \infty} x(t) - x(0^+) = \lim_{s \to 0} [s\mathcal{L}\{x(t)\}] - x(0^+)$, from which the desired result is derived

$$\lim_{s \to 0}\left[s\mathcal{L}\{x(t)\}\right] = \lim_{t \to \infty} x(t) \equiv x(\infty).$$

It is proven that the power of the initial value theorem *presupposes the absence of an impulsive function at the beginning* $t = 0$. As for the final value theorem, it holds when the transform $\mathcal{L}\{x(t)\}$ has a simple pole at the beginning $s = 0$, *but not when it has a higher-order pole or alternatively a pole on the imaginary semi-axis or on the right half-plane*. Finally, let us note that these two theorems apply only to causal functions that satisfy the condition $x(t) = 0$ for $t < 0$ and therefore they are actually properties of the unilateral Laplace transform.

Problem 4.6.1 *Using the initial value theorem, determine the value of the continuous-time signal* $x(t) = e^{-2t}u(t) + e^{-t}\cos(3t)\,u(t)$ *at the time* $t = 0$ *(Oppenheim, Willsky, Nawab, Example 9.4).*

Solution To apply the initial value theorem we will determine the Laplace transform of the above signal. Applying the property of the *linearity* of the Laplace transform, we get

$$\mathcal{L}\{x(t)\} = \mathcal{L}\{e^{-2t}u(t)\} + \mathcal{L}\{e^{-t}\cos(3t)u(t)\}.$$

Referring to tables of elementary transforms, we find

$$\mathcal{L}\{e^{-\alpha t}u(t)\} = \frac{1}{s+\alpha} \quad \text{and} \quad \mathcal{L}\{e^{-\alpha t}\cos(\omega t)u(t)\} = \frac{s+\alpha}{(s+\alpha)^2+\omega^2}$$

and therefore we have

$$\mathcal{L}\{e^{-2t}u(t)\} = \frac{1}{s+2} \quad \text{and} \quad \mathcal{L}\{e^{-t}\cos(3t)\} = \frac{s+1}{(s+1)^2+9}.$$

Therefore, we have

$$\mathcal{L}\{x(t)\} = \frac{1}{s+2} + \frac{s+1}{s^2+2s+10} = \frac{2s^2+5s+12}{(s^2+2s+10)(s+2)},$$

and applying the initial value theorem, we get

$$x(0) = \lim_{s\to\infty}[s\mathcal{L}\{x(t)\}] = \frac{2s^3+5s^2+12s}{s^3+4s^2+14s+20} = \lim_{s\to\infty}\frac{6s^2+10s+12}{3s^2+8s+14} = \lim_{s\to\infty}\frac{12s+10}{6s+8} = \frac{12}{6} = 2,$$

where we applied the *L'Hospital's rule* twice. The result we found is in agreement with that resulting from the defining equation of the signal $x(t)$, since for $t = 0$, we will have $x(0) = 2$.

Problem 4.6.2 *Apply the final value theorem to the functions*

$$\mathcal{L}\{x_1(t)\} = \frac{s+\alpha}{(s+\alpha)^2+\beta^2} \quad \textit{and} \quad \mathcal{L}\{x_2(t)\} = \frac{s}{s^2+\beta^2}.$$

Solution Applying the final value theorem for the two functions, we get

$$\lim_{t\to\infty} x_1(t) = \lim_{s\to 0}[s\mathcal{L}\{x_1(t)\}] = \lim_{s\to 0}\frac{s(s+\alpha)}{(s+\alpha)^2+\beta^2} = 0 \quad \text{and} \quad \lim_{t\to\infty} x_2(t) = \lim_{s\to 0}[s\mathcal{L}\{x_2(t)\}] = \lim_{s\to 0}\frac{s^2}{s^2+\beta^2}.$$

We observe, however, that the function $s\mathcal{L}\{x_2(t)\}$ has poles on the imaginary semi-axis at the points $s_{1,2} = \pm j\beta$ and therefore the final value theorem *cannot be applied*.

4.7 The Inverse Laplace Transform

Considering a *continuous differentiable signal $x(t)$ of exponential order* in the interval $[0, \infty)$ that has a Laplace transform $\mathcal{X}(s) \equiv \mathcal{L}\{x(t)\}$, we define the *inverse Laplace transform* as an *operator* which, when applied to the complex function $\mathcal{X}(s)$, returns the signal $x(t)$ from which it originated; in other words, this means that *the inverse of the Laplace transform of a function is the function itself.* The above property is mathematically written as

$$x(t) = \mathcal{L}^{-1}\{\mathcal{X}(s)\} \tag{4.11}$$

In practice, the equations of the direct and inverse transform appear in pairs, so that we can write the relations

$$\mathcal{X}(s) = \mathcal{L}\{x(t)\} \quad \text{and} \quad x(t) = \mathcal{L}^{-1}\{\mathcal{X}(s)\}.$$

To derive the defining equation of the inverse transform, we consider a piecewise continuous and differentiable signal $x(t)$ such that for $t < 0$ it is $x(t) = 0$ (this means that the signal is *causal*). In this case, the complex representation according to Fourier, the function $x(t)\,e^{-\sigma t}$, will have the form

$$x(t)\,e^{-\sigma t} = \frac{1}{2\pi}\int_{-\infty}^{+\infty} e^{j\omega t}\left\{\int_0^{\infty} x(\xi)\,e^{-\sigma\xi}\,e^{-j\omega\xi}d\xi\right\}d\omega,$$

where we silently assumed that the integral

$$I = \int_0^{\infty} |x(\xi)\,e^{-(\sigma+j\omega)\xi}|d\xi = \int_0^{\infty} |x(\xi)|e^{-\sigma\xi}d\xi,$$

converges so that the above function can be defined. Now, multiplying both sides of the above function by the quantity $e^{\sigma t}$, we get

$$x(t) = \frac{1}{2\pi} \int_{-\infty}^{+\infty} e^{(\sigma+j\omega)t} \left\{ \int_{0}^{\infty} x(\xi)\, e^{-(\sigma+j\omega)\xi} d\xi \right\} d\omega.$$

In the next step of the process we proceed to change the variable of integration from ω to $s = \sigma + j\omega$, so by differentiation we get $d\omega = ds/j$. As for the new integration limits, for $\omega = -\infty$, $s = \sigma - j\infty$, whereas for $\omega = +\infty$, $s = \sigma - j\infty$ and the above integral will take the form

$$x(t) = \frac{1}{2\pi j} \int_{\sigma-j\infty}^{\sigma+j\infty} e^{st} \left\{ \int_{0}^{\infty} x(\xi) e^{-s\xi} d\xi \right\} ds.$$

But the inner integral is recognized as the Laplace transform of the function $x(t)$,

$$\mathcal{X}(s) \equiv \mathcal{L}\{x(t)\} = \int_{0}^{\infty} x(t)\, e^{-st} dx,$$

and therefore we finally end up with the relation

$$x(t) = \frac{1}{2\pi i} \int_{\sigma-j\infty}^{\sigma+j\infty} e^{st} \mathcal{X}(s) ds, \tag{4.12}$$

which is also the defining equation of the *inverse Laplace transform.*

The integral of the last relation is known as the *Fourier-Mellin integral* or *Bromwich integral*. It is a *contour integral* calculated along a straight line $\Re(s) = c$ parallel to the imaginary axis of the complex plane with the value of c chosen large enough to ensure the existence of the integral. It is proven that *the convergence of the integral of the inverse transform is ensured, when the above straight line encloses all the poles of the transform, that is, when the value of the constant c is greater than the real part of all the poles of the function* $\mathcal{X}(s)$. This means that all the poles of the transform will be *to the left of the aforementioned straight line* and therefore the path along which the integral will be calculated must close to the left, so as to include these poles.[3]

There are many and different ways to calculate the inverse of the Laplace transform, which are presented in detail in the following pages.

4.7.1 The Method of Complex Integration

This method of calculating the inverse Laplace transform allows the finding of the signal $x(t)$ from the direct calculation of the above contour integral in the complex plane, along a *suitably chosen path*. Let us assume for this purpose that the function $\mathcal{X}(s)$ is a *rational function* of the form $P(s)/Q(s)$ (where $P(s)$ and $Q(s)$ are polynomial functions of the complex variable s, of degrees m and n respectively, with $n > m$), and has at most n poles at the points $s_1, s_2, \ldots, s_n$. Let us also assume that the path Γ_R along which we will calculate the contour integral, consists (a) of the vertical line $L_R : \{\Re(s) = \sigma_0\}$, which has been chosen so that the poles $s_1, s_2, \ldots, s_n$ are to its left and (b) from the left semicircle $C_R : \{s = \sigma_0 + Re^{j\vartheta}, [\pi/2 \le \vartheta \le (3\pi/2)]\}$ of radius R; this path is shown in picture (a) of Fig. 4.5. In this case, the calculation of the integral of the inverse transform leads to the result

$$\lim_{R\to\infty} \oint_{\Gamma_R} e^{st} \mathcal{X}(s)\, ds = \lim_{R\to\infty} \int_{C_R} \frac{P(s)}{Q(s)} e^{st} ds + \lim_{R\to\infty} \int_{L_R} \frac{P(s)}{Q(s)} e^{st} ds.$$

[3] According to the theory of complex analysis *contour integrals on the complex plane are calculated along closed paths*, that is, routes the beginning and end of which coincide. It is obvious that the straight line $\Re(s) = c$ is not a closed path and for this reason it must be incorporated into a closed path Γ_R along which the integral will be calculated.

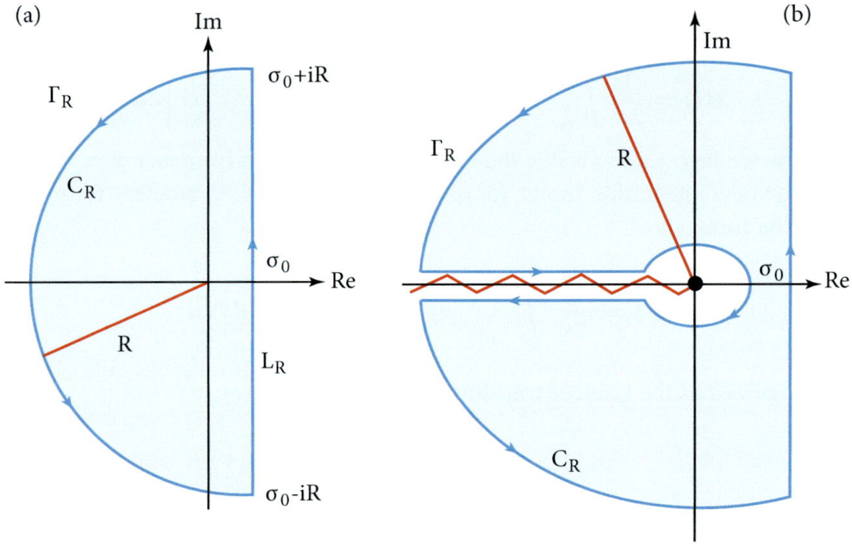

Fig. 4.5 The path Γ_R along which we calculate the contour integral of the inverse Laplace transform

However, as it emerges from the *Jordan's lemma of Jordan*[4] (the way of using this lemma will be demonstrated in the following exercises), *the calculation of the integral along the semicircle C_R leads to a value equal to zero*—thus,

$$\lim_{R\to\infty}\int_{C_R}\frac{P(s)}{Q(s)}\,e^{st}\,ds=0$$

and therefore

$$\lim_{R\to\infty}\oint_{\Gamma_R}e^{st}\mathcal{X}(s)\,ds=\lim_{R\to\infty}\int_{L_R}\frac{P(s)}{Q(s)}\,e^{st}\,ds=\lim_{R\to\infty}\int_{\sigma_0-jR}^{\sigma_0+jR}\frac{P(s)}{Q(s)}\,e^{st}\,ds$$

In the next step of the process we use *the basic theorem of residue calculus* according to which *the contour integral along the path Γ_R will be equal to the sum of the residues of the function $F(s)\,e^{st}$ calculated at the positions of its poles*—since all the poles are enclosed by the path Γ_R—*multiplied by the quantity* $(2\pi j)$. Therefore, we have

$$\lim_{R\to\infty}\int_{\sigma_0-jR}^{\sigma_0+jR}\frac{P(s)}{Q(s)}\,e^{st}\,ds=\int_{\sigma-j\infty}^{\sigma+j\infty}e^{st}\mathcal{X}(s)\,ds=\sum_{k=1}^{n}\text{Res}\left[\mathcal{X}(s)\,e^{st}\right]_{s=s_k}$$

(for the value $\sigma=\sigma_0$) and combining the above equation with the defining equation of the inverse Laplace transform, we will get

$$x(t)\equiv\mathcal{L}^{-1}\{\mathcal{X}(s)\}=\sum_{k=1}^{n}\text{Res}\left[\mathcal{X}(s)\,e^{st}\right]_{s=s_k}, \tag{4.13}$$

which is also our final result. This method of calculating the inverse transform, is demonstrated in an analytical way in the examples that follow.

Note Let us mention at this point that the choice of the closed path Γ_R in the way we presented previously applies to simple cases of complex functions with a *finite number of poles*. Conversely, if the function has an infinite number of poles, which, however, are located to the left of the line L_R, we can use as a path *a number of circular arcs*, the union of which will form a closed curve that contains all the poles. In this case, the inverse Laplace transform $x(t)$ will be calculated *as a series of*

[4] See Footnote 13 (page 264) of the homonymous section of the third chapter.

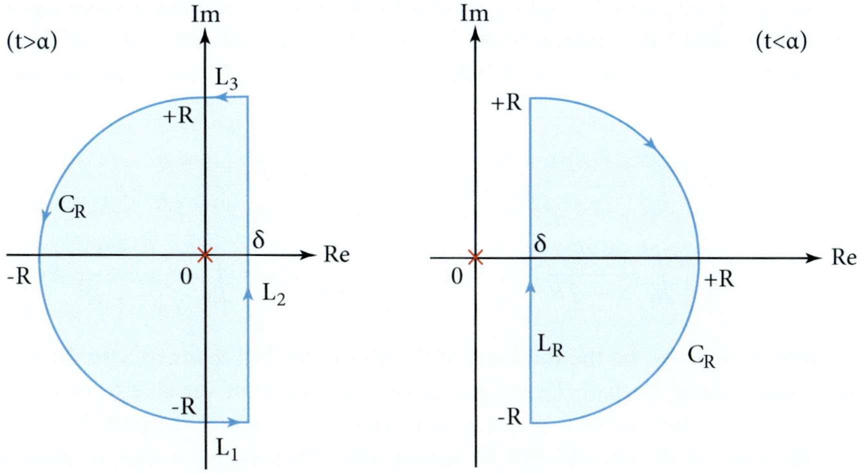

Fig. 4.6 Calculation of the contour integral for finding the inverse transform of the function $F(s) = e^{-\alpha s}/s$ (see Exercise 4.7.1)

infinite terms. Finally, if the function $\mathcal{X}(s)$ is a *multivalued function*, the use of a *cut*[5] and the choice of a path similar to the one presented in image (b) of Fig. 4.5. is required.

Problem 4.7.1 *Using the method of complex integration, determine the inverse Laplace transform of the function* $\mathcal{X}(s) = e^{-\alpha s}/s$.

Solution According to the method of *complex integration*, the inverse transform of the function $\mathcal{X}(s)$ will be calculated from the integral

$$x(t) = \frac{1}{2\pi j}\int_{\sigma-j\infty}^{\sigma+j\infty} e^{st}\mathcal{X}(s)\,ds = \frac{1}{2\pi j}\int_{\sigma-j\infty}^{\sigma+j\infty} \frac{e^{(t-\alpha)s}}{s}\,ds$$

In this case, the function

$$\mathcal{X}(s)\,e^{st} = \frac{e^{-\alpha s}}{s}\,e^{st} = \frac{e^{(t-\alpha)\,s}}{s}$$

has *a simple pole* at the position $s = 0$. To solve the problem, we will consider the cases $t > \alpha$, $t < \alpha$ and $t = \alpha$. In detail, we have the following.

Case 1: $t > \alpha$ In this case, we calculate the contour integral along the path Γ_R shown in the left image of Fig. 4.6 and which consists of

- The line L_1 defined by the points $s_1 = -jR$ and $s_2 = \delta - jR$
- The line L_2 defined by the points $s_1 = \delta - jR$ and $s_2 = \delta + jR$
- The line L_3 defined by the points $s_1 = \delta + jR$ and $s_2 = jR$.
- The left semicircle defined by the equation $s = \delta + Re^{j\vartheta}$ with the angle ϑ taking values in the interval $\pi/2 \le \vartheta \le 3\pi/2$.

Therefore, we have

$$\oint_{\Gamma_R} \frac{e^{(t-\alpha)\,s}}{s}ds = \int_{-jR}^{\delta-jR} \frac{e^{(t-\alpha)\,s}}{s}ds + \int_{\delta-jR}^{\delta+jR} \frac{e^{(t-\alpha)\,s}}{s}ds + \int_{\delta+jR}^{jR} \frac{e^{(t-\alpha)\,s}}{s}ds + \int_{C_R} \frac{e^{(t-\alpha)\,s}}{s}ds.$$

[5] *Cut* of an analytic multivalued complex function is a curve on the complex plane along which the function presents *discontinuities*.

It is proven that the integrals along the paths L_1 and L_3 and for the limit $R \to \infty$ have a value equal to zero. Starting from the first integral, we proceed to change the variable from s to $s = y - jR$; thus, by differentiating, we get $ds = dy$. As for the new limits of the integral, for $s = -jR$, $y = 0$, whereas for $s = \delta - jR$, $y = \delta$ and the modulus of the integral is calculated as

$$\left| \int_{-jR}^{\delta - jR} \frac{e^{(t-\alpha)s}}{s} ds \right| = \left| \int_0^{\delta} \frac{e^{(t-\alpha)(y-jR)}}{y - jR} dy \right| = \left| e^{-(t-\alpha)jR} \int_0^{\delta} \frac{e^{(t-\alpha)y}}{y - jR} dy \right|$$

$$= \left| \int_0^{\delta} \frac{e^{(t-\alpha)y}}{y - jR} dy \right| \leq \int_0^{\delta} \left| \frac{e^{(t-\alpha)y}}{y - jR} \right| dy = \int_0^{\delta} \frac{e^{(t-\alpha)y}}{\sqrt{y^2 + R^2}} dy \leq \frac{\delta e^{(t-\alpha)\delta}}{R}.$$

It is obvious that for the limit case $R \to \infty$ the modulus of the above integral *tends to zero*. In a completely similar way, for the calculation of the integral along the path L_3 we proceed to the change of variable from s to $s = y + jR$, so we get $ds = dy$. For $s = \delta + jR$ $y = \delta$, whereas for $s = jR$ $y = 0$. However, because the path L_3 is traversed *in the negative direction of the real axis*, the value of the integral will be accompanied by a *negative sign*, or equivalently the integral will be calculated for the interval from 0 to δ. Following the methodology we used previously, we get

$$\left| \int_{\delta + jR}^{jR} \frac{e^{(t-\alpha)s}}{s} ds \right| = \left| \int_0^{\delta} \frac{e^{(t-\alpha)(y+jR)}}{y + jR} dy \right| = \left| e^{(t-\alpha)jR} \int_0^{\delta} \frac{e^{(t-\alpha)y}}{y + jR} dy \right| =$$

$$= \left| \int_0^{\delta} \frac{e^{(t-\alpha)y}}{y + jR} dy \right| \leq \int_0^{\delta} \left| \frac{e^{(t-\alpha)y}}{y + jR} \right| dy = \int_0^{\delta} \frac{e^{(t-\alpha)y}}{\sqrt{y^2 + R^2}} dy \leq \frac{\delta e^{(t-\alpha)\delta}}{R},$$

and as before, in the limit $R \to \infty$ the measure of the integral will tend to *zero*.

On the other hand, the measure of the integral along the path C_R is calculated as follows: differentiating the defining equation of this path, which is $s = \delta + Re^{j\vartheta}$, we get $ds = jRe^{j\vartheta} d\vartheta$ with the angle ϑ taking values in the interval $\pi/2 \leq \vartheta \leq 3\pi/2$. Therefore, we have

$$\left| \int_{C_R} \frac{e^{(t-\alpha)s}}{s} ds \right| = \left| \int_{\pi/2}^{3\pi/2} \frac{e^{(t-\alpha)Re^{j\vartheta}}}{Re^{j\vartheta}} jRe^{j\vartheta}\, d\vartheta \right| \leq \int_{\pi/2}^{3\pi/2} |e^{(t-\alpha)Re^{j\vartheta}}| d\vartheta = \int_{\pi/2}^{3\pi/2} e^{(t-\alpha)R\cos\vartheta}\, d\vartheta,$$

since, as is well known, $|e^{(t-\alpha)Re^{j\vartheta}}| = |e^{(t-\alpha)R(\cos\vartheta + j\sin\vartheta)}| = |e^{(t-\alpha)R\cos\vartheta}| \cdot |e^{j(t-\alpha)R\sin\vartheta}| = e^{(t-\alpha)R\cos\vartheta}$. In the last expression, the quantity of the exponent $(t - \alpha)R\cos\vartheta$ is negative, since for the range of values $\pi/2 \leq \vartheta \leq 3\pi/2$, $\cos\vartheta < 0$, whereas we will still have $t - \alpha > 0$ (since we are examining the case $t > \alpha$). According to the *Jordan's lemma*, the above integral for the limit $R \to \infty$ will have a value equal to zero.

Combining the above results, we get

$$\oint_{\Gamma_R} \frac{e^{(t-\alpha)\, s}}{s} ds = \int_{\delta - jR}^{\delta + jR} \frac{e^{(t-\alpha)s}}{s} ds = 2\pi j \,\text{Res}\left[\mathcal{X}(s)\, e^{st}\right]_{s=0} = 2\pi j,$$

where we used *the basic theorem of residue calculus* and took into account the fact that the function $\mathcal{X}(s)\, e^{st}$ has a simple pole at the position $s = 0$ with a residual value

$$\text{Res}\left[\mathcal{X}(s)\, e^{st}\right]_{s=0} = \lim_{s\to 0}\left[s\mathcal{X}(s)\, e^{st}\right] = \lim_{s\to 0}\left[e^{(t-\alpha)s}\right] = 1$$

and from the defining equation of the inverse transform we will get

$$x(t) = \frac{1}{2\pi j} \int_{\delta - jR}^{\delta + jR} \frac{e^{(t-\alpha)s}}{s} ds = 1.$$

Case 2: $t < \alpha$ In this case, we calculate the contour integral along the path Γ_R depicted in the right image of Fig. 4.6 and which consists of the straight line L_R defined by the points $s_1 = \delta - jR$ and $s_2 = \delta + jR$, and from the right semicircle C_R, described by the equation $s = \delta + Re^{j\vartheta}$ with the angle ϑ, taking values within the range $\pi/2 \leq \vartheta \leq -\pi/2$. Since this path *does not contain poles of the function* $\mathcal{X}(s)\, e^{st}$, the curvilinear integral along the aforementioned path will be, according to the *Cauchy theorem*, equal to zero, or in mathematical writing

$$\oint_{\Gamma_R} \frac{e^{(t-\alpha)\,s}}{s}\, ds = \int_{\delta-jR}^{\delta+jR} \frac{e^{(t-\alpha)\,s}}{s}\, ds + \int_{C_R} \frac{e^{(t-\alpha)\,s}}{s} ds = 0.$$

Following the methodology of the previous case, the measure of the integral along the path C_R is calculated as

$$\left| \int_{C_R} \frac{e^{(t-\alpha)s}}{s} ds \right| = \left| \int_{+\pi/2}^{-\pi/2} \frac{e^{(t-\alpha)(\delta+Re^{j\vartheta})}}{\delta + Re^{j\vartheta}} jRe^{j\vartheta}\, d\vartheta \right|$$

$$\leq Re^{(t-\alpha)\delta} \int_{+\pi/2}^{-\pi/2} \left| \frac{e^{(t-\alpha)Re^{j\vartheta}}}{\delta + Re^{j\vartheta}} \right| d\vartheta \leq \frac{Re^{(t-\alpha)\delta}}{R-\delta} \int_{+\pi/2}^{-\pi/2} e^{(t-\alpha)R\cos\vartheta}\, d\vartheta.$$

In full analogy with the previous case, the exponent of the last expression is negative—since it is $t - \alpha < 0$ and $\cos\vartheta > 0$ for $\pi/2 \leq \vartheta \leq -\pi/2$, and therefore the measure of the integral along the path C_R will tend to zero in the limit $R \to \infty$. But this means that

$$\int_{\delta-jR}^{\delta+jR} \frac{e^{(t-\alpha)\,s}}{s}\, ds \quad \text{and therefore} \quad x(t) = \frac{1}{2\pi j} \int_{\delta-jR}^{\delta+jR} \frac{e^{(t-\alpha)\,s}}{s}\, ds = 0.$$

Case 3: $t = \alpha$ In this case, the inverse Laplace transform is calculated as

$$\mathcal{L}^{-1}\{\mathcal{X}(s)\} = \frac{1}{2\pi j} \lim_{R\to\infty} \int_{\delta-jR}^{\delta+jR} \frac{ds}{s}$$

with the requirement of the *analytic nature* of the function $1/s$ to be satisfied for every $\delta > 0$. From the basic theory, it is known that the *antiderivative* of this function is the *branch of the logarithm* (we denote this as $Log(s)$), which must be chosen in such a way that its *cut* does not intersect with the curve along which the above integral will be calculated. Considering the range of values $-\pi < arg(s) < \pi$, from the basic theory of complex analysis we have[6]

$$\int_{\delta-jR}^{\delta+jR} \frac{ds}{s} = \text{Log}(\delta + jR) - \text{Log}(\delta - jR) = \ln\left|\frac{\delta + jR}{\delta - jR}\right| + j\left[\arg(\delta + jR) - \arg(\delta - jR)\right].$$

However,

$$\lim_{R\to\infty}\left\{\ln\left|\frac{\delta + jR}{\delta - jR}\right|\right\} = \ln(-1) = 0, \qquad \lim_{R\to\infty}\left\{arg(\delta + jR)\right\} = \frac{\pi}{2}, \qquad \lim_{R\to\infty}\left\{arg(\delta - jR)\right\} = -\frac{\pi}{2},$$

from which it follows that

$$\lim_{R\to\infty}\int_{\delta-jR}^{\delta+jR} \frac{ds}{s} = j\left(\frac{\pi}{2} - \frac{-\pi}{2}\right) = j\pi \qquad \text{and finally} \qquad \mathcal{L}^{-1}\{\mathcal{X}(s)\} = \frac{1}{2\pi j}\lim_{R\to\infty}\int_{\delta-jR}^{\delta+jR}\frac{ds}{s} = \frac{j\pi}{2\pi j} = \frac{1}{2}.$$

[6] From complex analysis it is known that the function of the logarithm is a *multivalued function*, since if $e^s = w$ we will have $s = \ln|w| + j\arg(w) + 2\pi jn$ $(n \in Z)$. Specifically, if G is an open set, then any continuous function $f(s)$ that satisfies the relation $e^{f(s)} = s$—from which it follows that $f(s) = \ln|z| + j\arg(s) + 2\pi jn$ $(n \in Z)$—is called a *branch of the logarithm in the set G*. The above function is defined for infinite values of the parameter n, whereas for the value $n = 0$ the function is known as the *principal branch*. It is proven that the branch of a logarithm is an *analytic function* with the derivative $1/s$. Using the *principal branch* of the function we have

$$\int_\alpha^\beta \frac{ds}{s} = \text{Log}(z)\Big|_\alpha^\beta = \Big[\ln|s| + j\arg(s)\Big]_\alpha^\beta = \ln|\alpha| + j\arg(\alpha) - \ln|b| - j\arg(\beta) = \ln\left|\frac{\alpha}{\beta}\right| + j\Big(\arg(\alpha) - \arg(\beta)\Big),$$

where α and β are points on the complex plane.

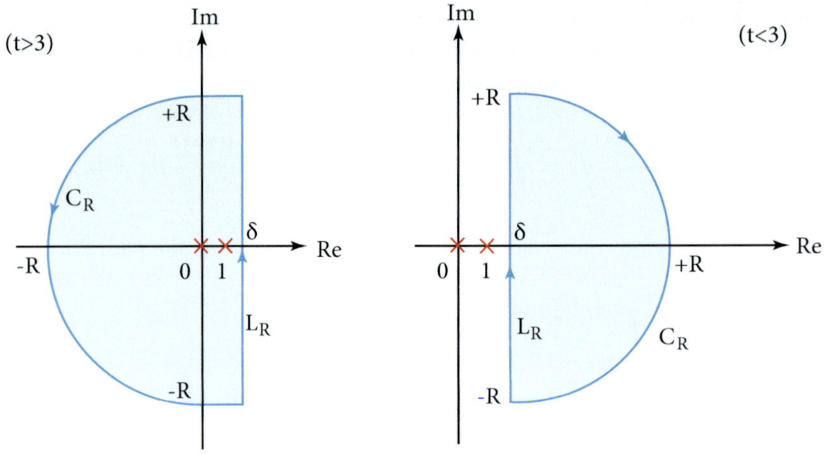

Fig. 4.7 Calculation of the contour integral for finding the inverse transform of the function $\mathcal{X}(s) = e^{-3s}/[s^2(s-1)]$ of Exercise 4.7.2

Combining the results obtained for the three above cases, we have

$$x(t) = \mathcal{L}^{-1}\{\mathcal{X}(s)\} = \begin{cases} 1 & \text{for } t > \alpha \\ 0 & \text{for } t < \alpha \\ 1/2 & \text{for } t = \alpha. \end{cases}$$

Problem 4.7.2 *Using the method of complex integration, determine the inverse Laplace transform of the function*

$$\mathcal{X}(s) = \frac{e^{-3s}}{s^2(s-1)}.$$

Solution In this case, the function to be integrated will have the form

$$\mathcal{X}(s)\, e^{st} = \frac{e^{-3s}}{s^2(s-1)}\, e^{st} = \frac{e^{(t-3)\,s}}{s^2(s-1)}$$

and therefore the equation of the *inverse Laplace transform* will take the form

$$x(t) = \frac{1}{2\pi i}\int_{\delta-i\infty}^{\delta+i\infty} \frac{e^{(t-3)s}}{s^2(s-1)}\, ds = \frac{1}{2\pi i}\oint_{\Gamma_R} \frac{e^{(t-3)s}}{s^2(s-1)}\, ds - \frac{1}{2\pi i}\int_{C_R} \frac{e^{(t-3)s}}{s^2(s-1)}\, ds,$$

where Γ_R is a closed path consisting of the straight line $L_R : (s_1, s_2) = (\delta - i\infty, \delta + i\infty)$ and the semicircle C_R radius $R \to \infty$, which will close *either to the left or to the right*, depending on the value of the parameter t. In this case, the function $\mathcal{X}(s)e^{st}$, has *a double pole* at the position $s = 0$ and *a simple pole* at the position $s = 1$; therefore, the position of the line L_R will be determined by the requirement to be $\delta > 1$. The path Γ_R that will be used for the calculation of the integral for the cases $t > 3$ and $t < 3$ is presented in Fig. 4.7.[7]

[7] In this case we considered for simplicity that the path Γ_R consists of the line L_R and from the semicircle C_R and *we did not take into account the existence of the paths L_1 and L_3 as we did in the previous exercise*. However, the proof does not change: in practice *the integrals along all paths except those of the straight line L_R tend to zero at the limit* $R \to \infty$ and the remaining integral is the one that appears in the defining equation of the inverse Laplace transform, thus leading to its final calculation.

Following the previous methodology, we consider the cases $t > 3$ and $t < 3$, whereas the case $t = 3$ is left as an exercise to the reader. In addition, we will not analyze the exercise in as great detail as before; however, the proof is more or less the same and in this sense the reader can use the previous exercise as an *example* for solving similar exercises, but we will use the findings that emerged from it. Starting from the case $t < 3$, which is simpler, the exponent of the expression $e^{(t-3)s}$ is negative and therefore according to *Jordan's lemma* we will use the *right semicircle* C_R presented in image (b) of Fig. 4.7. In this case, the path Γ_R *does not contain poles of the function* $\mathcal{X}(s)e^{st}$ with the result that both the integral along the straight line L_R and the function $x(t)$ have a value equal to *zero* in this area.

On the other hand, for $t > 3$ the exponent of the expression $e^{(t-3)s}$ is positive and according to *Jordan's lemma* we should choose the *left semicircle* C_R shown in image (a) of Fig. 4.7. In this case, the poles, $s = 0$ and $s = 1$ *are contained* within the path Γ_R and from the application of the residue theorem we get

$$x(t) = \text{Res}\left[\frac{e^{(t-3)s}}{s^2(s-1)}\right]_{s=0} + \text{Res}\left[\frac{e^{(t-3)s}}{s^2(s-1)}\right]_{s=1},$$

where

$$\text{Res}\left[\frac{e^{(t-3)s}}{s^2(s-1)}\right]_{s=0} = \lim_{s\to 0}\left\{\frac{d}{ds}\left[s^2\frac{e^{t-3}s}{s^s(s-1)}\right]\right\} = \lim_{s\to 0}\left[\frac{(t-3)e^{(t-3)s}(s-1) - e^{(t-3)s}}{(s-1)^2}\right] = 2 - t$$

and

$$\text{Res}\left[\frac{e^{(t-3)s}}{s^2(s-1)}\right]_{s=1} = \lim_{s\to 1}\left[(s-1)\frac{e^{(t-3)s}}{s^2(s-1)}\right] = \lim_{s\to 1}\left[\frac{e^{(t-3)s}}{s^2}\right] = e^{t-3}$$

Therefore, for the range of values $t > 3$ the function $x(t)$ will be given by the relation

$$x(t) = 2 - t + e^{t-3}.$$

Finally, as already mentioned, the case $t = 3$ is left as an exercise for the reader.

Problem 4.7.3 *Using the method of complex integration, determine the inverse Laplace transform of the function*

$$\mathcal{X}(s) = \frac{1}{s\sinh(\alpha s)}, \qquad (\alpha \in R).$$

Solution Using the representation of the function $\sinh(\alpha s)$ as an *infinite product of terms* of the form

$$\sinh(\alpha s) = \alpha s \prod_{n=1}^{\infty}\left[1 + \left(\frac{\alpha s}{\pi n}\right)^2\right]$$

the function to be integrated can take the form

$$\mathcal{F}(s)\,e^{st} = \frac{e^{st}}{s\sinh(\alpha s)} = \frac{e^{st}}{\alpha s^2 \prod_{n=1}^{\infty}\left[1 + \left(\frac{\alpha s}{\pi n}\right)^2\right]} = \frac{e^{st}}{\alpha s^2\left(1 + \frac{\alpha^2 s^2}{\pi^2}\right)\left(1 + \frac{\alpha^2 s^2}{4\pi^2}\right)\ldots}$$

from which we find that the function in question has a *second-order pole* at the position $s = 0$ and *infinite simple poles* at the positions $s_n = \pm(nj\pi/\alpha)$ $(n = 0, 1, 2, \ldots)$. Applying the residue theorem, we choose as a circular path C_R, to close the line Γ_R, a semicircle toward the left half-plane with a radius $R \to \infty$. In this way, the closed path Γ_R that will result will enclose all the poles of the function, as shown in Fig. 4.8 and the application of the residue theorem will lead to the equation

Fig. 4.8 The path Γ_R

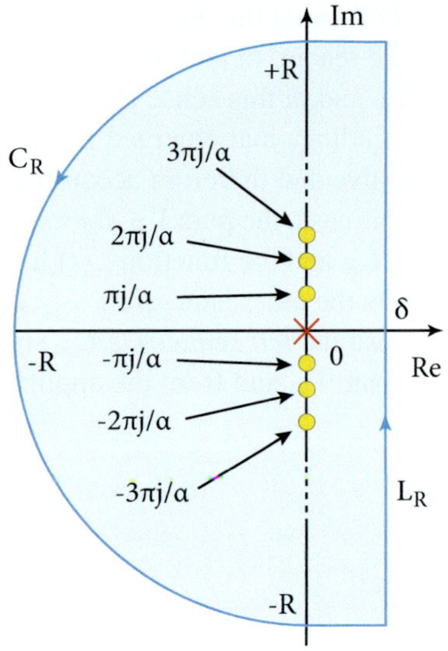

$$x(t) = \text{Res}\left[\frac{e^{st}}{s\sinh(\alpha s)}\right]_{s=0} + \sum_{n=1}^{\infty}\text{Res}\left[\frac{e^{st}}{s\sinh(\alpha s)}\right]_{s=s_n}$$

$$\text{Res}\left[\frac{e^{st}}{s\sinh(\alpha s)}\right]_{s=0} = \lim_{s\to 0}\left\{\frac{d}{ds}\left[\frac{s^2e^{st}}{s\sinh(\alpha s)}\right]\right\} = \lim_{s\to 0}\left\{\frac{d}{ds}\left[\frac{se^{st}}{\sinh(\alpha s)}\right]\right\}$$
$$= \lim_{s\to 0}\left[\frac{(e^{st}+ste^{st})\sinh(\alpha s) - \alpha s\cosh(\alpha t)e^{st}}{\sinh^2(\alpha s)}\right] = \frac{t}{\alpha}$$

$$\text{Res}\left[\frac{e^{st}}{s\sinh(\alpha s)}\right]_{s=s_n} = \lim_{s\to s_n}\frac{(s-s_n)\,e^{st}}{s\sinh(\alpha s)} = \lim_{s\to s_n}\frac{e^{st}}{\sinh(\alpha s)+\alpha s\cosh(\alpha s)}$$
$$= \frac{e^{\pm(n\pi jt/\alpha)}}{\sinh(\pm n\pi j)+(\pm n\pi j)\cosh(\pm n\pi j)} = \frac{e^{\pm(n\pi jt/\alpha)}}{(-1)^n(\pm n\pi j)}$$

,where we used the *L'Hospital's rule* and the properties $\sinh(\pm n\pi j) = 0$ and $\cosh(\pm n\pi j) = (-1)^n$. Therefore, combining the above relations, we get

$$x(t) = \frac{t}{\alpha} + \sum_{n=1}^{\infty}\frac{(-1)^n e^{n\pi jt/\alpha}}{n\pi j} - \sum_{n=1}^{\infty}\frac{(-1)^n e^{-n\pi jt/\alpha}}{n\pi j}$$
$$= \frac{t}{\alpha} + \frac{2}{\pi}\sum_{n=1}^{\infty}\frac{(-1)^n}{n}\frac{e^{n\pi jt/\alpha} - e^{-n\pi jt/\alpha}}{2j} = \frac{t}{\alpha} + \frac{2}{\pi}\sum_{n=1}^{\infty}\frac{(-1)^n}{n}\sin\left(\frac{n\pi t}{\alpha}\right),$$

which is the final result. To derive the final result, we silently assumed that the integral along the path C_R is zero, which is a consequence of the power of *Jordan's lemma*.

Problem 4.7.4 *Using the method of complex integration, determine the inverse Laplace transform of the function*

$$\mathcal{X}(s) = \frac{s}{s^2-k^2}.$$

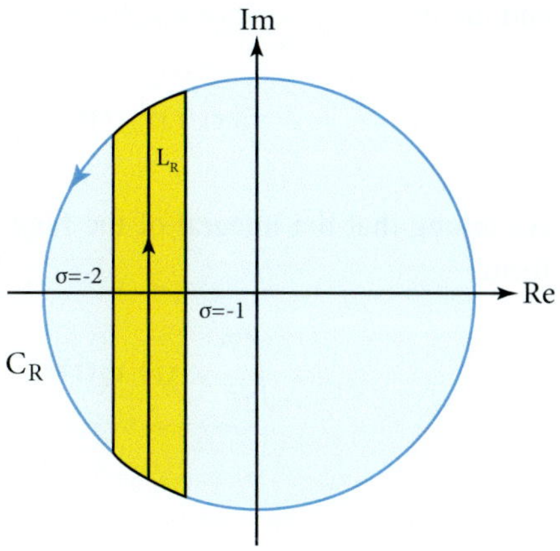

Fig. 4.9 The path Γ_R (Exercise 4.7.4)

Solution The function

$$\mathcal{X}(s)\, e^{st} = \frac{s e^{st}}{s^2 - k^2} = \frac{s e^{st}}{(s+k)(s-k)}$$

has *two simple poles* at the positions $s = k$ and $s = -k$. Considering as the integration path Γ_R the curve consisting of the straight line $L_R : \Re(s) = \delta > k$ and the circular arc C_R of the radius $R \to \infty$, which closes to the left, *so as to enclose the two poles* (see Fig. 4.9), the application of *Jordan's lemma* leads to the annulment of the integral along the curve C_R and therefore the inverse Laplace transform will be given by the relation

$$x(t) = \frac{1}{2\pi j} \int_{\delta - j\infty}^{\delta + j\infty} \frac{s e^{st}}{(s+k)(s-k)} ds = \text{Res}\left[\frac{s e^{st}}{s^2 - k^2}\right]_{s=-k} + \text{Res}\left[\frac{s e^{st}}{s^2 - k^2}\right]_{s=k}$$

as it results from the application of the residue theorem. However,

$$\text{Res}\left[\frac{s e^{st}}{s^2 - k^2}\right]_{s=-k} = \lim_{s \to -k}\left[(s+k)\frac{s e^{st}}{(s+k)(s-k)}\right] = \lim_{s \to -k}\left[\frac{s e^{st}}{(s-k)}\right] = -\frac{e^{-kt}}{2}$$

$$\text{Res}\left[\frac{s e^{st}}{s^2 - k^2}\right]_{s=+k} = \lim_{s \to +k}\left[(s-k)\frac{s e^{st}}{(s+k)(s-k)}\right] = \lim_{s \to -k}\left[\frac{s e^{st}}{(s+k)}\right] = \frac{e^{kt}}{2},$$

from which it follows that $x(t) = \dfrac{e^{kt} - e^{-kt}}{2} = \cosh(kt)$, which is also the desired result. ■

Convolution in the Complex Plane

Using the concept of the *inverse Laplace transform*, we can prove another useful property of it, which refers to the operation *of convolution in the complex plane*. According to this property, the Laplace transform of the product of two piecewise continuous-time signals $x_1(t)$ and $x_2(t)$ is given by the relation

$$\mathcal{L}\{x_1(t)x_2(t)\} = \frac{1}{2\pi j}\left[\mathcal{X}_1(s) * \mathcal{X}_2(s)\right], \tag{4.14}$$

where $\mathcal{X}_1(s) \equiv \mathcal{L}\{x_1(t)\}$ and $\mathcal{X}_2(s) \equiv \mathcal{L}\{x_2(t)\}$. In the following description, we assume that the regions of convergence of the transform $\mathcal{X}_1(s)$ and $\mathcal{X}_2(s)$ are described by the equations $\Re(s) > \sigma_1$ and $\Re(s) > \sigma_2$ with $\sigma_1 > \sigma_2$.

To prove the above property, we express the signal $x_2(t)$ as the *inverse Laplace transform* of the function $\mathcal{X}_2(s)$; Therefore,

$$x_2(t) = \frac{1}{2\pi j} \int_{C_2} \mathcal{X}_2(z)\, e^{zt}\, dz, \quad C_2 : \Re(z) = \delta > \sigma_2$$

and therefore the Laplace transform of the product of the signals $x_1(t)$ and $x_2(t)$ will take the form

$$\mathcal{L}\{x_1(t)x_2(t)\} = \int_0^{\infty} x_1(t)\, x_2(t)e^{-st}\, ds = \frac{1}{2\pi j}\int_0^{\infty} x_1(t)\, dt \int_{C_2} \mathcal{X}_2(z)\, e^{(z-s)t}\, dz.$$

Assuming that the integral of the function $\mathcal{X}_2(z)$ *converges along the path of integration*, the above equation can take the form

$$\int_0^{\infty} x_1(t)\, x_2(t)e^{-st}dt = \frac{1}{2\pi j}\int_{\sigma_2-j\infty}^{\sigma_2+j\infty} \mathcal{X}_2(z)\, dz \int_0^{\infty} x_1(t)\, e^{-(s-z)t}\, dt$$

$$= \frac{1}{2\pi j}\int_{\sigma_2-j\infty}^{\sigma_2+j\infty} \mathcal{X}_2(z)\mathcal{X}_1(s-z)\, dz = \frac{1}{2\pi j}\Big[\mathcal{X}_1(s) * \mathcal{X}_2(s)\Big],$$

which is also the desired property. It is proven that the region of convergence of the transform $\mathcal{L}\{x_1(t)x_2(t)\}$ is described by the equation $\Re(s) = \sigma > \sigma_1 + \sigma_2$. Finally, regarding the operation of *integration in the complex plane*, the semicircle C_R that we must add to the integration path to form the closed path Γ_R must close either to the left or to the right, so that the functions $\mathcal{X}_1(s)$ and $\mathcal{X}_2(s)$ tend to zero for the limit $s \to \infty$.

Problem 4.7.5 *Using the convolution method in the complex plane calculate the Laplace transform of the signal* $x(t) = e^{-t}e^{-2t}u(t)$.

Solution We set $x_1(t) = e^{-t}$ and $x_2(t) = e^{-2t}$ and referring to tables of elementary transforms, we find that

$$\mathcal{X}_1(s) = \frac{1}{s+1} \quad \text{with region of convergence} \quad \sigma_1 > -1 \quad \text{and} \quad \mathcal{X}_2(s) = \frac{1}{s+2} \quad \text{with region of convergence} \quad \sigma_2 > -2.$$

Therefore, we have

$$\mathcal{X}_1(z)\mathcal{X}_2(s-z) = \left(\frac{1}{z+2}\right)\left(\frac{1}{s-z+1}\right).$$

To calculate the integral, we will use the closed path depicted in Fig. 4.9. It is not difficult to see that the function $\mathcal{X}_1(z)\mathcal{X}_2(s-z)$ has two poles at positions $s = -2$ and $z = s+1$ of which *only the first one is within the closed path of integration*. Using *Jordan's lemma*, we find that the integral along the circular path C_R is equal to zero and from the application of the *residue theorem* we will get

$$\mathcal{X}(s) = \frac{1}{2\pi j}\int_{L_R} \mathcal{X}_2(z)\mathcal{X}_1(s-z)dz = \text{Res}\left[\left(\frac{1}{z+2}\right)\left(\frac{1}{s-z+1}\right)\right] = \lim_{z\to -2}\left[\frac{1}{s-z+1}\right] = \frac{1}{s+3},$$

which is also the desired result. ■.

The Property of Complex Conjugation

Another useful property that can be proven with the help of the inverse Laplace transform is the *property of complex conjugation*, according to which, if a signal $x(t)$ of real values has a Laplace transform $\mathcal{X}(s) = \mathcal{L}\{x(t)\}$, then $\mathcal{X}(s) = \mathcal{X}^*(s^*)$. In order to prove this property, we apply the operator of complex conjugation to the defining equation of the inverse transform, leading to the result

$$x^*(t) = -\frac{1}{2\pi j}\int_{\sigma+j\infty}^{\sigma-j\infty} \mathcal{X}^*(s)\, e^{s^*t}ds = x(t), \tag{4.15}$$

where we used the property $x^*(t) = x(t)$ that characterizes a *real function*.

Assuming that the complex quantity s has the general form $s = \sigma + j\omega$, we proceed to the substitution $p = s^* = \sigma - j\omega$. Differentiating the above relations (and assuming that the real part σ is a constant so that $d\sigma = 0$), we get $ds = jd\omega$ and $dp = -jd\omega = -ds$ and replacing in the last relation, it will take the form

$$x(t) = -\frac{1}{2\pi j}\int_{\sigma+j\omega}^{\sigma-j\omega} \mathcal{X}^*(p^*)\, e^{pt} dp = \frac{1}{2\pi j}\int_{\sigma-j\omega}^{\sigma+j\omega} \mathcal{X}^*(p^*)\, e^{pt} dp = \frac{1}{2\pi j}\int_{\sigma-j\omega}^{\sigma+j\omega} \mathcal{X}^*(s^*)\, e^{st} ds,$$

where in the last equation we replaced the dummy variable p with the variable s. If we compare the last relation with the equation definition of the inverse transform, it turns out that $\mathcal{X}(s) = \mathcal{L}\{x(t)\} = \mathcal{X}^*(s^*)$, which is also the desired result.

Note Using the above property we can formulate the proposition that *the poles and zero values of the Laplace transform of a real signal $x(t)$ appear in pairs of complex conjugates*. In order to prove this assertion, let us assume that the function $\mathcal{X}(s)$ has a zero at the position $s = s_1$, so it will satisfy the property $\mathcal{X}(s_1) = 0$. However, according to the previous property, it will also be $\mathcal{X}^*(s_1^*) = 0$, which in turn, and using again the above property, leads to the relation $\mathcal{X}(s_1^*) = 0$. Therefore, if the function $\mathcal{X}(s)$ has a zero at the position $s = s_1$, it will also have a zero at the position $s = s_1^*$, that is, *the zeros of the Laplace transform appear in pairs of complex conjugates*. Using similar arguments, we can prove that the same applies to the poles of the Laplace transform of the signal $x(t)$.

4.7.2 The Method of Expansion in Power Series

According to this calculation method, if the Laplace transform $\mathcal{X}(s)$ can be written as *an infinite term power series* of the form

$$\mathcal{X}(s) = \frac{\alpha_0}{s} + \frac{\alpha_1}{s^2} + \frac{\alpha_2}{s^3} + \frac{\alpha_3}{s^4} + \cdots = \sum_{n=0}^{\infty} \frac{\alpha_n}{s^{n+1}} \tag{4.16}$$

then the function of the inverse transform $x(t)$ is calculated as

$$x(t) \equiv \mathcal{L}^{-1}\{\mathcal{X}(s)\} = \alpha_0 + \alpha_1 t + \frac{\alpha_2}{2!}t^2 + \frac{\alpha_3}{3!}t^3 + \cdots = \sum_{n=0}^{\infty} \frac{\alpha_n t^n}{n!} \tag{4.17}$$

The above property is based on the *linearity* of the Laplace transform and the property

$$\mathcal{L}\{t^n\} = \frac{n!}{s^{n+1}} \quad \text{from where it follows that} \quad \mathcal{L}^{-1}\left\{\frac{1}{s^{n+1}}\right\} = \frac{t^n}{n!}.$$

Therefore, we have

$$x(t) = \mathcal{L}^{-1}\{\mathcal{X}(s)\} = \mathcal{L}^{-1}\left\{\sum_{n=0}^{\infty} \frac{\alpha_n}{s^{n+1}}\right\} = \sum_{n=0}^{\infty} \alpha_n \mathcal{L}^{-1}\left\{\frac{1}{s^{n+1}}\right\} = \sum_{n=0}^{\infty} \frac{\alpha_n t^n}{n!}. \tag{4.18}$$

Example In order to demonstrate in practice the application of this method, let us calculate the inverse Laplace transform of the function $\mathcal{X}(s) = (e^{-1/s})/s$. Developing the above function *into an infinite sum of terms*, we will get

$$\frac{e^{-1/s}}{s} = \frac{1}{s}\left\{1 - \frac{1}{s} + \frac{1}{2!\, s^2} - \frac{1}{3!\, s^3} + \ldots\right\} = \frac{1}{s} - \frac{1}{s^2} + \frac{1}{2!\, s^3} - \frac{1}{3!\, s^4} + \ldots.$$

Therefore, we have

$$x(t) = \mathcal{L}^{-1}\left\{\frac{1}{s}\right\} - \mathcal{L}^{-1}\left\{\frac{1}{s^2}\right\} + \frac{1}{2!}\mathcal{L}^{-1}\left\{\frac{1}{s^3}\right\} - \frac{1}{3!}\mathcal{L}^{-1}\left\{\frac{1}{s^4}\right\} + \cdots = 1 - t + \frac{t^2}{(2!)^2} - \frac{t^3}{(3!)^2} + \ldots,$$

which is the desired result. ■

4.7.3 The Method of Partial Fraction Expansion

The method of *partial fraction expansion* is the most widespread method of calculating the inverse Laplace transform and finds particular application in cases of transforms that can be expressed *in explicit form*, that is, as a *quotient of two polynomial functions*. This method is based on the property of *linearity* of the Laplace transform and the writing of the function $\mathcal{X}(s) \equiv \mathcal{L}\{x(t)\}$ as a *linear combination of elementary transforms*

$$\mathcal{X}(s) = \alpha_1 \mathcal{X}_1(s) + \alpha_2 \mathcal{X}_2(s) + \cdots + \alpha_n \mathcal{X}_n(s) = \sum_{i=1}^{n} \alpha_i \mathcal{X}_i(s). \tag{4.19}$$

In the above equation, the functions $\mathcal{X}_i(s)$ $(i = 1, 2, \ldots, n)$ are considered as the Laplace transforms of elementary functions of the form $x_1(t), x_2(t), \ldots, x_n(t)$, which are very likely to be found in *tables of elementary transforms* that have been created specifically for this purpose. If we manage to determine these functions, we can then determine the signal $x(t)$, the Laplace transform of which is the function $\mathcal{X}(s)$, taking advantage of the property of *linearity* of the inverse transform and using the equation

$$x(t) = \alpha_1 x_1(t) + \alpha_2 x_2(t) + \cdots + \alpha_k x_k(t) = \sum_{i=1}^{k} \alpha_i x_i(t). \tag{4.20}$$

The formulation of the transform $\mathcal{X}(s)$ into a linear combination of simpler transforms presupposes that this is an *explicit function*—something that, of course, applies in almost all cases—and is based on the application of the method of developing a fraction into a series of partial fractions, the description of which can be found in Appendix A (see Sect. A.6 on page 1487). The solved exercises that follow demonstrate the application of this method.

Problem 4.7.6 *Using the method of partial fraction expansion, determine the inverse Laplace transform of the function*

$$\mathcal{X}(s) = \frac{s-3}{s^2+5s+6}.$$

(Poularikas, The Transforms and Applications Handbook, *Example 5.4.1).*

Solution Applying the method of partial fraction expansion, we find that the Laplace transform function can be written in the form

$$\mathcal{X}(s) = \frac{s-3}{s^2+5s+6} = \frac{s-3}{(s+2)(s+3)} = \frac{A}{s+2} + \frac{B}{s+3} = \frac{(A+B)s + (3A+2B)}{(s+2)(s+3)}.$$

Equating the coefficients of the powers of s in the first and last fractions, we get the system of equations $A + B = 1$ and $2A + 3B = -3$ from which the values $A = -5$ and $B = 6$ emerge. Therefore,

$$\mathcal{X}(s) = \frac{s-3}{(s+2)(s+3)} = -\frac{5}{s+2} + \frac{6}{s+3},$$

and therefore the inverse transform is calculated as

$$x(t) = -5\mathcal{L}^{-1}\left\{\frac{1}{s+2}\right\} + 6\mathcal{L}^{-1}\left\{\frac{1}{s+3}\right\} = -5e^{-2t}u(t) + 6e^{-3t}u(t),$$

as it emerges by referring to tables of elementary transforms.

Problem 4.7.7 *Using the method of partial fraction expansion, determine the inverse Laplace transform of the function*

$$\mathcal{X}(s) = \frac{s^2+2s-4}{s^4+2s^3+s^2}$$

Solution Expressing the Laplace transform function in the form

$$\mathcal{X}(s) = \frac{s^2+2s-4}{s^4+2s^3+s^2} = \frac{s^2+2s-4}{s^2(s^2+2s+1)} = \frac{s^2+2s-4}{s^2(s+1)^2},$$

this can be expanded in the series of partial fractions

$$\mathcal{X}(s) = \frac{s^2+2s-4}{s^4+2s^3+s^2} = \frac{A}{s} + \frac{B}{s^2} + \frac{C}{s+1} + \frac{D}{(s+1)^2} = \frac{(A+C)s^3 + (2A+B+C+D)s^2 + (A+2B)s + B}{s^2(s+1)^2}.$$

Equating the coefficients of the powers of s in the first and last fractions leads to the system of equations $A + C = 0$, $2A + B + C + D = 1$, $A + 2B = 2$, and $B = -4$, from which the values $(A, B, C, D) = (10, -4, -10, -5)$ emerge, and finally the equation

$$\mathcal{X}(s) = \frac{s^2+2s-4}{s^4+2s^3+s^2} = \frac{10}{s} - \frac{4}{s^2} - \frac{10}{s+1} - \frac{5}{(s+1)^2}.$$

Utilizing the property of *linearity* of the inverse transform, we can write the desired function $x(t)$ in the form

$$x(t) = \mathcal{L}^{-1}\{\mathcal{X}(s)\} = 10\mathcal{L}^{-1}\left\{\frac{1}{s}\right\} - 4\mathcal{L}^{-1}\left\{\frac{1}{s^2}\right\} - 10\mathcal{L}^{-1}\left\{\frac{1}{s+1}\right\} - 5\mathcal{L}^{-1}\left\{\frac{1}{(s+1)^2}\right\}.$$

Referring to tables of elementary transforms, we find

$$\mathcal{L}^{-1}\left\{\frac{1}{s}\right\} = u(t), \quad \mathcal{L}^{-1}\left\{\frac{1}{s^2}\right\} = tu(t), \quad \mathcal{L}^{-1}\left\{\frac{1}{s+1}\right\} = e^{-t}u(t),$$

so

$$x(t) = (10 - 4t - 10e^{-t} - 5te^{-t})u(t).$$

Problem 4.7.8 *Using the method of partial fraction expansion, determine the inverse Laplace transform of the function*

$$\mathcal{X}(s) = \frac{s+1}{(s+3)(s^2+4s+5)}.$$

Solution The function of the Laplace transform is written in the form

$$\mathcal{X}(s) = \frac{s+1}{(s+3)(s^2+4s+5)} = \frac{A}{s+3} + \frac{B}{s^2+4s+5} = \frac{(A+B)s^2 + (4A+3B+C)s + (5A+3C)}{(s+3)(s^2+4s+5)}$$

By equating the coefficients of the powers of s in the first and last fractions, we end up with the system of equations $A+B=0$ $4A+3B+C=1$ and $5A+3C=1$ with the solution $A=-1$ $B=1$ and $C=2$. Therefore,

$$\mathcal{X}(s) = \frac{s+1}{(s+3)(s^2+4s+5)} = -\frac{1}{s+3} + \frac{s+2}{s^2+4s+5} = -\frac{1}{s+3} + \frac{s+2}{(s+2)^2+1},$$

and, referring to tables of elementary transforms, it turns out that

$$x(t) = -\mathcal{L}^{-1}\left\{\frac{1}{s+3}\right\} + \mathcal{L}^{-1}\left\{\frac{s+2}{(s+2)^2+1}\right\} = \left(e^{-3t} + e^{-2t}\cos(t)\right) u(t).$$

Note Although in this case the inverse Laplace transform was found in a table with a suitable conversion of the rational function $\mathcal{X}(s)$, in practice it may not always be possible to perform this conversion. Generally speaking, in cases where the denominator of a rational Laplace transform is a second-degree polynomial of the form $\alpha s^2 + \beta s + \gamma$, we can guess the form of the direct transform—that is, the signal $x(t)$—by examining the sign of the discriminant $\Delta = \beta^2 - 4\alpha\gamma$. We distinguish three cases:

- $\boldsymbol{\Delta < 0}$: The polynomial *does not have real roots* and the signal $x(t)$ will have one of the forms

$$x(t) \sim \cos(kt) \qquad x(t) \sim e^{-kt}\cos(\lambda t) \qquad x(t) \sim \sin(kt) \qquad x(t) \sim e^{-kt}\sin(\lambda t).$$

- $\boldsymbol{\Delta = 0}$:The polynomial has *one double real root* and the signal $x(t)$ will have the form $x(t) \sim te^{-kt}$.
- $\boldsymbol{\Delta > 0}$: In this case the polynomial has *two real roots* and the signal $x(t)$ will be described by the equation $x(t) \sim ke^{-\mu t} + \lambda e^{-\nu t}$.

Problem 4.7.9 *Using the method of partial fraction expansion, determine the inverse Laplace transform of the function*

$$\mathcal{X}(s) = \frac{s^2+2s+5}{(s+3)(s+5)^2}.$$

(Hsu, Signals and Systems, *Solved Exercise 3.19).*

Solution The function of the Laplace transform is written in the form

$$\mathcal{X}(s) = \frac{s^2+2s+5}{(s+3)(s+5)^2} = \frac{A}{s+3} + \frac{B}{s+5} + \frac{C}{(s+5)^2} = \frac{(A+B)s^2 + (10A+8B+C)s + (25A+15B+3C)}{(s+3)(s+5)^2}.$$

By equating the coefficients of the corresponding powers of s in the first and last fractions, we are led to the system of equations $A+B=1$ $10A+8B+C=2$ and $25A+15B+3C=5$ with solution $A=2$ $B=-1$ and $C=-10$. Therefore, we have

$$\mathcal{X}(s) = \frac{s^2+2s+5}{(s+3)(s+5)^2} = \frac{2}{s+3} - \frac{1}{s+5} - \frac{10}{(s+5)^2},$$

and by referring to tables of elementary transforms, we finally arrive at the equation

$$x(t) = 2\mathcal{L}^{-1}\left\{\frac{1}{s+3}\right\} - \mathcal{L}^{-1}\left\{\frac{1}{s+5}\right\} - 10\mathcal{L}^{-1}\left\{\frac{1}{(s+5)^2}\right\} = 2e^{-3t}u(t) - e^{-5t}u(t) - 10te^{-5t}u(t).$$

Note In cases of rational transforms where *the degree of the polynomial of the numerator is greater than or equal to the degree of the polynomial of the denominator*, we should, *before the development of the transformation into partial fractions, apply the method of polynomial division between the polynomials of the numerator and the denominator*. The next two exercises demonstrate the application of this method.

Problem 4.7.10 *Using the method of partial fraction expansion, determine the inverse Laplace transform of the function*

$$\mathcal{X}(s) = \frac{s^2 + 6s + 7}{s^2 + 3s + 2}$$

(Hsu, Signals and Systems, *Solved Exercise 3.20β).*

Solution Applying the method of polynomial division between the polynomials of the numerator and the denominator, we find

$$\begin{array}{r|lll} & 1 & & \\ \hline s^2 + 3s + 2 & s^2 & +6s & +7 \\ & s^2 & +3s & +2 \\ \hline & & +3s & +5 \end{array}$$

and by expanding the rational function that emerged into a sum of partial fractions, it turns out that

$$\frac{3s+5}{s^2+3s+2} = \frac{3s+5}{(s+1)(s+2)} = \frac{A}{s+5} + \frac{B}{s+2} = \frac{(A+B)s + (2A+B)}{s^2+3s+2}$$

By equating the coefficients of the corresponding powers of s in the first and last fractions, we will get the system of equations $A + B = 3$ and $2A + B = 5$ with the solution $A = 2$ and $B = 1$. Therefore, finally, the function $\mathcal{X}(s)$ is written in the form

$$\mathcal{X}(s) = 1 + \frac{2}{s+1} + \frac{1}{s+2}.$$

Therefore, referring to tables of elementary transforms, we arrive at the expression

$$x(t) = \mathcal{L}^{-1}\{1\} + 2\mathcal{L}^{-1}\left\{\frac{1}{s+1}\right\} + \mathcal{L}^{-1}\left\{\frac{1}{s+2}\right\} = \delta(t) + 2e^{-t}u(t) + e^{-2t}u(t),$$

which is also the desired result.

Problem 4.7.11 *Using the method of partial fraction expansion, determine the inverse Laplace transform of the function*

$$\mathcal{X}(s) = \frac{s^3 + 2s^2 + 6}{s^2 + 3s}.$$

(Hsu, Signals and Systems, *Solved Exercise 3.20γ).*

Solution Applying the method of polynomial division between the polynomials of the numerator and the denominator, we find

$$\begin{array}{r|llll} & s & -1 & & \\ \hline s^2+3s & s^3 & +2s^2 & & +6 \\ & s^3 & +3s^2 & & \\ \hline & & -1s^2 & & +6 \\ & & -1s^2 & -3s & \\ \hline & & & +3s & +6 \end{array}$$

so

$$\frac{s^3+2s^2+6}{s^2+3s} = s-1+\frac{3s+6}{s^2+3s}.$$

In the next step of the process, we expand the fraction that emerged into a partial fraction expansion. Thus,

$$\frac{3s+6}{s^2+3s} = \frac{3s+6}{s(s+3)} = \frac{A}{s}+\frac{B}{s+3} = \frac{(A+B)s+3A}{s^2+3s}.$$

Equating the coefficients of the corresponding powers of s in the first and last fractions, we get the system of equations $A+B=3$ and $3A=6$ with the solution $A=2$ and $B=1$. Combining all the above, the function $\mathcal{X}(s)$ can take the form

$$F(s) = \frac{s^3+2s^2+6}{s^2+3s} = s-1+\frac{2}{s}+\frac{1}{s+3}$$

and referring to tables of elementary transforms, it turns out that

$$x(t) = \mathcal{L}^{-1}\{s\} - \mathcal{L}^{-1}\{1\} + 2\mathcal{L}^{-1}\left\{\frac{1}{s}\right\} + \mathcal{L}^{-1}\left\{\frac{1}{s+3}\right\} = \delta'(t) - \delta(t) + 2u(t) + e^{-3t}u(t).$$

Note In some cases of finding the inverse transform, *knowledge of the convergence area allows us to draw useful conclusions about the form of the signal $x(t)$, which cannot be derived without this information.* This case is examined below.

Problem 4.7.12 *Using the method of partial fraction expansion, determine the inverse Laplace transform of the function*

$$\mathcal{X}(s) = \frac{s+2}{s^2+7s+12} \quad \textit{with convergence area} \quad -4 < \Re(s) < -3.$$

Solution If we did not know information about the convergence area of the function $\mathcal{X}(s)$, the expansion of the function $\mathcal{X}(s)$ in a series of partial fractions would lead us to the result

$$\mathcal{X}(s) = \frac{s+2}{s^2+7s+12} = \frac{s+2}{(s+4)(s+3)} = \frac{A}{s+4}+\frac{B}{s+3} = \frac{(A+B)s+(3A+4B)}{s^2+7s+12}.$$

By equating the coefficients of the corresponding powers of s in the first and last fractions, we get the system of equations $A+B=1$ and $3A+4B=2$ with the solution $A=2$ and $B=-1$, and therefore the function $\mathcal{X}(s)$ can be written in the form

$$\mathcal{X}(s) = \frac{s+2}{s^2+7s+12} = \frac{2}{s+4} - \frac{1}{s+3},$$

from which it follows that

$$x(t) = 2\mathcal{L}^{-1}\left\{\frac{1}{s+4}\right\} - \mathcal{L}^{-1}\left\{\frac{1}{s+3}\right\} = 2e^{-4t}u(t) - e^{-3t}u(t).$$

However, from the form of the region of convergence, it is concluded that the signal $x(t)$ should be defined *both for positive and for negative values of* t (see the section on the properties of the region of convergence) and therefore in the mathematical expression that describes it, both the function $u(t)$ and the function $u(-t)$ should exist. But this means that the second fraction in the function $\mathcal{X}(s)$ will not be treated as

$$\mathcal{L}^{-1}\left\{\frac{1}{s+3}\right\} = e^{-3t}u(t) \quad \text{but as} \quad \mathcal{L}^{-1}\left\{\frac{1}{s+3}\right\} = -e^{-3t}u(-t).$$

Therefore, the mathematical expression of the signal $x(t)$ will take the form $x(t) = 2e^{-4t}u(t) + e^{-3t}u(-t)$, which *is compatible* with the form of the region of convergence that has been defined.

Problem 4.7.13 *Using the method of partial fraction expansion, determine the inverse Laplace transform of the function*

$$\mathcal{X}(s) = \frac{s+1}{s^2+5s+6}$$

with the region of convergence $-4 < \Re(s) < -3$.

Solution Following the methodology of the previous exercise, we expand the function $\mathcal{X}(s)$ in a series of partial fractions. Therefore, we have

$$\mathcal{X}(s) = \frac{s+1}{s^2+5s+6} = \frac{s+1}{(s+2)(s+3)} = \frac{A}{s+2} + \frac{B}{s+3} = \frac{(A+B)s+(3A+2B)}{s^2+5s+6}.$$

By equating the coefficients of the corresponding powers of s in the first and last fractions, we are led to the system of equations $A + B = 1$ and $3A + 2B = 1$ with the solution $A = -1$ and $B = 2$, and therefore, the function $\mathcal{X}(s)$ can be written in the form

$$\mathcal{X}(s) = \frac{2}{s+3} - \frac{1}{s+2}.$$

Referring then to tables of elementary transforms, we find that

$$\mathcal{L}^{-1}\left\{\frac{1}{s+3}\right\} = e^{-3t}u(t) \quad \text{and} \quad \mathcal{L}^{-1}\left\{\frac{1}{s+2}\right\} = -e^{-2t}u(-t),$$

where for the derivation of the second relation we relied on the property of the signal $x(t)$, to be defined for both positive and negative time moments, as it emerges from the form of the region of convergence that has been given. Therefore, we have

$$x(t) = 2e^{-3t}u(t) + e^{-2t}u(-t),$$

which is also the desired result.

4.7.4 Heaviside's Theorem

In the special case where the transform $\mathcal{X}(s)$ is a rational function of the form

$$\mathcal{X}(s) = \frac{P(s)}{Q(s)} = \frac{P(s)}{(s-s_1)(s-s_2)\ldots(s-s_n)} \tag{4.21}$$

with the polynomial $Q(s)$ being of a higher degree than the polynomial $P(s)$ and having n distinct roots $s_1, s_2, \ldots, s_n$, it is proven that the *inverse transform* $x(t)$ can be calculated from the equation

$$x(t) \equiv \mathcal{L}^{-1}\{\mathcal{X}(s)\} = \mathcal{L}^{-1}\left\{\frac{P(s)}{Q(s)}\right\} = \sum_{i=1}^{n} \frac{P(s_i)}{Q'(s_i)} e^{s_i t}. \tag{4.22}$$

The above theorem—given without proof—is known as *Heaviside's theorem* and was formulated by the English mathematician *Oliver Heaviside* (1850-1925). Examples of the application of this theorem to the calculation of the inverse Laplace transform are presented below.

Problem 4.7.14 *Using Heaviside's theorem, find the inverse Laplace transform of the function*

$$\mathcal{X}(s) = \frac{s-3}{s^2+5s+6}.$$

Solution The function $\mathcal{X}(s)$ is a rational function of the form $P(s)/Q(s)$ with the polynomials $P(s)$ and $Q(s)$ being defined as $P(s) = s - 3$ and $Q(s) = s^2 + 5s + 6$ or equivalently, $Q(s) = (s+2)(s+3)$. We observe that the polynomial $Q(s)$ has two simple roots with values $s = -2$, $s = -3$, whereas its derivative is equal to $Q'(s) = 2s + 5$. Therefore, applying the theorem of Heaviside, we will have

$$x(t) = \sum_{i=1}^{2} \frac{P(s_i)}{Q'(s_i)} e^{s_i t} = \frac{P(-2)}{Q'(-2)} e^{-2t} + \frac{P(-3)}{Q'(-3)} e^{-3t} = -5e^{-2t} + 6e^{-3t}.$$

Problem 4.7.15 *Using the theorem of Heaviside to find the inverse Laplace transform of the function*

$$\mathcal{X}(s) = \frac{2s+3}{s^2+4s+7}.$$

Solution The function $\mathcal{X}(s)$ is a rational function of the form $P(s)/Q(s)$ with the polynomials $P(s)$ and $Q(s)$ defined as $P(s) = 2s + 3$ and $Q(s) = s^2 + 4s + 7$ or equivalently $Q(s) = (s + 2 + j\sqrt{3})(s + 2 - j\sqrt{3})$. In this case, the polynomial $Q(s)$ has two simple roots with values $s_1 = -2 - j\sqrt{3}$ and $s_2 = -2 + j\sqrt{3}$, whereas its derivative is equal to $Q'(s) = 2s + 4$. It is not difficult to see that $P(s_1) = -1 - j2\sqrt{3}$, $P(s_2) = -1 + j2\sqrt{3}$, $Q'(s_1) = -j2\sqrt{3}$ and $Q'(s_2) = j2\sqrt{3}$, and from Heaviside's theorem we get

$$x(t) = \frac{-1 - j2\sqrt{3}}{-j2\sqrt{3}} e^{(-2-j2\sqrt{3})t} + \frac{-1 + j2\sqrt{3}}{j2\sqrt{3}} e^{(-2+2\sqrt{3})t} = e^{-2t}\left[\frac{-1 - j2\sqrt{3}}{-j2\sqrt{3}} e^{-j2\sqrt{3}t} + \frac{-1 + j2\sqrt{3}}{j2\sqrt{3}} e^{2\sqrt{3}t}\right]$$

$$= e^{-2t}\left[\frac{e^{-j2\sqrt{3}t} - e^{j2\sqrt{3}t}}{j2\sqrt{3}} + \left(e^{-j2\sqrt{3}} + e^{-j2\sqrt{3}}\right)\right] = e^{-2t}\left[2\cos(2\sqrt{3}t) - \frac{1}{\sqrt{3}}\sin(2\sqrt{3}t)\right],$$

which is also the desired result.

4.8 Solving Differential Equations

Using the properties of *linearity* and the *differentiation* of the Laplace transform, we can very easily determine the solution $x(t)$ *of linear differential equations with constant coefficients*. To do this, we calculate the Laplace transforms of all the terms of the equation and end up with an algebraic equation in terms of the function $\mathcal{X}(s) \equiv \mathcal{L}\{x(t)\}$, from which we then determine the desired solution $x(t)$, calculating the inverse transform.

In a more detailed description, let us consider a *complete linear differential equation of order n with constant coefficients*, which, as is well known, has the form

$$\sum_{i=0}^{n} \alpha_i x^{(i)}(t) = \alpha_0 x(t) + \alpha_1 x'(t) + \alpha_2 x''(t) + \cdots + \alpha_{n-1} x^{(n-1)}(t) + \alpha_n x^{(n)}(t) = \varphi(t)$$

with solution the function $x = x(t)$. Taking the Laplace transform of the above equation and using the property of *linearity*, we have

$$\mathcal{L}\left\{\sum_{i=0}^{n} \alpha_i x^{(i)}(t)\right\} = \sum_{i=0}^{n} \alpha_i \mathcal{L}\{x^{(i)}(t)\} = \mathcal{L}\{\varphi(t)\}. \tag{4.23}$$

In the next step of the process, we use the *property of differentiation* according to which the Laplace transform of derivatives $x'(t)$ and $x''(t)$ is given by the relations

$$\mathcal{L}\{x'(t)\} = s\mathcal{L}\{x(t)\} - x(0) \qquad \text{and} \qquad \mathcal{L}\{x''(t)\} = s^2\mathcal{L}\{x(t)\} - sx(0) - x'(0),$$

whereas in the general case the following property applies:

$$\mathcal{L}\{x^{(n)}(t)\} = s^n\mathcal{L}\{x(t)\} - s^{n-1}x(0) - s^{n-2}x'(0) - \cdots - sx^{(n-2)}(0) - x^{(n-1)}(0).$$

Therefore, we have

$$\begin{aligned}
\sum_{i=1}^{n} \alpha_i \mathcal{L}\{x^{(i)}(t)\} &= \alpha_0\mathcal{L}\{x(t)\} + \alpha_1\mathcal{L}\{x'(t)\} + \alpha_2\mathcal{L}\{x''(t)\} + \cdots + \alpha_n\mathcal{L}\{x^{(n)}(t)\} \\
&= \alpha_0\mathcal{L}\{x(t)\} + \alpha_1 s\mathcal{L}\{x(t)\} - \alpha_1 x(0) + \alpha_2 s^2\mathcal{L}\{x(t)\} - \alpha_2 s x(0) - \alpha_2 x'(0) + \ldots \\
&= \left(\sum_{i=0}^{n} \alpha_i s^i\right) \mathcal{F}\{x(t)\} - \sum_{i=1}^{n} \alpha_i \left(\sum_{j=1}^{i} s^{i-j} x^{(i-j)}(0)\right)
\end{aligned}$$

and replacing in our original equation we get

$$\left(\sum_{i=0}^{n} \alpha_i s^i\right) \mathcal{F}\{x(t)\} - \sum_{i=1}^{n} \alpha_i \left(\sum_{j=1}^{i} s^{i-j} x^{(i-j)}(0)\right) = \mathcal{L}\{\varphi(t)\},$$

from where it finally turns out that

$$\mathcal{L}\{x(t)\} = \left\{\mathcal{L}\{\varphi(t)\} + \sum_{i=1}^{n} \alpha_i \left(\sum_{j=1}^{i} s^{i-j} x^{(i-j)(0)}\right)\right\} \Big/ \left\{\sum_{i=0}^{n} \alpha_i s^i\right\}. \tag{4.24}$$

Having determined the function $\mathcal{X}(s) \equiv \mathcal{L}\{x(t)\}$, we can then continue to find the solution of the differential equation $x(t)$ by calculating the *inverse transform*, a process that can be carried out using some of the methods we presented in the previous section. This methodology of solving differential equations is presented in the examples that follow.

Problem 4.8.1 *Using the method of the Laplace transform, solve the differential equation*

$$\frac{dx(t)}{dt} + \alpha x(t) = g(t)$$

with the function $g(t)$ defined as $g(t) = u(t)$ and for an initial condition of the form $x(0^+) = 0$.

Solution Taking the Laplace transform of the above equation, we are led to the relation

$$\mathcal{L}\{x'(t)\} + \alpha\mathcal{L}\{x(t)\} = \mathcal{L}\{g(t)\} \Rightarrow s\mathcal{L}\{x(t)\} - x(0^+) + \alpha\mathcal{L}\{x(t)\} = \mathcal{L}\{g(t)\} \text{ and finally } \mathcal{L}\{x(t)\} = \frac{\mathcal{L}\{g(t)\}}{s+\alpha} + \frac{x(0^+)}{s+\alpha}.$$

From a previous exercise we know that the Laplace transform of the step function is given by the equation $\mathcal{L}\{g(t)\} = 1/s$. Using this expression in the above equation, it is formulated as

$$\mathcal{X}(s) \equiv \mathcal{L}\{x(t)\} = \frac{1}{s(s+\alpha)}.$$

In order to determine the above transform, we will use the method of partial fraction expansion. Therefore,

$$\frac{1}{s(s+\alpha)} = \frac{A}{s} + \frac{B}{s+\alpha} = \frac{(A+B)s + \alpha A}{s(s+\alpha)}.$$

Equating the coefficients of the corresponding powers of s results in the system of equations $A + B = 0$ and $A\alpha = 1$ from which the values $A = 1/\alpha$ and $B = -1/\alpha$ emerge. Therefore,

$$\mathcal{X}(s) \equiv \mathcal{L}\{x(t)\} = \frac{1}{s\,(s+\alpha)} = \frac{1}{\alpha s} - \frac{1}{\alpha(s+\alpha)}$$

and by taking the inverse Laplace transform, we arrive at the desired function

$$x(t) = \mathcal{L}^{-1}\{\mathcal{X}(s)\} = \frac{1}{\alpha}\mathcal{L}^{-1}\left\{\frac{1}{\alpha}\right\} - \frac{1}{\alpha}\mathcal{L}^{-1}\left\{\frac{1}{s+\alpha}\right\} = \frac{1}{\alpha}(1 - e^{-\alpha t})u(t)$$

Problem 4.8.2 *Using the method of the Laplace transform, solve the differential equation*

$$\frac{d^2x(t)}{dt^2} + 5\frac{dx(t)}{dt} + 4x(t) = 10$$

with initial conditions $x(0) = x'(0) = 0$ (Poularikas, The Transforms and Applications Handbook, *Example 5.5.2).*

Solution Taking the Laplace transform of the derivatives of the above equation we will have

$$\mathcal{L}\{x''(t)\} = s^2\mathcal{L}\{x(t)\} - sx(0) - x'(0) = s^2\mathcal{L}\{x(t)\} \quad \text{and} \quad \mathcal{L}\{x'(t)\} = s\mathcal{L}\{x(t)\} - f(0) = s\mathcal{L}\{x(t)\}.$$

Therefore,

$$\mathcal{L}\{x''(t)\} + 5\mathcal{L}\{x'(t)\} + 4\mathcal{L}\{x(t)\} = 10\mathcal{L}\{1\}$$

or equivalently

$$s^2\mathcal{L}\{x(t)\} + 5s\mathcal{L}\{x(t)\} + 4\mathcal{L}\{x(t)\} = (s^2 + 5s + 4)\mathcal{L}\{x(t)\} = \frac{10}{s}$$

and finally

$$\mathcal{L}\{x(t)\} = \frac{10}{s\,(s^2 + 5s + 4)} = \frac{10}{s\,(s+1)\,(s+4)}$$

Developing the last function into a series of partial fractions we get

$$\frac{10}{s(s^2 + 5s + 4)} = \frac{10}{s(s+1)(s+4)} = \frac{A}{s} + \frac{B}{s+1} + \frac{C}{s+4} = \frac{(A+B+C)s^2 + (5A+4B+C)s + 4A}{s(s^2 + 5s + 4)}.$$

Equating the coefficients of the corresponding powers of s we get the system $A + B + C = 0$, $5A + 4B + C = 0$ and $4A = 10$ from which the values $A = 5/2$, $B = -10/3$ and $C = 5/6$ emerge. Therefore, we have

$$\mathcal{X}(s) \equiv \mathcal{L}\{x(t)\} = \frac{5}{2}\frac{1}{s} - \frac{10}{3}\frac{1}{s+1} + \frac{5}{6}\frac{1}{s+4}$$

and therefore

$$x(t) = \mathcal{L}^{-1}\{\mathcal{X}(s)\} = \frac{5}{2}\mathcal{L}^{-1}\left\{\frac{1}{s}\right\} - \frac{10}{3}\mathcal{L}^{-1}\left\{\frac{1}{s+1}\right\} + \frac{5}{6}\mathcal{L}^{-1}\left\{\frac{1}{s+4}\right\} = \frac{5}{2}u(t) - \frac{10}{3}e^{-t}u(t) + \frac{5}{6}e^{-4t}u(t)$$

as it emerges by referring to tables of elementary transforms.

Problem 4.8.3 *Using the method of the Laplace transform, solve the differential equation*

$$\frac{d^2x(t)}{dt^2} + 5\frac{dx(t)}{dt} + 6x(t) = g(t)$$

with initial conditions $x(0) = 2$, $x'(0) = 1$ *and* $g(t) = e^{-t}u(t)$.

Solution Taking the Laplace transform of the above equation we get

$$\mathcal{L}\{x''(t)\} = s^2\mathcal{L}\{x(t)\} - sx(0) - x'(0) = s^2\mathcal{L}\{x(t)\} - 2s - 1$$
$$\mathcal{L}\{x'(t)\} = s\mathcal{L}\{x(t)\} - x(0) = s\mathcal{L}\{x(t)\} - 2 \quad \text{and}$$
$$\mathcal{L}\{g(t)\} = \mathcal{L}\{e^{-t}u(t)\} = 1/(s+1)$$

Therefore, we have

$$s^2\mathcal{L}\{x(t)\} - 2s - 1 + 5\left(s\mathcal{L}\{x(t)\} - 2\right) + 6\mathcal{L}\{x(t)\} = \frac{1}{s+1}$$

and solving the above equation for $\mathcal{L}\{f(t)\}$ it turns out that

$$\mathcal{X}(s) \equiv \mathcal{L}\{x(t)\} = \frac{2s^2 + 13s + 12}{(s+1)(s^2 + 5s + 6)}.$$

In the next step of the process, we calculate the inverse transform $x(t)$ by applying the method of expansion in partial fractions. Working as before, we have

$$\frac{2s^2+13s+12}{(s+1)(s^2+5s+6)} = \frac{2s^2+13s+12}{(s+1)(s+2)(s+3)} = \frac{(A+B+C)s^2+(5A+4B+3C)s+(6A+3B+2C)}{(s+1)(s^2+5s+6)}.$$

Equating the coefficients of the corresponding powers of s results in the system of equations $A+B+C=2$, $5A+4B+3C=13$ and $6A+3B+2C=12$ with solution $A=1/2$, $B=6$ and $C=-9/2$. Therefore, we have

$$\mathcal{X}(s) = \mathcal{L}\{x(t)\} = \frac{2s^2+13s+12}{(s+1)(s^2+5s+6)} = \frac{1}{2}\frac{1}{s+1} + \frac{6}{s+2} - \frac{9}{2}\frac{1}{s+3}$$

and referring to tables of elementary transforms, it turns out that

$$x(t) = \mathcal{L}^{-1}\{\mathcal{X}(s)\} = \frac{1}{2}\mathcal{L}^{-1}\left\{\frac{1}{s+1}\right\} + \mathcal{L}^{-1}\left\{\frac{6}{s+2}\right\} - \frac{9}{2}\mathcal{L}^{-1}\left\{\frac{1}{s+3}\right\} = \frac{1}{2}e^{-t}u(t) + 6e^{-2t}u(t) - \frac{9}{2}e^{-3t}u(t).$$

Problem 4.8.4 *Using the method of the Laplace transform to solve the differential equation*

$$\frac{d^2x(t)}{dt^2} + 4x(t) = \sin(2t)$$

with initial conditions $x(0) = x'(0) = 0$.

Solution Taking the Laplace transform of the differential equation we will have

$$\mathcal{L}\{x''(t)\} + 4\mathcal{L}\{x(t)\} = \mathcal{L}\{\sin(2t)\} \qquad \text{where}$$

$$\mathcal{L}\{x''(t)\} = s^2\mathcal{L}\{f(t)\} - sx(0) - x'(0) = s^2\mathcal{L}\{x(t)\} \qquad \text{and} \qquad \mathcal{L}\{\sin(2t)\} = \frac{2}{s^2+4}$$

as it turns out from tables of elementary transforms. Replacing the above expressions in our initial equation we get

$$s^2\mathcal{L}\{x(t)\} + 4\mathcal{L}\{x(t)\} = \frac{2}{s^2+4}$$

and solving for the function $\mathcal{L}\{x(t)\}$ we get

$$\mathcal{X}(s) \equiv \mathcal{L}\{x(t)\} = \frac{2}{(s^2+4)^2} = \frac{2}{[(s+2i)(s-2i)]^2} = \frac{2}{(s+2i)^2(s-2i)^2}.$$

In order to determine the inverse transform of the above relation, let us use the *method of complex integration* and calculate the contour integral of the function $\mathcal{X}(s)e^{st}$ along a closed path Γ_R, which consists of the straight line $L_R : \Re(s) = \sigma > 0$ and the left semicircle C_R with a radius $R \to \infty$. From the application of *Jordan's lemma* it is known that *the integral along the semicircle* C_R *is equal to zero* and therefore the inverse Laplace transform will be calculated from the equation

$$x(t) = \frac{1}{2\pi j}\int_{\sigma-j\infty}^{\sigma+j\infty} \mathcal{X}(s)\, e^{st} ds = \sum_{i=1}^{2} \text{Res}\left[\mathcal{X}(s)\, e^{st}\right]_{s=s_j},$$

where $s_1 = 2j$ and $s_2 = -2j$ are the two second-order poles of the function $\mathcal{X}(s)e^{st}$ with residue values at them

$$\text{Res}\left[\mathcal{X}(s)e^{st}\right]_{s=2j} = \lim_{s\to 2j}\frac{d}{ds}\left[(s-2j)^2\frac{2e^{st}}{(s+2j)^2(s-2j)^2}\right] = \lim_{s\to 2j}\frac{d}{ds}\left[\frac{2e^{st}}{(s+2j)^2}\right]$$

$$= \lim_{s\to 2j}\frac{2te^{st}(s+2j)-4e^{st}}{(s+2j)^3} = -\frac{te^{2jt}}{8}+\frac{e^{2jt}}{16j}$$

$$\text{Res}\left[\mathcal{X}(s)e^{st}\right]_{s=-2j} = \lim_{s\to -2j}\frac{d}{ds}\left[(s+2j)^2\frac{2e^{st}}{(s+2j)^2(s-2j)^2}\right] = \lim_{s\to -2j}\frac{d}{ds}\left[\frac{2e^{st}}{(s-2j)^2}\right]$$

$$= \lim_{s\to =2j}\frac{2te^{st}(s-2j)-4e^{st}}{(s-2j)^3} = -\frac{te^{-2jt}}{8}-\frac{e^{-2jt}}{16j}.$$

Substituting the above expressions into the defining equation of the inverse transform, we get

$$x(t) = -\frac{t}{8}e^{2jt}+\frac{1}{16j}e^{2jt}-\frac{t}{8}e^{-2jt}-\frac{1}{16j}e^{-2jt} = \frac{1}{8}\frac{e^{2jt}-e^{-2jt}}{2j}-\left(\frac{t}{4}\frac{e^{2jt}+e^{-2jt}}{2}\right) = \frac{1}{8}\sin(2t)-\frac{t}{4}\cos(2t),$$

which is the desired result.

Problem 4.8.5 *Using the method of the Laplace transform, solve the system of first-order differential equations*

$$\frac{dx(t)}{dt}-3x(t)+4y(t)=t^2 \quad \textit{and} \quad \frac{dy(t)}{dt}+x(t)-5y(t)=e^t$$

with initial conditions $x(0)=1$ *and* $y(0)=-1$.

Solution In this exercise, the solution of *a system of differential equations* is requested, but the methodology we presented does not change. Taking the Laplace transform of all the terms of the two equations, we finally end up with an algebraic system of two equations with unknown functions $\mathcal{X}(s)\equiv\mathcal{L}\{x(t)\}$ and $\mathcal{Y}(s)\equiv\mathcal{L}\{y(t)\}$. In the last step of the process, we determine the desired functions $x(t)$ and $y(t)$ as the inverse Laplace transforms of the functions $\mathcal{X}(s)$ and $\mathcal{Y}(s)$. The exact same process applies to *any system of differential equations of arbitrary order and complexity degree*.

In this case, applying the Laplace transform to each member of the above equation will lead to the result

$$\mathcal{L}\{x'(t)\}-3\mathcal{L}\{x(t)\}+4\mathcal{L}\{y(t)\}=\mathcal{L}\{t^2\} \quad \text{and} \quad \mathcal{L}\{y'(t)\}+\mathcal{L}\{x(t)\}-5\mathcal{L}\{y(t)\}=\mathcal{L}\{e^t\}$$

with the transforms of the functions $x'(t)$ and $y'(t)$ being calculated as

$$\mathcal{L}\{x'(t)\}=s\mathcal{L}\{x(t)\}-x(0)=s\mathcal{L}\{x(t)\}-1 \quad \text{and} \quad \mathcal{L}\{y'(t)\}=s\mathcal{L}\{y(t)\}-y(0)=s\mathcal{L}\{y(t)\}+1.$$

Furthermore, referring to tables of elementary transforms, we find that

$$\mathcal{L}\{t^2\}=\frac{2}{s^3} \quad \text{and} \quad \mathcal{L}\{e^t\}=\frac{1}{s+1}.$$

Replacing the above quantities in our original system and performing the operations, the algebraic system of equations finally arises

$$(s-3)\mathcal{L}\{x(t)\}+4\mathcal{L}\{y(t)\}=\frac{s^3+2}{s^3} \quad \text{and} \quad \mathcal{L}\{x(t)\}+(s-5)\mathcal{L}\{y(t)\}=-\frac{s}{s+1}$$

the solution of which will give the functions $\mathcal{X}(s)\equiv\mathcal{L}\{x(t)\}$ and $\mathcal{Y}(s)\equiv\mathcal{L}\{y(t)\}$. The determination of the analytical form of these functions and the finding of the inverse transforms $x(t)$ and $y(t)$ is left as an exercise to the reader.

4.8.1 Continuous First- and Second-Order Systems

As an example of the application of the Laplace transform for solving differential equations, let us present the way in which the study of the behavior of systems described by *linear differential equations with constant coefficients* takes place. In the following pages, we study the behavior of systems described by first- and second-order equations, whereas the study of the behavior of higher-order systems can be found in the literature.

Let us start our study with the *simple continuous first-order systems* described by a differential equation of the form

$$\alpha_1 \frac{dy(t)}{dt} + \alpha_0 y(t) = \beta_0 x(t) \qquad \text{or equivalently} \qquad \tau \frac{dy(t)}{dt} + y(t) = \kappa x(t)$$

where $\tau = \alpha_1/\alpha_0$ and $\kappa = \beta_0/\alpha_0$. Taking the Laplace transform of the last relation, we get

$$\tau \mathcal{L}\left\{\frac{dy(t)}{dt}\right\} + \mathcal{L}\{y(t)\} = \kappa \mathcal{L}\{x(t)\}, \tag{4.25}$$

where $\mathcal{L}\{y(t)\} = \mathcal{Y}(s)$, $\mathcal{L}\{y'(t)\} = s\mathcal{Y}(s)$—according to the differentiation theorem of the Laplace transform—and $\mathcal{L}\{x(t)\} = \mathcal{X}(s)$. Replacing these relations in the above equation, it will take the form

$$\tau s \mathcal{Y}(s) + \mathcal{Y}(s) = \kappa \mathcal{X}(s) \qquad \text{and finally} \qquad \mathcal{Y}(s) = \frac{\kappa \mathcal{X}(s)}{\tau s + 1}$$

with a region of convergence of the form $\Re(s) > -1/\tau$.

In the last equation, the parameter τ is known as the *time constant* and constitutes a *measure of the speed at which the system responds to changes that occur at its input*, whereas the parameter κ is known as the *static gain* and expresses the *sensitivity of the system's output to the respective input signal*. If we use the function $x(t) = \alpha u(t)$ as such a signal, where α is a known constant and $u(t)$ is the *step function*, the transform $\mathcal{X}(s)$ is calculated as

$$\mathcal{X}(s) = \mathcal{L}\{x(t)\} = \alpha \mathcal{L}\{u(t)\} = \frac{\alpha}{s} \tag{4.26}$$

and therefore the Laplace transform of the system's output (which, of course, is none other than the *step response*) will take the form

$$\mathcal{Y}(s) = \frac{\kappa}{\tau s + 1} \frac{\alpha}{s}. \tag{4.27}$$

At this point, we can determine the system's output $y(t)$ from the inverse Laplace transform of the function $\mathcal{Y}(s)$. Using the method of partial fraction expansion, we get

$$\frac{\kappa}{\tau s + 1} \frac{\alpha}{s} = \frac{A}{\tau s + 1} + \frac{B}{s} = \frac{(A + B\tau)s + B}{s(\tau s + 1)}.$$

By equating the coefficients of the corresponding powers of s we are led to the system of equations $B = \alpha\kappa$ and $A + B\tau = 0$ with solution $B = \alpha\kappa$ and $A = -\alpha\kappa\tau$. Therefore, the Laplace transform $Y(s)$ can be written in the form

$$\mathcal{Y}(s) = \mathcal{L}\{y(t)\} = -\alpha\kappa \frac{\tau}{\tau s + 1} + \alpha\kappa \frac{1}{s} = -\alpha\kappa \frac{1}{s + (1/\tau)} + \alpha\kappa \frac{1}{s}$$

from which it finally follows that

$$y(t) = -\alpha\kappa\mathcal{L}^{-1}\left\{\frac{1}{s+(1/\tau)}\right\} + \alpha\kappa\mathcal{L}^{-1}\left\{\frac{1}{s}\right\} = \alpha\kappa e^{-t/\tau}u(t) + \alpha\kappa u(t) \tag{4.28}$$

From the above analysis, the following conclusions emerge regarding the behavior of a *continuous linear first-order system*:

- The system's response to each *step change* made at its input is the approximation of a *new steady state*.
- The *maximum value* of the system's output $y(t)$ is $y(t) = \alpha\kappa$, which is approached at the limit $t \to \infty$.
- The system's output at time $t = \tau$ will have a value

$$y(\tau) = \alpha\kappa\left(1 - \frac{1}{e}\right) = 0.632\alpha\kappa.$$

  Therefore, *the time constant τ is defined as the time period required for the system's output to approach 63.2% of its maximum value*.
- The smaller the value of the time constant τ, the more abrupt will be the *initial response* of the system's input.
- The larger the value of the gain κ, the larger will be *the value of the steady state* of the output $y(t)$ for the same change in the input signal.

Using the equation we derived for the Laplace transform of the system output $\mathcal{Y}(s)$, we can calculate the *impulse response* $h(t)$ as the *inverse Laplace transform of the system's transfer function*, which has the form[8]

$$\mathcal{H}(s) = \frac{\mathcal{Y}(s)}{\mathcal{X}(s)} = \frac{\kappa}{\tau s + 1} = \frac{\kappa}{\tau}\left\{\frac{1}{s+(1/\tau)}\right\}. \tag{4.29}$$

Thus,

$$h(t) = \mathcal{L}^{-1}\{\mathcal{H}(s)\} = \frac{\kappa}{\tau}\mathcal{L}^{-1}\left\{\frac{1}{s+(1/\tau)}\right\} = \frac{\kappa}{\tau}e^{-t/\tau}u(t) \tag{4.30}$$

The change in the system output $y(t)$ and its impulse response $h(t)$ for the simple case $\alpha = \kappa = 1$ is presented diagrammatically in Fig. 4.10.

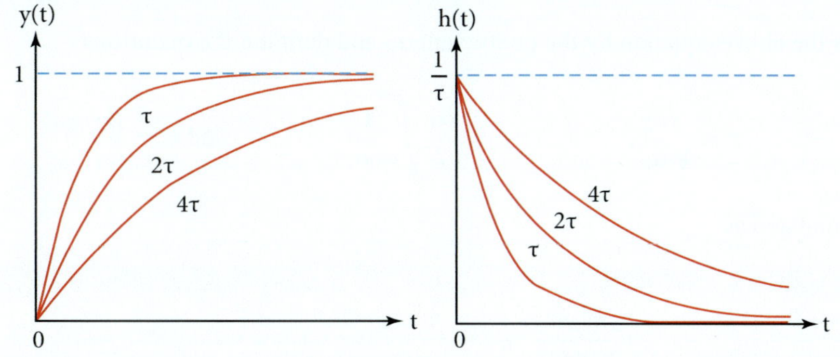

Fig. 4.10 The temporal change of the output $y(t)$ and the impulse response $h(t)$ of a continuous linear first-order system for parameter values $\alpha = \kappa = 1$

[8] The transfer function of linear time-invariant systems is presented in great detail in the next chapter.

In the special case where the constant α_0 appearing in the differential equation of the system *has a value equal to zero*—this means that the function $y(t)$ *does not appear in the above equation*—the situation is completely differentiated. In this case, the differential equation of the system takes the form

$$\alpha_1 \frac{dy(t)}{dt} = \beta_0 x(t) \qquad \text{or equivalently} \qquad \frac{dy(t)}{dt} = \lambda x(t),$$

where now the constant λ is defined as $\lambda = \beta_0/\alpha_1$. Taking, as before, the Laplace transform of the last relation, we get

$$\mathcal{L}\left\{\frac{dy(t)}{dt}\right\} = \lambda \mathcal{L}\{x(t)\} \qquad \text{or equivalently} \qquad s\mathcal{Y}(s) = \lambda \mathcal{X}(s),$$

and finally

$$\mathcal{Y}(s) \equiv \mathcal{L}\{y(t)\} = \frac{\lambda}{s}\mathcal{X}(s), \tag{4.31}$$

where $\mathcal{X}(s) \equiv \mathcal{L}\{x(t)\}$ is the Laplace transform of the input signal. If we consider $x(t) = \alpha u(t)$ as such a signal, as before, $\mathcal{X}(s) = \alpha/s$, and lastly

$$\mathcal{Y}(s) = \frac{\alpha\lambda}{s^2} \qquad \text{from which it follows that} \qquad y(t) = \alpha\lambda\mathcal{L}^{-1}\left\{\frac{1}{s^2}\right\} = \alpha\lambda t u(t).$$

We observe, therefore, that in this case the output of the system varies *proportionally as a function of time* and takes an *infinite value* at the limit $t \to \infty$.

Last of all, the impulse response of the system for this case arises from the relation

$$h(t) = \mathcal{L}^{-1}\{\mathcal{H}(s)\} = \lambda\mathcal{L}^{-1}\left\{\frac{1}{s}\right\} = \lambda u(t) \tag{4.32}$$

and therefore it is independent of time and equal to λ. ■

After the presentation of the behavior of continuous linear systems of the first order, let us now move on to the study of *second-order systems*, which are described by a *second-order linear differential equation with constant coefficients* of the form

$$\alpha_2 \frac{d^2y(t)}{dt^2} + \alpha_1 \frac{dy(t)}{dt} + \alpha_0 y(t) = \beta_0 x(t). \tag{4.33}$$

Dividing all the terms of the above equation by the coefficient α_0 and defining the quantities

$$\tau = \sqrt{\frac{\alpha_2}{\alpha_0}}, \qquad \zeta = \frac{\alpha_1}{2}\sqrt{\frac{1}{\alpha_0\alpha_2}} \qquad \text{and } \kappa = \frac{\beta_0}{\alpha_0}$$

the above relation is formulated as

$$\tau^2 \frac{d^2y(t)}{dt^2} + 2\tau\zeta \frac{dy(t)}{dt} + y(t) = \kappa x(t) \tag{4.34}$$

with the quantities τ, ζ and κ being known as *time constant*, *damping ratio*, and *static gain* respectively.

Taking the Laplace transform of the two terms of the above relation, we have

$$\tau^2\mathcal{L}\left\{\frac{d^2y(t)}{dt^2}\right\} + 2\tau\zeta\mathcal{L}\left\{\frac{dy(t)}{dt}\right\} + \mathcal{L}\{y(t)\} = \kappa\mathcal{L}\{x(t)\}, \tag{4.35}$$

where $\mathcal{L}\{x(t)\} = \mathcal{X}(s)$, $\mathcal{L}\{y(t)\} = \mathcal{Y}(s)$, whereas from the differentiation theorem of the Laplace transform we get, as is well known, $\mathcal{L}\{y''(t)\} = s^2\mathcal{Y}(s)$ and $\mathcal{L}\{y'(t)\} = s\mathcal{Y}(s)$.

Substituting the above quantities into our differential equation and solving for the function $\mathcal{Y}(s)$, it turns out that

$$\tau^2 s^2 Y(s) + 2\tau\zeta s Y(s) + Y(s) = \kappa\mathcal{X}(s)$$

and finally

$$\mathcal{Y}(s) = \frac{\kappa}{\tau^2 s^2 + 2\tau\zeta s + 1}\mathcal{X}(s).$$

Let us now calculate the *step response* of such a system and assume that we transmit to the system input the continuous-time signal $x(t) = \alpha u(t)$, where $u(t)$ is the continuous *step function*. Then, $\mathcal{X}(s) = \alpha/s$ and the Laplace transform $\mathcal{Y}(s)$ of the system's step response is written

$$\mathcal{Y}(s) = \frac{\kappa}{\tau^2 s^2 + 2\tau\zeta s + 1}\,\frac{\alpha}{s} = \frac{\kappa\alpha}{\tau^2}\,\frac{1}{s\left(s^2 + \frac{2\zeta}{\tau}s + \frac{1}{\tau^2}\right)}. \tag{4.36}$$

It is not difficult to see that the quadratic polynomial of the denominator has discriminant

$$\Delta = b^2 - 4ac = \frac{4\zeta^2}{\tau^2} - \frac{4}{\tau^2} = \frac{4}{\tau^2}(\zeta^2 - 1)$$

and roots

$$s_1 = \frac{-b + \sqrt{\Delta}}{2a} = -\frac{\zeta}{\tau} + \frac{\sqrt{\zeta^2 - 1}}{\tau} \quad \text{and} \quad s_2 = \frac{-b - \sqrt{\Delta}}{2a} = -\frac{\zeta}{\tau} - \frac{\sqrt{\zeta^2 - 1}}{\tau}.$$

These roots constitute two of the three poles of the Laplace transform $\mathcal{Y}(s)$—the third pole obviously is the one that corresponds to the value $s = 0$—which can now be written in the form

$$\mathcal{Y}(s) = \frac{\kappa\alpha}{\tau^2}\,\frac{1}{s(s - s_1)(s - s_2)}. \tag{4.37}$$

From the above analysis, it becomes obvious that *the functional form and the temporal evolution of the system's step response $y(t)$ depend on the value of the damping ratio ζ, which can be less than, greater than, or equal to one*. We distinguish, therefore, the following three cases.

Case 1: ($\zeta > 1$), Systems with High Damping In this case, the roots s_1 and s_2 are *real and different from each other*. In order to determine the output of the system $y(t)$, let us calculate the inverse Laplace transform of the function $\mathcal{Y}(s)$ by applying *the method of partial fraction expansion*. In this case, because the issue we are dealing with is purely theoretical, we present in detail all the intermediate steps of the method of partial fraction expansion, so that the reader can fully understand the way in which it is applied. Therefore, we have

$$\frac{1}{s(s - s_1)(s - s_2)} = \frac{A}{s} + \frac{B}{s - s_1} + \frac{C}{s - s_2} = \frac{(A + B + C)\,s^2 + (As_1 + As_2 + Bs_2 + Cs_1)\,s + As_1 s_2}{s(s - s_1)(s - s_2)}.$$

By equating the coefficients of the corresponding powers of s in the first and last fractions, we will get the system

$$A + B + C = 0, \quad As_1 + As_2 + Bs_2 + Cs_1 = 0, \quad As_1 s_2 = 1$$

with solution

$$A = \frac{1}{s_1 s_2} = \tau^2, \qquad B = \frac{1}{s_1(s_1 - s_2)} = -\frac{\tau^2}{2\sqrt{\zeta^2-1}(\zeta - \sqrt{\zeta^2-1})}$$

$$C = -\frac{1}{s_2(s_1 - s_2)} = \frac{\tau^2}{2\sqrt{\zeta^2-1}(\zeta + \sqrt{\zeta^2-1})}.$$

On the other hand, the inverse Laplace transform of the function $\mathcal{Y}(s)$ will be given by the relation

$$y(t) = \frac{\kappa\alpha}{\tau^2}\left[A\mathcal{L}^{-1}\left\{\frac{1}{s}\right\} + B\mathcal{L}^{-1}\left\{\frac{1}{s-s_1}\right\} + C\mathcal{L}^{-1}\left\{\frac{1}{s-s_2}\right\}\right]$$

with the inverse transforms of the above relation being calculated as

$$\mathcal{L}^{-1}\left\{\frac{1}{s}\right\} = u(t), \quad \mathcal{L}^{-1}\left\{\frac{1}{s-s_1}\right\} = e^{s_1 t}u(t) = e^{-\frac{\zeta}{\tau}t}e^{\frac{\sqrt{\zeta^2-1}}{\tau}t}u(t)$$

$$\mathcal{L}^{-1}\left\{\frac{1}{s-s_2}\right\} = e^{s_2 t}u(t) = e^{-\frac{\zeta}{\tau}t}e^{-\frac{\sqrt{\zeta^2-1}}{\tau}t}u(t).$$

Combining the above results, we arrive at the equation

$$y(t) = \frac{\kappa\alpha}{\tau^2}\left[\tau^2 - \frac{\tau^2 e^{-\frac{\zeta}{\tau}t}e^{\frac{\sqrt{\zeta^2-1}}{\tau}t}}{2\sqrt{\zeta^2-1}(\zeta - \sqrt{\zeta^2-1})} + \frac{\tau^2 e^{-\frac{\zeta}{\tau}t}e^{-\frac{\sqrt{\zeta^2-1}}{\tau}t}}{2\sqrt{\zeta^2-1}(\zeta + \sqrt{\zeta^2-1})}\right]u(t)$$

$$= \kappa\alpha\left[1 - \frac{e^{-\frac{\zeta}{\tau}t}}{2\sqrt{\zeta^2-1}}\left(\frac{e^{\frac{\sqrt{\zeta^2-1}}{\tau}t}}{\zeta - \sqrt{\zeta^2-1}} - \frac{e^{-\frac{\sqrt{\zeta^2-1}}{\tau}t}}{\zeta + \sqrt{\zeta^2-1}}\right)\right]$$

The expression within the parentheses is calculated as

$$\frac{e^{\frac{\sqrt{\zeta^2-1}}{\tau}t}}{\zeta - \sqrt{\zeta^2-1}} - \frac{e^{-\frac{\sqrt{\zeta^2-1}}{\tau}t}}{\zeta + \sqrt{\zeta^2-1}} = 2\sqrt{\zeta^2-1}\cosh\left(\frac{\sqrt{\zeta^2-1}}{\tau}t\right) + 2\zeta\sinh\left(\frac{\sqrt{\zeta^2-1}}{\tau}t\right)$$

and substituting in the above relation, we finally get

$$y(t) = \kappa\alpha\left\{1 - e^{-\frac{\zeta}{\tau}t}\left[\cosh\left(\frac{\sqrt{\zeta^2-1}}{\tau}t\right) + \frac{\zeta}{\sqrt{\zeta^2-1}}\sinh\left(\frac{\sqrt{\zeta^2-1}}{\tau}t\right)\right]\right\}u(t), \tag{4.38}$$

which is the output of the system for the value $\zeta > 1$. From the above equation we observe that *the output of the system changes relatively slowly in relation to time with the degree of slowness being proportional to the value of the parameter* ζ. In this sense, this parameter acts as *resistance*, since the larger its value, the slower the temporal evolution of the system.

Case 2: ($\zeta = 1$), Critically Damped Systems In this case, the polynomial of the denominator of the transform $Y(s)$ has a double root, the value of which is calculated as

$$s = s_p = -\frac{b}{2a} = -\frac{\zeta}{\tau} = -\frac{1}{\tau}.$$

Therefore, the Laplace transform is written in the form

$$\mathcal{Y}(s) = \frac{\kappa\alpha}{\tau^2}\frac{1}{(s-s_p)^2}. \tag{4.39}$$

Using again the method of partial fraction expansion, the above equation is written as

$$\frac{\kappa\alpha}{\tau^2}\frac{1}{(s-s_p)^2} = \frac{A}{s} + \frac{B}{s-s_p} + \frac{C}{(s-s_p)^2} = \frac{(A+B)s^2 + (C - Bs_p - 2As_p)s + As_p^2}{s(s-s_p)^2}.$$

Therefore, $A + B = 0$ $C - Bs_p - 2As_p = 0$ and $As_p^2 = 1$, from which it follows that $A = \tau^2$, $B = -\tau^2$, $C = -\tau$.
On the other hand, the inverse Laplace transform of the function $\mathcal{Y}(s)$ is

$$y(t) = \frac{\kappa\alpha}{\tau^2}\left[A\mathcal{L}^{-1}\left\{\frac{1}{s}\right\} + B\mathcal{L}^{-1}\left\{\frac{1}{s-s_p}\right\} + C\mathcal{L}^{-1}\left\{\frac{1}{(s-s_p)^2}\right\}\right].$$

Referring to tables of elementary transforms, it turns out that

$$\mathcal{L}^{-1}\left\{\frac{1}{s-s_p}\right\} = e^{s_p t}u(t) = e^{-\frac{t}{\tau}}u(t) \quad \text{and} \quad \mathcal{L}^{-1}\left\{\frac{1}{(s-s_p)^2}\right\} = te^{s_p t}u(t) = te^{-\frac{t}{\tau}}u(t),$$

whereas, as is well known, $\mathcal{L}\{1/s\} = u(t)$. Combining the above results, we get

$$y(t) = \frac{\kappa\alpha}{\tau^2}\left[\tau^2 - \tau^2 e^{-\frac{t}{\tau}} - \tau t e^{-\frac{t}{\tau}}\right]u(t) = \kappa\alpha\left[1 - \left(1 + \frac{t}{\tau}\right)e^{-\frac{t}{\tau}}\right]u(t), \tag{4.40}$$

an equation that expresses the *step response of the system* for $\zeta = 1$. It is proven that the time evolution of the system in this case *is faster* than that associated with the value $\zeta > 1$.

Case 3: ($\zeta < 1$), Systems of Small Damping In this case, the discriminant of the quadratic polynomial of the denominator of the function $\mathcal{Y}(s)$ has a *negative value* and therefore this polynomial has two conjugate complex roots, which are given by the relations

$$s_1 = -\frac{\zeta}{\tau} + \frac{j}{\tau}\sqrt{1-\zeta^2} \quad \text{and} \quad s_2 = -\frac{\zeta}{\tau} - \frac{j}{\tau}\sqrt{1-\zeta^2},$$

as can be proven. By expanding the function $\mathcal{Y}(s)$ in a partial fraction expansion, we obviously end up with the same system of equations for the coefficients A, B, C as the one we determined in Case 1, which however in this case are calculated as

$$A = \frac{1}{s_1 s_2} = \tau^2, \quad B = \frac{1}{s_1(s_1 - s_2)} = -\frac{\tau^2}{2\sqrt{1-\zeta^2}(j\zeta + \sqrt{1-\zeta^2})}$$

$$C = -\frac{1}{s_2(s_1 - s_2)} = \frac{\tau^2}{2\sqrt{1-\zeta^2}(j\zeta - \sqrt{1-\zeta^2})}$$

.The determination of the *step response* $y(t)$ will be done again through the calculation of the inverse Laplace transform of the function $\mathcal{Y}(s)$

$$y(t) = \frac{\kappa\alpha}{\tau^2}\left[A\mathcal{L}^{-1}\left\{\frac{1}{s}\right\} + B\mathcal{L}^{-1}\left\{\frac{1}{s-s_1}\right\} + C\mathcal{L}^{-1}\left\{\frac{1}{s-s_2}\right\}\right],$$

where this time the relations that will be used will be

$$\mathcal{L}^{-1}\left\{\frac{1}{s-s_1}\right\} = e^{s_1 t}u(t) = e^{-\frac{\zeta}{\tau}t}e^{\frac{j\sqrt{1-\zeta^2}}{\tau}t}u(t) \quad \text{and} \quad \mathcal{L}^{-1}\left\{\frac{1}{s-s_2}\right\} = e^{s_2 t}u(t) = e^{-\frac{\zeta}{\tau}t}e^{-j\frac{\sqrt{1-\zeta^2}}{\tau}t}u(t).$$

Combining the above results, we get

$$y(t) = \frac{\kappa\alpha}{\tau^2}\left[\tau^2 - \frac{\tau^2 e^{-\frac{\zeta}{\tau}t}e^{j\sqrt{\frac{1-\zeta^2}{\tau}}t}}{2\sqrt{1-\zeta^2}(j\zeta+\sqrt{1-\zeta^2})} + \frac{\tau^2 e^{-\frac{\zeta}{\tau}t}e^{-j\sqrt{\frac{1-\zeta^2}{\tau}}t}}{2\sqrt{1-\zeta^2}(j\zeta-\sqrt{1-\zeta^2})}\right]u(t)$$

$$= \kappa\alpha\left[1 - \frac{e^{-\frac{\zeta}{\tau}t}}{2\sqrt{1-\zeta^2}}\left(\frac{e^{j\frac{\sqrt{1-\zeta^2}}{\tau}t}}{j\zeta+\sqrt{1-\zeta^2}} - \frac{e^{-j\frac{\sqrt{1-\zeta^2}}{\tau}t}}{j\zeta-\sqrt{1-\zeta^2}}\right)\right]u(t).$$

The expression within the parentheses is calculated as

$$\frac{e^{j\frac{\sqrt{1-\zeta^2}}{\tau}t}}{j\zeta+\sqrt{1-\zeta^2}} - \frac{e^{-j\frac{\sqrt{1-\zeta^2}}{\tau}t}}{j\zeta-\sqrt{1-\zeta^2}} = -2\left[\sqrt{1-\zeta^2}\cos\left(\frac{\sqrt{1-\zeta^2}}{\tau}t\right) + \zeta\sin\left(\frac{\sqrt{1-\zeta^2}}{\tau}t\right)\right]$$

and therefore, finally, we get

$$y(t) = \kappa\alpha\left\{1 - \frac{e^{-\frac{\zeta}{\tau}t}}{\sqrt{1-\zeta^2}}\left[\sqrt{1-\zeta^2}\cos\left(\frac{\sqrt{1-\zeta^2}}{\tau}t\right) + \zeta\sin\left(\frac{\sqrt{1-\zeta^2}}{\tau}t\right)\right]\right\}u(t).$$

Defining the angle φ as $\sin\varphi = \sqrt{1-\zeta^2}$ and $\cos\varphi = \zeta$ [from which it follows that $\varphi = \tan^{-1}(\sqrt{1-\zeta^2}/\zeta)$] and the quantity $\omega = \sqrt{1-\zeta^2}/\tau$, the expression within the brackets is written as

$$\mathcal{G} = \sqrt{1-\zeta^2}\cos\left(\frac{\sqrt{1-\zeta^2}}{\tau}t\right) + \zeta\sin\left(\frac{\sqrt{1-\zeta^2}}{\tau}t\right) = \sin\varphi\cos(\omega t) + \cos\varphi\sin(\omega t) = \sin(\omega t + \varphi)$$

and the system's output takes the form

$$y(t) = \kappa\alpha\left[1 - \frac{e^{-\frac{\zeta}{\tau}t}}{\sqrt{1-\zeta^2}}\sin(\omega t + \varphi)\right]u(t). \tag{4.41}$$

From the last relation, it is concluded that in the case $0 < \zeta < 1$ the system's output performs *oscillation with circular frequency* ω and period with value $T = 2\pi/\omega = 2\pi\tau/\sqrt{1-\zeta^2}$.

The variation of the system's step response $y(t)$ for the three above cases, that is, for the values $\zeta > 1$, $\zeta = 1$, and $0 < \zeta < 1$, is presented in a diagrammatic way in Fig. 4.11.

Continuous second-order linear systems are characterized by certain response indicators, which in general are the following:

- ***Maximum overshoot M_p***: Defined as *the maximum deviation of the system's output from the step input during the transition* and is a measure of the system's stability.
- ***Delay time τ***: Defined as *the time period required for* the step response approaches 50% of its final value for the first time.
- ***Rise Time or Overshoot τ_r***: For the case of $\zeta > 1$ (i.e., for heavily damped systems) it is defined as *the time interval required for the step response to change from* 10% *to* 90% *or from* 0% *to* 100% *of its maximum value*. Conversely, for the

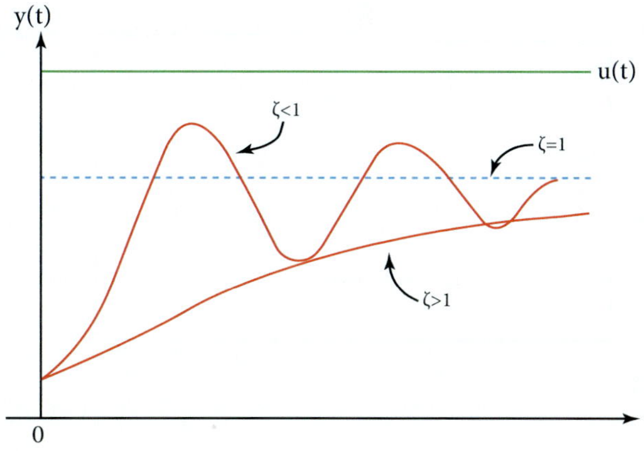

Fig. 4.11 Variation of the signal $y(t)$ for $\zeta > 1$, $\zeta = 1$, and $0 < \zeta < 1$

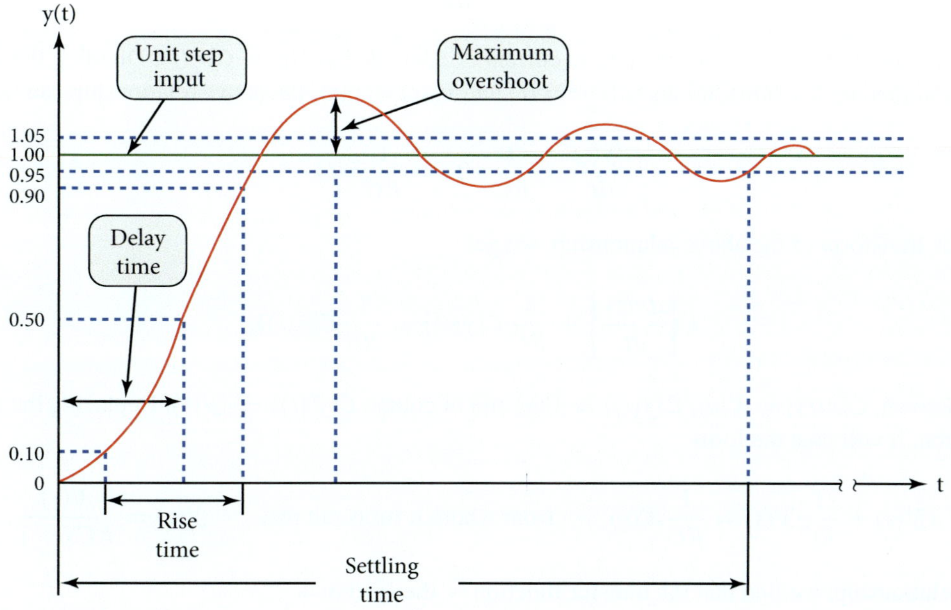

Fig. 4.12 The most important response indicators of a continuous linear second-order system

case of $\zeta < 1$ (i.e., for lightly damped systems) it is defined as *the time interval required for the step response to approach its maximum value for the first time.*

- ***Settling Time*** τ_s: Defined as *the time interval required for the step response to fall below a certain percentage—* approximately $2 - 5\%$*—of its final value* and stay there.
- ***Peak Time*** τ_p: Defined as *the time interval required for the system's step response to approach its first peak of overshoot.*

The most important of the above response indicators are presented in Fig. 4.12.

Problem 4.8.6 *Determine the transfer function* $\mathcal{H}(s)$ *and the impulse response* $h(t)$ *of the electrical circuit of Fig. 4.13.*

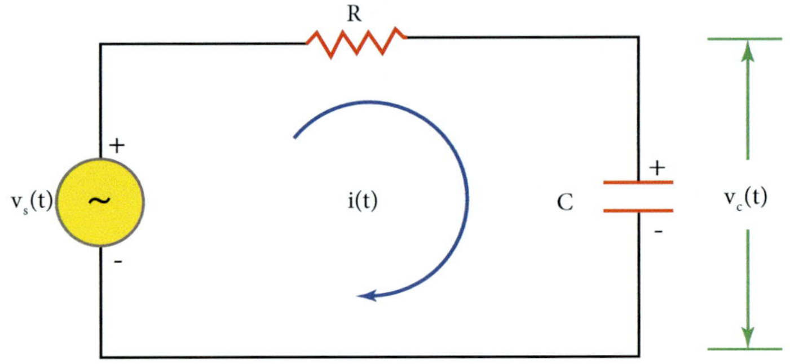

Fig. 4.13 The *RC* circuit of Exercise 4.8.6

Solution Applying *Kirchhoff's law* for the circuit of Fig. 4.13, we are led to the equation

$$Ri(t) + \upsilon_c(t) = RC\frac{\upsilon_c(t)}{dt} + \upsilon_c(t) = \upsilon_s(t),$$

since from the basic theory it is known that the circuit current is calculated as $i(t) = C\upsilon_c'(t)$. Dividing the above equation by the quantity RC and making the substitutions $\upsilon_s(t) = x(t)$ and $\upsilon_c(t) = y(t)$, the above relationship can take the form[9]

$$\frac{dy(t)}{dt} + \frac{1}{RC}y(t) = \frac{1}{RC}x(t)$$

Taking the Laplace transform of the above relationship, we get

$$\mathcal{L}\left\{\frac{dy(t)}{dt}\right\} + \frac{1}{RC}\mathcal{L}\{y(t)\} = \frac{1}{RC}\mathcal{L}\{x(t)\},$$

where, as is well known, $\mathcal{L}\{x(t)\} = \mathcal{X}(s)$, $\mathcal{L}\{y(t)\} = \mathcal{Y}(s)$ and of course $\mathcal{L}\{y'(t)\} = s\mathcal{Y}(s)$. Replacing the above expressions in our final equation, it will take the form

$$s\mathcal{Y}(s) + \frac{1}{RC}\mathcal{Y}(s) = \frac{1}{RC}\mathcal{X}(s) \qquad \text{from where it turns out that} \qquad \mathcal{Y}(s) = \frac{\mathcal{X}(s)}{RCs+1}.$$

From the above relationship, we find that the transfer function of the system is

$$\mathcal{H}(s) = \frac{\mathcal{Y}(s)}{\mathcal{X}(s)} = \frac{1}{RCs+1} = \frac{1}{RC}\frac{1}{s+(1/RC)},$$

whereas the impulse response of the system is calculated as

$$h(t) = \mathcal{L}^{-1}\{\mathcal{H}(s)\} = \frac{1}{RC}\mathcal{L}^{-1}\left\{\frac{1}{s+(1/RC)}\right\} = \frac{1}{RC}e^{-t/RC}u(t),$$

as emerges from tables of elementary transforms.

Problem 4.8.7 *Let us consider the circuit of image (a) in Fig. 4.14 and let us assume that the capacitor C has an initial voltage υ_c with an initial value $\upsilon_c(0^-) = \upsilon_0$. If at time $t = 0$ we lower the switch, the circuit will start to leak current, whereas the voltage $\upsilon_c(t)$ will start to change in function of time. Using the method of Laplace transform, determine (a) the current $i(t)$ and (b) the voltage $\upsilon_c(t)$ along the capacitor.*

[9] For the description of RC circuits see also the section on real filters in the previous chapter.

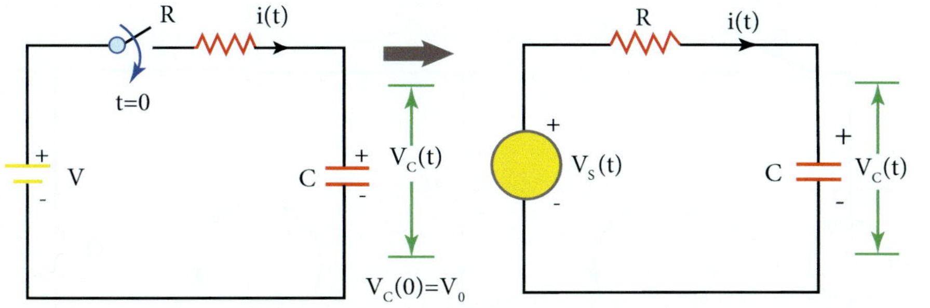

Fig. 4.14 The RC circuit of Exercise 4.8.7

Solution (a) If at time $t = 0$ we lower the switch of the circuit, then its structure coincides with the one presented in image (b) in Fig. 4.14, where now the voltage $\upsilon_s(t)$ is defined as $\upsilon_s(t) = Vu(t)$. In this case, and as emerges from the application of *Kirchhoff's law*, the temporal evolution of the circuit will be determined by the differential equation

$$Ri(t) + \frac{1}{C}\int_{-\infty}^{t} i(t)\, d\tau = \upsilon_s(t) = Vu(t).$$

Taking the *unilateral Laplace transform* of the above equation, we are led to the equation

$$R\mathcal{L}^{+}\{i(t)\} + \frac{1}{C}\mathcal{L}^{+}\left\{\int_{-\infty}^{t} i(t)\, d\tau\right\} = V\mathcal{L}^{+}\{u(t)\},$$

where, as is well known, we will have $\mathcal{L}^{+}\{i(t)\} = \mathcal{I}(s)$, $\mathcal{L}^{+}\{u(t)\} = 1/s$ and

$$\mathcal{L}^{+}\left\{\int_{-\infty}^{t} i(\tau)\, d\tau\right\} = \frac{1}{s}\mathcal{I}(s) + \frac{1}{s}\int_{-\infty}^{0^-} i(\tau)\, d\tau,$$

as emerges from the properties of the unilateral Laplace transform.[10] However, taking into account the definition of the current $i(t)$ as the first derivative of the function $\upsilon_c(t)$ and the description of the problem, it is not difficult to realize that

$$\upsilon_c(t) = \frac{1}{C}\int_{-\infty}^{t} i(\tau)\, d\tau \quad\longrightarrow\quad \upsilon_c(0^-) = \frac{1}{C}\int_{-\infty}^{0^-} i(\tau)\, d\tau = \upsilon_0.$$

Replacing the above quantities in our final equation, it will take the form

$$R\mathcal{I}(s) + \frac{1}{C}\left[\frac{1}{s}\mathcal{I}(s) + \frac{C}{s}\upsilon_0\right] = \frac{V}{s}$$

from where, solving for $\mathcal{I}(s)$, it turns out that

$$\mathcal{I}(s) = \frac{V - \upsilon_0}{s}\,\frac{1}{R + (1/Cs)} = \frac{V - \upsilon_0}{s}\,\frac{1}{s + (1/RC)}.$$

Therefore, taking the inverse Laplace transform of the above relationship we find that

[10] The proof of the above equation is relatively simple by writing the left side of the above equation in the form

$$\int_{-\infty}^{t} i(\tau)\, d\tau = \int_{-\infty}^{0^-} i(\tau)\, d\tau + \int_{0^-}^{t} i(\tau)\, d\tau$$

and using the *integration* property of the Laplace transform along with the fact that the first integral of the second term is actually constant.

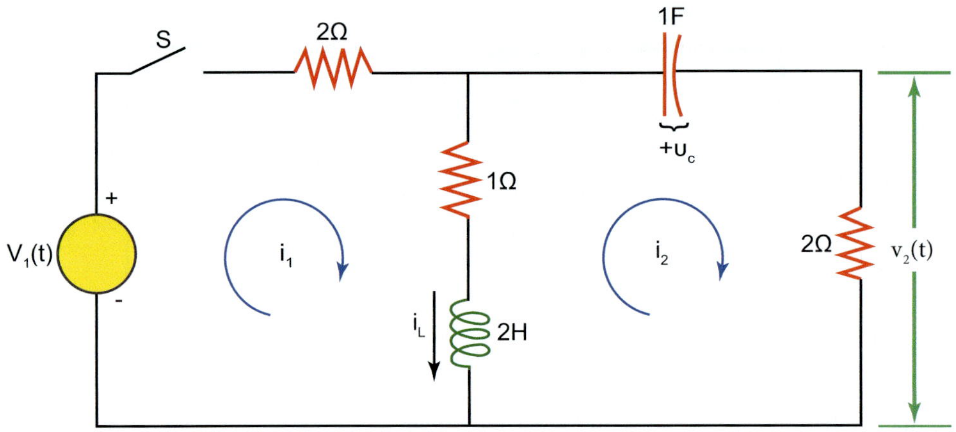

Fig. 4.15 The electrical circuit *RLC* of Exercise 4.8.8 (Poularikas, *The Transforms and Applications Handbook*, Example 5.5.4)

$$i(t) = \mathcal{L}^{-1}\{\mathcal{I}(s)\} = \frac{V - \upsilon_0}{s}\mathcal{L}^{-1}\left\{\frac{1}{s + (1/RC)}\right\} = \frac{V - \upsilon_0}{s}e^{-t/RC}u(t).$$

(b) Assuming—according to the problem description—that the input function is $\upsilon_s(t)$, whereas the output function is the $\upsilon_c(t)$, the differential equation governing the behavior of the circuit will be given by the relationship

$$\frac{d\upsilon_c(t)}{dt} + \frac{1}{RC}\upsilon_c(t) = \frac{1}{RC}\upsilon_s(t) = \frac{V}{RC}u(t).$$

Taking the *unilateral Laplace transform* of the above equation, we have

$$\mathcal{L}^+\left\{\frac{d\upsilon_c(t)}{dt}\right\} + \frac{1}{RC}\mathcal{L}^+\{\upsilon_c(t)\} = \frac{V}{RC}\mathcal{L}^+\{u(t)\},$$

where, as is well known, $\mathcal{L}^+\{\upsilon_c(t)\} = \mathcal{V}_c(s)$, $\mathcal{L}^+\{\upsilon_c'(t)\} = s\mathcal{V}_c(s) - \upsilon_c(0^-) = s\mathcal{V}_c(s) - \upsilon_0$ [since $\upsilon_c(0^-) = \upsilon_c(0^+) = \upsilon_0$] and $\mathcal{L}^+\{u(t)\} = 1/s$. Replacing the above expressions in the last equation, it will take the form

$$s\mathcal{V}_c(s) - \upsilon_0 + \frac{1}{RC}\mathcal{V}_c(s) = \frac{V}{RCs} \quad \text{or equivalently} \quad \left(s + \frac{1}{RC}\right)\mathcal{V}_c(s) = \frac{V}{RCs} + \upsilon_0$$

and finally

$$\mathcal{V}_c(s) = V\left(\frac{1}{s} - \frac{1}{s + (1/RC)}\right) + \frac{\upsilon_0}{s + (1/RC)}.$$

Taking the inverse Laplace transform of the above equation, we get

$$\upsilon_c(t) = V\left(\mathcal{L}^{-1}\left\{\frac{1}{s}\right\} - \mathcal{L}^{-1}\left\{\frac{1}{s + (1/RC)}\right\}\right) + \upsilon_0\mathcal{L}^{-1}\left\{\frac{1}{s + (1/RC)}\right\} = V(1 - e^{-t/RC})\,u(t) + \upsilon_0 e^{-t/RC}u(t).$$

Problem 4.8.8 *Given the RLC circuit of Fig. 4.15 and using the method of the Laplace transform, determine the analytical form of the voltage $\upsilon_2(t)$, assuming that the function $\upsilon_1(t)$ and the initial conditions $i_2(0^-)$ and $\upsilon_c(0^-)$ are known.*

Solution If at time $t = 0$ we turn off the switch of the circuit, it will start to be drained by the current. Therefore, applying the *laws of Kirchhoff* for the two closed loops defined in it, we get

$$\left(3i_1(t) + 2\frac{di_1(t)}{dt}\right) - \left(i_2(t) + 2\frac{di_2(t)}{dt}\right) = \upsilon_1(t)$$

$$\left(i_1(t) + 2\frac{di_1(t)}{dt}\right) + \left(3i_2(t) + 2\frac{di_2(t)}{dt} + \int_{-\infty}^{t} i_2(\tau)\, d\tau\right) = 0.$$

Taking the unilateral Laplace transform of the above equations, we get

$$\left(3\mathcal{L}\{i_1(t)\} + 2\mathcal{L}\left\{\frac{di_1(t)}{dt}\right\}\right) - \left(\mathcal{L}\{i_2(t)\} + 2\mathcal{L}\left\{\frac{di_2(t)}{dt}\right\}\right) = \mathcal{L}\{\upsilon_1(t)\}$$

$$\left(\mathcal{L}\{i_1(t)\} + 2\mathcal{L}\left\{\frac{di_1(t)}{dt}\right\}\right) + \left(3\mathcal{L}\{i_2(t)\} + 2\mathcal{L}\left\{\frac{di_2(t)}{dt}\right\} + \mathcal{L}\left\{\int_{-\infty}^{t} i_2(\tau)\, d\tau\right\}\right) = 0$$

where for reasons of brevity we used the symbol $\mathcal{L}$ to express the unilateral transform. However, $\mathcal{L}\{i_1(t)\}=\mathcal{I}_1(s)$, $\mathcal{L}\{i_2(t)\}=\mathcal{I}_2(s)$, $\mathcal{L}\{i_1'(t)\}=s\mathcal{I}_1(s) - i_1(0^+)$, $\mathcal{L}\{i_2'(t)\}=s\mathcal{I}_2(s) - i_2(0^+)$, $\mathcal{L}\{\upsilon_1(t)\} = U_1(s)$ and

$$\mathcal{L}\left\{\int_{\infty}^{t} i_2(t)\, dt\right\} = \frac{1}{s}\mathcal{I}_2(s) + \frac{1}{s}\int_{-\infty}^{0^-} i_2(\tau)\, d\tau = \frac{1}{s}\mathcal{I}_2(s) + \frac{C}{s}\upsilon_2(0^+) = \frac{1}{s}\mathcal{I}_2(s) + \frac{q_2(0^+)}{s}$$

(as we proved in the previous exercise), where for the derivation of the last relation we used the known equation $q(t) = C\upsilon(t)$ that correlates the charge with the voltage at the ends of a capacitor. Replacing the above quantities in the last relations and making simple mathematical operations, it turns out that

$$(3 + 2s)\mathcal{I}_1(s) - (1 + 2s)\mathcal{I}_2(s) = \mathcal{U}_1(s) + 2[i_1(0^+) - i_2(0^+)]$$

$$-(1 + 2s)\mathcal{I}_1(s) + (3 + 2s + \frac{1}{s})\mathcal{I}_2(s) = 2[-i_1(0^+) + i_2(0^+)] - \frac{q_2(0^+)}{s}$$

From the circuit's connectivity, it is not difficult to ascertain that the currents i_L, i_1 and i_2 are connected by the relation $i_L(t) = i_1(t) - i_2(t)$ and therefore at the time $t = 0^+$ $i_L(0^+) = i_1(0^+) - i_2(0^+)$. Also, from the relation

$$\frac{q_2(t)}{C} = \frac{1}{C}\int_{-\infty}^{t} i_2(\tau)d\tau = \frac{1}{C}\int_{-\infty}^{0} i_2(t)dt + \frac{1}{C}\lim_{t\to 0^+}\int_{0}^{t} i_2(\tau)d\tau = 0 + \upsilon_c(0^-) = \upsilon_c(0^-)$$

we have

$$\frac{q_2(0^+)}{C} \triangleq \upsilon_c(0^+) = \upsilon_c(0^-) = i_2^{(-1)}(0) = q_2(0^+).$$

Substituting the above relations into the system of equations, it will take the form

$$(3 + 2s)\mathcal{I}_1(s) - (1 + 2s)\mathcal{I}_2(s) = \mathcal{U}_1(s) + 2i_L(0^+)$$

$$(1 + 2s)\mathcal{I}_1(s) - (3 + 2s + \frac{1}{s})\mathcal{I}_2(s) = 2i_L(0^+) + \frac{\upsilon_c(0^+)}{s}.$$

The above system constitutes an algebraic system of two equations with two unknowns, the functions $\mathcal{I}_1(s)$ and $\mathcal{I}_2(s)$. Solving the system, it is not difficult to ascertain after operations that the transform $\mathcal{I}_2(s)$ is calculated as

$$\mathcal{I}_2(s) = \frac{-(2s^2 + 3s)\upsilon_c(0^+) - 4si_L(0^+) + (2s^2 + s)\mathcal{U}_1(s)}{8s^2 + 10s + 3}.$$

In order to determine the required function $\upsilon_2(t)$, we observe that it is related to the current $i_2(t)$ through the relation $\upsilon_2(t) = 2i_2(t)$, as can be verified by observing the diagram. Taking the Laplace transform of the two sides, we get $\mathcal{U}_2(s) = 2\mathcal{I}_2(s)$, and therefore the function $\upsilon_2(t)$ can be calculated from the relation

$$\upsilon_2(t) = \mathcal{L}^{-1}\{\mathcal{U}_2(s)\} = 2\mathcal{L}^{-1}\{\mathcal{I}_2(s)\} = 2\mathcal{L}^{-1}\left\{\frac{-(2s^2 + 3s)\upsilon_c(0^+) - 4si_L(0^+) + (2s^2 + s)\mathcal{U}_1(s)}{8s^2 + 10s + 3}\right\},$$

which is also the final result. From the last relation, we observe that the function $\upsilon_2(t)$ depends on the analytical form of the function $\upsilon_1(t)$—through the transform $\mathcal{U}_1(s)$—something that, of course, is expected.

In order to arrive at a more specific result, let us assume that the voltage $\upsilon_1(t)$ is the step function $u(t)$—so $\mathcal{U}_1(s) = \mathcal{L}\{u(t)\} = 1/s$—and that at the time $t = 0$ the circuit does not contain accumulated energy, which means that $i_L(0^+) = \upsilon_c(0^+) = 0$. In this case, the function $\upsilon_2(t)$ will take the simplest form

$$\upsilon_2(t) = 2\mathcal{L}^{-1}\left\{\frac{2s^2+s}{s(8s^2+10s+3)}\right\} = 2\mathcal{L}^{-1}\left\{\frac{2s+1}{8s^2+10s+3}\right\}$$

The polynomial of the denominator after factorization is written in the form $8s^2+10s+3 = (2s+1)(4s+3)$ and the above relation will take the form

$$\upsilon_2(t) = 2\mathcal{L}^{-1}\left\{\frac{1}{4s+3}\right\} = \frac{1}{2}\mathcal{L}^{-1}\left\{\frac{1}{s+(3/4)}\right\} = \frac{1}{2}e^{-3t/4}u(t),$$

which is also the desired result for the assumptions of the problem that have been set.

4.9 Problems

1. Determine the Laplace transform of the continuous-time signal $x(t) = \sin(\omega t)u(t)$.
2. Using the property of periodicity, determine the Laplace transform of the signals $x(t) = \sin(\omega t)$ and $x(t) = \cos(\omega t)$ and verify that the result obtained is the one we end up with, applying the defining equation of the transform.
3. Prove the following results regarding the process of calculating Laplace transforms.

$$\mathcal{L}\{e^{\alpha t}\sin(\beta t)\} = \frac{\beta}{(s-\alpha)^2+\beta^2} \quad \text{and} \quad \mathcal{L}\{te^{\alpha t}\sin(\beta t)\} = \frac{2\beta(s-\alpha)}{[(s-\alpha)^2+\beta^2]^2}.$$

4. Using Euler's formula $\sin(\omega t) = (e^{j\omega t} - e^{-j\omega t})/2j$, prove that

$$\mathcal{L}\left\{e^{-\alpha t}\sin(\omega t)u(t)\right\} = \frac{\omega}{\omega^2+(s+\alpha)^2}, \qquad \sigma \equiv \Re(s) > -\alpha.$$

5. Prove the following results regarding the process of calculating Laplace transforms.

$$\mathcal{L}\{\sinh(\alpha t)\} = \frac{\alpha}{s^2-\alpha^2} \quad \text{and} \quad \mathcal{L}\{t\sinh(\alpha t)\} = \frac{2\alpha s}{(s^2-\alpha^2)^2}.$$

6. Determine the Laplace transform of the signal $x(t) = e^{-6t}u(t-2)$ and its region of convergence.
7. Find the Laplace transform of the signal

$$x(t) = \delta(t) - \frac{5}{3}e^{-2t}u(t) + \frac{2}{3}e^{3t}u(t).$$

8. Considering the signal $x(t) = e^{-5t}u(t) + e^{-\alpha t}u(t)$, if the region of convergence is $\Re(s) > -4$ determine the restrictions concerning the real and imaginary parts of α.
9. Let us consider the signal $y(t) = x(t) + \alpha x(-t)$, where $x(t) = \beta e^{-t}u(t)$. If the Laplace transform of the signal $y(t)$ has the form $\mathcal{Y}(s) = s/(s^2-1)$ with the region of convergence $-1 < \Re(s) < 1$ determine the values of α and β.
10. Show that

$$\mathcal{L}\{t\sin(\omega t)\} = \frac{2s\omega}{(s^2+\omega^2)^2} \quad \text{and} \quad \mathcal{L}\{t^2\sin(\omega t)\} = \frac{6s^2\omega-\omega^3}{(s^2+\omega^2)^3}.$$

11. Using the result of Exercise 4.5.6 (refer to page 329) show that

$$\mathcal{L}\left\{\frac{\sin(\omega t)+\sin(|\omega t|)}{2}\right\} = \frac{\omega}{\omega^2+s^2}\left[1+\cosh\left(\frac{s\pi}{2\omega}\right)\right].$$

12. Determine the Laplace transform of the signals

$$x(t) = \sin(t)\sin(3t), \quad x(t) = \cos(3t)\sin(7t), \quad x(t) = \sin(t)\cos(t), \quad x(t) = \cos(t+4).$$

13. For each of the following continuous-time signals determine the Laplace transform and the region of convergence and draw the pole-zero diagram.

$$x_1(t) = e^{-6t}u(t) + e^{-5t}\sin(2t)u(t)$$
$$x_2(t) = te^{-4|t|}$$
$$x_3(t) = |t|e^{3t}u(-t)$$
$$x_4(t) = \delta(4t) + u(2t)$$
$$x_5(t) = e^{-2t}[u(t) - u(t-5)]$$
$$x_6(t) = \frac{t}{4}\sin(2t)u(t)$$
$$x_7(t) = A\cos(\omega t + \varphi)$$
$$x_8(t) = e^{\alpha t}\sinh(\beta t)$$
$$x_9(t) = t^2\sin(\alpha t)$$

14. Calculate the Laplace transform of the continuous-time signals

$$x(t) = u(t) - e^{-t}u(t) - te^{-t}u(t)$$
$$y(t) = -u(-t) - (1+t)e^{-t}u(t)$$
$$z(t) = -u(-t) + e^{-t}u(-t) + te^{-t}u(-t)$$
$$h(t) = (t/4)\sin(2t)u(t)$$

15. Determine the Laplace transform of the continuous-time signals

$$x(t) = \begin{cases} 0 & 0 \le t \le 1 \\ 1 & 1 \le t \le 2 \\ 0 & t > 2 \end{cases} \quad \text{and} \quad y(t) = \begin{cases} t^2 & 0 \le t \le 1 \\ 3-t & 1 \le t \le 2 \\ 1 & t > 2. \end{cases}$$

16. Determine the Laplace transform of the continuous-time signals

$$x(t) = \begin{cases} t & 0 \le t \le 2 \\ \sin t & t > 2 \end{cases} \qquad x(t) = \sin^2(t+3), \qquad x(t) = t^2e^{-2t} + t\sin(t).$$

17. Determine the Laplace transform of the continuous-time signals of Fig. 4.16.
18. Determine the Laplace transform of the functions

$$\varphi_n(t) = e^{-t/2}L_n(t)u(t), \quad n = 0, 1, 2, \ldots \quad \text{where}$$

$$L_n(t) = \frac{e^t}{n!}\frac{d^n}{dt^n}(t^n e^{-t})$$

These functions are known as *Laguerre polynomials*.

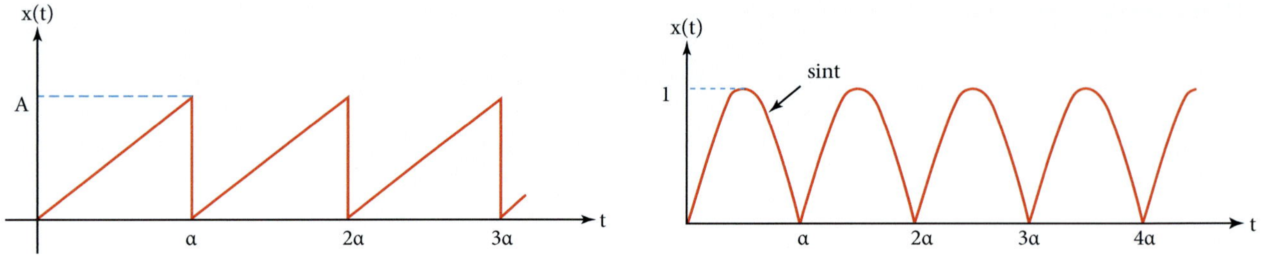

Fig. 4.16 The continuous-time signals of Problem 17

19. Apply the theorems of initial and final values for the signals

$$x_1(t) = e^{4t}\sinh(2t)u(t) + e^{-t}\cos(3t)u(t) \quad \text{and} \quad x_2(t) = 5t\cot(2t)u(t) + e^{-t}u(t).$$

20. According to the initial value theorem, if a causal signal $x(t)$ has a Laplace transform $\mathcal{X}(s)$, then its value at the time $t = 0^+$ is calculated from the relation $x(0^+) = \lim_{s\to\infty}[s\mathcal{X}(s)]$. Using the Taylor expansion of the signal $x(t)$ around the point $t = 0^+$

$$x(t) = \left[x(0^+) + x'(0^+)t + x''(0^+)\frac{t^2}{2!} + \cdots + x^{(n)}(0^+)\frac{t^n}{n!} + \ldots\right]u(t)$$

determine the Laplace transform of the quantity

$$y(t) = x^{(n)}(0^+)\frac{t^n}{n!}u(t)$$

and prove that the Laplace transform $X(s)$ can be calculated from the relation

$$\mathcal{X}(s) = \sum_{n=0}^{\infty} x^{(n)}(0^+)\frac{1}{s^{n+1}}.$$

21. Determine the inverse Laplace transform of the function

$$\mathcal{X}(s) = \frac{e^{-\alpha s} - e^{-\beta s}}{s},$$

where α and β are positive real constants with $\beta > \alpha$.

22. Using the method of partial fraction expansion, calculate the inverse Laplace transform of the function

$$\mathcal{X}(s) = \frac{2 + 2se^{-2s} + 4e^{-s}}{s^2 + 4s + 3} \quad \text{with the region of convergence} \quad \Re(s) > -1$$

(*Hint:* To solve the exercise, use the property of time shifting).

23. Determine the inverse Laplace transform of the functions

$$\mathcal{X}(s) = \frac{s^2}{s^2 + 4s + 5}, \qquad \mathcal{X}(s) = e^{-2s}\frac{1}{s(s^2 - 8s + 7)}, \mathcal{X}(s) = e^{-4s} + 1 + \frac{1}{s}.$$

24. Determine the Laplace transform of the function

$$\mathcal{X}(s) = \frac{s^3 + As^2 + Bs + C}{[(s-\alpha)^2 + \beta^2]^2},$$

where A, B, C, α, and β are constant values.

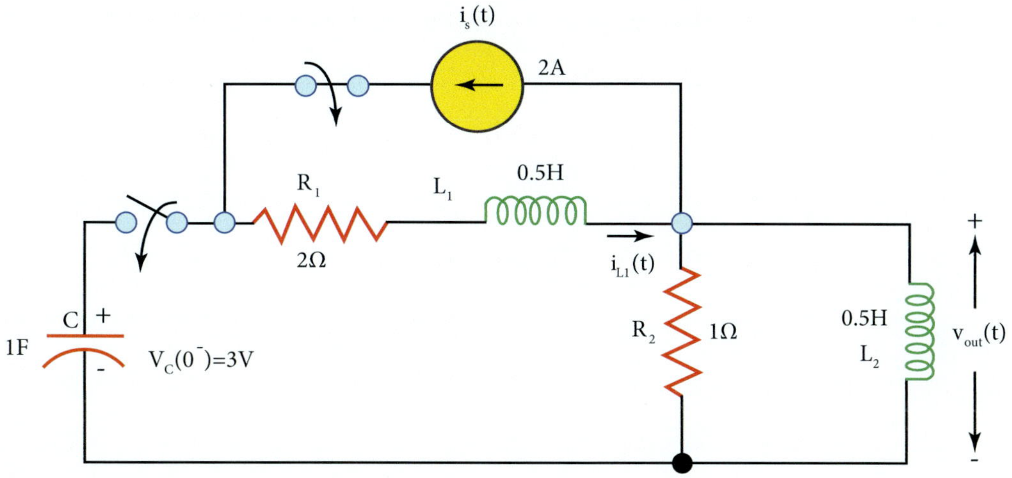

Fig. 4.17 The electrical circuit RLC of Problem 30

25. Using the method of partial fraction expansion, calculate the inverse Laplace transform of the functions

$$\mathcal{X}_1(s) = \frac{s+1}{s^2+2s+10} \quad \text{with the region of convergence} \quad \Re(s) < -1$$

$$\mathcal{X}_2(s) = \frac{(s+1)^2}{s^2-s+1} \quad \text{with the region of convergence} \quad \Re(s) > \frac{1}{2}$$

$$\mathcal{X}_3(s) = \frac{s^2-s+1}{(s+1)^2} \quad \text{with the region of convergence} \quad \Re(s) > -1$$

$$\mathcal{X}_4(s) = \frac{s}{s^3+2s^2+9s+18} \quad \text{with the region of convergence} \quad \Re(s) > -2.$$

26. Determine the signals $x(t)$ characterized by Laplace transforms of the form

$$\mathcal{X}(s) = \frac{1}{s^2+9}, \quad \Re(s) > 0, \qquad \mathcal{X}(s) = \frac{s+2}{s^2+7s+12}, \quad -4 < \Re(s) < -30, \qquad \mathcal{X}(s) = \frac{s^2-s+1}{(s+1)^2}, \quad \Re(s) > -1.$$

27. Using Heaviside's theorem, find the inverse transforms of all rational functions $\mathcal{X}(s)$ of the previous exercises that meet the prerequisites for the application of this theorem.
28. Referring to bibliographic sources, record and prove the defining equations of the response indices of a linear continuous second-order system.
29. Using the method of Laplace transform to solve the differential equations

$$\frac{d^2y(t)}{dt^2} + y(t) = \delta(t-2) \text{ with the initial conditions} y(0) = 1 \text{ and } y'(0) = 2$$

$$\frac{d^2y(t)}{dt^2} - 2\frac{dy(t)}{dt} + y(t) = 0 \text{ with the initial conditions} y(0) = 2 \text{ and } y'(0) = 1$$

$$\frac{d^3y(t)}{dt^3} + 3t\frac{dy(t)}{dt} - 6y(t) = 1 \text{ with the initial conditions} y(0) = 0 \text{ and } y'(0) = 0.$$

30. Given the RLC circuit of Fig. 4.17 and using the method of Laplace transform, determine the analytical form of the voltage V_{out} depicted.

System Analysis in the Frequency Domain 5

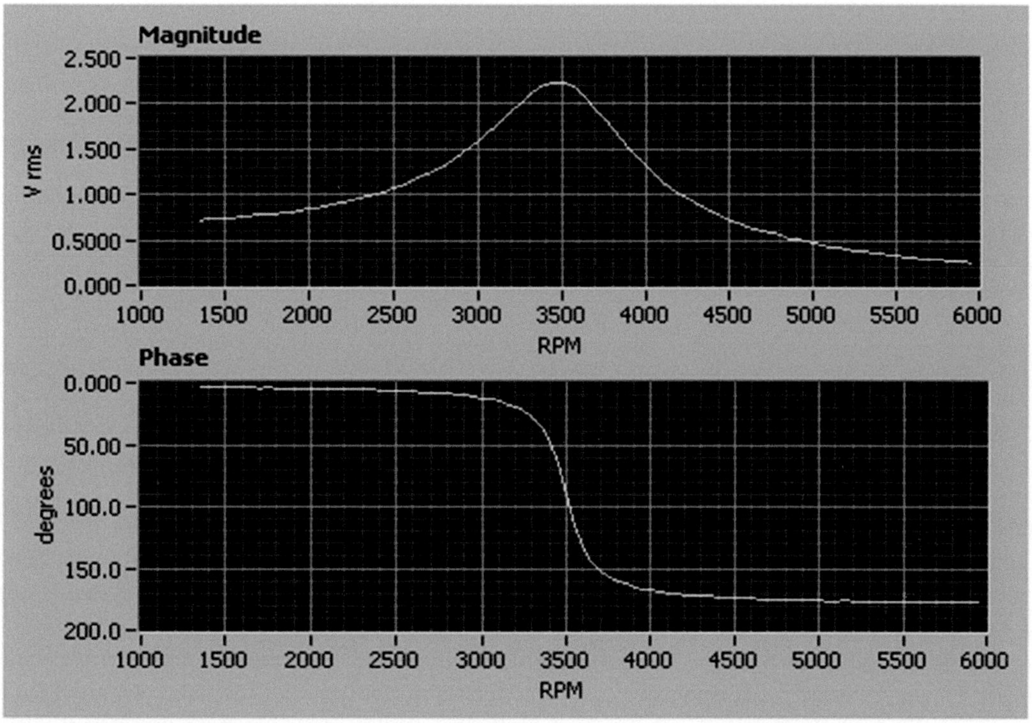

Bode diagrams depict the amplitude and phase responses of a continuous-time LTI system as a function of frequency ω

5.1 Introduction

Having provided an analytical description of the Laplace transform and all the characteristics that describe it, let us now move on to the way this transform is used for the study of continuous-time systems in the complex plane. The most important parameter related to this analysis process is the *transfer function* of a system, which, as mentioned in the chapter on continuous-time systems, is defined as the *Laplace transform of its impulse response*. In the general case of systems described by *linear differential equations with constant coefficients*, this function is a *rational expression*, that is, a ratio of two polynomials, such that the degree M of the numerator polynomial is less than the degree N of the denominator polynomial; such a function is known as a *proper function*. In other words, this means that, generally speaking, the *number of zeros of the function is less than the number of its poles*. The construction of a *pole-zero diagram* and the determination of the position of these points on the complex plane is a crucial stage in the study of a system, as it makes it possible to draw conclusions about many of its properties, most notably its stability.

A. I. Margaris, *Continuous and Discrete-Time Signals and Systems*,
https://doi.org/10.1007/978-3-031-81209-5_5

The most important techniques to be presented in this chapter concern the *determination of the transfer function from the differential equation of a system* and vice versa, the *characterization of a system's stability*, and *interconnecting simple systems to construct more complex structures*. Subsequently, we address the *study of systems in the state space* with the help of the Laplace transform, with emphasis given to the *construction of a transfer function from dynamic equations and vice versa, the solution of the state equation using the Laplace transform, and the study of the characteristics of controllability and observability*. The description of all these techniques, concepts, and methods is the focus of the following pages.

5.2 Transfer Function of Continuous-Time Systems

Let us consider a *continuous-time LTI system* whose impulse response is described by the function $h(t)$. From basic theory it is known that if we transmit a *causal* signal $x(t)$ at the input of such a system, its output $y(t)$ will be given by the equation of *convolution*:

$$y(t) = h(t) * x(t) = \int_0^{\infty} x(t)\, h(t-\tau)\, d\tau \tag{5.1}$$

If we take the Laplace transform of the two terms of the foregoing equation and use the *convolution theorem* that characterizes this transform, we find that

$$\mathcal{Y}(s) \equiv \mathcal{L}\{y(t)\} = \mathcal{L}\{x(t) * h(t)\} = \mathcal{L}\{h(t)\}\mathcal{L}\{x(t)\} = \mathcal{H}(s)\mathcal{X}(s) \tag{5.2}$$

where $\mathcal{Y}(s)$, $\mathcal{H}(s)$, and $\mathcal{X}(s)$ are the Laplace transforms of the functions $y(t)$, $h(t)$, and $x(t)$, respectively. From the last relation it follows that the Laplace transform of a continuous-time LTI system is equal to the product of the Laplace transforms of its impulse response and the input that is transmitted to it.

In the special case where the input transmitted to the system has the form $x(t) = e^{st}$, with the parameter s belonging to the region of convergence of the function $\mathcal{H}(s)$, it is proven that $\mathcal{Y}(s) = \mathcal{H}(s)\, e^{st}$, meaning the aforementioned input signal constitutes an eigenfunction of the system with the eigenvalue being the Laplace transform of the system's impulse response $\mathcal{H}(s)$ (see also Sect. 2.5.4, page 103).

Generally speaking, the function $\mathcal{H}(s) = \mathcal{Y}(s)/\mathcal{X}(s)$, which is known as a *system function* or *transfer function*, can be expressed in *explicit form*, meaning it can be written *as the quotient of two polynomial functions* of s provided that the system is described by *linear differential equations with constant coefficients* and is in *steady state*, meaning it is characterized by *zero initial conditions*.[1] To prove this assertion, let us consider the general case of a continuous-time LTI system described by a differential equation of the form

$$\begin{aligned}\alpha_N \frac{d^N y(t)}{dt^N} + \alpha_{N-1}\frac{d^{N-1}y(t)}{dt^{N-1}} + \cdots + \alpha_2 \frac{d^2y(t)}{dt^2} + \alpha_1 \frac{dy(t)}{dt} + \alpha_0 y(t)\\ = \beta_M \frac{d^M x(t)}{dt^M} + \beta_{M-1}\frac{d^{M-1}x(t)}{dt^{M-1}} + \cdots + \beta_2 \frac{d^2x(t)}{dt^2} + \beta_1 \frac{dx(t)}{dt} + \beta_0 x(t)\end{aligned}$$

or in condensed form

$$\sum_{i=0}^{N} \alpha_i \frac{d^i y(t)}{dt^i} = \sum_{j=0}^{M} \beta_j \frac{d^j x(t)}{dt^j} \tag{5.3}$$

[1] Note that this characteristic constitutes *a prerequisite* for the calculation of the transfer function: This function is defined *only* for LTI systems, while it does not exist for time-varying systems and in its calculation *all the initial conditions of the problem are considered equal to zero*.

with a solution of the form $y = y(t)$. If we take the Laplace transform of the foregoing equation and use the property of *linearity* of that transform, we will get

$$\sum_{i=0}^{N} \alpha_i \mathcal{L}\left\{\frac{d^i y(t)}{dt^i}\right\} = \sum_{j=0}^{M} \beta_j \mathcal{L}\left\{\frac{d^j x(t)}{dx^j}\right\} \tag{5.4}$$

then, using the *property of differentiation*, we end up with the expression

$$\left(\sum_{i=0}^{N} \alpha_i s^i\right) \mathcal{L}\{y(t)\} - \sum_{i=1}^{N} \alpha_i \left(\sum_{k=1}^{i} s^{i-k} y^{(i-k)}(0)\right) = \left(\sum_{j=0}^{M} \beta_j s^j\right) \mathcal{L}\{x(t)\} - \sum_{j=1}^{M} \beta_j \left(\sum_{\ell=1}^{j} s^{j-\ell} x^{(j-\ell)}(0)\right)$$

Relying on the assumption that the system under study is in a state of rest and is mathematically formulated as $y^{(i)}(0) = 0$ $(i = 1, 2, \ldots, N)$—this property will also characterize the input signal $x(t)$ for which we can therefore write that $x^{(j)}(0)$ $(j = 1, 2, \ldots, M)$—the foregoing relationship is written

$$\left(\sum_{i=0}^{N} \alpha_i s^i\right) \mathcal{L}\{y(t)\} = \left(\sum_{j=0}^{M} \beta_j s^j\right) \mathcal{L}\{x(t)\} \tag{5.5}$$

and therefore the system function $\mathcal{H}(s) \equiv \mathcal{L}\{h(t)\}$ can be expressed in the explicit form

$$\mathcal{H}(s) = \frac{\mathcal{Y}(s)}{\mathcal{X}(s)} = \left\{\sum_{j=0}^{M} \beta_j s^j\right\} \Big/ \left\{\sum_{i=0}^{N} \alpha_i s^i\right\} = \frac{\beta_0 + \beta_1 s + \beta_2 s^2 + \cdots + \beta_{M-1} s^{M-1} + \beta_M s^M}{\alpha_0 + \alpha_1 s + \alpha_2 s^2 + \cdots + \alpha_{N-1} s^{N-1} + \alpha_N s^N} \tag{5.6}$$

where $\mathcal{Y}(s) \equiv \mathcal{L}\{y(t)\}$ and $\mathcal{X}(s) \equiv \mathcal{L}\{x(t)\}$. From the last relation we observe that the transfer function of the system *depends only on the values of its parameters*—that is, the coefficients that appear in the differential equation defining it—*while it is independent of the input signal applied to it* (which is certainly expected).

In cases where the transfer function is a *rational function*, that is, it can be expressed as the quotient of two polynomial functions of the complex variable s, it must necessarily be that the *degree M of the numerator polynomial is smaller than or equal to the degree N of the denominator polynomial* [this of course is equivalent to the observation made in a previous chapter that the maximum order of the derivatives of the output $y(t)$ appearing in the differential equation of the system should be greater than or least equal to the maximum order of the derivatives of the input function $x(t)$], since in the opposite case the system is *BIBO (Bounded Input Bounded Output) stable*. Indeed, if we assume that $M > N$ holds, then the application of the method of *polynomial division* allows us to write the function $\mathcal{H}(s)$ as

$$\mathcal{H}(s) = \frac{N(s)}{D(s)} = \overbrace{c_k s^k + c_{k-1} s^{k-1} + \cdots + c_1 s + c_0}^{R(s)} + \frac{Q(s)}{D(s)} \tag{5.7}$$

with the degree k of the polynomial $R(s)$ being equal to $M - N$ and the degree d of the polynomial $Q(s)$ obviously being less than the degree N of the polynomial $D(s)$. Let us now assume that we transmit to the system the bounded input $x(t) = u(t)$ whose Laplace transform is given by the relation $\mathcal{X}(s) = 1/s$. In this case, the Laplace transform of the system's output will be calculated as

$$\mathcal{Y}(s) = \mathcal{H}(s)\mathcal{X}(s) = \frac{\mathcal{H}(s)}{s} = c_k s^{k-1} + c_{k-1} s^{k-2} + \cdots + c_1 + \frac{c_0}{s} + \frac{Q(s)}{sD(s)}$$

and therefore the system's output will be calculated from the inverse transform

$$y(t) = \mathcal{L}^{-1}\{\mathcal{Y}(s)\} = c_k\mathcal{L}^{-1}\{s^{k-1}\} + c_{k-1}\mathcal{L}^{-1}\{s^{k-2}\} + \cdots + \mathcal{L}^{-1}\{c_1\} + c_0\mathcal{L}^{-1}\left\{\frac{1}{s}\right\} + \mathcal{L}^{-1}\left\{\frac{Q(s)}{sD(s)}\right\}$$

However, we observe that $\mathcal{L}^{-1}\{c_1\} = \mathcal{L}^{-1}\{c_1 \cdot 1\} = c_1\mathcal{L}^{-1}\{1\} = c_1\delta(t)$, and therefore the output $y(t)$ will contain the *impulse function* $\delta(t)$ as one of its terms. But this function, as is well known, *is not bounded*, and therefore the system *is not BIBO stable*, since the transmission to the system of a bounded input resulted in the production of an unblocked output. For this reason, in the *transfer function* of the system, the *degree of the numerator polynomial should always be less than the degree of the denominator polynomial*. A rational function characterized by this property is known as a *proper or normal function*.

We conclude this brief introduction to the transfer function with the observation that since it is a complex function of a complex variable of the form $\mathcal{H}(s) = \Re\{\mathcal{H}(s)\} + j\Im\{\mathcal{H}(s)\}$, it can be expressed in exponential form, as is commonly known:

$$\mathcal{H}(s) = |\mathcal{H}(s)|e^{j\angle\mathcal{H}(s)}$$

where the magnitude and phase of the transfer function are defined by the relations

$$|\mathcal{H}(s)| = \sqrt{[\Re\{\mathcal{H}(s)\}]^2 + [\Im\{\mathcal{H}(s)\}]^2} \quad \text{and} \quad \angle\mathcal{H}(s) = \tan^{-1}\left\{\frac{\Im\{\mathcal{H}(s)\}}{\Re\{\mathcal{H}(s)\}}\right\}$$

If we assume that the complex variable $s = \sigma + j\omega$ has a *zero real part* $\sigma = 0$, that is, it is a purely imaginary number of the form $s = j\omega$, then the *transfer function transforms into our well-known frequency response function* $\mathcal{H}(j\omega)$, which is known to be defined as the Fourier transform of a system's impulse response.[2] From this perspective, we can consider the Laplace transforms $\mathcal{X}(s)$, $\mathcal{Y}(s)$, and $\mathcal{H}(s)$ as equivalent representations of the input $x(t)$, the output $y(t)$, and the impulse response $h(t)$ of a continuous-time LTI system in the frequency domain and vice versa. Note that because the study of a system in the frequency domain is generally simpler compared to its study in the continuous-time domain, it is quite common to study problems in this space and then, using an inverse transform, to express the results in relation to continuous-time. The application of this method for calculating the output of a system will be presented in future exercises.

Problem 5.2.1 Determine the transfer function of the continuous-time systems

$$\frac{d^2y(t)}{dt^2} - \frac{dy(t)}{dt} - 2y(t) = x(t) \quad \text{and} \quad \frac{d^3y(t)}{dt^3} + 6\frac{d^2y(t)}{dt^2} + 11\frac{dy(t)}{dt} + 6y(t) = x(t).$$

(Oppenheim, Schaffer and Nawab, Exercises 9.31α and 9.40).

Solution (a) The differential equation of the system is of the form

$$\alpha_2 y''(t) + \alpha_1 y'(t) + \alpha_0 y(t) = \beta_0 x(t)$$

with $\alpha_2 = 1$, $\alpha_1 = -1$, $\alpha_0 = -2$, and $\beta_0 = 1$. Therefore, we have

$$\mathcal{H}(s) = \frac{\beta_0}{\alpha_2 s^2 + \alpha_1 s + \alpha_0} = \frac{1}{s^2 - s - 2}$$

[2] Recall that the *Laplace transform* is a *generalization* of the Fourier transform, with the latter resulting from the former by setting $\Re\{s\} = \sigma = 0$; this is also why the variable s is known as a *complex frequency*.

(b) In a completely analogous way, the differential equation of the system is of the form

$$\alpha_3 y'''(t) + \alpha_2 y''(t) + \alpha_1 y'(t) + \alpha_0 y(t) = \beta_0 x(t)$$

with parameter values $\alpha_3 = 1, \alpha_2 = 6, \alpha_1 = 11, \alpha_0 = 1$, and $\beta_0 = 1$, from which it follows that

$$\mathcal{H}(s) = \frac{\beta_0}{\alpha_3 s^3 + \alpha_2 s^2 + \alpha_1 s + \alpha_0} = \frac{1}{s^3 + 6s^2 + 11s + 6}$$

Problem 5.2.2 Determine the transfer function and the impulse response of the system

$$\frac{d^2y(t)}{dt^2} + 5\frac{dy(t)}{dt} + 6y(t) = \frac{dx(t)}{dt} + 4x(t)$$

Solution Although the problem can be solved in the way we did in the previous exercise, let us apply directly the methodology that led to the construction of the defining equation of the transfer function $\mathcal{H}(s)$. If we take the Laplace transform of both terms of the differential equation, use the property of differentiation, and assume that the assumptions that were formulated during the presentation of the method are valid, we will get

$$\mathcal{L}\left\{\frac{d^2y(t)}{dt^2}\right\} + 5\mathcal{L}\left\{\frac{dy(t)}{dt}\right\} + 6\mathcal{L}\{y(t)\} = \mathcal{L}\left\{\frac{dx(t)}{dt}\right\} + 4\mathcal{L}\{x(t)\}$$

or, equivalently,

$$s^2\mathcal{Y}(s) + 5s\mathcal{Y}(s) + 6\mathcal{Y}(s) = s\mathcal{X}(s) + 4\mathcal{X}(s)$$

and finally

$$\mathcal{Y}(s)\left[s^2 + 5s + 6\right] = \mathcal{X}(s)\left(s + 4\right)$$

from which it follows that

$$\mathcal{H}(s) = \frac{\mathcal{Y}(s)}{\mathcal{X}(s)} = \frac{s+4}{s^2 + 5s + 6} = \frac{s+4}{(s+2)(s+3)}$$

which is also the transfer function of the system. We observe that the function has *one zero value* at the position $s = -4$ and *two simple poles* at the positions $s_1 = 2$ and $s_2 = 3$.

The impulse response of the system will be determined as the *inverse Laplace transform of the function* $\mathcal{H}(s)$. Using the method of partial fraction expansion in the way we did in the previous sections we find that

$$h(t) = \mathcal{L}^{-1}\{\mathcal{H}(s)\} = \mathcal{L}^{-1}\left\{\frac{s+4}{(s+2)(s+3)}\right\} = 2e^{-2t}u(t) - e^{-3t}u(t)$$

Note In some cases, the problem is formulated in the opposite way, that is, we know the transfer function of the system and we are asked to determine the differential equation that describes it. In this case, we work in the following way[3] (in this example we will start from the transfer function we just determined and construct the differential equation from which it emerges):

[3] See also the corresponding exercises of the third chapter.

$$\text{We have} \quad \mathcal{H}(s) = \frac{\mathcal{Y}(s)}{\mathcal{X}(s)} = \frac{s+4}{s^2+5s+6} \quad \text{and therefore } \mathcal{Y}(s)(s^2+5s+6) = \mathcal{X}(s)(s+4)$$

$$\text{or, equivalently,} \quad s^2\mathcal{Y}(s) + 5s\mathcal{Y}(s) + 6\mathcal{Y}(s) = s\mathcal{X}(s) + 4\mathcal{X}(s)$$

Taking the inverse Laplace transform of the last relation, it turns out that

$$\mathcal{L}^{-1}\{s^2\mathcal{Y}(s)\} + 5\mathcal{L}^{-1}\{s\mathcal{Y}(s)\} + 6\mathcal{L}^{-1}\{\mathcal{Y}(s)\} = \mathcal{L}^{-1}\{s\mathcal{X}(s)\} + 4\mathcal{L}^{-1}\{\mathcal{X}(s)\}$$

and finally

$$\frac{d^2y(t)}{dt^2} + 5\frac{dy(t)}{dt} + 6y(t) = \frac{dx(t)}{dt} + 4x(t)$$

which is also the differential equation that describes the system. We observe that the *order of the differential equation is equal to the value of the highest exponent of the variable s found in the polynomial of the denominator of the transfer function*, and for this reason it is also called the *order of the transfer function*.

The preceding exercises are examples of the process of calculating a *system's transfer function* when the *differential equation* that describes it is known. Alternatively, this calculation can be performed when we know the system's input $x(t)$, as well as the output $y(t)$ produced by it. From basic theory it is known that if a system's impulse response is the signal $h(t)$, then its output is calculated from the convolution equation $y(t) = h(t) * x(t)$; in this case, from the *convolution theorem* of the Laplace transform we will have $\mathcal{Y}(s) = \mathcal{H}(s)\mathcal{X}(s)$. Therefore, to determine the transfer function $\mathcal{H}(s)$, we will calculate the Laplace transforms of the input and output of the system $\mathcal{X}(s)$ and $\mathcal{Y}(s)$, respectively, and then the transfer function from the relation $\mathcal{H}(s) = \mathcal{Y}(s)/\mathcal{X}(s)$. The next exercise demonstrates the application of this method.

Problem 5.2.3 Consider a continuous-time LTI system that, when it receives as input the signal $x(t) = e^{-t}u(t)$, generates as output the signal $y(t) = 10e^{-t}\cos(4t)\,u(t)$. Determine the transfer function of the system and its impulse response.

Solution The Laplace transform of the input signal is calculated as

$$\mathcal{X}(s) = \int_{-\infty}^{+\infty} x(t)e^{-st}dt = \int_0^{\infty} e^{-t}e^{-st}dt = \int_0^{\infty} e^{-(s+1)t}dt = -\frac{e^{-(s+1)t}}{s+1}\bigg|_0^{\infty} = \frac{1}{s+1}$$

On the other hand, in Exercise 4.3.6 (page 313) we proved that

$$\mathcal{L}\left\{e^{-\alpha t}\cos(\omega t)u(t)\right\} = \frac{s+\alpha}{(s+\alpha)^2+\omega^2}$$

Using the foregoing equation for parameter values $\alpha = -1$ and $\omega = 4$, we find that

$$\mathcal{Y}(s) = 10\mathcal{L}\left\{e^{-t}\cos(4t)\,u(t)\right\} = 10\frac{s+\alpha}{(s+\alpha)^2+\omega^2}\bigg|_{\substack{\alpha=-1\\ \omega=4}} = \frac{10\,(s+1)}{(s+1)^2+16}$$

and applying the convolution theorem $y(t) = h(t) * x(t) \Leftrightarrow \mathcal{Y}(s) = \mathcal{H}(s)\mathcal{X}(t)$, we conclude that the transfer function of the system has the form

$$\mathcal{H}(s) = \frac{\mathcal{Y}(s)}{\mathcal{X}(s)} = \frac{10(s+1)^2}{(s+1)^2+16} = 10\,\frac{s^2+2s+1}{s^2+2s+17}$$

The function $\mathcal{H}(s)$ has a *double zero* at position $s = -1$ and *two simple poles* at positions $s_1 = -1 + 4j$ and $s_2 = -1 - 4j$, as can be verified.

Let us now move on to calculating the *impulse response* of the system $h(t)$, which is defined as the *inverse Laplace transform of the transfer function* $\mathcal{H}(s)$. Formulating this function as

$$\mathcal{H}(s) = \frac{\mathcal{Y}(s)}{\mathcal{X}(s)} = \frac{10\,(s+1)^2}{(s+1)^2+16} = 10 - 40\,\frac{4}{(s+1)^2+4^2}$$

it immediately follows that

$$h(t) = \mathcal{L}^{-1}\{\mathcal{H}(s)\} = 10\mathcal{L}^{-1}\{1\} - 40\mathcal{L}^{-1}\left\{\frac{4}{(s+1)^2+4^2}\right\} = 10\delta(t) - 40e^{-4}\sin(4t)u(t)$$

which is the desired result. To derive the last equation, we used the direct-inverse transform relations

$$\mathcal{L}\{\delta(t)\} = 1 \quad \Leftrightarrow \quad \mathcal{L}^{-1}\{1\} = \delta(t)$$

$$\mathcal{L}\left\{e^{-\alpha t}\sin(\omega t)u(t)\right\} = \frac{\omega}{(s+\alpha)^2+\omega^2} \quad \Leftrightarrow \quad \mathcal{L}^{-1}\left\{\frac{\omega}{(s+\alpha)^2+\omega^2}\right\} = e^{-\alpha t}\sin(\omega t)u(t)$$

(for the values $\alpha = -1$ and $\omega = 4$), which can be found in tables of elementary transforms. ■

The property of the convolution of the Laplace transform can also be used for the calculation of the system's output $y(t)$ *when we know the impulse response $h(t)$ and the input signal $x(t)$*. Although this calculation can be done directly using the relation $y(t) = h(t) * x(t)$, it is nonetheless possible that the operation of the *convolution* will be characterized by *high computational complexity*. In this case, we can alternatively apply the following procedure. We calculate the Laplace transforms $\mathcal{H}(s)$ and $\mathcal{X}(s)$ of the signals $h(t)$ and $x(t)$, respectively, and then determine their product $\mathcal{Y}(s)$. Relying on the property of convolution, we know that this product will constitute the Laplace transform of the system's output $y(t)$, which, therefore, can be determined as the inverse transform $\mathcal{L}^{-1}\{\mathcal{Y}(s)\}$. This method of solution is demonstrated in the next exercise.

Problem 5.2.4 Consider a continuous-time LTI system with an impulse response of the form $h(t) = e^{-2t}u(t)$ that receives at its input the signal $x(t) = e^{-t}u(t)$.

- Determine the Laplace transforms $\mathcal{H}(s)$ and $\mathcal{X}(s)$ of the signals $h(t)$ and $x(t)$.
- Using the convolution theorem, find the Laplace transform $\mathcal{Y}(s)$ of the output $y(t)$.
- Determine the output of the system $y(t)$ as the inverse transform of the function $\mathcal{Y}(s)$.
- Determine the output of the system from the convolution equation $y(t) = h(t) * x(t)$, thereby verifying the result of the previous question.

Solution

(a) From the defining equation of the Laplace transform we will have

$$\mathcal{H}(s) = \int_{-\infty}^{+\infty} h(t)\,e^{-st}\,dt = \int_0^{\infty} e^{-(s+2)t}u(t)\,dt = \left[\frac{e^{-(s+2)t}}{-(s+2)}\right]_0^{\infty} = \frac{1}{s+2} \quad \text{and}$$

$$\mathcal{X}(s) = \int_{-\infty}^{+\infty} x(t)\,e^{-st}\,dt = \int_0^{\infty} e^{-(s+1)t}u(t)\,dt = \left[\frac{e^{-(s+1)t}}{-(s+1)}\right]_0^{\infty} = \frac{1}{s+1}$$

with regions of convergence $\sigma \equiv \Re(s) > -2$ and $\sigma \equiv \Re(s) > -1$, respectively.

(b) The output of the system is calculated from the *convolution* between the impulse response of the system and the input signal or, in mathematical notation, $y(t) = h(t) * x(t)$. Therefore, applying the *convolution theorem* we have

$$\mathcal{Y}(s) = \mathcal{H}(s)\,\mathcal{X}(s) = \frac{1}{(s+1)(s+2)}$$

with the region of convergence being the intersection of the regions of convergence of the transforms $\mathcal{H}(s)$ and $\mathcal{X}(s)$, which is described by the equation $\sigma \equiv \Re(s) > -1$.

(c) To calculate the system's output $y(t)$, we need to determine the *inverse Laplace transform* of the function $\mathcal{Y}(s)$, which has two simple poles at positions $s_1 = -1$ and $s_2 = -2$. Therefore, applying the method of complex integration, we obtain

$$y(t) = \int_{\sigma-j\infty}^{\sigma+j\infty} \mathcal{Y}(s)\, e^{st}\, ds = \sum_{k=1}^{2} \text{Res}\left[\mathcal{Y}(s)\, e^{st}\right]_{s=s_k} \qquad \text{where}$$

$$\text{Res}\left[\mathcal{Y}(s)\, e^{st}\right]_{s=-1} = \lim_{s\to-1}\left[(s+1)\frac{e^{st}}{(s+1)(s+2)}\right] = \lim_{s\to-1}\left[\frac{e^{st}}{s+2}\right] = e^{-t}$$

$$\text{Res}\left[\mathcal{Y}(s)\, e^{st}\right]_{s=-2} = \lim_{s\to-2}\left[(s+2)\frac{e^{st}}{(s+1)(s+2)}\right] = \lim_{s\to-2}\left[\frac{e^{st}}{s+1}\right] = -e^{-2t}$$

and consequently, the output signal has the form $y(t) = (e^{-t} - e^{-2t})\, u(t)$.

(d) Calculating the system's output from the convolution equation, we will have

$$y(t) = h(t) * x(t) = \int_{-\infty}^{+\infty} h(t-\tau)\, x(\tau)\, d\tau = \int_{-\infty}^{+\infty} e^{-2(t-\tau)} u(t-\tau)\, e^{-t} u(\tau) d\tau =$$

$$= \int_0^t e^{-2t} e^{2\tau} e^{-\tau} d\tau = e^{-2t}\int_0^t e^{\tau} d\tau = e^{-2t}\left[e^{\tau}\right]_0^t = e^{-2t}(e^t - 1)$$

and finally $y(t) = (e^{-t} - e^{-2t})\, u(t)$, which is the same result as we concluded previously.

Problem 5.2.5 Consider a continuous-time LTI system whose response to the continuous input signal $x(t) = u(t)$ takes the form $y(t) = 2e^{-3t}u(t)$.

- Determine the impulse response of the system $h(t)$.
- Determine the output of the system when the continuous-time signal $x(t) = e^{-t}u(t)$ is used at its input.

Solution (a) Referring to tables of elementary transforms, we find that

$$\mathcal{X}(s) \equiv \mathcal{L}\{x(t)\} = \mathcal{L}\{u(t)\} = \frac{1}{s} \qquad \text{and} \qquad \mathcal{Y}(s) \equiv \mathcal{L}\{y(t)\} = 2\mathcal{L}\{e^{-3t}\} = \frac{2}{s+3}$$

and therefore the *transfer function* of the system will be calculated as

$$\mathcal{H}(s) = \frac{\mathcal{Y}(s)}{\mathcal{X}(s)} = \frac{2s}{s+3} = \frac{2s+6-6}{s+3} = \frac{2\,(s+3)-6}{s+3} = 2 - \frac{6}{s+3}$$

Taking, therefore, the *inverse Laplace transform* of the foregoing equation, we obtain

$$h(t) \equiv \mathcal{L}^{-1}\{\mathcal{H}(s)\} = 2\mathcal{L}^{-1}\{1\} - 6\mathcal{L}^{-1}\left\{\frac{1}{s+3}\right\} = 2\delta(t) - 6e^{-3t}u(t)$$

which is the required *impulse response* of the system.

(b) The output of the system will be calculated *from the convolution between the impulse response of the system and the input signal*. Therefore, it will hold that

$$y(t) = \int_{-\infty}^{+\infty} x(\tau)\, h(t-\tau)\, d\tau = \int_{-\infty}^{+\infty} e^{-\tau} u(\tau) \left[2\delta(t-\tau) - 6e^{-3(t-\tau)} u(t-\tau) \right] d\tau =$$

$$= 2\int_{-\infty}^{+\infty} \delta(t-\tau)\, e^{-\tau} d\tau - 6e^{-3t} \int_{-\infty}^{+\infty} e^{2\tau}\, u(\tau)\, u(t-\tau)\, d\tau$$

Using the properties of the function $\delta(t)$, the first integral is calculated as

$$\int_{-\infty}^{+\infty} \delta(t-\tau)\, e^{-\tau} d\tau = \int_{-\infty}^{+\infty} \delta(\tau - t)\, e^{-\tau} d\tau = e^{-t}$$

On the other hand, the second integral will take the form

$$\int_{-\infty}^{+\infty} e^{2\tau} u(\tau)\, u(t-\tau)\, d\tau = \int_{0}^{t} e^{2\tau} d\tau = \left[\frac{e^{2\tau}}{2} \right]_0^{\tau} = \frac{1}{2}(e^{2t} - 1)$$

Substituting the foregoing expressions into the last equation, the desired output is calculated as

$$y(t) = 2e^{-t} - 6e^{-3t} \left[\frac{1}{2}(e^{2t} - 1) \right] = (-e^{-t} + 3e^{-3t})\, u(t)$$

Problem 5.2.6 Let us consider a causal continuous-time LTI system with impulse response $h(t)$ that satisfies the following properties:

- When the input to the system is $x(t) = e^{2t}$, the output has the form $y(t) = (1/6)e^{2t}$.
- The impulse response of the system satisfies the differential equation (where β is an unknown constant)

$$\frac{dh(t)}{dt} + 2h(t) = e^{-4t} u(t) + \beta u(t)$$

Based on the foregoing information, determine the transfer function of the system (Oppenheim, Willsky and Nawab, Exercise for solution 9.3.2).

Solution Taking the Laplace transform of the two terms of the differential equation we get

$$\mathcal{L}\left\{ \frac{dh(t)}{dt} \right\} + 2\mathcal{L}\{h(t)\} = \mathcal{L}\{e^{-4t} u(t)\} + \beta \mathcal{L}\{u(t)\}$$

However, $\mathcal{L}\{h(t)\} = \mathcal{H}(s)$, $\mathcal{L}\{h'(t)\} = s\mathcal{H}(s)$ (as derived from the property of *differentiation*), while referring to tables of elementary transforms we find that

$$\mathcal{L}\{e^{-4t} u(t)\} = \frac{1}{s+4} \quad \text{and} \quad \mathcal{L}\{u(t)\} = \frac{1}{s}$$

Therefore, it will hold that

$$s\mathcal{H}(s) + \mathcal{H}(s) = \frac{1}{s+4} + \frac{\beta}{s} \quad \text{and consequently} \quad \mathcal{H}(s) = \frac{s + \beta(s+4)}{s\,(s+2)\,(s+4)}$$

as can be verified. To determine the value of the unknown constant β, we use the property of the system according to which, when the signal $x(t) = e^{2t}$ is transmitted to its input, the output signal produced by it is $y(t) = (1/6)e^{2t}$. But this signal is just an exponential signal of the form $x(t) = e^{st}$ for the value $s = 2$, which, as is well known, constitutes an *eigenfunction* of the system with, as a corresponding *eigenvalue*, the function $\mathcal{H}(s)$. Therefore, it will hold that $y(s) = \mathcal{H}(2)e^{2t} = (1/6)e^{2t}$, from which it immediately follows that $\mathcal{H}(2) = 1/6$. Replacing in the equation we determined for the function $\mathcal{H}(s)$ the value $s = 2$ and using the foregoing result we have

$$\mathcal{H}(2) = \frac{2+6\beta}{2(2+4)(2+2)} = \frac{2+6\beta}{48} = \frac{1}{6}$$

from which it follows that $\beta = 1$. Therefore, it will hold that

$$\mathcal{H}(s) = \frac{2s+4}{s\,(s+4)\,(s+2)} = \frac{2(s+2)}{s\,(s+4)\,(s+2)} = \frac{2}{s\,(s+4)}$$

Problem 5.2.7 Determine the transfer function of an RLC circuit in which the resistance R, the inductance L, and the capacitance C are connected in series.

Solution Let $\upsilon(t)$ be the *voltage* applied to the circuit and $i(t)$ be the *current intensity that flows through it*. In this case, we can treat the voltage $\upsilon(t)$ as the input signal of the system $x(t)$ and the current $i(t)$ as the output of the system $y(t)$. Applying *Kirchhoff's law* to the circuit, we get

$$\upsilon(t) = Ri(t) + L\frac{di(t)}{dt} + \frac{1}{C}\int_{-\infty}^{t} i(t)\,dt$$

If we now take the Laplace transform of the two terms of the preceding equation and use the properties of differentiation and integration

$$\mathcal{L}\left\{\frac{dx(t)}{dt}\right\} = s\mathcal{L}\{x(t)\} = s\mathcal{X}(s) \qquad \text{and} \qquad \mathcal{L}\left\{\int_{-\infty}^{t} x(t)\,dt\right\} = \frac{1}{s}\mathcal{L}\{x(t)\} = \frac{\mathcal{X}(s)}{s}$$

then the preceding relationship will take the form

$$\mathcal{L}\{\upsilon(t)\} = R\mathcal{L}\{i(t)\} + L\mathcal{L}\left\{\frac{di(t)}{dt}\right\} + \frac{1}{C}\mathcal{L}\left\{\int_{-\infty}^{t} i(t)\,dt\right\}$$

or, equivalently,

$$\mathcal{V}(s) = R\mathcal{I}(s) + Ls\mathcal{I}(s) + \frac{1}{Cs}\mathcal{I}(s) = \left(R + Ls + \frac{1}{Cs}\right)\mathcal{I}(s) = \frac{LCs^2 + RCs + 1}{Cs}\mathcal{I}(s)$$

From the last equation we see that the transfer function of the system will take the form

$$\mathcal{H}(s) = \frac{\mathcal{I}(s)}{\mathcal{V}(s)} = \frac{Cs}{LCs^2 + RCs + 1}$$

which is the desired result. Other examples of applying the Laplace transform in circuit study processes can be found in the previous chapter (Exercises 4.8.6, on page 365, 4.8.7 on page 366, and 4.8.8 on page 368).

5.3 Poles and Zeros of the Transfer Function

According to the discussion in the previous section, the transfer function of a continuous-time LTI system is a rational function of the form

$$\mathcal{H}(s) = \frac{\beta_0 + \beta_1 s + \beta_2 s^2 + \cdots + \beta_{M-1} s^{M-1} + \beta_M s^M}{\alpha_0 + \alpha_1 s + \alpha_2 s^2 + \cdots + \alpha_{N-1} s^{N-1} + \alpha_N s^N} \tag{5.8}$$

which, as can be seen, can also be written

$$\mathcal{H}(s) = \frac{\beta_M}{\alpha_N} \left\{ \frac{s^M + \frac{\beta_{M-1}}{\beta_M} s^{M-1} + \cdots + \frac{\beta_1}{\beta_M} s + \frac{\beta_0}{\beta_M}}{s^N + \frac{\alpha_{N-1}}{\alpha_N} s^{N-1} + \cdots + \frac{\alpha_1}{\alpha_N} s + \frac{\alpha_0}{\alpha_N}} \right\} = K \frac{N(s)}{D(s)} \tag{5.9}$$

In the foregoing equation, the parameter K is defined as $K = \beta_M/\alpha_N$, while the polynomials $N(s)$ and $D(s)$ are defined as

$$N(s) = s^M + \frac{\beta_{M-1}}{\beta_M} s^{M-1} + \cdots + \frac{\beta_1}{\beta_M} s + \frac{\beta_0}{\beta_M} \quad \text{and} \quad D(s) = s^N + \frac{\alpha_{N-1}}{\alpha_N} s^{N-1} + \cdots + \frac{\alpha_1}{\alpha_N} s + \frac{\alpha_0}{\alpha_N}$$

If we factorize the polynomials $N(s)$ and $D(s)$ by writing them in the form

$$N(s) = \prod_{i=1}^{M} (s - z_i) = (s - z_1)(s - z_2)(s - z_3)\ldots(s - z_M) \text{ and } D(s) = \prod_{j=1}^{N} (s - p_j) = (s - p_1)(s - p_2)(s - p_3)\ldots(s - p_N)$$

the transfer function of the system will be written[4,5]

$$\mathcal{H}(s) = \frac{\beta_M}{\alpha_N} \left\{ \frac{(s - z_1)(s - z_2)(s - z_3)\ldots(s - z_M)}{(s - p_1)(s - p_2)(s - p_3)\ldots(s - p_N)} \right\} \tag{5.10}$$

Having formulated the transfer function of the system in this form, we can define the *zeros of the function* as the roots $s = z_i$ ($i = 1, 2, \ldots, M$) of the equation $N(s) = 0$ and the *poles of the function* as the roots $s = p_j$ ($j = 1, 2, \ldots, N$) of the equation $D(s) = 0$. The function $\mathcal{H}(s)$ *becomes zero at the positions of the zeros, while it becomes infinite at the positions of the poles*, so that we can write

[4] In this description, we assume for simplicity that the *roots of the numerator and denominator polynomials are all distinct from each other* (i.e., they have a multiplicity value equal to one) and that *there is no zero whose value coincides with the value of a pole*.

[5] Having expressed the transfer function in this form, we can expand it into partial fractions as

$$\mathcal{H}(s) = K \times \left\{ N(s) \div \prod_{i=1}^{N} (s - p_n) \right\} = \sum_{i=1}^{n} \frac{A_i}{s - p_i} = \frac{A_1}{s - p_1} + \frac{A_2}{s - p_2} + \cdots + \frac{A_N}{s - p_N}$$

and from there to calculate the *impulse response* of the system as the inverse Laplace transform

$$h(t) = \mathcal{L}^{-1}\mathcal{H}(s) = A_1 \mathcal{L}^{-1}\left\{\frac{1}{s - p_1}\right\} + A_2 \mathcal{L}^{-1}\left\{\frac{1}{s - p_2}\right\} + \cdots + A_N \mathcal{L}^{-1}\left\{\frac{1}{s - p_N}\right\} = A_1 e^{p_1 t} + A_2 e^{p_2 t} + \cdots + A_N e^{p_N t} = \sum_{i=1}^{n} A_i e^{p_i t}$$

$$\lim_{s \to z_i} \mathcal{H}(s) = 0 \quad (i = 1, 2, \ldots, M) \qquad \text{and} \qquad \lim_{s \to p_j} \mathcal{H}(s) = \infty \quad (j = 1, 2, \ldots, N)$$

Taking into account that the variable s is a *complex quantity*, while the coefficients of the polynomials $N(s)$ and $D(s)$ are *real numbers*, we can conclude that the poles and zeros of the transfer function will be *either real or complex numbers* which, indeed, will appear in pairs of complex conjugates, since the appearance of a complex number but not its conjugate would lead to complex coefficients of the polynomials $N(s)$ and $D(s)$. Having determined the poles and zeros of the function $\mathcal{H}(s)$, we can plot them on the complex plane using the symbol $\times$ for the poles and the symbol $\circ$ for the zero values. The diagram that emerges in this way is known as the *pole-zero plot*, and its construction is considered very important, as it provides a diagrammatic way of representing the characteristics of the system in terms of its response to the respective input signal. Examples of such diagrams have been presented in the chapter of the *Laplace transform* (see Exercises 4.4.1 on page 317 and 4.4.2 on page 318), while they will also be given in the corresponding chapter for the *Z transform* so it is not deemed necessary to present them again at this point.

The knowledge of the poles of the transfer function is very important, since they appear in the equation of the zero-input response of the system and therefore determine the *temporal evolution of the components included in it*. Indeed, if we set the input $x(t)$ and all its derivatives to the zero value, in the differential equation that describes the system, it will take the form

$$\alpha_N \frac{d^N y(t)}{dt^N} + \alpha_{N-1} \frac{d^{N-1} y(t)}{dt^{N-1}} + \cdots + \alpha_2 \frac{d^2 y(t)}{dt^2} + \alpha_1 \frac{dy(t)}{dt} + \alpha_0 y(t) = 0 \tag{5.11}$$

Considering exponential solutions $y(t) = e^{\lambda t}$, we end up with the known *characteristic polynomial*

$$\alpha_N \lambda^N + \alpha_{N-1} \lambda^{N-1} + \cdots + \alpha_1 \lambda + \alpha_0 = 0 \tag{5.12}$$

with roots $\lambda_1, \lambda_2, \ldots \lambda_N$, which for simplicity are considered *distinct from each other*. Therefore, the foregoing equation has as solutions the functions $y_i(t) = e^{\lambda_i t}$ $(i = 1, 2, \ldots, N)$, while its general solution —which also expresses the *zero-input response*—will be

$$y(t) = \sum_{i=1}^{N} C_i y_i(t) = \sum_{i=1}^{N} C_i e^{\lambda_i t} \tag{5.13}$$

with the constants C_i calculated from the initial conditions. If we now compare the characteristic polynomial of the foregoing homogeneous differential equation with the equation

$$\alpha_0 + \alpha_1 s + \alpha_2 s^2 + \cdots + \alpha_{N-1} s^{N-1} + \alpha_N s^N = 0$$

we will find that the two equations are exactly the same, and obviously they will have the same roots.[6] Therefore, it holds that $\lambda_i = p_i$ $(i = 1, 2, \ldots, N)$, namely, the *poles of the system coincide with the roots of the characteristic polynomial*, so that we can write

[6] In reality, the polynomial of the denominator of the transfer function coincides with the characteristic polynomial of the system *only when the polynomials of the numerator and the denominator of the function $\mathcal{H}(s)$ do not have common roots* so that the cancellation of these common terms can take place. This situation is presented in a subsequent section.

$$y(t) = \sum_{i=1}^{N} C_i y_i(t) = \sum_{i=1}^{N} C_i e^{\lambda_i t} = \sum_{i=1}^{N} C_i e^{p_i t} \tag{5.14}$$

Finally, taking into account that the *roots of the characteristic polynomial coincide with the eigenvalues of the matrix* $\boldsymbol{A}$ *of the system's dynamic equations in the state space*, it is not difficult to understand that these three sets, that is, the poles of the transfer function, the roots of the characteristic polynomial, and the eigenvalues of the system matrix $\boldsymbol{A}$, express exactly the same thing.

Let us now see how the values of the poles of the transfer function affect the form of the *zero-input response* (or *natural response*) of the system. We distinguish the following cases:

- *Pole with zero value*: if one of the system's poles p_k has a value equal to zero, that is, it coincides with the origin O of the complex plane, then the component of the system's output corresponding to this will be $y_k(t) = C_k$. Therefore, we will have

$$y(t) = C_k + \sum_{\substack{i=1 \\ i \neq k}}^{N} C_i e^{p_i t}$$

  and so the system's output will have *one component of constant width* with value C_k.
- *Real pole with negative value*: If the transfer function has a real pole $p_k = -\xi$, then the corresponding component of the system's output, $y_k(t) = C_k e^{-\xi t}$, will *decrease* exponentially in relation to time *with the rate of decrease depending on the position of the pole on the real axis*. If the pole is far from the origin of the complex plane—that is, the value of ξ is quite large—then the damping of this component will be rapid, whereas, conversely, for a pole near the origin (which corresponds to a small value of ξ), the rate of damping will be relatively small.
- *Real pole with positive value*: If the transfer function has a real pole $p_k = \xi$, then the corresponding component of the system's output, which will be $y_k(t) = C_k e^{\xi t}$, will *increase* exponentially in relation *to time with the rate of increase also depending again on the position of the pole on the real axis*. It is obvious that such a system will be *unstable*, since obviously it will hold that $\lim y(t) = \infty$ in the limit $t \to \infty$.
- *Complex pole*: If the transfer function has a complex pole of the form $p_k = \sigma + j\omega$, then there must necessarily also be the conjugate complex pole $p_{k'} = \sigma - j\omega$, with the contribution of these two poles to the output of the system being calculated as

$$\begin{aligned} y_{k,k'}(t) &= C_k e^{p_k t} + C_{k'} e^{p_{k'} t} = C_k e^{(\sigma + j\omega)t} + C_{k'} e^{(\sigma - j\omega)t} = C_k e^{\sigma t} e^{j\omega t} \\ &+ C_{k'} e^{\sigma t} e^{-j\omega t} = C_k e^{\sigma t}[\cos(\omega t) + j\sin(\omega t)] + C_{k'} e^{\sigma t}[\cos(\omega t) - j\sin(\omega t)] \\ &= e^{\sigma t}\cos(\omega t)(C_k + C_{k'}) + e^{\sigma t}\sin(\omega t)[j(C_k - C_{k'})] \end{aligned}$$

  where for the derivation of the last relation, the known formulas of Euler were used:

$$e^{j\omega t} = \cos(\omega t) + j\sin(\omega t) \qquad \text{and} \qquad e^{-j\omega t} = \cos(\omega t) - j\sin(\omega t)$$

  Defining the new coefficients $A = C_k + C_{k'}$ and $B = j(C_k - C_{k'})$, the foregoing relation is written

$$y_{k,k'}(t) = Ae^{\sigma t}\cos(\omega t) + Be^{\sigma t}\sin(\omega t) = \sqrt{A^2 + B^2}\left(\frac{A}{\sqrt{A^2 + B^2}}\cos(\omega t) + \frac{B}{\sqrt{A^2 + B^2}}\sin(\omega t)\right)e^{\sigma t} =$$

$$= Ce^{\sigma t}\Big[\sin\varphi\cos(\omega t) + \cos\varphi\sin(\omega t)\Big] = Ce^{\sigma t}\sin(\omega t + \varphi)$$

where we have defined the quantities

$$C = \sqrt{A^2 + B^2}, \qquad \sin\varphi = \frac{A}{\sqrt{A^2 + B^2}}, \qquad \cos\varphi = \frac{B}{\sqrt{A^2 + B^2}}$$

and we used the known trigonometric identity $\sin(\alpha + \beta) = \sin\alpha\cos\beta + \cos\alpha\sin\beta$. We are led, therefore, to the conclusion that the *contribution of a pair of conjugate complex poles to the zero-input response is a sinusoidal oscillation of frequency* ω, *phase* φ, *and variable amplitude* $Ce^{\sigma t}$, *which in the limit* $t \to \infty$ *tends to zero if* $\sigma \equiv \Re\{s\} < 0$ *and to infinity if* $\sigma \equiv \Re\{s\} > 0$. It is obvious that in the first case the system is *stable*, while in the second it is *unstable*. The constants C and φ that appear in the foregoing relations can be calculated using the initial conditions.

- *Imaginary pole*: If the transfer function has a purely imaginary pole of the form $p_k = e^{j\omega}$, then the conjugate with respect to this pole $p_{k'} = e^{-j\omega}$ will necessarily appear. This case is a partial case of the previous one for the value $\sigma = 0$, which therefore leads to a contribution of the form

$$y_{k,k'} = C\sin(\omega t + \varphi)$$

We observe, therefore, that the *contribution of a pair of purely imaginary conjugate poles to the zero-input response of the system is a sinusoidal oscillation of constant amplitude C and phase* φ, whose values are calculated again using the initial conditions.

Problem 5.3.1 Determine the poles and the zero values of the transfer function

$$\mathcal{H}(s) = \frac{s^2 - 7s + 12}{s^4 + 6s^3 + 16s^2 + 26s + 5}$$

and then characterize the system in terms of the zero-input response.

Solution Proceeding with successive factorizations of the polynomials of the numerator and the denominator, the transfer function is formulated in the form

$$\mathcal{H}(s) = \frac{(s-4)(s-3)}{(s^2 + 4s + 3)(s^2 + 2s + 5)} = \frac{(s-4)(s-3)}{(s+1)(s+3)(s+1+2j)(s+1-2j)}$$

from which it follows that the function has *two zeros* at the positions $z_1 = 4$ and $z_2 = 3$ and *four poles* at the positions $p_1 = -1$, $p_2 = -3$, $p_3 = -1 + 2j$, and $p_4 = -1 - 2j$. According to the theory presented earlier, the poles p_1 and p_2 will contribute to the system's output the components $y_1(t) = C_1e^{-t}$ and $y_2(t) = C_2e^{-3t}$, while the conjugate complex poles p_3 and p_4 will give the component $y_3(t) = C_3e^{-t}\sin(2t + \varphi)$. Therefore, the zero-input response of the system will be

$$y(t) = C_1e^{-t} + C_2e^{-3t} + C_3e^{-t}\sin(2t + \varphi)$$

with the constants C_1, C_2, C_3, and φ being calculated from the initial conditions. We observe that the system described by the foregoing transfer function is *stable*, since all three terms of the zero-input response tend to zero in the limit $t \to \infty$. More specifically, the terms e^{-t} and e^{-3t} tend exponentially to zero, with the term e^{-3t} obviously undergoing a faster damping compared to the term e^{-t}, while the third term expresses a sinusoidal curve whose amplitude decreases exponentially over time as e^{-t}.

Taking into account that the variable $s = \sigma + j\omega$ appearing in the transfer function has the concept of *complex frequency*, since it is a generalization of the real frequency ω in the complex plane, we can consider that the *poles and zeros of the system's transfer function express such frequencies*. In a more detailed description, we can consider the frequency corresponding to a pole p_k as *a frequency generated by the system*, in the sense that a component of the form $Ae^{p_k t}$ will always appear at the system's output for $t \geq 0$ and for some value of the coefficient C_k, *even if this frequency does not appear in the system's input signal*. On the other hand, a zero value z_k represents a frequency that is "absorbed" by the system, in the sense that this frequency will never appear at the system's output, *even if it is included in the input signal*. This observation is in full agreement with the conclusion we drew in the description of systems in state space (where we stated that the eigenvalues λ_i $(i = 1, 2, \ldots, N)$ of the system matrix $\boldsymbol{A}$ constitute the *natural frequencies* of the system, with the exponential functions $e^{\lambda_i t}$ constituting the corresponding *eigenfunctions*), since we proved that the poles of the transfer function p_i coincide with the eigenvalues λ_i and that the zero-input response is a linear combination of the corresponding eigenfunctions.

5.3.1 Poles and Zeros of Higher Order

Although in the foregoing description we considered for reasons of simplicity only *distinct poles and zeros*, the multiplicity of which was equal to one, the appearance of higher-order poles and zeros is not prohibited. For example, the transfer function

$$\mathcal{H}(s) = \frac{(s-1)(s+2)^2}{s\,(s+3)\,(s+5)^4}$$

is characterized by the existence of *two zeros* at the positions $z_1 = 1$ and $z_2 = -2$ with multiplicity values $n_1 = 1$ and $n_2 = 2$, respectively, as well as by the existence of *three poles* at the positions $p_1 = 0$, $p_2 = -3$, and $p_3 = -5$ with corresponding multiplicity values $n_1 = n_2 = 1$ and $n_3 = 4$.

Limiting ourselves only to the poles of the transfer function, which are those that mainly characterize the system in terms of its stability and the rate of change of its response, let us consider a function $\mathcal{H}(s)$ that has N distinct poles $p_1, p_2, \ldots, p_N$ with multiplicity values $m_1, m_2, \ldots, m_N$. In this case, the total number of the system's poles that also express the order of the transfer function will obviously be equal to

$$P = \sum_{k=1}^{N} m_k$$

Assuming that the sets of poles and zeros *do not have common elements* (in a subsequent section we will see the consequences to which such a situation leads), we can express the rational transfer function using an equation of the form

$$\mathcal{H}(s) = N(s) \Bigg/ \left\{ \prod_{k=1}^{N} (s - p_k)^{m_k} \right\} \tag{5.15}$$

where $N(s)$ is the numerator polynomial, whose degree should generally be less than the degree of the denominator polynomial.[7] If we proceed to the expansion of the transfer function in a series of partial fractions, it is expressed as a sum of the form

[7] This is obviously a consequence of the fact that $N > M$ must hold, where N and M are the maximum orders of the derivatives of the output $y(t)$ and the input $x(t)$ in the differential equation of the system.

$$\mathcal{H}(s) = \sum_{k=1}^{N}\sum_{m=1}^{m_k} \frac{c_{km}}{(s-p_k)^m} \qquad \text{where} \qquad c_{km} = \frac{1}{(m_k-m)!}\lim_{s\to p_k}\frac{d^{(m_k-m)}}{ds^{(m_k-m)}}\left\{N(s)\Big/\prod_{\substack{r=1\\ r\neq k}}^{N}(s-p_k)^{m_r}\right\} \tag{5.16}$$

with the value of the residue corresponding to the pole p_k with multiplicity value m_k. Consequently, *a pole p_k with multiplicity value m_k contributes to the transfer function $\mathcal{H}(s)$ a sum of m_k terms of the form*

$$\frac{c_{k1}}{s-p_k} + \frac{c_{k2}}{(s-p_k)^2} + \frac{c_{k3}}{(s-p_k)^3} + \cdots + \frac{c_{km_k}}{(s-p_k)^{m_k}}$$

with the values of the coefficients $c_{k1}, c_{k2}, \ldots, c_{km_k}$ given by the foregoing general expression.

Having determined the transfer function, we can then calculate the *impulse response* of the system as the *inverse Laplace transform of the function $\mathcal{H}(s)$*. Therefore, it will hold that

$$h(t) = \mathcal{L}^{-1}\{\mathcal{H}(s)\} = \sum_{k=1}^{N}\sum_{m=1}^{m_k}\mathcal{L}^{-1}\left\{\frac{1}{(s-p_k)^m}\right\} = \sum_{k=1}^{N}\sum_{m=1}^{m_k}\frac{c_{km}}{(m-1)!}e^{p_k t}t^{m-1}$$

Therefore, the contribution of the pole p_k with multiplicity value m_k to the impulse response of the system will be the sum of the m_k terms

$$c_{k1}e^{p_k t} + c_{k2}te^{p_k t} + \frac{c_{k3}}{2!}t^2e^{p_k t} + \frac{c_{k4}}{3!}t^3e^{p_k t} + \cdots + \frac{c_{km_k}}{(m_k-1)!}t^{m_k-1}e^{p_k t}$$

5.4 Causality and Stability of Continuous-Time Systems

A continuous-time LTI system is called *causal* when its impulse response is a *causal signal*, that is, it satisfies the condition $h(t) = 0$ for $t < 0$. But this means that the function $h(t)$ is a *right-sided signal*, and according to theory, the region of convergence of the function $\mathcal{H}(s)$ will be a *right half-plane*. In the special case where the transfer function can be expressed in rational form, the property of causality is ensured if the region of convergence is described by the equation $\sigma > \sigma_{\max}$, where $\sigma_{\max}$ is the *maximum real part* of all the poles of the function $\mathcal{H}(s)$.

On the other hand, a continuous-time LTI system is called *anticausal* if its impulse response is an *anticausal signal*, that is, it satisfies the property $h(t) = 0$ for $t > 0$. In other words, the impulse response $h(t)$ is a *left-sided signal*, and according to theory, the region of convergence of the function $\mathcal{H}(s)$ will be a *left half-plane*. In the special case where the transfer function can be expressed in explicit form, the property of anticausality is ensured if the region of convergence is described by the equation $\sigma < \sigma_{\min}$, where $\sigma_{\min}$ is the *minimum real part* of all the poles of the function $\mathcal{H}(s)$.

A second important property that should characterize all realizable continuous-time LTI systems is *stability*. As already mentioned, such a system is characterized as stable *when its response to a bounded input signal is a bounded output signal*, and this type of stability is known as BIBO stability. Considering a system that accepts at its input the bounded causal input signal $x(t)$, which, as is widely known, should satisfy the condition $|x(t)| \le M$ for every time t, the measure of the system's output, as given by the convolution equation, will have the form

$$|y(t)| = |h(t) * x(t)| = \left|\int_0^{\infty} h(\tau)\,x(t-\tau)\,d\tau\right| \le \int_0^{\infty}|h(\tau)|\cdot|x(t-\tau)|d\tau \le M\int_0^{\infty}|h(\tau)|d\tau$$

But the last relationship is satisfied, only when the impulse response $h(t)$ is *absolutely integrable*, that is, it satisfies the condition

$$\int_0^{\infty} |h(\tau)|d\tau < \infty \tag{5.17}$$

which is to say that the condition of *absolute integrability* is a prerequisite for characterizing a continuous-time LTI system as a stable system.

The foregoing condition is one of the three *Dirichlet conditions* that the system's impulse response must meet in order for the Fourier transform to exist, and essentially its power is the one that determines the existence or nonexistence of the aforementioned transform, since the other two conditions are met by almost all functions that have practical interest. Taking into account that the Fourier and Laplace transforms coincide for the points of the imaginary axis $s = j\omega$ for which the relation $\sigma \equiv \Re(s) = 0$ holds, it is not difficult to find that the property of absolute integrability of the function $h(t)$—which ensures the stability of the system—is satisfied *only if the region of convergence of the transfer function includes the imaginary axis* $s = j\omega$.

In the special case where the function $\mathcal{H}(s)$ is expressed *in rational form* and the system it describes is a causal continuous-time system, the process of characterizing it in terms of its stability is significantly simplified, as it relies exclusively *on determining the positions of the poles of the transfer function on the complex plane*. As already mentioned, the convergence area of such a system is described by an equation of the form $\sigma > \sigma_{max}$, where σ_{max} is the *maximum real part of all the poles of the function*. It is obvious that for a convergence area of this form to contain the imaginary semi-axis, it must be to the right of the straight line $\sigma = \sigma_{max}$, a property that applies in the case where $\sigma_{max} < 0$. Therefore, a causal continuous-time LTI system described by a rational transfer function is stable *when the real parts of all the poles of the transfer function are negative numbers*.

It is proven that the foregoing requirement for the stability of the system according to BIBO constitutes a condition that is *both sufficient and necessary*. To prove the sufficiency of the condition, let us consider the equation of the impulse response for the case of a transfer function that contains N distinct poles $p_1, p_2, \ldots, p_N$ with multiplicity values $m_1, m_2, \ldots, m_N$ (see the previous section), and let us replace it in the equation of the system's output measure $y(t)$. It will then hold that

$$|y(t)| \leq M \int_0^{\infty} |h(t)|dt \equiv M \left\{ \lim_{t\to\infty} \int_0^t \left| \sum_{k=0}^{N} \sum_{m=1}^{m_k} \frac{c_{km}}{(m-1)!} t^{m-1} e^{p_k t} \right| dt \right\}$$

The integral in the last equation is formulated as

$$\int_0^t \left| \sum_{k=0}^{N} \sum_{m=1}^{m_k} \frac{c_{km}}{(m-1)!} t^{m-1} e^{p_k t} \right| dt = \sum_{k=0}^{N} \sum_{m=1}^{m_k} \frac{c_{km}}{(m-1)!} \int_0^t t^{m-1} |e^{p_k t}| dt = \sum_{k=0}^{N} \sum_{m=1}^{m_k} \frac{c_{km}}{(m-1)!} \int_0^t t^{m-1} e^{\Re\{p_k\} t} dt$$

since the pole p_k is a complex variable of the form $p_k = \Re\{p_k\} + j\Im\{p_k\}$ $(k = 1, 2, \ldots, N)$, while, as is well known,

$$|e^{p_k t}| = |e^{(\Re\{p_k\} + j\Im\{p_k\})t}| = |e^{\Re\{p_k\} t}||e^{j\Im\{p_k\} t}| = e^{\Re\{p_k\} t}$$

The value of the foregoing expression at the limit $t \to \infty$ is finite if and only if $\Re\{p_k\} < 0$, and it is proven that it is given by the equation

$$\int_0^{\infty} |h(t)|dt = \sum_{k=0}^{N} \sum_{m=1}^{m_k} \frac{c_{km}}{(m-1)!} \int_0^{\infty} t^{m-1} e^{\Re\{p_k\} t} dt = \sum_{k=0}^{N} \sum_{m=1}^{m_k} \frac{c_{mk}}{(-\Re\{p_k\})^m} < \infty$$

Therefore, the condition of absolute integrability holds only *when all the poles of the transfer function have negative real parts*.

On the other hand, to prove the necessity of the condition, let us assume that even though not all the poles of the system have negative real parts, the system is BIBO stable, and let us try to determine some bounded input signals that refute this claim. It is proven that if the transfer function has *simple or multiple poles on the imaginary axis*, then the response of the system is not bounded when the bounded input $x(t) = \cos(\omega t)$ is transmitted to the system. Also, it has been demonstrated that if the transfer function has *simple or multiple poles with positive real parts*—that is, poles that are located to the right of the imaginary axis—or, alternatively, *multiple poles on the imaginary axis*, then the response of the system becomes infinite when we transmit to it the bounded step function. The proof of some of these observations is presented in the following exercises.

Problem 5.4.1 Prove that if the transfer function $\mathcal{H}(s)$ contains poles that lie on the imaginary axis, the system it describes is BIBO unstable.

Solution Suppose that the transfer function $\mathcal{H}(s)$ contains a purely imaginary pole of the form $p_1 = j\omega$; in this case, the function will necessarily also contain the conjugate to this pole $p_2 = -j\omega$ and therefore will have the form

$$\mathcal{H}(s) = \frac{K(s-z_1)(s-z_2)\dots(s-z_M)}{(s-j\omega)(s+j\omega)(s-p_1)\dots(s-p_{N-2})} = \frac{K(s-z_1)(s-z_2)\dots(s-z_M)}{(s^2+\omega^2)(s-p_1)\dots(s-p_{N-2})}$$

where we assume that the function has a total of N poles. If we transmit the blocked input signal $x(t) = \cos(\omega t)$ to the system with a Laplace transform of the form

$$\mathcal{X}(s) = \mathcal{L}\{x(s)\} = \mathcal{L}\{\cos(\omega t)\} = \frac{s}{s^2+\omega^2}$$

then the Laplace transform of the system output will have the form

$$\mathcal{Y}(s) = \mathcal{H}(s)\mathcal{X}(s) = \frac{Ks(s-z_1)(s-z_2)\dots(s-z_M)}{(s^2+\omega^2)^2(s-p_1)\dots(s-p_{N-2})}$$

We observe that in the denominator of the function $\mathcal{Y}(s)$, the term $(s^2+\omega^2)^2$ appears, which, if we expand the function in question into a sum of partial fractions, will contribute the sum

$$\frac{As+B}{s^2+\omega^2} + \frac{Cs+D}{(s^2+\omega^2)^2}$$

with the values of the coefficients (A, B, C, D) depending on the coefficients of the polynomial of the numerator of the function $\mathcal{Y}(s)$. Therefore, in the equation of the output $y(t)$—which will result from the inverse Laplace transform of the function $\mathcal{Y}(s)$—the inverse transforms of the foregoing two terms will appear, and their calculation follows next.

The inverse transform of the first term is easily calculated as

$$\mathcal{L}^{-1}\left\{\frac{As+B}{s^2+\omega^2}\right\} = A\mathcal{L}^{-1}\left\{\frac{s}{s^2+\omega^2}\right\} + B\mathcal{L}^{-1}\left\{\frac{1}{s^2+\omega^2}\right\} = A\cos(\omega t) + \frac{B}{\omega}\sin(\omega t) = K_1 \sin(\omega t + \varphi)$$

where we have defined the parameters $K_1 = 1/\omega$, $\sin\varphi = A$, and $\cos\varphi = B/\omega$.

On the other hand, to calculate the inverse transform of the second term, we observe that since

$$\frac{d}{ds}\left(\frac{As+B}{s^2+\omega^2}\right) = -A\frac{s^2-\omega^2}{(s^2+\omega^2)^2} - 2B\frac{s}{(s^2+\omega^2)^2}$$

we can formulate the claim that the second term can be written as a sum in the form

$$\frac{Cs+D}{(s^2+\omega^2)^2} = \frac{d}{ds}\left(\frac{Ks+L}{s^2+\omega^2}\right) + M\frac{s^2-\omega^2}{(s^2+\omega^2)^2}$$

with the values of the coefficients that can be positive or negative depending on the form of the transfer function. It will hold, then, that

$$\mathcal{L}^{-1}\left\{\frac{Cs+D}{(s^2+\omega^2)^2}\right\} = \mathcal{L}^{-1}\left\{\frac{d}{ds}\left(\frac{Ks+L}{s^2+\omega^2}\right)\right\} + M\mathcal{L}^{-1}\left\{\frac{s^2-\omega^2}{(s^2+\omega^2)^2}\right\}$$

From the properties of the Laplace transform, we know that if $F(s) = \mathcal{L}\{f(t)\}$, then it will also be the case that

$$\mathcal{L}^{-1}\left\{\frac{dF(s)}{ds}\right\} = -tf(t)$$

Then we will have

$$\mathcal{L}\left\{A\sin(\omega t+\varphi)\right\} = \frac{As+B}{s^2+\omega^2} \quad \Leftrightarrow \quad \mathcal{L}^{-1}\left\{\frac{d}{ds}\left(\frac{Ks+L}{s^2+\omega^2}\right)\right\} = -tK_1\sin(\omega t+\varphi)$$

while from tables of elementary transforms we find that

$$\mathcal{L}^{-1}\left\{\frac{s^2-\omega^2}{(s^2+\omega^2)^2}\right\} = t\cos(\omega t)$$

Therefore, the inverse Laplace transform of the second term will have the form

$$\mathcal{L}^{-1}\left\{\frac{Cs+D}{(s^2+\omega^2)^2}\right\} = -tK_1\sin(\omega t+\varphi) + Mt\cos(\omega t) = K_2 t\sin(\omega t+\varphi)$$

as it can be verified via mathematical operations.

Combining the preceding results, we conclude that the total contribution of the term $1/[(s^2+\omega^2)^2]$ to the output of the system will be the expression

$$K_1\sin(\omega t+\varphi) + K_2 t\sin(\omega t+\varphi) = (K_1+K_2 t)\sin(\omega t+\varphi)$$

We observe that this expression tends to infinity in the limit $t\to\infty$, and therefore the system is unstable according to BIBO, as it creates an unbounded output when the bounded input $x(t) = \cos(\omega t)$ is transmitted to it. We conclude, therefore, that the existence of poles of the transfer function on the imaginary axis leads to the creation of an unstable system.

Problem 5.4.2 Show that if the transfer function $\mathcal{H}(s)$ contains a pole located at the origin of the coordinate system, the system it describes is unstable according to BIBO.

Solution Given that the transfer function contains a pole at the position $s = 0$, it will have the form

$$\mathcal{H}(s) = K\frac{(s-z_1)(s-z_2)\ldots(s-z_M)}{s(s-p_1)(s-p_2)\ldots(s-p_{N-1})}$$

and therefore if we transmit to it the blocked step function $x(t) = u(t)$, whose Laplace transform is known to be equal to $\mathcal{X}(s) = \mathcal{L}\{x(t)\} = \mathcal{L}\{u(t)\} = 1/s$, then the Laplace transform of the system output will have the form

$$\mathcal{Y}(s) = \mathcal{H}(s)\mathcal{X}(s) = K\frac{(s-z_1)(s-z_2)\ldots(s-z_M)}{s^2(s-p_1)(s-p_2)\ldots(s-p_{N-1})}$$

In this case, we observe that the denominator of the function $\mathcal{Y}(s)$ contains the term s^2, which in the calculation of the inverse Laplace transform will contribute to the system output a term of the form Kt (which can be easily verified by referring to tables of elementary transforms). It is obvious that in the limit $t \to \infty$, this term becomes infinite, and therefore the system *is not BIBO stable* since it produces an unbounded output in response to a bounded input signal. ■

From the foregoing examples we conclude that a necessary and sufficient condition for a system to be BIBO stable is to have a transfer function *all of whose poles have negative real parts*.[8]

Problem 5.4.3 Characterize in terms of causality and stability the systems

$$\mathcal{H}_1(s) = \frac{s-1}{(s+2)(s^2+2s+2)}, \quad \Re(s) > -0.5 \quad \text{and} \quad \mathcal{H}_2(s) = \frac{s+2}{(s-1)(s^2-2s+2)}, \quad \Re(s) < 0.5$$

Solution

(a) The transfer function of the system has a simple zero value at the position $s = 1$ and three simple poles at the positions $s_1 = -2$, $s_2 = -1 + j$, and $s_3 = -1 - j$. Taking into account that the region of convergence of the system contains the *imaginary semi-axis* and that all poles are located to the left of this region, it is concluded that the system is *causal and stable*.

(b) The transfer function of the system has a simple zero value at the position $s = -2$ and three simple poles at the positions $s_1 = 1$, $s_2 = 1 + j$, and $s_3 = 1 - j$. Taking into account that the region of convergence contains the *imaginary semi-axis* and all poles are located to the right of this area, it follows that the system is *causal and stable*.

Problem 5.4.4 Let us consider a causal continuous-time LTI system with impulse response $h(t)$ whose input $x(t)$ and output $y(t)$ satisfy a differential equation of the form

$$\frac{d^3y(t)}{dy^3} + (1+\xi)\frac{d^2y(t)}{dt^2} + \xi(1+\xi)\frac{dy(t)}{dt} + \xi^2 y(t) = x(t)$$

Determine the range of values of ξ so that the considered system is stable (Oppenheim, Willsky & Nawab, Exercise for solution 9.16β).

Solution (a) The differential equation of the system has the form

$$\alpha_3\frac{d^3y(t)}{dt^3} + \alpha_2\frac{d^2y(t)}{dt^2} + \alpha_1\frac{dy(t)}{dt} + \alpha_0 y(t) = \beta_0 x(t)$$

from which it follows that $\alpha_3 = 1$, $\alpha_2 = 1 + \xi$, $\alpha_1 = \xi(1 + \xi)$, $\alpha_0 = \xi^2$, and $\beta_0 = 1$. Therefore,

$$\mathcal{H}(s) = \frac{\mathcal{Y}(s)}{\mathcal{X}(s)} = \frac{\beta_0}{\alpha_3 s^3 + \alpha_2 s^2 + \alpha_1 s + \alpha_0} = \frac{1}{s^3 + (1+\xi)\,s^2 + \xi(1+\xi)\,s + \xi^2}$$

The polynomial of the denominator after the factorization process is written

$$s^3 + (1+\xi)\,s^2 + \xi(1+\xi)\,s + \xi^2 = (s+1)\left(s + \frac{\xi}{2} - j\xi\frac{\sqrt{3}}{2}\right)\left(s + \frac{\xi}{2} + j\xi\frac{\sqrt{3}}{2}\right)$$

[8] This stability criterion allows us to define stable/unstable poles and the zero values of a transfer function, in the following way: A zero value $z_k = \alpha_k + j\beta_k$ (for which $\mathcal{H}(z_k) = 0$) is called *stable* if $\Re\{z_k\} = \alpha < 0$ and *unstable* if $\Re\{z_k\} = \alpha > 0$. In a completely analogous way, a pole $p_k = \gamma + j\delta$ (for which $\lim \mathcal{H}(s) = \infty$ at the limit $s \to p_k$) is called *stable* if $\Re\{p_k\} = \gamma < 0$ and *unstable* if $\Re\{p_k\} = \gamma > 0$.

with roots $s_1 = -1$, $s_2 = -(\xi/2) + i\xi(\sqrt{3}/2)$, and $s_2 = -(\xi/2) - i\xi(\sqrt{3}/2)$, which also constitute the *poles* of the system's transfer function. According to the theory presented in the previous pages, in order for the system to be stable, the real parts of all the system's poles must be negative numbers. Therefore, it must be $-(\xi/2) < 0$, from which it follows that $\xi > 0$, which is also the desired result.

Problem 5.4.5 Let us consider a real function $f(t)$ with a Laplace transform $F(s) \equiv \mathcal{L}\{f(t)\}$ for which we know the following information:

- The function $F(s)$ has exactly two poles, one of which is $s = -1 + j$, and has no zeros.
- The function $g(t) = e^{2t} f(t)$ is not absolutely integrable.
- The function $F(s)$ is such that for $s = 0$, it holds that $F(0) = 8$

Based on this information, determine the analytical form of the function $F(s)$ as well as the region of convergence (Oppenheim, Willsky and Nawab, Exercise 9.27).

Solution Since the function $F(s)$ does not have zeros and has exactly two poles, it is obvious that it will be described by a relation of the form

$$F(s) = \frac{A}{(s+\alpha)(s+\beta)}$$

Furthermore, theory says that, as long as the function $f(s)$ is real, the poles of its transform $F(s)$ will appear in pairs of complex conjugate quantities, and therefore, since one pole is $s_1 = -1 + j$, the other pole will be $s_2 = -1 - j$, that is, the transform $F(s)$ will take the form

$$F(s) = \frac{A}{(s+1-j)(s+1+j)} = \frac{A}{s^2+2s+2}$$

As for the value of the constant A, this will result from the condition $F(0) = 8$. Replacing the value $s = 0$ in the foregoing equation, we get

$$F(0) = \frac{A}{2} = 8 \text{from which it follows that} A = 16$$

At this point we have determined the analytical form of the function $F(s)$, and we need to determine its region of convergence. From the position of the poles of this function on the complex plane it is obvious that there are two different choices for the region of convergence: to be described by (a) the equation $\sigma \equiv \Re(s) > -1$ and (b) the equation $\sigma \equiv \Re(s) < -1$. This situation is diagrammatically described in Fig. 5.1.

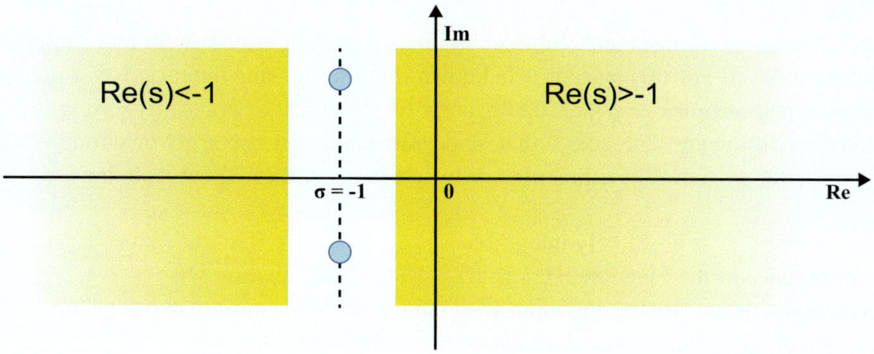

Fig. 5.1 The two choices for the region of convergence of the function $f(t)$ of Exercise 5.4.5

To decide which of the two regions of convergence is the one associated with the function $F(s)$, we will apply the second assumption of the exercise according to which the function $g(t) = e^{2t} f(t)$ is not absolutely integrable. Taking the Laplace transform of the function $g(t)$ and using the property of displacement in the complex plane, we are led, as is well known, to the result

$$g(t) = e^{2t} f(t) \leftrightarrows G(s) = F(s-2)$$

from which it follows that the region of convergence of the transform $G(s)$ arises from the displacement of the region of convergence of the transform $F(s)$ by two positions to the right along the real axis. However, since the function $e^{2t} f(t)$ is not absolutely integrable, it is obvious that the region of convergence of the transform $G(s)$ should not include the imaginary axis. Combining the last two properties, it follows that they are simultaneously satisfied only for the case $\Re(s) > -1$, an equation that also describes the region of convergence of the transform $F(s)$.

Problem 5.4.6 Considering a continuous-time LTI system with impulse response $h(t)$ and explicit transfer function $\mathcal{H}(s)$, examine whether the systems with impulse responses

$$g(t) = \frac{dh(t)}{dt} \quad \text{and} \quad r(t) = \int_{-\infty}^{t} h(\tau)d\tau$$

Solution From basic theory it is known that if a continuous LTI system is *causal and stable*, then all the poles of the (rational) transfer function have a negative real part. Regarding the two given systems, they are characterized in the following way:

(a) From the property of *differentiation* of the Laplace transform

$$g(t) = \frac{dh(t)}{dt} \leftrightarrows \mathcal{G}(s) = s\mathcal{H}(s)$$

it follows that the function $\mathcal{G}(s)$ will have the same poles and the same region of convergence as the function $\mathcal{H}(s)$, that is, the system with impulse response $g(t)$ is also *causal and stable*.

(b) From the property of integration of the Laplace transform

$$r(t) = \int_{-\infty}^{t} h(\tau)d\tau \leftrightarrows \mathcal{R}(s) = \frac{\mathcal{H}(s)}{s}$$

it follows that the function $\mathcal{R}(s)$ will have a pole at $s = 0$, only when $\mathcal{H}(s)$ does not have a zero value at the same position (as in the opposite case we would have mutual cancellation). Generally speaking, we cannot guarantee that the functions $\mathcal{H}(s)$ and $\mathcal{R}(s)$ will always have the same poles and the same region of convergence, that is, the system with impulse response $r(s)$ *is not generally causal and stable*.

Problem 5.4.7 Let us consider a causal and stable continuous-time LTI system with real impulse response $h(t)$ and rational transfer function $H(s)$. For this system it is known that the transfer function has a pole at the position $s = -1 + j$, a zero value at the position $s = 3 + j$, and exactly two zero values at infinity. Using this information, check the validity of each of the following sentences—that is, decide whether it is correct or wrong—unless you judge that the information given for the system is insufficient to draw a safe conclusion about its behavior.

- The function $g(t) = h(t)e^{-3t}$ is absolutely integrable.
- The region of convergence of the function $H(s)$ is described by the equation $\Re(s) > -1$.
- The differential equation that correlates the input with the output of the system contains only real coefficients.
- The transfer function of the system has a value $H(s) = 1$ at the limit $s \to \infty$.

(continued)

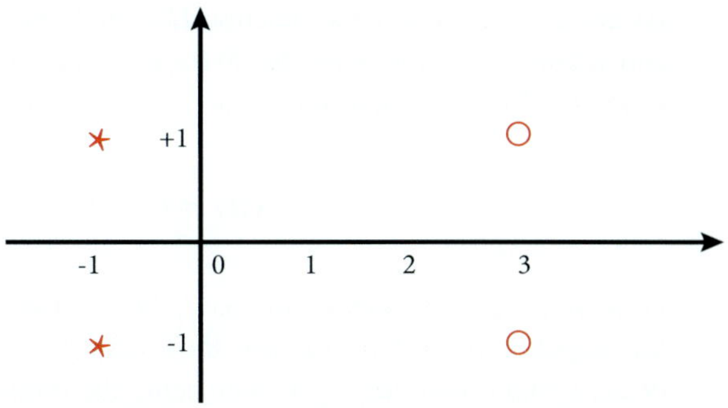

Fig. 5.2 The pole-zero diagram for the transfer function $H(s)$ of the system of Exercise 5.4.7

- The transfer function of the system has at least four poles.
- The system's response to the input signal $x(t) = e^{3t} \sin t$ has the form $y(t) = e^{3t} \cos t$.

Solution From the problem description we can formulate the following conclusions regarding the properties of the system with the transfer function $H(s) = \mathcal{L}\{h(t)\}$.

- Since the impulse response $h(t)$ is a real signal, the poles and zeros of the transfer function $H(s)$ will appear in complex conjugate pairs. Therefore, the pole-zero diagram regarding the pole $s = -1 + j$ and the zero value $s = 3 + j$ will have the form of Fig. 5.2. From this figure, we observe that the existence of the pole $s = -1 + j$ results in the appearance of the pole $s = -1 - j$, while the existence of the zero value $s = 3 + j$ results in the existence of the zero value $s = 3 - j$.
- Since the transfer function $H(s)$ has exactly two zero values at the limit $s \to \infty$, it will have at least two additional finite poles. In the case where the function $H(s)$ has more poles, then for each of them there will be an additional zero value at some position on the complex plane.
- Since the system is causal and stable, all the poles of the function $H(s)$ will belong to the left half-plane—that is, they will have negative real parts—while additionally the region of convergence of the function $H(s)$ will include the imaginary semi-axis $s = j\omega$.

Based on these properties of the considered continuous-time system, the propositions that have been formulated are evaluated in the following way:

(a) From basic theory it is known that for a functio $g(t)$ to be absolutely integrable, the region of convergence of $F(s)$ must contain the imaginary axis. In this case, using the property of shifting on the complex plane

$$g(t) = h(t)e^{-3t} \leftrightarrows G(s) = H(s + 3)$$

it is not difficult to ascertain that the region of convergence of the transform $G(s)$ will result from the region of convergence of the transform $H(s)$ if the latter is shifted by three positions to the left. It is obvious that since the region of convergence of the function $H(s)$ contains the imaginary axis, the same will apply to the region of convergence of the function $G(s)$, that is, the function $g(t) = h(t)e^{-3t}$ is absolutely integrable and proposition (a) is correct.

(b) The proposition cannot be verified or rejected because we do not have sufficient information about the system. Indeed, the validity of the equation $\Re(s) > -1$ presupposes that the pole of the function $H(s)$ with the largest real part is $s = -1 + j\omega$, something that, however, cannot be guaranteed since the function has other additional poles whose position we do not know.

(c) Since the function $H(s)$ is rational, it can be expressed as the ratio of two polynomial functions of s whose coefficients will be real numbers since the impulse response $h(t)$ is a real signal. If we take into account that these coefficients coincide with the coefficients of the differential equation of the system that arises from the inverse Laplace transform of the equation $Y(s) = H(s)X(s)$—which arises as a consequence of the explicit nature of the function $H(s)$—it is not difficult to understand that proposition (γ) is correct.

(d) Since the function $H(s)$ has zero values at infinity, $H(s) = 0$ at the limit $s \to \infty$, and therefore the proposition is wrong.

(e) As already mentioned, the function $H(s)$ has two complex conjugate poles at the positions $s = -1 + j$ and $s = -1 - j$ and at least two finite poles due to the existence of zero values at the limit $s \to \infty$. Therefore, proposition (ε) is correct.

(f) Using the Euler relation $\sin t = (e^{jt} - e^{-jt})/2j$ we can write the input signal $x(t)$ in the form

$$x(t) = e^{3t} \sin t = e^{3t} \frac{e^{jt} - e^{-jt}}{2j} = \frac{1}{2j} e^{(3+j)t} - \frac{1}{2j} e^{(3-j)t}$$

However, as is well known from basic theory, if in a continuous-time LTI system the input signal $x(t) = e^{st}$ is transmitted, the output of the system will have the form $y(t) = H(s)e^{st}$, that is, the exponential signal e^{st} constitutes an eigenfunction of the system, with the eigenvalue being the transfer function $H(s)$. Applying this property to the specific input signal, the output of the system will take the form

$$y(t) = \frac{1}{2j} H(3 + j) e^{(3+j)t} - \frac{1}{2j} H(3 - j) e^{(3-j)t}$$

However, as we proved in a previous proposition, the transfer function of the system $H(s)$ has two zero values at the positions $s = 3 + j$ and $s = 3 - j$. Therefore, $H(3 + j) = H(3 - j) = 0$ and, consequently, $y(t) = 0$. Thus, this statement is not correct.

Problem 5.4.8 Let us consider a causal and stable continuous-time LTI system with impulse response $h(t)$ and rational transfer function $H(s)$ for which the following information is known:

- The value of the function $H(s)$ for $s = 1$ is equal to 0.2.
- When the input is the step function $x(t) = u(t)$, the output of the system is absolutely integrable.
- When the input is the function $g(t) = tu(t)$, the output of the system is not absolutely integrable.
- The signal $r(t) = h''(t) + 2h'(t) + 2h(t)$ has finite duration.
- The function $H(s)$ has exactly one zero value at infinity.

Based on this information, determine the transfer function of the system $H(s)$ as well as its region of convergence (Oppenheim, Willsky and Nawab, Exercise for solution 9.33).

Solution From the application of the convolution theorem of the Laplace transform

$$y(t) = h(t) * x(t) \leftrightarrows Y(s) = H(s)X(s)$$

the following conclusions emerge:

The Laplace transform of the system's output for the input signal $x(t) = u(t)$ will have the form

$$Y(s) \equiv \mathcal{L}\{y(t)\} = H(s)\mathcal{L}\{u(t)\} = \frac{H(s)}{s}$$

Given that it is known that for the input signal $x(t) = u(t)$ the system's output is absolutely integrable, it is obvious that the region of convergence of the function $Y(s)$ should contain the imaginary axis. However, from the mathematical form of the function $Y(s)$ it is not difficult to see that a necessary condition for this to happen is that the system's transfer function $H(s)$ has a zero value at the position $s = 0$, which will cancel the pole located at the same position. To determine whether the zero value at this position is unique or there is more than one such value, we will use the third of the information we know about the system, according to which the system's response to the input $g(t) = tu(t)$ is not absolutely integrable. Using the convolution theorem again for the input $g(t)$ we get

$$Y(s) \equiv \mathcal{L}\{y(t)\} = H(s)\mathcal{L}\{tu(t)\} = \frac{H(s)}{s^2}$$

However, in this case, the system's output is not absolutely integrable, meaning that the function $H(s)$ does not have two zero values at the position $s = 0$, which would cancel the two poles at the same position, leading to the condition of absolute integrability. From the foregoing analysis, therefore, it is concluded that the function $H(s)$ has exactly one zero value at the position $s = 0$.

To determine the form of the explicit function $H(s)$, we will use the equation

$$r(x) = \frac{d^2h(t)}{dt^2} + 2\frac{dh(t)}{dt} + 2h(t)$$

Taking the Laplace transform of the preceding equation, we will have

$$\mathcal{L}\{r(x)\} = \mathcal{L}\left\{\frac{d^2h(t)}{dt^2}\right\} + 2\mathcal{L}\left\{\frac{dh(t)}{dt}\right\} + 2\mathcal{L}\{h(t)\}$$

or, equivalently,

$$R(s) = s^2H(s) + 2sH(s) + 2H(s) = (s^2 + 2s + 2)H(s)$$

and finally

$$H(s) = \frac{R(s)}{s^2 + 2s + 2}$$

where $R(s) \equiv \mathcal{L}\{r(t)\}$, $H(s) \equiv \mathcal{L}\{h(t)\}$, and we used the theorem of the differentiation of the Laplace transform according to which $\mathcal{L}\{h''(t)\} = s^2H(s)$ and $\mathcal{L}\{h'(t)\} = sH(s)$. It is not difficult to see that the function $R(s)$ is not explicit but polynomial: Since the signal $r(t)$ has finite duration, the region of convergence of the transform $R(s)$ will be the entire complex plane according to the theory. In other words, this transform will not have poles in the finite complex plane and therefore will have the general polynomial form

$$R(s) = A\prod_{i=1}^{N}(s - s_i) = A(s - s_1)(s - s_2)\ldots(s - s_N)$$

where A is an unknown constant whose value must be determined, N is the degree of the polynomial $R(s)$, and $s_1, s_2, \ldots, s_N$ are its zero values. Replacing the last relation in the equation of the transfer function of this system will take the form

$$H(s) = \frac{A\prod_{j=1}^{N}(s - s_i)}{s^2 + 2s + 2} = \frac{A(s - s_1)(s - s_2)\ldots(s - s_N)}{s^2 + 2s + 2}$$

To determine the degree N of the numerator polynomial, we will use the last piece of information we know about the system, according to which the function $H(s)$ has exactly one zero value at infinity. However, for this to happen, the degree of the denominator polynomial must be one unit greater than the degree of the numerator polynomial, and since the denominator polynomial is quadratic, it immediately follows that $N = 1$. Therefore, the function $R(s)$ will have the simple form $R(s) = As$, and the transfer function will finally take the form

$$H(s) = \frac{As}{s^2 + 2s + 2} = \frac{s}{s^2 + 2s + 2}$$

with the last relation resulting from the first property ($H(1) = 0.2$), the application of which immediately leads to the value $A = 1$.

It is not difficult to see that the function $H(s)$ has two simple poles at the positions $s_1 = -1 + i$ and $s_2 = -1 - i$, with real parts $\sigma_1 = \sigma_2 = -1$. Taking into account that the considered system is stable and causal, it immediately follows that the region of convergence of the function $H(s)$ will be described by the equation $\Re(s) > -1$, a finding that completes the solution of the exercise.

Problem 5.4.9 Considering a continuous LTI system that accepts at its input the continuous-time signal $x(t) = 2e^{-t}u(t)$ and produces as its output the continuous-time signal $y(t) = e^{-2t}\sin(2t)u(t) - te^{-t}u(t)$, determine the transfer function and its impulse response and characterize it in terms of its causality and stability.

Solution Referring to tables of elementary transforms, we find that

$$\mathcal{X}(s) = \frac{2}{s+1} \quad \Re(s) > -1 \qquad \text{and} \qquad \mathcal{Y}(s) = \frac{s^2-6}{(s^2+4s+8)(s+1)^2} \quad \Re(s) > -1$$

and therefore the transfer function of the system will have the form

$$\mathcal{H}(s) = \frac{\mathcal{Y}(s)}{\mathcal{X}(s)} = \frac{s^2-6}{2(s^2+4s+8)(s+1)}$$

The preceding function has two simple zero values at positions $s_{12} = \sqrt{6}$ and three simple poles with values $s_1 = -1$, $s_2 = -2+2i$, and $s_3 = -2-2i$. Therefore, there will be three different regions of convergence defined by the equations $\Re(s) < -2$, $-2 < \Re(s) < -1$, and $\Re(s) > -1$. To determine the one we will use for the analysis of the system, we should observe that since the relation $\mathcal{Y}(s) = \mathcal{H}(s)\mathcal{X}(s)$ holds, the region of convergence of the transform $\mathcal{Y}(s)$ should result from the intersection of the regions of convergence of the transforms $\mathcal{H}(s)$ and $\mathcal{X}(s)$. Taking into account that the region of convergence of the transforms $\mathcal{Y}(s)$ and $\mathcal{X}(s)$ is $\Re(s) > -1$, it is obvious that the only one of the three regions of convergence of the function $\mathcal{H}(s)$ that satisfies this property is the one described by the equation $-2 < \Re(s) < -1$, which also constitutes the region of convergence of the system's transfer function. If we take into account that this region describes a *right half-plane* and that all the poles of the transfer function—which is regular—have negative real parts, it immediately follows that the system is *causal and stable*.

5.4.1 Cancellation of Common Poles of Zero Values

In some cases, it is possible that in a rational transfer function $\mathcal{H}(s) = \mathcal{Y}(s)/\mathcal{X}(s)$, the polynomials $\mathcal{X}(s)$ and $\mathcal{Y}(s)$ are characterized by the existence of a common root, with a value, for example, $s = \alpha$. In other words, the function $\mathcal{H}(s)$ *will have a pole and a zero value at the same position* or, equivalently, both in the numerator and the denominator there will be the common term $(s-\alpha)$.[9] Although it may seem logical to cancel the common term from the numerator and the denominator, to simplify the transfer function, however, this could lead to erroneous conclusions regarding the stability of the system.

To understand the problem that arises in such a case, let us consider an explicit transfer function of the form

$$\mathcal{H}(s) = \frac{\mathcal{Y}(s)}{\mathcal{X}(s)} = \frac{s-2}{s^2+2s-8}$$

Observing that the polynomial of the denominator factorizes as $s^2+2s-8 = (s-2)(s+4)$, one might think that it would be good to eliminate the factor $s-2$ from the numerator and the denominator so as to express the function in its simplest form,

$$H(s) = \frac{s-2}{(s+4)(s-2)} = \frac{1}{s+4}$$

[9] Recall that the sign of the variable α determines the stability of the pole/zero value, which will be *stable* for negative α and *unstable* for positive α.

and then to assert that, since the function has a simple pole with a negative real part at the position $s = -4$, the system is stable. But this is wrong, since *in reality the system is unstable*. To understand the derivation of this conclusion, let us construct the differential equation of the system that corresponds to the original transfer function and that, as can be proven, is given by the relation

$$\frac{d^2y(t)}{dt^2} + 2\frac{dy(t)}{dt} - 8y(t) = \frac{dx(t)}{dt} - 2x(t)$$

If we take the Laplace transform of all the terms of the foregoing equation using the initial conditions of the problem $y(0)$, $\dot{y}(0) \equiv y'(0)$, and $x(0)$, we will be led to the equation

$$s^2\mathcal{Y}(s) - sy(0) - \dot{y}(0) + 2s\mathcal{Y}(s) - 2y(0) - 8\mathcal{Y}(s) = s\mathcal{X}(s) - x(0) - 2\mathcal{X}(s)$$

or, equivalently,

$$\mathcal{Y}(s) = \frac{s-2}{s^2+2s-8}\mathcal{X}(s) + \frac{s+2}{s^2+2s-8}y(0) + \frac{\dot{y}(0)}{s^2+2s-8} - \frac{x(0)}{s^2+2s-8}$$

Assuming for simplicity that $\dot{y}(0) = x(0) = 0$, the foregoing equation will take the form

$$\mathcal{Y}(s) = \frac{s-2}{s^2+2s-8}\mathcal{X}(s) + \frac{s+2}{s^2+2s-8}y(0)$$

and therefore the output of the system will be calculated from the inverse Laplace transform

$$y(t) = \mathcal{L}^{-1}\{\mathcal{Y}(s)\} = \mathcal{L}^{-1}\left\{\frac{s-2}{s^2+2s-8}\mathcal{X}(s)\right\} + \mathcal{L}^{-1}\left\{\frac{s+2}{s^2+2s-8}\right\}y(0)$$

We observe, then, that the two terms of the numerator and the denominator that were not simplified *led to the appearance of the second term in the foregoing equation*, which, developing the rational function into partial fractions, turns out to be equal to

$$\mathcal{L}^{-1}\left\{\frac{s+2}{s^2+2s-8}\right\} = \frac{2}{3}\mathcal{L}^{-1}\left\{\frac{1}{s-2}\right\} + \frac{1}{3}\mathcal{L}^{-1}\left\{\frac{1}{s+4}\right\} = \frac{2}{3}e^{2t}u(t) + \frac{1}{3}e^{-4t}u(t)$$

Taking into account that these two terms will appear in the output equation of the system $y(t)$ (along with those that will result from the first inverse transform), it is obvious that the system is *unstable* (since the term e^{2t} tends to infinity as $t \to \infty$), that is, the conclusion about the stable nature of the system to which we were led by simplifying the transfer function $H(s)$ is incorrect.

But why does this happen? First of all, let us note that the expression we calculated in the last equation is the *natural response of the system*, which, as we proved in a previous section, is written as a linear combination of the form

$$y(t) = \sum_{i=1}^{N} C_i e^{p_i t} = C_1 e^{p_1 t} + C_2 e^{p_2 t} + \cdots + C_N e^{p_N t}$$

where $p_1, p_2, \ldots, p_N$ are the N poles of the transfer function (for simplicity we consider *N distinct poles*, but obviously the situation does not differ if we consider poles with multiplicity greater than one). This means that even if a pole p_k is cancelled by a zero $z_k = p_k$ that is in the same position as it, so that it does not appear in the new simplified transfer function, the term $C_k e^{p_k t}$, due to its existence, *is not going to be removed from the natural response of the system*. This fact is, of course, expected, since in the derivation of the last equation we used the *complete characteristic polynomial of the system regardless of what was in the numerator of the transfer function*. This conclusion is confirmed and by the preceding example, in which, despite the fact that we simplified the function $\mathcal{H}(s)$ by cancelling the unstable pole at the position $s = 2$ due to the existence of a zero in the same position, in the natural response of the system, the term e^{2t} appeared, which is precisely due to the existence of this pole. We are led, therefore, to the conclusion that *if a transfer function has unstable poles, the system is unstable, since in its natural response unbounded terms appear that tend to infinity in the limit* $t \to \infty$. If we proceed to remove an unstable pole, the system does not become stable, since the preceding unstable terms are not removed but simply "*hide*," thereby leading to erroneous conclusions regarding the characterization of the system in terms of its stability.

In a previous section, we pointed out that the transfer function of a LTI system *is not sufficient for its full characterization* since it focuses exclusively on input–output characteristics without delving into details concerning its internal structure. The reason we made this claim is precisely the one we just described: If there are hidden poles that were removed from the transfer function, due to their cancellation by zeros located in the same position, the information associated with them *has been lost from the function* and cannot be retrieved in any way. Therefore, the only type of stability for which we can draw safe conclusions based on the transfer function is the *BIBO stability*, while for the internal stability of the system we resort to other methods, such as the *description of the state space*.

Are there cases where the transfer function can be used to characterize the system *in terms of its internal stability*? The answer is affirmative, as long as the polynomials of the numerator and the denominator *do not have common roots*, so that this process of mutual cancellation of poles and zero values cannot be carried out. In such a case, the polynomial of the denominator of the transfer function fully coincides with the characteristic polynomial of the system, and therefore we can, based on this, characterize the system in terms of its internal stability, leading to one of the following three conclusions:

- The system is characterized as *asymptotically stable* if and only if all the poles of the transfer function, regardless of their multiplicity, have negative real parts.
- The system is characterized as *unstable* if and only if at least one or both of the following conditions are met: (a) there is at least one pole with a positive real part and (b) there are repeated poles on the imaginary axis.
- The system is characterized as *marginally stable* if and only if there are no poles with a positive real part and there are distinct poles on the imaginary axis.

The final conclusion that emerges from the foregoing analysis is that *when analyzing the stability of a transfer function, one should avoid canceling common poles and zeros, particularly if these are known to be unstable.*

5.4.2 The Stability Criterion of Routh–Hurwitz

Although the stability of a system described by a rational transfer function *is fully determined by the sign of the real part of its poles*, in several cases the polynomial of the denominator is not easy to factorize, so as to determine its roots that also constitute the poles of the function. In this case, the stability of the system can be determined using a criterion known as the *Ruth–Hurwitz criterion*. This method provides us with information about the stability of the system, relying *only on the knowledge of the coefficients of the polynomial of the denominator of the transfer function*, and additionally informs us about the number of poles that belong to the positive half-plane of the complex plane or lie on the imaginary axis. A polynomial all of whose roots have negative real parts, that is, they belong to the left complex half-plane, is known as a *Hurwitz polynomial*.

Considering a polynomial of degree N of the form

$$P(s) = \alpha_0 + \alpha_1 s + \alpha_2 s^2 + \alpha_3 s^3 + \cdots + \alpha_{N-1} s^{N-1} + \alpha_N s^N$$

we can define the so-called Hurwitz matrix $H(P) \in R^{N\times N}$ from an equation of the form

$$\boldsymbol{H}(p) = \begin{bmatrix} \alpha_{N-1} & \alpha_N & 0 & 0 & \dots & 0 \\ \alpha_{N-3} & \alpha_{N-2} & \alpha_{N-1} & \alpha_N & \dots & 0 \\ \alpha_{N-5} & \alpha_{N-4} & \alpha_{N-3} & \alpha_{N-2} & \dots & 0 \\ \vdots & \vdots & \vdots & \vdots & \ddots & \vdots \\ 0 & 0 & 0 & 0 & \dots & 0 \end{bmatrix} \tag{5.18}$$

with the elements of the matrix that are not defined having a zero value. Having defined the matrix $\boldsymbol{H}(P)$, we can then define the submatrices $\boldsymbol{H}_k(P) \in R^{k\times k}$ $(k = 1, 2, \ldots, n)$ as $\boldsymbol{H}_k(P) = \{\boldsymbol{H}(P)_{ij}\}$ $(i, j = 1, 2, \ldots, k)$. In other words, the matrix $\boldsymbol{H}_k(P)$ is defined as the $k \times k$ submatrix starting from the top left element of the matrix $\boldsymbol{H}(p)$. Let us also symbolize by $D_k = \det[\boldsymbol{H}_k(P)]$ $(k = 1, 2, \ldots, n)$ the determinants of the matrices $\boldsymbol{H}_k(p)$; these determinants are known as *Hurwitz determinants*.

According to the *Hurwitz theorem*,[10] a polynomial is *Hurwitz type* if and only if for coefficient value $\alpha_n > 0$ all Hurwitz determinants that are defined as

$$D_1 = \alpha_{N-1}$$

$$D_2 = \begin{vmatrix} \alpha_{N-1} & \alpha_N \\ \alpha_{N-3} & \alpha_{N-2} \end{vmatrix} |$$

$$D_3 = \begin{vmatrix} \alpha_{N-1} & \alpha_N & 0 & 0 & \ldots & 0 \\ \alpha_{N-3} & \alpha_{N-2} & \alpha_{N-1} & \alpha_N & \ldots & 0 \\ \vdots & \vdots & \vdots & \vdots & \ddots & \vdots \\ 0 & 0 & 0 & 0 & \ldots & 0 \end{vmatrix}$$

$$\vdots$$

$$D_N = \alpha_N D_{N-1}$$

have values *greater than zero*. Using the preceding criterion, we can draw conclusions about the stability of a polynomial without having to calculate its roots.

Example 5.4.1 Let the polynomial $P(s) = s^5 + 3s^4 + 2s^3 + s^2 + 7s + 12$, which is of degree $N = 5$ with coefficient values $\alpha_0 = 12$, $\alpha_1 = 7$, $\alpha_2 = 1$, $\alpha_3 = 2$, $\alpha_4 = 3$, and $\alpha_5 = 1$. Applying the *Hurwitz criterion*, we find that

$$D_1 = \alpha_4 = 3 > 0$$

$$D_2 = \begin{vmatrix} \alpha_4 & \alpha_5 \\ \alpha_2 & \alpha_3 \end{vmatrix} = \begin{vmatrix} 3 & 1 \\ 1 & 2 \end{vmatrix} = 4 > 0$$

$$D_3 = \begin{vmatrix} \alpha_4 & \alpha_5 & 0 \\ \alpha_2 & \alpha_3 & \alpha_4 \\ \alpha_0 & \alpha_1 & \alpha_2 \end{vmatrix} = \begin{vmatrix} 3 & 1 & 0 \\ 1 & 2 & 3 \\ 12 & 7 & 1 \end{vmatrix} = -22 < 0$$

We observe that since the determinant D_3 was found to have a negative value, the polynomial is *unstable*, and there is no reason to continue the process. In the opposite case, we would have to calculate the determinants D_4 and D_5 and determine the stability of the polynomial according to the signs of the values that would emerge.

The main disadvantage of the stability criterion of Hurwitz is its *computational complexity*, since for polynomials of sufficiently high degree the dimensions of the determinants that will need to be calculated may be quite large. However, this criterion can be applied due to the work of *Routh*, who modified the process in such a way as not to need to calculate all these determinants but only one table known as *Routh's table* and whose construction is relatively easy. For this reason, this stability criterion in the international literature is known as the *Routh–Hurwitz stability criterion*.

To understand the application of the *Routh method*, let us consider two sets of numbers,

$$(\alpha_{11}, \alpha_{12}, \ldots, \alpha_{1k}, \ldots, \alpha_{1N}) \qquad \text{and} \qquad (\alpha_{21}, \alpha_{22}, \ldots, \alpha_{2k}, \ldots, \alpha_{2M})$$

[10] The proof of this and all related theorems can be found in the book by Andrew D. Lewis titled *A Mathematical Approach to Classical Control*, which is freely distributed in PDF form on the Internet.

with which we can construct a table of coefficients in the following way. In the first two lines of the table, we place the preceding sets of numbers under one another; then we will write

$$\begin{array}{l} \alpha_{11}\ \alpha_{12}\ \alpha_{13}\ \alpha_{14}\ \alpha_{15}\ \ldots\ \alpha_{1k}\ \ldots\ \alpha_{1N} \\ \alpha_{21}\ \alpha_{22}\ \alpha_{23}\ \alpha_{24}\ \alpha_{25}\ \ldots\ \alpha_{2k}\ \ldots\ \alpha_{2M} \end{array}$$

and subsequently the elements of the third line will be calculated using a relationship of the form

$$\alpha_{3i} = -\frac{\begin{vmatrix} \alpha_{11} & \alpha_{1,i+1} \\ \alpha_{21} & \alpha_{2,i+1} \end{vmatrix}}{\alpha_{21}} = \frac{\alpha_{21}\alpha_{1,i+1} - \alpha_{11}\alpha_{2,i+1}}{\alpha_{21}}$$

leading to the result

$$\begin{array}{cccc} \alpha_{11} & \alpha_{12} & \alpha_{13} & \alpha_{14} \\ \alpha_{21} & \alpha_{22} & \alpha_{23} & \alpha_{24} \\ \dfrac{\alpha_{21}\alpha_{12}-\alpha_{11}\alpha_{22}}{\alpha_{21}} & \dfrac{\alpha_{21}\alpha_{13}-\alpha_{11}\alpha_{23}}{\alpha_{21}} & \dfrac{\alpha_{21}\alpha_{14}-\alpha_{11}\alpha_{24}}{\alpha_{21}} & \dfrac{\alpha_{21}\alpha_{15}-\alpha_{11}\alpha_{25}}{\alpha_{21}} \end{array}$$

In the preceding table, we have retained *only the first four elements of each row* to demonstrate the application method. This process continues in exactly the same way. Therefore, using the elements of the second and third rows, we can calculate the elements of the fourth row, which, together with those of the third row, will allow us to calculate the elements of the fifth row, and so on. If the two sets of numbers do not have the same length or in general if empty positions arise in the table during the application of the method, these are filled with *zero values*.

How can we apply the foregoing process to the coefficients of a polynomial? Let us consider a polynomial of the form

$$P(s) = \alpha_0 + \alpha_1 s + \alpha_2 s^2 + \alpha_3 s^3 + \cdots + \alpha_{N-1} s^{N-1} + \alpha_N s^N$$

whose roots are, of course, the solutions of the equation $P(s) = 0$. It can be proven that this equation *does not have roots with positive real parts* if the following two conditions are met:

1. All coefficients of the polynomial are different from zero, that is, they appear in the equation of the polynomial $P(s)$.
2. All coefficients of the equation $P(s) = 0$ have the same sign.

However, let us note that these two conditions *are necessary but not sufficient* in the sense that even if all the coefficients have the same sign, it is still possible for a polynomial to have roots with positive real parts.

To proceed with the application of this method, let us calculate the *even and odd components* of the polynomial function $P(s)$, which obviously will be

$$P_e(s) = \alpha_0 + \alpha_2 s^2 + \alpha_4 s^4 + \alpha_6 s^6 + \ldots \quad \text{and} \quad P_o(s) = \alpha_1 s + \alpha_3 s^3 + \alpha_5 s^5 + \alpha_7 s^7 + \ldots$$

Alternatively, we can define these polynomials as

$$P_+(s) = \alpha_0 + \alpha_2 s + \alpha_4 s^2 + \alpha_6 s^3 + \ldots \quad \text{and} \quad P_-(s) = \alpha_1 + \alpha_3 s + \alpha_5 s^2 + \alpha_7 s^3 + \ldots$$

so the polynomial $P(s)$ will be written $P(s) = P_+(s^2) + sP_-(s^2)$. In this case, we can apply the *Routh method* using as the two sets of numbers we mentioned previously, the *coefficients of the polynomials $P_+(s)$ and $P_-(s)$*, which are the

$$(\alpha_0, \alpha_2, \alpha_4, \alpha_6, \alpha_8, \ldots) \quad \text{and} \quad (\alpha_1, \alpha_3, \alpha_5, \alpha_7, \alpha_9, \ldots)$$

concluding in a result of the form

$$\begin{array}{ccccc}
\alpha_0 & \alpha_2 & \alpha_4 & \cdots & \alpha_{2k} \\
\alpha_1 & \alpha_3 & \alpha_5 & \cdots & \alpha_{2k+1} \\
\dfrac{\alpha_1\alpha_2-\alpha_0\alpha_3}{\alpha_1} & \dfrac{\alpha_1\alpha_4-\alpha_0\alpha_5}{\alpha_1} & \dfrac{\alpha_1\alpha_6-\alpha_0\alpha_7}{\alpha_1} & \cdots & \dfrac{\alpha_1\alpha_{2k+2}-\alpha_0\alpha_{2k+3}}{\alpha_1}
\end{array}$$

Note that the order in which the coefficients are written does not matter, in the sense that if we wrote the polynomial in the form

$$P(s) = \alpha_N s^N + \alpha_{N-1} s^{N-1} + \cdots + \alpha_3 s^3 + \alpha_2 s^2 + \alpha_1 s + \alpha_0$$

the contents of the two sets of numbers would be $(\ldots, \alpha_6, \alpha_4, \alpha_2, \alpha_0)$ and $(\ldots, \alpha_7, \alpha_5, \alpha_4, \alpha_1)$. The result of applying the foregoing procedure for the general case of a polynomial of degree n is

$$\begin{array}{c|cccccc}
s^N & \alpha_N & \alpha_{N-2} & \alpha_{N-4} & \alpha_{N-6} & \cdots & \alpha_2 \; \alpha_0 \\
s^{N-1} & \alpha_{N-1} & \alpha_{N-3} & \alpha_{N-5} & \alpha_{N-7} & \cdots & \alpha_1 \; 0 \\
s^{N-2} & \beta_1 & \beta_2 & \beta_3 & \beta_4 & \cdots & \\
s^{N-3} & \gamma_1 & \gamma_2 & \gamma_3 & \cdots & & \\
s^{N-4} & \delta_1 & \delta_2 & \cdots & & & \\
\vdots & & & & & & \\
s^1 & & & & & & \\
s^0 & & & & & &
\end{array}$$

with the coefficients $\beta_1, \beta_2, \beta_3, \beta_4, \ldots, \gamma_1, \gamma_2, \gamma_3, \ldots, \delta_1, \delta_2, \ldots$ being calculated as

$$\beta_1 = \frac{\alpha_{N-2}\alpha_{N-1} - \alpha_N\alpha_{N-3}}{\alpha_{N-1}}, \; \beta_2 = \frac{\alpha_{N-1}\alpha_{N-4} - \alpha_N\alpha_{N-5}}{\alpha_{N-1}}$$

$$\beta_3 = \frac{\alpha_{N-1}\alpha_{N-6} - \alpha_N\alpha_{N-7}}{\alpha_{N-1}}, \; \beta_4 = \frac{\alpha_{N-1}\alpha_{N-8} - \alpha_n\alpha_{N-9}}{\alpha_{n-1}}$$

$$\gamma_1 = \frac{\beta_1\alpha_{N-3} - \beta_2\alpha_{N-1}}{\beta_1}, \quad \gamma_2 = \frac{\beta_1\alpha_{N-5} - \beta_3\alpha_{N-1}}{\beta_1}, \quad \gamma_3 = \frac{\beta_1\alpha_{N-7} - \beta_4\alpha_{N-1}}{\beta_1}$$

$$\delta_1 = \frac{\gamma_1\beta_2 - \beta_1\gamma_2}{\gamma_1}, \quad \delta_2 = \frac{\gamma_1\beta_3 = \beta_1\gamma_3}{\gamma_1}$$

and so forth. From the preceding table we observe that because the odd coefficients are one less than the even coefficients, a zero value was placed in the second row and immediately after the element α_1. We also observe that a helper column was placed in front of the first column of coefficients, which contains as labels the powers of the complex variable s and which is, of course, ignored. However, its role is quite significant, as it informs us about how many rows remain to be calculated, while also informing us about the way in which the various coefficients are related to the various powers of s. For example, the row

$$s^{N-1} \,\big|\, \alpha_{N-1} \; \alpha_{N-3} \; \alpha_{N-5} \; \alpha_{N-7} \; \cdots \; \alpha_1 \; \alpha_0$$

informs us that the coefficient α_{N-1} corresponds to the term s^{N-1}, the coefficient α_{N-3} corresponds to the term s^{N-3}, the coefficient α_{N-5} corresponds to the term s^{N-5}, and so forth, while the line

$$s^{N-3} \,\big|\, \gamma_1 \; \gamma_2 \; \gamma_3 \; \cdots$$

informs us that the coefficient γ_1 corresponds to the term s^{N-3}, the coefficient γ_2 corresponds to the term s^{N-5}, the coefficient γ_3 corresponds to the term s^{N-7}, and so forth.[11] In other words, this table consists of a number of lines that alternately refer

[11] Since the preceding lines alternately refer to the even and odd powers of the complex variable s, it is obvious that the powers associated with successive coefficients will differ from each other *by two units*.

to even and odd powers of s, with these powers being reduced by one as we go down the lines of the table until we reach the last line, which corresponds to the power s^0 and which, of course, will contain the constant term. This process for a polynomial of degree $s = 6$ is presented as follows:

$$
\begin{array}{c|cccc}
s^6 & \alpha_6 & \alpha_4 & \alpha_2 & \alpha_0 \\
s^5 & \alpha_5 & \alpha_3 & \alpha_1 & 0 \\
s^4 & \underbrace{\dfrac{\alpha_5\alpha_4 - \alpha_6\alpha_3}{\alpha_5}}_{A} & \underbrace{\dfrac{\alpha_5\alpha_2 - \alpha_6\alpha_1}{\alpha_5}}_{B} & \dfrac{\alpha_5\alpha_0}{\alpha_5} \dashrightarrow \alpha_0 & 0 \\
s^3 & \underbrace{\dfrac{A\alpha_3 - \alpha_5 B}{A}}_{C} & \underbrace{\dfrac{A\alpha_1 - \alpha_5\alpha_0}{A}}_{D} & 0 & 0 \\
s^2 & \underbrace{\dfrac{BC - AD}{C}}_{E} & \dfrac{C\alpha_0}{C} \dashrightarrow \alpha_0 & 0 & 0 \\
s^1 & \underbrace{\dfrac{ED - C\alpha_0}{E}}_{F} & 0 & 0 & 0 \\
s^0 & \dfrac{F\alpha_0}{F} \dashrightarrow \alpha_0 & 0 & 0 & 0
\end{array}
$$

It is not difficult to notice that the first column of coefficients of this table contains at most $(n + 1)$ elements that have values different from zero. This column is known as the *Routh table*. Having understood the structure and the method of constructing this table, we can now formulate the next theorem:

Theorem 5.4.1 *A polynomial of the form*

$$P(s) = \alpha_N s^N + \alpha_{N-1} s^{N-1} + \cdots + \alpha_3 s^3 + \alpha_2 s^2 + \alpha_1 s + \alpha_0$$

is of the Hurwitz type—that is, all its roots have negative real parts and therefore it is stable—if and only if all the elements of the Routh table are positive. Conversely, if the Routh table contains sign changes—that is, some of its elements have a negative sign—then the number of these changes is equal to the number of roots of the polynomial that have a positive real part.

To understand the application of the method, let us consider the simple case of the polynomial $P(s) = s^2 + \alpha s + \beta$ for which the application of the described procedure leads to the table

$$
\begin{array}{c|cc}
s^2 & 1 & \beta \\
s^1 & \alpha & 0 \\
s^0 & \alpha\beta & 0
\end{array}
\qquad \text{or, alternatively, to the table} \qquad
\begin{array}{c|cc}
s^2 & \beta & 1 \\
s^1 & \alpha & 0 \\
s^0 & \alpha & 0
\end{array}
$$

if the above polynomial had been rewritten in the form $P(s) = \beta + \alpha s + s^2$. The corresponding Routh tables for these two cases are $[\ 1\ \ \alpha\ \ \alpha\beta\ ]$ and $[\ \beta\ \ \alpha\ \ \alpha\ ]$,[12] both of which lead to the conclusion that for the polynomial $P(s)$ to be stable, $\alpha > 0$ and $\beta > 0$ must be true.

Let us try to verify the foregoing conclusion by calculating the roots of the polynomial and looking for the conditions that must be met for them to have negative real parts. These roots will obviously be given by the relations

$$s_1 = -\frac{\alpha}{2} + \frac{1}{2}\sqrt{\alpha^2 - 4\beta} \qquad \text{and} \qquad s_2 = -\frac{\alpha}{2} - \frac{1}{2}\sqrt{\alpha^2 - 4\beta}$$

[12] Because the degree of the polynomial is $n = 2$, the element of the Routh table that corresponds to the term s^0 is not equal to β; this is something that applies to degrees $n \geq 3$.

so we need to investigate the following three cases:

- $\alpha^2 - 4\beta < 0$: In this case, the equation $P(s) = 0$ has *conjugate complex roots* with real part $-\alpha$, which, to be in the left complex half-plane, must be $\alpha > 0$, as well as $\beta > \alpha^2/4$, from which it follows that $\beta > 0$ and, therefore, $\alpha\beta > 0$.
- $\alpha^2 - 4\beta < 0$: In this case, there is *one double root*, $s_{12} = -\alpha/2$, and in order for it to be in the left complex half-plane, it must hold that $\alpha > 0$. As for the parameter β, it is equal to $\alpha^2/4$, and since $\alpha > 0$, it must also hold that $\beta > 0$ and, therefore, $\alpha\beta > 0$.
- $\alpha^2 - 4\beta > 0$: In this case, we distinguish two subcases: if $\alpha > 0$, the roots will belong to the left complex half-plane if and only if $\alpha^2 - 4\beta < \alpha^2$, from which it follows that $\beta > 0$ and, therefore, $\alpha\beta > 0$. On the other hand, if $\alpha < 0$, then it will also hold that $-\alpha + \sqrt{\alpha^2 - 4\beta} > 0$, and there is no chance of having a root in the left complex half-plane.

We observe, therefore, that the results to which the Routh method led are similar to those we get by investigating the roots of the equation in the usual way. Therefore, this method is an excellent tool for studying the stability of a polynomial in cases where it is impossible to calculate its roots due to its exceptional complexity.

Problem 5.4.10 Use the Routh method to characterize the stability of the polynomial

$$P(s) = s^3 + 3s^2 + 4s + 2$$

Solution The degree of the polynomial is $N = 3$, and the coefficients included in it are $\alpha_0 = 2$, $\alpha_1 = 4$, $\alpha_2 = 3$, and $\alpha_3 = 1$. Applying the Routh method, it turns out that

$$\begin{array}{c|cc} s^3 & 1 & 4 \\ s^2 & 3 & 2 \\ s^1 & 10/3 & 0 \\ s^0 & 2 & 0 \end{array}$$

We observe that in the first column of this table, all elements are positive, there are no sign changes, and, therefore, *all the roots of the polynomial have negative real parts*. Indeed, if we calculate the roots of the polynomial $P(s)$, these are found to be equal to $p_1 = -1$ and $p_{2,3} = -1 \pm j$, a finding that confirms our conclusion.

Problem 5.4.11 Using the Routh method, characterize the stability of the polynomial

$$P(s) = s^6 + 4s^5 + 3s^4 + 2s^3 + s^2 + 4s + 4$$

Solution The degree of the polynomial is $N = 6$, and the coefficients included in it are $\alpha_0 = 4$, $\alpha_1 = 4$, $\alpha_2 = 1$, $\alpha_3 = 2$, $\alpha_4 = 3$, $\alpha_5 = 4$, and $\alpha_6 = 1$. Applying the Routh method, it turns out that

$$\begin{array}{c|cccc} s^6 & 1 & 3 & 1 & 4 \\ s^5 & 4 & 2 & 4 & 0 \\ s^4 & 5/2 & 0 & 4 & 0 \\ s^3 & 2 & -12/5 & 0 & 0 \\ s^2 & 3 & 4 & 0 & 0 \\ s^1 & -76/15 & 0 & 0 & 0 \\ s^0 & 4 & 0 & 0 & 0 \end{array}$$

In this case, we observe that there are two sign changes and, more specifically, $(s^2 \to s^1) = (3 \to -76/15)$ and $(s^1 \to s^0) = (-76/15 \to 4)$. Therefore, the polynomial has two poles with positive real parts, and the system it describes is unstable. This conclusion is verified by the detailed calculation of the system's poles, which leads to the following results: $p_1 = -3.2644$, $p_2 = -0.8858$, $p_{3,4} = -0.6046 \pm 0.9935j$, and $p_{5,6} = 0.6797 \pm 0.7488j$.

During the application of the *Routh method* in the manner described, there is a chance that *two unique situations* may occur that involve the appearance of lines in which either the *first element has a zero value* or *all the elements of the line are equal to zero*. Our approach to dealing with these cases is discussed in what follows.

First, we should note that if the first element of a line has a zero value, the method cannot continue to be applied. Considering the general case of a polynomial of degree N and assuming that the first element of the third line that corresponds to the term s^{N-2} has a value of $\beta_1 = 0$, it is obvious that the coefficients $\gamma_1, \gamma_2, \gamma_3, \ldots$ will take an infinite value since the parameter β_1 is in the denominator of the equations that are used for their calculation. If this situation occurs, we replace the zero value with a small positive constant ε, which is assumed to tend to zero, and we continue the application of the method in the manner described. Two versions are likely to occur:

- If a change in the sign is observed from the line preceding the one that contains the ε to the line located right below it, this means that there is a pole with a positive real part, as happens, for example, when

$$\begin{array}{l|l} s^{i+1} & 5 \ldots \\ s^{i} & \varepsilon \ldots \\ s^{i-1} & -2 \ldots \end{array} \quad \text{or, alternatively, when} \quad \begin{array}{l|r} s^{i+1} & -1 \ldots \\ s^{i} & \varepsilon \ldots \\ s^{i-1} & 6 \ldots \end{array}$$

- If such a change does not take place, this implies the existence of complex conjugate roots, as happens, for example, when

$$\begin{array}{l|l} s^{i+1} & 5 \ldots \\ s^{i} & \varepsilon \ldots \\ s^{i-1} & 6 \ldots \end{array} \quad \text{or, alternatively, when} \quad \begin{array}{l|r} s^{i+1} & -3 \ldots \\ s^{i} & \varepsilon \ldots \\ s^{i-1} & -4 \ldots \end{array}$$

Example 5.4.2 Consider the polynomial $P(s) = s^4 + s^3 + 2s^2 + 2s + 3$, which is of degree $N = 4$, with coefficient values equal to $\alpha_0 = 3$, $\alpha_1 = 2$, $\alpha_2 = 2$, $\alpha_3 = 1$, and $\alpha_4 = 1$. Applying the Routh method, we find that

$$\begin{array}{l|l} s^4 & 1\ 2\ 3 \\ s^3 & 1\ 2\ 0 \\ s^2 & 0\ 3\ 0 \\ s^1 & \uparrow \\ s^0 & \text{zero} \end{array} \quad \begin{array}{c} \text{To proceed,} \\ \text{we make the substitution} \\ 0 \Rightarrow \varepsilon \\ \text{and continue taking} \end{array} \quad \begin{array}{l|cc} s^4 & 1 & 2\ 3 \\ s^3 & 1 & 2\ 0 \\ s^2 & \varepsilon & 3\ 0 \\ s^1 & \dfrac{2\varepsilon - 3}{\varepsilon} & 0\ 0 \\ s^0 & 3 & 0\ 0 \end{array}$$

As the positive quantity ε tends to zero, the quantity $(2\varepsilon - 3)/\varepsilon$ tends to the value $-3/\varepsilon$, which is negative, since we have assumed that $\varepsilon > 0$. We therefore observe that we have two changes in the sign, $\varepsilon > 0 \to (-3/\varepsilon) < 0$ and $(-3/\varepsilon) < 0 \to 3 > 0$. Therefore, the system *has two poles with positive real parts, that is, it is unstable*. Indeed, if we calculate the roots of the equation $P(s) = 0$, we find that these are $p_{1,2} = 0.4057 \pm 1.2928j$ and $p_{3,4} = -0.9057 \pm 0.9020j$.

The second problematic case that may appear is one in which *all the elements of a line have zero values*; it is proven that this line always corresponds to an *odd power* of s. The appearance of such a situation indicates that the polynomial has *symmetrical real, imaginary, or complex roots with respect to the origin of the coordinate system*, expressed in the form $p_{12} = \pm\alpha$ $p_{12} = \pm j\beta$ or $p_{12} = \pm\alpha \mp j\beta$ or $p_{12} = \pm\alpha \pm j\beta$. If such a situation arises, we proceed as follows:

- We record the coefficients of the line that is *right above the line of zero elements*, and with their help we construct the so-called *auxiliary polynomial* $A(s)$, which always has *even symmetry* and in which the maximum degree is determined by the label associated with this line. The equation $A(s) = 0$ that arises is known as the *auxiliary equation*.
- We derive the auxiliary equation with respect to s and retain coefficients of the equation $A'(s) = 0$.
- We replace the row of zero elements with a new row that contains the coefficients we recorded in the previous step and continue the process.

In this case, it is proven that the *number of sign changes counted from the new row to the end of the table is equal to the number of roots of the auxiliary polynomial that have positive real parts*. Taking into account the symmetry that characterizes the position of the roots on the complex plane in this case, it is obvious that there must be an equal number of roots in the left complex half-plane, while if the polynomial has other roots, they must necessarily be on the imaginary axis.

Example 5.4.3 Consider the polynomial $P(s) = s^5 + 4s^4 + 8s^3 + 8s^2 + 7s + 4$, which is of degree $N = 5$, with coefficient values $\alpha_0 = 4$, $\alpha_7 = 2$, $\alpha_2 = 8$, $\alpha_3 = 8$, $\alpha_4 = 4$, and $\alpha_5 = 1$. Applying the *Routh method*, we find that

$$\begin{array}{c|ccc} s^5 & 1 & 8 & 7 \\ s^4 & 4 & 8 & 4 \\ s^3 & 6 & 6 & 0 \\ s^2 & 4 & 4 & 0 \\ s^1 & 0 & 0 & 0 \\ s^0 & & & \end{array}$$

We observe that the application of the method led to the appearance of a row of zero elements corresponding to the term s^1. Applying the method we presented earlier, we record the coefficients in the row right above, which are $[\,4\ \ 4\ \ 0\,]$. Considering that this line corresponds to the term s^2, the auxiliary equation will have the form $A(s) = 4s^2 + 4s = 0$, with roots $p_{1,2} = \pm j$. Differentiating, we get $A'(s) = 8s$, so the new coefficients are $[\,8\ \ 0\ \ 0\,]$. Therefore, replacing the zero line with this new line and continuing the process, we arrive at the result

$$\begin{array}{c|ccc} s^5 & 1 & 8 & 7 \\ s^4 & 4 & 8 & 4 \\ s^3 & 6 & 6 & 0 \\ s^2 & 4 & 4 & 0 \\ s^1 & 8 & 0 & 0 \\ s^0 & 4 & 0 & 0 \end{array}$$

We observe that after the line of zero elements no changes in the sign are observed. However, the polynomial cannot be characterized as stable, *only as marginally stable*, because it has roots on the imaginary axis, and, according to theory, the polynomials that have this property are characterized by this type of stability.

If the application of the foregoing process leads again to the appearance of a line of zero elements, this is repeated again in the same exact way until finally the line corresponding to the last term, which is always s^0, is completed.

5.5 The Response of Systems in the Frequency Domain

As we saw in several of the previous exercises—and of course we know from the study of systems in the field of continuous-time—the response of a LTI system consists of two components: the *zero-state response*, which is *independent of the initial conditions of the problem* and is calculated using the transfer function and the corresponding input signal, and the *zero-input response* or *natural response* of the system, which depends on the *initial conditions* of the problem. In this section, we will present a complete and systematic description of these characteristics expressed in the frequency domain, and then we will use the presented method to calculate the system's response for certain characteristic input signals. Let us consider, then, once again the continuous-time LTI system described by the differential equation

$$\alpha_N \frac{d^N y(t)}{dt^N} + \alpha_{N-1} \frac{d^{N-1} y(t)}{dt^{N-1}} + \cdots + \alpha_2 \frac{d^2 y(t)}{dt^2} + \alpha_1 \frac{dy(t)}{dt} + \alpha_0 y(t) =$$

$$= \beta_M \frac{d^M x(t)}{dt^M} + \beta_{M-1} \frac{d^{M-1} x(t)}{dt^{M-1}} + \cdots + \beta_2 \frac{d^2 x(t)}{dt^2} + \beta_1 \frac{dx(t)}{dt} + \beta_0 x(t)$$

and let us assume that the output $y(t)$ is characterized by a set of nonzero initial conditions of the form

$$y(0),\, y'(0),\, y''(0),\, \ldots,\, y^{(N-1)}(0)$$

while in contrast, the initial conditions corresponding to the input signal $x(t)$ are all equal to zero, so that we can write that $x(0) = x'(0) = x''(0) = \cdots = x^{(M-1)}(0) = 0$.

Taking the Laplace transform of the two terms of the foregoing equation, we will have

$$\alpha_N\left[s^n\mathcal{Y}(s) - s^{N-1}y(0) - s^{N-2}\frac{dy(0)}{dt} - \cdots - s\frac{d^{(N-2)}y(0)}{dt^{N-2}} - \frac{d^{(N-1)}y(0)}{dt^{N-1}}\right]$$
$$+\alpha_{N-1}\left[s^{N-1}\mathcal{Y}(s) - s^{N-2}y(0) - \cdots - s\frac{d^{(N-3)}y(0)}{dt^{N-3}} - \frac{d^{(N-2)}y(0)}{dt^{N-2}}\right]$$
$$+\alpha_1 s\mathcal{Y}(s) - \alpha_1 y(0) + \alpha_0\mathcal{Y}(s) = \beta_M s^M\mathcal{X}(s) + \beta_{M-1}s^{M-1}\mathcal{X}(s) + \cdots + \beta_1 s\mathcal{X}(s) + \beta_0\mathcal{X}(s)$$

The preceding relationship can be written more concisely in the form

$$\left(\sum_{i=1}^{N}\alpha_i s^i\right)\mathcal{Y}(s) - \sum_{i=1}^{N}\alpha_i\left(\sum_{j=1}^{i}s^{i-j}\frac{d^{(i-j)}y(0)}{dt^{i-j}}\right) = \left(\sum_{i=1}^{M}\beta_i s^i\right)\mathcal{X}(s) \tag{5.19}$$

and therefore the Laplace transform of the system output $y(t)$ is calculated as

$$\mathcal{Y}(s) = \left\{\sum_{i=1}^{M}\beta_i s^i\right\}\Bigg/\left\{\sum_{i=1}^{N}\alpha_i s^i\right\}\mathcal{X}(s) + \left\{\sum_{i=1}^{N}\alpha_i\left(\sum_{j=1}^{i}s^{i-j}\frac{d^{(i-j)}y(0)}{dt^{i-j}}\right)\right\}\Bigg/\left\{\sum_{i=1}^{N}\alpha_i s^i\right\} \tag{5.20}$$

But the explicit function that constitutes the coefficient of the Laplace transform of the input $\mathcal{X}(s)$ is recognized as the *system transfer function* $\mathcal{H}(s)$. Therefore, defining the polynomial functions

$$\mathcal{C}(s) = \sum_{i=1}^{N}\alpha_i\left(\sum_{j=1}^{i}s^{i-j}\frac{d^{(i-j)}y(0)}{dt^{i-j}}\right) \quad \text{and} \quad \mathcal{D}(s) = \sum_{i=1}^{N}\alpha_i s^i$$

where $\mathcal{C}(s)$ *depends on the initial conditions of the problem*, while $\mathcal{D}(s)$ is the *characteristic polynomial of the system*, the transform of the system's output $\mathcal{Y}(s)$ will take the form

$$\mathcal{Y}(s) = \mathcal{H}(s)\mathcal{X}(s) + \frac{\mathcal{C}(s)}{\mathcal{D}(s)}$$

It is not difficult to see that the term $\mathcal{H}(s)\mathcal{X}(s)$ is independent of the initial conditions of the problem and depends only on the input signal of the system through its Laplace transform $\mathcal{X}(s)$, while, in contrast, the term $\mathcal{C}(s)/\mathcal{D}(s)$ is *independent of the input* and depends only on the *initial conditions* that have been set. We can therefore formulate the assertion that in the preceding equation the first term expresses the *Laplace transform of the zero-state response* $y_{zs}(t)$, while the second term expresses the *Laplace transform of the zero-input response* $y_{zi}(t)$. Therefore, the output of the system $y(t)$ will be calculated from the inverse transform

$$y(t) = \mathcal{L}^{-1}\{\mathcal{Y}(s)\} = y_{sz}(t) + y_{zi}(t)$$

with the functions $y_{zs}(t)$ and $y_{zi}(t)$ being calculated as

$$y_{zs}(t) = \mathcal{L}^{-1}\{\mathcal{H}(s)\mathcal{X}(s)\} = \mathcal{L}^{-1}\left\{\frac{\sum_{i=1}^{M}\beta_i s^i}{\sum_{i=1}^{N}\alpha_i s^i}X(s)\right\} \quad \text{and}$$

$$y_{zi}(t) = \mathcal{L}^{-1}\left\{\frac{\mathcal{C}(s)}{\mathcal{D}(s)}\right\} = \mathcal{L}^{-1}\left\{\frac{\sum_{i=1}^{N}\alpha_i\left(\sum_{j=1}^{i}s^{i-j}\frac{d^{(i-j)}y(0)}{dt^{i-j}}\right)}{\sum_{i=1}^{N}\alpha_i s^i}\right\}$$

In the case of a *stable* LTI system whose transfer function contains poles with *negative real parts*, the foregoing output can be treated as the sum of the *transient or natural response* $y_{tr}(t)$ (which appears immediately after the start of the system operation and lasts for a short period of time, undergoing *damping* over time) and the *forced response* or *steady-state response* $y_{ss}(t)$, which does not diminish over time and after the damping of the natural response becomes the dominant term. For example, in output of the form

$$y(t) = C_1e^{-t} + C_2e^{-4t} + C_3u(t)$$

the first two terms diminish over time and vanish at the limit $t \to \infty$, while the third term will continue to exist forever, so that we can generally write that

$$y(t) = y_{tr}(t) + y_{ss}(t)$$

with the transient response satisfying the condition

$$\lim_{t\to\infty} y_{tr}(t) = 0$$

Therefore, and according to our earlier discussion, the first two terms in the preceding example constitute the *transient response of the system*, while the third term refers to the *steady-state response*. This characteristic of the system is demonstrated in the following two subsections, where we calculate the *step* and *frequency response* of a continuous-time LTI system, that is, its output when the input signal is (a) the step function $x(t) = u(t)$ and (b) the elementary trigonometric function $x(t) = A\cos(\omega t + \varphi)$.

5.5.1 Calculation of Step Response Function

Let us consider a stable continuous LTI system described by a transfer function $\mathcal{H}(s)$, which can be expanded into a sum of partial fractions of the form

$$\mathcal{H}(s) = \frac{C_1}{s-p_1} + \frac{C_2}{s-p_2} + \cdots + \frac{C_N}{s-p_n} = \sum_{i=1}^{N}\frac{C_i}{s-p_i} \tag{5.21}$$

where $p_1, p_2, \ldots, p_N$ are the N *distinct* poles that characterize it, and let us assume that we transmit the step function $x(t) = u(t)$ to its input, whose Laplace transform has the form

$$\mathcal{X}(s) = \mathcal{L}\{x(t)\} = \mathcal{L}\{u(t)\} = \frac{1}{s} \tag{5.22}$$

In this case, the Laplace transform of the output of the system will be given by the relation

$$\mathcal{Y}(s) = \mathcal{H}(s)\mathcal{X}(s) = \frac{\mathcal{H}(s)}{s} = \frac{1}{s}\sum_{i=1}^{N} \frac{C_i}{s - p_i} = \frac{A}{s} + \sum_{i=1}^{N} \frac{C_i}{s - p_i} \tag{5.23}$$

where for the derivation of the last relation we developed the complex function $\mathcal{Y}(s)$ into a sum of partial fractions. Therefore, the output of the system $y(t)$ will emerge through the calculation of the inverse Laplace transform of the preceding relation. It will then hold that

$$y(t) = \mathcal{L}^{-1}\{\mathcal{Y}(s)\} = A\mathcal{L}^{-1}\left\{\frac{1}{s}\right\} + \sum_{i=1}^{N} C_i\mathcal{L}^{-1}\left\{\frac{1}{s - p_i}\right\} = Au(t) + \sum_{i=1}^{N} C_i e^{p_i t} \tag{5.24}$$

From the last relation—which emerged based on the property of *linearity* of the Laplace transform and referring to tables of elementary transforms—it immediately follows that the transient response of the system and the steady-state response will be given by the relations[13]

$$y_{tr}(t) = \sum_{i=1}^{N} C_i\, e^{p_i t} \qquad \text{and} \qquad y_{ss}(t) = Au(t)$$

respectively. If we are interested *only in the steady-state response*, this can be calculated directly from the transfer function $\mathcal{H}(s)$, without having to calculate the output of the system $y(t)$. Indeed, observing that the functions $\mathcal{H}(s)$ and $\mathcal{Y}(s)$ are related through the relationship $\mathcal{H}(s) = s\mathcal{Y}(s)$, the application of the initial value theorem of the Laplace transform leads us to the equation

$$\lim_{s\to 0} [s\mathcal{Y}(s)] = \lim_{s\to 0} \mathcal{H}(s) = \mathcal{H}(0)$$

In other words, the *steady-state component of the step response is equal to the value of the transfer function for the argument value* $s = 0$, a conclusion that holds for all stable continuous-time LTI systems, *regardless of their order*.

Problem 5.5.1 Use the method of the Laplace transform to determine the step response of the system described by the transfer function

$$\mathcal{H}(s) = \frac{s+5}{s^2+3s+2}$$

Solution According to the previous section, when we transmit the step function $x(t) = u(t)$ to the input, the Laplace transform of the system's output will have the form

$$\mathcal{Y}(s) = \frac{\mathcal{H}(s)}{s} = \frac{s+5}{s(s+1)(s+2)} = \frac{5}{2}\left\{\frac{1}{s}\right\} - \frac{4}{s+1} + \frac{3}{2}\left\{\frac{1}{s+2}\right\}$$

with the last equation resulting from the expansion of the function $\mathcal{Y}(s)$ as a series of partial fractions. Therefore, the *step response* of the system will be calculated from the inverse Laplace transform of the above relationship, which is found to have the form

[13] For this claim to be valid, the system must be stable so that all poles of the transfer function have negative real parts.

$$y(t) = \mathcal{L}^{-1}\{\mathcal{Y}(s)\} = \frac{5}{2}\mathcal{L}^{-1}\left\{\frac{1}{s}\right\} - 4\mathcal{L}^{-1}\left\{\frac{1}{s+1}\right\} + \frac{3}{2}\mathcal{L}^{-1}\left\{\frac{1}{s+2}\right\} = \frac{5}{2}u(t) - 4e^{-t}u(t) + \frac{3}{2}e^{-2t}u(t)$$

From the last equation we conclude that the transient response and the steady-state response of the system will be given respectively by the relations

$$y_{tr}(t) = -4e^{-t}u(t) + \frac{3}{2}e^{-2t}u(t) \qquad \text{and} \qquad y_{ss}(t) = \frac{5}{2}u(t)$$

which is the desired result. If we are only interested in the steady-state response, we could use the initial value theorem to get

$$y_{ss}(t) = \lim_{s\to 0} \mathcal{H}(s) = \lim_{s\to 0} \frac{s+5}{s^2+3s+2} = \frac{5}{2} \dashrightarrow \frac{5}{2}u(t)$$

which is the same as what we obtained through the application of the Laplace transform method.

5.5.2 Calculation of Frequency Response Function

To calculate the *frequency response* of the system, we will use the methodology of the previous subsection. However, in this particular case, we will apply to the input of the system the elementary trigonometric signal $x(t) = A\cos(\omega t + \varphi)$, whose Laplace transform is

$$\mathcal{X}(s) = \mathcal{L}\{x(t)\} = \mathcal{L}\{A\cos(\omega t + \varphi)\} = \mathcal{L}\{A\cos(\omega)\cos\varphi - A\sin(\omega t)\sin\varphi\} =$$

$$= A\cos\varphi\mathcal{L}\{\cos(\omega t)\} - A\sin\varphi\mathcal{L}\{\sin(\omega t)\} = A\cos\varphi\frac{s}{s^2+\omega^2} - A\sin\varphi\frac{\omega}{s^2+\omega^2} = \frac{A(s\cos\varphi - \omega\sin\varphi)}{s^2+\omega^2}$$

Assuming, as before, that the transfer function $\mathcal{H}(s)$ has N distinct poles $p_1, p_2, \ldots, p_N$, so that it can be developed into a sum of partial fractions of the form

$$\mathcal{H}(s) = \frac{C_1}{s-p_1} + \frac{C_2}{s-p_2} + \cdots + \frac{C_N}{s-p_n} = \sum_{i=1}^{N} \frac{C_i}{s-p_i} \tag{5.25}$$

the Laplace transform of the system's output will have the form

$$\mathcal{Y}(s) = \mathcal{H}(s)\mathcal{X}(s) = \frac{A\mathcal{H}(s)(s\cos\varphi - \omega\sin\varphi)}{s^2+\omega^2} = \sum_{i=1}^{N}\left\{\frac{A(s\cos\varphi - \omega\sin\varphi)}{s^2+\omega^2} \times \frac{C_i}{s-p_i}\right\}$$

and by developing the last expression into partial fractions, we obtain

$$\mathcal{Y}(s) = \frac{C}{s-j\omega} + \frac{C^*}{s+j\omega} + \sum_{i=1}^{N}\frac{C_i}{s-p_i} \tag{5.26}$$

Therefore, the system's response to the trigonometric input signal $x(t) = A\cos(\omega t + \varphi)$, which is known as the *frequency response*, will be calculated from the inverse transform

$$y(t) = \mathcal{L}^{-1}\{\mathcal{Y}(s)\} = C\mathcal{L}^{-1}\left\{\frac{1}{s-j\omega}\right\} + C^*\mathcal{L}^{-1}\left\{\frac{1}{s+j\omega}\right\} + \sum_{i=1}^{N} C_i\mathcal{L}^{-1}\left\{\frac{1}{s-p_i}\right\}$$

Expressing the complex coefficient C in its exponential form $C = |C|e^{j\measuredangle C}$, it will obviously be the case that $C^* = |C|e^{-j\measuredangle C}$—where $|C|$ and $\measuredangle C$ are the magnitude and phase of the coefficient C—and the first two terms of the preceding equation are calculated as

$$C\mathcal{L}^{-1}\left\{\frac{1}{s-j\omega}\right\} + C^*\mathcal{L}^{-1}\left\{\frac{1}{s+j\omega}\right\} = |C|e^{j\measuredangle C}e^{j\omega t}|C|e^{-j\measuredangle C}e^{-j\omega t}$$

$$= |C|\left\{e^{j\omega t+\measuredangle C} + e^{-j(\omega t+\measuredangle C)}\right\} = 2|C|\cos(\omega t + \measuredangle C)$$

with the coefficient C being calculated as the residue of the complex function $\mathcal{Y}(s)$ at the position of the simple pole $s = j\omega$.[14] We will have, therefore,

$$C = \lim_{s\to j\omega}[(s-j\omega)\mathcal{Y}(s)] = \lim_{s\to j\omega}\left[(s-j\omega)\frac{A\mathcal{H}(s)(s\cos\varphi - \omega\sin\varphi)}{s^2+\omega^2}\right] = \lim_{s\to j\omega}\left[\frac{A\mathcal{H}(s)(s\cos\varphi - \omega\sin\varphi)}{s+j\omega}\right] =$$

$$= \frac{A\mathcal{H}(j\omega)(j\omega\cos\varphi - \omega\sin\varphi)}{2j\omega} = \frac{j\omega}{2j\omega}\mathcal{H}(j\omega)(\cos\varphi + j\sin\varphi) = \frac{A}{2}\mathcal{H}(j\omega)e^{j\varphi} = \frac{A}{2}|\mathcal{H}(j\omega)|e^{j(\varphi+\measuredangle\mathcal{H}(j\omega))}$$

where for the derivation of the last relation we formulated the transfer function $\mathcal{H}(s)$ calculated at the position $s = j\omega$, in its exponential form $\mathcal{H}(j\omega) = |\mathcal{H}(j\omega)|\exp[j\measuredangle\mathcal{H}(j\omega)]$, from which it immediately follows that

$$|C| = \frac{A}{2}|\mathcal{H}(j\omega)| \qquad \text{and} \qquad \measuredangle C = \varphi + \measuredangle\mathcal{H}(j\omega)$$

Replacing the foregoing expressions in the expression of the inverse Laplace transform of the first two terms of the function $\mathcal{Y}(s)$, it takes the form

$$\cdots = 2|C|\cos(\omega t + \measuredangle C) = A|\mathcal{H}(j\omega)|\cos(\omega t + \varphi + \measuredangle\mathcal{H}(j\omega))$$

On the other hand, the Laplace transform of the third term is calculated as

$$\sum_{i=1}^{N} C_i\mathcal{L}^{-1}\left\{\frac{1}{s-p_i}\right\} = \sum_{i=1}^{N} C_1 e^{p_i t}$$

that is, the frequency response of the system will be given by the relation

$$y(t) = A|\mathcal{H}(j\omega)|\cos(\omega t + \varphi + \measuredangle\mathcal{H}(j\omega)) + \sum_{i=1}^{N} C_1 e^{p_i t}$$

From the last relation we observe that the second term vanishes in the limit $t \to \infty$[15] and therefore corresponds to the transient response of the system, while the term

[14] The fact that the residue of the function $\mathcal{Y}(s)$ is calculated at the position $s = j\omega$ is another indication of the fact that what we say concerns only systems that are *BIBO stable* since only for these doeos the *region of convergence include the imaginary axis* so that we can set $s = j\omega$.

[15] Since the system is considered BIBO stable.

$$y_{ss}(t) = A|\mathcal{H}(j\omega)|\cos(\omega t + \varphi + \measuredangle\mathcal{H}(j\omega))$$

will continue to exist forever—since it is a cosine term—and which in the long term and as soon as the system's transient response weakens will become the *dominant term*. We find, therefore, that when we transmit as input to a stable linear and time-invariant system the elementary trigonometric signal $x(t) = A\cos(\omega t + \varphi)$, its output is also a trigonometric signal with the following characteristics:

- Its frequency ω is the same as the frequency ω of the input signal.
- Its amplitude $A|\mathcal{H}(j\omega)|$ is equal to the amplitude A of the input signal multiplied by the magnitude of the transfer function $\mathcal{H}(j\omega)$ calculated for the value of the frequency ω of the input signal.
- Its phase $\varphi + \measuredangle\mathcal{H}(j\omega)$ is equal to the phase φ of the input signal increased by the phase $\measuredangle\mathcal{H}(j\omega)$ of the transfer function $\mathcal{H}(j\omega)$ calculated for the value of the frequency ω of the input signal.

These conclusions are, of course, the same as those we reached in a previous chapter using other methods.

Since the amplitude of the input signal is multiplied by the factor $|\mathcal{H}(j\omega)|$ as it passes through the system, it is obvious that this factor, which is known as the amplitude response, will express the gain of the system. In a completely similar way, the factor $\measuredangle\mathcal{H}(j\omega)$, which is known as the phase response, describes the way in which the system modifies the phase of the input signal. Therefore, the graphical representations of the quantities $|\mathcal{H}(j\omega)|$ and $\measuredangle\mathcal{H}(j\omega)$ as a function of the frequency ω fully characterize the way in which the system responds to each input signal. Considering that these quantities were calculated at the position $s = j\omega$, it is obvious that they simply express the *magnitude and phase of the frequency response function of the system, which is commonly defined as the Fourier transform of the impulse response*. If these graphical representations are plotted on a logarithmic scale, they constitute a special type of diagram known as a *Bode diagram*; these diagrams will be studied in a subsequent section.

Problem 5.5.2 Determine and plot the amplitude and phase response of the system

$$\mathcal{H}(s) = \frac{s+3}{s+5}$$

and then calculate its response to the input signal $x(t) = 2\cos(20t - 30°)$.

Solution Proceeding with the substitution $s = j\omega$, the transfer function $\mathcal{H}(s)$ takes the form

$$\mathcal{H}(j\omega) = \frac{\mathcal{N}(j\omega)}{\mathcal{D}(j\omega)} = \frac{3 + j\omega}{5 + j\omega}$$

Expressing the functions $\mathcal{N}(j\omega)$ and $\mathcal{D}(j\omega)$ in exponential form, we get

$$\mathcal{N}(j\omega) = \sqrt{\omega^2 + 9}\,\exp\left[j\tan^{-1}\left(\frac{\omega}{3}\right)\right] \quad \text{and} \quad \mathcal{D}(j\omega) = \sqrt{\omega^2 + 25}\,\exp\left[j\tan^{-1}\left(\frac{\omega}{5}\right)\right]$$

Therefore, the function $\mathcal{H}(j\omega)$ is formulated as

$$\mathcal{H}(j\omega) = |\mathcal{H}(j\omega)|e^{j\measuredangle\mathcal{H}(j\omega)} = \sqrt{\frac{\omega^2 + 9}{\omega^2 + 25}}\,\exp\left\{j\left[\tan^{-1}\left(\frac{\omega}{3}\right) - \tan^{-1}\left(\frac{\omega}{5}\right)\right]\right\}$$

that is, the *amplitude response* and the *phase response* of the system in question will be given by the relations

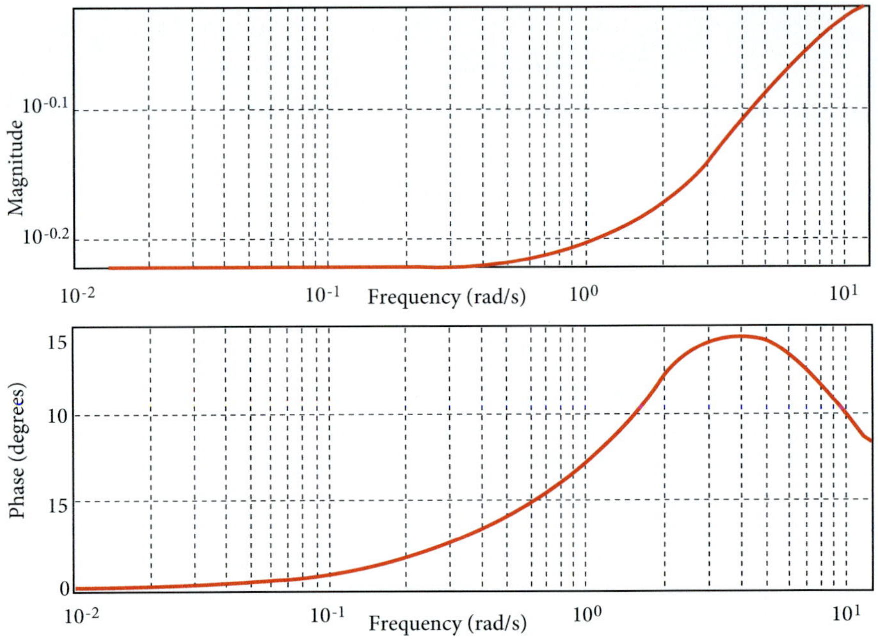

Fig. 5.3 Amplitude and phase response of system from Exercise 5.5.2

$$|\mathcal{H}(j\omega)| = \sqrt{\frac{\omega^2+9}{\omega^2+25}} \quad \text{and} \quad \measuredangle\mathcal{H}(j\omega) = \tan^{-1}\left(\frac{\omega}{3}\right) - \tan^{-1}\left(\frac{\omega}{5}\right)$$

The graphical representation of these two functions with respect to the frequency ω is presented in Fig. 5.3.

If we transmit the sinusoidal input signal $x(t) = 2\cos(20t - 30°)$, which is of the form $x(t) = A\cos(\omega t + \varphi)$, with parameter values $A = 2$, $\omega = 20$, and $\varphi = -30°$, the functions $|\mathcal{H}(j\omega)|$ and $\measuredangle\mathcal{H}(j\omega)$ for the value $\omega = 20$ are calculated as

$$|\mathcal{H}(20j)| = \sqrt{\frac{20^2+9}{20^2+25}} = \sqrt{\frac{409}{425}} = 0.980995 \quad \text{and}$$

$$\measuredangle\mathcal{H}(20j) = \tan^{-1}\left(\frac{20}{3}\right) - \tan^{-1}\left(\frac{20}{5}\right) = 81.469234° - 75.963756° = 5.505477°$$

Therefore, the system output $y(t)$ for this specific input signal will be a sinusoidal signal with the same frequency $\omega = 20$, amplitude equal to $2|\mathcal{H}(20j)| = 1.96199$, and phase $(-30 + 5.505477)° = -24.494523°$ and will be described by the equation

$$y(t) = 1.96199\cos(20t - 24.494523°)$$

Problem 5.5.3 Determine and design the amplitude and phase response of the system

$$\mathcal{H}(s) = \frac{s-2}{s^2+5s+6}$$

and then calculate its output when the signal $x(t) = 12\cos(15t + 34°)$ is transmitted to its input.

Solution When the substitution $s = j\omega$ is made, the transfer function $\mathcal{H}(s)$ takes the form

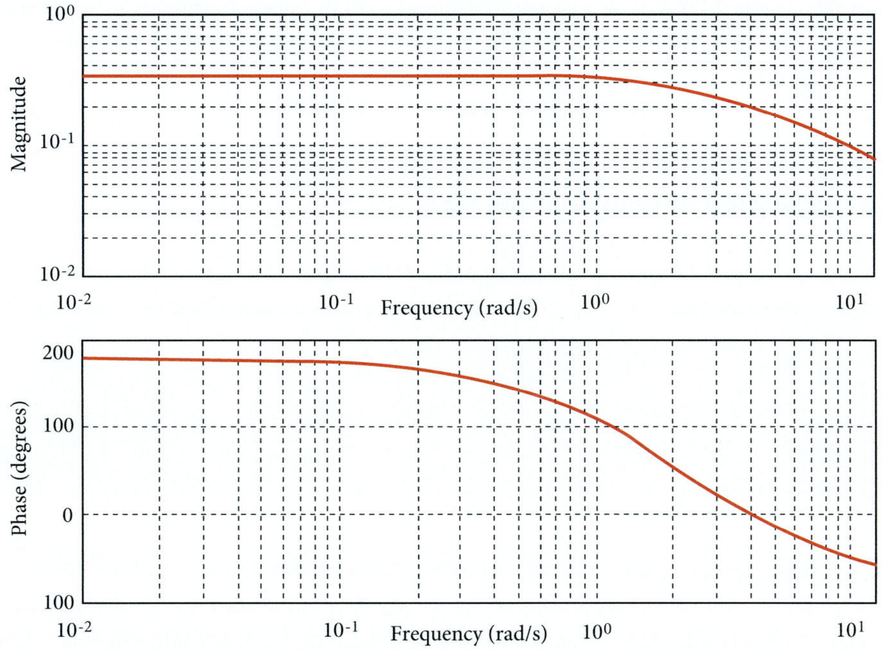

Fig. 5.4 Amplitude and phase response of system from Exercise 5.5.3

$$\mathcal{H}(j\omega) = \frac{\mathcal{N}(j\omega)}{\mathcal{D}(j\omega)} = \frac{j\omega - 2}{-\omega^2 + 5j\omega + 6} = \frac{-2 + j\omega}{(6-\omega^2) + 5j\omega}$$

Expressing the functions $\mathcal{N}(j\omega)$ and $\mathcal{D}(j\omega)$ in exponential form yields

$$\mathcal{N}(j\omega) = \sqrt{\omega^2 + 4}\, \exp\left[j \tan^{-1}\left(-\frac{\omega}{2}\right)\right] \quad \text{and} \quad \mathcal{D}(j\omega) = \sqrt{\omega^4 + 13\omega^2 + 36}\, \exp\left[j \tan^{-1}\left(\frac{5\omega}{6-\omega^2}\right)\right]$$

Therefore, the function $\mathcal{H}(j\omega)$ will take the form

$$\mathcal{H}(j\omega) = \sqrt{\frac{\omega^2 + 4}{\omega^4 + 13\omega^2 + 36}}\, \exp\left\{j\left[\tan^{-1}\left(-\frac{\omega}{2}\right) - \tan^{-1}\left(\frac{5\omega}{6-\omega^2}\right)\right]\right\}$$

That is, the *amplitude response* and the *phase response* of the system in question will be given by the relations

$$|\mathcal{H}(j\omega)| = \sqrt{\frac{\omega^2 + 4}{\omega^4 + 13\omega^2 + 36}} \quad \text{and} \quad \measuredangle\mathcal{H}(j\omega) = \tan^{-1}\left(-\frac{\omega}{2}\right) - \tan^{-1}\left(\frac{5\omega}{6-\omega^2}\right)$$

respectively. The variation of these functions with frequency ω is shown in Fig. 5.4.

If we transmit the sinusoidal input signal $x(t) = 12\cos(15t + 34°)$, which is of the form $x(t) = A\cos(\omega t + \varphi)$, with parameter values $A = 12$, $\omega = 15$, and $\varphi = 34°$, the functions $|\mathcal{H}(j\omega)|$ and $\measuredangle\mathcal{H}(j\omega)$ for the value $\omega = 15$ are calculated as

$$|\mathcal{H}(15j)| = \sqrt{\frac{15^2 + 4}{15^4 + 13(15)^2 + 36}} = \sqrt{\frac{229}{53586}} = 0.065372 \quad \text{and}$$

$$\measuredangle\mathcal{H}(20j) = \tan^{-1}\left(-\frac{15}{2}\right) - \tan^{-1}\left(-\frac{75}{219}\right) = -82.405356° - 18.904575° = -101.309931°$$

Therefore, the output of the system $y(t)$ will be a sinusoidal signal with the same frequency $\omega = 20$, amplitude $2|\mathcal{H}(15j)| = 0.784464$, and phase $(34 - 101.309931)^\circ = -67.309931^\circ$ and will be described by the equation

$$y(t) = 0.784464 \cos\left(15t - 67.309931^\circ\right)$$

5.6 Geometric Calculation of Transfer Function

Let us consider a *continuous-time LTI system* whose transfer function is characterized by the existence of N poles p_n $(n = 1, 2, \ldots, N)$ and M zero values z_m $(m = 1, 2, \ldots, M)$ so that it can be described by a rational expression of the form

$$\mathcal{H}(s) = K\frac{(s - z_1)(s - z_2)(s - z_3)\ldots(z - z_M)}{(s - p_1)(s - p_2)(s - p_3)\ldots(s - p_N)} = K\left\{\prod_{m=1}^{M}(s - z_m)\right\}\Big/\left\{\prod_{n=1}^{N}(s - p_n)\right\} \quad (5.27)$$

To facilitate our description, let us assume that the M zero points of the function, which are described by the vectors $\boldsymbol{z}_m = (\kappa_m, \lambda_m)$ $(m = 1, 2, \ldots, M)$, are represented as $z_m = \kappa_m + j\lambda_m$, where $\kappa_m = \Re\{z_m\}$ and $\lambda_m = \Im\{z_m\}$ $(m = 1, 2, \ldots, M)$, while in a completely analogous way the N poles of the transfer function, which are described by the vectors $\boldsymbol{p}_n = (\mu_n, \nu_n)$ $(n = 1, 2, \ldots, N)$, are represented as $p_n = \mu_n + j\nu_n$, where $\mu_n = \Re\{p_n\}$ and $\nu_n = \Im\{p_n\}$ $(n = 1, 2, \ldots, N)$, respectively. Having defined the preceding quantities and observing that *each quantity of the numerator and the denominator of the transfer function expresses a complex number*, we can formulate the proposition that if we wish to calculate the stated function at some point $s = \sigma + j\omega$ of the complex plane, which is also described by the vector $\boldsymbol{s} = (\sigma, \omega)$, we should use M vectors of the form $\boldsymbol{\chi}_m = \boldsymbol{s} - \boldsymbol{z}_m = r_m \exp(j\vartheta_m)$ $(m = 1, 2, \ldots, M)$ for the M representations of the numerator and N vectors of the form $\boldsymbol{\xi}_n = \boldsymbol{s} - \boldsymbol{p}_n = q_n \exp(j\varphi_n)$ $(n = 1, 2, \ldots, N)$ for the N representations of the denominator,[16] where

$$r_m = |\chi_m| = |\boldsymbol{s} - \boldsymbol{z}_m| = \sqrt{(\sigma - \kappa_m)^2 + (\omega - \lambda_m)^2}$$

$$\vartheta_m = \measuredangle\chi_m = \measuredangle(\boldsymbol{s} - \boldsymbol{z}_m) = \tan^{-1}\left\{\frac{\omega - \lambda_m}{\sigma - \kappa_m}\right\}$$

$$q_n = |\xi_n| = |\boldsymbol{s} - \boldsymbol{p}_n| = \sqrt{(\sigma - \mu_n)^2 + (\omega - \nu_n)^2}$$

$$\varphi_n = \measuredangle\xi_n = \measuredangle(\boldsymbol{s} - \boldsymbol{p}_n) = \tan^{-1}\left\{\frac{\omega - \nu_n}{\sigma - \mu_n}\right\}$$

the magnitude and phase of the quantities $s - z_m$ $(m = 1, 2, \ldots, M)$ and $s - p_i$ $(i = 1, 2, \ldots, N)$, respectively. Replacing these representations in the defining equation of the transfer function, it is written

$$\mathcal{H}(s) = K\frac{r_1 e^{j\vartheta_1} r_2 e^{j\vartheta_2}\ldots r_M e^{j\vartheta_M}}{q_1 e^{j\varphi_1} q_2 e^{j\varphi_2}\ldots q_N e^{j\varphi_N}} = K\frac{(r_1 r_2 \ldots r_M)\, e^{j(\vartheta_1+\vartheta_2+\cdots+\vartheta_M)}}{(q_1 q_2 \ldots q_N)\, e^{j(\varphi_1+\varphi_2+\cdots+\varphi_N)}} =$$

$$= \frac{\prod_{m=1}^{M} r_m}{\prod_{n=1}^{N} q_n} \exp\left[j\left(\sum_{m=1}^{M}\vartheta_m - \sum_{n=1}^{N}\varphi_n\right)\right] = |\mathcal{H}(s)|\exp[j\measuredangle\mathcal{H}(s)] \quad (5.28)$$

with the magnitude and phase of the transfer function defined as

[16] This description is based on the assumption that all *poles and zeros are distinct from each other* and that the sets of pole and zero values *do not share common elements*.

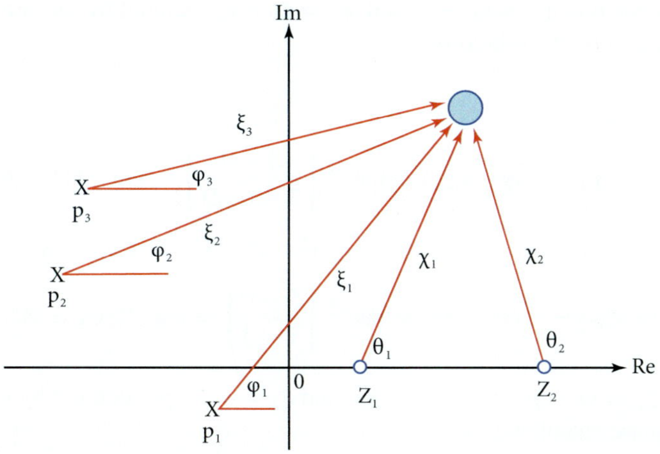

Fig. 5.5 Geometric calculation of transfer function of LTI system

$$|\mathcal{H}(s)| = \left\{\prod_{m=1}^{M} \sqrt{(\sigma-\kappa_m)^2+(\omega-\lambda_m)^2}\right\} \Big/ \left\{\prod_{n=1}^{N} \sqrt{(\sigma-\mu_n)^2+(\omega-\nu_n)^2}\right\} \tag{5.29}$$

$$\measuredangle\mathcal{H}(s) = \sum_{m=1}^{M}\left(\tan^{-1}\left\{\frac{\omega-\lambda_m}{\sigma-\kappa_m}\right\}\right) - \sum_{n=1}^{N}\left(\tan^{-1}\left\{\frac{\omega-\nu_n}{\sigma-\mu_n}\right\}\right) \tag{5.30}$$

where we have replaced the $q_n, \varphi_n, r_m, \vartheta_m$ $(n = 1, \ldots, N \ \& \ m = 1, \ldots, M)$ with their equivalent expressions.

Using the last two equations, we can calculate the magnitude and phase of the transfer function of the system at a given point in the complex plane $s = \sigma + j\omega$ (but *not* the value of the proportionality coefficient K) when we know the positions of the poles and zeros that describe it. We observe that the magnitude of the transfer function *is proportional to the product of the magnitudes of the geometric distances of each zero point from the considered point s divided by the product of the magnitudes of the geometric distances of the poles from the considered point*, while the phase of the transfer function *is equal to the sum of the angles of the vectors associated with the zero values minus the sum of the angles of the vectors associated with the positions of the poles*. The process of geometric calculation of a transfer function characterized by the existence of three poles and two zero values is presented diagrammatically in Fig. 5.5.

Problem 5.6.1 Use the geometric method to calculate the transfer function

$$\mathcal{H}(s) = \frac{s^2 - s - 6}{s^3 + 8s^2 + 29s + 52}$$

at positions (a) $s_1 = 3 + 5j$ and (b) $s_2 = -4 + j$.

Solution Factoring the polynomials of the numerator and denominator, we may write the function $\mathcal{H}(s)$ as

$$\mathcal{H}(s) = \frac{s^2 - s - 6}{s^3 + 8s^2 + 29s + 52} = \frac{(s+2)(s-3)}{(s^2+4s+13)(s+4)} = \frac{(s+2)(s-3)}{(s+2-3j)(s+2+3j)(s+4)}$$

from which it is observed that it has two zero values at the positions $z_1 = -2$ and $z_2 = 3$ and three poles at the positions $p_1 = -2 + 3j$, $p_2 = -2 - 3j$, and $p_3 = -4$. The details of the geometric calculation of the function $\mathcal{H}(s)$ at the positions $s_1 = 3 + 5j$ and $s_2 = -4 + j$ are represented diagrammatically in Fig. 5.6a,b, while the analytical process is as follows:

For case (a) $s_1 = 3 + 5j$, the vectors $\boldsymbol{\chi}_1 = s_1 - z_1$ and $\boldsymbol{\chi}_2 = s_1 - z_2$, defined by the zero values z_1 and z_2 and the point s_1, have magnitude and phase given by the relations

$$r_1 = |\boldsymbol{\chi}_1| = |s_1 - z_1| = \sqrt{[3-(-2)]^2 + (5-0)^2} = \sqrt{50} = 5\sqrt{2}$$

$$\vartheta_1 = \measuredangle \boldsymbol{\chi}_1 = \measuredangle (s_1 - z_1) = \tan^{-1}\left\{\frac{5-0}{3-(-2)}\right\} = \tan^{-1}\{1\} = 45^\circ$$

$$r_2 = |\boldsymbol{\chi}_2| = |s_1 - z_2| = \sqrt{(3-3)^2 + (5-0)^2} = \sqrt{25} = 5$$

$$\vartheta_2 = \measuredangle \boldsymbol{\chi}_2 = \measuredangle (s_1 - z_2) = \tan^{-1}\left\{\frac{5-0}{3-3}\right\} = \tan^{-1}\{\infty\} = 90^\circ$$

On the other hand, the vectors $\boldsymbol{\xi}_1 = s_1 - \boldsymbol{p}_1$, $\boldsymbol{\xi}_2 = s_1 - \boldsymbol{p}_2$, and $\boldsymbol{\xi}_3 = s_1 - \boldsymbol{p}_3$, defined by the poles $\boldsymbol{p}_1$, $\boldsymbol{p}_2$, and $\boldsymbol{p}_3$ and the point s_1, have magnitude and phase calculated as

$$q_1 = |\boldsymbol{\xi}_1| = |s_1 - \boldsymbol{p}_1| = \sqrt{[3-(-2)]^2 + (5-3)^2} = \sqrt{25+4} = \sqrt{29}$$

$$\varphi_1 = \measuredangle \boldsymbol{\xi}_1 = \measuredangle (s_1 - \boldsymbol{p}_1) = \tan^{-1}\left\{\frac{5-3}{3-(-2)}\right\} = \tan^{-1}\left\{\frac{2}{5}\right\} = 21.801409^\circ$$

$$q_2 = |\boldsymbol{\xi}_2| = |s_1 - \boldsymbol{p}_2| = \sqrt{[3-(-2)]^2 + [5-(-3)]^2} = \sqrt{25+64} = \sqrt{89}$$

$$\varphi_2 = \measuredangle \boldsymbol{\xi}_2 = \measuredangle (s_1 - \boldsymbol{p}_2) = \tan^{-1}\left\{\frac{5-(-3)}{3-(-2)}\right\} = \tan^{-1}\left\{\frac{8}{5}\right\} = 57.994616^\circ$$

$$q_3 = |\boldsymbol{\xi}_3| = |s_1 - \boldsymbol{p}_3| = \sqrt{[3-(-4)]^2 + (5-0)^2} = \sqrt{49+25} = \sqrt{74}$$

$$\varphi_3 = \measuredangle \boldsymbol{\xi}_3 = \measuredangle (s_1 - \boldsymbol{p}_3) = \tan^{-1}\left\{\frac{5-0}{3-(-4)}\right\} = \tan^{-1}\left\{\frac{5}{7}\right\} = 35.537677^\circ$$

Substituting the preceding expressions into the equation for calculating the transfer function using the geometric method, we get

$$\begin{aligned}
\mathcal{H}(s_1) = \mathcal{H}(3+5j) &= \frac{\prod_{m=1}^{2} r_m}{\prod_{n=1}^{3} q_n} \exp\left[j\left(\sum_{m=1}^{2}\vartheta_m - \sum_{n=1}^{3}\varphi_n\right)\right] \\
&= \frac{r_1 r_2}{q_1 q_2 q_3} \exp\left[j\left(\vartheta_1 + \vartheta_2 - \varphi_1 - \varphi_2 - \varphi_3\right)\right] \\
&= \frac{(5\sqrt{2})5}{\sqrt{29}\sqrt{89}\sqrt{74}} \times \exp\Big[j\,(45^\circ + 90^\circ - 21.801409^\circ \\
&\quad - 57.994616^\circ - 35.537677^\circ)\Big] \\
&= \frac{25\sqrt{2}}{\sqrt{190994}} e^{j19.666298^\circ} = 0.080899 e^{j19.666298^\circ}
\end{aligned}$$

This result is correct, since if we calculate directly the value of the expression $\mathcal{H}(3+5j)$ by replacing the value $s = 3+5j$ in the defining equation of the function $\mathcal{H}(s)$, we will get

$$\begin{aligned}
\mathcal{H}(3+5j) &= \frac{(3+5j)^2 - (3+5j) - 6}{(3+5j)^3 + 8(3+5j)^2 + 29(3+5j) + 52} = \frac{-25+25j}{-187+395j} \\
&= \frac{25\sqrt{2}e^{-j45^\circ}}{\sqrt{190994}e^{-j64.666295}} = 0.080899 e^{j19.666295^\circ}
\end{aligned}$$

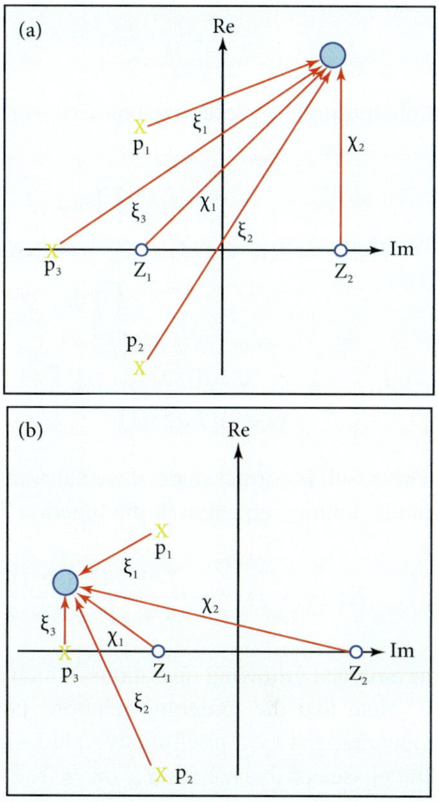

Fig. 5.6 Geometric calculation of function $\mathcal{H}(s)$ (Exercise 5.6.1) at positions $s_1 = 3 + 5j$ and $s_2 = -4 + j$

as emerges following operations, which is the same result as the one we obtained through the application of the *geometric method*. The difference in the sixth decimal digit of the angle in the final result has to do with the accuracy of the calculations and, of course, is considered negligible.

For case (b) $s_2 = -4 + j$, the vectors $\boldsymbol{\chi}_1 = s_2 - z_1$ and $\boldsymbol{\chi}_2 = s_2 - z_2$, defined by the zero values z_1 and z_2 and the point s_2, have magnitude and phase given by the relations

$$r_1 = |\boldsymbol{\chi}_1| = |s_2 - z_1| = \sqrt{[-4-(-2)]^2 + (1-0)^2} = \sqrt{4+1} = \sqrt{5}$$

$$\vartheta_1 = \measuredangle \boldsymbol{\chi}_1 = \measuredangle(s_2 - z_1) = \tan^{-1}\left\{\frac{1-0}{-4-(-2)}\right\} = \tan^{-1}\left\{-\frac{1}{2}\right\} = -26.565051^\circ$$

$$r_2 = |\boldsymbol{\chi}_2| = |s_2 - z_2| = \sqrt{(-4-3)^2 + (1-0)^2} = \sqrt{49+1} = \sqrt{50} = 5\sqrt{2}$$

$$\vartheta_2 = \measuredangle \boldsymbol{\chi}_2 = \measuredangle(s_2 - z_2) = \tan^{-1}\left\{\frac{1-0}{-4-3}\right\} = \tan^{-1}\left\{-\frac{1}{7}\right\} = -8.130102^\circ$$

while the vectors $\boldsymbol{\xi}_1 = s_1 - \boldsymbol{p}_1$, $\boldsymbol{\xi}_2 = s_1 - \boldsymbol{p}_2$, and $\boldsymbol{\xi}_3 = s_1 - \boldsymbol{p}_3$, defined by the poles $\boldsymbol{p}_1$, $\boldsymbol{p}_2$, and $\boldsymbol{p}_3$ and the corresponding point, have magnitude and phase that are calculated as

$$q_1 = |\boldsymbol{\xi}_1| = |s_2 - \boldsymbol{p}_1| = \sqrt{[-4-(-2)]^2 + (1-3)^2} = \sqrt{4+4} = 2\sqrt{2}$$

$$\varphi_1 = \measuredangle \boldsymbol{\xi}_1 = \measuredangle(s_2 - \boldsymbol{p}_1) = \tan^{-1}\left\{\frac{1-3}{-4-(-2)}\right\} = \tan^{-1}\{1\} = 45^\circ$$

$$q_2 = |\boldsymbol{\xi}_2| = |s_2 - \boldsymbol{p}_2| = \sqrt{[-4-(-2)]^2 + [1-(-3)]^2} = \sqrt{4+16} = 2\sqrt{5}$$

$$\varphi_2 = \measuredangle \boldsymbol{\xi}_2 = \measuredangle(s_2 - \boldsymbol{p}_2) = \tan^{-1}\left\{\frac{1-(-3)}{-4-(-2)}\right\} = \tan^{-1}\{-2\} = -63.434948^\circ$$

$$q_3 = |\boldsymbol{\xi}_3| = |s_2 - \boldsymbol{p}_3| = \sqrt{[-4-(-4)]^2 + (1-0)^2} = \sqrt{1} = 1$$

$$\varphi_3 = \measuredangle \boldsymbol{\xi}_3 = \measuredangle(s_2 - \boldsymbol{p}_3) = \tan^{-1}\left\{\frac{1-0}{-4-(-4)}\right\} = \tan^{-1}\{\infty\} = 90^\circ$$

Substituting the preceding expressions into the transfer function calculation equation, we get

$$\mathcal{H}(s_2) = \mathcal{H}(-4+j) = \frac{\prod_{m=1}^{2} r_m}{\prod_{n=1}^{3} q_n} \exp\left[j\left(\sum_{m=1}^{2}\vartheta_m - \sum_{n=1}^{3}\varphi_n\right)\right] = \frac{r_1 r_2}{q_1 q_2 q_3}\exp\left[j\,(\vartheta_1 + \vartheta_2 - \varphi_1 - \varphi_2 - \varphi_3)\right]$$

$$= \frac{\sqrt{5}(5\sqrt{2})}{(2\sqrt{2})(2\sqrt{5})(1)} \exp\left[j(-26.565^\circ - 8.130^\circ - 45^\circ + 63.434^\circ - 90^\circ)\right] = \frac{5}{4}e^{-j106.260^\circ} = 1.25e^{-j106.260^\circ}$$

This result is correct since, if we calculate directly the value of the expression $\mathcal{H}(3+5j)$ by substituting the value $s = -4+j$ in the defining equation of the function $\mathcal{H}(s)$, we obtain

$$\mathcal{H}(-4+j) = \frac{(-4+j)^2 - (-4+j) - 6}{(-4+j)^3 + 8(-4+j)^2 + 29(-4+j) + 52} = \frac{13-9j}{4+12j} = \frac{5\sqrt{10}e^{-j34.695^\circ}}{4\sqrt{10}e^{j71.565}} = 1.25e^{-j106.260^\circ}$$

as emerges following operations, which is the same result as the one obtained with the *geometric method*.

Note that the foregoing relations hold for poles and zero values that are *distinct* from each other—that is, they are characterized by a multiplicity value *equal to one*—while for larger multiplicity values we should count the measures and the phases of the vectors $\boldsymbol{\chi}_m$ $(m = 1, 2, \ldots, M)$ and $\boldsymbol{\xi}_n$ $(n = 1, 2, \ldots, N)$ as many times as is dictated by the value of the corresponding multiplicity. Considering, for example, a transfer function with *two zeros* at the positions z_1, z_2, and z_3 with multiplicity values $m_1 = 1$, $m_2 = 3$, and $m_3 = 2$ and *two poles* at the positions p_1 and p_2 with multiplicity values $n_1 = 2$ and $n_2 = 3$, the transfer function through the application of the geometric method is calculated as

$$\begin{aligned}\mathcal{H}(s) &= K\frac{r_1(r_2r_2r_2)r_3}{(q_1q_1)(q_2q_2q_2)}\exp[\,j\,(\vartheta_1 + \underbrace{\vartheta_2 + \vartheta_2 + \vartheta_2} + \vartheta_3 - \underbrace{\varphi_1 + \varphi_1} + \underbrace{\varphi_2 + \varphi_2 + \varphi_2})\,] \\ &= K\frac{r_1 r_2^3 r_3}{q_1^2 + q_2^3}\exp[\,j\,(\vartheta_1 + 3\vartheta_2 + \vartheta_3) - (2\varphi_1 + 3\varphi_2)\,]\end{aligned}$$

while for the general case of a transfer function with M zero values with multiplicities $m_1, m_2, \ldots, m_M$ and N poles with multiplicity values $n_1, n_2, \ldots, n_N$ the magnitude and phase of the transfer function are given by the relations

$$|\mathcal{H}(s)| = \prod_{j=1}^{M}(r_j)^{m_j} \Big/ \prod_{i=1}^{N}(q_i)^{n_i} \qquad \text{and} \qquad \measuredangle\mathcal{H}(s) = \sum_{j=1}^{M} m_j\vartheta_j - \sum_{i=1}^{N} n_i\varphi_i$$

5.6.1 The Case of the Function $\mathcal{H}(j\omega)$

The geometric method of calculating the transfer function $\mathcal{H}(s)$ presented in the previous section can be used without any modification for the calculation of the *frequency response function* of the system, since this is calculated on points of the *imaginary axis* $s = j\omega$[17] and therefore arises from the function $\mathcal{H}(s)$ if in this we set $\sigma = \Re\{s\} = 0$. In this case, the transfer function takes the form

[17] Recall that the transfer function and the frequency response function are defined as the *Laplace and Fourier transforms of the system's impulse response* and that the Fourier transform arises from the Laplace transform, if in the latter we set $s = j\omega$ or, equivalently, $\sigma = \Re\{s\} = 0$.

$$\mathcal{H}(j\omega) = K\frac{(j\omega - z_1)(j\omega - z_2)(j\omega - z_3)\ldots(j\omega - z_M)}{(j\omega - p_1)(j\omega - p_2)(j\omega - p_3)\ldots(j\omega - p_N)} = K\left\{\prod_{m=1}^{M}(j\omega - z_m)\right\}\Big/\left\{\prod_{n=1}^{N}(j\omega - p_n)\right\} \quad (5.31)$$

and refers to the *frequency response function of the system*, while the equations for calculating the measure and phase of the function of the positions of the poles and the zero values are formulated as

$$|\mathcal{H}(s)| = \left\{\prod_{m=1}^{M}\sqrt{\kappa_m^2 + (\omega - \lambda_m)^2}\right\}\Big/\left\{\prod_{n=1}^{N}\sqrt{\mu_n^2 + (\omega - \nu_n)^2}\right\} \quad (5.32)$$

$$\measuredangle\mathcal{H}(s) = \sum_{m=1}^{M}\left(\tan^{-1}\left\{\frac{\omega - \lambda_m}{-\kappa_m}\right\}\right) - \sum_{n=1}^{N}\left(\tan^{-1}\left\{\frac{\omega - \nu_n}{-\mu_n}\right\}\right) \quad (5.33)$$

Recall that the measure of the transfer function given by the foregoing equation should be multiplied by the value of the gain K, which unfortunately cannot be calculated by applying this geometric method. However, the *knowledge of the positions of the poles and the zero values of the Fourier transform of the impulse response allows us to draw qualitative conclusions about the characteristics of this magnitude*. The diagrammatic representation of this technique for the case of the function $\mathcal{H}(j\omega)$ will be presented through the next exercises; however, it is not difficult to see that since this function is defined at points of the form $s = j\omega$, all vectors carried from the positions of the poles and the zero values will end *at some point on the imaginary axis*.

Problem 5.6.2 Using the geometric method, calculate in any case point $s = j\omega$ of the imaginary axis the transfer function

$$\mathcal{H}(s) = \frac{s^2 - s - 1}{s^2 + s + 1}$$

Solution Factoring the polynomials of the numerator and the denominator, we find that the function $\mathcal{H}(s)$ has two zero values at the positions

$$z_1 = \kappa_1 + j\lambda_1 = \frac{1}{2} + j\frac{\sqrt{3}}{2} \quad \text{and} \quad z_2 = \kappa_2 + j\lambda_2 = \frac{1}{2} - j\frac{\sqrt{3}}{2}$$

and two poles at the positions

$$p_1 = \mu_1 + j\nu_1 = -\frac{1}{2} + j\frac{\sqrt{3}}{2} \quad \text{and} \quad p_2 = \mu_2 + j\nu_2 = -\frac{1}{2} - j\frac{\sqrt{3}}{2}$$

which are depicted in Fig. 5.7, along with the vectors $\boldsymbol{\chi}_1, \boldsymbol{\chi}_2$ and $\boldsymbol{\xi}_1, \boldsymbol{\xi}_2$ defined by them. By substituting the preceding values of the quantities $\kappa_m, \lambda_m, \mu_n, \nu_n$ $(i, j = 1, 2)$ into the equations defining the magnitude and phase of the frequency response, it turns out that

$$|\mathcal{H}(j\omega)| = \frac{\prod_{m=1}^{2}\sqrt{k_m^2 + (\omega - \lambda_m^2)}}{\prod_{n=1}^{2}\sqrt{\mu_n^2 + (\omega - \nu_n)^2}} = \frac{\sqrt{\frac{1}{4} + \left(\omega - \frac{\sqrt{3}}{2}\right)^2}\sqrt{\frac{1}{4} + \left(\omega + \frac{\sqrt{3}}{2}\right)^2}}{\sqrt{\frac{1}{4} + \left(\omega - \frac{\sqrt{3}}{2}\right)^2}\sqrt{\frac{1}{4} + \left(\omega + \frac{\sqrt{3}}{2}\right)^2}} = 1 \quad \text{and}$$

$$\measuredangle \mathcal{H}(j\omega) = \sum_{m=1}^{2} \left(\tan^{-1} \left\{ \frac{\omega - \lambda_m}{-\kappa_m} \right\} \right) + \sum_{n=1}^{2} \left(\tan^{-1} \left\{ \frac{\omega - \nu_n}{-\mu_n} \right\} \right) \Bigg] = \tan^{-1} \left[-2 \left(\omega - \frac{\sqrt{3}}{2} \right) \right]$$

$$+ \tan^{-1} \left[-2 \left(\omega + \frac{\sqrt{3}}{2} \right) \right] - \tan^{-1} \left[2 \left(\omega - \frac{\sqrt{3}}{2} \right) \right] - \tan^{-1} \left[2 \left(\omega + \frac{\sqrt{3}}{2} \right) \right]$$

$$= 2 \tan^{-1} \left[-2 \left(\omega - \frac{\sqrt{3}}{2} \right) \right] + 2 \tan^{-1} \left[-2 \left(\omega + \frac{\sqrt{3}}{2} \right) \right]$$

From mathematical analysis, we know that

$$\tan^{-1} x = \frac{j}{2} \ln \frac{1 - jx}{1 + jx} \quad \text{or, equivalently,} \quad 2 \tan^{-1} x = j \ln \frac{1 - jx}{1 + jx}$$

and the preceding relation is written

$$\measuredangle \mathcal{H}(j\omega) = j \ln \frac{1 + 2j \left(\omega - \frac{\sqrt{3}}{2} \right)}{1 - 2j \left(\omega - \frac{\sqrt{3}}{2} \right)} + j \ln \frac{1 + 2j \left(\omega + \frac{\sqrt{3}}{2} \right)}{1 - 2j \left(\omega + \frac{\sqrt{3}}{2} \right)} = j \ln \frac{\left[1 + 2j \left(\omega - \frac{\sqrt{3}}{2} \right) \right] \left[1 + 2j \left(\omega + \frac{\sqrt{3}}{2} \right) \right]}{\left[1 - 2j \left(\omega - \frac{\sqrt{3}}{2} \right) \right] \left[1 - 2j \left(\omega + \frac{\sqrt{3}}{2} \right) \right]}$$

$$= j \ln \frac{1 + j\omega - \omega^2}{1 - j\omega - \omega^2} = j \ln \frac{(1 - \omega^2) + j\omega}{(1 - \omega^2) - i\omega}$$

This result is correct since if we substitute the magnitude and the phase we calculated previously into the function $\mathcal{H}(j\omega)$, it will be written as

$$\mathcal{H}(j\omega) = |\mathcal{H}(j\omega)| e^{j \measuredangle \mathcal{H}(j\omega)} = \exp \left[-\ln \frac{(1 - \omega^2) + j\omega}{(1 - \omega^2) - j\omega} \right] = \frac{(1 - \omega^2) - j\omega}{(1 - \omega^2) + j\omega}$$

a result that is the same as the one we would get if we replaced the value $s = j\omega$ in the defining equation of the function $\mathcal{H}(s)$.

Problem 5.6.3 Use the geometric method to calculate, at the position $s = j\omega$, the magnitude and phase of

$$\mathcal{H}(s) = \frac{s^2 - s - 6}{s^3 + 8s^2 + 29s + 52}$$

Solution As we found in a previous exercise (Exercise 5.6.1, page 419), the function $\mathcal{H}(s)$ has two zero values and three poles at the positions

$$z_1 = \kappa_1 + i\lambda_1 = -2, \quad z_2 = \kappa_2 + i\lambda_2 = 3, \quad p_1 = \mu_1 + i\nu_1 = -2 + 3i, \quad p_2 = \mu_2 + i\nu_2 = -2 - 3i, \quad p_3 = \mu_3 + i\nu_3 = -4$$

which are graphically depicted in Fig. 5.8.

By substituting the values of the foregoing parameters into the equation of the magnitude of the function $\mathcal{H}(j\omega)$ and performing the operations, it turns out that

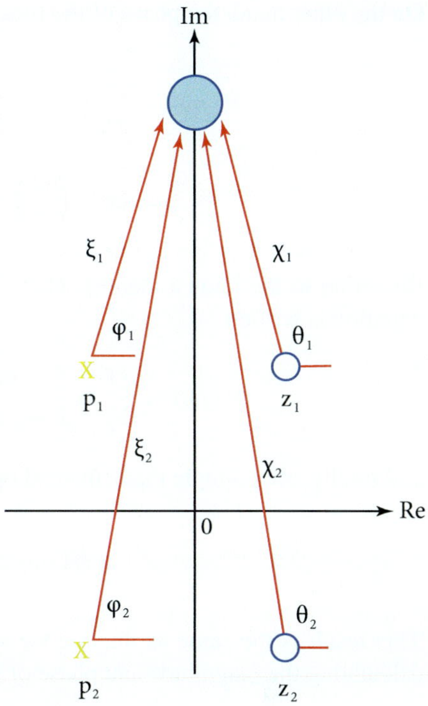

Fig. 5.7 Geometric calculation of transfer function $\mathcal{H}(s)$ from Exercise 5.6.2 at position $s = i\omega$

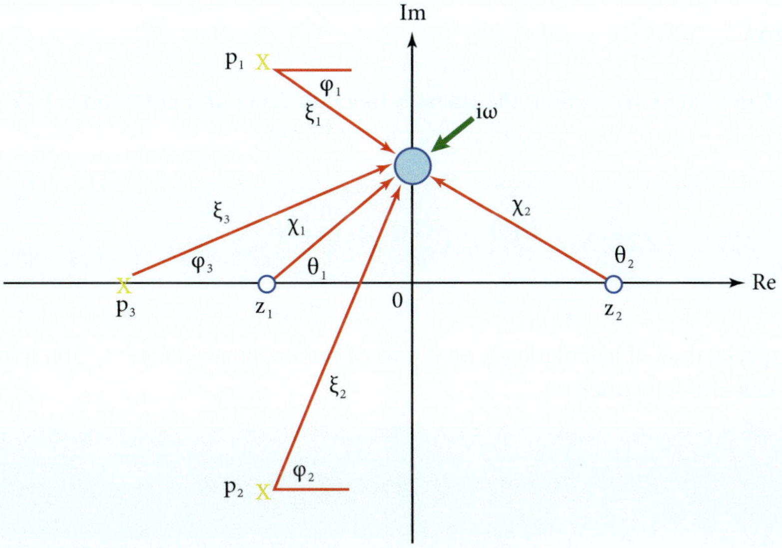

Fig. 5.8 Geometric calculation of function $\mathcal{H}(s)$ from Exercise 5.6.3 at position $s = j\omega$

$$|\mathcal{H}(j\omega)| = \frac{\prod_{m=1}^{2} \sqrt{\kappa_m^2 + (\omega - \lambda_m)^2}}{\prod_{n=1}^{3} \sqrt{\mu_n^2 + (\omega - \nu_n)^2}} = \frac{\sqrt{\kappa_1^2 + (\omega - \lambda_1)^2}\sqrt{\kappa_2^2 + (\omega - \lambda_2)^2}}{\sqrt{\mu_1^2 + (\omega - \nu_1)^2}\sqrt{\mu_2^2 + (\omega - \nu_2)^2}\sqrt{\mu_3^2 + (\omega - \nu_3)^2}} =$$

$$= \frac{\sqrt{4 + \omega^2}\sqrt{9 + \omega^2}}{\sqrt{4 + (\omega - 3)^2}\sqrt{4 + (\omega + 3)^2}\sqrt{16 + \omega^2}} = \sqrt{\frac{\omega^4 + 13\omega^2 + 36}{\omega^6 + 6\omega^4 + 9\omega^2 + 2704}}$$

On the other hand, the phase of the function $\mathcal{H}(j\omega)$ is calculated as

$$\begin{aligned}\measuredangle\mathcal{H}(j\omega) &= \sum_{j=1}^{2}\left(\tan^{-1}\left\{\frac{\omega-\lambda_j}{-\kappa_j}\right\}\right) - \sum_{i=1}^{3}\left(\tan^{-1}\left\{\frac{\omega-\nu_i}{-\mu_i}\right\}\right)\\ &= \tan^{-1}\left(\frac{\omega}{2}\right) + \tan^{-1}\left(-\frac{\omega}{3}\right) - \tan^{-1}\left(\frac{\omega-3}{2}\right) - \tan^{-1}\left(\frac{\omega+3}{2}\right) - \tan^{-1}\left(\frac{\omega}{4}\right)\end{aligned}$$

Resorting to the known identity $2\tan^{-1}(x) = j\ln[(1-jx)/(1+jx)]$ (see also Exercise 5.6.2, page 423) the preceding equation is written

$$\mathcal{H}(j\omega) = \frac{j}{2}\left[\ln\frac{2-j\omega}{2+j\omega} + \ln\frac{3+j\omega}{3-j\omega} - \ln\frac{2-j\omega+3j}{2+j\omega-3j} + \ln\frac{2-j\omega-3j}{2+j\omega+3j} - \ln\frac{4-j\omega}{4+j\omega}\right]$$

and finally, after simple mathematical operations,

$$\measuredangle\mathcal{H}(j\omega) = \frac{j}{2}\ln\left\{\frac{[(6+\omega^2)-j\omega][(52-8\omega^2)-j(\omega^3-29)]}{[(6+\omega^2)+j\omega][(52-8\omega^2)+j(\omega^3-29)]}\right\}$$

This result is the same as the one we would obtain by directly substituting the value $s = j\omega$ into the function $\mathcal{H}(s)$ and calculating the magnitude and phase of the complex expression that will result.

5.7 Bode Diagrams

As was already mentioned in a previous section, the transfer function $\mathcal{H}(s)$ of a continuous LTI system—being a complex function of a complex variable—can be expressed in its exponential form

$$\mathcal{H}(s) = |\mathcal{H}(s)|\exp[j\measuredangle\mathcal{H}(s)] \tag{5.34}$$

This function, for the special case of calculating it on points of the imaginary axis $s = j\omega$, transforms into the *frequency response function*, which can also be written as

$$\mathcal{H}(j\omega) = |\mathcal{H}(j\omega)|\exp[j\measuredangle\mathcal{H}(j\omega)] \tag{5.35}$$

In this expression, the functions $|\mathcal{H}(j\omega)|$ and $\measuredangle\mathcal{H}(j\omega)$, which are calculated either by directly substituting the value $s = j\omega$ into the defining equation of the transfer function $\mathcal{H}(s)$ or using the *pole-zero diagram* and in the manner described in the previous section, express the *amplitude and phase response* of the system, and their graphical representation in relation to the frequency ω of the input signal constitutes a special category of diagrams known as *Bode diagrams*.

Definition 5.7.1 Bode diagrams are graph structures depicting the steady-state response of a stable, continuous-time LTI system, when a basic trigonometric signal characterized by a certain cyclic frequency ω_0 is transmitted to its input.

The construction of the Bode diagram of a frequency response function $\mathcal{H}(j\omega)$ requires the formulation of the expression associated *with each pole or zero value of the function*—which we assume is located at some position $s = \alpha$ of the complex plane—as a term of the form

$$\left(1+\frac{s}{\alpha}\right)$$

To follow the established symbolism, we consider that the transfer function has M zero values at the positions $-z_1, -z_2, \ldots, -z_M$ and N poles at the positions $-p_1, -p_2, \ldots, -p_N$, so that it can be written in the form

$$\mathcal{H}(s) = K\,\frac{(s+z_1)(s+z_2)(s+z_3)\ldots(s+z_M)}{(s+p_1)(s+p_2)(s+p_3)\ldots(s+p_N)}$$

$$= K\,\frac{z_1\left(1+\frac{s}{z_1}\right)z_2\left(1+\frac{s}{z_2}\right)z_3\left(1+\frac{s}{z_3}\right)\ldots z_M\left(1+\frac{s}{z_M}\right)}{p_1\left(1+\frac{s}{p_1}\right)p_2\left(1+\frac{s}{p_2}\right)p_3\left(1+\frac{s}{p_3}\right)\ldots p_N\left(1+\frac{s}{p_N}\right)} =$$

$$= K\,\frac{z_1z_2z_3\ldots z_M}{p_1p_2p_3\ldots p_N}\,\frac{\left(1+\frac{s}{z_1}\right)\left(1+\frac{s}{z_2}\right)\left(1+\frac{s}{z_3}\right)\ldots\left(1+\frac{s}{z_M}\right)}{\left(1+\frac{s}{p_1}\right)\left(1+\frac{s}{p_2}\right)\left(1+\frac{s}{p_3}\right)\ldots\left(1+\frac{s}{p_N}\right)}$$

$$= K\left\{\frac{\prod_{m=1}^{M} z_m}{\prod_{n=1}^{N} p_n}\right\}\left\{\frac{\prod_{m=1}^{M}\left(1+\frac{s}{z_m}\right)}{\prod_{n=1}^{N}\left(1+\frac{s}{p_n}\right)}\right\} = K'\left\{\frac{\prod_{m=1}^{M}\left(1+\frac{s}{z_m}\right)}{\prod_{n=1}^{N}\left(1+\frac{s}{p_n}\right)}\right\}$$

where, of course, the new constant K' is defined as

$$K' = K\left\{\left(\prod_{m=1}^{M} z_m\right)\Big/\left(\prod_{n=1}^{N} p_n\right)\right\}$$

If we formulate the function $\mathcal{H}(s)$ in its exponential form for the value $s = j\omega$ so as to construct the frequency response function, we will get $\mathcal{H}(j\omega) = |\mathcal{H}(j\omega)|\exp[j\measuredangle\mathcal{H}(j\omega)]$, with the magnitude and phase of this function being defined as

$$|\mathcal{H}(j\omega)| = \left|K'\left\{\frac{\prod_{m=1}^{M}\left(1+\frac{j\omega}{z_m}\right)}{\prod_{n=1}^{N}\left(1+\frac{j\omega}{p_n}\right)}\right\}\right| = |K'|\frac{\prod_{m=1}^{M}\left|1+\frac{j\omega}{z_m}\right|}{\prod_{n=1}^{N}\left|1+\frac{j\omega}{p_n}\right|} \tag{5.36}$$

$$\measuredangle\mathcal{H}(j\omega) = \measuredangle K' + \sum_{m=1}^{M}\measuredangle\left(1+\frac{j\omega}{z_m}\right) - \sum_{n=1}^{N}\measuredangle\left(1+\frac{j\omega}{p_n}\right) \tag{5.37}$$

respectively. We are led, therefore, to the conclusion that the magnitude and phase of the function $\mathcal{H}(j\omega)$ expressed in this form consist of a constant term—*the magnitude and phase of the factor K', respectively*—as well as from other $M+N$ terms that are expressed as a function of the positions of the poles and the zero values of the function on the complex plane. Having described all the preceding quantities, we can initially define the Bode diagram as a set of two graphs of the form

$$(x, y) = \left\{\omega, |\mathcal{H}(j\omega)|\right\} \quad \text{and} \quad (x, y) = \left\{\omega, \measuredangle\mathcal{H}(j\omega)\right\}$$

of which the first describes the *amplitude response*, that is, the way the measure of the frequency response function changes with the frequency ω of the input signal, and the second describes the *phase response*, that is, the corresponding change in the phase of the function $\mathcal{H}(j\omega)$. In other words, this means that both these diagrams contain the frequency values ω on the horizontal axis, which, as we will see later, is calibrated on a *logarithmic scale*.

Starting with the phase response, which is the simplest as it simply contains sums and differences, it can be constructed very easily, since all we have to do is *draw each elementary phase term separately and then add all these terms together*. However, this cannot be done for the amplitude $|\mathcal{H}(j\omega)|$, as it contains *products instead of sums*. We know, however, how to convert a product into a sum: we just have to take the *logarithm of the product expression*, since from mathematical analysis it is known that

$$\log\left(\prod_{i=1}^{N} x_i\right) = \log(x_1 x_2 \ldots x_N) = \log(x_1) + \log(x_2) + \cdots + \log(x_N) = \sum_{i=1}^{N} \log(x_i)$$

In reality, for the construction of the Bode diagram regarding the amplitude response, we do take not the simple logarithm of each quantity X but its representation in units of dB (decibels), which is known to be defined as $Y = 20\log_{10}(X)$. Using this relationship with the defining equation of the measure of the function $\mathcal{H}(j\omega)$, we get the equation

$$20\log_{10}(|\mathcal{H}(j\omega)|) = 20\log_{10}\left\{|K'|\frac{\prod_{m=1}^{M}\left|1+\frac{j\omega}{z_m}\right|}{\prod_{n=1}^{N}\left|1+\frac{j\omega}{p_n}\right|}\right\}$$

$$= 20\log_{10}(|K'|) + 20\log_{10}\left\{\prod_{m=1}^{M}\left|1+\frac{j\omega}{z_m}\right|\right\} - 20\log_{10}\left\{\prod_{n=1}^{N}\left|1+\frac{j\omega}{p_n}\right|\right\}$$

and finally

$$20\log_{10}(|\mathcal{H}(j\omega)|) = 20\log_{10}(|K'|) + \sum_{m=1}^{M} 20\log_{10}\left(\left|1+\frac{j\omega}{z_m}\right|\right) - \sum_{n=1}^{N} 20\log_{10}\left(\left|1+\frac{j\omega}{p_n}\right|\right)$$

In this way, the equation of the measure of the function $\mathcal{H}(j\omega)$ is also expressed using a set of simple sums and differences, which allows for the calculation of the measure of each elementary term and then the addition of all these values in order to calculate the value of the expression $|\mathcal{H}(j\omega)|$. Based on all the points made in the preceding discussion, we can redefine the Bode diagrams as diagrams of the form

$$(x, y) = \left\{\log_{10}\omega, 20\log_{10}|\mathcal{H}(j\omega)|\right\} \quad \text{and} \quad (x, y) = \left\{\log_{10}\omega, \measuredangle\mathcal{H}(j\omega)\right\}$$

We observe, therefore, that the amplitude diagram is of the log-log type in the sense that *both axes use the logarithmic scale*, while the phase diagram is of the log-lin type, since the *horizontal axis uses the logarithmic while the vertical axis uses the linear scale*. As for the units of measurement, the horizontal logarithmic axis of the circular frequency ω is expressed in radians per second (rad/s), the vertical axis of the logarithmic amplitude is measured in decibels, and the vertical axis of the phase is graduated in degrees.

It should be noted that this necessity of converting products into sums also arises in other cases, such as, for example, the calculation of the output of a LTI system, as well as the determination of the *frequency response function* of a *composite* system resulting from the *series interconnection of simpler systems*. Starting from the first case, we know that the output $y(t)$ of such a system is given by the relation $y(t) = h(t) * x(t)$, where $h(t)$ *is its impulse response* and $x(t)$ the input signal, and that according to convolution theorem it will hold that $\mathcal{Y}(j\omega) = \mathcal{H}(j\omega)\mathcal{X}(j\omega)$. Therefore, the magnitude and phase of the function $\mathcal{Y}(j\omega)$ will be calculated as $|\mathcal{Y}(j\omega)| = |\mathcal{H}(j\omega)||\mathcal{X}(j\omega)|$ and $\measuredangle\mathcal{Y}(j\omega) = \measuredangle\mathcal{H}(j\omega) + \measuredangle\mathcal{X}(j\omega)$, a case that is similar to the one we are dealing with now. In other words, the phase diagram of the function $\mathcal{Y}(j\omega)$ will result *by adding the phase diagrams* of the functions $\mathcal{H}(j\omega)$ and $\mathcal{X}(j\omega)$, while the magnitude of the function in question is easily derived using the *logarithmic scale*, since in this case we will get

$$\log_{10}|\mathcal{Y}(j\omega)| = \log_{10}|\mathcal{H}(j\omega)| + \log_{10}|\mathcal{X}(j\omega)|$$

and therefore we calculate again *sums instead of products*. Regarding the series connection of two simple systems with frequency response functions $\mathcal{H}_1(j\omega) = |\mathcal{H}_1(j\omega)|e^{j\measuredangle\mathcal{H}_1(j\omega)}$ and $\mathcal{H}_2(j\omega) = |\mathcal{H}_2(j\omega)|e^{\measuredangle\mathcal{H}_2(j\omega)}$, this will lead to a composite system with a transfer function equal to $\mathcal{H}_1(j\omega)\mathcal{H}_2(j\omega)$ whose magnitude and phase can be calculated in exactly the same way.

The use of the logarithmic scale, apart from its practical usefulness, since it allows for the conversion of products into sums, is also indicated for another reason: *It gives us the ability to design functions whose free variable extends over a very large range of values*. The most characteristic example of such a parameter is the frequency, which in many applications ranges from values close to zero to values on the order of 10^{10} or even larger. It is obvious that designing such representations on a linear scale would smother many of the details of the phenomenon being described. For this reason, the logarithmic scale is used very often in science, with the classic example of its application being the *Richter logarithmic scale*, which is used, as is well known, for measuring *seismic activity*.

5.7.1 Basic Bode Diagrams

According to our discussion in the previous section, the construction of Bode diagrams consists in *separating the function $\mathcal{H}(j\omega)$ into its basic constituent structural units, calculating and designing the amplitude and phase response for each component separately and adding all these quantities to construct the diagram for the entire function*. Each of these basic terms can be associated with some *constant value*, with some *real pole or zero point* with multiplicity *greater than or equal to one* and with a value equal to or different from zero, and with some *pair of conjugate complex poles or zero values*. For each of these different types of components, the process of constructing the diagram is different, and the way in which this is done is presented in the following pages.

Case 1η: Constant Term

If the transfer function contains a constant term with a value equal to K, which happens in almost all cases, the situation is extremely simple, since *both the amplitude response and the phase response are constant and independent of the frequency ω*. If the constant K has a positive value, we can write that $K = |K|e^{j2\kappa\pi}$, from which it follows that[18]

$$f(\omega) = |K| \quad \text{and} \quad g(\omega) = 2\kappa\pi \quad (\kappa = 0, 1, 2, \dots)$$

On the other hand, if the value of K is negative, we will have $K = |K|e^{j(2\kappa+1)\pi}$, from which it follows

$$f(\omega) = |K| \quad \text{and} \quad g(\omega) = (2\kappa + 1)\pi \quad (\kappa = 0, 1, 2, \dots)$$

The Bode diagram corresponding to the constant value $K = 30$ is depicted in Fig. 5.9.

Case 2η: Real Pole or Zero Value of First Order

If the transfer function is characterized by the existence of a *simple pole* at the position $s = -\alpha$, so that it can be formulated as

$$\mathcal{H}(s) \sim 1\Big/\left\{1 + \frac{s}{\alpha}\right\}$$

and the magnitude of this term for the value $s = j\omega$ will be equal to

$$\left|1\Big/\left\{1 + \frac{j\omega}{\alpha}\right\}\right| = 1\Big/\left\{\left|1 + \frac{j\omega}{\alpha}\right|\right\}$$

[18] In this description, we will use the auxiliary functions $f(\omega)$ and $g(\omega)$ to describe the response *amplitude* and *phase*, respectively, where such a thing is necessary.

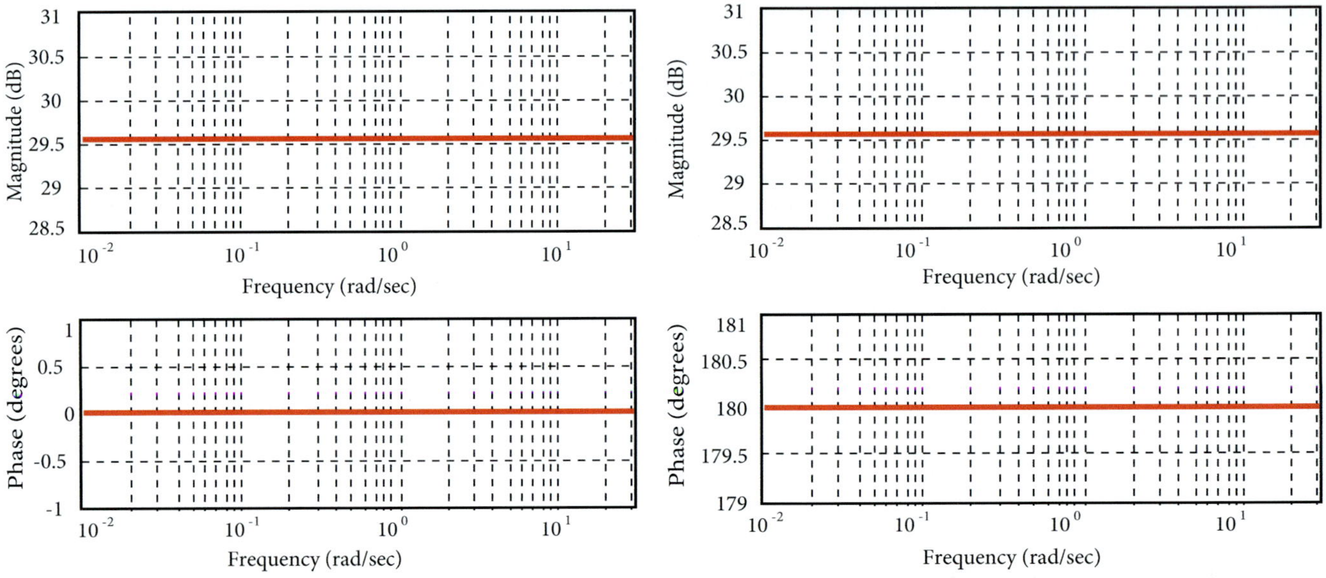

Fig. 5.9 Bode diagram of constant function of the form $\mathcal{H}(i\omega) = \pm K$ for the value $K = 30$

expression, which (expressed in dB) is formulated as

$$20\log_{10}\left\{1\Big/\left|1+\frac{j\omega}{\alpha}\right|\right\} = 20\log_{10}\left\{1\Big/\sqrt{1+\left(\frac{\omega}{\alpha}\right)^2}\right\} = -20\log_{10}\left(\sqrt{1+\left(\frac{\omega}{\alpha}\right)^2}\right) \text{ dB}$$

The preceding expression can be simplified *at the limit of low and high frequencies* by making the following approximations:

(a) Frequency range $\omega \gg \alpha$: For this range we can approximately assume that $(\omega/\alpha) \approx 0$ and the preceding relationship will take the form

$$-20\log_{10}\left(\sqrt{1+\left(\frac{\omega}{\alpha}\right)^2}\right) \approx -20\log_{10}(1) = 0$$

Therefore, for frequencies $\omega \gg \alpha$, the *function of logarithmic magnitude is equal to zero.*

(b) Frequency range $\omega \ll \alpha$: In this case, we will have $(\omega/\alpha) \gg 1$, and making the approximation $(\omega/\alpha) + 1 \approx \omega/\alpha$, the preceding relationship becomes

$$-20\log_{10}\left(\sqrt{1+\left(\frac{\omega}{\alpha}\right)^2}\right) \approx -20\log_{10}\left(\sqrt{\left(\frac{\omega}{\alpha}\right)^2}\right) = -20\log_{10}\left(\frac{\omega}{\alpha}\right) = -20\log\omega + 20\log\alpha = -20u + 20\log\alpha$$

where $u = \log\omega$. If we plot the graph of the preceding expression as a function of the parameter u, it is obviously nothing more than *a straight line with a negative slope equal to* -20 dB/decade *or* -8 dB/octave. It is not difficult to find that the above expression is zero for the value $\omega = \alpha$, that is, this straight line *intersects the horizontal axis at this point*.

The foregoing diagram of logarithmic amplitude is characterized as *asymptotic*, as it emerged from the approximations we adopted for the area of high and low frequencies, while to get the real curve we need to plot the real function

$$f(\omega) = -20\log_{10}\sqrt{1+\frac{\omega^2}{\alpha^2}} = -10\log_{10}\left(1+\frac{\omega^2}{\alpha^2}\right)$$

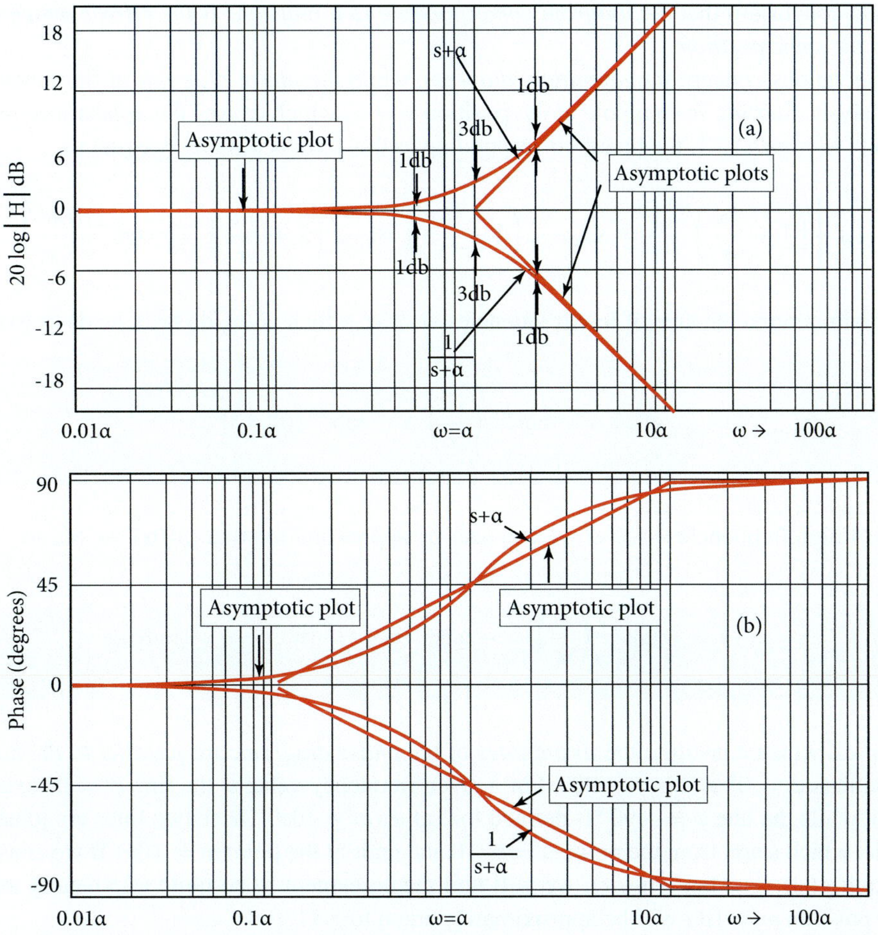

Fig. 5.10 Amplitude and phase response for simple pole and zero value at position $s = \alpha$

which is depicted in Fig. 5.10 along with the foregoing *asymptotic curves*. From this figure we observe that the agreement between the real and the asymptotic diagram is quite satisfactory and that the maximum approximation error that is observed at the position $\omega = \alpha$ is equal to -3 dB, since for this frequency value it turns out that

$$-20\log_{10}\sqrt{1+\frac{\omega^2}{\alpha^2}}\Bigg|_{\omega=\alpha} = -20\log_{10}(\sqrt{2}) = -3.01 \approx -3\text{ dB}$$

The preceding frequency value is known as the *slope frequency* and is characterized by the *maximum approximation error value*, while at any other point of the diagram this error is less than -3 dB.

In the case where the function has a zero value at the position $s = \alpha$, the corresponding Bode diagram is the *mirror image* of that corresponding to the pole in the same position. Therefore, at the limit of low frequencies, the curve of the logarithmic amplitude will coincide with the horizontal axis, at the limit of high frequencies it will be characterized by a *positive slope* with a value of 20 dB/decade or 8 dB/octave, while the maximum approximation error will appear at the position of the *slope frequency* $\omega = \alpha$ with a value equal to 3 dB. Indeed, in this case, the term $1 + (j\omega/\alpha)$ will appear in the numerator and not in the denominator of the transfer function, and the formulation of the measure of the expression (in dB) gives us

$$20\log_{10}\left|1+\frac{j\omega}{\alpha}\right| = 20\log_{10}\left[\sqrt{1+\left(\frac{\omega}{\alpha}\right)^2}\right]$$

We observe that in this case, the sign of this expression is positive and not negative—since the expression is in the *numerator* and not in the denominator, so that during logarithmization the sign changes as previously—and the application of a similar analysis leads us to the conclusion that the *diagram corresponding to a real zero is the mirror image of that resulting from considering a pole in the same position*.

The foregoing observations concern the *amplitude diagram*, while the *phase diagram* of the function $\mathcal{H}(j\omega)$ results by applying a similar analysis. Starting from a pole at the position $s = \alpha$, which causes the appearance in the transfer function of a term of the form $[1 + (j\omega/\alpha)]^{-1}$, the phase corresponding to this term is obviously equal to

$$\measuredangle\left\{\left(1+\frac{j\omega}{\alpha}\right)^{-1}\right\} = \measuredangle 1 - \measuredangle\left(1+\frac{j\omega}{\alpha}\right) = -\measuredangle\left(1+\frac{j\omega}{\alpha}\right) = -\tan^{-1}\left(\frac{\omega}{\alpha}\right)$$

In this case, the approximation in the area of low frequencies $\omega \ll \alpha$, which is $(\omega/\alpha) \approx 0$, leads us to the value

$$g(\omega) = -\tan^{-1}\left(\frac{\omega}{\alpha}\right) \approx -\tan^{-1}(0) = 0^\circ$$

while for the area of high frequencies $\omega \gg \alpha$ we consider by approximation that $(\omega/\alpha) \to \infty$, so

$$g(\omega) = -\tan^{-1}\left(\frac{\omega}{\alpha}\right) \approx -\tan^{-1}(\infty) = -90^\circ = -\frac{\pi}{2} \text{ radians}$$

The preceding curves, which constitute the *asymptotes of the phase diagram*, are *parallel to the horizontal axis* and are described by the equations $\varphi = 0^\circ$ and $\varphi = -90^\circ$. For a given frequency value α the line $\varphi = 0^\circ$ extends in the frequency area $0 \le \omega \le (\alpha/10)$, while the line $\varphi = -90^\circ$ is defined for values $\omega \ge 10\alpha$. These two lines are joined by a third line with a slope -45°/decade, which starts from the point $\omega = \alpha/10$ and ends at the point $\omega = 10\alpha$. If we construct the actual curve by plotting the function $g(\omega) = -\tan^{-1}(\omega/\alpha)$, we will find the agreement to be quite satisfactory and that the maximum error observed at the position $\omega = 10\alpha$ will be approximately equal to 5.71°.

The corresponding *phase diagram* for a zero value at the position $s = \alpha$ is the *mirror image of that associated with a simple pole in the same position*, a conclusion that arises using the same argumentation we developed previously.

The amplitude and phase diagrams for the case of a simple pole and zero value are also presented in Fig. 5.10, while the corresponding plots for the approximation error are presented in Fig. 5.11.

A typical example of a Bode diagram for the case of a pole and zero value at the position $s = 10$ is presented in Fig. 5.12, in which the theoretical analysis done in the preceding pages is illustrated. For example, in the phase diagram, the asymptotic line $\varphi = 0^\circ$ extends almost to the position $\omega = \alpha/10 = 1$, while the asymptotic line $\varphi = -90^\circ$ starts approximately from the position $\omega = 10\alpha = 100$. The aforementioned characteristics refer to the diagrams related to the pole at the position $s = 10$ are discussed, while those related to the zero value at the same position have corresponding representations that are *mirror images*.

In the special case where the transfer function has a pole or zero value at position $s = 0$, things are significantly simplified. Starting from the case of the pole, the term of the transfer function corresponding to it has the form

$$\left.\frac{1}{s/\alpha}\right|_{s=j\omega} = \frac{1}{j\omega/\alpha} = -j\frac{\alpha}{\omega}$$

with magnitude and phase given by the relations

$$f(\omega) = \left|-j\frac{\alpha}{\omega}\right| = \frac{\alpha}{\omega} = -20\log_{10}\left(\frac{\omega}{\alpha}\right) \text{ dB} \quad \text{and} \quad g(\omega) = \measuredangle\left(-j\frac{\alpha}{\omega}\right) = \tan^{-1}\left(\frac{-\alpha/\omega}{0}\right) = \tan^{-1}(-\infty) = -90^\circ$$

We observe that the logarithmic amplitude function is represented by a straight line with a slope of -20 dB/decade that intersects the horizontal axis at position $\omega = \alpha$, while the phase is constant and equal to $\varphi = -90^\circ$.

The situation for the zero value at position $s = 0$ is similar, with the difference that the slope of the logarithmic amplitude line is constant and equal to 20 dB/decade.

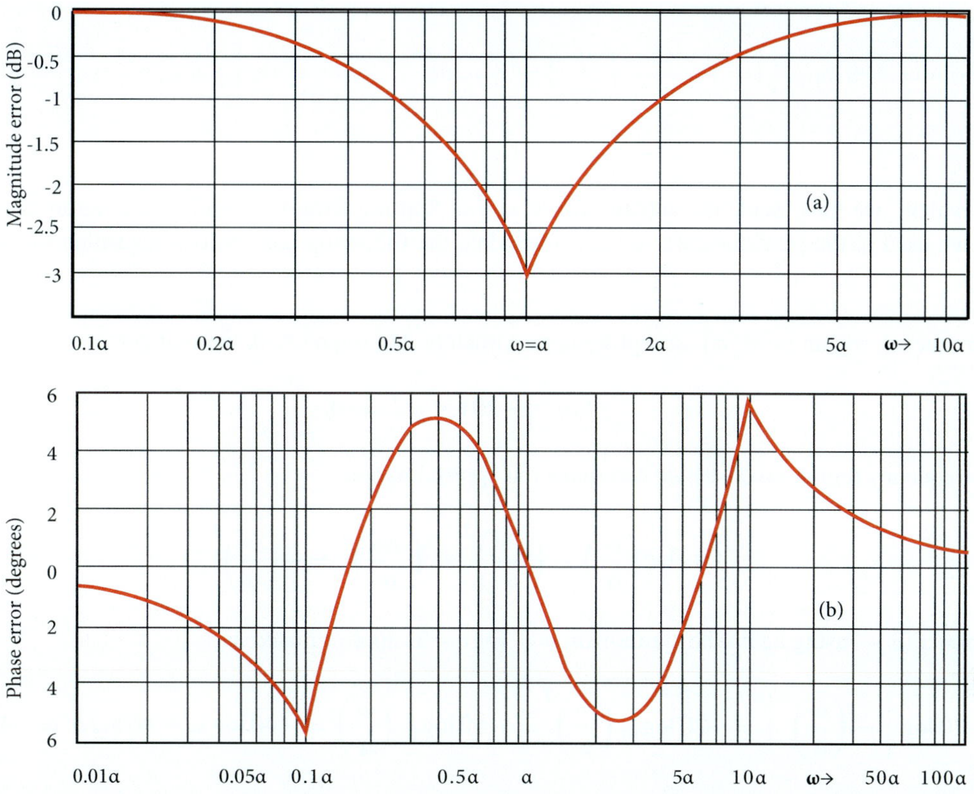

Fig. 5.11 Change in approximation error of amplitude and phase diagrams from their asymptotic versions for the case of a simple pole and zero value at position $s = \alpha$

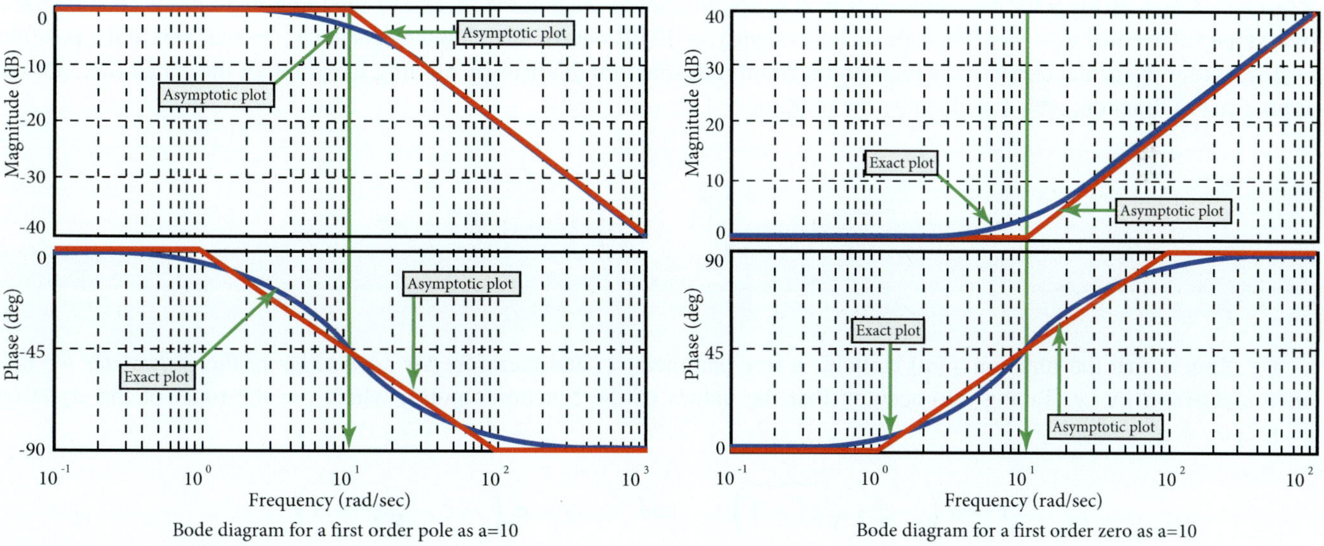

Fig. 5.12 Bode diagram for case of a pole and zero value located at position $s = 10$

Case 3$^{\eta}$: Second-Order Poles and Zero Values

If in the denominator of the transfer function there is some expression of the form

$$s^2 + 2\zeta\alpha s + \alpha^2$$

then the logarithmic amplitude and phase diagrams will be constructed using the functions

$$f(\omega) = -20\log_{10}\left|1 + 2j\zeta\frac{\omega}{\alpha} + \left(\frac{j\omega}{\alpha}\right)^2\right| \quad \text{and} \quad g(\omega) = -\measuredangle\left[1 + 2j\zeta\frac{\omega}{\alpha} + \left(\frac{j\omega}{\alpha}\right)^2\right]$$

where, as is evident, we have made the substitution $s = j\omega$. Starting from the amplitude diagram and resorting to the approximations we used in the previous case, we can formulate the following conclusions regarding the asymptotes of the diagram:

- For the low-frequency region ($\omega \ll \alpha$), accepting approximately that $(\omega/\alpha) \approx 0$, we will get

$$f(\omega) \approx -20\log_{10}(1) = 0.$$

- For the high-frequency region ($\omega \gg \alpha$) we can make the approximation

$$1 + 2i\zeta\left(\frac{\omega}{\alpha}\right) + \left(\frac{j\omega}{\alpha}\right)^2 \approx \left(\frac{j\omega}{\alpha}\right)^2 = -\left(\frac{\omega}{\alpha}\right)^2$$

  and the function $f(\omega)$—using again the symbolism $u = \log\omega$—is approximated as

$$f(\omega) \approx -20\log\left|-\left(\frac{\omega}{\alpha}\right)^2\right| = -20\log_{10}\left(\frac{\omega}{\alpha}\right)^2 = -40\log_{10}\left(\frac{\omega}{\alpha}\right) = -40\log\omega + 40\log\alpha = -40u + 40\log\alpha$$

From the foregoing analysis we observe that in the case of a *second-order pole*, the Bode diagram for the logarithmic amplitude is again characterized by the existence of *two asymptotic lines*, with equations $f(\omega) = 0$ for the region $\omega < \alpha$ and $f(\omega) = -40u + 40\log\alpha$ for the region $\omega > \alpha$. It is not difficult to see that the second of the foregoing lines is characterized by a slope value equal to -40 dB/decade or, equivalently, -12 dB/octave, which intersects the horizontal axis at the position $\omega = \alpha$. As for the actual curve of the logarithmic amplitude diagram, it arises by plotting the graph of the actual (i.e., not the approximate) function, which is given by the relation

$$f(\omega) = -20\log_{10}\sqrt{\left[1 - \left(\frac{\omega}{\alpha}\right)^2\right]^2 + 4\zeta^2\left(\frac{\omega}{\alpha}\right)^2} \text{ dB}$$

We observe that the function $f(\omega)$ contains a free parameter ζ, and therefore *for each value of this parameter we will have a different curve*. Taking into account that the values of the function's poles, which are the roots of the equation $s^2 + 2\zeta\alpha s + \alpha^2 = 0$ are

$$s_1 = \alpha\left(-\zeta + \sqrt{\zeta^2 - 1}\right) \quad \text{and} \quad s_2 = \alpha\left(-\zeta - \sqrt{\zeta^2 - 1}\right)$$

it is not difficult to see that if the parameter ζ has values in the interval $0 < \zeta < 1$, the function has *two conjugate complex poles*, whereas if it is $\zeta \geq 1$, the function has *two distinct real poles* and the branch of the Bode diagram corresponding to these will be constructed following the methodology we presented in the previous section. The Bode diagram regarding the logarithmic amplitude of the *frequency response function* for various values of the parameter ζ is shown in Fig. 5.13a.

Observing Fig. 5.13, we notice in it the existence of a *peak*, whose position and value are derived by calculating the frequency ω_r, which is the solution of the equation $|\mathcal{H}(j\omega)|' = 0$, and then determining the value of the expression $|\mathcal{H}(j\omega_r)|$. If we differentiate the magnitude of the function $|\mathcal{H}(j\omega)|$ and set the resulting expression to the zero value, we find that

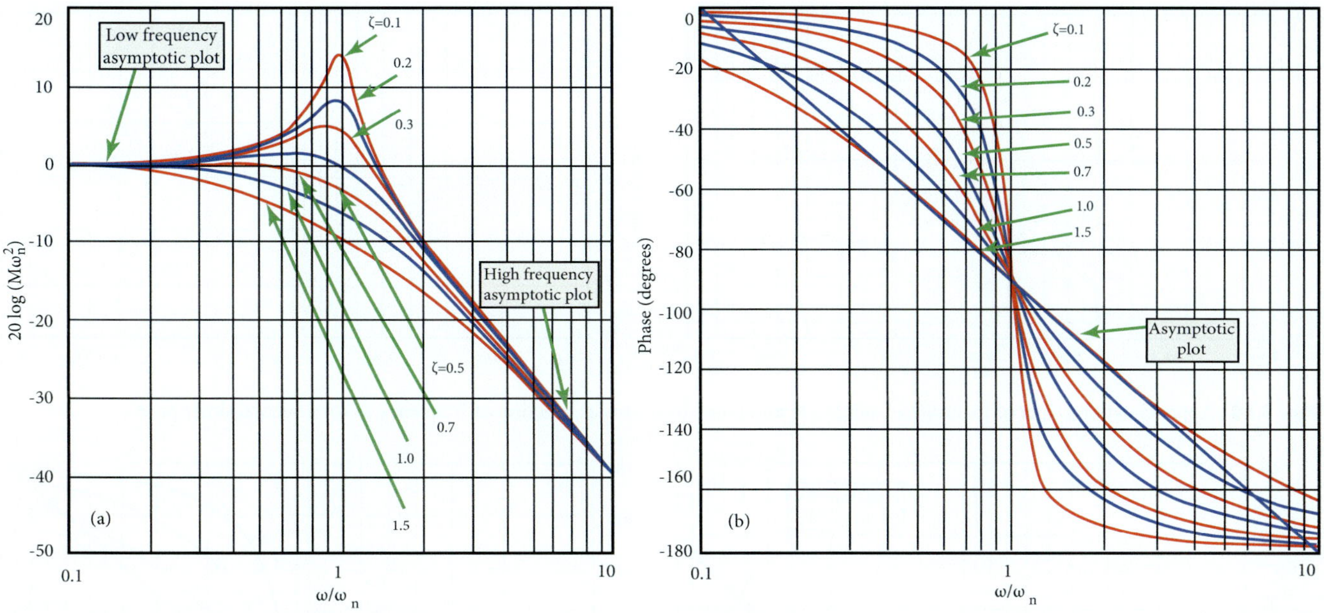

Fig. 5.13 Bode diagram of (**a**) logarithmic amplitude and (**b**) phase of a second-order pole

$$\frac{d|\mathcal{H}(j\omega)|}{d\omega} = \frac{2\omega}{\alpha^2} \times \left\{1 - \frac{\omega^2}{\alpha^2} - 2\zeta^2\right\} \Big/ \sqrt{\left[\left(1 - \frac{\omega^2}{\alpha^2}\right)^2 + 4\zeta^2\frac{\omega^2}{\alpha^2}\right]^3} = 0$$

In the next step, we solve for ω the equation

$$1 - \frac{\omega^2}{\alpha^2} - 2\zeta^2 = 0 \quad \text{obtaining the solution} \quad \omega_r = \alpha\sqrt{1 - 2\zeta^2}$$

which, if substituted into the function $|\mathcal{H}(j\omega)|$, leads to a value equal to

$$|\mathcal{H}(j\omega_r)| = -20\log_{10}\left\{\sqrt{\left(1 - \frac{\alpha^2(1 - 2\zeta^2)}{\alpha^2}\right)^2 + \frac{4\zeta^2}{\alpha^2}\left[\alpha^2(1 - 2\zeta^2)\right]}\right\} = 2\zeta\sqrt{1 - \zeta^2}$$

The foregoing analysis leads us to the conclusion that for every value of the parameter ζ the peak of the diagram appears at the frequency $\omega_r = \alpha\sqrt{1 - 2\zeta^2}$, while its value is equal to $|\mathcal{H}(j\omega_r)| = 2\zeta\sqrt{1 - \zeta^2}$ and therefore varies inversely proportional to the value of ζ—see Fig. 5.13a. It is obvious that for this peak to exist the frequency ω_r must be a real number, which happens when $1 - 2\zeta^2 > 0$ or, equivalently, when $\zeta < 1/\sqrt{2} \approx 0.707106$. Therefore, the peak appears when the parameter ζ takes values in the range $0 < \zeta < 0.707106$. For this range of ζ values, the peak frequency ω_r is quite close to the pole frequency α and coincides with it when $\zeta = 0$.

On the other hand, the *phase* of the term corresponding to the second-order pole is given by the relation

$$g(\omega) = -\measuredangle\left[\left(1 - \frac{\omega^2}{\alpha^2}\right) + j\left(\frac{2\zeta\omega}{\alpha}\right)\right] = -\tan^{-1}\frac{2\zeta\omega/\alpha}{1 - (\omega^2/\alpha^2)}$$

In this case, for the low-frequency limit $\omega \ll \alpha$ the function $g(\omega)$ approximates as $g(\omega) \approx 0°$, while the approximation for the high-frequency region is $g(\omega) \approx -180°$. These straight lines constitute the asymptotic curves of the phase diagram, along with a curve that connects these straight lines starting from the frequency $\omega_1 = (\alpha/2)\log_{10}(2/\zeta)$ and ending at the frequency $\omega_2 = 2\alpha/\log_{10}(2\zeta)$. As for the actual curve, it is constructed by graphing the phase response function defined as

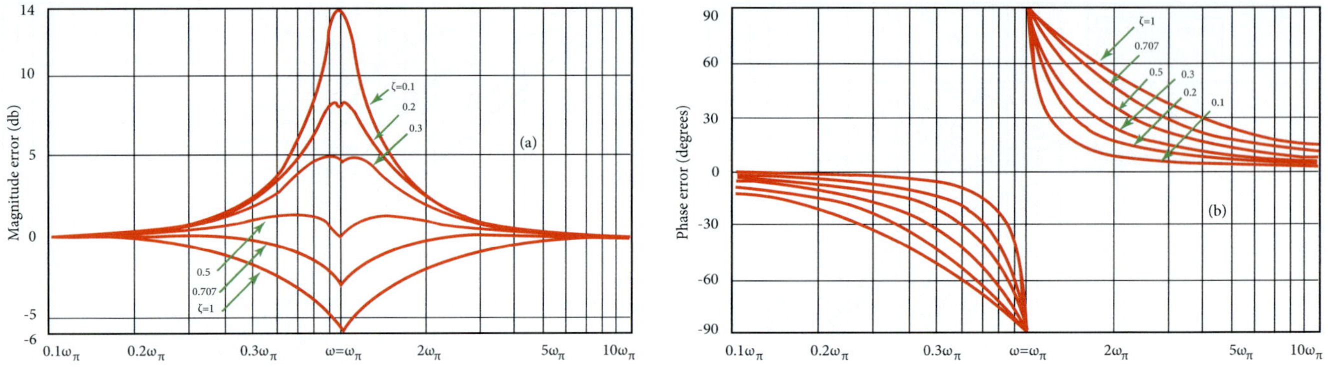

Fig. 5.14 Approximation errors of logarithmic amplitude and phase diagrams as function of frequency ω for a second-order pole

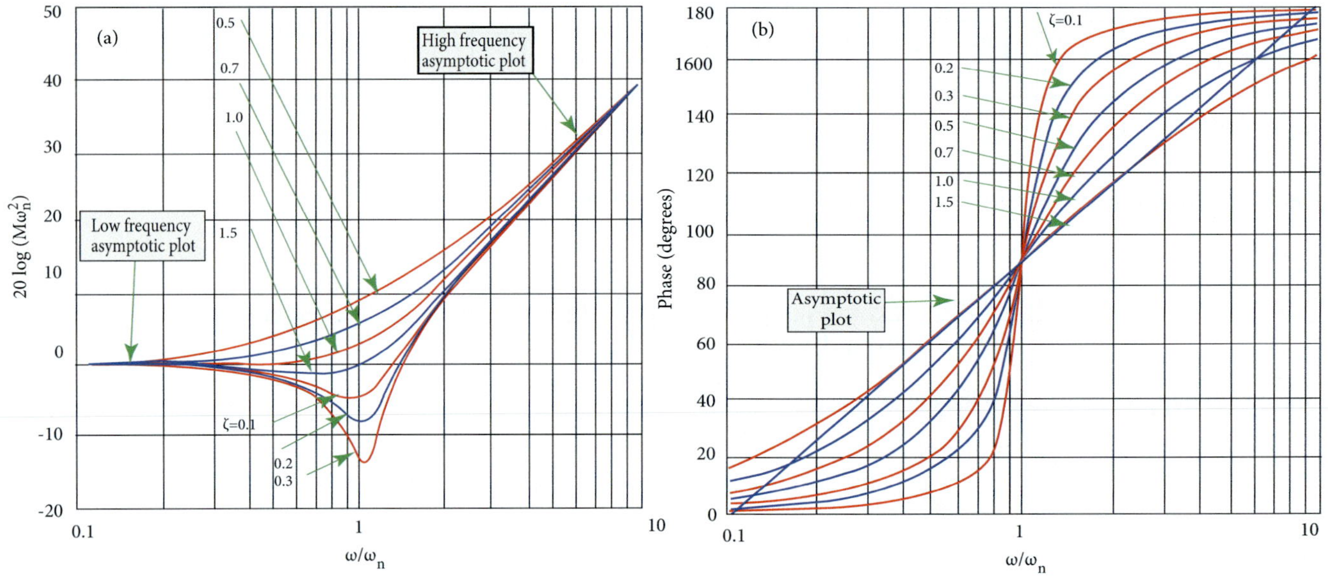

Fig. 5.15 Bode diagram of (**a**) logarithmic amplitude and (**b**) phase for a second-order zero value

$$g(\omega) = -\tan^{-1}\frac{2\zeta\omega/\alpha}{1-(\omega^2/\alpha^2)}$$

We observe that, in full analogy with the amplitude diagram, this function is characterized by the existence of the free parameter ζ, and therefore *for each distinct value of this parameter we will have a distinct curve*. The phase diagram for the case of the second-order pole is depicted in Fig. 5.13b, from which we observe that at the position of the frequency $\omega = \alpha$ the value of the function $g(\omega)$ is equal to $-90°$. The graphical representations of the approximation error for the logarithmic amplitude and phase diagrams are presented in Fig. 5.14. The foregoing analysis refers to a second-order pole, while in the case of a zero value, we observe that the logarithmic amplitude and phase diagrams have a similar shape, with the difference being that they are *mirror images with respect to the horizontal axis* of those related to the pole. This means that the amplitude diagram is characterized by the presence, not of a peak, but of a *trough*, while the slope of the asymptotic straight line is not negative but *positive* and, of course, equal to 40 dB/decade or, equivalently, 12 dB/octave. These diagrams are presented in Fig. 5.15a, b.

Figure 5.16 shows an example of a Bode diagram for a second-order pole and zero value for a frequency value of $\alpha = 4$ and parameter values of $\zeta = 0.1, 0.2, 0.3, 0.4, 0.5, 0.6, 0.7$.

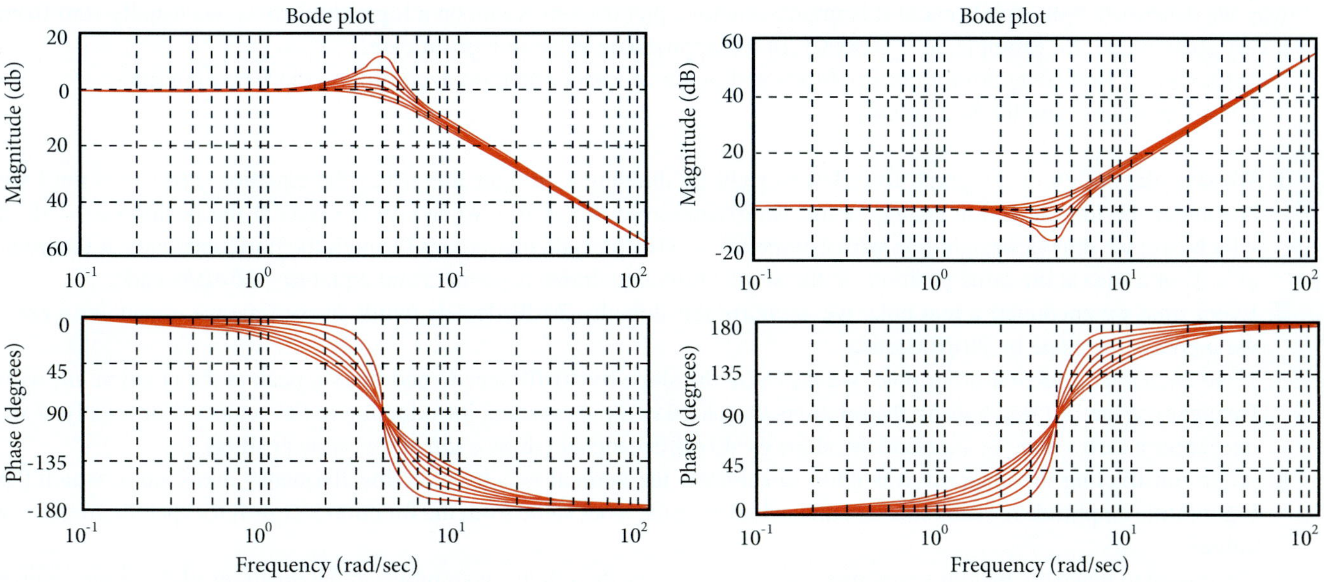

Bode diagram for a second order pole

Bode diagram for a second order zero

Fig. 5.16 Bode diagram for a second-order pole and zero value for a frequency value of $\alpha = 4$ and parameter values of $\zeta = 0.1, 0.2, 0.3, 0.4, 0.5, 0.6, 0.7$

Note According to the preceding discussion and analysis, the amplitude diagram of a first-order pole is characterized by an asymptote with a slope of -20 dB/decade, while the high-frequency asymptote in the corresponding phase diagram is at the position $\varphi = -90°$. We also observed that if the pole is of the second order, then the slope of the asymptote in the amplitude diagram is equal to -40 dB/decade, while the corresponding asymptote of the phase diagram is located at the position $\varphi = -180°$. If we follow the foregoing analysis for a pole of order n, we will find that the slope of the asymptote in the amplitude diagram is equal to $-20n$ dB/decade, while the asymptote in the phase diagram is located at the position $\varphi = -90n°$. These also apply for a zero value of order n with the values of the foregoing quantities obviously accompanied by a plus sign.

5.7.2 The Process of Constructing Bode Diagrams

Following the analytical description of how to create elementary Bode diagrams, let us now move on to the process of creating a diagram for the frequency response function $\mathcal{H}(j\omega)$; The general rules for doing so are as follows:

- Transformation of the *frequency response function*, so that it comes in the form

$$\mathcal{H}(s) = K' \left\{ \frac{z_1 z_2 z_3 \ldots z_M}{p_1 p_2 p_3 \ldots p_N} \right\} \times \frac{\left(1 + \frac{s}{z_1}\right)\left(1 + \frac{s}{z_2}\right)\left(1 + \frac{s}{z_3}\right) \ldots \left(1 + \frac{s}{z_M}\right)}{\left(1 + \frac{s}{p_1}\right)\left(1 + \frac{s}{p_2}\right)\left(1 + \frac{s}{p_3}\right) \ldots \left(1 + \frac{s}{p_N}\right)} \tag{5.38}$$

- Separation of the function expressed in its new form into the *basic structural units of which it is composed* and characterization of these as constant terms or real/conjugate complex poles or zero values.
- Design of the diagram axes and depiction on it *of the poles and zero values* on the horizontal frequency axis. For better clarity, it is customary to draw dashed vertical lines that start from the position of the poles or zero values and run upward, since at these positions, which are also known as *inflection change points*, the slope of asymptotic lines changes in the

way we described. Note that because it is impossible to depict the zero value on a logarithmic axis, we usually start from a very small value, for example, $\omega = 0.00001$, or the appropriate value in a given case.

- Construction of Bode diagram for the amplitude and phase of each basic term according to the following rules:
  For a logarithmic amplitude diagram:

1. We start the design of the graph from left to right by drawing a straight line using the equation $f(\omega) = 0$ until we encounter the first inflection change point. An exception to this rule is when a pole exists at the position $\omega = 0$, in which case we draw a straight line with a slope of -20 dB/decade that passes through the horizontal axis at the point $\omega = 1$, or a zero at the same position, so the slope of the line drawn is positive and equal to $+20$ dB/decade.
2. Every time we encounter a real pole, we *decrease* the slope by 20 dB/decade, while every time we encounter a zero, we *increase* the slope by 20 dB/decade.
3. If we encounter a second-order pole, we *decrease* the slope by 40 dB/decade and draw a peak at the position ω_r, with frequency value $|\mathcal{H}(j\omega_r)|$, using the relations discussed in the theoretical introduction to the subject. Conversely, if we encounter a zero value, we *increase* the slope by 40 dB/decade and draw a *dip* at the same position.
4. After constructing the diagram, it is good to calibrate the vertical axis by selecting frequency areas along which the logarithmic amplitude remains constant and calculating the label value from the function $|\mathcal{H}(j\omega)|$ for specific frequency values.
5. If we wish to make the diagram even more accurate, we can smooth the asymptotes at the positions of the slope change points, without exceeding the permissible error limits (e.g., the ± 3 dB associated with the real zeros and first-order poles).

For the phase diagram:

1. We start designing the graph from left to right by drawing a straight line using the equation $g(\omega) = 0$ until we encounter the first inflection point. An exception to this rule is when a pole exists at the position $\omega = 0$, where we draw a straight line using the equation $g(\omega) = -90°$, or a zero at the same position, where the equation of the line drawn is $g(\omega) = +90°$.
2. Every time we encounter a *pole* $\omega = \alpha$, *we reduce* the phase by 90° for the frequency range $[0.1\alpha, 10\alpha]$, while, conversely, for a *zero* at the same position, *we increase* the phase by 90° for the same range of values. Care should be taken so that at the position of the pole/zero value, the phase of the function has been reduced/increased by 45°, respectively.
3. Every time we encounter a *second-order pole* we should reduce the slope by 180°, while in the case of a *zero value* of the same order, we should add to the phase of the function the preceding value.
4. If we wish to improve the quality of the diagram, we can smooth the asymptotes by about 6° in the value range $[0.1\alpha, 10\alpha]$, so that the graphical representation resembles more the function of the tangent arc.

At this point we have completed the presentation of the basic theory related to *Bode diagrams*, and many exemplary solved exercises follow, so that the reader can fully understand how they are constructed.

Problem 5.7.1 Construct a Bode diagram of the transfer function $\mathcal{H}(s) = \dfrac{1}{s+5}$.

Solution Expressing the transfer function in the form

$$\mathcal{H}(j\omega) = \frac{1}{j\omega + 5} = \frac{1}{5}\left\{\frac{1}{(j\omega/5)+1}\right\} = 0.2\,\frac{1}{(j\omega/5)+1}$$

for the value $s = j\omega$, we find that it is characterized by the existence of *a constant term* with a value of $K = 0.2$ and a *pole* at the position $s = 5$, which corresponds to a slope frequency value of $\omega = 5$ rad/s. Therefore, the Bode diagram of the function $\mathcal{H}(s)$ will result from the superposition of the elementary Bode diagrams associated with the preceding two terms that are generally constructed as follows.

The *amplitude diagram* of the *constant term* $K = 0.2$ consists of a straight line described by the equation $f(\omega) = 20\log_{10}(0.2) = -13.979400$. Therefore, the amplitude value is constant for all frequency values ω. This property also characterizes the *phase diagram*, which is also described by an equation of the form $g(\omega) = 0°$ and corresponds to a phase value equal to zero for every frequency value.

On the other hand, the Bode diagram for the *simple pole* at the position $s = -5$ is characterized by the following properties. Starting from the *amplitude diagram*, it consists of a curve that starts from the value $\omega = 0$ and coincides with the horizontal extends to the position $\omega = 5$ rad/s, after which it begins to descend with a constant slope reduction equal to -20 dB/decade. As for the *phase diagram*, it starts from the position $\omega = 0$ and coincides with the horizontal axis up to one-tenth of the frequency slope, that is, up to 0.5 rad/s, while from there on a slope reduction is observed until the phase reaches $-90°$, a value that it approaches for a frequency value 10 times greater than the frequency slope, that is, at 50 rad/s.

The foregoing description refers to the asymptotic lines of the phase and logarithmic amplitude diagrams, while to get the real curves we need to plot the functions

$$f(\omega) = -10\log_{10}\left(1 + \frac{\omega^2}{25}\right) \quad \text{and} \quad g(\omega) = -\tan^{-1}\left(\frac{\omega}{5}\right)$$

The asymptotic and real Bode diagrams of the preceding terms are depicted in Fig. 5.17.

After constructing the foregoing elementary diagrams, we can construct the phase and logarithmic amplitude diagrams of the function $\mathcal{H}(s)$ from the superposition of the corresponding diagrams of the elementary terms. By the term *diagram superposition* we mean the addition of the corresponding curves in the way we add two continuous-time signals together, which consists of *adding their corresponding instantaneous values*. The result of this process is also presented in a diagrammatic way in Fig. 5.17.

Problem 5.7.2 Construct a Bode diagram of the transfer function

$$\mathcal{H}(s) = -100\frac{s}{s^3 + 9s^2 + 24s + 20}$$

Solution Factoring the polynomial of the denominator, we write the transfer function in the form

$$\mathcal{H}(s) = -\frac{100s}{s^3 + 9s^2 + 24s + 20} = -\frac{100s}{(s^2 + 4s + 4)(s + 5)} = -\frac{-100s}{(s + 2)^2(s + 5)}$$

or in a form suitable for the construction of the Bode diagram

$$\mathcal{H}(s) = -100\frac{s}{[2(s/2) + 1]^2 \times 5[(s/5) + 1]} = -5\frac{s}{[(s/2) + 1]^2[(s/5) + 1]}$$

From the last relation we observe that the function $\mathcal{H}(s)$ consists of a *constant term* with value $K = -5$, a *zero value* at the position $s = 0$, a *simple pole* at the position $s = -5$, and a *double pole* at the position $s = -2$. Therefore, according to the methodology presented earlier, we should construct *four elementary Bode diagrams*, one for each elementary term, from whose superposition the diagram of the transfer function $H(s)$ will result for the value $s = j\omega$.

In detail, then, we have the following: For the constant term $K = -5$, the *amplitude diagram* is a straight line with equation $f(\omega) = 20\log_{10}(-5) = 13.9794$. Therefore, the width is the same for all frequencies, while the same happens with the phase, which is described by the equation $\varphi = 180°$. The *amplitude* and *phase* diagrams for the constant term $K = -5$ are presented in Fig. 5.18a.

On the other hand, the *amplitude diagram* for the simple pole at the position $s = -5$ is a curve that starts from the position $\omega = 0$, continues on the horizontal axis (i.e., the amplitude value is equal to 0 dB) until the frequency of 5 rad/s, and then descends with a constant slope reduction equal to -20 dB/decade. As for the *phase diagram*, it is a curve that starts from the position $\omega = 0$ and coincides with the horizontal axis until 1/10 of the frequency of 5 rad/s (i.e., until 0.5 rad/s). Then

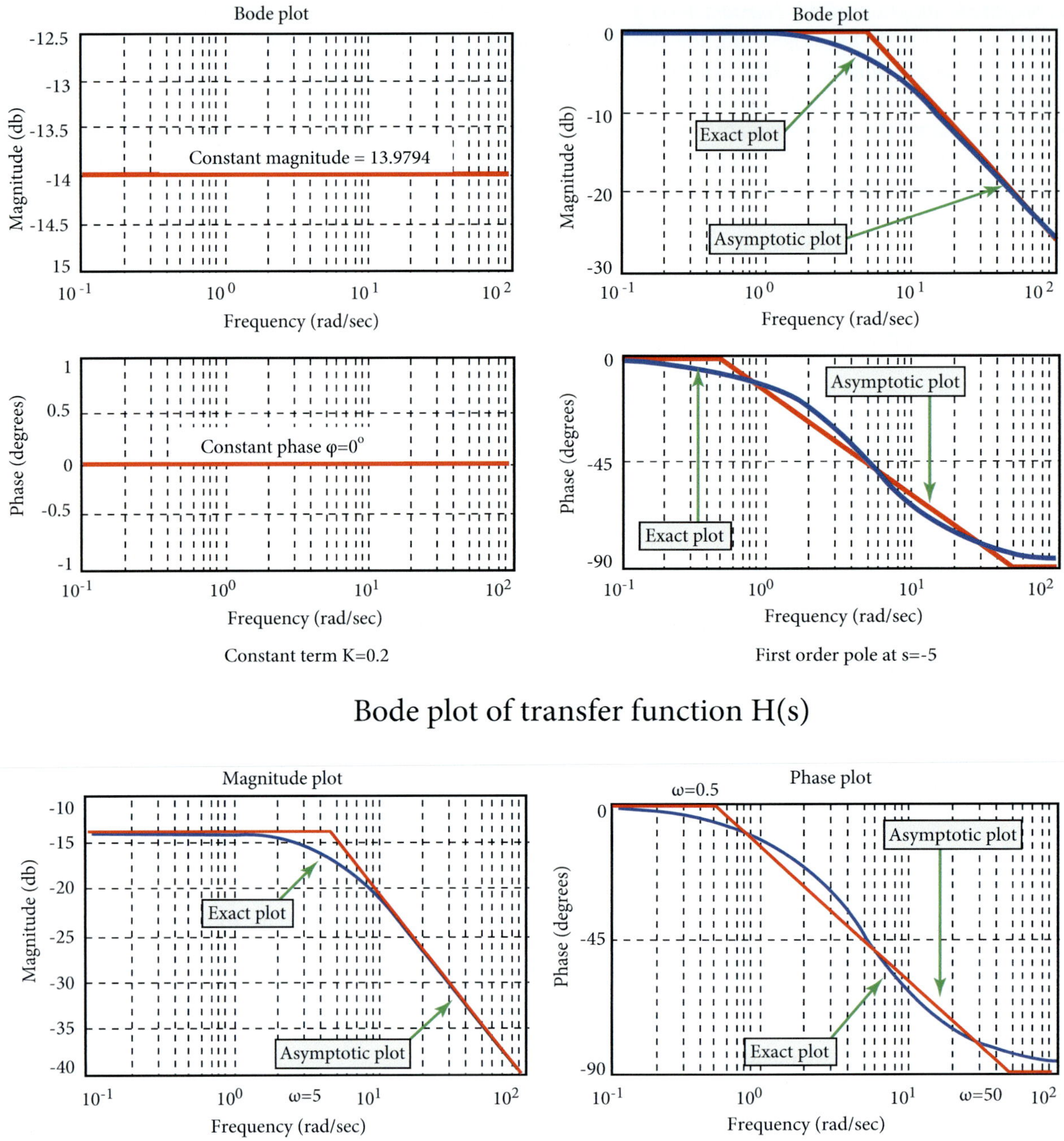

Fig. 5.17 Construction of Bode diagram of function $\mathcal{H}(s) = 1/(s+5)$ from Exercise 5.7.1

the curve descends until it approaches $-90°$ for a frequency value 10 times that of 5 rad/s (i.e., up to 50 rad/s), and from there on it remains constant at this value. The amplitude and phase diagrams for the simple pole at the position $s = -5$ are presented in Fig. 5.18b.

The next curve refers to the *double pole* at the position $s = -2$ for which, as we have mentioned, the reduction of the slope in the phase and logarithmic amplitude diagrams *is double in relation to that observed in diagrams related to simple poles*. Thus, the *amplitude diagram* starts from the position $\omega = 0$, coincides with the horizontal axis until the frequency value of 2 rad/s, and then descends with a slope reduction equal to -40 dB/decade. As for the *phase diagram*, it starts from $0°$, and from the position $\omega = 0$ coincides with the horizontal axis until the frequency of $2/10 = 0.2$ rad/s, and then descends by $180°$, a value that approaches the position $\omega = 2 \cdot 10 = 20$ rad/s. The amplitude and phase diagrams for the double pole at the position $s = -2$ are presented in Fig. 5.19a.

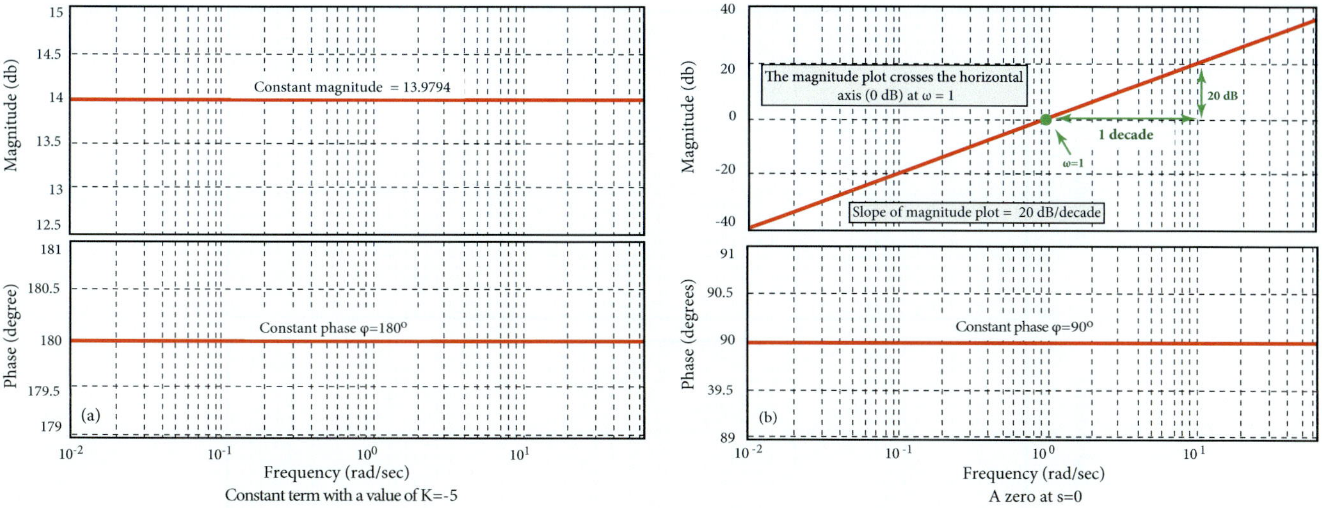

Fig. 5.18 Amplitude and phase diagrams for constant term $K = -5$ and for simple pole at position $s = -5$ of transfer function $\mathcal{H}(s)$ from Exercise 5.7.2

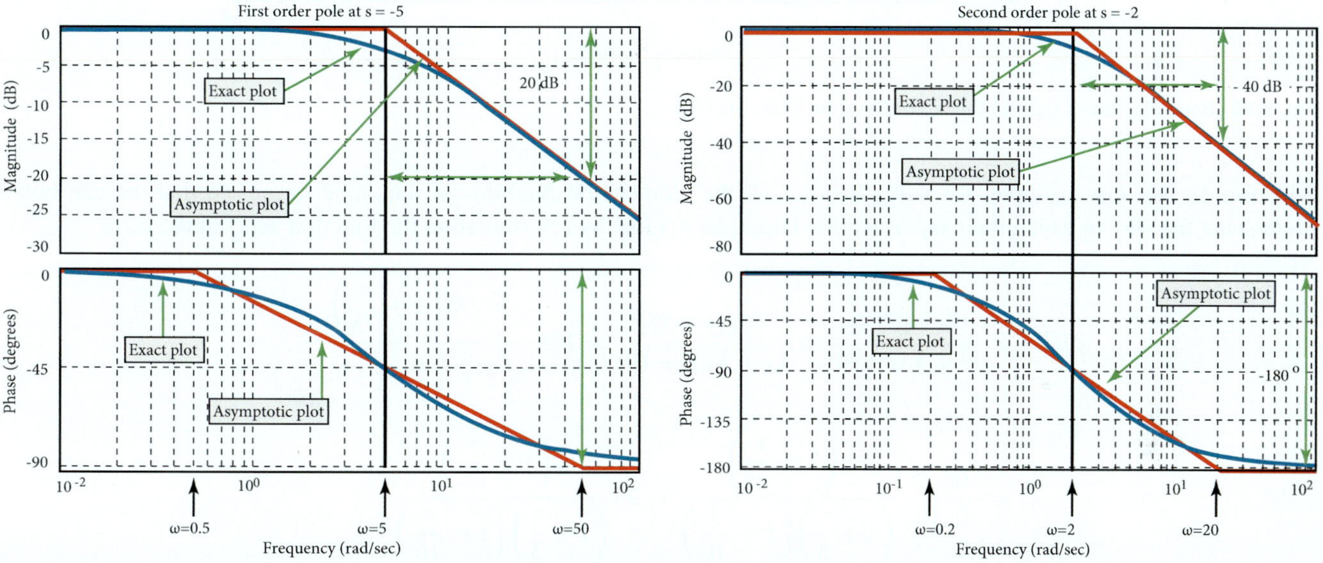

Fig. 5.19 Amplitude and phase diagrams for zero value at position $s = 0$ and for double pole at position $s = -2$ of transfer function $\mathcal{H}(s)$ from Exercise 5.7.2

Finally, the *amplitude diagram* for the zero value at the position $s = 0$ starts with a slope increase equal to 20 dB/decade, and so as to pass through the position of 0 dB for a frequency value of $\omega = 1$ rad/s, while on the other hand, the phase associated with this term is constant and equal to 90°. The amplitude and phase diagrams for the zero value at the position $s = 0$ are presented in Fig. 5.19b.

The foregoing diagrams depict both the *asymptotic lines* for the low- and high-frequency regions as well as the *actual curves* that result from the design of the functions $f(\omega)$ and $g(\omega)$ suitable for each case. As for the Bode diagram of the function $\mathcal{H}(j\omega)$, it results from the superposition of the four preceding elementary graphs in the manner described and is depicted in Fig. 5.20.

Problem 5.7.3 Construct a Bode diagram of the transfer function

$$\mathcal{H}(s) = \frac{100s^2 + 750s}{s^2 + 13s + 30}$$

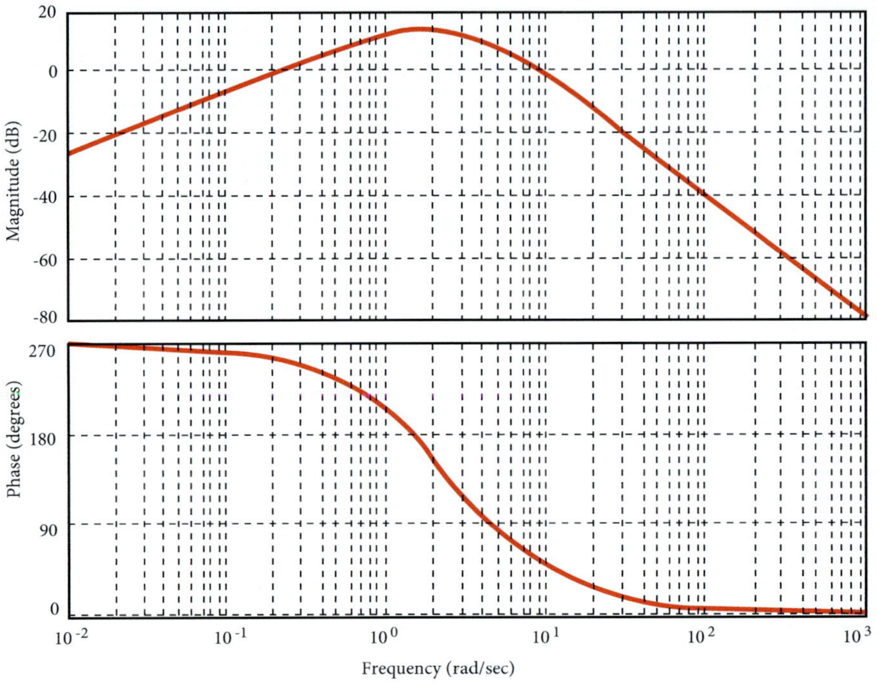

Fig. 5.20 Bode diagram of transfer function $\mathcal{H}(s)$ from Exercise 5.7.2

Solution In this last example, we will construct all the elementary diagrams in the same figure, so that we can better understand the way in which their *superposition* takes place. First, by reformulating the function $\mathcal{H}(s)$ in the form

$$\mathcal{H}(s) = \frac{10s^2 + 750s}{s^2 + 13s + 30} = \frac{10s(s+75)}{(s+3)(s+10)} = 10s\frac{75\left(1+\dfrac{s}{75}\right)}{3\left(1+\dfrac{s}{3}\right)10\left(1+\dfrac{s}{10}\right)} =$$

$$= \frac{10\cdot 75}{3\cdot 10}\frac{s\left(1+\dfrac{s}{75}\right)}{\left(1+\dfrac{s}{3}\right)\left(1+\dfrac{s}{10}\right)} = 25\frac{s\left(1+\dfrac{s}{75}\right)}{\left(1+\dfrac{s}{3}\right)\left(1+\dfrac{s}{10}\right)}$$

we find that it is characterized by the existence of a constant term with value $K = 25$, two simple zeros at the positions $s = 0$ and $s = -75$, and two simple poles at the positions $s = -3$ and $s = -10$. Therefore, in the Bode diagram there will be *three slope frequencies*, with values $\omega_1 = 3$ rad/s, $\omega_2 = 10$ rad/s, and $\omega_3 = 75$ rad/s.

Starting from the *magnitude plot* of the function $\mathcal{H}(s)$, the contribution of the constant term $K = 25$ has a value equal to $20\log_{10}(25) = 27.958800$. Since during the superposition of all curves this value will be added to each point of the diagram, thereby raising the entire curve by the aforementioned amount, we can simply shift the horizontal axis upward, so that it coincides with the line $f(\omega) = 20\log_{10}(25) = 27.958800$ dB. As for the contribution of the remaining terms, it is generally as follows:

- For the zero value at the position $s = 0$ we draw a line with *positive slope* $+20$ dB/decade, which passes through the point $\omega = 1$ rad/s of the shifted horizontal axis [line (a) in the asymptotic diagram of Fig. 5.21].

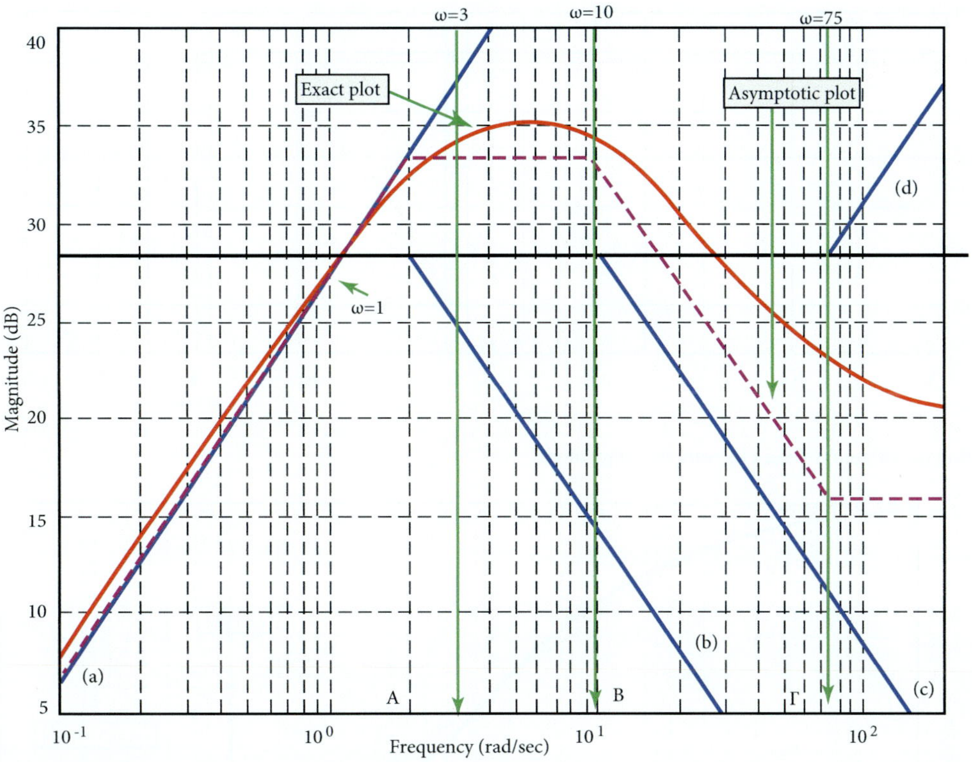

Fig. 5.21 Bode diagram for amplitude of function $\mathcal{H}(s)$ from Exercise 5.7.3

- For the simple pole at the position $s = -3$ we draw a line with *negative slope* $-$ 20 dB/decade that intersects the shifted horizontal axis at the point $\omega = 2$ rad/s [line (b) in the asymptotic diagram of Fig. 5.21].
- For the simple pole at the position $s = -10$ we draw a line with *negative slope* $-$ 20 dB/decade that intersects the shifted horizontal axis at the point $\omega = 10$ rad/s (the line (c) in the asymptotic diagram of Fig. 5.21).
- For the zero value at the position $s = -75$ we draw a line with *positive slope* +20 dB/decade that intersects the shifted horizontal axis at the point $\omega = 75$ rad/s (the line (d) in the asymptotic diagram of Fig. 5.21).

After constructing the foregoing asymptotic curves, we should add them together to construct a Bode diagram for the entire function. In this particular case, the process is relatively simple: Since all poles and zeros have a multiplicity value equal to one, the slope of all lines will be either +20 dB/decade or $-$ 20 dB/decade and therefore *in the areas where we add lines with equal and opposite slopes, the final result will be a horizontal line with a slope value equal to zero*, as is the case, for example, from point A to point B of the diagram, as well as after point C. In this diagram, the asymptotic Bode diagram is described by the dashed line, while the actual Bode diagram is described by the blue curve. Although the actual Bode diagram was constructed using MATLAB, it is possible to somewhat improve the approximate diagram by making some corrections. First, in the positions of the slope frequencies, we can *smooth* the curves by lowering the pole curves by $-$ 3 dB and raising the curve at the position $\omega = 75$ by the same amount. Corrections also need to be made *due to the existence of the rest of the frequencies that affect the shape of the curve that corresponds, for example, to a pole if it was studied separately from the others*. To determine the correction value, we should use the error change curve of the diagram's approximation from its asymptotic version, which for the frequency value $\omega = 3$ rad/s is shown in Fig. 5.22. From this figure we observe that the approximation error corresponding to the frequency $\omega = 10$ rad/s is equal to around $-$ 0.4 dB, while the one corresponding to the frequency of 75 rad/s can be considered negligible. Therefore, the total correction concerning the frequency of 3 rad/s is equal to $-(3 + 0.4)$ dB $= -3.4$ dB. Similar corrections can be made for the remaining frequencies and for some intermediate points, so that ultimately we get the real correct curve shown in Fig. 5.21.

On the other hand, the phase diagram will also be constructed based on the now known rules. Thus, the zero value at the position $s = 0$ will cause an *increase* in phase by 90°, while the constant term $K = 25$ will not cause the slightest change in the diagram, as it corresponds to a phase value equal to zero degrees. The pole at the position $s = -2$ will contribute *two*

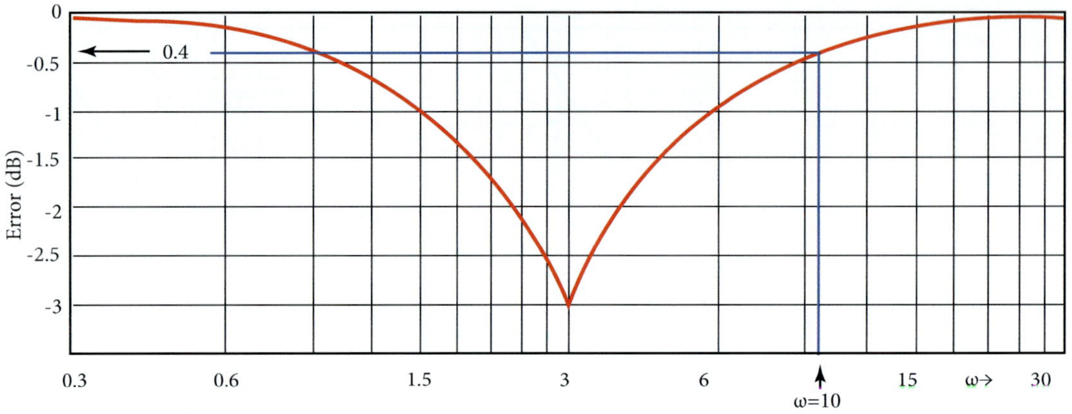

Fig. 5.22 Calculation of correction values for various frequencies

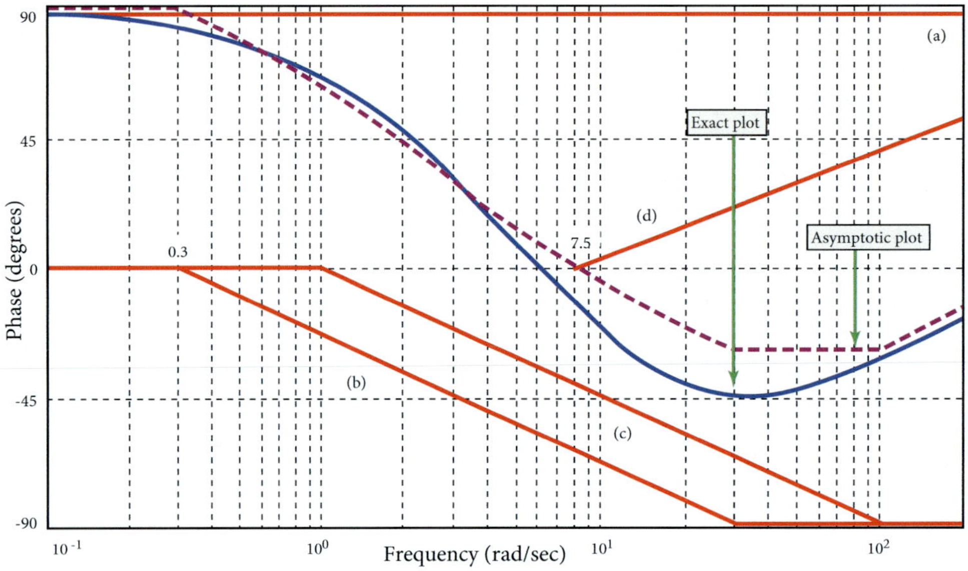

Fig. 5.23 Construction of Bode diagram for phase of function $\mathcal{H}(s)$ from Exercise 5.7.3

asymptotes in the areas $\omega < 0.2$ and $\omega > 20$ with phase values equal to $0°$ and $-90°$, respectively, which are connected by a straight line with a constant phase drop of -45 dB/decade, while the pole at the position $s = -10$ will contribute another *two asymptotes* defined in the areas $\omega < 1$ and $\omega > 100$ with phase values equal to $0°$ and $-90°$, respectively, that are connected by a straight line similar to the previous one. Finally, the zero value at the position $s = -75$ will give *two asymptotes* that have a zero value up to the frequency of 7.5 rad/s and a value equal to $90°$ for frequencies greater than 750 rad/s, respectively, that are connected by a straight line with a positive slope equal to $+45$ dB/decade. The asymptotic and the actual phase diagram are depicted by the black dashed and blue curve of Fig. 5.23, with the actual diagram resulting from the asymptotic approximate diagram by making corrections with the help of the approximation error diagram for the phase and in the manner presented during the construction of the amplitude diagram.

5.8 Interconnection of Continuous-Time Systems and Inverse Systems

One of the basic properties of continuous-time systems is their ability to interconnect *in series or in parallel*, so as to form more extensive and complex structures and processing arrangements. These two forms of interconnection are diagrammatically presented in Fig. 5.24. In this figure, the two systems are characterized by impulse responses $h(t)$ and $g(t)$ or, equivalently, by transfer functions $\mathcal{H}(s)$ and $\mathcal{G}(s)$.

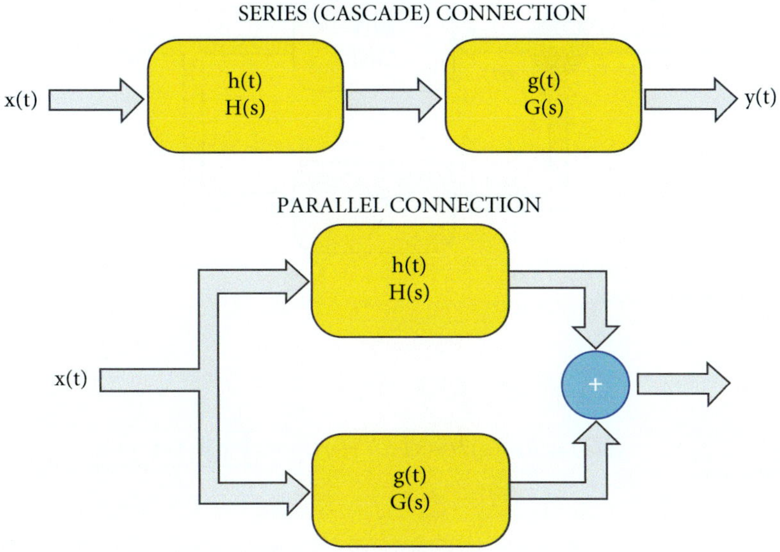

Fig. 5.24 Interconnection of continuous-time systems in series and in parallel

In the case of connecting the systems in series, the output of the first system $z(t) = x(t) * h(t)$ becomes the input to the second system, whose output can therefore be expressed in the form

$$y(t) = z(t) * g(t) = x(t) * [h(t) * g(t)] = x(t) * h_s(t) \tag{5.39}$$

where $h_s(t) = h(t) * g(t)$. Taking the *Laplace transform* of the last equation, we get

$$\mathcal{H}_s(s) = \mathcal{L}\{h_s(t)\} = \mathcal{L}\{h(t) * g(t)\} = \mathcal{L}\{h(t)\}\mathcal{L}\{g(t)\} = \mathcal{H}(s)\mathcal{G}(s) \tag{5.40}$$

(as derived from the *convolution theorem*), that is, in series connection, the *transfer function of the equivalent system results from the product of the functions of the basic systems*. On the other hand, in parallel connection, the output of the composite system arises from the *sum of the outputs of the basic systems*, or, in mathematical notation,

$$y(t) = x(t) * h(t) + x(t) * g(t) = x(t) * h_p(t) \tag{5.41}$$

where $h_p(t) = h(t) + g(t)$. Taking the Laplace transform of the last equation, we get

$$\begin{aligned} \mathcal{H}_p(s) &= \mathcal{L}\{h_p(t)\} = \mathcal{L}\{h(t) + g(t)\} = \\ &= \mathcal{L}\{h(t)\} + \mathcal{L}\{g(t)\} = \mathcal{H}(s) + \mathcal{G}(s) \end{aligned} \tag{5.42}$$

as emerges from the property of *linearity* of the Laplace transform, that is, the *transfer function of the composite system in parallel connection equals the sum of the transfer functions of the basic systems*.

The foregoing conclusions can be generalized for an *arbitrary number of linear and time-invariant systems* that are connected *in series or in parallel*. Considering N such systems with transfer functions of the form $\mathcal{H}_1(s), \mathcal{H}_2(s), \ldots, \mathcal{H}_N(s)$, the transfer function of the equivalent system resulting from their series connection will be given by the relation

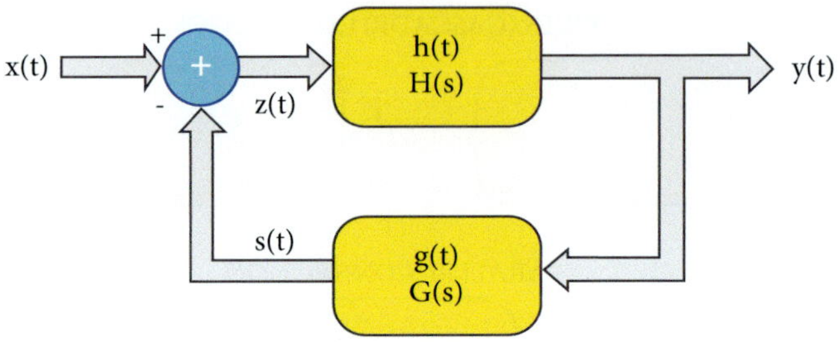

Fig. 5.25 Typical system structure with feedback

$$\mathcal{H}(s) = \mathcal{H}_1(s)\mathcal{H}_2(s)\ldots\mathcal{H}_N(s) = \prod_{i=1}^{N}\mathcal{H}_i(s) \tag{5.43}$$

while the corresponding expression for the parallel connection has the form

$$\mathcal{H}(s) = \mathcal{H}_1(s) + \mathcal{H}_2(s) + \cdots + \mathcal{H}_N(s) = \sum_{i=1}^{N}\mathcal{H}_i(s) \tag{5.44}$$

The third method of system interconnection allows for the construction of the so-called *feedback systems*, whose output *feeds back* their input. This situation is diagrammatically represented in Fig. 5.25 and refers to the case of two systems with impulse responses $h(t)$ and $g(t)$ and the corresponding transfer functions $\mathcal{H}(s) = \mathcal{L}\{h(t)\}$ and $\mathcal{G}(s) = \mathcal{L}\{g(t)\}$.

If we symbolize by $s(t)$ the output of the system $\mathcal{G}(s)$, this feeds back the system and is subtracted[19] from the input $x(t)$, to give the signal $z(t) = x(t) - s(t)$, which in turn is used as input to the system $\mathcal{H}(s)$. Taking into account that the systems $\mathcal{H}(s)$ and $\mathcal{G}(s)$ are *LTI*, their operation will be described by the operation of *convolution*. We will then have $y(t) = z(t) * h(t)$, from which it follows that $\mathcal{Y}(s) = \mathcal{Z}(s)\,\mathcal{H}(s)$, and $s(t) = y(t) * g(t)$, from which it follows that $\mathcal{S}(s) = \mathcal{Y}(s)\,\mathcal{G}(s)$. If we take the Laplace transform of the relation $z(t) = x(t) - s(t)$, we will get $\mathcal{Z}(s) = \mathcal{X}(s) - \mathcal{S}(s)$, and therefore

$$\mathcal{Y}(s) = \mathcal{H}(s)\,\mathcal{Z}(s) = \mathcal{H}(s)\Big[\mathcal{X}(s) - \mathcal{S}(s)\Big] = \mathcal{H}(s)\Big[\mathcal{X}(s) - \mathcal{Y}(s)\,\mathcal{G}(s)\Big]$$

from which it follows that the transfer function of the composite system will be given by the relation

$$\mathcal{H}(s) = \frac{\mathcal{Y}(s)}{\mathcal{X}(s)} = \frac{\mathcal{H}(s)}{1 + \mathcal{H}(s)\,\mathcal{G}(s)} \tag{5.45}$$

Using the interconnection of continuous-time systems in series, we have the ability to define the *inverse system* of a system $\mathcal{H}(s)$ as a system with transfer function $\mathcal{G}(s)$, which, if connected in series with $\mathcal{H}(s)$, leads to the creation of a composite system whose output $y(t)$ will coincide with its input $x(t)$. This situation is presented diagrammatically in Fig. 5.26.

To determine the mathematical relationship that exists between the functions $\mathcal{H}(s)$ and $\mathcal{G}(s)$, we will use the methodology presented earlier, according to which the output of the composite system is given by the relation

$$y(t) = x(t) * [h(t) * g(t)] = x(t) \tag{5.46}$$

[19] It is obvious that if instead of *subtraction* we consider the operation of *addition*, the equations change accordingly, and in reality there are two systems of this type characterized by *positive feedback* for the case of *addition* and by *negative feedback* for the case of *subtraction*.

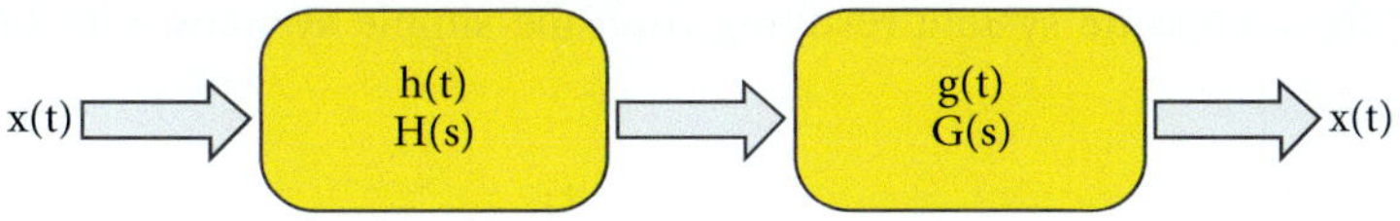

Fig. 5.26 The concept of the inverse system

If we take the *Laplace transform* of the foregoing equation and use the property

$$f(t) * [g(t) * h(t)] \leftrightarrows \mathcal{L}\{f(t)\}\mathcal{L}\{g(t)\}\mathcal{L}\{h(t)\}$$

it immediately follows that $\mathcal{L}\{y(t)\} = \mathcal{L}\{x(t)\}\mathcal{L}\{h(t)\}\mathcal{L}\{g(t)\} = \mathcal{L}\{x(t)\}$; therefore,

$$\mathcal{H}_s(t) = \mathcal{L}\{h(t)\}\mathcal{L}\{g(t)\} = \mathcal{H}(s)\,\mathcal{G}(s) = 1 \tag{5.47}$$

and therefore $\mathcal{G}(s) = 1/\mathcal{H}(s)$, an equation that constitutes the desired mathematical relationship between the functions $\mathcal{H}(s)$ and $\mathcal{G}(s)$. From the last relationship it is not difficult to see that the *zeros and poles of the function* $\mathcal{H}(s)$ *will become poles and zeros, respectively, of the function* $\mathcal{G}(s)$. Therefore, if we know the diagram of poles-zero values of the function $\mathcal{H}(s)$, we can very easily construct the corresponding diagram of the function $\mathcal{G}(s)$ *simply by swapping the positions of the zeros and the poles on the complex plane*.

A special category of continuous-time LTI systems is *minimum phase systems*. These systems are defined as *causal and stable systems whose inverse system exists and is also causal and stable*. In this case, it is not difficult to ascertain that *both the poles and the zeros of such a system will belong to the left half-plane of the complex plane, that is, they will have a negative real part*. Indeed, since the system is causal and stable, its poles will be characterized by the aforementioned property, as we have already proven, while obviously the same will apply to its zeros, which, being poles of the inverse system—which is also *causal* and *stable*—will also encounter this property. For more information about this special category of continuous-time systems, refer to footnote 14 on page 982.

Problem 5.8.1 Consider the following LTI systems with transfer functions

$$\mathcal{H}_1(s) = \frac{2}{s+3}, \qquad \mathcal{H}_2(s) = \frac{2s+5}{s^2-5s+7}, \qquad \mathcal{H}_3(s) = \frac{s-2}{s^3-2s^2+25s-50}$$

(a) Determine the transfer function of the composite system resulting from the series and parallel connection of the foregoing simple systems. (b) Determine the transfer function of the system with negative feedback resulting from the interconnection of the systems with transfer functions $\mathcal{H}_1(s)$ and $\mathcal{H}_2(s)$.

Solution (a) The transfer function of the system resulting from the series connection of these three systems is calculated as

$$G_1(s) = H_1(s)H_2(s)H_3(s) = \frac{2(2s+5)(s-2)}{(s+3)(s^2-5s+7)(s^3-2s^2+25s-50)}$$
$$= \frac{4s^2+2s-20}{s^6-4s^5+21s^4-63s^3-142s^2+925s-1050}$$

as derived following operations, while the corresponding expression for the parallel connection is calculated as

$$\mathcal{G}_2(s) = \mathcal{H}_1(s) + \mathcal{H}_2(s) + \mathcal{H}_3(s) = \frac{4s^5-6s^4+123s^3-237s^2+712s-1492}{s^6-4s^5+21s^4-63s^3-142s^2+925s-1050}$$

(b) The transfer function of the composite system resulting from the simple systems with functions $\mathcal{H}_1(s)$ and $\mathcal{H}_2(s)$ is calculated as

$$\mathcal{G}(s) = \frac{H_1(s)}{1 + H_1(s)\, H_2(s)} = \frac{2/(s+3)}{1 + [2/(s+3)] \times [(2s+5)/(s^2 - 5s + 7)]} = \frac{2s^2 - 10s + 14}{s^3 - 2s^2 - 4s + 31}$$

Note In the processes of calculating the transfer function of composite systems that result from the interconnection of simpler systems, it is possible to encounter the phenomenon of *mutual cancellation of common poles-zero values* with all the implications this entails and which we analyzed in a previous section. Let us consider, for example, two systems, S_1 and S_2, with transfer functions $\mathcal{H}_1(s)$ and $\mathcal{H}_2(s)$, such that the function $\mathcal{H}_1(s)$ has a zero at the position $s = \alpha$, while the function $\mathcal{H}_2(s)$ has a pole at the same position. It is obvious that if these two systems are connected in series, the transfer function of the composite system will be calculated as $\mathcal{H}(s) = \mathcal{H}_1(s)\mathcal{H}_2(s)$; in this case, the term $(s - \alpha)$ will appear *in both the numerator and the denominator* of the function $\mathcal{H}(s)$, and one might think that its simplification will lead to the construction of a simpler system. However, as was mentioned earlier, if the pole-zero pair that is removed is unstable, that is, it has a *positive real part*, then our conclusions regarding the stability of the system will not be correct.

To demonstrate this situation in practice, let us assume that the two systems we consider are described by the differential equations

$$\frac{dy(t)}{dt} + 4y(t) = \frac{dx(t)}{dt} - 2x(t) \quad \text{and} \quad \frac{d^2y(t)}{dt^2} - \frac{dy(t)}{dt} - 2y(t) = \frac{dx(t)}{dt} - 3x(t)$$

from which the corresponding transfer functions arise:

$$\mathcal{H}_1(s) = \frac{s-2}{s+4} \quad \text{and} \quad \mathcal{H}_2(s) = \frac{s-3}{s^2 - s - 2}$$

Therefore, if the two systems are connected in series, the transfer function of the composite system is

$$\mathcal{H}(s) = \mathcal{H}_1(s)\, \mathcal{H}_2(s) = \frac{(s-2)\,(s-3)}{(s+4)\,(s^2 - s - 2)}$$

Taking into account that the polynomial $s^2 - s - 2$ factors as $s^2 - s - 2 = (s-2)(s+1)$, one could simplify the composite function $\mathcal{H}(s)$ by writing it in the form

$$\mathcal{H}(s) = \frac{(s-2)\,(s-3)}{(s+4)\,(s^2 - s - 2)} = \frac{(s-2)\,(s-3)}{(s+4)\,(s-2)\,(s+1)} = \frac{s-3}{(s+4)\,(s+1)}$$

and then formulate the claim that the system described by the function $\mathcal{H}(s)$ is stable since it is characterized by the existence of two stable poles at the positions $s = -1$ and $s = -4$. Let us study the system a little more thoroughly to determine whether things are really like this.

Since systems S_1 and S_2 are connected in series, it is obvious that the output $y_1(t)$ of the first system will be transmitted as input to the second system, and therefore we will have $x(t) = y_1(t)$. Taking the Laplace transform of the differential equation of the first system, we find that

$$\mathcal{Y}_1(s) = \frac{s-2}{s+4}\mathcal{X}(s) + \frac{y_1(0)}{s+4}$$

where $\mathcal{Y}_1(s) = \mathcal{L}\{y_1(t)\}$ and $y_1(0)$ is the initial condition for the output $y_1(t)$ (for simplicity we have set $x(0) = 0$, although the use of a nonzero value for $x(0)$ does not change anything, but it does increase the mathematical operations). In a completely similar way, the application of the *Laplace transform* to the differential equation of the second system will lead to the result

$$\mathcal{Y}(s) = \frac{s-3}{s^2 - s - 2}\mathcal{Y}_1(s) + \frac{\dot{y}(0) + (s-1)y(0) - y_1(0)}{s^2 - s - 2} =$$

$$= \frac{(s-3)\,(s-2)}{(s^2-s-2)\,(s+4)}\mathcal{X}(s) - \frac{7y_1(0)}{(s^2-s-2)\,(s+4)} = \mathcal{H}(s)\,\mathcal{X}(s) - \frac{7}{(s-2)(s+1)(s+4)}$$

where for the derivation of the last relation we assumed that $x(t) = y_1(t)$—so we will also have $\mathcal{X}(s) = \mathcal{Y}_1(s)$—and we used the expression $\mathcal{Y}_1(s)$ that we constructed earlier, and for simplicity we assumed that $y(0) = \dot{y}(0) = 0$ and $y_1(0) = 1$. Therefore, to calculate the output $y(t)$ of the second system—which of course will also be the output of the *composite system* resulting from the series connection of the two systems—we must take the *inverse Laplace transform* of the preceding relation. Therefore, we obtain

$$y(t) = \mathcal{L}^{-1}\{Y(s)\} = \mathcal{L}^{-1}\{\mathcal{H}(s)\,\mathcal{X}(s)\} - 7\mathcal{L}^{-1}\left\{\frac{1}{(s+4)\,(s-2)\,(s+1)}\right\}$$

Using the convolution theorem, the first term is formulated as

$$\mathcal{L}^{-1}\{\mathcal{H}(s)\mathcal{X}(s)\} = h(t) * x(t) = [h_1(t) * h_2(t)] * x(t)$$

where $h_1(t) = \mathcal{L}^{-1}\{\mathcal{H}_1(s)\}$ and $h_2(t) = \mathcal{L}^{-1}\{\mathcal{H}_2(s)\}$ are the impulse responses of systems S_1 and S_2. On the other hand, using the method of partial fraction expansion we find that

$$\frac{1}{(s+4)(s-2)(s+1)} = \frac{1}{18}\frac{1}{s+4} + \frac{1}{18}\frac{1}{s-2} - \frac{1}{9}\frac{1}{s+1}$$

and the inverse Laplace transform of this term will take the form

$$\mathcal{L}^{-1}\left\{\frac{1}{(s+4)(s-2)(s+1)}\right\} = \frac{1}{18}\mathcal{L}^{-1}\left\{\frac{1}{s+4}\right\} + \frac{1}{18}\mathcal{L}^{-1}\left\{\frac{1}{s-2}\right\}$$
$$- \frac{1}{9}\mathcal{L}^{-1}\left\{\frac{1}{s+1}\right\} = \frac{1}{18}e^{-4t}u(t) + \frac{1}{18}e^{2t}u(t) - \frac{1}{9}e^{-t}u(t)$$

Therefore, the system output will be given by the relation

$$y(t) = \underbrace{[\,h_1(t) * h_2(t)\,] * x(t)}_{\text{zero-state response}} + \underbrace{\frac{1}{18}e^{-4t}u(t) + \frac{1}{18}e^{2t}u(t) - \frac{1}{9}e^{-t}u(t)}_{\text{zero-input response}}$$

We observe, then, once again that the unbounded exponential term e^{2t} appears in the system output whose existence is due to the unstable pole located at $s = 2$. Therefore, simplification of the term $(s-2)$ from the numerator and the denominator of the function $\mathcal{H}(s)$ does not remove the foregoing term, making the system stable, but simply *hides it from the equation*. Once again, then, we conclude that the *study of a transfer function in terms of its stability should be done without removing poles from it, especially if these poles are unstable*. Similar situations may occur in cases of systems connected in parallel or through feedback connections.

5.9 Representation with Schematic Diagrams

As we mentioned in the chapter on continuous-time systems, a simple system described by a simple linear differential equation can be represented using schematic diagrams that are created from the interconnections of elementary structural units, such as differentiators and integrators. These types of diagrams can also be used at a different level of representation to describe a complex system in a simpler and more abstract way.

Let us consider, for example, the continuous-time system described by the simple differential equation

$$2\frac{d^2y(t)}{dt^2} + 5\frac{dy(t)}{dt} + 4y(t) = x(t)$$

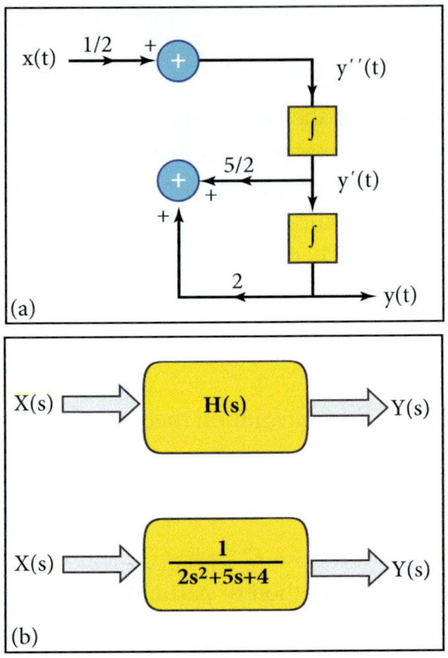

Fig. 5.27 System representation in time and frequency domains

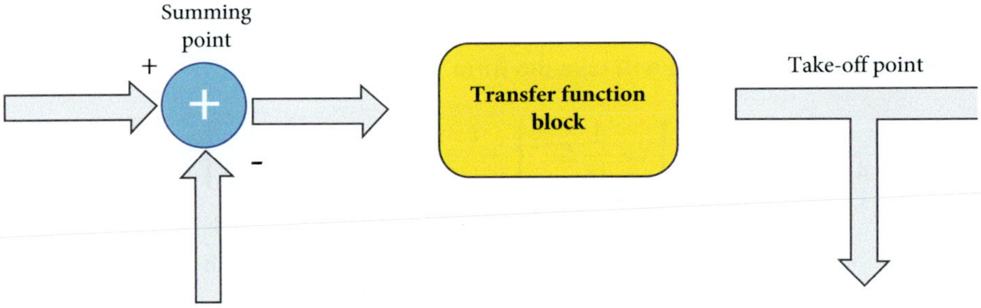

Fig. 5.28 The three fundamental building blocks of a schematic diagram

that, as was mentioned earlier, is represented by the schematic diagram in Fig. 5.27a. However, if this system is going to be used as a subsystem within a much larger and extensive system, it is usually not represented in this way but as shown in image (b) of the same figure. This figure represents the system with the help of its transfer function, which is proven to be

$$\mathcal{H}(s) = \frac{\mathcal{Y}(s)}{\mathcal{X}(s)} = \frac{1}{2s^2 + 5s + 4}$$

and describes it as a black box that accepts the input signal $\mathcal{X}(s)$ and generates the output signal $\mathcal{Y}(s)$. This representation is completely analogous to the one that arises from the equation $y(t) = \mathcal{T}\{x(t)\}$, only now it is not defined in the time domain but in the complex plane s. We observe that for the symbolism of the system, either the symbol $\mathcal{H}(s)$ or the equation of the transfer function itself can be used, provided, of course, that it is not too large.

Three fundamental building blocks can be used to construct schematic diagrams depicted in Fig. 5.28. The first of these is our well-known signal adder, which takes two signals as input and produces their sum or difference as its output. The signal adder is mainly encountered in feedback systems, which is considered positive for the case of the sum and negative for the case of the difference of signals; in the latter case, the difference between the two signals is known as the error signal. The second element is a rectangular parallelogram that represents the transfer function of a system and contains as a label either the symbol of the transfer function or the equation defining it, while the third building block is the so-called signal branching point, essentially a branching point in the diagram, through which the signal can be redirected to some other destination.

In the most common cases, complex schematic diagrams can be simplified, so that we can construct simpler structures that, however, will represent the same—or an equivalent—continuous-time system. We already encountered this simplification in the previous section, where we saw how a multitude of systems connected in series or in parallel could be replaced by an equivalent system whose transfer function was equal to the product or the sum respectively of the transfer functions of the

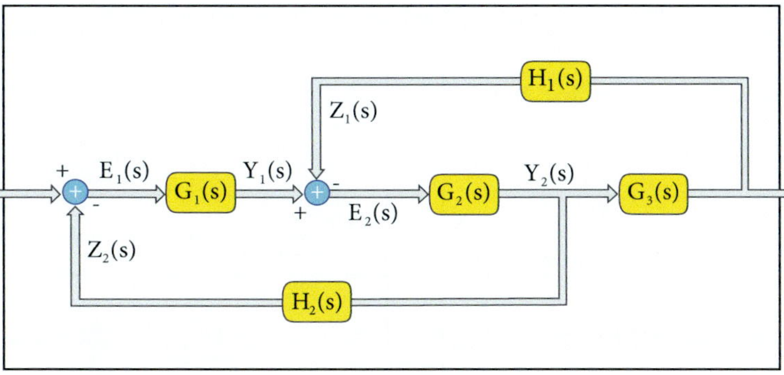

Fig. 5.29 Schematic diagram to be simplified from the example of Sect. 5.9

elementary systems. To understand this process, let us look at the way in which we can simplify the schematic diagram of Fig. 5.29 and construct the transfer function of the equivalent system that takes the signal $\mathcal{X}(s)$ as input and generates the signal $\mathcal{Y}(s)$ as its output.

From the figure we observe that the outputs of the two adders are described by the equations $e_1(t) = x(t) - z_2(t)$ and $e_2(t) = y_1(t) - z_1(t)$; therefore, taking the Laplace transform of the foregoing equations, it turns out that $\mathcal{E}_1(s) = \mathcal{X}(s) - \mathcal{Z}_2(s)$ and $\mathcal{E}_2(s) = \mathcal{Y}_1(s) - \mathcal{Z}_1(s)$.

On the other hand, the five elementary subsystems that appear in the figure are LTI systems and therefore their operation will be characterized by the act of convolution. Applying, then, the property of the convolution of the Laplace transform, it turns out that:

- For the subsystem $\mathcal{G}_1(s)$: $y_1(t) = g_1(t) * e_1(t) \Leftrightarrow \mathcal{Y}_1(s) = \mathcal{G}_1(s)\,\mathcal{E}_1(s)$;
- For the subsystem $\mathcal{G}_2(s)$: $y_2(t) = g_2(t) * e_2(t) \Leftrightarrow \mathcal{Y}_2(s) = \mathcal{G}_2(s)\,\mathcal{E}_2(s)$;
- For the subsystem $\mathcal{G}_3(s)$: $y(t) = g_3(t) * y_2(t) \Leftrightarrow \mathcal{Y}(s) = \mathcal{G}_3(s)\,\mathcal{Y}_2(s)$;
- For the subsystem $\mathcal{H}_1(s)$: $z_1(t) = h_1(t) * y(t) \Leftrightarrow \mathcal{Z}_1(s) = \mathcal{H}_1(s)\,\mathcal{Y}(s)$;
- For the subsystem $\mathcal{H}_2(s)$: $z_2(t) = h_2(t) * y_2(t) \Leftrightarrow \mathcal{Z}_2(s) = \mathcal{H}_2(s)\,\mathcal{Y}_2(s)$,

where obviously the impulse responses of the five subsystems are defined as

$$g_1(t) = \mathcal{L}^{-1}\{\mathcal{G}_1(s)\}, \quad g_2(t) = \mathcal{L}^{-1}\{\mathcal{G}_2(s)\}, \quad g_3(t) = \mathcal{L}^{-1}\{\mathcal{G}_3(s)\}, \quad h_1(t) = \mathcal{L}^{-1}\{\mathcal{H}_1(s)\}, \quad h_2(t) = \mathcal{L}^{-1}\{\mathcal{H}_2(s)\}$$

Based now on the foregoing relations, it turns out that

$$\begin{aligned}\mathcal{Y}(s) &= \mathcal{G}_3(s)\mathcal{Y}_2(s) = \mathcal{G}_3(s)\{\mathcal{G}_2(s)\mathcal{E}_2(s)\} = \mathcal{G}_3(s)\mathcal{G}_2(s)[\mathcal{Y}_1(s) - \mathcal{Z}_1(s)] = \mathcal{G}_3(s)\mathcal{G}_2(s)\mathcal{Y}_1(s) - \mathcal{G}_3(s)\mathcal{G}_2(s)\mathcal{Z}_1(s)\\ &= \mathcal{G}_3(s)\mathcal{G}_2(s)\mathcal{Y}_1(s) - \mathcal{G}_3(s)\mathcal{G}_2(s)\{\mathcal{H}_1(s)\mathcal{Y}(s)\} = \mathcal{G}_3(s)\mathcal{G}_2(s)\{\mathcal{G}_1(s)\mathcal{E}_1(s)\} - \mathcal{G}_3(s)\mathcal{G}_2(s)\mathcal{H}_1(s)\mathcal{Y}(s)\\ &= \mathcal{G}_3(s)\mathcal{G}_2(s)\mathcal{G}_1(s)[\mathcal{X}(s) - \mathcal{Z}_2(s)] - \mathcal{G}_3(s)\mathcal{G}_2(s)\mathcal{H}_1(s)\mathcal{Y}(s) = \mathcal{G}_3(s)\mathcal{G}_2(s)\mathcal{G}_1(s)\mathcal{X}(s)\\ &\quad - \mathcal{G}_3(s)\mathcal{G}_2(s)\mathcal{G}_1(s)\mathcal{Z}_2(s) - \mathcal{G}_3(s)\mathcal{G}_2(s)\mathcal{H}_1(s)\mathcal{Y}(s)\\ &= \mathcal{G}_3(s)\mathcal{G}_2(s)\mathcal{G}_1(s)\mathcal{X}(s) - \mathcal{G}_2(s)\mathcal{G}_1(s)\mathcal{H}_2(s)\mathcal{Y}(s) - \mathcal{G}_3(s)\mathcal{G}_2(s)\mathcal{H}_1(s)\mathcal{Y}(s)\end{aligned}$$

Therefore, we obtain

$$\mathcal{Y}(s)\left[1 + \mathcal{G}_1(s)\mathcal{G}_2(s)\mathcal{H}_2(s) + \mathcal{G}_2(s)\mathcal{G}_3(s)\mathcal{H}_1(s)\right] = \left[\mathcal{G}_1(s)\mathcal{G}_2(s)\mathcal{G}_3(s)\right] X(s)$$

and therefore the transfer function of the equivalent system will be given by the relation

$$\mathcal{H}(s) = \frac{\mathcal{Y}(s)}{\mathcal{X}(s)} = \frac{\mathcal{G}_1(s)\mathcal{G}_2(s)\mathcal{G}_3(s)}{1 + \mathcal{G}_1(s)\mathcal{G}_2(s)\mathcal{H}_2(s) + \mathcal{G}_2(s)\mathcal{G}_3(s)\mathcal{H}_1(s)}$$

The simplification of any schematic diagram takes place in the same way.

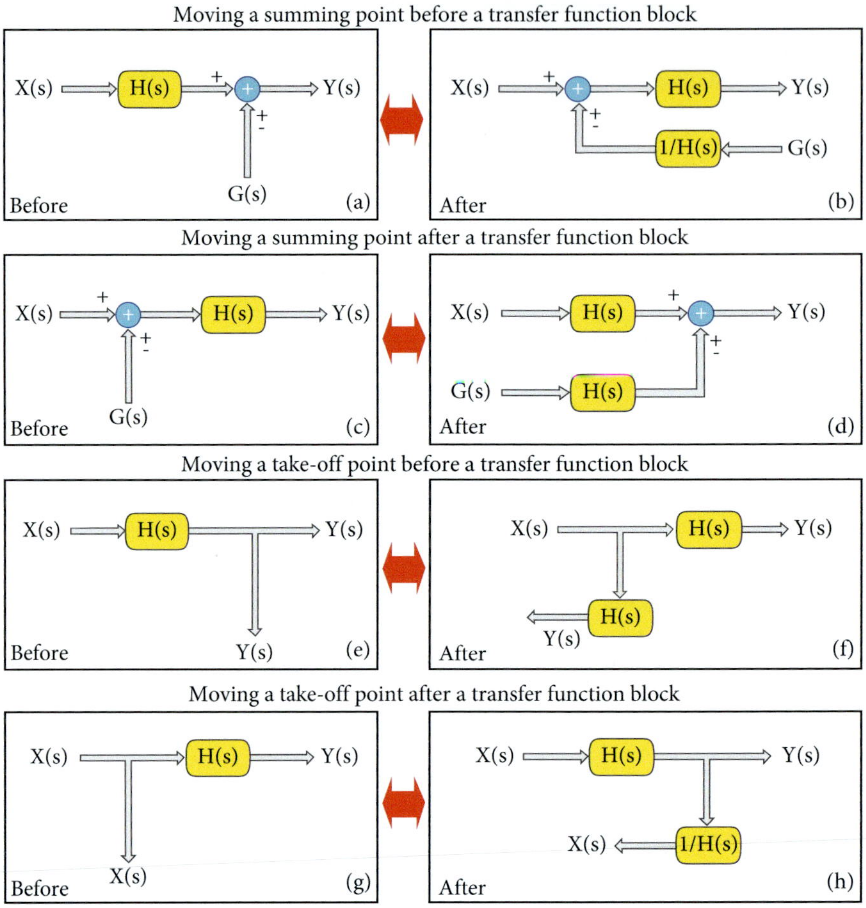

Fig. 5.30 Schematic diagram simplification rules

Although the foregoing process is direct and leads directly to the final result, in the case of complex diagrams, it can become quite complex. For this reason, in cases of large and extensive systems, their simplification is based on the use of some rules that determine a set of elementary operations regarding the movement of elements within a diagram, in order for these to be subsequently replaced by others. These rules are presented diagrammatically in Figs. 5.30 and 5.31—it is obvious that the circuits before and after the application of the process are equivalent to each other—and the operations determined by them are as follows:

- *Moving the signal adder in front of the transfer function element*. The image of the circuit before and after the process is presented in Fig. 5.30α, β, while the equation describing the circuit is in the form

$$\mathcal{Y}(s) = \mathcal{H}(s)\mathcal{X}(s) \pm \mathcal{G}(s) \tag{5.48}$$

- *Moving the signal adder behind the transfer function element*. The image of the circuit before and after the process is presented in Fig. 5.30γ, δ, while the equation describing the circuit is in the form

$$\mathcal{Y}(s) = \mathcal{H}(s)[\mathcal{X}(s) \pm \mathcal{G}(s)] \tag{5.49}$$

- *Moving the branching point in front of the transfer function element*. The image of the circuit before and after the process is presented in Fig. 5.30ε, $\sigma\tau$, while the equation describing the circuit is in the form

$$\mathcal{Y}(s) = \mathcal{H}(s)\mathcal{X}(s) \tag{5.50}$$

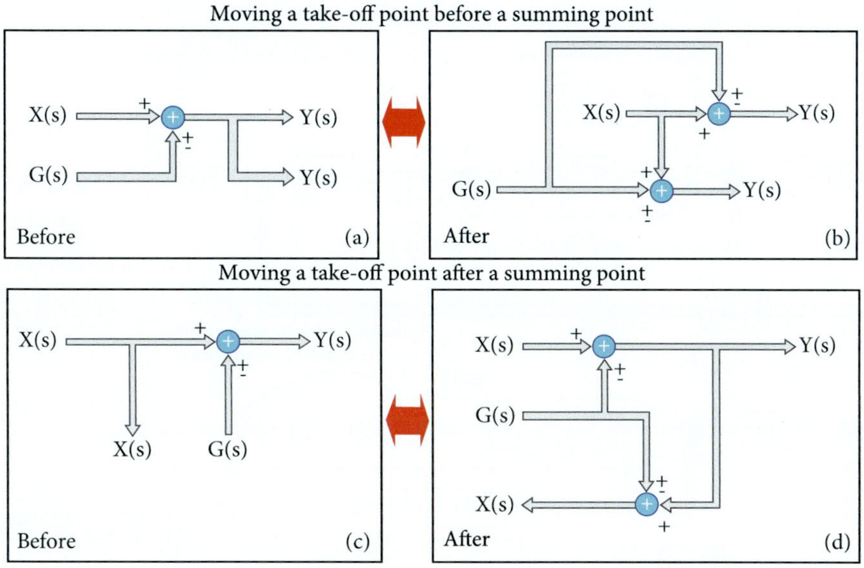

Fig. 5.31 Schematic diagram simplification rules (continued)

- *Moving the branching point behind a transfer function element*. The image of the circuit before and after the process is presented in Fig. 5.30z, h, while the equation describing the circuit is of the form

$$\mathcal{Y}(s) = \mathcal{H}(s)\mathcal{X}(s) \tag{5.51}$$

- *Moving the branching point in front of a signal adder*. The image of the circuit before and after the process is presented in Fig. 5.31a, b, while the equation describing the circuit is of the form

$$\mathcal{Y}(s) = \mathcal{X}(s) \pm \mathcal{G}(s) \tag{5.52}$$

- *Moving the branching point behind a signal adder*. The image of the circuit before and after the process is presented in Fig. 5.31g, d, while the equation describing the circuit is of the form

$$\mathcal{Y}(s) = \mathcal{X}(s) \pm \mathcal{G}(s) \tag{5.53}$$

The preceding simplification processes are particularly useful in representing control systems characterized by such structures.

An example of simplifying a schematic diagram using the foregoing rules is presented below. The simplification process applied is diagrammatically presented in Fig. 5.32 and consists of six steps, as follows:

- Systems $\mathcal{G}_2(s)$ and $\mathcal{G}_3(s)$ are characterized by a parallel connection and therefore can be replaced by a simple system with transfer function $\mathcal{G}_2(s) + \mathcal{G}_3(s)$.
- The branching point between systems $\mathcal{G}_2(s)$ and $\mathcal{G}_4(s)$ is transferred after system $\mathcal{G}_4(s)$, so it can be connected in series with the equivalent system of $\mathcal{G}_2(s)$ and $\mathcal{G}_3(s)$. This movement of the branching point causes the replacement of system $\mathcal{H}_1(s)$ with system $\mathcal{H}_1(s)/\mathcal{G}_4(s)$.
- Systems $\mathcal{G}_2(s) + \mathcal{G}_3(s)$ and $\mathcal{G}_4(s)$ are connected in series and can be replaced by an equivalent system with the transfer function $[\mathcal{G}_2(s) + \mathcal{G}_3(s)]\mathcal{G}_4(s)$.
- Systems $[\mathcal{G}_2(s) + \mathcal{G}_3(s)]\mathcal{G}_4(s)$ and $\mathcal{H}_1(s)/\mathcal{G}_4(s)$ are characterized by a connection with negative feedback and can be replaced by an equivalent system with the transfer function

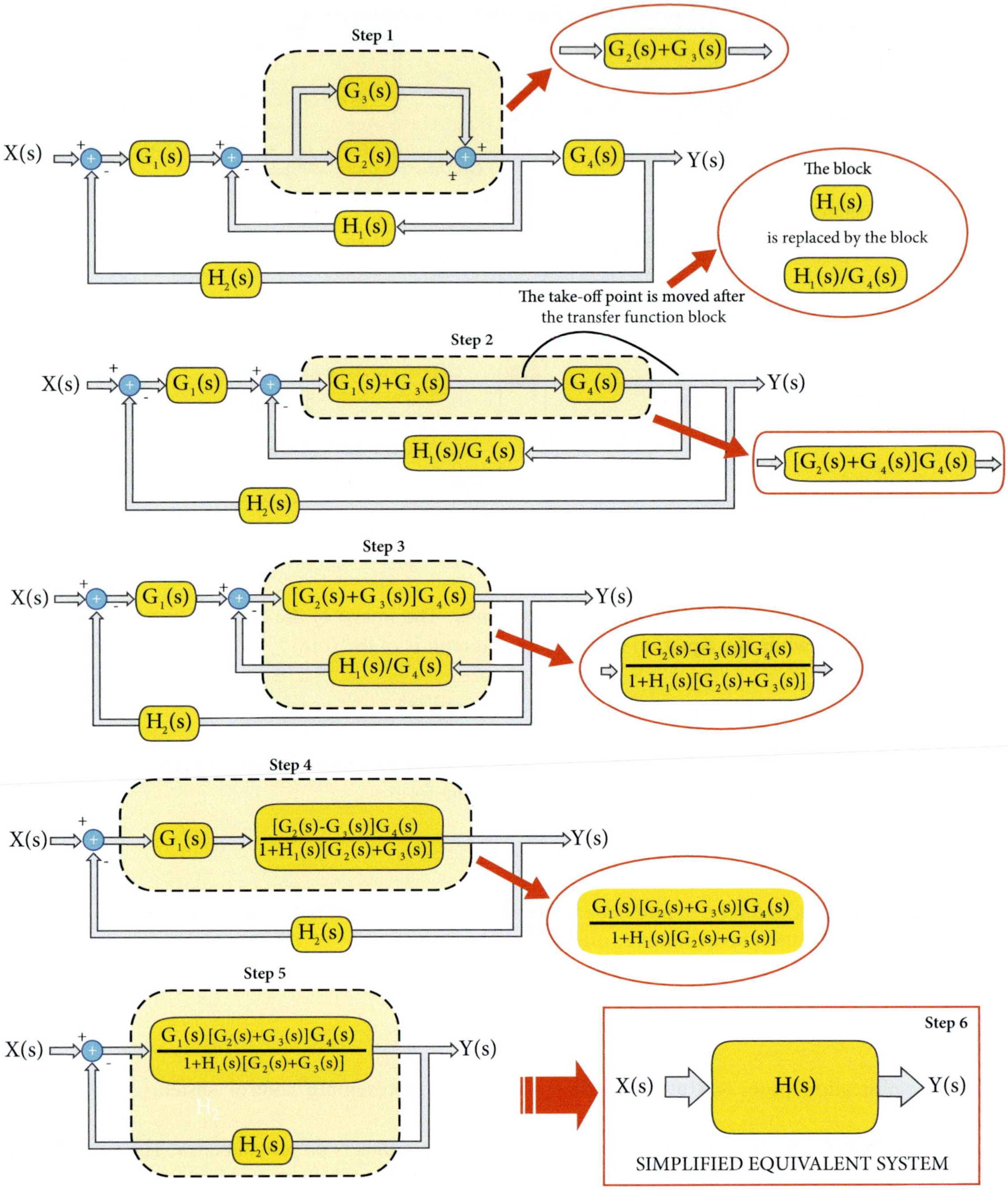

Fig. 5.32 Example of schematic diagram simplification

$$\mathcal{G}_5(s) = \frac{[\mathcal{G}_2(s) + \mathcal{G}_3(s)]\mathcal{G}_4(s)}{1 + \mathcal{H}_1(s)[\mathcal{G}_2(s) + \mathcal{G}_3(s)]}$$

- Systems $\mathcal{G}_1(s)$ and $\mathcal{G}_5(s)$ are connected in series and can be replaced by an equivalent system with the transfer function

$$\mathcal{G}_6(s) = \mathcal{G}_1(s)\mathcal{G}_5(s) = \frac{\mathcal{G}_1(s)[\mathcal{G}_2(s) + \mathcal{G}_3(s)]\mathcal{G}_4(s)}{1 + \mathcal{H}_1(s)[\mathcal{G}_2(s) + \mathcal{G}_3(s)]}$$

- Systems $\mathcal{G}_6(s)$ and $\mathcal{H}_2(s)$ are characterized by a negative feedback connection and can be replaced by an equivalent system with a transfer function

$$\mathcal{H}(s) = \frac{\mathcal{G}_6(s)}{1+\mathcal{H}_2(s)\mathcal{G}_6(s)} = \left\{\frac{\mathcal{G}_1(s)[\mathcal{G}_2(s)+\mathcal{G}_3(s)]\mathcal{G}_4(s)}{1+\mathcal{H}_1(s)[\mathcal{G}_2(s)+\mathcal{G}_3(s)]}\right\} \Big/ \left\{1+\frac{\mathcal{H}_2(s)\mathcal{G}_1(s)[\mathcal{G}_2(s)+\mathcal{G}_3(s)]\mathcal{G}_4(s)}{1+\mathcal{H}_1(s)[\mathcal{G}_2(s)+\mathcal{G}_3(s)]}\right\}$$

  which also constitutes the transfer function of the simple system, which is equivalent to the original complex system.

5.10 Signal Flow Graphs

Signal flow graphs, invented by Mason in 1953, are an alternative way of describing continuous-time linear systems and are particularly useful for determining the transfer function and describing systems in state space. Unlike schematic diagrams consisting of transfer function elements, signal adders, and branching points, the structure of signal flow graphs is similar to that of a network and is characterized by the existence of a set of nodes representing signals that are interconnected using oriented branches that represent systems. In this way, a signal can only move in one specific direction within a circuit. The most important properties of flow diagrams are as follows:

- The nodes of the graph represent the system's variables, describe the signals that are circulated in the system, and are arranged in such a way that they obey the cause–effect principle (since the systems described are obviously causal).
- Each branch of the graph corresponds to a weight known as gain and expresses the transfer function $\mathcal{H}(s)$ of the system described by the branch. This gain is multiplied by the signal passing through the branch (Fig. 5.33a), so as to calculate the output that is generated by it. This situation is entirely analogous to the one we encountered in schematic diagrams where the output of the system $\mathcal{Y}(s)$ resulted from the multiplication of the input signal $X(s)$ by the transfer function of the system $\mathcal{H}(s)$.
- The value of the variable, that is, the signal, associated with a node is transmitted to all branches of the circuit that start from it (Fig. 5.33b). This node is known as a source node.
- The value of the variable, that is, the signal, associated with a node is equal to the sum of the signals entering it (Fig. 5.33c). This node is known as the destination node.
- The passage of a signal through the same node more than once is not allowed, unless the signal starts and ends at the same node, forming a closed loop. In this case, the loop gain is defined as the product of the gains of the system branches traversed exactly once, in the direction of signal flow, until returning to the originating node. With reference to Fig. 5.34, there are four loop gains: $\mathcal{G}_2(s)\mathcal{H}_1(s)$, $\mathcal{G}_4(s)\mathcal{H}_2(s)$, $\mathcal{G}_4(s)\mathcal{G}_5(s)\mathcal{H}_3(s)$, and $\mathcal{G}_4(s)\mathcal{G}_6(s)\mathcal{H}_3(s)$. If within the same graph two or more loops do not share common nodes, then these loops are called foreign to each other. In Fig. 5.34, the loop $\mathcal{G}_2(s)\mathcal{H}_1(s)$ is foreign to the loops $\mathcal{G}_4(s)\mathcal{H}_2(s)$, $\mathcal{G}_4(s)\mathcal{G}_5(s)\mathcal{H}_3(s)$, and $\mathcal{G}_4(s)\mathcal{G}_6(s)\mathcal{H}_3(s)$, as it does not share common nodes with them.
- The value of the gain associated with some open, that is, not closed, path within a graph is equal to the product of the values of the gains associated with its individual branches and is described by the term path gain. Fig. 5.34 has two such path gains whose values are calculated as

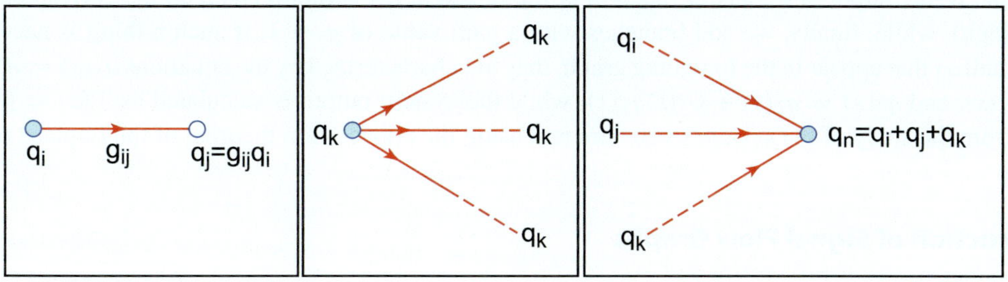

Fig. 5.33 Properties of signal flow graphs

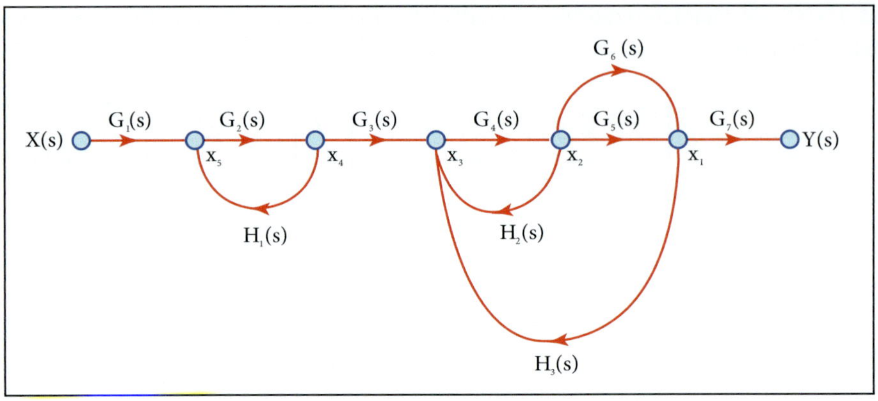

Fig. 5.34 Example of open and closed paths in a signal flow graph

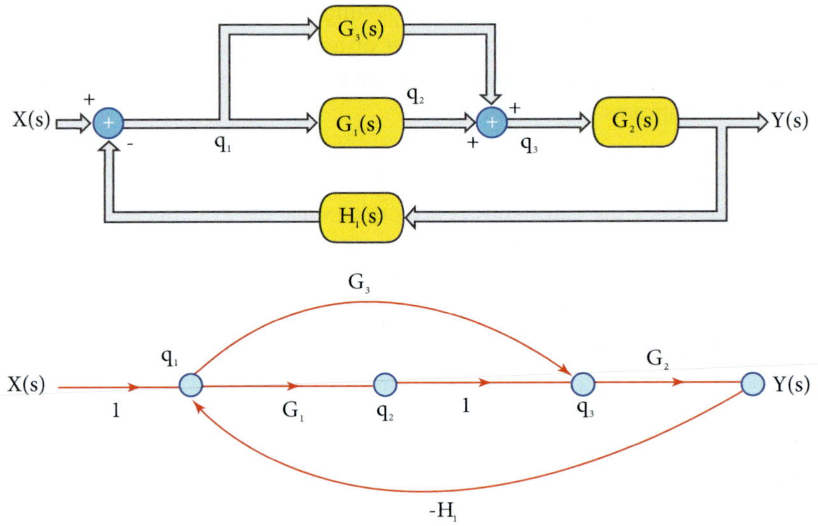

Fig. 5.35 Typical example of a schematic diagram and its corresponding signal flow graph

$$\mathcal{G}_1(s)\mathcal{G}_2(s)\mathcal{G}_3(s)\mathcal{G}_4(s)\mathcal{G}_5(s)\mathcal{G}_7(s) \quad \text{and} \quad \mathcal{G}_1(s)\mathcal{G}_2(s)\mathcal{G}_3(s)\mathcal{G}_4(s)\mathcal{G}_6(s)\mathcal{G}_7(s)$$

A typical example of a schematic diagram and its corresponding signal flow graph is presented in Fig. 5.35, from which the equivalence between these two different ways of representing a system becomes apparent. According to the analysis presented earlier, to convert a schematic diagram into a signal flow graph, we must replace the transfer function elements with oriented branches that represent the signals, which, in addition to the signal adders and branching points, in turn will be replaced by an equal number of nodes. In the next step of the process, we must place the appropriate labels on the system's branches (note that if the output of a transfer function is subtracted at the addition element, the corresponding label should bear a negative sign), while, finally, we add branches with a gain value of $g = 1$, if such a thing is necessary. As for the variables q_1, q_2, and q_3 that appear in the foregoing graph, they are characterized by the equations $q_1(s) = \mathcal{X}(s) - \mathcal{H}_1(s)\mathcal{Y}(s)$, $q_2(s) = \mathcal{G}_1(s)q_1(s)$, and $q_3(s) = q_2(s) + \mathcal{G}_3(s)q_1(s)$, while the system output is calculated as $\mathcal{Y}(s) = \mathcal{G}_2(s)q_3(s)$. These equations arise from the study of the system's structure and using the convolution theorem of the Laplace transform.

5.10.1 Construction of Signal Flow Graphs

In cases where the considered system is described not by a schematic diagram but by an equivalent system of equations that arise from it, the signal flow graph can be constructed gradually and in a way that reflects the structure of each of the

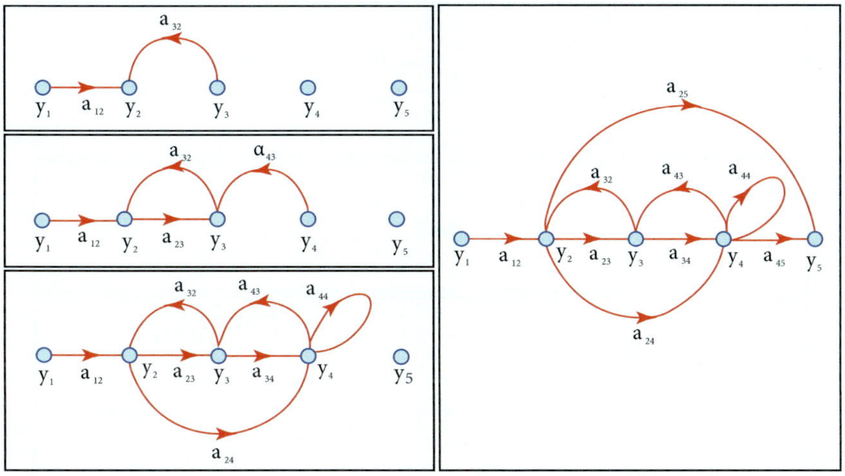

Fig. 5.36 Gradual construction of a signal flow graph

system's equations. Generally speaking, any linear algebraic system of equations can be described by such a graph, which therefore has many more applications than those we present in this brief description. Considering, for example, the system of algebraic equations

$$y_2 = \alpha_{12}y_1 + \alpha_{32}y_3$$
$$y_3 = \alpha_{23}y_2 + \alpha_{43}y_4$$
$$y_4 = \alpha_{24}y_2 + \alpha_{34}y_3 + \alpha_{44}y_4$$
$$y_5 = \alpha_{25}y_2 + \alpha_{45}y_4$$

the equivalent signal flow graph is constructed in the way described by Fig. 5.36.

From Fig. 5.36 we observe that initially we construct as many nodes as there are variables in the system, and then we connect the nodes with oriented curves, whose direction reveals the way in which the variables are correlated. For example, in the first image, the first of the system's equations is represented by the variable y_2, which is equal to the sum of the quantities $\alpha_{12}y_1$ and $\alpha_{13}y_3$. In the second image, the diagram has grown further, so as to include the second equation, and in the third image incorporates the third equation, while the complete diagram that describes the entire system is depicted in the fourth image. We observe, therefore, that for a continuous-time system we have the ability to construct a graph that describes it either by converting the schematic diagram or by using the equivalent algebraic equations.

5.10.2 Simplification of Signal Flow Graphs

In full analogy with the schematic diagrams, the signal flow graphs can be simplified in such a way that the final graph is related only to the input and output of the system, thereby allowing the calculation of the transfer function that describes it. There are six fundamental rules for simplifying graphs of this type, which are represented diagrammatically in Fig. 5.37 and are generally as follows:

- If the two nodes q_i and q_j are connected by the two different branches with gain values g_α and g_β, then these two branches can be replaced by a single branch with a weight value equal to $g_\alpha + g_\beta$ [image (1) of Fig. 5.37].
- A set of three nodes, q_i, q_j, and q_k, which are connected in series through two branches with gain values g_{ij} and g_{jk}, respectively, is equivalent to a simple graph that includes only the two extreme nodes q_i and q_k, which are connected by a simple branch with a gain value equal to the product $g_{ij}g_{jk}$ [image (2) of Fig. 5.37].
- If the two nodes q_i and q_j are connected in parallel to the node q_k with gain values g_{ik} and g_{jk}, which in turn is connected to the node q_n with a weight value equal to g_{kn}, then this set of four nodes is equivalent to a simple graph that only

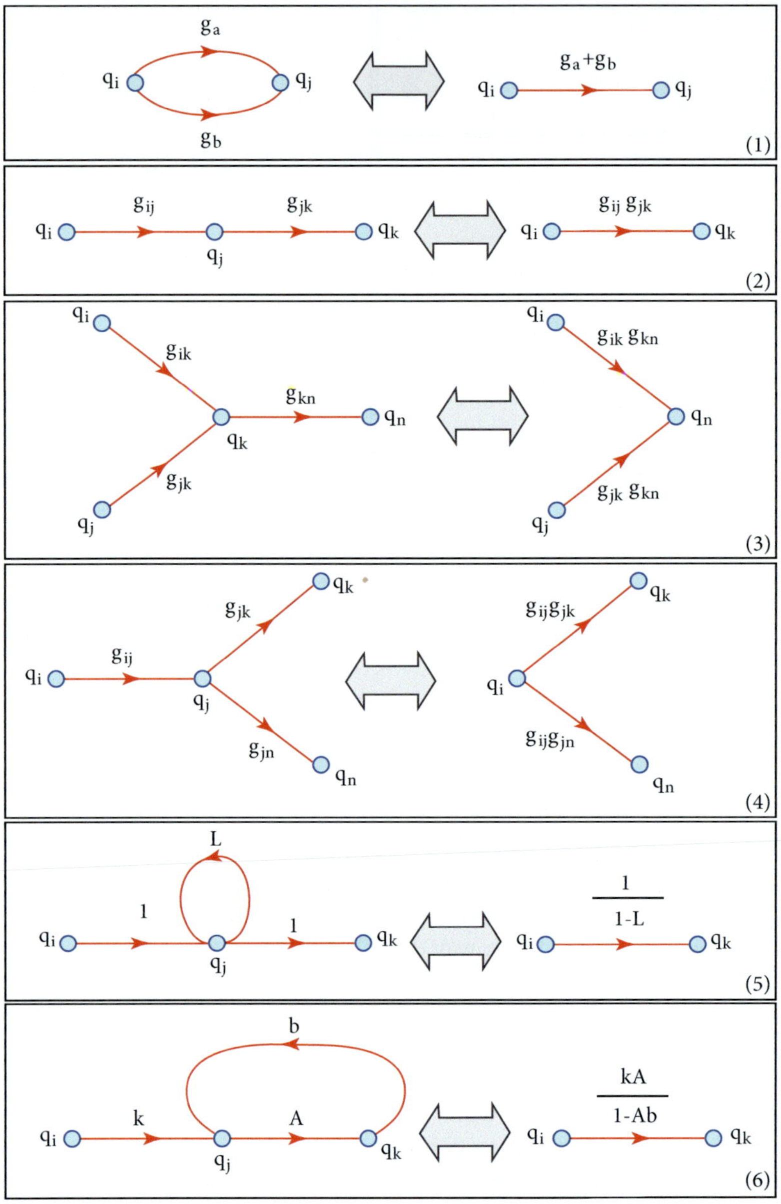

Fig. 5.37 Rules for simplifying signal flow graphs

contains the nodes q_i, q_j, and q_k, with the nodes q_i and q_j connected in parallel to the node q_k with gain values $g_{ik}g_{kn}$ and $g_{jk}g_{kn}$, respectively [image (3) of Fig. 5.37].

- A set of four nodes, q_i, q_j, q_k, and q_n, of which q_i is connected to q_j with a gain value of g_{ij}, which in turn transfers its output to the nodes q_k and q_n with gain values q_{jk} and q_{jn}, respectively, is equivalent to a simple graph in which the node q_j is connected to the nodes q_k and q_n with gain values $g_{ij}g_{jk}$ and $g_{ij}g_{jn}$, respectively [image (4) of Fig. 5.37].
- If the three nodes q_i, q_j, and q_k are connected in sequence with gain values $g_{ij} = g_{jk} = 1$, while the node q_j is connected to itself with a branch of loop shape with a gain value of $g_{jj} = L$, then this connection is equivalent to a simple graph that contains only the nodes q_i and q_k, which are connected through a branch with a gain value of $1/(L-1)$ [image (5) of Fig. 5.37].
- If the three nodes q_i, q_j, and q_k are interconnected with gain values $g_{ij} = k$ and $g_{jk} = A$ and there is also a recursive branch that connects the node q_k to the node q_j with a gain value of $g_{kj} = b$, then this connection is equivalent to a simple graph that contains only the nodes q_i and q_k, which are interconnected with a gain value equal to $kA/(1-Ab)$ [image (6) of Fig. 5.37].

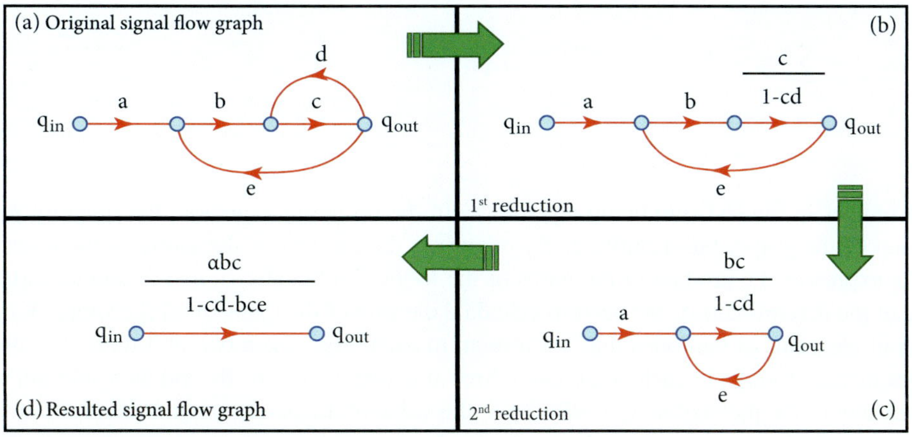

Fig. 5.38 Example of applying the signal flow graph simplification process

A typical example of applying the foregoing rules for simplifying a signal flow graph is presented in Fig. 5.38. In this figure, images (a) and (d) present the initial and final shapes of the diagram, respectively, while images (b) and (c) present the two simplifications that were needed for the foregoing conversion. The reader is invited as an exercise to determine the rules that were used in this process, as well as the way in which the final result emerged.

5.10.3 Mason's Rule

The calculation of the transfer function of the system that corresponds to some signal flow graph requires the simplification of the graph in the way presented in the previous example. After the graph is brought to a form that only contains the input and the output of the system connected with some branch, the transfer function of the equivalent simplified system (or in another formulation the total gain $\mathcal{G}$) is equal to the gain associated with this branch. If in the example from the previous section we assume that $\alpha = \mathcal{G}_1(s)$, $b = \mathcal{G}_2(s)$, $c = \mathcal{G}_3(s)$, $d = \mathcal{H}_1(s)$, and $e = \mathcal{H}_2(s)$, then the transfer function of the system is calculated as

$$\mathcal{H}(s) = \frac{\mathcal{Y}(s)}{\mathcal{X}(s)} \equiv \frac{q_{\text{out}}}{q_{\text{in}}} = \frac{\alpha bc}{1 - cd - bce} = \frac{\mathcal{G}_1(s)\mathcal{G}_2(s)\mathcal{G}_3(s)}{1 - \mathcal{G}_3(s)\mathcal{H}_1(s) - \mathcal{G}_2(s)\mathcal{G}_3(s)\mathcal{H}_2(s)}$$

Observing carefully the last relation and taking into account the definitions we presented on previous pages, it is not difficult to understand that the expression of the numerator expresses the total gain of the one and only forward path of the graph which is equal to $\mathcal{G}_1(s)\mathcal{G}_2(s)\mathcal{G}_3(s)$, while the denominator arises if we subtract from unity the sum of the gains of all simple loops included in it. Although in simple graphs we can calculate the transfer function using a simplified diagram, in complex situations, this function can be calculated using a mathematical relationship known as *Mason's rule* and is given by the equation

$$\mathcal{G} = \mathcal{H}(s) = \frac{\mathcal{Y}(s)}{\mathcal{X}(s)} = \frac{q_{\text{out}}}{q_{\text{in}}} = \frac{1}{\Delta}\sum_{k=1}^{N} \mathcal{G}_k \Delta_k \tag{5.54}$$

($k = 1, 2, 3, \ldots, N$). In the foregoing equation, the parameter N expresses the *number of simple branches of the graph in the forward direction*, while $\mathcal{G}_k$ is the *gain* associated with the branch number k of the graph. Particular emphasis should be given to the quantity Δ, which is known as the *determinant of the graph* and is given by a relationship of the form

$$\Delta = 1 - \sum_{i} L_i + \sum_{i,j} L_i L_j - \sum_{i,j,k} L_i L_j L_k + \cdots + (-1)^n \sum_{i,j,k,\ldots,} L_i L_j L_k \ldots \cdots + \ldots \tag{5.55}$$

with the number of indices in the last sum being equal to n. In the foregoing equation, the quantity L_i expresses the gain of the ith simple loop of the graph, the quantity $L_i L_j$ expresses the product of the gains of the loops numbered i and j, the quantity $L_i L_j L_k$ expresses the product of the gains of the loops numbered i, j, and k, and so forth. In other words, to determine the value of the determinant Δ, we need to calculate the sum of the gains of all the simple loops of the circuit, the sum of the products of the gains of the loops that are foreign to each other taken two at a time, the sum of the products of the gains of the loops that are foreign to each other taken three at a time, and so forth, and then subtract all these values from the unit. Finally, the value of the parameter Δ_k results from the value of the determinant Δ if we remove from the graph the branch numbered k or, equivalently, remove all the loops that have common nodes with that branch. Studying carefully all this analysis, we realize that the simple equation that gives the transfer function of the preceding graph is simply the result of the application of *Mason's rule*. Another conclusion that follows from the preceding description is that *two graphs are characterized by the same value for the total gain $\mathcal{G}$ when they have the same forward branches and the same loops*.

From the foregoing analysis it becomes obvious that the application of Mason's rule for solving a signal flow diagram requires the implementation of the following procedures:

- Determination of the forward branches of the graph and the gains associated with them.
- Determination of the loops of the graph and the gains associated with them.
- Determination of the nontouching loops and the sets that arise by taking them two by two, three by three, and so forth, and calculating the product of the gains for each group of such loops.
- Determination of the value of the determinant Δ and the values of the parameters Δ_k $(k = 1, 2, \ldots, N)$.
- Substitution of the preceding values into the Mason equation and calculation of the value of the total gain $\mathcal{G}$.

As an example of the application of Mason's rule, let us calculate the total gain—or, equivalently, the transfer function—of the graph of Fig. 5.39. In the first step of the process, let us determine the forward paths and their corresponding gains. Careful examination of the figure reveals that there are three such paths: $P_1 : y_1 \to y_2 \to y_3 \to y_4 \to y_5$, $P_2 : y_1 \to y_2 \to y_4 \to y_5$, and $P_3 : y_1 \to y_2 \to y_5$, with corresponding gains $\mathcal{G}_1 = \alpha_{12}\alpha_{23}\alpha_{34}\alpha_{45}$, $\mathcal{G}_2 = \alpha_{12}\alpha_{24}\alpha_{45}$, and $\mathcal{G}_3 = \alpha_{12}\alpha_{25}$. Next, let us determine the loops of the graph and their corresponding gains. Examination of the graph reveals four loops: $y_2 \to y_3 \to y_2$, $y_3 \to y_4 \to y_3$, $y_2 \to y_4 \to y_3 \to y_2$, and $y_4 \to y_4$, with corresponding gains $L_1 = \alpha_{23}\alpha_{32}$, $L_2 = \alpha_{34}\alpha_{43}$, $L_3 = \alpha_{24}\alpha_{43}\alpha_{32}$, and $L_4 = \alpha_{44}$. Among these loops, the first and fourth are foreign to each other—meaning they do not have common nodes—and therefore the product of their gains will be equal to $L_1 L_4 = \alpha_{23}\alpha_{32}\alpha_{44}$. With all the foregoing expressions calculated, the determinant of the graph is calculated as

$$\Delta = 1 - (L_1 + L_2 + L_3 + L_4) + L_1 L_4$$

In the final step of the process, we need to calculate the quantities Δ_1, Δ_2, and Δ_3, corresponding to the paths P_1, P_2, and P_3. These expressions will be calculated from the determinant of the graph Δ if we set to zero the gains of all the loops that have common nodes with each path. Studying the graph, we observe that all the loops have common nodes with the paths P_1 and P_2. Therefore, in the expression of the determinant Δ we should set all L to zero, from which it follows that $\Delta_1 = \Delta_2 = 1$. On the other hand, for the path P_3, we observe that the loops $y_2 \to y_3 \to y_2$ and $y_2 \to y_4 \to y_3 \to y_2$ have common nodes with the stated path, something that does not happen with the loops $y_3 \to y_4 \to y_3$ and $y_4 \to y_4$. Setting $L_1 = L_3 = 0$ in the equation of the determinant Δ, it follows that $\Delta_3 = 1 - L_2 - L_4 = 1 - \alpha_{34}\alpha_{43} - \alpha_{44}$.

Substituting the foregoing expressions in the Mason formula we get

$$G = \frac{y_5}{y_1} = \frac{\mathcal{G}_1\Delta_1 + \mathcal{G}_2\Delta_2 + \mathcal{G}_3\Delta_3}{\Delta} = \frac{(\alpha_{12}\alpha_{23}\alpha_{34}\alpha_{45}) + (\alpha_{12}\alpha_{24}\alpha_{45}) + (\alpha_{12}\alpha_{25})(1 - \alpha_{34}\alpha_{43} - \alpha_{44})}{1 - (\alpha_{23}\alpha_{32} + \alpha_{34}\alpha_{43} + \alpha_{24}\alpha_{32}\alpha_{43} + \alpha_{44}) + \alpha_{23}\alpha_{32}\alpha_{44}}$$

which constitutes the total gain or, equivalently, the transfer function of the system.

Fig. 5.39 Example of a signal flow graph

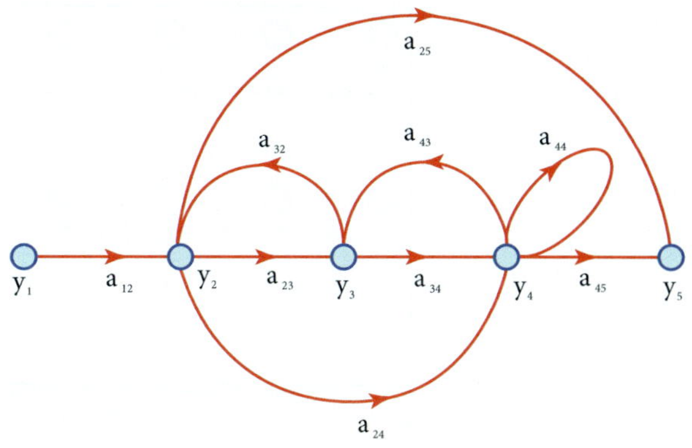

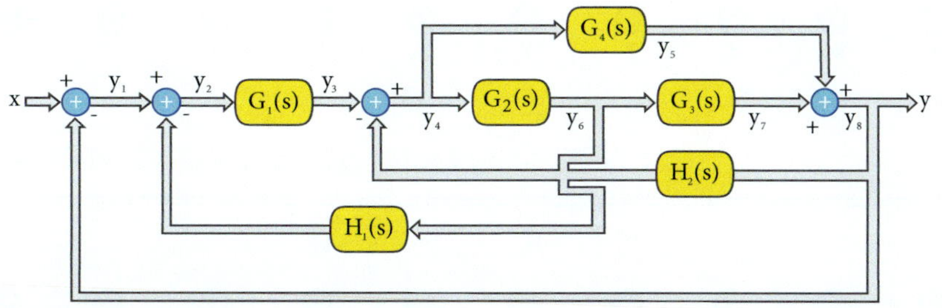

Fig. 5.40 Schematic diagram of Exercise 5.10.1

Problem 5.10.1 Construct an equivalent signal flow graph for the schematic diagram of Fig. 5.40 and calculate the transfer function using the Mason rule.

Solution From the schematic diagram of the exercise we find that the system is characterized by one input x and one output y, while there are also eight auxiliary variables denoted by y_i $(i = 1, 2, 3, \ldots, 8)$. Examination of the structure of the diagram readily reveals that these 10 parameters are related to each other through the equations

$$
\begin{array}{lllll}
y_1 = x - y & y_4 = y_3 - \mathcal{H}_2 y & y_7 = \mathcal{G}_3 y6 & y_2 = y_1 - \mathcal{H}_1 y_6 & y_5 = \mathcal{G}_4 y_4 \\
y_8 = y_5 + y_7 & y_3 = \mathcal{G}_1 y_2 & y_6 = \mathcal{G}_2 y_4 & y = y_9 &
\end{array}
$$

The gradual construction of the graph in which the preceding equations are incorporated one after the other and in the order in which they are written is done in the manner presented in the corresponding section with the process being presented in detail in Figs. 5.41 and 5.42. For the construction of the graph, we replace the 10 variables with an equal number of nodes, which we connect with branches, thereby creating the system equations. For the branches that correspond to transfer functions, the appropriate function is used as gain, while the branches that correspond to signal adders correspond to a gain value equal to one. The final graph that describes the system as a whole is depicted in the last image of Fig. 5.42.

After the construction of the flow graph that corresponds to the schematic diagram of the exercise, let us calculate the transfer function of the system using *Mason's rule*. Following the methodology of the previous example, we determine the forward paths and the loops of the graph and calculate their corresponding gains. Starting with the forward paths, there are two of them, specifically $P_1 : y_1 \to y_2 \to y_3 \to y_4 \to y_5 \to y_8 \to y$ and P_2: $y_1 \to y_2 \to y_3 \to y_4 \to y_6 \to y_7 \to y_8 \to y$, with corresponding values of gains $\mathcal{G}_1(s)\mathcal{G}_4(s)$ and $\mathcal{G}_1(s)\mathcal{G}_2(s)\mathcal{G}_3(s)$. On the other hand, careful study of the graph reveals the existence of five loops defined as $y_1 \to y_2 \to y_3 \to y_4 \to y_6 \to y_1$, $y_4 \to y_5 \to y_7 \to y_8 \to y_4$, $y_1 \to y_2 \to y_3 \to y_4 \to y_6 \to y_7 \to y_8 \to y_1$, $y_4 \to y_5 \to y_8 \to y_4$ and $y_1 \to y_2 \to y_3 \to y_6 \to y_7 \to y_8 \to y_1$, with corresponding values of gains $L_1 = -\mathcal{G}_1(s)\mathcal{G}_2(s)\mathcal{H}_1(s)$, $L_2 = -\mathcal{G}_2(s)\mathcal{G}_3(s)\mathcal{H}_2(s)$, $L_3 = -\mathcal{G}_1(s)\mathcal{G}_2(s)\mathcal{G}_3(s)$, $L_4 = -\mathcal{G}_4(s)\mathcal{H}_2(s)$, and $L_5 = -\mathcal{G}_1(s)\mathcal{G}_4(s)$. In addition, the circuit is not characterized by the existence of loops foreign to each other, and therefore

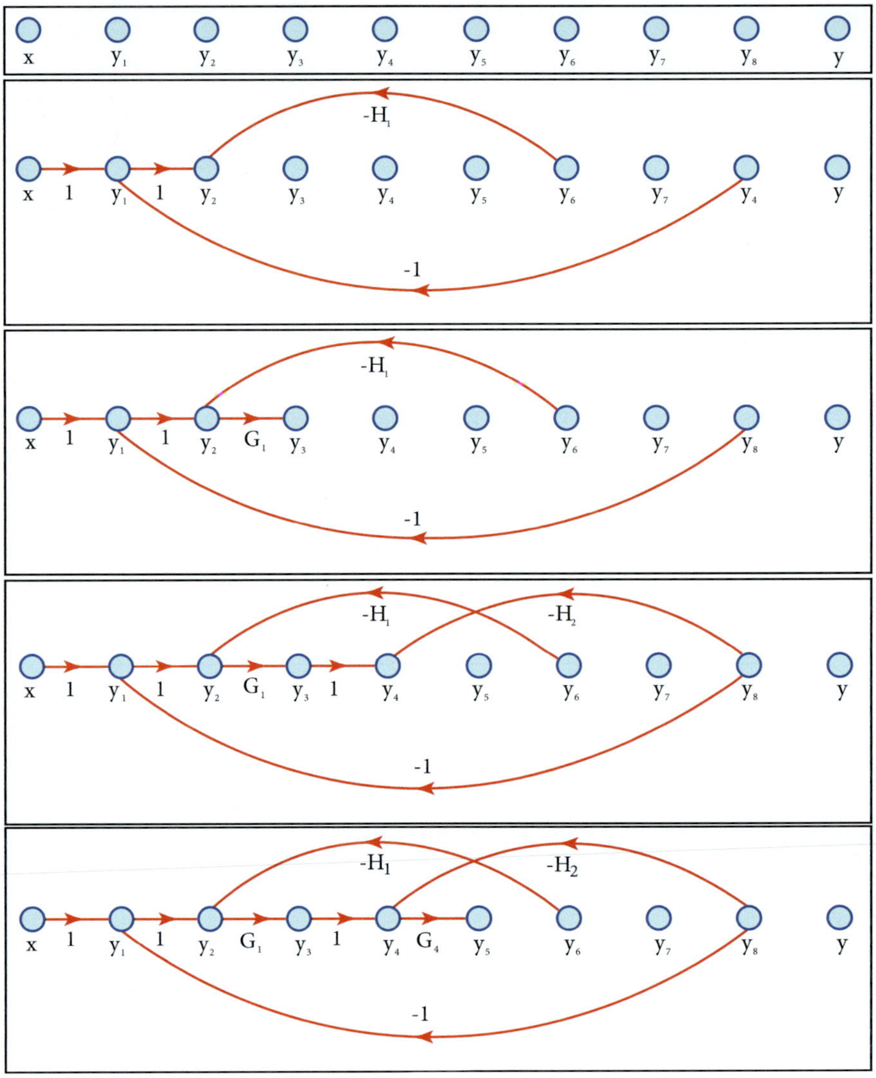

Fig. 5.41 Gradual construction of graph from Exercise 5.10.1 (part A)

the determinant of the circuit is calculated from the relation

$$\begin{aligned}\Delta &= 1-(L_1+L_2+L_3+L_4+L_5)\\ &= 1+\mathcal{G}_1(s)\mathcal{G}_2(s)\mathcal{H}_1(s)+\mathcal{G}_2(s)\mathcal{G}_3(s)\mathcal{H}_2(s)+\mathcal{G}_1(s)\mathcal{G}_2(s)\mathcal{G}_3(s)+\mathcal{G}_4(s)\mathcal{H}_2(s)+\mathcal{G}_1(s)\mathcal{G}_4(s)\end{aligned}$$

Finally, observing that all the aforementioned loops share common nodes with paths P_1 and P_2, we find that $\Delta_1 = \Delta_2 = 1$. Replacing the preceding expressions in Mason's formula, we conclude that the transfer function of the system is given by the relation

$$H(s) = \frac{\mathcal{G}_1(s)\mathcal{G}_4(s)+\mathcal{G}_1(s)\mathcal{G}_2(s)\mathcal{G}_3(s)}{1+\mathcal{G}_1(s)\mathcal{G}_2(s)\mathcal{H}_1(s)+\mathcal{G}_2(s)\mathcal{G}_3(s)\mathcal{H}_2(s)+\mathcal{G}_1(s)\mathcal{G}_2(s)\mathcal{G}_3(s)+\mathcal{G}_4(s)\mathcal{H}_2(s)+\mathcal{G}_1(s)\mathcal{G}_4(s)}$$

5.11 Description of Systems in State Space

As already mentioned, the description of a system in the so-called *state space* offers significant advantages over the *input–output description*, as it allows the study of the *internal structure* of a system and the derivation of conclusions concerning,

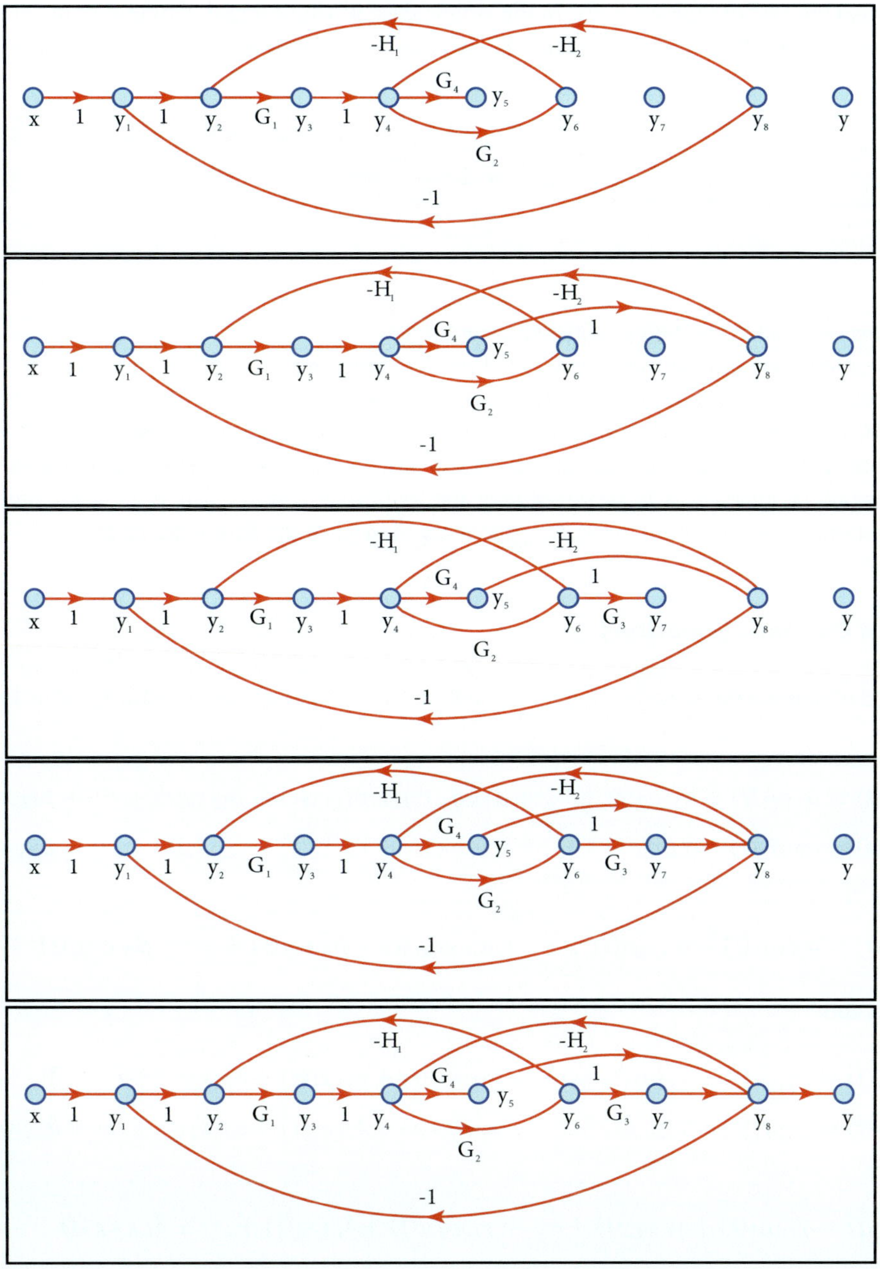

Fig. 5.42 Gradual construction of graph from Exercise 5.10.1 (part B)

for example, its stability. Generally speaking, the behavior of a linear multiple-input, multiple-output (MIMO) system in this space is described using a pair of equations known as *dynamic equations* and having the form

$$\begin{aligned}\dot{\boldsymbol{q}}(t) &= \boldsymbol{A}\boldsymbol{q}(t) + \boldsymbol{B}\boldsymbol{x}(t) \\ \boldsymbol{y}(t) &= \boldsymbol{C}\boldsymbol{q}(t) + \boldsymbol{D}\boldsymbol{x}(t)\end{aligned} \tag{5.56}$$

In the preceding equations, the parameter $\boldsymbol{q}(t) = [q_1(t)\ q_2(t)\ \ldots\ q_N(t)]^T$ constitutes the vector of the N *state variables of the system*, the quantities $\boldsymbol{x}(t) = [x_1(t)\ x_2(t)\ \ldots\ x_K(t)]^T$ and $\boldsymbol{y}(t) = [y_1(t)\ y_2(t)\ \ldots\ y_L(t)]^T$ express the *input and output vectors* that contain K and L elements, respectively, and the symbols $\boldsymbol{A}_{N\times N}$, $\boldsymbol{B}_{N\times K}$, $\boldsymbol{C}_{L\times N}$, and $\boldsymbol{D}_{L\times K}$ refer to the *system, input, measurement, and output matrices*. In the simple case of single-input, single-output (SISO) systems that have one input and one output—and for which, therefore, $K = L = 1$—the *dynamic equations* of the system take the simple form

$$\begin{aligned}\dot{\boldsymbol{q}}(t) &= \boldsymbol{A}\boldsymbol{q}(t) + \boldsymbol{B}x(t)\\ y(t) &= \boldsymbol{C}\boldsymbol{q}(t) + \boldsymbol{D}x(t)\end{aligned} \tag{5.57}$$

with the dimensions of the matrices being adjusted accordingly.

In the chapter on continuous-time systems and based on the foregoing description, we presented the theory related to this whole thematic unit—that is, the *solution of dynamic equations, the calculation of the exponential function* e^{At}, *and the study of the characteristics of controllability and observability*—using techniques defined in the time domain. However, there is also an alternative way of presenting this theory that is based on the use of the *Laplace transform* and the properties that describe it and which in some cases is preferred over the traditional form in which we have already presented it. This alternative way of describing the system in the state space is the subject of the following pages.

5.11.1 Solution of Dynamic Equations

To present the method of solving the system's dynamic equations using the Laplace transform, let us formulate them in their analytical form:

$$\dot{q}_1(t) = \alpha_{11}q_1(t) + \alpha_{12}q_2(t) + \cdots + \alpha_{1N}q_N(t) + \beta_{11}x_1(t) + \beta_{12}x_2(t) + \cdots + \beta_{1K}x_K(t)$$
$$\dot{q}_2(t) = \alpha_{21}q_1(t) + \alpha_{22}q_2(t) + \cdots + \alpha_{2N}q_N(t) + \beta_{21}x_1(t) + \beta_{22}x_2(t) + \cdots + \beta_{2K}x_K(t)$$
$$\cdots\cdots\cdots\cdots\cdots\cdots\cdots\cdots\cdots\cdots\cdots\cdots\cdots\cdots\cdots\cdots$$
$$\dot{q}_N(t) = \alpha_{N1}q_1(t) + \alpha_{N2}q_2(t) + \cdots + \alpha_{NN}q_N(t) + \beta_{N1}x_1(t) + \cdots + \beta_{NK}x_K(t)$$

for the *state equation* and

$$y_1(t) = \gamma_{11}q_1(t) + \gamma_{12}q_2(t) + \cdots + \gamma_{1N}q_N(t) + \delta_{11}x_1(t) + \delta_{12}x_2(t) + \cdots + \delta_{1K}x_K(t)$$
$$y_2(t) = \gamma_{21}q_1(t) + \gamma_{22}q_2(t) + \cdots + \gamma_{2N}q_N(t) + \delta_{21}x_1(t) + \delta_{22}x_2(t) + \cdots + \delta_{2K}x_K(t)$$
$$\cdots\cdots\cdots\cdots\cdots\cdots\cdots\cdots\cdots\cdots\cdots\cdots\cdots\cdots\cdots\cdots$$
$$y_L(t) = \gamma_{L1}q_1(t) + \gamma_{L2}q_2(t) + \cdots + \gamma_{LN}q_N(t) + \delta_{L1}x_1(t) + \cdots + \delta_{LK}x_K(t)$$

for the *output equation*. Defining the quantities $\mathcal{Q}_i(s) = \mathcal{L}\{q_i(t)\}$ $(i = 1, 2, \ldots, N)$ and $\mathcal{X}_j(s) = \mathcal{L}\{x_j(t)\}$ $(j = 1, 2, \ldots, K)$ and using the property of the *differentiation* of the Laplace transform described by the equation

$$\mathcal{L}\{\dot{q}(t)\} = s\mathcal{Q}(s) - q(0)$$

the application of the Laplace transform to each of the N state equations will lead to a result of the form

$$s\mathcal{Q}_1(s) - q_1(0) = \alpha_{11}\mathcal{Q}_1(s) + \cdots + \alpha_{1N}\mathcal{Q}_N(s) + \beta_{11}\mathcal{X}_1(s) + \cdots + \beta_{1K}\mathcal{X}_K(s)$$
$$s\mathcal{Q}_2(s) - q_2(0) = \alpha_{21}\mathcal{Q}_1(s) + \cdots + \alpha_{2N}\mathcal{Q}_N(s) + \beta_{21}\mathcal{X}_1(s) + \cdots + \beta_{2K}\mathcal{X}_K(s)$$
$$\cdots\cdots\cdots\cdots\cdots\cdots\cdots\cdots\cdots\cdots\cdots\cdots\cdots\cdots\cdots\cdots$$
$$s\mathcal{Q}_N(s) - q_N(0) = \alpha_{N1}\mathcal{Q}_1(s) + \cdots + \alpha_{NN}\mathcal{Q}_N(s) + \beta_{N1}\mathcal{X}_1(s) + \cdots + \beta_{NK}\mathcal{X}_K(s)$$

which, in matrix form, is restated as

$$s\begin{bmatrix} \mathcal{Q}_1(s) \\ \mathcal{Q}_2(s) \\ \vdots \\ \mathcal{Q}_N(s) \end{bmatrix} - \begin{bmatrix} q_1(0) \\ q_2(0) \\ \vdots \\ q_N(0) \end{bmatrix} = \begin{bmatrix} \alpha_{11} & \alpha_{12} & \dots & \alpha_{1N} \\ \alpha_{21} & \alpha_{22} & \dots & \alpha_{2N} \\ \vdots & \vdots & \ddots & \vdots \\ \alpha_{N1} & \alpha_{N2} & \dots & \alpha_{NN} \end{bmatrix} \begin{bmatrix} \mathcal{Q}_1(s) \\ \mathcal{Q}_2(s) \\ \vdots \\ \mathcal{Q}_N(s) \end{bmatrix} + \begin{bmatrix} \beta_{11} & \beta_{12} & \dots & \beta_{1K} \\ \beta_{21} & \beta_{22} & \dots & \beta_{2K} \\ \vdots & \vdots & \ddots & \vdots \\ \beta_{N1} & \beta_{N2} & \dots & \beta_{NK} \end{bmatrix} \begin{bmatrix} \mathcal{X}_1(s) \\ \mathcal{X}_2(s) \\ \vdots \\ \mathcal{X}_K(s) \end{bmatrix}$$

Defining the matrices $\boldsymbol{A}$ and $\boldsymbol{B}$ as in the chapter on continuous-time systems and the vectors

$$\boldsymbol{\mathcal{Q}}(s) = \left[\mathcal{Q}_1(s) \dots \mathcal{Q}_N(s)\right]^T$$

$$\boldsymbol{q}(0) = \left[q_1(0)\ q_2(0) \dots\ q_N(0)\right]^T$$

$$\boldsymbol{\mathcal{X}}(s) = \left[\mathcal{X}_1(s)\ \mathcal{X}_2(s) \dots \mathcal{X}_K(s)\right]^T$$

the preceding relation will take the form

$$s\boldsymbol{\mathcal{Q}}(s) - \boldsymbol{q}(0) = \boldsymbol{A}\boldsymbol{\mathcal{Q}}(s) + \boldsymbol{B}\boldsymbol{\mathcal{X}}(s) \tag{5.58}$$

which can also be written

$$s\boldsymbol{\mathcal{Q}}(s) - \boldsymbol{A}\boldsymbol{\mathcal{Q}}(s) = \boldsymbol{q}(0) + \boldsymbol{B}\boldsymbol{\mathcal{X}}(s) \qquad \text{or, equivalently, as} \qquad (s\boldsymbol{I} - \boldsymbol{A})\boldsymbol{\mathcal{Q}}(s) = \boldsymbol{q}(0) + \boldsymbol{B}\boldsymbol{\mathcal{X}}(s)$$

where $\boldsymbol{I}$ is the unit matrix of dimensions $N \times N$. Therefore, by multiplying both sides of the preceding equation by the matrix $\boldsymbol{\Phi}(s) = (s\boldsymbol{I} - \boldsymbol{A})^{-1}$, the last equation is formulated as

$$\boldsymbol{\mathcal{Q}}(s) = (s\boldsymbol{I} - \boldsymbol{A})^{-1}\Big[\boldsymbol{q}(0) + \boldsymbol{B}\boldsymbol{\mathcal{X}}(s)\Big] = \boldsymbol{\Phi}(s)\boldsymbol{q}(0) + \boldsymbol{\Phi}(s)\boldsymbol{B}\boldsymbol{\mathcal{X}}(s) \tag{5.59}$$

and constitutes the defining equation of the function $\boldsymbol{\mathcal{Q}}(s)$.

So how will we calculate the state vector $\boldsymbol{q}(t)$, which is also the desired solution of the state equation? Quite simply, by observing that the function $\boldsymbol{\mathcal{Q}}(s)$ is simply the *Laplace transform of the state vector* $\boldsymbol{q}(t)$, since it has been defined as

$$\boldsymbol{\mathcal{Q}}(s) = \left[\mathcal{Q}_1(s) \dots \mathcal{Q}_N(s)\right]^T = \left[\mathcal{L}\{q_1(t)\} \dots \mathcal{L}\{q_N(t)\}\right]^T = \mathcal{L}\Big\{\left[q_1(t) \dots q_N(t)\right]^T\Big\} \tag{5.60}$$

or, in summary notation, $\boldsymbol{\mathcal{Q}}(s) = \mathcal{L}\{\boldsymbol{q}(t)\}$. Thus, to calculate the *state vector* $\boldsymbol{q}(t)$, we should take the *inverse Laplace transform* of the preceding equation. Therefore, we will have

$$\boldsymbol{q}(t) = \mathcal{L}^{-1}\{\boldsymbol{\mathcal{Q}}(s)\} = \mathcal{L}^{-1}\{\boldsymbol{\Phi}(s)\}\boldsymbol{q}(0) + \mathcal{L}^{-1}\{\boldsymbol{\Phi}(s)\boldsymbol{B}\boldsymbol{\mathcal{X}}(s)\} \tag{5.61}$$

an equation that constitutes an alternative way of calculating the state vector $\boldsymbol{q}(t)$ that is based on the use of the *Laplace transform.*

On the other hand, the foregoing process can be used also for the calculation of the *output* of the system $\boldsymbol{y}(t)$. Defining the functions $\mathcal{Y}_i(s) = \mathcal{L}\{y_i(t)\}$ $(i = 1, 2, \dots, L)$ and taking the Laplace transform of the L output equations, we are led to a system of equations of the form

$$\mathcal{Y}_1(s) = \gamma_{11}\mathcal{Q}_1(s) + \gamma_{12}\mathcal{Q}_2(s) + \dots + \gamma_{1N}\mathcal{Q}_N(s) + \delta_{11}\mathcal{X}_1(s) + \dots + \delta_{1K}\mathcal{X}_K(s)$$

$$\mathcal{Y}_2(s) = \gamma_{21}\mathcal{Q}_1(s) + \gamma_{22}\mathcal{Q}_2(s) + \dots + \gamma_{2N}\mathcal{Q}_N(s) + \delta_{21}\mathcal{X}_1(s) + \dots + \delta_{2K}\mathcal{X}_K(s)$$

$$\dots\dots\dots\dots\dots\dots\dots\dots\dots\dots\dots\dots\dots\dots\dots\dots\dots\dots\dots$$

$$\mathcal{Y}_L(s) = \gamma_{L1}\mathcal{Q}_1(s) + \gamma_{L2}\mathcal{Q}_2(s) + \dots + \gamma_{LN}\mathcal{Q}_N(s) + \delta_{L1}X_1(s) + \dots + \delta_{LK}X_K(s)$$

which in matrix form is expressed as

$$\begin{bmatrix} \mathcal{Y}_1(s) \\ \mathcal{Y}_2(s) \\ \vdots \\ \mathcal{Y}_L(s) \end{bmatrix} = \begin{bmatrix} \gamma_{11} & \gamma_{12} & \cdots & \gamma_{1N} \\ \gamma_{21} & \gamma_{22} & \cdots & \gamma_{2N} \\ \vdots & \vdots & \ddots & \vdots \\ \gamma_{L1} & \gamma_{L2} & \cdots & \gamma_{LN} \end{bmatrix} \begin{bmatrix} \mathcal{Q}_1(s) \\ \mathcal{Q}_2(s) \\ \vdots \\ \mathcal{Q}_N(s) \end{bmatrix} + \begin{bmatrix} \delta_{11} & \delta_{12} & \cdots & \delta_{1K} \\ \delta_{21} & \delta_{22} & \cdots & \delta_{2K} \\ \vdots & \vdots & \ddots & \vdots \\ \delta_{L1} & \delta_{L2} & \cdots & \delta_{LK} \end{bmatrix} \begin{bmatrix} \mathcal{X}_1(s) \\ \mathcal{X}_2(s) \\ \vdots \\ \mathcal{X}_K(s) \end{bmatrix}$$

If we define the matrices $\boldsymbol{C}$ and $\boldsymbol{D}$ in the way we did in the chapter on continuous-time systems and consider the vector $\boldsymbol{\mathcal{Y}}(s) = [\mathcal{Y}_1(s)\ \mathcal{Y}_2(s)\ \ldots\ \mathcal{Y}_L(s)]^T$, the preceding relationship is written

$$\boldsymbol{\mathcal{Y}}(s) = \boldsymbol{C}\boldsymbol{\mathcal{Q}}(s) + \boldsymbol{D}\boldsymbol{\mathcal{X}}(s) \tag{5.62}$$

and replacing the function $\boldsymbol{Q}(s)$ from the equation $\boldsymbol{\mathcal{Q}}(s) = \boldsymbol{\Phi}(s)\boldsymbol{q}(0) + \boldsymbol{\Phi}(s)\boldsymbol{B}\boldsymbol{\mathcal{X}}(s)$, we will get

$$\boldsymbol{\mathcal{Y}}(s) = \boldsymbol{C}\Big[\boldsymbol{\Phi}(s)\boldsymbol{q}(0) + \boldsymbol{\Phi}(s)\boldsymbol{B}\boldsymbol{\mathcal{X}}(s)\Big] + \boldsymbol{D}\boldsymbol{\mathcal{X}}(s) = \boldsymbol{C}\boldsymbol{\Phi}(s)\boldsymbol{q}(0) + \Big[\boldsymbol{C}\boldsymbol{\Phi}(s)\boldsymbol{B} + \boldsymbol{D}\Big]\boldsymbol{\mathcal{X}}(s) \tag{5.63}$$

The preceding relationship constitutes the defining equation of the system's output expressed in the frequency domain. We observe that in full analogy with the corresponding output $\boldsymbol{y}(t)$ for the continuous-time domain, the output $\boldsymbol{\mathcal{Y}}(s)$ has been formulated *as a sum of two terms* of which the term $\boldsymbol{\mathcal{Y}}_{\text{zi}}(s) = \boldsymbol{C}\boldsymbol{\Phi}(s)\boldsymbol{q}(0)$ constitutes the *response to the zero input*, while the term $\boldsymbol{\mathcal{Y}}_{\text{ZS}}(s) = [\boldsymbol{C}\boldsymbol{\Phi}(s)\boldsymbol{B}+\boldsymbol{D}]\boldsymbol{\mathcal{X}}(s)$ constitutes the *response to the zero state* and arises from the output $\boldsymbol{\mathcal{Y}}(s)$ by setting it to $\boldsymbol{q}(0) = 0$. In the latter case, defining the function $\boldsymbol{\mathcal{H}}(s) = \boldsymbol{C}\boldsymbol{\Phi}(s)\boldsymbol{B} + \boldsymbol{D}$, the response to the zero state will take the form $\boldsymbol{\mathcal{Y}}_{\text{ZS}}(s) = \boldsymbol{\mathcal{H}}(s)\boldsymbol{\mathcal{X}}(s)$, from which it is not difficult to ascertain that the function $\boldsymbol{\mathcal{H}}(s)$ should constitute the Laplace transform of the *system's impulse response matrix*, that is, should be defined as $\boldsymbol{\mathcal{H}}(s) = \mathcal{L}\{\boldsymbol{h}(t)\}$, since if we take the *inverse Laplace transform* and use the *convolution theorem*, we end up with the known relation $\boldsymbol{y}(t) = \boldsymbol{h}(t) * \boldsymbol{x}(t)$ (later we will see that indeed things are so). Finally, to get the output of the system $\boldsymbol{y}(t)$, we must take the inverse Laplace transform of the function $\boldsymbol{\mathcal{Y}}(s)$, thus ending up with an equation of the form

$$\boldsymbol{y}(t) = \mathcal{L}^{-1}\{\boldsymbol{\mathcal{Y}}(s)\} = \mathcal{L}^{-1}\{\boldsymbol{C}\boldsymbol{\Phi}(s)\}\boldsymbol{q}(0) + \mathcal{L}^{-1}\Big\{\Big[\boldsymbol{C}\boldsymbol{\Phi}(s)\boldsymbol{B} + \boldsymbol{D}\Big]\boldsymbol{\mathcal{X}}(s)\Big\} \tag{5.64}$$

which constitutes an alternative way of calculating the output of the system using the Laplace transform.

The application of the foregoing method is demonstrated in the exercises that follow. The reader is invited as an exercise to solve the following exercises using the techniques presented in the homonymous section of Chap. 2, naturally ending up with the same results.

Problem 5.11.1 Solve the state equation $\dot{\boldsymbol{q}}(t) = \boldsymbol{A}\boldsymbol{q}(t) + \boldsymbol{B}x(t)$ for parameter values

$$\boldsymbol{A} = \begin{bmatrix} 0 & 2 \\ -1 & -3 \end{bmatrix}, \quad \boldsymbol{B} = \begin{bmatrix} 0 \\ 1 \end{bmatrix}, \quad \boldsymbol{q}(0) = \begin{bmatrix} 2 \\ 1 \end{bmatrix}, \quad x(t) = 0.$$

(Lathi, *Linear Systems and Signals*, Problem 10.3-1).

Solution Initially, let us calculate the matrix $s\boldsymbol{I} - \boldsymbol{A}$, which turns out to be equal to

$$s\boldsymbol{I} - \boldsymbol{A} = s\begin{bmatrix} 1 & 0 \\ 0 & 1 \end{bmatrix} - \begin{bmatrix} 0 & 2 \\ -1 & -3 \end{bmatrix} = \begin{bmatrix} s & 0 \\ 0 & s \end{bmatrix} - \begin{bmatrix} 0 & 2 \\ -1 & -3 \end{bmatrix} = \begin{bmatrix} s & -2 \\ 1 & s+3 \end{bmatrix}$$

Therefore, we will have

$$\boldsymbol{\Phi}(s) = (s\boldsymbol{I} - \boldsymbol{A})^{-1} = \begin{bmatrix} s & -2 \\ 1 & s+3 \end{bmatrix}^{-1} = \frac{1}{(s+2)(s+3)}\begin{bmatrix} s+3 & 2 \\ -1 & s \end{bmatrix} = \begin{bmatrix} \dfrac{s+3}{(s+1)(s+2)} & \dfrac{2}{(s+1)(s+2)} \\ \dfrac{-1}{(s+1)(s+2)} & \dfrac{s}{(s+1)(s+2)} \end{bmatrix}$$

and therefore

$$\boldsymbol{\Phi}(s)\boldsymbol{q}(0) = \begin{bmatrix} \dfrac{s+3}{(s+1)(s+2)} & \dfrac{2}{(s+1)(s+2)} \\ \dfrac{-1}{(s+1)(s+2)} & \dfrac{s}{(s+1)(s+2)} \end{bmatrix}\begin{bmatrix} 2 \\ 1 \end{bmatrix} = \begin{bmatrix} \dfrac{2s+8}{(s+1)(s+2)} \\ \dfrac{s-2}{(s+1)(s+2)} \end{bmatrix}$$

On the other hand, the system's input $x(t)$ is zero, and therefore, $\mathcal{X}(s) = \mathcal{L}\{x(t)\} = 0$. Replacing the preceding quantities in the defining equation of the function $\boldsymbol{\mathcal{Q}}(s)$, it will take the form

$$\boldsymbol{\mathcal{Q}}(s) = \boldsymbol{\Phi}(s)\boldsymbol{q}(0) + \boldsymbol{\Phi}(s)\boldsymbol{B}\boldsymbol{\mathcal{X}}(s) = \boldsymbol{\Phi}(s)\boldsymbol{q}(0) = \begin{bmatrix} \dfrac{2s+8}{(s+1)(s+2)} \\ \dfrac{s-2}{(s+1)(s+2)} \end{bmatrix}$$

and therefore the desired solution of the state equation is calculated as

$$\boldsymbol{q}(t) = \mathcal{L}^{-1}\{\boldsymbol{\mathcal{Q}}(s)\} = \mathcal{L}^{-1}\left\{\begin{bmatrix} \dfrac{2s+8}{(s+1)(s+2)} \\ \dfrac{s-2}{(s+1)(s+2)} \end{bmatrix}\right\} = \begin{bmatrix} \mathcal{L}^{-1}\left\{\dfrac{2s+8}{(s+1)(s+2)}\right\} \\ \mathcal{L}^{-1}\left\{\dfrac{s-2}{(s+1)(s+2)}\right\} \end{bmatrix}$$

To calculate the foregoing inverse transforms, we can use the method of partial fraction expansion, which leads to the result

$$\frac{2s+8}{(s+1)(s+2)} = \frac{6}{s+1} - \frac{4}{s+2} \quad \text{and} \quad \frac{s-2}{(s+1)(s+2)} = -\frac{3}{s+1} + \frac{4}{s+2}$$

It will therefore hold that

$$\mathcal{L}^{-1}\left\{\frac{2s+8}{(s+1)(s+2)}\right\} = 6\mathcal{L}^{-1}\left\{\frac{1}{s+1}\right\} - 4\mathcal{L}^{-1}\left\{\frac{1}{s+2}\right\} = 6e^{-t}u(t) - 4e^{-2t}u(t)$$

$$\mathcal{L}^{-1}\left\{\frac{2s+8}{(s+1)(s+2)}\right\} = 4\mathcal{L}^{-1}\left\{\frac{1}{s+2}\right\} - 3\mathcal{L}^{-1}\left\{\frac{1}{s+1}\right\} = 4e^{-2t}u(t) - 3e^{-t}u(t)$$

Thus, the solution of the state equation will have the form

$$\boldsymbol{q}(t) = \begin{bmatrix} q_1(t) \\ q_2(t) \end{bmatrix}\begin{bmatrix} 6e^{-t}u(t) - 4e^{-2t}u(t) \\ 4e^{-2t}u(t) - 3e^{-t}u(t) \end{bmatrix}$$

Problem 5.11.2 Solve the state equation $\dot{\boldsymbol{q}}(t) = \boldsymbol{A}\boldsymbol{q}(t) + \boldsymbol{B}x(t)$ for parameter values

$$\boldsymbol{A} = \begin{bmatrix} -2 & 0 \\ 1 & -1 \end{bmatrix}, \quad \boldsymbol{B} = \begin{bmatrix} 1 \\ 0 \end{bmatrix}, \quad \boldsymbol{q}(0) = \begin{bmatrix} 0 \\ -1 \end{bmatrix}, \quad x(t) = u(t).$$

(Lathi, *Linear Systems and Signals*, Problem 10.3-3).

Solution In this case, the matrix $s\boldsymbol{I} - \boldsymbol{A}$ is calculated as

$$s\boldsymbol{I} - \boldsymbol{A} = s\begin{bmatrix} 1 & 0 \\ 0 & 1 \end{bmatrix} - \begin{bmatrix} -2 & 0 \\ 1 & -1 \end{bmatrix} = \begin{bmatrix} s & 0 \\ 0 & s \end{bmatrix} - \begin{bmatrix} -2 & 0 \\ 1 & -1 \end{bmatrix} = \begin{bmatrix} s+2 & 0 \\ -1 & s+1 \end{bmatrix}$$

Therefore, we will have

$$\boldsymbol{\Phi}(s) = (s\boldsymbol{I} - \boldsymbol{A})^{-1} = \begin{bmatrix} s+2 & 0 \\ -1 & s+1 \end{bmatrix}^{-1} = \frac{1}{(s+1)(s+2)} \begin{bmatrix} s+1 & 0 \\ 1 & s+2 \end{bmatrix} = \begin{bmatrix} \dfrac{1}{s+2} & 0 \\ \dfrac{1}{(s+1)(s+2)} & \dfrac{1}{s+1} \end{bmatrix}$$

and therefore

$$\boldsymbol{\Phi}(s)\,\boldsymbol{q}(0) = \begin{bmatrix} \dfrac{1}{s+2} & 0 \\ \dfrac{1}{(s+1)(s+2)} & \dfrac{1}{s+1} \end{bmatrix} \begin{bmatrix} 0 \\ -1 \end{bmatrix} = \begin{bmatrix} 0 \\ -\dfrac{1}{s+1} \end{bmatrix}$$

On the other hand, the Laplace transform of the system input $x(t) = u(t)$ is calculated as

$$\mathcal{X}(s) = \mathcal{L}\{x(t)\} = \mathcal{L}\{u(t)\} = \frac{1}{s}$$

Using this result, we are led to the equations

$$\boldsymbol{B}\mathcal{X}(s) = \begin{bmatrix} 1 \\ 0 \end{bmatrix} \frac{1}{s} = \begin{bmatrix} \dfrac{1}{s} \\ 0 \end{bmatrix} \quad \text{and} \quad \boldsymbol{\Phi}(s)\boldsymbol{B}X(s) = \begin{bmatrix} \dfrac{1}{s+2} & 0 \\ \dfrac{1}{(s+1)(s+2)} & \dfrac{1}{s+1} \end{bmatrix} \begin{bmatrix} \dfrac{1}{s} \\ 0 \end{bmatrix} = \begin{bmatrix} \dfrac{1}{s(s+2)} \\ \dfrac{1}{s(s+1)(s+2)} \end{bmatrix}$$

and substituting the foregoing expressions into the defining equation of the function $\boldsymbol{Q}(s)$, we get the result

$$\boldsymbol{Q}(s) = \boldsymbol{\Phi}(s)\boldsymbol{q}(0) + \boldsymbol{\Phi}(s)\boldsymbol{B}X(s) = \begin{bmatrix} 0 \\ -\dfrac{1}{s+1} \end{bmatrix} + \begin{bmatrix} \dfrac{1}{s(s+2)} \\ \dfrac{1}{s(s+1)(s+2)} \end{bmatrix} = \begin{bmatrix} \dfrac{1}{s(s+2)} \\ \dfrac{-s^2 - 2s + 1}{s(s+1)(s+2)} \end{bmatrix}$$

Therefore, the solution $\boldsymbol{q}(t)$ of the state equation is calculated as

$$\boldsymbol{q}(t)=\mathcal{L}^{-1}\{\boldsymbol{Q}(s)\}=\mathcal{L}^{-1}\left\{\begin{bmatrix}\dfrac{1}{s(s+2)}\\[2ex] \dfrac{-s^2-2s+1}{s(s+1)(s+2)}\end{bmatrix}\right\}=\begin{bmatrix}\mathcal{L}^{-1}\left\{\dfrac{1}{s(s+2)}\right\}\\[2ex] \mathcal{L}^{-1}\left\{\dfrac{-s^2-2s+1}{s(s+1)(s+2)}\right\}\end{bmatrix}$$

For the calculation of the preceding inverse Laplace transforms, we use, as before, the method of partial fraction expansion, leading to the results

$$\frac{1}{s(s+2)}=\frac{1}{2}\frac{1}{s}-\frac{1}{2}\frac{1}{s+2}\qquad\text{and}\qquad\frac{-s^2-2s+1}{s(s+1)(s+2)}=\frac{1}{2}\frac{1}{s}-2\frac{1}{s+1}+\frac{1}{2}\frac{1}{s+2}$$

and referring to tables of elementary transforms, we find that

$$\mathcal{L}^{-1}\left\{\frac{1}{s\,(s+2)}\right\}=\frac{1}{2}\mathcal{L}^{-1}\left\{\frac{1}{s}\right\}-\frac{1}{2}\mathcal{L}^{-1}\left\{\frac{1}{s+2}\right\}=\frac{1}{2}u(t)-\frac{1}{2}e^{-2t}u(t)$$

$$\mathcal{L}^{-1}\left\{\frac{-s^2-2s+1}{s\,(s+1)\,s+2)}\right\}=\frac{1}{2}\mathcal{L}^{-1}\left\{\frac{1}{s}\right\}-2\mathcal{L}^{-1}\left\{\frac{1}{s+1}\right\}+\frac{1}{2}\mathcal{L}^{-1}\left\{\frac{1}{s+2}\right\}=\frac{1}{2}u(t)-2e^{-t}u(t)+\frac{1}{2}e^{-2t}u(t)$$

Therefore, the solution of the state equation will have the form

$$\boldsymbol{q}(t)=\begin{bmatrix}q_1(t)\\ q_2(t)\end{bmatrix}=\begin{bmatrix}\frac{1}{2}u(t)-\frac{1}{2}e^{-2t}u(t)\\[2ex] \frac{1}{2}u(t)-2e^{-t}u(t)+\frac{1}{2}e^{-2t}u(t)\end{bmatrix}$$

Problem 5.11.3 Solve the state equation $\dot{\boldsymbol{q}}(t)=\boldsymbol{A}\boldsymbol{q}(t)+\boldsymbol{B}x(t)$ for parameter values

$$\boldsymbol{A}=\begin{bmatrix}-1&1\\0&-2\end{bmatrix},\quad \boldsymbol{B}=\begin{bmatrix}1&1\\0&1\end{bmatrix},\quad \boldsymbol{q}(0)=\begin{bmatrix}1\\2\end{bmatrix},\quad \boldsymbol{x}(t)=\begin{bmatrix}u(t)\\ \delta(t)\end{bmatrix}$$

(Lathi, *Linear Systems and Signals*, Problem 10.3-4).

Solution Following the methodology of the previous exercises, we calculate the matrix

$$s\boldsymbol{I}-\boldsymbol{A}=s\begin{bmatrix}1&0\\0&1\end{bmatrix}-\begin{bmatrix}-1&1\\0&-2\end{bmatrix}=\begin{bmatrix}s&0\\0&s\end{bmatrix}-\begin{bmatrix}-1&1\\0&-2\end{bmatrix}=\begin{bmatrix}s+1&-1\\0&s+2\end{bmatrix}$$

and therefore we will have

$$\boldsymbol{\Phi}(s)=(s\boldsymbol{I}-\boldsymbol{A})^{-1}=\begin{bmatrix}s+1&-1\\0&s+2\end{bmatrix}^{-1}=\frac{1}{(s+1)(s+2)}\begin{bmatrix}s+2&1\\0&s+1\end{bmatrix}=\begin{bmatrix}\dfrac{1}{s+1}&\dfrac{1}{(s+1)(s+2)}\\[2ex] 0&\dfrac{1}{s+2}\end{bmatrix}\quad\text{and}$$

$$\boldsymbol{\Phi}(s)\,\boldsymbol{q}(0)=\begin{bmatrix}\dfrac{1}{s+1}&\dfrac{1}{(s+1)(s+2)}\\[2ex] 0&\dfrac{1}{s+2}\end{bmatrix}\begin{bmatrix}1\\2\end{bmatrix}=\begin{bmatrix}\dfrac{s+4}{(s+1)(s+2)}\\[2ex] \dfrac{2}{s+2}\end{bmatrix}$$

The Laplace transform of the input vector is calculated as

$$\boldsymbol{\mathcal{X}}(s) = \mathcal{L}\{x(t)\} = \mathcal{L}\left\{\begin{bmatrix} u(t) \\ \delta(t) \end{bmatrix}\right\} = \begin{bmatrix} \mathcal{L}\{u(t)\} \\ \mathcal{L}\{\delta(t)\} \end{bmatrix} = \begin{bmatrix} \frac{1}{s} \\ 1 \end{bmatrix}$$

from which it follows that

$$\boldsymbol{B}\boldsymbol{X}(s) = \begin{bmatrix} 1 & 1 \\ 0 & 1 \end{bmatrix}\begin{bmatrix} \frac{1}{s} \\ 1 \end{bmatrix} = \begin{bmatrix} \frac{s+1}{s} \\ 1 \end{bmatrix} \qquad \text{and}$$

$$\boldsymbol{\Phi}(s)\boldsymbol{B}\boldsymbol{X}(s) = \begin{bmatrix} \frac{1}{s+1} & \frac{1}{(s+1)(s+2)} \\ 0 & \frac{1}{s+2} \end{bmatrix}\begin{bmatrix} \frac{s+1}{s} \\ 1 \end{bmatrix} = \begin{bmatrix} \frac{s^2+4s+2}{s(s+1)(s+2)} \\ \frac{1}{s+2} \end{bmatrix}$$

Replacing the preceding quantities in the defining equation of the function $\boldsymbol{\mathcal{Q}}(s)$ we get

$$\boldsymbol{\mathcal{Q}}(s) = \boldsymbol{\Phi}(s)\boldsymbol{q}(0) + \boldsymbol{\Phi}(s)\boldsymbol{B}\boldsymbol{\mathcal{X}}(s) = \begin{bmatrix} \frac{s+4}{(s+1)(s+2)} \\ \frac{2}{s+2} \end{bmatrix} + \begin{bmatrix} \frac{s^2+4s+2}{s(s+1)(s+2)} \\ \frac{1}{s+2} \end{bmatrix} = \begin{bmatrix} \frac{2s^2+8s+2}{s(s+1)(s+2)} \\ \frac{3}{s+2} \end{bmatrix}$$

and therefore, the solution of the state equation $\boldsymbol{q}(t)$ will be calculated as the inverse transform

$$\boldsymbol{q}(t) = \mathcal{L}^{-1}\{\boldsymbol{\mathcal{Q}}(s)\} = \mathcal{L}^{-1}\left\{\begin{bmatrix} \frac{2s^2+8s+2}{s(s+1)(s+2)} \\ \frac{3}{s+2} \end{bmatrix}\right\} = \begin{bmatrix} \mathcal{L}^{-1}\left\{\frac{2s^2+8s+2}{s(s+1)(s+2)}\right\} \\ \mathcal{L}^{-1}\left\{\frac{3}{s+2}\right\} \end{bmatrix}$$

Using the method of partial fraction expansion we find that

$$\frac{2s^2+8s+2}{s(s+1)(s+2)} = \frac{1}{s} + \frac{4}{s+1} - \frac{3}{s+2}$$

and therefore we will have

$$\mathcal{L}^{-1}\left\{\frac{2s^2+8s+2}{s(s+1)(s+2)}\right\} = \mathcal{L}^{-1}\left\{\frac{1}{s}\right\} + 4\mathcal{L}^{-1}\left\{\frac{1}{s+1}\right\} - 3\mathcal{L}^{-1}\left\{\frac{1}{s+2}\right\} = u(t) + 4e^{-t}u(t) - 3e^{-2t}u(t)$$

while we still have

$$\mathcal{L}^{-1}\left\{\frac{3}{s+2}\right\} = 3\mathcal{L}^{-1}\left\{\frac{1}{s+2}\right\} = 3e^{-2t}u(t)$$

Therefore, the solution of the state equation will have the form

$$q_1(t) = u(t) + 4e^{-t}u(t) - 3e^{-2t}u(t) \qquad \text{and} \qquad q_2(t) = 3e^{-2t}u(t)$$

a result that is the same as the one we found working in the field of continuous-time.

Problem 5.11.4 Determine the output and the impulse response of the system

$$\boldsymbol{A}=\begin{bmatrix}0 & 2\\ -1 & -3\end{bmatrix},\qquad \boldsymbol{B}=\begin{bmatrix}0\\ 1\end{bmatrix},\qquad \boldsymbol{C}=\begin{bmatrix}2 \; 3\end{bmatrix},\qquad \boldsymbol{D}=\begin{bmatrix}-1\end{bmatrix},\qquad \boldsymbol{q}(0)=\begin{bmatrix}2\\ 1\end{bmatrix}$$

characterized by an input value $x(t)=0$.

Solution Following the methodology of the previous exercises, it turns out that

$$s\boldsymbol{I}-\boldsymbol{A}=\begin{bmatrix}s & -2\\ 1 & s+3\end{bmatrix}\qquad \boldsymbol{\Phi}(s)=\begin{bmatrix}\dfrac{s+3}{(s+1)(s+2)} & \dfrac{2}{(s+1)(s+2)}\\[2ex] \dfrac{-1}{(s+1)(s+2)} & \dfrac{s}{(s+1)(s+2)}\end{bmatrix}$$

$$\boldsymbol{\Phi}(s)\,\boldsymbol{q}(0)=\begin{bmatrix}\dfrac{2s+8}{(s+1)(s+2)}\\[2ex] \dfrac{s-2}{(s+1)(s+2)}\end{bmatrix}\qquad \text{and}\qquad \boldsymbol{C}\boldsymbol{\Phi}(s)\,\boldsymbol{q}(0)=\frac{7s+10}{(s+1)(s+2)}$$

while as for the other expressions required for the calculation of the output and the impulse response, these are calculated as

$$\boldsymbol{\Phi}(s)\boldsymbol{B}=\begin{bmatrix}\dfrac{2}{(s+1)(s+2)}\\[2ex] \dfrac{s}{(s+1)(s+2)}\end{bmatrix}\qquad \text{and}\qquad \mathcal{H}(s)=\boldsymbol{C}\boldsymbol{\Phi}(s)\boldsymbol{B}+\boldsymbol{D}=\frac{-s^2+2}{(s+1)(s+2)}$$

On the other hand, the system's input $x(t)$ is equal to zero, and therefore, $\mathcal{X}(s)=\mathcal{L}\{x(t)\}=0$. Replacing the foregoing expressions in the defining equation of the function $\mathcal{Y}(s)$, it turns out that

$$\mathcal{Y}(s)=\boldsymbol{C}\boldsymbol{\Phi}(s)\boldsymbol{q}(0)+\Big[\boldsymbol{C}\boldsymbol{\Phi}(s)\boldsymbol{B}+\boldsymbol{D}\Big]\mathcal{X}(s)=\boldsymbol{C}\boldsymbol{\Phi}(s)\boldsymbol{q}(0)=\frac{7s+10}{(s+1)(s+2)}$$

and therefore the desired system output will be calculated as the inverse transform

$$y(t)=\mathcal{L}^{-1}\{Y(s)\}=\mathcal{L}^{-1}\left\{\frac{7s+10}{(s+1)(s+2)}\right\}=\mathcal{L}^{-1}\left\{\frac{3}{s+1}+\frac{4}{s+2}\right\}$$
$$=3\mathcal{L}^{-1}\left\{\frac{1}{s+1}\right\}+4\mathcal{L}^{-1}\left\{\frac{1}{s+2}\right\}=3e^{-t}u(t)+4e^{-2t}u(t)$$

Similarly, the impulse response $h(t)$ will be calculated as the inverse Laplace transform of the function $\mathcal{H}(s)$—the careful reader will have already realized that this is precisely the transfer function of the system, something that will be commented on in detail in a subsequent section—and therefore we will have

$$h(t)=\mathcal{L}^{-1}\{H(s)\}=\mathcal{L}^{-1}\{\boldsymbol{C}\boldsymbol{\Phi}(s)\boldsymbol{B}+\boldsymbol{D}\}=\mathcal{L}^{-1}\left\{\frac{-s^2+2}{(s+1)(s+2)}\right\}$$
$$=\mathcal{L}^{-1}\left\{\frac{1}{s+1}-\frac{s}{s+2}\right\}=\mathcal{L}^{-1}\left\{\frac{1}{s+1}-1+\frac{2}{s+2}\right\}$$
$$=\mathcal{L}^{-1}\left\{\frac{1}{s+1}\right\}-\mathcal{L}^{-1}\{1\}+2\mathcal{L}^{-1}\left\{\frac{1}{s+2}\right\}=e^{-t}u(t)-\delta(t)+2e^{-2t}u(t)$$

Problem 5.11.5 Use the method of the Laplace transform to determine the output and the impulse response of the system described by the matrices

$$\boldsymbol{A} = \begin{bmatrix} -1 & 1 \\ 0 & -2 \end{bmatrix}, \quad \boldsymbol{B} = \begin{bmatrix} 1 & 1 \\ 0 & 1 \end{bmatrix}, \quad \boldsymbol{C} = \begin{bmatrix} 5 & -6 \\ 2 & 4 \end{bmatrix}, \quad \boldsymbol{D} = \begin{bmatrix} 1 & -1 \\ 0 & 1 \end{bmatrix}$$

while the input of $\boldsymbol{x}(t)$ and the initial value of the state vector $\boldsymbol{q}(0)$ have the form

$$\boldsymbol{x}(t) = \begin{bmatrix} u(t) \\ \delta(t) \end{bmatrix} \quad \text{and} \quad \boldsymbol{q}(0) = \begin{bmatrix} 1 \\ 2 \end{bmatrix}$$

Solution In this case, the matrices of the method are calculated as

$$s\boldsymbol{I} - \boldsymbol{A} = \begin{bmatrix} s+1 & -1 \\ 0 & s+2 \end{bmatrix} \qquad \boldsymbol{\Phi}(s) = \begin{bmatrix} \frac{1}{s+1} & \frac{1}{(s+1)(s+2)} \\ 0 & \frac{1}{s+2} \end{bmatrix}$$

$$\boldsymbol{\Phi}(s)\boldsymbol{q}(0) = \begin{bmatrix} \frac{s+4}{(s+1)(s+2)} \\ \frac{2}{s+2} \end{bmatrix} \qquad \boldsymbol{C}\boldsymbol{\Phi}(s)\boldsymbol{q}(0) = \begin{bmatrix} \frac{-7s+8}{(s+1)(s+2)} \\ \frac{10s+16}{(s+1)(s+2)} \end{bmatrix}$$

while in the same way we find that

$$\boldsymbol{\Phi}(s)\boldsymbol{B} = \begin{bmatrix} \frac{1}{s+1} & \frac{s+3}{(s+1)(s+2)} \\ 0 & \frac{1}{s+2} \end{bmatrix} \qquad \boldsymbol{C}\boldsymbol{\Phi}(s)\boldsymbol{B} = \begin{bmatrix} \frac{5}{s+1} & \frac{-s+9}{(s+1)(s+2)} \\ \frac{2}{s+1} & \frac{6s+10}{(s+1)(s+2)} \end{bmatrix}$$

and

$$\boldsymbol{H}(s) = \boldsymbol{C}\boldsymbol{\Phi}(s)\boldsymbol{B} + \boldsymbol{D} = \begin{bmatrix} \frac{s+6}{s+1} & \frac{-s^2-4s+7}{(s+1)(s+2)} \\ \frac{2}{s+1} & \frac{s^2+9s+12}{(s+1)(s+2)} \end{bmatrix}$$

with the function $\boldsymbol{H}(s)$ obviously expressing the function system transfer. On the other hand, the Laplace transform of the input vector will have the form

$$\boldsymbol{X}(s) = \mathcal{L}\{x(t)\} = \mathcal{L}\left\{\begin{bmatrix} u(t) \\ \delta(t) \end{bmatrix}\right\} = \begin{bmatrix} \mathcal{L}\{u(t)\} \\ \mathcal{L}\{\delta(t)\} \end{bmatrix} = \begin{bmatrix} \frac{1}{s} \\ 1 \end{bmatrix}$$

from which it follows that

$$\left[C\Phi(s)B + D\right]X(s) = \begin{bmatrix} \dfrac{-s^3 - 3s^2 + 15s + 12}{s(s+1)(s+2)} \\[2ex] \dfrac{s^3 + 9s^2 + 14s + 4}{s+2} \end{bmatrix}$$

Replacing the preceding expressions in the defining equation of the function $Y(s)$ we get the result

$$Y(s) = C\Phi(s)q(0) + \left[C\Phi(s)B + D\right]X(s) = \begin{bmatrix} \dfrac{-s^3 - 10s^2 + 23s + 12}{s(s+1)(s+2)} \\[2ex] \dfrac{s^3 + 19s^2 + 30s + 4}{s(s+1)(s+2)} \end{bmatrix}$$

and therefore the output vector of the system will be calculated as the inverse Laplace transform of the last expression. Using the method of partial fraction expansion we find

$$\frac{-s^3 - 10s^2 + 23s + 12}{s(s+1)(s+2)} = \frac{6}{s} + \frac{20}{s+1} - \frac{s+35}{s+2} = \frac{6}{s} + \frac{20}{s+1} - 1 - \frac{33}{s+2} \quad \text{and}$$

$$\frac{s^3 + 19s^2 + 30s + 4}{s(s+1)(s+2)} = \frac{2}{s} + \frac{8}{s+1} + \frac{s+8}{s+2} = \frac{2}{s} + \frac{8}{s+1} + 1 + \frac{6}{s+2}$$

and taking the inverse transform of the preceding expressions, we arrive at the result

$$y_1(t) = 6\mathcal{L}^{-1}\left\{\frac{1}{s}\right\} + 20\mathcal{L}^{-1}\left\{\frac{1}{s+1}\right\} - \mathcal{L}^{-1}\{1\} - 33\mathcal{L}^{-1}\left\{\frac{1}{s+2}\right\} = 6u(t) + 20e^{-t}u(t) - 33e^{-2t}u(t) - \delta(t)$$

$$y_2(t) = 2\mathcal{L}^{-1}\left\{\frac{1}{s}\right\} + 8\mathcal{L}^{-1}\left\{\frac{1}{s+1}\right\} + \mathcal{L}^{-1}\{1\} + 6\mathcal{L}^{-1}\left\{\frac{1}{s+2}\right\} = 2u(t) + 8e^{-t}u(t) + \delta(t) + 6e^{-2t}u(t)$$

On the other hand, the impulse response of the system will be calculated as the inverse transform of the transfer function $H(s)$, so we will have

$$h(t) = \mathcal{L}^{-1}\{H(s)\} = \mathcal{L}^{-1}\left\{\begin{bmatrix} \dfrac{s+6}{s+1} & \dfrac{-s^2 - 4s + 7}{(s+1)(s+2)} \\[2ex] \dfrac{2}{s+1} & \dfrac{s^2 + 9s + 12}{(s+1)(s+2)} \end{bmatrix}\right\} = \begin{bmatrix} \mathcal{L}^{-1}\left\{\dfrac{s+6}{s+1}\right\} & \mathcal{L}^{-1}\left\{\dfrac{-s^2 - 4s + 7}{(s+1)(s+2)}\right\} \\[2ex] \mathcal{L}^{-1}\left\{\dfrac{2}{s+1}\right\} & \mathcal{L}^{-1}\left\{\dfrac{s^2 + 9s + 12}{(s+1)(s+2)}\right\} \end{bmatrix}$$

It is not difficult to see that the fractions that appear in the preceding table develop as

$$\frac{s+6}{s+1} = 1 + \frac{5}{s+1}, \qquad \frac{-s^2 - 4s + 7}{(s+1)(s+2)} = \frac{10}{s+1} - 1 - \frac{11}{s+2}, \qquad \frac{s^2 + 9s + 12}{(s+1)(s+2)} = \frac{4}{s+1} + 1 + \frac{2}{s+2}$$

and taking the inverse transform of the preceding quantities, we arrive at the result

$$h(t) = \begin{bmatrix} 5e^{-t}u(t) + \delta(t) & 10e^{-t}u(t) - 11e^{-2t}u(t) - \delta(t) \\ 2e^{-t}u(t) & 4e^{-t}u(t) + 2e^{-2t}u(t) + \delta(t) \end{bmatrix}$$

which is the same as the one we had arrived at in the field of continuous-time.

5.11.2 Calculation of Exponential Function

As was already mentioned in a previous section, the solution of the state equation of a continuous LTI system is given by the expression

$$\boldsymbol{q}(t) = e^{\boldsymbol{A}t}\boldsymbol{q}(0) + e^{\boldsymbol{A}t} * \boldsymbol{Bx}(t) \tag{5.65}$$

where for simplicity we considered the initial time to be $t_0 = 0$. On the other hand, the formulation of the solution of this equation in the frequency domain has the form

$$\boldsymbol{\mathcal{Q}}(s) = \boldsymbol{\Phi}(s)\boldsymbol{q}(0) + \boldsymbol{\Phi}(s)\boldsymbol{B}\boldsymbol{\mathcal{X}}(s) \tag{5.66}$$

If we take into account that $\boldsymbol{\mathcal{Q}}(s) = \mathcal{L}\{\boldsymbol{q}(t)\}$ and $\mathcal{L}\{e^{\boldsymbol{A}t} * \boldsymbol{Bx}(t)\} = \mathcal{L}\{e^{\boldsymbol{A}t}\}\boldsymbol{B}\boldsymbol{\mathcal{X}}(s)$ (according to the convolution theorem) it immediately follows that the equation defining the function $\boldsymbol{\mathcal{Q}}(s)$ is the result of applying the Laplace transform on the equation defining the solution $\boldsymbol{q}(t)$ and that furthermore the function $\boldsymbol{\Phi}(s) = (s\boldsymbol{I} - \boldsymbol{A})^{-1}$ must correspond to the Laplace transform of the exponential function $e^{\boldsymbol{A}t}$, so that we can write that $\boldsymbol{\Phi}(s) = \mathcal{L}\{e^{\boldsymbol{A}t}\}$. Let us see, then, if things are indeed so.

Starting from the defining equation of the exponential function, which, as is well known, has the form

$$e^{\boldsymbol{A}t} = \sum_{k=0}^{\infty} \frac{t^k}{k!}\boldsymbol{A}^k \tag{5.67}$$

the *application of the Laplace transform* to this equation will lead to the expression

$$\mathcal{L}\{e^{\boldsymbol{A}t}\} = \mathcal{L}\left\{\sum_{k=0}^{\infty} \frac{t^k}{k!}\boldsymbol{A}^k\right\} = \sum_{k=0}^{\infty} \mathcal{L}\left\{\frac{t^k}{k!}\right\}\boldsymbol{A}^k \tag{5.68}$$

Referring to tables of elementary transforms, we find that

$$\mathcal{L}\{t^k\} = \frac{k!}{s^{k+1}} \qquad \text{or, equivalently, that} \qquad \mathcal{L}\left\{\frac{t^k}{k!}\right\} = \frac{1}{s^{k+1}}$$

due to the linearity property of the Laplace transform. Substituting the last relation into the *Laplace transform* of the exponential function, it will take the form

$$\mathcal{L}\{e^{\boldsymbol{A}t}\} = \sum_{k=0}^{\infty} \frac{\boldsymbol{A}^k}{s^{k+1}} = \sum_{k=0}^{\infty} \boldsymbol{A}^k s^{-(k+1)} = s^{-1}\sum_{k=0}^{\infty} \boldsymbol{A}^k s^{-k} = s^{-1}\sum_{k=0}^{\infty}(s^{-1}\boldsymbol{A})^k \tag{5.69}$$

Now relying on a theorem of linear algebra, according to which, *if the eigenvalues of the matrix* $\boldsymbol{A}$ *are in absolute value less than one, then the preceding power series of infinite terms converges within the space* $R^{N\times N}$ *on which it is defined—where* N *is, of course, the order of the matrix* $\boldsymbol{A}$*—to the matrix* $(\boldsymbol{I} - s^{-1}\boldsymbol{A})^{-1}$ (this relation is completely analogous to that which applies to the case of a simple real variable), the Laplace transform of the exponential function $e^{\boldsymbol{A}t}$ will take the form

$$\mathcal{L}\{e^{At}\} = s^{-1}(\boldsymbol{I} - s^{-1}\boldsymbol{A})^{-1} = [s(\boldsymbol{I} - s^{-1}\boldsymbol{A})]^{-1} = (s\boldsymbol{I} - ss^{-1}\boldsymbol{A})^{-1} = (s\boldsymbol{I} - \boldsymbol{A})^{-1}$$

that is, the function $\boldsymbol{\Phi}(s)$ indeed constitutes the *Laplace transform of the state transition matrix*, so that we can write that

$$\boldsymbol{\Phi}(s) = \mathcal{L}\{e^{At}\} = (s\boldsymbol{I} - \boldsymbol{A})^{-1} \tag{5.70}$$

The last relation is particularly useful, as it offers us another method of calculating the exponential function e^{At}. As already mentioned, the direct calculation of this function from its defining equation is particularly difficult, and for this reason we usually resort to the use of some alternative methods, some of which were presented in detail in the second chapter and used to solve exercises. The way in which we can calculate the matrix e^{At} using the Laplace transform works as follows. Initially, knowing the system matrix $\boldsymbol{A}$, we can calculate the function $\boldsymbol{\Phi}(s) = (s\boldsymbol{I} - \boldsymbol{A})^{-1}$ (something we have already done in all previous exercises), and then we can determine the exponential function e^{At} as an *inverse Laplace transform* of the form

$$e^{At} = \mathcal{L}^{-1}\{\boldsymbol{\Phi}(s)\}$$

This method of calculating the matrix e^{At} is demonstrated in the following exercises. The reader is asked to verify that the results obtained by applying this method are the same as those we get by calculating the exponential function e^{At} in the way we saw in Chap. 2.

Problem 5.11.6 Use the Laplace transform method to calculate the exponential function e^{At}, where the matrix $\boldsymbol{A}$ is given by the relation

$$\boldsymbol{A} = \begin{bmatrix} 0 & 2 \\ -1 & -3 \end{bmatrix}$$

Solution The function $\boldsymbol{\Phi}(s)$ is calculated as

$$\boldsymbol{\Phi}(s) = (s\boldsymbol{I} - \boldsymbol{A})^{-1} = \left\{ s\begin{bmatrix} 1 & 0 \\ 0 & 1 \end{bmatrix} - \begin{bmatrix} 0 & 2 \\ -1 & -3 \end{bmatrix} \right\}^{-1} = \begin{bmatrix} s & -2 \\ 1 & s+3 \end{bmatrix}^{-1}$$

$$= \frac{1}{s^2 + 3s + 2}\begin{bmatrix} s+3 & 2 \\ -1 & s \end{bmatrix} = \begin{bmatrix} \dfrac{s+3}{(s+1)(s+2)} & \dfrac{2}{(s+1)(s+2)} \\ \dfrac{-1}{(s+1)(s+2)} & \dfrac{s}{(s+1)(s+2)} \end{bmatrix}$$

and therefore we will have

$$e^{At} = \mathcal{L}^{-1}\{\boldsymbol{\Phi}(s)\} = \begin{bmatrix} \mathcal{L}^{-1}\left\{\dfrac{s+3}{(s+1)(s+2)}\right\} & \mathcal{L}^{-1}\left\{\dfrac{2}{(s+1)(s+2)}\right\} \\ \mathcal{L}^{-1}\left\{\dfrac{-1}{(s+1)(s+2)}\right\} & \mathcal{L}^{-1}\left\{\dfrac{s}{(s+1)(s+2)}\right\} \end{bmatrix}$$

Using the method of partial fraction expansion, we find that

$$\frac{s+3}{(s+1)(s+2)} = -\frac{1}{s+2} + \frac{2}{s+1} \qquad \frac{2}{(s+1)(s+2)} = -\frac{2}{s+2} + \frac{2}{s+1}$$

$$\frac{-1}{(s+1)(s+2)} = -\frac{1}{s+1} + \frac{1}{s+2} \qquad \frac{s}{(s+1)(s+2)} = -\frac{1}{s+1} + \frac{2}{s+2}$$

and referring back to tables of elementary transforms, we arrive at the expressions

$$\mathcal{L}^{-1}\left\{\frac{s+3}{(s+1)(s+2)}\right\} = -\mathcal{L}^{-1}\left\{\frac{1}{s+2}\right\} + 2\mathcal{L}^{-1}\left\{\frac{1}{s+1}\right\} = -e^{-2t}u(t) + 2e^{-t}u(t)$$

$$\mathcal{L}^{-1}\left\{\frac{2}{(s+1)(s+2)}\right\} = -2\mathcal{L}^{-1}\left\{\frac{1}{s+2}\right\} + 2\mathcal{L}^{-1}\left\{\frac{1}{s+1}\right\} = -2e^{-2t}u(t) + 2e^{-t}u(t)$$

$$\mathcal{L}^{-1}\left\{\frac{-1}{(s+1)(s+2)}\right\} = -\mathcal{L}^{-1}\left\{\frac{1}{s+1}\right\} + \mathcal{L}^{-1}\left\{\frac{1}{s+2}\right\} = -e^{-t}u(t) + e^{-2t}u(t)$$

$$\mathcal{L}^{-1}\left\{\frac{s}{(s+1)(s+2)}\right\} = -\mathcal{L}^{-1}\left\{\frac{1}{s+1}\right\} + 2\mathcal{L}^{-1}\left\{\frac{1}{s+2}\right\} = -e^{-t}u(t) + 2e^{-2t}u(t)$$

Therefore, the exponential function of the matrix e^{At} is given by the relation

$$e^{At} = \begin{bmatrix} -e^{-2t}u(t) + 2e^{-t}u(t) & -2e^{-2t}u(t) + 2e^{-t}u(t) \\ -e^{-t}u(t) + e^{-2t}u(t) & -e^{-t}u(t) + 2e^{-2t}u(t) \end{bmatrix}$$

Problem 5.11.7 Using the method of Laplace transform, calculate the exponential function e^{At}, where the matrix $\boldsymbol{A}$ is given by the relation

$$\boldsymbol{A} = \begin{bmatrix} 5 & -6 \\ 4 & -5 \end{bmatrix}$$

Solution In this case, the function $\boldsymbol{\Phi}(s)$ is calculated as

$$\boldsymbol{\Phi}(s) = (s\boldsymbol{I} - \boldsymbol{A})^{-1} = \left\{s\begin{bmatrix} 1 & 0 \\ 0 & 1 \end{bmatrix} - \begin{bmatrix} 5 & -6 \\ 4 & -5 \end{bmatrix}\right\}^{-1} = \begin{bmatrix} s-5 & 6 \\ -4 & s+5 \end{bmatrix}^{-1}$$

$$= \frac{1}{s^2-1}\begin{bmatrix} s+5 & -6 \\ 4 & s-5 \end{bmatrix} \begin{bmatrix} \dfrac{s+5}{(s+1)(s-1)} & \dfrac{-6}{(s+1)(s-1)} \\ \dfrac{4}{(s+1)(s-1)} & \dfrac{s-5}{(s+1)(s-1)} \end{bmatrix}$$

and therefore we will have

$$e^{At} = \mathcal{L}^{-1}\{\boldsymbol{\Phi}(s)\} = \begin{bmatrix} \mathcal{L}^{-1}\left\{\dfrac{s+5}{(s+1)(s-1)}\right\} & \mathcal{L}^{-1}\left\{\dfrac{-6}{(s+1)(s-1)}\right\} \\ \mathcal{L}^{-1}\left\{\dfrac{4}{(s+1)(s-1)}\right\} & \mathcal{L}^{-1}\left\{\dfrac{s-5}{(s+1)(s-1)}\right\} \end{bmatrix}$$

Using the method of partial fraction expansion, we arrive at the results

$$\frac{s+5}{(s+1)(s-1)} = -\frac{2}{s+1} + \frac{3}{s-1} \qquad \frac{-6}{(s+1)(s-1)} = -\frac{3}{s-1} + \frac{3}{s+1}$$

$$\frac{4}{(s+1)(s-1)} = -\frac{2}{s+1} + \frac{2}{s-1} \qquad \frac{s-5}{(s+1)(s-1)} = -\frac{2}{s-1} + \frac{3}{s+1}$$

and referring to tables of elementary transforms, we are led to the expressions

$$\mathcal{L}^{-1}\left\{\frac{s+5}{(s+1)(s-1)}\right\} = -2\mathcal{L}^{-1}\left\{\frac{1}{s+1}\right\} + 3\mathcal{L}^{-1}\left\{\frac{1}{s-1}\right\} = -2e^{-t}u(t) + 3e^{t}u(t)$$

$$\mathcal{L}^{-1}\left\{\frac{-6}{(s+1)(s-1)}\right\} = -3\mathcal{L}^{-1}\left\{\frac{1}{s-1}\right\} + 3\mathcal{L}^{-1}\left\{\frac{1}{s+1}\right\} = -3e^{t}u(t) + 3e^{-t}u(t)$$

$$\mathcal{L}^{-1}\left\{\frac{4}{(s+1)(s-1)}\right\} = -2\mathcal{L}^{-1}\left\{\frac{1}{s+1}\right\} + 2\mathcal{L}^{-1}\left\{\frac{1}{s-1}\right\} = -2e^{-t}u(t) + 2e^{t}u(t)$$

$$\mathcal{L}^{-1}\left\{\frac{s-5}{(s+1)(s-1)}\right\} = -2\mathcal{L}^{-1}\left\{\frac{1}{s-1}\right\} + 3\mathcal{L}^{-1}\left\{\frac{1}{s+1}\right\} = -2e^{t}u(t) + 3e^{-t}u(t)$$

Therefore, the exponential function of the matrix $e^{\boldsymbol{A}t}$ is calculated as

$$e^{\boldsymbol{A}t} = \begin{bmatrix} -2e^{-t}u(t) + 3e^{t}u(t) & -3e^{t}u(t) + 3e^{-t}u(t) \\ -2e^{-t}u(t) + 2e^{t}u(t) & -2e^{t}u(t) + 3e^{-t}u(t) \end{bmatrix}$$

a result that is again the same as the one we arrived at using the other two methods.

5.11.3 Transfer Function from Dynamic Equations

In a previous section, we proved that if we take the Laplace transform of the dynamic equations and set the vector of initial conditions $\boldsymbol{q}(0)$ to zero, the input and output of the system expressed in the frequency domain will be connected by a relationship of the form[20]

$$\mathcal{Y}(s) = \Big[\boldsymbol{C}\boldsymbol{\Phi}(s)\boldsymbol{B} + \boldsymbol{D}\Big]\mathcal{X}(s) = \mathcal{H}(s)\mathcal{X}(s) \tag{5.71}$$

from which it directly follows that the function

$$\mathcal{H}(s) = \frac{\mathcal{Y}(s)}{\mathcal{X}(s)} = \boldsymbol{C}\boldsymbol{\Phi}(s)\boldsymbol{B} + \boldsymbol{D} = \boldsymbol{C}\mathcal{L}\{e^{\boldsymbol{A}t}\}\boldsymbol{B} + \boldsymbol{D} = \boldsymbol{C}(s\boldsymbol{I} - \boldsymbol{A})^{-1}\boldsymbol{B} + \boldsymbol{D} \tag{5.72}$$

should constitute the *transfer function of the system*, which, therefore, can be calculated from the last relation, if we know the matrices $\boldsymbol{A}$, $\boldsymbol{B}$, $\boldsymbol{C}$, and $\boldsymbol{D}$ that appear in its dynamic equations. This conclusion is confirmed by the fact that since the transfer function is by definition the *Laplace transform of the system's impulse response*, this response should be able to be calculated from the relation $h(t) = \mathcal{L}^{-1}\{\mathcal{H}(s)\}$. Taking the inverse Laplace transform of the preceding relation, we end up with the equation

[20] Since the transfer function is going to be defined as the ratio $\mathcal{Y}(s)/\mathcal{X}(s)$, we consider the simple case of a SISO system for which the functions $\mathcal{X}(s)$ and $\mathcal{Y}(s)$ are *polynomials of the variable* s, and therefore their ratio will constitute a *rational function*. Conversely, for the case of MIMO systems, the functions $\boldsymbol{\mathcal{Y}}(s)$ and $\boldsymbol{\mathcal{X}}(s)$ are *vectors*, and it makes no sense to take the ratio of two vectors; however, even in this case, the transfer function continues to be given by the same equation.

$$h(t) = \mathcal{L}^{-1}\{\mathcal{H}(s)\} = \mathcal{L}^{-1}\{\boldsymbol{C}\boldsymbol{\Phi}(s)\boldsymbol{B} + \boldsymbol{D}\} = \boldsymbol{C}\mathcal{L}^{-1}\{\boldsymbol{\Phi}(s)\}\boldsymbol{B} + \boldsymbol{D}\mathcal{L}^{-1}\{1\} = \boldsymbol{C}e^{\boldsymbol{A}t}\boldsymbol{B} + \boldsymbol{D}\delta(t) \tag{5.73}$$

which was proven in the chapter on continuous-time systems, a conclusion that verifies our claim.

Definition 5.11.1 (Transfer Function Model) A state-space model described by the matrices $(\boldsymbol{A}, \boldsymbol{B}, \boldsymbol{C}, \boldsymbol{D})$ is aforementioned to implement a transfer function $\mathcal{H}(s)$ if this function can be expressed as

$$\mathcal{H}(s) = \boldsymbol{C}(s\boldsymbol{I} - \boldsymbol{A})^{-1}\boldsymbol{B} + \boldsymbol{D} = \left[\begin{array}{c|c} \boldsymbol{A} & \boldsymbol{B} \\ \hline \boldsymbol{C} & \boldsymbol{D} \end{array}\right]$$

On the other hand, if the matrix $(s\boldsymbol{I} - \boldsymbol{A})^{-1}$ does not exist, then the system described by the preceding matrices is considered nonimplementable.

If the matrix $(s\boldsymbol{I} - \boldsymbol{A})^{-1}$ is expanded into an infinite series

$$(s\boldsymbol{I} - \boldsymbol{A})^{-1} = s^{-1} + \boldsymbol{A}s^{-2} + \boldsymbol{A}^2s^{-3} + \boldsymbol{A}^3s^{-4} + \cdots = \sum_{k=0}^{\infty} \boldsymbol{A}^k s^{-(k+1)}$$

then the transfer function can be expressed in the form

$$\begin{aligned} \mathcal{H}(s) &= \boldsymbol{C}(s\boldsymbol{I} - \boldsymbol{A})^{-1}\boldsymbol{B} + \boldsymbol{D} = \boldsymbol{C}\left(\sum_{k=0}^{\infty} \boldsymbol{A}^k s^{-(k+1)}\right)\boldsymbol{B} + \boldsymbol{D} \\ &= \boldsymbol{D} + \boldsymbol{C}\boldsymbol{B}s^{-1} + \boldsymbol{C}\boldsymbol{A}\boldsymbol{B}s^{-2} + \boldsymbol{C}\boldsymbol{A}^2\boldsymbol{B}s^{-3} + \cdots = \mathcal{H}_0 + \mathcal{H}_1 s^{-1} + \mathcal{H}_2 s^{-2} + \mathcal{H}_3 s^{-3} + \ldots \end{aligned}$$

If we consider the matrices $\mathcal{H}_k \in C^{L\times K}$ that are defined as $\mathcal{H}_0 = \boldsymbol{D}$ and $\mathcal{H}_k = \boldsymbol{C}\boldsymbol{A}^{k-1}\boldsymbol{B}$ $(k \geq 1)$, we observe that these depend *only on the transfer function and not on the particular implementation of the system*; these matrices are known as *Markov parameters*.

Using a theorem of matrix algebra according to which the *inverse of a matrix* $\boldsymbol{A}$ *is equal to the adjoint matrix*[21] *Adj*$(\boldsymbol{A})$ *divided by its determinant det*$(\boldsymbol{A})$ or, in mathematical notation,

$$\boldsymbol{A}^{-1} = \frac{\mathrm{Adj}(\boldsymbol{A})}{\det(\boldsymbol{A})} \tag{5.74}$$

the defining equation of the transfer function is reformulated as

$$\mathcal{H}(s) = \frac{\mathcal{Y}(s)}{\mathcal{X}(s)} = \frac{\boldsymbol{C}\left[\mathrm{Adj}(s\boldsymbol{I} - \boldsymbol{A})\right]\boldsymbol{B}}{\det(s\boldsymbol{I} - \boldsymbol{A})} + \boldsymbol{D} \tag{5.75}$$

What is the physical meaning of the preceding relationship? Considering for a moment a system for which the matrix $\boldsymbol{D}$ is *equal to zero*, it is not difficult to realize that both the numerator $\boldsymbol{C}[\mathrm{Adj}(s\boldsymbol{I} - \boldsymbol{A})]\boldsymbol{B}$ and the denominator $\det(s\boldsymbol{I} - \boldsymbol{A})$

[21] From matrix algebra it is known that if from a square matrix $\boldsymbol{A}$ of order N we subtract the row number i and column number j, the determinant $|M_{ij}|$ of the remaining matrix is called the *minor determinant* of the element α_{ij}, while the determinant $A_{ij} = (-1)^{i+j}|M_{ij}|$ is known as the *algebraic complement* of the element α_{ij} of the matrix $\boldsymbol{A}$. Having defined the algebraic complement in this way, we can define the *adjoint matrix* Adj$(\boldsymbol{A})$ *as the inverse matrix that results from* $\boldsymbol{A}$ *if we replace each element with its algebraic complement.*

constitute *polynomial functions of the independent variable s with the degree of the numerator polynomial is always less than the degree of the denominator polynomial*. Although this observation is reasonable since the complement of the matrix $s\boldsymbol{I} - \boldsymbol{A}$ contains *minor determinants*, while the denominator contains the *full determinant* of the aforementioned matrix, let us see it through an example, and let us consider for this reason the system described by the dynamic equations

$$\dot{\boldsymbol{q}}(t) = \begin{bmatrix} -1 & 0 & 0 \\ 0 & -3 & 0 \\ 0 & 0 & -2 \end{bmatrix} \boldsymbol{q}(t) + \begin{bmatrix} 1 \\ 1 \\ 1 \end{bmatrix} x(t) \quad \text{and} \quad y(t) = \begin{bmatrix} 1 & 2 & 1 \end{bmatrix} \boldsymbol{q}(t)$$

and in which the matrix $\boldsymbol{A}$ for simplicity reasons is a diagonal matrix. Starting from the calculation of the matrix $s\boldsymbol{I} - \boldsymbol{A}$, it is found to have the form

$$s\boldsymbol{I} - \boldsymbol{A} = s \begin{bmatrix} 1 & 0 & 0 \\ 0 & 1 & 0 \\ 0 & 0 & 1 \end{bmatrix} - \begin{bmatrix} -1 & 0 & 0 \\ 0 & -3 & 0 \\ 0 & 0 & -2 \end{bmatrix} = \begin{bmatrix} s+1 & 0 & 0 \\ 0 & s+3 & 0 \\ 0 & 0 & s+2 \end{bmatrix}$$

In the next step of the process, let us calculate the algebraic complements C_{ij} of the elements $(s\boldsymbol{I} - \boldsymbol{A})_{ij}$ $(i, j = 1, 2, 3)$. Using the defining equation of the algebraic complement, we find that

$$C_{11} = (-1)^{1+1} \begin{vmatrix} s+2 & 0 \\ 0 & s+3 \end{vmatrix} = (s+2)(s+3)$$

$$C_{22} = (-1)^{2+2} \begin{vmatrix} s+1 & 0 \\ 0 & s+2 \end{vmatrix} = (s+1)(s+2)$$

$$C_{33} = (-1)^{3+3} \begin{vmatrix} s+1 & 0 \\ 0 & s+3 \end{vmatrix} = (s+1)(s+3)$$

while of course we will have $C_{12} = C_{13} = C_{21} = C_{23} = C_{31} = C_{32} = 0$, since the minor determinants of the nondiagonal elements will contain zero elements in both lines. Therefore, the adjoint matrix of the matrix $s\boldsymbol{I} - \boldsymbol{A}$ will have the form

$$\text{Adj}(s\boldsymbol{I} - \boldsymbol{A}) = \begin{bmatrix} C_{11} & C_{12} & C_{13} \\ C_{21} & C_{22} & C_{23} \\ C_{31} & C_{32} & C_{33} \end{bmatrix} = \begin{bmatrix} (s+2)(s+3) & 0 & 0 \\ 0 & (s+1)(s+2) & 0 \\ 0 & 0 & (s+1)(s+3) \end{bmatrix}$$

from which it follows that

$$\boldsymbol{C} \left[\text{Adj}\,(s\boldsymbol{I} - \boldsymbol{A}) \right] \boldsymbol{B} \begin{bmatrix} 1 & 2 & 1 \end{bmatrix} \begin{bmatrix} (s+2)(s+3) & 0 & 0 \\ 0 & (s+1)(s+2) & 0 \\ 0 & 0 & (s+1)(s+3) \end{bmatrix} \begin{bmatrix} 1 \\ 1 \\ 1 \end{bmatrix}$$

and finally

$$\boldsymbol{C} \left[\text{Adj}\,(s\boldsymbol{I} - \boldsymbol{A}) \right] \boldsymbol{B} = 5s^2 + 18s + 15$$

that is, the numerator of the transfer function is a polynomial function of s of the second degree. On the other hand, the denominator of this function is clearly a polynomial of the third degree, since it is found that

$$\det\,(s\boldsymbol{I} - \boldsymbol{A}) = (s+1)(s+2)(s+3) = s^3 + 6s^2 + 11s + 6$$

Therefore, in this equation defining the transfer function the *degree of the polynomial of the numerator is always less than the degree of the polynomial of the denominator*.

In a previous section, we mentioned that transfer functions that have this characteristic are called *proper or canonical rational functions* and represent systems that are *physically realizable*. Let us note that this is something that applies even in cases where the numerator and the denominator of the transfer function are *polynomials of the same degree* and that can appear *when the matrix* $\boldsymbol{D}$ *is different from zero*. If, for example, in the system of the previous example we assume that the matrix in question has the value $\boldsymbol{D} = [3]$, then the transfer function will take the form

$$\mathcal{H}(s) = \frac{\boldsymbol{C}\left[\mathrm{Adj}(s\boldsymbol{I} - \boldsymbol{A})\right]\boldsymbol{B}}{\det\left(s\boldsymbol{I} - \boldsymbol{A}\right)} + \boldsymbol{D} = \frac{5s^2 + 18s + 15}{s^3 + 6s^2 + 11s + 6} + 3 = \frac{3s^3 + 23s^2 + 51s + 33}{s^3 + 6s^2 + 11s + 6}$$

a result that verifies our claim. If, on the contrary, the degree of the polynomial of the numerator is greater than the degree of the polynomial of the denominator, then the system described by this function *cannot be realized*.

The second important conclusion that arises from the formulation of the system's transfer function in this way is *that the poles of the system coincide with the roots of the characteristic polynomial* $\det(s\boldsymbol{I} - \boldsymbol{A}) = 0$, *which are just the eigenvalues of the matrix* $\boldsymbol{A}$; once again, therefore, we see the close relationship that exists between these two quantities. Note that this conclusion does not mean that the set of poles of the system necessarily and in every case coincides with the set of eigenvalues of the system matrix, in the sense that these poles may simply be a *subset* of the set of eigenvalues of the matrix $\boldsymbol{A}$. The only case in which this identification exists between the two sets is *minimum dimension systems*, which will be studied in the next section.

Minimum Dimension Implementations

In a previous section, we mentioned that the *dimension of a system in the state space is equal to the number of state variables that are used to describe it* or, equivalently, with the value of the order N of the system matrix $\boldsymbol{A}$. We also mentioned that for a specific continuous-time LTI system *its representation in the state space is not unique*, and in this sense *many equivalent transfer functions describe the same system*.

To demonstrate this concept of *dimension*, let us consider the system

$$\mathcal{H}(s) = \frac{s + 3}{s^3 + 6s^2 + 11s + 6}$$

If we construct the state equations of the system using some of the techniques presented earlier, we find that a possible implementation of the system could be

$$\begin{bmatrix} \dot{q}_1(t) \\ \dot{q}_2(t) \\ \dot{q}_3(t) \end{bmatrix} = \begin{bmatrix} -6 & -11 & -6 \\ 1 & 0 & 0 \\ 0 & 1 & 0 \end{bmatrix} \begin{bmatrix} q_1(t) \\ q_2(t) \\ q_3(t) \end{bmatrix} + \begin{bmatrix} 1 \\ 0 \\ 0 \end{bmatrix} x(t) \quad \text{and} \quad y(t) = \begin{bmatrix} 0 & 1 & 3 \end{bmatrix} \begin{bmatrix} q_1(t) \\ q_2(t) \\ q_3(t) \end{bmatrix}$$

In the foregoing equation, the matrix $\boldsymbol{A}$ of the system, which is

$$\boldsymbol{A} = \begin{bmatrix} -6 & -11 & -6 \\ 1 & 0 & 0 \\ 0 & 1 & 0 \end{bmatrix}$$

is a 3×3 dimension matrix, and therefore the dimension of the system has a value of $N = 3$. The question that arises now is the following: *Is there another way to represent the same system that is characterized by a smaller dimension value*? It is obvious that if there is such a way, it will be preferred over the previous one, since it will be characterized by *fewer state variables* and therefore it will be *easier and less costly* to implement. The answer is yes, there is such a way, since if we try to factorize the polynomial of the denominator, we will see that it develops as

$$s^3 + 6s^2 + 11s + 6 = (s + 1)(s + 2)(s + 3)$$

and therefore the transfer function will be simplified—or, in an equivalent formulation, will be reduced—since it will take the form

$$\mathcal{H}(s) = \frac{s + 3}{s^3 + 6s^2 + 11s + 6} = \frac{s + 3}{(s + 1)(s + 2)(s + 3)} = \frac{1}{(s + 1)(s + 2)} = \frac{1}{s^2 + 3s + 2}$$

Although the preceding transfer function continues to describe exactly the same system, the implementation that represents it is simpler, since now the system's matrix has an order value of $N = 2$. Indeed, if we construct the *dynamic equations* that correspond to the new *simplified* transfer function, we find that a possible implementation of this system is

$$\begin{bmatrix} \dot{q}_1(t) \\ \dot{q}_2(t) \end{bmatrix} = \begin{bmatrix} -2 & -3 \\ 1 & 0 \end{bmatrix} \begin{bmatrix} q_1(t) \\ q_2(t) \end{bmatrix} + \begin{bmatrix} 1 \\ 0 \end{bmatrix} x(t) \quad \text{and} \quad y(t) = \begin{bmatrix} 0 & 1 \end{bmatrix} \begin{bmatrix} q_1(t) \\ q_2(t) \end{bmatrix}$$

which is obviously characterized by a dimension value of $N = 2$.

Closely related to minimum dimension implementations is the following theorem:

Theorem 5.11.1 *A continuous-time LTI system described in the state space by a minimum dimension implementation is* simultaneously *controllable and observable.*

Proof Let us consider a continuous-time LTI system described by dynamic equations of the form

$$\dot{\boldsymbol{q}}(t) = \boldsymbol{A}\boldsymbol{q}(t) + \boldsymbol{B}x(t)y(t) = \boldsymbol{C}\boldsymbol{q}(t) + \boldsymbol{D}x(t)$$

and let us assume for simplicity that the eigenvalues $\lambda_1, \lambda_2, \ldots, \lambda_N$ of the matrix $\boldsymbol{A}$ are all distinct from each other. If we proceed to the diagonalization of the matrix $\boldsymbol{A}$ by choosing the appropriate similarity transform described by the matrix $\boldsymbol{U}$, this can be formulated in the form

$$\dot{\boldsymbol{p}}(t) = \boldsymbol{\Lambda}\boldsymbol{p}(t) + \widetilde{\boldsymbol{B}}x(t)y(t) = \widetilde{\boldsymbol{C}}\boldsymbol{p}(t) + \boldsymbol{D}x(t)$$

where $\boldsymbol{\Lambda} = \boldsymbol{U}^{-1}\boldsymbol{A}\boldsymbol{U}$ is a diagonal matrix that contains the eigenvalues of the matrix $\boldsymbol{A}$, while the matrices $\widetilde{\boldsymbol{B}}$ and $\widetilde{\boldsymbol{C}}$ are defined as $\widetilde{\boldsymbol{B}} = \boldsymbol{U}^{-1}\boldsymbol{B} = \{\tilde{\beta}_i\}$ and $\widetilde{\boldsymbol{C}} = \boldsymbol{C}\boldsymbol{U} = \{\tilde{\gamma}_i\}$ $(i = 1, 2, \ldots, N)$ (for simplicity, we consider the case of a SISO system with a number of inputs and outputs $K = L = 1$). Developing the state equation expressed in matrix form of the diagonalized system in the N equations from which it consists, we will get

$$\dot{p}_i(t) = \lambda_i p_i(t) + \tilde{\beta}_i x(t), \quad (i = 1, 2, \ldots, N)$$

Taking the Laplace transform of each of the foregoing equations under zero initial conditions, we will be led to the equation

$$\mathcal{L}\{\dot{p}_i(t)\} = \lambda_i \mathcal{L}\{p_i(t)\} + \tilde{\beta}_i \mathcal{L}\{x(t)\}, \quad (i = 1, 2, \ldots, N)$$

or, equivalently,

$$s\mathcal{P}_i(s) = \lambda_i \mathcal{P}_i(s) + \tilde{\beta}_i \mathcal{X}(s), \quad (i = 1, 2, \ldots, N)$$

and finally

$$\mathcal{P}_i(s) = \frac{\tilde{\beta}_i}{s - \lambda_i}\mathcal{X}(s), \quad (i = 1, 2, \ldots, N)$$

On the other hand, expressing the output of the diagonalized system in the form

$$y(t) = \tilde{\gamma}_1 p_1(t) + \tilde{\gamma}_2 p_2(t) + \cdots + \tilde{\gamma}_N p_N(t) + dx(t) = \sum_{i=1}^{N} \tilde{\gamma}_i p_i(t) + dx(t)$$

where $\boldsymbol{D} = [d]$—and taking the Laplace transform of the preceding relation, it immediately follows that

$$\mathcal{Y}(s) = \sum_{i=1}^{N} \tilde{\gamma}_i \mathcal{P}_i(s) + d\mathcal{X}(s)$$

Finally, replacing the function $\mathcal{P}_i(s)$ from the expression constructed earlier, the function $\mathcal{Y}(s)$ will take the form

$$\mathcal{Y}(s) = \sum_{i=1}^{N} \frac{\tilde{\beta}_i \tilde{\gamma}_i}{s - \lambda_i} \mathcal{X}(s) + d\mathcal{X}(s) = \left(\sum_{i=1}^{N} \frac{\tilde{\beta}_i \tilde{\gamma}_i}{s - \lambda_i} + d \right) \mathcal{X}(s)$$

Therefore, the transfer function of the diagonalized system will be given by the relation

$$\mathcal{H}(s) = \frac{\mathcal{Y}(s)}{\mathcal{X}(s)} = \sum_{i=1}^{N} \frac{\tilde{\beta}_i \tilde{\gamma}_i}{s - \lambda_i} + d$$

In the next step of the process, let us formulate the function $\mathcal{H}(s)$ using the matrices that describe the system in the state space. We will then have

$$\mathcal{H}(s) = \frac{\widetilde{\boldsymbol{C}}\,[\,\mathrm{Adj}\,(s\boldsymbol{I} - \boldsymbol{\Lambda})\,]\,\widetilde{\boldsymbol{B}}}{\det\,(s\boldsymbol{I} - \boldsymbol{\Lambda})} + \boldsymbol{D} = \frac{\widetilde{\boldsymbol{C}}\,[\,\mathrm{Adj}\,(s\boldsymbol{I} - \boldsymbol{\Lambda})]\,\widetilde{\boldsymbol{B}}}{(s - \lambda_1)(s - \lambda_2)\ldots(s - \lambda_N)} + d = \sum_{i=1}^{N} \frac{G_i}{s - \lambda_i} + d$$

where for the derivation of the last relation we developed the first term of the preceding expression into a sum of partial fractions. Now comparing these two equivalent formulations of the transfer function of the diagonalized system, it immediately follows that

$$G_i = \tilde{\beta}_i \tilde{\gamma}_i, \quad (i = 1, 2, \ldots, N)$$

Let us now assume that the implementation of the considered system is not a minimum dimension implementation and therefore there are some poles λ_k that are neutralized by zero values of the transfer function located in the very same position. It is obvious that in such a case, the expansion of the function into partial fractions will not contain the terms corresponding to these poles, and therefore $G_k = \tilde{\beta}_k \tilde{\gamma}_k = 0$ must be an equation that presupposes that $\tilde{\beta}_k = 0$ or $\tilde{\gamma}_k = 0$. However, as we found in Sect. 2.9.9 (refer to page 153), if even one $\tilde{\beta}_k$ is equal to zero, then the system is not controllable, while if even one $\tilde{\gamma}_k$ is equal to zero, then the system is not observable. We are led, therefore, to the conclusion that a minimum dimension system is both controllable and observable, a finding of great importance in the field of continuous control systems. The proof of the converse proposition can also be given; it can be found in books and papers dealing with the theory of control systems (see, for example, the paper by J.M. Davis, I.A. Gravagne, B.J. Jackson & R.J. Marks II, "Controllability, Observability, Realizability and Stability of Dynamic Linear Systems," *Electronic Journal of Differential Equations*, Vol. 2009, No. 37, pp.1–32, 2009) and it is based to the use of a minimum dimension implementation that we assume is not controllable and observable, an assumption that ultimately leads to an absurd result.

Equivalent Systems

An interesting property of the transfer function, expressed using the matrices that appear in the dynamic equations, is that it remains *invariant* under *similarity transforms*, which as we mentioned in the chapter on continuous-time systems correlate the state vector $\boldsymbol{q}(t)$ with a new state vector $\boldsymbol{p}(t)$ through a relation of the form $\boldsymbol{p}(t) = \boldsymbol{W}\boldsymbol{q}(t)$, where $\boldsymbol{W}$ is the *transformation matrix* (this description is supposed to refer to *linear* systems, so that the foregoing matrix equation is equivalent to a system of N linear equations where N is the order of the matrix $\boldsymbol{A}$). As we proved in that section, the dynamic equations of the system

$$\begin{aligned} \dot{\boldsymbol{q}}(t) &= \boldsymbol{A}\boldsymbol{q}(t) + \boldsymbol{B}\boldsymbol{x}(t) \\ \boldsymbol{y}(t) &= \boldsymbol{C}\boldsymbol{q}(t) + \boldsymbol{D}\boldsymbol{x}(t) \end{aligned}$$

are transformed into the form

$$\begin{aligned} \dot{\boldsymbol{p}}(t) &= \widetilde{\boldsymbol{A}}\boldsymbol{p}(t) + \widetilde{\boldsymbol{B}}\boldsymbol{x}(t) \\ \boldsymbol{y}(t) &= \widetilde{\boldsymbol{C}}\boldsymbol{q}(t) + \widetilde{\boldsymbol{D}}\boldsymbol{x}(t) \end{aligned}$$

with the matrices of the new representation defined as $\widetilde{A} = WA^{-1}W$, $\widetilde{B} = WB$, and $\widetilde{C} = CW^{-1}$. If we calculate the transfer function $\widetilde{H}(s)$ of the system in this representation, we find that

$$\begin{aligned}\widetilde{\mathcal{H}}(s) &= \widetilde{C}(sI - \widetilde{A})^{-1}\widetilde{B}+D = CW^{-1}(sI - WAW^{-1})^{-1}WB+DCW^{-1}(WsIW^{-1}-WAW^{-1})^{-1}WB + D\\ &= CW^{-1}[W(sI - A)W^{-1}]^{-1}WB + D = CW^{-1}[(W^{-1})^{-1}(sI - A)^{-1}W^{-1}]WB + D\\ &= C(W^{-1}W)(sI - A)(W^{-1}W)B + D = C(sI - A)^{-1}B + D = \mathcal{H}(s)\end{aligned}$$

that is, the transfer functions

$$\mathcal{H}(s) = \left[\begin{array}{c|c} A & B \\ \hline C & D \end{array}\right] \quad \text{and} \quad \widetilde{\mathcal{H}}(s) = \left[\begin{array}{c|c} WAW^{-1} & WB \\ \hline CW^{-1} & D \end{array}\right]$$

are equivalent to each other. This conclusion is very important, as it tells us that *for every transfer function $\mathcal{H}(s)$ of a system described by the matrices (A, B, C, D) there are infinite equivalent implementations in the state space* that are defined as

$$(\widetilde{A}, \widetilde{B}, \widetilde{C}, \widetilde{B}) = (WAW^{-1}, WB, CW^{-1}, D) \quad \forall W \in R^{N\times N} : \det(W) \neq 0$$

Problem 5.11.8 Consider the system described by the matrices

$$A = \begin{bmatrix} 0 & 2 \\ -1 & -3 \end{bmatrix}, \quad B = \begin{bmatrix} 0 \\ 1 \end{bmatrix}, \quad C = \begin{bmatrix} 2 & 3 \end{bmatrix}, \quad D = \begin{bmatrix} -1 \end{bmatrix}$$

and whose transfer function is equal to $\mathcal{H}(s) = \dfrac{-s^2+2}{(s+1)(s+2)}$, as we proved in a previous exercise. Considering the matrices

$$W_1 = \begin{bmatrix} 5 & 1 \\ 0 & 6 \end{bmatrix}, \quad W_2 = \begin{bmatrix} -4 & 0 \\ 8 & 6 \end{bmatrix}, \quad W_3 = \begin{bmatrix} 5 & 0 \\ 15 & 3 \end{bmatrix}$$

prove the invariance of the transfer function under similarity transformations.

Solution First, let us calculate the inverses of the transformation matrices that are proven to be given by the relations

$$W_1^{-1} = \begin{bmatrix} 1/5 & -1/30 \\ 0 & 1/6 \end{bmatrix}, \quad W_2^{-1} = \begin{bmatrix} -1/4 & 0 \\ 1/3 & 1/6 \end{bmatrix}, \quad W_3^{-1} = \begin{bmatrix} 1/5 & 0 \\ -1 & 1/3 \end{bmatrix}$$

If we denote by $\widetilde{\mathcal{H}}_1(s)$, $\widetilde{\mathcal{H}}_2(s)$, and $\widetilde{\mathcal{H}}_3(s)$ the transfer functions of the systems resulting from the application of the transforms W_1, W_2, and W_3, respectively, in our initial system, we obviously need to show that $\widetilde{\mathcal{H}}_1(s) = \widetilde{\mathcal{H}}_2(s) = \widetilde{\mathcal{H}}_3(s) = \mathcal{H}(s)$. In detail, then, we have:

- For the transform W_1: The matrices $\widetilde{A}_1$, $\widetilde{B}_1$, and $\widetilde{C}_1$ are calculated as

$$\widetilde{A}_1 = W_1AW_1^{-1} = \begin{bmatrix} -1/5 & 6/5 \\ -6/5 & -14/5 \end{bmatrix}, \quad \widetilde{B}_1 = W_1B = \begin{bmatrix} 1 \\ 6 \end{bmatrix}, \quad \widetilde{C}_1 = CW_1^{-1} = \begin{bmatrix} 2/5 & 13/30 \end{bmatrix}$$

while we will also have

$$s\boldsymbol{I}-\widetilde{\boldsymbol{A}}_1=\begin{bmatrix}\dfrac{5s+1}{5} & -\dfrac{6}{5}\\[2ex] \dfrac{6}{5} & \dfrac{5s+14}{5}\end{bmatrix} \quad \text{and} \quad (s\boldsymbol{I}-\widetilde{\boldsymbol{A}}_1)^{-1}=\frac{1}{(s+1)(s+2)}\begin{bmatrix}\dfrac{5s+14}{5} & \dfrac{6}{5}\\[2ex] -\dfrac{6}{5} & \dfrac{5s+1}{5}\end{bmatrix}$$

Therefore, the transfer function $\widetilde{\mathcal{H}}_1(s)$ corresponding to the transform $\boldsymbol{W}_1$ is

$$\widetilde{\mathcal{H}}_1(s)=\widetilde{\boldsymbol{C}}_1(s\boldsymbol{I}-\widetilde{\boldsymbol{A}}_1)^{-1}\widetilde{\boldsymbol{B}}_1+\boldsymbol{D}=\frac{1}{(s+1)(s+2)}\begin{bmatrix}\dfrac{2}{5} & \dfrac{13}{50}\end{bmatrix}\begin{bmatrix}\dfrac{5s+14}{5} & \dfrac{6}{5}\\[2ex] -\dfrac{6}{5} & \dfrac{5s+1}{5}\end{bmatrix}\begin{bmatrix}1\\6\end{bmatrix}-1$$

$$=\frac{75s+100}{25(s+1)(s+2)}-1=\frac{-s^2+2}{(s+1)(s+2)}$$

- For the transform $\boldsymbol{W}_2$: The matrices $\widetilde{\boldsymbol{A}}_2$, $\widetilde{\boldsymbol{B}}_2$, and $\widetilde{\boldsymbol{C}}_2$ are calculated as

$$\widetilde{\boldsymbol{A}}_2=\boldsymbol{W}_2\boldsymbol{A}\boldsymbol{W}_2^{-1}=\begin{bmatrix}-8/3 & -4/3\\ 5/6 & -1/3\end{bmatrix}, \qquad \widetilde{\boldsymbol{B}}_2=\boldsymbol{W}_2\boldsymbol{B}=\begin{bmatrix}0\\6\end{bmatrix}, \qquad \widetilde{\boldsymbol{C}}_2=\boldsymbol{C}\boldsymbol{W}_2^{-1}=\begin{bmatrix}1/2 & 1/2\end{bmatrix}$$

while we will still have

$$s\boldsymbol{I}-\widetilde{\boldsymbol{A}}_2=\begin{bmatrix}\dfrac{3s+8}{3} & \dfrac{4}{3}\\[2ex] -\dfrac{5}{6} & \dfrac{3s+1}{3}\end{bmatrix} \quad \text{and} \quad (s\boldsymbol{I}-\widetilde{\boldsymbol{A}}_2)^{-1}=\frac{1}{(s+1)(s+2)}\begin{bmatrix}\dfrac{3s+1}{3} & -\dfrac{4}{3}\\[2ex] \dfrac{5}{6} & \dfrac{3s+8}{3}\end{bmatrix}$$

Therefore, the transfer function $\widetilde{\mathcal{H}}_2(s)$ that corresponds to the transform $\boldsymbol{W}_2$ is

$$\widetilde{\mathcal{H}}_2(s)=\widetilde{\boldsymbol{C}}_2(s\boldsymbol{I}-\widetilde{\boldsymbol{A}}_2)^{-1}\widetilde{\boldsymbol{B}}_2+\boldsymbol{D}=\frac{1}{(s+1)(s+2)}\begin{bmatrix}\dfrac{1}{2} & \dfrac{1}{2}\end{bmatrix}\begin{bmatrix}\dfrac{3s+1}{3} & -\dfrac{4}{3}\\[2ex] \dfrac{5}{6} & \dfrac{3s+8}{3}\end{bmatrix}\begin{bmatrix}0\\6\end{bmatrix}-1$$

$$=\frac{3s+4}{(s+1)(s+2)}-1=\frac{-s^2+2}{(s+1)(s+2)}$$

- For the transform $\boldsymbol{W}_3$: The matrices $\widetilde{\boldsymbol{A}}_3$, $\widetilde{\boldsymbol{B}}_3$, and $\widetilde{\boldsymbol{C}}_3$ are calculated as

$$\widetilde{\boldsymbol{A}}_3=\boldsymbol{W}_3\boldsymbol{A}\boldsymbol{W}_3^{-1}=\begin{bmatrix}-10 & 10/3\\ -108/5 & 7\end{bmatrix}, \quad \widetilde{\boldsymbol{B}}_3=\boldsymbol{W}_3\boldsymbol{B}=\begin{bmatrix}0\\3\end{bmatrix}, \quad \widetilde{\boldsymbol{C}}_3=\boldsymbol{C}\boldsymbol{W}_3^{-1}=\begin{bmatrix}-13/5 & 1\end{bmatrix}$$

while we will still have

$$s\boldsymbol{I}-\widetilde{\boldsymbol{A}}_3=\begin{bmatrix}s+10 & -\dfrac{10}{3}\\[2ex] \dfrac{108}{5} & s-7\end{bmatrix} \quad \text{and} \quad (s\boldsymbol{I}-\widetilde{\boldsymbol{A}}_3)^{-1}=\frac{1}{(s+1)(s+2)}\begin{bmatrix}s-7 & \dfrac{10}{3}\\[2ex] -\dfrac{108}{5} & s+10\end{bmatrix}$$

Therefore, the transfer function $\widetilde{\mathcal{H}}_3(s)$ that corresponds to the transform $\boldsymbol{W}_3$ is

$$\widetilde{\mathcal{H}}_3(s) = \widetilde{\boldsymbol{C}}_3(s\boldsymbol{I} - \widetilde{\boldsymbol{A}}_3)^{-1}\widetilde{\boldsymbol{B}}_3 + \boldsymbol{D} = \frac{1}{(s+1)(s+2)}\begin{bmatrix} -\frac{13}{5} & 1 \end{bmatrix}\begin{bmatrix} s-7 & \frac{10}{3} \\ -\frac{108}{5} & s+10 \end{bmatrix}\begin{bmatrix} 0 \\ 3 \end{bmatrix} - 1 =$$

$$= \frac{15s+20}{5(s+1)(s+2)} - 1 = \frac{-s^2+2}{(s+1)(s+2)}$$

We observe, then, that $\widetilde{\mathcal{H}}_1(s) = \widetilde{\mathcal{H}}_2(s) = \widetilde{\mathcal{H}}_3(s) = \mathcal{H}(s)$, that is, the *invariance* of the transfer function under similarity transformations indeed exists. ∎

It is obvious that since the impulse response of the system is defined as the *inverse Laplace transform of the transfer function*, this property of invariance will characterize the impulse response, so that we can claim that the *impulse response of a continuous-time LTI system remains invariant under similarity transformations*. To demonstrate in practice the truth of this claim, let us consider the example of the system from the previous exercise, whose impulse response is equal to

$$h(t) = \mathcal{L}^{-1}\{\mathcal{H}(s)\} = \mathcal{L}^{-1}\left\{\frac{-s^2+2}{(s+1)(s+2)}\right\} = e^{-t}u(t) - \delta(t) + 2e^{-2t}u(t)$$

and let us check whether this remains unchanged under the application of the transform $\boldsymbol{W}_1$. We already know from theory that the impulse response as a function of the matrices of the dynamic equations is given by the relation

$$h(t) = \boldsymbol{C}e^{\boldsymbol{A}t}\boldsymbol{B} + \boldsymbol{D}\delta(t)$$

and therefore what we need to prove is the validity of the equation $h(t) = \widetilde{h}(t) = \widetilde{\boldsymbol{C}}e^{\widetilde{\boldsymbol{A}}t}\widetilde{\boldsymbol{B}} + \boldsymbol{D}\delta(t)$, where the matrices $\widetilde{\boldsymbol{A}}, \widetilde{\boldsymbol{B}}$ and $\widetilde{\boldsymbol{C}}$ are defined in the way we have presented them in the theoretical analysis. As we proved in the previous exercise, these matrices for this specific system and for this transformation are calculated as

$$\widetilde{\boldsymbol{A}}_1 = \boldsymbol{W}_1\boldsymbol{A}\boldsymbol{W}_1^{-1} = \begin{bmatrix} -1/5 & 6/5 \\ -6/5 & -14/5 \end{bmatrix}, \quad \widetilde{\boldsymbol{B}}_1 = \boldsymbol{W}_1\boldsymbol{B} = \begin{bmatrix} 1 \\ 6 \end{bmatrix}, \quad \widetilde{\boldsymbol{C}}_1 = \boldsymbol{C}\boldsymbol{W}_1^{-1} = [2/5 \;\; 13/30]$$

while we will still have

$$(s\boldsymbol{I} - \widetilde{\boldsymbol{A}}_1)^{-1} = \frac{1}{(s+1)(s+2)}\begin{bmatrix} \frac{5s+14}{5} & \frac{6}{5} \\ -\frac{6}{5} & \frac{5s+1}{5} \end{bmatrix}$$

We will therefore have

$$e^{\widetilde{A}t} = \mathcal{L}^{-1}\left\{(s\boldsymbol{I} - \widetilde{\boldsymbol{A}}_1)^{-1}\right\} = \begin{bmatrix} \frac{1}{5}\mathcal{L}^{-1}\left\{\frac{5s+14}{(s+1)(s+2)}\right\} & \frac{6}{5}\mathcal{L}^{-1}\left\{\frac{1}{(s+1)(s+2)}\right\} \\ -\frac{6}{5}\mathcal{L}^{-1}\left\{\frac{1}{(s+1)(s+2)}\right\} & \frac{1}{5}\mathcal{L}^{-1}\left\{\frac{5s+1}{(s+1)(s+2)}\right\} \end{bmatrix}$$

or, equivalently,

$$e^{\widetilde{A}t} = \frac{1}{5}\begin{bmatrix} 9e^{-t}u(t) - 4e^{-2t}u(t) & 6e^{-t}u(t) - 6e^{-2t}u(t) \\ 6e^{-2t}u(t) - 6e^{-t}u(t) & 9e^{-2t}u(t) - 4e^{-t}u(t) \end{bmatrix}$$

as can be proven by applying the method of the Laplace transform. Substituting the foregoing expressions into the defining equation of the system's impulse response expressed in its new representation, we will get

$$\widetilde{h}(t) = \widetilde{C}e^{\widetilde{A}t}\widetilde{B} + D\delta(t) = \frac{1}{5}\begin{bmatrix}2/5 & 13/30\end{bmatrix}\begin{bmatrix}9e^{-t}u(t) - 4e^{-2t}u(t) & 6e^{-t}u(t) - 6e^{-2t}u(t)\\ 6e^{-2t}u(t) - 6e^{-t}u(t) & 9e^{-2t}u(t) - 4e^{-t}u(t)\end{bmatrix}\begin{bmatrix}1\\ 6\end{bmatrix} - \delta(t)$$

$$= e^{-t}u(t) - \delta(t) + 2e^{-2t}u(t) = h(t)$$

(as emerges after operations), that is, the *property of invariance under similarity transformations applies not only to the transfer function but also to the system's impulse response*.

5.11.4 Dynamic Equations from Transfer Function

In the previous section, we proved that if we know the dynamic equations of a continuous-time LTI system—and more specifically the values of the matrices $\boldsymbol{A}$, $\boldsymbol{B}$, $\boldsymbol{C}$, and $\boldsymbol{D}$ of the system's description in the state space—we can determine the transfer function of the system from the equation

$$\mathcal{H}(s) = \boldsymbol{C}(s\boldsymbol{I} - \boldsymbol{A})^{-1}\boldsymbol{B} + \boldsymbol{D} \tag{5.76}$$

where $\boldsymbol{I}$ is the *unit matrix of order N*. In this section, we will try to solve the inverse problem, which is formulated as follows. *If we know the transfer function of a system, can we determine the dynamic equations that correspond to it*? The answer to this question is yes; if we use the property of the invariance of the transfer function under similarity transformations, we can formulate the claim that there are infinitely many equivalent implementations of the system that correspond to a specific transfer function, with the two most important of these implementations being known as the *controllability scheme* and the *observability scheme* (for a detailed description of the characteristics of controllability and observability of continuous LTI systems see Sect. 2.9.9 on page 153).

Let us assume, then, that for given values of the matrices $\boldsymbol{A}$, $\boldsymbol{B}$, $\boldsymbol{C}$, and $\boldsymbol{D}$ of the considered system, the transfer function that resulted from the application of the foregoing relationship has the general rational form

$$\mathcal{H}(s) = \frac{\mathcal{Y}(s)}{\mathcal{X}(s)} = \frac{\beta_{N-1}s^{N-1} + \beta_{N-2}s^{N-2} + \cdots + \beta_1 s + \beta_0}{s^N + \alpha_{N-1}s^{N-1} + \alpha_{N-2}s^{N-2} + \cdots + \alpha_1 s + \alpha_0}$$

where for simplicity we assumed that $\alpha_N = 1$, we set $M = N$, and we considered that the degree of the polynomial of the numerator of the foregoing rational function is one unit less than the degree of the polynomial of the denominator for the reasons we explained in the previous section. If we now use the properties of the interconnection of continuous-time LTI systems, we can formulate the claim that the considered system can result from the series interconnection of two simpler systems of the form $z(t) = \mathcal{T}_1\{x(t)\}$ and $y(t) = \mathcal{T}_2\{z(t)\} = \mathcal{T}_2\{\mathcal{T}_1\{x(t)\}\}$, with transfer functions

$$\mathcal{H}_1(s) = \frac{\mathcal{Z}(s)}{\mathcal{X}(s)} = \frac{1}{s^N + \alpha_{N-1}s^{N-1} + \alpha_{N-2}s^{N-2} + \cdots + \alpha_1 s + \alpha_0} \quad \text{and}$$

$$\mathcal{H}_2(s) = \frac{\mathcal{Y}(s)}{\mathcal{Z}(s)} = \beta_{N-1}s^{N-1} + \beta_{N-2}s^{N-2} + \cdots + \beta_1 s + \beta_0$$

In this case, the transfer function of the composite system will result from the relationship

$$\mathcal{H}(s) = \mathcal{H}_1(s)\mathcal{H}_2(s) = \frac{\mathcal{Y}(s)}{\mathcal{X}(s)} = \frac{\beta_{N-1}s^{N-1} + \beta_{N-2}s^{N-2} + \cdots + \beta_1 s + \beta_0}{s^N + \alpha_{N-1}s^{N-1} + \alpha_{N-2}s^{N-2} + \cdots + \alpha_1 s + \alpha_0}$$

which is simply the equation we formulated previously. If we now construct the differential equations that describe the preceding subsystems in the way we described in the previous sections, it immediately follows that

$$\frac{d^N z(t)}{dt^N} + \alpha_{N-1}\frac{d^{N-1}z(t)}{dt^{N-1}} + \cdots + \alpha_1 \frac{dz(t)}{dt} + \alpha_0 z(t) = x(t) \quad \text{and}$$

$$y(t) = \beta_{N-1}\frac{d^{N-1}z(t)}{dt^{N-1}} + \beta_{N-2}\frac{d^{N-2}z(t)}{dt^{N-2}} + \cdots + \beta_1 \frac{dz(t)}{dt} + \beta_0 z(t)$$

To construct the controllability scheme (which, as we will find, corresponds to the *normal form of the controller*) from the above transfer function of the system, we define the N number of components of the state vector from the relations

$$\begin{aligned}
q_1(t) &= z(t)\\
q_2(t) &= \frac{dq_1(t)}{dt} = \frac{dz(t)}{dt}\\
q_3(t) &= \frac{dq_2(t)}{dt} = \frac{d^2z(t)}{dt^2}\\
&\cdots\cdots\cdots\cdots\cdots\cdots\\
q_N(t) &= \frac{dq_{N-1}(t)}{dt} = \frac{d^{N-1}z(t)}{dt^{N-1}}
\end{aligned}$$

Differentiating the last relation and replacing the derivative $z^{(N)}(t)$ from the first of the two foregoing differential equations, it follows that

$$\begin{aligned}
\frac{dq_N(t)}{dt} = \frac{d^N z(t)}{dt^N} &= x(t) - \alpha_{N-1}\frac{d^{N-1}z(t)}{dt^{N-1}} - \cdots - \alpha_1\frac{dz(t)}{dt} - \alpha_0 z(t) =\\
&= x(t) - \alpha_{N-1}q_N(t) - \alpha_{N-2}q_{N-1}(t) - \cdots - \alpha_1 q_2(t) - \alpha_0 q_1(t)
\end{aligned}$$

We end up, therefore, with the system of differential equations

$$\begin{aligned}
\dot{q}_1(t) &= q_2(t)\\
\dot{q}_2(t) &= q_3(t)\\
&\vdots\\
\dot{q}_{N-1}(t) &= q_N(t)\\
\dot{q}_N(t) &= -\alpha_0 q_1(t) - \alpha_1 q_2(t) - \cdots - \alpha_{N-1}q_N(t) + x(t)
\end{aligned}$$

which can be expressed by a matrix equation of the form

$$\begin{bmatrix} \dot{q}_1(t)\\ \dot{q}_2(t)\\ \dot{q}_3(t)\\ \vdots\\ \dot{q}_N(t) \end{bmatrix} = \begin{bmatrix} 0 & 1 & 0 & \dots & 0\\ 0 & 0 & 1 & \dots & 0\\ 0 & 0 & 0 & \dots & 0\\ \vdots & \vdots & \vdots & \ddots & \vdots\\ -\alpha_0 & -\alpha_1 & -\alpha_2 & \dots & -\alpha_{N-1} \end{bmatrix} \begin{bmatrix} q_1(t)\\ q_2(t)\\ q_3(t)\\ \vdots\\ q_N(t) \end{bmatrix} + \begin{bmatrix} 0\\ 0\\ 0\\ \vdots\\ 1 \end{bmatrix} x(t)$$

On the other hand, replacing the preceding state variables in the second of the foregoing differential equations, we get

$$y(t) = \beta_{N-1}q_N(t) + \beta_{N-2}q_{N-1}(t) + \cdots + \beta_1 q_2(t) + \beta_0 q_1(t)$$

or in matrix form

$$y(t)=\begin{bmatrix}\beta_0\ \beta_1\ \dots\ \beta_{N-2}\ \beta_{N-1}\end{bmatrix}\begin{bmatrix}q_1(t)\\ q_2(t)\\ \vdots\\ q_{N-1}(t)\\ q_N(t)\end{bmatrix}$$

The last two matrix equations constitute the so-called *controllability form* and are recognized as the *state equation and the output equation of the controller's normal form* (see also Theorem 2.9.2 in the chapter on s).

On the other hand, for the construction of the *observability scheme* we will use the *differential equation* of the system that arises from our initial function $\mathcal{H}(s)$ as presented in a previous section and which for educational reasons we repeat here. Reformulating the defining equation of the function $\mathcal{H}(s)$ in the form

$$s^n\mathcal{Y}(s)+\alpha_{N-1}s^{N-1}\mathcal{Y}(s)+\cdots+\alpha_1 s\mathcal{Y}(s)+\alpha_0\mathcal{Y}(s)=$$
$$\beta_{N-1}s^{N-1}\mathcal{X}(s)+\beta_{N-2}s^{N-2}\mathcal{X}(s)+\cdots+\beta_1 s\mathcal{X}(s)+\beta_0\mathcal{X}(s)$$

and taking the inverse Laplace transform, we arrive at the expression

$$\mathcal{L}^{-1}\{s^n\mathcal{Y}(s)\}+\alpha_{N-1}\mathcal{L}^{-1}\{s^{N-1}\mathcal{Y}(s)\}+\cdots++\alpha_0\mathcal{L}^{-1}\{\mathcal{Y}(s)\}=$$
$$\beta_{N-1}\mathcal{L}^{-1}\{s^{N-1}\mathcal{X}(s)\}+\beta_{N-2}\mathcal{L}^{-1}\{s^{N-2}\mathcal{X}(s)\}+\cdots+\beta_0\mathcal{L}^{-1}\{\mathcal{X}(s)\}$$

Therefore, from the property of differentiation expressed in the form[22] $\mathcal{L}^{-1}\{s^i\mathcal{Y}(s)\}=y^{(i)}(t)$ $(i=1,2,\dots,N)$, we end up with the differential equation of the system, which is

$$\frac{d^N y(t)}{dt^N}+\alpha_{N-1}\frac{d^{N-1}y(t)}{dt^{N-1}}+\cdots+\alpha_1\frac{dy(t)}{dt}+\alpha_0 y(t)=\beta_{N-1}\frac{d^{N-1}x(t)}{dt^{N-1}}+\cdots+\beta_1\frac{dx(t)}{dt}+\beta_0 x(t)$$

In this case, the state variables of the system should be chosen as

$$q_{N-1}(t)=\frac{dq_N(t)}{dt}+\alpha_{N-1}q_N(t)-\beta_{N-1}x(t)$$
$$q_{N-2}(t)=\frac{dq_{N-1}(t)}{dt}+\alpha_{N-2}q_N(t)-\beta_{N-2}x(t)$$
$$\dots\dots\dots\dots\dots\dots\dots\dots\dots\dots$$
$$q_2(t)=\frac{dq_3(t)}{dt}+\alpha_2 q_N(t)-\beta_2 x(t)$$
$$q_1(t)=\frac{dq_2(t)}{dt}+\alpha_1 q_N(t)-\beta_1 x(t)$$

Combining the foregoing relations, the last equation is written

$$q_1(t)=\frac{d^{N-1}y(t)}{dt^{N-1}}+\alpha_{N-1}\frac{d^{N-2}y(t)}{dt^{N-2}}+\cdots+\alpha_1 y(t)-\beta_{N-1}\frac{d^{N-2}x(t)}{dt^{N-2}}-\beta_{N-2}\frac{d^{N-3}x(t)}{dt^{N-3}}-\cdots-\beta_1 x(t)$$

which, after differentiation and using the differential equation we constructed previously, is reduced to the form

[22] The following relationship holds provided that the system is in a state of rest, so that $y^{(i)}(0)=0$ $(i=1,2,\dots,N)$.

$$\frac{dq_1(t)}{dt} = \frac{d^N y(t)}{dt^N} + \alpha_{N-1}\frac{d^{N-1}y(t)}{dt^{N-1}} + \cdots + \alpha_1\frac{dy(t)}{dt}$$
$$- \beta_{N-1}\frac{d^{N-1}x(t)}{dt^{N-1}} - \cdots - \beta_1\frac{dx(t)}{dt} = \beta_0 x(t) - \alpha_0 y(t) = \beta_0 x(t) - \alpha_0 q_N(t)$$

We end up, therefore, with the system of equations

$$\dot{q}_1(t) = \beta_0 x(t) - \alpha_0 q_N(t)$$
$$\dot{q}_2(t) = q_1(t) - \alpha_1 q_N(t) + \beta_1 x(t)$$
$$\dot{q}_3(t) = q_2(t) - \alpha_2 q_N(t) + \beta_2 x(t)$$
$$\cdots\cdots\cdots\cdots\cdots\cdots\cdots\cdots\cdots\cdots$$
$$\dot{q}_N(t) = q_{N-1}(t) - \alpha_{N-1} q_N(t) + \beta_{N-1} x(t)$$

which in matrix form is expressed as

$$\begin{bmatrix} \dot{q}_1(t) \\ \dot{q}_2(t) \\ \dot{q}_3(t) \\ \vdots \\ \dot{q}_N(t) \end{bmatrix} = \begin{bmatrix} 0\,0\,0 \ldots & -\alpha_0 \\ 1\,0\,0 \ldots & -\alpha_1 \\ 0\,1\,0 \ldots & -\alpha_2 \\ \vdots\,\vdots\,\vdots\,\ddots & \vdots \\ 0\,0\,0 \ldots & -\alpha_{N-1} \end{bmatrix} \begin{bmatrix} q_1(t) \\ q_2(t) \\ q_3(t) \\ \vdots \\ q_N(t) \end{bmatrix} + \begin{bmatrix} \beta_0 \\ \beta_1 \\ \beta_2 \\ \vdots \\ \beta_{N-1} \end{bmatrix} x(t)$$

On the other hand, the system's output equation is formulated as

$$y(t) = q_N(t) = \begin{bmatrix} 0\,0 \ldots 0\,0 \end{bmatrix} \begin{bmatrix} q_1(t) \\ q_2(t) \\ \vdots \\ q_{N-1}(t) \\ q_N(t) \end{bmatrix}$$

The last two equations constitute the so-called *observability scheme* and are recognized as the *state equation and the output equation of the observer's normal form* (see also Theorem 2.9.4 in the chapter on continuous-time systems).

5.12 Problems

1. Determine the differential equation of the systems described by the transfer functions

$$\mathcal{H}_1(s) = \frac{s+2}{s^2+3s+3}, \qquad \mathcal{H}_2(s) = \frac{s^2+5s+6}{s^3+5s^2+3s+4}, \qquad \mathcal{H}_3(s) = \frac{5s+3}{s^3+3s^2+s-2}$$

2. Determine the transfer function and the impulse response of the continuous-time systems described by the differential equations

$$\frac{dy(t)}{dt} + \alpha y(t) = \frac{dx(t)}{dt} - 2x(t) \;\; (\alpha > 0)$$

$$\frac{d^2y(t)}{dt^2} + 3\frac{dy(t)}{dt} + 2y(t) = x(t)$$

$$\frac{d^2y(t)}{dt^2} + 3\frac{dy(t)}{dt} + y(t) = \frac{dx(t)}{dt}$$

and verify that the result obtained is the same as what we obtain by applying the techniques described in the chapter on continuous-time systems.

3. Determine the transfer function of a system with impulse response $h(t) = e^{-3t}u(t)$, and calculate its output when the input signal is $x(t) = \cos(4t)$.
4. Determine the poles and zeros of the system

$$\frac{d^2y(t)}{dt^2} + \omega\sqrt{2}\frac{dy(t)}{dt} + \omega^2 y(t) = \frac{d^2x(t)}{dt^2} + \omega\sqrt{2}\frac{dx(t)}{dt}$$

5. Prove that the transfer function of a mass-spring system with damping is

$$\mathcal{H}(s) = \frac{\mathcal{X}(s)}{\mathcal{F}(s)} = \frac{1}{ms^2 + \beta s + k}$$

(for a description of the system see Example 3 of Sect. 2.9 of Chap. 2, page 127).

6. Consider a continuous-time LTI system described by the differential equation

$$\frac{d^2y(t)}{dt^2} + 4\frac{dy(t)}{dt} + 4y(t) = \frac{dx(t)}{dt} + x(t)$$

with initial conditions $y(0^-) = 1$ and $\dot{y}(0^-) = -1$ to which we transmit the input signal $x(t) = e^{-3t}u(t)$.

- Using the Laplace transform method, prove that the system's response to the zero state and the zero input is given respectively by the relations

$$y_{zs}(t) = e^{-2t}(2 - t)u(t) - 2e^{-3t}u(t) \qquad \text{and} \qquad y_{zi}(t) = e^{-2t}(1 + t)u(t)$$

Then calculate the total response of the system $y(t)$, and verify that it is equal to the sum $y_{zs}(t) + y_{zi}(t)$.

- Determine the zeros and poles of the transfer functions

$$\mathcal{H}_1(s) = \frac{s+2}{s^2+3s+3}, \qquad \mathcal{H}_2(s) = \frac{s^2+5s+6}{s^3+4s^2+6s+4}, \qquad \mathcal{H}_3(s) = \frac{5s+3}{s^3+4s^2+13s}$$

and then characterize the corresponding system in terms of the zero-input response.

- Use the initial and final value theorem to prove that $y(0^+) = 1$ and $y(\infty) = 0$.
- Determine the transfer function $\mathcal{H}(s)$ and the impulse response of the system $h(t)$.

7. Prove that the Laplace transform of the zero-state and zero-input responses of a system described by the differential equation

$$\frac{d^2y(t)}{dt^2} + 3\frac{dy(t)}{dt} + 2y(t) = \frac{dx(t)}{dt} + 3x(t)$$

is given by the relations

$$\mathcal{Y}_{zs}(s) = \frac{s+3}{s^3+3s+2}X(s) \qquad \text{and} \qquad \mathcal{Y}_{zi}(s) = \frac{(s+3)y(0)+y'(0)}{s^2+3s+2}$$

where $y(0)$ and $y'(0)$ are the initial conditions of the problem. If these conditions have values $y(0) = y'(0) = 1$, find the output of the system $y(t)$, when the input signal is the step function $u(t)$.

8. Determine the output $y(t)$ of a LTI system with impulse response $h(t) = e^{-2t}u(t)$ for the inputs $x_1(t) = e^{-t}u(t)$ and $x_2(t) = e^{-t}u(-t)$.
9. The step response of a LTI system is given by the relation $y_1(t) = (1 - e^{-t})u(t)$. Additionally, it is known that the response of the system to some unknown continuous input signal $x(t)$ is $y_2(t) = (2 - 3e^{-t} + e^{-3t})u(t)$. Based on this information, determine the unknown input $x(t)$.
10. From basic theory it is known that the autocorrelation function of a continuous-time signal $x(t)$ is given by the equation

$$\varphi_{xx}(\tau) = \int_{-\infty}^{+\infty} x(t)\, x(t+\tau)\, dt$$

- Determine the impulse response $h(t)$ of a LTI system whose response to the input signal $x(t)$ is the function $\varphi_{xx}(\tau)$.
- Determine the Laplace transform $\mathcal{L}\{\varphi_{xx}(\tau)\}$ of the autocorrelation function $\varphi_{xx}(\tau)$.

11. Characterize in terms of causality and stability systems with the transfer function

$$\mathcal{H}_1(s) = \frac{s-1}{(s+2)(s^2+2s+2)}, \quad \Re(s) > 1 \quad \text{and} \quad \mathcal{H}_2(s) = \frac{s+2}{(s-1)(s^2-2s+2)}, \quad \Re(s) < -0.5$$

12. Using the Routh method, characterize in terms of stability the polynomials

$$P_1(s) = s^5 + 7s^4 + 12s^3 + 6s^2 + 3s + 7$$

$$P_2(s) = 7s^7 + 6s^6 + s^5 + s^4 + 2s^3 + s^2 + 7s + 12$$

$$P_3(s) = s^4 + 12s^2 + 3$$

$$P_4(s) = 3s^6 + s^5 + 2s^3 + s^2 + 5s + 1$$

13. Considering a stable and causal continuous LTI system with impulse response $h(t)$ and rational transfer function $\mathcal{H}(s)$, prove that the function $g(s) = \Re[h(t)]$ also constitutes the impulse response of such a system.
14. Let us consider a continuous LTI system described by the equation

$$\frac{d^2y(t)}{dt^2} - \frac{dy(t)}{dt} - 2y(t) = x(t)$$

- Determine the transfer function $\mathcal{H}(s)$ as the ratio of two polynomial functions of s, and draw the pole-zero plot.
- Determine the impulse response of the system $h(t)$ so that (a) the system is causal, (b) the system is stable, and (c) the system is neither stable nor causal.

15. For each of the systems described by the transfer function

$$\mathcal{H}_1(s) = \frac{2s+3}{s^2+2s+5}, \quad \mathcal{H}_2(s) = \frac{2s}{s^2+4}, \quad \mathcal{H}_3(s) = \frac{s+2}{s^2+5s+6}$$

determine the step response and verify that the steady-state response determined by the use of the Laplace transform method is the same as that resulting from the application of the initial value theorem.
16. For each of the systems described by the differential equations

$$\frac{d^2y(t)}{dt^2} + 4\frac{dy(t)}{dt} + 20y(t) = \frac{d^2x(t)}{dt^2} - \frac{dx(t)}{dt} + 2x(t) \quad \text{and} \quad \frac{1}{4}\frac{d^2y(t)}{dt^2} + \frac{1}{\sqrt{2}}\frac{dy(t)}{dt} + y(t) = \frac{dx(t)}{dt} + x(t)$$

determine the step response and verify that the steady-state response determined by the use of the Laplace transform method is the same as that which arises from the application of the initial value theorem.
17. Determine the step response of a system described by the differential equation $y'(t) + y(t) = x(t)$.

18. Determine the impulse response of a system described by the differential equation $5y''(t) + 20y'(t) + y(t) = x(t)$.
19. Determine the impulse and step response of a system described by the differential equation $y''(t)+4y'(t)+4y(t) = x(t)$.
20. Determine the impulse and step response of a system that is described by the differential equation $y''(t)-5y'(t)+7y(t) = 2x(t)$.
21. Determine the step response of a system that is described by the differential equation $y'(t) + 5y(t) = 2x(t)$, and then calculate the value $s(\infty)$.
22. For each of the systems described by the transfer functions

$$\mathcal{H}_1(s) = \frac{s+5}{s^2+2s}, \quad \mathcal{H}_2(s) = \frac{s^2+1}{s^3+3s^2+5}, \quad \mathcal{H}_3(s) = \frac{s(s+2)}{(s+1)(s+3)(s+4)}$$

determine and plot the *amplitude and phase responses* and then calculate the system's output for each of the trigonometric input signals

$$x(t) = 5\cos(17t - 27°), \quad x(t) = \sin(t - 20°), \quad x(t) = 17\cos(23t + 17°)$$

23. Considering a stable continuous-time LTI system described by the transfer function

$$\mathcal{H}(s) = \frac{s+4}{(s+3)^2}$$

determine the steady-state response for each of the input signals

$$x(t) = 15u(t), \quad x(t) = \cos(2t + 60°)u(t), \quad x(t) = e^{i5t}u(t)$$

24. Calculate the output of a system described by the differential equation $y''(t) + 3y(t) = x(t)$ when the input signal $x(t) = 5\sin(4t)$ is transmitted.
25. Calculate the output of a system described by the differential equation $y'''(t) + 5y''(t) + 6y'(t) + y(t) = x(t)$ when the input signal $x(t) = \cos(2t)$ is transmitted.
26. Let us consider a LTI system described by the equation

$$\frac{d^3y(t)}{dt^3} + 9\frac{d^2y(t)}{dt^2} + 5y(t) = x(t)$$

with initial conditions $y(0) = y''(0) = -1$ and $y'(0) = 1$. (a) Determine the zero-input response and the zero-state response when the input signal $x(t) = e^{-5t}u(t)$ is transmitted. (b) Determine and plot the amplitude and phase responses of the system[23] and to calculate its output, when the elementary trigonometric signal $x(t) = 17\sin(35t - 45°)$ is used as the input.

[23] For the design of the amplitude and phase responses, you can use the function *freqs* of *MATLAB*, which is called

freqs (B,A,W).

where B and A are the vectors of the coefficients of the polynomials of the numerator and the denominator of the transfer function and W is the frequency vector, for each of which we want to calculate and design the aforementioned functions. Note that vectors B and A should have the *same number of elements*—this means that if some power of s is missing from the numerator or the denominator, the corresponding term in the appropriate vector will be set to zero—while the coefficients of the polynomials will correspond to continuously decreasing powers of s. If, for example, we want to design the amplitude and phase response of a system described by the transfer function

$$\mathcal{H}(s) = \frac{s^2 - 2s + 5}{s^3 + 3s^2 + 4s + 6}$$

vectors B and A will be defined as $B = [\ 0\ 1\ \text{-}2\ 5\ ]$ and $A = [\ 1\ 3\ 4\ 6\ ]$, while if we want the design of the aforementioned responses for the frequency range $[0, 2\pi]$ with a step change of 0.01, we should define vector W as $W = [\ 0 : 0.01 : 2 * \text{pi}]$. After defining the preceding vectors, we can design the two responses by calling the function freqs as mentioned earlier.

27. Considering the frequency response functions

$$\mathcal{H}(j\omega) = \frac{1}{j\omega + 1} \quad \text{and} \quad \mathcal{H}(j\omega) = \frac{1}{(j\omega + 1)(j\omega + 2)}$$

prove that these are characterized by a nonlinear phase by calculating the value of the group delay at the positions of two selected frequencies.

28. Use the geometric method to calculate the value of the transfer function

$$\mathcal{H}(s) = \frac{s^2 - 7s + 12}{s^4 + 6s^3 + 16s^2 + 26s + 5}$$

at the positions (a) $s_1 = 3 - 2j$, (b) $s_2 = -4 + 7j$, (c) $s_3 = 1 - j$, and (d) $s_4 = 7 + 2j$, and then verify the correctness of the results by substituting the values s_i $(i = 1, 2, 3, 4)$ into the defining equation of $\mathcal{H}(s)$.

29. Use the geometric method to calculate the value of the transfer functions

$$\mathcal{H}_1(s) = \frac{s - 1}{(s + 2)(s^2 + 2s + 2)} \quad \text{and} \quad \mathcal{H}_2(s) = \frac{s + 2}{(s - 1)(s^2 - 2s + 2)}$$

at the positions $s_1 = -6 + 3i$, $s_2 = 4 - 6i$, $s_3 = -2 - i$, and $s_4 = 2 + 5i$, and then verify the correctness of the results by replacing the values s_i $(i = 1, 2, 3, 4)$ in the defining equations of $\mathcal{H}_1(s)$ and $\mathcal{H}_2(s)$.

30. Use the geometric method to calculate the value of the transfer functions

$$\mathcal{H}_1(s) = \frac{s - 1}{(s + 2)(s^2 + 2s + 2)} \quad \text{and} \quad \mathcal{H}_2(s) = \frac{s + 2}{(s - 1)(s^2 - 2s + 2)}$$

at the position $s = j\omega$, and then verify the correctness of the results by substituting the respective value of s into the defining equations of the functions $\mathcal{H}_1(s)$ and $\mathcal{H}_2(s)$.

31. Construct Bode diagrams of the transfer functions

$$\mathcal{H}_1(s) = \frac{(s + 1)(s + 5)}{(s + 2)(s + 8)}$$

$$\mathcal{H}_2(s) = \frac{s(s + 15)}{s^2(s + 3)}$$

$$\mathcal{H}_3(s) = \frac{(s + 2)(s + 3)}{(s + 4)^2(s + 6)}$$

32. Construct Bode diagrams of the transfer functions

$$\mathcal{H}_1(s) = \frac{s^2}{(s + 1)(s^2 + 4s + 16)}$$

$$\mathcal{H}_2(s) = \frac{s + 6}{s^2 + 7s + 12}$$

33. Construct Bode diagrams of the transfer functions

$$\mathcal{H}_1(s) = \frac{100(s + 30)}{(s + 3)(s + 10)^2}$$

$$\mathcal{H}_2(s) = \frac{s + 10}{(s + 5)(s + 1)^2}$$

$$\mathcal{H}_3(s) = \frac{s^3}{(s + 20)^2}$$

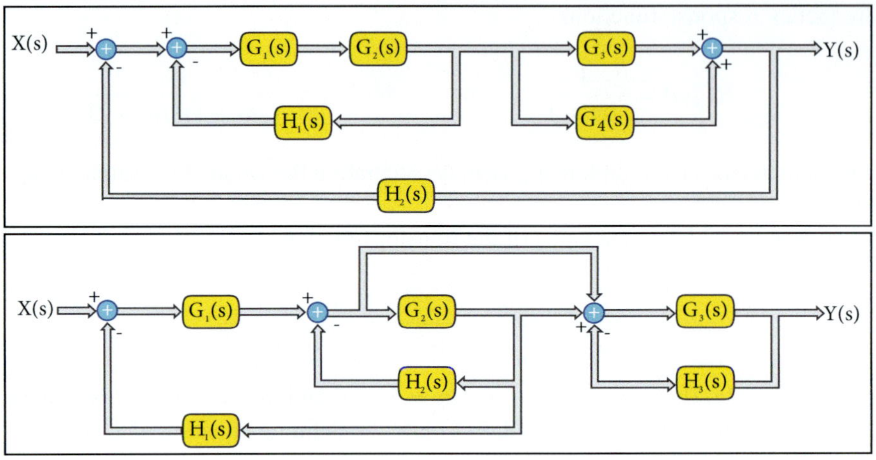

Fig. 5.43 Schematic diagrams of Problem 37

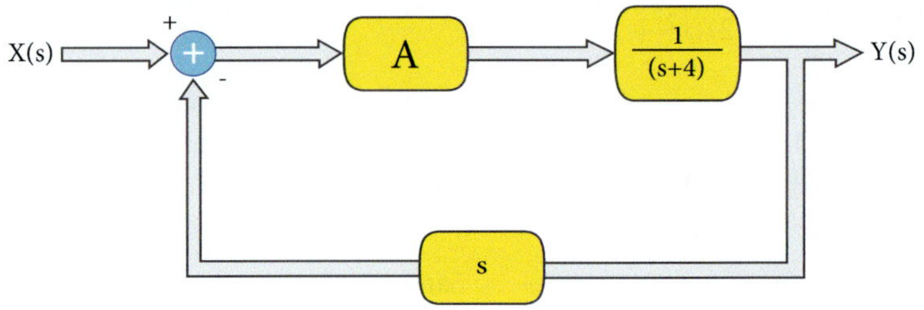

Fig. 5.44 Schematic diagram of Problem 38

34. Construct Bode diagrams of the transfer functions

$$\mathcal{H}_1(s) = \frac{100}{\left(1+\frac{s}{100}\right)^2}, \qquad \mathcal{H}_2(s) = \frac{s^2+2s+100}{100}, \qquad \mathcal{H}_3(s) = (100+0.1s)^2$$

35. Repeat Exercise 34 for systems with transfer functions

$$\mathcal{H}_1(s) = \frac{2s+1}{s^2+3}, \qquad \mathcal{H}_2(s) = \frac{s-4}{s^2-2s+2}, \qquad \mathcal{H}_3(s) = \frac{s^2}{2s^2+3s+1}$$

36. Calculate the transfer function of an equivalent circuit of Fig. 5.29 (refer to page 451) for the transfer functions

$$\mathcal{H}_1(s) = \frac{1}{s+3}, \quad \mathcal{H}_2(s) = \frac{1}{s}, \quad \mathcal{G}_1(s) = \frac{5s+2}{s^2+5s+5}, \quad \mathcal{G}_2(s) = G_3(s) = \frac{1}{s+5}$$

37. For each of the schematic diagrams of Fig. 5.43, proceed to its simplification and the determination of the transfer function of an equivalent circuit.
38. Determine the unknown transfer function A of the schematic diagram of Fig. 5.44.
39. For each of the schematic diagrams presented in this chapter, construct an equivalent signal flow graph and determine its transfer function using the Mason rule.
40. Using the Laplace transform method, prove that

$$e^{\mathbf{A}t} = \begin{bmatrix} -2e^{-2t}u(t)+3e^{-3t}u(t) & -6e^{-2t}u(t)+6e^{-3t}u(t) \\ e^{-2t}u(t)-e^{-3t}u(t) & 3e^{-2t}u(t)-2e^{-3t}u(t) \end{bmatrix} \quad \text{where} \quad \mathbf{A} = \begin{bmatrix} -5 & -6 \\ 1 & 0 \end{bmatrix}$$

41. Using the Laplace transform method, prove that

$$e^{At} = \begin{bmatrix} e^{-2t}u(t) & 0 \\ e^{-t}u(t) - e^{-2t}u(t) & e^{-t}u(t) \end{bmatrix} \quad \text{where} \quad A = \begin{bmatrix} -2 & 0 \\ 1 & -1 \end{bmatrix}$$

42. Using the Laplace transform method, prove that

$$e^{At} = \begin{bmatrix} e^{-t}u(t) & e^{-t}u(t) - e^{-2t}u(t) \\ 0 & e^{-2t}u(t) \end{bmatrix} \quad \text{where} \quad A = \begin{bmatrix} -1 & 1 \\ 0 & -2 \end{bmatrix}$$

43. Consider the system described by the matrices

$$\boldsymbol{A} = \begin{bmatrix} -1 & 1 \\ 0 & -2 \end{bmatrix}, \quad \boldsymbol{B} = \begin{bmatrix} 1 & 1 \\ 0 & 1 \end{bmatrix}, \quad \boldsymbol{C} = \begin{bmatrix} 5 & -6 \\ 2 & 4 \end{bmatrix}, \quad \boldsymbol{D} = \begin{bmatrix} 1 & -1 \\ 0 & 1 \end{bmatrix}$$

and whose transfer function is given by the relation

$$\boldsymbol{\mathcal{H}}(s) = \boldsymbol{C}\boldsymbol{\Phi}(s)\boldsymbol{B} + \boldsymbol{D} = \begin{bmatrix} \dfrac{s+6}{s+1} & \dfrac{-s^2 - 4s + 7}{(s+1)(s+2)} \\[2ex] \dfrac{2}{s+1} & \dfrac{s^2 + 9s + 12}{(s+1)(s+2)} \end{bmatrix}$$

Choose three arbitrary transformation matrices $\boldsymbol{W}_1$, $\boldsymbol{W}_2$, and $\boldsymbol{W}_3$ of dimensions 2×2 with nonzero determinant—so that the inverse matrix exists—and then show the invariance of the transfer function under their application.

44. For each of the systems described by the transfer functions

$$\mathcal{H}_1(s) = \frac{s^3 - 7s^2 + 8s + 12}{s^4 + 6s^3 - 16s^2 + 3s + 5} \quad \text{and} \quad \mathcal{H}_2(s) = \frac{s^2 + 5s + 6}{s^3 + 4s^2 + 6s + 4}$$

construct the dynamic equations of the state space that correspond to the normal form of the controller and the observer.

45. For each of the systems described by the transfer functions

$$\mathcal{H}_1(s) = \frac{s^2 - 17s + 10}{(s^3 + s^2 + s + 1)(s - 5)} \quad \text{and} \quad \mathcal{H}_2(s) = \frac{12s^2 + 15s + 3}{10s^3 - 7s^2 + 16s + 1}$$

construct state space dynamic equations that correspond to the normal form of the controller and the observer.

Discrete-Time Signals

6

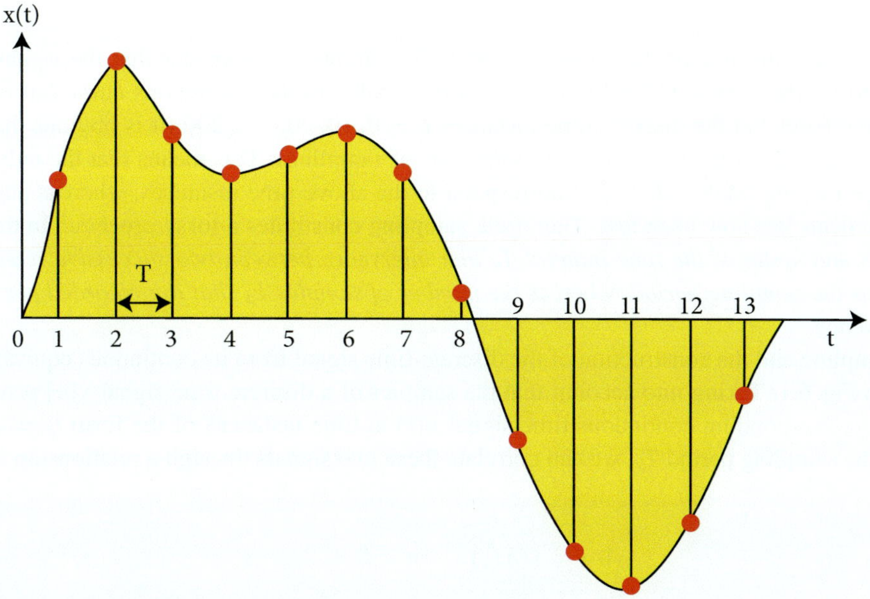

Discrete-time signals arise from continuous-time signals through the application of a process known as sampling

6.1 Introduction to Discrete-Time Signals

After the analytical study of signals and the systems that process them in the field of *continuous-time*, let us now turn our attention to the corresponding concepts as they are defined in the field of discrete-time. Starting our description from the so-called *discrete-time signals*, or *discrete-time signals*, these, in full analogy with their continuous equivalent, are defined as quantities that vary in function with time, which, however, is no longer *continuous* but *discrete*. This means that *the value of the signal is no longer defined at any time* but only at specific time instances, which are not necessarily equidistant from each other. From the mathematical side of the subject, a discrete-time signal is described by a function of the form $x[n]$ in which the independent variable n—which obviously expresses the *discrete-time*—takes only integer values in the interval $(-\infty, +\infty)$.

The creation of a discrete-time signal from its continuous equivalent is carried out by applying a process known as *sampling* and consists of *recording the amplitude of the continuous-time signal at discrete-time instances*. To simplify our description, let us consider the case of *uniform sampling*, during which the time instances of recording the signal samples are equidistant from each other. Let us consider, then, a device recording the values of some physical quantity and let us assume that at regular time intervals we record the value of the voltage appearing at some specific part of the circuit. Assuming that this recording process takes place every 10 ms, then after 200 ms the result would be what is included in Table 6.1.

A. I. Margaris, *Continuous and Discrete-Time Signals and Systems*,
https://doi.org/10.1007/978-3-031-81209-5_6

Table 6.1 Sampling of an analog signal

Sampling time instance (ms)	Signal amplitude (volts)	Sampling time instance (ms)	Signal amplitude (volts)
000	+ 0.437	110	− 0.098
010	− 0.430	120	− 0.040
020	+ 1.235	130	+ 0.115
030	− 0.495	140	+ 0.217
040	− 0.402	150	− 0.501
050	+ 0.645	160	+ 0.998
060	− 1.003	170	+ 0.017
070	+ 1.017	180	− 0.195
080	+ 0.112	190	+ 0.603
090	− 0.631	200	+ 0.342

What is the physical significance of the above process? Very simply, we consider that the signal we are dealing with is no longer described by the values of the infinite points that make up the continuous curve but only by the amplitude values that have been recorded at the discrete-time instances $t = 0, 10, 20, \ldots, 200$ (it is obvious that as the beginning of time measurement we considered the time moment of the start of sampling). This means that the only information we now have for the signal are the amplitude values that correspond to the above time instances, whereas anything concerning the intermediate time moments has now been lost. Therefore, sampling constitutes a lossy process.[1] In the terminology related to the field of signals and systems, *the time interval T_s that intervenes between two successive recordings of the signal's amplitude* is known as the *sampling period*, whereas *the number of samples F_s that are recorded per unit of time* is known as the *sampling frequency*.

The process of sampling and the construction of the discrete-time signal from its continuous equivalent are presented in a diagrammatic way in Fig. 6.1. Taking into account that the samples of a discrete-time signal $y[n]$ result from the recording of the samples of the corresponding continuous-time signal $x(t)$ at time instances of the form $t_n = nT_s$, which constitute integer multiples of the sampling period T_s, we can correlate these two signals through a relationship of the form

$$y[n] = x(nT_s), \quad (n = 0, 1, 2, \ldots). \tag{6.1}$$

The way of describing the discrete-time signal from the last image of Fig. 6.1 constitutes the so-called *diagrammatic representation* of the signal, as it allows the depiction of the content of the signal in a special type of graph in which the horizontal axis represents the *discrete-time*, whereas the vertical axis contains *the values of the signal's amplitude*. In reality, the discrete-time signal is described *only by the dots of the diagram*, whereas the vertical lines connecting these dots to the horizontal axis have been added *not so much for clarity, but to prevent the reader from connecting these dots by drawing lines from one to the other, as this would be incorrect*: time is now discrete and the signal is defined only for the time instances that correspond to these dots; under no circumstances, therefore, should we connect them together. Alternatively, we can represent the signal as a sequence of numbers using a symbolism of the form

$$\begin{aligned} x(n) &= \{+\mathbf{0.437}, -0.430, 1.235, -0.495, -0.402, 0.645, -1.003, 1.017, 0.112, \\ &= -0.631, -0.435, -0.098, -0.040, 0.115, 0.217, -0.501, 0.998, 0.017, -0.195, 0.603, 0.342, 0.961\}, \end{aligned}$$

which is known as the *sequential representation of the signal*. If we resort to the use of this symbolism, we must, in addition to the contents of the signal, also determine that value of the signal that corresponds to the time instance $n = 0$; otherwise, the description of the signal is not complete, for reasons that we will explain in a subsequent section. This is something we do *either by drawing exactly below this value an arrow pointing upward or by writing it in bold*. In this chapter, we use the

[1] In reality, this is not accurate; provided that certain conditions are met regarding the sampling process, it is possible to fully reconstruct the continuous-time signal from its discrete equivalent. Therefore, by the term lossy process in this description, we simply mean the absence of recording of the intermediate values of the signal.

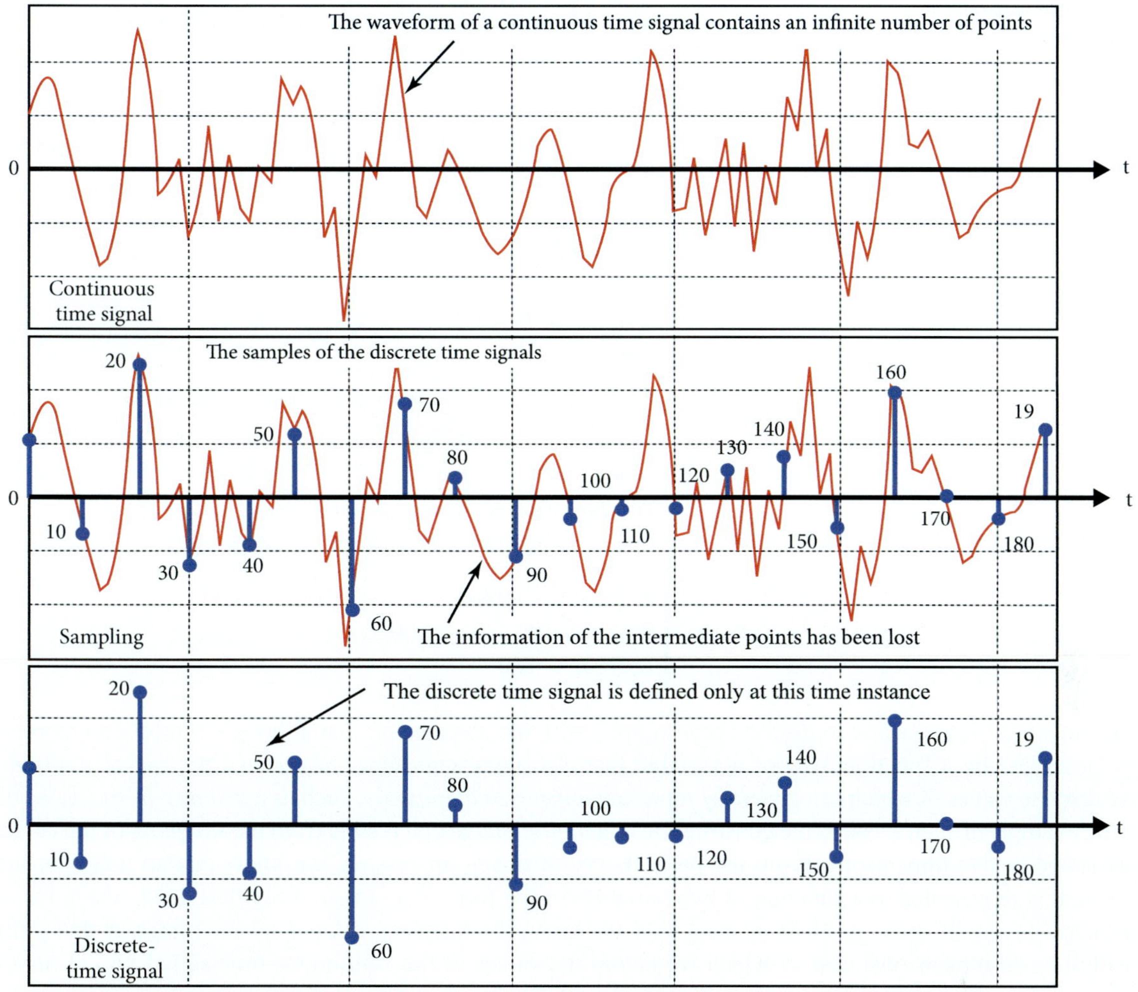

Fig. 6.1 Example of a continuous-time signal and its equivalent discrete-time signal

second method of symbolism; the fact, therefore, that the value +0.437 is written in bold characters means that this value corresponds to the time instance $n = 0$.

If the length of the discrete-time signal is *finite*, the signal can also be written in *vector form* as $\boldsymbol{x}[n] = \{x[0], x[1], x[2], \ldots,$
$x[N]\}$, where N is the number of data of the signal. Although generally the values that are included in a discrete-time signal are real, the use of complex discrete-time signals is not prohibited. These signals will obviously be defined by an equation of the form $z[n] = \alpha[n] + j\beta[n] = \Re\{z[n]\} + j\Im\{z[n]\}$ or equivalently, $z[n] = |z(n)| \exp(j\varphi\{z[n]\})$, where, as is well known, the amplitude and phase of complex quantities for each value of the variable n are defined as

$$|z(n)| = \sqrt{\{\Re\,[z(n)]\}^2 + \{\Im\,[z(n)]\}^2} \qquad \text{and} \qquad \varphi\,[z(n)] = \tan^{-1}\left(\frac{\Im\,[z(n)]}{\Re\,[z(n)]}\right)$$

respectively. On the other hand, if the discrete-time signal has *infinite length*, the two above forms of representation are not able to describe the signal (since we cannot construct a diagram with infinite dots or record a sequence with an infinite number of terms); in these cases we use the so-called *functional representation*, in which we determine the values of the signal for all value ranges of the independent variable n, as shown in the following example:

Table 6.2 Useful expressions of sums in closed form

Closed-form expressions for some well-known finite series	
$\sum_{k=\ell}^{n} \alpha^k = \frac{\alpha^\ell - \alpha^{n+1}}{1-\alpha} \quad (n \geq \ell)$	$\sum_{k=0}^{n} k\alpha^k = \frac{\alpha}{(1-\alpha)^2}\left[1-(n+1)\alpha^n + n\alpha^{n+1}\right]$
$\sum_{k=0}^{n} k^2\alpha^k = \frac{\alpha}{(1-\alpha)^3}\left[(1+\alpha)-(n+1)^2\alpha^n + (2n^2+2n-1)\alpha^{n+1} - n^2\alpha^{n+2}\right]$	
The above expressions are valid for parameter values $\alpha \neq 1$	
$\sum_{k=0}^{n} k = \frac{n(n+1)}{2}$	$\sum_{k=0}^{n} k^2 = \frac{n(n+1)(2n+1)}{6}$
$\sum_{k=0}^{n} k^3 = \frac{n^2(n+1)^2}{4}$	$\sum_{k=0}^{n} k^4 = \frac{n}{30}(n+1)(2n+1)(3n^2+3n-1)$
Closed-form expressions for some well-known infinite series	
$\sum_{k=0}^{\infty} \alpha^k = \frac{1}{1-\alpha} \quad \sum_{k=0}^{\infty} k\alpha^k = \frac{\alpha}{(1-\alpha)^2} \quad \sum_{k=0}^{\infty} k^2\alpha^k = \frac{\alpha^2+\alpha}{(1-\alpha)^3}$	
The above expressions are valid for parameter values $\|\alpha\| < 1$	

$$x[n] = \begin{cases} 1, \text{ for } n < -100 \\ 4, \text{ for } -100 \leq n \leq 100 \\ 0, \text{ for } n > 100 \end{cases}.$$

We conclude this brief introduction to *discrete-time signals* with the observation that although in most common cases these signals are described by a function $x[n]$ of a variable that also represents time, in practice the use of multidimensional signals, the discrete values of which can generally represent *any physical quantity*, such as a *distance in space*, is common. A characteristic example of such a case is the construction of a *digital image* that results from the *sampling* of the corresponding continuous image in two dimensions. From the basic theory of image processing, we know that an image of dimensions $M \times N$—which is represented as a function of two variables of the form $f(x, y)$—is fully described, when, for each point with coordinates (x, y) $(0 \leq x \leq M, 0 \leq y \leq N)$, we know the intensity value I of the image at that point. In full analogy with the one-dimensional case in which we record the values of the continuous-time signal $x(t)$ at time instances $t = nT_s$, in two-dimensional sampling we record the value of the function $f(x, y)$ for the coordinate values $x_{n_1} = n_1T_{s1}$ and $y_{n_2} = n_2T_{s2}$, where T_{s1} and T_{s2} are the *sampling periods* along the two dimensions of the image, which in this case express the *spatial distance* between two successive recordings of the image intensity toward each dimension. Based on this description, we can correlate the discrete image $g[n_1, n_2]$ with the continuous image $f(x, y)$ from which it originated through the relation

$$g(n_1, n_2) = f(n_1T_{s1}, n_2T_{s2}), \quad (n_1, n_2 = 0, 1, 2, \ldots). \tag{6.2}$$

In the remainder of this chapter, we present the basic theory of discrete-time signals. The structure of the chapter is similar to that of the corresponding chapter dealing with *continuous-time signals*, with the difference that in the following description *time will not be continuous but discrete*. Although in many cases our description is entirely analogous—in the sense that many of the continuous equations immediately transform into their corresponding discrete expressions simply by replacing *integrals* with *sums*—we will find that in some other cases this analogy does not exist and the continuous-time signals are entirely different from their discrete equivalent.

At several points in the text, we will need to calculate *sums of finite and infinite terms and of specific functional form*. Since the analytical calculation of these sums is certainly beyond the purpose of writing this book, in Table 6.2 we have collected in the form of a typology some characteristic identities, which we use directly whenever we need them, so that it is not necessary to proceed with their calculation. So, every time we refer to the *mathematical typology* we mean this table.

6.2 Sampling in the Time Domain

Returning to what we mentioned in Sect. 6.1 (refer to page 497), the basic difference that exists between a continuous and a discrete-time signal has to do, as we commented, with the fact that the continuous-time signal $x_\alpha(t)$ is defined *for every value of the time variable* t, whereas the discrete-time signal $x[n]$ is defined *only at specific time instances that are described by the integer variable* n, and the construction of a discrete-time signal from its continuous equivalent consists *in recording the values of the continuous-time signal at distinct time instances*, a process known as *sampling*. If the time instances of recording the successive values of the signal are equidistant from each other, the sampling process is characterized as *uniform*, the time interval T_s that intervenes between two successive recordings is known as the *sampling period*, whereas the number F_s of samples recorded per unit of time is known as the *sampling frequency*. From the mathematical side of the subject, the continuous-time signal $x_\alpha(t)$ and the discrete-time signal $x[n]$ that results from it are related, as we saw, through the expression

$$x[n] = x_\alpha(nT_s), \qquad -\infty < n < \infty. \tag{6.3}$$

For a quality demonstration of the sampling process, refer to the contents of Table 6.1.

It is not difficult to ascertain that as long as the distance between two successive recordings of the continuous-time signal values is equal to T_s, the variables of continuous and discrete-time will satisfy the equation

$$t = nT_s = n/F_s, \tag{6.4}$$

since the period and the sampling frequency satisfy a relationship of inverse proportionality (for example, considering the case $T_s = 10\,\text{ms}$, the first four samples of the discrete-time signal will be $x[0] = x_\alpha(0)$, $x[1] = x_\alpha(10)$, $x[2] = x_\alpha(20)$ and $x[3] = x_\alpha(30)$). Therefore, the discrete-time signal $x[n]$ that will result from the sampling of a continuous-time signal $x(t)$ with a sampling frequency F_s can be determined, if in the equation defining the continuous-time signal we make the substitution

$$t \Longrightarrow n/F_s$$

Considering, for example, that the continuous-time signal being sampled is the simple cosine

$$x(t) = A\cos(2\pi Ft + \varphi), \tag{6.5}$$

the discrete-time signal that will result from the sampling of the above signal with a sampling period $T_s = 1/F_s$ will be

$$x[n] = A\cos\left(2\pi\frac{F}{F_s}n + \varphi\right) = A\cos(2\pi fn + \varphi), \tag{6.6}$$

with the magnitude $f = F/F_s$ expressing the frequency of the discrete-time signal. The signal that will emerge is known as *relative* or *normalized* frequency. We find, therefore, that the frequency of the discrete-time signal depends *both on the frequency F of the continuous-time signal that is sampled and the sampling frequency* F_s, and in reality it is *the ratio of these two frequencies*. This conclusion is certainly expected, since if we sample the same continuous-time signal recording different numbers of samples in the unit of time, we will construct different discrete-time signals. Therefore, *a sampling process is fully defined when we specify both the continuous-time signal that is sampled and the number of samples that are recorded in the unit of time.*

Problem 6.2.1 Considering the continuous-time signal $x(t) = 5\cos(200\pi t)$, determine the discrete-time signal that will result from the sampling of the above signal with frequencies $F_{s1} = 300\,\text{Hz}$ and $F_{s2} = 800\,\text{Hz}$.

Solution If we compare the equation of the signal $x(t)$ with the equation of the sinusoidal signal

$$x(t) = A\cos(2\pi Ft + \varphi)$$

it immediately follows that $A = 5$, $\varphi = 0°$ and $F = 100\,\text{Hz}$. Therefore, the discrete-time signal that will result from a sampling process with frequency $F_{s1} = 300\,\text{Hz}$ will have the form

$$x[n] = 5\cos\left(2\pi\frac{100}{300}n\right) = 5\cos\left(\frac{2\pi n}{3}\right),$$

whereas the discrete-time signal that will result from sampling at a frequency $F_{s2} = 800\,\text{Hz}$ will be

$$x[n] = 5\cos\left(2\pi\frac{100}{800}n\right) = 5\cos\left(\frac{\pi n}{4}\right)$$

Folded Frequencies

The process of converting a continuous-time signal to its discrete equivalent must be carried out with particular care, since continuous and discrete-time signals have several differences, regarding the magnitude of the frequency (see Sect. 6.7.1 on page 520). First, the frequency F of the continuous-time signal (or equivalently the angular frequency $\Omega = 2\pi F$) can take *any value* in the interval $(-\infty, +\infty)$, in contrast to the frequency f of the equivalent discrete-time signal, which, as we will prove in Sect. 6.7.1, takes values only in the area $(-1/2, +1/2)$—if we use the angular frequency $\omega = 2\pi f$ of the discrete-time signals, this obviously takes values in the interval $(-\pi, +\pi)$. Second, in contrast to continuous-time signals, which, if they have different frequencies, *can always be distinguished from each other*, two discrete-time signals with frequencies f_1 and f_2, such that $f_2 - f_1 = m$ (or equivalently $\omega_2 - \omega_1 = 2\pi m$), where m is an integer variable, it is proven that *they are equivalent and cannot be distinguished from each other*. Moreover, as we will see in Sect. 6.7.1, if the frequency f of a discrete-time signal belongs to the value range $-\pi < f < \pi$, then it is *unique*, (in the sense that *there is no equivalent frequency that belongs to the same interval*), whereas, on the contrary, if it lies outside the boundaries of that interval, *there are infinite frequencies equivalent to it that are separated on the frequency axis by a distance that is an integer multiple of* 2π. Of all these infinite frequencies, the one that belongs to the interval $(-\pi, +\pi)$ is known as *true frequency*, whereas the infinite number of frequencies equivalent to it that lie outside this interval are copies of it and are known as *folded frequencies*.

But why does the existence of folded frequencies create a problem? In practice, the processing of a signal is based on the use of digital methods, which are more accurate and effective than their analog counterparts. For this reason, it is quite common *to convert a continuous-time signal to its discrete form, undergo the appropriate processing techniques in each case, and then convert it back to its analog representation*. If, however, the frequency of the resulting discrete-time signal has a value of $\omega > \pi$, then, according to what we have aforementioned, are there *infinite frequencies equivalent to it* (i.e., *infinite equivalent discrete-time signals*) that are at a distance of $2\pi m$ from it and therefore *infinite candidate analog signals*, the sampling of which would lead to the creation of the discrete-time signal we are considering? We do not know, therefore, which of these to choose. Therefore, the above conversion process would only be possible if there were a one-to-one correspondence between the frequencies of the continuous and the corresponding discrete-time signal.

In order to understand the above problem from the mathematical perspective of the subject, let us consider the sampling of a basic cosine signal of the form

$$x(t) = A\cos(2\pi F t)$$

with a sampling period T_s and sampling frequency $F_s = 1/T_s$. In this case, taking into account the correlation $t = nT_s$ and the fact that the cosine function is periodic with a period of 2π, the discrete-time signal that will result from the above process will have the form

$$x[n] = A\cos(2\pi F t) = A\cos(2\pi F n T_s) = A\cos(2\pi F n T_s + 2\pi m) = A\cos\left[2\pi\left(F + \frac{m}{nT_s}\right)nT_s\right] = A\cos[2\pi(F + kF_s)nT_s],$$

where $k = m/n$ (at this point we have silently assumed that m is an integer multiple of n, so that the value of k is an integer).

The above equation is very important, as it leads us to a very interesting conclusion: *considering a discrete sequence* $x[n]$ *which has resulted from the sampling of a simple cosine (or sine) waveform of frequency* F, *then the exact same sequence will result from the sampling of any sinusoidal waveform, the frequency of which is equal to* $F_k = F + kF_s$ *where* F_s *is the sampling frequency, whereas the integer variable* k *takes the values* $k = 0, \pm 1, \pm 2, \pm 3, \ldots$. This situation is graphically presented in Fig. 6.2.

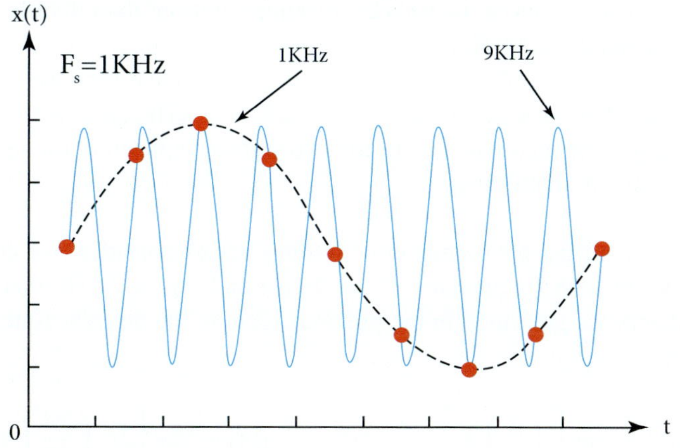

Fig. 6.2 Ambiguity of discrete-time signal representation

In Fig. 6.2 the discrete sequence has resulted from the sampling of a sinusoidal waveform of frequency $F = 1$ KHz—this curve is designed with a dashed line - whereas the sampling frequency is also equal to $F_s = 1$ KHz. However, we observe that this sequence *does not uniquely determine the continuous-time signal from which it originated*, since from the same points (as emerges from the figure) the sinusoidal waveform frequency $F' = F + 8F_s = 9KHz$ also passes. Therefore, *if we do not know information about the parameters of the sampling and the only thing we have is the discrete-time series that has resulted from it, we cannot in any way distinguish whether it has originated from the continuous-time signal of frequency F or from the continuous-time signal frequency F'*, whereas the same obviously happens for the infinite frequencies that correspond to the various values of the parameter k. This undesirable result in the international terminology related to the field of digital signal processing is known as frequency folding (aliasing).

Combining the conclusions we formulated in the previous discussion, it is not difficult to understand *that in order for the correlation of the discrete sampled sequence with a single continuous sinusoidal waveform to be possible, so as to eliminate the aforementioned ambiguity, the frequency of the sampling must be such that the frequency of the discrete-time signal that will result from it is a true frequency—that is, it belongs to the range of values $(-1/2, +1/2)$—since only then is it unique in the sense we described previously*. The last conclusion in mathematical form is expressed as

$$-\frac{1}{2} \leq f \leq \frac{1}{2} \Rightarrow -\frac{1}{2} \leq \frac{F}{F_s} \leq \frac{1}{2} - \frac{1}{2T_s} = -\frac{F_s}{2} \leq F \leq \frac{F_s}{2} = \frac{1}{2T_s}$$

or equivalently as

$$-\frac{1}{2T_s} = -\frac{F_s}{2} \leq F \leq \frac{F_s}{2} = \frac{1}{2T_s}, \tag{6.7}$$

where we have used the simple frequency F, whereas if the *cyclic frequency* is used, we get

$$-\frac{\pi}{T_s} = -\pi F_s \leq \Omega \leq \pi F_s = \frac{\pi}{T_s}. \tag{6.8}$$

The above equations determine the correct relationship that must exist between the sampling frequency F_s and the frequency of the signal to be sampled F in order to avoid displaying the phenomenon of frequency folding. *For a given sampling frequency F_s, the frequency of the continuous-time signal should be at most equal to $F_s/2$. Conversely, for a given signal frequency F, the sampling frequency should be at least equal to $2F$, so that the sampling is carried out in the correct way*. As we will see in the following pages, this minimum frequency value, which prevents the appearance of unwanted folded or alias frequencies, is known as the Nyquist frequency. A typical application of this rule is found in applications of human speech recording, which is carried out at 44,100 Hz, a frequency more than double the maximum frequency that is perceived by the human ear and is about 20,000 Hz.

Returning to Fig. 6.2, we find that the sampling frequency $F_s = 1$ KHz is less than the minimum value it should have, in order to successfully apply the sampling process, which is $F_s = 2F = 2$ KHz, and for this reason the discrete-time signal that has resulted cannot be uniquely correlated with any of the two signals depicted in the figure (since it could have resulted from either one with exactly the same probabilities).

Example 6.2.1 In order to understand the phenomenon of folding, let us consider two continuous-time signals $x_1(t) = \cos(20\pi t)$ and $x_2(t) = \cos(100\pi t)$ with frequencies $F_1 = 10$ Hz and $F_2 = 50$ Hz that are sampled at a frequency of $F_s = 40$ Hz. According to the theory we presented in the previous section, the discrete-time signals that will result from the above process will have the form

$$x_1[n] \cos\left[2\pi\left(\frac{10}{40}\right)n\right] = \cos\left[\left(\frac{\pi}{2}\right)n\right] \quad \text{and} \quad x_2[n] = \cos\left[2\pi\left(\frac{50}{40}\right)n\right] = \cos\left[\left(\frac{5\pi}{2}\right)n\right].$$

However, we observe that

$$\cos\left(\frac{5n\pi}{2}\right) = \cos\left(\frac{n\pi}{2} + 2\pi n\right) = \cos\left(\frac{n\pi}{2}\right)$$

from which it follows that $x_1[n] = x_2[n]$. In other words, the two signals are identical and cannot be distinguished from each other, which means that the signal $x_2[n]$ is an *alias* of the signal $x_1[n]$. This happens because the maximum frequency of the signal that can be used in a sampling process of frequency $F_s = 40$ Hz is $F_{max} = F_2/2 = 20$ Hz, whereas the frequency of the signal $x_2(t)$ is $F_2 = 50\,\text{Hz} > F_{max}$.

Problem 6.2.2 Consider the analog signal $x(t) = 3\cos(100\pi t)$ (Proakis & Manolakis, Example 1.4.2).

(a) What is the minimum value of the sampling frequency that should be used so that the phenomenon of frequency folding does not appear?

(b) Assuming that the signal is sampled at a frequency of $F_s = 200$ Hz, determine the discrete-time signal that will result. from this process.

(c) Considering that the signal is sampled at a frequency $F_s = 75$ Hz, determine the discrete-time signal that will result from this process.

(d) Determine the frequency F $(0 < F < F_s/2)$ of the sinusoidal signal, the sampling of which at a frequency $F_s = 75$ Hz will lead to the creation of the same discrete-time signal as that of the original.

Solution (a) The analog input signal has the form $x(t) = 3\cos(2\pi F t) = 3\cos(100\pi t)$ from which it follows that its frequency is equal to $F = 50$ Hz. Therefore, the minimum sampling frequency that should be used, to avoid the phenomenon of aliasing, is $F_s^{min} = 2F = 100$ Hz.

(b) If the above signal undergoes sampling at a frequency $F_s = 200$ Hz, the discrete-time signal that will result from this process will be given, according to the theory, by the relation

$$x[n] = 3\cos\left(\frac{100}{200}\pi n\right) = 3\cos\left(\frac{n\pi}{2}\right).$$

(c) In a completely similar way, if the above signal undergoes sampling at a frequency $F_s = 75$ Hz, the discrete-time signal that will result from this process will be given, according to the theory, by the relation

$$x[n] = 3\cos\left(\frac{100}{75}\pi n\right) = 3\cos\left(\frac{4n\pi}{3}\right) = 3\cos\left(-\frac{2n\pi}{3} + 2n\pi\right) = 3\cos\left(-\frac{2n\pi}{3}\right) = 3\cos\left(\frac{2n\pi}{3}\right)$$

(d) Taking into account that the discrete-time signal resulting from the sampling process takes the form

$$x[n] = A\cos(2\pi f_0 n)$$

the last equation is written as

$$x[n] = 3\cos\left[2\pi\left(\frac{1}{3}\right)n\right] = 3\cos(2\pi f n),$$

from which it follows that the relative or normalized frequency is equal to $f = F/F_s = 1/3$. Therefore, the frequency F of the sinusoidal signal, the sampling of which will lead to the same discrete-time signal as that of the original signal, will be equal to $F = f * F_s = 75/3 = 25$ Hz, whereas the mathematical representation of this signal will be $x(t) = 3\cos(2\pi Ft) = 3\cos(50\pi t)$. We observe that the analog signals with frequencies $F_1 = 50$ Hz and $F_2 = 25$ Hz give the same exact discrete-time signal for a sampling frequency $F_s = 75$ Hz, meaning that the signal of frequency F_1 is considered an *alias* of the signal of frequency F_2.

Example 6.2.2 Let us calculate the Nyquist frequency for the analog signal given by the equation $x(t) = 5\cos(80\pi t) + 3\sin(600\pi t) + \cos(100\pi t)$. If we write the signal in the form

$$x(t) = A_1\cos(2\pi F_1 t) + A_2\sin(2\pi F_2 t) + A_3\cos(2\pi F_3 t) = 3\cos[2\pi(40)t] + 10\sin[2\pi(300)t] - \cos[2\pi(50t)]$$

we find that this results from the parallelism of three sinusoidal signals with amplitudes $A_1 = 5$, $A_2 = 3$, $A_3 = 1$ and frequencies $F_1 = 40$ Hz, $F_2 = 300$ Hz and $F_3 = 50$ Hz. Therefore, the maximum frequency of the signal is $F_{\max} = F_2 = 300$ Hz and the Nyquist frequency will be equal to $F_N = 2F_{\max} = 2F_2 = 600$ Hz. This should also be the frequency at which the sampling of the signal should be carried out, in order for the full reconstruction of the original analog signal from the sampled values to be possible.

The Sampling Theorem

According to what has been mentioned in the previous section, *the sampling of a simple sinusoidal signal of frequency F should be done at a frequency at least equal to $2F$, so as not to present the undesirable phenomenon of frequency folding.* In practice, of course, continuous analog signals are not simple sinusoidal waveforms, but, according to Fourier theory, they can be broken down into a sum of such waveforms of the form

$$x(t) = \sum_{i=1}^{N} A_i\cos(2\pi F_i t + \varphi_i),$$

where the symbols A_i, F_i and φ_i $(i = 1, 2, 3, \ldots, N)$ refer to the *amplitude*, the *frequency*, and the *phase* of the component numbered i, whereas the parameter N expresses *the number of frequency components used.* In this case, the sampling frequency that should be used so as not to present the phenomenon of folding should satisfy the condition $F_s > 2F_{\max}$, where $F_{\max} = \max(F_1, F_2, \ldots, F_N)$ is *the maximum frequency that is contained in the analog signal for sampling.* It is obvious that the choice of the sampling frequency in the above way results in the representation of each sinusoidal component of frequency $|F_i| < F_{\max}$ $(i = 1, 2, \ldots, N)$ of the analog signal in a unique discrete sinusoidal component, the frequency of which will belong to the interval $[-1/2, 1/2]$ (for the usual frequency) or in the interval $[-\pi, \pi]$ (for the circular frequency $\omega_i = 2\pi f_i$). In this way, the phenomenon of aliasing is not going to appear, whereas additionally *the complete reconstruction of the original analog signal from its discrete equivalent through the use of interpolation methods* is possible. There are many interpolation methods that are used in practice, with the most ideal of them being dictated by the *sampling theorem*, which is formulated in the following way:

Theorem 6.2.1 *If the maximum frequency value contained in an analog signal $x_\alpha(t)$ is F_{max} and the sampling of the signal is done with a frequency value $F_s > 2F_{max}$, then the waveform $x_\alpha(t)$ can be fully reconstructed from the recorded values, using the interpolation function*

(continued)

$$g(t) = \frac{sin(2\pi Fmaxt)}{2\pi Fmaxt}$$

and by using an equation of the form

$$x_\alpha(t) = \sum_{n=-\infty}^{\infty} x_\alpha\left(\frac{n}{F_s}\right) g\left(t - \frac{n}{F_s}\right),$$

where $x_\alpha(n/F_s) = x_\alpha(nT_s) \equiv x[n]$ *are the samples of the analog waveform* $x_\alpha(t)$*. In the special case where the equality* $F_s = 2F_{max}$ *holds, the above relation takes the form*

$$x_\alpha(t) = \sum_{n=-\infty}^{\infty} x_\alpha\left(\frac{n}{2F_{max}}\right) \frac{\sin\{2\pi F_{max}[t - (n/2F_{max})]\}}{2\pi F_{max}[(t - (n/2F_{max}))]}.$$

In the terminology of digital signal processing, the frequency $F_N = 2F_{max}$ *is known as the* Nyquist frequency.

It should be noted that the reconstruction of the analog signal in the above way is quite difficult and in practice the calculation of the intermediate values of the signal amplitude is done with simpler methods.

Problem 6.2.3 Let the analog signal $x(t) = 3\cos(2000\,\pi t) + 5\sin(6000\,\pi t) + 10\cos(12000\,\pi t)$.

(a) What is the Nyquist frequency of the analog signal?
(b) What is the discrete-time signal that will result from sampling the above signal with a frequency of $F_s = 5000\,\text{Hz}$?
(c) What is the analog signal that will be reconstructed from the above discrete-time signal?

(Proakis and Manolakis, Example 1.4.4).

Answer: (a) By formulating the analog signal in the form

$$x(t) = A_1\cos(2\pi F_1 t) + A_2\sin(2\pi F_2 t) + A_3\cos(2\pi F_3 t) = 3\cos[2\pi(1000t] + 5\sin[2\pi(3000)t] + 10\cos[2\pi(6000t)]$$

we find that it results from the composition of three sine waves with frequencies $F_1 = 1000\,\text{Hz}$, $F_2 = 3000\,\text{Hz}$ and $F_3 = 6000\,\text{Hz}$. We find, therefore, that the maximum frequency contained in the signal is $F_{\text{max}} = F_3 = 6000\,\text{Hz}$ and therefore the Nyquist frequency will be equal to $F_n = 2F_{\text{max}} = 2F_3 = 12{,}000\,\text{Hz}$.

(b) The discrete-time signal that will result from the sampling of the analog signal is

$$\begin{aligned}
x[n] &= 3\cos\left[\left(\frac{1000}{5000}\right)2\pi n\right] + 5\sin\left[\left(\frac{3000}{5000}\right)2\pi n\right] + 10\cos\left[\left(\frac{6000}{5000}\right)2\pi n\right] \\
&= 3\cos\left[\left(\frac{1}{5}\right)2\pi n\right] + 5\sin\left[\left(\frac{3}{5}\right)2\pi n\right] + 10\cos\left[\left(\frac{6}{5}\right)2\pi n\right] \\
&= 3\cos\left[\left(\frac{1}{5}\right)2\pi n\right] + 5\sin\left[\left(1-\frac{2}{5}\right)2\pi n\right] + 10\cos\left[\left(1+\frac{1}{5}\right)2\pi n\right] \\
&= 3\cos\left[\left(\frac{1}{5}\right)2\pi n\right] + 5\sin\left[\left(-\frac{2}{5}\right)2\pi n\right] + 10\cos\left[\left(+\frac{1}{5}\right)2\pi n\right] = 13\cos\left[\left(\frac{1}{5}\right)2\pi n\right] - 5\sin\left[\left(\frac{2}{5}\right)2\pi n\right].
\end{aligned}$$

(c) The discrete-time signal that resulted from the sampling consists of two elementary waveforms whose frequencies are $F_1 = F_s f_0 = 5000\,\text{Hz} * (1/5) = 1000\,\text{Hz}$ and $F_2 = F_s f_0 = 5000\,\text{Hz} * (2/5) = 2000\,\text{Hz}$; therefore, the analog signal corresponding to this is

$$x(t) = 13\cos[2\pi(1000)t] - 5\sin[2\pi(2000\,t)] = 13\cos(2000\,\pi t) - 5\sin(4000\,t).$$

We observe, therefore, that the reconstructed signal is not the same as the original, with the distortion observed being due to the phenomenon of aliasing. Indeed, since the sampling frequency is $F_s = 5000\,\text{Hz}$, the maximum frequency that is fully reconstructed from the discrete sampled signal is $F_{\text{max}} = 2500\,\text{Hz}$, whereas the original analog signal has higher frequencies.

Problem 6.2.4 Let us assume an analog signal $x(t)$ that contains frequencies up to 10 kHz.

(a) For which frequency range of sampling is it possible to fully reconstruct the analog signal from its discrete samples?
(b) Let us assume that we sample the signal with a sampling frequency $F_s = 8$ KHz. Explain what will happen with the fundamental sinusoidal waveforms of frequencies $F_1 = 5$ KHz and $F_2 = 9$ KHz.

Solution (a) Since the maximum frequency contained in the signal is $F_{\text{max}} = 10$ KHz, the frequency range of sampling for which it is possible to fully reconstruct the analog signal is $F_s \geq 2F_{\text{max}} = 20$ KHz.

(b) For a value of the sampling frequency $F_s = 8$ KHz, the maximum frequency that can be fully reconstructed from the discrete samples will be $F = 4$ KHz. Therefore, the frequencies $F_1 = 5$ KHz and $F_2 = 9$ KHz constitute *aliases of lower frequencies* that belong to the interval [0 – 4] KHz. In order to determine these lower frequencies, we work in the following way:

- For the frequency $F_1 = 5$ KHz we have:

$$x[n] = \cos\left[2\pi\left(\frac{5000}{8000}\right)n\right] = \cos\left[2\pi\left(\frac{5}{8}\right)n\right] = \cos\left(\frac{5\pi n}{4}\right)$$
$$= \cos\left(-\frac{3n\pi}{4} + 2n\pi\right) = \cos\left(-\frac{3n\pi}{4}\right) = \cos\left(\frac{3n\pi}{4}\right)$$

  Therefore, the signal of frequency $F_1 = 5$ KHz is an alias of the signal of frequency $F = 3$ KHz.
- For the frequency $F_2 = 9$ KHz we have:

$$x[n] = \cos\left[2\pi\left(\frac{9000}{8000}\right)n\right] = \cos\left[2\pi\left(\frac{9}{8}\right)n\right] = \cos\left(\frac{9\pi n}{4}\right) = \cos\left(\frac{n\pi}{4} + 2n\pi\right) = \cos\left(\frac{n\pi}{4}\right)$$

  Therefore, the signal of frequency $F_2 = 5$ KHz is an alias of the signal of frequency $F = 1$ KHz.

Problem 6.2.5 Considering an analog signal of the form (Proakis and Manolakis, Problem 1.9).

$$x_\alpha(t) = \sin(480\,\pi t) + 3\sin(720\,\pi t),$$

which is sampled at a frequency $F_s = 600\,\text{Hz}$, answer the following questions:

(a) What is the value of the Nyquist frequency of the analog signal?
(b) What is the maximum frequency that can be can be represented in a unique way for the specific sampling frequency?

(continued)

(c) What are the frequencies (in radians) of the discrete-time signal $x[n]$?
(d) If the discrete-time signal $x[n]$ is converted to its analog form, what will be the resulting signal?

Solution (a) If we write the analog signal in the form

$$x_\alpha(t) = \sin(2\pi F_0 t) + 3\sin(2\pi F_1 t) = \sin[2\pi(240)t] + 3\sin[2\pi(360)t]$$

it is not difficult to see that it contains the simple frequencies $F_0 = 240$ Hz and $F_1 = 360$ Hz. We observe that the maximum frequency of the signal is $F_{\text{max}} = F_2 = 360$ Hz and therefore the Nyquist frequency of the signal will be equal to $F_N = 2F_{\text{max}} = 2F_1 = 720$ Hz.

(b) Since the sampling frequency of the signal is $F_s = 600$ Hz, the maximum frequency that can be represented in a unique way for this specific sampling frequency is $F = F_s/2 = 300$ Hz.

(c) The signal resulting from the sampling of the analog signal with frequency $F_s = 600$ Hz will be given by the relation

$$x[n] = \sin\left(2\pi\frac{240}{600}n\right) + 3\sin\left(2\pi\frac{360}{600}n\right) = -2\sin\left(\frac{4\pi n}{5}\right)$$

(as it results after simple mathematical operations) and therefore the frequency of the discrete-time signal (in radians) will be equal to $\omega = 4\pi/5$. As for the normalized frequency f_0, this will be calculated by the relation $f_0 = \omega/2\pi = (1/2\pi)(4\pi/5) = (4/10) = (2/5)$.

(d) The frequency of the analog signal that will result from the discrete-time signal will be given by the relation $F = f_0 F_s = (2/5) * 600 = 240$ Hz. Therefore, the reconstructed analog signal will have the form

$$x_\alpha(t) = -2\sin[2\pi(240)t] = -2\sin(480\pi t),$$

which is also the desired result. ■

In order to prove the sampling theorem, let us try to determine the relationship that exists *between the spectrum of the continuous-time signal that is sampled and the spectrum of the discrete-time signal that results from its sampling*. In the description that follows we use the variables Ω and ω to describe the frequencies in the field of continuous and discrete-time. It is not difficult to see that if we define the *sampling function* from the relation

$$s(t) = \sum_{n=-\infty}^{\infty} \delta(t - nT_s),$$

which describes a periodic train of impulse functions with a fundamental period T_s equal to the sampling period and a fundamental frequency equal to $\Omega_s = 2\pi F_s = 2\pi/T_s$ (it is the well-known *comb function* depicted in Fig. 1.13—refer to page 30), the signal that will result from the sampling of the continuous-time signal $x_\alpha(t)$ with period T_s will be a second train of impulse functions of the form

$$x_s(t) = x_\alpha(t)\,s(t) = x_\alpha(t)\sum_{n=-\infty}^{\infty}\delta(t - nT_s) = \sum_{n=-\infty}^{\infty} x_\alpha(t)\delta(t - nT_s). \tag{6.9}$$

Using the property of the *impulse function* $x(t)\delta(t - t_0) = x(t_0)\delta(t - t_0)$, we find that $x_\alpha(t)\delta(t - nT_s) = x_\alpha(nT_s)\delta(t - nT_s)$ and the above relationship will take the form

$$x_s(t) = \sum_{n=-\infty}^{\infty} x_\alpha(nT_s)\delta(t - nT_s) = \sum_{n=-\infty}^{\infty} x[n]\delta(t - nT_s), \tag{6.10}$$

where for the derivation of the last expression we used the equation $x[n] = x_\alpha(nT_s)$.[2] The above relation describes the correlation that exists between the continuous-time signal that is sampled and the signal that results as a consequence of this process, whereas to determine the type of correlation between the corresponding spectra, let us take the Fourier transform of Eq. (6.10). Therefore, we have

$$\mathcal{F}\{x_s(t)\} = \mathcal{F}\{x_\alpha(t)\, s(t)\} \qquad \text{or equivalently} \qquad \mathcal{X}_s(j\Omega) = \frac{1}{2\pi}\Big\{\mathcal{X}_\alpha(j\Omega) * \mathcal{S}(j\Omega)\Big\},$$

where we have defined the functions $\mathcal{X}_s(j\Omega) = \mathcal{F}\{x_s(t)\}$, $\mathcal{X}_\alpha(j\Omega) = \mathcal{F}\{x_\alpha(t)\}$ and $\mathcal{S}(j\Omega) = \mathcal{F}\{s(t)\}$, and we used *the property of multiplication* of the Fourier transform, according to which *the Fourier transform of the product of two signals is proportional to the convolution of the transforms of the aforementioned signals*. However, as we saw in Exercise 3.3.21 (refer to page 252), the Fourier transform of the function $s(t)$ is

$$\mathcal{S}(j\Omega) = \frac{2\pi}{T_s} \sum_{k=-\infty}^{\infty} \delta\left(\Omega - \frac{2k\pi}{T_s}\right) \tag{6.11}$$

and it also constitutes *a periodic train of impulse functions in the frequency domain with a period equal to* $\Omega_s = 2\pi/T_s$. By substituting the above relation into our final expression, it will take the form

$$\mathcal{X}_s(j\Omega) = \frac{1}{2\pi}\mathcal{X}_\alpha(j\Omega) * \left\{\frac{2\pi}{T_s} \sum_{k=-\infty}^{\infty} \delta\left(\Omega - \frac{2k\pi}{T_s}\right)\right\} = \frac{1}{T_s} \sum_{k=-\infty}^{\infty} \mathcal{X}_\alpha(j\Omega) * \delta\left(\Omega - \frac{2k\pi}{T_s}\right).$$

However, we know that *the convolution between a signal and the impulse function causes the displacement of the signal within the space where it is defined* and therefore we have

$$\mathcal{X}(j\Omega) * \delta(\Omega - \Omega_0) = \mathcal{X}[j(\Omega - \Omega_0)].$$

Using the above property in the last equation, it will take the form[3]

$$\mathcal{X}_s(j\Omega) = \frac{1}{T_s} \sum_{k=-\infty}^{\infty} \mathcal{X}_\alpha\left[j\left(\Omega - \frac{2k\pi}{T_s}\right)\right]. \tag{6.12}$$

The physical meaning of this equation is that *the function* $\mathcal{X}_s(j\Omega)$ *expressing the spectrum of the signal resulting from sampling and is a periodic function of frequency with period* $\Omega_s = 2\pi F_s = 2\pi/T_s$, *results from the periodic repetition of*

---

[2] The difference that exists between the signals $x_s(t)$ and $x[n]$ is quite subtle: although both signals describe *the same discrete sequence of recorded samples*, the signal $x_s(t)$ actually constitutes a *continuous-time signal* that is everywhere equal to zero except at the time instances $t = nT_s$, whereas the signal $x[n]$ constitutes a *discrete-time signal* that has undergone *normalization* with respect to its discrete-time variable (in the sense that it has been indexed with the help of the integer variable n) and *does not contain the slightest information that would allow us to recover details regarding the period or the sampling frequency*. For example, a sequence of the form $\{x_s(0), x_s(20), x_s(40), x_s(60), \ldots\}$ that allows us to guess that the sampling period has a value $T_s = 20$ is a conclusion that cannot be formulated in any way using the sequence $\{x[0], x[1], x[2], x[3], \ldots\}$.

[3] For a different proof of this relationship based on the use of the inverse Fourier transform, see Proakis and Manolakis, *Digital Signal Processing*, 3rd Edition, Section 4.2.9.

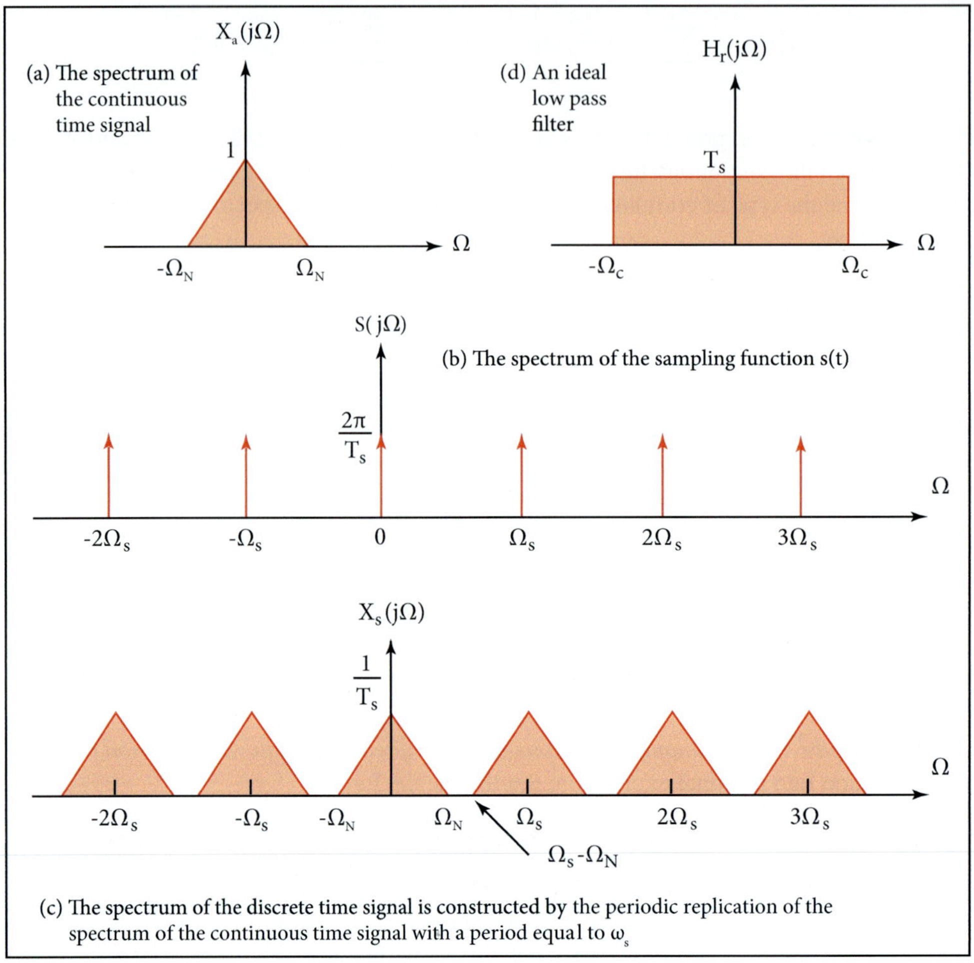

Fig. 6.3 The relationship between the spectra of the continuous-time signal and the corresponding discrete-time signal

frequency-shifted copies of the spectrum $\mathcal{X}_\alpha(j\Omega)$ of the signal $x_\alpha(t)$ that is sampled, which have also been scaled by the factor $F_s = 1/T_s$.

The above correlation between the spectra of the signals before and after the sampling process was important to determine, as it dictates the way in which under certain conditions the *recovery*—that is, the *reconstruction*—of the original continuous-time signal is possible, and indeed *without the slightest loss of information*. To understand this reconstruction process, let us assume that the continuous-time signal $x_\alpha(t)$ being sampled is a *band-limited signal*; this means that the Fourier transform $\mathcal{X}_\alpha(j\Omega)$ has a nonzero value only in the frequency range $-\omega_N \leq \omega \leq \omega_N$, and therefore the maximum frequency included in this signal is $\omega = \omega_N$ (this spectrum is shown in image (a) of Fig. 6.3). On the other hand, image (b) of the same figure depicts the spectrum $\mathcal{S}(j\Omega)$ of the sampling function, which is itself a *periodic train of impulse functions* in the frequency domain, whereas image (c) shows the spectrum $\mathcal{X}_s(j\Omega)$ resulting from the convolution of the spectra $\mathcal{X}_\alpha(j\Omega)$ and $\mathcal{S}(j\Omega)$.

How can we reconstruct the continuous-time signal $x_\alpha(t)$? Comparing images (a) and (c) of Fig. 6.3, we find that *all we have to do is keep only the part of the spectrum $\mathcal{X}_s(j\Omega)$ with frequencies belonging to the range* $[0, \Omega_c]$—where Ω_c is any frequency belonging to the range of values $\Omega_N < \Omega_c < \Omega_s - \Omega_N$ (see image (c) of Fig. 6.3)—something that can be achieved using an *ideal low-pass filter* with the frequency response

$$\mathcal{H}_r(j\Omega) = \begin{cases} T_s & |\Omega| \leq \Omega_c \\ 0 & |\Omega| > \Omega_c \end{cases}. \tag{6.13}$$

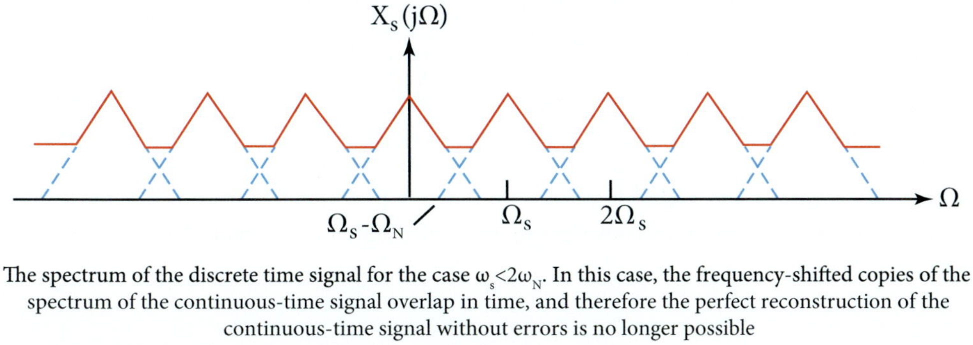

Fig. 6.4 The spectrum of the discrete-time signal for sampling frequency $\omega_s < 2\omega_N$

In this case, the filter output spectrum will have the form

$$\mathcal{X}_r(j\Omega) = \mathcal{H}_r(j\Omega)\,\mathcal{X}_s(j\Omega) = T_s\,\mathcal{X}_s(j\Omega) = \mathcal{X}_\alpha(j\Omega).$$

In other words, *the spectrum of the signal that will appear at the filter output will be the same as the spectrum of the continuous-time signal* $x_\alpha(t)$, *which means that the aforementioned signal can be fully reconstructed from its samples* (in a way that we will present later) and even *without the slightest distortion*. However, in order for the above process to be possible, the successive repetitions of the spectrum $\mathcal{X}_\alpha(j\Omega)$ *should not overlap with each other*, a property that is mathematically formulated as $\Omega_s - \Omega_N > \Omega_N$ or equivalently $\Omega_s > 2\Omega_N$ (see image (c) of Fig. 6.3). According to this property, *the full reconstruction of the original signal is possible only when the sampling frequency is greater than twice the maximum frequency contained in the signal being sampled*. Conversely, if $\Omega_s - \Omega_N \le \Omega_N$ or equivalently $\Omega_s \le 2\Omega_N$, the successive repetitions of the spectrum $\mathcal{X}_\alpha(j\Omega)$ *overlap with each other* (see Fig. 6.4) resulting in *it not being possible to reconstruct the original signal without errors*. In this case, the reconstructed signal $x_r(t)$ is a distorted version[4] of the original signal $x_\alpha(t)$. Examples of this distortion have been demonstrated in the previous pages (see, for example, Exercise 6.2.3 on page 506, in which the reconstructed continuous-time signal is different from the original).

In cases where the maximum frequency Ω_N of the continuous-time signal satisfies the condition $\Omega_N < \Omega_s/2$, so that *there is no overlap, as we described earlier*, the spectra of the signals before and after sampling satisfy, as we have seen, the condition

$$\mathcal{X}_r(j\Omega) = \begin{cases} T_s\,\mathcal{X}_s(j\Omega) & |\Omega| \le \Omega_s/2 \\ 0 & |\Omega| > \Omega_s/2 \end{cases} \tag{6.14}$$

and it is possible to fully reconstruct the continuous-time signal from these examples. Before, however, we describe the way in which this process is carried out, let us complete the analysis that we have started in the previous paragraphs and define the correlation that exists between the spectrum $\mathcal{X}_\alpha(j\omega)$ of the continuous-time signal $x_\alpha(t)$, which is sampled, and the spectrum $\mathcal{X}(e^{j\omega})$ of the discrete-time signal $x[n]$ that results from its sampling. Taking the Fourier transform of continuous-time of Eq. (6.10) (let us limit ourselves only to the first sum) and using the property of the selection of the time-shifted impulse function for the value $t_0 = nT_s$ and the function $\varphi(t) = e^{-i\Omega t}$ (see Eq. 1.23), it turns out that

$$\mathcal{X}_s(j\Omega) = \int_{-\infty}^{+\infty} x_s(t)\,e^{-j\Omega t} = \int_{-\infty}^{+\infty}\left\{\sum_{n=-\infty}^{\infty} x_\alpha(nT_s)\delta(t-nT_s)\right\} e^{-j\Omega t} = \sum_{n=-\infty}^{+\infty} x_\alpha(nT_s)\int_{-\infty}^{+\infty}\delta(t-nT_s)\,e^{-j\Omega t}dt$$

$$= \sum_{n=-\infty}^{+\infty} x_\alpha(nT_s)\,e^{-j\Omega(nT_s)} = \sum_{n=-\infty}^{+\infty} x_\alpha(nT_s)\exp\left(-j\frac{2\pi F}{F_s}n\right).$$

On the other hand, the Fourier transform of the discrete-time of the discrete sequence $x[n]$ is given, as known, by the relation

[4] In order to avoid the phenomenon of *frequency folding* or in any case to minimize the effects arising from this, it is customary before the application of the sampling process to transmit the continuous-time signal to a special filter (called an anti-aliasing filter), so that the frequency components $F \ge F_{\max}$ weaken as much as possible.

$$\mathcal{X}(e^{j\omega}) = \sum_{n=-\infty}^{+\infty} x[n]\, e^{-j\omega n}.$$

Comparing the last two relations and taking into account that $x[n] = x_{\alpha}(nT_s)$ it immediately follows that $\mathcal{X}_s(j\Omega) = \mathcal{X}(e^{j\omega})_{\omega=\Omega T_s} = \mathcal{X}(e^{j\Omega T_s})$. This relation correlates the spectra of the signals $x_s(t)$ and $x[n]$, whereas if combined with Eq. (9.81) it will give us

$$\mathcal{X}(e^{j\omega}) = \frac{1}{T_s} \sum_{k=-\infty}^{\infty} \mathcal{X}_{\alpha}\Big[j\Big(\frac{\omega}{T_s} - \frac{2k\pi}{T_s}\Big)\Big]. \tag{6.15}$$

This equation correlates the spectra $\mathcal{X}_{\alpha}(j\Omega)$ and $\mathcal{X}(e^{j\omega})$ of the continuous-time signal that is sampled and the corresponding discrete-time signal. From this relationship we observe that the spectrum of the discrete-time signal arises from the spectrum of the continuous-time signal through a frequency scaling process with the ratio coefficient being the sampling period T_s. This frequency scaling can be considered a process of normalizing the frequency axis, so that the frequency $\Omega = \Omega_s$ of the continuous-time signal frequency domain corresponds to the frequency $\omega = 2\pi$ of the discrete-time signal frequency domain. It is not difficult to understand that this frequency normalization that takes place during the transform $\mathcal{X}_s(j\Omega) \to \mathcal{X}(e^{j\omega})$ is the result of a corresponding time normalization that takes place during the transformation $x_s(t) \to x[n]$, since, as we have mentioned (see Footnote 2 (page 509) of this chapter), the distance between two consecutive samples of the discrete-time signal $x[n]$ is equal to one, in contrast to the distance between two consecutive samples of the equivalent continuous-time signal $x_s(t)$, which is equal to the sampling period T_s.

Returning now to the issue of reconstructing the original continuous-time signal, the reconstructed signal $x_r(t)$ will be calculated as the *inverse Fourier transform of the function* $\mathcal{X}_r(j\Omega)$; it will therefore be,

$$\begin{aligned} x_r(t) &= \frac{1}{2\pi}\int_{-\Omega_s/2}^{+\Omega_s/2} \mathcal{X}_r(j\Omega)\, e^{i\Omega t} d\Omega = \frac{1}{2\pi}\int_{-\Omega_s/2}^{+\Omega_s/2} T_s\, \mathcal{X}_s(j\Omega)\, e^{j\Omega t} d\Omega \\ &= \int_{-F_s/2}^{+F_s/2} T_s \Big[\sum_{n=-\infty}^{\infty} x_{\alpha}(nT_s) \exp\Big(-j\frac{2\pi F}{F_s} n\Big)\Big] e^{j2\pi F t} dF \\ &= T_s \sum_{n=-\infty}^{\infty} x_{\alpha}(nT_s) \Big\{\int_{-F_s/2}^{+F_s/2} \exp\Big[j2\pi F\Big(t - \frac{n}{F_s}\Big)\Big] dF\Big\}, \end{aligned}$$

where to derive the last equation we have alternated the operations of summation and integration and used the expression for the transform $\mathcal{X}_s(j\Omega)$ that we constructed previously.

It is not difficult to prove that the value of the above integral is calculated as

$$\int_{-F_s/2}^{+F_s/2} \exp\Big[j2\pi F\Big(t - \frac{n}{F_s}\Big)\Big] dF = \frac{1}{\pi(t - nT_s)} \sin\Big[\frac{\pi}{T_s}(t - nT_s)\Big]$$

and therefore the above equation will take the form

$$x_r(t) = \sum_{n=-\infty}^{\infty} x_{\alpha}(nT_s) \Big\{\sin\Big[\frac{\pi}{T_s}\Big(t - nT_s\Big)\Big]\Big\} \Big/ \Big\{\frac{\pi}{T_s}(t - nT_s)\Big\}, \tag{6.16}$$

which is also the final result. We observe that in this equation—which is known as the *interpolation equation*—the familiar function sinc appears, which is defined as

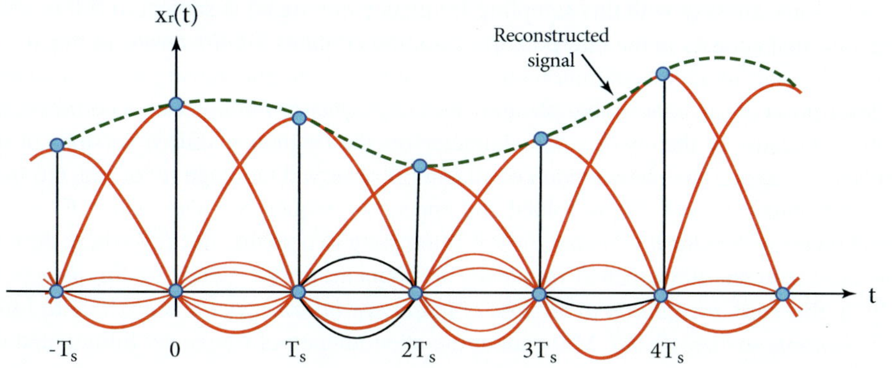

Fig. 6.5 The reconstruction of the original continuous-time signal through the application of the interpolation technique

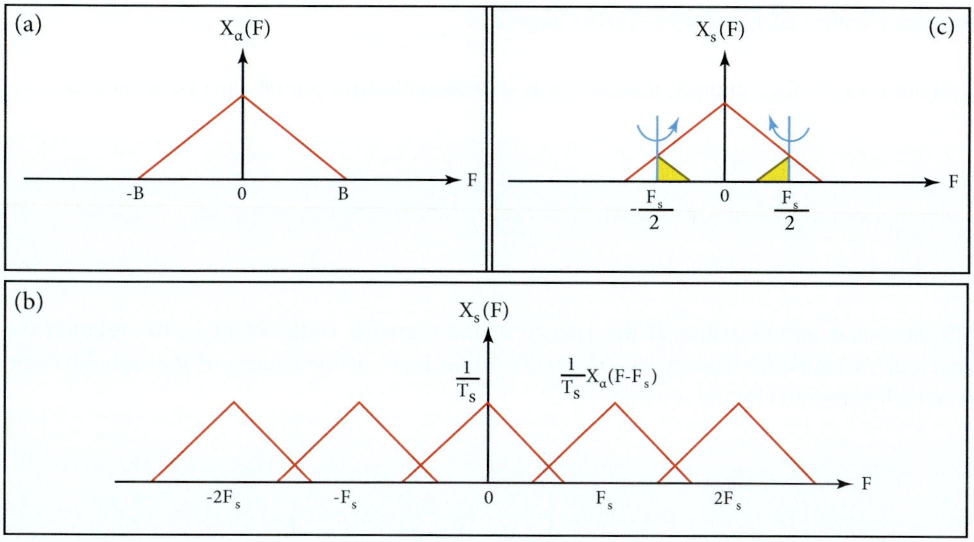

Fig. 6.6 Demonstration of the aliasing phenomenon

$$g(t) = \text{sinc}(t) = \sin(\pi t)/(\pi t) \tag{6.17}$$

shifted by the quantities nT_s ($n = 0, \pm 1, \pm 2, \dots$) and multiplied by the corresponding values of the continuous-time signal, which are $x_\alpha(nT_s)$ (for the graphical representation of this function refer to Fig. 1.12γ on page 30 and to Fig. 3.7 on page 211). We observe that at the time $t = kT_s$ the function $g(t - nT_s)$ has everywhere a value equal to *zero*, except at the position $k = n$, for which its value is equal to one, as results from the application of the *L'Hôpital's rule*. Therefore, the value of the function $x_r(t)$ at the position $t = kT_s$ is nothing but the sample $x_\alpha(kT_s)$, whereas for all the other times the weighted sum of the time-shifted versions of the interpolation function will give us exactly all the values of the signal $x_\alpha(t)$. This situation is presented diagrammatically in Fig. 6.5.

If the maximum frequency of the continuous-time signal that is assumed a limited spectrum signal is equal to $F_{\max}$, then the interpolation function takes the simple form

$$g(t) = \frac{\sin(2\pi F_{\max} t)}{2\pi F_{\max} t},$$

a relationship that also completes the proof of the theorem. Let us also note that *the sampling theorem holds true for the point of equality in the equation $\omega_s \geq 2\omega_N$, but only if the continuous-time signal $x_\alpha(t)$ does not contain a sinusoidal component of frequency ω_N*. The frequency value $\omega_s = 2\omega_N$ is called, as we have already mentioned, the *Nyquist frequency* and constitutes the *lowest possible limit, as far as the sampling frequency is concerned, so that it is possible to reconstruct a continuous limited spectrum signal from its recorded samples without errors*. This means that *we gain absolutely nothing by* *using higher frequencies*, since anyway with this sampling frequency our signal is subject to full reconstruction. Let us also note that although the sum that appears in the interpolation equation contains *infinite terms*, in practice only a finite number of signal samples are used, resulting in the continuous-time signal we reconstruct having *finite duration*.

We conclude the description of the undesirable phenomenon of *frequency folding* with a comment about the origin of its name. This name is given because of the *overlap* that characterizes the frequency-shifted versions of the spectrum $\mathcal{X}_\alpha(j\Omega)$ for sampling frequencies $\omega < 2\omega_N$. The phenomenon of aliasing is observed in image (c) of Fig. 6.6 from which we observe that high frequencies with values $F > F_s/2$ are folded and appear as frequencies with values $F < F_s$; for this reason, the frequency $F_s/2$ is also known as the *aliasing frequency*. It is interesting to note that in cases where the continuous-time signal we consider contains more than one fundamental frequency component, the frequency folding is applied to each frequency component separately. In this case, *in addition to the loss of frequency information, it is possible to lose information related to the amplitude of the continuous-time signal*. More about the phenomenon of frequency folding and the sampling theorem can be found in the literature.

6.3 Energy and Power of Discrete-Time Signals

In full analogy with continuous-time signals, the *energy* E of a discrete-time signal $x[n]$ is defined as

$$E = \sum_{n=-\infty}^{\infty} |x[n]|^2 \tag{6.18}$$

and can have both *finite* and *infinite* value. If the energy of the signal is finite—that is, the relation $0 < E < \infty$—the discrete-time signal $x[n]$ is called the *energy signal*. On the other hand, if the energy of the signal has an *infinite value* but its average power, which is defined by the relation

$$P = \lim_{N\to\infty}\left(\frac{1}{2N+1}\sum_{n=-N}^{N} |x(n)|^2\right) \tag{6.19}$$

is finite, the discrete-time signal $x[n]$ is known as the *power signal*.

Defining the energy of the signal over the finite range of values $-N \leq n \leq N$ as

$$E_N = \sum_{n=-N}^{N} |x[n]|^2 \tag{6.20}$$

it is obvious that the total energy will be given by the relation $E = \lim_{N\to\infty} E_N$, whereas the average power will be equal to

$$P = \lim_{N\to\infty}\left(\frac{1}{2N+1}E_N\right). \tag{6.21}$$

From the last relation, it immediately follows that if the energy E of the signal is *finite*, its average power P, will be equal to zero. On the other hand, if the energy E becomes infinite, the power P can have both finite and infinite value. We observe, therefore, that as far as the concepts of energy and power are concerned, discrete-time signals are characterized by those properties that also describe continuous-time signals.

Problem 6.3.1 Calculate the energy of the trigonometric signals

$$x_\kappa[n] = A\cos\left(\frac{2\kappa\pi n}{N}\right), \quad \kappa, n = 0, 1, 2, \ldots, N-1.$$

Solution From the defining equation of the energy of a discrete-time signal we have

$$E = \sum_{n=0}^{N-1} |x[n]|^2 = A^2 \sum_{n=0}^{N-1} \cos^2\left(\frac{2\kappa\pi n}{N}\right) = \frac{A^2}{2}\left\{\sum_{n=0}^{N-1}\left[1+\cos\left(\frac{4\kappa\pi n}{N}\right)\right]\right\} = \frac{A^2 N}{2} + \frac{A^2}{2}\sum_{n=0}^{N-1}\cos\left(\frac{4\kappa\pi n}{N}\right) = \frac{A^2 N}{2},$$

where we used the known trigonometric identity $\cos(2\alpha) = 2\cos^2\alpha - 1$ and took into account that

$$\sum_{n=0}^{N-1}\cos\left(\frac{4\kappa\pi n}{N}\right) = \Re\left\{\sum_{n=0}^{N-1}\exp\left(j\frac{4\kappa\pi n}{N}\right)\right\} = \Re\left\{\frac{e^{j(4\kappa\pi)}-1}{\exp[j(4\kappa\pi)/N]-1}\right\} = \Re\{0\} = 0.$$

Problem 6.3.2 Calculate the energy of the signal

$$x[n] = \begin{cases} e^{-n} & n \geq 0 \\ 0 & n < 0 \end{cases}.$$

Solution From the defining equation of the energy of the signal and using the appropriate expression from the mathematical typology, it turns out that

$$E = \sum_{n=-\infty}^{\infty} |x[n]|^2 = \sum_{n=0}^{\infty}(e^{-n})^2 = \sum_{n=0}^{\infty}(e^{-2})^n = \sum_{n=0}^{\infty}\left(\frac{1}{e^2}\right)^n = \frac{e^2}{e^2-1}.$$

6.4 Periodic Signals

A discrete-time signal $x[n]$ is called *periodic* with period N, if and only if it satisfies the condition

$$x[n+N] = x[n] \tag{6.22}$$

for every value of the discrete variable n. In this case, the smallest value of the parameter N for which the above relationship is satisfied is called the *fundamental period* of the signal. Conversely, if there is no value of N that satisfies this relationship, the signal is called *nonperiodic* or *aperiodic*. In full analogy with the continuous-time signals all periodic signals are power signals.

It is obvious that if the period of a signal is equal to N, then the signal in question has as periods the quantities $M = mN$ $(m = 1, 2, 3, \ldots)$ with the smallest of these being the *fundamental period.* If we consider two periodic signals $x_1[n]$ and $x_2[n]$ with period values N_1 and N_2 respectively, then the signals $x_1[n]+x_2[n]$ and $x_1[n]\cdot x_2[n]$ are also periodic with period value equal to $N = (N_1 N_2)/MK\Delta(N_1, N_2)$, where $MK\Delta(N_1, N_2)$ is the greatest common divisor of the integer numbers N_1 and N_2.

Taking into account that the value of N essentially expresses *the number of samples after which the periodic signal repeats itself again*, it is not difficult to realize that the value of N must necessarily be an *integer number* (for example, a periodic signal with a period of $N = 5$ means that the group of samples that is continuously repeated consists of five samples. This is certainly something that does not apply to continuous periodic signals, the period of which can be anything. Although this difference may not seem so significant, it leads to completely different characteristics of discrete-time signals than their continuous counterparts in terms of frequency size, which will be presented in a subsequent section.

We conclude our brief reference to *discrete periodic signals* with the observation that in full analogy with their continuous equivalents, they are characterized by *infinite energy*—since they have infinite duration—but by finite average power, which is defined as *the average energy of the signal included in a period interval.* If the value of this period is equal to N, then the average power of the periodic signal is given by the relation

$$P = \frac{1}{N} \sum_{n=\kappa}^{\kappa+N-1} |x[n]|^2 = \frac{1}{N} \sum_{n=<N>} |x[n]|^2 \tag{6.23}$$

with the symbol $< N >$ indicating any interval of length N.

Problem 6.4.1 The following discrete-time signals are given (Hayes, Problem 1.1, questions α and β)

(a) $x[n] = \cos\left(\frac{n\pi}{8}\right)$ (b) $x[n] = Re\left\{\exp\left(j\frac{n\pi}{12}\right)\right\} + \Im\left\{\exp\left(j\frac{n\pi}{18}\right)\right\}$.

For each of these, it should be determined whether it is periodic and if so, the value of its period should be calculated.

Solution The above exercises are solved by taking into account the fact that the functions of the sine and cosine are periodic with a period value of $T = 2\pi$ and that they satisfy the relations $\cos(x) = \cos(x + 2\pi)$ and $\sin(x) = \sin(x + 2\pi)$. In detail, then, we have:

(a) The discrete-time signal $x[n]$ is a cosine function—it will therefore be

$$(a)\ x[n] = \cos\left(\frac{n\pi}{8}\right) = \cos\left(\frac{n\pi}{8} + 2\pi\right) = \cos\left(\frac{n\pi}{8} + \frac{16\pi}{8}\right) = \cos\left[\frac{\pi}{8}(n + 16)\right] = x[n + 16]$$

and therefore, the signal is periodic with a period of $N = 16$.

(b) Using Euler's formula, we find that

$$\Re\left\{\exp\left(j\frac{n\pi}{12}\right)\right\} = \cos\left(\frac{n\pi}{12}\right) \quad \text{and} \quad \Im\left\{\exp\left(j\frac{n\pi}{18}\right)\right\} = \sin\left(\frac{n\pi}{18}\right)$$

and therefore the signal $x[n]$ can be expressed in the form

$$x[n] = x_1[n] + x_2[n] = \cos\left(\frac{n\pi}{12}\right) + \sin\left(\frac{n\pi}{18}\right).$$

In order, then, to determine the period of this signal, we must calculate the periods of the components $x_1[n]$ and $x_2[n]$ by applying the method we used in the previous case. In detail, then, we have:

$$(1)\ x_1[n] = \cos\left(\frac{n\pi}{12}\right) = \cos\left(\frac{n\pi}{12} + 2\pi\right) = \cos\left(\frac{n\pi}{12} + \frac{24\pi}{12}\right) = \cos\left[\frac{\pi}{12}(n + 24)\right] = x_1[n + 24]$$

$$(2)\ x_2[n] = \sin\left(\frac{n\pi}{18}\right) = \sin\left(\frac{n\pi}{18} + 2\pi\right) = \sin\left(\frac{n\pi}{18} + \frac{36\pi}{18}\right) = \sin\left[\frac{\pi}{18}(n + 36)\right] = x_2[n + 36],$$

that is, the components $x_1[n]$ and $x_2[n]$ are periodic discrete-time signals with period values $N_1 = 24$ and $N_2 = 36$ respectively. Therefore, according to the theory, the signal $x[n]$ will also be periodic with a period value

$$N = \frac{N_1 * N_2}{MK\Delta(N_1, N_2)} = \frac{24 * 36}{12} = 72.$$

Problem 6.4.2 The following discrete-time signals are given (Hayes, Solved Exercise 1.1, questions c and d)

$$\text{(a)} \quad x[n] = \sin(\pi + 0.2n), \quad \text{(b)} \quad x[n] = \exp\left(j\frac{n\pi}{16}\right)\cos\left(\frac{n\pi}{17}\right)$$

For each of these, it should be determined whether it is periodic and if so, the value of its period should be calculated.

Solution (a) For the signal $x[n]$ to be periodic, there must be a positive integer N such that $x[n] = x[n+N]$ or equivalently

$$\sin(\pi + 0.2n) = \sin[\pi + 0.2(n+N)] = \sin(\pi + 0.2n + 0.2N).$$

Taking into account that the sine function is periodic with period $T = 2\pi$, it should be $\sin(\pi + 0.2n) = \sin(\pi + 0.2n + 2m\pi)$ ($m = 1, 2, 3, \dots$). Comparing this relationship with the one we previously concluded leads to the conclusion that for the above identity to hold, the quantity $0.2N$ must be an integer multiple of 2π. Therefore, $0.2N = 2m\pi$, from which it follows that $N = 10m\pi$. However, we know that the constant π is an irrational number; therefore, there is no positive integer N that satisfies the above property, that is, the discrete-time signal $x[n] = \sin(\pi + 0.2n)$ is not periodic.

(b) The discrete-time signal $x[n]$ is a product of the form $x[n] = x_1[n] * x_2[n]$ with the signals $x_1[n]$ and $x_2[n]$ being given by the relations

$$x_1[n] = \exp\left[j\left(\frac{n\pi}{16}\right)\right] \quad \text{and} \quad x_2[n] = \cos\left(\frac{n\pi}{17}\right).$$

Therefore, in order to calculate its period, we need to determine the periods N_1 and N_2 of the signals $x_1[n]$ and $x_2[n]$ respectively. In detail, then, we have:

$$x_1[n] = \exp\left(j\frac{n\pi}{16}\right) = \cos\left(\frac{n\pi}{16}\right) + j\sin\left(\frac{n\pi}{16}\right) = \cos\left(\frac{n\pi}{16} + 2\pi\right) + j\sin\left(\frac{n\pi}{16} + 2\pi\right) = \cos\left(\frac{n\pi}{16} + \frac{32\pi}{16}\right)$$
$$+ j\sin\left(\frac{n\pi}{16} + \frac{32\pi}{16}\right) = \cos\left[\frac{\pi}{16}(n+32)\right] + j\sin\left[\frac{\pi}{16}(n+32)\right] = \exp\left[j\frac{(n+32)\pi}{16}\right] = x_1[n+32].$$

On the other hand, for the signal $x_2[n]$ applying the previous process we have

$$x_2[n] = \cos\left(\frac{n\pi}{17}\right) = \cos\left[\frac{\pi}{17}(n+34)\right] = x_2[n+34].$$

From the above analysis, we find that the components $x_1[n]$ and $x_2[n]$ are periodic discrete-time signals with period values $N_1 = 32$ and $N_2 = 34$ respectively. Therefore, according to the theory, the signal $x[n]$ will also be periodic with a period value

$$N = \frac{N_1 * N_2}{MK\Delta(N_1, N_2)} = \frac{32 * 34}{2} = 544.$$

Problem 6.4.3 Calculate the average power of the periodic signal of period N, which is defined as

$$x[n] = A\cos(\omega n + \varphi) = A\cos(2\pi f n + \varphi) = A\cos\left(\frac{2\pi n}{N} + \varphi\right).$$

Solution A periodic signal has *infinite duration* and *infinite energy*, whereas its average power, which is finite, is calculated for a duration of one period. Working as in Exercise 6.3.1 (refer to page 515) we find that

$$P_{\text{average}} = \frac{1}{N}\sum_{n=0}^{N-1} A^2 \cos^2\left(\frac{2\pi n}{N}+\varphi\right) = \frac{A^2}{2N}\sum_{n=0}^{N-1}\left[1+\cos\left(\frac{4\pi n}{N}+\varphi\right)\right] = \frac{A^2}{2}.$$

6.5 Even and Odd Signals

A discrete-time signal $x[n]$ of real values is called *even* or *symmetric*, if it satisfies the relation

$$x[-n] = x[n] \tag{6.24}$$

for every value of the discrete variable n. On the other hand, if the signal $x[n]$ is such that

$$x[-n] = -x[n] \tag{6.25}$$

the signal is called *odd* or *antisymmetric*. If a signal is characterized by odd symmetry, it immediately follows that $x[0] = 0$.

In full analogy with continuous-time signals, *each discrete-time signal* $x[n]$ *can be written as the sum of an even and an odd component*, or in mathematical notation

$$x[n] = x_e[n] + x_o[n] \tag{6.26}$$

with the signals $x_e[n]$ and $x_o[n]$ characterized by even and odd symmetry and defined as

$$x_e[n] = \frac{1}{2}\Big(x[n] + x[-n]\Big) \quad \text{and} \quad x_o(n) = \frac{1}{2}\Big(x[n] - x[-n]\Big)$$

(for a full description of the theory of even and odd signals refer to Sect. 1.4 on page 12, since the situation is completely analogous to that describing continuous-time signals). A typical example of an even (symmetric) and odd (antisymmetric) signal is presented in Fig. 6.7. From this figure we find that *even signals are symmetric with respect to the vertical axis, whereas odd signals are symmetric with respect to the origin of the axes*, a fact that is expected.

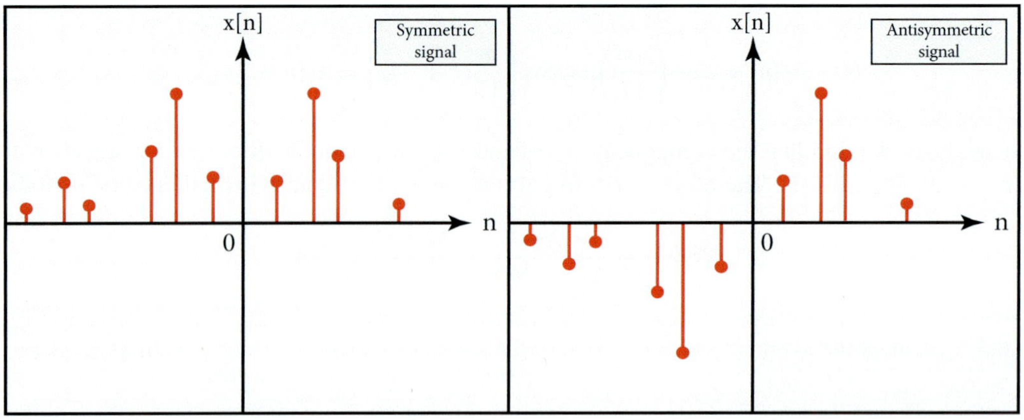

Fig. 6.7 Example of an even and an odd discrete-time signal

Examples of calculating the even and odd components of elementary discrete-time signals are given in a subsequent section.

Problem 6.5.1 Prove that the energy of a discrete-time signal $x[n]$ is equal to the sum of the energies of its even and odd components.

Solution Taking into account that the discrete-time signal $x[n]$ is defined as the sum $x[n] = x_e[n] + x_o[n]$, the equation defining the energy of the signal will take the form

$$E = \sum_{n=-\infty}^{\infty} x^2[n] = \sum_{n=-\infty}^{\infty} \left(x_e[n] + x_o[n]\right)^2 = \sum_{n=-\infty}^{\infty} x_e^2[n] + \sum_{n=-\infty}^{\infty} x_o^2[n] + 2\sum_{n=-\infty}^{\infty} x_e[n]x_o[n]$$

However, since the signal $x_e[n]$ is characterized by even symmetry, whereas the signal $x_o[n]$ is characterized by odd symmetry, their product $x_e[n]x_o[n]$ will be an odd sequence[5] and according to the basic theory

$$\sum_{n=-\infty}^{\infty} x_e[n]x_o[n] = 0$$

Therefore, the above equation will take the form

$$E = \sum_{n=-\infty}^{\infty} x^2[n] = \sum_{n=-\infty}^{\infty} x_e^2[n] + \sum_{n=-\infty}^{\infty} x_o^2[n] = E_e + E_o,$$

that is, *the total energy of a signal is equal to the sum of the energies of its even and odd components* (note that a similar property characterizes continuous-time signals as well).

Problem 6.5.2 Find the value of the energy of the odd component of a discrete-time signal $x[n]$ if its even component is given by the relation $x_e[n] = (1/2)^{|n|}$, whereas the energy of the signal $x[n]$ has a value of $E = 5$.

Solution Based on the result of the previous exercise, the energy of the odd component $x_o[n]$ is calculated as $E_o = E - E_e = 5 - E_e$, where E_e is the energy of the even component of the signal, which is given by the relation

$$E_e = \sum_{n=-\infty}^{\infty} x_e^2[n] = \sum_{n=-\infty}^{\infty} \left(\frac{1}{2}\right)^{2|n|} = -1 + 2\sum_{n=0}^{\infty} \left(\frac{1}{4}\right)^n = -1 + 2\frac{1}{1-(1/4)} = \frac{5}{3}.$$

Therefore, $E_o = E - E_e = 5 - E_e = 5 - (5/3) = 10/3$, which is the desired result.

6.6 Random and Stochastic Signals

Although what we have mentioned so far exclusively concerns deterministic signals whose values are assumed to arise in a specific way, such as the use of a mathematical expression, in practice, there are situations in which the processes that generate the signals used in each case are excessively complex, to the extent that the exact description of the signal

[5] See Sect. 1.4 on page 12 that deals with the symmetry properties of continuous-time signals.

is extremely difficult or even impossible. Examples of such situations appear in meteorology, where the fluctuations of temperature and air pressure are impossible to describe through an equation and are essentially considered random, as well as the noise that appears in a set of applications, such as telecommunications. For this reason, in order to study the signals related to these situations, we resort to the use of statistical methods, in exactly the same way as in thermodynamics, in which the huge number of gas molecules that exist in a container makes possible the use of any other method.

In cases where the signals we use are characterized by some degree of uncertainty, which makes it impossible to predict the values of the signal samples, these are characterized as stochastic. Stochastic signals are often also characterized as random signals, although in some cases these signals differ from each other based on the degree of uncertainty that characterizes them, which in random signals is absolute. The existence of this uncertainty makes it impossible to process the signals with the known tools we use in usual cases and which, therefore, is carried out in a different way and more specifically, with the help of statistical methods such as statistical averages and the probability density function, both in the time domain and in the frequency domain. This means that the amplitude of the signal at a certain time is not exactly known but is determined with the help of some probabilities. Therefore, each sample $x[n]$ of the considered signal is associated with some random stochastic variable x_n, with the set of all these variables constituting a stochastic process. Such variables, for example, are the temperatures recorded in a region during the day, as well as the values of thermal noise produced by the resistors of a circuit.

Note that although in most common cases stochastic signals do not have a Fourier transform as they are not absolutely or square summable, many of the properties of these signals, but not all, can be described using autocorrelation sequences for which, often, the Fourier transform can be calculated. For more details about these types of signals and the properties that describe them, see Chap. 13 of the textbook.

6.7 The Elementary Signals of Discrete-Time

In full analogy with the continuous-time signals presented in the first chapter, discrete-time signals can also be expressed with the help of *elementary* or *fundamental discrete-time signals*, the most important of which include the *elementary trigonometric and exponential signals*, the *unit step function*, and the *unit impulse function*. The analytical description of these elementary signals is the subject of the following pages.

6.7.1 Elementary Trigonometric and Exponential Signals

The *elementary trigonometric signal of discrete-time* is given by the relation

$$x[n] = A\cos(\omega n + \varphi) = A\cos(2\pi f n + \varphi) = A\cos\left(\frac{2\pi n}{N} + \varphi\right), \tag{6.27}$$

where A is the amplitude of the signal, f its frequency, $\omega = 2\pi f$ its circular frequency, φ its phase, and N its period, which must necessarily be an integer. The above relation refers to the *cosine* signal, whereas the corresponding *sine* signal has the form

$$x[n] = A\sin(\omega n + \varphi) = A\sin(2\pi f n + \varphi) = A\sin\left(\frac{2\pi n}{N} + \varphi\right). \tag{6.28}$$

The elementary discrete-time signals of the sine and cosine are presented in Fig. 6.8.

On the other hand, the general form of an *elementary exponential discrete-time signal* is

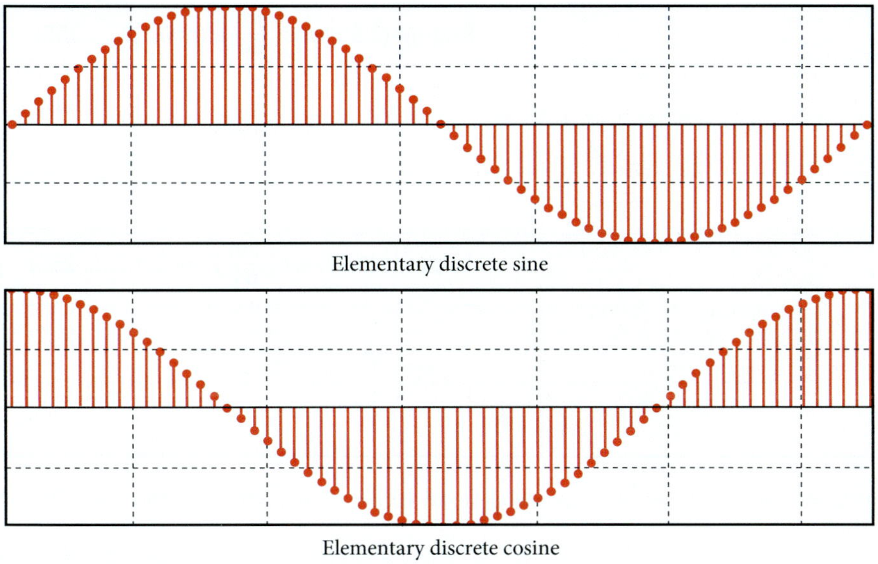

Fig. 6.8 The elementary discrete trigonometric signals

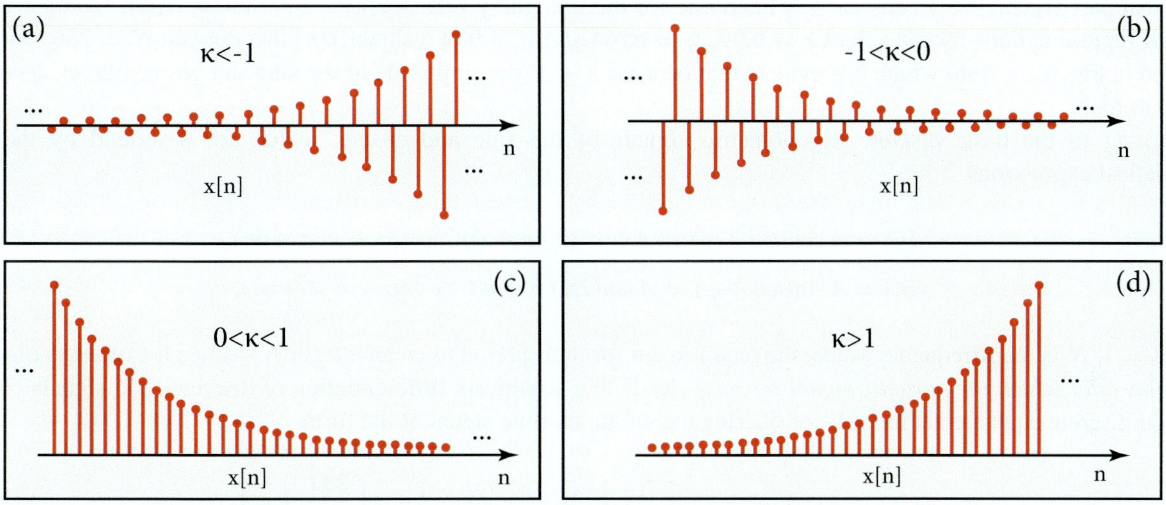

Fig. 6.9 The fundamental exponential signal $x[n] = \kappa^n$ for the various values of the real parameter κ

$$x[n] = A\kappa^n \tag{6.29}$$

with the exponent n naturally taking the integer values $n = 0, 1, 2, \ldots$ and the parameters A and κ being in the general case *complex numbers*. Expressing these complex quantities in their exponential form $A = \alpha e^{j\varphi}$ and $\kappa = re^{j\omega}$, the discrete-time signal $x[n]$ is formulated as

$$x[n] = \alpha e^{j\varphi}(re^{j\omega})^n = (\alpha r^n)\, e^{j(\omega n+\varphi)} = (\alpha r^n)\cos(\omega n + \varphi) + j(\alpha r^n)\sin(\omega n + \varphi), \tag{6.30}$$

where for the derivation of the last relation we used the Euler formula.

In the special case where $A = 1$, the discrete-time signal takes the form $x[n] = \kappa^n$ and its graphical representation for the cases (a) $\kappa < -1$, (b) $-1 < \kappa < 0$, (c) $0 < \kappa < 1$, and (d) $\kappa > 1$ is presented in Fig. 6.9, from which we observe that depending on the value of the constant κ the signal will *decrease* or *increase* in relation to time with the successive

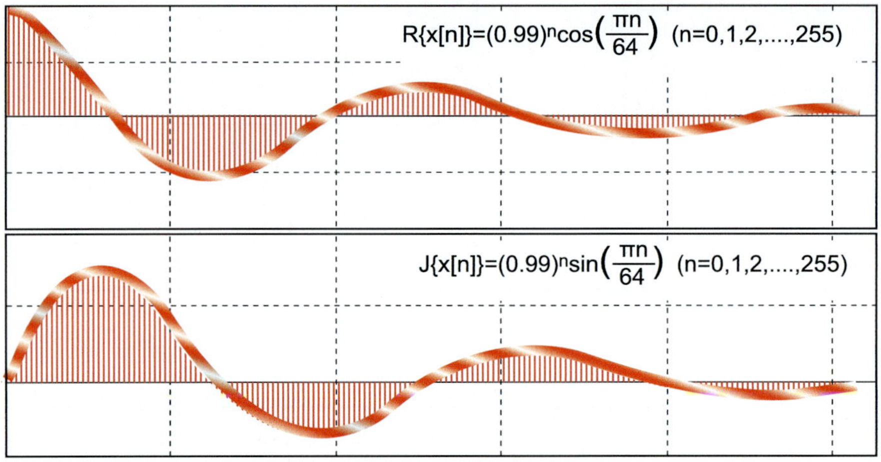

Fig. 6.10 The real and imaginary parts of the complex exponential signal $x[n] = \kappa^n$ for parameter values $r = 0.99$, $\omega = \pi/64$, and $\varphi = 0$

values having *the same sign* or characterized by an *alternation of signs*. In the above figure, the parameter κ is considered a *real number*, whereas for the case of a complex quantity we obviously need to construct two graphical representations, one for the real part $\Re\{x[n]\} = r^n \cos(\omega n + \varphi)$ and one for the imaginary part $\Im\{x[n]\} = r^n \sin(\omega n + \varphi)$. Examples of such graphical representations for the values $r = 0.99$, $\omega = \pi/64$ and $\varphi = 0$ of a signal $x[n]$ that contain $N = 256$ samples are presented in Fig. 6.10, from which it is evident that because $r < 1$, the amplitude of the sine and cosine curves *decreases in relation to time*.

Returning to the basic discrete trigonometric signals of the sine and cosine, which are described by the simple mathematical expressions

$$x[n] = A\cos(\omega n + \varphi) = A\cos(2\pi f n + \varphi) \quad (-\infty < n < +\infty)$$
$$y[n] = A\sin(\omega n + \varphi) = A\sin(2\pi f n + \varphi) \quad (-\infty < n < +\infty),$$

where $f = 1/N$ is their frequency value, the requirement for their period to be an integer N—since it expresses *the number of samples after which the periodic signal repeats*—leads to a significant differentiation of discrete-time signals compared with their discrete equivalents. Indeed, considering a continuous-time signal of the form

$$x(t) = A\cos(\omega t + \varphi) = A\cos(2\pi f t + \varphi) + A\cos\left(\frac{2\pi t}{T} + \varphi\right) \tag{6.31}$$

it is characterized by the following properties regarding its frequency:

- *For every value of the frequency f, the signal $x(t)$ is periodic, with a period value of $T = 1/f$*. This property is proven using the inverse proportionality relation $fT = 1$ and taking into account that the cosine is a *periodic function* with a period equal to 2π. Therefore, we have

$$x(t+T) = A\cos[2\pi f(t+T)+\varphi] = A\cos(2\pi f t + 2\pi f T + \varphi) = A\cos[(2\pi f t + \varphi) + 2\pi] = A\cos(2\pi f t + \varphi) = x(t)$$

  The proof of the above property was based on the observation that since the period T of a periodic signal is a continuous quantity that can take any real value, then whatever the value of the frequency f, *the value $T = 1/f$ will always be defined* and there will always be a continuous periodic signal with this period value.
- Considering two continuous sinusoidal signals with frequencies f_1 and f_2, these in all cases are *different and distinct from each other*. This property is a consequence of the fact that the frequency of a signal expresses the number of repetitions of the signal per unit of time and obviously two signals that repeat at different rates can always be distinguished from each other.

- *The increase in frequency f leads to an increase in the oscillation rate of the signal*, as the number of full repetitions per unit of time increases. This property arises directly from the definition of frequency, as well as from the previous observation.

Let us now try using the same argument to reformulate the above properties for the case of *discrete-time signals*, checking their validity at the same time. Starting from the first property, let us consider a discrete sinusoidal signal of the form $x[n] = A\cos(\omega n + \varphi)$ and see under what conditions provided the relation $x[n] = x[n+N]$ holds, so that this can be characterized as *periodic*. Following the methodology of continuous-time signals, we can write that

$$x[n+N] = A\cos[2\pi f(n+N) + \varphi] = A\cos(2\pi f n + 2\pi f N + \varphi).$$

At this point, can we assert that for the current value of the frequency f there is a period $N = 1/f$, so that $fN = 1$ and we can continue the analysis as before? In the general case *no, since N expresses the number of samples that are repeated, it must necessarily have an integer value*. Are there, of course, frequency values for which this requirement is met? For example, for the frequency $f = 1/8$ it is $N = 1/f = 8$ and the requirement is met. Generally, however, this is not the case and in order to be able to eliminate the term $2\pi f N$ from the above relation, we must resort to some other approach. It is not difficult to see that since the cosine is a *periodic function* with a period of $2\kappa\pi$, the above term is eliminated if it is set to this value. Indeed, if we set $2\pi f N = 2\kappa\pi$ (or equivalently $f = \kappa/N$), the above relation takes the form

$$\begin{aligned} x[n+N] &= A\cos(2\pi f n + 2\pi f N + \varphi) = A\cos(2\pi f n + 2\kappa\pi + \varphi) \\ &= A\cos[(2\pi f n + \varphi) + 2k\pi] = A\cos(2\pi f n + \varphi) = x[n], \end{aligned}$$

thus allowing the signal to be characterized as *periodic*. Taking into account that the parameters κ and N are integers, we can formulate the claim that *a discrete-time signal is periodic when its frequency f is a rational number, that is, it can be written as the quotient of two integers*.

Let us now move on to the second property of continuous-time signals according to which *two signals with different frequencies f_1 and f_2 can always be distinguished from each other*. Does this hold for discrete-time signals? The answer is *not always*. In order to verify this claim, let us consider two discrete-time signals with frequencies f_1 and f_2 of the form

$$x[n] = A\cos(\omega_1 n + \varphi) = A\cos(2\pi f_1 n + \varphi) \qquad \text{and} \qquad y[n] = A\cos(\omega_2 n + \varphi) = A\cos(2\pi f_2 n + \varphi)$$

and let us assume that the frequencies f_1 and f_2 differ by an integer value, so that we can write that $f_2 = f_1 + m$ or equivalently

$$\omega_2 = 2\pi f_2 = 2\pi(f_1 + m) = 2\pi f_1 + 2\pi m = \omega_1 + 2\pi m.$$

In this case, we find that

$$\begin{aligned} y[n] &= A\cos(\omega_2 n + \varphi) = A\cos[(\omega_1 + 2\pi m)n + \varphi] = A\cos[\omega_1 n + 2\pi(mn) + \varphi] \\ &= A\cos(\omega_1 n + 2\kappa\pi + \varphi) = A\cos(\omega_1 n + \varphi) = x[n] \end{aligned}$$

(where we have set $\kappa = mn$). We are led, therefore, to the conclusion that *two discrete periodic signals whose cyclic frequencies are a distance equal to an integer multiple of 2π are equivalent and cannot be distinguished from each other*. This is certainly something that does not characterize continuous-time signals, which can always be distinguished and separated, as long as they are characterized by different frequency values.

Finally, let us see what happens with the third property, according to which *the higher the frequency of a continuous-time signal, the more oscillations characterize its temporal evolution*. Does something like this apply to discrete-time signals? The answer is no, since according to the previous property two signals with frequencies differing by an integer multiple of 2π are identical and cannot be distinguished from each other. This situation is demonstrated diagrammatically in Fig. 6.11, in which discrete-time signals with steadily increasing frequency values are presented. Starting from the smallest possible frequency value, which is $\omega = 0$, the signal corresponding to this does not show any oscillations; indeed, if our signal has the form $x[n] = A\cos(\omega n)$ (where for simplicity we have considered the case of zero phase), the signal that results by setting $\omega = 0$ is $x[n] = A$, a signal of constant values that is not characterized by the existence of *oscillations*, that is, *alternations*

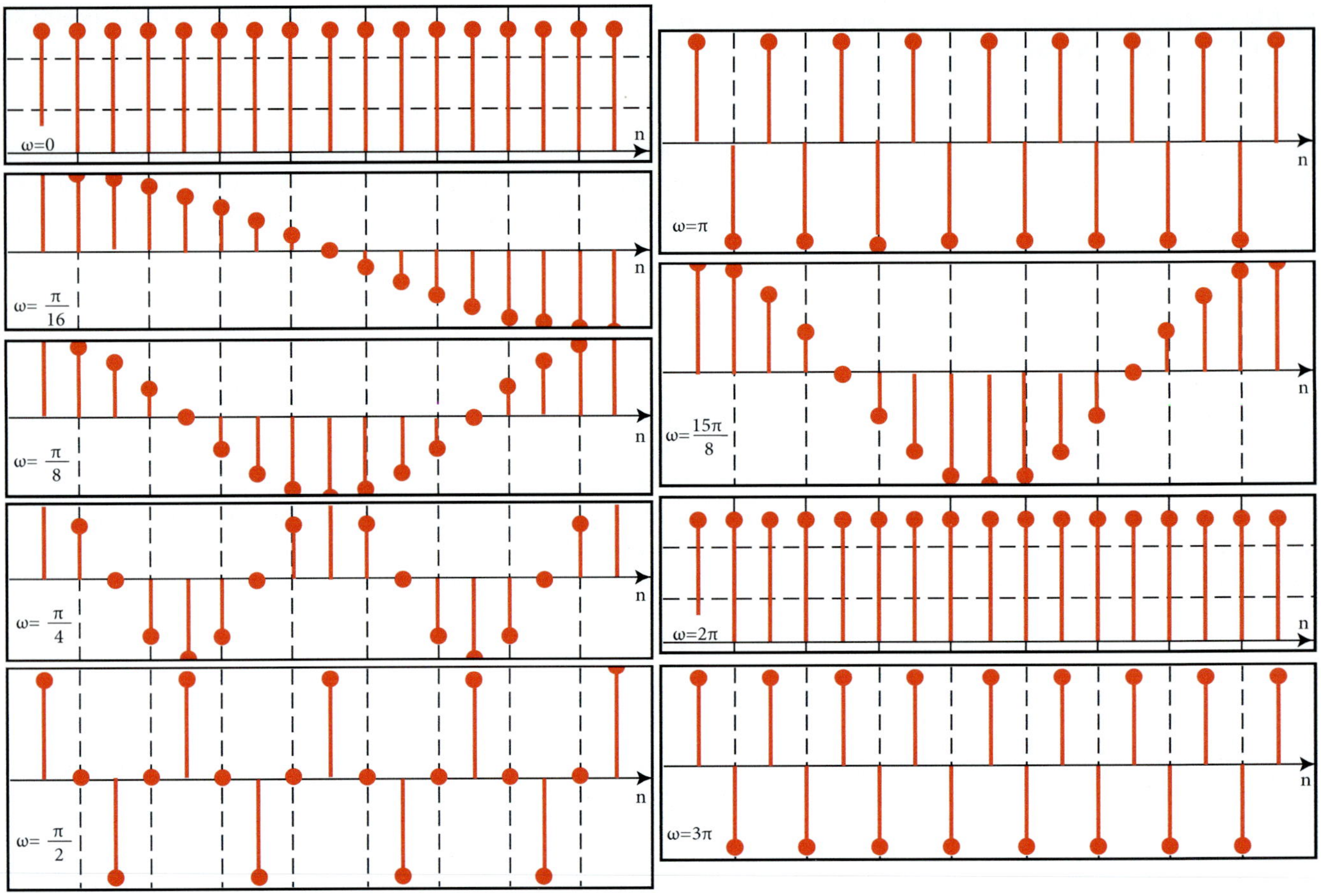

Fig. 6.11 Discrete trigonometric signals with increasing frequency value

in sign. As the frequency begins to increase, the number of oscillations also increases, as shown by the discrete-time signals of the figure corresponding to the frequencies $\omega{=}\pi/16$, $\omega{=}\pi/8$, $\omega{=}\pi/4$, $\omega{=}\pi/2$, and $\omega{=}\pi$.

For which frequency value ω does the oscillation rate of the signal reach its maximum value? It is obvious that the discrete-time signal that oscillates at the maximum possible rate is the one in which any two consecutive samples are characterized by a sign change (i.e., positive value, negative value, positive value, negative value, etc. There is nothing faster than this). But this means that since the group of repeating samples contains two samples, the minimum period of the signal will be equal to $N = 2$ (we remind the reader that *the value* $N = 1$ *does not constitute a valid period value, as it refers to a signal of constant values that can have any period*) and the maximum frequency corresponding to this minimum period is $\omega = 2\pi/N = \pi$; this is also the *maximum frequency* that a discrete-time signal can have.

On the other hand, if the frequency of the signal falls within the range of values $\pi \leq \omega \leq 2\pi$, the number of oscillations *decreases*. In order to prove the truth of this claim, let us consider two discrete-time signals $x[n]$ and $y[n]$ with frequencies $\omega_1 = \omega$ and $\omega_2 = 2\pi - \omega$; it is obvious that the two frequencies are such that $\omega_2 < \omega_1$ with $\omega_2 \in [0, \pi]$. It is not difficult to see that the two signals are of the same sign, as it is found that

$$y[n] = A\cos(\omega_2 n) = A\cos[(2\pi - \omega)n] = A\cos(-\omega n) = A\cos(\omega n) = A\cos(\omega_1 n) = x[n].$$

Therefore, for frequency values in the range $[\pi, 2\pi]$ the number of oscillations decreases, with its maximum value corresponding to the frequency value $\omega = \pi$.

What happens in cases where the signal frequency has a value $\omega > 2\pi$? The situation is now much simpler: for each such frequency there is a signal with the frequency $\omega' = \omega - 2\pi$ (which is within the range $0 \leq \omega \leq \pi$) that is *identical* to the considered signal, since their frequencies are 2π apart. This situation is presented in Fig. 6.11, from which we observe that signals with frequencies $\omega = \pi/8$ and $\omega = 17\pi/8$ are *exactly the same*. This observation can be generalized for arbitrarily large frequency values, so that we are able to state the claim that *all sinusoidal or cosinusoidal discrete sequences* $x_k(n) = A\cos(\omega_k n + \varphi)$ $(k = 0, 1, 2, \ldots)$ *where* $\omega_k = \omega_0 + 2k\pi$ $(-\pi \leq \omega_0 \leq \pi)$ *are identical and equivalent to the discrete-time signal of frequency* ω_0, *so that they cannot be distinguished from each other*. Taking into account that the minimum frequency value is $\omega = 0$, whereas the maximum frequency value is $\omega = \pi$, and using the above property, we can state the proposition that *the frequency axis for the case of discrete-time signals consists of low-frequency areas with frequencies adjacent to the values* $\omega_n = 0, 2\pi, 4\pi, \ldots, 2\kappa\pi$, *which alternate with high-frequency regions with frequencies adjacent to the values* $\omega = \pi, 3\pi, 5\pi, \ldots, (2\kappa + 1)\pi$ (for the values $k = 0, 1, 2, \ldots$). All the above frequency values are equivalent to each other and equivalent to some frequency that belongs to the interval $[0, \pi]$. Conversely, the frequencies that belong to this interval are *discrete* and *different from each other*, since, owing to the width of this interval, it is not possible to be at a distance from each other equal to 2π, so as to be equivalent to each other. For this reason, the frequencies that belong to the interval $[0, \pi]$, and that are unique, are considered *true* or *genuine frequencies*, whereas for each of these genuine frequencies the set of an infinite number of equivalent frequencies that are outside the above interval and are at a distance equal to some integer multiple of 2π consists of the *pseudo-epigraphic frequencies*, in the sense that these constitute *false copies of the genuine frequencies*.

If we extend the above description to the semi-axis of negative frequencies (which are defined through the exponential signals of the form $e^{-j\omega n}$ and are taken into account in the description of a signal in the frequency domain), the image we get will be the *mirror* of the one we presented, so that we can say that the *true* or *genuine* frequencies satisfy the property $|\omega| \leq \pi$ or equivalently $-\pi \leq \omega \leq \pi$, whereas all other frequencies are their aliases. Finally, if we use the simple frequency f for the circular frequency, it is not difficult to find that the genuine frequencies are those that belong to the value range $-\pi \leq 2\pi f \leq \pi$ or equivalently $-1/2 \leq f \leq 1/2$.

Note 1 We conclude our reference to the different nature of continuous and discrete-time signals regarding the concept of frequency,[6] with the observation that the requirement of rendering integer values to the period N of a periodic signal makes nonperiodic, signals that in their continuous form are *classic examples of periodic signals*. Let us consider, for example, the discrete-time signal $x[n] = \cos(n)$. For this signal to be periodic with a period value equal to N, there must be an integer variable N such that $x[n] = x[n + N]$ or equivalently $\cos(n) = \cos(n + N)$. It is obvious that there is no integer N that satisfies the above property and therefore the discrete-time signal $x[n] = \cos(n)$ is not periodic, something that applies to its continuous equivalent, which is known to be a periodic function with a period of 2π.

Note 2 The combination of the requirement of the *explicit nature* of the frequency of a discrete periodic signal with the property of *equivalence* that characterizes two periodic discrete-time signals, the frequencies of which differ by an integer multiple of 2π, leads us to a very interesting conclusion: considering any integer value N, *there are equal discrete frequency values*

$$\omega_\kappa = 2\pi \frac{\kappa}{N} \quad (\kappa = 0, 1, 2, \ldots, N - 1) \tag{6.32}$$

for which the corresponding discrete-time signals

[6] Although the proof of the above properties was based on the use of trigonometric signals, things are exactly the same for the *elementary complex exponential signals*. Considering, for example, such a signal of the form $x[n] = e^{j\omega n}$, it is periodic with period N only when $\omega N = 2\kappa\pi$ or equivalently $fN = \kappa$, since only then will we have

$$x[n + N] = e^{j\omega(n+N)} = e^{j\omega n}e^{j\omega N} = e^{j\omega n}e^{j(2\kappa\pi)} = e^{j\omega n} = x[n].$$

In a completely analogous way, if we consider two exponential signals of the form $x[n] = e^{j\omega_1 n}$ and $y[n] = e^{j\omega_2 n}$ with frequencies $\omega_2 = \omega_1 + 2\pi m$, they are identical since we have

$$y[n] = e^{j\omega_2 n} = e^{j(\omega_1 + 2\pi m)n} = e^{j\omega_1 n}e^{j(2\pi m)n} = e^{j\omega_1 n} = x[n].$$

$$x_\kappa[n] = A\cos(\omega_\kappa n + \varphi) \quad \text{or} \quad x_\kappa[n] = A\sin(\omega_\kappa n + \varphi) \quad (k = 0, 1, 2, \ldots, N-1)$$

are periodic with a common period value equal to N. Considering, for example, the parameter values $N = 8$, $A = 1$ and $\varphi = 0$, the eight discrete-time signals defined by the above relations with frequency values

$$\omega = 0, \frac{\pi}{4}, \frac{\pi}{2}, \frac{3\pi}{4}, \pi, \frac{5\pi}{4}, \frac{6\pi}{4}, \frac{7\pi}{4}$$

are *periodic*, with a period of $N = 8$, since, for example

$$x_3[n] = \cos\left(\frac{3\pi n}{4}\right) = \cos\left(\frac{3\pi n}{4} + 6\pi\right) = \cos\left[\frac{3\pi}{n}(n+8)\right] = x_3[n+8]$$

$$x_7[n] = \cos\left(\frac{7\pi n}{4}\right) = \cos\left(\frac{7\pi n}{4} + 14\pi\right) = \cos\left[\frac{7\pi}{n}(n+8)\right] = x_7[n+8],$$

etc. The above property also applies to *elementary complex exponential signals*, so that we can formulate the claim that *for every integer value N there are N exponential complex signals*

$$x_\kappa[n] = e^{j\omega_\kappa n} = \exp\left(j\frac{2\pi\kappa n}{N}\right) \quad (\kappa = 0, 1, 2, \ldots, N-1), \tag{6.33}$$

which all have a common period N. These properties of complex exponentials and trigonometric signals form the basis of *Fourier analysis* in the field of discrete-time, which will be presented in detail in a subsequent chapter.

Problem 6.7.1 Characterize in terms of their energy and power the elementary complex trigonometric signals of the form $x[n] = Ae^{j\omega n}$.

Solution From the equation defining the energy of the signal and taking into account that $|e^{j\omega n}| = 1$, we have

$$E = \sum_{n=-\infty}^{\infty} |x[n]|^2 = \sum_{n=-\infty}^{\infty} |A|^2 |e^{j\omega n}|^2 = \sum_{n=-\infty}^{\infty} A^2 \to \infty.$$

We observe that the energy of these signals *is not finite* and therefore they cannot be characterized as *energy signals*. However, it is not difficult to find that these signals are *power signals*, since they have *finite power*, as derived from the equation defining their power:

$$P = \lim_{N\to\infty}\left(\frac{1}{2N+1}\sum_{n=-N}^{+N} |Ae^{j\omega n}|^2\right) = \lim_{N\to\infty}\left(\frac{1}{2N+1}\sum_{n=-N}^{N} A^2\right) = \lim_{N\to\infty}\left[\frac{1}{2N+1}A^2(2N+1)\right] = A^2$$

Problem 6.7.2 Prove that the set of complex exponential signals

$$x_\kappa[n] = \exp\left(j\kappa\frac{2\pi}{N}n\right)$$

constitute an *orthogonal basis of functions* within any interval of length N.

Solution In order to solve the exercise, we must prove that *the complex exponential signals* $x_\kappa[n]$ *satisfy the orthogonality condition*. From the basic theory it is known that the orthogonality condition of functions within any interval $[N_1, N_2]$ for the case of discrete-time is described by the equation

$$P = < x_m[n], x_\kappa[n] > = \sum_{n=N_1}^{N_2} x_m[n]x_\kappa^*[n] = \begin{cases} M & m = \kappa \\ 0 & m \neq \kappa \end{cases}, \tag{6.34}$$

where $< x_m[n], x_\kappa[n] >$ is the *inner product* of the functions $x_m[n]$ and $x_\kappa[n]$, M is a constant different from zero, whereas the operator "*" refers to the operation of *complex conjugation*. In this case, the limits of this interval are $N_1 = 0$ and $N_2 = N - 1$, whereas the functions $x_m[n]$ and $x_\kappa^*[n]$ are defined as

$$x_m[n] = \exp\left(jm\frac{2\pi}{N}n\right) \quad \text{and} \quad x_\kappa^*[n] = \exp\left(-j\kappa\frac{2\pi}{N}n\right).$$

Therefore, the inner product of the *complex exponential signals* takes the form

$$P = \sum_{n=0}^{N-1} \exp\left(jm\frac{2\pi}{N}n\right)\exp\left(-j\kappa\frac{2\pi}{N}n\right) = \sum_{n=0}^{N-1} \exp\left[j(m-\kappa)\frac{2\pi}{N}n\right]$$

$$= \sum_{n=0}^{N-1} \exp\left[j(m-\kappa)\frac{2\pi}{N}\right]^n = \frac{1 - \left\{\exp\left[j(m-\kappa)\frac{2\pi}{N}\right]\right\}^N}{1 - \exp\left[j(m-\kappa)\frac{2\pi}{N}\right]} = \frac{1 - \exp\left[j(m-\kappa)2\pi\right]}{1 - \exp\left[j(m-\kappa)\frac{2\pi}{N}\right]},$$

where we used the appropriate expression from the typology to derive the final result.

It is not difficult to see that for values $m \neq \kappa$ the above relation becomes zero; indeed, if we set $m - \kappa = \ell$, $\exp[j(m - \kappa)2\pi] = \exp[j(2\pi\ell)] = 1$, and the value of the numerator becomes zero. On the other hand, if $m = \kappa$, the above relation reduces to the indeterminate form 0/0; therefore, applying *L'Hôpital's rule*, we will get

$$P = \lim_{m\to\kappa}\left\{\frac{\frac{d}{dm}\left\{1 - \exp\left[j(m-\kappa)2\pi\right]\right\}}{\frac{d}{dm}\left\{1 - \exp\left[j(m-\kappa)\frac{2\pi}{N}\right]\right\}}\right\} = \lim_{m\to\kappa}\left\{\frac{-j(2\pi)\exp\left[j(m-\kappa)2\pi\right]}{-j\frac{2\pi}{N}\exp\left[j(m-\kappa)\frac{2\pi}{N}\right]}\right\} = N.$$

Combining the above results, we arrive at the expression

$$P = < x_m[n], x_\kappa[n] > = \begin{cases} N & m = \kappa \\ 0 & m \neq \kappa \end{cases}. \tag{6.35}$$

We observe, therefore, that *the inner product of the elementary complex exponential signals* $x_\kappa[n]$ *satisfies the orthogonality condition* and therefore these signals constitute an *orthogonal basis of functions* in the interval $[0, N-1]$ and more generally in any interval of length N, since, as we previously proved, these signals are periodic functions with period N. Combining the above orthogonality condition with the fact that there are only N different such signals, we can express any discrete-time signal $x[n]$ as a linear combination of the above functions of the form

$$x[n] = \sum_{\kappa=0}^{N-1} \alpha_\kappa x_\kappa[n] = \sum_{\kappa=0}^{N-1} \alpha_\kappa \exp\left(j\kappa\frac{2\pi}{N}n\right), \tag{6.36}$$

a property that is different from the corresponding property of continuous-time signals in which the above sum contains infinite terms. The above relation constitutes the *Fourier expansion* of the discrete-time signal $x[n]$, with the coefficients α_κ, which appear in this, known as *Fourier coefficients*. We return to this topic later.

6.7.2 The Functions $u[n]$, $\delta[n]$ and $r[n]$

Apart from the elementary trigonometric and exponential functions of discrete-time that were presented in the previous section, the functions of the *unit step function* $u[n]$ and the *unit impulse function* or *unit pulse* $\delta[n]$ play a significant role in the analysis of discrete-time signals. These functions are the discrete equivalents of the homonymous continuous functions and are defined as

$$u[n] = \begin{cases} 1 & n \geq 0 \\ 0 & n < 0 \end{cases} \quad \text{and} \quad \delta[n] = \begin{cases} 1 & n = 0 \\ 0 & n \neq 0 \end{cases} \tag{6.37}$$

respectively. We observe that although the *step function* is defined in essentially the same way as the continuous step function $u(t)$, the definition of the discrete function $\delta[n]$ is simple and clear and is not characterized by the mathematical complexities associated with its continuous equivalent, which forced us to resort to the use of *generalized functions* or *distributions*. Using the discrete impulse function, we can define the discrete version of the *comb function* as

$$\text{comb}_N[n] = \sum_{\kappa=-\infty}^{+\infty} \delta[n - \kappa N]. \tag{6.38}$$

In the above expression, a *temporally shifted version* of the impulse function $\delta[n]$ is used, which also characterizes the step function $u[n]$. The temporally shifted versions of these two functions are given by the relations

$$u[n - n_0] = \begin{cases} 1 & n \geq n_0 \\ 0 & n < n_0 \end{cases} = \sum_{\kappa=-\infty}^{n} \delta[n - n_0 - \kappa] \quad \text{and} \quad \delta[n - \kappa] = \begin{cases} 1 & n = \kappa \\ 0 & n \neq \kappa \end{cases}$$

and are particularly useful in many cases, as we will see in subsequent sections. Specifically, the *discrete step function*, in full analogy with its continuous equivalent, allows the creation of *causal signals*, which have nonzero values only in the area $n \geq 0$. In order to construct the causal version of a signal, we simply multiply it by the step function, so as to zero the values of the samples that correspond to negative time instances.

It is proven that the two above elementary functions are related to each other and can arise from each other through the relations

$$\delta[n] = u[n] - u[n-1] \quad \text{and} \quad u[n] = \sum_{k=-\infty}^{n} \delta[k] = \sum_{k=0}^{\infty} \delta[n-k],$$

as is analytically proven in the next exercise.

Problem 6.7.3 Prove that $\delta[n] = u[n] - u[n-1]$ and $u[n] = \sum_{k=-\infty}^{n} \delta[k] = \sum_{k=0}^{\infty} \delta[n-k]$

Solution From the definition of the *unit step function* it is obvious that

$$u[n] = \begin{cases} 1 & n > 0 \\ 1 & n = 0 \\ 0 & n < 0 \end{cases} \text{ and } u[n-1] = \begin{cases} 1 & n-1 > 0 \Rightarrow n > 1 \\ 1 & n-1 = 0 \Rightarrow n = 1 \\ 0 & n-1 < 0 \Rightarrow n < 1 \end{cases} \text{ or equivalently } u[n-1] = \begin{cases} 1 & n > 0 \\ 0 & n = 0 \\ 0 & n < 0 \end{cases}.$$

Subtracting the two equations term by term, we get

$$u[n]-u[n-1]=\begin{cases}1 & n>0\\ 1 & n=0\\ 0 & n<0\end{cases}-\begin{cases}1 & n>0\\ 0 & n=0\\ 0 & n<0\end{cases}=\begin{cases}0 & n>0\\ 1 & n=0\\ 0 & n<0\end{cases}=\begin{cases}1 & n=0\\ 0 & n\neq 0\end{cases}=\delta[n].$$

The second equation arises from the property of the function $\delta[n]$ to have a value equal to one for the value $n=0$ and zero value anywhere else. Therefore,

$$\sum_{k=-\infty}^{n}\delta[k]=\begin{cases}0 & n<0\\ 1 & n\geq 0\end{cases}=u(n)\quad\text{and similarly}\quad\sum_{k=0}^{\infty}\delta[n-k]=\begin{cases}0 & n<0\\ 1 & n\geq 0\end{cases}=u(n),$$

which is the desired result. The last two proofs will become more easily understood when we present *the operation of summing* between two different signals. ■

The *unit step function* is characterized by the following properties:[7]

(1) $u[-n]=1-u[n]$ from which it follows that $u[n]+u[-n]=1$

(2) $u[+\infty]=1$ and $u[-\infty]=0$

(3) $$\int_0^n u[s]ds=\begin{cases}0 & n<0\\ \int_0^n ds=n & n\geq 0\end{cases}=nu[n].$$

The last useful elementary discrete function to be presented in this section is the *unit ramp function*, which in full analogy with its continuous equivalent is defined as

$$r[n]=\begin{cases}n & n\geq 0\\ 0 & n<0\end{cases}\tag{6.39}$$

and is related to the *unit step function* through the relations

$$r[n]=\sum_{\kappa=-\infty}^{n}u[\kappa-1]=nu[n],\tag{6.40}$$

the proof of which comes directly from the way in which the function is defined. The unit ramp function also has its time-shifted version, which is defined as

$$r[n-k]=(n-k)u[n-k].\tag{6.41}$$

The graphical representations of these elementary discrete functions are presented in Fig. 6.12.
The exercises that follow demonstrate some of the properties of the above discrete functions.

Problem 6.7.4 Characterize the signals $\delta[n]$, $u[n]$ and $r[n]$ in terms of their energy and power.

Solution From the definition of the *discrete impulse function* it immediately follows that this is an *energy signal*, since its energy is finite and equal to

$$E_\delta=\sum_{n=-\infty}^{+\infty}\delta^2[n]=\delta^2[0]=1,$$

[7] Note that the first of these two properties is based on the alternative definition of the unit step function according to which $u[n]=1$ for $n>0$, $u[n]=1/2$ for $n=0$, and $u[n]=0$ for $n<0$. Conversely, it does not hold for the definition we give here since for $n=0$ it is $u[n]+u[-n]=2$, unless we add to the expression the quantity $u[-n-1]$.

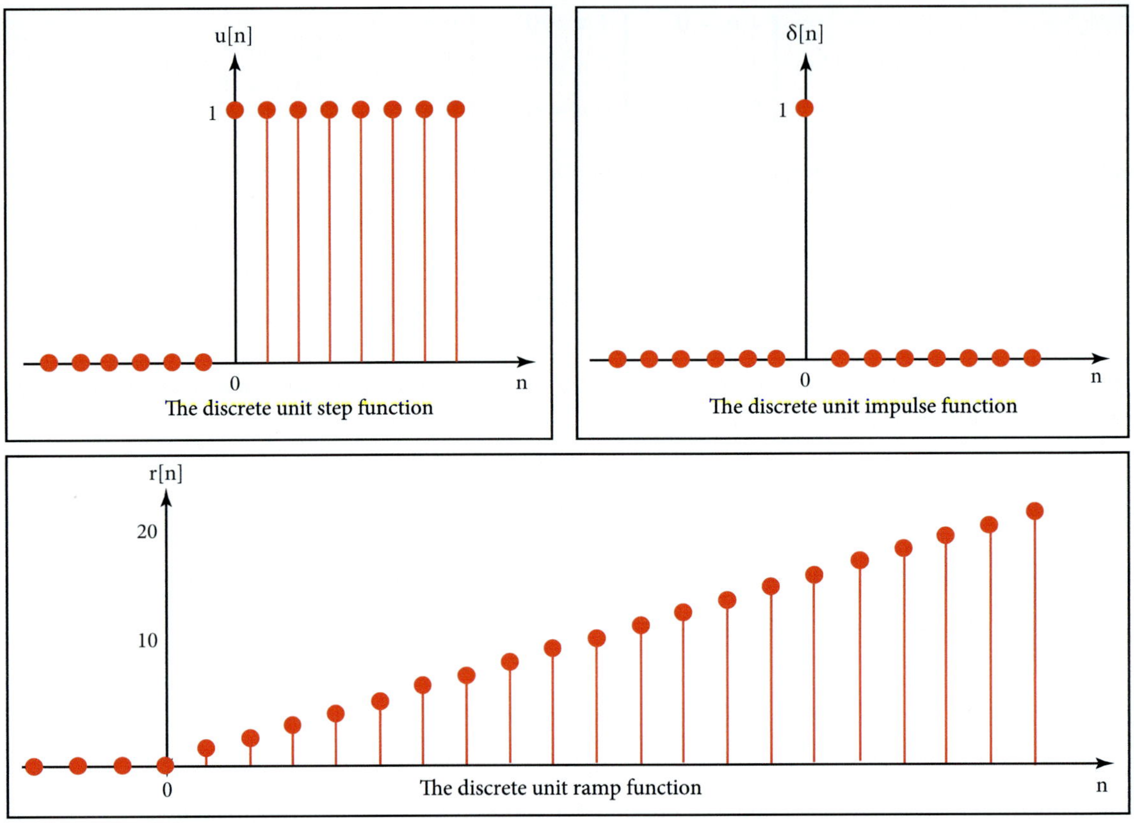

Fig. 6.12 The unit step function $u[n]$, the unit impulse function $\delta[n]$, and the unit ramp function $r[n]$

whereas, on the contrary, the *discrete step function* is characterized by infinite energy, since

$$E_u = \sum_{n=-\infty}^{+\infty} u^2[n] = \sum_{n=0}^{\infty} 1 \to \infty.$$

In reality, the discrete step function has finite power and therefore it is a *power signal*; indeed, from the definition of the power of a discrete-time signal, this is found to be equal to

$$P_u = \lim_{N\to\infty}\left(\frac{1}{2N+1}\sum_{n=-N}^{+N} u^2[n]\right) = \lim_{N\to\infty}\left(\frac{1}{2N+1}\sum_{n=0}^{N} 1\right) = \lim_{N\to\infty}\frac{N+1}{2N+1} = \frac{1}{2}.$$

Finally, as for the *unit ramp function*, it has *both infinite energy and infinite average power*, since it is proven that

$$E_r = \sum_{n=-\infty}^{+\infty} r^2[n] = \sum_{n=0}^{\infty} n^2 \to \infty \quad \text{and}$$

$$P_r = \lim_{N\to\infty}\left(\frac{1}{2N+1}\sum_{n=-N}^{+N} r^2[n]\right) = \lim_{N\to\infty}\left(\frac{1}{2N+1}\sum_{n=0}^{N} n^2\right) = \lim_{N\to\infty}\frac{N(N+1)}{6} \to \infty.$$

Therefore, the *ramp function* is neither an energy signal nor a power signal.

Problem 6.7.5 Given the discrete-time signals $x[n] = u[n]$ and $y[n] = \alpha^n u[n]$, determine their even and odd components.

Solution From the analytical form of the *unit step function* it is not difficult to see that

$$u[n] = \begin{cases} 1 & n > 0 \\ 1 & n = 0 \\ 0 & n < 0 \end{cases} \quad \text{and} \quad u[-n] = \begin{cases} 0 & n > 0 \\ 1 & n = 0 \\ 1 & n < 0 \end{cases}.$$

Therefore, the even component of the signal $x[n] = u[n]$ is given by the relation

$$x_e[n] = \frac{1}{2}\Big(x[n] + x[-n]\Big) = \frac{1}{2}\Big(u[n] + u[-n]\Big) = \frac{1}{2}\begin{cases} 1 & n > 0 \\ 1 & n = 0 \\ 0 & n < 0 \end{cases} + \frac{1}{2}\begin{cases} 0 & n > 0 \\ 1 & n = 0 \\ 1 & n < 0 \end{cases} = \begin{cases} 1/2 & n > 0 \\ 1 & n = 0 \\ 1/2 & n < 0 \end{cases}$$

or in equivalent notation

$$x_e[n] = \begin{cases} 1 & n = 0 \\ 1/2 & n \neq 0 \end{cases} = \frac{1}{2} + \frac{1}{2}\begin{cases} 1 & n = 0 \\ 0 & n \neq 0 \end{cases} = \frac{1}{2} + \frac{1}{2}\delta[n].$$

In a completely analogous way, the odd component of the signal $x[n] = u[n]$ is calculated as

$$x_o[n] = \frac{1}{2}\Big(x[n] - x[-n]\Big) = \frac{1}{2}\Big(u[n] - u[-n]\Big) = \frac{1}{2}\begin{cases} 1 & n > 0 \\ 1 & n = 0 \\ 0 & n < 0 \end{cases} - \frac{1}{2}\begin{cases} 0 & n > 0 \\ 1 & n = 0 \\ 1 & n < 0 \end{cases} = \frac{1}{2}\begin{cases} +1 & n > 0 \\ 0 & n = 0 \\ -1 & n < 0 \end{cases}$$

or equivalently $x_o[n] = (1/2)\ \mathrm{sgn}[n]$, where $sgn[n]$ is the sign function, which returns the value +1 for a positive argument, the value 0 for a zero argument and the value -1 for a negative argument.

Similarly, we use the same methodology for the case of the signal $y[n] = \alpha^n u[n]$. Thus, the even component is given by the relation

$$y_e[n] = \frac{1}{2}\Big(\alpha^n u[n] + \alpha^{-n} u[-n]\Big)$$

or equivalently

$$y_e[n] = \frac{1}{2}\alpha^n \begin{cases} 1 & n > 0 \\ 1 & n = 0 \\ 0 & n < 0 \end{cases} + \frac{1}{2}\alpha^{-n}\begin{cases} 0 & n > 0 \\ 1 & n = 0 \\ 1 & n < 0 \end{cases} = \begin{cases} \alpha^n/2 & n > 0 \\ 1 & n = 0 \\ \alpha^{-n}/2 & n < 0 \end{cases}$$

(since for $n = 0$ it is $y_e[0] = (\alpha^0 + \alpha^{-0}/2) = (1 + 1)/2 = 1]$ or in equivalent notation

$$y_e[n] = \begin{cases} 1 & n = 0 \\ \alpha^\nu/2 & n \neq 0 \end{cases} = \frac{1}{2}\alpha^\nu + \begin{cases} 1/2 & n = 0 \\ 0 & n \neq 0 \end{cases} = \frac{1}{2}\alpha^\nu + \frac{1}{2}\begin{cases} 1 & n = 0 \\ 0 & n \neq 0 \end{cases} = \frac{1}{2}\alpha^\nu + \frac{1}{2}\delta[n],$$

where $\nu = |n|$. Finally, as for the odd component, it is calculated as

$$y_o[n] = \frac{1}{2}\Big(\alpha^n u[n] - \alpha^{-n} u[-n]\Big)$$

or equivalently

$$y_o[n] = \frac{1}{2}\alpha^n \begin{cases} 1 & n > 0 \\ 1 & n = 0 \\ 0 & n < 0 \end{cases} - \frac{1}{2}\alpha^{-n} \begin{cases} 0 & n > 0 \\ 1 & n = 0 \\ 1 & n < 0 \end{cases} = \frac{1}{2}\alpha^{|n|} \begin{cases} +1 & n > 0 \\ 0 & n = 0 \\ -1 & n < 0 \end{cases},$$

and finally $y_o[n] = (1/2)\,\alpha^{|n|}\,\mathrm{sgn}[n]$.

Problem 6.7.6 Given the discrete-time signal $x[n] = (3/2)^n u[-n]$, calculate the values

$$A = \sum_{n=-\infty}^{\infty} x[n] \quad \text{and} \quad P = \sum_{n=-\infty}^{\infty} x^2[n].$$

Solution The calculation of the requested quantities leads to the following results:

$$A = \sum_{n=-\infty}^{\infty} x[n] = \sum_{n=-\infty}^{\infty} \left(\frac{3}{2}\right)^n u[-n] = \sum_{n=-\infty}^{0} \left(\frac{3}{2}\right)^n = \sum_{n=0}^{\infty} \left(\frac{3}{2}\right)^{-n} = \sum_{n=0}^{\infty} \left(\frac{2}{3}\right)^n = \frac{1}{1-(2/3)} = 3$$

$$P = \sum_{n=-\infty}^{\infty} x^2[n] = \sum_{n=-\infty}^{\infty} \left(\frac{3}{2}\right)^{2n} u[-n] = \sum_{n=-\infty}^{0} \left(\frac{3}{2}\right)^{2n} = \sum_{n=0}^{\infty} \left(\frac{3}{2}\right)^{-2n} = \sum_{n=0}^{\infty} \left(\frac{2}{3}\right)^{2n} = \frac{1}{1-(4/9)} = \frac{9}{5}.$$

6.7.3 The Functions $\Pi_D[n]$, $\Lambda_D[n]$ and sinc(ωn)

Two more useful discrete-time elementary signals that are used quite often in practice are the *square pulse* $\Pi_D[n]$, which finds application in digital filter design processes, and the *triangular pulse* $\Lambda_D[n]$. These two elementary signals are described by the equations

$$\Pi_D[n] = \begin{cases} 1 & 0 \le n < D \\ 0 & \text{elsewhere} \end{cases} = u[n] - u[n-D] \tag{6.42}$$

$$\Lambda_D[n] = \begin{cases} 1 - \dfrac{|n+1-D|}{D} & |n+1-D| \le D-1 \\ 0 & \text{elsewhere} \end{cases} \tag{6.43}$$

respectively and their graphical representations are depicted in Fig. 6.13. The well-known *sinc function* is also depicted in the same figure, which for the discrete case is defined as

$$sinc(\omega n) = \frac{\sin(\pi\omega n)}{\pi\omega n} \tag{6.44}$$

6.8 Basic Transformations of Discrete-Time Signals

Having created a discrete-time signal $x[n]$, we can then process it in a way that meets the requirements that exist in each case. In full analogy with continuous-time signals, there are many forms of transformations that we can apply to a discrete-time signal, the most characteristic of which include *time shifting* and *reversal*, as well as *time scaling*. In a more detailed

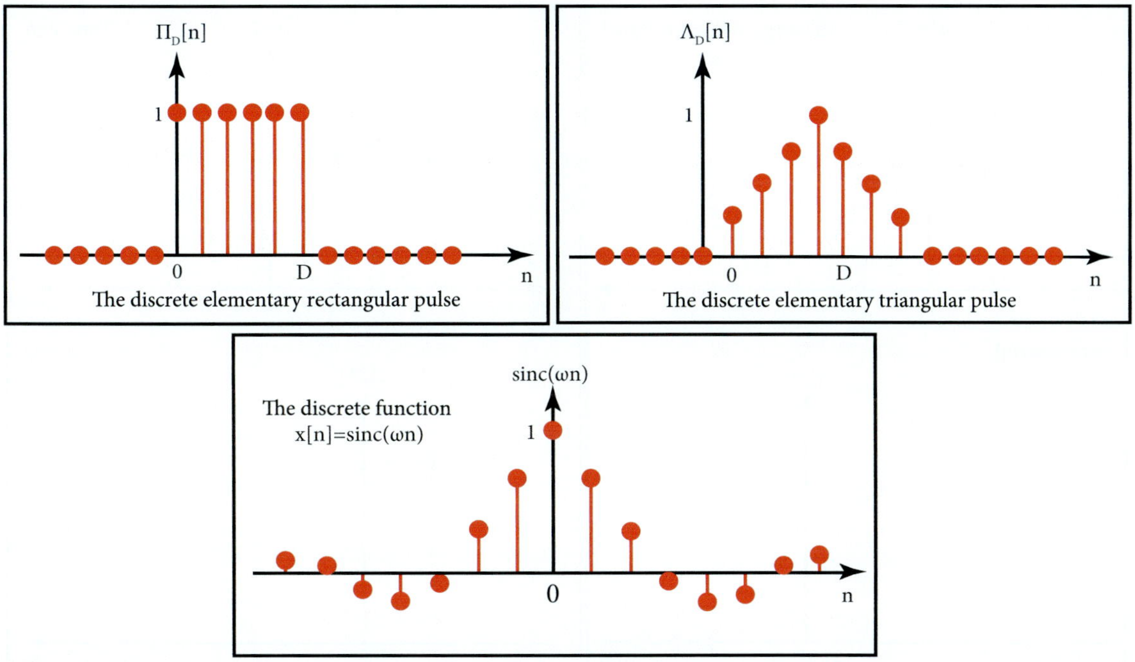

Fig. 6.13 The basic discrete-time signals $\Pi_D[n]$ and $\Lambda_D[n]$ and the function $\mathrm{sinc}(\omega n)$

description, the basic types of transformations that we can apply to the independent variable n of a discrete-time signal $x[n]$ are the following:

- *Time shifting:* a discrete-time signal $x[n]$ can be *shifted* in time by replacing the current value of the variable n with the value $(n - k)$, where k is an *integer* quantity. If the value of k is *positive*, this time shift corresponds to a *time delay* of the signal by k time units, whereas if k takes *negative values*, the process of time shifting becomes *forward* and the signal advances in time by $|k|$ time units. It is obvious that the process of *time advancement* can only be applied to signals that have already been recorded and are stored in memory or in a file of a computer system. Conversely, if the signal we consider is created *in real time* as the result of recording the value of a measurement, the above process is essentially inapplicable, because it involves predicting the value of future measurements, which of course is impossible.
- *Time reversal:* this type of transformation involves replacing the value of the independent variable n with the value $-n$, a process that results in the formation of the *mirror discrete-time signal* with respect to the origin $n = 0$ of the axis system.
- *Time scaling:* in this case, the independent variable n is replaced by the variable $n' = \alpha n$ with the coefficient α being of the form $\alpha = k$ or $\alpha = 1/k$, where k is an integer parameter.

Examples of applying these three basic transformations are presented in Fig. 6.14.

The above basic transformations can be combined in such a way as to lead to the construction of more complex signals, as demonstrated in detail in the next exercise.

Problem 6.8.1 Given the discrete-time signal $x[n] = (6 - n)(u[n] - u[n - 6])$, where $u[n]$ is the unit step function, design the discrete sequences (a) $y_1[n] = x[4 - n]$ and (b) $y_2[n] = x[2n - 3]$ (Hayes, Solved exercise 1.7, questions a and b).

Solution (a) In the first step of the process let us determine the content of the discrete sequence by applying the above relation for the various values of n. In detail, then, we have:

From the contents of the above table, we find that the signal $x[n]$ has a value equal to zero for the values $n \in (-\infty, -1]$ (since the arguments of the unit step function are negative and therefore its value is equal to zero), it takes a nonzero value

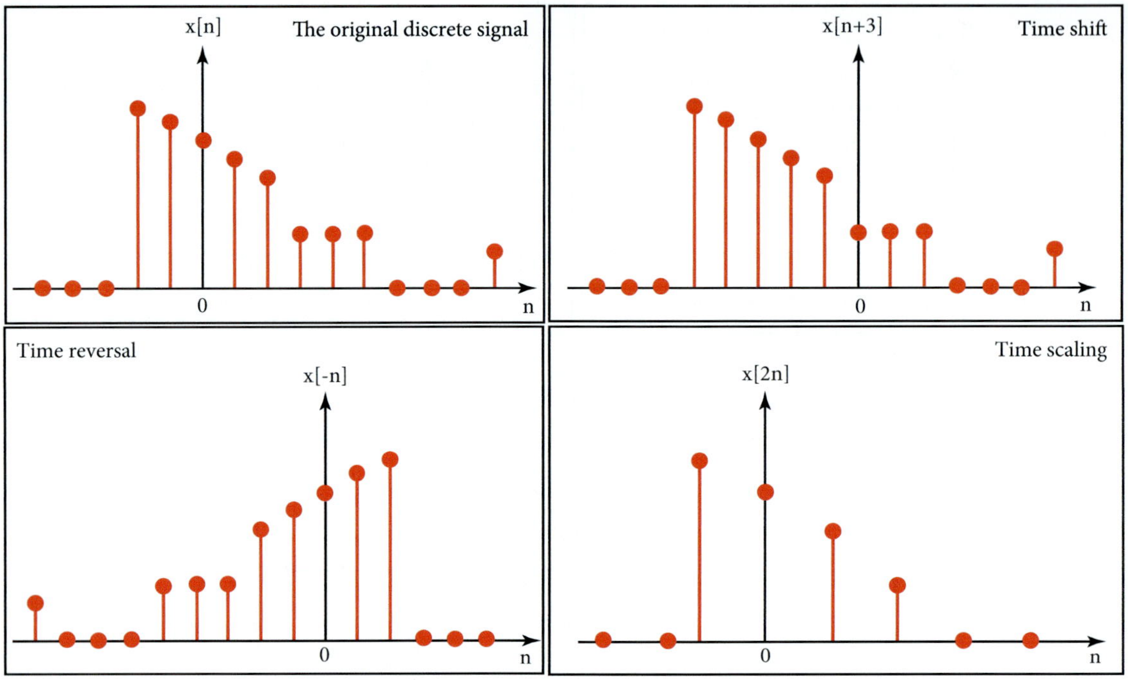

Fig. 6.14 The transformations of time shift, reversal, and scaling

....

for $n = -3 : x[n] = (6 + 3)(u[-3] - u[-9]) = 0$ for $n = -2 : x[n] = (6 + 2)(u[-2] - u[-8]) = 0$

for $n = -1 : x[n] = (6 + 1)(u[-1] - u[-7]) = 0$ for $n = +0 : x[n] = (6 + 0)(u[+0] - u[-6]) = 6$

for $n = +1 : x[n] = (6 - 1)(u[+1] - u[-5]) = 5$ for $n = +2 : x[n] = (6 - 2)(u[+2] - u[-4]) = 4$

for $n = +3 : x[n] = (6 - 3)(u[+3] - u[-3]) = 3$ for $n = +4 : x[n] = (6 - 4)(u[+4] - u[-2]) = 2$

for $n = +5 : x[n] = (6 - 5)(u[+5] - u[-1]) = 1$ for $n = +6 : x[n] = (6 - 6)(u[+6] - u[+0]) = 0$

for $n = +7 : x[n] = (6 - 7)(u[+7] - u[+1]) = 0$ for $n = +8 : x[n] = (6 - 8)(u[+8] - u[+2]) = 0$

for $n = +9 : x[n] = (6 - 9)(u[+9] - u[+3]) = 0$ ....

....

for the values $n \in [0, 5]$, whereas it has a value equal to zero for the values $n \in [6, +\infty)$ (since the arguments of the unit step function are positive and this function for each of them has a value equal to the unit). The graphical representation of the discrete-time signal $x[n]$ with the contents we determined above is shown in Fig. 6.15.

In order to determine the discrete-time signal $y_1[n]$ we should observe that this results from the signal $x[n]$ by applying two transformations to it: (a) *a reversal with respect to time*, so as to get the mirror discrete-time signal with respect to the origin $n = 0$ and (b) *a shift by four positions to the right*, in order to get the final output signal (we would end up with the same result, if we first applied the shift by four units to the left followed by the reversal with respect to time). This situation is presented diagrammatically in Fig. 6.16.

On the other hand, the creation of the discrete-time signal $y_2[n]$ is based on the implementation of three processes, the most interesting of which is *the halving of the input signal frequency*. Indeed, ignoring the existence of the term (-3), which refers to a time shift and calculating the output samples for the various values of n, we will get the correspondences $y[0] \to x[0]$, $y[1] \to x[2]$, $y[2] \to x[4]$, $y[3] \to x[6]$, etc. This, in other words, means that to construct the output signal, we will use *only the even samples of the original signal* ignoring the odd samples. In other words, we will use only half the samples of the input sequence, a fact that leads to the *halving* of the discrete-time signal frequency. After performing this process, all we have to do to get the final output signal is to shift the signal three positions to the right in order to achieve the desired time delay. The result of this process is depicted in Fig. 6.17.

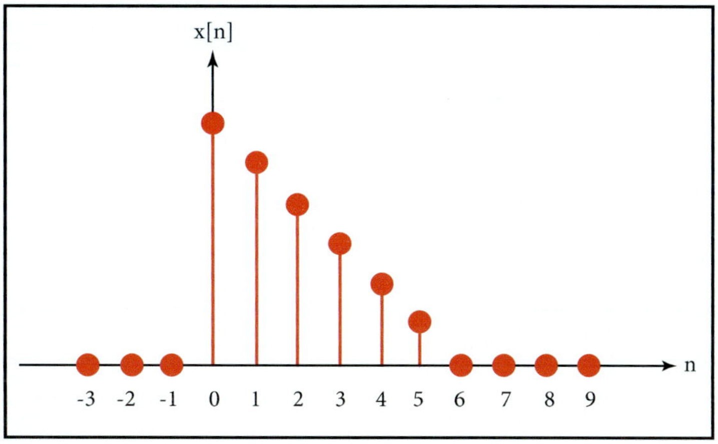

Fig. 6.15 The signal of Exercise 6.8.1

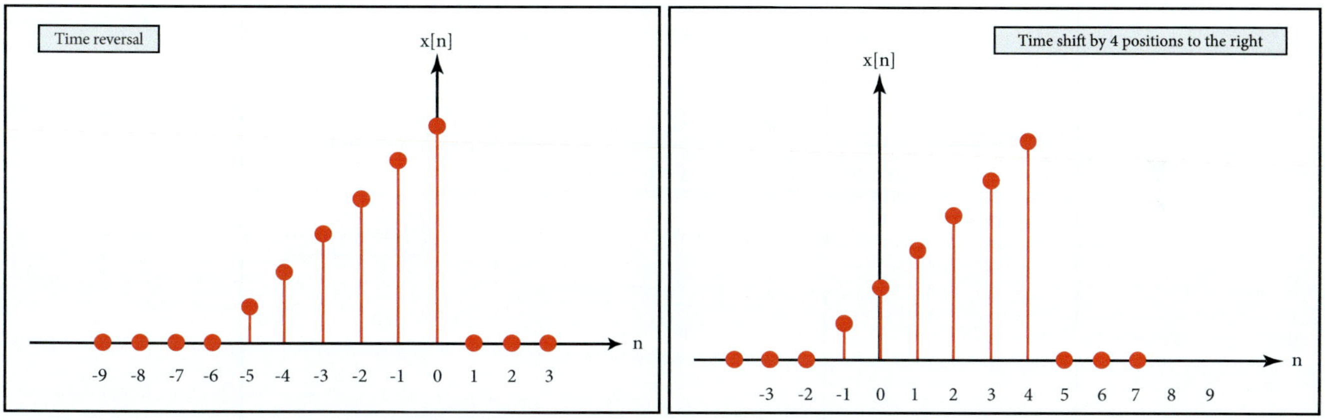

Fig. 6.16 Creation of the signal $y_1[n]$ of Exercise 6.8.1

6.9 Representation with Elementary Functions

In full analogy with their continuous counterparts, discrete-time signals can be formulated with the help of the *elementary discrete-time signals* that we presented in a previous section. This is something we have already observed for the *elementary complex exponential signals*, which, being an *orthogonal basis of functions* over any interval of length N, allow the formulation of any *periodic* discrete-time signal with the same period value as a sum of the form

$$x[n] = \frac{1}{N}\sum_{k=0}^{N-1} \alpha_k \exp\left(jk\frac{2\pi}{N}n\right). \tag{6.45}$$

The above sum is known as the *Fourier expansion* of discrete periodic signals, whereas the coefficients α_k appearing in it are known as *Fourier coefficients*. The analytical description of Fourier's theory for the field of discrete-time is presented in Chap. 10.

A very useful way of representing discrete-time signals is based on the use of the *elementary impulse function* $\delta[n]$. It is proven that *any discrete sequence* $x[n]$ *of nonzero elements of infinite length can be formulated as a linear combination of infinitely many time-shifted impulse functions of the form*

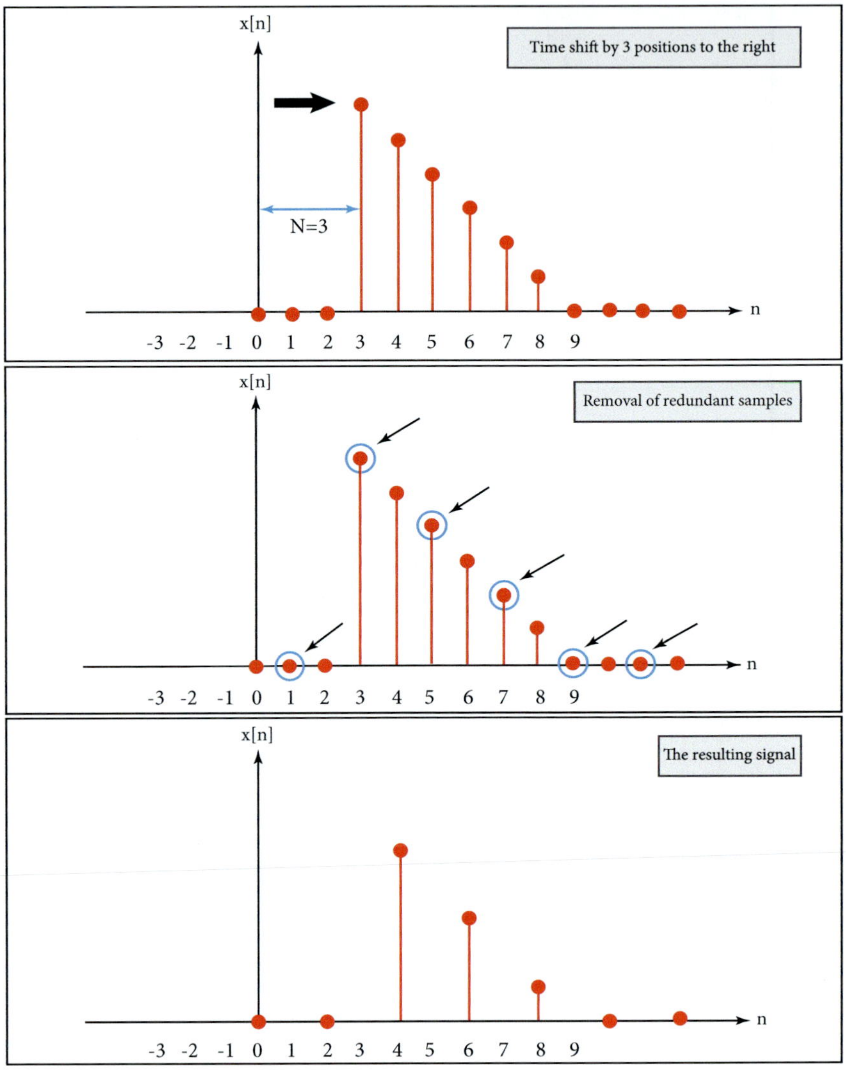

Fig. 6.17 The process of creating the signal $y_2(n)$ of Exercise 6.8.1

$$x[n] = \sum_{k=-\infty}^{\infty} x[k]\delta[n-k]. \tag{6.46}$$

In order to prove the truth of this proposition, let us rewrite the definition of the *temporally displaced impulse function* given by the equation

$$\delta[n-k] = \begin{cases} 1 & n = \kappa \\ 0 & n \neq \kappa \end{cases} \quad \text{or equivalently} \quad \alpha\delta[n-k] = \begin{cases} \alpha & n = \kappa \\ 0 & n \neq \kappa \end{cases},$$

where in the second equation we have multiplied the function $\delta[n-k]$ by a constant coefficient α. Taking into account that a discrete sequence $x[n]$ consists of a multitude of samples with constant and specific values, we can reformulate the last relation as

$$x[n]\delta[n-k] = \begin{cases} x[n] & k = n \\ 0 & k \neq n \end{cases}, \quad k \in (-\infty, +\infty). \tag{6.47}$$

Observing the last equation it is not difficult to ascertain that the multiplication of a discrete-time signal $x[n]$ with the temporally displaced function $\delta[n-k]$ will return us the value of the signal $x[n]$ at the time k, which is nothing but the sample $x[k]$. This fact in mathematical form is expressed as $x[n]\delta[n-k] = x[k]\delta[n-k]$, an equation that is the discrete equivalent of the corresponding property $x(t)\delta(t-t_0) = x(t_0)\delta(t-t_0)$ that characterizes the continuous-time signals. Generalizing the above conclusion, we can formulate the proposition that *if we multiply the considered discrete-time signal with a temporally displaced impulse function, we will we retrieve the signal value at the time of the shift*. Therefore, the application of this process for all possible shift values k in the interval $(-\infty, +\infty)$ will return *all the samples of the signal*, the overlap of which will ultimately give us the entire signal, exactly as dictated by our original equation.

In order to understand the above process, let us consider a signal with length $N = 6$ samples—the generalization to signals of infinite length is obvious—which is defined as

$$x[n] = \{-3, \quad 1, \quad \mathbf{4}, \quad -2, \quad 3, \quad 1\}.$$

We observe that the sample corresponding to the time $n = 0$ is $x[0] = 4$ and therefore the variable k used as an index in the sum

$$x[n] = \sum_{k=-\infty}^{\infty} y_k[n] = \sum_{k=-\infty}^{\infty} x[k]\delta[n-k]$$

will take the values $k = -2, -1, 0, 1, 2, 3$ in succession. In detail, we have:

$$\text{for } k=-2: y_{-2}[n] = x[-2]\delta[n+2] = \begin{cases} x[-2] = -3 & n = -2 \\ 0 & n \neq -2 \end{cases}$$

$$\text{for } k=-1: y_{-1}[n] = x[-1]\delta[n+1] = \begin{cases} x[-1] = 1 & n = -1 \\ 0 & n \neq -1 \end{cases}$$

$$\text{for } k=0: \quad y_0[n] = x[0]\delta[n] \quad = \begin{cases} x[0] = 4 & n = 0 \\ 0 & n \neq 0 \end{cases}$$

$$\text{for } k=1: \quad y_1[1] = x[1]\delta[n-1] \quad = \begin{cases} x[1] = -2 & n = 1 \\ 0 & n \neq 1 \end{cases}$$

$$\text{for } k=2: \quad y_2[n] = x[2]\delta[n-2] \quad = \begin{cases} x[2] = 3 & n = 2 \\ 0 & n \neq 2 \end{cases}$$

$$\text{for } k=3: \quad y_3[n] = x[3]\delta[n-3] \quad = \begin{cases} x[3] = 1 & n = 3 \\ 0 & n \neq 3 \end{cases}.$$

It is not difficult to see that the signals $y_k[n] = x[k]\delta[n-k]$ $(k = -2, -1, 0, 1, 2, 3)$ contain *six samples each* of which *the only one that has a value different than zero is the one that corresponds to the time instance* $n = k$. Therefore, by adding these six different signals, we will get our original signal $x[n]$, since it will obviously be[8]

n	−2	−1	0	1	2	3
$y_{-2}[n]$	−3	+0	+0	+0	+0	+0
+ $y_{-1}[n]$	+0	+1	+0	+0	+0	+0
+ $y_0[n]$	+0	+0	+4	+0	+0	+0
+ $y_1[n]$	+0	+0	+0	−2	+0	+0
+ $y_2[n]$	+0	+0	+0	+0	+3	+0
+ $y_3[n]$	+0	+0	+0	+0	+0	+1
= $\boldsymbol{x[n]}$	= −3	= 1	= 4	= −2	= 3	= 1

The above process of adding signals is mathematically formulated as

[8] The addition of discrete-time signals will be presented in the next section; however, it is not difficult to guess that this should be done by *adding the values of the corresponding samples of the considered signals*.

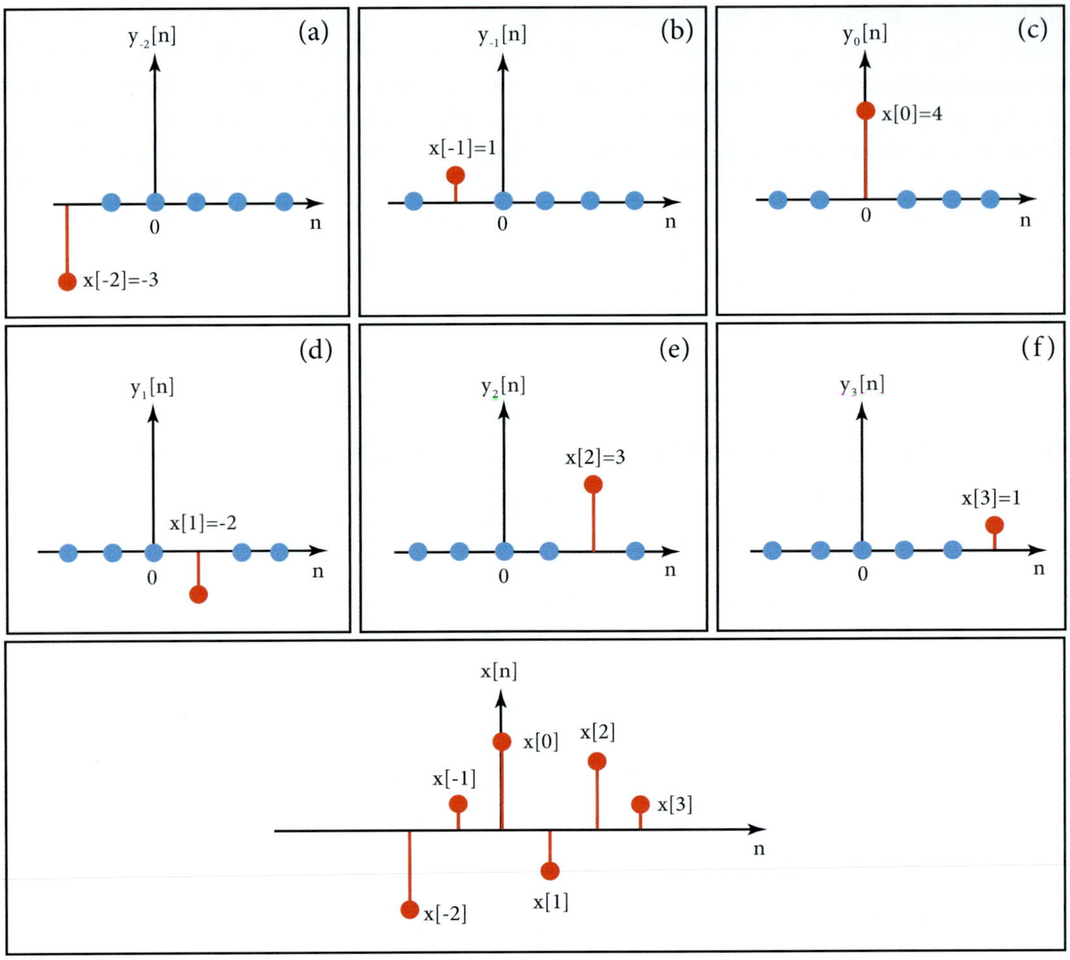

Fig. 6.18 The process of constructing a discrete-time signal as a linear combination of time-shifted impulse functions

$$x[n] = y_{-2}[n] + y_{-1}[n] + y_0[n] + y_1[n] + y_2[n] + y_3[n] = x[-2]\delta[n+2]+$$
$$+x[-1]\delta[n+1] + x[0]\delta[n] + x[1]\delta[n-1] + x[2]\delta[n-2] + x[3]\delta[n-3]$$

or more briefly, in the form of a sum, as

$$x[n] = \sum_{k=-2}^{3} x[k]\delta[n-k]. \tag{6.48}$$

The above relation holds as it is also for the general case of discrete-time signals with *an infinite number of elements*, with the difference that the index k of the sum takes all integer values in the interval $(-\infty, +\infty)$. The process of constructing the discrete-time signal $x[n] = \{-3, 1, \mathbf{4}, -2, 3, 1\}$ in the way we presented above is depicted diagrammatically in Fig. 6.18. We observe that the coefficients of the sum in the above relation are none other than the samples of the signal $x[k]$, which, if replaced by their numerical values, will lead to the final result, which has the form

$$x[n] = -3\delta[n+2] + \delta[n+1] + 4\delta[n] - 2\delta[n-1] + 3\delta[n-2] + \delta[n-3].$$

Problem 6.9.1 Given the discrete-time signal $x[n] = \{2, \mathbf{4}, 0, 3\}$. It should be formulated as a linear combination of discrete-time-shifted functions $\delta[n]$.

Solution With direct application of the relationship we proved in the previous section we will get

$$x[n] = \sum_{k=-1}^{2} x[k]\delta[n-k] = x[-1]\delta[n+1] + x[0]\delta[n]$$
$$+ x[1]\delta[n-1] + x[2]\delta[n-2] = 2\delta[n+1] + 4\delta[n] + 3\delta[n-2],$$

which is the desired result. ■

The last way of representing discrete-time signals that we study in this section concerns the representation with the help of the *unit step function*. Returning to the relation

$$x[n] = \sum_{k=-\infty}^{\infty} x[k]\delta[n-k] \tag{6.49}$$

and replacing in it the time-shifted impulse function $\delta[n-k]$ from the known expression $\delta[n-k] = u[n-k] - u[n-k-1]$ it immediately follows that

$$x[n] = \sum_{k=-\infty}^{\infty} x[k]\,\delta[n-k] = \sum_{k=-\infty}^{\infty} x[k]\left(u[n-k] - u[n-k-1]\right). \tag{6.50}$$

The above relation constitutes the discrete equivalent of Eq. (1.26) (see Sect. 1.9.1 on page 32) and the way of using it is demonstrated in the next exercise.

Problem 6.9.2 Write the discrete-time signal $x[n]$ as a linear combination of time-shifted unit step functions, which is defined as (Hayes, Solved exercise 1.11)

$$x[n] = \begin{cases} 1 \; n = 0 \\ 2 \; n = 1 \\ 3 \; n = 2 \\ 0 \text{ elsewhere} \end{cases}.$$

Solution With direct application of the relation we proved in the previous section we get

$$x[n] = \sum_{k=0}^{\infty} x[k]\delta[n-k] = x[0]\delta[n] + x[1]\delta[n-1] + x[2]\delta[n-2]$$
$$= \delta[n] + 2\delta[n-1] + 3\delta[n-2] = u[n] - u[n-1] + 2\left(u[n-1] - u[n-2]\right)$$
$$+ 3\left(u[n-2] - u[n-3]\right) = u[n] + u[n-1] + u[n-2] - 3u[n-3],$$

which is the desired result.

6.10 Fundamental Operations of Discrete-Time Signals

The fundamental operations that we can apply between discrete-time signals are the *scaling of the signal amplitude*, the *addition* and the *multiplication* of signals, as well as the operations of *convolution* and *correlation*. These operations are defined in a way similar to what we saw in continuous-time signals and their analytical description is presented in the following pages.

6.10.1 Scaling, Addition and Multiplication of Signals

The *scaling of the amplitude* of a discrete-time signal $x[n]$ by a constant A takes place *by multiplying the value of each sample of the signal by the value of the constant A*. The new signal that results from this is therefore given by the relation

$$y[n] = Ax[n] \quad -\infty < n < \infty.$$

On the other hand, the *addition* and the *multiplication* of two discrete-time signals $x[n]$ and $y[n]$ lead to the creation of a new signal $z[n]$ *each sample of which results from the addition or the multiplication of the corresponding samples* of the signals $x[n]$ and $y[n]$, so that we can write that

$$z[n] = x[n] + y[n] \qquad \text{and} \qquad z[n] = x[n] \cdot y[n]$$

$(n = \ldots, -2, -1, 0, 1, 2, \ldots)$ respectively. By the term *corresponding samples* we mean samples that correspond *to the same value of the discrete variable n* (this is also the reason why, apart from the signal samples, we should also specify which of these corresponds to the value $n = 0$), something that means that if the two signals are not "aligned" in time, we should take special care, so that the addition or multiplication of samples is done in the right way, as shown in the following exercises.

Problem 6.10.1 Find the sum of the discrete-time signals

$$x[n] = \{\mathbf{5}, \quad 2, \quad -1, \quad 4, \quad 6\} \qquad \text{and} \qquad y[n] = \{\mathbf{2}, \quad 1, \quad 3, \quad 5\}.$$

Solution The signals $x[n]$ and $y[n]$ are *aligned in time*, in the sense that the samples that correspond to the time value $n = 0$ are in the same position; therefore, all we have to do is add the corresponding samples. We also observe that the signal $y[n]$ has one sample less than the signal $y[n]$ and for this reason in the position of the missing sample we place the zero value. So,

n	0	1	2	3	4
x [n]	5	2	− 1	4	6
y [n]	2	1	3	5	0
z [n]	7	3	2	9	6

Therefore, the signal $z[n]$ will have the form $z[n] = \{\mathbf{7}, \quad 3, \quad 2, \quad 9, \quad 6\}$, which is the desired result.

Problem 6.10.2 Find the sum of discrete-time signals

$$x[n] = \{5, \quad 2, \quad \mathbf{-1}, \quad 4, \quad 6\} \quad \text{and} \quad y[n] = \{\mathbf{2}, \quad 1, \quad 3, \quad 5\}.$$

Solution In this case, the two signals, although they have the same samples as the previous exercise, *are not aligned in time*, since the samples corresponding to the time instance $n = 0$ are not in the same position. For this reason, before the addition operation, we should *align* the signals, placing the corresponding samples one under the other and adding zeros in the empty positions that arise from this process (these additional zeros that have been added for the alignment of the two signals are depicted in blue). The result of the addition of the two signals for this case is presented below.

n	−2	−1	0	1	2	3
x [n]	5	2	− 1	4	6	0
y [n]	0	0	2	1	3	5
z [n]	5	2	1	5	9	5

The above exercises concern the operation of addition, but of course the multiplication of signals is performed in exactly the same way.

6.10.2 Convolution of Discrete-Time Signals

The operation of *convolution* between two discrete-time signals $x[n]$ and $y[n]$ is given by an equation that is similar to that concerning continuous-time signals, under the understanding that *the symbol of the integral is simply replaced by the symbol of the sum*. Considering, then, that the two signals have infinite duration, their convolution is a third discrete-time signal $z[n]$ given by the relation

$$z[n] = x[n] * y[n] = \sum_{k=-\infty}^{\infty} x[k]y[n-k]. \tag{6.51}$$

In the case where the discrete-time signals $x[n]$ and $y[n]$ have *finite length* M, their convolution is defined by the relation

$$z[n] = x[n] * y[n] = \sum_{k=-(M/2)+1}^{M/2} x[k]y[n-k]. \tag{6.52}$$

The operation of convolution defined in the above way is known as *linear convolution*, so as to differentiate it from circular convolution, which is presented in Chap. 10.

Having defined the operation of convolution, we can now recognize the defining equation of a discrete-time signal $x[n]$ with the help of the impulse function as *the convolution between the signals* $x[n]$ *and* $\delta[n]$, so that we can write that

$$x[n] = \sum_{k=-\infty}^{+\infty} x[k]\delta[n-k] = x[n] * \delta[n]. \tag{6.53}$$

In other words, the convolution of a discrete-time signal with the function $\delta[n]$ returns the same signal, that is *the impulse function is the neutral element of the operation of convolution in the space of signals*, a property that, as we have seen, characterizes continuous-time signals.

In the special case where the signals $x[n]$ and $y[n]$ are *periodic* with a common period N, the sum of the convolution

$$z[n] = \sum_{k=-\infty}^{\infty} x[k]y[n-k] \tag{6.54}$$

does not converge. In this case, we use the periodic convolution, which is defined as

$$z[n] = x[n] \otimes y[n] = \sum_{n=0}^{N-1} x[k]y[n-k]. \tag{6.55}$$

It is not difficult to prove that the *periodic convolution* $z[n]$ is a periodic discrete-time signal with period N; indeed, since the signal $y[n]$ is periodic with period N, $y[n-k] = y[n-k+N]$ and therefore

$$z[n+N] = \sum_{n=0}^{N-1} x[k]y[(n+N)-k] = \sum_{n=0}^{N-1} x[k]y[n-k+N] = \sum_{n=0}^{N-1} x[k]y[n-k] = z[n],$$

from which the desired result follows. The following pages contain examples of calculating the convolution of discrete-time signals based on their defining equation, so that the reader can fully familiarize themselves with its use.

Problem 6.10.3 Calculate the convolution between the discrete-time signals

$$x[n] = \alpha^n u[n] \qquad \text{and} \qquad y[n] = \beta^n u[n]$$

for the cases $\alpha = \beta$ and $\alpha \neq \beta$ (Oppenheim, Willsky and Nawab, Exercise for solution 2.21 and Hsu, Exercise 3.29a).

Solution Using the defining equation of convolution, we get

$$z[n] = \sum_{k=-\infty}^{\infty} x[k]y[n-k] = \sum_{k=-\infty}^{\infty} \alpha^k u[k]\beta^{n-k} u[n-k].$$

From the above relationship, we observe that for $k < 0$ $u[k] = 0$, whereas for $k > n$, $u[n-k] = 0$. Therefore, the convolution of the two signals will be defined as

$$z[n] = \sum_{k=0}^{n} \alpha^k \beta^{n-k} = \sum_{k=0}^{n} \alpha^k \beta^n \beta^{-k} = \beta^n \sum_{k=0}^{n} \alpha^k \beta^{-k} = \beta^n \sum_{k=0}^{n} (\alpha\beta^{-1})^k$$

We distinguish two cases:

- Case 1, $\alpha = \beta$. $\alpha\beta^{-1} = \alpha\alpha^{-1} = 1$ and therefore

$$z[n] = \beta^n \sum_{k=0}^{n} 1^k = \beta^n (n+1) = \alpha^n (n+1).$$

- Case 2, $\alpha \neq \beta$. In this case, using the appropriate expression from the mathematical typology, we get

$$z[n] = \beta^n \frac{1-(\alpha\beta^{-1})^{n+1}}{1-(\alpha\beta^{-1})} = \beta^n \frac{1-\left(\dfrac{\alpha}{\beta}\right)^{n+1}}{1-\left(\dfrac{\alpha}{\beta}\right)} = \frac{\beta^{n+1}-\alpha^{n+1}}{\beta-\alpha}.$$

Therefore, finally, the convolution between the discrete-time signals $x[n]$ and $y[n]$ will be given by the relationship

$$z[n] = x[n] * y[n] = \begin{cases} \beta^n(n+1) & \alpha = \beta \\ \dfrac{\beta^{n+1}-\alpha^{n+1}}{\beta-\alpha} & \alpha \neq \beta \end{cases}$$

Problem 6.10.4 Calculate the convolution between the discrete-time signals

$$x[n] = \alpha^n u[n] = \begin{cases} \alpha^n & n \geq 0 \\ 0 & n < 0 \end{cases} \quad \text{and} \quad y[n] = u[n]$$

Solution Although the solution to the exercise can arise directly from the previous one by setting $\beta = 1$, let us repeat the process one more time so that we can familiarize ourselves with the process of calculating this quantity. From the convolution defining equation we have

$$z[n] = x[n] * y[n] = \sum_{k=-\infty}^{\infty} x[k]y[n-k] = \sum_{k=-\infty}^{\infty} \alpha^k u[k]u[n-k].$$

From the above relationship we find that for $k < 0$ the above sum is zero, since $u[k] = 0$, whereas the same applies for $k > n$, since then $u[n-k] = 0$. On the other hand, for $0 < k < n$, $u[k] = u[n-k] = 1$ and the above relationship takes the form

$$z[n] = \sum_{k=0}^{n} \alpha^k = \frac{1-\alpha^{n+1}}{1-\alpha}.$$

Combining the above cases we get the equation

$$z(n) = \begin{cases} \dfrac{1-\alpha^{n+1}}{1-\alpha} & n \geq 0 \\ 0 & n < 0 \end{cases} = \frac{1-\alpha^{n+1}}{1-\alpha} u(n).$$

Problem 6.10.5 Calculate the convolution between the discrete-time signals

$$x[n] = \left(\frac{1}{2}\right)^{n-4} u[n] \quad \text{and} \quad y[n] = \left(\frac{1}{3}\right)^n u[n-2].$$

Solution From the convolution defining equation we have

$$z[n] = \sum_{k=-\infty}^{\infty} x[k]y[n-k] = \sum_{k=-\infty}^{\infty} \left(\frac{1}{2}\right)^{k-4} u[k]\left(\frac{1}{3}\right)^{n-k} u[n-k-2].$$

In this case, for $k < 0$, $u[k] = 0$, whereas for $k > n-2$, $u[n-k-2] = 0$. On the other hand, for $0 \le k \le n-2$, $u[k] = u[n-k-2] = 1$ and the above relation will take the form

$$z[n] = \sum_{k=0}^{n-2} \left(\frac{1}{2}\right)^{k-4} \left(\frac{1}{3}\right)^{n-k} = \left(\frac{1}{2}\right)^{-4} \left(\frac{1}{3}\right)^{n} \sum_{k=0}^{n-2} \left(\frac{1}{2}\right)^{k} \left(\frac{1}{3}\right)^{-k}$$

$$= 16 \left(\frac{1}{3}\right)^{n} \sum_{k=0}^{n-2} \left(\frac{3}{2}\right)^{k} = 16 \left(\frac{1}{3}\right)^{n} \frac{1-(3/2)^{n-1}}{1-(3/2)} = 32 \left(\frac{1}{3}\right)^{n} \left[\left(\frac{3}{2}\right)^{n-1} - 1\right]$$

for the range of values $n \ge 2$, which is also the desired result.

Problem 6.10.6 Calculate the convolution between the discrete-time signals

$$x[n] = (0.5)^n u[n] \qquad \text{and} \qquad y(n) = nu[n].$$

(adapted from Hayes, Solved exercise 1.33).

Solution From the convolution defining equation we have

$$z[n] = x[n] * h[n] = \sum_{n=-\infty}^{\infty} x[k]h[n-k] = \sum_{k=-\infty}^{\infty} (0.5)^k u[k](n-k)u[n-k].$$

From the above relation we observe that for $k < 0$, $u[k] = 0$, whereas for $k > n$ it is $u[n-k] = 0$. Therefore, the above relation is written in the form

$$z[n] = \sum_{k=0}^{n} (n-k)(0.5)^k = n \sum_{k=0}^{n} (0.5)^k - \sum_{k=0}^{n} k(0.5)^k.$$

Referring to the mathematical typology, we find that

$$\sum_{k=0}^{n} (0.5)^k = 2\left[1-(0.5)^{n+1}\right] \qquad \text{and} \qquad \sum_{k=0}^{n} k(0.5)^k = 4\left[n(0.5)^{n+2} - (n+1)(0.5)^{n+1} + 0.5\right]$$

Substituting the above expression into the convolution equation, after operations, it turns out that

$$z[n] = 2\left[n + (0.5)^n - 1\right].$$

Problem 6.10.7 Calculate the convolution between the discrete-time signals

$$x[n] = \left(\frac{1}{3}\right)^{n} (u[n] - u[n-51]) \qquad \text{and} \qquad y[n] = \left(\frac{1}{5}\right)^{n} u[n]$$

(adapted from Hayes, Solved exercise 1.34).

Solution From the defining equation of convolution we have

$$z[n] = \sum_{k=-\infty}^{\infty} x[k]y[n-k] = \sum_{k=-\infty}^{\infty} \left(\frac{1}{3}\right)^k \Big(u[k] - u[k-51]\Big) \left(\frac{1}{5}\right)^{n-k} u[n-k].$$

From the above relation we observe that for $k < 0$ it is $u[k] = 0$ whereas for $k > 50$ it is $u[k-51] = 0$. Therefore, the system output will take nonzero values only in the interval $0 \le k \le 50$ and will be given by the relation

$$z(n) = \sum_{k=0}^{50} \left(\frac{1}{3}\right)^k \left(\frac{1}{5}\right)^{n-k} u[n-k].$$

The last relation contains the step function $u[n-k]$, which has a value of $u = 0$ or $u = 1$ depending on the range of values we refer to. We distinguish, therefore, three cases:

- $n < 0$: in this range of values, $n - k < 0$ (since $0 \le k \le 50$) and therefore the sum is zero.
- $0 \le n \le 50$: in this range of values, the function $u[n-k]$ is equal to one for $k \le n$, whereas it has a zero value for $k > n$. Therefore, we have

$$z[n] = \sum_{k=0}^{n} \left(\frac{1}{3}\right)^k \left(\frac{1}{5}\right)^{n-k} = \left(\frac{1}{5}\right)^n \sum_{k=0}^{n} \left(\frac{5}{3}\right)^k = \left(\frac{1}{2}\right)^n \frac{1-(5/3)^{n+1}}{1-(5/3)} = \frac{3}{2}\left(\frac{1}{5}\right)^n \left[\left(\frac{5}{3}\right)^{n+1} - 1\right].$$

- $n > 50$: in this range of values, the function $u[n-k]$ has a value equal to one for every k in the range $[0, 50]$ and therefore the system output is given by the relation

$$z[n] = \sum_{k=0}^{50} \left(\frac{1}{3}\right)^k \left(\frac{1}{5}\right)^{n-k} = \left(\frac{1}{5}\right)^n \sum_{k=0}^{50} \left(\frac{5}{3}\right)^k = \left(\frac{1}{5}\right) n \frac{1-(5/3)^{51}}{1-(5/3)} = \frac{3}{2}\left(\frac{1}{5}\right)^n \left[\left(\frac{5}{3}\right)^{51} - 1\right].$$

Combining the last three equations, we arrive at the equation

$$z[n] = \begin{cases} 0 & n < 0 \\ \frac{3}{2}\left(\frac{1}{5}\right)^n \left[\left(\frac{5}{3}\right)^{n+1} - 1\right] & 0 \le n \le 50 \\ \frac{3}{2}\left(\frac{1}{5}\right)^n \left[\left(\frac{5}{3}\right)^{51} - 1\right] & n > 50 \end{cases}.$$

Problem 6.10.8 Calculate the convolution between the discrete-time signals

$$x[n] = \begin{cases} 1 & 0 \le n \le 4 \\ 0 & \text{anywhere else} \end{cases} \quad \text{and} \quad y[n] = \begin{cases} \alpha^n & -3 \le n \le 5 \\ 0 & \text{anywhere else} \end{cases}$$

(Proakis & Manolakis, Exercise 2.19).

Solution In this case, we find that the two signals *do not have infinite length*, but on the contrary have nonzero values for specific ranges of the independent variable n; indeed, expressing the signals in their sequential representation, they are written as $x[n] = \{\mathbf{1}, 1, 1, 1, 1\}$ and $y[n] = \{\alpha^{-3}, \alpha^{-2}, \alpha^{-1}, \mathbf{1}, \alpha, \alpha^2, \alpha^3, \alpha^4, \alpha^5\}$, with the sample written in bold corresponding to the known value $n = 0$. Taking into account the defining equation of the convolution

$$z[n] = \sum_{k=-\infty}^{\infty} x[k]y[n-k],$$

we can conclude that the limits of the above sum do not extend to the infinite range of values $(-\infty, +\infty)$, but will range from $k_{\text{min}} = 0$ to $k_{\text{max}} = 4$, *since the signal only in this range of values* $x[k]$ *has nonzero values*. As for the limits of the range of values of n for which the signal $y[n]$ also has nonzero values, so that the calculation of the convolution is possible, these arise from the relations

$$n_1 - k_{\text{min}} = -3 \Rightarrow n_1 = -3 + k_{\text{min}} = -3 \quad \text{and} \quad n_2 - k_{\text{max}} = 5 \Rightarrow n_2 = 5 + k_{\text{max}} = 5 + 4 = 9.$$

Therefore, ultimately the convolution of the two discrete-time signals will have the form

$$z[n] = \begin{cases} \sum_{k=0}^{4} y[n-k] & -3 \le n \le 9 \\ 0 & \text{anywhere else} \end{cases},$$

(since for $0 \le k \le 4$ it is $x[k] = 1$). Using this relation we can calculate one by one all the elements of the discrete sequence $z[n]$; this calculation is quite easy and is left as an exercise to the reader.

Problem 6.10.9 Calculate the convolution between the discrete-time signals

$$x[n] = \begin{cases} 1 & n = -2, 0, 1 \\ 2 & n = -1 \\ 0 & \text{anywhere else} \end{cases} \quad \text{and}$$

$$y[n] = \delta[n] - \delta[n-1] + \delta[n-4] + \delta[n-5].$$

(Proakis & Manolakis, Exercise for solution 2.21).

Solution From the mathematical expression of the signal $x[n]$ it is not difficult to see that this signal in the form of a discrete sequence will be formulated as $x[n] = \{1, 2, \mathbf{1}, 1\}$. On the other hand, the signal $y[n]$ has the form

$$\begin{aligned} y[n] &= \delta[n] - \delta[n-1] + \delta[n-4] + \delta[n-5] \\ &= 1 \times \delta[n] + (-1) \times \delta[n-1] + 0 \times \delta[n-2] + 0 \times \delta[n-3] + 1 \times \delta[n-4] + 1 \times \delta[n-5] \end{aligned}$$

and taking into account that *each discrete-time signal is analyzed into a linear combination of temporally displaced discrete impulse functions, the coefficients of which also constitute the samples of the discrete sequence*, we can write the signal $y[n]$ in the form $y[n] = \{\mathbf{1}, -1, 0, 0, 1, 1\}$. Therefore, finally the two signals are written as

$$x[n] = \{1, 2, \mathbf{1}, 1\} \quad \text{and} \quad y[n] = \{\mathbf{1}, -1, 0, 0, 1, 1\}$$

and their convolution will be calculated according to the method we described in the previous exercise.

Problem 6.10.10 Calculate the convolution between the discrete-time signals

$$x[n] = -n3^n u[-n] \quad \text{and} \quad y[n] = u[-n-1].$$

(Hayes, Solved exercise 1.28).

Table 6.3 Convolution of known discrete-time signals

$x[n]$	$y[n]$	$z[n]=x[n]*y[n]$
$x[n]$	$\delta[n-N]$	$x[n-N]$
$e^{\lambda n}u[n]$	$u[n]$	$\frac{e^{\lambda n}-1}{\lambda}u[n]$
$u[n]$	u[n]	$nu[n]$
$e^{\lambda_1 n}u[n]$	$e^{\lambda_2 n}u[n]$	$\frac{e^{\lambda_1 n}-e^{\lambda_2 n}}{\lambda_1-\lambda_2}u[n] \quad (\lambda_1\neq\lambda_2)$
$e^{\lambda n}u[n]$	$e^{\lambda n}u[n]$	$ne^{\lambda n}u[n]$
$ne^{\lambda n}u[n]$	$e^{\lambda n}u[n]$	$(1/2)n^2e^{\lambda n}u[n]$
$n^Nu[n]$	$e^{\lambda n}u[n]$	$\frac{N\,!e^{\lambda n}}{\lambda^{N+1}}u[n]-\sum_{k=0}^{N}\frac{N\,!n^{N-k}}{\lambda^{k+1}(N-k)!}u[n]$
$n^Mu[n]$	$n^Nu[n]$	$\frac{M\,!N!}{(N+M+1)!}n^{M+N+1}\,u[n]$
$ne^{\lambda_1 n}u[n]$	$e^{\lambda_2 n}u[n]$	$\frac{e^{\lambda_2 n}-e^{\lambda_1 n}+(\lambda_1-\lambda_2)ne^{\lambda_1 n}}{(\lambda_1-\lambda_2)^2}u[n]$
$n^Me^{\lambda n}u[n]$	$n^Ne^{\lambda n}u[n]$	$\frac{M\,!N!}{(N+M+1)!}n^{M+N+1}\,e^{\lambda n}u[n]$
$n^Me^{\lambda_1 n}u[n]$	$n^Ne^{\lambda_2 n}u[n]$	$\sum_{k=0}^{M}\frac{(-1)^kM\,!(N+k)!t^{M-k^{\lambda_1 n}}}{k!(M-k)!\,(\lambda_1-\lambda_2)^{M+k+1}}u[n]+$
		$\sum_{k=0}^{\infty}\frac{(-1)^kN\,!(M+k)!t^{N-k_e\lambda_2 n}}{k!(N-k)!\,(\lambda_2-\lambda_1)^{N+k+1}}\,u[n]$
$e^{-\alpha n}\cos(\beta n+\vartheta)u[n]$	$e^{\lambda n}u[n]$	$\frac{\cos(\vartheta-\phi)e^{\lambda n}-e^{\alpha n}\cos(\beta n+\vartheta-\phi)}{\sqrt{(\alpha+\lambda)^2+\beta^2}}u[n]$

Solution From the convolution defining equation we have

$$z[n]=\sum_{k=-\infty}^{\infty}x[k]y[n-k]=\sum_{k=-\infty}^{\infty}-k3^ku[-k]u[-n-1-k].$$

From the above equation we observe that the function $u[-k]$ has zero value for $k>0$. The same applies to the function $u[-n-1-k]$ and for the value range $k<n+1$. Therefore, the output of the system will have a nonzero value in the area $n+1\leq k\leq 0\ (n\leq -2)$ for which $u[-k]=u[-n-1-k]=1$ and will be given by the relation

$$z[n]=\sum_{k=n+1}^{0}-k3^k=\sum_{m=0}^{-n-1}m\left(\frac{1}{3}\right)^m=\frac{3}{4}+(2n-1)\frac{3^{n+1}}{4}\quad (n\leq 2),$$

where we proceeded with the change of the dummy index from $-k$ to m and calculated the sum of the equation using the appropriate relation from the mathematical typology. ■

Table 6.3 contains the convolution between known discrete-time signals that are used quite often in practice. The calculation of the convolutions of this table is done in a way similar to the one we presented in the previous exercises and is left as an exercise to the reader.[9] Let us note that because in practice we are interested only in *causal signals*, which have nonzero values only in the area $n\geq 0$, the result of the convolution is multiplied by the discrete step function $u[n]$, in full analogy with the continuous-time signals.

[9] For the way of defining the angle φ in the last line of Table 6.3, see the corresponding equation of Table 1.1 on page 39.

Graphical Calculation of the Convolution

Observing the defining equation of the convolution of two discrete-time signals $x[n]$ and $y[n]$, which has the form

$$z[n] = \sum_{k=-\infty}^{\infty} x[k]y[n-k]. \tag{6.56}$$

It is not difficult to ascertain that the process of calculating each sample of the signal $z[n]$ requires the execution of the following procedures:

- *Inversion of the signal* $y[k]$ *with respect to time*, so that we obtain the reversed sequence $y[-k]$.
- *Temporal shift of the reversed sequence by n time units* to the right (for positive values of n) or to the left (for negative values of n), so as to create the signal $y[n-k]$.
- *Multiplication of the discrete-time signals* $x[k]$ *and* $y[n-k]$ *sample by sample*. In this way, we obtain a new discrete sequence $\upsilon_n[k]$ each sample of which will result as the product of the corresponding samples of the discrete-time signals $x[k]$ and $y[n-k]$. Particular attention should be given to the determination of the *corresponding samples*, if we are dealing with sequences of finite length in which the samples corresponding to the discrete-time instance $n = 0$ are not in the same position.
- *Addition of the values of the elements of the discrete sequence* $\upsilon_n[k]$, so as to calculate the value of the desired output of the system $y[n]$.

The above process allows the calculation of the sample of the signal $z[n]$ corresponding to the time instance n, whereas if we wish to calculate the values of all the samples of the signal, we will need to apply the above process for each time instance n in the interval $(-\infty, \infty)$.

In order to understand the above process, let us calculate diagrammatically the convolution between the discrete-time signals $x[n] = \{\mathbf{1}, 2, 3, 1\}$ and $y[n] = \{1, \mathbf{2}, 1, -1\}$. In this case, the application of the above process takes place as follows:

- Calculation of the value $z[0]$ corresponding to the value $n = 0$:

  1. We reverse the signal $y[-k]$ in time, so that it takes the form $y[-k]=\{-1, 1, \mathbf{2}, 1\}$.
  2. We shift the signal $y[-k]$ by $n = 0$ time units, resulting in the signal $y[-k]=\{-1, 1, \mathbf{2}, 1\}$.
  3. We multiply the signals $x[k]$ and $y[k-0]$. The result of this process is the construction of the signal $\upsilon_0 = \{1 * 2, 2 * 1, 3 * 0, 1 * 0\} = \{2, 2, 0, 0\}$.
  4. We add the elements of the sequence υ_0, to calculate the value $z[0]$. So, $z[0] = 2 + 2 + 0 + 0 = 4$.

- Calculation of the value $z[1]$ corresponding to the value $n = 1$:

  1. We reverse the signal $y[-k]$ in time, so that it takes the form $y[-k]=\{-1, 1, \mathbf{2}, 1\}$.
  2. We shift the signal $y[-k]$ by $n=1$ time units, resulting in the signal $y[1-k]=\{-1, \mathbf{1}, 2, 1\}$.
  3. We multiply the signals $x[k]$ and $y[k-1]$. The result of this process is the construction of the signal $\upsilon_1 = \{1 * 1, 2 * 2, 3 * 1, 1 * 0\} = \{1, 4, 3, 0\}$.
  4. We add the elements of the sequence υ_1, in order to calculate the value $z[1]$. Therefore, $z[1] = 1 + 4 + 3 + 0 = 8$.

- Calculation of the value $z[2]$ corresponding to the value $n = 2$:

  1. We reverse the signal $y[-k]$ in time, so that it takes the form $y[-k]=\{-1, 1, \mathbf{2}, 1\}$.
  2. We shift the signal $y[-k]$ by $n = 2$ units, obtaining the signal $y[2-k] = \{-\mathbf{1}, 1, 2, 1\}$.
  3. We multiply the signals $x[k]$ and $y[k-2]$. The result of this process is the construction of the signal $\upsilon_2 = \{(-1) * 1 + 1 * 2 + 2 * 3 + 1 * 1\} = \{-1, 2, 6, 1\}$.
  4. We add the elements of the sequence υ_2, in order to calculate the value $z[2]$. Therefore, $z[2] = -1 + 2 + 6 + 1 = 8$.

- Calculation of the value $z[3]$ corresponding to the value $n = 3$:

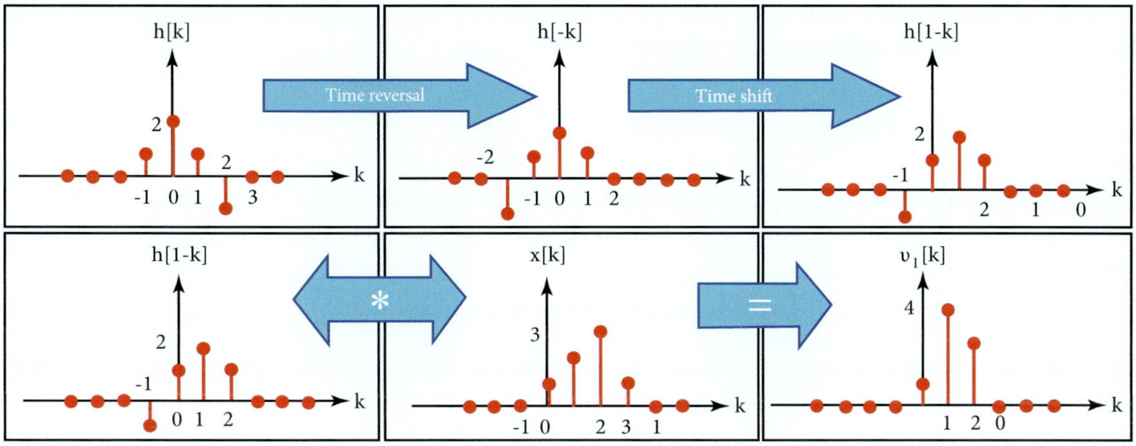

$$\upsilon_1[k] = \{ ..., 0, 1, 4, 3, 0, 0, 0, 0, ... \} \rightarrow y[1] = 1 + 4 + 3 = 8$$

Fig. 6.19 Diagrammatic calculation of the value $z[n]$ of the convolution of the discrete-time signals $x[n]$ and $y[n]$ with content $x[n] = \{\mathbf{1}, 2, 3, 1\}$ and $y[n] = \{1, \mathbf{2}, 1, -1\}$

1. We reverse the signal $y[-k]$ in time, so that it takes the form $y[-k]=\{-1, 1, \mathbf{2}, 1\}$.
2. We shift the signal $y[-k]$ by $n=3$ time units, obtaining the signal $y[3-k]=\{\mathbf{0}, -1, 1, 2\}$.
3. We multiply the signals $x[k]$ and $y[k-3]$. The result of this process is the construction of the signal $\upsilon_3=\{1*0, (-1)*2, 1*3, 2*1\}=\{0, -2, 3, 2\}$.
4. We add the elements of the sequence υ_3, in order to calculate the value $z[3]$. Therefore, $z[3] = 0 - 2 + 3 + 2 = 3$.

Applying the same process for other values of n, we calculate the values $z[4] = -2$ and $z[5] = -1$. We also find that for values $n > 5$ the two signals *are foreign to each other* and therefore all terms of the sequence υ_n (and obviously their sum) have zero value. Therefore, for $n > 5$, $z[n] = 0$.

The above process can also be applied to calculate the output of the system for negative values of n; in this case, as already mentioned, the shift of the impulse response sequence will be to the left. It is not difficult to find that for $n = -1$ the value of the signal $z[n]$ is $z[-1] = 1$, whereas for the value range $n < -1$ it is $z[n] = 0$. Therefore, the output of the system for the entire value range $(-\infty < n < +\infty)$ will be

$$z[n] = \{..., 0, 0, 0, 1, \mathbf{4}, 8, 8, 3, -2, -1, 0, 0, 0, 0...\}.$$

The way to calculate the value $z[n]$ of the convolution of the signals $x[n]$ and $y[n]$ is shown in Fig. 6.19.

Properties of the Convolution of Discrete-Time Signals

In full analogy with its continuous equivalent, the convolution of discrete-time signals is characterized by a set of useful properties, which in general are the following.

- **Commutative property:** according to the *commutative property*, the order in which the operation of convolution is performed between two discrete-time signals $x[n]$ and $y[n]$ does not matter, since in both cases we arrive at the same result. This property in mathematical form is expressed as

$$x[n] * y[n] = y[n] * x[n]$$

Proof Starting from the defining equation of convolution

$$z[n] = x[n] * y[n] = \sum_{k=-\infty}^{\infty} x[k]y[n-k]$$

let us change the dummy index of the sum from k to m with the two indices related through the relation $m = n - k$. In this case, $k = n - m$ and the above relation will take the form

$$z[n] = \sum_{m=-\infty}^{\infty} x[n-m]y[m] = \sum_{k=-\infty}^{\infty} x[n-k]y[k] = y[n] * x[n],$$

where in the second sum we have changed the dummy index of the sum from m to k.

- **Associative property:** considering three discrete-time signals $x[n]$, $y[n]$ and $z[n]$ it is proven that these satisfy the condition

$$\Big(x[n] * y[n]\Big) * z[n] = x[n] * \Big(y[n] * z[n]\Big).$$

Proof In order to prove the *associative property*, let us write the expressions that appear in the two sides using the convolution defining equation. Starting from the left member, let us set

$$x[n] * y[n] = r[n] = \sum_{k=-\infty}^{\infty} x[k]y[n-k].$$

It will then be

$$\Big(x[n] * y[n]\Big) * z[n] = r[n] * z[n] = \sum_{\ell=-\infty}^{\infty} r[\ell]z[n-\ell] = \sum_{\ell=-\infty}^{\infty} \Big(\sum_{k=-\infty}^{\infty} x[k]y[\ell-k]\Big)z[n-\ell].$$

In a completely analogous way, for the right member we will have

$$y[n] * z[n] = t[n] = \sum_{k=-\infty}^{\infty} y[k]z[n-k] \quad \text{and}$$

$$x[n] * \Big(y[n] * z[n]\Big) = x[n] * t[n] = \sum_{\ell=-\infty}^{\infty} x[\ell]t[n-\ell] = \sum_{\ell=-\infty}^{\infty} x[\ell]\Big(\sum_{k=-\infty}^{\infty} y[k]z[n-k-\ell]\Big).$$

In order to prove the validity of the *associative property* we need to show that the relations we derived for the two sides coincide with each other, or equivalently that

$$\sum_{\ell=-\infty}^{\infty} x[\ell]\Big(\sum_{k=-\infty}^{\infty} y[k]z[n-k-\ell]\Big) = \sum_{\ell=-\infty}^{\infty} \Big(\sum_{k=-\infty}^{\infty} x[k]y[\ell-k]\Big)\, z[n-\ell].$$

To do this, let us proceed to change the dummy index from k to q, setting $q = l + k$. It will then be

$$\begin{aligned}
\sum_{\ell=-\infty}^{\infty} x[\ell]\Big(\sum_{k=-\infty}^{\infty} y[k]z[n-k-\ell]\Big) &= \sum_{\ell=-\infty}^{\infty} x[\ell]\Big(\sum_{q=-\infty}^{\infty} y[q-\ell]z[n-q]\Big) \\
&= \sum_{\ell=-\infty}^{\infty}\sum_{q=-\infty}^{\infty} x[\ell]y[q-l]z[n-q] = \sum_{q=-\infty}^{\infty}\sum_{\ell=-\infty}^{\infty} x[\ell]y[q-\ell]z[n-q] \\
&= \sum_{q=-\infty}^{\infty}\Big(\sum_{\ell=-\infty}^{\infty} x[\ell]y[q-\ell]\Big)\, z[n-q].
\end{aligned}$$

If in the last sum we proceed to a simultaneous change of the indices from q to l and from l to k, we get

$$\sum_{l=-\infty}^{\infty} x[\ell]\Big(\sum_{k=-\infty}^{\infty} y[k]z[n-k-\ell]\Big) = \sum_{\ell=-\infty}^{\infty}\Big(\sum_{k=-\infty}^{\infty} x[k]y[\ell-k]\Big)z[n-\ell]$$

or equivalently $([x[n] * y[n]) * z[n] = x[n] * (y[n] * z[n])$, which is the desired result.

- **Distributive property:** considering three discrete-time signals $x[n]$, $y[n]$ and $z[n]$, it is proven that they satisfy the condition

$$x[n] * \Big(y[n] + z[n] \Big) = x[n] * y[n] + x[n] * z[n].$$

Proof the *distributive property* is proven from the defining equation of the convolution as follows:

$$\begin{aligned} x[n] * \Big(y[n] + z[n] \Big) &= \sum_{k=-\infty}^{\infty} x[k] \Big(y[n-k] + z[n-k] \Big) \\ &= \sum_{k=-\infty}^{\infty} x[k]y[n-k] + \sum_{k=-\infty}^{\infty} x[k]z[n-k] = x[n] * y[n] + x[n] * z[n] \end{aligned}$$

- **Homogeneity:** considering two discrete-time signals $x[n]$ and $y[n]$ it is proven that

$$\Big(\alpha x[n] \Big) * y[n] = \alpha \Big(x[n] * y[n] \Big) = x[n] * \Big(\alpha y[n] \Big).$$

Proof from the defining equation of the convolution it immediately follows that

$$\begin{aligned} \Big(\alpha x[n] \Big) * y[n] &= \sum_{k=-\infty}^{\infty} \Big(\alpha x[k] \Big) y[n-k] = \alpha \sum_{k=-\infty}^{\infty} x[k]y[n-k] = \alpha \Big(x[n] * y[n] \Big) \\ &= \sum_{k=-\infty}^{\infty} x[n] \Big(\alpha y[n-k] \Big) = x[n] * \Big(\alpha y[n] \Big). \end{aligned}$$

- **Linearity:** considering two discrete-time signals $z_1[n] = x_1[n] * y[n]$ and $z_2[n] = x_2[n] * y[n]$, it is proven that for the signals $x[n]$ and $z[n]$ defined as $x[n] = \alpha x_1[n] + \beta x_2[n]$ and $z[n] = \alpha z_1[n] + \beta z_2[n]$, $x[n] * y[n] = z[n]$.

Proof combining the properties of homogeneity and linearity, we get

$$x[n] * y[n] = \Big[\alpha x_1[n] + \beta x_2[n] \Big] * y[n] = \alpha \Big(x_1[n] * y[n] \Big) + \beta \Big(x_2[n] * y[n] \Big) = \alpha z_1[n] + \beta z_2[n] = z[n]$$

- **Time shift:** if the signals $x[n]$, $y[n]$ and $z[n]$ are connected between them through the relation

$$z[n] = x[n] * y[n] = \sum_{k=-\infty}^{+\infty} x[k]y[n-k],$$

then it will also be $z[n-\ell] = x[n] * y[n-\ell] = x[n-\ell] * y[n]$.

Proof from the *defining equation of the convolution* it immediately follows that

$$z[n-\ell] = \sum_{k=-\infty}^{+\infty} x[k]y[(n-\ell)-k] = x[n] * y[n-\ell].$$

Also, from the *commutative property* of the convolution we have

$$z[n-\ell] = \sum_{k=-\infty}^{+\infty} y[k]x[(n-\ell)-k] = y[n] * x[n-\ell] = x[n-\ell] * y[n].$$

From the equation of all the above, the desired result follows.

- **First difference:** if the signals $x[n]$, $y[n]$ and $z[n]$ are interconnected through the relation

$$z[n] = x[n] * y[n] = \sum_{k=-\infty}^{+\infty} x[k]y[n-k],$$

then it will also be $z[n] - z[n-1] = x[n] * \Big(y[n] - y[n-1]\Big)$.

Proof from the *property of time shift* for $\ell = 1$ it follows that

$$z[n-1] = x[n] * y[n-1].$$

Subtracting term by term the above relation from the equation defining the convolution and using the *distributive property*, we arrive at the desired result as

$$z[n] - z[n-1] = x[n] * y[n] - x[n] * y[n-1] = x[n] * \Big(y[n] - y[n-1]\Big).$$

- **Sum:** if the discrete-time signals $x[n]$, $y[n]$, and $z[n]$ are interconnected through the relation

$$z[n] = x[n] * y[n] = \sum_{k=-\infty}^{\infty} x[k]y[n-k],$$

then it will also be $S_z = S_x S_y$, with the quantities S_z, S_x, and S_y being defined as

$$S_z = \sum_{n=-\infty}^{\infty} z[n], \quad S_x = \sum_{k=-\infty}^{\infty} x[k], \quad S_y = \sum_{n=-\infty}^{\infty} y[n-k].$$

Proof from the equation defining the convolution we have

$$z[n] = \sum_{k=-\infty}^{\infty} x[k]y[n-k],$$

and therefore, we have

$$S_z = \sum_{n=-\infty}^{\infty} z[n] = \sum_{n=-\infty}^{\infty} \left(\sum_{k=-\infty}^{\infty} x[k]y[n-k]\right) = \left(\sum_{k=-\infty}^{\infty} x[k]\right)\left(\sum_{n=-\infty}^{\infty} y[n-k]\right) = S_x S_y,$$

where for the extraction of the final result we have changed the order of the sums.

Problem 6.10.11 Prove the commutative property of the convolution considering the discrete-time signals

$$x[n] = \left(\frac{1}{2}\right)^n u[n] \quad \text{and} \quad y[n] = \left(\frac{1}{4}\right)^n u[n].$$

Solution In order to prove the commutative property of convolution we need to show that $h_1(n) * h_2(n) = h_2(n) * h_1(n)$. In detail, then, we have:

$$h_1[n] * h_2[n] = \sum_{k=-\infty}^{\infty} h_1[k]h_2[n-k] = \sum_{k=-\infty}^{\infty} \left(\frac{1}{2}\right)^k u[k]\left(\frac{1}{4}\right)^{n-k} u[n-k].$$

From the above equation we observe that for $k < 0$ it is $u[k] = 0$, whereas for $k > n$ it is $u[n-k] = 0$. Therefore, the above equation will take the form

$$h_1[n] * h_2[n] = \sum_{k=0}^{n} \left(\frac{1}{2}\right)^k \left(\frac{1}{4}\right)^{n-k} = \left(\frac{1}{4}\right)^n \sum_{k=0}^{n} 2^k = \left(\frac{1}{4}\right)^n (2^{n+1} - 1) = \left(\frac{1}{2}\right)^n \left[2 - \left(\frac{1}{2}\right)^n\right] \quad n \geq 0.$$

In exactly the same way we have

$$h_2[n] * h_1[n] = \sum_{k=-\infty}^{\infty} h_2[k] h_1[n-k] = \sum_{k=-\infty}^{\infty} \left(\frac{1}{4}\right)^k u[k] \left(\frac{1}{2}\right)^{n-k} u[n-k].$$

Arguing as before, we can write the last relation as

$$\begin{aligned} h_2[n] * h_1[n] &= \sum_{k=0}^{n} \left(\frac{1}{4}\right)^k \left(\frac{1}{2}\right)^{n-k} = \left(\frac{1}{2}\right)^n \sum_{k=0}^{n} \left(\frac{1}{2}\right)^k \left(\frac{1}{2}\right)^n \left[\frac{1 - (1/2)^{n+1}}{1 - (1/2)}\right] = 2\left(\frac{1}{2}\right)^n \left[1 - \left(\frac{1}{2}\right)^{n+1}\right] \\ &= \left(\frac{1}{2}\right)^n \left[2 - 2\left(\frac{1}{2}\right)^{n+1}\right] = \left(\frac{1}{2}\right)^n \left[2 - \left(\frac{1}{2}\right)^{-1} \left(\frac{1}{2}\right)^{n+1}\right] = \left(\frac{1}{2}\right)^n \left[2 - \left(\frac{1}{2}\right)^n\right] \quad n \geq 0 \end{aligned}$$

and by comparing the two results, the desired result is obtained.

6.10.3 Cross-Correlation and Autocorrelation of Discrete-Time Signals

Let us consider two sequences of real numbers $x[n]$ and $y[n]$, which correspond to *discrete-time signals of finite energy*. In full analogy with the continuous case, the degree of similarity of the two signals is quantitatively described by the magnitude of the *cross-correlation* or simply *correlation*, which is defined as

$$r_{xy}[\ell] = \sum_{n=-\infty}^{\infty} x[n] y[n-\ell], \quad (\ell = 0, \pm 1, \pm 2, \ldots) \tag{6.57}$$

or equivalently

$$r_{xy}[\ell] = \sum_{n=-\infty}^{\infty} x[n+\ell] y[n], \quad (\ell = 0, \pm 1, \pm 2, \ldots). \tag{6.58}$$

In the above notation, the index ℓ expresses a *shift in the discrete-time*, whereas the indices x and y indicate the discrete-time signals that we correlate. The order in which these indices are written determines the "direction" in which one sequence is shifted in relation to the other. It should be noted that reversing the roles of the signals $x[n]$ and $y[n]$ does not change any information regarding the degree of correlation of the two signals, since, in this case, the above relations will take the form

$$r_{yx}[\ell] = \sum_{n=-\infty}^{\infty} y[n] x[n-\ell] \quad \text{and} \quad r_{yx}[\ell] = \sum_{n=-\infty}^{\infty} y[n+\ell] x[n]$$

from which it follows that $r_{xy}[\ell] = r_{yx}[-\ell]$.

In the special case where the two signals *coincide*—that is, the relation $x[n] = y[n]$ holds—the above quantity is called *autocorrelation* and is given by the relation

$$r_{xx}[\ell] = \sum_{n=-\infty}^{\infty} x[n]x[n-\ell] \quad \text{or equivalently} \quad r_{xx}[\ell] = \sum_{n=-\infty}^{\infty} x[n+\ell]x[n].$$

If the length of the sequences $x[n]$ and $y[n]$ is finite and equal to N (this means that the relation $x[n] = y[n] = 0$ for $n < 0$ and $n \geq N$) the above relations take the form

$$r_{xy}[\ell] = \sum_{n=i}^{N-|k|-1} x[n]y[n-\ell] \quad \text{and} \quad r_{xx}[\ell] = \sum_{n=i}^{N-|k|-1} x[n]x[n-\ell],$$

where for $\ell \geq 0$ it is $i = \ell$ and $k = 0$, whereas for $\ell < 0$ it is $i = 0$ and $k = \ell$.

The above equations are used in the case of *energy signals*; on the other hand, if the discrete-time signals we are dealing with are *power signals*, characterized by *infinite energy and finite power*, the quantities of *cross-correlation* and *autocorrelation* are defined as

$$r_{xy}[\ell] = \lim_{N\to\infty} \left(\frac{1}{2N+1} \sum_{n=-N}^{N} x[n]y[n-\ell] \right) \tag{6.59}$$

$$r_{xx}[\ell] = \lim_{N\to\infty} \left(\frac{1}{2N+1} \sum_{n=-N}^{N} x[n]x[n-\ell] \right) \tag{6.60}$$

respectively. Specifically, if the sequences $x[n]$ and $y[n]$ are *periodic discrete-time signals with period* N, the above averages for the interval $(-\infty, +\infty)$ coincide with those calculated *for a period interval* and the above relations take the form

$$r_{xy}[\ell] = \frac{1}{N} \sum_{n=0}^{N-1} x[n]y[n-l] \quad \text{and} \quad r_{xx}[\ell] = \frac{1}{N} \sum_{n=0}^{N-1} x[n]x[n-l].$$

It is proven that in the case of *periodic signals* both the cross-correlation and the autocorrelation are *periodic sequences* with a period equal to the period of the signals $x[n]$ and $y[n]$. The proof of this property is carried out in exactly the same way as for the discrete-time signal of the *periodic convolution* and is left as an exercise to the reader.

Properties of Cross-Correlation and Autocorrelation

In order to present the properties of the sequences regarding cross-correlation and autocorrelation, let us consider two finite energy sequences $x[n]$ and $y[n]$ and a linear combination of them in the form $\alpha x[n] + \beta y[n - \ell]$, where α and β are arbitrary constants and ℓ is the symbol of the time shift. In this case, the energy of the signal resulting from the above linear combination is given, as known, by the relation

$$E = \sum_{n=-\infty}^{\infty} \left(\alpha x[n] + \beta y[n-\ell] \right)^2 = \alpha^2 \sum_{n=-\infty}^{\infty} x^2[n] + \beta^2 \sum_{n=-\infty}^{\infty} y^2[n-\ell] + 2\alpha\beta \sum_{n=-\infty}^{\infty} x[n]y[n-\ell].$$

However, we observe that

$$\sum_{n=-\infty}^{\infty} x^2[n] = \sum_{n=-\infty}^{\infty} x[n]x[n] = \sum_{n=-\infty}^{\infty} x[n]x[n-0] = r_{xx}[0] = E_x,$$

whereas in a similar way we have

$$\sum_{n=-\infty}^{\infty} y^2[n-\ell] = \sum_{n=-\infty}^{\infty} y[n-\ell]y[n-\ell] = \sum_{n'=-\infty}^{\infty} y[n']y[n']) = \sum_{n'=-\infty}^{\infty} y[n']y[n'-0] = r_{yy}[0] = E_y.$$

In the above equations, the quantities E_x and E_y refer to the energies of the signals $x[n]$ and $y[n]$ respectively, whereas in the second equation the index was changed from $(n-\ell)$ to n' for better understanding. Therefore, the energy of the linear combination $\alpha x[n] + \beta y[n-\ell]$ will satisfy the property

$$E = \alpha^2 r_{xx}[0] + \beta^2 r_{yy}[0] + 2\alpha\beta r_{xy}[\ell] \geq 0,$$

which can also be formulated as

$$E = \alpha^2 E_x + \beta^2 E_y + 2\alpha\beta r_{xy}[\ell] \geq 0.$$

Assuming that the parameter β is *different than zero*, we can divide both sides of the above inequality by the quantity β^2. In this case, we get

$$r_{xx}[0]\left(\frac{\alpha}{\beta}\right)^2 + 2r_{xy}[\ell]\left(\frac{\alpha}{\beta}\right) + r_{yy}[0] \geq 0. \tag{6.61}$$

Setting $\xi = \alpha/\beta$, we are led to the quadratic inequality

$$r_{xx}[0]\xi^2 + 2r_{xy}[\ell]\xi + r_{yy}[0] \geq 0 \tag{6.62}$$

with discriminant $\Delta = \beta^2 - 4\alpha\gamma = 4r_{xy}^2[\ell] - 4r_{xx}[0]r_{yy}[0] \leq 0$ from which it follows that

$$4\left(r_{xy}^2[\ell] - r_{xx}[0]r_{yy}[0]\right) \leq 0.$$

Therefore, the correlation sequence $r_{xy}[\ell]$ satisfies the condition

$$|r_{xy}[\ell]| \leq \sqrt{r_{xx}[0]r_{yy}[0]} \tag{6.63}$$

or equivalently

$$|r_{xy}[\ell]| \leq \sqrt{E_x E_y}. \tag{6.64}$$

The above relation and for the case $x[n] = y[n]$ leads to the conclusion that *the autocorrelation $r_{xx}[\ell]$ takes the maximum value (which is equal to the energy of the signal E), for the value of the time shift $\ell = 0$*. This is of course expected, because for this value of ℓ the signals $x[n]$ and $x[n-\ell]$ are *identical* and therefore characterized by the *maximum degree of similarity*.

It is also important to mention at this point that if there is a *scaling* of one (or both) of the sequences $x[n]$ and $y[n]$, *the shape of the correlation sequence does not change* and only the amplitudes scale accordingly. Since this scaling does not serve any purpose (as the information conveyed by the signal is the same), it is common practice to normalize the values of the correlation and autocorrelation sequences, dividing them by the largest of them. In this way, a reduction of the values of these sequences takes place, in the range $[-1, +1]$. According to what we have mentioned in the previous sections, the maximum value of the correlation sequence is $r_{xy}^{\max} = \sqrt{r_{xx}[0]r_{yy}[0]}$, whereas the maximum value of the autocorrelation sequence is $r_{xx}^{\max} = r_{xx}[0]$. Therefore, the *normalized sequences* of *cross-correlation* and *autocorrelation* are given by the equations

$$\varrho_{xx}[\ell] = \frac{r_{xx}[\ell]}{r_{xx}[0]} \quad \text{and} \quad \varrho_{xy}[\ell] = \frac{r_{xy}[\ell]}{\sqrt{r_{xx}[0]r_{yy}[0]}}.$$

Finally, as we have already mentioned, the correlation function $r_{xy}[\ell]$ satisfies the condition $r_{xy}[\ell] = r_{yx}[-\ell]$. For $x[n] = y[n]$ the above relationship takes the form $r_{xx}[\ell] = r_{xx}[-\ell]$, from which follows the conclusion that *the autocorrelation function is an even function*. This means that to calculate the autocorrelation sequence, it is enough to calculate the values of the function $r_{xx}[\ell]$ for the area $\ell \geq 0$.

Example of Calculating the Correlation Sequence

In order to understand the process of calculating the values of the correlation sequence, let us consider the discrete-time signals

$$x[n] = \{\ldots, 0, 0, 2, -1, 3, +7, \mathbf{1}, +2, -3, 0, 0, \ldots\} \quad \text{and}$$
$$y[n] = \{\ldots, 0, 0, 1, -1, 2, -2, \mathbf{4}, 1, -2, 5, 0, 0, \ldots\},$$

where the values written in bold correspond to the index value $n = 0$ (this convention is also used in the remainder of the example). The calculation of the sequence values $r_{xy}[\ell]$ takes place based on the defining equation of the correlation magnitude. For $\ell = 0$ this equation will take the form

$$r_{xy}[0] = \sum_{n=-\infty}^{\infty} x[n]y[n].$$

The result of this operation is the sequence of values

$$\upsilon_0[n] = \{\ldots, 0, 0, 2, 1, 6, -14, \mathbf{4}, 2, 6, 0, 0, \ldots\},$$

the sum of which is equal to $r_{xx}[0] = 7$.

For *positive* values of the time shift ℓ ($\ell > 0$) *we shift the signal* $y[n]$ *to the right* in relation to the signal $x[n]$ by ℓ time units, we calculate the sequence of products $\upsilon_\ell[n] = x[n]y[n-\ell]$ and then sum these values. Applying this process, we arrive at the following results, regarding the values of the correlation of the signals $x[n]$ and $y[n]$ that correspond to positive values of the parameter ℓ:

$$r_{xy}[1] = +13 \; r_{xy}[2] = -18 \; r_{xy}[3] = +16 \quad r_{xy}[4] = -07$$
$$r_{xy}[5] = +05 \; r_{xy}[6] = -03 \; r_{xy}[l] = 0 \text{ for } l \geq 7.$$

On the other hand, for *negative* values of the time shift ℓ ($\ell > 0$), *we shift the signal* $y[n]$ *to the left* in relation to the signal $x[n]$ by ℓ time units, we calculate again the sequence of products $\upsilon_\ell[n]$, and then we add these values. Applying this process, we arrive at the following results regarding the values of the correlation of the signals $x[n]$ and $y[n]$ that correspond to negative values of the parameter l:

$$r_{xy}[-1) = +00 \; r_{xy}[-2] = +33 \; r_{xy}[-3] = -14$$
$$r_{xy}[-4] = +36 \; r_{xy}[-5] = +19 \; r_{xy}[-6] = -09$$
$$r_{xy}[-7] = +10 \; r_{xy}[\ell] = 0 \qquad \text{for } \ell \leq -8.$$

Therefore, the final correlation sequence of the signals $x[n]$ and $y[n]$ includes the values

$$r_{xy}[\ell] = \{10, -9, 19, 36, -14, 33, 0, \mathbf{7}, 13, -18, 16, -7, 5, -3\} \quad (-8 < \ell < 7).$$

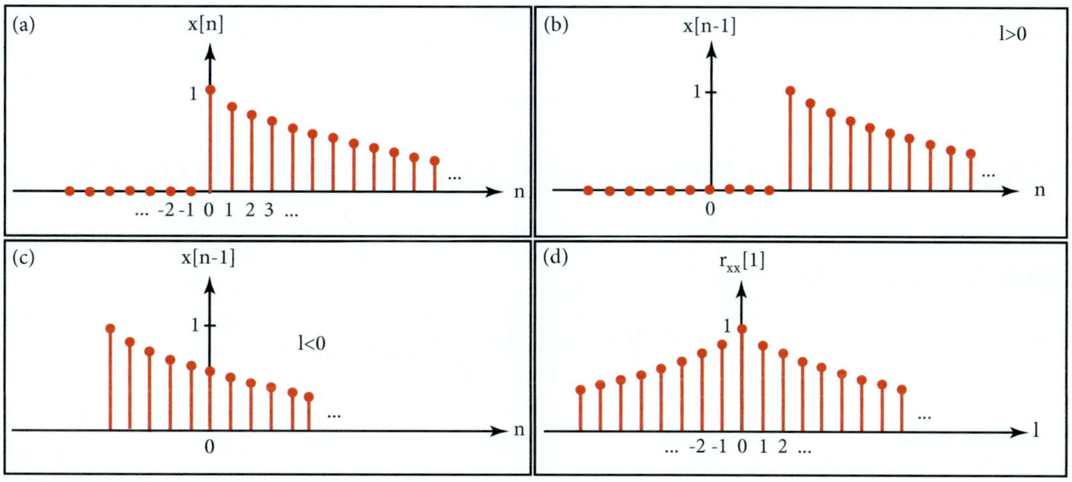

Fig. 6.20 Calculation of the autocorrelation of the signal $x[n] = \alpha^n u[n]$ of Exercise 6.10.12

Problem 6.10.12 Calculate the autocorrelation of the signal $x[n] = \alpha^n u[n]$—$(0 < \alpha < 1$ and $-\infty < n < \infty)$— depicted in image (a) of Fig. 6.20 (Proakis and Manolakis, Example 2.6.2).

Solution For the calculation of the autocorrelation sequence we consider the cases $\ell \geq 0$ and $\ell < 0$ and the form of the corresponding discrete-time signals $x[n-\ell]$, as depicted in Fig. 6.15. In detail, then, we have:

- Case 1 ($\ell \geq 0$). From image (b) of Fig. 6.20 we find that the first nonzero value of the right-shifted signal $x[n-\ell]$ corresponds to the position $n = \ell$, which will also be the lower limit of the sum in the equation defining autocorrelation. It will therefore be

$$r_{xx}[\ell] = \sum_{n=\ell}^{\infty} x[n]x[n-\ell] = \sum_{n=\ell}^{\infty} \alpha^n \alpha^{n-l} = \sum_{n=\ell}^{\infty} \alpha^n \alpha^n \alpha^{-\ell} = \alpha^{-\ell} \sum_{n=\ell}^{\infty} (\alpha^2)^n.$$

Since the parameter α is defined in the interval $0 < \alpha < 1$ the above series converges to a finite value. Referring to a mathematical formula book we find that

$$\sum_{n=\ell}^{\infty} (\alpha^2)^n = \frac{\alpha^{2\ell}}{1-\alpha^2}$$

and therefore the above relation is written

$$r_{xx}[\ell] = \alpha^{-\ell} \sum_{n=\ell}^{\infty} (\alpha^2)^n = \alpha^{-\ell} \frac{1}{1-\alpha^2} a^{2\ell} = \frac{\alpha^{\ell}}{1-\alpha^2}.$$

- Case 2 ($\ell < 0$). From image (c) of Fig. 6.20, we find that the first nonzero value of the expression $x[n-\ell]$ corresponds to the value $n = 0$, which is also the lower limit of the sum in the equation defining autocorrelation. It will therefore be

$$r_{xx}[\ell] = \sum_{n=0}^{\infty} x[n]x[n-\ell] = \sum_{n=0}^{\infty} \alpha^n \alpha^{n-\ell} = \sum_{n=0}^{\infty} \alpha^n \alpha^n \alpha^{-\ell} = \alpha^{-\ell} \sum_{n=0}^{\infty} (\alpha^2)^n = \frac{\alpha^{-\ell}}{1-\alpha^2}$$

(the last sum results as a special case of the previous one for the value $\ell = 0$).

Combining the results we obtained for the above two cases, we have

$$r_{xx}[\ell] = \frac{\alpha^\lambda}{1-\alpha^2} \quad (-\infty < \ell < \infty) \quad (0 < \alpha < 1),$$

where $\lambda = |\ell|$. The discrete-time signal of the autocorrelation $r_{xx}[\ell]$ is depicted in image (d) of Fig. 6.20.

From the last relation, we find that for the value $\ell = 0$, $r_{xx}[0] = 1/(1-\alpha^2)$. Therefore, the normalized sequence of the signal's autocorrelation $x[n] = \alpha^n u[n]$ is calculated as

$$\rho_{xx}[\ell] = \frac{r_{xx}[\ell]}{r_{xx}[0]} = \frac{\alpha^\lambda/(1-\alpha^2)}{1/(1-\alpha^2)} = \alpha^\lambda = \alpha^{|\ell|},$$

which is also the final result.

6.11 Problems

1. Formulate the discrete-time signals $x_1[n] = e^{-n/2}$, $x_2[n] = (-2)^n$, $x_3[n] = (0.5)^n$ and $x_4[n] = (-2)^n$ defined in the value range $-10 \le n \le 10$, in their diagrammatic and sequential representation.
2. Formulate in their diagrammatic and sequential representation each of the following discrete-time signals

$$x[n] = \cos\left(\frac{3\pi n}{7} + \frac{\pi}{16}\right) \qquad -5 \le n \le 15, \quad y[n] = 5n + 3^{-n} \quad -5 \le n \le 5$$

$$z[n] = \cos\left(\frac{2\pi n}{8}\right) + \sin\left(\frac{\pi n}{2}\right) \quad -10 \le n \le 10, \; t[n] = n^2\, 5^{-|n|} \quad -15 \le n \le 15.$$

3. Considering the continuous-time signal $x(t) = 5\cos(15000\,\pi t)$, determine the discrete-time signal that will result from the sampling of this continuous-time signal with frequencies $F_{s1} = 20$ KHz and $F_{s2} = 35$ KHz.
4. Considering a continuous-time signal of the form

$$x(t) = 5\cos(15000\,\pi t) - 15\sin(12000\,\pi t) + 3\cos(100\pi t)$$

determine the discrete-time signal that results from the sampling of the above signal with frequencies $F_{s1} = 25$ KHz and $F_{s2} = 45$ KHz.

5. Considering the analog signal

$$x_\alpha(t) = 3\sin(1000\,\pi t)$$

draw the waveform that describes it and then answer the following questions:

(a) Assuming that the signal is sampled at a frequency of $F_s = 300$ Hz, determine the frequency of the discrete-time signal $x[n] = x_\alpha(nT_s)$ $(T_s = 1/F_s)$ and prove that it is periodic.
(b) Calculate the samples of the discrete-time signal $x[n]$ for a period of time and draw the discrete-time signal in contrast to the continuous-time signal from which it originated.
(c) Determine the value of the sampling frequency F_s such that the discrete-time signal approaches its maximum amplitude value, which is the value $A = 3$. What is the minimum value for this frequency?

6. An analog signal originating from an electrocardiogram contains frequencies up to 100 Hz. (a) What is the Nyquist frequency? (b) Suppose that the above signal is sampled at a sampling frequency of $F_s = 250$ KHz. What is the highest frequency of the signal that can be uniquely represented for this sampling frequency?
7. Consider the continuous-time signal $x_\alpha(t) = \sin(20\,\pi t) + \cos(40\,\pi t)$, the sampling of which with period T_s led to the creation of the discrete-time signal

$$x[n] = \sin\left(\frac{\pi n}{5}\right) + \cos\left(\frac{2\pi n}{5}\right).$$

Determine the sampling period T_s and comment on whether or not its value is unique. In the latter case, determine another value of the sampling period that leads to the creation of the same discrete-time signal.

8. Repeat Exercise 7 for the pair of signals

$$x_\alpha(t) = \cos(4000\pi t) \quad \text{and} \quad x[n] = \cos\left(\frac{\pi n}{3}\right).$$

9. Provide an example of sampling a periodic analog signal that leads to a nonperiodic discrete-time signal.
10. Considering a continuous-time signal $x_\alpha(t)$ with the Nyquist frequency ω_0, find the Nyquist frequency of the signals

$$y_1(t)=x_\alpha(t)+x_\alpha(t-1), \quad y_2(t)=\frac{dx_\alpha(t)}{dt}, \quad y_3(t)=x_\alpha^2(t), \quad y_4(t)=x_\alpha(t)\cos(\omega_0 t).$$

11. Let $x_\alpha(t)$ be a real periodic continuous-time signal characterized by *even symmetry* and its Fourier expansion has the form

$$x_\alpha(t) = \sum_{k=0}^{5}\left(\frac{1}{2}\right)^k \sin(k\pi t).$$

If $x_s(t)$ is the signal resulting from the sampling of the above signal with a period of $T_s = 0.2$ (according to Eqs. 6.10 and 6.11), decide whether or not the phenomenon of aliasing will occur and determine the Fourier expansion of the resulting signal, if we transmit the signal $x_s(t)$ through an ideal low-pass filter with a cutoff frequency equal to π/T_s and a gain value equal to T_s.

12. Considering the continuous-time signal $x_\alpha(t) = \cos(2\pi Ft + \varphi)$, which is sampled at a frequency of $F_s = 1000\,\text{Hz}$, determine the distorted (owing to aliasing) waveforms that will result, if the frequency of the signal $x_\alpha(t)$ has values (a) $F_s = 600\,\text{Hz}$, (b) $F_s = 1000\,\text{Hz}$, and (c) $F_s = 2400\,\text{Hz}$.
13. Determine the Nyquist frequency of the continuous-time signals

$$x_\alpha^1(t) = \text{sinc}^2(100\,\pi t) \quad \text{and} \quad x_\alpha^2(t) = \text{sinc}(50\,\pi t)\text{sinc}(100\,\pi t).$$

14. Let the analog signal $x_\alpha(t) = 15\text{sinc}^2(5\pi t) + \cos(50\pi t)$ be sampled at a frequency of $F_s = 10\,\text{Hz}$. Determine the spectrum $\mathcal{X}_s(e^{j\omega})$ and find out if the original signal $x_\alpha(t)$ can be reconstructed from this spectrum by using a low-pass filter.
15. Repeat Exercise 14 for a sampling frequency of $F_s = 20\,\text{Hz}$.
16. Determine the energy of the discrete-time signals for the range of values $n > 0$.

$$(a) \quad x[n] = (0.1)^n\cos(2\pi n), \quad (b) \quad x[n] = 2n(0.7)^n, \quad (c) \quad x[n] = \frac{4}{n}.$$

17. The following discrete-time signals are given:

$$x_1[n] = \cos(0.01\,\pi n), \quad x_2[n] = \cos\left(\frac{30\,\pi n}{105}\right), \quad x_3[n] = \cos(3\,\pi n), \quad x_4[n] = \sin(3n)$$

$$x_5[n] = 3\cos\left[5n + \left(\frac{\pi}{6}\right)\right], \quad x_6[n] = 2\exp\left\{i\left[\frac{n}{6} - \pi\right]\right\}, \quad x_7[n] = \cos\left(\frac{n}{8}\right)\cos\left(\frac{n\pi}{8}\right)$$

$$x_8[n] = \cos\left(\frac{n\pi}{2}\right) - \sin\left(\frac{n\pi}{8}\right) + 3\cos\left[\left(\frac{n\pi}{4}\right) + \left(\frac{\pi}{3}\right)\right].$$

For each signal, determine whether it is periodic and if so, determine its fundamental period.

18. The following discrete-time signals are given:

$$x_1[n] = \cos\left(\frac{40\,\pi n}{115}\right), \quad x_2[n] = \cos(4\,\pi n), \quad x_3[n] = \cos\left(\frac{n\pi}{2}\right) - \sin\left(\frac{n\pi}{8}\right) + 3\cos\left[\left(\frac{n\pi}{4}\right) + \left(\frac{\pi}{3}\right)\right].$$

For each signal, determine whether it is periodic and if so, determine its fundamental period.

19. Show that the fundamental period N_p of the discrete-time signals of the form $x_k[n] = \exp(j2\pi kn/N)$ $(k = 0, 1, 2, \ldots)$ has the form $N_p = N/MK\Delta(N, k)$, where $MK\Delta(N, k)$ is the greatest common divisor of the numbers N and k. Then calculate the fundamental period of the above group of signals for $N = 7$ and $N = 16$.
20. Determine the even and odd components of the discrete-time signals

$$x[n] = nu[n] \qquad \text{and} \qquad y[n] = \cos\left(\frac{\pi n}{6}\right).$$

21. Determine the even and odd components of the next signal and verify that the sum of their energies is equal to the energy of the discrete-time signal $x[n]$.

$$x[n] = \begin{cases} (0.65)^n & n \geq 0 \\ 0 & n < 0 \end{cases}$$

22. Prove that for a real signal $x[n]$ the energies of the even and odd components are equal to each other and equal to half the energy of the signal $x[n]$.
23. Calculate the even and odd components of the discrete-time signal $x[n] = e^{i\omega n}$.
24. For each of the following discrete-time signals

$$x[n] = \sin(5n) + \cos\left(\frac{2\pi n}{3}\right) \qquad \text{and} \qquad y[n] = \exp\left(-\frac{5\pi n}{4}\right) + \cos\left(\frac{4\pi n}{5} + \pi\right)$$

examine whether they are even or odd. If you find that any of these signals are neither even nor odd, determine their even and odd components. Repeat the same process for the discrete-time signal $z[n] = (-1)^n$, which is defined in the range of values $n \geq 0$.
25. Calculate the period of the signal $x(n) = \cos(n\pi/4)$. Then increase its frequency from $\omega_1 = \pi/4 = 2\pi/8$ to $\omega_2 = 3\pi/8$ and calculate the new period of the signal. Comment on the result.
26. Prove that the sum of two discrete sinusoidal signals with the same frequency is also a sinusoidal signal and find its amplitude and initial phase.
27. Find the energy of the following discrete-time signals for the value range $-10 \leq n \leq 10$.

$$(a) \quad x[n] = \cos(2\pi n), \quad (b) \quad x[n] = u[n]\delta[n], \quad (c) \quad x[n] = u[n]u[-n]$$

28. Prove that

$$\left(u[n] + (-1)^{n+1}u[n]\right)\cos\left(\frac{\pi n}{2}\right) = 0.$$

29. Construct and plot the signals $x[4 - n]$, $x[2n - 1]$ and $x[4 - 5n]$ with the discrete-time signal $x[n]$ defined as $x[n] = \{\ldots, 0, 3, 1, \mathbf{2}, 1, 0, 2, \ldots\}$.
30. Considering a discrete-time signal with energy E, calculate the energy of the discrete-time signals $y[n] = x[-n]$, $z[n] = x[n - m]$, $h[n] = x[m - n]$ and $t[n] = Ax[n]$.
31. Calculate the values of the sums

$$\sum_{k=-\infty}^{\infty} \delta[k-4]u[k], \quad \sum_{k=-\infty}^{\infty} r[k]\delta[k+5], \quad \sum_{k=-\infty}^{\infty} \cos\left(\frac{k\pi}{12}\right)\delta[k-3].$$

32. Based on the definition of the discrete impulse function, prove that

$$\sum_{\kappa=-\infty}^{\infty} \delta[\kappa] = \sum_{\kappa=-\infty}^{\infty} \delta[-\kappa] = 1$$

33. Given the discrete-time signal

$$x[n] = \begin{cases} 1 + (n/5) & -3 \leq n \leq -1 \\ 1 & 0 \leq n \leq 5 \\ 0 & \text{anywhere else} \end{cases}.$$

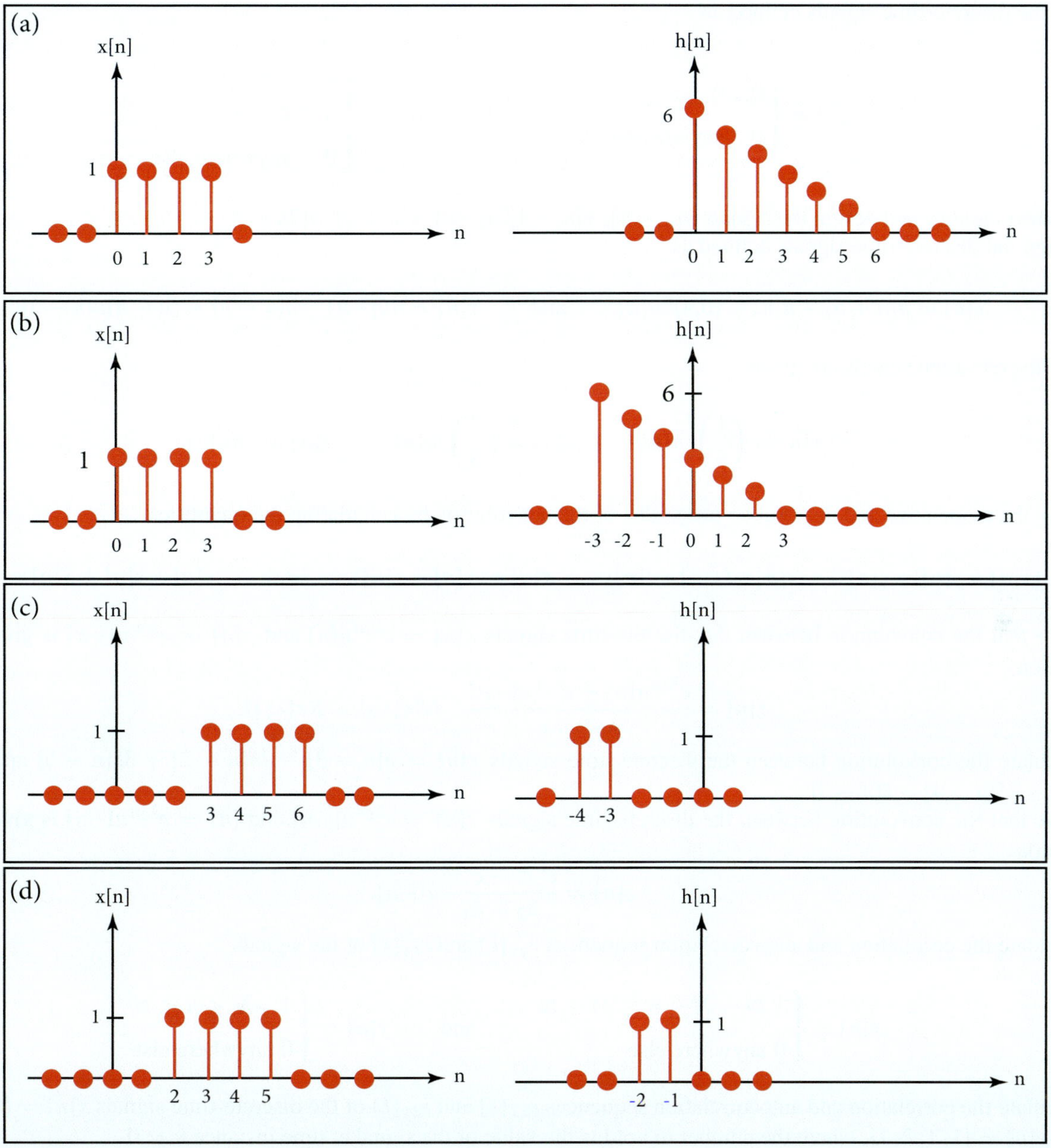

Fig. 6.21 The pairs of discrete-time signals of Problem 34

(a) Calculate the values of the signal for the range of values $-5 \leq n \leq 5$ and then plot the signal. (b) Formulate the above discrete-time signal as a function of the elementary discrete-time signals $\delta[n]$ and $u[n]$. (c) Plot the signal $y[n] = x[2n - 4]$.

34. Calculate the convolution between the signals $x[n]$ and $y[n]$ depicted in images (a), (b), (c), and (d) of Fig. 6.21.
35. Plot the discrete-time signals defined as

$$x[n] = |n|\Big(u[n+4] - u[n-4]\Big) \qquad \text{and} \qquad y[n] = n\Big(u[n-3] - u[n+3]\Big)$$

and then determine and plot the signals $x[3-n]$, $x[4-2n]$, $y[5n]$, $x[2n] + y[2n]$, $x[2n]y[-n]$.

36. Plot the discrete-time signals $x[4-n]$, $x[2n-4]$ and $x[n^2 + n - 1]$ where $x[n]$ is the discrete-time signal defined as $x[n] = (5-n)[u[n] - u[n-3]]$.

37. Plot the discrete-time signals defined as

$$x[n] = \begin{cases} 0.2 & 0 \le n \le 5 \\ 0 & \text{anywhere else} \end{cases} \quad \text{and} \quad y[n] = \begin{cases} 2 & -4 \le n \le -1 \\ n/7 & 0 \le n \le 4 \\ 0 & \text{anywhere else} \end{cases}$$

and then identify and design the signals $x[n-4]$, $x[n+3]$, $y[-n]$, $y[n+2]$, $y[2n]$.

38. Design the discrete-time signals defined as

$$x[n] = u[n+5] - u[n] + (0.4)^n u[n] \quad \text{and} \quad y[n] = nu[-n] - \delta[n-3] + (n-8)u[n-3].$$

39. The discrete-time signals are given

$$x[n] = \left(\frac{1}{4}\right)^n u[n], \qquad y[n] = \left(\frac{1}{8}\right)^n u[n], \qquad z[n] = 5nu[n].$$

Prove the associative and distributive properties of the convolution by calculating the quantities

$$(x[n] * y[n]) * z[n] \qquad x[n] * (y[n] * z[n]) \qquad x[n] * (y[n] + z[n]) \quad \text{and} \quad x[n] * y[n] + x[n] * z[n].$$

40. Prove that the convolution between the discrete-time signals $x[n] = e^{\lambda_1 n} u[n]$ and $y[n] = e^{\lambda_2 n} u[-n]$ is given by the relation

$$z[n] = \frac{e^{\lambda_1 n} u[n] + e^{\lambda_2 n} u[-n]}{\lambda_2 - \lambda_1} \quad (Re\{\lambda_2\} > Re\{\lambda_1\}).$$

41. Calculate the convolution between the discrete-time signals $x[n] = \delta[n-3] - 2\delta[n-2] + 3\delta[n-5]$ and $y[n] = \delta[n] + 2\delta[n+4] - \delta[n-3]$.

42. Prove that the convolution between the discrete-time signals $x[n] = e^{\lambda_1 n} u[-n]$ and $[n] = e^{\lambda_2 n} u[-n]$ is given by the relation

$$z[n] = \frac{e^{\lambda_1 n} - e^{\lambda_2 n}}{\lambda_2 - \lambda_1} u[-n].$$

43. Calculate the correlation and autocorrelation sequences $r_{xx}[\ell]$ and $r_{xy}[\ell]$ of the signals

$$x[n] = \begin{cases} 1 & M - N \le n \le M + N \\ 0 & \text{anywhere else} \end{cases} \quad \text{and} \quad y[n] = \begin{cases} 1 & -N \le n \le N \\ 0 & \text{anywhere else} \end{cases}.$$

44. Calculate the correlation and autocorrelation sequences $r_{xx}[\ell]$ and $r_{xy}[\ell]$ of the discrete-time signals $x[n] = \{\mathbf{1}, 2, 1, 1\}$ and $y[n] = \{\mathbf{1}, 1, 2, 1\}$, where the number in bold is the value of the signal at time instance $n = 0$.

45. Determine the normalized autocorrelation sequence of the signal

$$x[n] = \begin{cases} 1 & -N \le n \le N \\ 0 & \text{anywhere else} \end{cases}$$

Discrete-Time Systems

7

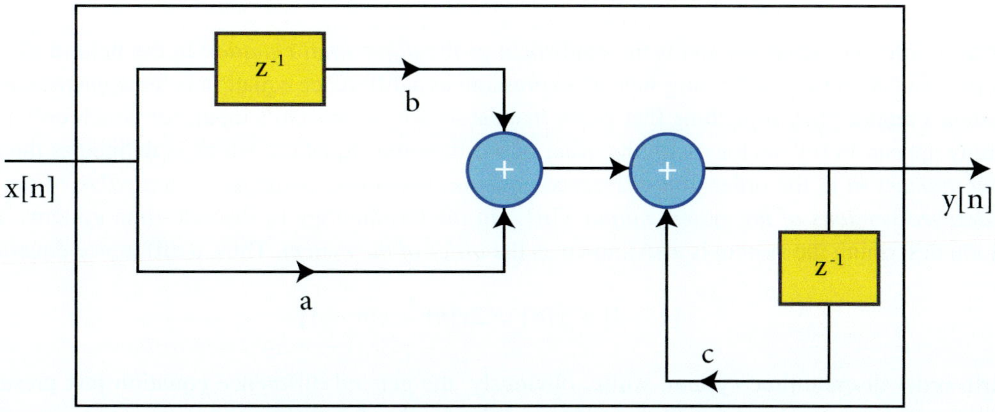

The structure of a typical recurrent system

In full analogy with their continuous-time counterparts, discrete-time systems are described using appropriate schematic diagrams

7.1 Introduction to Discrete-Time Systems

In full analogy with the continuous-time systems we studied in the second chapter, *discrete-time systems* or *discrete-time systems* are defined as *special hardware or software units* (i.e., *arrangements* or *algorithms*) that perform specific forms of processing on a discrete-time signal. The discrete-time signal on which the transformation is applied is called the *input* or *excitation* of the system, while the signal that results from this process is known as the *output* or *response*. If the input and output signals of the system are described by discrete sequences $x[n]$ and $y[n]$, we can, as in continuous-time systems, correlate these two signals through the relation

$$y[n] = \mathcal{T}\{x[n]\} \qquad \text{or alternatively} \qquad x[n] \rightarrowtail y[n]$$

where the symbol $\mathcal{T}$, which is simply an *operator*, symbolizes the *transformation* applied to the input signal, so as to create the output signal of the system. This mode of operation of the system is described by Fig. 2.1 (see page 68) that we had quoted for continuous-time signals, with the difference that in this case the signals x and y are *discrete quantities*. With the exception of this differentiation, all the discussion of Sect. 2.1 (see page 67) applies without any modification to discrete-time systems, and for this reason it is not deemed necessary to be repeated.

The mathematical description of a discrete-time system includes, among other things, the determination of the mathematical expression or rule that correlates the input of the system with the output resulting from it. This type of description—which is known as *input/output description*—is the only information we need to know about a system in order to use it, while the details of its internal structure are usually unknown or simply ignored. In other words, the system operates as a black box that accepts some input and produces some output, without our knowing anything at all about its internal

A. I. Margaris, *Continuous and Discrete-Time Signals and Systems*,
https://doi.org/10.1007/978-3-031-81209-5_7

mode of operation. However, as we saw in the chapter on continuous-time systems, it is possible to study the system using its internal structure, through its representation in the so-called *state space*.

The general form of the input/output equation of a discrete-time system[1] is described by an expression of the form

$$\alpha_0 y[n] + \alpha_1 y[n-1] + \alpha_2 y[n-2] + \cdots + \alpha_N y[n-N] = \beta_0 x[n] + \beta_1 x[n-1] + \beta_2 x[n-2] + \cdots + \beta_M x[n-M]$$

or, in condensed form[2]

$$\sum_{k=0}^{N} \alpha_k y[n-k] = \sum_{k=0}^{M} \beta_k x[n-k] \tag{7.1}$$

which is known as a *difference equation* and is the equivalent of the *differential equation* in the field of discrete-time. Note that a necessary prerequisite for characterizing such an expression as a difference equation is the *appearance of time-delayed versions of a system's output* $y[n]$, something that is not mandatory for the system's input, for which only the current value of $x[n]$ may simply appear. In full analogy with the order of a differential equation, which is defined as the *maximum order of derivative that appears in it*, the order of a difference equation is defined as the *maximum value of the time delay that appears in the delayed versions of the system output* $y[n]$.[3] In the terminology of discrete-time systems, the order of the difference equation describing the system is also known as the *order of the system*. Thus, a difference equation of the form

$$y[n-4] + y[n] = 2x[n] - x[n-1]$$

describes a fourth-order discrete-time system, while, obviously, the general difference equation just presented refers to a discrete-time system of order N. Equations of this type are solved either by applying specific techniques and methods that will be presented later or through *recursive processes*, which are based on solving the equation with respect to the output $y[n]$, which, as can be seen, is given by the relation

$$y[n] = \frac{1}{\alpha_0}\left(\sum_{k=0}^{M} \beta_k x[n-k] - \sum_{k=1}^{N} \alpha_k y[n-k]\right) \tag{7.2}$$

Finally, regarding the coefficients $\{\alpha_i\}_0^N$ and $\{\beta_i\}_0^M$ that appear in the general difference equation, these are always *real numbers*, some of which may have zero value (e.g., in the difference equation $y[n-4] + y[n] = 2x[n] - x[n-1]$ presented earlier, the values of the coefficients that appear in it are $\alpha_4 = 1$, $\alpha_3 = \alpha_2 = \alpha_1 = 0$, $\alpha_0 = 1$, and $\beta_0 = 2$, $\beta_1 = -1$). Usually, without loss of generality, we assume that $\alpha_N \neq 0$.

The preceding description applies to the general case of recursive discrete-time systems. However, if the input/output relationship does not involve any time-delayed versions of the system output $y[n]y[n]$—as in the case of a system such as $y[n] = x[n] - 2x[n-1]y[n] = x[n] - 2x[n-1]$—then the system is clearly non-recursive, and the analysis becomes significantly simpler[4] Recursive systems are particularly useful in cases involving an infinite number of previous inputs,

[1] In reality, this type of difference equation corresponds to the so-called *linear difference equations with constant coefficients*, describes only linear time-invariant (LTI) discrete-time systems that are also those used in practice.

[2] In full analogy with continuous-time systems and for reasons related to the principle of causality, we always impose the requirement $M \leq N$.

[3] Some books define the order of a difference equation as the *maximum of the parameters M and N*; however, in this description, we will consider the value of the parameter N as the order of the equation.

[4] In this case, we will obviously have $N = 1$, and the system output will be calculated from the relation

$$y[n] = \frac{1}{\alpha_0} \sum_{k=0}^{M} \beta_k x[n-k].$$

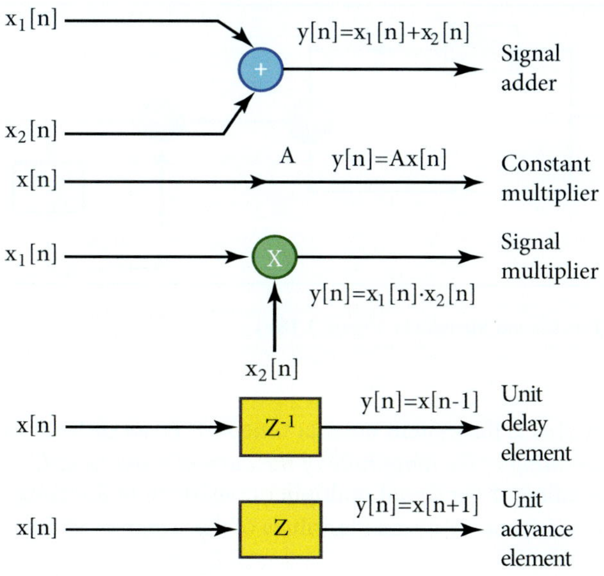

Fig. 7.1 Structural units of a discrete-time system

which in certain cases can be replaced by time-delayed outputs of the system. Considering, for example, the system of the *accumulator*, the input/output correlation that describes it is formulated as

$$y[n] = \sum_{k=-\infty}^{n} x[k] = \sum_{k=-\infty}^{n-1} x[k] + x[n] = y[n-1] + x[n] \tag{7.3}$$

The basic function of the accumulator is to add the current input value to its previous output value, a process that results in the calculation of the sum of all input values transmitted to it. We observe that in making our system *recursive*, we render its implementation very simple. Recursive systems are studied in great detail in subsequent sections.

An alternative way of describing the structure and function of a discrete-time system is with so-called schematic diagrams, which are used to describe the structural units of such a system. These structural units are the equivalents of the units used for the construction of continuous-time systems (depicted in Fig. 2.10, page 113) for the field of discrete-time and perform the basic transformations and forms of processing on a discrete input signal that were presented in Sects. 6.8 (see page 532) and 6.10 (see page 540). The fundamental structural units of a discrete-time system are presented in Fig. 7.1, and their operation, in general terms, is as follows:

- *Signal adder*: The basic operation of the adder consists in *adding the values of the two discrete-time signals it receives at its input*. If $x_1[n]$ and $x_2[n]$ are the values of these signals at time n, then the output of the adder will obviously be the signal

$$y[n] = x_1[n] + x_2[n] \tag{7.4}$$

- *Signal multiplier with constant*: The basic operation of this structural unit is the *multiplication of the current value of the input signal by the value of the constant that was determined during the design stage*. Therefore, the output of this unit will be the discrete-time signal

$$y[n] = Ax[n] \tag{7.5}$$

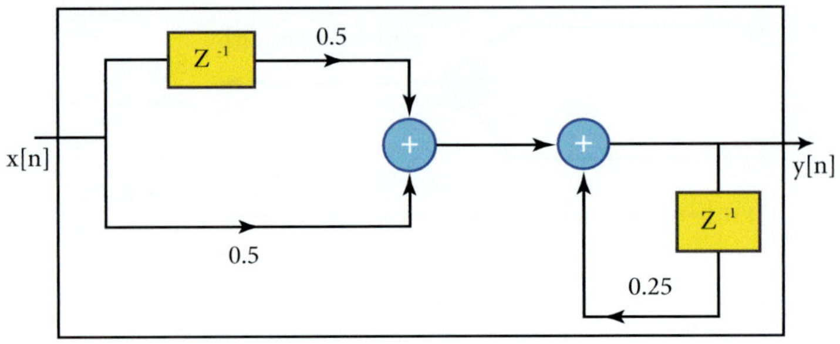

Fig. 7.2 Example schematic diagram (Proakis and Manolakis, Figure 2.18α)

where A is a constant value and $x[n]$ is the system input at time n. It is not difficult to see that the operation of this type of multiplier corresponds to the *scaling of the amplitude of the current input signal.*

- *Signal multiplier*: The basic operation of the signal multiplier consists in *multiplying the values of the two discrete-time signals it receives at its input*. If $x_1[n]$ and $x_2[n]$ are the values of these signals at time n, then the output of the multiplier will be the signal

$$y[n] = x_1[n] \cdot x_2[n] \tag{7.6}$$

- *Unit delay element*: The basic function of this element is the *temporal shift of the input signal by one time unit*. In other words, if we transmit the sample $x[n]$ as the input of the system, then the previous sample $x[n-1]$ will appear as its output. Since this shift is *backward*, the operation performed by this system is a *time delay*. Therefore, the output of the system at time n will have the form

$$y[n] = x[n-1] \tag{7.7}$$

- *Unit advance element*: The basic function of this element is also the *temporal shift of the input signal by one time unit*. However, in this case, this shift is *forward*, and for this reason it is considered a *time advance*. Thus, if we transmit the sample $x[n]$ at the input of the system, then the previous sample $x[n+1]$ will appear as its output. Therefore, the output of the system at time n will be

$$y[n] = x[n+1] \tag{7.8}$$

In full analogy with continuous-time signals, this type of time advancement can be applied *only to discrete-time signals that have already been recorded and stored*. In real-time acquisition scenarios, such an operation is practically unfeasible.

It is obvious that for every input/output representation of a discrete system corresponds to a schematic representation that describes the operation of the system using elementary structural units. For example, it is the schematic diagram of the discrete system described by the equation

$$y[n] = \frac{1}{4}y[n-1] + \frac{1}{2}x[n] + \frac{1}{2}x[n-1]$$

that is represented in Fig. 7.2.

7.2 Types and Categories of Discrete-Time Systems

Depending on their nature and the characteristics that describe their operation, discrete-time systems can be divided into the same categories as those described for continuous-time systems. The most important of these categories are briefly presented in the following sections.

7.2.1 Static and Dynamic Systems

A system is called *static* if its output $y[n]$ at time n depends *only on the current input of the system* $x[n]$ and not on inputs corresponding to earlier or later times; otherwise, the system is called *dynamic*. If the output of the system at time n depends on its N previous inputs, the system is characterized *by the existence of a memory of time duration N*. It is obvious that for static systems the relation $N = 0$ holds, and therefore systems of this type are characterized by an *absence of memory*. If the parameter N takes finite values (i.e., the relation $0 < N <$ holds), the system is characterized as a *dynamic system with finite memory*, while in the opposite case the system has *infinite memory*.

Examples of static systems are systems described by the equations $y[n] = \alpha x[n]$ and $y[n] = 5nx[n] + \beta x^2[n]$, where α and β are multiplicative coefficients with specific constant values. In these systems, the system's output at time n depends only on the value of the corresponding input. In contrast, the systems

$$y[n] = x[n] + 5x[n-1], \quad y_2[n] = \sum_{k=1}^{n} x[n-k], \quad y_3[n] = \sum_{k=0}^{\infty} x[n-k]$$

are characterized as dynamic systems, because their output at time n depends also on inputs corresponding to earlier times. Of these dynamic systems, the first two are characterized by the existence of *finite memory*—with duration $N = 1$ and $N = n$, respectively—while the third system is characterized as a dynamic system with *infinite memory*.

7.2.2 Time-Varying and Time-Invariant Systems

A discrete-time system is characterized as *time-invariant* when the correlation that exists between its input and output *does not change over time*; in the opposite case, the system is characterized as *time-varying*. In a more strict mathematical definition, if $\mathcal{T}$ is a transformation implemented by a system, then it is characterized as time-invariant *if and only if the validity of the relation* $y[n] = \mathcal{T}\{x[n]\}$ *implies the validity of the relation* $y[n-k] = \mathcal{T}\{x[n-k]\}$ *for every discrete-time signal* $x[n]$ *and for every integer number k.*

To understand this way of characterizing discrete-time systems, let us consider the *accumulator* system presented previously and described by the equation

$$y[n] = \mathcal{T}\{x[n]\} = \sum_{k=-\infty}^{n} x[k] \tag{7.9}$$

For this system to be time-invariant, the correlation that exists between its input and output should not change in relation to time. This, in other words, means that if we go back n_0 time units,[5] its output at that point in time, which is the function $y[n-n_0]$, should be given by the relation $y[n-n_0] = \mathcal{T}\{x[n-n_0]\}$. It is not difficult to observe that this is something that indeed holds, as it is found that[6]

[5] It is obvious that the characterization of the system can be carried out moving forward in time, since the mathematics do not change; essentially what we need to prove is that when we observe the system at any two different points in time, it always applies the same transform $\mathcal{T}$. The time shift backwards is done to ensure our compatibility with the principle of causality.

[6] In this example we use the variable n_0 as a time shift variable to avoid confusion with the variable k, which is used as a dummy index in the defining equation of the accumulator.

$$y[n-n_0] = \sum_{k=-\infty}^{n-n_0} x[k] = \mathcal{T}\{x[n-n_0]\} \tag{7.10}$$

Therefore, the accumulator system is a time-invariant discrete-time system.

On the other hand, considering the system of the *compressor*, whose operation involves ignoring the first $(M-1)$ samples of a group of M samples and storing only the Mth sample, leading to an input/output equation of the form

$$y[n] = x[Mn] \quad (-\infty < n < \infty)$$

it is not difficult to see that this system is *time-varying* since[7]

$$T\{x[n-k]\} = x[Mn-k] \qquad \text{and} \qquad y[n-k] = x[M(n-k)] \neq T\{x[n-k]\}$$

We observe that in this case, the correlation that exists between the input and the output is not constant in relation to time, and therefore the discrete-time system of the compressor is a *time-varying discrete-time system.*

Problem 7.2.1 For each of the following discrete-time systems, determine whether it is time-varying or time-invariant (Proakis and Manolakis, Example 2.2.4):

(a) $y[n] = x[n] - x[n-1]$ (b) $y[n] = nx[n]$ (c) $y[n] = x[-n]$ (d) $y[n] = x[n]\cos(\omega n)$.

Solution To determine the nature of each of the foregoing systems, we must transmit to the system an input $x[n]$, to receive the corresponding output $y[n]$, and then calculate the output $y(n,k) = \mathcal{T}\{x[n-k]\}$ resulting from the input $x[n-k]$, that is, the one characterized by a time delay k units in relation to the previous one. If the relation $y(n,k) = y[n-k]$ holds, the system will be *time-invariant*, while in the opposite case it will be *time-varying*. Applying this process for each of the discrete-time systems given, we will get the following results:

(a) For the system $y[n] = x[n] - x[n-1]$ we will have

$$y(n,k) = \mathcal{T}\{x[n-k]\} = x[n-k] - x[n-k-1] \qquad \text{and} \qquad y[n-k] = x[n-k] - x[n-k-1] = y(n,k)$$

and therefore the system is time-invariant.

(b) For the system $y[n] = nx[n]$ we will have

$$y(n,k) = \mathcal{T}\{x[n-k]\} = nx[n-k] \qquad \text{and} \qquad y[n-k] = (n-k)x[n-k] \neq y(n,k)$$

and therefore the system is time-varying.

(c) For the system $y[n] = x[-n]$ we will have

$$y(n,k) = \mathcal{T}\{x[n-k]\} = x[-n-k] \qquad \text{and} \qquad y[n-k] = x[-(n-k)] = x[k-n] \neq y(n,k)$$

and therefore the system is time-varying.

[7] In full analogy with continuous-time signals, the determination of the value $y[n-k]$ consists in the replacement of the discrete variable n with the quantity $n-k$, while, conversely, to calculate the expression $\mathcal{T}\{x[n-k]\}$, we must replace the function $x[n]$ with the function $x[n-k]$.

(d) For the system $y[n] = x[n]\cos(\omega n)$ we will have

$$y(n,k) = \mathcal{T}\{x[n-k]\} = x[n-k]\cos(\omega n) \qquad \text{and} \qquad y[n-k] = x[n-k]\cos[\omega(n-k)] \neq y(n,k)$$

and therefore the system is time-varying.

7.2.3 Linear and Nonlinear Systems

A discrete-time system is characterized as *linear* if and only if it obeys the *principle of superposition*, according to which (Sect. 2.2.5) the *response of a system to an input signal that is the linear combination of a set of elementary signals is equal to the linear combination of the system's responses to each elementary signal separately*. Considering two input signals $x_1[n]$ and $x_2[n]$, the principle of superposition is formulated in the form

$$\mathcal{T}\{\alpha_1 x_1[n] + \alpha_2 x_2[n]\} = \alpha_1 \mathcal{T}\{x_1[n]\} + \alpha_2 \mathcal{T}\{x_2[n]\} \tag{7.11}$$

for any two input signals $x_1[n]$ and $x_2[n]$ and for any two constants α_1 and α_2. Conversely, if the foregoing property is not satisfied, the system is characterized as nonlinear.

A simple example of a *linear discrete-time system* is the previously described *accumulator*. According to the analysis presented in the previous section, if we transmit any two inputs $x_1[n]$ and $x_2[n]$ to this discrete-time system, the system's response to these will be the outputs $y_1[n]$ and $y_2[n]$, which are defined as

$$y_1[n] = \sum_{k=-\infty}^{n} x_1[n] \qquad \text{and} \qquad y_2[n] = \sum_{k=-\infty}^{n} x_2[n]$$

respectively. To prove that the system is linear, we must show that the system's response to the input $x[n] = \alpha x_1[n] + \beta x_2[n]$ is the output $y[n] = \alpha y_1[n] + \beta y_2[n]$ for any values of the arbitrary constants α and β. We will therefore have

$$y[n] = \sum_{k=-\infty}^{n} x[n] = \sum_{k=-\infty}^{n} \left\{\alpha x_1[n] + \beta x_2[n]\right\} = \alpha \sum_{k=-\infty}^{n} x_1[n] + \beta \sum_{k=-\infty}^{n} x_2[n] = \alpha y_1[n] + \beta y_2[n]$$

We observe that this system is characterized by the power of the principle of superposition, and therefore it is a *discrete linear system*.

On the other hand, an example of a *nonlinear system* is the discrete-time system $y[n] = \log_{10}(|x[n]|)$. To prove the nonlinear nature of this system, we must identify at least one set of inputs/outputs that is *not compatible* with the principle of superposition. As such a set let us consider the inputs $x_1[n] = 10$ and $x_2[n] = 100$, for which the corresponding outputs of the system are $y_1[n] = \log_{10}(|10|) = 1$ and $y_2[n] = \log_{10}(|100|) = 2$. Observing that these two inputs are connected by the relation $x_2[n] = 10x_1[n]$, the corresponding outputs of the system, if it were linear, should satisfy the equation $y_2[n] = 10y_1[n]$. However, this is obviously not the case, meaning that this system is a *nonlinear discrete-time system*.

Problem 7.2.2 For each of the following systems, check whether it is linear or not.

(a) $y[n] = nx[n]$ (b) $y[n] = x[n^2]$ (c) $y(n) = x^2[n]$ (d) $y[n] = Ax[n] + B$ (e) $y(n) = e^{x[n]}$

(Proakis and Manolakis, Example 2.2.5).

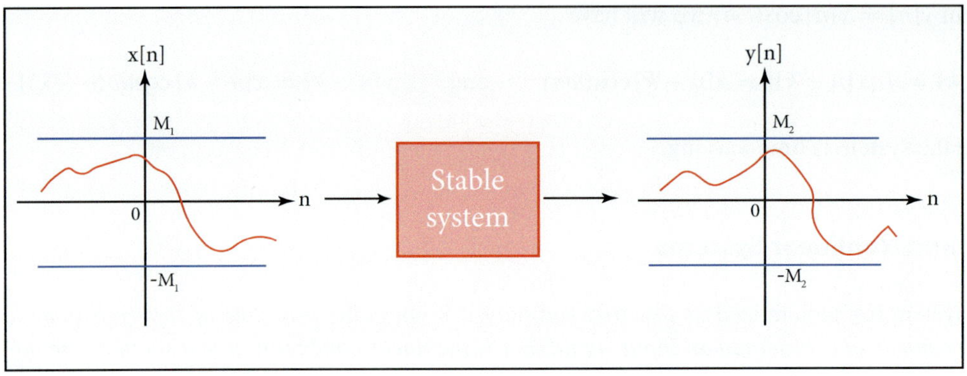

Fig. 7.3 The transmission of a bounded input to a stable system leads to the production of a bounded output

Solution As already mentioned, the criterion for characterizing a discrete-time system as linear or nonlinear is the *power or not of the principle of superposition*, according to which the system's response to the input $\alpha x_1[n] + \beta x_2[n]$ is the output $\alpha y_1[n] + \beta y_2[n]$. In detail, then, we have:

(a) The system's response to the inputs $x_1[n]$ and $x_2[n]$ are the outputs $y_1[n]=nx_1[n]$ and $y_2[n]=nx_2[n]$. Considering the combined input $x[n] = \alpha x_1[n] + \beta x_2[n]$, the system's response to this is

$$y[n] = nx[n] = n\left\{\alpha x_1[n] + \beta x_2[n]\right\} = \alpha\left\{nx_1[n]\right\} + \beta\left\{nx_2[n]\right\} = \alpha y_1[n] + \beta y_2[n]$$

Therefore, our system is a *linear discrete-time system.*

(b) The system's response to the inputs $x_1[n^2]$ and $x_2[n^2]$ are the outputs $y_1 = x_1[n^2]$ and $y_2 = x_2[n^2]$. Considering the combined input $x[n^2] = \alpha x_1[n^2] + \beta x_2[n^2]$, the system's response to this is

$$y[n] = x_3[n^2] = \alpha x_1[n^2] + \beta x_2[n^2] = \alpha y_1[n] + \beta y_2[n]$$

Therefore, our system is a *linear discrete-time system.*

(c) The system's response to the inputs $x_1[n]$ and $x_2[n]$ are the outputs $y_1[n] = x_1^2[n]$ and $y_2[n] = x_2^2[n]$. Considering the combined input $x[n] = \alpha x_1[n] + \beta x_2[n]$, the system's response to this is

$$y[n] = x^2[n] = \left\{\alpha x_1[n] + \beta x_2[n]\right\}^2 = \alpha^2 x_1^2[n] + \beta^2 x_2^2[n] + 2\alpha\beta x_1[n]x_2[n] = \alpha^2 y_1[n] + \beta^2 y_2[n] + 2\alpha\beta x_1[n]x_2[n]$$

Therefore, our system *is not linear*, because it *does not obey the principle of superposition.*

(d) The system's response to inputs $x_1[n]$ and $x_2[n]$ is the outputs $y_1[n] = Ax_1[n]+B$ and $y_2[n] = Ax_2[n]+B$. Considering the combined input of the form $x[n] = \alpha x_1[n] + \beta x_2[n]$, the system's response to this is the output

$$y[n] = Ax[n] + B = A\left\{\alpha x_1[n] + \beta x_2[n]\right\} + B = A\alpha x_1[n] + A\beta x_2[n] + B$$

Therefore, our system *is not linear*, because it *does not obey the principle of superposition.*

(e) The system's response to inputs $x_1[n]$ and $x_2[n]$ is the outputs $y_1[n] = e^{x_1[n]}$ and $y_2[n] = e^{x_2[n]}$. Considering the combined input $x[n] = \alpha x_1[n] + \beta x_2[n]$, the system's response to this is

$$y[n] = e^{x[n]} = \exp\left\{\left(\alpha x_1[n] + \beta x_2[n]\right)\right\} = e^{\alpha x_1[n]}\, e^{\beta x_2[n]} = [e^{x_1[n]}]^{\alpha}\, [e^{x_2[n]}]^{\beta} = \{y_1[n]\}^{\alpha}\, \{y_2[n]\}^{\beta}$$

Therefore, our system *is not linear*, because it *does not obey the principle of superposition.*

7.2.4 Stable and Unstable Systems

In full analogy with continuous-time systems, a discrete-time system is considered *BIBO stable* when it is *bounded* both in terms of input and output, that is, when *for every bounded input that is transmitted to it, the output produced by the system is always bounded.* As already mentioned, the input $x[n]$ and the output $y[n]$ are bounded signals, when there are finite numbers M and N, such that $|x[n]| \leq M < \infty$ and $|y[n]| \leq N < \infty$ (for every value of the parameter n). If our system has this property, it is bounded, and therefore *stable*, while in the opposite case the system is characterized as *unstable*.

It is important to mention at this point that the concept of stability plays a crucial role in the proper operation of discrete-time systems: An unstable system is usually characterized by problematic and extreme behavior, which makes it unsuitable for use in almost all cases. The behavior of a stable discrete-time system is presented in Fig. 7.3, in which, although the signal for reasons of clarity is described by a continuous curve, it is clear that it refers to a discrete quantity.

7.2.5 Causal and Noncausal Systems

A discrete-time system is called *causal* when the output $y[n]$ of the system at each time n depends *only on the current input of the system* $x[n]$, as well as on those inputs that correspond to earlier times $\{x[n-1], x[n-2], x[n-3], \ldots\}$, but not on the future inputs of the system $\{x[n+1], x[n+2], x[n+3], \ldots\}$. Therefore, in mathematical notation, the output of a causal system has the form (for some arbitrary function F)

$$y[n] = F(x[n], x[n-1], x[n-2], x[n-3], \ldots) \tag{7.12}$$

If the system does not satisfy the foregoing property, it is called *noncausal*. The output of a noncausal system depends not only on its previous inputs but also on its future inputs. It is obvious that systems of this kind cannot be used in a real-time signal processing application because it is impossible to predict the values of the measurements that will be made in the future. Conversely, if a signal has already been recorded and stored in a file, it is possible to create and use noncausal systems, as all the signal values are available.

Problem 7.2.3 Consider the discrete-time systems (Proakis and Manolakis, Solved Exercise 2.7)

$$(a) \quad y[n] = x[n] - x[n-1], \quad (b) \quad y[n] = \sum_{n=-\infty}^{n} x[k], \quad (c) \quad y[n] = \alpha x[n]$$

$$(d) \quad y[n] = x[n] + 3x[n+4], \quad (e) \quad y[n] = x[n^2] \quad (f) \quad y[n] = x[2n], \quad (g) \quad y[n] = x[-n]$$

For each of these, determine whether it is a causal or noncausal discrete-time system.

Solution From the definition of the *causal discrete-time system*, (a), (b), and (c) are causal systems—since their output at any time n is a function of the corresponding current input as well as previous inputs—while systems (d), (e), and (f) are noncausal, since the current output is a function of subsequent inputs. Finally, the discrete-time system (g) is also noncausal, since for negative values of n, it behaves like the aforementioned noncausal systems (e.g., for $n = -1$ it produces $y[-1] = x[1]$, that is, the current output depends on a subsequent system input).

Problem 7.2.4 Given the discrete-time systems

$$\text{(a)}\quad y[n] = [\cos(\pi n)]x[n] \qquad \text{and} \qquad \text{(b)}\quad y[n] = \sum_{k=n-1}^{\infty} x[n]$$

determine, for each system, whether it is stable, causal, linear, or time-invariant.

Solution (a) The system $y(n) = [\cos(\pi n)]x(n)$ is characterized by the following properties:

- The input/output relationship of the system can be written in the form

$$y[n] = [\cos(\pi n)]x[n] = \begin{cases} +x[n] \text{ for } n = 2k \\ -x[n] \text{ for } n = 2k+1 \end{cases} = (-1)^n x[n]$$

  We observe that the effect of the system on the input signal is the *change of its sign only, not its magnitude*. But this means that if the system's input is blocked, the same will happen with its output, that is, the system is *stable*.
- The system is *causal*, because the system's output depends *only on the current input*, not on subsequent inputs (for the same reason, the system is also characterized as a *static system*).
- Considering two inputs $x_1[n]$ and $x_2[n]$, the system's outputs are respectively $y_1[n] = [\cos(\pi n)]x_1[n]$ and $y_2[n] = [\cos(\pi n)]x_2[n]$. If we transmit the input $x[n] = \alpha x_1[n] + \beta x_2[n]$ to the system, the corresponding system output will be

$$y[n]=[\cos(\pi n)]x[n]=[\cos(\pi n)]\left(\alpha x_1[n]+\beta x_2[n]\right)=\alpha\,[\cos(\pi n)]x_1[n]+\beta\,[\cos(\pi n)]x_2[n]=\alpha y_1[n]+\beta y_2[n]$$

  that is, our system is *linear*.
- We have $y[n-k] = \{\cos[\pi(n-k)]\}x[n-k]$ and $y(n,k) = T\{x[n-k]\} = [\cos(\pi n)]x[n-k] \neq y(n-k)$, that is, our system is a *time-varying discrete-time system*.

(b) The system $y(n) = \sum_{k=n-1}^{\infty} x(k)$ is characterized by the following properties:

- First of all, the system is *not stable* since, if we transmit the step function $u[n]$, which is a bounded signal, its output becomes infinite since $y[n] \to \infty$ for every $n > 1$.
- The system is *not causal* because the system's output at the current time depends on subsequent inputs.
- Considering two inputs $x_1[n]$ and $x_2[n]$, the corresponding outputs of the system are

$$y_1[n] = \sum_{k=n-1}^{\infty} x_1[k] \qquad \text{and} \qquad y_2[n] = \sum_{k=n-1}^{\infty} x_2[k]$$

  If we transmit the input $x[n] = \alpha x_1[n] + \beta x_2[n]$ to the system, the output of the system will be

$$y[n] = \sum_{k=n-1}^{\infty} x[k] = \sum_{k=n-1}^{\infty}\left(\alpha x_1[k] + \beta x_2[k]\right) = \alpha \sum_{k=n-1}^{\infty} x_1[k] + \beta \sum_{k=n-1}^{\infty} x_2[k] = \alpha y_1[k] + \beta y_2[k]$$

  that is, our system is *linear*.
- Finally, it is not difficult to ascertain that

$$y[n-\ell] = \sum_{k=n-\ell-1}^{\infty} x[k] \qquad \text{and} \qquad y(n,\ell) = T\{x[n-\ell]\} = \sum_{k=n-\ell-1}^{\infty} x[k] = y[n-\ell]$$

that is, the system is a *time-invariant discrete-time system*.

Problem 7.2.5 Consider the discrete-time systems (Hayes, Solved Exercises 1.14 and 1.21)

$$\text{(a)}\quad y[n]=x^2[n],\qquad \text{(b)}\quad y[n]=\frac{e^{x[n]}}{x[n-1]},\qquad \text{(c)}\quad y[n]=\cos(x[n])$$

$$\text{(d)}\quad y[n]=\sum_{k=-\infty}^{n}x[k],\qquad \text{(e)}\quad y[n]=\log(1+|x[n]|)$$

For each of these, determine whether it is a stable or unstable discrete-time system.

Solution According to the theory presented earlier, a discrete-time system is *stable* if the transmission to it of a blocked input results in the appearance of a blocked output and *unstable* in the opposite case. If we apply this criterion to each of the foregoing discrete-time systems, we will get the following results:

(a) If any blocked input $x[n]$ is applied as the system's input such that $|x(n)| < M$, the system's output will also be blocked since it turns out that

$$|y[n]| = |x[n]|^2 < M^2$$

Therefore, the system is stable.

(b) Suppose we transmit the *discrete impulse function* $\delta[n]$ as input to the system. This function has a value equal to unity at the position $n = 0$ and zero value at all other positions. If this function passes through the system, its output will become infinite for all values except for $n = 1$ (as one can verify), and therefore the system is *unstable*.

(c) The system is *stable* since from trigonometry it is known that

$$|\cos(x[n])| \le 1 \quad (\forall x[n])$$

(d) If we transmit the *unit step function* $u[n]$ to the system for values $n < 0$, the system's output will have the form

$$y[n] = \sum_{k=-\infty}^{n} u[k] = n + 1$$

We observe that, although the system's input is blocked, its output does not satisfy this property (since at the limit $n \to \infty$ there will be $y[n] \to \infty$), and therefore the system is *unstable*.

(e) To solve the problem, we will use the mathematical relation $\log(1+x) \le x$, which characterizes the logarithmic function. If we transmit any bounded input $x[n]$ such that $|x[n]| \le M$, the output of the system will also be bounded since

$$|y(n)| = |\log(1 + |x[n]|)| \le 1 + |x[n]| < 1 + M$$

Therefore, our system is *stable*.

7.3 Impulse, Step, and Frequency Responses

The *impulse*, *step*, and *frequency responses* of discrete-time systems are described by the same definitions presented in the chapter on continuous-time systems, with the difference that the signals that appear in these will now be discrete. In detail, we will have:

Definition 7.3.1 (Impulse Response) The *impulse response* $h[n]$ of a linear discrete-time system is called the output produced by the system when the unit discrete impulse function is transmitted to its input, or, in mathematical notation,

$$h[n] = \mathcal{T}\{\delta[n]\}. \tag{7.13}$$

Definition 7.3.2 (Step Response) The *step response* $s[n]$ of a discrete-time linear system is called the output produced by the system when the unit discrete step function is transmitted to its input, or, in mathematical notation,

$$s[n] = \mathcal{T}\{u[n]\}. \tag{7.14}$$

Definition 7.3.3 (Frequency Response) The *frequency response* $\mathcal{H}(e^{j\omega})$ of a discrete-time linear system is called the ratio of the output signal produced by the system when an elementary discrete complex exponential signal with amplitude equal to A and arbitrary frequency ω is transmitted to its input, or, in mathematical notation,[8]

$$\mathcal{H}(e^{j\omega}) = \frac{\mathcal{T}\{Ae^{j\omega n}\}}{Ae^{j\omega n}}. \tag{7.15}$$

In almost all cases, the frequency response is a complex function, and if it is formulated in its polar form, $\mathcal{H}(e^{j\omega}) = |\mathcal{H}(e^{j\omega})| \exp[j\measuredangle\mathcal{H}(e^{j\omega})]$, then its magnitude $|\mathcal{H}(e^{j\omega})|$ is known as the amplitude response, while its phase $\measuredangle\mathcal{H}(e^{j\omega})$ is known as the phase response.

It should be noted that the foregoing definitions are valid *only for time-invariant systems*, whose transform $\mathcal{T}$ does not change over time, while in the opposite case the equations that describe the foregoing systems are the discrete equivalents of those presented in Sect. 2.4 (page 81). The analytical description of discrete-time LTI systems is the subject of the following pages.

7.4 LTI Systems

A special category of discrete-time systems that are very important and used in almost all cases are those systems that satisfy *both the property of linearity and the property of time invariance*. Recall from basic theory that a system is characterized as *linear* when it obeys the principle of superposition and as *time-invariant* when the transform $\mathcal{T}$ applied by the system to an input signal so as to create an output signal does not change over time. In full analogy with their continuous equivalents, the *output of a LTI system results from the convolution between the impulse response of the system and the input signal transmitted to it*.

To prove the truth of this claim, recall that any discrete-time signal $x[n]$ can be written as a linear combination of infinitely many time-shifted impulse functions of the form

$$x[n] = \sum_{k=-\infty}^{+\infty} x[k]\,\delta[n-k] \tag{7.16}$$

an equation that, as mentioned previously, expresses the convolution between the discrete-time signals $x[n]$ and $\delta[n]$. If we now transmit this signal as input to some discrete-time system $\mathcal{T}$, its output will be

$$y[n] = \mathcal{T}\{x[n]\} = \mathcal{T}\left\{ \sum_{k=-\infty}^{+\infty} x[k]\,\delta[n-k] \right\} \tag{7.17}$$

Assuming that the system is linear—and therefore that its operation will be characterized by the principle of superposition—the second term of the foregoing equation is formulated as[9]

$$\mathcal{T}\left\{ \sum_{k=-\infty}^{\infty} x[k]\,\delta[n-k] \right\} = \sum_{k=-\infty}^{\infty} x[k]\,\mathcal{T}\left\{\delta[n-k]\right\} \tag{7.18}$$

[8] In full analogy with continuous-time systems in which the frequency response is symbolized not by $\mathcal{H}(\omega)$ but by $\mathcal{H}(j\omega)$, the frequency response of discrete-time systems is symbolized by $\mathcal{H}(e^{j\omega})$ for reasons that will be explained later.

[9] On how to derive this equation, see Sect. 2.2.5, page 72.

If we further accept that the system is also time-invariant and take into account that the output of the system, when we transmit the impulse function $\delta[n]$ to it, is the impulse response $h[n]$—or, in mathematical notation, $\mathcal{T}\{\delta[n]\} = h[n]$—then it is not difficult to see that the expression $\mathcal{T}\{\delta[n-k]\}$ is equal to the time-delayed version of the impulse response $h[n]$, that is, the signal $h[n-k]$. Combining all of the preceding points, we can write the expression[10]

$$\underbrace{\underbrace{\underbrace{y[n] = \mathcal{T}\left\{ \sum_{k=-\infty}^{\infty} x[k]\,\delta[n-k] \right\}}_{\text{valid for all systems}} = \sum_{k=-\infty}^{\infty} x[k]\,\mathcal{T}\left\{\delta[n-k]\right\}}_{\text{valid only for linear systems}} = \sum_{k=-\infty}^{\infty} x[k]\,h[n-k] = h[n] * x[n]}_{\text{applies only to LTI systems}}$$

But the last expression is recognized as the *convolution between the discrete-time signals* $x[n]$ *and* $h[n]$. We are led, therefore, to the very important and extremely interesting conclusion that the *response* $y[n]$ *of a discrete-time LTI system to any input signal* $x[n]$ *arises from the convolution between the system's impulse response* $h[n]$ *and the input signal* $x[n]$, in full analogy with our discussion on continuous-time systems.

Depending on the length of the discrete impulse response signal, we can classify discrete LTI systems into two categories. More specifically, if an impulse response consists of a finite number of elements, our system will be characterized as a system *of finite impulse response*, while in the opposite case our system will be called a system *of infinite impulse response*. Therefore, the outputs of these two types of systems will be

$$y[n] = \sum_{k=0}^{N-1} h[k]x[n-k] \qquad \text{and} \qquad y[n] = \sum_{k=0}^{\infty} h[k]x[n-k]$$

respectively, where N is the number of elements of the finite impulse response of the first category of systems, while we assume that our systems are *causal*, and therefore the lower limit of the foregoing sums has the value $k_{\min} = 0$. Although systems of infinite impulse response at first glance seem impossible to implement, since the number of additions and multiplications required to calculate their output is practically *infinite*, their implementation is possible through the use of recursive discrete-time systems, whose current output does not depend only on the current input and previous inputs but also on *previous outputs*, which were created by the system at previous times. Systems of this type are described using difference equations, and approach to solving them will be presented in a subsequent section.

Problem 7.4.1 Consider a discrete-time LTI system whose impulse response has the form $h[n] = \alpha^n u[n]$. Determine the output of the system when we transmit the discrete-time signal $x[n] = \beta^n u[n]$ to its input.

Solution In Exercise 6.10.3 (page 542), we proved that the convolution between the signals $\alpha^n u[n]$ and $\beta^n u[n]$ is

$$\left\{\alpha^n u[n]\right\} * \left\{\beta^n u[n]\right\} = \begin{cases} \beta^n (n+1) & \alpha = \beta \\ \dfrac{\beta^{n+1} - \alpha^{n+1}}{\beta - \alpha} & \alpha \neq \beta \end{cases}$$

If, therefore, we consider the first of the two signals as the *impulse response* of the system $h[n]$ and the second as the *input signal* $x[n]$, then the output of the system for the cases $\alpha \neq \beta$ and $\alpha = \beta$ will be given by the corresponding expression of

[10] In full analogy with continuous-time systems and as noted in Sect. 2.2.5, the *total output of the system results from the sum of the response to the zero input, which depends only on the initial conditions of the problem and is calculated by setting all the expressions concerning the input signal to zero and the response to the zero state, which depends only on the input signal* $x[n]$ *and is calculated assuming that the system is at rest, that is, all the initial conditions of the problem have a zero value*. Therefore, the output $y[n]$ of the system, which results from the convolution of the signals $h[n]$ and $x[n]$ in the manner described in what follows, is actually the *response to the zero state*, while the total response of the system will result by adding to this the response to the zero input, provided the system is characterized by nonzero initial conditions.

the right-hand side of the foregoing equation. Generalizing this conclusion, we can formulate the assertion that *if all pairs of signals whose convolution we calculated in the previous exercises are considered the impulse response and the input of the system or vice versa (as dictated by the commutative property of convolution), then the signal resulting from their convolution can be considered the output of the system.* ■

In full analogy with continuous-time systems, the discrete-time systems we are dealing with are considered *causal*, which means that their output is calculated only for time instances $n \geq 0$. For this reason, if we want to be more meticulous and accurate in our description, we should write the output of the system of the previous exercise not as[11]

$$y_{zs}[n] = \begin{cases} \beta^n(n+1) & \alpha = \beta \\ \dfrac{\beta^{n+1} - \alpha^{n+1}}{\beta - \alpha} & \alpha \neq \beta \end{cases} \quad \text{but as} \quad y_{zs}[n] = \begin{cases} \beta^n(n+1)u[n] & \alpha = \beta \\ \dfrac{\beta^{n+1} - \alpha^{n+1}}{\beta - \alpha}u[n] & \alpha \neq \beta \end{cases}$$

where multiplication by the elementary step function $u[n]$ indicates the fact that the system output has nonzero values only for time instances $n \neq 0$.

7.4.1 The Step Response of LTI Systems

As already mentioned, the step response of a discrete-time system is defined as the *output of the system s[n] when the elementary step function u[n] is transmitted to its input*. In the special case where the considered system is *LTI*, its step response will result from the convolution between the step input function $u[n]$ and the system's impulse response $h[n]$; we will therefore have

$$s[n] = h[n] * u[n] = \sum_{k=-\infty}^{\infty} h[k]\, u[n-k] \tag{7.19}$$

From the definition of the step function it follows that

$$u[n-k] = \begin{cases} 1 \text{ for } n-k \geq 0 \\ 0 \text{ for } n-k < 0 \end{cases} \quad \text{or, equivalently,} \quad u[n-k] = \begin{cases} 1 \text{ for } k \leq n \\ 0 \text{ for } k > n \end{cases}$$

and the foregoing relation takes the form

$$s[n] = \sum_{k=-\infty}^{n} h[k] = \sum_{k=-\infty}^{n-1} h[k] + h[n] = s[n-1] + h[n] \tag{7.20}$$

We are led, therefore, to the conclusion that the step and impulse responses of a discrete-time LTI system are related to each other through the relations

$$s[n] = \sum_{k=-\infty}^{n} h[k] \quad \text{and} \quad h[n] = s[n] - s[n-1]$$

relations that constitute the discrete equivalents of the equations

[11] To emphasize the fact that the output of the system resulting from the convolution between the signals $h[n]$ and $x[n]$ is actually the *response of the system to the zero state*, we use in the symbolism of the output $y[n]$ the index zs (zero state).

$$s(t) = \int_{-\infty}^{+\infty} h(\tau)\, d\tau \qquad \text{and} \qquad h(t) = \frac{ds(t)}{dt}$$

which we proved for continuous-time signals.

Problem 7.4.2 Calculate the step response of a discrete-time LTI system whose impulse response is given by the relation $h[n] = \alpha^n u[n]$.

Solution If in the result of Exercise 7.4.1 (page 575) we set $\beta = 1$, it immediately follows that

$$y[n] = h[n] * u[n] = \frac{1 - \alpha^{n+1}}{1 - \alpha} u[n]$$

which is the desired result. Recall that the impulse function $u[n]$ in the expression of $y[n]$ does not result from the convolution operation but is added afterwards, so as to make the system output $y[n]$ a *causal signal.*

7.4.2 The Impulse Response of LTI Systems

Let us consider a discrete-time *LTI* system of order N described by a general difference equation of the form

$$\sum_{k=0}^{N} \alpha_k y[n-k] = \sum_{k=0}^{M} \beta_k x[n-k] \tag{7.21}$$

for which we assume that $N \geq M$. Following the same line of argumentation that we formulated for continuous-time systems and relying on the properties of *linearity* and *commutativity* of LTI systems (Footnote 9 (page 90) of the second chapter), we can assert that this system consists of two LTI subsystems, A and B, that implement the right and left sides of the difference equation, respectively, so that we can write that

$$w[n] = \mathcal{T}_1\{x[n]\} = \sum_{k=0}^{M} \beta_k x[n-k] \qquad \text{and} \qquad \sum_{k=0}^{N} \alpha_k y[n-k] = w[n]$$

By alternating the order of the two systems (a process that will not make the slightest change to the considered system) and passing to subsystem B the *discrete impulse function* $\delta[n]$, the output of subsystem A will be the impulse response $h[n]$, so that we can write

$$\delta[n] = \sum_{k=0}^{N} \alpha_k h_\beta[n-k] \qquad \text{and} \qquad h[n] \equiv h_\alpha[n] = \sum_{k=0}^{M} \beta_k h_\beta[n-k]$$

(for a full understanding of the method, see Fig. 2.8 on page 90, which refers to *continuous-time systems* but also describes well the situation related to the field of discrete-time). In full analogy, then, with continuous-time systems, we can determine the *impulse response* of a discrete-time LTI system by applying the following procedure:

1. Assuming that at time $n = 0$ the system is in a state of rest, so that we can write that

$$y[-1] = y[-2] = y[-3] = \cdots = y[N-1] = y[-N] = 0$$

we replace the right side of our original difference equation with the *discrete impulse function* $\delta[n]$. In this case, the output of subsystem B will be the impulse response $h_\beta[n]$, so that we can write

$$\sum_{k=0}^{N} \alpha_k h_\beta[n-k] = \delta[n]$$

2. The foregoing difference equation must be solved for the function $h_\beta[n]$, so that it can then be used to calculate the *impulse response* $h[n]$. Observing that the difference equation we consider is of order N, the full determination of its solution requires the specification of N initial conditions. Developing the foregoing relationship, we will get

$$\alpha_0 h_\beta[n] + \alpha_1 h_\beta[n-1] + \alpha_2 h_\beta[n-2] + \cdots + \alpha_N h_\beta[n-N] = \delta[n]$$

which for $n = 0$ takes the simple form

$$\alpha_0 h_\beta[0] + \alpha_1 h_\beta[-1] + \alpha_2 h_\beta[-2] + \cdots + \alpha_N h_\beta[-N] = \delta[0] = 1$$

Taking into account that the considered system is in a *state of rest* and assuming that it is *causal*, so its output—which is the impulse function $\delta[n]$—will depend only on the function $h_\beta[n]$, it follows that

$$h_\beta[-1] = h_\beta[-2] = h_\beta[-3] = \cdots = h_\beta[-N+1] = 0 \quad \text{and} \quad h_\beta[0] = \frac{1}{\alpha_0}$$

relations that also constitute the initial conditions of the problem.

3. After determining the initial conditions, we can solve the homogeneous difference equation

$$\sum_{k=0}^{N} \alpha_k h_\beta[n-k] = 0$$

determining the unknown function $h_\beta[n]$. Assuming, as is well known, an exponential solution of the form $h_\beta[n] = A\lambda^n$ $(n > 0)$, its substitution in the foregoing difference equation will lead to the equation

$$\alpha_0 A\lambda^n + \alpha_1 A\lambda^{n-1} + \alpha_2 A\lambda^{n-2} + \cdots + \alpha_N A\lambda^{n-N} = 0$$

or, equivalently,

$$A\lambda^{n-N}\left(\alpha_N + \alpha_{N-1}\lambda + \alpha_{N-2}\lambda^2 + \cdots + \alpha_1\lambda^{N-1} + \alpha_0\lambda^N\right) = 0$$

and finally

$$\alpha_N + \alpha_{N-1}\lambda + \alpha_{N-2}\lambda^2 + \cdots + \alpha_1\lambda^{N-1} + \alpha_0\lambda^N = \sum_{k=0}^{N} \alpha_k \lambda^{N-k} = 0$$

an equation that is also the *characteristic polynomial* of the considered difference equation. Taking into account that this polynomial will have *at most N distinct roots* of the form $\lambda_1, \lambda_2 \ldots, \lambda_N$, the foregoing equation will have N *elementary solutions* formulated as $h_\beta^k[n] = A_k\lambda_k^n$ $(k = 1, 2, \ldots, N)$, while its general solution will result from the linear combination of the foregoing elementary solutions, so that we can write that

$$h_\beta[n] = \sum_{k=1}^{N} A_k\lambda_k^n = A_1\lambda_1^n + A_2\lambda_2^n + \cdots + A_k\lambda_k^n$$

with the coefficients $A_1, A_2, \ldots, A_N$ resulting with the help of the initial conditions of the problem. Applying the N initial conditions that we determined in the previous step to the foregoing equation of the function $h_\beta[n]$, we will end up with a system of N equations with N unknowns of the form

$$h_\beta[0] = \sum_{k=1}^{N} A_k \lambda_k^n \bigg|_{n=0} = \sum_{k=1}^{N} A_k = \frac{1}{\alpha_0}$$

$$h_\beta[-1] = \sum_{k=1}^{N} A_k \lambda_k^n \bigg|_{n=-1} = \sum_{k=1}^{N} A_k \lambda_k^{-1} = 0$$

$$h_\beta[-2] = \sum_{k=1}^{N} A_k \lambda_k^n \bigg|_{n=-2} = \sum_{k=1}^{N} A_k \lambda_k^{-2} = 0$$

....

$$h_\beta[-N+1] = \sum_{k=1}^{N} A_k \lambda_k^n \bigg|_{n=-N+1} = \sum_{k=1}^{N} A_k \lambda_k^{-N+1} = 0$$

which in matrix form is given as

$$\begin{bmatrix} 1/\alpha_0 \\ 0 \\ 0 \\ \vdots \\ 0 \end{bmatrix} = \begin{bmatrix} 1 & 1 & \dots & 1 \\ \lambda_1^{-1} & \lambda_2^{-1} & \dots & \lambda_N^{-1} \\ \lambda_1^{-2} & \lambda_2^{-2} & \dots & \lambda_N^{-2} \\ \dots & \dots & \dots & \dots \\ \lambda_1^{1-N} & \lambda_2^{1-N} & \dots & \lambda_N^{1-N} \end{bmatrix} \begin{bmatrix} A_1 \\ A_2 \\ A_3 \\ \vdots \\ A_N \end{bmatrix} \tag{7.22}$$

In the foregoing equation, the $N \times N$ dimension matrix is known as the *Vandermonde matrix*; this matrix, as mentioned in the chapter on continuous-time systems, is always *invertible* so that its inverse exists and it is always possible to calculate the coefficients A_k $(k = 1, 2, \dots, N)$ and, consequently, the function $h_\beta[n]$.

4. In the last step of the process, we calculate the required *impulse response* from the relation

$$h[n] \equiv h_\alpha[n] = \sum_{k=0}^{M} \beta_k h_\beta[n-k] \tag{7.23}$$

The following solved exercises are examples of the method's application.

Problem 7.4.3 Determine the impulse response of a system described by the equation

$$y[n] - y[n-1] - 2y[n-2] = x[n-1].$$

Solution The difference equation of the system is a special case of the general expression, with parameter values $N = 2$, $M = 1$, $\alpha_0 = 1$, $\alpha_1 = -1$, $\alpha_2 = -2$, $\beta_0 = 0$, and $\beta_1 = 1$. Applying the process described earlier, we set the right side of the foregoing relation to the *discrete impulse function* $\delta[n]$, so that it takes the form

$$h_\beta[n] - h_\beta[n-1] - 2h_\beta[n-2] = \delta[n]$$

and then we solve the corresponding *homogeneous equation*

$$h_\beta[n] - h_\beta[n-1] - 2h_\beta[n-2] = 0$$

with initial conditions $h_\beta[0] = 1/\alpha_0 = 1$ and $h_\beta[-1] = 0$ for time instances $n > 0$. Considering the well-known exponential solution of the form $h_\beta[n] = A\lambda^n$ $(n > 0)$, we will have $h_\beta[n-1] = A\lambda^{n-1}$ and $h_\beta[n-2] = A\lambda^{n-2}$, and by substituting in the homogeneous difference equation we get

$$A\lambda^n + A\lambda^{n-1} + A\lambda^{n-2} = 0 \qquad \text{or equivalently} \qquad A\lambda^{n-2}(\lambda^2 - \lambda - 2) = 0$$

and finally $\lambda^2 - \lambda - 2 = 0$, an equation that also constitutes the characteristic polynomial of the difference equation. It is not difficult to understand that the roots of this polynomial are $\lambda_1 = 2$ and $\lambda_2 = -1$; therefore, the solutions of this equation are the functions $h_\beta^1[n] = (2)^n u[n]$ and $h_\beta^2[n] = (-1)^n u[n]$, while its general solution will be a linear combination of the form

$$h_\beta[n] = A_1(2)^n u[n] + A_2(-1)^n u[n]$$

with the values of the constants A_1 and A_2 being determined using the initial conditions. The use of these conditions leads to a system of two equations with two unknowns of the form

$$h_\beta[0] = A_1 + A_2 = 1 \qquad \text{and} \qquad h_\beta[-1] = \frac{A_1}{2} - A_2 = 0$$

with solution $(A_1, A_2) = (2/3, 1/3)$, and therefore the function $h_\beta[n]$ will have the form

$$h_\beta[n] = \left[\frac{2}{3}(2)^n + \frac{1}{3}(-1)^n\right]u[n]$$

Having calculated the function $h_\beta[n]$, we can calculate the function $h[n]$, which also expresses the impulse response of the system from the expression

$$h[n] = \sum_{k=0}^{1} \beta_k h_\beta[n-k] = \beta_0 h_\beta[n] + \beta_1 h_\beta[n-1] = \left[\frac{2}{3}(2)^{n-1} + \frac{1}{3}(-1)^{n-1}\right]u[n-1]$$

Problem 7.4.4 Determine the impulse response of a system described by the equation

$$y[n] - \frac{3}{5}y[n-1] + \frac{2}{25}y[n-2] = x[n-1] + \frac{1}{2}x[n-2]$$

Solution In this case, the values of the parameters appearing in the general equation are $N = M = 2$, $\alpha_0 = 1$, $\alpha_1 = -3/5$, $\alpha_2 = 2/25$, $\beta_0 = 0$, $\beta_1 = 1$, and $\beta_2 = 1/2$. Equating the second member of the system's difference equation with the discrete impulse function $\delta[n]$, we arrive at the equation

$$h_\beta[n] - \frac{3}{5}h_\beta[n-1] + \frac{2}{25}h_\beta[n-2] = \delta[n]$$

and then we solve the homogeneous difference equation

$$h_\beta[n] - \frac{3}{5}h_\beta[n-1] + \frac{2}{25}h_\beta[n-2] = \delta[n]$$

Considering again an exponential solution of the form $h_\beta[n] = A\lambda^n$, substituting it into the homogeneous difference equation will lead to the characteristic polynomial

$$\lambda^2 - \frac{3}{5}\lambda + \frac{2}{25} = 0$$

with roots $\lambda_1 = 2/5$ and $\lambda_2 = 1/5$. Therefore, the solutions of the equation in question will be $h_\beta^1[n] = (2/5)^n u[n]$ and $h_\beta^2[n] = (1/5)^n u[n]$, while its general solution will be given by the expression

$$h_\beta[n] = \left[A_1\left(\frac{2}{5}\right)^n + A_2\left(\frac{1}{5}\right)^n\right]u[n]$$

with the values of the constants A_1 and A_2 being determined again by the initial conditions. In this case, these conditions lead to a system of two equations with two unknowns

$$h_\beta[0] = A_1 + A_2 = 1 \qquad \text{and} \qquad h_\beta[-1] = \frac{5A_1}{2} + A_2 = 0$$

with solution $(A_1, A_2) = (2, -1)$, and therefore the function $h_\beta[n]$ will have the form

$$h_\beta[n] = \left[2\left(\frac{2}{5}\right)^n - 1\left(\frac{1}{5}\right)^n\right]u[n]$$

Therefore, the impulse response of the system $h[n]$ is calculated as

$$\begin{aligned} h[n] &= \sum_{k=0}^{2} \beta_k h_\beta[n-k] = \beta_0 h_\beta[n] + \beta_1 h_\beta[n-1] + \beta_2 h_\beta[n-2] \\ &= \left[2\left(\frac{2}{5}\right)^{n-1} - \left(\frac{1}{5}\right)^{n-1}\right]u[n-1] + \frac{1}{2}\left[2\left(\frac{2}{5}\right)^{n-2} - \left(\frac{1}{5}\right)^{n-2}\right]u[n-2] \\ &= \left[2\left(\frac{2}{5}\right)^{n-1} - \left(\frac{1}{5}\right)^{n-1}\right]u[n-1] + \frac{5}{2}\left[\left(\frac{2}{5}\right)^{n-1} - \left(\frac{1}{5}\right)^{n-1}\right]u[n-2] \end{aligned}$$

Problem 7.4.5 Determine the impulse response of a system described by the equation

$$y[n] + \sqrt{3}y[n-1] + y[n-2] = x[n-1] + x[n-2]$$

Solution Replacing the second term of the difference equation with the *impulse function* $\delta[n]$ and solving the corresponding homogeneous equation, which is

$$h_\beta[n] + \sqrt{3}h_\beta[n-1] + h_\beta[n-2] = 0$$

we end up with the characteristic polynomial $\lambda^2 + \sqrt{3} + 1 = 0$ with roots

$$\lambda_1 = -\frac{\sqrt{3}}{2} + \frac{1}{2}j = \exp\left(-j\frac{5\pi}{6}\right) \qquad \text{and} \qquad \lambda_2 = -\frac{\sqrt{3}}{2} - \frac{1}{2}j = \exp\left(j\frac{5\pi}{6}\right)$$

Therefore, the solutions of the homogeneous difference equation will be the functions

$$h_\beta^1[n] = \exp\left(-j\frac{5\pi n}{6}\right)u[n] \qquad \text{and} \qquad h_\beta^2[n] = \exp\left(j\frac{5\pi n}{6}\right)u[n]$$

while its general solution will be a linear combination of the form

$$h_\beta[n] = \left\{ A_1 \exp\left(-j\frac{5\pi n}{6} \right) + A_2 \exp\left(j\frac{5\pi n}{6} \right) \right\} u[n]$$

with the constants A_1 and A_2 resulting from the initial conditions. In this case, the initial conditions of the problem lead to a system of two equations with two unknowns of the form

$$h_\beta[0] = A_1 + A_2 = 1 \quad \text{and} \quad h_\beta[-1] = A_1 \exp\left(j\frac{5\pi}{6} \right) + A_2 \exp\left(-j\frac{5\pi}{6} \right) = 0$$

with a solution of the form $(A_1, A_2) = (e^{j\pi/3}, e^{-j\pi/3})$. Therefore, the function $h_\beta[n]$ will have the form

$$h_\beta[n] = \left\{ \exp\left(j\frac{5\pi n}{6} + \frac{\pi}{3} \right) + \exp\left[-j\left(\frac{5\pi n}{6} + \frac{\pi}{3} \right) \right] \right\} u[n] = 2\cos\left(\frac{5\pi n}{6} + \frac{\pi}{3} \right) u[n]$$

while the impulse response of the system will be calculated as

$$h[n] = \sum_{k=0}^{2} \beta_k h_\beta[n-k] = h_\beta[n-1] + h_\beta[n-2] = 2\cos\left(\frac{5\pi(n-1)}{6} + \frac{\pi}{3} \right) u[n-1] + 2\cos\left(\frac{5\pi(n-2)}{6} + \frac{\pi}{3} \right) u[n-2]$$

which is also the desired result.

7.4.3 The Frequency Response of LTI systems

Let us consider a *discrete-time LTI system* and suppose that the elementary discrete exponential sequence $x[n] = Ae^{j\omega_0 n}$ $(-\infty < n < \infty)$ with cyclic frequency $\omega_0 = 2\pi f_0$ is used as input to this system. In this case, the output of the discrete-time system will be given, as is well known, from the *convolution* between the impulse response of the system $h[n]$ and the input signal $x[n]$—we will therefore have

$$y[n] = h[n] * x[n] = \sum_{k=-\infty}^{\infty} h[k]\, x[n-k] = A \sum_{k=-\infty}^{\infty} h[k] e^{j\omega(n-k)} = Ae^{i\omega_0 n}\left(\sum_{k=-\infty}^{\infty} h[k] e^{-j\omega_0 k} \right)$$

Defining the function $\mathcal{H}(e^{j\omega})$ from the equation

$$\mathcal{H}(e^{j\omega}) = \sum_{k=-\infty}^{\infty} h[k] e^{-j\omega k} \tag{7.24}$$

the output of the system will take the form

$$y[n] = A\mathcal{H}(e^{j\omega_0}) e^{j\omega n} = \mathcal{H}(e^{j\omega_0}) x[n] \tag{7.25}$$

that is, the *exponential signal* $e^{j\omega_0 n}$ *constitutes an eigenfunction of the system with the eigenvalue being the function* $\mathcal{H}(e^{j\omega})$*, calculated at the position of the frequency* ω_0 *of the input signal*. This function describes the change in the complex amplitude of the exponential complex input signal as a function of the cyclic frequency ω and is called the *frequency response of the system*. It is not difficult to prove that this function is periodic with respect to the cyclic frequency ω, with period $T = 2\pi$, since (for simplicity we set $A = 1$)

$$\mathcal{H}\left(e^{j(\omega+2\pi)}\right) = \sum_{k=-\infty}^{\infty} h[k]\, e^{-j(\omega+2\pi)k} = \sum_{k=-\infty}^{\infty} h[k]\, e^{-j\omega k} e^{-j2\pi k} = \sum_{k=-\infty}^{\infty} h[k]\, e^{-j\omega k} = \mathcal{H}(e^{j\omega})$$

Generalizing the foregoing conclusion and using the *principle of superposition*, we can formulate the proposition that the response of the system to the input signal

$$x[n] = \sum_k \alpha_k\, e^{j\omega_k n} \quad \text{will have the form} \quad y[n] = \sum_k \alpha_k \mathcal{H}(e^{j\omega_k})\, e^{j\omega_k n}$$

Therefore, if we can express the random discrete-time signal $x[n]$ as a *linear combination of elementary complex exponential sequences*, we have the ability, as long as we know the *frequency response of the system*, to determine its output. More details about the frequency response of discrete-time systems will be given in the next chapter.

In full analogy with continuous-time systems, the *frequency response function* $\mathcal{H}(e^{j\omega})$ is a complex function of a complex variable, and if it is written in its exponential form

$$\mathcal{H}(e^{j\omega}) = |\mathcal{H}(e^{j\omega})| e^{j\measuredangle\mathcal{H}(e^{j\omega})} \tag{7.26}$$

the output of the system will be formulated as

$$y[n] = \mathcal{H}(e^{j\omega_0})\, x[n] = \left(|\mathcal{H}(e^{j\omega_0})| e^{i\measuredangle\mathcal{H}(e^{j\omega_0})}\right)\left(Ae^{j\omega_0 n}\right) = \left[A|\mathcal{H}(e^{j\omega_0})|\right] e^{j[\omega_0 n + \measuredangle\mathcal{H}(e^{j\omega_0})]}$$

We observe, therefore, that the effect of a discrete-time LTI system on the elementary exponential input signal is the *scaling of its amplitude by the magnitude of the frequency response function and its phase shift by the phase of the frequency response function, with this function being calculated at the position of the frequency* ω_0 *of the input signal*. This result is valid only for discrete LTI systems that are *asymptotically stable* or *BIBO stable*, while it does not make sense for systems that are *BIBO unstable* (or those that are characterized as *marginally stable* or *asymptotically unstable*, in terms of their internal stability).

Problem 7.4.6 Determine the output $y[n]$ of a discrete-time LTI system when the trigonometric signal $x[n] = A\cos(\omega_0 n + \varphi)$ is transmitted to its input.

Solution Using Euler's formula, the discrete input signal $x[n]$ is expressed as

$$x[n] = A\cos(\omega_0 n + \varphi) = A\,\frac{e^{j(\omega_0 n+\varphi)} + e^{-j(\omega_0 n+\varphi)}}{2} = \frac{1}{2}Ae^{j\varphi}e^{j\omega_0 n} + \frac{1}{2}Ae^{-j\varphi}e^{-j\omega_0 n}$$

and, according to the theory formulated earlier, the system output will be

$$y[n] = \frac{1}{2}Ae^{j\varphi}\mathcal{H}(e^{j\omega_0})\, e^{j\omega_0 n} + \frac{1}{2}Ae^{-j\varphi}\mathcal{H}(e^{-j\omega_0})\, e^{-j\omega_0 n}$$

with the functions $\mathcal{H}(e^{j\omega_0})$ and $\mathcal{H}(e^{-j\omega_0})$ being defined as

$$\mathcal{H}(e^{j\omega_0}) = \sum_{k=-\infty}^{\infty} h[k]\, e^{-j\omega_0 k} \quad \text{and} \quad \mathcal{H}(e^{-j\omega_0}) = \sum_{k=-\infty}^{\infty} h[k]\, e^{j\omega_0 k}$$

Assuming that the impulse response $h[n]$ of the system is *a signal of real values*—as is usually accepted in practice—we will have $h[n] = h^*[n]$ (where the symbol "*" refers to the operator of complex conjugation), and the foregoing equation will take the form

$$\mathcal{H}(e^{j\omega_0}) = \sum_{k=-\infty}^{\infty} h[k]\, e^{-j\omega_0 k} = \sum_{k=-\infty}^{\infty} h^*[k]\, e^{-j\omega_0 k} = \left(\sum_{k=-\infty}^{\infty} h[k]\, e^{j\omega_0 k} \right)^* = \mathcal{H}^*(e^{-j\omega_0})$$

or, equivalently, $\mathcal{H}(e^{-j\omega_0}) = \mathcal{H}^*(e^{j\omega_0})$. Therefore, the output of the system is formulated as

$$y[n] = \frac{1}{2} A e^{j\varphi} \mathcal{H}(e^{j\omega_0})\, e^{j\omega_0 n} + \frac{1}{2} A e^{-j\varphi} \mathcal{H}^*(e^{j\omega_0})\, e^{-j\omega_0 n}$$

Expressing the functions $\mathcal{H}(e^{j\omega_0})$ and $\mathcal{H}^*(e^{j\omega_0})$ in their exponential form, we will have

$$\mathcal{H}(e^{j\omega_0}) = |\mathcal{H}(e^{j\omega_0})| e^{j\measuredangle \mathcal{H}(e^{j\omega_0})} \quad \text{and} \quad \mathcal{H}^*(e^{j\omega_0}) = |\mathcal{H}(e^{j\omega_0})| e^{-j\measuredangle \mathcal{H}(e^{j\omega_0})}$$

and therefore the output of the system will be formulated as

$$\begin{aligned} y[n] &= \frac{1}{2} A |\mathcal{H}(e^{j\omega_0})| e^{j\varphi} e^{j\measuredangle \mathcal{H}(e^{j\omega_0})} e^{j\omega_0 n} + \frac{1}{2} A |\mathcal{H}(e^{j\omega_0})| e^{-j\varphi} e^{-j\measuredangle \mathcal{H}(e^{j\omega_0})} e^{-j\omega_0 n} \\ &= A\, \frac{e^{j[\omega_0 n + \varphi + \measuredangle \mathcal{H}(e^{j\omega_0})]} + e^{-j[\omega_0 n + \varphi + \measuredangle \mathcal{H}(e^{j\omega_0})]}}{2} = A |\mathcal{H}(e^{j\omega_0})| \cos\left[\omega_0 + \varphi + \measuredangle \mathcal{H}(e^{j\omega_0})\right] \end{aligned}$$

which is the desired result. From the last relation it follows that the effect of a discrete-time LTI system on a basic trigonometric input signal is the *scaling of its amplitude by the magnitude of the system's frequency response calculated at the position of the input signal's frequency, as well as the phase shift of the frequency response calculated at the same value*. This conclusion is very interesting: the *effect of a LTI system on a basic trigonometric signal is the same as that experienced by a basic complex exponential signal with the same frequency value, just as happens with continuous-time systems*.

Problem 7.4.7 Determine the frequency response of the ideal time delay subsystem described by an input/output equation of the form $y[n] = x[n - N]$, where N is an integer value.

Solution If in the considered discrete-time system we transmit the basic exponential signal $x[n] = e^{j\omega_0 n}$, the output will have the form

$$y[n] = x[n - N] = e^{j\omega_0 (n-N)} = e^{-j\omega_0 N} e^{j\omega_0 n} = e^{-\omega_0 N} x[n] = \mathcal{H}(e^{j\omega_0}) x[n]$$

Therefore, the frequency response of this system has the form $\mathcal{H}(e^{j\omega}) = e^{-j\omega N}$. Expressing the aforementioned complex function in its exponential form, it immediately follows that $|\mathcal{H}(e^{j\omega})| = 1$ and $\measuredangle \mathcal{H}(e^{j\omega n}) = -\omega N$, and therefore this system is a *linear phase system*.

Problem 7.4.8 Determine the magnitude and phase of the frequency response of the system described by the difference equation $y[n] = x[n] + 2x[n - 1] + x[n - 2]$.

Solution From the equation defining the frequency response, it is not difficult to see that the calculation of this function requires *finding the impulse response of the system* $h[n]$. Based on the properties of the time-shifted impulse function, it immediately follows that

$$x[n] = x[n] * \delta[n], \quad x[n-1] = x[n] * \delta[n-1], \quad x[n-2] = x[n] * \delta[n-2]$$

Substituting the foregoing expressions into the system's difference equation, it is written

$$y[n]=x[n]*\delta[n]+2x[n]*\delta[n-1]+x[n]*\delta[n-2]=x[n]*\Big(\delta[n]+2\delta[n-1]+\delta[n-2]\Big)=x[n]*h[n]$$

Consequently, the impulse response of the system is given by the relation

$$h[n]=\delta[n]+2\delta[n-1]+\delta[n-2]$$

while its frequency response will be calculated as

$$\mathcal{H}(e^{j\omega})=\sum_{k=-\infty}^{\infty}h[n]\,e^{-j\omega k}=\sum_{k=-\infty}^{\infty}\delta[k]\,e^{-j\omega k}+2\sum_{k=-\infty}^{\infty}\delta[k-1]\,e^{-j\omega k}+\sum_{k=-\infty}^{\infty}\delta[k-2]\,e^{-j\omega k}$$

From the definition, however, of the discrete impulse function it immediately follows that

$$\sum_{k=-\infty}^{\infty}\delta[k]\,e^{-j\omega k}=e^{-j\omega k}\Big|_{k=0}=1$$

$$\sum_{k=-\infty}^{\infty}\delta[k-1]\,e^{-j\omega k}=e^{-j\omega k}\Big|_{k=1}=e^{-j\omega}$$

$$\sum_{k=-\infty}^{\infty}\delta[k-2]\,e^{-j\omega k}=e^{-j\omega k}\Big|_{k=2}=e^{-j2\omega}$$

and the foregoing equation is written

$$\mathcal{H}(e^{j\omega})=1+2e^{-j\omega}+e^{-j2\omega}=2e^{-j\omega}\left(\frac{1}{2}e^{j\omega}+1+\frac{1}{2}e^{-j\omega}\right)=2e^{-j\omega}\Big[\cos(\omega)+1\Big]$$

Therefore, the magnitude and phase of the system's frequency response will be given by the relations

$$|\mathcal{H}(e^{j\omega})|=2\,[\cos(\omega)+1]\qquad\text{and}\qquad\measuredangle\mathcal{H}(e^{j\omega})=-\omega$$

7.4.4 The Output of LTI Systems

The description of discrete-time LTI systems presented in the previous sections was based on the knowledge of the step and, especially, the impulse response functions, whose convolution with the respective input signal allows the calculation of the corresponding system output. However, if the function $h[n]$ is not known and all we have is the system's difference equation and the input signal $x[n]$, the only way to determine the output $y[n]$ is to solve directly that difference equation, following the known methods for solving equations of this type. These methods were silently applied during the calculation of the impulse response from the system's difference equation, and in the following pages a detailed and systematic description of them is presented that covers all cases of interest.

As we saw in the first section of this chapter the general difference equation of a discrete-time system of order N is a *linear difference equation with constant coefficients* of the form

$$\sum_{k=0}^{N}\alpha_k y[n-k]=\sum_{k=0}^{M}\beta_k x[n-k]\qquad(7.27)$$

with the coefficients $\{\alpha_k\}_0^N$ and $\{\beta_k\}_0^M$ taking only real values (some of which may be zero), while generally we impose the requirement that $M \leq N$. Taking into account that *difference equations are the discrete equivalents of differential equations*, can we guess that the methods of solving them will be similar? Indeed, they will. For example, to solve such an equation, we must know the initial conditions of the problem, that is, the values of all the delayed versions of the function $y[n]$ at the time instance $n = 0$. Consider, for example, the difference equation

$$y[n-3] + 2y[n-2] - y[n-1] + y[n] = x[n-2] + 3x[n-1] + 5x[n]$$

The initial conditions of the problem are the values of the expressions $y[-3]$, $y[-2]$, and $y[-1]$, which must be known. On the contrary, for the input signal this requirement does not exist: This signal is considered to be causal, that is, it has *zero values* for time instances $n < 0$; therefore, in this case we will have $x[-1] = x[-2] = 0$, while generally it will hold that $x[-n] = 0$ (for values $n > 0$).

In full analogy with continuous-time systems that are described from differential equations, the response of a LTI discrete-time system consists of two components: the *system response due to the existence of initial conditions, known as a zero-input response, and the system response to the input signal transmitted to it, known as a zero-state response*. The analytical method of calculating the foregoing two types of outputs along with example solved exercises are presented in detail in the following sections.

Recursive Calculation of System Output

The simplest way to solve a difference equation is the recursive method in which we solve the equation in question for $y[n]$, so that it is transformed into the relation

$$y[n] = \frac{1}{\alpha_0}\left(\sum_{k=0}^{M} \beta_k x[n-k] - \sum_{k=1}^{N} \alpha_k y[n-k]\right) \tag{7.28}$$

and then, knowing the input signal of the system and the initial conditions of the problem, we calculate, one after another, the values of the system output $y[n]$. For example, for a second-order equation, we can use the values of the initial conditions $y[-1]$ and $y[-2]$ and the appropriate inputs to calculate the value of the function $y[0]$, which then together with the value $y[-1]$ will allow us to calculate the value of the function $y[1]$, and so forth. This recursive way of solving a difference equation is demonstrated in the next two exercises.

Problem 7.4.9 Determine the output $y[n]$ of the discrete-time LTI system described by the difference equation $y[n] = \alpha y[n-1] + x[n]$ (Proakis and Manolakis, Section 2.4.2).

Solution In this case, solving the difference equation for $y[n]$ is not necessary, since this function already appears on the left side of the foregoing equation. Following the methodology presented earlier, for each value of the discrete-time k we calculate successively the values of the system's output $y[k]$. In detail, we have:

- For $\boldsymbol{k = 0}$: $y[0] = \alpha y[-1] + x[0]$
- For $\boldsymbol{k = 1}$: $y[1] = \alpha y[0] + x[1] = \alpha^2 y[-1] + \alpha x[0] + x[1]$
- For $\boldsymbol{k = 2}$: $y[2] = \alpha y[1] + x[2] = \alpha^3 y[-1] + \alpha^2 x[0] + \alpha x[1] + x[2]$
- For $\boldsymbol{k = n}$: Generalizing the above process, it turns out that

$$y[n] = \alpha y[n-1] + x[n] = \alpha^{n+1} y[-1] + \alpha^n x[0] + \alpha^{n-1} x[1] + \cdots + \alpha x[n-1] + x[n]$$

an equation that, in more concise notation, is formulated as

$$y[n] = \alpha^{n+1} y[-1] + \sum_{k=0}^{n} \alpha^k x[n-k] \quad (n \geq 0)$$

In this case, the system's difference equation is first-order, and therefore the determination of the system's output through the application of the recursive method requires the determination of the value $y[-1]$. Assuming, for example, that this value is equal to $y[-1] = 16$, while the input signal has the value $x[n] = n^2$, the application of the foregoing process will lead to output of the form

$$y[n] = \{\mathbf{8}, \quad 5, \quad 6.5, \quad 12.15, \quad 22.125, \quad \ldots\}$$

Problem 7.4.10 Determine the first 10 samples of the output $y[n]$ of the discrete-time LTI system described by the difference equation

$$y[n] = y[n-1] - 0.24y[n-2] + x[n] - 2x[n-1]$$

for initial conditions $y[-1] = 2$, $y[-2] = 1$ and an input signal of the form $x[n] = n$.

Solution Following the methodology presented earlier, we successively have:

$$\boldsymbol{n=0}: \quad y[0] = y[-1] - 0.24y[-2] + x[0] - 2x[-1] = 1.76$$

$$\boldsymbol{n=1}: \quad y[1] = y[0] - 0.24y[-1] + x[1] - 2x[0] = 2.28$$

$$\boldsymbol{n=2}: \quad y[2] = y[1] - 0.24y[0] + x[2] - 2x[1] = 1.8576$$

$$\boldsymbol{n=3}: \quad y[3] = y[2] - 0.24y[1] + x[3] - 2x[2] = 0.3104$$

$$\boldsymbol{n=4}: \quad y[4] = y[3] - 0.24y[2] + x[4] - 2x[3] = -2.1354$$

$$\boldsymbol{n=5}: \quad y[5] = y[4] - 0.24y[3] + x[5] - 2x[4] = -5.2098$$

$$\boldsymbol{n=6}: \quad y[6] = y[5] - 0.24y[4] + x[6] - 2x[5] = -8.6974$$

$$\boldsymbol{n=7}: \quad y[7] = y[6] - 0.24y[5] + x[7] - 2x[6] = -12.4470$$

$$\boldsymbol{n=8}: \quad y[8] = y[7] - 0.24y[6] + x[8] - 2x[7] = -16.3597$$

$$\boldsymbol{n=9}: \quad y[9] = y[8] - 0.24y[7] + x[9] - 2x[8] = -20.3724$$

The Response to Zero Input or Natural Response

As already mentioned, a system's *response to zero input* is due to the existence of initial conditions and consists in solving the homogeneous difference equation that arises from the full equation if we set all expressions related to the system's input $x[n]$ to zero.[12] In this case, the general difference equation of order N will take the form

$$\sum_{k=0}^{N} \alpha_k y[n-k] = 0 \tag{7.29}$$

or, equivalently, by expanding the N terms of the sum, as

$$\alpha_0 y[n] + \alpha_1 y[n-1] + \alpha_2 y[n-2] + \cdots + \alpha_N y[n-N] = 0 \tag{7.30}$$

and its solution is carried out in the following way.

Adopting the methodology of solving *linear differential equations with constant coefficients*, according to which the solutions of these equations will have the exponential form $x(t) = e^{\lambda t}$ where the parameter λ in the general case is a complex number, we consider that the desired solution $y[n]$ will be an exponential function of discrete-time n, so that we can

[12] This conclusion is very interesting: The system will generate some output component, due to the existence of *nonzero initial conditions*, even if no signal is transmitted to its input.

express it as $y[n] = \lambda^n$, where the value of λ will again be a *real*, *imaginary*, or *complex* quantity. The use of this assumption makes similar in their functional form both the function $y[n]$ itself and all its time-delayed versions, since they will obviously be

$$y[n-N] = \lambda^{n-N}, \quad y[n-N+1] = \lambda^{n-N+1} = \lambda\lambda^{n-N}, \quad \ldots$$
$$y[n-2] = \lambda^{n-2} = \lambda^{N-2}\lambda^{n-N}, \quad y[n-1] = \lambda^{n-1} = \lambda^{N-1}\lambda^{n-N}, \quad y[n] = \lambda^n = \lambda^N\lambda^{n-N}$$

But this fact significantly simplifies the situation, since all the foregoing expressions contain as a common factor the expression λ^{n-N}. Therefore, by substituting the foregoing expressions into the homogeneous difference equation and taking into account that the exponential factor in question is always different from zero, it immediately follows that[13]

$$Q(\lambda) = \lambda^N + \alpha_1\lambda^{N-1} + \alpha_2\lambda^{N-2} + \cdots + \alpha_{N-2}\lambda^2 + \alpha_{N-1}\lambda + \alpha_N = 0 \qquad (7.31)$$

The polynomial $Q(\lambda)$ that appears in the foregoing equation, which is known as the *characteristic equation* of the system, is known as a *characteristic polynomial*, and the nature of its roots determines the form of the desired function $y[n]$, which also constitutes the zero-input response of the system. We distinguish the following cases:

- The polynomial $Q(\lambda)$ has *N distinct roots* $\lambda_1, \lambda_2, \ldots, \lambda_N$, and therefore there will be N linearly independent functions of the form $y_k[n] = \lambda_k^n$ (where $k = 1, 2, \ldots, N$), which will satisfy the homogeneous difference equation. Taking into account that this equation is a *linear differential equation with constant coefficients*, it follows that its general solution will be a linear combination of the form

$$y_{zi}[n] = \sum_{k=1}^{N} C_k\lambda_k^n = C_1\lambda_1^n + C_2\lambda_2^n + \cdots + C_N\lambda_N^n$$

  with the constants $C_1, C_2, \ldots, C_N$ being determined using the *initial conditions* of the problem.
- The polynomial $Q(\lambda)$ has *repeated roots*, that is, roots whose values are equal to each other. Considering a root λ_ξ with a multiplicity value of q, this means that there will be *q equal roots with a value equal to* λ_ξ, so that we can write that $\lambda_1 = \lambda_2 = \cdots = \lambda_{q-1} = \lambda_\xi$ and factorize the polynomial $Q(\lambda)$ by writing it in the form

$$Q(\lambda) = (\lambda - \lambda_\xi)^q(\lambda - \lambda_{q+1})(\lambda - \lambda_{q+2})\ldots(\lambda - \lambda_N)$$

  In this case and according to our known theory of difference equations, the $N - q$ number of discrete roots $\lambda_{q+1}, \lambda_{q+2}, \ldots, \lambda_N$ will create, as before, the functions $\lambda_{q+1}^n, \lambda_{q+2}^n, \ldots, \lambda_N^n$, while the root λ_ξ with multiplicity q will create q solutions of the form $\lambda_\xi, n\lambda_\xi, n^2\lambda_\xi, \ldots n^{q-1}\lambda_\xi$. As for the general solution of the characteristic equation $Q(\lambda) = 0$, it will have the form

$$y_{zi}[n] = (C_1 + C_2n + C_3n^2 + \cdots + C_qn^{q-1})\lambda_\xi^n + C_{q+1}\lambda_{q+1}^n + C_{q+2}\lambda_{q+2}^n + \cdots + C_N\lambda_N^n$$

  with the constants appearing in the foregoing relation being calculated from the *initial conditions*.

[13] In constructing the last relationship, we silently assumed that the coefficient α_0 was equal to one; in the opposite case, we simply divide all the coefficients by the value of α_0.

- The polynomial $Q(\lambda)$ has complex roots. In this case, since the coefficients appearing in it are *real numbers*, as we have seen in similar cases, these roots should appear in *pairs of complex conjugates*. This means that if a polynomial has the root $\lambda = |\lambda|e^{j\vartheta}$, then it must necessarily also have the conjugate complex root $\lambda^* = |\lambda|e^{-j\vartheta}$. In this case, the contribution of these two roots to the *zero-input response* of the system will have the form

$$C_1\lambda^n + C_2(\lambda^*)^n = C_1|\lambda|^n e^{in\vartheta} + C_2|\lambda|^n e^{-jn\vartheta}$$

If the discrete-time system we are dealing with creates *real outputs* $y[n]$, the coefficients C_1 and C_2 (which are generally of a complex nature), should be related by a complex conjugation relation. For the sake of simplicity, we set

$$C_1 = \frac{A}{2}e^{j\varphi} \qquad \text{and} \qquad C_2 = \frac{A}{2}e^{-j\varphi}$$

the foregoing expression will take the form

$$C_1|\lambda|^n e^{jn\vartheta} + C_2|\lambda|^n e^{-jn\vartheta} = A|\lambda|^n \frac{e^{j(n\vartheta+\varphi)} + e^{-j(n\vartheta+\varphi)}}{2} = A|\lambda|^n \cos(n\vartheta + \varphi)$$

with the constants C_1 and C_2 or, equivalently, the constants A and φ being determined again using the initial conditions of the problem.

The calculation of the *zero-input response*—or *natural response*, as it is often called—of representative discrete-time LTI systems is presented in the following exercises.[14]

Problem 7.4.11 Find the natural response of a discrete-time system described by the equation

$$y[n] - \alpha y[n-1] = x[n].$$

(Proakis and Manolakis, Example 2.44)

Solution To determine the natural response of the foregoing discrete-time system, we need to find the general solution of the homogeneous difference equation that describes its operation that is $y[n] + \alpha y[n-1] = 0$. Considering an exponential solution of the form $y[n] = \lambda^n$, then $y[n-1] = \lambda^{n-1}$, and substituting this into the homogeneous equation, it turns out that $\lambda - \alpha = 0$. From the last equation we find that the characteristic polynomial $Q(\lambda)$ is of the first degree with a root of the form $\lambda = \alpha$. Therefore, the system's response to the zero input will be given by the relation $y[n] = C\alpha^n$, with the constant C being calculated using the initial condition, which is assumed to be known. To calculate the value of this constant, we need to determine the value of the expression $y[0]$, both from the difference equation and from the solution we determined, and then equate these two quantities. In detail, then, we have:

- From the homogeneous equation: If we set $n = 0$ in the relation $y[n] + \alpha y[n-1] = 0$, we will get $y[0] = \alpha y[-1]$.
- From the general solution: If we set $n = 0$ in the relation $y[n] = C\alpha^n$, then we will get $y[0] = C$.

Equating, then, the two foregoing expressions that refer to the same quantity $y[0]$, we will get $C = \alpha y[-1]$. Therefore, the general solution of the homogeneous difference equation will take the form

[14] To emphasize the fact that the output calculated is the *response to zero input*, we use in the symbolism of the output $y[n]$ the index zi (zero input).

$$y_{zi}[n] = C\alpha^n = (\alpha y[-1]) \times \alpha^n = \alpha^{n+1} y[-1]$$

which is also the desired result. We observe that the expression we ended up with coincides with the first term of the solution of the difference equation of Exercise 7.4.9 (refer to the page 586), a finding that confirms our claim that the general solution of the equation arises *from the sum of the responses to the zero input and the zero state*. As can be proven, the second term of the solution $y[n]$ of Exercise 7.4.9 indeed coincides with the system's response *to the zero state*.

Problem 7.4.12 Find the natural response of a discrete-time system described by the equation

$$y[n] - 3y[n-1] - 4y[n-2] = 0.$$

(Proakis and Manolakis, Example 2.45)

Solution The system's difference equation is homogeneous, and therefore its solution will also give us the desired natural response. Considering a solution of the form $y[n] = \lambda^n$, the characteristic polynomial of the equation will have the form $\lambda^2 - 3\lambda - 4 = 0$ with roots $\lambda_1 = 4$ and $\lambda_2 = -1$, that is, the general solution of the homogeneous equation will be

$$y[n] = C_1\lambda_1^n + C_2\lambda_2^n = C_1(4)^n + C_2(-1)^n$$

with the constants C_1 and C_2 being determined by the initial conditions. Note that because in this case the order of the system is $N = 2$, we should know the values of two initial conditions corresponding to the values $y[-1]$ and $y[-2]$. Following the methodology applied previously, we will calculate the values of the quantities $y[0]$ and $y[1]$ from both the initial difference equation and the general solution that was determined, and then we will equate the corresponding quantities. In detail, then, we will have:

- From the homogeneous equation: If we set $n = 0$ and $n = 1$ in the difference equation of the system, we will get

$$y[0] = 3y[-1] + 4y[-2] \quad \text{and} \quad y[1] = 3y[0] + 4y[-1] = 13y[-1] + 12y[-2]$$

- From the general solution: If we set $n=0$ and $n=1$ in the relation $y[n]=C_1(4)^n+C_2(-1)^n$, we will get

$$y[0] = C_1 + C_2 \quad \text{and} \quad y[1] = 4C_1 - C_2$$

Equating, then, the corresponding quantities,[15] we will get the system of the two equations

$$C_1 + C_2 = 3y[-1] + 4y[-2] \quad \text{and} \quad 4C_1 - C_2 = 13y[-1] + 12y[-2]$$

with the two constants C_1 and C_2 as unknowns. The solution of this system is

$$C_1 = \frac{16}{5}y[-1] + \frac{16}{5}y[-2] \quad \text{and} \quad C_2 = -\frac{1}{5}y[-1] + \frac{4}{5}y[-2]$$

[15] This method of calculating the constants C_k is directly generalized for *linear difference equations of higher order with constant coefficients*. Considering a difference equation of order N, whose characteristic polynomial has (for simplicity) N distinct roots, the N constants C_k ($k = 0, 1, 2, \ldots, N$) will result from the N initial conditions of the problem $y[-n]$ ($n = 1, 2, \ldots, N$) in the following way. Initially, we will set the values $n = 0, 1, 2, \ldots, N-1$ in the difference equation of the system, calculating the values of the expressions $y[-n]$ ($n = 0, 1, \ldots, N-1$) as a function of the initial conditions. Then we will successively replace the values $n = 0, 1, 2, \ldots, N-1$ in the solution $y[n]$ we found, calculating the values of the expressions $y[-n]$ ($n = 0, 1, \ldots, N-1$) as a function of the constants C_k ($k = 0, 1, 2, \ldots, N-1$). By equating at this point the corresponding quantities, we will construct a system of N equations with N unknowns, whose solution will give us the values of the constants C_k ($k = 0, 1, 2, \ldots, N$). This is the exact method used in the two previous exercises for values of order of difference equations $N = 1$ and $N = 2$, respectively.

and therefore the natural response of the considered discrete-time system will be given by the relation

$$y_{zi}[n] = \left[\frac{16}{5}y(-1) + \frac{16}{5}y(-2)\right](4)^n + \left[-\frac{1}{5}y(-1) + \frac{4}{5}y(-2)\right](-1)^n$$

We observe that the function $y[n]$ depends on the values of the initial conditions $y[-1]$ and $y[-2]$ that are used in each case. If, for example, we use the values $y[-1] = 5$ and $y[-2] = 0$, then $C_1 = 16$ and $C_2 = -1$, and the natural response of the system will take the form

$$y_{zi}[n] = 16(4)^n + (-1)(-1)^n = (4)^2(4)^n + (-1)(-1)^n = (4)^{n+2} + (-1)^{n+1}$$

Problem 7.4.13 Find the natural response of a discrete-time system described by the equation

$$y[n] - 8y[n-1] + 16y[n-2] = x[n] - 2x[n-1]$$

with initial conditions $y[-1] = y[-2] = 1/8$.

Solution The natural response of the system will result from the solution of the homogeneous difference equation

$$y[n] - 8y[n-1] + 16y[n-2] = 0$$

Considering the exponential solution $y[n] = \lambda^n$, we are led to the characteristic equation $Q(\lambda) = \lambda^2 - 8\lambda + 16 = 0$, which has the double root $\lambda_1 = \lambda_2 = 4$. Therefore, according to the prevailing theory, this difference equation will have the simple solutions $y_1[n] = (4)^n$ and $y_2[n] = n(4)^n$, while its general solution will be

$$y[n] = C_1(4)^n + nC_2(4)^n = (C_1 + nC_2)(4)^n$$

with constants C_1 and C_2 resulting from the initial conditions.

Although in this case the root of the characteristic equation has multiplicity $p = 2$, the methodology we will apply will be the one presented earlier. By substituting the values $n = 0$ and $n = 1$ into the system's difference equation, we will get

$$y[0] = 8y[-1] - 16y[-2] \qquad \text{and} \qquad y[1] = 48y[-1] - 128y[-2]$$

while the substitution of the same values into the general solution we determined earlier will give

$$y[0] = C_1 \qquad \text{and} \qquad y[1] = 4C_1 + 4C_2$$

Equating the corresponding quantities, we end up with a system of two equations with the constants C_1 and C_2 as unknowns, whose value is calculated to be

$$C_1 = 8y[-1] - 16y[-2] = -1 \qquad \text{and} \qquad C_2 = 16y[-1] - 64y[-2] = -6$$

where for the extraction of the final values we used the values of the initial conditions of the problem. Substituting the preceding constants into our general solution, we end up with the final result

$$y_{zi}[n] = \left[(-1) + n(-6)\right](4)^n = -(1 + 6n)(4)^n$$

Problem 7.4.14 Find the natural response of the discrete-time system described by the equation

$$y[n] - 2y[n-1] + 5y[n-2] = x[n] - x[n-1]$$

Solution In this case, the solution of the homogeneous difference equation using an exponential solution of the form $y[n] = \lambda^n$ leads us to the characteristic polynomial $Q(\lambda) = \lambda^2 - 2\lambda + 5$, which has the conjugate complex roots $\lambda_1 = 1 + 2j = \sqrt{5}e^{j(63.434°)}$ and $\lambda_2 = 1 - 2j = \sqrt{5}e^{-j(63.434°)}$. Therefore, according to the theory presented in the previous section and considering that the outputs generated by the system will be real quantities, the general solution of the homogeneous difference equation will have the form

$$y[n] = C\lambda_1^n + C^*\lambda_2^n = C(5)^{n/2}e^{jn(63.434°)} + C^*(5)^{n/2}e^{-jn(63.434°)}$$

with the constant C being determined using the initial conditions. Applying the process described earlier, we are led to the system of the two equations

$$C + C^* = 2y[-1] - 5y[-2] \quad \text{and} \quad \sqrt{5}Ce^{i(63.434°)} + \sqrt{5}C^*e^{-i(63.434°)} = -y[-1] - 10y[-2]$$

from which we can determine the values of the parameters C and C^*, provided we know the initial conditions of the problem $y[-1]$ and $y[-2]$. The reader is asked to calculate these parameter values and complete the solution of the exercise for the pairs of initial conditions $(y[-1] = 1, y[-2] = 0)$, $(y[-1] = y[-2] = 1)$ and $(y[-1] = 3/4, y[-2] = -1/2)$.

The Zero-State Response or Forced Response

The second component of the output of a discrete-time LTI system is the *zero-state response*, which is defined as the response of the system to the input $x[n]$, assuming that the *initial conditions of the problem have zero values*. In reality, the zero-state response is simply the result of applying the convolution operation that is performed between the impulse response of the system $h[n]$ and the input signal $x[n]$, so we have

$$y[n] = h[n] * x[n] = \sum_{k=-\infty}^{\infty} x[k]y[n-k] \tag{7.32}$$

Taking into account that in almost all cases the input signals transmitted to the various discrete-time systems are causal, that is, they are defined only for time instances $n \geq 0$, the lower limit in the foregoing sum is set in almost all cases *to the zero value*.

Examples of calculating the forced response of discrete LTI systems are *all the examples involving convolution calculations from the previous chapter* if one of the two signals is treated as the *impulse response* of the system $h[n]$ and the other as the *system's input* $x[n]$. In the case where we do not know the impulse response of the system but only the difference equation, we must first calculate it by applying the methodology presented in a previous section and then calculate the convolution of the two signals by applying the now known method. It should be noted that for the special category of recursive discrete LTI systems, the impulse response is defined as the *response of the zero state* $y_{zs}[n]$ *when the unit impulse function is used as input to the system, with the system, of course, being in a state of rest*. Considering, for example, a system with zero-state response

$$y_{zs}[n] = \sum_{k=0}^{n} \alpha^k x[n-k]$$

Table 7.1 Known input signals and their corresponding partial solutions of difference equations

Input $[x]$	Partial solution $y_p[n]$
C	C_1
Cn	$C_1 n + C_2$
$C\alpha^n$	$C_1\alpha^n$
$C\cos(\omega n)$	$C_1\cos(\omega n) + C_2\sin(\omega n)$
$C\sin(\omega n)$	$C_1\cos(\omega n) + C_2\sin(\omega n)$
$C\alpha^n\cos(\omega n)$	$C_1\alpha^n\cos(\omega n) + C_2\alpha^n\sin(\omega n)$
C	$\delta[n]$

the transmission to its input of the discrete-time signal $x[n] = \delta[n]$ will lead to the output

$$y_{zs}[n] = \sum_{k=0}^{n} \alpha^k \delta[n-k] = \alpha^n \quad (n \geq 0)$$

and therefore its impulse response will be the discrete-time signal $h[n] = \alpha^n u[n]$.

The Total Response of LTI Systems

As already mentioned, the total response of a discrete LTI system is defined as the *sum of the zero-state response and the zero-input response*, which must be calculated separately and then added together. Alternatively, *we can solve the full difference equation using the input signal given and the initial conditions in each case*. In full analogy with the differential equations related to the field of continuous-time, the *general solution of a full difference equation equals the sum of the general solution of the homogeneous equation, as well as a partial solution of the full equation*. The general solution of the homogeneous equation is simply the response to the zero input of the system (since, to make an equation homogeneous, we zero all the expressions related to its input) and is calculated as in previous exercises. On the other hand, the determination of a partial solution of a full equation is not described by any specific process but is, among other things, a function of our experience and our familiarity with the use and solution of difference equations. However, it is worth mentioning that if some specific inputs with known forms are transmitted to the system, the required partial solution will also have a form similar to that of the input signal. Table 7.1 contains a number of known input signals and the corresponding partial solution that corresponds to them.[16]

Having determined the general solution of the homogeneous equation $y_h[n]$ and the partial solution of the full equation $y_p[n]$, the general solution of the full equation $y[n]$ arises, as we have already mentioned, from the sum of these two solutions, or, in mathematical notation, $y[n] = y_h[n] + y_p[n]$. Taking into account that the total output $y[n]$ of a discrete LTI system corresponding to the general solution of the full difference equation consists of the sum of the responses to the zero state and the zero input and how the general solution of the homogeneous equation is the response to the zero input, we can surmise that the *partial solution of the full difference equation will correspond to the response to the zero state*. As we will see in the next exercise, this is exactly how things are.

Problem 7.4.15 Determine the general solution of the full second-order difference equation

$$y[n] = \frac{5}{6}y[n-1] - \frac{1}{6}y[n-2] + x[n]$$

(continued)

[16] Note that while the general solution of the homogeneous difference equation contains linear combinations of the characteristic functions $y_k[n] = \lambda_k^n$, where $\lambda_k (k = 1, 2, \ldots, N)$ are the roots of the characteristic polynomial, the partial solution of the full equation *cannot by definition contain such terms*.

with the function $x[n]$ defined as $x[n] = (2)^n u[n]$, and then verify that the solution you found coincides with the sum of the responses to the zero state and the zero input of the discrete LTI system described by the foregoing equation (Proakis and Manolakis, Example 2.4.7 and Exercise for solution 2.26).

Solution As already mentioned, the general solution of a full difference equation equals the sum of the general solution of the homogeneous equation and a partial solution of the full equation. Therefore, in the first step of the process, we must solve the homogeneous difference equation that has the form

$$y[n] - \frac{5}{6}y[n-1] + \frac{1}{6}y[n-2] = 0$$

Considering a solution of the form $y[n] = \lambda^n$, we are led to the characteristic equation $\lambda^2 - (5/6)\lambda + (1/6) = 0$ with roots $\lambda_1 = 1/2$ and $\lambda_2 = 1/3$, so the general solution of the homogeneous equation will be

$$y[n] = C_1\left(\frac{1}{2}\right)^n + C_2\left(\frac{1}{3}\right)^n$$

with the constants C_1 and C_2 being determined using the initial conditions $y[-1]$ and $y[-2]$, which, of course, must be known.

In the next step of the process, we must seek a partial solution of the full difference equation. From the mathematical form of the function $x[n] = (2)^n u[n]$ and referring to Table 7.1, we find that the required partial solution should have the form $y_p[n] = k(2)^n u[n]$, with the value of the constant k arising from the requirement that the partial solution $y_p(n)$ satisfies the full difference equation. Therefore, replacing the function $y_p[n]$ in the full difference equation, we write

$$k(2)^n u[n] - \frac{5}{6}k(2)^{n-1}u[n-1] + \frac{1}{6}k(2)^{n-2}u[n-2] = (2)^n u[n]$$

For $n = 2$ the foregoing equation is written

$$k(2)^2 u[2] - \frac{5}{6}k(2)u[1] + \frac{1}{6}ku[0] = (2)^2 u[2]$$

from which the value $k = 8/5$ arises. Therefore, the partial solution of the full equation will be $y_p[n] = (8/5)(2)^n u[n]$, while its general solution will have the form

$$y[n] = \frac{8}{5}(2)^n u[n] + C_1\left(\frac{1}{2}\right)^n u[n] + C_2\left(\frac{1}{3}\right)^n u[n]$$

with the constants C_1 and C_2 arising from the initial conditions. Since these conditions are not known, we will determine the values of the constants C_1 and C_2 as a function of the values of the parameters $y[-1]$ and $y[-2]$. Following the previous methodology, we substitute the values $n = 0, 1$ into the difference equation of the system and into the general solution we have determined, resulting in the system

$$C_1 + C_2 = \frac{5}{6}y[-1] - \frac{1}{6}y[-2] + x[0] - \frac{8}{5}$$

$$\frac{C_1}{2} + \frac{C_2}{3} = \frac{19}{36}y[-1] - \frac{5}{36}y[-2] + \frac{5}{6}x[0] + x[1] - \frac{16}{5}$$

which if solved for C_1 and C_2, and after first substituting the values of the samples $x[0]$ and $x[1]$ of the signal $x[n]$, will lead us to the result

$$C_1 = \frac{3}{2}y[-1] - \frac{1}{2}y[-2] - 1 \quad \text{and} \quad C_2 = -\frac{2}{3}y[-1] + \frac{1}{3}y[-2] + \frac{2}{5}$$

Therefore, the general solution of the full difference equation will have the form

$$y[n] = \frac{8}{5}(2)^n u[n] + \left[\frac{3}{2}y[-1] - \frac{1}{2}y[-2] - 1\right]\left(\frac{1}{2}\right)^n u[n] + \left[-\frac{2}{3}y[-1] + \frac{1}{3}y[-2] + \frac{2}{5}\right]\left(\frac{1}{3}\right)^n u[n]$$

Let us now assume that the foregoing full difference equation describes the operation of a discrete LTI system, and let us examine *whether the sum of the responses in the zero state and the zero input is given by the foregoing equation.* Starting from the response to the zero input, this, as is well known, will result from the solution of a homogeneous equation that has the form

$$y[n] = C_1\left(\frac{1}{2}\right)^n + C_2\left(\frac{1}{3}\right)^n$$

with the constants C_1 and C_2 being determined by the initial conditions. Applying the methodology for finding the initial conditions, we find that the *response to zero input* is given by the relation

$$y_{zi}[n] = \left[\frac{3}{2}y[-1] - \frac{1}{2}y[-2]\right]\left(\frac{1}{2}\right)^n + \left[-\frac{2}{3}y[-1] + \frac{1}{3}y[-2]\right]\left(\frac{1}{3}\right)^n$$

On the other hand, the system's response to the zero state will arise, as is well known, from the calculation of the convolution between the impulse response of the system $h[n]$ and the input signal $x[n] = (2)^n u[n]$; initially, therefore, let us calculate the impulse response $h[n]$ from the system's difference equation in the way described in previous sections. Replacing the second term of the difference equation with the impulse function $\delta[n]$, we end up with the equation

$$h_\beta[n] - \frac{5}{6}h_\beta[n-1] + \frac{1}{6}h_\beta[n-2] = \delta[n]$$

the homogeneous version of which has the form

$$h_\beta[n] - \frac{5}{6}h_\beta[n-1] + \frac{1}{6}h_\beta[n-2] = 0$$

But this equation was solved previously. In this case, the constants C_1 and C_2 will be calculated with the help of the initial conditions, which are $h_\beta[n] = 1/\alpha_0 = 1$ and $h_\beta[-1] = 0$ and which lead to the system

$$C_1 + C_2 = 0 \quad \text{and} \quad 2C_1 + 3C_2 = 0$$

with the solution $(C_1, C_2) = (3, -2)$, and therefore the impulse response of the system will have the form

$$h[n] = \left[3\left(\frac{1}{2}\right)^n - 2\left(\frac{1}{3}\right)^n\right]u[n]$$

Having calculated the impulse response of the system, we can now calculate the zero-state response from the convolution between the discrete-time signals $h[n]$ and $x[n]$. We will therefore have

$$y_{zs}[n] = \sum_{k=-\infty}^{\infty} h[k]\,x[n-k] = \sum_{k=-\infty}^{\infty}\left[3\left(\frac{1}{2}\right)^k - 2\left(\frac{1}{3}\right)^k\right]u[k]\,(2)^{n-k}\,u[n-k]$$

$$= 3\cdot 2^n \sum_{k=0}^{n}\left(\frac{1}{4}\right)^k - 2\cdot 2^n \sum_{k=0}^{n}\left(\frac{1}{6}\right)^k = 3\cdot 2^n \frac{1-(1/4)^{n+1}}{1-(1/4)} - 2\cdot 2^n \frac{1-(1/6)^{n+1}}{1-(1/6)}$$

$$= 4\cdot 2^n\left[1 - \frac{1}{4}\frac{2^{-n}}{2^n}\right] - \frac{12}{5}\cdot 2^n\left[1 - \frac{1}{6}\frac{2^{-n}}{3^n}\right] = 4\cdot 2^n - \left(\frac{1}{2}\right)^n - \frac{12}{5}\cdot 2^n + \frac{2}{5}\left(\frac{1}{3}\right)^n = \frac{8}{5}(2)^n + \frac{2}{5}\left(\frac{1}{3}\right)^n - \left(\frac{1}{2}\right)^n$$

Adding the two foregoing responses, it turns out that

$$y[n]=\left[\frac{3}{2}y[-1]-\frac{1}{2}y[-2]\right]\left(\frac{1}{2}\right)^n+\left[-\frac{2}{3}y[-1]+\frac{1}{3}y[-2]\right]\left(\frac{1}{3}\right)^n+\frac{8}{5}(2)^n+\frac{2}{5}\left(\frac{1}{3}\right)^n-\left(\frac{1}{2}\right)^n$$

and finally

$$y[n]=\left[\frac{3}{2}y[-1]-\frac{1}{2}y[-2]-1\right]\left(\frac{1}{2}\right)^n u[n]+\left[-\frac{2}{3}y[-1]+\frac{1}{3}y[-2]+\frac{2}{5}\right]\left(\frac{1}{3}\right)^n u[n]$$

which is the same result as the one obtained previously. We arrive, therefore, at the very interesting conclusion that *to calculate the overall response of a discrete LTI system, we must either calculate the responses in the zero state and the zero input and then add them together or solve directly the full difference equation using the input signal and the initial conditions that were set in each case.*

Problem 7.4.16 Determine the overall response $y[n]$ $(n \geq 0)$ of the discrete LTI system described by the second-order difference equation

$$y[n]-3y[n-1]-4y[n-2]=x[n]+2x[n-1]$$

when the input $x[n]=(4)^n u[n]$ is transmitted to the system (Proakis and Manolakis, Exercise 2.27).

Solution Following the methodology for solving full difference equations presented previously, we will first solve the homogeneous equation that has the form $y[n]-3y[n-1]-4y[n-2]=0$. Considering a solution of the form $y[n]=\lambda^n$ and substituting it into the foregoing equation, we end up with the characteristic equation $\lambda^2-3\lambda-4=0$ with roots $\lambda_1=4$ and $\lambda_2=-1$, that is, the general solution of the homogeneous difference equation has the form $y[n]=C_1(4)^n+C_1(-1)^n$, with constants C_1 and C_2 to be calculated from the initial conditions, which are assumed to be known.

On the other hand, to determine a partial solution of the full difference equation, we find that the function $x[n]$ has the form $x[n]=\alpha^n x[n]$, with the value $\alpha=4$ being a root of the characteristic polynomial of multiplicity $q=1$. Therefore, according to theory, the required partial solution will have the form $y_p[n]=kn(4)^n u[n]$, with the value of the parameter k being determined by the requirement to satisfy the aforementioned partial solution of the non-homogenous difference equation. Substituting, therefore, the partial solution into of the full difference equation, we will get

$$kn(4)^n u[n]-3k(n-1)(4)^{n-1}u[n-1]-4k[n-2](4)^{n-2}u[n-2]=4^n u[n]+2*4^{n-1}u[n-1]$$

Setting $n=2$ and solving for the constant k, it turns out that $k=6/5$. Therefore, the general solution of the full equation will have the form

$$y[n]=\left[\frac{6}{5}n(4)^n+C_1(4)^n+C_2(-1)^n\right]u[n]$$

with the value of the constants C_1 and C_2 being calculated from the initial conditions. The calculation of these constants as a function of the initial conditions $y[-1]$ and $y[-2]$ and the final determination of the function $y[n]$ is left as an exercise to the reader.

Problem 7.4.17 Solve the full second-order difference equation

$$y[n]-\frac{3}{4}y[n-1]+\frac{1}{8}y[n-2]=\frac{5}{8}\sin\left(\frac{\pi n}{2}\right)u[n]$$

with initial conditions $y[-1]=y[-2]=1$ (adapted from Hayes, Exercise for resolution 1.38).

Solution Applying the previous methodology, we find that the homogeneous equation

$$y[n] - \frac{3}{4}y[n-1] + \frac{1}{8}y[n-2] = 0$$

has a characteristic equation of the form $\lambda^2 - (3/4)\lambda + (1/8) = 0$, with roots $\lambda_1 = (1/2)$ and $\lambda_2 = (1/4)$, and therefore its general solution will have the form

$$y_{zi}[n] = C_1\left(\frac{1}{2}\right)^n + C_2\left(\frac{1}{3}\right)^n$$

with the constants C_1 and C_2 resulting from the initial conditions. On the other hand, observing the form of the input signal $x[n]$ and referring to Table 7.1 (page 593), it is not difficult to find that the particular solution of the full differential equation should have the form

$$y_{zs}[n] = C_3\cos\left(\frac{\pi n}{2}\right) + C_4\sin\left(\frac{\pi n}{2}\right)$$

Using the known trigonometric identities

$$\cos(\alpha-\beta) = \cos\alpha\cos\beta + \sin\alpha\sin\beta \quad \text{and} \quad \sin(\alpha-\beta) = \sin\alpha\cos\beta - \cos\alpha\sin\beta$$

it turns out that

$$y_{zs}[n-1] = C_3\cos\left[\frac{\pi(n-1)}{2}\right] + C_4\sin\left[\frac{\pi(n-1)}{2}\right] = +C_3\sin\left(\frac{\pi n}{2}\right) - C_4\cos\left(\frac{\pi n}{2}\right)$$
$$y_{zs}[n-2] = C_3\cos\left[\frac{\pi(n-2)}{2}\right] + C_4\sin\left[\frac{\pi(n-2)}{2}\right] = -C_3\cos\left(\frac{\pi n}{2}\right) - C_4\sin\left(\frac{\pi n}{2}\right)$$

Replacing the expressions of the functions $y_{zs}[n]$, $y_{zs}[n-1]$ and $y_{zs}[n-2]$ in the full equation, we get

$$\left\{C_3\cos\left(\frac{\pi n}{2} + C_4\sin\left(\frac{\pi n}{2}\right)\right)\right\} - \frac{3}{4}\left\{C_3\sin\left(\frac{\pi n}{2} - C_4\cos\left(\frac{\pi n}{2}\right)\right)\right\}$$
$$+\frac{1}{8}\left\{-C_3\cos\left(\frac{\pi n}{2} - C_4\sin\left(\frac{\pi n}{2}\right)\right)\right\} = \frac{5}{8}\sin\left(\frac{\pi n}{2}\right)$$

and equating the corresponding quantities results in the system of two equations

$$C_3 - \frac{3C_4}{4} - \frac{C_3}{8} = 0 \quad \text{and} \quad C_4 - \frac{3C_3}{4} - \frac{C_4}{8} = 0$$

with unknowns the constants C_3 and C_4, whose values are $C_3 = -6/17$ and $C_4 = 7/17$, as can be proven. Therefore, it will hold that

$$y_{zs}[n] = -\frac{6}{17}\cos\left(\frac{\pi n}{2}\right) + \frac{7}{17}\sin\left(\frac{\pi n}{2}\right)$$

while the general solution of the equation will be

$$y[n] = y_{zi}[n] + y_{zs}[n] = C_1\left(\frac{1}{2}\right)^n + C_2\left(\frac{1}{3}\right)^n - \frac{6}{17}\cos\left(\frac{\pi n}{2}\right) + \frac{7}{17}\sin\left(\frac{\pi n}{2}\right)$$

with the values of the constants C_1 and C_2 resulting from the initial conditions of the problem. Following the now known process, we determine the values $C_1 = 5/4$ and $C_2 = -37/136$, so the general solution of the difference equation will take the form

$$y[n] = \frac{5}{4}\left(\frac{1}{2}\right)^n - \frac{37}{136}\left(\frac{1}{3}\right)^n - \frac{6}{17}\cos\left(\frac{\pi n}{2}\right) + \frac{7}{17}\sin\left(\frac{\pi n}{2}\right)$$

which is also the desired result. ■

Note Observing the foregoing output, we can draw the conclusion that the function $y[n]$ can be treated as the sum of two terms, of which the term

$$\frac{5}{4}\left(\frac{1}{2}\right)^n - \frac{37}{136}\left(\frac{1}{3}\right)^n$$

weakens exponentially over time and tends to zero in the limit $n \to \infty$, while the term

$$-\frac{6}{17}\cos\left(\frac{\pi n}{2}\right) + \frac{7}{17}\sin\left(\frac{\pi n}{2}\right)$$

lasts for infinite time, since it consists of a sum of periodic trigonometric functions. In the terminology of systems of continuous and discrete-time, the first term is known as the *transient response* (and obviously refers to our known *natural response*), while the second term is known as the *steady-state response*, since it remains for infinite time. If the transient response tends to zero over time, as is the case with us, the system is characterized as *stable*, while in the opposite case we are dealing with an *unstable* discrete-time system. As one can observe and as we will examine in detail in a subsequent section, these types of systems are stable *only when all the roots of the characteristic polynomial are in absolute value smaller than one*. Conversely, if even one of these roots has a value greater than one, the corresponding term will tend to infinity in the limit $n \to \infty$, and the system is characterized as *unstable*. This behavior of the system regarding the transient response and the steady-state response always appears when as the input of the system we transmit *periodic sinusoidal or cosinusoidal signals*. More about the stability of recursive discrete-time LTI systems will be presented in a subsequent section.

7.5 Causal and Stable Discrete LTI Systems

As already mentioned in a previous section, the output of a causal system at time n depends only on the current input, as well as on inputs that have been transmitted to the system at previous times. If additionally the system is LTI, then its output at time n will be given as known from the relation

$$y[n] = \sum_{k=-\infty}^{\infty} h[k]x[n-k] = \sum_{k=-\infty}^{-1} h[k]x[n-k] + \sum_{k=0}^{\infty} h[k]x[n-k] \tag{7.32}$$

However, because our system is causal, its output should be given *only by the second sum that contains the current and previous inputs*. Conversely, the first sum should not exist, as the values $x[n-k]$ corresponding to negative values of the index k refer to future inputs of the system whose existence makes the system noncausal. It is obvious that to remove the first sum from the foregoing equation, we must set $h[k] = 0$ for $k < 0$, a condition that constitutes the basic prerequisite for a system to be causal. In this case, the system's output at each time n will take the form

$$y[n] = \sum_{k=0}^{\infty} h[k]x[n-k] = \sum_{k=-\infty}^{n} x[k]h[n-k] \tag{7.33}$$

If a causal signal, i.e., a discrete sequence $x[n]$ with nonzero values for $n > 0$ and zero values for $n < 0$, is used as input to a causal system, the limits in the foregoing sums will be further restricted, and the preceding relationship will take the form

$$y[n] = \sum_{k=0}^{n} h[k]x[n-k] = \sum_{k=0}^{n} x[k]h[n-k] \tag{7.34}$$

Having described the causal LTI systems, let us now describe the behavior of a discrete LTI system that is characterized as stable. As mentioned previously, in such a system the transmission of a bounded input will lead to the appearance of a bounded output, a property that in mathematical notation is described by the expression $|x[n]| \leq M_x < \infty \Rightarrow |y[n]| \leq M_y < \infty$. If the stable system is a linear system invariant with respect to time, then its output will be given, as is well known, by the relation

$$y[n] = \sum_{k=-\infty}^{\infty} h[k]x[n-k] \tag{7.35}$$

and therefore its measure will satisfy the property

$$|y[n]|| = \left| \sum_{k=-\infty}^{\infty} h[k]x[n-k] \right| \leq \sum_{k=-\infty}^{\infty} |h(k)||x[n-k]| \leq M_x \sum_{k=-\infty}^{\infty} |h(k)|$$

(since its input is bounded, and therefore it will hold that $|x[n]| \leq M_x \ \forall n \in N$). Therefore, a discrete-time LTI system is stable *when the sum of the absolute values of the elements of its impulse response is finite*. It is not difficult to prove that this condition, which in mathematical notation is formulated as

$$S_h \equiv \sum_{k=-\infty}^{\infty} |h(k)| < \infty \tag{7.36}$$

is not only sufficient but also necessary, since if $S_h = \infty$, then there is *at least one bounded input for which the system's output is not bounded*. As such an input let us consider the signal

$$x[n] = \begin{cases} \dfrac{h^*[n]}{|h(n)|} & h[n] \neq 0 \\ 0 & h[n] = 0 \end{cases}$$

where $h^*[n]$ is the complex conjugate sequence of the impulse response $h[n]$. It is not difficult to ascertain that at the time instances $n = 0$, the system's output is not blocked, as we will have

$$y[0] = \sum_{k=-\infty}^{\infty} x[-k]\, h[k] = \sum_{k=-\infty}^{\infty} \frac{|h(k)|^2}{|h(k)|} = \sum_{k=-\infty}^{\infty} |h(k)| = S_h \rightarrow \infty$$

and therefore our system *is not stable*.

Problem 7.5.1 Determine the range of values of the parameter α that ensures stability of a discrete-time LTI system with an impulse response of the form $h[n] = \alpha^n u[n]$ (Hayes, Example 1.35).

Solution To prove that the system is stable, we must show that

$$S_h = \sum_{k=-\infty}^{\infty} |h[k]| < \infty$$

However, since our system is causal (since the current output depends only on the current input), the lower limit in the foregoing sum will have a value of $k_{\min} = 0$, and therefore the function $u[n]$ will have a value equal to one. Therefore, the stability condition will take the form

$$S_h = \sum_{k=0}^{\infty} |\alpha^k u[n]| = \sum_{k=0}^{\infty} |\alpha^k| = \sum_{k=0}^{\infty} |\alpha|^k = \frac{1}{1-|\alpha|}$$

However, we know that the foregoing series converges only when the relation $|\alpha| < 1$ holds. Therefore, our system is *stable* when the condition $|\alpha| < 1$ applies and *unstable* in the opposite case.

Problem 7.5.2 Determine the range of values of the parameters α and β for which a discrete-time LTI system with impulse response

$$h[n] = \begin{cases} \alpha^n & n \geq 0 \\ \beta^n & n < 0. \end{cases}$$

Solution In this case, the system is not causal and the stability condition is written

$$S_h = \sum_{k=-\infty}^{\infty} |h(k)| = \sum_{k=-\infty}^{-1} |\beta|^n + \sum_{k=0}^{\infty} |\alpha|^n$$

To determine the range of values of the variable β for which the first sum converges, we can write this sum in the form

$$\sum_{k=-\infty}^{-1} |\beta|^n = \sum_{k=1}^{\infty} |\beta|^{-n} = \sum_{k=1}^{\infty} \frac{1}{|\beta|^n} = \frac{1}{|\beta|}\left(1 + \frac{1}{|\beta|} + \frac{1}{|\beta|^2} + \frac{1}{|\beta|^3} + \ldots\right) = \gamma(1+\gamma+\gamma^2+\gamma^3+\ldots) = \frac{\gamma}{1-\gamma}.$$

as emerges using the appropriate expression from the mathematical typology. In order for the preceding series to converge, it must hold that $\gamma = (1/|\beta|) < 1$ or, equivalently, $|\beta| > 1$. As for the second sum, it was used in a previous exercise where we saw that the sequence resulting from this converges for the range of values $|\alpha| < 1$. Therefore, for our system to be stable, the conditions $|\alpha| < 1$ and $|\beta| > 1$ must be satisfied simultaneously.

7.6 Discrete Recursive LTI Systems

In the special case where the discrete-time system we consider is *recursive*, that is, it is described by the appropriate in each case *difference equation*, its characterization as a *linear system* requires the validity of the following three conditions (in this description the system is assumed to be time invariant, too).

- The overall response of the system should be equal to the sum of the response in the zero state and the response to the zero input—it will hold, therefore, that $y[n] = y_{zs}[n] + y_{zi}[n]$.
- The response in the zero state should be characterized by the validity of the principle of superposition; in other words, if the input signal $x[n] = c_1x_1[n] + c_2x_2[n]$ is transmitted to the system, its output should have the form $y_{zs}[n] = c_1 y_{zs}^{(1)}[n] + c_2 y_{zs}^{(2)}[n]$.
- The response to the zero input should be characterized by the power of the principle of superposition, that is, if the input signal $x[n] = c_1x_1[n]+c_2x_2[n]$ is transmitted to the system, its output should be in the form $y_{zi}[n] = c_1 y_{zi}^{(1)}[n]+c_2 y_{zi}^{(2)}[n]$.

If even one of the foregoing properties does not hold, the system is characterized as *nonlinear*. This process of characterizing a recursive system is presented in detail in the following exercises.

Problem 7.6.1 Determine whether the following discrete LTI system is linear and stable:

$$y[n] = \alpha y[n-1] + x[n]$$

Solution In a previous exercise we proved that the total output of this system is

$$y[n] = \alpha^{n+1} y[-1] + \sum_{k=0}^{n} \alpha^k x[n-k]$$

and therefore the responses of the system to the zero state and to the zero input will be defined as

$$y_{zs}[n] = \sum_{k=0}^{n} \alpha_k x[n-k] \qquad \text{and} \qquad y_{zi}[n] = \alpha^{k+1} y[-1]$$

as derived from the foregoing equation by setting $y[-1] = 0$ and $x[n] = 0$. We observe, therefore, that the first property is satisfied, that is, the relation $y[n] = y_{zs}[n] + y_{zi}[n]$ holds.

To prove the second property, let us input into the system the combined input $x[n] = c_1 x_1[n] + c_2 x_2[n]$. In this case, the zero-state response will take the form

$$y_{zs}[n] = \sum_{k=0}^{n} \alpha^k x[n-k] = \sum_{k=0}^{n} \alpha^k \Big[c_1 x_1[n-k] + c_2 x_2[n-k] \Big]$$

$$= c_1 \sum_{k=0}^{\infty} \alpha^k x_1[n-k] + c_2 \sum_{k=0}^{n} \alpha^k x_2[n-k] = c_1 y_{zs}^{(1)}[n] + c_2 y_{zs}^{(1)}[n]$$

and therefore the zero-state response obeys the *principle of superposition*. In a completely analogous way, inputting the combined input into the system

$$y[-1] = c_1 y_1[-1] + c_2 y_2[-1]$$

will lead to a zero-input response of the form

$$y_{zi}[n] = \alpha^{n+1} y[-1] = \alpha^{n+1} \Big[c_1 y_1[-1] + c_2 y_2[-1] \Big] = c_1 \alpha^{n+1} y_1[-1] + c_2 \alpha^{n+1} y_2[-1] = c_1 y_{zi}^{(1)}[n] + c_2 y_{zi}^{(2)}[n]$$

and therefore the zero-input response also obeys the principle of superposition. We observe, therefore, that our system obeys all three properties that characterize a linear system, and therefore it is *linear*.

On the other hand, to determine whether our system is stable, let us input into it the bounded input signal $x[n]$, which satisfies the condition $|x[n]| \leq M_x < \infty$ ($\forall n \geq 0$). In this case, the measure of the system's output will be given by the relation

$$|y[n]| \leq |\alpha^{n+1} y(-1)| + \left| \sum_{k=0}^{n} \alpha^k x[n-k] \right| \leq |\alpha|^{n+1} |y[-1]| + M_x \sum_{k=0}^{n} |\alpha|^k \leq |\alpha|^{n+1} |y[-1]| + M_x \frac{1 - |\alpha|^{n+1}}{1 - |\alpha|} = M_y$$

for values $n \geq 0$. We observe, therefore, that if the value of n is finite, the quantity M_y is also finite and the system is stable, as it gives a bounded output. However, if the parameter n takes infinite values, the system is stable only for values $|\alpha| < 1$ (since for this area, $|\alpha|^n \to 0$ as $n \to \infty$)—in this case, the output is bounded by the quantity

$$M_y = M_x \left[1 - \frac{|\alpha|^{n+1}}{1 - |\alpha|} \right]$$

Problem 7.6.2 Consider the discrete-time system described by the equation

$$y[n] = n y[n-1] + x[n] \quad (n \geq 0)$$

which is assumed to be at rest (i.e., the relation $y[-1] = 0$ holds). Check if it is linear, time-invariant, and stable.

Solution If we transmit the inputs $x_1[n]$ and $x_2[n]$ to the system, the corresponding outputs will have the form

$$y_1[n] = ny_1[n-1] + x_1[n] \qquad \text{and} \qquad y_2[n] = ny_2[n-1] + x_2[n]$$

If we transmit the combined input $x[n] = \alpha x_1[n] + \beta x_2[n]$, we will have

$$\begin{aligned}
\alpha y_1[n] + \beta y_2[n] &= \alpha\Big[ny_1[n-1] + x_1[n]\Big] + \beta\Big[ny_2[n-1] + \beta x_2[n]\Big] \\
&= \alpha n y_1[n-1] + \alpha x_1[n] + \beta n y_2[n-1] + \beta x_2[n] \\
&= n\Big[\alpha y_1[n-1] + \beta y_2[n-1]\Big] + \alpha x_1[n] + \beta x_2[n] = ny[n-1] + x[n] = y[n]
\end{aligned}$$

and therefore the system is *linear*.

To determine if our system is *time-invariant*, let us transmit the input $x[n-k]$. Thus,

$$y(n,k) = T\{x[n-k]\} = ny[n-k-1] + x[n-k]$$

but on the other hand, we will have

$$y[n-k] = (n-k)y[n-k-1] + x[n-k]$$

We observe that $y(n,k) \neq y[n-k]$, and therefore our system *is not time-invariant*.

Finally, the stability check for the system consists in transmitting a bounded input that should, if the system is stable, lead to a bounded output. However, we observe that this is not true because if we transmit the input $x[n] = u[n]$, which is bounded since $|x[n]| \leq 1$, the same does not happen with its output. Indeed, by repeatedly applying the difference equation of the system we have

$$\begin{aligned}
y[0] &= 0 * y[-1] + x[0] = 1 \\
y[1] &= 1 * y[0] + x[1] = 1 * 1 + 1 = 2 \\
y[2] &= 2 * y[1] + x[2] = 2 * 2 + 1 = 4 + 1 = 5 \\
y[3] &= 3 * y[2] + x[3] = 3 * 5 + 1 = 15 + 1 = 16 \\
&\dots\dots\dots\dots\dots\dots\dots\dots\dots\dots
\end{aligned}$$

Therefore, the system's output is not bounded, that is, our system *is not stable*.

7.6.1 Stability in Discrete Recursive LTI Systems

As already mentioned, a discrete recursive system is *stable* when its impulse response $h[n]$ is *absolutely summable*, that is, it satisfies the condition

$$S_h = \sum_{n=-\infty}^{\infty} |h[n]| < \infty \tag{7.37}$$

However, taking into account that the impulse response $h[n]$ is defined as

$$h[n] = \sum_{k=0}^{M} \beta_k h_\beta[n-k] \tag{7.38}$$

where $h_\beta[n]$ is the solution of the homogeneous version of the equation

$$\sum_{k=0}^{N} \alpha_k h_\beta[n-k] = \delta[n] \tag{7.39}$$

which is solved under zero initial conditions (see Sect. 7.4.2, page 577 on the calculation of the impulse response), the sum of the equation defining the quantity S_h takes the form

$$S_h = \sum_{n=0}^{\infty} |h[n]| = \sum_{n=0}^{\infty} \left| \sum_{k=0}^{M} \beta_k h_\beta[n-k] \right| \le \sum_{n=0}^{\infty}\sum_{k=0}^{M} |\beta_k||h_\beta[n-k]| = \sum_{k=0}^{M}\sum_{n=k}^{\infty} |\beta_k||h_\beta[n-k]|$$

In the preceding equation, the lower limit in the sum defining the parameter S_h was set to zero because we assumed that the considered recursive system was causal (that is, that $h[n] = 0$ for the range of values $n < 0$), while for the same reason the internal sum in the the last representation has as a lower limit the value $n = k$. However, considering that the function $h_\beta[n]$ is the solution of the homogeneous equation mentioned earlier, it will be known to have the form

$$h_\beta[n] = \sum_{m=1}^{N} C_m \lambda_m^n \tag{7.40}$$

where λ_m $(m = 1, 2, \ldots, N)$ are the *roots of the characteristic polynomial of the homogeneous differential equation.* If in the last expression we make the substitution $n \to n-k$ and use the expression of the function $h_\beta[n-k]$ that will result in the sum of the quantity S_h, we will successively get

$$\sum_{n=0}^{\infty} |h[n]| = \sum_{k=0}^{M}\sum_{n=k}^{\infty} |\beta_k| \left| \sum_{m=1}^{N} C_m \lambda_m^{n-k} \right| \le \sum_{k=0}^{M}\sum_{n=k}^{\infty} |\beta_k| \sum_{m=1}^{N} |C_m||\lambda_m^{n-k}|$$
$$= \sum_{k=0}^{M} |\beta_k| \sum_{m=1}^{N} |C_m| \sum_{n=k}^{\infty} |\lambda_m|^{n-k} = \sum_{k=0}^{M} |\beta_k| \sum_{m=1}^{N} |C_m| \sum_{r=0}^{\infty} |\lambda_m|^r$$

where in the last sum we changed the silent index from n to $r = n - k$, adjusting appropriately the values of the upper and lower limits. But the foregoing sum is finite if and only if $|\lambda_m| < 1$ $(m = 1, 2, \ldots, N)$. We are led, therefore, to the conclusion that *a causal recursive discrete-time system is stable only when all the roots of the system's characteristic equation are absolutely less than one*. This type of system stability that is based on the measure of the roots of the characteristic equation that describes it is known as *internal* or *asymptotic stability*; therefore, a recursive discrete LTI system is characterized as asymptotically stable when all the characteristic roots of the system have a measure less than one. Note that this condition, in addition to being sufficient, is also necessary.

Taking into account that in the general case the characteristic roots of a recursive system are *complex numbers* that, moreover, appear *in pairs* (for the case of real coefficients in the difference equation), the characterization of the system in terms of its stability is based on *determining the position of the roots on the complex plane*. Thus, the preceding formulation in terms of asymptotic stability is equivalent to the assertion that a *system is asymptotically stable when all the roots of the characteristic polynomials are within the unit circle* (this conclusion applies regardless of whether the roots are distinct from each other or repeated). On the other hand, if even one of these roots is outside the unit circle, the system is *unstable* since the characteristic term corresponding to this root tends to infinity as $n \to \infty$. If there are no roots of the characteristic polynomial that are outside the unit circle, but there are some *distinct* roots among them that are located *on the circumference of this circle*, the system is characterized as *marginally stable*. Conversely, if there are roots on the circumference of this circle that *coincide with each other*, then the system is proven to be *unstable*. These positions of the characteristic roots on the complex plane and the temporal variation of the characteristic terms corresponding to these are presented diagrammatically

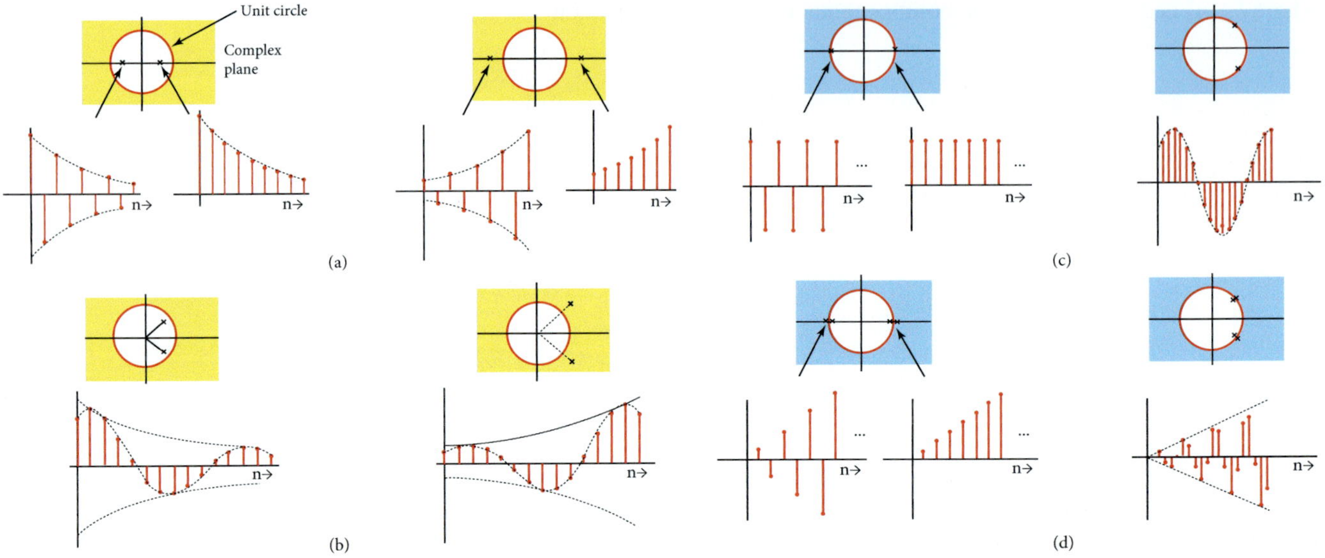

Fig. 7.4 The position of the characteristic roots on the complex plane and the temporal variation of the corresponding characteristic terms

in Fig. 7.4. As for the proof of these assertions, it is done in a similar way to what we presented for continuous-time systems and requires the use of the $\mathcal{Z}$ *transform*, which will be presented in a subsequent chapter.

In full analogy with continuous-time systems, *if a discrete recursive system is asymptotically stable, then it will automatically be BIBO stable*. Indeed, considering a root λ_k $(k = 1, 2, \ldots, N)$ located within the unit circle, that is, it has a modulus less than one, the corresponding characteristic function $\lambda_k^n u[n]$ will be absolutely summable since, as is well known, we will have

$$\sum_{n=-\infty}^{\infty} |\lambda_k|^n u[n] = \sum_{n=0}^{\infty} |\lambda_k|^n = \frac{1}{1-|\lambda_k|}, \quad (|\lambda_k| < 1) \tag{7.41}$$

If $|\lambda_k| > 1$, the foregoing sum does not converge and the system *is not BIBO stable*. Also, *a marginally stable or asymptotically unstable system is BIBO unstable*, while the reverse of this proposition is not necessarily true. As in continuous-time systems, a discrete-time system that is BIBO stable is not necessarily stable in terms of its internal stability.

Problem 7.6.3 Consider a discrete-time system described by the difference equation

$$y[n] + 7y[n-1] + 31y[n-2] + 25y[n-3] = 3x[n-1]$$

Characterize it in terms of BIBO stability and its asymptotic stability.

Solution The characteristic equation of the homogeneous difference equation

$$y[n] + 7y[n-1] + 31y[n-2] + 25y[n-3] = 0$$

is

$$\lambda^3 + 7\lambda^2 + 31\lambda + 25 = (\lambda+1)(\lambda^2 + 6\lambda + 25) = (\lambda+1)(\lambda+3-4j)(\lambda+3+4j) = 0$$

with roots $\lambda_1 = -1$, $\lambda_2 = -3 + 4j$, and $\lambda_3 = -3 - 4j$. The magnitudes of these roots are calculated as $|\lambda_1| = 1$ and $|\lambda_2| = |\lambda_3| = 5$, and since there are roots that are outside the unit circle, the system is *asymptotically unstable* and, therefore, also *unstable according to BIBO*.

Problem 7.6.4 Consider a discrete-time system described by the difference equation

$$y[n] - 1.5y[n-1] + 0.5y[n-3] = x[n] + 2x[n-1] + 3x[n-2]$$

Characterize it in terms of BIBO stability and its asymptotic stability.

Solution The characteristic equation of the homogeneous difference equation

$$y[n] - 1.5y[n-1] + 0.5y[n-3] = 0$$

is

$$\lambda^3 - 1.5\lambda^2 + 0.5 = (\lambda^2 - 2\lambda + 1)(\lambda + 0.5) = (\lambda - 1)^2(\lambda + 0.5) = 0$$

with roots $\lambda_1 = \lambda_2 = 1$ (double root) and $\lambda_3 = -0.5$. The measures of these roots are calculated as $|\lambda_1| = |\lambda_2| = 1$ and $|\lambda_3| = 0.5$, and since the equation has a multiple root that lies on the circumference of the unit circle, the system is *asymptotically unstable* and, therefore, also *unstable according to BIBO*.

Problem 7.6.5 Consider a discrete-time system described by the difference equation

$$y[n] - y[n-1] + 0.21y[n-2] = 2x[n-1] + 3x[n-2]$$

Characterize it in terms of BIBO stability and its asymptotic stability.

Solution The characteristic equation of the homogeneous difference equation

$$y[n] - y[n-1] + 0.21y[n-2] = 0$$

is

$$\lambda^2 - \lambda + 0.21 = (\lambda - 0.3)(\lambda - 0.7) = 0$$

with roots $\lambda_1 = 0.3$ and $\lambda_2 = 0.7$. The measures of these roots are calculated as $|\lambda_1| = 0.3$ and $|\lambda_2| = 0.7$, and since both are less than one, the system is asymptotically *stable* and, therefore, also *BIBO stable*.

7.7 Interconnection of Discrete-Time Systems

In cases where we use more than one discrete-time system to implement the desired processes in each case, there is the possibility of connecting them together, so as to create larger and more complex systems. In full analogy with their continuous equivalents, there are three basic ways of interconnecting discrete-time systems: (a) *series connection*, in which the systems are connected in such a way that the output of one system becomes the input to the other system; (b) *parallel connection*, in which the two systems receive the same input and operate in parallel without interacting, producing their own separate outputs, which are ultimately added together; and (c) *connection with feedback*, in which the output of the first system feeds its input after first passing through the second system. In this section, we will deal with the first two different ways of connecting systems, which are presented diagrammatically in Fig. 7.5. Following the argument we formulated in Sect. 2.6 (page 112) for continuous-time systems that can be applied here without any modification, it is not difficult to realize that the output of a system resulting from the series and parallel connections of elementary systems $\mathcal{T}_1$ and $\mathcal{T}_2$ will be given respectively by the relations

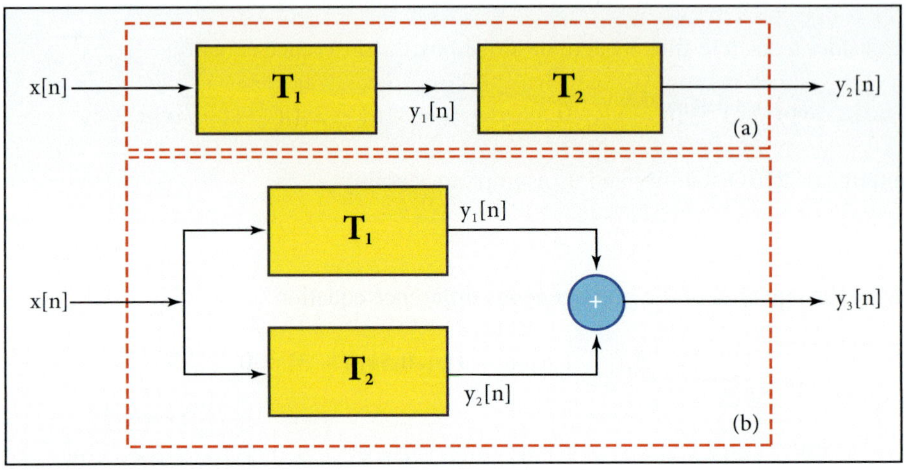

Fig. 7.5 Interconnection of discrete-time systems (**a**) in series and (**b**) in parallel

$$\underbrace{y[n] = \mathcal{T}_2\Big\{\mathcal{T}_1\{x[n]\}\Big\}}_{\text{series connection}} \quad \text{and} \quad \underbrace{y[n] = \mathcal{T}_1\{x[n]\} + \mathcal{T}_2\{x[n]\}}_{\text{parallel connection}}$$

In the special case of discrete LTI systems, whose output results from the convolution between their input and impulse response, the power of the commutative, associative, and distributive properties of this operation between signals allows for the formulation of the same conclusions reached for continuous-time systems. The analytical argument we formulated then and that applies without modifications also in the field of discrete-time can be found in Sect. 2.5.1 (page 88), while the conclusions we reached in this case are formulated as follows:

1. *A discrete LTI system with impulse response* $h[n]$ *that receives the input signal* $x[n]$ *is fully equivalent to a system with impulse response* $x[n]$ *that receives the input signal* $h[n]$.
2. *Two discrete LTI systems with impulse responses* $h_1[n]$ *and* $h_2[n]$ *that are connected in series are equivalent to a simple system, whose impulse response* $h[n]$ *results from the convolution of the impulse responses of the two systems, or, in mathematical notation,* $h[n] = h_1[n] * h_2[n]$.
3. *Two discrete LTI systems with impulse responses* $h_1[n]$ *and* $h_2[n]$ *that are connected in parallel are equivalent to a simple system whose impulse response results from the sum of the impulse responses of the two systems, or, in mathematical notation,* $h[n] = h_1[n] + h_2[n]$.

The diagrammatic representation of the foregoing conclusions can arise from Fig. 2.7 (page 89) if we replace all continuous-time signals with their discrete equivalents.

7.8 Inverse Discrete-Time Systems LTI

In full analogy with continuous-time systems, two discrete LTI systems are called *inverse to each other* when their series connection creates an *identity system*, which copies the input to its output, thereby implementing a transformation of the form $y[n] = x[n]$.

If the impulse responses of the two systems are $h_1[n]$ and $h_2[n]$, then the impulse response of the composite system will arise, as is well known, from the relation

$$h[n] = h_1[n] * h_2[n] \tag{7.42}$$

The output $y[n]$ of the composite system will therefore have the form

$$y[n] = h[n] * x[n] = \Big(h_1[n] * h_2[n]\Big) * x[n] = x[n] = \delta[n] * x[n] \tag{7.43}$$

and therefore

$$h_1[n] * h_2[n] = \delta[n] \tag{7.44}$$

since, as is well known, the impulse function $\delta[n]$ constitutes a neutral element of the convolution operation in the signal space. This relation, which connects the impulse responses $h_1[n]$ and $h_2[n]$ of the direct and the inverse system, is the discrete equivalent of the relation

$$h_1(t) * h_2(t) = \delta(t) \tag{7.45}$$

which applies to the case of continuous-time systems.

Problem 7.8.1 Prove that discrete-time systems LTI described by the relations

$$y_1[n] = \sum_{k=-\infty}^{n} x[k] \qquad \text{and} \qquad y_2[n] = x[n] - x[n-1]$$

constitute a pair of direct/inverse systems.

Solution To solve the problem, we need to determine the impulse responses of the two systems and to prove that their convolution equals the impulse function $\delta[n]$. Taking into account that the impulse response $h[n]$ is defined as the output $y[n]$ of a system, when the discrete-time signal $x[n] = \delta[n]$ is transmitted to its input, from the defining equations of the two systems, it turns out that

$$h_1[n] = \sum_{k=-\infty}^{n} \delta[k] = u[n] \qquad \text{and} \qquad h_2[n] = \delta[n] - \delta[n-1]$$

If we now calculate the convolution between the discrete-time signals $h_1[n]$ and $h_2[n]$ and take into account the properties of the impulse and step function, as well as their delayed versions, it immediately follows that

$$h_1[n] * h_2[n] = u[n] * \Big(\delta[n] - \delta[n-1]\Big) = u[n] * \delta[n] - u[n] * \delta[n-1] = u[n] - u[n-1] = \delta[n]$$

and therefore the discrete LTI systems described by the relations

$$y_1[n] = \sum_{k=-\infty}^{n} x[k] \qquad \text{and} \qquad y_2[n] = x[n] - x[n-1]$$

constitute a *direct/inverse system pair*. For your information, the first system, known as the *accumulator*, is the discrete analog of the elementary system of the *integrator* that we know from the field of continuous-time, while the second system is the discrete analog of the continuous-time system of the *differentiator*.

7.9 Eigenvalues and Eigenfunctions: The Transfer Function

Let us consider a discrete-time LTI system with impulse response $h[n]$. If the elementary discrete-time signal $x[n] = z^n$, where z is a complex quantity, is transmitted to this system, then the system output $y[n]$ will be given, as is well known, from the convolution between the impulse response $h[n]$ and the input $x[n]$, or in, mathematical notation,

$$y[n] = x[n] * h[n] = \sum_{k=-\infty}^{\infty} h[k]x[n-k] = \sum_{k=-\infty}^{\infty} h[k]z^{n-k} = z^n \sum_{k=-\infty}^{\infty} h[k]z^{-k} = \mathcal{H}(z)z^n$$

with the function $\mathcal{H}(z)$ defined as

$$\mathcal{H}(z) = \sum_{k=-\infty}^{\infty} h[k]z^{-k} \tag{7.46}$$

(it is obvious that if the considered system is causal—that is, it has an impulse response with values $h[n] = 0$ for $n < 0$—then the lower limit in the foregoing sum should be set to zero). Let us note that the preceding relation is valid *only for those values of z for which the sum of the infinite terms that appear in the defining equation of the function $\mathcal{H}(z)$ converges*. The set of points z in the complex plane for which the preceding sum converges is known as the *region of convergence*, while the preceding relationship defines a new type of transform that is known as the *$\mathcal{Z}$ transform*; it will be studied in detail in the next chapter.

From the foregoing analysis we observe that the *exponential complex signal $z[n] = z^n$ is the eigenfunction of any LTI system, with the eigenvalue being the value of the function $\mathcal{H}(z)$, calculated at the position of each value of the complex variable z*. This situation is completely analogous to the one encountered in continuous LTI systems, whose eigenfunctions had the form $x(t) = e^{st}$ with the eigenvalue being the *transfer function* of the system $\mathcal{H}(s)$. This analogy allows us to define the *transfer function of a discrete LTI system* in a similar way and, more specifically, as

$$\mathcal{H}(z) = \left.\frac{y[n]}{x[n]}\right|_{x[n]=z^n} = \frac{\mathcal{T}\{z^n\}}{z^n} \tag{7.47}$$

in full analogy with the definitions given for continuous-time systems. Recall that from these systems this function is defined *only for LTI systems and not for nonlinear or time-varying systems*. The detailed description of the $\mathcal{Z}$ transform, as well as the way it is used for the study of discrete LTI systems, is presented in the next chapter.

7.10 Implementation of Discrete-Time Systems

After the theoretical presentation of discrete-time LTI systems, let us now move on to the presentation of the basic structures used for their implementation and especially those described by an explicit transfer function. In full analogy with continuous-time systems, this implementation is based on the combined use of a *finite number of elementary subsystems* that perform the fundamental operations between discrete-time signals (for an analytical description of these operations, see Sect. 6.10 on page 540). The most important of these subsystems are presented in Fig. 7.6 (also see Fig. 7.1 on page 565, which depicts all the available subsystems and the related discussion) and constitute the *discrete equivalents* of the elementary continuous-time subsystems depicted in Fig. 2.10 on page 113.

In this section we will present *direct* or *immediate* implementations in which the schematic diagram that describes the structure of a system arises *directly from the system's function or from the equivalent difference equation*, a process that justifies their name. As will be noted in the chapter on *digital filters*, these implementations usually *are not* used, since,

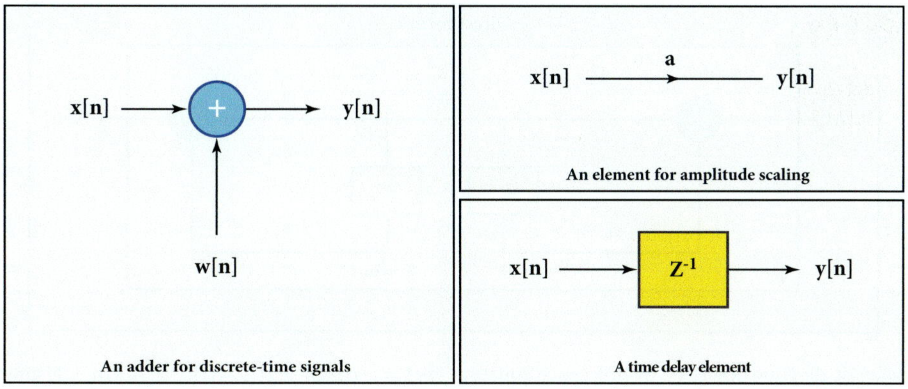

Fig. 7.6 The fundamental subsystems of discrete-time signal processing

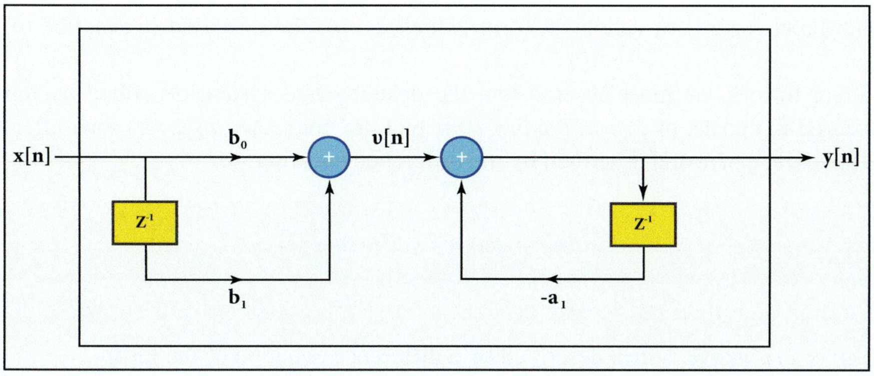

Fig. 7.7 Schematic diagram of recursive discrete-time system described by difference equation $y[n] = -\alpha_1 y[n-1] + \beta_0 x[n] + \beta_1 x[n-1]$, known as direct form I or noncanonical form

although they are characterized by the *exact same* requirements regarding storage requirements (N unit time delay elements) and the number of numerical operations ($2N + 1$ multiplications and $2N$ additions per sampling period) with the serial form (which arises by expressing the transfer function as the product of simpler functions—see Sect. 12.15.1 on page 1358) and the parallel form (which arises by expressing the transfer function as the sum of simpler functions—see Sect. 12.15.2 on page 1361), they are characterized by *very stringent restrictions and requirements*, regarding the accuracy of the filter coefficient values, so their use generally *is not recommended*. To understand how to construct these schematic diagrams, let us start from a simple case and consider a *discrete recursive LTI system* described by a difference equation of the form

$$y[n] = -\alpha_1 y[n-1] + \beta_0 x[n] + \beta_1 x[n-1]$$

which corresponds to a transfer function

$$\mathcal{H}(z) = \frac{\beta_0 + \beta_1 z^{-1}}{1 + \alpha_1 z^{-1}}$$

with the schematic diagram that describes it being the one depicted in Fig. 7.7. It is not difficult to understand how this diagram was constructed after careful study of the foregoing difference equation, and in reality, it is simply a direct implementation of it. For example, given that the system uses the values $x[n-1]$ and $y[n-1]$ of the input and output signals, *two unit time delay elements* are required, while for the implementation of the sums that appear in the difference

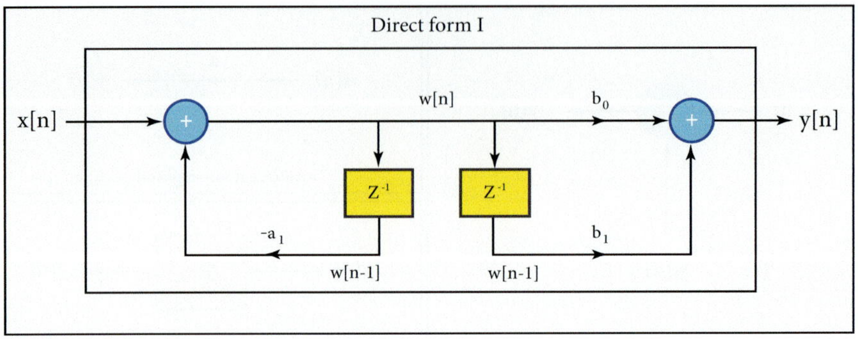

Fig. 7.8 Equivalent schematic diagram described by difference equation $y[n] = -\alpha_1 y[n-1] + \beta_0 x[n] + \beta_1 x[n-1]$ and resulting from previous one by alternating connection order of the two subsystems

equation of the system *two adders* are required. In reality, these direct forms appear in two variations, direct form I or noncanonical form and direct form II or canonical form, which is also the one used in practice for reasons that we will explain shortly.

To construct the direct form I, we must observe that the discrete-time system described by the preceding difference equation can be considered to consist of *two LTI subsystems* that are connected *in series*, one after the other. The first of these two subsystems is *nonrecursive* and described by the difference equation

$$v[n] = \beta_0 x[n] + \beta_1 x[n-1]$$

while the second system is a *recursive system* described by a difference equation of the form

$$y[n] = -\alpha_1 y[n-1] + v[n]$$

Therefore, if we connect these two subsystems in series, we construct the schematic diagram of Fig. 7.7, which represents the direct form I or noncanonical form of a system that is described by the foregoing transfer function $\mathcal{H}(z)$.

On the other hand, to construct the direct form II or canonical form, we must resort to a fundamental property of discrete-time LTI systems, according to which *if two such systems are connected in series to form a more complex system, the order in which these two systems are connected does not matter at all*. In other words, *if we change the order of connection of these two systems, the response of the composite system does not change*. Therefore, by making this exchange and constructing the schematic diagram of the composite system in the way we did previously, we obtain the structure depicted in Fig. 7.8.

Now relying on this new schematic diagram and determining the difference equations that describe the operation of the two subsystems that constitute the new discrete-time system, it is not difficult to see that they take the form

$$w[n] = -\alpha_1 w[n-1] + x[n] \qquad \text{and} \qquad y[n] = \beta_0 w[n] + \beta_1 w[n-1]$$

and that, considered as a system of equations, they are equivalent to the initial difference equation of the system

$$y[n] = -\alpha_1 y[n-1] + \beta_0 x[n] + \beta_1 x[n-1]$$

Furthermore, from the preceding diagram we find that the *two elements of time delay receive the same input and produce the same output*. Therefore, just as we did in continuous-time systems, we can proceed to the *merging* of these two elements and keep only one of them. The final schematic diagram that results in this way, as presented in Fig. 7.9, constitutes the so-called direct form II or *canonical form* and is one of the basic tools used in the design and implementation of discrete-time systems.

The preceding structures, which for reasons of understanding the subject are related to a very simple first-order difference equation, can be generalized to describe the general case of a recursive LTI system described by the difference equation

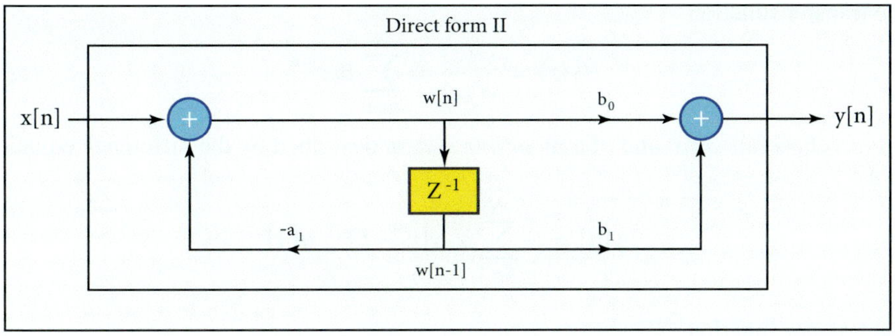

Fig. 7.9 Implementation of discrete-time system described by difference equation of the form $y[n] = -\alpha_1 y[n-1] + \beta_0 x[n] + \beta_1 x[n-1]$, which is based on the so-called direct form II or *canonical form*

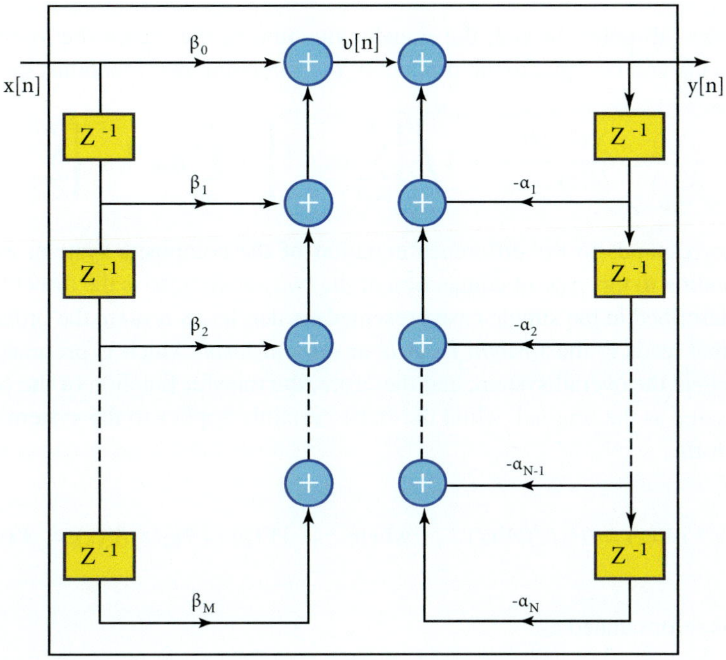

Fig. 7.10 *Noncanonical form* (direct form I realization) of a discrete recursive system LTI

$$y[n] = -\sum_{k=1}^{N} \alpha_k y[n-k] + \sum_{k=0}^{N} \beta_k x[n-k] \tag{7.48}$$

The schematic diagram corresponding to this general case for the representation of the system in noncanonical form is presented in Fig. 7.10.

It is not difficult to see that the calculation of the output of this generalized system requires the use of $M + N$ elements of time delay and the performance of $M + N + 1$ multiplications. In full analogy with the previous simple case, this system can be considered the union of a *nonrecursive system* that is described by a difference equation of the form

$$u[n] = \sum_{k=0}^{M} \beta_k x[n-k]$$

and corresponds to the transfer function

$$\mathcal{H}_1(z) = \frac{\mathcal{U}(z)}{\mathcal{X}(z)} = \sum_{k=0}^{N} \beta_k z^{-k}$$

which contains *only zero values* (all-zero) and of a *recursive system* described by the difference equation

$$y[n] = -\sum_{k=1}^{N} \alpha_k y[n-k] + u[n]$$

and corresponds to the transfer function

$$\mathcal{H}_2(z) = \frac{\mathcal{Y}(z)}{\mathcal{U}(z)} = \frac{1}{\sum_{k=1}^{N} \alpha_k z^{-k}}$$

and which contains *only poles* (all-pole). Indeed, the transfer function of the composite system resulting from the series connection of these two systems will be equal to the product of the foregoing two functions, that is,

$$\mathcal{H}(z) = \mathcal{H}_2(z)\mathcal{H}_1(z) = \frac{\mathcal{Y}(z)}{\mathcal{X}(z)} = \frac{\mathcal{Y}(z)\,\mathcal{U}(z)}{\mathcal{U}(z)\,\mathcal{X}(z)} = 1\Big/\left\{\sum_{k=1}^{N} \alpha_k z^{-k}\right\} \times \sum_{k=0}^{N} \beta_k z^{-k} = \left\{\sum_{k=0}^{N} \beta_k z^{-k}\right\}\Big/\left\{\sum_{k=1}^{N} \alpha_k z^{-k}\right\}$$

But this transfer function corresponds to the difference equation of the composite system, as can easily be proven. The schematic diagram corresponding to this type of connection of the two subsystems is the direct form I or noncanonical form.

Working in the way we described in the simple case presented earlier, let us *reverse* the order of interconnection of these two subsystems, a process that leads to the *straight form II* or *normal form*, which is presented in Fig. 7.11. This reversal process, as noted, *does not affect* the overall system, and therefore, the transfer function of the composite system that is now $\mathcal{H}(z) = \mathcal{H}_1(z)\mathcal{H}_2(z)$ is the same as the original, while the same certainly applies to the system's output, the $\mathcal{Z}$ transform of which can be written in the form

$$\mathcal{Y}(z) = \mathcal{H}_1(z)\mathcal{H}_2(z)\mathcal{X}(z) = \mathcal{H}_1(z)\mathcal{W}(z) \qquad \text{where} \qquad \mathcal{W}(z) = \mathcal{H}_2(z)\mathcal{X}(z) = \mathcal{X}(z)/\left\{\sum_{k=1}^{N} \alpha_k z^{-k}\right\}$$

The preceding relationship is reformulated as

$$\mathcal{W}(z) + \sum_{k=1}^{N} \alpha_k z^{-k}\mathcal{W}(z) = \mathcal{X}(z)$$

and taking the *inverse $\mathcal{Z}$ transform* we are led to the difference equation

$$w[n] = -\sum_{k=1}^{N} \alpha_k w[n-k] + x[n]$$

which describes a *recursive* system, while the $\mathcal{Z}$ transform of the output of the composite system is given by the equation

$$\mathcal{Y}(z) = \sum_{k=0}^{M} \beta_k z^{-k}\mathcal{W}(z)$$

whose inversion gives us

$$y[n] = \sum_{k=0}^{M} \beta_k w[n-k]$$

an expression that describes a nonrecursive system. The result of this reversal process is a composite system that consists of the *sequential connection of a nonrecursive and a recursive system*, described in the foregoing manner. The schematic

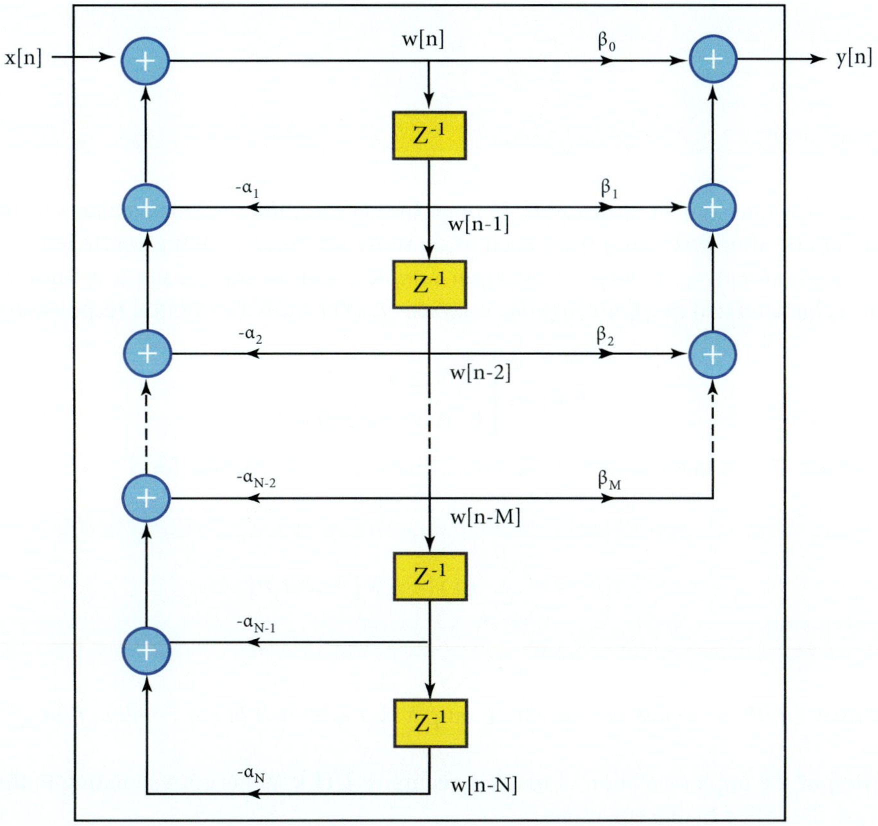

Fig. 7.11 Normal form (direct form II realization) of a discrete recursive LTI system

diagram of this system after merging the delay elements according to our previous discussion is known as direct form II or normal form and is depicted in Fig. 7.11.

It is interesting to note that in this second equivalent and simpler type of representation, the *number of delay elements required for the implementation of the system is the minimum possible* and, more specifically, is equal to max M, N, as can be inferred from the study of the diagram. However, the number of additions and multiplications is *the same* in these two implementations, since both structures require the execution of $M + N$ additions and $M + N + 1$ multiplications. It should also be noted that, apart from the number of delay elements, another significant difference between these two equivalent forms of representation is that in the nonnormal form, the zero values of the function $\mathcal{H}(z)$ represented by the function $\mathcal{H}_1(z)$ *are implemented first*, followed by the poles of the function $\mathcal{H}(z)$, which are represented by the transfer function $\mathcal{H}_2(z)$. Conversely, in its normal form, the order is reversed after the *poles are first implemented and then the zero values*. Although, as mentioned, these two forms are completely equivalent and, theoretically speaking, the order in which the poles and zero values are implemented does not matter, in practice when the implementation of a difference equation takes place using finite numerical accuracy, there may be a *significant differentiation* between these two systems, which are only theoretically equivalent, that is, only if we assume that numerical operations are performed with *infinite accuracy*. In reality, these structures, because they are proven to be extremely sensitive to the quantization of the filter coefficient values (see Sect. 12.9.1, page 1162), their use is not recommended in practical applications, which rely on the use of other structures for the implementation of filters.

From the preceding structures, which are also the most general that can appear, we can construct simpler and more specialized system descriptions. For example, if in the general equation

$$y[n] = -\sum_{k=1}^{N} \alpha_k y[n-k] + \sum_{k=0}^{M} \beta_k x[n-k]$$

we set $\alpha_k = 0$, it simplifies to

$$y(n) = \sum_{k=0}^{M} \beta_k x[n-k] \tag{7.49}$$

and refers to a *linear discrete-time system invariant as to time*. In this case, the system calculates its output with the help of its $M + 1$ most recent inputs, after first multiplying each of these by the corresponding coefficient β_k. In other words, this system calculates the *weighted moving average of the input samples*, and for this reason it is known as a *moving average system*. Such a system is characterized as a *finite impulse response system*, with the impulse response equal to the coefficients β_k, that is,

$$h[k] = \begin{cases} \beta_k & 0 \le k \le M \\ 0 & \text{anywhere else} \end{cases}$$

On the other hand, if we set $M = 0$ in the general difference equation, it will take the form

$$y[n] = -\sum_{k=1}^{N} \alpha_k y[n-k] + \beta_0 x[n] \tag{7.50}$$

This equation describes a *fully recursive system* whose output is *a weighted linear combination of its N previous outputs and its current input*.

The nonstandard form of the implementation of discrete recursive LTI systems corresponding to the values $M = N = 1$ and $M = N = 2$ that are described by the equations

$$\alpha_0 y[n] + \alpha_1 y[n-1] = \beta_0 x[n] + \beta_1 x[n-1] \quad \text{and}$$
$$\alpha_0 y[n] + \alpha_1 y[n-1] + \alpha_2 y[n-2] = \beta_0 x[n] + \beta_1 x[n-1] + \alpha_2 x[n-2]$$

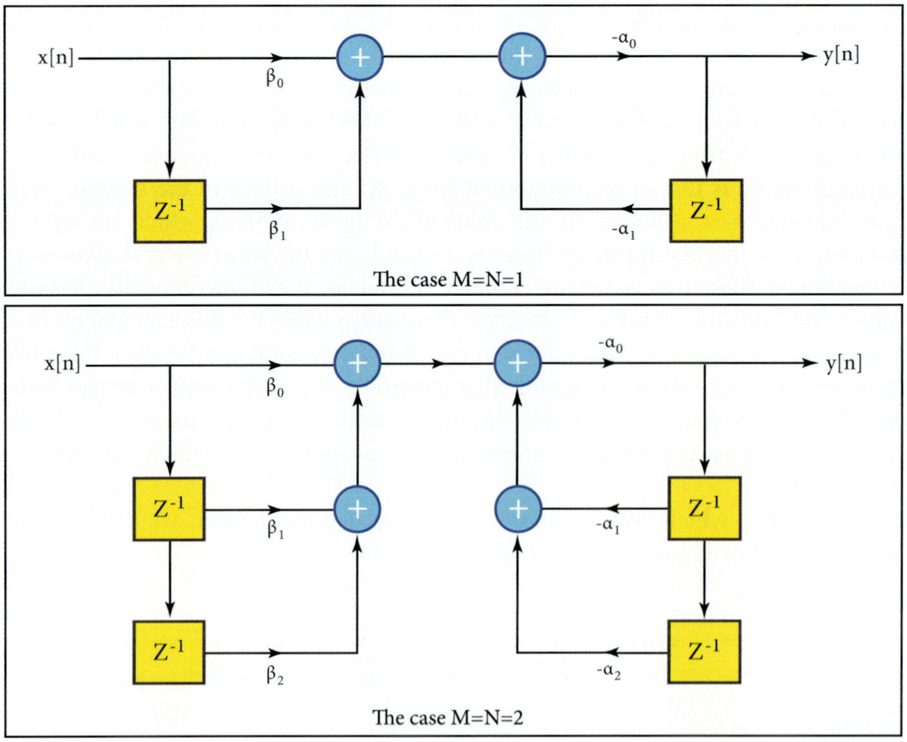

Fig. 7.12 Nonnormal form of recursive discrete LTI systems corresponding to parameter values $M = N = 1$ and $M = N = 2$

are presented in Fig. 7.12, while the construction of the normal form, according to the earlier discussion, is left as an exercise to the reader.

7.11 Description of Systems in State Space

The description of discrete-time systems in the *state space* with the help of appropriate *state variables* is governed by the same rules and the same principles that characterize the homonymous description of continuous-time systems. For this reason, the reader is encouraged, before studying the following material, to read the introductory paragraphs of Sect. 2.9 (page 124), in which the necessity of describing the systems from this perspective is analyzed, the concepts of *state*, *state variables*, and *dynamic equations* are defined, and indicative examples of system descriptions in this space are presented. The presentation in that description applies here as is, with the difference that time, instead of being *continuous*, is *discrete*, the symbols of *integrals* are replaced by *sums*, and the elementary *integrators* that appear in the nonnormal and in the normal forms of the diagrammatic representation of systems are replaced by *time delay elements*.

7.11.1 Constructing Dynamic Equations

In the general case of a *recursive discrete-time system of order N* described by the equation

$$\sum_{k=0}^{N} \alpha_k y[n-k] = \sum_{k=0}^{M} \beta_k x[n-k] \tag{7.51}$$

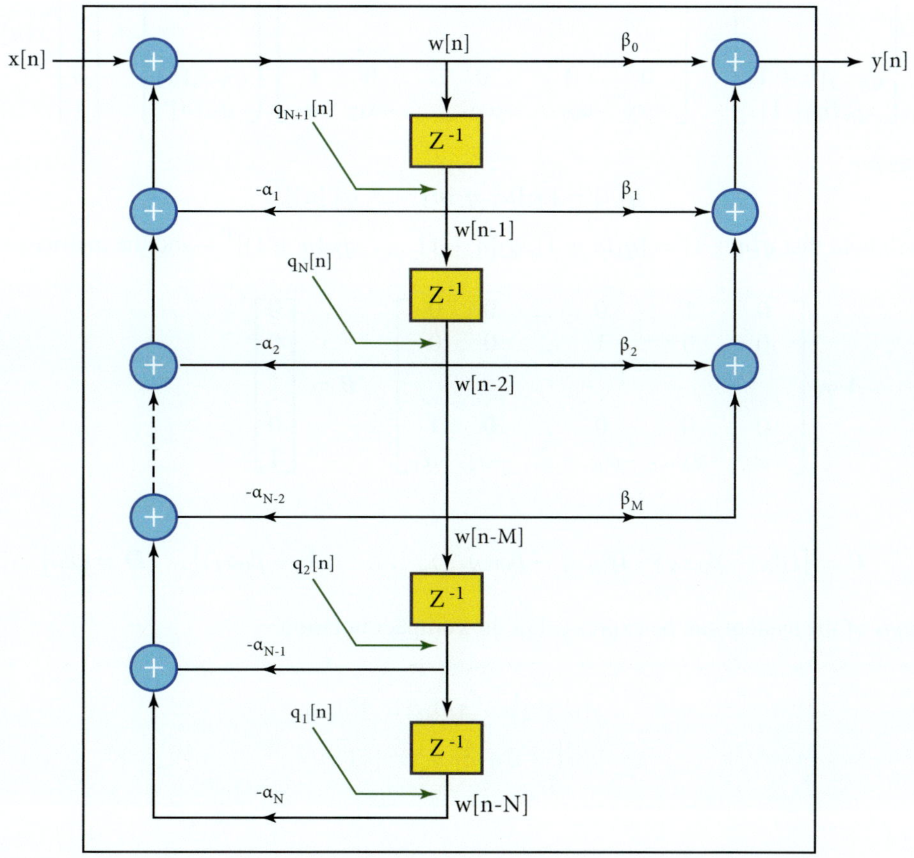

Fig. 7.13 Definition of system state variables from its normal form

we can construct the *dynamic equations* of the system based on both its *normal* and *nonnormal* forms. Starting from the first form, we choose as N state variables of the system the outputs of the N time delay elements, in the way presented in Fig. 7.13. Observing the diagram and following the methodology we developed for continuous-time systems, we find that these variables are related to each other through the relations

$$\begin{aligned} q_1[n+1] &= q_2[n] \\ q_2[n+1] &= q_3[n] \\ q_3[n+1] &= q_4[n] \\ &\cdots\cdots\cdots\cdots \\ q_{N-1}[n+1] &= q_N[n] \\ q_N[n+1] &= \beta_n q_1[n] + \beta_{N-1} q_2[n] + \cdots + \beta_1 q_N[n] + \beta_0 q_{N+1}[n] \end{aligned}$$

On the other hand, the output $y[n]$ of the system is calculated as

$$y[n] = \beta_N q_1[n] + \beta_{N-1} q_2[n] + \cdots + \beta_1 q_N[n] + \beta_0 q_{N+1}[n]$$

When we eliminate the term $q_{N+1}[n]$ from the last two equations, the system output defining equation will take the form

$$y[n] = (\beta_N - \beta_0 \alpha_N) q_1[n] + (\beta_{N-1} - \beta_0 \alpha_{N-1}) q_2[n] + \cdots + (\beta_1 - \beta_0 \alpha_1) q_N[n] + \beta_0 x[n]$$

On the other hand, the N number of equations that correlate among them the *state variables* of the system, which are the $q_1[n], q_2[n], \ldots, q_N[n]$, can be combined into the simple equation

$$\begin{bmatrix} q_1[n+1] \\ q_2[n+1] \\ \vdots \\ q_{N-1}[n+1] \\ q_N[n+1] \end{bmatrix} = \begin{bmatrix} 0 & 1 & 0 & \ldots & 0 & 0 \\ 0 & 0 & 1 & \ldots & 0 & 0 \\ \ldots & \ldots & \ldots & \ddots & \ldots & \ldots \\ 0 & 0 & 0 & \ldots & 0 & 1 \\ -\alpha_N & -\alpha_{N-1} & -\alpha_{N-2} & \ldots & -\alpha_2 & -\alpha_1 \end{bmatrix} \begin{bmatrix} q_1[n] \\ q_2[n] \\ \vdots \\ q_{N-1}[n] \\ q_N[n] \end{bmatrix} + \begin{bmatrix} 0 \\ 0 \\ \vdots \\ 0 \\ 1 \end{bmatrix} x[n]$$

Defining the *state vector*

$$\boldsymbol{q}[n] = [q_1[n], q_2[n], \ldots, q_N[n]]^T$$

—so obviously it will hold that $\boldsymbol{q}[n+1] = [q_1[n+1], q_2[n+1], \ldots, q_N[n+1]]^T$—and the matrices

$$\boldsymbol{A} = \begin{bmatrix} 0 & 1 & 0 & \ldots & 0 & 0 \\ 0 & 0 & 1 & \ldots & 0 & 0 \\ \ldots & \ldots & \ldots & \ddots & \ldots & \ldots \\ 0 & 0 & 0 & \ldots & 0 & 1 \\ -\alpha_N & -\alpha_{N-1} & -\alpha_{N-2} & \ldots & -\alpha_2 & -\alpha_1 \end{bmatrix} \quad \boldsymbol{B} = \begin{bmatrix} 0 \\ 0 \\ \vdots \\ 0 \\ 1 \end{bmatrix}$$

$$\boldsymbol{C} = \begin{bmatrix} (\beta_N - \beta_0 \alpha_N) & (\beta_{N-1} - \beta_0 \alpha_{N-1}) & \ldots & (\beta_1 - \beta_0 \alpha_1) \end{bmatrix}, \quad \boldsymbol{D} = \begin{bmatrix} \beta_0 \end{bmatrix}$$

the *dynamic equations* of the system can be expressed in the compact notation

$$\begin{aligned} \boldsymbol{q}[n+1] &= \boldsymbol{A}\boldsymbol{q}[n] + \boldsymbol{B}x[n] \\ y[n] &= \boldsymbol{C}\boldsymbol{q}[n] + \boldsymbol{D}x[n] \end{aligned} \tag{7.52}$$

and describe the behavior of the system in the state space based on the *canonical form* of its representation with schematic diagrams. In full analogy with continuous-time systems, the first of the preceding equations is known as the *state equation*

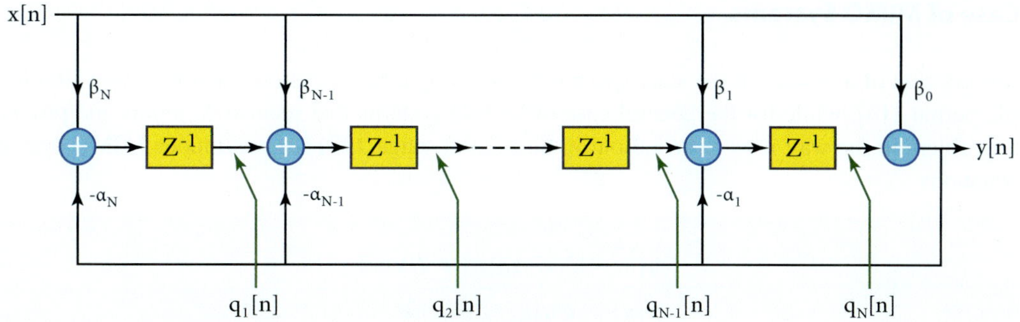

Fig. 7.14 Definition of system state variables from its nonnormal form

of the system, while the second is called the *output equation*. Finally, regarding the matrices that appear in these equations, they are characterized by the following properties:

- The matrix $\boldsymbol{A}$ has dimensions $N \times N$ and is called the *system matrix*.
- The matrix $\boldsymbol{B}$ has dimensions $N \times 1$ and is called the *input matrix*.
- The matrix $\boldsymbol{C}$ has dimensions $1 \times N$ and is called the *measurement matrix*.
- The matrix $\boldsymbol{D}$ has dimensions 1×1 and is called the *output matrix*.

On the other hand, the construction of the dynamic equations from the *nonnormal form* of the system's representation consists in the selection of the state variables in the way presented in Fig. 7.14. In this case, these variables are related to each other through the relations

$$\begin{aligned} q_1[n+1] &= -\alpha_N q_N[n] + (\beta_N - \alpha_N \beta_0)x[n] \\ q_2[n+1] &= q_1[n] - \alpha_{N-1} q_N[n] + (\beta_{N-1} - \alpha_{N-1}\beta_0)x[n] \\ q_3[n+1] &= q_2[n] - \alpha_{N-2} q_N[n] + (\beta_{N-2} - \alpha_{N-2}\beta_0)x[n] \\ &\cdots\cdots\cdots\cdots\cdots\cdots \\ q_N[n+1] &= q_{N-1}[n] - \alpha_1 q_N[n] + (\beta_1 - \alpha_1\beta_0)x[n] \end{aligned}$$

while the output of the system will have the form

$$y[n] = \beta_0 x[n] + (\beta_1 x[n-1] - \alpha_1 y[n-1]) + (\beta_2 x[n-2] - \alpha_2 y[n-2]) + \cdots + (\beta_N x[n-N] - \alpha_N y[n-N])$$

In this case and working as before, we can again arrive at the dynamic equations

$$\boldsymbol{q}[n+1] = \boldsymbol{A}\boldsymbol{q}[n] + \boldsymbol{B}x[n] \qquad \text{and} \qquad y[n] = \boldsymbol{C}\boldsymbol{q}[n] + \boldsymbol{D}x[n]$$

with the matrices $\boldsymbol{A}$, $\boldsymbol{B}$, $\boldsymbol{C}$, and $\boldsymbol{D}$ being defined as

$$\boldsymbol{A} = \begin{bmatrix} 0 & 0 & 0 & \dots & 0 & -\alpha_N \\ 1 & 0 & 0 & \dots & 0 & -\alpha_{N-1} \\ \dots & \dots & \dots & \ddots & \dots & \dots \\ 0 & 0 & 0 & \dots & 0 & -\alpha_2 \\ 0 & 0 & 0 & \dots & 1 & -\alpha_1 \end{bmatrix} \qquad \boldsymbol{B} = \begin{bmatrix} \beta_N - \beta_0\alpha_N \\ \beta_{N-1} - \beta_0\alpha_{N-1} \\ \vdots \\ \beta_2 - \beta_0\alpha_2 \\ \beta_1 - \beta_0\alpha_1 \end{bmatrix} \qquad \boldsymbol{C} = \begin{bmatrix} 0\,0\,0 \dots 1 \end{bmatrix}, \qquad \boldsymbol{D} = \begin{bmatrix} \beta_0 \end{bmatrix}$$

In full analogy with the previous case, we assumed that $\alpha_0 = 1$; if the use of this parameter is desired, all the foregoing expressions should simply be divided by this value.

7.11.2 The Case of MIMO Systems

The preceding description of a system in the state space refers to simple SISO systems that receive a simple input $x[n]$ and produce a simple output $y[n]$, while for the general case of MIMO systems that receive K inputs and produce L outputs[17] the study of the problem in the way it was carried out in Sect. 2.9.3 (page 133) for continuous MIMO systems will lead to the system of equations

$$\begin{aligned} \boldsymbol{q}[n+1] &= \boldsymbol{A}\boldsymbol{q}[n] + \boldsymbol{B}\boldsymbol{x}[n] \\ \boldsymbol{y}[n] &= \boldsymbol{C}\boldsymbol{q}[n] + \boldsymbol{D}\boldsymbol{x}[n] \end{aligned} \tag{7.53}$$

with the matrices $\boldsymbol{A}$, $\boldsymbol{B}$, $\boldsymbol{C}$, and $\boldsymbol{D}$ being defined as

$$\boldsymbol{A} = \begin{bmatrix} \alpha_{11} & \alpha_{12} & \cdots & \alpha_{1N} \\ \alpha_{21} & \alpha_{22} & \cdots & \alpha_{2N} \\ \vdots & \vdots & \ddots & \vdots \\ \alpha_{N1} & \alpha_{N2} & \cdots & \alpha_{NN} \end{bmatrix} \quad \boldsymbol{B} = \begin{bmatrix} \beta_{11} & \beta_{12} & \cdots & \beta_{1K} \\ \beta_{21} & \beta_{22} & \cdots & \beta_{2K} \\ \vdots & \vdots & \ddots & \vdots \\ \beta_{N1} & \beta_{N2} & \cdots & \beta_{NK} \end{bmatrix}$$

$$\boldsymbol{C} = \begin{bmatrix} \gamma_{11} & \gamma_{12} & \cdots & \gamma_{1N} \\ \gamma_{21} & \gamma_{22} & \cdots & \gamma_{2N} \\ \vdots & \vdots & \ddots & \vdots \\ \gamma_{L1} & \gamma_{L2} & \cdots & \gamma_{LN} \end{bmatrix} \quad \boldsymbol{D} = \begin{bmatrix} \delta_{11} & \delta_{12} & \cdots & \delta_{1K} \\ \delta_{21} & \delta_{22} & \cdots & \delta_{2K} \\ \vdots & \vdots & \ddots & \vdots \\ \delta_{L1} & \delta_{L2} & \cdots & \delta_{LK} \end{bmatrix}$$

and the vectors $\boldsymbol{x}[n] = [x_1[n], x_2[n], \ldots, x_K[n]]^T$ and $\boldsymbol{y}[n] = [y_1[n], y_2[n], \ldots, y_L[n]]^T$ containing K and L elements, respectively. In this case, the matrices that appear in the dynamic equations of the system are characterized by the following properties:

- The matrix $\boldsymbol{A}$ has dimensions $N \times N$ and is called the *system matrix*.
- The matrix $\boldsymbol{B}$ has dimensions $N \times K$ and is called the *input matrix*.
- The matrix $\boldsymbol{C}$ has dimensions $L \times N$ and is called the *measurement matrix*.
- The matrix $\boldsymbol{D}$ has dimensions $L \times K$ and is called the *output matrix*.

After the presentation of the basic theory, example solved exercises follow so that the reader can gain a deep understanding of how it is applied to the description of systems in the state space and to the field of discrete-time.

Problem 7.11.1 Determine the dynamic equations of a discrete recursive LTI system described by the difference equation

$$y[n] + 0.6y[n-1] - 0.16y[n-2] = -2x[n] + x[n-1]$$

based both on its normal and nonnormal form.

Solution The difference equation of the aforementioned discrete-time system arises from the general difference equation for parameter values $N = 2$, $M = 1$, $\alpha_0 = 1$, $\alpha_1 = 0.6$, $\alpha_2 = -0.16$, $\beta_0 = -2$, $\beta_1 = 1$, and $\beta_2 = 0$. Starting from the *normal form* of the system, the matrices $\boldsymbol{A}$, $\boldsymbol{B}$, $\boldsymbol{C}$, and $\boldsymbol{D}$ are

$$\boldsymbol{A} = \begin{bmatrix} 0 & 1 \\ -\alpha_2 & -\alpha_1 \end{bmatrix} = \begin{bmatrix} 0 & 1 \\ 0.16 & -0.6 \end{bmatrix}, \quad \boldsymbol{B} = \begin{bmatrix} 0 \\ 1 \end{bmatrix}$$

$$C = \begin{bmatrix} \beta_2 - \beta_0\alpha_2 & \beta_1 - \beta_0\alpha_1 \end{bmatrix} = \begin{bmatrix} 0.32 & 2.20 \end{bmatrix}, \quad \boldsymbol{D} = \begin{bmatrix} \beta_0 \end{bmatrix} = \begin{bmatrix} -2 \end{bmatrix}$$

[17] For the categorization of systems regarding the number of their inputs and outputs, see Sect. 2.2.2 on page 69. This section discusses *continuous-time systems*; however, the situation is completely analogous for *discrete-time systems*.

and therefore the dynamic equations that describe it will have the form

$$\boldsymbol{q}[n+1] = \boldsymbol{A}\boldsymbol{q}[n] + \boldsymbol{B}x[n] = \begin{bmatrix} 0 & 1 \\ 0.16 & -0.6 \end{bmatrix} \begin{bmatrix} q_1[n] \\ q_2[n] \end{bmatrix} + \begin{bmatrix} 0 \\ 1 \end{bmatrix} x[n]$$

$$y[n] = \boldsymbol{C}\boldsymbol{q}[n] + \boldsymbol{D}x[n] = \begin{bmatrix} 0.32 & 2.20 \end{bmatrix} \begin{bmatrix} q_1[n] \\ q_2[n] \end{bmatrix} - 2x[n]$$

In equivalent notation, the state equations of the system are reformulated as

$$q_1[n+1] = q_2[n] \quad \text{and} \quad q_2[n+1] = 0.16q_1[n] - 0.6q_2[n] + x[n]$$

while the output equation of the system will take the form

$$y[n] = 0.32q_1[n] + 2.20q_2[n] - 2x[n]$$

On the other hand, using the *nonnormal form* of the system's representation, the matrices $\boldsymbol{A}$, $\boldsymbol{B}$, $\boldsymbol{C}$, and $\boldsymbol{D}$ are calculated as

$$\boldsymbol{A} = \begin{bmatrix} 0 & -\alpha_2 \\ 1 & -\alpha_1 \end{bmatrix} = \begin{bmatrix} 0 & 0.16 \\ 1 & -0.6 \end{bmatrix}, \quad \boldsymbol{B} = \begin{bmatrix} \beta_2 - \beta_0\alpha_2 \\ \beta_1 - \beta_0\alpha_1 \end{bmatrix} = \begin{bmatrix} 0.32 \\ 2.20 \end{bmatrix}$$

$$C = \begin{bmatrix} 0 & 1 \end{bmatrix}, \quad \boldsymbol{D} = \begin{bmatrix} \beta_0 \end{bmatrix} = \begin{bmatrix} -2 \end{bmatrix}$$

and the dynamic equations of the system will take the form

$$\boldsymbol{q}[n+1] = \boldsymbol{A}\boldsymbol{q}[n] + \boldsymbol{B}x[n] = \begin{bmatrix} 0 & 0.16 \\ 1 & -0.6 \end{bmatrix} \begin{bmatrix} q_1[n] \\ q_2[n] \end{bmatrix} + \begin{bmatrix} 0.32 \\ 2.20 \end{bmatrix} x[n]$$

$$y[n] = \boldsymbol{C}\boldsymbol{q}[n] + \boldsymbol{D}x[n] = \begin{bmatrix} 0 & 1 \end{bmatrix} \begin{bmatrix} q_1[n] \\ q_2[n] \end{bmatrix} - 2x[n]$$

or, equivalently,

$$\left.\begin{aligned} q_1[n+1] &= 0.16q_2[n] + 0.32x[n] \\ q_2[n+1] &= q_1[n] - 0.6q_2[n] + 2.20x[n] \end{aligned}\right\} \quad \text{and} \quad y[n] = q_2[n] - 2x[n]$$

Problem 7.11.2 Determine the dynamic equations of a discrete recursive LTI system described by the difference equation

$$y[n] + 3y[n-1] + 2y[n-2] - y[n-3] = 5x[n] - 4x[n-1] + 12x[n-2]$$

based both on its normal and nonnormal form.

Solution The difference equation of the aforementioned discrete-time system arises from the general difference equation for parameter values $N = 3$, $M = 2$, $\alpha_0 = 1$, $\alpha_1 = 3$, $\alpha_2 = 2$, $\alpha_3 = -1$, $\beta_0 = 5$, $\beta_1 = -4$, $\beta_2 = 12$, and $\beta_3 = 0$. Starting from the *normal form* of the system, the matrices $\boldsymbol{A}$, $\boldsymbol{B}$, $\boldsymbol{C}$, and $\boldsymbol{D}$ are calculated as

$$\boldsymbol{A} = \begin{bmatrix} 0 & 1 & 0 \\ 0 & 0 & 1 \\ -\alpha_3 & -\alpha_2 & -\alpha_1 \end{bmatrix} = \begin{bmatrix} 0 & 1 & 0 \\ 0 & 0 & 1 \\ 1 & -2 & -3 \end{bmatrix}, \quad \boldsymbol{B} = \begin{bmatrix} 0 \\ 0 \\ 1 \end{bmatrix}$$

$$C = \begin{bmatrix} \beta_3 - \beta_0\alpha_3 & \beta_2 - \beta_0\alpha_2 & \beta_1 - \beta_0\alpha_1 \end{bmatrix} = \begin{bmatrix} 5 & 2 & -19 \end{bmatrix}, \quad \boldsymbol{D} = \begin{bmatrix} \beta_0 \end{bmatrix} = \begin{bmatrix} 5 \end{bmatrix}$$

Therefore, the state equation of the system will have the form

$$\boldsymbol{q}[n+1] = \boldsymbol{A}\boldsymbol{q}[n] + \boldsymbol{B}x[n] = \begin{bmatrix} 0 & 1 & 0 \\ 0 & 0 & 1 \\ 1 & -2 & -3 \end{bmatrix} \begin{bmatrix} q_1[n] \\ q_2[n] \\ q_3[n] \end{bmatrix} + \begin{bmatrix} 0 \\ 0 \\ 1 \end{bmatrix} x[n]$$

from which it follows that

$$q_1[n+1] = q_2[n]$$

$$q_2[n+1] = q_3[n]$$

$$q_3[n+1] = q_1[n] - 2q_2[n] - 3q_3[n] + x[n]$$

while the system's output equation is calculated as

$$y[n] = \boldsymbol{C}\boldsymbol{q}[n] + \boldsymbol{D}x[n] = \begin{bmatrix} 5 & 2 & -19 \end{bmatrix} \begin{bmatrix} q_1[n] \\ q_2[n] \\ q_3[n] \end{bmatrix} + 5x[n] = 5q_1[n] + 2q_2[n] - 19q_3[n] + 5x[n]$$

On the other hand, using the *nonnormal form* of the system's representation, the matrices $\boldsymbol{A}$, $\boldsymbol{B}$, $\boldsymbol{C}$, and $\boldsymbol{D}$ are calculated as

$$\boldsymbol{A} = \begin{bmatrix} 0 & 0 & -\alpha_3 \\ 1 & 0 & -\alpha_2 \\ 0 & 1 & -\alpha_1 \end{bmatrix} = \begin{bmatrix} 0 & 0 & 1 \\ 1 & 0 & -2 \\ 0 & 1 & -3 \end{bmatrix}, \quad \boldsymbol{B} = \begin{bmatrix} \beta_3 - \beta_0\alpha_3 \\ \beta_2 - \beta_0\alpha_2 \\ \beta_1 - \beta_0\alpha_1 \end{bmatrix} = \begin{bmatrix} 5 \\ 2 \\ -19 \end{bmatrix}$$

$$C = \begin{bmatrix} 0 & 0 & 1 \end{bmatrix}, \quad \boldsymbol{D} = \begin{bmatrix} \beta_0 \end{bmatrix} = \begin{bmatrix} 5 \end{bmatrix}$$

Therefore, the state equation of the system will have the form

$$\boldsymbol{q}[n+1] = \boldsymbol{A}\boldsymbol{q}[n] + \boldsymbol{B}x[n] = \begin{bmatrix} 0 & 0 & 1 \\ 1 & 0 & -2 \\ 0 & 1 & -3 \end{bmatrix} \begin{bmatrix} q_1[n] \\ q_2[n] \\ q_3[n] \end{bmatrix} + \begin{bmatrix} 5 \\ 2 \\ -19 \end{bmatrix} x[n]$$

from which it follows that

$$q_1[n+1] = q_3[n] + 5x[n]$$

$$q_2[n+1] = q_1[n] - 2q_3[n] + 2x[n]$$

$$q_3[n+1] = q_2[n] - 3q_3[n] - 19x[n]$$

while the system's output equation is calculated as

$$y[n] = \boldsymbol{C}\boldsymbol{q}[n] + \boldsymbol{D}x[n] = \begin{bmatrix} 0 & 0 & 1 \end{bmatrix} \begin{bmatrix} q_1[n] \\ q_2[n] \\ q_3[n] \end{bmatrix} + 5x[n] = q_3[n] + 5x[n]$$

7.11.3 Solving the State Equation

The solution of the state equation

$$\boldsymbol{q}[n+1] = \boldsymbol{A}\boldsymbol{q}[n] + \boldsymbol{B}x[n] \tag{7.54}$$

takes place through the application of the *recursive method* and assuming that the initial state $\boldsymbol{q}[0]$ corresponding to the time $n = 0$ is known. In this case, by setting in the foregoing equation the value $n = 0$, we will get

$$\boldsymbol{q}[1] = \boldsymbol{A}\boldsymbol{q}[0] + \boldsymbol{B}x[0]$$

while for $n = 1$ the foregoing relation will take the form

$$\boldsymbol{q}[2] = \boldsymbol{A}\boldsymbol{q}[1] + \boldsymbol{B}x[1]$$

Replacing the expression $\boldsymbol{q}[1]$ from the defining equation the last relation is written

$$\boldsymbol{q}[2] = \boldsymbol{A}\Big(\boldsymbol{A}\boldsymbol{q}[0] + \boldsymbol{B}x[0]\Big) + \boldsymbol{B}x[1] = \boldsymbol{A}^2\boldsymbol{q}[0] + \boldsymbol{A}\boldsymbol{B}x[0] + \boldsymbol{B}x[1]$$

Continuing in the same way, it turns out that

$$\boldsymbol{q}[3] = \boldsymbol{A}\boldsymbol{q}[2] + \boldsymbol{B}x[2] = \boldsymbol{A}^3\boldsymbol{q}[0] + \boldsymbol{A}^2\boldsymbol{B}x[0] + \boldsymbol{A}\boldsymbol{B}x[1] + \boldsymbol{B}x[2]$$

$$\boldsymbol{q}[4] = \boldsymbol{A}\boldsymbol{q}[3] + \boldsymbol{B}x[3] = \boldsymbol{A}^4\boldsymbol{q}[0] + \boldsymbol{A}^3\boldsymbol{B}x[0] + \boldsymbol{A}^2\boldsymbol{B}x[1] + \boldsymbol{A}\boldsymbol{B}x[2] + \boldsymbol{B}x[3]$$

and in the general case

$$\boldsymbol{q}[n] = \boldsymbol{A}^n\boldsymbol{q}[0] + \boldsymbol{A}^{n-1}\boldsymbol{B}x[0] + \boldsymbol{A}^{n-2}\boldsymbol{B}x[1] + \cdots + \boldsymbol{B}x[n-1]$$

or, equivalently,[18]

$$\boldsymbol{q}[n] = \boldsymbol{A}^n\boldsymbol{q}[0] + \sum_{k=0}^{n-1} \boldsymbol{A}^{n-1-k}\boldsymbol{B}x[k] \tag{7.55}$$

It is not difficult to see that the second term in the last equation expresses the convolution[19] between the matrices $\boldsymbol{A}^{n-1}u[n-1]$ and $\boldsymbol{B}x[n]$; thus,

$$\sum_{k=0}^{n-1} \boldsymbol{A}^{n-1-k}\boldsymbol{B}x[k] = \boldsymbol{A}^{n-1}u[n-1] * \boldsymbol{B}x[n]$$

and the foregoing equation takes the form

$$\boldsymbol{q}[n] = \boldsymbol{A}^n\boldsymbol{q}[0] + \boldsymbol{A}^{n-1}u[n-1] * \boldsymbol{B}x[n] \tag{7.56}$$

Of the two terms of the preceding sum, the first is independent of the system's input and is known as the *zero-input component*, while the second is independent of the initial state vector and is known as the *zero-state component*. The matrix

$$\boldsymbol{A}^n = \underbrace{\boldsymbol{A} \cdot \boldsymbol{A} \cdot \boldsymbol{A} \cdot \boldsymbol{A} \ldots \cdots \boldsymbol{A}}_{n \text{ times}}$$

[18] In this description, we assume that the beginning of time measurement is chosen at the time $n = 0$, while generally speaking, if the initial state of the system is some vector of the form $\boldsymbol{q}[n_0]$, while the input signal $x[n]$ is defined only for time instants $n \geq n_0$, the defining equation of the state vector $\boldsymbol{q}[n]$ is formulated as

$$\boldsymbol{q}[n] = \boldsymbol{A}^{n-n_0}\boldsymbol{q}[n_0] + \sum_{k=n_0}^{n-1} \boldsymbol{A}^{n-1-k}\boldsymbol{B}x[n_0 + k] \quad (n > n_0).$$

[19] For the definition of the convolution of matrices, see Footnote 26 (page 137) of the second chapter on continuous-time systems.

that appears in the preceding relations is known as the *state transition matrix* and has the same meaning as its continuous equivalent, and the way in which it is calculated will be presented in the next section.

If we replace the expression of the state vector $\boldsymbol{q}[n]$ previously calculated in the output equation of the considered discrete-time system, it will take the form

$$\begin{aligned} y[n] &= \boldsymbol{C}\boldsymbol{q}[n] + \boldsymbol{D}x[n] = \boldsymbol{C}\boldsymbol{A}^n\boldsymbol{q}[0] + \sum_{k=0}^{n-1} \boldsymbol{C}\boldsymbol{A}^{n-1-k}\boldsymbol{B}x[k] + \boldsymbol{D}x[n] \\ &= \boldsymbol{C}\boldsymbol{A}^n\boldsymbol{q}[0] + \boldsymbol{C}\boldsymbol{A}^{n-1}u[n-1] * \boldsymbol{B}x[n] + \boldsymbol{D}x[n] \end{aligned} \tag{7.57}$$

In full analogy with continuous-time systems, the term $\boldsymbol{C}\boldsymbol{A}^n\boldsymbol{q}[0]$ expresses the *system response to zero input*, while the sum of the terms $\boldsymbol{C}\boldsymbol{A}^{n-1}u[n-1] * \boldsymbol{B}x[n] + \boldsymbol{D}x[n]$ constitutes the *system response to the zero state*.

The foregoing equations concern the case of *discrete-time LTI systems*, while for the time-varying case and considering as the beginning of the discrete-time measurement some time $n = n_0$, the foregoing equations are formulated as

$$\boldsymbol{q}[n] = \boldsymbol{\Phi}[n, n_0]\boldsymbol{q}[n_0] + \sum_{k=n_0}^{n-1} \boldsymbol{\Phi}[n, k+1]\boldsymbol{B}[k]x[k] \tag{7.58}$$

$$y[n] = \boldsymbol{C}[n]\boldsymbol{\Phi}[n, n_0]\boldsymbol{q}[n_0] + \boldsymbol{C}[n] \sum_{k=n_0}^{n-1} \boldsymbol{\Phi}[n, k+1]\boldsymbol{B}[k]x[k] + \boldsymbol{D}[n]x[n] \tag{7.59}$$

with the matrix $\boldsymbol{\Phi}[n, n_0]$, which is the time-varying version of the *state transition matrix* to be the solution of the equation $\boldsymbol{\Phi}[n+1, n_0] = \boldsymbol{A}[n]\boldsymbol{\Phi}[n, n_0]$ under the restriction that it must be $\boldsymbol{\Phi}[n_0, n_0] = \boldsymbol{I}$ (where $\boldsymbol{I}$ is the unit matrix of order N) and given by the relation

$$\boldsymbol{\Phi}[n, n_0] = \prod_{k=1}^{n-n_0} \boldsymbol{A}[n-k] \tag{7.60}$$

More details about the time-varying case can be found in the literature.

7.11.4 Calculation of Exponential Function A^n

The calculation of the exponential function $\boldsymbol{A}^n$ that appears in the solution of the state equation, which, as we proved, is given by the relation

$$\boldsymbol{q}[n] = \boldsymbol{A}^n\boldsymbol{q}[n] + \boldsymbol{A}^{n-1}u[n-1] * \boldsymbol{B}x[n] \tag{7.61}$$

is carried out by applying techniques that are almost identical to those we used to calculate the exponential function $e^{\boldsymbol{A}t}$ in the case of continuous-time systems. For this reason, in the following section we will briefly present the methods for calculating this expression, and we encourage the reader to study in detail the material of Sect. 2.9.5 (page 138) of the chapter on continuous-time systems.

Let us start our description with the first of the two methods that relies on the use of the *eigenvalues* $\lambda_1, \lambda_2, \ldots, \lambda_N$ of the matrix $\boldsymbol{A}$ (which is supposed to have dimensions $N \times N$), which, as is well known, are the roots of the *characteristic equation*

$$Q(\lambda) = \lambda^N + \alpha_{N-1}\lambda^{N-1} + \cdots + \alpha_2\lambda^2 + \alpha_1\lambda + \alpha_0 = 0 \tag{7.62}$$

Using the *Cayley-Hamilton theorem* and the method described in the chapter on continuous-time systems, it is proven that the desired matrix $\boldsymbol{A}^n$ satisfies the equation

$$\boldsymbol{A}^n = \beta_0\boldsymbol{I} + \beta_1\boldsymbol{A} + \beta_2\boldsymbol{A}^2 + \cdots + \beta_{N-1}\boldsymbol{A}^{N-1} \tag{7.63}$$

from which it can be calculated. In the foregoing relation, the matrix $\boldsymbol{I}$ is the unit matrix of dimensions $N \times N$, while the coefficients $\beta_0, \beta_1, \beta_2, \ldots, \beta_{N-1}$ appearing in the preceding relation are calculated with the help of the eigenvalues of the matrix $\boldsymbol{A}$ as the solution of the system of equations

$$\begin{aligned}
\beta_0 + \beta_1\lambda_1 + \beta_2\lambda_1^2 + \cdots + \beta_{N-1}\lambda_1^{N-1} &= f(\lambda_1) = \lambda_1^n \\
\beta_0 + \beta_1\lambda_2 + \beta_2\lambda_2^2 + \cdots + \beta_{N-1}\lambda_2^{N-1} &= f(\lambda_2) = \lambda_2^n \\
\beta_0 + \beta_1\lambda_3 + \beta_2\lambda_3^2 + \cdots + \beta_{N-1}\lambda_3^{N-1} &= f(\lambda_3) = \lambda_3^n \\
&\cdots\cdots\cdots\cdots \\
\beta_0 + \beta_1\lambda_N + \beta_2\lambda_N^2 + \cdots + \beta_{N-1}\lambda_N^{N-1} &= f(\lambda_N) = \lambda_N^n
\end{aligned}$$

which in matrix form is formulated as

$$\begin{bmatrix} 1 & \lambda_1 & \lambda_1^2 & \ldots & \lambda_1^{N-1} \\ 1 & \lambda_2 & \lambda_2^2 & \ldots & \lambda_2^{N-1} \\ 1 & \lambda_3 & \lambda_3^2 & \ldots & \lambda_3^{N-1} \\ \vdots & \vdots & \vdots & \ddots & \vdots \\ 1 & \lambda_N & \lambda_N^2 & \ldots & \lambda_N^{N-1} \end{bmatrix} \begin{bmatrix} \beta_0 \\ \beta_1 \\ \beta_2 \\ \vdots \\ \beta_{N-1} \end{bmatrix} = \begin{bmatrix} \lambda_1^n \\ \lambda_2^n \\ \lambda_3^n \\ \vdots \\ \lambda_N^n \end{bmatrix} \tag{7.64}$$

In the case where the characteristic polynomial of the system has *repeated roots*, the problem is addressed in exactly the same way as presented in Sect. 2.9.5.

The second way of calculating the matrix $\boldsymbol{A}^n$ is based on the *diagonalization* of the matrix $\boldsymbol{A}$, which, as we have proven, is formulated as $\boldsymbol{A} = \boldsymbol{P}\boldsymbol{\Lambda}\boldsymbol{P}^{-1}$, where

$$\boldsymbol{\Lambda} = \begin{bmatrix} \lambda_1 & 0 & 0 & \ldots & 0 \\ 0 & \lambda_2 & 0 & \ldots & 0 \\ 0 & 0 & \lambda_3 & \ldots & 0 \\ \vdots & \vdots & \vdots & \ddots & \vdots \\ 0 & 0 & 0 & \ldots & \lambda_N \end{bmatrix}$$

is the *eigenvalue matrix* , $\boldsymbol{P} = [\boldsymbol{x}_1\ \boldsymbol{x}_2\ \boldsymbol{x}_3\ \ldots\ \boldsymbol{x}_N]$ is the *diagonalization matrix*, which contains as elements the eigenvectors of the matrix $\boldsymbol{A}$, and $\boldsymbol{P}^{-1}$ is the inverse of this matrix. Having calculated these matrices, it is proven, as we have seen, that the desired matrix $\boldsymbol{A}^n$ is calculated as

$$\boldsymbol{A}^n = \boldsymbol{P} \begin{bmatrix} \lambda_1^n & 0 & 0 & \ldots & 0 \\ 0 & \lambda_2^n & 0 & \ldots & 0 \\ 0 & 0 & \lambda_3^n & \ldots & 0 \\ \vdots & \vdots & \vdots & \ddots & \vdots \\ 0 & 0 & 0 & \ldots & \lambda_N^n \end{bmatrix} \boldsymbol{P}^{-1}$$

Finally, according to the theory of *spectral analysis of matrices*, if the matrix $\boldsymbol{A}$ is developed as the sum

$$\boldsymbol{A}=\sum_{i=1}^{N}\lambda_i\boldsymbol{E}_i=\lambda_1\boldsymbol{E}_1+\lambda_2\boldsymbol{E}_2+\cdots+\lambda_N\boldsymbol{E}_N$$

with the matrices $\boldsymbol{E}_i$ ($i=1,2,3,\ldots,N$) defined in the way described in Footnote 28 (page 143) of Chap. 2, it is proven that the desired matrix $\boldsymbol{A}^n$ can be calculated from the sum

$$\boldsymbol{A}^n=\sum_{i=1}^{N}\lambda_i^n\boldsymbol{E}_i=\lambda_1^n\boldsymbol{E}_1+\lambda_2^n\boldsymbol{E}_2+\cdots+\lambda_N^n\boldsymbol{E}_N \tag{7.65}$$

Let us also mention at this point that the matrix $\boldsymbol{A}^n$ can be calculated using the $\mathcal{Z}$ *transform* and applying a method that constitutes the *discrete equivalent of the method of calculating the exponential function* $e^{\boldsymbol{A}t}$ *with the method of the Laplace transform*; this method will be presented in a future chapter. The exemplary solved exercises that follow are examples of calculating the matrix $\boldsymbol{A}^n$ in the manner described in the preceding pages.

Problem 7.11.3 Calculate matrix $\boldsymbol{A}^n$, with matrix $\boldsymbol{A}$ having dimensions 2×2 being

$$\boldsymbol{A}=\begin{bmatrix}1/2 & 0\\ 1/4 & 1/4\end{bmatrix}$$

Solution To solve the problem, we will use the first of the methods presented in the previous section. The eigenvalues of matrix $\boldsymbol{A}$ will emerge as the roots of the equation

$$\det(\boldsymbol{A}-\lambda\boldsymbol{I})=\begin{vmatrix}(1/2)-\lambda & 0\\ 1/4 & (1/4)-\lambda\end{vmatrix}=\left(\frac{1}{2}-\lambda\right)\left(\frac{1}{4}-\lambda\right)=0$$

which, of course, are $\lambda_1=1/2$ and $\lambda_2=1/4$. Taking into account that the dimensions of matrix $\boldsymbol{A}$ are 2×2, the coefficients of the equation for calculating the matrix according to the *Cayley-Hamilton theorem* will be β_0 and β_1, which will be calculated as the solution of the system

$$\begin{aligned}\beta_0+\beta_1\lambda_1&=\lambda_1^n\\ \beta_0+\beta_1\lambda_2&=\lambda_2^n\end{aligned}\qquad\text{or, equivalently,}\qquad\begin{aligned}\beta_0+\frac{\beta_1}{2}&=\left(\frac{1}{2}\right)^n\\ \beta_0+\frac{\beta_1}{4}&=\left(\frac{1}{4}\right)^n\end{aligned}$$

The solution of the foregoing system leads to the expressions

$$\beta_0=2\left(\frac{1}{4}\right)^n-\left(\frac{1}{2}\right)^n\qquad\text{and}\qquad\beta_1=4\left[\left(\frac{1}{2}\right)^n-\left(\frac{1}{4}\right)^n\right]$$

which if substituted into the *Cayley-Hamilton equation* will lead to the expression

$$\boldsymbol{A}^n=\beta_0\boldsymbol{I}+\beta_1\boldsymbol{A}=\beta_0\begin{bmatrix}1 & 0\\ 0 & 1\end{bmatrix}+\beta_1\begin{bmatrix}1/2 & 0\\ 1/4 & 1/4\end{bmatrix}=\begin{bmatrix}\beta_0+\dfrac{\beta_1}{2} & 0\\ \dfrac{\beta_1}{4} & \beta_0+\dfrac{\beta_1}{4}\end{bmatrix}$$

Replacing the values of β_0 and β_1 in the foregoing relation results in

$$A^n = \begin{bmatrix} \left(\frac{1}{2}\right)^n & 0 \\ \left(\frac{1}{2}\right)^n - \left(\frac{1}{4}\right)^n & \left(\frac{1}{4}\right)^n \end{bmatrix}$$

On the other hand, the calculation of matrix A^n through the application of the *diagonalization method* of matrix A, requires, as is well known, the calculation of the *eigenvectors* $x_1 = \begin{bmatrix} x_{11} & x_{12} \end{bmatrix}^T$ and $x_2 = \begin{bmatrix} x_{21} & x_{22} \end{bmatrix}^T$ that correspond to the eigenvalues λ_1 and λ_2. Applying the usual methodology for calculating eigenvectors (e.g., Exercise 2.9.3, page 147), we find that $x_1 = \begin{bmatrix} 1 & 1 \end{bmatrix}^T$ and $x_1 = \begin{bmatrix} 0 & 1 \end{bmatrix}^T$, and therefore the matrices P and P^{-1} are calculated as

$$P = \begin{bmatrix} 1 & 0 \\ 1 & 1 \end{bmatrix} \quad \text{and} \quad P^{-1} = \begin{bmatrix} 1 & 0 \\ -1 & 1 \end{bmatrix}$$

Therefore, substituting the foregoing matrices in the defining equation of the matrix A^n that results from this method, we get

$$A^n = P \begin{bmatrix} \lambda_1^n & 0 \\ 0 & \lambda_2^n \end{bmatrix} P^{-1} = \begin{bmatrix} 1 & 0 \\ 1 & 1 \end{bmatrix} \begin{bmatrix} (1/2)^n & 0 \\ 0 & (1/4)^n \end{bmatrix} \begin{bmatrix} 1 & 0 \\ -1 & 1 \end{bmatrix}$$

Performing the operations, we end up with the same result as obtained with the first method.

Problem 7.11.4 Calculate matrix A^n with matrix A of dimensions 3×3 defined as

$$A = \begin{bmatrix} 3 & 0 & 0 \\ 0 & -2 & 1 \\ 0 & 4 & 1 \end{bmatrix}$$

(Hsu, *Signals and Systems*, Exercise for solution 7.60).

Solution In this case, the eigenvalue equation of matrix A is

$$\det(A - \lambda I) = \begin{vmatrix} 3-\lambda & 0 & 0 \\ 0 & -2-\lambda & 1 \\ 0 & 4 & 1-\lambda \end{vmatrix} = (3-\lambda)(\lambda-2)(\lambda+3) = 0$$

and therefore the eigenvalues of this matrix are $\lambda_1 = 3$, $\lambda_2 = 2$, and $\lambda_3 = -3$. Taking into account that the dimensions of matrix A are 3×3, we should calculate the values of the coefficients β_0, β_1, and β_2 of the *Cayley-Hamilton equation* as the solution of a system of three equations with three unknowns of the form

$$\begin{aligned} \beta_0 + \beta_1\lambda_1 + \beta_2\lambda_1^2 &= \lambda_1^n \\ \beta_0 + \beta_1\lambda_2 + \beta_2\lambda_2^2 &= \lambda_2^n \\ \beta_0 + \beta_1\lambda_3 + \beta_2\lambda_3^2 &= \lambda_3^n \end{aligned} \quad \text{or equivalently} \quad \begin{aligned} \beta_0 + 3\beta_1 + 9\beta_2 &= 3^n \\ \beta_0 + 2\beta_1 + 4\beta_2 &= 2^n \\ \beta_0 - 3\beta_1 + 9\beta_2 &= (-3)^n \end{aligned}$$

The solution of the preceding system of equations leads to the values

$$\beta_0 = -3^n + \frac{1}{5}(-3)^n + \frac{9}{5}2^n, \quad \beta_1 = \frac{1}{6}3^n - \frac{1}{6}(-3)^n, \quad \beta_2 = \frac{1}{6}3^n - \frac{1}{5}2^n + \frac{1}{30}(-3)^n$$

If these values are substituted into the *Cayley-Hamilton equation* of matrix $\boldsymbol{A}^n$ which is formulated as

$$\boldsymbol{A}^n = \beta_0 \boldsymbol{I} + \beta_1 \boldsymbol{A} + \beta_2 \boldsymbol{A}^2$$

it can be easily shown from the operations that.

$$\boldsymbol{A}^n = \begin{bmatrix} (3)^n & 0 & 0 \\ 0 & \frac{1}{5}(2)^n + \frac{4}{5}(-3)^n & \frac{1}{5}(2)^n - \frac{1}{5}(-3)^n \\ 0 & \frac{4}{5}(2)^n - \frac{4}{5}(-3)^n & \frac{4}{5}(2)^n + \frac{1}{5}(-3)^n \end{bmatrix}$$

On the other hand, the application of the diagonalization method leads to the calculation of the eigenvectors $\boldsymbol{x}_1 = [1 \quad 0 \quad 0]^T$ for the eigenvalue $\lambda_1 = 3$, $\boldsymbol{x}_2 = [\ 0 \quad 1 \quad 4\ ]^T$ for the eigenvalue $\lambda_2 = 2$, and $\boldsymbol{x}_3 = [\ 0 \quad 1 \quad -1\ ]^T$ for the eigenvalue $\lambda_3 = -3$. Therefore, the matrices $\boldsymbol{P}$ and $\boldsymbol{P}^{-1}$ are calculated as

$$\boldsymbol{P} = \begin{bmatrix} 1 & 0 & 0 \\ 0 & 1 & 1 \\ 0 & 4 & -1 \end{bmatrix} \quad \text{and} \quad \boldsymbol{P}^{-1} = \begin{bmatrix} 1 & 0 & 0 \\ 0 & 1/5 & 1/5 \\ 0 & 4/5 & -1/5 \end{bmatrix}$$

Replacing the preceding expressions in the defining equation of the matrix $\boldsymbol{A}^n$, we get

$$\boldsymbol{A}^n = \boldsymbol{P}\boldsymbol{\Lambda}^n\boldsymbol{P}^{-1} = \begin{bmatrix} 1 & 0 & 0 \\ 0 & 1 & 1 \\ 0 & 4 & -1 \end{bmatrix} \begin{bmatrix} (3)^n & 0 & 0 \\ 0 & (2)^n & 0 \\ 0 & 0 & (-3)^n \end{bmatrix} \begin{bmatrix} 1 & 0 & 0 \\ 0 & 1/5 & 1/5 \\ 0 & 4/5 & -1/5 \end{bmatrix}$$

and performing the operations, we arrive at the same result as that calculated previously.

In the following solved exercises dealing with the solution of the state equation in the field of discrete-time, we will use directly the value of the matrix $\boldsymbol{A}^n$ directly? It is obvious that the result we will present arises through the application of one of the calculation methods presented earlier, and readers are encouraged to verify the result so as to familiarize themselves with the process of calculating this type of matrix.

Problem 7.11.5 Solve the state equation $\boldsymbol{q}[n+1] = \boldsymbol{A}\boldsymbol{q}[n] + \boldsymbol{B}x[n]$ for parameter values

$$\boldsymbol{A} = \begin{bmatrix} 0 & 1 \\ -1/6 & 5/6 \end{bmatrix}, \quad \boldsymbol{B} = \begin{bmatrix} 0 \\ 1 \end{bmatrix}, \quad x[n] = u[n]$$

and initial state vector $\boldsymbol{q}[0] = \begin{bmatrix} 2 & 3 \end{bmatrix}^T$, and then calculate the output of the system if the matrices $\boldsymbol{C}$ and $\boldsymbol{D}$ have the form $\boldsymbol{C} = \begin{bmatrix} -1 & 5 \end{bmatrix}$ and $\boldsymbol{D} = \begin{bmatrix} 0 \end{bmatrix}$. (Lathi, *Linear systems and signals*, Example 10.12).

Solution The required solution $\boldsymbol{q}[n]$ of the system's state equation will arise from the relation

$$\boldsymbol{q}[n] = \boldsymbol{A}^n\boldsymbol{q}[0] + \sum_{k=0}^{n-1} \boldsymbol{A}^{n-1-k}\boldsymbol{B}x[k]$$

In the first step of the process, we need to calculate the exponential function of the matrix $\boldsymbol{A}^n$. Using the first of the methods for calculating this function presented earlier (the reader is asked to verify the result by applying the other two methods), we find that the eigenvalues of matrix $\boldsymbol{A}$ are $\lambda_1 = 1/3$ and $\lambda_2 = 1/2$, while the constants β_0 and β_1 that appear in the *Cayley-Hamilton equation* are

$$\beta_0 = 3\left(\frac{1}{3}\right)^n - 2\left(\frac{1}{2}\right)^n \quad \text{and} \quad \beta_1 = -6\left(\frac{1}{3}\right)^n + 6\left(\frac{1}{2}\right)^n$$

leading to a result in the form

$$\boldsymbol{A}^n = \begin{bmatrix} 3\left(\frac{1}{3}\right)^n - 2\left(\frac{1}{2}\right)^n & -6\left(\frac{1}{3}\right)^n + 6\left(\frac{1}{2}\right)^n \\ \left(\frac{1}{3}\right)^n - \left(\frac{1}{2}\right)^n & -2\left(\frac{1}{3}\right)^n + 3\left(\frac{1}{2}\right)^n \end{bmatrix}$$

It will still hold that

$$\boldsymbol{A}^n \boldsymbol{q}[0] = \begin{bmatrix} 3\left(\frac{1}{3}\right)^n - 2\left(\frac{1}{2}\right)^n & -6\left(\frac{1}{3}\right)^n + 6\left(\frac{1}{2}\right)^n \\ \left(\frac{1}{3}\right)^n - \left(\frac{1}{2}\right)^n & -2\left(\frac{1}{3}\right)^n + 3\left(\frac{1}{2}\right)^n \end{bmatrix} \begin{bmatrix} 2 \\ 3 \end{bmatrix} = \begin{bmatrix} -12\left(\frac{1}{3}\right)^n + 14\left(\frac{1}{2}\right)^n \\ -4\left(\frac{1}{3}\right)^n + 7\left(\frac{1}{2}\right)^n \end{bmatrix}$$

and therefore the system's response to the zero input is calculated as

$$y_{zi}[n] = \boldsymbol{C}\boldsymbol{A}^n \boldsymbol{q}[0] = \begin{bmatrix} -1 & 5 \end{bmatrix} \begin{bmatrix} -12\left(\frac{1}{3}\right)^n + 14\left(\frac{1}{2}\right)^n \\ -4\left(\frac{1}{3}\right)^n + 7\left(\frac{1}{2}\right)^n \end{bmatrix} = -8\left(\frac{1}{3}\right)^n + 21\left(\frac{1}{2}\right)^n$$

On the other hand, calculation of the term $\boldsymbol{A}^{n-1}u[n-1] * \boldsymbol{B}x[n]$ will be performed using the property of time shift of convolution (Sect. 6.10.2, page 541), which is formulated as

$$\boldsymbol{A}^{n-1}u[n-1] * \boldsymbol{B}x[n] = \boldsymbol{A}^n u[n] * \boldsymbol{B}x[n]\Big|_{n \to n-1}$$

Using the value of the matrix function $\boldsymbol{A}^n$ that we determined previously, we have

$$\boldsymbol{A}^n u[n] * \boldsymbol{B}x[n] = \begin{bmatrix} 3\left(\frac{1}{3}\right)^n u[n] - 2\left(\frac{1}{2}\right)^n u[n] & -6\left(\frac{1}{3}\right)^n u[n] + 6\left(\frac{1}{2}\right)^n u[n] \\ \left(\frac{1}{3}\right)^n u[n] - \left(\frac{1}{2}\right)^n u[n] & -2\left(\frac{1}{3}\right)^n u[n] + 3\left(\frac{1}{2}\right)^n u[n] \end{bmatrix} * \begin{bmatrix} 0 \\ u[n] \end{bmatrix}$$

$$= \begin{bmatrix} -6\left[\left(\frac{1}{3}\right)^n u[n] * u[n]\right] + 6\left[\left(\frac{1}{2}\right)^n u[n] * u[n]\right] \\ -2\left[\left(\frac{1}{3}\right)^n u[n] * u[n]\right] + 3\left[\left(\frac{1}{2}\right)^n u[n] * u[n]\right] \end{bmatrix}$$

On the other hand, from the defining equation of convolution we find that

$$\left(\frac{1}{3}\right)^n u[n] * u[n] = \frac{3}{2} - \frac{1}{2}\left(\frac{1}{3}\right)^n \quad \text{and} \quad \left(\frac{1}{2}\right)^n u[n] * u[n] = 2 - \left(\frac{1}{2}\right)^n$$

and substituting the preceding expressions into the last equation results in the form

$$\boldsymbol{A}^n u[n] * \boldsymbol{B}x[n] = \begin{bmatrix} 3\left(\frac{1}{3}\right)^n - 6\left(\frac{1}{2}\right)^n + 3 \\ \left(\frac{1}{3}\right)^n - 3\left(\frac{1}{2}\right)^n + 3 \end{bmatrix}$$

If we make the substitution $n \to n-1$ in the preceding relation, we get

$$\boldsymbol{A}^{n-1} u[n-1] * \boldsymbol{B}x[n] = \underbrace{\begin{bmatrix} 3\left(\frac{1}{3}\right)^n - 6\left(\frac{1}{2}\right)^n + 3 \\ \left(\frac{1}{3}\right)^n - 3\left(\frac{1}{2}\right)^n + 3 \end{bmatrix}}_{n \to n-1} = \begin{bmatrix} 9\left(\frac{1}{3}\right)^n - 12\left(\frac{1}{2}\right)^n + 3 \\ 3\left(\frac{1}{3}\right)^n - 6\left(\frac{1}{2}\right)^n + 3 \end{bmatrix}$$

By substituting the values of the expressions $\boldsymbol{A}^n \boldsymbol{q}[0]$ and $\boldsymbol{A}^{n-1} u[n-1] * \boldsymbol{B}x[n]$ that we calculated in the defining equation of the state vector $\boldsymbol{q}[n]$, it turns out that

$$\boldsymbol{q}[n] = \boldsymbol{A}^n \boldsymbol{q}[0] + \boldsymbol{A}^{n-1} u[n-1] * \boldsymbol{B}x[n] = \begin{bmatrix} -3\left(\frac{1}{3}\right)^n + 2\left(\frac{1}{2}\right)^n + 3 \\ -\left(\frac{1}{3}\right)^n + \left(\frac{1}{2}\right)^n + 3 \end{bmatrix}$$

which is also the desired solution of the system's state equation. The system's response in the zero state is calculated as

$$y_{zs}[n] = \boldsymbol{C}\boldsymbol{A}^{n-1} u[n-1] * \boldsymbol{B}x[n] + \boldsymbol{D}x[n] = \begin{bmatrix} -1 & 5 \end{bmatrix} \begin{bmatrix} 9\left(\frac{1}{3}\right)^n - 12\left(\frac{1}{2}\right)^n + 3 \\ 3\left(\frac{1}{3}\right)^n - 6\left(\frac{1}{2}\right)^n + 3 \end{bmatrix}$$

or, equivalently,

$$y_{zs}[n] = 6\left(\frac{1}{3}\right)^n - 18\left(\frac{1}{2}\right)^n + 12$$

while the total output of the system will have the form

$$y[n] = y_{zi}[n] + y_{zs}[n] = \left[-2\left(\frac{1}{3}\right)^n + 3\left(\frac{1}{2}\right)^n + 12\right] u[n]$$

with the step function $u[n]$ being used as a known to indicate that the system output $y[n]$ constitutes a causal signal.

7.11.5 The Impulse Response in the State Space

If we reformulate the defining equation of the response in the zero state in the form

$$y_{zs}[n] = \boldsymbol{C}\boldsymbol{A}^{n-1} u[n-1] * \boldsymbol{B}x[n] + \boldsymbol{D}(\delta[n] * x[n]) = \left(\boldsymbol{C}\boldsymbol{A}^{n-1} u[n-1]\boldsymbol{B} + \boldsymbol{D}\delta[n]\right) * x[n] = h[n] * x[n]$$

we can formulate the claim that the impulse response of the system $h[n]$ can be calculated from the matrices of the system representation in the state space from the equation

$$h[n] = \boldsymbol{C}\boldsymbol{A}^{n-1}u[n-1]\boldsymbol{B} + \boldsymbol{D}\delta[n] \tag{7.66}$$

In this case, the calculation of the preceding relation leads to the result

$$h[n] = \left[-12\left(\frac{1}{3}\right)^n + 18\left(\frac{1}{2}\right)^n\right]u[n-1]$$

as can be proven by making the substitution $n \to n-1$ in the defining equation of matrix $\boldsymbol{A}^n$ and using the expressions of matrices $\boldsymbol{B}$, $\boldsymbol{C}$, and $\boldsymbol{D}$ given in the statement of the exercise. To prove that the foregoing result is correct, let us calculate the convolution between the signals $h[n]$ and $x[n]$, whose value should coincide with the response of the system in the zero state. From the defining equation of the convolution of discrete-time signals we will have

$$h[n] * x[n] = \sum_{k=-\infty}^{\infty} h[k]x[n-k] = \sum_{k=-\infty}^{\infty}\left[-12\left(\frac{1}{3}\right)^k + 18\left(\frac{1}{2}\right)^k\right]u[k-1]u[n-k] = -12\sum_{k=1}^{n}\left(\frac{1}{3}\right)^k + 18\sum_{k=1}^{n}\left(\frac{1}{2}\right)^k$$

Changing the dummy index of the sums from k to $\xi = k-1$ and using the appropriate expression from the mathematical typology, it turns out that

$$\sum_{k=1}^{\infty}\left(\frac{1}{3}\right)^k = \frac{1}{2} - \frac{1}{2}\left(\frac{1}{3}\right)^n \quad \text{and} \quad \sum_{k=1}^{\infty}\left(\frac{1}{2}\right)^k = 1 - \left(\frac{1}{2}\right)^n$$

Substituting the preceding expressions into the defining equation of the convolution and performing the operations, we end up with the same expression for the zero-state response that we previously calculated, which confirms our claim.

Problem 7.11.6 Solve the state equation $\boldsymbol{q}[n+1] = \boldsymbol{A}\boldsymbol{q}[n] + \boldsymbol{B}x[n]$ for the parameter values

$$\boldsymbol{A} = \begin{bmatrix} 0 & 1 \\ -1/8 & 3/4 \end{bmatrix}, \quad \boldsymbol{B} = \begin{bmatrix} 0 \\ 1 \end{bmatrix}, \quad x[n] = \left(\frac{1}{2}\right)^n u[n]$$

and initial state vector $\boldsymbol{q}[0] = \begin{bmatrix} 4 \; 2 \end{bmatrix}^T$, and then calculate the output of the system if matrices $\boldsymbol{C}$ and $\boldsymbol{D}$ have the form $\boldsymbol{C} = \begin{bmatrix} -1/8 \; 3/4 \end{bmatrix}$ and $\boldsymbol{D} = \begin{bmatrix} 1 \end{bmatrix}$. (Adapted from Hsu, *Signals and Systems*, Solved exercise 7.29 (see also Exercise 7.7)).

Solution In this case, matrix $\boldsymbol{A}$ has the eigenvalues $\lambda_1 = 1/4$ and $\lambda_2 = 1/2$. The application of the *Cayley-Hamilton* method leads to the parameter values

$$\beta_0 = 2\left(\frac{1}{4}\right)^n - \left(\frac{1}{2}\right)^n \quad \text{and} \quad \beta_1 = 4\left(\frac{1}{2}\right)^n - 4\left(\frac{1}{4}\right)^n$$

while matrix $\boldsymbol{A}^n$ is calculated as

$$\boldsymbol{A}^n = \begin{bmatrix} -\left(\frac{1}{2}\right)^n + 2\left(\frac{1}{4}\right)^n & 4\left(\frac{1}{2}\right)^n - 4\left(\frac{1}{4}\right)^n \\ -\frac{1}{2}\left(\frac{1}{2}\right)^n + \frac{1}{2}\left(\frac{1}{4}\right)^n & 2\left(\frac{1}{2}\right)^n - \left(\frac{1}{4}\right)^n \end{bmatrix}$$

It will therefore hold that

$$A^n q[0] = \begin{bmatrix} -\left(\frac{1}{2}\right)^n + 2\left(\frac{1}{4}\right)^n & 4\left(\frac{1}{2}\right)^n - 4\left(\frac{1}{4}\right)^n \\ -\frac{1}{2}\left(\frac{1}{2}\right)^n + \frac{1}{2}\left(\frac{1}{4}\right)^n & 2\left(\frac{1}{2}\right)^n - \left(\frac{1}{4}\right)^n \end{bmatrix} \begin{bmatrix} 4 \\ 2 \end{bmatrix} = \begin{bmatrix} 4\left(\frac{1}{2}\right)^n \\ 2\left(\frac{1}{2}\right)^n \end{bmatrix}$$

while the system's response to the zero input is calculated as

$$y_{zi} = C A^n q[0] = \begin{bmatrix} -\frac{1}{8} & \frac{3}{4} \end{bmatrix} \begin{bmatrix} 4\left(\frac{1}{2}\right)^n \\ 2\left(\frac{1}{2}\right)^n \end{bmatrix} = \left(\frac{1}{2}\right)^n$$

On the other hand, to calculate the sum that appears in the defining equation of the state vector $q[n]$, let us not use the property of the temporal shift of the convolution as we did previously, but let us perform the mathematical operations directly. Using the expressions of matrices A^n and B, it immediately follows that

$$A^n B = \begin{bmatrix} 4\left(\frac{1}{2}\right)^n - 4\left(\frac{1}{4}\right)^n \\ 2\left(\frac{1}{2}\right)^n - \left(\frac{1}{4}\right)^n \end{bmatrix} \qquad \text{or equivalently} \qquad A^{n-1-k} B = \begin{bmatrix} 8\left(\frac{1}{2}\right)^n 2^k - 16\left(\frac{1}{4}\right)^n 4^k \\ 4\left(\frac{1}{2}\right)^n 2^k - 4\left(\frac{1}{4}\right)^n 4^k \end{bmatrix}$$

where for the derivation of the last relation we proceeded to the substitution $n \to n - 1 - k$. Therefore, we will have

$$A^{n-1-k} B\, x[k] = \begin{bmatrix} 8\left(\frac{1}{2}\right)^n 2^k \left(\frac{1}{2}\right)^k - 16\left(\frac{1}{4}\right)^n 4^k \left(\frac{1}{2}\right)^k \\ 4\left(\frac{1}{2}\right)^n 2^k \left(\frac{1}{2}\right)^k - 4\left(\frac{1}{4}\right)^n 4^k \left(\frac{1}{2}\right)^k \end{bmatrix} = \begin{bmatrix} 8\left(\frac{1}{2}\right)^n - 16\left(\frac{1}{4}\right)^n 2^k \\ 4\left(\frac{1}{2}\right)^n - 4\left(\frac{1}{4}\right)^n 2^k \end{bmatrix}$$

from which it follows that

$$\sum_{k=0}^{n-1-k} A^{n-1-k} B\, x[k] = \begin{bmatrix} 8n\left(\frac{1}{2}\right)^n - 16\left(\frac{1}{4}\right)^n \sum_{k=0}^{n-1} 2^k \\ 4n\left(\frac{1}{2}\right)^n - 4\left(\frac{1}{4}\right)^n \sum_{k=0}^{n-1} 2^k \end{bmatrix} = \begin{bmatrix} 8n\left(\frac{1}{2}\right)^n - 16\left(\frac{1}{2}\right)^n + 16\left(\frac{1}{4}\right)^n \\ 4n\left(\frac{1}{2}\right)^n - 4\left(\frac{1}{2}\right)^n + 4\left(\frac{1}{4}\right)^n \end{bmatrix}$$

The last equation was derived using the expression

$$\sum_{k=0}^{n-1} 2^k = \frac{1 - 2^n}{1 - 2} = -(1 - 2^n) = 2^n - 1$$

(see the appropriate relation in the mathematical typology) and performing simple algebraic operations. Using the value of the preceding expression along with the values of matrices C and D and the input signal $x[n]$, the system's response in the zero state is calculated as

$$y_{zi} = C \sum_{k=0}^{n-1-k} A^{n-1-k} B x[k] + D x[n] = 2n\left(\frac{1}{2}\right)^n + \left(\frac{1}{4}\right)^n$$

Therefore, the total response of the system will have the form

$$y[n] = y_{zs} + y_{zi} = \left[(2n+1)\left(\frac{1}{2}\right)^n + \left(\frac{1}{4}\right)^n\right]u[n]$$

while the state vector $\boldsymbol{q}[n]$ will have a value equal to

$$\boldsymbol{q}[n] = \boldsymbol{A}^n\boldsymbol{q}[0] + \sum_{k=0}^{n-1-k} \boldsymbol{A}^{n-1-k}\boldsymbol{B}\,x[k] = \begin{bmatrix} 8n\left(\frac{1}{2}\right)^n - 12\left(\frac{1}{2}\right)^n + 16\left(\frac{1}{4}\right)^n \\ 4n\left(\frac{1}{2}\right)^n - 2\left(\frac{1}{2}\right)^n + 4\left(\frac{1}{4}\right)^n \end{bmatrix}$$

which are also the desired results. The reader is encouraged to determine the impulse response of the system following the methodology presented in the previous section and then to verify the result by proving that the convolution between the impulse response $h[n]$ and the input signal $x[n]$ is equal to the response of the system in the zero state.

7.11.6 Similarity Transformations

In full analogy with continuous-time systems, the choice of the state variables to be used in each case *is not unique*, but there are *infinitely equivalent state vectors* that are related through suitable invertible *transformation matrices*. Assuming that the number of state variables of a system is equal to N and symbolizing by $\boldsymbol{q}[n] = \{q_1[n], q_2[n], \ldots, q_N[n]\}^T$ the current state vector, there is a square transformation matrix $\boldsymbol{W}$ of order N that relates this vector to another state vector, $\boldsymbol{p}[n] = \{p_1[n], p_2[n], \ldots, p_N[n]\}^T$, through the relation

$$\begin{bmatrix} p_1[n] \\ p_2[n] \\ p_3[n] \\ \vdots \\ p_N[n] \end{bmatrix} = \begin{bmatrix} w_{11} & w_{12} & w_{13} & \ldots & w_{1N} \\ w_{21} & w_{22} & w_{23} & \ldots & w_{2N} \\ w_{31} & w_{32} & w_{33} & \ldots & w_{3N} \\ \vdots & \vdots & \vdots & \ddots & \vdots \\ w_{N1} & w_{N2} & w_{N3} & \ldots & w_{NN} \end{bmatrix} \begin{bmatrix} q_1[n] \\ q_2[n] \\ q_3[n] \\ \vdots \\ q_N[n] \end{bmatrix} \tag{7.67}$$

or, in summary notation,

$$\boldsymbol{p}[n] = \boldsymbol{W}\boldsymbol{q}[n] \tag{7.68}$$

where $\boldsymbol{W} = \{w_{ij}\} : (i, j = 1, 2, \ldots, N)$. Because this matrix is *invertible*—which means it should be chosen such that its determinant is characterized by a nonzero value—there is the inverse matrix $\boldsymbol{W}^{-1}$, so that we can write that

$$\boldsymbol{q}[n] = \boldsymbol{W}^{-1}\boldsymbol{p}[n] \tag{7.69}$$

In this case, using the dynamic system equations, we find that

$$\boldsymbol{p}[n+1] = \boldsymbol{w}\boldsymbol{q}[n+1] = \boldsymbol{W}\Big(\boldsymbol{A}\boldsymbol{q}[n] + \boldsymbol{B}x[n]\Big) = (\boldsymbol{W}\boldsymbol{A}\boldsymbol{W}^{-1})\boldsymbol{p}[n] + (\boldsymbol{W}\boldsymbol{B})x[n]$$

$$y[n] = \boldsymbol{C}\boldsymbol{q}[n] + \boldsymbol{D}x[n] = (\boldsymbol{C}\boldsymbol{W}^{-1})\boldsymbol{p}[n] + \boldsymbol{D}x[n]$$

or, equivalently,

$$\begin{aligned} \boldsymbol{p}[n+1] &= \widetilde{\boldsymbol{A}}\boldsymbol{p}[n] + \widetilde{\boldsymbol{B}}x[n] \\ y[n] &= \widetilde{\boldsymbol{C}}p[n] + \widetilde{\boldsymbol{D}}x[n] \end{aligned} \tag{7.70}$$

where $\widetilde{A} = WAW^{-1}$, $\widetilde{B} = WB$, $\widetilde{C} = CW^{-1}$, and $\widetilde{D} = D$. The foregoing expressions constitute the dynamic equations of the system expressed in terms of the new state vector $\boldsymbol{p}[n]$.

It is proven that the *eigenvalues of the system matrix* **A** *are independent of the choice of the state vector that will be used to describe the system and therefore are invariant under similarity transformations*, as these transition processes from one state vector $\boldsymbol{q}[n]$ to a second state vector $\boldsymbol{p}[n]$ are typically called. This proof is based on the same argument that we developed for continuous-time systems and can be found in Sect. 2.9.8 (page 145).

7.11.7 Controllability and Observability

A basic theoretical introduction regarding the concepts of controllability and observability was presented in Sect. 2.9.9 (refer to page 153)—our discussion there applies also to discrete-time systems. Thus, a discrete LTI system described by the equation

$$\boldsymbol{q}[n+1] = \boldsymbol{A}\boldsymbol{q}[n] + \boldsymbol{B}x[n] \tag{7.71}$$

is characterized as *fully controllable* if we have the ability to start from an initial state $\boldsymbol{q}[0] = \boldsymbol{q}_0$ to transfer the system to any desired final state in any case $\boldsymbol{q}[n_1] = \boldsymbol{q}_1$ within a finite number of steps. Conversely, if we do not have the ability to control some or even all of the system's state variables, then this is called *partially controllable* and *uncontrollable*, respectively.

On the other hand, a discrete LTI system is characterized as *fully observable* within a finite time period of duration N if for every time instance n we have the ability to determine the initial state of the system $\boldsymbol{q}[0] = \boldsymbol{q}_0$ knowing only the system's input $x[n]$ and the corresponding output $y[n]$ for the corresponding time period. Conversely, if something like this does not happen, the system is characterized as *nonobservable*.

It is proven that the *controllability* and the *observability* of a LTI system can be described using criteria similar to those we formulated for continuous-time systems. Starting from the case of controllability, let us consider a discrete-time LTI system that is described by the state equation

$$\boldsymbol{q}[n+1] = \boldsymbol{A}\boldsymbol{q}[n] + \boldsymbol{B}x[n]$$

If the vector of the initial state is $\boldsymbol{q}[0]$, then the substitution of the values $n = 0, 1, 2, \ldots, N$ into the foregoing expression will lead us to a set of equations of the form

$$\begin{aligned}
\boldsymbol{q}[1] &= \boldsymbol{A}\boldsymbol{q}[0] + \boldsymbol{B}x[0] \\
\boldsymbol{q}[2] &= \boldsymbol{A}\boldsymbol{q}[1] + \boldsymbol{B}x[1] = \boldsymbol{A}^2 q[0] + \boldsymbol{A}\boldsymbol{B}x[0] + \boldsymbol{B}x[1] \\
\boldsymbol{q}[3] &= \boldsymbol{A}\boldsymbol{q}[2] + \boldsymbol{B}x[2] = \boldsymbol{A}^3 q[0] + \boldsymbol{A}^2\boldsymbol{B}x[0] + \boldsymbol{A}\boldsymbol{B}x[1] + \boldsymbol{B}x[2] \\
&\cdots\cdots\cdots\cdots\cdots\cdots\cdots\cdots\cdots\cdots \\
\boldsymbol{q}[N] &= \boldsymbol{A}^N q[0] + \boldsymbol{A}^{N-1}\boldsymbol{B}x[0] + \cdots + \boldsymbol{A}\boldsymbol{B}x[N-2] + \boldsymbol{B}x[N-1]
\end{aligned}$$

or, equivalently, in matrix form:

$$q[N] - \boldsymbol{A}^N \boldsymbol{q}[0] = \begin{bmatrix} \boldsymbol{B} & \boldsymbol{A}\boldsymbol{B} & \boldsymbol{A}^2\boldsymbol{B} & \ldots & \boldsymbol{A}^{N-1}\boldsymbol{B} \end{bmatrix} \begin{bmatrix} x[N-1] \\ x[N-2] \\ \vdots \\ x[1] \\ x[0] \end{bmatrix} \tag{7.72}$$

For the simple case of the SISO systems we examine, the matrix

$$S_c \begin{bmatrix} \boldsymbol{B} & \boldsymbol{A}\boldsymbol{B} & \boldsymbol{A}^2\boldsymbol{B} & \ldots & \boldsymbol{A}^{N-1}\boldsymbol{B} \end{bmatrix} \tag{7.73}$$

is a square matrix of order N known as the *controllability matrix*. If this matrix is *invertible*, we can determine the sequence of inputs that will cause the transition from the initial to the desired state of the system as the solution of the equation

$$\begin{bmatrix} x[N-1] \\ x[N-2] \\ \vdots \\ x[1] \\ x[0] \end{bmatrix} = \boldsymbol{S_c}^{-1}\left(\boldsymbol{q}[N] - \boldsymbol{A}^N\boldsymbol{q}[0]\right) \tag{7.74}$$

Conversely, in the case of MIMO systems with K inputs, the controllability matrix $\boldsymbol{S_c}$ has dimensions $N \times NK$, while the column matrix of inputs located on the left side of the last equation has dimensions $NK \times 1$. In full analogy with the continuous case, a discrete-time system of order N is fully controllable *if the order of the controllability matrix coincides with the order of the system*, or, in mathematical notation,

$$\text{rank } \boldsymbol{S_c} = N \tag{7.75}$$

(for the definition of matrix order, as well as comments regarding the characteristic of controllability, read Footnote 35 (page 155) of Chap. 2, as well as the discussion in the same section).

On the other hand, for the formulation of the *observability* criterion we apply the same process, which, however, this time is applied to the output equation of the system. Substituting successively into this equation the values $n = 0, 1, 2, \ldots, N-1$, we will get

$$\begin{aligned}
y[0] &= \boldsymbol{Cq}[0] + \boldsymbol{D}x[0] \\
y[1] &= \boldsymbol{Cq}[1] + \boldsymbol{D}x[1] = \boldsymbol{CAq}[0] + \boldsymbol{CB}x[0] + \boldsymbol{D}x[1] \\
y[2] &= \boldsymbol{Cq}[2] + \boldsymbol{D}x[2] = \boldsymbol{CA}^2\boldsymbol{q}[0] + \boldsymbol{CAB}x[0] + \boldsymbol{CB}x[1] + \boldsymbol{D}x[2] \\
&\cdots\cdots\cdots\cdots\cdots\cdots\cdots\cdots\cdots\cdots \\
y[N-1] &= \boldsymbol{CA}^{N-1}\boldsymbol{q}[0] + \boldsymbol{CA}^{N-2}\boldsymbol{B}x[0] + \cdots + \boldsymbol{CAB}x[n-3] + \boldsymbol{CB}x[n-2] + \boldsymbol{D}x[n-1]
\end{aligned}$$

or, equivalently, in matrix form:

$$\begin{bmatrix} y[0] \\ y[1] \\ y[2] \\ \vdots \\ y[N-1] \end{bmatrix} = \begin{bmatrix} \boldsymbol{C} \\ \boldsymbol{CA} \\ \boldsymbol{CA}^2 \\ \vdots \\ \boldsymbol{CA}^{N-1} \end{bmatrix} \boldsymbol{q}[n] + \begin{bmatrix} \boldsymbol{D} & 0 & 0 & \ldots & 0 & 0 \\ \boldsymbol{CB} & \boldsymbol{D} & 0 & \ldots & 0 & 0 \\ \boldsymbol{CAB} & \boldsymbol{CB} & \boldsymbol{D} & \ldots & 0 & 0 \\ \vdots & \vdots & \vdots & \ddots & \vdots & \vdots \\ \boldsymbol{CA}^{N-2} & \boldsymbol{CA}^{N-3} & \boldsymbol{CA}^{N-4} & \ldots & \boldsymbol{CB} & \boldsymbol{D} \end{bmatrix} \begin{bmatrix} x[0] \\ x[1] \\ x[2] \\ \vdots \\ x[N-1] \end{bmatrix}$$

An equation that is analogous to the one constructed for continuous-time systems. Following the argument we formulated then, we will call the matrix

$$\boldsymbol{S_o} = \begin{bmatrix} \boldsymbol{C} \\ \boldsymbol{CA} \\ \boldsymbol{CA}^2 \\ \vdots \\ \boldsymbol{CA}^{N-1} \end{bmatrix} \tag{7.76}$$

—which for simple SISO systems is a square matrix of order N, while for MIMO systems with L outputs it has dimensions $NL \times N$—an *observability matrix of the system*. It is proven that a discrete LTI system is fully observable *when the order of this matrix is equal to N*, or, in mathematical notation,

$$\text{rank}\, \boldsymbol{S_o} = N \tag{7.77}$$

The foregoing analysis refers to the general case of system matrices with arbitrary eigenvalues and corresponding eigenvectors, while for the special cases of matrices that can be diagonalized, the observations we formulated in Sect. 2.9.9 apply. As for the concept of *duality* between controllability and observability, as well as the normal forms of controllability (*controllable form* and *controller form*) and observability (*observable form* and *observer form*), the situation is completely analogous to that described for continuous-time systems and does not need to be described again. To familiarize themselves with the application of these concepts in discrete-time system description processes, readers are invited to study the exercises following Sect. 2.9.9, since the characterization of a system in terms of its controllability and observability depends exclusively on the ranks of the controllability and observability matrices and has nothing to do with whether the system we are dealing with is defined in the field of continuous or discrete-time.

7.11.8 Stability in State Space

The theory that describes the stability of continuous LTI systems in state space applies to discrete-time systems and so need not be repeated here; the reader is encouraged to review the material in Sect. 2.9.10 on page 173 and the exercise that follows in order to understand in depth how to characterize a system in terms of its stability. At this point we simply recall for the sake of the text's autonomy that *a discrete LTI system is characterized as asymptotically stable when all the eigenvalues of the system matrix* $\boldsymbol{A}$ *have negative real parts*. We will revisit the issue of stability of discrete LTI systems in a later chapter, where we will examine this characteristic of discrete-time systems using the $\mathcal{Z}$ transform.

7.12 Problems

1. Determine the input/output equation of the discrete-time systems depicted in Fig. 7.15a,b. For any of these equations that you determine to be difference equations, specify their order and the coefficients that appear in them.

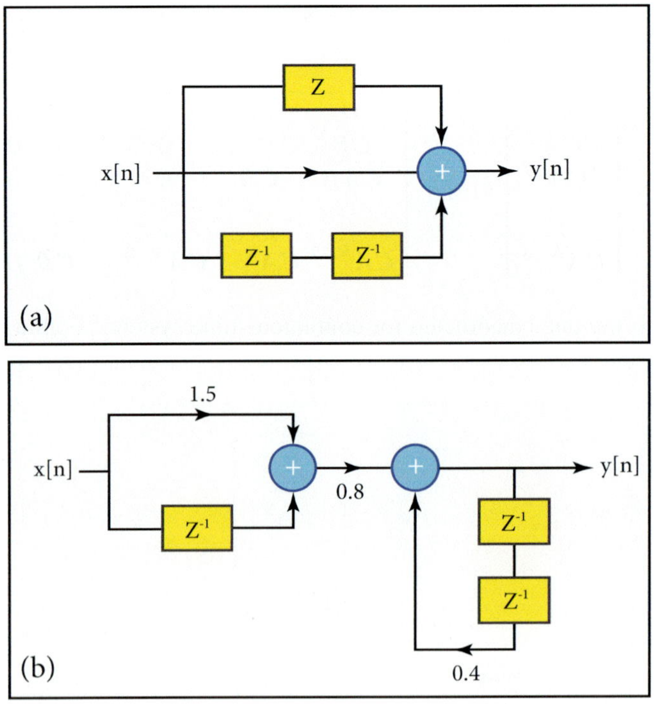

Fig. 7.15 Schematic diagrams of Problem 1

2. For each of the following equations, determine the values of the coefficients α_k and β_k, as well as the values of the parameters M and N, and decide whether or not it is a difference equation. Then draw the schematic diagrams of the discrete-time systems described by them.

$$y[n] = [n-1] + x[n-2] + x[n-3] + x[n-4]$$
$$y[n] = 5x[n] + 3x[n-1] + y[n-2]$$
$$y[n] = x^2[n] + x[n-1] + y[n-1]$$

3. Consider a discrete-time system described by the input/output equation

$$y[n] = \mathcal{T}\{x[n]\} = x[n^2]$$

(a) Examine whether the system is time-invariant. To substantiate the previous result, use as input to the system the discrete-time signal

$$x[n] = \begin{cases} 1\, 0 \leq n \leq 3 \\ 0 \text{ anywhereelse} \end{cases}$$

and perform the following actions:
(1) Design the signal $x[n]$.
(2) Determine and design the signal $y[n] = \mathcal{T}\{x[n]\}$.
(3) Design the signal $y_2'[n] = y[n-2]$.
(4) Determine and design the signal $x_2[n] = x[n-2]$.
(5) Determine and design the signal $y_2[n] = \mathcal{T}\{x_2[n]\}$.
(6) Compare the signals $y_2[n]$ and $y[n-2]$. What do you observe?
(b) Repeat the foregoing process for the discrete-time systems

$$y[n] = \mathcal{T}\{x[n]\} = x[n] - x[n-1]$$
$$y[n] = \mathcal{T}\{x[n]\} = nx[n]$$

4. The following discrete-time systems are given:

$$\text{(a)} y[n] = x^3[n], \quad \text{(b)} y[n] = \frac{e^{2x[n]}}{x[n-1]}, \quad ;\text{(c)} y[n] = \sin(x[n]), \quad ;\text{(d)} y[n] = \sum_{k=-\infty}^{n} x[k], \quad ;\text{(e)} y[n] = \log(1-|x[n]|)$$

Determine whether each of these system is a stable or unstable discrete-time system.

5. The following discrete-time systems are given:

$$y[n] = x[2+n] \quad y[n] = [x[n]] \quad y[n] = [x[n]] + 1$$
$$y[n] = |x[n]| \quad y[n] = x(n)u[n] \quad y[n] = x[n] + n^2 x[n+1]$$
$$y[n] = x[3n] \quad y[n] = x[-2n] \quad y[n] = sgn(x[n])$$
$$y[n] = \cos(x[n]) \quad y[n] = \sum_{k=-\infty}^{n+1} x[k] \quad y[n] = x[n]\cos(\omega n)$$

For each of the foregoing systems, examine whether it is (a) static or dynamic, (b) time-varying or time-invariant, (c) linear or nonlinear, (d) causal or noncausal, and (e) stable or unstable.

6. The following discrete-time systems are given:

$$y[n] = \frac{1}{2} \sum_{k=-\infty}^{\infty} x[k]\Big(\delta[n-k] + \delta[n+k]\Big), \quad y[n] = \frac{x[n-1]}{x[n]}, \quad y[n] = x[n]r[n]$$

[in the third system, the discrete-time signal $r[n]$ is the ramp function]. For each of these systems, determine whether it is (a) static or dynamic, time-varying or time-invariant, (c) linear or nonlinear, (d) causal or noncausal, and (e) stable or unstable.

7. For every possible combination of input signal/impulse response pairs, determine the output of the discrete-time LTI system associated with them:

$$x[n] = 2u[n] \qquad h[n] = \alpha^n u[-n-2], \quad \alpha > 1$$
$$x[n] = u[n-4] \qquad h[n] = 2^n u[-n-1]$$
$$x[n] = u[n] \qquad h[n] = 0.8(0.1)^n u[-n]$$
$$x[n] = u[n] - u[n-10] \; h[n] = 2^n u[-n]$$

8. Considering a discrete-time LTI system with an impulse response of the form $h[n] = u[n]$, determine its output for the input signal

$$x[n] = \begin{cases} 0 & n < 0 \\ \alpha^n & 0 \le n \le N_1 \\ 0 & N_1 < n < N_2 \\ \alpha^{n-N_2} & N_2 \le n \le N_1 + N_2 \\ 0 & n > N_1 + N_2 \end{cases}$$

9. Considering a discrete-time LTI system with impulse response $h[n]$, prove that if a periodic signal $x[n]$ with period N is used at its input, then its output $y[n]$ is also a periodic signal characterized by the same period value.
10. Considering a discrete-time LTI system with impulse response $h[n] = (1/2)^n u[n]$, calculate its output when the signals $x_1[n] = 2^{n+1}u[n]$ and $x_2[n] = u[-n-2]$ are used at its input.
11. Determine the impulse response of a second-order discrete-time LTI system described by the difference equation

$$y[n] - 6y[n-1] + 25y[n-2] = 2x[n] - 4x[n-1]$$

12. Determine the impulse response of a LTI nonrecursive discrete-time system described by the difference equation

$$y[n] = 3x[n] - 5x[n-1] - 2x[n-3]$$

13. Determine the impulse response of a second-order discrete-time LTI system described by the difference equation

$$y[n] - 3y[n-1] - 4y[n-2] = x[n] + 2x[n-1].$$

14. Prove that the impulse response of a discrete-time LTI system described by the difference equation

$$y[n] + 0.4y[n-1] = x[n] + x[n-1] \qquad \text{has the form} \qquad h[n] = (-0.4)^n u[n] + (-0.4)^{n-1} u[n-1].$$

15. Prove that the impulse response of a discrete-time LTI system described by the difference equation

$$y[n] - 0.9y[n-1] + 0.2y[n-2] = x[n-1] \qquad \text{has the form} \qquad h[n] = 10\left[(0.5)^n - (0.4)^n\right]u[n-1].$$

16. Prove that the impulse response of a discrete-time LTI system described by the difference equation

$$y[n] - 0.6y[n-1] - 0.16y[n-2] = 5x[n] \qquad \text{has the form} \qquad h[n] = \left[(-0.2)^n + 4(0.8)^n\right]u[n]$$

17. Applying the theory presented earlier, determine the output of the system described by the frequency response function

$$\mathcal{H}(e^{j\omega}) = e^{-[j\omega-(\pi/4)]}\left(\frac{1+e^{-j3\omega}+4e^{-j2\omega}}{1+2e^{-j\omega}}\right), \quad (-\pi < \omega \leq \pi)$$

when the discrete trigonometric signal $x[n] = 5\cos(\pi n/8)$ is used at its input.

18. Determine the frequency response of discrete recursive systems described by the difference equations

$$S_1: \quad y[n] - 6y[n-1] + 5y[n-2] = 2x[n] - 4x[n-1]$$
$$S_2: \quad y[n] - y[n-1] - 15y[n-2] = 3x[n] - 5x[n-1] - 5x[n-2]$$
$$S_3: \quad y[n] - 3.7y[n-1] + 0.2y[n-2] = x[n-1]$$

and then determine their output when signals are used as their input

$$x_1[n] = -6\cos\left(\frac{\pi n}{12}\right) \quad \text{and} \quad x_n[2] = 15\cos\left(\frac{\pi n}{8}+\frac{\pi}{2}\right)$$

19. Determine the output $y[n]$ of a discrete-time system LTI with frequency response

$$\mathcal{H}(e^{j\omega}) = e^{-j\omega/4}, \quad -\pi < \omega \leq \pi$$

when the elementary trigonometric signal is used at its input

$$x[n] = \cos\left(\frac{5\pi n}{7}\right)$$

20. Determine the output $y[n]$ of a discrete-time system LTI with frequency response

$$\mathcal{H}(e^{i\omega}) = \begin{cases} e^{-i\omega} & |\omega| < \frac{2\pi}{16}\left(\frac{3}{2}\right) \\ 0 & \frac{2\pi}{16}\left(\frac{3}{2}\right) \leq |\omega| \leq \pi \end{cases}$$

when the next discrete-time signal characterized by a period value $k = 16$ is used at its input.

$$x[n] = \sum_{k=-\infty}^{\infty} \delta[n+16k]$$

21. Determine the first five samples of the output $y[n]$ of a discrete-time LTI system described by the difference equation

$$5y[n] + 2y[n-1] - 5y[n-2] = 3x[n] - 5x[n-1]$$

for initial conditions $y[-1] = 3$ and $y[-2] = 3$ and input signal $x[n] = (5)^n u[n]$.

22. Determine the first 15 samples of the output $y[n]$ of a discrete-time LTI system described by the difference equation

$$y[n] + 3y[n-1] - 5y[n-2] = x[n] + 3x[n-1] - x[n-2]$$

for initial condition $y[-1] = 5$ and input signal of the form $x[n] = e^{-n}u[n]$.

23. Determine the first 10 samples of the output $y[n]$ of a discrete-time LTI system described by the difference equation

$$y[n] + 3y[n-1] - 5y[n-2] = x[n] + 3x[n-1] - x[n-2]$$

for initial conditions $y[-1] = -1$ and $y[-2] = 2$ and input signal $x[n] = (-3)^n u[n]$.

24. Determine the natural response of second-order discrete-time LTI systems defined as

$$y[n] - 1.56y[n-1] + 0.81y[n-2] = x[n] + 3x[n-1] \quad \text{and} \quad y[n] + 4y[n-2] = 2x[n]$$

25. Prove that the natural response of a second-order discrete-time LTI system described by the equation

$$y[n] - 1.56y[n-1] + 0.81y[n-2] = x[n] + 3x[n-1]$$

with initial conditions $y[-1] = 2$ and $y[-2] = 1$ has the form

$$y_{zi}[n] = 2.34(0.9)^n \cos\left(\frac{\pi n}{6} - 0.17\right) \quad (n \geq 0)$$

26. Prove that the natural response of a second-order discrete-time LTI system described by the equation

$$y[n] + 4y[n-2] = 2x[n]$$

with initial conditions $y[-1] = -1/(2\sqrt{2})$ and $y[-2] = 1/(4\sqrt{2})$ takes the form

$$y_{zi}[n] = (2)^n \cos\left(\frac{\pi n}{2} - \frac{3\pi}{4}\right) \quad (n \geq 0)$$

27. Assuming that the difference equation of Exercise 26 describes the operation of a discrete LTI system, determine the zero-state response and the zero-input response and verify that their sum, which also expresses the total output of the system, is given by the same relation as the one you previously calculated.

28. Solve the full second-order difference equation

$$y[n] - 0.8y[n-1] = x[n-1]$$

for an input signal of the form $x[n] = \cos[(\pi n/6) - 0.2]$ and arbitrary initial conditions.

29. Solve the full second-order difference equation

$$y[n] + 3y[n-1] + 2y[n-2] = 2^n u[n-1]$$

for initial conditions $y[-1] = 0$ and $y[-2] = 1$.

30. Find the response of a recursive LTI system described by the equation

$$y[n] - 0.8y[n-1] = x[n-1]$$

to the input signals

$$x_1[n] = \cos\left(\frac{\pi n}{2} + \frac{\pi}{3}\right)u[n] \quad \text{and} \quad x[n] = \sin\left(\frac{3\pi n}{5}\right)u[n]$$

and for initial conditions $y[-1] = y[-2 = -1$.

31. Assuming that the difference equation of Exercise 30 describes the operation of a discrete LTI system to determine the zero-state response and the zero-input response and to verify that their sum, which expresses the total output of the system, is given by the same relation as the one you previously calculated.

32. Find the general solution of the full difference equation

$$y[n] - 5y[n-1] + 36y[n-2] = n^2 - 1$$

Hint: Use the quadratic polynomial function as a particular solution of the full equation

$$y_p[n] = An^2 + Bn + C$$

33. Determine the impulse and step response of discrete LTI systems described by the equations

$$y[n] = 0.6y[n-1] - 0.08y[n-2] + x[n]$$

$$y[n] = 0.7y[n-1] - 0.1y[n-2] + 2x[n] - x[n-2]$$

Hint: To calculate the step response $s[n]$, solve the full difference equation for input signal $x[n] = u[n]$, and consider a particular solution of the form $y[n] = C$, then you can calculate the impulse response $h[n]$ from the relation $h[n] = s[n] - s[n-1]$. You can verify the correctness of the result you arrived at by calculating the impulse response in the manner presented in detail in a previous section and comparing the results.

34. Determine the impulse and step response of a discrete LTI system described by the difference equation

$$y[n] + 3y[n-1] = x[n]$$

35. If the step response $s[n]$ of the discrete LTI system described by the equation

$$y[n] - \frac{3}{4}y[n-1] + \frac{1}{8}y[n-2] = x[n]$$

is given by the relation

$$s[n] = \left[\frac{8}{3} - 2\left(\frac{1}{2}\right)^n + \frac{1}{3}\left(\frac{1}{4}\right)^n\right]u[n]$$

determine the initial conditions of the problem.

36. Considering the discrete-time LTI system of second-order

$$x[n] - 3y[n-1] - 4y[n-2] = 0$$

prove (using the recursive method) that the system's response to zero input is

$$y_{zi}[n] = \left(-\frac{4}{3}\right)^{n+2} y[-2]$$

37. Determine the zero-state response of the discrete LTI system

$$y[n] + \frac{1}{2}y[n-1] = x[n] + 2x[n-2]$$

when the signal $x(n) = \{1, 2, \mathbf{3}, 4, 2, 1\}$ is used as the input. The exercise should be solved by recursively using the system's difference equation.

38. Determine the response $y[n]$ $(n \geq 0)$ of the discrete LTI system

$$y[n] - 4y[n-1] + 4y[n-2] = x[n] - x[n-1]$$

to the input $x[n] = (-1)^n u(n)$ and for initial conditions $y[-1] = y[-2] = 0$.

39. Assume a discrete LTI system described by the equation

$$y[n] - \frac{1}{2}y[n-1] = x[n]$$

receives at its input the signal $x[n] = (1/3)^n u[n]$. (a) Prove that the solution of the homogeneous equation is $y[n] = A(1/2)^n u[n]$. (b) If $y_p[n] = B(1/3)^n u[n]$ is a partial solution of the system's difference equation such that

$$y_p[n] - \frac{1}{2}y_p[n-1] = \left(\frac{1}{3}\right)^n u[n]$$

determine the value of the constant B.

40. Find the impulse response of the LTI system described by the equation

$$y[n]-4y[n-1]+4y[n-2]=x[n]-x[n-1]$$

41. Considering a LTI system described by the equation

$$y[n]-\frac{3}{4}y[n-1]+\frac{1}{8}y[n-2]=2x[n-1]$$

determine the output $y[n]$ for time instants $n \geq 0$ if $x[n]=\delta[n]$ and $y[n]=0$ for $n<0$.

42. Determine the natural, impulse, and step responses of a causal discrete LTI system described by the difference equation

$$y[n]-5y[n-1]+6y[n-2]=2x[n-1]$$

43. Consider a discrete LTI system described by the difference equation

$$y[n]-\frac{5}{6}y[n-1]+\frac{1}{6}y[n-2]=\frac{1}{3}x[n-1]$$

- Find its natural, impulse, and step responses.
- Find the general form of the solution of a homogeneous system (i.e., a system described by the homogeneous difference equation).
- If $y[0]=y[1]=1$, find the system response $y[n]$ to the input $x[n]=\delta[n]$.

44. Determine the natural response of a LTI second-order discrete-time system described by the equation

$$y[n]+10y[n-1]+25y[n-2]=x[n]+x[n-2]$$

and express it in terms of the initial conditions $y[-1]$ and $y[-2]$.

45. Determine the natural response of a LTI third-order discrete-time system described by the equation

$$y[n]-6y[n-1]+12y[n-2]-8y[n-3]=5x[n]+2x[n-1]-4x[n-2]$$

and express it in terms of the initial conditions $y[-1]$, $y[-2]$, and $y[-3]$.

46. Consider a discrete-time systems LTI described by the difference equations

$$y[n]-\frac{1}{4}y[n-1]-\frac{1}{8}y[n-2]=3x[n]$$

$$y[n]+\frac{1}{15}y[n-1]-\frac{2}{15}y[n-2]=x[n]$$

Find their response $y[n]$ when the discrete-time signals $x[n]=(1/2)^n u[n]$ and $x[n]=(3/5)^n u[n]$ are transmitted to their input.

47. Consider a discrete-time system described by the equation

$$y[n]=ny[n-1]+x[n] \quad (n \geq 0)$$

which is assumed to be at rest (i.e., the relation $y[-1]=0$ holds). Check whether it is linear, time-invariant, and stable.

48. Examine whether the discrete LTI system with impulse response

$$h[n]=\delta[n]+\left(\frac{1}{6}\right)^n u[n-2]$$

is causal and stable, and then determine the system's response in the zero state when the discrete-time signal $x[n] = u[n-5] - u[n+5]$ is used at its input.

49. Consider recursive LTI systems described by the difference equations

$$S_1: \quad y[n] + 0.5y[n-1] - 0.15y[n-2] = x[n-1] - 2x[n]$$
$$S_2: \quad y[n] + 7y[n-1] + 2y[n-2] = x[n-1] + 2x[n-2]$$
$$S_3: \quad y[n] + y[n-1] - 8y[n-2] = x[n] + 2x[n-1]$$

(a) For each system, determine the type of asymptotic stability and BIBO stability.
(b) Determine their impulse response $h[n]$.
(c) Determine the total output $y[n]$ when the input signals are used at its input

$$x_1[n] = u[n], \qquad x_2[n] = 3^n u[n], \qquad x_3[n] = \sin\left(\frac{5\pi n}{9} + \frac{\pi}{4}\right)u[n]$$

for initial conditions $y[-1] = 1$ and $y[-2] = 2$, both with the classic method of solving full difference equations and using the appropriate partial solution from Table 7.1 (page 593), as well as by calculating the response of the systems in the zero state and at the zero input.

50. Show that the step response of a discrete-time LTI system with an impulse response of the form

$$h[n] = (n+1)\alpha^n u[n], \quad |\alpha| < 1$$

is given by the equation

$$s[n] = \left[\frac{1}{(\alpha-1)^2} - \frac{\alpha}{(\alpha-1)^2}\alpha^n + \frac{\alpha}{(\alpha-1)}(n+1)\alpha^n\right]u[n]$$

51. Let there be two discrete-time LTI systems, $\mathcal{T}_1$ and $\mathcal{T}_2$, which are connected in series according to the scheme presented earlier. Based on the theory that describes their connection method, characterize each of the following statements as correct or incorrect.

- If the systems $\mathcal{T}_1$ and $\mathcal{T}_2$ are linear, then the system resulting from their connection is also a linear discrete-time system.
- If the systems $\mathcal{T}_1$ and $\mathcal{T}_2$ are time-invariant, then the system resulting from their connection is also a time-invariant discrete-time system.
- If the systems $\mathcal{T}_1$ and $\mathcal{T}_2$ are causal, then the system resulting from their connection is also a causal discrete-time system.
- If the systems $\mathcal{T}_1$ and $\mathcal{T}_2$ are LTI, then the system resulting from their connection is also a discrete-time LTI system.
- If the systems $\mathcal{T}_1$ and $\mathcal{T}_2$ are nonlinear, then the system resulting from their connection is also a nonlinear discrete-time system.
- If the systems $\mathcal{T}_1$ and $\mathcal{T}_2$ are stable, then the system resulting from their connection is also a stable discrete-time system.

52. Considering three discrete-time LTI systems with impulse responses $h_1[n] = \delta[n] - \delta[n-1]$, $h_2[n] = u[n]$, and $h_3[n] = \alpha^n u[n]$, find the impulse response of the composite system that will result if (a) the three systems are connected in series, (b) the three systems are connected in parallel, and (c) the systems $h_1[n]$ and $h_2[n]$ are connected in parallel, while the composite system resulting from their connection is connected in series with the system $h_3[n]$.

53. Let us consider the interconnection of the discrete LTI systems of Fig. 7.16.

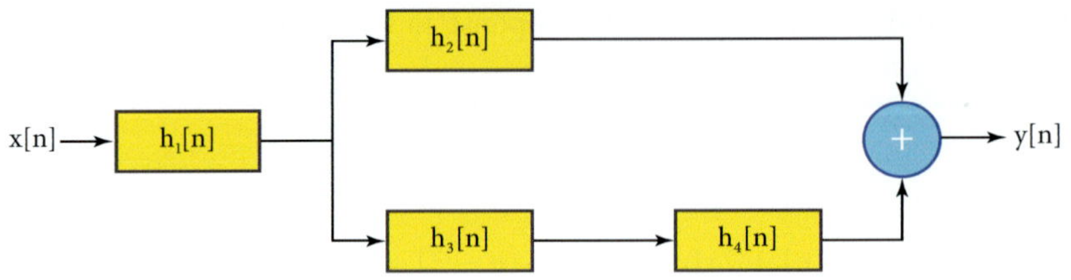

Fig. 7.16 Example of interconnection of LTI systems

- Express the impulse response of a composite system as a function of the impulse responses $h_1[n]$, $h_2[n]$, $h_3[n]$, and $h_4[n]$.
- Find the impulse response $h[n]$ when the impulse responses of the three elementary systems are

$$h_1[n] = \left\{\frac{1}{3}, \frac{1}{5}, \frac{1}{3}\right\}, \qquad h_2[n] = h_3[n] = (n+1)u[n], \qquad h_4[n] = \delta[n-3]$$

- Determine the response of a composite system when the discrete-time signal is transmitted to its input $x[n] = \delta[n+2] + 3\delta[n-3] - 4\delta[n+3]$.

54. Considering two discrete LTI systems with impulse responses

$$h_1[n] = 3^n u[n] \qquad \text{and} \qquad h_2[n] = \delta[n] - \delta[n-2]$$

characterize in terms of BIBO stability and its asymptotic stability the composite system that will result from the interconnection of the two systems (a) in series and (b) in parallel.

55. For each of the recursive discrete LTI systems described by the equations

$$\begin{aligned} S_1 &: \quad y[n] - 6y[n-1] + 25y[n-2] = 2x[n] - 4x[n-1] \\ S_2 &: \quad y[n] - 0.9y[n-1] + 0.2y[n-2] + y[n-4] = x[n] - 3x[n-3] \\ S_3 &: \quad y[n] + 0.2y[n-1] + 0.7y[n-2] - 9y[n-3] = x[n] + 3x[n-1] - 12x[n-3] \end{aligned}$$

Determine their dynamic equations based on both their normal and nonnormal form.

56. Construct the normal and nonnormal form of the schematic diagrams of discrete recursive LTI systems described by the difference equations

$$\begin{aligned} S_1 &: \quad y[n] - 6y[n-1] + 25y[n-2] = 2x[n] - 4x[n-1] \\ S_2 &: \quad y[n] - y[n-1] - 5y[n-2] = 3x[n] - 5x[n-1] - 2x[n-2] \\ S_3 &: \quad y[n] - 0.9y[n-1] + 0.2y[n-2] = x[n-1] \end{aligned}$$

57. Prove that

$$\boldsymbol{A}^n = \begin{bmatrix} 3(-1)^n - 2(-2)^n & -2(-1)^n + 2(-2)^n \\ 3(-1)^n - 3(-2)^n & -2(-1)^n + 3(-2)^n \end{bmatrix} \quad \text{where} \quad \begin{bmatrix} 1 & -2 \\ 3 & -4 \end{bmatrix}$$

58. Calculate matrix $\boldsymbol{A}^n$, with matrix $\boldsymbol{A}$ having dimensions 3×3 being the

$$\boldsymbol{A} = \begin{bmatrix} -1 & 2 & 2 \\ 2 & -1 & 2 \\ 2 & 2 & -1 \end{bmatrix}$$

59. Prove that the impulse response of a system described by the equations

$$\boldsymbol{q}[n+1] = \begin{bmatrix} 0 & 1 \\ 1 & 1 \end{bmatrix} \boldsymbol{q}[n] + \begin{bmatrix} 0 \\ 1 \end{bmatrix} x[n] \quad \text{and} \quad y[n] = \begin{bmatrix} 1 & 1 \end{bmatrix} \boldsymbol{q}[n] + x[n]$$

is given by the equation

$$h[n] = \frac{1}{\sqrt{5}} \left[\left(\frac{3+\sqrt{5}}{2}\right)\left(\frac{1+\sqrt{5}}{2}\right)^{n-1} - \left(\frac{3-\sqrt{5}}{2}\right)\left(\frac{1-\sqrt{5}}{2}\right)^{n-1} \right] u[n-1] + \delta[n]$$

This system is used to construct the well-known Fibonacci sequence, each number of which is derived as the sum of the two previous elements.

60. For each of the discrete LTI systems described by the equations

(a) $$\begin{cases} \boldsymbol{q}[n+1] = \\ \quad ! \begin{bmatrix} 0 & 1 \\ -0.81 & 1 \end{bmatrix} \boldsymbol{q}[n] + \begin{bmatrix} 0 \\ 1 \end{bmatrix} x[n] \\ y[n] = \quad \begin{bmatrix} -0.81 & 1 \end{bmatrix} \boldsymbol{q}[n] + x[n] \end{cases}$$

(b) $$\begin{cases} \boldsymbol{q}[n+1] = \begin{bmatrix} 0 & 1 \\ -2 & -3 \end{bmatrix} \boldsymbol{q}[n] + \begin{bmatrix} 0 \\ 1 \end{bmatrix} x[n] \\ y[n] = \begin{bmatrix} 1 & 0 \end{bmatrix} \boldsymbol{q}[n] + x[n] \end{cases}$$

(c) $$\begin{cases} \boldsymbol{q}[n+1] = \begin{bmatrix} -0.3 & 0.4 \\ 0.4 & -0.3 \end{bmatrix} \boldsymbol{q}[n] + \begin{bmatrix} 0 \\ 1 \end{bmatrix} x[n] \\ y[n] = \begin{bmatrix} 1 & 1 \end{bmatrix} \boldsymbol{q}[n] + x[n] \end{cases}$$

(a) determine the characteristic polynomial, the eigenvalues of the system, and the value of matrix $\boldsymbol{A}^n$ and (b) considering a step input of the form $x[n] = u[n]$, determine the state vector and the overall response of the system if the value of the initial state vector is $\boldsymbol{q}[0] = \begin{bmatrix} 2 & -3 \end{bmatrix}$.

8 The Z Transform

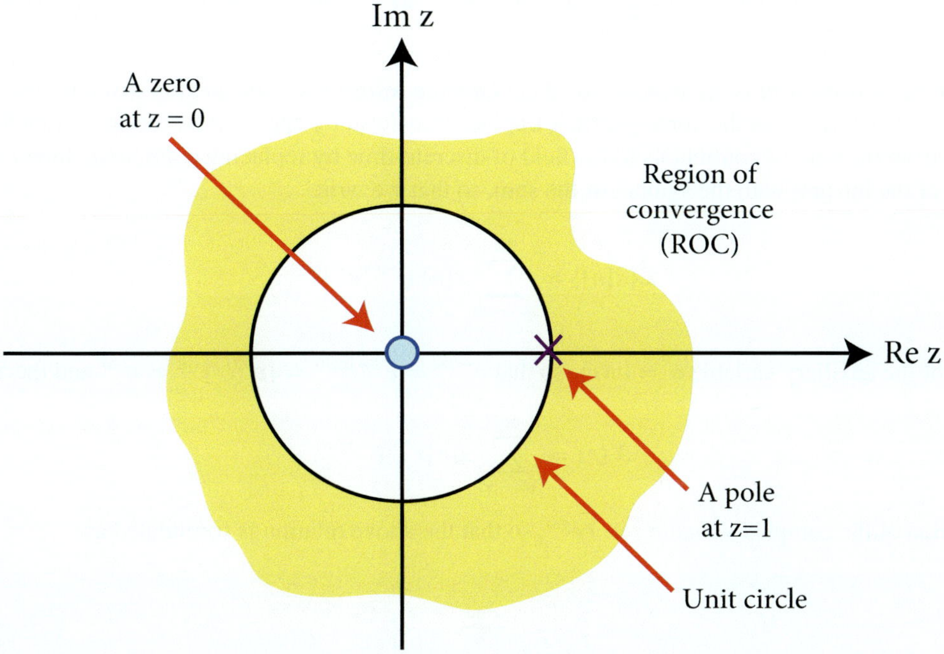

The causality and stability of a discrete-time signal or system depends on the positions of the poles of the Z transform of the system's impulse response in the complex plane, in relation to the circumference of the unit circle

8.1 Introduction

One of the fundamental tools for studying signals and systems in the field of discrete-time is the $\mathcal{Z}$ *transform*; this transform is applied to a discrete sequence $x[n]$ and is defined as

$$\mathcal{X}(z) \equiv \sum_{n=-\infty}^{\infty} x[n]z^{-n}, \tag{8.1}$$

an equation that describes an *infinite series of terms* of the *complex quantity* z. The above transform can also be considered an *operator* that converts the *discrete sequence* $x[n]$ into the *complex function* $\mathcal{X}(z)$ and in this sense it can be written in the form

A. I. Margaris, *Continuous and Discrete-Time Signals and Systems*,
https://doi.org/10.1007/978-3-031-81209-5_8

$$\mathcal{Z}\{x(n)\} = \sum_{n=-\infty}^{\infty} x[n]z^{-n} = \mathcal{X}(z). \tag{8.2}$$

As for the correlation between the discrete scheme $x[n]$ and the transform $\mathcal{X}[z]$, this is symbolized as $x[n] \leftrightarrows \mathcal{X}(z)$. In full analogy with the *Fourier* and Laplace transforms that were presented in previous chapters, this symbolism implies that *the $\mathcal{Z}$ transform of the discrete-time signal $x[n]$ is the function $\mathcal{X}(z)$, whereas the inverse $\mathcal{Z}$ transform of the function $\mathcal{X}(z)$ is the signal $x[n]$*.

If we compare the defining equations of the Laplace and $\mathcal{Z}$ transforms

$$\mathcal{L}\{x(t)\} = \int_{-\infty}^{+\infty} x(t)\, e^{-st} dt = \int_{-\infty}^{+\infty} x(t)\, e^{-\sigma t} e^{-j\omega t} dt \qquad \text{and} \qquad \mathcal{Z}\{x(n)\} = \sum_{n=-\infty}^{\infty} x[n]z^{-n},$$

we can assert that *the $\mathcal{Z}$ transform is the equivalent of the Laplace transform in the field of discrete-time*, and in reality we can construct the $\mathcal{Z}$ transform from the corresponding Laplace transform by applying a process that includes the following steps: (a) moving from the field of continuous to the field of discrete-time by replacing continuous-time t with discrete-time n and the symbol of the integral with the symbol of the sum, so that we write

$$\mathcal{L}\{x[n]\} = \sum_{n=-\infty}^{\infty} x[n]\, e^{-\sigma n} e^{-j\omega n}; \tag{8.3}$$

(b) the definition of the auxiliary variable $\sigma = \ln(r)$, so that $e^{-\sigma n} = e^{-n\ln(r)} = [e^{\ln(r)}]^{-n} = r^{-n}$ and the above relation will take the form

$$\mathcal{X}(r) = \sum_{n=-\infty}^{\infty} x[n]r^{-n} e^{-j\omega n}; \tag{8.4}$$

and (c) the definition of the complex variable $z = re^{j\omega}$, so that the above relation is formulated as

$$\mathcal{X}(z) \equiv \sum_{n=-\infty}^{\infty} x[n]z^{-n}. \tag{8.5}$$

The basic consequence of the above transformation of one transform to another is the shape of the *region of convergence* of the $\mathcal{Z}$ transform. In full analogy with the Laplace transform, the $\mathcal{Z}$ transform will also have its own region of convergence, which is defined as *the set of points z of the complex plane for which the power series appearing in the defining equation converges to some finite value*. For the case of the Laplace transform in which the complex variable s is defined in Cartesian coordinates as $s = \sigma + j\omega$, the region of convergence consists of *vertical regions of the complex plane*, the boundaries of which are *lines parallel to the vertical axis*. Conversely, for the case of the $\mathcal{Z}$ transform, the complex variable describing it is expressed in polar form $z = re^{-j\omega}$ from which it follows that the region of convergence of the transform should take the form *of a circle's circumference*. From the relation $\sigma = -\ln(r)$ it is not difficult to realize that the case $\sigma = 0$ corresponds to the case $r = 1$, and therefore *the unit circle* (described by the last relation) *plays the same role for the $\mathcal{Z}$ transform as the vertical axis of the complex plane in the Laplace transform*. In a completely analogous way, the vertical lines that in the Laplace transform belong to the left or right half-plane, in the $\mathcal{Z}$ transform correspond to circumferences that are located inside or outside the unit circle respectively. This observation allows us to draw qualitative conclusions regarding the stability of discrete-time systems, extending the corresponding conclusions we had formulated for the continuous-time systems according to the above analogy. We return to this issue later.

The above conclusions regarding the shape and characteristics of the region of convergence of the $\mathcal{Z}$ transform can arise directly from the defining equation. From the fundamental theory of power series we know that *the convergence of a power series requires its characterization to be absolutely summable*, a property that is mathematically formulated as

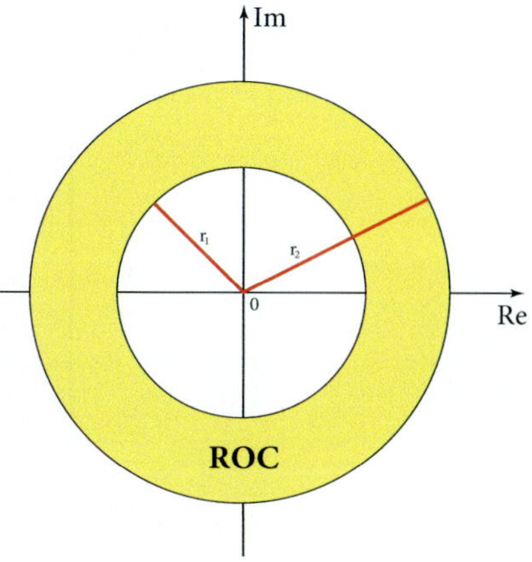

Fig. 8.1 The region of convergence of the $\mathcal{Z}$ transform of the sequence $x[n]$ is a ring of the complex plane with an inner radius r_1 and an outer radius r_2. ROC region of convergence

$$\sum_{n-\infty}^{\infty} |x[n]z^{-n}| < \infty \quad \text{or equivalently} \quad \sum_{n-\infty}^{\infty} |x(n)||z|^{-n} < \infty.$$

According to the above property, *the convergence of the power series of the $\mathcal{Z}$ transform will depend on the absolute value $|z|$ of the complex quantity z*. In other words, this means that if there is some variable $z = z_1$ for which the above power series converges to a finite value, then this convergence will be achieved for all complex quantities z for which the relation $z = |z_1|$ holds. But the last relation is nothing more than *the equation of a circle on the complex plane*, with a radius $r = |z_1|$. Generalizing the above conclusion, we can formulate the proposition that *the region of convergence of the $\mathcal{Z}$ transform of a random discrete sequence $x[n]$ will always be a ring defined on the complex plane with the center at the origin $z = 0$ and radii $r_1 = |z_1|$ for the inner circle and $r_2 = |z_2|$ for the outer circle*. A typical example of a region of convergence for the $\mathcal{Z}$ transform of a random discrete sequence $x[n]$ is shown in Fig. 8.1. It is important to mention at this point that depending on the nature and values of the discrete sequence $x[n]$ the outer circle can extend up to the point $z = \infty$ (or even include it), whereas the inner circle can include the origin of the system $z = 0$. As we prove in the next section, *if the region of convergence of the $\mathcal{Z}$ transform of the discrete sequence $x[n]$ includes the unit circle described by the equation $|z| = 1$*, then the *Fourier transform* of the aforementioned discrete sequence exists, as it is defined as the *$\mathcal{Z}$ transform calculated on the circumference of the unit circle*.

Example 8.1.1 The $\mathcal{Z}$ transform of the discrete-time signal with contents $x[n] = \{\mathbf{1}, 2, 5, 7, 0, 1\}$ will be according to its definition the power series

$$\mathcal{X}(z) = \sum_{n=0}^{5} x[n]z^{-n} = 1 + \frac{2}{z} + \frac{5}{z^2} + \frac{7}{z^3} + \frac{1}{z^5}.$$

The region of convergence of this transform is the entire complex plane, except for the point $z = 0$, as for this value the above expression becomes infinite.

Example 8.1.2 Let us consider the discrete-time signal

$$x[n] = \alpha^n u[n] = \begin{cases} \alpha^n & n \geq 0 \\ 0 & n < 0. \end{cases}$$

Applying the defining equation of the $\mathcal{Z}$ transform, it is given by the relation

$$\mathcal{X}(z) = \sum_{n=-\infty}^{\infty} x[n]z^{-n} = \sum_{n=-\infty}^{0} x[n]z^{-n} + \sum_{n=0}^{\infty} x[n]z^{-n} = \sum_{n=-\infty}^{0} 0z^{-n} + \sum_{n=0}^{\infty} \alpha^n z^{-n} = \sum_{n=0}^{\infty} (\alpha z^{-1})^n.$$

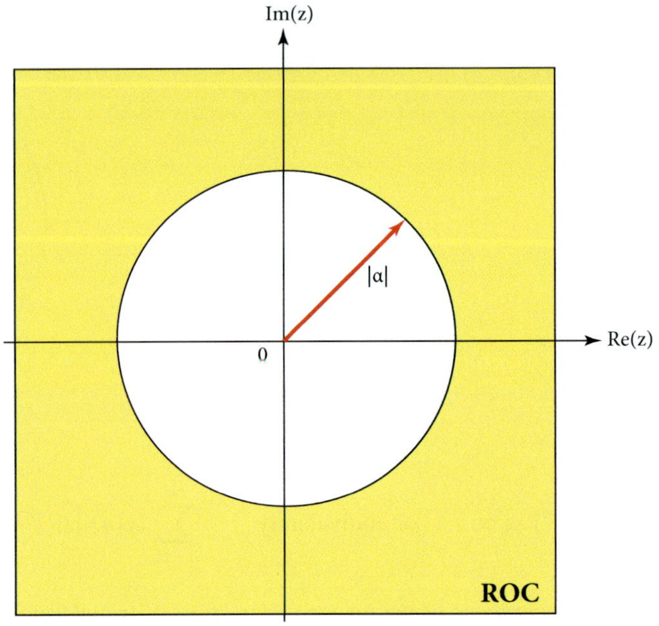

Fig. 8.2 The region of convergence of the $\mathcal{Z}$ transform for the discrete-time signal $x[n] = \alpha^n u[n]$. ROC region of convergence

The convergence of the above sequence is possible only when the condition relation $|\alpha z^{-1}| < 1$, or equivalently when $|z| > |\alpha|$. The region of the complex plane defined by the above relation is the complement of a circle that has its center at the point $z = 0$ and radius $r = |\alpha|$. This area is depicted in Fig. 8.2.

It is important to note that the defining equation of the region of convergence of the transform $\mathcal{X}(z)$ allows the derivation of useful conclusions about the discrete-time signal $x[n]$. In the previous example, we saw that if this equation takes the form $|z| > |\alpha|$—which means that the region of convergence is the complement of the circle with a radius $r = |\alpha|$—then the powers of z in the homonymous transform are *negative* and therefore the discrete-time signal $x[n]$ is a *causal signal* (i.e., it has nonzero values for the region $n \geq 0$ and zero values for the region $n < 0$). Conversely, if the region of convergence of the Z transform is described by the equation $|z| < |\alpha|$—i.e., it is confined to the interior of the circle with a radius $r = |\alpha|$, then our signal is *anti-causal* (i.e., it has nonzero values for the region $n < 0$ and zero values for the region $n \geq 0$) and the $\mathcal{Z}$ transform contains *positive powers* of z. These properties apply *only for discrete-time signals with an infinite number of elements*; as we have already mentioned, for *finite discrete-time signals* the region of convergence of the $\mathcal{Z}$ transform is *the entire complex plane* except for possibly from the points $z = 0$ and $z = \infty$.

Problem 8.1.1 Find the $\mathcal{Z}$ transform of the discrete-time signal

$$x[n] = \{3, 0, 0, 0, 0, \mathbf{6}, 1, -4\}$$

(Proakis and Manolakis, Exercise for solution 3.1).

Solution From the defining equation of the $\mathcal{Z}$ transform we get

$$X(z) = \sum_{n=-\infty}^{\infty} x[n]z^{-n} = \sum_{n=-5}^{2} x[n]z^{-n} = x[-5]z^5 + x[0]z^0 + x[1]z^{-1} + x[2]z^{-2}$$

$$= 3z^5 + 6 + z^{-1} - 4z^{-2} = 3z^5 + 6 + \frac{1}{z} - \frac{4}{z^2}$$

with the region of convergence of the transform being defined by an equation of the form $0 < |z| < \infty$.

Problem 8.1.2 Find the $\mathcal{Z}$ transform of the discrete-time signal $x[n] = (1/2)^n u[n-5]$ (adapted from Oppenheim, Willsky, and Nawab, Exercise for solution 10.2).

Solution From the defining equation of the $\mathcal{Z}$ transform we get

$$\mathcal{X}(z) = \sum_{n=-\infty}^{\infty} x[n]z^{-n} = \sum_{n=5}^{\infty} x[n]z^{-n} = \sum_{n=5}^{\infty} \left(\frac{1}{2z}\right)^n.$$

In order to set the lower limit of the above sum to zero, we proceed to change the dummy index from n to $m = n - 5$. Then $n = m + 5$, whereas for $n = 5$, $m = 0$. By replacing these parameters in the above equation and using the appropriate expression from the mathematical typology, we get

$$\mathcal{X}(z) = \sum_{m=0}^{\infty} \left(\frac{1}{2z}\right)^{m+5} = \left(\frac{1}{2z}\right)^5 \sum_{m=0}^{\infty} \left(\frac{1}{2z}\right)^m = \left(\frac{1}{2z}\right)^5 \frac{1}{1-(1/2z)} = \frac{1}{(2z)^4}\frac{1}{2z-1}.$$

In order for the above sequence to converge to the calculated value, it must be $|1/2z| < 1$ or equivalently $|z| > 1/2$, which is also the defining equation of the region of convergence of this transform.

8.2 Poles and Zeros of the Z Transform

The concept of poles and zeros of a rational complex function has been detailed in Sects. 4.4 (refer to page 316) and 5.3 (refer to page 385). At this point, we remind the reader that the *zeros* of the transform $\mathcal{X}(z)$ are the values of the complex variable z that satisfy the equation $\mathcal{X}(z) = 0$, whereas the *poles* of the transform are those of z for which the function $\mathcal{X}(z)$ diverges, so we can write that $\mathcal{X}(z) \to \infty$ (hence, the poles constitute *roots of the polynomial of the denominator* of the rational function $\mathcal{X}(z)$). If the $\mathcal{Z}$ transform is formulated as a rational complex function

$$\mathcal{X}(z) = \frac{N(z)}{D(z)} = \frac{\beta_0 + \beta_1 z^{-1} + \beta_2 z^{-2} + \cdots + \beta_M z^{-M}}{\alpha_0 + \alpha_1 z^{-1} + \alpha_2 z^{-2} + \cdots + \alpha_N z^{-N}} = \left\{\sum_{k=0}^{M} \beta_k z^{-k}\right\} \Big/ \left\{\sum_{k=0}^{N} \alpha_k z^{-k}\right\} \tag{8.6}$$

and we assume that $\alpha_0 \neq 0$ and $\beta_0 \neq 0$, we can find its *poles* and *zero values*, if we write it in the form

$$\mathcal{X}(z) = \frac{N(z)}{D(z)} = \frac{\beta_0 z^{-M}}{\alpha_0 z^{-N}} \left\{ \frac{z^M + \left(\frac{\beta_1}{\beta_0}\right) z^{M-1} + \left(\frac{\beta_2}{\beta_0}\right) z^{M-2} + \cdots + \left(\frac{\beta_M}{\beta_0}\right)}{z^N + \left(\frac{\alpha_1}{\alpha_0}\right) z^{N-1} + \left(\frac{\alpha_2}{\alpha_0}\right) z^{N-2} + \cdots + \left(\frac{\alpha_N}{\alpha_0}\right)} \right\} \tag{8.7}$$

or equivalently

$$\mathcal{X}(z) = \frac{\beta_0}{\alpha_0} \frac{(z-z_1)(z-z_2)\ldots(z-z_M)}{(z-p_1)(z-p_2)\ldots(z-p_N)} z^{-M+N} = G \times \left\{\prod_{k=1}^{M}(z-z_k)\right\} \Big/ \left\{\prod_{k=1}^{N}(z-p_k)\right\} \times z^{N-M}, \tag{8.8}$$

where $G = \beta_0/\alpha_0$, whereas we have proceeded to factorize the polynomials of the numerator and the denominator. Therefore, the transform $\mathcal{X}(z)$ has M *finite zero values* at the positions $z = z_1, z_2, \ldots, z_M$ (the roots of the numerator), N *finite poles* at positions $z = p_1, p_2, \ldots, p_N$ (the roots of the denominator) and $|N - M|$ *zero values* (if $N > M$) or *poles* (if $N < M$) at the origin of the complex plane $z = 0$. Let it be noted that the poles and the zero values can also appear for the value $z \to \infty$: if for the value $z \to \infty$ the relation $X(\infty) = 0$ holds, this value corresponds to a zero value, whereas if for this value it is $X(\infty) \to \infty$, this value corresponds to a pole.

An interesting process related to the $\mathcal{Z}$ transform is its representation in a special diagram that displays on the complex plane the positions of the zero values and the poles of the transform. This diagram is called the *pole-zero plot*. It has been presented in detail in a previous chapter (see Sect. 4.4 on page 316 and the Exercises 4.4.1 on page 317 and 4.4.2 on page 318), and indicative examples of its construction are presented in the following exercises.

Problem 8.2.1 Draw the pole-zero plot of the $\mathcal{Z}$ transform of the discrete-time signal $x[n] = \alpha^n u[n]$, for a parameter value $\alpha > 0$ (Oppenheim, Willsky, and Nawab, Example 10.1; Proakis and Manolakis, Example 3.1.3).

Solution The $\mathcal{Z}$ transform of the signal $x[n]$ is given by the last sum of Example 8.1.2, which, referring to the mathematical typology, is calculated as

$$X(z) = \frac{1}{1 - \alpha z^{-1}} = \frac{z}{z - \alpha}$$

with the region of convergence being the area of the complex plane for which the relation $|z| > \alpha$ holds. From the equation defining the poles and zeros, it is not difficult to realize that this transform has a zero value at the position $z_1 = 0$ and a pole at the position $p_1 = \alpha$. The diagram of poles-zeros for the $\mathcal{Z}$ transform of the signal $x[n] = \alpha^n u[n]$ is presented in Fig. 8.3.

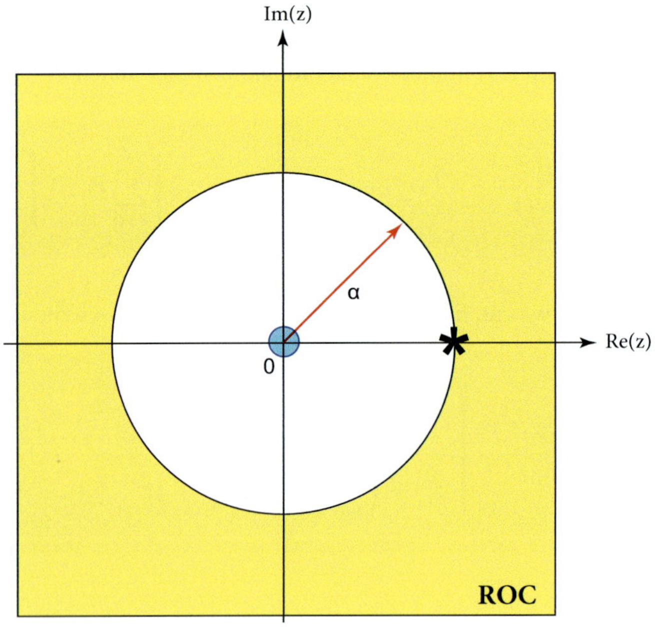

Fig. 8.3 The pole-zero plot for the $\mathcal{Z}$ transform of the discrete-time signal $x[n] = \alpha^n u[n]$ ($\alpha > 0$). ROC region of convergence

Problem 8.2.2 Draw the diagram of poles-zeros of the $\mathcal{Z}$ transform of the signal

$$x[n] = \begin{cases} \alpha^n & 0 \leq n \leq M-1 \\ 0 & \text{anywhere else} \end{cases}$$

(Oppenheim, Willsky, and Nawab, Example 10.6).

Solution The $\mathcal{Z}$ transform of the above discrete-time signal is given by the relation

$$\mathcal{X}(z) = \sum_{n=0}^{M-1} (\alpha z^{-1})^n = \frac{1-(\alpha z^{-1})^M}{1-\alpha z^{-1}} = \frac{z^M - \alpha^n}{z^{M-1}(z-\alpha)}$$

as can be proven by referring to the mathematical typology. Assuming that the relation $\alpha > 0$ holds, the equation $z^M = \alpha^M$ has M roots at the positions[1]

$$z_k = \alpha \exp\left(\frac{j2\pi k}{M}\right) \quad (k = 0, 1, 2, \ldots, M-1).$$

It is not difficult to see that the root corresponding to the value $k = 0$ is $z_0 = \alpha$. Taking into account that the function $\mathcal{X}(z)$ also has a pole in the same position, there will be, as we have seen in a previous chapter (see Sect. 5.4.1 on page 400) *mutual cancellation of the pole and the zero in this position*. But this means that the $\mathcal{Z}$ transform of the above discrete-time signal can be written as a ratio of polynomials of the form

$$\mathcal{X}(z) = \frac{(z-z_1)(z-z_2)\ldots(z-z_{M-1})}{z^{M-1}}$$

and therefore is characterized by the existence of a pole with a multiplicity value of $q = M-1$ at the position $z = 0$ and $M-1$ zero values, the positions of which on the circumference of a circle of radius α for the value $M = 8$, are depicted in Fig. 8.4.

Note it is not difficult to ascertain that for the value $\alpha = 1$ the discrete-time signal $x[n]$ is written as

$$x_1[n] = \begin{cases} 1 & 0 \leq n \leq M-1 \\ 0 & \text{anywhere else,} \end{cases}$$

which describes the *square discrete pulse* $\Pi_{N-1}[n]$ (see Sect. 6.7.3 on page 532). Replacing the value $\alpha = 1$ in the equation of the $\mathcal{Z}$ transform that we have concluded, it will take the form

$$\mathcal{X}(z) = \sum_{n=0}^{M-1} (z^{-1})^n = \frac{1-(z^{-1})^M}{1-z^{-1}} = \frac{z^M - 1}{z^{M-1}(z-1)},$$

which is a special case of the previous one. ■

An interesting topic worth mentioning is *the correlation that exists between the type of temporal evolution of a discrete-time signal and the positions of the poles and zero values of the $\mathcal{Z}$ transform of the aforementioned signal on the complex plane.* Considering the unit circle centered at the origin of the complex plane $z = 0$ and with a radius equal to one, a pole can be located in three different areas on the complex plane: (a) in the *interior* of the unit circle, (b) in the *exterior* of the unit circle,

[1] For the solution of this equation refer to a basic book of complex analysis that deals with the calculation of the M roots of unity.

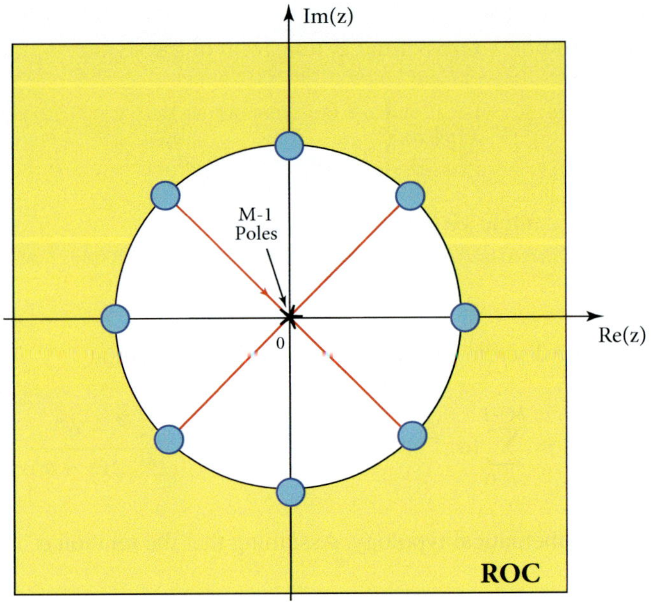

Fig. 8.4 The pole-zero plot for the $\mathcal{Z}$ transform of the signal $x[n] = \alpha^n$ $(0 \leq n \leq M - 1)$ for values $\alpha > 0$. ROC region of convergence

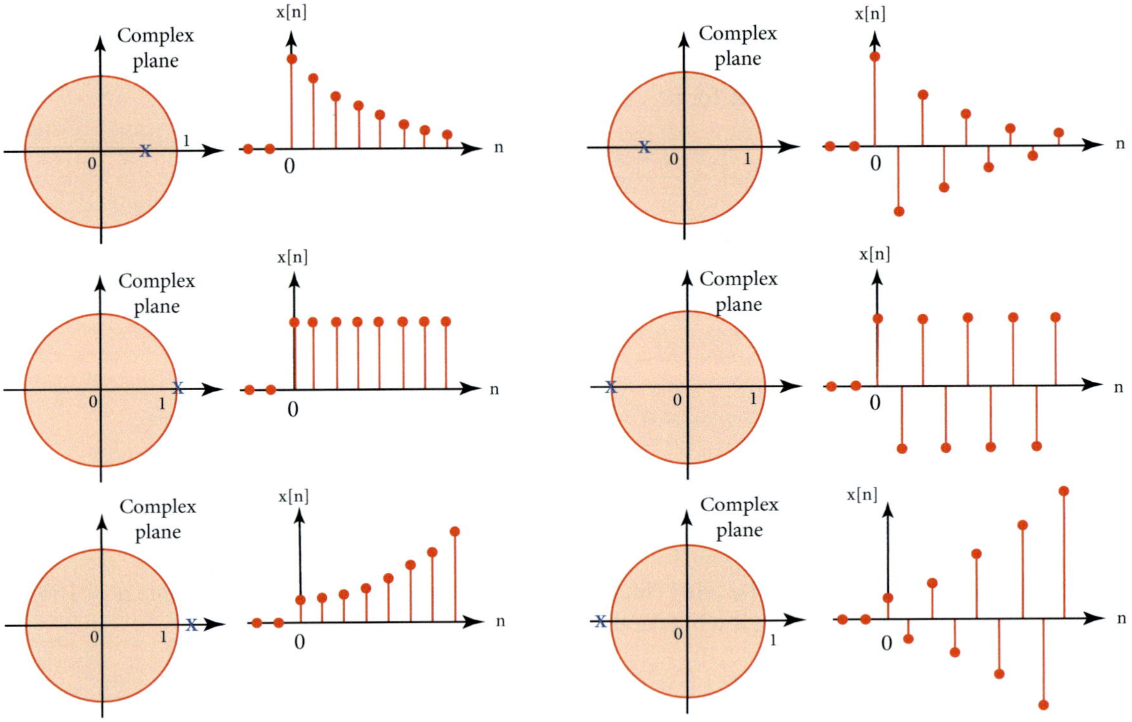

Fig. 8.5 Typical positions of the poles of the $\mathcal{Z}$ transform of a discrete-time signal on the complex plane and the corresponding temporal evolution of the discrete-time signal with which it is associated

and (c) on the *periphery* of the unit circle. These three different positions of a pole are presented in Fig. 8.5, which also refers to the entirety of the description presented in the following paragraphs.

The simplest case we can study is that of the signal $x[n] = \alpha^n u[n]$, the $\mathcal{Z}$ transform of which is given by the relation $\mathcal{X}(z) = (1 - \alpha z^{-1})^{-1}$ with a region of convergence of $|z| > |\alpha|$. This transform has a zero value at the beginning of the complex plane $z = 0$ and a pole at the position $p = \alpha$, which lies *on the positive real axis*. In this case, *the temporal evolution of the signal depends on the position of the above pole on the complex plane*. More specifically, *if the pole is inside the unit circle, the signal weakens over time; if the pole is on the circumference of the unit circle, the amplitude of the signal remains constant; if the pole is outside the unit circle, the amplitude of the signal increases indefinitely and ceases to be bounded.*

The same behavior is observed in cases where the pole of the $\mathcal{Z}$ transform is on the negative real axis. The difference in this case is that the sign of the signal alternates at each time repetition, with the result that *two consecutive samples are characterized by different signs*. These cases are also presented in Fig. 8.5.

Apart from this simple case of the one and only real pole there are more complex cases, each of which has its own special characteristics. Generalizing the above description, we can formulate the claim that *a causal discrete-time signal of real values that has simple real poles or pairs of conjugate complex poles that are located inside the unit circle of the complex plane is always bounded as far as the value of its amplitude is concerned*. In addition, a signal that has a pole that is very close to the origin of the complex plane weakens *much faster* during its evolution over time compared with another signal, whose pole is at a greater distance from the origin (but always, of course, *within the limits of the unit circle*). From the above discussion, it follows that *the temporal evolution of a causal discrete-time signal depends to a large extent on the positions of the poles of the $\mathcal{Z}$ transform of the signal on the complex plane*. There is, of course, also a dependence on the positions of the zero values of the transform, but this *is not as strong* as the dependence on the positions of the poles. Finally, it is necessary to mention that the above properties that apply to the causal discrete-time signals also apply to the linear and time-invariant ones, since the impulse response of these systems is considered to be described by a causal discrete-time signal. This means that *if any of the poles of the system are outside the limits of the unit circle, the impulse response of the system is not blocked and therefore the system is unstable*. We return to this issue later.

8.3 Examples of $\mathcal{Z}$ Transform Calculations

After the brief theoretical introduction of the previous pages, examples of solved exercises are presented, so that the reader can understand the basic features of the $\mathcal{Z}$ transform and its calculation methods. To solve these exercises, extensive use will be made of the sums in closed form that are included in the mathematical typology of Sect. 6.1 on page 497 (see Table 6.2 on page 500).

Problem 8.3.1 Find the $\mathcal{Z}$ transform and design the pole-zero diagram of the discrete-time signal $x[n] = (1+n)u[n]$ (Proakis and Manolakis, Exercise for solution 3.2α).

Solution The Z transform of the signal will be given according to the defining equation from the relation

$$\mathcal{X}(z) = \sum_{n=-\infty}^{\infty} (1+n)\, u[n] z^{-n} = \sum_{n=0}^{\infty} (1+n)\, z^{-n} = \sum_{n=0}^{\infty} z^{-n} + \sum_{n=0}^{\infty} n z^{-n}.$$

It is not difficult to ascertain that the two sums of the above equation have a common convergence area, which is described by the equation $|z| > 1$, whereas the sums themselves are calculated as

$$\sum_{n=0}^{\infty} z^{-n} = \frac{1}{1-z^{-1}} = \frac{1}{1-(1/z)} = \frac{z}{z-1}$$

and

$$\sum_{n=0}^{\infty} n z^{-n} = \sum_{n=0}^{\infty} n \left(\frac{1}{z}\right)^n = \frac{z^{-1}}{(1-z^{-1})^2} = \frac{1/z}{[1-(1/z)]^2} = \frac{z}{(z-1)^2}.$$

Therefore, the $\mathcal{Z}$ transform of the discrete-time signal will be given by the equation

$$\mathcal{X}(z) = \frac{z}{z-1} + \frac{z}{(z-1)^2} = \frac{z^2}{(z-1)^2},$$

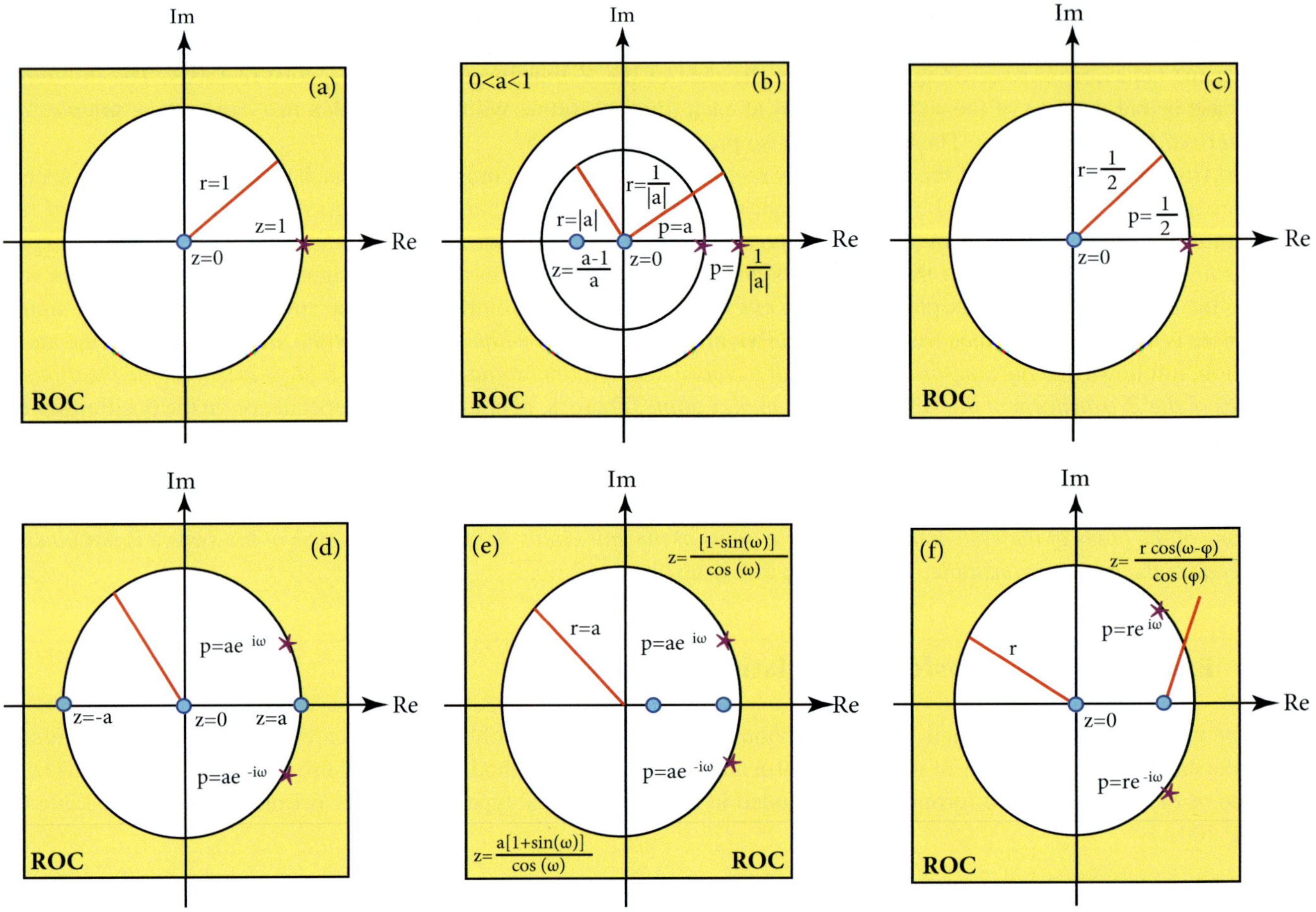

Fig. 8.6 From left to right, the pole-zero diagram for the $\mathcal{Z}$ transform of Exercises 8.3.1, 8.3.2 and 8.3.3 (the three top images), as well as the corresponding diagram for the exercises 8.3.4, 8.3.5 and 8.3.6 (the three bottom images) are depicted. Note that in several cases the positions of the zeros and poles are indicative, as they depend on the values of the problem parameters. The shaded area also represents the region of convergence for each transform. ROC region of convergence

from which we ascertain that the transform has a double zero value at the position $z = 0$ and a double pole at the position $p = 1$. The convergence area and the pole-zero diagram of the transform are depicted in image (a) of Fig. 8.6.

Problem 8.3.2 Find the $\mathcal{Z}$ transform and draw the pole-zero diagram of the discrete-time signal $x[n] = (\alpha^n + \alpha^{-n})u[n]$ ($\alpha \in R$) (Proakis and Manolakis, Exercise for solution 3.2β).

Solution The $\mathcal{Z}$ transform of the signal will be given according to the defining equation from the relation

$$\mathcal{X}(z) = \sum_{n=-\infty}^{\infty} x[n]z^{-n} = \sum_{n=-\infty}^{\infty} (\alpha^n + \alpha^{-n})\, u[n]z^{-n} = \sum_{n=0}^{\infty}(\alpha^n + \alpha^{-n})\, z^{-n} = \sum_{n=0}^{\infty}\alpha^n z^{-n} + \sum_{n=0}^{\infty}\alpha^{-n} z^{-n}.$$

The first of the two sums is calculated as

$$\sum_{n=0}^{\infty}\alpha^n z^{-n} = \sum_{n=0}^{\infty}\left(\frac{\alpha}{z}\right)^n = \frac{1}{1-(\alpha/z)} = \frac{z}{z-\alpha},$$

with the series converging for values of the variable z, such that $|z| > |\alpha|$. In a completely similar way, the second sum will take the form

$$\sum_{n=0}^{\infty} \alpha^{-n} z^{-n} = \sum_{n=0}^{\infty} \left(\frac{2}{\alpha z}\right)^n = \frac{1}{1-(1/\alpha z)} = \frac{\alpha z}{\alpha z - 1},$$

with the series converging for values of the variable z, such that $|z| > 1/|\alpha|$. Therefore, the $\mathcal{Z}$ transform of the discrete-time signal $x[n]$ is calculated as

$$\mathcal{X}(z) = \frac{z}{z-\alpha} + \frac{\alpha z}{\alpha z - 1} = \frac{z\,(2\alpha z - \alpha^2 - 1)}{(z-\alpha)\,(\alpha z - 1)},$$

with the region of convergence $|z| > \max(|\alpha|, 1/|\alpha|)$. From the last equation, it is not difficult to see that the transform $\mathcal{X}(z)$ has two zero values at the points $z_1 = 0$ and $z_2 = (\alpha^2 + 1)/2\alpha$ and two poles at the points $p_1 = \alpha$ and $p_2 = 1/\alpha$. The region of convergence and the pole-zero diagram of this transform are depicted in image (b) of Fig. 8.6.

Problem 8.3.3 Find the $\mathcal{Z}$ transform and design the pole-zero diagram of the discrete-time signal $x[n] = (-1)^n (2)^{-n} u[n]$ (Proakis and Manolakis, Exercise for solution 3.2γ).

Solution Using the equation of the $\mathcal{Z}$ transform for the signal $x[n] = (-1)^n (2)^{-n} u[n]$ we get

$$X(z) = \sum_{n=-\infty}^{\infty} (-1)^n (2)^{-n} u[n] z^{-n} = \sum_{n=0}^{\infty} \left(-\frac{1}{2z}\right)^n = \frac{2z}{2z+1}$$

with the region of convergence $|z| > 1/2$. From the above equation, we observe that the transform $X(z)$ has a zero value at the position $z = 0$ and a simple pole at the position $p = 1/2$. The region of convergence and the pole-zero diagram of the transform are depicted in image (c) of Fig. 8.6.

Problem 8.3.4 Find the $\mathcal{Z}$ transform and design the pole-zero diagram of the discrete-time signal $x[n] = n\alpha^n \sin(\omega n) u[n]$ (Proakis and Manolakis, Exercise for solution 3.2δ).

Solution Using the equation of the definition of the $\mathcal{Z}$ transform for the discrete-time signal $x[n]$, we get

$$\begin{aligned}
\mathcal{X}(z) &= \sum_{n=-\infty}^{\infty} x[n] z^{-n} = \sum_{n=-\infty}^{\infty} n\alpha^n \sin(\omega n) u[n] z^{-n} = \sum_{n=0}^{\infty} n\alpha^n \sin(\omega n) z^{-n} \\
&= \sum_{n=0}^{\infty} n\alpha^n \left[\frac{e^{j\omega n} - e^{-j\omega n}}{2j}\right] z^{-n} = \frac{1}{2j}\left[\sum_{n=0}^{\infty} n\alpha^n e^{j\omega n} z^{-n} - \sum_{n=0}^{\infty} n\alpha^n e^{-j\omega n} z^{-n}\right],
\end{aligned}$$

where for the derivation of the last relation we replaced the sine function from the Euler type. From the two sums of the above relation, the first sum is calculated as

$$\sum_{n=0}^{\infty} n\alpha^n e^{j\omega n} z^{-n} = \sum_{n=0}^{\infty} n \left(\frac{\alpha}{z} e^{j\omega}\right)^n = \frac{\alpha e^{j\omega}/z}{\left(1 - \dfrac{\alpha}{z} e^{j\omega}\right)^2} = \frac{\alpha e^{j\omega}/z}{\dfrac{(z - \alpha e^{j\omega})^2}{z^2}} = \frac{\alpha z^2 e^{j\omega}}{z(z-\alpha e^{j\omega})^2} = \frac{\alpha z e^{j\omega}}{(z-\alpha e^{j\omega})^2},$$

with the series converging for values of the complex variable z, such that $|z| > |\alpha|$, whereas, in a completely analogous way, for the second sum we have

$$\sum_{n=0}^{\infty} n\alpha^n e^{-j\omega n} z^{-n} = \sum_{n=0}^{\infty} n\left(\frac{\alpha}{z}e^{-j\omega}\right)^n = \frac{\alpha e^{-j\omega}/z}{\left(1-\frac{\alpha}{z}e^{-j\omega}\right)^2} = \frac{\alpha e^{-j\omega}/z}{\frac{(z-\alpha e^{-j\omega})^2}{z^2}} = \frac{\alpha z^2 e^{-j\omega}}{z(z-\alpha e^{-j\omega})^2} = \frac{\alpha z e^{-j\omega}}{(z-\alpha e^{-j\omega})^2},$$

with the region of convergence being the same as the previous one. Replacing the above expressions in the defining equation of the $\mathcal{Z}$ transform for the discrete-time signal in question, it will take the form

$$\mathcal{X}(z) = \frac{1}{2j}\left[\frac{\alpha z e^{j\omega}}{(z-\alpha e^{j\omega})^2} - \frac{\alpha z e^{-j\omega}}{(z-\alpha e^{-j\omega})^2}\right] = \frac{\alpha z}{2j}\left[\frac{e^{j\omega}}{(z-\alpha e^{j\omega})^2} - \frac{e^{-i\omega}}{(z-\alpha e^{-j\omega})^2}\right],$$

with the expression within the brackets being calculated as

$$\begin{aligned} A &= \frac{e^{j\omega}}{(z-\alpha e^{j\omega})^2} - \frac{e^{-j\omega}}{(z-\alpha e^{-j\omega})^2} = \frac{(z-\alpha e^{-j\omega})^2 e^{j\omega} - (z-\alpha e^{j\omega})^2)\, e^{-j\omega}}{(z-\alpha e^{j\omega})^2(z-\alpha e^{-j\omega})^2} \\ &= \frac{(z^2 - 2\alpha z e^{-j\omega} + \alpha^2 e^{-2j\omega})\, e^{j\omega} - (z^2 - 2\alpha z e^{j\omega} + \alpha^2 e^{2j\omega})\, e^{-j\omega}}{[(z-\alpha e^{j\omega})(z-\alpha e^{-j\omega})]^2} \\ &= \frac{z^2 e^{j\omega} - 2\alpha z + \alpha^2 e^{-j\omega} - z^2 e^{-j\omega} + 2\alpha z - \alpha^2 e^{j\omega}}{(z^2 - \alpha z e^{-j\omega} - \alpha z e^{j\omega} + \alpha^2)^2} \\ &= \frac{z^2(e^{j\omega} - e^{-j\omega}) - \alpha^2(e^{j\omega} - e^{-j\omega})}{[z^2 - \alpha z(e^{j\omega} + e^{-j\omega}) + \alpha^2]^2} = \frac{(z^2-\alpha^2)(e^{j\omega} - e^{-j\omega})}{[z^2 - \alpha z(e^{j\omega} + e^{-j\omega}) + \alpha^2]^2}, \end{aligned}$$

and finally

$$A = \frac{(z^2-\alpha^2)2j\sin(\omega)}{[z^2 - 2\alpha z\cos(\omega) + \alpha^2]^2},$$

where we used again our known equations $e^{jx} + e^{-jx} = 2\cos(x)$ and $e^{jx} - e^{-jx} = 2j\sin(x)$. Substituting the above expression into the defining equation of the transform $X(z)$, we get

$$\mathcal{X}(z) = \frac{\alpha z}{2j}\frac{(z^2-\alpha^2)2j\sin(\omega)}{[z^2 - 2\alpha z\cos(\omega) + \alpha^2]^2} = \frac{\alpha z(z^2-\alpha^2)\sin(\omega)}{[z^2 - 2\alpha z\cos(\omega) + \alpha^2]^2},$$

with the convergence area $|z| > |\alpha|$, which is also the desired result. From the last equation it is not difficult to see that the complex function $\mathcal{X}(z)$ has three zero values at the positions $z_1 = 0$ and $z_{2,3} = \pm\alpha$. In order to determine the poles of the function we must solve the equation $z^2 - 2\alpha z\cos(\omega) + \alpha^2 = 0$, which is a quadratic algebraic equation with respect to z with discriminant $\Delta = -4\alpha^2\sin^2(\omega)$ and roots $z_1 = \alpha e^{i\omega}$ and $z_2 = \alpha e^{-i\omega}$.

$$\begin{aligned} z_1 &= \frac{2\alpha\cos(\omega) + 2\alpha j\sin(\omega)}{2} = \alpha[\cos(\omega) + j\sin(\omega)] = \alpha e^{j\omega} \quad \text{and} \\ z_2 &= \frac{2\alpha\cos(\omega) - 2\alpha j\sin(\omega)}{2} = \alpha[\cos(\omega) - j\sin(\omega)] = \alpha e^{-j\omega}. \end{aligned}$$

Since the algebraic equation in question appears in the denominator of the function $\mathcal{X}(z)$ squared, the function $\mathcal{X}(z)$ has two double poles at the points $p_1 = \alpha e^{i\omega}$ and $p_2 = \alpha e^{-i\omega}$. The region of convergence and the pole-zero diagram of the transform $\mathcal{X}(z)$ are depicted in image (d) of Fig. 8.6.

Problem 8.3.5 Find the $\mathcal{Z}$ transform and draw the pole-zero diagram of the discrete-time signal $x[n] = n\alpha^n\cos(\omega n)u[n]$ (Proakis and Manolakis, Exercise for solution 3.2e).

Solution Following the methodology of the previous exercise, the use of the defining equation of the $\mathcal{Z}$ transform leads us to the equation

$$X(z) = \sum_{n=-\infty}^{\infty} x[n]z^{-n} = \sum_{n=-\infty}^{\infty} n\alpha^n \cos(\omega n)u[n]z^{-n} = \sum_{n=0}^{\infty} n\alpha^n \cos(\omega n)z^{-n}$$

$$= \sum_{n=0}^{\infty} n\alpha^n \left[\frac{e^{i\omega n} + e^{-i\omega n}}{2}\right] z^{-n} = \frac{1}{2}\left[\sum_{n=0}^{\infty} n\alpha^n e^{i\omega n} z^{-n} + \sum_{n=0}^{\infty} n\alpha^n e^{-i\omega n} z^{-n}\right],$$

with the two sums of the last expression being calculated as

$$\sum_{n=0}^{\infty} n\alpha^n e^{+i\omega n} z^{-n} = \sum_{n=0}^{\infty} n\left(\frac{\alpha}{z}e^{i\omega}\right)^n = \frac{\alpha e^{i\omega}/z}{\left(1-\frac{\alpha}{z}e^{i\omega}\right)^2} = \frac{\alpha e^{i\omega}/z}{\frac{(z-\alpha e^{i\omega})^2}{z^2}} = \frac{\alpha z^2 e^{i\omega}}{z(z-\alpha e^{i\omega})^2} = \frac{\alpha z e^{i\omega}}{(z-\alpha e^{i\omega})^2}$$

$$\sum_{n=0}^{\infty} n\alpha^n e^{-i\omega n} z^{-n} \sum_{n=0}^{\infty} n\left(\frac{\alpha}{z}e^{-i\omega}\right)^n = \frac{\alpha e^{-i\omega}/z}{\left(1-\frac{\alpha}{z}e^{-i\omega}\right)^2} = \frac{\alpha e^{-i\omega}/z}{\frac{(z-\alpha e^{-i\omega})^2}{z^2}} = \frac{\alpha z^2 e^{-i\omega}}{z(z-\alpha e^{-i\omega})^2} = \frac{\alpha z e^{-i\omega}}{(z-\alpha e^{-i\omega})^2},$$

and to have a common region of convergence, which is described by the equation $|z| > |\alpha|$. Substituting the above expressions into the defining equation of the $\mathcal{Z}$ transform we get

$$X(z) = \frac{1}{2}\left[\frac{\alpha z e^{i\omega}}{(z-\alpha e^{i\omega})^2} + \frac{\alpha z e^{-i\omega}}{(z-\alpha e^{-i\omega})^2}\right] = \frac{\alpha z}{2}\left[\frac{e^{i\omega}}{(z-\alpha e^{i\omega})^2} + \frac{e^{-i\omega}}{(z-\alpha e^{-i\omega})^2}\right]$$

with the expression inside the brackets being calculated as

$$A = \frac{e^{i\omega}}{(z-\alpha e^{i\omega})^2} + \frac{e^{-i\omega}}{(z-\alpha e^{-i\omega})^2} = \frac{(z-\alpha e^{-i\omega})^2 e^{i\omega} + (z-\alpha e^{i\omega})^2) e^{-i\omega}}{(z-\alpha e^{i\omega})^2 (z-\alpha e^{-i\omega})^2}$$

$$= \frac{(z^2 - 2\alpha z e^{-i\omega} + \alpha^2 e^{-2i\omega})e^{i\omega} + (z^2 - 2\alpha z e^{i\omega} + \alpha^2 e^{2i\omega})e^{-i\omega}}{[(z-\alpha e^{i\omega})(z-\alpha e^{-i\omega})]^2}$$

$$= \frac{z^2 e^{i\omega} - 2\alpha z + \alpha^2 e^{-i\omega} + z^2 e^{-i\omega} - 2\alpha z + \alpha^2 e^{i\omega}}{(z^2 - \alpha z e^{-i\omega} - \alpha z e^{i\omega} + \alpha^2)^2} = \frac{z^2(e^{i\omega} + e^{-i\omega}) - 4\alpha z + \alpha^2(e^{i\omega} + e^{-i\omega})}{(z^2 - \alpha z(e^{i\omega} + e^{-i\omega}) + \alpha^2)^2}$$

$$= \frac{(z^2 + \alpha^2)(e^{i\omega} + e^{-i\omega}) - 4\alpha z}{(z^2 - \alpha z(e^{i\omega} + e^{-i\omega}) + \alpha^2)^2} = \frac{2(z^2 + \alpha^2)\cos(\omega) - 4\alpha z}{(z^2 - 2\alpha z\cos(\omega) + \alpha^2)^2},$$

where we used again the well-known expression $e^{ix} + e^{ix} = 2\cos(x)$. Substituting the above expression into the defining equation of the transform $X(z)$ we finally get

$$X(z) = \frac{\alpha z}{2}\left\{\frac{2(z^2 + \alpha^2)\cos(\omega) - 4\alpha z}{(z^2 - 2\alpha z\cos(\omega) + \alpha^2)^2}\right\} = \frac{az[(z^2 + \alpha^2)\cos(\omega) - 2\alpha z]}{(z^2 - 2\alpha z\cos(\omega) + \alpha^2)^2},$$

with the convergence area $|z| > |\alpha|$, which is also the desired result. From the last equation it is not difficult to realize that the expression of the denominator is the same as that in the previous exercise and therefore the complex function $X(z)$ has two double poles at the points $p_1 = \alpha e^{i\omega}$ and $p_2 = \alpha e^{-i\omega}$. As for its zero values, the function $X(z)$ has one zero value at the position $z = 0$ and two zero values at the positions z_2 and z_3, which are roots of the equation $(z^2 + \alpha^2)\cos(\omega) - 2\alpha z = 0$. The above equation is reduced to the form $\cos(\omega)z^2 - 2\alpha z + \alpha^2\cos(\omega) = 0$ and is recognized as a quadratic algebraic equation as to z with a discriminant equal to $\Delta = 4\alpha^2\sin^2\omega$ and roots

$$z_1 = \frac{2\alpha + 2\alpha\sin(\omega)}{2\cos(\omega)} = \frac{\alpha + \alpha\sin(\omega)}{\cos(\omega)} = \frac{\alpha[1 + \sin(\omega)]}{\cos(\omega)}$$

$$z_2 = \frac{2\alpha - 2\alpha\sin(\omega)}{2\cos(\omega)} = \frac{\alpha - \alpha\sin(\omega)}{\cos(\omega)} = \frac{\alpha[1-\sin(\omega)]}{\cos(\omega)},$$

which are also the other two zero values of the function. The convergence area and the pole-zero diagram of the transform $X(z)$ are depicted in image (e) of Fig. 8.6.

Problem 8.3.6 Find the $\mathcal{Z}$ transform and draw the pole-zero diagram of the discrete-time signal $x[n] = Ar^n\cos(\omega n + \varphi)u[n]$ (for the range of values $0 < r < 1$) (Proakis and Manolakis, Exercise for solution 3.2st).

Solution Starting from the defining equation of the $\mathcal{Z}$ transform for the discrete-time signal $x[n]$ and replacing the cosine function from the Euler formula, we get

$$\begin{aligned}\mathcal{X}(z) &= \sum_{n=-\infty}^{\infty} x[n]z^{-n} = \sum_{n=-\infty}^{\infty} Ar^n\cos(\omega n+\varphi)u[n]z^{-n} = A\sum_{n=0}^{\infty} r^n\cos(\omega n+\varphi)z^{-n}\\ &= A\sum_{n=0}^{\infty} r^n\left[\frac{e^{j(\omega n+\varphi)}+e^{-j(\omega n+\varphi)}}{2}\right]z^{-n} = \frac{A}{2}\sum_{n=0}^{\infty} r^n(e^{j\omega n}e^{j\varphi}+e^{-j\omega n}e^{-j\varphi})z^{-n}\\ &= \frac{A}{2}\left\{\sum_{n=0}^{\infty} r^n e^{j\omega n}e^{j\varphi}z^{-n} + \sum_{n=0}^{\infty} r^n e^{-j\omega n}e^{-j\varphi}z^{-n}\right\}.\end{aligned}$$

Then, using the appropriate expressions from the mathematical typology, we find that the two sums of the above equation are calculated as

$$\sum_{n=0}^{\infty} r^n e^{+j\omega n}e^{+j\varphi}z^{-n} = e^{+j\varphi}\sum_{n=0}^{\infty}\left(\frac{r}{z}e^{+j\omega}\right)^n = \frac{e^{j\varphi}}{1-\frac{r}{z}e^{j\omega}} = \frac{e^{j\varphi}}{\frac{z-re^{j\omega}}{z}} = \frac{ze^{j\varphi}}{z-re^{j\omega}}$$

$$\sum_{n=0}^{\infty} r^n e^{-j\omega n}e^{-j\varphi}z^{-n} = e^{-j\varphi}\sum_{n=0}^{\infty}\left(\frac{r}{z}e^{-j\omega}\right)^n = \frac{e^{-j\varphi}}{1-\frac{r}{z}e^{-j\omega}} = \frac{e^{-j\varphi}}{\frac{z-re^{-j\omega}}{z}} = \frac{ze^{-j\varphi}}{z-re^{-j\omega}}$$

and they have the same region of convergence, which is given by the equation $|z| > r$. Replacing the above sums in our initial equation, it will take the form

$$\mathcal{X}(z) = \frac{A}{2}\left(\frac{ze^{j\varphi}}{z-re^{j\omega}} + \frac{ze^{-j\varphi}}{z-re^{-j\omega}}\right) = \frac{Az}{2}\left(\frac{e^{j\varphi}}{z-re^{j\omega}} + \frac{e^{-j\varphi}}{z-re^{-j\omega}}\right),$$

with the expression inside the parentheses being calculated as

$$\begin{aligned}A &= \frac{e^{j\varphi}}{z-re^{j\omega}} + \frac{e^{-j\varphi}}{z-re^{-j\omega}} = \frac{(z-re^{-j\omega})\,e^{j\varphi} + (z-re^{j\omega})\,e^{-j\varphi}}{(z-re^{j\omega})(z-re^{-j\omega})} = \frac{ze^{j\varphi} - re^{-j\omega+j\varphi} + ze^{-j\varphi} - re^{j\omega}e^{-j\varphi}}{z^2 - zre^{-j\omega} - zre^{j\omega} + r^2}\\ &= \frac{z(e^{j\varphi}+e^{-j\varphi}) - r[e^{j(\omega-\varphi)} + e^{-j(\omega-\varphi)}]}{z^2 - zr(e^{j\omega}+e^{-j\omega}) + r^2} = \frac{2z\cos(\varphi) - 2r\cos(\omega-\varphi)}{z^2 - 2rz\cos(\omega) + r^2}.\end{aligned}$$

Therefore, the defining equation of the $\mathcal{Z}$ transform for the discrete-time signal $x[n]$ will take the form

$$\mathcal{X}(z) = \frac{Az}{2}\left\{\frac{2z\cos(\varphi) - 2r\cos(\omega-\varphi)}{z^2 - 2rz\cos(\omega) + r^2}\right\} = \frac{Az[z\cos(\varphi) - r\cos(\omega-\varphi)]}{z^2 - 2rz\cos(\omega) + r^2},$$

which is also the desired result. From the above equation we find that the complex function $\mathcal{X}(z)$ has two simple zero values at the positions $z = 0$ and $z = r\cos(\omega-\varphi)/\cos(\varphi)$. In order to calculate the poles of the function $\mathcal{X}(z)$ we need to determine the roots of the equation $z^2 - 2rz\cos(\omega) + r^2 = 0$, which is recognized as a quadratic algebraic equation with respect to z with discriminant $\Delta = -4r^2\sin^2(\omega)$ and roots

$$z_1 = \frac{2r\cos(\omega) + j\,2r\sin(\omega)}{2} = r\,[\cos(\omega) + j\sin(\omega)] = re^{j\omega}$$

$$z_2 = \frac{2r\cos(\omega) - j\,2r\sin(\omega)}{2} = r\,[\cos(\omega) - j\sin(\omega)] = re^{-j\omega}.$$

Therefore, the function $\mathcal{X}(z)$ has two simple poles at the positions $p_1 = re^{j\omega}$ and $p_2 = re^{-j\omega}$. The convergence area and the pole-zero diagram of the transform $\mathcal{X}(z)$ are depicted in the image (f) of Fig. 8.6.

Problem 8.3.7 Find the $\mathcal{Z}$ transform and draw the pole-zero diagram of the discrete-time signal $x[n] = \alpha^{|n|}$ $(0 < \alpha < 1)$ (Hayes, Exercise 4.4d and Hsu, Exercise 4.7).

Solution The absolute value function is defined as known from the relation

$$|n| = \begin{cases} +n \text{ for } n \geq 0 \\ -n \text{ for } n < 0 \end{cases}$$

and therefore the $\mathcal{Z}$ transform of the signal $x[n]$ will have the form

$$\mathcal{X}(z) = \sum_{n=-\infty}^{\infty} x[n]\,z^{-n} = \sum_{n=-\infty}^{\infty} \alpha^{|n|}\,z^{-n} = \sum_{n=-\infty}^{-1} \alpha^{|n|}\,z^{-n} + \sum_{n=0}^{\infty} \alpha^{|n|}\,z^{-n} = \sum_{n=-\infty}^{-1} \alpha^{-n} z^{-n} + \sum_{n=0}^{\infty} \alpha^{n} z^{-n}.$$

The first sum of the above equation is formulated as

$$\sum_{n=-\infty}^{-1} \alpha^{-n} z^{-n} = -\sum_{n=-1}^{-\infty} \alpha^{-n} z^{-n} = \sum_{n=1}^{\infty} \alpha^{n} z^{n} = \sum_{n=1}^{\infty} (\alpha z)^n$$

- where for the extraction of the last expression we proceeded to the substitution $-n \longrightarrow n$—and finally

$$\sum_{n=1}^{\infty} (\alpha z)^n = \sum_{n=0}^{\infty} (\alpha z)^n - (\alpha z)^0 = \frac{1}{1-\alpha z} - 1 = \frac{\alpha z}{1-\alpha z}.$$

The region of convergence of the above sum will result from the requirement that the inequality $|\alpha z| < 1$ holds, from which it follows that $|z| < 1/|\alpha|$.

On the other hand, the second sum is calculated as

$$\sum_{n=0}^{\infty} \alpha^n z^{-n} = \sum_{n=0}^{\infty} \left(\frac{\alpha}{z}\right)^n = \frac{1}{1-(\alpha/z)} = \frac{z}{z-\alpha},$$

with a region of convergence of $|z| > |\alpha|$, as can be proven.

Therefore, the transform $\mathcal{X}(z)$ will be calculated as the sum

$$\mathcal{X}(z) = \frac{\alpha z}{1-\alpha z} + \frac{z}{z-\alpha} = \frac{\alpha z\,(z-\alpha) + z(1-\alpha z)}{(1-\alpha z)(z-\alpha)} = \frac{z\,(1-\alpha^2)}{(1-\alpha z)(z-\alpha)},$$

Fig. 8.7 Exercise 8.3.7

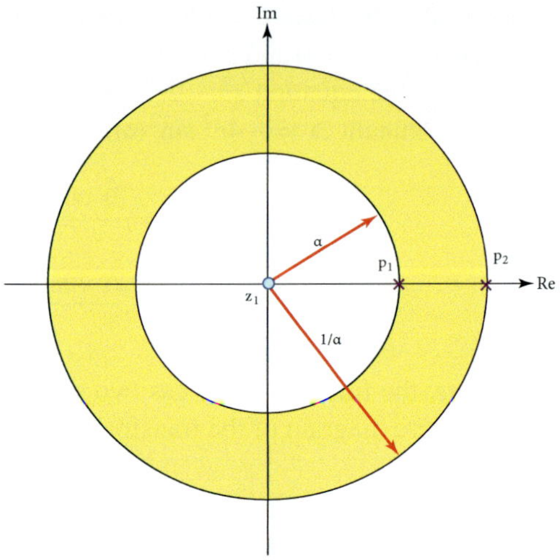

with a region of convergence of $|\alpha| < |z| < (1/|\alpha|)$, which is the desired result. From the last equation, it is concluded that the transform $\mathcal{X}(z)$ has a zero at the position $z = 0$ and two poles at the positions $z = \alpha$ and $z = 1/\alpha$. The region of convergence of this transform along with the positions of the poles and zeros on the complex plane are shown in Fig. 8.7.

Problem 8.3.8 Find the $\mathcal{Z}$ transform of the discrete-time signal

$$x[n] = \begin{cases} n & \text{for } 0 \leq n \leq N \\ 2N - n & \text{for } N + 1 \leq n \leq 2N \\ 0 & \text{elsewhere} \end{cases}$$

(Oppenheim, Schafer, and Buck, Exercise for solution 3.3β).

Solution From the defining equation of the Z transform and taking into account the functional form of the discrete-time signal $x[n]$ we have

$$X(z) = \sum_{n=-\infty}^{\infty} x[n]z^{-n} = \sum_{n=0}^{N} nz^{-n} + \sum_{n=N+1}^{2N} (2N - n)z^{-n} = S_1 + S_2,$$

where

$$S_1 = \sum_{n=0}^{N} nz^{-n} = \frac{Nz^{-N-2} - (N+1)z^{-N-1} + z^{-1}}{(1 - z^{-1})^2} = \frac{z^{-N-2}[N - (N+1)z + z^{N+1}]}{\frac{(z-1)^2}{z^2}}$$

$$= \frac{z^2}{z^{N+2}} \left(\frac{N - (N+1)z + z^{N+1}}{(z-1)^2} \right) = \frac{N - (N+1)z + z^{N+1}}{z^N (z-1)^2} \quad \text{and}$$

$$S_2 = \sum_{n=N+1}^{2N} (2N - n)z^{-n} = 2N \sum_{n=N+1}^{2N} z^{-n} - \sum_{n=N+1}^{2N} nz^{-n} = S_{21} + S_{22}.$$

In order to calculate the first of the above sums we proceed to change the dummy index from n to $n' = n - (N + 1)$ from where it turns out that $n = n' + (N + 1)$. For $n = N + 1$, $n' = 0$, whereas for $n = 2N$, $n' = 2N - (N + 1) = N - 1$. We have, therefore,

$$S_{21}=2N\sum_{n=N+1}^{2N}z^{-n}=2N\sum_{n'=0}^{N-1}(1/z)^{n'+(N+1)}=\frac{2N}{z^{N+1}}\sum_{n'=0}^{N-1}(1/z)^{n'}=\frac{2N}{z^{N+1}}\left(\frac{1-z^{-N}}{1-z^{-1}}\right)=\frac{2N}{z^{2N}}\left(\frac{z^N-1}{z-1}\right).$$

In a completely similar way, the second sum is calculated as

$$\begin{aligned}S_{22}&=\sum_{n=N+1}^{2N}n(1/z)^n=\sum_{n'=0}^{N-1}[n'+(N+1)](1/z)^{n'+(N+1)}=\frac{1}{z^{N+1}}\sum_{n'=0}^{N-1}n'(1/z)^{n'}+\frac{N+1}{z^{N+1}}\sum_{n'=0}^{N-1}(1/z)^{n'}\\&=\frac{1}{z^{N+1}}\left(\frac{(N-1)\,z^{-N-1}-Nz^{-N}+z^{-1}}{(1-z^{-1})^2}\right)+\frac{N+1}{z^{N+1}}\left(\frac{1-z^{-N}}{1-z^{-1}}\right)\\&=\frac{1}{z^{2N}}\left(\frac{z^N-Nz+N-1}{(z-1)^2}\right)+\frac{N+1}{z^{2N}}\left(\frac{z^N-1}{z-1}\right)\end{aligned}$$

and therefore we have

$$\begin{aligned}S_2&=\frac{2N}{z^{2N}}\left(\frac{z^N-1}{z-1}\right)+\frac{1}{z^{2N}}\left(\frac{z^N-Nz+N-1}{(z-1)^2}\right)+\frac{N+1}{z^{2N}}\left(\frac{z^N-1}{z-1}\right)\\&=\frac{1}{z^{2N}(z-1)}\left[(N-1)(z^N-1)-\frac{z^N-Nz+N-1}{z-1}\right]=\frac{(N-1)z-N+z^{1-N}}{z^N(z-1)^2},\end{aligned}$$

as it emerges after operations. Therefore,

$$X(z)=S_1+S_2=\frac{N-(N+1)z+z^{N+1}}{z^N(z-1)^2}+\frac{(N-1)z-N+z^{1-N}}{z^N(z-1)^2}=\frac{1}{z^{2N-1}}\left(\frac{z^N-1}{z-1}\right)^2$$

—as can be proven—which is also the desired result.

Problem 8.3.9 Find the $\mathcal{Z}$ transform of the discrete-time signal

$$x[n]=\alpha^n u[n]+\beta^n u[n]+\gamma^n u[-n-1],\quad |\alpha|<|\beta|<|\gamma|$$

(Oppenheim, Schafer, and Buck, Problem 3.31α).

Solution From the defining equation of the Z transform we have

$$X(z)=\sum_{n=-\infty}^{\infty}\alpha^n u[n]z^{-n}+\sum_{n=-\infty}^{\infty}\beta^n u[n]z^{-n}+\sum_{n=-\infty}^{\infty}\gamma^n u[-n-1]z^{-n}=\sum_{n=0}^{\infty}\left(\frac{\alpha}{z}\right)^n+\sum_{n=0}^{\infty}\left(\frac{\beta}{z}\right)^n+\sum_{n=-\infty}^{-1}\left(\frac{\gamma}{z}\right)^n,$$

with the three sums being calculated as

$$\sum_{n=0}^{\infty}\left(\frac{\alpha}{z}\right)^n=\frac{1}{1-(\alpha/z)}=\frac{z}{z-\alpha},\quad \text{ROC: } |z|>|\alpha|$$

$$\sum_{n=0}^{\infty}\left(\frac{\beta}{z}\right)^n=\frac{1}{1-(\beta/z)}=\frac{z}{z-\beta},\quad \text{ROC: } |z|>|\beta|,$$

and

$$\sum_{n=-\infty}^{-1}\left(\frac{\gamma}{z}\right)^{n}=\sum_{n'=1}^{\infty}\left(\frac{z}{\gamma}\right)^{n'}=\sum_{n'=0}^{\infty}\left(\frac{z}{\gamma}\right)^{n'}-1=\frac{1}{1-(z/\gamma)}-1=\frac{z}{\gamma-z},$$

with the region of convergence of the last sum defined by the equation $|z| < |\gamma|$. Therefore, the Z transform of the discrete-time signal $x[n]$ will be given from the relation

$$X(z)=\frac{z}{z-\alpha}+\frac{z}{z-\beta}+\frac{z}{\gamma-z}=\frac{z[z^2-2\gamma z-(\alpha\beta-\beta\gamma-\alpha\gamma)]}{(z-\alpha)(z-\beta)(z-\gamma)}$$

—as can be verified after operations—with the region of convergence $|\beta| < |z| < |\gamma|$. The design of the region of convergence and the pole-zero diagram is left as an exercise to the reader.

Problem 8.3.10 Find the $\mathcal{Z}$ transform of the discrete-time signal

$$x[n]=\sum_{k=-\infty}^{\infty}\delta(n-4k)$$

Solution From the defining equation of the Z transform we get

$$X(z)=\sum_{n=-\infty}^{\infty}x[n]\,z^{-n}=\sum_{n=-\infty}^{\infty}\sum_{k=-\infty}^{\infty}\delta[n-4k]\,z^{-n}.$$

However, among the infinite values of the index n in the above sum, the only value for which the function δ is nonzero, is the value $n = 4k$—for which we have $\delta[n-4k] = \delta[0] = 1$. Therefore, we have

$$X(z)=\sum_{k=-\infty}^{\infty}z^{-4k},$$

which is also the desired result.

8.4 Properties of the Region of Convergence

As already mentioned, *the region of convergence of the $\mathcal{Z}$ transform is defined as the region of the complex plane for which the power series of infinite terms appearing in the defining equation converges to a finite value*. There are many interesting properties that characterize the region of convergence of this transform, the most important of which are the following:

- *The shape of the region of convergence of the $\mathcal{Z}$ transform of a discrete sequence $x[n]$ depends on its length*. More specifically, if the sequence $x[n]$ has finite length, that is, it has zero values everywhere except for a range of indices $-\infty < N_1 \leq n \leq N_2 < \infty$, then the region of convergence of the $\mathcal{Z}$ transform will extend *to the entire complex plane*, possibly except for the points $z = 0$ and $z = \infty$. On the other hand, if the sequence $x[n]$ has infinite length, the shape of the region of convergence depends on the nature and characteristics of the discrete sequence.

  In a more detailed description, if the sequence is *two-sided* (i.e., it has nonzero values for both positive and negative values of the index n), the region of convergence will have, as already mentioned, a *ring* shape with the inner and outer circle limited by *poles*, i.e., points on the complex plane, at which the function $\mathcal{X}(z)$ takes infinite value. If the sequence $x[n]$ is a *left-sided sequence* (i.e., it has nonzero values only for the index region $n > N_2 > -\infty$), then the region of convergence will have a shape of *circle* with the center at the point $z = 0$ (which may be included in it) and the radius equal to the measure *of the smallest (in measure) nonzero pole*. Finally, if the sequence $x[n]$ is a *right-sided sequence*

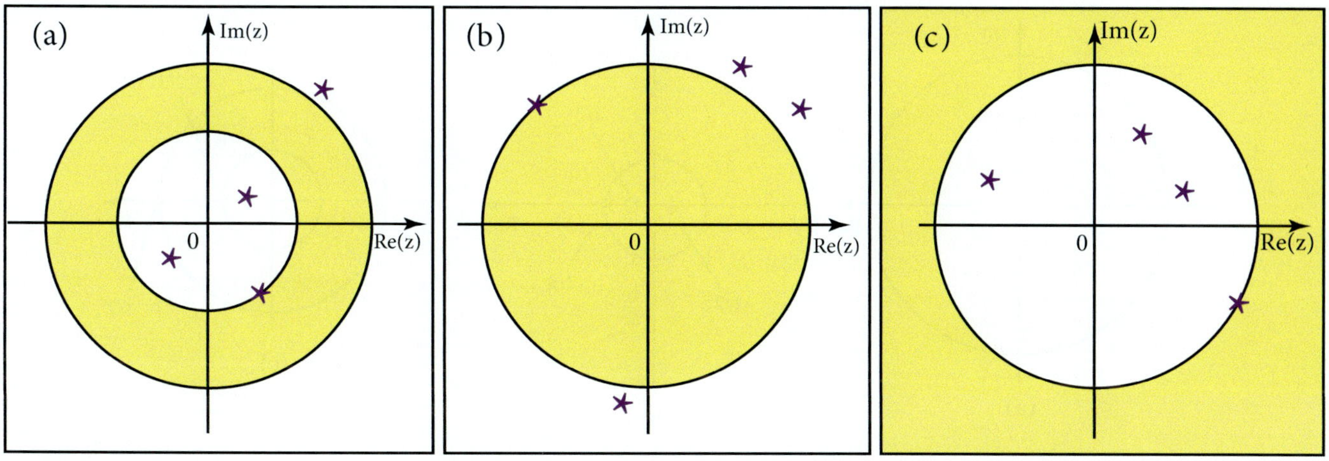

Fig. 8.8 The region of convergence of the $\mathcal{Z}$ transform for a two-sided sequence and for a left-sided and right-sided sequence

(i.e., it has nonzero values only for the index region $n < N_1 < \infty$), then the region of convergence will be defined as *the complement on the complex plane of a circle with the radius equal to the measure of the largest (in measure) nonzero pole* that may possibly extend to the point $z \to \infty$. These three cases are depicted diagrammatically in images (a), (b), and (c) respectively of Fig. 8.8.

- The *Fourier transform* of the discrete sequence $x[n]$ converges in absolute value, *if and only if the region of convergence of the $\mathcal{Z}$ transform of this sequence contains the unit circle*. As we will see in the next section, the power of this property is based on the fact that *the $\mathcal{Z}$ transform of a discrete sequence is reduced to the Fourier transform, if calculated for those values of the complex variable z for which the relation $|z| = 1$ holds*—which, as is well known, is also the defining equation *of the unit circle* on the complex plane.
- The region of convergence of the $\mathcal{Z}$ transform of a discrete sequence $x[n]$ cannot contain points of the complex plane that correspond to *pole positions*. This property is a direct consequence of the definition of the region of convergence of the $\mathcal{Z}$ transform, as if this region contained poles, then the function $\mathcal{X}(z)$ at these positions would take an infinite value and therefore the power series appearing in the defining equation of the $\mathcal{Z}$ transform would not converge.

Problem 8.4.1 For each of the following Z transforms, draw the pole-zero diagram and determine the region of convergence

$$\mathcal{X}_1(z) = \frac{1 - 0.5z^{-1}}{1 + 2z^{-1}}$$

$$\mathcal{X}_2(z) = \frac{1 - (1/3)z^{-1}}{(1 + 0.5z^{-1})[1 - (2/3)z^{-1}]}$$

$$\mathcal{X}_3(z) = \frac{1 + z^{-1} - 2z^{-2}}{1 - (13/6)z^{-1} + z^{-2}},$$

taking into account that the signal $x_2[n]$ associated with the transform $\mathcal{X}_2(z)$ is causal, whereas the signal $x_3[n]$ associated with the transform $\mathcal{X}_3(z)$ is absolutely summable.

Solution For each of the above transforms $\mathcal{Z}$ the following conclusions apply:

(a) The transform $\mathcal{X}_1(z)$ has a zero value at the position $z = 1/2$ and a pole at the position $z = -2$ and its region of convergence is described by the equation $|z| < 2$. The positions of the poles and the zero values of the transform $\mathcal{X}_1(z)$, as well as its region of convergence, are depicted in Fig. 8.9a.

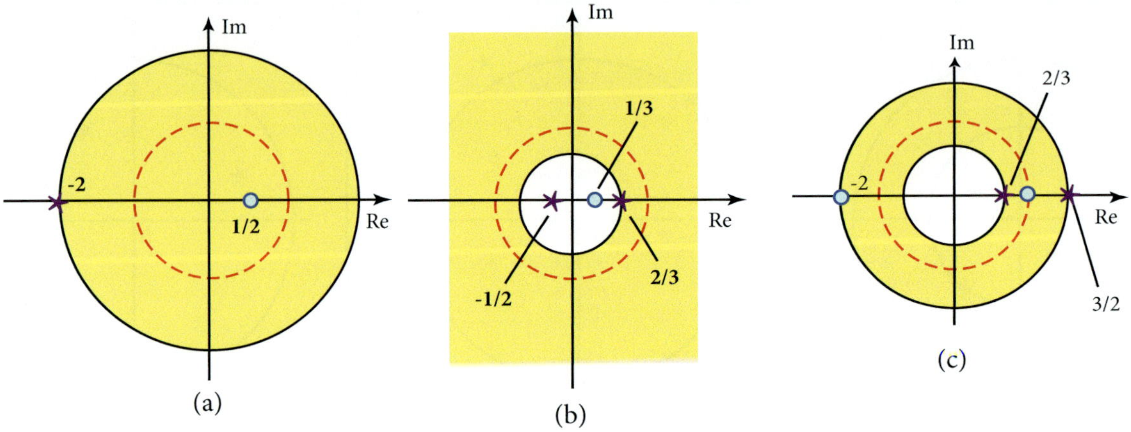

Fig. 8.9 The pole-zero diagram and the region of convergence for the $\mathcal{Z}$ transforms of Exercise 8.4.1. In the above images, the dashed line represents the unit circle

(b) The transform $\mathcal{X}_2(z)$ has a zero value at the position $z = 1/3$ and two poles at the positions $p_1 = -1/2$ and $p_2 = 2/3$. Since the signal $x_2[n]$ is causal, the region of convergence is *the complement of the circle*, the radius of which is equal to the measure of the most distant pole—that is, the pole $p_2 = 2/3$—and therefore is described by the equation $|z| > 2/3$. The positions of the poles and the zero values of the transform $\mathcal{X}_2(z)$, as well as its region of convergence, are depicted in Fig. 8.9b.

(c) The transform $\mathcal{X}_3(z)$ has two zero values at the positions $z_1 = 1$ and $z_2 = -2$ and two poles at the positions $p_1 = 3/2$ and $p_2 = 2/3$. Since the signal $x_3[n]$ is absolutely summable, the region of convergence of the transform will contain the unit circle and therefore is described by the equation $2/3 < |z| < 3/2$. The positions of the poles and the zero values of the transform $\mathcal{X}_3(z)$, as well as its region of convergence, are depicted in Fig. 8.9γ.

8.5 Initial and Final Value Theorems

In some cases, it is useful to be able to predict the initial value of the discrete-time signal $x[0]$ and especially its final value, which is defined as $x[\infty] \equiv \lim_{n\to\infty} x[n]$ with the help of the $\mathcal{Z}$ transform. The calculation of these values can be carried out with the help of the known theorems *initial and final value*, the formulation and proof of which are presented in this section.

Theorem 8.5.1 *Let a discrete-time signal be $x[n]$, the $\mathcal{Z}$ transform of which is the function $\mathcal{X}(z)$. In this case, if the signal is causal (i.e., the relation $x[n] = 0$ for $n < 0$ holds), it is proven that*

$$x[0] = \lim_{z\to\infty} \mathcal{X}(z).$$

On the other hand, if the signal is anti-causal (i.e., the relationship $x[n] = 0$ for $n > 0$ holds), it follows that

$$x[0] = \lim_{z\to 0} \mathcal{X}(z).$$

Proof From the defining equation of the $\mathcal{Z}$ transform of a causal signal we have

$$\lim_{z\to\infty} \mathcal{X}(z) = \lim_{z\to\infty} \sum_{n=-\infty}^{\infty} x[n]z^{-n} = \lim_{z\to\infty} \sum_{n=0}^{\infty} x[n]z^{-n}$$

(since for $n < 0$ it is $x[n] = 0$). Developing the term of the sum for the value $n = 0$, we get

$$\lim_{z\to\infty} \mathcal{X}(z) = x[0] + \lim_{z\to\infty} \sum_{n=1}^{\infty} x[n]z^{-n} = x[0].$$

In a completely analogous way, for an anti-causal signal $x[n]$ we have

$$\lim_{z\to 0} \mathcal{X}(z) = \lim_{z\to 0} \sum_{n=-\infty}^{\infty} x[n]\, z^{-n} = \lim_{z\to 0} \sum_{n=-\infty}^{0} x[n]z^{-n}.$$

Changing the dummy index from n to $m = -n$, the above relationship is written as

$$\lim_{z\to 0} \mathcal{X}(z) = \lim_{z\to 0} \sum_{m=0}^{\infty} x[-m]\, z^{m} = x[0] + \lim_{z\to 0} \sum_{m=1}^{\infty} x[-m]\, z^{m} = x[0].$$

Theorem 8.5.2 *Let a causal discrete-time signal $x[n]$ with $\mathcal{Z}$ transform be the function $\mathcal{X}(z)$. In this case, it is proven that if the region of convergence of the complex function is $(z-1)\mathcal{X}(z)$ it contains the unit circle and in addition all its poles are within this circle—except perhaps for a simple pole that may be located on the circumference of the unit circle and at the position $z = 1$; the relation holds*

$$x[\infty] \equiv \lim_{n\to\infty} x[n] = \lim_{z\to 1} \left[\left(1 - \frac{1}{z}\right)\mathcal{X}(z)\right]$$

Proof From the defining equation of the $\mathcal{Z}$ transform we have

$$\mathcal{Z}\{x[n]\} = \mathcal{X}(z) = \sum_{n=0}^{\infty} x[n]z^{-n} \quad \text{and} \quad \mathcal{Z}\{x[n-1]\} = z^{-1}\mathcal{X}(z) = \sum_{n=0}^{\infty} x[n-1]z^{-n},$$

where the lower limit of the sum has been set to zero, since the discrete-time signal $x[n]$ is considered causal. Subtracting the second equation from the first, we get

$$\sum_{n=0}^{\infty} x[n]z^{-n} - \sum_{n=0}^{\infty} x[n-1]z^{-n} = \mathcal{X}(z) - z^{-1}\mathcal{X}(z) = (1 - z^{-1})\mathcal{X}(z)$$

and taking the limit of the above equation for the value $z \to 1$, we arrive at the equation

$$\lim_{z\to 1} \left[\sum_{n=0}^{\infty} x[n]z^{-n} - \sum_{n=0}^{\infty} x[n-1]z^{-n}\right] = \lim_{z\to 1} [(1 - z^{-1})\mathcal{X}(z)].$$

On the other hand, the limit of the left term of the above equation is calculated as

$$\lim_{z\to 1} \left[\sum_{n=0}^{\infty} x[n]z^{-n} - \sum_{n=0}^{\infty} x[n-1]z^{-n}\right] = \sum_{n=0}^{\infty} (x[n] - x[n-1])$$
$$= (x[0] - x[-1]) + (x[1] - x[0]) + (x[2] - x[1]) + \cdots = x[\infty] = \lim_{n\to\infty} x[n].$$

Of course, for the last relation to hold, the value $x[\infty]$ must be *finite*, which is ensured *if the conditions regarding the positions of the poles* that were formulated in the statement of the theorem are met.

By combining the last equations, we finally get

$$x[\infty] \equiv \lim_{n\to\infty} x[n] = \lim_{z\to 1}\left[\left(1 - \frac{1}{z}\right)\mathcal{X}(z)\right]. \tag{8.9}$$

The exercises that follow allow understanding of the theorems of initial and final values for specific cases of causal and anti-causal signals.

Problem 8.5.1 Let the $\mathcal{Z}$ transform be described by the equation

$$\mathcal{X}(z) = \frac{1}{1-\alpha z^{-1}}.$$

Determine the value $x[0]$ assuming that the discrete-time signal $x[n]$ is causal (Hayes, Solved Exercise 4.15).

Solution From the initial value theorem for causal signals we have

$$x[0] = \lim_{z\to\infty} \mathcal{X}(z) = \lim_{z\to\infty} \frac{1}{1-\alpha z^{-1}} = 1.$$

Problem 8.5.2 Let the $\mathcal{Z}$ transform be described by the equation

$$\mathcal{X}(z) = \frac{3z^{-1} + 2z^{-2}}{3 - z^{-1} + z^{-2}}.$$

Determine the value $x[0]$ assuming that the discrete-time signal $x[n]$ is anti-causal (Hayes, Solved Exercise 4.18).

Solution From the initial value theorem for anti-causal signals we have

$$x[0] = \lim_{z\to 0} \mathcal{X}(z) = \lim_{z\to 0} \frac{3z^{-1} + 2z^{-2}}{3 - z^{-1} + z^{-2}} = \lim_{z\to 0} \frac{3z+2}{3z^2 - z + 1} = 2.$$

Problem 8.5.3 Let the $\mathcal{Z}$ transform of the causal discrete-time signal $x[n]$ be defined as

$$\mathcal{X}(z) = \frac{5z^2 + 4z}{(z-1)(6z-1)^2}.$$

Using the final value theorem, determine the value $x[\infty]$.

Solution Applying the final value theorem, we get

$$x[\infty] = \lim_{n\to\infty} x[n] = \lim_{z\to 1}[(1 - z^{-1})\mathcal{X}(z)] = \lim_{z\to 1}\left[\left(\frac{z-1}{z}\right)\frac{5z^2+4z}{(z-1)(6z-1)^2}\right] = \lim_{z\to 1}\left[\frac{5z^2+4z}{z(6z-1)^2}\right] = \frac{9}{25}.$$

8.6 Properties of the $\mathcal{Z}$ Transform

In full analogy with the Fourier and Laplace transforms presented in previous chapters, the $\mathcal{Z}$ transform is also characterized by the existence of a set of corresponding properties, which significantly simplify the calculation process; these properties are presented in detail in the following pages:

- **Linearity:** if the transforms $\mathcal{Z}$ of the discrete-time signals $x_1[n]$ and $x_2[n]$ are the functions $\mathcal{X}_1(z)$ and $\mathcal{X}_2(z)$, then the $\mathcal{Z}$ transform of the signal $x[n] = \alpha x_1[n] + \beta x_2[n]$ will have the form $\mathcal{X}(z) = \alpha\mathcal{X}_1(z) + \beta\mathcal{X}_2(z)$ for every value of the constants α and β.

  Proof considering the linear combination
$$x[n] = \alpha x_1[n] + \beta x_2[n]$$
  the transform $\mathcal{X}(z)$ of this signal is given by definition from the relation
$$\mathcal{X}(z) = \sum_{n=-\infty}^{\infty} x[n]z^{-n} = \sum_{n=-\infty}^{\infty} \Big(\alpha x_1[n] + \beta x_2[n]\Big) z^{-n}$$
$$= \alpha \sum_{n=-\infty}^{\infty} x_1[n]z^{-n} + \beta \sum_{n=-\infty}^{\infty} x_2[n]z^{-n} = \alpha\mathcal{X}_1(z) + \beta\mathcal{X}_2(z),$$
  an equation that expresses the *property of linearity*. According to this property, *the $\mathcal{Z}$ transform of the linear combination of a number of signals coincides with the linear combination of the transforms $\mathcal{Z}$ of the signals separately*. As for the region of convergence of the transform $\mathcal{X}(z)$, this is at least *the intersection of the regions of convergence of the transforms $\mathcal{X}_1(z)$ and $\mathcal{X}_2(z)$*. More specifically, for signals whose $\mathcal{Z}$ transform is a *rational function* of the form $\mathcal{B}(z)/\mathcal{A}(z)$, if the poles of the function $\mathcal{X}(z)$ contain all the poles of the transforms $\mathcal{X}_1(z)$ and $\mathcal{X}_2(x)$, the region of convergence is *precisely* the intersection of the regions of convergence of the two transforms. Conversely, if in the transform $\mathcal{X}(z)$ there are *zeros*, which *cancel* one or more of the above poles, the region of convergence may be *even larger*.

  It is important to mention at this point that the property of *linearity* of the $\mathcal{Z}$ transform is the principle upon which the process of calculating the inverse $\mathcal{Z}$ transform through the application of the *method of partial fraction expansion* is based, which is presented in a subsequent section.
- **Time shift:** if the $\mathcal{Z}$ transform of the signal $x[n]$ is the function $\mathcal{X}(z)$, then the $\mathcal{Z}$ transform of the time-delayed signal $x[n-k]$ will be $z^{-k}\mathcal{X}(z)$.

  Proof considering the discrete-time signal $y[n] = x[n-k]$, the $\mathcal{Z}$ transform of this signal is the
$$\mathcal{Y}(z) = \sum_{n=-\infty}^{\infty} x[n-k]z^{-n} = \sum_{m=-\infty}^{\infty} x[m]z^{-m-k} = z^{-k} \sum_{m=-\infty}^{\infty} x[m]z^{-m} = z^{-k}\mathcal{X}(z),$$
  where for the transition from the second to the third sum, a change of the dummy index from n to $m = n - k$ was made. The region of convergence of the $\mathcal{Y}(z)$ transform is the same as that of the $\mathcal{X}(z)$ transform with the possible addition of the point $z = 0$, if $k > 0$ or $z \to \infty$, if $k < 0$.
- **Scaling:** If the $\mathcal{Z}$ transform of the discrete-time signal $x[n]$ is the function $\mathcal{X}(z)$ with the region of convergence $r_1 < |z| < r_2$, then the $\mathcal{Z}$ transform of the signal $\alpha^n x[n]$ is the function $\mathcal{X}(\alpha^{-1}z)$ with the region of convergence $|\alpha|r_1 < |z| < |\alpha|r_2$, for any real or complex value of the parameter α.

  Proof considering the discrete-time signal $y[n] = \alpha^n x[n]$, the $\mathcal{Z}$ transform of this signal is
$$\mathcal{Y}(z) = \sum_{n=-\infty}^{\infty} y[n]z^{-n} = \sum_{n=-\infty}^{\infty} x[n]\,\alpha^n z^{-n} = \sum_{n=-\infty}^{\infty} x[n](\alpha^{-1}z)^{-n} = \sum_{n=-\infty}^{\infty} x[n]\left(\frac{z}{\alpha}\right)^{-n} = \mathcal{X}\left(\frac{z}{\alpha}\right),$$

an equation that also expresses the *property of scaling*. The basic consequence of this property is *the multiplication of the positions of the poles and zeros by the value of the parameter* α, a fact that arises from the observation that if the $\mathcal{X}(z)$ transform has a pole at the position $z = p_1$, then the $\mathcal{X}(\alpha^{-1}z)$ transform will have a pole at the position $z = \alpha p_1$. If the value of α is real, this scaling can be interpreted as *contraction or expansion of the complex plane with the poles and zeros moving along radial lines, that is, straight lines passing through the point* $z = 0$. Conversely, if α is a complex number with a modulus equal to one—that is, given by the relation $\alpha = e^{j\omega}$—the scaling corresponds *to a rotation on the complex plane with the poles and zeros to shift along the peripheries of cycles centered at the point* $z = 0$. The latter fact is interpreted in turn as a *frequency shift process*, which arises by multiplying the discrete-time signal $x[n]$ with the complex exponential component $e^{-i\omega n}$.

- **Time reversal:** if the $\mathcal{Z}$ transform of the discrete-time signal $x[n]$ is the function $\mathcal{X}(z)$ with the region of convergence $r_1 < |z| < r_2$, then the $\mathcal{Z}$ transform of the signal $x[-n]$ is the function $\mathcal{X}(1/z)$ with the region of convergence $(1/r_2) < |z| < (1/r_1)$.

  Proof considering the discrete-time signal $y(n) = x[-n]$, the $\mathcal{Z}$ transform of the signal is expressed as

$$\mathcal{Y}(z) = \sum_{n=-\infty}^{\infty} x[-n]z^{-n} = \sum_{n=-\infty}^{\infty} x[n]z^{n} = \sum_{n=-\infty}^{\infty} x[n](z^{-1})^{-n} = X(z^{-1}) = \mathcal{X}\,(1/z).$$

  The above relation holds when the sequence $x[n]$ is real, whereas for a signal of complex values it is similarly proven that $x^*[-n] \rightleftarrows \mathcal{X}^*(1/z^*)$, where "$*$" is the operator of complex conjugation.
- **Conjugation:** if the $\mathcal{Z}$ transform of the discrete-time signal $x[n]$ is the function $\mathcal{X}(z)$ with region of convergence $r_1 < |z| < r_2$, then the $\mathcal{Z}$ transform of the discrete-time signal $x^*[-n]$ will be the function $\mathcal{X}^*(z^*)$ with the same region of convergence.

  Proof From the defining equation of the $\mathcal{Z}$ transform it immediately follows that

$$\mathcal{Z}\{x^*[n]\} = \sum_{n=-\infty}^{\infty} x^*[n]z^{-n} = \left(\sum_{n=-\infty}^{\infty} x[n](z^*)^{-n}\right)^* = \mathcal{X}^*(z^*).$$

- **Differentiation:** if the $\mathcal{Z}$ transform of the discrete-time signal $x[n]$ is the function $\mathcal{X}(z)$, then the $\mathcal{Z}$ transform of the discrete-time signal $nx[n]$ will be the function $-z[d\mathcal{X}(z)/dz]$ with the two functions being characterized by the same region of convergence.
  Proof differentiating the defining equation of the $\mathcal{Z}$ transform, we get

$$\frac{d\mathcal{X}(z)}{dz} = \sum_{n=-\infty}^{\infty} (-n)x[n]z^{-n-1},$$

  from which it follows that

$$-z\frac{d\mathcal{X}(z)}{dz} = -z\sum_{n=-\infty}^{\infty} (-n)x[n]z^{-n-1} = \sum_{n=-\infty}^{\infty} \Big(nx[n]\Big)z^{-n} = \mathcal{Z}\{nx[n]\}.$$

- **Convolution:** if the $\mathcal{Z}$ transforms of the discrete-time signals $x_1[n]$ and $x_2[n]$ are the functions $\mathcal{X}_1(z)$ and $\mathcal{X}_2(z)$, then the $\mathcal{Z}$ transform of the signal $x[n] = x_1[n] * x_2[n]$ will have the form $\mathcal{X}(z) = \mathcal{X}_1(z)\mathcal{X}_2(z)$. As for the region of convergence of the $X(z)$ transform, it will be *at least the intersection of the regions of convergence of the* $\mathcal{X}_1(z)$ *and* $\mathcal{X}_2(z)$ transforms, whereas if there are zeros inside the region of convergence of the $\mathcal{X}_2(z)$ transform that cancel poles that are on the boundary of the region of convergence of the $\mathcal{X}_1(z)$ transform or conversely, the region of convergence of the $\mathcal{X}(z)$ transform may be *even larger*. ***Proof*** considering the discrete-time signal

$$y[n] = x_1[n] * x_2[n] = \sum_{k=-\infty}^{\infty} x_1[k]x_2[n-k]$$

Table 8.1 The properties of the $\mathcal{Z}$ transform

Property	Discrete-time signal $x[n]$	Z transform	Region of convergence
Linearity	$\alpha_1 x_1[n] + \alpha_2 x_2[n]\alpha_1$	$X_1(z) + \alpha_2 X_2(z)$	$R_1 \cap R_2$
Time shift	$x[n-k]$	$z^{-k}X(z)$	$R - \begin{cases} \{0\}(k>0) \\ \{\infty\}(k<0) \end{cases}$
Scaling	$\alpha^n x[n]$	$X\left(\alpha^{-1}z\right)$	αR
Time reversal	$x[-n]$	$X(1/z)$	$\frac{1}{r_1} < \lvert z\rvert < \frac{1}{r_2}$
Conjugation	$x^*[n]$	$X^*(z^*)$	R
Differentiation	$nx[n]$	$-z\frac{dX(z)}{dz}$	R
Convolution	$x_1[n] * x_2[n]$	$X_1(z)X_2(z)$	$R_1 \cap R_2$
Correlation	$r_{x_1x_2}[\ell]$	$X_1(z)X_2\left(z^{-1}\right)$	$R_1 \cap R_2$
Notation	$Z\{x[n]\} = X(z),\ Z\{x_1[n]\} = X_1(z),\ Z\{x_2[n]\} = X_2(z)$ with regions of convergence R : $r_2 < \lvert z\rvert < r_1,\ R_1, R_2$		

the $\mathcal{Z}$ transform of this signal will be given according to what is known from the relation

$$\mathcal{Y}(z) = \sum_{n=-\infty}^{\infty} y[n]z^{-n} = \sum_{n=-\infty}^{\infty} \left\{ \sum_{k=-\infty}^{\infty} x_1[k]x_2[n-k] \right\} z^{-n} = \sum_{k=-\infty}^{\infty} x_1[k] \left\{ \sum_{n=-\infty}^{\infty} x_2[n-k]z^{-n} \right\},$$

with the second sum resulting from the first *by changing the order of the sums*. Finally, changing the dummy index of the second sum from n to $m = n - k$, we get

$$\mathcal{Y}(z) = \sum_{k=-\infty}^{\infty} x_1[k] \sum_{m=-\infty}^{\infty} x_2[m]z^{-m-k} = \left(\sum_{k=-\infty}^{\infty} x_1[k]z^{-k} \right) \left(\sum_{m=-\infty}^{\infty} x_2[m]z^{-m} \right) = \mathcal{X}_1(z)\mathcal{X}_2(z).$$

- **Correlation of sequences:** if the transforms $\mathcal{Z}$ of the discrete-time signals $x_1[n]$ and $x_2[n]$ are the functions $\mathcal{X}_1(z)$ and $\mathcal{X}_2(z)$, then the $\mathcal{Z}$ transform of the discrete-time signal $r_{x_1x_2}[\ell]$ will be the function $R_{x_1x_2}(z)$ with the functions $r_{x_1x_2}[\ell]$ and $R_{x_1x_2}(z)$ defined as

$$r_{x_1x_2}[\ell] = \sum_{n=-\infty}^{\infty} x_1[n]x_2[n-\ell] \quad \text{and} \quad R_{x_1x_2}(z) = \mathcal{X}_1(z)\mathcal{X}_2(z^{-1}).$$

As for the region of convergence of the transform $\mathcal{X}(z)$, it will be at least the intersection of the regions of convergence of the transforms $\mathcal{X}_1(z)$ and $\mathcal{X}_2(z)$.

Proof this property is proven using the properties of *convolution* and *time reversal* and taking into account that the correlation of two sequences is characterized by the property $r_{x_1x_2}[\ell] = x_1[\ell] * x_2[-\ell]$. Therefore, we have

$$\mathcal{Z}\{r_{x_1x_2[\ell]}\} = \mathcal{Z}\{x_1[\ell]\}\mathcal{Z}\{x_2[-\ell]\} = \mathcal{X}_1(z)\mathcal{X}_2(z^{-1}).$$

The above properties of the $\mathcal{Z}$ transform are summarized in Table 8.1.

8.6.1 Examples of the Application of the Properties of the Transformation

In full analogy with the *Fourier* and *Laplace* transforms, the use of the properties of the $\mathcal{Z}$ transform significantly simplifies the process of calculating the aforementioned transformation. The following examples confirm the truth of this assertion.

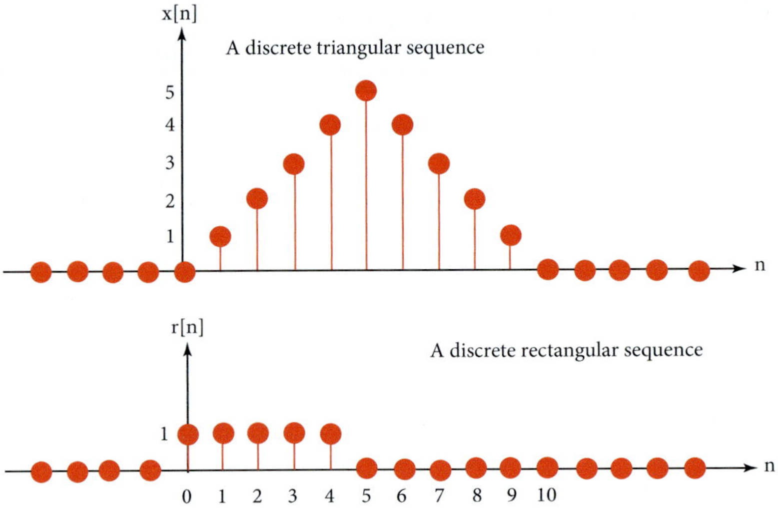

Fig. 8.10 The rectangular and the triangular discrete sequences

Example 8.6.1 Let us consider the discrete-time signal

$$x[n] = \begin{cases} n & \text{for } 0 \leq n \leq N \\ 2N - n & \text{for } N+1 \leq n \leq 2N \\ 0 & \text{elsewhere,} \end{cases}$$

the $\mathcal{Z}$ transform of which it is proven to be given by the relation

$$X(z) = \frac{1}{z^{2N-1}} \left(\frac{z^N - 1}{z - 1} \right)^2.$$

This result arises by directly using the defining equation of the $\mathcal{Z}$ transform, which constitutes the subject of Exercise 8.3.8 (refer to page 660) and arises after many mathematical operations. However, the above process can be significantly simplified, if we observe that the discrete-time signal $x[n]$ is nothing but the *triangular discrete sequence* $\Lambda[n]$, which is proven to be given by the relation[2] $\Lambda[n] = \Pi_{N-1}[n-1] * \Pi_{N-1}[n]$, where $\Pi_{N-1}[n]$ is the *discrete rectangular sequence*, which is defined as

$$\Pi_{N-1}[n] = \begin{cases} 1 \, 0 \leq n \leq N-1 \\ 0 \text{ anywhere else,} \end{cases}$$

whereas the symbol $*$ refers to the operator of convolution. The triangular and the rectangular sequences for a parameter value of $N = 5$ are presented in Fig. 8.10. According to the property of the convolution of the $\mathcal{Z}$ transform, if $z[n] = x[n] * y[n]$, then it will also be $\mathcal{Z}\{z[n]\} = \mathcal{Z}\{x[n]\}\mathcal{Z}\{y[n]\}$. Therefore, the $\mathcal{Z}$ transform of the triangular sequence defined as $\Lambda[n] = \Pi_{N-1}[n-1] * \Pi_{N-1}[n]$ is given by the relation

$$\mathcal{Z}\{x[n]\} = \mathcal{Z}\{\Pi_{N-1}[n-1]\}\mathcal{Z}\{\Pi_{N-1}[n]\} = \left(z^{-1}\mathcal{Z}\{\Pi_{N-1}[n]\} \right) \mathcal{Z}\{\Pi_{N-1}[n]\} = \frac{1}{z}\left[\mathcal{Z}\{\Pi_{N-1}[n]\} \right]^2,$$

where to derive the last relation we used the property of time shifting $x[n-k] \leftrightarrows z^{-k}X(z)$ for the value $k = 1$. Taking into account that the $X(z)$ transform of the rectangular sequence is $X(z) = (z^{N-1})/[z^{N-1}(z-1)]$ (see the note of Sect. 8.2 on page 649) and replacing this function in the last relation, we arrive at the result we presented at the beginning of this example, performing far fewer operations and spending much less time.

Example 8.6.2 In Exercise 8.3.4 (refer to page 655) we proved that the $\mathcal{Z}$ transform of the discrete-time signal defined as $y[n] = n\alpha^n \sin(\omega n)u[n]$ is given by the relation

$$\mathcal{Y}(z) = \frac{\alpha z(z^2 - \alpha^2)\sin(\omega)}{(z^2 - 2\alpha z\cos(\omega) + \alpha^2)^2}.$$

The last relation can be proven much more easily, if we take into account that the considered discrete-time signal is of the form $y[n] = nx[n]$, where $x[n] = \alpha^n \sin(\omega n)u[n]$ and we use the property $nx[n] \leftrightarrows -z\mathcal{X}'(z)$, where $\mathcal{X}(z)$ is the $\mathcal{Z}$ transform of the signal $x[n]$. From tables of elementary transforms—or even with direct calculation—it turns out that

$$\mathcal{X}(z) = \frac{\alpha z^{-1}\sin(\omega)}{1 - 2\alpha z^{-1}\cos(\omega) + \alpha^2 z^{-2}}.$$

Differentiating the above equation, we get

$$\frac{d\mathcal{X}(z)}{dz} = \frac{\alpha^3 z^{-4}\sin(\omega) - \alpha z^{-2}\sin(\omega)}{(1 - 2\alpha z^{-1}\cos(\omega) + \alpha^2 z^{-2})^2} = \frac{\alpha^3\sin(\omega) - \alpha z^2\sin(\omega)}{(z^2 - 2\alpha z\cos(\omega) + \alpha^2)^2}.$$

Therefore, we have

$$\mathcal{Y}(z) = -z\frac{d\mathcal{X}(z)}{dz} = -z\frac{\alpha^3\sin(\omega) - \alpha z^2\sin(\omega)}{(z^2 - 2\alpha z\cos(\omega) + \alpha^2)^2} = \frac{\alpha z(z^2 - \alpha^2)\sin(\omega)}{(z^2 - 2\alpha z\cos(\omega) + \alpha^2)^2},$$

which is the same result as the one we concluded by using the $\mathcal{Z}$ transform equation.

Example 8.6.3 Let us assume that we want to calculate the $\mathcal{Z}$ transform of the discrete-time signal $x[n] = n^2\alpha^n u[n]$. If we start from the defining equation of this transform, we get

$$\mathcal{X}(z) = \sum_{n=-\infty}^{\infty} x[n]z^{-n} = \sum_{n=-\infty}^{\infty} n^2\alpha^n u[n]z^{-n} = \sum_{n=0}^{\infty} n^2\left(\frac{\alpha}{z}\right)^n$$

We observe that the last sum we ended up with contains the term n^2 and is not included in any of the tables of the mathematical typology—without of course meaning that it is not possible to find or calculate it in closed form. Therefore, in order to calculate the $\mathcal{Z}$ transform of the discrete-time signal $x[n]$, we need to use a different method. Indeed, setting $y[n] = \alpha^n u[n]$, we find that our initial signal can be written in the form $x[n] = ny'[n]$, where $y'[n] = ny[n]$. Using the property $ny[n] \leftrightarrows -z[d\mathcal{Y}(z)/dz]$ and taking into account that the transform $\mathcal{Y}(z)$ of the signal $y[n]$ is given by the relation $\mathcal{Y}(z) = z/(z - \alpha)$, it is not difficult to find that the $\mathcal{Z}$ transform of the signal $y'[n]$ will be calculated from the relation

$$\mathcal{Y}'(z) = -z\frac{d\mathcal{Y}(z)}{dz} = -z\frac{d}{dz}\left(\frac{z}{z - \alpha}\right) = \frac{\alpha z}{(1 - \alpha)^2}.$$

Using again the above property for the signal $x[n] = ny'[n] = n^2\alpha^n u[n]$, we get

$$\mathcal{X}(z) = -z\frac{d\mathcal{Y}'(z)}{dz} = -z\frac{d}{dz}\left(\frac{\alpha z}{(1 - \alpha)^2}\right) = \frac{\alpha z\,(z + \alpha)}{(z - \alpha)^3}.$$

8.7 The Inverse $\mathcal{Z}$ Transform

Following the same mode of operation as the other inverse transforms, the *inverse $\mathcal{Z}$ transform* transforms the discrete-time signal $x[n]$ from the representation on the complex plane to the field of discrete-time n. In order to construct the defining equation of the inverse transform, we start from the equation of the direct transformation

[2] For the proof of the continuous equivalent of this equation, see Exercise 1.10.9 on page 51.

$$\mathcal{X}(z) \equiv \sum_{k=-\infty}^{\infty} x[k]z^{-k} \tag{8.10}$$

and we multiply both terms by the quantity z^{n-1}. In the next step of the process, we *calculate the contour integral of both terms along a closed curve C on the complex plane, which is located within the region of convergence of the power series of the infinite terms of the transform and contains the origin* $z = 0$. Therefore, we have

$$\oint_C \mathcal{X}(z)z^{n-1}dz = \oint_C \sum_{k=-\infty}^{\infty} x[k]z^{n-k-1}dz. \tag{8.11}$$

The calculation of the integral along the closed curve C in the above equation is done *in the counterclockwise direction*. Since the curve in question is *within the region of convergence of the power series of the infinite terms*, it follows that the power series in question *converges*, and therefore we can change the order of the operations of summation and integration in the above equation, formulating it as follows:

$$\oint_C \mathcal{X}(z)z^{n-1}dz = \sum_{k=-\infty}^{\infty} \oint_C x[k]z^{n-k-1}dz. \tag{8.12}$$

In the final step of the process we use the *Cauchy's integral theorem*, which is given by the mathematical relation

$$\frac{1}{2\pi j} \oint_C z^{n-1-k}dz = \begin{cases} 1 & k = n \\ 0 & k \neq n, \end{cases} \tag{8.13}$$

where the curve C can be *any closed curve that contains the origin of the complex plane* $z = 0$. In this case, the above equation is written

$$2\pi j\, x[n] = \oint_C \mathcal{X}(z)z^{n-1}dz, \tag{8.14}$$

that is,

$$x[n] = \frac{1}{2\pi j} \oint_C \mathcal{X}(z)z^{n-1}dz = \frac{1}{2\pi j} \oint_C \frac{\mathcal{X}(z)}{z} z^n dz. \tag{8.15}$$

The last equation also constitutes the defining equation of the inverse $\mathcal{Z}$.

In full analogy with the inverse Laplace transform, there are three different ways of calculating the inverse $\mathcal{Z}$ transform: (a) the *direct calculation of the contour integral* of its defining equation, (b) the method of *power series*, and (c) the method of *partial fractions*. The analytical description of the three methods and examples of their application are given in the following pages.

8.7.1 The Method of Complex Integration

The calculation of the contour integral in the above equation can be performed using the *residue theorem* of complex analysis, which is known to be formulated as follows.

Theorem 8.7.1 *If C is a closed path on the complex plane and $f(z)$ is a function that is analytic both along the path C and within the area D enclosed by it, except for a finite number of isolated singular points $z_1, z_2, \ldots, z_n$ of the region D, then*

$$\oint_C f(z)dz = 2\pi j \sum_{k=1}^{n} Res[f(z_k)].$$

Observing the equation of the inverse $\mathcal{Z}$ transform, we find that the contour integral appearing in it can be calculated with the help of the residue theorem if *as a complex function $f(z)$ we use the function $\mathcal{Y}(z) = [\mathcal{X}(z)/z]z^n$ and as path C we choose any closed curve on the complex plane that contains both the origin $z = 0$ and all the poles of the function $\mathcal{Y}(z)$*. Therefore, the equation of the inverse $\mathcal{Z}$ transform will take the form

$$x[n] = \frac{1}{2\pi j}\oint_C \mathcal{Y}(z)dz = \frac{1}{2\pi j}\oint_C \frac{\mathcal{X}(z)}{z} z^n dz = \sum_{k=1}^{n} Res[\mathcal{Y}(z_k)] \tag{8.16}$$

with the residues of the function $Y(z)$ at its poles being calculated in the known way (see Sect. 4.7.1 on page 335 and Exercises 4.7.1 on page 337, 4.7.2 on page 340, and 4.7.3 on page 341 dealing with the same problem for the case of the inverse Laplace transform).[3] The following are some examples of calculating the inverse $\mathcal{Z}$ transform using the method of *complex integration*, so that the reader can fully understand its method of application.

Problem 8.7.1 Find the inverse of the $\mathcal{Z}$ transform that is described by the equation

$$\mathcal{X}(z) = \frac{z^2}{(z-0.8)(z-0.9)(z-1)}.$$

Solution The complex function

$$\mathcal{Y}(z) = \frac{\mathcal{X}(z)}{z} z^n = \frac{z^{n+1}}{(z-0.8)(z-0.9)(z-1)}$$

has *three simple poles* at the points $p_1 = 0.8$, $p_2 = 0.9$ and $p_3 = 1$ with residue values at these calculated as

$$Res[Y(0.8)] = \lim_{z\to 0.8}\left[(z-0.8)\frac{z^{n+1}}{(z-0.8)(z-0.9)(z-1)}\right] = \lim_{z\to 0.8}\left[\frac{z^{n+1}}{(z-0.9)(z-1)}\right] = \frac{(0.8)^{n+1}}{(-0.1)(-0.2)} = 40(0.8)^n$$

$$Res[Y(0.9)] = \lim_{z\to 0.9}\left[(z-0.9)\frac{z^{n+1}}{(z-0.8)(z-0.9)(z-1)}\right] = \lim_{z\to 0.9}\left[\frac{z^{n+1}}{(z-0.8)(z-1)}\right] = \frac{(0.9)^{n+1}}{(0.1)(-0.1)} = -90(0.9)^n$$

$$Res[Y(1)] = \lim_{z\to 1}\left[(z-1)\frac{z^{n+1}}{(z-0.8)(z-0.9)(z-1)}\right] = \lim_{z\to 1}\left[\frac{z^{n+1}}{(z-0.8)(z-0.9)}\right] = \frac{(1)^{n+1}}{(0.2)(0.1)} = 50.$$

[3] From the theory of complex analysis it is known that the residue of the complex function $f(z)$ at the point $z = z_0$, which is a pole of the function of order m, is given by the relation

$$Res[f(z_0)] = \begin{cases} \lim\limits_{z\to z_0}[(z-z_0)f(z)] & \text{if } m = 1 \\ \dfrac{1}{(m-1)!}\lim\limits_{z\to z_0}\dfrac{d^{m-1}}{dz^{m-1}}\left[(z-z_0)^m f(z)\right] & \text{if } m > 1 \end{cases}$$

Therefore, we have

$$x[n] = \oint_C \mathcal{Y}(z)dz = 40(0.8)^n - 90(0.9)^n + 50.$$

Problem 8.7.2 Find the inverse of the $\mathcal{Z}$ transform described by the equation

$$X(z) = \frac{z - 1.5}{z^2 - 0.7z + 0.1}.$$

Solution The equation $z^2 - 0.7z + 0.1 = 0$ is a *quadratic algebraic equation* with respect to z with discriminant $\Delta = 0.09$ and roots $z_1 = 0.5$ and $z_2 = 0.2$. Therefore, the function

$$Y(z) = \frac{X(z)}{z}z^n = \frac{(z - 1.5)z^n}{z(z^2 - 0.7z + 0.1)} = \frac{(z - 1.5)z^n}{z(z - 0.2)(z - 0.5)}$$

has *three simple poles* at the points $p_1 = 0.2$, $p_2 = 0.5$, and $p_3 = 0$ with residue values at these

$$Res[Y(0.2)] = \lim_{z\to 0.2}\left[(z - 0.2)\frac{(z - 1.5)z^n}{z(z - 0.5)(z - 0.2)}\right] = \lim_{z\to 0.2}\left[\frac{(z - 1.5)z^n}{z(z - 0.5)}\right] = \frac{(0.2 - 1.5)(0.2)^n}{(0.2)(0.2 - 0.5)} = +21.66(0.2)^n$$

$$Res[Y(0.5)] = \lim_{z\to 0.5}\left[(z - 0.5)\frac{(z - 1.5)z^n}{z(z - 0.5)(z - 0.2)}\right] = \lim_{z\to 0.5}\left[\frac{(z - 1.5)z^n}{z(z - 0.2)}\right] = \frac{(0.5 - 1.5)(0.5)^n}{(0.5)(0.5 - 0.2)} = -06.66(0.5)^n$$

$$Res[Y(0)] = \lim_{z\to 0}\left[z\frac{(z - 1.5)z^n}{z(z - 0.5)(z - 0.2)}\right] = \lim_{z\to 0}\left[\frac{(z - 1.5)z^n}{(z - 0.5)(z - 0.2)}\right] = \frac{-1.5}{(-0.5)(-0.2)} = -15.$$

Therefore, we have

$$x[n] = \oint_C Y(z)dz = -15 + 21.666(0.2)^n - 6.666(0.5)^n.$$

Problem 8.7.3 Find the inverse of the $\mathcal{Z}$ transform, which is described by the equation

$$\mathcal{X}(z) = \frac{z}{(z + 0.2)(z^2 - 0.4z + 0.04)}.$$

Solution The equation $z^2 - 0.4z + 0.04 = 0$ is a *quadratic algebraic equation* with respect to z, with discriminant $\Delta = 0$ and a double root $z_1 = z_2 = 0.2$. Therefore, the function

$$\mathcal{Y}(z) = \frac{\mathcal{X}(z)}{z}z^n = \frac{z^{n+1}}{z(z + 0.2)(z^2 - 0.4z + 0.04)} = \frac{z^n}{(z + 0.2)(z - 0.2)^2}$$

has a *simple pole* at the position $z = -0.2$ and a *double pole* at the position $z = 0.2$, with residue values at these

$$Res[Y(-0.2)] = \lim_{z\to -0.2}\left[(z + 0.2)\frac{z^n}{(z + 0.2)(z - 0.2)^2}\right] = \lim_{z\to -0.2}\left[\frac{z^n}{(z - 0.2)^2}\right] = \frac{(-0.2)^n}{(-0.4)^2} = 6.25(-0.2)^n$$

$$Res[Y(+0.2)] = \frac{1}{(2 - 1)!}\lim_{z\to +0.2}\frac{d^{2-1}}{dz^{2-1}}\left[(z - 0.2)^2\frac{z^n}{(z + 0.2)(z - 0.2)^2}\right]$$

$$= \lim_{z\to +0.2}\frac{d}{dz}\left[\frac{z^n}{(z + 0.2)}\right]\lim_{z\to +0.2}\frac{nz^{n-1}(z + 0.2) - z^n}{(z + 0.2)^2} = \frac{n(0.2)^{n-1}0.4 - (0.2)^n}{(0.4)^2} = 6.25[(0.2)^n(2n - 1)]$$

as derived after operations. Therefore, we have

$$x[n] = \oint_C \mathcal{Y}(z)dz = 6.25[(-0.2)^n + (2n-1)(0.2)^n].$$

8.7.2 The Method of Power Series Expansion

The central idea of this method is to write the function $\mathcal{X}(z)$ as a power series of infinite terms

$$\mathcal{X}(z) = \sum_{n=-\infty}^{\infty} c_n z^{-n}, \tag{8.17}$$

which *must necessarily converge in the (known) region of convergence of the transform*. If we combine this relationship with the defining equation of the $\mathcal{Z}$ transform, it immediately follows that

$$\mathcal{X}(z) = \sum_{n=-\infty}^{\infty} x[n]\, z^{-n} = \sum_{n=-\infty}^{\infty} c_n z^{-n} \Rightarrow \sum_{n=-\infty}^{\infty} x[n]\, z^{-n} - \sum_{n=-\infty}^{\infty} c_n z^{-n} = 0,$$

and finally

$$\sum_{n=-\infty}^{\infty} [x(n) - c_n]\, z^{-n} = 0. \tag{8.18}$$

In the next step of the process, we use the *uniqueness theorem*, according to which *the representation of a function by a series of infinite terms is unique*. The application of this theorem in our case imposes the requirement that all the coefficients of the quantity z^{-n} of the above expansion are equal to zero. Therefore, $x[n] - c_n = 0$ and consequently $x[n] = c_n$ ($\forall n \in N$), an equation from which we can calculate the values of the discrete-time signal $x[n]$. The way of applying this method is demonstrated in the next two exercises.

Problem 8.7.4 Find the inverse of the $\mathcal{Z}$ transform that is described by the equation

$$\mathcal{X}(z) = \log(1 + \alpha z^{-1}) \quad (|z| > |\alpha|)$$

using the method of power series expansion (Proakis and Manolakis, Example 4.3, Hayes, Example 4.4.2, and Oppenheim, Schafer, and Buck, Example 3.11).

Solution We know that the series expansion of the function $f(x) = \log(1 + x)$ $(-1 \le x \le 1)$ has the form

$$\log(1+x) = x - \frac{1}{2}x^2 + \frac{1}{3}x^3 - \frac{1}{4}x^4 + \cdots = \sum_{n=1}^{\infty} \frac{(-1)^{n+1}}{n} x^n.$$

If in the above equation we set $x = \alpha z^{-1}$ the $\mathcal{Z}$ transform of the exercise takes the form

$$\mathcal{X}(z) = \log(1 + \alpha z^{-1}) = \sum_{n=1}^{\infty} \frac{(-1)^{n+1}}{n}(\alpha z^{-1})^n = \sum_{n=1}^{\infty} \left[\frac{(-1)^{n+1}}{n}\alpha^n\right] z^{-n}.$$

If we compare the above equation with the equation

$$\mathcal{X}(z) = \sum_{n=-\infty}^{\infty} x[n]z^{-n}$$

it is not difficult to see that for $n \geq 1$, $x[n] = [(-1)^{n+1}/n]\alpha^n$, whereas for $n < 1$, $x[n] = 0$. Therefore, the discrete-time signal $x[n]$ is given by the relation

$$x[n] = \begin{cases} \dfrac{(-1)^{n+1}}{n}\alpha^n & \text{for } n \geq 1 \\ 0 & \text{for } n < 1 \end{cases} = \frac{(-1)^{n+1}}{n}\alpha^n u[n-1]$$

Problem 8.7.5 Find the inverse of the $\mathcal{Z}$ transform described by the equation

$$\mathcal{X}(z) = \sin(z).$$

Using the method of expansion in power series (Hayes, Solved exercise 4.26 and Oppenheim, Schafer, and Buck, Problem 3.28).

Solution Following the methodology of the previous exercise, we develop the sine function in *infinite term power series*. From the basic theory, we know that this function has the form

$$\sin(z) = z - \frac{z^3}{3!} + \frac{z^5}{5!} - \frac{z^7}{7!} + \cdots = \sum_{n=0}^{\infty}(-1)^n \frac{z^{2n+1}}{(2n+1)!}.$$

Therefore, we have

$$\mathcal{X}(z) = \sin(z) = \sum_{n=-\infty}^{\infty} x[n]z^{-n} = \sum_{n=0}^{\infty}(-1)^n \frac{z^{2n+1}}{(2n+1)!},$$

from which it follows that $x[n] = (-1)^n/(2|n|+1)!$ $(n = -1, -3, -5, \ldots)$, which is the desired result.

Problem 8.7.6 Using the method of expansion in power series, find the inverse of the transforms $\mathcal{Z}$ defined as

$$\mathcal{X}(z) = e^z + e^{1/z} \quad (z \neq 0) \qquad \text{and} \qquad \mathcal{X}(z) = \log\left[2\left(\frac{1}{2} - z\right)\right] \ \left(|z| < \frac{1}{2}\right).$$

(Question a: Proakis and Manolakis, Exercise for solution 3.2.3 and Oppenheim, Schafer, and Buck, Exercise for solution 3.29; question b: Oppenheim, Schafer, and Buck, Exercise for solution 3.30).

Solution (a) From the basic theory, it is known that the function e^x can be written as an *infinite power series sum* of the form

$$e^x = 1 + x + \frac{x^2}{2!} + \frac{x^3}{3!} + \frac{x^4}{4!} + \frac{x^5}{5!} + \cdots = \sum_{n=0}^{\infty} \frac{x^n}{n!}.$$

Therefore,

$$e^z = \sum_{n=0}^{\infty} \frac{z^n}{n!} \quad \text{and} \quad e^{1/z} = \sum_{n=0}^{\infty} \frac{1}{n!}\left(\frac{1}{z}\right)^n = \sum_{n=0}^{\infty} \frac{z^{-n}}{n!}$$

and thus the $\mathcal{Z}$ transform will take the form

$$\mathcal{X}(z) = \sum_{n=0}^{\infty} \frac{z^n}{n!} + \sum_{n=0}^{\infty} \frac{z^{-n}}{n!} = \sum_{n=-\infty}^{0} \frac{z^{-n}}{(-n)!} + \sum_{n=0}^{\infty} \frac{z^{-n}}{n!},$$

where in the first sum of the last equation we proceeded to change the dummy index from n to $-n$. Now, using the absolute value function and developing the term of the second sum for the value $n = 0$—this term appears in both sums—the above equation is written as

$$\mathcal{X}(z) = \sum_{n=-\infty}^{0} \frac{z^{-n}}{(-n)!} + \sum_{n=1}^{\infty} \frac{z^{-n}}{n!} + 1 = \sum_{n=-\infty}^{\infty} \frac{z^{-n}}{|n|!} + 1.$$

Taking the *inverse transform* of the above equation, it immediately follows that

$$x[n] = \frac{1}{|n|!} + \delta[n].$$

(b) To solve the exercise we use the property of the logarithm function to develop into an *infinite power series sum* of the form

$$\log(1 - x) = -\sum_{n=1}^{\infty} \frac{x^n}{n} \quad (|x| < 1).$$

Using the above property, the $\mathcal{Z}$ transform can be written in the form

$$\mathcal{X}(z) = \log\left[2\left(\frac{1}{2} - z\right)\right] = \log(1 - 2z) = -\sum_{n=1}^{\infty} \frac{(2z)^n}{n} = -\sum_{n'=-\infty}^{-1} \frac{(2z)^{-n'}}{-n'}$$

$$= \sum_{n'=-\infty}^{-1} \frac{1}{n'}\left(\frac{1}{2}\right)^{n'} z^{-n'} = \sum_{n=-\infty}^{1} \frac{-1}{n}\left(\frac{1}{2}\right)^{n} z^{-n},$$

where initially we proceeded to change the dummy index from n to $n' = -n$ and in the end we set $n' = n$.

From the above relationship, it follows that the signal $x[n]$ is defined *only for the range of values* $(-\infty, -1]$ and therefore it should be multiplied by the elementary discrete-time signal $u[-n-1]$ whose value is equal to one for $n \le -1$ and zero anywhere else. Using, therefore, the defining equation of the $\mathcal{Z}$ transform and the above conclusion, it is not difficult to find that the discrete-time signal $x[n]$ will be given by the relationship

$$x[n] = \frac{1}{n}\left(\frac{1}{2}\right)^n u[-n-1].$$

Problem 8.7.7 Find the inverse of the Z transforms

$$\mathcal{X}(z) = \frac{1}{1 + 0.5z^{-1}} \quad \text{and} \quad \mathcal{X}(z) = \frac{1 - 0.5z^{-1}}{1 + (3/4)z^{-1} + (1/8)z^{-2}}$$

using the method of expansion in power series (Question a: Oppenheim, Schafer, and Buck, Exercise for solution 3.6α, β, question β: Oppenheim, Schafer, and Buck, Exercise for resolution 3.6.γ). (Hint: for the construction of the power series use the method of polynomial long division).

Solution In this case, we cannot develop the function of the transform into a closed form series as in the previous exercise, and for this reason we resort to the application of the method of polynomial long division, which is described in detail in the homonymous section of the mathematical appendix. In detail, therefore, we have:

$$
\begin{array}{r|llll}
 & 1 & -\frac{1}{2}z^{-1} & +\frac{1}{4}z^{-2} & \ldots \\
\hline
1+\frac{1}{2}z^{-1} & 1 & & & \\
 & 1 & +\frac{1}{2}z^{-1} & & \\
\cline{2-5}
 & & -\frac{1}{2}z^{-1} & & \\
 & & -\frac{1}{2}z^{-1} & -\frac{1}{4}z^{-2} & \\
\cline{3-5}
 & & & +\frac{1}{4}z^{-2} & \\
 & & & +\frac{1}{4}z^{-2} & +\frac{1}{8}z^{-3} \\
\cline{4-5}
 & & & \ldots\ldots & \ldots\ldots
\end{array}
$$

From the above operation it immediately follows that

$$
\frac{1}{1+0.5z^{-1}} = 1-\frac{1}{2}z^{-1}+\frac{1}{4}z^{-2}-\cdots = \left(-\frac{1}{2}\right)^{0} z^{-0} + \left(-\frac{1}{2}\right)^{1} z^{-1} + \left(-\frac{1}{2}\right)^{2} z^{-2} = \sum_{n=0}^{\infty}\left(-\frac{1}{2}\right)^{n} z^{-n}
$$

and therefore the discrete-time signal $x[n]$ will be given by the relation $x[n] = (1/2)^n u[n]$, which is also the desired result.

In a completely analogous way, the application of the method of polynomial long division for the second transform will be carried out as follows:

$$
\begin{array}{r|lllll}
 & 1 & -\frac{5}{4}z^{-1} & +\frac{13}{16}z^{-2} & +\frac{29}{64}z^{-3} & +\ldots \\
\hline
1+\frac{3}{4}z^{-1}+\frac{1}{8}z^{-2} & 1 & -\frac{1}{2}z^{-1} & & & \\
 & 1 & +\frac{3}{4}z^{-1} & +\frac{01}{08}z^{-2} & & \\
\cline{2-6}
 & & +\frac{5}{4}z^{-1} & +\frac{01}{08}z^{-2} & & \\
 & & +\frac{5}{4}z^{-1} & +\frac{15}{16}z^{-2} & +\frac{05}{32}z^{-3} & \\
\cline{3-6}
 & & & +\frac{13}{16}z^{-2} & +\frac{05}{32}z^{-3} & \\
 & & & +\frac{13}{16}z^{-2} & +\frac{39}{64}z^{-3} & +\frac{13}{108}z^{-4} \\
\cline{4-6}
 & & & & +\frac{29}{64}z^{-3} & +\frac{13}{108}z^{-4} \\
\cline{5-6}
 & & & & \ldots\ldots & \ldots\ldots
\end{array}
$$

Therefore,

$$
\frac{1-0.5z^{-1}}{1+(3/4)z^{-1}+(1/8)z^{-2}} = 1-\frac{5}{4}z^{-1}+\frac{13}{16}z^{-2}+\frac{29}{64}z^{-3}+\ldots
$$

from which it immediately follows that

$$
x[n] = \left\{1, -\frac{5}{4}, \frac{13}{16}, \frac{29}{64}, \ldots\right\}.
$$

8.7.3 The Method of Expansion into Partial Fractions

According to this method, which has been described and used in previous chapters, if we express the $\mathcal{Z}$ transform as a linear combination of elementary transforms of the form

Table 8.2 Examples of elementary $\mathcal{Z}$ transforms

Discrete-time signal $x[n]$	Z transform	Region of convergence
$\delta[n]$	1	All z
$u[n]$	$\frac{1}{1-z^{-1}}$	$\lvert z\rvert > 1$
$\alpha^n u[n]$	$\frac{1}{1-\alpha z^{-1}}$	$\lvert z\rvert > \lvert\alpha\rvert$
$n\alpha^n u[n]$	$\frac{\alpha z^{-1}}{(1-\alpha z^{-1})^2}$	$\lvert z\rvert > \lvert\alpha\rvert$
$-\alpha^n u[-n-1]$	$\frac{1}{1-\alpha z^{-1}}$	$\lvert z\rvert < \lvert\alpha\rvert$
$-n\alpha^n u[-n-1]$	$\frac{\alpha z^{-1}}{(1-\alpha z^{-1})^2}$	$\lvert z\rvert < \lvert\alpha\rvert$
$\cos(\omega_0 n)\, u[n]$	$\frac{1-z^{-1}\cos(\omega_0)}{1-2z^{-1}\cos(\omega_0)+z^{-2}}$	$\lvert z\rvert > 1$
$\sin(\omega_0 n)\, u[n]$	$\frac{z^{-1}\sin(\omega_0)}{1-2z^{-1}\cos(\omega_0)+z^{-2}}$	$\lvert z\rvert > 1$
$\alpha^n \cos(\omega_0 n) u[n]$	$\frac{1-\alpha z^{-1}\cos(\omega_0)}{1-2\alpha z^{-1}\cos(\omega_0)+\alpha^2 z^{-2}}$	$\lvert z\rvert > \lvert\alpha\rvert$
$\alpha^n \sin(\omega_0 n) u[n]$	$\frac{\alpha z^{-1}\sin(\omega_0)}{1-2\alpha z^{-1}\cos(\omega_0)+\alpha^2 z^{-2}}$	$\lvert z\rvert > \lvert\alpha\rvert$

$$\mathcal{X}(z) = \alpha_1 \mathcal{X}_1(z) + \alpha_2 \mathcal{X}_2(z) + \cdots + \alpha_k \mathcal{X}_k(z) = \sum_{i=1}^{k} \alpha_i \mathcal{X}_i(z) \tag{8.19}$$

we can use the property of *linearity* of this transform to calculate the inverse $x[n]$ transform from the relation

$$x[n] = \alpha_1 x_1[n] + \alpha_2 x_2[n] + \cdots + \alpha_k x_k[n] = \sum_{i=1}^{k} \alpha_i x_i[n] \tag{8.20}$$

with the inverse transforms $x_i[n] = \mathcal{Z}^{-1}\{\mathcal{X}_i(z)\}$ $(i = 1, 2, \ldots, k)$ being able to be calculated or even found in *tables of elementary transforms*, such as Table 8.2.[4]

The calculation of the *inverse $\mathcal{Z}$ transform* using the method of partial fraction expansion is an efficient process in cases where the function $\mathcal{X}(z)$ is a *rational function*, that is, it can be written as the quotient of two polynomial functions $f(z)$ and $g(z)$. In this case, the determination of the elementary functions $\mathcal{X}_1(z), \mathcal{X}_2(z), \ldots, \mathcal{X}_k(z)$ can be achieved *by expanding the function $\mathcal{X}(z)$ into a series of partial fractions*, using the techniques and methods provided by the relevant theory. In order to understand the application of this method for the calculation of the inverse $\mathcal{Z}$ transform, examples of solved exercises are presented that fully clarify its use. The reader is encouraged to study the exercises on the calculation of the *inverse transforms of Laplace and Fourier* that rely on the use of the same method.

[4] The pairs of elementary transforms of the table and the accompanying exercises as well as many others can be found in the book E. I. Jury: *Theory and Applications of the Z transform Method*, New York, John Wiley and Sons, Inc, 1964.

Problem 8.7.8 Find the inverse of the $\mathcal{Z}$ transform that is described by the equation

$$\mathcal{X}(z) = \frac{1}{1+3z^{-1}+2z^{-2}}$$

using the method of partial fraction expansion.

Solution For simplicity, we set $z^{-1} = y$. Observing that the denominator of the explicit function $\mathcal{X}(y) = 1/(1+3y+2y^2)$ is written as a product of first-degree polynomials of the form $1+3y+2y^2 = (2y+1)(y+1)$, we can formulate the function $\mathcal{X}(y)$ as a sum of partial fractions. Relying on the theory of expansion into partial fractions, we can write that

$$\mathcal{X}(y) = \frac{1}{1+3y+2y^2} = \frac{A}{2y+1} + \frac{B}{y+1} = \frac{(A+2B)y+(A+B)}{1+3y+2y^2}.$$

It will therefore be $(A+2B)y+(A+B) = 1$, from which the system of equations $A+2B = 0$ and $A+B = 1$ with solution $A = 2$ and $B = 1$ arises. Therefore, the function $\mathcal{X}(y)$ is formulated as

$$\mathcal{X}(y) = \frac{1}{1+3y+2y^2} = \frac{2}{2y+1} - \frac{1}{y+1}$$

and returning to our original variable, we get

$$\mathcal{X}(z) = 2\left\{\frac{1}{1+2z^{-1}}\right\} - 1\left\{\frac{1}{1+z^{-1}}\right\}.$$

At this point we have developed the transform $\mathcal{X}(z)$ as the linear combination

$$\mathcal{X}(z) = \alpha\mathcal{X}_1(z) + \beta\mathcal{X}_2(z)$$

for parameter values $\alpha = 2$, $\beta = -1$ and for the functions $\mathcal{X}_1(z) = (1+2z^{-1})^{-1}$ and $\mathcal{X}_2(z) = (1+z^{-1})^{-1}$. In the next step of the process we check *if these simple functions constitute the transforms $\mathcal{Z}$ of some elementary discrete-time signals that can be found in tables of elementary transforms*. Indeed, observing the contents of Table 8.2, we find that:

- The function $\mathcal{X}_1(z) = (1+2z^{-1})^{-1}$ arises from the simple function $\mathcal{X}(z) = (1-\alpha z^{-1})^{-1}$ for the parameter value $\alpha = -2$. However, as emerges from the contents of Table 8.2, the latter function constitutes the $\mathcal{Z}$ transform of the discrete-time signal $x[n] = \alpha^n u[n]$. Therefore, the inverse transform of the function $\mathcal{X}_1(z) = (1+2z^{-1})^{-1}$ is the discrete-time signal $x_1[n] = (-2)^n u[n]$.
- The function $\mathcal{X}_2(z) = (1+z^{-1})^{-1}$ arises from the simple function $\mathcal{X}(z) = (1-\alpha z^{-1})^{-1}$ for the parameter value $\alpha = -1$. However, as it emerges from the contents of Table 8.2, the latter function constitutes the $\mathcal{Z}$ transform of the discrete-time signal $x[n] = \alpha^n u[n]$. Therefore, the inverse transform of the function $\mathcal{X}_2(z) = (1+z^{-1})^{-1}$ is the discrete-time signal $x_1[n] = (-1)^n u[n]$.

Therefore, using the property of *linearity* that characterizes the $\mathcal{Z}$ transform, we conclude that the inverse of the transform $\mathcal{X}(z) = 1/(1+3z^{-1}+2z^{-2})$ is the signal

$$x[n] = \alpha x_1[n] + \beta x_2[n] = 2(-2)^n u[n] - (-1)^n u[n] = \left[2(-2)^n - (-1)^n\right]u[n].$$

Problem 8.7.9 Find the inverse of the $\mathcal{Z}$ transform described by the equation

$$\mathcal{X}(z) = \left\{1 + \frac{1}{4}z^{-1}\right\} \Big/ \left\{\left(1 - \frac{1}{2}z^{-1}\right)^2\right\}$$

using the method of partial fraction expansion (Hayes, Solved exercise 4.24).

Solution For simplicity, we set $y = z^{-1}$ and following the methodology of the previous exercise, we can write that

$$\mathcal{X}(y) = \frac{1 + (y/4)}{[1 - (y/2)]^2} = \frac{A}{1 - (y/2)} + \frac{B}{[1 - (y/2)]^2} = \frac{A[1 - (y/2)] + B}{[1 - (y/2)]^2} = \frac{-(A/2)y + (A + B)}{[1 - (y/2)]^2}.$$

Therefore, $-(A/2)y + (A + B) = (1/4)y + 1$, from which the system $-(A/2) = (1/4)$ and $A + B = 1$ is derived with solution $A = -1/2$ and $B = 3/2$. Therefore, the function $\mathcal{X}(y)$ will take the form

$$\mathcal{X}(y) = -\frac{1}{2[1 - (y/2)]} + \frac{3}{2[1 - (y/2)]^2}$$

or equivalently—returning to our original variable—as

$$\mathcal{X}(z) = -\frac{1}{2}\underbrace{\left\{\frac{1}{(1 - 0.5z^{-1})}\right\}}_{\mathcal{X}_1(z)} + \frac{3}{2}\underbrace{\left\{\frac{1}{(1 - 0.5z^{-1})^2}\right\}}_{\mathcal{X}_2(z)}.$$

Referring to Table 8.2, we find that

$$x_1[n] = \mathcal{Z}^{-1}\{\mathcal{X}_1(z)\} = \mathcal{Z}^{-1}\left\{\frac{1}{1 - 0.5z^{-1}}\right\} = \left(\frac{1}{2}\right)^n u[n].$$

On the other hand, the function $\mathcal{X}_2(z)$ is not included in Table 8.2 and therefore its inverse transform should be calculated in a different way. In order to proceed with this calculation, we formulate the function $\mathcal{X}_2(z)$ with the form

$$\mathcal{X}_2(z) = \frac{1}{(1 - 0.5z^{-1})^2} = \mathcal{X}_{2\alpha}(z) + \mathcal{X}_{2\beta}(z) = \frac{0.5z^{-1}}{(1 - 0.5z^{-1})^2} + \frac{1}{1 - 0.5z^{-1}},$$

and referring back to Table 8.2, we now find that $x_{2\alpha}[n] = n(1/2)^n u[n]$ and $x_{2\beta}[n] = (1/2)^n u(n)$. Therefore, $x_2[n] = (n + 1)(1/2)^n u[n]$, and finally the inverse transform of the given function $\mathcal{X}(z)$ is the discrete sequence

$$x[n] = \left[3(n + 1) - 1\right]\left(\frac{1}{2}\right)^{n+1} u[n]$$

(as emerges after simple mathematical operations), which is also the desired result.

Problem 8.7.10 Find the inverse of the $\mathcal{Z}$ transform, which is described by the equation

$$X(z) = \frac{1}{1 + 0.5z^{-1}} \quad \left(|z| > \frac{1}{2}\right)$$

using the methods (a) complex integration and (b) expansion in partial fraction series.

Solution (a) According to the *method of complex integration*, the inverse transform is defined as

$$x[n] = \frac{1}{2\pi i}\oint_C \frac{X(z)}{z} z^n dz = \sum_{k=1}^{n} Res[Y(z_k)],$$

where z_k $(k = 1, 2, \ldots, n)$ are the poles of the function $Y(z) = X(z)z^{n-1}$. In this case, the function $Y(z)$ is written in the form

$$Y(z) = \frac{X(z)}{z} z^n = \frac{1}{1+0.5z^{-1}} z^{n-1} = \frac{z^n}{z+0.5}$$

and has a simple pole at the position $z = -0.5$. The residue of the function $Y(z)$ at the pole position is

$$Res[Y(-0.5)] = \lim_{z\to -0.5}\left[(z+0.5)\frac{z^n}{z+0.5}\right] = \lim_{z\to -0.5} z^n = \left(-\frac{1}{2}\right)^n$$

and therefore the discrete-time signal $x[n]$ is given by the relation $x[n] = (-1/2)^n u[n]$, which is the desired result.

(b) As for the application of the method of partial fraction expansion, in this case it cannot (and does not need to) be applied because the function $X(z)$ cannot be further decomposed. Referring to Table 8.2, we immediately find that the inverse $\mathcal{Z}$ transform of the function $X(z)$ is the function $x[n] = (-1/2)^n u[n]$.

Problem 8.7.11 Find the inverse of the $\mathcal{Z}$ transform described by the equation

$$X(z) = \frac{1}{(1-z^{-1})(1-z^{-2})}$$

using the method of partial fraction expansion (Hayes, Solved Exercise 4.23d).

Solution For simplicity, we set $z^{-1} = y$ and observing that the denominator of the rational function $X(y) = 1/(1-y)(1-y^2)$ is expressed as the product of the first-degree of the polynomial $(1+y)(1-y)^2$, we expand the function $X(y)$ as a sum of partial fractions of the form[5]

$$\frac{1}{(1+y)(1-y)^2} = \frac{A}{1+y} + \frac{B}{1-y} + \frac{C}{(1-y)^2} = \frac{A(1-y^2) + B(1+y)(1-y) + C(1+y)}{(1+y)(1-y^2)}$$

$$= \frac{A(1-2y+y^2) + B(1-y^2) + C(1+y)}{(1+y)(1-y^2)} = \frac{(A-B)y^2 + (C-2A)y + (A+B+C)}{(1-y)(1-y^2)}.$$

By equating the coefficients of the corresponding powers of z we end up with the system $A - B = 0$, $C - 2A = 0$ and $A + B + C = 1$ with the solution $A = B = 1/4$ and $C = 1/2$. Therefore,

$$\mathcal{X}(y) = \frac{1}{(1-y)(1-y^2)} = \frac{1}{4(1+y)} + \frac{1}{4(1-y)} + \frac{1}{2(1-y)^2}$$

or, returning to our original variables,

$$\mathcal{X}(z) = \frac{1}{(1-z^{-1})(1-z^{-2})} = \frac{1}{4}\left\{\frac{1}{1+z^{-1}}\right\} + \frac{1}{4}\left\{\frac{1}{1-z^{-1}}\right\} + \frac{1}{2}\left\{\frac{1}{(1-z^{-1})^2}\right\}.$$

From the above analysis we find that the desired transform has been written as a linear combination of the form

$$\mathcal{X}(z) = \alpha\mathcal{X}_1(z) + \beta\mathcal{X}_2(z) + \gamma\mathcal{X}_3(z),$$

[5] For an introduction to the theory of expansion in partial fractions, see the corresponding section of the Appendix.

with $\alpha = 1/4$, $\beta = 1/4$, $\gamma = 1/2$, and $\mathcal{X}_1(z) = (1+z^{-1})^{-1}$, $\mathcal{X}_2(z) = (1-z^{-1})^{-1}$, $\mathcal{X}_3(z) = [(1-z^{-1})^2]^{-1}$. Referring then to Table 8.2 of elementary transformation pairs, we find that:

- The function $\mathcal{X}_1(z) = (1+z^{-1})^{-1}$ is of the form $\mathcal{X}(z) = (1-\alpha z^{-1})^{-1}$ for parameter value $\alpha = -1$ and therefore, its inverse transform is the discrete sequence $x_1[n] = (-1)^n u(n)$.
- The inverse transform of the function $\mathcal{X}_2(z) = (1-z^{-1})^{-1}$ is the discrete-time signal $x_2[n] = u[n]$.
- The function $\mathcal{X}_3(z) = [(1-z^2)^2]^{-1}$ is not included in Table 8.2 and therefore its inverse transform must be calculated in a different way.

In order to calculate the third of the above inverse transforms, we observe that if we expand the term $\mathcal{X}_3(z)$ as

$$\mathcal{X}_3(z) = \frac{1}{(1-z^{-1})^2} = \mathcal{X}_{3\alpha}(z) + \mathcal{X}_{3\beta}(z) = \frac{z^{-1}}{(1-z^{-1})^2} + \frac{1}{1-z^{-1}}$$

and resort to the property of linearity of the $\mathcal{Z}$ transform, we can formulate the claim that if we determine the inverse transforms $x_{3\alpha}[n]$ and $x_{3\beta}[n]$ of the functions $X_{3\alpha}(z)$ and $X_{3\beta}(x)$, then obviously the relation $x_3[n] = x_{3\alpha}[n] + x_{3\beta}[n]$ will hold. But the function $X_{3\beta}(z)$ is included in Table 8.2, according to which its inverse transform is the discrete-time signal $x_{3\beta}[n] = u[n]$. As for the function $X_{3\alpha}(z)$, if we formulate as

$$X_{3\alpha}(z) = \frac{z^{-1}}{(1-z^{-1})^2} = (-z)\left\{-\frac{z^{-2}}{(1-z^{-1})^2}\right\} = -z\frac{d}{dz}\left\{\frac{1}{1-z^{-1}}\right\}$$

and we use the *property of differentiation* of the Z transform according to which if it is $x[n] \rightleftarrows X(z)$ then it will also be $nx[n] \rightleftarrows [-zX'(z)]$, it follows that

$$nu(n) \rightleftarrows -z\frac{d}{dz}\left\{\frac{1}{1-z^{-1}}\right\}.$$

So, $x_{3\alpha}[n] = nu[n]$, from which it follows that $x_3[n] = (n+1)u[n]$. and finally the discrete-time signal $x[n]$ is calculated as

$$x[n] = \frac{1}{4}\Big[(-1)^n + 1 + 2(n+1)\Big]u(n),$$

which is also the desired result.

Problem 8.7.12 Find the inverse of the $\mathcal{Z}$ transform, which is described by the equation

$$\mathcal{X}(z) = \frac{1-0.5z^{-1}}{1+(3/4)z^{-1}+(1/8)z^{-2}} \quad \left(|z| > \frac{1}{2}\right),$$

using (a) the method of complex integration and (b) the method of expansion in a series of partial fractions (Oppenheim, Schafer, and Buck, Exercise for solution 3.6.γ).

Solution (a) If we factorize the function of the denominator writing it in the form

$$1+\frac{3}{4}z^{-1}+\frac{1}{8}z^{-2} = \frac{1}{8}(z^{-1}+2)(z^{-1}+4)$$

the function $\mathcal{Y}(z)$ required for the application of the method of complex integration is calculated as

$$\mathcal{Y}(z) = \frac{\mathcal{X}(z)}{z}z^n = \frac{8(1-0.5z^{-1})}{(z^{-1}+2)(z^{-1}+4)}\frac{z^n}{z} = \frac{8z-4}{(2z+1)(4z+1)}z^n.$$

The above function has *two simple poles* at the positions $z = -1/2$ and $z = -1/4$ with residue values

$$Res[Y(-1/2)] = \lim_{z \to -1/2} \left[\left(z + \frac{1}{2}\right) \frac{(8z-4)z^n}{2\left(z+\frac{1}{2}\right)(4z+1)} \right]$$

$$= \lim_{z \to -1/2} \left[\frac{(8z-4)z^n}{2(4z+1)} \right] = \frac{[8(-1/2)-4](-1/2)^n}{2[4(-1/2)+1]} = 4\left(-\frac{1}{2}\right)^n$$

$$Res[Y(-1/4)] = \lim_{z \to -1/4} \left[\left(z + \frac{1}{4}\right) \frac{(8z-4)z^n}{4\left(z+\frac{1}{4}\right)(2z+1)} \right]$$

$$= \lim_{z \to -1/4} \left[\frac{(8z-4)z^n}{4(2z+1)} \right] = \frac{[8(-1/4)-4](-1/4)^n}{4[2(-1/4)+1]} = -3\left(-\frac{1}{4}\right)^n.$$

Therefore, we have

$$x[n] = 4\left(-\frac{1}{2}\right)^n u[n] - 3\left(-\frac{1}{4}\right)^n u[n].$$

(b) Using the result of the factorization of the denominator and applying the method of partial fraction expansion, the function $\mathcal{X}(y)$—where $y = z^{-1}$—is formulated as

$$\mathcal{X}(y) = \frac{1-0.5y}{(1/8)(y+2)(y+4)} = \frac{8-4y}{(y+2)(y+4)} = \frac{\alpha}{y+2} + \frac{\beta}{y+4} = \frac{\alpha(y+4)+\beta(y+2)}{(y+2)(y+4)} = \frac{(\alpha+\beta)y + (4\alpha+2\beta)}{(y+2)(y+4)}.$$

Therefore, we have $\alpha + \beta = -4$ and $4\alpha + 2\beta = 8$, from which the values $\alpha = 8$ and $\beta = -12$ are derived, and, returning to our original variables, we finally get

$$\mathcal{X}(z) = \frac{8}{2+z^{-1}} - \frac{12}{4+z^{-1}} = 4\frac{1}{1+(1/2)z^{-1}} - 3\frac{1}{1+(1/4)z^{-1}}.$$

Referring to tables of elementary transforms, it turns out that

$$\frac{1}{1+(1/2)z^{-1}} \leftrightarrows \left(-\frac{1}{2}\right)^n u(n) \quad \text{and} \quad \frac{1}{1+(1/4)z^{-1}} \leftrightarrows \left(-\frac{1}{4}\right)^n u(n),$$

from which we get the result we also got with the method of complex integration.

Problem 8.7.13 Find the inverse of the $\mathcal{Z}$ transform that is described by the equation

$$X(z) = \frac{1-0.5z^{-1}}{1-0.25z^{-2}} \quad \left(|z| > \frac{1}{2}\right).$$

using (a) the method of complex integration and (b) the method of expansion in a series of partial fractions.

Solution (a) The function $Y(z)$ used in the method of complex integration

$$Y(z) = \frac{X(z)}{z} z^n = \frac{1-0.5z^{-1}}{(1-0.5z^{-1})(1+0.5z^{-1})} z^{n-1} = \frac{2z}{2z+1} z^{n-1} = \frac{2z^n}{2z+1}$$

has *a simple pole* at the position $z = -1/2$ with a residue at this position calculated as

$$Res[Y(-1/2)] = \lim_{z\to -1/2}\left[\left(z+\frac{1}{2}\right)\frac{2z^n}{2\left(z+\frac{1}{2}\right)}\right] = \lim_{z\to -1/2} z^n = \left(-\frac{1}{2}\right)^n$$

from which it follows that $x[n] = (-1/2)^n u[n]$.

(b) Expressing the function $X(z)$ in the form

$$X(z) = \frac{1-0.5z^{-1}}{1-0.25z^{-2}} = \frac{1-0.5z^{-1}}{(1-0.5z^{-1})(1+0.5z^{-1})} = \frac{1}{1+0.5z^{-1}}$$

and referring to tables of elementary transforms, we find that $x[n] = (-1/2)^n u[n]$.

Problem 8.7.14 Find the inverse of the $\mathcal{Z}$ transform, which is described by the equation

$$X(z) = \frac{1-\alpha z^{-1}}{z^{-1}-\alpha}\quad\left(|z| > \frac{1}{|\alpha|}\right)$$

using the method of partial fraction expansion (Proakis and Manolakis, Exercise 3.14(ij) and Oppenheim, Schafer, Buck, Exercise for solution 3.6e).

Solution For simplicity, let $y = z^{-1}$ and expand the function $X(y)$ in the form

$$X(y) = \frac{1-\alpha y}{y-\alpha} = \frac{1-\alpha y}{\alpha[(y/\alpha)-1]} = \frac{A}{\alpha} + \frac{B}{(y/\alpha)-1} = \frac{A[(y/\alpha)-1]+\alpha B}{\alpha[(y/\alpha)-1]} = \frac{(A/\alpha)y + (\alpha B - A)}{\alpha[(y/\alpha)-1]}.$$

We are led to the system of equations $(A/\alpha) = -\alpha$ and $\alpha B - A = 1$, with solution $A = -\alpha^2$ and $B = (1-\alpha^2)/\alpha$, and, returning to our original variables, we get

$$X(z) = -\alpha\cdot 1 - \frac{1-\alpha^2}{\alpha}\left[\frac{1}{1-(1/\alpha)z^{-1}}\right].$$

Referring to tables of elementary transforms, we find that

$$1 \leftrightarrows \delta[n] \qquad \text{and} \qquad \frac{1}{1-(1/\alpha)z^{-1}} \leftrightarrows \left(\frac{1}{\alpha}\right)^n u[n] = \alpha^{-n}u[n].$$

Therefore, we have

$$x[n] = -\alpha\delta[n] - \frac{1-\alpha^2}{\alpha}\alpha^{-n}u[n] = -\alpha\delta[n] - (1-\alpha^2)\alpha^{-(n+1)}u[n].$$

8.7.4 Product Transformation/Parseval's Identity

After the analytical description of the inverse $\mathcal{Z}$ transform, we are now in a position to prove an additional property that concerns the transform of the product of two discrete sequences and is formulated in the following way.

If the transforms $\mathcal{Z}$ of the discrete sequences $x_1[n]$ and $x_2[n]$ are the functions $\mathcal{X}_1(z)$ and $\mathcal{X}_2(z)$, then the Z transform of the discrete-time signal $x[n] = x_1[n]x_2[n]$ will be given by the relation

$$\mathcal{X}(z) = \frac{1}{2\pi j}\oint_C \mathcal{X}_1(\upsilon)\mathcal{X}_2\left(\frac{z}{\upsilon}\right)\upsilon^{-1}d\upsilon$$

with the contour integral appearing in this calculation along a closed curve C that is *at the intersection of the regions of convergence of the transforms* $\mathcal{X}_1(\upsilon)$ *and* $\mathcal{X}_2(1/\upsilon)$ *and contains the origin of the complex plane* $z = 0$.

Proof considering the sequence $x[n] = x_1[n]x_2[n]$, the $\mathcal{Z}$ transform of this is calculated as

$$\mathcal{X}(z) = \sum_{n=-\infty}^{\infty} x[n]z^{-n} = \sum_{n=-\infty}^{\infty} x_1[n]x_2[n]z^{-n}.$$

If $\mathcal{X}_1(\upsilon)$ is the $\mathcal{Z}$ transform of the signal $x_1[n]$, then the aforementioned signal, being the inverse transform of this function, will be given by the relation

$$x_1[n] = \frac{1}{2\pi j}\oint_C \frac{\mathcal{X}_1(\upsilon)}{\upsilon}\upsilon^n d\upsilon = \frac{1}{2\pi j}\oint_C \mathcal{X}_1(\upsilon)\upsilon^{n-1}d\upsilon.$$

By substituting this expression of $x_1[n]$ into our initial equation, it will take the form

$$\mathcal{X}(z) = \sum_{n=-\infty}^{\infty}\left\{\frac{1}{2\pi j}\oint_C \mathcal{X}_1(\upsilon)\upsilon^{n-1}d\upsilon\right\}x_2[n]z^{-n}.$$

By interchanging the terms of the summation and the integration into the last equation, it is written

$$\mathcal{X}_1(z) = \frac{1}{2\pi j}\oint_C \mathcal{X}_1(\upsilon)\left\{\sum_{n=-\infty}^{\infty} x_2[n]\left(\frac{z}{\upsilon}\right)^{-n}\right\}\upsilon^{-1}d\upsilon.$$

But the term between the brackets is nothing other than the transform $\mathcal{X}_2(z/\upsilon)$

$$\mathcal{X}_2\left(\frac{z}{\upsilon}\right) = \sum_{n=-\infty}^{\infty} x_2[n]\left(\frac{z}{\upsilon}\right)^{-n}.$$

By substituting the last expression into the previous equation, it will take the form.

$$\mathcal{X}(z) = \frac{1}{2\pi j}\oint_C \mathcal{X}_1(\upsilon)\mathcal{X}_2\left(\frac{z}{\upsilon}\right)\upsilon^{-1}d\upsilon. \tag{8.18}$$

But what is the region of convergence of the transform $\mathcal{X}(z)$? To determine it, we must observe how if the transform $\mathcal{X}_1(\upsilon)$ converges in the region $r_{1\ell} < |\upsilon| < r_{1u}$ and the transform $\mathcal{X}_2(z)$ converges in the region $r_{2\ell} < |z| < r_{2u}$, the transform $\mathcal{X}(z/\upsilon)$ will converge in the region $r_{2\ell} < |z/\upsilon| < r_{2u}$. Therefore, the region of convergence of the transform $\mathcal{X}(z)$ will be at least $r_{1\ell}r_{2\ell} < |z| < r_{1u}r_{2u}$.

Finally, working in the same way, we can prove that if the sequences $x_1[n]$ and $x_2[n]$ are *complex*, then their product will be defined as $x[n] = x_1[n]x_2^*[n]$ (with the symbol "*" referring to the operator of complex conjugation), the $\mathcal{Z}$ transform of this product will have the form

$$\mathcal{X}(z) = \frac{1}{2\pi j}\oint_C \mathcal{X}_1(\upsilon)\mathcal{X}_2^*\left(\frac{z^*}{\upsilon^*}\right)\upsilon^{-1}d\upsilon, \tag{8.19}$$

with the closed path C defined as before. ∎

Another interesting property that is proven with the help of the inverse $\mathcal{Z}$ transform, is the *Parseval'sidentity*, according to which, if $x_1[n]$ and $x_2[n]$ are two sequences of complex values, then the relation holds

$$\sum_{n=-\infty}^{\infty} x_1[n]\, x_2^*[n] = \frac{1}{2\pi j} \oint_C \mathcal{X}_1(v)\mathcal{X}_2^*(\frac{1}{v^*})v^{-1}dv. \tag{8.20}$$

The Parseval's identity arises from the last equation, if we set in it $z = 1$; in this case, we get

$$\mathcal{X}(z)_{z=1} = \left[\sum_{n=-\infty}^{\infty} x_1[n]x_2^*[n]z^{-n}\right]_{z=1} = \sum_{n=-\infty}^{\infty} x_1[n]x_2^*[n] = \frac{1}{2\pi j} \oint_C \mathcal{X}_1(v)\mathcal{X}_2^*\left(\frac{1}{v^*}\right)v^{-1}dv,$$

from which the desired result follows.

8.8 The Unilateral $\mathcal{Z}$ Transform

The unilateral (or one-sided) $\mathcal{Z}$ transform is a special case of the $\mathcal{Z}$ transform that has been studied in the previous sections, in the sense that *the lower limit of the sum in the defining equation is set to zero* and therefore *the discrete-time variable n takes only positive and not negative values*; in other words, this transform, for which we use the symbol $\mathcal{Z}^+$, is given by the relation

$$\mathcal{X}^+(z) = \sum_{n=0}^{\infty} x[n]z^{-n}. \tag{8.21}$$

The most important characteristics of the $\mathcal{Z}^+$ transform are the following:

- The $\mathcal{Z}^+$ transform does not contain information about the values of the discrete-time signal $x[n]$ that correspond to negative values of the discrete-time n.
- The $\mathcal{Z}^+$ transform is unique *only for causal signals*, because only for these signals is $x[n] = 0$ for $n < 0$.
- The $\mathcal{Z}^+$ transform of the signal $x[n]$ is identical to the $\mathcal{Z}$ transform of the signal $x[n]u[n]$. Because this signal is a causal signal, the region of convergence of the $\mathcal{Z}$ transform of this signal—and therefore also the region of convergence of the $\mathcal{Z}^+$ transform of the signal $x[n]$—*is always the exterior of the unit circle*. For this reason, *when we define the $\mathcal{Z}^+$ transform of a signal, it is not necessary to define its region of convergence as well.*

The properties of the $\mathcal{Z}^+$ transform are the same as those of the $\mathcal{Z}$ transform, except for the property of *time shift*, which is formulated *differently*. This property is analyzed in the *property of time delay* and in the *property of time advance*. According to the first property, if the $\mathcal{Z}^+$ transform of the discrete-time signal $x[n]$ is the function $\mathcal{X}^+(z)$, then the $\mathcal{Z}^+$ transform of the time-delayed signal $x[n-k]$ will be the function

$$\mathcal{Z}^+\{x[n-k]\} = z^{-k}\left[\mathcal{X}^+(z) + \sum_{n=1}^{k} x[-n]z^n\right] \quad (k > 0). \tag{8.22}$$

To prove this property, we start from the defining equation of the unilateral $\mathcal{Z}$ transform of the signal $x[n-k]$, which is formulated as

$$\mathcal{Z}^+\{x[n-k]\} = \sum_{n=0}^{\infty} x[n-k]z^{-n}.$$

Proceeding with the substitution $n \longrightarrow m = n - k$, for $n = 0$, $m = -k$, and we have

$$\sum_{n=0}^{\infty} x[n-k]z^{-n} = \sum_{m=-k}^{\infty} x[m]z^{-(m+k)} = z^{-k}\left\{\sum_{m=-k}^{-1} x[m]z^{-m} + \sum_{m=0}^{\infty} x[m]z^{-m}\right\}.$$

But the second sum expresses the *unilateral* $\mathcal{X}^{+}(z)$ *transform* of the signal $x[n]$. On the other hand, if in the first sum we make the substitution $m = -n$, it is written as

$$\sum_{m=-k}^{-1} x[m]z^{-n} = \sum_{n=1}^{k} x[-n]z^{n}.$$

Therefore, we have

$$\mathcal{Z}^{+}\{x[n-k]\} = z^{-k}\left[\mathcal{X}^{+}(z) + \sum_{n=1}^{k} x[-n]z^{n}\right] \quad (k > 0),$$

an equation that expresses the *property of time delay* of the unilateral $\mathcal{Z}$ transform.

In the special case where the signal $x[n]$ is causal, the above relationship is written as

$$\mathcal{Z}^{+}\{x[n-k]\} = z^{-k}\mathcal{X}^{+}(z). \tag{8.23}$$

Finally, according to the second property, if the $\mathcal{Z}^{+}$ transform of the discrete-time signal $x[n]$ is the function $\mathcal{X}^{+}(z)$, then the $\mathcal{Z}^{+}$ transform of the time advanced signal $x[n+k]$ will be the function

$$\mathcal{Z}^{+}\{x[n+k]\} = z^{k}\left[\mathcal{X}^{+}(z) - \sum_{n=0}^{k-1} x[n]z^{-n}\right] \quad (k > 0). \tag{8.24}$$

In full analogy with the previous case, we start again from the defining equation of the unilateral $\mathcal{Z}$ transform of the signal $x[n+k]$, which is formulated as

$$\mathcal{Z}^{+}\{x[n+k]\} = \sum_{n=0}^{\infty} x[n+k]z^{-n}.$$

Proceeding to change the dummy index from n to $m = n + k$, for $n = 0$, $m = k$, and the above sum is written

$$\sum_{n=0}^{\infty} x[n+k]z^{-n} \sum_{m=k}^{\infty} x[m]z^{-(m-k)} = z^{k}\sum_{m=k}^{\infty} x[m]z^{-m}z^{k}\left\{\sum_{m=0}^{\infty} x[m]z^{-m} - \sum_{m=0}^{k-1} x[m]z^{-m}\right\}.$$

But the first sum is nothing but the unilateral $\mathcal{X}^{+}(z)$ transform of the discrete-time signal $x[n]$ and finally the above relation will take the form

$$\mathcal{Z}^{+}\{x[n+k]\} = z^{k}\left[\mathcal{X}^{+}(z) - \sum_{n=0}^{k-1} x[n]z^{-n}\right],$$

where in the second sum we have changed the dummy index from m to n.

A very important application of the unilateral $\mathcal{Z}$ transform is the *solution of difference equations* that describe *causal discrete recursive systems*. The methodology followed is relatively simple and includes (a) *the calculation of the unilateral $\mathcal{Z}$ transform of each expression included in the difference equation*, (b) *the solution of the expression that arises in terms* *of the quantity* $\mathcal{Y}^+(z)$, and (c) *the calculation of the function* $y[n]$ *as the inverse transform of the function* $\mathcal{Y}^+(z)$. This methodology will be used several times in the next chapter and for this reason we simply demonstrate a typical example of its use.

Suppose we want to solve the difference equation

$$y[n] + \frac{1}{2}y[n-1] - \frac{1}{4}y[n-2] = 0$$

for the range of values $n \geq 0$ and with initial conditions $y[-1] = y[-2] = 1$. Taking the unilateral $\mathcal{Z}$ transform of the above equation, we have[6]

$$\mathcal{Y}^+(z) + \frac{1}{2}\mathcal{Z}^+\{y[n-1]\} - \frac{1}{4}\mathcal{Z}^+\{y[n-2]\} = 0.$$

Applying the property of time shifting of the unilateral $\mathcal{Z}$ transform, we get

$$\mathcal{Z}^+\{y[n-1]\} = z^{-1}[\mathcal{Y}^+(z) + y[-1]z] = z^{-1}\mathcal{Y}^+(z) + y[-1] = z^{-1}\mathcal{Y}^+(z) + 1$$

$$\mathcal{Z}^+\{y[n-2]\} = z^{-2}[\mathcal{Y}^+(z) + y[-1]z + y[-2]z^2] = z^{-2}\mathcal{Y}^+(z) + z^{-1}y[-1] + y[-2] = z^{-2}\mathcal{Y}^+(z) + z^{-1} + 1.$$

Substituting the above expressions into the unilateral $\mathcal{Z}$ transform of the equation, we get

$$\mathcal{Y}^+(z) + \frac{1}{2}\left(z^{-1}\mathcal{Y}^+(z) + 1\right) - \frac{1}{4}\left(z^{-2}\mathcal{Y}^+(z) + z^{-1} + 1\right) \quad \text{or equivalently} \quad \mathcal{Y}^+(z) = \frac{(1/4)z^{-1} - (1/4)}{1 + (1/2)z^{-1} - (1/4)z^{-2}}.$$

In the last step of the process we need to determine the function $y[n]$ as the *inverse transform of the above relation*, a problem left as an exercise to the reader.

8.9 Problems

1. Using the property $nx(n) \leftrightarrows -z\mathcal{X}'(z)$ and taking into account that

$$\mathcal{Z}\left\{\alpha^n \cos(\omega n)u[n]\right\} = \frac{\alpha z^{-1}\cos(\omega)}{1 - 2\alpha z^{-1}\cos(\omega) + \alpha^2 z^{-2}},$$

   determine the $\mathcal{Z}$ transform of the discrete sequence $y[n] = n\alpha^n \cos(\omega n)u[n]$.
2. Using the defining equation of the $\mathcal{Z}$ transform, prove the validity of the pairs of elementary transforms of Table 8.2 (refer to page 679).
3. Using the defining equation of the $\mathcal{Z}$ transform, prove the validity of the following pairs of elementary transforms:

$$u[n-k] \leftrightarrows \frac{z^{-k+1}}{z-1}, \quad n(n-1)(n-2)\dots(n-k+1) \leftrightarrows k!\,\frac{z}{(z-1)^{k+1}}$$

$$\frac{1}{n}\,(n>0) \leftrightarrows \ln\frac{z}{z-1}, \quad e^{-\alpha n}x[n] \leftrightarrows \mathcal{X}(e^{\alpha z}), \quad \sum_{m=1}^{n}\frac{1}{m} \leftrightarrows \frac{z}{z-1}\ln\frac{z}{z-1}.$$

4. Using the defining equation of the $\mathcal{Z}$ transform, prove the validity of the following pairs of elementary transforms:

$$\frac{c^n}{n!} \leftrightarrows e^{c/z}, \quad \frac{(\ln c)^n}{n!} \leftrightarrows c^{1/z}, \quad \frac{1-e^{\alpha n}}{n} \leftrightarrows \alpha + \ln\frac{z-e^{-\alpha}}{z-1}\quad (\alpha > 0).$$

[6] It is not difficult to realize that this method is the discrete equivalent of the method of solving differential equations by using the *Laplace transform*.

5. Using the defining equation of the $\mathcal{Z}$ transform, prove the validity of the following pairs of elementary transforms:

$$\frac{\sin(\alpha n)}{n} \leftrightarrows \alpha + \tan^{-1}\frac{\sin(\alpha)}{z-\cos(\alpha)} \ (\alpha > 0), \qquad \frac{sin(\alpha n)}{n!} \leftrightarrows e^{\cos(\alpha)/z}\sin\left(\frac{\sin(a)}{z}\right)$$

$$\frac{\cos(\alpha n)}{n}\ (n>0) \leftrightarrows \ln\frac{z}{\sqrt{z^2-2z\cos(\alpha)+1}}, \qquad \frac{\cos(\alpha n)}{n!} \leftrightarrows e^{\cos(\alpha)/z}\sin\left(\frac{\cos(a)}{z}\right).$$

6. Using the defining equation of the $\mathcal{Z}$ transform, prove the validity of the following pairs of elementary transforms:

$$\frac{1}{m(m+1)(m+2)\dots(m+n)} \leftrightarrows (m-1)!z^m\left[e^{1/z}-\sum_{k=0}^{m-1}\frac{1}{k!z^k}\right]$$

$$\alpha^n\cos(\pi n) \leftrightarrows \frac{z}{z+\alpha}, \qquad \frac{\alpha^n+(-\alpha)^n}{2\alpha^2} \leftrightarrows \frac{1}{\alpha^2}\frac{z^2}{z^2-\alpha^2}, \qquad \frac{\alpha^n-\beta^n}{\alpha-\beta} \leftrightarrows \frac{z}{(z-\alpha)(z-\beta)}$$

$$\frac{1}{n}\sin\left(\frac{\pi n}{2}\right) \leftrightarrows \frac{\pi}{2}+\tan^{-1}\frac{1}{z}, \qquad e^{-\alpha n}\cos(\alpha n) \leftrightarrows \frac{z(z-e^{-\alpha}\cos(\alpha))}{z^2-2ze^{-\alpha}\cos(\alpha)+e^{-2\alpha}}.$$

7. Using the defining equation of the $\mathcal{Z}$ transform, prove the validity of the following pairs of elementary transforms:

$$\frac{\cos[\alpha(2n-1)]}{2n-1} \leftrightarrows \frac{1}{4\sqrt{z}}\ln\frac{z+2\sqrt{z}\cos(\alpha)+1}{z-2\sqrt{z}\cos(\alpha)+1}, \qquad -n\alpha^n\frac{1+\cos(\pi n)}{2} \leftrightarrows \frac{2\alpha^2z^2}{(z^2-\alpha^2)^2}$$

$$e^{-\alpha n}\sinh(\alpha n+\xi) \leftrightarrows \frac{z^2\sinh(\xi)+ze^{-\alpha}\sinh(\alpha-\xi)}{z^2-2ze^{-\alpha}\cosh(\alpha)+e^{-2\alpha}}.$$

8. Using the defining equation of the $\mathcal{Z}$ transform, prove the validity of the following pairs of elementary transforms:

$$-n\alpha^n\cos\left(\frac{\pi n}{2}\right) \leftrightarrows \frac{2\alpha^2z^2}{(z^2+\alpha^2)^2}, \qquad \alpha^n\sin\left(\frac{n\pi}{4}\right)\frac{1+\cos(\pi n)}{2} \leftrightarrows \frac{\alpha^2z^2}{z^4+\alpha^4}$$

$$\alpha^n\left[\frac{1+\cos(\pi n)}{2}-\cos\left(\frac{\pi n}{2}\right)\right] \leftrightarrows \frac{\alpha^2z^2}{z^4-\alpha^4}, \qquad \alpha^n\left[\frac{1+\cos(\pi n)}{2}-\cos\left(\frac{\pi n}{2}\right)\right] \leftrightarrows \frac{2z^4}{z^4-\alpha^4}$$

$$(-1)^n\frac{\cos(\alpha n)}{n}\ (n>0) \leftrightarrows \ln\frac{z}{\sqrt{z^2+2z\cos(\alpha)+1}} \qquad \frac{n\alpha^n}{(n+1)(n+2)} \leftrightarrows \frac{z(\alpha-2z)}{\alpha^2}\ln\left(1-\frac{\alpha}{z}\right)-\frac{2z}{\alpha}.$$

9. Show that the $\mathcal{Z}$ transform of the signal

$$x[n] = \sum_{k=-n}^{n}\alpha^{|k|}u[n], \qquad |\alpha|<1$$

has the form

$$X(z) = \frac{1+(1+\alpha)z^{-1}-\alpha z^{-2}}{(1-\alpha z^{-1})(1-z^{-1})}, \qquad |z|>1.$$

10. Find the $\mathcal{Z}$ transform of the signal

$$y[n] = \sum_{k=0}^{n}k\left(\frac{2}{3}\right)^k u[n]$$

if it is known that

$$\mathcal{Y}(z) = -\frac{z^2}{z-1}\frac{d}{dz}\mathcal{X}(z).$$

11. Determine the causal signals $x[n]$ for each of the $\mathcal{Z}$ transforms

$$\mathcal{X}_1(z)=\frac{1+3.5z^{-1}}{1+3z^{-1}+5z^{-2}}\quad \mathcal{X}_2(z)=\frac{1}{1+z^{-1}-0.5z^{-4}}\quad \mathcal{X}_3(z)=\frac{z^{-3}}{[1-0.95z^{-1}]^2}$$

$$\mathcal{X}_4(z)=\frac{z^{-6}+z^{-8}}{1+z^{-1}}\quad \mathcal{X}_5(z)=\frac{1-(1/3)z^{-1}}{1+(1/3)z^{-1}}\quad \mathcal{X}_6(z)=\frac{1+2z^{-2}}{1+z^{-3}}$$

$$\mathcal{X}_7(z)=\frac{1+6z^{-1}+z^{-2}}{(1-0.5z^{-1})}\quad \mathcal{X}_8(z)=\frac{2-1.5z^{-1}}{1-1.5z^{-1}-0.5z^{-2}}\quad \mathcal{X}_9(z)=\frac{2+6.5z^{-2}}{1+4.5z^{-1}-6z^{-2}}$$

using any of the methods for calculating the inverse transform. The use of the properties of the $\mathcal{Z}$ transform may prove useful in some cases.

12. Determine the discrete-time signal $x[n]$ whose $\mathcal{Z}$ transform has the form $\mathcal{X}(z)=2e^{z}+2e^{1/z}$ with $|z|\neq 0$.
13. Find the $\mathcal{X}(z)$ transform of the convolution of the signals (a) $x[n]=3^n u[-n]$, (b) $y[n]=(1/2)^n u[n]$, and (c) $x[n]=u[n]$ and $y[n]=\delta[n]+(1/2)^n u[n]$.
14. Find the $\mathcal{Z}$ transform and draw the pole-zero diagram of the discrete-time signal $x[n]=n\alpha^n\cos(\omega n)u[n]$.
15. Suppose a real signal with a rational $\mathcal{Z}$ transform, which has two poles and two zeros. If one of these zeros is $z=\sqrt{2}\exp(j\pi/4)$ whereas one of these poles is $p=\exp(-j\pi/8)$, find the $\mathcal{Z}$ transform of this signal.
16. Given the discrete-time signals

$$x_1[n]=(1/3)(n^2+n)(1/2)^{n-1}u[n-1]\quad x_2[n]=(1/5)^n(u[n]-u[n-12])\quad x_3[n]=n(-1)^n u[n]$$

$$x_4[n]=n^2u[n]\quad x_5[n]=-n\alpha^n u[-n-5]\quad x_6[n]=(-1)^n\cos(n\pi/5)u[n]$$

for each of these, the function $\mathcal{X}(z)$ should be calculated and the pole-zero plot and the region of convergence should be drawn.

17. The pole-zero plot and the region of convergence for the transform should be drawn

$$X_1(z)=\frac{1+\frac{1}{2}z^{-1}}{1+3z^{-1}},\quad |a|<3,\qquad X_2(z)=\frac{1+\frac{1}{3}z^{-1}}{1+\frac{3}{4}z^{-1}+\frac{1}{8}z^{-2}},\qquad x_2[n]\ \text{causal}$$

18. The pole-zero plot and the region of convergence for the transform should be drawn

$$X_3(z)=\frac{1+\frac{1}{2}z^{-1}}{1-\frac{5}{3}z^{-1}+\frac{1}{4}z^{-2}},\qquad |X_3(z)(e^{j\omega})|<\infty.$$

19. The $\mathcal{Z}$ transform of the signals $x[n]=-n(1/2)^n u[n]$ and $y[n]=(1/2)(n^2+n)(-1/4)^{n-1}u[n-1]$ should be found.
20. The $\mathcal{Z}$ transform of the convolution of the signals $x[n]=(1/2)^n u[n]$ and $y[n]=(-1/3)^n u[n]$ should be found.
21. The $\mathcal{Z}$ transform of the signal should be found

$$x[n]=\sum_{k=-\infty}^{n}\alpha^k\cos(\omega_0 k)u[k].$$

22. Using the initial and final value theorems, determine the values $x[0]$ and $x[\infty]$ for the $\mathcal{Z}$ transforms

$$\mathcal{X}(z)=\frac{2z(z-0.75)}{(z-0.5)(z-0.25)},\quad |z|>\frac{1}{2}\qquad \text{and}\qquad \mathcal{X}(z)=\frac{z}{2z^2+3z+1},\quad |z|>1.$$

23. Determine the value of $x[0]$ when it is known that

$$\mathcal{X}(z)=\frac{z}{\left(1-\frac{1}{2}z^{-1}\right)\left(1-\frac{1}{3}z^{-2}\right)},\qquad |z|>\frac{1}{2}.$$

24. Find the limit of the signal $x[n]$ when $n \to \infty$ for the discrete-time signal

$$x[n] = \left[\left(\frac{1}{3}\right)^n - \frac{1}{3}\right] u[n].$$

25. Using the method of complex integration, calculate the inverse $\mathcal{Z}$ transform of the following transforms

$$X(z) = \frac{1+3z^{-1}}{1+3z^{-1}+2z^{-2}}, \quad X(z) = \frac{1+2z^{-2}}{1+z^{-2}}, \quad X(z) = \frac{1-\alpha z^{-1}}{z^{-1}-\alpha}.$$

26. Using the method of complex integration, calculate the inverse $\mathcal{Z}$ transform of the following transforms

$$\mathcal{X}_1(z) = \frac{1}{(1-2z^{-1})(1-z^{-1})^2}, \quad \mathcal{X}_2(z) = \frac{1}{1+1.5z^{-1}-0.5z^{-2}}, \quad \mathcal{X}_3(z) = \frac{1}{1-0.5z^{-1}+0.25z^{-2}}.$$

27. Find the inverse $\mathcal{Z}$ transform of the following transforms $X(z) = \ln(1+\alpha z^{-1})$ $|z| > |\alpha|$ and $X(z) = \cos(z)$.
28. Find the inverse $\mathcal{Z}$ transform of the following transforms

$$X(z) = \frac{1}{1-\alpha z^{-1}}, \quad |z| > |\alpha| \quad \text{and} \quad X(z) = \frac{z+z^{-1}}{1-\frac{1}{2}z^{-1}}, \quad |z| > \frac{1}{2}.$$

29. Find the inverse $\mathcal{Z}$ transform of the following transforms

$$X(z) = \frac{1}{\left(1-\frac{1}{2}z^{-1}\right)\left(1-\frac{1}{2}z^{-1}\right)}, \quad |z| > \frac{1}{2} \quad \text{and} \quad X(z) = \frac{1}{\left(1-\frac{1}{2}z^{-1}\right)(1-z^{-1})}, \quad |z| > 1.$$

30. Show that the inverse $\mathcal{Z}$ transform of the following transforms

$$\mathcal{X}(z) = \frac{1-z^{-1}+z^{-2}}{\left(1-\frac{1}{2}z^{-1}\right)(1-2z^{-1})(1-z^{-1})} \quad \text{and} \quad \mathcal{X}(z) = \frac{1-10z^{-1}-4z^{-2}+4z^{-3}}{1-z^{-1}-2z^{-2}}, \quad |z| < 1$$

are the discrete-time signals

$$x[n] = \left(\frac{1}{2}\right)^n u[n] - 2^{n+1}u[-n-1] - 2u[n] \quad \text{and} \quad x[n] = -2\delta[n-1] + 3\delta[n] - (-1)^n u[-n-1] + 3(2)^n u[-n-1]$$

respectively.

Systems in the Z Domain: Digital Processing Techniques

9

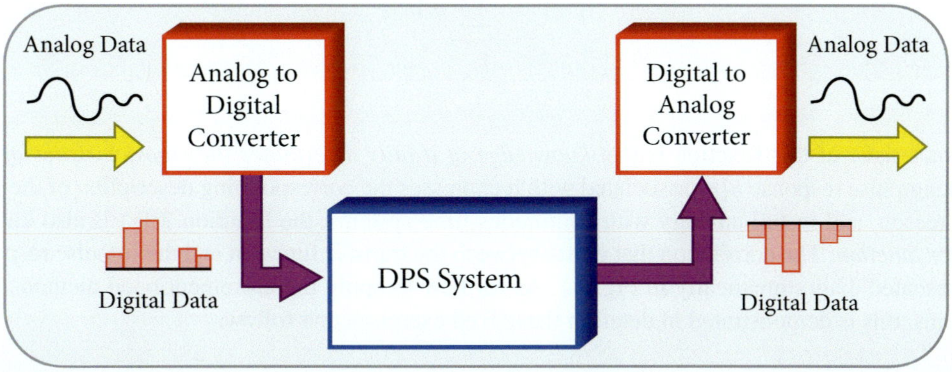

The causality and stability of a discrete-time signal or system depend on the positions of the poles of the $\mathcal{Z}$ transform of the system's impulse response in the complex plane, in relation to the circumference of the unit circle

9.1 The Transfer Function of Discrete-Time Systems

As already mentioned, the response $y[n]$ of a discrete, linear time-invariant (LTI) system with impulse response $h[n]$ to a corresponding input signal $x[n]$ is given by the equation of the *convolution*

$$y[n] = h[n] * x[n] = \sum_{k=-\infty}^{\infty} h[k]x[n-k] \tag{9.1}$$

If the $\mathcal{Z}$ transforms of the discrete sequences $y[n]$, $h[n]$, and $x[n]$ are the functions $\mathcal{Y}(z)$, $\mathcal{H}(z)$, and $\mathcal{X}(z)$, respectively, then, according to the property of the convolution of the $\mathcal{Z}$ transform, the following relation will hold:

$$\mathcal{Y}(z) = \mathcal{H}(z)\mathcal{X}(z) \tag{9.2}$$

Therefore, if we know the impulse response of the system $h[n]$ and the input sequence of the system $x[n]$, we can determine the output $y[n]$ produced by it *if we calculate the transforms* $\mathcal{H}(z)$ *and* $\mathcal{X}(z)$*, determine their product* $\mathcal{Y}(z) = \mathcal{H}(z)\mathcal{X}(z)$*, and then calculate the inverse transform* $\mathcal{Z}$ *of the function* $\mathcal{Y}(z)$. Alternatively, if we know the input signal $x[n]$

A. I. Margaris, *Continuous and Discrete-Time Signals and Systems*,
https://doi.org/10.1007/978-3-031-81209-5_9

and the output of the system $y[n]$, we can find the function of the impulse response, $h[n]$, *if we calculate the inverse $\mathcal{Z}$ transform* of the function.[1]

$$\mathcal{H}(z) = \frac{\mathcal{Y}(z)}{\mathcal{X}(z)} \tag{9.3}$$

which is defined as the ratio of the transforms $\mathcal{Y}(z)$ and $\mathcal{X}(z)$ and which, based on the defining equation of the $\mathcal{Z}$ transform, will obviously be given by the relation

$$\mathcal{H}(z) = \sum_{n=-\infty}^{\infty} h[n]z^{-n} \tag{9.4}$$

The main characteristic of this function is that *knowledge of it fully determines the characteristics of the system in the z domain* since the impulse response $h[n]$ associated with it expresses the corresponding description of the system in the time domain. For this reason, and in full analogy with continuous-time systems, the function $\mathcal{H}(z)$ is also known as the *system function* or *transfer function*. The correlation that exists between the transfer function and the impulse response of a discrete-time system is presented diagrammatically in Fig. 9.1. As for how to apply the aforementioned methods for characterizing discrete LTI systems, this is demonstrated in detail in the solved exercises that follow.

Problem 9.1.1 Considering a discrete LTI system with transfer function

$$\mathcal{H}(z) = \frac{1 + 2z^{-1} + z^{-2}}{(1 + 0.5z^{-1})(1 - z^{-1})}$$

determine its impulse response $h[n]$.

Solution The impulse response of the system will be calculated from the inverse $\mathcal{Z}$ transform of the function $\mathcal{H}(z)$. Setting for the sake of simplicity $y = z^{-1}$, the expansion of the function $\mathcal{H}(y)$ in partial fractions will take the form

$$\mathcal{H}(y) = \frac{1 + 2y + y^2}{(1 + 0.5y)(1 - y)} = \frac{\alpha y + \beta}{1 + 0.5y} + \frac{\gamma}{1 - y}$$
$$= \frac{(\alpha y + \beta)(1 - y) + \gamma(1 + 0.5y)}{(1 + 0.5y)(1 - y)} = \frac{-\alpha y^2 + (\alpha - \beta + 0.5\gamma)y + (\beta + \gamma)}{(1 + 0.5y)(1 - y)}$$

Equating the coefficients of the corresponding powers of z, we get $-\alpha = 1$, $\alpha - \beta + 0.5\gamma = 2$, and $\beta + \gamma = 1$, from which the values $\alpha = -1$, $\beta = -5/3$, and $\gamma = 8/3$ emerge. Therefore, the transfer function of the system can be written in the form

$$\mathcal{H}(y) = \frac{-y - (5/3)}{1 + 0.5y} + \frac{8/3}{1 - y} = \frac{-y - 2 + (1/3)}{1 + 0.5y} = -2 + \frac{1/3}{1 + 0.5y} + \frac{8/3}{1 - y}$$

[1] For a description of the corresponding processes in the domain of continuous-time see Sect. 3.5.1 on page 271 and Exercises 3.5.2 (page 272), 3.5.3 (page 273), 3.5.4 (page 275), 3.5.5 (page 276), and 3.5.6 (page 277) (Fourier transform), as well as Sect. 5.2 on page 376, along with Exercises 5.2.4 (page 381), 5.2.5 (page 382), and 5.2.6 (page 383) (Laplace transform).

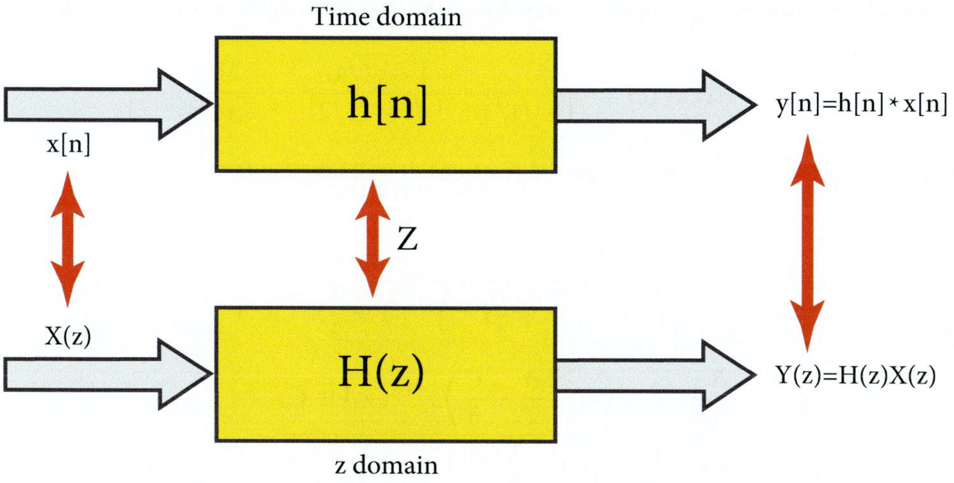

Fig. 9.1 Correlation between *impulse response* $h[n]$ and *transfer function* $\mathcal{H}(z)$ of discrete-time LTI system

or, returning to our original variables,

$$\mathcal{H}(z) = -2 \cdot (1) + \frac{1}{3}\left(\frac{1}{1+0.5z^{-1}}\right) + \frac{8}{3}\left(\frac{1}{1-z^{-1}}\right)$$

Referring back to Table 8.2 (page 679) we find that

$$\mathcal{Z}^{-1}\{1\} = \delta[n], \quad \mathcal{Z}^{-1}\left\{\frac{1}{1+0.5z^{-1}}\right\} = \left(-\frac{1}{2}\right)^n u[n], \quad \mathcal{Z}^{-1}\left\{\frac{1}{1-z^{-1}}\right\} = u[n]$$

that is, the inverse $\mathcal{Z}$ transform of the function $\mathcal{H}(z)$ will have the form

$$h[n] = -2\delta[n] + \frac{1}{3}\left(-\frac{1}{2}\right)^n u[n] + \frac{8}{3}u[n]$$

which is also the impulse response of the considered discrete-time system.

Problem 9.1.2 Use the $\mathcal{Z}$ transform method to calculate the output of a discrete-time system with impulse response $h[n] = (1/3)^n u[n]$ when the signal $x[n] = (1/2)^n \cos(\pi n/3)u[n]$ (Proakis and Manolakis, Problem 3.38α) is transmitted to its input.

Solution From basic theory it is known that the system function $\mathcal{H}(z)$ is defined as the $\mathcal{Z}$ transform of the system's impulse response. Therefore,

$$\mathcal{H}(z) = \sum_{n=-\infty}^{\infty} h[n]z^{-n} = \sum_{n=-\infty}^{\infty}\left(\frac{1}{3}\right)^n u[n]z^{-n} = \sum_{n=0}^{\infty}\left(\frac{1}{3}z^{-1}\right)^n = \frac{1}{1-(1/3)z^{-1}}$$

On the other hand, referring to Table 8.2, it is not difficult to see that the $\mathcal{Z}$ transform of the signal $x[n] = (1/2)^n \cos(\pi n/3)u[n]$ has the form (the following equation arises from the penultimate line of Table 8.2 for parameter values $\alpha = 1/2$ and $\omega_0 = \pi/3$)

$$\mathcal{X}(z) = \frac{1-(1/4)z^{-1}}{1-(1/2)z^{-1}+(1/4)z^{-2}}$$

Therefore, using the *property of convolution* of the $\mathcal{Z}$ transform, the transform of the system's output is calculated as

$$\mathcal{Y}(z) = \mathcal{H}(z)\mathcal{X}(z) = \frac{1-(1/4)z^{-1}}{(1-(1/3)z^{-1})[1-(1/2)z^{-1}+(1/4)z^{-2}]}$$

Expanding the foregoing expression into a sum of partial fractions, it will take the form

$$\mathcal{Y}(z) = \frac{1-\frac{1}{4}z^{-1}}{\left(1-\frac{1}{3}z^{-1}\right)\left(1-\frac{1}{2}z^{-1}+\frac{1}{4}z^{-2}\right)} = \frac{A}{1-\frac{1}{3}z^{-1}} + \frac{Bz^{-1}+C}{1-\frac{1}{2}z^{-1}+\frac{1}{4}z^{-2}}$$

$$= \frac{\left(\frac{A}{4}-\frac{B}{3}\right)z^{-2} + \left(B-\frac{A}{2}-\frac{C}{3}\right)z^{-1} + (A+C)}{\left(1-\frac{1}{3}z^{-1}\right)\left(1-\frac{1}{2}z^{-1}+\frac{1}{4}z^{-2}\right)}$$

Equating the coefficients of the corresponding powers of z, we are led to the following system of three equations:

$$\frac{A}{4}-\frac{B}{3}=0, \quad B-\frac{A}{2}-\frac{C}{3}=-\frac{1}{4}, \quad A+C=1$$

with the solution $(A, B, C) = (1/7, 3/28, 6/7)$. Therefore, the function $\mathcal{Y}(z)$ is written

$$\mathcal{Y}(z) = \frac{1/7}{1-(1/3)z^{-1}} + \frac{(3/28)z^{-1}+(6/7)}{1-(1/2)z^{-1}+(1/4)z^{-2}}$$

Referring back to Table 8.2, we find that

$$\mathcal{Z}^{-1}\left\{\frac{1/7}{1-(1/3)z^{-1}}\right\} = \frac{1}{7}\mathcal{Z}^{-1}\left\{\frac{1}{1-(1/3)z^{-1}}\right\} = \frac{1}{7}\left(\frac{1}{3}\right)^n u[n]$$

On the other hand, the second term of our final equation is not included in the table of elementary transforms. However, we observe that if we write it in the form[2]

$$\frac{(3/28)z^{-1}+(6/7)}{1-(1/2)z^{-1}+(1/4)z^{-2}} = \frac{6}{7}\frac{1-(1/4)z^{-1}}{1-(1/2)z^{-1}+(1/4)z^{-2}} + \frac{3\sqrt{3}}{7}\frac{(\sqrt{3}/4)z^{-1}}{1-(1/2)z^{-1}+(1/4)z^{-2}}$$

then from Table 8.2 we will get

$$\mathcal{Z}^{-1}\left\{\frac{1-(1/4)z^{-1}}{1-(1/2)z^{-1}+(1/4)z^{-2}}\right\} = \left(\frac{1}{2}\right)^n \cos\left(\frac{\pi n}{3}\right) u[n]$$

$$\mathcal{Z}^{-1}\left\{\frac{(\sqrt{3}/4)z^{-1}}{1-(1/2)z^{-1}+(1/4)z^{-2}}\right\} = \left(\frac{1}{2}\right)^n \sin\left(\frac{\pi n}{3}\right) u[n]$$

Combining all the foregoing expressions, it follows that the desired output signal of the system $y[n]$ is given by the relation

$$y[n] = \left[\frac{1}{7}\left(\frac{1}{3}\right)^n + \frac{6}{7}\left(\frac{1}{2}\right)^n \cos\left(\frac{\pi n}{3}\right) + \frac{3\sqrt{3}}{7}\left(\frac{1}{2}\right)^n \sin\left(\frac{\pi n}{3}\right)\right]u[n].$$

[2] This way of writing the second term *is not so obvious*: we must have enough experience to realize that since the input of the system is a signal of the form $\alpha^n \cos(\omega_0 n)$, its output will contain the linear combination of the functions $\alpha^n \cos(\omega_0 n)$ and $\alpha^n \sin(\omega_0 n)$ and therefore in the following formulation of this term, the transforms $\mathcal{Z}$ of the last two lines of Table 8.2 corresponding to the parameter values $\alpha = 1/2$ and $\omega_0 = \pi/3$ should appear.

Problem 9.1.3 Use the method of the $\mathcal{Z}$ transform to calculate the output of a discrete-time system with impulse response $h[n] = (1/2)^n u[n]$ when the signal is transmitted to its input

$$x[n] = (1/3)^n u[n] + (1/2)^{-n} u[-n-1]$$

(Proakis and Manolakis, Problem 3.38β).

Solution Following the previous methodology, the transfer function $\mathcal{H}(z)$ is calculated as

$$\mathcal{H}(z) = \sum_{n=-\infty}^{\infty} h[n] z^{-n} = \sum_{n=-\infty}^{\infty} \left(\frac{1}{2}\right)^n u[n] z^{-n} = \sum_{n=0}^{\infty} \left(\frac{1}{2} z^{-1}\right)^n = \frac{1}{1-(1/2)z^{-1}}$$

On the other hand, referring to tables of elementary transforms, it is proven that the $\mathcal{Z}$ transform of the discrete-time signal $x[n] = (1/3)^n u[n] + (1/2)^{-n} u[-n-1]$ is the

$$\mathcal{X}(z) = \frac{1}{1-(1/3)z^{-1}} - \frac{1}{1-2z^{-1}} = \frac{-(5/3)z^{-1}}{[1-(1/3)z^{-1}][1-2z^{-1}]}$$

Therefore, using the property of convolution we will get

$$\mathcal{Y}(z) = \mathcal{H}(z)\mathcal{X}(z) = \frac{-(5/3)z^{-1}}{[1-(1/2)z^{-1}][1-(1/3)z^{-1}][1-2z^{-1}]} = \frac{10/3}{1-(1/2)z^{-1}} + \frac{-2}{1-(1/3)z^{-1}} + \frac{-(4/3)}{1-2z^{-1}}$$

where for the calculation of the last expression, there took place an expansion of the function $\mathcal{Y}(z)$ in a series of partial fractions. Calculating the inverse $\mathcal{Z}$ transform of the function $Y(z)$ by referring to tables of elementary transforms, it turns out that

$$y[n] = \left[\frac{10}{3}\left(\frac{1}{2}\right)^n - 2\left(\frac{1}{3}\right)^n\right] u[n] + \frac{4}{3}(2)^n u[-n-1]$$

which is also the desired result. The reader is encouraged to verify the correctness of this result by calculating the convolution between the impulse response and the input signal of the system.

Problem 9.1.4 Let us consider a causal discrete LTI system that receives at its input the discrete-time signal

$$x[n] = u[-n-1] + \left(\frac{1}{2}\right)^n u[n].$$

If the $\mathcal{Z}$ transform of the system's output has the form

$$\mathcal{Y}(z) = -\frac{0.5z^{-1}}{(1-0.5z^{-1})(1+z^{-1})}$$

find the transfer function of the system $\mathcal{H}(z)$ and the output of the system $y[n]$ (Hayes, Problem 5.38).

Solution In this case, the transfer function of the system will be calculated as the ratio of the transforms $\mathcal{Y}(z)/\mathcal{X}(z)$, with the transform $\mathcal{X}(z)$ of the input signal $x[n]$ being calculated as

$$\mathcal{X}(z) = \sum_{n=-\infty}^{\infty} x[n] z^{-n} = \sum_{n=-\infty}^{\infty} u[-n-1] z^{-n} + \sum_{n=-\infty}^{\infty} \left(\frac{1}{2}\right)^n u[n] z^{-n}$$

with the step functions $u[n]$ and $u[-n-1]$ being defined as known

$$u[n] = \begin{cases} 1 \text{ for } n \geq 0 \\ 0 \text{ for } n < 0 \end{cases} \quad \text{and} \quad u[-n-1] = \begin{cases} 1 \text{ for } n \leq -1 \\ 0 \text{ for } n > -1 \end{cases}$$

The first of the two sums is calculated as

$$\sum_{n=-\infty}^{\infty} u[-n-1]z^{-n} = \sum_{n=-\infty}^{-1} z^{-n} = \sum_{n'=1}^{\infty} z^{n'} = \sum_{n'=0}^{\infty} z^{n'} - 1 = \frac{1}{1-z} - 1 = \frac{z}{1-z}$$

—where we changed the dummy index from n to $n' = -n$—with a region of convergence of $|z| < 1$, as can be verified. In a completely analogous way for the second sum, we have

$$\sum_{n=-\infty}^{\infty} \left(\frac{1}{2}\right)^n u[n]z^{-n} = \sum_{n=0}^{\infty} \left(\frac{1}{2z}\right)^n = \frac{1}{1-(1/2z)} = \frac{2z}{2z-1}$$

with a region of convergence of $|z| > 1/2$. Therefore, the $\mathcal{Z}$ transform of the input signal will be

$$\mathcal{X}(z) = \frac{z}{1-z} + \frac{2z}{2z-1} = \frac{z(2z-1)+2z(1-z)}{(1-z)(2z-1)} = \frac{z}{(1-z)(2z-1)}$$

and taking into account that the regions of convergence of the foregoing sums are $|z| < 1$ and $|z| > 1/2$, it is obvious that the region of convergence of the transform $\mathcal{X}(z)$ will be the intersection of the preceding two regions, which is described by the equation $(1/2) < |z| < 1$. Now using the defining equation of the transform $\mathcal{Y}(z)$, it turns out that

$$\mathcal{H}(z) = \frac{\mathcal{Y}(z)}{\mathcal{X}(z)} = -\frac{z}{(2z-1)(z+1)} \frac{(1-z)(2z-1)}{z} = \frac{z-1}{z+1}$$

with a region of convergence of $|z| > 1$, since the system is a causal system.

On the other hand, the calculation of the system output consists in finding the inverse transform of the function $\mathcal{Y}(z)$. Expressing this function in *sum of partial fractions*, it turns out that

$$\mathcal{Y}(z) = -\frac{0.5z^{-1}}{(1-0.5z^{-1})(1-z^{-1})} = -\frac{1}{3}\left[\frac{1}{1-(1/2)z^{-1}}\right] + \frac{1}{3}\left[\frac{1}{1+z^{-1}}\right]$$

and then referring to Table 8.2 (page 679), we find that

$$\mathcal{Z}^{-1}\left\{\frac{1}{1-(1/2)z^{-1}}\right\} = \left(\frac{1}{2}\right)^n u[n] \quad \text{and} \quad \mathcal{Z}^{-1}\left\{\frac{1}{1+z^{-1}}\right\} = (-1)^n u[n]$$

We will then have

$$y[n] = -\frac{1}{3}\left(\frac{1}{2}\right)^n u[n] + \frac{1}{3}(-1)^n u[n]$$

Problem 9.1.5 Assuming that the $\mathcal{Z}$ transform of the discrete-time signal $h[n] = A_1\alpha_1^n u[n] + A_2\alpha_2^n u[n]$ is the function $\mathcal{H}(z) = 1/[1-(1/4)z^{-2}]$, determine the values of the parameters α_1, α_2 and A_1, A_2.

Solution To determine the values of the aforementioned parameters, we must calculate the inverse transform of the function $\mathcal{H}(z)$ and then compare the result that will emerge with the discrete-time signal $h[n]$. Using the method of expansion in partial fractions, the function $\mathcal{H}(z)$ is written

$$\frac{1}{1-\frac{z^{-2}}{4}} = \frac{1}{\left(1-\frac{z^{-1}}{2}\right)\left(1+\frac{z^{-1}}{2}\right)} = \frac{A}{1-\frac{z^{-1}}{2}} + \frac{B}{1+\frac{z^{-1}}{2}}$$

$$= \frac{A\left(1+\frac{z^{-1}}{2}\right)+B\left(1-\frac{z^{-1}}{2}\right)}{\left(1+\frac{z^{-1}}{2}\right)\left(1-\frac{z^{-1}}{2}\right)} = \frac{\frac{1}{2}(A-B)y+(A+B)}{1-\frac{y^2}{4}}$$

where, as before, we have defined for simplicity the variable $y = z^{-1}$.

By equating the coefficients of the corresponding powers of y, it turns out that $[(A-B)/2] = 0$ and $A+B = 1$; therefore, $A = B = (1/2)$, meaning the function $H(z)$ is written in the form

$$\mathcal{H}(z) = \frac{1}{2}\left[\frac{1}{1-(1/2)z^{-1}}\right] + \frac{1}{2}\left[\frac{1}{1+(1/2)z^{-1}}\right]$$

Referring back to Table 8.2, we find that

$$\mathcal{Z}^{-1}\left\{\frac{1}{1-(1/2)z^{-1}}\right\} = \left(\frac{1}{2}\right)^n u[n] \qquad \text{and} \qquad \mathcal{Z}^{-1}\left\{\frac{1}{1+(1/2)z^{-1}}\right\} = \left(-\frac{1}{2}\right)^n u[n]$$

meaning the inverse transform of the function $\mathcal{H}(z)$ will be given by the relation

$$h[n] = \frac{1}{2}\left(\frac{1}{2}\right)^n u[n] + \frac{1}{2}\left(-\frac{1}{2}\right)^n u[n]$$

If we compare the last relation with the relation $h[n] = A_1\alpha_1^n u[n] + A_2\alpha_2^n u[n]$, it easily follows that $A_1 = \alpha_1 = A_2 = (1/2)$ and $\alpha_2 = -(1/2)$, which is also the desired result.

9.2 Transfer Functions and Difference Equations

Let us consider a recursive discrete LTI system of order N described by the general linear difference equation with constant coefficients

$$y[n] = -\sum_{k=1}^{N} \alpha_k y[n-k] + \sum_{k=0}^{M} \beta_k x[n-k] \tag{9.5}$$

Taking the $\mathcal{Z}$ transform of both sides of the preceding equation, it is written

$$\mathcal{Z}\{y[n]\} = -\sum_{k=1}^{N} \alpha_k \mathcal{Z}\{y[n-k]\} + \sum_{k=0}^{M} \beta_k \mathcal{Z}\{x[n-k]\} \tag{9.6}$$

where to derive the foregoing relation, we used the *property of linearity* of the $\mathcal{Z}$ transform, which is formulated as

$$\mathcal{Z}\left\{\sum_{k=\alpha}^{\beta} \xi_k x[k]\right\} = \sum_{k=\alpha}^{\beta} \mathcal{Z}\{\xi_k x[k]\} = \sum_{k=\alpha}^{\beta} \xi_k \mathcal{Z}\{x[k]\} \tag{9.7}$$

In the last step of the process, we use the *property of time shifting* of the $\mathcal{Z}$ transform, according to which

$$\mathcal{Z}\{x[n-k]\} = z^{-k}\mathcal{X}(z) \qquad \text{and} \qquad \mathcal{Z}\{y[n-k]\} = z^{-k}\mathcal{Y}(z)$$

and the foregoing relation is written

$$\mathcal{Y}(z) = -\sum_{k=1}^{N} \alpha_k \mathcal{Y}(z) z^{-k} + \sum_{k=0}^{M} \beta_k \mathcal{X}(z) z^{-k} \tag{9.8}$$

or, equivalently,

$$\mathcal{Y}(z)\left(1 + \sum_{k=1}^{N} \alpha_k z^{-k}\right) = \mathcal{X}(z)\left(\sum_{k=0}^{M} \beta_k z^{-k}\right) \tag{9.9}$$

and finally

$$\mathcal{H}(z) = \frac{\mathcal{Y}(z)}{\mathcal{X}(z)} = \left\{\sum_{k=0}^{M} \beta_k z^{-k}\right\} \Big/ \left\{1 + \sum_{k=1}^{N} \alpha_k z^{-k}\right\} \tag{9.10}$$

an equation that allows for the calculation of a system's transfer function when the difference equation that describes it is known. We are led, therefore, to the conclusion that recursive discrete LTI systems are characterized by a *rational transfer function*.[3]

The foregoing equation is the most general form of the transfer function, from which we can construct more simplified forms that are of particular interest. The first of these forms concerns the case in which the coefficients α_k of the foregoing equation take zero values in the range $1 \leq k \leq N$, and therefore the system described by it *is not recursive*. In this case, the system's transfer function is given by the relation

$$\mathcal{H}(z) = \sum_{k=0}^{M} \beta_k z^{-k} = \frac{1}{z^M} \sum_{k=0}^{M} \beta_k z^{M-k} \tag{9.11}$$

and, as can be observed, it is characterized by the existence of M zero values (whose values are determined by the system parameters β_k) and a pole of order M located at the origin of the zero level $z = 0$. It is proven that such a system is characterized by a finite impulse response, and for this reason it is known as a *finite impulse response system* or *moving average system*.

On the other hand, if the relation $\beta_k = 0$ holds (for the range of values $1 \leq k \leq M$), the transfer function of the system that is now formulated as

$$\mathcal{H}(z) = \beta_0 \Big/ \left\{1 + \sum_{k=1}^{N} \alpha_k z^{-k}\right\} = \left\{\beta_0 z^N\right\} \Big/ \left\{\sum_{k=0}^{N} \alpha_k z^{N-k}\right\} \quad (\alpha_0 = 1) \tag{9.12}$$

[3] This process we just described is the discrete equivalent of the one we presented in Sect. 5.2 (page 376) for calculating the transfer function of a continuous LTI system from the system's differential equation using the *Laplace transform*. The reader is also encouraged to study Sect. 5.3 on page 385, which deals with the description of the poles and zeros of the function $\mathcal{H}(s)$, as the situation is completely analogous for the function $\mathcal{H}(z)$.

will contain N poles with values determined by the system parameters α_k, as well as a zero value of order N, at the position $z = 0$. In this case, the impulse response of the system will be characterized by infinite duration (due to the existence of these poles), that is, the system will be characterized as a system of *infinite impulse response*.

The process of calculating the transfer function of a discrete LTI system from the difference equation that describes it is demonstrated through the example solved exercises that follow.

Problem 9.2.1 Let there be a recursive discrete-time LTI system, whose output, when the discrete-time signal is transmitted to its input

$$x[n] = \left(\frac{1}{3}\right)^n u[n] + 2^n u[-n-1] \quad \text{is given by the relation} \quad y[n] = 5\left(\frac{1}{3}\right)^n u[n] - 5\left(\frac{2}{3}\right)^n u[n]$$

Determine the transfer function and the impulse response of the system and construct the difference equation that describes its operation (adapted from Hayes, Solved Exercise 5.1).

Solution The system's transfer function will be calculated from the ratio of the $\mathcal{Z}$ transforms of the input and output signals. Starting with the determination of the function $\mathcal{X}(z)$, from the defining equation of the transform we have

$$\mathcal{X}(z) = \sum_{n=-\infty}^{\infty} x[n]z^{-n} = \sum_{n=-\infty}^{\infty} \left(\frac{1}{3}\right)^n u[n]z^{-n} + \sum_{n=-\infty}^{\infty} 2^n u[-n-1]$$

The first of the above sums is calculated as

$$\sum_{n=-\infty}^{+\infty} \left(\frac{1}{3}\right)^n u(n)z^{-n} = \sum_{n=0}^{\infty} \left(\frac{1}{3z}\right)^n = \frac{1}{1-(1/3z)} = \frac{3z}{3z-1}$$

with a convergence area of $\|z\| > 1/3$. In a completely analogous way, for the second sum we will have

$$\sum_{n=-\infty}^{\infty} 2^n u(-n-1) = \sum_{n=-\infty}^{-1} \left(\frac{z}{2}\right)^{-n} = \sum_{n'=1}^{\infty} \left(\frac{z}{2}\right)^{n'} = \sum_{n'=0}^{\infty} \left(\frac{z}{2}\right)^{n'} - 1 = \frac{1}{1-(z/2)} - 1 = \frac{2}{2-z} - 1 = \frac{z}{2-z}$$

—where we proceeded to change the dummy index from n to $n' = -n$—with region of convergence $\|z\| < 2$. Therefore, the $\mathcal{Z}$ transform of the input signal $x[n]$ will have the form

$$X(z) = \frac{3z}{3z-1} + \frac{z}{2-z} = \frac{3z(2-z)+z(3z-1)}{(3z-1)(2-z)} = \frac{5z}{(3z-1)(2-z)}$$

with region of convergence $(1/3) < \|z\| < 2$, as can be easily verified.

On the other hand, the $\mathcal{Z}$ transform of the output signal is calculated as

$$\mathcal{Y}(z) = \sum_{n=-\infty}^{\infty} y(n)z^{-n} = 5\sum_{n=0}^{\infty}\left(\frac{1}{3z}\right)^n - 5\sum_{n=0}^{\infty}\left(\frac{2}{3z}\right)^n = \frac{5}{1-(1/3z)} - \frac{5}{1-(2/3z)} =$$
$$= \frac{15z(3z-2) - 15z(3z-1)}{(3z-1)(3z-2)} = \frac{15z}{(3z-1)(2-3z)}$$

with region of convergence $|z| > (2/3)$, as can be proven. Having calculated the functions $\mathcal{X}(z)$ and $\mathcal{Y}(z)$, we can calculate the transfer function of the system from the relation

$$\mathcal{H}(z) = \frac{\mathcal{Y}(z)}{\mathcal{X}(z)} = \left(\frac{15z}{(3z-1)(2-3z)}\right) \Big/ \left(\frac{5z}{(3z-1)(2-z)}\right) = \frac{6-3z}{2-3z}$$

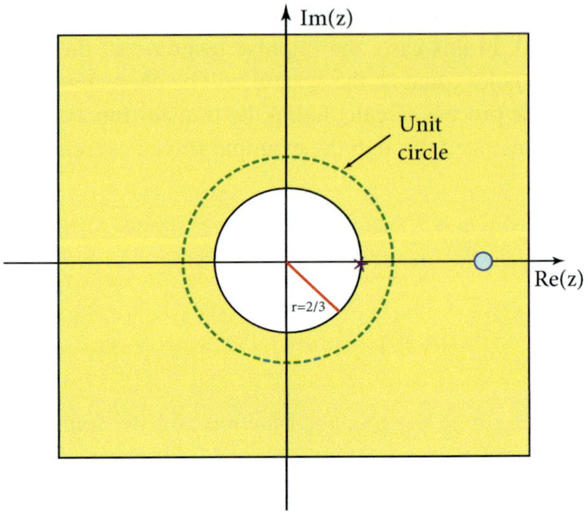

Fig. 9.2 Pole-zero diagram and region of convergence of function $\mathcal{H}(z)$ from Exercise 9.2.1

with region of convergence $|z| > 2/3$. We observe that the region of convergence of the system function includes the unit circle; therefore, as we will see later, this system is *stable*.

From the mathematical form of the function $\mathcal{H}(z)$, it is not difficult to see that it has a zero at the position $z = 2$ and a pole at the position $z = 2/3$. The pole-zero diagram and the region of convergence of this function are presented in Fig. 9.2.

After calculating the transfer function we can now determine its impulse response, as the inverse $\mathcal{Z}$ transform of the function $\mathcal{H}(z)$. Formulating this function in the form

$$\mathcal{H}(z) = \frac{6-3z}{2-3z} = \frac{(6/z)-3}{(2/z)-3} = \frac{1}{1-(2/3z)} - 2z^{-1}\frac{1}{1-(2/3z)}$$

and referring to Table 8.2, we find that

$$\mathcal{Z}^{-1}\left\{\frac{1}{1-(2/3z)}\right\} = \left(\frac{2}{3}\right)^n u[n] \qquad \text{and}$$

$$\mathcal{Z}^{-1}\left\{z^{-1}\frac{1}{1-(2/3z)}\right\} = \left(\frac{2}{3}\right)^{n-1} u[n-1]$$

where for the construction of the last relation using the property of time shift of the transform $\mathcal{Z}\{x[n-k]\} = z^{-k}X(z)$ for the value $k = 1$. Combining the foregoing results, the impulse response of the system is calculated as

$$h[n] = \left(\frac{2}{3}\right)^n u[n] - 2\left(\frac{2}{3}\right)^{n-1} u[n-1] = \left(\frac{2}{3}\right)^n \Big(u[n] - 3u[n-1]\Big)$$

which is also the desired result. We observe that for $n < 0$ we have $h[n] = 0$, and therefore the considered discrete-time system is causal.

Finally, from the defining equation of the function $\mathcal{H}(z)$ we will get

$$\mathcal{H}(z) = \frac{\mathcal{Y}(z)}{\mathcal{X}(z)} = \frac{-3+6z^{-1}}{-3+2z^{-1}} \Rightarrow \mathcal{Y}(z)(-3+2z^{-1}) = \mathcal{X}(z)(-3+6z^{-1})$$

and finally $-3\mathcal{Y}(z) + 2z^{-1}\mathcal{Y}(z) = -3\mathcal{X}(z) + 6z^{-1}\mathcal{X}(z)$. If we take the inverse $\mathcal{Z}$ transform of the two terms and use the property of time shifting, we end up with the expression

$$y[n] - \frac{2}{3}y[n-1] = x[n] - 2x[n-1]$$

which also constitutes the difference equation of the considered discrete-time system.

Problem 9.2.2 Considering a recursive discrete LTI system with transfer function

$$\mathcal{H}(z) = \frac{1 - 0.5z^{-2}}{(1 - 0.5z^{-1})(1 - 0.25z^{-1})}$$

determine its impulse response $h[n]$ and the difference equation of the system.

Solution The impulse response of the system will be determined as the inverse $\mathcal{Z}$ transform of the transfer function that describes it. Setting for simplicity reasons $y = z^{-1}$, the function $\mathcal{H}(y)$ expands into a sum of partial fractions of the form

$$\mathcal{H}(y) = \frac{\alpha y + \beta}{1 - 0.5y} + \frac{\gamma}{1 - 0.25y} = \frac{-0.25\alpha y^2 + (\alpha - 0.25\beta - 0.5\gamma)y + (\beta + \gamma)}{(1 - 0.5y)(1 - 0.25y)}$$

By equating the coefficients of the corresponding powers of z, we end up with the system of equations $-0.25\alpha = -0.5$, $\alpha - 0.25\beta - 0.5\gamma = 0$, and $\beta + \gamma = 1$, with solution $\alpha = 2$, $\beta = -6$, and $\gamma = 7$, and returning to our original variables, we can write the transfer function of the system in the form

$$\mathcal{H}(z) = \frac{2z^{-1} - 6}{1 - 0.5z^{-1}} + \frac{7}{1 - 0.25z^{-1}} = -4 \cdot (1) - 2\left(\frac{1}{1 - 0.5z^{-1}}\right) + 7\left(\frac{1}{1 - 0.25z^{1}}\right)$$

From the contents of Table 8.2 we find that

$$\mathcal{Z}^{-1}\{1\} = \delta[n], \quad \mathcal{Z}^{-1}\left\{\frac{1}{1 - 0.5z^{-1}}\right\} = \left(\frac{1}{2}\right)^n u[n], \quad \mathcal{Z}^{-1}\left\{\frac{1}{1 - 0.25z^{-1}}\right\} = \left(\frac{1}{4}\right)^n u[n]$$

and therefore the impulse response of the system is calculated as

$$h[n] = -4\delta[n] - 2\left(\frac{1}{2}\right)^n u[n] + 7\left(\frac{1}{4}\right)^n u[n]$$

On the other hand, to determine the difference equation, we will apply the method described earlier. Starting from the defining equation of the transfer function we will have

$$\mathcal{H}(z) = \frac{\mathcal{Y}(z)}{\mathcal{X}(z)} = \frac{1 - 0.5z^{-2}}{(1 - 0.5z^{-1})(1 - 0.25z^{-1})} \quad \text{or, equivalently,} \quad \mathcal{Y}(z)\left(1 - \frac{1}{2}z^{-1}\right)\left(1 - \frac{1}{4}z^{-1}\right) = \mathcal{X}(z)\left(1 - \frac{1}{2}z^{-2}\right)$$

and finally, after simple algebraic operations,

$$\mathcal{Y}(z) - \frac{3}{4}z^{-1}\mathcal{Y}(z) + \frac{1}{8}z^{-2}\mathcal{Y}(z) = \mathcal{X}(z) - \frac{1}{2}z^{-2}\mathcal{X}(z)$$

If we take the inverse $\mathcal{Z}$ transform of the preceding expressions and use the property of time shift, which is formulated as $\mathcal{Z}^{-1}\{x[n-k]\} = z^{-k}\mathcal{X}(z)$, we will get the equation

$$y[n] - \frac{3}{4}y[n-1] + \frac{1}{8}y[n-2] = x[n] - \frac{1}{2}x[n-2]$$

which is also the required difference equation of the system.

9.3 Stability and Causality of Discrete-Time Systems

In full analogy with their continuous equivalents, the discrete LTI systems used in practice must fulfill the properties of *causality* and *BIBO stability*. It is obvious that a discrete-time system of this type is *causal* when its impulse response $h[n]$ is a *causal sequence*—or, equivalently, *right-sided sequence*—that is, it satisfies the condition $h[n] = 0$ for $n < 0$. In this case, and as was mentioned in a previous section, *the region of convergence of the system's transfer function will be the complement of a circle on the complex plane with an equation of the form* $|z| > r$ (with the value of r being determined by the nature of the system), and therefore *all the poles of the function* $\mathcal{H}(z)$ *will be located inside—or even on the periphery—of this circle*. On the other hand, if the system is *anticausal*, that is, its impulse response is a *left-sided sequence* and satisfies the relation $h[n] = 0$ for $n \geq 0$, then *the region of convergence of the system's transfer function will be the interior of a circle* and will be described, therefore, by an equation of the form $|z| < r$—in this case, all the poles of the function $\mathcal{H}(z)$ should be located *outside or even on the periphery of this circle*.

On the other hand, the sufficient and necessary condition that allows the characterization of a LTI system as *BIBO stable* is *the finite value of the sum of the absolute values of the elements of the impulse response*.[4] This condition—which has been presented in detail in the introductory chapter of discrete-time systems—is mathematically rendered by an equation of the form

$$S = \sum_{n=-\infty}^{\infty} |h[n]| < \infty \tag{9.13}$$

and its power is shown to affect the shape of the region of convergence of the system's transfer function, *which must necessarily contain the unit circle*. To prove the truth of this claim, let us start from the definition of the transfer function as the $\mathcal{Z}$ transform of the impulse response, and let us take the measure of this equation. We will then have

$$|\mathcal{H}(z)| \leq \sum_{n=-\infty}^{\infty} |h[n]z^{-n}| = \sum_{n=-\infty}^{\infty} |h[n]||z^{-n}| \tag{9.14}$$

Calculating the foregoing equation for the circumference of the unit circle ($|z| = 1$), we will get

$$|\mathcal{H}(z)| \leq \sum_{n=-\infty}^{\infty} |h[n]| \tag{9.15}$$

from which it follows that *the system is BIBO stable when the region of convergence of the transfer function* $\mathcal{H}(z)$ *contains the unit circle, a condition that, besides being sufficient, is shown to be necessary*. It is interesting to note at this point that in the general case, *the validity of causality does not presuppose the simultaneous validity of stability, and vice versa*. For example, *a stable system can be both causal and noncausal*, while in a completely similar way *a causal system can be both stable and unstable*. However, it is shown that *if a system is simultaneously both causal and stable, then all the poles of the system's transfer function will be located inside the unit circle, which obviously will belong to the region of convergence of the function* $\mathcal{H}(z)$.

In the general case, the transfer function of recursive discrete LTI systems can be expressed in the explicit form $\mathcal{H}(z) = \mathcal{B}(z)/\mathcal{A}(z)$, and therefore its poles will constitute the roots of the denominator polynomial, which has the general form[5]

---

[4] This condition constitutes the discrete equivalent of the condition of *absolute integrability*, the integral of which due to the discrete nature of the problem has been replaced by sum.

[5] This polynomial is simply the characteristic polynomial of the system related to the homogeneous difference equation that describes it, and the requirement set by the stability criterion presented in this section is the same as the one we formulated in the chapter on discrete-time systems regarding the roots of the characteristic polynomial.

$$\mathcal{A}_m(z) = 1 + \alpha_1^m z^{-1} + \alpha_2^m z^{-2} + \cdots + \alpha_m^m z^{-m} = \sum_{k=0}^{m} \alpha_k z^{-k} \quad (\alpha_0^m \equiv 1) \tag{9.16}$$

with the symbol m indicating the degree of the polynomial $\mathcal{A}(z)$. In the following description, we will assume that the recursive system is of order N—that is, its current output $y[n]$ depends on the output $y[n-N]$—so the polynomial $\mathcal{A}(z)$ will also be a polynomial of degree N; for this reason, let us symbolize it by $\mathcal{A}_N(z)$.

In the international literature, there are many stability criteria by which we can check whether the roots of the polynomial $\mathcal{A}_N(z)$ are found *within the unit circle*, and in this section we will present the *stability criterion according to Schur-Cohn*, which is applied in the following way:

1. In the first step of the process, we initialize the algorithm by setting $m = N$, $\zeta_N = \alpha_N^N$, and we use as initial polynomial $\mathcal{A}_N(z)$.
2. If the relation $m \geq 1$ holds, we perform the following actions:

   - We determine the polynomial of degree $m-1$ from the relation

   $$\mathcal{A}_{m-1}(z) = \frac{\mathcal{A}_m(z) - \zeta_m \mathcal{A}_m^*(z)}{1 - \zeta_m^2} \tag{9.17}$$

   where $\mathcal{A}_m^*(z)$ is the inverse polynomial[6] of $\mathcal{A}_m(z)$.
   - We determine the value of the *reflection coefficient* ζ_m from the relation $\zeta_m = \alpha_m^m$.

3. We decrease the value of m by one unit.

In this case, since the polynomial $\mathcal{A}(z)$ is a polynomial of negative powers of z, the reverse polynomial $\mathcal{A}_m^*(z)$ is calculated from the relation

[6] Considering a polynomial $P_n(\xi)$ of degree n of the complex variable ξ with complex coefficients of the form

$$P_n(\xi) = \alpha_0^n + \alpha_1^n \xi + \alpha_2^n \xi^2 + \cdots + \alpha_n^n \xi^n = \sum_{k=0}^{n} \alpha_k^n \xi^k$$

the *inverse polynomial* is defined as

$$P_n^*(\xi) = \overline{\alpha_n} + \overline{\alpha_{n-1}}\xi + \overline{\alpha_{n-2}}\xi^2 + \cdots + \overline{\alpha_2}\xi^{n-2} + \overline{\alpha_1}\xi^{n-1} + \overline{\alpha_0}\xi^n = \xi^n \overline{P(1/\overline{\xi})}$$

where the symbol $\overline{\xi}$ refers to the complex conjugate. It is proven that in the special case where the coefficients α_k^n $(k = 0, 1, \ldots, n)$ are real numbers, they will appear in the reverse polynomial in the reverse order—that is, the relation $\alpha_k^n = \alpha_{n-k}^n$ will hold. In this case, the reverse polynomial is called a *palindromic polynomial* and is given by the relation

$$P_n^*(\xi) = \xi^n \overline{P(1/\overline{\xi})} = \xi^n \sum_{k=0}^{n} \overline{\alpha_{n-k}^n} \xi^k$$

Using the polynomials $P_n(\xi)$ and $P_n^*(\xi)$, we can define the *reflection coefficients* ζ_k—which are also known as Schur-Szegö parameters or *PARCOR (partial correlation coefficients) coefficients*—from the relation

$$\xi P_{n-1}(\xi) = \frac{P_n(\xi) - \zeta_k P_n(\xi)}{1 - |\zeta_k|^2}$$

where $\zeta_k = \alpha_0^k$. From the last relation it is proven that

$$P_n(\xi) = \xi P_{n-1}(\xi) + \zeta_k P_{n-1}^*(\xi)$$

According to a fundamental theorem of polynomial theory, a sufficient and necessary condition for all the roots of the polynomial to be within the unit circle is $|\zeta_k| < 1$ $(k = 1, 2, \ldots, n)$.

$$\mathcal{A}_m^*(z) = Z^{-m}\mathcal{A}_m(z^{-1}) = \sum_{k=0}^{m} \alpha_{m-k}^m z^{-k} \tag{9.18}$$

where the notation α_{m-k}^m refers to a *palindromic polynomial*, since the coefficients of the polynomial $A_m(z)$ are considered real numbers.

By repeatedly applying the foregoing process, we determine the reflection coefficients ζ_k $(k = 1, 2, \ldots, N)$. If for each reflection coefficient the relation $|\zeta_k| < 1$ holds, it is proven that *all the roots of the polynomial $\mathcal{A}(z)$ are within the unit circle*, and therefore the system with a transfer function of the form $\mathcal{B}(z)/\mathcal{A}(z)$ is *simultaneously causal and stable*.

Examples using the stability check according to *Schur-Cohn* are presented in the following exercises.

Problem 9.3.1 Using the Schur-Cohn stability criterion, check whether a discrete-time system described by the transfer function

$$\mathcal{H}(z) = \frac{1 + 2z^{-1} + 2z^{-2} + z^{-3}}{1 - (1/2)z^{-1} + (3/4)z^{-2} - z^{-3} + (1/4)z^{-4}}$$

is causal and stable.

Solution In this case, the polynomial of the denominator of the transfer function is

$$\mathcal{A}(z) = \mathcal{A}_4(z) = 1 - \frac{1}{2}z^{-1} + \frac{3}{4}z^{-2} - z^{-3} + \frac{1}{4}z^{-4}$$

from which it follows that $\zeta_4 = \alpha_4^4 = 1/4$. We observe that $|\zeta_4| < 1$, and we continue to the next step of the method. The inverse polynomial of $\mathcal{A}_4(z)$ has the form

$$\mathcal{A}_4^*(z) = z^{-4} - \frac{1}{2}z^{-3} + \frac{3}{4}z^{-2} - z^{-1} + \frac{1}{4}$$

Therefore, we will have

$$A_3(z) = \frac{A_4(z) - \zeta_4 A_4^*(z)}{1 - (\zeta_4)^2}$$

$$= \frac{1 - \frac{1}{2}z^{-1} + \frac{3}{4}z^{-2} - z^{-3} + \frac{1}{4}z^{-4} - \frac{1}{4}\left(z^{-4} - \frac{1}{2}z^{-3} + \frac{3}{4}z^{-2} - z^{-1} + \frac{1}{4}\right)}{1 - \left(\frac{1}{4}\right)^2} = 1 - \frac{4}{15}z^{-1} + \frac{9}{15}z^{-2} - \frac{14}{15}z^{-3}$$

From the last equation we observe that $\zeta_3 = \alpha_3^3 = -14/15$, and because $|\zeta_3| < 1$, we proceed to the next step. The inverse polynomial of the function $\mathcal{A}_3(z)$ is given by the equation

$$\mathcal{A}_3^*(z) = z^{-3} - \frac{4}{15}z^{-2} + \frac{9}{15}z^{-1} - \frac{14}{15}$$

and therefore the polynomial $\mathcal{A}_2(z)$ will have the form

$$A_2(z) = \frac{A_3(z) - \zeta_3 A_3^*(z)}{1 - (\zeta_3)^2}$$

$$= \frac{1 - \frac{4}{15}z^{-1} + \frac{9}{15}z^{-2} - \frac{14}{15}z^{-3} + \frac{14}{15}\left(z^{-3} - \frac{4}{15}z^{-2} + \frac{9}{15}z^{-1} - \frac{14}{15}\right)}{1 - \left(-\frac{14}{15}\right)^2} = 1 + \frac{69}{29}z^{-1} + \frac{79}{29}z^{-2}$$

Therefore, $\zeta_2 = \alpha_2^2 = 79/29$, and because $|\zeta_2| > 1$, the system is proven to be *not causal and stable*, and the method terminates.

Problem 9.3.2 Using the Schur-Cohn stability criterion, check whether the discrete-time system described by the transfer function

$$\mathcal{H}(z) = \frac{1}{1 - (7/4)z^{-1} - (1/2)z^{-2}}$$

is causal and stable (Proakis and Manolakis, Example 3.6.7).

Solution The polynomial of the denominator of the transfer function has the form

$$\mathcal{A}(z) \equiv \mathcal{A}_2(z) = 1 - \frac{7}{4}z^{-1} - \frac{1}{2}z^{-2}$$

from which it follows that $\zeta_2 = \alpha_2^2 = -1/2$. We observe that $|\zeta_2| < 1$ and continue to the next step of the method. The inverse polynomial of the function $A_2(z)$ is

$$\mathcal{A}_2^*(z) = z^{-2} - \frac{7}{4}z^{-1} - \frac{1}{2}$$

and therefore the polynomial $\mathcal{A}_1(z)$, is calculated as

$$\mathcal{A}_1(z) = \frac{\mathcal{A}_2(z) - \zeta_2\mathcal{A}_2^*(z)}{1 - \zeta_2^2} = \frac{1 - \frac{7}{4}z^{-1} - \frac{1}{2}z^{-2} + \frac{1}{2}\left(-\frac{1}{2} - \frac{7}{4}z^{-1} + z^{-2}\right)}{1 - \left(-\frac{1}{2}\right)^2} = 1 - \frac{7}{2}z^{-1}$$

We observe that $\zeta_1 = -7/2$, and because $|\zeta_1| > 1$, the system is *unstable*.

From the foregoing examples, we conclude that during the application of the Schur-Cohn method, it is not necessary to calculate all the reflection coefficients of the polynomial $\mathcal{A}(z)$: *If even one coefficient is found with a modulus greater than one, the system is characterized as unstable and causal, and the method terminates.*

Problem 9.3.3 Use the Schur-Cohn stability criterion to theoretically analyze the stability of a typical recursive discrete LTI system of second order described by a difference equation of the form

$$y[n] = -\alpha_1 y[n-1] - \alpha_2 y[n-2] + \beta_0 x[n].$$

(Proakis and Manolakis, Section 3.6.8).

Solution In the first step of the process, we need to find the transfer function of the system using the $\mathcal{Z}$ transform and the way we did it in previous exercises. Therefore, taking the $\mathcal{Z}$ transform of all the terms of the difference equation and using the property of time shift formulated as $\mathcal{Z}\{x[n-k]\} = z^{-k}\mathcal{X}(z)$, we will get

$$\mathcal{Y}(z) = -\alpha_1 z^{-1}\mathcal{Y}(z) - \alpha_2 z^{-2}\mathcal{Y}(z) + \beta_0\mathcal{X}(z) \quad \text{or, equivalently,} \quad \mathcal{Y}(z)(1 + \alpha_1 z^{-1} + \alpha_2 z^{-2}) = \beta_0\mathcal{X}(z)$$

and finally

$$\mathcal{H}(z) = \frac{\mathcal{Y}(z)}{\mathcal{X}(z)} = \frac{\beta_0}{1 + \alpha_1 z^{-1} + \alpha_2 z^{-2}}$$

In order for this system to be *causal and stable*, the polynomial of the denominator

$$\mathcal{A}(z) \equiv \mathcal{A}_2(z) = 1 + \alpha_1 z^{-1} + \alpha_2 z^{-2}$$

must successfully pass the *Schur-Cohn* test. From the foregoing relation we directly observe that the reflection coefficient ζ_2 is given by the relation $\zeta_2 = \alpha_2$, and therefore a first stability condition will be $|\alpha_2| < 1$. In the next step of the process, we must calculate the coefficient ζ_1 by calculating the polynomial $\mathcal{A}_1(z)$. Applying the method we presented in the previous section, we define the *inverse polynomial* of $\mathcal{A}_2(z)$ from the relation

$$\mathcal{A}_2^*(z) = z^{-2} + \alpha_1 z^{-1} + \alpha_2$$

and then we calculate the polynomial $\mathcal{A}_1(z)$ as

$$\begin{aligned}\mathcal{A}_1(z) &= \frac{\mathcal{A}_2(z) - \zeta_2 \mathcal{A}_2^*(z)}{1 - (\zeta_2)^2} = \frac{1 + \alpha_1 z^{-1} + \alpha_2 z^{-2} - \alpha_2(z^{-2} + \alpha_1 z^{-1} + \alpha_2)}{1 - \alpha_2^2} \\ &= \frac{1}{1 - \alpha_2^2}\left[(1 - \alpha_2^2) + \alpha_1(1 - \alpha_2)z^{-1}\right] = 1 + \frac{1 - \alpha_2}{1 - \alpha_2^2} z^{-1} = 1 + \frac{\alpha_1}{1 + \alpha_2} z^{-1}\end{aligned}$$

From the last relation we observe that the reflection coefficient ζ_1 is given by the relation $\zeta_1 = \alpha_1/(1 + \alpha_2)$, and therefore a second stability condition is

$$\left|\frac{\alpha_1}{1 + \alpha_2}\right| < 1 \Rightarrow -1 < \frac{\alpha_1}{1 + \alpha_2} < 1 \Rightarrow \begin{cases} \alpha_1 < 1 + \alpha_2 \\ \alpha_1 > -1 - \alpha_2 \end{cases}$$

Note An alternative way to solve the problem is to write the transfer function as

$$\mathcal{H}(z) = \frac{\beta_0}{1 + \alpha_1 z^{-1} + \alpha_2 z^{-2}} = \frac{\beta_0 z^2}{z^2 + \alpha_1 z + \alpha_2} = \frac{\beta_0 z^2}{(z - p_1)(z - p_2)}$$

with the poles of the system p_1 and p_2 being the roots of the quadratic equation of the denominator and given by the known relations

$$p_1 = -\frac{\alpha_1}{2} + \sqrt{\frac{\alpha_1^2 - 4\alpha_2}{4}}, \qquad p_2 = -\frac{\alpha_1}{2} - \sqrt{\frac{\alpha_1^2 - 4\alpha_2}{4}}$$

If we solve the system of the foregoing equations with respect to the parameters α_1 and α_2, we will get $\alpha_1 = -(p_1 + p_2)$ and $\alpha_2 = p_1 p_2$, and using the conditions $|p_1| < 1$ and $|p_2| < 1$, which ensure that the poles p_1 and p_2 are *within the unit circle*, we end up with stability conditions compatible with those we determined using the *Schur-Cohn method*. These stability conditions define a region in the plane (α_1, α_2), which is *triangular in shape*, is called a *stability region*, and is defined as *the set of points whose coordinates* (α_1, α_2) *correspond to values of the parameters* α_1 *and* α_2 *for which the system is stable*. This stability region can be subdivided into three smaller regions for which the poles of the system are *conjugate complex values* or *real values equal to or different from each other*, depending on the values of the system parameters and that are presented diagrammatically in Fig. 9.3.

9.4 Minimum and Maximum Phase Systems

A special category of LTI systems are the *minimum phase systems*. These systems, defined both in the continuous- and discrete-time domain, constitute *causal and stable systems whose inverse exists and is also causal and stable*. Taking into account that the poles and zeros of the inverse of a discrete LTI system coincide respectively with the zeros and poles of the direct system (in other words, to construct a pole-zero diagram for the inverse system, we can simply construct a pole-zero

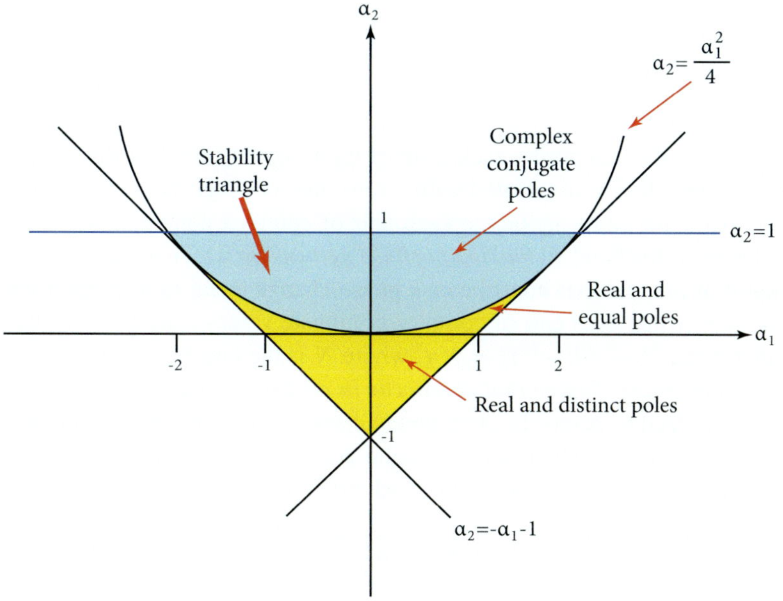

Fig. 9.3 Stability region for second-order recursive discrete LTI system

diagram of the direct system and then interchange the positions of the zeros and poles),[7] it is not difficult to understand that *both the poles and the zero values of such a system, for a continuous-time system, will belong to the left complex half-plane, that is, they will have a negative real part, while correspondingly, for a discrete-time system, they will be located in the interior of the unit circle*. Indeed, as long as the system is causal and stable, its poles, as we have shown, will be characterized by the foregoing property, while obviously the same will apply to its zero values, which, being poles of the inverse system—which is also causal and stable—will also have this property. In other words, a minimum phase system is defined as a system in which *both the poles and the zero values of its transfer function meet the appropriate stability criterion* (i.e., they are located in the left complex half-plane or in the interior of the unit circle of the complex plane, depending on whether the system is defined in the domain of continuous or discrete-time; in what follows, we will of course limit ourselves to the second case). The necessity of defining minimum phase systems stems from the fact that the frequency response of a LTI system with a rational transfer function *does not uniquely determine the system* since, although the characterization of the system as causal and stable imposes the foregoing restriction for its poles, this does not apply to its zero values, which can be found anywhere. Conversely, if we impose the requirement for the inverse system to also be a causal and stable system then the transfer function $\mathcal{H}(z)$ *will be determined in a unique way* and will contain the poles and zero values of the function $\mathcal{C}(z) = \mathcal{H}(z)\mathcal{H}^*(1/z^*)$, where the symbol $*$ indicates complex conjugation.

Minimum phase systems are at one end of the spectrum of different types of systems, with the criterion being the phase response, while at the other end of this spectrum are the so-called *maximum phase systems*, in which *all the zero values of the transfer function are located in the exterior of the unit circle* (or, equivalently, in the right half-plane if the system is defined in the continuous-time domain). In other words, a system is characterized as a maximum phase system *if the inverse system exists and is causal but not stable* as in minimum phase systems, but in contrast, *unstable*. All the other systems are located between these two extremes of the spectrum, where the set of zero values is divided into two subsets, with the zeros of the first subset being inside and the zeros of the second subset being outside of the unit circle.

The graphical representation of the phase response $\measuredangle\mathcal{H}(e^{j\omega})$ for these two types of systems reveals that in minimum phase systems, the change in phase between the frequency values $\omega = 0$ and $\omega = \pi$ has a minimum value equal to zero, while, in contrast, in systems of maximum phase, this change has a maximum value equal to π. Conversely, in all other systems, the foregoing phase change has a value that lies between these two extreme values. Indeed, considering a digital finite impulse response (FIR) filter of length $N + 1$ that contains N zero values and, hence, is characterized by a frequency response of the

[7] See Sect. 9.5 on page 714 for discrete-time systems and Sect. 5.8 on page 444 for continuous-time systems, as mentioned previously, since, according to what we mentioned earlier, exactly the same holds true for the domain of continuous-time.

form

$$\mathcal{H}(e^{j\omega}) = H_0 \prod_{i=0}^{N}(1 - z_i e^{-j\omega}) = H_0(1 - z_1 e^{-j\omega})(1 - z_2 e^{-j\omega}) \cdots (1 - z_N e^{-j\omega})$$

when all the zero values are inside the unit circle, each term of the foregoing product that corresponds to a zero value that lies on the real axis of the complex plane is associated with a zero phase change as the frequency ω changes from the value $\omega = 0$ to the value $\omega = \pi$, while the same applies to each pair of complex conjugate values. Therefore, it will hold that $\measuredangle\mathcal{H}(e^{j\pi} - \measuredangle\mathcal{H}(e^{j0})) = 0$. On the other hand, in *maximum phase systems*, in which all zero values are outside the unit circle, each zero value that is a point of the real axis contributes a phase change equal to π as the frequency ω changes from the value $\omega = 0$ to the value $\omega = \pi$, while each pair of complex conjugate values contributes a change amount equal to 2π. It will therefore be the case that $\measuredangle\mathcal{H}(e^{j\pi} - \measuredangle\mathcal{H}(e^{j0})) = N\pi$, where N is the number of zeros of the transfer function, a value that also constitutes *the maximum phase change that can occur in an FIR filter with M zero values*. This is also why these two types of systems are characterized respectively as *minimum phase systems* and *maximum phase systems*, while of course it will hold that $\measuredangle\mathcal{H}_{\max}(e^{j\omega}) \leq \measuredangle\mathcal{H}_{\min}(e^{j\omega})$, where $\mathcal{H}_{\max}(e^{j\omega})$ and $\mathcal{H}_{\min}(e^{j\omega})$ are the frequency responses of the maximum and minimum phase systems, respectively, which are related to each other according to the expression

$$\mathcal{H}_{\max}(z) = z^{-N}\mathcal{H}^*_{\min}(1/z^*) = \alpha^*_N + \alpha^*_{N-1}z^{-1} + \cdots + z^{-N}$$

where

$$\mathcal{H}_{\min}(z) = 1 + \alpha_1 z^{-1} + \alpha_2 z^{-2} + \cdots + \alpha_N z^{-N}$$

from which it follows that the function $\mathcal{H}_{\max}(z)$ is the inverse polynomial of the function $\mathcal{H}_{\min}(z)$ (see also the discussion in Footnote 6 on page 705) (from this relationship it follows that every zero z of the function $\mathcal{H}_{\min}(z)$ that is located inside the unit circle is mapped to the position $1/z$, which is located outside the unit circle, so as to become a zero value of the function $\mathcal{H}_{\max}(z)$—this property will be used later. Conversely, all other systems that fall into the intermediate area with this characteristic of phase change are known as *nonminimum phase systems* or *mixed phase systems*.

According to a very important theorem related to the space of signals and systems, a transfer function of a system that is not a minimum phase system can be written as a product of the form

$$\mathcal{H}(z) = \mathcal{H}_{\min}(z)\mathcal{H}_{\mathrm{ap}}(z)$$

where $\mathcal{H}_{\min}(z)$ is the transfer function of a *minimum phase system* and $\mathcal{H}_{\mathrm{ap}}(z)$ is the transfer function of an *all-pass system*. This in an equivalent formulation and means that *a nonminimum phase system arises from the series connection of a minimum phase system with an all-pass system*. To prove this property, let us consider a system with a transfer function $\mathcal{H}(z)$, which is not a minimum phase system and therefore contains a zero that is located outside of the unit circle and at the position $z = 1/c*$, where $|c| < 1$, while all other zero values and all the poles are located inside the unit circle. In this case, the transfer function of the system $\mathcal{H}(z)$ will be written

$$\mathcal{H}(z) = \mathcal{H}_1(z)(z^{-1} - c^*)$$

with the function $\mathcal{H}_1(z)$ being by definition a *minimum phase system function*. The foregoing relationship can also be written in the form

$$\mathcal{H}(z) = \mathcal{H}_1(z)(1 - cz^{-1})\frac{z^{-1} - c^*}{1 - cz^{-1}}$$

We now observe that, since $|c| < 1$, the function $\mathcal{H}_1(z)(1 - cz^{-1})$ will also be a minimum phase system transfer function and will differ from the function $\mathcal{H}(z)$ only in that the zero value of the function $\mathcal{H}(z)$, which was outside the unit circle and at the position $z = 1/c*$, is mapped inside the unit circle and at the conjugate reciprocal position $z = c$ (recall that in the general case, this mapping has the form $z \mapsto (1/z^*)$, and consequently, the value $z = 1/c^*$ will be mapped to the value $1/z^* = 1/((1/c^*))^* = (c^*)^* = c$, which is inside the unit circle since $|c| < 1$). On the other hand, the term $(z^{-1} - c^*)/(1 - cz^{-1})$ refers to an *all-pass filter*. Indeed, the transfer function of an all-pass filter with unit gain, which allows the passage of all frequency components, in the general case has the form

$$\mathcal{H}_{\text{ap}}(z) = \prod_{k=1}^{N} \frac{(z^{-1} - c_k^*)}{(1 - c_k z^{-1})}$$

and, as it turns out, after operations it holds that $|\mathcal{H}_{\text{ap}}(z)| = 1$. Replacing the preceding transfer functions from their representation in amplitude and phase we will have

$$|\mathcal{H}(z)| = |\mathcal{H}_{\text{min}}(z)||\mathcal{H}_{\text{ap}}(z)| \quad \text{and} \quad \measuredangle\mathcal{H}(s) = \measuredangle\mathcal{H}_{\text{min}}(s) + \measuredangle\mathcal{H}_{\text{ap}}(s)$$

and since as we noted that $|\mathcal{H}_{\text{ap}}(z)| = 1$, finally we arrive at the conclusion

$$|\mathcal{H}(z)| = |\mathcal{H}_{\text{min}}(z)|$$

Consequently, the functions $\mathcal{H}(z)$ and $\mathcal{H}_{\text{min}}(z)$ will be characterized *by the same amplitude response*. However, we should not ignore the fact that an all-pass filter, though characterized by unit gain, has a positive (i.e., nonzero) phase response and, therefore, causes a *phase shift* from 0° to − 180°. We end up, therefore, with the very interesting conclusion that *a minimum phase system has the same amplitude response as a nonminimum phase system and the smallest possible phase value*, from where it also derives its characterization as a minimum phase system.

According to the foregoing analysis, *for a given transfer function that is not of minimum phase, we can construct a corresponding minimum phase function by determining the zero values of the original function that belong to the right half-plane and associating with each of these zero values an all-pass transfer function, adding a zero in the left half-plane and in a mirror position with respect to the imaginary axis*. In the foregoing proof, we considered a single zero value located outside the unit circle, but the generalization of this process to include an arbitrary number of zero values of this type is straightforward. If we differentiate the relation $\measuredangle\mathcal{H}(z) = \measuredangle\mathcal{H}_{\text{min}}(z) + \measuredangle\mathcal{H}_{\text{ap}}(z)$, we find that a similar property characterizes the magnitude of group delay (since this is the negative derivative of the phase response), and for this reason, minimum phase filters are also known as minimum group delay systems.

Example 9.4.1 To demonstrate this process, let us consider the causal and stable system (Oppenheim, Schafer, Buck, Example 5.14)

$$\mathcal{H}_1(z) = \frac{1 + 3z^{-1}}{1 + \frac{1}{2}z^{-1}}$$

whose transfer function has a pole inside the unit circle at the position $z = -1/2$ and a zero value outside the unit circle at the position $z = -3$, which, according to our earlier discussion, corresponds to the value $c = 1/z^* = -1/3$. Therefore, the transfer function of the all-pass filter is

$$\mathcal{H}_{\text{ap}}(z) = \frac{z^{-1} + \frac{1}{3}}{1 + \frac{1}{3}z^{-1}}$$

while the minimum phase transfer function will be given by the expression

$$\mathcal{H}_{\text{min}}(z) = 3\frac{1 + \frac{1}{3}z^{-1}}{1 + \frac{1}{2}z^{-1}}$$

Combining the foregoing results we have

$$\mathcal{H}_1(z) = \left(3\frac{1 + \frac{1}{3}z^{-1}}{1 + \frac{1}{2}z^{-1}}\right)\left(\frac{z^{-1} + \frac{1}{3}}{1 + \frac{1}{3}z^{-1}}\right)$$

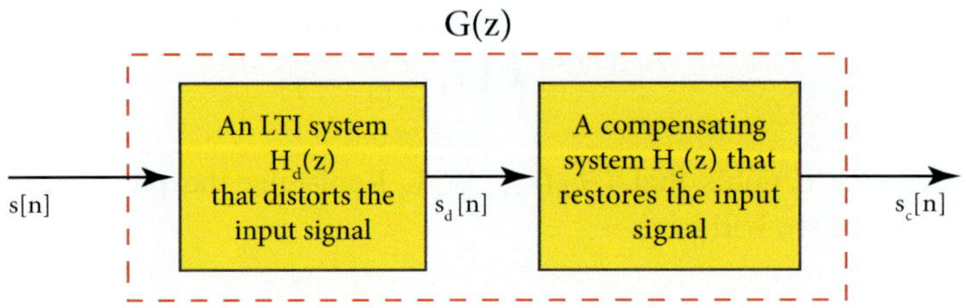

Fig. 9.4 A signal distortion compensation system

An interesting application of minimum phase systems is their use as *compensation systems* in cases where a distortion of an input signal $s[n]$ from a LTI system whose frequency response does not have the desired characteristics, as can sometimes happen *during the passage of a signal through a communication channel*. This situation is depicted in Fig. 9.4. If the addition of this compensation system, which is connected in series to the LTI system so as to receive its output $s_d[n]$ in order to restore it, leads to *perfect compensation*, then the output of the system $s_c[n]$ will coincide with the original input signal $s[n]$. In this case, it is obvious that *the combination of these two systems is the identity system* (since it receives as input the signal $s[n]$ and produces as its output the same signal), and consequently, the frequency response of the system $\mathcal{H}_c(z)$ will be the *inverse* of the frequency response of the system $\mathcal{H}_d(z)$. It is not difficult to understand that if the LTI system that distorts the signal is *stable and causal*, while this requirement must also be satisfied by the compensation system, then according to our earlier discussion, the system described by the transfer function $\mathcal{H}_d(z)$ will be a *minimum phase system*. On the other hand, if this is not the case, then we can convert this minimum phase system in the way mentioned previously, that is, *by mapping all the zeros that are outside the unit circle of the complex plane to the inverse conjugate complex number that is inside the unit circle*, thereby constructing the function $\mathcal{H}_{d_{\min}}(z)$. As mentioned previously, the functions $\mathcal{H}_d(z)$ and $\mathcal{H}_{d_{\min}}(z)$ are characterized by *the same frequency response* and will be related to each other through the transfer function of an all-pass system as

$$\mathcal{H}_d(z) = \mathcal{H}_{d_{\min}}(z)\mathcal{H}_{\text{ap}}(z)$$

If we choose as the frequency response for the compensation system the function

$$\mathcal{H}_c(z) = \frac{1}{\mathcal{H}_{d_{\min}}(z)}$$

we find that the transfer function of the composite system that maps the signal $s[n]$ to the signal $s_c[n]$ will be

$$\mathcal{G}(z) = \mathcal{H}_d(z)\mathcal{H}_c(z) = \mathcal{H}_{\text{ap}}(z)$$

and therefore, the function $\mathcal{G}(z)$ will correspond to an *all-pass system*, as required (all we want is *the restoration of distortion, without cutting off any frequency*, and therefore, the system should allow the passage of all frequencies). The measure of the frequency response is fully compensated, while the phase response is modified by the factor $\measuredangle H_{\text{ap}}(e^{j\omega})$.

Let us complete our reference to minimum phase systems by presenting three significant and interesting properties that characterize them.

- *The property of minimum phase delay* . From the defining equation of the transfer function of any system, as the product of the transfer function of a minimum phase system and the transfer function of an all-pass system, it immediately follows that

$$\arg[\mathcal{H}(e^{j\omega})] = \arg[\mathcal{H}_{min}(e^{j\omega})] + \arg[\mathcal{H}_{ap}(e^{j\omega})]$$

  Therefore, the continuous phase that corresponds to the phase of the primary value (see the note at the end of this section) is *the sum of the unwrapped phase (unwrapped phase) associated with the minimum phase function and the unwrapped phase of the all-pass system.* Taking into account that the unwrapped phase of an all-pass system is negative for the range of values $0 \leq \omega \leq \pi$, we are led to the conclusion that the mapping of the zero values of the function $\mathcal{H}_{\min}(z)$ from the interior of the unit circle to the conjugate reciprocal positions located outside the unit circle in the complex plane leads to a reduction of the nonfolded phase or an increase in the negative value of the phase, which is understood as phase delay.

We therefore conclude that a causal and stable system with an amplitude response $\mathcal{H}_{\min}(z)$ whose poles and zero values are all located inside the unit circle has the minimum phase delay compared to all systems characterized by the same amplitude response.

- *The property of minimum group delay*. From the defining equation of the transfer function of any system, as the product of the transfer function of a minimum phase system and the transfer function of an all-pass system, it immediately follows that

$$grd[\mathcal{H}(e^{j\omega})] = grd[\mathcal{H}_{\min}(e^{j\omega})] + grd[\mathcal{H}_{ap}(e^{j\omega})]$$

This equation tells us that the group delay (grd) of a minimum phase system *is always less than the group delay of a nonminimum phase system* since the all-pass system that transforms the nonminimum phase system into a minimum phase system has a positive group delay. We can reach the same conclusion with the help of the equations of Exercise 10.4.8 (page 845), in which expressions are constructed for the magnitude and phase of the frequency response as well as for the group delay, depending on the magnitude and phase of the poles and zero values with which it is associated. Considering a zero value of the form $z = re^{j\vartheta}$ that is within the unit circle, its contribution to the phase response and group delay is given respectively by the expressions

$$-\tan^{-1}\frac{\sin(\omega-\vartheta)}{(1/r)-\cos(\omega-\vartheta)} \quad \text{and} \quad \frac{r-\cos(\omega-\vartheta)}{[r+(1/r)]-2\cos(\omega-\vartheta)}$$

as can be easily verified. If in the foregoing expressions we replace the zero value z with the value $1/z^* = (1/r)e^{j\vartheta}$, which is outside the unit circle, this will lead to an *increase* in group delay. Consequently, *a minimum phase system is indeed characterized by the minimum group delay among all systems that have the same amplitude response*. Note that despite the fact that this property of the minimum refers to the group delay but also to the phase delay, in the literature, these systems are characterized as minimum phase systems.

- *The property of minimum energy delay*. Studying the problem of minimum phase, we find that the minimum phase sequence seems to have higher sample values at its left end in relation to all the other sequences. In this case, it is proven that

$$|h[0]| \le |h_{\min}[0]|$$

for every causally stable sequence $h[n]$, for which it holds that

$$|\mathcal{H}(e^{j\omega})| = |\mathcal{H}_{\min}(e^{j\omega})|$$

Resorting to the *theorem of Parseval* we can prove that *all impulse responses whose amplitude response is equal to* $|\mathcal{H}_{\min}(e^{j\omega})|$ *have the same total energy as the impulse response* $h_{\min}[0]$ since, as we can ascertain, it holds that

$$\sum_{n=0}^{\infty}|h[n]|^2 = \frac{1}{2\pi}\int_{-\pi}^{\pi}|\mathcal{H}(e^{j\omega})|^2 d\omega = \frac{1}{2\pi}\int_{-\pi}^{\pi}|\mathcal{H}_{min}(e^{j\omega})|^2 d\omega = \sum_{n=0}^{\infty}|h_{min}[n]|^2$$

If we define the partial energy of the impulse response as

$$E[n] = \sum_{m=0}^{n}|h[m]|^2$$

then, as Oppenheim and Schafer note, it is proven that

$$\sum_{m=0}^{n}|h[m]|^2 \le \sum_{m=0}^{n}|h_{min}[m]|^2$$

for all impulse responses $h[n]$ of the family of systems for which the following property holds:

$$|\mathcal{H}(e^{j\omega})| = |\mathcal{H}_{min}(e^{j\omega})|$$

The foregoing relation tells us that *the partial energy of a minimum phase system is more concentrated around the value* $n = 0$ *or, in an equivalent formulation, that the energy of the minimum phase system is characterized by the smallest possible delay compared to all systems characterized by the same amplitude response*. For this reason, minimum phase delay systems are also characterized as *systems of minimum energy delay* or simply as *minimum delay systems*.

Note Based on complex analysis, the exponential form of a complex number $z = a + ib$ is

$$z = re^{i\varphi} = r(\cos\varphi + i\sin\varphi)$$

(the last expression is known as the *polar form*), where $r = \sqrt{a^2 + b^2}$ is the modulus of the complex number and $\varphi = \tan^{-1}(b/a)$ its phase. Observing that $re^{i\varphi} = re^{i(\varphi+2k\pi)}$, it is not difficult to see that *the phase of a complex number is not uniquely defined* since any angle of the form $\varphi_k = \varphi + 2k\pi$ determines the same point in the complex plane. Conversely, the value of the angle φ is *unique* only within the interval $0 \leq \varphi \leq 2\pi$, although in general we can use any other interval of width 2π, such as $-\pi \leq \varphi \leq \pi$. To remove this *indeterminacy* regarding which phase value we will use, we choose in advance an interval of width 2π (such as, for example, the interval $0 \leq \varphi \leq 2\pi$), and we characterize it as a *primary interval*, while the value of the angle φ corresponding to this interval is known as the *primary value*.

However, in practice, the primary value is defined in the range of values $-\pi \leq \varphi \leq \pi$, and the reason for this is that the function of the inverse tangent used to calculate the angle φ—remember that $\varphi = \tan^{-1}(b/a)$—always returns a value in the interval $-\pi \leq \varphi \leq \pi$. This exact value is characterized as a *primary phase value* and is symbolized by $\arg(z)$. Having calculated this angle value, any other phase for the considered complex quantity can be calculated, as previously noted, as $\varphi_k = \arg(z) + 2k\pi$ $(k \in \mathbb{Z})$. In the special case of the frequency response $\mathcal{H}(e^{j\omega})$ of a discrete-time LTI system, which as we know is a complex function with magnitude $|\mathcal{H}(e^{j\omega})|$ and primary phase $-\pi \leq \arg(\mathcal{H}(e^{j\omega})) \leq \pi$, we can replace the variable k with a function $k(\omega) \in \mathbb{Z}$ and determine any phase value of the considered function from the equation

$$\varphi_{\mathcal{H}}(e^{j\omega}) = \measuredangle\mathcal{H}(e^{j\omega}) = \arg(\mathcal{H}(e^{j\omega})) + 2\pi k(\omega)$$

This process of calculating the continuous phase function from the primary phase is known as *phase unwrapping* (phase unwrapping) and the nonunwrapping (or wrapped) phases referred to here specifically concern phase functions calculated in the aforementioned manner (the term "wrapping" arises from the fact that the angle $\varphi_k = \varphi + 2k\pi$ is equal to the angle φ, in the sense that if we start from the point of the complex plane determined by the angle φ and travel a distance that is an integer multiple of the circumference of the circle—recall that we do something like this when we wrap a thread around a spool—we return to the point from which we started). This process of "unwrapping" the phase response is often necessary when we use the magnitude of the *phase delay* to ensure that *in the phase response of the system, all appropriate integer multiples of* 2π *have been taken into account*, since if these multiples are ignored, the knowledge of the primary phase is not sufficient for the calculation of a phase response that can be used for the correct calculation of time delays. Indeed, if we ignore these multiples, *the phase delay is altered by multiples of the period of the sinusoidal signal*, and since the analysis of LTI systems relies on the use of sinusoids that have no beginning or end, *it is impossible, by studying the output of the system for a specific frequency value, to perceive whether a phase delay is a real phase delay or a delay due to a time-delayed sinusoidal signal*.

9.5 Interconnection of Systems/Inverse Systems

The theory of interconnection of discrete LTI systems was described in Sect. 7.7 (page 605) and is summarized in Exercise 9.5.1 (page 716). As for the so-called *inverse systems* (see Sect. 7.8 on page 606), these, in full analogy with their continuous counterparts, satisfy the condition $\mathcal{H}_1(z)\mathcal{H}_2(z) = 1$, where $\mathcal{H}_1(z)$ and $\mathcal{H}_2(z)$ are the *transfer functions* that describe them. Taking into account that the product $\mathcal{G}(z) = \mathcal{H}_1(z)\mathcal{H}_2(z)$ refers to the *transfer function* of the system that arises *from the connection of the systems* $\mathcal{H}_1(z)$ *and* $\mathcal{H}_2(z)$ *in series*, it is obvious that if we connect two inverse systems in the aforementioned manner, we will construct the *identity system*, whose output $y[n]$ at each time instant n coincides with its input $x[n]$. Indeed, if we take the inverse $\mathcal{Z}$ transform of the relation $\mathcal{G}(z) = \mathcal{H}_1(z)\mathcal{H}_2(z) = 1$ and use *the property of convolution* and the relation $\mathcal{Z}^{-1}(1) = \delta[n]$, we will get $h[n] = h_1[n] * h_2[n] = \delta[n]$, where $h_1[n] = \mathcal{Z}^{-1}\{\mathcal{H}_1(z)\}$ and $h_2[n] = \mathcal{Z}^{-1}\{\mathcal{H}_2(z)\}$ are the impulse responses of the two systems and $h[n] = \mathcal{Z}^{-1}\{\mathcal{G}(z)\}$ *the impulse response of the composite system*. The response of this composite system to the input signal $x[n]$ is given by the relation

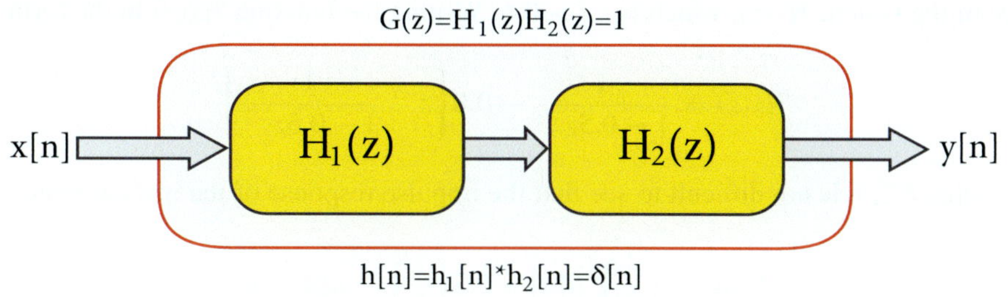

Fig. 9.5 The identity discrete-time system resulting from the connection in series of two inverse discrete-time systems

$$y[n] = x[n] * h[n] = x[n] * \delta[n] = \sum_{n=-\infty}^{\infty} x[k]\delta[n-k] = x[n] \tag{9.19}$$

that is, the system $\mathcal{G}(z)$ is indeed the *identity system*. The connection of two inverse systems in series and the equivalent identity system that results from this are presented in Fig. 9.5.

Generally speaking, the existence of the inverse of a system *is not given* and there are many examples of discrete-time systems whose inverse system *does not exist*. A particularly interesting category of systems that have an inverse system are those whose transfer function is a rational function of the form

$$\mathcal{H}_1(z) = \left\{\sum_{k=0}^{M} \beta_k z^{-k}\right\} \Big/ \left\{\sum_{k=0}^{N} \alpha_k z^{-k}\right\} = \frac{\beta_0}{\alpha_0}\left\{\prod_{k=1}^{M}(z-z_k)\right\} \Big/ \left\{\prod_{k=1}^{N}(z-p_k)\right\} z^{N-M} \tag{9.20}$$

It is not difficult to see that the transfer function of the inverse system is

$$\mathcal{H}_2(z) = \frac{1}{\mathcal{H}_1(z)} = \left\{\sum_{k=0}^{N} \alpha_k z^{-k}\right\} \Big/ \left\{\sum_{k=0}^{M} \beta_k z^{-k}\right\} = \frac{\alpha_0}{\beta_0}\left\{\prod_{k=1}^{N}(z-p_k)\right\} \Big/ \left\{\prod_{k=1}^{M}(z-z_k)\right\} z^{M-N} \tag{9.21}$$

that is, *the poles and zeros of the function $\mathcal{H}_1(z)$ become zeros and poles of the function $\mathcal{H}_2(z)$, respectively*. As for the *region of convergence* of the inverse system, this will result from the fact that the regions of convergence of the two systems must *overlap*, as required by the *convolution property*, which correlates the impulse responses of the two systems.

Example 9.5.1 Let us consider the LTI system with transfer function (Hayes, Example 2.5.1)

$$\mathcal{H}_1(z) = \frac{1-0.5z^{-1}}{1-0.8z^{-1}}, \quad |z| > 0.8$$

In this case, the inverse system will be defined as

$$\mathcal{H}_2(z) = \frac{1}{\mathcal{H}_1(z)} = \frac{1-0.8z^{-1}}{1-0.5z^{-1}}$$

If $h_2[n]$ is the impulse response of the system $\mathcal{H}_2(z)$, there are *two candidate regions of convergence* for the function $\mathcal{H}_2(z) = \mathcal{Z}\{h_2[n]\}$, of which the first is $|z| > 1/2$ and the second is $|z| < 1/2$. Of these regions of convergence, the one that will be associated with the function $\mathcal{H}_2(z)$ is $|z| > 1/2$ since the other region of convergence *does not overlap* with the region of convergence of the system $\mathcal{H}_1(z)$, which is $|z| > 0.8$. Writing the function $\mathcal{H}_2(z)$ in the form

$$\mathcal{H}_2(z) = \frac{1}{1-0.5z^{-1}} - 0.8\left\{z^{-1}\frac{1}{1-0.5z^{-1}}\right\}$$

and referring back to Table 8.2, it is not difficult to see that the impulse response of the inverse system will be given by the relation

$$h_2[n] = \left(\frac{1}{2}\right)^n u[n] - 0.8\left(\frac{1}{2}\right)^{n-1} u[n-1]$$

where for the derivation of the second term we used the property of time shifting. ■

In the foregoing example, the function $\mathcal{H}_1(z)$ has a zero value at the position $z_1 = 0.5$, which is located *within the region of convergence of the system* and constitutes the *pole of the inverse system* $\mathcal{H}_2(z)$. In the general case where the function $\mathcal{H}_1(z)$ has N zero values at the positions z_k $(k = 1, 2, \ldots, N)$, it is proven that the inverse system will be causal if the region of convergence that will be associated with the function $\mathcal{H}_2(z)$ is described by the equation $|z| > \max_{(k)} |z_k|$. If we want the inverse system to be not only causal but also stable, its region of convergence should contain the unit circle, and therefore, the relation $\max_{(k)} |z_k| < 1$ $(k = 1, 2, \ldots, N)$ should hold.

From the preceding discussion follows the conclusion that *a discrete LTI system is causal and stable and has an inverse system that is also causal and stable if and only if the zeros and poles of the function* $\mathcal{H}(z)$ *are within the unit circle*. These types of systems are known as *minimum phase systems*.

Problem 9.5.1 Considering two discrete LTI systems with impulse responses $h_1[n]$ and $h_2[n]$, determine the transfer function of the system that will result from the interconnection of the foregoing systems (a) in series and (b) in parallel.

Solution Let $\mathcal{H}_1(z)$ and $\mathcal{H}_2(z)$ be the transfer functions of the two systems, which are calculated as

$$\mathcal{H}_1(z) = \mathcal{Z}\{h_1[n]\} = \sum_{n=-\infty}^{\infty} h_1[n]z^{-n} \quad \text{and} \quad \mathcal{H}_2(z) = \mathcal{Z}\{h_2[n]\} = \sum_{n=-\infty}^{\infty} h_2[n]z^{-n}$$

In a previous chapter, we proved that the convolution of the system resulting from the connection of the systems (a) in series and (b) in parallel will be given by the relations $h[n] = h_1[n] * h_2[n]$ and $h[n] = h_1[n] + h_2[n]$, respectively. Therefore, the transfer function of the system resulting from the connection of the two systems in series will be

$$\mathcal{H}(z) = \mathcal{Z}\{h[n]\} = \mathcal{Z}\{h_1[n] * h_2[n]\} = \mathcal{H}_1(z)\mathcal{H}_2(z)$$

—as derived from the convolution theorem of the $\mathcal{Z}$ transform—while for the parallel connection

$$\mathcal{H}(z) = \sum_{n=-\infty}^{\infty} h[n]z^{-n} = \sum_{n=-\infty}^{\infty} h_1[n]z^{-n} + \sum_{n=-\infty}^{\infty} h_2[n]z^{-n} = \mathcal{H}_1(z) + \mathcal{H}_2(z)$$

thereby leading to the same conclusions as those we had derived for continuous-time systems.

9.6 The Output of LTI Systems

Let us consider a *discrete LTI system* described by the general difference equation with constant coefficients

$$y[n] = -\sum_{k=1}^{N} \alpha_k y[n-k] + \sum_{k=0}^{M} \beta_k x[n-k] \tag{9.22}$$

which, as is well known, leads to a transfer function of the form

$$\mathcal{H}(z) = \left\{ \sum_{k=0}^{M} \beta_k z^{-k} \right\} \Bigg/ \left\{ 1 + \sum_{k=1}^{N} \alpha_k z^{-k} \right\} \tag{9.23}$$

As mentioned previously, this function can be formulated as the *ratio of two polynomials* $\mathcal{B}(z)/\mathcal{A}(z)$, where $\mathcal{B}(z)$ is the numerator polynomial containing the *zero values* of the system function, while $\mathcal{A}(z)$ is the denominator polynomial containing the *poles* of this function. Let us now assume that in the preceding system we transmit a discrete input signal $x[n]$, whose $\mathcal{Z}$ transform can also be written as a ratio of two polynomials of the form $\mathcal{N}(z)/\mathcal{Q}(z)$.

Assuming that initially the system is in a *state of rest* and therefore the initial conditions of the problem $y[-1], y[-2], \ldots, y[-N]$ will have zero values, the $\mathcal{Z}$ transform of the system's output can be written in the form

$$\mathcal{Y}(z) = \mathcal{H}(z)\mathcal{X}(z) = \frac{\mathcal{B}(z)\mathcal{N}(z)}{\mathcal{A}(z)\mathcal{Q}(z)} \tag{9.24}$$

as emerges from the *property of convolution* of the aforementioned transform.[8] The preceding relation holds under the condition that the relation $p_k \neq q_m$ applies to every $k = 1, 2, \ldots, N$ and $m = 1, 2, \ldots, L$, where $p_1, p_2, \ldots, p_N$ are the poles of the system[9] and $q_1, q_2, \ldots, q_L$ are the poles of the $\mathcal{Z}$ transform of the input signal, while we also assume that the zero values of the polynomials of the numerator $\mathcal{B}(z)$ and $\mathcal{N}(z)$ do not coincide with the poles $\{p_k\}$ and $\{q_k\}$ so that a *mutual cancellation of poles and zero values* takes place. Developing the function $\mathcal{Y}(z)$ in a series of partial fractions, it is formulated as

$$\mathcal{Y}(z) = \sum_{k=1}^{N} \frac{A_k}{1 - p_k z^{-1}} + \sum_{k=1}^{L} \frac{Q_k}{1 - q_k z^{-1}} \tag{9.25}$$

Referring to Table 8.2, we find that

$$\mathcal{Z}^{-1}\left\{\frac{1}{1 - p_k z^{-1}}\right\} = (p_k)^n u[n] \qquad \text{and} \qquad \mathcal{Z}^{-1}\left\{\frac{1}{1 - q_k z^{-1}}\right\} = (q_k)^n u[n]$$

that is, the output of the system will be calculated from an expression of the form

$$y[n] = \sum_{k=1}^{N} A_k (p_k)^n u[n] + \sum_{k=1}^{L} Q_k (q_k)^n u[n] \tag{9.26}$$

[8] In reality, the transform $\mathcal{Y}(z)$ defined by the foregoing relation refers to the *zero-state response* $y[n] = h[n] * x[n]$, the calculation of which takes place under zero initial conditions.

[9] For the effect of the nature of the poles on the mathematical form of the system's output, read also Sect. 5.3 (page 385), which refers to continuous-time systems, since the situation in general terms is more or less the same. For example, in the case of discrete poles, the output $y[n]$ has the form described earlier, while if a pole p of the function is of complex form, then the conjugate pole p^* will necessarily appear since we assume that the coefficients of the difference equation are real numbers. Following the same argumentation with continuous-time systems, it is proven that the contribution of the pair of complex conjugate poles (p, p^*) to the sum of partial fractions will be the expression

$$\frac{A}{1 - pz^{-1}} + \frac{A^*}{1 - p^* z^{-1}}$$

while the inverse $\mathcal{Z}$ transform corresponding to this expression has the form

$$\mathcal{Z}^{-1}\left\{\frac{A}{1 - pz^{-1}} + \frac{A^*}{1 - p^* z^{-1}}\right\} = 2|A||p|^n \cos(n\measuredangle p + \measuredangle A)u[n]$$

where we used the notation $A = |A|e^{j\measuredangle A}$ and $p = |p|e^{j\measuredangle p}$.

From the foregoing analysis, it is concluded that the output of the system's $y[n]$ can be written *as a sum of two terms.* Of these terms, the first is a function of the system's poles $\{p_k\}$ $(k = 1, 2, \ldots, N)$ and is called the *zero-input response* or *natural response* of the system, while the second is a function of the poles $\{q_k\}$ $(k = 1, 2, \ldots, L)$ of the $\mathcal{Z}$ transform of the input signal and is called the *zero-state response* or *forced response* of the system. The effect of these two terms on the output $y[n]$ is realized through the coefficients A_k and Q_k, which are functions of both the poles $\{p_k\}$ and the poles $\{q_k\}$. It is not difficult to see that the results we obtained are the same as those arrived at by studying the system in the discrete-time domain.

Now suppose that the initial conditions of the system's difference equation $y[-1], y[-2], \ldots, y[-N]$ *are not zero.* In this case, the natural response of the system *changes*. To prove the truth of this assertion, let us assume that at time $n = 0$, we transmit a causal signal $x[n]$ to the system. In this case, to calculate the system's output, we must take the *unilateral* $\mathcal{Z}^+$ *transform* of the difference equation that describes it. Using the property of *time shift* in the case of a single transform, we find that

$$\mathcal{Y}^+(z) = -\sum_{k=1}^{N} \alpha_k z^{-k}\left[\mathcal{Y}^+(z) + \sum_{n=1}^{k} y[-n]z^n\right] + \sum_{k=0}^{M} \beta_k z^{-k}\mathcal{X}^+(z) \tag{9.27}$$

However, since the input signal is causal (and therefore satisfies the condition $x[n] = 0$ for $n < 0$), we can make the substitution $\mathcal{X}^+(z) \to \mathcal{X}(z)$. In this case, the transform $\mathcal{Y}^+(z)$ is written

$$\mathcal{Y}^+(z) = \left\{\sum_{k=0}^{M} \beta_k z^{-k}\right\} \Big/ \left\{1 + \sum_{k=1}^{N} \alpha_k z^{-k}\right\} \mathcal{X}(z) - \left\{\sum_{k=1}^{N} \alpha_k z^{-k}\left(\sum_{n=1}^{k} y[-n]z^n\right)\right\} \Big/ \left\{1 + \sum_{k=1}^{N} \alpha_k z^{-k}\right\}$$

or, equivalently,

$$\mathcal{Y}^+(z) = \mathcal{H}(z)\mathcal{X}(z) + \frac{\mathcal{D}(z)}{\mathcal{A}(z)}$$

where

$$\mathcal{D}(z) = -\sum_{k=1}^{N}\left(\alpha_k z^{-k}\sum_{n=1}^{k} y[-n]z^n\right) \tag{9.28}$$

We are led, therefore, to the conclusion that in this case the $\mathcal{Z}$ transform of the system's output can be written as a *sum of two terms*, the first of which constitutes the *transform of the system's response in the zero state* and is formulated as

$$\mathcal{Y}_{zs}(z) = \mathcal{H}(z)\mathcal{X}(z) \tag{9.29}$$

while the second constitutes the *transform of the system's response to zero input* due to the existence of *nonzero initial conditions* and is described by the equation

$$\mathcal{Y}_{zi}^+(z) = \frac{\mathcal{D}(z)}{\mathcal{A}(z)} \tag{9.30}$$

Therefore, the calculation of the total response of the system expressed in the time domain requires the determination of the inverse transform of the foregoing expressions and the addition of the representations that will arise, with the result of this process being given by the equation

$$y[n] = y_{zs}[n] + y_{zi}[n] \tag{9.31}$$

Since the denominator of the transform $\mathcal{Y}_{zi}^{+}$ is the polynomial $\mathcal{A}(z)$, it follows that the poles of this transform will be $p_1, p_2, \ldots, p_N$. Therefore, the zero-input response will have the form

$$\mathcal{Y}_{zi}^{+}(z) = \sum_{k=1}^{N} \mathcal{C}_k (p_k)^n u[n] \tag{9.32}$$

while the total output of the system will be described by the relation

$$y[n] = \sum_{k=1}^{N} \mathcal{P}_k (p_k)^n u[n] + \sum_{k=1}^{L} \mathcal{Q}_k (q_k)^n u[n] \tag{9.33}$$

where $\mathcal{P}_k = \mathcal{A}_k + \mathcal{C}_k$. We observe, therefore, that the result of the existence of nonzero initial conditions in our system is *the change of the natural response of the system*, while, in contrast, the forced response of the system remains unchanged.[10]

A final issue worth commenting on is related to the *measure* of the poles $\{p_k\}$ of a discrete-time system. If this measure satisfies the inequality $|p_k| < 1$, then *the natural response of the system tends to zero as time n tends to infinity*. In this case, the natural response of the system is characterized as a *transient or temporary response*. As for the rate of this reduction, it *depends on the position of the poles* $\{p_k\}$ *on the complex plane*. If all the poles are *very close to the origin of the complex plane* (which corresponds to the point $z = 0$), the signal attenuation rate is *very large*. Conversely, if the poles are *close to the circumference of the unit circle*, the attenuation of the natural response is *much slower*, and this transient term lasts for a much longer period of time.

On the other hand, the *forced response* of a system is a function of the coefficients $\mathcal{Q}_k$ and the poles $\{q_k\}$ of the *input signal*. If all these poles are within the unit circle, the forced response *will also tend to zero*, like the natural response. This fact is expected, as, in this case, the input signal is a *transient signal* whose width *decreases* in relation to time. Conversely, if the input signal is a *sine wave*, the poles of the signal are *on the circumference of the unit circle*, and the forced response is also a *sine wave* defined for time values $n \geq 0$. In this case, the system's response is known by the name *steady-state response*.

Problem 9.6.1 Determine the step response of a recursive discrete LTI system described by the difference equation

$$y[n] = 0.9y[n-1] - 0.81y[n-2] + x[n]$$

for initial conditions (a) $y[-1] = y[-2] = 0$ and (b) $y[-1] = y[-2] = 1$ (Proakis and Manolakis, Example 3.6.1).

Solution

(a) If we compare the system's difference equation

$$y[n] = 0.9y[n-1] - 0.81y[n-2] + x[n]$$

with the general recursive equation that has the form

[10] This result is expected: The initial conditions constitute *a characteristic* of the system *and not of the corresponding input signal*. Therefore, the only component that is influenced by these will be one that exists independently of the existence of an input signal, that is, the *natural response*.

$$y[n] = -\sum_{k=1}^{N} \alpha_k y[n-k] + \sum_{k=0}^{M} \beta_k x[n-k]$$

it is not difficult to ascertain that the discrete-time system in question corresponds to parameter values $N = 2$ and $M = 0$, while the system's coefficients will be $\alpha_1 = -0.9$, $\alpha_2 = 0.81$, and $\beta_0 = 1$. Therefore, the system's transfer function is calculated as

$$\mathcal{H}(z) = \left\{\sum_{k=0}^{0} \beta_k z^{-k}\right\} \Big/ \left\{1 + \sum_{k=1}^{2} \alpha_k z^{-k}\right\} = \frac{\beta_0}{1 + \alpha_1 z^{-1} + \alpha_2 z^{-2}} = \frac{1}{1 - 0.9z^{-1} + 0.81z^{-2}}$$

Setting for simplicity's sake $z^{-1} = y$, the poles of the transfer function will emerge as the roots of the quadratic equation $0.81y^2 - 0.9y + 1 = 0$, which has a discriminant with root $\sqrt{\Delta} = 0.9j\sqrt{3}$ and roots

$$y_{1,2} = \frac{0.9 \pm 0.9j\sqrt{3}}{2 \cdot (0.81)} = \frac{1}{2 \cdot (0.9)}(1 \pm j\sqrt{3}) = \frac{1}{2 \cdot (0.9)}\left[2e^{\pm j\pi/3}\right] = \frac{1}{0.9}e^{\pm j\pi/3}$$

Therefore, $z_{1,2} = 0.9e^{\mp j\pi/3}$, meaning that the transfer function will be characterized by the existence of two simple poles at the positions $p_1 = 0.9e^{j\pi/3}$ and $p_2 = 0.9e^{-j\pi/3}$, so it can be written in the form

$$\mathcal{H}(z) = \frac{1}{(1 - 0.9e^{j\pi/3}z^{-1})(1 - 0.9e^{-j\pi/3}z^{-1})}$$

On the other hand, referring to tables of elementary transforms, we find that the $\mathcal{Z}$ transform of the input signal, which is the step function $x[n] = u[n]$, has the simple form

$$\mathcal{X}(z) = \frac{1}{1 - z^{-1}}$$

and therefore, according to the convolution theorem, the $\mathcal{Z}$ transform of the output has the form

$$\begin{aligned} Y(z) = H(z)X(z) &= \frac{1}{(1 - z^{-1})(1 - 0.9\exp(j\pi/3)z^{-1})(1 - 0.9\exp(-j\pi/3)z^{-1})} \\ &= \frac{0.542 - 0.049j}{1 - 0.9\exp(j\pi/3)z^{-1}} + \frac{0.542 + 0.049j}{1 - 0.9\exp(-j\pi/3)z^{-1}} + \frac{1.099}{1 - z^{-1}} \end{aligned}$$

Developing the function $\mathcal{Y}(z)$ into a sum of partial fractions, it turns out that[11]

$$\begin{aligned} \mathcal{Y}(z) &= \frac{A}{1 - 0.9e^{j\pi/3}z^{-1}} + \frac{B}{1 - 0.9e^{-j\pi/3}z^{-1}} + \frac{C}{1 - z^{-1}} \\ &= \frac{(0.9e^{-j\pi/3}A + 0.9e^{j\pi/3}B + 0.81C)z^{-2} - [(1 + 0.9e^{-j\pi/3})A + (1 + 0.9e^{j\pi/3})B + 0.9C]z^{-1} + (A+B+C)}{(1 - z^{-1})(1 - 0.9\exp(j\pi/3)z^{-1})(1 - 0.9\exp(-j\pi/3)z^{-1})} \end{aligned}$$

Equating the coefficients of the corresponding powers of z in the last two equations, we end up with the system of the three equations

$$\begin{aligned} &0.9e^{-j\pi/3}A + 0.9e^{j\pi/3}B + 0.81C = 0 \\ &[(1 + 0.9e^{-j\pi/3})A + (1 + 0.9e^{j\pi/3})B + 0.9C] = 0 \\ &A + B + C = 1 \end{aligned}$$

[11] In many cases, because of space constraints, only the final results are presented, with their analytical derivation being left as an exercise to the reader.

with a solution of the form

$$A = -0.0495 - 0.5425j, \quad B = -0.0495 + 0.5425j, \quad C = 1.0989$$

and therefore the $\mathcal{Z}$ transform of the system's output will be given by the equation

$$\mathcal{Y}(z) = \frac{-0.0495 - 0.5425j}{1 - 0.9e^{j\pi/3}z^{-1}} + \frac{-0.0495 + 0.5425j}{1 - 0.9e^{-j\pi/3}z^{-1}} + \frac{1.0989}{1 - z^{-1}}$$

Referring back to Table 8.2, we immediately find that

$$\mathcal{Z}^{-1}\left\{\frac{1.0989}{1 - z^{-1}}\right\} = 1.0989u[n]$$

On the other hand, the sum of the remaining two terms is of the form

$$\frac{A}{1 - pz^{-1}} + \frac{A^*}{1 - p^*z^{-1}}$$

for the values $A = -0.0495 - 0.5425j = 0.544753e^{j87.786526^\circ}$ and $p = 0.9e^{j\pi/3}$ and, according to our discussion in Footnote 9 (page 717), the inverse transform of this will be

$$2|A||p|^n \cos(n\measuredangle p + \measuredangle A)u[n] = 1.089(0.9)^n \cos\left(\frac{\pi n}{3} + 87.786526^\circ\right)u[n]$$

Combining the foregoing expressions, the output of the system $y[n]$ takes the form

$$y[n] = \left[1.0989 + 1.089(0.9)^n \cos\left(\frac{\pi n}{3} + 87.786526^\circ\right)\right]u[n]$$

(b) In this case, the existence of nonzero initial conditions $y[-1] = y[-2] = 1$ causes the appearance of the additional term $\mathcal{D}(z)/\mathcal{A}(z)$ in the equation of the unilateral $\mathcal{Z}^+$ transform of the system output $y[n]$, with the polynomial $\mathcal{D}(z)$ calculated as

$$\mathcal{D}(z) = -\sum_{k=1}^{2} \alpha_k z^{-k}\left(\sum_{n=1}^{k} y[-n]z^n\right) = -\alpha_1 z^{-1}(y[-1]z) - \alpha_2 z^{-2}(y[-1]z + y[-2]z^2)$$

$$= -\alpha_1 y[-1] - \alpha_2 y[-2]z^{-1} - \alpha_2 y[-2] = 0.9 - 0.81z^{-1} - 0.81 = 0.09 - 0.81z^{-1}$$

Expanding the expression of the additional term into partial fractions, we get

$$\frac{\mathcal{D}(z)}{\mathcal{A}(z)} = \frac{0.09 - 0.81z^{-1}}{1 - 0.9z^{-1} + 0.81z^{-2}} = \frac{(A + B) - (0.9e^{-j\pi/3}A + 0.9e^{j\pi/3}B)z^{-1}}{1 - 0.9z^{-1} + 0.81z^{-2}}$$

Equating the coefficients of the corresponding powers of z, we arrive at the system

$$e^{-j\pi/3}A + e^{j\pi/3}B = 0.9 \qquad \text{and} \qquad A + B = 0.09$$

whose solution has the form

$$A = 0.045 + 0.4936j = 0.495647e^{j84.790919^\circ} \qquad \text{and} \qquad B = A^* = 0.045 - 0.4936j = 0.495647e^{-j84.790919^\circ}$$

Therefore, we have

$$\frac{0.09-0.81z^{-1}}{1-0.9z^{-1}+0.81z^{-2}}=\frac{0.045+0.4936j}{1-0.9e^{j\pi/3}z^{-1}}+\frac{0.045-0.4936j}{1-0.9e^{-j\pi/3}z^{-1}}$$

and following the previous methodology, we arrive at the inverse transform

$$y'[n]=2|A||p|^n\cos(n\measuredangle p+\measuredangle A)u[n]=0.991294(0.9)^n\cos\left(\frac{\pi n}{3}+84.790919^\circ\right)u[n]$$

Based on the theory described in the previous sections, it is obvious that the output $y[n]$ we calculated in question (a) is actually the zero-state response, while the additional term we calculated in question (b) is the zero-input response. Therefore, to determine the total output of the system, we need to add these two components together. We can arrive at the same result, if we add together the two rational functions of the variable z that we determined in the two subquestions and then determine the inverse transform of the overall function that will arise.

Problem 9.6.2 Determine the output of a recursive discrete-time system

$$y[n]-0.5y[n-1]=x[n]$$

and is characterized by the existence of the initial conditions $y[-1]=y[-2]=0$, when the signal $x[n]=10\cos(n\pi/4)u[n]$ is used as input (Proakis and Manolakis, Example 3.6.2).

Solution Following the methodology for solving the previous exercise—that is, comparing the difference equation of a system with the general difference equation and determining in this way the number of inputs and outputs, as well as the values of the coefficients—it turns out that the transfer function of the system is

$$\mathcal{H}(z)=\frac{\beta_0}{1+\alpha_1 z^{-1}}=\frac{1}{1-0.5z^{-1}}$$

and has a simple pole at the position $p=0.5$. On the other hand, referring to tables of elementary transforms, it is not difficult to find that the $\mathcal{Z}$ transform of the sinusoidal input signal has the form

$$\mathcal{X}(z)=\frac{10[1-z^{-1}\cos(\pi/4)]}{1-2z^{-1}\cos(\pi/4)+z^{-2}}=\frac{10[(1-(z^{-1}/\sqrt{2}))]}{1-\sqrt{2}z^{-1}+z^{-2}}$$

and has two simple poles at the positions $p_1=\exp(j\pi/4)$ and $p_2=\exp(-j\pi/4)$. Therefore, the $\mathcal{Z}$ transform of the output of the system will be given by the relation

$$\mathcal{Y}(z)=\mathcal{H}(z)\mathcal{X}(z)=\frac{10[(1-(z^{-1}/\sqrt{2}))]}{(1-0.5z^{-1})(1-\exp(j\pi/4)z^{-1})(1-\exp(-j\pi/4)z^{-1})}$$
$$=\frac{6.3}{1-0.5z^{-1}}+\frac{6.78\exp(-j28.7^\circ)}{1-\exp(j\pi/4)z^{-1}}+\frac{6.78\exp(j28.7^\circ)}{1-\exp(-j\pi/4)z^{-1}}$$

Applying the method of partial fraction expansion, that is, expressing the last expression as a sum of the form

$$\mathcal{Y}(z)=\frac{A}{1-0.5z^{-1}}+\frac{B}{1-e^{j\pi/4}z^{-1}}+\frac{C}{1-e^{-j\pi/4}z^{-1}}$$

results in the system of three equations after operations

$$A + 0.5e^{j\pi/4}B + 0.5e^{-j\pi/4}C = 0$$

$$A\sqrt{2} + (0.5 + e^{j\pi/4})B + (0.5 + e^{-j\pi/4})C = \frac{10}{\sqrt{2}}$$

$$A + B + C = 10$$

with solution

$$A = -1.9074$$

$$B = 5.9537 + 3.2562j = 6.785830e^{j28.675148°}$$

$$C = 5.9537 - 3.2562j = 6.785830e^{-j28.675148°}$$

Calculating the inverse transform of the final expression that arises and applying the argumentation we used in the previous exercise, it turns out that

$$y[n] = \left[-1.9074(0.5)^n + 13.56\cos\left(\frac{n\pi}{4} - 28.7°\right)\right]u[n]$$

which is the desired result. In the last equation, the first term refers to the *transient output of the system*, which tends to zero over time, while the second term refers to the *steady-state response*, which does not decay with time, but it lasts indefinitely.

Problem 9.6.3 Determine the zero-input response of the recursive discrete-time system

$$y[n] + \frac{1}{2}y[n-1] - \frac{1}{4}y[n-2] = 0$$

which is characterized by the existence of the initial conditions $y[-1] = y[-2] = 1$.

Solution Taking the unilateral $\mathcal{Z}$ transform of the two sides and using the property

$$\mathcal{Z}^+\left\{y[n-k]\right\} = z^{-k}\left[\mathcal{Y}^+(z) + \sum_{n=1}^{k} y[-n]z^n\right]$$

we will get

$$\mathcal{Z}^+\left\{y[n]\right\} + \frac{1}{2}\mathcal{Z}^+\left\{y[n-1]\right\} - \frac{1}{4}\mathcal{Z}^+\left\{y[n-2]\right\} = 0$$

or, equivalently,

$$\mathcal{Y}^+(z) + \frac{1}{2}z^{-1}\left[\mathcal{Y}^+(z) + y[-1]z\right] - \frac{1}{4}z^{-2}\left[\mathcal{Y}^+(z) + y[-1]z + y[-2]z^2\right] = 0$$

and finally

$$\mathcal{Y}^+(z) = \frac{(z^{-1}/4) - (1/4)}{1 + (z^{-1}/2) - (z^{-2}/4)} = \frac{0.154}{1 - 0.31z^{-1}} - \frac{0.404}{1 + 0.81z^{-1}}$$

where we solved for $\mathcal{Y}^+(z)$, after having substituted the values of the initial conditions that were given.

To calculate the inverse transform of the last relation, we will use *the method of partial fraction expansion*. For simplicity's sake, setting $z^{-1} = y$, we find that the foregoing relationship is written

$$\mathcal{Y}^+(y) = \frac{(y/4) - (1/4)}{1 + (y/2) - (y^2/4)} = \frac{1 - y}{y^2 - 2y - 4} = \frac{0.5}{3.2361 - y} - \frac{0.5}{1.2361 + y}$$

where $y_1 = 1 + \sqrt{5} = 3.2361$ and $y_2 = 1 - \sqrt{5} = -1.2361$ are the two roots of the polynomial of the denominator. Returning to our original variable and reformulating the preceding fractions in a form that can be looked up in tables of elementary transforms, we will get

$$\mathcal{Y}^+(z) = \frac{0.5}{3.2361 - z^{-1}} - \frac{0.5}{1.2361 + z^{-1}} = \frac{0.1545}{1 - 0.309z^{-1}} - \frac{0.4045}{1 + 0.809z^{-1}}$$

From the contents of Table 8.2 we find that

$$\mathcal{Z}^{-1}\left\{\frac{1}{1 - 0.309z^{-1}}\right\} = (0.309)^n u[n] \quad \text{and} \quad \mathcal{Z}^{-1}\left\{\frac{1}{1 + 0.809z^{-1}}\right\} = (-0.809)^n u[n]$$

and therefore $y[n] = \mathcal{Z}^{-1}\{\mathcal{Y}^+(z)\} = \left[0.1545(0.309)^n - 0.4045(-0.809)^n\right]u[n]$.

Problem 9.6.4 Using the methodology of the previous exercise, prove that the zero-input response of the recursive discrete-time system described by the difference equation

$$y[n] - \frac{3}{2}y[n-1] + \frac{1}{2}y[n-2] = 0$$

and characterized by the existence of the initial conditions $y[-1] = 1$, $y[-2] = 0$ is given by the relationship

$$y[n] = \left[2 - (0.5)^{n+1}\right]u[n]$$

Solution Taking the unilateral $\mathcal{Z}$ transform of the two sides and using the property

$$y[n-k] \rightleftarrows z^{-k}\left[Y^+(z) + \sum_{n=1}^{k} y(-n)z^n\right]$$

we end up with the equation

$$Y^+(z) - \frac{3}{2}\left[z^{-1}Y^+(z) + y(-1)\right] + \frac{1}{2}\left[z^{-2}Y^+(z) + z^{-1}y(-1)\right] = 0$$

from which, by substituting the initial conditions and solving for $Y^+(z)$ we find

$$Y^+(z) = \frac{(3/2) - (1/2)z^{-1}}{1 - (3/2)z^{-1} + (1/2)z^{-2}} = \frac{2}{1 - z^{-1}} - \frac{0.5}{1 - 0.5z^{-1}}$$

where for the calculation of the last expression expansion of the function $Y^+(z)$ took place in a series of partial fractions. Calculating the inverse $\mathcal{Z}$ transform of the function $Y^+(z)$ through our reference to tables of elementary transforms, it turns out that

$$y[n] = \Big[2 - 0.5(0.5)^n\Big]u(n) = \Big[2 - (0.5)^{n+1}\Big]u[n]$$

Problem 9.6.5 Considering the discrete-time system described by the transfer function

$$\mathcal{H}(z) = \frac{z^{-1} + 0.5z^{-2}}{1 - (3/5)z^{-1} + (2/25)z^{-2}}$$

determine (a) the impulse response $h[n]$, (b) the zero-state response to the input signal $x[n] = u[n]$, and (c) the step response for the initial conditions $y[-1] = 1$ and $y[-2] = 2$ (Proakis and Manolakis, Problem 3.45).

Solution

(a) The determination of the system's impulse response is based on the calculation of the *inverse $\mathcal{Z}$ transform* of the transfer function $\mathcal{H}(z)$. Setting for simplicity's sake $y = z^{-1}$, the polynomial of the denominator is factored as

$$f(y) = \frac{2}{25}y^2 - \frac{3}{5}y + 1 = \frac{2}{25}\left(y^2 - \frac{15}{2}y + \frac{25}{2}\right) = \frac{(y-5)(2y-5)}{25}$$

and, returning to our original variables, the system's transfer function is formulated as

$$\mathcal{H}(z) = \frac{25(z^{-1} + 0.5z^{-2})}{(z^{-1} - 5)(2z^{-1} - 5)} = \frac{25}{2}\frac{2z+1}{(1-5z)(2-5z)}$$

Using the method of complex integration[12] for the calculation of the inverse transform, we define as known the function

$$\mathcal{Y}(z) = \frac{\mathcal{H}(z)}{z}z^n = \frac{25}{2}\frac{(2z+1)z^{n-1}}{(1-5z)(2-5z)} = z^{-1}\left\{\frac{25}{2}\frac{(2z+1)z^n}{(1-5z)(2-5z)}\right\} = z^{-1}\mathcal{Y}_1(z)$$

The function $\mathcal{Y}_1(z)$ has two simple poles at the positions $z_1 = 1/5$ and $z_2 = 2/5$ with remainder values

$$\text{Res}[Y(1/5)] = \frac{25}{2}\lim_{z\to 1/5}\left[\left(z - \frac{1}{5}\right)\frac{(2z+1)z^n}{5\left(\frac{1}{5} - z\right)(2-5z)}\right] = -\frac{5}{2}\lim_{z\to 1/5}\left[\frac{(2z+1)z^n}{2-5z}\right] = -\frac{7}{2}\left(\frac{1}{5}\right)^n$$

$$\text{Res}[Y(2/5)] = \frac{25}{2}\lim_{z\to 2/5}\left[\left(z - \frac{2}{5}\right)\frac{(2z+1)z^n}{5(1-5z)\left(\frac{2}{5} - z\right)}\right] = -\frac{5}{2}\lim_{z\to 2/5}\left[\frac{(2z+1)z^n}{1-5z}\right] = +\frac{9}{2}\left(\frac{2}{5}\right)^n$$

Therefore, the inverse $\mathcal{Z}$ transform of the function $\mathcal{Y}_1(z)$ will be given by the relation

$$h_1[n] = \left[\frac{9}{2}\left(\frac{2}{5}\right)^n - \frac{7}{2}\left(\frac{1}{5}\right)^n\right]u(n)$$

and using the property of time shift $\mathcal{Z}\{x[n-k]\} = z^{-k}\mathcal{Y}(z)$ (for $k = 1$) we have

$$h[n] = \left[\frac{9}{2}\left(\frac{2}{5}\right)^{n-1} - \frac{7}{2}\left(\frac{1}{5}\right)^{n-1}\right]u[n-1]$$

[12] The inverse $\mathcal{Z}$ transforms of this and the next exercise will be calculated using the method of complex integration, so that readers can familiarize themselves with the use of this method. However, in each case, they can verify the result we get using the method of partial fraction expansion.

a relation that expresses the *impulse response* of the considered discrete-time system.

(b) The output of the system, when the step function $x[n] = u[n]$ is used at its input, will be given by the convolution equation

$$y[n] = h[n] * x[n] = \frac{9}{2}\sum_{k=-\infty}^{\infty}\left(\frac{2}{5}\right)^{k-1} u[k-1]u[n-k] - \frac{7}{2}\sum_{k=-\infty}^{\infty}\left(\frac{1}{5}\right)^{k-1} u[k-1]u[n-k]$$

and finally

$$y[n] = h[n] * x[n] = \frac{9}{2}\sum_{k=1}^{n}\left(\frac{2}{5}\right)^{k-1} - \frac{7}{2}\sum_{k=1}^{n}\left(\frac{1}{5}\right)^{k-1}$$

Changing the dummy index from k to $m = k - 1$, the preceding equation will take the form

$$y[n] = \frac{9}{2}\sum_{m=0}^{n-1}\left(\frac{2}{5}\right)^{m} - \frac{7}{2}\sum_{m=0}^{n-1}\left(\frac{1}{5}\right)^{m} = \frac{9}{2}\left[\frac{1-(2/5)^n}{1-(2/5)}\right] - \frac{7}{2}\left[\frac{1-(1/5)^n}{1-(1/5)}\right] = \frac{35}{8}\left(\frac{1}{5}\right)^{n} - \frac{15}{2}\left(\frac{2}{5}\right)^{n} + \frac{25}{8}$$

or considering that our system is causal,

$$y[n] = \frac{9}{2}\sum_{m=0}^{n-1}\left(\frac{2}{5}\right)^{m} - \frac{7}{2}\sum_{m=0}^{n-1}\left(\frac{1}{5}\right)^{m} = \left[\frac{35}{8}\left(\frac{1}{5}\right)^{n} - \frac{15}{2}\left(\frac{2}{5}\right)^{n} + \frac{25}{8}\right]u(n)$$

Let us note that the last equation can also arise through the use of the $\mathcal{Z}$ transform: Indeed, using this transform for the step function in combination with the convolution property that characterizes it, we will get

$$\mathcal{Y}(z) = \mathcal{H}(z)\mathcal{X}(z) = \frac{25}{2}\frac{(2z+1)z}{(1-5z)(2-5z)(z-1)}$$

Using once again the method of *complex integration*, we define the function

$$\mathcal{Y}'(z) = \frac{\mathcal{Y}(z)}{z}z^n = \frac{25}{2}\frac{(2z+1)z^n}{(1-5z)(2-5z)(z-1)}$$

which has three simple poles at the positions $z_1 = 1/5$, $z_2 = 2/5$, and $z_3 = 1$. Following the usual methodology for calculating residues, it is not difficult to prove that

$$\mathrm{Res}[\mathcal{Y}'(1/5)] = \frac{35}{8}\left(\frac{1}{5}\right)^{n}$$

$$\mathrm{Res}[\mathcal{Y}'(2/5)] = -\frac{15}{2}\left(\frac{2}{5}\right)^{n}$$

$$\mathrm{Res}[\mathcal{Y}'(1)] = \frac{25}{8}$$

thereby arriving at the same result as the one we obtained using the convolution method. It is obvious that since the initial conditions of the system, which were considered equal to zero, were not used at all in the calculation of this response, the result we arrived at is simply the *zero-state response*. If we compare the equation of the output of the system that we determined previously with the corresponding equation that arises from the theoretical analysis we presented in the previous pages

$$y[n] = \sum_{k=1}^{N} A_k(p_k)^n u[n] + \sum_{k=1}^{L} Q_k(q_k)^n u[n]$$

we find that the first two terms are due to the $N = 2$ poles of the system $p_1 = 1/5$ and $p_2 = 2/5$, while the third is due to the $L = 1$ pole of the $\mathcal{Z}$ transform of the input signal, which is $q_1 = 1$.

(c) In this case and applying the theory presented in the previous pages, the total response of the system will arise *from the sum of the zero-state response*—which we calculated previously—*and the zero-input response*, which depends *only on the poles of the system* and is *independent* of the poles of the signal. To determine the zero-input response, we will use the system's difference equation. Following the methodology used in previous exercises, we will write as known[13]

$$\mathcal{H}(z) = \frac{\mathcal{Y}(z)}{\mathcal{X}(z)} = \frac{z^{-1} + 0.5z^{-2}}{1 - (3/5)z^{-1} + (2/25)z^{-2}} \quad \text{or, equivalently,} \quad \mathcal{Y}(z)\left(1-\frac{3}{5}z^{-1}+\frac{2}{25}z^{-2}\right) = \mathcal{X}(z)\left(z^{-1}+\frac{1}{2}z^{-2}\right)$$

and finally

$$\mathcal{Y}(z) - \frac{3}{5}z^{-1}\mathcal{Y}(z) + \frac{2}{25}z^{-2}\mathcal{Y}(z) = z^{-1}\mathcal{X}(z) + \frac{1}{2}z^{-2}\mathcal{X}(z)$$

If we take the inverse $\mathcal{Z}$ transform of the foregoing relation and use the property of time shifting $\mathcal{Z}\{x[n-k]\} = z^{-k}\mathcal{X}(z)$ for the values $k = 1$ and $k = 2$, we will end up with the system's difference equation, which has the form

$$y[n] - \frac{3}{5}y[n-1] + \frac{2}{25}y[n-2] = x[n-1] + \frac{1}{2}x[n-2]$$

In the next step of the process, we will set $x[n-1] = x[n-2] = 0$ (since we are interested in the zero-input response) and, as before, calculate *the unilateral $\mathcal{Z}$ transform of the homogeneous difference equation*, so as to incorporate it into the problem and the *initial conditions* that were given. Using the *property of time shift* of the unilateral $\mathcal{Z}$ transform, which is known to be given by the equation

$$\mathcal{Z}^{+}\{y[n-k]\} = z^{-k}\left[\mathcal{Y}^{+}(z) + \sum_{n=1}^{k} y[-n]z^{n}\right]$$

and the initial conditions $y[-1] = 1$ and $y[-2] = 2$, we will get

$$\mathcal{Z}^{+}\{y[n-1]\} = z^{-1}(\mathcal{Y}^{+}(z) + y[-1]z) + z^{-1}\mathcal{Y}^{+}(z) + y[-1] = z^{-1}\mathcal{Y}^{+}(z) + 1$$

$$\mathcal{Z}^{+}\{y[n-2]\} = z^{-2}(\mathcal{Y}^{+}(z) + y[-1]z + y[-2]z^{2}) = z^{-2}\mathcal{Y}^{+}(z) + z^{-1}y[-1] + y[-2] = z^{-2}\mathcal{Y}^{+}(z) + z^{-1} + 2$$

that is, the application of the *unilateral $\mathcal{Z}$* transform to the homogeneous difference equation of the system will lead to the result

$$\mathcal{Y}^{+}(z) - \frac{3}{5}(z^{-1}\mathcal{Y}^{+}(z) + 1) + \frac{2}{25}(z^{-2}\mathcal{Y}^{+}(z) + z^{-1} + 2)$$

from which it finally follows that

$$\mathcal{Y}^{+}(z) = \frac{-(2/25)z^{-1} + (11/25)}{(2/25)z^{-2} - (3/5)z^{-1} + 1}$$

Therefore, the zero-input response will be calculated as *the inverse $\mathcal{Z}$ transform of the function $\mathcal{Y}^{+}(z)$*. Taking into account the result of the factorization of question (a) of the exercise

$$\frac{1}{(2/25)z^{-2} - (3/5)z^{-1} + 1} = \frac{25}{(z^{-1} - 5)(2z^{-1} - 5)}$$

[13] This process is a demonstration of the way we can determine *the system's difference equation*, when the *transfer function* that describes it is known.

we can write the function $\mathcal{Y}^+(z)$ in the form

$$\mathcal{Y}^+(z) = \frac{25[-(2/25)z^{-1} + (11/25)]}{(z^{-1} - 5)(2z^{-1} - 5)} = \frac{11 - 2z^{-1}}{(z^{-1} - 5)(2z^{-1} - 5)} = \frac{(11z - 2)z}{(1 - 5z)(2 - 5z)}$$

Using the method of *complex integration*, we define the function as known

$$\mathcal{A}(z) = \frac{\mathcal{Y}^+(z)}{z} z^n = \frac{(11z - 2)z^n}{(1 - 5z)(2 - 5z)}$$

which has *two simple poles* at the points $z_1 = 1/5$ and $z_2 = 2/5$, with residue values at these

$$\text{Res}[\mathcal{A}(1/5)] = \lim_{z \to 1/5} \left[\left(z - \frac{1}{5} \right) \frac{(11z - 2)z^n}{5\left(\frac{1}{5} - z\right)(2 - 5z)} \right] = -\frac{1}{5} \lim_{z \to 1/5} \left[\frac{(11z - 2)z^n}{2 - 5z} \right] = -\frac{1}{25} \left(\frac{1}{5} \right)^n$$

$$\text{Res}[\mathcal{A}(2/5)] = \lim_{z \to 2/5} \left[\left(z - \frac{2}{5} \right) \frac{(11z - 2)z^n}{5\left(\frac{2}{5} - z\right)(1 - 5z)} \right] = -\frac{1}{5} \lim_{z \to 2/5} \left[\frac{(11z - 2)z^n}{1 - 5z} \right] = +\frac{12}{25} \left(\frac{2}{5} \right)^n$$

Therefore, the *zero-input response* will be given by the relation

$$y[n] = \left[-\frac{1}{25} \left(\frac{1}{5} \right)^n + \frac{12}{25} \left(\frac{2}{5} \right)^n \right] u[n]$$

We observe that this response depends *only on the poles of the system and not on the poles of the signal* in full application of the corresponding theory. Finally, the total response of the system will be equal to the sum of the zero-state response that we calculated in question (b) and the zero-input response that we calculated previously. It will therefore be the case that

$$y[n] = \left[\left(\frac{35}{8} - \frac{1}{25} \right) \left(\frac{1}{5} \right)^n + \left(\frac{12}{25} - \frac{15}{2} \right) \left(\frac{2}{5} \right)^n + \frac{25}{8} \right] u[n] = \left[\frac{867}{200} \left(\frac{1}{5} \right)^n - \frac{351}{50} \left(\frac{2}{5} \right)^n + \frac{25}{8} \right] u[n]$$

Problem 9.6.6 Considering the discrete-time system described by the difference equation

$$y[n] = -\alpha_1 y[n - 1] + \beta_0 x[n] + \beta_1 x[n - 1]$$

determine (a) its impulse response $h[n]$, (b) the zero-state step response, (c) the step response for an initial condition with value $y[-1] = A \neq 0$, and (d) the system's response to the input signal $x[n] = \cos(\omega_0 n)$ $(0 \leq n < \infty)$ (Proakis and Manolakis, Problem 3.47).

Solution

(a) Determination of the system's impulse response is based on the well-known *calculation of the inverse $\mathcal{Z}$ transform of the transfer function*, which in turn can be determined from the *difference equation* that correlates the input with its output. Taking the $\mathcal{Z}$ transform of this difference equation we get

$$\mathcal{Y}(z) = -\alpha_1 \mathcal{Z}\{y[n - 1]\} + \beta_0 \mathcal{X}(z) + \beta_1 \mathcal{Z}\{x[n - 1]\} = -\alpha_1 z^{-1} \mathcal{Y}(z) + \beta_0 \mathcal{X}(z) + \beta_1 z^{-1} \mathcal{X}(z)$$

from which we get

$$\mathcal{Y}(z)(1+\alpha_1 z^{-1}) = \mathcal{X}(z)(\beta_0+\beta_1 z^{-1})$$

and finally

$$\mathcal{H}(z) = \frac{\mathcal{Y}(z)}{\mathcal{X}(z)} = \frac{\beta_0+\beta_1 z^{-1}}{1+\alpha_1 z^{-1}}$$

If we write the preceding function in the form

$$\mathcal{H}(z) = \beta_0 \frac{1}{1+\alpha_1 z^{-1}} + \beta_1 \left\{ z^{-1} \frac{1}{1+\alpha_1 z^{-1}} \right\}$$

and calculate the inverse $\mathcal{Z}$ transform of the last relation, it turns out that

$$h[n] = \beta_0(-\alpha_1)^n u[n] + \beta_1(-\alpha_1)^{n-1} u[n-1]$$

an equation that also expresses the *impulse response* of the considered discrete-time system.

(b) The *step response* of the system, that is, its response to the input step $x[n] = u[n]$, can be determined by the *convolution equation*; it will therefore be the case that

$$y[n] = h[n] * x[n] = \sum_{k=-\infty}^{+\infty} h[k]x[n-k]$$

$$= \beta_0 \sum_{k=-\infty}^{+\infty} (-\alpha_1)^k u[k]u[n-k] + \beta_1 \sum_{k=-\infty}^{+\infty} (-\alpha_1)^{k-1} u[k-1]u[n-k] = \beta_0 \sum_{k=0}^{n} (-\alpha_1)^k + \beta_1 \sum_{k=1}^{n} (-\alpha_1)^{k-1}$$

The sums of the foregoing equation are calculated as

$$\sum_{k=0}^{n} (-\alpha_1)^k = \frac{1-(-\alpha_1)^{n+1}}{1-(-\alpha_1)} = \frac{1+\alpha_1(-\alpha_1)^n}{1+\alpha_1} \quad \text{and} \quad \sum_{k=1}^{n} (-\alpha_1)^{k-1} = \sum_{m=0}^{n-1} (-\alpha_1)^m = \frac{1-(-\alpha_1)^n}{1+\alpha_1}$$

where for the calculation of the second sum we proceeded to the substitution $k \longrightarrow m = k-1$. Using the foregoing sums in the equation of the step response, we write

$$y[n] = h[n] * x[n] = \sum_{k=-\infty}^{+\infty} h[k]x[n-k] = \beta_0 \sum_{k=0}^{n} (-\alpha_1)^k + \beta_1 \sum_{k=1}^{n} (-\alpha_1)^{k-1}$$

$$= \left[\beta_0 \frac{1+\alpha_1(-\alpha_1)^n}{1+\alpha_1} + \beta_1 \frac{1-(-\alpha_1)^n}{1+\alpha_1} \right] u[n] = \left[\frac{\beta_0+\beta_1}{1+\alpha_1} + \frac{\beta_0\alpha_1-\beta_1}{1+\alpha_1}(-\alpha_1)^n \right] u[n]$$

which is also the desired result. The reader is asked to verify the preceding result by calculating the output $y[n]$ through the use of the $\mathcal{Z}$ transform method.

(c) In the case where our system is characterized by the existence of the initial condition $y[-1] = A \neq 0$, we should, in the *zero-state response* that we calculated previously, add the *zero-input response*, which will result from the application of the unilateral $\mathcal{Z}$ transform to the homogeneous difference equation. It will therefore be the case that

$$\mathcal{Y}^+(z) = -\alpha_1 \mathcal{Z}^+ \Big\{ y[n-1] \Big\}$$

with the unilateral transform of the right-hand side being calculated as

$$\mathcal{Z}^+ \Big\{ y[n-1] \Big\} = z^{-1} \Big[\mathcal{Y}^+(z) + y[-1]z \Big] = z^{-1}\mathcal{Y}^+(z) + y[-1] = z^{-1}\mathcal{Y}^+(z) + A$$

Combining the foregoing expressions, it immediately follows that

$$\mathcal{Y}^+(z) = -\alpha_1 z^{-1}\mathcal{Y}^+(z) - \alpha_1 A \qquad \text{or, equivalently,} \qquad \mathcal{Y}^+(z) = -\alpha_1 A\left\{\frac{1}{1+\alpha_1 z}\right\}$$

Referring to tables of elementary transforms, it is not difficult to find that

$$y^+[n] = -\alpha_1 A(-\alpha_1)^n u[n]$$

and therefore the total output of the system will be calculated as

$$y[n] = \left\{\frac{\beta_0+\beta_1}{1+\alpha_1} + \left[\frac{\beta_0\alpha_1-\beta_1}{1+\alpha_1} - \alpha_1 A\right](-\alpha_1)^n\right\}u[n]$$

(d) Referring to Table 8.2, we find that the $\mathcal{Z}$ transform of the elementary signal $x[n] = \cos(\omega_0 n)u[n]$ takes the form

$$\mathcal{X}(z) = \frac{1-z^{-1}\cos(\omega_0)}{1-2z^{-1}\cos(\omega_0)+z^{-2}} = \frac{z[z-\cos(\omega_0)]}{z^2-2z\cos(\omega_0)+1}$$

On the other hand, the transfer function of the considered discrete-time system is formulated as

$$\mathcal{H}(z) = \frac{\beta_0+\beta_1 z^{-1}}{1+\alpha_1 z^{-1}} = \frac{\beta_0 z+\beta_1}{z+\alpha_1}$$

Applying, therefore, the *convolution theorem* of the $\mathcal{Z}$ transform, we can calculate the output of the system $y[n]$ as the *inverse $\mathcal{Z}$ transform* of the function

$$\mathcal{Y}(z) = \mathcal{H}(z)\mathcal{X}(z) = \frac{z[z-\cos(\omega_0)](\beta_0 z+\beta_1)}{(z^2-2z\cos(\omega_0)+1)(z+\alpha_1)}$$

To solve the problem, let us use the *method of complex integration*, and let us define for this purpose the function

$$\mathcal{A}(z) = \frac{\mathcal{Y}(z)}{z}z^n = \frac{[z-\cos(\omega_0)](\beta_0 z+\beta_1)z^n}{(z^2-2z\cos(\omega_0)+1)(z+\alpha_1)} = \frac{[z-\cos(\omega_0)](\beta_0 z+\beta_1)z^n}{(z-e^{i\omega_0})(z-e^{-i\omega_0})(z+\alpha_1)}$$

We observe that the function $\mathcal{A}(z)$ has three simple poles at the positions $p_1 = -\alpha_1$, $p_2 = e^{i\omega_0}$, and $p_3 = e^{-i\omega_0}$, with residue values at these

$$\begin{aligned}
\text{Res}[A(-\alpha_1)] &= \lim_{z\to-\alpha_1}\left[(z+\alpha_1)\frac{[z-\cos(\omega_0)](\beta_0 z+\beta_1)z^n}{(z-e^{i\omega_0})(z-e^{-i\omega_0})(z+\alpha_1)}\right] = \\
&= \lim_{z\to-\alpha_1}\left[\frac{[z-\cos(\omega_0)](\beta_0 z+\beta_1)z^n}{(z-e^{i\omega_0})(z-e^{-i\omega_0})}\right] = \frac{[\alpha_1+\cos(\omega_0)](\beta_0\alpha_1-\beta_1)(-\alpha_1)^n}{\alpha_1^2+2\alpha_1\cos(\omega_0)+1} \\
\text{Res}[A(e^{i\omega_0})] &= \lim_{z\to e^{i\omega_0}}\left[(z-e^{i\omega_0})\frac{[z-\cos(\omega_0)](\beta_0 z+\beta_1)z^n}{(z-e^{i\omega_0})(z-e^{-i\omega_0})(z+\alpha_1)}\right] = \lim_{z\to e^{i\omega_0}}\left[\frac{[z-\cos(\omega_0)](\beta_0 z+\beta_1)z^n}{(z-e^{-i\omega_0})(z+\alpha_1)}\right] \\
&= \frac{[e^{i\omega_0}-\cos(\omega_0)](\beta_0 e^{i\omega_0}+\beta_1)e^{i\omega_0 n}}{(e^{i\omega_0}-e^{-i\omega_0})(e^{i\omega_0}+\alpha_1)} = \frac{1}{2}\left(\frac{\beta_0 e^{i\omega_0}+\beta_1}{\alpha_1+e^{i\omega_0}}\right)e^{i\omega_0 n} \\
\text{Res}[A(e^{-i\omega_0})] &= \lim_{z\to e^{-i\omega_0}}\left[(z-e^{-i\omega_0})\frac{[z-\cos(\omega_0)](\beta_0 z+\beta_1)z^n}{(z-e^{i\omega_0})(z-e^{-i\omega_0})(z+\alpha_1)}\right] = \lim_{z\to e^{-i\omega_0}}\left[\frac{[z-\cos(\omega_0)](\beta_0 z+\beta_1)z^n}{(z-e^{i\omega_0})(z+\alpha_1)}\right] \\
&= \frac{[e^{-i\omega_0}-\cos(\omega_0)](\beta_0 e^{-i\omega_0}+\beta_1)e^{-i\omega_0 n}}{(e^{-i\omega_0}-e^{i\omega_0})(e^{-i\omega_0}+\alpha_1)} = \frac{1}{2}\left(\frac{\beta_0 e^{-i\omega_0}+\beta_1}{\alpha_1+e^{-i\omega_0}}\right)e^{-i\omega_0 n}
\end{aligned}$$

meaning that the system's response to the input signal $x[n] = \cos(\omega_n)nu[n]$ will have the form

$$y[n] = \left\{ \frac{[\alpha_1 + \cos(\omega_0)](\beta_0\alpha_1 - \beta_1)(-\alpha_1)^n}{\alpha_1^2 + 2\alpha_1\cos(\omega_0) + 1} + \frac{1}{2}\left(\frac{\beta_0 e^{i\omega_0} + \beta_1}{\alpha_1 + e^{i\omega_0}}\right)e^{i\omega_0 n} + \frac{1}{2}\left(\frac{\beta_0 e^{-i\omega_0} + \beta_1}{\alpha_1 + e^{-i\omega_0}}\right)e^{-i\omega_0 n} \right\} u[n]$$

which is also the desired result. The preceding equation may be subject to improvements and simplifications if we replace the exponential quantities with the functions of sine and cosine as they are related through the use of Euler's equations $e^{\pm ix} = \cos x \pm \sin x$ $(x = \omega_0 n)$. This process is left as an exercise to the reader, who is also asked to verify the correctness of the preceding result by calculating the output of the system from the convolution equation, as well as the preceding inverse transform through the application of the partial fraction expansion method.

Problem 9.6.7 Determine the zero-state response of the system

$$y[n] = \frac{1}{2}y[n-1] + 4x[n] + 3x[n-1]$$

when the signal $x[n] = e^{i\omega_0 n}u[n]$ is used at its input. What is the steady-state response? (Proakis and Manolakis, Problem 3.48)

Solution

(a) In the first step of the process, we need to determine the impulse response of the system from its transfer function $H(z)$. By taking the $\mathcal{Z}$ transform of the system's difference equation and using the properties of this transform, we will get

$$Y(z) - \frac{1}{2}z^{-1}Y(z) = 4X(z) + 3z^{-1}X(z) \quad \Rightarrow \quad Y(z)\left(1 - \frac{1}{2}z^{-1}\right) = X(z)(4 + 3z^{-1})$$

and finally

$$H(z) = \frac{Y(z)}{X(z)} = \frac{4 + 3z^{-1}}{1 - 0.5z^{-1}} = 4\left\{\frac{1}{1 - 0.5z^{-1}}\right\} + 3\left\{z^{-1}\frac{1}{1 - 0.5z^{-1}}\right\}$$

from which it follows that

$$h[n] = 4\left(\frac{1}{2}\right)^n u[n] + 3\left(\frac{1}{2}\right)^{n-1} u[n-1]$$

Therefore, the system's response to the input signal $x[n] = e^{i\omega_0 n}u[n]$ is calculated as

$$\begin{aligned} y[n] &= h[n] * x[n] = \sum_{k=\infty}^{\infty} x[k]h[n-k] \\ &= 4\sum_{k=-\infty}^{\infty}\left(\frac{1}{2}\right)^{n-k} u[n-k]e^{i\omega_0 k}u[k] + 3\sum_{k=-\infty}^{\infty}\left(\frac{1}{2}\right)^{n-k-1} u[n-k-1]e^{i\omega_0 k}u[k] \\ &= 4\left(\frac{1}{2}\right)^n \sum_{k=0}^{n}(2e^{i\omega_0})^k + 3\left(\frac{1}{2}\right)^{n-1}\sum_{k=0}^{n-1}(2e^{i\omega_0})^k = 4\left(\frac{1}{2}\right)^n\left[\frac{1 - (2e^{i\omega_0})^{n+1}}{1 - 2e^{i\omega_0}}\right] + 6\left(\frac{1}{2}\right)^n\left[\frac{1 - (2e^{i\omega_0})^n}{1 - 2e^{i\omega_0}}\right] \end{aligned}$$

or, equivalently, after operations

$$y[n] = h[n] * x[n] = 4\left(\frac{1}{2}\right)^n \left[\frac{1-(2e^{i\omega_0})^{n+1}}{1-2e^{i\omega_0}}\right] + 6\left(\frac{1}{2}\right)^n \left[\frac{1-(2e^{i\omega_0})^{n}}{1-2e^{i\omega_0}}\right] =$$

$$= \left\{\frac{10}{1-2e^{i\omega_0}}\left(\frac{1}{2}\right)^n - \frac{8e^{i\omega_0}+6}{1-2e^{i\omega_0}} e^{i\omega_0 n}\right\} u[n]$$

where we used the appropriate expressions from the mathematical typology and considered that the discrete-time system in question was causal.

On the other hand, to determine the *steady-state response* of the system, we must calculate the value of the foregoing relationship at the limit $n \to \infty$. It will therefore be the case that

$$y_s[n] = \lim_{n\to\infty} y[n] = \lim_{n\to\infty} \left\{\frac{10}{1-2e^{i\omega_0}}\left(\frac{1}{2}\right)^n - \frac{8e^{i\omega_0}+6}{1-2e^{i\omega_0}} e^{i\omega_0 n}\right\}$$

Of the two preceding terms, the first obviously tends to zero, while the second does not weaken over time, since the functions $\cos(\omega_0 n)$ and $\sin(\omega_0 n)$ of the quantity $e^{i\omega_0 n} = \cos(\omega_0 n) + i\,\sin(\omega_0 n)$ are bounded in the interval $[-1, +1]$ for any value of the index n. Therefore, the steady-state response of the system will be given by the relationship

$$y_s[n] = -\frac{8e^{i\omega_0}+6}{1-2e^{i\omega_0}} e^{i\omega_0 n} = \frac{8e^{i\omega_0}+6}{2e^{i\omega_0}-1} e^{i\omega_0 n}$$

9.7 Description of Systems in State Space

As already mentioned, discrete-time systems, in full analogy with their continuous-time counterparts, are described in the so-called *state space* with the help of two equations known as *dynamic equations* and have the form

$$\begin{aligned} \boldsymbol{q}[n+1] &= \boldsymbol{A}\boldsymbol{q}[n] + \boldsymbol{B}\boldsymbol{x}[n] \\ \boldsymbol{y}[n] &= \boldsymbol{C}\boldsymbol{q}[n] + \boldsymbol{D}\boldsymbol{x}[n] \end{aligned} \tag{9.34}$$

where $\boldsymbol{x}[n] = [x_1[n], x_2[n], \ldots, x_K[n]]^T$ and $\boldsymbol{y}[n] = [y_1[n], y_2[n], \ldots, y_L[n]]^T$ are the vectors of the K inputs and L outputs, $\boldsymbol{q}[n] = [q_1[n], q_2[n], \ldots, q_N[n]]^T$ is the vector of the N *state variables*, and $\boldsymbol{A}_{N\times N}$, $\boldsymbol{B}_{N\times K}$, $\boldsymbol{C}_{L\times N}$, and $\boldsymbol{D}_{L\times K}$ are the matrices of the *system*, *input*, *measurement*, and *output*, respectively. Using this representation, we studied *recursive LTI systems* in the state space solving the dynamic equations, calculating the matrix $\boldsymbol{A}^n$, and examining significant features, such as *similarity transformations*, *controllability*, and *observability*. Taking into account that in the case of continuous-time systems, the aforementioned features can be studied alternatively with the help of the *Laplace transform*, whose discrete version can be considered to be the $\mathcal{Z}$ *transform*, we can surmise (which is, of course, true) that the description of discrete-time systems in the state space can also be carried out with the help of this transform. In the following pages, we will describe this method of study.

9.7.1 Solution of Dynamic Equations

The solution of the dynamic equations of discrete-time systems using the $\mathcal{Z}$ transform is obtained in the way described for the case of continuous-time systems. Starting from the state equation and taking the unilateral $\mathcal{Z}^+$ transform of its two sides, we have

$$\mathcal{Z}^+\{\boldsymbol{q}[n+1]\} = \mathcal{Z}^+\{\boldsymbol{A}\boldsymbol{q}[n] + \boldsymbol{B}\boldsymbol{x}[n]\} = \boldsymbol{A}\mathcal{Z}^+\{\boldsymbol{q}[n]\} + \boldsymbol{B}\mathcal{Z}^+\{\boldsymbol{x}[n]\} \tag{9.35}$$

where to derive the last expression, we used the *property of linearity* of the unilateral transform. Now defining the vectors

$$\boldsymbol{\mathcal{Q}}^{+}(z) = \mathcal{Z}^{+}\{\boldsymbol{q}[n]\}[\mathcal{Z}^{+}\{q_1[n]\}, \mathcal{Z}^{+}\{q_2[n]\}, \ldots, \mathcal{Z}^{+}\{q_N[n]\}]^T = [\mathcal{Q}_1(z), \mathcal{Q}_2(z), \ldots, \mathcal{Q}_N(z)]^T \quad \text{and}$$

$$\boldsymbol{\mathcal{X}}^{+}(z) = \mathcal{Z}^{+}\{\boldsymbol{x}[n]\} = [\mathcal{Z}^{+}\{x_1[n]\}, \mathcal{Z}^{+}\{x_2[n]\}, \ldots, \mathcal{Z}^{+}\{q_L[n]\}]^T = [\mathcal{X}_1(z), \mathcal{X}_2(z), \ldots, \mathcal{X}_L(z)]^T$$

and using the property of time shift of the unilateral transform

$$\mathcal{Z}^{+}\{q[n+k]\} = z^k\left[\mathcal{X}^{+}(z) - \sum_{n=0}^{k-1} q[n]z^{-n}\right] \quad (k > 1) \tag{9.36}$$

which for the value $k = 1$ will give us

$$\mathcal{Z}^{+}\{q_i[n+1]\} = z\left[\mathcal{Q}_i(z) - q_i[0]\right] \quad (i = 1, 2, \ldots, N) \tag{9.37}$$

or, in vector notation,

$$\mathcal{Z}^{+}\{\boldsymbol{q}[n+1]\} = z\boldsymbol{\mathcal{Q}}^{+}(z) - z\boldsymbol{q}[0] \tag{9.38}$$

the state equation transformed by $\mathcal{Z}^{+}$ will take the form

$$z\boldsymbol{\mathcal{Q}}^{+}(z) - z\boldsymbol{q}[0] = \boldsymbol{A}\boldsymbol{\mathcal{Q}}^{+}(z) + \boldsymbol{B}\boldsymbol{\mathcal{X}}^{+}(z) \tag{9.39}$$

or, equivalently,

$$(z\boldsymbol{I} - \mathrm{A})\boldsymbol{\mathcal{Q}}^{+}(z) = z\boldsymbol{q}[0] + \boldsymbol{B}\boldsymbol{\mathcal{X}}^{+}(z) \tag{9.40}$$

and finally

$$\boldsymbol{\mathcal{Q}}^{+}(z) = (z\boldsymbol{I} - \boldsymbol{A})^{-1}z\boldsymbol{q}[0] + z(\boldsymbol{I} - \boldsymbol{A})^{-1}\boldsymbol{B}\boldsymbol{\mathcal{X}}^{+}(z) = (\boldsymbol{I} - z^{-1}\boldsymbol{A})\boldsymbol{q}[0] + (z\boldsymbol{I} - \boldsymbol{A})^{-1}\boldsymbol{B}\boldsymbol{\mathcal{X}}^{+}(z) \tag{9.41}$$

where $\boldsymbol{I}$ is the *unit matrix of dimensions* $N \times N$. Having arrived at the last relation, we can take the state vector $\boldsymbol{q}[n]$, which constitutes the desired solution of the state equation by calculating *its inverse unilateral transform*, a process that mathematically is formulated as

$$\boldsymbol{q}[n] = \underbrace{(\mathcal{Z}^{+})^{-1}\left[(\boldsymbol{I} - z^{-1}\boldsymbol{A})^{-1}\boldsymbol{q}[0]\right]}_{\text{zero-input component}} + \underbrace{(\mathcal{Z}^{+})^{-1}\left[(z\boldsymbol{I} - \boldsymbol{A})^{-1}\boldsymbol{B}\boldsymbol{\mathcal{X}}^{+}(z)\right]}_{\text{zero-state component}} \tag{9.42}$$

In a completely analogous way, by taking the unilateral $\mathcal{Z}$ transform of the system's output equation and replacing the function $\boldsymbol{\mathcal{Q}}^{+}(z)$ from the preceding relation, it turns out that

$$\mathcal{Z}^{+}\{\boldsymbol{y}[n]\} = \boldsymbol{C}\mathcal{Z}^{+}\{\boldsymbol{q}[n]\} + \boldsymbol{D}\mathcal{Z}^{+}\{\boldsymbol{x}[n]\} = \boldsymbol{C}\boldsymbol{\mathcal{Q}}^{+}(z) + \boldsymbol{D}\boldsymbol{\mathcal{X}}^{+}(z) \tag{9.43}$$

or, equivalently,

$$\begin{aligned}\boldsymbol{\mathcal{Y}}^{+}(z) &= \boldsymbol{C}(\boldsymbol{I} - z^{-1}\boldsymbol{A})\boldsymbol{q}[0] + \boldsymbol{C}(z\boldsymbol{I} - \boldsymbol{A})^{-1}\boldsymbol{B}\boldsymbol{\mathcal{X}}^{+}(z) + \boldsymbol{D}\boldsymbol{\mathcal{X}}^{+}(z) = \\ &= \boldsymbol{C}(\boldsymbol{I} - z^{-1}\boldsymbol{A})\boldsymbol{q}[0] + \left[\boldsymbol{C}(z\boldsymbol{I} - \boldsymbol{A})^{-1}\boldsymbol{B} + \boldsymbol{D}\right]\boldsymbol{\mathcal{X}}^{+}(z)\end{aligned} \tag{9.44}$$

From the inverse unilateral transform of the last relation, the output $\boldsymbol{y}[n]$ is calculated as

$$\boldsymbol{y}[n] = \underbrace{(\mathcal{Z}^{+})^{-1}\left[\boldsymbol{C}(\boldsymbol{I} - z^{-1}\boldsymbol{A})\boldsymbol{q}[0]\right]}_{\text{zero-input response}} + \underbrace{(\mathcal{Z}^{+})^{-1}\left\{\left[\boldsymbol{C}(z\boldsymbol{I} - \boldsymbol{A})^{-1}\boldsymbol{B} + \boldsymbol{D}\right]\boldsymbol{\mathcal{X}}^{+}(z)\right\}}_{\text{zero-state response}} \tag{9.45}$$

The following solved exercises demonstrate the application of the method.

Problem 9.7.1 Using the $\mathcal{Z}$ transform method, solve the state equation

$$\boldsymbol{q}[n+1] = \boldsymbol{A}\boldsymbol{q}[n] + \boldsymbol{B}x[n]$$

for parameter values

$$\boldsymbol{A} = \begin{bmatrix} 0 & 1 \\ -1/6 & 5/6 \end{bmatrix}, \quad \boldsymbol{B} = \begin{bmatrix} 0 \\ 1 \end{bmatrix}, \quad x[n] = u[n]$$

and initial state vector $\boldsymbol{q}[0] = [\ 2 \quad 3\ ]^T$, and then calculate the system output if the matrices $\boldsymbol{C}$ and $\boldsymbol{D}$ have the forms $\boldsymbol{C} = [\ -1 \quad 5\ ]$ and $\boldsymbol{D} = [0]$.

Solution The desired solution of the state equation will be calculated from the relation

$$\boldsymbol{q}[n] = (\mathcal{Z}^{+})^{-1}\left\{(\boldsymbol{I} - z^{-1}\boldsymbol{A})^{-1}\boldsymbol{q}[0]\right\} + (\mathcal{Z}^{+})^{-1}\left\{(z\boldsymbol{I} - \boldsymbol{A})^{-1}\boldsymbol{B}\mathcal{X}^{+}(z)\right\}$$

with the values of the expressions included in this being calculated subsequently.

Using the value of the system matrix $\boldsymbol{A}$, it turns out that

$$\boldsymbol{I} - z^{-1}\boldsymbol{A} = \begin{bmatrix} 1 & -\dfrac{1}{z} \\ \dfrac{1}{6z} & \dfrac{6z-5}{6z} \end{bmatrix}$$

with the determinant

$$\text{Det}(\boldsymbol{I} - z^{-1}\boldsymbol{A}) = \frac{6z^2 - 5z + 1}{6z^2}$$

and therefore the inverse matrix $(\boldsymbol{I} - z^{-1}\boldsymbol{A})^{-1}$ is calculated as

$$(\boldsymbol{I} - z^{-1}\boldsymbol{A})^{-1} = \begin{bmatrix} \dfrac{6z^2 - 5z}{6z^2 - 5z + 1} & \dfrac{6z}{6z^2 - 5z + 1} \\ \dfrac{-z}{6z^2 - 5z + 1} & \dfrac{6z^2}{6z^2 - 5z + 1} \end{bmatrix}$$

We will therefore have

$$(\boldsymbol{I} - z^{-1}\boldsymbol{A})^{-1}\boldsymbol{q}[0] = \begin{bmatrix} \dfrac{12z^2 + 8z}{6z^2 - 5z + 1} \\ \\ \dfrac{18z^2 - 2z}{6z^2 - 5z + 1} \end{bmatrix}$$

and the *zero-input component* of the state vector will be calculated from the equation

$$(\mathcal{Z}^+)^{-1}\left\{(\boldsymbol{I} - z^{-1}\boldsymbol{A})^{-1}\boldsymbol{q}[0]\right\} = \begin{bmatrix} (\mathcal{Z}^+)^{-1}\left\{\dfrac{12z^2 + 8z}{6z^2 - 5z + 1}\right\} \\ \\ (\mathcal{Z}^+)^{-1}\left\{\dfrac{18z^2 - 2z}{6z^2 - 5z + 1}\right\} \end{bmatrix}$$

Taking into account that the common polynomial of the denominator is factored as

$$6z^2 - 5z + 1 = 6\left(z - \frac{1}{2}\right)\left(z - \frac{1}{3}\right)$$

we find that these functions have two simple poles at positions $z = 1/2$ and $z = 1/3$, and the application of the *complex integration* method will lead to the following results:

For the first expression, the auxiliary function $\mathcal{Y}(z)$ is defined as

$$\mathcal{Y}(z) = \frac{12z^2 + 8z}{6z^2 - 5z + 1}z^{n-1}$$

with residue values at the pole positions

$$\text{Res}\,[\mathcal{Y}(1/2)] = \lim_{z\to 1/2}\left[\left(z - \frac{1}{2}\right)\frac{12z^2 + 8z}{6\left(z - \frac{1}{2}\right)\left(z - \frac{1}{3}\right)}z^{n-1}\right] = +14\left(\frac{1}{2}\right)^n$$

$$\text{Res}\,[\mathcal{Y}(1/3)] = \lim_{z\to 1/3}\left[\left(z - \frac{1}{3}\right)\frac{12z^2 + 8z}{6\left(z - \frac{1}{2}\right)\left(z - \frac{1}{3}\right)}z^{n-1}\right] = -12\left(\frac{1}{3}\right)^n$$

Therefore, we will have

$$(\mathcal{Z}^+)^{-1}\left\{\frac{12z^2 + 8z}{6z^2 - 5z + 1}\right\} = 14\left(\frac{1}{2}\right)^n - 12\left(\frac{1}{3}\right)^n$$

For the second expression, the auxiliary function $\mathcal{Y}(z)$ is defined as

$$\mathcal{Y}(z) = \frac{18z^2 - 2z}{6z^2 - 5z + 1}z^{n-1}$$

with residue values at the pole positions

$$\text{Res}\,[\mathcal{Y}(1/2)] = \lim_{z\to 1/2}\left[\left(z - \frac{1}{2}\right)\frac{18z^2 - 2z}{6\left(z - \frac{1}{2}\right)\left(z - \frac{1}{3}\right)}z^{n-1}\right] = +7\left(\frac{1}{2}\right)^n$$

$$\text{Res}\,[\mathcal{Y}(1/3)] = \lim_{z\to 1/3}\left[\left(z-\frac{1}{3}\right)\frac{18z^2-2z}{6\left(z-\frac{1}{2}\right)\left(z-\frac{1}{3}\right)}z^{n-1}\right] = -4\left(\frac{1}{3}\right)^n$$

We will therefore have

$$(\mathcal{Z}^+)^{-1}\left\{\frac{18z^2-2z}{6z^2-5z+1}\right\} = 7\left(\frac{1}{2}\right)^n - 4\left(\frac{1}{3}\right)^n$$

and finally the *zero-input component* of the state vector will have the form

$$(\mathcal{Z}^+)^{-1}\left\{(\boldsymbol{I}-z^{-1}\boldsymbol{A})^{-1}\boldsymbol{q}[0]\right\} = \begin{bmatrix} 14\left(\frac{1}{2}\right)^n - 12\left(\frac{1}{3}\right)^n \\ 7\left(\frac{1}{2}\right)^n - 4\left(\frac{1}{3}\right)^n \end{bmatrix}$$

On the other hand, to calculate the *zero-state component*, we need the unilateral transform $\mathcal{Z}$ of the input signal $x[n] = u[n]$, which is given by the relation $\mathcal{X}^+(z) = z/(z-1)$. We will still have

$$z\boldsymbol{I} - \boldsymbol{A} = \begin{bmatrix} z & -1 \\ \frac{1}{6} & \frac{6z-5}{6} \end{bmatrix}$$

with the determinant

$$\text{Det}(z\boldsymbol{I}-\boldsymbol{A}) = \frac{6z^2-5z+1}{6}$$

and therefore the inverse matrix $(z\boldsymbol{I}-\boldsymbol{A})^{-1}$ is calculated as

$$(z\boldsymbol{I}-\boldsymbol{A})^{-1} = \begin{bmatrix} \frac{6z-5}{6z^2-5z+1} & \frac{6}{6z^2-5z+1} \\ \frac{-1}{6z^2-5z+1} & \frac{6z}{6z^2-5z+1} \end{bmatrix}$$

By substituting the foregoing expressions into the argument of the inverse unilateral transform of the second term of the defining equation of the state vector, it turns out that

$$(z\boldsymbol{I}-\boldsymbol{A})^{-1}\boldsymbol{B}\mathcal{X}^+(z) = \begin{bmatrix} \frac{6z}{(6z^2-5z+1)(z-1)} \\ \frac{6z^2}{(6z^2-5z+1)(z-1)} \end{bmatrix}$$

that is, the *zero-state component* of the desired state vector will be calculated from the inverse transform

$$(\mathcal{Z}^+)^{-1}\left\{(z\boldsymbol{I}-\boldsymbol{A})^{-1}\boldsymbol{B}\mathcal{X}^+(z)\right\} = \begin{bmatrix} (\mathcal{Z}^+)^{-1}\left\{\frac{6z}{(6z^2-5z+1)(z-1)}\right\} \\ (\mathcal{Z}^+)^{-1}\left\{\frac{6z^2}{(6z^2-5z+1)(z-1)}\right\} \end{bmatrix}$$

In this case, the polynomial of the denominator has three roots with values $z_1 = 1/2$, $z_2 = 1/3$, and $z_3 = 1$, which are also the poles of the functions to be inverted. Starting from the first expression, the auxiliary function $\mathcal{Y}(z)$ is defined as

$$\mathcal{Y}(z) = \frac{6z^n}{(6z^2 - 5z + 1)(z - 1)}$$

with residue values at the positions of its poles equal to

$$\text{Res}\,[\mathcal{Y}(1/2)] = -12\left(\frac{1}{2}\right)^n, \quad \text{Res}\,[\mathcal{Y}(1/3)] = 9\left(\frac{1}{3}\right)^n, \quad \text{Res}\,[\mathcal{Y}(1)] = 3$$

On the other hand, the auxiliary function for the second expression is

$$\mathcal{Y}(z) = \frac{6z^{n+1}}{(6z^2 - 5z + 1)(z - 1)}$$

with residue values at the poles equal to

$$\text{Res}\,[\mathcal{Y}(1/2)] = -6\left(\frac{1}{2}\right)^n, \quad \text{Res}\,[\mathcal{Y}(1/3)] = 3\left(\frac{1}{3}\right)^n, \quad \text{Res}\,[\mathcal{Y}(1)] = 3$$

Combining the foregoing results, we get

$$(\mathcal{Z}^+)^{-1}\left\{\frac{6z}{(6z^2 - 5z + 1)(z - 1)}\right\} = -12\left(\frac{1}{2}\right)^n + 9\left(\frac{1}{3}\right)^n + 3, \quad \text{and}$$

$$(\mathcal{Z}^+)^{-1}\left\{\frac{6z^2}{(6z^2 - 5z + 1)(z - 1)}\right\} = -6\left(\frac{1}{2}\right)^n + 3\left(\frac{1}{3}\right)^n + 3$$

that is, the zero-state component of the state vector will be given by the equation

$$(\mathcal{Z}^+)^{-1}\left\{(z\boldsymbol{I} - \boldsymbol{A})^{-1}\boldsymbol{B}\mathcal{X}^+(z)\right\} = \begin{bmatrix} 9\left(\frac{1}{3}\right)^n - 12\left(\frac{1}{2}\right)^n + 3 \\ 3\left(\frac{1}{3}\right)^n - 6\left(\frac{1}{2}\right)^n + 3 \end{bmatrix}$$

Therefore, the desired state vector will emerge by *adding together the zero-input component and the zero-state component*. This process will lead to the result

$$\boldsymbol{q}[n] = \begin{bmatrix} 2\left(\frac{1}{2}\right)^n - 3\left(\frac{1}{3}\right)^n + 3 \\ \left(\frac{1}{2}\right)^n - \left(\frac{1}{3}\right)^n + 3 \end{bmatrix}$$

On the other hand, the output of the system through the application of the $\mathcal{Z}$ transform will be calculated from the equation

$$y[n] = (\mathcal{Z}^+)^{-1}\left\{\boldsymbol{C}(\boldsymbol{I} - z^{-1}\boldsymbol{A})^{-1}\boldsymbol{q}[0]\right\} + (\mathcal{Z}^+)^{-1}\left\{\left[\boldsymbol{C}(z\boldsymbol{I} - \boldsymbol{A})^{-1}\boldsymbol{B} + \boldsymbol{D}\right]\mathcal{X}^+(z)\right\}$$

Using the result of the previous question, we find

$$\boldsymbol{C}(\boldsymbol{I} - z^{-1}\boldsymbol{A})^{-1}\boldsymbol{q}[0] = \begin{bmatrix} -1 & 5 \end{bmatrix}\begin{bmatrix} \dfrac{12z^2 + 8z}{6z^2 - 5z + 1} \\ \dfrac{18z^2 - 2z}{6z^2 - 5z + 1} \end{bmatrix} = \frac{78z^2 - 18z}{6z^2 - 5z + 1}$$

Applying the method of complex integration and using the auxiliary function

$$\mathcal{Y}(z) = \frac{78z^2 - 18z}{6z^2 - 5z + 1} z^{n-1}$$

we find that the residues at the positions of its poles have the values

$$\text{Res}\,[\mathcal{Y}(1/2)] = 21\left(\frac{1}{2}\right)^n \quad \text{and} \quad \text{Res}\,[\mathcal{Y}(1/2)] = -8\left(\frac{1}{3}\right)^n$$

that is, the *zero-input response* of the system will have the form

$$y_{zi}[n] = 21\left(\frac{1}{2}\right)^n - 8\left(\frac{1}{3}\right)^n$$

In the next step of the process, we will need to calculate the *system response to the zero state*. It is not difficult to ascertain that the argument of the unilateral inverse $\mathcal{Z}$ transform of the second term of the output equation $y[n]$ has a value equal to

$$\left[\boldsymbol{C}(z\boldsymbol{I} - \boldsymbol{A})^{-1}\boldsymbol{B} + \boldsymbol{D}\right]\mathcal{X}^+(z) = \begin{bmatrix}-1 & 5\end{bmatrix}\begin{bmatrix}\dfrac{6z}{(6z^2 - 5z + 1)(z-1)} \\ \dfrac{6z^2}{(6z^2 - 5z + 1)(z-1)}\end{bmatrix} = \frac{30z^2 - 6z}{(6z^2 - 5z + 1)(z-1)}$$

Applying again the method of complex integration for the auxiliary function

$$\mathcal{Y}(z) = \frac{30z^2 - 6z}{(6z^2 - 5z + 1)(z-1)} z^{n-1} = \frac{30z^2 - 6z}{6\,[z - (1/2)]\,[z - (1/3)](z-1)} z^{n-1}$$

we find that the values of the residues at the positions of the poles are given by the relations

$$\text{Res}\,[\mathcal{Y}(1/2)] = -18\left(\frac{1}{2}\right)^n, \quad \text{Res}\,[\mathcal{Y}(1/3)] = 6\left(\frac{1}{3}\right)^n, \quad \text{Res}\,[\mathcal{Y}(1)] = 12$$

Therefore, the *zero-state response* of the system will be

$$y_{zs}[n] = -18\left(\frac{1}{2}\right)^n + 6\left(\frac{1}{3}\right)^n + 12$$

while the *total response* of the system is calculated as

$$y[n] = y_{zi}[n] + y_{zs}[n] = \left[3\left(\frac{1}{2}\right)^n - 2\left(\frac{1}{3}\right)^n + 12\right] u[n]$$

with the step function $u[n]$ being added, as usual, to make the system output a causal signal.

Problem 9.7.2 Use the method of the $\mathcal{Z}$ transform to solve the state equation

$$\boldsymbol{q}[n+1] = \boldsymbol{A}\boldsymbol{q}[n] + \boldsymbol{B}x[n]$$

for parameter values

$$\boldsymbol{A} = \begin{bmatrix} 0 & 1 \\ -1/8 & 3/4 \end{bmatrix}, \quad \boldsymbol{B} = \begin{bmatrix} 0 \\ 1 \end{bmatrix}, \quad x[n] = \left(\frac{1}{2}\right)^n u[n]$$

and initial state vector $\boldsymbol{q}[0] = \begin{bmatrix}4 & 2\end{bmatrix}^T$.

Solution Using the methodology of the preceding exercise we find that

$$(\boldsymbol{I} - z^{-1}\boldsymbol{A})^{-1} = \begin{bmatrix} \dfrac{8z^2 - 6z}{8z^2 - 6z + 1} & \dfrac{8z}{8z^2 - 6z + 1} \\ \dfrac{-z}{8z^2 - 6z + 1} & \dfrac{8z^2}{8z^2 - 6z + 1} \end{bmatrix}$$

so the *zero-input component* of the state vector will be calculated from the inverse transform

$$(\mathcal{Z}^+)^{-1}\left\{(\boldsymbol{I} - z^{-1}\boldsymbol{A})^{-1}\boldsymbol{q}[0]\right\} = \begin{bmatrix} (\mathcal{Z}^+)^{-1}\left\{\dfrac{32z^2 - 8z}{8z^2 - 6z + 1}\right\} \\ (\mathcal{Z}^+)^{-1}\left\{\dfrac{16z^2 - 4z}{8z^2 - 6z + 1}\right\} \end{bmatrix}$$

The functions to be inverted have common poles at positions $z_1 = 1/2$ and $z_2 = 1/4$. Applying the method of the complex integration, the auxiliary function for the first expression is

$$\mathcal{Y}(z) = \frac{32z^2 - 8z}{8z^2 - 6z + 1} z^{n-1}$$

with residue values at the positions of its poles

$$\text{Res}\,[\mathcal{Y}(1/2)] = 4\left(\frac{1}{2}\right)^n \qquad \text{and} \qquad \text{Res}\,[\mathcal{Y}(1/4)] = 0$$

while the auxiliary function for the second expression is

$$\mathcal{Y}(z) = \frac{16z^2 - 4z}{8z^2 - 6z + 1} z^{n-1}$$

with residue values at the positions of its poles

$$\text{Res}\,[\mathcal{Y}(1/2)] = 2\left(\frac{1}{2}\right)^n \qquad \text{and} \qquad \text{Res}\,[\mathcal{Y}(1/4)] = 0$$

Combining the foregoing results, we conclude that the *zero-input component* of the state vector $\boldsymbol{q}[n]$ will have the form

$$\boldsymbol{q}_{zi}[n] = \left[4\left(\frac{1}{2}\right)^n \; 2\left(\frac{1}{2}\right)^n\right]^T$$

In the next step of the process, we need to calculate the unilateral $\mathcal{Z}$ transform of the input signal $x[n] = (1/2)^n u[n]$, which has the form $\mathcal{X}^+(z) = 2z/(2z - 1)$, as we easily find referring to tables of elementary transforms. It is also proven that

$$(z\boldsymbol{I} - \boldsymbol{A})^{-1} = \begin{bmatrix} \dfrac{8z - 6}{8z^2 - 6z + 1} & \dfrac{8}{8z^2 - 6z + 1} \\ \dfrac{-1}{8z^2 - 6z + 1} & \dfrac{8z}{8z^2 - 6z + 1} \end{bmatrix}$$

and therefore the *zero-state component* of the state vector will result from the inverse transform

$$(\mathcal{Z}^+)^{-1}\left\{(z\boldsymbol{I}-\boldsymbol{A})^{-1}\boldsymbol{B}\mathcal{X}^+(z)\right\} = \begin{bmatrix} (\mathcal{Z}^+)^{-1}\left\{\dfrac{16z}{(8z^2-6z+1)(2z-1)}\right\} \\ (\mathcal{Z}^+)^{-1}\left\{\dfrac{16z^2}{(8z^2-6z+1)(2z-1)}\right\} \end{bmatrix}$$

In this case, the functions to be inverted have a simple pole at the position $z = 1/4$ and a double pole[14] at the position $z = 1/2$. To invert the first expression, we will use the auxiliary function

$$\mathcal{Y}(s) = \frac{16z^n}{(8z^2-6z+1)(2z-1)}$$

with residue values at the positions of its poles

$$\text{Res}\,[\mathcal{Y}(1/4)] = 16\left(\frac{1}{4}\right)^n \quad \text{and} \quad \text{Res}\,[\mathcal{Y}(1/2)] = 8n\left(\frac{1}{2}\right)^n - 16\left(\frac{1}{2}\right)^n$$

while to invert the second expression, we will use the auxiliary function

$$\mathcal{Y}(s) = \frac{16z^{n+1}}{(8z^2-6z+1)(2z-1)}$$

with residue values at the positions of its poles

$$\text{Res}\,[\mathcal{Y}(1/4)] = 4\left(\frac{1}{4}\right)^n \quad \text{and} \quad \text{Res}\,[\mathcal{Y}(1/2)] = 4n\left(\frac{1}{2}\right)^n - 4\left(\frac{1}{2}\right)^n$$

as can be proven. Combining the foregoing results leads to the conclusion that the *zero-state component* of the state vector will be given by the relation

$$\boldsymbol{q}_{zs}[n] = \begin{bmatrix} 16\left(\frac{1}{4}\right)^n + 8n\left(\frac{1}{2}\right)^n - 16\left(\frac{1}{2}\right)^n \\ 4\left(\frac{1}{4}\right)^n - 4\left(\frac{1}{2}\right)^n + 4n\left(\frac{1}{2}\right)^n \end{bmatrix}$$

while the state vector itself will emerge as the sum

$$\boldsymbol{q}[n] = \boldsymbol{q}_{zi}[n] + \boldsymbol{q}_{zs}[n] = \begin{bmatrix} 16\left(\frac{1}{4}\right)^n + 8n\left(\frac{1}{2}\right)^n - 12\left(\frac{1}{2}\right)^n \\ 4\left(\frac{1}{4}\right)^n - 2\left(\frac{1}{2}\right)^n + 4n\left(\frac{1}{2}\right)^n \end{bmatrix}$$

This result is the same as the one we calculated in Exercise 7.11.6 (page 629) by solving the state equation using the methodology presented in the chapter of discrete-time systems. The reader is encouraged to determine the system's output if the matrices $\boldsymbol{C}$ and $\boldsymbol{D}$ have the forms $\boldsymbol{C} = \begin{bmatrix}-1/8 & 3/4\end{bmatrix}$ and $\boldsymbol{D} = \begin{bmatrix}1\end{bmatrix}$ and to verify that this is the same as the one calculated in the earlier exercise by applying the method presented in the previous chapter.

[14] To see how to calculate the residue of a complex function at the position of a higher-order pole, refer to the footnote in Sect. 8.7.1 on page 672.

9.7.2 Calculation of Exponential Matrix A^n

If we compare the equation defining the zero-input component of the state vector, as was calculated in the traditional way (Eq. 7.55) and which has the form

$$\boldsymbol{q}_{zi}[n] = \boldsymbol{A}^n \boldsymbol{q}[0] \tag{9.46}$$

with the corresponding relationship that arises through the application of the $\mathcal{Z}$ transform method and which is formulated as

$$\boldsymbol{q}_{zi}[n] = (\mathcal{Z}^+)^{-1}\left[(\boldsymbol{I} - z^{-1}\boldsymbol{A})^{-1}\boldsymbol{q}[0]\right] = (\mathcal{Z}^+)^{-1}\left[(\boldsymbol{I} - z^{-1}\boldsymbol{A})^{-1}\right]\boldsymbol{q}[0] \tag{9.47}$$

(where for the derivation of the last relationship we used the property of the linearity of the unilateral transform), it immediately follows that

$$\boldsymbol{A}^n = (\mathcal{Z}^+)^{-1}\left[(\boldsymbol{I} - z^{-1}\boldsymbol{A})^{-1}\right] \tag{9.48}$$

The last relationship offers another alternative method of calculating the exponential function $\boldsymbol{A}^n$, which is based on the use of the unilateral $\mathcal{Z}$ transform; the way it is applied is trivial since examples of calculating such expressions were presented in previous exercises, and readers can familiarize themselves with its use by solving the next exercise.

9.7.3 Transfer Function and Dynamic Equations

Using the argument developed in the fifth chapter regarding the method of calculating the transfer function of continuous-time systems from their state model, we can formulate entirely similar conclusions for discrete-time systems. Therefore, considering a discrete-time system described by the dynamic equations

$$\boldsymbol{q}[n+1] = \boldsymbol{A}\boldsymbol{q}[n] + \boldsymbol{B}x[n]$$
$$y[n] = \boldsymbol{C}\boldsymbol{q}[n] + \boldsymbol{D}x[n]$$

(for simplicity's sake we consider a SISO system) we can calculate its explicit *transfer function* from the relation

$$\mathcal{H}(z) = \boldsymbol{C}(z\boldsymbol{I} - \boldsymbol{A})^{-1}\boldsymbol{B} + \boldsymbol{D} = \frac{\boldsymbol{C}[\mathrm{Adj}(z\boldsymbol{I} - \boldsymbol{A})]\boldsymbol{B}}{\det(z\boldsymbol{I} - \boldsymbol{A})} + \boldsymbol{D} \tag{9.49}$$

For the method of deriving the foregoing relation and the discussion related to it, the reader is referred to Sect. 5.11.3 on page 477. On the other hand, if we know the *transfer function* of the system—which we can calculate either from the *difference equation* that describes it or from the *$\mathcal{Z}$ transform of its impulse response*—we can construct the *dynamic equations of the system in the state space* by appropriately selecting the *state variables* of the problem. This process is left as an exercise to the reader, who will have to work through the method used in Sect. 5.11.4 (page 486). In order, for example, to construct the *controller's normal form*, we need to formulate the system's transfer function

$$\mathcal{H}(z) = \frac{\mathcal{Y}(z)}{\mathcal{X}(z)} = \frac{\beta_0 + \beta_1 z^{-1} + \beta_2 z^{-2} + \cdots + \beta_{N-1} z^{-N+1}}{1 + \alpha_1 z^{-1} + \alpha_2 z^{-2} + \cdots + \alpha_N z^{-N}} \tag{9.50}$$

(analogously to the case of continuous-time systems we consider the case $M = N - 1$) as the product

$$\mathcal{H}(z) = \underbrace{\left\{\frac{1}{1+\alpha_1 z^{-1} + \alpha_2 z^{-2} + \cdots + \alpha_N z^{-N}}\right\}}_{\mathcal{H}_1(z)} \underbrace{\left\{\beta_0 + \beta_1 z^{-1} + \beta_2 z^{-2} + \cdots + \beta_{N-1} z^{-N+1}\right\}}_{\mathcal{H}_2(z)}$$

and then choose the state variables in the way we did then; similar conclusions apply to the *observer's normal form*. Finally, regarding the so-called *minimum dimension implementations*, similar conclusions apply to those formulated for continuous-time systems, and more specifically:

(a) *A system implementation is characterized as a minimum dimension implementation when it is both controllable and observable.*
(b) *All minimum dimension implementations that describe a specific transfer function are related to each other through similarity transformations.*

To prove the foregoing properties, the reader is referred to the bibliography, although they can be proven by analogy applying the arguments formulated for continuous-time systems.

9.8 Digital Processing of Analog Signals

Taking into account that in the general case, processing methods in the domain of discrete-time are *more accurate and effective* compared to the corresponding methods in the domain of continuous-time, it is common practice to transform an analog signal $x_\alpha(t)$ into its discrete equivalent $x[n]$, to carry out on it the appropriate forms of processing in each case and then to *transform* the processed digital scheme $y[n] = \mathcal{T}\{x[n]\}$ (where $\mathcal{T}$ is the transformation implemented by the discrete-time system on the input signal $x[n]$) back to the analog signal $x_\alpha(t)$ from which it originated. Both of these transformation processes were described in previous sections, with the process of converting an analog signal to digital based on the periodic sampling of the analog signal with a sampling period T_s, leading to the discrete-time signal $x[n] = x_\alpha(nT_s)$, and the process of converting a digital signal to analog based on the reconstruction of the analog signal from its samples, which, as we have shown, is based on the use of the interpolation equation

$$x_r(t) = \sum_{n=-\infty}^{\infty} x_\alpha(nT_s) \left\{\sin\left[\frac{\pi}{T_s}\left(t - nT_s\right)\right]\right\} \Big/ \left\{\frac{\pi}{T_s}(t - nT_s)\right\}$$

with the signal $x_r(t)$ being identical to the *original* analog signal $x_\alpha(t)$, if the undesirable phenomenon of frequency folding does not occur, or it constitutes a modified version of it, in the opposite case. During the application of this process, it is particularly important to choose in advance the *bandwidth* of the signals we are interested in processing, since this bandwidth determines, as is well known, the maximum sampling frequency. The aforementioned processes are supposed to represent the ideal situation, with the conversion arrangement of a continuous to a discrete-time signal being characterized as *ideal converter C/D (continuous/discrete)* and the conversion arrangement of the discrete back to a continuous-time signal being characterized as *ideal converter D/C (discrete/continuous)* (Fig. 9.6). The characterization *ideal* is due to the assumption that the analog signal being sampled is a signal with *strictly limited bandwidth*, while the low-pass filter used in the process of reconstructing the continuous-time signal from its samples is an *ideal filter*.

Let us spend some time examining the ideal system of Fig. 9.6. It is obvious that the foregoing interpolation equation holds as it is *only for the case of the identity system* described by the difference equation $y[n] = x[n]$. In the general case it holds that $y[n] = \mathcal{T}\{x[n]\} = \mathcal{T}\{x_\alpha(nT_s)\}$, and consequently, the output of the system will be, not $x_r(t)$, but some other signal $y_r(t)$. If we symbolize by $\mathcal{H}_r(j\omega)$ the frequency response of the converter D/C and use the symbolisms $\mathcal{Y}_r(j\Omega) = \mathcal{F}\{y_r(t)\}$, $\mathcal{X}(e^{j\Omega}) = \mathcal{F}\{x[n]\}$, and $\mathcal{Y}(e^{j\Omega}) = \mathcal{F}\{y[n]\}$ (where $\mathcal{F}$ is the Fourier transform of continuous or discrete-time, depending on the case), then, as we saw in the proof of the *sampling theorem* (Sect. 6.2, page 501), we will have the equation (suitably adapted for the case)

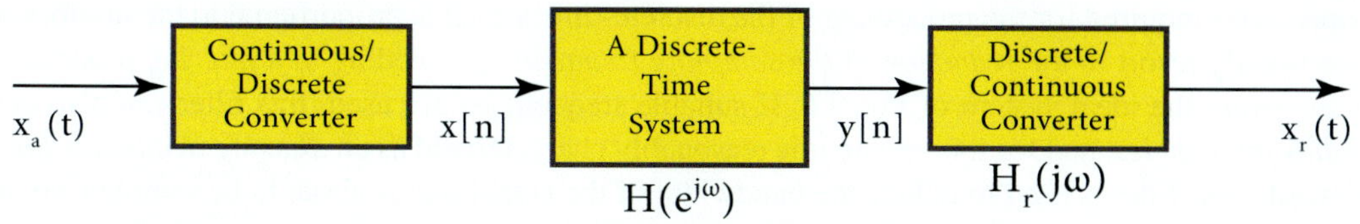

Fig. 9.6 Processing of continuous-time signals in domain of discrete-time. Note that although a discrete-time system is included in the above diagram, the composite system that arises in the aforementioned way is a continuous-time system (since both its input and its output are continuous-time signals) whose behavior depends on the choice of the discrete-time system and the sampling frequency

$$\mathcal{Y}_r(j\Omega) = \mathcal{H}_r(j\Omega)\mathcal{Y}(e^{j\Omega T_s}) = \begin{cases} T_s\mathcal{Y}(e^{j\Omega T_s}) & |\Omega| < \Omega_c \\ 0 & \text{in the opposite case} \end{cases}$$

where $\Omega_c = \omega/T_s = \pi/T_s$ (where we use the maximum frequency of a discrete-time signal $\omega = \pi$).

Assuming that the discrete-time system that is interposed between the C/D and D/C converters of Fig. 9.6 is a LTI system with frequency response $\mathcal{H}(e^{j\omega})$, the application of the *convolution theorem* will give us

$$\mathcal{Y}(e^{j\omega}) = \mathcal{H}(e^{j\omega})\mathcal{X}(e^{j\omega}) = \mathcal{H}(e^{j\Omega T_s})\mathcal{X}(e^{j\Omega T_s})$$

(since $\omega = \Omega T_s$) and the foregoing relation becomes

$$\mathcal{Y}_r(j\Omega) = \mathcal{H}_r(j\Omega)\mathcal{Y}(e^{j\Omega T_s}) = \mathcal{H}_r(j\Omega)\mathcal{H}(e^{j\Omega T_s})\mathcal{X}(e^{j\Omega T_s}) = \mathcal{H}_r(j\Omega)\mathcal{H}(e^{j\Omega T_s})\frac{1}{T_s}\sum_{k=-\infty}^{\infty}\mathcal{X}_\alpha\left[j\left(\Omega - \frac{2k\pi}{T_s}\right)\right]$$

(where we used Eq. 6.15—page 512—for $\Omega = \omega/T_s$). At this point we make the very important observation that if $\mathcal{X}_\alpha(j\Omega) = 0$ for $|\Omega| \geq \pi/T_s$, then from the preceding sum of infinite terms, the only term that remains is the one that corresponds to the value $k = 0$. In addition, because for $|\Omega| < \pi/T_s$ it holds that $\mathcal{H}_r(j\Omega) = T_s$, we will have $\mathcal{H}_r(j\Omega)(1/T_s) = 1$ and the foregoing relation is written

$$\mathcal{Y}_r(j\Omega) = \begin{cases} \mathcal{H}(e^{j\Omega T_s})\mathcal{X}_\alpha(j\Omega) & |\Omega| < \pi/T_s \\ 0 & \text{in the opposite case} \end{cases}$$

Consequently, if the Fourier transform $\mathcal{X}_\alpha(j\Omega)$ is of limited bandwidth, while the sampling of the analog signal is performed at a frequency greater than or equal to the *Nyquist frequency*, then the output of the composite system (which includes the two converters and the discrete-time system) is related to the input of the system in the frequency domain, according to the relation

$$\mathcal{Y}_r(j\Omega) = \mathcal{H}_{\text{eff}}(j\Omega)\mathcal{X}_\alpha(j\Omega)$$

where

$$\mathcal{H}_{\text{eff}}(j\Omega) = \begin{cases} \mathcal{H}(e^{j\Omega T_s}) & |\Omega| < \pi/T_s \\ 0 & |\Omega| \geq \pi/T_s \end{cases}$$

In other words, the ideal system that allows the processing of analog signals in the domain of discrete-time can be considered a *LTI system* whose frequency response is given by the last equation.

However, it is not difficult to understand that the foregoing two conditions that appear in the definition of the ideal system *cannot be satisfied in practice* since, on the one hand, real signals *do not have strictly limited bandwidth* and, on the other hand, ideal filters *are not implementable systems* since, as is well known, they lead to noncausal impulse responses. For this reason, in practice we resort to the use of *approximate* systems. Specifically, the ideal C/D system is approximated by an analog-to-digital signal conversion arrangement *(A/D converter)*, while the ideal D/C system is approximated by a digital-to-analog signal conversion arrangement *(D/A converter)*.

Another common problem encountered in practice is the appearance of *alias frequencies* if the periodic sampling is performed at a frequency that does not satisfy the Nyquist criterion, although, as noted, this is a problem that is easily addressed by choosing the appropriate value for this frequency. Generally speaking, in every case and because the number of mathematical operations required for the processing of the discrete-time signal is proportional to the number of samples that are recorded, we usually resort to the *reduction of the sampling frequency* proceeding to a *filtering* process, which consists of placing a filter before the ideal system of Fig. 9.6. If suitable frequencies are used, this filter can also contribute to the *elimination of alias frequencies*, and for this reason it is reasonably characterized as an *aliasing avoidance filter* (antialiasing filter). In other words, this filter is used to reduce the bandwidth of the signal that is about to be sampled, so that it is limited to the desired frequency range, resulting in the *minimization of distortion* caused by the cutoff of frequency components with values greater than $F_s/2$ (where F_s is the sampling frequency), while it is also used for the *limitation of the spectrum of additive noise*, which is usually characterized by a large bandwidth that exceeds the bandwidth of the signal. Therefore, using this filter, there is a reduction in the power of the additive noise, so that it is limited to the bandwidth of the signal. As for the operating characteristics of this filter, these depend on the type of signal and the application in which it is used. For example, if the application includes the *digital transmission of a voice signal* through a telephone channel, this filter can be designed to be characterized by a passband extending up to 3000 Hz, a transition bandwith a range of 400–500 Hz and a cutoff band for the elimination of frequencies with a value greater than 3500 Hz. The signal can be sampled at a frequency of 8000 Hz, which leads to a folding frequency equal to 4000 Hz. In this case, the undesirable effects of the phenomenon of the frequency folding is reduced to such an extent that it is considered negligible.

The system that arises from the addition of this additional filter is depicted in Fig. 9.7, with the subsystem that is enclosed within the frame easily recognized as the *ideal system* referred to earlier, before which, as we aforementioned, is placed in series the aforementioned filter. In other words, before the analog signal is sent to the ideal C/D converter, it passes through that filter, which ensures that its output, i.e., the signal that will enter the ideal system, is characterized by a *limited bandwidth* with frequencies less than the value π/T. Considering that the frequency response of this filter, in the ideal case, will be given by the relation

$$\mathcal{H}_{\text{aa}}(j\Omega) = \begin{cases} 1, \ |\Omega| < \Omega_c \le \pi/T, \\ 0, \ |\Omega| \ge \Omega_c \end{cases}$$

(in other words, we consider it an ideal filter) and symbolizing by $\mathcal{H}_{\text{cdc}}(j\Omega)$ the frequency response of the ideal system that is within the frame, the frequency response of the composite system of Fig. 9.7 according to the *convolution theorem* and according to our discussion in the section on series connections of LTI systems will be defined as the product

$$\begin{aligned}\mathcal{H}_{\text{eff}}(j\Omega) &= \mathcal{H}_{\text{aa}}(j\Omega) \times \mathcal{H}_{\text{cdc}}(j\Omega)\\ &= \begin{cases} 1, \ |\Omega| < \Omega_c \le \pi/T, \\ 0, \ |\Omega| \ge \Omega_c \end{cases} \times \begin{cases} \mathcal{H}(e^{j\Omega T_s}) & |\Omega| < \pi/T_s \\ 0 & |\Omega| \ge \pi/T_s \end{cases} = \begin{cases} \mathcal{H}(e^{j\Omega T_s}) & |\Omega| < \Omega_c \\ 0 & |\Omega| \ge \Omega_c \end{cases}\end{aligned}$$

where we used for the frequency response $\mathcal{H}_{\text{cdc}}(j\Omega)$ of the ideal system the expression for the quantity $\mathcal{H}_{\text{eff}}(j\Omega)$ constructed previously. This conclusion is particularly important because it tells us that this new system to which the ideal filter has been added to avoid the phenomenon of aliasing behaves as a *LTI system*, despite the fact that the input signal is not a strictly band-limited signal. Of course, in practice, there are no ideal filters and it is impossible to apply $|\mathcal{H}_{\text{aa}}(j\Omega)| = 0$ for $|\Omega| > \pi/T$. However, we can design a filter with very small values of the measure $|\mathcal{H}_{\text{aa}}(j\Omega)|$ for the aforementioned frequency range, so as to minimize to a great extent the impact of the aliasing phenomenon. Of course, in this case, the equality of the product of the frequency responses will apply *only approximately*, so that we can write that

$$H_{\text{eff}}(j\Omega) \approx H_{\text{aa}}(j\Omega)\mathcal{H}(e^{j\Omega T})$$

The previous discussion is based on the silent assumption that the filter connected before the ideal system to avoid the appearance of the aliasing phenomenon is characterized by a *very sharp transition at the cutoff frequency position*, while it *does not cause distortion due to time delay* within the passband. The fact that, in reality, this filter is analog, since the signal on which it is applied is an analog signal, constitutes a significant cost factor; indeed, these filters are manufactured using *active elements* and *integrated circuits*, which are much more expensive than the considerably cheaper digital processors used in digital filters. Apart from the high cost, these analog filters, with the sharp transition at the cutoff frequency position, are characterized by a *phase response with a high degree of nonlinearity*, especially in the passband. For all these reasons, it is particularly important to somehow eliminate the need for the use of these filters or, if such a thing is not possible, at least to resort to *compromises* regarding the requirements of the characteristics and the quality of their operation. Such an

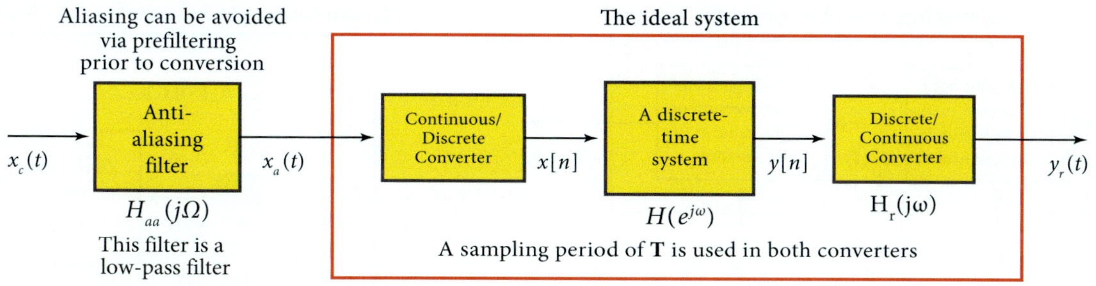

Fig. 9.7 Using a filter before the input of the analog signal to the system to avoid the phenomenon of aliasing. In some cases, the use of a second filter after the ideal D/C system is possible for the implementation of the appropriate processes in each case, such as, for example, the reconstruction of the output signal

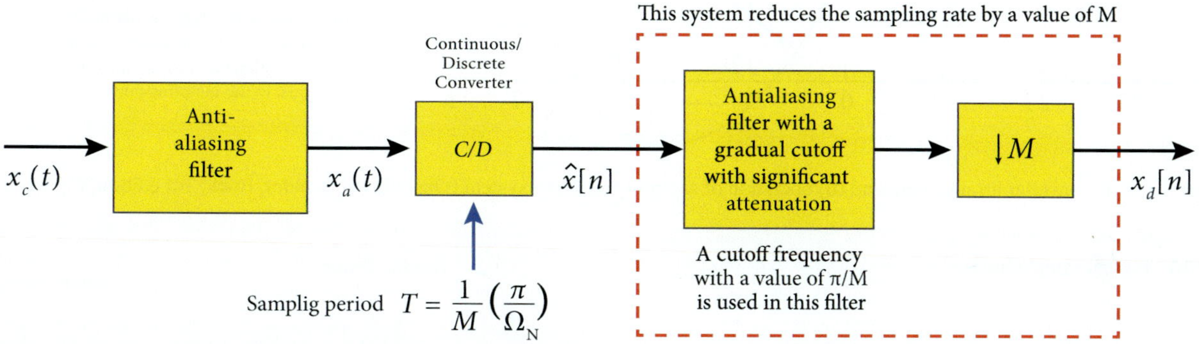

Fig. 9.8 A more compromising lower-cost solution

approach, which is depicted in Fig. 9.8, consists in the use of a quite simple and cheap analog filter to avoid the phenomenon of aliasing characterized by a *smoother and gradual transition* in the cutoff zone, so as to cause significant attenuation at the frequency $M\Omega_N$, where Ω_N is the highest frequency component that we wish to be preserved after the passage of the original analog signal through this filter. Then the process of converting an analog to a digital signal takes place, with a sampling rate much greater than the value $2\Omega_N$—for example, using the value $2M\Omega_N$—while subsequently a *subsampling* system is used (which includes a low-pass filter and a compressor)—so as to reduce the sampling frequency and to enable the processing of the signal in the discrete-time domain, with a relatively reduced number of necessary calculations. This process of changing the sampling rate is carried out in the way described in the previous sections.

9.8.1 Conversion of Analog to Digital Signal

After documenting the existence of the aliasing filter, let us now move on to the practical implementations of ideal continuous-to-discrete-time signal (C/D) and discrete-to-continuous-time signal (D/C) conversion arrangements. It is self-evident that *the use of a discrete-time signal, as defined in the previous sections, cannot occur in practice* because the sampling process, despite the fact that it takes place at specific time instances, is assumed to be carried out with *infinite accuracy* in terms of the value of the recorded samples, in the sense that the value of each sample of the discrete-time signal $x[n]$ is characterized by the presence of *an infinite number of decimal digits*. In practice, of course, this is not possible, but on the contrary, the whole process is carried out with *finite accuracy*. In other words, the result of the sampling is a *discrete sequence of samples of finite accuracy*, otherwise known, for reasons that we will explain later, as *quantized* samples. This sequence of quantized samples is known as a *digital signal*. The digital signal is therefore defined as a *discrete-time signal the values of whose samples are known not with infinite but finite accuracy*. The conversion of an analog signal into digital is carried out with the help of an *A/D (analog-to-digital) device* , that is, a physical, real device that can be implemented with the appropriate circuits and be used in practice. This device accepts at its input a value that expresses the amplitude of a voltage or a current (since analog signals appear either as voltages or as currents flowing through circuits) and transforms it in the appropriate way into a *sequence of binary numbers*, which constitutes the binary representation of the quantized value

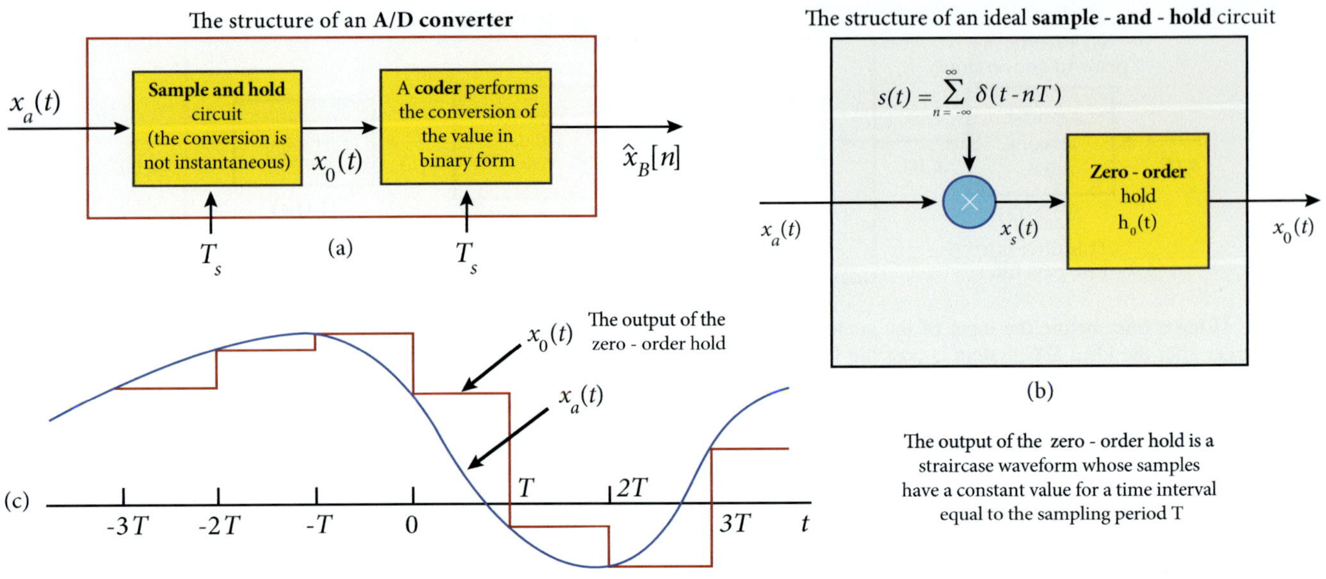

Fig. 9.9 (**a**) A/D converter arrangement and (**b**) operation of sampling–holding system (Oppenheim, Schafer, Buck, 3rd edition, Figures 4.51 and 4.52)

Fig. 9.10 Typical sample/hold circuit

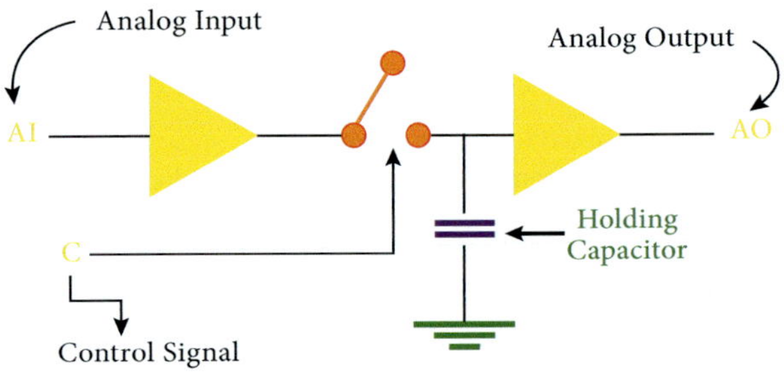

of the amplitude, that is, the value of the amplitude that has resulted from sampling that took place with past accuracy. The structure of such a device is depicted in Fig. 9.9a.

9.8.2 Returning to Sampling of Continuous-Time Signal

Observing Fig. 9.9, we find that *the A/D converter consists of a sampling device that records the value of the current sample and an encoder device that reads this value at its input and converts it into a binary number*, so that eventually the digital signal is produced. Note that because this conversion process *is not instantaneous* but requires some *very small but nonzero* time interval, after the recording of the sample during the sampling process, it is maintained for the appropriate time interval required in order to convert the sample value into binary code. The arrangement that performs this combination of sampling and temporary retention of the sample, for as long as such a thing is necessary, is characterized as a *sample-and-hold arrangement* and constitutes a digitally controlled analog circuit that uses a capacitor for the retention of the charge representing the recorded sample. Usually this sampling/hold circuit, a form of which is presented in Fig. 9.10, is *embedded* in the A/D arrangement system, the operation of which is characterized by the successive alternation from the sampling operation state to the hold operation state, with the first state being related to the recording of the next sample and the second state maintaining it for the time period required in order to construct its binary representation. The use of this hold circuit gives the A/D converter the ability to operate at a lower speed compared to that required for the recording of the sample, a fact that makes its existence an *essential prerequisite* in processes of converting an analog into a digital signal that use *high resolution* during the quantization process and are applied to signals characterized by *rapid time changes*, which refer to *wide bandwidth*. Although in the ideal case the sampling/hold circuit should not introduce the slightest degree of distortion and be modeled as an ideal sampling system, in practice, distortions and errors related to *the periodicity of the conversion*

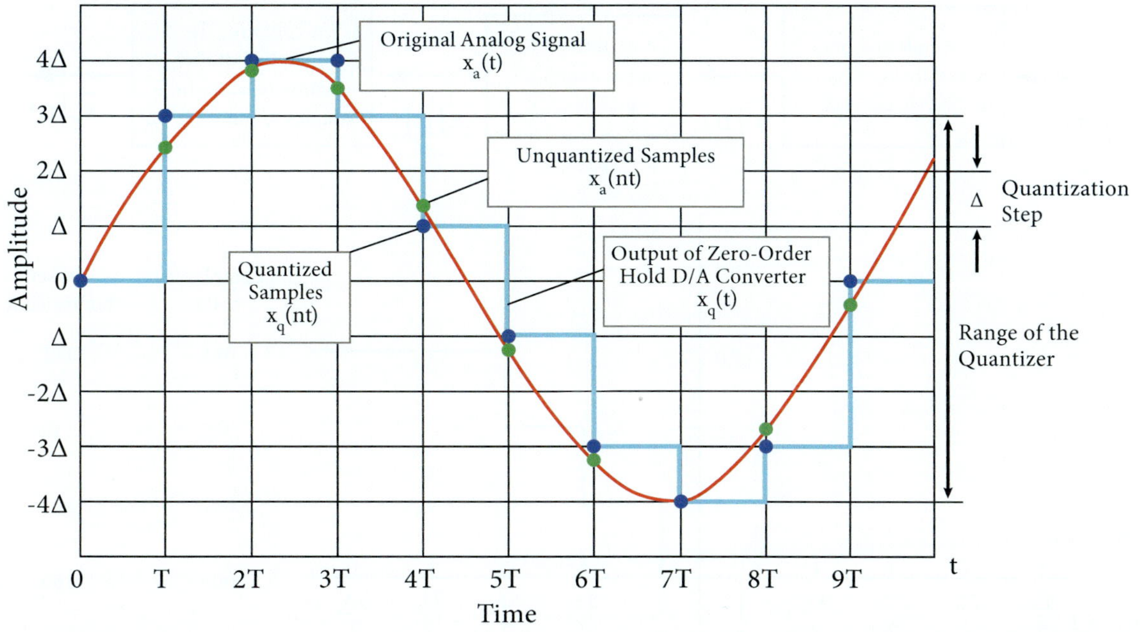

Fig. 9.11 Details of sampling and quantization processes taking place within conversion of an analog to digital signal. The scheme refers to the application of processes for the case of a simple sinusoidal signal (Proakis and Manolakis, Figure 1.21)

process, nonlinear changes during sampling, and the instability of the voltage value, which is maintained during the hold stage. However, in the description that follows, these errors are considered negligible.

From the mathematical side of the subject, the operation of the ideal sampling/hold arrangement is described by the equation

$$x_0(t) = \sum_{n=-\infty}^{\infty} x[n]h_0(t - nT)$$

where $x[n] = x_\alpha(nT_s)$ are the ideal samples of the signal $x_\alpha(t)$ undergoing sampling, while the signal $h_0(t)$, which is defined as

$$h_0(t) = \begin{cases} 1, \; 0 < t < T, \\ 0 \;\; \text{in the opposite case} \end{cases}$$

constitutes the impulse response of the so-called *zero-order sample holding system*, which only retains the current sample undergoing sampling. The defining equation of the signal $x_0(t)$ can also take the equivalent form

$$x_0(t) = h_0(t) * \sum_{n=-\infty}^{\infty} x_a(nT)\delta(t - nT)$$

from which we observe that the ideal sampling process and sample holding are equivalent to the *modulation* of the signal with the help of a *train of impulse functions*, which is accompanied by filtering performed by the sampling and holding system. The output of this system, along with the broader process of sampling and quantization presented later, is depicted in Fig. 9.11 and, as can be easily seen, takes the form of a *staircase waveform*, with the value of each sample remaining constant during the sampling period. In this way, the input transmitted to the encoder in order for it to perform the conversion to a binary number, which as we stated expresses a momentary voltage or current value, is constant for the time period required for this conversion.

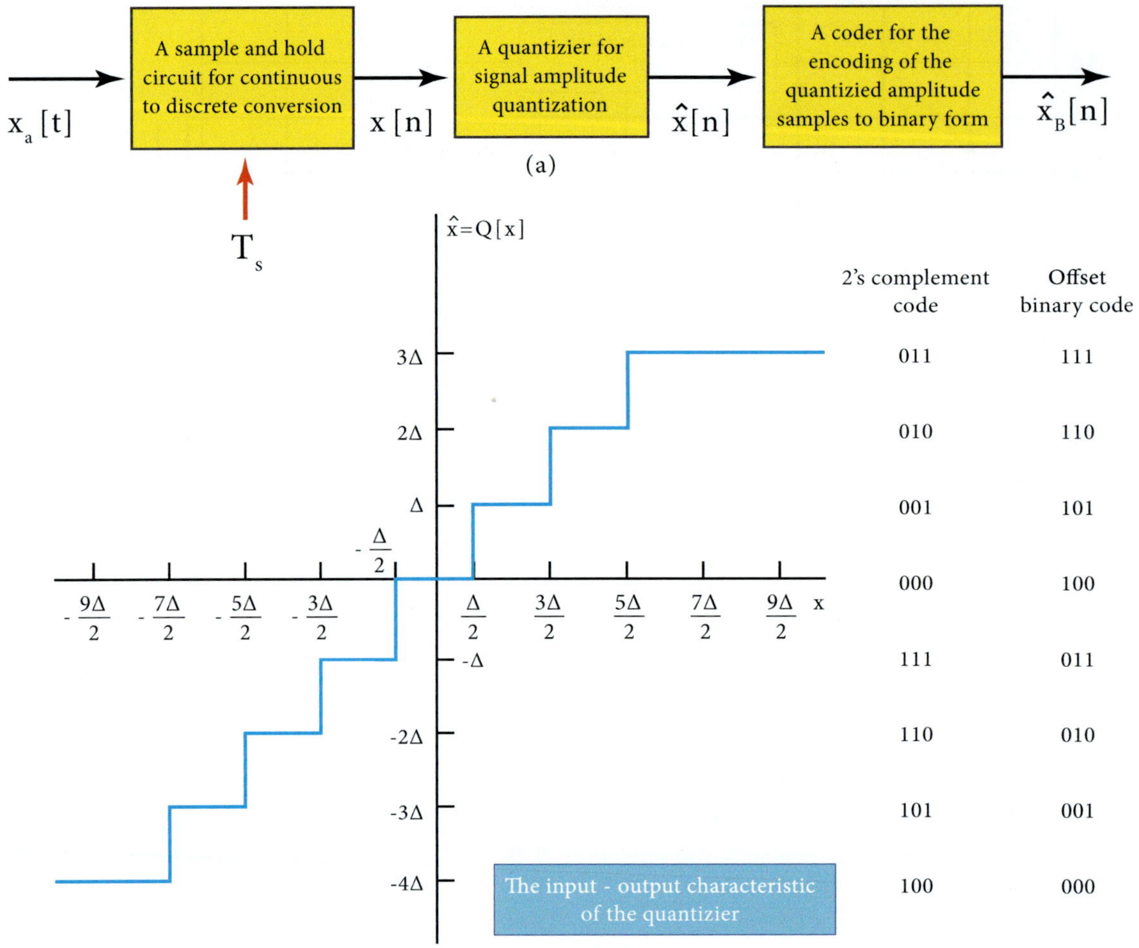

Fig. 9.12 (**a**) A/D converter layout with a quantizer between the sampling–holding system and the encoder and characteristic curve of a uniform eight-level quantizer that supports both positive and negative values

9.8.3 Quantization of Discrete-Time Signal

The conversion of the samples of the discrete-time signals that arise during the sampling process and are theoretically characterized by *infinite accuracy* to the samples of digital signals that are characterized, as we mentioned, by *finite accuracy* is known as *quantization*. This word has its roots in physics and, more specifically, in quantum mechanics, according to which the energy levels in the Bohr model for the hydrogen atom *have no energy values but only discrete and specific values*. In exactly the same way, the samples of digital signals have no values but values that belong to a set of specific values that are known in advance. The process of quantization is performed by the system of a *quantizer*, which we can consider as being either integrated into the encoder or located before it and immediately after the sampling–holding arrangement, as depicted in image (a) of Fig. 9.12. The operation of the quantizer is described by the equation

$$\tilde{x}[n] = Q(x[n])$$

where $x[n]$ is the nth sample of the discrete-time signal received at its input, which is characterized by *infinite accuracy*, and $\tilde{x}[n]$ is the quantized sample of the digital signal appearing as its output, which is characterized by *finite accuracy*. During the quantization process, the value of the input sample $x[n]$ *is rounded up or down* so as to coincide with one of the values of the set of predetermined values that have been defined during the design of the quantizer and that correspond to the permissible energy levels of the hydrogen atom. Since these values are entirely specific, the number of decimal places used to describe them is *finite*, and therefore, according to our earlier discussion, the same will apply to the value of the sample that is quantized, resulting in the creation of a *digital signal*, that is, a signal whose values belong to a *finite* set of values. These values are known as quantization levels. If these quantization levels are equidistant from each other, the quantizer is characterized as uniform or linear, while in the opposite case, it is characterized as nonuniform. In the general case and provided that the samples of the digital signal are to be used in numerical calculations, we resort for simplicity to the use of uniform quantizers for simplicity.

The characteristic curve of a uniform quantizer with input x and output $\widetilde{x}$ is depicted in Fig. 9.12b. This quantizer allows the encoding of a discrete-time signal that contains *both positive and negative values*. If it is known in advance that all samples of the signal have positive values, a different characteristic curve is used. If we want to define a quantizer that has a quantization level with a value equal to zero and at the same time an equal number of positive and negative quantization levels, the number of these levels must necessarily be an *odd* number. Due to the subsequent encoding in the binary system in which a specific number of bits is used, usually *the number of quantization levels is a power of 2*. In other words, if for the binary system encoding of each value of the digital signal N bits are required, which as is well known, allow the definition of 2^N different values, then a quantizer with 2^N quantization levels is used, with each level corresponding to one of the foregoing 2^N values. Therefore, during the operation of the quantizer, if the value of a sample of the digital signal is depicted at some quantization level, then its encoding in the binary system consists simply *in replacing it with the corresponding binary value*. In Fig. 9.12, the considered quantizer has $2^3 = 8$ quantization levels, and consequently, each level is described by a binary number of length 3 bits, with the numbering of levels starting from the most negative level corresponding to the value 000 and ending at the most positive level that corresponds to the value 001. Notice that, apart from the usual binary value, its representation according to the *2's complement* can be used (Sect. 12.5.1, page 1136), which is the most common encoding scheme used in most computers and microprocessors, but generally, *any scheme that involves assigning a binary sequence* or, as it is typically called, a *codeword* (codeword), can be used for the values of the quantization levels. Of course, in digital signal processing, for purely practical reasons, it is desirable to use an encoding scheme that allows the *direct use of the binary sequences* that arise in the appropriate numerical calculations in each case, and this scheme is simply that which is based on the use of the complement with respect to 2.

In the case of a *uniform quantizer*, the distance between any two successive quantization levels, as we noted earlier, is constant, and let us say that its value is equal to Δ (Fig. 9.12). It is not difficult to understand that the value of this constant depends *on the range of values of the signal being sampled*, as well as *on the number of quantization levels* or, equivalently, according to our earlier discussion, *on the number of bits used for encoding the quantized values*. Assuming for symmetry reasons that the minimum and maximum values of the analog signal being sampled are equal to A and $-A$, respectively (so the same will apply to the values of a discrete-time signal that results from the sampling), as happens, for example, with a sinusoidal signal of the form $x_\alpha(t) = A\sin(\omega t + \varphi)$, the range of signal values is equal to $E = A - (-A) = A + A = 2A$. If we assume that for the encoding of the quantized samples, k bits are used, this means that the number of different values supported by this number of bits, and therefore the number of different quantization levels is equal to 2^k. In other words, the value range $2A$ will be divided into 2^k quantization levels *equidistant from each other*, and therefore, the distance Δ between two successive quantization levels will be equal to

$$\Delta = \frac{2A}{2^k} = \frac{A}{2^{k-1}}$$

Note that the smaller quantization levels that have values $\pm\Delta$ correspond to the *least significant bit* of the codeword produced by the encoder, while the numerical correlation between the codewords and the quantized samples has the form $\widetilde{x}[n] = A\widetilde{x}_B[n]$ since for the case of binary representation according to the complement to 2 usually used in practice we consider that the quantity $\widetilde{x}_B[n]$ constitutes a binary number, such that $-1 < \widetilde{x}_B[n] < +1$. Of course, in practice, we consider that the used signal is *normalized*, that is, $A = 1$, so that $\widetilde{x}[n] = \widetilde{x}_B[n]$.

9.8.4 Quantization Error

Taking into account that during the quantization process each sample of the discrete-time signal *is rounded up or down*, so as to coincide with the nearest quantization level to it, it is obvious that the quantization process constitutes a *lossy process* (i.e., a process that leads to *information loss*), since during its implementation there is a *reduction in the accuracy* with which we know the values of the signal samples. Considering for simplicity's sake, a uniform quantizer characterized by the existence of eight quantization levels with integer values $\{0, 1, 2, 3, 4, 5, 6, 7\}$ and two samples with values $x[i] = 2.17$ and $x[j] = 3.89$ for some values of i and j, the application of the quantization process will lead to new samples with values $\widetilde{x}[i] = 2$ and $\widetilde{x}[j] = 4$. In other words, the foregoing process is characterized by error values $e[i] = \widetilde{x}[i] - x[i] = 2 - 2.17 = -0.17$ and $e[j] = \widetilde{x}[j] - x[j] = 4 - 3.89 = 0.11$, the first of which is negative, because the rounding was done downward, while the second is positive, because the rounding was done upward. If we apply this process to all the samples that are transmitted to the quantizer, an error value will emerge for each of them, with the set of all these values defining a *sequence* or a *discrete quantization error signal* defined as

$$e[n] = \widetilde{x}[n] - x[n]$$

Studying Fig. 9.12 and considering a sample with value $\Delta/2 < x[n] < 3\Delta/2$, it will obviously be rounded to the value Δ. That is, $\widetilde{x}[n] = \Delta$, and therefore we will have $e[n] = \Delta - x[n]$. We now observe that for $x[n] = \Delta/2$ we will have $e[n] = \Delta - (\Delta/2) = \Delta/2$, while for $x[n] = 3\Delta/2$, we will have $e[n] = \Delta - (3\Delta/2) = -\Delta/2$. Therefore, the range of error values will be

$$-\Delta/2 \leq e[n] < \Delta/2$$

In other words, the instantaneous quantization error *cannot become larger than half the quantization step*. Δ, which expresses the distance between two successive quantization levels.

Of course, for the preceding relationship to hold, the samples of the discrete-time signal must take values in the operating region of the quantizer, which in the case of Fig. 9.12 is the[15]

$$-9\Delta/2 \leq x[n] \leq 7\Delta/2$$

In the general case of a quantizer that uses $k + 1$ bits to describe the quantization levels and is characterized by a distance Δ between the successive quantization levels given by the foregoing relationship, the preceding operating region is

$$(-A - \Delta/2) < x[n] \leq (A - \Delta/2)$$

If the samples of the signal $x[n]$ are outside this region, there is a chance that the quantization error $e[n]$ may have a value greater than $\Delta/2$. In this case, the input samples are *clipped*, i.e., they do not appear at the output of the quantizer, with the additional error due to this clipping being characterized as *overload distortion* or as *overload noise* (in other words, the quantizer is characterized as *overloaded* because it has received at its input a sample with a value that is outside the range of values it supports and uses for its operation).

An example of quantization of a typical input signal is shown in Fig. 9.13. At the top of the figure, the original continuous-time signal (the dashed curve) is shown along with the quantized signal in the form of a *step function*, while at the bottom of the figure, the quantization error curve is shown. This error can be considered the application of a *modulation* process that arises from the application of the frequency components of the original signal to the nonlinear input/output characteristic of the quantizer.

In the general case, the quantization error signal $e[n]$ *is not known* or *cannot be calculated*, and its analysis relies on techniques that fall within the domain of *statistical signal processing* since the dependence of the quantization error on the characteristics of the input signal and on the nonlinear nature of the quantizer (for the general case) make it *impossible* to carry out a *deterministic analysis*, except for some simple cases. In other words, the quantization error is considered to be a *random signal*, and as such, it can be modeled as a signal noise, which is added to the original unquantized signal. Particularly problematic is the case where the dynamic range of the input signal, defined as the difference between the maximum and minimum values, is *greater* than the operating range of the quantizer, during which, as noted, there is a cutoff of the samples at the output of the quantizer and the appearance of overload noise. If we ignore this peculiar case and limit ourselves to the usual cases, we can proceed to a statistical analysis of the noise $e[n]$, accepting that it (a) is characterized by *uniform distribution* in the range of values $(-\Delta/2, \Delta/2)$, (b) constitutes a *stationary sequence* of white noise, and (c) is *uncorrelated with the input sequence* $x[n]$, for which we also accept that it is stationary and characterized by a zero mean. Although these assumptions do not hold in the general case, as is the case, for example, when the step function is used as a signal, they hold with good accuracy when the quantization step Δ is quite small, while the distance between two successive samples of the input signal, extends over several quantization levels, as is the case with vocal signals.

In cases where the prerequisites for the application of the techniques of statistical signal processing are met, the signal of the quantization error $e[n]$ can be considered as a sequence of *uniformly distributed white noise*, with a mean value equal to zero and variance given by the relation

[15] Note that in Fig. 9.12 the range of values on the vertical axis is not symmetrical but extends from -4Δ to 3Δ.

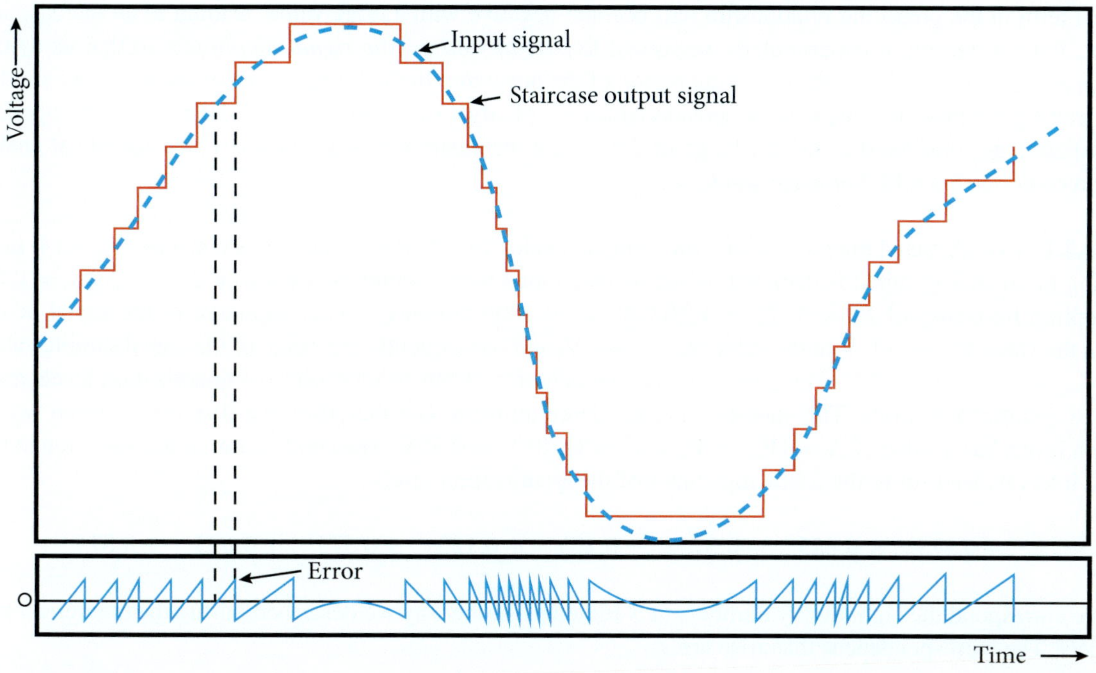

Fig. 9.13 The original input signal to the quantizer, along with the quantized signal that is created (top image) and the corresponding quantization error (bottom image) (Bennett, Figure 2)

$$\sigma_e^2 = \int_{-\Delta/2}^{\Delta/2} e^2 \frac{1}{\Delta} de = \frac{\Delta^2}{12}$$

For a uniform quantizer that uses $k + 1$ bits for the description of quantization levels and with a full-scale range equal to $2A$, the *variance* or *power* of the noise is given by the relation

$$\sigma_e^2 = \frac{2^{-2k} A^2}{12}$$

while the *spectral power density* is described by the equation

$$P_{ee}(e^{j\omega}) = \sigma_e^2 = \frac{2^{-2k} A^2}{12} \qquad |\omega| \leq \pi$$

Closely related to the foregoing concepts is also the concept of the *signal-to-noise ratio* (SNR). This concept allows the quantitative determination of the degradation of signal quality from the existence of white additive noise and is defined as the *ratio of signal power to noise power*. This ratio for a uniform quantizer that uses $k + 1$ bits for the quantization levels, expressed in decibels (dB), is given by the relation

$$\mathrm{SNR}_Q = 10 \log_{10}\left(\frac{\sigma_x^2}{\sigma_e^2}\right) = 10 \log_{10}\left(\frac{12 \cdot 2^{2k} \sigma_x^2}{A^2}\right) = 6.02k + 16.81 - 20 \log_{10}\left(\frac{A}{\sigma_x}\right)$$

The last term in the preceding relation in the general case depends on the operating region of the quantizer and on its input signal statistics, and its basic consequence is that the *signal-to-noise ratio increases approximately by 6 dB for each bit added to the word length of the quantized samples*. The quantity σ_x, which constitutes the root mean square value of the signal amplitude, must necessarily be *less than the maximum value* of this amplitude, while if, in contrast, it takes very large values, then the signal amplitude will take a value that is outside the operating region of the quantizer, resulting in the appearance of the distortions commented on previously. On the other hand, however, if this size takes a very small value, then the last term in the preceding relationship will become negative with a large value, leading to an increase in the value of the SNR. For this reason, it is particularly important to *carefully adjust the signal amplitude*, so that its dynamic range corresponds as much as possible to the operating region of the quantizer. In the foregoing analysis, where we used the simple sine wave as a signal, these two areas were considered for simplicity's sake, equal to each other, since the operating region of the quantizer was considered to have a range of $2A$, as can be easily verified. For a detailed statistical analysis of the quantization error, see Sect. 13.7 on page 1443.

Example 9.8.1 To understand the process of converting an analog to a digital signal, let us consider Fig. 9.14. In this figure, the sampling of an analog signal is depicted, which is performed with a sampling period of $T_s = 0.5$ ms $= 0.0005$ s, i.e., with a sampling frequency of $F_s = 1/T_s = 1/(0.0005\text{ s}) = 2000$ samples/s. The amplitude of the signal being sampled varies from the value $V_{\min} = -1$ V to the value $V_{\max} = +1$ V, and consequently the range of the signal sample values is equal to $V_{\max} - V_{\min} = 1 - (-1) = 2$ V. The quantizer arrangement uses 3 bits to represent the quantization levels and therefore uses $2^3 = 8$ quantization levels. The quantizer is considered uniform, and therefore, the distance between two successive quantization levels has a value of $\Delta = (V_{\max} - V_{\min})/2^3 = (2/8)\text{ V} = 0.25$V. The subdivision of the value range $[-1, 1]$ into eight equal intervals leads us to the following values of the quantization levels

$$-0.875 \quad -0.625 \quad -0.375 \quad -0.125 \quad +0.125 \quad +0.375 \quad +0.625 \quad +0.875$$

to which we correspond the eight binary numbers of 3-bit length in the way we described during the theoretical presentation of the subject. The correspondences that arise are

$$-0.875 \longrightarrow \mathbf{000} \quad -0.625 \longrightarrow \mathbf{001} \quad -0.375 \longrightarrow \mathbf{010} \quad -0.125 \longrightarrow \mathbf{011}$$

$$+0.125 \longrightarrow \mathbf{100} \quad +0.375 \longrightarrow \mathbf{101} \quad +0.625 \longrightarrow \mathbf{110} \quad +0.875 \longrightarrow \mathbf{111}$$

The details of the quantization process are included in Table 9.1. The first column of the table includes the samples of the discrete-time signal $x[n]$ (in Fig. 9.14 the values of these samples are in parentheses), the second column includes the values of the quantized samples of the digital signal $\tilde{x}[n]$ (note that each of these values corresponds to the value of some of the quantization levels), the third column includes the binary number or code word of length 3 bits corresponding to the quantization level of each value, while the fourth column includes the quantization error. After performing this process, the only thing we have to do is replace each sample of the digital sequence with its corresponding binary value and, thus, construct the binary digital signal. The result of this process is the binary sequence of 0 and 1 located at the bottom of the figure. Note that $\Delta/2 = 0.125$ and that $-0.125 < e[n] < 0.125$, in full agreement with what we mentioned during the theoretical presentation of the subject.

Table 9.1 Results of quantization process of Example 9.8.1

$x[n]$	$\tilde{x}[n]$	Binary	$e[n]$
+0.613000	+0.625	110	+0.01200
+0.391700	+0.375	101	−0.01670
+0.005200	+0.125	100	+0.11980
−0.214000	−0.125	011	+0.08900
−0.170100	−0.125	011	+0.04510
+0.230000	+0.125	100	−0.10500
+0.589210	+0.625	110	+0.03579
+0.653100	+0.625	110	−0.02810
+0.198000	+0.125	100	−0.07300
−0.360000	−0.375	010	−0.01500
−0.762000	−0.875	000	−0.11300
−0.873400	−0.875	000	−0.00160
−0.660000	−0.625	001	+0.00350

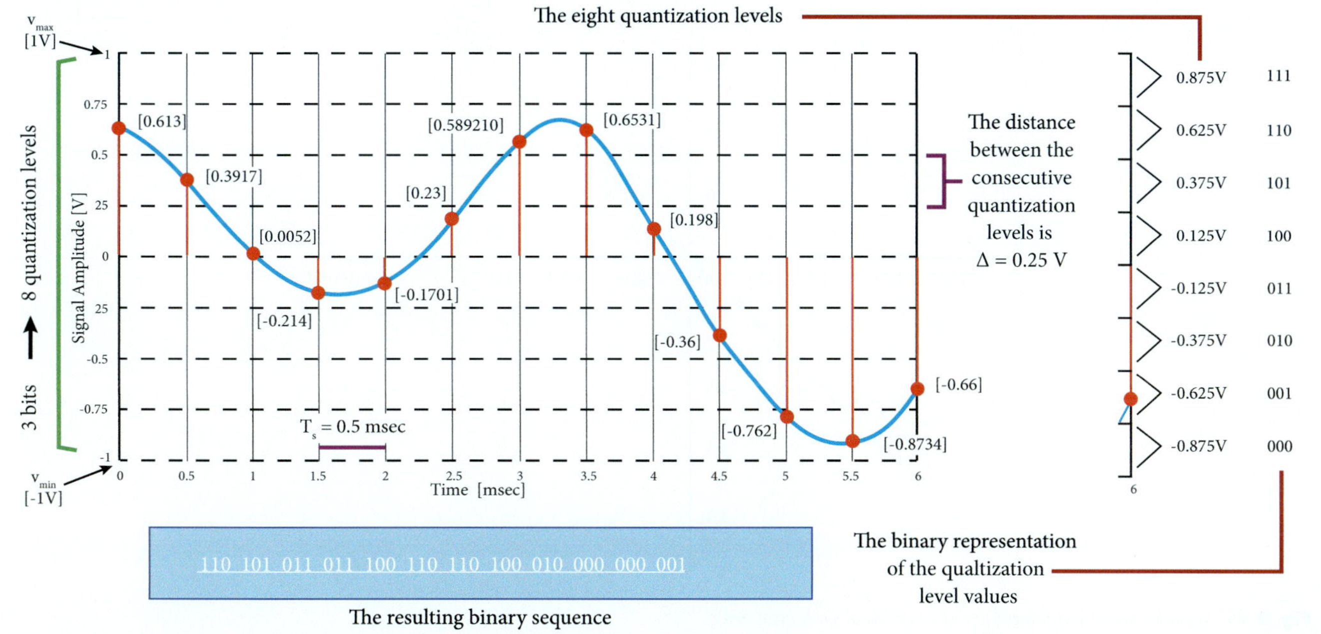

Fig. 9.14 Demonstration of quantization process with uniform eight-level quantizer

9.8.5 Conversion of Digital Signal to Analog Signal

In the proof of the sampling theorem, we explained the way in which a continuous-time signal of limited bandwidth can be reconstructed from a sequence of samples, using low-pass filtering. This *reconstruction process in the frequency domain* is described by the equation $\mathcal{X}_r(j\Omega) = \mathcal{X}(e^{j\Omega T})\mathcal{H}_r(j\Omega)$, where $\mathcal{X}(e^{j\omega})$ is the *Fourier transform of discrete-time* of the sequence of samples, while $\mathcal{X}_r(j\Omega)$ is the *Fourier transform of the reconstructed continuous-time signal*. As for the ideal low-pass filter of the reconstruction, which in a previous section we characterized as the ideal C/D converter, this is characterized by a frequency response of the form

$$H_r(j\Omega) = \begin{cases} T, & |\Omega| < \pi/T, \\ 0, & |\Omega| \geq \pi T \end{cases}$$

with the correlation between the reconstructed continuous-time signal and the sequence of samples from which it originated having the form

$$x_r(t) = \sum_{n=-\infty}^{\infty} x_\alpha(nT_s)\left\{\sin\left[\frac{\pi}{T_s}\left(t - nT_s\right)\right]\right\} \Big/ \left\{\frac{\pi}{T_s}(t - nT_s)\right\}$$

As we have already proven, this expression arises from the linear combination of an infinite number of time-shifted copies of the function

$$g(t) = \frac{\sin(\pi t/T_s)}{\pi t/T_s}$$

which is considered the *ideal interpolation function*, with the coefficients of this linear combination, which are treated as weight functions, being the samples $x_\alpha(nT_s)$ that have arisen from the sampling of the original analog signal, with a sampling period T_s. This ideal reconstruction filter characterized by an impulse response of the form $h(t) = g(t)$, where $g(t)$ is the foregoing function, receives as input the discrete-time signal $x[n]$ expressed as a linear combination of time-shifted impulse functions of the form

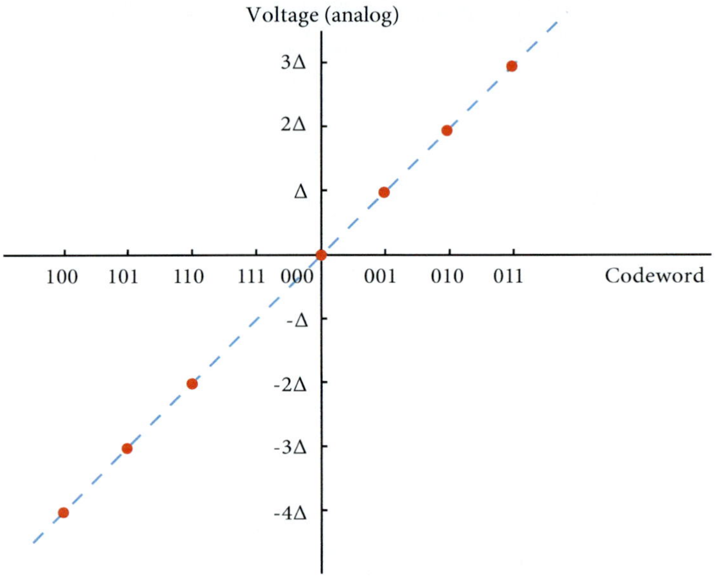

Fig. 9.15 Input/output characteristic curve of ideal D/A converter

$$x[n] = \sum_{n=-\infty}^{\infty} x_\alpha(nT_s)\delta(t - nT_s)$$

and produces as its output the reconstructed analog signal $x_r(t)$. This ideal D/C converter, as we have already noted, *cannot be constructed in practice*, and despite the fact that it can be approximated to some extent by introducing some delay, it does not constitute a practical solution for most applications that involve digital-to-analog signal conversion. In contrast, the circuit used in practice that constitutes an *approximate version of the ideal D/C converter* is a digital-to-analog signal conversion arrangement [D/A (digital-to-analog) converter], which is used in combination with a *sampling/holding circuit* followed by a low-pass filter. During its operation, the D/A converter receives at its input electrical signals that represent one or more codewords $\tilde{x}_B$ $[n]$ (i.e., strings of bits that are supposed to have arisen through the use of an encoder in the way explained in the previous section) and produces as its output a voltage or a current that corresponds to a continuous-time signal, which is proportional to the value of the corresponding codeword, which therefore undergoes *scaling* by a proportionality factor A. In other words, in the ideal case at least, the input/output characteristic curve of this system would be a straight line with a slope A, which passes through the origin (examples of such a characteristic curve for three-length codewords bits are depicted in Fig. 9.15), although in practical arrangements, the characteristic curve is not a straight line, due to all kinds of errors involved in the process.

In order for the sampling/holding circuit to maintain the stable output of the D/A system until the next sample appears during the next sampling period, the analog signal is approximated as *a series of orthogonal pulses* whose height is proportional to the corresponding value of the pulse signal. In other words, the output of the sampling/holding circuit, which is the signal $x_{\mathrm{DA}}(t)$, is approximated by a *step function* that takes as input a sample of the signal from the D/A converter and keeps its value stable for a time period of duration T_s seconds. After the passage of this time period, the next sample appears at the input of the function, which is read by the function and remains stable for the next T_s seconds, and so forth. From the mathematical side of the subject, this behavior requires the modeling of the sampling holding arrangement as a filter with impulse response

$$h_0(t) = \begin{cases} 1\ 0 < t < T \\ 0 \text{ anywhere else} \end{cases}$$

and frequency response

$$\mathcal{H}_0(j\omega) = T\frac{\sin(\omega T/2)}{\omega T/2}\exp\left(-j\frac{\omega T}{2}\right)$$

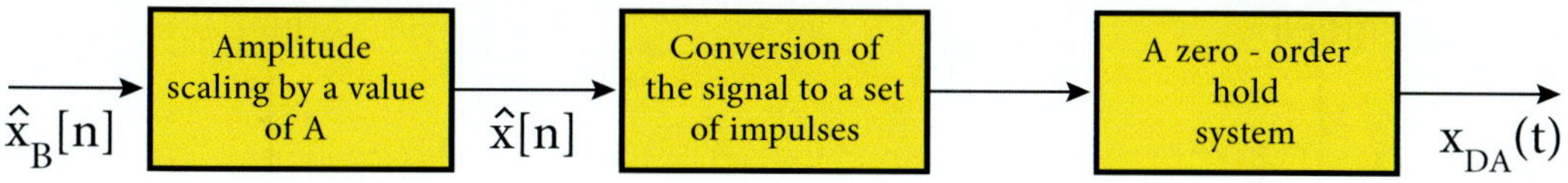

Fig. 9.16 Schematic diagram of digital-to-analog signal converter (D/A)

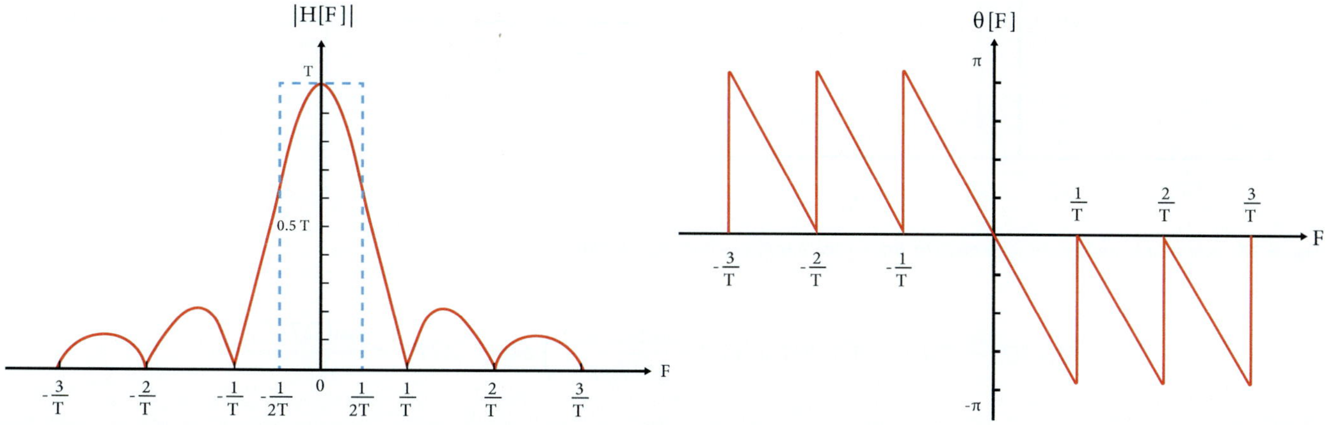

Fig. 9.17 Amplitude and phase responses of zero-order hold system

whose output has the form

$$x_{\text{DA}}(t) = \sum_{n=-\infty}^{\infty} A\tilde{x}_B\,[n]h_0(t-nT) = \sum_{n=-\infty}^{\infty} \tilde{x}\,[n]h_0(t-nT)$$

with the index o indicating that the sampling/holding circuit is a *zero-order circuit*, which, as mentioned, maintains only the value of the current sample. The operation of the foregoing system is depicted diagrammatically in Fig. 9.16, while the amplitude and phase responses of the zero-order sampling hold system are depicted in Fig. 9.17. From this figure we find that this zero-order sampling/holding circuit *does not offer a sharp transition from the passband to the cutoff band*, and therefore, at the output of this system, *unwanted spurious frequencies* will appear, as dictated by the phenomenon of folding. To smooth out this problem, we can add a low-pass filter to the output of this system to attenuate the frequency components with a value less than $F_s/2$.

It is interesting to note that, apart from the foregoing simple zero-order system, whose performance is not satisfactory, as the signal generated as its output is characterized by *discontinuities*, in practice, other more efficient techniques are used, which are based on the *method of linear interpolation* and collectively described by the name higher-order hold systems. The simplest of these systems is the *first-order hold system*, in which the mathematical correlation between the output signal and the samples of the input signal has the form

$$x_{\text{DA}}(t) = x(nT_s) + \beta\frac{x(nT_s) - x(nT_s - Ts)}{T_s}(t - nT_s)$$

for $nT_s \leq t < (n+1)T_s$. If this first-order hold system is treated as a filter, it is characterized by an impulse response of the form

$$h_1(t) = \begin{cases} 1+\beta(t/T) \; 0 \leq t \leq T \\ 1-\beta(t/T) \; T \leq t \leq 2T \\ 0 \qquad\qquad \text{anywhereelse} \end{cases}$$

and frequency response

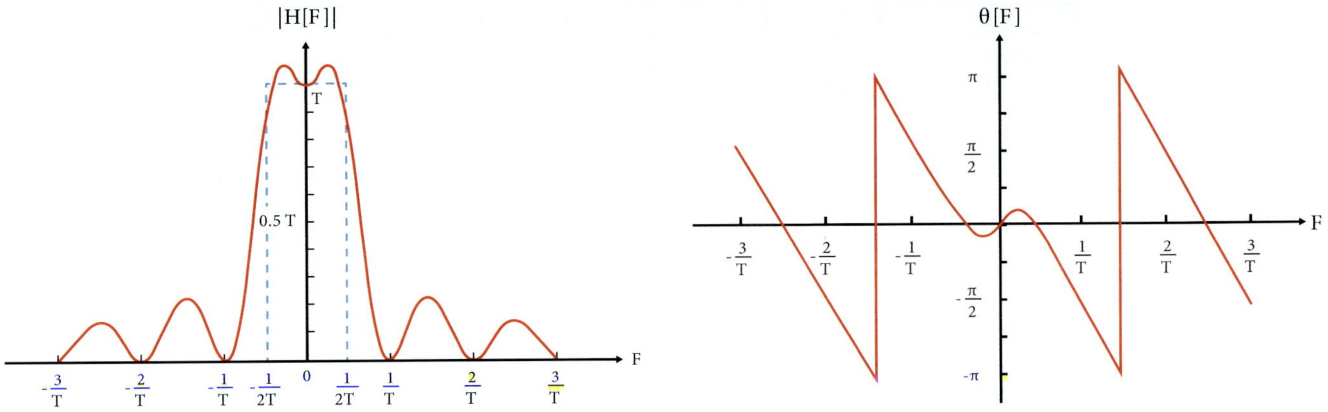

Fig. 9.18 Amplitude and phase responses of first-order sampling–holding system

$$\mathcal{H}_1(j\Omega) = T\left(1-\beta+\beta(1+j\Omega T)\frac{\sin(\Omega T/2)}{\Omega T/2}\right)\exp(-j\Omega T)\times\frac{\sin(\Omega T/2)}{\Omega T/2}$$

as can be proven after simple operations. The parameter β in the preceding expressions expresses *the slope of a straight line segment that connects two consecutive samples* and is particularly useful in cases where we want to modify the frequency response of the system, eliminating the peaks that appear in the amplitude response for the frequency range $|F_s|$, if such a thing is desirable. The mathematical analysis of the problem reveals that these peaks are relatively small for the value $\beta = 0.3$, while they disappear completely for the value $\beta = 0.1$. In the common first-order hold system, we set $\beta = 1$. In this case, it is proven after simple operations that the frequency response of the system has the simplest form

$$\mathcal{H}_1(j\Omega) = T\sqrt{1+\Omega^2T^2}\left[\frac{\sin(\Omega T/2)}{\Omega T/2}\right]^2\exp[j(\tan^{-1}(\Omega T)-\Omega T)]$$

The amplitude and phase responses of the first-order hold sampling system are presented in Fig. 9.18. It is worth noting that, although this system is clearly superior to the simple zero-order hold system, it also leads *to the appearance of folded (or aliased) frequencies at the output of the system* with values greater than $F_s/2$, and for this reason, it is required before the use of a low-pass filter for the removal of these frequencies. Let us also mention that in reality, the equation that describes the operation of the first-order hold system is not an interpolation equation but *an equation of deviation*, as it tries to calculate, or rather predict, the value of the next sample based on the value of the two previous samples, which leads to the appearance of *discontinuity jumps*. To improve the situation, we can introduce into the process of signal reconstruction a delay of one sample, and therefore, this process will be based, not on the use of the previously mentioned equation, but, correspondingly, on the equation

$$x_{\text{DA}}(t) = x(nT_s - T_s) + \frac{x(nT_s) - x(nT_s - Ts)}{T_s}(t - nT_s)$$

(where for simplicity we have set $\beta = 1$) for the same range of values of t ($nT_s \le t < (n+1)T_s$). In this case, the impulse response of the system is the

$$h_1^d(t) = \begin{cases} t/T & 0 \le t \le T \\ 2-(t/T) & T \le t \le 2T \\ 0 & \text{anywhere else} \end{cases} \quad \text{with frequency response} \quad \mathcal{H}_1^d(j\Omega) = T\left[\frac{\sin(\Omega T/2)}{\Omega T/2}\right]^2\exp\left(-j\frac{\omega T}{2}\right)$$

If we resort to the design of the amplitude response of this system, we will find that this function decreases *more sharply* and contains smaller sidelobes for frequency values greater than the frequency F_s, while the phase response is linear due to the delay T. The use of a low-pass filter at the output of this system leads as before *to the removal of unwanted spurious frequencies*. Taking into account that the quantization process introduces a *quantization error* $e[n]$, such that $\widetilde{x}[n] = x[n] +$

$e[n]$, the substitution of this expression at the output of the zero-order hold system and the analysis of the equation in two sums, one for the term $x[n]$ and one for the term $e[n]$, gives us

$$x_{\text{DA}}(t) = \sum_{n=-\infty}^{\infty} x[n]h_0(t-nT) + \sum_{n=-\infty}^{\infty} e[n]h_0(t-nT) = x_0(t) + e_0(t)$$

where for simplicity's sake we defined the quantities

$$x_0(t) = \sum_{n=-\infty}^{\infty} x[n]h_0(t-nT) \qquad \text{and} \qquad e_0(t) = \sum_{n=-\infty}^{\infty} e[n]h_0(t-nT)$$

The Fourier transform of the first of the preceding equations related to the input signal $x_\alpha(t)$ [since $x[n] = x_\alpha(nT_s)$] has the form

$$\mathcal{X}_0(j\Omega) = \sum_{n=-\infty}^{\infty} x[n]\mathcal{H}_0(j\Omega)e^{-j\Omega T_s n} = \left(\sum_{n=-\infty}^{\infty} x[n]e^{-j\Omega T_s n}\right)\mathcal{H}_0(j\Omega) = \mathcal{X}(e^{j\Omega T_s})\mathcal{H}_0(j\Omega)$$

where $\mathcal{H}_0(j\omega) = \mathcal{F}\{h_0(t)\}$ is the frequency response of the zero-order hold system. Observing now that

$$\mathcal{X}(e^{j\Omega T_s}) = \frac{1}{T}\sum_{k=-\infty}^{\infty} X_\alpha\left(j\left(\Omega - \frac{2\pi k}{T_s}\right)\right)$$

we will have

$$\mathcal{X}_0(j\Omega) = \left[\frac{1}{T_s}\sum_{k=-\infty}^{\infty} X_\alpha\left(j\left(\Omega - \frac{2\pi k}{T_s}\right)\right)\right]\mathcal{H}_0(j\Omega)$$

If the Fourier transform $\mathcal{X}_\alpha(j\Omega)$ is limited to frequencies less than the value π/T, then the shifted copies of the spectrum $\mathcal{X}_\alpha(j\Omega)$ in the preceding equation *do not overlap*, thereby defining a *compensation filter* of the form

$$\widetilde{\mathcal{H}}_r(j\Omega) = \frac{\mathcal{H}_r(j\Omega)}{\mathcal{H}_0(j\Omega)}$$

In this case, the output of the filter will be the signal $x_\alpha(t)$ if its input is the signal $x_0(t)$. Using the mathematical expression for the previously mentioned function $\mathcal{H}_0(j\omega)$, the preceding relation becomes

$$\widetilde{\mathcal{H}}_r(j\Omega) = \begin{cases} \dfrac{\Omega T/2}{\sin(\Omega T/2)} e^{j\Omega T/2}, & |\Omega| < \pi/T \\ 0, & |\Omega| \geq \pi/T \end{cases}$$

This *compensation filter* is connected to the output of the D/A converter, and according to our earlier discussion in the proof of the sampling theorem, the output of this composite system depicted in Fig. 9.19 will be

$$\begin{aligned}\widetilde{x}_r(t) &= \sum_{n=-\infty}^{\infty} \widetilde{x}[n]\left\{\sin\left[\frac{\pi}{T_s}\left(t - nT_s\right)\right]\right\} \Big/ \left\{\frac{\pi}{T_s}(t-nT_s)\right\} \\ &= \sum_{n=-\infty}^{\infty} x[n]\left\{\sin\left[\frac{\pi}{T_s}\left(t - nT_s\right)\right]\right\} \Big/ \left\{\frac{\pi}{T_s}(t-nT_s)\right\} + \sum_{n=-\infty}^{\infty} e[n]\left\{\sin\left[\frac{\pi}{T_s}\left(t - nT_s\right)\right]\right\} \Big/ \left\{\frac{\pi}{T_s}(t-nT_s)\right\}\end{aligned}$$

Consequently, the analog signal that results from this digital-to-analog signal conversion process will be given by the equation

$$\widetilde{x}_r(t) = x_\alpha(t) + e_\alpha(t)$$

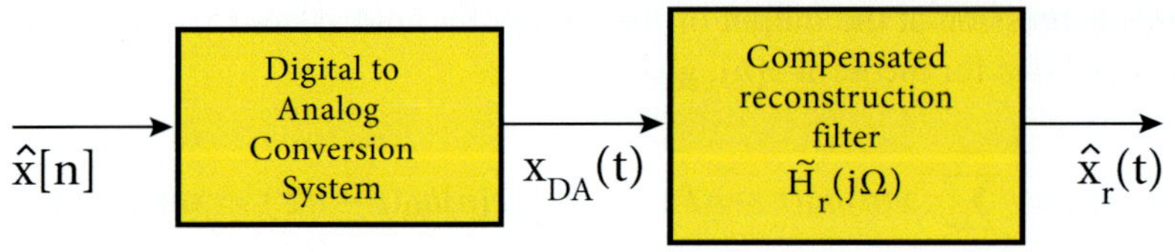

Fig. 9.19 Structure of practical digital-to-analog signal conversion system

where $e_\alpha(t)$ is a band-limited signal that describes *white noise*. The foregoing analysis refers to the simple case of using a *zero-order hold system*, while similar conclusions can be formulated for *higher-order hold systems*.

9.8.6 The Structure of a Practical Digital Signal Processing System of an Analog Signal

Combining the entirety of what we presented in detail on the preceding pages, the detailed structure of a system that allows for the digital processing of an analog signal is depicted in Fig. 9.20. This system consists of *a subsystem for converting an analog signal to digital, as well as a subsystem for converting a digital signal to analog*, while between these two subsystems, there is a discrete-time system that applies to the digital signal created in the above way, the appropriate forms of digital signal processing for each case.

Considering that the output of the aliasing filter is limited to frequencies less than the value of π/T_s, that the spectrum $\widetilde{H}_r(j\omega)$ is *band-limited*, and that the discrete-time system performing the digital signal processing is *LTI*, then the output of this system is given by the equation

$$\widetilde{y}_r(t) = y_a(t) + e_a(t)$$

where

$$T\mathcal{Y}_\alpha(j\Omega) = \widetilde{\mathcal{H}}_r(j\Omega)\mathcal{H}_0(j\Omega)\mathcal{H}(e^{j\Omega T})\mathcal{H}_{aa}(j\Omega)\mathcal{X}_c(j\Omega)$$

with the quantities appearing here having been explained in the previous sections. Similarly, assuming that the quantization noise added to the signal by the quantizer arrangement is white noise with variance $\sigma_e^2 = \Delta^2/12$, it is proven that the power spectrum of the noise at the output is given by the equation

$$P_{e_\alpha}(j\Omega) = |\widetilde{\mathcal{H}}_r(j\Omega)\mathcal{H}_0(j\Omega)\mathcal{H}(e^{j\Omega T})|^2\sigma_e^2$$

Finally, relying on the foregoing model for quantization noise and considering that the effects of the aliasing phenomenon are negligible, the frequency response of the system of Fig. 9.20 that receives the signal $x_c(t)$ as input and produces the signal $\widehat{y}_r(t)$ as its output has the form

$$T\mathcal{H}_{eff}(j\Omega) = \widetilde{\mathcal{H}}_r(j\Omega)\mathcal{H}_0(j\Omega)\mathcal{H}(e^{j\Omega T})\mathcal{H}_{aa}(j\Omega)$$

9.9 Multirate Signal Processing

Multirate signal processing emerged as a natural consequence of the necessity of data processing using *more than one* value for the sampling rate. In practice, there are two basic techniques to which we resort for the implementation of this process. The first of these techniques is the *decimation*, which leads to a *reduction* of the sampling rate and, consequently, the number of signal samples, thereby leading to signal compression. Note, however, that this process is carried out in such a way as to maintain the desired information in each case. On the other hand, the second technique is well known from numerical analysis, *interpolation*, which leads to an *increase* of the sampling rate, calculating and interpolating new samples between the existing signal samples, which are calculated using a filter specially designed for this purpose; this is quite reasonably

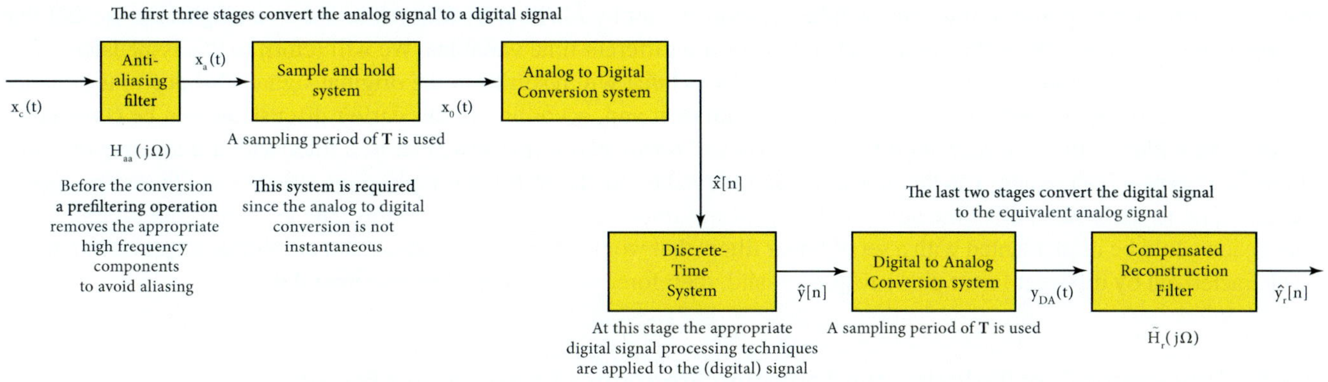

Fig. 9.20 Structure of practical digital processing system of analog signal

characterized as *filter interpolation*. It is important to mention that the aforementioned techniques are applied directly to a digital signal since, although it is clear how the change in the sampling rate could be carried out by converting the digital signal to an analog one and proceeding to its redigitization using a different sampling frequency, this would lead to the appearance of all the errors involved in the digitization of an analog signal (such as *quantization errors*) and subsequent *degradation* of the signal quality. In addition, even if we had chosen to work in this way, the new sampling frequency to be used would not have been related in any way to the original sampling frequency, something that is especially desirable but is, however, chosen arbitrarily. For all these reasons, the whole process is carried out digitally, and the signal is converted to its analog form only when it is about to be sent to an analog device, such as a speaker.

The use of polyrhythmic signal processing makes the processing of signals characterized by *different bandwidths* particularly efficient, such as those used in telecommunications applications, where sending and receiving signals of different types (such as fax signals, voice signals, and video signals), each of which has its own characteristic bandwidth and therefore should be sampled at different frequencies. More specifically, the ability to change the sampling rate makes the use of analog filters unnecessary, such as those used to remove the consequences of the phenomenon of frequency folding, which are characterized by high cost. The reason why this happens is that if we proceed to the sampling of an analog signal at a frequency much higher than that dictated by the sampling theorem, a much simpler analog filter of the aforementioned type should be used before digitizing the signal. These techniques of polyrhythmic processing are widely used in *applications of voice signal processing* to reduce the storage requirements of voice data as well as the frequency at which they are transmitted, while they also find applications in the processes of converting analog signals to digital, as it has been proven that the oversampling of quantization noise causes its dispersion in a wider frequency range. In this way, the *in-band noise* or otherwise *intraband noise* is significantly reduced, something that leads to an increase in the number of bits used in the processes of converting an analog to a digital signal. In the following pages, we will discuss how to implement the two basic techniques of polyrhythmic signal processing—*subsampling* (in which the technique of decimation is used) and *oversampling* (in which the technique of interpolation is used).

Generally speaking, the change in sampling rate that is carried out in the time domain, whether it refers to undersampling or oversampling, can be considered a *linear filtering process*, which is carried out by a filter that takes as input the digital signal $x[n] = x_\alpha(nT_s)$, which results after the sampling of some analog signal $x_\alpha(t)$ at a frequency of $F_s = 1/T_s$, where T_s is the *sampling period* and produces as its output the signal $x'[n] = x_\alpha(nT_s')$, which, although, as we will see, arises directly from the digital signal $x[n]$, is considered to coincide with the signal that would result from the sampling of the original analog signal $x_\alpha(t)$ at a different sampling frequency, $F_s' = 1/T_s'$, corresponding to a sampling period T_s'. In the description that follows, the ratio of the sampling frequencies F_s'/F_s is considered to be a *rational* number, that is, it equals the ratio of two integers. If $T_s' > T_s$, then the signal $x'[n]$ will obviously have fewer samples than the signal $x[n]$, while if it is $T_s' < T_s$, then the signal $x'[n]$ will obviously have more samples than the signal $x[n]$. Indeed, in the first case, the distances between successive samples will be larger than the distances of the samples in the original signal, while in the second case, the samples are characterized by shorter distances. These two situations are described, as noted earlier, by the terms *undersampling* and *oversampling*, and both constitute the two alternative forms of the so-called resampling, which is defined as the sampling of the original signal at a different period or sampling frequency. As will be discussed later, the digital filter that implements the aforementioned processes is linear but nevertheless constitutes not a time-invariant but a time-varying system. For this reason, the output of this filter will be symbolized by $y[m]$—and not by $y[n]$ due to the time variability—while for the same reason, the impulse response of this filter will be symbolized not by $h[n]$ but by $h[n, m]$ in order to emphasize the fact that the input and output signals of this filter, $x[n]$ and $y[m]$, use different time variables. We will return to this issue later.

It is not difficult to understand that the digital signal $y[m]$ that will result from the original signal $x[n]$ due to the change in the sampling frequency will be a *time-delayed* version of it and, according to our earlier discussion, can be constructed by applying a filter with a *constant amplitude response* and *linear phase response*, that is, a filter with a frequency response of the form $\exp(-j\omega t)$, where τ is the time delay introduced by the filter. If the sampling rates that create these two signals are not equal to each other, this time delay will vary from sample to sample, and therefore, the system of the sampling rate conversion could be implemented with a set of linear filters that would share the same constant amplitude response but would be characterized by different linear phase responses and, therefore, would create different time delays.

9.9.1 Downsampling: Reducing the Sampling Frequency by an Integer Factor

In reality, the process of *downsampling* is very simple, since all that is required is the resampling of the discrete-time signal $x[n] = x_\alpha(nT_s)$ that resulted from the original sampling of the analog signal $x_\alpha(t)$, so as to retain only some of the samples. For example, we can consider only the samples that are in the *even* positions of the discrete sequence $x[n]$, thereby constructing the signal $y = [m] = x_d[n] = x[2n]$ $(n = 0, 1, 2, \dots)$, that is, considering only one sample for each group of $M = 2$ samples. In the general case, for each value of M, which is a parameter of the process, our signal will have the form $y[m] = x_d[n] = x[Mn]$ $(M = 1, 2, 3, \dots)$ (note that the two variables describing the discrete-time for the input and output signals are related as $m = Mn$), and therefore, the new signal $y[m]$ arises by considering one sample for each group of M samples, that is, ignoring the remaining $M - 1$ samples (it is obvious that for the value $M = 1$ we will have $m = n$ and $y[m] = x[n]$, since we consider all the samples, so at least we should have $M = 2$); because $x[n] = x_\alpha(nT_s)$, we will have $y[m] = x_d[n] = x_\alpha(nMT_s)$. In other words, the subsampling of the discrete-time signal $x[n]$, which resulted from the sampling of the continuous-time signal $x_\alpha(t)$ with a sampling period T_s, leads to a discrete-time signal $y[m]$, which is the same as the one we would get if we sampled the analog signal $x_\alpha(t)$ with a sampling period $T_d = MT_s$ or, equivalently, with a sampling frequency $F_d = F_s/M$. If we assume that this subsampling process is performed by a system that receives the discrete-time signal $x[n]$ at its input and produces the discrete-time signal $y[m]$ as its output, this system could quite reasonably be described by the name *compressor* (Fig. 9.21), since it compresses the input signal, in the sense that the signal that emerges as its output has M times fewer samples than the original signal, which are moreover closer together in time compared to the samples of the original discrete-time signal (Fig. 9.22) (as we saw in Sect. 7.2.2, page 567, the compressor system is a time-varying system, in full agreement with what we mentioned earlier). As is well known from the *sampling theorem* and the related discussion, if the analog signal $x_\alpha(t)$ is characterized by a finite bandwidth, that is, the continuous-time Fourier transform $\mathcal{X}_\alpha(j\omega)$ is equal to zero for frequencies $|\Omega| \geq \Omega_N$ (where Ω_N is the limit of the bandwidth), then if the condition $\pi/T_d = \pi/(MT) \geq \Omega_N$ is satisfied, the discrete-time signal $x_d[n]$ is an exact representation of the continuous-time signal $x_\alpha(t)$ since the phenomenon of *frequency folding* will not appear.

Adopting the approach we followed for the proof of the sampling theorem, let us determine the correlation that exists between the discrete-time Fourier transforms of the discrete-time signals $x[n]$ and $y[m] = x_d[n] = x[nM]$, which constitute the input and output of the compressor system. As we analytically proved during the presentation of the foregoing theorem, the aforementioned transformation for the signal $x[n]$, which is the $\mathcal{X}\alpha(e^{j\omega})$, is related to the continuous-time Fourier transform $\mathcal{X}(j\omega)$ of the analog signal $x_\alpha(t)$ through the equation

$$\mathcal{X}(e^{j\omega}) = \frac{1}{T}\sum_{k=-\infty}^{\infty} \mathcal{X}_\alpha\left[j\left(\frac{\omega}{T} - \frac{2\pi K}{T}\right)\right] \tag{9.51}$$

while for the corresponding transformation of the discrete-time signal $y[m] = x_d[n] = x[nM] = x_\alpha(nMT_s) = x_\alpha(nT_d)$, we will have

$$\mathcal{X}_d(e^{j\omega}) = \frac{1}{T_d}\sum_{r=-\infty}^{\infty} \mathcal{X}_\alpha\left[j\left(\frac{\omega}{T_d} - \frac{2\pi r}{T_d}\right)\right] = \frac{1}{MT}\sum_{r=-\infty}^{\infty} \mathcal{X}_\alpha\left[j\left(\frac{\omega}{MT} - \frac{2\pi r}{MT}\right)\right] \tag{9.52}$$

Expressing the index r in the second equation as $r = i + kM$, where k and i are integer indices, such that $-\infty < k < \infty$ and $0 \leq i \leq M - 1$, the last expression takes the form

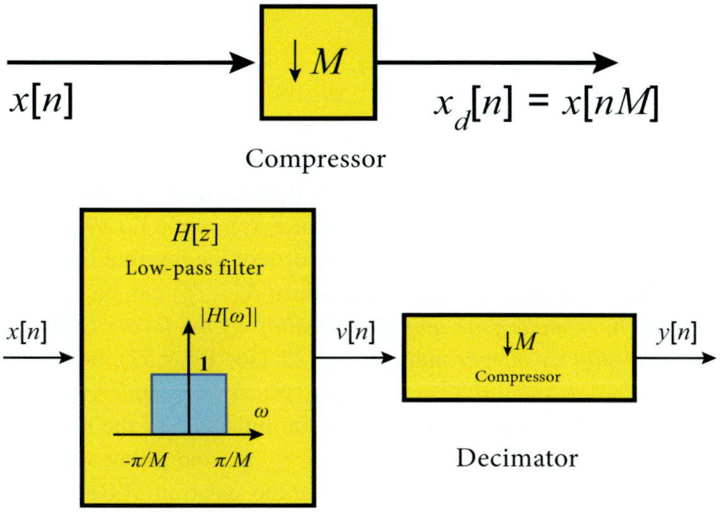

Fig. 9.21 The process of subsampling a discrete-time signal is performed by the compressor system. This system is symbolized by a rectangle containing an arrow pointing down (since the sampling frequency is reduced) and the coefficient M of the decimation process

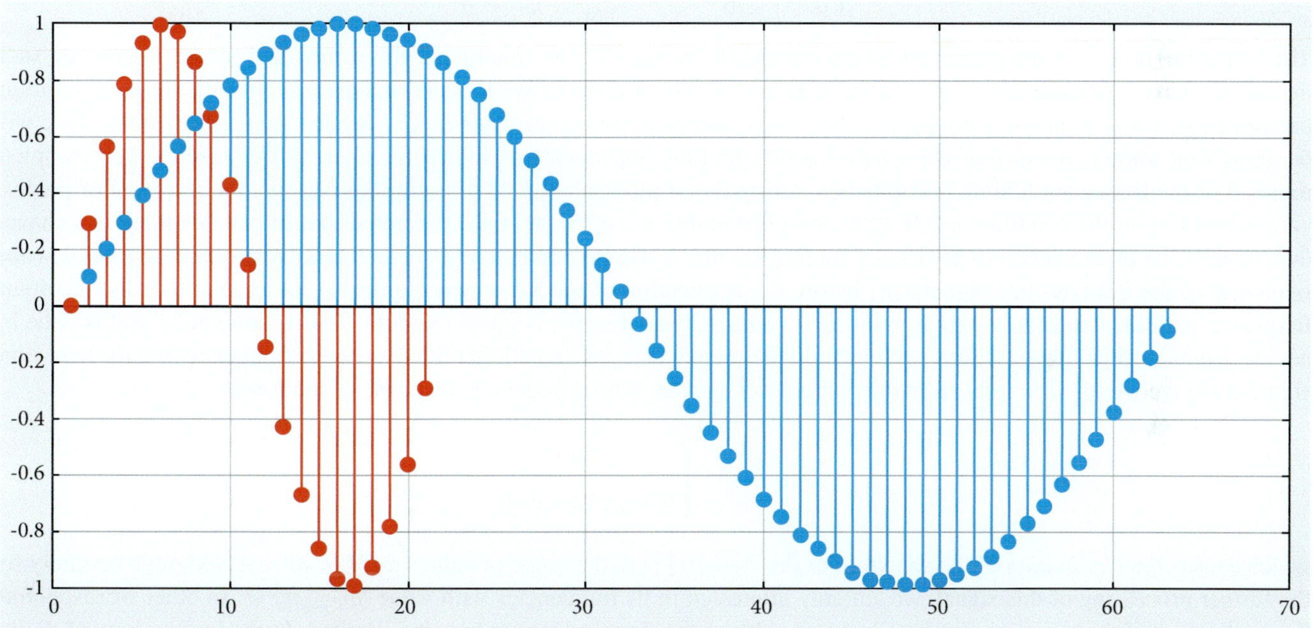

Fig. 9.22 Subsampling of a discrete sinusoidal signal (blue curve) with parameter value $M = 3$ (which means that for each group of three samples we consider only the first sample while ignoring the next two samples). Notice that if we plot the signal resulting from this process (red curve), it is compressed in time, since the samples of the original signal between which two samples were interpolated now become consecutive

$$\mathcal{X}_d(e^{j\omega}) = \frac{1}{M}\sum_{i=0}^{M-1}\left\{\frac{1}{T}\sum_{k=-\infty}^{\infty}\mathcal{X}_\alpha\left[j\left(\frac{\omega}{MT} - \frac{2\pi k}{T} - \frac{2\pi i}{MT}\right)\right]\right\}$$

But the term inside the brackets of the last expression is simply the transform

$$\mathcal{X}(e^{j(\omega-2\pi i)/M}) = \frac{1}{T}\sum_{k=-\infty}^{\infty}\mathcal{X}_\alpha\left[j\left(\frac{\omega - 2\pi i}{MT} - \frac{2\pi k}{T}\right)\right]$$

and consequently, we finally get

$$\mathcal{X}_d(e^{j\omega}) = \frac{1}{M}\sum_{i=0}^{M-1} \mathcal{X}(e^{j(\omega/M - 2\pi i/M)}) \tag{9.53}$$

From the preceding two equations, Eq. (9.51) expresses the Fourier transform of the discrete-time signal $x[n]$ as a function of the continuous-time Fourier transform $\mathcal{X}_\alpha(j\omega)$ of the analog signal $x_\alpha(t)$, while Eq. (9.53) expresses the Fourier transform of the discrete-time signal $x_d[n]$ as a function of the Fourier transform of the discrete-time signal $x[n]$. Careful study of the last relationship leads us to the conclusion that the Fourier transform $\mathcal{X}_d(e^{j\omega})$ can be considered either *as a set of infinite copies of the spectrum $\mathcal{X}_\alpha(j\omega)$ that have undergone amplitude scaling by the factor $1/T_d$, frequency scaling by the factor T_d (since it is $\omega = \Omega T_d$) and phase shift by integer multiples of 2π* (see Eq. 9.52), or, alternatively, *as a set of M copies of the spectrum $\mathcal{X}(e^{j\omega})$ of the original discrete-time signal $x[n]$, which have undergone amplitude scaling by the factor $1/M$, frequency scaling by the factor M* (this means that under the influence of the sampling process the frequency area $0 \leq |\omega_x| \leq \pi/M$ is depicted in the frequency region $0 \leq |\omega_y| \leq \pi$), *and phase shift by integer multiples of 2π* (see Eq. 9.52). Both of our representations lead us to the conclusion that the spectrum $\mathcal{X}_d(e^{j\omega})$ of the discrete-time signal $x_d[n]$ is periodic with period 2π (as naturally happens with all Fourier transforms of discrete-time) and that the phenomenon of frequency folding can be avoided if we ensure that the spectrum $\mathcal{X}(e^{j\omega})$ is characterized by *finite bandwidth*, and therefore, the following condition applies:

$$\mathcal{X}(e^{j\omega}) = 0 \qquad \omega_N \leq |\omega| \leq \pi$$

The foregoing processes are presented diagrammatically in Fig. 9.23. In this figure, the parameters of the process are such that *no* overlap of the successive copies of the spectrum occurs, and therefore, there is *no appearance* of the frequency folding phenomenon, while if this does not happen, but on the contrary, these parameters have not been selected correctly, there will be, as is well known, *overlap* of the spectral copies and the appearance of folded (or aliased) frequencies. The complete removal of the frequency folding phenomenon during the implementation of subsampling requires the selection of process parameters so that the condition $\omega_N M \leq \pi$ or, equivalently, $\omega_N \leq \pi/M$. Let it be noted that in practice there is a chance that, even if the phenomenon of frequency folding occurs, it can be addressed to some extent by proceeding to *bandwidth reduction* of the discrete-time signal $x[n]$ before the application of the subsampling process, so as to reduce its maximum frequency with the expectation that it will take a value that satisfies the Nyquist theorem. This is something that is done by adding a digital ideal low-pass filter with a cutoff frequency of π/M (recall that the maximum frequency of a discrete-time signal is the frequency $\omega = \pi$), before the compressor system, with a frequency response of the form

$$\mathcal{H}_M(\omega) = \begin{cases} 1 & |\omega| \leq \pi/M \\ 0 & \text{anywhere else} \end{cases}$$

which causes the *elimination* of the spectrum of the signal $x[n]$ in the range of values $\pi/M < \omega < \pi$, although certainly for the further processing of this signal, we are only interested in its frequencies with value $|\omega| \leq \pi/M$. In other words, before the application of the sampling process, the bandwidth of the signal spectrum becomes limited. In the terminology of digital signal processing, this arrangement, resulting from the series connection of this ideal low-pass filter with the compressor system, is known as a *decimator* and is depicted in Fig. 9.24. If $h[n]$ is the impulse response of the aforementioned filter, output will be in the form $v[n] = h[n] * x[n]$, and consequently, the output of the compressor that receives the output of the filter $v[n]$ at its input will be in the form

$$y[m] = v[mD] = \sum_{k=0}^{\infty} x(mM - k)$$

By determining the $\mathcal{Z}$ transform of the signal $y[m]$ and then calculating it on the perimeter of the unit circle so as to calculate the Fourier transform of discrete-time, we can reach the same conclusions as those formulated in the earlier analysis regarding the characteristics of the spectrum of the signal $y[n]$ and the way in which it is related to the spectrum of the original digital signal $x[n]$.

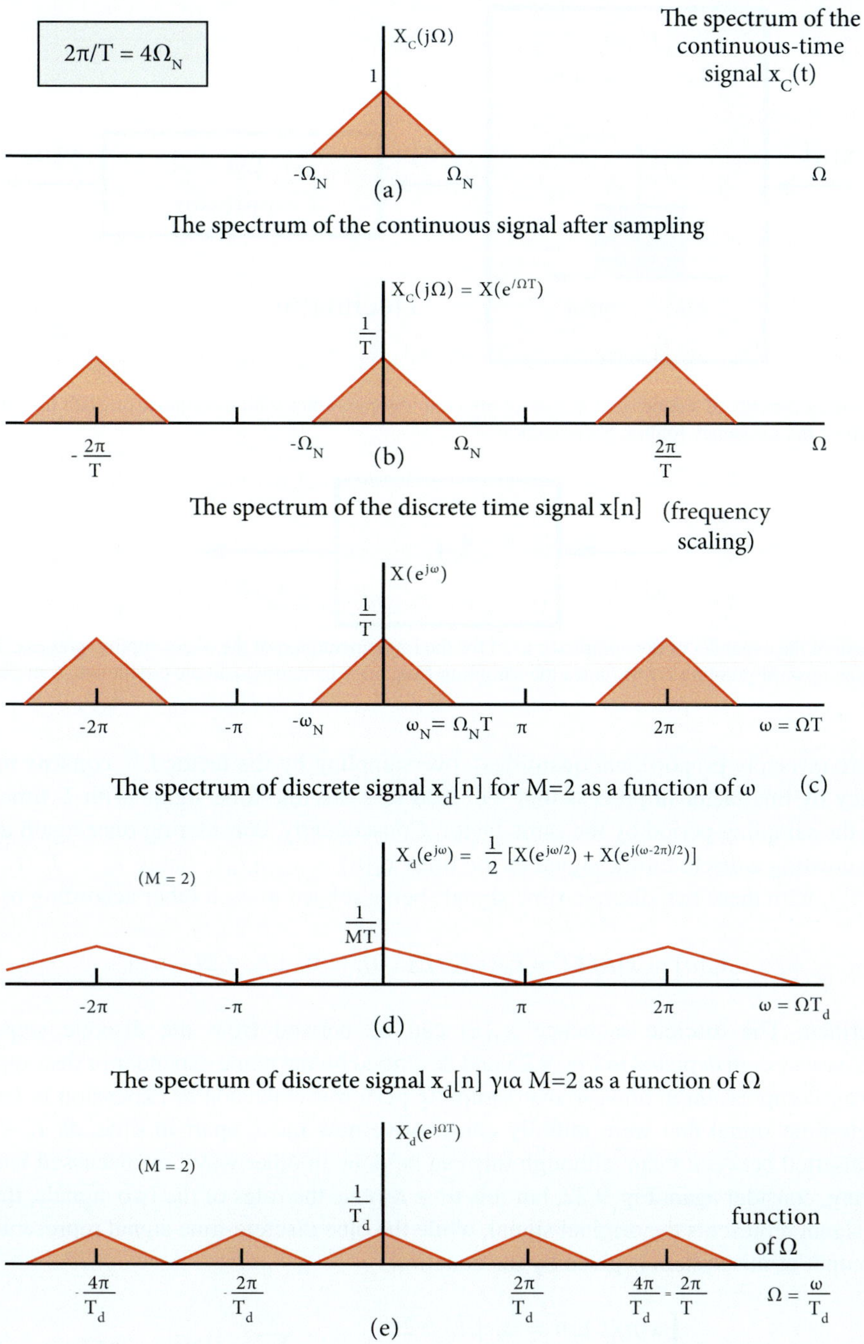

Fig. 9.23 Subsampling of discrete-time signal without appearance of aliasing phenomenon (Oppenheim, Schafer, Buck, 3rd edition, Figure 4.20)

9.9.2 Oversampling: Increasing the Sampling Frequency by an Integer Factor

Oversampling, as a process, is the *inverse* of subsampling, under the concept that while the latter consists in reducing the number of samples of the discrete-time signal that is sampled, the former consists in increasing this number of samples. Taking into account that *the reduction in the sampling rate by a factor M consists in multiplying the sampling period by this factor (and, consequently, in dividing the sampling frequency $F = 1/T_s$ by the same factor*, since the period and the

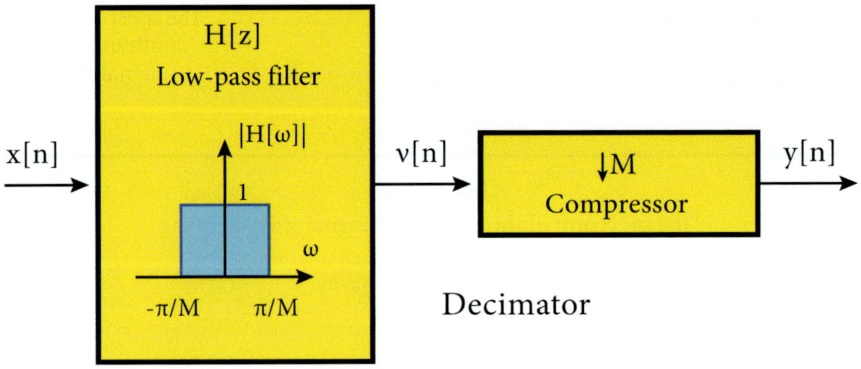

Fig. 9.24 A decimator system consists of a series connection of an ideal low-pass filter with a compressor, with the addition of the filter leading to a reduction in the effects of the frequency folding phenomenon

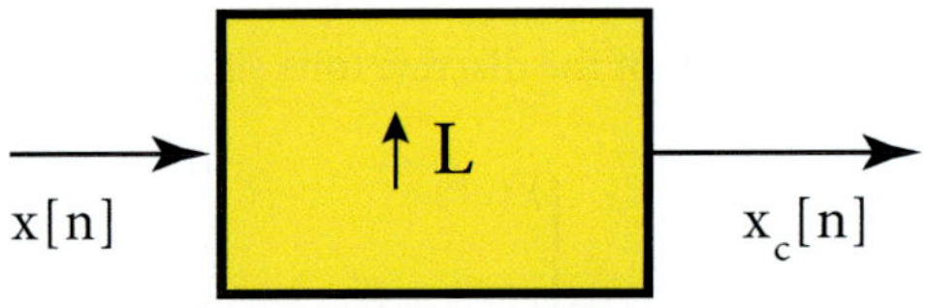

Fig. 9.25 The arrangement of the expander or decompressor used for the implementation of the oversampling process. This system is symbolized by a rectangle containing an upward-pointing arrow (since the sampling frequency increases) and the coefficient L of the interpolation process

sampling frequency are inversely proportional quantities), oversampling by the factor L[16] consists in the *multiplication* of the sampling frequency by this factor (a process that will lead to a discrete-time signal with L times more samples) and, therefore, in dividing the sampling period by the same factor. Consequently, considering once again an analog signal $x_\alpha(t)$, *oversampling* aims at creating a discrete-time signal of the form $x_u[n] = x_\alpha(nT_u)$, where $T_u = T_s/L$, from the sequence of samples $x[n] = x_\alpha(nT_s)$, with these two discrete-time signals being related to each other according to the relation

$$x_u[n] = x[n/L] = x_\alpha(nT_s/L) \quad (n = 0, \pm L, \pm 2L, \ldots)$$

as can easily be verified. The discrete sequence $x_u[n]$ can be derived from the discrete sequence $x[n]$ using the oversampling rate increase system depicted in Fig. 9.25 and described by the name expander or decompressor (indeed, just as undersampling performs compression in time, so oversampling performs extension or expansion in time, since two samples of the original discrete-time signal that were initially consecutive now move apart in time, as $L - 1$ additional samples with zero values are inserted between them, although this can be done in other ways, as discussed later). To understand the process of oversampling, consider again Fig. 9.22, but this time reverse the roles of the two signals, that is, consider that the orange discrete-time signal represents the original signal, while the blue discrete-time signal represents what results from its oversampling). The output of this system is given by the equation

$$x_e[n] = \begin{cases} x[n/L], \ n = 0, \pm L, \pm 2L \ldots \\ 0 \quad \text{otherwise} \end{cases} = \sum_{k=-\infty}^{\infty} x[k]\delta[n - kL]$$

The low-pass filter included in the preceding arrangement is used in the way we saw during the reconstruction of the analog signal from its samples. Specifically, the process of oversampling, which can be considered in some sense the *digital equivalent of the process of converting a digital to an analog signal* (although in this case only the calculation of some specific sample values takes place) is a two-step process, during which in the first step we create the discrete-time signal $x_e[n]$ from the discrete-time signal $x[n]$ (note that these two signals constitute the input and output of a LTI system, so they are related to each other by the convolution equation). In the second step, we use the samples of the discrete-time signal $x_e[n]$ that was created in the first stage for the reconstruction of the signal $x_u[n]$ through the use of the low-pass filter, whose impulse and frequency responses have the forms

[16] The reason we use different symbols is to make clear which process we are referring to in each case.

$$h_u[n] = \frac{\sin(\pi n/L)}{\pi n/L} \qquad \text{and} \qquad \mathcal{H}_L(\omega) = \begin{cases} L & 0 \le |\omega| \le \pi/L \\ 0 & \text{anywhere else} \end{cases}$$

and then this signal is used for the *removal* of the imaginary frequencies that have appeared as a result of the increase in the sampling frequency. From the study of the preceding equation, it is not difficult to see that $h_u[0] = 1$ and $h_u[n] = 0$ $(n = \pm L, \pm 2L, \dots)$. Therefore, the output of the filter $x_u[n]$ will be given by the expression

$$x_u[n] = h_u[n] * x_e[n] = \sum_{k=-\infty}^{\infty} x[k] \frac{\sin[\pi(n-kL)/L]}{\pi(n-kL)/L}$$

and, using the foregoing properties, it immediately follows that

$$x_u[n] = x[n/L] = x_\alpha(nT_s/L) = x_\alpha(nT_u), \qquad n = 0, \pm L, 2L, \dots$$

which is exactly what we expect. Taking into account that the oversampling process is simply an *interpolation process*, since we calculate intermediate samples of an initial sample sequence, this ideal low-pass filter is quite reasonably characterized as an *interpolation filter*. The system resulting from the series connection of the expander and the low-pass filter is known as an *interpolator*. The reason the frequency response $\mathcal{H}_L(\omega)$ has a value of L for the frequency range $0 \le |\omega| \le \pi/L$ is justified by the fact that the interpolation of $L - 1$ zero values between the samples of the original signal for the implementation of the oversampling process results in the *dispersion of the energy* of each sample into a group of L input samples, i.e., the attenuation of each sample by a factor of L. To compensate for this attenuation, we must multiply the value of each sample by the factor L.

To describe the foregoing process in the frequency of as we did in the case of sampling, let us calculate the *discrete-time Fourier transform* of the discrete sequence $x_e[n]$, which according to the defining equation will have the form

$$\mathcal{X}_e(e^{j\omega}) = \sum_{n=-\infty}^{\infty} \left(\sum_{k=-\infty}^{\infty} x[k]\delta[n-kL] \right) e^{-j\omega n} = \sum_{k=-\infty}^{\infty} x[k]e^{-j\omega Lk} = \mathcal{X}(e^{j\omega L}) \tag{9.54}$$

As can be verified, this expression follows from the discrete-time Fourier transform $\mathcal{X}(e^{j\omega})$ of the signal $x[n]$, provided that a frequency scaling is applied, in which ω is replaced by the frequency $\omega' = \omega L = \omega T/T_u$, since $T_u = T_s/L$ and therefore $L = T_s/T_u$. However, since it holds that $\omega' = \Omega T_s$, a process of *normalization of frequency* must be implemented that will take the form $\omega = \Omega T_u$, since then the expression becomes $\omega' = \Omega T$, as required. The foregoing process as described in the frequency domain is depicted diagrammatically in Fig. 9.26 for the case of $L = 2$. In this figure, the spectrum $\mathcal{X}_u(e^{j\omega})$ results from the spectrum $\mathcal{X}_e(e^{j\omega})$ by changing the amplitude scaling factor from $1/T$ to $1/T_i$ and removing all frequency-scaled copies of the spectrum $\mathcal{X}_\alpha(j\omega)$, except for those located at frequencies with a value that is an integer multiple of 2π. This process is performed using a low-pass filter with the gain value equal to 2 and the cutoff frequency equal to $\pi/2$. The system that arises in this way, that is, from the series connection of an expander and an ideal low-pass filter, is known as an *interpolator* and is depicted in Fig. 9.27. The foregoing discussion applies to the value $L = 2$, while in the general case the values of the gain and the cutoff frequency of the filter are obviously equal to L and π/L, respectively. Note that, in contrast to the case of subsampling, in which the ideal low-pass filter is placed to the left of the *compressor* system, so that the filter output becomes the compressor input, in oversampling this filter is placed after the expander, so that the expander output becomes the filter input. This particular way of connection is especially useful when we combine an interpolator with a decimator to change the sampling frequency value by a noninteger factor in a way we will describe later.

Returning to the aforementioned *interpolation filter*, this, as an ideal filter, is *impossible* to implement, and therefore, in practice, we resort to the use of approximate filters. From numerical analysis we know that there are many different types of interpolation, such as *linear interpolation* (in which we define two points, which are supposed to belong to the same straight line, and by constructing the equation of the line from these two points, we can then calculate the coordinates of the intermediate points that belong to this line), as well as *cubic* and, in the general case, the *polynomial* interpolation, which operate in the same way, with the difference that they use more complex equations whose graphical representations are not straight lines but polynomial curves. Let us examine the simplest of the aforementioned interpolation methods and, more specifically, linear interpolation. This interpolation process can be carried out using a filter, which, quite logically, is called a *linear interpolation filter*. This filter is characterized by a discrete impulse response of *triangular* shape defined by the

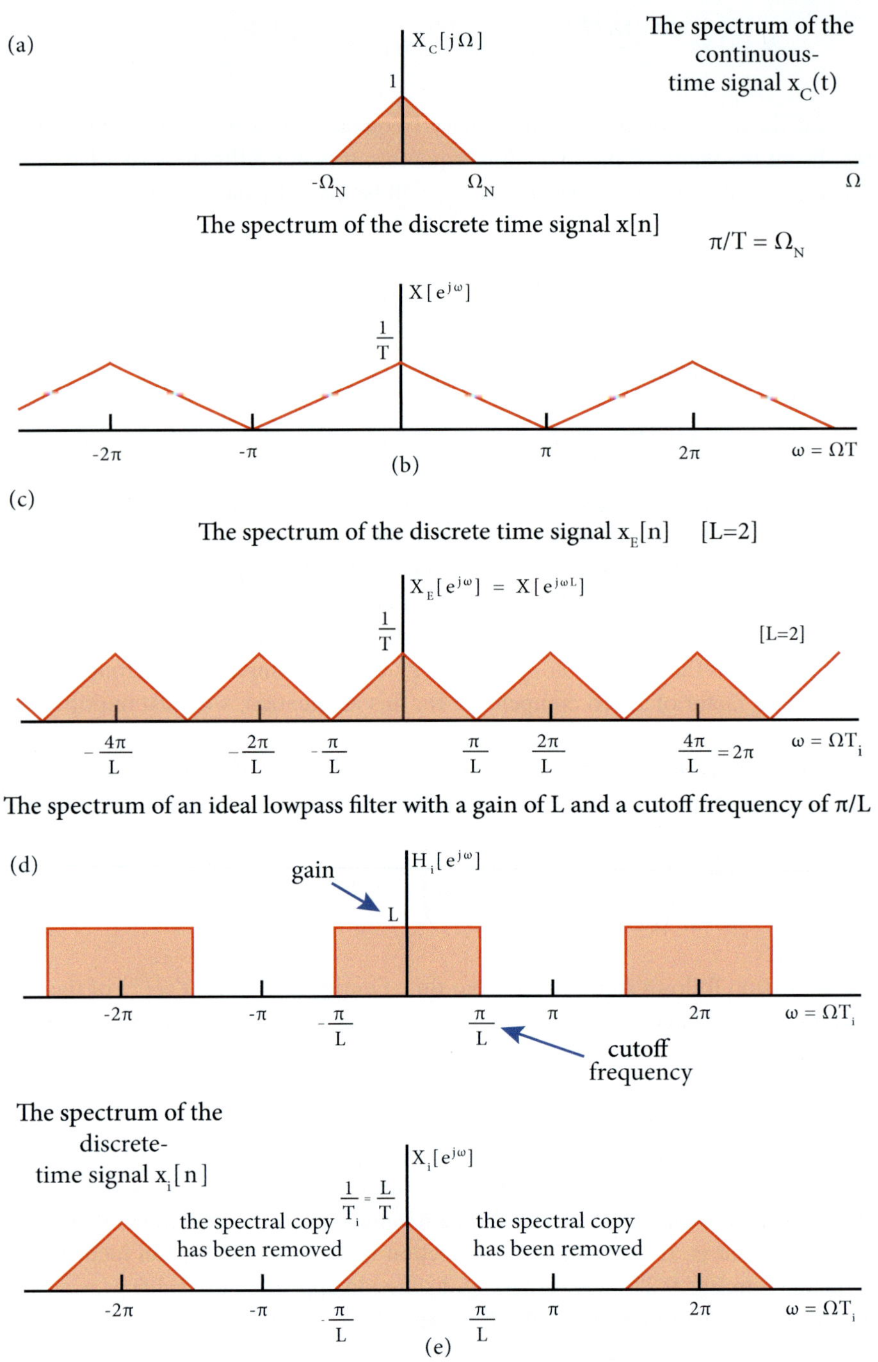

Fig. 9.26 Demonstration of oversampling process (which simply an interpolation process in the frequency domain) (Oppenheim, Schafer, Buck, 3rd edition, Figure 4.24)

equation $h[n] = 1 - |n|/L$ for the value range $|n| \leq L$, while it has a zero value for all other values of n and is characterized by the frequency response

$$H_{lin}(e^{j\omega}) = \frac{1}{L}\left[\frac{\sin(\omega L/2)}{\sin(\omega/2)}\right]^2$$

as can easily be verified by the reader. The use of this filter involves the method of linear interpolation in the sense that its output is what we would get if we connected two consecutive points of the original signal and then calculated the $L-1$ intermediate points that are on the straight line connecting these two points, exactly as dictated by the operation of this

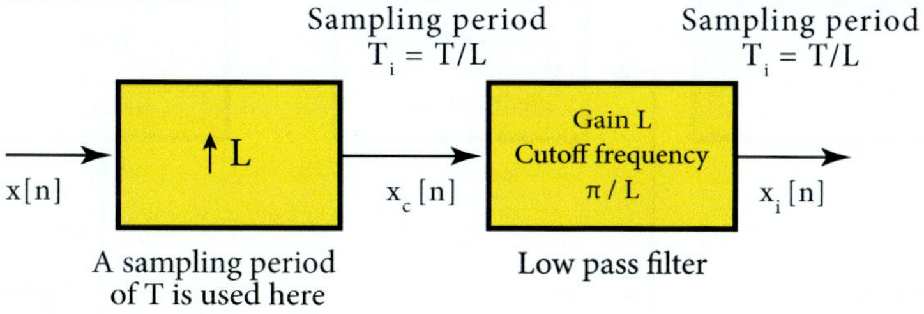

Fig. 9.27 The system of the interpolator that arises from the series connection of an expander and an ideal low-pass filter

method. The main advantage of using this filter, which, however, to function satisfactorily must be used at a higher frequency than the Nyquist frequency (since, in the opposite case, the filter output in the frequency range $\pi/L \leq |\omega| \leq \pi$ will contain a significant amount of energy due to the existence of the repeating copies of the spectrum of the continuous-time signal at the positions $2\pi k/L$, which are not removed by the linear interpolation filter), is that the process of calculating the intermediate samples can be performed using *only two samples and not all the samples of the signal* as happens in the case of the ideal filter, although it is clear that the use of more samples leads to better results. For more details on this topic, the reader is referred to the extensive bibliography.

9.9.3 Change of Sampling Frequency by a Specific Coefficient

According to our discussion in the preceding pages, the use of a decimator allows for the subsampling of a signal, that is, the reduction of the sampling rate by an integer coefficient, while the use of an interpolator allows for oversampling of a signal, that is, the increase of the sampling rate by an integer coefficient. If we combine these two systems and, more specifically, if we connect in series an interpolator with a decimator so that the output of the interpolator becomes input to the decimator (the reason why these two systems are connected in this specific way is to preserve the spectral characteristics of the signal), we can construct a *hybrid system* that changes the sampling frequency by a *noninteger* coefficient. This capability is extremely useful in several cases, including, among others, the exchange of data between two systems that use different sampling frequencies. This situation is presented in Fig. 9.28. In this figure, the output of system A, which has resulted from sampling with some frequency F_x, is transmitted to an interpolator, which causes an increase in the value of the frequency by L and then forwards its output, which corresponds to a sampling frequency LF_x, to a digital system *of sampling/holding*, which forwards this output to system B, maintaining the same frequency LF_x. In turn, system B reads the signals sent at frequency DF_y. As shown in Fig. 9.28, the sampling frequencies F_x and F_y are determined by two clocks that impose different operating rates for the two systems. Other cases in which techniques and arrangements *of polyrhythmic processing* are used include the design *of phase shift systems*, which, as is well known, corresponds to a delay in the time domain, in the implementation of filter array spectrum analysis, and signal synthesis, and the subzone encoding that is inextricably linked to the domain of digital processing of voice signals. For more cases related to the characteristics and the method of implementing these techniques, the reader is referred to the very rich bibliography.

The structure of a typical frequency sampling change system by a rational coefficient is depicted in Fig. 9.29. In this figure, the interpolator system causes *reduction of the signal period* from the value T_s to the value T_s/L, while subsequently the output of this system becomes the input to the decimator system, which causes an *increase in the signal period* by the coefficient M. The combined use of these interconnected systems leads to an output that has undergone sampling with a period of TM/L. If we choose the values of the parameters M and L in the appropriate way, we can achieve practically any ratio of sampling periods we desire. If $M > L$, then the final result will be an increase in the sampling period, while if $M < L$, then the final result will be a reduction in the sampling period.

From Fig. 9.29 we observe that the way of interconnecting these two systems leads to the *cascade* connection of the two low-pass filters included in them, which can then, according to our earlier discussion in a previous chapter, be combined into a low-pass filter with gain equal to L and cutoff frequency equal to the smaller of the values π/L and π/M. Consequently, if $M > L$, then the dominant cutoff frequency is the frequency π/M, and the final result will be the *reduction of the sampling rate*, while if $M < L$, then the dominant cutoff frequency will be the frequency π/L and the pure result will be the *increase in the sampling frequency*. To emphasize the fact that the systems involved in the foregoing process are not time invariant, but time-varying, we use different variables for the discrete-time, with the variable k related to the output of the expander,

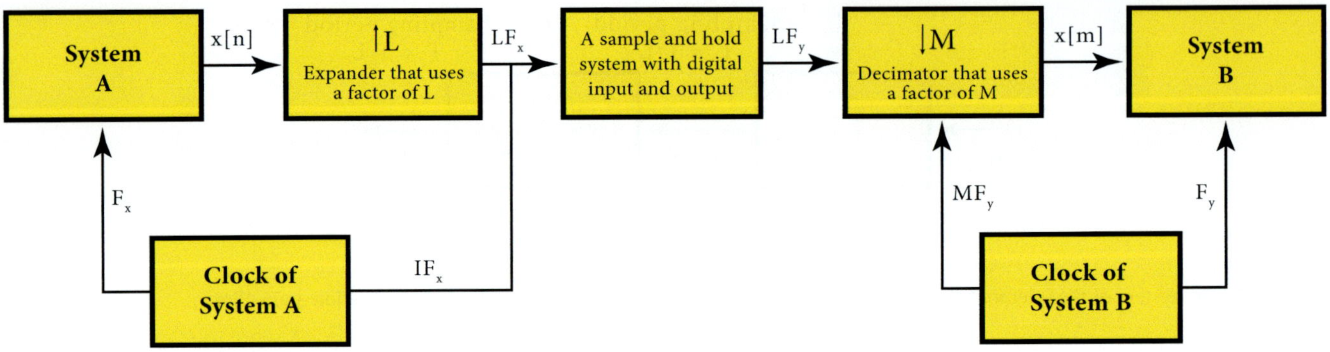

Fig. 9.28 Signal exchange between two systems using different sampling rates. A typical example of such a case is a digital radio station that reproduces music originating from a CD, with the audio signal sampling taking place at, say, 35 kHz, and the digital CD recording is known to take place at 44.1 kHz. Therefore, in order for this process to take place, a polyrhythmic processing system must be used that is capable of handling both of these rates and converting the second frequency value to the first

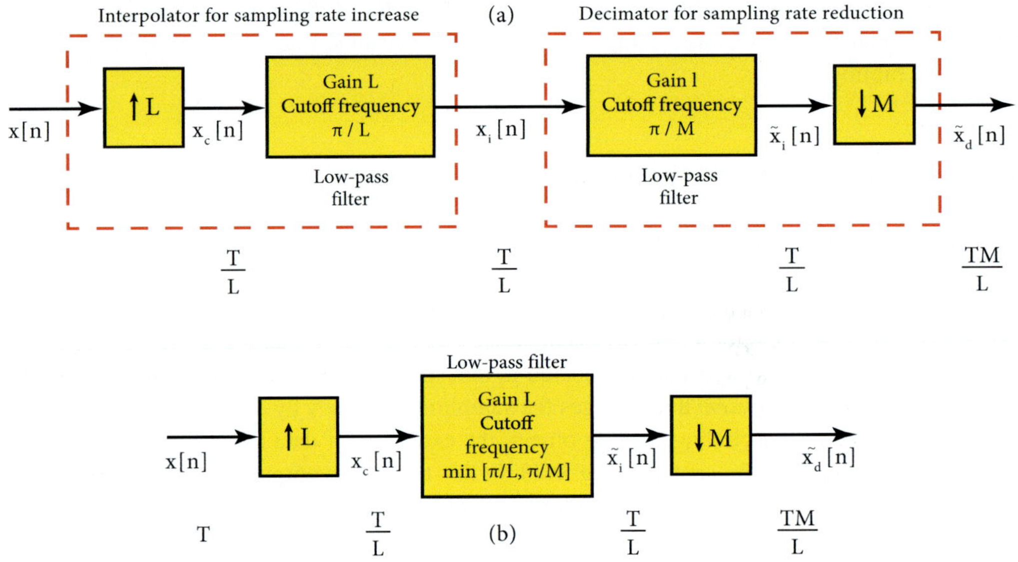

Fig. 9.29 Sampling of a signal by a noninteger coefficient is carried out with the help of a composite system that arises from the cascade connection of a supersampling system and a subsampling system

the variable ℓ to the output of the intermediate filter, and the variable m to the output of the compressor. The output of the composite system consisting of these three subsystems is the digital signal $y[m]$.

From the mathematical side of the subject, let $h_u[n]$ and $h_d[n]$ be the impulse responses of the interpolator and decimator filters, respectively, which operate at the same frequency LF_s, where F_s is the sampling frequency of the original digital signal $x[n]$. Because these two filters are connected in series and operate under the same conditions, they can be replaced, as mentioned previously, by an equivalent filter with impulse response $h[n] = h_u[n] * h_d[n]$. According to our previous discussion, the frequency response of the composite filter that combines the processes of subsampling and oversampling will have the form

$$\mathcal{H}(e^{j\omega_v}) = \begin{cases} L \text{ for } 0 \le |\omega_v| \le \min(\pi/M, \pi/L) \\ 0 \text{ otherwise} \end{cases}$$

with the frequency ω_v being related to the frequency ω_x of the original digital signal $x[n]$ through the relation $\omega_v = \omega_x/L$. The output of the expander will have the known form

$$v[\ell] = \begin{cases} x[\ell/L], \ \ell = 0, \pm L, \pm 2L \dots \\ 0 \quad \text{otherwise} \end{cases}$$

and consequently, according to the *convolution theorem*, the output of the intermediate low-pass filter that receives the signal $v[\ell]$ as its input will be

$$w[\ell] = \sum_{k=-\infty}^{\infty} h[\ell - k]v[k] = \sum_{k=-\infty}^{\infty} h[\ell - kL]x[k]$$

with the last expression resulting from the fact that according to the definition of the signal $v[\ell]$, the only nonzero terms of the foregoing sum are those that correspond to values of the index k, which are *integer multiples* of L. Finally, the output $w[\ell]$ of the composite filter is used as input to the decimator system, whose output, as mentioned earlier, will be the

$$y[m] = w[mM] = \sum_{k=-\infty}^{\infty} h[mM - kL]x[k]$$

with the final form of the output being justified by resorting to similar arguments. This system, as mentioned earlier, *is not a time-invariant but, on the contrary, a time-varying system*, since its impulse response depends not only on the discrete-time n but is a function of the time variables m and n. To prove the truth of this claim, let us change the variable $k = \lfloor mM/L \rfloor - n$, where $\lfloor x \rfloor$ is the largest integer that has a value less than x. In this case, the index appearing in the impulse response of the last expression can be written in the form $mM - kL = mM - \lfloor mM/L \rfloor L + nL = ((mM))_L + nL$, where $((x))_y \equiv x$ MOD y is the remainder of the division of x by y. Given the preceding definitions of terms, the previous relation is written

$$y[m] = \sum_{k=-\infty}^{+\infty} h[((mM))_L + nL]x[\lfloor mM/L \rfloor - n] = \sum_{k=-\infty}^{+\infty} g(n, m)x[\lfloor mM/L \rfloor - n]$$

where $g(n, m) = h[((mM))_L + nL]$ $(-\infty < n, m < \infty)$ is the impulse response of a time-varying system that receives at its input a signal that uses the variable of discrete-time n and produces as its output a signal that uses the discrete-time variable m. This impulse response is periodic with respect to the variable m, since it satisfies the property $g(n, m + kL) = g(n, m)$, as is easily proven from the defining equation.

If we combine the results we arrived at in the sections on subsampling and oversampling, we can calculate the expression of the spectrum for the output sequence $y[n]$ of the aforementioned rate change system by a noninteger coefficient. Specifically, we will have

$$\mathcal{V}(e^{j\omega_v}) = \mathcal{H}(e^{j\omega_v})\mathcal{X}(e^{j\omega_v L}) = L\mathcal{X}(e^{j\omega_v L})$$

for the range of values $0 \leq |\omega_v| \leq \min(\pi/M, \pi/L)$ and

$$\mathcal{Y}(e^{j\omega_y}) = \frac{1}{M}\sum_{k=0}^{M-1}\sum_{k=0}^{M-1} \mathcal{V}(e^{j(\omega_y/M - 2\pi k/M)})$$

where $\omega_y = M\omega_v = (M/L)\omega_x$. Combining these two equations, it turns out that

$$\mathcal{Y}(e^{j\omega_y}) = \frac{L}{M}\mathcal{X}(e^{j(\omega_y/M)})$$

for the range of values $0 \leq |\omega_y| \leq \min(\pi, \pi M/L)$.

Finally, regarding the implementation of the ideal low-pass filter with finite impulse response used in the foregoing arrangement and interposed between the arrangements of the expander and the compressor, the simplest possible approach is to implement it in its normal form, as was presented in the relevant chapter (Sect. 7.10, page 608) and in such a way that it is characterized by linear phase and by specific values for the variance in the passband and the attenuation in the cutoff zone. This filter can be implemented using any of the techniques presented in this book, such as the *window method* (Sect. 12.12.1, page 1233) or the *frequency sampling method* (Sect. 12.12.4, page 1247). However, the implementation of this filter in this way is highly inefficient, since the addition of the $L-1$ zero samples during the oversampling phase and for large values of the coefficient L results in most of the filter coefficients having a zero value, thereby leading to a zero result for the majority of the

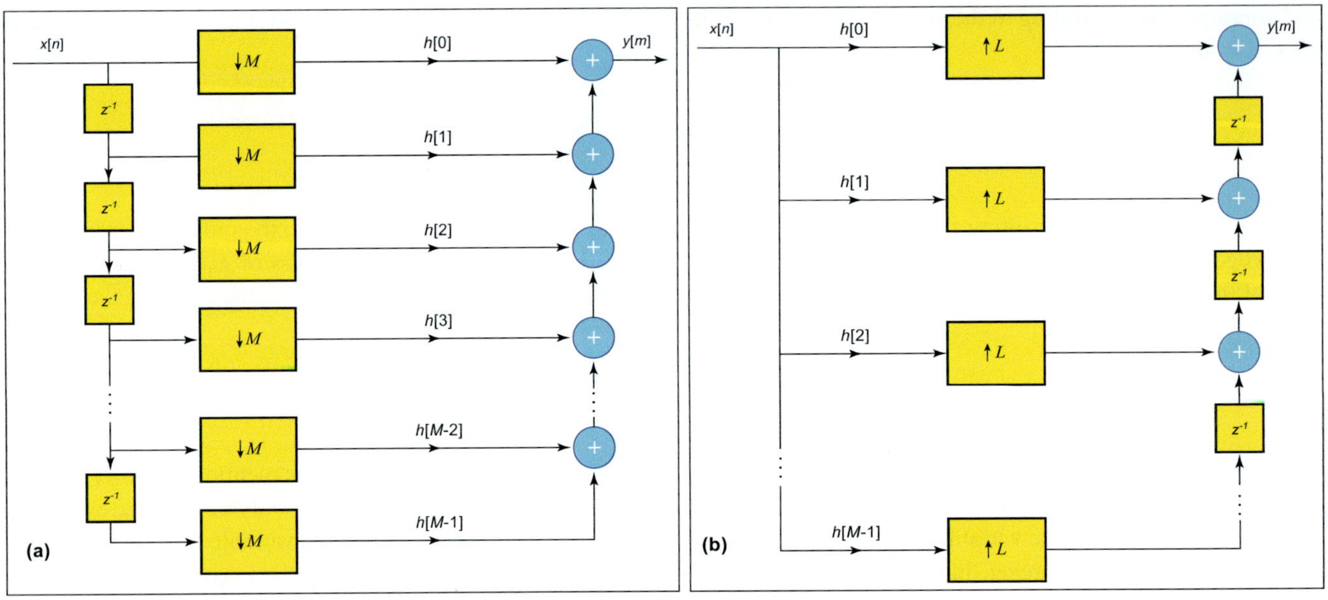

Fig. 9.30 More efficient implementations of decimator and interpolator systems. Note that these two structures are inverses of each other (left image: Proakis and Manolakis, Figure 10.9, right image: Proakis and Manolakis, Figure 10.12)

involved additions and multiplications. In addition, the use of this filter will obviously lead to the calculation of all the samples of the output of the composite system, despite the fact that, due to the subsequent undersampling applied, we only need the value of one sample for each group of M samples. For this reason, we can adopt a *more intelligent design*, incorporating the undersampling process into the aforementioned linear filter, resulting in a significant reduction in the number of necessary calculations and the increase in the performance of the filter. In the same way, for the case of the interpolator, we can transform the filter in the appropriate way, which includes the incorporation of the oversampling process. Finally, in some cases, we can further reduce the complexity of the filter, taking advantage of the characteristics of the *symmetry of the impulse response* if it indeed has such characteristics. These efficient implementations of the decimator and the interpolator are presented in the left and right images of Fig. 9.30, and it is not difficult to see that these two implementations are inverses of each other. In other words, *the inverse structure of a decimator is an interpolator, and vice versa*. An alternative way to construct filters that are characterized by the same efficiency as those presented here is to use *polyphase analysis filters*, which are presented in the next section (Sect. 9.9.5, page 772).

9.9.4 Multiple Stage Downsampling and Oversampling

Although the combination of oversampling processes and the sampling method in the way described in the previous section makes it possible to change the sampling rate by any integer or rational factor, the direct implementation of this technique is characterized by very high computational cost, because the number of operations that need to be performed for each sample of the output signal is quite large. For this reason, more efficient techniques have appeared in practice in which the systems that perform the previously discussed processes are *complex multistage systems* (i.e., they consist of the interconnection of simpler subsystems), which lead to a significant reduction in the volume of calculations, especially in cases where the values of the parameters M and L are very large. The construction of these systems is based on the observation that *if we alternate the positions of the filter and the compressor in the decimator system, the system that will result from this alternation will be equivalent to the original system*, while the same exact situation appears if we alternate the positions of the filter and the expander in the interpolator system. Although this property is known to characterize the series connection of two LTI systems, it is proven to apply in this case as well, despite the fact that some of the involved systems are not time-invariant but, on the contrary, time-varying systems. This situation is demonstrated in Fig. 9.31, in which the left image refers to the process of undersampling, while the right image refers to the process of oversampling. In each of these images, the top diagram describes the system presented in previous sections, while the bottom diagram describes the equivalent system that arises by alternating the two subsystems. This equivalence is demonstrated using the spectra of the input and output signals

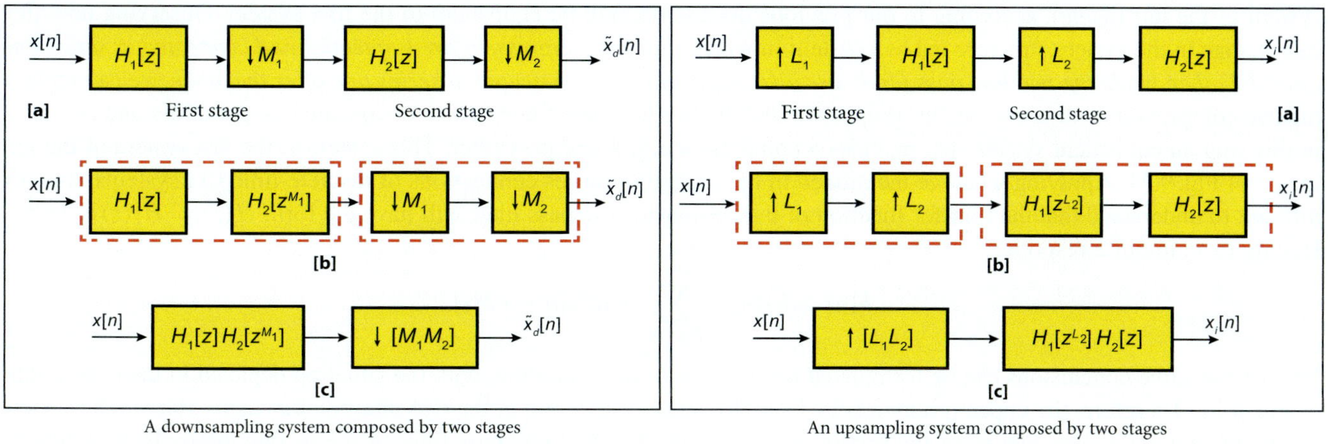

Fig. 9.31 Two-stage subsampling and oversampling systems

of the subsystems in the following way. Starting from the bottom diagram of the left image, the digital filter $\mathcal{H}(z^M)$ accepts as input the discrete-time signal $x[n]$ and produces as its output the discrete-time signal $x_b[n]$. From the convolution theorem we know that the spectra of these two signals will be related through the relation $\mathcal{X}_b(e^{j\omega}) = \mathcal{H}(e^{j\omega M})\mathcal{X}(e^{j\omega})$. From Eq. (9.53) (page 762) we know, however, that

$$\mathcal{Y}(e^{j\omega}) = \frac{1}{M}\sum_{i=0}^{M-1}\mathcal{X}_b(e^{j(\omega/M-2\pi i/M)})$$

and combining the two preceding equations we get

$$\mathcal{Y}(e^{j\omega}) = \frac{1}{M}\sum_{i=0}^{M-1}\mathcal{X}(e^{j(\omega/M-2\pi i/M)})\mathcal{H}(e^{j(\omega-2\pi i)})$$

or, equivalently,

$$\mathcal{Y}(e^{j\omega}) = \mathcal{H}(e^{j\omega})\frac{1}{M}\sum_{i=0}^{M-1}\mathcal{X}(e^{j(\omega/M-2\pi i/M)}) = \mathcal{H}(e^{j\omega})\mathcal{X}_a(e^{j\omega})$$

(since $\mathcal{H}(e^{j\omega-2\pi i}) = \mathcal{H}(e^{j\omega})$). This equation is simply a description of the system in the left image, expressed in the frequency domain. We are therefore led to the conclusion that the two systems of the left image are equivalent and, thus, *the interchange of the two subsystems of the decimator leads to an equivalent system.* In exactly the same way, the equivalence of the two systems of the right image is proven, as the reader can easily verify.

How can we exploit the preceding equivalence to construct more efficient arrangements of *polyrhythmic signal processing*? We can simply combine simpler arrangements, an approach that, on the one hand, leads to equivalent, in terms of their operation, complex arrangements, and on the other hand, to the same result, performing significantly fewer calculations. To understand this, let us limit ourselves to subsampling and assume that we want to construct a system that performs subsampling of a discrete symbol by a factor M, which can be expressed as the product $M = M_1M_2$, where M_1 and M_2 are integer quantities. The use of a simple decimation system, according to our previous discussion, would require the use of a *compressor* and an *ideal low-pass filter* with a cutoff frequency $\pi/M = \pi/(M_1M_2)$. However, the foregoing equivalence of systems allows us to construct such a system *by connecting in series two decimators* that implement subsampling by the factors M_1 and M_2, and therefore, the low-pass filters used in them will be characterized by cutoff frequencies π/M_1 and π/M_2, respectively. In the terminology of analog and digital filters, each of these two subsystems is characterized as a *stage*, while, quite reasonably, the composite system resulting from their series connection is characterized as a *two-stage system.* This situation is depicted in the upper system of the left image of Fig. 9.31. If we now proceed to swap the positions of the compressor of the first decimator and the ideal low-pass filter of the second decimator, the system that will result (the middle system of the left image), according to our previous discussion, will be equivalent to the first system. Observing now that this change in the system structure led to a system in which *the two compressors are sequential and connected in series one after the other, while the two low-pass filters are also sequential and connected in series one after the other*, we can replace the two compressors with a composite compressor and the two low-pass filters with a composite low-pass filter and construct in this way an equivalent system that includes a compressor and a low-pass filter. This system is the last system of the left image of Fig. 9.31. Applying what we mentioned in the section on series connections of discrete-time LTI systems, it is not difficult to understand that the transfer function of the composite low-pass filter will have the form $H(z) = H_1(z)H_2(z^{M_1})$, leading to an impulse response

$$h[n] = h_1[n] * \sum_{k=-\infty}^{\infty} h_2[k]\delta[n - kM_1]$$

Exactly the same conclusions can be formulated for the case of oversampling, with the situation depicted in the right image of Fig. 9.31. Therefore, the upper system results from the series connection of the two interpolator stages, the middle system results by alternating the damper of the first interpolator with the ideal low-pass filter of the second interpolator, while the third system results by combining the two dampers and the two low-pass filters.

The foregoing process refers to the connection of two stages, while in the general case, we can combine K numbers of decimation stages with coefficients $M_1, M_2, \ldots, M_K$ for the construction of an equivalent composite system with coefficient $M = M_1M_2 \cdots M_K$ and K numbers of oversampling stages with coefficients $L_1, L_2, \ldots, L_K$ for the construction of an equivalent composite system with coefficient $L = L_1L_2 \cdots L_K$. As already mentioned, the use of this approach *leads to a significant reduction in the volume of calculations* in cases where the values of the parameters M and L are very large, and therefore, to approach satisfactorily the ideal low-pass filters, we must use approximate filters with impulse responses with a very large number of samples.

9.9.5 The Technique of Polyphase Analysis

As already mentioned, in the case of decimation by a coefficient M, we retain one and only one sample for each group of M samples, ignoring all the other samples. However, although simple in its implementation, this technique is not particularly efficient, since although the majority of the output signal samples are ignored, the technique requires the calculation of all samples. This filter behavior can be improved using the technique of *polyphase analysis*.

To understand the principles and characteristics of this technique, let us consider any discrete-time signal $h[n]$ that, for the purposes of our description, expresses the impulse response of some filter. It is not difficult to understand that this signal can be represented as the *overlap* M of discrete sequences $h_k[n]$ ($k = 0, 1, 2, \ldots, M - 1$), each of which results from the successive subsampling of the signal $h[n]$ with a coefficient M, after the process of unit time delay has been applied to it. These k discrete sequences have the form

$$h_k[n] = \begin{cases} h[n+k] & n = \text{integer multiple of } M\text{, i.e., } n = \ell M \ (\ell \in \mathbb{I}) \\ 0 & \text{in the opposite case} \end{cases}$$

($k = 0, 1, 2, \ldots, M - 1$) and allow the reconstruction of the original signal $h[n]$ according to the equation

$$h[n] = \sum_{k=0}^{M-1} h_k[n - k]$$

It is interesting to note that the analysis of the impulse response $h[n]$ in this way is compatible with the theoretical analysis of the frequency change system by a noninteger coefficient that we presented earlier. Recall that, according to this analysis, this change process is equivalent to the passage of the original digital signal $x[n]$ through a time-varying linear filter with impulse response $g[n, m] = h[(mM)_L + nL]$, which is periodic with period L. This gives us the ability to use *different filter coefficient sets*, to construct the L samples of the output signal $y[n]$.

The foregoing process is depicted in the schematic diagrams of Fig. 9.32a,b, which are in reality equivalent to each other, since their only difference is that each element of time advancement (time delay) by N time units of the first diagram is replaced in the second diagram by a chain of N elements of time advancement (time delay), each of which causes an advancement (delay) by one time unit each time. The sequences $e_k[n]$ ($k = 0, 1, 2, \ldots, M - 1$) that appear in these diagrams

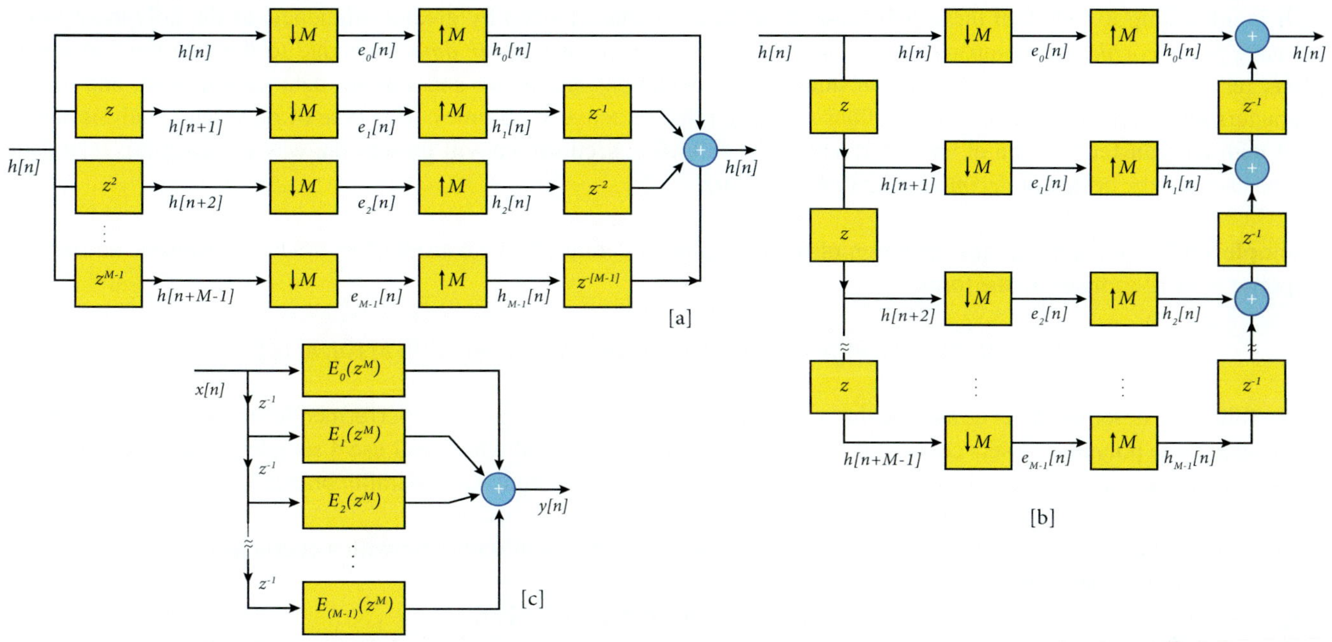

Fig. 9.32 The demonstration of the polyphase analysis technique and the filter structure resulting from it

are defined as $e_k[n] = h[nM + k] = h_k[nM]$ and constitute the so-called *polyphase components* of the discrete-time signal $h[n]$ with respect to the integer number M. These diagrams present the way in which a filter can be decomposed into a set of M simpler filters that are connected in parallel. If we consider these sequences as the impulse responses of elementary digital filters that are quite reasonably characterized as *polyphase filters* under number k ($k = 0, 1, 2, \ldots, M - 1$), then the transfer functions of these filters will have the form

$$E_k(z) = \mathcal{Z}\{e_k[n]\} = \sum_{n=-\infty}^{\infty} e_k[n]z^{-n} = \sum_{n=-\infty}^{\infty} h[nM + k]z^{-n}$$

To determine the correlation between the transfer function $\mathcal{H}(z)$ of the filter with a shock response $h[n]$ and the transfer functions $E_k(z)$ of multiphase filters, let us express the integer index n ($-\infty < n < \infty$) in its modulo representation M (according to the remainder of division theorem) as $n = Mr + k$ for values $-\infty < r < \infty$ and $0 \le k \le M - 1$. We will then have

$$\mathcal{H}(z) = \sum_{n=-\infty}^{\infty} h[n]z^{-n} = \sum_{k=0}^{M-1} \sum_{r=-\infty}^{\infty} h[Mr + k]z^{-(Mr+k)} = \sum_{k=0}^{M-1} \left(\sum_{r=-\infty}^{\infty} h[Mr + k](z^M)^{-r} \right) z^{-k} = \sum_{k=0}^{M-1} E_k(z^M)z^{-k}$$

or, in matrix representation,

$$\mathcal{H}(z) = \left[1 \; z^{-1} \; z^{-2} \ldots z^{-(M-1)}\right] \begin{bmatrix} E_0(z^M) \\ E_1(z^M) \\ E_2(z^M) \\ \vdots \\ E_{M-1}(z^M) \end{bmatrix}$$

The functions $E_k(z)$ ($k = 0, 1, 2, \ldots, M - 1$) are characterized as the *polyphase components* of the transfer function $\mathcal{H}(z)$ in the complex frequency domain, while this function's representation in the preceding manner leads to the implementation of the digital filter depicted in Fig. 9.32c. The reason why the simpler filters described by these functions are characterized as *polyphase* is that they constitute *all-pass filters* that mainly differ in terms of their phase response characteristics.

It is interesting to note that these polyphase structures are characterized by greater efficiency, as the polyphase filters (a) have *fewer coefficients* compared to the original filter $\mathcal{H}(z)$, (b) use *lower data rates* due to the subsampling taking place, resulting in the execution of a larger number of calculations and easier and simpler data transmission, and (c) are characterized by *greater efficiency*, as the filter structure remains active at all times. This type of polyphase analysis can be used with the transfer function of any filter, and in the following sections we will present the way in which this technique can be used for the design of *decimators* and *interpolators*.

Example 9.9.1 To understand the operation of the foregoing structures, let us consider Fig. 9.32a and suppose an impulse response with 12 samples, in the form

$$h[n] = \{h[0], h[1], h[2], h[3], h[4], h[5], h[6], h[7], h[8], h[9], h[10], h[11]\}$$

In the following description, let us assume that subsampling and oversampling will be performed for the value $M = 3$, which means that three polyphase filters connected in parallel will be used, each of which will use 4 samples of the original signal. These samples are calculated in the following way (see the first three horizontal and parallel branches of Fig. 9.32a):

- In the first filter (see the first branch), the signal $h[n]$ undergoes direct subsampling with a coefficient of $M = 3$, leading to the creation of the signal

$$\boldsymbol{e_0[n] = \{h[0], h[3], h[6], h[9]\}}$$

- In the second filter (see the second branch), the signal $h[n]$ before its sampling passes through a unit time advance element (i.e., time advance by one sample) described by the equation $y[n] = x[n + 1]$. The output of this filter is the signal

$$h[n] = \{h[1], h[2], h[3], h[4], h[5], h[6], h[7], h[8], h[9], h[10], h[11], 0\}$$

  which, if it undergoes subsampling with a coefficient of $M = 3$, will lead to the creation of the signal

$$\boldsymbol{e_1[n] = \{h[1], h[4], h[7], h[10]\}}$$

- In the third filter (see the third branch), the signal $h[n]$ before its sampling passes through a time advance element by two samples described by the equation $y[n] = x[n + 2]$. The output of this filter is the signal

$$h[n] = \{h[2], h[3], h[4], h[5], h[6], h[7], h[8], h[9], h[11], h[11], 0, 0\}$$

  which, if it undergoes subsampling with a coefficient of $M = 3$, will lead to the creation of the signal

$$\boldsymbol{e_2[n] = \{h[2], h[5], h[8], h[11]\}}$$

The signals $e_0[n]$, $e_1[n]$, and $e_2[n]$ constitute the polyphase components of the discrete-time signal $h[n]$. Subsequently, each of these signals will undergo oversampling with a value of $M = 3$, which in this case consists of the interpolation of two zero values between two successive samples, so that the number of their samples is ultimately tripled. From the study of Fig. 9.32a we find that the results of this process are the discrete sequences $h_0[n]$, $h_1[n]$, and $h_2[n]$:

$$h_0[n] = \{h[0], 0, 0, h[3], 0, 0, h[6], 0, 0, h[9], 0, 0\}$$

$$h_1[n] = \{h[1], 0, 0, h[4], 0, 0, h[7], 0, 0, h[10], 0, 0\}$$

$$h_2[n] = \{h[2], 0, 0, h[5], 0, 0, h[8], 0, 0, h[11], 0, 0\}$$

Examination of the diagram reveals that while the operation of the first filter has been completed, the second filter requires the passage of the signal $h_1[n]$ through a unit time delay system described by the difference equation $y[n] = x[n - 1]$ and leads to the signal

$$h_1[n] = \{0, h[1], 0, 0, h[4], 0, 0, h[7], 0, 0, h[10], 0\}$$

while the second filter requires the passage of the signal $h_2[n]$ through a system that causes a delay of two time units, described by the difference equation $y[n] = x[n-2]$, and leads to the signal

$$h_2[n] = 0, 0, h[2], 0, 0, h[5], 0, 0, h[8], 0, 0, h[11]$$

The three remaining signals that reach the input of the adder are

$$\begin{array}{rl} h_0[n] = \{ & h[0] \quad 0 \quad 0 \quad h[3] \quad 0 \quad 0 \quad h[6] \quad 0 \quad 0 \quad h[9] \quad 0 \quad 0 \quad \} \\ h_1[n] = \{ & 0 \quad h[1] \quad 0 \quad 0 \quad h[4] \quad 0 \quad 0 \quad h[7] \quad 0 \quad 0 \quad h[10] \quad 0 \quad \} \\ h_2[n] = \{ & 0 \quad 0 \quad h[2] \quad 0 \quad 0 \quad h[5] \quad 0 \quad 0 \quad h[8] \quad 0 \quad 0 \quad h[11] \} \\ \hline h[n] = \{ & h[0]\; h[1]\; h[2]\; h[3]\; h[4]\; h[5]\; h[6]\; h[7]\; h[8]\; h[9]\; h[10]\; h[11] \} \end{array}$$

and if we add them together, adding the corresponding samples, we will get the original signal $h[n]$, as can be easily verified and which, therefore, can be analyzed in the preceding way.

Example 9.9.2 Considering the transfer function $\mathcal{H}(z) = (1-2z^{-1})/(1+3z^{-1})$ and the value $M = 2$ we observe that

$$\mathcal{H}(z) = \frac{(1-2z^{-1})(1-3z^{-1})}{(1+3z^{-1})(1-3z^{-1})} = \frac{1-5z^{-1}+6z^{-2}}{1-9z^{-2}} = \frac{1+6z^{-2}}{1-9z^{-2}} + \frac{-5z^{-1}}{1-9z^{-2}}$$

We will therefore have $\mathcal{H}(z) = E_0(z^2) + z^{-1}E_1(z^2)$, where

$$E_0(z^2) = \frac{1+6z^{-2}}{1-9z^{-2}} \quad \text{and} \quad E_1(z^2) = \frac{-5}{1-9z^{-2}}$$

The foregoing techniques for designing a digital filter find immediate application in the arrangements of *subsampling* and *oversampling*. Recall that a decimator (which performs subsampling) consists of a low-pass filter connected in series with a compressor, while an interpolator (which performs oversampling) consists *of an expander connected in series with a low-pass filter*. If we replace the filters of these arrangements with the structure shown in the left image of Fig. 9.32, we can construct *more advanced* decimator and interpolator structures, which are depicted in the left images of Figs. 9.33 and 9.34. On the other hand, the right images of these figures depict the equivalent circuits that arise by alternating the two subsystems of the decimator and the interpolator, respectively, according to our earlier discussion in a previous section. Similar structures can be constructed using the alternative equivalent diagram of Fig. 9.32.

To understand why these structures are *more efficient* compared to traditional structures that operate in the usual way, let us assume that the input signal $x[n]$ is *synchronized*, so that a sample appears per unit of time and that the transfer function $\mathcal{H}(z)$ describes a finite impulse response filter, of length N samples. In this case, as Oppenheim and Schafer report, while in the traditional implementation, N multiplications and $N-1$ additions per unit of time are required, in contrast, in the new, more efficient implementation, each of the polyphase filters $E_k(z)$ is characterized by an impulse response of length

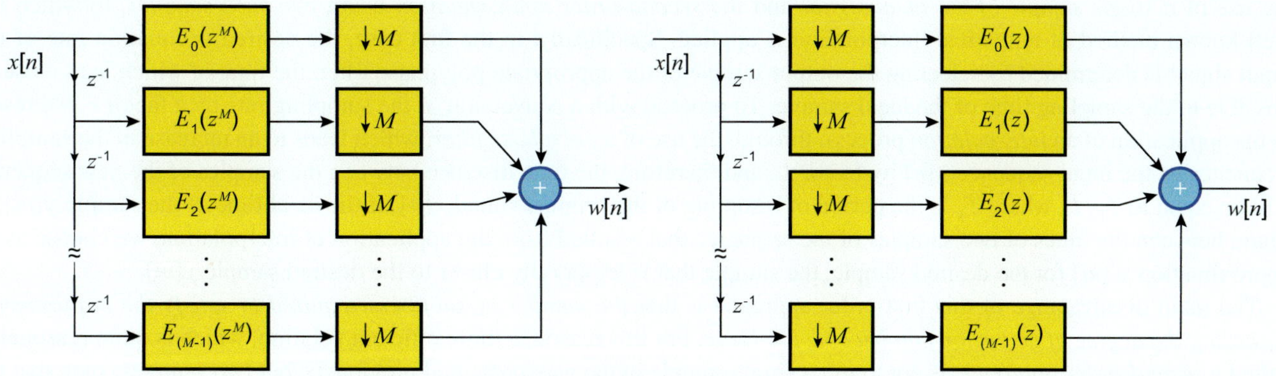

Fig. 9.33 Implementations of subsampling filter with polyphase analysis technique (Oppenheim, Schafer, Buck, Figures 4.36 and 4.37)

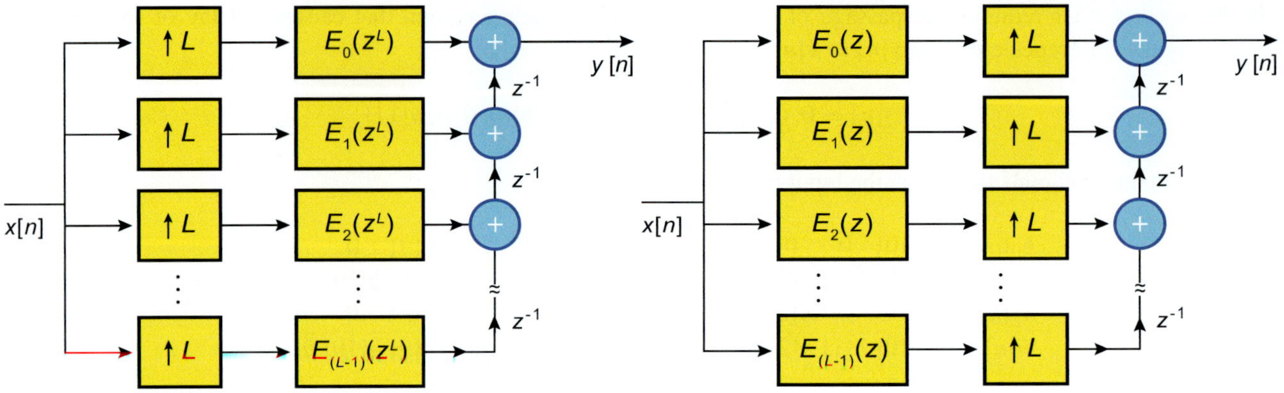

Fig. 9.34 Implementations of oversampling filter with polyphase analysis technique

N/M, while its input is synchronized, so that a sample appears every M units of time. Consequently, each filter requires $(1/M)(N/M)$ multiplications and $(1/M)[(N/M) - 1]$ additions per unit of time, and since there are M such filters in total, the system as a whole requires (N/M) multiplications and $[(N/M) - 1] + (M - 1)$ additions (the additional $M - 1$ additions are related to the operation of the adder that gives the final output of the filter) per unit of time, with this number of operations being significantly less than the previous one. The situation is similar in the case of oversampling. If we assume again that the input $x[n]$ is synchronized, so that a sample appears again per unit of time, then in the traditional implementation, at the output of the filter L values appear per unit of time, while the operation of the filter, if it is considered as a finite impulse response filter of length N samples, requires NL multiplications and $(NL - 1)$ additions per unit of time. Conversely, in the new efficient implementation, $L(N/L)$ multiplications and $L[(N/L) - 1]$ additions are required per unit of time, and moreover $L - 1$ additions for the calculation of its output. Again, this number of operations is significantly less than what is required by the traditional approach.

9.9.6 Change of Sampling Frequency by Arbitrary Coefficient

Although in most common cases the conversion of the sampling frequency is performed by an *integer* or *rational* coefficient whose value is exactly known; however, sometimes this process cannot be carried out, either because the number of polyphase filters used is *too large* or because the conversion rate *is not known* during the design of the system or, alternatively, it changes continuously during the conversion process. In these cases, it is clear that the sample we will receive at the output of this system is not the same as what we would get in the ideal case, and therefore, the output of the system will be characterized by some *degree of distortion*, which, in order for the result we will deem acceptable, must be less than some predetermined threshold value or error tolerance. In an equivalent formulation, the sample $y'[m]$ that we will receive at the output of the system at each time m will constitute only an *approximation* of the ideal sample $y[m]$. Generally speaking, there are two approximate techniques to which we can resort to calculate the sample $y'[m]$, with the *first-order approximation* based on the use of *a single sample of the ideal signal* and the *second-order approximation* using *two such samples*, to which the well-known method of numerical interpolation is applied. Specifically, in the first case, the desired decimation rate of the input signal is determined by selecting the output sample of the appropriate polyphase filter, the time of which is as close as possible to the sampling time of the ideal sample. To proceed with a conversion of the sampling rate by a factor r, we resort to the application of an *interpolation* process, through the use of a *polyphase filter*, which leads to an increase in the sampling frequency of the input sequence $x[n]$ by factor L, and therefore, the time distance between the samples of the new sequence will be equal to T_x/L, where T_x is the period of sampling of the original signal $x[n]$. If the ideal time of the sample $y[m]$, is found between the times of two samples of the sequence that resulted from the application of interpolation, we choose as an approximation $y'[m]$ for the desired sample, the sample that is temporally closer to the desired sample $y[m]$.

The main disadvantage of this first-order approach is that *the number of subfilters required to satisfy the requirements regarding the degree of distortion can become too large*. For this reason, a more efficient solution, which is quite reasonably called a *second-order approach*, is not to use a single sample in the way presented previously, but two samples, such that the ideal sampling time is located between the temporal moments time instances that correspond to these two samples. All we have to do next to calculate the approximation $y'[m]$ of the desired sample is resort to a process of *numerical interpolation* using the equation

$$y[m] = (1 - \alpha_m)y_1[m] + \alpha_2 y_2[m]$$

where $y_1[m]$ and $y_2[m]$ are these two samples, t_m is the ideal sampling time, and $\alpha_m = Lt_m$ $(0 \leq \alpha_m \leq 1)$. More details on this topic, such as the calculation of the approximation error and the way in which this depends on the number of subfilters used, can be found in the literature.

9.10 Problems

1. Solve Exercise 7.11.3 on page 624 and Problems 57 and 58 on page 642, which deal with the calculation of the exponential matrix $\boldsymbol{A}^n$, using the $\mathcal{Z}$ transform method.
2. Using the $\mathcal{Z}$ transform method, determine the transfer function and the impulse response of the discrete LTI system of Exercise 7.11.6 (page 629).
3. Using the $\mathcal{Z}$ transform method, determine the transfer function and the impulse response of the discrete LTI system of Exercise 9.7.1 (page 734).
4. Using the $\mathcal{Z}$ transform method, determine the transfer function and the impulse response of the three discrete-time systems of Exercise 32 (page 780).
5. Using the $\mathcal{Z}$ transform method, construct the dynamic equations of the normal form of the controller and the observer for the discrete LTI system of Exercise 9.6.5 (page 725).
6. Using the property $x(n) * y(n) \leftrightarrows X(z)Y(z)$ and the techniques for calculating the inverse $\mathcal{Z}$ transform, compute the convolution between the discrete-time signals

$$x(n) = \begin{cases} \left(\frac{1}{3}\right)^n & n \geq 0 \\ \left(\frac{1}{4}\right)^{-n} & n < 0 \end{cases} \quad \text{and} \quad y(n) = \left(\frac{1}{8}\right)^n u(n)$$

7. Determine the impulse response of the system, whose transfer function is given by the relation

$$\mathcal{H}(z) = \frac{1 - 2z^{-1} + 2z^{-2} - z^{-3}}{(1 - z^{-1})(1 - 0.5z^{-1})(1 - 0.4z^{-1})}$$

8. Verify the result of Exercise 9.1.2 (page 695) by directly calculating the output of the system, that is, as the convolution of the impulse response signal $h[n] = (1/3)^n u[n]$ and the input signal $x[n] = (1/2)^n \cos(\pi n/3)u[n]$.
9. Consider the recursive discrete LTI system characterized by the impulse response

$$h[n] = \left(\frac{1}{4}\right)^n u[n]$$

Calculate the output of the system $y[n]$ when the discrete-time signal is transmitted to its input

$$x[n] = 5\cos\left(\frac{\pi n}{3}\right)u[n]$$

(a) from the convolution equation $y[n] = h[n] * x[n]$ and (b) through the application of the $\mathcal{Z}$ transform method and the relation

$$y[n] = \mathcal{Z}^{-1}\{\mathcal{Y}(z)\} = \mathcal{Z}^{-1}\{\mathcal{H}(z)\mathcal{X}(z)\} = \mathcal{Z}^{-1}\left\{\mathcal{Z}\{h[n]\}\mathcal{Z}\{x[n]\}\right\}$$

What is the difference equation of the system?

10. Determine the transfer function and the impulse response of the system described by the difference equation

$$y[n] + \frac{1}{2}y[n-1] - \frac{1}{4}y[n-2] = x[n] + \frac{1}{2}x[n-1] - 2x[n-2]$$

and then calculate, using the method of $\mathcal{Z}$ transform, the response of the system to the discrete input signal

$$x[n] = (n+1)\left(\frac{1}{3}\right)^n u[n]$$

11. Construct a pole-zero plot and calculate impulse and step responses of a system described by the transfer function

$$\mathcal{H}(z) = \frac{1 - 2z^{-1} + 2z^{-2} - z^{-3}}{(1 - z^{-1})(1 - 0.5z^{-1})(1 - 0.2z^{-1})}$$

with a region of convergence described by the equation $0.5 < |z| < 1$. What is the difference equation that describes the operation of the system?
12. Determine the transfer function and the impulse and step responses of a system described by the difference equation

$$y[n] - \frac{3}{4}y[n-1] + \frac{1}{8}y[n-2] = x[n]$$

13. Prove that the transfer function and the impulse response of a system described by the difference equation

$$y[n] - \frac{5}{6}y[n-1] + \frac{1}{6}y[n-2] = x[n]$$

have the form

$$\mathcal{H}(z) = \frac{z^2}{[z-(1/2)][z-(1/3)]} \quad |z| > \frac{1}{2} \quad \text{and} \quad h[n] = \left[3\left(\frac{1}{2}\right)^n - 2\left(\frac{1}{3}\right)^n\right]u[n]$$

respectively. What is the step response of the system?
14. For each of the recursive discrete LTI systems described by the equations

$$y[n] - 0.6y[n-1] + 0.08y[n-2] = x[n]$$
$$y[n] - 0.7y[n-1] + 0.10y[n-2] = 2x[n] - x[n-2]$$

determine the type of its stability using the Schur-Cohn stability criterion.
15. Let us consider a discrete LTI system whose response to the input signal

$$x[n] = -\frac{1}{3}\left(\frac{1}{5}\right)^n u[n] - \frac{5}{3}(3)^n u[-n-1]$$

is characterized by a $\mathcal{Z}$ transform of the form

$$\mathcal{Y}(z) = \frac{1 + 0.8z^{-1}}{(1 - z^{-1})(1 + 0.5z^{-1})(1 - 2z^{-1})}$$

Determine the transfer function, the impulse response, and the difference equation of the system and characterize it in terms of its stability by applying the Schur-Cohn criterion.
16. Considering a discrete-time system described by the difference equation

$$y[n] = x[n] - 3x[n-1] + 2x[n-2]$$

(a) determine the impulse response of the system, (b) express the output of the system as the convolution $y[n] = x[n] * x[n]$, and (c) calculate the step response of the system.

17. Calculate the impulse response of the discrete-time systems

$$y[n] = \sum_{k=0}^{\infty} \alpha_k x[n-k] \qquad \text{and} \qquad y[n] = \frac{1}{2}y[n-1] + x[n]$$

18. Determine the response of the recursive discrete-time system described by the difference equation

$$y[n] = 0.85y[n-1] + 0.18y[n-2] + 3x[n-1] - x[n-2]$$

when the discrete-time signal $x[n] = n^2u[n]$ is transmitted as its input.

19. Considering a discrete LTI system described by the equation

$$y[n] - \frac{5}{8}y[n-1] + y[n-2] = x[n] - 3x[n-1]$$

construct the pole-zero plot of the transfer function $\mathcal{H}(z)$ and determine the impulse response $h[n]$.

20. Calculate the output of a system with impulse response

$$h[n] = -\delta[n] + 2\delta[n-1] - \delta[n-2]$$

when the discrete-time signal

$$x[n] = \delta[n+2] + 2\delta[n+1] + 3\delta[n] + 2\delta[n-1] + \delta[n-2]$$

21. Calculate the output of a system with impulse response $h[n] = (1/3)^n u[n]$ when the signals $x[n] = (1/4)^n$ for $n \geq 0$ and $x[n] = 4^n$ for $n < 0$ are transmitted to its input. What is the transfer function and the difference equation of the system?

22. Considering a discrete LTI system, whose response to the input signal

$$x[n] = \left(\frac{1}{4}\right)^n u[n] + (-3)^n u[-n-1] \qquad \text{is the discrete-time signal} \qquad y[n] = 6\left(\frac{1}{2}\right)^n u[n] - 6\left(\frac{3}{4}\right)^n u[n]$$

determine the transfer function $\mathcal{H}(z)$ and the impulse response $h[n]$ and characterize the system in terms of its stability and causality properties.

23. Considering a discrete LTI system whose response to the input signal

$$x[n] = -\frac{1}{3}\left(\frac{1}{2}\right)^n u[n] - \frac{4}{3}(2)^n u[-n-1]$$

is characterized by a $\mathcal{Z}$ transform of the form

$$\mathcal{Y}(z) = \frac{1 - z^{-2}}{(1 - 0.5z^{-1})(1 - 2z^{-1})}$$

determine the transfer function $\mathcal{H}(z)$ and the impulse response $h[n]$.

24. Considering a discrete LTI system with a transfer function

$$\mathcal{H}(z) = \frac{1 - \alpha^{-1}z^{-1}}{1 - \alpha z^{-1}} \qquad (\alpha \in R)$$

(a) determine the system's difference equation and the range of values of the parameter α for which the system is stable and (b) determine the system's impulse response.

25. Considering a system described by the difference equation

$$y[n] = 0.25x[n-1] + 0.5x[n] + 0.25x[n+1]$$

(a) determine its impulse response and (b) calculate and plot its frequency response $\mathcal{H}(e^{j\omega})$.

26. Considering a discrete LTI system with a transfer function

$$\mathcal{H}(z) = \frac{1 - 2z^{-1} + 2z^{-2} - z^{-3}}{(1 - z^{-1})(1 - 0.5z^{-1})(1 - 0.2z^{-1})} \quad (0.5 < |z| < 1)$$

(a) determine the system's impulse response and (b) plot the system's pole-zero diagram and characterize it in terms of its stability.

27. Determine the step response of the system described by the equation

$$y[n] = 0.7y[n-1] - 0.12y[n-2] + x[n-1] + x[n-2]$$

28. Considering a discrete LTI system with transfer function

$$\mathcal{H}(z) = \frac{z(z-4)}{(z-1)(z-2)(z-3)}$$

determine all possible region of convergences and examine for which of these the impulse response $h[n]$ is a causal discrete-time signal.

29. Considering a discrete LTI system with transfer function

$$\mathcal{H}(z) = \frac{z}{z - 3.5} \quad (|z| > 0.5)$$

determine the step response of the system, as well as its response to the signal $x[n] = nu[n]$.

30. Suppose we want to design a discrete-time LTI system, such that its output $y[n]$ at each time n is the average of the current and the M previous inputs corresponding to the times $n, n-1, n-2, \ldots, n-M+1$. (a) Construct the difference equation of the system, (b) determine the transfer function $\mathcal{H}(z)$ and draw the pole-zero diagram for the value $M = 3$, and (c) determine the impulse response of the system $h[n]$.

31. Let us consider a discrete-time LTI system whose output to the input signal $x[n] = (-2)^n$ is equal to zero, while in contrast, when the signal is transmitted to the system $x[n] = (1/2)^n u[n]$, the output takes the form $y[n] = \delta[n] + \alpha(1/4)^n u[n]$. The value of the parameter α, as well as the output $y[n]$, should be determined when the input signal $x[n] = 1$ $(\forall n \in N)$ is transmitted to the system.

32. Using the $\mathcal{Z}$ transform method and considering the discrete LTI systems that are described by the dynamic equations

$$\text{(a)} \quad \begin{cases} \boldsymbol{q}[n+1] = \begin{bmatrix} 0 & 1 \\ -0.81 & 1 \end{bmatrix} \boldsymbol{q}[n] + \begin{bmatrix} 0 \\ 1 \end{bmatrix} x[n] \\ y[n] \quad = \begin{bmatrix} -0.81 & 1 \end{bmatrix} \boldsymbol{q}[n] + x[n] \end{cases}$$

$$\text{(b)} \quad \begin{cases} \boldsymbol{q}[n+1] = \begin{bmatrix} 0 & 1 \\ -2 & -3 \end{bmatrix} \boldsymbol{q}[n] + \begin{bmatrix} 0 \\ 1 \end{bmatrix} x[n] \\ y[n] = \begin{bmatrix} 1 & 0 \end{bmatrix} \boldsymbol{q}[n] + x[n] \end{cases}$$

$$\text{(c)} \quad \begin{cases} \boldsymbol{q}[n+1] = \begin{bmatrix} -0.3 & 0.4 \\ 0.4 & -0.3 \end{bmatrix} \boldsymbol{q}[n] + \begin{bmatrix} 0 \\ 1 \end{bmatrix} x[n] \\ y[n] = \begin{bmatrix} 1 & 1 \end{bmatrix} \boldsymbol{q}[n] + x[n] \end{cases}$$

determine for each of them the value of the matrix $\boldsymbol{A}^n$, the state vector $\boldsymbol{q}[n]$, and the total response $y[n]$ if the vector of the initial state is $\boldsymbol{q}[0] = [\ 2 \quad -3\ ]^T$. The results you will arrive at should be the same as those of Exercise 60 (page 643) in which the foregoing systems are studied by applying alternative methods.

33. Using the $\mathcal{Z}$ transform method, prove that the impulse response of the system described by the dynamic equations

$$\boldsymbol{q}[n+1] = \begin{bmatrix} 0 & 1 \\ 1 & 1 \end{bmatrix} \boldsymbol{q}[n] + \begin{bmatrix} 0 \\ 1 \end{bmatrix} x[n] \quad \text{and} \quad \boldsymbol{y}[n] = \begin{bmatrix} 1 & 1 \end{bmatrix} \boldsymbol{q}[n] + x[n]$$

is given by the equation

$$h[n] = \frac{1}{\sqrt{5}} \left[\left(\frac{3+\sqrt{5}}{2} \right) \left(\frac{1+\sqrt{5}}{2} \right)^{n-1} - \left(\frac{3-\sqrt{5}}{2} \right) \left(\frac{1-\sqrt{5}}{2} \right)^{n-1} \right] u[n-1] + \delta[n]$$

This system is used to construct the well-known Fibonacci sequence, each number of which arises as the sum of the two previous numbers.

34. Let us assume that a discrete-time signal $x[n]$ originating from some source is read by a system at a rate of $1/T_1$ samples per second, and let us say we want to proceed to resample this signal and create a second discrete-time signal $y[n]$, which is read at a rate of $1/T_2$ samples per second, where $T_2/T_1 = 5/7$. Design the structure of the required resampling system, and for each subsystem included in this system determine the input/output correlation in the frequency domain.
35. Let $h(t) = e^{-\alpha t} u(t)$ be the impulse response of a continuous-time filter, where α is a positive real constant, and let $h[n] = Th(nT)$ be the response of a discrete-time filter. For a given value of the sampling period T, design the amplitude response of the discrete-time filter as a function of the parameter α.
36. Let us assume that an analog-to-digital conversion channel that uses 5 bits for encoding receives an analog signal with amplitude values from 0 to 5 V. (a) What are the quantization levels of the signal? (b) What is the value of the quantization step Δ? (c) What is the value of the quantization step if, in contrast, the amplitude of the signal is equal to 3.5 V? (d) What is the binary code that will result from the use of this system? (e) What is the quantization error?
37. What is the binary sequence that will result if the samples $\{1.2, 4.3, 7.6, 0.9, 4.8\}$ of a continuous-time signal $x(t)$ are transmitted to the quantizer of Problem 36?
38. Let us consider an analog signal of the form $x(t) = 10 \sin(2000\pi t)$, which is about to undergo a quantization process with 6 bits. (a) What is the number of quantization levels? (b) What is the value of the quantization step Δ? (c) What is the value of the SNR?
39. If the continuous-time signal to be quantized has the form

$$x(t) = A_1 \cos(\omega_1 t + \varphi_1) \times A_2 \cos(\omega_2 t + \varphi_2)$$

determine the value of the SNR if the quantizer uses 5 bits for encoding.

40. Let us assume that a continuous-time signal $x(t)$ with a dynamic range of $[-5, 5]$ V is about to undergo quantization by a type of quantizer midriser that uses for encoding 5 bits. (a) Design the transfer function of the system. (b) What are the quantized values of the values 0.25 V and -3.15 V?

Fourier Analysis in the Discrete-Time Domain 10

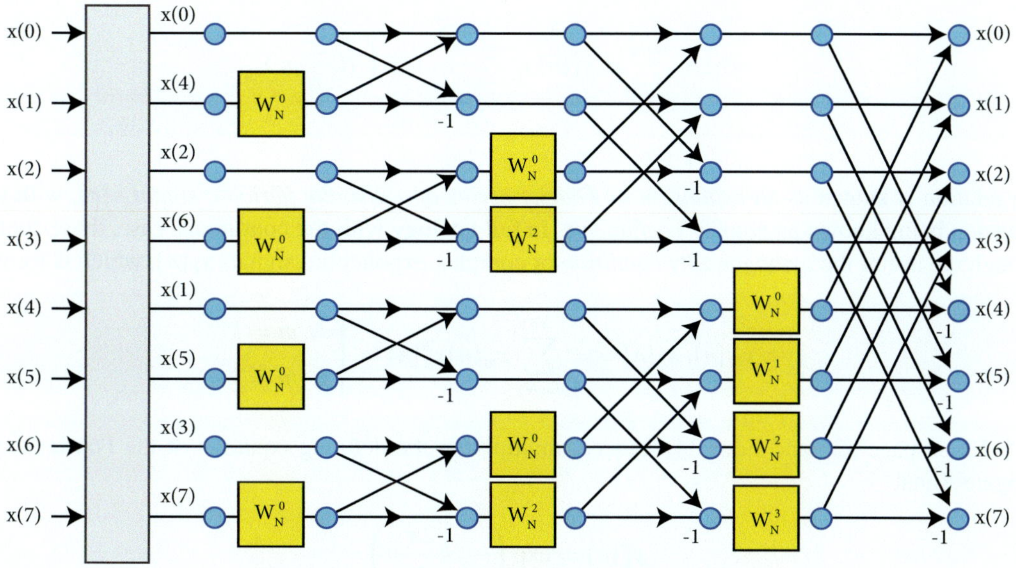

Example of an 8-point FFT butterfly scheme

Process of calculating spectrum of finite discrete sequence using fast Fourier transform described using special diagrams known as butterfly diagrams

10.1 Introduction

This chapter covers Fourier theory and how it is applied in the domain of discrete-time. The most basic concepts included in this are (a) *the Fourier development of discrete-time signals* and the properties that describe it, (b) the continuous *Fourier transform* of discrete-time signals, (c) the *discrete Fourier transform* (DFT) and the way in which it is implemented and used for the frequency analysis of signals and systems in the field of discrete-time, and (d) the short-time system analysis applied in the case of rapidly changing signals, based on the use of the time-dependent Fourier transform.

10.2 Development of Fourier Discrete-Time Signals

As already mentioned in a previous chapter, elementary complex exponential signals of the form

A. I. Margaris, *Continuous and Discrete-Time Signals and Systems*,
https://doi.org/10.1007/978-3-031-81209-5_10

$$x_k[n] = \exp\left(j\frac{2\pi kn}{N}\right) = \exp(jk\omega_0 n) \quad (k = \ldots, -2, -1, 0, 1, 2, \ldots) \tag{10.1}$$

are *periodic signals* with a common period $N = 2\pi/\omega_0$ (in reality, as we have proven, there are N discrete sequences that satisfy this property, and for this reason the index k can only be limited to the interval $[0, N-1]$). Taking into account that these N discrete sequences constitute an *orthogonal basis of functions* within any interval of duration N (for the proof of this property see Exercise 6.7.2 on page 526), we can formulate the claim that *every*[1] *periodic discrete-time signal* $x[n] = x[n+rN](r = 0, \pm1, \pm2)$ *of period N can be expressed as a linear combination* of the N complex exponential signals $x_k[n]$ $(k = 0, 1, 2, \ldots, N-1)$ of the form

$$x[n] = \sum_{k=0}^{N-1} \alpha_k x_k[n] = \sum_{k=0}^{N-1} \alpha_k \exp\left(jk\frac{2\pi}{N}n\right) \tag{10.2}$$

The preceding relation is known as an *expansion or Fourier series of a discrete periodic signal* $x[n]$, with the coefficients α_k of this expansion being known as *Fourier coefficients*. In full analogy with the continuous case, these coefficients can be calculated through the use of the *orthogonality condition* of complex exponential signals $x_k[n]$, which is formulated as

$$< x_m[n], x_k[n] >= \sum_{n=N_1}^{N_2} x_m[n]x_k^*[n] = \begin{cases} N & m = k \\ 0 & m \neq k \end{cases} \tag{10.3}$$

(see Exercise 6.7.2 on page 526 and Eq. 6.34). Indeed, multiplying the defining equation of the Fourier expansion by the complex conjugate signal

$$x_k^*[n] = \exp\left(-jk\frac{2\pi}{N}n\right)$$

[1] In reality, for the aforementioned relationship to apply *for every periodic discrete-time signal* $x[n]$, we need to prove that the correlation that exists between the quantities $x[n]$ and α_k is *reversible*, i.e., that the Fourier coefficients $\alpha_1, \alpha_2, \ldots, \alpha_{N-1}$ corresponding to a given periodic discrete-time signal $x[n]$ are *unique*, and vice versa. But this is something that holds true since, if we develop an equation defining the coefficients α_k, we will end up with the system of equations

$$\alpha_0 = \frac{1}{N}\left(x[0] + x[1] + x[2] + \cdots + x[N-1]\right)$$

$$\alpha_1 = \frac{1}{N}\left(x[0] + x[1]\,e^{-j\frac{2\pi}{N}} + x[2]\,e^{-j\frac{2\pi}{N}2} + \cdots + x[N-1]\,e^{-j\frac{2\pi}{N}(N-1)}\right)$$

$$\cdots\cdots\cdots\cdots\cdots\cdots\cdots\cdots\cdots\cdots\cdots\cdots$$

$$\alpha_{N-1} = \frac{1}{N}\left(x[0] + x[1]\,e^{-j(N-1)\frac{2\pi}{N}} + x[2]\,e^{-j(N-1)\frac{2\pi}{N}2} + \cdots + x[N-1]\,e^{-j(N-1)\frac{2\pi}{N}(N-1)}\right)$$

which in matrix form is expressed as

$$\begin{bmatrix} \alpha_0 \\ \alpha_1 \\ \vdots \\ \alpha_{N-1} \end{bmatrix} = \frac{1}{N}\begin{bmatrix} 1 & 1 & 1 & \ldots & 1 \\ 1 & e^{j\frac{2\pi}{N}} & e^{j\frac{2\pi}{N}2} & \ldots & e^{j\frac{2\pi}{N}(N-1)} \\ \ldots & \ldots & \ldots & \ldots & \ldots \\ 1 & e^{-j(N-1)\frac{2\pi}{N}} & e^{-j(N-1)\frac{2\pi}{N}2} & \ldots & e^{-j(N-1)\frac{2\pi}{N}(N-1)} \end{bmatrix}\begin{bmatrix} x[0] \\ x[1] \\ \vdots \\ x[N-1] \end{bmatrix}$$

But the foregoing matrix is proven to be invertible. Therefore, each discrete periodic signal corresponds to a unique set of Fourier coefficients, that is, the Fourier expansion exists for every periodic signal.

and summing for the entire duration N, we get

$$\sum_{n=<N>} x[n]x_k^*[n] = \sum_{n=<N>} x[n]\exp\left(-jk\frac{2\pi}{N}n\right) = \sum_{n=<n>} \underbrace{\left\{\sum_{m=<N>} \alpha_m \exp\left(jm\frac{2\pi}{N}n\right)\right\}}_{x[n]} \exp\left(-jk\frac{2\pi}{N}n\right) = N\alpha_n$$

from which it immediately follows that

$$\alpha_k = \frac{1}{N}\sum_{n=0}^{N-1} x[n]\exp\left(-jk\frac{2\pi}{N}n\right) \qquad (10.4)$$

(for the derivation of the last relation we proceeded to replace the dummy index from n to k, while as duration N we considered the range of values $[0, N-1]$).

It is not difficult to see that *the Fourier coefficients* α_k *are periodic sequences with period N*. Indeed, from the foregoing equation it follows that[2]

$$\alpha_{k+N} = \sum_{n=0}^{N-1} x[n]e^{-j\frac{2\pi}{N}(k+N)n} = \left(\sum_{n=0}^{N-1} x[n]e^{-j\frac{2\pi}{N}kn}\right)e^{-j2\pi n} = \alpha_k \quad (\forall k \in N) \qquad (10.5)$$

The defining equations of the quantities $x[n]$ and α_k usually appear *in pairs*, and one is considered *the inverse of the other*. From these two equations, the defining equation of the coefficients α_k is called the *analysis equation*, while the defining equation of the discrete-time signal $x[n]$ is called the *synthesis equation*. Defining the quantity $W_N = e^{-j\frac{2\pi}{N}}$, the preceding equations are written in the form of a pair as

$$\alpha_k = \frac{1}{N}\sum_{n=0}^{N-1} x[n]W_N^{kn} \Longleftrightarrow x[n] = \sum_{k=0}^{N-1} \alpha_k W_N^{-kn}$$

We conclude this brief introduction regarding the *Fourier expansion of discrete-time signals* with the observation that, *provided the sum appearing in the defining equation is finite, it always converges to some finite value*; this is something that contrasts with the sum of the infinite terms of the Fourier expansion of continuous-time signals formulated earlier and the appropriate convergence conditions. For the same reason, *the Gibbs phenomenon that characterizes the Fourier development of continuous-time signals (Sect. 3.2.5 on page 200) does not appear in the discrete case.*

10.2.1 Trigonometric Form of Fourier Expansion

In the special case where the periodic discrete-time signal $x[n]$ is *a signal of real values*, its Fourier expansion can be formulated in a trigonometric form, thereby allowing the formulation of the signal *as a linear combination of simple sines and cosines*. Indeed, starting with the defining equation of the exponential Fourier expansion and developing the term that corresponds to the value $k = 0$, we get

---

[2] If the signal $x[n]$ is a real-valued signal, $x[n] = x^*[n]$, and therefore it will hold that

$$\alpha_k^* = \left\{\frac{1}{N}\sum_{n=0}^{N-1} x[n]\exp\left(-jk\frac{2\pi}{N}n\right)\right\}^* = \frac{1}{N}\sum_{n=0}^{N-1} x^*[n]\exp\left(jk\frac{2\pi}{N}n\right) = \frac{1}{N}\sum_{n=0}^{N-1} x[n]\exp\left(-j\frac{2\pi}{N}(-k)n\right) = \alpha_{-k}.$$

$$x[n]=\sum_{k=0}^{N-1}\alpha_k\exp\left(-j\frac{2\pi}{N}kn\right)=\alpha_0+\sum_{k=1}^{N-1}\alpha_k\exp\left(-j\frac{2\pi}{N}kn\right)$$

At this point we distinguish two cases. If the period N of the signal is *an odd number*, the expression $N-1$ will be an even number, and the preceding equation can be written in the form

$$x[n]=\alpha_0+\sum_{k=1}^{(N-1)/2}\left\{\alpha_k\exp\left(-j\frac{2\pi}{N}kn\right)+\alpha_{N-k}\exp\left[-j\frac{2\pi}{N}(N-k)n\right]\right\}$$

But from the defining equation of the Fourier coefficients it follows that $\alpha_{N-k}=\alpha_k^*$, while we will also have

$$\exp[j(N-k)(2\pi/N)n]=\exp[-jk(2\pi/N)n]$$

Therefore, it will hold that

$$x[n]=\alpha_0+\sum_{k=1}^{(N-1)/2}\left[\alpha_k\exp\left(j\frac{2\pi}{N}kn\right)+\alpha_k^*\exp\left(-j\frac{2\pi}{N}kn\right)\right]=$$

$$=\alpha_0+2\sum_{k=1}^{(N-1)/2}\Re\left\{\alpha_k\exp\left(j\frac{2\pi}{N}kn\right)\right\}=\alpha_0+2\sum_{k=1}^{(N-1)/2}\left\{\Re\{\alpha_k\}\cos\left(k\frac{2\pi}{N}n\right)-\Im\{\alpha_k\}\sin\left(k\frac{2\pi}{N}n\right)\right\}$$

The preceding equation expresses the trigonometric form of a real periodic discrete-time signal with *odd period value*. However, if the signal's period is *even*, we can formulate the signal $x[n]$ in the form

$$x[n]=\alpha_0+\sum_{k=1}^{(N-2)/2}\left\{\alpha_k\exp\left(-j\frac{2\pi}{N}kn\right)+\alpha_{N-k}\exp\left[-j\frac{2\pi}{N}(N-k)n\right]\right\}+\alpha_{N/2}\exp\left(j\frac{2\pi}{N}n\frac{N}{2}\right)$$

Performing the analysis carried out earlier and observing that

$$\exp\left(j\frac{2\pi}{N}n\frac{N}{2}\right)=e^{j\pi n}=(-1)^n$$

we can formulate the signal $x[n]$ in the form

$$x[n]=\alpha_0+(-1)^n\alpha_{N/2}+2\sum_{k=1}^{(N-1)/2}\left\{\Re\{\alpha_k\}\cos\left(k\frac{2\pi}{N}n\right)-\Im\{\alpha_k\}\sin\left(k\frac{2\pi}{N}n\right)\right\}$$

which constitutes the trigonometric form of a real signal with *even period value*.

Problem 10.2.1 Find the Fourier expansion of the periodic signal depicted in Fig. 10.1 (Boulet, Solved Exercise 2.1).

Solution Observing the graphical representation of the signal, we find that its period has a value of $N=6$, while the group of repeating samples is $\{\mathbf{2},1,-1,-2,-1,1\}$. Therefore, from the defining equation of the Fourier coefficients we will have

$$\alpha_k=\frac{1}{6}\sum_{n=0}^{5}x[n]\exp\left(-j\frac{2\pi}{6}kn\right)=\frac{1}{6}\left\{x[0]+x[1]\exp\left(-j\frac{2\pi}{6}k\right)+x[2]\exp\left(-j\frac{4\pi}{6}k\right)+x[3]\exp\left(-j\frac{6\pi}{6}k\right)\right.$$

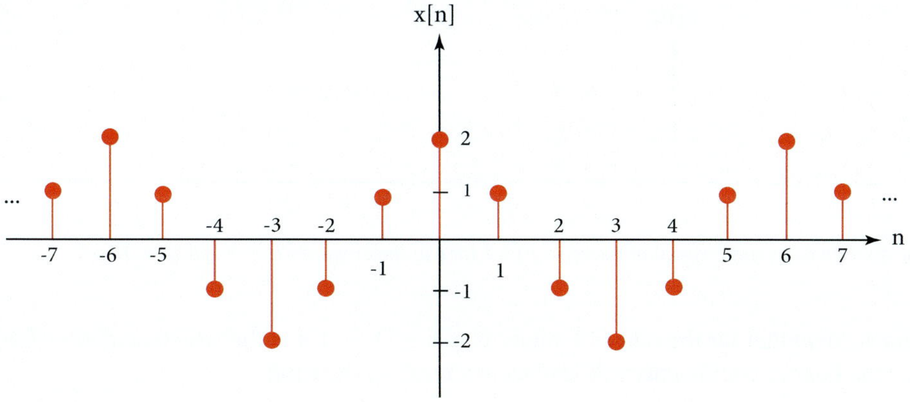

Fig. 10.1 The discrete-time signal of Exercise 10.2.1

$$+ x[4] \exp\left(- j\frac{8\pi}{6}k \right) + x[5] \exp\left(- j\frac{10\pi}{6}k \right) \Big\} = \frac{1}{6} \Big\{ 2 + \exp\left(- j\frac{\pi}{3}k \right) - \exp\left(- j\frac{2\pi}{3}k \right)$$
$$- \exp(-j\pi k) - \exp\left(- j\frac{4\pi}{3}k \right) + \exp\left(- j\frac{5\pi}{3}k \right) \Big\}$$

It is not difficult to see that the exponential expressions of the preceding equation are written as $\exp(-j\pi k) = (-1)^k$, $\exp(-j4\pi k/3) = (-1)^k \exp(-j\pi k/3)$, and $\exp(-j5\pi k/3) = (-1)^k \exp(-j2\pi k/3)$, and therefore the last expression can be transformed into

$$\alpha_k = \frac{1-(-1)^k}{6}\left[2 + \exp\left(- j\frac{\pi k}{3} \right) - \exp\left(- j\frac{2\pi k}{3} \right) \right]$$

Substituting in the foregoing equation the values $k = 0, 1, 2, 3, 4, 5$, we find that $\alpha_0 = \alpha_2 = \alpha_3 = \alpha_4 = 0$ and $\alpha_1 = \alpha_5 = 1$. Therefore, the discrete periodic signal $x[n]$ will take the form

$$x[n] = \sum_{k=0}^{5} \alpha_k \exp\left(j\frac{2\pi}{6}kn \right) = \exp\left(j\frac{\pi n}{3} \right) + \exp\left(j\frac{5\pi n}{3} \right)$$

Problem 10.2.2 Find the Fourier expansion of the elementary trigonometric signal $x[n] = \cos(n)$.

Solution As already mentioned in the previous section, a necessary prerequisite for the expansion of a discrete-time signal $x[n]$ in a Fourier series is the characterization of the signal as *periodic* with some period value equal to N. However, as we found in the chapter on discrete-time signals (Note 1 on page 525), the signal $x[n] = \cos(n)$ *is not periodic*, and therefore its expansion in a Fourier series is not possible. This observation naturally applies to any aperiodic signal.

Problem 10.2.3 Find the Fourier expansion of the discrete periodic signal

$$x[n] = \sum_{m=-\infty}^{\infty} \delta[n - mN] = \begin{cases} 1 \;\; n = mN, \quad (m \in N) \\ 0 \text{ in the opposite case} \end{cases}$$

with period N, where $\delta[n]$ is the discrete impulse function.

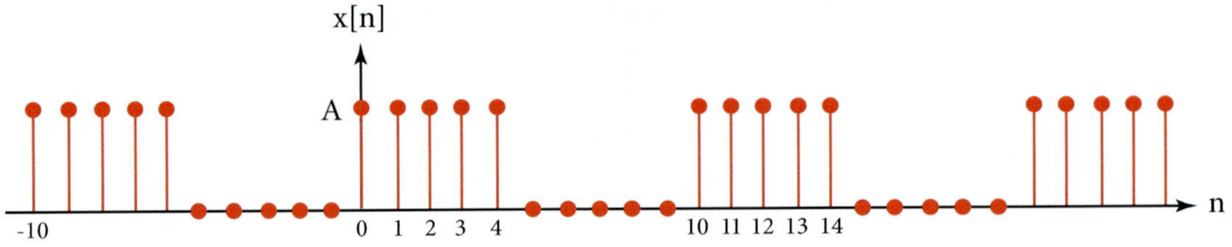

Fig. 10.2 The periodic discrete rectangular signal of Exercise 10.2.4 for parameter values $L = 5$ and $N = 10$

Solution Taking into account that for the range of values $0 \leq n \leq N - 1$ it is the case that $x[n] = \delta[n]$, the application of the equation defining the Fourier coefficients will lead us to a result of the form

$$\alpha_k = \sum_{n=0}^{N-1} x[n] \exp\left(-j\frac{2\pi}{N}kn\right) = \sum_{n=0}^{N-1} \delta[n] \exp\left(-j\frac{2\pi}{N}kn\right) = 1$$

since the impulse function $\delta[n]$ has a *nonzero value* equal to unity *only* for the parameter $k = 0$. Therefore, the signal $x[n]$ is formulated as

$$x[n] = \sum_{m=-\infty}^{\infty} \delta[n - mN] = \frac{1}{N} \sum_{k=0}^{N-1} \exp\left(j\frac{2\pi}{N}kn\right)$$

Problem 10.2.4 Find the coefficients of the Fourier expansion of the periodic rectangular signal

$$y[n] = \sum_{r=-\infty}^{\infty} x[n + rN] \quad \text{with the discrete-time signal } x[n] \text{ defined as} \quad x[n] = \begin{cases} A & 0 \leq n \leq L - 1 \\ 0 & L < n \leq N \end{cases}$$

This discrete-time signal is depicted in Fig. 10.2 for parameter values $L = 5$ and $N = 10$ (Proakis and Manolakis, Example 4.2.2, part A).

Solution From the defining equation of the Fourier coefficients we will have

$$\alpha_k = \frac{1}{N} \sum_{n=0}^{N-1} x[n] \exp\left(-j\frac{2\pi}{N}kn\right) = \frac{A}{N} \sum_{n=0}^{L-1} \exp\left(-j\frac{2\pi}{N}kn\right)$$

It is not difficult to see that for $k = 0$ it will hold that

$$\alpha_0 = \frac{A}{N} \sum_{k=0}^{L-1} 1 = \frac{AL}{N}$$

while for the range of values $0 < k \leq L - 1$ and using the appropriate expression from mathematical typology (see Table 6.2 on page 500) we will have

$$\alpha_k = \frac{A}{N} \sum_{n=0}^{L-1} \left\{\exp\left(-j\frac{2\pi}{N}k\right)\right\}^n$$

or, equivalently,

$$\alpha_k = \frac{A}{N} \left\{\frac{1 - \exp\left(-j\frac{2\pi}{N}kL\right)}{1 - \exp\left(-j\frac{2\pi}{N}k\right)}\right\} = \frac{A}{N} \frac{\exp\left(-j\frac{\pi kL}{N}\right)}{\exp\left(-j\frac{\pi k}{N}\right)} \left\{\frac{\exp\left(j\frac{\pi kL}{N}\right) - \exp\left(-j\frac{\pi kL}{N}\right)}{\exp\left(j\frac{\pi k}{N}\right) - \exp\left(-j\frac{\pi k}{N}\right)}\right\}$$

and finally

$$\alpha_k = \frac{A}{N}\exp\left[-j\frac{\pi k(L-1)}{N}\right]\left\{\sin\left(\frac{\pi kL}{N}\right)\right\}\Big/\left\{\sin\left(\frac{\pi k}{N}\right)\right\},\quad k=1,2,\ldots,N-1$$

10.2.2 Properties of the Fourier Expansion

The most important properties of the Fourier expansion of discrete periodic signals are presented in Table 10.1. Following the methodology of Sect. 3.2.6 (refer to page 204), we will use for the Fourier coefficients α_k ($k = 0, 1, 2\ldots, N-1$) of the discrete periodic signals $x[n]$ and $y[n]$ the symbols x_k and y_k, respectively, and therefore the two signals will have the form

$$x[n] = \sum_{k=0}^{N-1} x_k \exp\left(jk\frac{2\pi}{N}n\right) \quad \text{and} \quad y[n] = \sum_{k=0}^{N-1} y_k \exp\left(jk\frac{2\pi}{N}n\right)$$

A detailed presentation of the properties of the Fourier series of discrete periodic signals is the subject of the following pages.

- **Linearity**: If the coefficients of a Fourier series of discrete periodic signals $x[n]$ and $y[n]$ that have a common period N are x_k and y_k, then the coefficients of a Fourier series of the linear combination $z[n] = \alpha x[n] + \beta y[n]$ will be $z_k = \alpha x_k + \beta y_k$.

Table 10.1 Basic properties of the Fourier series of discrete-time signals

Property	Discrete-time sequence $X[n]$	Fourier coefficient
Homogeneity	$\alpha x[n]$	$a x_k$
Addition	$x[n] + y[n]$	$x_k + y_k$
Linearity	$\alpha x[n] + \beta y[n]$	$\alpha x_k + \beta y_k$
Multiplication	$x[n]y[n]$	
Periodic convolution	$\sum_{m=0}^{N-1} x[m]y[n-m]$	$N x_k y_k$
Time shift	$x[n-m]$	$\exp\left(\frac{-2j\pi \text{km}}{N}\right) x_k$
Frequency shift	$\left(\frac{2j\pi \text{nm}}{N}\right) x[n]$	x_{k-m}
Time reversal	$x[-n]$	x_{-k}
Time scaling	$x[n/m]$	x_k/m
Symmetry and conjugation	$x^{\boxtimes}[n]$	$x_{-k}^{\boxtimes}$
	$x^{\boxtimes}[-n]$	$x_k^{\boxtimes}$
	$\Re\{x[n]\}$	$x_k^e = \frac{x_k + x_{-k}}{2}$
	$j\Im\{x[n]\}$	$x_k^o = \frac{x_k - x_{-k}}{2}$
	$\frac{x[n] + x[-n]}{2}$	$\Re\{x_k\}$
	$\frac{x[n] - x[-n]}{2}$	$j\Im\{x_k\}$
Dualism	x_k	$x[-n]/N$
Difference	$x[n] - x[n-1]$	$\left[1 - \exp\left(-j\frac{2\pi}{N}k\right)\right] x_k$
Accumulation	$\sum_{m=-\infty}^{n} x[m]$	$\frac{x_k}{1 - \exp\left(-j\frac{2\pi}{N}k\right)}$

Proof From the defining equation of the Fourier coefficients of the discrete-time signal $z[n]$ we will have

$$z_k = \frac{1}{N}\sum_{n=0}^{N-1} z[n]\exp\left(-jk\frac{2\pi}{N}n\right) = \frac{1}{N}\sum_{n=0}^{N-1}\Big(\alpha x[n] + \beta y[n]\Big)\exp\left(-jk\frac{2\pi}{N}n\right)$$

$$= \alpha\left\{\frac{1}{N}\sum_{n=0}^{N-1} x[n]\exp\left(-jk\frac{2\pi}{N}n\right)\right\} + \beta\left\{\frac{1}{N}\sum_{n=0}^{N-1} y[n]\exp\left(-jk\frac{2\pi}{N}n\right)\right\} = \alpha x_k + \beta y_k$$

which is the desired result. The property of linearity can arise from the combination of the properties of addition and homogeneity, which are proven in exactly the same way.

- **Time shift**: If the coefficients of the Fourier expansion of the periodic signal $x[n]$ with period N are x_k, then the coefficients of the Fourier expansion of the time-shifted signal $y[n] = x[n-m]$ will be $y_k = W_N^{km}x_k = \exp(-2j\pi mk/N)x_k$.
Proof From the defining equation of the Fourier coefficients of the discrete-time signal $y[n]$ we will have

$$y_k = \frac{1}{N}\sum_{n=0}^{N-1} y[n]\exp\left(-j\frac{2\pi}{N}kn\right) = \frac{1}{N}\sum_{n=0}^{N-1} x[n-m]\exp\left(-j\frac{2\pi}{N}kn\right)$$

In the next step of the process, let us proceed to change the dummy index of the foregoing sum from n to $r = n - m$. For $n = 0$ we will have $r = -m$, while for $n = N - 1$ we will have $r = N - 1 - m$, and the last relation is written

$$y_k = \frac{1}{N}\sum_{r=-m}^{N-1-m} x[r]\exp\left[-j\frac{2\pi}{N}k(r+m)\right] = \frac{1}{N}\sum_{r=-m}^{N-1-m} x[r]\exp\left(-jk\frac{2\pi}{N}m\right)\exp\left(-jk\frac{2\pi}{N}r\right)$$

$$= \exp\left(-jk\frac{2\pi}{N}m\right)\left\{\frac{1}{N}\sum_{r=-m}^{N-1-m} x[r]\exp\left(-jk\frac{2\pi}{N}r\right)\right\}$$

But the last sum of the foregoing relation is expressed as the sum of smaller sums of the form

$$\sum_{r=-m}^{N-1-m} x[r]\exp\left(-jk\frac{2\pi}{N}r\right) = \sum_{r=-m}^{-1} x[r]\exp\left(-jk\frac{2\pi}{N}r\right) + \sum_{r=0}^{N-1-m} x[r]\exp\left(-jk\frac{2\pi}{N}r\right)$$

However, taking into account that the discrete-time signal $x[r]$ and the exponential expression W_N^{kr} are periodic functions[3] with a period of N, $x[r] = x[r+N]$ and, similarly, $W_N^{kr} = W_N^{k(r+N)}$, and the first subsum is expressed as

$$\sum_{r=-m}^{-1} x[r]\exp\left(-jk\frac{2\pi}{N}r\right) = \sum_{r=-m}^{-1} x[r+N]\exp\left[-jk\frac{2\pi}{N}(r+n)\right]$$

Changing the dummy index of the sum from r to $\ell = r + N$ for $r = -m$ yields $\ell = N - m$, while for $r = -1$ we get $\ell = N - 1$, and the preceding relation is written

[3] The foregoing assumption is a necessary prerequisite regarding the signal $x[n]$, while it is proven that the property of periodicity also applies to the exponential expression W_N^{kr} since

$$W_N^{k(r+N)} = \exp\left(-j\frac{2\pi}{N}k(r+N)\right) = \exp\left(-j\frac{2\pi}{N}r\right)\exp(-j2\pi k) = \exp\left(-j\frac{2\pi}{N}kr\right) = W_N^{kr}.$$

$$\sum_{r=-m}^{-1} x[r]\exp\left(-jk\frac{2\pi}{N}r\right)=\sum_{\ell=N-m}^{N-1} x[\ell]\exp\left(-jk\frac{2\pi}{N}\ell\right)$$

Therefore, using the foregoing subsum (expressed in its final form and using r as the index instead of ℓ), we will get

$$\sum_{r=-m}^{N-1-m} x[r]\exp\left(-jk\frac{2\pi}{N}r\right)=\sum_{r=0}^{N-1-m} x[r]\exp\left(-jk\frac{2\pi}{N}r\right)$$
$$+\sum_{r=N-m}^{N-1} x[r]\exp\left(-jk\frac{2\pi}{N}r\right)=\sum_{r=0}^{N-1} x[r]\exp\left(-jk\frac{2\pi}{N}r\right)$$

Therefore, the Fourier coefficients of the periodic discrete-time signal $y[n]$ will take their final form

$$y_k=\exp\left(-jk\frac{2\pi}{N}m\right)\left\{\frac{1}{N}\sum_{r=0}^{N-1} x[r]\exp\left(-jk\frac{2\pi}{N}r\right)\right\}=\exp\left(-jk\frac{2\pi}{N}m\right)x_k$$

- **Multiplication**: If the coefficients of the Fourier expansion of the discrete periodic signals $x[n]$ and $y[n]$ that have a common period N are x_k and y_k, then the coefficients of the Fourier expansion of the signal $z[n]=x[n]y[n]$ will be given by the relation

$$z_k=\sum_{\ell=0}^{N-1} x_\ell y_{\ell-k}$$

with the sum that appears in this expressing the so-called *periodic convolution*.
Proof From the defining equation of the Fourier coefficients z_k we have

$$z_k=\frac{1}{N}\sum_{n=0}^{N-1} z[n]\exp\left(-j\frac{2\pi}{N}kn\right)=\frac{1}{N}\sum_{n=0}^{N-1} x[n]y[n]\exp\left(-j\frac{2\pi}{N}kn\right)$$

Replacing the discrete-time signal $x[n]$ from its Fourier expansion, which has the form

$$x[n]=\sum_{\ell=0}^{N-1} x_\ell\exp\left(j\frac{2\pi}{N}\ell n\right)$$

the preceding relation is formulated as

$$z_k=\frac{1}{N}\sum_{n=0}^{N-1}\left\{\sum_{\ell=0}^{N-1} x_\ell\exp\left(j\frac{2\pi}{N}\ell n\right)\right\}y[n]\exp\left(-j\frac{2\pi}{N}kn\right)=\sum_{m=0}^{N-1} x_\ell\left\{\frac{1}{N}\sum_{n=0}^{N-1} y[n]\exp\left[-j\frac{2\pi}{N}(\ell-k)\right]\right\}$$

and finally

$$z_k = \sum_{\ell=0}^{N-1} x_\ell y_{\ell-k}$$

which is the desired result.

- **Periodic convolution**: If the coefficients of the Fourier expansion of the discrete periodic signals $x[n]$ and $y[n]$ that have a common period N are x_k and y_k, then the coefficients of the Fourier expansion of the discrete periodic signal defined by the relation

$$z[n] = \sum_{m=0}^{N-1} x[m]y[n-m]$$

will be calculated as $z_k = Nx_k y_k$.
Proof From the defining equation of the coefficients z_k we get

$$z_k = \frac{1}{N}\sum_{n=0}^{N-1} z[n]\exp\left(-j\frac{2\pi}{N}nk\right) = \frac{1}{N}\sum_{n=0}^{N-1}\left\{\sum_{m=0}^{N-1} x[m]y[n-m]\right\}\exp\left(-j\frac{2\pi}{N}nk\right)$$
$$= \sum_{m=0}^{N-1} x[m]\left\{\frac{1}{N}\sum_{n=0}^{N-1} y[n-m]\exp\left(-j\frac{2\pi}{N}nk\right)\right\}$$

Using the property of time shift, we find

$$\frac{1}{N}\sum_{n=0}^{N-1} y[n-m]\exp\left(-j\frac{2\pi}{N}nk\right) = \exp\left(-j\frac{2\pi}{N}mk\right)y_k$$

and substituting in the defining equation of the coefficients z_k, this will take the form

$$z_k = y_k\left\{\sum_{m=0}^{N-1} x[m]\exp\left(-j\frac{2\pi}{N}mk\right)\right\} = Nx_k y_k$$

- **Frequency shift**: If the coefficients of the Fourier expansion of the discrete periodic signal $x[n]$ with period N are the x_k, then the coefficients of the Fourier expansion of the discrete periodic signal defined by the relation

$$y[n] = \exp\left(j\frac{2\pi}{N}nm\right)x[n]$$

will have the form $y_k = x_{k-m}$.
Proof From the defining equation of the Fourier coefficients it immediately follows that

$$y_k = \frac{1}{N}\sum_{n=0}^{N-1} y[n]\exp\left(-j\frac{2\pi}{N}kn\right) = \frac{1}{N}\sum_{n=0}^{N-1} x[n]\exp\left[-j\frac{2\pi}{N}(k-m)n\right] = x_{k-m}$$

- **Symmetry and conjugation**: If the coefficients of the Fourier expansion of the discrete complex periodic signal $x[n]$ with period N are the x_k, then the following symmetry properties apply:

1. The Fourier coefficients of the complex conjugate signal $y[n] = x^*[n]$ are $y_k = x^*_{-k}$.
2. The Fourier coefficients of the signal $y[n] = x^*[-n]$ are $y_k = x^*_k$.
3. The Fourier coefficients of the real signal $y[n] = \Re\{x[n]\}$ are $y_k = x^e_k$, where x^e_k is the even component of the coefficients x_k, while the symbol $\Re$ denotes the operator of the real part.
4. The Fourier coefficients of the imaginary signal $y[n] = i\Im\{x[n]\}$ are $y_k = x^o_k$, where x^o_k is the odd component of the coefficients x_k, while the symbol $\Im$ denotes the operator of the imaginary part.

In the special case of a periodic signal $x[n]$ with *real values*, it is proven that

$$\Re\{x_k\} = \Re\{x_{-k}\}, \quad \Im\{x_k\} = -\Im\{x_{-k}\}, \quad |x_k| = |x_{-k}|, \quad \measuredangle x_k = -\measuredangle x_{-k}$$

where $|x_k|$ and $\measuredangle x_k$ are the *magnitude* and *phase* of the coefficients x_k, which are generally *complex numbers*.
Proof The foregoing properties are proven analytically in the following way:

(1) From the defining equation of the Fourier coefficients of the signal $y[n]$ it immediately follows that

$$y_k = \frac{1}{N}\sum_{n=0}^{N-1} y[n]\exp\left(-j\frac{2\pi}{N}kn\right) = \frac{1}{N}\sum_{n=0}^{N-1} x^*[n]\exp\left(-j\frac{2\pi}{N}kn\right)$$

$$= \left\{\frac{1}{N}\sum_{n=0}^{N-1} x[n]\exp\left[-j\frac{2\pi}{N}(-k)n\right]\right\}^* = x^*_{-k}$$

(2) from the defining equation of the Fourier coefficients of the signal $y[n]$ we will have

$$y_k = \frac{1}{N}\sum_{n=0}^{N-1} y[n]\exp\left(-j\frac{2\pi}{N}kn\right) = \frac{1}{N}\sum_{n=0}^{N-1} x^*[-n]\exp\left(-j\frac{2\pi}{N}kn\right)$$

In the next step of the process, we change the dummy index of the sum from n to $m = -n$. In this case, for $n = 0$, $m = 0$, while for $n = N - 1$, $m = 1 - N$, and the foregoing expression will take the form[4]

$$y_k = \frac{1}{N}\sum_{m=1-N}^{0} x^*[m]\exp\left(j\frac{2\pi}{N}km\right) = \left\{\frac{1}{N}\sum_{m=1-N}^{0} x[m]\exp\left(-j\frac{2\pi}{N}km\right)\right\}^* = x^*_k$$

which is the desired result. To derive the final conclusion we relied on the assumption *that since the discrete-time signal $x[n]$ is periodic with period N, we can use any N consecutive samples of it, such as those belonging to the range of values* $[1 - N, 0]$.

(3) Since the signal $y[n]$ expresses the *real part* of the complex signal $x[n]$, it will have the form

$$y[n] = \Re\{x[n]\} = \frac{x[n] + x^*[n]}{2}$$

and therefore, the coefficients of the Fourier expansion y_k will be calculated as

$$y_k = \frac{1}{N}\sum_{n=0}^{N-1} y[n]\exp\left(-j\frac{2\pi}{N}kn\right) = \frac{1}{N}\sum_{n=0}^{N-1} \frac{x[n] + x^*[n]}{2}\exp\left(-j\frac{2\pi}{N}kn\right)$$

[4] The change in the limits in the last sum is due to the fact that the value of the difference $1 - N$ is less than zero.

$$= \frac{1}{2}\left\{ \frac{1}{N}\sum_{n=0}^{N-1} x[n]\exp\left(-j\frac{2\pi}{N}kn\right) + \frac{1}{N}\sum_{n=0}^{N-1} x^*[n]\exp\left(-j\frac{2\pi}{N}kn\right)\right\} = \frac{x_k + x^*_{-k}}{2} = x^e_k$$

(4) Since the signal $y[n]$ expresses the *imaginary part* of the complex signal $x[n]$, it will have the form

$$y[n] = \Im\{x[n]\} = \frac{x[n] - x^*[n]}{2j}$$

and therefore, the coefficients of the Fourier expansion of the signal $iy[n]$ will be calculated as

$$y_k = \frac{1}{N}\sum_{n=0}^{N-1} jy[n]\exp\left(-j\frac{2\pi}{N}kn\right) = \frac{1}{N}\sum_{n=0}^{N-1} \frac{x[n] - x^*[n]}{2}\exp\left(-j\frac{2\pi}{N}kn\right)$$

$$= \frac{1}{2}\left\{ \frac{1}{N}\sum_{n=0}^{N-1} x[n]\exp\left(-j\frac{2\pi}{N}kn\right) - \frac{1}{N}\sum_{n=0}^{N-1} x^*[n]\exp\left(-j\frac{2\pi}{N}kn\right)\right\} = \frac{x_k - x^*_{-k}}{2} = x^o_k$$

On the other hand, if the signal $x[n]$ is a signal of real values, it will hold that $x[n] = x^*[n]$ and $x[-n] = x^*[-n]$. Using the results of subquestions 1 and 2, which are in the form $x^*[n] \Leftrightarrow x^*_{-k}$ and $x^*[-n] \Leftrightarrow x^*_k$, as well as the property of time reversal, which is immediately proven afterwards—but also arises from the defining equation of the Fourier coefficients—we find that

$$\left.\begin{array}{l} x[n] \Leftrightarrow x_k \\ x^*[n] \Leftrightarrow x^*_{-k} \\ x[n] = x^*[n] \end{array}\right\} \Leftrightarrow x_k = x^*_{-k} \quad \text{and} \quad \left.\begin{array}{l} x[-n] \Leftrightarrow x_{-k} \\ x^*[-n] \Leftrightarrow x^*_k \\ x[-n] = x^*[-n] \end{array}\right\} \Leftrightarrow x^*_k = x_{-k}$$

Therefore, the real part of the Fourier coefficients of the signal $x[n]$ will have the form

$$\Re\{x_k\} = \frac{x_k + x^*_k}{2} = \frac{x_{-k} + x^*_{-k}}{2} = \Re\{x_{-k}\}$$

while for the imaginary part we will have

$$\Im\{x_k\} = \frac{x_k - x^*_k}{2} = \frac{x^*_{-k} - x_{-k}}{2} = -\frac{x_{-k} - x^*_{-k}}{2} = -\Im\{x_{-k}\}$$

From the last two equations we find that *the real and imaginary parts of the Fourier coefficients constitute even and odd functions of the variable k, respectively*. Finally, using the definitions of the magnitude and phase of a complex quantity, we will have

$$|x_k| = \sqrt{x_k x^*_k} = \sqrt{x^*_{-k}x_{-k}} = |x_{-k}| \quad \text{and} \quad \measuredangle x_k = \arctan\left(\frac{\Im\{x_k\}}{\Re\{x_k\}}\right) = \arctan\left(-\frac{\Im\{x_{-k}\}}{\Re\{x_k\}}\right) = -\measuredangle x_{-k}$$

That is, *the magnitude and phase of the Fourier expansion coefficients constitute even and odd functions of the variable k, respectively.*

Note that the foregoing properties also characterize the Fourier expansion coefficients of continuous-time signals.

- **Time reversal**: If the coefficients of the Fourier expansion of the discrete periodic signal $x[n]$ with period N are the x_k, then the coefficients of the Fourier expansion of the discrete periodic signal $y[n] = x[-n]$ will have the form $y_k = x_{-k}$.
  Proof From the defining equation of the Fourier coefficients of the signal $y[n]$ we will have

$$y_k = \frac{1}{N}\sum_{n=0}^{N-1} y[n]\exp\left(-j\frac{2\pi}{N}kn\right) = \frac{1}{N}\sum_{n=0}^{N-1} x[-n]\exp\left(-j\frac{2\pi}{N}kn\right)$$

Changing the dummy index of the sum from n to $m = -n$ and working as in the second property of symmetry/conjugation (see the previous proof), it immediately follows that

$$y_k = \frac{1}{N}\sum_{m=1-N}^{0} x[m]\exp\left(j\frac{2\pi}{N}km\right) = \frac{1}{N}\sum_{m=1-N}^{0} x[m]\exp\left[-j\frac{2\pi}{N}(-k)m\right] = x_{-k}$$

which is the desired result. Relying on this property and using the argument used previously, we can prove that *the coefficients of the Fourier expansion of an even or odd periodic discrete-time signals are respectively even or odd functions of the variable k.*

- **Time scaling**: Considering a periodic signal $x[n]$ of period N, we define the *downsampled* signal $x_{\downarrow m}[n]$ from the relation $x_{\downarrow m} = x[mn]$ and the *upsampled* signal $x_{\uparrow m}[n]$ from the relation

$$x_{\uparrow m}[n] = \begin{cases} x[n/m] \text{ for values of } n \text{ that are integer} \\ 0 \quad \text{multiples of } m \text{ in the opposite case} \end{cases}$$

where m is a positive integer. From these two signals, the signal $x_{\downarrow m}[n]$ is created by the operation of *downsampling*, which consists in considering *only the samples of the signal $x[n]$ whose positions are integer multiples of the parameter m* (i.e., the samples $x[0], x[m], x[2m], \ldots$). On the other hand, the signal $x_{\uparrow m}[n]$ is created by the operation of *upsampling*, which consists in *adding* $(m-1)$ *zero values between any two successive samples of the signal* $x[n]$. It is proven that the signals $x_{\downarrow m}[n]$ and $x_{\uparrow m}[n]$ are periodic, and therefore, their Fourier expansion exists.
Indeed, starting from the signal $x_{\downarrow m}[n]$, it is periodic with period M if there is a value M such that

$$x[nm] = x[m(n+M)] = x[nm + mM]$$

This condition is satisfied if there are numbers $p, M \in Z$ such that $mM = pN$, a property that is obviously satisfied if we set $p = m$ and $M = N$. Considering the minimum value of p, which is $p = m = 1$, $N = mM$, that is, *the fundamental period of the signal $x_{\downarrow m}[n]$ is $M = N/m$.* On the other hand, the oversampled signal $x_{\uparrow m}[n]$ is also periodic with fundamental period $M = mN$ since it turns out that

$$x_{\uparrow m}[n+mN] = \begin{cases} x\left[\frac{n+mN}{m}\right] = x\left[\frac{n}{m}+N\right] = x\left[\frac{n}{m}\right] & \text{if } n \text{ MOD } m = 0 \\ 0 & \text{if } n \text{ MOD } m \neq 0 \end{cases}$$

where (n MOD m) is the remainder of the division n/m.

Calculation of the Fourier coefficients of the signal $x_{\downarrow m}[n]$ requires knowledge of elements from the theory of sampling of discrete-time signals and will not be presented at this point; as for the calculation of the Fourier coefficients of the signal $y[n] = x_{\uparrow m}[n]$, these arise from their defining equation and taking into account that the signal in question is periodic with a period $M = mN$, it will therefore be the case that

$$y_k = \frac{1}{mN}\sum_{n=0}^{mN-1} x_{\uparrow m}[n]\exp\left(-j\frac{2\pi}{mN}nk\right) = \frac{1}{mN}\sum_{n=km\,k=0,1,\ldots,N-1} x\left[\frac{n}{m}\right]\exp\left(-j\frac{2\pi}{mN}nk\right)$$

Changing the dummy index of the sum from n to $\ell = n/m$, the preceding equation is written

$$y_k = \frac{1}{m}\left\{\frac{1}{N}\sum_{\ell=0}^{N-1} x[\ell]\exp\left(-j\frac{2\pi}{N}\ell k\right)\right\} = \frac{1}{m}x_k$$

which is also the final result. Based on the fundamental theory of the Fourier coefficients presented at the beginning of this chapter, we can assert that *the coefficients x_k/m are periodic functions with period mN.*

- **Duality**: If the Fourier coefficients of the discrete periodic signal $x[n]$ with period N are the x_k, then the Fourier coefficients of the coefficients x_k will be equal to $x[-n]/N$.
Proof From the defining equation of the Fourier coefficients of the periodic signal $x[n]$ we have

$$x_k = \frac{1}{N}\sum_{n=0}^{N-1} x[n]\exp\left(-j\frac{2\pi}{N}kn\right) = \sum_{n=0}^{N-1}\frac{x[n]}{N}\exp\left(-j\frac{2\pi}{N}kn\right)$$

By changing the dummy index of the sum from n to $m = -n$, the preceding relation will take the form

$$x_k = \sum_{m=1-N}^{0}\frac{x[-m]}{N}\exp\left(j\frac{2\pi}{N}km\right) = \sum_{m=<N>}\frac{x[-m]}{N}\exp\left(j\frac{2\pi}{N}km\right)$$

If we compare the last relation with the defining equation of the Fourier coefficients, we can assert that *the values of the signal $x[-m]/N$ constitute the coefficients of the Fourier expansion of the representations x_k*, so that we can write the correspondence

$$x_k \Longleftrightarrow x[-n]/N$$

where in the last term we simply changed the dummy index from m to n. The foregoing result is very important, as it leads us to the conclusion that *between the analysis equation and the synthesis equation related to the development of a discrete periodic signal according to Fourier, there is a kind of duality, similar to that encountered in continuous-time signals.* However, let us note that this duality *does not extend to the discrete-time Fourier transform (DTFT)* since aperiodic discrete-time signals and their associated transforms according to Fourier are described by completely different types of functions. We have already encountered this kind of duality when we proved that *the Fourier coefficients of the product $x[n]y[n]$ are given by the periodic convolution of the Fourier coefficients x_k and y_k, while the Fourier coefficients of the periodic convolution $x[n] * y[n]$ are given by the product of the coefficients x_k and y_k.*

- **Difference**: If the Fourier coefficients of the periodic discrete-time signal $x[n]$ with period N are the x_k, then the Fourier coefficients of the signal $y[n] = x[n] - x[n-1]$ will have the form

$$y_k = \left[1 - \exp\left(-j\frac{2\pi}{N}k\right)\right]x_k$$

Proof Combining the properties of linearity and time shift for the value $m = 1$, we have

$$y_k = x_k - \exp\left(-j\frac{2\pi}{N}k\right)x_k = \left[1 - \exp\left(-j\frac{2\pi}{N}k\right)\right]x_k$$

- **Parseval's equation**: If the coefficients of the Fourier development of the periodic discrete-time signal $x[n]$ with a period value equal to N being the x_k, it is proven that the following relation holds:

$$\frac{1}{N}\sum_{n=0}^{N-1}|x[n]|^2 = \sum_{k=0}^{N-1}|x_k|^2$$

Proof From the defining equation of the Fourier series expansion, we will have

$$x[n] = \sum_{k=0}^{N-1} x_k \exp\left(j\frac{2\pi}{N}kn\right) \quad \text{and} \quad x^*[n] = \sum_{k=0}^{N-1} x_k^* \exp\left(-j\frac{2\pi}{N}kn\right)$$

and the left side of the preceding equation is formulated as

$$\frac{1}{N}\sum_{n=0}^{N-1}|x[n]|^2 = \frac{1}{N}\sum_{n=0}^{N-1}x[n]\cdot x^*[n] = \frac{1}{N}\sum_{n=0}^{N-1}x[n]\left\{\sum_{k=0}^{N-1}x_k^*\exp\left(-j\frac{2\pi}{N}kn\right)\right\}$$

Interchanging the order of the foregoing sums we will get

$$\frac{1}{N}\sum_{n=0}^{N-1}|x[n]|^2 = \sum_{n=0}^{N-1}x_k^*\left\{\frac{1}{N}\sum_{k=0}^{N-1}x[n]\exp\left(-j\frac{2\pi}{N}kn\right)\right\} = \sum_{k=0}^{N-1}x_k^*x_k = \sum_{k=0}^{N-1}|x_k|^2$$

which is the desired result. This relation is known as Parseval's theorem or identity, which is the discrete equivalent of the relation

$$\frac{1}{T}\int_T |x(t)|^2 = \sum_{n=-\infty}^{\infty}|x_n|^2$$

which we proved for continuous-time signals, and its physical significance will be given in a subsequent section.

10.2.3 Physical Significance of Fourier Series Expansion

In full analogy with their continuous counterparts, discrete periodic signals can be developed in a *Fourier series* in order to highlight their frequency content, that is, to reveal the simple frequency components contained in them. However, unlike continuous-time signals, which contain *infinite frequencies* and whose expansion exists *only when certain convergence conditions are met* (see Sect. 3.2.4 on page 198), the frequencies of discrete periodic signals are unique only in the value range $-\pi < \omega < \pi$ or, equivalently, $-1/2 < f < 1/2$ (see Sect. 6.7.1 on page 520). For this reason, and as we have already proven, *their Fourier expansion will contain at most N harmonically correlated frequency components*, with the distance between two consecutive such components being equal to the fundamental frequency, which expressed in radians is equal to $\omega = 2\pi/N$, and expressed in cycles is equal to $f = 1/N$, where N is the period of the discrete-time signal. If in the previous analysis we used the cyclic frequency $\omega = 2\pi f = 2\pi/N$ as a variable, then the expansion of the periodic discrete-time signal in Fourier series will be given by

$$x[n] = \sum_{k=0}^{N-1}\alpha_k x_k[n] = \sum_{k=0}^{N-1}\alpha_k e^{j\omega kn} \tag{10.6}$$

which is also very often used in practice.

What is the physical meaning of the Fourier development? According to this development, *any periodic discrete-time signal $x[n]$ with period N consists of N frequency components* of the form

$$\alpha_0, \quad \alpha_1 e^{j\omega n}, \quad \alpha_2 e^{2j\omega n}, \quad \ldots, \quad \alpha_{N-1} e^{(N-1)j\omega n}$$

with frequencies $\omega_k = (2\pi/N)\,k = \omega k$ $(k = 0, 1, 2, \ldots, N-1)$, which constitute *integer multiples* of the fundamental frequency $\omega = 2\pi/N$. If we plot the coefficients α_k as a function of the index k or the frequency ω, we will construct the so-called *Fourier spectrum*, which shows diagrammatically *the contribution of elementary harmonic components to the discrete periodic signal* $x[n]$.

Taking into account that the Fourier coefficients are *complex numbers* of the form $\alpha_k = |\alpha_k| e^{j\measuredangle\alpha_k}$, the graphical representation of the measures $|\alpha_k|$ as a function of the index k is known as the *amplitude spectrum*, while the graphical representation of the phase $\measuredangle\alpha_k$ as a function of the index k is known as the *phase spectrum*. If we know these two spectra—which together constitute the so-called *frequency representation of the signal*—then we can *reconstruct* or, equivalently, *compose* the discrete periodic signal $x[n]$ using the previously mentioned *synthesis equation*. Taking into account that the coefficients of the Fourier series are *periodic functions of the variable k with period N*, we can in this reconstruction process to use *any N consecutive Fourier coefficients*, a property that has been used without mention to prove some of the properties of the expansion. For the same reason, if we plot the amplitude spectrum for all available values of k and not just for the first N values, we will find that it constitutes a *periodic structure* with a period value equal to N.

We conclude this concise account of the Fourier expansion of periodic discrete-time signals with the observation that the sequence of values $|\alpha_k|^2$ $(k = 0, 1, 2, \ldots, N-1)$ describes *the power distribution of the signal as a function of frequency* and constitutes the so-called *power spectral density* of the periodic discrete-time signal $x[n]$. In full analogy with the continuous case, this spectrum contains absolutely no information about the phase of the Fourier coefficients, while it itself constitutes a discrete periodic structure with a period value equal to N. Taking into account that the average power of a discrete periodic signal with a period of N is defined as

$$P = \frac{1}{N}\sum_{n=0}^{N-1} |x[n]|^2 \tag{10.7}$$

the power of the Parseval identity, which as we have proven has the form

$$\frac{1}{N}\sum_{n=0}^{N-1} |x[n]|^2 = \sum_{k=0}^{N-1} |\alpha_k|^2 \tag{10.8}$$

allows us to formulate the claim that *the average power of a discrete periodic signal is equal to the sum of the powers of all elementary harmonic components of which it consists*, a property that, as we have already pointed out, also characterizes signals defined in the domain of continuous-time.

Problem 10.2.5 Design the amplitude and phase spectra, as well as the power spectral density of the rectangular discrete-time signal of Exercise 10.2.4 (see page 788) for parameter values $A = 1$, $L = 5$, and $N = 10$, and then verify the Parseval identity (Proakis and Manolakis, Example 4.2.2, part B).

Solution The coefficients of the Fourier series of the rectangular periodic signal of Exercise 10.2.4 are given, as we have proven, by the relation

$$\alpha_k = \frac{A}{N}\exp\left[-j\frac{\pi k(L-1)}{N}\right]\left\{\sin\left(\frac{\pi k L}{N}\right)\Big/\sin\left(\frac{\pi k}{N}\right)\right\}, \quad k = 1, 2, \ldots, N-1$$

which for parameter values $A = 1$, $L = 5$, and $L = 10$ simplifies to

Table 10.2 Coefficients of Fourier expansion of rectangular signal of Exercise 10.2.4 calculated in Exercise 10.2.5

k	α_k	$\lvert\alpha_k\rvert$	$\measuredangle\alpha_k$	$\lvert\alpha_k\rvert^2$
0	0.5	0.5000	0.0000	0.2500
1	$0.1-0.3078j$	0.3236	−1.2566	0.1047
2	0.0	0.0000	0.0000	0.0000
3	$0.1-0.0727j$	0.1236	−0.6283	0.0153
4	0.0	0.0000	0.0000	0.0000
5	0.1	0.1000	0.0000	0.0100
6	0.0	0.0000	0.0000	0.0000
7	$0.1+0.0727j$	0.1236	0.6283	0.0153
8	0.0	0.0000	0.0000	0.0000
9	$0.1+0.3078j$	0.3236	1.2566	0.1047

$$\alpha_k = \frac{1}{10}\exp\left(-j\frac{2\pi k}{5}\right)\left\{\sin\left(\frac{\pi k}{2}\right)\Big/\sin\left(\frac{\pi k}{10}\right)\right\}, \quad k = 1, 2, \ldots 9$$

(for the value $k = 0$ we have $\alpha_0 = AL/N = 0.5$). By calculating the foregoing relation for the values of k in the range [0, 9], we can determine the coefficients $\alpha_0, \alpha_1, \ldots, \alpha_9$, which are presented in Table 10.2. The phases of the Fourier coefficients included in the fourth column of this table are expressed in radians and are therefore defined in the interval $(-\pi, +\pi)$.

The amplitude and phase spectra, as well as the power spectral density of the rectangular periodic signal, result from the graphical representation of the columns $(k, |\alpha_k|)$, $(k, \measuredangle\alpha_k)$, and $(k, |\alpha_k|^2)$ and are presented in Fig. 10.3a–c.

Finally, to verify *Parseval's identity*, we must substitute in both sides of the expression that describes it, the nonzero elements of the discrete periodic signal, as well as the nonzero measures of the corresponding Fourier coefficients. From the equation defining the discrete-time signal and the contents of the preceding table, it follows that

$$\frac{1}{N}\sum_{n=0}^{N-1}|x[n]|^2 = \frac{1}{10}\sum_{n=0}^{9}|x[n]|^2 = \frac{1}{10}\left\{x[0]^2 + x[1]^2 + x[2]^2 + x[3]^2 + x[4]^2\right\}$$

$$= \frac{1+1+1+1+1}{10} = \frac{5}{10} = 0.5 \quad \text{and}$$

$$\sum_{k=0}^{N-1}|\alpha_k|^2 = \sum_{k=0}^{9}|\alpha_k|^2 = |\alpha_0|^2 + |\alpha_1|^2 + |\alpha_3|^2 + |\alpha_5|^2 + |\alpha_7|^2 + |\alpha_9|^2 =$$

$$= 0.25 + 0.1047 + 0.0153 + 0.01 + 0.0153 + 0.1047 = 0.5$$

We observe that the preceding expressions lead to the same result, a conclusion that also verifies the validity of the Parseval identity.

10.2.4 Fourier Expansion and LTI Systems

As mentioned in Sect. 7.9 (page 608), a complex exponential signal described by the equation

$$x[n] = z^n \tag{10.9}$$

It is an eigenfunction of every discrete-time LTI system, whose eigenvalue is the system's transfer function

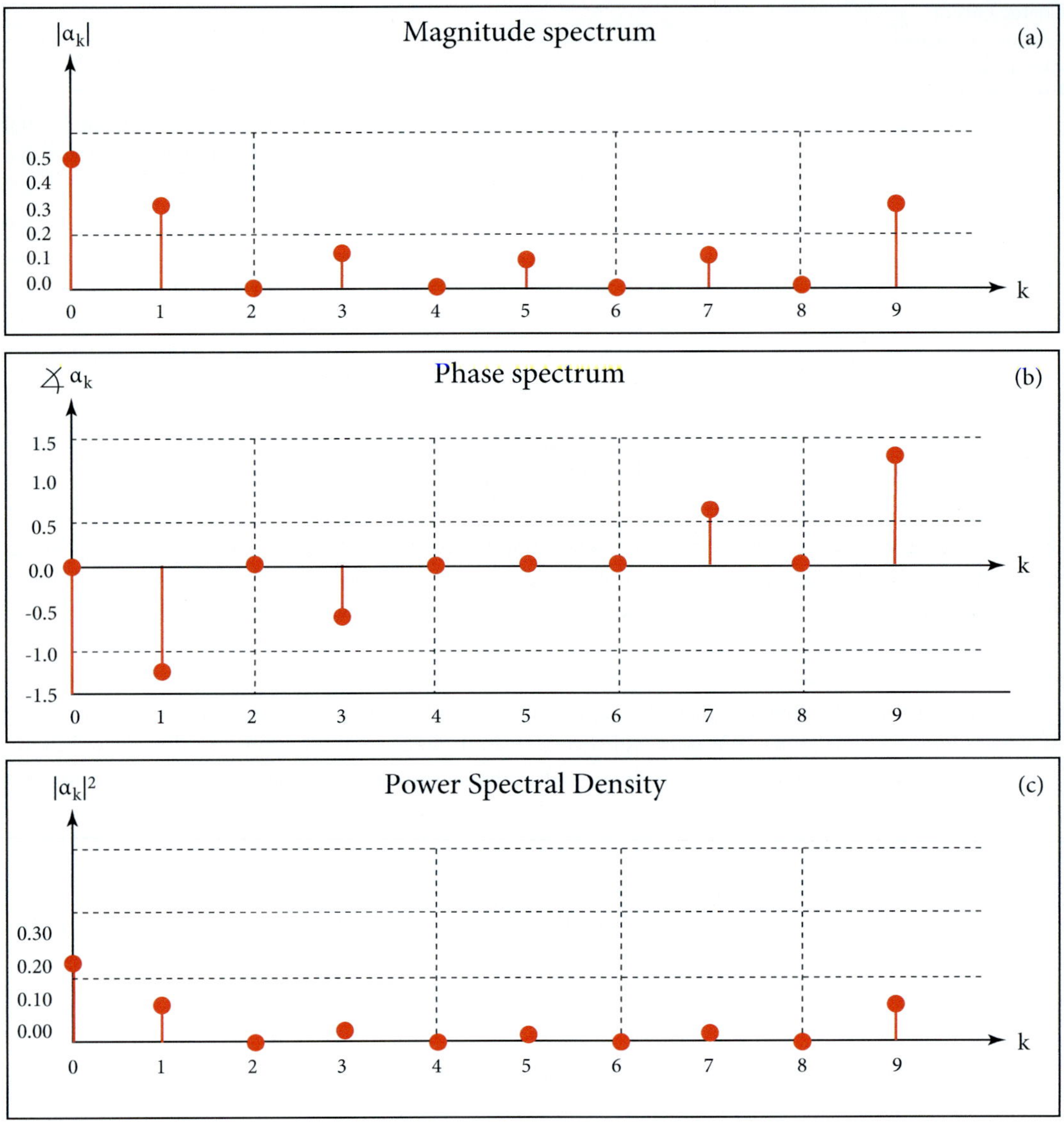

Fig. 10.3 Amplitude and phase spectra, as well as power spectral density of discrete rectangular signal of Exercise 10.2.5

$$\mathcal{H}(z) = \sum_{k=-\infty}^{\infty} h[k]z^{-k} \tag{10.10}$$

calculated for the value of z of the input signal; in other words, this means that the output of the system will have the form

$$y[n] = \mathcal{H}(z)x[n] = \mathcal{H}(z)z^n \tag{10.11}$$

If the discrete input signal is periodic with period N and is expanded in a Fourier series as

$$x[n] = \sum_{k=0}^{N-1} \alpha_k \exp\left(j\frac{2\pi}{N}kn\right) = \sum_{k=0}^{N-1} \alpha_k \left\{\exp\left(j\frac{2\pi}{N}k\right)\right\}^n = \sum_{k=0}^{N-1} \alpha_k z_k^n \tag{10.12}$$

then the response of the system to the elementary signal

$$x_k[n] = z_k^n = \left\{ \exp\left(j\frac{2\pi}{N}k \right) \right\}^n \tag{10.13}$$

will be given, according to the previous description, by the *eigenvalue equation*

$$y_k[n] = \mathcal{H}(z_k)z_k^n = \mathcal{H}\Big[\exp\left(j\frac{2\pi}{N}k \right) \Big] \exp\left(j\frac{2\pi}{N}kn \right) \tag{10.14}$$

while according to the *principle of superposition*—which is satisfied, since the considered discrete-time system is linear—the total output of the system will have the form

$$y[n] = \sum_{k=0}^{N-1} \alpha_k \mathcal{H}\Big[\exp\left(j\frac{2\pi}{N}k \right) \Big] \exp\left(j\frac{2\pi}{N}kn \right) \tag{10.15}$$

Alternatively, taking into account that *the Fourier coefficients and the exponential quantities of the foregoing relation are periodic with period N*, we can calculate the preceding sum using *any N consecutive values of the index k*, so we can write that

$$y[n] = \sum_{k=<N>} \alpha_k \mathcal{H}\Big[\exp\left(j\frac{2\pi}{N}k \right) \Big] \exp\left(j\frac{2\pi}{N}kn \right) \tag{10.16}$$

Therefore, *the output* $y[n]$ *of a discrete LTI system is also a periodic function with period N*; moreover, if we take its Fourier series

$$y[n] = \sum_{k=<N>} y_k \exp\left(j\frac{2\pi}{N}kn \right) \tag{10.17}$$

and compare the last two relations, it immediately follows that

$$y_k = \mathcal{H}\Big[\exp\left(j\frac{2\pi}{N}k \right) \Big] \alpha_k \tag{10.18}$$

that is, *the Fourier coefficient number k of the output signal results from the product of the Fourier coefficient number k of the input signal, multiplied by the value of the frequency response function of the system* $\mathcal{H}[\exp(j2\pi k/N)]$ *calculated at the position of each frequency k*. This conclusion is analogous to the one formulated for continuous-time systems.

Problem 10.2.6 Let us consider a discrete-time recursive LTI system described by the difference equation

$$y[n] - \frac{1}{4}y[n-1] = x[n].$$

Determine the Fourier expansion of the output signal $y[n]$ when the following discrete periodic signals are transmitted at its input

$$x_1[n] = \sin\left(\frac{3\pi n}{4} \right) \quad \text{and} \quad x_2[n] = \cos\left(\frac{\pi n}{4} \right) + 2\cos\left(\frac{\pi n}{2} \right).$$

Solution As already mentioned, the Fourier expansion of the system's output $y[n]$ is calculated as

$$y[n] = \sum_{k=<N>} \alpha_k \mathcal{H}\Big[\exp\Big(j\frac{2\pi}{N}k\Big)\Big]\exp\Big(j\frac{2\pi}{N}kn\Big)$$

where $\mathcal{H}(e^{j\omega})$ is the *frequency response function* of the system, which therefore must be calculated. Following the methodology of Sect. 7.4.3 (page 582), we transmit to the system a discrete exponential input signal of the form $x[n] = e^{j\omega n}$, so, according to the theory, the output of the system will have the form $y[n] = \mathcal{H}(e^{j\omega})e^{j\omega n}$. It is clear that

$$y[n-1] = \mathcal{H}(e^{j\omega})e^{j\omega(n-1)} = e^{-j\omega}\mathcal{H}(e^{j\omega})\, e^{j\omega n}$$

and replacing $y[n]y[n]$ and $y[n-1]y[n-1]$ with their respective expressions in the difference equation of the system it turns out that

$$\mathcal{H}(e^{j\omega}) = \frac{1}{1-(1/4)\, e^{-j\omega}}$$

Replacing the preceding relation in the defining equation of the system output $y[n]$, we will have

$$y[n] = \sum_{k=<N>} \frac{\alpha_k}{1-\frac{1}{4}\exp\Big(-j\frac{2\pi}{N}k\Big)}\exp\Big(j\frac{2\pi}{N}kn\Big)$$

where α_k $(k = 0, 1, 2, \ldots, N-1)$ are the coefficients of the Fourier expansion of the discrete-time signal $x[n]$, which is assumed to be periodic with period N. In detail, then, we will have:

(a) For the signal $x_1[n]$: The signal is periodic with period $N = 8$, since it is proven that

$$x_1[n] = \sin\Big(\frac{3\pi n}{4}\Big) = \sin\Big(\frac{3\pi n}{4}+6\pi\Big) = \sin\Big(\frac{3\pi n+24\pi}{4}\Big) = \sin\Big[\frac{3\pi(n+8)}{4}\Big] = x_1[n+8]$$

and therefore its Fourier expansion will have the form

$$x_1[n] = \sum_{k=<8>} \alpha_k \exp\Big(j\frac{2\pi}{8}kn\Big)$$

On the other hand, the above signal can be expressed, using Euler's formula, as follows

$$x_1[n] = \sin\Big(\frac{3\pi n}{4}\Big) = \frac{1}{2j}\Big[\exp\Big(j\frac{3\pi n}{4}\Big)-\exp\Big(-j\frac{3\pi n}{4}\Big)\Big] = \frac{1}{2j}\exp\Big(j\frac{2\pi}{8}3n\Big) - \frac{1}{2i}\exp\Big[j\frac{2\pi n}{8}(-3)n\Big]$$

Comparing the last two relations, we find that the only nonzero coefficients of the Fourier expansion of the signal $x_1[n]$ are

$$\alpha_3 = \frac{1}{2j} \quad \text{and} \quad \alpha_{-3} = -\frac{1}{2j}$$

Therefore, the nonzero coefficients of the Fourier expansion of the system output $y[n]$ are

$$y_3 = \mathcal{H}\Big\{\exp\Big(j\frac{3\pi}{4}\Big)\Big\}\alpha_3 = \frac{1}{2j}\cdot\frac{1}{1-\frac{1}{4}\exp\Big(-j\frac{3\pi}{4}\Big)} \quad \text{and}$$

$$y_{-3} = \mathcal{H}\Big\{\exp\Big(-j\frac{3\pi}{4}\Big)\Big\}\alpha_{-3} = -\frac{1}{2j}\cdot\frac{1}{1-\frac{1}{4}\exp\Big(j\frac{3\pi}{4}\Big)}$$

while the Fourier expansion of the system output will have the form

$$y[n] = y_3 \exp\left(j\frac{3\pi n}{4}\right) + y_{-3}\exp\left(-j\frac{3\pi n}{4}\right) = \frac{1}{2j}\left\{\frac{\exp\left(j\frac{3\pi n}{4}\right)}{1-\frac{1}{4}\exp\left(-j\frac{3\pi}{4}\right)} - \frac{\exp\left(-j\frac{3\pi n}{4}\right)}{1-\frac{1}{4}\exp\left(j\frac{3\pi}{4}\right)}\right\}$$

The reader is asked to simplify this result and prove that it is the same as that obtained by directly calculating the system output.

(b) The signal $x_2[n]$ consists of two simple harmonic components with periods $N_1 = 8$ and $N_2 = 4$. Therefore, its period is equal to $N = 8$ (Sect. 6.3, page 514), which allows us to express it as a Fourier series in the form

$$x_2[n] = \sum_{k=<8>} \alpha_k \exp\left(j\frac{2\pi}{8}kn\right)$$

On the other hand, using Euler's formula for the cosine function, we get

$$\begin{aligned} x_2[n] &= \frac{1}{2}\exp\left(j\frac{\pi n}{4}\right) + \frac{1}{2}\exp\left(-j\frac{\pi n}{4}\right) + \exp\left(j\frac{\pi n}{2}\right) + \exp\left(-j\frac{\pi n}{2}\right) \\ &= \frac{1}{2}\exp\left(j\frac{2\pi}{8}(1)n\right) + \frac{1}{2}\exp\left(j\frac{2\pi}{8}(-1)n\right) + \exp\left(j\frac{2\pi}{8}(2)n\right) + \exp\left(j\frac{2\pi}{8}(-2)n\right) \end{aligned}$$

Comparing the last two equations, we find that the only nonzero coefficients of the Fourier series of the discrete periodic signal $x[n]$ are

$$\alpha_1 = \frac{1}{2}, \quad \alpha_{-1} = \frac{1}{2}, \quad \alpha_2 = 1, \quad \alpha_{-2} = 1$$

that is, the corresponding nonzero Fourier coefficients of the output signal $y[n]$ are calculated as

$$y_1 = \mathcal{H}\left\{\exp\left(j\frac{\pi}{4}\right)\right\}\alpha_1 = \frac{1}{2}\cdot\frac{1}{1-\frac{1}{4}\exp\left(-j\frac{\pi}{4}\right)} \qquad y_{-1} = \mathcal{H}\left\{\exp\left(-j\frac{\pi}{4}\right)\right\}\alpha_{-1} = \frac{1}{2}\cdot\frac{1}{1-\frac{1}{4}\exp\left(j\frac{\pi}{4}\right)}$$

$$y_2 = \mathcal{H}\left\{\exp\left(j\frac{\pi}{2}\right)\right\}\alpha_2 = \frac{1}{1-\frac{1}{4}\exp\left(-j\frac{\pi}{2}\right)} \qquad y_{-2} = \mathcal{H}\left\{\exp\left(-j\frac{\pi}{2}\right)\right\}\alpha_{-2} = \frac{1}{1-\frac{1}{4}\exp\left(j\frac{\pi}{2}\right)}$$

Therefore, the Fourier expansion of the system output will be given by the equation

$$y[n] = y_1\exp\left(j\frac{\pi n}{4}\right) + y_{-1}\exp\left(-j\frac{\pi n}{4}\right) + y_2\exp\left(j\frac{\pi n}{2}\right) + y_{-2}\exp\left(-j\frac{\pi n}{2}\right)$$

and by substituting the values we get

$$y[n] = \frac{1}{2}\left\{\frac{\exp\left(j\frac{\pi n}{4}\right)}{1-\frac{1}{4}\exp\left(-j\frac{\pi}{4}\right)} + \frac{\exp\left(-j\frac{\pi n}{4}\right)}{1-\frac{1}{4}\exp\left(j\frac{\pi}{4}\right)}\right\} + \frac{\exp\left(j\frac{\pi n}{2}\right)}{1-\frac{1}{4}\exp\left(-j\frac{\pi}{2}\right)} + \frac{\exp\left(-j\frac{\pi n}{2}\right)}{1-\frac{1}{4}\exp\left(j\frac{\pi}{2}\right)}$$

which is the desired result. The reader is again asked to simplify the preceding expression and verify that the final result is the one we arrive at by other methods.

10.3 Discrete-Time Fourier Transform

In full analogy with the continuous case in which *periodic signals are represented by the Fourier expansion, while aperiodic signals are described by the Fourier transform*, an aperiodic signal defined in the domain of discrete-time can be represented by the homonymous transform whose construction is based on the same arguments and using the same methodology that we used in the continuous case. As we saw in detail in Sect. 3.3 (page 219), *considering a nonperiodic signal $x(t)$ of finite duration $T < T_0$, we could construct from it a periodic signal $x_p(t)$by continuously repeating the nonperiodic signal $x(t)$ every T_0 time units.* By the same logic, let us consider a discrete nonperiodic signal $x[n]$ that has nonzero values only in the area $-N_1 \leq n \leq N_2$, and let us construct a periodic signal $x_p[n]$ by continuously repeating the nonperiodic signal $x[n]$ in the way presented in Fig. 10.4. In this figure, the period N expresses the period of the periodic signal $x_p[n]$, which coincides with the nonperiodic signal $x[n]$ at the limit $N \to \infty$, a property that is mathematically formulated as

$$x[n] = \lim_{N\to\infty} x_p[n] \tag{10.19}$$

Taking into account that the signal $x_p[n]$ is *periodic* with period N, we can develop it in a Fourier series; we will then have

$$x_p[n] = \sum_{k=<N>} \alpha_k \exp\left(j\frac{2\pi}{N}kn\right) \tag{10.20}$$

with the coefficients of the Fourier expansion defined as

$$\alpha_k = \frac{1}{N}\sum_{n=<N>} x_p[n]\exp\left(-j\frac{2\pi}{N}kn\right) \tag{10.21}$$

Considering as a range of width N that which includes the range values $-N_1 \leq n \leq N_2$ and taking into account that in that area the two signals coincide, so we can write that $x_p[n] = x[n]$, the preceding relationship takes the form

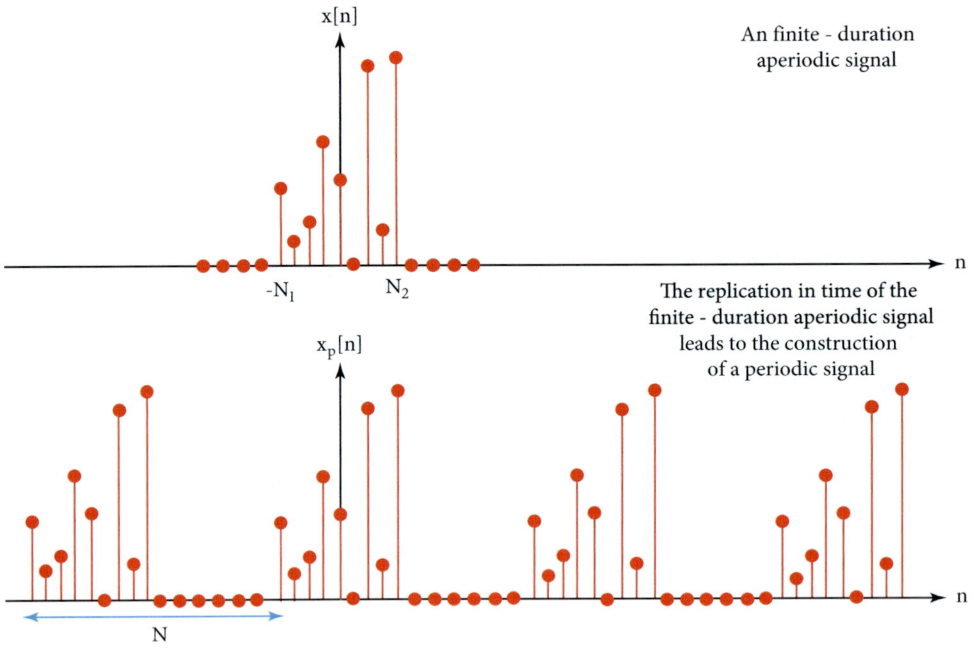

Fig. 10.4 The nonperiodic discrete-time signal $x[n]$ and the periodic signal $x_p[n]$ that arises from it by its periodic repetition with period N

$$\alpha_k = \frac{1}{N}\sum_{n=-N_1}^{N_2} x[n]\exp\left(-j\frac{2\pi}{N}kn\right) = \frac{1}{N}\sum_{n=-\infty}^{\infty} x[n]\exp\left(-j\frac{2\pi}{N}kn\right) \tag{10.22}$$

where for the construction of the last expression we extended the sum over the entire range of values $(-\infty, \infty)$ since the signal $x[n]$ has zero values outside the interval $[-N_1, N_2]$, so nothing changes. Using for simplicity the cyclic frequency $\omega_k = 2\pi k/N = k\omega_0$ (where $\omega_0 = 2\pi/N$ is the fundamental frequency) and defining the function

$$\mathcal{X}(e^{j\omega}) \equiv \mathcal{X}(e^{j\omega}) = \sum_{n=-\infty}^{\infty} x[n]e^{-j\omega n} \tag{10.23}$$

which is known as the *DTFT*[5] of the discrete-time signal $x[n]$, the coefficients of the Fourier expansion are calculated as

$$\alpha_k = \frac{1}{N}\mathcal{X}(e^{j\omega_k}) = \frac{1}{N}\mathcal{X}(e^{jk\omega_0}) \tag{10.24}$$

or, equivalently,

$$N\alpha_k = \mathcal{X}(e^{jk\omega_0}) \tag{10.25}$$

Therefore, *the coefficients of the Fourier expansion of the discrete periodic signal $x_p[n]$ can be considered as samples of the function $\mathcal{X}(e^{j\omega})$ corresponding to frequencies $\omega_k = k\omega_0$, which are integer multiples of the fundamental frequency ω_0* (this conclusion applies exactly to continuous-time signals, as we proved in Sect. 3.3). Replacing this expression in the Fourier expansion of the signal $x_p[n]$, it will take the form

$$x_p[n] = \sum_{k=<N>} \frac{1}{N}\mathcal{X}(e^{jk\omega_0})\exp\left(j\frac{2\pi}{N}kn\right) = \frac{1}{2\pi}\sum_{k=<N>} \mathcal{X}(e^{jk\omega_0})\, e^{jk\omega_0 n}\omega_0 \tag{10.26}$$

where for the derivation of the last relation we used the identity $\omega_0 = 2\pi/N$, from which it follows that $1/N = \omega_0/2\pi$.

The preceding relation refers to *the periodic discrete-time signal $x_p[n]$*, while in order to construct the aperiodic signal $x[n]$, we must, as we have already mentioned, calculate the preceding relation in the limit $N \to \infty$. In this case, it turns out that

$$x[n] = \frac{1}{2\pi}\lim_{N\to\infty}\left\{\sum_{k=<N>} \mathcal{X}(e^{jk\omega_0})\, e^{jk\omega_0 n}\omega_0\right\} = \frac{1}{2\pi}\int_{2\pi} \mathcal{X}(e^{j\omega})\, e^{j\omega n}d\omega \tag{10.27}$$

with the conversion of the sum into an integral due to the expression $\omega_0 = 2\pi/N$, from which it follows that in the limit $N \to \infty$ will be $\omega_0 \to 0$. We also observe that because the expression $\mathcal{X}(e^{j\omega})e^{j\omega n}$ is *periodic* with a period of 2π, as can be proven from its defining equation, the preceding integral can be calculated *in any interval of width* 2π. Therefore, applying the foregoing analysis, we arrive at the equations

$$\mathcal{X}(e^{j\omega}) \equiv \mathcal{F}\{x[n]\} = \sum_{n=-\infty}^{\infty} x[n]e^{-j\omega n} \tag{10.28}$$

$$x[n] \equiv \mathcal{F}^{-1}\{\mathcal{X}(e^{j\omega})\} = \frac{1}{2\pi}\int_{2\pi} \mathcal{X}(e^{j\omega})\, e^{j\omega n}d\omega \tag{10.29}$$

[5] We observe that, in contrast to the continuous case in which the Fourier transform is defined by the use of an *integral*, in the discrete case this integral is replaced by a *sum*, which, of course, is expected.

which constitute the discrete versions of Eqs. (3.25) and (3.26) and express *the pair of direct-inverse Fourier transform of aperiodic discrete-time signals*, with the defining equation of the function $\mathcal{X}(e^{j\omega})$ known as the *analysis equation* and the defining equation of the discrete-time signal $x[n]$ known as the *synthesis equation*. In full analogy with the continuous case, *the Fourier transform of a discrete aperiodic signal represents the frequency content of the signal* and in this sense describes *the analysis of the signal into the frequency components of which it consists*.

It is proven that the Fourier transform in the domain of discrete-time is a *periodic function with a period of* 2π, a property that was used without comment to derive the inverse transform, *does not characterize the continuous case*, and is proven by its defining equation, since it will hold that

$$\mathcal{X}(e^{j(\omega+2k\pi)}) = \sum_{n=-\infty}^{\infty} x[n]\, e^{-j(\omega+2k\pi)n} = \sum_{n=-\infty}^{\infty} x[n]\, e^{-j\omega n} e^{-j2k\pi n} = \sum_{n=-\infty}^{\infty} x[n]\, e^{-j\omega n} = \mathcal{X}(e^{j\omega})$$

It is not difficult to see that *this property of periodicity is a consequence of the fact that the frequencies of discrete-time signals have unique and distinct values only in the range* $(-\pi, +\pi)$, while any frequency outside this range is always equivalent to a frequency that belongs to the preceding range. Using this property of periodicity, we can develop the periodic function $\mathcal{X}(e^{j\omega})$ into a Fourier series whose coefficients are proven to be the samples of the discrete-time signal $x[n]$. Alternatively, we can prove that the functions $x[n]$ and $\mathcal{X}(e^{j\omega})$ appearing in the synthesis equation and in the analysis equation are connected by a forward-inverse transform relationship, replacing the function $\mathcal{X}(e^{j\omega})$ in the synthesis equation and checking if the signal resulting from this is the discrete-time signal $x[n]$, which, of course, is proven to be true.[6]

10.3.1 Physical Significance of DTFT

As already mentioned, the Fourier transform $\mathcal{X}(e^{j\omega})$ of a discrete aperiodic signal $x[n]$ describes the analysis of the signal into the frequency components of which it is composed and allows its formulation as a linear combination of complex exponential functions whose frequencies are infinitesimally apart on the frequency axis, while their amplitudes are of the form

$$\frac{1}{2\pi}\mathcal{X}(e^{j\omega})\, e^{j\omega n}$$

Taking into account that the frequencies of discrete-time signals are unique *only in the interval* $(-\pi, \pi)$, we can use *either the preceding interval or any other interval of width* 2π for the integral of the synthesis equation. In full analogy with the continuous case, the Fourier transform $\mathcal{X}(e^{j\omega})$ is known as the *spectrum* of the discrete-time signal $x[n]$ with which it is associated.

Generally speaking, the function $\mathcal{X}(e^{j\omega})$ is a *complex function of the real frequency* ω and can be formulated as

[6] Indeed, if we replace the function $\mathcal{X}(e^{j\omega})$ in the synthesis equation and assume that the signal created by this is, let us say $\tilde{x}[n]$, we will have

$$\tilde{x}[n] = \frac{1}{2\pi}\int_{2\pi}\left(\sum_{m=-\infty}^{\infty} x[m]\, e^{-j\omega m}\right) e^{j\omega n} d\omega = \sum_{m=-\infty}^{\infty} x[n]\left(\frac{1}{2\pi}\int_{2\pi} e^{j\omega(m-n)} d\omega\right)$$

where in the last expression we have interchanged the operations of summation and integration. Now observing that

$$\frac{1}{2\pi}\int_{2\pi} e^{j\omega(m-n)} d\omega = \frac{\sin\left[\pi(m-n)\right]}{\pi(m-n)} = \begin{cases} 1 & m = n \\ 0 & m \neq n \end{cases} = \delta[n-m]$$

the preceding relation takes the form

$$\tilde{x}[n] = \sum_{m=-\infty}^{\infty} x[m]\delta[n-m] = x[n]$$

as derived from the convolution equation. We therefore find that $\tilde{x}[n] = x[n]$, that is, *the functions* $x[n]$ *and* $\mathcal{X}(e^{j\omega})$ *are inverses of each other and therefore are connected by a forward-inverse transform relationship.*

$$\mathcal{X}(e^{j\omega}) = \Re\{\mathcal{X}(e^{j\omega})\} + i\Im\{\mathcal{X}(e^{j\omega})\} = |\mathcal{X}(e^{j\omega})|e^{i\measuredangle\mathcal{X}(e^{j\omega})}$$

with the representations $|\mathcal{X}(e^{j\omega})|$ and $\measuredangle\mathcal{X}(e^{j\omega})$ expressing the *amplitude* and *phase* functions. Note that because the function $\measuredangle\mathcal{X}(e^{j\omega})$ is periodic with a period of 2π, *the phase of the transform is not uniquely determined*, and for this reason we use the notation arg $[\mathcal{X}(e^{j\omega})]$ to refer to the *principal* (or *main*) *value* for which the phase $\measuredangle\mathcal{X}(e^{j\omega})$ is limited to the interval $(-\pi, +\pi)$. In full analogy with the continuous case, the function $S_{xx}(\omega) = |\mathcal{X}(e^{j\omega})|^2$ defined as the square of the magnitude of the Fourier transform is known as the *energy spectral density* of the discrete-time signal $x[n]$. This function does not contain the slightest information regarding the phase of the transform and expressed as

$$\frac{1}{2\pi}|\mathcal{X}(e^{j\omega})|^2 d\omega$$

describes *the energy of the signal $x[n]$ that is included in a frequency band of width $d\omega$*. Taking into account that the frequencies of discrete-time signals have unique values only within the interval $(-\pi, +\pi)$, we can formulate the conclusion that the total energy of the signal expressed in the frequency domain will be given by the relation

$$E_x = \frac{1}{2\pi}\int_{-\pi}^{+\pi} |\mathcal{X}(e^{j\omega})|^2 d\omega \tag{10.30}$$

Closely connected with the foregoing expression is the *Parseval identity*, which expresses the *principle of energy conservation*, according to which *the energy of a signal in the time domain is equal to the energy of that signal in the frequency domain*. For the proof of this identity, let us consider the complex conjugate expression of the synthesis equation

$$x^*[n] = \frac{1}{2\pi}\int_{-\pi}^{+\pi} \mathcal{X}^*(e^{j\omega})\, e^{-j\omega n} d\omega$$

Replacing the foregoing relation in the defining equation of the energy of a discrete-time signal, we have

$$E_x = \sum_{n=-\infty}^{\infty} |x[n]|^2 = \sum_{n=\infty}^{\infty} x[n]x^*[n] = \sum_{n=-\infty}^{\infty} x[n]\left(\frac{1}{2\pi}\int_{-\pi}^{+\pi} \mathcal{X}^*(e^{j\omega})\, e^{-j\omega n} d\omega\right)$$

Alternating the operations of addition and integration and using the analysis equation, the last relation is formulated as

$$E_x = \frac{1}{2\pi}\int_{-\pi}^{+\pi} \mathcal{X}^*(e^{j\omega})\left[\sum_{n=-\infty}^{\infty} x[n]\, e^{-j\omega n}\right] d\omega = \frac{1}{2\pi}\int_{-\pi}^{+\pi} \mathcal{X}^*(e^{j\omega})\, \mathcal{X}(e^{j\omega})\, d\omega = \frac{1}{2\pi}\int_{-\pi}^{+\pi} |\mathcal{X}(e^{j\omega})|^2 d\omega$$

and combining the foregoing relations, we finally get

$$E_x = \sum_{n=-\infty}^{\infty} |x[n]|^2 = \frac{1}{2\pi}\int_{-\pi}^{+\pi} |\mathcal{X}(e^{j\omega})|^2 d\omega \tag{10.31}$$

which constitutes *Parseval's theorem for the Fourier transform of discrete aperiodic signals* and obviously applies only to *energy signals* whose energy E_x is finite.

Examples of amplitude, phase, and energy spectral density calculations of discrete-time signals are presented in detail in the following exercises.

10.3.2 Convergence Conditions of DTFT

Based on our discussions in previous sections, it is obvious that *the Fourier transform of a discrete-time signal $x[n]$ exists only when the sum that appears in the defining equation converges uniformly to a finite value for each frequency ω*, a property that is mathematically formulated as

$$|\mathcal{X}(e^{j\omega})| = \left| \sum_{n=-\infty}^{\infty} x[n]\, e^{j\omega n} \right| < \infty \quad \text{for every } \omega \tag{10.32}$$

By the term uniform convergence, we mean that if we express the Fourier transform as

$$\mathcal{X}(e^{j\omega}) = \lim_{N\to\infty} \mathcal{X}_N(\omega) = \lim_{N\to\infty} \sum_{n=-N}^{N} x[n]\, e^{-j\omega n} \tag{10.33}$$

then, for every value of the frequency ω the condition

$$\lim_{N\to\infty} |\mathcal{X}(e^{j\omega}) - \mathcal{X}_N(\omega)| = 0 \tag{10.34}$$

must be satisfied. It is proven that a sufficient condition for the convergence of the Fourier transform of a discrete-time signal is the characterization of the signal as *absolutely summable*, a property that is mathematically expressed as[7,8]

$$\sum_{n=-\infty}^{\infty} |x[n]| < \infty \tag{10.35}$$

since in this case it will hold that

$$|\mathcal{X}(e^{j\omega})| = \left| \sum_{n=-\infty}^{\infty} x[n]\, e^{-j\omega n} \right| \le \sum_{n=-\infty}^{\infty} |x[n]||e^{-j\omega n}| \le \sum_{n=-\infty}^{\infty} |x[n]| < \infty$$

A second convergence condition that, let it be noted, is not as strong as the condition of *absolute summability* is the condition of *square summability*, which is not satisfied by the measure of the discrete-time signal $x[n]$ but by the square of its measure and is formulated as

$$\sum_{n=-\infty}^{\infty} |x[n]|^2 < \infty \tag{10.36}$$

Taking into account that the foregoing sum expresses the *energy* of the discrete-time signal (Sect. 6.3, page 514 and, more specifically, Eq. 6.3), this condition refers to *discrete-time signals of finite energy*, which in this way are included in the category of discrete-time signals that have a Fourier transform. In this case, taking into account the analysis equation

[7] This condition constitutes *the discrete equivalent of the first condition of Dirichlet* that characterizes the convergence of the Fourier transform of continuous-time (see Sect. 3.3.3 on page 225).

[8] It is obvious that this condition is automatically satisfied by all sequences of *finite* length that, therefore, always have a Fourier transform.

$$\mathcal{X}(e^{j\omega}) = \sum_{n=-\infty}^{\infty} x[n]\, e^{-j\omega n} \qquad \text{and defining the function} \qquad \mathcal{X}_N(\omega) = \sum_{n=-N}^{N} x[n]\, e^{-j\omega n}$$

which in the limit $N \to \infty$ coincides with the Fourier transform, the convergence of the transform requires the validity of the condition

$$\lim_{N\to\infty} \int_{-\pi}^{+\pi} \left| \mathcal{X}(e^{j\omega}) - \mathcal{X}_N(\omega) \right|^2 d\omega = \lim_{N\to\infty} \int_{-\pi}^{+\pi} \left| \mathcal{X}(e^{j\omega}) - \sum_{n=-\infty}^{\infty} x[n]\, e^{-j\omega n} \right|^2 d\omega = 0$$

which imposes the *zeroing of the error energy* $|\mathcal{X}(e^{j\omega}) - \mathcal{X}_N(\omega)|$ in the limit $N \to \infty$, but this does not necessarily apply to its measure.

10.3.3 The Relationship Between Fourier and $\mathcal{Z}$ Transforms

As we mentioned in Chap. 8 (Eq. 8.1), the $\mathcal{Z}$ of a discrete input sequence $x[n]$ is defined by a relationship of the form

$$\mathcal{X}(z) = \sum_{n=-\infty}^{\infty} x[n]\, z^{-n} \tag{10.37}$$

with a region of convergence being a *subset* of the complex plane that satisfies the relationship $r_2 < |z| < r_1$. Expressing the complex variable z in *polar coordinates*, it will be written in the form $z = r\exp(j\omega)$ where r and ω are the magnitude and phase, respectively, of the complex variable z. Now replacing this new form of the variable z in the defining equation of the $\mathcal{Z}$ transform, it is written as

$$\mathcal{X}(z)\Big|_{z=r\exp(j\omega)} = \sum_{n=-\infty}^{\infty} \left(x[n]\, r^{-n} \right) e^{-j\omega n} \tag{10.38}$$

From the preceding relationship it follows that *the transform $\mathcal{X}(z)$ can be interpreted as the Fourier transform of the sequence* $x[n]z^{-n}$. The factor r^{-n} increases with the variable n if the relationship $r < 1$ holds and decreases with the variable n if $r > 1$. Alternatively, we can state the claim that if the $\mathcal{Z}$ transform converges for complex variables that satisfy the condition $|z| = 1$ (this relationship is known to be the defining equation of the unit circle centered at the origin of the complex at level $z = 0$ and radius equal to unity), then we will have[9]

$$\mathcal{X}(z)\Big|_{z=\exp(i\omega)} = \mathcal{X}(e^{j\omega}) = \sum_{n=-\infty}^{\infty} x[n]\, e^{-j\omega n} \tag{10.39}$$

Therefore, the Fourier transform can be defined as *the $\mathcal{Z}$ transform of the discrete-time signal we consider in each case*, calculated on the circumference of the unit circle. If the transform $\mathcal{X}(z)$ *does not converge* in the area $|z| = 1$—which means that the unit circle *does not belong to the convergence area of the $\mathcal{Z}$ transform*—then the Fourier transform $\mathcal{X}(e^{j\omega})$ *does not exist*.

---

[9] This is exactly why the DTFT is symbolized as $\mathcal{X}(e^{j\omega})$ and not as $\mathcal{X}(\omega)$. This situation is analogous to the corresponding continuous transform, which, by analogy, is not symbolized as $\mathcal{X}(\omega)$ but as $\mathcal{X}(j\omega)$. This notation reminds us that the continuous-time Fourier transform arises from the Laplace transform for values $s = j\omega$, while the DTFT arises from the $\mathcal{Z}$ transform for values $z = e^{j\omega}$.

10.3.4 Examples of DTFT Calculation

After the analytical presentation of the basic theory related to the Fourier transform of aperiodic discrete-time signals, some examples of solved exercises follow to allow readers to fully familiarize themselves with the calculation process. In this process, will use, where necessary, the appropriate identities from the mathematical typology included in Table 6.2 (page 500).

Problem 10.3.1 Determine the Fourier transform and design the amplitude, phase, and energy spectral density of the discrete sequence $x[n] = \alpha^n u[n]$ with parameter value $|\alpha| < 1$ (Hayes, Example 2.5.1—also, Oppenheim, Schafer, Buck, Example 2.21 and Proakis and Manolakis, Example 4.2.3).

Solution First of all, we observe that the foregoing sequence is *absolutely summable* for the preceding range of parameter values α, and therefore the Fourier transform exists. From the defining equation of this transform we will have

$$\mathcal{X}(e^{j\omega}) = \sum_{n=-\infty}^{\infty} x[n]\,e^{-j\omega n} = \sum_{n=-\infty}^{\infty} \alpha^n u[n]\,e^{-j\omega n} = \sum_{n=0}^{\infty}(\alpha e^{-j\omega})^n = \frac{1}{1-\alpha e^{-j\omega}}$$

(where we used the appropriate expression from mathematical typology), which is the required result. Expressing the foregoing complex function in its Cartesian form, it turns out that

$$\mathcal{X}(e^{j\omega}) = \frac{1}{(1-\alpha\cos\omega)+j\alpha\sin\omega} = \frac{1-\alpha\cos\omega}{1-2\alpha\cos\omega+\alpha^2} - j\frac{\alpha\sin\omega}{1-2\alpha\cos\omega+\alpha^2}$$

Therefore, the *magnitude* and *phase* of the transform will be given by the relations

$$|\mathcal{X}(e^{j\omega})| = \frac{1}{\sqrt{1-2\alpha\cos\omega+\alpha^2}} \quad \text{and} \quad \measuredangle\mathcal{X}(e^{j\omega}) = \tan^{-1}\left(\frac{\alpha\sin\omega}{1-\alpha\cos\omega}\right)$$

while the energy density function will have the form

$$S_{xx}(\omega) = |\mathcal{X}(e^{j\omega})|^2 = \frac{1}{1-2\alpha\cos\omega+\alpha^2}$$

The amplitude and phase spectra, as well as the energy spectral density, of the discrete-time signal in question are presented in images (a) (amplitude and phase spectra for values $\alpha > 0$), (b) (amplitude and phase spectra for values $\alpha < 0$), and (c) (energy spectral density for values $\alpha > 0$ and $\alpha < 0$) of Fig. 10.5.

Problem 10.3.2 Determine the Fourier transform and design the amplitude and phase spectra of the discrete sequence $x[n]$, which is defined as

$$x[n] = \begin{cases} A & 0 \le n \le L-1 \\ 0 & \text{in the opposite case} \end{cases}$$

(Proakis and Manolakis, Example 4.2.4).

Solution Taking into account that the discrete-time signal $x[n]$ is a *finite-length sequence*, the required Fourier transform exists and is calculated as

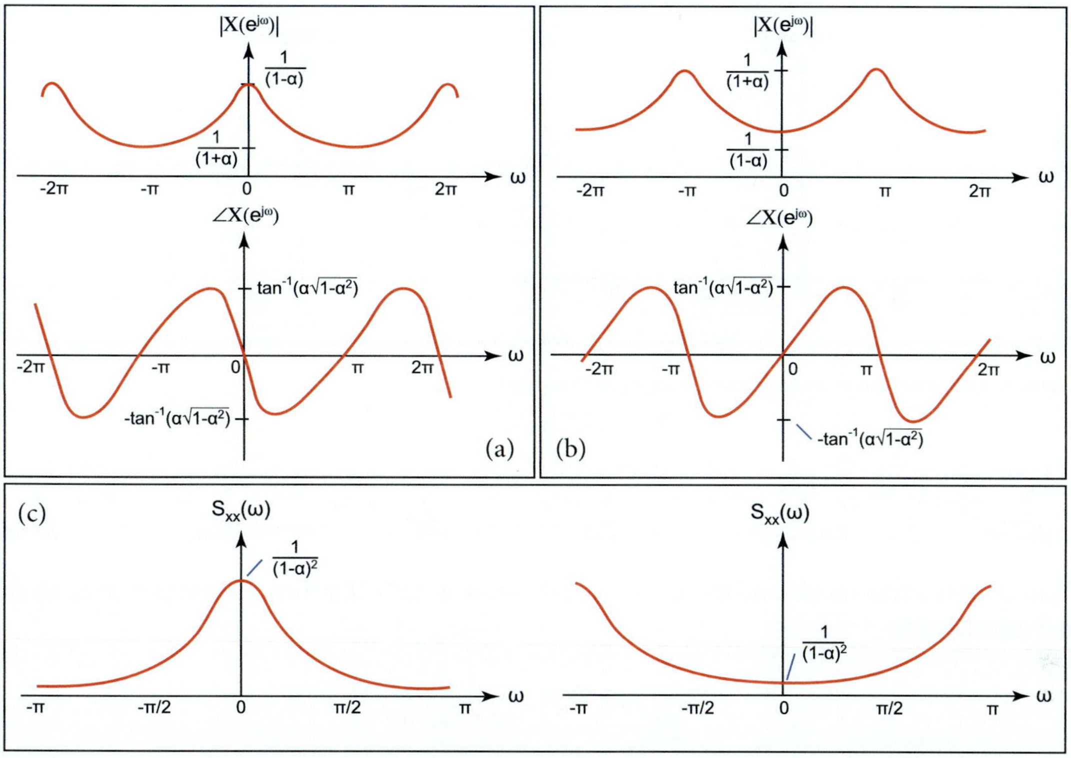

Fig. 10.5 The amplitude, phase, and energy spectral density of the discrete-time signal $x[n] = \alpha^n u[n]$ of Exercise 10.3.1

$$\mathcal{X}(e^{j\omega}) = \sum_{n=-\infty}^{\infty} x[n]\, e^{-j\omega n} = \sum_{n=0}^{L-1} A e^{-j\omega n} = A \sum_{n=0}^{L-1} (e^{-j\omega})^n = \frac{A(1-e^{-j\omega L})}{1-e^{-j\omega}}$$

Working as in Exercise 10.2.5, the preceding relationship is written

$$\mathcal{X}(e^{j\omega}) = A\left\{\frac{e^{-j\omega L/2}}{e^{-j\omega/2}}\right\}\left\{\frac{e^{j\omega L/2}-e^{-j\omega L/2}}{e^{j\omega/2}-e^{-j\omega/2}}\right\} = A\frac{\sin(\omega L/2)}{\sin(\omega/2)}\exp\left[-j\frac{\omega(L-1)}{2}\right]$$

Therefore, the magnitude and phase of the function $\mathcal{X}(e^{j\omega})$ are calculated as

$$|\mathcal{X}(e^{j\omega})| = |A|\left|\frac{\sin(\omega L/2)}{\sin(\omega/2)}\right| \quad \text{and} \quad \measuredangle\mathcal{X}(e^{j\omega}) = -\frac{\omega(L-1)}{2}$$

It is not difficult to see that for the value $\omega = 0$, the foregoing relationships take the simple form $|\mathcal{X}(e^{j\omega})| = |A|L$ and $\measuredangle\mathcal{X}(e^{j\omega}) = 0$. The function $|\mathcal{X}(e^{j\omega})|$ for the value $A = 1$ is known as the Dirichlet kernel. The graphical representations of the functions $|\mathcal{X}(e^{j\omega})|$ and $\measuredangle\mathcal{X}(e^{j\omega})$ are presented in Fig. 10.6 (see also Proakis and Manolakis, Example 4.3.3).

Problem 10.3.3 Determine the Fourier transform and draw the amplitude and phase spectra of the discrete sequence $x[n]$, which is defined as $x[n] = \alpha^{|n|}, \quad |\alpha| < 1$.

Solution The Fourier transform of the preceding discrete-time signal exists, due to the condition $|\alpha| < 1$, and is given by the relation

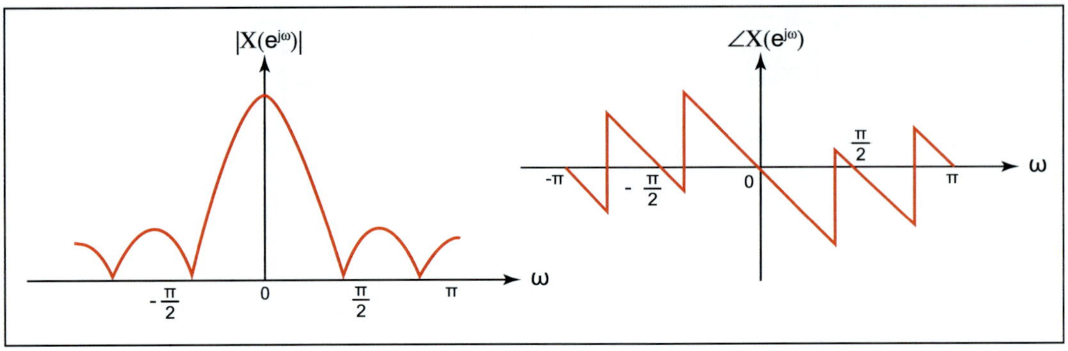

Fig. 10.6 Amplitude and phase spectra of discrete-time signal of Exercise 10.3.2

$$\mathcal{X}(e^{j\omega}) = \sum_{n=-\infty}^{\infty} x[n]\, e^{-j\omega n} = \sum_{n=-\infty}^{\infty} \alpha^{|n|} e^{-j\omega n} = \sum_{n=0}^{\infty} \alpha^n e^{-j\omega n} + \sum_{n=-\infty}^{-1} \alpha^{-n} e^{-j\omega n} = \sum_{n=0}^{\infty} (\alpha e^{-j\omega})^n + \sum_{n=-\infty}^{-1} (\alpha e^{j\omega})^{-n}$$

where to construct the last relation we used the definition of the *absolute value* function. Referring to mathematical typology, the first sum is calculated as

$$\sum_{n=0}^{\infty} (\alpha e^{-j\omega})^n = \frac{1}{1 - \alpha e^{-j\omega}}$$

On the other hand, to calculate the second sum, we proceed to the replacement $n \longrightarrow m = -n$. For $n = -\infty$ we have $m = \infty$, while for $n = -1$ we have $m = 1$, and the preceding sum is written

$$\sum_{n=-\infty}^{-1} (\alpha e^{j\omega})^{-n} = \sum_{m=1}^{\infty} (\alpha e^{j\omega})^m = \sum_{m=0}^{\infty} (\alpha e^{j\omega})^m - (\alpha e^{j\omega})^m \Big|_{m=0} = \frac{1}{1 - \alpha e^{j\omega}} - 1 = \frac{\alpha e^{j\omega}}{1 - \alpha e^{j\omega}}$$

Combining the foregoing results, the Fourier transform of the signal $x[n]$ is calculated as

$$\mathcal{X}(e^{j\omega}) = \frac{1}{1 - \alpha e^{-j\omega}} + \frac{\alpha e^{j\omega}}{1 - \alpha e^{j\omega}} = \frac{1 - \alpha^2}{1 - 2\alpha \cos\omega + \alpha^2}$$

as emerges after simple algebraic operations, which is also the desired result. We observe that, in this case, the function $\mathcal{X}(e^{j\omega})$ is a *real function* whose graphical representation for parameter values $0 < \alpha < 1$ is presented in Fig. 10.7 (see also Proakis and Manolakis, Example 4.3.3).[10]

Problem 10.3.4 Determine the Fourier transform of the discrete sequence $x[n] = n\alpha^n u[n]$, $|\alpha| < 1$.

Solution The signal $x[n]$ has a Fourier transform—since $|\alpha| < 1$—in the form

$$\mathcal{X}(e^{j\omega}) = \sum_{n=-\infty}^{\infty} x[n]\, e^{-j\omega n} = \sum_{n=-\infty}^{\infty} n\alpha^n e^{-j\omega n} u[n] = \sum_{n=0}^{\infty} n(\alpha e^{-j\omega})^n = \frac{\alpha e^{-j\omega}}{(1 - \alpha e^{-j\omega})^2} = \frac{\alpha e^{j\omega}}{(e^{j\omega} - \alpha)^2}$$

[10] In the following exercises, to economize on space, we will not present the graphical representations of the transform but only the final equation that emerges. Readers are encouraged, once they have understood how it is calculated, to proceed with the construction of the corresponding graphical representations using the appropriate software applications (like MATLAB).

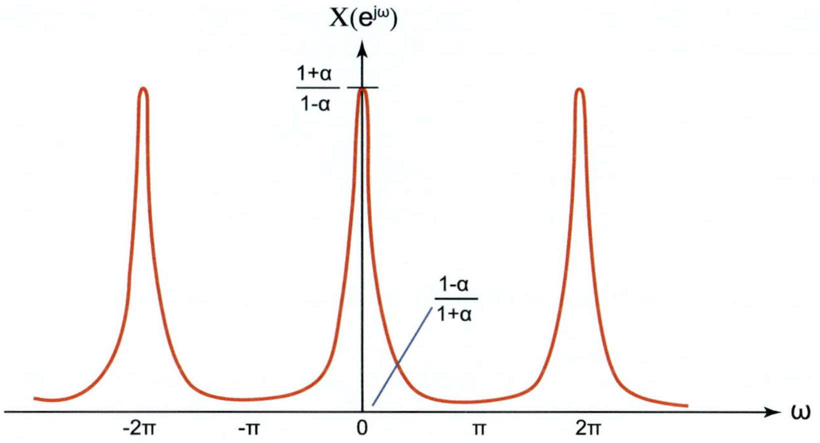

Fig. 10.7 The Fourier transform of the discrete-time signal $x[n] = \alpha^{|n|}$ of Exercise 10.3.3

as it arises after simple mathematical operations and using the appropriate expression from mathematical typology (Table 6.2, page 500).

Problem 10.3.5 Determine the Fourier transform of the discrete sequence

$$x[n] = -\alpha^n u[-(n+1)], \quad |\alpha| < 1$$

(Hsu, Problem 6.11).

Solution From the defining equation of the Fourier transform we will have

$$\mathcal{X}(e^{j\omega}) = \sum_{n=-\infty}^{\infty} x[n]\, e^{-j\omega n} = -\sum_{n=-\infty}^{\infty} \alpha^n e^{-j\omega n} u[-(n+1)]$$

From the defining equation of the step function it turns out that

$$u[-(n+1)] = \begin{cases} 1 \; n \leq -1 \\ 0 \; n > -1 \end{cases}$$

and the preceding relation will take the form

$$\mathcal{X}(e^{j\omega}) = -\sum_{n=-\infty}^{-1} \alpha^n e^{-j\omega n} = -\sum_{n=-1}^{-\infty} (\alpha e^{-j\omega})^n$$

In the next step of the process we proceed to change the index of the sum from n to $m = -n$. For $n = -1$ we will have $m = 1$, while for $n = -\infty$ we will have $m = \infty$, and the preceding relation becomes

$$\mathcal{X}(e^{j\omega}) = -\sum_{m=1}^{\infty} (\alpha e^{-j\omega})^{-m} = -\sum_{m=1}^{\infty} \left(\frac{e^{j\omega}}{\alpha}\right)^m = -\sum_{m=0}^{\infty} \left(\frac{e^{j\omega}}{\alpha}\right)^m + \left(\frac{e^{j\omega}}{\alpha}\right)^m \Bigg|_{m=0}$$

$$= -\frac{1}{1-(e^{j\omega}/\alpha)} + 1 = -\frac{e^{j\omega}}{\alpha - e^{j\omega}} = \frac{1}{1-\alpha e^{-j\omega}}$$

Problem 10.3.6 Determine the Fourier transform of the discrete sequence

$$x[n] = [\alpha^n \sin(\omega_0 n) + \beta^n]u[n], \quad |\alpha| < 1, \ |\beta| < 1$$

(Boulet, Problem 12.3).

Solution Replacing the sine function with the Euler formula, the signal $x[n]$ is written

$$x[n] = \left(\alpha^n \frac{e^{j\omega_0 n} - e^{-j\omega_0 n}}{2i} + \beta^n\right)u[n] = \left[\frac{1}{2i}(\alpha e^{i\omega_0})^n - \frac{1}{2i}(\alpha e^{-i\omega_0})^n + \beta^n\right]u[n]$$

and from the defining equation of the Fourier transform we will have

$$\begin{aligned}\mathcal{X}(e^{j\omega}) &= \sum_{n=-\infty}^{\infty} x[n]e^{-j\omega n} = \frac{1}{2i}\sum_{n=-\infty}^{\infty}(\alpha e^{i\omega_0})^n e^{-j\omega n}u[n] - \frac{1}{2i}\sum_{n=-\infty}^{\infty}(\alpha e^{-i\omega_0})^n e^{-j\omega n}u[n] + \sum_{n=-\infty}^{\infty}\beta^n u[n] \\ &= \frac{1}{2i}\sum_{n=0}^{\infty}\left[\alpha e^{-i(\omega-\omega_0)n}\right] - \frac{1}{2i}\sum_{n=0}^{\infty}\left[\alpha e^{-i(\omega+\omega_0)n}\right]^n + \sum_{n=0}^{\infty}(\beta e^{-j\omega})^n\end{aligned}$$

Using the appropriate expression from mathematical typology, it turns out that

$$\sum_{n=0}^{\infty}\left[\alpha e^{-i(\omega-\omega_0)n}\right]^n = \frac{1}{1-\alpha e^{-i(\omega-\omega_0)}}, \quad \sum_{n=0}^{\infty}\left[\alpha e^{-i(\omega+\omega_0)n}\right]^n = \frac{1}{1-\alpha e^{-i(\omega+\omega_0)}}$$

$$\sum_{n=0}^{\infty}(\beta e^{-j\omega})^n = \frac{1}{1-\beta e^{-j\omega}}$$

and replacing the preceding relations in the Fourier transform equation we get

$$\mathcal{X}(e^{j\omega}) = \frac{1}{2i}\left\{\frac{1}{1-\alpha e^{-i(\omega-\omega_0)}} - \frac{1}{1-\alpha e^{-i(\omega+\omega_0)}}\right\} + \frac{1}{1-\beta e^{-j\omega}} \frac{1-\alpha(\sin\omega_0+2\cos\omega_0)e^{-j\omega}+\alpha(\alpha+\beta\sin\omega_0)e^{-2i\omega}}{(1-2\alpha\cos\omega_0 e^{-j\omega}+\alpha^2 e^{-2i\omega})(1-\beta e^{-j\omega})}$$

Problem 10.3.7 Determine the Fourier transform of the impulse function

$$\delta[n] = \begin{cases} 1 & n = 0 \\ 0 & n \neq 0. \end{cases}$$

(Oppenheim, Willsky, Nawab, Example 5.4 and Mitra, Example 3.1).

Solution From the defining equation of the Fourier transform and the discrete impulse function we have

$$\mathcal{X}(e^{j\omega}) = \sum_{n=-\infty}^{\infty} x[n]\, e^{-j\omega n} = \sum_{n=-\infty}^{\infty} \delta[n]\, e^{-j\omega n} = \delta[n]\, e^{-j\omega n}\Big|_{n=0} = 1$$

Therefore, *the Fourier transform of the discrete impulse function has a unit amplitude and zero phase for all frequencies that belong to the interval* $(-\pi, +\pi)$.

Problem 10.3.8 Determine the Fourier transform of the time-shifted impulse function

$$\delta[n-n_0]=\begin{cases}1 & n=n_0\\ 0 & n\neq n_0\end{cases}$$

Solution In full analogy with the previous case we have

$$\mathcal{X}(e^{j\omega})=\sum_{n=-\infty}^{\infty}x[n]\,e^{-j\omega n}=\sum_{n=-\infty}^{\infty}\delta[n-n_0]\,e^{-j\omega n}=\delta[n]\,e^{-j\omega n}\Big|_{n=n_0}=e^{-j\omega n_0}$$

Therefore, *the magnitude of the transform is, as before, equal to unity, but its phase is now nonzero and proportional to the frequency for all frequencies that belong to the interval* $(-\pi,+\pi)$.

10.3.5 The Fourier Transform of Periodic Signals

In full analogy with continuous periodic signals, *their discrete equivalents do not meet the conditions of existence of their Fourier transform*, since *they do not tend to zero* at the limit $n\rightarrow\infty$, and therefore *they cannot be characterized as absolutely or square summable*. Nevertheless, following the argumentation formulated in Sect. 3.3.6 (page 250), we can extend the concept of Fourier transform also to these signals, *using impulse functions in the frequency domain and interpreting that transform as a train of discrete impulse functions defined in this space whose values are proportional to the coefficients of the Fourier expansion of the discrete periodic signal under consideration*.

To prove the truth of this claim, let us calculate the Fourier transform of the discrete complex of the exponential signal $x[n]=e^{j\omega_0 n}$. Having calculated this transform and taking into account that any periodic discrete-time signal can be developed into a Fourier series, that is, into a linear combination of elementary exponential complex signals, we can, taking advantage of the *linear nature* of the transform,[11] calculate the transformation of the periodic discrete-time signal $x[n]$. So let us start with the simple case of the exponential signal

$$x[n]=e^{j\omega_0 n} \tag{10.40}$$

which, as is well known, is *periodic* with a period of 2π. From the defining equation of the Fourier transform of this periodic signal, we will get

$$\mathcal{X}(e^{j\omega})=\sum_{n=-\infty}^{\infty}x[n]\,e^{-j\omega n}=\sum_{n=-\infty}^{\infty}e^{j\omega_0 n}e^{-j\omega n}=\sum_{n=-\infty}^{\infty}e^{-j(\omega-\omega_0)n} \tag{10.41}$$

It is obvious that the preceding sum in the form in which it has been formulated does not converge to any conventional function. However, the preceding relationship is the discrete equivalent of the expression

$$\mathcal{F}\{e^{jn\omega_0 t}\}=\int_{-\infty}^{+\infty}e^{j(n\omega_0-\omega)t}dt=\left[\frac{e^{j(n\omega_0-\omega)t}}{j(n\omega_0-\omega)}\right]_{-\infty}^{+\infty} \tag{10.42}$$

which we encountered in Sect. 3.3.6 and which was characterized by the same convergence problem. To bypass this problem, we used the result $\mathcal{F}\{1\}=\delta(t)$, along with the frequency shifting property of the Fourier transform, leading to the result

[11] The properties of the DTFT will be studied and proven analytically in a subsequent section.

$$\mathcal{F}\{e^{jn\omega_0 t}\} = 2\pi\delta(\omega - n\omega_0) \tag{10.43}$$

Following the same line of thought, we can formulate the claim that *the sum that appears in the defining equation of the Fourier transform of the exponential discrete-time signal* $x[n] = e^{j\omega_0 n}$ *does not converge to a typical, usual function, but to a generalized function or distribution*. Indeed, considering the finite sum[12]

$$\mathcal{X}_N(\omega) = \sum_{n=-N}^{N} e^{-j(\omega-\omega_0)n} = \sum_{m=0}^{2N} e^{-j(\omega-\omega_0)m} = \frac{\sin[(N+1/2)(\omega-\omega_0)]}{\sin[(\omega-\omega_0)/2]} \tag{10.44}$$

we observe that this constitutes the Fourier transform of a rectangular signal of finite length[13] but shifted on the frequency axis by the value of the frequency ω_0. Considering that the height of the mainlobe of the transform (see the amplitude spectrum in Fig. 10.6) is proportional to N, while its width is inversely proportional to N, with the area enclosed by it approaching the value 2π, it is not difficult to see that in the limit $N \to \infty$, this expression tends to the *impulse function* $\delta(\omega - \omega_0)$, so we can write that

$$\mathcal{X}(e^{j\omega}) = \lim_{N\to\infty} \mathcal{X}_N(\omega) = 2\pi\delta(\omega - \omega_0) \tag{10.45}$$

In reality, of course, the preceding relationship is not entirely correct, since we know that the Fourier transform of discrete-time signals is a *periodic function with a period of* 2π and therefore should contain impulse functions at the positions $\omega = \omega_0, \omega_0 \pm 2\pi, \omega_0 \pm 4\pi, \ldots,$ and so forth. Thus, the Fourier transform of the discrete sequence $x[n] = e^{j\omega_0 n}$ has the form

$$\mathcal{X}(e^{j\omega}) = \mathcal{F}\{e^{j\omega_0 n}\} = \sum_{k=-\infty}^{\infty} 2\pi\delta(\omega - \omega_0 - 2k\pi) \tag{10.46}$$

which is the desired result. To verify whether this result is correct, let us check if the inverse transform of the function $\mathcal{X}(e^{j\omega})$ is indeed the signal $x[n] = e^{j\omega_0 n}$. Substituting the foregoing relationship into the defining equation of the inverse transform, we will get

$$\frac{1}{2\pi}\int_{2\pi} \mathcal{X}(e^{j\omega})\, e^{j\omega n} d\omega = \int_{2\pi} \left\{ \sum_{k=-\infty}^{\infty} \delta(\omega - \omega_0 - 2k\pi) \right\} e^{j\omega n} d\omega \tag{10.47}$$

It is not difficult to see that within the integration interval, which has a range of 2π, there is *only one impulse function*. Therefore, if we choose as such an interval the one that includes the impulse function at the position $\omega = \omega_0 + 2\pi m$, the foregoing relation will be calculated as

$$\int_{2\pi} \mathcal{X}(e^{j\omega})\, e^{j\omega n} d\omega = e^{j(\omega_0+2\pi m)n} = e^{j\omega_0 n} = x[n] \tag{10.48}$$

from which the desired result follows.

Having calculated the Fourier transform of the elementary discrete-time signal $x[n] = e^{j\omega_0 n}$, we can now calculate the Fourier transform *of any periodic discrete-time signal of period N*, developing the signal in a Fourier series of the form

[12] For the derivation of the final form of this sum we proceeded to change the dummy index from n to $m = n+N$ and worked as in Exercises 10.2.4 (page 788) and 10.3.2 (page 810).

[13] See Exercise 10.3.2 on page 810 and the amplitude spectrum of Fig. 10.6 on page 812.

$$x[n] = \sum_{k=<N>} \alpha_k e^{j\omega_0 kn} = \sum_{k=<N>} \alpha_k \exp\left(j\frac{2\pi}{N}kn\right) \tag{10.49}$$

and taking advantage of the linear nature of that transform. We will therefore have

$$\begin{aligned}\mathcal{X}(e^{j\omega}) &= \mathcal{F}\{x[n]\} = \mathcal{F}\left\{\sum_{k=<N>} \alpha_k \exp\left(j\frac{2\pi}{N}kn\right)\right\}\\ &= \sum_{k=<N>} \alpha_k \mathcal{F}\left\{\exp\left(j\frac{2\pi}{N}kn\right)\right\} = \sum_{k=<N>} \alpha_k \sum_{\ell=-\infty}^{\infty} 2\pi\delta\left(\omega - k\frac{2\pi}{N} - 2\pi\ell\right)\end{aligned}$$

and finally

$$\mathcal{X}(e^{j\omega}) = \sum_{k=-\infty}^{\infty} 2\pi\alpha_k\delta\left(\omega - k\frac{2\pi}{N}\right)$$

which is our final result.[14] The last relationship is very important because it dictates the way we can calculate the Fourier transform of a periodic signal directly from its Fourier expansion coefficients.

Problem 10.3.9 Determine the Fourier transform of the impulse train

$$x[n] = \sum_{k=-\infty}^{\infty} \delta[n - kN].$$

(Oppenheim, Willsky, Nawab, Example 5.6).

---

[14] Following the argumentation formulated for the *elementary exponential signal* $x[n] = e^{j\omega_0 n}$, we can prove that this result is correct, replacing the expression for the transforming $\mathcal{X}(e^{j\omega})$ into the defining equation of the inverse transform and checking if the expression that will emerge is identical to the discrete periodic signal $x[n]$. We will therefore have

$$\frac{1}{2\pi}\int_{0-\varepsilon}^{2\pi-\varepsilon} \mathcal{X}(e^{j\omega})\, e^{j\omega n} d\omega = \frac{1}{2\pi}\int_{0-\varepsilon}^{2\pi-\varepsilon} \left\{\sum_{k=-\infty}^{\infty} 2\pi\alpha_k\delta\left(\omega - k\frac{2\pi}{N}\right)\right\} e^{j\omega n} d\omega$$

In the preceding equation, the parameter ε has a value in the interval $(0, 2\pi/N)$ and has been chosen so as to include in the integral the impulse function located at $\omega = 0$, excluding at the same time the one located at $\omega = 2\pi$. By alternating the order of the operations of summation and integration in the second term of the preceding equation, we will get

$$\frac{1}{2\pi}\int_{0-\varepsilon}^{2\pi-\varepsilon} \mathcal{X}(e^{j\omega})\, e^{j\omega n} d\omega = \sum_{k=-\infty}^{\infty} \alpha_k \int_{0-\varepsilon}^{2\pi-\varepsilon} \delta\left(\omega - k\frac{2\pi}{N}\right) e^{j\omega n} d\omega$$

However, taking into account that the integration interval from $0 - \varepsilon$ to $2\pi - \varepsilon$ includes only those impulse responses that correspond to the positions $k = 0, 1, 2, \ldots, N-1$ and using the defining equation of the time-shifted impulse function, it immediately follows that

$$\sum_{k=-\infty}^{\infty} \alpha_k \int_{0-\varepsilon}^{2\pi-\varepsilon} \delta\left(\omega - k\frac{2\pi}{N}\right) e^{j\omega n} d\omega = \sum_{k=0}^{N-1} \alpha_k \exp\left(j\frac{2\pi}{N}kn\right) = x[n]$$

—since the last relationship is simply the *Fourier expansion of the discrete periodic signal* $x[n]$—a finding that validates the correctness of the result we arrived at for the Fourier transform $\mathcal{X}(e^{j\omega})$.

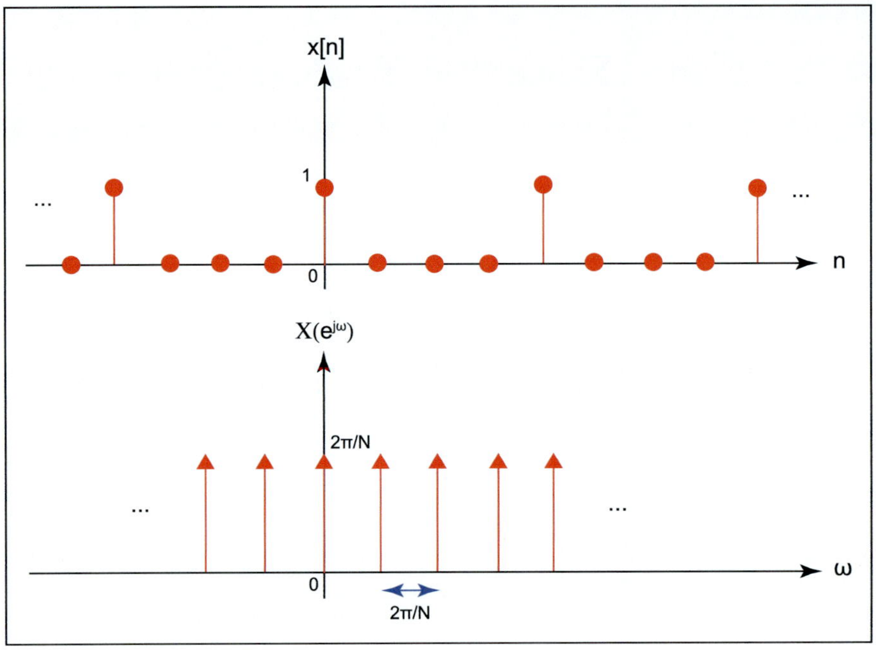

Fig. 10.8 Periodic train of impulse functions and its Fourier transform calculated in Exercise 10.3.9

Solution From the defining equation of the discrete-time signal $x[n]$ we find that it is periodic with period N with its Fourier expansion coefficients calculated as

$$\alpha_k = \frac{1}{N} \sum_{n=<N>} x[n] \exp\left(-jk\frac{2\pi}{N}\right) = \sum_{n=<N>} \left\{ \sum_{k=-\infty}^{\infty} \delta[n-kN] \right\} \exp\left(-jk\frac{2\pi}{N}\right)$$

Choosing the length interval N as the value range $[0, N-1]$, from all the terms of the infinite sum with respect to k, *only the one* corresponding to the value $k = 0$ will remain, thereby leading to the equation

$$\alpha_k = \frac{1}{N} \sum_{n=0}^{N-1} \delta[n] \exp\left(-jk\frac{2\pi}{N}\right) = \frac{1}{N}\left[\delta[n] \exp\left(-jk\frac{2\pi}{N}\right)\right]_{n=0} = \frac{1}{N}$$

Therefore, according to the theory presented earlier, the Fourier transform of the discrete-time signal $x[n]$ will have the form

$$\mathcal{X}(e^{j\omega}) = \sum_{k=0}^{N-1} \frac{2\pi}{N} \sum_{\ell=-\infty}^{\infty} \delta\left(\omega - \frac{2\pi}{N} - 2\pi\ell\right) = \frac{2\pi}{N} \sum_{k=-\infty}^{\infty} \delta\left(\omega - \frac{2\pi}{N}\right)$$

which is also the final result. From the last equation we observe that *the Fourier transform of a periodic train of impulse functions is also an impulse train defined in the frequency domain, with the impulse functions included in it being equidistant from each other with distances equal to* $2\pi/N$. The signal $x[n]$ and its Fourier transform $\mathcal{X}(e^{j\omega})$ are presented in Fig. 10.8.

10.3.6 Properties of Discrete-Time Fourier Transform

The Fourier transform of discrete-time signals is characterized by properties similar to those that characterize its continuous equivalent. These properties are presented in Table 10.3, and their analytical proof follows in the continuation.

Table 10.3 Basic properties of Fourier transform

Property	Discrete-time sequence $x[n]$	Fourier transform $X(e^{j\omega})$
Periodicity	$x[n]$	$X(e^{j\omega}) = X(e^{j(\omega \pm 2k\pi)}),\ k \in N$
Homogeneity	$\alpha x\ [n]$	$\alpha X(e^{j\omega})$
Addition	$x[n] + y[n]$	$X(e^{j\omega}) + Y(e^{j\omega})$
Linearity	$\alpha x\ [n]\ +\ \beta y[n]$	$\alpha X(e^{j\omega})\ +\ \beta Y(e^{j\omega})$
Differentiation in frequency	$nx[n]$	$jX'(e^{j\omega})$
Integration	$\int_t^{-\infty} x(\tau)\,d\tau$	$\frac{X(e^{j\omega})}{j\omega} + \pi X(0)\delta(\omega)$
Frequency shifting	$\exp(j\omega_0 n)\, x[n]$	$X(e^{j(\omega-\omega_0)})$
Modulation (sine)	$x[n]\cos(\omega_0 n)$	$\frac{1}{2}\left[X(e^{j(\omega+\omega_0)}) + X(e^{j(\omega-\omega_0)})\right]$
Modulation (cosine)	$x[n]\sin(\omega_0 n)$	$\frac{j}{2}\left[X(e^{j(\omega+\omega_0)}) - X(e^{j(\omega-\omega_0)})\right]$
Time shifting	$x\,[n - n_0\,]$	$\exp(-j\omega n_0)X(e^{j\omega})$
Time reversal	$x[-n]$	$X(e^{-j\omega})$
Convolution	$x[n] * y[n]$	$X(e^{j\omega})Y(e^{j\omega})$
Difference	$x[n] - x[n-1]$	$(1 - e^{-j\omega})X(e^{j\omega})$
Accumulation	$\sum_{k=-\infty}^{n} x[k]$	$\frac{X(e^{j\omega})}{1 - e^{-j\omega}}$
Multiplication	$x[n]y[n]$	$\frac{1}{2\pi}\int_{2\pi} X(e^{j\vartheta})Y(e^{j}(\omega - \vartheta))d\vartheta$
Scaling	$x_k[n]$	$X(e^{jk\omega})$
Symmetry	$x[n] \in R$	$\begin{cases} X(e^{j\omega}) = X^*(e^{-j\omega}) \\ \Re\{\mathcal{X}(e^{j\omega})\} = \Re\{\mathcal{X}(e^{-j\omega})\} \\ \Im\{\mathcal{X}(e^{j\omega})\} = -\Im\{\mathcal{X}(e^{-j\omega})\} \\ \lvert X(e^{j\omega})\rvert = \lvert X(e^{-j\omega})\rvert \\ \measuredangle X(e^{j\omega}) = -X(e^{-j\omega}) \end{cases}$
Conjugation	$x^*[n]$	$X^*(e^{-j\omega})$
Correlation	$x[n] * y[n]$	$X(e^{j\omega})Y(e^{-j\omega})$
Autocorrelation	$x[n] * x[n]$	$\lvert X(e^{j\omega})\rvert^2$
Odd part	$\mathcal{E}v\{x[n]\}$	$\Re\{X(e^{j\omega})\}$
Even part	$Od\{x[n]\}$	$i\Im\{\mathcal{X}(e^{j\omega})\}$

- **Periodicity**: The Fourier transform of a discrete sequence $x[n]$ is always a *periodic function of frequency* ω *with a period value equal to* 2π, so we can write that

$$\mathcal{X}(e^{j\omega}) = \mathcal{X}(e^{j(\omega \pm 2\pi m)}) \quad m = 1, 2, \ldots$$

  Note that this property *does not apply to continuous-time signals* whose Fourier transform *does not constitute a periodic function.*
- **Linearity**: If the Fourier transforms of discrete sequences $x[n]$ and $y[n]$ are the functions $\mathcal{X}(e^{j\omega})$ and $\mathcal{Y}(e^{j\omega})$, then the Fourier transform of the linear combination $z[n] = \alpha x[n] + \beta y[n]$ will have the form $\mathcal{Z}(e^{j\omega}) = \alpha\mathcal{X}(e^{j\omega}) + \beta\mathcal{Y}(e^{j\omega})$ for every value of the constants α and β.
  Proof From the defining equation of the Fourier transform of the signal $z[n]$ it immediately follows that

$$\mathcal{Z}(e^{j\omega}) = \sum_{n=-\infty}^{\infty} z[n]\, e^{-j\omega n} = \sum_{n=-\infty}^{\infty} \Big(\alpha x[n] + \beta y[n]\Big)\, e^{-j\omega n}$$

$$= \alpha \sum_{n=-\infty}^{\infty} x[n]\, e^{-j\omega n} + \beta \sum_{n=-\infty}^{\infty} y[n]\, e^{-j\omega n} = \alpha \mathcal{X}(e^{j\omega}) + \beta \mathcal{Y}(e^{j\omega})$$

which is the desired result. The property of linearity can arise from the combination of the properties of *addition* and *homogeneity*, which are proven in exactly the same way.

- **Time shift**: If the Fourier transform of the discrete-time signal $x[n]$ is the function $\mathcal{X}(e^{j\omega})$, then the Fourier transform of the time-delayed signal $y[n] = x[n - n_0]$ is the function $\mathcal{Y}(e^{j\omega}) = e^{-j\omega n_0} \mathcal{X}(e^{j\omega})$.
  Proof From the defining equation of the Fourier transform of the signal $y[n]$ we will have

$$\mathcal{Y}(e^{j\omega}) = \sum_{n=-\infty}^{\infty} y[n]\, e^{-j\omega n} = \sum_{n=-\infty}^{\infty} x[n - n_0]\, e^{-j\omega n}$$

  Proceeding to change the dummy index of the sum from n to $m = n - n_0$, we will have $n = m + n_0$ (the limits of the sum, being infinite, obviously do not change), and the preceding relation will take the form

$$\mathcal{Y}(e^{j\omega}) = \sum_{m=-\infty}^{\infty} x[m]\, e^{-j\omega(m+n_0)} = e^{-j\omega n_0} \sum_{m=-\infty}^{\infty} x[m]\, e^{-j\omega m} = e^{-i\omega n_0} \mathcal{X}(e^{j\omega})$$

  which is the desired result. Using the same argument and considering the discrete-time signal $y[n] = x[n + n_0]$, we will have

$$\mathcal{Y}(e^{j\omega}) = \sum_{n=-\infty}^{\infty} y[n]\, e^{-j\omega n} = \sum_{n=-\infty}^{\infty} x[n + n_0]\, e^{-j\omega n}$$

$$= \sum_{m=-\infty}^{\infty} x[m]\, e^{-j\omega(m-n_0)} = e^{j\omega n_0} \sum_{n=-\infty}^{\infty} x[m]\, e^{-j\omega m} = e^{j\omega n_0} \mathcal{X}(e^{j\omega})$$

- **Frequency shift**: If the Fourier transform of the signal $x[n]$ is the function $\mathcal{X}(e^{j\omega})$, then the Fourier transform of the signal $y[n] = e^{j\omega_0 n} x[n]$ will have the form $\mathcal{Y}(e^{j\omega}) = \mathcal{X}(e^{j(\omega-\omega_0)})$.
  Proof From the defining equation of the Fourier transform of the signal $y[n]$ we will have

$$\mathcal{Y}(e^{j\omega}) = \sum_{n=-\infty}^{\infty} y[n]\, e^{-j\omega n} = \sum_{n=-\infty}^{\infty} e^{j\omega_0 n} x[n]\, e^{-j\omega n} = \sum_{n=-\infty}^{\infty} x[n]\, e^{-j(\omega-\omega_0)n} = \mathcal{X}(e^{j(\omega-\omega_0)})$$

  which is the desired result. Applying the same method, it is proven that $\mathcal{F}\{e^{-j\omega_0 n} x[n]\} = \mathcal{X}(e^{j(\omega+\omega_0)})$.

- **Conjugation and symmetry**: If the Fourier transform of the signal $x[n]$ is the function $\mathcal{X}(e^{j\omega})$, then the Fourier transform of the conjugate signal $y[n] = x^*[n]$ will be $\mathcal{Y}(e^{j\omega}) = \mathcal{X}^*(e^{-j\omega})$.
  Proof From the defining equation of the Fourier transform of the signal $y[n]$ it immediately follows that

$$\mathcal{Y}(e^{j\omega}) = \sum_{n=-\infty}^{\infty} y[n]\, e^{-j\omega n} = \sum_{n=-\infty}^{\infty} x^*[n]\, e^{-j\omega n} = \left(\sum_{n=-\infty}^{\infty} x[n]\, e^{j\omega n} \right)^* = \left(\sum_{n=-\infty}^{\infty} x[n]\, e^{-j(-\omega)n} \right)^* = \mathcal{X}^*(e^{-j\omega})$$

  This property has some interesting consequences. First, if the signal $x[n]$ is a *real-valued signal*, it holds that $x[n] = x^*[n]$, and from the preceding property it follows that

$$\mathcal{X}(e^{j\omega}) = \mathcal{X}^*(e^{-j\omega})$$

  that is, *the Fourier transform of a real-valued signal is characterized by Hermitian symmetry*. Expressing the preceding functions in their Cartesian form, the foregoing relationship becomes

$$\Re\{\mathcal{X}(e^{j\omega})\} + i\Im\{\mathcal{X}(e^{j\omega})\} = \Re\{\mathcal{X}(e^{-j\omega})\} - i\Im\{\mathcal{X}(e^{j\omega})\}$$

or, equivalently,

$$\Re\{\mathcal{X}(e^{j\omega})\} = \Re\{\mathcal{X}(e^{-j\omega})\} \qquad \text{and} \qquad \Im\{\mathcal{X}(e^{j\omega})\} = -\Im\{\mathcal{X}(e^{j\omega})\}$$

Therefore, *the real part of the Fourier transform of a real periodic signal is an even function of the frequency* ω*, while its imaginary part is an odd function of that frequency*. Finally, expressing the functions $\mathcal{X}(e^{j\omega})$ and $\mathcal{X}^*(e^{-j\omega})$ in their exponential form, the relationship $\mathcal{X}(e^{j\omega}) = \mathcal{X}^*(e^{-j\omega})$ is formulated as

$$|\mathcal{X}(e^{j\omega})|e^{j\measuredangle\mathcal{X}(e^{j\omega})} = |\mathcal{X}(e^{-j\omega})|e^{-j\measuredangle\mathcal{X}(e^{-j\omega})}$$

or, equivalently,

$$|\mathcal{X}(e^{j\omega})| = |\mathcal{X}(e^{-j\omega})| \qquad \text{and} \qquad \measuredangle\mathcal{X}(e^{j\omega}) = -\measuredangle\mathcal{X}(e^{-j\omega})$$

This result is also very interesting because it leads us to the conclusion that *the amplitude spectrum of a periodic discrete-time signal is an even function of the frequency* ω*, while, in contrast, the phase of the signal in question* is an odd function of frequency.

- **Difference**: If the Fourier transform of the signal $x[n]$ is the function $\mathcal{X}(e^{j\omega})$, then the Fourier transform of the discrete-time signal $y[n] = x[n] - x[n-1]$ will be $\mathcal{Y}(e^{j\omega}) = (1 - e^{-j\omega})\mathcal{X}(e^{j\omega})$.
Proof From the property of time shift for the value $k = 1$ we will have

$$\mathcal{F}\{x[n-1]\} = e^{-j\omega}\mathcal{X}(e^{j\omega})$$

and therefore, using the property of the linearity of the transform, we get

$$\mathcal{Y}(e^{j\omega}) = \mathcal{F}\{x[n]\} - \mathcal{F}\{x[n-1]\} = \mathcal{X}(e^{j\omega}) - e^{-j\omega}\mathcal{X}(e^{j\omega}) = (1 - e^{-j\omega})\mathcal{X}(e^{j\omega})$$

- **Time reversal**: If the Fourier transform of the discrete-time signal $x[n]$ is the function $\mathcal{X}(e^{j\omega})$, then the Fourier transform of the time-reversed discrete-time signal $y[n] = x[-n]$ will be the function $\mathcal{Y}(e^{j\omega}) = \mathcal{X}(e^{-j\omega})$.
Proof From the defining equation of the Fourier transform of the signal $y[n]$ we will get

$$\mathcal{Y}(e^{j\omega}) = \sum_{n=-\infty}^{\infty} y[n]\, e^{-j\omega n} = \sum_{n=-\infty}^{\infty} x[-n]\, e^{-j\omega n}$$

Proceeding to change the dummy index of the sum from n to $m = -n$, the limits of the sum do not change, and the preceding relation will take the form

$$\mathcal{Y}(e^{j\omega}) = \sum_{m=-\infty}^{\infty} x[m]\, e^{j\omega m} = \sum_{m=-\infty}^{\infty} x[m]\, e^{-j(-\omega)m} = \mathcal{X}(e^{-j\omega})$$

- **Time scaling**: The property of time scaling of a DFT presents some differences compared to the corresponding property of the continuous transform. Indeed, if we try to define this property in the way we did in the domain of continuous-time where we had ended up with a relationship of the form

$$\mathcal{F}\{x(\alpha t)\} = \frac{1}{|\alpha|}\mathcal{X}\left(\frac{j\omega}{\alpha}\right)$$

we will encounter problems in cases where the value of α is not an integer. For example, it is not possible to *slow down* the signal considering values $\alpha < 1$. On the other hand, considering values $\alpha > 1$, such as the value $\alpha = 2$, the signal $x[2n]$ is not the accelerated version of the original signal, since we simply retrieve its even samples.

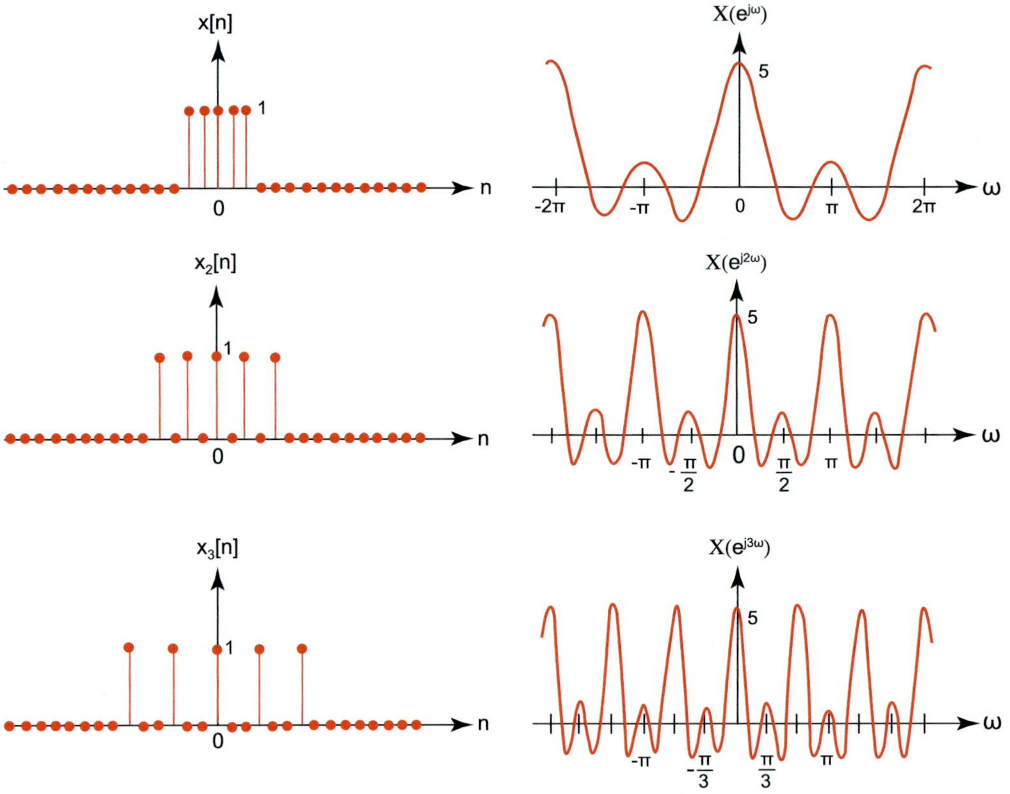

Fig. 10.9 Change in signal $x_k[n]$ and its Fourier transform $\mathcal{X}_k(e^{j\omega})$ for different values of parameter k

To address the foregoing problem, we define the signal

$$x_k[n] = \begin{cases} x[n/k] & \text{for values of } n \text{ that are integer multiples of } k \\ 0 & \text{in the opposite case} \end{cases}$$

Taking into account that the discrete-time signal is nonzero *only for the values of n for which* $n/k = r$ or, equivalently, $n = rk$, the Fourier transform of the discrete-time signal $x_k[n]$ is calculated as

$$\mathcal{X}_k(e^{j\omega}) = \sum_{k=-\infty}^{\infty} x_k[n]\, e^{-j\omega n} = \sum_{r=-\infty}^{\infty} x_k[rk]\, e^{-j\omega rk}$$

Taking into account that $x_k[rk] = x[r]$, the foregoing equation will take the form

$$\mathcal{X}_k(e^{j\omega}) = \sum_{r=-\infty}^{\infty} x[r]\, e^{-j(k\omega)r} = \mathcal{X}(e^{jk\omega})$$

which is also the final result. If we construct the graphs of the functions $x[n]$ and $\mathcal{X}(e^{jk\omega})$ for continuously increasing values of k, we will find that the signal *extends in the time domain*, while its Fourier transform *compresses in the frequency domain*, as depicted in Fig. 10.9.

- **Differentiation in the frequency domain**: If the Fourier transform of the discrete-time signal $x[n]$ is the function $\mathcal{X}(e^{j\omega})$, then the Fourier transform of the discrete-time signal $y[n] = nx[n]$ is the function $\mathcal{Y}(e^{j\omega}) = i\mathcal{X}'(e^{j\omega})$.

Proof Differentiating the defining equation of the Fourier transform

$$\mathcal{X}(e^{j\omega}) = \sum_{n=-\infty}^{\infty} x[n]\, e^{-j\omega n}$$

and interchanging the operations of summation and differentiation, we will get

$$\frac{d\mathcal{X}(e^{j\omega})}{d\omega} = \frac{d}{d\omega}\left\{\sum_{n=-\infty}^{\infty} x[n]\, e^{-j\omega n}\right\} = \sum_{n=-\infty}^{\infty} x[n]\frac{d}{d\omega}(e^{-j\omega n}) = -j\sum_{n=-\infty}^{\infty} nx[n]\, e^{-j\omega n}$$

or, equivalently,

$$\sum_{n=-\infty}^{\infty} (nx[n])\, e^{-j\omega n} = j\frac{d\mathcal{X}(e^{j\omega})}{d\omega}$$

where for the derivation of the last relation we multiplied both members sides of the previous equation by the imaginary unit j. But the left side of the preceding equation is simply the Fourier transform of the discrete-time signal $y[n] = nx[n]$, from which the desired result arises.

- **Convolution**: If the Fourier transforms of the discrete sequences $x[n]$ and $y[n]$ are the functions $\mathcal{X}(e^{j\omega})$ and $\mathcal{Y}(e^{j\omega})$, then the Fourier transform of the convolution of the two signals defined as

$$z[n] = x[n] * y[n] = \sum_{k=-\infty}^{\infty} x[k]\, y[n-k]$$

will have the form $\mathcal{Z}(e^{j\omega}) = \mathcal{X}(e^{j\omega})\mathcal{Y}(e^{j\omega})$.

Proof From the equation defining the Fourier transform of the signal $z[n]$ we have

$$\mathcal{Z}(e^{j\omega}) = \sum_{n=-\infty}^{\infty} z[n]\, e^{-j\omega n} = \sum_{n=-\infty}^{\infty}\left(\sum_{k=-\infty}^{\infty} x[k]\, y[n-k]\right) e^{-j\omega n} = \sum_{k=-\infty}^{\infty} x[k]\left(\sum_{n=-\infty}^{\infty} y[n-k]\, e^{-j\omega n}\right)$$

where in the last expression we have simply changed the order of the sums. But the expression in parentheses is recognized as the Fourier transform of the signal $y[n-k]$, which according to the *property of time shift*, can be given in the form

$$\sum_{n=-\infty}^{\infty} y[n-k]\, e^{-j\omega k} = e^{-j\omega k}\mathcal{Y}(e^{j\omega})$$

By substituting the preceding expression into the defining equation of the function $\mathcal{Z}(e^{j\omega})$, we will get

$$\mathcal{Z}(e^{j\omega}) = \sum_{k=-\infty}^{\infty} x[k]\, e^{-j\omega k}\mathcal{Y}(e^{j\omega}) = \mathcal{Y}(e^{j\omega})\left\{\sum_{k=-\infty}^{\infty} x[k]\, e^{-j\omega k}\right\} = \mathcal{X}(e^{j\omega})\mathcal{Y}(e^{j\omega})$$

- **Multiplication**: If the Fourier transforms of the sequences $x[n]$ and $y[n]$ are the functions $\mathcal{X}(e^{j\omega})$ and $\mathcal{Y}(e^{j\omega})$, then the Fourier transform of the product $z[n] = x[n]y[n]$ will be

$$\mathcal{Z}(e^{j\omega}) = \frac{1}{2\pi}\int_{2\pi} \mathcal{X}(e^{j\vartheta})\mathcal{Y}(e^{j(\omega-\vartheta)})d\vartheta$$

Proof From the defining equation of the Fourier transform of the signal $z[n]$ we have

$$\mathcal{Z}(e^{j\omega}) = \sum_{n=-\infty}^{\infty} z[n]\, e^{-j\omega n} = \sum_{n=-\infty}^{\infty} x[n]\, y[n]\, e^{-j\omega n}$$

By substituting the sequence $x[n]$ from the defining equation of the inverse transform

$$x[n] = \frac{1}{2\pi}\int_{2\pi} \mathcal{X}(e^{j\vartheta})\, e^{jn\vartheta}\, d\theta$$

the foregoing relation is written

$$\mathcal{Z}(e^{j\omega}) = \sum_{n=-\infty}^{\infty} y[n]\left\{\frac{1}{2\pi}\int_{2\pi} \mathcal{X}(e^{j\vartheta})\, e^{jn\vartheta}\, d\theta\right\} e^{-j\omega n} = \frac{1}{2\pi}\int_{2\pi} \mathcal{X}(e^{j\vartheta})\left[\sum_{n=-\infty}^{\infty} y[n]\, e^{-j(\omega-\vartheta)n}\right] d\vartheta$$

where for the derivation of the last expression we alternated the operations of summation and integration. But the expression in brackets is recognized as the transform

$$\mathcal{Y}(e^{j(\omega-\vartheta)}) = \sum_{n=-\infty}^{\infty} y[n]\, e^{-j(\omega-\vartheta)n}$$

and consequently we arrive at the equation

$$\mathcal{Z}(e^{j\omega}) = \frac{1}{2\pi}\int_{2\pi} \mathcal{X}(e^{j\vartheta})\, \mathcal{Y}(e^{j(\omega-\vartheta)})\, d\vartheta$$

which is the desired result.
The type of convolution that appears in the preceding expression constitutes the *convolution of two periodic functions with a common period value* and is known as *periodic convolution*.[15] It can be calculated using *any completion interval with a period range*, which is equal to 2π.

- **Duality**: In contrast to the *analysis* and *synthesis* equations of the continuous Fourier transform

$$\mathcal{X}(j\omega) = \int_{-\infty}^{+\infty} x(t)\, e^{-j\omega t}\, dt \qquad \text{and} \qquad x(t) = \frac{1}{2\pi}\int_{-\infty}^{+\infty} \mathcal{X}(j\omega)\, e^{j\omega t}\, d\omega$$

which are characterized by a *duality* of the form $\mathcal{F}\{\mathcal{X}(t)\} = 2\pi x(-\omega)$ or, equivalently, $\mathcal{F}\{\mathcal{X}(t)\} = x(-f)$, this duality does not characterize the forward/inverse transform relations

$$\mathcal{X}(e^{j\omega}) = \sum_{n=-\infty}^{\infty} x[n]\, e^{-j\omega n} \qquad \text{and} \qquad x[n] = \frac{1}{2\pi}\int_{2\pi} \mathcal{X}(e^{j\omega})\, e^{j\omega n}\, d\omega$$

in the *discrete-time* domain (incidentally, such a duality correlation *exists* in the Fourier expansion of discrete-time signals, where we have proven that $x_k \rightleftarrows x[-n]/N$ [Sect. 10.2.2, page 789]). However, a duality correlation exists between the Fourier transform of discrete-time signals and the Fourier expansion of continuous-time signals. Indeed, if we compare the inverse transform of a discrete-time signal $x[n]$ with the defining equation of the Fourier expansion of a continuous-time signal $x(t)$, which have the form

[15] Due to the *periodicity* of the Fourier transform of discrete-time signals, *there is no perfect duality* between the time and frequency domains regarding convolution, as is the case in the continuous-time domain. Thus, the convolution in the time domain (which is a *nonperiodic operation*) is equivalent to the multiplication of two continuous periodic Fourier transforms; however, *the multiplication of two aperiodic sequences is equivalent to the periodic convolution of their Fourier transforms.*

$$x[n] = \frac{1}{2\pi}\int_{2\pi} \mathcal{X}(e^{j\omega})\, e^{j\omega n} d\omega \qquad \text{and} \qquad \alpha_k = \frac{1}{T}\int_T x(t)\, e^{-j\omega_0 nt} dt$$

as well as the pair of equations

$$\mathcal{X}(e^{j\omega}) = \sum_{n=-\infty}^{\infty} x[n]\, e^{-j\omega n} \qquad \text{and} \qquad x(t) = \sum_{n=-\infty}^{\infty} \alpha_n\, e^{j\omega_0 nt}$$

then we find that they are characterized by a high degree of similarity. In reality, we can interpret the defining equations of the functions $x[n]$ and $\mathcal{X}(e^{j\omega})$ *as a representation in a Fourier series of the function* $\mathcal{X}(e^{j\omega})$, *which is periodic with respect to the frequency* ω. In a more detailed description, taking into account that the $\mathcal{X}(e^{j\omega})$ transform is a periodic function of frequency with a period of 2π, it can be represented by a Fourier series having the form of a *weighted sum of harmonically correlated periodic exponential functions of frequency* ω*, all of which have a common period equal to* 2π.

- **Correlation**: If the Fourier transforms of the discrete sequences $x[n]$ and $y[n]$ are the functions $\mathcal{X}(e^{j\omega})$ and $\mathcal{Y}(e^{j\omega})$, then the Fourier transform of the correlation of the two signals defined as

$$r_{xy}[n] = \sum_{k=-\infty}^{\infty} x[k]\, y[k-n]$$

will be given by the relation $\mathcal{R}_{xy}(e^{j\omega}) = \mathcal{X}(e^{j\omega})\mathcal{Y}(e^{-j\omega})$.
Proof From the defining equation of the Fourier transform of the discrete-time signal $r_{xy}[n]$ we have

$$\mathcal{R}_{xy}(e^{j\omega}) = \sum_{n=-\infty}^{\infty} r_{xy}[n]\, e^{-j\omega n} = \sum_{n=-\infty}^{\infty}\left\{\sum_{k=-\infty}^{\infty} x[k]\, y[k-n]\right\} e^{-j\omega n} = \sum_{k=-\infty}^{\infty} x[k]\left\{\sum_{n=-\infty}^{\infty} y[k-n]\, e^{-j\omega n}\right\}$$

where for the derivation of the last expression we alternated the order of the two sums. Proceeding now to change the dummy index of the second sum from n to $m = k - n$ will be $n = k - m$ (the limits of the sum, being infinite, obviously do not change), and the preceding relation takes the form

$$\mathcal{R}_{xy}(e^{j\omega}) = \sum_{k=-\infty}^{\infty} x[k] \sum_{m=-\infty}^{\infty} y[m]\, e^{-j\omega(k-m)} = \sum_{k=-\infty}^{\infty} x[k] \sum_{m=-\infty}^{\infty} y[m]\, e^{-j\omega k} e^{j\omega m}$$

$$= \left\{\sum_{k=-\infty}^{\infty} x[k]\, e^{-j\omega k}\right\}\left\{\sum_{m=-\infty}^{\infty} x[k]\, e^{-j(-\omega)m}\right\} = \mathcal{X}(e^{j\omega})\, \mathcal{Y}(e^{-j\omega})$$

The function $\mathcal{R}_{xy}(e^{j\omega})$ is called the spectrum density of energy of the discrete sequences $x[n]$ and $y[n]$.

In the case of real signals, such that $x[n] = y[n]$, the preceding relation is written $\mathcal{R}_{xy}(e^{j\omega}) = \mathcal{X}(e^{j\omega})\mathcal{X}(e^{-j\omega})$, and because for real signals $\mathcal{X}(e^{-j\omega}) = \mathcal{X}^*(e^{j\omega})$, we will have $\mathcal{R}_{xy}(e^{j\omega}) = \mathcal{X}(e^{j\omega})\mathcal{X}^*(e^{j\omega}) = |\mathcal{X}(e^{j\omega})|^2 = S_{xx}(\omega)$. This result is very important because it tells us that *the energy spectral density of an energy signal is equal to the Fourier transform of the autocorrelation function.* This property also characterizes continuous-time signals (see Sect. 3.3.5 on page 235) and is known as the *Wiener–Khinchin theorem.*

- **Modulation**: If the Fourier transform of the discrete-time signal $x[n]$ is the function $\mathcal{X}(e^{j\omega})$, then it is proven that

$$\mathcal{F}\left\{x[n]\cos(\omega_0 n)\right\} = \frac{1}{2}\left[\mathcal{X}(e^{j(\omega+\omega_0)}) + \mathcal{X}(e^{j(\omega-\omega_0)})\right]$$

Proof Replacing the cosine function with Euler's formula $\cos(\omega_n) = (e^{j\omega_0 n} - e^{-j\omega_0 n})/2$ and using the property of linearity, we get

$$\mathcal{F}\left\{x[n]\cos(\omega_0 n)\right\} = \frac{1}{2}\mathcal{F}\left\{e^{j\omega_0 n}\right\} + \frac{1}{2}\mathcal{F}\left\{e^{-j\omega_0 n}\right\}$$

But from the property of frequency shift it follows that

$$\mathcal{F}\left\{e^{j\omega_0 n}\right\} = \mathcal{X}(e^{j(\omega-\omega_0)}) \qquad \text{and} \qquad \mathcal{F}\left\{e^{-j\omega_0 n}\right\} = \mathcal{X}(e^{j(\omega+\omega_0)})$$

and replacing the foregoing expressions in our final expression results in the desired result. In a completely analogous way and replacing the sine function with Euler's formula, it immediately follows that

$$\mathcal{F}\left\{x[n]\sin(\omega_0 n)\right\} = \frac{j}{2}\left[\mathcal{X}(e^{j(\omega+\omega_0)}) - \mathcal{X}(e^{j(\omega-\omega_0)})\right]$$

Using the properties of the DTFT, the calculation process is significantly simplified in some cases, as demonstrated in the following exercises.

Problem 10.3.10 Determine the Fourier transform of the signals $x_1[n] = \cos(\omega_0 n)$ and $x_2[n] = \sin(\omega_0 n)$.

Solution In Sect. 10.3.5 (page 815), we proved that

$$\mathcal{F}\{e^{j\omega_0 n}\} = \sum_{k=-\infty}^{\infty} 2\pi\delta(\omega - \omega_0 - 2k\pi)$$

while, using the same argument, we arrive at the expression

$$\mathcal{F}\{e^{-j\omega_0 n}\} = \sum_{k=-\infty}^{\infty} 2\pi\delta(\omega + \omega_0 - 2k\pi)$$

Therefore, by replacing the cosine function from Euler's formula $\cos(\omega_n) = (e^{j\omega_0 n} - e^{-j\omega_0 n})/2$ and using the linearity property of the transform, it follows that

$$\mathcal{F}\{\cos(\omega_0 n)\} = \frac{1}{2}\mathcal{F}\{e^{j\omega_0 n}\} + \frac{1}{2}\mathcal{F}\{e^{-j\omega_0 n}\} = \pi\sum_{k=-\infty}^{\infty}\Big[\delta(\omega - \omega_0 - 2k\pi) + \delta(\omega + \omega_0 - 2k\pi)\Big]$$

Applying the same process for the sine function, we arrive at the result

$$\mathcal{F}\{\sin(\omega_0 n)\} = \frac{1}{2j}\mathcal{F}\{e^{j\omega_0 n}\} - \frac{1}{2j}\mathcal{F}\{e^{-j\omega_0 n}\} = j\pi\sum_{k=-\infty}^{\infty}\Big[\delta(\omega + \omega_0 - 2k\pi) - \delta(\omega - \omega_0 - 2k\pi)\Big]$$

Problem 10.3.11 Define the Fourier transform of the discrete-time signal $x[n] = 1$.

Solution The discrete-time signal $x[n] = 1$ arises from the discrete exponential sequence $x[n] = e^{j\omega_0 n}$ for parameter value $\omega_0 = 0$. Therefore, its Fourier transform will result by making the preceding substitution in the function

$$\mathcal{F}\{e^{j\omega_0 n}\} = \sum_{k=-\infty}^{\infty} 2\pi\,\delta(\omega - \omega_0 - 2k\pi)$$

from which the desired result emerges in the form

$$\mathcal{F}\{x[n]\} = \mathcal{F}\{1\} = \mathcal{F}\{e^{j\omega_0 n}\}\Big|_{\omega_0=0} = \sum_{k=-\infty}^{\infty} 2\pi\,\delta(\omega - 2k\pi)$$

Problem 10.3.12 Define the Fourier transform of the step function $u[n]$ (Proakis and Manolakis, Example 4.2.5α).

Solution We know that the even and odd components of the step function $u[n]$ are given by the relations

$$u_e[n] = \frac{1}{2} + \frac{1}{2}\delta[n] \qquad \text{and} \qquad u_o[n] = \frac{1}{2}\text{sgn}[n]$$

where $\delta[n]$ and $\text{sgn}[n]$ are the discrete impulse function and the sign function, respectively. Therefore, applying the property of addition of the Fourier transform we get

$$\mathcal{U}(e^{j\omega}) = \mathcal{F}\{u[n]\} = \mathcal{F}\{u_e[n]\} + \mathcal{F}\{u_o[n]\}$$

Using the results of Exercises 10.3.7 (page 814) and 10.3.11 (page 826), it follows that

$$\mathcal{F}\{u_e[n]\} = \mathcal{F}\left\{\frac{1}{2} + \frac{1}{2}\delta[n]\right\} = \frac{1}{2}\mathcal{F}\{\delta[n]\} + \frac{1}{2}\mathcal{F}\{1\} = \frac{1}{2} + \sum_{k=-\infty}^{\infty} \pi\,\delta(\omega - 2k\pi)$$

while from the definition of the sign function (see Exercise 6.7.5 on page 531) we have

$$\mathcal{F}\{u_o[n]\} = \frac{1}{2}\left(\sum_{n=1}^{\infty}(+1)\,e^{-j\omega n} + \sum_{n=-\infty}^{-1}(-1)\,e^{-j\omega n}\right)$$

If we proceed to change the dummy index from n to $-n$ in the second sum, its limits coincide with those of the first sum, and referring to mathematical typology, it follows that

$$\mathcal{F}\{u_o[n]\} = \frac{1}{2}\sum_{n=1}^{\infty}\left(e^{-j\omega n} - e^{j\omega n}\right) = \frac{1}{2}\frac{e^{-j\omega} - e^{j\omega}}{(1 - e^{-j\omega})(1 - e^{j\omega})}$$

Replacing the preceding expressions in the defining equation of the function $\mathcal{U}(\omega)$, we end up after simple mathematical operations with the desired expression

$$\mathcal{U}(e^{j\omega}) = \frac{1}{1 - e^{-j\omega}} + \pi\sum_{k=-\infty}^{\infty}\delta(\omega - 2k\pi)$$

Table 10.4 contains some pairs of elementary transforms of the form $\{x[n], \mathcal{X}(e^{j\omega})\}$, which are encountered quite often in practice. Several of the foregoing correspondences were already proven in previous exercises, and readers are asked to determine as many more than the remaining transform can, in order to familiarize themselves with the process of calculating this function.

Table 10.4 Examples of elementary discrete-time Fourier transforms

Discrete-time sequence $x[n]$	Fourier transform $X\left(e^{j\omega}\right)$
$\delta[n]$	1
$\delta\left[n-n_0\right]$	$e^{-j\omega n_0}$
$\sum_{k=-\infty}^{\infty} \delta[n-kN]$	$\frac{2\pi}{N} \sum_{k=-\infty}^{\infty} \delta\left(\omega-\frac{2\pi}{N}\right)$
$u[n]$	$\frac{1}{1-e^{-j\omega}}+\pi \sum_{k=-\infty}^{\infty} \delta(\omega-2k\pi)$
$e^{j\omega_0 n}$	$\sum_{k=-\infty}^{\infty} 2\pi\delta\left(\omega-\omega_0-2k\pi\right)$
$\cos\left(\omega_0 n\right)$	$\pi \sum_{k=-\infty}^{\infty}\left[\delta\left(\omega-\omega_0-2k\pi\right)+\delta\left(\omega+\omega_0-2k\pi\right)\right]$
$\sin\left(\omega_0 n\right)$	$j\pi \sum_{k=-\infty}^{\infty}\left[\delta\left(\omega+\omega_0-2k\pi\right)-\delta\left(\omega-\omega_0-2k\pi\right)\right]$
$\alpha^n u[n], \|\alpha\|<1$	$\frac{1}{1-\alpha e^{-j\omega}}$
$\sum_{k=<N>} \alpha_k \exp\left(jk\frac{2\pi}{N}n\right)$	$2\pi \sum_{k=-\infty}^{\infty} \alpha_k\delta\left(\omega-\frac{2\pi k}{N}\right)$
$\begin{cases}1 \|n\| \leq N \\ 0 \|n\|>N\end{cases}$	$\frac{\sin[\omega(N+1/2)]}{\sin(\omega/2)}$
$\frac{W}{\pi} sinc\left(\frac{Wn}{\pi}\right)$	$\begin{cases}10 \leq \|\omega\| \leq W \\ W \leq \|\omega\| \leq \pi\end{cases}$
$\frac{(n+r-1)!}{n!(r-1)!}\alpha^n u[n], \|\alpha\|<1$	$\frac{1}{\left(1-\alpha^{-j\omega}\right)^r}$

10.3.7 The Inverse Fourier Transform

In full analogy with its continuous equivalent, the *inverse Fourier transform* in the discrete-time domain can be calculated either by directly using the synthesis equation

$$x[n]=\frac{1}{2\pi} \int_{2\pi} \mathcal{X}\left(e^{j\omega}\right) e^{j\omega n} d\omega \tag{10.50}$$

or by separating the function to be inverted into smaller representations, which are included in tables of elementary transforms, and calculating the final result using the linearity property, as well as the appropriate properties remaining in each case. The next two exercises demonstrate how to apply these methods, while for more examples the reader can refer to Sect. 3.4 (page 259) and the following exercises, which deal with the same problem for the continuous case.

Problem 10.3.13 Determine the inverse Fourier transform of the function

$$\mathcal{X}\left(e^{j\omega}\right)=\begin{cases}1 \ |\omega| \leq W \\ 0 \ W<|\omega| \leq \pi .\end{cases}$$

Solution From the synthesis equation of the Fourier transform it immediately follows that

$$x[n]=\frac{1}{2\pi} \int_{-\pi}^{+\pi} \mathcal{X}\left(e^{j\omega}\right) e^{j\omega n}=\frac{1}{2\pi} \int_{-W}^{+W} e^{j\omega n} d\omega=\frac{1}{2\pi}\left[\frac{e^{jWn}-e^{-jWn}}{in}\right]=\frac{1}{\pi n} \frac{e^{jWn}-e^{-jWn}}{2j}=\frac{\sin(Wn)}{\pi n}$$

Problem 10.3.14 Determine the inverse Fourier transform of the function $\mathcal{X}(e^{j\omega}) = \cos^2\omega$ (Hayes, Solved Exercise 2.2.6).

Solution Replacing the cosine function with Euler's formula, we get

$$\mathcal{X}(e^{j\omega}) = \cos^2\omega = \left(\frac{e^{j\omega}+e^{-j\omega}}{2}\right)^2 = \frac{1}{4}e^{2j\omega} + \frac{1}{4}e^{-2j\omega} + \frac{1}{2}$$

and applying the linearity property that also characterizes the inverse transform, we find

$$x[n] = \mathcal{F}^{-1}\{\mathcal{X}(e^{j\omega})\} = \frac{1}{4}\mathcal{F}^{-1}\{e^{2j\omega}\} + \frac{1}{4}\mathcal{F}^{-1}\{e^{-2j\omega}\} + \mathcal{F}^{-1}\left\{\frac{1}{2}\right\}$$

But from the time-shift property for the values $n_0 = \pm 2$ we will have

$$\mathcal{F}^{-1}\{e^{2j\omega}\} = e^{2j\omega}\mathcal{F}\{1\} = \delta[n]\Big|_{n\to n+2} = \delta[n+2] \quad \text{and} \quad \mathcal{F}^{-1}\{e^{-2j\omega}\} = e^{-2j\omega}\mathcal{F}\{1\} = \delta[n]\Big|_{n\to n-2} = \delta[n-2]$$

while it will still hold that

$$\mathcal{F}^{-1}\left\{\frac{1}{2}\right\} = \frac{1}{2}\mathcal{F}^{-1}\{1\} = \frac{1}{2}\delta[n]$$

Replacing the preceding expressions in the defining equation of the function $x[n]$, it turns out that

$$x[n] = \mathcal{F}^{-1}\{\cos^2\omega\} = \frac{1}{2}\delta[n] + \frac{1}{4}\delta[n+2] + \frac{1}{4}\delta[n-2]$$

10.4 Frequency Analysis of Discrete-Time Systems

In the following pages, we will describe techniques for studying discrete-time systems in the frequency domain with the help of the DTFT presented in previous sections. In full analogy with the continuous case, we will consider the special category of *discrete-time LTI systems* that accept a discrete-time signal $x[n]$ as their input and produce another discrete-time signal $y[n]$ as their output, with these two signals being related through a *recursive difference equation* of order N of the form

$$\sum_{k=0}^{N}\alpha_k y[n-k] = \sum_{k=0}^{M}\beta_k x[n-k] \tag{10.51}$$

Following the approach to studying the corresponding continuous-time systems, we can determine the *frequency response* of a system in two different ways, one of which is based on the *convolution theorem* and the other on the property of elementary exponential complex signals to constitute *eigenfunctions* of discrete LTI systems. Starting with the first method, if we have the input and output of the system at each time n, we know that they are related by the operation of convolution, so we can write that $y[n] = h[n] * x[n]$. In this case, applying the property of the convolution of the Fourier transform, we will get $\mathcal{Y}(e^{j\omega}) = \mathcal{H}(e^{j\omega})\mathcal{X}(e^{j\omega})$, from which we can calculate the *frequency response function* as a ratio of Fourier transforms of the form

$$\mathcal{H}(e^{j\omega}) = \frac{\mathcal{Y}(e^{j\omega})}{\mathcal{X}(e^{j\omega})} \tag{10.52}$$

On the other hand, if we do not know the pair of inputs/outputs ($x[n]$, $y[n]$) but only the *difference equation* of the system, we can calculate the preceding function by applying the Fourier transform to each of its sides and using the properties of *linearity* and *time shift* of that transform. We will therefore have

$$\mathcal{F}\left\{\sum_{k=0}^{N} \alpha_k y[n-k]\right\} = \mathcal{F}\left\{\sum_{k=0}^{M} \beta_k x[n-k]\right\} \tag{10.53}$$

or, equivalently,

$$\sum_{k=0}^{N} \alpha_k \mathcal{F}\{y[n-k]\} = \sum_{k=0}^{M} \beta_k \mathcal{F}\{x[n-k]\} \tag{10.54}$$

and since, according to the *time shift* property, it holds that

$$\mathcal{F}\{y[n-k]\} = e^{-j\omega k}\mathcal{Y}(e^{j\omega}) \qquad \text{and} \qquad \mathcal{F}\{x[n-k]\} = e^{-j\omega k}\mathcal{X}(e^{j\omega})$$

the preceding relationship will take the form

$$\left(\sum_{k=0}^{N} \alpha_k e^{-j\omega k}\right)\mathcal{Y}(e^{j\omega}) = \left(\sum_{k=0}^{M} \beta_k e^{-j\omega k}\right)\mathcal{X}(e^{j\omega}) \tag{10.55}$$

and therefore, the *frequency response function* will be calculated as the ratio

$$\mathcal{H}(e^{j\omega}) = \frac{\mathcal{Y}(e^{j\omega})}{\mathcal{X}(e^{j\omega})} = \left\{\sum_{k=0}^{M} \beta_k e^{-j\omega k}\right\} \Big/ \left\{\sum_{k=0}^{N} \alpha_k e^{-j\omega k}\right\} \tag{10.56}$$

We are thus once again led to a rational frequency response function, for which all properties discussed earlier in the case of continuous-time systems apply exactly–regarding the poles and zeros, the corresponding plot, and the relationship between the degrees of the numerator and denominator polynomials in terms of system realization.

The second method for determining the system's frequency response is based on the property of complex exponential signals to serve as eigenfunctions of a discrete-time LTI system, with the corresponding eigenvalue being the frequency response of the system evaluated at ω_0 — the frequency of the input signal. Indeed, considering an input signal of the form $x[n] = e^{j\omega_0 n}$, the output of the system, according to the convolution equation, will have the form

$$y[n] = x[n] * h[n] = \sum_{k=-\infty}^{\infty} h[k]\, x[n-k] = \sum_{k=-\infty}^{\infty} h[k]\, e^{j\omega_0(n-k)} = e^{j\omega_0 n} \sum_{k=-\infty}^{\infty} h[k]\, e^{-j\omega_0 k} = \mathcal{H}(\omega_0)x[n]$$

with this $\mathcal{H}(e^{j\omega})$ being defined as

$$\mathcal{H}(e^{j\omega}) = \sum_{k=-\infty}^{\infty} h[k]\, e^{-j\omega k} \tag{10.57}$$

But this relationship is recognized as the *DTFT* of the system's impulse response, and therefore the frequency response function is defined, as in the continuous case, as the aforementioned transform.

Taking into account that the function $\mathcal{H}(e^{j\omega})$ in the general case is of a complex nature, it can be formulated in its exponential form[16]

$$\mathcal{H}(e^{j\omega}) = |\mathcal{H}(e^{j\omega})|e^{j\measuredangle\mathcal{H}(e^{j\omega})}$$

with the functions $|\mathcal{H}(e^{j\omega})|$ and $\measuredangle\mathcal{H}(e^{j\omega})$ known as the *amplitude response* and the *phase response* of the discrete-time system under consideration. In this case, formulating the Fourier transforms of the input and output signals as

$$\mathcal{Y}(e^{j\omega}) = |\mathcal{Y}(e^{j\omega})|e^{j\measuredangle\mathcal{Y}(e^{j\omega})} \quad \text{and} \quad \mathcal{X}(e^{j\omega}) = |\mathcal{X}(e^{j\omega})|e^{j\measuredangle\mathcal{X}(e^{j\omega})}$$

we observe that

$$|\mathcal{Y}(e^{j\omega})| = |\mathcal{H}(e^{j\omega})||\mathcal{X}(e^{j\omega})| \quad \text{and} \quad \measuredangle\mathcal{Y}(e^{j\omega}) = \measuredangle\mathcal{H}(e^{j\omega}) + \measuredangle\mathcal{X}(e^{j\omega})$$

which are the same equations we constructed for continuous-time systems (Eqs. 3.53 and 3.54). We are led, therefore, to the conclusion that *the effect of a discrete LTI system on an aperiodic input signal, whose frequencies belong to a specific area, is the amplification of the amplitude of certain frequency components and the attenuation of the amplitude of others, which is accompanied by the appropriate phase shift in each case*. We also observe that *the output of a discrete LTI system cannot contain frequencies that are not included in the input signal*; if we want the existence of such frequencies, we must use a *nonlinear system* or a *linear system* that, however, *varies* in relation to time.

The exemplary solved exercises that follow demonstrate how to use the preceding functions in the four different approaches presented in Sect. 3.5 (page 270) that are used as is in the domain of discrete-time.

Problem 10.4.1 To determine the response of a discrete LTI system to elementary trigonometric signals $x_1[n] = A\cos(\omega n)$ and $x_2[n] = A\sin(\omega n)$.

---

[16] Alternatively, using *Euler's formula* we can express the function as

$$\mathcal{H}(e^{j\omega}) = \sum_{n=-\infty}^{\infty} h[n]e^{-j\omega n} = \sum_{n=-\infty}^{\infty} h[n]\cos(\omega n) - j\sum_{n=-\infty}^{\infty} h[n]\sin(\omega n) = \mathcal{H}_R(\omega) + i\mathcal{H}_I(\omega)$$

with the *real* and *imaginary* parts being defined as

$$\mathcal{H}_R(\omega) = \sum_{n=-\infty}^{\infty} h[n]\cos(\omega n) \quad \text{and} \quad \mathcal{H}_I(\omega) = -\sum_{n=-\infty}^{\infty} h[n]\sin(\omega n)$$

and related to the *magnitude* and *phase* of the frequency response function by the known relationships

$$|\mathcal{H}(e^{j\omega})| = \sqrt{\mathcal{H}_R^2(\omega) + \mathcal{H}_I^2(\omega)} \quad \text{and} \quad \measuredangle\mathcal{H}(e^{j\omega}) = \tan^{-1}\left\{\frac{\mathcal{H}_I(\omega)}{\mathcal{H}_R(\omega)}\right\}$$

Considering that the impulse response $h[n]$ is the *inverse Fourier transform* of the function $\mathcal{H}(e^{j\omega})$ so we can write that

$$h[n] = \frac{1}{2\pi}\int_{-\pi}^{+\pi} \mathcal{H}(e^{j\omega})\, e^{j\omega n} d\omega$$

and based on the properties of the Fourier transform that we proved in a previous section, we can state that *if a discrete LTI system has a real impulse response, then it is characterized by the symmetry properties $\mathcal{H}_R(e^{-j\omega}) = \mathcal{H}_R(e^{j\omega})$ and $\mathcal{H}_I(e^{-j\omega}) = -\mathcal{H}_I(e^{j\omega})$, that is, the real and imaginary parts of the frequency response function are even and odd functions of the frequency, respectively.*

Solution To solve the problem, we will use *the property of eigenvalues* along with the *principle of parallelism*. According to the first property, the responses of a LTI system to the elementary exponential signals $\kappa_1[n] = Ae^{j\omega n}$ and $\kappa_2[n] = Ae^{-j\omega n}$ will have the form

$$\lambda_1[n] = \mathcal{H}(e^{j\omega})\kappa_1[n] = A|\mathcal{H}(e^{j\omega})|e^{j\omega n}e^{j\measuredangle\mathcal{H}(e^{j\omega})} \qquad \text{and} \qquad \lambda_2[n] = \mathcal{H}(e^{j\omega})\kappa_2[n] = A|\mathcal{H}(e^{j\omega})|e^{-j\omega n}e^{j\measuredangle\mathcal{H}(e^{j\omega})}$$

where we replaced the frequency response function from its exponential form and made use of the symmetry properties $|\mathcal{H}(e^{-j\omega})| = |\mathcal{H}(e^{j\omega})|$ and $\measuredangle\mathcal{H}(e^{-j\omega}) = -\measuredangle\mathcal{H}(e^{j\omega})$. Using now the *principle of parallelism* we can claim that the response of the system to the linear combination

$$x_1[n] = \cos(\omega n) = \frac{1}{2}e^{j\omega n} + \frac{1}{2}e^{-j\omega n} = \frac{1}{2}\kappa_1[n] + \frac{1}{2}\kappa_2[n]$$

will be the signal

$$y_1[n] = \frac{1}{2}\lambda_1[n] + \frac{1}{2}\lambda_2[n] = \frac{1}{2}A|\mathcal{H}(e^{j\omega})|e^{j\omega n}e^{j\measuredangle\mathcal{H}(e^{j\omega})} + \frac{1}{2}A|\mathcal{H}(e^{j\omega})|e^{-j\omega n}e^{j\measuredangle\mathcal{H}(e^{j\omega})} =$$

$$= A|\mathcal{H}(e^{j\omega})|\frac{e^{j[\omega n+\measuredangle\mathcal{H}(e^{j\omega})]} + e^{-j[\omega n+\measuredangle\mathcal{H}(e^{j\omega})]}}{2} = A|\mathcal{H}(e^{j\omega})|\cos(\omega n+\measuredangle\mathcal{H}(e^{j\omega}))$$

while, correspondingly, the system's response to the linear combination

$$x_2[n] = \sin(\omega n) = \frac{1}{2j}e^{j\omega n} - \frac{1}{2j}e^{-j\omega n} = \frac{1}{2}\kappa_1[n] - \frac{1}{2}\kappa_2[n]$$

will be the signal

$$y_2[n] = \frac{1}{2j}\lambda_1[n] - \frac{1}{2j}\lambda_2[n] = \frac{1}{2j}A|\mathcal{H}(e^{j\omega})|e^{j\omega n}e^{j\measuredangle\mathcal{H}(e^{j\omega})} - \frac{1}{2j}A|\mathcal{H}(e^{j\omega})|e^{-j\omega n}e^{j\measuredangle\mathcal{H}(e^{j\omega})} =$$

$$= A|\mathcal{H}(e^{j\omega})|\frac{e^{j[\omega n+\measuredangle\mathcal{H}(e^{j\omega})]} - e^{-j[\omega n+\measuredangle\mathcal{H}(e^{j\omega})]}}{2j} = A|\mathcal{H}(e^{j\omega})|\sin(\omega n+\measuredangle\mathcal{H}(e^{j\omega}))$$

which is also the desired result. We are therefore led to the conclusion that the passage of a simple sinusoidal or cosinusoidal signal through a discrete-time LTI system results in the creation of a corresponding sinusoidal or cosinusoidal signal, whose amplitude is amplified if $|\mathcal{H}(e^{j\omega})| > 1|\mathcal{H}(e^{j\omega})| > 1$ or attenuated if $|\mathcal{H}(e^{j\omega})| < 1|\mathcal{H}(e^{j\omega})| < 1$, while its phase is shifted by the amount $\measuredangle\mathcal{H}(e^{j\omega})\measuredangle\mathcal{H}(e^{j\omega})$. On the other hand, if the input signal is more complex and consists of multiple sinusoidal and cosinusoidal components, the aforementioned effect is exerted on each individual component separately, and the overall system output results from the sum of all these modified components, as dictated by the superposition principle.[17]

[17] Following the argumentation applied in a previous chapter, we will characterize the response of the system to the elementary trigonometric signals of the sine and cosine as a *steady-state response*, since, being periodic, it lasts for an infinite amount of time. Recall (see Footnote 7, page 7, in the first chapter, which also applies to discrete-time signals) that periodic signals that obviously have infinite duration always start at time $n = -\infty$, which is considered the time of their application to the discrete LTI system under consideration. On the other hand, if elementary exponential and trigonometric signals (which are equivalent in terms of how they are treated by the system) are applied to the system at a specific time (for example, at time $n = 0$), then the output of the system consists, as is well known, *of the sum of the steady-state response* $y_{ss}[n]$ *and the transient response* $y_{tr}[n]$, of which the first lasts indefinitely and is the one that ultimately remains, while the second weakens over time and after a certain period ceases to exist. Considering, for example, the simple discrete-time system

$$y[n] = \alpha y[n-1] + x[n]$$

the response to the input signal $x[n]$ will, as we have proven (Exercise 7.4.9, page 586), have the form

$$y[n] = \alpha^{n+1}y[-1] + \sum_{k=0}^{n}\alpha^k x[n-k]$$

where $y[-1]$ is the initial condition of the problem. If in the preceding system we transmit the signal $x[n] = Ae^{j\omega n}$ from the foregoing relation, it follows that

Problem 10.4.2 Determine the frequency response of a discrete recursive LTI system described by the difference equation

$$y[n] - \frac{1}{2}y[n-1] = x[n] + 2x[n-1] + x[n-2].$$

Solution Taking the *Fourier transform* of the two sides of the foregoing equation, we will get

$$\mathcal{F}\{y[n]\} - \frac{1}{2}\mathcal{F}\{y[n-1]\} = \mathcal{F}\{x[n]\} + 2\mathcal{F}\{x[n-1]\} + \mathcal{F}\{x[n-2]\}$$

However, $\mathcal{F}\{y[n]\} = \mathcal{Y}(e^{j\omega})$ and $\mathcal{F}\{x[n]\} = \mathcal{X}(e^{j\omega})$, while from the *property of time shift* of the transformation it follows that

$$\mathcal{F}\{y[n-1]\} = e^{-j\omega}y[n], \quad \mathcal{F}\{x[n-1]\} = e^{-j\omega}x[n], \quad \mathcal{F}\{x[n-2]\} = e^{-2j\omega}x[n]$$

Substituting these expressions into our original equation, it will take the form

$$\mathcal{Y}(e^{j\omega}) - \frac{1}{2}e^{-j\omega}\mathcal{Y}(e^{j\omega}) = \mathcal{X}(e^{j\omega}) + 2e^{-j\omega}\mathcal{X}(e^{j\omega}) + e^{-2j\omega}\mathcal{X}(e^{j\omega})$$

from which we finally get

$$\mathcal{H}(e^{j\omega}) = \frac{\mathcal{Y}(e^{j\omega})}{\mathcal{X}(e^{j\omega})} = \frac{1 + 2e^{-j\omega} + e^{-2j\omega}}{1 - (1/2)\,e^{-j\omega}}$$

which is also the result (for the continuous analog of this method see Exercise 3.5.6 on page 277).

Problem 10.4.3 Determine the frequency and impulse response of a discrete recursive LTI system described by the equation differences

$$y[n] - \frac{1}{6}y[n-1] - \frac{1}{6}y[n-2] = x[n].$$

Solution Taking the *Fourier transform* of the two terms of the difference equation and using the *property of time shift*, the foregoing relation is written as

$$\mathcal{Y}(e^{j\omega}) - \frac{1}{6}e^{-j\omega}\mathcal{Y}(e^{j\omega}) - \frac{1}{6}e^{-2j\omega}\mathcal{Y}(e^{j\omega}) = \mathcal{X}(e^{j\omega})$$

and therefore the frequency response of the system will have the form

---

$$y[n] = \alpha^{n+1}y[-1] + A\sum_{k=0}^{n}\alpha^k e^{j\omega(n-k)} = \alpha^{n+1}y[-1] - \frac{A\alpha^{n+1}e^{-j\omega(n+1)}}{1-\alpha e^{-j\omega}} + \frac{Ae^{j\omega n}}{1-\alpha e^{-j\omega}}$$

(for values $n \geq 0$). Taking into account that for values $|\alpha| < 1$ (for which the system is BIBO stable) the first two terms containing the term α^{n+1} tend to zero in the limit $n \to \infty$, it is obvious that they constitute the *transient response of the system*, while the third term that finally remains is the *steady-state response*, which is calculated as

$$y_{ss}[n] = \lim_{n\to\infty} y[n] = \frac{A}{1-\alpha e^{j\omega n}}e^{j\omega n} = A\mathcal{H}(e^{j\omega})\,e^{j\omega n}$$

This behavior characterizes *all BIBO-stable discrete LTI systems* that receive as their input an elementary *exponential* or *trigonometric* signal at time $n = 0$ or some later time.

$$\mathcal{H}(e^{j\omega}) = \frac{\mathcal{Y}(e^{j\omega})}{\mathcal{X}(e^{j\omega})} = \frac{1}{1-(1/6)\,e^{-j\omega}-(1/6)\,e^{-2j\omega}}$$

To determine the *impulse response* of the system, we need to determine the *inverse Fourier transform* of the preceding relation. Expanding the function $\mathcal{H}(e^{j\omega})$ into a sum of partial fractions, it turns out that

$$\mathcal{H}(e^{j\omega}) = \frac{1}{1-(1/6)\,e^{-j\omega}-(1/6)\,e^{-2j\omega}} = \frac{1}{1-(1/2)\,e^{-j\omega}} \times \frac{1}{1+(1/3)\,e^{-2j\omega}}$$
$$= \frac{3}{5}\left\{\frac{1}{1-(1/2)\,e^{-j\omega}}\right\} + \frac{2}{5}\left\{\frac{1}{1+(1/3)\,e^{-j\omega}}\right\}$$

Taking the inverse Fourier transform of the preceding relation and referring to tables of elementary transforms, the *impulse response* of the considered LTI system is calculated as

$$h[n] = \frac{3}{5}\mathcal{F}^{-1}\left\{\frac{1}{1-(1/2)\,e^{-j\omega}}\right\} + \frac{2}{5}\mathcal{F}^{-1}\left\{\frac{1}{1+(1/3)\,e^{-j\omega}}\right\} = \frac{2}{5}\left(\frac{1}{2}\right)^n + \frac{2}{5}\left(-\frac{1}{3}\right)^n u[n]$$

Problem 10.4.4 Determine the difference equation of a discrete recursive LTI system whose frequency response is given by the relation

$$\mathcal{H}(e^{j\omega}) = \frac{1-(1/2)\,e^{-j\omega}+e^{-3j\omega}}{1+(1/2)\,e^{-j\omega}+(3/4)\,e^{-2j\omega}}.$$

(Hayes, Solved Exercise 2.41)

Solution In this case, the methodology we follow is *the reverse of the previous one*. Formulating the transfer function of the system as

$$\mathcal{H}(e^{j\omega}) = \frac{\mathcal{Y}(e^{j\omega})}{\mathcal{X}(e^{j\omega})} = \frac{1-(1/2)\,e^{-j\omega}+e^{-3j\omega}}{1+(1/2)\,e^{-j\omega}+(3/4)\,e^{-2j\omega}}$$

it immediately follows that

$$\mathcal{Y}(e^{j\omega}) + \frac{1}{2}e^{-j\omega}\mathcal{Y}(e^{j\omega}) + \frac{3}{4}e^{-2j\omega}\mathcal{Y}(e^{j\omega}) = \mathcal{X}(e^{j\omega}) - \frac{1}{2}e^{-j\omega}\mathcal{X}(e^{j\omega}) + e^{-3j\omega}\mathcal{X}(e^{j\omega})$$

Taking the *inverse Fourier transform* of the two terms of the preceding relation, it is written as

$$\mathcal{F}^{-1}\{\mathcal{Y}(e^{j\omega})\} + \frac{1}{2}\mathcal{F}^{-1}\{e^{-j\omega}\mathcal{Y}(e^{j\omega})\} + \frac{3}{4}\mathcal{F}^{-1}\{e^{-2j\omega}\mathcal{Y}(e^{j\omega})\} =$$
$$\mathcal{F}^{-1}\{\mathcal{X}(e^{j\omega})\} - \frac{1}{2}\mathcal{F}^{-1}\{e^{-j\omega}\mathcal{X}(e^{j\omega})\} + \mathcal{F}^{-1}\{e^{-3j\omega}\mathcal{X}(e^{j\omega})\}$$

However, we have that $\mathcal{F}^{-1}\{\mathcal{Y}(e^{j\omega})\} = y[n]$ and $\mathcal{F}^{-1}\{\mathcal{X}(e^{j\omega})\} = x[n]$, while from the property of the time shift of the transform we will get

$$\mathcal{F}^{-1}\{e^{-j\omega}\mathcal{Y}(e^{j\omega})\} = y[n-1], \quad \mathcal{F}^{-1}\{e^{-2j\omega}\mathcal{Y}(e^{j\omega})\} = y[n-2],$$
$$\mathcal{F}^{-1}\{e^{-j\omega}\mathcal{X}(e^{j\omega})\} = x[n-1], \quad \mathcal{F}^{-1}\{e^{-3j\omega}\mathcal{Y}(e^{j\omega})\} = x[n-3]$$

Substituting these expressions into the last equation, it is written as

$$y[n]+\frac{1}{2}y[n-1]+\frac{3}{4}y[n-2]=x[n]-\frac{1}{2}x[n-1]+x[n-3]$$

which is the difference equation of the system (for the continuous analog of the method see Exercise 3.5.5 on page 276).

Problem 10.4.5 Using the method of the Fourier transform, determine the output of the system $y[n]$ with impulse response $h[n]=5(-1/2)^n u[n]$ when the discrete-time signal $x[n]=(1/3)^n u[n]$ is transmitted to its input.

Solution To solve the problem, we will determine the *Fourier transform* of the system's output, which, according to the *convolution theorem*, will have the form $\mathcal{Y}(e^{j\omega})=\mathcal{H}(e^{j\omega})\mathcal{X}(e^{j\omega})$ and then the desired output will be the inverse transform $y[n]=\mathcal{F}^{-1}\{\mathcal{Y}(e^{j\omega})\}$. Initially, let us calculate the Fourier transforms of the signals $h[n]$ and $x[n]$, which have the form

$$\mathcal{H}(e^{j\omega})=\sum_{n=-\infty}^{\infty}5\left(\frac{1}{2}\right)^n u[n]\,e^{-j\omega n}=5\sum_{n=0}^{\infty}\left(-\frac{1}{2}e^{-j\omega}\right)^n=\frac{5}{1+(1/2)\,e^{-j\omega}}$$

$$\mathcal{X}(e^{j\omega})=\sum_{n=-\infty}^{\infty}\left(\frac{1}{3}\right)^n u[n]\,e^{-j\omega n}=\sum_{n=0}^{\infty}\left(\frac{1}{3}e^{-j\omega}\right)^n=\frac{1}{1-(1/3)\,e^{-j\omega}}$$

Therefore, the Fourier transform of the output will be given, according to the *convolution theorem*, by the relation

$$\mathcal{Y}(e^{j\omega})=\mathcal{H}(e^{j\omega})\mathcal{X}(e^{j\omega})=\frac{5}{[1+(1/2)\,e^{-j\omega}][1-(1/3)\,e^{-j\omega}]}=\frac{3}{1+(1/2)\,e^{-j\omega}}+\frac{2}{1-(1/3)\,e^{-j\omega}}$$

where for the derivation of the last expression we developed the function $\mathcal{Y}(e^{j\omega})$ in *partial fraction series*. Taking the inverse transform of the last relation, we end up with the expression

$$y[n]=3\left(-\frac{1}{2}\right)^n u[n]+2\left(\frac{1}{3}\right)^n u[n]$$

—as emerges by referring to tables of elementary transforms—which is also the desired result (for the continuous analog of this method see Exercise 3.5.3 on page 273).

Problem 10.4.6 Using the method of the Fourier transform, determine the output of the system $y[n]$ with an impulse response of the form

$$h[n]=\left(\frac{1}{2}\right)^n\cos\left(\frac{\pi n}{2}\right)u[n]$$

when the discrete-time signal $x[n]=\cos(\pi n/2)$ is transmitted to its input (Oppenheim, Willsky, Nawab, Exercise for solution 5.29β(ii)).

Solution In this case, instead of following the methodology of the previous exercise, we will rely on the fact that *complex exponential functions constitute eigenvalues of a discrete LTI system*. Initially, we will calculate the frequency response of the system. Replacing the cosine function with Euler's formula, the impulse response of the system takes the form

$$h[n]=\frac{1}{2}\left(\frac{1}{2}\right)^n(e^{i\pi n/2}+e^{-i\pi n/2})u[n]$$

and the *frequency response function* is calculated as

$$\mathcal{H}(e^{j\omega}) = \sum_{n=-\infty}^{\infty} h[n]e^{-j\omega n} = \frac{1}{2}\sum_{n=0}^{\infty}\left(\frac{1}{2}e^{i\pi/2}e^{-j\omega}\right)^n + \frac{1}{2}\sum_{n=0}^{\infty}\left(\frac{1}{2}e^{-i\pi/2}e^{-j\omega}\right)^n$$
$$= \frac{1}{2}\left\{\frac{1}{1-(1/2)e^{i\pi/2}e^{-j\omega}}\right\} + \frac{1}{2}\left\{\frac{1}{1-(1/2)e^{-i\pi/2}e^{-j\omega}}\right\}$$

In the next step of the process, using Euler's formula for the cosine function, we can present the input signal $x[n]$ in the form

$$x[n] = \cos\left(\frac{\pi n}{2}\right) = \frac{1}{2}e^{i\pi n/2} + \frac{1}{2}e^{-i\pi n/2} = \frac{1}{2}x_1[n] + \frac{1}{2}x_2[n]$$

that is, as a linear combination of complex exponential signals

$$x_1[n] = e^{i\omega_1 n} = e^{i\pi n/2} \qquad \text{and} \qquad x_2[n] = e^{i\omega_2 n} = e^{-i\pi n/2}$$

with frequency values $\omega_1 = \pi/2$ and $\omega_2 = -\pi/2$. In this case, according to the theory of *eigenfunctions* and *eigenvalues* that we briefly presented in Sect. 10.4 (page 829), the responses of the system to the elementary inputs $x_1[n]$ and $x_2[n]$ will have the form

$$y_1[n] = \mathcal{H}(\omega_1)e^{i\omega_1 n} = \frac{4}{3}e^{i\pi n/2} \qquad \text{and} \qquad y_2[n] = \mathcal{H}(\omega_2)e^{i\omega_2 n} = \frac{4}{3}e^{-i\pi n/2}$$

—since it is proven that $\mathcal{H}(\omega_1) = \mathcal{H}(\omega_2) = 4/3$—and according to the principle of superposition that characterizes the considered linear system, its output is calculated as

$$y[n] = \frac{1}{2}y_1[n] + \frac{1}{2}y_2[n] = \frac{1}{2}\left(\frac{4}{3}e^{i\pi n/2} + \frac{4}{3}e^{-i\pi n/2}\right) = \frac{4}{3}\frac{e^{i\pi n/2} + e^{-i\pi n/2}}{2} = \frac{4}{3}\cos\left(\frac{\pi n}{2}\right)$$

which is also the desired result.

10.4.1 Frequency Selection Filters

According our analysis in previous sections, in full analogy with continuous-time systems, *the effect of a discrete LTI system on an elementary discrete input signal is the multiplication of the signal's amplitude by the magnitude of the frequency response function and the phase shift of the signal by the phase of the frequency response function, with that function being calculated at the frequency position* ω_0 *of the input signal*. On the other hand, if the input signal $x[n]$ contains many frequencies, then the foregoing effect—which obviously introduces some types of *distortion*—is applied *to each frequency component separately* and is either desirable or undesirable, depending on the requirements of each case. In reality, the described behavior of LTI systems allows us to construct systems that operate as *frequency selection filters* whose frequency response has a measure with a relatively low value at the positions of the frequencies of the signal we wish to remove. Following the discussion of Sect. 3.6 (page 282), we can define the *ideal low-pass filter* as a discrete LTI system that allows the passage of only the frequencies of the signal that have a value *lower* than the cutoff frequency $\omega = \omega_c$. Therefore, the frequency response function of this filter will have the form

$$\mathcal{H}^{lp}(e^{j\omega}) = \begin{cases} 1 \text{ for } |\omega| \leq \omega_c \\ 0 \text{ for } \omega_c \leq \omega \leq \pi \end{cases} \tag{10.58}$$

while its impulse response will arise through the calculation of the *inverse Fourier transform* of the preceding relation. It will therefore be the case that

$$h^{lp}[n] = \frac{1}{2\pi}\int_{-\pi}^{+\pi} \mathcal{H}^{lp}(e^{j\omega})\, e^{j\omega n} d\omega = \frac{1}{2\pi}\int_{-\omega_c}^{+\omega_c} e^{j\omega n} d\omega = \frac{\omega_c}{\pi}\frac{\sin(\omega_c n)}{\omega_c n}$$

as can be proven. The preceding relationship holds for $n \neq 0$, while for the case $n = 0$ it is proven by applying L'Hôpital's rule that $h^{lp}[n] = \omega_c/\pi$. As for the impulse and frequency response of the remaining types of ideal filters—which are the *ideal high-pass*, the *ideal band-pass*, and the *ideal band-stop filters*—which are presented in the next exercise.

Problem 10.4.7 Using the methodology for calculating the impulse response of the ideal low-pass filter, show how the impulse responses of the ideal high-pass, band-pass, and band-stop filters are defined in the following way:

- For the ideal high-pass filter

$$h_d^{hp}[n] = \begin{cases} 1 - \dfrac{\omega_c}{\pi} & n = 0 \\ -\dfrac{\omega_c}{\pi}\dfrac{\sin(n\omega_c)}{n\omega_c} & n \neq 0 \end{cases}$$

- For the ideal band-pass filter

$$h_d^{bp}[n] = \begin{cases} \dfrac{\omega_{c_2}}{\pi} - \dfrac{\omega_{c_1}}{\pi} & n = 0 \\ \dfrac{\omega_{c_2}}{\pi}\dfrac{\sin(n\omega_{c_2})}{n\omega_{c_2}} - \dfrac{\omega_{c_1}}{\pi}\dfrac{\sin(n\omega_{c_1})}{n\omega_{c_1}} & n \neq 0 \end{cases}$$

- For the ideal band-stop filter

$$h_d^{bs}[n] = \begin{cases} 1 + \dfrac{\omega_{c_1}}{\pi} - \dfrac{\omega_{c_2}}{\pi} & n = 0 \\ \dfrac{\omega_{c_1}}{\pi}\dfrac{\sin(n\omega_{c_1})}{n\omega_{c_1}} - \dfrac{\omega_{c_2}}{\pi}\dfrac{\sin(n\omega_{c_2})}{n\omega_{c_2}} & n \neq 0 \end{cases}$$

where ω_c is the cutoff frequency of the ideal high-pass filter, while the frequencies $(\omega_{c_1}, \omega_{c_2})$ define the pass and cutoff region for the ideal band-pass and band-stop filter, respectively.

Solution

(a) The frequency response function of the ideal high-pass filter that allows the passage of frequencies belonging to the value region $\omega_c \leq |\omega| \leq \pi$ is given by the relation

$$\mathcal{H}^{hp}(e^{j\omega}) = \begin{cases} 0 \text{ for } |\omega| \leq \omega_c \\ 1 \text{ for } \omega_c \leq \omega \leq \pi \end{cases} = 1 - \begin{cases} 1 \text{ for } |\omega| \leq \omega_c \\ 0 \text{ for } \omega_c \leq \omega \leq \pi \end{cases} = 1 - \mathcal{H}^{lp}(e^{j\omega})$$

where $\mathcal{H}^{lp}(e^{j\omega})$ is the *frequency response function of the ideal low-pass filter*. Taking the inverse Fourier transform of the preceding relation leads us to the equation

$$\mathcal{F}^{-1}\{\mathcal{H}^{hp}(e^{j\omega})\} = \mathcal{F}^{-1}\{1\} - \mathcal{F}^{-1}\{\mathcal{H}^{lp}(e^{j\omega})\}$$

From basic theory, however, it is known that $\mathcal{F}\{\delta[n]\} = 1$ and $\mathcal{F}\{h[n]\} = \mathcal{H}(e^{j\omega})$. Therefore, $\mathcal{F}^{-1}\{1\} = \delta[n]$ and $\mathcal{F}^{-1}\{\mathcal{H}(e^{j\omega})\} = h[n]$, and the preceding relationship is written

$$h^{hp}[n] = \delta[n] - h^{lp}[n] = \delta[n] - \frac{\omega_c}{\pi}\frac{\sin(\omega_c n)}{\omega_c n}$$

For $n \neq 0$ we write $\delta[n] = 0$, and the first term of the preceding relationship is zeroed. On the other hand, for $n = 0$ it is $\delta[n] = 1$, while the other term is proven, by applying L'Hôpital's rule, to be equal to ω_c/π. Combining the foregoing results, we get

$$h_d^{hp}[n] = \begin{cases} 1 - \dfrac{\omega_c}{\pi} & n = 0 \\ -\dfrac{\omega_c}{\pi}\dfrac{\sin(n\omega_c)}{n\omega_c} & n \neq 0 \end{cases}$$

(b) The frequency response function of the *ideal band-pass filter* that allows the passage of frequencies $\omega_{c_1} \leq \omega \leq \omega_{c_2}$ and cuts off all the remaining ones has the form

$$\mathcal{H}^{bp}(e^{j\omega}) = \begin{cases} 0 \text{ for } 0 \leq |\omega| \leq \omega_{c_1} \\ 1 \text{ for } \omega_{c_1} \leq |\omega| \leq \omega_{c_2} \\ 0 \text{ for } \omega_{c_2} \leq |\omega| \leq \pi \end{cases}$$

Therefore, to determine the *impulse response* of the ideal band-pass filter, we must calculate the inverse Fourier transform of the preceding relationship. In detail, therefore, we will have

$$\begin{aligned} h_d^{bp}[n] &= \frac{1}{2\pi}\int_{-\pi}^{+\pi} \mathcal{H}_d(e^{j\omega})\, e^{j\omega n} d\omega = \frac{1}{2\pi}\int_{-\omega_{c_2}}^{-\omega_{c_1}} e^{j\omega n} d\omega + \frac{1}{2\pi}\int_{+\omega_{c_1}}^{+\omega_{c_2}} e^{j\omega n} d\omega \\ &= \frac{1}{2\pi jn}\Big[e^{j\omega_n}\Big]_{\omega=-\omega_{c_2}}^{\omega=-\omega_{c_1}} + \frac{1}{2\pi jn}\Big[e^{j\omega_n}\Big]_{\omega=+\omega_{c_1}}^{\omega=+\omega_{c_2}} = \frac{1}{n\pi}\frac{e^{-j\omega_{c_1}n} - e^{-j\omega_{c_2}n}}{2j} + \frac{1}{n\pi}\frac{e^{j\omega_{c_2}n} - e^{j\omega_{c_1}n}}{2j} \\ &= \frac{1}{n\pi}\frac{e^{j\omega_{c_2}n} - e^{-j\omega_{c_2}n}}{2j} - \frac{1}{n\pi}\frac{e^{j\omega_{c_1}n} - e^{-j\omega_{c_1}n}}{2j} = \frac{\sin(\omega_{c_2}n)}{n\pi} - \frac{\sin(\omega_{c_1}n)}{n\pi} = \frac{\omega_{c2}}{\pi}\frac{\sin(n\omega_{c_2})}{n\omega_{c_2}} - \frac{\omega_{c1}}{\pi}\frac{\sin(n\omega_{c_1})}{n\omega_{c_1}} \end{aligned}$$

The preceding relation holds for $n \neq 0$, while for $n = 0$ the value takes the indeterminate form $0/0$. Applying L'Hôpital's rule, we get

$$\begin{aligned} h^{bp}[0] &= \lim_{n\to 0}\frac{\sin(\omega_{c_2}n)}{n\pi} - \lim_{n\to 0}\frac{\sin(\omega_{c_1}n)}{n\pi} = \lim_{n\to 0}\frac{(\sin(\omega_{c_2}n))'}{(n\pi)'} - \lim_{n\to 0}\frac{(\sin(\omega_{c_1}n))'}{(n\pi)'} \\ &= \lim_{n\to 0}\frac{\omega_{c_2}\cos(\omega_{c_2}n)}{\pi} - \lim_{n\to 0}\frac{\omega_{c_1}(\sin(\omega_{c_1}n))'}{\pi} = \frac{\omega_{c2}}{\pi} - \frac{\omega_{c1}}{\pi} \end{aligned}$$

Combining the foregoing results, we conclude that the impulse response of the ideal band-pass filter will be given by the relation

$$h^{bp}[n] = \begin{cases} \dfrac{\omega_{c2}}{\pi} - \dfrac{\omega_{c1}}{\pi} & n = 0 \\ \dfrac{\omega_{c2}}{\pi}\dfrac{\sin(n\omega_{c_2})}{n\omega_{c_2}} - \dfrac{\omega_{c1}}{\pi}\dfrac{\sin(n\omega_{c_1})}{n\omega_{c_1}} & n \neq 0 \end{cases}$$

(c) The frequency response of the *ideal band-stop filter* that allows the passage of all frequencies except those belonging to the region $\omega_{c_1} \le \omega \le \omega_{c_2}$ that it cuts off is

$$\mathcal{H}^{bs}(e^{j\omega}) = \begin{cases} 1 \text{ for } 0 \le |\omega| \le \omega_{c_1} \\ 0 \text{ for } \omega_{c_1} \le |\omega| \le \omega_{c_2} \\ 1 \text{ for } \omega_{c_2} \le |\omega| \le \pi \end{cases}$$

which is alternatively formulated as

$$\mathcal{H}^{bs}(e^{j\omega}) = 1 - \begin{cases} 0 \text{ for } 0 \le |\omega| \le \omega_{c_1} \\ 1 \text{ for } \omega_{c_1} \le |\omega| \le \omega_{c_2} \\ 0 \text{ for } \omega_{c_2} \le |\omega| \le \pi \end{cases} = 1 - \mathcal{H}^{bp}(e^{j\omega})$$

where $\mathcal{H}^{bp}(e^{j\omega})$ is the *frequency response function of the ideal band-pass filter*. Taking the inverse Fourier transform of the foregoing relation, we are led to the equation

$$\mathcal{F}^{-1}\{\mathcal{H}^{bs}(e^{j\omega})\} = \mathcal{F}^{-1}\{1\} - \mathcal{F}^{-1}\{\mathcal{H}^{bp}(e^{j\omega})\}$$

Following the known methodology, we have $\mathcal{F}\{\delta[n]\} = 1$ and $\mathcal{F}\{h[n]\} = \mathcal{H}(e^{j\omega})$. Therefore, $\mathcal{F}^{-1}\{1\} = \delta[n]$ and $\mathcal{F}^{-1}\{\mathcal{H}(e^{j\omega})\} = h[n]$, and the foregoing relation is written

$$h^{bs}[n] = \delta[n] - h^{bp}[n] = \delta[n] + \frac{\omega_{c1}}{\pi}\frac{\sin(n\omega_{c1})}{n\omega_{c1}} - \frac{\omega_{c2}}{\pi}\frac{\sin(n\omega_{c2})}{n\omega_{c2}}$$

For $n \ne 0$ it holds that $\delta[n] = 0$, and the first term of the preceding equation is zeroed. On the other hand, for $n = 0$ we have $\delta[0] = 1$, while the remaining term (which we shall denote by L) takes the indeterminate form $0/0$. Applying L'Hôpital's rule, we get

$$\begin{aligned} L &= \lim_{n\to 0}\frac{\sin(\omega_{c_1}n)}{n\pi} - \lim_{n\to 0}\frac{\sin(\omega_{c_2}n)}{n\pi} = \lim_{n\to 0}\frac{(\sin(\omega_{c_1}n))'}{(n\pi)'} - \lim_{n\to 0}\frac{(\sin(\omega_{c_2}n))'}{(n\pi)'} \\ &= \lim_{n\to 0}\frac{\omega_{c_1}\cos(\omega_{c_1}n)}{\pi} - \lim_{n\to 0}\frac{\omega_{c_2}(\sin(\omega_{c_2}n))'}{\pi} = \frac{\omega_{c1}}{\pi} - \frac{\omega_{c2}}{\pi} \end{aligned}$$

and therefore for $n = 0$ the impulse response of the ideal band-pass filter will be

$$h[0] = 1 + \frac{\omega_{c1}}{\pi} - \frac{\omega_{c2}}{\pi}$$

Combining the foregoing results, we are led to our final result, which has the form

$$h_d^{bs}[n] = \begin{cases} 1 + \frac{\omega_{c1}}{\pi} - \frac{\omega_{c2}}{\pi} & n = 0 \\ \frac{\omega_{c1}}{\pi}\frac{\sin(n\omega_{c1})}{n\omega_{c1}} - \frac{\omega_{c2}}{\pi}\frac{\sin(n\omega_{c2})}{n\omega_{c2}} & n \ne 0 \end{cases}$$

In full analogy with their continuous equivalents, the ideal discrete-time filters are *noncausal* and therefore cannot be implemented in practice; for this reason, they merely constitute *idealized models of the corresponding real-world filters*, which are the ones actually used in all practical applications. Additionally, they are aforementioned to be *zero phase filters*, since their frequency response has a magnitude equal to unity—for the passband—and phase equal to zero. Recall from Sect. 3.5.2 (page 278) that *the existence of a phase component in a frequency response function causes distortion in the input signal only when it constitutes a nonlinear function of frequency*, the existence of linear phase has the sole effect of introducing a time shift (delay) to the signal. The simplest discrete-time system characterized by the linear phase is the element of the *ideal time delay*, which is described by an impulse response of the form $h[n] = \delta[n - n_0]$ and by a function frequency response that is calculated as

$$\mathcal{H}(e^{j\omega}) = \mathcal{F}\{h[n]\} = \mathcal{F}\{\delta[n - n_0]\} = e^{-j\omega_0 n} \tag{10.59}$$

from which it follows that $|\mathcal{H}(e^{j\omega})| = 1$ and $\measuredangle\mathcal{H}(e^{j\omega}) = -\omega n_0$ $(|\omega| < \pi)$. On the other hand, the *ideal low-pass filter* with a linear phase response is characterized by the *frequency response function*

$$\mathcal{H}^{lp}(e^{j\omega}) = \begin{cases} e^{\ j\omega n_0} & \text{for } |\omega| \leq \omega_c \\ 0 & \text{for } \omega_c \leq \omega \leq \pi \end{cases}$$

while its *impulse response*, following the methodology used previously, is shown to have the form

$$h_d^{lp}[n] = \frac{\sin[\omega_c(n - n_0)]}{\pi(n - n_0)} \tag{10.60}$$

It is obvious that the situation for the remaining types of ideal discrete-time filters is entirely analogous. However, let it be noted that even though such a filter more closely approximates a real filter compared to a zero-phase filter, it continues to be *noncausal* and, hence, *nonimplementable*.

Finally, in full analogy with continuous-time filters, the *group delay* is related *to the effects of phase on signals characterized by limited bandwidth*[18] and is defined as *the negative slope of the phase component of the system's frequency response* (for the definition and characteristics of this size see Eq. 3.60 and the observations that follow).

10.4.2 Correlation of Functions $\mathcal{H}(z)$ and $\mathcal{H}(e^{j\omega})$

As mentioned earlier, the transfer function of a discrete recursive LTI system described by a difference equation of the form

$$\sum_{k=0}^{N} \alpha_k y[n-k] = \sum_{k=0}^{M} \beta_k x[n-k]$$

is given by the relation

[18] In full analogy with continuous-time signals (see Exercise 3.3.12 on page 233 and Sect. 3.3.7 on page 255), *the bandwidth of a discrete-time signal is defined as the area of frequencies $\omega_1 \leq \omega \leq \omega_2$ in which the majority of the energy or power of the signal is concentrated.* For example, if 98% of a signal's energy is concentrated in the aforementioned area, then the interval $\omega_2 - \omega_1$ is known as the 98% bandwidth, and so forth. Depending on the values of the frequencies that belong to this area, the signal can be characterized as a *low- or high-frequency signal* if its energy is concentrated around the frequency $\omega = 0$ or around some higher frequency, respectively, while, finally, if its energy is concentrated at frequencies that belong to some intermediate areas, the signal is aforementioned to be a *midfrequency signal* or *band-pass signal*, respectively. In the last case, the signal is aforementioned to be a *limited-bandwidth signal* if the value of the difference $\omega_2 - \omega_1$ is much smaller (e.g., by an order of magnitude or more) than the central frequency of this area, which is $(\omega_1 + \omega_2)/2$, and as a signal of extended bandwidth in the opposite case. A signal whose spectrum is everywhere zero, except for a frequency range $\omega_0 \leq |\omega| < \pi$, is known as a *limited-spectrum signal*.

$$\mathcal{H}(z) = \left\{\sum_{k=0}^{M} \beta_k z^{-k}\right\} \Big/ \left\{\sum_{k=0}^{N} \alpha_k z^{-k}\right\} = \left\{\sum_{k=0}^{M} \beta_k z^{-k}\right\} \left\{1 + \sum_{k=1}^{N} \alpha_k z^{-k}\right\}$$

with the last relation, which is also often used in practice, corresponding to a coefficient value $\alpha_0 = 1$ (Eq. 9.8). Taking into account that *the Fourier transform of a discrete sequence arises from the $\mathcal{Z}$ transform by making the substitution $z = e^{j\omega}$* (see Sect. 10.3.3 on page 809), we can assert that as long as the region of convergence of the transfer function of the system $\mathcal{H}(z)$ includes the unit circle, so that the frequency response function exists, the latter function can be calculated as[19]

$$\mathcal{H}(e^{j\omega}) = \mathcal{H}(z)\Big|_{z=e^{j\omega}} = \left\{\sum_{k=0}^{M} \beta_k e^{-j\omega k}\right\} \Big/ \left\{\sum_{k=0}^{N} \alpha_k e^{-j\omega k}\right\} \tag{10.61}$$

with the coefficients $\{\alpha_k\}$ and $\{\beta_k\}$ being real numbers.

In many cases, it is useful to express the frequency response function using its *poles* $\{p_k\}$ and its *zero values* $\{z_k\}$ from an equation of the form

$$\mathcal{H}(e^{j\omega}) = \frac{\beta_0}{\alpha_0} \times \left\{\prod_{k=1}^{M}(1 - z_k e^{-j\omega})\right\} \Big/ \left\{\prod_{k=1}^{N}(1 - p_k e^{-j\omega})\right\} \tag{10.62}$$

which is equivalent to Eq. (8.8) for $z = e^{j\omega}$. Therefore, the magnitude of the system's frequency response will have the form

$$|\mathcal{H}(e^{j\omega})| = \left|\frac{\beta_0}{\alpha_0}\right| \times \left\{\prod_{k=1}^{M}\left|1 - z_k e^{-j\omega}\right|\right\} \Big/ \left\{\prod_{k=1}^{N}\left|1 - p_k e^{-j\omega}\right|\right\} \tag{10.63}$$

while the square of the measure, which is also used in several cases, will be calculated from the equation

$$|\mathcal{H}(e^{j\omega})|^2 = \mathcal{H}(e^{j\omega})\mathcal{H}^*(e^{j\omega}) \tag{10.64}$$

Taking into account that the function $\mathcal{H}^*(e^{j\omega})$ can be calculated from the function

$$\mathcal{H}^*(1/z^*) = \frac{\beta_0}{\alpha_0} \times \left\{\prod_{k=1}^{M}(1 - z_k^* z)\right\} \Big/ \left\{\prod_{k=1}^{N}(1 - p_k^* z)\right\} \tag{10.65}$$

for values $z = e^{j\omega}$ and that for systems that have a *real impulse response*—or, equivalently, real coefficients $\{\alpha_k\}$ and $\{\beta_k\}$ in their difference equation—*the poles and zeros of the transfer function, if they are complex quantities, always appear in pairs*, we will have $\mathcal{H}^*(1/z^*) = \mathcal{H}(z^{-1})$ and therefore $\mathcal{H}^*(e^{j\omega}) = \mathcal{H}(e^{-j\omega})$; thus, we will have

$$|\mathcal{H}(e^{j\omega})|^2 = \mathcal{H}(e^{j\omega})\mathcal{H}(e^{-j\omega}) = \left(\frac{\beta_0}{\alpha_0}\right)^2 \times \left\{\prod_{k=1}^{M}(1 - z_k e^{-j\omega})(1 - z_k^* e^{j\omega})\right\} \Big/ \left\{\prod_{k=1}^{N}(1 - p_k e^{-j\omega})(1 - p_k^* e^{j\omega})\right\} \tag{10.66}$$

---

[19] In this way, it is possible to calculate the frequency response of a discrete recursive LTI system from the coefficients of its difference equation, provided, of course, the convergence conditions we mentioned above are met.

From the last relation we conclude *that the measure of a system's frequency response is calculated as the product of the measures of a set of factors that contain the zero values of the function divided by the product of the measures of another group of factors that contain the poles of the frequency response of the system.* In full analogy with continuous-time systems, it is common practice to convert the foregoing products into sums by logarithmizing the last relation and using decibels (dB) as the unit of measurement (Sect. 5.7, page 426). By logarithmizing, therefore, Eq. (10.67) and multiplying all its terms by 20—so they can be expressed in decibels—we will get

$$20\log_{10}|\mathcal{H}(e^{j\omega})| = 20\log_{10}\left|\frac{\beta_0}{\alpha_0}\right| + \sum_{k=1}^{M} 20\log_{10}|1 - z_k e^{-j\omega}| - \sum_{k=1}^{N} 20\log_{10}|1 - p_k e^{-j\omega}| \tag{10.67}$$

The function $20\log_{10}|\mathcal{H}(e^{j\omega})|$ is known as the *logarithmic amplitude of the frequency response of a system* and is expressed in decibels. In several cases this quantity is known as *gain*, while its negative value is known as *attenuation*. On the other hand, the *phase response* of the frequency response function will be given according to Equation 10.67 by the relation

$$\measuredangle\mathcal{H}(e^{j\omega}) = \measuredangle\left(\frac{\beta_0}{\alpha_0}\right) + \sum_{k=1}^{M}\measuredangle\left[1 - z_k e^{-j\omega}\right] - \sum_{k=1}^{N}\measuredangle\left[1 - p_k e^{-j\omega}\right] \tag{10.68}$$

from which it immediately follows that

$$\tau(\omega) = -\frac{d}{d\omega}\left\{\measuredangle\mathcal{H}(e^{j\omega})\right\} = \sum_{k=1}^{N}\frac{d}{d\omega}\measuredangle\left(1 - p_k e^{-j\omega}\right) - \sum_{k=1}^{M}\frac{d}{d\omega}\measuredangle\left(1 - z_k e^{-j\omega}\right) \tag{10.69}$$

It is proven that the *group delay* can alternatively be calculated from the relation

$$\tau(\omega) = \sum_{k=1}^{N}\frac{|p_k|^2 - \Re\{p_k e^{-j\omega}\}}{1 + |p_k|^2 - 2\Re\{p_k e^{-j\omega}\}} - \sum_{k=1}^{M}\frac{|z_k|^2 - \Re\{z_k e^{-j\omega}\}}{1 + |z_k|^2 - 2\Re\{z_k e^{-j\omega}\}} \tag{10.70}$$

10.4.3 Geometric Calculation of Function $\mathcal{H}(e^{j\omega})$

Following the methodology of calculating the measure and phase of the frequency response function of continuous LTI systems from the positions of the zero values and the poles (Sect. 5.6, page 418), let us reformulate equation (10.63) by writing it in the equivalent form

$$\mathcal{H}(e^{j\omega}) = \frac{\beta_0}{\alpha_0}\, e^{j\omega(N-M)} \left\{\prod_{k=1}^{M}(e^{j\omega} - z_k)\right\} \Big/ \left\{\prod_{k=1}^{N}(e^{j\omega} - p_k)\right\} \tag{10.71}$$

Taking into account that the factors $e^{j\omega} - z_k$ $(k = 1, 2, \ldots, M)$ and $e^{j\omega} - p_k$ $(k = 1, 2, \ldots, N)$ are of a complex nature, let us formulate them in their polar form

$$e^{j\omega} - z_k = V_k(\omega)\, e^{j\Theta_k(\omega)} \quad (k = 1, 2, \ldots, M) \qquad \text{and} \qquad e^{j\omega} - p_k = U_k(\omega)\, e^{j\Phi_k(\omega)} \quad (k = 1, 2, \ldots, N)$$

with the functions appearing in these expressions being defined as

$$V_k(\omega) = |e^{j\omega} - z_k|, \quad \Theta_k(\omega) = \measuredangle(e^{j\omega - z_k}) \quad (k = 1, 2, \ldots, M)$$

$$U_k(\omega) = |e^{j\omega} - p_k|, \quad \Phi_k(\omega) = \measuredangle(e^{j\omega - p_k}) \quad (k = 1, 2, \ldots, N)$$

Substituting the preceding expressions into the defining equation of the frequency response $\mathcal{H}(e^{j\omega})$ and formulating the quantities α_0 and β_0 in their exponential forms[20] $\alpha_0 = |\alpha_0|e^{j\measuredangle\alpha_0}$ and $\beta_0 = |\beta_0|e^{j\measuredangle\beta_0}$, we get

$$\begin{aligned}\mathcal{H}(e^{j\omega}) &= \frac{|\beta_0|e^{j\measuredangle\beta_0}}{|\alpha_0|e^{j\measuredangle\alpha_0}} e^{j\omega(N-M)} \left\{\prod_{k=1}^{M} V_k(\omega)e^{j\Theta_k(\omega)}\right\} \Big/ \left\{\prod_{k=1}^{N} U_k(\omega)e^{j\Phi_k(\omega)}\right\} \\ &= \frac{|\beta_0|}{|\alpha_0|} \left\{\prod_{i=1}^{M} V_k(\omega)\right\} \Big/ \left\{\prod_{k=1}^{N} U_k(\omega)\right\} \times \exp\left\{j\left[\omega(N-M) + \measuredangle\beta_0 - \measuredangle\alpha_0 + \sum_{k=1}^{M}\Theta_k(\omega) - \sum_{k=1}^{N}\Phi_k(\omega)\right]\right\}\end{aligned}$$

Therefore, the magnitude and phase of the frequency response function will be calculated as

$$|\mathcal{H}(e^{j\omega})| = \left|\frac{\beta_0}{\alpha_0}\right| \frac{\prod_{i=1}^{M} V_k(\omega)}{\prod_{k=1}^{N} U_k(\omega)} = \left|\frac{\beta_0}{\alpha_0}\right| \frac{V_1(\omega)V_2(\omega)\ldots V_M(\omega)}{U_1(\omega)U_2(\omega)\ldots U_N(\omega)} \quad \text{and}$$

$$\begin{aligned}\measuredangle\mathcal{H}(e^{j\omega}) &= \omega(N-M) + \measuredangle\beta_0 - \measuredangle\alpha_0 + \sum_{k=1}^{M}\Theta_k(\omega) - \sum_{k=1}^{N}\Phi_k(\omega) \\ &= \omega(N-M) + \measuredangle\beta_0 - \measuredangle\alpha_0 + [\Theta_1(\omega) + \Theta_2(\omega) + \cdots + \Theta_M(\omega)] - [\Phi_1(\omega) + \Phi_2(\omega) + \cdots + \Phi_N(\omega)]\end{aligned}$$

From the last two equations we can calculate the frequency response function of a discrete LTI system knowing the positions of the zeros and poles.

The methodology for solving problems calculating the frequency response function using geometric methods is similar to that presented in the chapter on continuous-time systems (see Exercise 5.6.1 on page 419), Exercise 5.6.2 on page 423, and Exercise 5.6.3 on page 423), and the reader is asked to apply it in the discrete-time domain by completing the following exercises.

To understand the foregoing process of calculating the *frequency response* $\mathcal{H}(e^{j\omega})$ from the positions of the zeros and poles, suppose we want to calculate the value of this function at the position of some frequency ω. Taking into account that this function results from the *transfer function* $\mathcal{H}(z)$ by setting $z = e^{j\omega}$, it is obvious that *the frequency* ω *under consideration corresponds to a point* $\boldsymbol{L}$ *that belongs to the circumference of the unit circle, such that the angle formed by the straight line segment* $\boldsymbol{OL}$ *and the real axis of the complex plane is equal to* ω (see Fig. 10.10a). But this means that the expression $e^{j\omega}$ will be uniquely determined by the straight line segment $\boldsymbol{OL}$. Considering additionally a pole p_k and a zero value z_k corresponding to the points $\boldsymbol{A}$ and $\boldsymbol{B}$ of the same figure and described by vectors $\boldsymbol{OA}$ and $\boldsymbol{OB}$, from the geometry of the figure we have

$$\boldsymbol{OL} = \boldsymbol{OA} + \boldsymbol{AL} \quad \text{and} \quad \boldsymbol{OL} = \boldsymbol{OB} + \boldsymbol{BL}$$

Taking into account that $\boldsymbol{OL} = e^{j\omega}$, $\boldsymbol{OA} = p_k$, and $\boldsymbol{OB} = z_k$, we finally arrive at the relations

$$e^{j\omega} - p_k = \boldsymbol{AL} = U_k(\omega)e^{j\Phi_k(\omega)}$$

$$e^{j\omega} - z_k = \boldsymbol{BL} = V_k(\omega)e^{j\Theta_k(\omega)}$$

[20] Taking into account that the coefficients α_0 and β_0 are *real numbers*, it is obvious that their phase will be equal to 0 if their value is positive or equal to π if their value is negative.

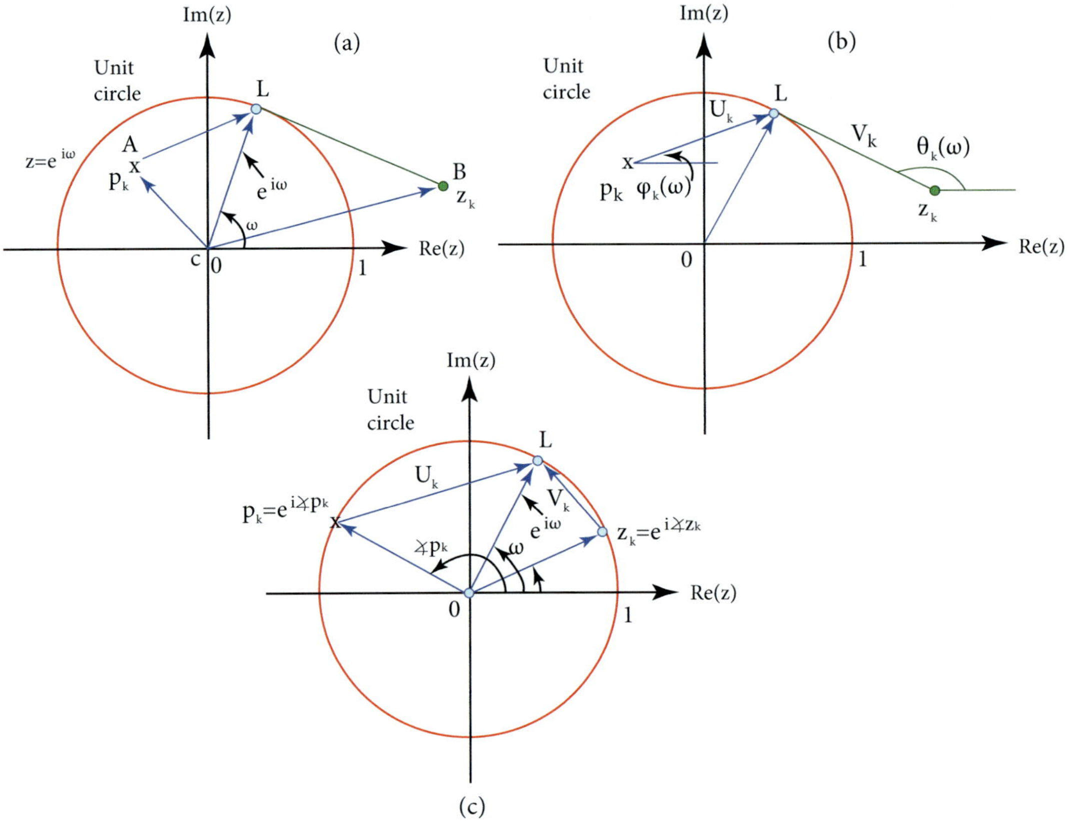

Fig. 10.10 Geometric calculation of frequency response from pole/zero positions (Proakis and Manolakis, Figures 4.40 and 4.41)

From the foregoing relations and Fig. 10.10 it follows that the function $U_k(\omega)$ expresses *the length of vector* $\boldsymbol{AL}$, which is *the distance of the pole* p_k *from the point* $\boldsymbol{L}$ *corresponding to the frequency of interest*—that is, the quantity $e^{j\omega}$—while the function $V_k(\omega)$ expresses *the length of vector* $\boldsymbol{BL}$, which is *the distance of the zero value* z_k *from that point*. Similarly, the functions $\Phi_k(\omega)$ and $\Theta_k(\omega)$ express the angles of vectors $\boldsymbol{AL}$ and $\boldsymbol{BL}$ with the horizontal axis (Fig. 10.10b).

Using the preceding geometric representation, we can formulate useful conclusions regarding the way in which the positions of the poles and the zero values of the frequency response function affect the magnitude and phase of this function (and, more generally, of the Fourier transform of any signal). Let us consider a pole p_k that lies on the circumference of the unit circle. In this case, it is not difficult to see that at the position $\omega = \measuredangle p_k$ the function $\mathcal{H}(e^{j\omega})$ *becomes infinite* since the length of vector $\boldsymbol{AL}$ described by the function $V_k(\omega)$ becomes equal to zero. In a completely analogous way, the existence of a zero z_k on the circumference of the unit circle causes the *zeroing* of the function $\mathcal{H}(e^{j\omega})$ at the position $\omega = \measuredangle z_k$ since in this case $U_k(\omega) = |\boldsymbol{BL}| = 0$. These two cases are presented in Fig. 10.10c.

From the preceding discussion we conclude that *the existence of a zero in the frequency response function that is located near the circumference of the unit circle leads to small amplitude values for this function at the frequency positions represented by points neighboring the point representing the zero value*. Conversely, *the existence of a pole in the aforementioned position leads to large amplitude values for those frequencies*. Thus, we observe how the existence of a zero value brings about the opposite result from that which the existence of a pole brings about, and in the special case in which these two points coincide, there is a *complete and mutual neutralization of their effects on the frequency response function* [the last conclusion is also mathematically substantiated: indeed, in the case where a pole is identified with a zero value, the expressions $e^{j\omega} - p_k$ and $e^{j\omega} - z_k$ become equal to each other and simplify, since one is in the numerator and the other in the denominator of the function $\mathcal{H}(e^{j\omega})$]. The positions of the poles and zero values of the frequency response play a decisive role in the process of designing digital filters.

Problem 10.4.8 Investigate theoretically the effect of a zero value, a pole, a pair of complex conjugate zero values, and a pair of complex conjugate poles on the frequency response of a LTI system and then construct the corresponding expressions for the amplitude response, the phase response, and the group delay.

Solution Let us consider a point z of the complex plane representing a complex value of the frequency response $\mathcal{H}(e^{j\omega})$. Expressing this complex quantity in its polar form $z = re^{j\vartheta}$, it is obvious that the frequency response function will necessarily contain a term of the form

$$1 - ze^{-j\omega} = 1 - re^{j\vartheta}e^{-j\omega} = 1 - re^{-j(\omega-\vartheta)} = [1 - r\cos(\omega - \vartheta)] + jr\sin(\omega - \vartheta)$$

(see Eq. 10.63). The square of the magnitude of this factor is equal to

$$|1 - re^{j\vartheta}e^{-j\omega}|^2 = (1 - re^{j\vartheta}e^{-j\omega})(1 - re^{j\vartheta}e^{-j\omega})^* = (1 - re^{j\vartheta}e^{-j\omega})(1 - re^{-j\vartheta}e^{j\omega}) = 1 + r^2 - 2r\cos(\omega - \vartheta)$$

and therefore the magnitude will have the form

$$|1 - re^{j\vartheta}e^{-j\omega}| = \sqrt{1 + r^2 - 2r\cos(\omega - \vartheta)}$$

or, expressed in decibels,

$$20\log_{10}|1 - re^{j\vartheta}e^{-j\omega}| = 10\log_{10}\left[1 + r^2 - 2r\cos(\omega - \vartheta)\right]$$

We observe that the maximum value of the logarithmic amplitude corresponds to the frequency $\omega = \pi + \vartheta$ and is equal to $20\log_{10}(1 + r)$, while the minimum value corresponds to the frequency $\omega = \vartheta$ and is equal to $20\log_{10}|1 - r|$. On the other hand, the phase of the term of the transfer function contributed by the zero value is equal to

$$\measuredangle(1 - re^{j\vartheta}e^{-j\omega}) = \tan^{-1}\left\{\frac{r\sin(\omega - \vartheta)}{1 - r\cos(\omega - \vartheta)}\right\}$$

—in reality, the preceding relation gives us the principal value of the phase, which takes values in the interval $(-\pi, +\pi]$—while differentiating, we find that the corresponding group delay has the form

$$\tau_\omega(1 - re^{j\vartheta}e^{-j\omega}) = -\frac{d}{d\omega}\measuredangle(1 - re^{j\vartheta}e^{-j\omega}) = \frac{r^2 - r\cos(\omega - \vartheta)}{1 + r^2 - 2r\cos(\omega - \vartheta)}$$

The foregoing analysis refers to the case of a simple zero value. In the case where the expression $(1 - re^{j\vartheta}e^{-j\omega})$ corresponds to a pole of the function $\mathcal{H}(e^{j\omega})$, it will obviously be in the denominator of the equation that describes it, resulting in the phase functions, logarithmic amplitude, and group delay being *the negatives of those we calculated earlier*, as can be proven.

In the next step of the process, let us look at the effect on the function frequency response of two complex conjugate zero values that have the forms $z_1 = re^{j\vartheta}$ and $z_2 = z_1^* = re^{-j\vartheta}$. In this case, the frequency response of the system will necessarily contain the product

$$(1 - re^{j\vartheta}e^{-j\omega})(1 - re^{-j\vartheta}e^{-j\omega}) = [1 - re^{-j(\omega-\vartheta)}][1 - re^{-j(\omega+\vartheta)}]$$

Following the methodology applied previously, we find that the logarithmic amplitude function will be equal to

$$10\log_{10}\left[1 - 2r\cos(\omega - \vartheta) + r^2\right] + 10\log_{10}\left[1 - 2r\cos(\omega + \vartheta) + r^2\right]$$

the phase that corresponds to this product has the value

$$\tan^{-1}\left\{\frac{r\sin(\omega-\vartheta)}{1-r\cos(\omega-\vartheta)}\right\}+\tan^{-1}\left\{\frac{r\sin(\omega+\vartheta)}{1-r\cos(\omega+\vartheta)}\right\}$$

while, finally, the group delay that constitutes the derivative of the foregoing relation with respect to the frequency ω is

$$\frac{r^2-r\cos(\omega-\vartheta)}{1+r^2-2r\cos(\omega-\vartheta)}+\frac{r^2-r\cos(\omega+\vartheta)}{1+r^2-2r\cos(\omega+\vartheta)}$$

Finally, if the quantities z_1 and z_2 refer to poles of the frequency response, the product

$$[1-re^{-j(\omega-\vartheta)}][1-re^{-j(\omega+\vartheta)}]$$

will be found in the denominator of the function in question and the phase, logarithmic amplitude, and group delay functions will be the *negative* of those previously calculated.

Considering that the poles and zeros of the transfer function have the form $p_k = r_k e^{j\varphi_k}$ $(k = 1, 2, \ldots, N)$ and $z_k = q_k e^{j\vartheta_k}$ $(k = 1, 2, \ldots, M)$ and replacing the preceding results in the previously constructed equations, we arrive at the expressions

$$|\mathcal{H}(e^{j\omega})| = \left|\frac{\beta_0}{\alpha_0}\right| \times \frac{\prod_{k=1}^{M}\sqrt{1+q_k^2-2q_k\cos(\omega-\vartheta_k)}}{\prod_{k=1}^{N}\sqrt{1+r_k^2-2r_k\cos(\omega-\varphi_k)}}$$

$$\measuredangle\mathcal{H}(e^{j\omega}) = \measuredangle\mathcal{H}\left(\frac{\beta_0}{\alpha_0}\right)+\sum_{k=1}^{M}\tan^{-1}\left(\frac{q_k\sin(\omega-\vartheta_k)}{1-q_k\cos(\omega-\vartheta_k)}\right)-\sum_{k=1}^{N}\tan^{-1}\left(\frac{r_k\sin(\omega-\varphi_k)}{1-r_k\cos(\omega-\varphi_k)}\right)$$

$$\tau(\omega) = \sum_{k=1}^{N}\frac{r_k^2-r_k\cos(\omega-\varphi_k)}{1+r_k^2-2r_k\cos(\omega-\varphi_k)}-\sum_{k=1}^{M}\frac{q_k^2-q_k\cos(\omega-\vartheta_k)}{1+q_k^2-2q_k\cos(\omega-\vartheta_k)}$$

10.5 Discrete Fourier Transform

In the previous sections, the theoretical background of discrete-time signal and system analysis techniques in the space of frequencies using *expansion* and, mainly, the *Fourier transform*, which covers both *periodic* and *aperiodic* discrete-time signals, was given. Following the logic of Laplace and $\mathcal{Z}$ transforms, as well as the Fourier transform defined in the continuous-time domain, the aforementioned techniques are based on the *convolution theorem*, which characterizes the operation of LTI systems and allows for *the conversion of the system's representation from the time domain to the frequency domain, the study of the system in this space, and a return to the time domain, if such a thing is necessary*. However, the DTFT, as important as it is, constitutes a *continuous function of frequency* ω and, therefore, is not an effective tool for the frequency analysis of signals and systems, since it cannot be implemented by a digital computational system. For this reason, in practice, a discrete sequence $x[n]$ is not studied using its continuous spectrum $\mathcal{X}(e^{j\omega})$, but instead a set of samples of this continuous spectrum is used, recorded periodically in the frequency domain, and having a constant frequency distance between them equal to $\delta\omega$. This process of *sampling in the frequency domain* leads to another type of transform known as the *discrete Fourier transform* (DFT) and can be implemented on a computer. A theoretical presentation of the DFT, its properties, and the way it is used in the study of discrete-time systems is detailed in the following pages, along with a detailed description of the sampling process in the frequency domain. The reader is encouraged, before moving on to the next section, to study and understand the section that deals with sampling in the time domain, as the concepts and techniques used in both cases are the same.

10.5.1 Sampling in Frequency Domain

As we mentioned in the introductory section on the DFT, the DTFT, being a continuous function of frequency ω, *cannot be represented using a digital computer*. For this reason, in practice, a discrete version of it is used, which can be calculated using *a finite number of samples*, which are naturally expressed, not in the time domain, but in the frequency domain. To obtain these samples, we must proceed to *uniform sampling of the DTFT*, which is known to be defined as

$$\mathcal{X}(e^{j\omega}) = \sum_{n=-\infty}^{\infty} x[n]\, e^{-j\omega n} \tag{10.72}$$

recording its value at frequency positions that are separated by a *constant distance equal to* $\delta\omega$ *radii*. Let us note that because the DTFT is a *periodic function of frequency with a period equal to* 2π, the foregoing sampling process can be limited only to the value range $0 \le \omega \le 2\pi$. Assuming that during the application of this process we record the values of N samples of the function $\mathcal{X}(e^{j\omega})$, the frequency values at the positions where this recording will take place will be at a distance from each other equal to $\delta\omega = 2\pi/N$ (this situation is depicted in Fig. 10.11a). Therefore, these values will have the form

$$\omega_k = \frac{2\pi k}{N}, \quad (k = 0, 1, 2, \ldots, N-1) \tag{10.73}$$

while the value of the function $\mathcal{X}(e^{j\omega})$ at these frequency positions will be equal to

$$\mathcal{X}(e^{j\omega_k}) = \mathcal{X}\left[\exp\left(j\frac{2\pi k}{N}\right)\right] = \sum_{n=-\infty}^{\infty} x[n] \exp\left(-j\frac{2\pi k}{N}n\right) \quad (k = 0, 1, 2, \ldots, N-1) \tag{10.74}$$

It is obvious that because the function $\mathcal{X}(e^{j\omega})$ is *periodic* with respect to frequency with a period of 2π, the sequence of samples $\mathcal{X}(e^{j\omega_k})$ $(k = 0, 1, 2, \ldots, N-1)$ that will result in the described way will also be a *periodic function of the variable* k with a period value equal to N. Taking into account that *the DTFT is the* $\mathcal{Z}$ *transform, calculated on the circumference of the unit circle*, we can claim that *the foregoing sampling process in the frequency domain is equivalent to determining N points of the complex plane that belong to the circumference of the unit circle and are at an angular distance equal to* $2\pi/N$—see Fig. 10.11b.

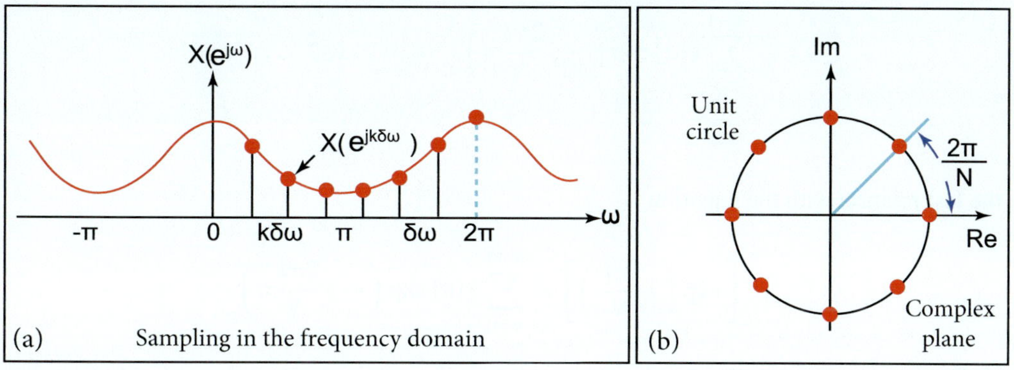

Fig. 10.11 Sampling in the frequency domain and equivalent representation in the complex plane (left image: Proakis and Manolakis, Figure 5.1, right image Oppenheim, Schafer, Buck, Figure 8.7)

In the next step of the process, let us divide the sum of the infinite terms of the preceding equation into infinite groups of N consecutive terms, putting it in the form[21]

$$\mathcal{X}\left[\exp\left(j\frac{2\pi k}{N}\right)\right] = \cdots + \sum_{n=-N}^{-1} x[n]\exp\left(-j\frac{2\pi k}{N}n\right) + \sum_{n=0}^{N-1}\exp\left(-j\frac{2\pi k}{N}n\right)$$
$$+ \sum_{n=N}^{2N-1}\exp\left(-j\frac{2\pi k}{N}n\right) + \cdots = \sum_{m=-\infty}^{\infty}\sum_{n=mN}^{mN+N-1}\exp\left(-j\frac{2\pi k}{N}n\right)$$

Proceeding with the index change $n \longrightarrow r = n - mN$, the new limits of the sum will be $r = 0$ and $r = N - 1$, while we still have $\exp(-j2k\pi n/N) = \exp(-j2k\pi r/N)$. Therefore, we will have

$$\mathcal{X}\left[\exp\left(j\frac{2\pi k}{N}\right)\right] = \sum_{m=-\infty}^{\infty}\sum_{r=0}^{N-1} x[r]\exp\left(-j\frac{2k\pi r}{N}\right)$$
$$= \sum_{r=0}^{N-1}\left[\sum_{m=-\infty}^{\infty} x[n-mN]\right]\exp\left(-j\frac{2\pi k}{N}r\right) = \sum_{n=0}^{N-1}\xi[n]\exp\left(-j\frac{2\pi k}{N}n\right)$$

$(k = 0, 1, 2, \ldots, N-1)$, where for the extraction of the last relation we proceeded to the interchange of the sums included in it and defined the discrete-time signal $\xi[n]$ from the relation

$$\xi[n] = \sum_{m=-\infty}^{\infty} x[n-mN] \tag{10.75}$$

making at the same time for simplicity's sake the replacement $r \to n$. Observing now that the signal $\xi[n]$ is periodic with fundamental period N (since it results from the periodic repetition of the signal $x[n]$ every N samples), we can develop it in a Fourier series of the form

$$\xi[n] = \sum_{k=0}^{N-1}\xi_k\exp\left(j\frac{2\pi k}{N}n\right) \quad (n = 0, 1, 2, \ldots, N-1) \tag{10.76}$$

with the Fourier coefficients, as is well known, to be defined as

$$\xi_k = \frac{1}{N}\sum_{n=0}^{N-1}\xi[n]\exp\left(-j\frac{2\pi k}{N}n\right) \quad (k = 0, 1, 2, \ldots, N-1) \tag{10.77}$$

If we compare the last relation with the equation

$$\mathcal{X}\left[\exp\left(j\frac{2\pi k}{N}\right)\right] = \sum_{n=0}^{N-1}\xi[n]\exp\left(-j\frac{2\pi k}{N}n\right) \tag{10.78}$$

[21] For an alternative derivation see A.Oppenheim, R.Schafer, J.Back, *Discrete-time Signal Processing*, Sect. 8.4; in that approach it is assumed that the periodic sequence $\mathcal{X}(e^{j\omega_k})$ consists of the coefficients of the Fourier series of an unknown signal, which is ultimately proven to be the signal $\xi[n]$ resulting from the aperiodic convolution between the signal $x[n]$ and a periodic train of impulse functions.

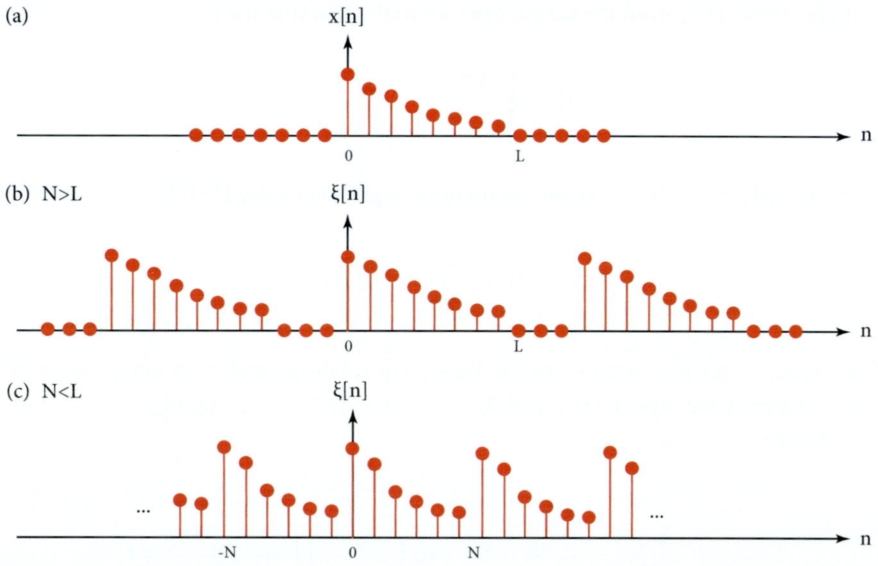

Fig. 10.12 Discrete nonperiodic sequence $x[n]$ (**a**) and its periodic repetition in absence of folding (**b**) and in the presence of folding (**c**) (Proakis and Manolakis, Figure 5.2)

it immediately follows that

$$\xi_k = \frac{1}{N}\sum_{k=0}^{N-1} \mathcal{X}\left[\exp\left(j\frac{2\pi k}{N}\right)\right] \quad (k = 0, 1, 2, \ldots, N-1) \tag{10.79}$$

and therefore the discrete periodic signal $\xi[n]$ will take the form

$$\xi[n] = \frac{1}{N}\sum_{k=0}^{N-1} \mathcal{X}\left[\exp\left(j\frac{2\pi k}{N}\right)\right]\exp\left(j\frac{2\pi k}{N}n\right) \quad (n = 0, 1, 2, \ldots, N-1) \tag{10.80}$$

Although the preceding equation allows the reconstruction of the periodic signal $\xi[n]$ from the spectrum $\mathcal{X}(e^{j\omega})$, *nothing guarantees that it will be possible to reconstruct the given spectrum from its recorded samples*. It is not difficult, however, to realize that since the signal $\xi[n]$ arises from the periodic repetition of the signal $x[n]$ every N samples, we can fully reconstruct the signal $x[n]$ from the signal $\xi[n]$ *when the phenomenon of aliasing does not occur*, that is, *when the length L of the signal $x[n]$ is less than the period N of the signal $\xi[n]$*—see Fig. 10.12b.[22] In contrast, when the phenomenon of aliasing is present, that is, when the relation $L > N$ holds (this means that there is *overlap* between the instances of the signal $x[n]$ that is repeated), the foregoing process cannot be applied (see Fig. 10.12c). We conclude, therefore, that we have the ability to fully reconstruct the spectrum $\mathcal{X}(e^{j\omega})$ of a nonperiodic discrete-time signal of duration L from its samples $\mathcal{X}(e^{j\omega_k}) = \mathcal{X}(e^{j2\pi k/N})$ $(k = 0, 1, 2, \ldots, N-1)$ by applying a three-step process that includes the following actions: (a) the calculation of the periodic signal $\xi[n]$ from the relation

$$\xi[n] = \frac{1}{N}\sum_{k=0}^{N-1} \mathcal{X}\left[\exp\left(j\frac{2\pi k}{N}\right)\right]\exp\left(j\frac{2\pi k}{N}n\right) \quad (n = 0, 1, 2, \ldots, N-1) \tag{10.81}$$

[22] In an equivalent formulation, this means that, *considering one period of the periodic signal $\xi[n]$, this is recognized as the signal $x[n]$*, an observation that agrees with the finding reported in the section on the development of Fourier discrete-time signals, where we saw that *the Fourier coefficients constituted samples of the DTFT recorded for a time interval of one period*.

(b) the determination of the signal $x[n]$ from the signal $\xi[n]$ through the expression

$$x[n] = \begin{cases} \xi[n] & 0 \leq n \leq N-1 \\ 0 & \text{anywhere else} \end{cases} \tag{10.82}$$

and (c) the calculation of the spectrum $\mathcal{X}(e^{j\omega})$ from the defining equation of the DTFT, which is known to have the form

$$\mathcal{X}(e^{j\omega}) = \sum_{n=-\infty}^{\infty} x[n]\, e^{-j\omega n} \tag{10.83}$$

In the last step of the process, let us combine all the foregoing relations trying to determine a correlation between the spectrum $\mathcal{X}(e^{j\omega})$ of the discrete-time signal $x[n]$ and its samples $\mathcal{X}(e^{j2\pi k/N})$. Taking into account that for the range of values $0 \leq n \leq N-1$ we have

$$x[n] = \xi[n] = \frac{1}{N} \sum_{k=0}^{N-1} \mathcal{X}\left[\exp\left(j\frac{2\pi k}{N}\right)\right] \exp\left(j\frac{2\pi k}{N}n\right) \tag{10.84}$$

the defining equation of the Fourier transform will take the form

$$\begin{aligned} \mathcal{X}(e^{j\omega}) &= \sum_{n=0}^{N-1} \left\{ \frac{1}{N} \sum_{k=0}^{N-1} \mathcal{X}\left[\exp\left(j\frac{2\pi k}{N}\right)\right] \exp\left(j\frac{2\pi k}{N}n\right) \right\} e^{-j\omega n} \\ &= \sum_{k=0}^{N-1} \mathcal{X}\left[\exp\left(j\frac{2\pi k}{N}\right)\right] \left\{ \frac{1}{N} \sum_{n=0}^{N-1} \exp\left[-j\left(\omega - \frac{2\pi k}{N}\right)\right] \right\} \end{aligned}$$

Defining the function

$$\mathcal{P}(e^{j\omega}) = \frac{1}{N} \sum_{n=0}^{N-1} e^{-j\omega n} = \frac{1}{N} \left\{ \frac{1 - e^{-j\omega N}}{1 - e^{-j\omega}} \right\} = \frac{\sin(\omega N/2)}{N \sin(\omega/2)} e^{-j\omega(N-1)/2} \tag{10.85}$$

(on deriving the last expression, see Exercise 10.3.2 on page 810), the preceding relationship is formulated as

$$\mathcal{X}(e^{j\omega}) = \sum_{k=0}^{N-1} \mathcal{X}\left[\exp\left(j\frac{2\pi k}{N}\right)\right] \mathcal{P}\left\{\exp\left[j\left(\omega - \frac{2\pi k}{N}\right)\right]\right\} \tag{10.86}$$

However, observing that

$$\mathcal{P}\left[\exp\left(j\frac{2\pi k}{N}\right)\right] = \begin{cases} 1 & k = 0 \\ 0 & k = 1, 2, \ldots, N-1 \end{cases} \tag{10.87}$$

we conclude that the preceding equation for the frequencies $\omega_k = 2\pi k/N$ gives *exactly* the values of the samples $\mathcal{X}(e^{j2\pi k/N})$, while for all intermediate values it calculates the values of the spectrum $\mathcal{X}(e^{j\omega})$ as a *weighted sum of the values of its samples*; it is, therefore, the correct interpolation equation that allows for *the complete reconstruction of the spectrum of the discrete-time signal* $x[n]$ *from its recorded samples.*

Summarizing the foregoing description, the sampling of the DTFT

$$\mathcal{X}(e^{j\omega}) = \sum_{n=0}^{L-1} x[n]\, e^{-j\omega n}, \quad (0 \le \omega \le 2\pi) \tag{10.88}$$

of a nonperiodic discrete-time signal $x[n]$ with length $L > N$, that is, the recording of its values at the positions of the frequencies $\omega = 2\pi k/N$ ($k = 0, 1, 2, \ldots, N-1$), leads to the discrete sequence

$$\mathcal{X}[k] \equiv \mathcal{X}\left[\exp\left(j\frac{2\pi k}{N}\right)\right] = \sum_{n=0}^{N-1} x[n] \exp\left(-j\frac{2\pi kn}{N}\right), \quad (k = 0, 1, 2, \ldots, N-1) \tag{10.89}$$

with the variable L in the upper limit of the sum having been replaced by the variable N, since we assume that for $n \ge L$ it holds that $x[n] = 0$. The preceding relation is known as the *DFT of the signal* $x[n]$ and allows for the transformation of a sequence of N discrete values $x[n]$ expressed in the domain of discrete-time into a sequence of equal number of values $\mathcal{X}[k]$ defined in the frequency domain. On the other hand, the relation

$$x[n] = \frac{1}{N} \sum_{k=0}^{N-1} \mathcal{X}[k] \exp\left(j\frac{2\pi kn}{N}\right) \quad (n = 0, 1, 2, \ldots, N-1) \tag{10.90}$$

is known as the *inverse DFT* and allows for the construction of the sequence $x[n]$ knowing the values of the spectrum samples $\mathcal{X}[k]$. In full analogy with the examples of transforms encountered in previous chapters, these two relations constitute a *pair of forward/inverse transforms*.

It is not difficult to see that for the calculation of each point of the discrete transform, N multiplications and $N-1$ additions are required, and therefore, for the calculation of a discrete transform with a length of N elements, N^2 multiplications and $N(N-1)$ additions are required, a number that makes the process of calculating the DFT quite time-consuming, especially if it has a large length. For this reason, over time more efficient ways of calculating the mentioned transform have been invented, with the most characteristic example of all being the *fast Fourier transform* (FFT), which will be presented in a subsequent section.

We conclude this brief introduction to the DFT with the observation that the N values $\omega_k = 2\pi k/N$ of the cyclic frequency at the positions where this transform is calculated concern the domain of discrete-time, while corresponding values of the continuous-time signal can arise if we know the sampling frequency F_s that was used to create the discrete sequence $x[n]$. Taking into account that the frequencies F and f of the continuous and corresponding discrete-time signals are related to each other through the sampling frequency as $f = F/F_s$ (see Sect. 6.2 on page 501), the corresponding relationship for the cyclic frequencies Ω and ω will be $\omega = 2\pi f = 2\pi F/F_s = \Omega/F_s = \Omega T_s$, where $T_s = 1/F_s$ is the sampling period. Taking into account that the N frequencies of the discrete-time signal on which the DFT is calculated are $\omega_k = 2\pi k/N$, the corresponding N frequencies of the continuous-time signal will result from the equation $\Omega_k T_s = 2\pi k/N$ and as it is found to be equal to $\Omega_k = 2\pi k/NT_s = 2\pi k F_s/N$. The preceding relationship for the simple (not cyclic) frequencies $F_k = \Omega_k/2\pi$ are $F_k = kF_s/N$ ($k = 0, 1, 2, \ldots, N-1$).

Therefore, if we apply the DFT to a set of N samples that resulted from the sampling of a continuous-time signal with a sampling frequency F_s, then the k frequencies of the continuous-time signal at positions where the DFT will be calculated are

$$F_k = k\frac{F_s}{N} = \frac{k}{NT_s}, \quad k = 0, 1, 2 \ldots, N-1$$

Adopting the argumentation developed during the description of the continuous-time Fourier transform, we can characterize the expression $1/(NT_s)$ as the fundamental frequency of the DFT and calculate the remaining frequencies F_k as integer multiples of this fundamental frequency. It is not difficult to see that the product NT_s describes the duration of sampling of the N samples of the original signal with a sampling period of T_s. These N discrete values of frequency F_k ranging from the value $F_0 = 0$ to the value $F_{N-1} = F_s$ divide the value range $[0, F_s]$ into N intervals of width F_s/N, with the interval number k having the frequency F_k at its center. This value of the parameter N is very important, because it affects the resolution of the transform (i.e., the minimum distance between two frequencies, so that they can be distinguished from each other), as well as the processing time and the volume of required calculations.

Example 10.5.1 Considering the sampling of a signal with a sampling frequency $F_s = 500$ samples/sec and the application on the resulting discrete sequence of a DFT with a length of $N = 16$ points, the fundamental frequency F_0 will be calculated as

$$F_0 = \frac{F_s}{N} = \frac{500 \text{ samples/sec}}{16 \text{ samples}} = 31.25 \text{ Hz}$$

while the remaining frequencies will have values $F_k = k \times (31.25 \text{ Hz})$ $(k = 1, 2, \ldots, 15)$. These are also the frequencies at the positions where the DFT will be calculated.

Problem 10.5.1 Prove that the DFT of the discrete sequence $x[n]$ and the Fourier coefficients x_k of its Fourier expansion are related through the expression $\mathcal{X}[k] = Nx_k$, where N is the number of points of the transform.

Solution Considering a periodic sequence $\xi[n]$ with period N, its Fourier expansion is

$$\xi[n] = \sum_{k=0}^{N-1} \xi_k \exp\left(j\frac{2\pi k}{N}n\right) \quad (-\infty < n < \infty)$$

with the Fourier coefficients defined as

$$\xi_k = \frac{1}{N}\sum_{n=0}^{N-1} \xi[n] \exp\left(-j\frac{2\pi k}{N}n\right)$$

If we define the discrete sequence $x[n] = \xi[n]$ for the region $0 \le n \le N-1$, then $x_k = \xi_k$. Comparing the foregoing relations with the defining equations of the direct and inverse DFT that we presented earlier, it immediately follows that $\mathcal{X}[k] = Nx_k$, which is the desired result.

Problem 10.5.2 Determine the correlation between the DFT and the $\mathcal{Z}$ transform.

Solution Theory reveals that the $\mathcal{Z}$ transform of a discrete sequence $x[n]$ is defined as

$$\mathcal{X}(z) = \sum_{n=-\infty}^{\infty} x[n]z^{-n}$$

and that for the DTFT of the aforementioned signal to exist, *the region of convergence of the $\mathcal{Z}$ transform must contain the unit circle*. As we have also mentioned in the previous section, *sampling the spectrum $\mathcal{X}(e^{j\omega})$ at frequency positions spaced by $\Delta\omega = 2\pi N/\Delta\omega = N/2\pi$ along the frequency axis is equivalent to sampling the $\mathcal{Z}$ transform, that is, selecting N points on the unit circle, spaced by the same angular distance*.

Assuming that the sequence $x[n]$ has a finite length N that is even less than or equal to the length of the points of the DFT, we know that this sequence *can be fully recovered from the N values of the aforementioned discrete transform*. But this means that also *the $\mathcal{Z}$ transform of the discrete sequence $x[n]$ will be recoverable from the same number of points*. Indeed, replacing the signal $x[n]$ from the equation of the inverse DFT in the defining equation of the $\mathcal{Z}$ transform, the latter relation will take the form

$$\mathcal{X}(z) = \sum_{n=0}^{N-1} x[n]z^{-n} = \sum_{n=0}^{N-1}\left[\frac{1}{N}\sum_{k=0}^{N-1}\mathcal{X}[k]\exp\left(j\frac{2\pi k}{N}n\right)\right]z^{-n} = \frac{1}{N}\sum_{k=0}^{N-1}\mathcal{X}[k]\left\{\sum_{n=0}^{N-1}\left[\frac{1}{z}\exp\left(j\frac{2\pi k}{N}\right)\right]^n\right\}$$

Referring to the mathematical typology, it immediately follows that

$$\sum_{n=0}^{N-1}\left[\frac{1}{z}\exp\left(j\frac{2\pi k}{N}\right)\right]^n = \left\{1-\left[\frac{1}{z}\exp\left(j\frac{2\pi k}{N}\right)\right]^N\right\}\Big/\left\{1-\frac{1}{z}\exp\left(j\frac{2\pi k}{N}\right)\right\} = \left\{1-z^{-N}\right\}\Big/\left\{1-\frac{1}{z}\exp\left(j\frac{2\pi k}{N}\right)\right\}$$

and the preceding equation takes the form

$$\mathcal{X}(z) = \frac{1-z^{-N}}{N}\sum_{k=0}^{N-1}\mathcal{X}[k]\Big/\left\{1-\frac{1}{z}\exp\left(j\frac{2\pi k}{N}\right)\right\}$$

which is also the final result. This equation describes the correlation that exists between the $\mathcal{Z}$ transform and its samples, and for the special case $z = e^{j\omega}$ it is reduced to the form

$$\mathcal{X}(e^{j\omega}) = \frac{1-e^{-j\omega N}}{N}\sum_{k=0}^{N-1}\mathcal{X}[k]\Big/\left\{1-\exp\left[-j\left(\omega-\frac{2\pi k}{N}\right)\right]\right\}$$

which is an alternative expression for the DTFT $\mathcal{X}(e^{j\omega})$.

Problem 10.5.3 Determine the correlation that exists between the DFT and the coefficients of the Fourier expansion of a continuous periodic signal.

Solution Considering a continuous periodic signal $x_\alpha(t)$ with a fundamental period of the form $T_0 = 1/F_0$, its Fourier expansion is known from the equation

$$x_\alpha(t) = \sum_{k=-\infty}^{\infty} x_k e^{j2\pi kF_0t}$$

with the quantities x_k being the coefficients of the Fourier expansion. Proceeding to uniform sampling with frequency $F_s = N/T_0 = 1/T_s$, we end up with the discrete-time sequence

$$x[n] = x_\alpha(nT_s) = \sum_{k=-\infty}^{\infty} x_k \exp(j2\pi kF_0nT_s)$$

(since $F_0T_s = T_s/T_0 = 1/N$). Now, following the methodology we used to derive the equation of the discrete transform (i.e., dividing the preceding infinite sum into infinite sums of N terms each), the foregoing relation takes the form

$$x[n] = \sum_{k=0}^{N-1}\left[\sum_{m=-\infty}^{\infty} x_{k-mN}\right]\exp\left(j\frac{2\pi}{N}kn\right)$$

But the last relation has the form of the inverse DFT equation, where

$$\mathcal{X}[k] = N \sum_{m=-\infty}^{\infty} x_{k-mN} \equiv N\tilde{x}_k \qquad \text{where} \qquad \tilde{x}_k = \sum_{m=-\infty}^{\infty} x_{k-mN}$$

an equation that also reveals the desired correlation. ■

Examples of the calculation of the DFT of indicative signals follow.

Problem 10.5.4 Determine the DFT with a length of N elements of the discrete impulse function (Hsu, Problem 6.49α).

Solution From the defining equation of the DFT and the discrete impulse function

$$\mathcal{X}[k] = \sum_{n=0}^{N-1} x[n] \exp\left(j\frac{2\pi k}{N}n\right) = \sum_{n=0}^{N-1} \delta[n] \exp\left(j\frac{2\pi k}{N}n\right) = 1$$

arises, which is also the desired result (see also Exercise 10.3.7 on page 814).

Problem 10.5.5 Determine the DFT with a length of N elements of a time-shifted discrete impulse function defined as $x[n] = \delta[n - n_0]$ $(0 < n_0 < N)$.

Solution From the defining equation of the discrete transform and the time-shifted discrete impulse function arises

$$\mathcal{X}[k] = \sum_{n=0}^{N-1} x[n] \exp\left(j\frac{2\pi k}{N}n\right) = \sum_{n=0}^{N-1} \delta[n] \exp\left(j\frac{2\pi k}{N}n\right) = \exp\left(j\frac{2\pi k}{N}n_0\right)$$

which is also the desired result (see also Exercise 10.3.8 on page 815).

Problem 10.5.6 Determine the DFT with a length of N elements of the rectangular discrete-time signal $x[n] = u[n] - u[n - N]$, where $u[n]$ is the discrete unit step function (Hsu, Solved Exercise 6.49β).

Solution From the definition of the discrete step function we will have

$$u[n] = \begin{cases} 1 & n \geq 0 \\ 0 & n < 0 \end{cases} \qquad \text{and} \qquad u[n-N] = \begin{cases} 1 & n \geq N \\ 0 & n < N \end{cases}$$

and from the defining equation of the DFT we get

$$\mathcal{X}[k] = \sum_{n=0}^{N-1} x[n] \exp\left(-j\frac{2\pi k}{N}n\right) = \frac{1 - e^{-j2\pi k}}{1 - \exp\left(-j\frac{2\pi k}{N}\right)} = 0$$

The preceding relation holds for $k \neq 0$, while for $k = 0$ it turns out that $\mathcal{X}[0] = N$.

Problem 10.5.7 Determine the DFT with a length of N elements of the discrete-time signal $x[n] = \alpha^n$ $(0 \le n \le N-1)$.

Solution From the defining equation of the DFT it immediately follows that

$$\mathcal{X}[k] = \sum_{n=0}^{N-1} x[n] \exp\left(-j\frac{2\pi k}{N} n\right) = \sum_{n=0}^{N-1} \left[\alpha \exp\left(-j\frac{2\pi k}{N}\right)\right]^n = \frac{1 - \left[\alpha \exp\left(-j\frac{2\pi k}{N}\right)\right]^N}{1 - \alpha \exp\left(-j\frac{2\pi k}{N}\right)}$$

and finally

$$\mathcal{X}[k] = \frac{1 - \alpha^N}{1 - \alpha \exp\left(-j\frac{2\pi k}{N}\right)}$$

Problem 10.5.8 Determine the DFT with a length of N elements of the signal

$$x[n] = \begin{cases} e^{j\omega_0 n} & 0 \le n \le N-1 \\ 0 & \text{elsewhere} \end{cases}$$

and to prove that this arises from the discrete-time Fourier transform $\mathcal{X}(e^{j\omega})$ for frequency values $\omega = 2\pi k/N$.

Solution From the defining equation of the DFT and using the methodology of Exercise 10.2.4 (refer to page 788), it turns out that

$$\mathcal{X}[k] = \sum_{n=0}^{N-1} x[n] \exp\left(-j\frac{2\pi k}{N} n\right) = \sum_{n=0}^{N-1} e^{j\omega_0 n} \exp\left(-j\frac{2\pi k}{N} n\right) = \sum_{n=0}^{N-1} \exp\left[-j\left(\frac{2k\pi}{N} - \omega_0\right) n\right]$$

$$= \frac{1 - \exp\left[-j\left(\frac{2k\pi}{N} - \omega_0\right) N\right]}{1 - \exp\left[-j\left(\frac{2k\pi}{N} - \omega_0\right)\right]} = \exp\left[-j\left(\frac{2k\pi}{N} - \omega_0\right)\left(\frac{N-1}{2}\right)\right] \left\{\frac{\sin\left[\left(\frac{2k\pi}{N} - \omega_0\right)\frac{N}{2}\right]}{\sin\left[\frac{1}{2}\left(\frac{2k\pi}{N} - \omega_0\right)\right]}\right\}$$

On the other hand, the DTFT of the signal $x[n]$ is calculated as

$$\mathcal{X}(e^{j\omega}) = \sum_{n=-\infty}^{\infty} x[n]\, e^{-j\omega_0 n} = \sum_{n=0}^{N-1} e^{j\omega_0 n} e^{-j\omega n} = \sum_{n=0}^{N-1} e^{-j(\omega-\omega_0)n} = \frac{1 - e^{-j(\omega-\omega_0)N}}{1 - e^{-j(\omega-\omega_0)}}$$

$$= \exp\left[-j(\omega - \omega_0)\left(\frac{N-1}{2}\right)\right] \left\{\sin\left[(\omega - \omega_0)\frac{N}{2}\right]\right\} \Big/ \left\{\sin\left(\frac{\omega - \omega_0}{2}\right)\right\}$$

Comparing the expressions of the $\mathcal{X}(e^{j\omega})$ and $\mathcal{X}[k]$ transforms, we immediately realize how the second function arises from the first for frequency values $\omega_k = 2\pi k/N (k = 0, 1, 2, \ldots, N-1)$, a finding that proves the desired result.

Problem 10.5.9 Determine the DFT with a length of N elements of the discrete-time signal $x[n] = \exp(j2\pi k_0 n/N)$ $(0 \le n \le N-1)$.

Solution From the defining equation of the DFT it immediately follows that

$$\mathcal{X}[k]=\sum_{n=0}^{N-1} x[n]\exp\left(-j\frac{2\pi k}{N}n\right)=\sum_{n=0}^{N-1}\exp\left[-j\frac{2\pi}{N}(k-k_0)n\right]=\frac{1-\exp[-j2\pi(k-k_0)]}{1-\exp\left[-j\frac{2\pi}{N}(k-k_0)\right]}$$

It is not difficult to see that for $k \neq k_0$ the exponential expression of the numerator is equal to unityand therefore we will have $\mathcal{X}[k]=0$, while for $k=k_0$ the foregoing relation takes the indeterminate form $0/0$. Therefore, applying L'Hôpital's rule, we get

$$\mathcal{X}[k]=\lim_{k\to k_0}\frac{d}{dk}\left\{\frac{1-\exp[-j\,2\pi(k-k_0)]}{1-\exp\left[-j\frac{2\pi}{N}(k-k_0)\right]}\right\}=\left.\frac{j2\pi\exp[-j\,2\pi(k-k_0)]}{\frac{j\,2\pi}{N}\exp\left[-j\frac{2\pi}{N}(k-k_0)\right]}\right|_{k=k_0}=N$$

Combining the preceding results, we will get

$$\mathcal{X}[k]=\begin{cases}0 & k\neq k_0\\ N & k=k_0\end{cases}=N\begin{cases}0 & k\neq k_0\\ 1 & k=k_0\end{cases}=N\delta[k-k_0]$$

10.5.2 Matrix Form of Discrete Transform

Defining for simplicity the quantity $W_N=\exp(-j2\pi/N)$, the equations of the direct and inverse DFT are formulated as

$$\mathcal{X}[k]=\sum_{n=0}^{N-1} x[n]\,W_N^{kn}\quad (k=0,1,2,\ldots,N-1) \tag{10.91}$$

$$x[n]=\frac{1}{N}\sum_{k=0}^{N-1}\mathcal{X}[k]\,W_N^{-kn}\quad (n=0,1,2\ldots,N-1) \tag{10.92}$$

Taking into account that the DFT is characterized by the property of *linearity*,[23] we can consider this transform as a mapping of linear nature of the vector of times samples $\boldsymbol{x}_N=\{x[0],x[1],x[2],\ldots,x[N-1]\}^T$ to the vector of frequency samples $\boldsymbol{\mathcal{X}}_N=\{\mathcal{X}[0],\mathcal{X}[1],\mathcal{X}[2],\ldots,\mathcal{X}[N-1]\}^T$ through a relationship of the form

$$\boldsymbol{\mathcal{X}}_N=\boldsymbol{W}_N\boldsymbol{x}_N \tag{10.93}$$

where

[23] Although the properties of the DFT will be studied in the next section, this conclusion arises as a direct consequence of the fact that the DTFT leading to the DFT is a linear transform.

$$\mathbf{W}_N = \begin{bmatrix} 1 & 1 & 1 & \dots & 1 \\ 1 & W_N & W_N^2 & \dots & W_N^{N-1} \\ 1 & W_N^2 & W_N^4 & \dots & W_N^{2(N-1)} \\ \dots & \dots & \dots & \ddots & \dots \\ 1 & W_N^{N-1} & W_N^{2(N-1)} & \dots & W_N^{(N-1)(N-1)} \end{bmatrix} \tag{10.94}$$

The matrix $\mathbf{W}_N$ appearing in the foregoing relation is known as the *matrix of the DFT of dimension N* and is a *symmetric matrix*, in the sense that it *coincides with its inverse*; therefore, $\mathbf{W}_N = (\mathbf{W}_N)^T$. Now, using this equation, we can formulate the inverse DFT using the matrix expression

$$\boldsymbol{x}_N = \mathbf{W}_N^{-1}\boldsymbol{\mathcal{X}}_N \tag{10.95}$$

where to derive the last expression we multiplied the expression of the forward transform matrix by the matrix $\mathbf{W}_N^{-1}$. On the other hand, however, if we apply the same logic to the original equation of the inverse discrete transform, it can be formulated as

$$\boldsymbol{x}_N = \frac{1}{N}\mathbf{W}_N^{*}\boldsymbol{\mathcal{X}}_N \tag{10.96}$$

with the symbol $*$ referring to the operation of *complex conjugation*. Comparing the last two equations, it immediately follows that $\mathbf{W}_N^{*} = N\mathbf{W}_N^{-1}$, an expression that constitutes an interesting property of the inverse transform matrix.[24]

In practice, the contents of matrices $\mathbf{W}_N$ for the various values of the length N of the transform are calculated in advance and used directly in matrix multiplications. For example, matrices $\mathbf{W}_2$ and $\mathbf{W}_3$ are proven to have the form

$$\mathbf{W}_2 = \begin{bmatrix} 1 & 1 \\ 1 & -1 \end{bmatrix}, \quad \mathbf{W}_3 = \begin{bmatrix} 1 & 1 & 1 \\ 1 & -\dfrac{1+j\sqrt{3}}{2} & -\dfrac{1-j\sqrt{3}}{2} \\ 1 & -\dfrac{1-j\sqrt{3}}{2} & -\dfrac{1+j\sqrt{3}}{2} \end{bmatrix}$$

while matrix $\mathbf{W}_4$, which allows the calculation of DFTs of length $N = 4$, is

$$\mathbf{W}_4 = \begin{bmatrix} 1 & 1 & 1 & 1 \\ 1 & -j & -1 & j \\ 1 & -1 & 1 & -1 \\ 1 & j & -1 & -j \end{bmatrix}$$

Using the preceding representation in matrix form of the DFT, we can directly calculate the aforementioned transform by performing simple multiplications between matrices, as demonstrated in the following exercises, along with their solutions.

---

[24] From the equation $\mathbf{W}_N^{*} = N\mathbf{W}_N^{-1}$ it immediately follows that $\mathbf{W}_N\mathbf{W}_N^{*} = N\boldsymbol{I}_N$, where I_N is *the unit matrix of order N*. Therefore, we will have $\boldsymbol{I}_N = N^{-1}\mathbf{W}_N\mathbf{W}_N^{*} = (N^{-1/2}\mathbf{W})(N^{-1/2}\mathbf{W}^{*}) = \boldsymbol{Q}\boldsymbol{Q}^{*}$, and consequently, the symmetric matrix $\boldsymbol{Q} = N^{-1/2}\mathbf{W}_N$, which is known as the *normalized matrix of the DFT*, will be an *orthonormal matrix*. Therefore, the columns of this matrix will constitute an *orthonormal basis in a complex vector space of N dimensions*, and if we symbolize these columns as $N^{-1/2}\boldsymbol{f}_{N,k}$ $(0 \le k \le N-1)$, then the equation of the inverse discrete transform is formulated as

$$\boldsymbol{x}_N = \frac{1}{N}\mathbf{W}_N^{*}\boldsymbol{\mathcal{X}}_N = \sum_{k=0}^{N-1}(N^{-1/2}\mathcal{X}[k])(N^{-1/2}\boldsymbol{f}_{N,k})$$

The physical significance of the foregoing relationship is that *the quantities $N^{-1/2}\mathcal{X}[k]$ constitute the coordinates of the vector $\boldsymbol{x}_N$ in this vector space*; for this reason, the quantities $\mathcal{X}[k]$ (with the approximation of the multiplicative constant $N^{-1/2}$) are often characterized as *the basis of the DFT of the discrete sequence $x[n]$*.

Problem 10.5.10 Using the representation in matrix form of the DFT, determine the transform of the discrete sequence $x[n] = [1, 3, 0, 2]$ corresponding to the value $N = 4$.

Solution Applying the foregoing procedure, it immediately follows that

$$\mathcal{X}_4 = \mathbf{W}_4 \boldsymbol{x}_4 = \begin{bmatrix} 1 & 1 & 1 & 1 \\ 1 & -j & -1 & j \\ 1 & -1 & 1 & -1 \\ 1 & j & -1 & -j \end{bmatrix} \begin{bmatrix} 1 \\ 3 \\ 0 \\ -2 \end{bmatrix} = \begin{bmatrix} 2 \\ 1-5j \\ 0 \\ 1+5j \end{bmatrix}$$

Problem 10.5.11 Using the representation in matrix form of the DFT, determine the transform of the discrete sequence $x[n] = [1, 2, 3, 1]$ corresponding to the value $N = 4$.

Solution It immediately follows the application of the foregoing process that

$$\mathcal{X}_4 = \mathbf{W}_4 \boldsymbol{x}_4 = \begin{bmatrix} 1 & 1 & 1 & 1 \\ 1 & -i & -1 & i \\ 1 & -1 & 1 & -1 \\ 1 & i & -1 & -i \end{bmatrix} \begin{bmatrix} 1 \\ 2 \\ 3 \\ 1 \end{bmatrix} = \begin{bmatrix} 7 \\ -2-j \\ 1 \\ -2+j \end{bmatrix}$$

Problem 10.5.12 Using the matrix representation of the DFT, determine the inverse transform of the discrete sequence $\mathcal{X}[k] = [6, -2+2j, -2, -2-2j]$ corresponding to the value $N = 4$.

Solution In this case, we will use matrix $\mathbf{W}_4^{-1}$, which is calculated as

$$\mathbf{W}_4^{-1} = \frac{1}{4}\mathbf{W}_4^{*} = \frac{1}{4} \begin{bmatrix} 1 & 1 & 1 & 1 \\ 1 & j & -1 & -j \\ 1 & -1 & 1 & -1 \\ 1 & -j & -1 & j \end{bmatrix}$$

Thus, it will hold that

$$\boldsymbol{x}_4 = \frac{1}{4}\mathbf{W}_4^{*}\mathcal{X}_4 = \frac{1}{4} \begin{bmatrix} 1 & 1 & 1 & 1 \\ 1 & j & -1 & -j \\ 1 & -1 & 1 & -1 \\ 1 & -j & -1 & j \end{bmatrix} \begin{bmatrix} 6 \\ -2+2j \\ -2 \\ -2-2j \end{bmatrix} = \begin{bmatrix} 0 \\ 1 \\ 2 \\ 3 \end{bmatrix}$$

10.5.3 Properties of DFT

The DFT of finite-length discrete-time signals is characterized by a set of useful properties that are presented in detail in the following pages. Taking into account the close relationship that exists between the aforementioned transform and the coefficients of the Fourier expansion of discrete-time signals (see Exercise 10.5.1 on page 852), several of these properties will be given *without proof*, since it is based on the same philosophy and methodology as the proofs presented so far; however, readers are encouraged to verify them, so as to familiarize themselves with their use. Let us note, incidentally, that the DFT presents several variations compared to the transforms we presented previously, mainly in terms of *the finite length of the sequences on which it is applied*, as well as the property of *circular convolution*; these differences will be emphasized at the appropriate degree and will be demonstrated through the presentation of exemplary solved exercises.

- **Linearity**: If the DFTs of the sequences $x[n]$ and $y[n]$ with lengths N_1 and N_2 samples are $\mathcal{X}[k]$ and $\mathcal{Y}[k]$, then the DFT of the linear combination $z[n] = \alpha x[n] + \beta y[n]$ will be the function $\mathcal{Z}[k] = \alpha\mathcal{X}[k] + \beta\mathcal{Y}[k]$. Taking into account that the length of the $\mathcal{Z}[k]$ transform will be equal to $N_3 = \max(N_1, N_2)$, it is obvious that the common length of the $\mathcal{X}[k]$ and $\mathcal{Y}[k]$ transforms should have a value $N \geq N_3$. Therefore, if the sequence $x[n]$ has a length $N_1 < N_2$, *it should be supplemented with* $N_2 - N_1$ *zero values*, so as to obtain the same length as the sequence $y[n]$. In this case, the $\mathcal{X}[k]$ and $\mathcal{Y}[k]$ transforms will be calculated from the relations

$$\left.\begin{aligned} \mathcal{X}[k] &= \sum_{n=0}^{N_1-1} x[n]W_{N_2}^{kn} \\ \mathcal{Y}[k] &= \sum_{n=0}^{N_2-1} y[n]W_{N_2}^{kn} \end{aligned}\right\} \quad (0 \leq k \leq N_2 - 1) \tag{10.97}$$

and, of course, they will have a length equal to N_2, while the $\mathcal{Z}[k]$ transform and the property of *linearity* that characterizes it arise directly from the equation defining the forward discrete transform.
- **Periodicity**: If the discrete sequences $x[n]$ and $\mathcal{X}[k]$, with a length of N samples, constitute a pair of direct/inverse DFT, it is directly proven from their defining equations that they constitute periodic functions of period N, or, in mathematical notation,

$$x[n+N] = x[n] \quad (\forall n) \qquad \text{and} \qquad \mathcal{X}[k+N] = \mathcal{X}[k] \quad (\forall k)$$

- **Symmetry**: The properties of *symmetry* of the DFT arise in the way we saw in the DTFT. Let there be two complex discrete sequences $x[n]$ and $\mathcal{X}[k]$ each with a length of N samples, which constitute a *pair of forward/inverse transforms.*

$$\begin{aligned} x[n] &= \Re\{x[n]\} + j\Im\{x[n]\} \quad (0 \leq n \leq N-1) \\ \mathcal{X}[k] &= \Re\{\mathcal{X}[k]\} + j\Im\{\mathcal{X}[k]\} \quad (0 \leq k \leq N-1) \end{aligned}$$

By substituting the preceding relations into the defining equations of the direct and inverse DFT, we can construct the equations that correlate the real and imaginary parts of the functions $x[n]$ and $\mathcal{X}[k]$ and that have the form

$$\Re\{\mathcal{X}[k]\} = \sum_{n=0}^{N-1} \left[\Re\{x[n]\}\cos\left(\frac{2\pi kn}{N}\right) + \Im\{x[n]\}\sin\left(\frac{2\pi kn}{N}\right)\right]$$

$$\Im\{\mathcal{X}[k]\} = \sum_{n=0}^{N-1} \left[\Im\{x[n]\}\cos\left(\frac{2\pi kn}{N}\right) - \Re\{x[n]\}\sin\left(\frac{2\pi kn}{N}\right)\right]$$

$$\Re\{x[n]\} = \frac{1}{N}\sum_{k=0}^{N-1}\left[\Re\{\mathcal{X}[k]\}\cos\left(\frac{2\pi kn}{N}\right) - \Im\{\mathcal{X}[k]\}\sin\left(\frac{2\pi kn}{N}\right)\right]$$

$$\Im\{x[n]\} = \frac{1}{N}\sum_{k=0}^{N-1}\left[\Re\{\mathcal{X}[k]\}\sin\left(\frac{2\pi kn}{N}\right) + \Im\{\mathcal{X}[k]\}\cos\left(\frac{2\pi kn}{N}\right)\right]$$

The foregoing properties apply to the case of complex discrete-time signals, while they can be formulated in a simpler form if the signals are *purely real or imaginary discrete sequences*. Assuming, for example, that the discrete sequence $x[n]$ is a real-valued signal, we will have

$$\mathcal{X}[N-k] = \mathcal{X}^*[k] = \mathcal{X}[-k]$$

and therefore we will have $|\mathcal{X}[N-k]| = |\mathcal{X}[k]|$ and $\measuredangle\mathcal{X}[N-k] = -\measuredangle\mathcal{X}[k]$. If the real signal $x[n]$ is characterized by *even* or *odd symmetry*, things are simplified even further. It is proven that for the case of an *even real signal of length N* the pair of forward/inverse transforms has the form

$$\mathcal{X}[k] = \sum_{n=0}^{N-1} x[n]\cos\left(\frac{2\pi kn}{N}\right) \quad (0 \le k \le N-1) \tag{10.98}$$

$$x[n] = \frac{1}{N}\sum_{k=0}^{N-1} \mathcal{X}[k]\cos\left(\frac{2\pi kn}{N}\right) \quad (0 \le n \le N-1) \tag{10.99}$$

while the corresponding equations for an *odd real signal* are

$$\mathcal{X}[k] = -j\sum_{n=0}^{N-1} x[n]\sin\left(\frac{2\pi kn}{N}\right) \quad (0 \le k \le N-1) \tag{10.100}$$

$$x[n] = j\frac{1}{N}\sum_{k=0}^{N-1} \mathcal{X}[k]\sin\left(\frac{2\pi kn}{N}\right) \quad (0 \le n \le N-1) \tag{10.101}$$

Finally, for a *purely imaginary signal* it is proven that the real and imaginary parts of the DFT have the form

$$\Re\{\mathcal{X}[k]\} = \sum_{n=0}^{N-1} \Im\{x[n]\}\sin\left(\frac{2\pi kn}{N}\right) \tag{10.102}$$

$$\Im\{\mathcal{X}[k]\} = \sum_{n=0}^{N-1} \Im\{x[n]\}\cos\left(\frac{2\pi kn}{N}\right) \tag{10.103}$$

that is, the functions $\Im\{\mathcal{X}[k]\}$ and $\Re\{\mathcal{X}[k]\}$ constitute *even and odd functions of the variable k*.
The basic consequence of the symmetry of a DFT applied to a signal of real values is that of the N terms of the result, only the $(N/2)+1$ terms are independent of each other, while the remaining terms can be calculated from the previous ones. Therefore, for the calculation of the DFT of a discrete sequence $x[n]$ we only need the first $(N/2)+1$ terms of the sequence $\mathcal{X}[k]$, while the terms from $\mathcal{X}[(N/2)+1]$ to $\mathcal{X}[N-1]$ do not provide any additional information regarding the spectrum of $x[n]$.

- **Cyclic time shift**: If the DTFT of the discrete aperiodic sequence $x[n]$ is the function $\mathcal{X}(e^{j\omega})$, then the transform of the *temporally shifted discrete sequence* $y[n] = x[n-\ell]$ will be the function $\mathcal{Y}(e^{j\omega}) = e^{-j\omega\ell}\mathcal{X}(e^{j\omega})$, while the coefficients of the corresponding *Fourier series expansion* will be calculated as $y_k = e^{-j2\pi k\ell/N}x_k$ (if, of course, the sequence $x[n]$ is periodic with period N). It is not difficult to see that if we try to extend the same property to the space of the DFT, we will encounter problems; in other words, trying to determine the discrete sequence $y[n]$ whose Fourier transform has the form

$$\mathcal{Y}[k] = \exp\left(-j\frac{2\pi k\ell}{N}\right)\mathcal{X}[k] \tag{10.104}$$

we cannot claim that it will result from a simple time shift, as we saw earlier, since the sequences $x[n]$ and $y[n]$ have finite length and therefore have *zero values* for index n values outside the range $0 \leq n \leq N-1$. So how should we deal with such a situation?

To address this issue, let us return to the introductory section on the DFT; there we saw *that the discrete transform of N points of a discrete sequence $x[n]$ of finite length $L \leq N$ was equivalent to the discrete transform of N points of a periodic sequence $\xi[n]$ of infinite length that arises from the periodic repetition of the discrete-time signal $x[n]$ every N samples* and therefore is defined as[25]

$$\xi[n] = \sum_{m=-\infty}^{\infty} x[n-mN] = x[(n \text{ modulo } N)] = x[((n))_N] \tag{10.105}$$

where, as is well known, the operation (n modulo N) returns the *remainder* of the division n/N, with the parameters n and N being integer numbers.

Based on this equivalence, let us apply the operation of *time shift* by ℓ time units, not to the discrete aperiodic sequence of finite length L, but to the *infinite discrete sequence* $\xi[n]$ with period N. If $\xi'[n]$ is the discrete-time signal resulting from this process, it will be defined as

$$\xi'[n] = \xi[n-\ell] = \sum_{m=-\infty}^{\infty} x[n-\ell-mN] \tag{10.106}$$

In this case, the discrete-time signal $x'[n]$ of finite length that results by considering the part of the signal $\xi'[n]$ for a duration of one period and therefore is defined as

$$x'[n] = \begin{cases} \xi'[n] & 0 \leq n \leq N-1 \\ 0 & \text{anywhere else} \end{cases} \tag{10.107}$$

will be related to the original discrete-time signal $x[n]$, which results in the same way from the periodic signal of infinite length $\xi[k]$, through a *cyclic shift*; this is also the way we deal with time shifts in *discrete sequences of finite length*.

Generalizing the process of the previous example for any value of time shift ℓ and for any period value N, we can formulate the time-shifted discrete sequence $x'[n]$ that arises in the previously described way as

[25] In reality, this conclusion arises from the combination of the following conclusions: (a) the coefficients ξ_k of the Fourier expansion of the discrete periodic signal $\xi[n]$ of period N arise from the *sampling* of the DTFT $\Xi(e^{j\omega}) = \mathcal{F}\{\xi[n]\}$ at frequency values $\omega_k = 2k\pi/N$, (b) the coefficients ξ_k constitute a *periodic function* of the parameter k with period N, and (c) the DFT of the discrete sequence $x[n]$ arises from the uniform sampling of the $\mathcal{X}(e^{j\omega})$ transform in the interval $[0, N]$ and at the positions of the frequencies ω_k defined earlier. Taking into account that the signal $x[n]$ coincides with the signal $\xi[n]$ *for a duration of one period*, we can formulate the claim that *the samples of the DFT of the signal $x[n]$ are simply the coefficients ξ_k of the Fourier expansion of the periodic signal $\xi[n]$ for the same time period*. Therefore, the quantities ξ_k and $\mathcal{X}[k]$ are related through the relations

$$\mathcal{X}[k] = \begin{cases} \xi_k & 0 \leq k \leq N-1 \\ 0 & \text{anywhere else} \end{cases} \quad \text{and} \quad \xi_k = \mathcal{X}[(k \text{ modulo } N)] = \mathcal{X}[((k))_N].$$

$$x'[n] = x[(n-\ell) \text{ modulo } N] = x[((n-\ell))_N] \tag{10.108}$$

Having understood the necessity and method of using the cyclic shift,[26] we can reformulate the property of the time shift of the DFT as follows: *If a DFT of length N of the N-element sequence* $x[n]$ *is the function* $\mathcal{X}[k]$*, then the discrete Fourier transform of the sequence* $y[n] = x[((n-\ell))_N]$ cyclically shifted by ℓ *elements will be the function* $\mathcal{Y}[k] = e^{-j2k\pi\ell/N}\mathcal{X}[k]$.

Proof From the defining equation of the DFT of the sequence $x[n]$ we will have

$$\mathcal{Y}[k] \sum_{n=0}^{N-1} y[n] \exp\left(-j\frac{2\pi kn}{N}\right) = \sum_{n=0}^{N-1} x[((n-\ell))_N] \exp\left(-j\frac{2\pi kn}{N}\right)$$
$$\sum_{n=0}^{\ell-1} x[((n-\ell))_N] \exp\left(-j\frac{2\pi kn}{N}\right) + \sum_{n=\ell}^{N-1} x[n-\ell] \exp\left(-j\frac{2\pi kn}{N}\right)$$

with the value of the shift ℓ having been chosen in the range $0 \le \ell \le N-1$. However, observing that

$$x[((n-\ell))_N] = x[N-\ell+n]$$

as we saw in the previous example, the first sum is formulated as

$$\sum_{n=0}^{\ell-1} x[((n-\ell))_N] \exp\left(-j\frac{2\pi kn}{N}\right) = \sum_{n=0}^{\ell-1} x[N-\ell+n] \exp\left(-j\frac{2\pi kn}{N}\right) = \sum_{m=N-\ell}^{N-1} x[m] \exp\left[-j\frac{2k\pi(m+\ell)}{N}\right]$$

where for the derivation of the last sum we proceeded to change the dummy index from n to $m = N-\ell+n$ and took into account that the exponential expression appearing in it is periodic with period N. On the other hand, the second sum is formulated as

---

[26] Using the concept of *cyclic shift* presented here, we can define a finite-length discrete sequence N as *cyclically symmetric* when it satisfies the condition $x[N-n] = x[n]$ $(1 \le n \le N-1)$ and as *cyclically antisymmetric* if the relation $x[N-n] = -x[n]$ holds for the same range of index values n. In addition, based on these concepts, we can alternatively define a signal as *even* if it is $x[n] = x^*[N-n]$ and as *odd* if it holds that $x[n] = -x^*[N-n]$. It is also useful to mention that in full analogy with the equations

$$\xi_e[n] = \frac{1}{2}\left\{\xi[n] + \xi^*[-n]\right\} \quad \text{and} \quad \xi_o[n] = \frac{1}{2}\left\{\xi[n] - \xi^*[-n]\right\}$$

which allow the formulation of a periodic discrete sequence as the sum of a conjugate symmetric and a conjugate antisymmetric periodic component, we can formulate in a similar way a finite-length discrete sequence $x[n]$ N *as the sum of two finite-length sequences N that correspond to a period of the conjugate symmetric and conjugate antisymmetric component*. Therefore, defining the periodic signal $\xi[n]$ from the relation $\xi[n] = x[((n))_N]$ and using the definitions of the sequences $\xi_e[n]$ and $\xi_o[n]$, we can define the sequences

$$x_e[n] = \xi_e[n] \quad (0 \le n \le N-1) \quad \text{and} \quad x_o[n] = \xi_o[n] \quad (0 \le n \le N-1)$$

or, in equivalent notation,

$$x_e[n] = \frac{1}{2}\left\{x[((n))_N] + x^*[((n))_N]\right\} \quad \text{and} \quad x_o[n] = \frac{1}{2}\left\{x[((n))_N] - x^*[((n))_N]\right\}$$

with the index n taking values in the interval $[0, N-1]$. It is proven that the discrete sequences $x_e[n]$ and $x_o[n]$ (which are known as the *periodic conjugate symmetric* and *periodic conjugate antisymmetric* components, respectively) *do not coincide* with the even and odd components of the signal $x[n]$, but in reality they are related to each other through the expressions

$$x_e[n] = \mathcal{E}\text{v}\{x[n]\} + \mathcal{E}\text{v}\{x[n-N]\} \quad \text{and} \quad x_o[n] = \mathcal{O}\text{d}\{x[n]\} + \mathcal{O}\text{d}\{x[n-N]\}$$

the proof of which is left as an exercise to the reader.

$$\sum_{n=\ell}^{N-1} x[n-\ell]\exp\Big(-j\frac{2\pi kn}{N}\Big) = \sum_{m=0}^{N-\ell-1} x[m]\exp\Big[-j\frac{2k\pi(m+\ell)}{N}\Big]$$

where in this case we proceeded to change the dummy index from n to $m = n - \ell$. Replacing the final expressions of the preceding sums in the equation of the function $\mathcal{Y}[k]$, this is written as

$$\begin{aligned}\mathcal{Y}[k] &\sum_{m=0}^{N-\ell-1} x[m]\exp\Big[-j\frac{2k\pi(m+\ell)}{N}\Big] + \sum_{m=N-\ell}^{N-1} x[m]\exp\Big[-j\frac{2k\pi(m+\ell)}{N}\Big]\\ &= \exp\Big(-j\frac{2\pi k\ell}{N}\Big)\sum_{m=0}^{N-1} x[m]\exp\Big(-j\frac{2\pi km}{N}\Big) = \exp\Big(-j\frac{2\pi k\ell}{N}\Big)\mathcal{X}[k]\end{aligned}$$

- **Cyclic frequency shift**: If a DFT with a length of N of the discrete sequence $x[n]$ of length N is the function $\mathcal{X}[k]$, then the DFT of the discrete sequence $y[n] = \exp(j2\pi\ell n/N)x[n]$ will be the function $\mathcal{X}[((k-\ell))_N]$, which results from the *cyclic frequency shift* of the function $\mathcal{X}[k]$ by ℓ samples.
  The proof of this property is exactly the same as the previous one: Starting from the defining equation of the inverse transform

$$y[n] = \frac{1}{N}\sum_{k=0}^{N-1} \mathcal{X}[((k-\ell))_N]\exp\Big(j\frac{2\pi kn}{N}\Big) \tag{10.109}$$

  and following the steps outlined in the preceding discussion, we find that $y[n] = \exp(j2\pi\ell n/N)x[n]$.
- **Time reversal**: If the DFT of length N of the discrete sequence $x[n]$ with a length of N elements is the function $\mathcal{X}[k]$, then the DFT of the discrete sequence $y[n] = x[((-n))_N] = x[N-n]$ is proven to have the form

$$\mathcal{Y}[k] = \mathcal{X}[((-k))_N] = \mathcal{X}[N-k]$$

  The physical significance of this property is that *the time reversal of a sequence of N points is equivalent to the reversal of the samples of the discrete transformation in the frequency domain.*
  Proof From the equation of the DFT of the sequence $y[n] = x[N-n]$ we will have

$$\mathcal{Y}[k] = \sum_{n=0}^{N-1} x[N-n]\exp\Big(-j\frac{2\pi kn}{N}\Big)$$

  Proceeding to change the dummy index from n to $m = n - N$, the preceding sum takes the form

$$\begin{aligned}\mathcal{Y}[k] &= \sum_{m=0}^{N-1} x[m]\exp\Big[-j\frac{2\pi k(N-m)}{N}\Big] = \sum_{m=0}^{N-1} x[m]\exp\Big(j\frac{2\pi km}{N}\Big)\\ &= \sum_{m=0}^{N-1} x[m]\exp\Big[-j\frac{2\pi m(N-k)}{N}\Big] = \mathcal{X}[n-k]\end{aligned}$$

- **Duality**: For the formulation and proof of the property of duality we can use the close correlation that exists *between the DFT of the discrete sequence* $x[n]$ *and the coefficients of the Fourier expansion* described in the property of time cyclic shift (see Footnote 25 on page 861). Let us consider, therefore, two sequences, $x[n]$ and $\mathcal{X}[k]$, of finite length N which form a pair of *forward/inverse transforms*, and let us construct the periodic sequences

$$\xi[n] = x[((n))_N] \qquad \text{and} \qquad \xi_k = \mathcal{X}[((k))_N]$$

such that *the time series* ξ_k *constitutes the coefficients of the Fourier expansion of the discrete periodic sequence* $\xi[n]$. In this case, using the property of *duality* that characterizes the coefficients of the Fourier expansion of discrete-time signals (see Sect. 10.2.2 on page 789), we know that the coefficients of the Fourier expansion of the discrete periodic sequence ξ_n are the $\xi[-k]/N$.

Now defining the periodic sequence ξ_n that results from the *periodic repetition* of the sequence ξ_n, the coefficients of its Fourier expansion will be the $\Xi_k = \xi[-k]/N$. Therefore, the DFT of the sequence Ξ_n will be defined as

$$\Xi[k] = \begin{cases} \xi[-k]/N & 0 \leq k \leq N-1 \\ 0 & \text{elsewhere} \end{cases}$$

or, equivalently, as

$$\Xi[k] = \begin{cases} x[((-k))_N]/N & 0 \leq k \leq N-1 \\ 0 & \text{elsewhere} \end{cases}$$

Based on all the foregoing points, we can formulate the property of the duality of the DFT in the following way: *If the DFT with a length of N elements of the discrete sequence* $x[n]$ *of length N is the function* $\mathcal{X}[k]$, *then the DFT of the discrete sequence* $\mathcal{X}[n]$ *will be the function* $x[((-k))_N]/N$.

- **Conjugation**: Combining the equations $\xi[n] = x[((n))_N]$ and $\xi_k = \mathcal{X}[((k))_N]$, which were formulated previously, along with the properties of *symmetry* that characterize the coefficients of the Fourier expansion of a periodic discrete-time signal, it is proven that the DFT of the discrete sequence $y[n] = x^*[n]$ is the function $\mathcal{Y}[k] = \mathcal{X}^*[((-k))_N] = \mathcal{X}^*[N-k]$, while the DFT of the discrete-time signal $y[n] = x^*[((-n))_N]$ is the function $\mathcal{Y}[k] = \mathcal{X}^*[k]$.
- **Circular convolution**: Let us consider two discrete sequences, $x[n]$ and $y[n]$, of finite length N whose DFTs are defined as

$$\mathcal{X}[k] = \sum_{n=0}^{N-1} x[n] \exp\left(-j\frac{2\pi kn}{N}\right) \quad (k = 0, 1, 2, \ldots, N-1) \tag{10.110}$$

$$\mathcal{Y}[k] = \sum_{n=0}^{N-1} y[n] \exp\left(-j\frac{2\pi kn}{N}\right) \quad (k = 0, 1, 2, \ldots, N-1) \tag{10.111}$$

and let us multiply them so as to construct the discrete transforms

$$\mathcal{Z}[k] = \mathcal{X}[k]\mathcal{Y}[k] \quad (k = 0, 1, 2, \ldots, N-1) \tag{10.112}$$

which will also consist of N samples. If we denote by $z[n]$ the discrete-time signal corresponding to the $\mathcal{Z}[k]$ transform, it is obvious that this will be related to the signals $x[n]$ and $y[n]$ that correspond to the $\mathcal{X}[k]$ and $\mathcal{Y}[k]$ transforms; let us try, then, to determine this correlation.

Starting from the defining equation of the inverse Fourier transform and using the expressions of the $\mathcal{X}[k]$ and $\mathcal{Y}[k]$ transforms given earlier, we will successively obtain

$$z[m] = \frac{1}{N}\sum_{k=0}^{N-1} \mathcal{Z}[k] \exp\left(j\frac{2\pi km}{N}\right) = \frac{1}{N}\sum_{k=0}^{N-1} \mathcal{X}[k]\,\mathcal{Y}[k] \exp\left(j\frac{2\pi km}{N}\right)$$

$$= \frac{1}{N}\sum_{k=0}^{N-1}\left[\sum_{n=0}^{N-1} x[n]\exp\left(-j\frac{2\pi kn}{N}\right)\right]\left[\sum_{\ell=0}^{N-1} y[\ell]\exp\left(-j\frac{2\pi k\ell}{N}\right)\right]\exp\left(j\frac{2\pi km}{N}\right)$$

$$= \frac{1}{N}\sum_{n=0}^{N-1} x[n]\sum_{\ell=0}^{N-1} y[\ell]\left\{\sum_{k=0}^{N-1}\exp\left[j\frac{2\pi k(m-n-\ell)}{N}\right]\right\}$$

Defining for simplicity the parameter $\alpha = \exp[2\pi j\,(m-n-\ell)/N]$ and referring to mathematical typology, the last sum of the preceding equation is written

$$\sum_{k=0}^{N-1}\exp\left[j\frac{2\pi k(m-n-\ell)}{N}\right] = \sum_{k=0}^{N-1}\alpha^k = \begin{cases} N & \text{for } \alpha = 1 \\ \dfrac{1-\alpha^N}{1-\alpha} & \text{for } \alpha \neq 1 \end{cases}$$

However, we observe that the parameter α is equal to one *only when the value of the quantity $m-n-\ell$ is an integer multiple of N* and therefore will have the form $m-n-\ell = pN$, where p is an *integer quantity*. Therefore, the preceding sum is reformulated as

$$\sum_{k=0}^{N-1}\exp\left[j\frac{2\pi k(m-n-\ell)}{N}\right] = \begin{cases} N\ \ell = m-n+pN = ((m-n))_N \quad (p \in N) \\ 0 \text{ in the opposite case} \end{cases}$$

and the foregoing relationship is written

$$z[m] = \sum_{n=0}^{N-1} x[n]\, y[((m-n))_N], \quad (m = 0, 1, 2, \ldots, N-1) \tag{10.113}$$

which is also the desired result. It is not difficult to realize that the operation defined in the foregoing way contains a *convolution sum* that, however, *does not contain the usual time shift that characterizes our known linear convolution*[27] *but, rather, a circular shift*; for this reason, the type of convolution that is defined in the foregoing way is known as *circular convolution*. Summarizing the preceding results we can formulate the proposition that *if the DFTs of the discrete sequences $x[n]$ and $y[n]$ of finite length N are the functions $\mathcal{X}[k]$ and $\mathcal{Y}[k]$, then the DFT of the discrete sequence $z[n]$ that results from the circular convolution of the sequences $x[n]$ and $y[n]$ will result from the product of the DFTs of the sequences $x[n]$ and $y[n]$*. In the international literature, the operation of circular convolution of two discrete sequences of finite length N is symbolized as

$$z[n] = x[n] \circledN y[n] = \sum_{m=0}^{N-1} x[m]\, y[((n-m))_N] \tag{10.114}$$

- **Circular correlation**: Considering two discrete complex sequences of finite length N whose N-point DFTs are the functions $\mathcal{X}[k]$ and $\mathcal{Y}[k]$, it is proven that the DFT of the circular correlation of the two signals defined as

[27] The difference that exists between the *linear* and *circular* convolution is that in the linear convolution, each sample of the convolutive signal arises from the multiplication of one sequence *by a time-reversed and linearly shifted version of the other sequence* and the addition of the values of the products $x[k]y[n-k]$ for all the values of the index k, while in the circular convolution the second sequence is *circularly time-reversed* and *circularly shifted* in relation to the first. Observing, also, that the discrete $\mathcal{X}[k]$ and $\mathcal{Y}[k]$ transforms satisfy the relation $\mathcal{X}[k]\mathcal{Y}[k] = \mathcal{Y}[k]\mathcal{X}[k]$, it immediately follows that *the operation of circular convolution satisfies the commutative property*, or, in mathematical notation, $x[n] \circledN y[n] = y[n] \circledN x[n]$.

$$\rho_{xy}[\ell] = \sum_{n=0}^{N-1} x[n]\, y^*[((n-\ell))_N] \tag{10.115}$$

is given by the relation $\mathcal{R}_{xy}[k] = \mathcal{X}[k]\mathcal{Y}^*[k]$.

For the proof of the exercise, we can formulate the *circular correlation* of the discrete sequences $x[n]$ and $y[n]$ as the *circular convolution* between the sequences $x[n]$ and $y^*[-n]$, that is, as

$$\rho_{xy}[\ell] = x[\ell] \circledN y^*[-\ell] \tag{10.116}$$

and then, using the properties of *circular convolution* and *complex conjugation*, the desired result is derived directly. In the special case where $x[n] = y[n]$, the aforementioned property is formulated as $\mathcal{R}_{xx}[k] = |\mathcal{X}[k]|^2$ and constitutes the equivalent of the properties of the other transforms presented in the previous chapters regarding the magnitude of *autocorrelation*.

- **Product**: If the DFTs with a length of N elements of the discrete sequences $x[n]$ and $y[n]$ that also contain N elements are the functions $\mathcal{X}[k]$ and $\mathcal{Y}[k]$, then the DFT of the discrete sequence $z[n] = x[n]y[n]$ will be the function

$$\mathcal{Z}[k] = \frac{1}{N}\left\{\mathcal{X}[k] \circledN \mathcal{Y}[k]\right\} \tag{10.117}$$

This property is proven by the property of circular convolution *by alternating the roles of time and frequency*, as imposed by the property of *duality*. The reader is asked to prove the aforementioned relationship by applying the proof methodology of the product property of other transforms in the way it was used in previous chapters.

- **Parseval's theorem**: If the DFTs with a length of N elements of the sequences $x[n]$ and $y[n]$ that also contain N elements are $\mathcal{X}[k]$ and $\mathcal{Y}[k]$, it is proven that

$$\sum_{n=0}^{N-1} x[n]\, y^*[n] = \frac{1}{N}\sum_{k=0}^{N-1} \mathcal{X}[k]\, \mathcal{Y}^*[k] \tag{10.118}$$

Proof From the aforementioned property of circular correlation and using the defining equation of the inverse DFT, we will have

$$r_{xy}[\ell] = \frac{1}{N}\sum_{k=0}^{N-1} \mathcal{R}_{xy}[k]\exp\left(j\frac{2k\pi\ell}{N}\right) = \frac{1}{N}\sum_{k=0}^{N-1} \mathcal{X}[k]\mathcal{Y}^*[k]\exp\left(j\frac{2k\pi\ell}{N}\right)$$

For $\ell = 0$ the preceding equation simplifies to

$$r_{xy}[0] = \frac{1}{N}\sum_{k=0}^{N-1} \mathcal{X}[k]\mathcal{Y}^*[k]$$

which, compared with the corresponding equation that arises from the defining equation and is the

$$r_{xy}[0] = r_{xy}[l]\Big|_{\ell=0} = \sum_{n=0}^{N-1} x[n]y^*[((n-\ell))_N]\Big|_{\ell=0} = \sum_{n=0}^{N-1} x[n]y^*[n]$$

Table 10.5 Basic properties of discrete Fourier transform

Property	Discrete-time sequence $x[n]$	Discrete Fourier transform $X[k]$
Periodicity	$x[n] = x[n+N]$	$X[k] = X[k+N]$
Homogeneity	$\alpha x\ [n]$	$\alpha X\ [k]$
Addition	$x[n] + y[n]$	$X[k] + Y[k]$
Linearity	$\alpha x[n]\ +\ \beta y[n]$	$\alpha X[k]\ +\ \beta Y[k]$
Frequency shifting	$\exp(i2\pi \ell n/N)x[n]$	$X\ [((k-\ell))_N]$
Time shifting	$x\ [((n-\ell))_N]$	$\exp(-i2\pi k\ell/N)X[k]$
Time reversal	$x[N-n]$	$X[N-k]$
Duality	$X[n]$	$x\ [((-k))_N]\ /N$
Convolution	$x[n] \textcircled{N} y[n]$	$X[k]Y[k]$
Multiplication	$x[n]y[n]$	$\frac{1}{N} X[k] \textcircled{N} Y[k]$
Symmetry	$x[n] \in R$	$\begin{cases} X[k] = X^*\ [((-k))_N] \\ \Re\{\mathcal{X}[k]\} = \Re\{\mathcal{X}\ [((-k))_N]\} \\ \Im\{\mathcal{X}[k]\} = -\Im\{\mathcal{X}\ [((-k))_N]\} \\ \lvert X[k]\rvert\ = \lvert X\ [((-k))_N]\rvert \\ \measuredangle X[k] = -X\ [((-k))_N] \end{cases}$
Conjugation	$x^*[n]$	$X^*\ [((-k))_N]$
	$x^*\ [((-n))_N]$	$X^*[k]$
Correlation	$x[n] \textcircled{N} y^*[-n]$	$X[k]Y^*[k]$
Autocorrelation	$x[n] \textcircled{N} x^*[-n]$	$\lvert X[k]\rvert^2$

gives us the desired result.

The most important properties of the DFT presented in the previous pages are summarized in Table 10.5.

Example 10.5.2 Let us consider the discrete-time signal $x[n] = \{\mathbf{1}, 2, 3, 4\}$ of finite length $N = 4$ and a segment of the periodic signal $\xi[n]$ that will result from the periodic repetition of the foregoing signal as

$$\xi[n] = \{\ldots, 1, 2, 3, 4, \underbrace{\mathbf{1}, 2, 3, 4}_{x[n]}, 1, 2, 3, 4, \ldots\}$$

If we shift the foregoing signal by $\ell = 2$ time units, we find that it will take the form

$$\xi[n-2] = \{\ldots, 1, 2, 3, 4, 1, 2, \underbrace{\mathbf{3}, 4, 1, 2}_{x'[n]}, 3, 4, \ldots\}$$

and considering only the part of the time-shifted signal $\xi[n-2]$ that corresponds to the value range $0 \leq n \leq 3$, we will get the discrete-time signal $x'[n] = \{3, 4, 1, 2\}$, which is indeed a *cyclically shifted version* of the original signal with a shift value equal to 2. Adopting the symbolism that is usually used to describe such shifts, we will formulate the signal $x'[n]$ as $x'[n] = x[((n-2))_4]$, to show that it has resulted from the shift, by two samples, of a signal with length $N = 4$.

Example 10.5.3 Let us consider the discrete-time signal $x[n] = \{\mathbf{6}, 5, 4, 3, 2, 1\}$. In this case, the application of the process described earlier will give us

$$x[((n-1))_6] = \{\mathbf{5}, 4, 3, 2, 1, 6\} x[((n+1))_6] = \{\mathbf{1}, 6, 5, 4, 3, 2\}$$

$$x[((n-2))_6] = \{\mathbf{4}, 3, 2, 1, 6, 5\} x[((n+2))_6] = \{\mathbf{2}, 1, 6, 5, 4, 3\}$$

$$x[((n-3))_6] = \{\mathbf{3}, 2, 1, 6, 5, 4\} x[((n+3))_6] = \{\mathbf{3}, 2, 1, 6, 5, 4\}$$

$$x[((n-4))_6] = \{\mathbf{2}, 1, 6, 5, 4, 3\} x[((n+4))_6] = \{\mathbf{4}, 3, 2, 1, 6, 5\}$$

$$x[((n-5))_6] = \{\mathbf{1}, 6, 5, 4, 3, 2\} x[((n+5))_6] = \{\mathbf{5}, 4, 3, 2, 1, 6\}$$

$$x[((n-6))_6] = \{\mathbf{6}, 5, 4, 3, 2, 1\} x[((n+6))_6] = \{\mathbf{6}, 5, 4, 3, 2, 1\}$$

These results are summarized in the relation $x[((n-\ell))_N] = x[N-\ell+n]$ $(1 \leq \ell \leq N)$. Although this property has been demonstrated on a purely practical level, it is also theoretically justified, observing that the expression W_N defined above satisfies the condition $W_N^{\ell k} = W_N^{-(N-\ell)k}$. This equality, however, means that *a cyclic shift of a group of N values in one direction by the distance ℓ is equivalent to a cyclic shift in the opposite direction by the quantity N − ℓ.*

Problem 10.5.13 Find the circular convolution between the signals $x[n] = \{\mathbf{2}, 1, 2, 1\}$ and $y[n] = \{\mathbf{1}, 2, 3, 4\}$ (Mitra, Example 3.15).

Solution We observe that the discrete sequences $x[n]$ and $y[n]$ have a length of $N = 4$ samples; therefore, applying the defining equation of the *circular convolution* for this value of N we get

$$z[n] = \sum_{m=0}^{3} x[m]y[((n-m))_4], \qquad n = 0, 1, 2, 3$$

Taking into account that the variable m takes the values 0, 1, 2, 3, it is obvious that we need to determine the contents of the temporally shifted discrete-time signals $y[((n-0))_4]$, $y[((n-1))_4]$, $y[((n-2))_4]$, and $y[((n-3))_4]$. Applying the methodology of Example 1.5.4, we find that

$$y[((n-0))_4] = \{\mathbf{1}, 2, 3, 4\} y[((n-1))_4] = \{\mathbf{2}, 3, 4, 1\}$$

$$y[((n-2))_4] = \{\mathbf{3}, 4, 1, 2\} y[((n-3))_4] = \{\mathbf{4}, 1, 2, 3\}$$

To calculate the values of $z[n]$, we need to multiply the signal $x[n]$ sample by sample by the four preceding discrete-time signals and add the individual products. For ease of notation, we will adopt the symbolism

$$\{x_1, x_2, x_3, x_4\} \times \{y_1, y_2, y_3, y_4\} = x_1y_1 + x_2y_2 + x_3y_3 + x_4y_4$$

In detail, then, we will have

$$\text{for } n = 0: \quad z[0] = \{2, 1, 2, 1\} \times \{1, 2, 3, 4\} = 2 + 2 + 6 + 4 = 14$$

$$\text{for } n = 1: \quad z[1] = \{2, 1, 2, 1\} \times \{2, 3, 4, 1\} = 4 + 3 + 8 + 1 = 16$$

$$\text{for } n = 2: \quad z[2] = \{2, 1, 2, 1\} \times \{3, 4, 1, 2\} = 6 + 4 + 2 + 2 = 14$$

$$\text{for } n = 3: \quad z[3] = \{2, 1, 2, 1\} \times \{4, 1, 2, 3\} = 8 + 1 + 4 + 3 = 16$$

Therefore, the circular convolution of the discrete-time signals $x[n]$ and $y[n]$ is the discrete-time signal $z[n] = \{\mathbf{14}, 16, 14, 16\}$, which is also the desired result. It is obvious that it makes no sense to continue the process for values $n > 3$, since we will constantly obtain the sequence of the four preceding numbers, which, of course, is expected.

Problem 10.5.14 Determine the circular convolution between the signals defined by the relations

$$x[n] = \cos\left(\frac{\pi n}{2}\right) \qquad \text{and} \qquad y[n] = 2^n \quad (n = 0, 1, 2, 3).$$

Solution Following the methodology of the preceding exercise, we will calculate the *circular convolution* from the defining equation for a length of discrete sequences $N = 4$. We will therefore have

$$z[n] = \sum_{m=0}^{3} x[m]y[((n-m))_4], \quad n = 0, 1, 2, 3$$

with the discrete sequences $x[n]$ and $y[n]$ having the values $x[n] = \{\mathbf{1}, 0, -1, 0\}$ and $y[n] = \{\mathbf{1}, 2, 4, 8\}$, as immediately results from substituting the values $n = 0, 1, 2, 3$ in their defining equations. In this case, the time-shifted sequences are calculated as

$$y[((n-0))_4] = \{\mathbf{1}, 2, 4, 8\} y[((n-1))_4] = \{\mathbf{2}, 4, 8, 1\}$$
$$y[((n-2))_4] = \{\mathbf{4}, 8, 1, 2\} y[((n-3))_4] = \{\mathbf{8}, 1, 2, 4\}$$

and therefore we will have

$$\text{for } n = 0: \quad z[0] = \{1, 0, -1, 0\} \times \{1, 2, 4, 8\} = 1 - 4 = -3$$
$$\text{for } n = 1: \quad z[1] = \{1, 0, -1, 0\} \times \{2, 4, 8, 1\} = 2 - 8 = -6$$
$$\text{for } n = 2: \quad z[2] = \{1, 0, -1, 0\} \times \{4, 8, 1, 2\} = 4 - 1 = +3$$
$$\text{for } n = 3: \quad z[3] = \{1, 0, -1, 0\} \times \{8, 1, 2, 4\} = 8 - 2 = +6$$

Therefore, the circular convolution of the signals $x[n]$ and $y[n]$ is the discrete-time signal $z[n] = \{\mathbf{-3}, -6, +3, +6\}$. This problem will be solved in the next section (Sect. 10.5.4) by the application of the convolution theorem of the discrete transform (see Exercise 10.5.15 on page 871).

10.5.4 The Relationship of Cyclic to Linear Convolution

As already mentioned, *the product of the DFT of two discrete sequences* $x[n]$ and $y[n]$ *of finite length N corresponds to their circular convolution* $x[n] \circledN y[n]$. Therefore, based on the *convolution theorem* of the discrete transform, we can calculate this quantity in full analogy with the other transforms in the following way:

- We calculate the DFT $\mathcal{X}[k]$ of the sequence $x[n]$ with length N elements.
- We calculate the DFT $\mathcal{Y}[k]$ of the sequence $y[n]$ with length N elements.
- We calculate *the product of the discrete transforms* $\mathcal{Z}[k] = \mathcal{X}[k]\mathcal{Y}[k]$

In this case, the convolution theorem tells us that *if we take the inverse DFT of the function* $\mathcal{Z}[k]$, *the resulting signal will be equal to the circular convolution* $z[n] = x[n] \circledN y[n]$.

Although the circular convolution is a useful and easily calculated quantity, in practice, we are mainly interested in the usual *linear convolution* that we encountered several times in previous chapters, as it is inextricably linked with *LTI continuous- and discrete-time systems*; let us see, then, if these two types of convolution are related, and if so, in what way. Taking into account that the circular convolution is related to sequences of *finite length*, let us first see if for these sequences we can define the *linear convolution* that is known to be described by the equation

$$z[n] = x[n] * y[n] = \sum_{k=-\infty}^{\infty} x[k]\, y[n-k] \tag{10.119}$$

Assuming that the discrete-time signals $x[n]$ and $y[n]$ contain N_1 and N_2 samples, respectively, it is not difficult to see that the product appearing in the foregoing relation will be characterized by a nonzero value only for the range of values

$0 \leq n \leq N_1 + N_2 - 2$, since outside this range one of the two signals—or *both*–will have a value equal to zero. We are led, then, to the conclusion that *the linear convolution is defined also for signals of finite length*; however, it itself constitutes a *finite signal* with its length being equal to *the sum of the lengths of the two signals reduced by one unit*.

It is proven that the circular convolution $x[n] \circledN y[n]$, which corresponds to the product of two DFTs $\mathcal{X}[k]\mathcal{Y}[k]$, is identical or not with the usual linear convolution of the two finite-length sequences $x[n]$ and $y[n]$, *depending on the correlation that exists between the length of the discrete transforms and the length of the considered discrete-time signals*. To investigate this situation, let us remember that if the DTFT $\mathcal{X}(e^{j\omega})$ of a sequence $x[n]$ has undergone *sampling* at the positions of the frequencies $\omega_k = 2\pi k/N$, then the sequence that will result will correspond *to the coefficients of the Fourier expansion of the discrete periodic sequence of infinite length*, defined as

$$\xi[n] = \sum_{m=-\infty}^{\infty} x[n - mN] \tag{10.120}$$

and that the sequence *of finite length* defined by the relation

$$\mathcal{X}[k] = \begin{cases} \mathcal{X}(e^{j2\pi k/N}) & 0 \leq k \leq N-1 \\ 0 & \text{in the opposite case} \end{cases} \tag{10.121}$$

constitutes *the DFT of a period of the signal* $\xi[n]$, which is defined as

$$x'[n] = \begin{cases} \xi[n] & 0 \leq n \leq N-1 \\ 0 & \text{in the opposite case} \end{cases} \tag{10.122}$$

From the foregoing description, it is not difficult to understand that if the length L of the discrete-time signal $x[n]$ is less than or equal to the length N of the DFT, *the phenomenon of folding does not appear* (see also Sect. 10.5.1 on page 847), and therefore, $x[n] = x'[n]$. However, if $L > N$, then the signals $x[n]$ and $x'[n]$ will not coincide for some or even all values of the index n, resulting in the emergence of the problems we described in previous sections.

What has been mentioned so far is already known; but how all this stuff is connected with the correlation that exists between the cyclic and linear convolutions? Let us return to the convolution $z[n] = x[n] * y[n]$ of the discrete-time signals $x[n]$ and $y[n]$ with lengths N_1 and N_2, which, as we have proven, constitutes a signal *of finite length* with $N_1 + N_2 - 1$ elements, and, following the methodology of Sect. 10.5.1, let us construct a periodic sequence $\zeta[n]$ *of infinite length* that is defined by the equation

$$\zeta[n] = \sum_{m=-\infty}^{\infty} z[n - mN]$$

If $\mathcal{X}(e^{j\omega})$, $\mathcal{Y}(e^{j\omega})$ and $\mathcal{Z}(e^{j\omega})$ are the DTFTs of the discrete sequences $x[n]$, $y[n]$, and $z[n]$, then from the convolution theorem we will have $\mathcal{Z}(e^{j\omega}) = \mathcal{X}(e^{j\omega})\mathcal{Y}(e^{j\omega})$, an equation that for frequency values $\omega = 2k\pi/N$ is reduced to $\mathcal{Z}[k] = \mathcal{X}[k]\mathcal{Y}[k]$. In this case, taking into account that *the inverse DFT of the function* $\mathcal{Z}[k]$ *is obviously the discrete-time signal* $z[n]$, from the convolution theorem of the discrete transform we will have $z[n] = x[n] \circledN y[n]$. On the other hand, the function $\mathcal{Z}[k]$ will constitute, according to the discussion in the previous paragraph, the DFT of a period of the periodic sequence $\zeta[n]$, which is defined as

$$z'[n] = \begin{cases} \zeta[n] & 0 \leq n \leq N-1 \\ 0 & \text{in the opposite case} \end{cases}$$

Applying, therefore, the theory we applied previously, we come to the conclusion that if the length N of the discrete transform $\mathcal{X}[k]$, $\mathcal{Y}[k]$, and $\mathcal{Z}[k]$ is greater than $N_1 + N_2 - 1$, then the phenomenon of folding does not appear; therefore, we will have

$z'[n] = z[n]$, and *the circular convolution of the discrete sequences $x[n]$ and $y[n]$ will coincide with their linear convolution.* Conversely, if $N < N_1 + N_2 - 1$, then the circular convolution of the considered sequences is equivalent to the linear convolution, which, however, has been distorted by the phenomenon of folding.

Problem 10.5.15 Using the method of the DFT, determine the circular convolution between the discrete-time signals defined by the relations

$$x[n] = \cos\left(\frac{\pi n}{2}\right) \quad \text{and} \quad y[n] = 2^n \quad (n = 0, 1, 2, 3).$$

Solution Initially, let us calculate the DFTs of the discrete sequences $x[n]$ and $y[n]$. Starting with the transform $\mathcal{X}[k]$, it is calculated as

$$\mathcal{X}[k] = \sum_{n=0}^{N-1} x[n] \exp\left(j\frac{2\pi kn}{N}\right) = \sum_{n=0}^{3} \cos\left(\frac{\pi n}{2}\right) \exp\left(j\frac{2\pi kn}{4}\right)$$

However, we observe that

$$x[1] = \cos\left(\frac{\pi}{2}\right) = x[3] = \cos\left(\frac{3\pi}{2}\right) = 0$$

while also $x[0] = \cos(0) = 1$ and $x[2] = \cos(\pi) = -1$. Therefore,

$$\mathcal{X}[k] = \sum_{n=0,2} \cos\left(\frac{\pi n}{2}\right) \exp\left(j\frac{\pi kn}{2}\right) = 1 - e^{jk\pi}$$

On the other hand, the function $\mathcal{Y}[k]$ is calculated as

$$\mathcal{Y}[k] = \sum_{n=0}^{N-1} y[n] \exp\left(j\frac{2\pi kn}{N}\right) = \sum_{n=0}^{3} 2^n \exp\left(j\frac{\pi kn}{2}\right) = 1 + 2\exp\left(j\frac{\pi k}{2}\right) + 4\exp(j\pi k) + 8\exp\left(j\frac{3\pi k}{2}\right)$$

The product of the preceding discrete transforms is given by the equation

$$\begin{aligned}\mathcal{Z}[k] = \mathcal{X}[k]\mathcal{Y}[k] &= (1 - e^{jk\pi}) \times \left[1 + 2\exp\left(j\frac{\pi k}{2}\right) + 4\exp(j\pi k) + 8\exp\left(j\frac{3\pi k}{2}\right)\right] \\ &= -3 - 6\exp\left(j\frac{\pi k}{2}\right) + 3\exp(j\pi k) + 6\exp\left(j\frac{3\pi k}{2}\right)\end{aligned}$$

and therefore, the required circular convolution will have the form

$$z[n] = x[n] \circledast y[n] = -3\mathcal{F}^{-1}\{1\} - 6\mathcal{F}^{-1}\left\{\exp\left(j\frac{\pi k}{2}\right)\right\} + 3\mathcal{F}^{-1}\{\exp(j\pi k)\} + 6\mathcal{F}^{-1}\left\{\exp\left(j\frac{3\pi k}{2}\right)\right\}$$

where, in this case, the symbol $\mathcal{F}$ is obviously used to denote the *DFT*. Taking into account that $\mathcal{F}\{\delta[n]\} = 1$ and using the property of *time shift*, it follows that

$$z[n] = x[n] \circledast y[n] = -3\delta[n] - 6\delta[n-1] + 3\delta[n-2] + 6\delta[n-3]$$

or, equivalently, $z[n] = \{-3, -6, +3, +6\}$, which is the same result as that of Exercise 10.5.14 (page 868).

10.5.5 Implementation of DFT

As already mentioned, the calculation of the N values of the Fourier transform of a discrete sequence of N samples requires the execution of N^2 *multiplications and* $N(N-1)$ *additions*, a number that is considered prohibitive for the use of the aforementioned transform in signal processing applications *in real time*.[28] It should be noted that in the case where the number of samples of the transform is *constant* and *known in advance*, the foregoing process can be significantly simplified since the N^2 number of exponential quantities

$$\exp\left(-j\frac{2\pi kn}{N}\right) \quad n, k = 0, 1, 2, \ldots, N-1 \tag{10.123}$$

is *independent of the samples of the discrete sequence* $x[n]$ and therefore can be calculated *from the beginning* once and only once and stored in auxiliary memory areas future use. In addition, observing that for a given length N the relation

$$\left[\exp\left(-j\frac{2\pi kn}{N}\right)\right]^N = 1 \tag{10.124}$$

always holds for every pair of values (n, k), which means that the N quantities

$$\exp\left(-j\frac{2\pi n}{N}\right) \quad n = 0, 1, 2, \ldots, N-1 \tag{10.125}$$

are the *Nth roots of unity*, we can calculate the DFT using only these representations, so the storage requirements of the transform calculation process are further reduced. Even though using these techniques means that the computational cost of calculating the discrete transform reduced significantly,[29] in recent decades, fast algorithms for calculating this function have been developed, with the most characteristic of all being the *FFT*, whose computational complexity is of the order $\mathcal{O}(N\log_2 N)$, much smaller than the order $\mathcal{O}(N^2)$ involved in the process of calculating the DFT directly from its defining equation.

Fast Fourier Transform

Although the term *fast Fourier transform* is usually used to describe a specific algorithm that is mainly used in practice, in reality it refers to an entire family of algorithms that are based on an algorithmic approach described by the term *divide and conquer*. According to this approach, the solution of an algorithmic problem is based on the application of the following process:

1. The problem to be solved is *broken down into smaller similar problems* that, ideally, are exactly the same size.
2. Each of the smaller problems that arise from the breakdown of the original problem *is solved separately from the rest* using mostly *recursive methods.*

---

[28] The preceding numbers concern *discrete sequences of real numbers*, while for the general case of *complex discrete sequences*, the situation is even more complicated; indeed, by separating the defining equation of the function $\mathcal{X}[k]$ *into its real and imaginary parts*, we will get the equations

$$\Re\{\mathcal{X}[k]\} = \sum_{n=0}^{N-1}\left[\Re\{x[n]\}\cos\left(\frac{2\pi nk}{N}\right) + \Im\{x[n]\}\sin\left(\frac{2\pi nk}{N}\right)\right]$$

$$\Im\{\mathcal{X}[k]\} = \sum_{n=0}^{N-1}\left[\Im\{x[n]\}\cos\left(\frac{2\pi nk}{N}\right) - \Re\{x[n]\}\sin\left(\frac{2\pi nk}{N}\right)\right]$$

from which we find that the calculation of the foregoing quantities requires the calculation of $2N^2$ trigonometric expressions, performing $4N^2$ real multiplications and $4N(N-1)$ real additions, as well as a large number of addressing and indexing processes.

[29] See also Goertzel G, "An Algorithm for the Evaluation of Finite Trigonometric Series," *Am. Math. Monthly*, Vol. 65, pp. 34–35, 1958 for a variation of the usual way of calculating the discrete transform that is based on the properties of the periodicity of the quantities W_N^{kn} and is characterized by lower computational complexity, although its order continues to be $\mathcal{O}(N^2)$.

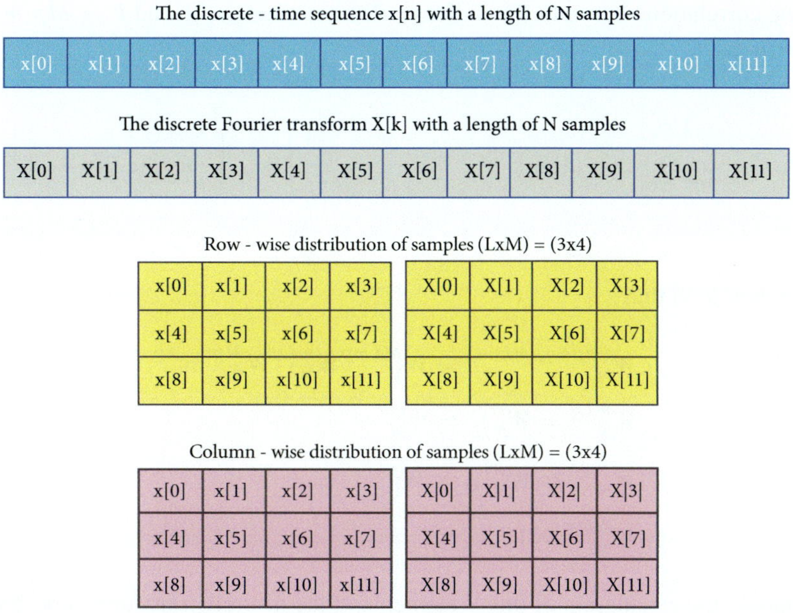

Fig. 10.13 Distribution of samples of discrete sequence and its DFT by rows and columns

3. The solutions that emerge in the previous step *are combined with each other*—if necessary—in order to obtain a solution to the original large problem.

How can this technique be applied to the problem of calculating the DFT? Generally speaking, this can happen *by separating the original discrete sequence of N elements into M sequences of L elements (or, equivalently, into L sequences of M elements) and calculating M discrete transforms of length L (or L discrete transforms of length M), which are then combined among themselves in order to give us the desired transform of length N.*[30] These two approaches are, of course, equivalent: The elements of the discrete sequence $x[n]$ are stored in a two-dimensional array of dimensions $L \times M$, and which of these will be used depends on whether the data of the discrete-time signal $x[n]$ are stored row-wise (for the first case) or column-wise (for the second case). From the algorithmic side of the issue, if the two-dimensional array of dimensions $L \times M$ that will host the data of the discrete-time signal is symbolized by ζ and is described using a pair of indices (ℓ, m) $(0 \leq \ell \leq L-1$ and $0 \leq m \leq M-1)$, with the index ℓ referring to the rows and the index m referring to the columns of the matrix ζ, then the preceding data rearrangement requires the determination of a correspondence of the form $x[n] \Rightarrow \zeta[\ell, m]$ $(0 \leq n \leq N-1)$, with the indices n, ℓ, and m being associated with the equation $n = M\ell + m$ for the storage *row-wise* and the equation $n = mL + \ell$ for the storage *column-wise*.

On the other hand, the management of the samples of the discrete transform $\mathcal{X}[k]$ with a length of N elements will be carried out in exactly the same way, and therefore, they will also be stored in a corresponding *two-dimensional array* $\mathcal{Z}$ of dimensions $L \times M$ using one of the two approaches. This means that if the elements of matrix $\mathcal{Z}$ are described using a corresponding pair of indices (p, q) $(0 \leq p \leq L-1$ and $0 \leq q \leq M-1)$, we must again determine a correspondence of the form $\mathcal{X}[k] \Rightarrow \mathcal{Z}[p, q]$, with the indices k $(0 \leq k \leq N-1)$ and (p, q) being connected with the relation $k = Mp + q$ for storage *row-wise* and $k = qL + p$ for storage *column-wise*. These two different ways of distributing by rows and columns are presented in Fig. 10.13 for the case $N = 12$, $L = 4$, and $M = 3$.

From the foregoing analysis it is not difficult to understand that for the calculation of the DFT we can adopt for the discrete sequence $x[n]$, as well as for its transform $\mathcal{X}[k]$, *either row-wise storage or column-wise storage*. In this case, the transform will be calculated as a *double sum* over the set of elements of the two-dimensional matrix ζ multiplied by the corresponding exponential factors. Considering, for example, that the elements of the discrete sequence are stored *by columns*, while the

[30] It is obvious that to make it possible to apply the aforementioned technique, the lengths L, M, and N must be connected by the relation $N = ML$; in the case where the length N is a prime number, we can *add zero elements* at the end of the sequence $x[n]$ so that the length of the increased sequence can be factorized in the described way.

DFT is stored *by rows*, the correlations between the indices will be $n = mL + \ell$ and $k = Mp + q$, and from the defining equation of the DFT we will get

$$\mathcal{Z}(p,q) = \sum_{m=0}^{M-1}\sum_{\ell=0}^{L-1} \zeta(\ell,m) W_N^{(Mp+q)(mL+\ell)} = \sum_{\ell=0}^{L-1}\left\{W_N^{\ell q}\left[\sum_{m=0}^{L-1} \zeta(\ell,m) W_M^{mq}\right]\right\} W_L^{\ell p}$$

as can easily be verified, it will hold that

$$\begin{aligned}
W_N^{(Mp+q)(mL+\ell)} &= W_N^{MLmp} W_N^{mLq} W_N^{Mp\ell} W_N^{\ell q} \\
W_N^{mLq} &= W_{N/L}^{mq} = W_M^{mq} \\
W_N^{Mp\ell} &= W_{N/M}^{p\ell} = W_L^{p\ell} \\
W_N^{Nmp} &= 1
\end{aligned}$$

According to the last equation, the calculation of the N elements of the discrete transform $\mathcal{X}[k]$ has been shaped as a three-step process:
(a) *Calculate the Fourier transform of M elements for each of the L rows of matrix* $\mathcal{Z}$:

$$\Gamma(\ell,q) = \sum_{m=0}^{M-1} \zeta(\ell,m)\, W_M^{mq}, \quad 0 \le q \le M-1 \tag{10.126}$$

(b) *Define the auxiliary two-dimensional matrix*

$$\mathcal{G}(\ell,q) = W_N^{\ell q}\,\Gamma(\ell,q) \quad \text{for the value range} \begin{cases} 0 \le q \le M-1 \\ 0 \le \ell \le L-1 \end{cases} \tag{10.127}$$

(c) *Calculate the DFT of L points for each of the M columns of matrix* $\mathcal{G}$, which is defined by the relation

$$\mathcal{Z}(p,q) = \sum_{\ell=0}^{L-1} \mathcal{G}(\ell,q)\, W_L^{\ell p} \tag{10.128}$$

It is proven that the calculation of the DFT in the way described above, requires the implementation of $N(M+L+1)$ multiplications and $N(M+L-2)$ additions between complex quantities and therefore is more efficient than the one based on the direct calculation of the this transform from the defining equation. In the preceding description, we assume that the samples of the signal $x[n]$ are stored in columns and the samples of the transform $\mathcal{X}[k]$ are stored in rows, while if we reverse the storage methods (i.e., we store the samples $x[n]$ in rows and the samples $\mathcal{X}[k]$ in columns), the corresponding equation that arises takes the form

$$\mathcal{Z}(p,q) = \sum_{m=0}^{M-1}\sum_{\ell=0}^{L-1} \zeta(\ell,m)\, W_N^{pm}\, W_L^{p\ell}\, W_M^{qm} = \sum_{m=0}^{M-1} W_M^{mq}\left[\sum_{\ell=0}^{L-1} \zeta(\ell,m) W_L^{\ell p}\right] W_N^{mp}$$

and the problem is solved in a completely analogous way.

Having proceeded to a first breakdown of the problem, we can, according to the method of "divide and conquer", break down each of the two small problems into even smaller pieces and each of these into even smaller problems until the length of each individual transform is a prime number, so that it cannot be further divided. Considering, for example, the problem of calculating a discrete transform with a length of $N = 15015$ elements, we can break it down into two problems of calculating transforms with a length of $L = 143$ and $M = 105$ elements. Of these two problems, the first can be divided into two even smaller problems of calculating transforms with a length of $A = 13$ and $B = 11$ elements (since the numbers 13 and 11 are prime, the problem does not allow further division), while the second can be divided into two smaller problems with a length of $C = 15$ and $D = 7$ elements. Finally, the problem of calculating the transform of 15 elements can be broken down one last time into two smaller problems of transforms of 3 and 5 elements, respectively. This subdivision is based on the fact that the number 15015 can be expressed as a product of the form

$$N = 15015 = [(3 \cdot 5) \cdot 7] \cdot (11 \cdot 13)$$

while, in the general case, the length N of the discrete transform is considered as a product of ν prime numbers of the form $N = r_1 r_2 \ldots r_\nu$. Of particular interest is the case $r_1 = r_2 = \cdots = r_\nu = r$, in which the length of the transform becomes an exponential expression of the form $N = r^\nu$, while additionally all the small transforms that arise *have the same length r*; this size is known as *the base of the discrete transform*.[31] Having defined the size, we are now able to say that the algorithm known to the general public as the FFT is simply an algorithm based on the so-called *divide and conquer* algorithmic approach for which the *base value is equal to 2*; this algorithm is presented in the next section.

Base $r = 2$ Algorithms

According to the discussion in the previous section, if the base of the algorithm implementing the DFT is equal to $r = 2$, then the total number of data of the discrete sequence must be a power of 2 in the form

$$N = \underbrace{2 \cdot 2 \cdot 2 \ldots 2}_{\nu \text{ times}} = 2^\nu$$

and the decomposition of the original problem into smaller problems involving the calculation of discrete transforms will be based on the successive subdivision of the original data table into *two equal parts*. Although it may seem logical that this subdivision should be done in the usual way according to which each subsequence that emerges contains $N/2$ consecutive samples of the original signal (for example, the decomposition of a sequence of 100 samples in this way will lead to the creation of two smaller sequences, the first of which will include samples from 1 to 50 and the second samples from 51 to 100), in practice this method is not followed (it should be noted that it is used in a variation of this algorithm presented subsequently): according to a lemma formulated by Danielson and Lanczos in 1942, *the DFT of a sequence of values of size N can be written as the sum of two DFTs, each of which has a length of* $N/2$. Of these two transforms, the first is calculated from the samples located *at even positions of the sequence of input data*, while the second arises from the samples located *at the odd positions of the sequence of input data*.[32] The proof of this lemma is relatively easy and is presented in what follows.

Considering two sequences, $g[n] = x[2n]$ and $h[n] = x[2n+1]$, for the range of values ($n = 1, 2, \ldots, N/2$) that contain the samples of the even and odd positions of the sequence of input data $x[n]$—which we assume consists of N samples—the *DFT* of the signal $x(n)$ can be written as a sum of two terms in the form

$$\begin{aligned} \mathcal{X}[k] &= \sum_{n=0}^{N-1} x[n] \exp\left(-j\frac{2\pi kn}{N}\right) = \sum_{n=0}^{N-1} x[n]\, W_N^{kn} \\ &= \sum_{n \text{ even}}^{N-1} x[n]\, W_N^{kn} + \sum_{n \text{ odd}}^{N-1} x[n]\, W_N^{kn} = \sum_{\ell=0}^{(N/2)-1} g[\ell]\, W_N^{k(2\ell)} + \sum_{\ell=0}^{(N/2)-1} h[\ell]\, W_N^{k(2\ell+1)} \end{aligned}$$

[31] Note that from time to time *hybrid algorithms* have been invented that during the stage of successive divisions use different base values in order to reduce as much as possible the computational cost. For example, the initial sequence could be divided into two subsequences of equal length, each of which would be subdivided into four smaller sequences, thus adopting an approach that relies on the use of base values $r = 2$ for the first subdivision and $r = 4$ for the second subdivision.

[32] It is obvious that this method of subdivision is a special case of the more general problem of subdivision presented in the previous section and corresponds to the values $M = N/2$ and $L = 2$.

with the quantity W_N defined as $W_N = \exp(-2\pi j/N)$. However, we observe that

$$W_N^{k(2\ell)} = \exp\left(-j\frac{2\pi}{N}\right)^{k(2\ell)} = \exp\left[-j\frac{2\pi}{N}k(2\ell)\right] = \exp\left(-j\frac{2\pi k\ell}{N/2}\right) = W_{N/2}^{k\ell}$$

Therefore, we will have $W_N^{k(2\ell+1)} = W_N^{2k\ell}W_N^k = W_{N/2}^{k\ell}W_N^k$ and the preceding sum is formulated as

$$\mathcal{X}[k] = \sum_{\ell=0}^{(N/2)-1} g[\ell]\, W_M^{k(2\ell)} + W_N^k \sum_{\ell=0}^{(N/2)-1} h[\ell]\, W_N^{2k\ell} = \sum_{\ell=0}^{(N/2)-1} g[\ell]\, W_{N/2}^{k\ell} + W_N^k \sum_{\ell=0}^{(N/2)-1} h[\ell]\, W_{N/2}^{k\ell} = \mathcal{G}(k) + W_N^k \mathcal{H}(k)$$

with the functions

$$\mathcal{G}[k] = \sum_{\ell=0}^{(N/2)-1} g[\ell]\, W_{N/2}^{kl} \quad \text{and} \quad \mathcal{H}[k] = \sum_{\ell=0}^{(N/2)-1} h[\ell]\, W_{N/2}^{kl}$$

constituting the *DFTs* of the signals $g[n]$ and $h[n]$, each of which has a length of $N/2$. The last relation also constitutes the proof of the theorem of Danielson and Lanczos.

It is proven that the functions $\mathcal{G}[k]$ and $\mathcal{H}[k]$ are *periodic functions with a period of* $N/2$, and therefore they will satisfy the relations

$$\mathcal{G}[k] = \mathcal{G}\left[k + \frac{N}{2}\right] \quad \text{and} \quad \mathcal{H}[k] = \mathcal{H}\left[k + \frac{N}{2}\right]$$

Since, as was proven, the relation holds

$$W_N^{k+(N/2)} = W_N^k W_N^{N/2} = -W_N^k \tag{10.129}$$

we will finally get

$$\begin{aligned} \mathcal{X}\left[k + \frac{N}{2}\right] &= \mathcal{G}\left[k + \frac{N}{2}\right] + W_N^{k+(N/2)}\mathcal{H}\left[k + \frac{N}{2}\right] \\ &= \mathcal{G}[k] - W_N^k \mathcal{H}[k] \end{aligned} \tag{10.130}$$

The coefficients W_N^k of the Fourier transform $\mathcal{H}(k)$ are called *twiddle factors*, and their calculation occupies a central position in the process of calculating the FFT.

The most important of the properties that characterize the lemma of Danielson and Lanczos is that it can be called *recursively*, as is also dictated by the so-called *divide and conquer* algorithmic approach: In other words, *having divided the transform* $\mathcal{X}[k]$ *into the transforms* $\mathcal{G}[k]$ *and* $\mathcal{H}[k]$*, each of which is characterized by the use of* $N/2$ *points of the input sequence* $x[n]$*, we can then use the same process to divide the transforms* $\mathcal{G}[k]$ *and* $\mathcal{H}[k]$ *into two smaller transforms, each of which will use a number of points equal to* $N/4$. Thus, the transform $\mathcal{G}[k]$ can be written as the sum

$$\mathcal{G}[k] = \sum_{n=0}^{(N/4)-1} g[2n]\, W_{N/4}^{nk} + W_{N/2}^k \sum_{n=0}^{(N/4)-1} g[2n+1]\, W_{N/4}^{nk} \tag{10.131}$$

while a similar relationship can be written for the transform $\mathcal{H}[k]$. Because in every application of the algorithm the number of used points *is halved*, it is obvious that this should be a *power of* 2, as mentioned previously, so that the quantity $\log_2 N$ is an integer. In the opposite case, the application of the FFT is not possible. In cases where the length of the data sequence is not a power of 2, we can resort to two solutions: *Either ignore some samples* keeping only those equal in number to some power of 2, *or add zeros at the end of the sequence* so that its length is rounded to the next power of 2. Which of the two solutions we will choose obviously depends on the length of the discrete sequence used in each case.

Applying the foregoing process $\log_2 N$ times, we will finally end up at an input data sequence length with a value equal to one. In this case, we will use *the termination condition of the recursive process*, according to which *the DFT of a data sequence that includes only one point* $x[0] = \xi$ *is this very point*. This property is proven very easily, since for a number of points $N = 1$ the DFT is calculated as

$$\mathcal{X}[k] = \sum_{n=0}^{1-1} x[n] \exp\left(-\frac{j2\pi n}{1}\right) = x[0] = \xi \tag{10.132}$$

Using mathematical notation, this property is expressed as

$$\mathcal{F}_k^{eoeeoeo...oee} = x[n] \quad \text{for some value of } n \tag{10.133}$$

where it is obvious that the number of e and o in the sequence $eoeeoeo \ldots oee$ is equal to $\log_2 N$. In the preceding notation, the symbol $\mathcal{F}$ is used to symbolize the Fourier transform of the current sequence of points that has resulted from the subdivision of the previous sequence into even (e) or odd (o) samples.

To identify the value of n that corresponds to some sequence $eoeeoeo \ldots oee$ for which the termination condition $\mathcal{F}[k] = \xi$ holds, *we must reverse the sequence and make the substitutions* $e = 0$ *and* $o = 1$. In this case, the result that will emerge will be the value of n expressed in the *binary system*.

Let us now move on to a detailed example of the application of the FFT for a discrete-time signal containing $N = 16$ samples. The application of the successive breakdown of the signal into even and odd samples is presented diagrammatically in Fig. 10.14.

From Fig. 10.14 we can see that the initial sequence $x[n]$ of the 16 points at the positions

$$\{00, 01, 02, 03, 04, 05, 06, 07, 08, 09, 10, 11, 12, 13, 14, 15\}$$

was split into the sequence of even points

$$x_e[n] = \{00, 02, 04, 06, 08, 10, 12, 14\}$$

and the sequence of odd points

$$x_o[n] = \{01, 03, 05, 07, 09, 11, 13, 15\}$$

In the next step of the process, the sequence $x_e[n]$ was split into the sequences

$$x_{ee}[n] = \{00, 04, 08, 12\} \qquad \text{and} \qquad x_{eo}[n] = \{02, 06, 10, 14\}$$

while in the same way the sequence $x_o[n]$ was split into the sequences

$$x_{oe}[n] = \{01, 05, 09, 13\} \qquad \text{and} \qquad x_{oo}[n] = \{03, 07, 11, 15\}$$

In the third stage of the split, the sequences $x_{ee}[n]$ and $x_{eo}[n]$ were split into the sequences

$$x_{eee}[n] = \{00, 08\} \quad x_{eeo}[n] = \{04, 12\}$$
$$x_{eoe}[n] = \{02, 10\} \quad x_{eoo}[n] = \{06, 14\}$$

while the sequences $x_{oe}[n]$ and $x_{oo}[n]$ were split into the sequences

$$x_{oee}[n] = \{01, 09\} \quad x_{oeo}[n] = \{05, 13\}$$
$$x_{ooe}[n] = \{03, 11\} \quad x_{ooo}[n] = \{07, 15\}$$

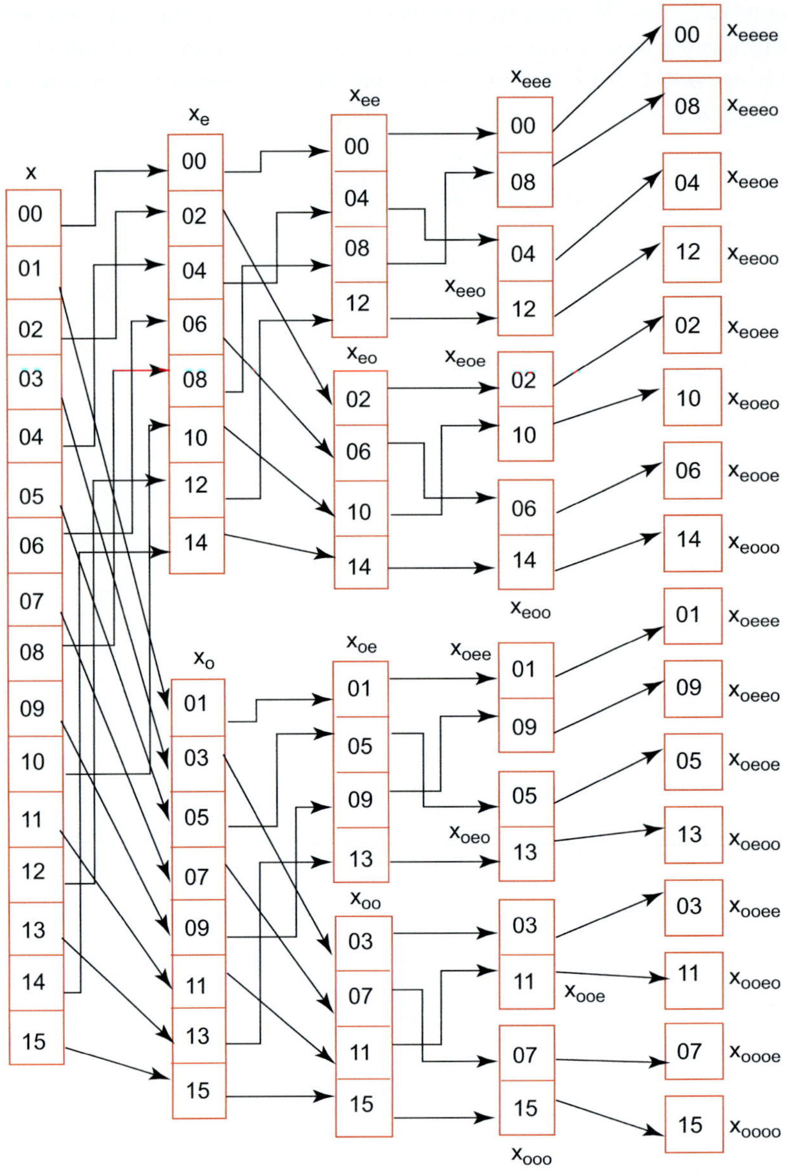

Fig. 10.14 Successive breakdown of a 16-sample signal into its even and odd samples

Finally, each of the eight two-point sequences was split into two sequences containing only one sample, and they are symbolized as

$$x_{eeee}[n] = \{00\} \quad x_{eeeo}[n] = \{08\} \quad x_{eeoe}(n) = \{04\} \quad x_{eeoo}[n] = \{12\}$$
$$x_{eoee}[n] = \{02\} \quad x_{eoeo}[n] = \{10\} \quad x_{eooe}[n] = \{06\} \quad x_{eooo}[n] = \{14\}$$
$$x_{oeee}[n] = \{01\} \quad x_{oeeo}[n] = \{09\} \quad x_{oeoe}[n] = \{05\} \quad x_{oeoo}[n] = \{13\}$$
$$x_{ooee}[n] = \{03\} \quad x_{ooeo}[n] = \{11\} \quad x_{oooe}[n] = \{07\} \quad x_{oooo}(n) = \{15\}$$

Observing carefully the implementation of the preceding process, it is not difficult to see that for the FFT to be possible on the sequence of 16 points, these must be *rearranged*, so that they appear in the order

$$\{00, 08, 04, 12, 02, 10, 06, 14, 01, 09, 05, 13, 03, 11, 07, 05\}$$

It is proven that this rearrangement can be easily and quickly done in the following way. For each sample $x[n]$ we convert its index, n, to the binary system—it is obvious that the length of this binary number will be equal to $\log_2 N$—and then we reverse the bits of this number, so as to form its mirror image (considering, for example, the number 1001111, it is obvious that its mirror image that will result from the reversal of the bits will be 1111001). This process is known as *bit reverse sorting*, and its application for a number of points $N = 16$ will lead to the following results:

Sample Value	Value Index	Binary Value	Mirror Value	Decimal Value
$X_{00} = 5.27$	00	0000	0000	00
$X_{01} = 8.48$	01	0001	1000	08
$X_{02} = 3.71$	02	0010	0100	04
$X_{03} = 0.33$	03	0011	1100	12
$X_{04} = 4.85$	04	0100	0010	02
$X_{05} = 1.70$	05	0101	1010	10
$X_{06} = 9.23$	06	0110	0110	06
$X_{07} = 0.45$	07	0111	1110	14
$X_{08} = 5.17$	08	1000	0001	01
$X_{09} = 3.07$	09	1001	1001	09
$X_{10} = 1.09$	10	1010	0101	05
$X_{11} = 3.12$	11	1011	1101	13
$X_{12} = 8.16$	12	1100	0011	03
$X_{13} = 9.96$	13	1101	1011	11
$X_{14} = 1.12$	14	1110	0111	07
$X_{15} = 6.85$	15	1111	1111	15

In the foregoing process, $b = 4$ bits were used for the conversion of the indices to the binary system, because the time series contains $N = 16$ samples, and therefore their positions are encoded with $b = \log_2(16) = 4$ digits. After calculating the new indices, we should *sort* the samples according to them, thereby causing their *rearrangement*. This rearrangement involves *the swapping of the appropriate samples in each case*, a process justified by the fact that *if a number k_2 is the reversed binary of a number k_1, then the number k_1 is also the reversed binary of k_2*. This swapping process is very important because *all we have to do to rearrange the input sequence is swap one or more pairs of samples*, a process that does not require additional memory and is applied directly to the vector itself.

Having constructed the new sequence of points in the described way, we can then use the algorithm of Danielson and Lanczos to calculate the DFT of the sequence of input data. To do this, we must take into account that *each of the points of the new, rearranged sequence can be considered as the discrete transform of itself*. Using, therefore, two consecutive points—one of which is in a position with an even index and the other in a position with an odd index—and applying the lemma of Danielson and Lanczos, we can calculate the Fourier transform of the sequence of these two points. After the calculation of these transforms, *we double the number of points*—combining consecutive pairs of points—so as to calculate the Fourier transform for each of the $N/4$ groups of consecutive points that have emerged. This process is continuously applied *until we calculate the Fourier transform of the sequence of N points from the two discrete transforms that have emerged from the two equal parts of the mentioned sequence, each of which includes $N/2$ samples*.

To better understand this synthesis process, let us consider two input data sequences, $\{\kappa, \lambda, \mu, \nu\}$ and $\{\pi, \rho, \sigma, \tau\}$, whose Fourier transforms are described by the sequences $\{K, L, M, N\}$ and $\{P, R, S, T\}$. According to the analysis we presented in the previous paragraphs, the composition of the two transforms will lead to the Fourier transform of the sequence $\{\kappa, \lambda, \mu, \nu, \pi, \rho, \sigma, \tau\}$ that results from the union of the two sequences. The way in which the preceding process takes place is presented diagrammatically in Fig. 10.15.

In Fig. 10.15, we assume that the values $\{\kappa, \lambda, \mu, \nu\}$ belong to the *odd* positions of the vector from which they originated, while the values $\{\pi, \rho, \sigma, \tau\}$ belong to its *even* positions (in this description, we assume that the first cell of the table is in an *odd* position and corresponds to index 1; in the opposite case, we simply need to swap the roles of the two input sequences). For this reason, we *double the length of the sequence of the four points* and place *the odd samples in the odd positions*, while in the cells that are interspersed between them, we place *zero values*. The result of this process is the creation of the vector $\{\kappa, 0, \lambda, 0, \mu, 0, \nu, 0\}$. The same process is carried out for the *even* samples, with the difference that after the creation of the

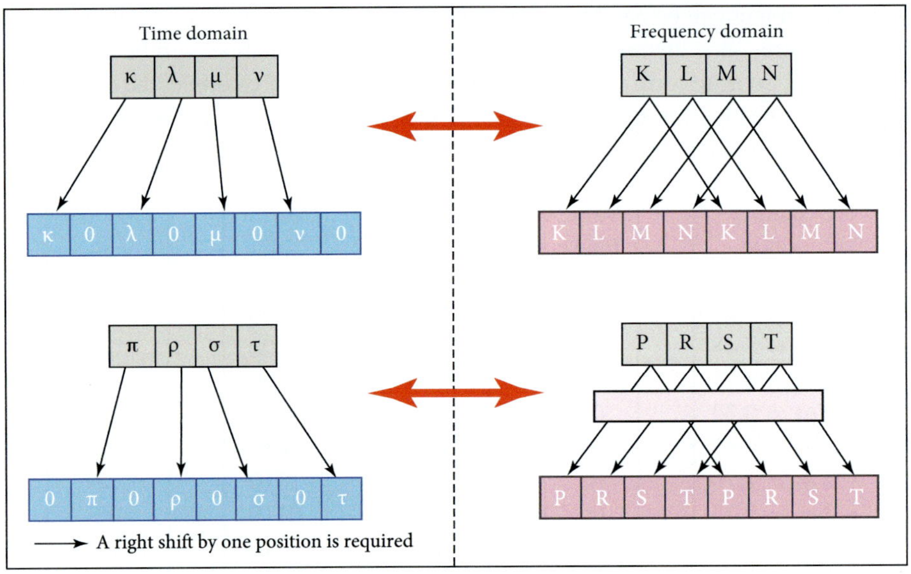

Fig. 10.15 Phase of FFT synthesis

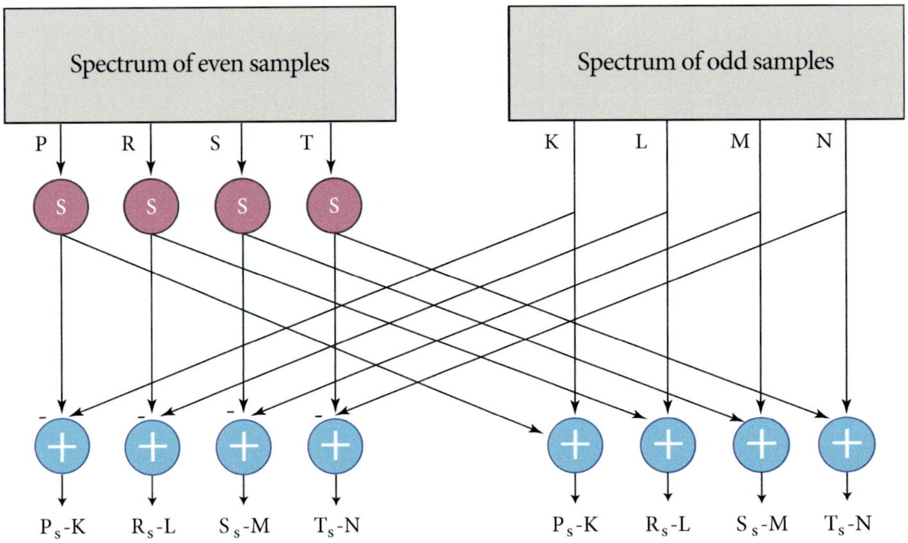

Fig. 10.16 Calculation of spectrum of a sequence of eight points from the spectra of two sequences of four points each

new vector its contents are *shifted one position to the right*—so as to move to the even positions—while in the first cell of the vector we place a zero value. In this way, the vector of even values will lead to the creation of the vector $\{0, \pi, 0, \rho, 0, \sigma, 0, \tau\}$. If we calculate the values of the Fourier transform coefficients of the discrete sequence $\{\kappa, 0, \lambda, 0, \nu, 0, \nu, 0\}$, we will find that they will have the values $\{K, L, M, N, K, L, M, N\}$, and therefore, *the process of expanding the vector to double its length with the simultaneous interpolation of zero values led to the doubling of the spectrum of the four-point signal*. The exact same process characterizes the Fourier coefficients of the sequence $\{0, \pi, 0, \rho, 0, \sigma, 0, \tau\}$, which will have the values $\{P, R, S, T, P, R, S, T\}$. However, in the latter case, the shift of the input signal points by one position to the right has caused the multiplication of the Fourier coefficients with a sinusoidal function, a process that arises from the properties of the transform presented in a previous section (see the property of *time shift*, which is equivalent to *the convolution of a signal with a time-shifted impulse function*, as well as the *convolution theorem*). In the last step of the process, the addition of the signals $\{\kappa, 0, \lambda, 0, \mu, 0, \nu, 0\}$ and $\{0, \pi, 0, \rho, 0, \sigma, 0, \tau\}$, so as to get the signal $\{\kappa, \lambda, \mu, \nu, \pi, \rho, \sigma, \tau\}$, will lead *to the addition of the spectra of the two signals*, as depicted in Fig. 10.16. We observe that the basic structural unit of this diagram is the one depicted in Fig. 10.17, which, due to its similarity to a butterfly, is known as the butterfly structure of the FFT (FFT butterfly structure). This structure for the calculation of the FFT of the sequence of 16 points is presented in Fig. 10.18.

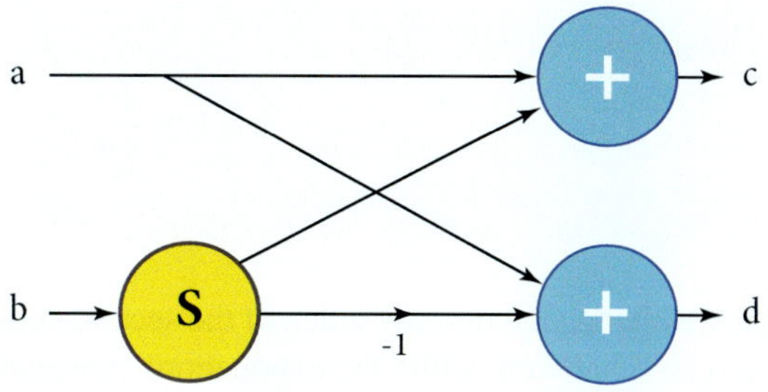

Fig. 10.17 Basic structural unit of butterfly-type diagrams

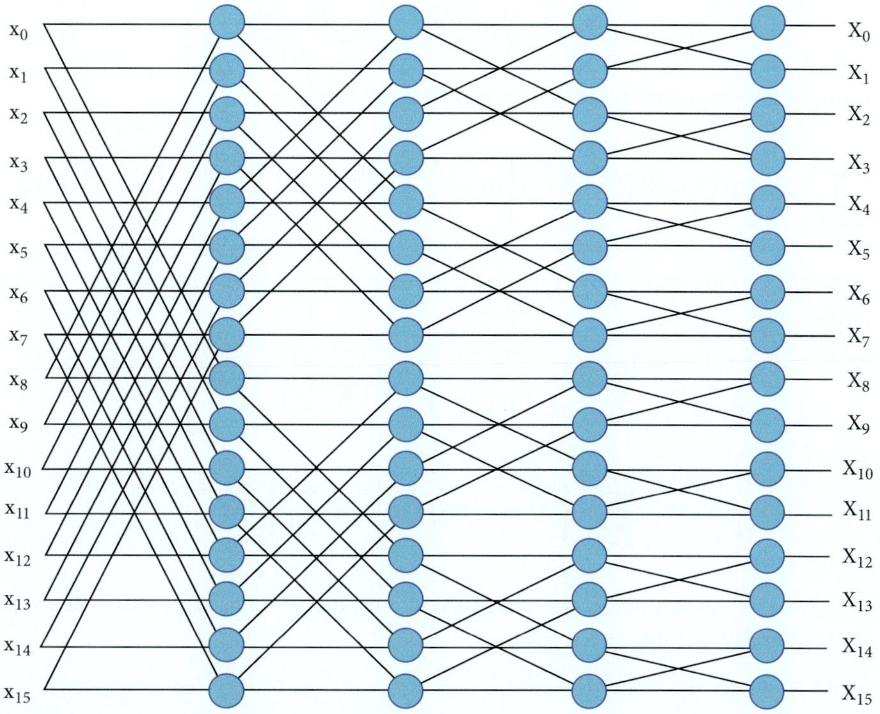

Fig. 10.18 Butterfly diagram for calculation of FFT of 16 points

From a programming point of view, the implementation of the Fourier transform is carried out *in two stages*. In the first stage, *the samples are rearranged in the way we described earlier*. This rearrangement is done very easily *by swapping the appropriate elements in each case*. The second part of the algorithm is characterized by the existence of a *double loop*. The outer loop is executed $\log_2 N$ times and calculates the Fourier transforms for sequences with a length of 2, 4, 8, …, N elements. Inside the outer loop are *two inner loops* that implement the Danielson and Lanczos algorithm , according to our discussion in the previous pages. To reduce the execution time of the calculations, *we can calculate the values of the sines and cosines of the algorithm in the outer loop*, so that these operations are performed $\log_2 N$ times.

This technique for calculating the DFT is known as *decimation in time*, since the process of successive decomposition is applied *to the discrete sequence* $x[n]$, which is defined in the time domain. Another interesting method of calculation based on $r = 2$ that falls into the algorithmic approach of "divide and conquer" is known as *decimation in frequency*, and, as one can see, it is applied to the discrete sequence $\mathcal{X}[k]$, which is obviously defined *in the frequency domain*. The application of this method consists *in the initial decomposition of the sequence* $x[n]$ *into two sequences of length* $N/2$*, of which the first includes the first* $N/2$ *samples and the second the last* $N/2$ *samples*[33] of the input sequence $x[n]$, and therefore the defining equation of the DFT will take the form

[33] This way of subdividing the original sequence is a special case of the more general problem of subdivision presented in the previous section and corresponds to the values $L = N/2$ and $M = 2$.

$$\mathcal{X}[k] = \sum_{n=0}^{N-1} x[n]\, W_N^{kn} = \sum_{n=0}^{(N/2)-1} x[n]\, W_N^{kn} + \sum_{n=N/2}^{N-1} x[n]\, W_N^{kn} \tag{10.134}$$

If we change the dummy index of the second sum from n to $m = n-(N/2)$, its new limits will be $m = 0$ and $m = (N/2)-1$, as can be proven, while the exponential expression that appears in this is reformulated as

$$W_N^{kn} = W_N^{k[m+(N/2)]} = W_N^{kN/2} W_N^{km}$$

Therefore, the second sum of the preceding relationship will take the form

$$\sum_{n=N/2}^{N-1} x[n]\, W_N^{kn} = W_N^{kN/2} \sum_{m=0}^{N-1} x\left[m + \frac{N}{2}\right] W_N^{km} = W_N^{kN/2} \sum_{n=0}^{N-1} x\left[n + \frac{N}{2}\right] W_N^{kn}$$

(with the third sum resulting from the second by restoring the dummy index from m to n), and the defining equation of the discrete transform is written as

$$\begin{aligned}\mathcal{X}[k] &= \sum_{n=0}^{(N/2)-1} x[n]\, W_N^{kn} + W_N^{kN/2} \sum_{n=0}^{N-1} x\left[n + \frac{N}{2}\right] W_N^{kn} \\ &= \sum_{n=0}^{(N/2)-1} \left\{ x[n] + (-1)^k x\left[n + \frac{n}{2}\right]\right\} W_N^{kn}\end{aligned} \tag{10.135}$$

since obviously $W_{kN/2} = (-1)^k$. At this point, we can proceed with the first decomposition of the sequence $\mathcal{X}[k]$ into its even- and odd-indexed samples, by defining the sequences

$$\mathcal{G}[k] = \mathcal{X}[2k] = \sum_{n=0}^{(N/2)-1} \left\{ x[n] + x\left[n + \frac{N}{2}\right]\right\} W_{N/2}^{kn} = \sum_{n=0}^{(N/2)-1} g[n]\, W_{N/2}^{kn} \tag{10.136}$$

$$\mathcal{H}[k] = \mathcal{X}[2k+1] = \sum_{n=0}^{(N/2)-1} \left\{\left(x[n] - x\left[n + \frac{N}{2}\right]\right) W_N^n \right\} W_{N/2}^{kn} \tag{10.137}$$

(for the values $0 \leq k \leq (N/2) - 1$) with the signals $g[h]$ and $h[n]$ defined as

$$\left.\begin{aligned} g[n] &= x[n] + x\left[n + \frac{N}{2}\right] \\ h[n] &= \left(x[n] - x\left[n + \frac{N}{2}\right]\right) W_N^n \end{aligned}\right\} \quad n = 0, 1, 2, \ldots, \frac{N}{2} - 1 \tag{10.138}$$

We can then proceed to further subdivide the sequences $\mathcal{G}[k]$ and $\mathcal{H}[k]$ into their even and odd samples, and so forth. The diagrammatic representation of this sequence of successive divisions and subsequent compositions is again highlighted through a butterfly-type structure in which $\nu = \log_2 N$ divisions are involved, that is, $(N/2)\log_2 N$ multiplications and

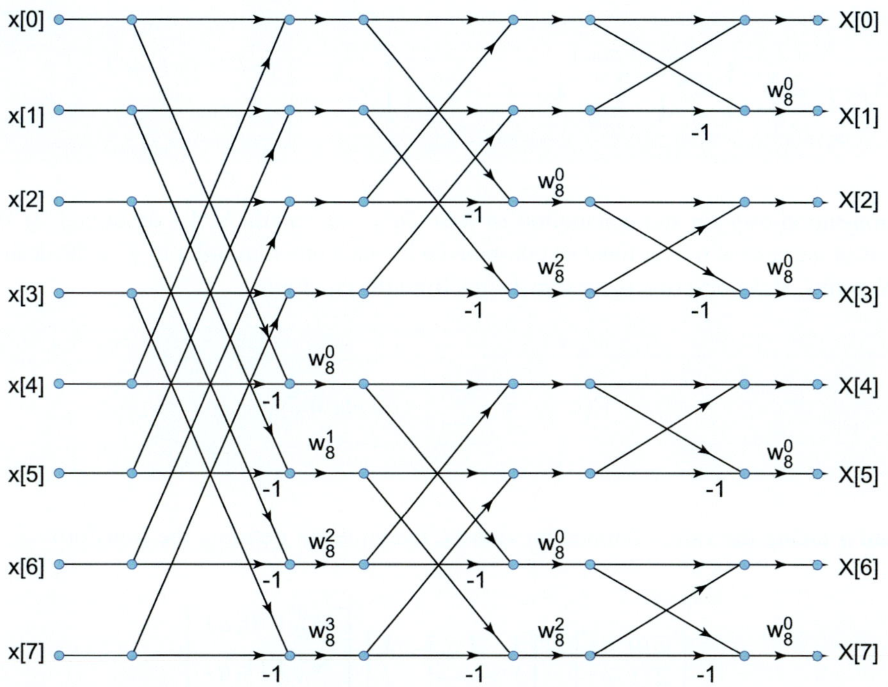

Fig. 10.19 Butterfly diagram for calculation of a DFT with a length of eight points using the frequency decimation method (Proakis and Manolakis, Figure 6.11)

$N \log_2 N$ additions between complex quantities. This structure for the calculation of a DFT with a length of $N = 8$ elements is presented in Fig. 10.19, from which we observe that the samples of the sequence $\mathcal{X}[k]$ are characterized not by the usual increase with respect to their position classification but one that results from the method of reversing the binary digits described previously.

Algorithms Based on $r = 4$

Following the methodology presented in previous sections, the application of an algorithm based on $r = 4$ requires a sequence length of the form $N = 4^\nu$ (note that since 4 is a power of 2, the calculation of the transform *can also be performed with an algorithm based on* $r = 2$, although the use of the algorithms described in this section is much more efficient). In this case, the separation of the data of the discrete sequence, according to the *divide and conquer* programming approach, is based on the values $L = 4$ and $M = L/4$; this means that the relationships between the indices of one-dimensional and two-dimensional arrays will be $n = 4m + \ell$ and $k = (N/4)p + q$, while the arrays themselves will be related through the equations

$$\zeta[\ell, m] = x[4m + \ell] \qquad \text{and} \qquad \mathcal{Z}[p, q] = \mathcal{X}\left[\frac{N}{4}p + q\right]$$

Therefore, the initial discrete sequence $x[n]$ with a length of N elements will be divided into the four sequences $e[n]$, $f[n]$, $g[n]$, and $h[n]$, which are defined as

$$\left.\begin{aligned} e[n] &= x[4n] \\ f[n] &= x[4n+1] \\ g[n] &= x[4n+2] \\ h[n] &= x[4n+3] \end{aligned}\right\} \quad n = 0, 1, 2, \ldots, \frac{N}{4} - 1 \tag{10.139}$$

while the application of an analysis similar to the preceding one will lead to the expression

$$\mathcal{Z}(p,q)=\sum_{\ell=0}^{3}\left\{W_N^{\ell q}\left[\sum_{m=0}^{(N/4)-1}\zeta(\ell,m)W_{N/4}^{mq}\right]\right\}W_4^{\ell p}\quad\begin{cases}p,\ell=0,1,2,3\\ q=0,1,2,\ldots,(N/4)-1\end{cases}$$

The foregoing relationship allows for the calculation of *four DFTs* of length $N/4$—described by the inner sum of this expression—which, as in the case of $r=2$, must be combined with each other, in order to give the desired discrete transform of the sequence of N points. If we describe these transforms from the equation

$$\Gamma(\ell,q)=\sum_{m=0}^{(N/4)-1}\zeta(\ell,m)\,W_{N/4}^{mq} \tag{10.140}$$

with the indices ℓ and q taking the values defined previously, the equation defining the transform $\mathcal{Z}(p,q)$ can be given in the form

$$\begin{bmatrix}\mathcal{Z}(0,q)\\ \mathcal{Z}(1,q)\\ \mathcal{Z}(2,q)\\ \mathcal{Z}(3,q)\end{bmatrix}=\begin{bmatrix}1&1&1&1\\ 1&-j&-1&j\\ 1&-1&1&-1\\ 1&j&-1&-j\end{bmatrix}\begin{bmatrix}W_N^0\,\Gamma(0,q)\\ W_N^q\,\Gamma(1,q)\\ W_N^{2q}\,\Gamma(2,q)\\ W_N^{3q}\,\Gamma(3,q)\end{bmatrix} \tag{10.141}$$

The first matrix on the right-hand side is the matrix $\boldsymbol{W}_4$ that we defined in Sect. 10.5.2 on page 856. The foregoing decomposition process *can be applied recursively* ν *times*, leading to a ν step algorithm with each step containing $N/4$ butterfly structures, that is, $(3N/8)\log_2 N$ multiplications and $(3N/2)\log_2 N$ additions between complex quantities. The last equation defines the butterfly structure depicted in Fig. 10.20, while, as for a schematic representation of the process of calculating a transform of 16 elements, this is depicted in Fig. 10.21.

The foregoing description refers to the method *of decimation in time*, while the method *of decimation in frequency* will be carried out in the way described in the previous case. For the implementation of this method, we should choose the values $L=N/4$ and $M=4$, while the indices are related to each other through the relations $n=(N/4)m+\ell$ and $k=4p+q$ and take the values $\ell,p=0,1,\ldots,(N/4)-1$ and $q=0,1,2,3$. Following the methodology used in the case $r=2$, we will formulate the defining equation of the discrete transform as a sum of four terms of the form

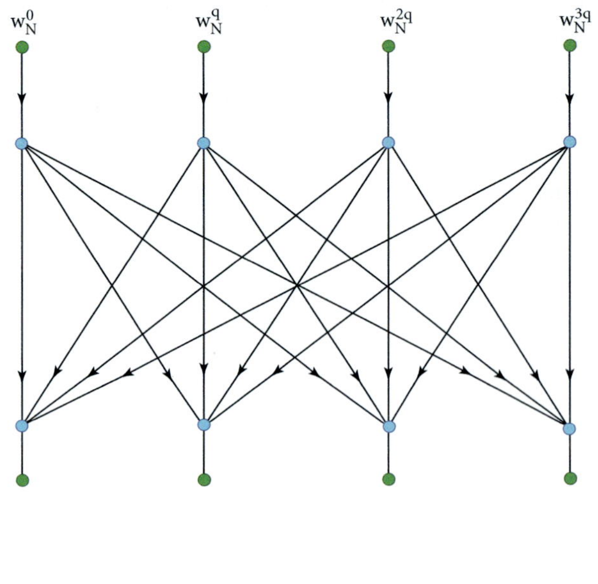

Fig. 10.20 Butterfly structure for decimation in time algorithm and for base value $r=4$

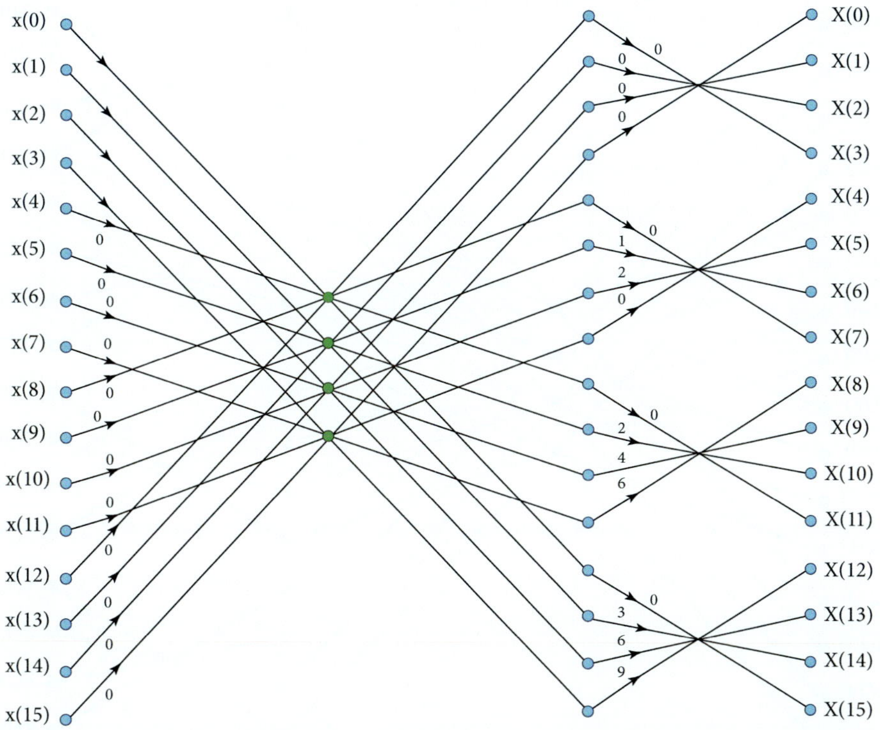

Fig. 10.21 Butterfly diagram of process of calculating DFT of length $N = 16$ for decimation in time method and for base value $r = 4$ (Proakis and Manolakis, Figure 6.13)

$$\mathcal{X}[k] = \sum_{n=0}^{(N/4)-1} x[n]\, W_N^{kn} + \sum_{n=N/4}^{(N/2)-1} x[n]\, W_N^{kn} + \sum_{n=N/2}^{(3N/4)-1} x[n]\, W_N^{kn} + \sum_{n=3N/4}^{N-1} x[n]\, W_N^{kn} = \sum_{n=0}^{(N/4)-1} x[n]\, W_N^{kn}$$

$$+ W_N^{kN/4} \sum_{n=0}^{(N/4)-1} x\left[n + \frac{N}{4}\right] W_N^{kn} + W_N^{kN/2} \sum_{n=0}^{(N/4)-1} x\left[n + \frac{N}{2}\right] W_N^{kn} + W_N^{3kN/4} \sum_{n=0}^{(N/4)-1} x\left[n + \frac{3N}{4}\right] W_N^{kn}$$

where for the construction of the last expression we dealt with the dummy indices of the sums in the way we did in the previous case. Taking into account that $W_N^{kN/4} = (-j)^k$, $W_N^{kN/2} = (-1)^k$, and $W_N^{3kN/4} = (+j)^k$, the defining equation of the DFT is formulated as

$$\mathcal{X}[k] = \sum_{n=0}^{(N/4)-1} \left\{ x[n] + (-j)^k x\left[n + \frac{N}{4}\right] + (-1)^k x\left[n + \frac{N}{2}\right] + (+j)^k x\left[n + \frac{3N}{4}\right] \right\} \qquad (10.142)$$

Having formulated the function $\mathcal{X}[k]$ in the previous way, we can now proceed to the *first subdivision* defining the sequences

$$\mathcal{X}[4k] = \sum_{n=0}^{(N/4)-1} \left\{ x[n] + x\left[n + \frac{N}{4}\right] + x\left[n + \frac{N}{2}\right] + x\left[n + \frac{3N}{4}\right] \right\} W_N^0\, W_{N/4}^{kn}$$

$$\mathcal{X}[4k+1] = \sum_{n=0}^{(N/4)-1} \left\{ x[n] - jx\left[n + \frac{N}{4}\right] - x\left[n + \frac{N}{2}\right] + jx\left[n + \frac{3N}{4}\right] \right\} W_N^n\, W_{N/4}^{kn}$$

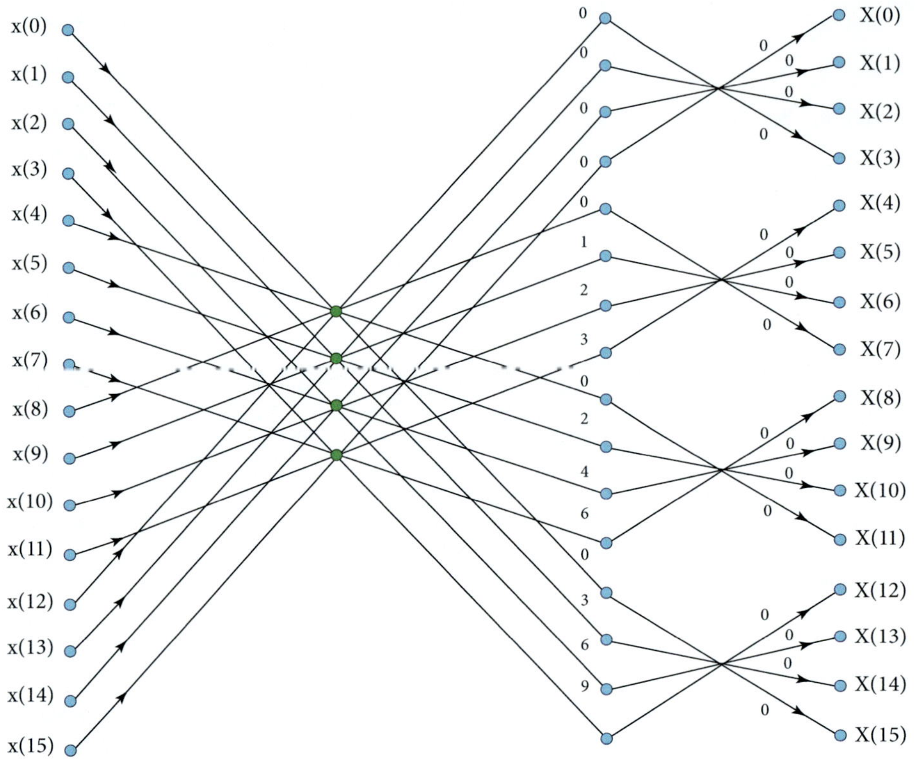

Fig. 10.22 Butterfly diagram of process of calculating DFT of length $N = 16$ for method of decimation in frequency and for base value $r = 4$ (Proakis and Manolakis, Figure 6.14)

$$\mathcal{X}[4k+2] = \sum_{n=0}^{(N/4)-1} \left\{ x[n] - x\left[n + \frac{N}{4}\right] + x\left[n + \frac{N}{2}\right] - x\left[n + \frac{3N}{4}\right] \right\} W_N^{2n} W_{N/4}^{kn}$$

$$\mathcal{X}[4k+3] = \sum_{n=0}^{(N/4)-1} \left\{ x[n] + jx\left[n + \frac{N}{4}\right] - x\left[n + \frac{N}{2}\right] - jx\left[n + \frac{3N}{4}\right] \right\} W_N^{3n} W_{N/4}^{kn}$$

where for the construction of the foregoing relations we took into account that $W_N^{4kn} = W_{N/4}^{kn}$. In the next step of the process, we can, as before, proceed to the subdivision of each of the foregoing sequences in the same way, a process that will be carried out *recursively* $\log_4 N$ times. This process of calculating the DFT is represented diagrammatically by the indicative butterfly diagram of Fig. 10.22.

10.5.6 Frequency Analysis of Signals and Systems

Although in the ideal case the frequency analysis of a signal or system is based on the use of the *DTFT*, in practice, the *DFT* is used, which is implemented by adopting some of the fast calculation methods presented in the previous section. However, it should be noted that this solution is only an *approximation*, as the discrete transform does not contain *only certain samples* of the DTFT and can also be applied *only to discrete-time signals of finite length*. For this reason, special care must be taken to ensure that the results we get *approximate as closely as possible the ideal case*. For example, we should ensure that the *circular convolution*—to which we are forced to resort, since we use finite sequences—*gives results identical to those given by the usual linear convolution* (see Sect. 10.5.4 on page 869), while the incompatibility between the requirement of finite length of the sequences used in the discrete transform and the infinite-length discrete-time signals used in the DTFT can be addressed using a well-known and widely used method based on *applying a window function to the infinite-length signal under consideration*. This method is presented in detail in the following pages.[34]

Construction of Discrete Sequences of Finite Length

As already mentioned in previous sections, the DFT accepts as input a discrete sequence $x[n]$ with a length of N samples that has resulted from the sampling of a continuous-time signal $x(t)$ with sampling frequency F_s and returns the values of N samples of the continuous spectrum that correspond to N discrete frequencies $F_k = kF_s/N$ $(k = 0, 1, 2, \ldots, N-1)$, which are integer multiples of the fundamental frequency $F_s/N = 1/T_sN$, where $T_s = 1/F_s$ is the sampling period. The question, therefore, that arises reasonably at this point is how the conversion of the sequence $x[n]$, which is assumed to have infinite length, into a finite sequence $y[n]$, which consists of N samples of the signal $x[n]$, happens.

The simplest way to perform such a conversion is to use a rectangular discrete pulse with a length of N samples in the form

$$w[n] = \begin{cases} 1 \text{ for } 0 \le n \le N-1 \\ 0 \text{ anywhere else} \end{cases} \tag{10.143}$$

and to construct the finite signal $y[n]$ by multiplying between the signals $x[n]$ and $w[n]$, leading to a result in the form

$$y[n] = x[n]w[n] = x[n]\begin{cases} 1 \text{ for } 0 \le n \le N-1 \\ 0 \text{ anywhere else} \end{cases} = \begin{cases} x[n] \text{ for } 0 \le n \le N-1 \\ 0 \quad \text{anywhere else} \end{cases}$$

In the terminology of digital signal processing, the *orthogonal discrete sequence* used in the way described here is known as a window function (since it allows the isolation of only a part of the original signal) and is a member of a family of windows that can be used alternatively depending on the requirements of each case; these windows will be presented in detail in a subsequent section. It is obvious that the aforementioned multiplication process must be carried out *before the calculation of the DFT*, which, therefore, will be applied to the finite-length sequence $y[n]$ that will produce the described result. This process for an orthogonal window with a length of $N = 15$ samples is graphically presented in Fig. 10.23.

Taking into account that the discrete finite-length signal $y[n]$ constitutes *an approach to the discrete-time signal $x[n]$ of infinite length*, it is not difficult to see that the use of the window function will make the spectrum $\mathcal{Y}(e^{j\omega}) = \mathcal{F}\{y[n]\}$ *an approximate version of the desired spectrum $\mathcal{X}(e^{j\omega}) = \mathcal{F}\{x[n]\}$*. It is proven that for a given length N of the window function, the rectangular window leads to the *smallest approximation error* compared to all windows of the same length. To verify the truth of this claim, let us consider the mean square approximation error of the function $\mathcal{X}(e^{j\omega})$ from the function $\mathcal{Y}(e^{j\omega})$, which is defined as

$$\varepsilon^2 = \frac{1}{2\pi}\int_{-\pi}^{+\pi} |\mathcal{X}(e^{j\omega}) - \mathcal{Y}(e^{j\omega})|^2 d\omega \tag{10.144}$$

and which, using Parseval's theorem, takes the form

$$\varepsilon^2 = \frac{1}{2\pi}\int_{-\pi}^{+\pi} |\mathcal{X}(e^{j\omega}) - \mathcal{Y}(e^{j\omega})|^2 d\omega = \sum_{n=-\infty}^{+\infty} |x[n] - y[n]|^2 = \sum_{n=0}^{N-1} |x[n] - y[n]|^2 + \sum_{n \in Z-[0,N]} |x[n] - y[n]|^2 \tag{10.145}$$

But from the last relation it immediately follows that *minimizing the error consists in nullifying the second sum of the foregoing relation, that is, in cutting off all elements of the sequence $x[n]$ that correspond to $n \ge N$ positions*, thereby

[34] Although in most textbooks the use of a window function is examined in conjunction with the processes of constructing digital filters with finite impulse response, it is a general method of transforming an infinite-length signal of an infinite sequence into a discrete-time signal of finite length, and for this reason it is presented in this section. In the next chapter we will study the consequences of this process on the frequency response function of a digital filter.

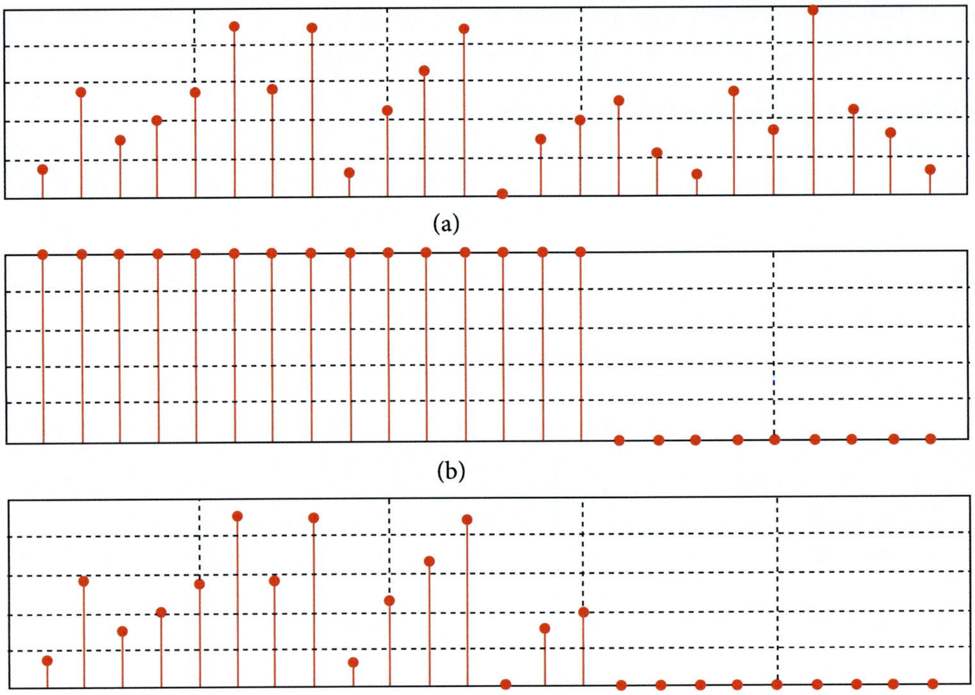

Fig. 10.23 Construction of a finite-length signal by multiplying an infinite-length signal by an orthogonal discrete sequence with a length of $N = 15$ samples. (**a**) The discrete-time signal $x[n]$ with a possibly infinite number of samples. (**b**) The discrete rectangular pulse $w[n]$ with a length of $N = 15$ samples. (**c**) The discrete-time signal $y[n] = x[n] * w[n]$ with a length of $N = 15$ samples

leading to the conclusion formulated previously. However, let us note that in digital filter design processes, this criterion of the mean square error approximation *is not necessarily the best possible choice* because it considers all frequencies *equivalent*, ascribing to them *the same statistical weight*. In practice, however, we are usually only interested in *specific frequency ranges* that are considered more important than the others, and for this reason advanced optimization algorithms use a *weight function* $\Omega(\omega)$ that allows the selection of the relative error size for each frequency separately.

In the following description, we will find that *the shape and characteristics of the window function spectrum greatly influence the spectrum of the finite signal that will result from the application of the described process*. For this reason, let us examine the characteristics of the spectrum of a discrete orthogonal sequence with a length of N samples given by the relation

$$\mathcal{W}(e^{j\omega}) = e^{-j\omega(N-1)/2}\left[\frac{\sin(\omega N/2)}{\sin(\omega/2)}\right]$$

whose measure is depicted in Fig. 10.24. From this figure we find that the function $|\mathcal{W}(e^{j\omega})|$ is characterized by the existence of a *central lobe* (whose range, as we will soon discover, is simply *the active range* $\Delta\omega_w$ of the window $w[n]$ that was defined previously), as well as a set of *sidelobes* whose height *decreases* at a relatively slow rate with increases in frequency; in reality, these sidelobes weaken at a rate proportional to $1/\omega$, and in the general case the effect of the window is considered *more efficient the smaller the height of the lobes and the greater the rate at which it weakens*. We will return to this issue later.

Spectral Diffusion and Leakage

The next question we must answer concerns the way in which the window function will affect the spectrum $\mathcal{X}(e^{j\omega})$ of the original discrete-time signal $x[n]$. Generally speaking, if we denote by $\mathcal{W}(e^{j\omega}) = \mathcal{F}\{w[n]\}$ the spectrum of the window $w[n]$ we use and make use of *the theorem of the product of the DTFT* (Sect. 10.3.6 on page 818), then we will discover the spectrum $\mathcal{Y}(e^{j\omega})$ of the finite-length discrete sequence $y[n]$ will be calculated as a convolution of two spectra of the form

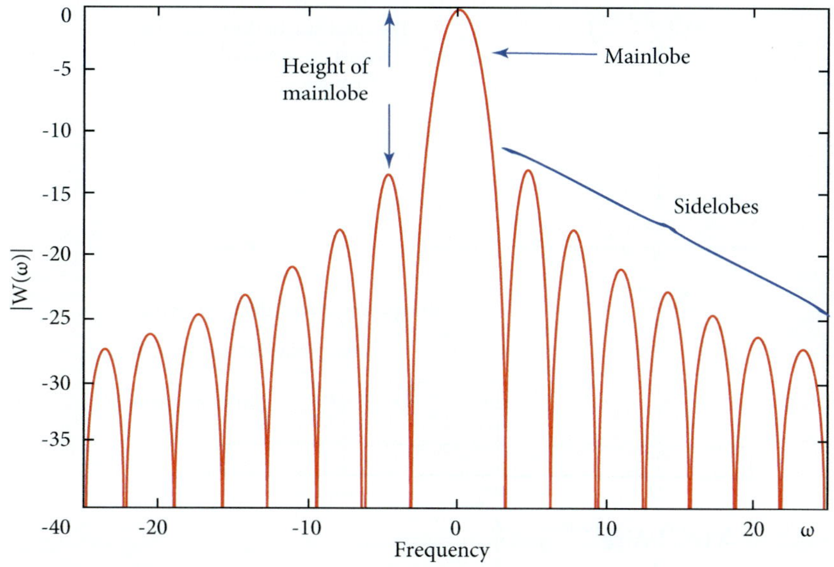

Fig. 10.24 The measure of the spectrum of a discrete orthogonal sequence of finite length

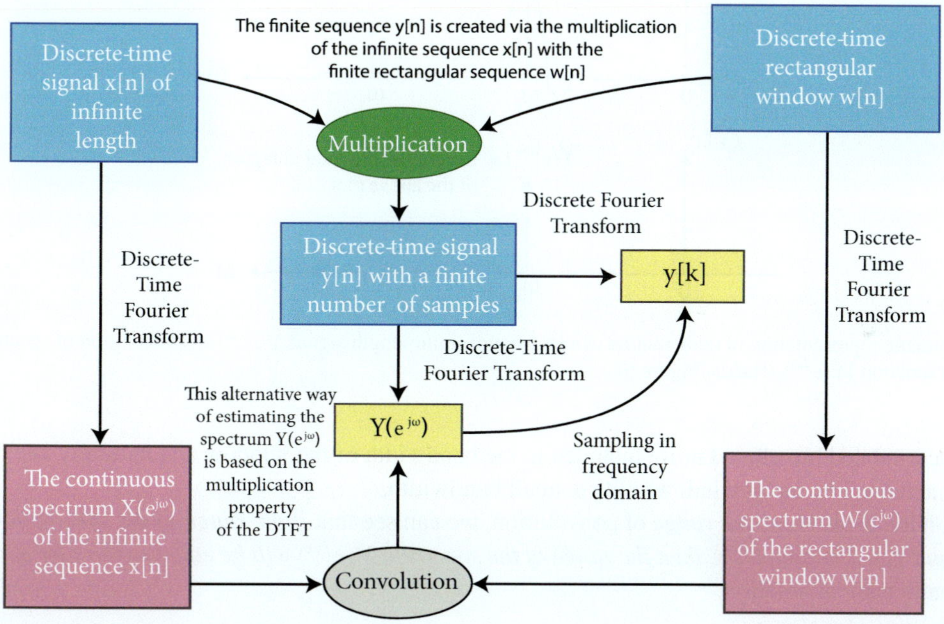

Fig. 10.25 Diagrammatic representation of correlations existing between sequences $x[n]$, $y[n]$, and $w[n]$ regarding the calculation of the DFT of a finite-length sequence

$$\mathcal{Y}(e^{j\omega}) = \mathcal{F}\{y[n]\} = \mathcal{F}\{w[n]\cdot x[n]\} = \frac{1}{2\pi}\int_{-\pi}^{+\pi} \mathcal{X}(e^{j\lambda})\,\mathcal{W}(e^{j(\omega-\lambda)})\,d\lambda \tag{10.146}$$

This calculation process is presented diagrammatically in Figs. 10.25 and 10.26, from which it is apparent that the window function operates as a weight function and that in the calculation of the spectrum value $\mathcal{Y}(e^{j\omega})$ at the position of a frequency of interest ω_0, all frequencies contribute, with the degree of contribution of each frequency ω being determined by the value of the frequency-shifted function $\mathcal{W}(e^{j(\omega_0-\omega)})$. From Fig. 10.26 we also observe that the amplitude of the spectrum at the position of a frequency—which, being an integral, will be equal to the area of the shaded region of the third graph—is

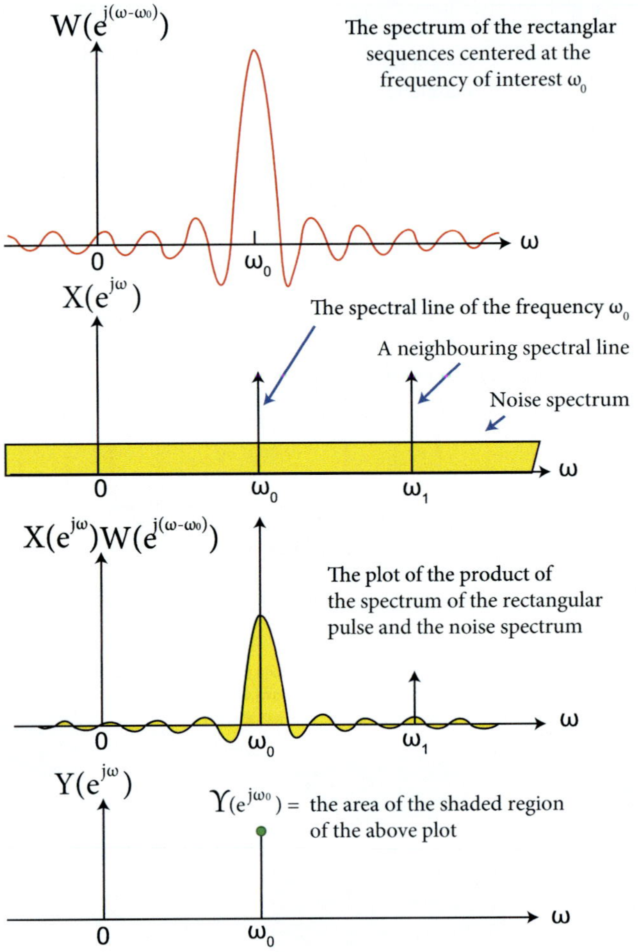

Fig. 10.26 Diagrammatic representation of computation of spectrum of finite-length signal $\mathcal{Y}(e^{j\omega})$ as convolution of spectrum of original signal $\mathcal{X}(e^{j\omega})$ and window function $\mathcal{W}(e^{j\omega})$ (Harris, Figure 6)

affected by the accumulated broadband noise included in the bandwidth of the window and naturally needs to be minimized, something that is achieved by using windows with a small bandwidth.

Now, relying on the *property of the range* of convolution, we can see that *if the range of the spectra* $\mathcal{X}(e^{j\omega})$ *and* $\mathcal{W}(e^{j\omega})$ *is equal to* $\Delta\omega_x$ *and* $\Delta\omega_w$, *respectively, then the range of the spectrum* $\mathcal{Y}(e^{j\omega})$ *will be equal to the sum of the two with higher values,*[35] or, in mathematical notation,

$$\Delta\omega_y = \Delta\omega_x + \Delta\omega_w \tag{10.147}$$

Taking into account that our description takes place in the frequency domain, this size that we describe as a range should actually be simply the *active bandwidth of the finite discrete sequence* $y[n]$, which, as is well known, is defined as the *frequency range outside of which the Fourier transform* $\mathcal{Y}(e^{j\omega})$ *has a very small value, so it is considered negligible*. We are led, therefore, to the very important conclusion that *the application of a window function* $w[n]$ *to an infinite discrete-time signal* $x[n]$ *in order for it to become finite causes the widening of the spectrum of the original signal by the active bandwidth of the spectrum of the window function that is applied to it*; this phenomenon, known as *spectral spreading*, is due to the central lobe of the spectrum of the rectangular discrete sequence and as we shall see in what follows, it impacts

[35] For a diagrammatic demonstration of the property of the range of convolution see Fig. 1.22 on page 47. This figure refers to the *time domain*, but of course the situation is completely analogous in the frequency domain.

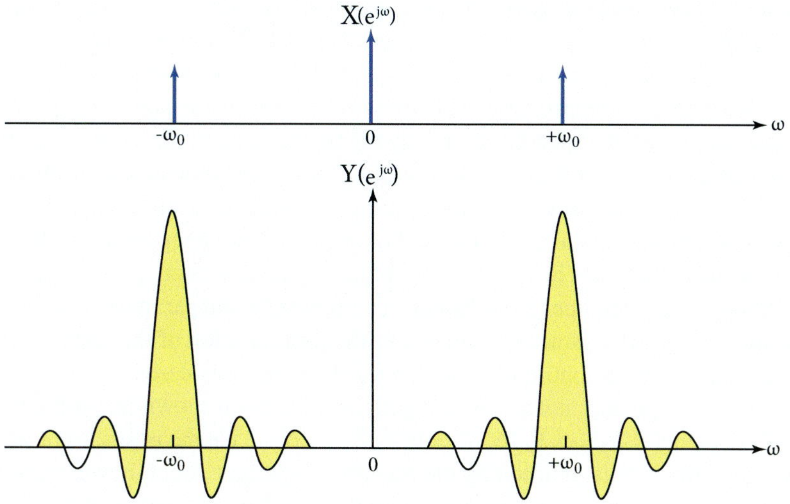

Fig. 10.27 Demonstration of spectral leakage and its effect on the spectrum of the simple sequence $x[n] = \cos(\omega_0 n)$

the frequency resolution with respect to the ability to distinguish between closely spaced frequencies in the spectrum of the discrete transform.

Example 10.5.4 To demonstrate the phenomenon of spectral diffusion and the type of distortion it causes in the spectrum of infinite-length signals, let us consider the discrete cosine

$$x[n] = A\cos(\omega_0 n), \quad -\infty < n < \infty$$

whose spectrum $\mathcal{X}(e^{j\omega})$, as is well known, has two spectral lines[36] at the frequency positions $\omega = +\omega_0$ and $\omega = -\omega_0$ (see Exercise 10.3.10 on page 826); this spectrum is depicted in the top image of Fig. 10.27. If we multiply the signal $x[n]$ by a rectangular window $w[n]$ of N elements defined in the value range $-N/2 \leq n \leq N/2$, then the spectrum $\mathcal{Y}(e^{j\omega})$ of the finite discrete-time signal $y[n]$ that will result will be calculated, as we have already mentioned, from the convolution between the spectrum $\mathcal{X}(e^{j\omega})$ of the signal $x[n]$ and the spectrum $\mathcal{W}(e^{j\omega})$ of the rectangular window $w[n]$, which is known to be the sinc function (see Exercise 10.3.2 on page 810). Taking into account (a) that the spectrum of the signal $x[n]$ contains two impulse functions located at the positions $\omega = \pm\omega_0$ and (b) that the convolution between any function and the impulse function is the same function shifted to the position at which the impulse function has been calculated (see Sect. 1.6.3 on page 24 and, more specifically, Eq. 1.23), we will find that the spectrum $\mathcal{Y}(e^{j\omega})$ has the form that is presented in the lower image of Fig. 10.27. From this figure we observe that the spectrum of the signal $y[n]$ has been extended around the frequencies $\omega = \pm\omega_0$ by the amount $2\pi/N$ in each direction, that is, in total in a region of width $4\pi/N$, which is also the width of the mainlobe of the spectrum of the rectangular window.

But how extensive is this diffusion? According to *bandwidth theorem* examined in Sect. 3.3.7 (page 255), *the active bandwidth* $\Delta\omega$ *and the active duration* Δt *of a signal are connected by a relationship of inverse proportionality* since they satisfy the inequality

$$\Delta\omega \cdot \Delta t \geq \gamma \tag{10.148}$$

where γ is a global constant. The basic consequence of the application of this theorem is that *the longer the length of the window we use, the smaller is its active bandwidth and, therefore, the smaller will be the degree of spectral diffusion we described*, while of course the *inverse* applies. Let us note that this is expected, since *the larger the length of the window*

[36] For simplicity, we limit ourselves to the basic frequency range $-\pi \leq \omega \leq \pi$.

$w[n]$ we use, the more samples of the original signal $x[n]$ we take into account, the greater the approximation of this signal from its finite equivalent and the smaller the deviation that exists between the corresponding spectra, which is expressed in the form of this diffusion. However, it should be noted that this spectral diffusion cannot be completely removed, as the degree to which it appears tends to zero only *asymptotically*. This means that even if the signal $x[n]$ is strictly *of finite bandwidth*, in the sense that its spectrum exists within a specific frequency range, the application of the window will cause *the extension of the spectrum to frequency areas where its measure was initially zero*; this undesirable phenomenon is known as *leakage* and is due to the existence of the sidelobes of the spectrum of the rectangular discrete sequence. The main consequence of spectral leakage is that the energy of the signal is not focused on a single frequency but extends to neighboring frequencies, thereby introducing unwanted noise that must be removed, as it hinders the detection of low-amplitude signals corresponding to adjacent frequencies. Although the phenomenon of leakage is affected by the parameters of the sampling process, it is not due to this or to the application of the DFT (since it is related to the past duration of the signal), while the degree to which it appears is proportional to the degree of discontinuity that characterizes the considered signal.

Note that, in reality, spectral leakage does not always happen, as it has been shown that if the frequency of an input signal coincides with one of the N frequencies $F_k = kF_s/N$ ($k = 0, 1, 2\ldots, N-1$) at the positions of which the DFT is calculated, then everything will go well and the results returned by the transform will be correct; however, this obviously constitutes an exception to the rule, since in the general case the frequency of the input signal can be anything. To prove the truth of this claim, let us consider an arbitrary sinusoidal signal of infinite length and frequency ω_0 of the form $x[n] = \cos(\omega_0 n)$, whose DTFT, as we proved in Exercise 10.3.10 (refer to page 826), is given by the relation

$$\mathcal{X}(e^{j\omega}) = \pi\delta(\omega-\omega_0) + \pi\delta(\omega+\omega_0)$$

(for the value $k = 0$, so as to limit ourselves to the fundamental frequency region). If we denote by $\mathcal{W}(e^{j\omega})$ the spectrum of the rectangular window $w[n]$ with a length of N elements by which we multiply the input signal $x[n]$ in order for it to become finite, then, according to our earlier discussion, the continuous spectrum (i.e., the DTFT) of the signal $y[n] = x[n]w[n]$ is calculated as

$$\begin{aligned}\mathcal{Y}(e^{j\omega}) &= \frac{1}{2\pi}\int_{-\pi}^{+\pi}\mathcal{X}(e^{j\lambda})\,\mathcal{W}(e^{j(\omega-\lambda)})\,d\lambda = \frac{1}{2\pi}\int_{-\pi}^{+\pi}\pi\Big[\delta(\lambda-\omega_0)+\delta(\lambda+\omega_0)\Big]\mathcal{W}(e^{j(\omega-\lambda)})\,d\lambda \\ &= \frac{1}{2}\int_{-\pi}^{+\pi}\delta(\lambda-\omega_0)\mathcal{W}(e^{j(\omega-\lambda)})\,d\lambda + \frac{1}{2}\int_{-\pi}^{+\pi}\delta(\lambda+\omega_0)\mathcal{W}(e^{j(\omega-\lambda)})\,d\lambda = \frac{1}{2}\mathcal{W}(e^{j(\omega-\omega_0)}) + \frac{1}{2}\mathcal{W}(e^{j(\omega+\omega_0)})\end{aligned}$$

where to derive the final expression we used the appropriate property of the impulse function (see Sect. 1.6.3 on page 24). Taking into account that the DTFT of the discrete orthogonal sequence with a length of N elements has the form

$$\mathcal{W}(e^{j\omega}) = e^{-j\omega(N-1)/2}\left[\frac{\sin(\omega N/2)}{\sin(\omega/2)}\right]$$

(see Exercise 10.3.2 on page 810), the foregoing relationship is written

$$\mathcal{Y}(e^{j\omega}) = \frac{1}{2}\exp\left[-j(\omega-\omega_0)\frac{N-1}{2}\right]\left\{\frac{\sin\left[\dfrac{(\omega-\omega_0)N}{2}\right]}{\sin\left(\dfrac{\omega-\omega_0}{2}\right)}\right\} + \frac{1}{2}\exp\left[-j(\omega+\omega_0)\frac{N-1}{2}\right]\left\{\frac{\sin\left[\dfrac{(\omega+\omega_0)N}{2}\right]}{\sin\left(\dfrac{\omega+\omega_0}{2}\right)}\right\}$$

Let us now assume that the sinusoidal input signal is periodic and characterized by a frequency equal to m, that is, by m full repetitions that extend to the set of N samples that make it up. As we analytically proved in Sect. 6.7.1 (page 520), for a signal to be periodic, its frequency f_0 must be a rational number of the form $f_0 = m/N$, leading to a cyclic frequency of the form $\omega_0 = 2\pi f_0 = 2\pi m/N$. Consequently, the input signal will be given by the relation $x[n] = \cos(2\pi mn/N)$, while the spectrum $\mathcal{Y}(e^{j\omega})$ of the finite discrete-time signal $y[n]$ will be described by the equation

$$\mathcal{Y}(e^{j\omega}) = \frac{1}{2}\exp\left[-j\left(\omega - \frac{2\pi m}{N}\right)\frac{N-1}{2}\right]\frac{\sin\left[\left(\omega - \frac{2\pi m}{N}\right)N/2\right]}{\sin\left[\left(\omega - \frac{2\pi m}{N}\right)/2\right]}$$

$$+\frac{1}{2}\exp\left[-j\left(\omega + \frac{2\pi m}{N}\right)\frac{N-1}{2}\right]\frac{\sin\left[\left(\omega + \frac{2\pi m}{N}\right)N/2\right]}{\sin\left[\left(\omega + \frac{2\pi m}{N}\right)/2\right]}$$

Finally, the DFT of the signal $y[n]$ will result from the sampling of the spectrum $\mathcal{Y}(e^{j\omega})$, which consists in recording the values of $\mathcal{Y}(e^{j\omega_k})$ at the positions of the frequencies $\omega_k = 2\pi k/N$. If in the preceding relation we make the substitution $\omega \to \omega_k$, this will take the form

$$\mathcal{Y}(e^{j\omega_k}) = \frac{1}{2}\exp\left[-j\frac{\pi}{N}(k-m)(N-1)\right]\frac{\sin[\pi(k-m)]}{\sin[\pi(k-m)/N]} + \frac{1}{2}\exp\left[-j\frac{\pi}{N}(k+m)(N-1)\right]\frac{\sin[\pi(k+m)]}{\sin[\pi(k+m)/N]}$$

Of the two terms in the above expression, the first corresponds to the region of positive frequencies, while the second corresponds to the region of negative frequencies, and its existence is due to the real nature of the discrete input sequence $x[n]$. Using the approximate expression $\sin\alpha \approx \alpha$, which holds for very small values of the parameter α, we can for quite large values of the length N and small values of the differences $k-m$ and $k-(-m) = k+m$ make the approximations

$$\sin\left[\frac{\pi}{N}(k-m)\right] \approx \frac{\pi}{N}(k-m) \quad \text{and} \quad \sin\left[\frac{\pi}{N}(k+m)\right] \approx \frac{\pi}{N}(k+m)$$

and formulate the discrete spectrum $\mathcal{Y}(e^{j\omega_k})$ in the form

$$\mathcal{Y}(e^{j\omega_k}) \approx \left\{\frac{N}{2}\frac{\sin[\pi(k-m)]}{\pi(k-m)}\right\}\exp\left[-j\frac{\pi}{N}(k-m)(N-1)\right] + \left\{\frac{N}{2}\frac{\sin[\pi(k+m)]}{\pi(k+m)}\right\}\exp\left[-j\frac{\pi}{N}(k+m)(N-1)\right]$$

Therefore, the magnitude of the discrete spectrum in the positive and negative frequency regions will be given by the equations

$$|\mathcal{Y}(e^{j\omega_k})|^{+} \approx \frac{N}{2}\frac{\sin[\pi(k-m)]}{\pi(k-m)} \quad \text{and} \quad |\mathcal{Y}(e^{j\omega_k})|^{-} \approx \frac{N}{2}\frac{\sin[\pi(k+m)]}{\pi(k+m)}$$

respectively, and as we can see, these are simply our known sinc function scaled by the factor $N/2$. Limiting ourselves for simplicity to the positive frequency region, it is not difficult to see that this function will be centered around the position $k = m$, where it also shows its maximum value is equal to $N/2$, while it intersects the horizontal axis at positions $(\ldots, m-3, m-2, m-1, m, m+1, m+2, m+3, \ldots)$. The graphical representation of the function $|\mathcal{Y}(e^{j\omega_k})|^{+}$ is presented in Fig. 10.28, while the function $|\mathcal{Y}(e^{j\omega_k})|^{-}$ is described by a similar curve, which of course is centered with respect to the position $k = -m$; therefore, the curves of the functions $|\mathcal{Y}(e^{j\omega_k})|^{+}$ and $|\mathcal{Y}(e^{j\omega_k})|^{-}$ will be symmetrical with respect to the vertical axis corresponding to the value $k = 0$.

Let us now assume that the frequency $\omega_0 = 2\pi m/N$ of the periodic input signal coincides with one of the k frequencies $\omega_k = 2\pi k/N$ at whose positions the DFT is calculated. We will therefore have $k = m$, while for reasons of symmetry we should also consider the case $k = -m$ for the area of negative frequencies. Considering the two terms of the foregoing sum separately, since they refer to different frequency regions, it is not difficult to see that for $k \neq m$ (and, correspondingly, for $k \neq -m$), the functions $\sin[\pi(k-m)]$ (and, correspondingly, $\sin[\pi(k+m)]$) have a value equal to zero and therefore will be $\mathcal{Y}(e^{j\omega_k})_{k\neq m} = \mathcal{Y}(e^{j\omega_k})_{k\neq -m} = 0$. On the other hand, for $k = m$ (and, correspondingly, for $k = -m$), the values of the exponential expressions are equal to unity, while the two ratios of the sines take the indeterminate form $0/0$. Therefore, applying L'Hôpital's rule, it turns out that

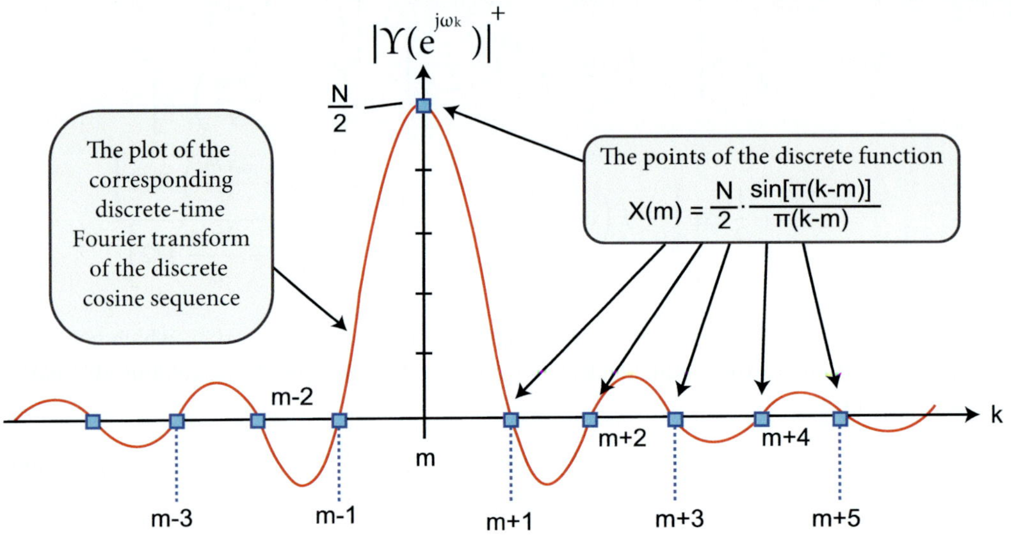

Fig. 10.28 Amplitude response of DFT with a length of N points of a periodic discrete sine wave of frequency m. This curve, which in reality is the same as that of the previous figure that demonstrated the characteristic of spectral diffusion, reveals the existence of a central lobe and a set of sidelobes that originate from the spectrum of the rectangular window that convolved with the spectrum of the original infinite-length discrete cosine (Lyons, Figure 3.9a)

$$\lim_{k\to m}\frac{\sin[\pi(k-m)]}{\sin\left[\frac{\pi}{N}(k-m)\right]}=\lim_{k\to m}\frac{\frac{d}{dk}\left\{\sin[\pi(k-m)]\right\}}{\frac{d}{dk}\left\{\sin\left[\frac{\pi}{N}(k-m)\right]\right\}}=\lim_{k\to m}\frac{\pi\cos[\pi(k-m)]}{\frac{\pi}{N}\cos\left[\frac{\pi}{N}(k-m)\right]}=N$$

$$\lim_{k\to -m}\frac{\sin[\pi(k+m)]}{\sin\left[\frac{\pi}{N}(k+m)\right]}=\lim_{k\to m}\frac{\frac{d}{dk}\left\{\sin[\pi(k+m)]\right\}}{\frac{d}{dk}\left\{\sin\left[\frac{\pi}{N}(k+m)\right]\right\}}=\lim_{k\to m}\frac{\pi\cos[\pi(k+m)]}{\frac{\pi}{N}\cos\left[\frac{\pi}{N}(k+m)\right]}=N$$

and therefore $\mathcal{Y}(e^{j\omega_k})_{k=m} = \mathcal{Y}(e^{j\omega_k})_{k=-m} = N/2$. A graphical representation of the spectrum's magnitude $\mathcal{Y}(e^{j\omega_k})$ (the set of discrete points) as well as the magnitude of the corresponding continuous spectrum $\mathcal{Y}(e^{j\omega})$ (the continuous curve) for the regions of positive and negative frequencies is given in Fig. 10.29. From this figure we can see that the spectrum contains only two spectral lines corresponding to the frequencies $\omega = \pm\omega_0$ (something that is, of course, expected), and therefore, when the frequency of the input signal coincides with some of the N frequencies of the transform, the results returned from it are correct.

On the other hand, when the foregoing condition does not exist, things change dramatically. In this case, none of the terms of the preceding expression is zeroed, so there will be spectral lines of nonzero height at the positions of all the N frequencies at which the DFT is calculated. But this means that the energy of the periodic input signal *is not concentrated at the frequencies associated with it* but has extended (or leaked) to all frequencies $F_k = kF_s/N$ $(k = 0, 1, 2, \ldots, N-1)$. This is also why this phenomenon, which is clearly undesirable, is known as leakage. This situation for a sinusoidal signal of frequency $m + \frac{1}{4}$ is shown in Fig. 10.30, which, as we can see, is shifted in relation to the previous one concerning the ideal case by $0.25F_s/N$ (i.e., by one quarter of the distance between two successive frequencies of the transform). In this case, we observe that this shift has established a correspondence between the N frequencies of the transform and the nonzero points of the curve of the continuous spectrum $\mathcal{Y}(e^{j\omega})$, resulting in the appearance of the phenomenon of leakage.

Use of the Window Function

Based on our discussion in the preceding pages, let us now try to lay down some rules that, if applied, will *minimize* the undesirable phenomena of *spectral diffusion* and *leakage* just described:

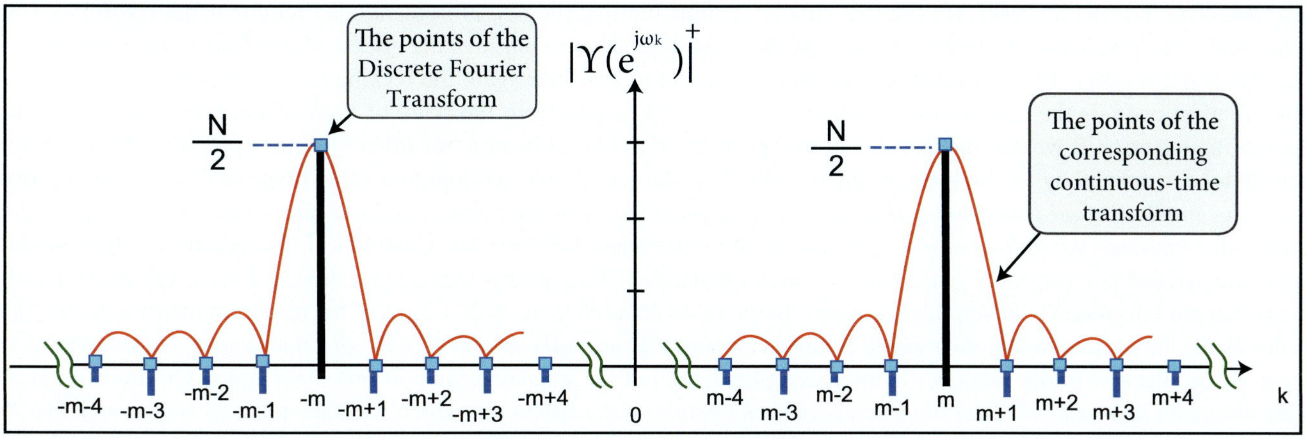

Fig. 10.29 Discrete and continuous spectrum of a discrete periodic sinusoidal signal whose frequency m belongs to the set of discrete frequencies for which the DFT is calculated. We observe that, in this case, the samples of the DFT coincide with the points at which the corresponding curve of the continuous transform intersects the horizontal axis (and, therefore, have zero value), with the exception of two points that have a value equal to $N/2$ (where N is the length of the transform) and correspond to the frequencies $\pm\omega_0$, at whose positions the energy of the signal is concentrated (Lyons, Figure 3.43)

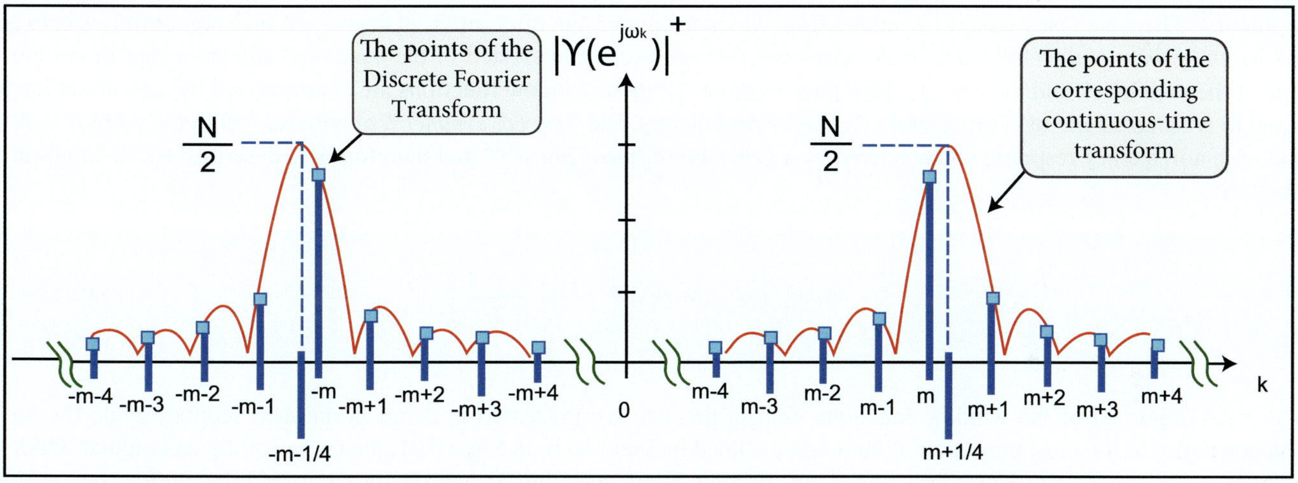

Fig. 10.30 Discrete and continuous spectrum of discrete periodic sinusoidal signal with frequency $m+\frac{1}{4}$ that does not belong to the set of discrete frequencies for which the DFT is calculated. We observe that, in this case, the energy of the signal leaks into all the frequencies at which the DFT is calculated, an undesirable phenomenon known as leakage (Lyons, Figure 3.44)

- Taking into account that *the size of a diffusion is equal to the active bandwidth of the spectrum* $\mathcal{W}(e^{j\omega})$ *of the applied window*, which according to bandwidth theorem is inversely proportional to the active duration of the signal $w[n]$, we can reduce the amount of diffusion by choosing *long windows*.
- To reduce the phenomenon of *leakage*, we should increase as much as possible *the degree of attenuation of the height of the sidelobes*. It is proven that the spectrum of a signal whose derivatives up to order $k-1$ are *continuous* while in contrast the k-order derivative is characterized by the existence of *discontinuity* is reduced with a frequency at a rate proportional to $1/\omega^{k+1}$. Taking into account that the number of continuous derivatives of a signal also determines *the degree of smoothness that characterizes it*, we can state that the smoother the window we use, *the greater will be the attenuation of the height of the sidelobes and the smaller will be the effects of the undesirable phenomenon of leakage*.

Taking into account that the rectangular window is characterized by a high degree of discontinuity at its two ends (since it includes abrupt transitions from 0 to 1 and vice versa), we can, instead, use other types of windows characterized by a gradual reduction in their sample values at their limits, so as to minimize as much as possible the appearance of leakage. By increasing the length N of the window, then, according to our earlier discussion, the size of the spectral diffusion will also decrease. Things, however, are not that simple: It turns out that the two aforementioned requirements *contradict each other* and cannot be satisfied simultaneously; on the contrary, *efforts to improve the effects of one lead to the worsening of the effects of the other*. In order for this to be understood, let us consider again the rectangular window, which, as we have already mentioned, is the *optimal* of all windows of the same length from the point of view of the criterion of minimum approximation error. If we plot the measure of the spectrum of this window as a function of frequency ω, we will find that its central lobe is defined by the frequencies $\omega_1 = -2\pi/N$ and $\omega_2 = 2\pi/N$, leading to a value range of $4\pi/N$, which proves to be *the smallest of all values related to the range of the other windows*. Therefore, the use of the rectangular window causes the *minimum possible spectral diffusion*. On the other hand, however, the sidelobes of the spectrum of this window are characterized *by a large size and relatively small attenuation*. Taking into account that these sidelobes are defined as the areas that are interposed between the successive roots $\omega_n = 2\pi n/N$ $(n = 1, 2, 3, \ldots)$ of the spectrum measure $\mathcal{W}(e^{j\omega})$, it is not difficult to ascertain that *their range is inversely proportional to the length N of the discrete sequence*. However, if we try to reduce the size of the sidelobes by increasing the length of the sequence, we will find that nothing will improve at all: *This reduction in length will lead to an increase in the height of the sidelobes*, which will take place in such a way that *the area defined by them and the horizontal axis will remain constant*.

On the other hand, if we use a window characterized by a *higher degree of smoothness* compared to the rectangular discrete sequence, we will find that *its spectrum is characterized by smaller sidelobes that weaken more quickly compared to those of the rectangular window, but the range of the central lobe is much larger, which leads to a higher degree of spectral diffusion*. We find, therefore, that in each case we are obliged to resort to some kind of *compromise* between the undesirable effects of *spectral diffusion* and *leakage* and to choose the type of window that is considered most suitable for each specific problem.[37] There are many types of windows from which to choose that offer different degrees of such compromises between the *width of the mainlobe*, the *size of the sidelobes* (we are usually interested in the *maximum sidelobes* that lie on either side of the mainlobe), and the *rate* at which they weaken. All these window functions are characterized by a common length equal to N samples and by a satisfactory degree of smoothness, and they are *symmetrical* with respect to the point $n = N/2$ and have a frequency response characterized by a *generalized linear phase*,[38] and therefore it is described by an equation of the form

$$\mathcal{W}(e^{j\omega}) = \mathcal{A}(\omega)\exp\left[-j\omega\frac{N-1}{2}\right] \tag{10.149}$$

The most important of the window functions used in practice are presented in detail in the next section, while the basic characteristics of the most important of these are included in Table 10.6. In Fig. 10.31, the curves of the rectangular window, as well as the windows of Hamming, von Hann, Bartlett, Blackman, and Kaiser (Fig. 10.31a) and the comparison of their spectra as regards (1) the width of the central lobe (Fig. 10.31b) and (2) the height of the first sidelobe (Fig. 10.31c). In practice, the characteristics and performance of a window function are compared with those of the rectangular window, which is considered, therefore, as a reference window. In this figure, the compromise between the phenomena of diffusion and leakage is evident. If, for example, we choose a rectangular window characterized by the minimum central lobe width but the maximum sidelobe height, we minimize spectral diffusion but maximize spectral leakage, while if we choose the Blackman window characterized by the maximum central lobe width but the minimum sidelobe height (among the windows depicted in the figure), we minimize spectral leakage but maximize spectral diffusion. It should be noted that in practice, a reduction in spectral leakage is considered much more significant than a reduction in spectral diffusion, as the advantages we gain from reducing leakage more than offset the disadvantage of reducing the discriminative ability caused by spectral diffusion.

Usually, the representation of the spectra of window functions using a linear scale does not allow for an easy preview of their characteristics, and for this reason this representation is done on a logarithmic scale and even under normalization, so

[37] Note that window functions, apart from their use in the way we describe for reducing the effects of spectral leakage, are also used for the design of filters with finite impulse response. In practice, the windows that are used in DFT applications have an even length, while those that are used in filter design have an odd length.

[38] The phase shift associated with this phase is due to the fact that the window function has been defined in the range of values $0 \le n \le N-1$. Conversely, if the window is symmetrical with respect to the origin of the horizontal axis, this exponential term does not exist and we will simply have $\mathcal{W}(\omega) = \mathcal{A}(\omega)$. The use of windows with this symmetry is quite common in the literature. Note, however, that symmetrical windows of this type with an even number of samples are generally problematic.

Table 10.6 Basic characteristics of known window functions. We observe that the rectangular window is characterized by the smallest width of the central lobe, but also by the smallest attenuation and the greatest height of the maximum sidelobe

Window function $w[n]$	Mainlobe width	Rate of attenuation (d/B octave)	Height of maximum sidelobe (dB)
Rectangular window	$4\pi/N$	–06	–13.3
Bartlett window	$8\pi/N$	–12	–26.5
von Hann window	$8\pi/N$	–18	–31.5
Hamming window	$8\pi/N$	–06	–42.7
Blackman window	$12\pi/N$	–18	–58.1

that the maximum value associated with the central lobe corresponds to 0 dB. This process is presented in Fig. 10.32 and mathematically described by the equation

$$|\mathcal{W}_{\text{dB}}| = 20\log_{10}\left(\frac{|\mathcal{W}[k]|}{|\mathcal{W}[0]|}\right)$$

A significant problem that the phenomenon of *spectral leakage* leads to is a *reduction in the discriminative ability*, which consists *in the ability to detect neighboring frequencies of the spectrum of the original signal* $\mathcal{X}(e^{j\omega})$, *within the spectrum of the finite signal resulting from the original through the application of the window function*. It is not difficult to see that *if the distance between two frequencies* ω_1 *and* ω_2 *of the spectrum of the original signal* $x[n]$ *is smaller than the width of the mainlobe of the spectrum of the used window, then these two frequencies cannot be distinguished from each other in the spectrum* $\mathcal{Y}(e^{j\omega})$ *of the finite signal* $y[n]$ since they will be drowned by the phenomenon of spectral leakage (in this sense, the *discriminative ability* of spectral analysis that relies on the use of windows is considered to be expressed *by the width of the mainlobe of the spectrum of the window function*). To understand this problem in practice, let us consider the discrete-time signal of infinite length $x[n] = A_1\cos(\omega_1 n) + A_2\cos(\omega_2 n)$ defined in the value range $-\infty < n < \infty$ and containing two simple frequencies with values $\omega = \omega_1$ and $\omega = \omega_2$. In line with our discussion in the previous example, the spectrum $\mathcal{X}(e^{j\omega})$ will contain four distinct spectral lines at $\omega = \pm\omega_1$ and $\omega = \pm\omega_2$, while the application of a window $w[n]$ will cause the spectrum to be modified in the manner shown in Fig. 10.33. Considering a rectangular window of length $N = 64$, plotting *of the measure* of the spectra $\mathcal{W}(e^{j\omega})$ and $\mathcal{X}(e^{j\omega})$ will lead to the first two images in Fig. 10.33, while the spectrum of the finite signal $\mathcal{Y}(e^{j\omega}) = \mathcal{F}\{y[n]\} = \mathcal{F}\{x[n] \cdot w[n]\}$ will be given by

$$\mathcal{Y}(e^{j\omega}) = \frac{A_1}{2}\mathcal{W}(e^{j(\omega-\omega_1)}) + \frac{A_1}{2}\mathcal{W}(e^{j(\omega+\omega_1)}) + \frac{A_2}{2}\mathcal{W}(e^{j(\omega-\omega_2)}) + \frac{A_2}{2}\mathcal{W}(e^{j(\omega+\omega_2)}) \tag{10.150}$$

as can be proven using the property of the *frequency shift* of the Fourier transform of discrete-time and replacing the cosine functions from *Euler's formula*.

Taking into account that the function $w[n]$ describes a rectangular window of length $N = 64$, the width of the central lobe of the spectrum $\mathcal{W}(e^{j\omega})$ will be equal to $4\pi/N = \pi/16$, and this value also expresses the *limit of the resolution in the spectrum* $\mathcal{Y}(e^{j\omega})$. Therefore, the frequencies ω_1 and ω_2 can be distinguished from each other only when their distance is equal to or greater than $\pi/16$. In Fig. 10.33b, the two frequencies have a value of $\omega_1 = 2\pi/6$ and $\omega_2 = 2\pi/3$. We observe that

$$|\Delta\omega| = |\omega_2 - \omega_1| = \left|\frac{2\pi}{3} - \frac{2\pi}{6}\right| = \frac{\pi}{3} > \frac{\pi}{16}$$

meaning the two frequencies can be distinguished from each other in the spectrum $\mathcal{Y}(e^{j\omega})$, as is also evident from the preceding image. In a completely analogous way, in Fig. 10.33c, the two frequencies are $\omega_1 = 2\pi/14$ and $\omega_2 = 4\pi/15$ with a distance $|\Delta\omega| = 13\pi/105 > \pi/16$, and therefore, even now their separation is possible. It should be noted, however, that although the positions of the frequencies can still be determined, the amplitude of the spectrum at frequency ω_0 is affected by the amplitude of the sinusoidal signal corresponding to the frequency ω_1 and vice versa, which is a consequence of the leakage phenomenon. If the frequencies of the two harmonic components approach even closer to each other, the situation will become even worse, and when this distance becomes less than $\pi/16$, the two frequencies become *indistinguishable*.

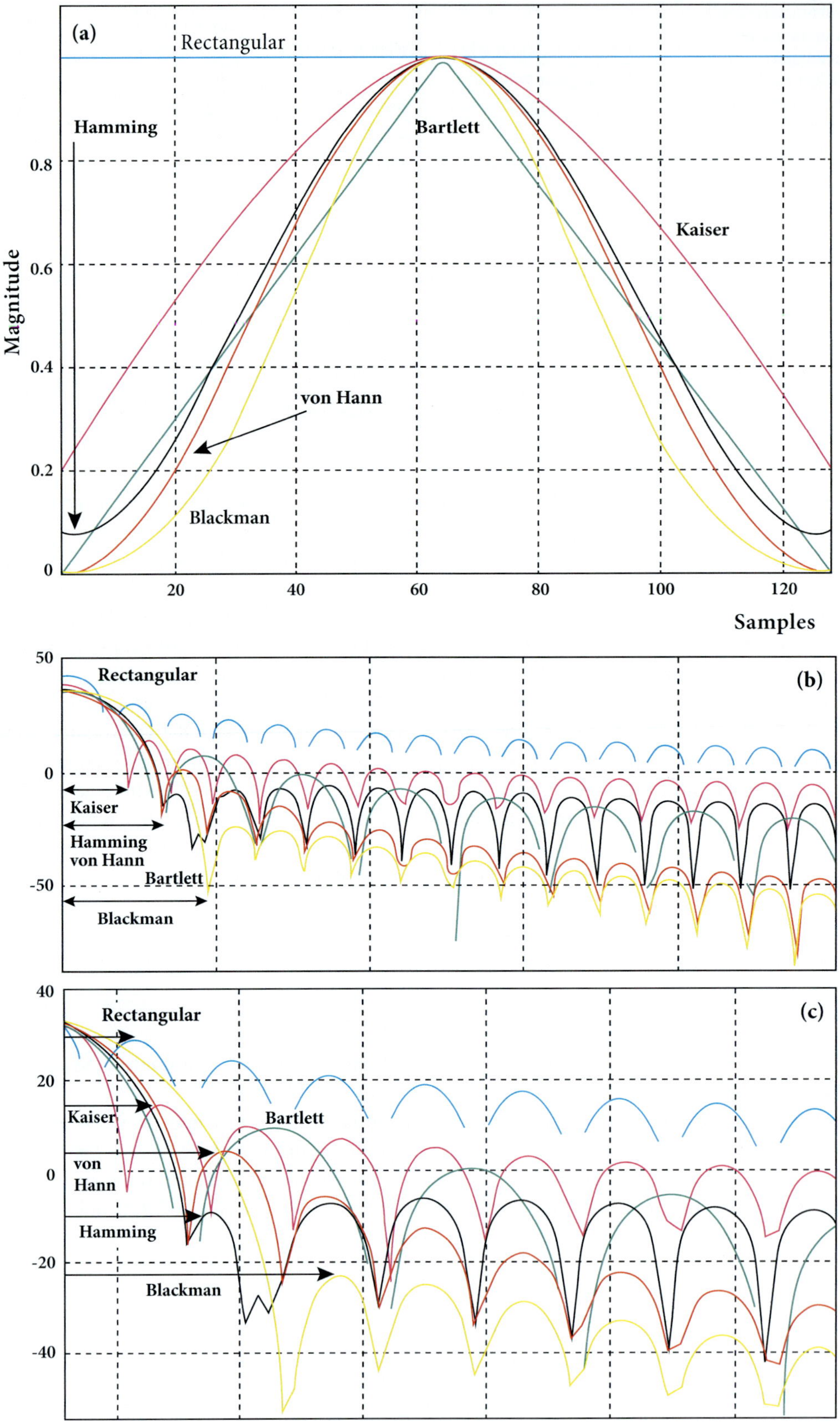

Fig. 10.31 Comparative presentation of rectangular window and Hamming, von Hann, Bartlett, Blackman, and Kaiser windows, in terms of their graphical representations (**a**), the width of the mainlobe (**b**), and the height of the first sidelobe (**c**)

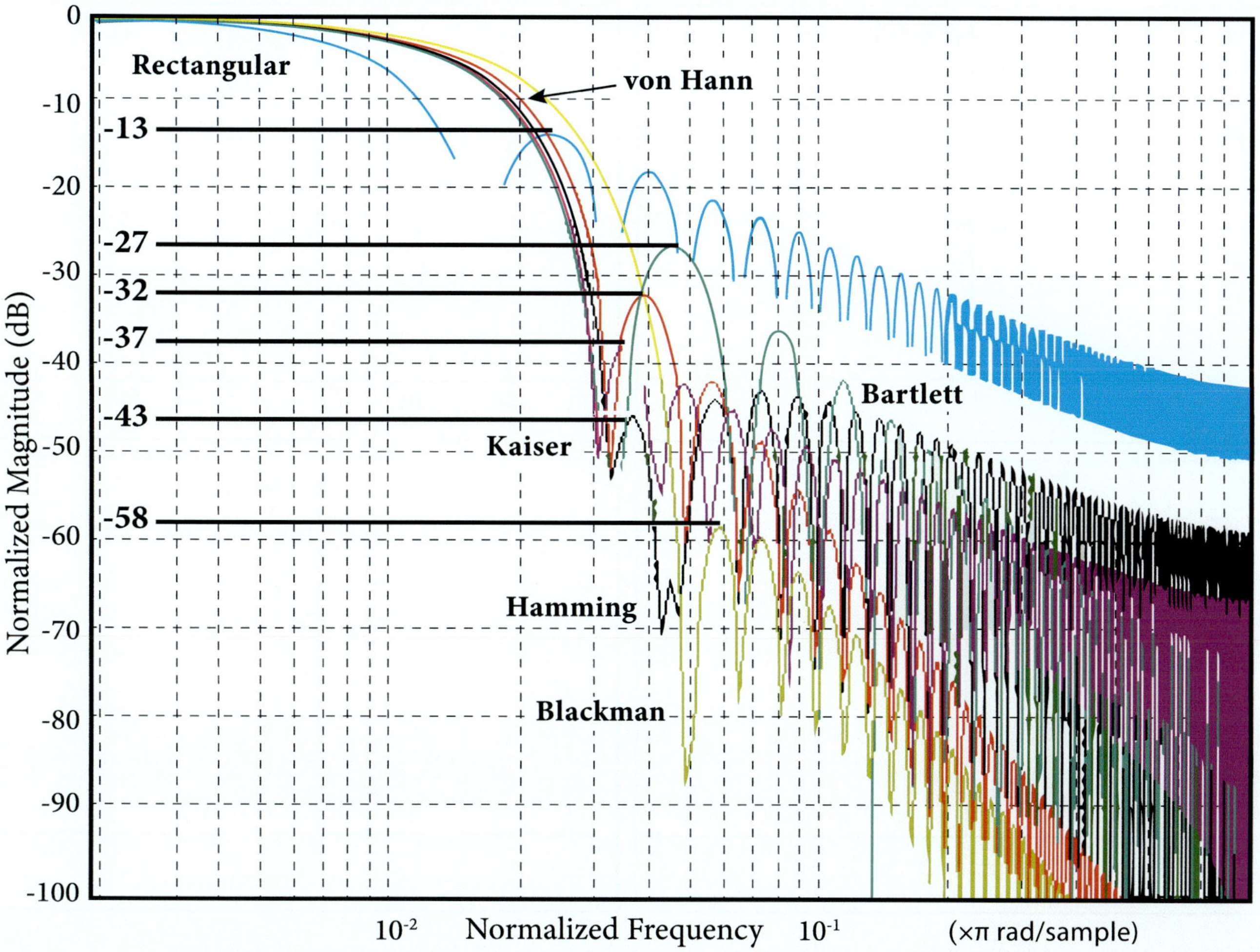

Fig. 10.32 Comparative presentation of spectra of rectangular window, as well as Hamming, von Hann, Bartlett, Blackman, and Kaiser windows on a normalized logarithmic scale

This situation is presented in Fig. 10.33d, in which the two frequencies have values $\omega_1 = 2\pi/14$ and $\omega_2 = 2\pi/25$ and the distance between them is equal to $|\Delta\omega| = 11\pi/175 < \pi/16$.

Closely related to the characteristic of resolution is the ability to observe features of the input signal that are related to specific frequency values. As we will describe in detail in a subsequent section, if these frequencies do not coincide with some of the N discrete frequencies ω_k $(k = 1, 2, \ldots, N)$ at which the DFT is calculated, they undergo attenuation to some extent, with the result that the detection of these features becomes problematic. One way to address this problem is to reduce the distance between the frequencies ω_k, which, being inversely proportional to the length N of the transform, can be reduced by increasing the value of N. If we do not have additional samples of the discrete input sequence at our disposal, this increase can be achieved by adding p zero values to the input signal, whose length now becomes equal to $L = N + p$. Subsequently, we must, of course, recalculate the DFT, which will now return L values. Although this addition of zero values improves the ability to observe features of interest, it does not lead to an increase in the resolution, as it was previously defined. If we want to increase the resolution, we must use L values, which, however, must be related to L and are real samples of the input signal, not including p imaginary zero values, as we saw earlier.

This process of adding zero values is presented diagrammatically in Fig. 10.34, which depicts a discrete sequence and the 16-point DFT. If we judge that the sampling of the spectrum does not lead to satisfactory results, then we can add the appropriate number of zero values at the end of the sequence and recalculate the DFT, thereby constructing more detailed discrete spectra. In this figure, the addition of 16, 48, and 112 zero values has taken place and the calculation of the DFT with lengths of 32, 64, and 128 points, respectively, resulting in the discrete spectrum that emerges to approximate more and more the corresponding continuous spectrum. Let us note that this process does not require a redefinition of the frequency

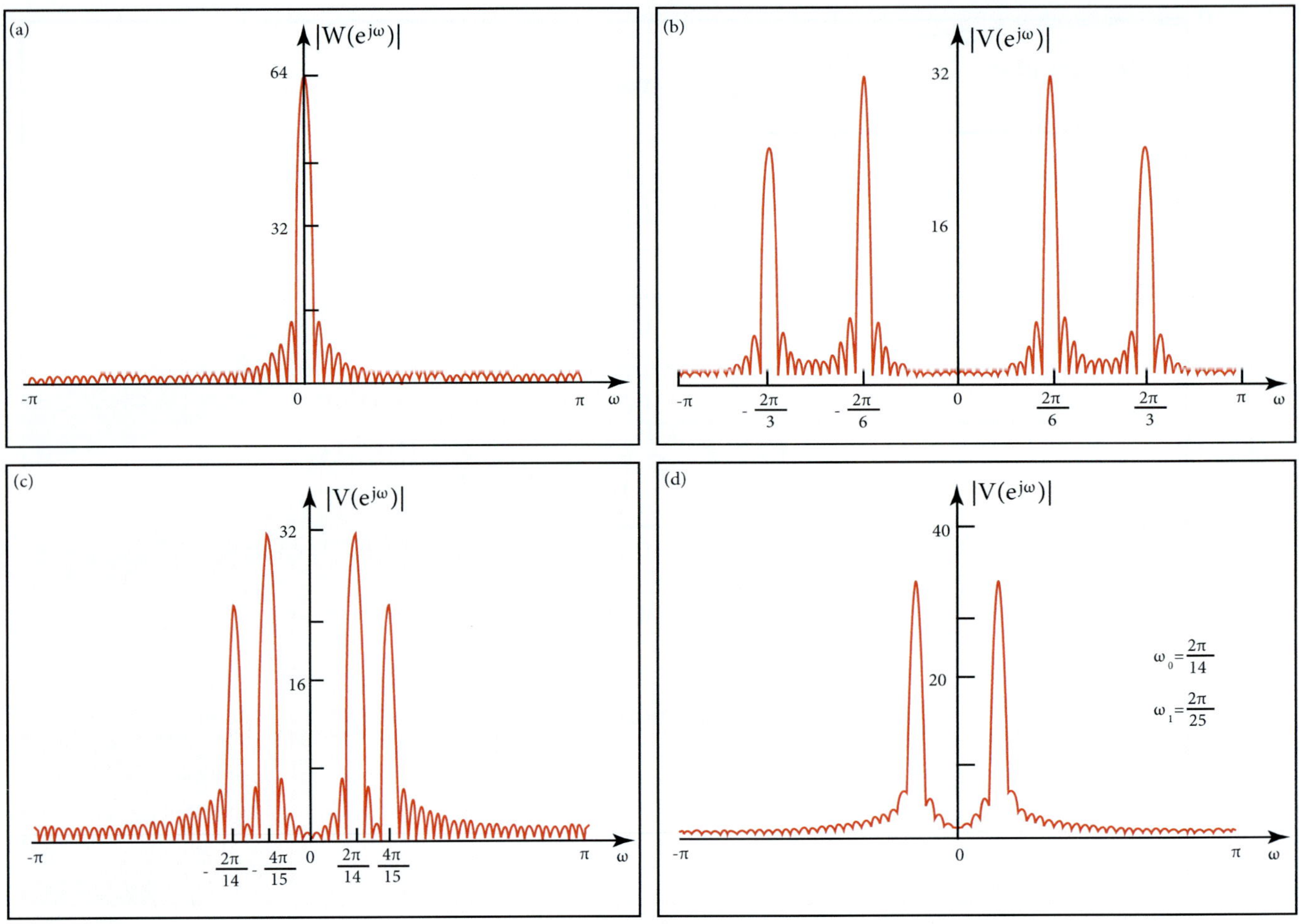

Fig. 10.33 Reduction of discrete ability as a result of spectral diffusion (Oppenheim, Schafer, Buck, Figures 10.3 and 10.4)

axis, since the frequency number m after the addition of zeros will continue to be mF_s/N. It is not difficult to see that if the problem to be solved is based on the use of windows, then these should be applied to the original signal before the addition of the zero values, as in the opposite case the results we would get would not be correct.

Window Functions

Having described the problems created by the use of the rectangular (ideal) window, let us now proceed to an analytical description of the most important of the window functions that are used in practice. The presentation of this material will be followed by a the description of a set of performance evaluation indicators of windows that are particularly useful in terms of choosing the appropriate window function in each case.

- **Von Hann and Hamming windows**: These window functions are described by the equation

$$w_H[n] = \begin{cases} \alpha - (1-\alpha)\cos\left(\dfrac{2\pi n}{N-1}\right) & \text{for } n = 0, 1, 2, \ldots, N-1 \\ 0 & \text{elsewhere} \end{cases} \tag{10.151}$$

with the value $\alpha = 0.5$, creating the *von Hann window*, which has the form

$$w_H[n]=\frac{1}{2}\left[1-\cos\left(\frac{2\pi n}{N}\right)\right]=\cos^2\left(\frac{n\pi}{N-1}\right) \quad n = 0, 1, 2, \ldots, N-1 \tag{10.152}$$

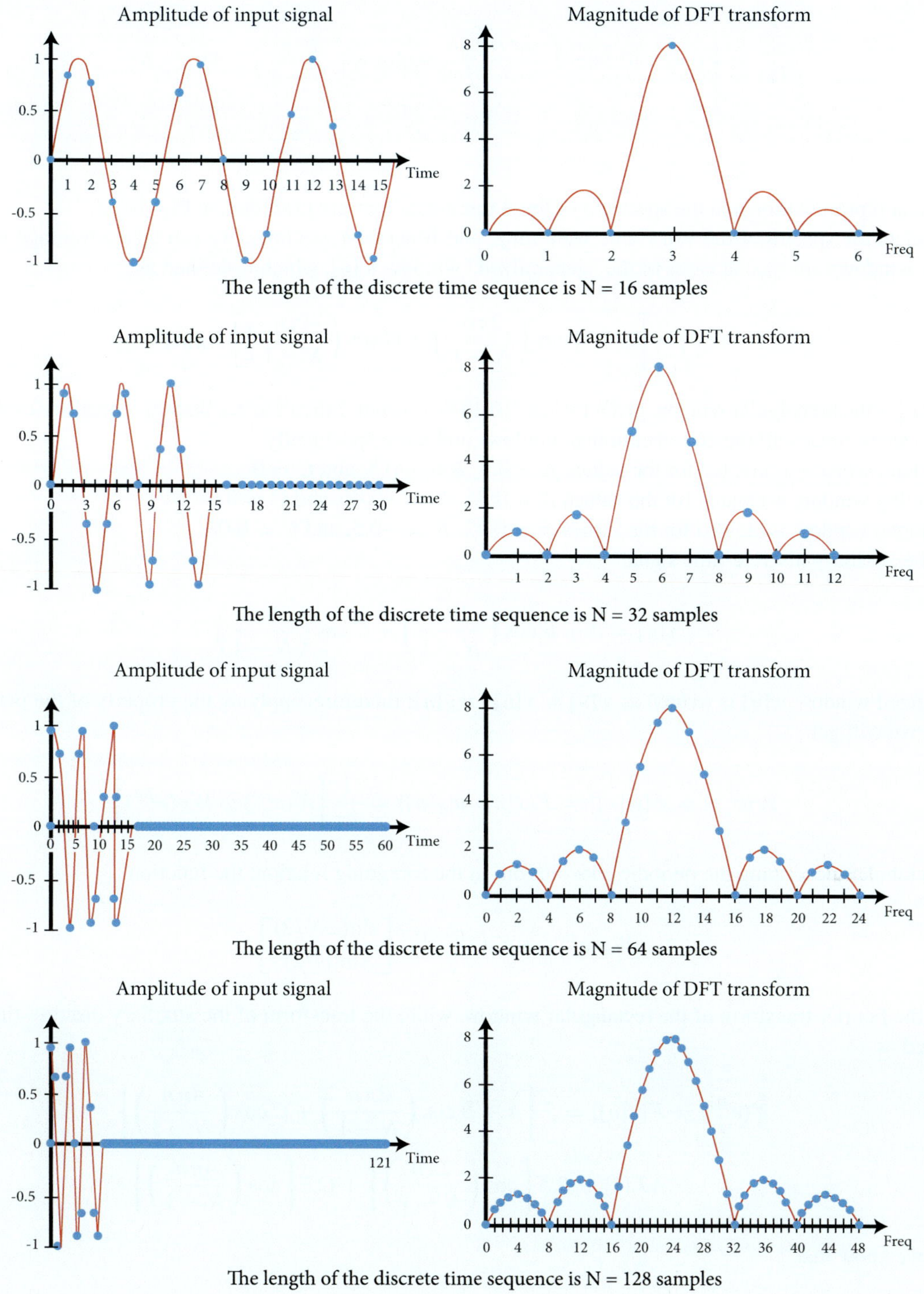

Fig. 10.34 Adding zero values to improve the results of the DFT (Lyons, Figure 3.21)

and the value $\alpha = 0.54$, creating the Hamming window, which is defined as

$$w_H[n] = 0.54 - 0.46\cos\left(\frac{2\pi n}{N-1}\right) \tag{10.153}$$

- **Blackman window**: The defining equation of the *Blackman window* has the form

$$w[n] = \begin{cases} 0.42 - 0.5\cos\left(\dfrac{2\pi n}{N-1}\right) + 0.08\cos\left(\dfrac{4\pi n}{N-1}\right) & 0 \le n \le N-1 \\ 0 & \text{elsewhere} \end{cases}$$

The graphical representation and the spectrum of these three windows are presented in Fig. 10.35.

To calculate the spectra of the von Hann, Hamming, and Blackman windows, we can take advantage of the fact that these three windows are special cases of the "generalized" window $w[n]$, which is defined as

$$w[n] = \left[A + B\cos\left(\frac{2\pi n}{N-1}\right) + C\cos\left(\frac{4\pi n}{N-1}\right)\right] w_R[n]$$

where $w_R[n]$ is the rectangular window given by Eq. (10.144). It is not difficult to see that this equation for the appropriate parameter values creates all the aforementioned windows and more specifically:

- A von Hann window is created for the values $A = 0.5$, $B = -0.5$, and $C = 0$.
- A Hamming window is created for the values $A = 0.54$, $B = -0.46$, and $C = 0$.
- A Blackman window is created for the values $A = 0.42$, $B = -0.5$, and $C = 0.08$.

Defining the auxiliary discrete-time signal

$$x[n] = A + B\cos\left(\frac{2\pi n}{N-1}\right) + C\cos\left(\frac{4\pi n}{N-1}\right)$$

the generalized window $w[n]$ is written as $w[n] = x[n] \cdot w_R[n]$; therefore, applying the property of the product of the of the DTFT, we will get

$$\mathcal{W}(e^{j\omega}) = \mathcal{F}\{w[n]\} = \mathcal{F}\{x[n] \cdot w_R[n]\} = \frac{1}{2\pi}\left\{\mathcal{X}(e^{j\omega}) \otimes W_R(e^{j\omega})\right\}$$

with the symbol $\otimes$ describing the periodic convolution. In the foregoing relation, the function

$$W_R(e^{j\omega}) = e^{-j\omega(N-1)/2}\left[\frac{\sin(\omega N/2)}{\sin(\omega/2)}\right]$$

expresses the Fourier transform of the rectangular window, while the transform of the auxiliary discrete-time signal $x[n]$ is calculated as

$$\begin{aligned} \mathcal{X}(e^{j\omega}) = \mathcal{F}\{x[n]\} &= \mathcal{F}\left\{A + B\cos\left(\frac{2\pi n}{N-1}\right) + C\cos\left(\frac{4\pi n}{N-1}\right)\right\} \\ &= A\mathcal{F}\{1\} + B\mathcal{F}\left\{\cos\left(\frac{2\pi n}{N-1}\right)\right\} + C\mathcal{F}\left\{\cos\left(\frac{4\pi n}{N-1}\right)\right\} \end{aligned}$$

However, we know that

$$\mathcal{F}\{1\} = 2\pi\delta(\omega) \qquad \text{and} \qquad \mathcal{F}\{\cos(\omega_0 n)\} = \pi\left[\delta(\omega + \omega_0) + \delta(\omega - \omega_0)\right]$$

(see Exercise 10.3.10 on page 826 and Exercise 10.3.11 on page 826, respectively, for the value $k = 0$ referring to the fundamental frequency region)—therefore, we will have

$$\mathcal{X}(e^{j\omega}) = 2\pi A\delta(\omega) + \pi B\left[\delta\left(\omega + \frac{2\pi}{N-1}\right) + \delta\left(\omega - \frac{2\pi}{N-1}\right)\right] + \pi C\left[\delta\left(\omega + \frac{4\pi}{N-1}\right) + \delta\left(\omega - \frac{4\pi}{N-1}\right)\right]$$

and therefore the Fourier transform of the window function $w[n]$ will have the form

$$\mathcal{W}(e^{j\omega}) = \frac{1}{2\pi}\left\{\mathcal{X}(e^{j\omega}) \otimes W_R(e^{j\omega})\right\} = A\{\delta(\omega) \otimes W_R(e^{j\omega})\}$$
$$+\frac{B}{2}\left\{\left[\delta\left(\omega+\frac{2\pi}{N-1}\right)\otimes W_R(e^{j\omega})\right]+\left[\delta\left(\omega-\frac{2\pi}{N-1}\right)\otimes W_R(e^{j\omega})\right]\right\}$$
$$+\frac{C}{2}\left\{\left[\delta\left(\omega+\frac{4\pi}{N-1}\right)\otimes W_R(e^{j\omega})\right]+\left[\delta\left(\omega-\frac{4\pi}{N-1}\right)\otimes W_R(e^{j\omega})\right]\right\}$$

Using the known properties of the impulse function

$$\delta(\omega)\otimes f(\omega) = f(\omega) \qquad \text{and} \qquad \delta(\omega+\omega_0)\otimes f(\omega) = f(\omega_0)$$

the preceding relationship is formulated as

$$\mathcal{X}(e^{j\omega}) = AW_R(e^{j\omega}) + \frac{B}{2}\left[W_R\left\{j\left[\left(\omega+\frac{2\pi}{N-1}\right)\right]\right\} + W_R\left\{j\left[\left(\omega+\frac{2\pi}{N-1}\right)\right]\right\}\right]$$
$$+\frac{C}{2}\left[W_R\left\{j\left[\left(\omega+\frac{2\pi}{N-1}\right)\right]\right\} + W_R\left\{j\left[\left(\omega+\frac{2\pi}{N-1}\right)\right]\right\}\right]$$

Using the analytical expression of the function $W_R(e^{j\omega})$ in the preceding equation, it turns out after operations that the Fourier transform $W(e^{j\omega})$ of the window $w[n]$ has the form dictated by Eq. (10.150), with the function $\mathcal{A}(\omega)$ having the form

$$\mathcal{A}(\omega) = A\frac{\sin\left(\frac{\omega N}{2}\right)}{\sin\left(\frac{\omega}{2}\right)} - \frac{B}{2}\left\{\frac{\sin\left(\frac{\omega N}{2}+\frac{N\pi}{N-1}\right)}{\sin\left(\frac{\omega}{2}+\frac{\pi}{N-1}\right)} + \frac{\sin\left(\frac{\omega N}{2}-\frac{N\pi}{N-1}\right)}{\sin\left(\frac{\omega}{2}-\frac{\pi}{N-1}\right)}\right\}$$
$$+\frac{C}{2}\left\{\frac{\sin\left(\frac{\omega N}{2}+\frac{2N\pi}{N-1}\right)}{\sin\left(\frac{\omega}{2}+\frac{2\pi}{N-1}\right)} + \frac{\sin\left(\frac{\omega N}{2}-\frac{2N\pi}{N-1}\right)}{\sin\left(\frac{\omega}{2}-\frac{2\pi}{N-1}\right)}\right\}$$

The foregoing equation defines the magnitude of the spectrum $\mathcal{A}(\omega)$ of the generalized window $w[n]$, while to calculate the magnitude of the spectrum of the von Hann, Hamming, and Blackman windows, we must replace the appropriate values of the constants A, B, and C. So we will obtain the following results:

– For the von Hann window:

$$\mathcal{A}(\omega) = 0.5\frac{\sin\left(\frac{\omega N}{2}\right)}{\sin\left(\frac{\omega}{2}\right)} + 0.25\left\{\frac{\sin\left(\frac{\omega N}{2}+\frac{N\pi}{N-1}\right)}{\sin\left(\frac{\omega}{2}+\frac{\pi}{N-1}\right)} + \frac{\sin\left(\frac{\omega N}{2}-\frac{N\pi}{N-1}\right)}{\sin\left(\frac{\omega}{2}-\frac{\pi}{N-1}\right)}\right\}$$

– For the Hamming window:

$$\mathcal{A}(\omega) = 0.54\frac{\sin\left(\frac{\omega N}{2}\right)}{\sin\left(\frac{\omega}{2}\right)} + 0.23\left\{\frac{\sin\left(\frac{\omega N}{2}+\frac{N\pi}{N-1}\right)}{\sin\left(\frac{\omega}{2}+\frac{\pi}{N-1}\right)} + \frac{\sin\left(\frac{\omega N}{2}-\frac{N\pi}{N-1}\right)}{\sin\left(\frac{\omega}{2}-\frac{\pi}{N-1}\right)}\right\}$$

– For the Blackman window:

$$\mathcal{A}(\omega) = 0.42 \frac{\sin\left(\frac{\omega N}{2}\right)}{\sin\left(\frac{\omega}{2}\right)} + 0.25 \left\{ \frac{\sin\left(\frac{\omega N}{2} + \frac{N\pi}{N-1}\right)}{\sin\left(\frac{\omega}{2} + \frac{\pi}{N-1}\right)} + \frac{\sin\left(\frac{\omega N}{2} - \frac{N\pi}{N-1}\right)}{\sin\left(\frac{\omega}{2} - \frac{\pi}{N-1}\right)} \right\}$$
$$+ 0.04 \left\{ \frac{\sin\left(\frac{\omega N}{2} + \frac{2N\pi}{N-1}\right)}{\sin\left(\frac{\omega}{2} + \frac{2\pi}{N-1}\right)} + \frac{\sin\left(\frac{\omega N}{2} - \frac{2N\pi}{N-1}\right)}{\sin\left(\frac{\omega}{2} - \frac{2\pi}{N-1}\right)} \right\}$$

It is not difficult to see that in order to create the spectra of the von Hann and Hamming windows, we must add two copies of the spectrum of the rectangular window function, the first of which is shifted by $2\pi/(N-1)$ to the left and the second by $2\pi/(N-1)$ to the right. If we construct the graphs of the three terms of the foregoing equation, we will find that *the second and third terms tend to cancel the two large sidelobes on either side of the central lobe, and for this reason the corresponding lobes of these two windows have a much smaller width.* The graph and the spectrum of the von Hann and Hamming windows are presented, as mentioned, in Fig. 10.35. It is proven that the height of the first sidelobe of the von Hann window is 31.5 dB lower compared to the height of the central lobe, while the corresponding reduction for the Hamming window is equal to 42.7. As for the corresponding attenuations for the first sidelobe, these are equal to 18 and 6 dB/octave, respectively. On the other hand, in the Blackman window, the presence of the additional term in the spectrum measure $\mathcal{A}(\omega)$ further improves the characteristics of the window, with the height of the first sidelobe being 58.1 dB lower than the height of the mainlobe. While its attenuation occurs at a rate of 18 dB/octave (however, the width of the mainlobe has increased to $12\pi/N$). These performances of this window were improved by *Harris*, who created two variations of the Blackman window known as *Blackman–Harris windows*. From these variations the first is given by the equation

$$w[n] = 0.44959 + 0.49364 \cos\left(\frac{2\pi n}{N}\right) + 0.05677 \cos\left(\frac{4\pi n}{N}\right)$$

—where $n = -(N-1)/2, \ldots, -1, 0, +1, \ldots, +(N-1)/2$ and is characterized by a maximum sidelobe height equal to -61 dB, while the second is defined by the equation

$$w[n] = 0.40217 + 0.49703 \cos\left(\frac{2\pi n}{N}\right) + 0.09892 \cos\left(\frac{4\pi n}{N}\right) + 0.00183 \cos\left(\frac{6\pi n}{N}\right)$$

and is characterized by a lobe height value equal to -74 dB.

- **Bartlett window**: The *Bartlett window* with a length of N elements is defined as

$$w_{\mathrm{BR}}[n] = 1 - \frac{2\left|n - \frac{N-1}{2}\right|}{N-1} \qquad n = 0, 1, 2, \ldots, N-1 \tag{10.154}$$

or, equivalently, as

$$w_{\mathrm{BR}}[n] = \begin{cases} \frac{2n}{N-1} & \text{for } 0 \le n \le \frac{N-1}{2} \\ 2 - \frac{2n}{N-1} & \text{for } \frac{N-1}{2} < n \le N-1 \\ 0 & \text{anywhere else.} \end{cases}$$

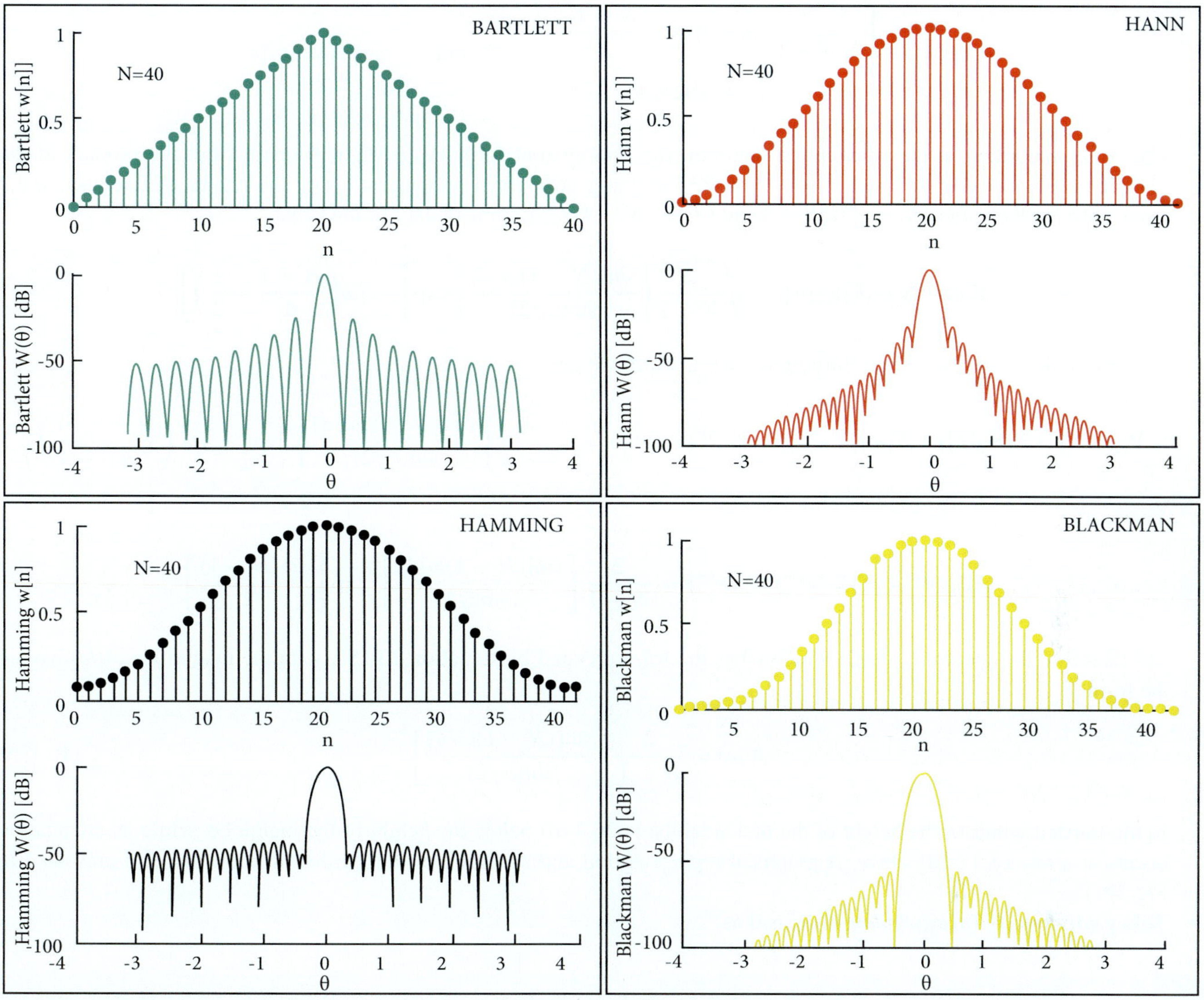

Fig. 10.35 Graphical representation and spectrum of Bartlett, von Hann, Hamming, and Blackman windows

In full analogy with the continuous case (see Exercise 1.10.9 on page 51), the Bartlett window of length N elements (which is simply the triangular window) can be computed as the convolution of two rectangular windows with a length of $N/2$ samples each,[39] which are defined as

---

[39] Following the methodology of Exercise 1.10.9, we find that the convolution of two orthogonal discrete pulses $x[n]$ and $y[n]$ with a length of N_1 and N_2 samples, respectively, with $N_2 < N_1$ is the discrete-time signal

$$z[n] = x[n] * y[n] = \begin{cases} n+1 & \gamma\iota\alpha 0 \le n < N_2 \\ N_2 & \gamma\iota\alpha N_2 \le n < N_1 \\ -n + (N_1 + N_2 - 1) & \gamma\iota\alpha N_1 \le n < N_1 + N_2 - 1 \\ 0 & \text{οπουδήποτε αλλού} \end{cases}$$

with a length of $N_1 + N_2 - 1$ samples and a maximum height equal to N_2. The reader is asked to draw the signal $z[n]$ for specific values of the parameters N_1 and N_2 and to find that the result is a trapezoidal signal. In the special case where the two square pulses have a common length $N_1 = N_2 = N$, the aforementioned trapezoidal signal degenerates into a triangular signal with a length of $2N - 1$ samples and height N. These two signals are the discrete equivalents of those depicted in Fig. 1.28 (page 53), which, however, are symmetric with respect to the origin O, since the square pulses used there were characterized by the same type of symmetry, while here they are defined only for values $n \ge 0$; however, the logic of the calculation is the same.

$$w_1[n] = \begin{cases} \sqrt{\dfrac{2}{N-1}} & \text{for } 0 \leq n \leq \dfrac{N-1}{2} - 1 \\ 0 & \text{anywhere else} \end{cases} \qquad \text{and} \qquad w_2[n] = w_1[n-1]$$

Therefore, according to the convolution theorem, the Fourier transform $\mathcal{W}_{\text{BR}}(e^{j\omega})$ will result from the product of the Fourier transforms $\mathcal{W}_1(e^{j\omega})$ and $\mathcal{W}_2(e^{j\omega})$ of the foregoing orthogonal sequences. Using the results of Exercise 10.3.2 (page 810) for the values $A = \sqrt{2/(N-1)}$ and $L = (N-1)/2$, it immediately follows that

$$\mathcal{W}_1(e^{j\omega}) = \mathcal{F}\{w_1[n]\} = \sqrt{\frac{2}{N-1}} \left\{ \frac{\sin[(N-1)\omega/4]}{\sin(\omega/2)} \right\} \exp\left[-j\omega\left(\frac{N-1}{4} - \frac{1}{2} \right) \right]$$

while from the property of time shifting of the transform we get

$$\mathcal{W}_2(e^{j\omega}) = \mathcal{F}\{w_2[n]\} = \mathcal{F}\{w_1[n-1]\} = e^{-j\omega}\mathcal{W}_1(e^{j\omega}) = \sqrt{\frac{2}{N-1}} \left\{ \frac{\sin[(N-1)\omega/4]}{\sin(\omega/2)} \right\} \exp\left[-j\omega\left(\frac{N-1}{4} + \frac{1}{2} \right) \right]$$

Thus,

$$\mathcal{W}_{\text{BR}}(e^{j\omega}) = \mathcal{W}_1(e^{j\omega})\mathcal{W}_2(e^{j\omega}) = \frac{2}{N-1} \left[\frac{\sin[(N-1)\omega/4]}{\sin(\omega/2)} \right]^2 \exp\left[j\frac{\omega(N-1)}{2} \right]$$

and therefore the spectrum of this window has the form dictated by Equation 10.150, with the magnitude $\mathcal{A}(\omega)$ having the form

$$\mathcal{A}(\omega) = \frac{2}{N-1} \left[\frac{\sin[(N-1)\omega/4]}{\sin(\omega/2)} \right]^2 \tag{10.155}$$

In the Bartlett window, the height of the first sidelobe is 26.5 dB below the height of the mainlobe, while its attenuation occurs at a rate of 12 dB/octave. A graphical representation and spectrum of the Bartlett window are also presented in Fig. 10.35.

- **Tukey window**: A *Tukey window* is defined as[40]

$$w[n] = \begin{cases} \dfrac{1}{2}\left\{ 1 + \cos\left[\pi\left(\dfrac{2n}{\alpha N} - 1 \right) \right] \right\} & \text{for } 0 \leq n < \dfrac{\alpha N}{2} \\ 1 & \text{for } \dfrac{\alpha N}{2} \leq n < N\left(1 - \dfrac{\alpha}{2}\right) \\ \dfrac{1}{2}\left\{ 1 + \cos\left[\pi\left(\dfrac{2n}{\alpha N} - \dfrac{2}{\alpha} + 1 \right) \right] \right\} & N\left(1 - \dfrac{\alpha}{2}\right) \leq n \leq N \end{cases}$$

It is proven that the Tukey window results from the convolution of a sine wave lobe of length $\alpha N/2$ with a square window of width $[1 - (\alpha/2)]N$. For $\alpha = 0$ the Tukey window degenerates into a rectangular window, while for $\alpha = 1$ it transforms into a von Hann window.

- **Lanczos window**: A *Lanczos window* has the form

[40] See Tukey, J., *An Introduction to the Calculations of Numerical Spectrum Analysis*, in Advanced Seminar on Spectral Analysis of Time Series, edited by B. Harris, pp. 25–46, Wiley, New York, 1967.

$$w[n] = \left(\frac{\sin[2\pi(n-L)(N-1)]}{2\pi(n-L)/L} \right)^{\ell} \quad \ell > 0 \tag{10.156}$$

where again we have set $L = (N-1)/2$.

- **Dolph–Chebyshev window**: The defining equation of a *Dolph–Chebyshev window* has the form

$$w[n] = \frac{1}{N}\left[\frac{1}{r} + 2\sum_{i=1}^{(N-1)/2} T_{N-1}\left[x_0 \cos\left(\frac{j\pi}{N}\right)\right] \cos\left(\frac{2n\pi j}{N}\right)\right] \tag{10.157}$$

In the preceding equation, the parameter r refers to the *residual ripple ratio*,[41] which is arbitrarily chosen and according to the given requirements, while the width of the mainlobe is determined by the number of window samples, N. We observe, therefore, that in contrast to the other windows in which the aforementioned ratio, being independent of N, was practically impossible to control, here this possibility exists, which makes this window more flexible and practical compared to the others. In the preceding equation, the parameter x_0 is defined as

$$x_0 = \cosh\left(\frac{1}{N-1}\cosh^{-1}\frac{1}{r}\right) \tag{10.158}$$

while the functions $T_k(x)$ are the *Chebyshev polynomials* defined as[42]

$$T_k(x) = \begin{cases} \cos\left(k\cos^{-1}x\right) & |x| \le 1 \\ \cosh\left(k\cosh^{-1}x\right) & |x| > 1 \end{cases} \tag{10.159}$$

A graphical representation and spectrum of the Dolph–Chebyshev window are presented in Fig. 10.36. It is proven that for specific values of the parameters N and r, the range of the central lobe of the spectrum of this window is the smallest

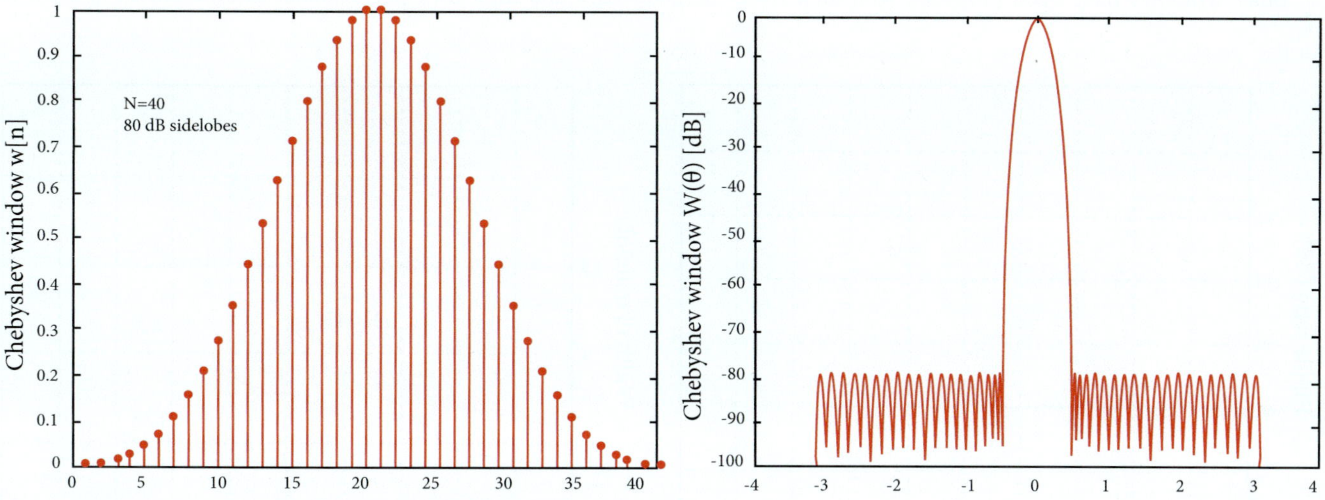

Fig. 10.36 Graphical representation and spectrum of Dolph–Chebyshev window

[41] This quantity is defined as the *ratio of the width of the maximum sidelobe to the width of the mainlobe expressed as a percentage* (see also the next chapter).

[42] See also the homonymous section in the mathematical appendix.

compared to all windows of the same length, while all the sidelobes have the same width, resulting in an approximation error characterized by a uniform distribution across all frequency bands.

- **Kaiser window**: The defining equation of a *Kaiser window* is in the form

$$w[n] = \frac{I_0(\alpha\sqrt{L^2 - (n-L)^2})}{I_0(\alpha L)} \tag{10.160}$$

where $L = (N-1)/2$. In the foregoing equation, the parameter α determines, among other things, the *shape of the spectrum* and usually takes values in the interval [0, 4], while the function I_0 is the *modified Bessel function of zero order of the first kind* calculated using the equation

$$I_0(x) = 1 + \sum_{k=1}^{\infty} \left[\frac{1}{k!} \left(\frac{x}{2} \right)^k \right]^2 \tag{10.161}$$

It is not difficult to see that for a parameter value of $\alpha = 0$, the *Kaiser window* transforms into the *rectangular window function*.

It is proven that the spectrum of a Kaiser window is given by the equation

$$\mathcal{W}(\omega) = \frac{2}{I_0(\alpha)} \frac{\sin[2\sqrt{\omega^2 - \omega_\alpha^2}]}{\sqrt{\omega^2 - \omega_\alpha^2}} \tag{10.162}$$

where $\omega_\alpha = \alpha/2$. Examination of the characteristics of this spectrum has shown that *a reduction in the parameter value* α—for a constant window length N—*causes a decrease in the height of the sidelobes* (and therefore also of the *ratio of remaining variance* that depends on it) and a simultaneous *increase in the width of the central lobe*. On the other hand, *the increase in the value of N*—for a constant value of the parameter α—*causes a reduction in the width of the central lobe, while the height of the sidelobes remains unchanged*. The change in the width of the central lobe and the ratio of the remaining variance for the Kaiser window are presented in Fig. 10.37, while a graphical representation of the function $w[n]$ and the spectrum $\mathcal{W}(\omega)$ are presented in Fig. 10.38.

- **Other types of windows**: In addition to these more well-known and widely used window functions, from time to time other windows have been proposed with their own characteristics and their own performance coefficients. For each of

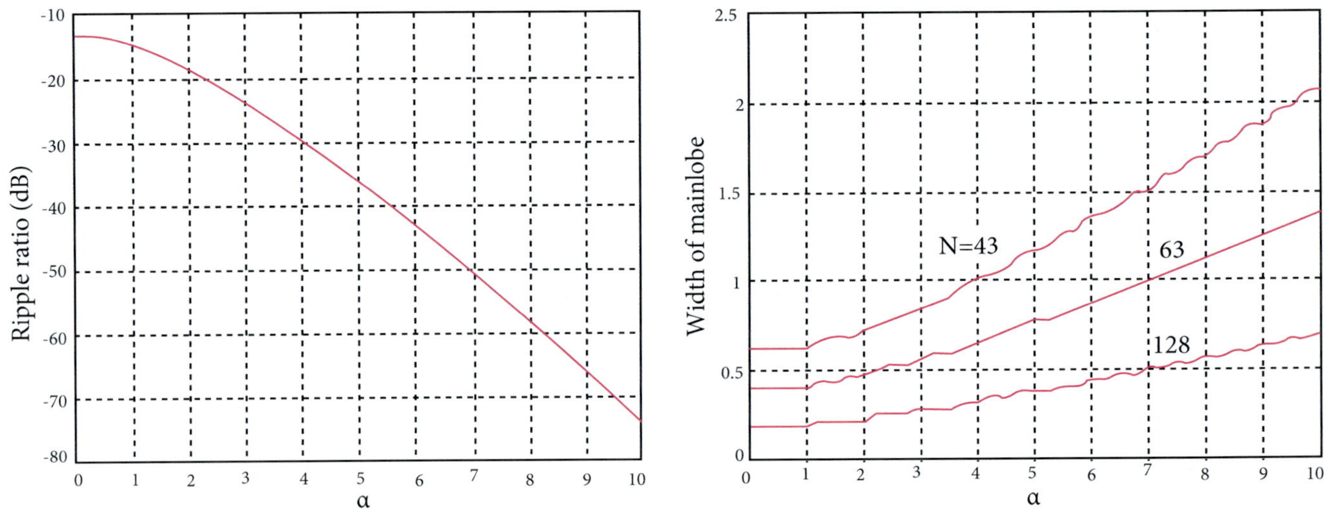

Fig. 10.37 Change in ratio of remaining variance and width of central lobe of Kaiser window as a function of the parameter α

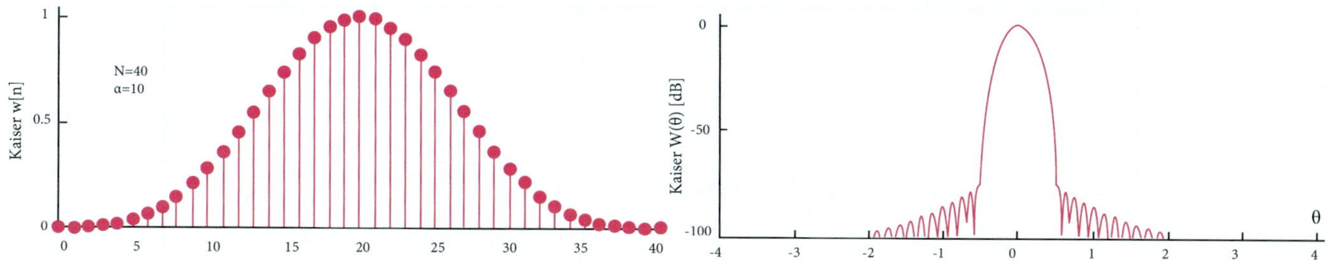

Fig. 10.38 Graphical representation and spectrum of Kaiser window

these windows, only their defining equation is given; for more details the reader is referred to the relevant literature. In the following equations, the windows have been defined so that they are symmetric with respect to the origin of the horizontal axis:

- de la Vallé–Pussin window $w[n] = \begin{cases} 1.0 - 6\left[\dfrac{n}{N/2}\right]^2\left[1.0 - \dfrac{|n|}{N/2}\right] & 0 \le |n| \le \dfrac{N}{2} \\ 2\left[1.0 - \dfrac{|n|}{N/2}\right]^3 & \dfrac{N}{4} \le |n| \le \dfrac{N}{2} \end{cases}$
- Bohman window $w[n] = \left[1.0 - \dfrac{|n|}{N/2}\right]\cos\left[\pi\dfrac{|n|}{N/2}\right] + \dfrac{1}{\pi}\sin\left[\pi\dfrac{|n|}{N/2}\right] \quad 0 \le |n| \le \dfrac{N}{2}$
- Poisson window $w[n] = \exp\left(-\alpha\dfrac{|n|}{N/2}\right) \quad 0 \le |n| \le \dfrac{N}{2}$
- Hanning–Poisson window $w[n] = 0.5\left[1.0 + \cos\left(\pi\dfrac{n}{N/2}\right)\right]\exp\left(-\alpha\dfrac{|n|}{N/2}\right) \quad 0 \le |n| \le \dfrac{N}{2}$
- Cauchy window $w[n] = \left[1.0 + \left(\alpha\dfrac{n}{N/2}\right)^2\right]^{-1} \quad 0 \le |n| \le \dfrac{N}{2}$
- Riesz window $w[n] = 1.0 - \left[\dfrac{n}{N/2}\right]^2 \quad 0 \le |n| \le \dfrac{N}{2}$
- Riemann window $w[n] = \sin\left[\dfrac{n}{N}2\pi\right] \Big/ \left(\dfrac{n}{N}2\pi\right) \quad 0 \le |n| \le \dfrac{N}{2}$

Example 10.5.5 Let there be a continuous-time signal of the form $x(t) = \cos(6000\pi t) + \cos(9000\pi t) + \cos(12000\pi t)$ that contains the frequencies $F_1 = 3000$ Hz, $F_2 = 4500$ Hz, and $F_3 = 6000$ Hz, whose sampling with frequency $F_s = 15{,}000$ Hz will lead, as is well known, to the discrete-time signal

$$x[n] = \cos\left(\frac{2\pi n}{5}\right) + \cos\left(\frac{3\pi n}{5}\right) + \cos\left(\frac{4\pi n}{5}\right)$$

(see Sect. 6.2 on page 501 and, more specifically, Exercise 6.2.1 on page 501). This discrete-time signal is presented in Fig. 10.39, while Fig. 10.40 shows the discrete spectrum of the signal on which the six best known window functions have been applied. In this figure, the spectral peaks at positions 0.4, 0.6, and 0.8 correspond to the frequencies $F_1 = 3000$ Hz, $F_2 = 4500$ Hz, and $F_3 = 6000$ Hz, while the maximum value of the horizontal axis corresponds to half the sampling frequency $F_s/2 = 7500$ Hz. From this figure all the findings we formulated in the previous section regarding the spectral characteristics of the applied windows are verified. We observe, for example, that the rectangular window led to the smallest central lobe but the largest sidelobes, while, on the other hand, the Blackman window led to the largest central lobe, but the smallest sidelobes. The length of the windows and the corresponding discrete transforms in all cases is $N = 128$.

Performance Coefficients of Discrete Window Functions

The window functions presented in the previous pages can be evaluated and compared with each other by calculating the values of some coefficients that are related to specific operating characteristics and generally are as follows.

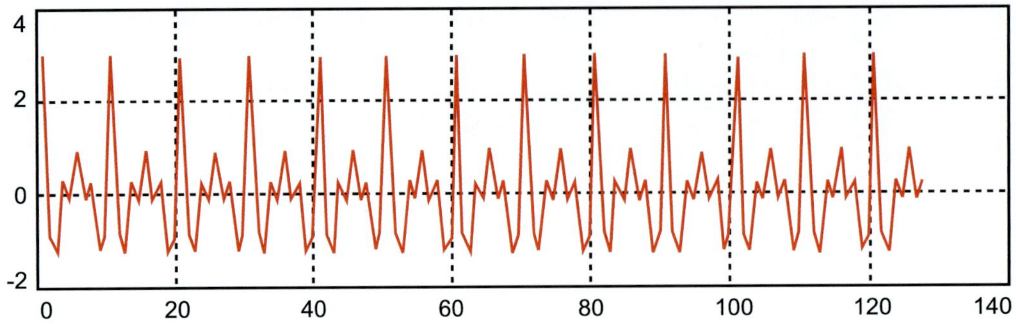

Fig. 10.39 Discrete-time signal $x[n]$ resulting from sampling of continuous-time signal defined as $x(t) = \cos(6000\pi t) + \cos(9000\pi t) + \cos(12000\pi t)$

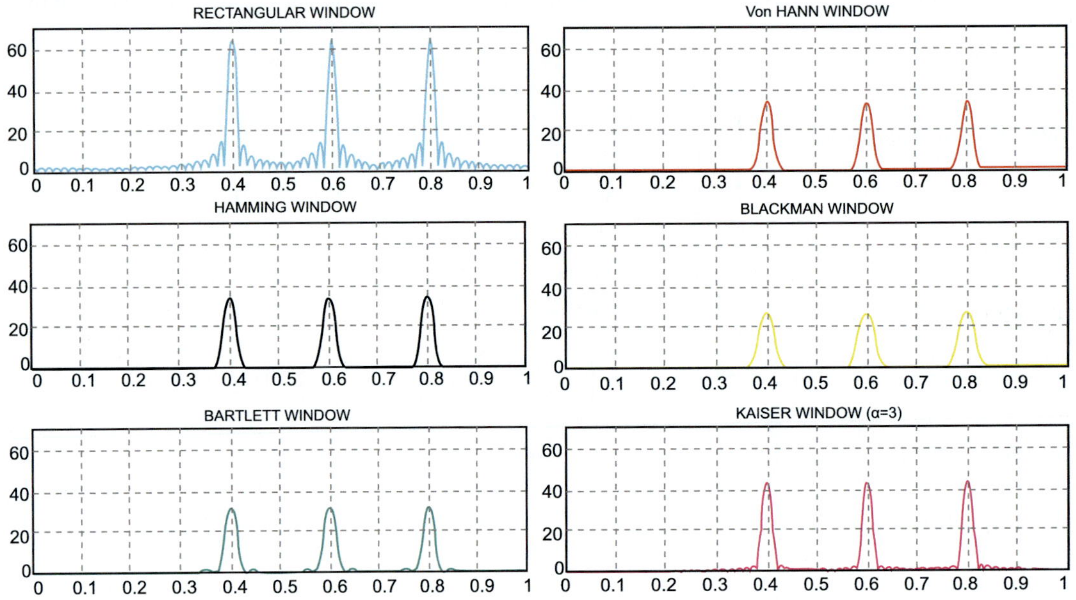

Fig. 10.40 Discrete spectra of signal $x[n]$ for six most significant discrete window functions

Central Lobe width W_{main}

This is the most important characteristic of each window function (since, as we have already mentioned, it determines the degree of spectral diffusion) and is inversely proportional to the length of the window. Significant characteristics also related to this parameter are the window width corresponding to a 6-dB attenuation, as well as the window width corresponding to a 3-dB attenuation. The first of these parameters is important because, as it turns out, in order for two frequencies to be distinguishable as the output of a transform, they must be at a distance greater than the window width, which corresponds to an attenuation of 6 dB. On the other hand, the range that corresponds to an attenuation of 3 dB can be used in combination with the equivalent noise bandwidth B_{eq} that is described in what follows as a performance index of the considered window. As Harris proved[43] by comparing the characteristics of 40 or more windows, the ratio of these two parameters for windows characterized by satisfactory performance falls within the range of values

$$1.04 \leq \frac{B_{3\,\mathrm{db}}}{B_{eq}} \leq 1.055$$

[43] See Harris F, "On the Use of Windows for Harmonic Analysis with the Discrete Fourier Transform," Proceedings of IEEE, Volume 66, pp.51–83, January 1978.

Conversely, windows with a ratio value less than 1.04 are characterized by a central lobe with a large range, leading to significant attenuation, while windows with a ratio greater than 1.055 have large sidelobes and therefore are characterized by reduced resolution.

Sidelobe Attenuation A_{side}

The attenuation associated with the sidelobes is described by the height of the first (and largest) sidelobe H_{side}, as well as by the attenuation of most of a remote sidelobe that is closer than all the lobes to the frequency $\omega = \pi$. If this attenuation does not have a high value, there will be a significant reduction in the discriminative ability regarding the separation of two neighboring frequencies and especially in cases where one signal is much stronger in relation to the other.

Equivalent Noise Bandwidth B_{eq}

As was already mentioned in the previous section, the width of the spectrum for a specific frequency value is affected by the accumulated broadband noise included in the spectrum of the window used, which should be minimized to the greatest possible extent (at least in problems involving the detection of harmonic components of input signals), something that is achieved by choosing windows with a small bandwidth. A handy way of describing this feature is the equivalent noise bandwidth that characterizes the performance of a window in terms of the presence of noise and is defined as the bandwidth of the ideal rectangular window that causes an accumulation of the same noise power (for the case of white noise) as the window function under consideration, for the same value of the maximum power gain. If the considered window is applied to a signal containing zero-mean white noise, then the total noise power will be given by the equation

$$P_w = \frac{N_0}{2}\int_{-1/2T}^{+1/2T} |\mathcal{W}(e^{j\omega})|^2 \frac{d\omega}{2\pi}$$

where N_0 is the noise power per unit bandwidth and T is the sampling period in the time domain. The concept of equivalent noise bandwidth is diagrammatically represented in Fig. 10.41, in which the area of the surface bounded by the rectangular window is equal to the area of the region bounded by the curve of the spectrum of the window under consideration. This rectangle has a height equal to the maximum height of the window curve and width equal to $2B_{\text{eq}}$, with the equivalent noise bandwidth $2B_{\text{eq}}$ defined as the *bandwidth for which the considered filter produces the same noise power as the rectangular filter*. Taking into account that this power for the ideal filter is equal to $P_F = N_0 B[\mathcal{W}(0)]^2$, the equation of the quantities P_w and P_F and the solution for $2B_{\text{eq}}$ gives us

$$2B_{\text{eq}} = \frac{1}{2[\mathcal{W}(0)]^2}\int_{-1/2T}^{+1/2T} |\mathcal{W}(\omega)|^2 \frac{d\omega}{2\pi} = \left\{\sum_n w^2[nT]\right\} \Big/ \left\{2\left(\sum_n w[nT]\right)^2\right\}$$

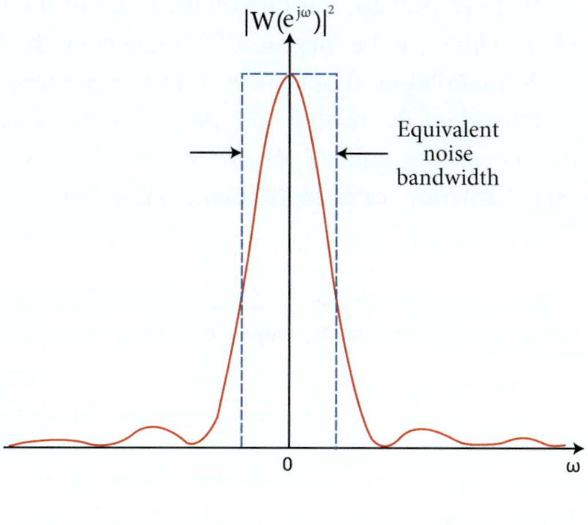

Fig. 10.41 Equivalent noise bandwidth (Harris, Figure 7)

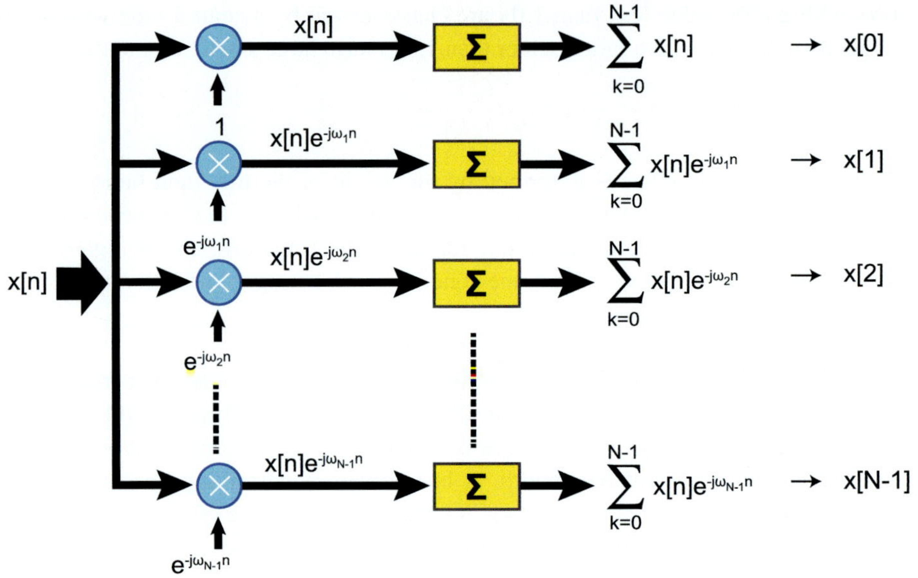

Fig. 10.42 Calculation of DFT from an array of band-pass filters with each band-pass filter being simulated by a modulator

with the last expression resulting from the application of the Parseval theorem. Note, that the range of values of n does not appear in the sums, since this index can vary depending on the application either from $-N/2$ to $N/2$ or from 0 to $N-1$, where N is the length of the window.

Gain and Attenuation Factors Due to Processing G_p and L_p

Considering the DFT of a finite sequence $x[n]$ with a length of N samples, which is defined as

$$\mathcal{X}[k] = \sum_{n=0}^{N-1} x[n] \exp\left(-j\frac{2k\pi n}{N}\right) = \sum_{n=0}^{N-1} x[n] e^{-j\omega_k n}, \quad \omega_k = \frac{2k\pi}{N}, k = 0, 1, 2, \ldots N-1$$

it is proven that this can be calculated exactly from an array of N band-pass filters with finite impulse response and with the central frequency of each filter coinciding with one of the N discrete frequencies ω_k. A variation of this situation is presented in Fig. 10.42 with the N modulators multiplying the discrete-time signal $x[n]$ by the exponential expression $e^{-j\omega_k n}$ ($k = 0, 1, 2 \ldots, N-1$) and the sum filter calculating the sum of the last N samples it receives at its input.[44] This fact allows us to describe each sample of the transform as the output of a band-pass filter and, consequently, to evaluate a window function using concepts from the theory of filters, such as gain and loss, which are described in detail in the next chapter.

Taking into account that the output of the DFT is described by a curve that refers to the sinc function, we can assert that its value is mainly influenced by the energy that is located under the central lobe or even in areas very close to it. Because, as we have proven, the maximum value of the DFT output is equal to $N/2$,[45] we can consider this value as the gain of the filter, which can be increased by increasing the length N of the transform.

A quantitative description of this processing gain can be given using the concept of *signal-to-noise ratio* (SNR), which is defined as *the ratio of the power of the signal S to the power of the noise N included in it*. In this case, we can define the processing gain G_pG_p *as the ratio of the SNR values at the input and output of the filter* that implements the DFT transformation, or in mathematical notation.

---

[44] This system is described by a causal impulse response of the form $h[n] = 1$ for $0 \le n \le N-1$ and $h[n] = 0$ for $n > N$ and calculates the sum of the values of the last N samples of the signal, since its output, as can be proven, has the form

$$y[n] = \sum_{k=0}^{\infty} h[k]x[n-k] = \sum_{k=0}^{N-1} x[n-k] = x[n] + x[n-1] + x[n-2] + \cdots + x[n-N+1].$$

[45] This result is based on the use of a sine wave of unit amplitude, while for the general case of amplitude A, this maximum value will obviously be equal to $AN/2$.

$$G_p = \frac{\mathrm{SNR}_{\mathrm{out}}}{\mathrm{SNR}_{\mathrm{in}}} = \frac{S_{\mathrm{out}}/N_{\mathrm{out}}}{S_{\mathrm{in}}/N_{\mathrm{in}}}$$

when as an input signal we consider a sine wave with a frequency of some of the frequencies ω_k, distorted by white noise of variability σ_p^2 of the form $x[n] = Ae^{j\omega_k nT} + \rho[nT]$, where A is the amplitude of the input signal, T is the sampling period in the time domain, and $\rho[nT]$ is a discrete sequence of white noise. It is not difficult to see that the power of the signal and the noise as the input of the filter will have the form $S_{\mathrm{in}} = A^2$ and $N_{\mathrm{out}} = \sigma_p^2$.

On the other hand, the contribution of the signal as the output of the filter is given by the equation

$$\mathcal{X}_s(e^{j\omega_k}) = \sum_{n=0}^{N-1} \left(w[nT] A e^{j\omega_k nT} \right) e^{-j\omega_k (nT)} = A \sum_{n=0}^{N-1} w[nT]$$

and therefore the value $\mathcal{X}_s(\omega_k)$ is proportional to the amplitude A of the input signal, with the proportionality coefficient

$$G_{\mathrm{coh}} = \left\{ A \left| \sum_{n=0}^{N-1} w[nT] \right| \right\} \Big/ |A| = \sum_{n=0}^{N-1} w[nT]$$

known as the coherent gain. This coefficient, which is calculated by adding together the values of all the samples of the window, has its maximum value—which is equal to N—for the rectangular window, while for all other window functions its value is smaller, due to the progressively smaller values of the samples of these windows as we approach their two ends. Using the foregoing expression for the function $\mathcal{X}_s(\omega_k)$, the power of the signal as the output of the filter is written

$$S_{\mathrm{out}} = |\mathcal{X}_s(e^{j\omega_k})|^2 = A^2 \left[\sum_{n=0}^{N-1} W[nT] \right]^2$$

Finally, the contribution of the noise as the output of the filter has the form

$$\mathcal{X}_N(\omega_k) = \sum_{n=0}^{N-1} w[nT] \rho[nT]\, e^{-j\omega_k nT}$$

leading to a noise power value as the output equal to

$$N_{\mathrm{out}} = \mathcal{E}\left\{ |\mathcal{X}_N(\omega_k)|^2 \right\} = \sigma_p^2 \sum_{n=0}^{N-1} \left[w[nT] \right]^2$$

with the operator $\mathcal{E}\{\dots\}$ referring to the magnitude of the expected value (see the relevant section). Having calculated all the foregoing quantities, the processing gain G_p will finally be given by the relation

$$G_p = \frac{S_{\mathrm{out}}/N_{\mathrm{out}}}{S_{\mathrm{in}}/N_{\mathrm{in}}} = \frac{A^2 \left[\sum_{n=0}^{N-1} w[nT] \right]^2 \Big/ \sigma_p^2 \sum_{n=0}^{N-1} \left[w[nT] \right]^2}{A^2/\sigma_p^2} = \left\{ \left[\sum_{n=0}^{N-1} w[nT] \right]^2 \right\} \Big/ \left\{ \sum_{n=0}^{N-1} \left[w[nT] \right]^2 \right\} = \frac{1}{2B_{\mathrm{eq}}}$$

where B_{eq} is the equivalent previously described noise bandwidth. We find, therefore, that there is a relationship of inverse proportionality between these two characteristic sizes of a window function, and therefore a large equivalent noise bandwidth leads to a small processing gain value. This result is certainly expected, as the larger the bandwidth, the greater the amount of noise that will enter the signal.

The inverse of the processing gain is known as processing attenuation and is given by the relation $L_p = 1/G_p = 2B_{eq}$.

Overlap Correlation Coefficient $c(r)$

In cases of processing long discrete sequences, it is common practice to divide the original signal $x[n]$ into shorter sequences with a length of N with this parameter value dictated by the desired resolution $\Delta f = \beta(f_s/N)$, where F_s is the sampling frequency that must satisfy the Nyquist criterion (note that the quantity f_s/N is the minimum resolution associated with a transform of N samples), and β is a proportionality coefficient that reflects the increase in bandwidth due to the use of the considered window in each case. Taking into account that the samples of the window are characterized by increasingly smaller values as we approach their two ends, we conclude that if the segments into which we divided the original signal do not overlap with each other, then the samples that are multiplied by the samples of the window that are located near the ends are essentially not taken into account, due to the very small value of the product of these two values. This situation is undesirable especially in cases where the cost of recording the samples of the signal is quite high and we want to extract from them the maximum amount of information. For this reason, in practice, the successive segments into which we divide the original signal overlap with each other, with the overlap percentage r having a value from 50% for large windows up to 75% for smaller windows. This process is diagrammatically depicted in Fig. 10.43. In the top image of the figure, a sequence of windows applied to successive nonoverlapping segments is depicted, while in the second case, the windows are applied to overlapping segments, so that the samples corresponding to the overlap area lead to nonnegligible contributions. In the case of overlapping segments of the original signal, the degree of correlation between successive segments is described by the correlation coefficient given by the relation

$$c(r) = \left\{\sum_{n=0}^{rN-1}\left\{w[n]w[n+(1-r)N]\right\}\right\} \Big/ \left\{\sum_{n=0}^{N-1} w^2[n]\right\}$$

Note that if the overlap of successive windows is very large, the results associated with spectral calculations will be characterized by a high degree of correlation, even if the input signal is random.

Taking into account that the application of a window function consists of sample-by-sample multiplication of the discrete-time signal $x[n]$ by the sequence of the discrete window, it is obvious that the sample number k of the finite-length discrete-time signal $y[n] = x[n]w[n]$ will result in the product

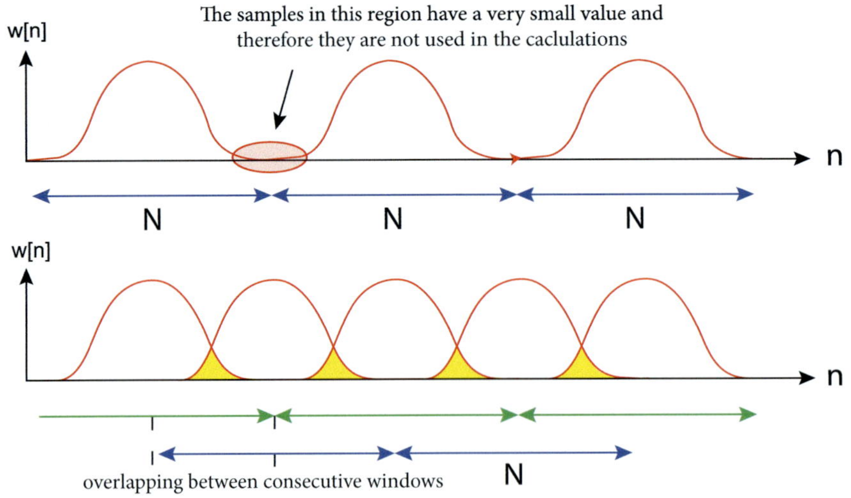

Fig. 10.43 Application of window function sequences to nonoverlapping parts of signal $x[n]$ (top image) and overlapping parts (bottom image)

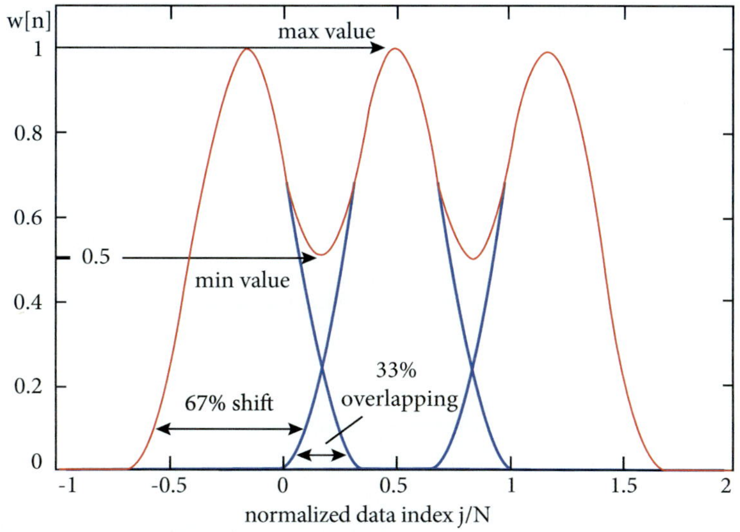

Fig. 10.44 Definition of AF coefficient for the case of successive Hamming windows with an overlap of 33% (Heinzel et al., Figure 10)

$$y[k] = w[k]x[k] \quad (k = 1, 2, \ldots, N - 1)$$

It is not difficult to understand that in the case of overlapping windows, the samples of the original signal $x[n]$ that correspond to the overlap area will be affected by all the windows that surround them. Considering for simplicity two successive windows, $w_1[n]$ and $w_2[n]$, of the same type (e.g., two Hamming windows) that overlap by $r = \alpha$ percent, a sample of the signal $x[n]$ that belongs to the overlap area will be multiplied by the corresponding samples of both windows, thereby leading to a result of the form

$$y[k] = w_1[k]x[k] + w_2[k]x[k] = \Big(w_1[k] + w_2[k]\Big)\, x[k] \quad (k = 1, 2, \ldots, N - 1)$$

If all samples of the original signal are equally important, the value of the sum $w_1[k] + w_2[k]w_1[k] + w_2[k]$, which can be treated as the weight coefficient of the sample $x[k]x[k]$, should be more or less the same for all samples. This feature is described by the amplitude flatness (AF), which is defined as the ratio of the minimum total weight that can be applied to any point to the maximum total weight, with the value of AF being calculated for each value of the overlap percentage under consideration. This situation for two overlapping Hamming-type windows and for an overlap percentage of 33% is presented in Fig. 10.44, from which it is clear that the value of the AF is equal to 0.5. For the special case of the Hamming window, the maximum value of AF, which is equal to one, corresponds to a 50% overlap, which is considered optimal and is almost always used. Conversely, for the other windows there is no corresponding value, although the value of AF tends very quickly to one for increasingly larger overlap percentages.

Closely related to the AF coefficient is the PF coefficient (Power Flatness), which is defined as the sum of the squares of the values of the samples $w_1[k]w_1[k]$ and $w_2[k]w_2[k]$. This quantity is primarily used with non-correlated signals, such as, for example, noise signals.

The variation of the coefficients AF, PF, and $C(r)$ as a function of the percentage overlap for the most well-known of the discrete window functions is shown in Fig. 10.45.

Additional Attenuation Factor L_s

As has already been mentioned in previous sections, the DFT returns results that concern a specific set of discrete frequencies ω_k, and in reality the concepts of equivalent bandwidth and gain and attenuation, due to processing, concern frequency components associated with some of these frequencies. But what happens to signals whose frequencies do not coincide with the aforementioned basic frequencies? Considering that the Fourier transform can be described as an array of N band-pass filters with central frequencies that coincide with the N basic frequencies kf_s/N, it is obvious that if signals with frequencies different from these pass through these filters. the basic ones, apart from the weakening due to processing, which

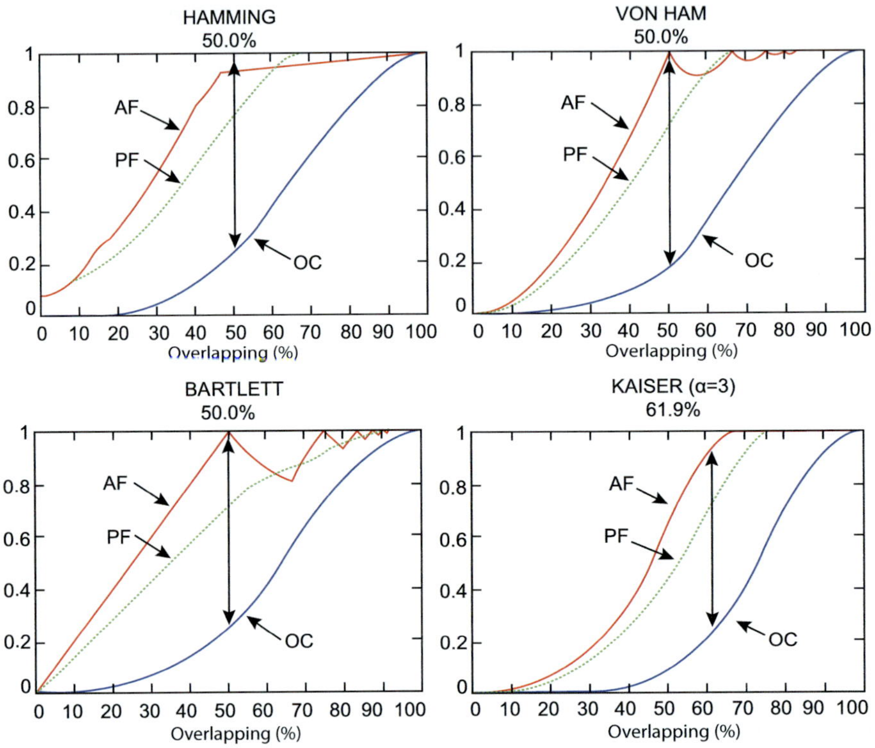

Fig. 10.45 Variation of coefficients AF, PF, and $C(r)$ as a function of percentage overlap for Hamming, von Hann, Bartlett, and Kaiser windows (Heinzel et al., Figures 16, 18, 19, and 28)

we described earlier, will also undergo additional weakening, the degree of which will depend on the distance of the signal frequency from these basic frequencies. Let us calculate, then, this additional weakening for the worst possible case in which the frequency of the input signal is exactly in the middle of two basic frequencies and, therefore, is of the form $(k+\frac{1}{2})f_s/N$ for some value of the parameter k.

Considering, as before, a sinusoidal signal distorted by noise of the form

$$x[n] = A\exp\left(j\omega_{(k+\frac{1}{2})}nT\right) + \rho[nT], \quad \omega_{(k+\frac{1}{2})} = \frac{2\pi}{NT}\left(k+\frac{1}{2}\right)$$

the contribution of the noise-free signal as the output of the filter will have the form

$$\mathcal{X}_s(e^{j\omega_{k+(1/2)}}) = \sum_{n=0}^{N-1} w[nT]A\exp\left(j\omega_{(k+\frac{1}{2})}nT\right)e^{-j\omega_k nT} = A\sum_{n=0}^{N-1} w[nT]\exp(j\pi n/N)$$

Using this result, we can define this coefficient of additional weakening as *the ratio of the agreed gain for an input signal with a frequency that belongs in the middle of the interval defined by the basic frequencies of the DFT to the agreed gain for an input signal whose frequency coincides with one of the basic frequencies of the DFT*, or in mathematical notation,[46]

[46] Alternatively, this maximum weakening for the frequency $(k+\frac{1}{2})f_s/N$ can be calculated as

$$\mathcal{X}\left[k+\frac{1}{2}\right] = \frac{N}{2}\times\left\{\sin\left[\pi\left(k+\frac{1}{2}-k\right)\right]\right\}\left\{\pi\left(m+\frac{1}{2}-k\right)\right\} = \frac{N}{2}\frac{\sin(\pi/2)}{\pi/2} = \frac{2}{\pi}\left(\frac{N}{2}\right)$$

Taking into account that the value $N/2$ expresses the maximum value of the central lobe, this maximum attenuation corresponds to a reduction in the spectrum amplitude to $2/\pi = 0.637$ of its maximum value.

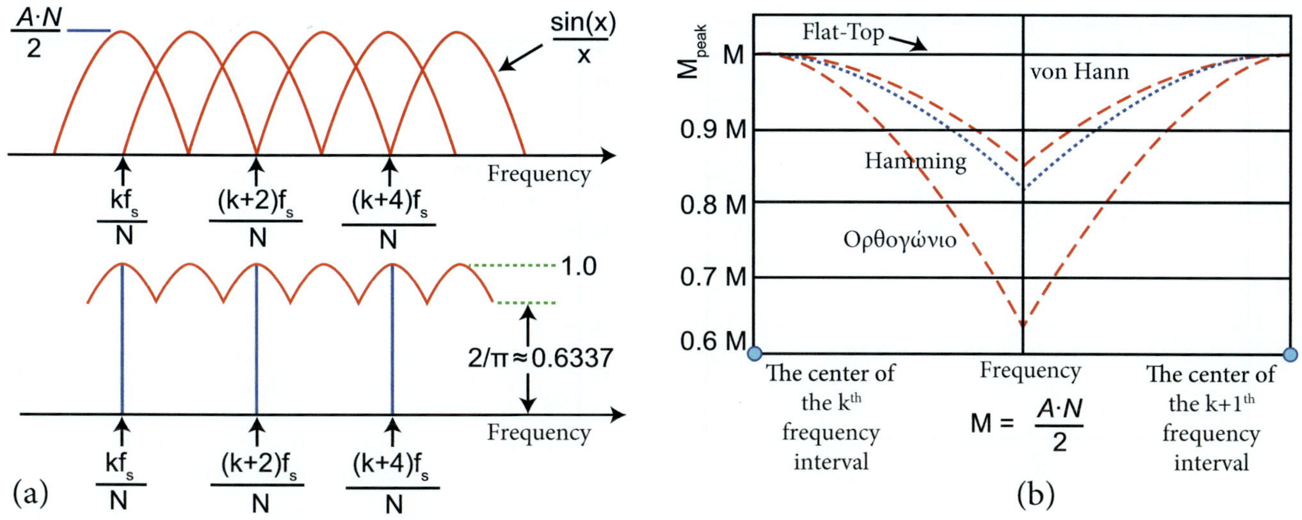

Fig. 10.46 Diagrammatic representation of additional attenuation undergone by frequency components of input signal whose frequency does not coincide with N frequencies, at whose positions the DFT is calculated (Lyons, Figure 3.19)

$$L_s = \frac{G_{\text{coh}}(\omega_{k+\frac{1}{2}})}{G_{\text{coh}}(\omega_k)} = \left\{\left|\sum_{n=0}^{N-1} w[nT]\exp\left(j\frac{\pi}{N}n\right)\right|\right\} \Big/ \left\{\sum_{k=0}^{N-1} w[nT]\right\} = \frac{\left|\mathcal{W}\left(\frac{\pi}{NT}\right)\right|}{\mathcal{W}(0)}$$

This characteristic of additional attenuation is presented diagrammatically in Fig. 10.46a. In this figure, for each of the frequencies F_k of the discrete transform, the curve of the amplitude response corresponding to it is depicted (for better representation, only the central lobe of each curve is shown, not the sidelobes, as the latter are not significant in this discussion). From this figure we observe that the maximum value of the amplitude response is equal to $N/2$ and corresponds to the frequency value kF_s/N (this means that the input signal with this frequency value does not undergo any such attenuation), while for frequency values on either side of the preceding value, the amplitude attenuates, with the value of the attenuation increasing as we move away from the central frequency and in both directions. It is not difficult to see that this behavior is due to the shape of the curve of the central lobe, which increases monotonically before the central frequency and decreases monotonically after the central frequency. The maximum value of this attenuation corresponds to a frequency value $(k + \frac{1}{2})F_s/N$, which is midway between the frequency values kF_s/N and $(k + 1)F_s/N$; the value of this maximum attenuation is equal to 0.637 of the maximum amplitude value, which, as we have proven, is equal to $N/2$.

The foregoing description refers to the use of a rectangular window whose central lobe has the smallest width compared to all the other windows, so its amplitude response decreases at the highest possible rate at frequency values on either side of the central lobe (since the amplitude responses for all windows have the same height). But this means that the rectangular window leads to the maximum degree of attenuation. This situation can be improved by using other windows whose lobes, being larger in relation to that of the rectangular window, are characterized by a higher degree of overlap, thereby leading to a lesser degree of attenuation.[47] This situation for the rectangular window, as well as the Hamming and von Hann windows, is presented in Fig. 10.46b, from which it can be seen that the attenuation caused by the Hamming and von Hann windows is significantly less compared to that caused by the rectangular window. Therefore, considering a sinusoidal input signal with amplitude A, the maximum value of the amplitude of the DFT—denoted by M—will belong to the range of values

$$0.637AN/2 \leq M \leq AN/2$$

depending on the frequency of this sinusoidal waveform.

[47] See Lyons R, "Reducing FFT Scalloping Loss Without Multiplication," *IEEE Signal Processing Magazine*, March 2012, pp. 112–116.

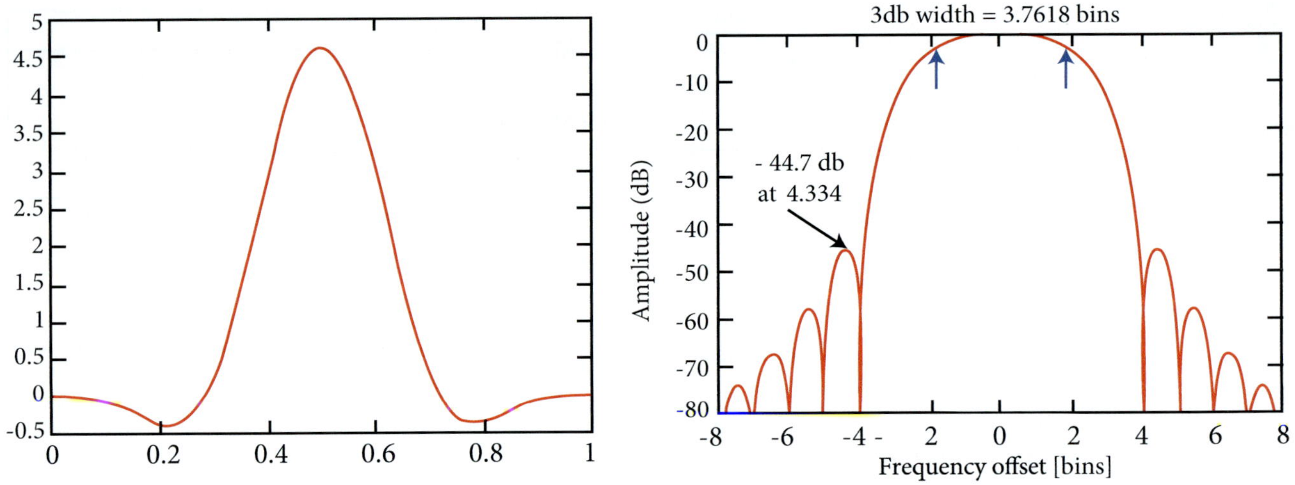

Fig. 10.47 Curve of window function and spectrum of typical flat-top window that leads to zero attenuation for frequencies near its center (Heinzel et al., Figure 32)

In cases where we wish to completely eliminate this additional attenuation, we can resort to the use of a special group of windows whose amplitude response is as flat as possible. In the international literature, these windows are known as flat-top windows and are defined like the Blackman–Harris windows, that is, as a sum of some cosine terms, and their bandwidth is about twice the bandwidth of the known windows, for the same level of suppression of the sidelobes. These windows are defined by a relation of the form

$$w[n] = \sum_{k=0}^{M} \alpha_k \cos\left(\frac{2\pi kn}{N}\right) = \alpha_0 + \sum_{k=1}^{M} \alpha_k \cos\left(\frac{2\pi kn}{N}\right)$$

with the number of terms M taking values in the region $3 \leq M \leq 5$. The curve and spectrum of a typical window of this category are presented in Fig. 10.47, from where it becomes apparent why these windows are aforementioned to be flat-top windows, while more details about them regarding their coefficients and their graphical representations can be found in the literature.[48]

We conclude this discussion on attenuation with the observation that the total attenuation caused by a window function on an input signal will obviously be equal to $L_p + L_s$ and expresses the total reduction of the value of the SNR at the filter output. The value of this quantity expressed in decibels for all window functions ranges between 3.0 and 4.3, although windows with a total attenuation value greater than 3.8 dB are considered very poor in terms of their performance and should not be used.

Table 10.7 contains the values of the most important performance coefficients for the most important windows used in practice, while Fig. 10.48 depicts the relative position of the windows on a two-axis diagram, where the horizontal axis contains the total attenuation $L_p + L_s$, while the vertical axis contains the maximum value of sidelobe attenuation. The windows characterized by good performance are those that have small values for these two quantities and therefore are located in the lower left area of this diagram.

Problem 10.5.16 Considering the problem of determining the spectrum of a discrete sequence of infinite length $x[n]$ on which a Hamming-type window function of length $L = 2^\nu$ is applied, determine the minimum value of ν for which the resolution is at most equal to $\pi/100$.

---

[48] See Heinzel G, Rüdiger A, and Schilling R, *Spectrum and Spectral Density Estimation by the Discrete Fourier Transform (DFT), Including a Comprehensive List of Window Functions and Some New Flat Top Windows*, Max-Planck-Institut für Gravitationsphysik, Document Id, 395068.0, 2002.

Table 10.7 Performance coefficients of best-known discrete window functions (Harris, Table 1)

Window function $w[n]$	H_{side} (dB)	A_{side} (dB/oct)	G_{coh}	B_{eq} (bins)	$B_{3\,\text{dB}}$ (bins)	L_s (dB)	$L_s + L_p$ (dB)	$B_{6\,\text{dB}}$ (bins)	c (0.75)	c (0.50)
Rectangle	–13	–06	1.00	1.00	0.89	3.92	3.92	1.21	75.0	50.0
Triangle	–27	–12	0.50	1.33	1.28	1.82	3.07	1.78	71.9	25.0
von Hann ($\alpha = 1.0$)	–23	–12	0.64	1.23	1.20	2.10	3.01	1.65	75.5	31.8
von Hann ($\alpha = 2.0$)	–32	–18	0.50	1.50	1.44	1.42	3.18	2.00	65.9	16.7
von Hann ($\alpha = 3.0$)	–39	–24	0.42	1.73	1.66	1.08	3.47	2.32	56.7	08.5
von Hann ($\alpha = 4.0$)	–47	–30	0.38	1.94	1.86	0.86	3.75	2.59	48.6	04.3
Hamming	–43	–06	0.54	1.36	1.30	1.78	3.10	1.81	70.7	23.5
Riesz	–21	–12	0.67	1.20	1.16	2.22	3.01	1.59	76.5	34.4
Riemann	–26	–12	0.59	1.30	1.26	1.89	3.03	1.74	73.4	27.4
de la Vallé-Poussin	–53	–24	0.38	1.92	1.82	0.90	3.72	2.55	49.3	05.0
Tukey ($\alpha = 0.25$)	–14	–18	0.88	1.10	1.01	2.96	3.39	1.38	74.1	44.4
Tukey ($\alpha = 0.50$)	–15	–18	0.75	1.22	1.15	2.24	3.11	1.57	72.7	36.4
Tukey ($\alpha = 0.75$)	–19	–18	0.63	1.36	1.31	1.73	3.07	1.80	70.5	25.1
Bohman	–46	–24	0.41	1.79	1.71	1.02	3.54	2.38	54.5	07.4
Poisson ($\alpha = 2.0$)	–19	–06	0.44	1.30	1.21	2.09	3.23	1.69	69.9	27.8
Poisson ($\alpha = 3.0$)	–24	–06	0.32	1.65	1.45	1.46	3.64	2.08	54.8	15.1
Poisson ($\alpha = 4.0$)	–31	–06	0.25	2.08	1.75	1.03	4.21	2.58	40.4	07.4
Hanning–Poisson ($\alpha = 0.5$)	–35	–18	0.43	1.61	1.54	1.26	3.33	2.14	61.3	12.6
Hanning–Poisson ($\alpha = 1.0$)	–39	–18	0.38	1.73	1.64	1.11	3.50	2.30	56.0	09.2
Hanning–Poisson ($\alpha = 2.0$)	–	–18	0.29	2.02	1.87	0.87	3.94	2.65	44.6	04.7
Cauchy ($\alpha = 3.0$)	–31	–06	0.42	1.48	1.34	1.71	3.40	1.90	61.6	202
Cauchy ($\alpha = 4.0$)	–35	–06	0.33	1.76	1.50	1.36	3.83	2.20	48.8	13.2
Cauchy ($\alpha = 50$)	–30	–06	0.28	2.06	1.68	1.13	4.28	2.53	38.3	09.0
Dolph–Chebyshev ($\alpha = 2.5$)	–50	+00	0.53	1.39	1.33	1.70	3.12	1.85	69.6	22.3
Dolph–Chebyshev ($\alpha = 3.0$)	–60	+00	0.48	1.51	1.44	1.44	3.23	2.01	64.7	16.3
Dolph–Chebyshev ($\alpha = 3.5$)	–70	+00	0.45	1.62	1.55	1.25	3.35	2.17	60.2	11.9
Dolph–Chebyshev ($\alpha = 4.0$)	–80	+00	0.42	1.73	1.65	1.10	3.48	2.31	55.9	8.07
Kaiser–Bessel ($\alpha = 2.0$)	–46	–06	0.49	150	1.43	1.46	320	1.99	65.7	16.9
Kaiser–Bessel ($\alpha = 2.5$)	–57	–06	0.44	1.65	1.57	1.20	3.38	2.20	59.5	11.2
Kaiser–Bessel ($\alpha = 3.0$)	–69	–06	0.40	1.80	1.71	1.02	3.56	2.39	53.9	07.4
Kaiser–Bessel ($\alpha = 3.5$)	–82	–06	0.37	1.93	1.83	0.89	3.74	2.57	48.8	04.8
Blackman	–58	–18	0.42	1.73	1.68	1.10	3.47	2.35	5.67	090

Solution From basic theory we know *that the resolution, as far as the spectrum of the finite-length sequence is concerned, is equal to the width of the central lobe of the spectrum of the window used.* Referring to Table 10.6 (page 897), we find that the width of the central lobe of the spectrum of the Hamming window is equal to $8\pi/L$. On the other hand, the upper limit of the resolution value is equal to $\pi/100$. Combining the foregoing results, it immediately follows that

$$\frac{8\pi}{L} \leq \frac{\pi}{100} \quad \text{or, equivalently,} \quad L \geq 800$$

However, since the length L is a power of 2, we will finally have $L_{\text{min}} = 1024$, from which it follows that $\nu = 10$, which is the desired result.

Problem 10.5.17 Suppose we want to determine the spectrum of an infinite-length discrete-time signal $x[n]$ with a resolution at least equal to $\pi/25$ and using a window with a length of $N = 256$ samples. Which of the windows in Table 10.6 is suitable for our purpose?

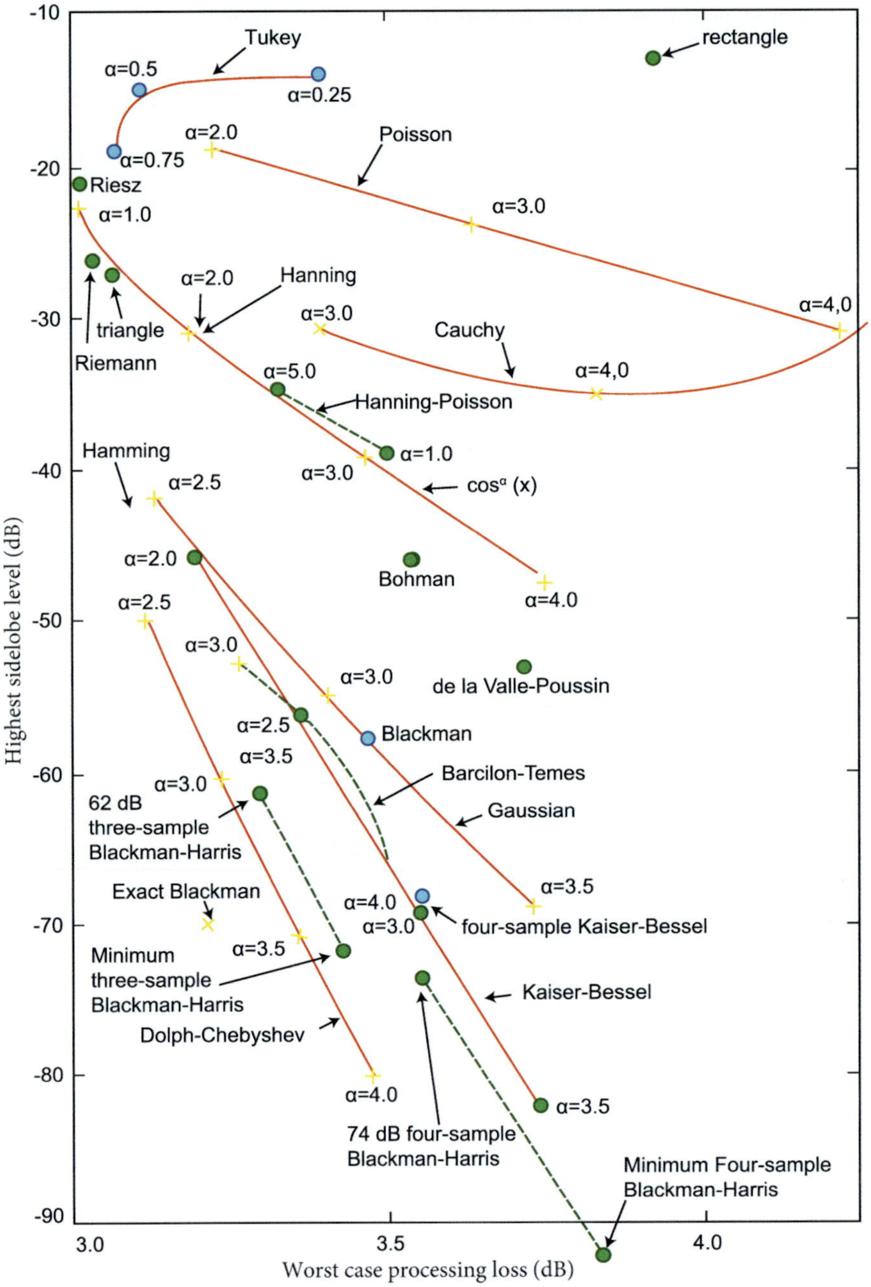

Fig. 10.48 Relative position of examined windows in a diagram (maximum sidelobe height, maximum total attenuation) (Harris, Figure 12)

Solution To solve the problem we must, *for each window in Table 10.6, determine the resolution that coincides with the width of the central lobe of its spectrum*, which corresponds to the length of the 256 samples. If the value that emerges is less than or equal to $\pi/25$, the window is considered suitable for our purpose, while in the opposite case, it is unsuitable for use, as it does not meet the criteria set by the exercise. In detail, then, we have:

- For the *rectangular window*: The width of the mainlobe is calculated as

$$\Delta\omega_w = \frac{4\pi}{N} = \frac{4\pi}{256} = \frac{\pi}{64} \leq \frac{\pi}{25}$$

- For the *Bartlett, Hamming, and von Hann windows*: The width of the mainlobe is calculated as

$$\Delta\omega_w = \frac{8\pi}{N} = \frac{8\pi}{256} = \frac{\pi}{32} \leq \frac{\pi}{25}$$

- For the *Blackman window*: The width of the mainlobe is calculated as

$$\Delta\omega_w = \frac{12\pi}{N} = \frac{12\pi}{256} = \frac{3\pi}{64} \nleq \frac{\pi}{25}$$

Therefore, except for the Blackman window, all the other windows are suitable for our purposes, since *the resolution they offer is within the limits of what is required.* ■

Although the construction of a window with a small width of the mainlobe and simultaneously small size of sidelobes is not theoretically possible; however, *the Kaiser window* offers a way to handle this incompatibility; in reality, a method known as *the Kaiser window method* is used *for the construction of digital filters that meet certain specifications.* Observing the defining equation of the Kaiser window, we find that it contains two parameters, the α and L, which can be determined with the help of empirical relations proposed by its inspirer, so that the spectrum will have the appropriate characteristics in each case. More details on the *Kaiser window method* and how it is used for the design *of digital filters* are given in the next chapter.

10.6 Short-Time Signal Analysis

In previous chapters, we constructed the defining equation of the DFT and saw how it was used for the analysis of a signal in the frequency domain. Recall that signals in practice *do not have infinite duration*, and to make a signal of infinite duration a finite-duration signal, we multiply it by a *window function*, being aware of all the consequences that this process brings to the spectral characteristics of the considered signal (such as the appearance of *spectral spreading* and *spectral leakage*). The basic assumption we resorted to in the earlier analysis was that the characteristics of the sinusoidal components of the considered signal and, more specifically, the amplitude, the frequency, and the phase *did not change over time* or, more specifically, over the duration dictated by the length of the applied window (recall that *the larger the length of this window, the better the resolution ability regarding the detection of neighboring frequencies*). However, in practice, this assumption *is not* always satisfied, and therefore, the classic discrete (and time-independent) Fourier transform *is not sufficient* for a spectral description of known signals that we encounter in practical applications, such as, for example, radar signals, human speech signals, and signals used in data transmission networks. On the contrary, in many cases of practical interest, the signals we consider are characterized by rapid changes in their characteristics, which is particularly true of voice signals, in which changes take place in intervals on the order of milliseconds. Of course, since we have no other tools in our toolbox beyond those we use in the general case, we will inevitably resort to their use trying, however, to apply them in the appropriate way and with the hope that the results we obtain for the case of rapidly changing signals will be reliable.

Although the type of analysis applied to rapidly changing signals consists, as before, in the multiplication of the signal by a window function, the logic this time is different since in this case *we have, not a static discrete sequence of infinite length with specific samples*, which through multiplication by a window function becomes a finite-length sequence, but a *rapidly changing* signal whose content *continuously changes*, as happens, for example, during the recording of a vocal signal in real time using a microphone. If we assumed that we could freeze time and consider the instantaneous content of a vocal signal at some point in time, then the situation would be the same as before, although of course such things do not happen. Time passes and new samples constantly appear. Consequently, the window function *is applied not to a specific static sequence as before, but, on the contrary, each time to a new group of samples*. It should be noted that *the faster the rate of change of signal characteristics, the shorter should be the duration of this window* . However, the very short duration of the window has the disadvantage that it leads to a reduction in discriminative ability, that is, our ability to distinguish different frequencies that are near each other, but on the other hand, it leads to an increase in our ability to distinguish changes that are made in the time domain. In other words, we must resort to a kind of *compromise* between the time and frequency domains in terms of discriminative ability.

Considering the case of a vocal signal, there are *two ways* in which the window function can be applied to the signal samples. If the signal has already been recorded in advance and we know all its samples (i.e., from the programmer's perspective, everything is stored in memory in a one-dimensional array of N elements), the window function, whose length M also defines the time duration during which we assume that the signal characteristics do not change, *slides over the signal from its beginning to its end* and *cuts off a different part of it each time*, and the appropriate type of processing is applied to each part. Conversely, if the signal is recorded *in real time*, and therefore, the samples are not known in advance but are continuously created, *the window remains stationary and the signal passes through it*, resulting in a *different* area being selected for processing each time. The situation is exactly the same as the detection of explosives and weapons carried out at airports, where there is a stationary detection device through which passengers' luggage, moving on a conveyor belt one behind the other, passes. At any given time, the system selects and examines a different piece of luggage (the one inside the detector). Two important parameters of this process are *the type of segmentation* of the signal into smaller pieces, which can either overlap or not (in practice, the first approach is chosen for reasons we will explain later) and *the type of window* used. A widely used window function is the *Hamming window*.

To mathematically describe this short-time analysis, as it is typically called, we need to invent and use new notation and define the related terminology. Up to this point, we have used the notation $x[n]$ to refer to any sample of the discrete-time signal $x[n]$, with the index n taking values in the range $(-\infty, \infty)$ for an infinite-length discrete sequence and in the range $[0, M-1]$ for a causal finite-length discrete sequence M. In addition, the quantities calculated so far generally took into account the entire set of signal samples, that is, *its entire duration*, while in this section, because we refer to rapidly changing signals, we must consider *only a very small part of them* and, more specifically, *a relatively small number of samples* located on either side of a sample numbered m, with the value of m changing continuously. To distinguish these two cases, we will characterize the quantities and the involved calculations for the first case as *long-term*, while correspondingly, the quantities and the related calculations for the second case will be characterized as *short-term*.

Based on all the foregoing considerations, let us use the notation $\mathcal{X}(\lambda)$ to describe a *long-term property* of a discrete-time signal $x[n]$ that depends on a parameter λ. This symbolism is very general and can describe any of the properties we have defined and used up to this point, with the parameter λ being replaced by the appropriate variable in each case and the property $\mathcal{X}(\lambda)$ being considered as resulting from the application to the discrete-time signal $x[n]$ of an operation $\mathcal{J}(\lambda)$, which leads to the quantity $\mathcal{J}(\lambda)\{x[n]\}$. Using this notation, we could, for example, describe the *DTFT* writing

$$\mathcal{X}(\omega) \equiv \mathcal{X}(e^{j\omega}) = \mathcal{J}(\omega)\{x[n]\} = \sum_{n=-\infty}^{\infty} x[n]e^{-j\omega n}$$

Apart from the application of the operation $\mathcal{J}(\lambda)$ to the discrete-time signal $x[n]$, the calculation of the property $\mathcal{X}(\lambda)$ may also include the application to the quantity $\mathcal{J}(\lambda)\{x[n]\}$ of an operator $\mathcal{L}$, which allows *the calculation of the average value over time of this magnitude*—it will therefore hold that$\mathcal{X}(\lambda) = \mathcal{L}\{\mathcal{J}(\lambda)\{x[n]\}\}$.

On the other hand, we use the notation $\mathcal{X}(\lambda; m)$ to describe a *short-term property* of the discrete-time signal $x[n]$, which depends on the parameter λ and is calculated from the values of a small set of samples of the signal, which are on either side of a point m and correspond to the values $m - M + 1 \leq n \leq M$, where N is the number of samples of this set. In the terminology of digital signal processing, this small part of the signal that is used at each stage of the study and is assumed to maintain constant the values of the signal's properties is known as a *frame*. As already mentioned, the creation of this frame is based on the application to the discrete-time signal $x[n]$ of a *window function* $w[n]$, which is performed in such a way that the resulting signal has nonzero values in the region $m - M + 1 \leq n \leq M$ and zero values anywhere else. If this window is a *unit height rectangular window*, then the samples of the frame coincide with the samples of the signal $x[n]$, although usually according to the discussion in the relevant section, in practice other window functions are used, such as the *Hamming window*, which leads to *better spectral characteristics* (for the selection criteria and the type of effect that the spectrum of a window has on the spectrum of the original signal, as well as for the parameters that characterize in general a window function, see Sect. 10.5.6 on page 886 and the related discussion). It is not difficult to understand that the part of the signal, that is, the frame, that is created in this way depends on the position m of the signal sample on either side of which the frame samples are located, and for this reason it is defined as $f(n; m) = x[n]w[m-n]$.

Based on the foregoing discussion, we can calculate the value of a short-term property $\mathcal{X}(\lambda; m)$ that approximates the corresponding long-term property $\mathcal{X}(\lambda)$ in the following way. Initially, we choose the appropriate frame of the discrete-time signal $f(n; m) = x[n]w[m-n]$ with a length of N samples by multiplying the signal by the appropriate window function, and then the value of the considered property is calculated as

$$\mathcal{X}(\lambda; m) = \frac{1}{N} \sum_{n=-\infty}^{\infty} \mathcal{J}(\lambda)\{f(n; m)\} = \frac{1}{N} \sum_{n=-\infty}^{\infty} \mathcal{J}(\lambda)\{x[n]w[m-n]\}$$

In other words, the calculation process simply consists of the *cutting*, or rather the selection from the signal, of a frame of length N and the application to it of the exact same process that we apply to this signal during the phase of long-term analysis. Usually, the frame is chosen in such a way that the last sample of the frame is the sample numbered m. Also, note that the application of the operator calculating the average value over time simply consists in dividing the value of the sum by the time duration N of the frame, that is, by the number of samples included in it. In practice, this division *is not particularly important* and sometimes is not performed, which is the rule for energy signals, for which the power value is known to be equal to zero.

A case of particular interest is one in which the operation $\mathcal{J}(\lambda)$ is *separable*, under the concept that the result it brings if applied to a product of two signals is equal to the product of the results of the application of the operation to each signal separately, a property that is formulated mathematically as

$$\mathcal{J}(\lambda)\{x[n]y[n]\} = \mathcal{J}(\lambda)\{x[n]\}\mathcal{J}(\lambda)\{y[n]\} = \mathcal{X}_{\mathcal{J}}(n; \lambda)\mathcal{Y}_{\mathcal{J}}(n; \lambda)$$

where $\mathcal{X}_{\mathcal{J}}(n; \lambda) = \mathcal{J}(\lambda)\{x[n]\}$ and $\mathcal{Y}_{\mathcal{J}}(n; \lambda) = \mathcal{J}(\lambda)\{y[n]\}$. In this special case, the equation for calculating the value of $\mathcal{X}(\lambda; m)$ takes the form

$$\mathcal{X}(\lambda; m) = \frac{1}{N} \sum_{n=-\infty}^{\infty} \mathcal{J}(\lambda)\{x[n]w[m-n]\} = \frac{1}{N} \sum_{n=-\infty}^{\infty} \mathcal{J}(\lambda)\{x[n]\}\mathcal{J}(\lambda)\{w[m-n]\} = \frac{1}{N} \sum_{k=-\infty}^{\infty} \mathcal{X}_{\mathcal{J}}(k; \lambda)\mathcal{W}_{\mathcal{J}}(m-k; \lambda)$$

and it is not difficult to understand that the calculation of this value involves the calculation of a *convolution*. We are therefore led to the conclusion that for each parameter of interest λ, the value $\mathcal{X}(\lambda; m)$ can be calculated as the convolution of the sequences $(1/N)\mathcal{X}_{\mathcal{J}}(n; \lambda)$ and $\mathcal{W}_{\mathcal{J}}(n; \lambda)$ calculated at the position $n = m$, while for the calculation of this value for each value of n we will have

$$\mathcal{X}(\lambda; n) = \frac{1}{N}\mathcal{X}_{\mathcal{J}}(n; \lambda) * \mathcal{W}_{\mathcal{J}}(n; \lambda)$$

Therefore, the quantity $\mathcal{X}(\lambda; n)$ can be considered the output of a filter with impulse response $\mathcal{W}_{\mathcal{J}}(n; \lambda)$ that receives at its input the signal $(1/N)\mathcal{X}_{\mathcal{J}}(n; \lambda)$.

Using this method of defining short-term values, we can construct short-term versions for all the features of the signal that we defined and used in previous chapters. For example, *short-term average power* is defined as

$$P(m) = \frac{1}{N} \sum_{n=-\infty}^{\infty} \{x[n]w[m-n]\}^2$$

while *short-term autocorrelation* is defined through the equation

$$r_{xx}(n; m) = \frac{1}{N} \sum_{n=-\infty}^{\infty} \mathcal{J}(\ell)\{x[n]w[m-n]\} = \frac{1}{N} \sum_{n=m-N+1}^{m} x[n]w[m-n]x[n-|\ell|]w[m-n+|\ell|]$$

We therefore observe that in the first case, due to the way average power is defined, *no parameter* λ is required, while in the second case, the parameter λ coincides with the time delay ℓ. The study of the foregoing relations reveals that the operation applied in each of the preceding two cases constitutes a separable operation as previously defined. Other quantities for which we can construct such short-term versions are the *covariance*, the *correlation*, and the *power spectral density*. The interested reader is encouraged to consult to the relevant literature for more details.

The main feature of the foregoing calculation process of the value $\mathcal{X}(\lambda; n)$ is that the operation $\mathcal{J}(\lambda)$ is applied not to the signal $x[n]$ but to the product $x[n]w[n]$, that is, in the context of the signal created by the application of the window $w[n]$ to the signal. Another alternative way of defining the value $\mathcal{X}(\lambda; n)$, which in some cases leads to the same result as the one we previously described, is to apply the operation *to the entire signal* and then apply the window function to the new signal that has emerged from this application of the operation, so as to cut off only one part of it. Working as before and using this time a short-term version for the size of the average of the signal defined as

$$\mu(m) = \mathcal{A}(m)\{x[n]\} = \frac{1}{N} \sum_{n=-\infty}^{\infty} x[n]w[n-m] = \frac{1}{N} x[n] * w[n]$$

we can calculate the value of the quantity $\mathcal{X}(\lambda; n)$ using the equation

$$\mathcal{X}(\lambda; n) = \mathcal{A}(m)\{\mathcal{J}(\lambda)\{x[n]\}\} = \mathcal{A}(m)\{\mathcal{X}_{\mathcal{J}}(n; \lambda)\} = \frac{1}{N} \sum_{n=-\infty}^{\infty} \mathcal{X}_{\mathcal{J}}(n; \lambda)w[n-m] = \frac{1}{N} \mathcal{X}_{\mathcal{J}}(n; \lambda) * w[n]$$

We end up again with a *convolution equation*, and therefore the value $\mathcal{X}(\lambda; n)$ can be considered *the output of a filter* with the impulse response being the window function $w[n]$ that receives as input the signal $(1/N)\mathcal{X}_{\mathcal{J}}(n; \lambda)$. The main difference between these two different methods of calculating the aforementioned value has to do with the fact that in the first case, the window function is applied to the signal $x[n]$ before the application of the operation $\mathcal{J}(\lambda)$, while in the second case, the window function is applied to the signal after the application of this operation. In other words, the window function is not affected by this operation and is always the same, regardless of the value of the parameter λ that is used, if such a thing is necessary. It is not difficult to understand that, precisely because of this characteristic, this second method of calculation is simpler compared to the first. In reality, all the algorithms for calculating short-term characteristics of signals that we encounter in the literature will belong to one or the other category.

10.6.1 Time-Dependent Fourier Transform

Although, as already mentioned, we can construct short-term versions for all the characteristics and properties of discrete-time signals, the most important of these is undoubtedly the one related to the discrete-time Fourier transform. The short-term version of this quantity is commonly referred to in the literature as the time-dependent Fourier transform. Based on our discussion in the previous section, this new time-dependent transform is defined as

$$\mathcal{X}'(n; \lambda) = \sum_{m=-\infty}^{\infty} x[n]w[n-m]e^{-j\lambda m}$$

In the foregoing expression, as we mentioned, a *shift* of the window takes place in relation to the signal, keeping the principle of time measurement constant, so that each time the window function is applied to a different frame of the signal, while another equivalent version is to have a shift of the signal in relation to the window, so the *time-dependent Fourier transform* is defined as

$$\mathcal{X}(n; \lambda) = \sum_{m=-\infty}^{\infty} x[n+m]w[m]e^{-j\lambda m}$$

and these two versions are related as $\mathcal{X}'(n; \lambda) = e^{-j\lambda n}\mathcal{X}(n; \lambda)$. In this description, we will consider the second version, which is also frequently encountered in the literature. In the preceding equations, the index n takes values from $-\infty$ to ∞, although obviously the calculations will be limited only to the values of n at positions where the window function has nonzero values, while note that we have removed the coefficient $1/N$, in full agreement with our previous comments. It is not difficult to understand that the parameter λ corresponds to the frequency ω. Regarding the range of the index in which the window is defined, if it is defined so that $w[n] \neq 0$ for $n > 0$, then it is aforementioned to be *noncausal*, since the calculation of the quantity $\mathcal{X}(n; \lambda)$ requires signal samples that are located after the current sample $x[n]$. The sequence $w[m]$ has *finite length* and is assumed to be centered around the value $m = 0$, so the transform $\mathcal{X}(n; \lambda)$ displays the frequency characteristics of the signal centered on the value n. It is not difficult to see that the time-dependent Fourier transform is *periodic* with respect to the frequency λ with a period of 2π, and for this reason, in its study and use, it is sufficient to consider only frequency values λ that belong to a frequency with a range of 2π such as the area $0 \leq \lambda < 2\pi$.

Adopting the presented method for the general case of short-term signal properties, we can describe the *time-dependent Fourier transform* as the output of a filter calculated through the use of a convolution. Indeed, by making the substitution $m' = n + m$ in the defining equation of the transform $\mathcal{X}(n; \lambda)$, so that $-m = n - m'$, it takes the equivalent form

$$\mathcal{X}(n; \lambda) = \sum_{m'=-\infty}^{\infty} x[m']w[-(n - m')]e^{j\lambda(n-m')}$$

which can be considered the convolution

$$\mathcal{X}(n; \lambda) = x[n] * h_\lambda[n] \qquad \text{where} \qquad h_\lambda[n] = w[-n]e^{j\lambda n}$$

Consequently, the time-dependent Fourier transform can be interpreted as *the output of a LTI filter*, with impulse response $h_\lambda[n]$ and frequency response function

$$\mathcal{H}_\lambda(e^{j\omega}) = \mathcal{W}(e^{j(\lambda-\omega)})$$

If $w[n] = 0$ for $n < 0$, then the impulse response $h_\lambda[n]$ is a *noncausal* signal, while if $w[n] \neq 0$ for $n \leq 0$, then the filter described by this is a *causal system*. The function $\mathcal{W}(e^{j\omega})$ used in the foregoing relationship is characterized by the *usual low-pass spectrum of window functions*, and therefore, the function $\mathcal{H}_\lambda(e^{j\omega})$ will describe a band-pass filter whose passband is centered around the frequency $\omega = \lambda$, with the width of this band being approximately equal to the width of the central lobe of the Fourier transform of the window. As for the degree of elimination of adjacent frequency components, this depends, as is well known, on the relative height of the sidelobe.

Closely related to the time-dependent Fourier transform is the so-called *spectrum*, which is a diagram whose horizontal axis is the time index n, while the vertical axis is the quantity $20 \log_{10} |\mathcal{X}(n; \lambda)|$.

As with all transformations used in the field of digital signal processing, the time-dependent Fourier transform is *reversible*, provided, of course, that the window function has at least one nonzero sample. The equation defining the inverse transform has the form

$$x[n + m]w[m] = \frac{1}{2\pi} \int_0^{2\pi} X(n; \lambda)e^{j\lambda m} d\lambda, \qquad -\infty < m < \infty$$

or, equivalently,

$$x[n + m] = \frac{1}{2\pi\, w[m]} \int_0^{2\pi} X(n; \lambda) d\lambda$$

and it looks very much like the equation of the DTFT, which is of course expected.

10.6.2 Sampling in Time and Frequency

According to our discussion in previous sections, the DTFT is a function of frequency ω, which, as is well known, is a continuous variable and, therefore, *cannot be used* for the calculation of the transform from an electronic computer. For this reason, we resorted to sampling the DTFT in the frequency domain, with the result of this process being the DFT, which is also the one used in practice. The situation is similar in the case of the time-dependent Fourier transform, since the frequency λ appearing in the defining equation constitutes a continuous variable, and therefore, to calculate the transform $\mathcal{X}(n; \lambda)$, we should once again proceed to *sampling in the frequency domain* λ. Assuming that the window function has a length of L samples, whose values start from the position $m = 0$, that is, defined as $w[m] = 0$ for $m < 0$ and $m > L - 1$, this sampling consists in the *recording of the transform value* $\mathcal{X}(n; \lambda)$ at the positions of the frequency values $\lambda_k = 2\pi k/N$, where $N \geq L$. In this case, we are in a position to recover the original sequence from the samples of the time-dependent Fourier transform. Defining the quantity

$$\mathcal{X}[n, k] = X(n, 2\pi k/N) = \sum_{m=0}^{L-1} x[n + m]w[m]e^{-j(2\pi/N)km} \qquad 0 \leq k \leq N - 1 \tag{10.163}$$

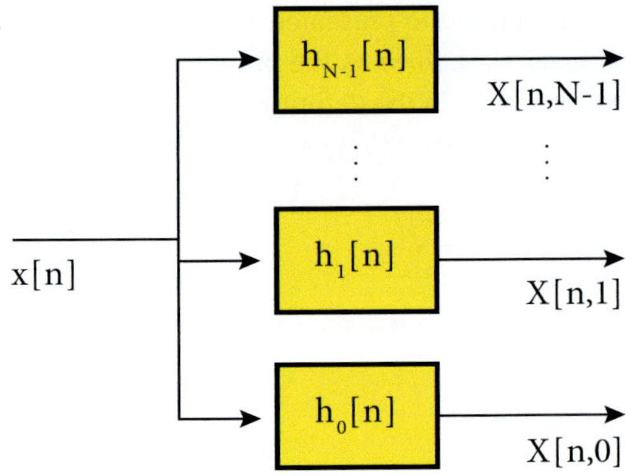

Fig. 10.49 Representation of the time-dependent Fourier transform by a filter array

this, for a fixed value of n, is recognized as the *DFT* of the sequence $x[n+m]w[m]$, that is, of the considered frame of the signal and from the defining equation of the inverse discrete transform, we will have

$$x[n+m]w[m] = \frac{1}{N}\sum_{k=0}^{N-1} X[n,k]e^{j(2\pi/N)km} \qquad 0 \le m \le L-1$$

Therefore, solving for $x[n+m]$, the samples of the discrete sequence can be recovered for the index value range from n to $(n+L-1)$ using the equation

$$x[n+m] = \frac{1}{Nw[m]}\sum_{k=0}^{N-1} X[n,k]e^{j(2\pi/N)km} \qquad 0 \le m \le L-1$$

The foregoing analysis allows us to implement the *time-dependent Fourier transform*, as a result of applying a sequence of filters whose structure is determined in the following way. Initially, we observe that Eq. (10.163) can also be written as

$$\mathcal{X}[n,k] = x[n] * h_k[n] \qquad (0 \le k \le N-1) \quad \text{where} \quad h_k[n] = w[-n]e^{j(2\pi/N)kn}$$

These equations can be considered to describe a sequence of N filters, which are connected in parallel, as dictated by the summation symbol of Eq. (10.163) and depicted in Fig. 10.49, with filter number k characterized by a frequency response of the form

$$\mathcal{H}_k(e^{j\omega}) = \mathcal{W}(e^{j[(2\pi k/N)-\omega]})$$

Note that the reconstruction of the signal $x[n]$ for the value range $-\infty < n < \infty$ can be achieved if the functions $\mathcal{X}(n;\lambda)$ and $\mathcal{X}[n,k]$ are sampled in the time domain. In this case, the corresponding function is

$$\mathcal{X}[rR,k] = \mathcal{X}(rR, 2\pi k/N) = \sum_{m=0}^{L-1} x[rR+m]w[m]e^{-j(2\pi/N)km}$$

where r and k are integer quantities with values $-\infty < r < \infty$ and $0 \le k \le N-1$. Defining the quantity

$$\mathcal{X}_r[k] = \mathcal{X}[rR,k] = \mathcal{X}[rR,\lambda_k), \qquad -\infty$$

where $\lambda_k = 2\pi k/N$, the sequence resulting from the sampling in the time domain of the time-dependent Fourier transform is simply the DFT of length N of the discrete sequences

$$x_r[m] = x[rR+m]w[m], \qquad -\infty < r < \infty \qquad 0 \le m \le L-1$$

with the position of the window moving in time, making jumps of length R. Note that in the defining equation of the function $\mathcal{X}[rR, k]$, all the significant parameters of the process appear, and more specifically, the length L of the window $w[m]$, the length N of the DFT, and the interval R of the sampling that takes place in the time domain. As for the correlation that exists between these quantities, if $L \leq N$, the reconstruction of the sequences $x_r[m]$ from the transforms $\mathcal{X}_r[k]$ is ensured. On the other hand, if $R < L$, then these sequences overlap, but if $R > L$, then some of the signal samples are not used, and therefore, they cannot be reconstructed from the transforms $\mathcal{X}_r[k]$. Finally, if $R \leq L \leq N$, we can recover R samples of $x[n]$ for each value of n from the transforms $\mathcal{X}_r[k]$, with each group of R samples represented by N complex numbers.

10.6.3 The Method of Overlap-Add Reconstruction

In the special case where the condition $R = L = N$ holds, the complete reconstruction of the signal from the samples of the time-dependent Fourier transform is possible. More specifically, first, we proceed to the recovery of the signal $x[rR+m]w[m]$ $(0 \leq m \leq N-1)$ through the calculation of the inverse DFT of the function $\mathcal{X}_r[k]$ and then the signal itself $x[n]$ for the range of values $rR \leq n \leq [(r+1)R - 1]$ through the equation

$$x[n] = \frac{X_r[n - rR]}{w[n - rR]} \qquad rR \leq n \leq [(r+1)R - 1]$$

However, this reconstruction *can only theoretically be achieved* since usually in practice the time-dependent Fourier transform *is constantly modified*, as happens, for example, in applications of audio signal encoding, resulting in a significant increase in error values at the ends of successive frames of the signal when we try to join them, since they do not exactly match each other. In this case, it is preferable to use for the parameter R a value *smaller* than those of the parameters L and N, a fact that leads to *the overlap of successive frames*. In this case, if we choose the appropriate window function, there is no reason later to remove it, as we did previously.

In a more detailed description, let us assume that the values of the parameters are such that $R \leq L \leq N$ applies. It will then hold that

$$x_r[m] = x[rR + m]w[m] = \frac{1}{N} \sum_{k=0}^{N-1} X_r[k]e^{j(2\pi k/N)m} \qquad 0 \leq m \leq L - 1$$

In this equation, the parts of the signal that have been recovered have shaped by the applied window, with the beginning of the time measurement *to coincide* with the beginning of the window. Alternatively, we can shift the aforementioned sections in time, so that they return to their original positions rR and then add them together, so the reconstructed signal will have the form

$$\hat{x}[n] = \sum_{r=-\infty}^{\infty} x_r[n - rR]$$

This result is particularly interesting since, if we manage to prove that the signal $\hat{x}[n]$ coincides with the signal $x[n]$ for all values of the parameter n, then the two preceding equations jointly constitute a method of *synthesis*, which allows for the perfect reconstruction of the signal $x[n]$. Combining the two equations, we get

$$\hat{x}[n] = \sum_{r=-\infty}^{\infty} x[rR + n - rR]w[n - rR] = x[n] \sum_{r=-\infty}^{\infty} w[n - rR]$$

and defining the sequence

$$\hat{w}[n] = \sum_{r=-\infty}^{\infty} w[n - rR]$$

the reconstructed signal given by the foregoing expression is formulated as

$$\hat{x}[n] = x[n]\hat{w}[n]$$

The *perfect reconstruction* of the signal becomes possible when the condition

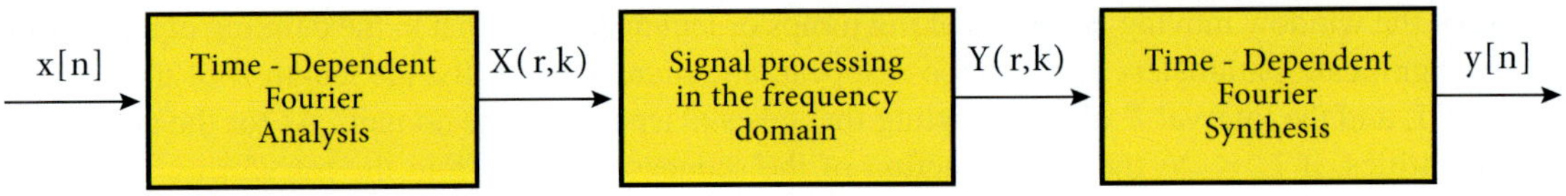

Fig. 10.50 A signal processing system that uses the time-dependent Fourier transform

$$\bar{w}[n] = \sum_{r=-\infty}^{\infty} w[n - rR] = C \qquad -\infty < n < \infty$$

is satisfied. Finally, regarding the window functions that are used in the foregoing process, two widely used functions are the *Bartlett and Hann windows* for the analytical description, on which the reader is referred to the relevant section. Specifically, it is proven that these windows with a length that is a power of 2 can lead to perfect reconstruction for values $R = M/2, M/4, \ldots, 1$ and with a corresponding reconstruction gain equal to $M/(2R)$. More details on this subject can be found in the bibliography.

10.6.4 Signal Processing Using the Time-Dependent Fourier Transform

The time-dependent Fourier transform can be used as a discrete-time signal analysis technique, in the same way as other transforms are used. A signal processing system based on the use of this transform is depicted in Fig. 10.50. This system consists of *three subsystems*, of which the first and third perform the *analysis* and *synthesis* using this new transform, while the middle subsystem performs the *processing in the frequency domain* as appropriate in each case.

Let us consider a causal discrete-time signal, and let us assume that we calculate the time-dependent Fourier transform using a rectangular window. In this case, the transform $\mathcal{X}_r[k]$ will be defined as a set of DFTs with a length of N samples of the sequence

$$x_r[m] = x[rL + m], \qquad 0 \le m \le L - 1$$

Since in this case all samples of the discrete sequence $x[n]$ are used, while the successive segments of the signal do not overlap, it follows that

$$x[n] = \sum_{r=0}^{\infty} x_r[n - rL]$$

Let us now define a *time-dependent Fourier transform* of the form

$$\mathcal{Y}_r[k] = \mathcal{H}[k]\mathcal{X}_r[k], \qquad 0 \le k \le N - 1$$

where $\mathcal{H}[k]$ is the DFT of a causal sequence of unit samples $h[n]$ of finite length P. If we calculate the inverse DFT of the preceding transform, we will get

$$y_r[m] = \frac{1}{N}\sum_{k=0}^{N-1} \mathcal{Y}_r[k]e^{j(2\pi/N)km} = \sum_{\ell=0}^{N-1} x_r[\ell]h[((m - \ell))_N]$$

In other words, the signal $y_r[m]$ is the *circular convolution* of the sequences $h[m]$ and $x_r[m]$, and according to our earlier discussion, if the signal $x_r[m]$ has a finite length equal to L samples, if $N \ge L+P-1$, then the signal $y_r[m] =$ will coincide with the linear convolution of the signals $h[m]$ and $x_r[m]$ in the value range $0 \le m \le L + P - 2$ and will have a value of zero anywhere else. In this case, the output signal

$$y[n] = \sum_{r=0}^{\infty} y_r[n - rL]$$

will be the output of a discrete LTI system with impulse response $h[n]$. This method corresponds exactly to the overlap-add method presented previously.

10.7 Problems

1. Following the methodology of Exercise 10.2.1 (page 786), determine the Fourier expansion of the signals $x[n] = \{\mathbf{3}, -2, 9, 7, 1, -1\}$ with period $N = 6$ and $y[n] = \{\mathbf{1}, 0, 1, -1\}$ with period $N = 4$.
2. Check if the discrete-time signal

$$x[n] = \cos\left(\frac{\pi n}{8}\right) + \cos\left(\frac{\pi n}{4} + \frac{\pi}{4}\right)$$

is periodic, and if it is, determine its period and its Fourier series expansion.

3. Determine the discrete periodic signals $x[n]$, $y[n]$, and $z[n]$ of period $N = 8$ whose Fourier expansion coefficients are given by the relations

$$\text{(a)} \quad x_k = \cos\left(\frac{k\pi}{4}\right) + \sin\left(\frac{5k\pi}{7}\right)$$

$$;(b) \quad y_k = \begin{cases} \sin\left(\frac{k\pi}{5}\right) & 0 \le k \le 8 \\ 0 & k = 9 \end{cases}$$

$$;(c) \quad z_k = \left\{ \ldots, \quad 0, \quad \frac{1}{4}, \quad \frac{1}{2}, \quad 1, \quad 2, \quad 1, \quad \frac{1}{2}, \quad \frac{1}{4}, \quad 0, \quad \ldots \right\}$$

4. Prove the property of a *sum* according to which, if the coefficients of the Fourier expansion of the discrete periodic signal $x[n]$ are x_k, then the coefficients of the Fourier expansion of the discrete-time signal

$$y[n] = \sum_{m=-\infty}^{n} x[n] \qquad \text{are given by the relation} \qquad y_k = \frac{x_k}{1 - \exp(-j\,k2\pi/N)}$$

(Hint: A necessary condition for the signal $y[n]$ to be periodic and bounded is the condition $x_0 = 0$, according to which the periodic discrete-time signal $x[n]$ does not have a continuous component [i.e., a component independent of time n]).

5. Considering the discrete-time signal $x[n]$ defined as

$$x[n] = \begin{cases} 1\, 0 \le n \le 7 \\ 0\, 8 \le n \le 9 \end{cases}$$

which is periodic with value $N = 10$, determine the coefficients of the Fourier expansion of the discrete-time signals defined as

$$y[n] = x[n] - x[n-1], \quad y[n] = x[n-5], \quad y[n] = \exp\left(i4\frac{\pi n}{5}\right)$$

6. Verify the validity of Parseval's theorem considering the pairs $[\xi[n], \xi_k]$ (where $\xi = x, y, z$) from Problem 3.
7. Repeat the calculations of Exercise 10.2.5 (refer to page 798) for the values ($L = 5$, $N = 40$) and ($L = 8$, $N = 100$) and comment on the result. As amplitude value use for simplicity $A = 1$. To observe in practice the symmetry properties that were commented on in the relevant section, extend your calculations to the area of negative values of the index k.
8. Let the discrete periodic signal $x[n]$ be given by the equation

$$x[n] = 1 + \sin\left(\frac{2\pi n}{7}\right) + \sqrt{5}\sin\left(\frac{\pi n}{5}\right) + \cos\left(\frac{4\pi n}{5} + \frac{\pi}{3}\right)$$

(1) What are the fundamental frequency and fundamental period of the signal? (2) Find the coefficients of the exponential Fourier expansion. (3) Draw the variation of the magnitude and phase of the complex Fourier coefficients that you will determine for a time interval of two periods. (4) Construct the power spectral density of the signal and verify Parseval's identity. (*Hint*: The problem can be solved directly, expressing the functions of the sines and cosines as a function of the complex exponential function with the help of Euler's formulas).

9. Repeat Exercise 8 for the signals of Exercise 10.2.1 (page 786) and Problems 1 and 2.
10. Calculate the coefficients of the Fourier expansion, draw the amplitude, phase, and power spectral density, and verify Parseval's identity for the periodic discrete-time signals

$$x_1[n] = \sin\left(\frac{2\pi n}{3}\right)\cos\left(\frac{\pi n}{2}\right) \quad \text{with period } N = 12$$

$$x_2[n] = 1 - \sin\left(\frac{\pi n}{4}\right) \quad (0 \le n \le 3) \quad \text{with period } N = 4 \quad \text{and}$$

$$x_3[n] = 1 - \sin\left(\frac{\pi n}{4}\right) \quad (0 \le n \le 11) \quad \text{with period } N = 12$$

11. Determine the Fourier expansion of the output $y[n]$ of a discrete LTI system with an impulse response in the form $h[n] = \alpha^{|n|}$ when the periodic discrete-time signal is transmitted to its input

$$x[n] = \sum_{k=-\infty}^{\infty} \delta[n - 5k]$$

12. Prove that the Fourier transform of the signal $x[n] = u[n] - u[n-4]$ is

$$\mathcal{X}(e^{j\omega}) = \frac{\sin(2\omega)}{\sin(\omega/2)} \exp(-j3\omega/2)$$

13. Prove that the Fourier transform of the discrete-time signals

$$x[n] = \alpha^n \cos(\omega_0 n)u[n] \quad \text{and} \quad x[n] = \alpha^n \sin(\omega_0 n)u[n]$$

for the range of values $|\alpha| < 1$ is given respectively by the relations

$$\mathcal{X}(e^{j\omega}) = \frac{1 - \alpha\cos(\omega_0)\, e^{-j\omega}}{1 - 2\alpha\cos(\omega_0)\, e^{-j\omega} + \alpha^2 e^{-2i\omega}} \quad \text{and} \quad \mathcal{X}(e^{j\omega}) = \frac{\alpha\sin(\omega_0)\, e^{-j\omega}}{1 - 2\alpha\cos(\omega_0)\, e^{-j\omega} + \alpha^2 e^{-2i\omega}}$$

14. Prove that the Fourier transform of the signal $x[n] = \alpha^n \sin(\omega_0 n)u[n]$ for the range of values $|\alpha| < 1$ is given by the relation

$$\mathcal{X}(e^{j\omega}) = \frac{\alpha\sin(\omega_0)\, e^{-j\omega}}{1 - 2\alpha\cos(\omega_0)\, e^{-j\omega} + \alpha^2 e^{-2i\omega}}$$

15. Prove that the Fourier transform of the signal $x[n] = (1/4)^n$ is given by the relation

$$\mathcal{X}(e^{j\omega}) = \frac{256 e^{4i\omega}}{1 - 0.25\, e^{-j\omega}}$$

16. Using the arguments of symmetry and conjugation, prove that the Fourier transform of an even real signal is an even real function, while the Fourier transform of an odd real signal is an odd imaginary function.
17. Using the property of convolution and the Fourier transform of the step function, prove the property of the sum of the transform, according to which if the Fourier transform of the discrete-time signal $x[n]$ is the function $\mathcal{X}(e^{j\omega})$, then the Fourier transform of the signal

$$y[n] = \sum_{k=-\infty}^{n} x[k] \quad \text{is the function} \quad \mathcal{Y}(e^{j\omega}) = \frac{\mathcal{X}(e^{j\omega})}{1 - e^{-j\omega}} + \pi \mathcal{X}(e^{j\omega}) \sum_{k=-\infty}^{\infty} \delta(\omega - 2k\pi)$$

(Hint: Formulate the discrete-time signal $y[n]$ as the convolution between the signals $x[n]$ and $u[n]$).

18. Determine the Fourier transform of the signal $x[n] = \sin(\omega_0 n)(u[n] - u[n - M])$, where $u[n]$ is the discrete step function.
19. Prove that the Fourier transform of the signals

$$x[n] = \left(\frac{1}{3}\right)^n u[n] + 2^n u[-n-1] \quad \text{and} \quad x[n] = e^{\tau|n|} \quad (\tau < 0)$$

is given respectively by the relations

$$\mathcal{X}(e^{j\omega}) = -\frac{7}{3}\frac{e^{-j\omega}}{[1 + (1/3)\, e^{-j\omega}][1 - 2e^{-j\omega}]} \quad \text{and} \quad \mathcal{X}(e^{j\omega}) = \frac{1 - e^{2\tau}}{1 - 2e^{\tau}\cos(\omega) + e^{2\tau}}$$

20. Prove that the Fourier transform of the signal $x[n] = e^{\tau|n|}$ $(\tau < 0)$ is

$$\mathcal{X}(e^{j\omega}) = \frac{1 - e^{2\tau}}{1 - 2e^{\tau}\cos(\omega) + e^{2\tau}}$$

21. Prove that the inverse Fourier transform of the function $\mathcal{X}(e^{j\omega})$ defined as $\mathcal{X}(e^{j\omega}) = \cos(2\omega) + j\sin(\omega)$ is given by the relation

$$x[n] = \frac{1}{2}\delta[n+2] + \frac{1}{2}\delta[n+1] - \frac{1}{2}\delta[n-1] + \frac{1}{2}\delta[n-2]$$

22. Prove that $\mathcal{F}^{-1}\{|\omega|\} = [(-1)^n - 1]/(\pi n^2)$.
23. Prove that

$$\mathcal{F}^{-1}\left\{\text{tri}\left(\frac{\omega}{2\pi A}\right)\right\} = A\text{sinc}^2(An) \quad \text{and} \quad \mathcal{F}\left\{\text{rect}\left(\frac{\omega}{2\pi B}\right) e^{j\omega n_0}\right\} = B\text{sinc}[B(n + n_0)]$$

(for parameter values $0 < A \leq 0.5$ and $0 < B \leq 1$) with the functions rect and tri defined as

$$\text{rect}(t) = \begin{cases} 0 & |t| > (1/2) \\ 1/2 & |t| = (1/2) \\ 1 & |t| < (1/2) \end{cases} \quad \text{and} \quad \text{tri}(t) = \begin{cases} 1 + t & -1 \leq t \leq 0 \\ 1 - t & 0 < t \leq 1 \\ 0 & \text{for all other values} \end{cases}$$

24. Using the properties of the Fourier transform and referring to tables of elementary transforms, find the inverse Fourier transform of the function

$$\mathcal{X}(e^{j\omega}) = \frac{1}{1 - e^{-j\omega}}\left[\frac{\sin(3\omega/2)}{\sin(\omega/2)}\right] + 5\pi\delta(\omega)$$

25. Determine the inverse Fourier transform of the function $\mathcal{X}(e^{j\omega})$ given by the relation $\mathcal{X}(e^{j\omega}) = |\pi - \omega|$, $(0 \leq \omega < 2\pi)$.
26. Determine the inverse Fourier transform of the functions

$$\mathcal{X}(e^{j\omega}) = \cos^2\omega + \sin^2(3\omega) \quad \text{and} \quad \mathcal{X}(e^{j\omega}) = \sum_{k=-\infty}^{\infty} (-1)^k \delta\left(\omega - \frac{\pi k}{2}\right)$$

27. Determine the inverse Fourier transform of the function $\mathcal{X}(e^{j\omega}) = |\omega|$, $(-\pi \leq \omega < \pi)$.

28. Determine the transient response and the steady-state response of the discrete LTI system described by the difference equation

$$y[n] = \frac{1}{2}\left(x[n] - x[n-3]\right)$$

when the discrete input signal is applied to its input

$$x[n] = 3 + 15\cos\left(\frac{\pi n}{2} + 60^\circ\right)$$

29. Determine the amplitude and phase response of a discrete LTI system described by the difference equation

$$y[n] = \alpha y[n-1] + (1-\alpha)x[n], \quad (n \geq 0)$$

for a parameter value of $\alpha = 0.9$ and initial condition $y[-1] = 0$ when the discrete-time signal $x[n] = 2\sin(2\pi f n)$, $0 \leq n \leq 100$ is transmitted to its input for frequency values $f = 1/2, 1/5, 1/10$.

30. Using the Fourier transform method, determine the output of the system $y[n]$ with an impulse response of the form

$$h[n] = \left(\frac{1}{2}\right)^n \cos\left(\frac{\pi n}{2}\right) u[n]$$

when the discrete-time signal $x[n] = \cos(\pi n/2)$ is transmitted to its input.

31. Determine the output $y[n]$ of a discrete recursive LTI system with an impulse response of the form

$$h[n] = \left(\frac{1}{4}\right)^n u[n]$$

when discrete-time signals are transmitted to its input

$$x_1[n] = \left(\frac{3}{5}\right)^n u[n], \quad x_2[n] = (n+1)\left(\frac{1}{5}\right)^n u[n], \quad x_3[n] = (-1)^n$$

32. Determine the output $y[n]$ of a discrete recursive LTI system with an impulse response of the form

$$h[n] = \sin\left(\frac{\pi n}{3}\right) u[n]$$

when discrete-time signals are transmitted to its input

$$x_1[n] = \sum_{k=-\infty}^{\infty} \delta[n-8k] \quad \text{and} \quad x_2[n] = \delta[n+1] + \delta[n-1]$$

33. Consider the discrete recursive LTI system described by the equation

$$y[n] + \frac{1}{2}y[n-1] = x[n]$$

Determine the frequency response function $\mathcal{H}(e^{j\omega})$, as well as the system's output, when discrete-time signals are transmitted to its input

$$x_1[n] = \left(\frac{1}{2}\right)^n u[n] \quad \text{and} \quad x_2[n] = \delta[n] + \frac{1}{2}\delta[n-1]$$

34. Considering an ideal band-pass filter, its frequency response is

$$\mathcal{H}(e^{j\omega}) = \begin{cases} 0 \text{ for frequency values } |\omega| \le \dfrac{\pi}{8} \\ 1 \text{ for frequency values } \dfrac{\pi}{8} \le |\omega| < \dfrac{3\pi}{8} \\ 0 \text{ for frequency values } \dfrac{3\pi}{8} \le |\omega| \le \pi \end{cases}$$

Determine its impulse response $h[n]$, and prove that this can be calculated as the product of the expression $\cos(\pi n/4)$ times the impulse response of an ideal low-pass filter.

35. Considering a discrete LTI system with frequency response $\mathcal{H}(e^{j\omega})$, prove that the group delay $\tau(\omega)$ can be calculated from the equation

$$\tau(\omega) = \frac{\Re\{\mathcal{H}(e^{j\omega})\}\Re\{\mathcal{G}(e^{j\omega})\} + \Im\{\mathcal{H}(e^{j\omega})\}\Im\{\mathcal{G}(e^{j\omega})\}}{|\mathcal{H}(e^{j\omega})|^2}$$

where $\mathcal{G}(e^{j\omega})$ is the Fourier transform of the discrete-time signal $g[n] = nh[n]$, with the discrete sequence $h[n]$ expressing the impulse response of the system. Then verify the validity of the preceding relationship considering a discrete LTI system with impulse response $h[n] = (1/3)^n u[n]$.

36. Prove that the group delay of discrete LTI systems with frequency response[49]

$$\mathcal{H}_1(\omega) = 1 - \alpha e^{-j\omega} \quad \text{and} \quad \mathcal{H}_2(\omega) = \frac{1}{1 - \alpha e^{-j\omega}}$$

has the form

$$\tau_1(\omega) = \frac{\alpha^2 - \alpha\cos\omega}{1 - 2\alpha\cos\omega + \alpha^2} \quad \text{and} \quad \tau_2(\omega) = \frac{\alpha\cos\omega - \alpha^2}{1 - 2\alpha\cos\omega + \alpha^2}$$

Then use the appropriate software application to design the functions $\tau_1(\omega)$ and $\tau_2(\omega)$ for the frequency range $-\pi \le \omega \le \pi$ and for the parameter values $\alpha = 1/2$ and $\alpha = 1/4$, and comment on the result.

37. Determine the group velocity of a discrete LTI system with frequency response

$$\mathcal{H}(e^{j\omega}) = \frac{1}{1 - 2\alpha\cos\vartheta e^{-j\omega} + \alpha^2 e^{-2i\omega}}$$

and construct its graph for the frequencies that belong to the value range $-\pi \le \omega \le \pi$ and Problems 36 and 37, and verify your solution by calculating the group delay from Eq. (10.71).

38. Considering a discrete LTI system whose frequency response has a double pole at the position $p_{1,2} = 0.5$ and two zero values at the positions $z_{1,2} = \exp(\pm j3\pi/4)$, determine, using geometric arguments, the value of its gain, so that $|\mathcal{H}(0)| = 1$.

39. Considering a discrete LTI system with impulse response $h[n] = [(1/3)^n \cos(\pi n/6)]u[n]$, determine the value of the magnitude of the frequency response $\mathcal{H}(e^{j\omega})$ from the pole–zero diagram, as well as the system's response to the input signal $x[n] = (1/4)^n u[n]$.

40. Performing the appropriate algebraic operations, prove that the expression

$$\mathcal{X}(e^{j\omega}) = \frac{1 - e^{-j\omega N}}{N}\sum_{k=0}^{N-1} \mathcal{X}[k] \Big/ \left\{1 - \exp\left[-j\left(\omega - \frac{2\pi k}{N}\right)\right]\right\}$$

---

[49] For the calculation of the function $\tau_2(\omega)$ you can rely on the fact that systems with frequency responses $\mathcal{H}_1(e^{j\omega})$ and $\mathcal{H}_2(e^{j\omega})$ are *inverse systems* (see Sect. 7.8 on page 606) since by analogy with the functions in their transfer, their frequency responses will be connected by a relationship of the form $\mathcal{H}_2(e^{j\omega}) = 1/\mathcal{H}_1(e^{j\omega})$. But this means that *the logarithmic amplitude, the phase, and the group delay of the inverse system will constitute the negatives of the corresponding functions of the original system.* Note that there are systems whose inverse does not exist, such as, for example, the ideal low-pass filter.

for the function $\mathcal{X}(e^{j\omega})$ is equivalent to the relation

$$\mathcal{X}(e^{j\omega}) = \sum_{k=0}^{N-1} \mathcal{X}\left[\exp\left(j\frac{2\pi k}{N}\right)\right]\mathcal{P}\left\{\exp\left[j\left(\omega - \frac{2\pi k}{N}\right)\right]\right\}$$

41. Calculate the DFT of the discrete-time signal

$$x[n] = \begin{cases} 1 \text{ for even values of } n \\ 0 \text{ for odd values of } n \end{cases} \qquad (0 \le n \le N-1)$$

42. Calculate the DFT of the discrete-time signal

$$x[n] = \begin{cases} 1 \; 0 \le n \le (N/2) - 1 \\ 0 \; (N/2) \le n \le N-1 \end{cases} \qquad (0 \le n \le N-1)$$

43. Calculate the DFT of the discrete-time signal

$$x[n] = 0.42 - 0.5\cos\left(\frac{2\pi n}{N-1}\right) + 0.08\cos\left(\frac{4\pi n}{N-1}\right) \qquad (0 \le n \le N-1)$$

44. If the DFT of the discrete-time signal $x[n]$ is the function $\mathcal{X}[k]$, determine the Fourier transforms of the discrete-time signals

$$\left.\begin{aligned} x_1[n] &= x[n]\cos\left(\frac{2k\pi n}{N}\right) \\ x_2[n] &= x[n]\sin\left(\frac{2k\pi n}{N}\right) \end{aligned}\right\}, \qquad (0 \le n \le N-1)$$

in terms of the transform $\mathcal{X}[k]$.

45. Calculate the DFT of the discrete sinusoidal waveform

$$x[n] = \sin\left(\frac{2k\pi n}{N} + \varphi\right) \qquad (n = 0, 1, 2, \ldots, N-1)$$

46. Using the defining equation of the DFT, prove that the stated transform of length $N = 4$ for the discrete-time signal $x[n] = \{1, 2, 2, 1\}$ is the discrete sequence $\mathcal{X}[k] = \{6, -1-i, 0, -1+i\}$.
47. Prove that the DFT of length $N = 8$ of the discrete-time signal $x[n] = 6\cos^2(\pi n/4)$ is the discrete sequence $\mathcal{X}[k] = \{24, 0, 12, 0, 0, 0, 12, 0\}$.
48. Prove that the DFT of the discrete-time signal given by the relation $x[n] = \delta[n] + 0.9\delta[n-6]$ has the form $\mathcal{X}[k] = 1 + 0.9\exp(-j\,3\pi k/2)$. What is the length of the transform and why?
49. Considering the DFT with a length of N samples of the form

$$\mathcal{X}[k] = \begin{cases} e^{+j\varphi} \; k = k_0 \\ e^{-j\varphi} \; k = N - k_0 \\ 0 \quad \text{anywhere else} \end{cases}, \qquad k_0 \in \left\{1, \ldots, \frac{N}{2} - 1, \frac{N}{2} + 1, \ldots, N-1\right\}$$

prove that the inverse transform has the form

$$x[n] = \frac{2}{N}\cos\left(\frac{2\pi}{N}k_0 n + \varphi\right), \qquad (n = 0, 1, 2, \ldots, N-1)$$

50. Prove that the DFT of the discrete cosine sequence $x[n] = \cos(\omega n + \varphi)$ $(0 \le n \le N-1)$ is given by the relation

$$\mathcal{X}[k] = \frac{e^{j\varphi}}{2} \times \frac{1-e^{j\omega N}}{1-\exp\left[-\left(\omega-\dfrac{2\pi k}{N}\right)\right]} + \frac{e^{-j\varphi}}{2} \times \frac{1-e^{-j\omega N}}{1-\exp\left[-\left(\omega+\dfrac{2\pi k}{N}\right)\right]}$$

What is the form of the function $\mathcal{X}[k]$ for frequencies of the form $\omega = 2\pi m/N$?

51. Let us consider two discrete-time signals, $x[n]$ and $y[n]$, whose Fourier transforms $\mathcal{X}(e^{j\omega})$ and $\mathcal{Y}(e^{j\omega})$ are related through the expression

$$\frac{1}{2\pi}\int_{-\pi}^{\pi} \mathcal{X}(e^{j\vartheta})\mathcal{Y}(e^{j(\omega-\vartheta)})d\vartheta = 1 + e^{j\omega}$$

If the signal $x[n]$ has the form $x[n] = \delta[n]$, determine the signal $y[n]$. What is your answer when the signal $x[n]$ is $x[n] = (1/2)^n u[n]$?

52. Find the impulse response and difference equation of a composite system resulting from the series connection of two discrete LTI systems with impulse responses

$$\mathcal{H}_1(e^{j\omega}) = \frac{3-e^{-j2\omega}}{1+(1/5)e^{-j\omega}} \quad \text{and} \quad \mathcal{H}_2(e^{j\omega}) = \frac{1}{1-(1/3)e^{-j\omega}+(3/4)e^{-j2\omega}}$$

53. Considering a discrete LTI system with input $x[n]$ and output $y[n]$ whose Fourier transforms are related to each other through the expression

$$\mathcal{Y}(e^{j\omega}) = 2\mathcal{X}(e^{j\omega}) + e^{-j\omega}\mathcal{X}(e^{j\omega}) - \frac{d}{d\omega}\mathcal{X}(e^{j\omega})$$

determine the output $y[n]$ when the input signal $x[n]$ is the impulse function $\delta[n]$. Is the considered system a discrete-time LTI system?

54. Considering the DFT with a length of N samples of the form

$$\mathcal{X}[k] = \begin{cases} +1 & k = 0 \\ -1 & k = N/2 \\ 0 & \text{elsewhere} \end{cases} \quad \text{(for even value of } N\text{)}$$

prove that its inverse transform is the discrete sequence $x[n] = [1-(-1)^n]/N$.

55. Calculate the matrices $\mathbf{W}_5$, $\mathbf{W}_6$, $\mathbf{W}_7$, and $\mathbf{W}_8$, as well as their inverses that are used in processes of calculating the direct and inverse DFT with five, six, seven, and eight elements, respectively.

56. Using the properties of the DFT, determine the discrete N-point transforms of the discrete sequences

$$\left.\begin{aligned} y[n] &= x[n]\cos\left(\frac{2k\pi n}{N}\right) \\ z[n] &= x[n]\sin\left(\frac{2k\pi n}{N}\right) \end{aligned}\right\} \quad (0 \le n \le N-1)$$

in terms of the DFT $\mathcal{X}[k]$ of the sequence $x[n]$.

57. Determine the DFT of the eight elements of the discrete sequence $x[n] = \{\mathbf{1}, 1, 1, 1, 1, 1, 0, 0\}$, and draw its magnitude and phase.

58. Determine the DTFT of the discrete sequence $x[n] = \{1, 2, \mathbf{3}, 2, 1, 0\}$ and the DFT of the six-element sequence $y[n] = \{\mathbf{3}, 2, 1, 0, 1, 2\}$, and determine whether they are related to each other, and if so, how.

59. Describe the indexing of the input sequence in a radix-3 FFT algorithm that uses the decimation in time technique.

60. If a radix-3 FFT algorithm uses the decimation in frequency technique, what is the number of complex multiplications required for its implementation?

61. Determine the way in which it is possible to calculate the DFT of a real sequence with a length of $2N$ points from a FFT algorithm with a length of N points.
62. Calculate the circular convolution of the discrete sequences

$$x[n] = \{\mathbf{1}, 2, 1, 1, 2, 1, 1, 2\} \qquad \text{and} \qquad y[n] = \{\mathbf{0}, 1, 3, 2, 0, 0, 0, 0\}$$

63. Considering the sequences $x[n] = \{0, \mathbf{1}, -2, -1, 3, 0, 0\}$ and $y[n] = \{0, \mathbf{0}, 2, 0, 0, -1, 1, 0, 0\}$, determine the smallest possible value of N for which the linear convolution of the two sequences coincides with their circular convolution, or, in mathematical notation, $x[n]\,\textcircled{\scriptsize N}\,y[n] = x[n] * y[n]$. What discrete-time signal results from the calculation of this convolution?
64. Considering a discrete sequence $x[n]$ of even length N to calculate the DFT of the sequences $x[n] + x[n - (N/2)]$, $x[n] - x[n - (N/2)]$, and $(-1)^n x[n]$ as a function of the discrete transform $\mathcal{X}[k]$ of the discrete-time signal $x[n]$.
65. Considering three discrete-time signals $x[n]$, $y[n]$, and $z[n]$ with even length N that are connected through a circular convolution equation of the form $z[n] = x[n]\,\textcircled{\scriptsize 8}\,y[n]$ to prove that

$$\sum_{n=0}^{N-1}(-1)^n z[n] = \left\{\sum_{m=0}^{N-1}(-1)^m x[m]\right\}\left\{\sum_{\ell=0}^{N-1}(-1)^\ell x[\ell]\right\}$$

66. Let there be two discrete sequences $x[n]$ and $\mathcal{X}[k]$ with an even length of N elements connected by a direct/inverse Fourier transform relation. Prove that if the sequence $x[n]$ is symmetric, that is, it satisfies the condition $x[n] = x[N-1-n]$, then $\mathcal{X}[N/2] = 0$, while if the sequence $x[n]$ is antisymmetric, that is, it satisfies the condition $x[n] = -x[N-1-n]$, then $\mathcal{X}[0] = 0$.
67. Determine the circular convolution of the same-sign orthogonal discrete pulses with a length of N elements defined as

$$x[n] = y[n] = \begin{cases} 1\; 0 \leq n \leq N-1 \\ 0 \text{ anywhere else} \end{cases}$$

68. Construct a butterfly diagram of the process calculating the DFT with a length of $N = 16$ elements for the case of decimation in time and for a base value of $r = 4$ in which both the signal samples $x[n]$ and the samples of the discrete sequence $\mathcal{X}[k]$ appear in the usual ascending order.
69. Construct a butterfly diagram of the hybrid algorithm described in Footnote 31 (page 875).
70. Suppose we want to determine the DFT of a sequence of 32 samples using the method of decimation in time and for a base value of $r = 2$. What is the order of appearance of the samples according to the bit reverse sorting method?

MIX
Papier aus verantwortungsvollen Quellen
Paper from responsible sources
FSC® C105338

If you have any concerns about our products,
you can contact us on
ProductSafety@springernature.com

In case Publisher is established outside the EU,
the EU authorized representative is:
Springer Nature Customer Service Center GmbH
Europaplatz 3, 69115 Heidelberg, Germany

Printed by Libri Plureos GmbH
in Hamburg, Germany

Continuous and Discrete-Time Signals and Systems

Athanasios I. Margaris

Continuous and Discrete-Time Signals and Systems

Volume 2

Springer

Athanasios I. Margaris
Trikala, Thessaly, Greece

ISBN 978-3-031-81208-8 ISBN 978-3-031-81209-5 (eBook)
https://doi.org/10.1007/978-3-031-81209-5

This Springer imprint is published by the registered company Springer Nature Switzerland AG
The registered company address is: Gewerbestrasse 11, 6330 Cham, Switzerland

In loving memory of my parents, Ioannis and Agoris

Preface

Digital signal processing (DSP) is a foundational discipline in modern engineering, powering advancements in telecommunications, audio and video processing, biomedical engineering, and numerous other fields. Despite the abundance of DSP literature, there remains a need for a comprehensive, well-structured resource that seamlessly blends theoretical rigor with practical application, accessible to students, educators, and professionals alike. This two-volume work aims to address this need, offering a systematic, authoritative exploration of DSP that serves as both a pedagogical textbook and a technical reference. By combining mathematical precision with real-world examples, it equips readers with the tools to tackle complex signal processing challenges while fostering a deep understanding of the field.

The book is meticulously organized to provide a logical progression from fundamental concepts to advanced techniques, catering to readers at various levels of expertise. Volume 1 establishes the core principles of continuous- and discrete-time signals and systems, while Volume 2 focuses on the design and implementation of analog and digital filters, concluding with an introduction to statistical signal processing. Each chapter is designed to be self-contained yet interconnected, supporting both linear study and selective reference. Extensive problem sets at the end of each chapter reinforce theoretical concepts through practical exercises, making the text ideal for classroom use, self-directed learning, and professional development.

Structure of Volume 1

Volume 1 lays the groundwork for DSP by introducing the essential principles of signal and system analysis. **Continuous-Time Signals** (Chap. 1) covers foundational concepts such as signal energy and power, periodicity, even and odd signals, and elementary functions like exponentials, sinusoids, and generalized functions (e.g., unit step, ramp, and impulse). It introduces key operations including convolution, correlation, and vector space representations, providing a robust basis for signal manipulation. **Continuous-Time Systems** (Chap. 2) builds on this foundation, exploring system classifications (e.g., linear vs. nonlinear, causal vs. noncausal, time-invariant vs. time-varying), stability, and responses (impulse, step, and frequency). It also addresses system interconnections, schematic representations, and state-space descriptions, emphasizing practical design considerations.

Fourier Analysis in the Continuous-Time Domain (Chap. 3) introduces frequency-domain analysis through Fourier series and transforms. It covers the exponential and trigonometric forms of Fourier series, convergence conditions, the Gibbs phenomenon, and their application to linear time-invariant (LTI) systems. The continuous-time Fourier transform is explored in depth, including its properties, convergence, and applications to periodic signals and frequency selection filters. **The Laplace Transform** (Chap. 4) provides a powerful tool for system analysis, detailing regions of convergence, poles and zeros, properties, and inverse transforms. It emphasizes solving differential equations for first- and second-order systems, bridging time- and frequency-domain techniques.

System Analysis in the Frequency Domain (Chap. 5) integrates these tools to analyze transfer functions, stability (e.g., Routh-Hurwitz criterion), and frequency responses. It includes practical methods like Bode diagrams, signal-flow graphs, and state-space solutions, equipping readers with design-oriented skills. The focus then shifts to discrete-time systems with **Discrete-Time Signals** (Chap. 6), which parallels Chap. 1 by addressing sampling, energy, periodicity, and operations like convolution and correlation. **Discrete-Time Systems** (Chap. 7) explores system classifications, LTI system responses, and state-space representations, with a focus on stability and recursive systems.

The Z Transform (Chap. 8) introduces a key tool for discrete-time system analysis, covering poles, zeros, convergence, and inverse transforms. **Systems in the Z Domain: Digital Processing Techniques** (Chap. 9) delves into digital signal processing, including transfer functions, difference equations, and multirate processing techniques like downsampling and oversampling. Volume 1 concludes with **Fourier Analysis in the Discrete-Time Domain** (Chap. 10), addressing discrete-time Fourier transforms (DTFT), discrete Fourier transforms (DFT), and their applications in frequency analysis and short-time signal processing.

Structure of Volume 2

Volume 2 builds on Volume 1's foundations, focusing on practical design and advanced topics. **Design of Analog Filters** (Chap. 11) provides a thorough exploration of filter types (e.g., low-pass, high-pass, band-pass), characteristics, and design methodologies, including Butterworth, Chebyshev, elliptic, and Bessel-Thomson approximations. It covers frequency transformations, circuit theory, and the Paley-Wiener theorem, ensuring a practical perspective on filter implementation. **Design of Digital Filters** (Chap. 12) contrasts analog and digital approaches, addressing finite and infinite impulse response (FIR and IIR) filters, quantization effects, and advanced design techniques such as the Parks-McClellan algorithm, Kaiser window method, and bilinear transformation. It emphasizes implementation through signal-flow graphs, hardware considerations, and error analysis (e.g., quantization and overflow).

Elements of Statistical Signal Processing (Chap. 13) introduces stochastic processes, power spectral density, and system responses to random inputs. It covers key concepts like the Wiener-Khinchin theorem, white noise, and quantization error analysis, providing a foundation for advanced signal processing applications. The book concludes with Appendix A, a comprehensive reference consolidating fundamental mathematical concepts, including complex numbers, difference equations, Chebyshev polynomials, elliptic functions, and probability theory. This appendix supports readers by providing quick access to essential mathematical tools used throughout the text.

Objectives and Audience

This book is designed for a diverse audience, including undergraduate and graduate students in electrical and computer engineering, practicing engineers, and researchers seeking a definitive DSP reference. Its balanced approach—combining theoretical depth with practical examples—makes it suitable for introductory courses, advanced study, and professional applications. The inclusion of contemporary topics, such as multirate processing, statistical signal analysis, and advanced filter design, ensures alignment with current industry and research trends.

By offering a unified treatment of continuous and discrete-time systems, analog and digital filters, and statistical processing within a cohesive pedagogical framework, this work fills a critical gap in DSP literature. Our goal is to inspire readers to explore the dynamic and evolving field of digital signal processing, fostering both intellectual curiosity and technical proficiency. We hope this book will serve as a trusted companion for learning, teaching, and applying DSP in real-world contexts.

Trikala, Thessaly, Greece
April 2025

Athanasios I. Margaris

Contents

Volume 1

11 Design of Analog Filters

Example of analog filter structure

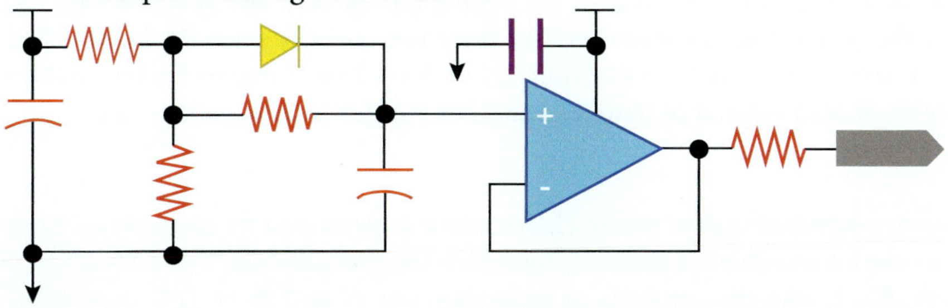

Analog and digital filters are used in many signal processing applications in the domains of continuous and discrete-time

11.1 Introduction to Analog Filters

In the following pages, we will describe the principles of operation of analog filters, which, as we mentioned in the third chapter, are linear time-invariant (LTI) systems that either select or shape frequencies. In this description, emphasis will be given to *frequency selection filters*, which allow the passage of only a band of frequencies of interest, cutting off all the remaining frequency bands. Recall that a LTI system *does not modify the frequency content of a signal* (i.e., it does not add new frequencies or change existing ones) but affects only the relative amplitudes and/or phases of the frequency components of the input signal through *scaling* and *shifting* processes, respectively. Depending on the values of the frequencies that belong to the passband, filters are divided into *low-pass, high-pass, band-pass,* and *band-stop* filters, whose operation is mainly based on knowledge of the cutoff frequency values and the desired degree of attenuation in each case. As discussed previously, in practice, filters are characterized by the existence of a *zone transition* between the pass and cutoff zones, as well as by fluctuations in the width of the frequency response function in these frequency regions. For a concise description of the basic operating characteristics of analog filters, as well as the properties of ideal and real filters, the reader is referred to Sect. 3.6.

11.2 Elements of Circuit Theory

Analog filters are implemented in the form of electrical circuits, and for this reason it is particularly useful and instructive to present some basic information about these structures. From basic theory we know that an electrical circuit is defined as a set of electrical elements that interact with each other to perform a function. In practice, we are primarily interested in the so-called *lumped circuits*, in which the dimensions of the electrical elements that form them are very small compared to the wavelength of the current that flows through them.[1] The simplest circuit elements are characterized by the presence of

[1] If this is not the case, the circuit is aforementioned to be *distributed* and its operation is characterized by the radiation of energy, as happens, for example, in an antenna.

A. I. Margaris, *Continuous and Discrete-Time Signals and Systems*,
https://doi.org/10.1007/978-3-031-81209-5_11

two terminals, and an electric current flows through such an element entering from one terminal and exiting from the other. Assuming that the speed of development and propagation of electrical phenomena is infinite, that is, a current entering from one terminal is at every instant equal to the current coming out of the other, we can describe the operation of an electrical circuit using the *Kirchhoff's laws*, which are limiting cases of the *Maxwell's equations* and are formulated as follows:

- *Law of intensities*: In a concentrated electrical circuit, the algebraic sum of the currents of all branches approaching a node of the circuit or moving away from it is at every instant equal to zero.
- *Law of voltages*: In a concentrated electrical circuit, the sum of the voltages of the branches forming loops is at every instant equal to zero.

Typical examples of electrical circuits and the application of Kirchhoff's laws are presented in Exercises 4.8.6 (page 365), 4.8.7 (page 366), and 4.8.8 (page 368).

The simplest elements of an electrical circuit are characterized by the existence of two terminals, and their operation can be fully described with knowledge of the intensity of the electric current $i(t)$ that flows through them and the voltage $\upsilon(t)$ developed at their ends. The correlation that exists between these two quantities defines the so-called characteristic $\upsilon - i$ of the element and is described by an expression of the form $i(t) = f(\upsilon)$. Depending on the type of the foregoing expression—that is, whether it is algebraic (polynomial or transcendental) or a differential equation—the basic electrical elements of a circuit are described as follows:

Resistor A resistor is a two-terminal system whose characteristic is described by an algebraic expression. If the resistor is LTI, then it simplifies to our known ohmic resistance, for which the characteristic $\upsilon - i$ is $i(t) = G\upsilon(t)$, while its inverse is formulated as $\upsilon(t) = Ri(t)$, with the constants of proportionality G and $R = 1/G$ expressing in the known way the *electrical conductivity* and the *electrical resistance*, respectively. The foregoing expressions constitute the mathematical formulation of the *Ohm's law*. If the preceding quantities vary with time, then $G = G(t)$ and $R = R(t)$, and the expressions $i(t) = G(t)\upsilon(t)$ and $\upsilon(t) = R(t)i(t)$ describe a linear *time-varying* resistor.

On the other hand, typical examples of nonlinear resistors are contact diodes and tunnel diodes, which are characterized by a more complex characteristic curve. For example, the characteristic of a contact diode is described by the equation

$$i(t) = I_s\left[\exp\left(\frac{q\upsilon(t)}{kT}\right) - 1\right]$$

where I_s is the electric current flowing through the circuit when $\upsilon(t) < 0$, q is the electron charge, k is Boltzmann's constant, and T is the absolute temperature.

Capacitor A capacitor is defined as a system of two terminals that allows the storage of electric charges and, consequently, electric energy in the form of an electric field. The relationship between the voltage and charge stored in a capacitor, known as the characteristic $\upsilon - q$ of the capacitor, takes the form $q(t) = f(\upsilon)$, and since it is known that $i(t) = \dot{q}(t) = dq(t)/dt$, it immediately follows that $i(t) = df(\upsilon)/dt$. Therefore, the characteristic $\upsilon - i$ of a capacitor is related not to an algebraic but to a *differential* equation.

The operation of a LTI capacitor is described by the relation $q(t) = C\upsilon(t)$, with the proportionality constant C referring to the capacitance of the capacitor, which is independent of charge $q(t)$, voltage $\upsilon(t)$, and time t. In this case, from the defining equation of the current we have

$$i(t) = \frac{dq(t)}{dt} = C\frac{d\upsilon(t)}{dt}$$

For continuous voltage, $\dot{\upsilon}(t) = 0$, and consequently, $i(t) = 0$. Therefore, a capacitor in a direct current circuit behaves as an *open circuit*.

Integrating the $\upsilon - i$ equation of a LTI capacitor for the time period from t_0 to t, it turns out that

$$v(t) = v(t_0) + \frac{1}{C}\int_{t_0}^{t} i(\tau)d\tau \qquad \text{and} \qquad q(t) = q(t_0) + \int_{t_0}^{t} i(\tau)d\tau$$

Finally, if the capacitor is linear but *time-varying*, $C = C(t)$; in this case, we will have $q(t) = C(t)v(t)$ and

$$i(t) = C(t)\frac{dv(t)}{dt} + v(t)\frac{dC(t)}{dt}$$

Coil The coil is defined as a two-terminal system that stores energy in the form of a magnetic field, and its operation is described by the characteristic $\phi - i$, with equation $\phi(t) = f(i)$, where $\phi = \phi(t)$ is our known magnetic field flow or magnetic flux. According to *Faraday's law of induction*, the voltage at the ends of the coil is a function of the time variation of the magnetic flux, that is,

$$v(t) = \frac{d\phi(t)}{dt} = \frac{df(i)}{dt}$$

For a LTI coil, we will have $\phi(t) = Li(t)$, where L is the *self-induction of the coil*, which is independent of $v(t)$ and $i(t)$; hence, we will have

$$v(t) = L\frac{di(t)}{dt}$$

For a time-invariant (i.e., continuous) electric current, we have $\dot{i}(t) = 0$ and, consequently, $v(t) = 0$. Therefore, in a continuous current circuit, the coil behaves as a short circuit. Integrating the foregoing relationship, we get

$$i(t) = i(t_0) + \frac{1}{L}\int_{t_0}^{t} v(\tau)d\tau$$

Voltage and Current Sources Voltage and current sources are elements of two terminals that cause the excitation of electrical circuits and in this sense constitute their inputs. Starting from the voltage source, it is aforementioned to be *independent* if the voltage developed at its ends is independent of the current flowing through it, while by analogy the independent source of current is characterized by a current that is independent of the voltage. If this independence does not exist, these elements are aforementioned to be *dependent*. Independent voltage and current sources are *nonlinear elements* since their characteristic does not pass through the origin of the axes. It is not difficult to find that a zero-voltage source is equivalent to a *short circuit*, while a zero-current source is equivalent to an open circuit. These two elements cannot exist in reality, and for this reason they are aforementioned to be ideal elements.

The circuit elements of the five aforementioned circuit elements are presented in Fig. 11.1. In this figure, the resistor, capacitor, and coil are treated as LTI elements.

Energy Distinction of Electrical Elements of Two Terminals The electrical elements of two terminals presented in the previous paragraphs are characterized by instantaneous power $p(t) = v(t)i(t)$, which is different from zero; this fact allows us to characterize them as *active elements*, so as to distinguish them from *nonactive elements*—such as short circuits, open circuits, switches, and ideal transformers—which are characterized by zero instantaneous power and neither absorb nor produce energy. Active elements are further divided into two categories: *passive elements*, whose energy has a positive value, and *active elements*, whose energy can have both positive and negative values. Typical examples of active elements are voltage and current sources. Passive elements are also divided into those that consume energy, such as *ohmic resistance*, and

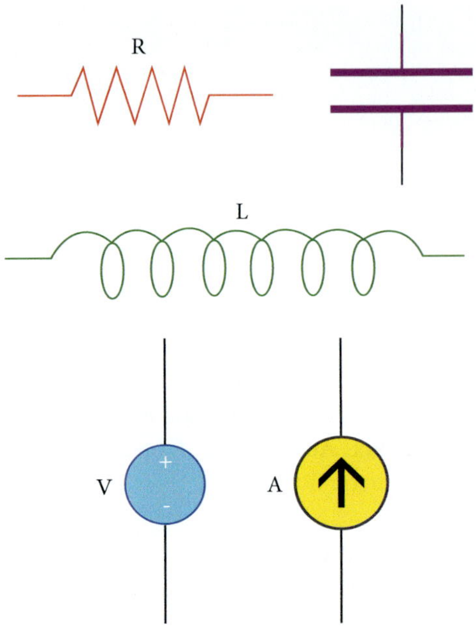

Fig. 11.1 Circuit symbols of the two basic terminal elements

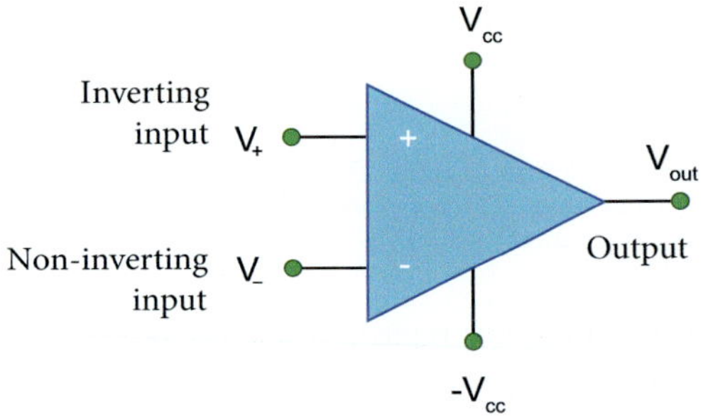

Fig. 11.2 Circuit symbol of operational amplifier

those that store energy, such as *capacitors* and *coils*. As is well known from the theory of electrical circuits, the instantaneous power of a LTI resistor, capacitor, and coil, respectively, is defined as

$$p_R(t) = i^2(t)R, \qquad p_C(t) = Cv(t)\frac{dv(t)}{dt}, \qquad p_L(t) = Li(t)\frac{di(t)}{dt}$$

Operational Amplifier An operational amplifier, whose circuit symbol is presented in Fig. 11.2, is characterized by two inputs and one output, and in this sense it constitutes an element of three terminals. In this diagram, the input denoted by $(-)$ is called an *inverting input*, as it inverts the polarity of the signal appearing as the output, while, in contrast, input denoted by $(+)$ maintains the polarity of the signal that is applied to it. The characterization of this amplifying arrangement as *operational* is due to its ability to perform mathematical operations, and for this reason one of its first applications was the analog computer.

The activation of the amplifier is carried out using continuous voltages V_{cc} and $-V_{cc}$, the typical value of which is ± 15 V. Considering the special case of the ideal operational amplifier, the composite input resistance $Z_i(\omega)$ is defined as the resistance between its inputs has magnitude with *infinite value*, while in contrast, the magnitude of the composite output resistance $Z_o(\omega)$ is equal to zero. In reality, however, the magnitudes of these resistances are 10^6–10^{13} Ω and 50–200 Ω, respectively. Although the ideal operational amplifier cannot exist in reality, it is approximated with satisfactory accuracy by the real amplifier when it operates in the region of low frequencies, that is, at frequencies of less than 10 KHz. The operational amplifier is an active electrical element.

The basic function of an operational amplifier is *to amplify any difference that might arise between its two inputs*. The ratio of input and output voltages is related to the so-called *gain factor* μ, which depends on the angular frequency of the input signal. For the ideal amplifier, the voltage gain has an infinite value, while for the real amplifier and for low frequencies less than 10 Hz the voltage gain ranges from $2 \cdot 10^5$–10^6. Although an operational amplifier is by design an *unstable circuit element*, in practice the output voltage cannot become infinite, since for input voltages beyond a certain specified value the operational amplifier enters a *saturation state* and the output voltage becomes equal to the saturation voltage V_{SAT}, which is always slightly less than the continuous supply voltage. The stability of an operational amplifier is ensured by the application of a *feedback* process (or *refeeding*), which is characterized as being positive or negative, depending on whether the sample returned from the output is added to or subtracted from the current input.

11.3 Types of Analog Filters

As was mentioned earlier, analog filters are simply electrical circuits that consist of both *passive* elements (*resistors, capacitors, and coils*) whose operation does not require an energy source, and *active* elements (such as an *operational amplifier* or a *transistor*), whose use requires some power source. The use or nonuse of the aforementioned types of elements allows us to characterize analog filters as *passive* or *active* filters, respectively, while special categories of the aforementioned filters are *integrated analog active filters* that are manufactured in the form of integrated circuits, *switched capacitor filters* that simulate high resistance values, and *microwave filters*.

Passive filters, whose operation is based on the phenomenon of *electrical resonance* that characterizes the series or parallel connection of a capacitor with a coil, are used in cases where continuous current passage is required with an intensity greater than 1 ME or the presence of a noise level significantly lower than the signal level. Although the filters of this category are characterized by low power consumption (without being able to offer gain, since they do not have a power source), they are quite bulky (especially when they have been designed to allow the passage of large currents), since the use of conductors with large cross-sectional area is required, and the coils themselves are characterized by a relatively large physical size. Commercial coils are characterized by their high cost and unsuitability as resistors and capacitors, especially in low-frequency applications, since their inductances are specific, making it practically difficult to find a coil with an inductance value deviating by an order of 10% from the desired one; however, their use is essential for the manifestation of the phenomenon of *electric resonance*. Passive filters work just as well in high-frequency areas, as they are free from the bandwidth limitations of operational amplifiers. However, they have the disadvantage that the complex input resistances may be smaller than desired, while the complex output resistances are larger than optimal. Passive filters are at a disadvantage in that if high accuracy or small size is required, then their manufacturing cost is prohibitive.

On the other hand, *active analog filters* rely on the use of operational amplifiers combined with resistors and capacitors to form a filter with the frequency response desired in each particular case. The main advantage of these filters is their small volume, since they do not use coils, and they can be further reduced if they are constructed in the form of *integrated circuits*. Filters of this type are suitable for applications that allow rapid change in the signal width as a result of their frequency change, although they should be able to handle large amplitude values for frequencies that are quite far from the passband. Active filters are easier to design than passive filters, while they are also characterized by high complex input impedance and low output impedance, and they can provide practically any gain value. The main disadvantage of these types of filters is *signal distortion due to noise* caused by the presence of operational amplifiers, the limitations imposed on the signal width due to the amplifier's response rate at its output, and the limitations on the filter's performance resulting from the use of operational amplifiers in terms of width and bandwidth.

A special category of active analog filters are those that are manufactured in the form of integrated circuits and that for this reason are known as *integrated active filters*. Filters of this category, which are clearly smaller than the common active filters and much smaller than passive analog filters, use *high-frequency amplification systems* and *integrated circuit elements*, thereby allowing frequencies up to 15 GHz, in contrast to the more common active filters that initially supported frequencies of up to 100 KHz, although later with the use of operational amplifiers with a wide bandwidth, they managed to support frequencies on the order of megahertz.

The operation of integrated active filters is based on the use of a multitude of operational amplifiers included in integrated circuits (which often incorporate resistors and capacitors), while the desired response of the filters is achieved by adding such passive elements outside the limits of the circuit.

The second type of active analog filters are the *switched capacitor filters* whose use is based on the rapid charging and discharging of capacitors (which in this way act as switches) for the simulation of high resistance values, which are generally difficult to implement in the form of an integrated circuit. The operation of these filters is guided by a clock with the input signal undergoing sampling and processing with *discrete* and not continuous-time techniques. The active value of the resistance simulated in this way depends on the switching speed of the charging and discharging cycles. These types of filters allow for the creation of signals with frequencies equal to the switching frequency (which, however, are characterized

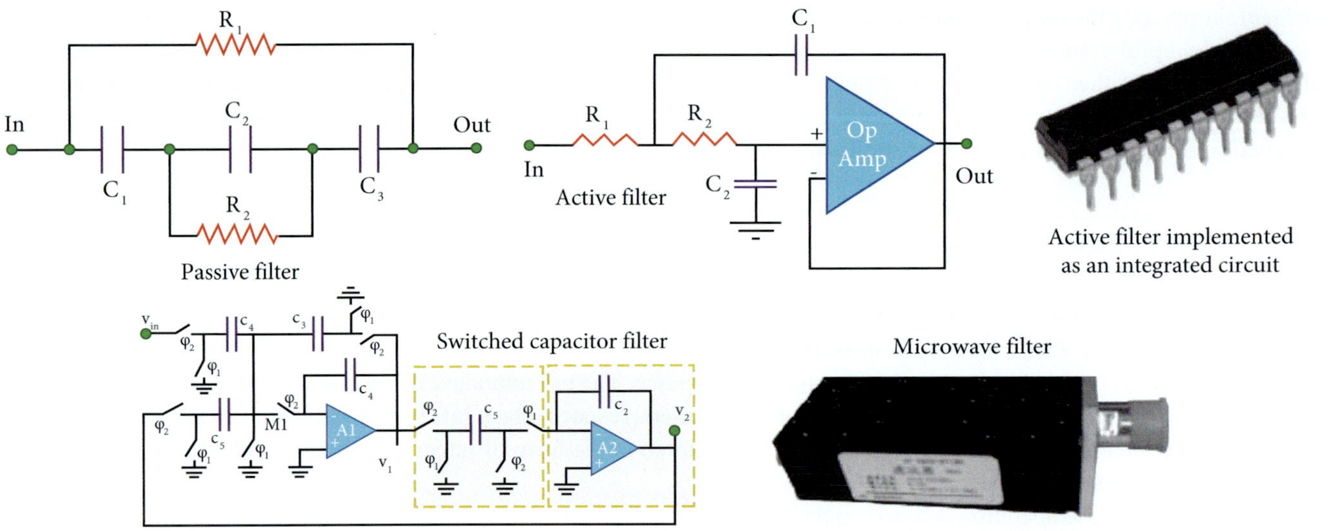

Fig. 11.3 Structures and arrangements of analog filters

by *high noise content*), while their cutoff frequency depends on the operating frequency of the clock and can be adjusted through it. Most filters of this type are low-pass filters, and the frequencies they support are below 100 kHz.

Finally, *microwave filters* are constructed by cascading components of microwave nature. Examples of such filters include *transverse electromagnetic wave transmission lines, waveguides, dielectric resonators, and surface acoustic wave devices*, which are used in applications with frequencies ranging from 0.5 to 500 GHz.

Examples of structures and arrangements for the aforementioned types of filters are presented in Fig. 11.3.

11.4 Applications of Analog Filters

Analog electronic filters are used in almost every type of electronic equipment, such as radios and televisions, stereo systems, signal generators, spectral analyzers, and devices for converting analog signals to digital to prevent the occurrence of spurious frequencies. Analog filters are also used in telephone and telecommunication systems for the implementation of important processes, such as *frequency division multiplexing* and *noise removal*. A more detailed description of the most classic applications of analog filters follows:

Radio and Television The selection of a radio station or a television channel is based on the tuning of the receiver of the radio or television to the desired frequency in each case, something that is done using a band-pass filter that rejects all other frequencies and allows only the passage of the frequency of interest. Band-pass filters are also used to remove unwanted noise due to sources such as *lightning*, *electric motors*, *power lines*, and *transformers*, and their frequency can generally be extended over a wide range of frequencies.

Telephony The basic application of analog filters in a telephone system is to implement the mechanism of *frequency division multiplexing*, which allows for the transmission of content of thousands of telephone conversations through the same communication channel. This type of multiplexing is based on the use of a *unique carrier signal* whose frequency is added to the frequency of each voice signal, a process that results in the shift of the frequency spectrum of each voice signal during the carrier frequency. Then all these different frequencies undergo modulation in the *transmitter* and are combined among them using a *signal adder*, forming a composite signal that is transmitted through the communication channel and is received by a *receiver*, which in turn extracts individual voice signals through demodulation and distributes them to interlocutors via telephone lines. The implementation of this process requires the use of band-pass filters whose number is equal to the number of voice signals being transmitted. The technique of frequency division multiplexing is presented diagrammatically in Fig. 11.4.

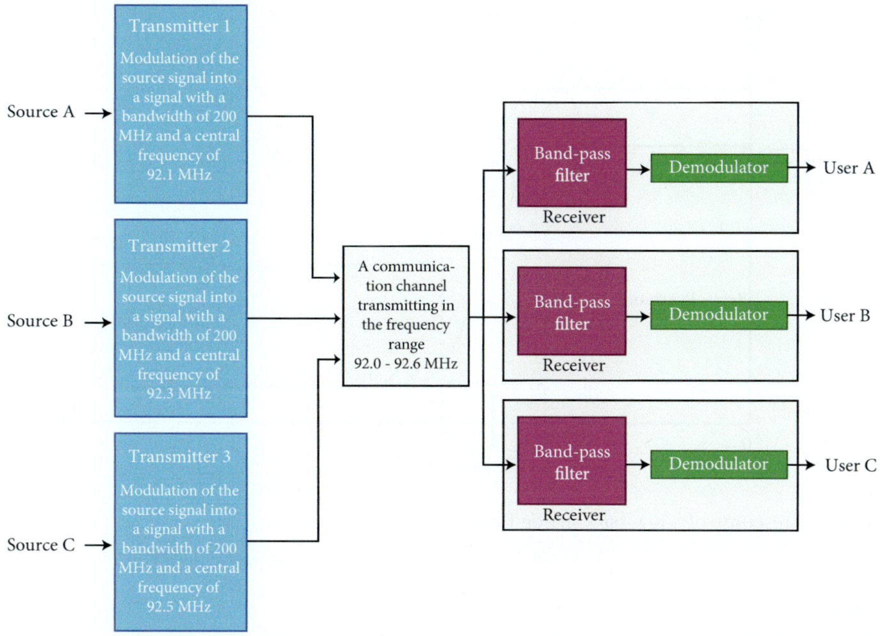

Fig. 11.4 The technique of frequency division multiplexing

Sampling As mentioned in the previous chapter, the sampling processes of analog signals should be carried out at a frequency that is at least twice the maximum frequency included in the signal being sampled, so as to avoid the occurrence of unwanted folded (or aliased) frequencies. Therefore, if the maximum frequency of a signal is greater than half the sampling frequency, we should appropriately limit the bandwidth of the signal, which can be done using a low-pass filter. Low-pass filters are also used, as we have seen, for the reconstruction of analog signals from their discrete equivalents (Eq. 6.16). *Sound reproduction systems:* Sound reproduction devices use equalizers that function as filters, restoring the spectrum of audio signals that are reproduced and which is distorted due to the mechanical resonances that are due to the enclosure or cabinets of speaker systems and which are superimposed on the audio signal.

11.5 Characteristics of Analog Filters

To describe the characteristics and operating parameters of analog filters, let us design once again the frequency response of a real low-pass filter, an equivalent version of which is given in Fig. 3.19 (page 288). In Fig. 11.5, the frequency response is aforementioned to be *normalized* because the maximum value of the measure of the frequency response in the passband is equal to the unit (that is, equal to 0 dB). Conversely, the minimum value of the ripple in the passband is equal to $1/\sqrt{1+\varepsilon^2}$, with the constant ε expressing the *ripple parameter* in this band. As for the stopband, the maximum value of the ripple in it is equal to $1/A$, where A is the so-called *stopband attenuation parameter*. The frequencies ω_p and ω_s express the frequencies at the limits of the zones of passage and cutoff, respectively, with the ratio $k = \omega_p/\omega_s$ of these frequencies known as the *transition ratio* or *selectivity parameter*. It is obvious that for a low-pass filter it will hold that $k < 1$. From the foregoing description, it is not difficult to understand that the range of values of the measure of the frequency response of the filter will be $1/\sqrt{1+\varepsilon^2} \leq |\mathcal{H}(j\omega)| \leq 1$ for the passband and $0 \leq |\mathcal{H}(j\omega)| \leq 1/A$ for the cutoff zone.

The foregoing description relates to the parameters ε and A, while if we use the nonnormalized graphical representation of Fig. 3.19 from Sect. 3.6.2 (page 287), we will have respectively $1-\delta_p \leq |\mathcal{H}(j\omega)| \leq 1+\delta_p$ for the passband and $0 \leq |\mathcal{H}(j\omega)| \leq \delta_s$ for the cutoff zone, where δ_p and δ_s are the *maximum ripple values* for the passband and cutoff zones. These two sets of parameters are related to each other as

$$\frac{1-\delta_p}{1+\delta_p} = \sqrt{\frac{1}{1+\varepsilon^2}} \quad \text{and} \quad \frac{\delta_s}{1+\delta_p} = \frac{1}{A}$$

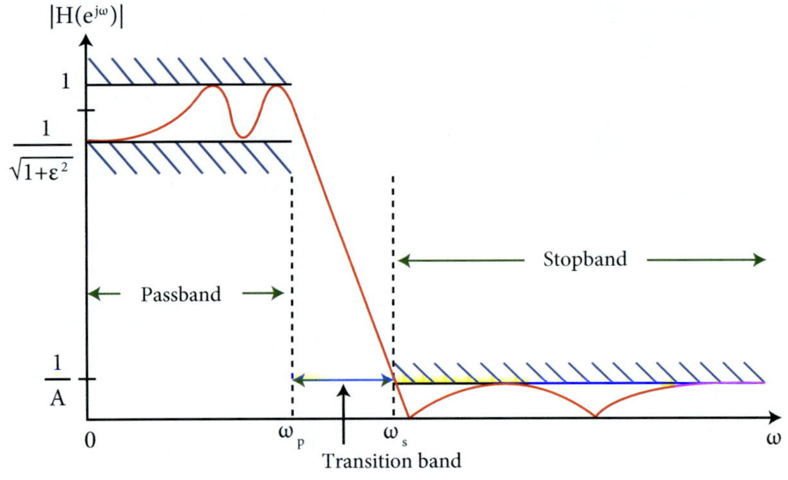

Fig. 11.5 Normalized frequency response of a low-pass analog filter

from which it follows that

$$\varepsilon = \frac{2\sqrt{\delta_p}}{1-\delta_p} \quad \text{and} \quad A = \frac{1+\delta_p}{\delta_s}$$

In practice, the characteristics of a filter are not given in the preceding form but are expressed in units of decibels and are defined as

$$A_p = -20\log\frac{1-\delta_p}{1+\delta_p} \quad \text{and} \quad A_s = -20\log\frac{\delta_s}{1+\delta_p} \tag{11.1}$$

Replacing the expressions of the deviations δ_p and δ_s as a function of ε and A, we get

$$A_p = -20\log\sqrt{\frac{1}{1+\varepsilon^2}} = -10\log\frac{1}{1+\varepsilon^2} \quad \text{and} \quad A_s = -20\log\frac{1}{A} \tag{11.2}$$

from which it follows that

$$\varepsilon = \sqrt{10^{R_p/10}-1} \quad \text{and} \quad A_s = 10^{A_s/20} \tag{11.3}$$

Another parameter used to describe the operation of an analog low-pass filter is the *selectivity parameter*, which is defined as $\alpha = \varepsilon/\sqrt{A^2-1}$ and takes values $\alpha \ll 1$.

Analog filters, being continuous-time LTI systems, are described with the help of all the concepts, methodologies, and tools we presented in detail in the first five chapters of the book. Thus, the operation of a filter is described by the general differential equation (i.e., Eq. 5.3), in which the coefficients α_i $(i = 1, 2, \ldots, N)$ and β_j $(j = 1, 2, \ldots, M)$ must necessarily be real numbers, as they depend on the values of the elements of the circuit that implements the filter (resistances, capacitances, and self-inductances), which, of course, are real quantities. On the other hand, the transfer function of the filter is known as the *ratio of the Laplace transforms of the input and output signals* (Eq. 5.6) and, expressed in the form of Eq. (5.10), allows for the calculation of the poles and zero values p and z. In practice, we are only interested in feasible transfer functions that describe causal and stable LTI systems; these functions, as we explained in Chap. 5, must meet the following requirements:

- The transfer function $\mathcal{H}(s)$ must be a rational function of the complex variable s with real coefficients.
- The poles of the transfer function $\mathcal{H}(s)$ must lie in the left half-plane of the complex plane.
- The degree of the numerator polynomial of the rational transfer function must be less than or equal to the degree of the denominator polynomial of this function.

For a detailed description of the relevant theory, see Sects. 5.2 (page 376), 5.3 (page 385), and 5.4 (page 390), as well as the discussion of Sect. 5.11.3 (page 477).

The response of an analog filter in the time domain is calculated either directly from the convolution equation—Eq. (5.1)—or indirectly through the calculation of the inverse Laplace transform of a system's output defined as $y(t) = \mathcal{L}^{-1}[\mathcal{H}(s)\mathcal{X}(s)]$, where $\mathcal{H}(s)$ and $\mathcal{X}(s)$ are the transfer function and the Laplace transform of the input signal, respectively (for a demonstration of the indirect calculation of the output of a LTI system, see Exercise 5.2.5 on page 382, while for the techniques of calculating the inverse Laplace transform, see Sect. 4.7 on page 334). As for the frequency response function of an analog filter, this can be calculated either algebraically, if in the transfer function $\mathcal{H}(s)$ we make the substitution $s = j\omega$, or geometrically from the positions of its poles and zero values, in the way we described in Sect. 5.6.1 and on page 422 (it is obvious that this process applies only when the imaginary axis of the complex plane belongs to the convergence area of the transfer function). The effect of the poles and zero values on the frequency response of a filter is described using Bode diagrams (Sect. 5.7, page 426). The construction of the transfer function of an analog filter is based on the use of electrical circuit analysis techniques, such as *Ohm's law* and the *voltage and current laws of Kirchhoff*, while the mathematical expression that describes it depends on the form of the circuit that implements the analog filter (see, for example, Exercise 5.2.7, page 384).

As we proved in detail in Sect. 5.5.2 (page 413), the response of a filter to the elementary trigonometric signal $x(t) = A\cos(\omega t + \varphi)$ has the form

$$y(t) = A|\mathcal{H}(j\omega)|\cos(\omega t + \varphi + \measuredangle\mathcal{H}(j\omega)) + \sum_{i=1}^{N} C_1 e^{jp_i t} \tag{11.4}$$

and consists of a transient term that weakens over time and vanishes in the limit $t \to \infty$ and from a periodic trigonometric term that constitutes the steady-state response and persists forever. This result allows us to characterize the magnitude of the frequency response function calculated at the frequency value desired in each particular case as *filter gain*. Since the value of this measure is always less than or equal to one, the value of the gain in decibels will always be less than or equal to zero. For this reason, the analog filters are often described in terms of the inverse gain expressed in decibels has a value greater than or equal to zero. This parameter is known as *filter attenuation* and expressed in decibels as

$$\mathcal{A}(\omega) = 20\log\frac{1}{|\mathcal{H}(j\omega)|} = 10\log\frac{1}{|\mathcal{H}(j\omega)|^2} = 10\log\frac{1}{\mathcal{H}(j\omega)\mathcal{H}(-j\omega)} = 10\log L(\omega^2) \tag{11.5}$$

where

$$L(\omega^2) = \frac{1}{\mathcal{H}(j\omega)\mathcal{H}(-j\omega)} \tag{11.6}$$

If we set $\omega = s/j$ in the foregoing equation, it immediately follows that

$$L(-s^2) = \frac{1}{\mathcal{H}(s)\mathcal{H}(-s)} = \frac{D(s)D(-s)}{N(s)N(-s)} = \frac{\prod_{i=1}^{N}(s-p_i)\prod_{i=1}^{N}(-s-p_i)}{\prod_{i=1}^{N}(s-z_i)\prod_{i=1}^{N}(-s-z_i)} = (-1)^{N-M} \times \left\{\frac{\prod_{i=1}^{N}(s-p_i)\prod_{i=1}^{N}[s-(-p_i)]}{\prod_{i=1}^{N}(s-z_i)\prod_{i=1}^{N}[s-(-z_i)]}\right\}$$

where $N(s)$ and $D(s)$ are the polynomials of the numerator and the denominator of the transfer function $\mathcal{H}(s)$, expressed in their factorized form. The function $L(-s^2)$ is known as the *loss function* of the filter and, as follows from the preceding equation, *its poles are the zeros of the transfer function and their negatives, while its zeros are the poles of the transfer function and their negatives*. The graphical representation of the attenuation $\mathcal{A}(\omega)$ as a function of the frequency ω is often referred to as the *loss characteristic* of the filter, and typical forms of it for the four categories of filters used in practice will be presented in a subsequent section.

The number of circuit elements with the ability to store energy (i.e., coils and capacitors) included in the circuit that implements an analog filter is known as the *order of the filter*. Considering that higher-order filters arise from a series connection of many simple first- and second-order filters (this process will be examined later) and how the transfer function for the series connection arises from the product of the transfer functions of the simple systems that are connected (Sect. 5.8, page 444), it is not difficult to see that in most cases, the value of the order of the filter is equal to the highest power of s that appears in the polynomial of the denominator of the transfer function. It is obvious that the order of a filter directly affects its size, cost, and design process: the higher the order of a filter, the larger its size (especially if it contains coils that occupy a large volume), the higher its manufacturing cost (especially if the filter is to be manufactured in large quantities), and the more complex its design (since it deals with large and complex characteristic polynomials and a large number of poles). However, higher-order filters are more effective than lower-order filters in distinguishing between signals characterized by different frequencies, while leading to a steeper attenuation in the transition zone, i.e., a smaller range for this zone, which is particularly useful when we want to reject frequencies that are very close to the cutoff zone.

The characterization of a filter as *low-pass, high-pass, band-pass,* or *band-stop* is dictated by the way the filter's frequency response changes as a function of frequency ω (Fig. 3.19, page 288). Although, as mentioned earlier, the maximum gain value is equal to one, in active filters this can become *even greater* through the use of *amplifying arrangements*. Depending on the type and characteristics of the filter, the gain change curve in the pass and cutoff zones can be *constant* for some frequency range, change *monotonically*—that is, the gain value increases or decreases continuously with increases in frequency, without the slope changing sign—or present *fluctuations* (in the latter case, the amplitude of the fluctuations or the range of values within which this should be limited is determined during the design of the filter). Note that we are usually only interested in fluctuations in the pass band, since in the cutoff band the signal is supposed to have weakened to the desired degree. It is obvious that the property of *monotonicity* characterizes only low-pass and high-pass filters, not band-pass or band-stop filters, which, however, can be monotonic in each of the frequency ranges located to the left and right of the two cutoff frequencies that describe their operation. Typical examples of these four types of second-order analog filters implemented in the form of passive circuits are shown in Fig. 11.6, which also contains the equation of the transfer function of the filter, as well as the amplitude and phase responses for each of them. In full analogy with the method of determining the cutoff frequency of low-pass and high-pass filters (Sect. 3.6.2, page 287), the two cutoff frequencies ω_{c1} and ω_{c2} for the band-pass filters are defined as the frequencies at whose positions the gain value is equal to 0.7071 of its maximum value, that is, it shows an attenuation of -3 dB. More specifically, the lower cutoff frequency ω_{c1} is the frequency at which the gain has approached 0.7071 of its maximum value as it increases approaching the maximum value, while the higher cutoff frequency ω_{c2} is the frequency at which the gain has approached 0.7071 of its maximum value as it decreases, moving away from the maximum value. This means that these frequencies can be calculated when given the change in the filter's gain as a function of frequency (although in other cases they are determined from the outset according the needs of the particular context), and the filter is designed in such a way as to behave in the desired way. Having defined these two frequencies, we can calculate two more parameters that can be used to describe the operation of a band-pass filter. These parameters are the *central frequency* ω_c, which is defined as *the geometric mean of the two cutoff frequencies*, and its value is equal to $\omega_c = \sqrt{\omega_{c1}\omega_{c2}}$ and the *quality factor* Q, which is calculated from the relation $Q = \omega_c/(\omega_{c2} - \omega_{c1})$ and constitutes a measure of the *sharpness* of the amplitude response curve.

It is interesting to note that although the quality factor Q, as defined earlier, makes sense *only for band-pass filters*, it can be calculated for all types of second-order filters if the denominator of the transfer function is formulated as

$$D(s) = s^2 + \frac{\omega_c}{Q}s + \omega_c^2$$

where ω_c is the central frequency of the filter. In some cases, instead of the factor Q, the parameters $\alpha = 1/Q$ and $\xi = 1/2\alpha$ are defined and used. Starting from the equation of the transfer function for the elementary low-pass, high-pass, band-pass, and band-stop second-order filters (Fig. 11.6), it is proven that this can be formulated as

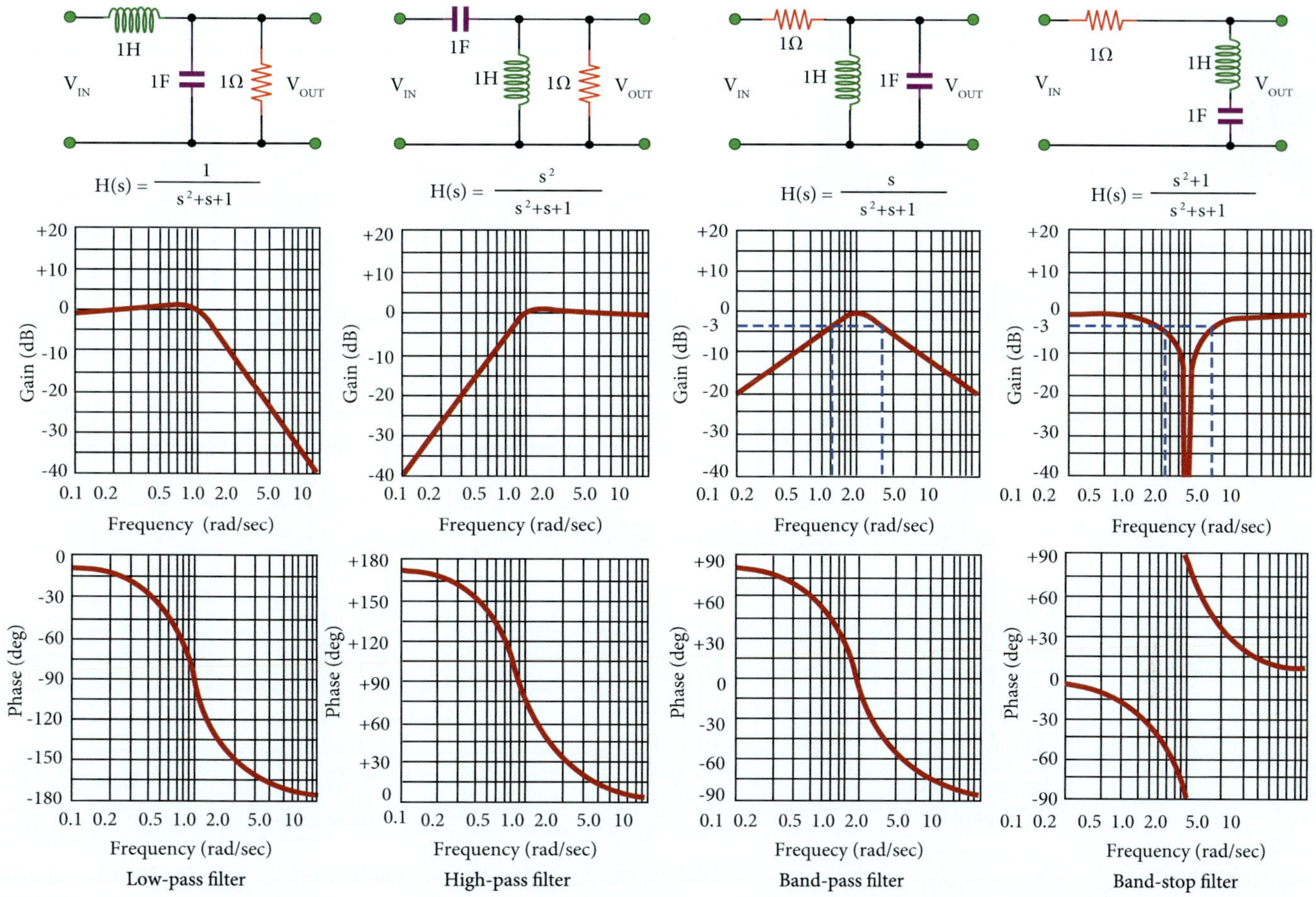

Fig. 11.6 Typical examples of low-pass, high-pass, band-pass, and band-stop second-order analog filters implemented in the form of a passive RLC circuit. For each type of filter, the transfer function and the amplitude and phase response as a function of frequency ω are presented

$$\mathcal{H}(s) = H_0 \left\{ Cs^2 + B\frac{\omega_c}{Q}s + A\omega_0^2 \right\} \Big/ \left\{ s^2 + \frac{\omega_c}{Q}s + \omega_c^2 \right\}$$

for the values $A = 1$, $B = C = 0$ for the low-pass filter, $A = B = 0$, $C = 1$ for the high-pass filter, $A = C = 0$, $B = 1$ for the band-pass filter, and $A = C = 1$, $B = 0$ for the band-stop filter. In the foregoing relation, the proportionality coefficient H_0 is the *gain* of the function. Replacing the preceding values in the last equation, we are led to the relations

$$\mathcal{H}(s) = H_0 \frac{\omega_c^2}{s^2 + \frac{\omega_c}{Q}s + \omega_c^2}, \qquad \mathcal{H}(s) = H_0 \frac{s^2}{s^2 + \frac{\omega_c}{Q}s + \omega_c^2}$$

$$\mathcal{H}(s) = H_0 \frac{\frac{\omega_c}{Q}s}{s^2 + \frac{\omega_c}{Q}s + \omega_c^2}, \qquad \mathcal{H}(s) = H_0 \frac{s^2 + \omega_c^2}{s^2 + \frac{\omega_c}{Q}s + \omega_c^2}$$

for the low-pass, high-pass, band-pass, and band-stop filters, respectively. These three parameters (i.e., the gain H_0, the frequency ω_c, and the quality factor Q), which, it should be noted, are related to observable and measurable quantities, constitute an alternative way of describing the operation of a filter, different from one based on a knowledge of the transfer function's characteristic polynomial coefficient values. Considering, for example, the simple case of a second-order low-pass filter, we can formulate its transfer function as

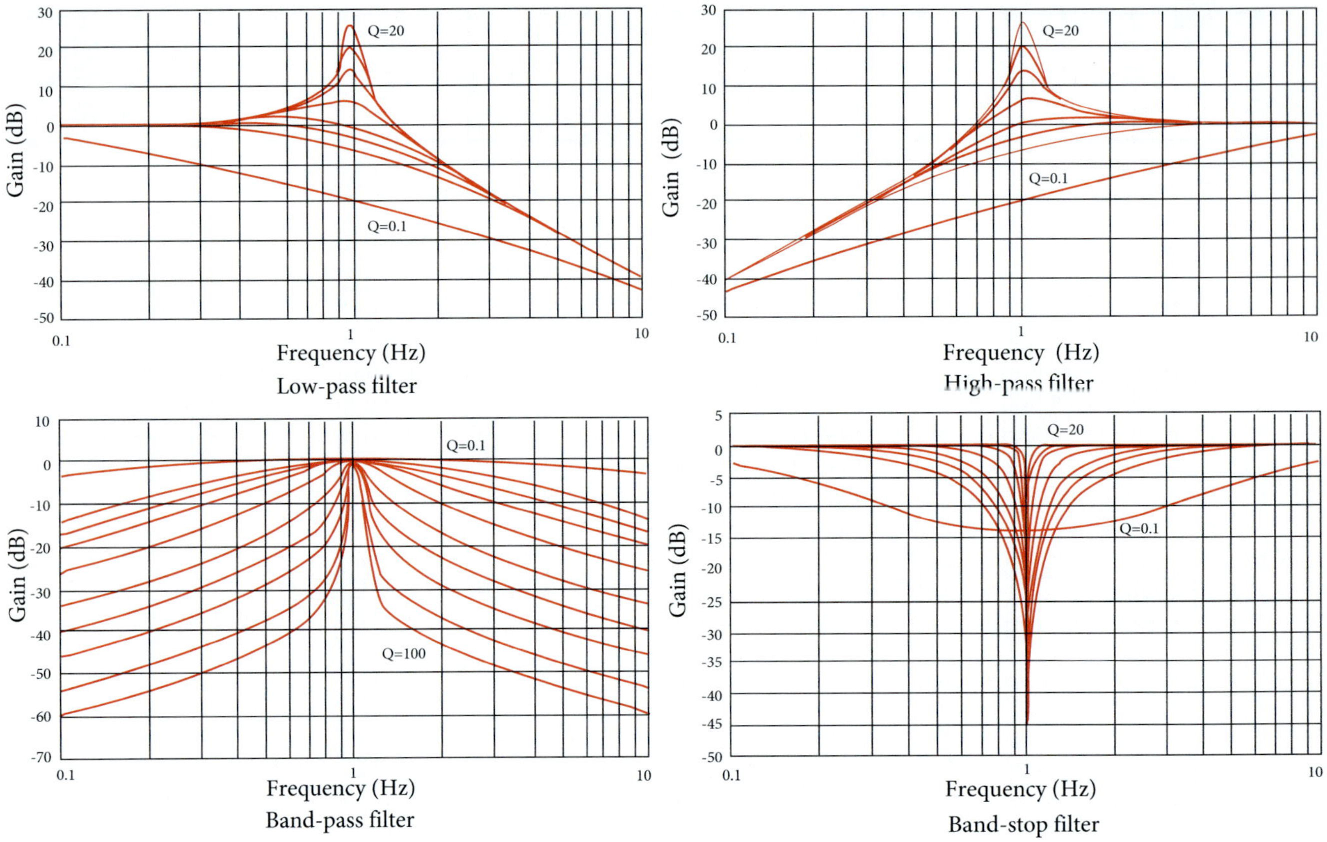

Fig. 11.7 Change in gain of low-pass, high-pass, band-pass, and band-stop second-order filters as a function of frequency ω for various values of quality factor Q

$$\mathcal{H}(s) = H_0 \frac{\alpha_0}{s^2 + \alpha_1 s + \alpha_0} = H_0 \frac{\omega_c^2}{s^2 + \dfrac{\omega_c}{Q} s + \omega_c^2}$$

from which it immediately follows that $\omega_c^2 = \alpha_0$ and $Q = \omega_c/\alpha_1 = \sqrt{\alpha_0}/\alpha_1$. Therefore, these two sets of parameters are related to each other.

The change in the filter's gain as a function of frequency for the four categories of second-order filters and for various values of the quality factor Q is presented diagrammatically in Fig. 11.7. From this figure we observe that when the value of Q exceeds a certain threshold, the gain in the low-pass and high-pass filters exhibits some *peak*, with the maximum corresponding to the position of the central frequency ω_c.

We conclude this brief summary of the basic properties of analog filters with a description of their phase characteristics and, more specifically, the *phase response*, *phase delay*, and *group delay*. Considering, therefore, a typical filter transfer function like the one described by Eq. (5.6), the frequency response function is obtained by substituting $s = j\omega$ into the transfer function, and it takes the form:

$$\mathcal{H}(j\omega) = \frac{\displaystyle\sum_{k=0}^{M} \beta_k (j\omega)^k}{\displaystyle\sum_{k=0}^{N} \alpha_k (j\omega)^k} = \frac{\displaystyle\sum_{k \text{ even}} \beta_k (-1)^{\frac{k}{2}} \omega^k + j \sum_{k \text{ odd}} \beta_k (-1)^{\frac{k-1}{2}} \omega^k}{\displaystyle\sum_{k \text{ even}} \alpha_k (-1)^{\frac{k}{2}} \omega^k + j \sum_{k \text{ odd}} \alpha_k (-1)^{\frac{k-1}{2}} \omega^k} \tag{11.7}$$

Defining the auxiliary functions

$$R_n(\omega) = \sum_{k \text{ even}} \beta_k (-1)^{\frac{k}{2}} \omega^k, \qquad I_n(\omega) = \sum_{k \text{ odd}} \beta_k (-1)^{\frac{k-1}{2}} \omega^k,$$

$$R_d(\omega) = \sum_{k \text{ even}} \alpha_k (-1)^{\frac{k}{2}} \omega^k, \qquad I_d(\omega) = \sum_{k \text{ odd}} \alpha_k (-1)^{\frac{k-1}{2}} \omega^k$$

the transfer function of the system is written in the form

$$\begin{aligned}\mathcal{H}(j\omega) &= \frac{R_n(\omega) + jI_n(\omega)}{R_d(\omega) + jI_d(\omega)} = \frac{[R_n(\omega) + jI_n(\omega)][R_d(\omega) - jI_d(\omega)]}{[R_d(\omega) + jI_d(\omega)][R_d(\omega) - jI_d(\omega)]} \\ &= \frac{R_n(\omega)R_d(\omega) + I_n(\omega)I_d(\omega)}{R_d^2(\omega) + I_d^2(\omega)} + j\frac{R_d(\omega)I_n(\omega) - R_n(\omega)I_d(\omega)}{R_d^2(\omega) + I_d^2(\omega)}\end{aligned}$$

and therefore, its phase response will be calculated from the equation

$$\begin{aligned}\measuredangle\mathcal{H}(j\omega) &= \tan^{-1}\frac{R_d(\omega)I_n(\omega) - R_n(\omega)I_d(\omega)}{R_n(\omega)R_d(\omega) + I_n(\omega)I_d(\omega)} \\ &= \frac{\sum_{k \text{ even}} \alpha_k (-1)^{\frac{k}{2}} \omega^k \sum_{k \text{ odd}} \beta_k (-1)^{\frac{k-1}{2}} \omega^k - \sum_{k \text{ even}} \beta_k (-1)^{\frac{k}{2}} \omega^k \sum_{k \text{ odd}} \alpha_k (-1)^{\frac{k-1}{2}} \omega^k}{\sum_{k \text{ even}} \beta_k (-1)^{\frac{k}{2}} \omega^k \sum_{k \text{ even}} \alpha_k (-1)^{\frac{k}{2}} \omega^k + \sum_{k \text{ odd}} \beta_k (-1)^{\frac{k-1}{2}} \omega^k \sum_{k \text{ odd}} \alpha_k (-1)^{\frac{k-1}{2}} \omega^k}\end{aligned} \tag{11.8}$$

Having calculated the phase response of the filter from the foregoing relation, we can calculate the *phase delay* and the *group delay* from the expressions

$$\tau_p(\omega) = -\frac{\measuredangle\mathcal{H}(j\omega)}{\omega} \quad \text{and} \quad \tau_g(\omega) = -\frac{d}{d\omega}\{\measuredangle\mathcal{H}(j\omega)\} = -\frac{d}{d\omega}\left\{\tan^{-1}\frac{I(\omega)}{R(\omega)}\right\} = \frac{I(\omega)\dfrac{dR(\omega)}{d\omega} - R(\omega)\dfrac{dI(\omega)}{d\omega}}{R^2(\omega) + I^2(\omega)}$$

where

$$R(\omega) = \Re\{\mathcal{H}(j\omega)\} = \frac{R_n(\omega)R_d(\omega) + I_n(\omega)I_d(\omega)}{R_d^2(\omega) + I_d^2(\omega)} \quad \text{and} \quad I(\omega) = \Im\{\mathcal{H}(j\omega)\} = \frac{R_d(\omega)I_n(\omega) - R_n(\omega)I_d(\omega)}{R_d^2(\omega) + I_d^2(\omega)}$$

are the real and imaginary parts of the frequency response function of the considered filter.

In the special case where the filter is characterized by ideal linear phase response, it will hold that $\measuredangle\mathcal{H}(j\omega) = -\tau_d\omega$, and therefore, $\tau_d(\omega) = \tau_g(\omega) = \tau_d$. Consequently, for such an ideal filter the phase delay is equal to the group delay, something that, of course, does not hold in the general case.

11.5.1 Characteristics of First-Order Passive and Active Filters

First-order filters can be implemented with both passive and active electrical circuits, and, depending on the frequency range they cut off, they will operate as either low-pass or high-pass analog filters. Using first-order circuits, we can also construct all-pass filters, while in contrast, *there are no first-order band-pass or band-stop filters*; these filters are constructed using two energy storage elements (i.e., two capacitors, two coils, or, alternatively, one capacitor and one coil), and therefore, their order must be at least equal to 2. There are two types of first-order passive analog filters: *RC type* filters consisting of a resistor and a capacitor and *RL type* filters consisting of a resistor and a coil. As we saw in Sect. 3.6.2, these filters operate either as low-pass or high-pass, depending on which of the two voltages developed at the terminals of the circuit elements is considered as the output voltage (this also applies to RL filters). The conversion of an RC or RL circuit from low-pass to high-pass and vice versa can be achieved by swapping the positions of the two elements of the circuit.

Although RC type filters were essentially studied in Sect. 3.6.2 and in Exercise 4.8.6 (page 365), let us repeat here our discussion from Chaps. 3 and 4, so that the description of first-order analog filters is concentrated in a simple section. Starting with the passive implementation of first-order filters, let us describe the operation of low-pass and high-pass RC filters

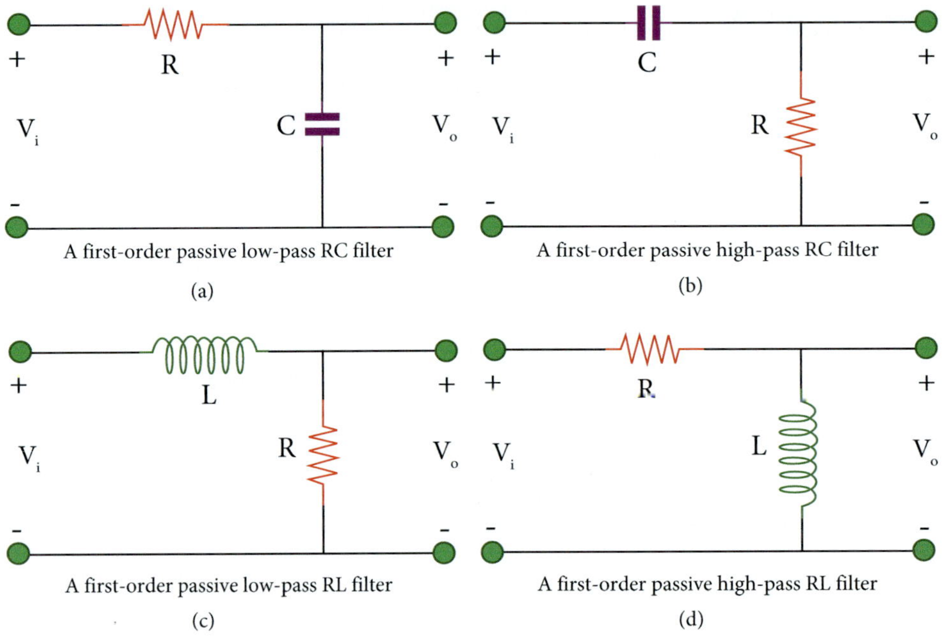

Fig. 11.8 Passive first-order low-pass and high-pass filters implemented in the form of RC and RL electrical circuits

whose circuit is depicted in Fig. 11.8a, b. Using the laws of Kirchhoff and the Laplace transform method as was done in Exercise 4.8.6 and taking into account that the capacitive resistance of a capacitor is $Z_c = 1/sC$, we find that

$$V_o(s) = \frac{\dfrac{1}{sC}}{R + \dfrac{1}{sC}} V_i(s) = \frac{1}{RCs+1} V_i(s) \quad \text{and} \quad V_o(s) = \frac{R}{R + \dfrac{1}{sC}} V_i(s) = \frac{RCs}{1+RCs} V_i(s)$$

for the low-pass and high-pass filter, respectively, where $V_o(s) = \mathcal{L}\{u_o(t)\}$ and $V_i(s) = \mathcal{L}\{u_i(t)\}$ are the Laplace transforms of the input and output voltages, respectively. Therefore, from the definition of the transfer function $\mathcal{H}(s) = V_o(s)/V_i(s)$ we will have

$$\mathcal{H}(s) = \begin{cases} \dfrac{1}{1+RCs} = \dfrac{1}{1+\tau s} & \text{for the low-pass RC filter} \\ \dfrac{RCs}{1+RCs} = \dfrac{\tau s}{1+\tau s} & \text{for the high-pass RC filter} \end{cases}$$

where $\tau = RC$ is the so-called time constant of the circuit.

To obtain the frequency response of these filters, we must set $s = j\omega$. Taking into account that the *cutoff frequency* of these two types of filters is equal to $\omega_c = 1/\tau = 1/RC$ (see Sect. 3.6.2 and discussion related to Fig. 3.20 on page 289), it immediately follows that

$$\mathcal{H}(j\omega) = \begin{cases} 1\Big/\left(1 + j\dfrac{\omega}{\omega_c}\right) & \text{for the low-pass filter RC} \\ 1\Big/\left(1 - j\dfrac{\omega_c}{\omega}\right) & \text{for the high-pass filter RC} \end{cases}$$

Expressing the preceding frequency response functions in exponential form, their magnitude and phase for these two categories of filter are found to be equal to

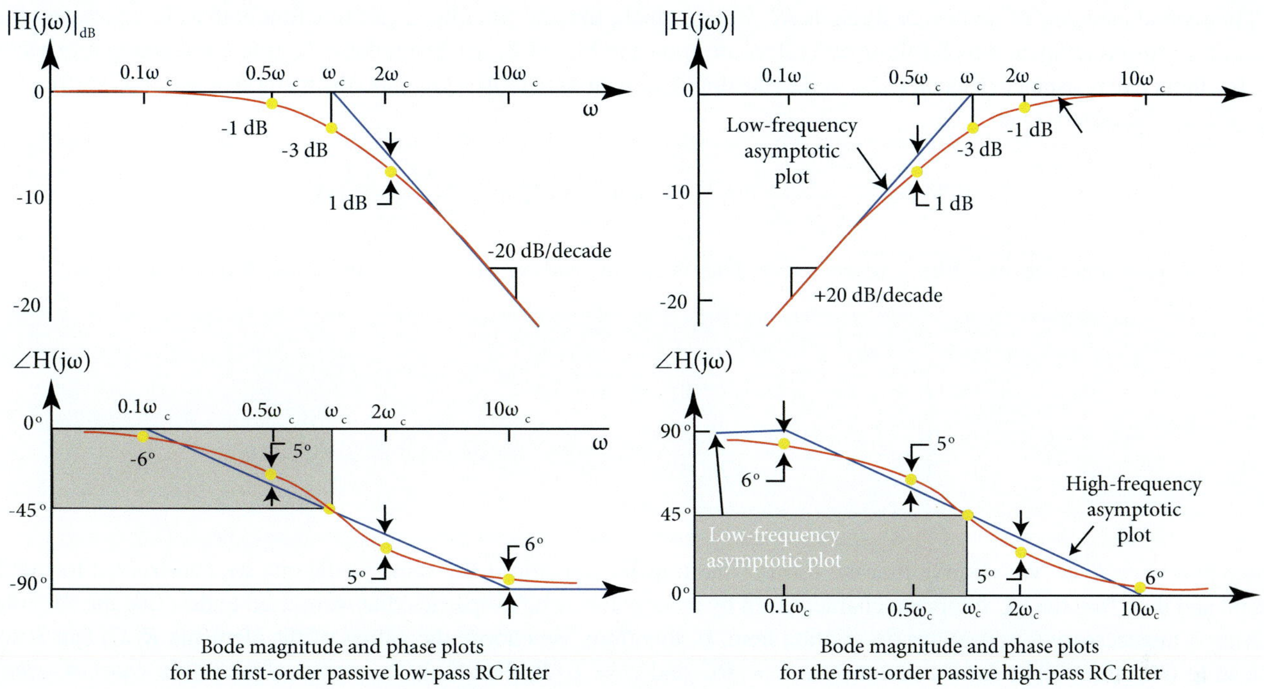

Fig. 11.9 Bode diagrams for amplitude and phase of first-order low-pass and high-pass RC filters

$$|\mathcal{H}(j\omega)| = \begin{cases} 1\Big/\sqrt{1+\left(\dfrac{\omega}{\omega_c}\right)^2} & \text{for the low-pass filter RC} \\ 1\Big/\sqrt{1+\left(\dfrac{\omega_c}{\omega}\right)^2} & \text{for the high-pass filter RC} \end{cases}$$

and

$$\measuredangle\mathcal{H}(j\omega) = \begin{cases} -\tan^{-1}\left(\dfrac{\omega}{\omega_c}\right) & \text{for the low-pass filter RC} \\ +\tan^{-1}\left(\dfrac{\omega_c}{\omega}\right) & \text{for the high-pass filter RC} \end{cases}$$

Bode amplitude and phase diagrams of low-pass and high-pass RC filters are presented in Fig. 11.9 (for the process of constructing these diagrams see Sect. 5.7.2 on page 437). From this figure we observe that for the amplitude response and for the low-pass filter in the range of values $\omega \gg \omega_c$, we have $|\mathcal{H}(j\omega)| \approx \omega_c/\omega$, and therefore, $|\mathcal{H}(j\omega)|_{dB} \approx 20\log(\omega_c/\omega) \approx 20\log\omega_c - 20\log\omega$ which is a straight line equation with a slope of -20 dB/decade. Conversely, for $\omega \ll \omega_c$ we will have $|\mathcal{H}(j\omega)| \approx 1$, an expression that also constitutes a straight line equation that intersects the previous one at $\omega = \omega_c$. As for the phase response, from the figure it is observed that for $\omega \ll \omega_c$ we have $\measuredangle\mathcal{H}(j\omega) \approx 0$, for $\omega \gg \omega_c$ we have $\measuredangle\mathcal{H}(j\omega) \approx -90°$, while at the cutoff frequency position we have $\measuredangle\mathcal{H}(\omega_c) \approx -45°$. On the other hand, in the case of the high-pass filter, we will correspondingly have $|\mathcal{H}(j\omega)| \approx \omega/\omega_c$ for $\omega \ll \omega_c$, from which it follows that $|\mathcal{H}(j\omega)|_{dB} \approx 20\log(\omega/\omega_c) \approx 20\log\omega - 20\log\omega_c$, and $|\mathcal{H}(j\omega)| \approx 1$ for $\omega \gg \omega_c$, while for the phase diagram we will have $\measuredangle\mathcal{H}(j\omega) \approx 90°$ and $\measuredangle\mathcal{H}(j\omega) \approx 0$, respectively.

The second category of first-order filters is *RL filters*, which, like the RC filters, can function both as low-pass and high-pass (the corresponding circuits for these two filters are shown in Fig. 11.8c, d). From these diagrams and taking into account that the inductive resistance of a coil is $Z_L = Ls$, the application of *Kirchhoff's laws* and the method of the *Laplace transform* will give us the expressions

$$V_o(s) = \frac{R}{R + sL} V_i(s) \quad \text{and} \quad V_o(s) = \frac{sL}{R + Ls} V_i(s)$$

for the low-pass and high-pass filters, respectively. Therefore, the *transfer function* for these two types of filters is

$$\mathcal{H}(s) = \begin{cases} \dfrac{R}{R + Ls} = \dfrac{1}{1 + (L/R)\,s} = \dfrac{1}{1 + \tau s} & \text{for the low-pass LC filter} \\[2ex] \dfrac{Ls}{R + Ls} = \dfrac{(L/R)\,s}{1 + (L/R)\,s} = \dfrac{\tau s}{1 + \tau s} & \text{for the high-pass RC filter} \end{cases}$$

where now $\tau = L/R$. We observe that the form of the transfer function is the same as the one we constructed for the RC filters, and therefore the LC filters are characterized by exactly the same properties that were used to describe the RC filters, making it unnecessary to repeat our description here. If, therefore, we choose the values of the elements R, C, and L so as to lead to the same values for the parameters $\tau = RC$ and $\tau = L/R$, then *these two types of filters are characterized by the same cutoff frequency and by the identical amplitude and phase responses*. This conclusion, of course, holds under the assumption that the passive elements of a circuit are characterized by ideal behavior, something that usually does not happen in practice.

On the other hand, first-order active filters do not use coils but instead consist of an RC circuit followed by an *operational amplifier* connected to the circuit using a *noninverting* or *inverting* connection. As with the case of passive RC filters, the corresponding active filter operates as a low-pass or high-pass filter, depending on which circuit element voltage is taken as the output voltage. The simplest case of an active filter includes a simple amplifier, leading to a gain value equal to one, while if we add two additional resistors to the circuit, we can achieve higher gain values. The two types of first-order active low-pass filters are depicted in Fig. 11.10a, b. Applying the usual circuit analysis techniques, it is proven that the transfer function for the noninverting and the inverting connection has the form

$$\mathcal{H}(s) = \frac{1 + (R_2/R_3)}{1 + \omega_c R_1 C_1 s} \quad \text{and} \quad \mathcal{H}(s) = -\frac{R_2/R_1}{1 + \omega_c R_1 C_1 s}$$

leading to gain values of

$$K = 1 + \frac{R_2}{R_3} \quad \text{and} \quad K = -\frac{R_2}{R_1}$$

respectively, where ω_c is the cutoff frequency of the filter. The negative sign in the inverting connection indicates that the output signal is characterized by a phase shift equal to 180°, due to the inverting operation of the operational amplifier.

Finally, *first-order active high-pass filters* arise, as in passive filters, *by swapping the positions of the capacitor and the resistor in the corresponding low-pass filter*, resulting in the circuits shown in Fig. 11.10c, d for noninverting and the inverting connections. In this case, the application of circuit analysis techniques leads us to transfer functions

$$\mathcal{H}(s) = \frac{1 + (R_2/R_3)}{1 + \dfrac{1}{\omega_c R_1 C_1} \cdot \dfrac{1}{s}} \quad \text{and} \quad \mathcal{H}(s) = \frac{R_2/R_1}{1 + \dfrac{1}{\omega_c R_1 C_1} \cdot \dfrac{1}{s}}$$

and therefore, to gain values of

$$K = 1 + \frac{R_2}{R_3} \quad \text{and} \quad K = -\frac{R_2}{R_1}$$

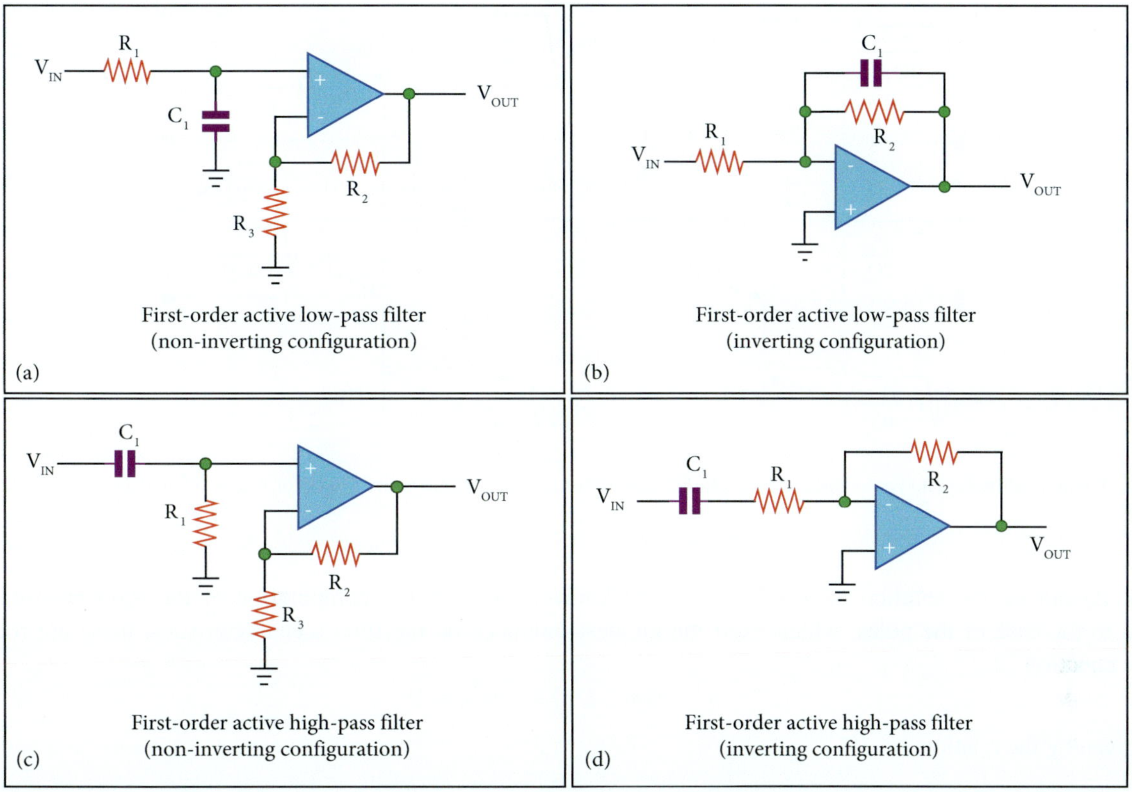

Fig. 11.10 First-order low-pass and high-pass active filters for noninverting and inverting connections

for the noninverting and inverting connections, respectively. The operating characteristics and amplitude and phase responses of these active filters are exactly the same as those of passive first-order filters and so need not be repeated here.

11.5.2 Characteristics of Second-Order Filters

In contrast to first-order filters that operate exclusively as low-pass or high-pass filters, second-order filters can also operate as *band-pass* or *band-stop* circuits, allowing or preventing the passage of a specific frequency band. These types of filters are described by a *quadratic transfer function* that can be formulated in the general form

$$\mathcal{H}(s) = \frac{\beta_0 + \beta_1 s + \beta_2 s^2}{\alpha_0 + \alpha_1 s + \alpha_2 s^2} = \frac{\beta_2}{\alpha_2}\left\{s^2 + \frac{\beta_1}{\beta_2}s + \frac{\beta_0}{\beta_2}\right\} \Big/ \left\{s^2 + \frac{\alpha_1}{\alpha_2}s + \frac{\alpha_0}{\alpha_2}\right\} \tag{11.9}$$

Defining the parameters $K = \beta_2/\alpha_2$, $\omega_{0z}^2 = \beta_0/\beta_2$, $Q_z = \omega_{0z}/(\beta_1/\beta_2)$, $\omega_{0p}^2 = \alpha_0/\alpha_2$, and $Q_p = \omega_{0p}/(\alpha_1/\alpha_2)$, the transfer function of the filter is formulated as

$$\mathcal{H}(s) = K\left\{s^2 + \frac{\omega_{0z}}{Q_z}s + \omega_{0z}^2\right\} \Big/ \left\{s^2 + \frac{\omega_{0p}}{Q_p}s + \omega_{0p}^2\right\} = K\frac{(s - z_1)(s - z_2)}{(s - p_1)(s - p_2)} \tag{11.10}$$

where z_1, z_2 and p_1, p_2 are the zeros and poles of the function, respectively, with the quantities related to these being described by the indices z and p. The numerator and denominator equations in the above expression are second-order

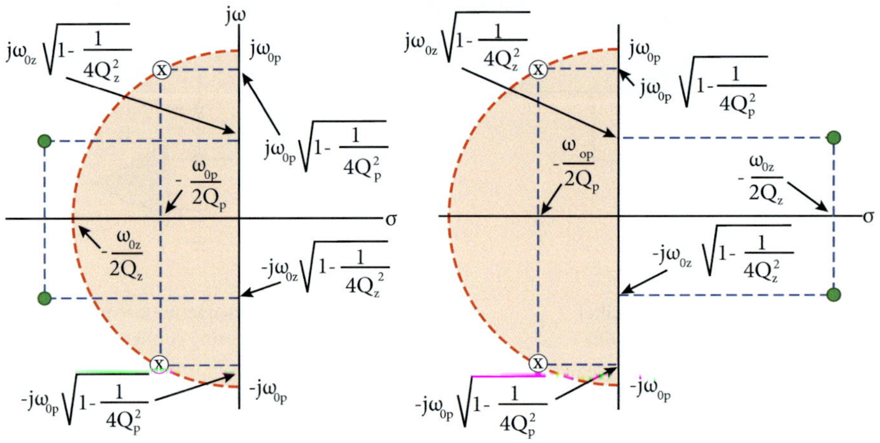

Fig. 11.11 Positions of poles and zero values of biquadratic transfer function of second-order filters for $Q_p > 0.5$ and $Q_z > 0.5$ in left and right half-planes

algebraic equations, the solution of which, as is well known, relies on the computation of the *discriminant*. Limiting ourselves to the case of the poles, which exert the greatest influence on the filter's characteristics, these are roots of the quadratic equation

$$s^2 + (\omega_{0p}/Q_p)\, s + \omega_{0p}^2 = 0$$

and are given by the relation

$$p_{1,2} = \frac{1}{2}\left(-\frac{\omega_{0p}}{Q_p} \pm \sqrt{\frac{\omega_{0p}^2}{Q_p^2} - 4\omega_{0p}^2}\right) = \omega_{0p}\left(-\frac{1}{2Q_p} \pm \sqrt{\left(\frac{1}{2Q_p}\right)^2 - 1}\right) \tag{11.11}$$

as can be verified. As for the values of the zeros of the transfer function, these are obviously given by a similar expression.

From the expressions of the poles of the transfer function it turns out that for $Q_p > 0.5$, these are of a complex nature, while for $Q < 0.5$ they are real numbers. Taking into account that the real part of a complex number expresses the distance from the imaginary axis and that in this case it is equal to $1/2Q_p$ in absolute value, it follows that for large values of Q_p *the poles are closer to the imaginary axis*, and as a result it exerts a greater influence on the behavior of the filter. However, passive RC circuits generally do not lead to high values of the quality factor Q_p, and for this reason, active RC circuits as well as passive RLC circuits are used in their place. The positions of the poles and zero values on the complex plane for values $Q_p > 0.5$ and $Q_z > 0.5$ are shown in Fig. 11.11.

The parameters Q_p and Q_z are known as *quality factors of poles and zero values*, respectively, with the quantities ω_{0p} and ω_{0z} expressing the corresponding *natural frequencies*. These two parameters allow for the full determination of the filter's behavior, and it is desirable to be able to adjust them independently of each other. The foregoing relationship refers to the general case of the transfer function of a second-order filter, from which arise as special cases the transfer functions of all types of filters used in practice, namely, low-pass, high-pass, band-pass, band-stop, and all-pass filters, which are described in the following pages.

Low-Pass Filters The transfer function of the low-pass filters has only poles, no zeros, and is described by the equation

$$\mathcal{H}(s) = \frac{A}{s^2 + \dfrac{\omega_0}{Q}s + \omega_0^2} \tag{11.12}$$

To determine the corresponding frequency response, we set $s = j\omega$; then we have

$$\mathcal{H}(j\omega) = \frac{A}{-\omega^2 + j\dfrac{\omega_0}{Q}\omega + \omega_0^2} = \frac{A(\omega_0^2 - \omega^2)}{\omega^4 - 2\omega_0^2\omega^2\left(1 - \dfrac{1}{2Q^2}\right) + \omega_0^4} - j\frac{A\dfrac{\omega_0}{Q}\omega}{\omega^4 - 2\omega_0^2\omega^2\left(1 - \dfrac{1}{2Q^2}\right) + \omega_0^4}$$

as can be proven after simple operations. Expressing the foregoing function in an *amplitude–phase representation*, it turns out that

$$G(\omega) = |\mathcal{H}(j\omega)| = \frac{A}{\sqrt{\omega^4 - 2\omega_0^2\omega^2\left(1 - \dfrac{1}{2Q^2}\right) + \omega_0^4}} \quad \text{and} \quad \measuredangle\mathcal{H}(j\omega) = -\tan^{-1}\frac{\dfrac{\omega_0}{Q}\omega}{\omega_0^2 - \omega^2}$$

From the preceding equation we find that $G_0 = G(0) = |\mathcal{H}(j\omega)|_{\omega=0} = A/\omega_0^2$ and $G_\infty = |\mathcal{H}(j\omega)|_{\omega\to\infty} = 0$. The graphical representation of the gain $|\mathcal{H}(j\omega)|$ as a function of the quality factor Q is shown in Fig. 11.7a, from where it is evident that the value of the gain *maximizes* at the position of some frequency $\omega_{\max}$. This frequency arises as the solution of the equation $\dot{G}(\omega) = 0$ and is found to be equal to

$$\omega_{\max} = \omega_0\sqrt{1 - \frac{1}{2Q^2}} \quad \text{leading to a maximum gain value} \quad G_{\max} = \frac{G_0 Q}{\sqrt{1 - \dfrac{1}{4Q^2}}} = \frac{AQ}{\omega_0^2\sqrt{1 - \dfrac{1}{4Q^2}}}$$

for values $Q \geq \sqrt{2}/2 \geq 0.707$, so that the frequency $\omega_{\max}$ is a real quantity. If Q takes large values, we can make the approximation $\omega_{\max} \approx \omega_0$ and, therefore, $G_{\max} \approx G_0 Q = AQ/\omega_0^2$.

The value of the frequency ω_c for which the gain value becomes equal to $\sqrt{2}/2 = 0.707$ of the maximum value of $G_0 = A/\omega_0^2$ is called the *cutoff frequency*. Substituting the foregoing values into the defining equation of the measure of the frequency response results in

$$\frac{\sqrt{2}}{2}\frac{A}{\omega_0^2} = \frac{A}{\sqrt{\omega_c^4 - 2\omega_0^2\omega_c^2\left(1 - \dfrac{1}{2Q^2}\right) + \omega_0^4}}$$

and solving for ω_0 we get

$$\omega_c = \omega_0\sqrt{\left(1 - \frac{1}{2Q^2}\right) + \sqrt{\left(1 - \frac{1}{2Q^2}\right)^2 + 1}}$$

The foregoing relationship for values $Q \gg 1$ is approximated by the equation $\omega_c \approx \omega_0\sqrt{1 + \sqrt{2}} \approx 1.544\omega_0 \approx 1.544\omega_{\max}$.

High-Pass Filters The transfer function of high-pass filters has the form

$$\mathcal{H}(s) = \frac{As^2}{s^2 + \dfrac{\omega_0}{Q}s + \omega_0^2} \tag{11.13}$$

and therefore differs from the corresponding function of low-pass filters in that it contains a second-order zero value at the position $s = 0$. Depending on the value of the quality factor Q, this function may exhibit either two real poles or a pair of complex conjugate poles. The frequency response function arises from the preceding function by making the substitution $s = j\omega$ and has the form

$$\mathcal{H}(j\omega) = \frac{-A\omega^2}{(\omega_0^2 - \omega^2) + j\dfrac{\omega_0}{Q}\omega} = -\frac{A\omega^2(\omega_0^2 - \omega^2)}{(\omega_0^2 - \omega^2)^2 + \dfrac{\omega_0^2}{Q^2}\omega^2} + j\frac{A\dfrac{\omega_0}{Q}\omega^3}{(\omega_0^2 - \omega^2)^2 + \dfrac{\omega_0^2}{Q^2}\omega^2}$$

Expressing the preceding function in an amplitude–phase representation, it turns out that

$$G(\omega) = |\mathcal{H}(j\omega)| = \frac{A}{\sqrt{\omega^4 - 2\omega_0^2\omega^2\left(1 - \dfrac{1}{2Q^2}\right) + \omega_0^4}} \quad \text{and} \quad \measuredangle\mathcal{H}(j\omega) = -\tan^{-1}\frac{\dfrac{\omega_0}{Q}\omega}{\omega_0^2 - \omega^2}$$

In this case, for $\omega = 0$ we have $G_0 = G(0) = |\mathcal{H}(j\omega)|_{\omega=0} = 0$ and $G_\infty = |\mathcal{H}(j\omega)|_{\omega\to\infty} = A$. The graphical representation of the gain $|\mathcal{H}(j\omega)|$ as a function of the quality factor Q is shown in Fig. 11.7b. Following the methodology we used to describe low-pass filters, it is proven that for values $Q \geq \sqrt{2}/2 \geq 0.707$ the value of the gain is maximized at the frequency position

$$\omega_{\text{max}} = \omega_0\sqrt{1 - \frac{1}{2Q^2}} \quad \text{with maximum value equal to} \quad G_{\text{max}} = \frac{G_\infty Q}{\sqrt{1 - \dfrac{1}{4Q^2}}} = \frac{AQ}{\sqrt{1 - \dfrac{1}{4Q^2}}}$$

Finally, the *cutoff frequency* ω_c is defined, as before, and performing the corresponding process we find

$$\omega_c = \omega_0\sqrt{\frac{1}{2Q^2} - 1 + \sqrt{\left(1 - \frac{1}{2Q^2}\right)^2 + 1}}$$

In this case and in the limit of large values of Q ($Q \gg 1$), the preceding frequency is approximated by the equation $\omega_c \approx 0.6436\omega_0 \approx 0.6436\omega_{\text{max}}$.

Band-pass Filters The transfer function of band-pass filters is given by the expression

$$\mathcal{H}(s) = \frac{As}{s^2 + \dfrac{\omega_0}{Q}s + \omega_0^2} \tag{11.14}$$

and therefore it is characterized by the existence of a simple zero value at the position $s = 0$ and two real poles or a pair of conjugate complex poles, depending on the value of the quality factor Q. The frequency response function arises from the function $\mathcal{H}(s)$ for $s = j\omega$ and has the form

$$\mathcal{H}(j\omega) = \frac{jA\omega}{(\omega_0^2 - \omega^2) + j\dfrac{\omega_0}{Q}\omega} = \frac{A\dfrac{\omega_0}{Q}\omega^2}{(\omega_0^2 - \omega^2) + \dfrac{\omega_0^2}{Q^2}\omega^2} + j\frac{A\omega(\omega_0^2 - \omega^2)}{(\omega_0^2 - \omega^2) + \dfrac{\omega_0^2}{Q^2}\omega^2}$$

Expressing the preceding function in an amplitude–phase representation, it turns out that

$$G(\omega) = |\mathcal{H}(j\omega)| = \frac{A\omega}{\sqrt{(\omega_0^2 - \omega^2)^2 + \dfrac{\omega_0^2}{Q^2}\omega^2}} = \frac{A\omega}{\sqrt{\omega^4 - 2\omega_0^2\omega^2\left(1 - \dfrac{1}{2Q^2}\right) + \omega_0^4}}$$

and

$$\measuredangle\mathcal{H}(j\omega) = -\tan^{-1}\frac{\omega_0^2 - \omega^2}{\dfrac{\omega_0}{Q}\omega}$$

It is not difficult to see that for $\omega = 0$ it holds that $G_0 = G(0) = |\mathcal{H}(j\omega)|_{\omega=0} = 0$. On the other hand, in the limit $\omega \to \infty$, the parameter G_∞ takes the indeterminate value ∞/∞. Applying *L'Hôpital's rule*, it turns out that $G_\infty = 0$, and therefore the function $G(\omega)$ will present a maximum value at the position of some frequency $\omega_{\max}$, which is calculated as the solution of the algebraic equation $\dot{G}(\omega) = 0$ and is found to be equal to $\omega_{\max} = \omega_0$. Note that the frequency $\omega_{\max}$ is *independent* of the quality factor Q, and therefore the gain function always presents a maximum, regardless of the value of the factor Q, in contrast to the corresponding functions of low-pass and high-pass filters, which present this maximum only for values $Q \geq 0.707$. Replacing the value of $\omega_{\max}$ in the defining equation of the function $G(\omega)$, it turns out that

$$G_{\max} = G(\omega_0) = \frac{A\omega_0}{\sqrt{2\omega_0^4 - 2\omega_0^4\left(1 - \dfrac{1}{2Q^2}\right)}} = \frac{A\omega_0}{\sqrt{2\omega_0^4\dfrac{1}{2Q^2}}} = \frac{AQ}{\omega_0}$$

In full analogy with previous cases, the cutoff frequency ω_c is defined as the frequency at whose position the value of the function $G(\omega)$ is equal to $(\sqrt{2}/2)\ G_0$. Using the defining equation of the measure $G(\omega)$ and the value of G_0 calculated earlier, we will obtain

$$G(\omega_c) = \frac{A\omega_c}{\sqrt{(\omega_0^2 - \omega_c^2)^2 + \dfrac{\omega_0^2}{Q^2}\omega_c^2}} = \frac{\sqrt{2}}{2}\frac{AQ}{\omega_0}$$

from which it easily follows that

$$\omega_c^2 \pm \frac{\omega_0}{Q}\omega_c - \omega_0^2 = 0$$

From the preceding quadratic equations, the one corresponding to the positive sign has discriminant

$$\Delta = (\omega_0^2/Q^2)(1 + 4Q^2)$$

and roots

$$\omega_{c1} = -\frac{\omega_0}{2Q}(1 - \sqrt{1 + 4Q^2}) \quad \text{and} \quad \omega_{c3} = -\frac{\omega_0}{2Q}(1 + \sqrt{1 + 4Q^2})$$

while the one corresponding to the negative sign has the same discriminant and roots

$$\omega_{c2} = \frac{\omega_0}{2Q}(1 + \sqrt{1 + 4Q^2}) \quad \text{and} \quad \omega_{c4} = \frac{\omega_0}{2Q}(1 - \sqrt{1 + 4Q^2})$$

Of these four roots, ω_{c2} and ω_{c4} are *rejected* since they are *negative*, so the two cutoff frequencies that are known to characterize the operation of a band-pass filter are

$$\omega_{c1} = -\frac{\omega_0}{2Q}(1 - \sqrt{1 + 4Q^2}) \quad \text{and} \quad \omega_{c2} = \frac{\omega_0}{2Q}(1 + \sqrt{1 + 4Q^2})$$

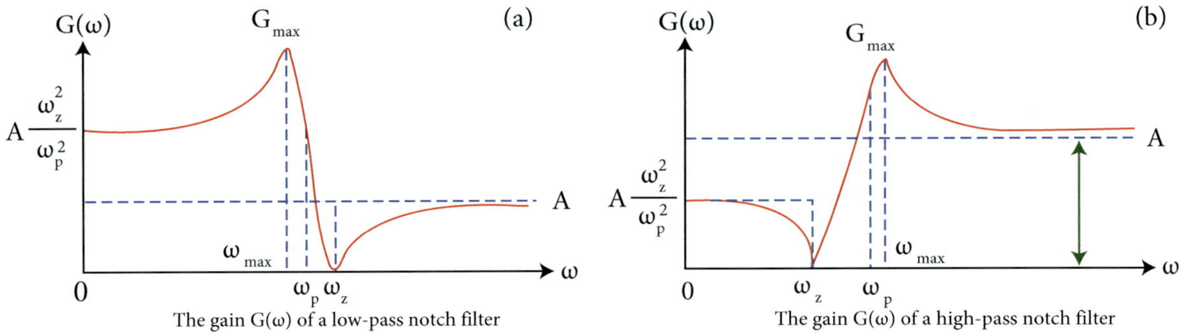

Fig. 11.12 Variation of gain $G(\omega)$ as a function of the frequency ω for low-pass and high-pass notch filters

From the last relations we observe that

$$\omega_{c1}\omega_{c2} = \left[-\frac{\omega_0}{2Q}(1-\sqrt{1+4Q^2})\right]\left[\frac{\omega_0}{2Q}(1+\sqrt{1+4Q^2})\right] = \omega_0^2 \qquad \text{and} \qquad \Delta\omega = \omega_{c2}-\omega_{c1} = \frac{\omega_0}{Q}$$

and consequently, *the bandwidth of a band-pass filter is inversely proportional to the quality factor Q*. This means that as the value of Q increases, the bandwidth value decreases and the amplitude response curve becomes increasingly sharper. We can therefore state that the *quality factor Q in a band-pass filter is a measure of the sharpness of the filter's frequency response curve*.

Band-pass filters can be examined in two ways. If we know the transfer function of the filter or we can calculate it from the electrical circuit that implements it, we can initially determine the values of Q and ω_0 and then the two cutoff frequencies ω_{c1} and ω_{c2} using the equations constructed earlier. Alternatively, we can find these frequencies first using the method applied earlier and then calculate the values of ω_0 and Q from the relations $\omega_0 = \sqrt{\omega_{c1}\omega_{c2}}$ and $Q = \omega_0/\Delta\omega$.

Band-stop Filters The *band-stop filters* (also known as *notch filters*) are described by the transfer function

$$\mathcal{H}(s) = \frac{A(s^2+\omega_z^2)}{s^2+\dfrac{\omega_p}{Q}s+\omega_p^2} \tag{11.15}$$

and are characterized by the existence of two conjugate imaginary zero values at the positions $s = \pm j\omega_z$ and two real poles or a pair of conjugate complex poles depending on the value of the coefficient of quality Q. The parameters ω_p and ω_z express the frequencies of the poles and zeros, respectively, while the frequency response function will emerge, as is well known, from the preceding function, setting $s = j\omega$. We will then have

$$\mathcal{H}(j\omega) = \frac{A(\omega_z^2-\omega^2)}{(\omega_p^2-\omega^2)+j\dfrac{\omega_p}{Q}\omega} = \frac{A(\omega_z^2-\omega^2)(\omega_p^2-\omega^2)}{(\omega_p^2-\omega^2)^2+\dfrac{\omega_p^2}{Q^2}\omega^2} - j\frac{A(\omega_z^2-\omega^2)\dfrac{\omega_p}{Q}\omega}{(\omega_p^2-\omega^2)^2+\dfrac{\omega_p^2}{Q^2}\omega^2}$$

Formulating the foregoing transfer function in an *amplitude–phase representation*, we will get

$$G(\omega) = |\mathcal{H}(j\omega)| = \frac{A|\omega_z^2-\omega^2|}{\sqrt{(\omega_p^2-\omega^2)^2+\dfrac{\omega_p^2}{Q^2}\omega^2}} \qquad \text{and} \qquad \measuredangle\mathcal{H}(j\omega) = -\tan^{-1}\frac{\dfrac{\omega_p}{Q}\omega}{\omega_p^2-\omega^2}$$

The graphical representations of the function $G(\omega)$ for $\omega_z > \omega_p$ and $\omega_z < \omega_p$ are shown in Fig. 11.12.

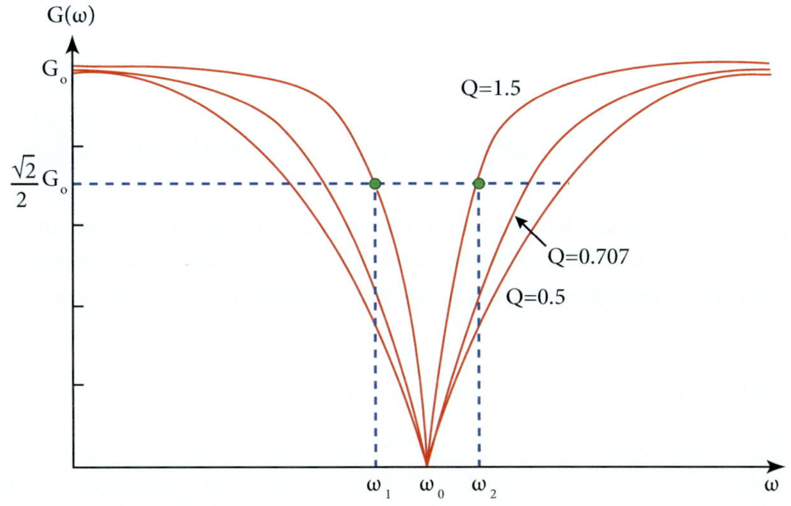

Fig. 11.13 Variation of gain with respect to frequency for symmetric notch filters

There are three types of notch filters, depending on the correlation that exists between the frequencies ω_p and ω_z of the poles and zeros. If $\omega_z > \omega_p$, the cutoff filter is aforementioned to be *low-pass* and is described by the amplitude response shown in Fig. 11.12a, while if $\omega_z < \omega_0$, the cutoff filter is aforementioned to be *high-pass* and is described by the amplitude response shown in Fig. 11.12b. In both cases, the gain value for the frequency $\omega = 0$ is equal to $G(0) = A\omega_z^2/\omega_p^2$, as can be proven, with the line $G(\omega) = G(0)$ being above or below the line $G(\omega) = A$, depending on whether the cutoff filter is *low-pass* or *high-pass*. On the other hand, if the two aforementioned frequencies are equal to each other, the filter is stb*symmetrical cutoff filter*, and its characteristics arise from the foregoing equations by setting $\omega_p = \omega_z = \omega_0$. The frequency response of the filter will therefore be given by the relation

$$\mathcal{H}(j\omega) = \frac{A(\omega_0^2 - \omega^2)}{(\omega_0^2 - \omega^2) + j\dfrac{\omega_0}{Q}\omega} = \frac{A(\omega_0^2 - \omega^2)^2}{(\omega_0^2 - \omega^2)^2 + \dfrac{\omega_0^2}{Q^2}\omega^2} - j\frac{A(\omega_0^2 - \omega^2)\dfrac{\omega_0}{Q}\omega}{(\omega_0^2 - \omega^2)^2 + \dfrac{\omega_0^2}{Q^2}\omega^2}$$

with amplitude and phase response

$$G(\omega) = |\mathcal{H}(j\omega)| = \frac{A|\omega_0^2 - \omega^2|}{\sqrt{(\omega_0^2 - \omega^2)^2 + \dfrac{\omega_0^2}{Q^2}\omega^2}} = \frac{A|\omega_0^2 - \omega^2|}{\sqrt{\omega^4 - 2\omega_0^2\omega^2\left(1 - \dfrac{1}{2Q^2}\right) + \omega_0^4}}$$

and

$$\measuredangle\mathcal{H}(j\omega) = -\tan^{-1}\frac{\dfrac{\omega_0}{Q}\omega}{\omega_0^2 - \omega^2}$$

In this case, for $\omega = \omega_0$ we have $G(\omega_0) = 0$, while for $\omega = 0$ we have $G(0) = A$, as can be proven. A graphical representation of the gain of the band-pass filter in relation to the frequency ω is shown in Fig. 11.13.

The cutoff frequency for band-pass filters is defined, as before, as the frequency value ω_c for which the corresponding gain function value $G(\omega_c)$ is equal to 0.707 of the value $G(0)$. In this case, this requirement is mathematically formulated as

$$G(\omega_c) = \frac{A|\omega_0^2 - \omega_c^2|}{\sqrt{(\omega_0^2 - \omega_c^2)^2 + \dfrac{\omega_0^2}{Q^2}\omega_c^2}} = \frac{\sqrt{2}}{2}A$$

from which it follows

$$\omega_c^2 \pm \frac{\omega_0}{Q}\omega_c - \omega_0^2 = 0$$

But these equations are the same as those obtained in the previous section for band-pass filters. Therefore, applying the previous analysis, we arrive at the same results as with the band-pass filters regarding the defining equations of the cutoff frequencies ω_{c1} and ω_{c2} and the expressions that correlate these frequencies with the *central frequency* ω_0 and *quality factor* Q.

All-Pass Filters All-pass filters are described by a transfer function of the form

$$\mathcal{H}(s) = \frac{s^2 - \dfrac{\omega_0}{Q}s + \omega_0^2}{s^2 + \dfrac{\omega_0}{Q}s + \omega_0^2} \tag{11.16}$$

which contains two poles and two zero values characterized by the same frequency and the same value for quality factor Q, but located in different half-planes. As before, depending on the value of the constant Q, these poles will be real or conjugate complex quantities, which in this case also applies to the two zero values. The frequency response function arises from the transfer function $\mathcal{H}(s)$ by setting $s = j\omega$ and has the form

$$\mathcal{H}(j\omega) = \frac{(\omega_0^2 - \omega^2) - j\dfrac{\omega_0}{Q}\omega}{(\omega_0^2 - \omega^2) + j\dfrac{\omega_0}{Q}\omega} = \frac{(\omega_0^2 - \omega^2)^2 - \dfrac{\omega_0^2}{Q^2}\omega^2}{(\omega_0^2 - \omega^2)^2 + \dfrac{\omega_0^2}{Q^2}\omega^2} - j\frac{2\dfrac{\omega_0}{Q}\omega(\omega_0^2 - \omega^2)}{(\omega_0^2 - \omega^2)^2 + \dfrac{\omega_0^2}{Q^2}\omega^2}$$

By formulating the preceding transfer function in an amplitude–phase representation, we will get

$$G(\omega) = |\mathcal{H}(j\omega)| = 1 \qquad \text{and} \qquad \measuredangle\mathcal{H}(j\omega) = -\tan^{-1}\frac{2\dfrac{\omega_0}{Q}\omega(\omega_0^2 - \omega^2)}{(\omega_0^2 - \omega^2)^2 - \dfrac{\omega_0^2}{Q^2}\omega^2}$$

The foregoing conclusion is very important, as it tells us that *filters of this category allow the passage of all frequencies*—which is also why they are called *all-pass* filters—maintaining the characteristics of the signal amplitude unchanged at their output. Therefore, their effect on input signals concerns *their phase characteristics and the time delay they cause to them* (recall from basic theory that the group delay $\tau(\omega)$ that a filter causes is defined as *the derivative of its phase response*). Graphical representations of the functions $G(\omega)$, $\measuredangle\mathcal{H}(j\omega)$, and $\tau(\omega)$ in relation to the frequency ω and for indicative values of the quality factor Q are presented in Fig. 11.14.

11.5.3 Passive and Active Second-Order Filters

In full analogy with first-order filters, second-order filters can be constructed from both passive circuits consisting of *capacitors*, *resistors*, and *coils* and active circuits that contain capacitors, resistors, and devices such as *operational amplifiers*. *Passive second-order filters* can take many different forms and be implemented by quite different electrical circuits that, however, have a common characteristic: They have at least two energy storage elements (i.e., capacitors and coils). Typical examples of passive second-order low-pass, high-pass, band-pass, and band-stop filters are presented in Fig. 11.15.

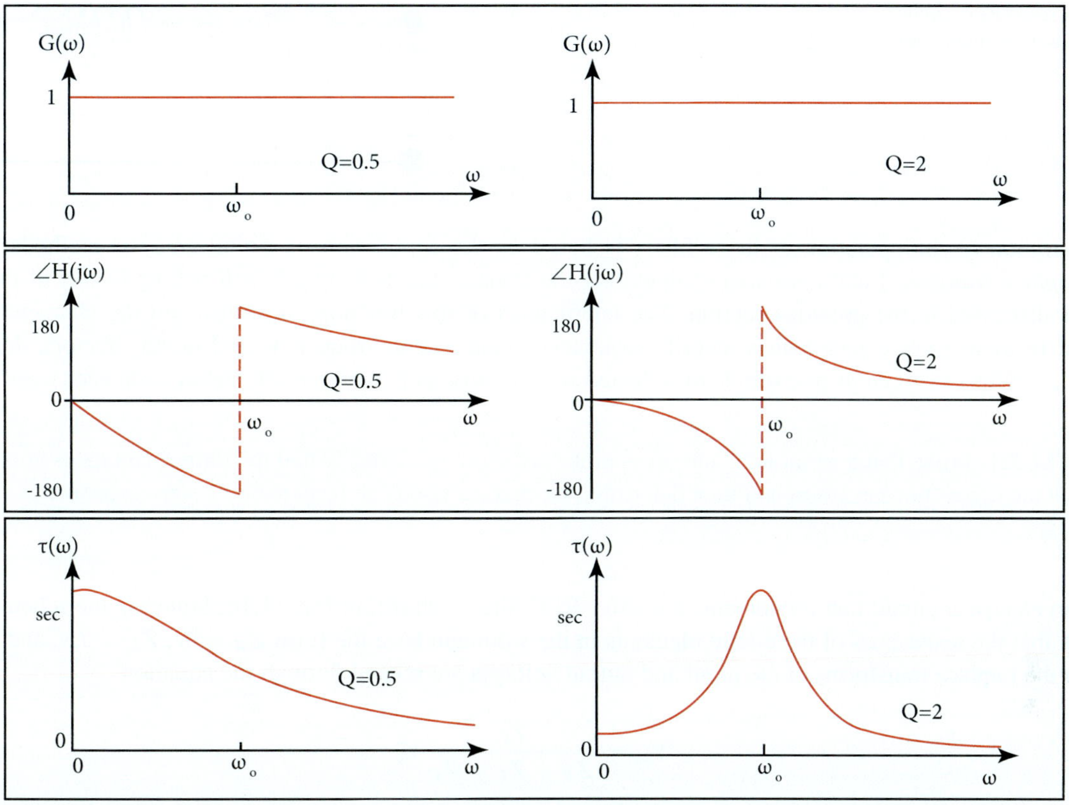

Fig. 11.14 Amplitude and phase response and group delay of a band-pass filter for various values of quality factor Q

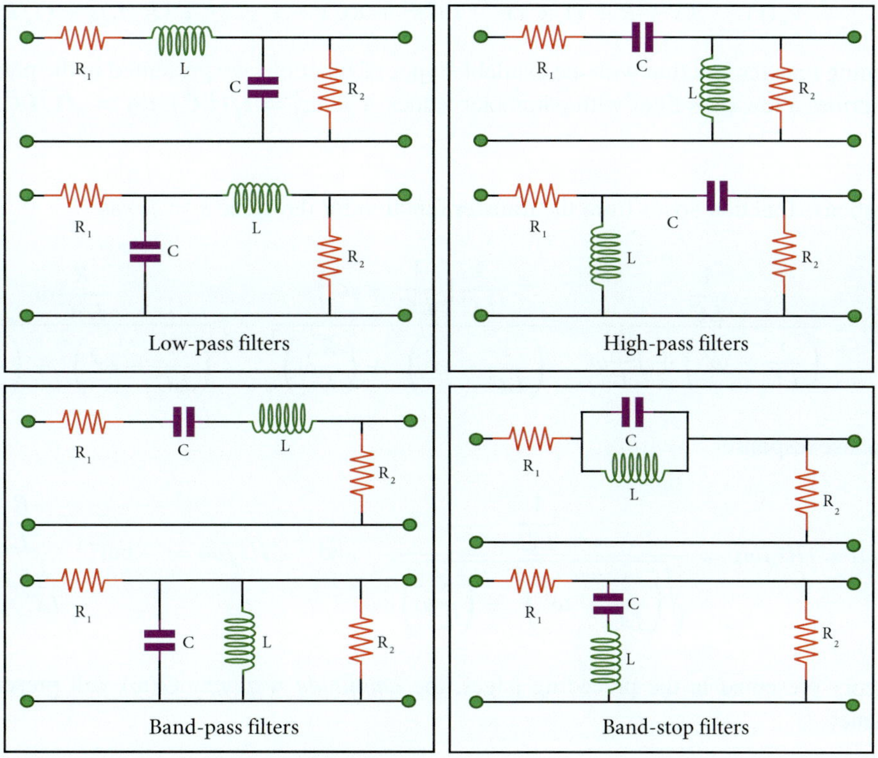

Fig. 11.15 Typical examples of passive second-order low-pass, high-pass, band-pass, and band-stop filters

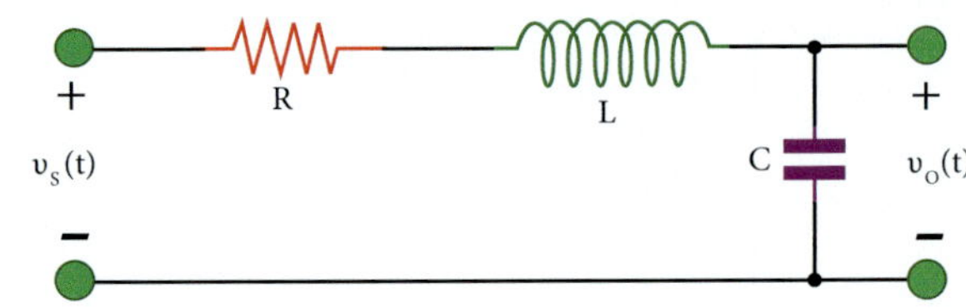

Fig. 11.16 A serial RLC circuit that implements a low-pass filter

The description of the operation of these filters is based on the application of circuit analysis methods (*Kirchhoff's laws* and *Laplace transform*) and consists in calculating the transfer function, which will belong to one of the categories of functions discussed in the previous section. The coefficients of this function, as well as all the characteristics of the filter related to them (poles, zero values, cutoff frequencies, gain, center frequency, and quality factor), depend on the characteristics of the circuit elements, which are selected so as to prompt the appropriate behavior in each case.

Problem 11.5.1 Prove that a serial RLC circuit operates as a low-pass filter when the output voltage is considered to be the voltage across the capacitor, and then determine its characteristics and calculate its step response.

Solution An electrical circuit that implements a passive RLC filter is shown in Fig. 11.16. Studying the scheme and taking into account that the resistances of the circuit elements in the s domain have the form $Z_R = R$, $Z_L = Ls$, and $Z_c = 1/sC$, we find that the Laplace transforms of the input and output voltages are related through the equation

$$V_{\text{out}} = \frac{Z_C}{Z_R + Z_L + Z_C} V_{\text{in}}$$

and therefore, the system's transfer function will be calculated from the defining equation as

$$\mathcal{H}(s) = \frac{V_{\text{out}}(s)}{V_{\text{in}}(s)} = \frac{1/(Cs)}{Ls + R + (1/Cs)} = \frac{1}{LCs^2 + RCs + 1} = \frac{1/(LC)}{s^2 + (R/L)s + (1/LC)}$$

Comparing the preceding transfer function with the available types of functions we presented in the previous section, we find that this function describes a low-pass filter with parameter values $A = \omega_0^2 = 1/(LC)$, $\omega_0 = \sqrt{1/LC}$, and $Q = L\omega_0/R = \sqrt{L/CR^2}$.

The frequency response function arises from the transfer function for the value $s = j\omega$ as

$$\mathcal{H}(j\omega) = \frac{\dfrac{1}{LC}}{\left(\dfrac{1}{LC} - \omega^2\right) + j\dfrac{R}{L}\omega} = \frac{\dfrac{1}{LC}\left(\dfrac{1}{LC} - \omega^2\right)}{\left(\dfrac{1}{LC} - \omega^2\right)^2 + \left(\dfrac{R}{L}\omega\right)^2} - j\frac{\dfrac{R}{L^2C}\omega}{\left(\dfrac{1}{LC} - \omega^2\right)^2 + \left(\dfrac{R}{L}\omega\right)^2}$$

with amplitude and phase responses

$$G(\omega) = |\mathcal{H}(j\omega)| = \frac{\dfrac{1}{LC}}{\sqrt{\left(\dfrac{1}{LC} - \omega^2\right)^2 + \left(\dfrac{R}{L}\omega\right)^2}} \quad \text{and} \quad \measuredangle\mathcal{H}(j\omega) = -\tan^{-1}\frac{\dfrac{R}{L}\omega}{\dfrac{1}{LC} - \omega^2}$$

According to the theory presented in the preceding pages, the *amplitude response* $G(\omega)$ will present a maximum at the position of the frequency

$$\omega_{\text{max}} = \omega_0\sqrt{1 - \frac{1}{2Q^2}} = \frac{1}{LC}\sqrt{1 - \frac{CR^2}{2L}} = \sqrt{\frac{2L - CR^2}{2C^2L^3}}$$

with a value equal to

$$G_{\max} = \frac{AQ}{\omega_0^2\sqrt{1-\dfrac{1}{4Q^2}}} = \frac{\dfrac{1}{LC}\sqrt{\dfrac{L}{CR^2}}}{\dfrac{1}{LC}\sqrt{1-\dfrac{CR^2}{4L}}} = \frac{2}{R}\sqrt{\frac{L}{C(4L-CR^2)}}$$

Finally, the cutoff frequency of the filter ω_c will be given by the relation

$$\omega_c = \omega_0\sqrt{\left(1-\frac{1}{2Q^2}\right)+\sqrt{\left(1-\frac{1}{2Q^2}\right)^2+1}} = \frac{1}{LC}\sqrt{\frac{2L-CR^2}{2L}+\sqrt{1+\left(\frac{2L-CR^2}{2L}\right)^2}}$$

To calculate the *step response* of the system, defined as the output of the system, when we transmit the step function $u(t)$ at its input, let us construct a differential equation that describes the operation of the filter, which obviously is a second-order differential equation with constant coefficients. Starting from the transfer function of the filter and setting $s = j\omega$ and $\omega_0/Q = R/L = 2\zeta\omega_0$, where $\zeta = 1/2Q$, our known ratio or damping factor, the transfer function of the system is formulated as

$$\mathcal{H}(s) = \frac{V_{\text{out}}(s)}{V_{\text{in}}(s)} = \frac{\omega_0^2}{(j\omega)^2 + j\omega(2\zeta\omega_0) + \omega_0^2}$$

from which it follows that

$$(j\omega)^2 V_{\text{out}}(s) + 2\zeta\omega_0(j\omega)V_{\text{out}}(s) + \omega_0^2 V_{\text{out}}(s) = \omega^2 V_{\text{in}}(s)$$

Taking into account that $V_{\text{out}}(s) = \mathcal{L}\{\upsilon_o(t)\}$ and $V_{\text{in}}(s) = \mathcal{L}\{\upsilon_i(t)\}$ and using the property of the differentiation of the Laplace transform—see Sect. 4.5 on page 319, as well as Exercises 4.8.6 (page 365), 4.8.7 (page 366), and 4.8.8 (page 368—the application of the inverse transform to the foregoing relation leads us to the differential equation

$$\frac{d^2\upsilon_o(t)}{dt^2} + 2\zeta\omega_0\frac{d\upsilon_o(t)}{dt} + \omega_0^2\upsilon_o(t) = \omega_0^2\upsilon_i(t)$$

which has the same form as Eq. (4.35) describing a LTI second-order system if we make the correspondences $x(t) \to \upsilon_i(t)$ and $y(t) \to \upsilon_o(t)$ and set $\omega_0 = 1/\tau$ and $\kappa = 1$. Therefore, applying the theory of Sect. 4.8.1 (page 358), the Laplace transform of the system's step response will have the form

$$V_{\text{out}}(s) = \frac{\omega_0^2}{s(s-s_1)(s-s_2)}$$

(as derived from Eq. (4.37) for $\kappa = \alpha = 1$), with the roots s_1 and s_2 calculated as

$$s_{1,2} = -\frac{\zeta}{\tau} \pm \frac{\sqrt{\zeta^2-1}}{\tau} = \omega_0(-\zeta \pm \sqrt{\zeta^2-1})$$

Therefore, the behavior of the system will depend on the value of ζ and, according to Sect. 4.8.1, will be as follows:

- Case 1 ($\zeta > 1$): The system is characterized by *high damping*, and, according to Eq. (4.38), its output will have the form

$$\upsilon_o(t) = \left\{1 - e^{-\omega_0\zeta t}\left[\cosh(\omega_0\sqrt{\zeta^2-1}t) + \frac{\zeta}{\sqrt{\zeta^2-1}}\sinh(\omega_0\sqrt{\zeta^2-1}t)\right]\right\}u(t)$$

- Case 2 ($\zeta = 1$): The filter is characterized as a *critically damped system*, and, according to Eq. (4.40), its output will have the form

$$\upsilon_o(t) = [1 - (1+\omega_0 t)e^{-\omega_0 t}]u(t)$$

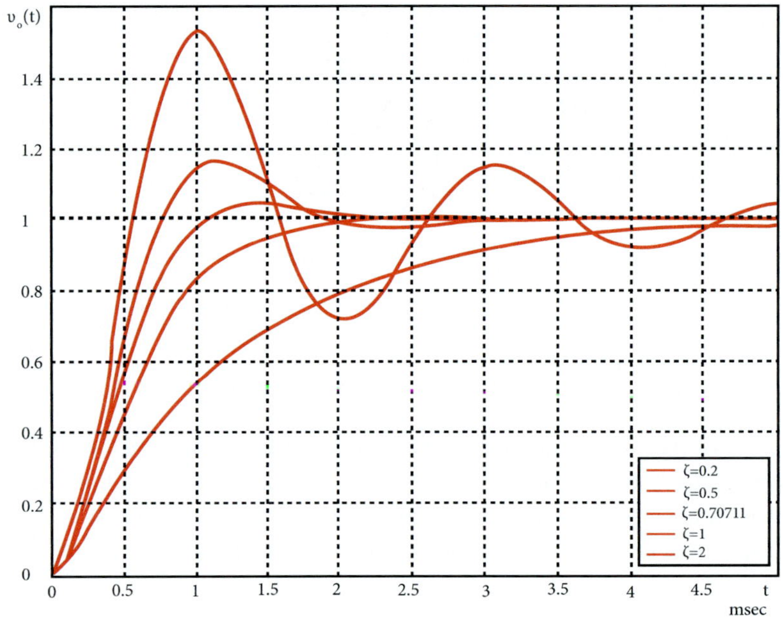

Fig. 11.17 Step response of low-pass RLC serial circuit

- Case 3 ($0 < \zeta < 1$): The system is characterized by *low damping*, and, according to Eq. (4.41), its output will have the form

$$v_o(t) = \left[1 - \frac{e^{-\omega_0 \zeta t}}{\sqrt{1-\zeta^2}} \sin(\omega t + \varphi)\right] u(t) \qquad \text{where} \qquad \varphi = \tan^{-1} \frac{\sqrt{\zeta^2 - 1}}{\zeta} \qquad \text{and} \qquad \omega = \omega_0 \sqrt{1-\zeta^2}$$

The change in the step response for the value $\omega_0 = 3141.5927 \equiv 100\pi$ is shown in Fig. 11.17.

Problem 11.5.2 Prove that a series RLC circuit operates as a band-pass filter when the output voltage is considered to be the voltage across the resistance and determine its characteristics.

Solution The electrical circuit that implements the passive RLC filter is shown in Fig. 11.18a. Studying the diagram and taking into account that the resistances of the circuit elements in the s domain have the form $Z_R = R$, $Z_L = Ls$, and $Z_c = 1/sC$, we find that the Laplace transforms of the input and output voltages are related through the equation

$$V_{\text{out}} = \frac{Z_R}{Z_R + Z_L + Z_C} V_{\text{in}}$$

and therefore, the transfer function of the system has the form

$$\mathcal{H}(s) = \frac{V_{\text{out}}}{V_{\text{in}}} = \frac{R}{R + Ls + \dfrac{1}{Cs}} = \frac{Rs}{Ls^2 + Rs + \dfrac{1}{C}} = \frac{\dfrac{R}{L}s}{s^2 + \dfrac{R}{L}s + \dfrac{1}{LC}}$$

It is not difficult to see that this function refers to a band-pass filter with parameter values

$$A = \frac{R}{L}, \qquad \omega_0 = \frac{1}{\sqrt{LC}}, \qquad Q = \frac{L}{R}\omega_0 = \frac{L}{R}\sqrt{\frac{1}{LC}} = \sqrt{\frac{L}{CR^2}}$$

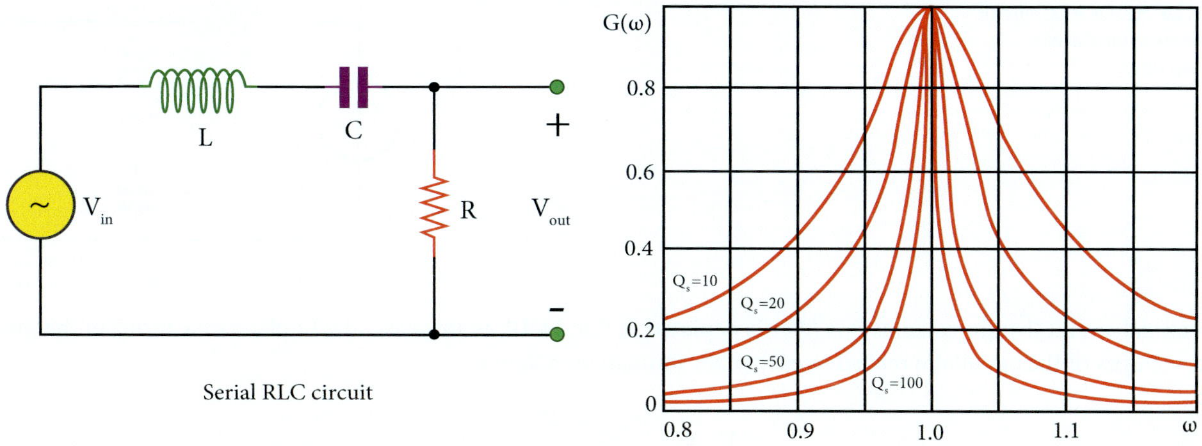

Fig. 11.18 Series RLC circuit and change in its frequency response measure $G(\omega)$ as function of quality factor Q

(this conclusion arises from a comparison of the foregoing transfer function with that of a general band-pass filter). The amplitude and phase response of the foregoing function are formulated as

$$G(\omega) = |\mathcal{H}(j\omega)| = \frac{A\omega}{\sqrt{(\omega_0^2 - \omega^2)^2 + \dfrac{\omega_0^2}{Q^2}\omega^2}} = \frac{\dfrac{R}{L}\omega}{\sqrt{\left(\dfrac{1}{LC} - \omega^2\right)^2 + \left(\dfrac{R}{L}\omega\right)^2}} \quad \text{and}$$

$$\measuredangle\mathcal{H}(j\omega) = -\tan^{-1}\left(\frac{\omega_0^2 - \omega^2}{\dfrac{\omega_0}{Q}\omega}\right) = -\tan^{-1}\left(\frac{\dfrac{1}{LC} - \omega^2}{\dfrac{R}{L}\omega}\right) = 90^\circ - \tan^{-1}\left(\frac{\dfrac{R}{L}\omega}{\dfrac{1}{LC} - \omega^2}\right)$$

The variation in the gain $G(\omega)$ of the RLC band-pass filter as a function of the frequency ω for various values of the quality factor Q is shown in Fig. 11.18b.

The cutoff frequencies ω_{c1} and ω_{c2} are calculated as

$$\omega_{c1} = -\frac{\omega_0}{2Q}(1 - \sqrt{1 + 4Q^2}) = -\frac{R}{2L}\left(1 - \sqrt{1 + 4\frac{L}{CR^2}}\right) = -\frac{R}{2L} + \sqrt{\left(\frac{R}{2L}\right)^2 + \frac{1}{LC}}$$

$$\omega_{c2} = +\frac{\omega_0}{2Q}(1 + \sqrt{1 + 4Q^2}) = +\frac{R}{2L}\left(1 + \sqrt{1 + 4\frac{L}{CR^2}}\right) = +\frac{R}{2L} + \sqrt{\left(\frac{R}{2L}\right)^2 + \frac{1}{LC}}$$

Therefore, the bandwidth of the filter is

$$\Delta\omega = \omega_{c2} - \omega_{c1} = \left\{\frac{R}{2L} + \sqrt{\left(\frac{R}{2L}\right)^2 + \frac{1}{LC}}\right\} - \left\{-\frac{R}{2L} + \sqrt{\left(\frac{R}{2L}\right)^2 + \frac{1}{LC}}\right\} = \frac{R}{L}$$

Problem 11.5.3 Prove that the series RLC circuit operates as a symmetrical band-stop filter when the output voltage is considered to be the voltage across the complex resistance resulting from the series connection of the capacitive and the inductive resistance.

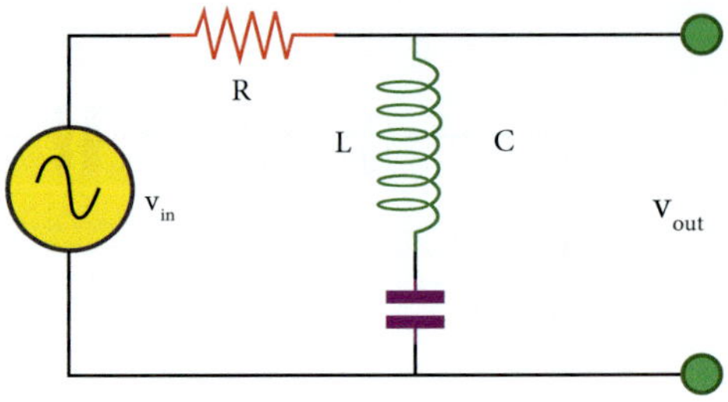

Fig. 11.19 Series RLC circuit implements symmetrical band-stop filters

An electrical circuit that implements this filter is depicted in Fig. 11.19. In this case, the Laplace transforms of the input and output voltages of the circuit are related to each other through the relation

$$V_{\text{out}} = \frac{Z_L + Z_c}{Z_R + Z_L + Z_C} V_{\text{in}}$$

leading to the transfer function

$$\mathcal{H}(s) = \frac{V_{\text{out}}}{V_{\text{in}}} = \frac{Ls + \dfrac{1}{Cs}}{R + Ls + \dfrac{1}{Cs}}$$

$$= \frac{LCs^2 + 1}{LCs^2 + RCs + 1} = \frac{s^2 + \dfrac{1}{LC}}{s^2 + \dfrac{R}{L}s + \dfrac{1}{LC}}$$

Comparing this function with the five types of functions described in the previous pages, we find that the aforementioned function describes a *symmetrical band-pass filter* with parameter values $A = 1$, $\omega_z = \omega_p = \omega_0 = 1/\sqrt{LC}$, and

$$Q = L\omega_0/R = \sqrt{L/(CR^2)}$$

The frequency response arises from the transfer function with the substitution $s = i\omega$ and has the form

$$\mathcal{H}(j\omega) = \frac{1 - \omega^2 LC}{1 - \omega^2 LC + j\omega RC} = \frac{(1 - \omega^2 LC)^2}{(1 - \omega^2 LC)^2 + (\omega RC)^2} - j\frac{(1 - \omega^2 LC)(\omega RC)}{(1 - \omega^2 LC)^2 + (\omega RC)^2}$$

with amplitude and phase responses

$$G(\omega) = \left\{\left|\frac{1}{LC} - \omega^2\right|\right\} \Big/ \left\{\sqrt{\left(\frac{1}{LC} - \omega^2\right)^2 + \left(\frac{R}{L}\omega\right)^2}\right\} \quad \text{and} \quad \measuredangle\mathcal{H}(j\omega) = -\tan^{-1}\left\{\frac{R}{L}\omega\right\} \Big/ \left\{\frac{1}{LC} - \omega^2\right\}$$

As for the cutoff frequencies, these are given by the same expressions we constructed for the case of the RLC band-pass filter presented in the previous exercise. ■

In cases where the values of the central frequency ω_0 and the quality factor Q are known, we should solve the circuit equations and express the values of the elements R, L, and C as functions of ω_0 and Q, so that for given values of these two parameters we can choose the appropriate passive elements that will allow the construction of a filter that will be characterized by the given values. For simplicity's sake, considering resistance values $R_2 = R$ and $R_1 = \kappa R$, the expressions of ω_0 and Q for the preceding filter are formulated as

$$\omega_0 = \sqrt{\frac{\kappa + 1}{LC}} \quad \text{and} \quad Q = \frac{\omega_0}{\dfrac{\kappa R}{L} + \dfrac{1}{CR}}$$

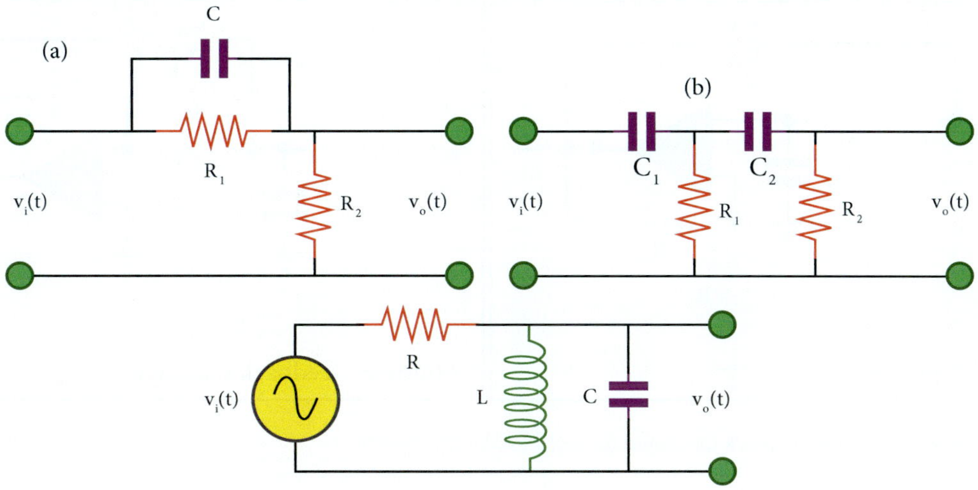

Fig. 11.20 Examples of electrical circuits implementing analog filters—an exercise

Solving the foregoing expressions for L and C, we find that

$$L = \frac{\kappa + 1}{C\omega_0^2} \quad \text{and} \quad C = \frac{\kappa + 1}{2\kappa R Q \omega_0}\left[1 \pm \sqrt{1 - \frac{4\kappa Q^2}{\kappa + 1}}\right]$$

with the restriction $\kappa \leq 1/(4Q^2 - 1)$, so that the capacitance of the capacitor is a real number. For simplicity's sake, choosing the value of κ that zeroes the square root in the expression of C [$\kappa = 1/(4Q^2 - 1)$], the capacitive and inductive resistances of the circuit are proven to have the form

$$C = \frac{\kappa + 1}{2\kappa R Q \omega_0} = \frac{2Q}{R\omega_0} \quad \text{and} \quad L = \frac{\kappa + 1}{C\omega_0^2}$$

Therefore, for given values of ω_0 and Q and for an arbitrarily chosen resistance value R, we can calculate the values of L and C that we should choose so that the filter that will be constructed from these will be characterized by the given values of ω_0 and Q. If these values are $\omega_0 = 10000\pi$ rad/s and $Q = 1$ and we arbitrarily choose $R = 1000\ \Omega$, then from the foregoing relations we find that $C = 63.6620$ nF, $L = 21.22$ mH, $k = 1/3$, $R_2 = R = 1000\ \Omega$, and $R_1 = \kappa R = 333.33\ \Omega$.
Here, the reader is asked, as an exercise, to determine the type, transfer function, and amplitude and phase response for the filters implemented by the electrical circuits depicted in Fig. 11.20a, b. If any of these filters are characterized as band-pass or band-stop, the reader is asked to calculate their cutoff frequencies, as well as the value of their quality factor Q.

Following the presentation of second-order passive filters, let us move on to a description of their active equivalents, which also use resistors and capacitors but instead of coils include active circuit elements, such as operational amplifiers. There are two basic topologies of second-order active filters, the first of which is known as *Sallen–Key topology*, while the second is called *multiple feedback or feedback topology*. The Sallen–Key topology, also known as voltage control voltage source (VCVS) topology, was proposed in 1955 by R.P. Sallen and E.L. Key of the Massachusetts Institute of Technology's Lincoln Labs.[2] The main advantage of this method is that the performance of the filter it describes depends as little as possible on the performance of the operational amplifier, while the values of ω_0 and Q are to some extent independent of each other but strongly depend on the gain of the filter. The main disadvantage of this topology is that these filters cannot be easily adjusted due to the interaction of the circuit element values with the values of ω_0 and Q. On the other hand, the *multiple feedback topology* uses the operational amplifier as an *integrator*, and for this reason the dependence of the transfer function of the filter on the characteristics of the operational amplifier is greater compared to the Sallen–Key method. The following pages present the electrical circuits that implement these two filter topologies for low-pass, high-pass, band-pass, and band-stop filters and provide the expressions for their *transfer function* $\mathcal{H}(s)$, the *center frequency* ω_0, and the *quality factor* Q. The

[2] See R.P. Sallen & E.L. Key, "A Practical Method of Designing RC Active Filters," *IRE Trans. Circuit Theory*, Vol. CT-2, 1955, pp. 74–85.

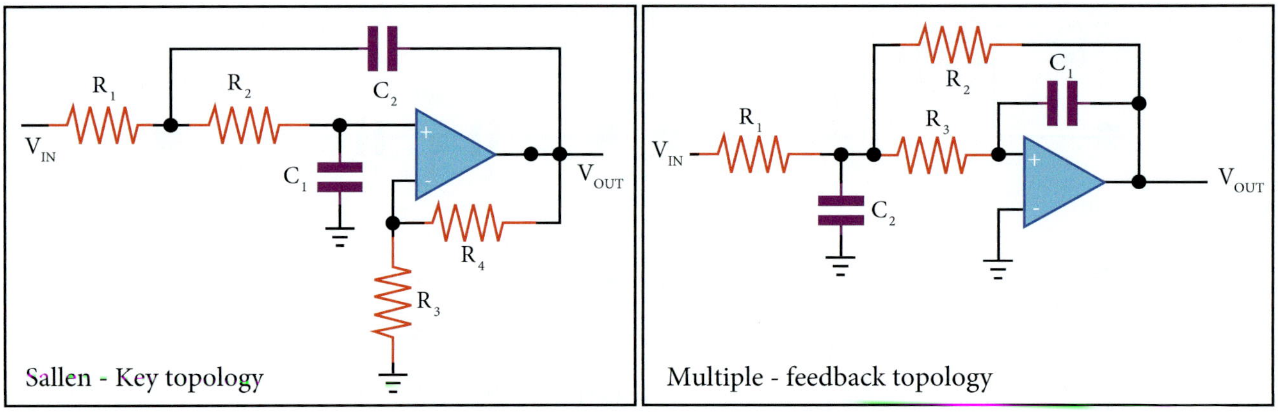

Fig. 11.21 Sallen–Key and multiple feedback topologies for second-order low-pass filters

proof of these relationships is conducted in the same manner as for the second-order passive filters and is left as an exercise to the reader. As for the theory governing the characteristics of the transfer function and the amplitude and phase responses, it is the same as that for passive filters and does not need to be repeated.

Active Second-Order Low-Pass Filter: Sallen–Key Topology The electrical circuit that implements active low-pass second-order filters, according to the *Sallen-Key topology*, is presented in Fig. 11.21a. The transfer function $\mathcal{H}(s)$ has the form

$$\mathcal{H}(s) = \left\{\frac{1+(R_4/R_3)}{R_1 R_2 C_1 C_2}\right\} \Big/ \left\{s^2 + \left(\frac{1}{R_1 C_2} + \frac{1}{R_2 C_2} - \frac{(R_4/R_3)}{R_2 C_1}\right) s + \frac{1}{R_1 R_2 C_1 C_2}\right\}$$

while the parameters ω_0 and Q are given by the equations

$$\omega_0 = \sqrt{\frac{1}{R_1 R_2 C_1 C_2}} \quad \text{and} \quad Q = \sqrt{\frac{1}{R_1 R_2 C_1 C_2}} \Big/ \left\{\frac{1}{R_1 C_2} + \frac{1}{R_2 C_2} - \frac{R_4/R_3}{R_2 C_1}\right\}$$

Active Second-Order Low-Pass Filter: Multiple Feedback Topology The electrical circuit that implements active low-pass second-order filters, according to the *multiple feedback topology*, is presented in Fig. 11.21b. The transfer function $\mathcal{H}(s)$ has the form

$$\mathcal{H}(s) = \left\{\frac{1}{R_1 R_3 C_1 C_2}\right\} \Big/ \left\{s^2 + \left(\frac{1}{R_1} + \frac{1}{R_2} + \frac{1}{R_3}\right)\frac{s}{C_2} + \frac{1}{R_2 R_3 C_1 C_2}\right\}$$

while the parameters ω_0 and Q are given by the equations

$$\omega_0 = \sqrt{\frac{1}{R_1 R_3 C_1 C_2}} \quad \text{and} \quad Q = \left\{\sqrt{\frac{C_2}{C_1}}\right\} \Big/ \left\{\sqrt{\frac{R_3}{R_2}} + \sqrt{\frac{R_2}{R_3}} + \frac{\sqrt{R_2 R_3}}{R_1}\right\}$$

Second-Order Active High-Pass Filter: Sallen–Key Topology The electrical circuit that implements second-order active high-pass filters, according to the *Sallen–Key topology*, is presented in Fig. 11.22a. The transfer function $\mathcal{H}(s)$ has the form

$$\mathcal{H}(s) = \left\{\left(1 + \frac{R_4}{R_3}\right) s^2\right\} \Big/ \left\{s^2 + \left(\frac{1}{R_1 C_1} + \frac{1}{R_1 C_2} - \frac{R_4/R_3}{R_1 C_1}\right) s + \frac{1}{R_1 R_2 C_1 C_2}\right\}$$

while the parameters ω_0 and Q are given by the equations

$$\omega_0 = \sqrt{\frac{1}{R_1 R_2 C_1 C_2}} \quad \text{and} \quad Q = \left\{\sqrt{\frac{1}{R_1 R_2 C_1 C_2}}\right\} \Big/ \left\{\frac{1}{R_1 C_1} + \frac{1}{R_1 C_2} - \frac{R_4/R_3}{R_1 C_1}\right\}$$

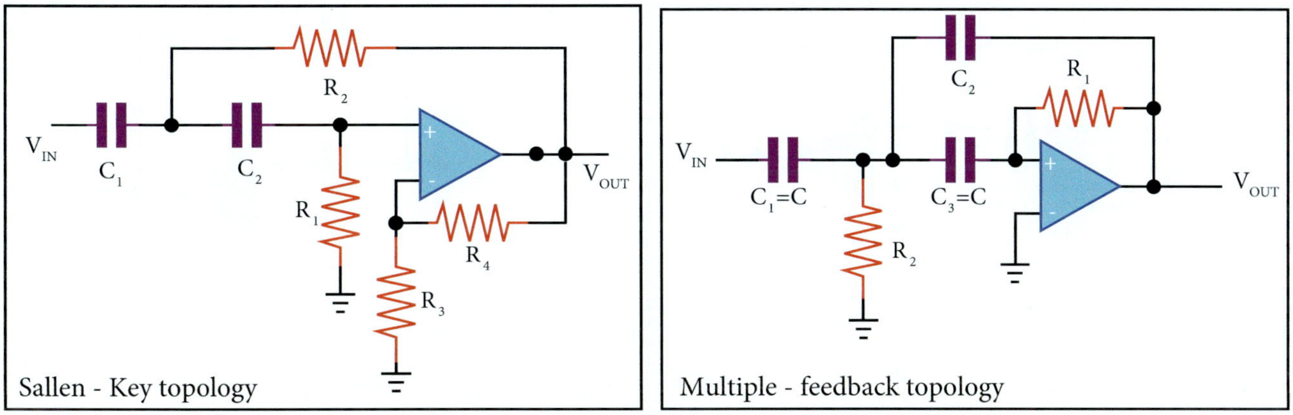

Fig. 11.22 Sallen–Key and multiple feedback topologies for second-order high-pass filters

Second-Order Active High-Pass Filter: Multiple Feedback Topology An electrical circuit that implements active second-order high-pass filters according to the *multiple feedback topology* is presented in Fig. 11.22b. The transfer function $\mathcal{H}(s)$ has the form

$$\mathcal{H}(s) = \left\{(C_1/C_2)\, s^2\right\} \Big/ \left\{s^2 + \left(\frac{C_1}{C_2 + C_3} + \frac{1}{C_2} + \frac{1}{C_3}\right)\frac{s}{R_1} + \frac{1}{R_1 R_2 C_2 C_3}\right\}$$

while the parameters ω_0 and Q are given by the equations

$$\omega_0 = \sqrt{\frac{1}{R_1 R_2 C_2 C_3}} \quad \text{and} \quad Q = \left\{\sqrt{\frac{R_1}{R_2}}\right\} \Big/ \left\{\sqrt{\frac{C_3}{C_2}} + \sqrt{\frac{C_2}{C_3}} + \frac{C_1}{\sqrt{C_2 C_3}}\right\}$$

Active Second-Order Band-Pass Filter: Sallen–Key Topology An electrical circuit that implements active second-order band-pass filters, according to the *Sallen–Key topology*, is presented in Fig. 11.23a. The transfer function $\mathcal{H}(s)$ has the form

$$\mathcal{H}(s) = \left\{\left(1 + \frac{R_4}{R_3}\right)\frac{s}{R_1 C_1}\right\} \Big/ \left\{s^2 + \left(\frac{1}{R_1 C_1} + \frac{1}{R_3 C_2} + \frac{1}{R_3 C_1} - \frac{R_4/R_3}{R_2 C_1}\right) s + \frac{R_1 + R_2}{R_1 R_2 R_3 C_1 C_2}\right\}$$

while the parameters ω_0 and Q are given by the equations

$$\omega_0 = \sqrt{\frac{R_1 + R_2}{R_1 R_2 R_3 C_1 C_2}} \quad \text{and} \quad Q = \left\{\sqrt{\frac{R_1 + R_2}{R_1 R_2 R_3 C_1 C_2}}\right\} \Big/ \left\{\frac{1}{R_1 C_1} + \frac{1}{R_3 C_2} + \frac{1}{R_3 C_1} - \frac{R_4/R_3}{R_2 C_1}\right\}$$

Second-Order Active Band-pass Filter: Multiple Feedback Topology An electrical circuit that implements second-order active band-pass filters, according to the *multiple feedback topology*, is presented in Fig. 11.23b. The transfer function $\mathcal{H}(s)$ has the form

$$\mathcal{H}(s) = -\left\{\frac{s}{R_1 C_2}\right\} \Big/ \left\{s^2 + \left(\frac{1}{C_1} + \frac{1}{C_2}\right)\frac{s}{R_3} + \frac{1}{R_3 C_1 C_2}\left(\frac{1}{R_1} + \frac{1}{R_2}\right)\right\}$$

while the parameters ω_0 and Q are given by the equations

$$\omega_0 = \sqrt{\frac{R_1 + R_2}{R_1 R_2 R_3 C_1 C_2}} \quad \text{and} \quad Q = \sqrt{\frac{R_1}{R_2}\left(1 + \frac{R_2}{R_1}\right)} \Big/ \left\{\sqrt{\frac{C_1}{C_2}} + \sqrt{\frac{C_2}{C_1}}\right\}$$

Second-Order Active Band-stop Filter: Sallen–Key Topology An electrical circuit that implements second-order active band-pass filters, according to the *Sallen–Key topology*, is presented in Fig. 11.24a. However, it should be noted that this

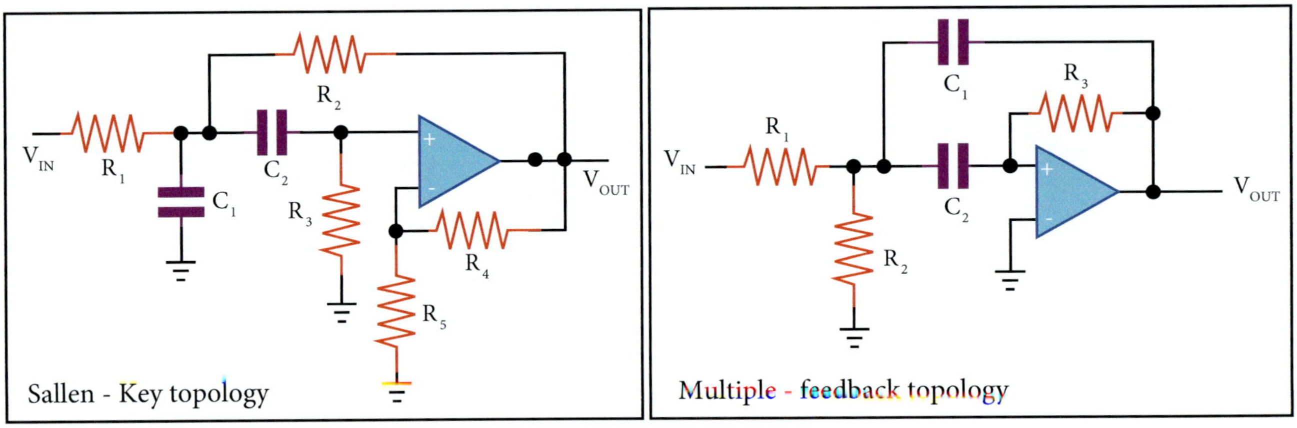

Fig. 11.23 Sallen–Key and multiple feedback topologies for second-order band-pass filters

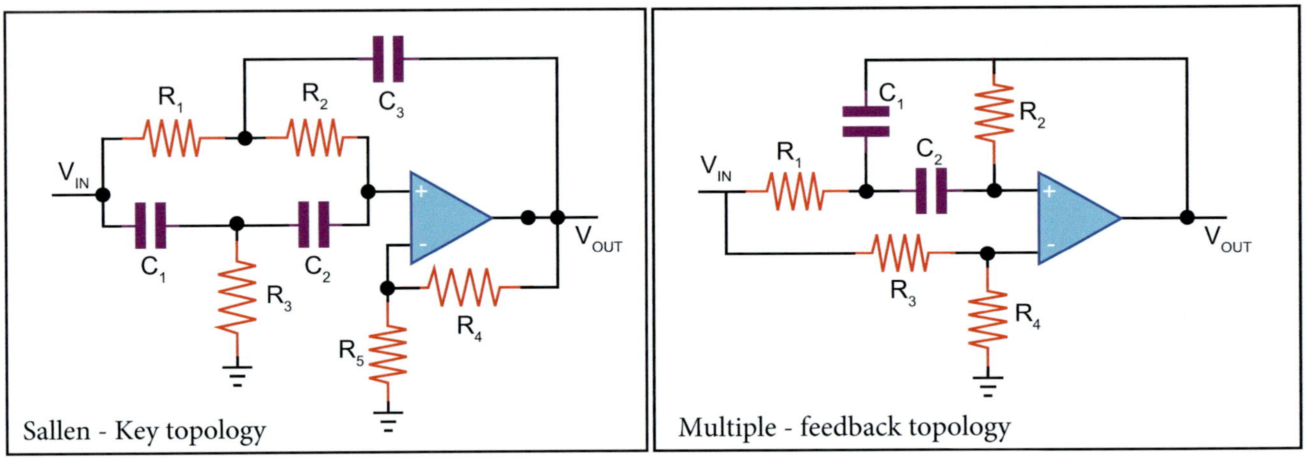

Fig. 11.24 Sallen–Key and multiple feedback topologies for second-order band-stop filters

topology is not often used, as it is characterized by many undesirable properties, such as *a high degree of difficulty in adjusting the frequency* ω_0. For this reason, alternative topologies are used in practice, which are presented subsequently.

Active Second-Order Band-pass Filter: Multiple Feedback Topology An electrical circuit that implements active second-order band-pass filters, according to the *multiple feedback topology*, is presented in Fig. 11.24b. The transfer function $\mathcal{H}(s)$ has the form

$$\mathcal{H}(s) = \frac{R_4}{R_3 + R_4}\left\{s^2 + \left(\frac{1}{R_2C_1} + \frac{1}{R_2C_2} - \frac{R_3}{R_1R_4C_1}\right)s + \frac{1}{R_1R_2C_1C_2}\right\} \Big/ \left\{s^2 + \left(\frac{1}{R_2C_1} + \frac{1}{R_2C_2}\right)s + \frac{1}{R_1R_2C_1C_2}\right\}$$

while the parameters ω_0 and Q are given by the equations

$$\omega_0 = \sqrt{\frac{1}{R_1R_2C_1C_2}} \quad \text{and} \quad Q = \sqrt{\frac{R_2}{R_1}} \Big/ \left\{\sqrt{\frac{C_2}{C_1}} + \sqrt{\frac{C_1}{C_2}}\right\}$$

Two popular electrical circuits that allow the creation of active second-order band-pass filters are the *twin-T circuit* and the *Wien–Robinson circuit* presented in Fig. 11.25. The main advantage of the twin-T filter is that it *allows the variation of the quality factor* Q *through the internal gain* $G = 1 + (R_2/R_1)$ *without modifying the central frequency* ω_0. On the other hand, the *Wien–Robinson filter* allows the modification of the gain in the transition zone *without affecting the quality factor* Q. The reader is asked to determine the transfer function of these two filters by applying Kirchoff's laws in the way we demonstrated previously.

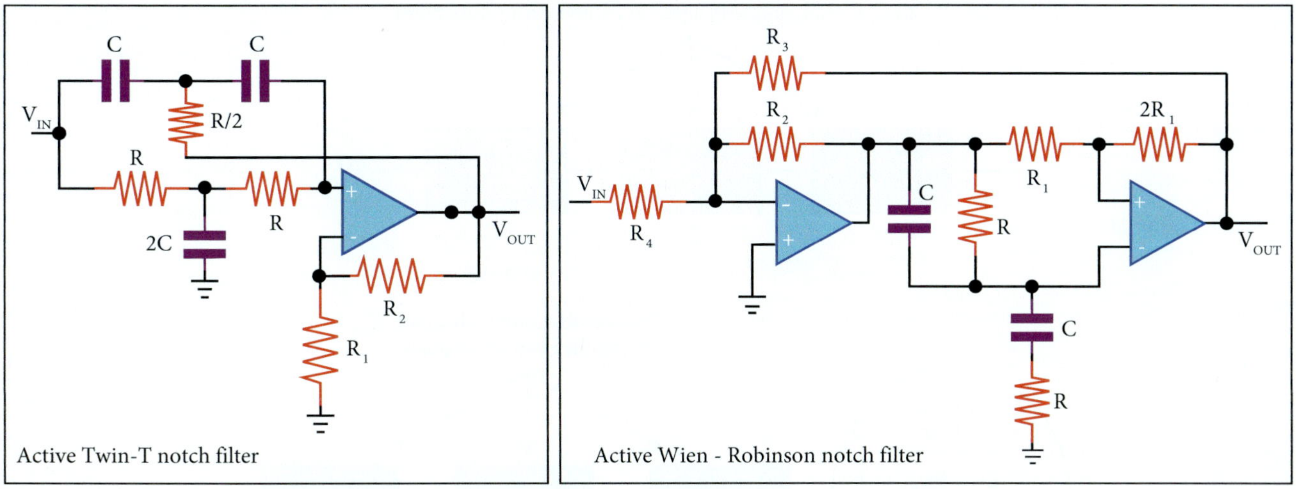

Fig. 11.25 Active second-order band-pass filters implemented by twin-T and Wien–Robinson circuits

At this point, we conclude our reference to second-order active filters with the observation that in addition to the aforementioned types of circuits, many others cannot be presented here due space limitation. Examples of such filters include *Deliyannis circuits*,[3] which allow the construction of band-pass filters, and *current generalized immittance converter (CGIC) circuits* of *Andreas Antonius*,[4] which allow the construction of all types of analog filters presented in the previous pages. More complex second-order circuits, such as *Tow–Thomas circuits* and *Kerwin–Huelsmann–Newcomb*, use three operational amplifiers. Detailed information about the characteristics and operation of these filters can be found in the literature.[5]

11.5.4 Higher-Order Filters

Using the first- and second-order filters presented in the previous sections as the *basic building units*, we can construct more complex higher-order filters by connecting the aforementioned simple filters. For example, by connecting in series a first-order low-pass filter with a second-order low-pass filter, we can construct a third-order low-pass filter. It is obvious that higher-order filters are characterized by their large size and degree of complexity, as well as higher cost. On the other hand, they offer *greater attenuation in the transition zone* and therefore are more effective in distinguishing signals characterized by different frequencies.

A method of connecting first- and second-order filters for the construction of higher-order filters is presented in Fig. 11.26. From this figure we observe that when the order N of a composite filter is an *even number*, it results from the series connection of $N/2$ second-order filters whose transfer function generally has a pair of complex conjugate poles. On the other hand, if the order of the filter is an *odd number*, then we should add to the aforementioned connection (which now will contain $(N-1)/2$ series connected second-order filters) a *first-order filter* whose transfer function has *a real pole*. As for the implementation of the transfer function, this can be done using both *Sallen–Key circuits* and *multiple feedback topologies* (or even some *combination* of these two). Although the order in which simple filters are connected does not matter at all, in practice, to avoid *saturation phenomena*, the filters are placed in *ascending order* in terms of the value of the quality factor Q, with the value of Q increasing from the input to the output of the filter. It is obvious that the gain of a composite filter is equal to the product of the gains of the simple filters from which it is composed.

From the mathematical side of the subject and taking into account that first- and second-order transfer functions have the general form

[3] See T. Deliyannis, "High-Q Factor Circuit with reduced sensitivity," *Electron. Lett.*, Vol. 4, p. 567, 1968.

[4] See Wai-Kai Chen (Ed), *Circuits and Filters Handbook*, 2nd Edition, ISBN 978-0849309120, 2002, Chapter 82, pp. 2495–2515.

[5] See, for example, (1) J. Tow, *IEEE Spectrum*, 6, 64 (1969), L. Thomas, *IEEE Trans. Circuit Theory*, CT 18, 350 (1971). and (2) W. Kerwin, L. Huelsman and R. Newcomb, *IEEE J. Solid State Circuit*, 2, 87 (1967).

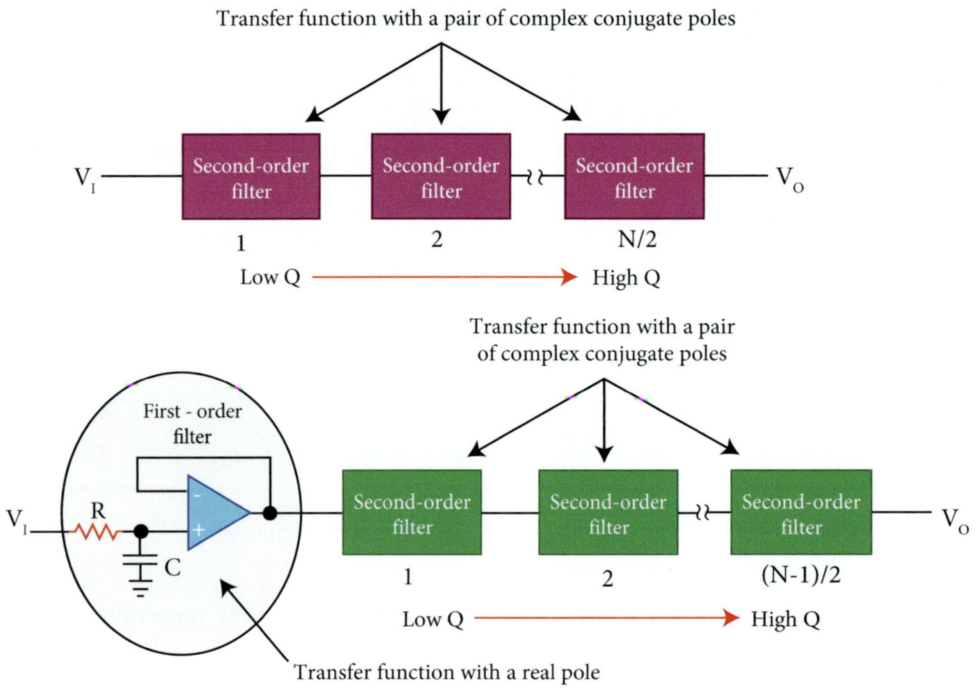

Fig. 11.26 Construction of higher-order filters from series connection of first- and second-order filters

$$\mathcal{H}(s) = K\frac{s+z}{s+p} \qquad \text{and} \qquad \mathcal{H}(s) = K\frac{s^2+\xi s+\zeta}{s^2+\mu s+\nu}$$

the general form of the transfer function of a higher-order filter with N poles and M zero values can arise as the *product of first- and second-order functions*

$$\mathcal{H}(s) = \mathcal{H}_1(s)\mathcal{H}_2(s)\ldots\mathcal{H}_{N_1+N_2}(s) = \frac{\sum_{k=0}^{M}\beta_k s^k}{\sum_{k=0}^{N}\alpha_k s^k} \sim \left\{\frac{\prod_{k=1}^{M_1}(s+z_k)}{\prod_{k=1}^{N_1}(p+s_k)}\right\}\left\{\frac{\prod_{k=1}^{M_2}(s^2+\xi_k s+\zeta_k)}{\prod_{k=1}^{N_2}(s^2+\mu_k s+\nu_k)}\right\} \tag{11.17}$$

where $M = M_1 + 2M_2$ and $N = N_1 + 2N_2$. Therefore, the amplitude and phase responses are formulated as

$$|\mathcal{H}(j\omega)| = |\mathcal{H}_1(j\omega)||\mathcal{H}_2(j\omega)|\ldots|\mathcal{H}_{N_1+N_2}(j\omega)| \qquad \text{and} \qquad \measuredangle\mathcal{H}(j\omega) = \measuredangle\mathcal{H}_1(j\omega) + \measuredangle\mathcal{H}_2(j\omega) + \cdots + \measuredangle\mathcal{H}_{N_1+N_2}(j\omega)$$

Taking into account that the gain of first-order filters is 20 dB/decade and that of second-order filters is 40 dB/decade (Sect. 5.7, page 426), it is evident that as the filter order increases, the behavior of the circuit more closely approximates that of an ideal filter. The amplitude and phase responses of the first- and second-order elementary filters used in the foregoing equations are calculated analytically in the next exercise.

Problem 11.5.4 Calculate the analytical expressions of amplitude and phase responses for first- and second-order elementary filters used for the construction of higher-order filters, and construct the corresponding expressions for the higher-order filters for even- and odd-order values.

Solution The transfer function of a second-order filter has the general explicit form

$$\mathcal{H}(s) = \frac{\beta_2 s^2 + \beta_1 s + \beta_0}{\alpha_2 s^2 + \alpha_1 s + \alpha_0} \tag{11.18}$$

which, by making the substitution $s = j\omega$, is formulated as

$$\mathcal{H}(j\omega) = \frac{-\beta_2\omega^2 + \beta_1(j\omega) + \beta_0}{-\alpha_2\omega^2 + \alpha_1(j\omega) + \alpha_0} = \frac{(\beta_0 - \beta_2\omega^2) + j\,(\beta_1\omega)}{(\alpha_0 - \alpha_2\omega^2) + j\,(\alpha_1\omega)}$$

Expressing the expressions of the numerator and the denominator in polar form, they take the form

$$(\beta_0 - \beta_2\omega^2) + j\,(\beta_1\omega) = \sqrt{(\beta_0 - \beta_2\omega^2)^2 + (\beta_1\omega)^2}\exp\left[j\tan^{-1}\left(\frac{\beta_1\omega}{\beta_0 - \beta_2\omega^2}\right)\right] \quad \text{and}$$

$$(\alpha_0 - \alpha_2\omega^2) + j\,(\alpha_1\omega) = \sqrt{(\alpha_0 - \alpha_2\omega^2)^2 + (\alpha_1\omega)^2}\exp\left[j\tan^{-1}\left(\frac{\alpha_1\omega}{\alpha_0 - \alpha_2\omega^2}\right)\right]$$

Therefore, we have

$$\mathcal{H}(j\omega) = |\mathcal{H}(j\omega)|e^{j\measuredangle\mathcal{H}(j\omega)} = \sqrt{\frac{(\beta_0 - \beta_2\omega^2)^2 + (\beta_1\omega)^2}{(\alpha_0 - \alpha_2\omega^2)^2 + (\alpha_1\omega)^2}}\exp\left\{j\left[\tan^{-1}\left(\frac{\beta_1\omega}{\beta_0 - \beta_2\omega^2}\right) - \tan^{-1}\left(\frac{\alpha_1\omega}{\alpha_0 - \alpha_2\omega^2}\right)\right]\right\} \tag{11.19}$$

and therefore the *magnitude* and *phase* of the transfer function are calculated as

$$|\mathcal{H}(j\omega)| = \sqrt{\frac{(\beta_0 - \beta_2\omega^2)^2 + (\beta_1\omega)^2}{(\alpha_0 - \alpha_2\omega^2)^2 + (\alpha_1\omega)^2}} \quad \text{and} \quad \measuredangle\mathcal{H}(j\omega) = \tan^{-1}\left(\frac{\beta_1\omega}{\beta_0 - \beta_2\omega^2}\right) - \tan^{-1}\left(\frac{\alpha_1\omega}{\alpha_0 - \alpha_2\omega^2}\right)$$

On the other hand, the transfer function for first-order filters arises from the transfer function of second-order filters by zeroing the appropriate coefficients. If the degree of the numerator polynomial of the rational transfer function is the same as the degree of the denominator polynomial, we must set $\beta_2 = \alpha_2 = 0$. In this case, we will have

$$\mathcal{H}(j\omega) = |\mathcal{H}(j\omega)|e^{j\measuredangle\mathcal{H}(j\omega)} = \sqrt{\frac{\beta_0^2 + (\beta_1\omega)^2}{\alpha_0^2 + (\alpha_1\omega)^2}}\exp\left\{j\left[\tan^{-1}\left(\frac{\beta_1\omega}{\beta_0}\right) - \tan^{-1}\left(\frac{\alpha_1\omega}{\alpha_0}\right)\right]\right\}$$

from which it follows that

$$|\mathcal{H}(j\omega)| = \sqrt{\frac{\beta_0^2 + (\beta_1\omega)^2}{\alpha_0^2 + (\alpha_1\omega)^2}} \quad \text{and} \quad \measuredangle\mathcal{H}(j\omega) = \tan^{-1}\left(\frac{\beta_1\omega}{\beta_0}\right) - \tan^{-1}\left(\frac{\alpha_1\omega}{\alpha_0}\right) \tag{11.20}$$

In practice, however, we are interested in transfer functions in which *the degree of the numerator polynomial is less than the degree of the denominator polynomial* (i.e., $M < N$). In this case, we should set $\beta_1 = \beta_2 = \alpha_2 = 0$, so the transfer function simplifies to

$$\mathcal{H}(j\omega) = |\mathcal{H}(j\omega)|e^{j\measuredangle\mathcal{H}(j\omega)} = \frac{\beta_0}{\sqrt{\alpha_0^2 + \alpha_1^2\omega^2}}\exp\left\{-j\tan^{-1}\left(\frac{\alpha_1\omega}{\alpha_0}\right)\right\} \tag{11.21}$$

with magnitude and phase given by the equations

$$|\mathcal{H}(j\omega)| = \frac{\beta_0}{\sqrt{\alpha_0^2 + \alpha_1^2\omega^2}} \quad \text{and} \quad \measuredangle\mathcal{H}(j\omega) = -\tan^{-1}\left(\frac{\alpha_1\omega}{\alpha_0}\right) \tag{11.22}$$

Taking into account that *an even-order filter N results from the series connection of N/2 elementary second-order filters*, while *an odd-order filter results from the series connection of an elementary first-order filter and N − 1 elementary second-order filters*, the amplitude responses for these two cases will have the form[6]

$$|\mathcal{H}(j\omega)| = \frac{\beta_0}{\sqrt{\alpha_0^2 + \alpha_1^2\omega^2}} \prod_{k=1}^{(N-1)/2} \sqrt{\frac{(\beta_{0k} - \beta_{2k}\omega^2)^2 + (\beta_{1k}\omega)^2}{(\alpha_{0k} - \alpha_{2k}\omega^2)^2 + (\alpha_{1k}\omega)^2}} \;\; \& \;\; |\mathcal{H}(j\omega)| = \prod_{k=1}^{N/2} \sqrt{\frac{(\beta_{0k} - \beta_{2k}\omega^2)^2 + (\beta_{1k}\omega)^2}{(\alpha_{0k} - \alpha_{2k}\omega^2)^2 + (\alpha_{1k}\omega)^2}} \tag{11.23}$$

while the corresponding phase responses are formulated as

$$\measuredangle\mathcal{H}(j\omega) = \sum_{k=1}^{N/2} \measuredangle\mathcal{H}_k(j\omega) = \sum_{k=1}^{N/2} \left\{ \tan^{-1}\left(\frac{\beta_{1k}\omega}{\beta_{0k} - \beta_{2k}\omega^2}\right) - \tan^{-1}\left(\frac{\alpha_{1k}\omega}{\alpha_{0k} - \alpha_{2k}\omega^2}\right) \right\} \tag{11.24}$$

$$\measuredangle\mathcal{H}(j\omega) = -\tan^{-1}\left(\frac{\alpha_{1k}\omega}{\alpha_{0k}}\right) + \sum_{k=1}^{(N-1)/2} \left\{ \tan^{-1}\left(\frac{\beta_{1k}\omega}{\beta_{0k} - \beta_{2k}\omega^2}\right) - \tan^{-1}\left(\frac{\alpha_{1k}\omega}{\alpha_{0k} - \alpha_{2k}\omega^2}\right) \right\} \tag{11.25}$$

The foregoing *additive property* will obviously characterize both phase and group delays—since they are expressed using the phase response—so we can write that

$$\tau_p(\omega) = \sum_{k=1}^{N} \tau_p^k(\omega) \quad \text{and} \quad \tau_d(\omega) = \sum_{k=1}^{N} \tau_d^k(\omega)$$

where $\tau_d^k(\omega)$ and $\tau_p^k(\omega)$ are the *phase and group delays* of the simple filter numbered k. We observe that the phase delay arises by dividing the phase response by the frequency ω and adding a negative sign, while the group delay is defined as the negative derivative of the phase response with respect to the frequency ω, with the result for even- and odd-order filter given respectively by the equations

$$\tau(\omega) = -\frac{d\measuredangle\mathcal{H}(j\omega)}{d\omega} = \sum_{k=1}^{N/2} \left\{ \frac{\alpha_{1k}\alpha_{0k} + \alpha_{1k}\alpha_{2k}\omega^2}{(\alpha_{0k} - \alpha_{2k}\omega^2)^2} - \frac{\beta_{1k}\beta_{0k} + \beta_{1k}\beta_{2k}\omega^2}{(\beta_{0k} - \beta_{2k}\omega^2)^2} \right\} \tag{11.26}$$

$$\tau(\omega) = -\frac{d\measuredangle\mathcal{H}(j\omega)}{d\omega} = \frac{\alpha_1\alpha_0}{\alpha_0^2 + \alpha_1^2\omega^2} + \sum_{k=1}^{(N-1)/2} \left\{ \frac{\alpha_{1k}\alpha_{0k} + \alpha_{1k}\alpha_{2k}\omega^2}{(\alpha_{0k} - \alpha_{2k}\omega^2)^2} - \frac{\beta_{1k}\beta_{0k} + \beta_{1k}\beta_{2k}\omega^2}{(\beta_{0k} - \beta_{2k}\omega^2)^2} \right\} \tag{11.27}$$

11.6 Design of Analog Filters

In practice, all analog filters are designed using a *normalized low-pass filter* whose frequency response is characterized by a passband whose upper limit is the frequency value $\omega = 1$ rad/s. However, the design process is different for passive and active filters. Starting with passive filter models, these are characterized by passive element values, which *are normalized*, so as to correspond to a load value equal to 1 Ω. This normalization method is very effective because it allows the construction of practically any filter structure using *graphs and tables of values*, which are then *scaled*, to give the values of the passive

[6] In the following equations, we use the simplified expressions corresponding to the case $M < N$, which is mainly used in practice.

elements for the desired filter structure in every case. This scaling is based on performing multiplications and divisions with the values of some constant coefficients, so that the frequency can be derived from the normalized value of 1 rad/s to its desired value and the complex resistance from the normalized value of 1 Ω to the desired load value.

The preceding design process refers to *low-pass analog filters*, while the corresponding high-pass filters are constructed from their analog equivalents, since the frequency response of a high-pass filter is the inverse of the frequency response of a low-pass filter, while the passive elements of a high-pass filter are the inverses of those of a low-pass filter. That is, the capacitors (coils) of a low-pass filter should be replaced by coils (capacitors) in the equivalent high-pass model. As for band-pass analog filters, these can be constructed from normalized low-pass models (although their construction process is more complex), while the response of band-stop filters is the *inverse* of that of band-pass filters.

On the other hand, the design process of *active analog filters* is not based on the use of tables with normalized element values but *on the positions of the poles and the zero values of the transfer function of the filter to be constructed on the complex plane*. These positions can be used to provide normalized values of elements for each type of active filter. As with passive filters, the frequency of the filter being designed is normalized to the value $\omega = 1$ rad/s and should be scaled so as to give the frequency value desired in each case. As for high-pass, band-pass, and band-stop active filters, these arise by scaling from the low-pass models, after first applying to them the appropriate *transformations*.

11.7 Paley–Wiener Theorem

As has already been mentioned (see Sect. 3.6.1 on page 283), the ideal analog filters have a noncausal impulse response $h(t)$ and therefore are not physically realizable. For this reason, in practice, we resort to their approximations by real filters, with each approximation offering its own distinct characteristics. Considering a transfer function that describes some filter, its characterization as realizable or not can be carried out using a fundamental theorem known as the *Paley–Wiener theorem*[7] and is formulated in the following way:

Theorem 11.7.1 *Considering a magnitude response function $|\mathcal{H}(j\omega)|$ that is square integrable, that is, it satisfies the condition*

$$\int_{-\infty}^{+\infty} |\mathcal{H}(j\omega)|^2 d\omega < \infty \tag{11.28}$$

the sufficient and necessary condition *for this function to constitute the amplitude response of a causal filter with impulse response $h(t)$ is given by the equation*

$$\int_{-\infty}^{+\infty} \frac{|\ln|\mathcal{H}(j\omega)||}{1+\omega^2} d\omega < \infty \tag{11.29}$$

The proof of the Paley–Wiener theorem is based on the verification of the validity of both the *sufficient* and *necessary* conditions. Starting with the necessary condition, we will prove that *if the impulse response of the filter $h(t)$ is a causal signal, then the condition dictated by the theorem applies*. To do this, let us assume that the magnitude and phase of the frequency response function have the form $|\mathcal{H}(j\omega)| = e^{\rho(\omega)}$ and $\measuredangle\mathcal{H}(j\omega) = \vartheta(\omega)$, so that the frequency response of the filter can be formulated as $\mathcal{H}(j\omega) = \exp\{\rho(\omega) + j\vartheta(\omega)\}$, with the functions $\rho(\omega)$ and $\vartheta(\omega)$ known as an *attenuation function* (or *loss function*) and a *phase function*, respectively. Without loss of generality, we assume that $|\mathcal{H}(j\omega)| \leq 1$.

Substituting the foregoing expression into the condition of the Paley–Wiener theorem, we write

$$\int_{-\infty}^{+\infty} \frac{|\ln|\mathcal{H}(j\omega)||}{1+\omega^2} d\omega = \int_{-\infty}^{+\infty} \frac{|\ln e^{\rho(\omega)}|}{1+\omega^2} d\omega = \int_{-\infty}^{+\infty} \frac{\rho(\omega)}{1+\omega^2} d\omega < \infty$$

[7] See Paley R. & Wiener N. (1934), *Fourier Transforms in the Complex Domain*, American Mathematical Society Colloquium Publications, Vol. 19, American Mathematical Society, New York.

Let us now consider the auxiliary function $f(s) = \rho(s/j)/(1+s)$. If, for simplicity's sake, we assume that the transfer function $\mathcal{H}(s)$ *does not have zero values that belong to the right half-plane or the imaginary axis*[8] and, of course, *does not contain poles in the right half-plane* (since we assume that the impulse response $h(t)$ is a causal signal), then the function $\rho(s/j)$—and, consequently, $f(s)$—will be *analytic* throughout the right complex half-plane, including the imaginary axis defined as the locus of points $s = j\omega$. In this case, the application of Gauss's formula (see the relevant section of the mathematical appendix) for the value $s = 1$ will give us

$$f(1) = \frac{1}{2\pi j}\oint_C \frac{f(s)}{s-1}ds = -\frac{1}{2\pi j}\oint_C \frac{f(s)}{1-s}ds = -\frac{1}{2\pi j}\oint_C \frac{\rho(s/j)}{(1-s)(1+s)}ds = -\frac{1}{2\pi j}\oint_C \frac{\rho(s/j)}{1-s^2}ds$$

However, from the definition of the function $f(s)$ it is obvious that $f(1) = \rho(1/j)/2$, and therefore,

$$\frac{1}{2\pi j}\oint_C \frac{\rho(1/j)}{1-s^2}ds = -\frac{1}{2}\rho(1/j)$$

If we reverse the direction in which we traverse the closed path C,[9] the foregoing relationship is written

$$\oint_C \frac{\rho(s/j)}{1-s^2}ds = j\pi\rho(1/j)$$

Let us now assume that as a path C we choose a closed curve that consists of (a) the *imaginary axis* $s = j\omega$, which crosses from $-\infty$ to $+\infty$, and (b) from the *right semicircle* C_R radius $R \to \infty$ described by the equation $s = Re^{i\varphi}$ with the angle φ varying in the range of values $[\pi/2, -\pi/2]$ (see the right image of Fig. 4.6 on page 337 for the value $\delta = 0$). Working as in Exercise 4.7.1 (page 337), it is not difficult to find that the curve integral along the path C_R is equal to zero, and therefore, the only contribution to the value of the integral along the closed path C will come from the integral of the function along the imaginary axis. Therefore, replacing the value $s = j\omega$ in the foregoing relationship, we get

$$\oint_C \frac{\rho(s/j)}{1-s^2}ds = \int_{-\infty}^{+\infty} \frac{\rho(j\omega/j)}{1-(j\omega)^2}(jd\omega) = j\int_{-\infty}^{+\infty} \frac{\rho(\omega)}{1+\omega^2}d\omega = j\pi\rho(1/j)$$

Combining the foregoing results, it turns out that

$$\int_{-\infty}^{+\infty} \frac{\rho(\omega)}{1+\omega^2}d\omega = \int_{-\infty}^{+\infty} \frac{|\ln|\mathcal{H}(j\omega)||}{1+\omega^2}d\omega = \pi\rho(1/j) = \pi \ln|\mathcal{H}(1)| < \infty$$

is an equation that proves the power of the *necessary* condition.

On the other hand, to prove the power of the *sufficient condition*, we must show that if the inequality of the Paley–Wiener theorem holds, then the impulse response $h(t)$ of the filter is *causal*, meaning that we can determine a function $\vartheta(\omega)$ for which the function $\mathcal{H}(j\omega) = |\mathcal{H}(j\omega)|e^{j\measuredangle\mathcal{H}(j\omega)}$ is analytic in the right complex half-plane. We recall from basic theory that the transfer function of a causal signal is *analytic in the aforementioned half-plane*, and therefore, if we prove the analytic nature of the transfer function in this half-plane, then the corresponding impulse response will be a causal signal. To do this, let us proceed with the variable change $\omega = -\tan(\zeta/2)$. It is not difficult to see that for $\omega = -\infty$, it will hold that $\zeta = -\pi$, while for $\omega = \infty$ it will hold that $\zeta = \pi$. Also, we will have $\ln|\mathcal{H}(j\omega)| = \ln[e^{\rho(\omega)}] = \rho(\omega) = \rho[-\tan(\zeta/2)] \triangleq \overline{\rho}(\zeta)$ and

$$d\omega = -d\left[\tan\left(\frac{\zeta}{2}\right)\right] = -\frac{1}{2}\frac{d\zeta}{\cos^2(\zeta/2)} = \frac{1+\omega^2}{2}d\zeta$$

Replacing the foregoing expressions in the condition of the Paley–Wiener theorem, it will take the form

[8] This assumption is made for convenience, as the theorem is still satisfied even if such zeros exist.

[9] Recall from basic theory that integrals defined in the complex plane are *line integrals* calculated along some curve and that if we traverse the curve in the opposite direction during their calculation, the sign of the result *is reversed*.

$$\int_{-\infty}^{+\infty} \frac{|\ln|\mathcal{H}(j\omega)||}{1+\omega^2} d\omega = \int_{-\pi}^{+\pi} \left[\frac{\overline{\rho}(\zeta)}{1+\omega^2}\right]\left(\frac{1+\omega^2}{2}\right) d\zeta = \frac{1}{2}\int_{-\pi}^{+\pi} |\overline{\rho}(\zeta)| d\zeta < \infty$$

But the last relationship means that the function $\overline{\rho}(\zeta)$ is *absolutely integrable* and therefore can be represented by a *convergent Fourier series* with period $T = 2\pi$ of the form (see Sect. 3.2.4 on page 198)

$$\overline{\rho}(\zeta) = \sum_{k=-\infty}^{+\infty} d_k \cos(k\zeta)$$

where d_k are the coefficients of the Fourier expansion. Now, defining the auxiliary function

$$F(w) = \sum_{k=-\infty}^{\infty} d_k w^k$$

it is not difficult to find that *this will also converge* for values $|w| < 1$ since it is proven that $F(1) = \overline{\rho}(0)$, while each w can be represented by an exponential quantity multiplied by a coefficient that is less than or equal to one. Therefore, the function $F(w)$ will be analytic *inside the unit circle*, and consequently, proceeding with the change in variable $w = (1-s)/(1+s)$, the function

$$F\left(\frac{1-s}{1+s}\right) \triangleq Z(s)$$

will be *analytic* throughout the right complex half-plane, since it is proven (see also Sect. 12.13.4 on page 1297) that the aforementioned change in variable *maps* the inside of the unit circle to the aforementioned half-plane. Defining, therefore, the transfer function of the filter as $\mathcal{H}(s) = e^{Z(s)}$ will lead to $\mathcal{H}(j\omega) = e^{Z(j\omega)}$ where

$$Z(j\omega) = F\left(\frac{1-j\omega}{1+j\omega}\right) = F\left(\frac{1 - j\tan\dfrac{\zeta}{2}}{1 + j\tan\dfrac{\zeta}{2}}\right) = F\left(\frac{1 - j\dfrac{\sin\frac{\zeta}{2}}{\cos\frac{\zeta}{2}}}{1 - j\dfrac{\sin\frac{\zeta}{2}}{\cos\frac{\zeta}{2}}}\right)$$

$$= F\left(\frac{\cos\dfrac{\zeta}{2} - j\sin\dfrac{\zeta}{2}}{\cos\dfrac{\zeta}{2} + j\sin\dfrac{\zeta}{2}}\right) = F\left(\frac{\exp\left[-j\dfrac{\zeta}{2}\right]}{\exp\left[+j\dfrac{\zeta}{2}\right]}\right) = F\left(e^{-i\zeta}\right) = \sum_{k=-\infty}^{+\infty} d_k e^{-ik\zeta}$$

But because the function $Z(s)$ is analytic throughout the right half-plane, the same will happen with $\mathcal{H}(s) = e^{Z(s)}$, and according to the previous argument, the impulse response of the filter $h(t) = \mathcal{L}^{-1}\{\mathcal{H}(s)\}$ will be a *causal* signal. Combining the expression $\mathcal{H}(j\omega) = e^{Z(j\omega)}$, the appropriate relations of complex analysis, as well as the constructed expressions, it immediately follows that

$$\ln|\mathcal{H}(j\omega)| = \Re\{Z(j\omega)\} = \Re\left\{\sum_{k=-\infty}^{+\infty} d_k e^{-ik\zeta}\right\} = \sum_{k=-\infty}^{+\infty} d_k \cos k\zeta = \overline{\rho}(\zeta) = \rho(\omega)$$

an equation that verifies the validity and the sufficient condition. We are led, therefore, to the conclusion that *if the condition of the Paley–Wiener theorem is satisfied, then this is a necessary and sufficient condition for the existence of a causal impulse response $h(t)$ that corresponds to the given function $|\mathcal{H}(j\omega)|$*. It is not difficult to understand that the Paley–Wiener theorem is *the equivalent of the causality condition in the frequency domain*.

There are many interesting conclusions related to the Paley–Wiener theorem, the most important of which are presented in what follows.

- Taking into account that high-pass and band-stop filters are characterized by an amplitude response that is *not square integrable*, it is obvious that they *do not satisfy the Paley–Wiener theorem*. However, these types of filters are characterized by a rational transfer function in which the degree of the numerator polynomial is equal to the degree of the denominator polynomial. Applying the method of polynomial long division to this function (see the corresponding section of the mathematical appendix), we can express it in the form $\mathcal{H}(s) = K_0 + \mathcal{H}_k(s)$, with the function $\mathcal{H}_k(s)$ being square integrable. Consequently, the Paley–Wiener theorem can be applied to these types of filters.
- The Paley–Wiener theorem does not guarantee that the function $\mathcal{H}(j\omega)$ that satisfies the condition of the theorem will have a *causal inverse function* $h(t)$. In reality, the theorem tells us that if the amplitude response $|\mathcal{H}(j\omega)|$ satisfies the condition of the theorem, then there will be a phase response $\measuredangle\mathcal{H}(j\omega)$ for which the function $\mathcal{H}(j\omega) = |\mathcal{H}(j\omega)|e^{\measuredangle\mathcal{H}(j\omega)}$ has a causal inverse function $h(t)$, but it does not provide us with information about this phase response.[10]
- If the function $|\mathcal{H}(j\omega)|$ is not square integrable, then the condition of the Paley–Wiener theorem is neither sufficient nor necessary.
- Although the condition of the Paley–Wiener theorem must be satisfied in order for the associated impulse response $h(t)$ to be a causal signal, in practice, there also exist other restrictions that are not dictated by the theorem, such as, for example, the requirement for the impulse response $h(t)$ to be a real signal, since it depends on the values of the circuit elements (such as, for example, the ohmic resistance R, the capacitor capacity C, and the self-induction of a coil L). However, for an impulse response to be a real signal, according to the symmetry properties of the Fourier transform, the amplitude response should be characterized by *even symmetry* and the phase response by *odd symmetry*.
- One requirement that must be met in order for the condition of the Paley–Wiener theorem to apply is for there *not to exist frequency zones for which the function* $|\mathcal{H}(j\omega)|$ *is everywhere zero* (something that is, however, allowed for *isolated points* of the imaginary axis. In this case, although the integrable function has an infinite value at these positions, the preceding integral can be finite). In a more detailed description, if there is a frequency zone $\omega_1 < \omega < \omega_2$ ($\omega_1 \neq \omega_2$) for which $|\mathcal{H}(j\omega)| = 0$, then $|\ln|\mathcal{H}(j\omega)|| = \infty$ and, consequently, also

$$\int_{-\infty}^{+\infty} \frac{|\ln|\mathcal{H}(j\omega)||}{1+\omega^2} d\omega = \infty$$

  In other words, the condition of the theorem is violated, and since this is a necessary condition, there is no corresponding *causal impulse response* $h(t)$. The same happens when the function $|\mathcal{H}(j\omega)|$ decreases exponentially (or at an even faster rate) as a function of the frequency ω. Relying on this property and observing the width response of ideal filters, we can see that their impulse responses are not causal signals, and for this reason ideal filters cannot be implemented in practice.
- An interesting consequence of the Paley–Wiener theorem is that *there is no continuous-time signal of finite energy* $h(t)$ *that is limited both in the time domain and in the frequency domain*, that is, it simultaneously satisfies the relations $h(t) = 0$ for $|t| > t_{\max}$ and $\mathcal{H}(j\omega) = \mathcal{F}\{h(t)\} = 0$ for $|\omega| > \omega_{\max}$, where $t_{\max}$ and $\omega_{\max}$ are specific time and frequency values. Indeed, if the signal $h(t)$ has finite energy, then the function $|\mathcal{H}(j\omega)|$ will be square integrable (see the Bessel equation of Sect. 3.3.5 on page 235) and for all causal signals that belong to this category, the Paley–Wiener theorem will apply. Consequently, if the signal is frequency-limited, that is, there is a frequency band $|\omega| > \omega_{\max}$ for which $\mathcal{H}(j\omega) = 0$, then, according to the previous observation, the condition of the Paley–Wiener theorem is violated, and therefore the signal will not be causal. Taking into account that every time-limited signal can be made causal, undergoing the appropriate time shift process in each case (which does not affect the function $|\mathcal{H}(j\omega)|$), it is obvious that a signal limited in frequency cannot simultaneously be limited in time. Conversely, if the signal is limited in time, then it will satisfy the condition of

[10] In reality, during the design of analog (and digital) filters, we only take into account the amplitude response characteristics, ignoring the phase response characteristics. This is mainly due to the fact that as a starting point for the design of the filter, we consider the power response $|\mathcal{H}(j\omega)|^2$ since it is real, characterized by zero phase, as demonstrated by the properties of the symmetry of the Fourier transform related to the real nature of the impulse response $h(t)$. As we will prove later, the magnitude and phase of a filter's transfer function are not independent but related through a Hilbert transform. Therefore, since the choice of a phase response for a given amplitude response is not a difficult process, the emphasis during the filter design process is on satisfying the requirements dictated by the form of the function $|\mathcal{H}(j\omega)|$.

the Paley–Wiener theorem (after a possible time shift to become causal) and therefore cannot simultaneously be limited in frequency.

11.8 Correlation of Amplitude and Phase Responses of Analog Filters

A very interesting property of analog filters described by a *real* and *causal* impulse response $h(t)$ is that their frequency response $\mathcal{H}(j\omega)$ can be fully determined *from only its real part* at any time t or *from only its imaginary part*, but only for times $t \neq 0$. To prove this property, let us express the impulse response $h(t)$ in terms of its even and odd components, which are known to be defined as

$$h_e(t) = \frac{h(t) + h(-t)}{2} \quad \text{and} \quad h_o(t) = \frac{h(t) - h(-t)}{2}$$

Starting with the formulation of the signal $h(t)$ in terms of its even component, it is not difficult to see that since the signal in question is causal, it will hold that $h(t) = 0$ for $t < 0$. But this means that for $t > 0$, $h(-t) = 0$, and therefore $h_e(t) = h(t)/2$, from which it follows that $h(t) = 2h_e(t)$. Finally, at the time $t = 0$ the nonzero parts of the signals $h(t)$ and $h(-t)$ will overlap and will be equal to each other; thus, we will have $h(t) = h_e(t)$. Combining the foregoing results, we can express the *real impulse response* of the filter as

$$h(t) = \begin{cases} 0 & \text{for } t < 0 \\ h_e(t) & \text{for } t = 0 \\ 2h_e(t) & \text{for } t > 0 \end{cases}$$

From the properties of the Fourier transform, we know that if a continuous-time signal $x(t)$ has a Fourier transform $\mathcal{X}(j\omega) = \mathcal{F}\{x(t)\}$, then the Fourier transform of the even component of the signal will be equal to the real part of the function $\mathcal{X}(j\omega)$, while the Fourier transform of the odd component of the signal will be equal to the imaginary part of the function $\mathcal{X}(j\omega)$, or, in mathematical notation, $\mathcal{F}\{\mathcal{E}\mathrm{v}(t)\} = \Re\{\mathcal{X}(j\omega)\}$ and $\mathcal{F}\{\mathcal{O}\mathrm{d}(t)\} = j\Im\{\mathcal{X}(j\omega)\}$. Therefore, if we only know the real part of the frequency response of a filter, we can determine the even component of its impulse response $h(t)$ and from this the impulse response $h(t)$ (provided it is a real causal signal, which is always the case in practice) by taking the inverse Fourier transform. After calculating the impulse response $h(t)$, we can calculate the *frequency response* of the system as $\mathcal{H}(j\omega) = \mathcal{F}\{h(t)\}$. Therefore, the frequency response of the filter is fully determined by knowing only the real part of the impulse response.

Let us now see what conclusions we reach by expressing the impulse response of a filter as a function of only its odd component. For $t < 0$ we will again have $h(t) = 0$, since the signal is causal, and therefore, for $t > 0$ it will hold that $h(-t) = 0$, from which it follows that $h(t) = 2h_o(t)$. However, in this case, for $t = 0$, the expression $h(t) - h(-t)$ that appears in the numerator of the equation defining the odd component is zeroed. But this means that the function $h(t)$ has been removed from the equation, and therefore, it is impossible to express the impulse response of the filter as a function of its odd component. The foregoing results are stated in summary form as

$$h(t) = \begin{cases} 0 & \text{for } t < 0 \\ \text{unknown} & \text{for } t = 0 \\ 2h_o(t) & \text{for } t > 0 \end{cases}$$

Following the argument presented earlier, we can assert that if we know *only the imaginary part of a transfer function*, we can recover the odd component of the impulse response $h_o(t)$ from the inverse Fourier transform of the function $j\Im\{\mathcal{X}(j\omega)\}$ and, from this, the impulse response $h(t)$ for all time points except the value $t = 0$. In the last step of the process, knowing the impulse response, we can calculate the frequency response function from the relationship $\mathcal{H}(j\omega) = \mathcal{F}\{h(t)\}$. Note that if the impulse response does not contain anomalies at the position $t = 0$, the function $\mathcal{H}(j\omega)$ can be calculated even if the value $h(0)$ is not known. Under these conditions, we find that *if we only have the imaginary part of the function $\mathcal{H}(j\omega)$, we can precisely determine the function in question.*

After proving the foregoing properties, let us move on to another issue that is also interesting: Is there any correlation between the real and imaginary parts of the frequency response of a causal filter? To answer this question, let us start with the impulse response of the filter, which should be a causal signal of real values, a property that mathematically is formulated as $h(t) = h(t)u(t)$, where $u(t)$ is our known step function (see Eq. 1.14 as well as the discussion in Sect. 1.6.1 on page 21 on causality). Taking the Fourier transform of the two sides and using the property of the product (or multiplication) of the Fourier transform (Sect. 3.3.5), we get

$$\mathcal{H}(j\omega) = \mathcal{F}\{h(t)\} = \mathcal{F}\{h(t)u(t)\} = \frac{1}{2\pi}\Big[\mathcal{H}(j\omega) * \mathcal{U}(j\omega)\Big]$$

where $\mathcal{H}(j\omega) = \mathcal{F}\{h(t)\}$ and $\mathcal{U}(j\omega) = \mathcal{F}\{u(t)\}$ are the Fourier transforms of the continuous-time signals $h(t)$ and $u(t)$. However, we know that the Fourier transform of the step function has the form

$$\mathcal{U}(j\omega) = \mathcal{F}\{u(t)\} = \frac{1}{j\omega} + \pi\delta(\omega)$$

(see the proof of the integration property in Sect. 3.3.5), and therefore it will hold that

$$\mathcal{H}(j\omega) = \frac{1}{2\pi}\left\{\mathcal{H}(j\omega) * \left[\frac{1}{j\omega} + \pi\delta(\omega)\right]\right\} = \frac{1}{2}\mathcal{H}(j\omega) * \delta(\omega) + \frac{1}{2\pi j}\left[\mathcal{H}(j\omega) * \frac{1}{\omega}\right] = \frac{1}{2}\mathcal{H}(j\omega) + \frac{1}{2\pi j}\left[\mathcal{H}(j\omega) * \frac{1}{\omega}\right]$$

since it is known that *the impulse function is the neutral element of a convolution operation*, and therefore it will hold that $\mathcal{H}(j\omega) * \delta(\omega) = \mathcal{H}(j\omega)$. Now defining the auxiliary function $f(\omega) = 1/\omega$ and using the defining equation of convolution, the last expression is written

$$\mathcal{H}(j\omega) * \frac{1}{\omega} = \mathcal{H}(j\omega) * f(\omega) = \int_{-\infty}^{+\infty} \mathcal{H}(j\eta) f(\omega - \eta) d\eta = \int_{-\infty}^{+\infty} \frac{\mathcal{H}(j\eta)}{\omega - \eta} d\eta$$

Therefore, we will get

$$\mathcal{H}(j\omega) = \frac{1}{2}\mathcal{H}(j\omega) + \frac{1}{2\pi j}\int_{-\infty}^{+\infty} \frac{\mathcal{H}(j\eta)}{\omega - \eta} d\eta \qquad \text{or, equivalently,} \qquad \mathcal{H}(j\omega) = \frac{1}{\pi j}\int_{-\infty}^{+\infty} \frac{\mathcal{H}(j\eta)}{\omega - \eta} d\eta$$

Taking into account that the function $\mathcal{H}(j\omega)$ is a complex function, let us express it in its Cartesian form, $\mathcal{H}(j\omega) = R(j\omega) + jI(j\omega)$, where $R(j\omega) = \Re\{\mathcal{H}(j\omega)\}$ and $I(j\omega) = \Im\{\mathcal{H}(j\omega)\}$ are its real and imaginary parts. Replacing this relation in the last expression, we write

$$\begin{aligned} R(j\omega) + jI(j\omega) &= \frac{1}{\pi j}\int_{-\infty}^{+\infty} \frac{R(j\eta) + j\,I(j\eta)}{\omega - \eta} d\eta = \frac{1}{\pi j}\int_{-\infty}^{+\infty} \frac{R(j\eta)}{\omega - \eta} d\eta + j\frac{1}{\pi j}\int_{-\infty}^{+\infty} \frac{I(j\eta)}{\omega - \eta} d\eta \\ &= \frac{1}{\pi}\int_{-\infty}^{+\infty} \frac{I(j\eta)}{\omega - \eta} d\eta - j\frac{1}{\pi}\int_{-\infty}^{+\infty} \frac{R(j\eta)}{\omega - \eta} d\eta \end{aligned}$$

Comparing the real and imaginary parts in the initial and final expressions, we find that[11]

$$R(j\omega) = \frac{1}{\pi}\int_{-\infty}^{+\infty} \frac{I(j\eta)}{\omega - \eta} d\eta = -\frac{1}{\pi}\int_{-\infty}^{+\infty} \frac{I(j\eta)}{\eta - \omega} d\eta \qquad \text{and} \qquad I(j\omega) = -\frac{1}{\pi}\int_{-\infty}^{+\infty} \frac{R(j\eta)}{\omega - \eta} d\eta = \frac{1}{\pi}\int_{-\infty}^{+\infty} \frac{R(j\eta)}{\eta - \omega} d\eta$$

[11] In reality, the foregoing relationship applies to low-pass and band-pass filters, not to high-pass and band-stop filters, whose impulse response includes an impulse function (see Sect. 3.6.1 on page 283). If we wish to include in the analysis the last two categories of filters, we must start from an impulse response of the form $h(t) = K\delta(t) + h(t)u(t)$, with the proportionality constant K being zero for low-pass and band-pass filters and different from zero for high-pass and band-stop filters. If we apply the same process and take into account that $\mathcal{F}\{\delta(t)\} = 1$, we will end up again with the same relationships, with the difference that in the expression of the function $R(j\omega)$ we should add the proportionality constant K. Let it be noted that this constant theoretically will appear also in the case where the magnitude of the frequency response of the low-pass filter has a nonzero value at the position $\omega = \infty$; however, its value is very small and can be considered negligible.

Although the foregoing results are valid only for causal systems (since the starting point was the expression $h(t) = h(t)u(t)$, which refers to a causal impulse response), in reality, the opposite is also true, that is, every frequency response whose real and imaginary parts are related to each other through the foregoing expressions constitutes the Fourier transform of a causal impulse response.

The integrals that appear in the defining equations of the functions $R(j\omega)$ and $I(j\omega)$ refer to a new generalized integral transform known as the *Hilbert transform,*[12] in honor of the mathematician David Hilbert who invented it in an attempt to solve problems related to the space of holomorphic functions. In full analogy with the other transforms presented in previous chapters, the Hilbert transform is accompanied by its inverse, with the pair of equations of direct and inverse transforms defined as[13]

$$f(t) = -\frac{1}{\pi}\int_{-\infty}^{+\infty}\frac{x(\zeta)}{t-\zeta}d\zeta = \frac{1}{\pi}\int_{-\infty}^{+\infty}\frac{x(\zeta)}{\zeta-t}d\zeta = x(t) * \left(\frac{1}{\pi t}\right) \tag{11.30}$$

$$x(t) = -\frac{1}{\pi}\int_{-\infty}^{+\infty}\frac{f(\zeta)}{\zeta-t}d\zeta = \frac{1}{\pi}\int_{-\infty}^{+\infty}\frac{f(\zeta)}{t-\zeta}d\zeta = f(t) * \left(\frac{1}{\pi t}\right) \tag{11.31}$$

If two functions are related to each other in the described way, then we will say that *the function $f(t)$ constitutes the Hilbert transform of the function $x(t)$, while the function $x(t)$ constitutes the inverse Hilbert transform of the function $f(t)$*. We are led, therefore, to the very important conclusion that *the real and the imaginary parts of the frequency response of a causal system*—that is, of a system described by a causal impulse response—*are related to each other through a Hilbert transform*, and therefore, *knowledge of one allows the calculation of the other*. What we do, then, in practice is determine the desired characteristics of a filter by constructing the appropriate amplitude response *and as long as it is feasible*, that is, as long as it meets the requirements of the Paley–Wiener theorem, then we calculate the corresponding phase response as the inverse Hilbert transform of the amplitude response that has been determined. The foregoing relations, which connect the real and imaginary parts of the transfer function via the Hilbert transform, are known as the *Kramers–Kronig* relations..

A very interesting property of the Hilbert transform that finds wide application in the field of continuous-time signals is that if two functions $f(t)$ and $x(t)$ are related to each other through that transform, then the Fourier transform of the complex function $h(t) = x(t) + if(t)$ is *one-sided* and, therefore, is *nonzero* only for frequency values $\omega > 0$. This is something that is easily proven if we take the Fourier transform of the two sides of the defining equation of the direct Hilbert transform, use the property of the *convolution* of the Fourier transform, and take into account that $\mathcal{F}(1/t) = -j\pi\,\mathrm{sgn}(\omega)$, where $\mathrm{sgn}(\omega)$ is the sign function (for the proof of the last relation see Exercise 3.3.17 on page 248). Thus, it will hold that

$$\mathcal{F}(\omega) = \mathcal{F}\{f(t)\} = \mathcal{F}\left\{x(t) * \left(\frac{1}{\pi t}\right)\right\} = \frac{1}{\pi}\mathcal{F}\{x(t)\}\mathcal{F}\left\{\frac{1}{t}\right\} = \frac{1}{\pi}\mathcal{X}(j\omega)[-j\pi\,\mathrm{sgn}(\omega)]$$

$$= -j\mathcal{X}(j\omega)\mathrm{sgn}(\omega) = \begin{cases} -j\mathcal{X}(j\omega) & \omega > 0 \\ +j\mathcal{X}(j\omega) & \omega < 0 \end{cases}$$

Now defining the complex function $h(t) = x(t) + if(t)$, its Fourier transform, according to the linearity property that characterizes it, will be calculated as

---

[12] In contrast to other transforms, such as the *Fourier transform*, which causes a change in the representation domain from t to ω, and the *Laplace transform*, which causes a change in the representation domain from t to s, the Hilbert transform does not change this space; in other words, two functions $f(t)$ and $g(t)$ that are related to each other through a Hilbert transform are both defined in the time domain t.

[13] Note that because the integral that appears in the equation defining the Hilbert transform is generalized (since it tends to infinity for the value $t = \zeta$), this transform is defined in terms of the *principal value* of the aforementioned integral. Therefore, a more accurate formulation of the direct transform is

$$f(t) = -\frac{1}{\pi}\mathcal{P}\int_{-\infty}^{+\infty}\frac{x(\zeta)}{t-\zeta}d\zeta = -\frac{1}{\pi}\lim_{\varepsilon\to 0}\underbrace{\left\{\int_{-A}^{-\varepsilon}\frac{x(\zeta)}{t-\zeta}d\zeta + \int_{\varepsilon}^{A}\frac{x(\zeta)}{t-\zeta}d\zeta\right\}}_{\lim A\to\infty}$$

while a similar formulation characterizes the equation defining the inverse transform.

$$\mathcal{H}(j\omega)=\mathcal{F}\{h(t)\}=\mathcal{F}\{x(t)+if(t)\}=\mathcal{F}\{x(t)\}+j\mathcal{F}\{f(t)\}$$

$$=\mathcal{X}(j\omega)+j\begin{cases}-j\mathcal{X}(j\omega) & \omega>0\\ +j\mathcal{X}(j\omega) & \omega<0\end{cases}=\mathcal{X}(j\omega)+\begin{cases}\mathcal{X}(j\omega) & \omega>0\\ -\mathcal{X}(j\omega) & \omega<0\end{cases}=\begin{cases}2\mathcal{X}(j\omega) & \omega>0\\ 0 & \omega<0\end{cases}=2\mathcal{X}(j\omega)u(\omega)$$

from which the desired result is derived. The complex signals that satisfy the foregoing property are known as *complex analytical signals*.

The foregoing correlation of the real and imaginary parts of the frequency response function through the Hilbert transform applies also in cases where this function is expressed in the amplitude–phase representation. In other words, *the magnitude and phase of a frequency response function of a causal continuous-time filter constitute a pair of Hilbert transforms*. However, in contrast to the previous case in which there are no restrictions regarding the phase characteristics of the function, the correlation of the magnitude and phase of the function $\mathcal{H}(j\omega)$ is valid only when the system implemented by the function is a *minimum phase system*.[14] If the considered system is not a minimum phase system, then it is proven that the frequency response function[15] can always be expressed as a product of the form $\mathcal{H}(j\omega)=\mathcal{H}_{\text{min}}(\omega)\mathcal{H}_{\text{ap}}(\omega)$, where the function $\mathcal{H}_{\text{min}}(\omega)=|\mathcal{H}_{\text{min}}(\omega)|e^{j\varphi(\omega)}=A_\varphi(\omega)+jB_\varphi(\omega)$ is a *minimum phase function*, while the function $\mathcal{H}_{\text{ap}}(\omega)$ is an *entire*

---

[14] As was already mentioned, a necessary condition for the stability of a LTI system is the appearance of all the poles of its transfer function in the negative complex half-plane. This restriction does not concern the zero values of the function, which can be located on both the right and left half-planes. In the special case where *all the zero values of a transfer function are also located in the left half-plane*, the corresponding system is known as a *minimum phase system*. It is proven that if the number of zero values is the same as the number of poles, then the considered system is *causal* and *stable* and has an inverse system that is also causal and stable; however, usually the number of zero values is less than the number of poles. It is also proven that a transfer function that is not of minimum phase, i.e., has zeros in the right half-plane (but not on the imaginary axis), can be written as a product of the form $\mathcal{H}(s)=\mathcal{H}_{\text{min}}(s)\mathcal{H}_{\text{ap}}(s)$, where $\mathcal{H}_{\text{min}}(s)$ is the transfer function of a minimum phase system and $\mathcal{H}_{\text{ap}}(s)$ is the transfer function of an all-pass system. This in an equivalent formulation means that *a nonminimum phase system arises from the series connection of a minimum phase system with an all-pass system*. Replacing the foregoing transfer functions from their amplitude and phase representation immediately results in $|\mathcal{H}(s)|=|\mathcal{H}_{\text{min}}(s)||\mathcal{H}_{\text{ap}}(s)|$ and $\measuredangle\mathcal{H}(s)=\measuredangle\mathcal{H}_{\text{min}}(s)+\measuredangle\mathcal{H}_{\text{ap}}(s)$. But the magnitude of the transfer function of an all-pass system is always equal to one—that is, $|\mathcal{H}_{\text{ap}}(s)|=1$—and therefore, for $s=j\omega$, it will hold that $|\mathcal{H}(j\omega)|=|\mathcal{H}_{\text{min}}(j\omega)|$. On the other hand, however, the all-pass system is characterized by a *positive phase response* and causes a phase shift from 0° to $-180°$. We end up, therefore, with the very interesting conclusion that *a minimum phase system has the same amplitude response as a nonminimum phase system and the lowest possible phase value*, from which its name also derives.

To prove the foregoing property, let us consider a nonminimum phase system described by a causal and stable transfer function, $\mathcal{H}(s)$, which contains N poles and $M=M_1+M_2$ zero values, of which M_1 belong to the left half-plane—let us denote them by ℓ_k $(k=1,2,\ldots,M_1)$—and the M_2 to the right half-plane—let us denote them by r_i $(i=1,2,\ldots,M_2)$. In this case, based on the definition of the system's transfer function we will have

$$\mathcal{H}(s)=A\left\{\prod_{k=1}^{M_1}(s-\ell_l)\prod_{i=1}^{M_2}(s-r_i)\right\}\Big/\left\{\prod_{n=1}^{N}(s-p_n)\right\}$$

Now expressing the second product of the numerator in the form

$$\prod_{i=1}^{M_2}(s-r_i)=\prod_{i=1}^{M_2}(s+r_i)\prod_{i=1}^{M_2}\frac{(s-r_i)}{(s+r_i)}$$

we find that since these zeros belong to the right half-plane, $\Re\{r_i\}>0$, and therefore, the first of the two products concerns zero values that belong to the left half-plane. On the other hand, the second product refers to the transfer function of an *all-pass filter*, which contains poles in the left half-plane and zeros in the mirror positions with respect to the imaginary axis of the poles in the right half-plane. Therefore, expressing the transfer function $\mathcal{H}(s)$ in the form

$$\mathcal{H}(s)=\underbrace{\left\{A\frac{\prod_{k=1}^{M_1}(s-\ell_k)\prod_{i=1}^{M_2}(s+r_i)}{\prod_{i=1}^{N}(s-p_n)}\right\}}_{\mathcal{H}_{\text{min}}(s)}\underbrace{\left\{\prod_{i=1}^{M_2}\frac{s-r_i}{s+r_i}\right\}}_{\mathcal{H}_{\text{ap}}(s)}$$

we prove the desired property, since the function $\mathcal{H}_{\text{min}}(s)$, having its poles and zeros only in the left half-plane, is a minimum phase function, while the function $\mathcal{H}_{\text{ap}}(s)$ is an entire function. Therefore, for a given transfer function that is not minimum phase, we can construct the corresponding minimum phase function *by determining the zero values of the initial function that belong to the right half-plane and associating with each of these zero values an entire transfer function by adding a zero in the left half-plane and in a mirror position with respect to the imaginary axis*. If we derive the relation $\measuredangle\mathcal{H}(s)=\measuredangle\mathcal{H}_{\text{min}}(s)+\measuredangle\mathcal{H}_{\text{ap}}(s)$, we find that a similar property characterizes the magnitude of group delay, and for this reason minimum phase filters are also known as *minimum group delay systems*.

[15] Note that all these observations apply provided the frequency response function satisfies the condition of the Paley–Wiener theorem.

function, so the correlation through the Hilbert transform will concern the real and imaginary parts of the $A_{\varphi}(\omega)$ and $B_{\varphi}(\omega)$, respectively. In this case, it is proven that the desired pair of equations takes the form

$$\measuredangle\mathcal{H}(j\omega) = \frac{1}{\pi}\int_{-\infty}^{+\infty}\frac{d}{du}\Big\{\ln|\mathcal{H}(j\omega)|\Big\}\ln\left(\coth\frac{|u|}{2}\right)du \tag{11.32}$$

and

$$\ln|\mathcal{H}(j\omega_0)| = \ln|\mathcal{H}(j\infty)| - \frac{1}{\pi\omega_0}\int_{-\infty}^{+\infty}\frac{d}{du}\Big\{\omega\measuredangle\mathcal{H}(j\omega)\Big\}\ln\left(\coth\frac{|u|}{2}\right)du \quad \text{or} \tag{11.33}$$

$$\ln|\mathcal{H}(j\omega_0)| = \ln|\mathcal{H}(0)| - \frac{\omega_0}{\pi}\int_{-\infty}^{+\infty}\frac{d}{du}\left\{\frac{\measuredangle\mathcal{H}(j\omega)}{\omega}\right\}\ln\left(\coth\frac{|u|}{2}\right)du \tag{11.34}$$

The preceding equations are known as *Bode integrals*, and their proof follows.

Following the methodology of the proof of the Paley–Wiener theorem, let us assume that the function $\mathcal{H}(j\omega)$ is a minimum phase function, let us express it in the form $|\mathcal{H}(j\omega) = \exp\{\rho(\omega) + j\vartheta(\omega)\}$, and let us define the so-called *propagation function* as the logarithm $\gamma(\omega) = \ln\{\mathcal{H}(j\omega)\} = \rho(\omega) + j\vartheta(\omega)$. Taking into account that this function is analytic in the entire right half-plane, we can correlate its real and imaginary parts with the help of the *Kramers–Kronig equations*—since $\rho(\omega) = \ln|\mathcal{H}(j\omega)|$ and $\vartheta(\omega) = \measuredangle\mathcal{H}(j\omega)$, we will have

$$\ln|\mathcal{H}(j\omega)| = \frac{1}{\pi}\int_{-\infty}^{+\infty}\frac{\measuredangle\mathcal{H}(j\eta)}{\omega-\eta}d\eta \qquad \text{and} \qquad \measuredangle\mathcal{H}(j\omega) = -\frac{1}{\pi}\int_{-\infty}^{+\infty}\frac{\ln|\mathcal{H}(j\eta)|}{\omega-\eta}d\eta \tag{11.35}$$

In practice, however, the foregoing relations are not used, but two special types of integrals known as the *Bode's first and second integrals*, which allow the correlation of the functions $|\mathcal{H}(j\omega)|$ and $\measuredangle\mathcal{H}(j\omega)$ calculated at the position of some frequency $\omega = \omega_0$. The process of constructing these integrals is presented in detail below.

It is not difficult to see that the propagation function $\gamma(\omega)$ is characterized by a real part with even symmetry, an imaginary part with odd symmetry [for real impulse response $h(t)$], while its analytical nature at the position $\omega = \infty$ allows us to write it in the form

$$\gamma(\omega) = \rho(\infty) + j\frac{\vartheta(\infty)}{\omega}$$

Although the preceding expansion contains other higher-order terms, these are ignored, because the analysis we will present is based on the assumption that the function $\gamma(\omega)$—as well as the functions that contain it—changes at most proportionally to $1/\omega$. The main reason we resort to this assumption has to do with the ability to calculate contour integrals on the complex plane along circular arcs of radius R. Considering an arc C of radius R defined by the points $z_1 = Re^{j\vartheta_1}$ and $z_2 = Re^{j\vartheta_2}$ (see Fig. 11.27a) and is a segment of a circle's circumference with radius R lying on the complex plane with the origin as its center, the contour integral of the function $1/z$ along path C is calculated as

$$\int_C\frac{dz}{z} = \int_{\vartheta_1}^{\vartheta_2}\frac{jRe^{j\vartheta}d\vartheta}{Re^{j\vartheta}} = \int_{\vartheta_1}^{\vartheta_2}jd\vartheta = -j(\vartheta_1-\vartheta_2)$$

since $z = Re^{j\vartheta}$, and therefore, $dz = jRe^{j\vartheta}d\vartheta$ (the type of sign will obviously depend on the direction in which we traverse path C). If, therefore, we impose the requirement for the function $\gamma(\omega)$ to change at most proportionally to $1/\omega$, then we can calculate its appropriate contour integrals in each case, using the foregoing relation. If this change is even greater—for example, proportional to $1/\omega^2$—then for a long path length the value of the function's contour integrals tends to zero. A final property that the function $\gamma(\omega)$ must satisfy is *not to have poles on the vertical axis*, which in this case *is true*, since it is a minimum phase function. Generally speaking, if there are poles on this axis, they are bypassed by appropriately choosing the closed path (see Fig. 11.27b).

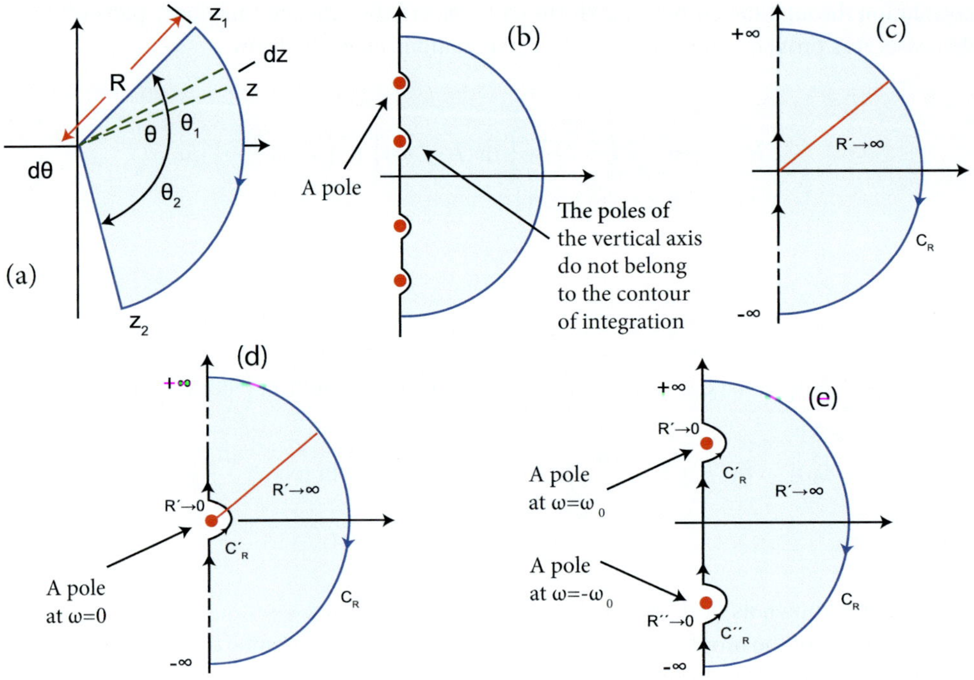

Fig. 11.27 (**a**) Calculation of contour integral of function $f(z) = 1/z$ along closed path C, (**b**) bypassing of the poles that lie along a closed path, (c), (d), and (e) the closed paths used for the calculation of the integrals of Bode

In the following description, we will use the *Cauchy's theorem* and *techniques of integrating functions on the complex plane* to construct expressions that correlate the functions $\rho(\omega)$ and $\vartheta(\omega)$, that is, essentially the amplitude and phase functions $|\mathcal{H}(j\omega)|$ and $\measuredangle\mathcal{H}(j\omega)$. Since the function $\gamma(\infty)$ is defined and is analytic at the position $\omega = \infty$, let us choose as the function to be integrated $\gamma(\omega) - \rho(\infty)$ and as a closed path the one we used for the proof of the Paley–Wiener theorem [this path consists of the vertical axis that is crossed from bottom to top and from $-\infty$ to $+\infty$ and from a semicircle C_R of radius $R \to \infty$ extending to the right half-plane (see image (c) of Fig. 11.27)]. Since within this closed path there are no poles, the contour integral along this will be equal to zero, that is,

$$\oint[\gamma(\omega) - \rho(\infty)]d\omega = \int_{-\infty}^{+\infty}[\gamma(\omega) - \rho(\infty)]d\omega + \int_{C_R}[\gamma(\omega) - \rho(\infty)]d\omega = 0$$

However, $\gamma(\omega) = \rho(\omega) + j\vartheta(\omega)$ and noting that $\gamma(\omega) - \rho(\infty) = j\vartheta(\infty)/\omega$, the above relation is written

$$\int_{-\infty}^{+\infty}[\rho(\omega) - \rho(\infty)]\,d\omega + j\int_{-\infty}^{+\infty}\vartheta(\omega)\,d\omega + j\int_{C_R}\frac{\vartheta(\infty)}{\omega}d\omega = 0$$

Taking into account that the functions $\rho(\omega)$ and $\vartheta(\omega)$ are characterized by even and odd symmetry, respectively, we will have $\rho(-\omega) = \rho(\omega)$ and $\vartheta(-\omega) = -\vartheta(\omega)$, and working as in Sect. 3.2.1 (page 190), we will get

$$\int_{-\infty}^{+\infty}[\rho(\omega) - \rho(\infty)]\,d\omega = 2\int_{0}^{+\infty}[\rho(\omega) - \rho(\infty)]\,d\omega \qquad \text{and} \qquad \int_{-\infty}^{+\infty}\vartheta(\omega)d\omega = 0$$

As for the third integral along the semicircular arc, it arises using the foregoing relationship for values $\vartheta_1 = \pi/2$ and $\vartheta_2 = -\pi/2$; therefore,

$$\int_{C_R}\frac{\vartheta(\infty)}{\omega}d\omega = \vartheta(\infty)\int_{-\frac{\pi}{2}}^{+\frac{\pi}{2}} j\,d\vartheta = j\pi$$

and finally

$$\int_0^{+\infty} [\rho(\omega) - \rho(\infty)]\, d\omega = -\frac{\pi}{2}\vartheta(\infty)$$

The last equation expresses the integral of the logarithm of the amplitude of the system's transfer function in relation to the behavior of the phase function at the position $\omega = \infty$, while there is also a similar relationship that expresses the integral of the phase component in relation to the behavior of the logarithm of the amplitude at high frequencies. To construct this relationship, we will use the procedure applied previously, but as an integrable function we will use, not $\gamma(\omega) - \rho(\infty)$, but $\gamma(\omega)/\omega$. However, in this case, we will not choose the closed path of image (c) of Fig. 11.27, but conversely, the closed path of image (d) of the same figure, since the function $\gamma(\omega)/\omega$ contains a pole at the position $\omega = 0$, which we must bypass, something that is done by adding the small semicircular path of radius $R' \to 0$ that circles the origin of the O axes. Applying *Cauchy's theorem* and taking into account that within this path there are no poles, we will have

$$\begin{aligned}\oint \frac{\gamma(\omega)}{\omega} d\omega &= \int_{-\infty}^{+\infty} \frac{\gamma(\omega)}{\omega} d\omega + \int_{C_R} \frac{\gamma(\omega)}{\omega} d\omega + \int_{C'_R} \frac{\gamma(\omega)}{\omega} d\omega \\ &= \int_{-\infty}^{+\infty} \frac{\rho(\omega)}{\omega} d\omega + j\int_{-\infty}^{+\infty} \frac{\vartheta(\omega)}{\omega} d\omega + \int_{C_R} \frac{\gamma(\omega)}{\omega} d\omega + \int_{C'_R} \frac{\gamma(\omega)}{\omega} d\omega = 0\end{aligned}$$

where we again made the substitution $\gamma(\omega) = \rho(\omega) + j\vartheta(\omega)$. In this case, however, things are different in terms of the symmetry exhibited by the integrable quantities: The function $\rho(\omega)/\omega$ shows odd symmetry, while the function $\vartheta(\omega)/\omega$ shows even symmetry, and therefore, we obtain

$$\int_{-\infty}^{+\infty} \frac{\rho(\omega)}{\omega} d\omega = 0 \qquad \text{and} \qquad \int_{-\infty}^{+\infty} \frac{\vartheta(\omega)}{\omega} d\omega = 2\int_0^{+\infty} \frac{\vartheta(\omega)}{\omega} d\omega$$

As for the last two integrals, they are calculated, as before, by replacing the function $\gamma(\omega)$ from the appropriate approximation for the boundary cases $R \to \infty$ for the path C_R and $R \to 0$ for the path C'_R. Starting with the first integral, the second term of the expansion of $\gamma(\omega)$ (and, of course, all higher-order terms) are zero for $\omega \to \infty$, so we can write that $\gamma(\omega) = \rho(\infty)$. On the other hand, considering that the function is *analytic* around the point $\omega = 0$, it can be described by a similar expansion of the form $\gamma(\omega) = \rho(0) + j\vartheta(0)\omega$, so we can make the approximation that $\gamma(\omega) = \rho(0)$. Replacing these expressions in the last two integrals and working as before, it turns out that

$$\int_{C_R} \frac{\gamma(\omega)}{\omega} d\omega = -j\pi\rho(\infty), \qquad \int_{C'_R} \frac{\gamma(\omega)}{\omega} d\omega = j\pi\rho(0)$$

and finally

$$\int_0^{\infty} \frac{\vartheta(\omega)}{\omega} d\omega = \frac{\pi}{2}[\rho(\infty) - \rho(0)]$$

The foregoing relationship is analogous to the previous one, with the difference that the integration is performed *on the logarithmic frequency scale*—since the expression $d\omega/\omega$ is recognized as the differential $d(\log\omega)$—while the right side includes the difference of the values of the amplitude component at the positions $\omega = 0$ and $\omega = \infty$. According to the foregoing relationship, *the area of the region located below the curve of the phase component, designed on the logarithmic frequency scale, depends on the difference in the values of the logarithmic amplitude at the positions $\omega = 0$ and $\omega = \infty$ and not on the way it changes between these values.*

The two foregoing results that define the so-called *amplitude and phase* can be used for the calculation of the functions $\rho(\omega)$ and $\vartheta(\omega)$ at any frequency value. Taking into account that the appearance of the term $\varphi(\infty)$ is due to the fact that the function $\gamma(\omega)$ has a residue at $\omega = \infty$, we can assert that to calculate the value of the function at a certain frequency $\omega = \omega_0$, we must ensure the existence of a residue at that position. But this, as is well known, means the placement of a pole at that position, something that can be achieved by dividing the function by the quantity $\omega - \omega_0$. To maintain symmetry, we must place a pole also at the mirror position to this, which means division by the quantity $\omega + \omega_0$. For reasons of convenience that are not so important to go into detail about here and are related to the simplification of the presentation of these principles, the function that is divided by the preceding quantities is not $\gamma(\omega)$, but $\gamma(\omega) - \rho(\omega_0)$, and therefore the application of Cauchy's theorem will lead to the expression

$$\oint\left(\frac{\gamma(\omega)-\rho(\omega_0)}{\omega-\omega_0}-\frac{\gamma(\omega)-\rho(\omega_0)}{\omega+\omega_0}\right)d\omega=\oint\frac{2\omega_0[\gamma(\omega)-\rho(\omega_0)]}{\omega^2-\omega_0^2}d\omega=0$$

In this case, the closed contour along which the preceding integral is evaluated consists, as before, of the vertical axis—traversed from bottom to top and extending from $-\infty$ to $\infty+\infty$—and of the semicircular arc C_R in the right half of the complex plane, with radius $R\to\infty$. However, in this case, the existence of the poles at the positions $\omega=\pm\omega_0$ forces us to bypass them as shown in Fig. 11.27e, surrounding them with two small circular arcs of radius $R'\to 0$ and $R''\to 0$. Replacing the function $\gamma(\omega)$ with the equivalent expression $\rho(\omega)+j\vartheta(\omega)$, the preceding integral is written

$$\oint\left(\frac{\gamma(\omega)-\rho(\omega_0)}{\omega-\omega_0}-\frac{\gamma(\omega)-\rho(\omega_0)}{\omega+\omega_0}\right)d\omega=2\omega_0\int_{-\infty}^{+\infty}\frac{\rho(\omega)-\rho(\omega_0)}{\omega^2-\omega_0^2}d\omega+2j\omega_0\int_{-\infty}^{+\infty}\frac{\vartheta(\omega)}{\omega^2-\omega_0^2}d\omega$$
$$+j\vartheta(\omega_0)\oint_{C'_R}\left(\frac{1}{\omega-\omega_0}-\frac{1}{\omega+\omega_0}\right)d\omega+j\vartheta(\omega_0)\oint_{C''_R}\left(\frac{1}{\omega-\omega_0}-\frac{1}{\omega+\omega_0}\right)d\omega=0$$

We observe that in the preceding equation, the integral does not appear along the path C_R, since the integrable function changes faster than $1/\omega$—and more specifically as $1/\omega^2$—and therefore, the contribution of this path is negligible. As for the small circular paths of radius $R'\to 0$ and $R''\to 0$ that bypass the poles $\omega=\omega_0$ and $\omega=-\omega_0$, we can assume that the function $\gamma(\omega)$ maintains a constant value along them equal to $\gamma(\omega_0)=\rho(\omega_0)+j\vartheta(\omega_0)$ and $\gamma(-\omega_0)=\rho(\omega_0)-j\vartheta(\omega_0)$, which allows the formulation of the preceding equation in the form in which it has been written. This equation simplifies further, the second integral from the left in the expression on the right-hand side evaluates to zero, due to the odd symmetry characterizing the function $\vartheta(\omega)/(\omega^2-\omega_0^2)$. The first integral from $-\infty$ to $+\infty$ is equal to twice the integral from 0 to ∞, due to the even symmetry that characterizes the integrable quantity. Finally, in the third and fourth integrals and for frequencies very close to the frequency ω_0 we can ignore the term $1/(\omega+\omega_0)$ in relation to the term $1/(\omega-\omega_0)$. The calculation of these integrals using the foregoing relation for the values $\vartheta_1=-\pi/2$ and $\vartheta_2=\pi/2$ leads for both to a value equal to $-\pi\vartheta(\omega_0)$. Combining the preceding results and solving the equation that arises for $\vartheta(\omega_0)$, it turns out that

$$\vartheta(\omega_0)=\frac{2\omega_0}{\pi}\int_{-\infty}^{+\infty}\frac{\rho(\omega)-\rho(\omega_0)}{\omega^2-\omega_0^2}d\omega=\frac{2\omega_0}{\pi}\int_0^{\infty}\frac{\rho(\omega)-\rho(\omega_0)}{[(\omega/\omega_0)-(\omega_0/\omega)]\omega_0\omega}=\frac{2}{\pi}\int_0^{\infty}\frac{\rho(\omega)-\rho(\omega_0)}{\omega/\omega_0-\omega_0/\omega}\times\frac{d\omega}{\omega}$$

Let us now proceed to the change of variable $u=\log(\omega/\omega_0)$. It is not difficult to see that for $\omega=0$, we have that $u=-\infty$, and for $\omega=+\infty$ we have $u=\infty$, while differentiating we get $du=d\omega/\omega$. Also, we will have $\omega/\omega_0=e^u$ and $\omega_0/\omega=e^{-u}$. Substituting these expressions into the last integral, we write

$$\vartheta(\omega_0)=\frac{2}{\pi}\int_{-\infty}^{+\infty}\frac{\rho(u)-\rho(\omega_0)}{e^u-e^{-u}}du=\frac{1}{\pi}\int_{-\infty}^{\infty}\frac{\rho(u)-\rho(\omega_0)}{\sinh u}du=\frac{1}{\pi}\left\{\int_{-\infty}^{0}\frac{\rho(u)-\rho(\omega)_0}{\sinh u}du+\int_0^{\infty}\frac{\rho(u)-\rho(\omega_0)}{\sinh u}du\right\}$$

Resorting to integration by parts and taking into account that $\int du/\sin hu=\ln[\coth(|u/2|)]$, the foregoing expression is written

$$\vartheta(\omega_0)=\frac{1}{\pi}\left[[\rho(u)-\rho(\omega_0)]\ln\left(\coth\frac{-u}{2}\right)\right]_{-\infty}^{0}+\frac{1}{\pi}\int_{-\infty}^{0}\frac{d\rho(u)}{du}\ln\left(\coth\frac{-u}{2}\right)du$$
$$-\frac{1}{\pi}\left[[\rho(u)-\rho(\omega_0)]\ln\left(\coth\frac{u}{2}\right)\right]_{0}^{\infty}+\frac{1}{\pi}\int_0^{\infty}\frac{d\rho(u)}{du}\ln\left(\coth\frac{u}{2}\right)du$$

Near the value $u=0$, the quantity $\rho(u)-\rho(\omega_0)$ is approximately proportional to u, while the expression $\ln[\coth(u/2)]$ will change as $-\ln(u/2)$. Therefore, the expressions in brackets will change approximately to $u\ln u$, an expression that tends to zero in the limit $u\to 0$. On the other hand, at the other limit, the term $\ln\coth$ approaches the values $2e^u=2\omega/\omega_c$ and $2e^{-u}=2\omega_c/\omega$ and vanishes at the limits $u\to-\infty$ and $u\to\infty$, respectively. Therefore, the expressions in brackets vanish and the preceding relation is written

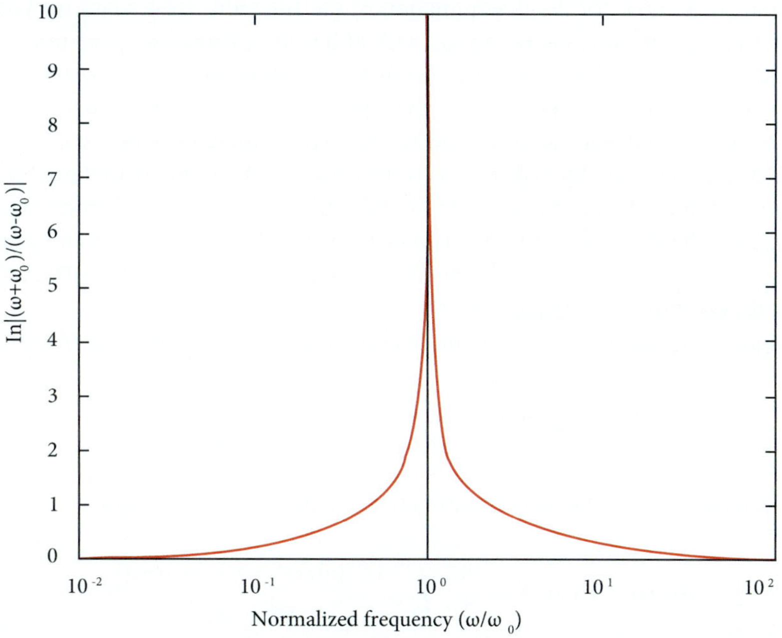

Fig. 11.28 Weighting coefficient of Bode integral for function $\vartheta(\omega_0)$

$$\vartheta(\omega_0) = \frac{1}{\pi}\int_{-\infty}^{0} \frac{dA(u)}{du}\ln\left(\coth\frac{-u}{2}\right)du + \frac{1}{\pi}\int_{0}^{\infty} \frac{d\rho(u)}{du}\ln\left(\coth\frac{u}{2}\right)du = \frac{1}{\pi}\int_{-\infty}^{\infty} \frac{d\rho(u)}{du}\ln\left(\coth\frac{|u|}{2}\right)du \quad (11.36)$$

We therefore arrive at the important conclusion *that the phase characteristics of a filter are proportional to the derivative of the logarithmic amplitude function, expressed in logarithmic form*. However, because the integration extends over the entire frequency spectrum, *the phase characteristics at each frequency depend on the slope of the function across all areas of the spectrum*. The relative weight of the slopes across all areas of the spectrum is determined by the term $\ln\coth|u/2|$, which therefore acts as a weighting coefficient and can also be written as $\ln|(\omega+\omega_0)/(\omega-\omega_0)|$. We observe that the value of this coefficient is greater in the vicinity of the frequency ω_0, while it becomes infinite at the position of this frequency, something that is naturally expected. For frequencies $\omega \gg \omega_0$ the value of this coefficient is approximately equal to $2\omega_0/\omega$, while for frequencies $\omega \ll \omega_0$ this value is equal to $2\omega/\omega_0$. The change of this coefficient in the range of values $[\omega_0/10,\ 10\omega_0]$—which corresponds to the area $-\ln 10 \le u \le \ln 10$—is presented diagrammatically in Fig. 11.28, from which we observe that the curve is symmetrical with respect to the frequency $\omega = \omega_0$ at the position of which it shows a maximum.

In the simple case where the amplitude response curve designed on a logarithmic scale has a constant slope $K = d\rho(u)/du$ for all frequencies, then it can be taken out of the integral. The value of the remaining integral is equal to $\pi^2/2$ (as can be ascertained by referring to integral tables), and therefore the foregoing relationship takes the simple form

$$\vartheta(\omega) = \frac{K}{\pi}\int_{-\infty}^{\infty} \ln\left(\coth\frac{|u|}{2}\right)du = \frac{K}{\pi}\left(\frac{\pi^2}{2}\right) = K\frac{\pi}{2} \quad (11.37)$$

Therefore, in this case, *the phase shift caused by the passage of an input signal through a causal and stable filter is an integer multiple of* 90°, *with the constant of proportionality equal to the slope constant of the characteristic attenuation, designed on the logarithmic scale of frequencies*.

Equation (11.36) allows us to calculate the phase of the frequency response at the position of some frequency ω_0 when we know the value of its amplitude, while the reverse process is also possible, which consists in calculating the amplitude of the frequency response for some frequency value when the value of its phase is known. Recall again that these calculations are applicable only when the considered frequency response function is a *minimum phase function*. Note also that while knowledge of the amplitude response uniquely determines the corresponding minimum phase, the opposite is not true, in the sense that if we determine some phase function, the corresponding amplitude response will be determined by approximating a constant that will usually be $\rho(0)$ or $\rho(\infty)$, that is, the value of the function $\rho(\omega)$ at the positions $\omega = 0$ and $\omega = \infty$.

To construct the equations that allow for the determination of the function $\rho(\omega)$ when we know the function $\vartheta(\omega)$, we must replace the function $\gamma(\omega)$ in all previous equations with either the expression $j\omega[\gamma(\omega) - \rho(\infty)]$ or the expression $[\gamma(\omega) - \rho(0)]/j\omega$. The reason we use the expression $j\omega$ in these substitutions is *to alternate the role of the real and imaginary components* at all stages of the analysis. For example, if we use the expression $j\omega[\gamma(\omega) - \rho(\infty)]$, then the real component will be $-\omega\vartheta(\omega)$ and will be characterized by even symmetry, while the imaginary component will be $j\omega[\rho(\omega)-\rho(\infty)]$ and will be characterized by odd symmetry. Proceeding, therefore, with the substitutions $\rho(\omega) \to -\omega\vartheta(\omega)$ and $\vartheta(\omega) \to \omega[\rho(\omega) - \rho(\infty)]$, the relationship that allows the determination of $\vartheta(\omega)$ from $\rho(\omega)$ can be used to calculate $\rho(\omega)$ from $\vartheta(\omega)$. If, on the other hand, we use the expression $[\gamma(\omega) - \rho(0)]/j\omega$, then the substitutions that will be made are $\rho(\omega) \to \vartheta(\omega)/\omega$ and $\vartheta(\omega) \to -[\gamma(\omega) - \rho(0)]/\omega$. In the preceding equations, the terms $\rho(0)$ and $\rho(\infty)$ are used to prevent the appearance of poles at the positions $\omega = 0$ and $\omega = \infty$.

Therefore, using the expression $j\omega[\gamma(\omega) - \rho(\infty)]$ and making the foregoing substitutions, we finally arrive at the result

$$\rho(\omega_0) - \rho(\infty) = -\frac{2}{\pi}\int_0^{\infty} \frac{\omega\vartheta(\omega) - \omega_0\vartheta(\omega_0)}{\omega^2 - \omega_0^2}\,d\omega$$

while if we use the substitutions related to the expression $[\gamma(\omega) - \rho(0)]/j\omega$, we will get

$$\rho(\omega_0) - \rho(0) = -\frac{2\omega_0^2}{\pi}\int_0^{\infty} \frac{[\vartheta(\omega)/\omega] - [\vartheta(\omega_0)/\omega_0]}{\omega^2 - \omega_0^2}\,d\omega$$

Making the foregoing substitutions in Eq. (10.33), we are led to the expressions

$$\rho(\omega_0) - \rho(\infty) = -\frac{1}{\pi\omega_0}\int_{-\infty}^{+\infty} \frac{d[\omega\vartheta(u)]}{du} \ln\left(\coth\frac{|u|}{2}\right) du \quad \text{and}$$

$$\rho(\omega_0) - \rho(0) = -\frac{\omega_0}{\pi}\int_{-\infty}^{+\infty} \frac{d[\vartheta(u)/\omega]}{du} \ln\left(\coth\frac{|u|}{2}\right) du$$

respectively. Finally, making the substitutions $\rho(\omega) = \ln|\mathcal{H}(j\omega)|$ and $\vartheta(\omega) = \measuredangle\mathcal{H}(j\omega)$, we get

$$\measuredangle\mathcal{H}(j\omega) = \frac{1}{\pi}\int_{-\infty}^{+\infty} \frac{d}{du}\left\{\ln|\mathcal{H}(j\omega)|\right\} \ln\left(\coth\frac{|u|}{2}\right) du \tag{11.38}$$

and

$$\ln|\mathcal{H}(j\omega_0)| = \ln|\mathcal{H}(j\infty)| - \frac{1}{\pi\omega_0}\int_{-\infty}^{+\infty} \frac{d}{du}\left\{\omega\measuredangle\mathcal{H}(j\omega)\right\} \ln\left(\coth\frac{|u|}{2}\right) du \quad \text{or} \tag{11.39}$$

$$\ln|\mathcal{H}(j\omega_0)| = \ln|\mathcal{H}(0)| - \frac{\omega_0}{\pi}\int_{-\infty}^{+\infty} \frac{d}{du}\left\{\frac{\measuredangle\mathcal{H}(j\omega)}{\omega}\right\} \ln\left(\coth\frac{|u|}{2}\right) du \tag{11.40}$$

which are the final results. The foregoing relations are known as *Bode integrals* and are widely used—mainly the first one—in electrical circuit analysis processes.

11.9 Approximations to Ideal Analog Filters

As mentioned in Sect. 3.6.2 (page 287), ideal analog filters are not causal and for this reason cannot be implemented in practice. Using the concepts defined in the previous section, we can formulate the claim that an ideal filter will be characterized by *zero attenuation* in the passbands and by *infinite attenuation* in the stopbands, while the transition from passband to stopband and vice versa will be *instantaneous*. This behavior for the four types of ideal filters is presented diagrammatically in Fig. 11.29, which is equivalent to Fig. 3.19 (page 288), which presents the variation of the filter's gain as a function of frequency. For this reason, ideal filters constitute *idealized models for the construction of real filters*, which are designed in such a way as to approximate the behavior of the corresponding ideal filters to the maximum possible extent. It is obvious that in a real filter, the attenuation in the passband *is not zero*, the attenuation in the stopband *is not infinite*, and the transition from passband to stopband *is not instantaneous but gradual*, thereby defining the so-called *transition zone*. In

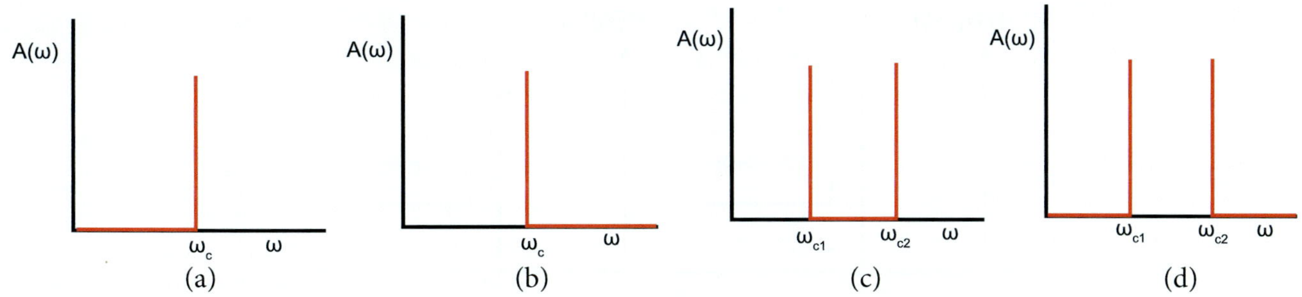

Fig. 11.29 Loss characteristics of the four types of ideal analog filters: (**a**) low-pass, (**b**) high-pass, (**c**) band-pass, and (**d**) band-stop

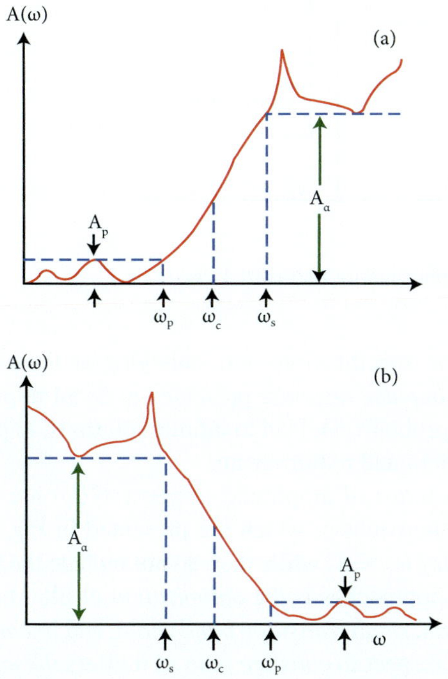

Fig. 11.30 Loss characteristics of the four types of real analog filters: (**a**) low-pass, (**b**) high-pass, (**c**) band-pass, and (**d**) band-stop (Antoniou, Figures 10.6 and 10.7)

addition, while in ideal filters the gain value in the passbands is *constant* for all frequencies, in practice this is not the case, resulting in the various frequency components of the input signal with frequencies within the passbands of the filter being amplified *to different degrees*. Therefore, the loss characteristic will not have the form of Fig. 11.29, which refers to ideal filters, but on the contrary will resemble the curve of Fig. 11.30. In this figure, the parameters ω_p and ω_s are the ends of the passbands and cutoff zones, A_p is *the maximum permissible attenuation in the passband*, and A_s is *the minimum permissible attenuation in the cutoff zone*.

The foregoing description refers to the signal's bandwidth response, while as for the *phase response*, for real filters it is not zero or linear, as would happen in the ideal case, but *nonlinear*, which results in a distortion of the input signal to some extent.

Speaking generally, the approach of an ideal filter by its realistic equivalent can be considered a mathematical process *mapping* specific bandwidth response specifications to transfer functions that can be implemented using passive or active electronic circuits. In all cases, the specifications given refer to *low-pass filters*, which then, under the transforms appropriate in each case, can be converted into *high-pass*, *band-pass*, and *band-stop* filters. For this reason, what follows concerns exclusively the case of low-pass filters. Let us note, also, that at this stage we are not dealing with the phase characteristics of a filter, but only with its *bandwidth response* since, as we mentioned in a previous section, for some specific bandwidth response, a suitable phase response can be found according to the *Paley–Wiener theorem* and the discussion on *Hilbert transforms* (Sects. 11.7 (page 975) and 11.8 (page 979). Therefore, in the following pages we describe the process of determining an appropriate *transfer function* $\mathcal{H}(s) = \mathcal{Y}(s)/\mathcal{X}(s)$ (by calculating its zeros and poles) for which the

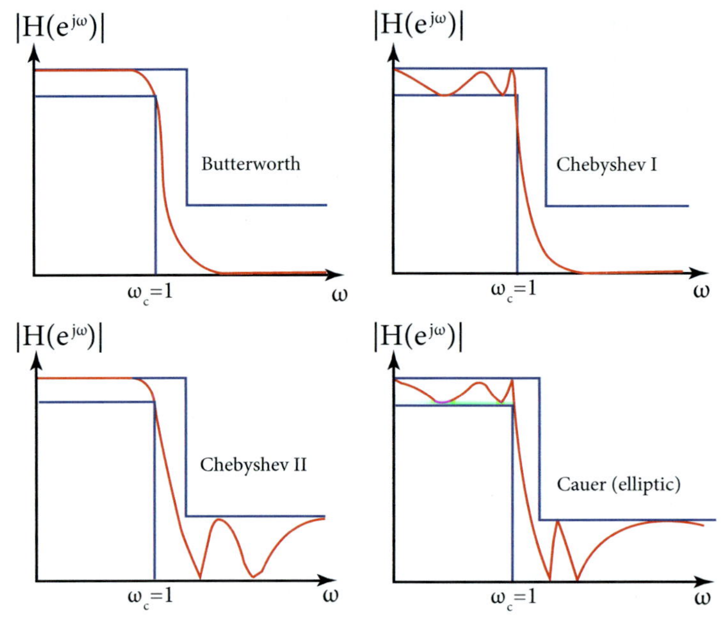

Fig. 11.31 Butterworth, Chebyshev I, Chebyshev II, and Cauer (or elliptic approach) for normalized case $\omega_c = 1$

corresponding *gain function* $G(\omega) \equiv |\mathcal{H}(s)|_{s=j\omega}$ must meet the specifications set, satisfying at the same time all the requirements about *causality*, *stability*, and *the real nature of the impulse response* presented in detail in previous chapters. For this reason, although theoretically solving this approximation problem can lead to infinite solutions, in practice, we must reject many of them, considering only those that meet the aforementioned requirements.

The most important of the low-pass ideal filter approaches in terms of amplitude response $G(\omega)$ are the *Butterworth, Chebyshev I, Chebyshev II, Bessel*, and *Cauer* (or *elliptic* filter approaches), which are presented in Fig. 11.31 (graphical representations of the figure concern the *normalized cutoff frequency* $\omega_c = 1$, while they do not include the Bessel approach, which is related to phase characteristics). The main goal of these approaches is the *optimization* of filter behavior from the perspective of various desirable characteristics, such as *constant gain*, *small transition band width*, and *linear phase response*. More specifically, the Butterworth approach optimizes filters with respect to *constant gain* as it offers *the maximum possible flat behavior in the passband*. The Chebyshev approach optimizes filters with respect to the *degree of attenuation*, offering a more *sharp* transition band (a feature also offered by the *Cauer approach*), while the Bessel approach addresses the issue of the *nonlinear phase response*, linearizing the phase of a filter for frequencies up to the cutoff frequency ω_c. Note that the Bessel approach addresses the phase response of the filter, and for this reason, it will be studied separately at the end, distinct from the others. From Fig. 11.31 we find that there exist approaches exhibiting monotonic behavior in the passband and stopband, as well as approaches with variations in the passband and/or the stopband, while in some cases the gain value is zeroed at specific frequency values, such as in the case of the elliptic approach. A detailed description of all these different approaches of the ideal low-pass analog filters will be presented in the following sections.

11.9.1 Polynomial and Rational Approaches

The approaches of ideal low-pass filters that were mentioned in the previous section can be grouped into two categories, *polynomial approaches* that lead to transfer functions with *only poles and no zero values* and *rational approaches* that allow the construction of transfer functions that contain *both zero values and poles*. Starting from the polynomial approaches, these lead to transfer functions

$$\mathcal{H}(s) = \frac{H_0}{D(s)} = \frac{H_0}{s^N + \alpha_{N-1}s^{N-1} + \alpha_{N-2}s^{N-2} + \cdots + \alpha_2 s^2 + \alpha_1 s + \alpha_0} \tag{11.41}$$

where $D(s)$ is a polynomial of degree N whose value determines the *order of the filter*. It is obvious that the poles of the transfer function will emerge as the roots of the polynomial equation $D(s) = 0$ and their value determines the form of the system output, based on our discussion in Sect. 5.3 (page 385). Adopting the conclusions formulated in the discussion on stability—see Sect. 5.4 on page 390, as well as Exercises 5.4.1 (page 392) and 5.4.2 (page 393)—we can assert that the only acceptable poles are those that have *negative real parts*, since, as is well known, poles that belong to the right complex half-plane lead to *unstable systems*, while poles that lie on the imaginary axis lead to *marginally stable systems* and to infinities of their gain at their positions. In the case where the order of the filter N is an even number, there are $N/2$ pairs of complex conjugate poles of the form $p_m = \sigma_m \pm j\omega_m$ ($m = 1, 2, \ldots, N/2$), each of which contributes to the denominator of the transfer function

$$(s - p_m)(s - p_m^*) = s^2 - (p_m + p_m^*)\, s + p_m p_m^* = s^2 - 2\sigma_m s + \sigma_m^2 + \omega_m^2$$

(which describes the contribution of the mth $N/2$ second-order elementary filters that will be connected in series, to form the considered filter of order N), and therefore the transfer function of the system can be given in the form

$$\mathcal{H}(s) = \frac{H_0}{D(s)} = \frac{H_0}{\prod_{m=1}^{N/2}\left[s^2 - 2\sigma_m s + \sigma_m^2 + \omega_m^2\right]} \tag{11.42}$$

On the other hand, if a filter is described by an odd-order value, it will have $N - 1$ pairs of *conjugate complex poles* with the foregoing contribution, as well as a negative real pole $p_0 = -\sigma_0$ ($\sigma_0 > 0$). This pole contributes to the denominator of the transfer function the term $s + \sigma_0$, which expresses the contribution of the first-order elementary filter that should be connected in series with $N - 1$ second-order elementary filters, to form an odd-order filter (see Sect. 11.5.4 on page 971). Therefore, in the case of filters with odd-order value, the transfer function of the filter has the form

$$\mathcal{H}(s) = \frac{H_0}{D(s)} = \frac{H_0}{(s + \sigma_0)\prod_{m=1}^{(N-1)/2}\left[s^2 - 2\sigma_m s + \sigma_m^2 + \omega_m^2\right]} \tag{11.43}$$

The constant H_0 is defined by the filter specifications, and, as we will see in the following sections, its value depends on the normalization requirements.

In almost all cases, the filter specifications concern the gain function

$$G(\omega) = |\mathcal{H}(j\omega)| = \frac{|H_0|}{|D(\omega)|} = \frac{|H_0|}{\sqrt{1 + \left[|D(\omega)|^2 - 1\right]}} = \frac{|H_0|}{\sqrt{1 + Q(\omega)}}$$

where $Q(\omega) = |D(\omega)|^2 - 1$. Taking into account that the gain function $G(\omega)$ must necessarily be characterized by *even symmetry* (since the impulse response of the filter is a signal of *real values*), the same thing must happen with the polynomial $Q(\omega)$, and therefore it will have the form

$$\begin{aligned} Q(\omega) &= \omega^{2N} + \beta_{2N-2}\omega^{2N-2} + \beta_{2N-4}\omega^{2N-4} + \cdots + \beta_4\omega_4 + \beta_2\omega_2 + \beta_0 \\ &= (\omega^2)^N + \beta_{2N-2}(\omega^2)^{N-1} + \beta_{2N-4}(\omega^2)^{N-2} + \cdots + \beta_4(\omega^2)^2 + \beta_2(\omega^2) + \beta_0 = P(\omega^2) \end{aligned}$$

However, in practice, the amplitude response of the filter is not given in this form but is formulated as

$$G(\omega) = |\mathcal{H}(j\omega)| = \frac{H_0}{\sqrt{1 + \gamma^2 P_N^2(\omega)}} \tag{11.44}$$

where $P_N(\omega)$ is an even or odd polynomial of degree N and γ is the so-called *design parameter* of the filter. Typical examples of polynomial approximations are the *Butterworth approach*, which uses the polynomial $P_N(\omega) = \omega^N$, and the *Chebyshev I approach*, which uses the homonymous polynomials (see the relevant section of the mathematical appendix). It should also be noted that since we refer to *low-pass filters*, the polynomial $P_N(\omega)$ should be characterized by small values for frequencies $0 \leq \omega \leq 1$ and by large values for $\omega > 1$ (for the normalized case $\omega_c = 1$). In practice, we usually accept that $P_N(1) = 1$.

The second category of approximations of ideal low-pass filters is the *rational approximations*, which are also described by a gain function of the form

$$G(\omega) = |\mathcal{H}(j\omega)| = \frac{H_0}{\sqrt{1 + \gamma^2 P^2(\omega)}}$$

with the difference that, in this case, the function $P(\omega)$ is not polynomial but a *rational function*, that is, a ratio of polynomial functions of degree N and M respectively and defined as

$$P(\omega) = \frac{N(\omega)}{D(\omega)} = \frac{\alpha_N\omega^N + \alpha_{N-1}\omega^{N-1} + \alpha_{N-2}\omega^{N-2} + \cdots + \alpha_2\omega_2 + \alpha_1\omega + \alpha_0}{\beta_M\omega^M + \beta_{M-1}\omega^{M-1} + \beta_{M-2}\omega^{M-2} + \cdots + \beta_2\omega_2 + \beta_1\omega + \beta_0} \tag{11.45}$$

Therefore, the gain function and its square are formulated as

$$G(\omega) = \frac{H_0}{\sqrt{1 + \gamma^2 \dfrac{N^2(\omega)}{D^2(\omega)}}} = \frac{H_0 D(\omega)}{\sqrt{D^2(\omega) + \gamma^2 N^2(\omega)}} \quad \text{and} \quad G^2(\omega) = \frac{H_0}{1 + \gamma^2 \dfrac{N^2(\omega)}{D^2(\omega)}} = \frac{H_0^2 D^2(\omega)}{D^2(\omega) + \gamma^2 N^2(\omega)}$$

Taking into account that the gain function should, as before, be characterized by *even symmetry*, we can formulate the claim that the rational function $P^2(\omega)$ should be an even function of ω. Note that the polynomial $D(\omega)$ appearing in the numerator of the foregoing relation must necessarily be an *even function*, since if it is an odd function, then, as is well known, it will be characterized by the property $D(0) = 0$, and therefore, there will be a zeroing of the filter gain at the position $\omega = 0$, which is not permissible for low-pass filters. Conversely, at the limit $\omega \to \infty$, we should have $G(\omega) = 0$, which is automatically ensured if $N > M$, while in the case $M = N$ (and for an even value of N), we will have $P(\infty) = \alpha_N/\beta_M = K$ and $G(\infty) = H_0/\sqrt{1 + \gamma^2 K^2}$, so we must ensure that the value of K is such that the filter specifications are met. Finally, the polynomial $N(\omega)$ can be either an even or an odd function of ω.

The main feature of the rational approximation is the *appearance of zero values in the gain function*, which, therefore, is zeroed at the position of some frequencies of the cutoff zone. We recall from basic theory that this does not constitute a problem, since the *Paley–Wiener theorem* does prohibit the vanishing of the gain over entire frequency bands, but allows it at specific discrete frequency values. As we will prove in the next section, these zero values appear in the form of conjugates of purely imaginary values of the form $z_m = \pm j\omega_m$, with each pair contributing to the numerator of the transfer function the term $(s - z_m)(s - z_m^*) = (s + j\omega_m)(s - j\omega_m) = s^2 + \omega_m^2$. Typical examples of approximations that behave in this way are the *Chebyshev II* approximation and the *elliptic approximation*.

11.9.2 Poles and Zero Values of Transfer Function

The calculation of the poles and zero values of the transfer function of real analog filters is based on the construction of the function $\mathcal{H}(s)\mathcal{H}(-s)$—which, as it turns out based on our discussion in the introductory section, is the inverse of the *loss function* of the filter $L(-s^2)$—given by the relation

$$\mathcal{H}(s)\mathcal{H}(-s) = |\mathcal{H}(j\omega)|^2_{\omega=s/j} = G^2(\omega)_{\omega^2=-s^2} \tag{11.46}$$

For a filter of order N this function has $2N$ poles, which, due to symmetry with respect to the origin of the complex plane, *are evenly distributed in the left and right half-planes*, with each half-plane containing N poles. Taking into account that the poles with positive real parts lead to unstable filters, it is obvious that for the construction of the transfer function we will use *only the N poles of the left half-plane*, ignoring the rest. Starting from the *polynomial* approximations, these have only poles and no zeros, and therefore their transfer function will have the form presented previously.

On the other hand, in *rational approximations*, the transfer function is constructed from the relation

$$\mathcal{H}(s)\mathcal{H}(-s) = |\mathcal{H}(j\omega)|^2_{\omega=s/j} = \frac{H_0^2 D^2(\omega)_{\omega^2=-s^2}}{D^2(\omega)_{\omega^2=-s^2} + \gamma^2 N^2(\omega)_{\omega^2=-s^2}} \tag{11.47}$$

with the condition $N > M$ ensuring that the transfer function is a *realizable function*. From the foregoing relation we observe that in this case the function $\mathcal{H}(s)$ will have not only poles, but also, and zeros, which are roots of the polynomial $D(\omega)_{\omega=s/j}$. Taking into account that the degree of this polynomial is equal to M, it is obvious that the polynomial $D(\omega)_{\omega=s/j} = D(\omega)_{\omega^2=-s^2}$ will be an *even* polynomial of degree $2M$ that will contribute to the transfer function M zeros z_m ($m = 1, 2, \ldots, M$). As for the representation of the denominator, the polynomials $D^2(\omega)_{\omega^2=-s^2}$ and $N^2(\omega)_{\omega^2=-s^2}$ are both *even* polynomials of degree $2N$ that will contribute to the transfer function N poles p_m ($m = 1, 2, \ldots, N$). Therefore, for values $N \geq M$ (with M being an *even* number, an assumption that will be silently adopted throughout the description that follows), the transfer function will have the general form

$$\mathcal{H}(s) = H_0 \left\{ \prod_{m=1}^{N} (s - z_m) \right\} \Big/ \left\{ \prod_{m=1}^{N} (s - p_m) \right\}$$

An interesting property that characterizes rational approximations is that the zero values of the even polynomial $D(\omega)$ can be *either purely real or purely imaginary, but not complex variables*. To prove the truth of this claim, let us assume that the polynomial $D(\omega)$ has zero values of the form $p_m = \alpha_m \pm j\beta_m$ (from complex analysis we know that a polynomial equation with real coefficients has conjugate complex roots). However, since the polynomial $D(\omega)$ is characterized by *even symmetry*, it is obvious that it will also have as roots the quantities $-p_m = -\alpha_m \mp j\beta_m$, and therefore, in its factored form the expression

$$(\omega^2 - 2\alpha_m\omega + \alpha_m^2 + \beta_m^2)(\omega^2 + 2\alpha_m\omega + \alpha_m^2 + \beta_m^2)$$

will appear, as we showed previously. If we make the substitution $\omega^2 = -s^2$, then in the equation of the function $\mathcal{H}(s)\mathcal{H}(-s)$—which contains the term $D^2(\omega)$—the expression

$$(-s^2 - 2j\alpha_m s + \alpha_m^2 + \beta_m^2)^2(-s^2 + 2j\alpha_m + \alpha_m^2 + \beta_m^2)^2$$

will appear, and consequently, in the defining equation of the transfer function $\mathcal{H}(s)$ will appear either the expression $(-s^2 - 2j\alpha_m s + \alpha_m^2 + \beta_m^2)^2$ or the expression $(-s^2 + 2j\alpha_m + \alpha_m^2 + \beta_m^2)^2$. However, we know that a rational transfer function is defined as the ratio of two polynomials with *real coefficients*, and therefore, it must necessarily be the case that $\alpha_m = 0$. Consequently, the zero values of the transfer function will either be of the form $p_m = \pm\omega_m$ or of the form $p_m = \pm j\omega_m$ (where we returned to our original symbolism) with the type of root depending on the nature of the respective polynomial (e.g., the polynomial $D(\omega) = \omega^2 - 4$ will lead to the real roots $\omega_{1,2} = \pm 2$, while the polynomial $D(\omega) = \omega^2 + 16$ will lead to the imaginary roots $\omega_{1,2} = \pm 4j$). In practice, all approximations of analog ideal filters lead to *real frequencies*, with the pair of roots $\omega_{1,2} = \pm\omega_m$ leading to the appearance of the term $(\omega - \omega_m)(\omega + \omega_m) = \omega^2 - \omega_m^2$ in the factorized form of the polynomial $D(\omega)$, which therefore, if all its $M/2$ roots are taken into account (for simplicity's sake we consider only discrete roots), the expression will be formulated as

$$D(\omega) = \beta_M \prod_{m=1}^{M/2} (\omega^2 - \omega_m^2)$$

By analogy, the contribution of the roots $\omega_{1,2} = \pm\omega_m$ to the function $\mathcal{H}(s)\mathcal{H}(-s)$ for the value $s = j\omega$ will be $(s - j\omega_m)(s + j\omega_m) = s^2 + \omega_m^2$ and therefore the aforementioned function will be described by the equation

$$\mathcal{H}(s)\mathcal{H}(-s) = \frac{\mathcal{Y}(s)}{\mathcal{X}(s)}\frac{\mathcal{Y}(-s)}{\mathcal{X}(-s)} = \frac{H_0^2\, D^2(\omega)_{\omega^2=-s^2}}{D^2(\omega)_{\omega^2=-s^2} + \gamma^2 N^2(\omega)_{\omega^2=-s^2}} = \frac{H_0^2\, \beta_M^2 \prod_{m=1}^{M/2}(s^2+\omega_m^2)^2}{D^2(\omega)_{\omega^2=-s^2} + \gamma^2 N^2(\omega)_{\omega^2=-s^2}} \tag{11.48}$$

where we silently assumed that the transfer function was expressed in the explicit form $\mathcal{H}(s) = \mathcal{Y}(s)/\mathcal{X}(s)$. From the foregoing relation it follows that the numerator of the transfer function is formulated as

$$\mathcal{Y}(s) = \mathcal{Y}(-s) = H_0|\beta_M| \prod_{m=1}^{M/2}(s^2+\omega_m^2)$$

while the function itself will be given by the expression

$$\mathcal{H}(s) = \frac{\mathcal{Y}(s)}{\mathcal{X}(s)} = \frac{H_0|\beta_M|}{B} \times \left\{\prod_{m=1}^{M/2}(s^2+\omega_m^2)\right\} \Big/ \left\{\prod_{m=1}^{N}(s-p_m)\right\}$$

In the foregoing relation, the proportionality constant B is related to the poles of the function, which will emerge as the roots of the equation

$$D^2(\omega)_{\omega^2=-s^2} + \gamma^2 N^2(\omega)_{\omega^2=-s^2} = \beta_M^2 \prod_{m=1}^{M/2}(s^2+\omega_m^2) + \gamma^2 N^2(\omega)_{\omega^2=-s^2} = 0$$

As we have already mentioned, this equation has $2N$ poles, which, due to symmetry, are equally distributed in the two half-planes and of which we will consider only the N of them that belong to the left half-plane, thereby leading to a *stable filter*. If $N > M$, then the coefficient of the term with the highest power will be the $\gamma^2\alpha_N^2$, while if $N = M$, then the value of this coefficient will be $\gamma^2\alpha_N + \beta_M^2$. Taking into account that the foregoing expression describing a polynomial of degree $2N$ refers to the function $\mathcal{H}(s)\mathcal{H}(-s)$, the proportionality constant B for the two preceding cases will be equal to $\gamma|\alpha_N|$ and $\sqrt{\gamma^2\alpha_N^2 + \beta_M^2}$, respectively, and therefore the transfer function will be given by the relation

$$\mathcal{H}(s) = H_0\, C \left\{\prod_{m=1}^{M/2}(s^2+\omega_m^2)\right\} \Big/ \left\{\prod_{m=1}^{N}(s-p_m)\right\} \quad \text{where } C = \begin{cases} \dfrac{|\beta_M|}{\gamma|\alpha_N|} & \text{for } N > M \\[2ex] \dfrac{|\beta_M|}{\sqrt{\gamma^2\alpha_N^2+\beta_M^2}} & \text{for } N = M \end{cases}$$

11.9.3 Selectivity and Spectral Efficiency Factor

To characterize a real filter as to the degree to which its behavior approximates that of the corresponding ideal filter, we can calculate the values of two parameters, the first of which is called the *selectivity* of the filter and the second *spectral efficiency factor*. The selectivity of a filter is defined as *the negative slope of the filter's amplitude response calculated at the cutoff frequency position*, or, in mathematical notation,

$$F_s = -\frac{d}{d\omega}\left\{|\mathcal{H}(j\omega)|\right\}_{\omega=\omega_c} \tag{11.49}$$

In other words, the *selectivity* of a filter is defined as *the slope of the amplitude response at the limit of the passband*, and the larger its value, the better the the filter's performance is considered to be. In the limiting case of the ideal low-pass filter, the value of this magnitude is infinite.

On the other hand, the *spectral efficiency factor* is defined as *the ratio of bandwidth values at specific attenuation levels* and is given by the equation

$$S_\alpha^\beta = \frac{\mathrm{BW}_\beta}{\mathrm{BW}_\alpha} \tag{11.50}$$

where β is the attenuation in decibels at the wider bandwidth, and α is the attenuation in decibels at the narrower bandwidth. In this case, *the smaller the value of this magnitude, the better the filter's performance is considered to be*. In the limiting case of the ideal low-pass filter, the value of the spectral efficiency factor is equal to one.

11.9.4 Butterworth Approximation

As we showed in Sect. 3.6.2 (page 287), the measure of the frequency response of a low-pass RC filter is given by the equation

$$|\mathcal{H}(j\omega)| = \frac{1}{\sqrt{1+\omega^2R^2C^2}} = \frac{1}{\sqrt{1+(\omega/\omega_c)^2}}$$

where $\omega_c = 1/RC$ is the *cutoff frequency* (see Eq. 3.70 and the following discussion). This filter is the simplest of a family of filters known as *Butterworth filters* and could be characterized as a Butterworth filter, first class, although this term is not used. As for the construction of higher-order Butterworth filters, this can be achieved by connecting first-class structures in series (in this case, as we saw in Sect. 5.8 on page 444, the transfer function of the composite system will result from the product of the transfer functions of the individual simple systems). These simple structures can be both passive (e.g., LC circuits consisting of alternating capacitors and coils) and active (i.e., structures containing capacitors, resistors, and operational amplifiers), and characteristic examples of them are depicted in Fig. 11.32. The study of these composite filters takes place in the manner presented in previous sections. Considering, for example, a second-order Butterworth low-pass filter that is a direct generalization of a simple RC filter and includes two ohmic resistors with values R_1 and R_2 and two capacitors with capacities C_1 and C_2, it is proven in a similar way that the measure of the frequency response function will be given by the relation

$$|\mathcal{H}(j\omega)| = \frac{1}{\sqrt{1+(\omega/\omega_c)^4}}$$

with the cutoff frequency defined as $\omega_c = 1/\sqrt{R_1C_1R_2C_2}$. Finally, for the general case of a low-pass Butterworth filter of order N, it is similarly proven that[16]

$$|\mathcal{H}(j\omega)| = \frac{1}{\sqrt{1+(\omega/\omega_c)^{2N}}} \quad \text{or alternatively} \quad |\mathcal{H}(j\omega)| = \frac{1}{\sqrt{1+\varepsilon^2(\omega/\omega_p)^{2N}}} \tag{11.51}$$

where ε is the *oscillation parameter in the passband*, which is related to the frequencies ω_p and ω_c through the expression $\varepsilon = (\omega_p/\omega_c)^N$. The change in the magnitude and phase of the frequency response of the low-pass Butterworth filter as a

[16] As already mentioned, the *order* of an analog filter is defined as the number of *energy storage elements* that form the filter, and its value affects the cost of its construction, size, and design process.

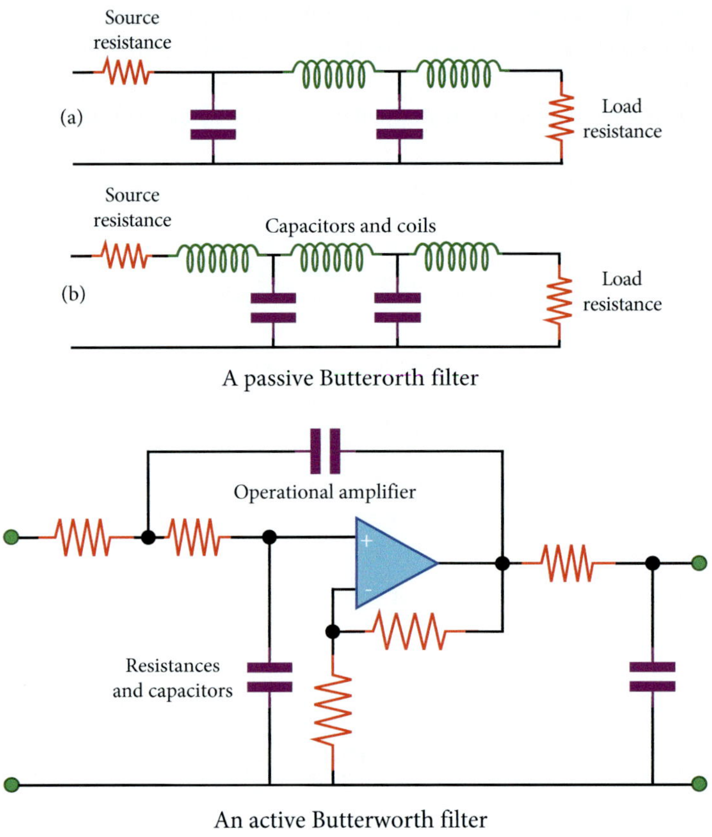

Fig. 11.32 Examples of passive and active higher-order Butterworth low-pass filters

function of frequency ω is shown in Fig. 11.33, in which the horizontal frequency axis is calibrated in the usual linear scale. On the other hand, in Fig. 11.34, the same curves are presented, but with the horizontal axis calibrated according to the logarithmic scale, an approach that is commonly used in practice.

By substituting the function $\mathcal{H}(j\omega)$ into the defining equation of the function $L(\omega^2)$, it immediately follows that

$$L(\omega^2) = \frac{1}{\mathcal{H}(i\omega)\mathcal{H}(-i\omega)} = \sqrt{1 + (\omega/\omega_c)^{2N}}\sqrt{1 + (-\omega/\omega_c)^{2N}} = 1 + (\omega/\omega_c)^{2N} \tag{11.52}$$

and therefore the attenuation of the filter will be calculated as

$$\mathcal{A}(\omega) = 10\log L(\omega^2) = 10\log\left[1 + (\omega/\omega_c)^{2N}\right] \tag{11.53}$$

The Butterworth approximation constitutes *the simplest approximation of a low-pass analog filter and leads to a strictly genuine transfer function with constant coefficients, with the maximum possible flat behavior in the passband.* To prove the truth of this claim, let us consider the square of a typical amplitude response—which is known as *power response*—of the form

$$|\mathcal{H}(j\omega)|^2 = K^2 \Big/ \left\{\sum_{k=0}^{N} \alpha_k \omega^{2k}\right\} = \left(\frac{K^2}{\alpha_0}\right)\left\{1 \Big/ \sum_{k=0}^{N} \beta_k \omega^{2k}\right\}$$

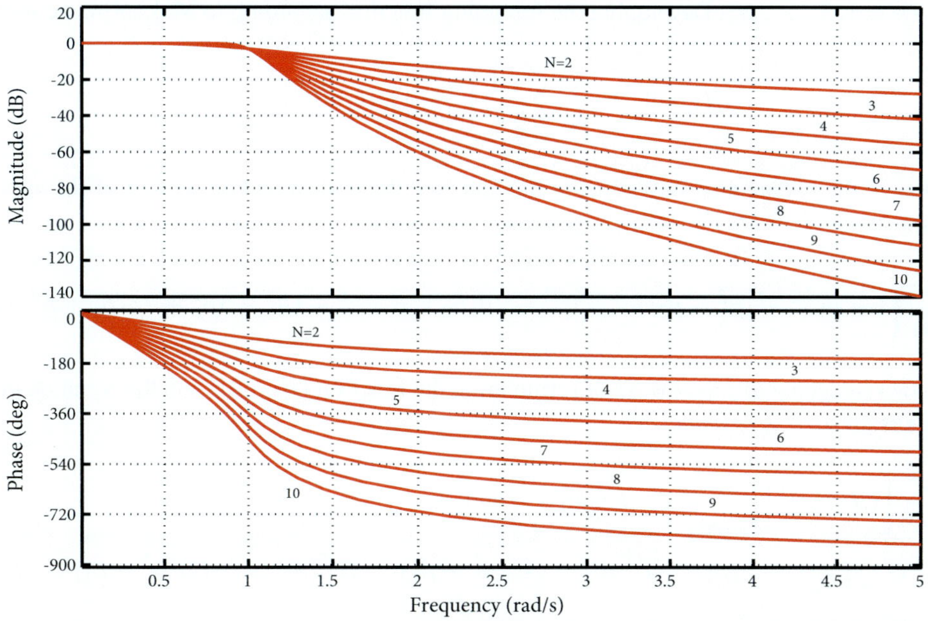

Fig. 11.33 Change in amplitude response and phase response of a low-pass Butterworth filter, with frequency axis calibrated in a linear scale, for filter order values $N = 2, 3, \ldots, 10$

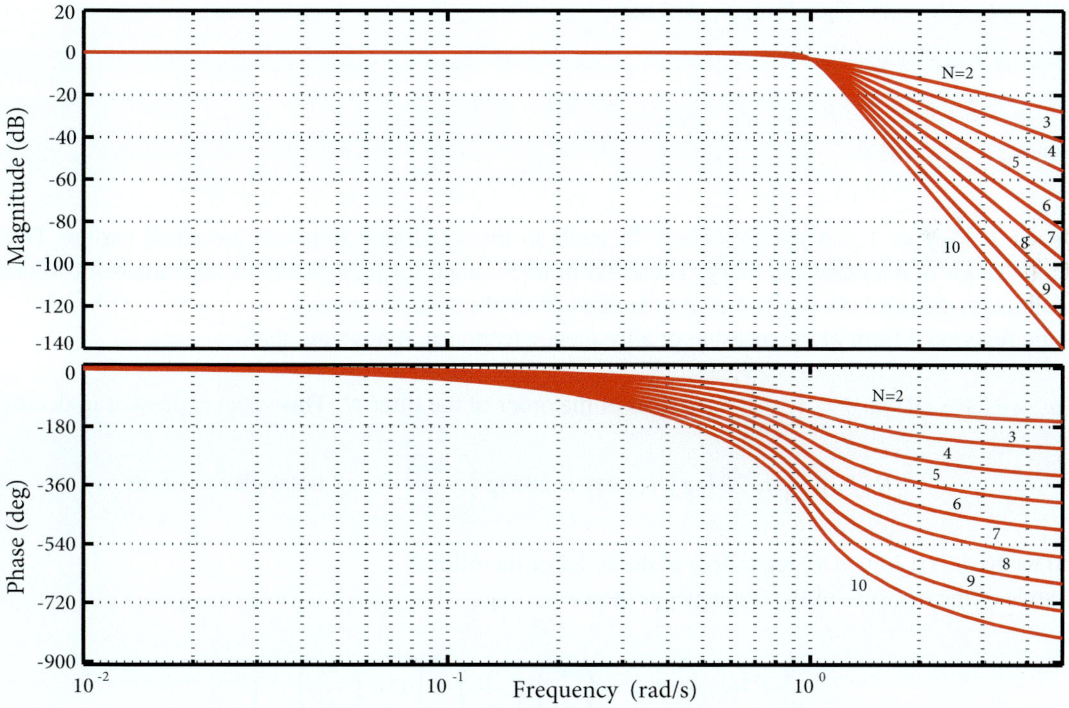

Fig. 11.34 Variation of amplitude response and phase response of low-pass Butterworth filter, with frequency axis graded on logarithmic scale, for filter order values $N = 2, 3, \ldots, 10$

where $\beta_k = \alpha_k/\alpha_0$ $(k = 1, 2, \ldots, N)$. Developing the foregoing function according to McLaurin, we will get

$$|\mathcal{H}(j\omega)|^2 = \frac{K^2}{\alpha_0}\sum_{k=0}^{\infty}\frac{1}{k!}\left\{\frac{d^k}{d\omega^k}\left[|\mathcal{H}(j\omega)|^2\right]\right\}_{\omega=0}\omega^k$$

while the application of the *polynomial long division* method (see the corresponding section of the appendix) allows us to present it in the form

$$|\mathcal{H}(j\omega)|^2 = \frac{K^2}{\alpha_0}\left\{1 - \beta_1\omega^2 + (\beta_1^2 - \beta_2)\,\omega^4 - (\beta_1^3 - 2\beta_1\beta_2 + \beta_3)\,\omega^6 + (\beta_1^4 - 3\beta_1^2 + 2\beta_1\beta_3 + \beta_2^2 - \beta_4)\,\omega^8 - \ldots\right\}$$

Comparison of the two foregoing different ways of formulating the function $|\mathcal{H}(j\omega)|^2$ and the equation of the corresponding terms leads us to the results

$$G^{(k)}(0) = \left\{\frac{d^k}{d\omega^k}\left[|\mathcal{H}(j\omega)|^2\right]\right\}_{\omega=0}\omega^k, \quad (k \text{ odd}) \quad \text{and}$$

$$G^{(0)}(0) = 1, \quad G^{(2)}(0) = (-2!)\beta_1, \quad G^{(4)}(0) = (4!)(\beta_1^2 - \beta_2), \quad G^{(6)}(0) = (-6!)(\beta_1^3 - \beta_1\beta_2 + \beta_3), \quad \ldots\ldots\ldots$$

In order for the power response and, consequently, the amplitude response to be characterized by the *maximum flat behavior*, we should set as many derivatives as possible to zero. It is obvious that if we set all the coefficients β_k $(1 \le k \le N)$ to zero, then the power response will become *independent* of the frequency ω. What we do, then, is to set to zero only the $N - 1$ coefficients β_k $(1 \le k \le N - 1)$, keeping nonzero only the coefficients $\beta_0 = 1$ and β_N, from which it follows that $G^{(k)}(0) = 0$ $(1 \le k \le N - 1)$. Therefore, we will have

$$|\mathcal{H}(j\omega)|^2 = \frac{K^2}{\alpha_0} \times \frac{1}{1 + \beta_N\omega^{2N}}$$

which, if we set $K^2/\alpha_0 = 1$ and $\beta_N = (1/\omega_c)^{2N}$, leads to the amplitude response presented earlier. This property of *maximum flat behavior in the passband* makes Butterworth filters ideal for preventing the appearance of folded (or aliased) frequencies, although the price we pay is *the large bandwidth of the transition zone*.

Studying the functional form of the measure of a frequency response, it turns out that:

- For $\omega = \omega_c$ we have $|\mathcal{H}(j\omega)| = 1/\sqrt{2}$, regardless of the order of the filter N. This value expressed in decibels is equal to

$$|\mathcal{H}(j\omega)|_{\text{dB}} = 20\log|\mathcal{H}(j\omega)| = 20\log\left(\frac{1}{\sqrt{2}}\right) = -3.0103 \text{ dB} \approx -3 \text{ dB}$$

- For $\omega = 0$ we have $|\mathcal{H}(j\omega)| = 0$, regardless of the order of the filter N.
- The derivative of the measure of the frequency response

$$\frac{d}{d\omega}(|\mathcal{H}(j\omega)|) = -\left(\frac{N}{\omega}\right)\left(\frac{\omega}{\omega_c}\right)^{2N}\left[1 + \left(\frac{\omega}{\omega_c}\right)^{2N}\right]^{-\frac{3}{2}}$$

  is always negative for positive values of ω, and therefore the frequency response of the Butterworth filter decreases monotonically with increases in the frequency value. This means that for two frequency values ω_1 and ω_2, such that $\omega_1 < \omega_2$, it will hold that $|\mathcal{H}(\omega_2)| < |\mathcal{H}(\omega_1)|$, while in the limit $\omega \to \infty$, it will hold that $|\mathcal{H}(j\omega)| \to 0$.

Poles and Zeros of Transfer Function

Applying the previous methodology, the function $\mathcal{H}(s)\mathcal{H}(-s)$ takes the form

$$\mathcal{H}(s)\mathcal{H}(-s) = |\mathcal{H}(j\omega)|^2\Big|_{\omega=\frac{s}{j}} = \frac{1}{1+[s/(j\omega_c)]^{2N}} = \frac{(j\omega_c)^{2N}}{s^{2N}+(j\omega_c)^{2N}} \tag{11.54}$$

and therefore the poles of the function $\mathcal{H}(s)\mathcal{H}(-s)$ will emerge as the roots of the equation $1+(s/j\omega_c)^{2N}=0$, which is formulated as

$$1+\left(\frac{s}{j\omega_c}\right)^{2N} = 1+(-j)^{2N}\left(\frac{s}{\omega_c}\right)^{2N} = 1+(-1)^N\left(\frac{s}{\omega_c}\right)^{2N} = 0 \tag{11.55}$$

We distinguish two cases regarding the value of the filter order N:

N Odd In this case, we have $(-1)^N=-1$ and the foregoing equation is written

$$1-\left(\frac{s}{\omega_c}\right)^{2N} = \left[1+\left(\frac{s}{\omega_c}\right)^N\right]\left[1+\left(-\frac{s}{\omega_c}\right)^N\right] = 0 \tag{11.56}$$

with the roots of the first expression constituting the N poles of the left half-plane and the roots of the second expression constituting the N poles of the right half-plane. From the preceding relation will we have

$$\left(\frac{s}{\omega_c}\right)^{2N} = 1 = e^{j2\pi m}$$

from which it follows that

$$s \equiv p_m = \omega_c \exp\left(j\frac{\pi m}{N}\right) = \omega_c\cos\left(\frac{\pi m}{N}\right) + j\omega_c\sin\left(\frac{\pi m}{N}\right)$$

for the values $m=0,1,2,\ldots,2N-1$, or, alternatively, using the preceding $2N$ values of m,

$$\begin{array}{c|l}
0 & p_0 = \omega_c e^0 = \omega_c \\
1 & p_1 = \omega_c \exp\left(j\frac{\pi}{N}\right) = \omega_c\cos\left(\frac{\pi}{N}\right) + j\omega_c\sin\left(\frac{\pi}{N}\right) \\
2 & p_2 = \omega_c \exp\left(j\frac{2\pi}{N}\right) = \omega_c\cos\left(\frac{2\pi}{N}\right) + j\omega_c\sin\left(\frac{2\pi}{N}\right) \\
3 & p_3 = \omega_c \exp\left(j\frac{3\pi}{N}\right) = \omega_c\cos\left(\frac{3\pi}{N}\right) + j\omega_c\sin\left(\frac{3\pi}{N}\right) \\
\ldots & \ldots\ldots\ldots\ldots\ldots\ldots \\
N-2 & p_{N-2} = \omega_c \exp\left(j\frac{(N-2)\pi}{N}\right) = \omega_c\cos\left(\frac{(N-2)\pi}{N}\right) + j\omega_c\sin\left(\frac{(N-2)\pi}{N}\right) \\
N-1 & p_{N-1} = \omega_c \exp\left(j\frac{(N-1)\pi}{N}\right) = \omega_c\cos\left(\frac{(N-1)\pi}{N}\right) + j\omega_c\sin\left(\frac{(N-1)\pi}{N}\right) \\
N & p_N = \omega_c \exp\left(j\frac{i\pi N}{N}\right) = \omega_c e^{j\pi} = -\omega_c \\
N+1 & p_{N+1} = \omega_c \exp\left(j\frac{\pi(N+1)}{N}\right) = -\omega_c\exp\left(j\frac{\pi}{N}\right) = -\omega_c\cos\left(\frac{\pi}{N}\right) - j\omega_c\sin\left(\frac{\pi}{N}\right) = -p_1 \\
N+2 & p_{N+2} = \omega_c \exp\left(j\frac{\pi(N+2)}{N}\right) = -\omega_c\exp\left(j\frac{2\pi}{N}\right) = -\omega_c\cos\left(\frac{2\pi}{N}\right) - j\omega_c\sin\left(\frac{2\pi}{N}\right) = -p_2 \\
N+3 & p_{N+3} = \omega_c \exp\left(j\frac{\pi(N+3)}{N}\right) = -\omega_c\exp\left(j\frac{3\pi}{N}\right) = -\omega_c\cos\left(\frac{3\pi}{N}\right) - j\omega_c\sin\left(\frac{3\pi}{N}\right) = -p_3 \\
\ldots & \ldots\ldots\ldots\ldots\ldots\ldots \\
2N-2 & p_{2N-2} = \omega_c \exp\left(j\frac{\pi[N+(N-2)]}{N}\right) = -\omega_c\exp\left(j\frac{(N-2)\pi}{N}\right) = -\omega_c\cos\left(\frac{(N-2)\pi}{N}\right) - j\omega_c\sin\left(\frac{(N-2)\pi}{N}\right) = -p_{N-2} \\
2N-1 & p_{2N-1} = \omega_c \exp\left(j\frac{\pi[N+(N-1)]}{N}\right) = -\omega_c\exp\left(j\frac{(N-1)\pi}{N}\right) = -\omega_c\cos\left(\frac{(N-1)\pi}{N}\right) - j\omega_c\sin\left(\frac{(N-1)\pi}{N}\right) = -p_{N-1}
\end{array}$$

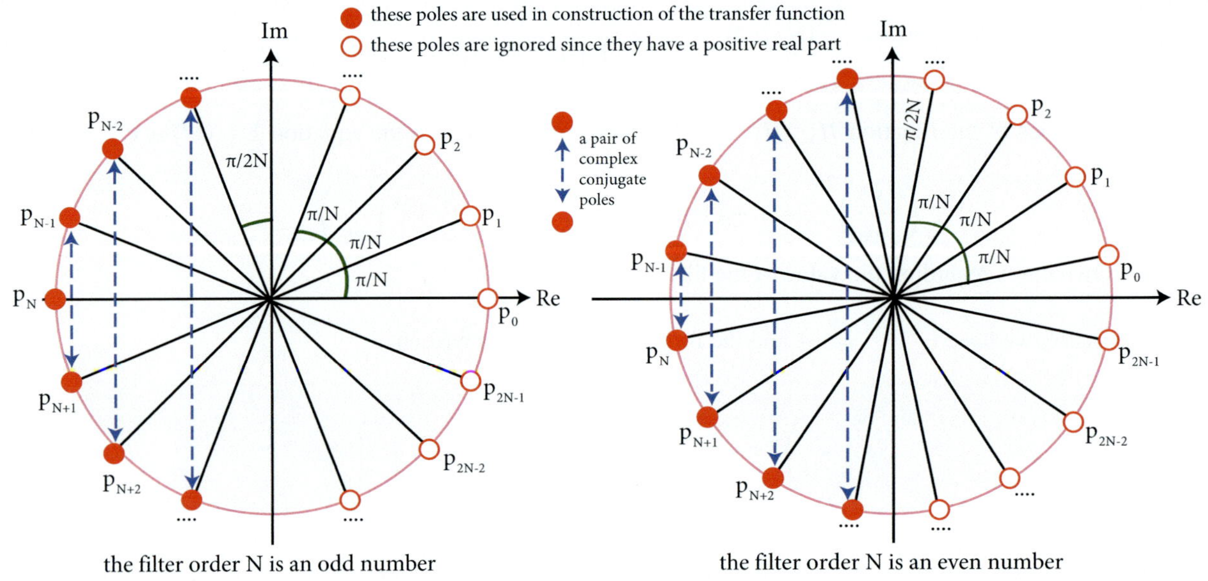

Fig. 11.35 The $2N$ poles of the function $\mathcal{H}(s)\mathcal{H}(-s)$ for a low-pass Butterworth filter and for even- and odd-order filters. The N poles used for the construction of the transfer function are denoted by filled circles; if they are of a complex nature, they appear in pairs of complex conjugate points of the left half-plane, so that the filter being constructed is a stable system

From the foregoing analysis we find that the function $\mathcal{H}(s)\mathcal{H}(-s)$ has *two real poles* with values $p_0 = \omega_c$ and $p_N = -\omega_c$ (corresponding to the values $m = 0$ and $m = N$) and $2N-2$ *complex poles* that form $N-1$ pairs of complex poles symmetric to each other with respect to the origin of the axes. Of all these $2N$ poles, we will keep only those that have negative real parts, that is, the real pole $p_N = -\omega_c$, as well as the $N-1$ complex poles of the left half-plane, which, as can be inferred from Fig. 11.35a, correspond to values of the index[17]

$$m = \frac{N+1}{2}, \frac{N+3}{2}, \frac{N+5}{2}, \ldots, \frac{3N-3}{2}, \frac{3N-1}{2}$$

Conversely, all other poles have positive real parts and are ignored. For simplicity's sake, let us denote the N poles of the left half-plane that will be used in the transfer function by p_k $(k = 1, 2, \ldots, N)$, with the index $k = 0$ corresponding to the *real pole*. Observing that the $N-1$ complex poles participating in the transfer function form pairs of conjugate complex poles (also depicted in the figure), there will be $(N-1)/2$ such pairs of the form (p_k, p_k^*) $(k = 1, 2, \ldots, N)$, with each pair contributing to the transfer function the expression

$$(s - p_k)(s - p_k^*) = s^2 - 2\Re\{p_k\} + |p_k|^2 = s^2 - 2\cos\left(\frac{\pi k}{N}\right) + \omega_c^2$$

(the signs in the preceding expressions should not cause confusion to the reader, since the signs of the values of the trigonometric functions depend on the quadrant in which they are defined, resulting in a positive sign actually corresponding to a negative value; see the next exercise). As for the real pole, it will contribute to the denominator of the transfer function the primary term $(s + \omega_c)$.

[17] The fact that solid circles were used to denote the poles in the transfer function and empty (unfilled) circles were used to denote the poles that were rejected exclusively for better design reasons and should not cause confusion to the reader, who knows that circles are used to represent zero values: Butterworth filters only have poles and no zero values. This observation also applies to Chebyshev I filters, which are presented immediately afterward.

Problem 11.9.1 Determine the poles of the transfer function of a low-pass Butterworth filter for order values $N = 3, 5$ and for normalized frequency $\omega_c = 1$.

Solution For the value $N = 3$ the values of index m are $0, 1, 2, 3, 4, 5$, and the poles of the function $\mathcal{H}(s)\mathcal{H}(-s)$ are

$$\left|\begin{array}{l|l} 0 & p_0 = 1 \\ 1 & p_1 = \cos\left(\frac{\pi}{3}\right) + j\sin\left(\frac{\pi}{3}\right) = 0.500000 + 0.866024j \\ 2 & p_2 = \cos\left(\frac{2\pi}{3}\right) + j\sin\left(\frac{2\pi}{3}\right) = -0.500000 + 0.866025j \\ 3 & p_3 = -1 \\ 4 & p_4 = -\cos\left(\frac{\pi}{3}\right) - j\sin\left(\frac{\pi}{3}\right) = -0.500000 - 0.866025j \\ 5 & p_5 = -\cos\left(\frac{2\pi}{3}\right) - j\sin\left(\frac{2\pi}{3}\right) = 0.500000 - 0.086025j \end{array}\right|$$

From the preceding poles in the transfer function, the real pole $p_3 = -1$ and the complex and conjugate poles $p_2 = -0.5 + 0.866025j$ and $p_4 = -0.5 - 0.866025j$ will be used, since only these have negative real parts.

On the other hand, for the value $N = 5$ the values of index m are $0, 1, 2, 3, 4, 5, 6, 7, 8, 9$, and the poles of the function $\mathcal{H}(s)\mathcal{H}(-s)$ are calculated as

$$\left|\begin{array}{l|l} 0 & p_0 = 1 \\ 1 & p_1 = \cos\left(\frac{\pi}{5}\right) + j\sin\left(\frac{\pi}{5}\right) = 0.809016 + 0.587785j \\ 2 & p_2 = \cos\left(\frac{2\pi}{5}\right) + j\sin\left(\frac{2\pi}{5}\right) = 0.309016 + 0.951056j \\ 3 & p_3 = \cos\left(\frac{3\pi}{5}\right) + j\sin\left(\frac{3\pi}{5}\right) = -0.309016 + 0.951056j \\ 4 & p_4 = \cos\left(\frac{4\pi}{5}\right) + j\sin\left(\frac{4\pi}{5}\right) = -0.809016 + 0.587784j \\ 5 & p_5 = -1 \\ 6 & p_6 = -\cos\left(\frac{\pi}{5}\right) - j\sin\left(\frac{\pi}{5}\right) = -0.809016 - 0.587785j \\ 7 & p_7 = -\cos\left(\frac{2\pi}{5}\right) - j\sin\left(\frac{2\pi}{5}\right) = -0.309016 - 0.951056j \\ 8 & p_8 = -\cos\left(\frac{3\pi}{5}\right) - j\sin\left(\frac{3\pi}{5}\right) = 0.309016 - 0.951056j \\ 9 & p_9 = -\cos\left(\frac{4\pi}{5}\right) - j\sin\left(\frac{4\pi}{5}\right) = 0.809016 - 0.587784j \end{array}\right|$$

From the foregoing poles in the transfer function, the real pole $p_5 = -1$ and the pairs of conjugate complex poles $(p_3, p_7) = (-0.309016 + 0.951056j, -0.309016 - 0.951056j)$ and $(p_4, p_6) = (-0.809016 + 0.587784j, -0.809016 - 0.587785j)$ will be used, since only these have negative real parts. The poles of the Butterworth filter transfer function for the order value $N = 3$ are presented in the left image of Fig. 11.36, while the construction of the corresponding diagram for an order value of $N = 5$ is left as an exercise for the reader.

From the foregoing analysis we find that the poles of the function $\mathcal{H}(s)\mathcal{H}(-s)$ are *uniformly distributed on the circumference of a circle in the complex plane* (this circle is known as the *Butterworth circle*) with radius $\omega_c = \omega_p(1/\varepsilon)^{1/N}$, where N is the order of the filter. We also observe that the radii connecting two successive poles with the center of the circle form an angle π/N, while the radius connecting the pole to the right of the vertical axis with the center of the circle forms

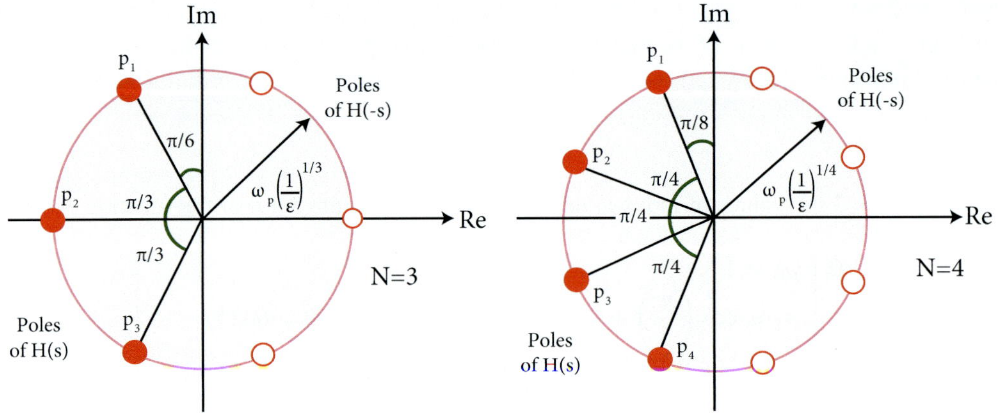

Fig. 11.36 The poles of the transfer function of a low-pass filter Butterworth for $N = 3, 4$

an angle $\pi/2N$ with this axis. Considering, as we mentioned earlier, *only the N poles of the left half-plane*, which ensure the stability of the filter, we can formulate the transfer function of the low-pass Butterworth filter for an odd-order value N in the form[18]

$$\mathcal{H}(s) = \frac{\omega_c^N}{(s+\omega_c)\prod_{k=1}^{(N-1)/2}(s-p_k)(s-p_k^*)} = \frac{\omega_c^N}{(s+\omega_c)\prod_{k=1}^{(N-1)/2}\left[s^2 - 2\cos\left(\frac{\pi k}{N}\right) + \omega_c^2\right]} \tag{11.57}$$

N **Even** In this case, it will hold that $(-1)^N = +1$, and the foregoing equation will be written

$$1+\left(\frac{s}{\omega_c}\right)^{2N} = \left[1+j\left(\frac{s}{\omega_c}\right)^N\right]\left[1-j\left(\frac{s}{\omega_c}\right)^N\right] = 0 \tag{11.58}$$

with the roots of the first expression constituting the N poles of the left half-plane and the roots of the second expression constituting the N poles of the right half-plane. From the foregoing relationship we will have

$$\left(\frac{s}{\omega_c}\right)^{2N} = -1 = e^{j2\pi m + j\pi}$$

or, equivalently,

$$s \equiv p_m = \omega_c \exp\left(j\pi\frac{2m+1}{2N}\right) = \omega_c\cos\left(\frac{\pi(2m+1)}{2N}\right) + j\omega_c\sin\left(\frac{\pi(2m+1)}{2N}\right)$$

for the values $m = 0, 1, 2, \ldots, 2N-1$, or, alternatively, using the preceding $2N$ values of m,

[18] In the general case of filters, the numerator of the function $\mathcal{H}(s)$ is multiplied by a constant coefficient K that expresses the required gain in each case.

0	$p_0 = \exp\left(j\dfrac{\pi}{2N}\right) = \omega_c \cos\left(\dfrac{\pi}{2N}\right) + j\omega_c \sin\left(\dfrac{\pi}{2N}\right)$
1	$p_1 = \exp\left(j\dfrac{3\pi}{2N}\right) = \omega_c \cos\left(\dfrac{3\pi}{2N}\right) + j\omega_c \sin\left(\dfrac{3\pi}{2N}\right)$
2	$p_2 = \exp\left(j\dfrac{5\pi}{2N}\right) = \omega_c \cos\left(\dfrac{5\pi}{2N}\right) + j\omega_c \sin\left(\dfrac{5\pi}{2N}\right)$
3	$p_3 = \exp\left(j\dfrac{7\pi}{2N}\right) = \omega_c \cos\left(\dfrac{7\pi}{2N}\right) + j\omega_c \sin\left(\dfrac{7\pi}{2N}\right)$
...	....
$N-2$	$p_{N-2} = \omega_c \exp\left(j\dfrac{\pi(2N-3)}{2N}\right) = \omega_c \cos\left(\dfrac{\pi(2N-3)}{2N}\right) + j\omega_c \sin\left(\dfrac{\pi(2N-3)}{2N}\right)$
$N-1$	$p_{N-1} = \omega_c \exp\left(j\dfrac{\pi(2N-1)}{2N}\right) = \omega_c \cos\left(\dfrac{\pi(2N-1)}{2N}\right) + j\omega_c \sin\left(\dfrac{\pi(2N-1)}{2N}\right)$
N	$p_N = \omega_c \exp\left(j\dfrac{\pi(2N+1)}{2N}\right) = \omega_c \cos\left(\dfrac{\pi(2N+1)}{2N}\right) + j\omega_c \sin\left(\dfrac{\pi(2N+1)}{2N}\right)$
$N+1$	$p_{N+1} = \omega_c \exp\left(j\dfrac{\pi(2N+3)}{2N}\right) = -\exp\left(j\dfrac{3\pi}{2N}\right) = -\omega_c \cos\left(\dfrac{3\pi}{2N}\right) - j\omega_c \sin\left(\dfrac{3\pi}{2N}\right) = -p_1$
$N+2$	$p_{N+2} = \omega_c \exp\left(j\dfrac{\pi(2N+5)}{2N}\right) = -\exp\left(j\dfrac{5\pi}{2N}\right) = -\omega_c \cos\left(\dfrac{5\pi}{2N}\right) - j\omega_c \sin\left(\dfrac{5\pi}{2N}\right) = -p_2$
$N+3$	$p_{N+3} = \omega_c \exp\left(j\dfrac{\pi(2N+7)}{2N}\right) = -\exp\left(j\dfrac{7\pi}{2N}\right) = -\omega_c \cos\left(\dfrac{7\pi}{2N}\right) - j\omega_c \sin\left(\dfrac{7\pi}{2N}\right) = -p_3$
...	....
$2N-2$	$p_{2N-2} = \omega_c \exp\left(j\dfrac{\pi[2(2N-2)+1]}{2N}\right) = -\omega_c \exp\left(j\dfrac{\pi(2N-3)}{2N}\right) = -\omega_c \cos\left(\dfrac{(2N-3)\pi}{2N}\right) - j\omega_c \sin\left(\dfrac{(2N-3)\pi}{2N}\right) = -p_{N-2}$
$2N-1$	$p_{2N-1} = \omega_c \exp\left(j\dfrac{\pi[2(2N-1)+1]}{2N}\right) = -\omega_c \exp\left(j\dfrac{\pi(2N-1)}{2N}\right) = -\omega_c \cos\left(\dfrac{(2N-1)\pi}{2N}\right) - j\omega_c \sin\left(\dfrac{(2N-1)\pi}{2N}\right) = -p_{N-1}$

We observe that in this case, *there are no real poles, but* $2N$ *complex poles that are pairwise symmetric with respect to the origin of the coordinate system*. This situation is shown in Fig. 11.35b, from where it can be seen that, as in the case of odd order, the rays that connect two successive poles with the center of the circle form an angle between them equal to π/N, while the radius that connects the first pole with the center of the circle forms an angle $\pi/2N$ with the vertical axis. In full analogy with the previous case, in the transfer function will be used *only the N poles of the left half-plane*, which correspond to index values

$$m = \frac{N+2}{2}, \frac{N+4}{2}, \frac{N+6}{2}, \ldots, \frac{3N-2}{2}, \frac{3N}{2}$$

and form $N/2$ pairs of conjugate complex poles of the form (p_k, p_k^*) $(k = 1, 2, \ldots, N)$, with each pair contributing to the transfer function the expression

$$(s - p_k)(s - p_k^*) = s^2 - 2\Re\{p_k\} + |p_k|^2 = s^2 - 2\cos\left(\frac{\pi k}{N}\right) + \omega_c^2$$

Consequently, the transfer function of a low-pass Butterworth filter of even order will have the form[19]

$$\mathcal{H}(s) = \frac{\omega_c^N}{\prod_{k=1}^{N/2}(s - p_k)(s - p_k^*)} = \frac{\omega_c^N}{\prod_{k=1}^{N/2}\left[s^2 - 2\cos\left(\frac{\pi k}{N}\right) + \omega_c^2\right]} \tag{11.59}$$

[19] For the order value $N = 2$ the transfer function of the filter is given by the equation

$$\mathcal{H}(s) = \frac{\omega_c^2}{s^2 + \sqrt{2}\omega_c s + \omega_c^2} = \frac{\omega_c^2}{s^2 + \frac{\omega_c}{Q}s + \omega_c^2}$$

from which it immediately follows that the quality factor of the Butterworth filter has a value of $Q = 1/\sqrt{2} = 0.707$. This is also the value of the quality factor that ensures *maximum flat behavior in the passband*, which, as mentioned earlier, characterizes low-pass Butterworth filters.

Problem 11.9.2 Determine the poles of the transfer function of a low-pass Butterworth filter for order values $N = 4, 6$ and for normalized frequency $\omega_c = 1$.

Solution For the value $N = 4$ the values of index m are 0, 1, 2, 3, 4, 5, 6, 7, and the poles of the function $\mathcal{H}(s)\mathcal{H}(-s)$ are

0	$p_0 = \cos\left(\frac{\pi}{8}\right) + j\sin\left(\frac{\pi}{8}\right) = 0.923879 + 0.382683j$
1	$p_1 = \cos\left(\frac{3\pi}{8}\right) + j\sin\left(\frac{3\pi}{8}\right) = 0.382683 + 0.923879j$
2	$p_2 = \cos\left(\frac{5\pi}{8}\right) + j\sin\left(\frac{5\pi}{8}\right) = -0.382683 + 0.923879j$
3	$p_3 = \cos\left(\frac{7\pi}{8}\right) + j\sin\left(\frac{7\pi}{8}\right) = -0.923879 + 0.382683j$
4	$p_4 = \cos\left(\frac{9\pi}{8}\right) + j\sin\left(\frac{9\pi}{8}\right) = -0.923879 - 0.382683j$
5	$p_5 = -\cos\left(\frac{3\pi}{8}\right) - j\sin\left(\frac{3\pi}{8}\right) = -0.382683 - 0.923879j$
6	$p_6 = -\cos\left(\frac{5\pi}{8}\right) - j\sin\left(\frac{5\pi}{8}\right) = 0.382683 - 0.923879j$
7	$p_7 = -\cos\left(\frac{7\pi}{8}\right) - j\sin\left(\frac{7\pi}{8}\right) = 0.923879 - 0.382683j$

Of these poles in the transfer function, the complex conjugate poles $(p_2, p_5) = (-0.382683 + 0.923879j, -0.382683 - 0.923879j)$ and $(p_3, p_4) = (-0.923879+0.382683j, -0.923879-0.382683j)$ will be used, since only these have negative real parts.

For the value $N = 6$ the values of index m are 0, 1, 2, 3, 4, 5, 6, 7, 8, 9, 10, 11, and the poles of the function $\mathcal{H}(s)\mathcal{H}(-s)$ are calculated as

0	$p_0 = \cos\left(\frac{\pi}{12}\right) + j\sin\left(\frac{\pi}{12}\right) = 0.965925 + 0.258819j$
1	$p_1 = \cos\left(\frac{3\pi}{12}\right) + j\sin\left(\frac{3\pi}{12}\right) = 0.707106 + 0.707106j$
2	$p_2 = \cos\left(\frac{5\pi}{12}\right) + j\sin\left(\frac{5\pi}{12}\right) = 0.258819 + 0.965925j$
3	$p_3 = \cos\left(\frac{7\pi}{12}\right) + j\sin\left(\frac{7\pi}{12}\right) = -0.258819 + 0.965925j$
4	$p_4 = \cos\left(\frac{9\pi}{12}\right) + j\sin\left(\frac{9\pi}{12}\right) = -0.707106 + 0.707106j$
5	$p_5 = \cos\left(\frac{11\pi}{12}\right) + j\sin\left(\frac{11\pi}{12}\right) = -0.965925 + 0.258819j$
6	$p_6 = \cos\left(\frac{13\pi}{12}\right) + j\sin\left(\frac{13\pi}{12}\right) = -0.965925 - 0.258819j$
7	$p_7 = -\cos\left(\frac{3\pi}{12}\right) - j\sin\left(\frac{3\pi}{12}\right) = -0.707106 - 0.707106j$
8	$p_8 = -\cos\left(\frac{5\pi}{12}\right) - j\sin\left(\frac{5\pi}{12}\right) = -0.258819 - 0.965925j$
9	$p_9 = -\cos\left(\frac{7\pi}{12}\right) - j\sin\left(\frac{7\pi}{12}\right) = 0.258819 - 0.965925j$
10	$p_{10} = -\cos\left(\frac{9\pi}{12}\right) - j\sin\left(\frac{9\pi}{12}\right) = 0.707106 - 0.707106j$
11	$p_{11} = -\cos\left(\frac{11\pi}{12}\right) - j\sin\left(\frac{11\pi}{12}\right) = 0.965925 - 0.258819j$

Among the above poles of the transfer function, only the complex conjugate pairs $(p_3, p_8) = (-0.258819 + 0.965925j, -0.258819 - 0.965925j)$, $(p_4, p_7) = (-0.707106 + 0.707106j, -0.707106 - 0.707106j)$, and $(p_5, p_6) = (-0.965925 + 0.258819j, -0.965925 - 0.258819j)$ will be used, as only these have negative real parts. The poles of the transfer function of a Butterworth filter for the order value $N = 4$ are presented in the right image of Fig. 11.36, while the construction of the corresponding diagram for the order value $N = 6$ is left as an exercise to the reader.

The foregoing results for even- and odd-order values can be combined in the expression

$$s_k = \exp\left(j\pi\frac{2k+N-1}{2N}\right) = \cos\left(\pi\frac{2k+N-1}{2N}\right) + j\sin\left(\pi\frac{2k+N-1}{2N}\right) \tag{11.60}$$

In this case, the normalized transfer function $\mathcal{H}(s)$ for the value $\omega_c = 1$ is formulated as $\mathcal{H}(s) = 1/B_N(s)$, where $B_N(s)$ are the so-called *Butterworth polynomials*, which are defined as

$$B_n(s) = \begin{cases} \displaystyle\prod_{k=1}^{N/2}\left[s^2 - 2s\cos\left(\frac{2k+N-1}{2N}\pi\right) + 1\right] & N \text{ even} \\ \displaystyle(s+1)\prod_{k=1}^{(N-1)/2}\left[s^2 - 2s\cos\left(\frac{2k+N-1}{2N}\pi\right) + 1\right] & N \text{ odd} \end{cases} \tag{11.61}$$

The factored form of the Butterworth polynomials and their corresponding roots, that is, the poles of the transfer function, for order values $N = 1, 2, \ldots, 10$ are included in Table 11.1. This table contains N poles that are characterized by a negative real part and therefore belong to the left half-plane, thereby leading to a stable and causal filter.

Phase Characteristics and Time-Domain Response

The *phase response* of a low-pass Butterworth filter arises from the equations of Exercise 11.5.4 (page 972) for parameter values $\alpha_1 = \alpha_0 = 1$, $\alpha_2 = 0$, $\alpha_{0k} = \omega_c^2$, $\alpha_{1k} = -2\sigma_k = -2\Re\{p_k\}$, $\alpha_{2k} = 1$, $\beta_{1k} = \beta_{2k} = 0$, and $\beta_{0k} = \omega_c^2$, and for the even and odd filter orders it is given by the relations

$$\measuredangle\mathcal{H}(j\omega) = -\sum_{k=1}^{N/2}\tan^{-1}\frac{-2\Re\{p_k\}\omega}{\omega_c^2 - \omega^2} \qquad \measuredangle\mathcal{H}(j\omega) = -\tan^{-1}\omega - \sum_{k=1}^{(N-1)/2}\tan^{-1}\frac{2\Re\{p_k\}\omega}{\omega^2 - \omega_c^2} \tag{11.62}$$

respectively. The *phase delay* is calculated as $\tau_{\text{ph}} = -\measuredangle\mathcal{H}(j\omega)/\omega$, while the *group delay*, which is defined as *the negative derivative of a phase response*, can be calculated from the equations for the foregoing parameter values. The result that arises for the even and odd filter orders has the form

$$\tau(\omega) = -\sum_{k=1}^{N/2}\frac{2\Re\{s_k\}(\omega_c^2 + \omega^2)}{(\omega_c^2 - \omega^2)^2 + 4\Re\{p_k\}^2\omega^2} \qquad \tau(\omega) = \frac{1}{1+\omega^2} - \sum_{k=1}^{(N-1)/2}\frac{2\Re\{s_k\}(\omega_c^2 + \omega^2)}{(\omega_c^2 - \omega^2)^2 + 4\Re\{p_k\}^2\omega^2} \tag{11.63}$$

On the other hand, the *impulse response* of a Butterworth low-pass filter will emerge, as is well known, from the inverse Laplace transform of the filter's transfer function, using methods of calculating the inverse transform, such as *the method of complex integration* (Sect. 4.7.1, page 335), *the method of partial fraction expansion* (Sect. 4.7.3, page 346), or *the application of Heaviside's theorem* (Sect. 4.7.4, page 351). In any case, the impulse response of a Butterworth low-pass filter is shown to be given by an expression of the form

$$h(t) = \mathcal{L}^{-1}\{\mathcal{H}(s)\} = \sum_{k=1}^{N} A_k e^{p_k t}$$

Table 11.1 Polynomial of denominator of transfer function and values of corresponding poles of normalized low-pass Butterworth filter for first 10 order values of filter

N	Characteristic polynomial of transfer function	Pole $s = \sigma + j\omega$
1	$s+1$	–1.0000
2	$s^2+1.414s+1$	$-0.7071 \pm 0.7071j$
3	$(s+1)(s^2+s+1)$	-1.0000
		$-0.5000 \pm 0.8660j$
4	$(s^2+0.7654s+1)(s^2+1.8478s+1)$	$-0.9239 \pm 0.3827j$
		$-0.3827 \pm 0.9239j$
5	$(s+1)(s^2+0.6810s+1)(s^2+1.6810+1)$	-1.0000
		$-0.8090 \pm 0.5878j$
		$-0.3090 \pm 0.9511j$
6	$(s^2+0.5176s+1)(s^2+1.4142s+1)(s^2+1.9319s+1)$	$-0.9659 \pm 0.2588j$
		$-0.7071 \pm 0.7071j$
		$-0.2588 \pm 0.9659j$
7	$(s+1)(s^2+0.4450s+1)(s^2+1.2470s+1)(s^2+1.8019s+1)$	-1.0000
		$-0.9010 \pm 0.4339j$
		$-0.6235 \pm 0.7818j$
		$-0.2225 \pm 0.9749j$
8	$(s^2+0.3902s+1)(s^2+1.1111s+1)(s^2+1.6629s+1)(s^2+1.9616s+1)$	$-0.9808 \pm 0.1951j$
		$-0.8315 \pm 0.5556j$
		$-0.5556 \pm 0.8315j$
		$-0.1951 \pm 0.9808j$
9	$(s+1)(s^2+0.3473s+1)(s^2+s+1)\times$	-1.0000
	$(s^2+1.5321s+1)(s^2+1.8794s+1)$	-0.9397 ± 0.3420
		$-0.7660 \pm 0.6428j$
		$-0.5000 \pm 0.8660j$
		$-0.1737 \pm 0.9848j$
10	$(s^2+0.3129s+1)(s^2+0.9080s+1)(s^2+1.4142s+1)\times$	$-0.9877 \pm 0.1564j$
	$(s^2+1.7820s+1)(s^2+1.9754s+1)$	$-0.8910 \pm 0.4540j$
		$-0.7071 \pm 0.7071j$
		$-0.4540 \pm 0.8910j$
		$-0.1564 \pm 0.9877j$

(see also Eq. 4.22) with the proportionality constants A_k being calculated as

$$A_k = \left. \frac{(s-p_k)}{(s-p_1)(s-p_2)\ldots(s-p_N)} \right|_{s=p_k}, \quad k = 1, 2, \ldots, N$$

In the general case, the quantities A_k and p_k are complex numbers, although for the case of the Butterworth low-pass filter the poles p_k $(k = 1, 2, \ldots, N)$ appear in pairs of complex conjugates. It is shown that when the order N of a filter is an *even number*, the foregoing relationship is formulated as

$$h(t) = \sum_{k=1}^{N/2} \left[2\Re\{A_k\}e^{\sigma_k t}\cos(\omega_k t) - 2\Im\{A_k\}e^{\sigma_k t}\sin(\omega_k t) \right] \tag{11.64}$$

where $p_k = \sigma_k + j\omega_k$, while the roots p_k are enumerated such that for $k = 1, 2, \ldots, N/2$, the points s_k belong to the same quadrant of the complex plane. On the other hand, when the order N of the filter is an *odd number*, the impulse response $h(t)$ is calculated from the equation

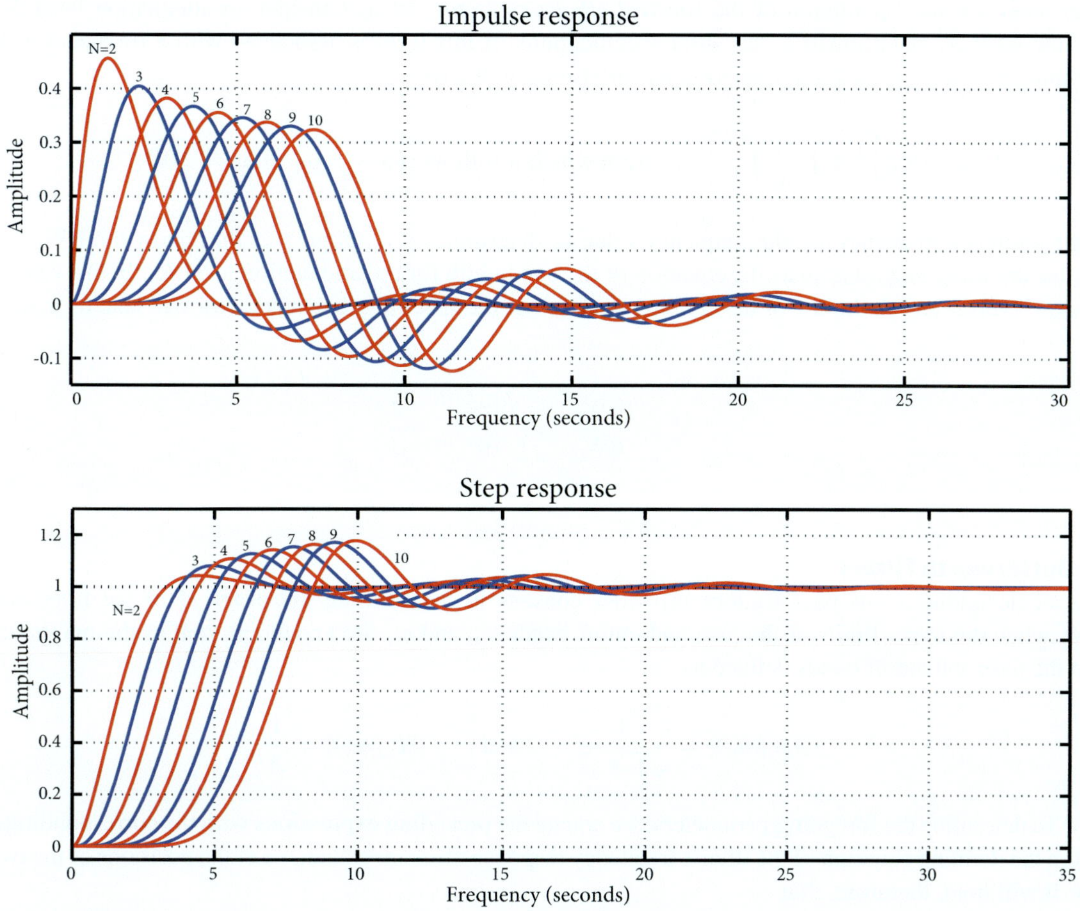

Fig. 11.37 Change in impulse and step response of low-pass Butterworth filter as a function of frequency for filter order values $N = 2, 3, \ldots, 10$

$$h(t) = Ke^{-t} + \sum_{k=1}^{(N-1)/2} \left[2\Re\{A_k\}e^{\sigma_k t}\cos(\omega_k t) - 2\Im\{A_k\}e^{\sigma_k t}\sin(\omega_k t) \right] \tag{11.65}$$

where in this case there are no pairs of roots that are related to each other by the property of complex conjugation. The change in the impulse response over time for filter order values Butterworth $N = 2, 3, \ldots, 10$ is shown in the top image of Fig. 11.37.

Once the impulse response of the filter has been computed, the step response can be obtained through *integration of the impulse response*, in accordance with standard procedure. The corresponding step response for the impulse responses of Fig. 11.37 is shown in the bottom image of this figure.

Selectivity and Spectral Efficiency Factor of Butterworth Filters

From the defining equation of *selectivity* it follows that

$$F_s = -\frac{d}{d\omega}\{|\mathcal{H}(j\omega)|\}_{\omega=\omega_c} = -\frac{d}{d\omega}\left\{\frac{1}{\sqrt{1+(\omega/\omega_p)^{2N}}}\right\}_{\omega=\omega_c} = \left\{\frac{N(\omega^{2N-1}/\omega_c^{2N})}{\sqrt[3]{[1+(\omega/\omega_c)^{2N}]^2}}\right\}_{\omega=\omega_c} = \frac{N}{(2\sqrt{2})\,\omega_c} \tag{11.66}$$

On the other hand, for the calculation of the *spectral efficiency factor*, let us consider an attenuation level R_ℓ dB that is greater than the level R_p determined by the filter specifications. If this level is associated with a frequency value ω_ℓ, then from the defining equation of the filter's amplitude response we will have

$$R_\ell = 10\log\left[1+\left(\frac{\omega_\ell}{\omega_c}\right)^{2N}\right] \qquad \text{from which it follows that} \qquad \omega_\ell = \omega_c\sqrt[2N]{10^{R_\ell/10}-1}$$

Considering a reference frequency ω_0, the frequency value ω_ℓ determines *in a unique way* the bandwidth $\mathrm{BW}_\ell = \omega_\ell - \omega_0$, so that we can use the foregoing relation as the equation of the bandwidth for the attenuation level R_ℓ. In this case, considering two attenuation levels α and β, the use of the defining equation of the spectral efficiency factor will lead us to the expression

$$S_\alpha^\beta = \frac{\mathrm{BW}_\beta}{\mathrm{BW}_\alpha} = \sqrt{\frac{10^{\beta/10}-1}{10^{\alpha/10}-1}} \tag{11.67}$$

Design of Butterworth Filters

The process of designing a low-pass Butterworth filter consists in determining the *order* N of the filter and the *cutoff frequency* ω_c when the upper limits of the pass and cutoff bands ω_p and ω_s are known, as well as the maximum/minimum deviation of the pass and cutoff bands defined as

$$|\mathcal{H}(j\omega_p)| = \frac{1}{\sqrt{1+\varepsilon^2}} \qquad \text{and} \qquad |\mathcal{H}(j\omega_s)| = \frac{1}{A}$$

respectively. To determine the foregoing parameters, we equate the preceding expressions with the corresponding expressions resulting from the defining equation of the filter's frequency response measure calculated at the positions of the two preceding frequencies. It will hold, therefore, that

$$|\mathcal{H}(j\omega_p)| = \frac{1}{\sqrt{1+(\omega_p/\omega_c)^{2N}}} = \frac{1}{\sqrt{1+\varepsilon^2}} \quad \text{from where it follows that} \quad \left(\frac{\omega_p}{\omega_c}\right)^{2N} = \varepsilon^2 \quad \text{and}$$

$$|\mathcal{H}(j\omega_s)| = \frac{1}{\sqrt{1+(\omega_s/\omega_c)^{2N}}} = \frac{1}{A} \quad \text{from which it follows that} \quad \left(\frac{\omega_s}{\omega_c}\right)^{2N} = A^2-1$$

Dividing the last two equations term by term, we get

$$\left(\frac{\omega_s}{\omega_p}\right)^{2N} = \frac{A^2-1}{\varepsilon^2} \qquad \text{or, equivalently,} \qquad 2N\log\left(\frac{\omega_s}{\omega_p}\right) = \log\left(\frac{A^2-1}{\varepsilon^2}\right)$$

and finally

$$N = \frac{1}{2}\frac{\log\left(\dfrac{A^2-1}{\varepsilon^2}\right)}{\log\left(\dfrac{\omega_s}{\omega_p}\right)} = \frac{\log\left(\dfrac{\sqrt{A^2-1}}{\varepsilon}\right)}{\log\left(\dfrac{\omega_s}{\omega_p}\right)} = \frac{1}{2}\frac{\log\left(\dfrac{10^{A_s/10}-1}{10^{A_p/10}-1}\right)}{\log\left(\dfrac{\omega_s}{\omega_p}\right)} = \frac{\log\alpha}{\log k} \tag{11.68}$$

where k and α are the parameters of *selectivity* and *discrimination* of the filter. From the last relation we can calculate the order of the filter N knowing the values of the parameters ω_s, ω_p, A_s and A_p. In almost all cases, the value of N that will be calculated in this way is a decimal number and should be rounded to the next largest integer, since the order of a filter is obviously an integer quantity.

Knowing the order N of the filter, we can then determine the corresponding Butterworth polynomial $B_N(s)$ assuming that it has the general form

$$B_N(s) = 1 + d_1 s + d_2 s^2 + \cdots + d_N s^N \tag{11.69}$$

with the Butterworth d_k ($k = 1, 2, \ldots, N$) coefficients calculated from the recursive relation

$$d_k = \frac{\cos[(k-1)\pi/2]}{\sin(k\pi/2N)}, \quad k = 1, 2, \ldots, N \tag{11.70}$$

starting with the value $d_0 = 1$. Having determined the polynomial $B_N(s)$, we calculate the transfer function of the filter as $H(s) = 1/B_N(s)$. However, for order values $N = 1, 2, \ldots, 10$, we can directly determine this function and its poles from Table 11.1 on page 1006.

Using the value of the order N of the filter, we can calculate the cutoff frequency ω_c either from the relation $(\omega_p/\omega_c)^{2N} = \varepsilon^2$ that uses the frequency ω_p or from the relation $(\omega_s/\omega_c)^{2N} = A^2 - 1$ that uses the frequency ω_s. Solving these two equations for ω_c, we get

$$\omega_c = \frac{\omega_p}{\sqrt[N]{\varepsilon}} = \frac{\omega_p}{\sqrt[2N]{10^{A_p/10} - 1}} \quad \text{and} \quad \omega_c = \frac{\omega_s}{\sqrt[2N]{A^2 - 1}} = \frac{\omega_p}{\sqrt[2N]{10^{A_s/10} - 1}}$$

The Butterworth filter that will be designed in this way will satisfy exactly *either the passband restriction or the cutoff band restriction*, depending on which of the foregoing expressions is used for the calculation of the cutoff frequency ω_c. Note that because the value of N that is used in the foregoing relation was rounded *to the next largest integer*, the frequency of the filter that will result and is related to the other restriction may eventually become larger than the one that has been specified due to this rounding.

Problem 11.9.3 Construct a normalized Butterworth filter with operating frequencies $\omega_p = 0.6$ rad/s and $\omega_s = 2.5$ rad/s and attenuation levels $A_p = 1.0$ dB and $A_s = 40$ dB (Antoniou, Exercise for solution 10.5).

Solution In the first step of the process, we need to calculate the minimum order of the filter. Using the values of the specifications that have been given, we will have

$$N = \frac{1}{2}\frac{\log\left(\dfrac{10^{A_s/10} - 1}{10^{A_p/10} - 1}\right)}{\log\left(\dfrac{\omega_s}{\omega_p}\right)} = \frac{1}{2}\frac{\log\left(\dfrac{10^{40/10} - 1}{10^{1/10} - 1}\right)}{\log\left(\dfrac{2.5}{0.6}\right)}$$

$$= \frac{1}{2}\frac{\log\left(\dfrac{9999}{0.258925}\right)}{\log\left(\dfrac{2.5}{0.6}\right)} = \frac{1}{2}\frac{\log(38617.298)}{\log(4.16666)} = \frac{1}{2}\frac{4.586781}{0.619781} = 3.668008$$

The resulting value is a decimal and should be rounded up to the next integer. Therefore, the minimum filter order is $N = 4$. The poles of the transfer function for a low-pass Butterworth filter with an order value of $N = 4$ were calculated in Exercise 11.9.1 (page 1001) as $p_2 = -0.382683 + 0.923879j$, $p_3 = -0.923879 + 0.382683j$, $p_4 = -0.923879 - 0.382683j$, and $p_5 = -0.382683 - 0.923879j$. The pair of conjugate complex poles (p_2, p_5) contributes to the denominator of the transfer function the term

$$s^2 - 2(-0.382683)\,s + (0.382683)^2 + (0.923879)^2 = s^2 + 0.765366s + 1$$

while the pair of conjugate complex poles (p_3, p_4) contributes analogously the term

$$s^2 - 2(-0.923879)\,s + (0.923879)^2 + (0.382683)^2 = s^2 + 1.847758s + 1$$

Therefore, the transfer function of the filter will have the form[20]

$$\mathcal{H}(s)=\frac{1}{(s^2+0.765366s+1)(s^2+1.847758s+1)}=\frac{1}{s^4+2.613124s^3+3.414211s^2+2.613124s+1}$$

After calculating the transfer function of the system $\mathcal{H}(s)$, we can calculate its frequency response, proceeding to the substitution $s=j\omega$. In this case, the frequency response of the filter is

$$\begin{aligned}\mathcal{H}(j\omega)&=\frac{1}{(j\omega)^4+2.613124(j\omega)^3+3.414211(j\omega)^2+2.613124(j\omega)+1}\\&=\frac{1}{(1-3.414211\omega^2+\omega^4)+j2.613124(\omega-\omega^3)}\\&=\frac{1-3.414211\omega^2+\omega^4}{(1-3.414211\omega^2+\omega^4)^2+6.828417(\omega-\omega^3)^2}-j\frac{2.613214(\omega-\omega^3)}{(1-3.414211\omega^2+\omega^4)^2+6.828417(\omega-\omega^3)^2}\end{aligned}$$

with magnitude and phase given respectively by the relations

$$|\mathcal{H}(j\omega)|=\frac{1}{\sqrt{1+\omega^8}}\quad\text{and}\quad\measuredangle\mathcal{H}(j\omega)=-\tan^{-1}\left[\frac{2.613124\,(\omega-\omega^3)}{1-3.414211\omega^2+\omega^4}\right]$$

as can be proven. The phase response can alternatively be calculated from the theoretical expression of the function $\measuredangle\mathcal{H}(j\omega)$ that we gave in the section on the phase characteristics of the filter and using the trigonometric identity $\tan^{-1}x+\tan^{-1}y=\tan^{-1}[(x+y)/(1-xy)]$. Taking into account that the square of the magnitude of all the poles of the filter is equal to unity, since they lie on the circumference of the unit circle (recall that the filter is considered *normalized*). The application of this equation will lead to the same result:

$$\begin{aligned}\measuredangle\mathcal{H}(j\omega)&=-\tan^{-1}\frac{2\times(0.382683)\,\omega}{1-\omega^2}-\tan^{-1}\frac{2\times(0.923879)\,\omega}{1-\omega^2}=\\&=-\tan^{-1}\frac{\dfrac{0.765366\omega}{1-\omega^2}+\dfrac{1.874758\omega}{1-\omega^2}}{1-\dfrac{0.765366\omega}{1-\omega^2}\cdot\dfrac{1.874758\omega}{1-\omega^2}}=-\tan^{-1}\left[\frac{2.613124\,(\omega-\omega^3)}{1-3.414211\omega^2+\omega^4}\right]\end{aligned}$$

The phase delay of the filter is calculated as $-\measuredangle\mathcal{H}(j\omega)/\omega$, while the group delay is calculated as the derivative of the foregoing equation and, as is proven following simple operations, it has the form

$$\tau(\omega)=\frac{2.613126+1.08239\omega^2+1.08239\omega^4+2.613126\omega^6}{1+\omega^2}$$

while the same result can be reached by the direct application of the theoretical expression. The reader is asked as an exercise to calculate the impulse response $h(t)$ as the inverse Laplace transform of the transfer function $\mathcal{H}(s)$ and the step response $s(t)$ as the integral of the impulse response (on how to perform these calculations, see the relevant exercises of previous chapters). The change in the amplitude and phase responses of this filter as a function of the frequency ω, in logarithmic scale, are presented in Fig. 11.38.

[20] In this exercise, for purely educational reasons, we calculate the transfer function analytically; however, in practice, in the denominator of that function we can directly use the Butterworth polynomial of the appropriate degree (and the corresponding poles) from Table 11.1 (page 1006).

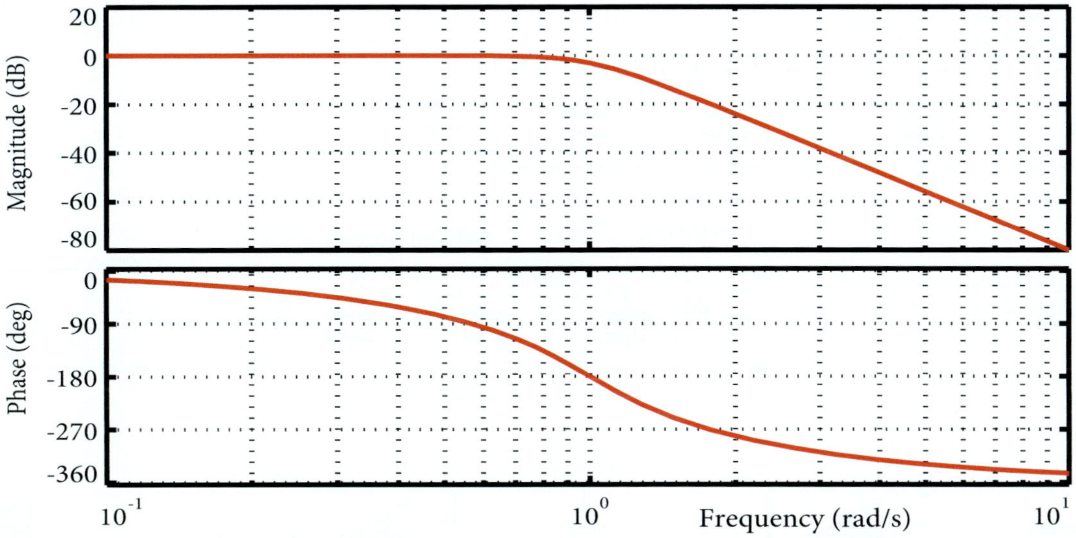

Fig. 11.38 Change in amplitude and phase responses as function of frequency ω in logarithmic scale for Butterworth filter from Exercise 11.9.3

11.9.5 Chebyshev Approach

In contrast to Butterworth filters, which ensure *maximum flat behavior in the passband*, Chebyshev filters are designed to offer an amplitude response with a relatively *sharp transition* from the passband to the cutoff band and therefore are characterized by a transition band with a relatively small width. However, the price we pay for this sharp transition is the *occurrence of equal amplitude fluctuations* in the amplitude response function. Since the transfer function of these filters is based on the use of *Chebyshev polynomials*, in Appendix A (Sect. A.9, page 1498), we briefly present the basic properties of these polynomials. Typical examples of passive and active circuits that implement Chebyshev filters are presented in Fig. 11.39. Historically, passive circuits that mainly include *a sequence of inductive and capacitive resistances with terminations at both ends* were the first to be developed to offer maximum power transfer for frequency values corresponding to the maxima of the gain function in the passband, and they continue to constitute *prototype structures* for both active and passive filters, since they are characterized by *low sensitivity* to errors related to the values of the circuit elements. Two types of analog filters are based on the Chebyshev approach: the first type of Chebyshev filters, or Chebyshev I filters, are examples of polynomial approximation, while the second type of Chebyshev filters, or Chebyshev II filters, are examples of rational approximation. A detailed presentation of these two types of filters is the subject of the following pages.

11.9.6 Chebyshev Type I Filters

Low-pass Chebyshev I filters are characterized by a transfer function that contains *only poles* and by an amplitude response characterized by the presence of *equal amplitude ripples* in the passband and by *monotonic attenuation* (i.e., absence of ripples) in the stopband (in reality, these filters exhibit the maximum possible flat behavior in the stopband since the gain function has $2N-1$ zeros in it, a property that is proven in the way we saw in Butterworth filters). Conversely, the transfer function of Chebyshev II filters contains, as we will see in the next section, *both poles and zeros* and is characterized by *monotonic attenuation* in the passband and by the appearance of *equal amplitude ripples* in the stopband. It is proven[21] that Chebyshev I filters ensure the maximum possible weakening in the cutoff zone A_s among all polynomial approximations of ideal filters characterized by the same value of A_p and the same filter order. The value of the factor Q for Chebyshev I filters is greater than the value $Q = 0.707$ that characterizes Butterworth filters and can be found along with other characteristics of the filters in special data tables.

The amplitude response of Chebyshev I filters has the form

[21] See de Queiroz, A.C.M. 1999, "Chebyshev Filters," *Wiley Encyclopedia of Electrical and Electronics Engineering*.

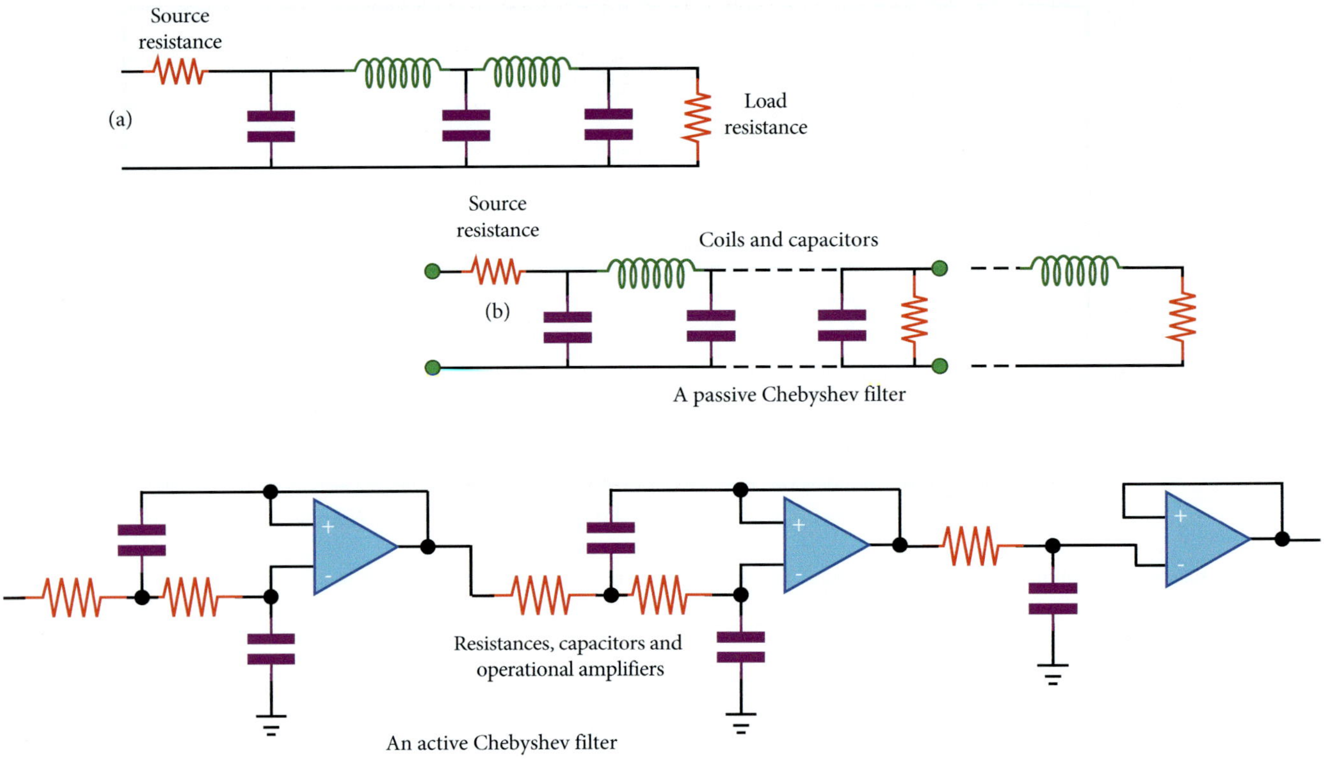

Fig. 11.39 Examples of passive and active higher-order Chebyshev filters

$$|\mathcal{H}(j\omega)| = \frac{1}{\sqrt{1+\varepsilon^2 T_N^2(\omega/\omega_p)}} \tag{11.71}$$

where ω_p is the *cutoff frequency of the passband* and ε is the *parameter of oscillations in the passband* that makes it possible to determine the amplitude of the fluctuations in this band. If we use a normalized implementation for which, as is well known, the cutoff frequency $\omega_p = 1$ is used, the foregoing relation simplifies to

$$|\mathcal{H}(j\omega)| = \frac{1}{\sqrt{1+\varepsilon^2 T_N^2(\omega)}} \tag{11.72}$$

Studying the foregoing expression, we can formulate the following conclusions regarding the normalized implementation of low-pass Chebyshev I filters (on defining the homonymous polynomials, see the relevant section of the mathematical appendix):

- For the frequency value $\omega = 0$, we have

$$T_N(0) = \cos\left(N\cos^{-1}0\right) = \cos\left[\frac{N(2k+1)\pi}{2}\right] = \begin{cases} \pm 1 & N \text{ even} \\ 0 & N \text{ odd} \end{cases}$$

and therefore

$$|\mathcal{H}(0)| = \frac{1}{\sqrt{1+\varepsilon^2 T_N^2(0)}} = \begin{cases} \dfrac{1}{\sqrt{1+\varepsilon^2}} & N \text{ even} \\ 1 & N \text{ odd} \end{cases}$$

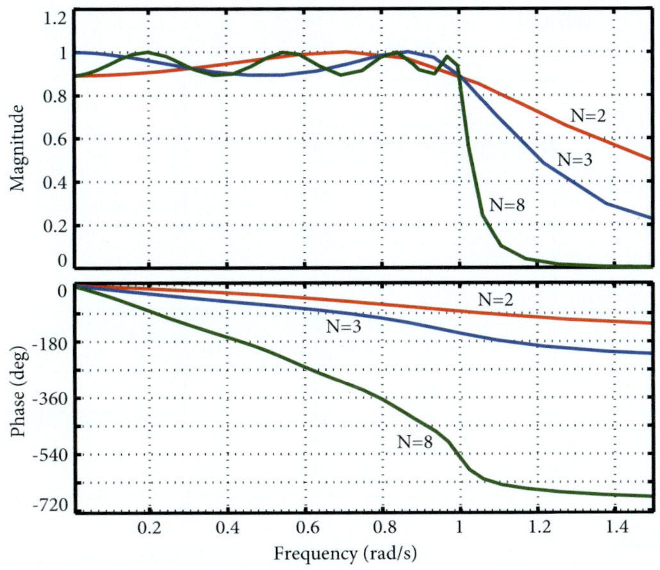

Fig. 11.40 Amplitude and phase responses of low-pass Chebyshev I filter for a cutoff frequency $\omega_p = 1$, ripple width 1 dB, and filter order values $N = 2, 3, 8$. Both axes of these diagrams are normalized in a linear scale

• For a frequency value of $\omega = 1$ rad/s, it will hold that

$$T_N(1) = \cos\left[N \cos^{-1}(1)\right] = \cos\left(2k\pi N\right) = 1$$

and therefore $|\mathcal{H}(1)| = 1/\sqrt{1+\varepsilon^2}$, a value that is independent of the order of the filter N.

Finally, regarding the high frequency values ω within the cutoff zone, we can make the approximation $\varepsilon T_N(\omega) \gg 1$, so we will have $|\mathcal{H}(j\omega)| \approx 1/[\varepsilon T_N(\omega)]$. Additionally, if $N \gg 1$ (i.e., if the filter has a sufficiently high order value), we can ignore the terms of the polynomial that correspond to small values of N and write $T_N(\omega) \approx 2^{N-1}\omega^N$, so we will also have $|\mathcal{H}(j\omega)| \approx 1/[2^{N-1}\varepsilon\omega^N]$.

The amplitude response of a low-pass Chebyshev I filter is shown in Fig. 11.40 for a cutoff frequency value of $\omega_p = 1$, ripple width with a deviation equal to 1 dB, and filter order values $N = 2, 3, 8$. From this figure we observe that *as the order of the filter N increases, the attenuation in the transition zone becomes increasingly steep* (and, consequently, *the width of the transition zone becomes increasingly smaller*), but at the cost of *increasing the amplitude of the deviations*, as can be observed. Taking into account that in the range of values $0 \le \omega < \omega_p$, the function $T_N^2(\omega/\omega_p)$ oscillates between 0 and 1, it is not difficult to see that the value of the measure $|\mathcal{H}(j\omega)|$ will oscillate between values 1 and $1/\sqrt{1+\varepsilon^2}$. Specifically, as we mentioned earlier, for filters with odd-order value, for $\omega = 0$ it will hold that $1 + T_N^2(\omega) = 1$, and therefore $G(0) = 1$, while for filters with even-order value, for $\omega = 0$ it will hold that $1 + T_N^2(\omega) = 1 + \varepsilon^2$, and therefore $G(0) = 1/\sqrt{1+\varepsilon^2} = 10^{-A_p/20} = 10^{-0.05A_p}$ (since, as we saw in the introductory section, the parameters ε and A_p are connected by the relation $A_p = -20\log(1/\sqrt{1+\varepsilon^2})$). This behavior of the filter, which is used to determine the constant ratio H_0, as we will see in the next section, is presented in a diagrammatic way in Fig. 11.41. If the amplitude response is designed expressed in decibels, then the range of deviations in the passband will be equal to

$$20\log\left[\frac{1}{1/\sqrt{1+\varepsilon^2}}\right] = 20\log\sqrt{1+\varepsilon^2} = 20\log(1+\varepsilon)^{1/2} = 10\log(1+\varepsilon^2) \tag{11.73}$$

and in practice a value around 1 dB is usually acceptable.

On the other hand, the frequency ω_s that signals the limit of the cutoff zone is determined, as for Butterworth filters, depending on the attenuation parameter A. More specifically, in the frequency range $\omega > \omega_p$, the amplitude response decreases monotonically as the frequency increases, with ω_s being defined as the frequency at which

$$\frac{1}{\sqrt{1+\varepsilon^2 T_N^2(\omega/\omega_p)}} < \frac{1}{A} \quad (\omega_p < \omega_s < \omega)$$

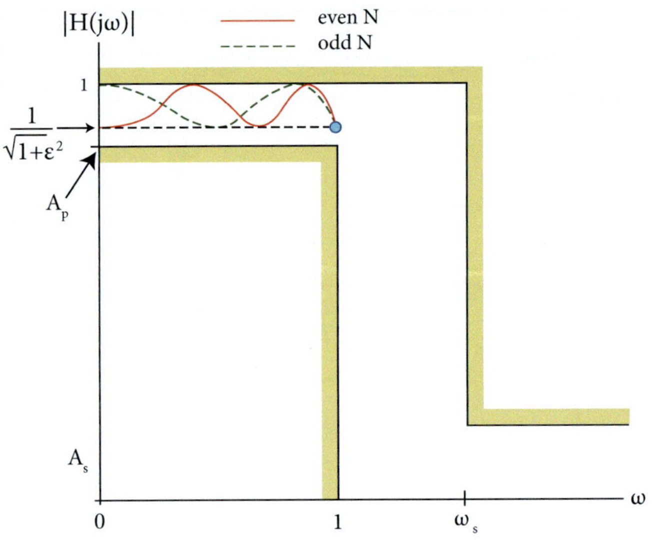

Fig. 11.41 Value of gain function $G(\omega) = |\mathcal{H}(j\omega)|$ for even- and odd-order Chebyshev I filters. The same figure also shows the effect of the ripple width ε, whose maximum allowable value is known to be equal to $\sqrt{10^{A_p/10} - 1}$, which will lead to the coincidence of the dashed line with the line $G(\omega) = A_p$

From the graphical representation of the gain function and the equation defining it, we find that it is maximized at frequencies for which $T_N^2(\omega/\omega_p) = 0$ and minimized at frequencies for which $T_N^2(\omega/\omega_p) = 1$. If we denote these frequencies by $\omega = \omega_{\max}$ and $\omega = \omega_{\min}$, then these will be determined as the roots of the equations

$$T_N^2\left(\frac{\omega_{\max}}{\omega_p}\right) = \cos^2\left[N\cos^{-1}\left(\frac{\omega_{\max}}{\omega_p}\right)\right] = 0 \quad \text{and} \quad T_N^2\left(\frac{\omega_{\max}}{\omega_p}\right) = \cos^2\left[N\cos^{-1}\left(\frac{\omega_{\min}}{\omega_p}\right)\right] = 1$$

correspondingly, where we employed the *trigonometric form* of the Chebyshev polynomials, as we refer to the transmission band where $\omega < \omega_p$, and thus $\omega/\omega_p = 1$. Using the properties of the cosine trigonometric function, this is nullified at the positions (see the relevant section of the mathematical appendix)

$$N\cos^{-1}\left(\frac{\omega_{\max}}{\omega_p}\right) = \frac{\pi}{2}, \frac{3\pi}{2}, \frac{5\pi}{2}, \ldots, \frac{k\pi}{2}, \quad k = 1, 2, 3, \ldots$$

from which it follows that

$$\omega_{\max}^k = \omega_p \cos\left[\frac{(2k-1)\pi}{2N}\right], \quad k = 1, 2, \ldots, L \tag{11.74}$$

where $L = (N+1)/2$ for odd-order and $L = N/2$ for even-order filters, while for the frequencies $\omega_{\min}$ in the same way we find

$$\omega_{\min}^k = \omega_p \cos\left[\frac{k\pi}{N}\right], \quad k = 1, 2, \ldots, L \tag{11.75}$$

with the value of the parameter L defined as before. From the preceding equations, but also from the graphical representation of the gain function we find that the maximum and minimum frequencies *are not equidistant*; however, the maximum and minimum values corresponding to these have *the same measure* between them.

Poles of the Transfer Function

Based on the defining equation of the function $|\mathcal{H}(j\omega)|$ of low-pass filters Chebyshev I and by the well-known methodology for calculating its poles, the function $\mathcal{H}(s)\mathcal{H}(-s)$ takes the form

$$\mathcal{H}(s)\mathcal{H}(-s) = |\mathcal{H}(j\omega)|^2_{\omega=s/j} = \frac{1}{1+\varepsilon^2 T_N^2[s/(j\omega_p)]} \tag{11.76}$$

and therefore the poles of the transfer function of these filters will be calculated as the values of the complex variable $s = j\omega$, which are roots of the equation

$$1+\varepsilon^2 T_N^2\left(\frac{s}{j\omega_p}\right) = 0 \qquad \text{which can also be written as} \qquad T_N\left(\frac{s}{j\omega_p}\right) = \cos\left[N\cos^{-1}\left(\frac{s}{j\omega_p}\right)\right] = \pm\frac{j}{\varepsilon}$$

Defining the auxiliary variable $\varphi = \cos^{-1}(s/j\omega_p) = u + jv$, the foregoing relation is reformulated as[22]

$$T_N(s/j\omega_p) = \cos[N\cos^{-1}(s/j\omega_p)] = \cos(N\varphi) = \cos[N(u+jv)] = \cos(Nu)\cosh(Nv) - j\sin(Nu)\sinh(Nv) = \pm\frac{j}{\varepsilon}$$

where for the derivation of the last relation we used the known trigonometric identity

$$\cos(\alpha + j\beta) = \cos\alpha\cosh\beta - j\sin\alpha\sinh\beta$$

Comparing the real and imaginary parts of the last two expressions, it immediately follows that

$$\cos(Nu)\cosh(Nv) = 0 \qquad \text{and} \qquad \sin(Nu)\sinh(Nv) = \mp\frac{1}{\varepsilon}$$

However, since the function of the hyperbolic cosine cannot take a zero value (which immediately follows from its defining equation), it must necessarily be $\cos(Nu) = 0$, from which it follows that $u = u_\kappa = [(2\kappa - 1)\pi]/(2N)$, $(k = 1, 2, \ldots, N)$. But for these values of u it will hold that $\sin(Nu) = \pm 1$, and therefore, we will have $\sinh(Nv) = \pm 1/\varepsilon$ or, equivalently,

$$v = \pm\frac{1}{N}\sinh^{-1}\left(\frac{1}{\varepsilon}\right)$$

Consequently, the auxiliary variable φ is formulated as

$$\varphi_k = u + jv = \frac{(2\kappa-1)\pi}{2N} \pm \frac{j}{N}\sinh^{-1}\left(\frac{1}{\varepsilon}\right)$$

that is, the poles of the transfer function of the low-pass Chebyshev I filter of order N will be

$$p_k = j\omega_k = j\omega_p\cos\varphi_k = j\omega_p\cos\left[\frac{(2\kappa-1)\pi}{2N} \pm \frac{j}{N}\sinh^{-1}\left(\frac{1}{\varepsilon}\right)\right]$$

or, equivalently,

$$p_k = \pm\sin\frac{(2\kappa-1)\pi}{2N}\sinh\left[\frac{1}{N}\sinh^{-1}\left(\frac{1}{\varepsilon}\right)\right]\omega_p + j\left\{\cos\frac{(2\kappa-1)\pi}{2N}\cosh\left[\frac{1}{N}\sinh^{-1}\left(\frac{1}{\varepsilon}\right)\right]\right\}\omega_p \tag{11.77}$$

[22] The use of the *trigonometric* form of the Chebyshev polynomials, $T_N(\omega) = \cos(N\cos^{-1}\omega)$, is based on the silent assumption that $\varepsilon > 1$. Conversely, if we accept that $\varepsilon < 1$, we should use the *hyperbolic form* of the Chebyshev polynomials $T_N(\omega) = \cosh(N\cosh^{-1}\omega)$; however, the methodology of the analysis and the results we will get are the same as those we will get here and, therefore, are valid for any value of the parameter ε. The repetition of the process presented here using the hyperbolic form of the Chebyshev polynomials is left as an exercise to the reader.

for the values $\kappa = 1, 2, \ldots, N$. Therefore, taking into account that the poles of the transfer function are complex quantities of the form $s_k = \sigma_k + j\omega_k$, it is not difficult to see that their real and imaginary parts have the form[23]

$$\sigma_k = \pm \sin \frac{(2\kappa - 1)\pi}{2N} \sinh\left[\frac{1}{N}\sinh^{-1}\left(\frac{1}{\varepsilon}\right)\right]\omega_p \quad \text{and} \quad \omega_k = \cos\frac{(2\kappa - 1)\pi}{2N}\cosh\left[\frac{1}{N}\sinh^{-1}\left(\frac{1}{\varepsilon}\right)\right]\omega_p \tag{11.78}$$

for the values $k = 1, 2, \ldots, 2N$. In the special case where the order of the filter N is an odd number, the values $(N + 1)/2$ and $(3N + 1)/2$ are even, and for these two values of the index k the trigonometric functions that appear in the equations of the parameters σ_k and ω_k will take the values

$$\sin\frac{(2\kappa - 1)\pi}{2N}\bigg|_{k=\frac{N+1}{2}} = 1, \qquad \cos\frac{(2\kappa - 1)\pi}{2N}\bigg|_{k=\frac{N+1}{2}} = 0$$

$$\sin\frac{(2\kappa - 1)\pi}{2N}\bigg|_{k=\frac{3N+1}{2}} = -1, \qquad \cos\frac{(2\kappa - 1)\pi}{2N}\bigg|_{k=\frac{3N+1}{2}} = 0$$

But this means that for these values of k the imaginary component of the pole *vanishes*, and therefore, for an odd value of order the filter has *two real poles* with a value of

$$\sigma = \pm \sinh\left[\frac{1}{N}\sinh^{-1}\left(\frac{1}{\varepsilon}\right)\right]\omega_p \tag{11.79}$$

from which we only keep the negative pole, which belongs to the left complex half-plane. If we write the foregoing relations as

$$\frac{\sigma_k}{\sinh\left[\frac{1}{N}\sinh^{-1}\left(\frac{1}{\varepsilon}\right)\right]\omega_p} = \pm\sin\frac{(2\kappa - 1)\pi}{2N} \quad \text{and} \quad \frac{\omega_k}{\cosh\left[\frac{1}{N}\sinh^{-1}\left(\frac{1}{\varepsilon}\right)\right]\omega_p} = \cos\frac{(2\kappa - 1)\pi}{2N}$$

and then square them and add them term by term, it turns out that

$$\left(\frac{\sigma_k}{\sinh\left[\frac{1}{N}\sinh^{-1}\left(\frac{1}{\varepsilon}\right)\right]\omega_p}\right)^2 + \left(\frac{\omega_k}{\cosh\left[\frac{1}{N}\sinh^{-1}\left(\frac{1}{\varepsilon}\right)\right]\omega_p}\right)^2 = \sin^2\frac{(2\kappa - 1)\pi}{2N} + \cos^2\frac{(2\kappa - 1)\pi}{2N} = 1 \tag{11.80}$$

which is an ellipse equation with minor and major semi-axis lengths

$$\alpha = \sinh\left[\frac{1}{N}\sinh^{-1}\left(\frac{1}{\varepsilon}\right)\right]\omega_p \quad \text{and} \quad \beta = \cosh\left[\frac{1}{N}\sinh^{-1}\left(\frac{1}{\varepsilon}\right)\right]\omega_p \tag{11.81}$$

We are led, therefore, to the conclusion that *the* $2N$ *poles of the function* $\mathcal{H}(s)\mathcal{H}(-s) = |\mathcal{H}(j\omega)|^2_{\omega=s/j}$ *lie on the circumference of an ellipse on the complex plane*, in contrast to the poles of the corresponding function of the low-pass Butterworth filter, which lie *on the circumference of a circle*. It is not difficult to see that this ellipse intersects the vertical axis at points with ordinates $\pm\beta$ and the horizontal axis at points with abscissas $\pm\alpha$. In full analogy with the Butterworth

[23] Alternatively, we can consider that the sine in the expression of σ_k is related to a positive sign and change the variable k in the range of values $1, 2, \ldots, 2N$, an approach that we will adopt in the exercises that follow.

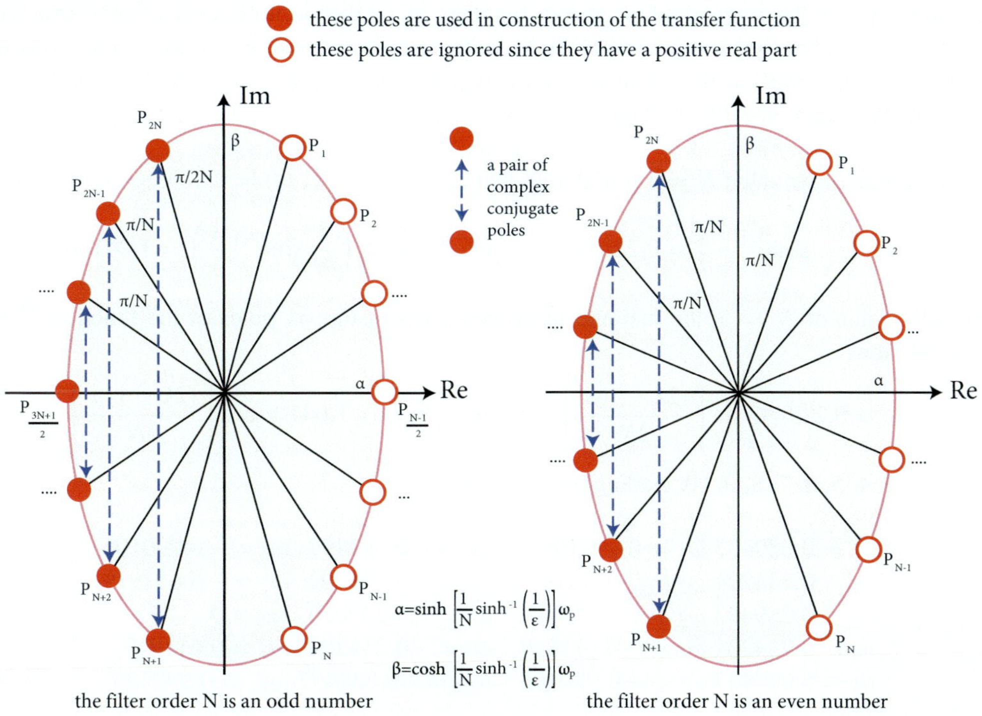

Fig. 11.42 $2N$ poles of function $\mathcal{H}(s)\mathcal{H}(-s)$ for low-pass Chebyshev I filter and for even- and odd-order filter. The N poles used for the construction of the transfer function are represented by blue circles, and if they are of a complex nature, they appear *in pairs of conjugate complex points of the left half-plane*, so that the filter being constructed is a stable system

filters, we use in the transfer function of the low-pass Chebyshev I filter *only the N poles that belong to the left complex half-plane*, so that the filter that will result is *causal* and *stable*. If the order of the filter is even, these N poles form $N/2$ pairs of conjugate complex poles, while for an odd-order filter there is a real pole with a negative value and $(N-1)/2$ pairs of conjugate complex poles. This situation is presented diagrammatically in Fig. 11.42.

Problem 11.9.4 Determine the poles of the transfer function of a normalized low-pass Chebyshev I filter for order values $N = 3, 5$. Assume that the maximum loss in the passband is equal to $A_p = 1.0$ dB.

Solution For this specific value of the loss A_p the ripple parameter is $\varepsilon = \sqrt{10^{0.1} - 1} = 0.508846$. Therefore, we have $x = 1/\varepsilon = 1.965227$ and $\sinh^{-1}(1/\varepsilon) = \sinh^{-1} x = \ln(x + \sqrt{x^2+1}) = 1.427975$.

Starting from the value $N = 3$ we will have $(1/N)\sinh^{-1}(1/\varepsilon) = 0.475991$, from which it follows that

$$\sinh\left(\frac{1}{N}\sinh^{-1}\frac{1}{\varepsilon}\right) = 0.494170 \quad \text{and} \quad \cosh\left(\frac{1}{N}\sinh^{-1}\frac{1}{\varepsilon}\right) = 1.115440$$

Therefore, for an order value of $N = 3$, the real and imaginary parts of the six poles of the function $\mathcal{H}(s)\mathcal{H}(-s)$ will be calculated from the relations

$$\sigma_k = 0.494170 \sin\left[\frac{(2k-1)\pi}{2N}\right] \quad \text{and} \quad \omega_k = 1.115440 \cos\left[\frac{(2k-1)\pi}{2N}\right]$$

for the values $k = 1, 2, 3, 4, 5, 6$, leading to the results

$\sigma_1 = 0.247085$ $\omega_1 = 0.965998$	$\sigma_4 = -0.247085$ $\omega_4 = -0.965998$
$\sigma_2 = 0.494170$ $\omega_2 = 0$	$\sigma_5 = -0.247085$ $\omega_5 = 0$
$\sigma_3 = 0.247085$ $\omega_3 = 0.965998$	$\sigma_6 = -0.247085$ $\omega_6 = 0.965998$

The design of the filter poles on the complex plane reveals that they lie on the circumference of an ellipse with lengths of the major and minor axes $\beta = 1.115440$ and $\alpha = 0.494170$. From these poles we will use in the transfer function only the real pole $p_5 = \sigma_5 = -0.494170$, as well as the conjugate complex poles $(p_4, p_6) = (-0.247085 - 0.965998j, -0.247085 + 0.965998j)$, since only these have negative real parts.

On the other hand, for an order value $N = 5$ it will hold that $(1/N)\sinh^{-1}(1/\varepsilon) = 0.285595$, from which it follows that

$$\sinh\left(\frac{1}{N}\sinh^{-1}\frac{1}{\varepsilon}\right) = 0.289493 \quad \text{and} \quad \cosh\left(\frac{1}{N}\sinh^{-1}\frac{1}{\varepsilon}\right) = 1.04106$$

Therefore, for an order value of $N = 5$, the real and imaginary parts of the six poles of the function $\mathcal{H}(s)\mathcal{H}(-s)$ will be calculated from the relations

$$\sigma_k = 0.289493\sin\left[\frac{(2k-1)\pi}{2N}\right] \quad \text{and} \quad \omega_k = 1.04106\cos\left[\frac{(2k-1)\pi}{2N}\right]$$

for the values $k = 1, 2, 3, 4, 5, 6, 7, 8, 9, 10$, leading to the results

$\sigma_1 = 0.089457$ $\omega_1 = 0.990106$	$\sigma_6 = -0.089457$ $\omega_6 = -0.990106$
$\sigma_2 = 0.234204$ $\omega_2 = 0.611919$	$\sigma_7 = -0.234204$ $\omega_7 = -0.611919$
$\sigma_3 = 0.289493$ $\omega_3 = 0$	$\sigma_8 = -0.289016$ $\omega_8 = 0$
$\sigma_4 = 0.234204$ $\omega_4 = -0.611919$	$\sigma_9 = -0.234204$ $\omega_9 = 0.611919$
$\sigma_5 = 0.089457$ $\omega_5 = -0.990106$	$\sigma_{10} = -0.089457$ $\omega_{10} = 0.990106$

The design of the filter poles on the complex plane reveals that they lie on an ellipse with lengths of the major and minor axes $\beta = 1.04106$ and $\alpha = 0.289493$. From these poles we will use in the transfer function only the real pole $p_8 = \sigma_5 = -0.289016$, as well as the pairs of conjugate complex poles $(p_6, p_{10}) = (-0.089457-0.990106j, -0.089457+0.990106j)$ and $(p_7, p_9) = (-0.234204 - 0.611919j, -0.234204 + 0.611919j)$, since only these have negative real parts.

Problem 11.9.5 Determine the poles of the transfer function of a normalized low-pass Chebyshev I filter for order values $N = 4, 6$. Assume that the maximum loss in the passband is equal to $A_p = 1.0$ dB.

Solution For the specific value of the loss A_p the variance parameter is $\varepsilon = \sqrt{10^{0.1} - 1} = 0.508846$. Therefore, $x = 1/\varepsilon = 1.965227$ and $\sinh^{-1}(1/\varepsilon) = \sinh^{-1}x = \ln(x + \sqrt{x^2 + 1}) = 1.427975$. Starting with the value $N = 4$, we will have $(1/N)\sinh^{-1}(1/\varepsilon) = 0.356993$ from which it follows that

$$\sinh\left(\frac{1}{N}\sinh^{-1}\frac{1}{\varepsilon}\right) = 0.364625 \quad \text{and} \quad \cosh\left(\frac{1}{N}\sinh^{-1}\frac{1}{\varepsilon}\right) = 1.064402$$

Therefore, for an order value $N = 4$ the real and imaginary parts of the six poles of the function $\mathcal{H}(s)\mathcal{H}(-s)$ will be calculated from the relations

$$\sigma_k = 0.364625\sin\left[\frac{(2k-1)\pi}{2N}\right] \quad \text{and} \quad \omega_k = 1.064402\cos\left[\frac{(2k-1)\pi}{2N}\right]$$

for the values $k = 1, 2, 3, 4, 5, 6, 7, 8$, leading to the results

$\sigma_1 = 0.139535$ $\omega_1 = 0.983378$	$\sigma_5 = -0.139535$ $\omega_5 = -0.983378$
$\sigma_2 = 0.336869$ $\omega_2 = 0.407328$	$\sigma_6 = -0.336869$ $\omega_6 = -0.407328$
$\sigma_3 = 0.336869$ $\omega_3 = -0.407328$	$\sigma_7 = -0.336869$ $\omega_7 = 0.407328$
$\sigma_4 = 0.139537$ $\omega_4 = -0.983378$	$\sigma_8 = -0.139537$ $\omega_8 = 0.983378$

The design of the filter poles on the complex plane reveals that they lie on the circumference of an ellipse with lengths of the major and minor semi-axes $\beta = 1.064402$ and $\alpha = 0.364625$. From these poles we will use in the transfer function the pairs of conjugate complex poles $(p_5, p_8) = (-0.139535 - 0.983378j, -0.139537 + 0.983378j)$ and $(p_6, p_7) = (-0.336869 - 0.407328j, -0.336869 + 0.407328j)$, since only these have negative real parts.

On the other hand, for an order value of $N = 6$ we will have $(1/N)\sinh^{-1}(1/\varepsilon) = 0.237995$, from which it follows that

$$\sinh\left(\frac{1}{N}\sinh^{-1}\frac{1}{\varepsilon}\right) = 0.240248 \qquad \text{and} \qquad \cosh\left(\frac{1}{N}\sinh^{-1}\frac{1}{\varepsilon}\right) = 1.028450$$

Therefore, for an order value of $N = 5$ the real and imaginary parts of the six poles of the function $\mathcal{H}(s)\mathcal{H}(-s)$ will be calculated from the relations

$$\sigma_k = 0.240248\sin\left[\frac{(2k-1)\pi}{2N}\right] \qquad \text{and} \qquad \omega_k = 1.028450\cos\left[\frac{(2k-1)\pi}{2N}\right]$$

for the values $k = 1, 2, 3, 4, 5, 6, 7, 8, 9, 10, 12$, leading to the results

$\sigma_1 = 0.062180$ $\omega_1 = 0.993405$	$\sigma_7 = -0.062180$ $\omega_7 = -0.993405$
$\sigma_2 = 0.169880$ $\omega_2 = 0.727223$	$\sigma_8 = -0.169880$ $\omega_8 = -0.727223$
$\sigma_3 = 0.232061$ $\omega_3 = 0.266182$	$\sigma_9 = -0.232061$ $\omega_9 = -0.266182$
$\sigma_4 = 0.232061$ $\omega_4 = -0.266182$	$\sigma_{10} = -0.232061$ $\omega_{10} = 0.266182$
$\sigma_5 = 0.169880$ $\omega_5 = -0.727223$	$\sigma_{11} = -0.169880$ $\omega_{11} = 0.727223$
$\sigma_6 = 0.062180$ $\omega_6 = -0.993405$	$\sigma_{12} = -0.062180$ $\omega_{12} = 0.993405$

The design of the filter poles on the complex plane reveals that they lie on an ellipse with lengths of the major and minor axes $\beta = 1.028450$ and $\alpha = 0.240248$. From these poles we will use in the transfer function the pairs of conjugate complex poles $(p_7, p_{12}) = (-0.062180 - 0.993405j, -0.062180 + 0.993405j)$, $(p_8, p_{11}) = (-0.169880 - 0.727223j, -0.169880 + 0.727223j)$ and $(p_9, p_{10}) = (-0.232061 - 0.266182j, -0.232061 + 0.266182j)$, since only these have negative real parts. ■

Having determined the N poles of the transfer function in the way previously described, we can then determine that function from its defining equation. For the case of a normalized Chebyshev I filter ($\omega_p = 1$) and for an even-order filter N, the transfer function contains, as is well known, $N/2$ pairs of complex conjugate poles and therefore will have the form

$$\mathcal{H}(s) = \frac{H_0}{\prod_{i=1}^{N/2}(s-p_i)(s-p_i^*)} = \frac{H_0}{\prod_{i=1}^{N/2}(s^2 - 2\Re\{p_i\}s + |p_i|^2)} \tag{11.82}$$

with the proportionality constant H_0 resulting from the requirement to be $G(0) = 10^{-0.05A_p}$ (see earlier discussion about the amplitude response). If, therefore, in the preceding relation we set $s = 0$, then the value of the gain for $\omega = 0$ will have the form

$$G(0) = |\mathcal{H}(0)| = \frac{H_0}{\prod_{i=1}^{N/2}|p_i|^2} = 10^{-0.05A_p} \qquad \text{from which it follows that} \qquad H_0 = 10^{-0.05A_p}\prod_{i=1}^{N/2}|p_i|^2$$

and therefore

$$\mathcal{H}(s) = \frac{10^{-0.05R_p}\prod_{i=1}^{N/2}|p_i|^2}{\prod_{i-1}^{N/2}(s-p_i)(s-p_i^*)} = \frac{10^{-0.05R_p}\prod_{i=1}^{N/2}|p_i|^2}{\prod_{i=1}^{N/2}(s^2 - 2\Re\{p_i\}s + |p_i|^2)} = \frac{10^{-0.05A_p}\prod_{i=1}^{N/2}(\sigma_i^2+\omega_i^2)}{\prod_{i=1}^{N/2}(s^2 - 2\sigma_i s + \sigma_i^2 + \omega_i^2)} \tag{11.83}$$

On the other hand, for an odd-order filter, the transfer function will have the form

$$\mathcal{H}(s) = \frac{H_0}{(s-p_0)\prod_{i=1}^{(N-1)/2}(s-p_i)(s-p_i^*)} = \frac{H_0}{(s+\sigma)\prod_{i=1}^{(N-1)/2}(s^2 - 2\Re\{p_i\}s + |p_i|^2)} \tag{11.84}$$

where $p_0 = -\sigma$ is the negative real pole we considered previously. In this case, the proportionality constant will arise from the requirement that $G(0) = |\mathcal{H}0| = 1$—so it will hold that

$$G(0) = |\mathcal{H}(0)| = \frac{H_0}{\sigma \prod_{i=1}^{(N-1)/2} |p_i|^2} = 1$$

from which it follows that

$$H_0 = \sigma \prod_{i=1}^{(N-1)/2} |p_i|^2 = \sinh\left[\frac{1}{N}\sinh^{-1}\left(\frac{1}{\varepsilon}\right)\right] \prod_{i=1}^{(N-1)/2} |p_i|^2$$

Therefore, for the case of an odd-order Chebyshev I filter N, the transfer function has the form

$$\begin{aligned}\mathcal{H}(s) &= \frac{\sinh\left[\frac{1}{N}\sinh^{-1}\left(\frac{1}{\varepsilon}\right)\right] \prod_{i=1}^{(N-1)/2} |p_i|^2}{\left\{s + \sinh\left[\frac{1}{N}\sinh^{-1}\left(\frac{1}{\varepsilon}\right)\right]\right\} \prod_{i=1}^{(N-1)/2} (s - p_i)(s - p_i^*)} \\ &= \frac{\sinh\left[\frac{1}{N}\sinh^{-1}\left(\frac{1}{\varepsilon}\right)\right] \prod_{i=1}^{(N-1)/2} |p_i|^2}{\left\{s + \sinh\left[\frac{1}{N}\sinh^{-1}\left(\frac{1}{\varepsilon}\right)\right]\right\} \prod_{i=1}^{(N-1)/2} (s^2 - 2\Re\{p_i\}s + |p_i|^2)}\end{aligned} \tag{11.85}$$

It should be noted that because the expressions of the poles p_i contain *hyperbolic functions* that are not very easy to handle, a formalism is usually adopted that is based on the use of *nonexcessive functions*. To express this formalism, let us use the mathematical identity $\sinh^{-1} x = \ln(x + \sqrt{x^2 + 1})$. In this case, it turns out that

$$\frac{1}{N}\sinh^{-1}\left(\frac{1}{\varepsilon}\right) = \frac{1}{N}\ln\left(\frac{1}{\varepsilon} + \sqrt{\frac{1}{\varepsilon^2} + 1}\right) = \frac{1}{N}\ln\left(\frac{1 + \sqrt{1+\varepsilon^2}}{\varepsilon}\right) = \ln\sqrt[N]{\frac{1 + \sqrt{1+\varepsilon^2}}{\varepsilon}} = \ln\xi$$

where $\xi = [(1 + \sqrt{1+\varepsilon^2})/\varepsilon]^{1/N}$. In this case, we will have

$$\cosh\left[\frac{1}{N}\sinh^{-1}\left(\frac{1}{\varepsilon}\right)\right] = \cosh(\ln\xi) = \frac{e^{\ln\xi} + e^{-\ln\xi}}{2} = \frac{\xi + \frac{1}{\xi}}{2} = \frac{\xi^2 + 1}{2\xi}$$

$$\sinh\left[\frac{1}{N}\sinh^{-1}\left(\frac{1}{\varepsilon}\right)\right] = \sinh(\ln\xi) = \frac{e^{\ln\xi} - e^{-\ln\xi}}{2} = \frac{\xi - \frac{1}{\xi}}{2} = \frac{\xi^2 - 1}{2\xi}$$

Therefore, the real and imaginary parts of the poles of the transfer function are expressed as

$$\sigma_k = -\sin\frac{(2\kappa+1)\pi}{2N}\left(\frac{\xi^2 - 1}{2\xi}\right)\omega_p \quad \text{and} \quad \omega_k = \cos\frac{(2\kappa+1)\pi}{2N}\left(\frac{\xi^2 + 1}{2\xi}\right)\omega_p \tag{11.86}$$

while the semi-axes of the ellipse on which the poles of the transfer function lie are

$$\alpha = \frac{\xi^2 - 1}{2\xi}\omega_p \quad \text{and} \quad \beta = \frac{\xi^2 + 1}{2\xi}\omega_p$$

Table 11.2 Positions of poles of Chebyshev I low-pass filter for order values $N = 2, 3, 4, 5, 6, 7, 8$ and for ripple range $\varepsilon = 0.1, 0.5$ and 1 dB

ε = 0.1 dB		ε = 0.5 dB		ε = 1.0 dB	
N	Pole position	N	Pole position	N	Pole position
2	$-1.186178 \pm 1.380948j$	2	$-0.712812 \pm 1.004020j$	2	$-0.548867 \pm 0.895129j$
3	–0.969406	3	–0.626457	3	–0.494171
	$-0.484703 \pm 1.206155j$		$-0.313228 \pm 1.021928j$		$-0.247085 \pm 0.965999j$
4	$-0.637730 \pm 0.465000j$	4	$-0.423340 \pm 0.420946j$	4	$-0.336870 \pm 0.407329j$
	$-0.264156 \pm 1.122610j$		$-0.175353 \pm 1.016253j$		$-0.139536 \pm 0.983379j$
5	–0.538914	5	–0.362320	5	–0.289493
	$-0.435991 \pm 0.667707j$		$-0.293123 \pm 0.625177j$		$-0.234205 \pm 0.611920j$
	$-0.166534 \pm 1.080372j$		$-0.111963 \pm 1.011557j$		$-0.089458 \pm 0.990107j$
6	$-0.428041 \pm 0.283093j$	6	$-0.289794 \pm 0.270216j$	6	$-0.232063 \pm 0.266184j$
	$-0.313348 \pm 0.773426j$		$-0.212144 \pm 0.738245j$		$-0.169882 \pm 0.727227j$
	$-0.114693 \pm 1.056519j$		$-0.077650 \pm 1.008461j$		$-0.062181 \pm 0.993411j$
7	–0.376778	7	–0.256170	7	–0.205414
	$-0.339465 \pm 0.463659j$		$-0.230801 \pm 0.447894j$		$-0.185072 \pm 0.442943j$
	$-0.234917 \pm 0.835485j$		$-0.159719 \pm 0.807077j$		$-0.128074 \pm 0.798156j$
	$-0.083841 \pm 1.041833j$		$-0.057003 \pm 1.006409j$		$-0.045709 \pm 0.995284j$
8	$-0.321650 \pm 0.205314j$	8	$-0.219293 \pm 0.199907j$	8	$-0.175998 \pm 0.198206j$
	$-0.272682 \pm 0.584684j$		$-0.185908 \pm 0.569288j$		$-0.149204 \pm 0.564444j$
	$-0.182200 \pm 0.875941j$		$-0.124219 \pm 0.852000j$		$-0.099695 \pm 0.844751j$
	$-0.063980 \pm 1.032181j$		$-0.043620 \pm 1.005002j$		$-0.035008 \pm 0.996451j$

From the foregoing analysis it is concluded that a low-pass Chebyshev I filter of order N has N poles, and it is characterized by the absence of zero values.[24] These N poles lie on an elliptical circumference with semi-axis lengths α and β, of which the major semi-axis coincides with the imaginary axis, while the minor semi-axis coincides with the real axis of the complex plane. We observe that the lengths of the semi-axes and, consequently, the positions of the poles of the transfer function on the complex plane depend on the value of the ripple parameter ε. It is proven that for the special case of zero ripple $\varepsilon = 0$ dB the Butterworth and Chebyshev filters *have exactly the same transfer function and the same frequency response characteristics*. The positions of the poles of the Chebyshev low-pass filter for order values $N = 2, 3, 4, 5, 6, 7, 8$ and for ripple range $\varepsilon = 0, 1, 0.5$ and 1 dB are included in Table 11.2.

Loss Characteristic

The *loss characteristic* of Chebyshev I filters is calculated from the equation

$$\mathcal{A}(\omega) = 10 \log L(\omega^2) = 10 \log \frac{1}{\mathcal{H}(j\omega)\mathcal{H}(-j\omega)} = 10 \log[1 + \varepsilon^2 T_N^2(\omega)] \tag{11.87}$$

(for the normalized frequency $\omega_p = 1$) and for a fourth-order normalized Chebyshev filter is represented diagrammatically in Fig. 11.43. From this figure we observe that the loss in the passband fluctuates between zero and the value of the parameter A_p, which is defined during the specification of the filter. For an alternative presentation of the filter characteristics starting from its loss characteristic, the reader is referred to the book by Andreas Antoniou (see the bibliography at the end of the book). In this book, the description starts from the desired properties that the loss characteristic should have, and through a detailed analysis we conclude that to achieve these properties, the function in question should be expressed in terms of Chebyshev polynomials. Having calculated the characteristic $\mathcal{A}(\omega)$, it is then possible to determine the transfer function and its poles, leading to the same results as those we obtained in the preceding pages.

[24] The mathematical form of the function $\mathcal{H}(s)$ should not confuse the reader since the complex quantities $s_1, s_2, \ldots, s_N$ that appear in the numerator of the function are its poles and not some zero values.

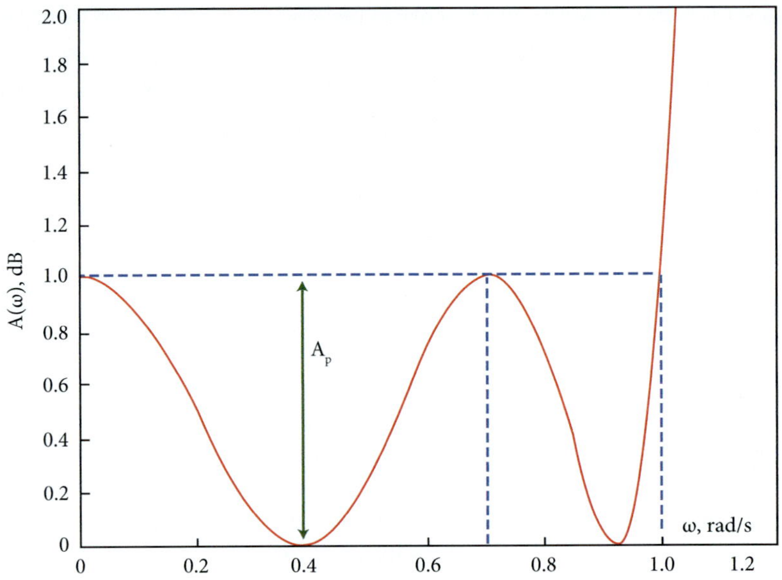

Fig. 11.43 Loss characteristic $\mathcal{A}(\omega)$ for fourth-order normalized Chebyshev I filter (Antoniou, Figure 10.9)

Phase Characteristics and Time-Domain Response

The *phase response* of the prototype low-pass Chebyshev I filter for normalized frequency $\omega_p = 1$, for attenuation $A_p = 1$ dB, and for filter order values $N = 2$ to $N = 10$ is presented in Fig. 11.44. In this case, the application of the general equations for calculating the phase response for even-order filter and odd-order filter leads to the results

$$\measuredangle\mathcal{H}(j\omega) = -\sum_{k=1}^{N/2} \tan^{-1} \frac{2\sigma_k \omega}{\omega^2 - \sigma_k^2 - \omega_k^2}$$

$$\measuredangle\mathcal{H}(j\omega) = -\tan^{-1}\left\{\frac{\omega}{\sinh\left[\frac{1}{N}\sinh^{-1}\left(\frac{1}{\varepsilon}\right)\right]}\right\} - \sum_{k=1}^{(N-1)/2} \tan^{-1}\frac{2\sigma_k\omega}{\omega^2 - \sigma_k^2 - \omega_k^2}$$

The *phase delay* is typically obtained by dividing the negative of the preceding expression by the frequency ω, whereas the *group delay* is defined as the negative derivative of the function $\measuredangle\mathcal{H}(j\omega)$; the last quantity, for the cases of even and odd filter orders, is given by the equations

$$\tau(\omega) = -\sum_{k=1}^{N/2} \frac{2\sigma_k(\sigma_k^2 + \omega_k^2 + \omega^2)}{(\sigma_k^2 + \omega_k^2 - \omega^2)^2} \tag{11.88}$$

$$\tau(\omega) = -\frac{\sinh\left[\frac{1}{N}\sinh^{-1}\left(\frac{1}{\varepsilon}\right)\right]}{\sinh^2\left[\frac{1}{N}\sinh^{-1}\left(\frac{1}{\varepsilon}\right)\right] + \omega^2} - \sum_{k=1}^{(N-1)/2} \frac{2\sigma_k(\sigma_k^2 + \omega_k^2 + \omega^2)}{(\sigma_k^2 + \omega_k^2 - \omega^2)^2} \tag{11.89}$$

The *impulse* and *step response* of the prototype low-pass Chebyshev I filter for the attenuation value $A_p = 1$ dB and for the filter order values $N = 2$ to $N = 10$ are presented in Fig. 11.45. From this figure we observe that over time the step response of the filter stabilizes at a value which for odd-order filters is equal to one, while for even-order filters it is equal to $1/\sqrt{1+\varepsilon^2}$. As for the impulse response, this can be calculated from the derivative of the step response. The process of

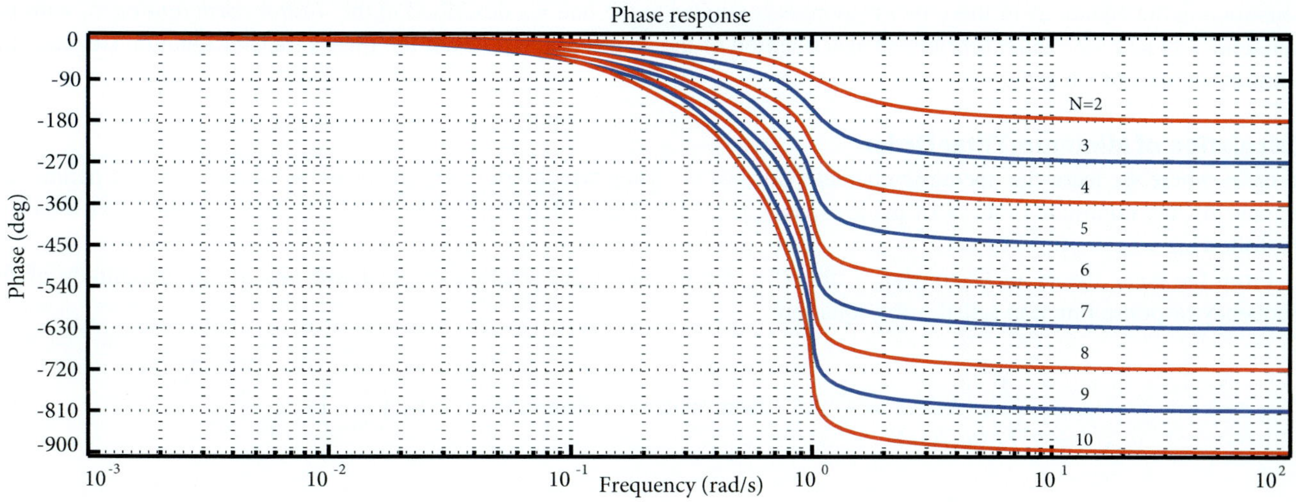

Fig. 11.44 Phase response of prototype low-pass Chebyshev I filter for values $\omega_p = 1$, $R_p = 1$ dB and for filter order values $N = 2$ to $N = 10$

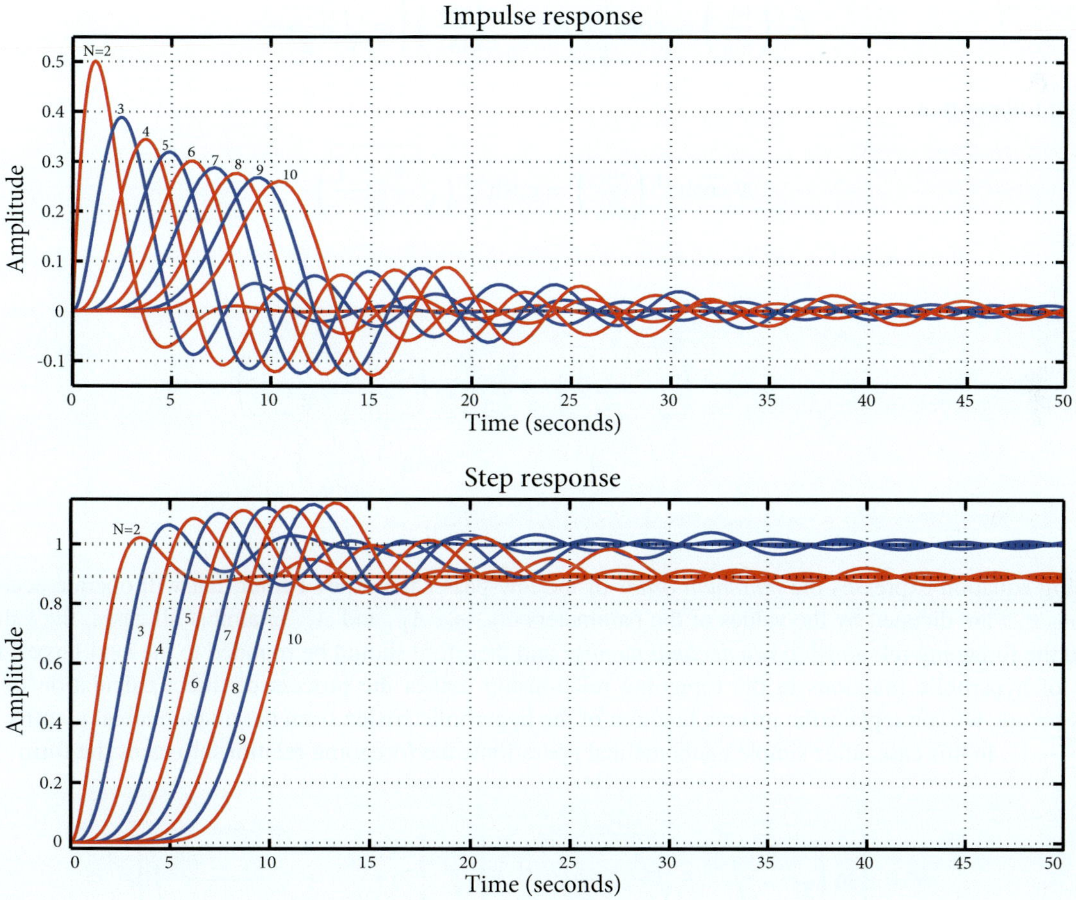

Fig. 11.45 Impulse and step responses of prototype low-pass Chebyshev I filter for the value $A_p = 1$ dB and for filter order values $N = 2 - 10$

constructing the equations of these two responses is similar to the one we described in the *Butterworth approach*, with the functions $s(t)$ and $h(t)$ being formulated as a function of the values of the poles of the filter's transfer function (see also the relevant sections of Chap.5).

Calculation of Minimum Filter Order

In cases where we learn the operating characteristics of the filter, which are the frequencies ω_p and ω_s and the parameters A_p and A_s, the minimum order N of the low-pass Chebyshev I filter that meets the foregoing specifications is calculated in the following way:

From the basic theory of Chebyshev I filters we know that at the position of the frequency ω_s the magnitude of the frequency response function satisfies the equation

$$\frac{1}{\sqrt{1+\varepsilon^2 T_N^2(\omega_s/\omega_p)}} = \frac{1}{A}, \qquad \text{from which it follows that} \qquad T_N\left(\frac{\omega_s}{\omega_p}\right) = \sqrt{\frac{A^2-1}{\varepsilon^2}}$$

where, as is well known, $\varepsilon = \sqrt{10^{A_p/10}-1}$ and $A = 10^{A_s/20}$ (see the introductory section on analog filters).

Using the definition of Chebyshev polynomials of the first kind, the foregoing relationship is written

$$T_N\left(\frac{\omega_s}{\omega_p}\right) = \cosh\left[N\cosh^{-1}\left(\frac{\omega_s}{\omega_p}\right)\right] = \sqrt{\frac{A^2-1}{\varepsilon^2}}$$

from which it follows that

$$N\cosh^{-1}\left(\frac{\omega_s}{\omega_p}\right) = \cosh^{-1}\left(\sqrt{\frac{A^2-1}{\varepsilon^2}}\right)$$

and finally

$$N = \frac{\cosh^{-1}\left(\sqrt{\frac{A^2-1}{\varepsilon^2}}\right)}{\cosh^{-1}\left(\frac{\omega_s}{\omega_p}\right)} = \frac{\cosh^{-1}\left(\frac{\sqrt{10^{A_s/10}-1}}{\sqrt{10^{A_p/10}-1}}\right)}{\cosh^{-1}\left(\frac{\omega_s}{\omega_p}\right)}$$

The foregoing equation expresses the *minimum order* of the low-pass Chebyshev I filter that will operate according to the specifications that are dictated by the values of the parameters ω_p, ω_s, A_p, and A_s. In almost all cases, the value of N that results from the foregoing relationship is a *decimal number* and therefore should be rounded to the next largest integer.

The use of hyperbolic functions in the foregoing relationship makes the process of direct calculation of N difficult. Working as before, we can replace the inverse function of the hyperbolic cosine from the mathematical identity $\cosh^{-1} x = \ln(x+\sqrt{x^2-1})$. In this case, after simple mathematical operations, the foregoing relationship takes the form

$$N = \left\{\ln\left(\sqrt{\frac{A^2-1}{\varepsilon^2}} + \sqrt{\frac{A^2-\varepsilon^2-1}{\varepsilon^2}}\right)\right\} \Big/ \left\{\ln\left(\frac{\omega_s}{\omega_p} + \sqrt{\frac{\omega_s^2-\omega_p^2}{\omega_p^2}}\right)\right\} \tag{11.90}$$

Selectivity and Spectral Efficiency Factor

The *selectivity* of the Chebyshev I filters is calculated directly from the defining equation as

$$F_s = -\frac{d}{d\omega}\{|\mathcal{H}(j\omega)|\}_{\omega=\omega_c} = -\frac{d}{d\omega}\left\{\frac{1}{\sqrt{1+\varepsilon^2 T_N^2(\omega/\omega_p)}}\right\}_{\omega=\omega_c} = \left\{\frac{\varepsilon^2 T_N(\omega/\omega_p)\frac{d}{d\omega}\{T_N(\omega/\omega_p)\}}{\sqrt[3]{[1+\varepsilon^2 T_N^2(\omega/\omega_p)]^2}}\right\}_{\omega=\omega_c}$$

Relying on the properties of the derivatives of hyperbolic functions, we find that

$$\frac{dT_N(\omega)}{d\omega} = \frac{d}{d\omega}\left[\cosh(N\cosh^{-1}\omega)\right] = \frac{N\sinh(N\cosh^{-1}\omega)}{\sqrt{\omega^2-1}} \quad \text{and therefore}$$

$$\frac{d}{d\omega}\left\{T_N\left(\frac{\omega}{\omega_p}\right)\right\} = \frac{N\sinh[N\cosh^{-1}(\omega/\omega_p)]}{\sqrt{(\omega/\omega_p)^2-1}}\frac{1}{\omega_p} = \frac{N\sinh[N\cosh^{-1}(\omega/\omega_p)]}{\sqrt{\omega^2-\omega_p^2}}$$

so by substitution we get

$$F_s = \left\{\frac{N\varepsilon^2}{\sqrt{\omega^2-\omega_p^2}}\frac{\cosh\left[N\cosh^{-1}\left(\frac{\omega}{\omega_p}\right)\right]\sinh\left[N\cosh^{-1}\left(\frac{\omega}{\omega_p}\right)\right]}{\sqrt[3]{\left[1+\varepsilon^2 T_N^2\left(\frac{\omega}{\omega_p}\right)\right]^2}}\right\}_{\omega=\omega_c}$$

$$= \frac{N\varepsilon^2}{\sqrt{\omega_c^2-\omega_p^2}}\frac{\cosh\left[N\cosh^{-1}\left(\frac{\omega_c}{\omega_p}\right)\right]\sinh\left[N\cosh^{-1}\left(\frac{\omega_c}{\omega_p}\right)\right]}{\sqrt[3]{\left\{1+\varepsilon^2\cosh^2\left[N\cosh^{-1}\left(\frac{\omega_c}{\omega_p}\right)\right]\right\}^2}}$$

Taking into account that at the cutoff frequency $\omega = \omega_c$ the magnitude of the frequency response becomes equal to $1/\sqrt{2}$ (a reduction value which in the logarithmic scale is known to be equal to -3 dB), we will have

$$|\mathcal{H}(j\omega_c)| = \frac{1}{\sqrt{1+\varepsilon^2 T_N^2\left(\frac{\omega_c}{\omega_p}\right)}} = \frac{1}{\sqrt{1+\varepsilon^2\cosh^2\left[N\cosh^{-1}\left(\frac{\omega_c}{\omega_p}\right)\right]}}$$

from which it follows that

$$\frac{\omega_c}{\omega_p} = \cosh\left[\frac{1}{N}\cosh^{-1}\left(\frac{1}{\varepsilon}\right)\right]$$

Substituting the foregoing frequency ratio into the equation of *selectivity*, it is written

$$F_s = \frac{N\varepsilon\cosh\left[\frac{1}{N}\cosh^{-1}\left(\frac{1}{\varepsilon}\right)\right]\sinh\left[\cosh^{-1}\left(\frac{1}{\varepsilon}\right)\right]}{(2\sqrt{2}\omega_c)\sinh\left[\frac{1}{N}\cosh^{-1}\left(\frac{1}{\varepsilon}\right)\right]} \tag{11.91}$$

On the other hand, to calculate the *spectral efficiency factor*, let us consider an attenuation level A_ℓ dB that is greater than the level A_p, which is determined by the filter specifications. If this level is associated with a frequency value ω_ℓ, then from the amplitude response defining equation of the filter we will have

$$A_\ell = 10\log\left[1+\varepsilon^2 T_N^2\left(\frac{\omega_\ell}{\omega_p}\right)\right]$$

Solving for ω_ℓ, and using the *hyperbolic form* of the Chebyshev polynomials—since $\omega_\ell > \omega_p$—it turns out that

$$\omega_\ell = \omega_p \cosh\left[\frac{1}{N}\cosh^{-1}\left(\frac{\sqrt{e^{A_\ell/10}-1}}{\varepsilon}\right)\right] \tag{11.92}$$

Considering a reference frequency ω_0, the frequency ω_ℓ uniquely determines the bandwidth $\mathrm{BW}_\ell = \omega_\ell - \omega_0$, so that we can use the foregoing relationship as the bandwidth equation that corresponds to the attenuation level A_ℓ. In this case, considering two attenuation levels α and β, the use of the spectral efficiency factor defining equation will lead us to the expression

$$S_\alpha^\beta = \frac{\mathrm{BW}_\beta}{\mathrm{BW}_\alpha} = \left\{\cosh\left[\frac{1}{N}\cosh^{-1}\left(\frac{\sqrt{e^{\beta/10}-1}}{\varepsilon}\right)\right]\right\} \Big/ \left\{\cosh\left[\frac{1}{N}\cosh^{-1}\left(\frac{\sqrt{e^{\alpha/10}-1}}{\varepsilon}\right)\right]\right\} \tag{11.93}$$

The Process of Designing Low-Pass Chebyshev I Filters

At this point we have completed the presentation of all the characteristics associated with the low-pass Chebyshev I filters, and we can move on to the description of their design process, which consists of the following steps:

1. Determine the values of the filter's operating parameters ω_p, ω_s, A_p, and A_s, so that the existing requirements in each case are met.
2. Calculate the values of the parameters $\varepsilon = \sqrt{10^{A_p/10}-1}$ and $A = 10^{A_s/20}$ and then the value of the *minimum order filter* from the relations

$$N = \left\{\cosh^{-1}\left(\sqrt{\frac{A^2-1}{\varepsilon^2}}\right)\right\} \Big/ \left\{\cosh^{-1}\left(\frac{\omega_s}{\omega_p}\right)\right\}$$
$$= \left\{\ln\left(\sqrt{\frac{A^2-1}{\varepsilon^2}}+\sqrt{\frac{A^2-\varepsilon^2-1}{\varepsilon^2}}\right)\right\} \Big/ \left\{\ln\left(\frac{\omega_s}{\omega_p}+\sqrt{\frac{\omega_s^2-\omega_p^2}{\omega_p^2}}\right)\right\}$$

   which is rounded to the next highest integer.
3. Knowing the values of ε and N, calculate the values of the real and imaginary parts of the poles $s_\kappa = \sigma_\kappa + j\omega_\kappa$ ($\kappa = 1, 2, \ldots, 2N$) from the equations

$$\sigma_k = -\sin\frac{(2\kappa+1)\pi}{2N}\sinh\left[\frac{1}{N}\sinh^{-1}\left(\frac{1}{\varepsilon}\right)\right] \quad \text{and} \quad \omega_k = \cos\frac{(2\kappa+1)\pi}{2N}\cosh\left[\frac{1}{N}\sinh^{-1}\left(\frac{1}{\varepsilon}\right)\right]$$

4. From the calculated $2N$ poles, consider only the N of them that belong to the *left complex half-plane*, and calculate the transfer function of the system from the relation

$$H(s) = H_0\frac{(-1)^N s_1 s_2 \ldots s_N}{[(s/\omega_p)-s_1][(s/\omega_p)-s_2]\ldots[(s/\omega_p)-s_N]}$$

   where H_0 is the value of the gain that is determined in the way presented in the relevant section. In the special case where the filter to be constructed is *normalized*, the transfer function is calculated from the relations

$$\mathcal{H}(s) = \frac{10^{-0.05A_p} \prod_{i=1}^{N/2} (\sigma_i^2 + \omega_i^2)}{\prod_{i=1}^{N/2} (s^2 - 2\sigma_i s + \sigma_i^2 + \omega_i^2)}$$

and

$$\mathcal{H}(s) = \frac{\sinh\left[\frac{1}{N} \sinh^{-1}\left(\frac{1}{\varepsilon}\right)\right] \prod_{i=1}^{(N-1)/2} |p_i|^2}{\left\{s + \sinh\left[\frac{1}{N} \sinh^{-1}\left(\frac{1}{\varepsilon}\right)\right]\right\} \prod_{i=1}^{(N-1)/2} (s^2 - 2\Re\{p_i\}s + |p_i|^2)}$$

for *even*- and *odd*-order filters, respectively.

5. We calculate the frequency response function by making the substitution $s = j\omega$ in the transfer function and from this the amplitude and phase responses $|\mathcal{H}(j\omega)|$ and $\measuredangle\mathcal{H}(j\omega)$.

The design process for low-pass Chebyshev I filters is demonstrated in detail in the next exercise.

Problem 11.9.6 A normalized Chebyshev filter should be constructed with operating frequencies $\omega_p = 1.0$ rad/s and $\omega_s = 2.2$ rad/s and attenuation levels $A_p = 0.2$ dB and $A_s = 40$ dB (Antoniou, Exercise for solution 10.9).

Solution Initially, we need to calculate the *minimum filter order* as a function of the values of the preceding four parameters. From the defining equation of the minimum order N it turns out that

$$N = \frac{\cosh^{-1}\left(\frac{\sqrt{10^{A_s/10} - 1}}{\sqrt{10^{A_p/10} - 1}}\right)}{\cosh^{-1}\left(\frac{\omega_s}{\omega_p}\right)} = \frac{\cosh^{-1}\left(\frac{\sqrt{10^{40/10} - 1}}{\sqrt{10^{0.2/10} - 1}}\right)}{\cosh^{-1}\left(\frac{2.2}{1.0}\right)} = \frac{\cosh^{-1}\left(\frac{\sqrt{9999}}{\sqrt{0.0471285}}\right)}{\cosh^{-1}(2.2)} = \frac{6.825704}{1.425416} = 4.788569$$

where for the calculation of the inverse hyperbolic cosine we used the known identity $\cosh^{-1} x = \ln(x + \sqrt{x^2 - 1})$. The foregoing value should be rounded to the next highest integer, and therefore the order of the filter that will satisfy the operating specifications is $N = 5$.

On the other hand, the ripple parameter ε corresponding to the attenuation level $A_p = 0.2$ dB is calculated as

$$\varepsilon = \sqrt{10^{0.2/10} - 1} = 0.217091$$

from which it follows that

$$\sinh^{-1}\left(\frac{1}{\varepsilon}\right) = \sinh^{-1}\left(\frac{1}{0.217091}\right) = \sinh^{-1}(4.606363)$$
$$= \ln(4.606363 + \sqrt{(4.606363)^2 + 1}) = \ln(9.32002) = 2.232164$$

Therefore,

$$\sinh\left[\frac{1}{N} \sinh\left(\frac{1}{\varepsilon}\right)\right] = \sinh\left(\frac{1}{5} \times 2.232164\right) = 0.461410$$

$$\cosh\left[\frac{1}{N}\sinh\left(\frac{1}{\varepsilon}\right)\right] = \cosh\left(\frac{1}{5}\times 2.232164\right) = 1.101320$$

and consequently, the real and imaginary parts of the poles of the transfer function are written

$$\sigma_k = 0.461410\sin\left[\frac{(2k-1)\pi}{2N}\right] \quad \text{and} \quad \omega_k = 1.101320\sin\left[\frac{(2k-1)\pi}{2N}\right]$$

for the values $k = 1, 2, 3, \ldots, 10$. Working as before, we choose for the transfer function only the poles that have negative real parts and that are the real pole $p_3 = -0.461410$ and the pairs of conjugate complex poles $p_{1,5} = -0.142583 \pm 1.047417j$ and $p_{2,4} = -0.373288 \pm 0.647339j$. Applying the theory we presented in previous sections, we find that the real pole contributes to the denominator of the transfer function the first degree term $(s + 0.416410)$, while the pairs of the poles (p_1, p_5) and (p_2, p_4) contribute respectively the terms $(s^2 + 0.285166s + 1.117411)$ and $(s^2 + 0.746576s + 0.558390)$. Therefore, the transfer function will have the form

$$\begin{aligned}\mathcal{H}(s) &= \frac{H_0}{(s+0.416410)(s^2+0.285166s+1.117411)(s^2+0.746576s+0.558390)} = \\ &= \frac{0.287897}{s^5+1.493152s^4+2.364755s^3+1.864929s^2+1.082345s+0.287897}\end{aligned}$$

with the constant H_0 being calculated, so that $G(0) = 1$. Having determined the transfer function, we can calculate the frequency response by making the substitution $s = j\omega$; then

$$\begin{aligned}\mathcal{H}(j\omega) &= \frac{0.287897}{(j\omega)^5+1.493152(j\omega)^4+2.364755(j\omega)^3+1.864929(j\omega)^2+1.082345(j\omega)+0.287897} \\ &= \frac{0.287897}{(0.287897-1.864929\omega^2+1.493152\omega^4)+j(1.082345\omega-2.364755\omega^3+\omega^5)} \\ &= \frac{(0.082884-0.536907\omega^2+0.429873\omega^4)-j(0.311603-0.680805\omega^3+0.287897\omega^5)}{\omega^{10}-2.500008\omega^8+2.187512\omega^6-0.781234\omega^4+0.097656\omega^2+0.082884}\end{aligned}$$

After calculating the transfer function and the frequency response of the filter we can apply the now known equations to calculate the *amplitude and phase responses*, the *phase and group delays*, and the *step* and *impulse* of the *response*, processes that are left as an exercise to the reader. The variations of the amplitude and phase responses as a function of the frequency ω, expressed in logarithmic scale for the filter of this exercise, are depicted in Fig. 11.46.

11.9.7 Chebyshev Type II Filters

After the description of Chebyshev I filters let us move on to Chebyshev II filters that are associated with the so-called *inverse Chebyshev approximation*.[25] This approach owes its name to the fact that it leads to a response of amplitude that is the *inverse* of that given by Chebyshev I filters. More specifically, while the amplitude response of Chebyshev I filters shows equal amplitude variations in the passband and monotonic behavior in the stopband, Chebyshev II filters show *monotonic behavior in the passband* and *equal amplitude variations in the stopband*. Of particular interest is the existence of zeros of the gain function in the stopband and at frequency values that are *the inverses of the roots of the corresponding Chebyshev polynomial*, while between these zeros the gain function presents maxima with a value equal to $-A_s$ (dB). On the other hand, in the passband and for a frequency value of $\omega = 0$, the gain function of the Chebyshev II filters shows *the maximum possible behavior*, just as we saw in the Butterworth approach. This similarity between these two approaches regarding their behavior in the passband gives the phase response of Chebyshev II filters *a higher degree of linearity* compared to that of Chebyshev I filters, while the maximum value of Q for the same gain specifications is significantly smaller compared to the corresponding value of the Chebyshev I filters. An example of a passive circuit that implements a Chebyshev II filter is

[25] The inverse Chebyshev polynomials on which this approach is based arise from the Chebyshev polynomials by making the substitution $x \to (1/x)$ and are characterized by the recursive relationship that the Chebyshev polynomials satisfy.

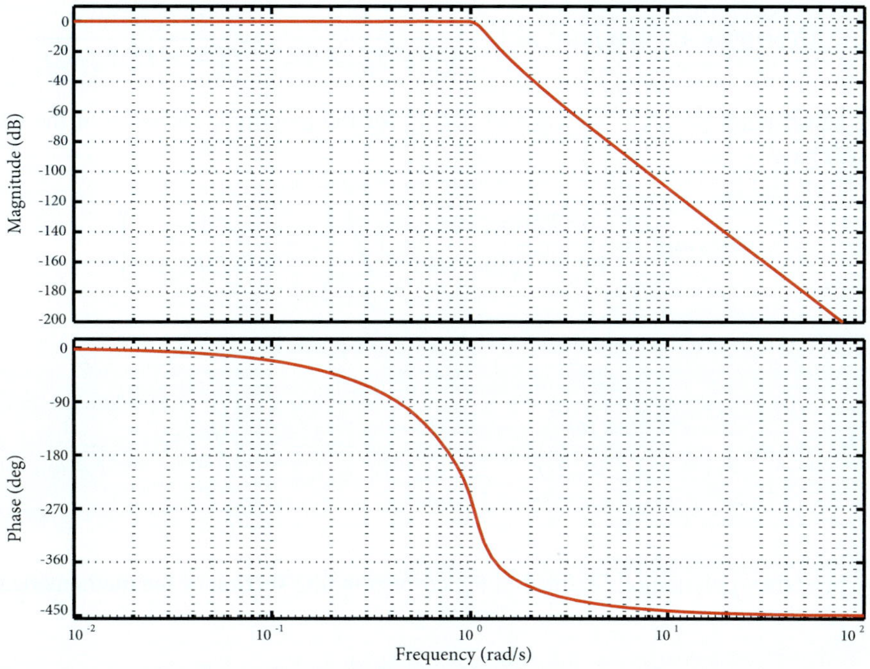

Fig. 11.46 Amplitude and phase responses of Chebyshev I filter of Exercise 11.9.6

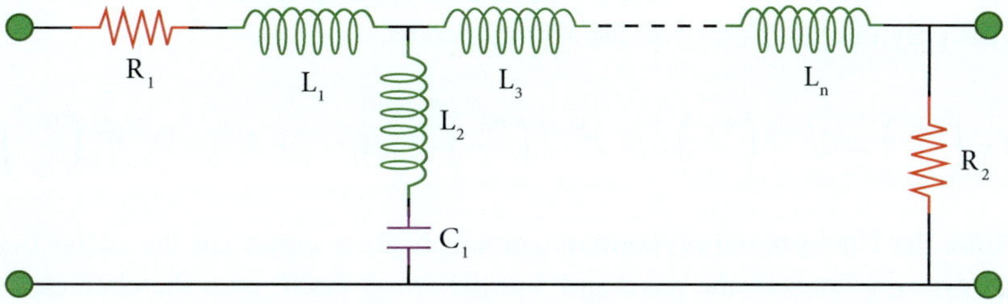

Fig. 11.47 Passive implementation of Chebyshev II filters

presented in Fig. 11.47. Note that this circuit allows for the implementation of *only odd-order filters*, since it cannot handle the appearance of constant gain at the position $\omega = \infty$ that characterizes even-order filters. The construction of even-order filters requires the application to the Chebyshev polynomials of the so-called *Moebius transform*, which has the form

$$x^2 \longrightarrow \frac{x^2 - z_{z1}^2}{1 - x_{z1}^2}, \quad x_{z1} = \cos\frac{k_{\max}\pi}{2N}$$

where $k_{\max}$ is the largest odd integer that is less than the order of the filter N. More information about this transform can be found in the bibliography.[26]

Considering the power response of Chebyshev I filters

$$|\mathcal{H}(j\omega)|^2 = \frac{1}{1 + \varepsilon^2 T_N^2(\omega/\omega_p)}$$

the amplitude response of Chebyshev II filters arises as follows: Initially we define the function[27]

[26] See, for example, (1) R.W. Daniels, *Approximation Methods for Electronic Filter Design*, McGraw-Hill, New York, 1974 and (2) G.C. Temes, J.W. LaPatra, *Circuit Synthesis and Design*, McGraw-Hill, Kogakusha, Tokyo, 1977.

[27] From this analysis it turns out that Chebyshev I and II filters are *complementary to each other in terms of power response* since the sum of these responses for these filters is equal to one.

$$|\mathcal{G}(j\omega)|^2 = 1 - |\mathcal{H}(j\omega)|^2 = 1 - \frac{1}{1+\varepsilon^2 T_N^2(\omega/\omega_p)} = \frac{\varepsilon^2 T_N^2(\omega/\omega_p)}{1+\varepsilon^2 T_N^2(\omega/\omega_p)}$$

and then we make the substitution $(\omega/\omega_p) \longrightarrow (\omega_s/\omega)$. Using for the amplitude response of the Chebyshev II filters the symbol $|\mathcal{H}(j\omega)|$, we will have

$$|\mathcal{H}(j\omega)|^2 = \frac{\varepsilon^2 T_N^2(\omega_s/\omega)}{1+\varepsilon^2 T_N^2(\omega_s/\omega)} = 1 \Big/ \left\{1 + \frac{1}{\varepsilon^2 T_N^2(\omega_s/\omega)}\right\}$$

and therefore

$$|\mathcal{H}(j\omega)| = \sqrt{\frac{\varepsilon^2 T_N^2(\omega_s/\omega)}{1+\varepsilon^2 T_N^2(\omega_s/\omega)}} = \frac{\varepsilon\,|\,T_N(\omega_s/\omega)\,|}{\sqrt{1+\varepsilon^2 T_N^2(\omega_s/\omega)}} = 1 \Big/ \sqrt{1 + \frac{1}{\varepsilon^2 T_N^2(\omega_s/\omega)}} \tag{11.94}$$

where $T_N(\omega_s/\omega)$ are Chebyshev polynomials defined in the corresponding section of the mathematical appendix. Using the relation

$$T_N(x) = c_N x^N + c_{N-2} x^{N-2} + \cdots + c_n x^n$$

where $C_N = 2^{N-1}$ and $c_n = N^n(-1)^{(N+3n)/2}$ (with the parameter n being equal to 0 for an even-order filter and equal to one for odd-order filter), the polynomials $T_N(\omega_s/\omega)$ are formulated as

$$T_N\left(\frac{\omega_s}{\omega}\right) = 2^{N-1}\left(\frac{\omega_s}{\omega}\right)^N + 2^{N-3}\left(\frac{\omega_s}{\omega}\right)^{N-2} + \cdots + N^n(-1)^{\frac{N+3n}{2}}\left(\frac{\omega_s}{\omega}\right)^n$$

We observe that this time the Chebyshev polynomials contain in their argument the *cutoff frequency* ω_s, and therefore their equiripple fluctuations do not concern the passband, but the *cutoff band*, with the characteristics of these fluctuations depending on the value of the ripple parameter ε. If we set $\omega_s = 1$ in the foregoing relation, we will get the *normalized version* of the Chebyshev II filter, whose amplitude response is defined as

$$|\mathcal{H}(j\omega)| = \sqrt{\frac{\varepsilon^2 T_N^2(1/\omega)}{1+\varepsilon^2 T_N^2(1/\omega)}} = \frac{\varepsilon\,|\,T_N(1/\omega)\,|}{\sqrt{1+\varepsilon^2 T_N^2(1/\omega)}} \tag{11.95}$$

It is not difficult to see that the foregoing gain function is an example of a rational approximation (Sect. 11.9.1, page 990) of the form

$$G(\omega) = |\mathcal{H}j\omega| = \frac{1}{\sqrt{1+\gamma^2 F^2(\omega)}} = \frac{1}{\sqrt{1+\gamma^2[N^2\omega/D^2(\omega)]}} = \frac{D(\omega)}{\sqrt{D^2(\omega)+\gamma^2 N^2(\omega)}}$$

for the parameter value $\gamma = 1/\varepsilon$ and for the rational function

$$F(\omega) = \frac{1}{T_N(\omega/\omega_s)} = \frac{\omega^N}{\omega^N T_N(\omega/\omega_s)} = \frac{\omega^N}{(N^n(-1)^{\frac{N+3n}{2}}\omega_s^n)\,\omega^{N-n} + \cdots + (2^{N-3}\omega_s^{N-2})\,\omega^2 + (2^{N-1}\omega_s^N)} = \frac{N(\omega)}{D(\omega)}$$

We observe that the degree $M = N - n$ of the polynomial $D(\omega)$ *always has an even value*, while also $N \geq M$, and therefore all the requirements for the occurrence of *low-pass behavior* presented in the chapter on explicit approaches are met. Taking into account that the roots of the Chebyshev polynomials that will emerge as the roots of the equation $T_N(\omega/\omega_s) = \cos[N\cos^{-1}(\omega_s/\omega)] = 0$ have the form

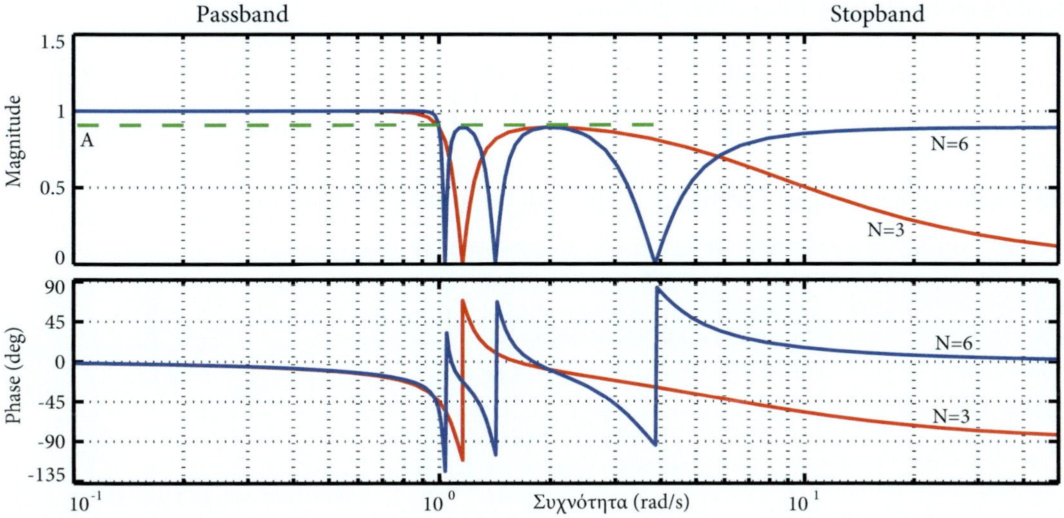

Fig. 11.48 Amplitude and phase responses of prototype low-pass Chebyshev II filter for even- and odd-order values. We observe that in the case of the odd-order filter, the gain function is zero at the position $\omega = \infty$. The dashed line that appears in the amplitude response diagram intersects the vertical axis at the position $A = 1/\sqrt{[1 + (1/\varepsilon^2)]}$

$$\omega_k^z = \pm \frac{\omega_s}{\cos\left[(2k-1)\pi/(2N)\right]} \quad \text{for the values } k = 1, 2, \ldots, \frac{N-n}{2} \tag{11.96}$$

it is obvious that at these frequency positions a *nullification of the gain function in the cutoff zone*[28] takes place (recall that the nullification of the gain function at specific discrete frequency values—but not in an entire frequency zone—is allowed by the *Paley–Wiener theorem*). If the filter order is an odd number, there are $N - 1$ discrete finite frequency values at which the filter gain is nullified and one additional such frequency at the position $\omega = \infty$ (this situation is presented in Fig. 11.48). Conversely, for even-order filters, the gain nullification appears for $N/2$ finite frequency values. On the other hand, if the relation $T_N^2(\omega_s/\omega) = 1$ holds, we will get $G(\omega) = \varepsilon/\sqrt{1+\varepsilon^2} = A_s$, and the frequencies for which the foregoing relation holds are

$$\omega_k^z = \pm \frac{\omega_s}{\cos(k\pi/N)} \quad \text{for values } k = 1, 2, \ldots, \frac{N-n}{2} \tag{11.97}$$

It is not difficult to ascertain that for $\omega = 0$, we have $|\mathcal{H}(j\omega)|^2_{\omega=0} = 1$, regardless of the order of the filter N, while for $\omega = \omega_c$, we have $|\mathcal{H}(j\omega)|^2_{\omega=\omega_c} = \varepsilon^2/(1+\varepsilon^2)$, or, equivalently, $10\log|\mathcal{H}(j\omega)|_{\omega=0} = 0$ and $10\log|\mathcal{H}(j\omega)|_{\omega=\omega_c} = 10\log[\varepsilon^2/(1+\varepsilon^2)]$ in decibels. Applying the theory related to the characteristics of analog filters, it is proven that the ripple parameter ε is given by the relation $\varepsilon = 1/\sqrt{10^{A_s/10}-1}$ where A_s is the *minimum attenuation in the stopband.*

Returning to the rational approximation and observing that the roots of the polynomial $T_N(\omega/\omega_s)$ will also be the roots of the even polynomial $D(\omega) = \omega^N T_N(\omega_s/\omega)$, we can formulate this polynomial in its factored form

$$D(\omega) = |c_n|\omega_s^n \prod_{k=1}^{(N-n)/2} [\omega^2 - (\omega_k^z)^2] = N^n(-1)^{(N+3n)/2}\omega_s^n \prod_{k=1}^{(N-n)/2} \left(\omega^2 - \frac{\omega_s^2}{\cos^2[(2k-1)\pi/(2N)]}\right)$$

[28] In this expression, the *trigonometric form* of Chebyshev polynomials is used since the fluctuations appear in the *cutoff zone* and therefore correspond to the frequencies $\omega_s < \omega$ or, equivalently, $(\omega_s/\omega) < 1$.

Consequently, the explicit function $F(\omega)$ is formulated as

$$F(\omega) = \frac{N(\omega)}{D(\omega)} = \omega^N \Bigg/ \left\{ N^n (-1)^{(N+3n)/2} \omega_s^n \prod_{k=1}^{(N-n)/2} \left(\omega^2 - \frac{\omega_s^2}{\cos^2[(2k-1)\pi/(2N)]} \right) \right\} \tag{11.98}$$

while the *gain function* of the low-pass Chebyshev II filter will be given by the equation

$$G(\omega) = |\mathcal{H}(j\omega)| = \frac{\varepsilon N^n (-1)^{(N+3n)/2} \omega_s^n \displaystyle\prod_{k=1}^{(N-n)/2} \left(\omega^2 - \frac{\omega_s^2}{\cos^2\left[\frac{(2k-1)\pi}{2N}\right]} \right)}{\sqrt{\omega^{2N} + N^{2n}(-1)^{(N+3n)} \omega_s^{2n} \left\{ \displaystyle\prod_{k=1}^{(N-n)/2} \left(\omega^2 - \frac{\omega_s^2}{\cos^2[(2k-1)\pi/(2N)]} \right) \right\}^2}} \tag{11.99}$$

It is proven that the foregoing function is characterized by *monotonic reduction* in the frequency range $0 \le \omega \le \omega_s$, while it has $(N-n)/2$ maxima in the cutoff zone with a value equal to $G(\omega_s)$.

Poles and Zero Values of Transfer Function

The determination of the poles and the zero values of the transfer function of Chebyshev II filters is based on the construction of the function $\mathcal{H}(s)\mathcal{H}(-s)$, which in this case is written

$$\mathcal{H}(s)\mathcal{H}(-s) = |\mathcal{H}(j\omega)|^2_{\omega=s/j} = \frac{\varepsilon^2 T_N^2(j\omega_s/s)}{1+\varepsilon^2 T_N^2(j\omega_s/s)} = \frac{\varepsilon^2 T_N^2(-\omega_s/js)}{1+\varepsilon^2 T_N^2(-\omega_s/js)}$$

Consequently, the zeros of the transfer function arise as the roots of the equation $T_N(j\omega_s/s) = 0$, while the poles of the function result as the roots of the equation $1 + \varepsilon^2 T_N^2(j\omega_s/s) = 0$. Starting with the zero values, these will obviously constitute the roots of Chebyshev I polynomials of order N. It should be noted that since the inverse hyperbolic cosine of zero does not exist, the trigonometric form of Chebyshev polynomials will be used. Taking into account that a Chebyshev polynomial $T_N(x)$ of the first kind of order N has in the interval $[-1, +1]$ N simple roots of the form

$$x_k = \pm\cos\left[(2k-1)\pi/(2N)\right] \quad k = 1, 2, \ldots, \frac{N-n}{2}$$

as we saw earlier, it turns out that

$$j\frac{\omega_s}{z_k} = \pm\cos\left[(2k-1)\pi/(2N)\right] \quad k = 1, 2, \ldots, \frac{N-n}{2}$$

and therefore the zero values of the transfer function will be calculated from the equation

$$z_k = \pm\frac{j\omega_s}{\cos\left[(2k-1)\pi/(2N)\right]} \quad k = 1, 2, \ldots, \frac{N-n}{2} \tag{11.100}$$

Repeating what we mentioned in the introductory section on this approach, filters with even order have $N/2$ pairs of complex (in reality, *purely imaginary*) conjugate zero values, while in odd-order filters there are $(N-1)/2$ such finite pairs and one additional zero value at the position $\omega = \infty$.

On the other hand, the poles of a transfer function constitute roots of the equation

$1 + \varepsilon^2 T_N^2(j\omega_s/s) =$ where comparison to the corresponding equation $1 + \varepsilon^2 T_N^2(s/j\omega_p) = 0$

related to the Chebyshev I polynomials reveals that *the roots of one equation are the inverses of the roots of the other* (something that, of course, is expected, since the transition from the Chebyshev I filters to the Chebyshev II filters is based, among other things, on the substitution $\omega/\omega_p \longrightarrow \omega_s/\omega$, which is equivalent to the substitution $s \longrightarrow \omega_s/s$). Applying, therefore, the same process as the one used for the previous type of filters (the reader can perform the analytical steps as an exercise), we end up with the relation

$$p_k = \frac{j\omega_s}{\cosh\left[\frac{1}{N}\sinh^{-1}\left(\frac{1}{\varepsilon}\right) + j\frac{(2k-1)\pi}{2N}\right]}$$
$$= \frac{\omega_s}{-\sinh\left[\frac{1}{N}\sinh^{-1}\left(\frac{1}{\varepsilon}\right)\right]\sin\frac{(2k-1)\pi}{2N} + j\cosh\left[\frac{1}{N}\sinh^{-1}\left(\frac{1}{\varepsilon}\right)\right]\cos\frac{(2k-1)\pi}{2N}}$$

Of the $2N$ poles of the transfer function, we again consider the N poles of the *left complex half-plane* characterized by *negative real parts* and have the form $p_k = \sigma_k + j\omega_k$, where

$$\sigma_k = \frac{-\omega_s \sinh\left[\frac{1}{N}\sinh^{-1}\left(\frac{1}{\varepsilon}\right)\right]\sin\frac{(2k-1)\pi}{2N}}{\sinh^2\left[\frac{1}{N}\sinh^{-1}\left(\frac{1}{\varepsilon}\right)\right]\sin^2\frac{(2k-1)\pi}{2N} + \cosh^2\left[\frac{1}{N}\sinh^{-1}\left(\frac{1}{\varepsilon}\right)\right]\cos^2\frac{(2k-1)\pi}{2N}} \qquad (11.101)$$

$$\omega_k = \frac{-\omega_s \cosh\left[\frac{1}{N}\sinh^{-1}\left(\frac{1}{\varepsilon}\right)\right]\cos\frac{(2k-1)\pi}{2N}}{\sinh^2\left[\frac{1}{N}\sinh^{-1}\left(\frac{1}{\varepsilon}\right)\right]\sin^2\frac{(2k-1)\pi}{2N} + \cosh^2\left[\frac{1}{N}\sinh^{-1}\left(\frac{1}{\varepsilon}\right)\right]\cos^2\frac{(2k-1)\pi}{2N}} \qquad (11.102)$$

Denoting the common denominator of the two foregoing expressions by $D(k)$ and defining the angle

$$\vartheta_k = -\tan^{-1}\left\{\frac{\cosh\left[\frac{1}{N}\sinh^{-1}\left(\frac{1}{\varepsilon}\right)\right]\cos\frac{(2k-1)\pi}{2N}}{\sinh\left[\frac{1}{N}\sinh^{-1}\left(\frac{1}{\varepsilon}\right)\right]\sin\frac{(2k-1)\pi}{2N}}\right\}$$

we find that the poles of the normalized transfer functions of Chebyshev I ($\omega_p = 1$) and Chebyshev II ($\omega_s = 1$) are given by the relations

$$p_k^I = \sqrt{D(k)}e^{j\vartheta_k} \quad \text{and} \quad p_k^{II} = \frac{1}{\sqrt{D(k)}}e^{-j\vartheta_k}$$

or, equivalently,

$$p_k^I = \alpha_k + j\beta_k \quad \text{and} \quad p_k^{II} = \frac{1}{p_k^I} = \frac{\omega_s\alpha_k}{\alpha_k^2+\beta_k^2} - j\frac{\omega_s\beta_k}{\alpha_k^2+\beta_k^2}$$

and therefore *the poles p_k^{II} are the mirror images of the poles p_k^I with respect to the circumference of the unit circle*. For this reason, we can determine the poles of the Chebyshev II transfer function by finding the poles of the Chebyshev I transfer function and then converting them in the foregoing way. However, it should be noted that the foregoing relationship does not have particular practical value, since we can calculate the poles of the Chebyshev II function directly from their defining equation.

The pole-zero plot of the Chebyshev II filter and the correlation of its poles with the poles of the Chebyshev I filter, from which the inverse proportionality relationship we presented earlier emerges, is shown in Fig. 11.49. Although for specific

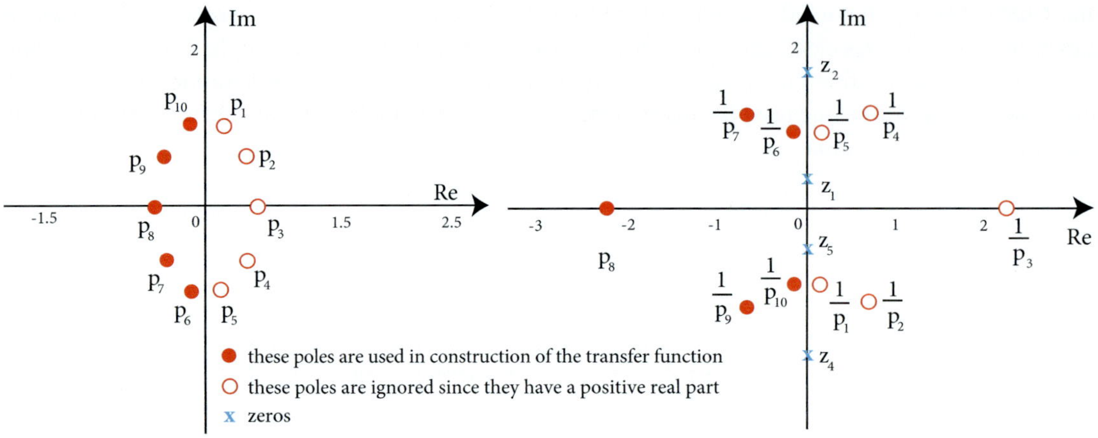

Fig. 11.49 Correlation of poles of Chebyshev I filters [image (a)] and Chebyshev II [image (b)] as well as the zero values of Chebyshev II filters for odd-order value (and, more specifically, for the value $N = 5$). For even-order filters, the image is similar, although there are no real poles, only pairs of complex conjugate poles. For better preview reasons, we maintain the symbolism of the blue and white circles for the used and unused poles and X for the zero values, keeping in mind the convention used in practice

values of the parameters ε and N the poles of the Chebyshev II filter seem to form an ellipse, in the general case their geometric locus is not an ellipse but a set of points that constitute a more complex geometric structure.[29]

Problem 11.9.7 Determine the poles and zeros of the transfer function of a normalized ($\omega_s = 1$) low-pass Chebyshev II filter for the parameter value $\varepsilon = 0.217344$ and order values $N = 5$ and $N = 8$.

Solution Following the methodology of previous exercises for the value $\varepsilon = 0.217344$, we have $x = 1/\varepsilon = 4.600984$ and $\sinh^{-1} x = \ln(x + \sqrt{x^2 + 1}) = 3.287283$. For the order value $N = 5$ we have $\sinh^{-1} x/N = 0.657456$, and therefore,

$$\sinh(\sinh^{-1} x/N) = 0.705854 \qquad \text{and} \qquad \cosh(\sinh^{-1} x/N) = 1.22402$$

Applying the pole defining equations for the values $k = 1, 2, 3, 4, 5, 6, 7, 8, 9, 10$, we arrive at the following results:

$$\begin{array}{ll}
\sigma_1 = 0.120565\ \omega_1 = 0.953287 & \sigma_6 = -0.120565\ \ \omega_6 = -0.953287 \\
\sigma_2 = 0.652345\ \omega_2 = 1.217630 & \sigma_7 = -0.652345\ \ \omega_7 = -1.217630 \\
\sigma_3 = 2.366460\ \omega_3 = 0 & \sigma_8 = -2.366460\ \ \omega_8 = 0 \\
\sigma_4 = 0.652345\ \omega_4 = -1.217630 & \sigma_9 = -0.652345\ \ \omega_9 = 1.217630 \\
\sigma_5 = 0.120565\ \omega_5 = -0.953287 & \sigma_{10} = -0.120565\ \omega_{10} = 0.953287
\end{array}$$

Of the 10 preceding poles, we consider for the transfer function only the conjugate complex poles $(p_6, p_{10}) = (-0.120565 - 0.953287j, -0.120565 + 0.953287j)$ and $(p_7, p_9) = (-0.652345 - 1.217630j, -0.652345 + 1.217630j)$ and the real pole $p_8 = -2.366460$ that have negative real parts, while the rest are ignored. As for the zero values, these are equal to $z_{1,5} = \pm 1.051462j$ and $z_{2,4} = \pm 1.701301j$, while there is also a zero value z_3 at infinity that cannot be depicted.

On the other hand, for $N = 8$ we have $\sinh^{-1} x/N = 0.410910$, and therefore $\sinh(\sinh^{-1} x/N) = 0.422572$ and $\cosh(\sinh^{-1} x/N) = 1.08562$. From the pole defining equations for values $k = 1, 2, 3, 4, 5, 6, 7, 8, 9, 10, 11, 12, 13, 14, 15, 16$ we arrive at the results

[29] Once again, to distinguish the poles that are used in the transfer function from those that are not used, we did not resort to the usual symbolism of the pole–zero plot, but we used the symbols in the way referred to in the figure.

$$\begin{array}{ll}
\sigma_1 = 0.072283 \; \omega_1 = 0.933582 & \sigma_9 = -0.072283 \;\; \omega_9 = -0.933582 \\
\sigma_2 = 0.269876 \; \omega_2 = 1.037645 & \sigma_{10} = -0.269876 \; \omega_{10} = -1.037645 \\
\sigma_3 = 0.721134 \; \omega_3 = 1.237900 & \sigma_{11} = -0.721134 \; \omega_{11} = -1.237900 \\
\sigma_4 = 1.913203 \; \omega_4 = 0.977687 & \sigma_{12} = -1.913203 \; \omega_{12} = -0.977687 \\
\sigma_5 = 1.913203 \; \omega_5 = -0.977687 & \sigma_{13} = -1.913203 \; \omega_{13} = 0.977687 \\
\sigma_6 = 0.721134 \; \omega_6 = -1.237900 & \sigma_{14} = -0.721134 \; \omega_{14} = 1.237900 \\
\sigma_7 = 0.269876 \; \omega_7 = -1.037645 & \sigma_{15} = -0.269876 \; \omega_{15} = 1.037645 \\
\sigma_8 = 0.072283 \; \omega_8 = -0.933582 & \sigma_{16} = -0.072283 \; \omega_{16} = 0.933582
\end{array}$$

In this case, due to the even order of the filter, there is no real pole, but only complex ones of which we consider only those characterized by a negative real part. Therefore, for the transfer function of the filter we will consider the pairs of complex conjugate poles $(p_9, p_{16}) = (-0.072283 - 0.933582j, -0.072283 + 0.933582j)$, $(p_{10}, p_{15}) = (-0.269876 - 1.037645j, -0.269876 + 1.037645j)$, $(p_{11}, p_{14}) = (-0.721134 - 1.237900j, -0.721134 + 1.237900j)$, and $(p_{12}, p_{13}) = (-1.913203 - 0.977687j, -1.913203 + 0.977687j)$, while all the others are ignored. As for the zero values for $N = 8$, these emerge equal to $z_{1,2} = \pm 1.019591j$, $z_{3,4} = \pm 1.202689j$, $z_{5,6} = \pm 1.799982j$, and $z_{7,8} = \pm 5.125830j$. The construction of the pole–zero diagram is a simple process and is left as an exercise to the reader. ■

After determining the poles and zeros of the transfer function, it can be calculated from its defining equation. For the case of an *even-order filter* we will have

$$\mathcal{H}(s) = H_0 \frac{\prod_{k=1}^{N/2} (s - z_k)(s - z_k^*)}{\prod_{k=1}^{N/2} (s - p_k)(s - p_k^*)} = H_0 \frac{\prod_{k=1}^{N/2} \left(s^2 + \frac{\omega_s^2}{\cos^2[(2k-1)\pi/(2N)]} \right)}{\prod_{k=1}^{N/2} (s^2 - 2\sigma_k s + \sigma_k^2 + \omega_k^2)} \tag{11.103}$$

with the expressions of σ_k and ω_k given by the foregoing equations. The multiplicative factor H_0 arises from the requirement that $G(0) = 1$ for $\omega = 0$. Therefore, setting $s = 0$, we get

$$G(0) = |\mathcal{H}(0)| = H_0 \frac{\prod_{k=1}^{N/2} \frac{\omega_s^2}{\cos^2[(2k-1)\pi/(2N)]}}{\prod_{k=1}^{N/2} (\sigma_k^2 \omega_k^2)} = 1$$

from which it follows that

$$H_0 = \frac{\prod_{k=1}^{N/2} (\sigma_k^2 + \omega_k^2)}{\prod_{k=1}^{N/2} \frac{\omega_s^2}{\cos^2[(2k-1)\pi/(2N)]}}$$

On the other hand, for an *odd-order filter*, the transfer function has a *real pole* at the position $p_0 = 1/\sinh[(1/N)\sinh^{-1}(1/\varepsilon)]$ and therefore will be given by the equation

$$\mathcal{H}(s) = H_0 \frac{\prod_{k=1}^{(N-1)/2} (s - z_k)(s - z_k^*)}{(s - p_0) \prod_{k=1}^{(N-1)/2} (s - p_k)(s - p_k^*)} = H_0 \frac{\prod_{k=1}^{(N-1)/2} \left(s^2 + \frac{\omega_s^2}{\cos^2[(2k-1)\pi/(2N)]} \right)}{\left(s - \frac{1}{\sinh[(1/N)\sinh^{-1}(1/\varepsilon)]} \right) \prod_{k=1}^{(N-1)/2} (s^2 - 2\sigma_k s + \sigma_k^2 + \omega_k^2)}$$

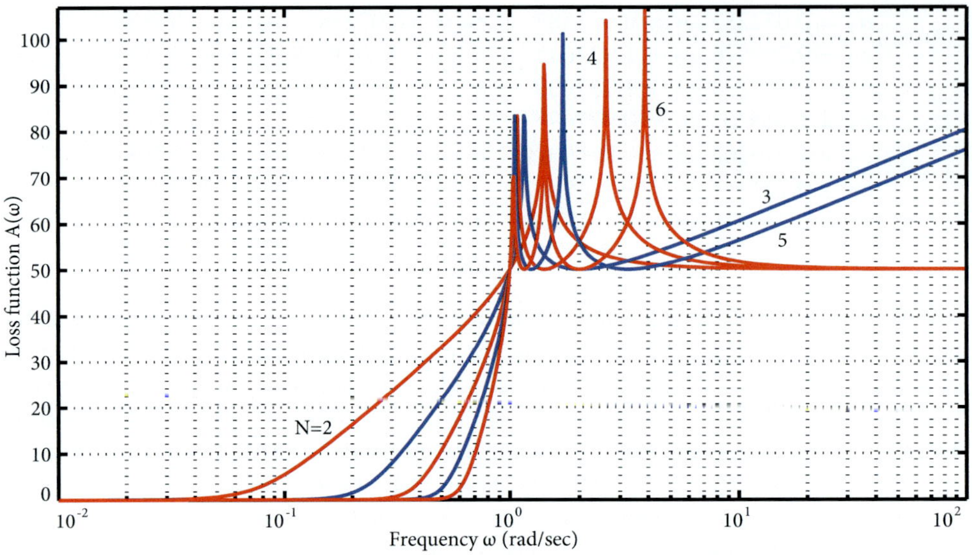

Fig. 11.50 Characteristic losses of prototype analog Chebyshev II filter and for the parameter value $A_s = 50$ dB and for filter orders $N = 2, 3, 4, 5, 6$

For $\omega = 0$ it will still be the case that $G(0) = 1$—thus,

$$H_0 \frac{\sinh\left[\frac{1}{N}\sinh^{-1}\left(\frac{1}{\varepsilon}\right)\right]\prod_{k=1}^{(N-1)/2}\left(\frac{\omega_s^2}{\cos^2\left[\frac{(2k-1)\pi}{2N}\right]}\right)}{-\prod_{k=1}^{(N-1)/2}(\sigma_k^2+\omega_k^2)} = 1$$

and therefore,

$$H_0 = -\frac{\prod_{k=1}^{(N-1)/2}(\sigma_k^2+\omega_k^2)}{\sinh\left[\frac{1}{N}\sinh^{-1}\left(\frac{1}{\varepsilon}\right)\right]\prod_{k=1}^{(N-1)/2}\frac{\omega_s^2}{\cos^2\left[\frac{(2k-1)\pi}{2N}\right]}}$$

Characteristic Loss

The characteristic loss of a Chebyshev II filter is given by the relation

$$A(\omega) = 10\log L(\omega^2) = 10\log\left[1 + \frac{1}{\varepsilon^2 T_N^2(\omega_s/\omega)}\right] \tag{11.104}$$

The *loss in the passband* looks very much like that of a Butterworth filter and is described by a monotonically increasing function of frequency ω. Conversely, the *loss in the cutoff zone* fluctuates, as we already noted, between infinity and the minimum attenuation value A_s, which is determined by the filter specifications. Typical examples of loss characteristics for various orders of Chebyshev II filter and for specific values of the parameters A_p and A_s are presented in Fig. 11.50.

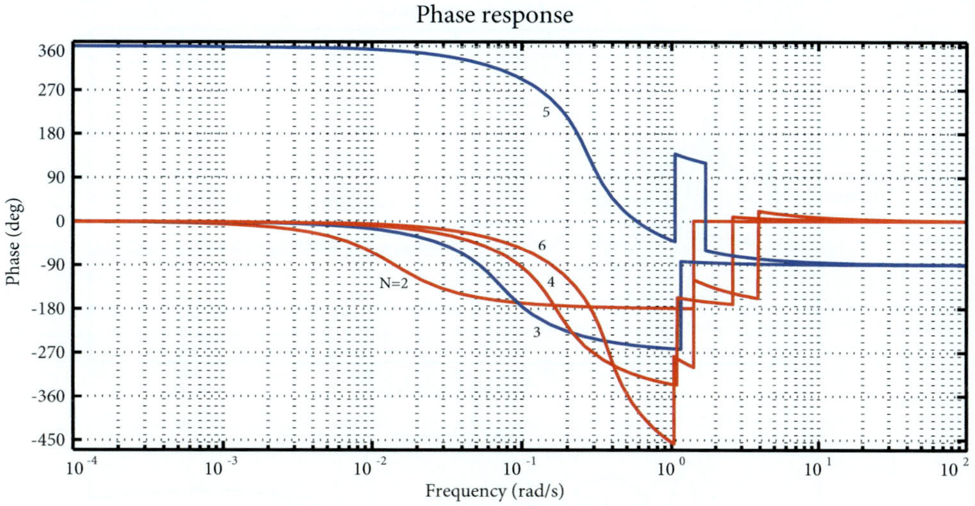

Fig. 11.51 Phase response of Chebyshev II filters for parameter values $\omega_c = 1$ and $A_s = 80$ dB and for filter orders $N = 2, 3, 4, 5, 6$

Phase Characteristics and Time-Domain Response

The *phase response* of Chebyshev II filters is calculated from the defining equation, and its variation as a function of frequency ω for filter order values from $N = 2$ to $N = 6$ is presented in Fig. 11.51 for parameter values $\omega_c = 1$ and $A_s = 80$ dB (which corresponds to a parameter value of deviations of $\varepsilon = 0.0001$). The analytical expressions of the phase response for even and odd filter orders arise from the general equations for parameter values $\alpha_{0k} = \sigma_k^2 + \omega_k^2$, $\alpha_{1k} = -2\sigma_k$, $\alpha_{2k} = 1$, $\beta_0 = 1$, $\beta_2 = \beta_1 = 0$, $\alpha_2 = \alpha_1 = 0$, $\alpha_0 = -1/\{\sinh[(1/N)\sinh^{-1}(1/\varepsilon)]\}$ and have the form

$$\measuredangle\mathcal{H}(j\omega) = -\sum_{k=1}^{N/2} \frac{-2\sigma_k}{\sigma_k^2 + \omega_k^2 - \omega^2}$$

and

$$\measuredangle\mathcal{H}(j\omega) = -\tan^{-1}\left\{-\sinh\left[\frac{1}{N}\sinh^{-1}\left(\frac{1}{\varepsilon}\right)\right]\right\} - \sum_{k=1}^{(N-1)/2} \frac{-2\sigma_k}{\sigma_k^2 + \omega_k^2 - \omega^2}$$

while the corresponding expressions for the *group delay* are formulated as

$$\tau(\omega) = -\sum_{N=1}^{N/2} \frac{2\sigma_k(\sigma_k^2 + \omega_k^2 + \omega^2)}{(\sigma_k^2 + \omega_k^2 - \omega^2)^2}$$

and

$$\tau(\omega) = -\frac{\sinh\left[\frac{1}{N}\sinh^{-1}\left(\frac{1}{\varepsilon}\right)\right]}{1 + \sinh\left[\frac{1}{N}\sinh^{-1}\left(\frac{1}{\varepsilon}\right)\right]\omega^2} - \sum_{N=1}^{(N-1)/2} \frac{2\sigma_k(\sigma_k^2 + \omega_k^2 + \omega^2)}{(\sigma_k^2 + \omega_k^2 - \omega^2)^2}$$

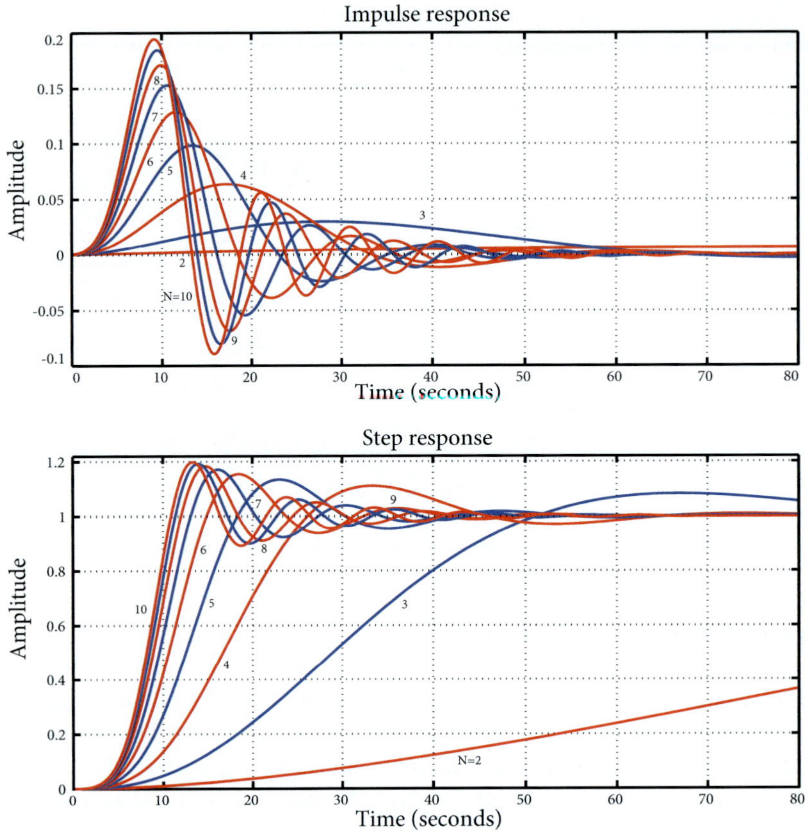

Fig. 11.52 Impulse and step responses of Chebyshev II filters for parameter values $N = 1 - 10$, $\omega_c = 1$ and $A_s = 80$ dB

In Fig. 11.52, the impulse and step responses of the Chebyshev II filter for parameter values $\omega_c = 1$ and $A_s = 80$ dB are presented.

Calculation of Minimum Filter Order

An interesting expression that allows the calculation of the *minimum order* of Chebyshev II filters (as well as selectivity coefficients and spectral performance) is the equation that correlates the cutoff and passband frequencies ω_p and ω_s. If we denote by A the attenuation level defined by the filter specifications, then, according to basic theory presented in a previous section, the frequency ω_p that corresponds to the value of A will satisfy the relation

$$|\mathcal{H}(j\omega_p)|^2 = \frac{\varepsilon^2 T_N^2(\omega_s/\omega_p)}{1+\varepsilon^2 T_N^2(\omega_s/\omega_p)} = \frac{1}{A^2}$$

from which it follows that

$$A = \sqrt{\frac{1+\varepsilon^2 T_N^2(\omega_s/\omega_p)}{\varepsilon^2 T_N^2(\omega_s/\omega_p)}} \quad \text{or, equivalently,} \quad A_p = (A)_{\text{dB}} = 10\log\frac{1+\varepsilon^2 T_N^2(\omega_s/\omega_p)}{\varepsilon^2 T_N^2(\omega_s/\omega_p)}$$

Solving the foregoing equation of the parameter A with respect to N, after first making the substitution $\varepsilon = 1/\sqrt{10^{A_s/10}-1}$ and using the *hyperbolic form* of the Chebyshev polynomials, it turns out, after simple mathematical operations, that the minimum order N of the filter can be calculated as

$$N = \left\{\cosh^{-1}\sqrt{\frac{10^{A_s/10}-1}{10^{A_p/10}-1}}\right\} \Big/ \cosh^{-1}(\omega_s/\omega_p) \tag{11.105}$$

when the values of A_s, A_p, ω_s, and ω_p are known. In almost all cases, the value resulting from the foregoing equation is a real number and should be rounded up to the next integer. On the other hand, if we solve the foregoing equation with respect to ω_p, we will get

$$\omega_p = \frac{\omega_s}{\cosh\left[\dfrac{1}{N}\cosh^{-1}\left(\dfrac{1}{\varepsilon\sqrt{10^{A_p/10}-1}}\right)\right]} \tag{11.106}$$

an equation that reveals the functional dependence that exists between the frequencies ω_p and ω_s.

Selectivity and Spectral Efficiency Factor

Using the defining equation of a filter's selectivity, it turns out that

$$F_s = -\frac{d}{d\omega}\{|\mathcal{H}(j\omega)|\}_{\omega=\omega_c} = \frac{\varepsilon N\omega_s}{\omega_c\sqrt{\omega_s^2-\omega_c^2}}\cdot\frac{\sinh[N\cosh^{-1}(\omega_s/\omega_c)]}{\sqrt[3]{\{1+\varepsilon^2\cosh^2[N\cosh^{-1}(\omega_s/\omega_c)]\}^2}} \tag{11.107}$$

The foregoing equation can be simplified by eliminating the frequency ω_s; to do this, we replace in Eq. (10.100) the value $A_p = 3$ dB, so by definition the frequency that corresponds to the preceding attenuation level is the cutoff frequency ω_c. If, therefore, we make the replacement $\omega_p \to \omega_c$ and combine the last two equations, we will get

$$F_s = \frac{N\varepsilon\cosh\left[\dfrac{1}{N}\cosh^{-1}\left(\dfrac{1}{\varepsilon}\right)\right]\sinh\left[\cosh^{-1}\left(\dfrac{1}{\varepsilon}\right)\right]}{(2\sqrt{2}\omega_c)\sinh\left[\dfrac{1}{N}\cosh^{-1}\left(\dfrac{1}{\varepsilon}\right)\right]}$$

which is the same equation as that characterizing Chebyshev I filters.

Finally, using the expression for the attenuation level A_p, the definition of the spectral efficiency factor, and the argumentation of the corresponding derivation for Chebyshev I filters, we will get

$$S_\alpha^\beta = \frac{\mathrm{BW}_\beta}{\mathrm{BW}_\alpha} = \frac{\cosh\left[\dfrac{1}{N}\cosh^{-1}\left(\dfrac{1}{\varepsilon\sqrt{10^{\alpha/10}-1}}\right)\right]}{\cosh\left[\dfrac{1}{N}\cosh^{-1}\left(\dfrac{1}{\varepsilon\sqrt{10^{\beta/10}-1}}\right)\right]}$$

The Design Process of Low-Pass Filters Chebyshev II

The design of low-pass Chebyshev II filters is done in a similar way to that used with other types of filters and generally includes:

- Determination of the values of the parameters ω_s, ω_p, A_s, and A_p that constitute the specifications of the filter.
- Calculation of the values of the variance parameter ε and the minimum order of the filter N as

$$\varepsilon = \frac{1}{\sqrt{10^{A_s/10}-1}} \quad \text{and} \quad N = \cosh^{-1}\sqrt{\frac{10^{A_s/10}-1}{10^{A_p/10}-1}}\Bigg/\left\{\cosh^{-1}(\omega_s/\omega_p)\right\}$$

- Determination of the zero values and the poles of the transfer function and then the construction of this function in the way we saw in the homonymous section.
- Construction of desired amplitude and phase responses that should satisfy the specifications of the filter.

Problem 11.9.8 Construct a Chebyshev II filter with operating frequencies $\omega_p = 0.5$ rad/s and $\omega_s = 1.0$ rad/s and attenuation levels $A_p = 0.5$ dB and $A_s = 30$ dB (Antoniou, Exercise for solution 10.14).

Solution In full analogy with the corresponding exercise for Chebyshev I filters, initially we need to calculate the minimum filter order as a function of the values of the four preceding parameters. From the defining equation of minimum order N it follows that

$$N = \frac{\cosh^{-1}\left(\frac{\sqrt{10^{A_s/10}-1}}{\sqrt{10^{A_p/10}-1}}\right)}{\cosh^{-1}\left(\frac{\omega_s}{\omega_p}\right)} = \frac{\cosh^{-1}\left(\frac{\sqrt{10^{30/10}-1}}{\sqrt{10^{0.5/10}-1}}\right)}{\cosh^{-1}\left(\frac{1.0}{0.5}\right)} = \frac{\cosh^{-1}\left(\frac{\sqrt{999}}{\sqrt{0.122018}}\right)}{\cosh^{-1}(2.0)} = \frac{5.198285}{1.316957} = 3.947191$$

where for the calculation of the inverse hyperbolic cosine we used the identity $\cosh^{-1} x = \ln(x + \sqrt{x^2 - 1})$. The preceding value should be rounded to the next highest integer, and therefore the order of the filter that will satisfy the operating specifications is $N = 4$.

On the other hand, the variance parameter ε that corresponds to the attenuation level $A_s = 30$ dB is calculated as

$$\varepsilon = 1/\sqrt{10^{30/10} - 1} = 0.031638$$

from which it follows that

$$\sinh^{-1}\left(\frac{1}{\varepsilon}\right) = \sinh^{-1}\left(\frac{1}{0.031638}\right) = \sinh^{-1}(31.607560)$$
$$= \ln(31.607560 + \sqrt{(31.607560)^2 + 1}) = \ln(63.230935) = 4.146793$$

and therefore

$$\sinh\left[\frac{1}{N}\sinh\left(\frac{1}{\varepsilon}\right)\right] = \sinh\left(\frac{1}{4} \times 4.146793\right) = 1.232633$$
$$\cosh\left[\frac{1}{N}\sinh\left(\frac{1}{\varepsilon}\right)\right] = \cosh\left(\frac{1}{4} \times 4.146793\right) = 1.587257$$

Applying the defining equations of the poles and zero values for Chebyshev II filters, we arrive at the following results regarding the poles and zeros of the transfer function, as well as the contributions of these parameters to the mathematical expression of the function $\mathcal{H}(s)$.

- Zeros of the transfer function
  - $z_{1,2} = \pm 2.613126j$, which contributes the term $(s^2 + 6.828427)$.
  - $z_{3,4} = \pm 1.082392j$, which contributes the term $(s^2 + 1.171572)$.
- Poles of the transfer function
  - $p_{1,4} = -0.198786 \pm 0.617982j$, which contributes the term $(s^2 + 0.397572s + 0.421417)$.
  - $p_{2,3} = -0.683625 \pm 0.364633j$, which contributes the term $(s^2 + 1.367250s + 0.600300)$.

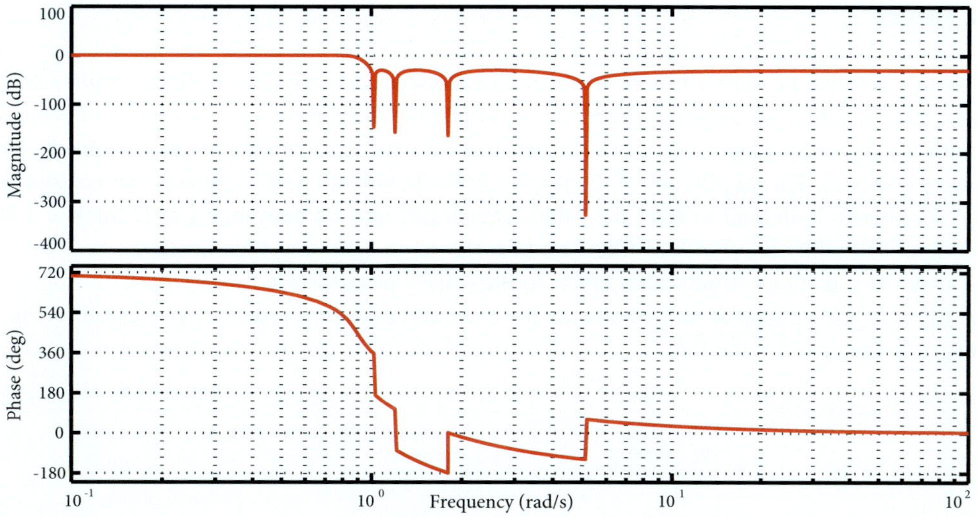

Fig. 11.53 Amplitude and phase responses of Chebyshev II filter from Exercise 11.9.8

On the other hand, the transfer function of the filter will be calculated from the equation

$$\mathcal{H}(s) = \frac{H_0(s^2 + 6.828427)(s^2 + 1.171572)}{(s^2 + 0.397572s + 0.421417)(s^2 + 1.367250s + 0.600300)}$$
$$= \frac{H_0(s^4 + 7.999999s^2 + 7.999993)}{s^4 + 1.764822s^3 + 1.565297s^2 + 0.814844s + 0.252976}$$

with the constant H_0 being calculated using the method described for the case of the even-order Chebyshev II filter and for the value $N = 4$ as

$$H_0 = \left\{ \prod_{k=1}^{2} (\sigma_k^2 + \omega_k^2) \right\} \Big/ \left\{ \prod_{k=1}^{2} \left[\omega_s^2 / \cos^2 \left[\frac{(2k-1)\pi}{8} \right] \right] \right\} = \frac{0.421417 \times 0.600300}{(1/0.853553) \times (1/0.146447)} = 0.031622$$

Consequently, the transfer function will take its final form

$$\mathcal{H}(s) = \frac{H_0(s^4 + 7.999999s^2 + 7.999993)}{s^4 + 1.764822s^3 + 1.565297s^2 + 0.814844s + 0.252976}$$
$$= \frac{0.031622s^4 + 0.252975s^2 + 0.252975}{s^4 + 1.764822s^3 + 1.565297s^2 + 0.814844s + 0.252976}$$

and proceeding with the substitution $s = j\omega$, we construct the frequency response function, which is given by the equation

$$\mathcal{H}(j\omega) = \frac{0.031622\omega^4 - 0.252975\omega^2 + 0.252975}{(\omega^4 - 1.565297\omega^2 + 0.252975) + j(0.814844\omega - 1.764822\omega^3)}$$

After calculating the frequency response function, we can construct the amplitude and phase responses whose variation, depending on the frequency in the logarithmic scale, is presented in Fig. 11.53. The reader is asked as an exercise to determine the analytical forms of these responses and then to calculate analytically the phase and group delays as well as the step and impulse response of the filter. The foregoing results are verified using the appropriate functions of MATLAB, which was used for the construction of the foregoing graphs.

11.9.8 Elliptic Filters

Elliptic filters,[30] which are also known as *Cauer filters*, are characterized by *equal amplitude ripples in the passband* like Chebyshev I filters and by *equal amplitude ripples in the stopband*, like Chebyshev II filters, offering simultaneously a transition zone with a very small range; however, the passband is characterized *by the maximum degree of nonlinearity* regarding their phase response. The amplitudes of the ripples in the passbands and stopbands can be adjusted independently of each other, and if either or both tend to zero, the filter degenerates into a Chebyshev I or II filter or a Butterworth filter, respectively. In comparison with other types of filters and for the same operating specifications, i.e., for the same values of the parameters A_p, A_s, ω_p, and ω_s, elliptic filters lead to the *lowest possible filter order* N. The ripples of the *amplitude response* of these filters in the passbands and stopbands for even and odd filter order are presented in Fig. 11.54, while the equation describing this function has the form

$$|\mathcal{H}(j\omega)| = \frac{1}{\sqrt{1 + \varepsilon_p^2 R_N^2(\omega, \omega_s, \omega_p, \varepsilon_s, \varepsilon_p)}} \tag{11.108}$$

where $R_N(\omega, \omega_s, \omega_p, \varepsilon_s, \varepsilon_p)$ is the *Jacobi elliptic function* of order N, which is defined as

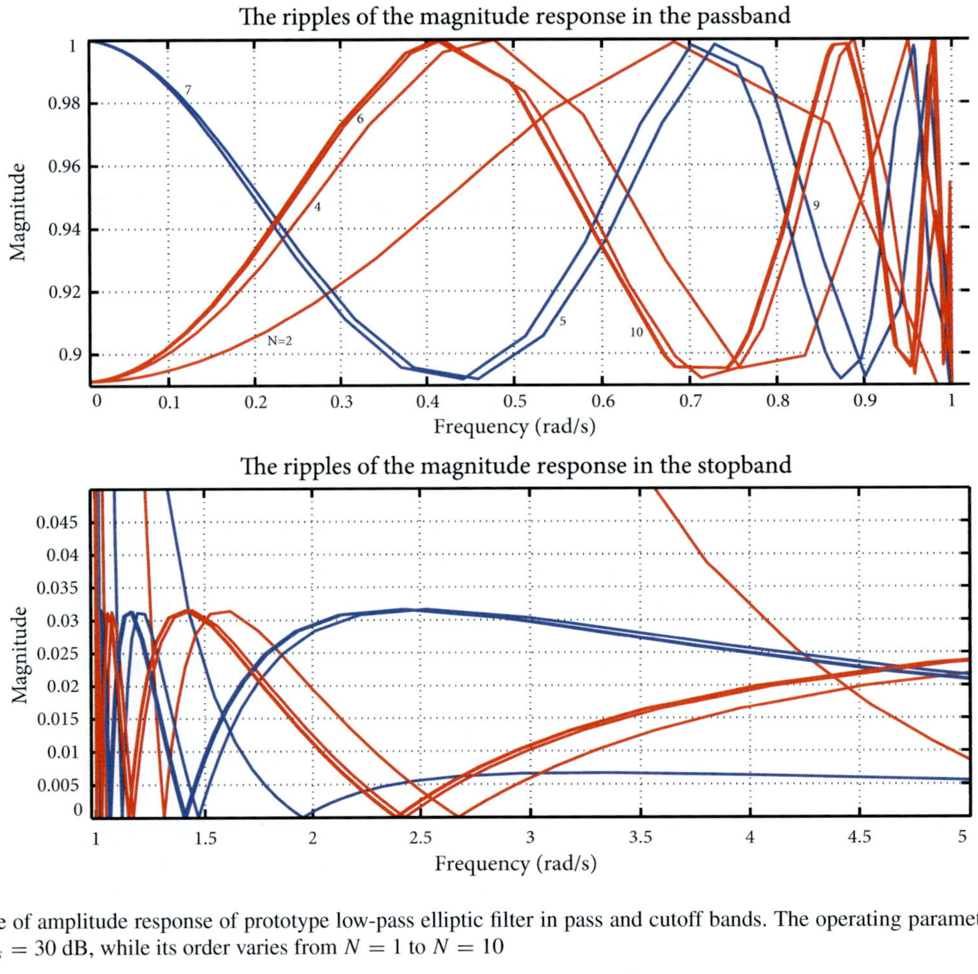

Fig. 11.54 Ripple of amplitude response of prototype low-pass elliptic filter in pass and cutoff bands. The operating parameters of the filter are $A_p = 1$ dB and $A_s = 30$ dB, while its order varies from $N = 1$ to $N = 10$

[30] For an understanding of the theory that follows, it is necessary to know the basic definitions and properties related to Jacobi elliptic integrals and elliptic functions, material that is included in the relevant section of the mathematical appendix.

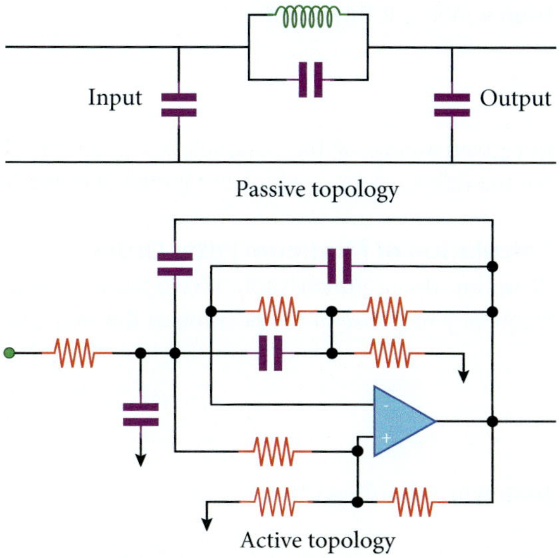

Fig. 11.55 Example of passive and active elliptic filter of third order

$$R_N(\omega, \omega_s, \omega_p, \varepsilon_s, \varepsilon_p) = \mathrm{sn}\left[\kappa\,\mathrm{sn}^{-1}\left(\frac{\omega}{\omega_p}, \frac{\omega_p}{\omega_s}\right) + q\boldsymbol{K}_1, \varepsilon_p\varepsilon_s\right] \tag{11.109}$$

In the foregoing equation, $\mathrm{sn}(x, k)$ is the *Jacobi elliptic sine function* of the parameter x with elliptic measure k, ω_p and ω_s are *the limits of the pass and cutoff bands*, κ is a parameter that depends on the order of the filter N and on the values of the elliptic integrals related to the filter, ε_p and ε_s are *the ripple amplitudes in the pass and cutoff bands*, respectively, q is a constant with a value of 0 for odd-order filters and 1 for even-order filters, and $\boldsymbol{K}_1$ is *the complete elliptic integral of the first kind* of the elliptic measure $\varepsilon_p\varepsilon_s$. Defining for convenience the elliptic measures $k_1 = \varepsilon_p\varepsilon_s$ and $k_2 = \omega_p/\omega_s$ and the parameters $\xi = \kappa\,\mathrm{sn}^{-1}(\omega/\omega_p, \omega_p/\omega_s)$ and $\zeta = \omega/\omega_p$, the preceding function takes the simpler form

$$R_N(\omega, \omega_s, \omega_p, \varepsilon_s, \varepsilon_p) = \mathrm{sn}[\kappa\,\mathrm{sn}^{-1}(\zeta, k_2) + q\boldsymbol{K}_1, k_1] = \mathrm{sn}(\xi + q\boldsymbol{K}_1, k_1) \tag{11.110}$$

Typical examples of passive and active elliptic filters are presented in Fig. 11.55.

Applying the analysis techniques of analog filters used in the previous sections, let us construct the equations that describe the operation of the filter correlating between them the appropriate parameters in each case. Starting with the passband, which presents *equal amplitude fluctuations* ε_p, this, by analogy with Chebyshev I filters, is described by the parameter A_p, which satisfies the condition $A_p = 10\log(1 + \varepsilon_p^2)$ (in decibels), from which it follows that

$$\varepsilon_p = \sqrt{10^{A_p/10} - 1}$$

On the other hand, the stopband is characterized by equal amplitude fluctuations ε_2, with the minimum width in this band being equal to $1/(\varepsilon_p\varepsilon_s) = 1/k_1$. The behavior of this band is described by the parameter A_s, which by analogy with Chebyshev II filters will satisfy the equation

$$\frac{1}{A_s} = \frac{\varepsilon_s}{\sqrt{1+\varepsilon_s^2}} = \frac{1}{\sqrt{1+(1/\varepsilon_s^2)}} \tag{11.111}$$

The foregoing relationship expressed in decibels is formulated as

$$A_s = 10\log\left(1 + \frac{1}{\varepsilon_s^2}\right)$$

from which it follows that

$$\varepsilon_s = \frac{1}{\sqrt{10^{A_s/10} - 1}}$$

For given values of the parameters A_p and A_s, from the foregoing relations, we can therefore calculate the values of the parameters ε_s and ε_p, which are included in the frequency response of the filter.

Calculation of Minimum Filter Order

Adopting the argumentation developed for the description of Chebyshev I filters, the square of the measure of the filter's frequency response at the position of the frequency $\omega = \omega_p$ will have a value

$$|\mathcal{H}(j\omega_p)|^2 = \frac{1}{1 + \varepsilon_p^2 R_N^2(\omega_p, \omega_s, \omega_p, \varepsilon_s, \varepsilon_p)} = \frac{1}{1 + \varepsilon_p^2} \tag{11.112}$$

from which it follows that

$$R_N^2(\omega_p, \omega_s, \omega_p, \varepsilon_s, \varepsilon_p) = \mathrm{sn}^2(\xi_{\omega=\omega_p} + q\boldsymbol{K}_1, k_1) = 1$$

or, equivalently,

$$\mathrm{sn}(\xi_{\omega=\omega_p} + q\boldsymbol{K}_1, k_1) = \pm 1 \tag{11.113}$$

However, from the properties of the Jacobi elliptic sine function (see also Fig. A.10 on page 1502), we know that $\mathrm{sn}[(2m + 1)\boldsymbol{K}, k] = \pm 1$. Therefore, we will have

$$\xi_{\omega=\omega_p} + q\boldsymbol{K}_1 = (2m + 1)\boldsymbol{K}_1$$

or, equivalently,

$$\kappa\, \mathrm{sn}^{-1}\left(\frac{\omega_p}{\omega_p}, \frac{\omega_p}{\omega_s}\right) = (2m + 1 - q)\boldsymbol{K}_1$$

and finally

$$\kappa = \frac{2m + 1 - q}{\mathrm{sn}^{-1}[1, (\omega_p/\omega_s)]}\boldsymbol{K}_1 = \frac{2m + 1 - q}{\mathrm{sn}^{-1}(1, k_2)}\boldsymbol{K}_1 = (2m + 1 - q)\frac{\boldsymbol{K}_1}{\boldsymbol{K}_2} \tag{11.114}$$

since from the defining equation of the *inverse Jacobi elliptic sine function*

$$\mathrm{sn}^{-1}(\sin\varphi, k) = \int_0^{\varphi} \frac{d\vartheta}{\sqrt{1 - k^2 \sin^2\vartheta}}$$

it follows that

$$\mathrm{sn}^{-1}(1, k_2) = \mathrm{sn}^{-1}\left[\sin\left(\frac{\pi}{2}\right), k_2\right] = \int_0^{\pi/2} \frac{d\vartheta}{\sqrt{1 - k_2^2 \sin^2\vartheta}} = \boldsymbol{K}(k_2) = \boldsymbol{K}_2$$

On the other hand, at the cutoff frequency $\omega = \omega_s$, the amplitude response will have the form

$$|\mathcal{H}(j\omega_s)| = \frac{1}{\sqrt{1 + \varepsilon_p^2 \mathrm{sn}^2\left[\kappa\, \mathrm{sn}^{-1}\left(\frac{\omega_s}{\omega_p}, \frac{\omega_p}{\omega_s}\right) + q\boldsymbol{K}_1, \varepsilon_p\varepsilon_s\right]}} = \frac{1}{A_s} = \frac{1}{\sqrt{1 + \frac{1}{\varepsilon_s^2}}} \tag{11.115}$$

From a comparison of the expressions found in the roots of the denominator it immediately follows that

$$\varepsilon_p^2 \mathrm{sn}^2\left[\kappa\,\mathrm{sn}^{-1}\left(\frac{\omega_s}{\omega_p}, \frac{\omega_p}{\omega_s}\right) + q\boldsymbol{K}_1, \varepsilon_p\varepsilon_s\right] = \frac{1}{\varepsilon_s^2} \qquad \text{or, equivalently,} \qquad \mathrm{sn}\left[\kappa\,\mathrm{sn}^{-1}\left(\frac{\omega_s}{\omega_p}, \frac{\omega_p}{\omega_s}\right) + q\boldsymbol{K}_1, \varepsilon_p\varepsilon_s\right] = \frac{1}{\varepsilon_s\varepsilon_p} = \frac{1}{k_1}$$

Comparing the last relation with the first of Equations (A.57) (see the relevant section of the mathematical appendix) for the values $n = 0$ and $\boldsymbol{K} = \boldsymbol{K}_1$, which is written

$$\mathrm{sn}[(2m+1)\boldsymbol{K}_1 + j\boldsymbol{K}_1', k_1] = \frac{1}{k_1}$$

we will get

$$\kappa\,\mathrm{sn}^{-1}\left(\frac{\omega_s}{\omega_p}, \frac{\omega_p}{\omega_s}\right) + q\boldsymbol{K}_1 = \kappa\,\mathrm{sn}^{-1}\left(\frac{1}{k_2}, k_2\right) + q\boldsymbol{K}_1 = (2m+1)\boldsymbol{K}_1 + j\boldsymbol{K}_1'$$

and therefore

$$\mathrm{sn}^{-1}\left(\frac{1}{k_2}, k_2\right) = \frac{2m+1-q}{\kappa}\boldsymbol{K}_1 + j\frac{\boldsymbol{K}_1'}{\kappa} = \boldsymbol{K}_2 + j\frac{\boldsymbol{K}_1'}{\kappa} \tag{11.116}$$

where for the derivation of the last relation we substituted the parameter κ from the expression we calculated earlier. Now using the equation[31]

$$\mathrm{sn}^{-1}\left(\frac{1}{k}, k\right) = \boldsymbol{K} + j\boldsymbol{K}' \qquad \text{the foregoing relation is written as} \qquad \boldsymbol{K}_2 + j\boldsymbol{K}_2' = \boldsymbol{K}_2 + j\frac{\boldsymbol{K}_1'}{\kappa}$$

from which we finally get

$$\kappa = (2m+1-q)\frac{\boldsymbol{K}_1}{\boldsymbol{K}_2} = \frac{\boldsymbol{K}_1'}{\boldsymbol{K}_2'} \tag{11.117}$$

The above relation is particularly important in the process of designing elliptical filters, as it dictates the correlation that must exist between the parameters ω_p, ω_s, ε_p, and ε_s, which is a consequence of the fact that the integrals $\boldsymbol{K}_1$ and $\boldsymbol{K}_1'$ depend on the amplitudes of fluctuations ε_p and ε_s, while the integrals $\boldsymbol{K}_2$ and $\boldsymbol{K}_2'$ depend on the frequencies ω_p and ω_s, as derived from their defining equations. The preceding expression contains the coefficient $2m+1-q$, which is even if $q = 1$ and odd if $q = 0$ and expresses the *order* N of the filter. Therefore, if we set $N = 2m+1-q$ in the last relation, it is written

$$\kappa = N\frac{\boldsymbol{K}_1}{\boldsymbol{K}_2} = \frac{\boldsymbol{K}_1'}{\boldsymbol{K}_2'} \tag{11.118}$$

from which it follows that the minimum order of the filter corresponding to the values of the four preceding parameters that have been given is calculated as

$$N = \frac{\boldsymbol{K}_2\,\boldsymbol{K}_1'}{\boldsymbol{K}_1\,\boldsymbol{K}_2'} \tag{11.119}$$

In most cases, the value resulting from this expression is a *real number* and as usual should be rounded *to the next highest integer*.

[31] This relation constitutes one of the properties of the inverse Jacobi elliptic sine function, whose verification is left as an exercise to the reader.

Although the foregoing relation allows the calculation of the minimum order of the filter, usually the value of N is calculated using a simpler expression that will be constructed later and that does not include the calculation of elliptic integrals (which, however, can be calculated using the *arithmetic-geometric mean*). The usefulness of the foregoing relationship is the definition of a *specification constraint* of the form $f_1(N, \omega_p/\omega_s) = f_2(A_p, A_s)$, which the values of the parameters ω_s, ω_p, A_s, and A_p must satisfy. Therefore, if we know the values of these parameters, then the value that the order of the filter N should have, to satisfy the above relationship, is calculated in the following way:

Starting from the representation of the Jacobi elliptic sine function using the functions $\vartheta(u, q)$ (see the relevant section of the mathematical appendix)

$$\mathrm{sn}[u,k] = \frac{1}{\sqrt{k}} \frac{\vartheta_1\left[\dfrac{u}{2\boldsymbol{K}}, q(k)\right]}{\vartheta_4\left[\dfrac{u}{2\boldsymbol{K}}, q(k)\right]} = \frac{2\sqrt[4]{q(k)}}{\sqrt{k}} \left\{ \frac{\displaystyle\sum_{n=0}^{\infty}(-1)^n q^{n(n+1)}(k)\sin\left[(2n+1)\frac{\pi u}{2\boldsymbol{K}}\right]}{\displaystyle 1+2\sum_{n=0}^{\infty}(-1)^n q^{n^2}(k)\cos\left(2m\frac{\pi u}{2\boldsymbol{K}}\right)} \right\}$$

and taking into account that for $u = \boldsymbol{K}_1$ it holds that $\mathrm{sn}(\boldsymbol{K}_1, k_1) = 1$ (as derived from the properties of the elliptic sine function) and that

$$\sin\left[(2n+1)\frac{\pi u}{2\boldsymbol{K}_1}\right]_{u=\boldsymbol{K}_1} = \sin\left[(2n+1)\frac{\pi}{2}\right] = \cos\left(2n\frac{\pi u}{2\boldsymbol{K}_1}\right)_{u=\boldsymbol{K}_1} = \cos(n\pi) = (-1)^n$$

the foregoing relationship is written

$$\frac{2\sqrt[4]{q(k_1)}}{\sqrt{k_1}} \left\{\sum_{n=0}^{\infty} q^{n(n+1)}(k_1)\right\} \Big/ \left\{1+2\sum_{n=1}^{\infty} q^{n^2}(k_1)\right\} = 1$$

or, equivalently,

$$\begin{aligned} k_1 &= 4\sqrt{q(k_1)} \times \left[\left\{\sum_{n=0}^{\infty} q^{n(n+1)}(k_1)\right\} \Big/ \left\{1+2\sum_{n=1}^{\infty} q^{n^2}(k_2)\right\}\right]^2 \\ &= 4\sqrt{q(k_1)} \times \left[\frac{1+q^2(k_1)+q^6(k_1)+q^{12}(k_1)+\ldots}{1+2q(k_1)+2q^4(k_1)+2q^9(k_1)+\ldots}\right]^2 \end{aligned}$$

where $q(k_1) = \exp(-\pi \boldsymbol{K}_1'/\boldsymbol{K}_1)$. Now, resorting to the simplifying assumptions $k_1 \approx 0$ and $k_1' \approx 1$, which generally hold in practice, $\boldsymbol{K}_1'/\boldsymbol{K}_1 \gg 1$ and $q(k_1) \ll 1$, so we can consider that the squared expression in square brackets is approximately equal to one. Therefore, $k_1^2 \approx 4\sqrt{q(k_1)}$, and consequently,

$$k_1^2 \approx 16q(k_1) = 16\exp(-\pi \boldsymbol{K}_1'/\boldsymbol{K}_1) = 16\exp(-\pi N\boldsymbol{K}_2'/\boldsymbol{K}_2) = 16\,[\exp(-\pi \boldsymbol{K}_2'/\boldsymbol{K}_2)]^N = 16q^N(k_2)$$

where for the derivation of the last expression we used the relation $N = \boldsymbol{K}_2\boldsymbol{K}_1'/\boldsymbol{K}_1\boldsymbol{K}_2'$, which we previously proved. Substituting this relation into the equation

$$k_1 = \varepsilon_p\varepsilon_s = \sqrt{\frac{10^{A_p/10}-1}{10^{A_s/10}-1}} = \frac{1}{\sqrt{D}} \quad \text{we get} \quad k_1^2 = \frac{10^{A_p/10}-1}{10^{A_s/10}-1} = \frac{1}{D} = 16q^N(k_2)$$

For given values of the design parameters ε_s, ε_s, A_p, and A_s, the *minimum filter order* is calculated by solving the foregoing expression for N, from which it follows that $N = \log(16D)/\log[1/q(k_2)]$, where $D = (10^{A_s/10}-1)/(10^{A_p/10}-1)$, which is also the desired result. Let us note that if we know the order of the filter N, as well as the values of A_p and k, then the value of the parameter A_s, so that the constraints of specifications are met, is calculated from the foregoing relationship as

$$A_s = 10\log\left(\frac{10^{A_p/10}-1}{16q^N(k_2)}+1\right) \tag{11.120}$$

Poles and Zeros of Elliptic Function, Maximum/Minimum Frequencies, and Cutoff Frequency

From the mathematical expression of the function $|\mathcal{H}(j\omega)|^2$ it is not difficult to find that it will take its maximum value $|\mathcal{H}(j\omega)|^2 = 1$ at the frequencies ω, for which we have $R_N(\omega, \omega_s, \omega_p, \varepsilon_s, \varepsilon_p) = 0$, and its minimum value for the frequencies ω, for which we have $R_N^2(\omega, \omega_s, \omega_p, \varepsilon_s, \varepsilon_p) = 1$ (recall that the Jacobi elliptic functions are defined in the range of values $[-1, +1]$). To determine the values of the frequency for which the function $|\mathcal{H}(j\omega)|^2$ presents maxima, we should calculate the roots of the equation

$$\mathrm{sn}\left[\kappa\,\mathrm{sn}^{-1}\left(\frac{\omega}{\omega_p}, \frac{\omega_p}{\omega_s}\right) + q\boldsymbol{K}_1, \varepsilon_p\varepsilon_s\right] = 0 \tag{11.121}$$

something that can be done taking into account that the function $\mathrm{sn}(u, k)$ is zero at the positions $u = 2m\boldsymbol{K}$ (see Fig. A.8 of the mathematical appendix). It will therefore hold that

$$\kappa\,\mathrm{sn}^{-1}\left(\frac{\omega}{\omega_p}, \frac{\omega_p}{\omega_s}\right) + q\boldsymbol{K}_1 = 2m\boldsymbol{K}_1$$

from which it follows that

$$\mathrm{sn}^{-1}\left(\frac{\omega}{\omega_p}, \frac{\omega_p}{\omega_s}\right) = \frac{(2m-q)\boldsymbol{K}_1}{k} = \frac{(2m-q)\boldsymbol{K}_2}{N}$$

where for the derivation of the last relation we used the expression $k = N\boldsymbol{K}_1/\boldsymbol{K}_2$. The solution to the preceding equation with respect to ω, which is based on the equivalence $\mathrm{sn}^{-1}(x, k) = \alpha \Leftrightarrow x = \mathrm{sn}(\alpha, k)$, leads us to the result

$$\omega_m^{\max} = \omega_p\,\mathrm{sn}\left[\frac{(2m-q)\boldsymbol{K}_2}{N}, \frac{\omega_p}{\omega_s}\right]$$

where $m = 0, 1, 2, \ldots, (N-1)/2$ for odd values of N and $m = 0, 1, 2, \ldots, N/2$ for even values of N. This relation gives us the frequencies for which the square of the magnitude of the frequency response function is maximized in the passband.

On the other hand, the frequencies corresponding to the minimum points are the roots of the equation

$$\mathrm{sn}^2\left[\kappa\,\mathrm{sn}^{-1}\left(\frac{\omega}{\omega_p}, \frac{\omega_p}{\omega_s}\right) + q\boldsymbol{K}_1, \varepsilon_p\varepsilon_s\right] = 1 \quad \text{or, equivalently,} \quad \mathrm{sn}\left[\kappa\,\mathrm{sn}^{-1}\left(\frac{\omega}{\omega_p}, \frac{\omega_p}{\omega_s}\right) + q\boldsymbol{K}_1, \varepsilon_p\varepsilon_s\right] = \pm 1$$

In this case, we should take into account that the function of the *Jacobi elliptic sine function* satisfies the condition $\mathrm{sn}(u, k) = \pm 1$ for the values $u = (2m+1)\boldsymbol{K}$. Working, as before, we get

$$\omega_m^{\min} = \omega_p\,\mathrm{sn}\left[\frac{(2m+1-q)\boldsymbol{K}_2}{N}, \frac{\omega_p}{\omega_s}\right] \tag{11.122}$$

where $m = 0, 1, 2, \ldots, (N-3)/2$ for odd values of N and $m = 0, 1, 2, \ldots, (N-1)/2$ for even values of N. This relationship gives us the frequencies for which the square of the magnitude of the frequency response function *is minimized* in the passband.

In the next step of the process, let us move on to calculating the *cutoff frequency* ω_c as a function of the values of the parameters ω_p, ω_s, ε_p, and ε_s. As is well known from basic theory, the cutoff frequency $\omega = \omega_c$ corresponds to an attenuation $A_p = 3$ dB, a property that in mathematical notation is formulated as

$$|\mathcal{H}(j\omega_c)| = \frac{1}{\sqrt{2}}|\mathcal{H}(j\omega_c)|_{\max} = \frac{1}{\sqrt{2}}$$

or, equivalently,

$$|\mathcal{H}(j\omega_c)|^2 = \frac{1}{1+\varepsilon_p^2\mathrm{sn}^2\left[\kappa\,\mathrm{sn}^{-1}\left(\frac{\omega_c}{\omega_p}, \frac{\omega_p}{\omega_s}\right) + q\boldsymbol{K}_1, \varepsilon_p\varepsilon_s\right]} = \frac{1}{2}$$

Therefore,

$$\varepsilon_p\mathrm{sn}\left[\kappa\,\mathrm{sn}^{-1}\left(\frac{\omega_c}{\omega_p}, \frac{\omega_p}{\omega_s}\right) + q\boldsymbol{K}_1, \varepsilon_p\varepsilon_s\right] = 1$$

from which it follows that

$$\mathrm{sn}^{-1}\left(\frac{\omega_c}{\omega_p}, \frac{\omega_p}{\omega_s}\right) = \left[\frac{\mathrm{sn}^{-1}(1/\varepsilon_p, \varepsilon_p\varepsilon_s) - q\boldsymbol{K}_1}{N\boldsymbol{K}_1}\right]\boldsymbol{K}_2$$

where for the derivation of the last relation we used again the equation $k = N\boldsymbol{K}_1/\boldsymbol{K}_2$. However, according to the properties of the *inverse Jacobi elliptic sine function*, for values of the frequency ω_c/ω_p, namely, for values $z = \omega_c/\omega_p > 1$ the function $\mathrm{sn}^{-1}(z, k_2)$ (where $k_2 = \omega_p/\omega_s$) will be of a complex nature and will have an imaginary part

$$\Im\left\{\mathrm{sn}^{-1}\left(\frac{\omega_c}{\omega_p}, \frac{\omega_p}{\omega_s}\right)\right\} = \Im\left\{\left[\frac{\mathrm{sn}^{-1}(1/\varepsilon_p, \varepsilon_p\varepsilon_s)}{N\boldsymbol{K}_1}\right]\boldsymbol{K}_2\right\}$$

which will be zero for $\omega_c = \omega_p$, namely, for the value $z = 1$ (this, on the contrary, is something that *does not hold* for values $z < 1$, or, equivalently, $\omega_c < \omega_p$ as shown by Fig. 11.56, which presents a change in the real and imaginary parts of the function $\mathrm{sn}^{-1}(z, k_2)$ for the values of elliptic measure $k_2 = 0.2, 0.3, 0.5, 0.7$). Taking into account that the real part of this function is equal to $\boldsymbol{K}_2$, as shown by the foregoing figure, we can write our initial function in the form

$$\mathrm{sn}^{-1}\left(\frac{\omega_c}{\omega_p}, \frac{\omega_p}{\omega_s}\right) = \boldsymbol{K}_2 + j\Im\left\{\left[\frac{\mathrm{sn}^{-1}(1/\varepsilon_p, \varepsilon_p\varepsilon_s) - q\boldsymbol{K}_1}{N\boldsymbol{K}_1}\right]\boldsymbol{K}_2\right\}$$

and solving for the cutoff frequency ω_c we end up with the expression

$$\omega_c = \omega_p\,\mathrm{sn}\left\{\boldsymbol{K}_2 + j\Im\left[\frac{\mathrm{sn}^{-1}(1/\varepsilon_p, \varepsilon_p\varepsilon_s) - q\boldsymbol{K}_1}{N\boldsymbol{K}_1}\right]\right\} \tag{11.123}$$

Finally, let us calculate the frequencies that lead to infinity of the function $R_N(\omega, \omega_s, \omega_p, \varepsilon_s, \varepsilon_p)$ and therefore constitute *its poles*. From the theory of elliptic functions we know that the function of the elliptic sine is zero every $2\boldsymbol{K}$. Therefore, it will hold that $\mathrm{sn}(2m\boldsymbol{K} + j\boldsymbol{K}', k) = \pm\infty$, and therefore

$$\kappa\,\mathrm{sn}^{-1}\left(\frac{\omega}{\omega_p}, \frac{\omega_p}{\omega_s}\right) + q\boldsymbol{K}_1 = 2m\boldsymbol{K}_1 + j\boldsymbol{K}_1'$$

Now, considering the relation

$$\mathrm{sn}^{-1}\left(\frac{1}{k_2}, k_2\right) = \boldsymbol{K}_2 + j\boldsymbol{K}_2'$$

where $k_2 = \omega_p/\omega_s$, it is obvious that this will be satisfied exactly for the frequency $\omega_0 = \omega_s/\omega_p$, while for the frequencies $\omega > \omega_0$ the value of this inverse elliptic function will be equal to $x + j\boldsymbol{K}_2'$, where $x < \boldsymbol{K}_2$ (see also Fig. 11.56). Therefore, for the frequencies $\omega > \omega_0$, it will hold that $k(x + j\boldsymbol{K}_2') + q\boldsymbol{K}_1 = 2m\boldsymbol{K}_1 + j\boldsymbol{K}_1'$. Replacing the value of k from the defining

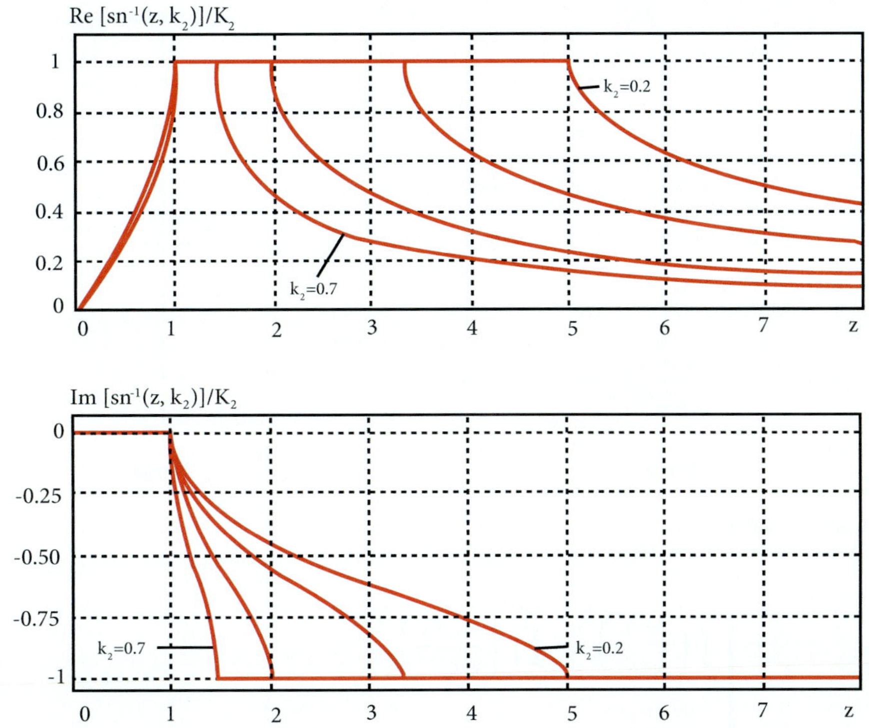

Fig. 11.56 Variation of real and imaginary parts of function $\mathrm{sn}^{-1}(z, k_2)$ for elliptic measure values $k_2 = 0.2, 0.3, 0.5, 0.7$

equation and solving for x, it turns out that

$$x = \frac{(2m-q)\boldsymbol{K}_2}{N} \quad \text{and consequently} \quad \mathrm{sn}^{-1}\left(\frac{\omega}{\omega_p}, \frac{\omega_p}{\omega_s}\right) = \frac{(2m-q)\boldsymbol{K}_2}{N} + j\boldsymbol{K}_2'$$

Thus, solving for ω we get

$$\omega_m^{\infty} = \omega_p \mathrm{sn}\left[\frac{(2m-q)\boldsymbol{K}_2}{N} + j\boldsymbol{K}_2', \frac{\omega_p}{\omega_s}\right] = \frac{\omega_s}{\mathrm{sn}[(2m-q)\boldsymbol{K}_2/N, \omega_p/\omega_s]} = \frac{\omega_p\omega_s}{\omega_m^{\max}} \tag{11.124}$$

where for the construction of the last relation we used the known property of the elliptic sine function $\mathrm{sn}[u + j\boldsymbol{K}', k] = 1/[k\mathrm{sn}(u, k)]$. In the foregoing expressions, the parameter m takes the values $m = 1, 2, \ldots, (N-1)/2$ for odd-order filters and $m = 1, 2, \ldots, N/2$ for even-order filters. Note that odd-order filters have one more such frequency at the position $\omega = \infty$.

From the last relation it follows that *the poles and zeros of the Jacobi elliptic function are related to each other by inverse proportionality*. A significant consequence of this property is that if zero values can be found for which the function R_N is characterized by equal amplitude fluctuations in the passband, then *automatically* it will be characterized by equal amplitude fluctuations in the stopband. Having calculated the poles $\omega_p^m = \omega_m^{\infty}$ and the zeros $\omega_z^m = \omega_m^{\max}$ of the elliptic function R_N, which in the literature is known as an *explicit Chebyshev function*, we can formulate it as

$$R_N = D\omega^{1-q} \times \left\{\prod_{m=1}^{L}[\omega^2 - (\omega_z^m)^2]\right\} \Big/ \left\{\prod_{m=1}^{L}[\omega^2 - (\omega_p^m)^2]\right\} = D\omega^{1-q} \times \left\{\prod_{m=1}^{L}[\omega^2 - (\omega_z^m)^2]\right\} \Big/ \left\{\prod_{m=1}^{L}\left[\omega^2 - \frac{\omega_p^2\omega_s^2}{(\omega_z^m)^2}\right]\right\}$$

with the parameter L being equal to $N/2$ for even-order filters and equal to $(N-1)/2$ for odd-order filters and the scaling factor D being defined as

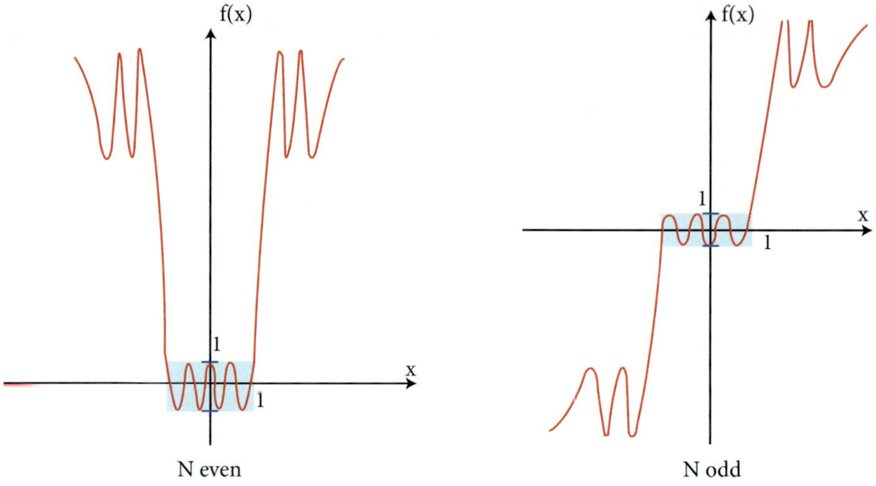

Fig. 11.57 Explicit Chebyshev function

$$D = \begin{cases} \prod_{m=1}^{N/2} \left(\frac{\omega_p^m}{\omega_z^m}\right)^2 = \prod_{m=1}^{N/2} \left[\frac{\omega_p^2 \omega_s^2}{(\omega_z^m)^4}\right] \\ \frac{\prod_{m=1}^{(N-1)/2} [(\omega_p^m)^2 - \omega_p^2]}{\omega_p \prod_{m=1}^{(N-1)/2} [\omega_p^2 - (\omega_z^m)^2]} = \omega_p \frac{\prod_{m=1}^{(N-1)/2} [\omega_s^2 - (\omega_z^m)^2]}{\prod_{m=1}^{(N-1)/2} (\omega_z^m)^2 [\omega_p^2 - (\omega_z^m)^2]} \end{cases}$$

for even- and odd-order filters, respectively. This constant is calculated in the same way as we calculated the normalization constant H_0 in the previous approaches, and the value that results will ensure that at the edge of the passband $\omega = \omega_p$ and at the position $\omega = 0$, the magnitude of the function R_N will be equal to one. It is proven that this function is characterized by oscillations of equal amplitude in the value range $[-1, +1]$, as well as in the areas $(-\infty, -1]$ and $[+1, +\infty)$, and its graphical representation for even- and odd-order value is shown in Fig. 11.57. The frequency ω that appears in the expression of R_N for odd-order filters due to the fact that for these orders the degree of the polynomial of the numerator is $N - 1$, and therefore a first-degree term should be added, so that a polynomial of degree N arises.

Note that in practice the poles and zero values of the elliptic function are calculated not from the foregoing relations but from the following equations, in which the elliptic sine function is expressed using the functions $\vartheta_1(u, q)$, $\vartheta_4(u, q)$, and $q(k)$

$$\omega_z^m = \frac{\omega_p}{\sqrt{k}} \frac{\vartheta_1\left(\frac{2m-q}{2N}, q(k)\right)}{\vartheta_4\left(\frac{2m-q}{2N}, q(k)\right)} = \frac{\omega_p}{\sqrt{k}} \frac{\vartheta_1\left[\frac{2m-q}{2N}, \exp\left(-\pi \frac{M(1,\sqrt{1-k^2})}{M(1,k)}\right)\right]}{\vartheta_4\left[\frac{2m-q}{2N}, \exp\left(-\pi \frac{M(1,\sqrt{1-k^2})}{M(1,k)}\right)\right]}$$

$$= \frac{2\omega_p \sqrt[4]{\exp\left(-\pi \frac{M(1,\sqrt{1-k^2})}{M(1,k)}\right)} \sum_{n=0}^{\infty} \left\{(-1)^n \left[\exp\left(-\pi \frac{M(1,\sqrt{1-k^2})}{M(1,k)}\right)\right]^{n(n+1)} \sin\left[(2n+1)\pi \frac{2m-q}{2N}\right]\right\}}{\sqrt{k} \; 1 + 2\sum_{n=1}^{\infty} \left\{(-1)^n \left[\exp\left(-\pi \frac{M(1,\sqrt{1-k^2})}{M(1,k)}\right)\right]^{n^2} \cos\left[2n\pi \frac{2m-q}{2N}\right]\right\}}$$

$$\omega_p^m = \frac{\omega_p}{\sqrt{k}} \frac{\vartheta_1\left(\frac{2m-q}{2N} + j\frac{\boldsymbol{K}_2'}{2\boldsymbol{K}_2}, q(k)\right)}{\vartheta_4\left(\frac{2m-q}{2N} + j\frac{\boldsymbol{K}_2'}{2\boldsymbol{K}_2}, q(k)\right)} = \frac{\omega_p}{\sqrt{k}} \frac{\vartheta_1\left[\frac{2m-q}{2N} + j\frac{\boldsymbol{K}_2'}{2\boldsymbol{K}_2}, \exp\left(-\pi \frac{M(1,\sqrt{1-k^2})}{M(1,k)}\right)\right]}{\vartheta_4\left[\frac{2m-q}{2N} + j\frac{\boldsymbol{K}_2'}{2\boldsymbol{K}_2}, \exp\left(-\pi \frac{M(1,\sqrt{1-k^2})}{M(1,k)}\right)\right]} =$$

$$= \frac{2\omega_p \sqrt[4]{\exp\left(-\pi\frac{M(1,\sqrt{1-k^2})}{M(1,k)}\right)}\sum_{n=0}^{\infty}\left\{(-1)^n\left[\exp\left(-\pi\frac{M(1,\sqrt{1-k^2})}{M(1,k)}\right)\right]^{n(n+1)}\sin\left[(2n+1)\pi\left(\frac{2m-q}{2N}+j\frac{\boldsymbol{K}_2'}{2\boldsymbol{K}_2}\right)\right]\right\}}{\sqrt{k}\left(1+2\sum_{n=1}^{\infty}\left\{(-1)^n\left[\exp\left(-\pi\frac{M(1,\sqrt{1-k^2})}{M(1,k)}\right)\right]^{n^2}\cos\left[2n\pi\left(\frac{2m-q}{2N}+j\frac{\boldsymbol{K}_2'}{2\boldsymbol{K}_2}\right)\right]\right\}\right)}$$

The foregoing relations *do not include elliptic functions* and allow for the easy calculation of the poles and roots of the elliptic function R_N, since the arithmetic-geometric mean of two numbers $M(x, y)$ is calculated very quickly even with a handheld calculator, while the functions $\vartheta_1(u, q)$, $\vartheta_4(u, q)$, and $q(k)$ are calculated from their defining equations, with an accurate determination of their value requiring the calculation of only a few terms.

Poles and Zero Values of Transfer Function

After calculating the elliptic function R_N, let us now move on to the calculation of the poles and zero values of the transfer function of the low-pass elliptic filter $\mathcal{H}(s)$, applying the methodology followed in the previous approaches to ideal filters. In this case, the function $\mathcal{H}(s)\mathcal{H}(-s)$ takes the form

$$\mathcal{H}(s)\mathcal{H}(-s) = |\mathcal{H}(j\omega)|^2_{\omega=s/j} = \frac{1}{1+\varepsilon_p^2 R_N^2(s/j,\omega_s,\omega_p,\varepsilon_s,\varepsilon_p)} = \frac{1}{1+\varepsilon_p^2 D^2(-1)^{q+1}s^{2(1-q)}\left\{\prod_{m=1}^{L}\frac{s^2+(\omega_z^m)^2}{s^2+(\omega_p^m)^2}\right\}}$$

$$= \frac{\prod_{m=1}^{L}[s^2+(\omega_p^m)^2]^2}{\prod_{m=1}^{L}[s^2+(\omega_p^m)^2]^2+\varepsilon_p^2 D^2(-1)^{q+1}s^{2(1-q)}\prod_{m=1}^{L}[s^2+(\omega_p^z)^2]^2} \tag{11.125}$$

Therefore, the *zero values* of the transfer function are determined as the roots of the algebraic equation $s^2+(\omega_p^m)^2=0$, from which it immediately follows that

$$z_m = \pm j\omega_p^m j\omega_p \mathrm{sn}\left[\frac{(2m-q)\boldsymbol{K}_2}{N}+j\boldsymbol{K}_2', \frac{\omega_p}{\omega_s}\right] = \pm j\omega_p\frac{\omega_s}{\omega_p \mathrm{sn}\left[\frac{(2m-q)\boldsymbol{K}_2}{N}, \frac{\omega_p}{\omega_s}\right]} = \pm j\frac{\omega_s\omega_p}{\omega_z^m} = \pm\frac{j\omega_p}{\Omega_m}$$

with the parameter Ω_m defined as

$$\Omega_m = \frac{1}{\omega_s}\mathrm{sn}\left[\frac{(2m-q)\boldsymbol{K}_2}{N}, \frac{\omega_p}{\omega_s}\right] \tag{11.126}$$

The foregoing equation for calculating the zero values of the transfer function applies *for both even- and odd-order values* for the values $q = 1$ and $q = 0$, respectively.

On the other hand, the poles of the transfer function will emerge as the roots of the equation

$$1+\varepsilon_p^2 R_N^2(s/j,\omega_s,\omega_p,\varepsilon_s,\varepsilon_p) = 0$$

which can take the equivalent form

$$\mathrm{sn}\left[\kappa\,\mathrm{sn}^{-1}\left(\frac{s}{j\omega_p}, \frac{\omega_p}{\omega_s}\right)+q\boldsymbol{K}_1, \varepsilon_p\varepsilon_s\right] = \frac{j}{\varepsilon_p}$$

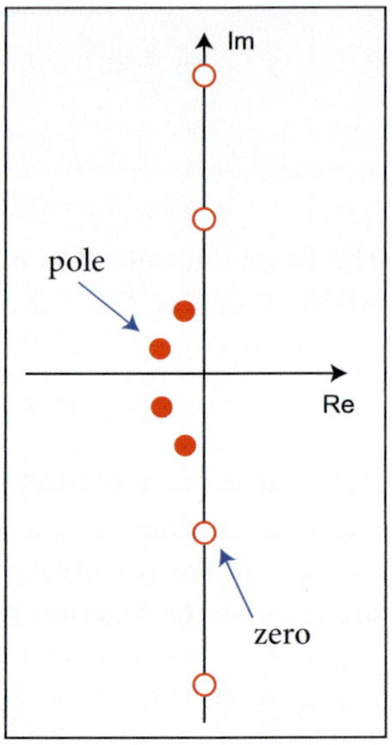

Fig. 11.58 Pole–zero diagram of a fourth-order low-pass elliptic filter

Taking into account that the Jacobi elliptic sine function is a *doubly periodic function* with real period $4m\boldsymbol{K}_1$ (see the relevant section of the mathematical appendix), the foregoing relationship is written

$$\mathrm{sn}\left[\kappa\,\mathrm{sn}^{-1}\left(\frac{s}{j\omega_p},\frac{\omega_p}{\omega_s}\right)+(q+4m)\boldsymbol{K}_1,\varepsilon_p\varepsilon_s\right]=\frac{j}{\varepsilon_p}$$

from which it follows that the poles of the transfer function are given by the relationship

$$p_m=j\omega_p\mathrm{sn}\left[\frac{\mathrm{sn}(j/\varepsilon_p,\varepsilon_p\varepsilon_s)-(q+4m)\boldsymbol{K}_1}{k},\frac{\omega_p}{\omega_s}\right] \tag{11.127}$$

for parameter values $m=1,2,\ldots,N$. A pole–zero values diagram of a fourth-order low-pass elliptic filter is presented in Fig. 11.58.

Approximate Pole Equations

The foregoing equation for calculating the poles of the transfer function can be simplified and expressed using *trigonometric* and *hyperbolic* functions, if we make the simplifying assumption $k_1=\varepsilon_p\varepsilon_s\approx 0$, which generally holds in practice. Now, using the property of the Jacobi elliptic sine $\mathrm{sn}(u,0)=\sin u$ and the result $\boldsymbol{K}_1(0)=\pi/2$, we will have

$$\kappa\ (k_1=0)=\frac{N}{\boldsymbol{K}_2}\boldsymbol{K}_1(0)=\frac{\pi N}{2\boldsymbol{K}_2}$$

and the pole calculation equation is given as

$$\mathrm{sn}\left[\frac{\pi N}{2\boldsymbol{K}_2}\mathrm{sn}^{-1}\left(\frac{s}{j\omega_p},\frac{\omega_p}{\omega_s}\right)+q\frac{\pi}{2},0\right]=\sin\left[\frac{\pi N}{2\boldsymbol{K}_2}\mathrm{sn}^{-1}\left(\frac{s}{j\omega_p},\frac{\omega_p}{\omega_s}\right)+q\frac{\pi}{2}\right]=\frac{j}{\varepsilon_p} \tag{11.128}$$

At this point we distinguish two cases:

N **Odd** In this case, $q=0$, and the last relation is written

$$\sin\left[\frac{\pi N}{2\boldsymbol{K}_2}\mathrm{sn}^{-1}\left(\frac{s}{j\omega_p},\frac{\omega_p}{\omega_s}\right)\right]=\frac{j}{\varepsilon_p}\quad\text{or, equivalently,}\quad -j\sin\left[\frac{\pi N}{2\boldsymbol{K}_2}\mathrm{sn}^{-1}\left(\frac{s}{j\omega_p},\frac{\omega_p}{\omega_s}\right)\right]=\frac{1}{\varepsilon_p}$$

From mathematical analysis we know, however, that $j \sin u = \sinh(ju)$, and therefore we will have

$$-\sinh\left[j\frac{\pi N}{2\boldsymbol{K}_2}\text{sn}^{-1}\left(\frac{s}{j\omega_p}, \frac{\omega_p}{\omega_s}\right)\right] = \frac{1}{\varepsilon_p} \quad \text{or, equivalently,} \quad j\frac{\pi N}{2\boldsymbol{K}_2}\text{sn}^{-1}\left(\frac{s}{j\omega_p}, \frac{\omega_p}{\omega_s}\right) = \sinh^{-1}\left(-\frac{1}{\varepsilon_p}\right) = -\sinh^{-1}\left(\frac{1}{\varepsilon_p}\right)$$

Using now the identity $\sinh^{-1} x = \ln(x + \sqrt{x^2+1})$, the second term is written

$$\sinh^{-1}\left(\frac{1}{\varepsilon_p}\right) = \ln\left(\frac{1}{\varepsilon_p} + \sqrt{\frac{1}{\varepsilon_p^2}+1}\right) = \ln\left(\frac{1}{\varepsilon_p} + \sqrt{\frac{1+\varepsilon_p^2}{\varepsilon_p^2}}\right) = \ln\frac{1+\sqrt{1+\varepsilon_p^2}}{\varepsilon_p}$$

and since $\varepsilon_p = \sqrt{10^{A_p/10}-1}$, we will have

$$\sinh^{-1}\left(\frac{1}{\varepsilon_p}\right) = \ln\frac{1+\sqrt{1+10^{A_p/10}-1}}{\sqrt{10^{A_p/10}-1}} = \ln\frac{1+10^{A_p/20}}{\sqrt{10^{A_p/20}+1}\sqrt{10^{A_p/20}-1}} = \ln\sqrt{\frac{10^{A_p/20}+1}{10^{A_p/20}-1}} = \frac{1}{2}\ln\frac{10^{A_p/20}+1}{10^{A_p/20}-1}$$

Therefore, we have

$$j\frac{\pi N}{2\boldsymbol{K}_2}\text{sn}^{-1}\left(\frac{s}{j\omega_p}, \frac{\omega_p}{\omega_s}\right) = -\frac{1}{2}\ln\frac{10^{A_p/20}+1}{10^{A_p/20}-1} \quad \text{or, equivalently,} \quad \text{sn}^{-1}\left(\frac{s}{j\omega_p}, \frac{\omega_p}{\omega_s}\right) = j\frac{\boldsymbol{K}_2}{N\pi}\ln\frac{10^{A_p/20}+1}{10^{A_p/20}-1}$$

and solving for s it turns out that

$$s = j\omega_p \text{sn}\left[j\frac{\boldsymbol{K}_2}{N\pi}\ln\frac{10^{A_p/20}+1}{10^{A_p/20}-1}, \frac{\omega_p}{\omega_s}\right] = j\omega_p\sqrt{\frac{\omega_s}{\omega_p}}\frac{2\sqrt[4]{q(k_2)}\sum_{n=0}^{\infty}(-1)^n q^{n(n+1)}(k_2)\sin\left[j(2n+1)\frac{\boldsymbol{K}_2}{N\pi}\pi\ln\frac{10^{A_p/20}+1}{10^{A_p/20}-1}\right]}{1+2\sum_{n=1}^{\infty}(-1)^n q^{n^2}(k_2)\cos\left[2\pi nj\frac{\boldsymbol{K}_2}{N\pi}\ln\frac{10^{A_p/20}+1}{10^{A_p/20}-1}\right]}$$

and finally

$$s = -\sqrt{\omega_s\omega_p}\frac{2\sqrt[4]{q(k_2)}\sum_{n=0}^{\infty}(-1)^n q^{n(n+1)}(k_2)\sinh\left[(2n+1)\frac{\boldsymbol{K}_2}{N}\ln\frac{10^{A_p/20}+1}{10^{A_p/20}-1}\right]}{1+2\sum_{n=1}^{\infty}(-1)^n q^{n^2}(k_2)\cosh\left[2n\frac{\boldsymbol{K}_2}{N}\ln\frac{10^{A_p/20}+1}{10^{A_p/20}-1}\right]} = p_0 \qquad (11.129)$$

where for the derivation of the last relation we used the identities $\sin ju = j\sinh u$ and $\cos(ju) = \cosh u$. The preceding value of the constant s is a *real number*, and therefore the elliptic low-pass filters of odd order have a *real pole* $s = p_0$, which, in full analogy with the previous approximations of ideal filters, does not appear in the filters of even order N (let us note, however, that the expression of p_0 *is used* in the equations of the poles for the even order, as we will see later).

On the other hand, to calculate the remaining poles of the function, we will use the *property of periodicity* of the elliptic sine function $\text{sn}(u + 4\boldsymbol{K}, k) = \text{sn}(u, k)$, from which it follows that the fundamental period of the function $\text{sn}(\alpha u, k)$ is $T = 4\boldsymbol{K}/\alpha$ (to verify this property, consider the case $k = 0$, so we end up with the corresponding known property of the trigonometric sine function). In this case, the multiplicative coefficient of the quantity $u = \text{sn}^{-1}(s/j\omega_p, \omega_p/\omega_s)$ is $\alpha = N\boldsymbol{K}_1/\boldsymbol{K}_2 = N\pi/2\boldsymbol{K}_2$ (where we used the result $\boldsymbol{K}_1(0) = \pi/2$), and therefore, the period of the elliptic function that appears in the equation of the filter poles will be equal to $T = 4\boldsymbol{K}_1/\alpha = 2\pi/(N\pi/2\boldsymbol{K}_2) = 4\boldsymbol{K}_2/N$, but this means that if the expression

$$z_0 = j\frac{\boldsymbol{K}_2}{N\pi}\ln\frac{10^{A_p/20}+1}{10^{A_p/20}-1}$$

is a solution of the pole defining equation of the function, then the same will apply to the quantities

$$z_m = z_0 + \frac{4m\boldsymbol{K}_2}{N}, \quad z \equiv \mathrm{sn}^{-1}\left(\frac{s}{j\omega_p}, \frac{\omega_p}{\omega_s}\right)$$

Hence, we will have

$$\mathrm{sn}^{-1}\left(\frac{s}{j\omega_p}, \frac{\omega_p}{\omega_s}\right) = j\frac{\boldsymbol{K}_2}{N\pi}\ln\frac{10^{A_p/20}+1}{10^{A_p/20}-1} + 4\frac{\boldsymbol{K}_2}{N}m$$

and therefore, the poles of the transfer function are calculated as

$$p_m = \sigma_m + j\omega_m = j\omega_p \mathrm{sn}\left(j\frac{\boldsymbol{K}_2}{N\pi}\ln\frac{10^{A_p/20}+1}{10^{A_p/20}-1} + 4\frac{\boldsymbol{K}_2}{N}m, \frac{\omega_p}{\omega_s}\right) \quad (11.130)$$

(for the values $m = 0, 1, 2, \ldots, N-1$), as can be proven, by solving the foregoing equation for s. It is not difficult to see that for $m = 0$, we will get the real pole p_0, while for the remaining values of m we will get the remaining $N-1$ poles. Using the properties $\mathrm{sn}(u+2\boldsymbol{K}, k) = -\mathrm{sn}(u, k)$ and $\mathrm{sn}(u+4\boldsymbol{K}, k) = \mathrm{sn}(u, k)$ (as one can verify looking at the graph of the function), the foregoing equation is written

$$p_m = \sigma_m + j\omega_m = j\omega_p(-1)^m \mathrm{sn}\left(j\frac{\boldsymbol{K}_2}{N\pi}\ln\frac{10^{A_p/20}+1}{10^{A_p/20}-1} \pm 2\frac{\boldsymbol{K}_2}{N}m, \frac{\omega_p}{\omega_s}\right) \quad (11.131)$$

where now the parameter m takes the values $m = 0, 1, 2\ldots, (N-1)/2$.

To calculate the foregoing elliptic function, let us use the property of the sum of the elliptic sine function, which is formulated as

$$\mathrm{sn}(u+\upsilon, k) = \frac{\mathrm{sn}(u,k)\mathrm{cn}(\upsilon,k)\mathrm{dn}(\upsilon,k) \pm \mathrm{sn}(\upsilon,k)\mathrm{cn}(u,k)\mathrm{dn}(u,k)}{1 - k^2\mathrm{sn}^2(u,k)\mathrm{sn}^2(\upsilon,k)}$$

for parameter values

$$u = j\frac{\boldsymbol{K}_2}{N\pi}\ln\frac{10^{A_p/20}+1}{10^{A_p/20}-1}$$

$$\upsilon = \frac{2\boldsymbol{K}_2}{N}m$$

$$k = k_2 = \frac{\omega_p}{\omega_s}$$

Starting with the calculation of the denominator, which is simpler, and taking into account the equation

$$p_0 = j\omega_p \mathrm{sn}\left(j\frac{\boldsymbol{K}_2}{N\pi}\ln\frac{10^{A_p/20}+1}{10^{A_p/20}-1}, \frac{\omega_p}{\omega_s}\right)$$

we will have

$$\begin{aligned}1 - k^2\mathrm{sn}^2(u,k)\mathrm{sn}^2(\upsilon,k) &= 1 - k^2\mathrm{sn}^2\left(j\frac{\boldsymbol{K}_2}{N\pi}\ln\frac{10^{A_p/20}+1}{10^{A_p/20}-1}, \frac{\omega_p}{\omega_s}\right)\mathrm{sn}^2\left(\frac{2\boldsymbol{K}_2}{N}m, \frac{\omega_p}{\omega_s}\right)\\ &= 1 + \frac{\omega_p^2}{\omega_s^2}\frac{p_0^2}{\omega_p^2}\mathrm{sn}^2\left(\frac{2\boldsymbol{K}_2}{N}m, \frac{\omega_p}{\omega_s}\right) = 1 + p_0^2\left[\frac{1}{\omega_s}\mathrm{sn}\left(\frac{2\boldsymbol{K}_2}{N}m, \frac{\omega_p}{\omega_s}\right)\right] = 1 + p_0^2\Omega_m^2\end{aligned}$$

with the quantity Ω_m defined as[32]

$$\Omega_m = \frac{1}{\omega_s}\operatorname{sn}\left[\frac{2\boldsymbol{K}_2}{N}m, \frac{\omega_p}{\omega_s}\right] = \frac{1}{\sqrt{\omega_s\omega_p}} \frac{2\sqrt[4]{q(k_2)}\sum_{n=0}^{\infty}(-1)^n q^{n(n+1)}(k_2)\sin\left[(2n+1)\frac{\pi m}{N}\right]}{1+2\sum_{n=1}^{\infty}(-1)^n q^{n^2}(k_2)\cos\left(\frac{2n\pi m}{N}\right)}, \; m = 1, 2, \ldots, \frac{N-1}{2} \tag{11.132}$$

To calculate the two expressions of the numerator, let us combine the properties

$$\operatorname{sn}^2(u,k) + \operatorname{cn}^2(u,k) = 1 \qquad \text{and} \qquad k^2\operatorname{sn}^2(u,k) + \operatorname{dn}^2(u,k) = 1$$

to construct the auxiliary function

$$\operatorname{cn}(u,k)\operatorname{dn}(u,k) = \sqrt{[1-\operatorname{sn}^2(u,k)][1-k^2\operatorname{sn}^2(u,k)]}$$

Therefore,

$$\operatorname{sn}\left(j\frac{\boldsymbol{K}_2}{N\pi}\ln\frac{10^{A_p/20}+1}{10^{A_p/20}-1}, \frac{\omega_p}{\omega_s}\right)\operatorname{cn}\left(\frac{2\boldsymbol{K}_2}{N}m, \frac{\omega_p}{\omega_s}\right)\operatorname{dn}\left(\frac{2\boldsymbol{K}_2}{N}m, \frac{\omega_p}{\omega_s}\right) =$$

$$\operatorname{sn}\left(j\frac{\boldsymbol{K}_2}{N\pi}\ln\frac{10^{A_p/20}+1}{10^{A_p/20}-1}, \frac{\omega_p}{\omega_s}\right)\sqrt{\left[1-\operatorname{sn}^2\left(\frac{2\boldsymbol{K}_2}{N}m, \frac{\omega_p}{\omega_s}\right)\right]\left[1-k^2\operatorname{sn}^2\left(\frac{2\boldsymbol{K}_2}{N}m, \frac{\omega_p}{\omega_s}\right)\right]}$$

and using the defining equations of the parameters p_0 and Ω_m, we get

$$\operatorname{sn}\left(j\frac{\boldsymbol{K}_2}{N\pi}\ln\frac{10^{A_p/20}+1}{10^{A_p/20}-1}, \frac{\omega_p}{\omega_s}\right)\operatorname{cn}\left(\frac{2\boldsymbol{K}_2}{N}m, \frac{\omega_p}{\omega_s}\right)\operatorname{dn}\left(\frac{2\boldsymbol{K}_2}{N}m, \frac{\omega_p}{\omega_s}\right) = \frac{p_0}{j\omega_p}\sqrt{(1-\omega_p^2\Omega_m^2)(1-\omega_s^2\Omega_m^2)} = \frac{p_0}{j\omega_p}V_m$$

with the quantity V_m defined as

$$V_m = \sqrt{(1-\omega_p^2\Omega_m^2)(1-\omega_s^2\Omega_m^2)}, \quad m = 1, 2, \ldots, \frac{N-1}{2} \tag{11.133}$$

For the second expression of the numerator, in a similar way we will have

$$\operatorname{sn}\left(\frac{2\boldsymbol{K}_2}{N}m, \frac{\omega_p}{\omega_s}\right)\operatorname{cn}\left(j\frac{\boldsymbol{K}_2}{N\pi}\ln\frac{10^{A_p/20}+1}{10^{A_p/20}-1}, \frac{\omega_p}{\omega_s}\right)\operatorname{dn}\left(j\frac{\boldsymbol{K}_2}{N\pi}\ln\frac{10^{A_p/20}+1}{10^{A_p/20}-1}, \frac{\omega_p}{\omega_s}\right) =$$

$$\operatorname{sn}\left(\frac{2\boldsymbol{K}_2}{N}m, \frac{\omega_p}{\omega_s}\right)\sqrt{\left[1-\operatorname{sn}^2\left(j\frac{\boldsymbol{K}_2}{N\pi}\ln\frac{10^{A_p/20}+1}{10^{A_p/20}-1}, \frac{\omega_p}{\omega_s}\right)\right]\left[1-k^2\operatorname{sn}^2\left(j\frac{\boldsymbol{K}_2}{N\pi}\ln\frac{10^{A_p/20}+1}{10^{A_p/20}-1}, \frac{\omega_p}{\omega_s}\right)\right]}$$

and using the expressions p_0 and Ω_m, we will get

$$\operatorname{sn}\left(\frac{2\boldsymbol{K}_2}{N}m, \frac{\omega_p}{\omega_s}\right)\operatorname{cn}\left(j\frac{\boldsymbol{K}_2}{N\pi}\ln\frac{10^{A_p/20}+1}{10^{A_p/20}-1}, \frac{\omega_p}{\omega_s}\right)\operatorname{dn}\left(j\frac{\boldsymbol{K}_2}{N\pi}\ln\frac{10^{A_p/20}+1}{10^{A_p/20}-1}, \frac{\omega_p}{\omega_s}\right)$$
$$= \omega_s\Omega_m\sqrt{\left(1+\frac{p_0^2}{\omega_p^2}\right)\left(1+\frac{p_0^2}{\omega_s^2}\right)} = \omega_s\Omega_m W$$

[32] This relationship also arises from the general expression of Ω_m for $q = 0$.

with the parameter W (which is independent of m) defined as

$$W = \sqrt{\left(1 + \frac{p_0^2}{\omega_p^2}\right)\left(1 + \frac{p_0^2}{\omega_s^2}\right)} \tag{11.134}$$

Replacing the three foregoing expressions in the pole equation p_m, this equation takes the form

$$p_m = \sigma_m + j\omega_m = j\omega_p(-1)^m \frac{(p_0/j\omega_p)V_m \pm \omega_s \Omega_m V}{1 + p_0^2 \Omega_m^2}$$

and finally

$$p_m = \frac{(-1)^m p_0 V_m \pm j\omega_p \omega_s (-1)^m \Omega_m W}{1 + p_0^2 \Omega_m^2} = \frac{(-1)^m p_0 V_m \pm j\omega_p \omega_s \Omega_m W}{1 + p_0^2 \Omega_m^2} \tag{11.135}$$

for the values $m = 1, 2, \ldots, N/2$ which is also our final result. Therefore, the transfer function of the system has a *negative real pole* at the position $s = p_0$, which contributes to the denominator the term $s + p_0$ (the sign $+$ is due to the existence of the negative sign in the defining equation of p_0) and $(N - 1)/2$ *pairs of conjugate complex poles* $p_m = \sigma_m + j\omega_m$ and $p_m^* = \sigma_m - j\omega_m$, which are given by the foregoing relation, with each pair of poles contributing to the transfer function the term

$$\frac{1}{(s - p_m)(s - p_m^*)} = \frac{1}{s^2 - (p_m + p_m^*)s + p_m p_m^*} = \frac{1}{s^2 - 2\sigma_m s + \sigma_m^2 + \omega_m^2}$$
$$= \frac{1}{s^2 + \dfrac{2p_0 V_m}{1 + p_0^2 \Omega_m^2} s + \dfrac{(p_0 V_m)^2 + (\omega_p \omega_s \Omega_m W)^2}{(1 + p_0^2 \Omega_m^2)^2}}$$

The transfer function also has *imaginary conjugate zero values* at the positions $z_m = \pm j\omega_p/\Omega_m$ (for the values $m = 1, 2, \ldots, (N - 1)/2$, with each such pair contributing to the aforementioned function the term

$$(s - z_m)(s - z_m^*) = \left(1 - \frac{j\omega_p}{\omega_m}\right)\left(1 + \frac{j\omega_p}{\omega_m}\right) = s^2 + \frac{\omega_p^2}{\Omega_m^2}$$

Combining all of the foregoing considerations, the transfer function of the odd-order low-pass elliptic filter is formulated as

$$\mathcal{H}(s) = \frac{H_0}{s + p_0} \prod_{m=1}^{(N-1)/2} \left\{ \left[s^2 + \frac{\omega_p^2}{\Omega_m^2} \right] \Big/ \left[s^2 + \frac{2p_0 V_m}{1 + p_0^2 \Omega_m^2} s + \frac{(p_0 V_m)^2 + (\omega_p \omega_s \Omega_m W)^2}{(1 + p_0^2 \Omega_m^2)^2} \right] \right\} \tag{11.136}$$

The proportionality constant H_0 will arise from the requirement that $R_N(\omega, \omega_p, \omega_s, \varepsilon_p, \varepsilon_s) = 1$ for an odd-order filter N, from which it follows that $\mathcal{H}(0) = 1$. From the foregoing relation it immediately follows that

$$\mathcal{H}(0) = \frac{H_0}{p_0} \prod_{m=1}^{(N-1)/2} \left\{ \frac{\dfrac{\omega_p^2}{\Omega_m^2}}{\dfrac{(p_0 V_m)^2 + (\omega_p \omega_s \Omega_m W)^2}{(1 + p_0^2 \Omega_m^2)^2}} \right\} = 1$$

from which it follows that

$$H_0 = p_0 \prod_{m=1}^{(N-1)/2} \left\{ \frac{\dfrac{(p_0 V_m)^2 + (\omega_p \omega_s \Omega_m W)^2}{(1+p_0^2\Omega_m^2)^2}}{\dfrac{\omega_p^2}{\Omega_m^2}} \right\}$$

N **Even** In this case, we will have $q = 1$, and therefore, we have

$$\sin\left[\frac{\pi N}{2\boldsymbol{K}_2}\mathrm{sn}^{-1}\left(\frac{s}{j\omega_p}, \frac{\omega_p}{\omega_s}\right) + \frac{\pi}{2}\right] = \frac{j}{\varepsilon_p}$$

and, working as before, we end up with the relation

$$\mathrm{sn}^{-1}\left(\frac{s}{j\omega_p}, \frac{\omega_p}{\omega_s}\right) = j\frac{\boldsymbol{K}_2}{N\pi}\ln\frac{10^{A_p/20}+1}{10^{A_p/20}-1} - \frac{\boldsymbol{K}_2}{N}$$

Taking into account that the function involved in the equation of the poles has, as we have proven, a fundamental period equal to $4\boldsymbol{K}_2/N$, we will have by analogy

$$z_m = z_0 \pm \frac{2m-1}{N}\boldsymbol{K}_2 \tag{11.137}$$

(see the discussion on defining z_m for the case of an odd order), and consequently, the poles of the transfer function will take the form

$$p_m = \sigma_m + j\omega_m = j\omega_p(-1)^m \mathrm{sn}\left(j\frac{\boldsymbol{K}_2}{N\pi}\ln\frac{10^{A_p/20}+1}{10^{A_p/20}-1} \pm \frac{2m-1}{N}\boldsymbol{K}_2, \frac{\omega_p}{\omega_s}\right) \tag{11.138}$$

We observe, therefore, that in this case, the parameter m cannot take a zero value, and therefore, *we do not have a real pole, but pairs of conjugate complex poles* that are calculated for the values $m = 1, 2, \ldots, N/2$.

From this point on, the analysis will be exactly the same as that described for the case of odd order. Developing the function of the elliptic sine in the way we did previously, we finally end up with the expression

$$p_m = \pm(\sigma_m + j\omega_m) \qquad \text{where} \qquad \sigma_m + j\omega_m = \frac{\pm[p_0 V_m + j\omega_p\omega_s(-1)^m\Omega_m W]}{1+p_0^2\Omega_m^2} \tag{11.139}$$

with the parameters V_m and W being defined as previously, and the parameter Ω_m being defined as

$$\Omega_m = \frac{1}{\omega_s}\mathrm{sn}\left[\frac{2m-1}{N}\boldsymbol{K}_2, \frac{\omega_p}{\omega_s}\right] \tag{11.140}$$

which results from the general expression for $q = 1$. Therefore, the transfer function for even-order filters will take the form

$$\mathcal{H}(s) = H_0 \prod_{m=1}^{N/2} \left\{ \left[s^2 + \frac{\omega_p^2}{\Omega_m^2}\right] \Big/ \left[s^2 + \frac{2p_0 V_m}{1+p_0^2\Omega_m^2}s + \frac{(p_0 V_m)^2 + (\omega_p\omega_s\Omega_m W)^2}{(1+p_0^2\Omega_m^2)^2}\right] \right\} \tag{11.141}$$

with the proportionality constant H_0 arising this time from the requirement to be $G(0) = 10^{-A_p/20}$ (see the relevant discussion in the section on the Chebyshev I filters since they present the same type of variations with elliptic filters). Therefore,

$$\mathcal{H}(0) = H_0 \prod_{m=1}^{(N-1)/2} \left\{ \frac{\dfrac{\omega_p^2}{\Omega_m^2}}{\dfrac{(p_0 V_m)^2 + (\omega_p \omega_s \Omega_m W)^2}{(1 + p_0^2 \Omega_m^2)^2}} \right\} = 10^{-0.05 A_p}$$

and hence, we will have

$$H_0 = 10^{-0.05 A_p} \prod_{m=1}^{(N-1)/2} \left\{ \frac{\dfrac{(p_0 V_m)^2 + (\omega_p \omega_s \Omega_m W)^2}{(1 + p_0^2 \Omega_m^2)^2}}{\dfrac{\omega_p^2}{\Omega_m^2}} \right\}$$

After constructing the transfer function for even- and odd-order filters, we can compare the calculation equation with Eq. (10.12), which defines this function in terms of the central frequency $\omega_c \equiv \omega_0$ and the quality factor Q to prove that these quantities are defined as

$$\omega_c^m = \frac{\sqrt{(p_0 V_m)^2 + (\omega_p \omega_s \Omega_m W)^2}}{1 + p_0^2 \Omega_m^2} \quad \text{and} \quad Q_m = \frac{1}{2}\sqrt{1 + \left(\frac{\omega_p \omega_s \Omega_m W}{p_0 V_m}\right)^2}$$

The difficulty of calculating values of elliptic functions makes difficult the analytical presentation of an exercise involving the calculation of poles and zero values of elliptic filters, as we did in the previous approaches. Although later we will present such a detailed calculation process for educational purposes only, in practice, the design of filters of this type is based on the use of specialized software and libraries of elliptic functions, such as, for example, those offered by the applications MATLAB and Mathematica.

Selectivity and Spectral Efficiency Factor

From the defining equation of a filter's *selectivity* it follows that

$$F_s = -\left.\frac{d\{|\mathcal{H}(j\omega)|\}}{d\omega}\right|_{\omega=\omega_c} = -\frac{d}{d\omega}\left\{\frac{1}{\sqrt{1 + \varepsilon_p^2 R_N^2(\omega, \omega_s, \omega_p, \varepsilon_s, \varepsilon_p)}}\right\}_{\omega=\omega_c}$$

$$= -\frac{d}{d\omega}\left\{\frac{1}{\sqrt{1 + \varepsilon_p^2 \mathrm{sn}\left[\kappa\, \mathrm{sn}^{-1}\left(\frac{\omega}{\omega_p}, \frac{\omega_p}{\omega_s}\right) + q\boldsymbol{K}_1, \varepsilon_p \varepsilon_s\right]}}\right\}_{\omega=\omega_c}$$

and finally

$$F_s = \frac{j\varepsilon_p k \mathrm{cn}\left[\kappa\, \mathrm{sn}^{-1}\left(\frac{\omega_c}{\omega_p}, \frac{\omega_p}{\omega_s}\right) + q\boldsymbol{K}_1, \varepsilon_p \varepsilon_s\right]\sqrt{1 - \varepsilon_p^2 \varepsilon_s^2 \mathrm{sn}^2\left[\kappa\, \mathrm{sn}^{-1}\left(\frac{\omega_c}{\omega_p}, \frac{\omega_p}{\omega_s}\right) + q\boldsymbol{K}_1, \varepsilon_p \varepsilon_s\right]}}{2\sqrt{2}\sqrt{1 - \left(\frac{\omega_p}{\omega_s}\right)^2}\cosh^2\left\{\Im\left[\sin^{-1}\left(\frac{\omega_c}{\omega_p}\right)\right]\right\}\sqrt{\omega_c^2 - \omega_p^2}} \tag{11.142}$$

The proof of the foregoing relationship is based on the properties of differentiation of the forward and inverse Jacobi elliptic functions and is left as an exercise to the reader.

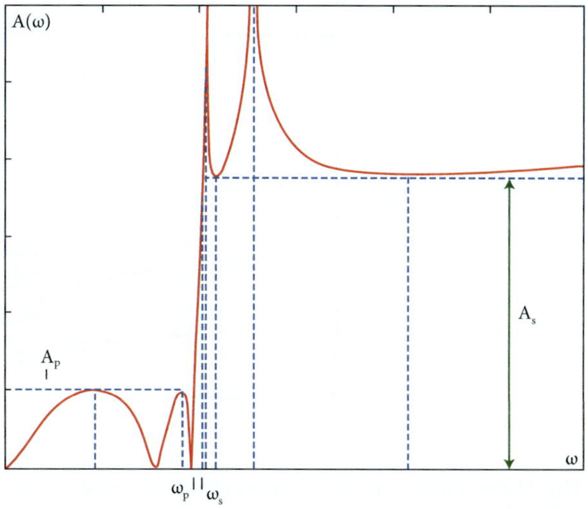

Fig. 11.59 Loss characteristic for fifth-order elliptic low-pass filter

On the other hand, for the calculation of the *spectral efficiency factor*, let us consider an attenuation plane A, expressed in decibels, that will be given from the relation

$$A = 10\log\left\{1 + \varepsilon_p^2 \mathrm{sn}^2\left[\kappa\,\mathrm{sn}^{-1}\left(\frac{\omega_s}{\omega_p}, \frac{\omega_p}{\omega_s}\right) + q\boldsymbol{K}_1, \varepsilon_p\varepsilon_s\right]\right\}$$

Using the expressions $\varepsilon_p = \sqrt{10^{A_p/10} - 1}$ and $k = N\boldsymbol{K}_1/\boldsymbol{K}_2$, the solution to the foregoing equation with respect to the frequency ω will lead us to the equation

$$\omega = \omega_p \mathrm{sn}\left\{\frac{\left[\mathrm{sn}^{-1}\left(\frac{\sqrt{10^{A/10} - 1}}{\sqrt{10^{A_p/10} - 1}}, \varepsilon_p\varepsilon_s\right) - q\boldsymbol{K}_1\right]\boldsymbol{K}_2}{N\boldsymbol{K}_1}, \frac{\omega_p}{\omega_s}\right\} \tag{11.143}$$

Assuming, as in previous approaches, that the preceding equation gives us the bandwidth BW, the spectral efficiency factor will be calculated from its defining equation $S_\alpha^\beta = \mathrm{BW}_\beta/\mathrm{BW}_\alpha$, with the bandwidth BW_α corresponding to the attenuation level $A = \alpha$ and the bandwidth BW_β corresponding to the attenuation level $A = \beta$.

Loss Characteristic

The loss characteristic of a *low-pass elliptical filter* has the form

$$A(\omega) = 10\log\left[1 + \varepsilon_p^2 R_N^2(s/j, \omega_s, \omega_p, \varepsilon_s, \varepsilon_p)\right] = 10\log\left\{1 + \varepsilon_p^2 \mathrm{sn}^2\left[\kappa\,\mathrm{sn}^{-1}\left(\frac{\omega}{\omega_p}, \frac{\omega_p}{\omega_s}\right) + q\boldsymbol{K}_1, \varepsilon_p\varepsilon_s\right]\right\} \tag{11.144}$$

A typical example of a loss characteristic for a fifth-order elliptic low-pass filter is shown in Fig. 11.59.

Phase Characteristics and Time-Domain Response

The *phase response* of a prototype low-pass elliptic filter for a cutoff frequency $\omega_c = 1$, parameter values $A_p = 1$ dB, $A_s = 30$ dB, and order values $N = 1$ to 10 is shown in Fig. 11.60. The defining equation of this quantity, as well as the

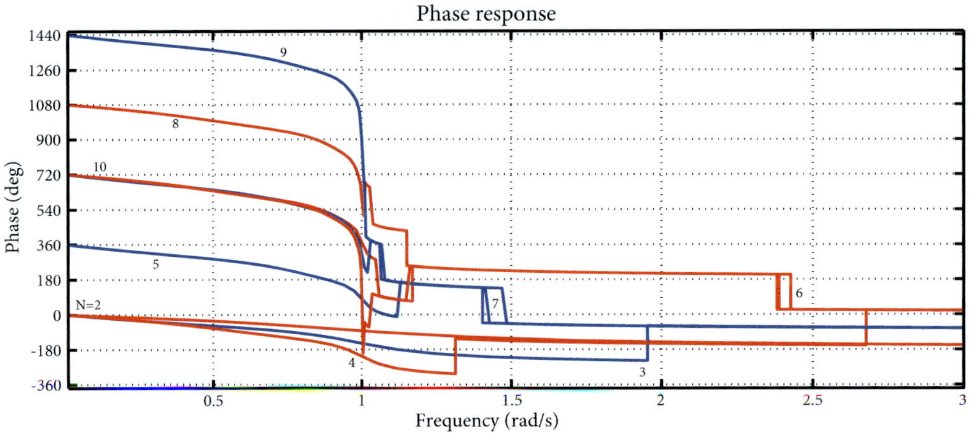

Fig. 11.60 Phase response of prototype low-pass elliptic filter for cutoff frequency $\omega_c = 1$, parameter values $A_p = 1$ dB, $A_s = 30$ dB, and order values $N = 1$ up to $N = 10$

equations for calculating the phase and group delays, results from the general equations for the parameter values $\alpha_0 = p_0$, $\alpha_1 = 1$, $\alpha_2 = 0$, $\beta_0 = 1$, $\beta_1 = \beta_2 = 0$, $\alpha_{2k} = \beta_{2k} = 1$, $\beta_{1k} = 0$, and

$$\beta_{0k} = \frac{\omega_p^2}{\Omega_k^2}, \quad \alpha_{0k} = \frac{(p_0 V_k)^2 + (\omega_p \omega_s \Omega_k W)^2}{(1 + p_0^2 \Omega_k^2)}, \quad \alpha_{1k} = \frac{2 p_0 V_k}{1 + p_0^2 \Omega_k^2}$$

The phase response of elliptic low-pass filters will, therefore, have the form

$$\measuredangle \mathcal{H}(j\omega) = -\sum_{k=1}^{N/2} \frac{2 p_0 V_k (1 + p_0^2 \Omega_k^2)\omega}{(p_0 V_k)^2 + (\omega_p \omega_s \Omega_k W)^2 - (1 + p_0^2 \Omega_k^2)\omega^2} \quad \text{(N even)} \tag{11.145}$$

$$\measuredangle \mathcal{H}(j\omega) = -\tan^{-1}\left(\frac{\omega}{p_0}\right) - \sum_{k=1}^{(N-1)/2} \frac{2 p_0 V_k (1 + p_0^2 \Omega_k^2)\omega}{(p_0 V_k)^2 + (\omega_p \omega_s \Omega_k W)^2 - (1 + p_0^2 \Omega_k^2)\omega^2} \quad \text{(N odd)} \tag{11.146}$$

respectively, while the corresponding expressions for the group delay are formulated as

$$\tau(\omega) = \sum_{k=1}^{N/2} \left\{ \frac{2 p_0 V_k}{1 + p_0^2 \Omega_k^2} \left[\frac{(p_0 V_k)^2 + (\omega_p \omega_s \Omega_k W)^2}{(1 + p_0^2 \Omega_k^2)^2} + \omega^2 \right] \right\} \Big/ \left\{ \left[\frac{(p_0 V_k)^2 + (\omega_p \omega_s \Omega_k W)^2}{(1 + p_0^2 \Omega_k^2)^2} - \omega^2 \right]^2 \right\}$$

$$\tau(\omega) = \sum_{k=1}^{(N-1)/2} \left\{ \frac{2 p_0 V_k}{1 + p_0^2 \Omega_k^2} \left[\frac{(p_0 V_k)^2 + (\omega_p \omega_s \Omega_k W)^2}{(1 + p_0^2 \Omega_k^2)^2} + \omega^2 \right] \right\} \Big/ \left\{ \left[\frac{(p_0 V_k)^2 + (\omega_p \omega_s \Omega_k W)^2}{(1 + p_0^2 \Omega_k^2)^2} - \omega^2 \right]^2 \right\} + \frac{p_0}{p_0^2 + \omega^2}$$

On the other hand, the *impulse* and *step responses* of the same filter are presented in the top and bottom images of Fig. 11.61. These characteristics are calculated by applying the equations we constructed in the corresponding chapters on continuous-time systems and those we used in the presentation of the corresponding characteristics in the previous approaches of ideal low-pass filters.

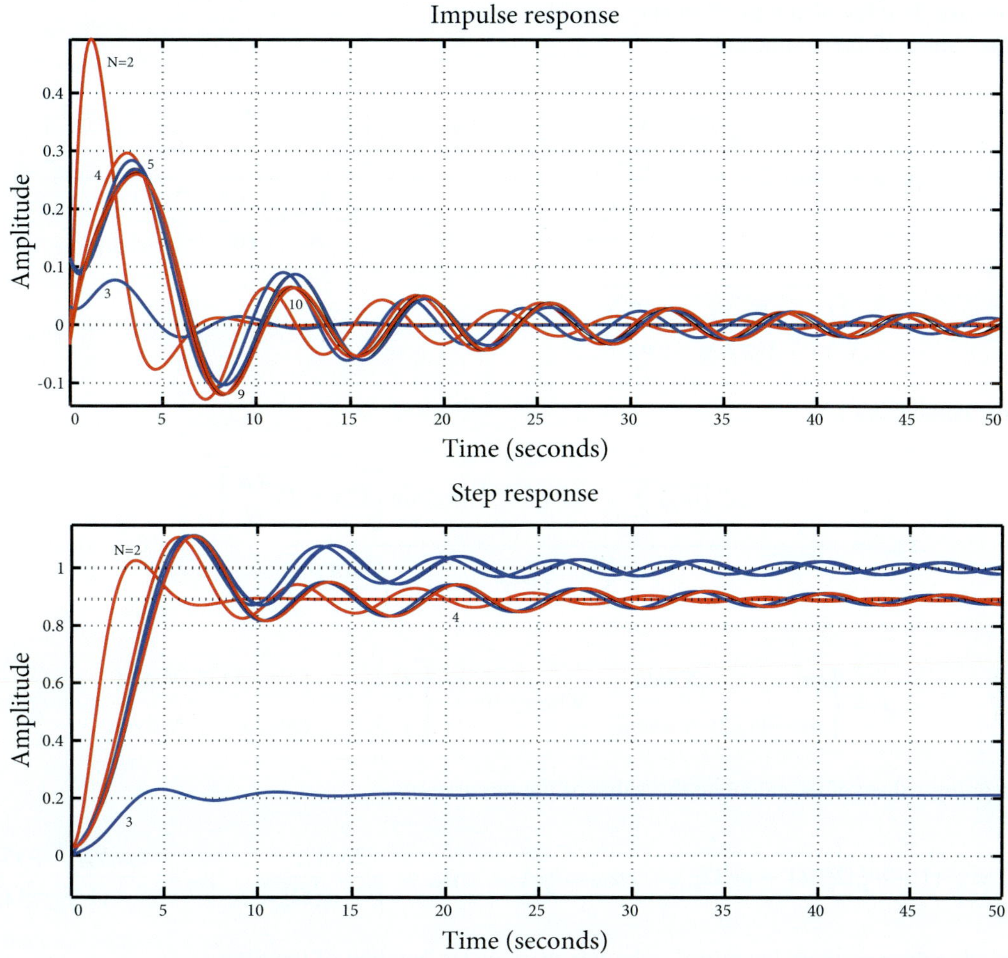

Fig. 11.61 Impulse and step responses of prototype low-pass elliptic filter for cutoff frequency $\omega_c = 1$, parameter values $A_p = 1$ dB, $A_s = 30$ dB, and order values $N = 1 - 10$

Design of Elliptic Filters

After the analytical description of all the foregoing processes we can now summarize the design process of low-pass elliptic filters, which includes the following steps:

1. Determine the values of the *maximum attenuation in the passband* A_p, the *minimum attenuation in the stopband* A_s, and the *pass and stop frequencies* ω_p and ω_s.
2. Calculate the *elliptic measure*

$$k_1 = \varepsilon_p \varepsilon_s = \sqrt{\frac{10^{A_p/10} - 1}{10^{A_s/10} - 1}} = \frac{1}{\sqrt{D}} \quad \text{where} \quad D = \frac{10^{A_s/10} - 1}{10^{A_p/10} - 1}$$

3. Calculate the *elliptic measure* $k_2 = \omega_p / \omega_s$ which, as is well known, expresses the *selectivity parameter* of the filter.
4. Calculate the *complementary elliptic measures* $k_1' = \sqrt{1 - k_1^2}$ and $k_2' = \sqrt{1 - k_2'^2}$ and the *elliptic integrals* $\boldsymbol{K}_1$, $\boldsymbol{K}_1'$, $\boldsymbol{K}_2$, and $\boldsymbol{K}_2'$ using the numerical geometric mean.
5. Calculate the *minimum filter order* from the relationship

$$N = \frac{\log(16D)}{\log(1/q(k_2))} \quad \text{or, alternatively, from the relationship} \quad N = \frac{\boldsymbol{K}_1' \, \boldsymbol{K}_2}{\boldsymbol{K}_1 \, \boldsymbol{K}_2'}$$

and round the result to the next highest integer.

6. Calculate the values of the parameters

$$p_0 = -\sqrt{\omega_s\omega_p}\frac{2\sqrt[4]{q(k_2)}\sum_{n=0}^{\infty}(-1)^n q^{n(n+1)}(k_2)\sinh\left[(2n+1)\frac{\boldsymbol{K}_2}{N}\ln\frac{10^{A_p/20}+1}{10^{A_p/20}-1}\right]}{1+2\sum_{n=1}^{\infty}(-1)^n q^{n^2}(k_2)\cosh\left[2n\frac{\boldsymbol{K}_2}{N}\ln\frac{10^{A_p/20}+1}{10^{A_p/20}-1}\right]}$$

$$W = \sqrt{\left(1+\frac{p_0^2}{\omega_p^2}\right)\left(1+\frac{p_0^2}{\omega_s^2}\right)}$$

7. Calculate the parameters

$$\Omega_m = \frac{1}{\sqrt{\omega_s\omega_p}}\frac{2\sqrt[4]{q(k_2)}\sum_{n=0}^{\infty}(-1)^n q^{n(n+1)}(k_2)\sin\left[(2n+1)\frac{\pi\mu}{N}\right]}{1+2\sum_{n=1}^{\infty}(-1)^n q^{n^2}(k_2)\cos\left(\frac{2n\pi\mu}{N}\right)} \quad \text{for the values}$$

$$\mu = \begin{cases} m & N \text{ odd} \\ m-(1/2) & N \text{ even} \end{cases} \quad \text{where} \quad m = \begin{cases} 1,2,\ldots,(N-1)/2 & N \text{ odd} \\ 1,2,\ldots,(N/2) & N \text{ even} \end{cases}$$

8. For the foregoing values of m, calculate the value of the parameters

$$V_m = \sqrt{(1-\omega_p^2\Omega_m^2)(1-\omega_s^2\Omega_m^2)}, \quad \alpha_m = \frac{\omega_p^2}{\Omega_m^2}, \quad \beta_m = \frac{2\sigma_0 V_m}{1+\sigma_0^2\Omega_m^2}, \quad \gamma_m = \frac{(p_0 V_m)^2+(\Omega_m W)^2}{(1+p_0^2\Omega_m^2)^2}$$

9. After calculating the preceding quantities, calculate the transfer function of the filter as

$$\mathcal{H}(s) = \frac{H_0}{D(s)}\prod_{m=1}^{L}\frac{s^2+\alpha_m}{s^2+\beta_m s+\gamma_m}$$

$$L = \begin{cases} (N-1)/2 & N \text{ odd} \\ N/2 & N \text{ even} \end{cases}, \quad D(s) = \begin{cases} s+p_0 & N \text{ odd} \\ 1 & N \text{ even} \end{cases}, \quad H_0 = \begin{cases} p_0\prod_{m=1}^{L}\frac{\gamma_m}{\alpha_m} & N \text{ odd} \\ 10^{-A_p/20}\prod_{m=1}^{L}\frac{\gamma_m}{\alpha_m} & N \text{ even} \end{cases}$$

10. Formulate the transfer function in the amplitude–phase representation and creates diagrams for the amplitude response and the phase response.

The foregoing elliptic filter design process is demonstrated in the following exercise. Note that, although in practice these filters are characterized by an increased degree of complexity and are always carried out with the help of specialized software, for educational purposes only, the exercise will be solved analytically so that the reader can understand this design process in depth.

Problem 11.9.9 Design an elliptic filter with selectivity value $k_2 = \omega_p/\omega_s = 0.95$ and attenuation levels $A_p = 0.3$ dB and $A_s = 60$ dB (Antoniou, Exercise for solution 10.18).

Solution Initially, we need to calculate the minimum filter order from the relation $N = \log(16D)/\log[1/q(k_2)]$. Using the values of the attenuation levels, it turns out that

$$D = \frac{10^{A_s/10} - 1}{10^{A_p/10} - 1} = \frac{10^{60/10} - 1}{10^{0.3/10} - 1} = \frac{999999}{0.071519} = 13982284.427914$$

On the other hand, the value of the function $q(k_2) = q(0.95)$ can be calculated in two ways, both of which will be presented for educational purposes. The first way is based on the defining equation of this function, which has the form $q(k_2) = \exp(-\pi \boldsymbol{K}'_2/\boldsymbol{K}_2)$. To calculate the elliptic integrals that appear in the foregoing relation, we can use the method of the *numerical geometric mean*, which is computationally as simple in its implementation as it is fast in terms of convergence time. We end up with the results $\boldsymbol{K}_2 = \boldsymbol{K}(k_2) = \boldsymbol{K}(0.95) = 2.590011$ and $\boldsymbol{K}'_2 = \boldsymbol{K}(k'_2) = \boldsymbol{K}(\sqrt{1-k_2^2}) = \boldsymbol{K}(0.312249) = 1.611337$, and therefore, we will have

$$q(k_2) = q(0.95) = \exp\left(-\pi\frac{\boldsymbol{K}'_2}{\boldsymbol{K}_2}\right) = \exp\left(-3.14159 \times \frac{1.611337}{2.590011}\right) = \exp(-1.954493) = 0.141636$$

The other method is to calculate the auxiliary quantity

$$q_0 = \frac{1}{2}\left(\frac{1-\sqrt{k'_2}}{1+\sqrt{k'_2}}\right) = \frac{1}{2}\left(\frac{1-\sqrt{0.312249}}{1+\sqrt{0.312249}}\right) = \frac{1}{2}\left(\frac{0.441207}{1.558972}\right) = 0.141522$$

and then calculate the required function from the approximate expression

$$q(k_2) = q(0.95) = q_0 + 2q_0^5 + 15q_0^9 = 0.141522 + 2(0.141522)^5 + 15(0.141522)^9 = 0.141636$$

Although both methods lead to the same result, it is obviously preferable to use the second method, since it *does not include calculations of elliptic integrals*. Having calculated the two foregoing functions, the minimum order of the filter is calculated as

$$N = \frac{\log(16D)}{\log[1/q(k_2)]} = \frac{\log(16 \times 13982284.427914)}{\log(1/0.141636)} = \frac{\log(223716550.846628)}{7.060351} = \frac{8.349698}{0.848826} = 9.836756$$

This value obviously needs to be rounded up to the nearest integer, and therefore the minimum order of the filter is $N = 10$.

Observing that the filter specifications do not include the frequencies ω_p and ω_s, but only the selectivity factor $k_2 = 0.75$, we can make the assumption $\omega_p = \sqrt{k_2} = \sqrt{0.75} = 0.974679$ and $\omega_s = 1/\sqrt{k_2} = 1/\sqrt{0.75} = 1.025978$—this assumption is often used in practice and allows the construction of normalized filters with cutoff frequency $\omega_c = \sqrt{\omega_p\omega_s} = 1$. Calculating the auxiliary quantity

$$j\frac{\boldsymbol{K}_2}{N\pi}\ln\frac{10^{A_p/20}+1}{10^{A_p/20}-1} = j\frac{2.590011}{10 \times 3.14159}\ln\frac{10^{0.3/20}+1}{10^{0.3/20}-1} = j\,0.0082442\ln\frac{10^{0.015}+1}{10^{0.015}-1} = 0.334365j$$

the real parameter p_0 is calculated as

$$p_0 = j\omega_p \mathrm{sn}\left[j\frac{\boldsymbol{K}_2}{N\pi}\ln\frac{10^{A_p/20}+1}{10^{A_p/20}-1}, \frac{\omega_p}{\omega_s}\right] = j\,0.974679\,\mathrm{sn}(j\,0.334625, 0.75)$$

At this point, we need to calculate the *elliptic sine function*, a process that is generally not easy. To do this, we can implement a mathematical function in some programming language that calculates this function from its infinite series expansion using the functions θ. In this case, for simplicity, we will use the function [SN,CN,DN]=ellipj(u,k) in MATLAB, which accepts the values of two parameters, u and k, and returns the values of the elliptic functions sn, cn and du. However, since this function only accepts real arguments, while we want to calculate the elliptic sine of an imaginary quantity, we will use the identity

$$\mathrm{sn}(ju, m) = j\,\mathrm{sc}(u, m_1) = j\frac{\mathrm{sn}(u, m_1)}{\mathrm{cn}(u, m_1)}$$

where $u = 0.334625$, $m = k_2^2 = (0.75)^2 = 0.9025$, and $m_1 = (k_2')^2 = 1 - k_2^2 = 1 - m_1 = 1 - 0.9025 = 0.0975$. Therefore, $\mathrm{sn}(0.334625, 0.0975) = 0.327852$ and $\mathrm{cn}(0.334625, 0.0975) = 0.944728$ and consequently,

$$\mathrm{sn}(j\,0.334625, 0.0975) = j\frac{0.327852}{0.944728} = j\,0.347033$$

Using this result, we find that

$$p_0 = j\omega_p(j\,0.347033) = j\,(0.974679)(j\,0.347033) = -0.338245$$

and then this value of p_0 is used for the calculation of the value of the auxiliary parameter W as

$$W = \sqrt{\left(1 + \frac{p_0^2}{\omega_p^2}\right)\left(1 + \frac{p_0^2}{\omega_s^2}\right)} = \sqrt{\left(1 + \frac{(-0.338245)^2}{(0.974679)^2}\right)\left(1 + \frac{(-0.338245)^2}{(1.025978)^2}\right)} = 1.114634$$

Having all the necessary parameters at our disposal, let us calculate the transfer function of the filter starting with the calculation of its zero values. Initially, we need to calculate the values of the quantities

$$\Omega_m = \frac{1}{\omega_s}\mathrm{sn}\left[\frac{2m-1}{N}\boldsymbol{K}_2, \frac{\omega_p}{\omega_s}\right] = 0.974679\mathrm{sn}[0.259001(2m-1), 0.75], \quad m = 1, 2, 3, 4, 5$$

Using again the function [SN,CN,DN]=ellipj(u,k), the foregoing quantities are calculated as

$$\Omega_1 = 0.247614, \quad \Omega_2 = 0.648846, \quad \Omega_3 = 0.868917, \quad \Omega_4 = 0.959913, \quad \Omega_5 = 0.970943$$

and therefore the zero values of the transfer function are

$$z_{1,2} = \pm\frac{j}{\Omega_1} = \pm\frac{j}{0.247614} = \pm 4.038539j \quad z_{3,4} = \pm\frac{j}{\Omega_2} = \pm\frac{j}{0.648846} = \pm 1.545961j$$

$$z_{5,6} = \pm\frac{j}{\Omega_3} = \pm\frac{j}{0.868917} = \pm 1.150857j \quad z_{7,8} = \pm\frac{j}{\Omega_4} = \pm\frac{j}{0.959913} = \pm 1.041760j$$

$$z_{9,10} \pm \frac{j}{\Omega_5} = \pm\frac{j}{0.970943} = \pm 1.029926j$$

On the other hand, to calculate the poles of the function, we will need the values

$$V_m = \sqrt{(1 - \omega_p^2\Omega_m^2)(1 - \omega_s^2\Omega_m^2)}, \quad m = 1, 2, 3, 4, 5$$

which, as can be proven after simple operations, are

$$V_1 = 0.938601, \quad V_2 = 0.578042, \quad V_3 = 0.240895, \quad V_4 = 0.061221, \quad V_5 = 0.028267$$

By substituting these values into the pole defining equations, we finally arrive at the results

$$p_{1,2} = -0.315265 \pm 0.275998j \quad p_{3,4} = -0.186534 \pm 0.689991j \quad p_{5,6} = -0.075002 \pm 0.891514j$$

$$p_{7,8} = -0.018732 \pm 0.967913j, \qquad p_{9,10} = -0.008630 \pm 0.976882j$$

The normalization constant H_0 is calculated from its defining equation for even-order elliptic filters and is found to be equal to $H_0 = 0.001197$.

To construct the transfer function of the filter, we need to determine the contributions of the poles and zeros to its mathematical formulation, which, as can be proven, are as follows:

- The zero values $z_{1,2}$ contribute the term $(s^2 + 16.309797)$.
- The zero values $z_{3,4}$ contribute the term $(s^2 + 2.389995)$.
- The zero values $z_{5,6}$ contribute the term $(s^2 + 1.324471)$.
- The zero values $z_{7,8}$ contribute the term $(s^2 + 1.085263)$.
- The zero values $z_{9,10}$ contribute the term $(s^2 + 1.060747)$.
- The poles $p_{1,2}$ contribute the term $s^2 + 0.630539s + 0.175566$.
- The poles $p_{3,4}$ contribute the term $s^2 + 0.373068s + 0.510881$.
- The poles $p_{5,6}$ contribute the term $s^2 + 0.150004s + 0.800422$.
- The poles $p_{7,8}$ contribute the term $s^2 + 0.037464s + 0.937205$.
- The poles $p_{9,10}$ contribute the term $s^2 + 0.017260s + 0.954372$.

Therefore, the numerator polynomial of the transfer function will result from the product of the five contributions of its zero values, while the denominator polynomial will result from the product of the five contributions of its poles. Both polynomials will be of degree $N = 10$, with the numerator polynomial containing only even powers of s and the denominator polynomial containing all powers between 0 and 10. To complete the construction of the transfer function, we must multiply the numerator polynomial by the normalization constant H_0. The result we arrive at (where, for brevity's sake, we go to only three decimal points) has the form

$$\mathcal{H}(s) = \frac{0.001197(s^{10} + 22.170s^8 + 1.078s^6 + 2.144s^4 + 1.841s^2 + 59.433)}{s^{10} + 1.208s^9 + 3.828s^8 + 3.666s^7 + 5.456s^6 + 4.016s^5 + 3.509s^4 + 1.851s^3 + 0.944s^2 + 0.293s + 0.0642}$$

The amplitude and phase responses of this filter are presented in Fig. 11.62. Note that the results we obtained are approximate, since we used approximate expressions, and it is certain that we will achieve greater accuracy if we construct the filter using specialized applications of calculation such as, for example, MATLAB. The implementation of this filter in MATLAB leads us to the results

$$z_{1,2} = \pm 4.076817j \quad p_{1,2} = -0.327805 \pm 0.285514j$$

$$z_{3,4} = \pm 1.583155j \quad p_{3,4} = -0.195119 \pm 0.705739j$$

$$z_{5,6} = \pm 1.195265j \quad p_{5,6} = -0.085341 \pm 0.902603j$$

$$z_{7,8} = \pm 1.084517j \quad p_{7,8} = -0.031622 \pm 0.977975j$$

$$z_{9,10} = \pm 1.051635j \, p_{9,10} = -0.007805 \pm 1.002470j$$

while the value of the gain is calculated as $H_0 = 9.9999 \times 10^{-4}$. Therefore, the accuracy we achieved is quite satisfactory. In practice, elliptic filters are constructed using specialized applications due to the complexity of their design and the elliptic functions that describe them.

11.9.9 Bessel–Thomson Approach

The *Bessel–Thomson* approach was devised by Kiyasu in 1943 and later by Thomson in 1952 and allows the construction of filters with a *linear phase*[33] and *maximum flat behavior of group delay at the position* $\omega = 0$, which ensures that all frequencies of the input signal passing through the filter undergo the same delay. *Bessel–Thomson filters* have only poles characterized by a very small value of the quality factor Q characteristic, which in combination with the aforementioned property leads to the *minimization of overshoot* that generally characterizes the transient step response of filters (see Sect. 4.8.1 on page 358, as well as Fig. 4.12 on page 365). Applying the same analysis as the one used to prove the corresponding property for the amplitude response of Butterworth low-pass filters, we can prove that all derivatives of the function $\tau(\omega)$ at the position $\omega = 0$ have *zero value*. In contrast to previous approaches that rely on the construction of a

[33] Recall from basic theory that an ideal filter is characterized by a linear phase response and, consequently, by constant group delay τ_d, so that its response to the input signal $x(t)$ is the signal $y(t) = x(x - \tau_d)$, which is a time-delayed version of the input signal without changing the characteristics of the waveform.

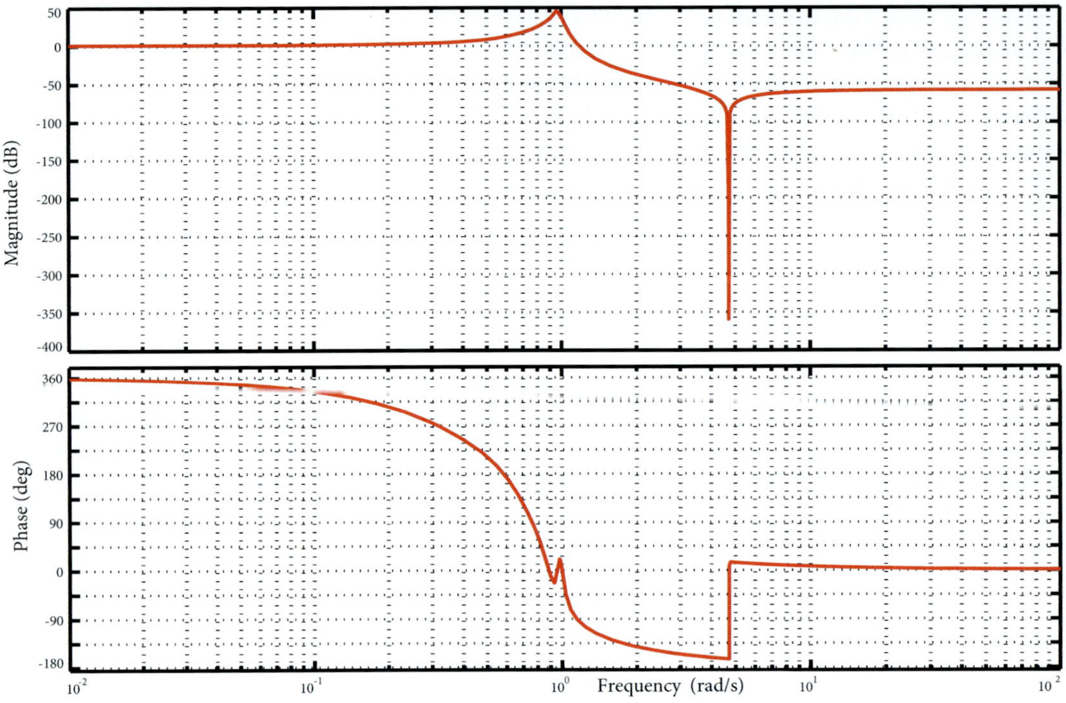

Fig. 11.62 Amplitude and phase responses of elliptic filter from Exercise 11.9.9

bandwidth response with the desired characteristics in each case, Bessel type filters are designed in such a way as to offer a constant time delay for a specific frequency range; in this sense, a Bessel filter can be considered a circuit that causes a *time delay*, although, as we will see, in reality they behave as *low-pass filters whose passband is chosen according to the bandwidth of the input signal*.

The *loss characteristic* and the *group delay* of a Bessel filter for order values $N = 1$ to 5 are presented in Fig. 11.63. Examination of these types of filters reveals a *monotonic reduction of the filter gain* with relatively small attenuation in the cutoff zone, as well as the existence of a transition zone with a *very wide range*; this is the price we pay for the constant group delay. From Fig. 11.63 we find that the limit of the passband depends on the order of the filter, while it is interesting to note that, in contrast to the previous approaches of ideal filters, there is no mathematical expression for calculating the minimum order of a filter so as to meet the existing specifications in each case, at the level of attenuation and the behavior of time delay. On the other hand, the change in time delay in relation to the frequency, which appears normalized so as to be $\tau(0) = 1$, reveals a constant group delay that extends to a frequency range whose width is proportional to the order N of the filter.

It is proven that *the product of the bandwidth times the group delay is constant*, so that we can achieve a large group delay over a small bandwidth and vice versa. In reality, the larger the value of this product, the larger the order of the filter we need to use.

The Transfer Function of Bessel–Thomson Filters

The transfer function of the Bessel–Thomson filters is constructed so that *at the limit of infinite filter order it approaches the transfer function of the ideal filter*, which has the form $\mathcal{H}(j\omega) = Ge^{-j\tau_d\omega}$, where $|\mathcal{H}(j\omega)| = G$ is the constant gain of the filter, which can be equal to the unit, and $\measuredangle\mathcal{H}(j\omega) = -\tau_d\omega$ is the linear phase response, which leads to constant group delay $\tau(\omega) = -\measuredangle\mathcal{H}'(j\omega) = \tau_d$. To construct this function, let us assume that it has the general form

$$\mathcal{H}(s) = 1\bigg/\left\{\sum_{k=0}^{N}\alpha_k s^k\right\} = \frac{1}{1 + \alpha_1 s + \alpha_2 s^2 + \cdots + \alpha_N s^N} \tag{11.147}$$

(for simplicity, we assume that $\beta_0 = 1$, while of course it will hold that $\alpha_N = 1$), which contains *only poles* and is characterized by the value $\alpha_0 = 1$. By making the substitution $s = j\omega$, we construct the frequency response function, which, according to our discussion in Sect. 11.5 (page 943), is formulated as

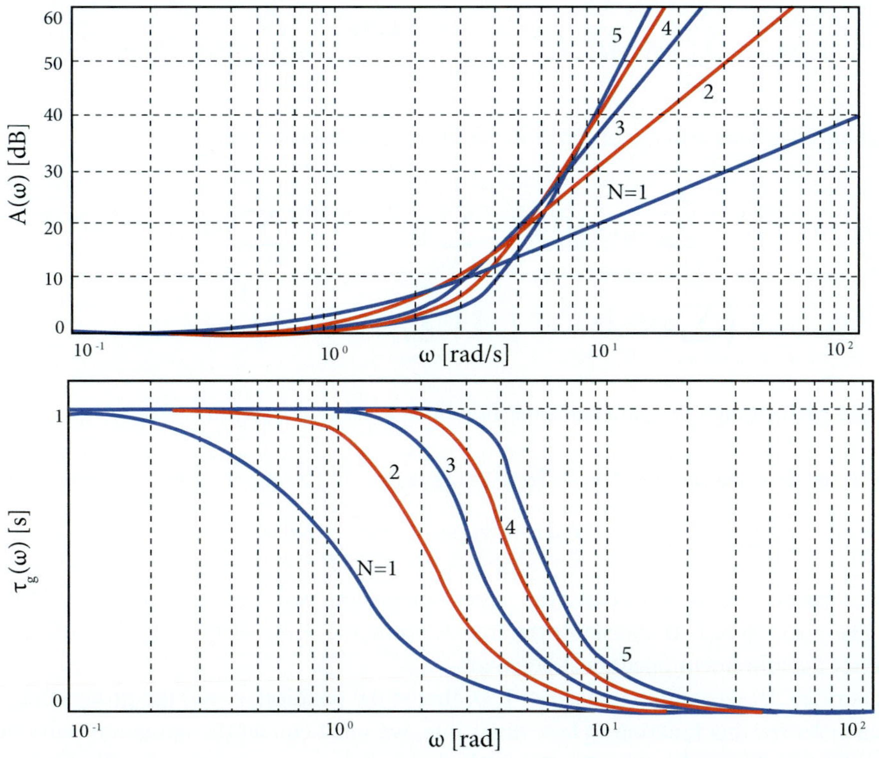

Fig. 11.63 Change in attenuation and group delay in relation to frequency for a Bessel–Thomson filter

$$\mathcal{H}(j\omega) = \frac{1}{\sum\limits_{k\ \text{even}}^{N} \alpha_k (-1)^{\frac{k}{2}} \omega^k + j \sum\limits_{k\ \text{odd}}^{N} \alpha_k (-1)^{\frac{k-1}{2}} \omega^k} \tag{11.148}$$

and leads to the phase response

$$\measuredangle \mathcal{H}(j\omega) = -\tan^{-1} \left\{ \left[\sum_{k\ \text{odd}}^{N} \alpha_k (-1)^{\frac{k-1}{2}} \omega^k \right] \Big/ \left[\sum_{k\ \text{even}}^{N} \alpha_k (-1)^{\frac{k}{2}} \omega^k \right] \right\} = -\tan^{-1} \frac{\xi(\omega)}{\zeta(\omega)} \tag{11.149}$$

and to the group delay

$$\tau(\omega) = -\frac{d}{d\omega} \left\{ \measuredangle \mathcal{H}(j\omega) \right\} = \frac{d}{d\omega} \left\{ \tan^{-1} \frac{\xi(\omega)}{\zeta(\omega)} \right\} = \frac{1}{\xi^2(\omega) + \zeta^2(\omega)} \left[\frac{d\xi(\omega)}{d\omega} - \xi(\omega) \frac{d\zeta(\omega)}{d\omega} \right] \tag{11.150}$$

where

$$\xi(\omega) = \sum_{k\ \text{odd}}^{N} \alpha_k (-1)^{\frac{k-1}{2}} \omega^k \qquad \frac{d\xi(\omega)}{d\omega} = \sum_{k\ \text{odd}}^{N} k \alpha_k (-1)^{\frac{k-1}{2}} \omega^{k-1}$$

$$\zeta(\omega)=\sum_{k\text{ even}}^{N}\alpha_k(-1)^{\frac{k}{2}}\omega^k \qquad \frac{d\zeta(\omega)}{d\omega}=\sum_{k\text{ even}}^{N}k\alpha_k(-1)^{\frac{k}{2}}\omega^{k-1}$$

Substituting the foregoing functions into the group delay defining equation, we get

$$\tau(\omega)=\frac{\sum_{k\text{ odd}}^{N}k\alpha_k(-1)^{\frac{k-1}{2}}\omega^{k-1}\sum_{m\text{ even}}^{N}\alpha_m(-1)^{\frac{m}{2}}\omega^m-\sum_{k\text{ odd}}^{N}\alpha_k(-1)^{\frac{k-1}{2}}\omega^k\sum_{m\text{ even}}^{N}m\alpha_m(-1)^{\frac{m}{2}}\omega^{m-1}}{\left(\sum_{k\text{ odd}}^{N}\alpha_k(-1)^{\frac{k-1}{2}}\omega^k\right)^2+\left(\sum_{k\text{ even}}^{N}\alpha_k(-1)^{\frac{k}{2}}\omega^k\right)^2}=\frac{\lambda(\omega)}{\mu(\omega)} \tag{11.151}$$

Expanding the sums and performing the operations, the functions $\lambda(\omega)$ and $\mu(\omega)$ are written

$$\lambda(\omega)=\alpha_1+(\alpha_1\alpha_2-3\alpha_3)\omega^2+(5\alpha_5-3\alpha_3\alpha_4+\alpha_2\alpha_3)\omega^4+(5\alpha_1\alpha_6+\alpha_3\alpha_4-3\alpha_2\alpha_5-7\alpha_7)\omega^6+\dots$$

$$\mu(\omega)=1+(\alpha_1^2-2\alpha_2)\omega^2+(\alpha_2^2-2\alpha_1\alpha_3+2\alpha_4)\omega^4+(\alpha_3^2+2\alpha_1\alpha_5-2\alpha_2\alpha_4-2\alpha_6)\omega^6+\dots$$

We observe that in the limit $\omega\to 0$, it holds that $\tau(\omega)=\alpha_1=$ constant, and therefore, for $\omega\to 0$ the behavior of the ideal filter is approached. The goal, then, is to construct the transfer function $\mathcal{H}(s)$ of the filter in such a way that under certain conditions, it approaches the aforementioned ideal behavior.

Without loss of generality, let us assume that $\alpha_1=1$, so that at the position $\omega=0$ the group delay is normalized, taking the value $\tau(0)=1$. In order for this function to take this value, we must equate the numerator with the denominator of the foregoing equation, which is achieved by equating the coefficients of the same powers of ω. In this case, we will get

$$\alpha_2-3\alpha_3=1-2\alpha_2$$

$$5\alpha_5-3\alpha_4+\alpha_2\alpha_3=\alpha_2^2-2\alpha_3+2\alpha_4$$

$$5\alpha_6+\alpha_3\alpha_4-3\alpha_2\alpha_5-7\alpha_7=\alpha_3^2+2\alpha_5-2\alpha_2\alpha_4-2\alpha_6$$

$$\dots\dots\dots\dots\dots\dots\dots\dots\dots\dots$$

$$\text{or, equivalently,}\quad \alpha_2-\alpha_3=\frac{1}{3},\quad 3(\alpha_5-\alpha_4)=\frac{1}{5},\quad 15(\alpha_6-\alpha_7)+\alpha_4-6\alpha_5=\frac{1}{105},\quad \dots\dots\dots$$

Continuing in this way, we find that in order for the group delay to be characterized by *maximum flat behavior* at the limit $\omega\to 0$ (for the normalized case $\tau(0)=1$), the successive coefficients of the transfer function must be related to each other as

$$\alpha_{k+1}=\frac{2(N-k)}{(2N-k)(k+1)}\alpha_k \quad\text{or, equivalently,}\quad \alpha_k=\frac{(2N-k)(k+1)}{2(N-k)}\alpha_{k+1}\quad (k=0,1,2,\dots,N-1)$$

with a starting value of $\alpha_N=1$ and the parameter N expressing the order of the filter. Using the preceding equation iteratively, successively we have

$$\alpha_{N-1}=\frac{N(N+1)}{1\cdot 2}\alpha_N=\frac{N(N+1)}{(2^1)(1!)}$$

$$\alpha_{N-2}=\frac{(N-1)(N+2)}{(2)(2)}\alpha_{N-1}=\frac{(N-1)N(N+1)(N+2)}{(2^2)(2!)}$$

$$\alpha_{N-3}=\frac{(N-2)(N+3)}{(2)(3)}\alpha_{N-2}=\frac{(N-2)(N-1)N(N+1)(N+2)(N+3)}{(2^3)(3!)}$$

$$\dots\dots=\dots\dots\dots\dots\dots\dots\dots\dots$$

$$\alpha_{N-m}=\frac{(N-m+1)(N-m+2)\dots N(N+1)\dots(N+m)}{2^m m!}$$

Proceeding with a change in the index from m to $k = N - m$, the last relation becomes

$$\alpha_k = \frac{(k+1)(k+2)\dots(2N-k)}{2^{N-k}(N-k)!} = \frac{(2N-k)!}{2^{N-k}k!(N-k)!}$$

since it will obviously be the case that

$$(k+1)(k+2)\dots(2N-k) = \frac{1\cdot 2\cdot 3\cdot k\ (k+1)\ (k+2)\cdots(2N-k)}{1\cdot 2\cdot 3\cdot k} = \frac{(2N-k)!}{k!}$$

and therefore the transfer function $\mathcal{H}(s)$ will take the form[34]

$$\mathcal{H}(s) = \frac{\beta_0}{\sum_{k=0}^{N}\alpha_s s^k} = \frac{\beta_0}{\sum_{k=0}^{N}\frac{(2N-k)!}{2^{N-k}k!(N-k)!}s^k} = \frac{\beta_0}{s^N B_N\left(\frac{1}{s}\right)} \tag{11.152}$$

where $B_N(s)$ are the Bessel polynomials, defined as[35]

$$B_N(s) = \sum_{k=0}^{N}\frac{(n+k)!}{(n-k)!\,k!}\left(\frac{s}{2}\right)^k = 1 + N(N+1)\frac{s}{2} + \dots + \frac{(2N-1)!}{(N-1)!}\left(\frac{s}{2}\right)^{N-1} + \frac{(2N)!}{N!}\left(\frac{s}{2}\right)^N \tag{11.153}$$

These polynomials constitute solutions of the differential equation

$$s^2\frac{d^2B_N(s)}{ds^2} + (2s+2)\frac{dB_N(s)}{dx} = N(N+1)B_N(s)$$

and can be calculated from the recursive relation

$$B_{N+1}(s) = (2N+1)sB_N(s) + B_{N-1}(s)$$

Using the foregoing relation, the transfer functions $\mathcal{H}_N(s)$ for filter order values $N = 1$ to 5 have the form[36]

$$\mathcal{H}_1(s) = \frac{1}{s+1},\quad \mathcal{H}_2(s) = \frac{3}{s^2+3s+3},\quad \mathcal{H}_3(s) = \frac{15}{s^3+6s^2+15s+15}$$

$$\mathcal{H}_4(s) = \frac{105}{s^4+10s^3+45s^2+105s+105},\quad \mathcal{H}_5(s) = \frac{945}{s^5+15s^4+105s^3+420s^2+945s+945}$$

while the values of these coefficients for filter order values $N = 1$ up to 8 are included in Table 11.3.

It is not difficult to see that for $k = 0$ it will hold that $\alpha_0 = (2N)!/(2^N)N!$, and consequently, the coefficients α_k and α_0 are related to each other through the expression

---

[34] At this point we reintroduce to the numerator the constant β_0, which we had removed for simplicity since it did not affect the calculations.

[35] See Krall H.L & O. Frink, *A New Class of Orthogonal Polynomials: The Bessel Polynomials*, Transactions of American Mathematical Society, Vol. 65, pp. 100–115, 1949.

[36] The constant β_0 in these relations, which also expresses the gain of the function, is calculated from the requirement to be $\mathcal{H}(0) = 1$.

Table 11.3 Polynomial of denominator of transfer function and values of corresponding poles of Bessel filter for order values $N = 2, 3, \ldots, 8$ and for unit time delay at position $\omega = 0$

N	Characteristic polynomial of transfer function	Pole $s = \sigma + j\omega$
2	$s^2 + 3s + 3$	$-1.5 \pm 0.8660j$
3	$s^3 + 6s^2 + 15s + 15$	-2.3222
		$-1.8390 \pm 1.7543j$
4	$s^4 + 10s^3 + 45s^2 + 105\,s + 105$	$-2.1039 \pm 2.6575j$
		$-2.8961 \pm 0.8672j$
5	$s^5 + 15\,s^4 + 105\,s^3 + 420\,s^2 + 945\,s + 945$	-3.6467
		$-2.3246 \pm 3.5710j$
6	$s^6 + 21\,s^5 + 210\,s^4 + 1260\,s^3 + 4725\,s^2 + 10395\,s + 10395$	$-2.5158 \pm 4.4927j$
		$-3.7357 \pm 2.6263j$
		$-4.2484 \pm 0.8675j$
7	$s^7 + 28\,s^6 + 378\,s^5 + 3150\,s^4 + 17325\,s^3 + 62370\,s^2$ $+ 135135\,s + 135135$	-4.9716
		$-2.6857 \pm 5.4206j$
		$-4.0701 \pm 3.5173j$
		$-4.7584 \pm 1.7393j$
8	$s^8 + 36\,s^7 + 630\,s^6 + 6930\,s^5 + 51975\,s^4 + 270270\,s^3$ $+ 945945\,s^2 + 2027025\,s + 2027025$	$-5.2049 \pm 2.6162j$
		$-4.3683 \pm 4.6162j$
		$-2.8388 \pm 6.3540j$
		$-5.5878 \pm 0.8676j$

$$\frac{\alpha_k}{\alpha_0} = \frac{(2N-k)!}{2^{N-k}k!(N-k)!} \times \frac{2^N N!}{(2N)!} = \frac{N!/(N-k)!}{2^{-k}(2N)!/(2N-k)!} \times \frac{1}{k!} = \frac{(N-k+1)(N-k+2)\ldots(N-1)N}{2^{-k}(2N-k+1)(2N-k+2)\ldots(2N-1)2N} \times \frac{1}{k!}$$

For the marginal case $N \to \infty$ we can make the approximations $N-k+1 \approx N,\ N-k+2 \approx N, \ldots, N-1 \approx N$ and, similarly, $2N-k+1 \approx 2N,\ 2N-k+2 \approx 2N, \ldots, 2N-1 \approx 2N$, so we will have

$$\lim_{N\to\infty} \frac{\alpha_k}{\alpha_0} = \frac{N^k}{2^{-k}(2N)^k}\frac{1}{k!} = \frac{1}{k!}$$

Therefore, formulating the transfer function in the form

$$\mathcal{H}(s) = \frac{\beta_0}{\sum\limits_{k=0}^{\infty} \alpha_k s^k} = \frac{\beta_0}{\alpha_0} \frac{1}{\sum\limits_{k=0}^{\infty} \frac{\alpha_k}{\alpha_0} s^k} \tag{11.154}$$

we observe that in the limit $N \to \infty$, we will have (with the approximation of the multiplicative constant β_0/α_0)

$$\lim_{N\to\infty} \mathcal{H}(s) \sim \frac{1}{\sum\limits_{k=0}^{\infty} \frac{\alpha_k}{\alpha_0} s^k} \equiv \frac{1}{1 + s + \frac{1}{2}s^2 + \frac{1}{6}s^3 + \frac{1}{24}s^4 + \ldots} \tag{11.155}$$

But the last expression is recognized as the *McLaurin expansion* of the function e^{-s}, and therefore we will have

$$\lim_{N\to\infty} \mathcal{H}(s) \sim e^{-s} \qquad \text{or, equivalently,} \qquad \lim_{N\to\infty} \mathcal{H}(j\omega) \sim e^{-j\omega}$$

which constitutes the frequency response of an ideal filter for gain and group delay values $G = \tau_d = 1$. We are led, therefore, to the conclusion that *as the order of the Bessel filter becomes increasingly larger, the behavior of the group delay approaches that of the ideal filter*, as indeed is verified by Fig. 11.63, where it appears that the frequency range within which the delay group takes a constant value increases with the increase in the order of the filter.

The foregoing analysis can be generalized for the *nonnormalized case* for which $\alpha_\tau(0) = \tau_0$, if we make the substitutions in the transfer function $s' = \tau_0 s$ and $\alpha'_k = \alpha_k/\tau_0^k$ and work as before.

Without being able at this point to proceed to detailed mathematical derivations, it is proven that the Bessel polynomials are related to our known Bessel functions through the relation

$$B_N\left(\frac{1}{ir}\right) = \frac{1}{j^N}\sqrt{\frac{\pi r}{2}} \times \left[(-1)^N J_{-N-\frac{1}{2}}(r) - jJ_{N+\frac{1}{2}}(r)\right]e^{ir} \tag{11.156}$$

with the *Bessel functions* defined as

$$J_{N+\frac{1}{2}}(r) = r^{N+\frac{1}{2}}\sum_{k=0}^{\infty}\frac{(-1)^k r^{2k}}{2^{2k+N+\frac{1}{2}}\,k!\,\Gamma\left(N+k+\frac{3}{2}\right)} \quad \text{where} \quad \Gamma(z) = \int_0^{\infty} t^{z-1}e^{-t}dt \quad \text{the function } \Gamma.$$

Proceeding now with the substitution $s = j\omega$ and using the foregoing relation, the *frequency response* of the Bessel filters is formulated as

$$\begin{aligned}\mathcal{H}(j\omega) &= \frac{\beta_0}{(j\omega)^n B\left(\frac{1}{j\omega}\right)} = \frac{\beta_0\sqrt{2}}{\sqrt{\pi}\omega^{N+\frac{1}{2}}\left[(-1)^N J_{-N-\frac{1}{2}}(\omega) - jJ_{N+\frac{1}{2}}(\omega)\right]}e^{-i\omega} \\ &= \beta_0\sqrt{\frac{2}{\pi\omega^{2N+1}}}\frac{(-1)^N J_{-N-\frac{1}{2}}(\omega) + jJ_{N+\frac{1}{2}}(\omega)}{J^2_{-N-\frac{1}{2}}(\omega) + J^2_{N+\frac{1}{2}}(\omega)}e^{-i\omega}\end{aligned}$$

and finally

$$\mathcal{H}(j\omega) = \frac{\beta_0\sqrt{\dfrac{2}{\pi\omega^{2N+1}}}}{\sqrt{J^2_{-N-\frac{1}{2}}(\omega) + J^2_{N+\frac{1}{2}}(\omega)}}\exp\left[j\left(-\omega + \tan^{-1}\frac{(-1)^N J_{N+\frac{1}{2}}(\omega)}{J_{-N-\frac{1}{2}}(\omega)}\right)\right] \tag{11.157}$$

as emerges after simple mathematical operations. From the last relation, it is concluded that the amplitude and phase of the frequency response of the Bessel filters have the form

$$|\mathcal{H}(j\omega)| = \frac{\beta_0\sqrt{\dfrac{2}{\pi\omega^{2N+1}}}}{\sqrt{J^2_{-N-\frac{1}{2}}(\omega) + J^2_{N+\frac{1}{2}}(\omega)}} \quad \text{and} \quad \measuredangle\mathcal{H}(j\omega) = -\omega + \tan^{-1}\frac{(-1)^N J_{N+\frac{1}{2}}(\omega)}{J_{-N-\frac{1}{2}}(\omega)} \tag{11.158}$$

The *group delay* of the Bessel filter will emerge, as is well known, as *a negative derivative of the phase response*. Differentiating the foregoing relation leads to

$$\tau(\omega) = -\frac{d\measuredangle\mathcal{H}(j\omega)}{d\omega} = 1 - \frac{(-1)^N\left[\dfrac{dJ_{N+\frac{1}{2}}(\omega)}{d\omega}J_{-N-\frac{1}{2}}(\omega) - J_{N+\frac{1}{2}}(\omega)\dfrac{dJ_{-N-\frac{1}{2}}(\omega)}{d\omega}\right]}{J^2_{-N-\frac{1}{2}}(\omega) + J^2_{N+\frac{1}{2}}(\omega)}$$

But the expression in square brackets in the numerator is recognized as the *Wronski determinant* of the functions $J_{-N-\frac{1}{2}}(\omega)$ and $J_{N+\frac{1}{2}}(\omega)$, which is proven[37] to have the form

$$J_{-m}(\omega)\frac{dJ_m(\omega)}{d\omega} - J_m(\omega)\frac{dJ_{-m}(\omega)}{d\omega} = \frac{2\sin(m\pi)}{\pi\omega}$$

The foregoing relation for $m = N + \frac{1}{2}$ takes the form

$$\frac{dJ_{N+\frac{1}{2}}(\omega)}{d\omega}J_{-N-\frac{1}{2}}(\omega) - J_{N+\frac{1}{2}}(\omega)\frac{dJ_{-N-\frac{1}{2}}(\omega)}{d\omega} = \frac{2}{\pi\omega}\sin\left[\left(N+\frac{1}{2}\right)\pi\right] = \frac{2}{\pi\omega}(-1)^N$$

or, equivalently,

$$(-1)^N\left[\frac{dJ_{N+\frac{1}{2}}(\omega)}{d\omega}J_{-N-\frac{1}{2}}(\omega) - J_{N+\frac{1}{2}}(\omega)\frac{dJ_{-N-\frac{1}{2}}(\omega)}{d\omega}\right] = \frac{2}{\pi\omega}$$

Substituting the last relation into the group delay calculation equation, we get

$$\tau(\omega) = 1 - \frac{2}{\pi\omega}\frac{1}{J^2_{-N-\frac{1}{2}}(\omega) + J^2_{N+\frac{1}{2}}(\omega)} = 1 - \frac{\omega^{2N}}{\beta_0^2}\frac{2\beta_0^2}{\pi\omega^{2N+1}\left[J^2_{-N-\frac{1}{2}}(\omega) + J^2_{N+\frac{1}{2}}(\omega)\right]} = 1 - \frac{\omega^{2N}}{\beta_0^2}|\mathcal{H}j\omega|^2 \quad (11.159)$$

In the next section, we will prove that the power response of a Bessel filter can be expressed as the *sum of infinite terms*

$$|\mathcal{H}(j\omega)|^2 = 1 - \frac{\omega^2}{2N-1} + \frac{2(N-1)\omega^4}{(2N-1)^2(2N-3)} + \ldots$$

From the foregoing relationship, we find that at the limit $\omega \to 0$, it will hold that $|\mathcal{H}(j\omega)| \to 1$ and $\tau(\omega) \to 1$. On the other hand, at the limit $\omega \to \infty$, it will hold that $|\mathcal{H}(j\omega)| \to 1/(j\omega)^N = 0$, and therefore, *the Bessel filter is an approximation of an ideal low-pass filter* that introduces a constant time delay in the input signal. This property is highlighted in Fig. 11.64, where the amplitude response of the filter in the pass and cutoff zones is presented for various filter orders. If we calculate the first $N-1$ derivatives of the function $\tau(\omega)$ with respect to the quantity ω^2, we will find that these are zero at the position $\omega = 0$, that is,[38]

$$\left.\frac{d\tau(\omega)}{d(\omega^2)}\right|_{\omega=0} = \left.\frac{d^2\tau(\omega)}{d(\omega^2)^2}\right|_{\omega=0} = \cdots = \left.\frac{d^{N-1}\tau(\omega)}{d(\omega^2)^{N-1}}\right|_{\omega=0} = 0$$

which is expected since the transfer function has been constructed in such a way as to offer the *maximum possible flat behavior* with regard to the group delay of the filter.

Phase Characteristics and Time-Domain Response

The change in the *phase response* of the prototype Bessel filter for order values from $N = 1$ to 6 are presented in Fig. 11.65, while Fig. 11.66 presents the corresponding graphical representations of the *impulse* and the *step response*. It is proven that the impulse response of these filters is *of a Gaussian type* and symmetric around the position τ_0, leading to a step response characterized by very small-amplitude variation compared to that of the other approximations of ideal analog filters.

Poles of Transfer Function

Applying the theory of Bessel polynomials, we can prove that the zero values of the expression $s^N B(1/s)$ are all located in the *left complex half-plane*, a fact that ensures the *stability* of Bessel filters, since these coincide with the poles of its transfer function. Unlike previous types of filters, for Bessel filters it is difficult to find pole calculation equations expressed in closed

[37] See N.G. Watson, *A Treatise on the Theory of Bessel Functions*, Cambridge University Press, 1995, ISBN 978-0-521-48391-9, p. 43.

[38] In this analysis we only consider the derivatives with respect to ω^2 because the group delay is an even function of ω.

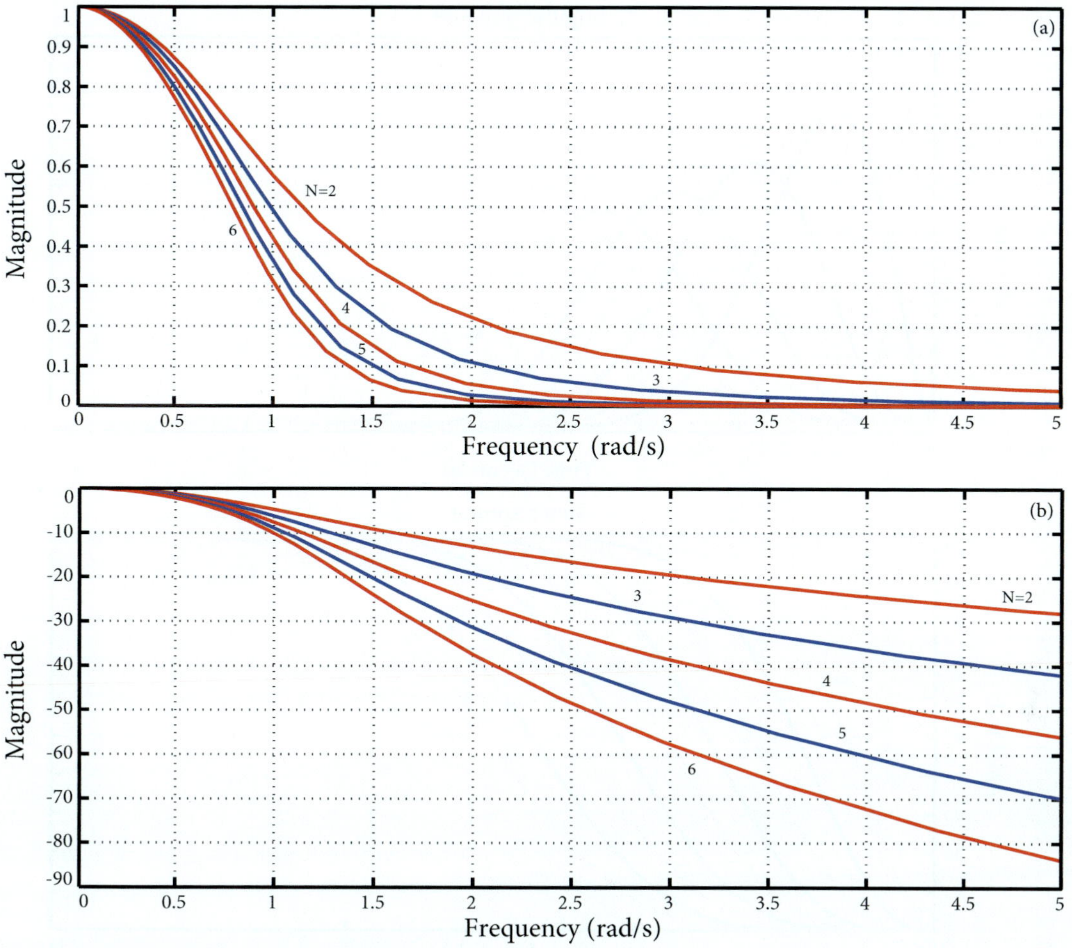

Fig. 11.64 Amplitude response of prototype Bessel filter for filter order values $N = 2, 3, 4, 5, 6$ plotted on a linear scale (**a**) and on a logarithmic scale (**b**). In both diagrams, the horizontal frequency axis is graduated on a linear scale

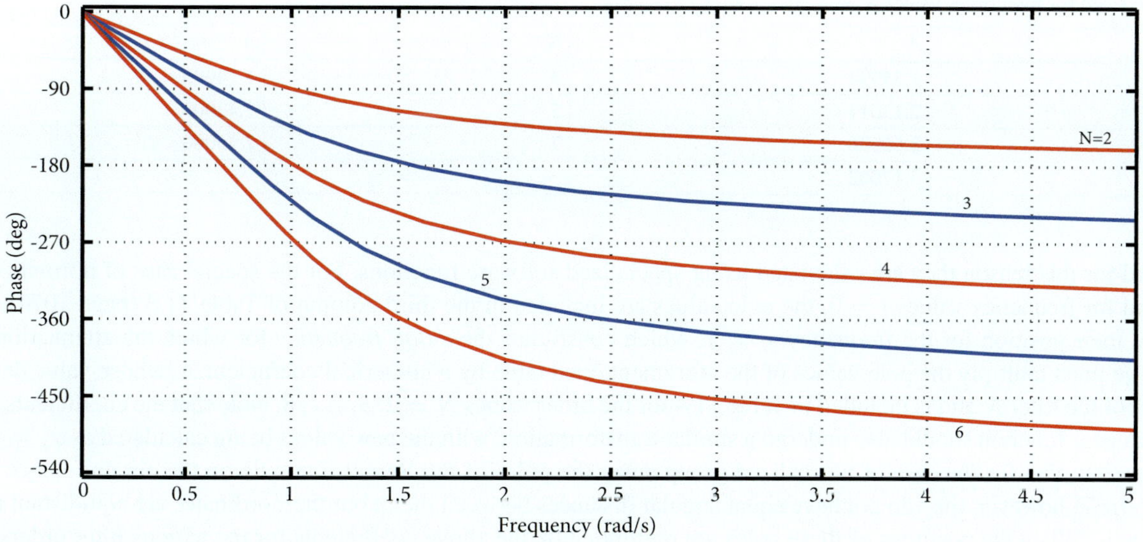

Fig. 11.65 Phase response of prototype Bessel filter for filter order values $N = 1$ to 6

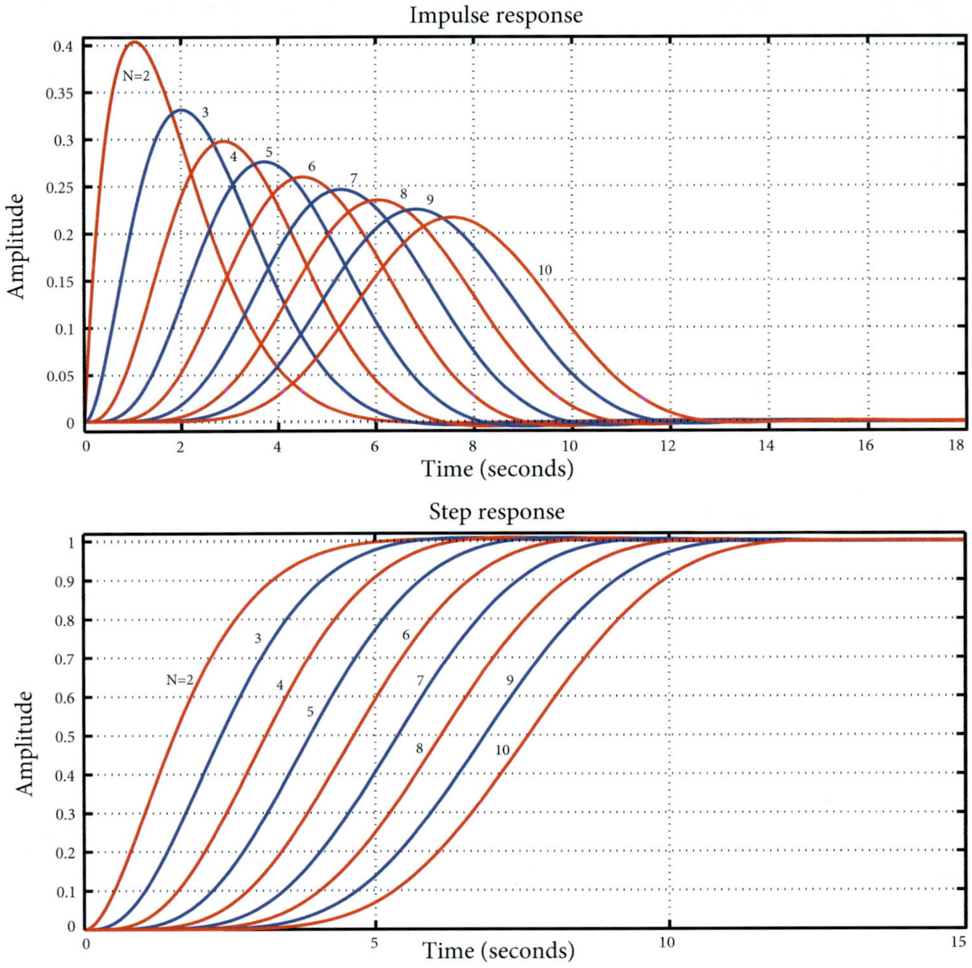

Fig. 11.66 Impulse response (**a**) and step response (**b**) of prototype Bessel filter for filter orders $N = 1$ to 10

Table 11.4 Renormalization coefficients of filter poles Bessel from *unit delay* for frequency value $\omega = 0$ in attenuation of 3 dB for frequency value $\omega = 1$ and for filter order values $N = 2, 3, \ldots, 8$

N	A	N	A
2	1.35994	3	1.74993
4	2.13011	5	2.42003
6	2.66996	7	2.95000
8	3.17002		

form, and for this reason they are calculated using specialized software functions. For the special case of normalized group delay and for frequency value $\omega = 0$, the pole values are included in the third column of Table 11.3 (page 1070), while to calculate their position for the frequency $\omega = 1$, which constitutes the *cutoff frequency* for which the attenuation of 3 dB occurs, we must multiply the pole values of the aforementioned table by a numerical coefficient A, whose value depends on the order of the filter N and is included in Table 11.4 for the order values $N = 2, 3, \ldots, 8$. Note that the coefficients α_k of the filter's transfer function should also undergo a similar transformation, with the new values being calculated as $\alpha'_k = A^{N-k}\alpha_k$.

It is proven that for the case of normalized group delay the poles of the Bessel filters lie on the circumference of the *of the unit circle*; however, they do not have equal angular distances between them, but their ordinates are equidistant along the vertical axis. When the positions of these poles are multiplied by the above coefficients for the various filter orders, so as to correlate with the frequency $\omega = 1$, the circles on which they belong no longer have a unit radius, while their center does not coincide with the origin of the coordinate system. In reality, after the above scaling, the poles belong to elliptical peripheries, one of which focus is near the origin of the coordinate system and the other at some point of the positive real semi-axis. This situation for filter orders $N = 1, 2, 3, \ldots, 10$ is diagrammatically presented in Fig. 11.67.

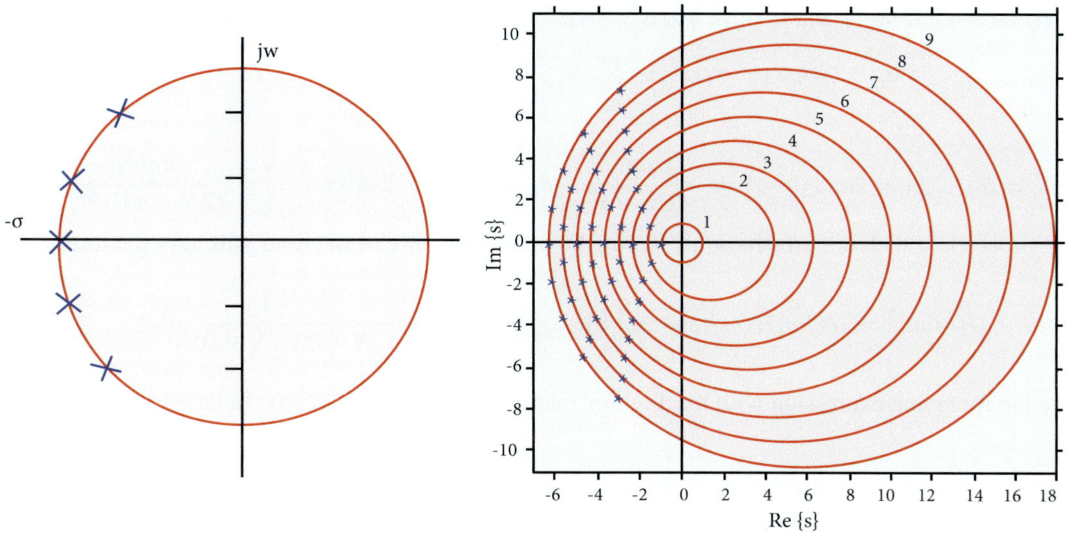

Fig. 11.67 Poles of Bessel filter for normalized and nonnormalized case. In the left image, the order of the filter is $N = 5$, while in the right image, it ranges from $N = 1$ to $N = 10$. In full analogy with the previous approaches to ideal filters, the even-order filters have $N/2$ pairs of conjugate complex poles, while the odd-order filters have one real negative pole and $(N-1)/2$ pairs of conjugate complex poles

Loss Characteristic

The *loss characteristic* of Bessel filters will result from the relation

$$A(\omega) = 10\log L(\omega^2) = 10\log \left.\frac{1}{\mathcal{H}'(s)\mathcal{H}'(-s)}\right|_{s=j\omega} = 10\log[\mathcal{H}(s)\mathcal{H}(-s)]_{s=j\omega} \tag{11.160}$$

where we use the function $\mathcal{H}'(s) = 1/\mathcal{H}(s)$ (this approach is often used in the case of Butterworth approximations). From the defining equation of the function $\mathcal{H}(s)$ we have

$$[\mathcal{H}(s)\mathcal{H}(-s)] = \left(\sum_{k=0}^{N}\alpha_k s^k\right)\left(\sum_{k=0}^{N}\alpha_k(-s)^k\right) = \left(\sum_{k=0}^{N}\alpha_k s^k\right)\left[\sum_{k=0}^{N}\alpha_k(-1)^k s^k\right] \tag{11.161}$$

To examine the characteristic losses, let us calculate some terms of the foregoing product. Taking into account that the successive coefficients of the Bessel polynomials are related to each other through the expression

$$\alpha_{k+1} = \frac{2(N-k)}{(k+1)(2N-k)}\alpha_k \qquad \text{successively we will have}$$

$$\alpha_1 = \frac{2N}{2N}\alpha_0 = \alpha_0$$

$$\alpha_2 = \frac{2(N-1)}{2(2N-1)}\alpha_1 = \frac{N-1}{2N-1}\alpha_1 = \frac{N-1}{2N-1}\alpha_0$$

$$\alpha_3 = \frac{2(N-2)}{3(2N-2)}\alpha_2 = \frac{2(N-2)(N-1)}{3(2N-2)(2N-1)}\alpha_0 = \frac{N-2}{3(2N-1)}\alpha_0$$

$$\alpha_4 = \frac{2(N-3)}{4(2N-3)}\alpha_3 = \frac{2(N-3)(N-2)}{4(2N-3)3(2N-1)}\alpha_0 = \frac{(N-2)(N-3)}{6(2N-1)(2N-3)}\alpha_0$$

and so forth. Therefore, the function $\mathcal{H}(s)\mathcal{H}(-s)$ will be a polynomial that will contain *only even powers* of s and given by the relation

$$\mathcal{H}(s)\mathcal{H}(-s) = (\alpha_0 + \alpha_1 s + \alpha_2 s^2 + \alpha_3 s^3 + \alpha_4 s^4 + \ldots)(\alpha_0 - \alpha_1 s + \alpha_2 s^2 - \alpha_3 s^3 + \alpha_4 s^4 - \ldots) =$$
$$= \alpha_0^2 + (2\alpha_0\alpha_2 - \alpha_1^2)s^2 + (2\alpha_0\alpha_4 - 2\alpha_1\alpha_3 + \alpha_2^2)s^4 + \cdots = \beta_0 + \beta_1 s^2 + \beta_2 s^4 + \ldots$$

where

$$\beta_0 = \alpha_0^2, \quad \beta_1 = 2\alpha_0\alpha_2 - \alpha_1^2 = -\frac{\alpha_0^2}{2N-1}, \quad \beta_2 = 2\alpha_0\alpha_4 - 2\alpha_1\alpha_3 + \alpha_2^2 = \frac{\alpha_0^2(N-1)}{(2N-1)^2(2N-3)}$$

Consequently, proceeding with the substitution $s = j\omega$, the power response of the filter will have the form

$$|\mathcal{H}(j\omega)|^2 = \mathcal{H}(j\omega)\mathcal{H}(-j\omega) = \alpha_0^2\left[1 - \frac{1}{2N-1}\omega^2 + \frac{N-1}{(2N-1)^2(N-3)}\omega^4 + \ldots\right] \tag{11.162}$$

If we compare the foregoing expression with the Taylor expansion of the Gaussian type function

$$f(\omega) = \alpha_0^2 \exp\left(\frac{\omega^2}{2N-1}\right), \qquad \text{which has the form} \qquad \alpha_0^2 \exp\left(\frac{\omega^2}{2N-1}\right) = \alpha_0^2\left[1 - \frac{1}{2N-1}\omega^2 + \frac{1}{2}\frac{\omega^4}{(2N-1)^2} + \ldots\right]$$

we will find that the relative difference between these two functions is

$$\frac{\alpha_0^2 \exp\left(\frac{\omega^2}{2N-1}\right) - \mathcal{H}(j\omega)\mathcal{H}(-j\omega)}{\alpha_0^2 \exp\left(\frac{\omega^2}{2N-1}\right)} = \exp\left(-\frac{\omega^2}{2N-1}\right) \times \left[-\frac{\omega^4}{2(2N-1)^2(2N-3)} + \mathcal{O}(\omega^6)\right] \tag{11.163}$$

as can be proven. In this case, for filter orders $N \geq 3$ and for small values of the frequency ω we can make the approximation

$$|\mathcal{H}(j\omega)|^2 = \mathcal{H}(j\omega)\mathcal{H}(-j\omega) \approx \alpha_0^2 \exp\left(\frac{\omega^2}{2N-1}\right) \tag{11.164}$$

and therefore, the loss characteristic is calculated as

$$A(\omega) = 10\log\left[\alpha_0^2 \exp\left(\frac{\omega^2}{2N-1}\right)\right] = 20\log\alpha_0 + 10\log\left[\exp\left(\frac{\omega^2}{2N-1}\right)\right] = 20\log\alpha_0 + \frac{10\log e}{2N-1}\omega^2 \tag{11.165}$$

The same approach can be applied to the *group delay*, which can be written

$$\tau(\omega) \approx 1 - \frac{\omega^{2N}}{\beta_0^2} \times \left\{\alpha_0^2 \exp\left(\frac{\omega^2}{2N-1}\right)\right\}$$

Using the loss function equation, we can calculate the frequency ω_c as the frequency associated with a loss level equal to 3 dB. Proceeding with the substitutions $\omega = \omega_c$ and $A = 3$, we will have

$$10\log\left[\alpha_0^2 \exp\left(\frac{\omega_c^2}{2N-1}\right)\right] = 3$$

from which it follows that

$$\omega_c = \sqrt{(2N-1)\ln\left(\frac{2}{\alpha_0^2}\right)}$$

Problem 11.9.10 Design the amplitude and phase responses of a Bessel filter of order $N = 8$ (Antoniou, Exercise for solution 10.19).

Solution Referring to the table of Bessel polynomials, we find that the transfer function of this filter will have the form

$$\mathcal{H}(s) = \frac{2027025}{s^8 + 36s^7 + 630s^6 + 6930s^5 + 51975s^4 + 270270s^3 + 945945s^2 + 2027025s + 2027025}$$

The poles of this function and the contribution of each of them to the denominator of the transfer function are presented as folows:

Poles $p_{1,2} = -5.2049 \pm 2.6162j$, which contribute the term $s^2 + 10.4096825s + 33.934741$

Poles $p_{3,4} = -4.3683 \pm 4.4146j$, which contribute the term $s^2 + 8.736578s + 38.569256$

Poles $p_{5,6} = -2.8388 \pm 6.3540j$, which contribute the term $s^2 + 5.677968s + 48.432015$

Poles $p_{9,10} = -5.5878 \pm 0.8676j$, which contribute the term $s^2 + 11.1757725 + 31.977224$

Proceeding with the substitution $s = j\omega$, the frequency response of the filter is calculated as

$$\mathcal{H}(j\omega) = \frac{2027025}{(\omega^8 - 630\omega^6 + 51975\omega^4 - 94594\omega^2 + 2027025) + j(2027025 - 270270\omega^3 + 6930\omega^5 - 36\omega^7)}$$

with magnitude and phase

$$|\mathcal{H}(j\omega)| = \frac{2027025}{\sqrt{(\omega^8 - 630\omega^6 + 51975\omega^4 - 94594\omega^2 + 2027025)^2 + (2027025 - 270270\omega^3 + 6930\omega^5 - 36\omega^7)^2}}$$

$$\measuredangle\mathcal{H}(j\omega) = -\tan^{-1}\frac{2027025 - 270270\omega^3 + 6930\omega^5 - 36\omega^7}{\omega^8 - 630\omega^6 + 51975\omega^4 - 94594\omega^2 + 2027025}$$

respectively, as can be verified. The variation of the amplitude and phase responses of this filter are presented in Fig. 11.68.

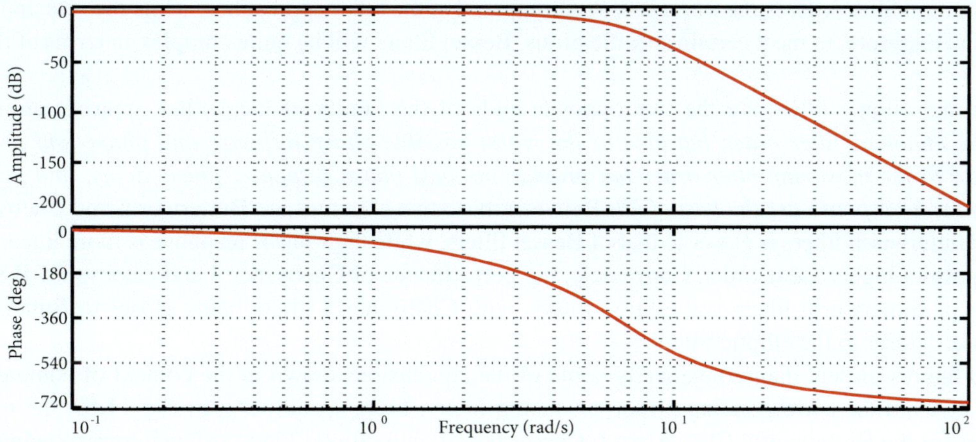

Fig. 11.68 Change in amplitude and phase responses of Bessel filter from Exercise 11.9.10

11.9.10 Comparison of Ideal Filter Approximations

After the analytical presentation of the characteristics of the ideal filter approximations, let us move on to a brief comparison of them to determine their advantages and disadvantages. Among the preceding approximations, the *Butterworth filters* are the simplest of all, as they offer flat behavior in the pass and cutoff bands. On the other hand, the order of Chebyshev I filters *will never be greater* than the order of the Butterworth filter for the same specifications, while for all practical values of the ripple range ε, the *phase factor* is always smaller than that of a Butterworth filter for the same filter order, in contrast to the value of *selectivity*, which is greater in Chebyshev I filters. Although the foregoing comparison may suggest that Chebyshev I filters are more suitable than Butterworth filters, the latter offer a phase response with *greater linearity* compared to that of Chebyshev I of the same filter order, an advantage that is reflected in the phase and group delays. The impulse and step responses of Butterworth filters are also better compared to those of Chebyshev I filters, as the latter present *ripples* that decrease at a slower rate compared to the corresponding responses of Butterworth filters. It should be noted that in reality, Butterworth filters can be treated as a special case of Chebyshev filters, since the pole defining equations of Chebyshev filters transform into those of Butterworth filters in the limit $\varepsilon \to 0$, as can be verified.

On the other hand, the minimum order of Chebyshev II filters for specific operating specifications is *the same* as that of Chebyshev I filters, which will never be greater than the order of Butterworth filters for the same specifications (although sometimes this is not the case due to the rounding of the filter order to the next highest integer). As far as *the response in the passband* is concerned, although it has not been designed to exhibit the maximum possible flat behavior, it is comparable to that of Butterworth filters. On the other hand, Butterworth filters offer a phase response with a higher degree of linearity, but the difference between the various responses is not particularly significant. The comparison between the impulse and step responses of the two filters reveals that there are no significant differences there, either.

If we compare Chebyshev II with Chebyshev I filters, we find that for the same set of specifications a Chebyshev II filter has a *more stable amplitude response* compared to a Chebyshev I filter of the same order, a *more linear* phase response, a *more stable* phase delay and group delay, and a *smaller* bandwidth variation. On the other hand, the value of the bandwidth variation ε for Chebyshev II filters is significantly smaller than the corresponding parameter of Chebyshev I filters, resulting in the selectivity F_s of the latter being much better.

The main characteristic of elliptic filters is that they offer the *minimum filter order* for the same specifications in relation to the other approaches of ideal filters. For example, for frequencies $f_p = 3000\,\text{Hz}$, $f_s = 10000\,\text{Hz}$, $A_p = 1$ dB, and $A_s = 70$ dB, the minimum filter order is $N = 8$ for a Butterworth filter, $N = 6$ for a Chebyshev I or Chebyshev II filer, and just $N = 4$ for an elliptic filter. However, in many cases, their implementation is characterized by a high degree of difficulty and maximum cost, especially for the components of filters (the elementary filters from which it is composed) that are characterized by large values of the quality factor Q. As for their phase response, elliptic filters have a response with *the highest degree of nonlinearity*—and therefore are the least desirable in terms of this characteristic—while the bandwidth variation in the step and impulse responses is comparable to (and perhaps slightly worse than) that of the Chebyshev I and Chebyshev II filters.

Finally, Bessel filters excel in terms of *phase response*, *group delay*, and *impulse* and *step response* compared to any other approach, a characteristic that does not apply to amplitude response, which is the least desirable compared to any other type of filter. The minimum filter order is *the worst possible* compared to the other approaches, while the selectivity factor value is not very good, and therefore, to meet certain specifications, Bessel filters will be quite complex in terms of their construction process.

Bessel filters and elliptic filters are the two opposite ends of the family of ideal filter approximations, since *elliptic filters lead to the minimum filter order, but also to the worst possible phase response and phase and group delay, while Bessel filters lead to the maximum filter order but provide the best phase response, phase delay, and group delay, while their impulse and step response are the best of all.* Between these two extremes are Butterworth and Chebyshev filters. The performance of Butterworth filters is closer to that of Bessel filters, while their phase response is more linear than Chebyshev filters, but it requires a higher value order. Conversely, the performance of Chebyshev I and Chebyshev II filters is between the performance of Butterworth filters and elliptic filters, with Chebyshev I filters being closer to Butterworth filters and Chebyshev II filters closer to the elliptic filters.

The next two figures present diagrammatically some of the conclusions drawn in the context of comparing analog filter approaches and concern the corresponding prototype analog filters. More specifically, in Fig. 11.69, the relative change of the gain function for the Butterworth, Chebyshev I, Chebyshev II, and elliptic filters (a) and corresponding attenuation (b)

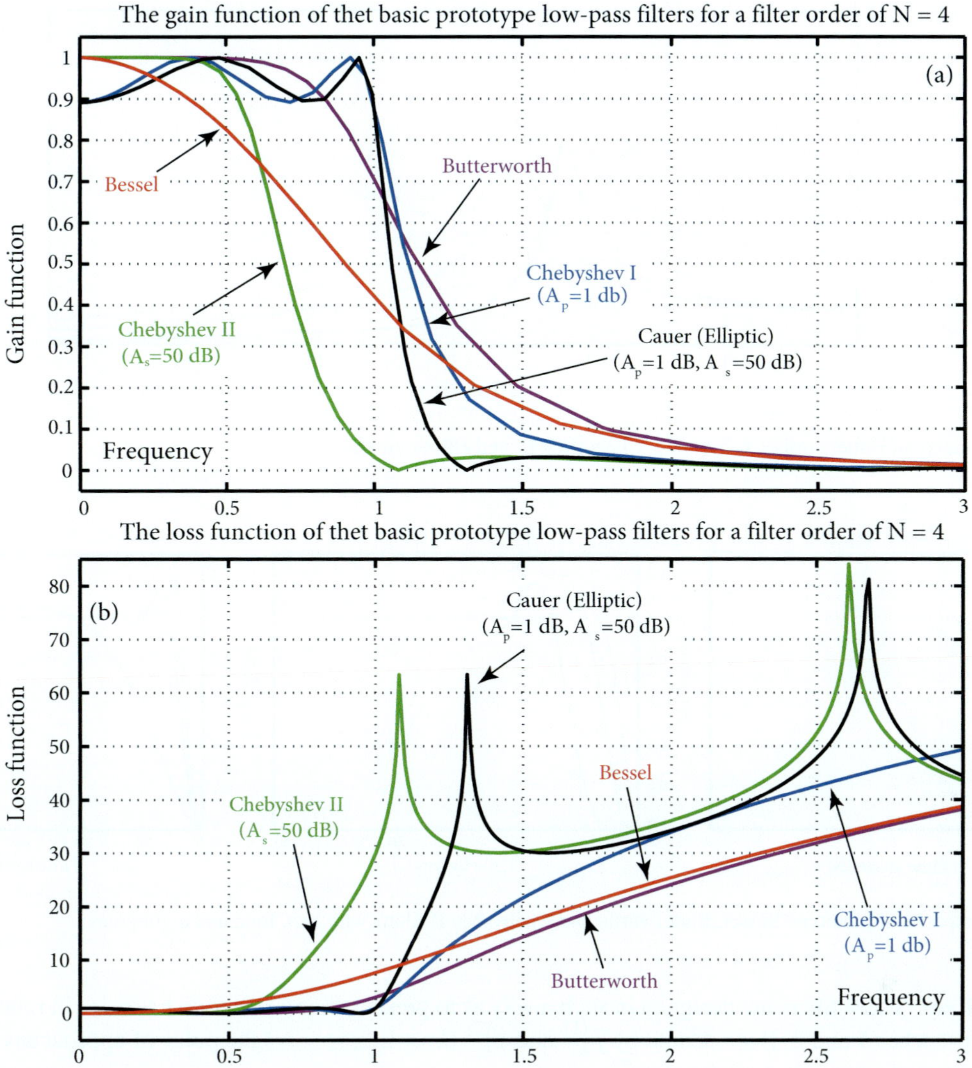

Fig. 11.69 Gain and loss functions of prototype Butterworth, Chebyshev I, Chebyshev II, and Cauer filters for a filter order value of $N = 4$

for some operating specifications and for order value $N = 4$ are depicted. From this figure we observe that Chebyshev II filters are characterized by a more gradual transition between pass and cutoff zones compared to Chebyshev I filters, despite the fact that the two filters correspond to the same order value. On the other hand, Fig. 11.70 depicts the group delay of the aforementioned types of filters, while, finally, Fig. 11.71 shows the step response of the prototype Bessel, Butterworth, and Chebyshev II filters with two, four, and eight poles, respectively, characterized by different wave amplitudes depending on the type of filter and the number of poles, with the Bessel filters being, as we have already mentioned, the best possible choice.

11.9.11 Other Approaches to Ideal Filters

Apart from the classic approaches of analog filters presented in previous sections, there are others that, although they have not been widely accepted and used, are nevertheless briefly presented for completeness. The reader who is interested in more information regarding these alternative approaches is referred to the relevant literature.

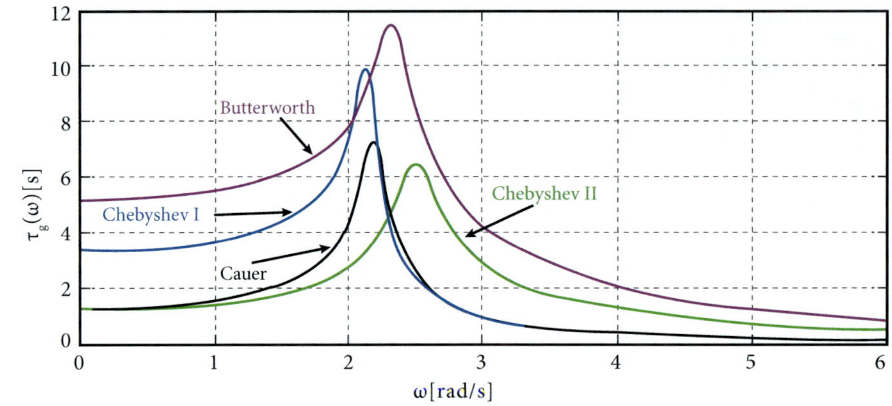

Fig. 11.70 Group delay of Butterworth, Chebyshev I, Chebyshev II, and elliptic filters

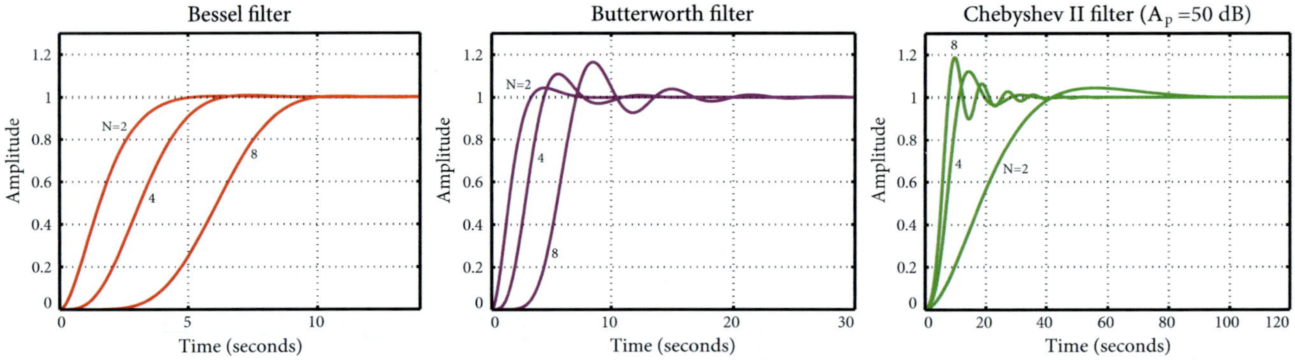

Fig. 11.71 Step response of prototype Bessel, Butterworth, and Chebyshev II filters with two, four, and eight poles

Transitional Filters These filters are designed in such a way as to present an *intermediate behavior* in relation to that which characterizes approaches 1 and 2. If we denote by $p_k^{(1)}$ and $p_k^{(2)}$ $(k = 1, 2, \ldots, N)$ the poles of approaches 1 and 2, then the poles of the transitional approach are defined as[39]

$$p_k = [p_k^{(2)}]^m \times [p_k^{(1)}]^{1-m} \quad (0 \le m \le 1) \tag{11.166}$$

From the foregoing equation we find that for $m = 0$, the preceding poles are reduced to the poles of approach 1, while for $m = 1$ we end up with the poles of approach 2. It is obvious that if the parameter m takes some intermediate value in the interval [0, 1], then the filter that will result will be characterized by some *intermediate behavior*. Using this approach of *transitional filters*, we can construct filters that present the advantages of one approach while simultaneously measuring the weaknesses of the other.

Gaussian Filters The ideal Gaussian filter is characterized by a power response of the form[40]

[39] See (1) Aiello G.L & Angelo P.M, "Transitional Legendre–Thomson Filters," *IEEE Transactions on Circuits and Systems*, Vol. CAS-21, No. 1, pp. 159–162, 1974, (2) Lindquist G.C, *Active Network Design with Signal Filtering Applications*, Steward & Sons, Long Beach, 1977, (3) Peles Y. & Murakami T., "Analysis and Synthesis of Transitional Butterworth–Thomson Filters and Band-pass Amplifiers," *RCA Review*, Vol. 18, No. 1, pp. 60–94, 1957.

[40] See Dishal M., "Gaussian-Response Filter Design," *Electrical Communication*, Vol. 36, No. 1, pp. 3–26, 1959.

$$|\mathcal{H}(j\omega)|^2 = \exp(-\gamma\omega^2) = \frac{1}{1 + \sum_{k=1}^{N} (\gamma^k/k!)\,\omega^{2k}}, \quad \gamma = \frac{\ln 2}{\omega_c^2} \tag{11.167}$$

with the last equation resulting from the *McLaurin expansion of the exponential function*. The parameter ω_c that appears in the foregoing equation obviously expresses the cutoff frequency of the filter. It is proven that*the frequency response, the phase response, the phase and group delays, and the step and impulse responses of Gaussian filters are very similar to the corresponding characteristics of Bessel filters*, which, of course, is expected since the power response of Bessel filters at the limit $N \to \infty$ is a Gaussian function, as we pointed out during the presentation of these filters.

Legendre Filters The power response of Legendre filters is given by the equation[41]

$$|\mathcal{H}(j\omega)|^2 = \frac{1}{1 + \varepsilon^2[P_N^{(m)}(\omega/\omega_p)]^2} \tag{11.168}$$

where $P_N^{(m)}(\omega/\omega_p)$ is the *Legendre polynomial attached to number m*, N is the filter order, ε is the variance parameter, and ω_p is the limit of the passband. The attached Legendre polynomials can be derived from the recursive relationship

$$P_{N+1}(\omega) = \frac{2N+1}{N+1}\omega P_N(\omega) - \frac{N}{N+1}P_{N-1}(\omega) \tag{11.169}$$

starting with the expressions $P_0(\omega) = 1$ and $P_1(\omega) = \omega$. The Legendre polynomial of order N attached to number m can be calculated from the Legendre polynomial of order N from the relationship

$$P_N^{(m)}(\omega) = \frac{d^m}{d\omega^m}P_{N+m}(\omega) \tag{11.170}$$

It is proven that all the characteristics (frequency and phase response, phase and group delay, impulse and step response) of Legendre filters are very similar to the corresponding characteristics of Chebyshev I filters.

Hyperspherical Filters The power response of hyperspherical filters is given by the equation[42]

$$|\mathcal{H}(j\omega)|^2 = \frac{1}{1 + \varepsilon^2[F_N^{(\alpha)}(\omega/\omega_p)]^2} \tag{11.171}$$

where N is the filter order, ω_p is the limit of the passband, ε is a parameter that controls attenuation at the position $\omega = \omega_p$, and α is a second parameter that controls the shape of the filter's response curve. The function $F_N^{(\alpha)}(\omega/\omega_p)$ is given by the relationship

[41] See Ku Y.H. & Drubin M, "Network Synthesis using Legendre and Hermite Polynomials," Journal of Franklin Institute, Vol. 273, No. 2, pp. 138–157, 1962.

[42] See Johnson D.E. & Johnson J.R., "Low Pass Filters Using Ultrashperical Polynomials," *IEEE Transactions on Circuit Theory*, Vol. CT-13, No. 4, pp. 364–369, 1966.

$$F_N^{(\alpha)}\left(\frac{\omega}{\omega_p}\right) = \frac{N!}{(1+\alpha)\cdot(2+\alpha)\cdot(3+\alpha)\cdots(N+\alpha)}\, P_N^{(\alpha,\alpha)}\left(\frac{\omega}{\omega_p}\right) \tag{11.172}$$

for values $\alpha > -1$. The function $P_N^{(\alpha,\alpha)}(\omega/\omega_p)$ is known as *hyperspherical polynomial* and is a special case ($\beta = \alpha$) of the *Jacobi polynomial*

$$P_N^{(\alpha,\beta)}(\omega) = \frac{\Gamma(2N+\alpha+\beta+1)}{N!\Gamma(N+\alpha+\beta+1)} G_N(\alpha+\beta+1, \beta+1, \frac{1}{2}(\omega+1))$$

where

$$G_n(p, q, \omega) \equiv \frac{N!\Gamma(N+p)}{\Gamma(2N+p)} P_N^{(p-q,q-1)}(2\omega-1)$$

and $\Gamma(z)$ is the Γ function. It is proven that for $\alpha = -0.5$ the hyperspherical filter reduces to a Chebyshev I filter, while for $\alpha = 0$ the hyperspherical filter reduces to a Legendre filter. If $\alpha = m$, where m is a positive integer, the filter response is the one associated with the modified Legendre polynomials of order m.

Papoulis Filters The power response of Papoulis filters is given by the equation [43]

$$|\mathcal{H}(j\omega)|^2 = \frac{1}{1+\varepsilon^2[L_N^2(\omega/\omega_p)]^2} \tag{11.173}$$

where N is the order of the filter, ω_p is the limit of the passband, ε is a parameter that controls attenuation at the position $\omega = \omega_p$, and $L_N(\omega)$ is a polynomial associated with the nonmodified Legendre polynomials $P_1(x), P_2(x), \ldots, P_N(x)$. This polynomial is given by the relation

$$L_N^2(\omega) = \int_1^{2\omega^2-1}\left[\sum_{k=0}^{(N-1)/2} \alpha_k P_k(x)\right]^2 dx, \quad \text{where} \quad \alpha_0 = \frac{\alpha_1}{3} = \frac{\alpha_2}{5} = \cdots = \frac{\alpha_r}{2r+1} = \frac{1}{\sqrt{2}(k+1)}$$

for the odd-order filter $N = 2k+1$ and from the relation

$$L_N^2(\omega) = L\int_1^{2\omega^2-1}(x+1)\left(\frac{dP_{k+1}(x)}{dx}\right)^2 dx$$

for the even-order filter $N = 2k+2$, where L is a normalization constant, so that $L_N^2(1) = 1$. The most characteristic property of Papoulis filters is that they have the maximum negative slope of the response curve at the position $\omega = \omega_p$, and therefore, the value of the quantity

$$-\left.\frac{d|\mathcal{H}(j\omega)|}{d\omega}\right|_{\omega=\omega_p}$$

of Papoulis filters is the maximum compared to all filters, whose frequency response is not characterized by a monotonic increase in function of the frequency ω.

Halpern Filters Halpern filters are closely related to Papoulis filters, and their power response is expressed as[44]

[43] See Papoulis A., "Optimum Filters with Monotonic Response," *Proceedings of IRE*, Vol. 46, No. 3, pp. 606–609, 1958 and Papoulis A., "On Monotonic Response Filters," *Proceedings of IRE*, Vol. 47, pp. 332–333, 1959.

[44] See Halpern P.H., "Optimum Monotonic Low-Pass Filters," *IEEE Transactions on Circuit Theory*, Vol. CT-16, No. 2, pp. 240–242, 1969.

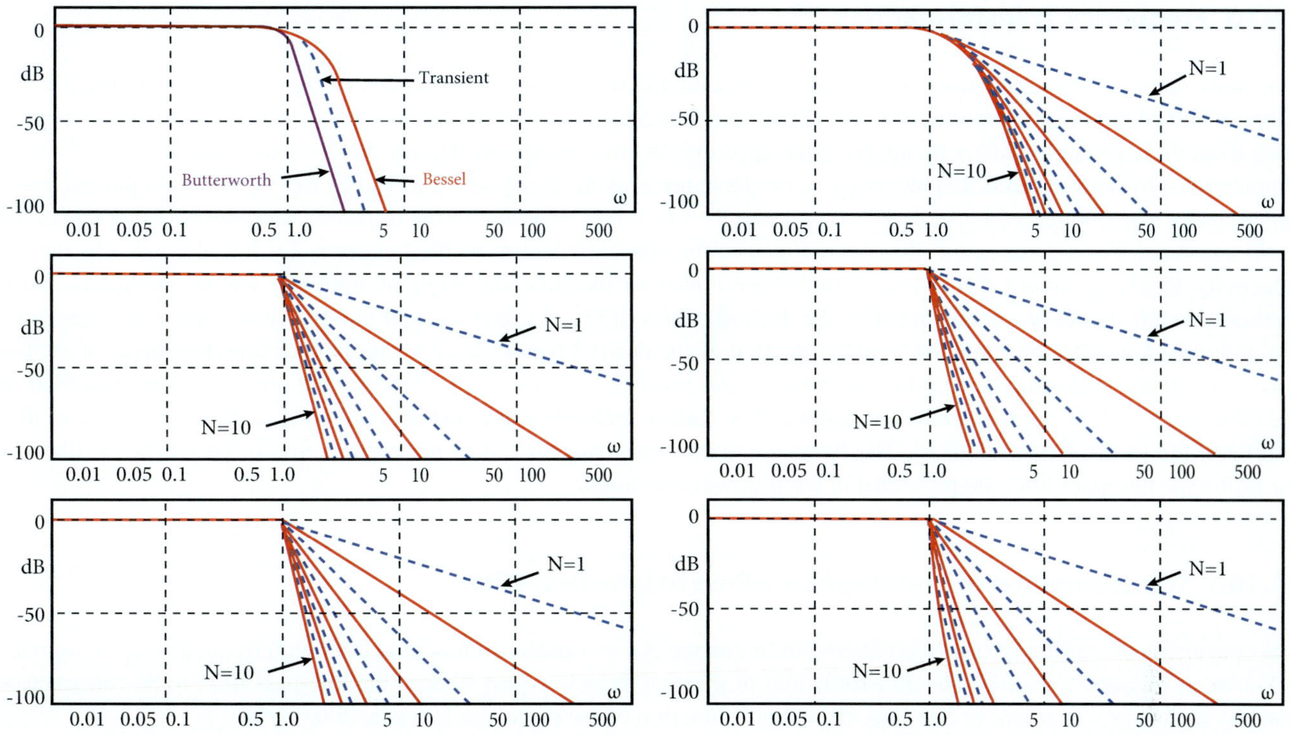

Fig. 11.72 Typical amplitude response curves for transitional filters (**a**), Gauss filters (**b**), Legendre filters (**c**), hyperspherical filters (**d**), Papoulis filters (**e**), and Halpern filters (**f**) and for various operating specifications

$$|\mathcal{H}(j\omega)|^2 = \frac{1}{1+\varepsilon^2[T_N^2(\omega/\omega_p)]^2} \tag{11.174}$$

where N is the order of the filter, ω_p is the limit of the passband, ε is a parameter that controls attenuation at the position $\omega = \omega_p$, and $T_N(\omega)$ is a polynomial related to the Jacobi polynomials and for the normalized frequency $\omega_p = 1$ is given by the relation

$$T_N^2(\omega) - \int_0^{\omega} x U_{N-1}^2(x)dx \tag{11.175}$$

with the polynomials $U(x)$ for even- and odd-order filters, respectively, defined as

$$U_{2k}(x) = \sqrt{4k+2}\sum_{m=0}^{k}(-1)^{k-m}x^{2m}\frac{(m+k)!}{(k-m)!(m!)^2} \tag{11.176}$$

$$U_{2k+1}(x) = 2\sqrt{k+1}\sum_{m=0}^{k}(-1)^{m}x^{2(k-m)}x^{2(k-m)+1}\frac{(2k+1-m)!}{m!(k+1-m)!(k-m)!} \tag{11.177}$$

Indicative amplitude response curves for the foregoing approximations of ideal filters and for various operating specifications are presented in Fig. 11.72.

11.10 Frequency Transformations

The construction of approximations of ideal filters presented in the previous pages refers to the so-called *prototype low-pass filters*, while all the other categories of filters (high-pass, band-pass, band-stop, and low-pass with any cutoff frequency) can arise from these prototype filters by applying the appropriate transformations. Recall that the prototype low-pass filters are characterized by a normalized frequency $\Omega_p = 1$ rad/s expressing the *passband limit* and by the values of the parameters A_p and A_s, which are expressed in decibels and describe the *maximum deviation in the passband* and *the minimum attenuation in the stopband*. For a given set of filter operating specifications, we determine the frequency Ω_s (or, alternatively, the filter selectivity factor, assuming that $\Omega_p = 1$ rad/s), we construct the function $\mathcal{H}(p)$ by applying one of the approximation methods, which should be characterized by the desired values of Ω_s, A_p, and A_s, and finally, we *transform* this function of the prototype low-pass filter into the transfer function of the desired type of filter by applying a *transformation* of the form $s = f(p)$. In the following description, we will use the variable $s = u + j\Omega$ to describe the prototype low-pass filter and the variable $p = \sigma + j\omega$ to describe the desired filter, and consequently, the transfer functions of these two types of filters will be of the forms $\mathcal{H}(s)$ and $\mathcal{H}(p)$. The frequency transformations that allow the construction of each type of filter from the prototype low-pass filter are presented in the following sections.

11.10.1 Transformation of Prototype Low-Pass to Low-Pass Filter

The conversion of a prototype low-pass filter with frequency $\Omega_p = 1$ rad/s to a low-pass filter with frequency $\omega_p = \omega_0\, rad/s$ is known as *frequency scaling*. The transformation of the prototype low-pass filter to the low-pass filter to be constructed is done by replacing $s = p/\omega_0$. In this case, the transfer function of the prototype low-pass filter, which is

$$\mathcal{H}_{\text{LPP}}(s) = \sum_{k=0}^{M} \beta_k s^k \Big/ \sum_{k=0}^{N} \alpha_k s^k \tag{11.178}$$

will be transformed into the function of the desired low-pass filter

$$\mathcal{H}_{\text{LP}}(p) = \frac{\sum_{k=0}^{M} \beta_k \left(\frac{p}{\omega_0}\right)^k}{\sum_{k=0}^{N} \alpha_k \left(\frac{p}{\omega_0}\right)^k} = \frac{\sum_{k=0}^{M} \beta_k p^k \omega_0^{-k}}{\sum_{k=0}^{N} \alpha_k p^k \omega_0^{-k}} = \frac{\omega_0^{-M} \sum_{k=0}^{M} \beta_k p^k \omega_0^{M-k}}{\omega_0^{-N} \sum_{k=0}^{N} \alpha_k p^k \omega_0^{N-k}} = \omega_0^{N-M} \frac{\sum_{k=0}^{M} \xi_k p^k}{\sum_{k=0}^{N} \zeta_k p^k} \tag{11.179}$$

where $\xi_k = \beta_k \omega_0^{M-k}$ $(k = 1, 2, \ldots, M)$ and $\zeta_k = \alpha_k \omega_0^{N-k}$ $(k = 1, 2, \ldots, N)$.

To see the effect of the foregoing transformation on the frequency of the prototype low-pass filter, let us replace in the equation $s = p/\omega_0$ the quantities $s = j\Omega$ and $p = j\omega$, from which it immediately follows that $\Omega = \omega/\omega_0$ or, equivalently, $\omega = \omega_0\Omega$. It is not difficult to see that the range of values $[0, \infty)$ of the Ω axis is mapped to the range of values $[0, \infty)$ of the ω axis, while a frequency with value Ω is depicted at the frequency with a value of $\omega_0\Omega$, with the corresponding amplitude response value remaining the same; therefore, we will have $\mathcal{H}_{\text{LP}}(j\omega) = \mathcal{H}_{\text{LPP}}(j\omega/\omega_0)$, from which it follows that $|\mathcal{H}_{\text{LP}}(j\omega)| = |\mathcal{H}_{\text{LPP}}(j\omega/\omega_0)|$ and $\measuredangle\mathcal{H}_{\text{LP}}(j\omega) = \measuredangle\mathcal{H}_{\text{LPP}}(j\omega/\omega_0)$. The limits of the pass and cutoff zones $\Omega_p = 1$ rad/s and Ω_s are depicted at the values $\omega_p = \omega_0$ and $\omega_s = \omega_0\Omega_s$, respectively. This situation is presented in Fig. 11.73, from which we see that the filter that results from the foregoing transformation is indeed a *low-pass filter*. Therefore, for a low-pass filter with specific frequency values ω_s and ω_p, the prototype low-pass filter that will be designed to then be transformed into the desired filter will be characterized by a frequency value of $\Omega_s = \omega_s/\omega_p$, with the parameter ω_0 corresponding to the cutoff frequency of the desired low-pass filter.

Poles and Zeros of Transfer Function

Formulating the transfer function $\mathcal{H}_{\text{LPP}}(s)$ in its factorized form

$$\mathcal{H}_{\text{LPP}}(s) = K_{\text{LPP}} \left\{\prod_{k=0}^{M} (s - s_{z_k})\right\} \Big/ \left\{\prod_{k=0}^{N} (s - s_{p_k})\right\} \tag{11.180}$$

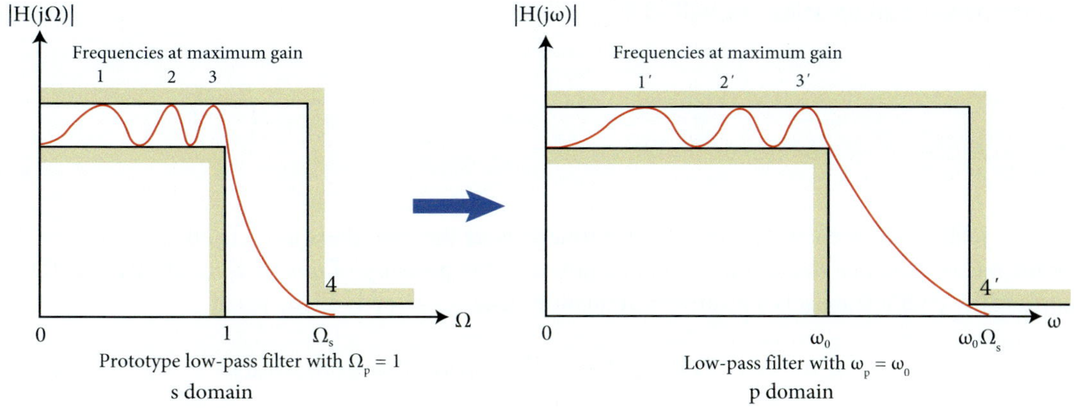

Fig. 11.73 Construction of low-pass filter from prototype low-pass filter (Chen, 2002, Figure 3.2)

where s_{z_k} $(k = 1, 2, \ldots, M)$ and s_{p_k} $(k = 1, 2, \ldots, N)$ are the zeros and poles of the function $\mathcal{H}_{\text{LPP}}(s)$, the substitution $s = p/\omega_0$, will lead to the factored function of the low-pass filter

$$\mathcal{H}_{\text{LP}}(p) = K_{\text{LPP}} \frac{\prod_{k=0}^{M}\left(\dfrac{p}{\omega_0} - s_{z_k}\right)}{\prod_{k=0}^{N}\left(\dfrac{p}{\omega_0} - s_{p_k}\right)} = K_{\text{LPP}}\omega_0^{N-M} \frac{\prod_{k=0}^{M}(p - \omega_0 s_{z_k})}{\prod_{k=0}^{N}(p - \omega_0 s_{p_k})} = K_{\text{LP}} \frac{\prod_{k=0}^{M}(p - p_{z_k})}{\prod_{k=0}^{N}(p - p_{p_k})} \tag{11.181}$$

where $K_{\text{LP}} = K_{\text{LPP}}\omega_0^{N-M}$ and $p_{z_k} = \omega_0 s_{z_k}$ $(k = 1, 2, \ldots, M)$, $p_{p_k} = \omega_0 s_{p_k}$ $(k = 1, 2, \ldots, N)$ are the zeros and the poles of the transfer function $\mathcal{H}_{\text{LP}}(p)$. We observe, therefore, that *the poles and zeros of the transfer function of the low-pass filter to be constructed arise from the multiplication of the corresponding zeros and poles of the transfer function of the prototype low-pass filter by the scaling factor* ω_0. Since the value of this factor is positive and real, this process causes a corresponding scaling in the *magnitudes* of the poles and zeros of the prototype low-pass filter, leaving the phases unaffected. By analogy, the pole–zero diagram of the low-pass filter that arises will be the same as that of the prototype low-pass filter, with the axes of the diagram having undergone the previously described process of *scaling*.

Selectivity and Spectral Efficiency Factor

For the calculation of the *selectivity* and the *spectral efficiency factor* of the transformed low-pass filter, we can directly use the equations constructed in the description of the various approximations of ideal filters, since these concern the *general case* and not the special case of normalized approximations.

Phase Characteristics and Time-Domain Response

Taking into account that the phase responses of the prototype and the transformed low-pass filter are related to each other through the expression

$$\measuredangle\mathcal{H}_{\text{LP}}(j\omega) = \measuredangle\mathcal{H}_{\text{LPP}}(j\Omega) = \measuredangle\mathcal{H}_{\text{LPP}}\left(j\frac{\omega}{\omega_0}\right) \tag{11.182}$$

we can see that the phase delays of the two filters will be related by the equation

$$\tau_{\text{LP}}^{\varphi}(\omega) = -\frac{\measuredangle\mathcal{H}_{\text{LP}}(j\omega)}{\omega} = -\frac{1}{\omega}\left[\measuredangle\mathcal{H}_{\text{LPP}}\left(j\frac{\omega}{\omega_0}\right)\right] = \frac{1}{\omega_0}\left[-\frac{\omega_0}{\omega}\measuredangle\mathcal{H}_{\text{LPP}}\left(j\frac{\omega}{\omega_0}\right)\right] = \frac{1}{\omega_0}\tau_{\text{LPP}}^{\varphi}\left(\frac{\omega}{\omega_0}\right) \tag{11.183}$$

while for the corresponding group delays we will have

$$\tau^g_{\text{LP}}(\omega) = -\frac{d\measuredangle\mathcal{H}_{\text{LP}}(j\omega)}{d\omega} = -\frac{d}{\omega}\left[\measuredangle\mathcal{H}_{\text{LPP}}\left(j\frac{\omega}{\omega_0}\right)\right] = \frac{1}{\omega_0}\tau^g_{\text{LPP}}\left(\frac{\omega}{\omega_0}\right) \tag{11.184}$$

On the other hand, taking into account that the transfer functions of the two filters are related as $\mathcal{H}_{\text{LP}}(p) = \mathcal{H}_{\text{LPP}}(\omega_0 s)$, the application of the *inverse Laplace transform* in combination with the property of *time scaling* of this transform and the fact that the proportionality coefficient ω_0 is a positive real quantity leads us directly to the result

$$h_{\text{LP}}(t) = \mathcal{L}^{-1}\{\mathcal{H}_{\text{LP}}(p)\} = \mathcal{L}^{-1}\{\mathcal{H}_{\text{LPP}}(\omega_0 s)\} = \omega_0 h_{\text{LPP}}(\omega_0 t) \tag{11.185}$$

We are led, therefore, to the conclusion that *the impulse response of the transformed low-pass filter is a version of the corresponding response of the original filter that is scaled both in time and amplitude*. If $\omega_0 > 1$, then the response $h_{\text{LP}}(t)$ will have a larger amplitude than $h_{\text{LPP}}(t)$ and will be *compressed* in time, while in the opposite case it will have a smaller amplitude and will be *extended* in time. Finally, the *step response* of the transformed filter will result from the integration of its impulse response, so it turns out that $s_{\text{LP}}(t) = h_{\text{LPP}}(\omega_0 t)$. Therefore, this response is *scaled in time* compared to the response of the original low-pass filter, but not in terms of amplitude.

Problem 11.10.1 Construct a low-pass filter with operating specifications $\omega_p = 2000$ rad/s, $\omega_s = 7000$ rad/s, $A_p = 0.4$ dB, and $A_s = 45$ dB using a Butterworth approximation as the prototype low-pass filter (Antoniou, Exercise for solution 10.29).

Solution Applying the theory presented in the previous section, let us calculate the poles of the transfer function of a prototype low-pass Butterworth filter with operating specifications $\Omega_p = 1$ rad/s, $\Omega_s = \omega_s/\omega_p = 7000/2000 = 3.5$ rad/s, $A_p = 0.4$ dB, and $A_s = 45$ dB. Applying the defining equation of the minimum order filter, it turns out that

$$N = \frac{1}{2}\frac{\log\left(\dfrac{10^{A_s/10}-1}{10^{R_p/10}-1}\right)}{\log\left(\dfrac{\Omega_s}{\Omega_p}\right)} = \frac{1}{2}\frac{\log\left(\dfrac{10^{45/10}-1}{10^{0.4/10}-1}\right)}{\log\left(\dfrac{3.5}{1.0}\right)} = \frac{1}{2}\frac{\log\left(\dfrac{31621}{0.09647}\right)}{\log(3.5)} = 5.068$$

The foregoing value should be rounded to the next highest integer, and therefore the minimum order of the Butterworth filter we will use is $N = 6$. However, this rounding will affect the operating frequency values of the prototype low-pass filter, which will no longer be $\Omega_p = 1$ rad/s and $\Omega_s = 3.5$ rad/s, but $\Omega_p = (10^{0.1A_p} - 1)^{1/2N} = (10^{0.4} - 1)^{1/12} = 0.822941$ rad/s and $\Omega_s = (10^{0.1A_s} - 1)^{1/2N} = (10^{4.5} - 1)^{1/12} = 2.371367$ rad/s. Therefore, the scaling factor ω_0 will be equal to $\omega_0 = \omega_p\Omega_p = 2000/0.822941 = 2430.307883$.

Referring back to Table 11.1 on page 1006, we find that the transfer function of the prototype low-pass Butterworth filter has three pairs of complex conjugate poles with values

$$s_{p_1} = -0.9659 \pm 0.2588j, \quad s_{p_2} = -0.7071 \pm 0.7071j, \quad s_{p_3} = -0.2588 \pm 0.9659j$$

and therefore, the corresponding poles of the low-pass filter to be constructed will be

$$p_{p_1} = 2430.307883 \times (-0.9659 \pm 0.2588j) = -2347.434383 \pm 628.963680j$$

$$p_{p_2} = 2430.307883 \times (-0.7071 \pm 0.7071j) = -1718.470704 \pm 1718.470704j$$

$$p_{p_3} = 2430.307883 \times (-0.2588 \pm 0.9659j) = -628.963680 \pm 2347.434384j$$

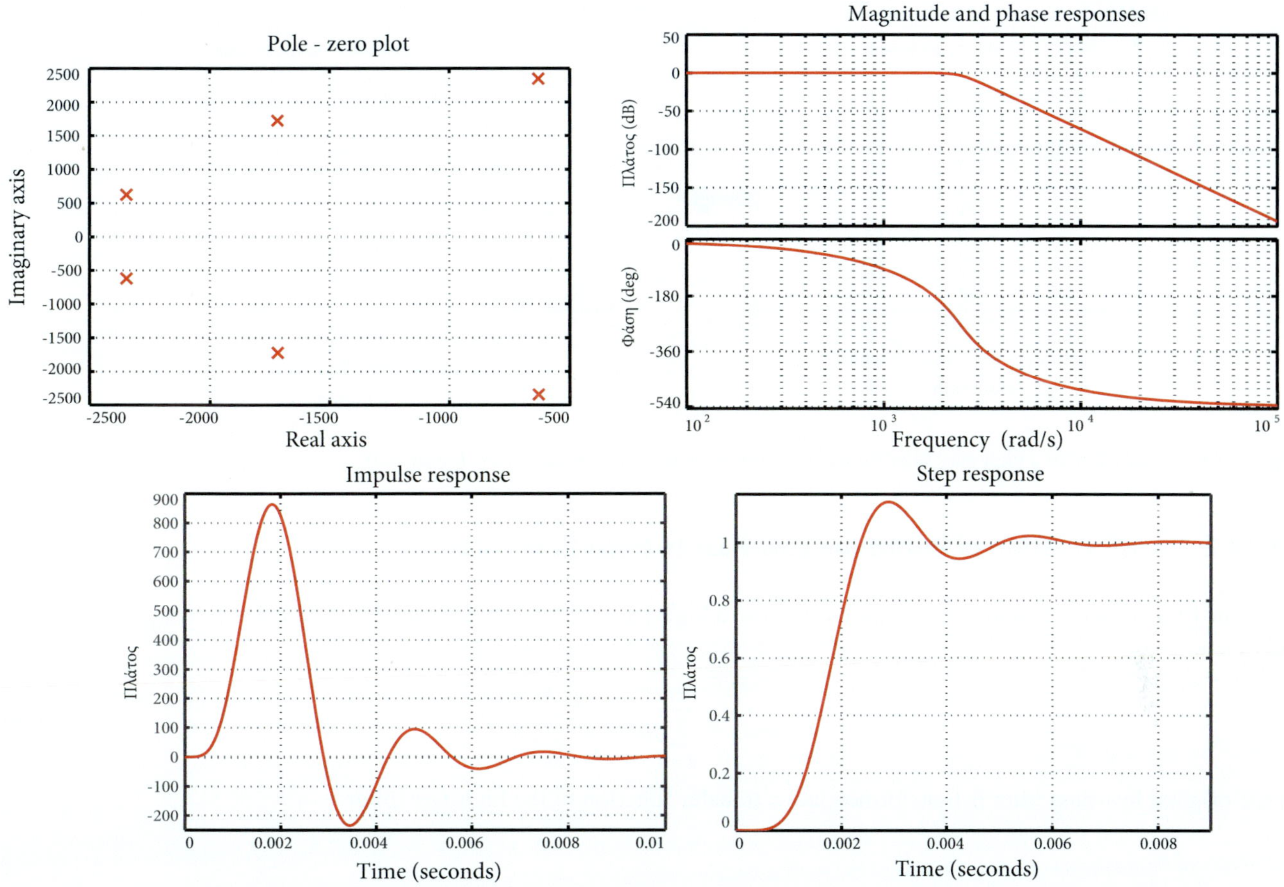

Fig. 11.74 Pole–zero diagram, amplitude and phase responses, and impulse and step responses of low-pass filter from Exercise 11.10.1

with contributions to the denominator of the transfer function $\mathcal{H}_{\text{LP}}(p)$ of the quadratic terms

$$p^2 + (4.694868 \times 10^3)\, p + 5.906043 \times 10^6$$

$$p^2 + (3.436941 \times 10^3)\, p + 5.906283 \times 10^6$$

$$p^2 + (1.257927 \times 10^3)\, p + 5.906043 \times 10^6$$

respectively. Therefore, the transfer function will have the form

$$\mathcal{H}_{\text{LP}}(p)$$

$$= \frac{H_0}{[p^2 + (4.694 \cdot 10^3)p + 5.906 \cdot 10^6] \times [p^2 + (3.436 \cdot 10^3)p + 5.906 \cdot 10^6] \times [p^2 + (1.257 \cdot 10^3)p + 5.906 \cdot 10^6]}$$

$$= \frac{2.060 \cdot 10^{20}}{p^6 + 9.389 \cdot 10^3 p^5 + 4.408 \cdot 10^7 s^4 + 1.312 \cdot 10^{11} s^3 + 2.603 \cdot 10^{14} s^2 + 3.275 \cdot 10^{17} s + 2.060 \cdot 10^{20}}$$

The pole–zero diagram, the amplitude and phase responses, and the impulse and step responses of the desired low-pass filter are presented in Fig. 11.74. These graphical representations were created using MATLAB functions.

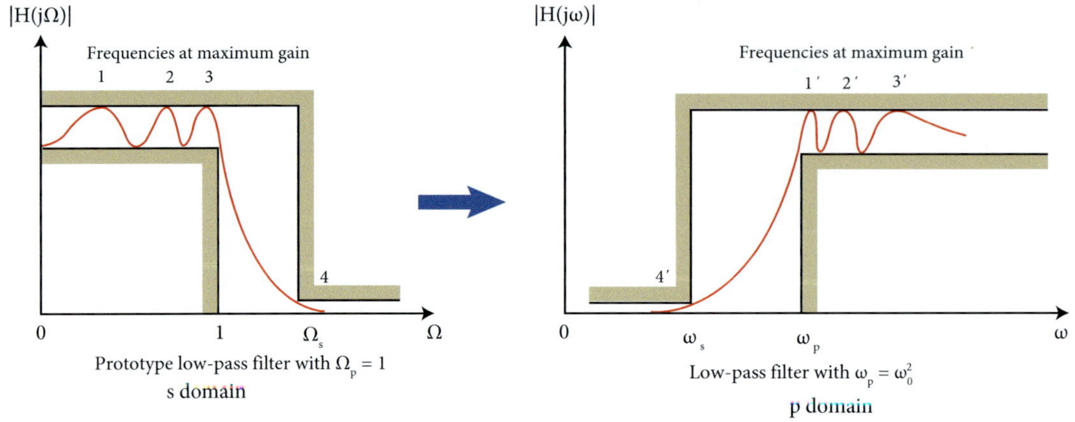

Fig. 11.75 Construction of a high-pass filter from a prototype low-pass filter (Chen, 2002, Figure 3.5)

11.10.2 Transformation of Prototype Low-Pass to High-Pass Filter

The transformation of a prototype low-pass filter into a high-pass filter is implemented by replacing $s = \omega_0^2/p$. In this case, the function

$$\mathcal{H}_{\text{LPP}}(s) = \sum_{k=0}^{M} \beta_k s^k \Big/ \sum_{k=0}^{N} \alpha_k s^k \tag{11.186}$$

of the original low-pass filter is transformed into a transfer function of the high-pass filter

$$\mathcal{H}_{\text{HP}}(p) = \frac{\sum_{k=0}^{M} \beta_k \left(\frac{\omega_0^2}{p}\right)^k}{\sum_{k=0}^{N} \alpha_k \left(\frac{\omega_0^2}{p}\right)^k} = \frac{\sum_{k=0}^{M} \beta_k \left(\frac{p}{\omega_0^2}\right)^{-k}}{\sum_{k=0}^{N} \alpha_k \left(\frac{p}{\omega_0^2}\right)^{-k}} = \frac{\left(\frac{p}{\omega_0^2}\right)^{-M} \sum_{k=0}^{M} \beta_k \left(\frac{p}{\omega_0^2}\right)^{M-k}}{\left(\frac{p}{\omega_0^2}\right)^{-N} \sum_{k=0}^{N} \alpha_k \left(\frac{p}{\omega_0^2}\right)^{N-k}} = \left(\frac{p}{\omega_0^2}\right)^{N-M} \frac{\sum_{k=0}^{M} \beta_k \left(\frac{p}{\omega_0^2}\right)^{M-k}}{\sum_{k=0}^{N} \alpha_k \left(\frac{p}{\omega_0^2}\right)^{N-k}} \tag{11.187}$$

To see exactly how the frequency transformation is carried out in the s and p spaces, let us make the substitutions $s = j\Omega$ and $p = j\omega$, so it immediately follows that $\omega = -\omega_0^2/\Omega$. From this relationship the conclusion is drawn that *the positive frequencies of the original low-pass filter are depicted in the inverse scaled frequencies of the high-pass filter*. More specifically, the value range $[0, \infty)$ of the Ω axis is depicted in the value range $(-\infty, 0]$ of the ω axis, while the value range $(-\infty, 0]$ of the Ω axis is depicted in the value range $[0, \infty)$ of the ω axis. Finally, the limit of the passband $\Omega_p = \pm 1$ rad/s and the cutoff limit $\pm\,\Omega_s$ of the prototype low-pass filter are depicted at the frequencies $\omega_p = \pm\omega_0^2$ and $\omega_s = \pm\omega_0^2/\Omega_s$ of the high-pass filter. This situation is presented diagrammatically in Fig. 11.75. In this figure, we see that the application of the transformation, which is characterized by a high degree of nonlinearity, caused a change in the characteristics of the amplitude response curve, since the peaks and valleys of the low-pass filter's fluctuations in the passband (which, as is well known, describe the Chebyshev I filters and the elliptical filters) have been distorted, so as to occupy the entire value range $[\omega_p, \infty)$ in the amplitude response of the transformed high-pass filter. It is not difficult to see that the frequency responses of the two filters are connected by the relation $\mathcal{H}_{\text{HP}}(j\omega) = \mathcal{H}_{\text{LPP}}(\omega_0^2/j\Omega)$, while it will also hold that $\mathcal{H}_{\text{HP}}(\infty) = \mathcal{H}_{\text{LPP}}(0)$ and $\mathcal{H}_{\text{HP}}(j\omega_0^2) = \mathcal{H}_{\text{LPP}}(j)$.

From the foregoing description it is concluded that to represent a high-pass filter with operating frequencies ω_p and ω_s in the *equivalent prototype low-pass filter*, we define the cutoff frequency of the prototype low-pass filter as $\Omega_s = \omega_p/\omega_s$, and we construct its transfer function using some of the approaches presented in the previous sections, so that it satisfies simultaneously the values of A_p and A_s dictated by the operating specifications of the filter. After completing this process, the corresponding high-pass filter is constructed through the application of the transformation $p = \omega_0^2/s$.

The calculation of the minimum order of the prototype low-pass filter that will be transformed into the high-pass filter to be constructed is carried out using the equations constructed in the previous sections, in which we must make the substitution $\Omega_s/\Omega_p \to \omega_p/\omega_s$, since, under the influence of the transformation, the frequencies $\pm\,\Omega_p$ and $\pm\,\Omega_s$ will be represented at the frequencies $\mp\,\omega_0^2/\Omega_p$ and $\mp\,\omega_0^2/\Omega_s$, and therefore, $\omega_p/\omega_s = \Omega_s/\Omega_p$.

Poles and Zeros of Transfer Function

Formulating the transfer function $\mathcal{H}_{\text{LPP}}(s)$ in its factorized form

$$\mathcal{H}_{\text{LPP}}(s) = K_{\text{LPP}} \left\{\prod_{k=0}^{M}(s - s_{z_k})\right\} \Big/ \left\{\prod_{k=0}^{N}(s - s_{p_k})\right\} \tag{11.188}$$

where s_{z_k} $(k = 1, 2, \ldots, M)$ and s_{p_k} $(k = 1, 2, \ldots, N)$ are the zero values and the poles of the function $\mathcal{H}_{\text{LPP}}(s)$, the substitution $s = \omega_0^2/p$ will lead to the factorized function of the high-pass filter

$$\begin{aligned}\mathcal{H}_{\text{HP}}(p) &= K_{\text{LPP}} \frac{\prod_{k=0}^{M}\left(\frac{\omega_0^2}{p} - s_{z_k}\right)}{\prod_{k=0}^{N}\left(\frac{\omega_0^2}{p} - s_{p_k}\right)} = K_{\text{LPP}} \frac{\prod_{k=1}^{M}\left[\frac{1}{p}(\omega_0^2 - p s_{z_k})\right]}{\prod_{k=1}^{N}\left[\frac{1}{p}(\omega_0^2 - p s_{p_k})\right]} = K_{\text{LPP}} \frac{\prod_{k=1}^{M}\left[\frac{s_{zk}}{p}\left(\frac{\omega_0^2}{s_{zk}} - p\right)\right]}{\prod_{k=1}^{N}\left[\frac{s_{pk}}{p}\left(\frac{\omega_0^2}{s_{pk}} - p\right)\right]} \\ &= \left\{K_{\text{LPP}}(-1)^{M-N} \frac{\prod_{k=1}^{M}(s_{z_k})}{\prod_{k=1}^{N}(s_{p_k})}\right\} p^{N-m} \frac{\prod_{k=1}^{M}\left(p - \frac{\omega_0^2}{s_{zk}}\right)}{\prod_{k=1}^{N}\left(p - \frac{\omega_0^2}{s_{pk}}\right)} = K_{\text{HP}} p^{N-M} \frac{\prod_{k=0}^{M}(p - p_{z_k})}{\prod_{k=0}^{N}(p - p_{p_k})}\end{aligned} \tag{11.189}$$

In the foregoing equation, the constant K_{HP} is defined as

$$K_{\text{HP}} = K_{\text{LPP}}(-1)^{M-N} \left\{\prod_{k=1}^{M}(s_{z_k})\right\} \Big/ \left\{\prod_{k=1}^{N}(s_{p_k})\right\}, \quad p_{z_k} = \frac{\omega_0^2}{s_{zk}}, \quad p_{p_k} = \frac{\omega_0^2}{s_{pk}} \tag{11.190}$$

where p_{z_k} and p_{p_k} are the poles and the zero values of the function $\mathcal{H}_{\text{HP}}(p)$ of the high-pass filter. We observe, therefore, that *the poles and the zeros of the function $\mathcal{H}_{HP}(p)$ are the reciprocals of the poles and zeros of the function $\mathcal{H}_{LPP}(s)$, having also undergone scaling by the factor ω_0^2.* It is interesting to note that the multiplication of the quantity p^{N-M} by the numerator polynomial of degree M leads to a numerator polynomial whose degree is equal to N, that is, equal to the degree of the denominator polynomial, something that is consistent with the already presented theory of high-pass filters.

Selectivity and Spectral Efficiency Factor

The selectivity of a high-pass filter is characterized by a *positive sign*, so as to be a *positive number*, and can be calculated using the selectivity of the corresponding prototype low-pass filter from the equation

$$F_s^{\text{HP}} = \left.\frac{d|\mathcal{H}_{\text{HP}}(j\omega)|}{d\omega}\right|_{\omega=\omega_c} = \left\{\frac{d}{d\omega}\left|\mathcal{H}_{\text{LPP}}\left(-j\frac{\omega_0}{\omega^2}\right)\right|\right\}_{\omega=\omega_c} = \frac{\omega_0^2}{\omega_c^2} F_s^{\text{LPP}} \tag{11.191}$$

with the selectivity F_s^{LPP} being calculated at the position of the frequency Ω_c, which corresponds to attenuation 3 dB. On the other hand, the *spectral efficiency factors of the two filters proves that they are equal to each other*, or, in mathematical notation, $S_{\text{HP}} = S_{\text{LPP}}$. However, let us note that because the passband of high-pass filters has *infinite width*, this size has no particular usefulness.

Phase Characteristics and Response in the Time Domain

The phase responses of the prototype low-pass filter and the transformed high-pass filter are related to each other through the expression

$$\measuredangle\mathcal{H}_{\text{HP}}(j\omega) = \measuredangle\mathcal{H}_{\text{LPP}}(j\Omega) = \measuredangle\mathcal{H}_{\text{LPP}}\Big(-j\frac{\omega_0^2}{\omega}\Big) \tag{11.192}$$

since the phase response is characterized by *odd symmetry*. Let us also note that in a graphical representation of the phase response, it is permissible to add a *constant phase shift* by an integer multiple of 2π, so that we can write that $\measuredangle\mathcal{H}_{\text{HP}}(j\omega) = \measuredangle\mathcal{H}_{\text{LPP}}(-j\omega_0^2/\omega) + 2\pi m$. From the foregoing relationship it follows that $\measuredangle\mathcal{H}_{\text{HP}}(0) = -\measuredangle\mathcal{H}_{\text{LPP}}(\infty)$, $\measuredangle\mathcal{H}_{\text{HP}}(\infty) = -\measuredangle\mathcal{H}_{\text{LPP}}(0)$ and $\measuredangle\mathcal{H}_{\text{HP}}(j\omega_0^2) = -\measuredangle\mathcal{H}_{\text{LPP}}(j)$.

On the other hand, the *phase delays* for these two filters are related to each other as

$$\tau_{\text{HP}}^{\varphi}(\omega) = -\frac{\measuredangle\mathcal{H}_{\text{HP}}(j\omega)}{\omega} = -\frac{1}{\omega}\Big[\measuredangle\mathcal{H}_{\text{LPP}}\Big(-j\frac{\omega_0^2}{\omega}\Big) + 2\pi m\Big] \tag{11.193}$$

From the foregoing relation it follows that $\tau_{\text{HP}}^{\varphi}(\infty) = 0$. Finally, the *group delay* of the high-pass filter can be calculated from the relation

$$\tau_{\text{HP}}^{g}(\omega) = -\frac{d\measuredangle\mathcal{H}_{\text{HP}}(j\omega)}{d\omega} = -\frac{d}{d\omega}\measuredangle\mathcal{H}_{\text{LPP}}\Big(-j\frac{\omega_0^2}{\omega}\Big) = -\frac{d}{d\omega}\Big(-\frac{\omega_0^2}{\omega}\Big)\tau_{\text{LPP}}^{g}\Big(-\frac{\omega_0^2}{\omega}\Big) = \frac{\omega_0^2}{\omega^2}\tau_{\text{LPP}}^{g}\Big(\frac{\omega_0^2}{\omega}\Big) \tag{11.194}$$

where to derive the final result we used the fact that the group delay is characterized by *even symmetry* with respect to the frequency ω.

To describe the *impulse* and the *step responses* of a high-pass filter, let us formulate its transfer function in the form

$$\mathcal{H}_{\text{HP}}(p) = K_{\text{HP}}\,p^{N-m}\frac{\prod_{k=1}^{M}\Big(p-\frac{\omega_0^2}{s_{zk}}\Big)}{\prod_{k=1}^{N}\Big(p-\frac{\omega_0^2}{s_{pk}}\Big)} = K_{\text{HP}} - \frac{K_{\text{HP}}\Big[\prod_{k=1}^{N}\Big(p-\frac{\omega_0^2}{s_{pk}}\Big) - p^{N-M}\prod_{k=1}^{M}\Big(p-\frac{\omega_0^2}{s_{zk}}\Big)\Big]}{\prod_{k=1}^{N}\Big(p-\frac{\omega_0^2}{s_{pk}}\Big)} \tag{11.195}$$

Taking into account that the products of the first-degree terms in the numerator and the denominator constitute the factorized form of the polynomials $N(s) = \beta_0 + \beta_1 p + \cdots + \beta_M p^M$ and $D(s) = \alpha_0 + \alpha_1 p + \cdots + \alpha_M p^M$ of the rational transfer function, we can distinguish two cases:

$\boldsymbol{M = 0}$ In this case, the foregoing equation is written

$$\mathcal{H}_{\text{HP}}(p) = K_{\text{HP}} - \Big\{K_{\text{HP}}\Big[\prod_{k=1}^{N}\Big(p-\frac{\omega_0^2}{s_{pk}}\Big) - p^N\Big]\Big/\prod_{k=1}^{N}\Big(p-\frac{\omega_0^2}{s_{pk}}\Big)\Big\} \tag{11.196}$$

Expanding the product in the numerator and taking into account that the term ρ^N that will emerge will be eliminated by the term ρ^N that exists at the end, it is not difficult to realize that the term with the highest power in the numerator will be $\alpha_{N-1}\rho^{N-1}$, so we can write that

$$\mathcal{H}_{\text{HP}}(p) = K_{\text{HP}} - \left\{ K_{\text{HP}}\alpha_{N-1} \prod_{k=1}^{N-1} \left(p - \frac{\omega_0^2}{s_{pk}}\right) \Big/ \prod_{k=1}^{N} \left(p - \frac{\omega_0^2}{s_{pk}}\right) \right\} \tag{11.197}$$

***M* > 0** In this case, the multiplication of the second term that expresses a factorized polynomial of degree M over p^{N-M} makes it a polynomial of degree N, and therefore

$$\mathcal{H}_{\text{HP}}(p) = K_{\text{HP}} - \left\{ K_{\text{HP}} \prod_{k=1}^{N} (p - \beta_k) \Big/ \prod_{k=1}^{N} \left(p - \frac{\omega_0^2}{s_{pk}}\right) \right\} \tag{11.198}$$

where β_k are suitably calculated coefficients. But the second term in both of the foregoing cases is recognized as a *low-pass filter transfer function* $\mathcal{H}_{\text{LP}}(s)$, so that we can write

$$\mathcal{H}_{\text{HP}}(p) = K_{\text{HP}} - \mathcal{H}_{\text{LP}}(p) \tag{11.199}$$

Therefore, taking the *inverse Laplace transform* of the foregoing relation and using the relation $h(t) = \mathcal{L}^{-1}\{\mathcal{H}(p)\}$, the impulse response of the high-pass filter is calculated as

$$h_{\text{HP}}(t) = K_{\text{HP}}\delta t - h_{\text{LP}}(t) \tag{11.200}$$

from which we find that it has an impulse function at the position $t = 0$, which, of course, is expected. Finally, the *step response* of the high-pass filter will be calculated from the integral of the impulse response, leading to a result of the form

$$s_{\text{HP}}(t) = \int_0^t h_{\text{HP}}(\tau)d\tau = K_{\text{HP}} \int_0^t \delta(\tau)d\tau - \int_0^t h_{\text{LP}}(\tau)d\tau = K_{\text{HP}} - s_{\text{LP}}(t) \tag{11.201}$$

Problem 11.10.2 Construct a high-pass filter with operating specifications $\omega_p = 2000$ rad/s, $\omega_s = 1000$ rad/s, $A_p = 0.5$ dB, and $A_s = 40$ dB using the Chebyshev I approximation as the prototype low-pass filter (Antoniou, Exercise for resolution 10.34).

Solution Applying the theory of the previous section, let us calculate the poles of the transfer function of a prototype low-pass Chebyshev filter with operating specifications $\Omega_p = 1$ rad/s, $\Omega_s = \omega_p/\omega_s = 2000/1000 = 2$ rad/s, $A_p = 0.5$ dB, and $A_s = 40$ dB. Applying the equation defining the minimum order of the filter results in

$$N = \frac{\cosh^{-1}\left(\dfrac{\sqrt{10^{A_s/10}-1}}{\sqrt{10^{R_p/10}-1}}\right)}{\cosh^{-1}\left(\dfrac{\Omega_s}{\Omega_p}\right)} = \frac{\cosh^{-1}\left(\dfrac{\sqrt{10^{40/10}-1}}{\sqrt{10^{0.5/10}-1}}\right)}{\cosh^{-1}\left(\dfrac{2}{1}\right)} = \frac{\cosh^{-1}(286.263201)}{\cosh^{-1}(2)} = \frac{6.350055}{1.316957} = 4.821760$$

where for the calculation of the inverse hyperbolic cosine we used the identity $\cosh^{-1} x = \ln(x + \sqrt{x^2 - 1})$. The foregoing value should be rounded to the next highest integer, and therefore the minimum value of order N is $N = 5$. For the value of the attenuation level $A_p = 0.5$ dB, successively we have $\varepsilon = \sqrt{10^{A_p/10-1}} = \sqrt{10^{0.05} - 1} = 0.349311$, $x = 1/\varepsilon = 2.862775$,

$\sinh^{-1}(/\varepsilon) = \sinh^{-1}(x) = \ln(x + \sqrt{x^2+1}) = 1.774135$, $(1/N)\sinh^{-1}(/\varepsilon) = 0.354827$, $\sinh[(1/N)\sinh^{-1}(/\varepsilon)] = 0.362320$, and $\cosh[(1/N)\sinh^{-1}(/\varepsilon)] = 1.063610$. Substituting the values into the pole defining equations of the transfer function of the low-pass Chebyshev I filter, we find that it has two pairs of conjugate complex poles $p_{1,5}$, $p_{2,4}$ and a negative real pole p_3 with values

$$s_{1,5} = -0.111963 \pm 1.011553j, \quad s_{2,4} = -0.293123 \pm 0.625174j, \quad s_3 = -0.362$$

and gain value $H_0 = K_{\text{LPP}} = |s_{1,5}|^2|s_{2,4}|^2|s_3| = 0.178920$.

On the other hand, assuming that the coefficient ω_0 has a positive value and that under the influence of the transformation the frequencies $\pm\Omega_p$ will correspond to the frequencies $\mp\omega_p$, it immediately follows that $\omega_0^2 = -(\pm\Omega_p)(\mp\omega_p) = \Omega_p\omega_p = 1 \times 2000 = 2000$. Therefore, according to the theory presented in the previous section, the poles of the transfer function of the high-pass filter are calculated as

$$p_{1,5} = \frac{\omega_0^2}{s_{1,5}} = \frac{2000}{-0.111963 \pm 1.011553j} = -216 \mp 1953.228875j, \qquad p_3 = \frac{\omega_0^2}{s_3} = -5519.982336$$

$$p_{2,4} = \frac{\omega_0^2}{s_{2,4}} = \frac{2000}{-0.293123 \pm 0.625174j} = -1229.636598 \mp 2622.574245j$$

and contribute to the denominator of their transfer function the terms

$$p^2 + 432.383p + 3.861841 \times 10^6, \quad p^2 + 2459.273p + 8.389901 \times 10^6, \quad p + 5519.982$$

while the proportionality constant, calculated according to its defining equation, is found to be equal to one. Therefore, the transfer function of the high-pass filter will have the form

$$\begin{aligned}\mathcal{H}_{\text{HP}}(p) &= \frac{p^5}{(p^2 + 432.383p + 3.861841 \times 10^6)(p^2 + 2459.273p + 8.389901 \times 10^6)(p + 5519.982)} \\ &= \frac{p^5}{p^5 + 8.411 \cdot 10^3 p^4 + 2.927 \cdot 10^7 p^3 + 8.662 \times 10^{10} p^2 + 1.048 \times 10^{14} p + 1.788 \times 10^{17}}\end{aligned}$$

The pole-zero diagram, the amplitude and phase responses, and the impulse and step response of the required low-pass filter are presented in Fig. 11.76. The construction of these graphs was again done using MATLAB functions.

11.10.3 Transformation of Prototype Low-Pass to Band-Pass Filter

The transformation of a prototype low-pass filter into a band-pass filter with central frequency ω_0 and bandwidth $B = \omega_{p2} - \omega_{p_1}$, where ω_{p1} and ω_{p2} are respectively the *lower* and *upper* limits of the passband, is achieved by applying to the transfer function of the prototype filter

$$\mathcal{H}_{\text{LPP}}(s) = \sum_{k=0}^{M} \beta_k s^k \Big/ \sum_{k=0}^{N} \alpha_k s^k \tag{11.202}$$

the transformation

$$s = \frac{p^2 + \omega_0^2}{p\,(\omega_{p2} - \omega_{p1})} = \frac{p^2 + \omega_0^2}{pB} \tag{11.203}$$

The transfer function of the band-pass filter constructed in this way has the form

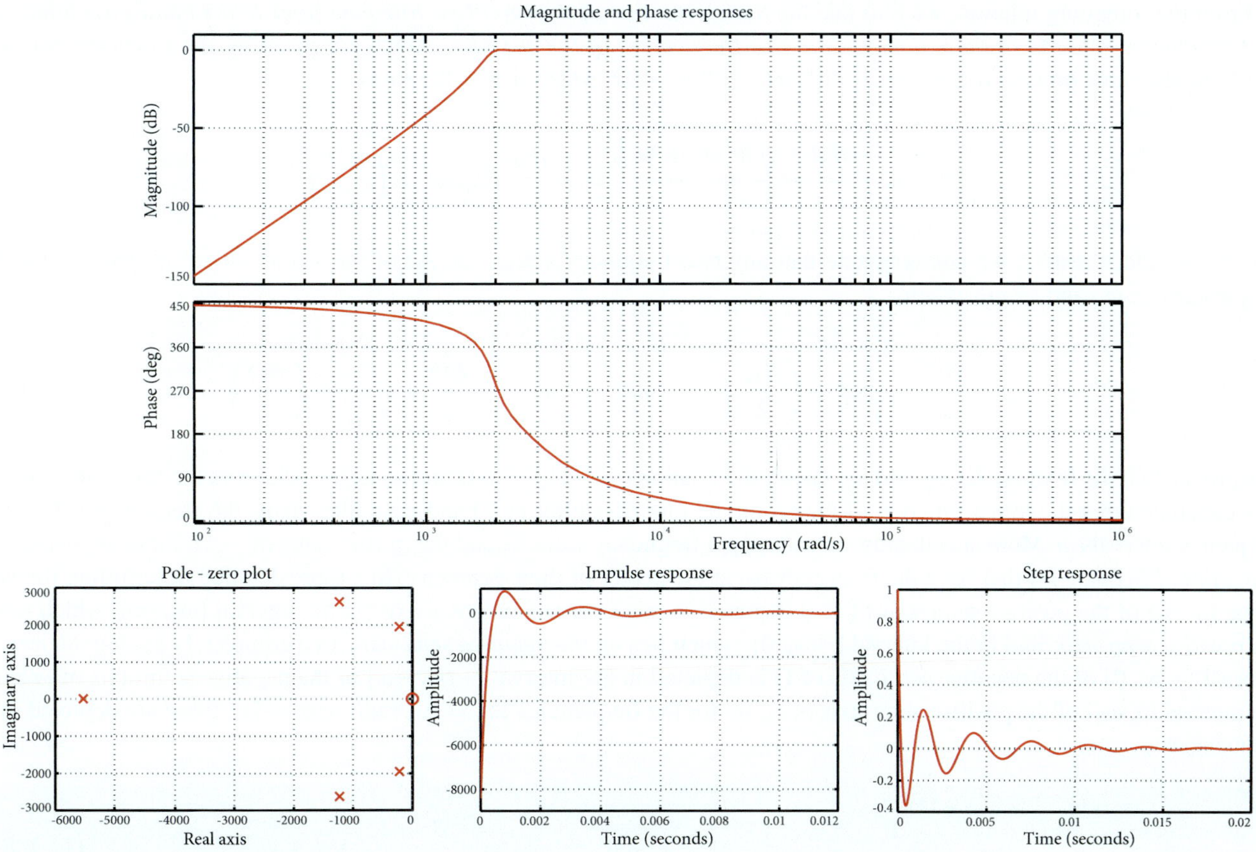

Fig. 11.76 Pole–zero diagram, amplitude and phase responses, and impulse and step responses of low-pass filter from Exercise 11.10.2

$$\mathcal{H}_{\text{BP}}(p) = \left\{ \sum_{k=0}^{M} \beta_k \left(\frac{p^2 + \omega_0^2}{pB} \right)^k \right\} \Big/ \left\{ \sum_{k=0}^{N} \alpha_k \left(\frac{p^2 + \omega_0^2}{pB} \right)^k \right\} \tag{11.204}$$

It is not difficult to see that *the order of the transfer function of the band-pass filter* $\mathcal{H}_{BP}(p)$ *is twice the order of the transfer function of the original low-pass filter* $\mathcal{H}(s)$. Furthermore, it can be proven that *the function* $\mathcal{H}(s)$ *can always be expressed as the product of a low-pass filter transfer function times a high-pass filter transfer function.* If the original low-pass filter exhibits variations in the cutoff zone, as happens, for example, with the Chebyshev II and Cauer filters, then the amplitude response of the band-pass filter will also exhibit very small-amplitude variations, while in the opposite case it will hold that $|\mathcal{H}_{\text{BP}}(0)| = |\mathcal{H}_{\text{BP}}(\infty)| = 0$.

To study the effect of this transformation on the frequency of the original low-pass filter, let us proceed with the substitutions $s = j\Omega$ and $p = j\omega$, so we will get

$$j\Omega = \frac{-\omega^2 + \omega_0^2}{j\omega B} \qquad \text{or, equivalently,} \qquad \omega^2 - (\Omega B)\omega - \omega_0^2 = 0$$

The last equation is a quadratic equation with respect to ω with discriminant $\Delta = B^2\Omega^2 + 4\omega_0^2$ and roots

$$\omega_{1,2} = \frac{B\Omega \pm \sqrt{B^2\Omega^2 + 4\omega_0^2}}{2} = \frac{B\Omega}{2} \pm \sqrt{\omega_0^2 + \left(\frac{B\Omega}{2}\right)^2} \tag{11.205}$$

From the foregoing relation, we find that *the transformation of the prototype low-pass filter to the band-pass filter maps each frequency value* Ω *of the* s *domain into two frequencies of the* p *space*, which are given by the foregoing relations. Applying the same process for the mirror frequency $s^* = -j\Omega$, we are led to the roots

$$\omega_{3,4} = \frac{-B\Omega \pm \sqrt{B^2\Omega^2 + 4\omega_0^2}}{2} = -\frac{B\Omega}{2} \pm \sqrt{\omega_0^2 + \left(\frac{B\Omega}{2}\right)^2} \tag{11.206}$$

Combining these results, we can conclude that any two frequency values $\pm j\Omega$ of the s domain are mapped to the four frequencies $\pm\omega_\alpha$ and $\pm\omega_\beta$, where

$$\omega_\alpha = \frac{B\Omega}{2} + \sqrt{\omega_0^2 + \left(\frac{B\Omega}{2}\right)^2} \quad \text{and} \quad \omega_\beta = -\frac{B\Omega}{2} + \sqrt{\omega_0^2 + \left(\frac{B\Omega}{2}\right)^2} \tag{11.207}$$

It is proven that the frequencies ω_α and ω_β satisfy the relations $\omega_\alpha - \omega_\beta = B\Omega$ and $\omega_\alpha\omega_\beta = \omega_0^2$. From the foregoing analysis, it is concluded that the transformation of the prototype low-pass filter to a band-pass filter maps the frequency values Ω to frequency intervals ω. More specifically, considering a frequency value Ω and the mirror value of $-\Omega$, these are depicted *in two pairs of frequencies* that have the frequency ω_0 at the center of their geometry. In a more detailed description, the value range $[0, \infty)$ of the positive semi-axis of Ω is depicted in two intervals on the ω axis: in the interval $[\omega_0, \infty)$, which lies *on the positive semi-axis*, and in the interval $[-\omega_0, 0]$, which lies *on the negative semi-axis*. In a completely analogous way, the interval $(-\infty, 0]$ of the negative semi-axis of Ω is depicted in the interval $(-\infty, -\omega]$ of the negative semi-axis of ω and in the interval $[0, \omega_0]$ of the positive semi-axis of ω. As for the frequencies $\Omega_p = \pm 1$ rad/s and $+\Omega_s$, these are depicted in the frequencies

$$\omega_{p_1} = -\frac{B}{2} + \frac{1}{2}\sqrt{4\omega_0^2 + B^2} \quad \text{and} \quad \omega_{p_2} = +\frac{B}{2} + \frac{1}{2}\sqrt{4\omega_0^2 + B^2} \tag{11.208}$$

which define *the limits of the passband of the band-pass filter* and in the

$$\omega_{s_1} = -\frac{B\Omega_s}{2} + \frac{1}{2}\sqrt{\omega_0^2 + \left(\frac{\Omega_s B}{2}\right)^2} \quad \text{and} \quad \omega_{s_2} = +\frac{B\Omega_s}{2} + \frac{1}{2}\sqrt{\omega_0^2 + \left(\frac{\Omega_s B}{2}\right)^2} \tag{11.209}$$

which define *the lower and upper limits of the two cutoff zones* and satisfy the relations $\omega_{p2} - \omega_{p1} = B$, $\omega_{s2} - \omega_{s2} = B\Omega_s$, and $\omega_{p1}\omega_{p2} = \omega_{s1}\omega_{s2} = \omega_0^2$. This situation is presented diagrammatically in Fig. 11.77.

Taking into account that the frequency response functions of the prototype low-pass filter and the band-pass filter are related through the expression

$$\mathcal{H}_{\text{BP}}(j\omega) = \mathcal{H}_{\text{LPP}}(j\Omega) = \mathcal{H}_{\text{LPP}}\left(\frac{-\omega^2 + \omega_0^2}{jB\omega}\right) \tag{11.210}$$

we can prove that

$$\begin{aligned}
|\mathcal{H}_{\text{BP}}(j\omega_{p_1})| &= |\mathcal{H}_{\text{LPP}}(j\Omega_p)| & \quad \measuredangle\mathcal{H}_{\text{BP}}(j\omega_{p_1}) &= -\measuredangle\mathcal{H}_{\text{LPP}}(j\Omega_p)\\
|\mathcal{H}_{\text{BP}}(j\omega_{p_2})| &= |\mathcal{H}_{\text{LPP}}(j\Omega_p)| & \quad \measuredangle\mathcal{H}_{\text{BP}}(j\omega_{p_2}) &= +\measuredangle\mathcal{H}_{\text{LPP}}(j\Omega_p)\\
|\mathcal{H}_{\text{BP}}(j\omega_{s_1})| &= |\mathcal{H}_{\text{LPP}}(j\Omega_s)| & \quad \measuredangle\mathcal{H}_{\text{BP}}(j\omega_{s_1}) &= -\measuredangle\mathcal{H}_{\text{LPP}}(j\Omega_s)\\
|\mathcal{H}_{\text{BP}}(j\omega_{s_2})| &= |\mathcal{H}_{\text{LPP}}(j\Omega_s)| & \quad \measuredangle\mathcal{H}_{\text{BP}}(j\omega_{s_2}) &= +\measuredangle\mathcal{H}_{\text{LPP}}(j\Omega_s)
\end{aligned}$$

In cases where the operating specifications of the band-pass filter do not satisfy the equation $\omega_{p1}\omega_{p2} = \omega_{s1}\omega_{s2} = \omega_0^2$, there should be a *redefinition of the preceding frequencies*, so as to satisfy the foregoing restriction. In a more detailed description, the determination of the frequency Ω_s of a prototype low-pass filter for a band-pass filter with operating specifications $\omega_{p1}, \omega_{p2}, \omega_{s1}, \omega_{s2}, A_s, A_p$ is performed as follows:

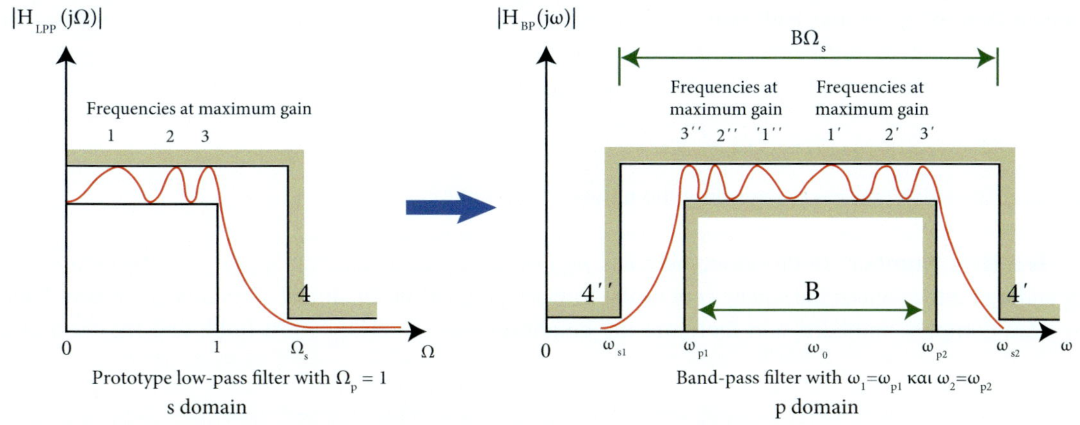

Fig. 11.77 Construction of band-pass filter from prototype low-pass filter (Chen, 2002, Figure 3.7)

1. Redefine the value of the frequency ω_{s_2} from the relation $\omega_{s_2} = \omega_{p_1}\omega_{p_2}/\omega_{s_1}$.
2. Calculate the parameter $\Omega_s^{(1)} = (\omega_{s_2} - \omega_{s_1})/(\omega_{p_2} - \omega_{p_1})$.
3. Redefine the frequency ω_{s_1} as $\omega_{s_1} = \omega_{p_1}\omega_{p_2}/\omega_{s_2}$ (using the initial value of ω_{s_2}).
4. Calculate the parameter $\Omega_s^{(2)} = (\omega_{s_2} - \omega_{s_1})/(\omega_{p_2} - \omega_{p_1})$ using the new values.
5. Select the value of the parameter Ω_s as $\Omega_s = \max(\Omega_s^{(1)}, \Omega_s^{(2)})$.

Finally, taking into account that the transformation of a prototype low-pass filter into a band-pass causes the mappings $\omega_p \to \omega_{p_2}$, $-\omega_p \to \omega_{p_1}$, $\omega_s \to \omega_{s_2}$, and $-\omega_s \to \omega_{s_1}$, we will have

$$\frac{B_s}{B_p} = \frac{\omega_{s_2} - \omega_{s_1}}{\omega_{p_2} - \omega_{p_1}} = \frac{2\omega_s}{2\omega_p} = \frac{\omega_s}{\omega_p}$$

Consequently, to calculate the *minimum order* of the prototype low-pass filter that we will transform into the desired band-pass filter, we can use the appropriate equation (depending on the approach of our choice) by making the substitution $\omega_s/\omega_p \to B_s/B_p$ in it.

Poles and Zero Values of Transfer Function

Formulating the transfer function of the prototype low-pass filter $\mathcal{H}_{\text{LPP}}(s)$ in the form

$$\mathcal{H}_{\text{LPP}}(s) = K\left\{\prod_{k=1}^{M}(s - s_{z_k})\right\}\Big/\left\{\prod_{k=1}^{M}(s - s_{p_k})\right\} \tag{11.211}$$

the substitution $s = (p^2 + \omega_0^2)/pB$ will lead to the function

$$\mathcal{H}_{\text{BP}}(p) = K\left\{\prod_{k=1}^{M}\left(\frac{p^2 + \omega_0^2}{pB} - s_{z_k}\right)\right\}\Big/\left\{\prod_{k=1}^{M}\left(\frac{p^2 + \omega_0^2}{pB} - s_{p_k}\right)\right\} \tag{11.212}$$

Therefore, the poles of the function $\mathcal{H}_{\text{BP}}(s)$ will emerge as the roots of the equation

$$\frac{p^2 + \omega_0^2}{pB} - s_{p_k} = 0 \qquad \text{or, equivalently,} \qquad p^2 - (Bs_{p_k})p + \omega_0^2 = 0$$

with discriminant $\Delta = B^2 s_{p_k}^2 - 4\omega_0^2$ and roots

$$p_{1,2} = \frac{B s_{p_k} \pm \sqrt{B^2 s_{p_k}^2 - 4\omega_0^2}}{2} \tag{11.213}$$

At this point, we distinguish two cases regarding the nature of the pole s_{p_k}.

The Pole s_{p_k} Is a Real Number In this case, the poles $p_{1,2}$ of the transfer function $\mathcal{H}_{\text{BP}}(p)$ will be *either real or complex numbers*, depending on the sign of the square root of the foregoing equation. In almost all practical cases of band-pass filters the poles $p_{1,2}$ are complex numbers, and therefore the condition $4\omega_0^2 > B^2 s_{p_k}^2$ must hold, and the foregoing relation is written

$$p_{1,2} = \frac{B s_{p_k}}{2} \pm \frac{j}{2}\sqrt{4\omega_0^2 - B^2 s_{p_k}^2} = \frac{1}{2}\left(\frac{\omega_0}{Q} \pm j\sqrt{4Q^2 - 1}\right) \tag{11.214}$$

where $Q = \omega_0 / B s_{p_k}$ is the *quality factor* of the poles. We find, therefore, that the pole s_{p_k} of the prototype low-pass filter leads to the creation of two *complex conjugate poles* in the transfer function of the band-pass filter with frequency ω_0 and quality factor $Q = \omega_0 / B s_{p_k}$, while if we take into account that for the value $p = 0$ the expression $(p^2 + \omega_0^2)/(Bp) - s_{p_k}$ takes an infinite value, we will also have a zero value at the position $p = 0$. In cases where $4\omega_0^2 \gg B^2 s_{p_k}^2$, we can make the approximation $4\omega_0^2 - B^2 s_{p_k}^2 \approx 4\omega_0^2$ (known as a *narrow-band approximation*), and consequently, the imaginary part of the complex conjugate poles of the function $\mathcal{H}_{\text{BP}}(s)$ is approximately equal to $\pm\,\omega_0$, as can be verified.

The Pole s_{p_k} Is a Complex Number of the Form $s_{p_k} = \xi_k + j\zeta_k$ In this case, we will have

$$s_{p_k}^2 = (\xi_k + j\zeta_k)^2 = (\xi_k^2 - \zeta_k^2) + j(2\xi_k\zeta_k)$$

and the defining equation of the poles $p_{1,2}$ takes the form

$$p_{1,2} = \frac{B s_{p_k}}{2} \pm \frac{1}{2}\sqrt{B^2 s_{p_k}^2 - 4\omega_0^2} = \frac{B}{2}(\xi_k + j\zeta_k) \pm \sqrt{[B^2(\xi_k^2 - \zeta_k^2) - 4\omega_0^2] + jB^2(2\xi_k\zeta_k)} \tag{11.215}$$

Observing that the expression of the radicand of the last relation is a complex number of the form

$$C_k = [B^2(\xi_k^2 - \zeta_k^2) - 4\omega_0^2] + jB^2(2\xi_k\zeta_k)$$

this can be formulated in its exponential form $C_k = |C_k| e^{j\varphi_k}$, where

$$|C_k| = \sqrt{[B^2(\xi_k^2 - \zeta_k^2) - 4\omega_0^2]^2 + 4B^4\xi_k^2\zeta_k^2} \qquad \text{and} \qquad \varphi_k = \tan^{-1}\frac{2B^2\xi_k\zeta_k}{B^2(\xi_k^2 - \zeta_k^2) - 4\omega_0^2}$$

and therefore the poles $p_{1,2}$ of the function $\mathcal{H}_{\text{BP}}(s)$ will take the form

$$\begin{aligned}
p_{1,2} &= \frac{B}{2}\xi_k + j\frac{B}{2}\zeta_k \pm \frac{1}{2}\sqrt{|C_k| e^{j\varphi_k}} = \frac{B}{2}\xi_k + j\frac{B}{2}\zeta_k \pm \frac{1}{2}\sqrt{|C_k|} e^{j\varphi_k/2} \\
&= \frac{B}{2}\xi_k + j\frac{B}{2}\zeta_k \pm \frac{1}{2}\sqrt{|C_k|}\left(\cos\frac{\varphi_k}{2} + j\sin\frac{\varphi_k}{2}\right) = \frac{B}{2}\xi_k + j\frac{B}{2}\zeta_k \pm \frac{1}{2}\sqrt{|C_k|}\cos\frac{\varphi_k}{2} \pm j\frac{1}{2}\sqrt{|C_k|}\sin\frac{\varphi_k}{2} \\
&= \frac{1}{2}\left(B\xi_k \pm \sqrt{|C_k|}\cos\frac{\varphi_k}{2}\right) + \frac{j}{2}\left(B\zeta_k \pm \sqrt{|C_k|}\sin\frac{\varphi_k}{2}\right)
\end{aligned}$$

Therefore, we will have

$$p_1 = \frac{1}{2}\left(B\xi_k + \sqrt{|C_k|}\cos\frac{\varphi_k}{2}\right) + \frac{j}{2}\left(B\zeta_k + \sqrt{|C_k|}\sin\frac{\varphi_k}{2}\right) \tag{11.216}$$

$$p_2 = \frac{1}{2}\left(B\xi_k - \sqrt{|C_k|}\cos\frac{\varphi_k}{2}\right) + \frac{j}{2}\left(B\zeta_k - \sqrt{|C_k|}\sin\frac{\varphi_k}{2}\right) \tag{11.217}$$

From basic theory, however, it is known that *if a transfer function has a complex pole, then the complex conjugate of this will also be a pole of the function*, and therefore, if the function $\mathcal{H}_{\text{LPP}}(s)$ has a pole at the position $s_{p_k} = \xi_k + j\zeta_k$, then it will also have a pole at the position $s^*_{p_k} = \xi_k - j\zeta_k$, which will lead to the creation of two additional poles of the transfer function $\mathcal{H}_{\text{BP}}(p)$ with values $p_3 = p_1^*$ and $p_4 = p_2^*$, as demonstrated by a similar analysis. Consequently, *for every pair of complex conjugate poles in the transfer function of the prototype low-pass filter, two pairs of complex conjugate poles will appear in the transfer function of the transformed band-pass filter as well as two zero values at the position* $p = 0$, for the reason given previously. For all practical band-pass filters that arise in this way, the imaginary parts of their transfer function poles are *centered* with respect to the central frequency ω_0 of the band-pass filter.

On the other hand, if the transfer function of the prototype low-pass filter has zero values, these, as proven in Sect. 11.9.2 (page 992) will be *purely imaginary numbers* of the form $s_{z_k} = j\nu_k$, where ν_k is a real variable. Consequently, the zero values of the transfer function $\mathcal{H}_{\text{BP}}(p)$ of the band-pass filter will arise as the roots of the equation

$$\frac{p^2 + \omega_0^2}{Bp} - j\nu_k = 0 \qquad \text{or, equivalently,} \qquad p^2 - j(B\nu_k)p + \omega_0^2 = 0$$

which, as it turns out, are given by the relation

$$p_{1,2} = \frac{jB\nu_k}{2} \pm \frac{j}{2}\sqrt{4\omega_0^2 + B^2\nu_k^2} \tag{11.218}$$

We find, therefore, that for each zero value of the function $\mathcal{H}_{\text{LPP}}(s)$ of the prototype low-pass filter, there are *two purely imaginary zero values* in the transfer function $\mathcal{H}_{\text{BP}}(s)$ of the transformed band-pass filter, which, however, *do not have the same magnitude* and, therefore, are not related to each other by the property of complex conjugation. Taking into account that the zero values in the function $\mathcal{H}_{\text{LPP}}(s)$ appear in *complex conjugate pairs*, we can assert that for the zero value $s_{z_k} = j\nu_k$ there will also be the complex conjugate value $s^*_{z_k} = -j\nu_k$, which will lead to the pair of zero values

$$p_{3,4} = -\frac{jB\nu_k}{2} \pm \frac{j}{2}\sqrt{4\omega_0^2 + B^2\nu_k^2} \tag{11.219}$$

Finally, observing that the expression $(s^2 + \omega_0^2)/(Bp) \pm j\nu_k$ diverges at the position $p = 0$, it is obvious that the zero values $\pm j\nu_k$ cause the appearance of two poles at the position $p = 0$. In the special case where $4\omega_0^2 \gg B^2\nu_k^2$, we can make the approximations $p_{1,2} = j\omega_0 \pm B\nu_k/2$ and $p_{3,4} = -j\omega_0 \pm jB\nu_k/2$.

Selectivity and Spectral Efficiency Factor

The selectivity of a band-pass filter as a function of the selectivity of the corresponding prototype low-pass filter is calculated as

$$F_s^{\text{BP}} = \left.\frac{d|\mathcal{H}_{\text{BP}}(j\omega)|}{d\omega}\right|_{\omega=\omega_c} = \left\{\frac{d}{d\omega}\left|\mathcal{H}_{\text{LPP}}\left(-j\frac{-\omega^2+\omega_0^2}{B\omega}\right)\right|\right\}_{\omega=\omega_c} = \left.\frac{d}{d\omega}\left(\frac{\omega^2-\omega_0^2}{B\omega}\right)\right|_{\omega=\omega_c} F_s^{\text{LPP}} = \frac{\omega_c^2+\omega_0^2}{B^2\omega_c^2} F_s^{\text{LPP}} \tag{11.220}$$

with the selectivity F_s^{LPP} being calculated at the position of the frequency $\Omega_c = (\omega_c^2 + \omega_0^2)/(B\omega_c)$, where the frequency ω_c can be either of the frequencies ω_{c_1} or ω_{c_2} of the band-pass filter. On the other hand, the spectral efficiency factors of the two filters are proven to be equal to each other, or, in mathematical notation, $S_{\text{BP}} = S_{\text{LPP}}$.

Phase Characteristics and Time-Domain Response

Using the correlation between the frequency responses of the prototype low-pass and the band-pass filter, it is proven that the *phase delay* of the band-pass filter has the form

$$\tau_{\text{BP}}^{\varphi}(\omega) = -\frac{1}{\omega}\measuredangle\mathcal{H}_{\text{BP}}(\omega) = -\frac{1}{\omega}\measuredangle\mathcal{H}_{\text{LPP}}\left(\frac{-\omega^2+\omega_0^2}{jB\omega}\right) = -\frac{1}{\omega}\measuredangle\mathcal{H}_{\text{LPP}}\left(j\frac{\omega^2-\omega_0^2}{B\omega}\right) \tag{11.221}$$

while the corresponding *group delay* is formulated as

$$\tau_{\text{BP}}^{g}(\omega) = -\frac{d}{d\omega}\measuredangle\mathcal{H}_{\text{BP}}(\omega) = -\frac{d}{d\omega}\measuredangle\mathcal{H}_{\text{LPP}}\left(j\frac{\omega^2-\omega_0^2}{B\omega}\right) = \frac{d}{d\omega}\left(\frac{\omega^2-\omega_0^2}{B\omega}\right)\tau_{\text{LPP}}^{g}\left(\frac{\omega^2-\omega_0^2}{B\omega}\right) = \frac{\omega^2+\omega_0^2}{B\omega^2}\tau_{\text{LPP}}^{g}\left(\frac{\omega^2-\omega_0^2}{B\omega}\right) \tag{11.222}$$

The reader is asked as an exercise to calculate the values of phase delay and group delay at the positions $\omega = \omega_0$, $\omega = \omega_{p_{1,2}}$, and $\omega = \omega_{s_{1,2}}$.

The *impulse response* of the band-pass filter will emerge from the *inverse Laplace transform* of the transfer function $\mathcal{H}_{\text{BP}}(p)$. Taking into account that the transfer function of the band-pass filter results from the multiplication of the transfer function of a low-pass filter by the transfer function of a high-pass filter, or, in mathematical notation, $\mathcal{H}_{\text{BP}}(p) = \mathcal{H}_{\text{LP}}(p)\mathcal{H}_{\text{HP}}(p)$, the application of the convolution theorem will lead us to the result $h_{\text{BP}}(t) = h_{\text{LP}}(t) * h_{\text{HP}}(t)$. After calculating the impulse response of the system, we can also calculate its step response as the integral of the impulse response.

Problem 11.10.3 Construct a band-pass filter with operating specifications $\omega_{p_1} = 9500$ rad/s, $\omega_{p_2} = 10500$ rad/s, $\omega_{s_1} = 5000$ rad/s, $\omega_{s_2} = 15000$ rad/s, $A_p = 1.0$ dB, and $A_s = 50$ dB using the Chebyshev II approximation as the prototype low-pass filter (Antoniou, Exercise for solution 10.40).

Solution First of all, we find that

$$\sqrt{\omega_{p_1}\omega_{p_2}} = \sqrt{9500 \times 10500}\ \text{rad/s} = 9987.492177 \quad \text{and}$$

$$\sqrt{\omega_{s_1}\omega_{s_2}} = \sqrt{5000 \times 15000}\ \text{rad/s} = 8660.254037\ \text{rad/s}$$

and since the foregoing values are different, a *redefinition* of the filter's operating frequencies is required. Applying the process described in the previous section, it turns out that

$$\begin{aligned}
\omega'_{s_2} &= \frac{\omega_{p_1}\omega_{p_2}}{\omega_{s_1}}\ \text{rad/s} = \frac{9500 \times 10500}{5000}\ \text{rad/s} = 19950\ \text{rad/s} \\
\Omega_s^{(1)} &= \frac{\omega'_{s_2} - \omega_{s_1}}{\omega_{p_2} - \omega_{p_1}}\ \text{rad/s} = \frac{19950 - 5000}{10500 - 9500}\ \text{rad/s} = 14.950\ \text{rad/s} \\
\omega'_{s_1} &= \frac{\omega_{p_1}\omega_{p_2}}{\omega_{s_2}}\ \text{rad/s} = \frac{9500 \times 10500}{15000}\ \text{rad/s} = 6650\ \text{rad/s} \\
\Omega_s^{(2)} &= \frac{\omega_{s_2} - \omega'_{s_1}}{\omega_{p_2} - \omega_{p_1}}\ \text{rad/s} = \frac{15000 - 6650}{10500 - 9500}\ \text{rad/s} = 8.350\ \text{rad/s}
\end{aligned}$$

Therefore, we will have $\Omega_s = \min(\Omega_s^{(1)}, \Omega_s^{(2)}) = \min(14.950, 8.350)$ rad/s $= 8.350$ rad/s, and consequently, the selectivity factor for $\Omega_p = 1$ rad/s is $k = \Omega_p/\Omega_s = 0.119760$.

Following the methodology of the previous examples, the *minimum order* of the Chebyshev II filter for the given parameter values is calculated as

$$N = \frac{\cosh^{-1}\sqrt{\dfrac{10^{A_s/10}-1}{10^{A_p/10}-1}}}{\cosh^{-1}(\Omega_s/\Omega_p)} = \frac{\cosh^{-1}\sqrt{\dfrac{10^{50/10}-1}{10^{1/10}-1}}}{\cosh^{-1}(8.350/1)} = \frac{\cosh^{-1}(621.456150)}{\cosh^{-1}(8.350)} = \frac{7.125211}{2.811803} = 2.534036$$

The foregoing value should be rounded to the next highest integer, so the Chebyshev II filter we will use will have an order value of $N = 3$. Considering the case of normalized approximation, we define $\Omega_s = 1$ rad/s, and therefore, the frequency Ω_p is calculated as

$$\Omega_p = \frac{1}{\cosh\left[\dfrac{1}{N}\cosh^{-1}\sqrt{\dfrac{10^{A_s/10}-1}{10^{A_p/10}-1}}\right]} = \frac{1}{\cosh(2.375070)} = \frac{1}{5.422390} = 0.184420 \text{ rad/s}$$

But the fact that the frequency Ω_p is different from unity will also affect the value of the bandwidth B, which will be equal not to $B = \omega_{p_2} - \omega_{p_1} = 1000$ but to

$$B = \frac{\omega_{p_2} - \omega_{p_1}}{\Omega_p} = \frac{10500 - 9500}{0.184420} = 5329.382484$$

The value of the constant ω_0 is equal to $\sqrt{\omega_{p_1}\omega_{p_2}} = \sqrt{9500 \times 10500}$ rad/s $= 9987.492177$ rad/s, and consequently, the actual cutoff frequencies of the real filter are

$$\omega_{s_1} = -\frac{5329.382484}{2} + \sqrt{(9987.492177)^2 + \left(\frac{5329.382484}{2}\right)^2} = 7.672164 \text{ rad/s} \quad \text{and}$$

$$\omega_{s_2} = +\frac{5329.382484}{2} + \sqrt{(9987.492177)^2 + \left(\frac{5329.382484}{2}\right)^2} = 13001.546634 \text{ rad/s}$$

We observe that the frequency ω_{s_1} is *greater* than the one that has been defined, while the frequency ω_{s_2} is *smaller* than the corresponding given value, and therefore the operating specifications of the filter regarding the cutoff frequencies are satisfied, and even more so.

Let us now move on to the calculation of the zero values and the poles of the transfer function of the prototype normalized Chebyshev II filter with operating specifications $\Omega_p = 0.184420$ rad/s, $\Omega_s = 1$ rad/s $A_p = 1.0$ dB, and $A_s = 50.0$ dB. Applying the methodology used in Exercise 11.9.8 (page 1040) successively we have

$$\varepsilon = 1/\sqrt{10^{A_s/10} - 1} = 1/\sqrt{10^5 - 1} = \sqrt{9999} = 0.003162$$

$$x = 1/\varepsilon = 316.226184$$

$$\sinh^{-1}(x) = 6.449607$$

$$\sinh[(1/3)\sinh^{-1}(x)] = 4.233617$$

$$\cosh[(1/3)\sinh^{-1}(x)] = 4.350117$$

Using the equations for calculating zero values and the poles of the transfer function we find that it has two *complex conjugate zero values* $z_{1,2} = \pm j/\cos(\pi/6) = \pm 1.154700j$, a real pole $p_3 = -0.236204$, and a pair of complex conjugate poles with a value of $p_{1,2} = -0.113335 \pm 0.201746j$. Regarding the value of the coefficient H_0, which turns out to be equal to 0.009486.

By substituting the foregoing values into the defining equations of the poles and zeros, it turns out that the pole $s_1 = -0.113335 + 0.201746j$ leads to the creation of the complex poles $p_1 = -285.763172 - 9459.811958j$ and $p_2 = -318.242391 + 10534.993556j$, while the pole $s_2 = -0.113335 - 0.201746j$ leads to the creation of the poles $p_3 = p_1^* = -285.763172 + 9459.811958j$ and $p_4 = p_2^* = -318.242391 - 10534.993556j$. It is not difficult to see that the pair of poles (p_1, p_3) contributes to the transfer function the term $p^2 + 571.526344p + 8.956970 \cdot 10^7$, while the term contributed by the pair of poles (p_2, p_4) is $p^2 + 636.484782p + 1.1108736 \cdot 10^8$. On the other hand, the real pole $s_3 = -0.236204$ leads to the creation of two complex conjugate poles $p_{5,6} = -629.410730 \pm 9967.639746j$ that contribute to the denominator of the transfer function the term $p^2 + 1258.821460p + 9,974999 \cdot 10^7$, as well as a zero value $p = 0$ that contributes to the numerator of the transfer function the term p. Finally, the zero value $s_1 = 1.154701j$ leads to the creation of two zero values with values $z_1 = 23978.351906j$ and $z_2 = -17824.508622j$, while the zero value $s_2 = s_1^* = -1.154701j$ leads to the creation of the zero values $z_3 = z_2^* = 17824.508622j$ and $z_4 = z_1^* = -23978.351906j$. These four zero values are two by two complex conjugate quantities, with the pair (z_1, z_4) contributing to the numerator of the transfer function the term $p^2 + 5.749613 \cdot 10^8$ and the pair (z_2, z_3) contributing the term $p^2 + 3.177130 \cdot 10^8$. Therefore, the transfer function of the band-pass filter to be designed will take the form

$$\mathcal{H}_{\text{BP}}(p) = \frac{p(p^2 + 5.749613 \cdot 10^8)(p^2 + 3.177130 \cdot 10^8)}{(p^2 + 571.526344p + 8.956970 \cdot 10^7)(p^2 + 636.484782p + 1.1108736 \cdot 10^8)(p^2 + 1258.821460p + 9.974999 \cdot 10^7)}$$

or, equivalently, $\mathcal{H}_{\text{BP}}(p) = N(p)/D(p)$, where

$$N(p) = p^5 + 8.926744 \cdot 10^8 p^3 + 1.826727 \cdot 10^{17} p$$

$$D(p) = p^6 + 2.466832 \cdot 10^3 p^5 + 3.022915 \cdot 10^8 p^4 + 4.940475 \cdot 10^{11} p^3 + 3.015357 \cdot 10^{16} p^2 + 2.454513 \cdot 10^{19} p + 9.925187 \cdot 10^{23}$$

The pole–zero diagram, the amplitude and phase responses, and the step and impulse responses of this band-pass filter are presented in Fig. 11.78.

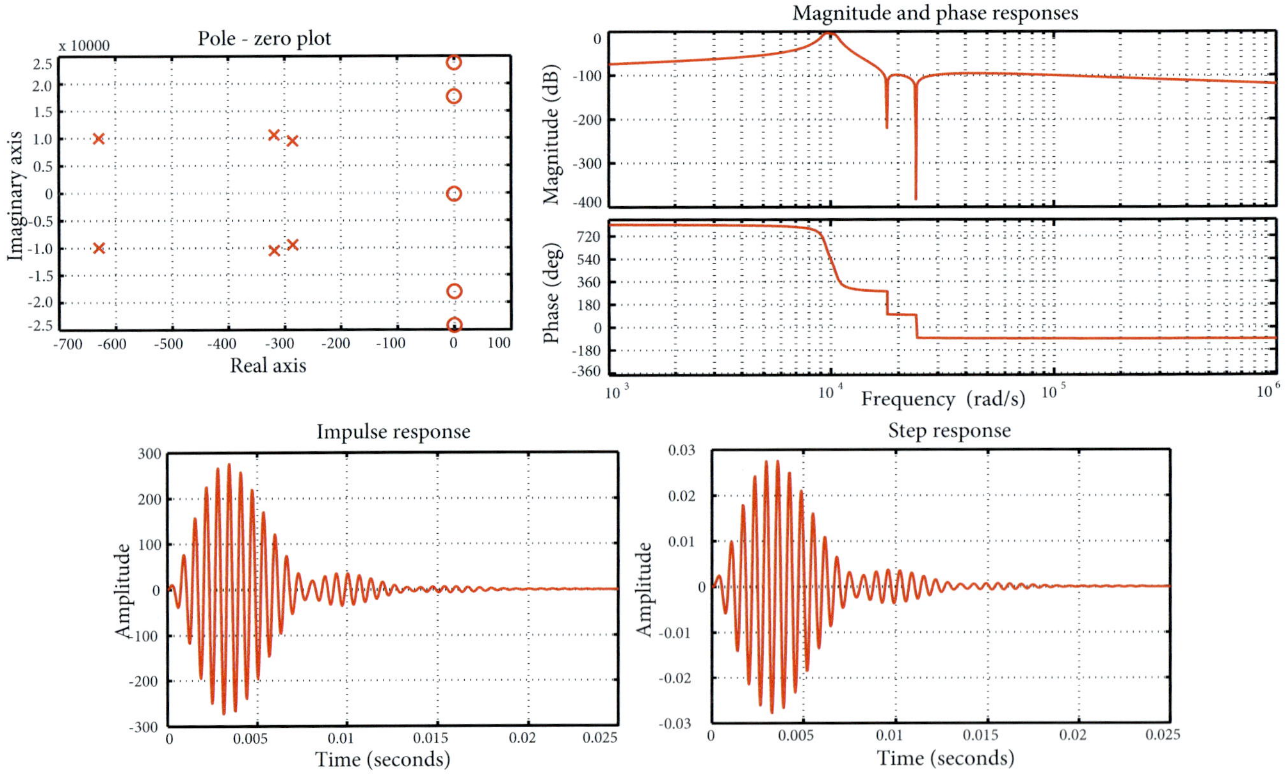

Fig. 11.78 Pole–zero diagram, amplitude and phase responses, and impulse and step responses of band-pass filter of Exercise 11.10.3

11.10.4 Transformation of a Prototype Low-Pass Filter into a Band-stop Filter

The transformation of a prototype low-pass filter into a band-stop filter with a central frequency ω_0 and bandwidth $B = \omega_{p2} - \omega_{p_1}$, where ω_{p1} and ω_{p2} are the *lower* and *upper* cutoff limits, is achieved by applying to the transfer function of the prototype filter

$$\mathcal{H}_{\text{LPP}}(s) = \sum_{k=0}^{M} \beta_k s^k \Big/ \sum_{k=0}^{N} \alpha_k s^k \tag{11.223}$$

the transformation

$$s = \frac{p\,(\omega_{p2} - \omega_{p1})}{p^2 + \omega_0^2} = \frac{pB}{p^2 + \omega_0^2} \tag{11.224}$$

This can be considered a transformation of a low-pass filter into a high-pass filter of the form $s = 1/p'$, which is followed by a transformation into a band-pass filter of the form $p' = (p^2 + \omega_0^2)/Bp$. In this case, the transfer function of the band-stop filter that is constructed in this way is formulated as

$$\mathcal{H}_{\text{BS}}(p) = \left\{\sum_{k=0}^{M} \beta_k \left(\frac{pB}{p^2 + \omega_0^2}\right)^k\right\} \Big/ \left\{\sum_{k=0}^{N} \alpha_k \left(\frac{pB}{p^2 + \omega_0^2}\right)^k\right\} \tag{11.225}$$

It is not difficult to see that *the order of the transfer function of the band-stop filter $\mathcal{H}_{BS}(p)$ is twice the order of the transfer function of the prototype low-pass filter $\mathcal{H}(s)$*. Moreover, it can be proven that *the function $\mathcal{H}_{\text{BS}}(p)$ can always be formulated as the sum of a low-pass filter transfer function and a high-pass filter transfer function.* If the prototype low-pass filter exhibits variations either in the passband or in the stopband, then the amplitude response of the band-pass filter will also present very small-amplitude fluctuations, while in the opposite case we will have $|\mathcal{H}_{\text{BS}}(0)| = |\mathcal{H}_{\text{BS}}(\infty)| = 1$.

To study the effect of the transformation on the frequency of the prototype low-pass filter, let us proceed with the substitutions $s = j\Omega$ and $p = j\omega$, so we get

$$j\Omega = \frac{j\omega B}{-\omega^2 + \omega_0^2} \quad \text{or, equivalently,} \quad \omega^2 + \frac{B}{\Omega}\omega - \omega_0^2 = 0$$

The last equation is a quadratic equation with respect to ω with discriminant $\Delta = (B^2/\Omega^2) + 4\omega_0^2$ and roots

$$\omega_{1,2} = -\frac{B}{2\Omega} \pm \frac{1}{2}\sqrt{\frac{B^2}{\Omega^2} + 4\omega_0^2} = -\frac{B}{2\Omega} \pm \sqrt{\omega_0^2 + \left(\frac{B}{2\Omega}\right)^2} \tag{11.226}$$

From the foregoing relation we find that *the transformation of the prototype low-pass to the band-pass filter maps each value of the frequency Ω of the s domain to two frequencies of the p space*, which are given by the foregoing relations. Applying the same process for the mirror frequency $s^* = -j\Omega$, we are led to the roots

$$\omega_{3,4} = \frac{B}{2\Omega} \pm \frac{1}{2}\sqrt{\frac{B^2}{\Omega^2} + 4\omega_0^2} = \frac{B}{2\Omega} \pm \sqrt{\omega_0^2 + \left(\frac{B}{2\Omega}\right)^2} \tag{11.227}$$

Combining these results, we can conclude that any two frequency values $\pm\, j\Omega$ of the s domain are mapped to the four frequencies $\pm\,\omega_\alpha$ and $\pm\,\omega_\beta$, where

$$\omega_\alpha = \frac{B}{2\Omega} + \sqrt{\omega_0^2 + \left(\frac{B}{2\Omega}\right)^2} \quad \text{and} \quad \omega_\beta = -\frac{B}{2\Omega} + \sqrt{\omega_0^2 + \left(\frac{B}{2\Omega}\right)^2} \tag{11.228}$$

It is not difficult to see that the frequencies ω_α and ω_β satisfy the equations $\omega_\alpha - \omega_\beta = B/\Omega$ and $\omega_\alpha\omega_\beta = \omega_0^2$.

A way to represent the frequency Ω under the influence of this transformation is diagrammatically presented in Fig. 11.79, in which the frequency response of the band-pass filter emerges as a result of the *synthesis* of the frequency responses of a low-pass and a high-pass filter. In the first step of the process, applying the transformation $p' = 1/s$, we construct the response of a *normalized high-pass filter* with a passband extending from the value $\Omega'_p = 1$ to the position $\Omega' = \infty$ and a cutoff zone defined by the frequency values $\Omega' = 0$ to $\Omega' = 1/\Omega_s$. Then we apply to the variable p' the *band-pass filter transformation*, so as to construct the frequency response function of the desired band-pass filter. In this case, the frequency range from $\Omega' = 0$ to ∞ is mapped to the ranges $[\omega_0, \infty)$ and $[-\omega_0, 0]$, while the frequency range $(-\infty, 0]$ is mapped to the ranges $[0, \omega_0]$ and $(-\infty, -\omega_0]$. Using the equations constructed previously, we can assert that the transformation of the band-pass filter, applied to the transfer function of a normalized high-pass filter, creates *two passbands* with variance related to the attenuation value A_p for the value ranges defined by the frequencies

$$\omega_{p_1} = -\frac{B\Omega'_p}{2} + \frac{1}{2}\sqrt{\omega_0^2 + \left(\frac{B\Omega'_p}{2}\right)} = -\frac{B}{2\Omega_p} + \frac{1}{2}\sqrt{\omega_0^2 + \left(\frac{B}{2\Omega_p}\right)} \tag{11.229}$$

$$\omega_{p_2} = +\frac{B\Omega'_p}{2} + \frac{1}{2}\sqrt{\omega_0^2 + \left(\frac{B\Omega'_p}{2}\right)} = +\frac{B}{2\Omega_p} + \frac{1}{2}\sqrt{\omega_0^2 + \left(\frac{B}{2\Omega_p}\right)} \tag{11.230}$$

and *one cutoff region* with variance related to the attenuation value A_s and for the value ranges defined by the frequencies

$$\omega_{s_1} = -\frac{B\Omega'_s}{2} + \frac{1}{2}\sqrt{\omega_0^2 + \left(\frac{B\Omega'_s}{2}\right)} = -\frac{B}{2\Omega_s} + \frac{1}{2}\sqrt{\omega_0^2 + \left(\frac{B}{2\Omega_s}\right)} \tag{11.231}$$

$$\omega_{s_2} = +\frac{B\Omega'_s}{2} + \frac{1}{2}\sqrt{\omega_0^2 + \left(\frac{B\Omega'_s}{2}\right)} = +\frac{B}{2\Omega_s} + \frac{1}{2}\sqrt{\omega_0^2 + \left(\frac{B}{2\Omega_s}\right)} \tag{11.232}$$

It is not difficult to prove that the foregoing frequencies satisfy the relations $\omega_{p2} - \omega_{p1} = B$, $\omega_{s2} - \omega_{s2} = B\Omega'_s = B/\Omega_s$, and $\omega_{p1}\omega_{p2} = \omega_{s1}\omega_{s2} = \omega_0^2$.

For given band-stop filter specifications ω_{s_1}, ω_{s_2}, ω_{p_1}, ω_{p_2}, A_p, and A_s, the parameter Ω_s of the equivalent prototype low-pass filter will be calculated from the relation $\Omega_s = (\omega_{p_2} - \omega_{p_1})/(\omega_{s_2} - \omega_{s_1})$, with the foregoing operating frequencies of the filter undergoing *readjustment*, if necessary, by applying the process described for band-pass filters.

Taking into account that the frequency response functions of the prototype low-pass and band-stop filters are related through the expression

$$\mathcal{H}_{\text{BP}}(j\omega) = \mathcal{H}_{\text{LPP}}(j\Omega) = \mathcal{H}_{\text{LPP}}\left(\frac{jB\omega}{-\omega^2 + \omega_0^2}\right) \tag{11.233}$$

we can prove that

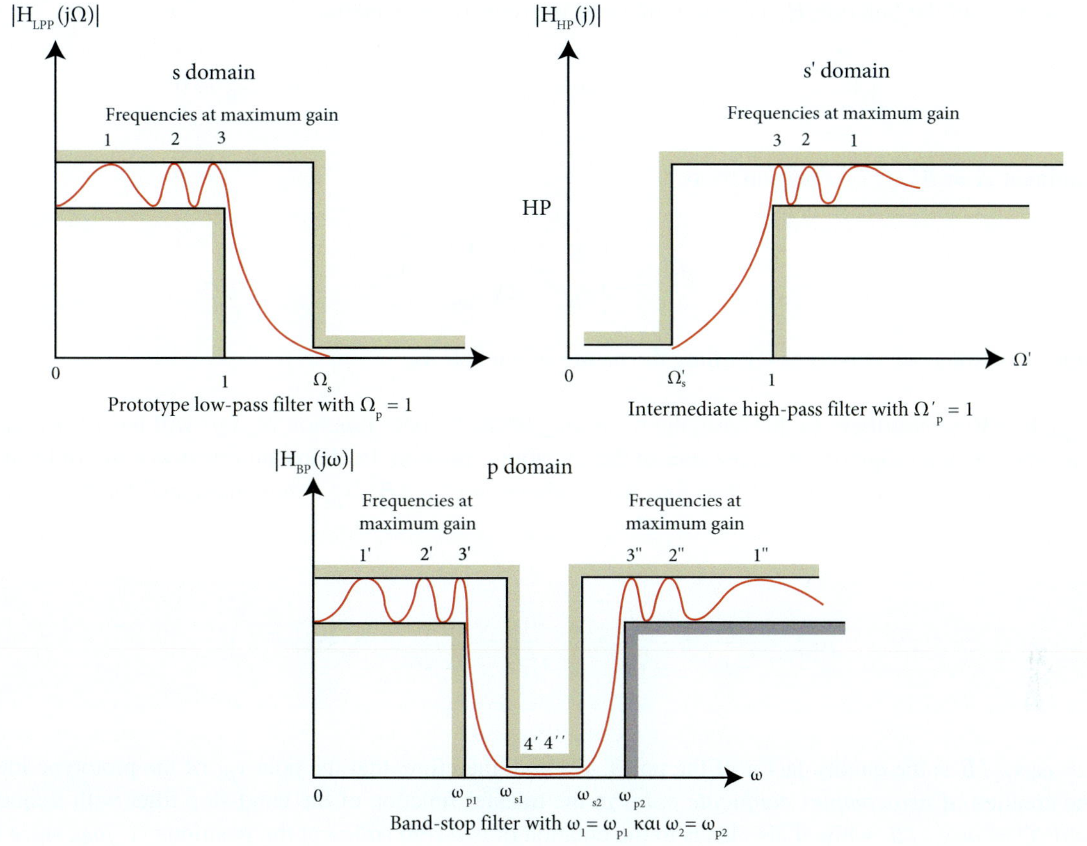

Fig. 11.79 Construction of a band-stop filter from a prototype low-pass filter (Chen, 2002, Figure 3.9)

$$|\mathcal{H}_{\text{BS}}(j\omega_{p_1})| = |\mathcal{H}_{\text{LPP}}(j\Omega_p)| \qquad \measuredangle\mathcal{H}_{\text{BS}}(j\omega_{p_1}) = +\measuredangle\mathcal{H}_{\text{LPP}}(j\Omega_p)$$
$$|\mathcal{H}_{\text{BS}}(j\omega_{p_2})| = |\mathcal{H}_{\text{LPP}}(j\Omega_p)| \qquad \measuredangle\mathcal{H}_{\text{BP}}(j\omega_{p_2}) = -\measuredangle\mathcal{H}_{\text{LPP}}(j\Omega_p)$$
$$|\mathcal{H}_{\text{BS}}(j\omega_{s_1})| = |\mathcal{H}_{\text{LPP}}(j\Omega_s)| \qquad \measuredangle\mathcal{H}_{\text{BP}}(j\omega_{s_1}) = +\measuredangle\mathcal{H}_{\text{LPP}}(j\Omega_s)$$
$$|\mathcal{H}_{\text{BS}}(j\omega_{s_2})| = |\mathcal{H}_{\text{LPP}}(j\Omega_s)| \qquad \measuredangle\mathcal{H}_{\text{BP}}(j\omega_{s_2}) = -\measuredangle\mathcal{H}_{\text{LPP}}(j\Omega_s)$$

Finally, following the argumentation presented in the previous section regarding the *minimum order* of the prototype low-pass filter that should be used for the construction of the band-stop filter, this can be calculated from the equations constructed in the approximations of ideal filters by simply making the substitution $\omega_s/\omega_p = B_p/B_s$, where $B_p = \omega_{p_2} - \omega_{p_1}$ and $B_s = \omega_{s_2} - \omega_{s_1}$.

Poles and Zeros of Transfer Function

Formulating the transfer function of the prototype low-pass filter $\mathcal{H}_{\text{LPP}}(s)$ as

$$\mathcal{H}_{\text{LPP}}(s) = K\left\{\prod_{k=1}^{M}(s - s_{z_k})\right\} \Big/ \left\{\prod_{k=1}^{M}(s - s_{p_k})\right\} \tag{11.234}$$

the substitution $s = (p^2 + \omega_0^2)/pB$ will lead to the function

$$\mathcal{H}_{\text{BS}}(p) = K\left\{\prod_{k=1}^{M}\left(\frac{pB}{p^2+\omega_0^2} - s_{z_k}\right)\right\} \Big/ \left\{\prod_{k=1}^{M}\left(\frac{pB}{p^2+\omega_0^2} - s_{p_k}\right)\right\} \tag{11.235}$$

Therefore, the poles of the function $\mathcal{H}_{\mathrm{BS}}(s)$ will emerge as the roots of the equation

$$\frac{pB}{p^2+\omega_0^2} - s_{p_k} = 0 \qquad \text{or, equivalently,} \qquad p^2 - \frac{B}{s_{p_k}}p + \omega_0^2 = 0$$

with discriminant $\Delta = B^2/s_{p_k}^2 - 4\omega_0^2$ and roots

$$p_{1,2} = \frac{1}{2}\frac{B}{s_{p_k}} \pm \frac{1}{2}\sqrt{\frac{B^2}{s_{p_k}^2} - 4\omega_0^2} \tag{11.236}$$

At this point, we distinguish two cases regarding the nature of the pole s_{p_k}.

The Pole s_{p_k} Is a Real Number In this case, the poles $p_{1,2}$ of the transfer function $\mathcal{H}_{\mathrm{BS}}(p)$ will be *either real or complex numbers*, depending on the sign of the square root of the foregoing equation. In almost all practical cases of band-stop filters, the poles $p_{1,2}$ are *complex numbers*, and therefore the condition $4\omega_0^2 > (B^2/s_{p_k}^2)$ must hold, and the foregoing relation is written

$$p_{1,2} = \frac{B}{2s_{p_k}} \pm \frac{j}{2}\sqrt{4\omega_0^2 - \frac{B^2}{s_{p_k}^2}} = \frac{1}{2}\left(\frac{\omega_0}{Q} \pm j\sqrt{4Q^2-1}\right) \tag{11.237}$$

where $Q = \omega_0 s_{p_k}/B$ is the quality factor of the poles. We find, therefore, that the pole s_{p_k} of the prototype low-pass filter leads to the creation of *two complex conjugate poles* in the transfer function of the band-stop filter with frequency ω_0 and quality factor $Q = \omega_0 s_{p_k}/B$, while it also leads to the creation of *two zero values* at the positions $\pm\, j\omega_0$, since to construct the second-order equation for calculating the poles, we multiplied the numerator and the denominator of the transfer function by the expression $(p^2+\omega_0^2)$. In cases where $4\omega_0^2 \gg B^2/s_{p_k}^2$, we can make the approximation $4\omega_0^2 - B^2/s_{p_k}^2 \approx 4\omega_0^2$, and consequently, the imaginary part of the complex conjugate poles of the function $\mathcal{H}_{\mathrm{BS}}(s)$ is approximately equal to $\pm\,\omega_0$, as can be verified.

The Pole s_{p_k} Is a Complex Number of the Form $s_{p_k} = \xi_k + j\zeta_k$ In this case, we will have

$$s_{p_k}^2 = (\xi_k + j\zeta_k)^2 = (\xi_k^2 - \zeta_k^2) + j(2\xi_k\zeta_k)$$

and the defining equation of the poles $p_{1,2}$ takes the form

$$p_{1,2} = \frac{B}{2s_{p_k}} \pm \frac{1}{2}\sqrt{\frac{B^2}{s_{p_k}^2} - 4\omega_0^2} = \frac{B}{2(\xi_k + j\zeta_k)} \pm \frac{1}{2}\sqrt{\frac{B^2}{(\xi_k^2-\zeta_k^2) + j(2\xi_k\zeta_k)} - 4\omega_0^2}$$

$$= \frac{B(\xi_k - j\zeta_k)}{2(\xi_k^2+\zeta_k^2)} \pm \frac{1}{2}\sqrt{\left[\frac{B^2(\xi_k^2-\zeta_k^2)}{(\xi_k^2+\zeta_k^2)^2} - 4\omega_0^2\right] - j\frac{2B^2\xi_k\zeta_k}{(\xi_k^2+\zeta_k^2)^2}} \tag{11.238}$$

Observing that the expression under the square root of the last relation is the complex number

$$C_k = \left[\frac{B^2(\xi_k^2-\zeta_k^2)}{(\xi_k^2+\zeta_k^2)^2} - 4\omega_0^2\right] - j\frac{2B^2\xi_k\zeta_k}{(\xi_k^2+\zeta_k^2)^2} \tag{11.239}$$

this can be formulated in its exponential form $C_k = |C_k|e^{j\varphi_k}$, where

$$|C_k| = \sqrt{\left[\frac{B^2(\xi_k^2-\zeta_k^2)}{(\xi_k^2+\zeta_k^2)^2} - 4\omega_0^2\right]^2 + \frac{4B^4\xi_k^2\zeta_k^2}{(\xi_k^2+\zeta_k^2)^4}} \qquad \text{and} \qquad \varphi_k = -\tan^{-1}\frac{2B^2\xi_k\zeta_k}{B^2(\xi_k^2-\zeta_k^2) - 4\omega_0^2(\xi_k^2+\zeta_k^2)^2}$$

and therefore, the poles $p_{1,2}$ of the function $\mathcal{H}_{\text{BS}}(p)$ will take the form

$$\begin{aligned}
p_{1,2} &= \frac{B\xi_k}{2(\xi_k^2+\zeta_k^2)} - j\frac{B\zeta_k}{2(\xi_k^2+\zeta_k^2)} \pm \frac{1}{2}\sqrt{|C_k|e^{j\varphi_k}} = \frac{B\xi_k}{2(\xi_k^2+\zeta_k^2)} - j\frac{B\zeta_k}{2(\xi_k^2+\zeta_k^2)} \pm \frac{1}{2}\sqrt{|C_k|}e^{j\varphi_k/2} \\
&= \frac{B\xi_k}{2(\xi_k^2+\zeta_k^2)} - j\frac{B\zeta_k}{2(\xi_k^2+\zeta_k^2)} \pm \frac{1}{2}\sqrt{|C_k|}\left(\cos\frac{\varphi_k}{2} + j\sin\frac{\varphi_k}{2}\right) \\
&= \frac{B\xi_k}{2(\xi_k^2+\zeta_k^2)} - j\frac{B\zeta_k}{2(\xi_k^2+\zeta_k^2)} \pm \frac{1}{2}\sqrt{|C_k|}\cos\frac{\varphi_k}{2} \pm j\frac{1}{2}\sqrt{|C_k|}\sin\frac{\varphi_k}{2} \\
&= \frac{1}{2}\left(\frac{B\xi_k}{\xi_k^2+\zeta_k^2} \pm \sqrt{|C_k|}\cos\frac{\varphi_k}{2}\right) - \frac{j}{2}\left(\frac{B\zeta_k}{\xi_k^2+\zeta_k^2} \mp \sqrt{|C_k|}\sin\frac{\varphi_k}{2}\right)
\end{aligned}$$

Therefore, we will have

$$p_1 = \frac{1}{2}\left(\frac{B\xi_k}{\xi_k^2+\zeta_k^2} + \sqrt{|C_k|}\cos\frac{\varphi_k}{2}\right) - \frac{j}{2}\left(\frac{B\zeta_k}{\xi_k^2+\zeta_k^2} - \sqrt{|C_k|}\sin\frac{\varphi_k}{2}\right) \tag{11.240}$$

$$p_2 = \frac{1}{2}\left(\frac{B\xi_k}{\xi_k^2+\zeta_k^2} - \sqrt{|C_k|}\cos\frac{\varphi_k}{2}\right) - \frac{j}{2}\left(\frac{B\zeta_k}{\xi_k^2+\zeta_k^2} + \sqrt{|C_k|}\sin\frac{\varphi_k}{2}\right) \tag{11.241}$$

From basic theory, however, it is known that *if a transfer function has a complex pole, then the complex conjugate of this pole will also be a pole of the function*, and therefore, if the function $\mathcal{H}_{\text{LPP}}(s)$ has a pole at the position $s_{p_k} = \xi_k + j\zeta_k$, then it will have a pole at the position $s_{p_k}^* = \xi_k - j\zeta_k$, which will lead to the creation of two additional poles of the transfer function $\mathcal{H}_{\text{BS}}(p)$ with values $p_3 = p_1^*$ and $p_4 = p_2^*$, as proven by performing a similar analysis. Consequently, *for each pair of complex-conjugate poles in the transfer function of the prototype low-pass filter, the transfer function of the resulting band-stop filter will exhibit two pairs of complex-conjugate poles, as well as two zeros located at* $p = \pm j\omega_0$, for the reason previously discussed. For all practical band-stop filters that arise in this way, the imaginary parts of the poles of their transfer function are *centered* with respect to the central frequency ω_0 of the band-stop filter.

On the other hand, if the transfer function of the prototype low-pass filter has zero values, then these, as was proven in Sect. 11.9.2 (page 992), will be purely imaginary numbers of the form $s_{z_k} = j\nu_k$, where ν_k is a real variable. Consequently, the zero values of the transfer function $\mathcal{H}_{\text{BS}}(p)$ of the band-stop filter will arise as the roots of the equation

$$\frac{Bp}{p^2+\omega_0^2} - j\nu_k = 0 \qquad \text{or, equivalently,} \qquad p^2 + j\frac{B}{\nu_k}p + \omega_0^2 = 0$$

which, as proven, are given by the relation

$$p_{1,2} = -\frac{j}{2}\frac{B}{\nu_k} \pm \frac{j}{2}\sqrt{4\omega_0^2 + (B^2/\nu_k^2)} \tag{11.242}$$

We therefore find that for each zero value of the function $\mathcal{H}_{\text{LPP}}(s)$ of the prototype low-pass filter, *two purely imaginary zero values* are created in the transfer function $\mathcal{H}_{\text{BS}}(p)$ of a transformed band-pass filter, which, however, *do not have the same measure* and therefore are not related to each other by the property of complex conjugation. Taking into account that the zero values in the function $\mathcal{H}_{\text{LPP}}(s)$ appear in complex conjugate pairs, we can formulate the assertion that for the zero value $s_{z_k} = j\nu_k$ there will also be the complex conjugate value $s_{z_k}^* = -j\nu_k$, which will lead to the pair of zero values

$$p_{3,4} = \frac{j}{2}\frac{B}{\nu_k} \pm \frac{j}{2}\sqrt{4\omega_0^2 - (B^2/\nu_k^2)} \tag{11.243}$$

Taking into account that the numerator and denominator of the function $\mathcal{H}_{\text{BS}}(p)$ have been multiplied by the expression $(p^2+\omega_0^2)$, so as to form the quadratic equation for calculating the zeros of the values, we can formulate the conclusion that each pair of conjugate zero values of the function $\mathcal{H}_{\text{LPP}}(s)$ leads to the creation of four poles of the function $\mathcal{H}_{\text{BS}}(p)$ at the positions $\pm j\omega_0$. However, two of these poles will be cancelled by the two zero values that exist in the same positions due to the existence of the poles (as we saw a little while ago), and consequently, the zero values $\pm j\nu_k$ will lead to the appearance of *only two such poles* in the transfer function of the band-pass filter.

Selectivity and Spectral Efficiency Factor

The *selectivity* of a band-stop filter, as a function of the selectivity of the corresponding prototype low-pass filter, is computed as follows:

$$F_s^{\text{BS}} = \frac{d|\mathcal{H}_{\text{BS}}(j\omega)|}{d\omega}\bigg|_{\omega=\omega_c} = \left\{\frac{d}{d\omega}|\mathcal{H}_{\text{LPP}}\left(j\frac{B\omega}{-\omega^2+\omega_0^2}\right)|\right\}_{\omega=\omega_c} = \frac{d}{d\omega}\left(\frac{B\omega}{\omega^2-\omega_0^2}\right)\bigg|_{\omega=\omega_c} F_s^{\text{LPP}} = \frac{B(\omega_c^2+\omega_0^2)}{(\omega_0^2-\omega_c^2)^2}F_s^{\text{LPP}} \tag{11.244}$$

with the selectivity F_s^{LPP} being calculated at the position of the frequency $\Omega_c = (B\omega_c)/(\omega_0^2-\omega_c^2)$, where the frequency ω_c can be any of the frequencies ω_{c_1} and ω_{c_2} of the band-stop filter. On the other hand, the *spectral efficiency factors* of the two filters are proven to be equal to each other, or, in mathematical notation, $S_{\text{BS}} = S_{\text{LPP}}$.

Phase Characteristics and Time-Domain Response

Using the correlation between the frequency responses of the prototype low-pass and band-stop filters, it is proven that the *phase delay* of the band-stop filter has the form

$$\tau_{\text{BS}}^{\varphi}(\omega) = -\frac{1}{\omega}\measuredangle\mathcal{H}_{\text{BS}}(\omega) = -\frac{1}{\omega}\measuredangle\mathcal{H}_{\text{LPP}}\left(j\frac{B\omega}{-\omega^2+\omega_0^2}\right) \tag{11.245}$$

while the corresponding *group delay* is formulated as

$$\begin{aligned}\tau_{\text{BS}}^{g}(\omega) &= -\frac{d}{d\omega}\measuredangle\mathcal{H}_{\text{BS}}(\omega) = -\frac{d}{d\omega}\measuredangle\mathcal{H}_{\text{LPP}}\left(j\frac{B\omega}{\omega^2-\omega_0^2}\right)\\ &= \frac{d}{d\omega}\left(\frac{B\omega}{\omega^2-\omega_0^2}\right)\tau_{\text{LPP}}^{g}\left(\frac{B\omega}{\omega^2-\omega_0^2}\right) = \frac{B(\omega^2+\omega_0^2)}{(\omega_0^2-\omega^2)^2}\tau_{\text{LPP}}^{g}\left(\frac{\omega^2-\omega_0^2}{B\omega}\right)\end{aligned} \tag{11.246}$$

The reader is invited as an exercise to calculate the values of phase delay and group delay at the positions $\omega=\omega_0$, $\omega=\omega_{p_{1,2}}$, and $\omega=\omega_{s_{1,2}}$.

The impulse response of a band-stop filter will emerge from the *inverse Laplace transform of the transfer function* $\mathcal{H}_{\text{BS}}(p)$. Taking into account that *the transfer function of the band-stop filter results from the sum of the transfer function of a low-pass filter and the transfer function of a high-pass filter*, or, in mathematical notation, $\mathcal{H}_{\text{BS}}(p) = \mathcal{H}_{\text{LP}}(p) + \mathcal{H}_{\text{HP}}(p)$, the application of the inverse Laplace transform will lead us to the result $h_{\text{BS}}(t) = h_{\text{LP}}(t) + h_{\text{HP}}(t)$. After calculating the impulse response of the system, we can also calculate its step response as the *integral of the impulse response*. Taking into account that the impulse response $h_{\text{HP}}(t)$ shows an impulse function at position $t=0$, the same thing will happen with the impulse response $h_{\text{BS}}(t)$, with the value of the latter function tending to unity at the limit $t\to\infty$, since at this limit $h_{\text{LP}}(t)\to 1$ and $h_{\text{HP}}(t)\to 0$.

Problem 11.10.4 Construct a band-stop filter with operating specifications $\omega_{p_1} = 20$ rad/s, $\omega_{p_2} = 80$ rad/s, $\omega_{s_1} = 48$ rad/s, $\omega_{s_2} = 52$ rad/s, $A_p = 1.0$ dB, and $A_s = 25.0$ dB using the Cauer approximation as the prototype low-pass filter (Antoniou, Exercise for solution 10.46).

Solution Initially, let us calculate the *minimum order* of the prototype elliptic filter we will use. Using the values of the attenuation levels A_p and A_s, it turns out that

$$D = \frac{10^{A_s/10} - 1}{10^{A_p/10} - 1} = \frac{10^{2.5} - 1}{10^{0.1} - 1} = 1217.446228$$

On the other hand, for the calculation of the *selectivity factor* we find that the values of the quantities

$$\sqrt{\omega_{p_1}\omega_{p_2}} = \sqrt{20 \times 80} \text{ rad/s} = 40 \text{ rad/s} \quad \text{and} \quad \sqrt{\omega_{s_1}\omega_{s_2}} = \sqrt{48 \times 52} \text{ rad/s} = 49.959983 \text{ rad/s}$$

are different from each other, so a *redefinition of the cutoff frequencies* is required. Let us apply, therefore, the process we used for the design of the band-pass filter in Exercise 11.10.3 (page 1098), but this time, because the filter operates *in reverse*, we will choose not the minimum but the *maximum* value, while the selectivity factor will be *the inverse of the previous one*. Therefore, we will have

$$\omega'_{s_2} = \frac{\omega_{p_1}\omega_{p_2}}{\omega_{s_1}} \text{ rad/s} = \frac{20 \times 80}{48} \text{ rad/s} = 33.333333 \text{ rad/s}$$

$$\Omega_s^{(1)} = \frac{\omega'_{s_2} - \omega_{s_1}}{\omega_{p_2} - \omega_{p_1}} \text{ rad/s} = \frac{33.333333 - 48}{80 - 20} \text{ rad/s} = -0.244444 \text{ rad/s}$$

$$\omega'_{s_1} = \frac{\omega_{p_1}\omega_{p_2}}{\omega_{s_2}} \text{ rad/s} = \frac{20 \times 80}{52} \text{ rad/s} = 30.769230 \text{ rad/s}$$

$$\Omega_s^{(2)} = \frac{\omega_{s_2} - \omega'_{s_1}}{\omega_{p_2} - \omega_{p_1}} \text{ rad/s} = \frac{52 - 30.769230}{80 - 20} \text{ rad/s} = 0.353846 \text{ rad/s}$$

Therefore, $\Omega_s = \max(\Omega_s^{(1)}, \Omega_s^{(2)}) = \max(-0.244444, 0.353846)$ rad/s $= 0.353846$ rad/s, so the selectivity factor will be equal to $k_2 = 0.353846$. Following the methodology developed in the section on elliptic filters, we define the frequencies Ω_p and Ω_s as $\Omega_p = \sqrt{k_2} = 0.594849$ rad/s and $\Omega_s = 1/\sqrt{k_2} = 1.681096$ rad/s.

Having calculated the values of D and k_2, we can now calculate the order N of the filter and then the poles and its zero values, working as in Exercise 11.9.9 (page 1062). Using the second method of calculating the function $q(k_2)$, which does not include the use of elliptic integrals, yields $k'_2 = \sqrt{1 - k_2^2} = \sqrt{1 - 0.353846^2} = 0.935303$, and therefore,

$$q_0 = \frac{1}{2}\left(\frac{1 - \sqrt{k'_2}}{1 + \sqrt{k'_2}}\right) = \frac{1}{2}\left(\frac{1 - \sqrt{0.935303}}{1 + \sqrt{0.935303}}\right) = \frac{1}{2}\left(\frac{0.032889}{1.967110}\right) = 0.08359$$

from which it follows that

$$q(k_2) = q(0.353846) = q_0 + 2q_0^5 + 15q_0^9 = 0.08359 + 2(0.08359)^5 + 15(0.08359)^9 = 0.08359$$

Having calculated the two foregoing functions, the *minimum filter order* is calculated as

$$N = \frac{\log(16D)}{\log[1/q(k_2)]} = \frac{log(16 \times 1217.446228)}{\log(1/0.08359)} = \frac{\log(19479.139653)}{119.621121} = \frac{9.877099}{4.784329} = 2.064468.$$

This value obviously needs to be rounded up to the nearest integer, and therefore the minimum filter order is $N = 3$.

Using the result $\boldsymbol{K}_2 = \boldsymbol{K}(k_2) = \boldsymbol{K}(0.353486) = 1.623644$ and calculating the quantity

$$j\frac{\boldsymbol{K}_2}{N\pi} \ln\frac{10^{A_p/20} + 1}{10^{A_p/20} - 1} = j\frac{1.623644}{3 \times 3.14159} \ln\frac{10^{1.0/20} + 1}{10^{1.0/20} - 1} = j\, 0.0082442 \ln\frac{10^{0.015} + 1}{10^{0.015} - 1} = 0.334365j$$

the real parameter p_0 is calculated as

$$p_0 = j\,\Omega_p \mathrm{sn}\left[j\frac{\boldsymbol{K}_2}{N\pi}\ln\frac{10^{A_p/20}+1}{10^{A_p/20}-1}, \frac{\omega_p}{\omega_s}\right] = j\,0.594849\ \mathrm{sn}\,(j\,0.492003, 0.353486)$$

$$= j\,0.594849\left(j\frac{\mathrm{sn}(0.935303)}{\mathrm{cn}(0.935303)}\right) = j\,0.594849\left(j\frac{0.456872}{0.889531}\right) = -0.305520$$

where for the calculation of the elliptic sine we used the appropriate function in MATLAB and the properties of this function (see Exercise 11.9.9 on page 1062). After the value of the constant p_0 is calculated, the value of the auxiliary parameter W is calculated as

$$W = \sqrt{\left(1+\frac{p_0^2}{\Omega_p^2}\right)\left(1+\frac{p_0^2}{\Omega_s^2}\right)} = \sqrt{\left(1+\frac{(-0.305520)^2}{(0.594849)^2}\right)\left(1+\frac{(-0.305520)^2}{(1.681096)^2}\right)} = 1.142600$$

Therefore,

$$\Omega_m = \frac{1}{\omega_s}\mathrm{sn}\left[\frac{2m}{N}\boldsymbol{K}_2, \frac{\omega_p}{\omega_s}\right] = 0.594850\ \mathrm{sn}\,[1.082429m, 0.353846] \quad (m=1)$$

and following performance of the operations, it turns out that

$$\Omega_1 = 0.5944850\mathrm{sn}[1.082429, 0.353846] = 0.594850 \cdot 0.861371 = 0.512386$$

Therefore, the transfer function of the prototype elliptic filter has a pair of complex conjugate zero values with values

$$z_{1,2} = \pm\frac{j}{\Omega_1} = \pm\frac{j}{0.512386} = \pm 1.951650j$$

With regard to the poles of the function, it is a function of *odd order* and therefore, it has the real pole $p_0 = -0.305520$ and a *pair of complex conjugate poles* for the calculation of which we need the value of the auxiliary parameter $V_m = \sqrt{(1-\omega_p^2\Omega_m^2)(1-\omega_s^2\Omega_m^2)}$ for the value $m = 1$. By substituting the values, it turns out that

$$V_1 = \sqrt{[1-(0.594849)^2(0.512386)^2][1-(1.681096)^2(0.512386)^2]} = 0.483807$$

and therefore the pair of complex conjugate poles is calculated as (the product $\Omega_p\Omega_s$ is omitted, since its value is equal to one)

$$p_{1,2} = \frac{p_0 V_1 \pm j\Omega_p\Omega_s\Omega_1 W}{1+p_0^2\Omega_1^2} = \frac{-0.305520\cdot 0.483808 \pm j\,0.512386\cdot 1.142600}{1+(-0.305520)^2(0.512386)^2} = -0.144277 \pm 0.571448j$$

The normalization constant H_0 is calculated from the defining equation for the odd-order elliptic filter and is found to be equal to

$$H_0 = \frac{0.305520 \times 0.347308}{3.808937} = 0.027858$$

By substituting the foregoing values of poles and zeros of the prototype elliptic filter and the operating parameters $\omega_0 = \sqrt{\omega_{p_1}\omega_{p_2}} = \sqrt{20 \times 80}$ rad/s $= 40$ rad/s and $B = \Omega_p(\omega_{p_1} - \Omega_{p_2}) = 35.690939$ of the band-stop filter to be constructed in the equations for calculating the poles and zeros of that filter, we arrive at the following results:

The complex pole $s_{p_1} = -0.144277 + 0.571448j$ leads to the poles $p_1 = -2.994372 + 19.901472j$ and $p_2 = -11.828599 - 78.616330j$, while the conjugate pole $s_{p_2} = -0.144277 - 0.571448j$ leads to the poles $p_3 = -2.994372 - 19.901472j$ and $p_4 = -11.828599 + 78.616330j$. We observe that the poles p_1 and p_3 are complex conjugate poles and contribute to the denominator of the transfer function the term $(p^2 + 5.988744p + 405.034885)$, while the poles p_2 and

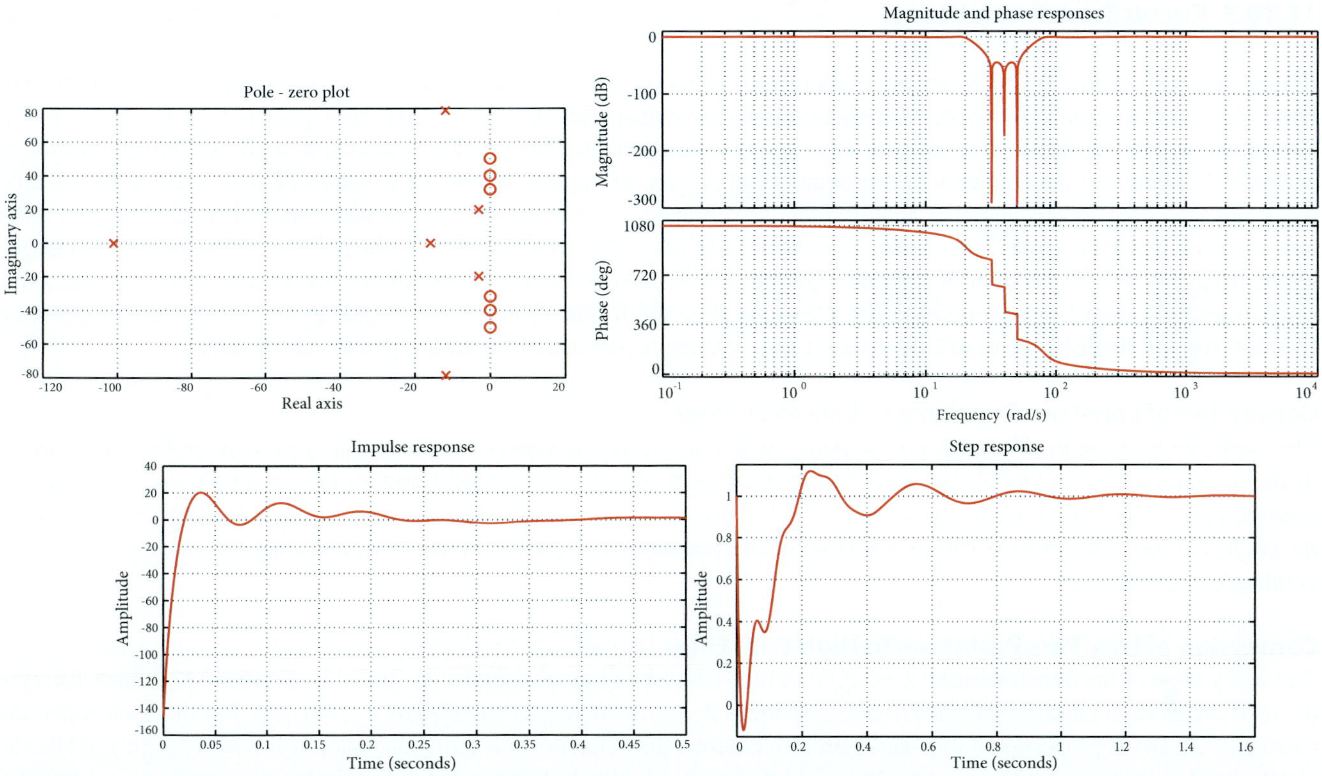

Fig. 11.80 Pole–zero plot, amplitude and phase responses, and impulse and step responses of band-stop filter of Exercise 11.10.4

p_4 are also complex conjugates of each other, contributing the term $(p^2 + 23.657199p + 6320.443230)$. Finally, the pole $s_{p_3} = -0.305520$ leads to the creation of the real poles $p_5 = -100.974758$ and $p_6 = -15.845544$, which contribute to the denominator of the transfer function the terms $(p + 100.974758)$ and $(p + 15.845544)$, respectively. This pole also leads to the creation of two zero values at the positions $z_{5,6} = \pm\omega_0 j = \pm 40j$, which contribute to the numerator of the transfer function the term $(p^2 + 1600)$.

On the other hand, the zero value $s_{z_1} = 1.951650j$ leads to the zero values of the band-stop filter $z_1 = 31.888016j$ and $z_2 = 50.175588j$, while its complex conjugate leads to the zero values $z_3 = -31.888016j$ and $z_4 = -50.175588j$. We observe that the values z_1 and z_3 are complex conjugates of each other and contribute to the numerator of the transfer function the term $(p^2 + 1016.845612)$, while the values z_2 and z_4 are also complex conjugates and contribute the term $(p^2 + 2517.589660)$. Combining all of the foregoing points, the transfer function $\boldsymbol{H}_{\text{BS}}(p)$ of the band-stop filter to be constructed will have the form

$$\boldsymbol{H}_{\text{BS}}(p) = \frac{(p^2 + 1600)(p^2 + 1016.845612)(p^2 + 2517.589660)}{(p^2 + 5.988744p + 405.034885)(p^2 + 23.657199p + 6320.443230)(p + 100.974758)(p + 15.845544)}$$

or, equivalently, $\boldsymbol{H}_{\text{BS}}(p) = N(p)/D(p)$, where

$$\begin{aligned} N(p) &= p^6 + 5.134435 \cdot 10^3 p^4 + 8.215096 \cdot 10^6 p^2 + 4.095999 \cdot 10^9 \\ D(p) &= p^6 + 1.464662 \times 10^2 p^5 + 1.193040 \times 10^4 p^4 + 8.970901 \cdot 10^5 p^3 \\ &\quad + 1.908864 \cdot 10^7 p^2 + 3.749535 \cdot 10^8 p + 4.095999 \cdot 10^9 \end{aligned}$$

The pole–zero plot, the amplitude and phase responses, and the step and impulse responses of this band-stop filter are presented in Fig. 11.80.

11.10.5 Circuit Transformations

Taking into account that analog filters are implemented using *passive or active electrical circuits*, the preceding transformations that will convert the prototype low-pass filter we have constructed into a *low-pass, high-pass, band-pass,* or *band-stop* filter must also be applied to the electrical circuit that implements it. This process is known as a *circuit transformation*, is performed directly on the elements of the circuit, and is easy to implement since the values of the elements for specific types and operating specifications of analog filters are grouped and recorded in *special data tables*. It is obvious that these transformations will be applied *only to the capacitors and coils of the circuit* (and not to the resistances), as only these depend on the frequency ω of the current flowing through the circuit. From basic theory it is known that the capacitive and inductive resistances for a capacitor and a coil in the s domain have the forms $Z_C(s) = 1/sC$ and $Z_L(s) = sL$. In the following sections, we will describe the transformations of these elements for each of the four types of analog filters.

Conversion of Low-Pass Prototype to Low-Pass Filter

The application of the transformation $s = p/\omega_0$ (which converts a low-pass prototype into a transformed low-pass filter) to the resistances $Z_C(s)$ and $Z_L(s)$ leads to the resistances $Z_C(p) = \omega_0/(pC)$ and $Z_L(p) = (pL)/\omega_0$. Therefore, to convert a prototype into a transformed low-pass filter, *we divide all the resistances of the circuit elements by the value of the coefficient* ω_0. This process causes a scaling of the frequency axis, without, however, changing the shape of the filter's amplitude response curve.

Conversion of Low-Pass Prototype to High-Pass Filter

The application of the transformation $s = \omega_0^2/p$ (which converts a low-pass prototype into a transformed high-pass filter) to the resistances $Z_C(s)$ and $Z_L(s)$ leads to the resistances $Z_C(p) = p/(\omega_0^2 C)$ and $Z_L(p) = (\omega_0^2 L)/p$. But this means that the resistance $Z_C(p)$ is proportional to p and therefore exhibits the behavior of a self-inductance coil $L = 1/(\omega_0^2 C)$, while the resistance $Z_L(p)$ is inversely proportional to p and therefore exhibits the behavior of a capacitor with capacity $C = 1/(\omega_0^2 L)$. Therefore, to transform a prototype low-pass filter into a high-pass filter, *we must replace all coils with capacitors and all capacitors with coils with the foregoing values of self-inductance and capacity.*

Conversion of Prototype Low-Pass to Band-Pass Filter

The application of the transformation $s = (p^2 + \omega_0^2)/pB$ (which converts a prototype low-pass into a transformed band-pass filter) to the resistance $Z_L(s)$ leads to a new resistance of the form

$$Z_L(p) = \frac{L(p^2+\omega_0^2)}{Bp} = \frac{Lp^2 + L\omega_0^2}{Bp} = \left(\frac{L}{B}\right)p + \left(\frac{L\omega_0^2}{B}\right)\frac{1}{p} \tag{11.247}$$

But the first term of the last expression corresponds to a resistance of a self-inductance coil equal to L/B, while the second term corresponds to a resistance of a capacitor with capacity $B/(L\omega_0^2)$. Therefore, *each coil of the circuit that implements the prototype low-pass filter should be replaced by a coil and a capacitor with the aforementioned characteristics, connected in series.*

On the other hand, the application of the above transformation to the resistance $Z_C(s)$ leads to a new resistance of the form $Z_C(p) = Bp/[C(p^2 + \omega_0^2)]$. However, we observe that

$$\frac{1}{Z_C(p)} = \frac{C(p^2+\omega_0^2)}{Bp} = \frac{CP}{B} + \frac{C\omega_0^2}{Bp} = \frac{1}{\left(\frac{1}{C/B}\right)\frac{1}{p}} + \frac{1}{\left(\frac{B}{C\omega_0^2}\right)p} \tag{11.248}$$

But the denominator of the first fraction corresponds to a capacitor resistance equal to C/B, while the second term corresponds to a coil self-induction resistance $B/(\omega_0^2 C)$. Therefore, *each capacitor in the circuit should be replaced by a capacitor and a coil with the foregoing characteristics, connected in parallel.*

Conversion of Low-Pass Prototype to Band-Stop Filter

The application of the transformation $s = pB/(p^2+\omega_0^2)$ (which converts a low-pass prototype into a transformed band-stop filter) to the resistance $Z_C(s)$ leads to a new resistance of the form

$$Z_C(p) = \frac{p^2+\omega_0^2}{BpC} = \left(\frac{1}{CB}\right)p + \left(\frac{\omega_0^2}{CB}\right)\frac{1}{p} \quad (11.249)$$

But the first term of the last expression corresponds to a coil self-induction resistance equal to $1/(CB)$, while the second term corresponds to a capacitor resistance with capacity CB/ω_0^2. Therefore, *each capacitor in the circuit that implements a low-pass filter prototype should be replaced by a coil and a capacitor with the foregoing characteristics, connected in series.*

On the other hand, the application of the foregoing transformation to the resistance $Z_L(s)$ leads to a new resistance of the form $Z_L(p) = LBp/(p^2+\omega_0^2)$. However, we observe that

$$\frac{1}{Z_L(p)} = \frac{p^2+\omega_0^2}{LBp} = \frac{P}{LB} + \frac{\omega_0^2}{LBp} = \frac{1}{(LB)\dfrac{1}{p}} + \frac{1}{\left(\dfrac{LB}{\omega_0^2}\right)p} \quad (11.250)$$

But the denominator of the first fraction corresponds to a capacitor resistance equal to $1/LB$, while the second term corresponds to an inductance coil resistance LB/ω_0^2. Therefore, *each coil of the circuit should be replaced by a capacitor and a coil with the foregoing characteristics, connected in parallel.*

The preceding transformations of circuit elements for these four conversion cases are presented in Fig. 11.81, while an example of their application for arbitrary operating parameters is depicted in Fig. 11.82.

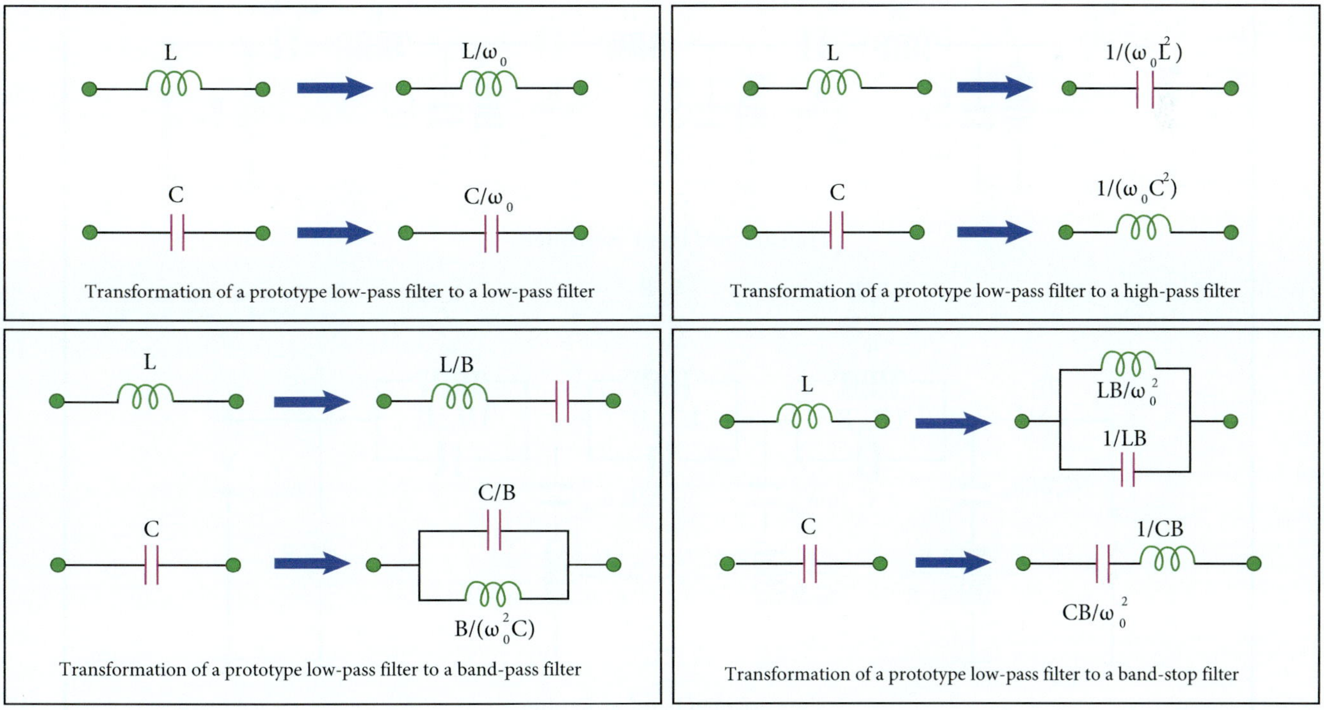

Fig. 11.81 Transformations of circuit elements for conversion of prototype low-pass filter into low-pass, high-pass, band-pass, and band-stop filters

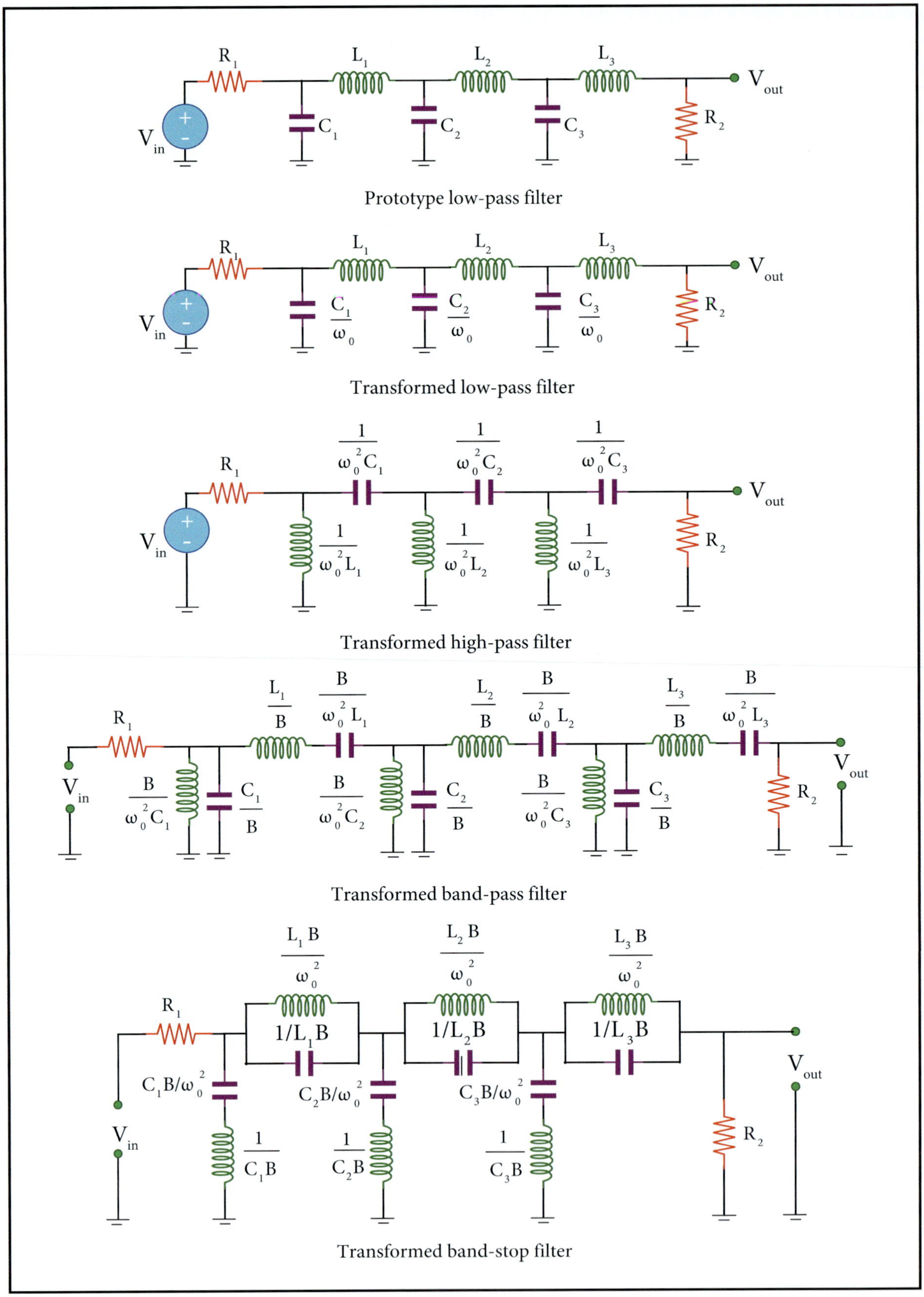

Fig. 11.82 Example of application of transformations of circuit elements for conversion of prototype low-pass filter into low-pass, high-pass, band-pass, and band-stop filters

11.11 Problems

1. Using the methodology of Exercise 10.5.1 (page 852), determine the step response of the band-pass filter of Exercise 10.5.2 (page 852).
2. Using the methodology of Exercise 10.5.1 (page 852), determine the step response of the band-stop filter of Exercise 10.5.3 (page 853).
3. The equations of the transfer function presented in this section for the structures of active filters arise from the application of Kirchoff's rules to the branches of the circuits that implement them. Considering, for example, the topology of Sallen–Key of a second-order band-stop filter and assuming that the voltages $v_1(t)$, $v_2(t)$, and $v_3(t)$ are those presented in Fig. 11.24α (page 970), the application of Kirchhoff's voltage law will lead to the equations

$$C_3s[V_{\text{out}} - V_3] + \frac{V_{\text{in}} - V_3}{R_1} + \frac{V_2 - V_3}{R_2} = 0$$

$$C_2s(V_1 - V_2) + \frac{V_3 - V_2}{R_2} = 0$$

$$\frac{V_1}{R_3} + C_1s[V_1 - V_{\text{in}}] + C_2s(V_1 - V_2) = 0$$

Considering that the resistances R_4 and R_5 have the forms $R_4 = (A-1)R_0$ and $R_5 = R_0$, the voltages V_{out} and V_2 will be related through the relationship $V_{\text{out}} = AV_2$. By combining all these equations, determine the transfer function of the filter as $\mathcal{H}(s) = V_{\text{out}}/V_{\text{in}}$, where $V_1 = V_1(s) = \mathcal{L}\{v_1(t)\}$, $V_2 = V_2(s) = \mathcal{L}\{v_2(t)\}$, and $V_3 = V_3(s) = \mathcal{L}\{v_3(t)\}$ and $V_{\text{out}} = V_{\text{out}}(s) = v_{\text{out}}(t)$, $V_{\text{in}} = V_{\text{in}}(s) = v_{\text{in}}(t)$.
4. Applying Kirchhoff's current law in the way presented in Problem 3, determine the transfer function and the expressions of the parameters ω_0 and Q for the Sallen–Key topology and multiple feedback topologies of an active second-order low-pass filter.
5. Applying Kirchhoff's current law as presented in Problem 3, determine the transfer function and the expressions of the parameters ω_0 and Q for the Sallen–Key topology and multiple feedback topologies of an active second-order high-pass filter.
6. Applying Kirchhoff's current law as presented in Problem 3, determine the transfer function and the expressions of the parameters ω_0 and Q for the Sallen–Key topology and multiple feedback topologies of an active second-order band-pass filter.
7. Determine the poles of the transfer function of a Butterworth low-pass filter for order values $N = 7, 9$ and for normalized frequency $\omega_c = 1$.
8. Determine the poles of the transfer function of a Butterworth low-pass filter for order values $N = 8, 10$ and for normalized frequency $\omega_c = 1$.
9. Repeat Exercise 11.9.3 (page 1009) for operating specifications $\omega_p = 0.7$ rad/s and $\omega_s = 2.0$ rad/s and attenuation levels $A_p = 0.5$ dB and $A_s = 30$ dB.
10. Repeat Exercise 11.9.3 (page 1009) for operating specifications $\omega_p = 1000$ Hz and $\omega_s = 5000$ Hz and attenuation levels $A_p = 2.5$ dB and $A_s = 60$ dB.
11. For each of the Problems 9 and 10 determine the frequency value that corresponds to a 3 dB attenuation, as well as the selectivity of the filter.
12. Determine the transfer function, the amplitude and phase responses, the phase and group delays, and the impulse and step responses of a low-pass Chebyshev I filter of order $N = 3$ with operating parameters $\omega_p = 1$ rad/s and $A_p = 1.0$ dB.
13. Considering a low-pass Chebyshev I filter of order $N = 6$ with operating parameters $\omega_c = 1000$ rad/s and $A_p = 1.5$ dB, determine the values of ε and ω_p, calculate the frequencies of the gain function's maxima and minima, and plot the amplitude response curve.
14. For each of the filters in Problems 12 and 13, calculate the filter's selectivity and the spectral efficiency factor.
15. Repeat Exercise 11.9.8 (page 1040) for a Chebyshev II filter with operating frequencies $\omega_p = 0.4$ rad/s and $\omega_s = 0.75$ rad/s and attenuation levels $A_p = 0.3$ dB and $A_s = 40$ dB.
16. For each of the filters in Problems 12, 13, and 15, calculate the selectivity of the filter, the spectral efficiency factor, and the value of the frequency ω that corresponds to an attenuation of 3 dB.

17. Repeat Exercise 11.9.9 (page 1062) for a selectivity value of $k = 0.6$ and attenuation levels $A_p = 0.6$ dB and $A_s = 50$ dB.
18. Design an elliptic filter with operating specifications $\omega_p = 0.948$ rad/s and $\omega_p = 1.054$ rad/s and attenuation levels $A_p = 0.3$ dB and $A_s = 60$ dB.
19. Returning to the filter of Exercise 11.9.10 (page 1077), calculate and design the phase delay and the group delay as a function of the frequency ω.
20. Repeat the process of Problem 19 for Bessel filters of order $N = 3, 4, 5, 6$, and prove that the group delay for these four cases has the form

$$\tau(\omega) = \frac{225 + 45\omega^2 + 6\omega^4}{225 + 45\omega^2 + 6\omega^4 + \omega^6}$$

$$\tau(\omega) = \frac{11025 + 1575\omega^2 + 135\omega^4 + 10\omega^6}{11025 + 1575\omega^2 + 135\omega^4 + 10\omega^6 + \omega^8}$$

$$\tau(\omega) = \frac{893025 + 99225\omega^2 + 6300\omega^4 + 315\omega^6 + 15\omega^8}{893025 + 99225\omega^2 + 6300\omega^4 + 315\omega^6 + 15\omega^8 + \omega^{10}}$$

$$\tau(\omega) = \frac{108056025 + 982327\omega^2 + 496125\omega^4 + 18900\omega^6 + 630\omega^8 + 21\omega^{10}}{108056025 + 9823275\omega^2 + 496125\omega^4 + 18900\omega^6 + 630\omega^8 + 21\omega^{10} + \omega^{12}}$$

21. The frequency response of an analog filter has the form

$$\mathcal{H}(j\omega) = \frac{1}{1 + (j\omega/1000)}$$

Determine the frequencies at the boundaries of the passband and cutoff band considering a deviation in the passband of 1 dB and attenuation in the cutoff band on the order of 40 dB.
22. Design the step response of a Chebyshev I filter with order value $N = 10$ and operating specifications $f_c = 1$ kHz AND $A_p = 1$ dB. What is the maximum group delay of this filter and to what frequency does it correspond? What is the maximum value of the impulse response and at what time does it appear?
23. Determine the transfer function expressed in polynomial form as well as the values of its poles for a Chebyshev I filter with operating specifications $\omega_p = 1$ and $A_p = 1.2$ dB and order value $N = 6$. What are the impulse and step responses of this filter? Design the amplitude and phase responses, as well as the phase delay and group delay.
24. Considering a Chebyshev II filter with operating specifications $\omega_c = 1000$ rad/s and $A_s = 60$ dB and order value $N = 6$, determine the values of the parameters ε and ω_s, the values of the frequency at which the amplitude response takes its minimum value, and the frequencies in the cutoff band that correspond to an attenuation value equal to A_s.
25. Suppose a Chebyshev II filter is required to be designed with operating specifications $f_c = 3.5$ kHz, $f_s = 7$ kHz. What should be the required filter order so that (a) $A_s = 40$ dB, (b) $A_s = 60$ dB, (c) $A_s = 80$ dB, and (d) $A_s = 100$ dB.
26. Repeat Problem 25 for a Butterworth filter.
27. Considering that the operating specifications for a low-pass Chebyshev I type filter are $A_p = 1.2$ dB, $A_s = 75$ dB, $\omega_p = 3500$ rad/s, and $\omega_s = 7000$ rad/s, determine the minimum order of the filter for which these specifications are met.
28. Repeat Problem 27 for operating specifications $A_p = 1.2$ dB, $A_s = 75$ dB, $f_p = 6.5$ KHz rad/s, and $f_s = 13$ kHz.
29. Determine the minimum order of an elliptic filter to meet the operating specifications $f_p = 5$ kHZ, $f_s = 75$ kHZ, $A_p = 1$ dB, and $A_s = 75dB$.
30. Determine the poles and zeros of an elliptic filter of order $N = 3$, with operating characteristics $\omega_p = 1$, $A_p = 3$ dB, $A_s = 40dB$.
31. Repeat Exercise 11.10.1 (page 1086) using a Chebyshev I type low-pass filter as a prototype.
32. Repeat Exercise 11.10.1 using a Chebyshev II type low-pass filter as a prototype.
33. Repeat Exercise 11.10.1 using an elliptic low-pass filter as a prototype.
34. Compare the results of Exercise 11.10.1 and Problems 31, 32, and 33, in terms of the minimum filter order and the characteristics of the transformed filter responses.
35. Repeat Exercise 11.10.2 (page 1091) using a Butterworth type low-pass filter as a prototype.
36. Repeat Exercise 11.10.2 using a Chebyshev II filter as a prototype low-pass filter.
37. Repeat Exercise 11.10.2 using an elliptical filter as a prototype low-pass filter.

38. Compare the results of Exercise 11.10.2 and Problems 35, 36, and 37 regarding the minimum order of the filter and the characteristics of the responses of the transformed filter.
39. Repeat Exercise 11.10.3 (page 1098) using a Butterworth filter as a prototype low-pass filter.
40. Repeat Exercise 11.10.3 using a Chebyshev I filter as a prototype low-pass filter.
41. Repeat Exercise 11.10.3 using an elliptical filter as a prototype low-pass filter.
42. Compare the results of Exercise 11.10.3 and Problems 39, 40, and 41 regarding the minimum order of the filter and the characteristics of the responses of the transformed filter.
43. Repeat Exercise 11.10.4 (page 1106) using a Butterworth filter as a prototype low-pass filter.
44. Repeat Exercise 11.10.4 using a Chebyshev I filter as a prototype low-pass filter.
45. Repeat Exercise 11.10.4 using a Chebyshev Type II filter as the prototype low-pass filter.
46. Compare the results of Exercise 11.10.4 and Problems 43, 44 and 45 regarding the minimum order of the filter and the characteristics of the responses of the transformed filter.
47. Let us assume that the equation

$$\mathcal{H}(s) = \frac{1}{(s+1)(s^2+s+1)}$$

represents the transfer function of a Butterworth low-pass filter of order $N = 3$ with a unit cutoff frequency. What is the transfer function of the corresponding high-pass filter?
48. If the transfer function of a Butterworth low-pass filter is

$$\mathcal{H}(s) = \frac{1}{(s+1)(s^2+\sqrt{2}s+1)}$$

what is the transfer function of the corresponding high-pass filter with cutoff frequency ω_c?
49. If the transfer function of a low-pass filter is

$$\mathcal{H}(s) = \frac{1}{s+1}$$

determine the transfer function of the corresponding band-pass filter with a unit cutoff frequency and bandwidth $B = 0.1$.
50. Let us assume that the polynomial of the denominator of the transfer function of a prototype Butterworth low-pass filter has the form

$$D(s) = (s+1)(s^2+0.618s+1)(s^2+1.618s+1)$$

Determine the transfer function of the corresponding high-pass filter with a unit cutoff frequency, and design the amplitude response of this filter.

12 Design of Digital Filters

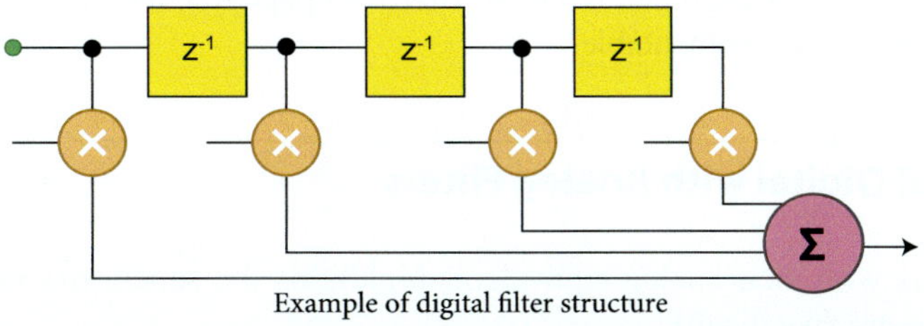

Example of digital filter structure

Analog and digital filters are used in many signal processing applications in the field of continuous and discrete-time

12.1 Introduction to Digital Filters

We use the term *digital filter* to describe a hardware or software system that accepts as input a *discrete-time signal* $x[n]$ and produces as its output a second *discrete-time signal* $y[n]$, by modifying the shape of the waveform, that is, the characteristics of *amplitude, frequency* and *phase* of the input signal. Although the term digital filter refers to a device that is applied on *digital signals*—that is, on *sequences of binary values*—what is mentioned in the following pages applies to *discrete-time filters*, which are applied to *discrete sequences of recorded data*, that is, to discrete-time signals. Digital filters have been used continuously over the past decades in a plethora of applications, such as *data compression*, *noise removal*, *encoding and decoding of voice signals*, and *digital image processing applications*, which are aimed both at *improving the quality of the signal* and at *extracting information from it*. In the following pages, we use the term digital filters to describe discrete-time filters that are applied to discrete sequences of data.

Speaking generally, the basic functions of a digital filter are *the alteration of spectral characteristics, the separation of frequencies included in the input signal, the removal of unwanted characteristics*, and *the extraction of certain others*, processes that are easily performed when *the frequency content of the signal carrying the useful information is quite different from the frequency content associated with the noise*. In other words, this means that the signal and the noise should not have common frequencies, so that their complete separation is possible. If this requirement is not met, we have to use other processing techniques that are related to the field of *stochastic signals*. As already mentioned (see Sect. 3.6 on page 282), filters can be divided into two categories: *shaping filters* that alter the spectral characteristics of frequency components and *selection filters* that allow the passage or cutoff of specific frequency ranges.

The quantitative assessment of the frequency separation ability of the filter used in each case is a measure of its quality and is based on the use of the concepts of *gain* and *attenuation*, which are used in the same way as we described for analog filters. We remember from the basic theory that the response of a discrete-time filter impulse $h[n]$ to an exponential signal $x[n] = Ae^{j\omega_0 n}$ is the signal

$$y[n] = \mathcal{H}(e^{j\omega_0})\, x[n] = (A|\mathcal{H}(e^{j\omega_0})|)\, e^{j\,[\omega_0 n + \measuredangle \mathcal{H}(e^{j\omega_0})]}, \tag{12.1}$$

A. I. Margaris, *Continuous and Discrete-Time Signals and Systems*,
https://doi.org/10.1007/978-3-031-81209-5_12

where $\mathcal{H}(e^{j\omega_0})$ is the frequency response of the filter—which is known to be defined as *the Fourier transform of its impulse response*—calculated at the position of the frequency of the input signal, $\omega = \omega_0$. In this case, if the amplitude of the output signal remains unchanged, that is, the relation $|\mathcal{H}(e^{j\omega_0})| = 1$ holds, the value of the filter's gain expressed in decibels is equal to zero. Conversely, in a filter characterized by an attenuation value equal, let us say, to -100 dB the amplitude of the output signal is five orders of magnitude smaller than the amplitude of the input signal, that is, it has been divided by the factor 10^5. The unit dB is, as is well known, a *logarithmic unit of measurement of the ratio of two values*. Considering, for example, an amplifier with power values P_i for the input signal and P_o for the output signal, the power gain in decibels is equal to $10 \log_{10}(P_o/P_i)$ dB. In the general case of a filter with frequency response function $\mathcal{H}(e^{j\omega})$, its gain expressed in dB is equal to $G = 20 \log_{10} |\mathcal{H}(e^{j\omega})|$.

In full analogy with the analog filters presented in the previous sections, real digital filters should be characterized by the properties of stability and causality, which requires the impulse response of the filter to be defined only for values $n \geq 0$. If the above property does not hold, the filter is characterized as noncausal and it is impossible to implement in practice. As we have proven in previous chapters (see, for example, Exercise 10.4.7 on page 837 and the following discussion), ideal digital filters are noncausal and therefore non-implementable.

12.2 The Comparison of Digital with Analog Filters

The comparison of digital filters with their analog equivalents highlights the superiority of digital over the corresponding analog filters at several points, more specifically:

- Digital filters have properties, such as, for example, *the completely linear phase response*, which cannot be provided by an analog filter.
- The behavior of digital filters *does not depend on and is not affected by environmental factors*, unlike analog filters, which require *very frequent calibration*, since the characteristics of their electronic components are affected by temperature and aging.
- The frequency response of digital filters can easily be adapted to the respective requirements, since they contain a *programmable processor* (in practice, it is possible to construct a digital filter with any characteristic curve). This is also the reason why these filters are used as *adaptive filters*.
- Digital filters allow the filtering of many input signals from the same filter and the storage of both the filtered and the unfiltered values for further processing.
- Digital filters are characterized by *much greater computational accuracy* than analog filters in which the achievement of such accuracy is quite limited and requires the use of complex circuits.
- The continuous and impressive developments of *very-large-scale integration technology VLSI* allow the construction of digital filters with ever smaller size, cost, and power consumption.

On the other hand, the main disadvantage of digital filters compared with analog ones is the *introduction of time delays* in the production of the output signal, owing to the use of digital processors, with the values of these delays mainly depending on the *processor speed* and *the degree of complexity* of the processing applied to the input signal, such as, for example, the number of arithmetic operations performed. These delays are not encountered in analog filters, which are implemented with the help of *classic analog circuits* and therefore offer direct processing of the input signals. It should be noted, incidentally, that certain types of filters, such as for example, reconstruction filters and filters for removing the effects of aliasing, only appear in analog form.

Typical examples of analog and digital filter structures are presented in Fig. 12.1.

12.3 Frequency Selection Filters

As already mentioned, the effect of a digital filter—described by a *linear and time-invariant discrete-time system*—on an input signal, results in the *attenuation* (or *complete cutoff* for the case of an ideal filter) of certain frequencies included in it. This filtering process depends on the characteristics of the *frequency response* of the system $\mathcal{H}(e^{j\omega})$, which in turn depends on the coefficients α_k and β_k of the difference equation that describes the system and is known to have the form

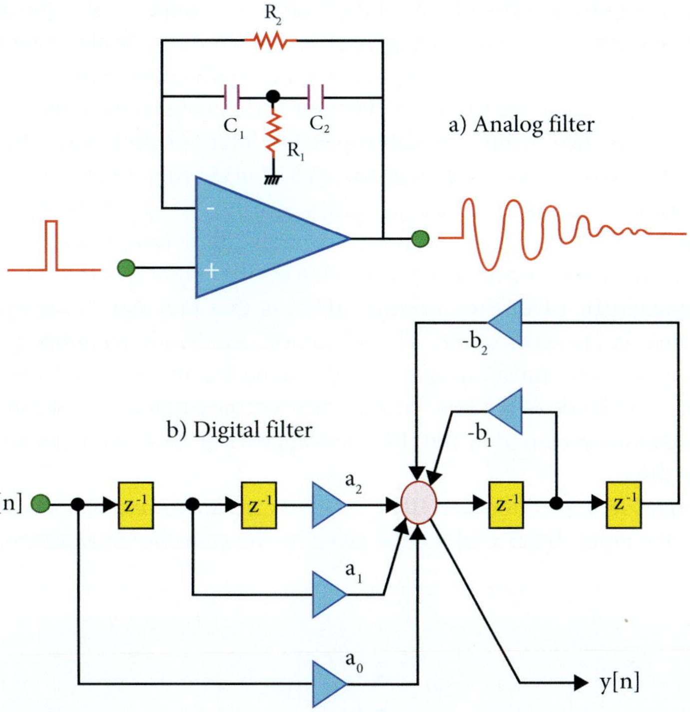

Fig. 12.1 Analog (**a**) and digital (**b**) filter structures

$$y[n] = -\sum_{k=1}^{N} \alpha_k y[n-k] + \sum_{k=0}^{M} \beta_k x[n-k] \tag{12.2}$$

(in the terminology of digital filters these are known as *filter coefficients*). Therefore, with a suitable choice of these coefficients we can construct digital filters that cut off the desired frequency range in each case, while simultaneously allowing passage to some others. Knowing the number of past inputs M and outputs N that appear in the difference equation of the filter, we can determine its order as the value $n = \max(N, M)$.

An alternative, but equivalent way of describing the structure and operation of a digital filter, is based on the use of the transfer function $\mathcal{H}(z)$ (see Sect. 9.1 on page 693), which is defined as *the $\mathcal{Z}$ transform of the discrete impulse response signal* $h[n]$. In this case, the order of the filter is determined *by the number of poles and zeros of the transfer function*, the position of which on the complex plane also determines, as is well known, the magnitude of the function $\mathcal{H}(e^{j\omega})$ (see Sect. 10.4.2 on page 840). As has been proven in Sect. 10.4.3 on page 842, a pole causes *amplification* of frequencies close to it, in contrast to a zero that causes *attenuation* of neighboring frequencies. Apart from the magnitude of the frequency response function, the positions of the poles and zeros also affect the *stability* of the filter, since, as has already been proven (see Sect. 9.3 on page 704), *the poles of a stable filter are all located inside the unit circle* described by the equation $z = e^{j\omega}$. The most classic and widely used type of stability is *BIBO stability*, according to which *the response of the filter, when a bounded input is applied to it, is the production of a bounded output* (see Sect. 7.2.4 on page 571). For the effect of the filter on the spectrum of the input signal, see Sect. 10.4.1 on page 836, whereas for the categories into which frequency selection filters are divided, see Sect. 3.6 on page 282, which refers to continuous-time filters, but nevertheless this categorization applies as is to digital filters.

12.3.1 Finite and Infinite Impulse Response Filters

We conclude our brief reference to the basic characteristics of frequency selection digital filters with the observation that they can be categorized in many different ways, the most well-known of which is based *on the length of the discrete-time signal of the impulse response* $h[n]$ that describes them. If the number of samples—or equivalently the length—of the signal $h[n]$ is *finite*, the filter is characterized as a *finite impulse response (FIR) filter*; on the contrary, if the impulse response has *infinite length*, the filter is characterized as an *infinite impulse response (IIR) filter*. FIR-type filters are also known as *moving average filters* or *zero-only filters*, because their transfer function $\mathcal{H}(z)$ contains only zeros and no poles (see Eq. 9.11). On the other hand, IIR filters appear in two forms—in *autoregressive* filters or pole-only filters—since the transfer function $\mathcal{H}(z)$ contains only poles and no zeros[1]—and in *autoregressive moving average filters*—or for short *ARMA-type filters*—the transfer function of which contains both poles and zero values (see Eq. 9.10). Of the two above categories of filters, the autoregressive filters and the ARMA-type filters are always recursive in nature, in contrast to the moving average filters, which are nonrecursive and always stable, since their poles, lying on the origin $z = 0$ of the complex plane, have a modulus less than one. The basic characteristic of moving average filters is that they can be designed in such a way as to cause *the minimum possible distortion on the input signal*. This distortion, as already mentioned, mainly concerns the amplitude characteristic, as the frequency response function cannot be the same for all frequency values; however, its existence does not create a problem, since the magnitude of the filter's frequency response that causes it can be controlled during the design phase. On the other hand, the distortion caused by the IIR filters appears in the form of a *time delay* that cannot be controlled during the design stage of the filter.

Taking into account that both FIR-type filters and IIR-type filters are *linear and time-invariant discrete-time systems*, it is obvious that their response to the input signal $x[n]$ will be given by the *convolution equation*, which for FIR-type filters will take the form

$$y[n] = \sum_{k=0}^{N-1} h[k]x[n-k], \tag{12.3}$$

where N is the number of samples of the FIR of the filter $h[n]$, whereas for IIR-type filters it will be given by the equation

$$y[n] = \sum_{k=0}^{\infty} h[k]x[n-k], \tag{12.4}$$

since in this case the impulse response has *infinite length*. In practice, of course, the output of the filter is not calculated from the above sum of infinite terms, but instead from the *difference equation* of the system (Eq. 9.5) or alternatively through the use of the *transfer function* $\mathcal{H}(z)$ (Eq. 9.10). As for the choice of the type of filter that we will use in each case, this depends on the relative advantages and disadvantages of one type of filter over the other. Speaking generally, the comparison of these two types of filters leads to the following conclusions:

- FIR-type filters are able to offer a *completely linear phase response* (this type of response characterizes ideal filters and simply introduces a *time delay* into the input signal without distorting its phase characteristics). On the contrary, IIR-type filters are characterized by *nonlinear phase responses*, specifically at the end of frequency bands, resulting in them being unsuitable for applications such as data transmission, biomedical signal analysis, and digital audio and image processing.
- FIR-type filters are implemented directly through the use of *nonrecursive structures* and, as we mentioned in the previous section, they are always *stable*. On the other hand, no one can guarantee in advance the stability of an IIR filter, which, like any system with *feedback*, is sensitive to disturbances of its input, which are likely to lead to instability and oscillations in its output. An IIR filter is causal and stable, *only when the poles of its transfer function are located in the interior of the unit circle*.
- In cases where we want the transition zones of the frequency response function to have *a small range*, that is, to be characterized by *sharp transitions* from the pass areas to the cutoff areas and vice versa, FIR-type filters require *more filter coefficients* than IIR filters. This in other words means that for a given filter specification regarding its amplitude

[1] The existence of the pole of order M in Eq. (9.11) and the zero value of order N in Eq. (9.12) should not confuse the reader, since their value is equal to zero and in practice we never refer to them.

response, the use of an FIR filter requires *more processing time* and *more storage space* than an IIR-type filter. However, the overall performance of a FIR filter can be significantly improved, if we take advantage of the computational speed *of the fast Fourier transform* and all related techniques and algorithms.

- In contrast to digital filters IIR for which the corresponding analog structures are defined and it is possible to convert an analog filter into a digital IIR filter, the FIR-type filters *do not have the corresponding analog equivalents*; however, they offer a greater degree of ease in terms of the synthesis of a filter characterized by any arbitrary frequency response.
- In cases where we use a small number of bits for the implementation of digital filters, the appearance of unwanted errors occurs, such as *errors due to the rounding of values*, as well as *quantization errors of the filter coefficients*. It turns out that these errors are much more intense in IIR filters than in FIR filters.

Based on all the above, we can formulate the following practical rules regarding the type of filter we will use in each case:

- If we want *abrupt transitions* from the pass bands to the cutoff bands and vice versa, as well as operation under a *high-performance regime*, we should resort to the use of IIR filters, since these use fewer coefficients in relation to the FIR filters. The IIR filters are very efficient and require many fewer calculations per output sample to compute their response and for this reason their implementation in hardware is used in *real-time systems*, which operate at much higher speeds than FIR filters.
- If the number of filter coefficients is not too large and we desire *zero or as little as possible distortion of the phase characteristics of the input signal*, we should resort to the use of FIR filters.

12.3.2 Ideal Digital Filters

An interesting category of digital filters is the *ideal digital filters*, which are of theoretical interest only, because they cannot exist in practice. These filters, in full analogy with their analog equivalents (see Sects. 3.5.2 (see page 278) and 3.6.1 (see page 283) have two significant properties: (a) their amplitude response leads to *the abrupt nullification of the value of the amplitude of the input signal at the cutoff frequencies* whereas it has a constant value in the passband and (b) *their phase response is a linear function of the frequency ω of the input signal*, since if we impose the requirement for the filter to be causal, this cannot be equal to zero. The amplitude response of the ideal digital filter for each of the four basic filter categories is presented in Fig. 12.2. We observe that this figure is the same as Fig. 3.16 (see page 285), which characterizes the ideal analog filters, with the difference that in this case the frequency takes values in the interval $-\pi < \omega < \pi$, a property that characterizes the discrete-time signals and was presented in detail in the homonymous chapter (see Sect. 6.7.1 on page 520).

In order to understand the concept of *linear phase response*, let us resort again to the argumentation of Sect. 3.5.2 (see page 278) concerning continuous linear time invariant (LTI) systems of linear phase and consider an input signal $x[n]$ with frequency in the interval $-\pi < \omega < \pi$ passing through a digital filter with frequency response

$$\mathcal{H}(e^{j\omega}) = \begin{cases} Ae^{-j\omega\alpha} & -\pi < \omega < \pi \\ 0 & \text{anywhere else,} \end{cases} \tag{12.5}$$

where $A = |\mathcal{H}(e^{j\omega})|$ and α are constants. In this case, the spectrum of the output signal will have the form

$$\mathcal{Y}(e^{j\omega}) = \mathcal{X}(e^{j\omega})\mathcal{H}(e^{j\omega}) = A\mathcal{X}(e^{j\omega})\, e^{-j\omega\alpha} \qquad (-\pi < \omega < \pi). \tag{12.6}$$

Applying the properties of *scaling* and *time shifting* that characterize the discrete-time Fourier transform, which are expressed as

$$\alpha x[n] \rightleftarrows \alpha\mathcal{X}(e^{j\omega}) \qquad \text{and} \qquad x[n-\alpha] \rightleftarrows \mathcal{X}(e^{j\omega})\, e^{-i\omega\alpha},$$

we can formulate the system output in the time domain as $y[n] = Ax[n-\alpha]$. We observe, therefore, that the system output is a *modified version of the input signal* characterized by *time delay* and *amplification or attenuation of its amplitude* (depending

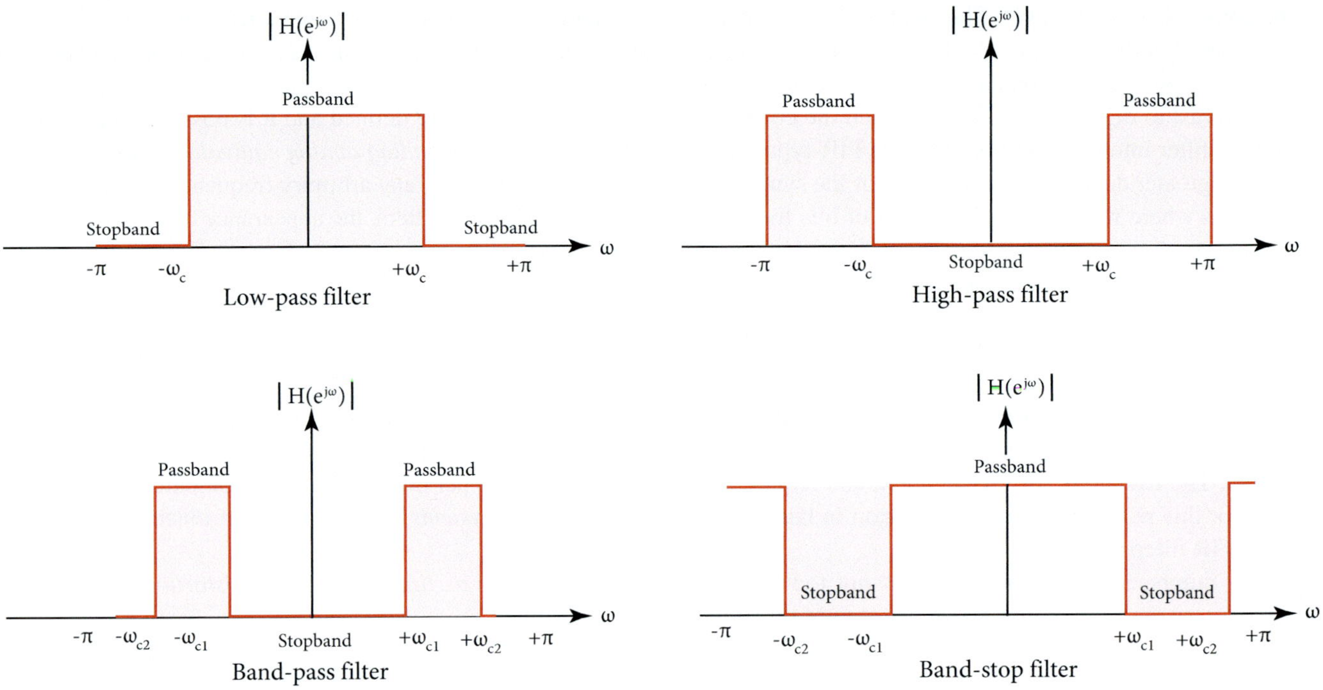

Fig. 12.2 Amplitude response of ideal low-pass, high-pass, band-pass, and band-stop digital filter

on the value of the constant A) compared with the original signal. However, it is not difficult to realize that this time delay that appeared as a result of the signal passing through the filter does not distort the signal, since it can be removed very easily, by transmitting the output of the filter to a simple system that applies the corresponding time advance to this signal. We are led, therefore, to the conclusion that the ideal filters are characterized by a *linear phase response curve* that will be described by an equation of the form $\measuredangle \mathcal{H}(e^{j\omega}) = -\omega\alpha$. Indeed, in this case, the filter output will be

$$y[n] = Ax[n-\alpha] = Ae^{j\,[\omega n + \measuredangle \mathcal{H}(e^{j\omega})]} = Ae^{j\omega(n-\alpha)}, \tag{12.7}$$

from which it appears that the frequency of the input signal *has not been modified* during its passage through the digital filter. The above results apply to the simple discrete exponential input signal $x[n] = e^{j\omega n}$, whereas in the general case of a signal $x[n]$, the application of the convolution theorem will give us

$$y[n] = Ax[n] * \frac{\sin \pi(n-\alpha)}{\pi(n-\alpha)} = A \sum_{k=-\infty}^{\infty} x[k] \frac{\sin \pi(n-k-\alpha)}{\pi(n-k-\alpha)},$$

where

$$h_{id}[n] = A\frac{\sin \pi(n-\alpha)}{\pi(n-\alpha)}, \qquad -\infty < n < \infty$$

is the impulse response corresponding to the frequency response of the ideal filter. Let it be noted that in order for all of the above to apply, that is, to get an output characterized by a time delay of α time units in relation to the input, the value of α must be an integer. Conversely, if this condition does not apply, this conclusion cannot be formulated, because it is not possible to shift a discrete-time signal by a number of samples that is not an integer. In this case we can use a continuous-time system with frequency response $\mathcal{H}(j\Omega) = e^{-j\Omega\alpha T_s}$, where T_s is the sampling period, so according to the property of time shift of the continuous-time Fourier transform, the output of this system will have the form $y_c(t) = x_c(t - \alpha T)$, where $x_c(t)$ is the continuous-time signal that has resulted from the discrete-time signal $x[n]$ in the way we have described in the relevant section and using the interpolation method. In this case, we can construct the discrete output signal $y[n]$, proceeding to sampling of the continuous-time signal $y_c(t)$, which is the output of the above continuous-time system. This situation is an example of processing a discrete-time signal with the help of a continuous-time system. The sampling period T_s in reality has no significance and it is common practice for reasons of simplicity to be considered equal to the unit, so as to simplify the analysis.

The derivative of the phase $\measuredangle\mathcal{H}(e^{j\omega})$ with respect to the frequency ω is expressed in *units of time delay*. Therefore, we can define the *delay* of the signal as a function of the frequency through a relationship of the form

$$\tau(\omega) = -\frac{d}{d\omega}\left\{\measuredangle\mathcal{H}(e^{j\omega})\right\}, \tag{12.8}$$

which is none other than Eq. (3.56), which applies to continuous-time filters. Following the analysis we presented then, we name the magnitude $\tau(\omega)$ appearing in the above relationship the *group delay* of the filter and for the case of the frequency response of the ideal filter we used previously, it follows that $\tau(\omega) = \alpha$. We are therefore led to the conclusion that in cases where the phase response $\measuredangle\mathcal{H}(e^{j\omega})$ of the filter is a *linear function* of the frequency ω, the group delay $\tau(\omega)$ is *independent* of the frequency ω, which means that all signals undergo the same delay from an ideal digital filter, regardless of their frequency.

The main disadvantage of ideal filters is that they *are not causal systems* and therefore *cannot be implemented in practice*, so as to be used in real applications. In order to confirm the truth of this claim, we must calculate their impulse response, a process that will lead us to the conclusion that this constitutes a *noncausal* signal (for the calculation of this magnitude for the four types of ideal digital filters, see Sect. 10.4.1 on page 836 and Exercise 10.4.7 on page 837). Therefore, ideal filters are merely *theoretical models*, the behavior of which is approximated to the best possible degree by real filters that must necessarily be *causal*. The basic techniques for designing real filters will be studied in a subsequent section. However, let us mention at this point that a special role in the construction of a digital filter is played by *the positions of the poles and the zero values of the filter on the complex plane*. As one can infer from the discussion of Sect. 10.4.3 (see page 842), the poles of the system should be placed *near the points of the unit circle corresponding to the desired frequency values in each case*, whereas on the contrary, the zero values should be placed *near the less significant frequencies*. In addition to the above rules, the following two restrictions must be imposed:

- All poles of the filter must be placed *inside the unit circle*, in order for the filter to be *causal* and *stable*. However, zero values can be placed *anywhere on the complex plane*.
- If the system is characterized by the existence of *complex poles and zero values*, these should appear in pairs of *conjugate quantities*, so that the coefficients of the filter are *real numbers*.

In the case of low-pass digital filters, the poles should be placed near the unit circle and at points corresponding to *low frequencies* (that is, near the frequency $\omega = 0$). On the other hand, zero values should be placed near or on the unit circle and at points corresponding to *high frequencies* (more specifically near the frequency $\omega = \pi$). The opposite applies to the case of high-pass filters. Finally, in the case of band-pass filters, there should be *one or more pairs of conjugate complex poles*, which are located near the unit circle and in the vicinity of the two frequencies that define the frequency band that is allowed to pass through the filter. Examples of pole-zero diagrams for typical cases of low-pass, high-pass and band-pass filters are shown in Fig. 12.3.

12.3.3 Generalized Linear Phase Systems

A generalization of linear phase systems with frequency response $\mathcal{H}(e^{j\omega}) = |\mathcal{H}(e^{j\omega})|e^{-j\omega\alpha}$ are the generalized linear phase systems whose frequency response has the form

$$\mathcal{H}(e^{j\omega}) = \mathcal{A}(e^{j\omega})e^{-j\alpha\omega+j\beta},$$

where α and β are both real constants, whereas $\mathcal{A}(e^{j\omega})$ is a real function of ω. These systems are characterized in this way because the phase of such a system consists of a constant term β, which is added to the linear function $-\omega\alpha$ with the quantity $-\omega\alpha + \beta$ being the equation of a straight line with slope α. If we ignore any discontinuity that arises from this addition of

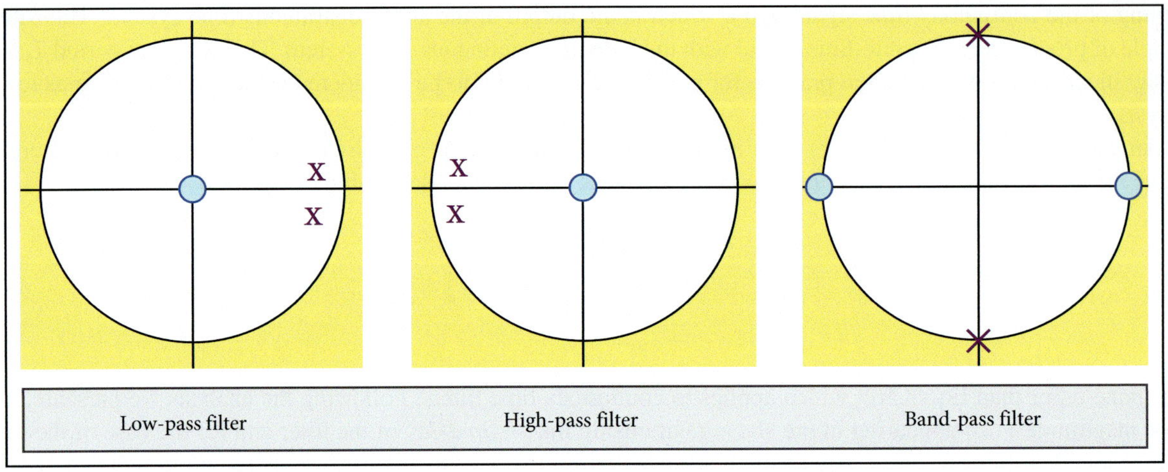

Fig. 12.3 Typical positions of poles and zero values on the complex plane for the cases of low-pass, high-pass, and band-pass filter

a constant phase value for all frequencies ω that fall within the region $|\omega| < \pi$, then this system is characterized by constant group delay. In other words, the class of systems for which it holds that $\tau(\omega) = \alpha$ are characterized by linear phase of the most general form

$$arg[\mathcal{H}(e^{j\omega})] = \beta - \omega\alpha, \qquad 0 < \omega < \pi, \quad \alpha, \beta \in \mathbb{R}.$$

Rewriting the above relation as

$$\mathcal{H}(e^{j\omega}) = A(e^{j\omega})e^{j(\beta-\alpha\omega)} = A(e^{j\omega})\cos(\beta - \omega\alpha) + jA(e^{j\omega})\sin(\beta - \omega\alpha)$$

and comparing it with the defining equation of the frequency response

$$\mathcal{H}(e^{j\omega}) = \sum_{n=-\infty}^{\infty} h[n]e^{-j\omega n} = \sum_{n=-\infty}^{\infty} h[n]\cos\omega n - j\sum_{n=-\infty}^{\infty} h[n]\sin\omega n$$

(which holds under the condition that the impulse response $h[n]$ is a real signal), we get

$$\cos(\beta - \omega\alpha) = \frac{1}{A(e^{j\omega})}\sum_{n=-\infty}^{\infty} h[n]\cos\omega n \quad \text{and} \quad \sin(\beta - \omega\alpha) = -\frac{1}{A(e^{j\omega})}\sum_{n=-\infty}^{\infty} h[n]\sin\omega n.$$

Consequently, the tangent of the phase angle of the function $\mathcal{H}(e^{j\omega})$ will be given by the expression

$$\tan(\beta - \omega\alpha) = \frac{\sin(\beta - \omega\alpha)}{\cos(\beta - \omega\alpha)} = \frac{-\sum_{n=-\infty}^{\infty} h[n]\sin\omega n}{\sum_{n=-\infty}^{\infty} h[n]\cos\omega n}.$$

Multiplying the fractions diagonally in the second and third terms and transferring all terms to the same side of the equation, it easily follows that

$$\sum_{n=-\infty}^{\infty} h[n]\cos\omega n\sin(\beta-\omega\alpha) + \sum_{n=-\infty}^{\infty} h[n]\sin\omega n\cos(\beta-\omega\alpha) = \sum_{n=-\infty}^{\infty} h[n]\big(\cos\omega n\sin(\beta-\omega\alpha) + \sin\omega n\cos(\beta-\omega\alpha)\big) = 0.$$

Resorting to the known trigonometric identity $\sin\vartheta\cos\varphi + \cos\vartheta\sin\varphi = \sin(\vartheta + \varphi)$ for the angles $\varphi = \beta - \omega\alpha$ and $\vartheta = \omega n$ it follows that $\vartheta + \varphi = \beta - \omega\alpha + \omega n = \beta + \omega(n - \alpha)$ and

$$\cos\omega n \sin(\beta - \omega\alpha) + \sin\omega n \cos(\beta - \omega\alpha) = \sin[\beta + \omega(n - \alpha)]$$

and the above relation becomes

$$\sum_{n=-\infty}^{\infty} h[n]\sin[\beta + \omega(n - \alpha)] = 0$$

for all frequencies ω. This relation constitutes the necessary (not sufficient though) condition that must be satisfied by the impulse response $h[n]$ and the real constants α and β so that the system is characterized by a constant group delay. Special cases are the systems with values $\beta = 0$ or $\beta = \pi$, $2\alpha = M =$ integer and $h[2\alpha - n] = h[n]$ for which the above equation becomes

$$\sum_{n=-\infty}^{\infty} h[n]\sin[\omega(n - \alpha)] = 0,$$

as well as the systems with values $\beta = \pi/2$ or $\beta = 3\pi/2$, $2\alpha = M =$ integer and $h[2\alpha - n] = -h[n]$ for which the above equation becomes

$$\sum_{n=-\infty}^{\infty} h[n]\cos[\omega(n - \alpha)] = 0.$$

Examples of causal systems with generalized linear phase are the FIR filters of types I, II, III, and IV, which are studied in a subsequent section of this chapter.

What is the correlation between generalized linear phase FIR filters and minimum phase systems? According to the relative analysis, when the impulse response of these filters is a discrete-time signal consisting of real values, the corresponding transfer function exhibits zeros that lie either on the unit circle, or in positions that are complex conjugates or reciprocals of each other. This property allows us to formulate the transfer function of a FIR filter, as the product of a minimum phase transfer function $\mathcal{H}_{\text{min}}(z)$ whose M_i zeros are all located inside the unit circle, a maximum phase transfer function $\mathcal{H}_{\text{max}}(z)$ whose M_i zeros are all located outside the unit circle, and a transfer function that contains M_o zero values, which are located on the circumference of the unit circle. It will therefore be

$$\mathcal{H}(z) = \mathcal{H}_{\text{min}}(z)\mathcal{H}_{\text{uc}}(z)\mathcal{H}_{\text{max}}(z) \quad \text{where} \quad \mathcal{H}_{\text{max}}(z) = \mathcal{H}_{\text{min}}(z^{-1})z^{-M_i}.$$

Consequently, the order of the transfer function $\mathcal{H}(z)$ is equal to $M = 2M_i + M_o$.

12.3.4 Real Digital Filters

The design processes of digital filters consist of *determining the desired characteristics that the frequency response function of the filter* $\mathcal{H}(e^{j\omega})$ should have, so that it behaves in the appropriate way in each case. Taking into account that the function $\mathcal{H}(e^{j\omega})$, being a complex function, is characterized by some *measure* $|\mathcal{H}(e^{j\omega})|$ and by some *phase* $\measuredangle\mathcal{H}(e^{j\omega})$, it is obvious that the correct design of the filter consists in the correct choice of the two above functions. This choice should, among other things, take into account that the filter being designed must be *causal*, so that it can be *practically implementable*. Therefore, this filter cannot be ideal, since, as we have already mentioned, filters of this category *are not causal*.

The basic characteristics of the frequency response function of digital filters are similar to those of analog filters, more specifically:

- The frequency response function of frequency selection filters can take zero value for certain frequency values, but not in every finite frequency band, as if this happened then the integral of the discrete version of the *Paley-Wiener theorem* would take infinite value (see Sect. 12.10 on page 1220).
- The *amplitude response* of the filter $|\mathcal{H}(e^{j\omega})|$ does not have a constant value along the *passband*, but presents *fluctuations*. If the width of these fluctuations is equal to δ_p, then the value of the amplitude of the function fluctuates between $1 \pm \delta_p$. In a completely analogous way, the value of the measure of the function in the *frequency cutoff region* is usually not

constant and equal to zero, but it also presents *fluctuations*, the width of which is equal to δ_s. Using the values of δ_p and δ_s we can, as we did in the case of analog filters, define the range of variation in the passband A_p and the attenuation in the cutoff zone A_s (expressed in units dB) from the relations

$$A_p = -20\log_{10}\frac{1-\delta_p}{1+\delta_p} \quad \text{and} \quad A_s = -20\log_{10}\frac{\delta_s}{1+\delta_p}.$$

- The transition from the passband to the cutoff zone is not abrupt and therefore no *discontinuity* is observed at the cutoff frequency, as happens in ideal filters. On the contrary, the fall of the measure $|\mathcal{H}(e^{j\omega})|$ is quite smooth and the actual zeroing of its value occurs at another frequency that also dictates the range of the *transition zone*. As for the range of the *passband*, this is known as the *bandwidth* of the filter. Therefore, the amplitude response of a real filter consists of the *passband*, the *transition zone*, and the *cutoff zone* defined by the frequency areas $[0, \omega_p]$, $[\omega_p, \omega_s]$ and $[\omega_s, \pi]$ respectively.
- The real and imaginary parts of the frequency response function of a digital causal filter are related through the *discrete Hilbert transform* (see Sect. 12.10). Bear in mind that the magnitude and phase of the frequency response function depend on the real and imaginary parts of the aforementioned function through the known relationships

$$|\mathcal{H}(e^{j\omega})| = \sqrt{\mathcal{H}_R^2(\omega) + \mathcal{H}_I^2(\omega)} \quad \text{and} \quad \measuredangle\{\mathcal{H}(e^{j\omega})\} = \tan^{-1}\left[\frac{\mathcal{H}_I(\omega)}{\mathcal{H}_R(\omega)}\right].$$

It is obvious that these two functions *cannot be chosen arbitrarily and independently of each other*. What usually happens is that the magnitude of the function $\mathcal{H}(e^{j\omega})$ is designed so that the filter behaves in the appropriate way in each case and then the appropriate *phase response function* $\measuredangle\{\mathcal{H}(e^{j\omega})\}$ is chosen.

The basic characteristics of the function $|\mathcal{H}(e^{j\omega})|$ for the case of real low-pass, high-pass, band-pass, and band-stop filters, with central frequency ω_c, are presented in the images (a), (b), (c), and (d) of Fig. 12.4. In this figure the quantities δ_p and δ_s refer to the width of the deviation of the function's value in the *pass* and *stop* regions, whereas the range of the transition, pass, and stop bands is determined by the quantities $\Delta\omega_\Gamma$, $\Delta\omega_p$ and $\Delta\omega_s$. The ratio $\delta_p/2\delta_s$ is characterized by some authors as *residual deviation ratio* and is usually defined in a different way, as we will see in a subsequent section.

In full analogy with their analog equivalents, digital filters can also be described by the four basic approaches we presented in the previous sections, which are the approaches Butterworth, Chebyshev I, Chebyshev II and Cauer (or elliptical approach). The way of constructing the digital versions of these four approaches from their analog versions is presented in subsequent sections. Figures 12.5 and 12.6 show the variation of the group delay for the filters Butterworth and Chebyshev I (Fig. 12.5), as well as for the Chebyshev II and Cauer approaches (Fig. 12.6) for various operating specifications. It is obvious that in the case of digital filters, the size of the group delay is expressed in the number of samples and not in seconds, as is the case with analog filters.

12.4 Design of Digital Filters

The goal of the process of designing a digital filter, consists *in determining the characteristics of the filter*, so that the specifications related to its frequency response function are approached *to the maximum possible extent*. This process can be described as a *sequence of five steps*, which, depending on the circumstances, may or may not be *independent of each other*, whereas the order in which they are carried out may or may not be the one listed below. In several cases, it is possible *to merge two or more steps into a composite and more complex step*, whereas in some cases it is possible *to return to a step more than once*, in order to make revisions to the design decisions, if such a thing is deemed necessary. Generally speaking, the five stages of the process of designing and constructing a digital filter are as follows:

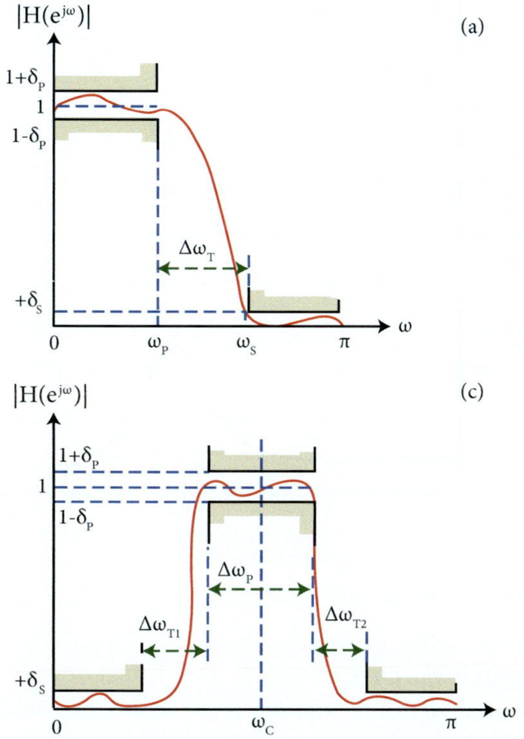

Fig. 12.4 Variation of the width of the function $|\mathcal{H}(e^{j\omega})|$ of real digital filters

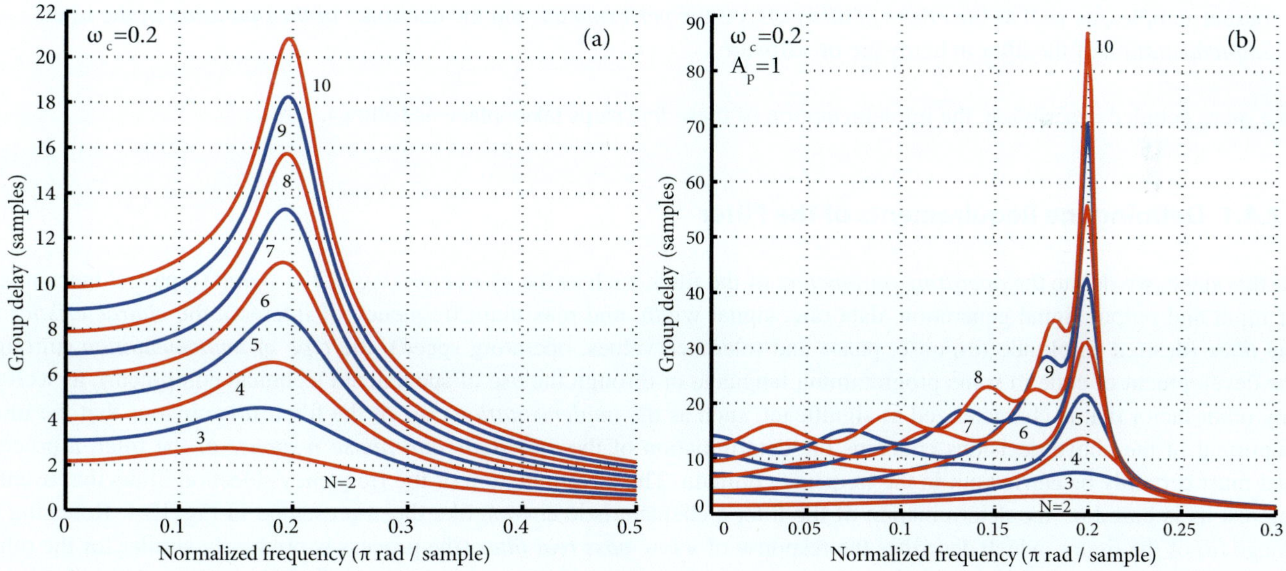

Fig. 12.5 The variation of the group delay for the digital filters Butterworth and Chebyshev I and for specific operating specifications

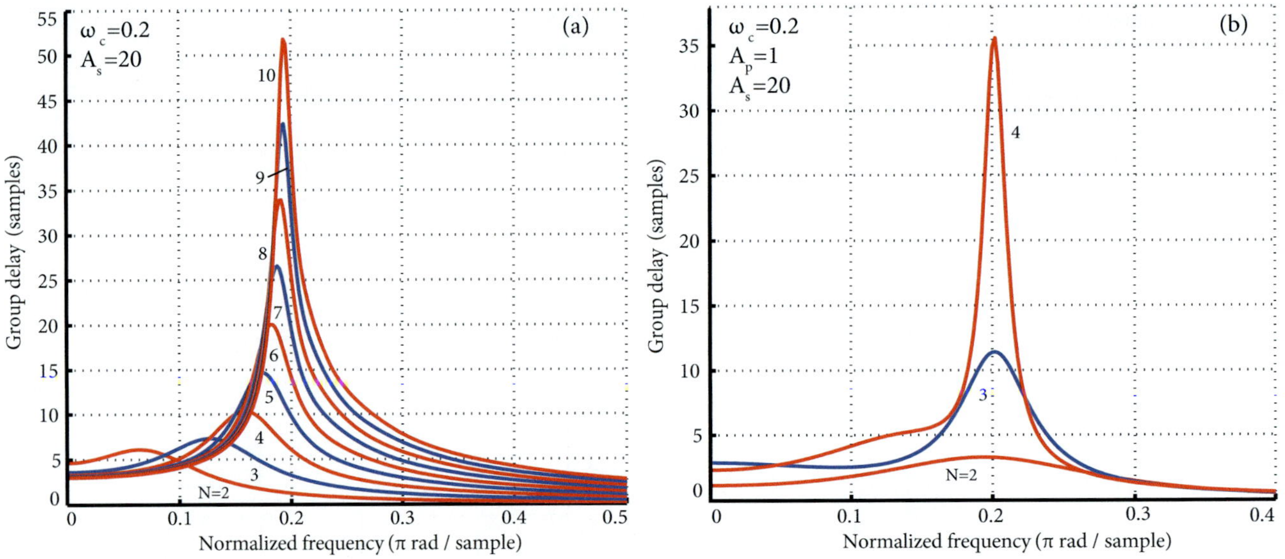

Fig. 12.6 The group delay variation for the digital filters Chebyshev II and Cauer and for specific operating specifications

- Determining the *specifications* and the *operational requirements of the filter*.
- Calculation of the appropriate *coefficients of the filter* in each case, so that the transfer function that they describe satisfies to the greatest possible extent the specifications that have been set regarding the amplitude and phase response, as well as the characteristics of the filter in the time domain.
- Representation of the filter by a structure of *schematic diagrams* or *signal flow graphs*.
- Estimation of the effects of the *finite length of the registers* used for storing the coefficients of the filter and the input and output samples, as well as the *errors of all kinds*, on the *performance* and the *accuracy of the operation* of the filter.
- *Implementation* of the filter in hardware or software.

In a more detailed description, the implementation of these five steps takes place as follows.

12.4.1 Defining the Requirements of the Filter

At this stage, we define the *operating parameters* of the filter, such as the *characteristics of the signal* (type and mechanism of input and output signal generation, data rate, signal width, and maximum frequency of interest), the *characteristics of the filter* (desired amplitude response, phase and tolerance values, operating speed), the *type of implementation* (through the development of code in some programming language or through the use of specialized technical equipment), as well as any other factor that is characterized as significant, such as the *implementation cost* of the filter. It is obvious that the most important of the above decisions concerns the determination of the *amplitude and phase response* of the filter, a process that must certainly be carried out in the frequency domain. This determination of the frequency selection filters that usually interest us is based on the determination of the filter's characteristic curves, like those presented in Fig. 12.4. Referring to image (a) of the figure, which describes the response of a *low-pass real filter* (the process is obviously similar for the other categories of real filters), the operating parameters that we need to define concern the desired width of the deviation δ_p and δ_s for the pass and cutoff zones, as well as the width of the transition zone, which is described by the frequencies ω_p and ω_s. The values of the parameters δ_p and δ_s can either be given as simple numbers or expressed in decibels, whereas the frequencies $f_p = \omega_p/2\pi$ and $f_s = \omega_s/2\pi$ are defined either in Hertz or as normalized (and therefore dimensionless) frequencies of the form f/F_s, where F_s is the *sampling frequency*.

12.4.2 Calculation of the Coefficients and the Transfer Function of the Filter

After determining the desired operating characteristics and more generally, the specifications of the filter, we must then determine the *transfer function* that it describes, so that, immediately after, we can proceed to its representation with the appropriate schematic diagram structure and its implementation. Speaking generally, a method of calculating the transfer function is characterized as *direct*, when the problem is solved directly in the z domain and as *indirect*, when we initially construct a *continuous-time* transfer function, which is then *transformed into its discrete equivalent*. If the problem is solved by applying only a few design steps and using a set of closed-form equations,[2] then the method we use is also characterized as a *closed-form method*, whereas alternatively we can resort to the use of *recursive methods*. The main feature of these methods is the solution of the problem starting from an initial solution, which, through the application of appropriate optimization methods, eventually leads to a sequence of progressively improved solutions, until the set criteria are finally met. In practice, nonrecursive-type filters are always designed with *direct methods*, whereas, recursive filters can be designed using *both direct and indirect methods*.

As we have mentioned in the chapter on the $\mathcal{Z}$ transform (see Sect. 9.1 on page 693), the transfer function of a discrete LTI system can be calculated both from the *impulse response* of the system (see Exercise 9.1.2 on page 695) and from the *coefficients* α_k and β_k of the *difference equation* that describes it (see Sect. 9.2 on page 699, as well as Exercises 9.6.1 on page 719 and 9.6.2 on page 722). For the case of FIR filters, the transfer function $\mathcal{H}(z)$ is determined by calculating the samples $h[n]$ of their impulse response, since it has a *finite length*. It is obvious that this process cannot be applied to IIR filters, whose impulse response has an *infinite length*; for this reason, for this category of filters, the determination of the transfer function consists in the calculation of the coefficients α_k and β_k of the difference equation that describes their operation.

Starting with FIR filters, the calculation of the values of the samples $h[n]$ can be carried out in many different ways, the most important of which are the *use of a window function*, the *sampling in the frequency domain*, and the *Parks-McClellan algorithm*. The window functions that were studied in detail in the previous chapter (see Sect. 10.5.6 on page 886) *allow the control of the length of the input signal while simultaneously changing its spectral characteristics* in the way we described in the corresponding section; however, *they do not offer sufficient control over the values of the filter parameters*. Nevertheless, their use is considered a very simple and flexible way of calculating the samples of the system's impulse response. On the other hand, the *frequency sampling method*, allows the construction of a computationally efficient recursive model of an FIR filter, without, however, offering great flexibility regarding the determination—or even the control—of the filter parameters. Finally, the Parks-McClellan method (which is also known as the optimal method), is the one that is mostly used in practice, as it is supported by many software applications that have been implemented for the construction of digital FIR filters.

On the other hand, the calculation of the coefficients α_k and β_k of an IIR filter, is mainly based *on the transformation of characteristics of analog filters to their digital equivalents*, with the methods usually used in practice being *the method of the invariant impulse response* and the *bilinear transformation*. The application of the first method consists in the *digitization* of an analog filter carried out in such a way that *the impulse response of the filter remains unchanged*, something, however, that does not happen with the characteristics of the filter in the frequency domain, which *are modified*. The main disadvantage of this method is that it is greatly affected by the undesirable phenomenon of *frequency folding*, a fact that makes it *unsuitable* for the creation of high-pass and band-stop IIR filters. On the other hand, the method of *bilinear transformation* is considered an effective and flexible method for calculating the coefficients of frequency selection filters of the IIR type and allows the design of digital filters, with known classical characteristics (such as those associated with the Butterworth and Chebyshev filters, as well as elliptical filters), which have the same amplitude response as their analog equivalents, but this does not apply to the characteristics associated with their representation in the time domain.

As a final method of calculating the coefficients of an IIR filter we mention the *method of placing the poles and zeros in the complex plane*. However, this method is not recommended for the design of high-specification filters, in terms of amplitude response, since it is based on the repetitive alternation of the positions of the poles and the zero values on the complex plane and the use of the trial and error method.

In practice, the method we will ultimately choose to use for the calculation of the filter's transfer function depends on the type of filter as well as the requirements it is intended to meet.

[2] By the term closed form of an equation that contains a sum of finite or infinite terms we mean an alternative formulation of this equation with the help of appropriate transformations, from which *the symbol of the sum is absent*.

12.4.3 Representation of the Filter with Schematic Diagrams

After determining the samples of the impulse response for the FIR-type filters or the coefficients α_k and β_k of the difference equation for the IIR filters, it is possible to calculate the transfer function of the filter, in the way we explained in the chapter on the $\mathcal{Z}$ transform, which can then be represented diagrammatically by constructing the corresponding *schematic diagrams* (see Sect. 7.10 on page 608) or *signal flow graphs* (see Sect. 12.14 on page 1350). In full analogy with the approximation methods we presented previously, the methods of representing the transfer function are characterized as *direct*, when they use directly the transfer function of the digital filter, or as *indirect*, when the structure of the filter is obtained through the use of an *equivalent prototype analog filter*. It is not difficult to understand that the choice of the type of representation we use will be different for the FIR- and IIR-type filters.

More specifically, for IIR-type filters, the *normal* or *non-normal* form is used, which offers a direct and straightforward representation of the filter's transfer function in its entirety. Alternatively, we can separate the transfer function into simpler functions and proceed to the interconnection of the subsystems that arise in this way, either in series or parallel, in the first case, the *factorization of the transfer function into a product of simpler functions* that describe *second-order* structures, which are connected in series. In the second case, the transfer function is developed into an *sum of partial fractions* representing transfer functions of second-order subsystems, which are connected in parallel (for the description of the way of interconnecting discrete-time systems LTI, see Sect. 7.7 on page 605 and Exercise 9.5.1 on page 716). Although all these types of diagrammatic representation, allow the description of the same IIR filter, in practice *the series and the parallel connection* are mainly used, since it is proven that they lead to simpler filtering algorithms, and they are more resistant to errors associated with the use of *finite length registers*, compared with the normal forms of the filter, a claim that is particularly true for filters with a high-order value.

On the other hand, the representation of FIR-type filters is mainly based on the *normal* and *non-normal* forms, as these are proven to be characterized by greater ease of implementation. Other structures used for the representation of these filters are based on *frequency sampling* and the use of *convolution techniques*. Of these two structures, the first is computationally more efficient, as it leads to fewer filter coefficients, although it is not so easy to implement, as it is characterized by high computational requirements. On the other hand, the technique of fast convolution, as it is characteristically called, uses the computational advantage of the *fast Fourier transform (FFT)* (see Sect. 10.5.5 on page 872) and is particularly attractive when, among other things, we want to calculate the power spectrum of the signal.

12.4.4 Analysis of the Effects of the Limited Length of the Registers

Although during the design and description phase of the filter, we assume that the accuracy of the calculations is *infinite* or at least *very large*, in practice, the digital material used—and by extension the process of calculation and use of the filter coefficients—is characterized by *finite accuracy*, which depends on the *length of the registers* used for storing the numbers, the *type of the representation* that has been adopted, and the type of arithmetic used (which can be fixed or floating point—see the following sections). The same limitations naturally characterize digital filter implementations that are based on software, as this will be executed on a computer that will perform the above calculations. In other words, this means that in reality, the general recursive difference equation

$$y[n] = -\sum_{k=1}^{N} \alpha_k y[n-k] + \sum_{k=0}^{M} \beta_k x[n-k] \tag{12.9}$$

that describes the operation of the filter *does not* describe the actual digital filter that will be used in practice, but only a discrete-time system (sampled-data filter) in which time is *discrete* but all other quantities are *continuous* (remember, the difference between a discrete-time signal that has arisen through sampling and a digital signal is that *in the digital signal, both time and the values of the signal samples are characterized by finite accuracy, whereas in the discrete-time signal, this property characterizes only the time variable, and the signal samples are assumed to be known with infinite accuracy*). This requirement for finite accuracy representation has *dramatic* consequences in the overall process of designing digital filters and in reality, constitutes *the main reason for existence* of all alternative but equivalent forms of implementing the transfer function of the underlying digital system. Indeed, if this requirement did not exist, the order of performing the additions and multiplications required to implement the above difference equation would not matter and we could simply implement the filter using the direct form I that arises directly from the system's difference equation. However, in practice, things are not like that, since it turns out that *the length of the registers used for storing the values of the involved quantities has a strong impact on the amount of material required for the construction of the filter* and is *different* for the various forms of implementation. For example, practice has shown that the implementation of a filter from the serial and parallel connection of second-order stages is much more efficient than the direct form, a fact that makes the second-order stage a popular structure, which has been particularly studied, since among other things the calculation of the coefficients of the filter for this stage is easily and quickly done, knowing the positions of the poles and the zero values of the transfer function.

A second very important consequence of the use of *finite precision representation* for the numerical values involved in the operation of a digital filter involves the occurrence of *errors* and *numerical inaccuracies*, the impact of which on the overall performance and accuracy of the filter's operation must be evaluated. Generally speaking, these errors are due to the *quantization* of three different types of numerical data and more specifically, (a) to the *quantization of the filter's coefficients*, (b) to the *quantization of the input signal values*, and (c) to the *quantization of the values of the signals circulating within the filter*. In a more detailed description, the impact of the quantization of these three types of parameters on the performance and operation of the filter is described in the following.

Quantization of the Filter's Coefficients

The *quantization of the filter's coefficients* (see Sect. 12.9.1 on page 1162), that is, the coefficients α_k $(k = 1, 2, 3, \ldots, N)$ and β_k $(k = 1, 2, 3, \ldots, M)$ of the difference equation that describes its operation, introduces *errors* in the positions of the poles and the zeros of the transfer function. These errors, obviously, are transferred to the amplitude and phase responses, resulting in them being different from those that were determined during the specification stage of the filter (note, however, that the linearity of the system *is not affected*, in the sense that despite the change in the positions of the poles and the zeros, the modified quantized system that results, continues to be (a linear system). It is interesting to note that this quantization of the filter coefficients, along with the *rounding noise* we comment on below, affect in different ways the various equivalent implementations of the used transfer function and therefore, this equivalence, in reality, characterizes *only the theoretical structures of infinite accuracy and not the structures that operate under a regime of finite accuracy representation*, which are also the ones used in practice. In reality, if we assume that the quantization of the coefficients was done correctly, in the sense that it leads to acceptable behavior, the choice of the implementation that will ultimately be used, depends largely on *the level of rounding noise at the output of the filter*. The errors due to the quantization of the filter coefficients are *deterministic* in nature and all that is required, as we will see, is the recalculation of the filter parameters, in order to take into account the small changes in the values of the coefficients resulting from their quantization.

As we comment in detail in Sect. 12.9.1, for the construction of the actual quantized system, we initially calculate the filter coefficients under a regime of *infinite accuracy* and then, by applying some method, such as *the method of trial and error*, we progressively change the length of the registers, essentially the accuracy of the representation, until we finally end up with a *minimum register length*, for which the filter that has emerged continues to meet the defined operating specifications. The work of Jackson, who theoretically calculates a *optimal statistical register length*, on which a minimization process is then applied, is of particular interest. Speaking generally, it is proven that in practice *different alternative implementations are characterized by several variations regarding this feature*, since practice has shown that implementations related to serial and parallel connection of simple stages, *outperform* direct implementation, as they require the same number of adders, multipliers and time-delay elements as it, but generally lead to filters, which are characterized by *smaller* register lengths for the same tolerance value, regarding the approximation of the ideal transfer function.

It is proven that the effects of the quantization of the filter coefficients, are *particularly severe* in those filters in which the poles of the transfer function are *close to the periphery of the unit circle* and therefore, the sensitivity of these filters to changes in the values of the coefficients, due to quantization, are particularly high. This means that *only small changes in the values of the coefficients can lead to large changes in the frequency response of the filter*, which, as we commented, may affect their stability. For this reason, second-order *low-sensitivity* stages have been invented, which can be combined both in series and in parallel, leading to structures with much lower sensitivity, compared to the usual direct implementations. Also, in many cases, the reduction of the length of the registers affects the minimum order of the filter that leads to a transfer function that satisfies the defined operating specifications and for this reason, we may need to resort to a kind of compromise between the order of the filter and the degree of accuracy of the representation.

Quantization of Input Signal Values

This quantization process is applied *to the samples of the input signal* $x[n]$, which appears in the system's difference equation. In the special case where the input signal results from the sampling of an analog signal, as often happens in practice, this process leads to the creation of a *noise signal* at the output of the filter (see Sect. 9.8.4 on page 749), with a mean square value equal to $\Delta^2/12$, where Δ is the width that corresponds to the *least significant bit of a data word*, which decreases by 6 dB with the increase in the length of the registers by one bit (see the relevant section). Let it be noted, by the way, that because a filter can be used for a wide variety of applications and not just for frequency filtering, the determination of the length of the registers *does not always constitute part of the design of the digital filter*, and nor is the associated cost part of the cost of the filter. In engineering terminology, the maximum values that can be represented within the finite-precision framework of a digital system—resulting from quantization—define what is known as the system's *dynamic range*. The exact definition of this range varies depending on the specific representation format used to encode numbers in the binary system. Specifically, for the case of a floating-point value, this range is determined *by the size of the exponent* and therefore, for the same word length, *a floating-point processor is characterized by a larger dynamic range compared with a fixed-point processor*, at the expense, however, of higher cost and *greater power consumption*. On the other hand, for the *same word length*, a fixed-point processor is characterized *by greater accuracy* than a mobile subprocessor (for a detailed description of these two types of representation, see Sect. 12.5 on page 1135). The use of finite-precision values affects the operation of the filter owing to the introduction of a small or large degree of *nonlinearity* in its behavior, whereas the deviation from the ideal linear behavior can be limited and become arbitrarily small, using the appropriate length for the registers of the digital system.

Quantization of the Values of the Signals Circulating in the Interior of the Filter

The signals that circulate *inside the filter*, on the values of the samples of which the process of *quantization* is applied, are all kinds of products and sums calculated in the context of implementing the difference equation of the system, such as the products $\alpha_k y[n-k]$ $(k = 1, 2, \ldots, N)$ and $\beta_k x[n-k]$ $(k = 0, 1, 2, \ldots, M)$. The calculation of these products is associated with the appearance of *numerical rounding errors* in cases during which the numerical results that emerge require the use of more bits than those dictated by the length of the registers, something that results in their *rounding* (see also Sect. 12.9.4 on page 1179). Closely connected to the calculation of these products is also the appearance of *phenomena of overflow*(see Sect. 12.7 on page 1154), which occur when the value of a number that results from an operation*is greater than the maximum value that can be encoded* using the word length of the digital system, resulting in the production of erroneous results at the filter output. In reality, this very overflow is what leads to *quantization and rounding errors*, since, as is well known, in the opposite case, the operations of addition and multiplication do not create any problem.[3] Problems of this type characterize particularly the realization of *multiplications*, since the number of digits of the product of two numbers equals the *sum of the number* of digits of the two numbers and therefore, it is very likely that the length of the registers is not sufficient for storing the value of the product. For this reason, and as we comment in the relevant section, the values of the products pass through quantizers in order to undergo the appropriate processing, which, owing to their nonlinear nature, are replaced by sources of linear noise, the handling of which usually relies on the use of techniques and methods from the field of *statistical signal processing*. Note that regardless of the way in which quantization is performed, that is, regardless of whether this is done by truncation or rounding, the resulting noise of quantization is described by the term rounding noise and as we comment in the relevant section, this noise *can be modeled in the same way as quantization error is modeled*, that is, as an *uncorrelated additive noise* with a mean square value equal to $\Delta^2/12$, where the parameter Δ, as we mentioned earlier, is the width corresponding to the least significant bit of a data word. In order to address the large increase in the noise component due to this error, which can lead to noise values that are even larger than the encoding error itself, we can proceed to a *temporary increase in the length of the registers* during the computation phase, which, however, leads to an increase in the complexity of the underlying digital system, such as the increase in memory capacity requirements, as well as the increase in the number of adders and multipliers.

Rounding errors, unlike quantization errors, differ in that the data to be quantized *are already expressed in digital form*, whereas the quantization of values is performed at many points inside the filter and not only at its input. For this reason, the rounding error is generally *greater than the quantization error* and is one of the most significant factors affecting the complexity of the design of the digital filter, especially in the case of using specialized hardware.

It is interesting to note that the level and nature of the rounding noise that arises during the operation of a digital filter is dictated by three factors, specifically, (a) *the number of bits used to represent the data in the digital system*, (b) *the type of arithmetic used* (fixed or floating point—see the relevant section), and (c) *the structure of the underlying electronic circuit*. Of

[3] See for example, Wilkinson J.H., *Rounding Effects in Algebraic Processes, Englewood Cliffs, NJ, Prentice Hall, 1963.*

these three parameters, the number of bits affects either the value of the quantization step or the dynamic range of the signal and we usually choose the latter option if we want all filters to have the same quantization step. Therefore, in this approach, the number of bits is considered not to affect the level of rounding noise, but on the contrary, *the maximum permissible value of the signal* and consequently, *the value of the signal-to-noise ratio (SNR)*. Generally speaking, in every case, as commented elsewhere, the appropriate scaling process is applied to the input signal, as happens for example when we want to maintain the maximum value of the SNR ratio, without simultaneously violating the constraints related to the dynamic range.

Experience has shown that the rounding noise at the input of the filter, when using floating-point arithmetic, is usually but not always, *smaller* than that of a filter that uses fixed-point arithmetic, owing to the automatic scaling supported by the first type of arithmetic.[4] Note that in practice, as commented in a subsequent section, there is also another representation, the *block floating-point representation*. This representation, which was proposed by Oppenheim in 1969 and offers a simplified form of automatic scaling, is characterized by performance that is between the performances of the two previous representations.[5]

On the other hand, the implementation of the underlying digital system, that is, the structure of the electronic circuit that describes it, affects the *spectrum of the rounding noise* at the output of the filter and together with the type of arithmetic (fixed or floating point), dictates *the number of bits* required, in order for the specifications related to performance to be met. As is well known, there are many different and equivalent implementations of the same transfer function. However, these implementations are equivalent *only theoretically*, as in practice they are affected to varying degrees by the *quantization* of the filter coefficients and the *rounding error*. Taking into account that the effects of the quantization of the coefficients are *deterministic* in nature and can be incorporated into the filter design, the choice between these alternative implementations is based on the level and character of the rounding noise at the filter output and it is interesting how the differences that exist between these implementations, as far as this characteristic is concerned, are *particularly evident*. In addition, practice has shown that the type and complexity of the analysis of rounding noise depends, among other things, *on whether or not there is a correlation between rounding errors*. If we assume that these errors are not related to each other, both from sample to sample and from multiplier to multiplier, then the analysis of the rounding error is relatively simple, with the results that emerge being *independent* of the characteristics of the input signal. However, in the opposite case, where such a correlation exists, the analysis is *more complex* and the results it leads to depend on the characteristics of the input signal. This assumption about the lack of correlation from sample to sample and from source to source is based on the intuitive observation, which is also confirmed by the experiment, that *if the dynamic range of the signal being circulated within the filter is quite large, then the relatively small (compared with the signal values) rounding error, created at some point in the system and/or at some point in time, will be related to a small extent only, with the corresponding errors that are created at other points in the filter and/or at different times*. This assumption of independence from sample to sample allows us to consider the rounding error as *white noise with uniform spectral power density* N_0, whereas considering uniformly distributed rounding errors with zero mean value, the dispersion of this noise from each source will be equal to $\Delta^2/12$, where Δ is the quantization step. In order to remove the sampling period T from the involved equations, we define $N_0 = \Delta^2/12$. On the other hand, the assumption of independence from source to source allows us to calculate the spectrum of the rounding noise at the output of the filter, *by adding together the errors that arise from all the noise sources* located inside the filter.

Another undesirable consequence, apart from rounding noise due to the quantization of product values, is the appearance of the so-called *limit cycles* (see Sect. 12.9.9 on page 1198), which manifest as *the appearance at the filter output of a stable periodic sequence*, in cases where the input to the filter is *zero* or has a *very small value*, as happens during the pauses between successive recordings of vocal signal. It is interesting to note that although the mechanism leading to the creation of these cycles is essentially the same as that which creates rounding noise, predicting the measure of the filter output exhibiting such behavior, which is considered unstable, is *generally difficult*, despite the fact that relationships have been developed that allow the theoretical prediction of an upper limit for this measure. Practice has shown that this measure has a relatively small value, which rarely exceeds the value of quantization noise. For this reason, these limit cycles are characterized as *small-scale limit cycles*, in order to distinguish them from another significant category of limit cycles, the so-called *large-scale limit cycles*, which are associated with the phenomenon of *overflow* and in which the amplitude of oscillations is very large. For more information about limit cycles, the reader is referred to the relevant section.

[4] See references Kaneko T. & Liu B., *Roundoff Error of Floating Point Digital Filters*, Proceedings of Sixth Annual Allerton Conference on Circuit and System Theory, Monticello, Illinois, October 1968, pp. 219–227 and Weinstein C. & Oppenheim A.V., *A Comparison of Roundoff Noise in Floating Point and Fixed Point Digital Filter Realizations*, Proceedings of IEEE, Vol 57, No. 6 (June 1969), pp. 1181–1183.

[5] See the reference Oppenheim A.V., *Block Floating Point Realization of Digital Filters*, MIT Lincoln Laboratory, Technical Note 1969–19 (March 20, 1969).

Quantization Noise and Rounding Noise

It is interesting to note that although numerical rounding errors are related, as are those that arise during the conversion of the analog signal to digital, with the appropriate quantization process in each case, they differ from these in two points: (a) *the numerical values to be quantized are already expressed in digital form* and (b) *the truncation and rounding processes* that constitute the two ways in which the quantization process is applied and are presented in the next section *can appear at any point within the structure of the digital filter* and not only at its input, as is the case with the quantization of the input signals. In order to distinguish between these two types of errors, we characterize the quantization error of the input signals as *quantization noise* and the quantization error due to truncation or rounding of the signal values inside the filter as *rounding noise*. It is proven that in practice this rounding noise is much more intense than the quantization noise of the input signal and its handling is one of the most important factors affecting the complexity of the design of a digital filter. Speaking generally, the characteristics and the level of this rounding noise depend on three significant factors, more specifically, (a) *from the length of the registers* of the underlying system, that is, from the number of bits used to represent the values expressed in the binary system and which determines either the quantization step or the dynamic range of the system; (b) *from the type of arithmetic used*, that is, whether it is fixed- or floating-point; and (c) *from the type of implementation chosen*. Usually, the length of the registers is used to determine the dynamic range, so that all candidate structures are characterized by the same quantization step and therefore, it does not affect the level of the rounding noise but only the value of the SNR. In practice, as we comment in the relevant section, there is a scaling of the input signal, if necessary, in order to maintain *the maximum value of the SNR*, without simultaneously violating the conditions imposed by the range of the dynamic range. Note also that although the rounding noise can appear at any point within the system, in practice we are interested *mainly in the noise of this type, which appears at the output of the system* and which usually (although not always), has a lower value in systems that they use floating-point arithmetic. However, these systems are characterized *by more complex structures and higher implementation costs* and for this reason, many digital systems perform their calculations using fixed-point arithmetic.

Finally, regarding the analysis of rounding noise, its complexity depends to a significant extent *on the degree of correlation that exists between these types of errors*. Specifically, if these errors are considered *uncorrelated*, both among themselves and from multiplier to multiplier, then this analysis is relatively *simple*, with the resulting outcome being *independent* of the nature of the input signal, something that does not hold true when we assume that correlations of this type exist. It is proven that if this correlation is not taken into account, the results that emerge are valid both for the case of truncation and for the case of rounding, whereas on the contrary, if we assume that the signals are characterized by a lesser degree of randomness, then this assumption about the lack of correlation begins to collapse more quickly for the case of truncation, compared with the case of rounding, with the first case therefore requiring more attention.

How can we avoid the occurrence of the above three types of error? As we comment elsewhere, this can be done by choosing the *appropriate word length* for the above parameters, the value of which can even be different for each of the above cases. Specifically, the word length of the filter coefficients is usually chosen so as to *satisfy the specifications of the frequency response function*, whereas the word length for the signal samples is chosen so as to *satisfy the specifications related to the (SNR)*. It is obvious that the larger register length leads to a higher implementation cost and in almost all cases, when implementing a specific filter structure, we must resort to some compromise, regarding the cost of implementation, the SNR and the accuracy of representation. Speaking generally and as we comment later, if the system is designed carefully, we can ensure that overflow is a rare occurrence and that the values of the quantization error are relatively small. In this case, the system that uses the quantized coefficients behaves very well as a linear system, with quantization errors appearing only at points where clipping and rounding techniques are applied. For a detailed presentation of the three above causes of reducing the accuracy, see the corresponding sections of the textbook.

12.4.5 Implementation

In the final stage of the process, after all the above steps have been taken, the only thing left is the *actual implementation of the filter*, that is, its physical construction, which can be done either by developing an application that will simulate the operation of the filter on a computer or integrated digital signal processing circuit, or by resorting to a hardware implementation. Specifically, in common nonreal-time applications, where often the entire set of data to be processed is known and available in advance, the implementation of the filter can be done in software (such as, for example, the MATLAB environment and the accompanying Simulink application) and using a simple computer. In this case, the implementation usually relies on the difference equation that describes the behavior of the system. In this type of implementation, emphasis should be placed on describing the filter to be implemented in the form of an algorithm, which then needs to be implemented, that is, converted into a computer program. In other words, design parameters related to hardware are not particularly critical, except for those elements that relate to any type of software application, such as memory and processing speed requirements as well as the type of input/output system.

On the other hand, if the filter is going to be used in *real-time applications* (such as, for example, *telecommunication systems*), it must support the fastest processing of the current input sample and the production of the corresponding output before the arrival of the next sample, a requirement that imposes *the implementation of the filter in hardware* and the use of fast equipment that implements fast algorithms, such as the FFT. There are many possibilities regarding the type of hardware that will be used for the implementation of the filter, such as the construction of a *specialized* arrangement using electronic components and integrated circuits, the use of *programmable arrangements* such as programmable logic devices) and field programmable gate arrays), as well as the *use of digital signal processors (DSPs)* for general use, which can be selected from a variety of such arrangements, differing in terms of the type of arithmetic they use (i.e., fixed- or floating-point arithmetic), the clock speed, and the amount of internal memory. In each case and from the study of the general recursive difference equation that describes its operation, it is not difficult to realize that the hardware that should be used includes *adders* and *multipliers* for the implementation of the necessary additions and multiplications, as well as elements of time delay. However, in addition to the above, the use of the appropriate amount of memory for storing the values of the filter coefficients and the values of the input and output samples is required. In cases where the filter coefficients *are never going to change*, but will always be the same, they can be stored in a *read-only memory (ROM)* for better protection after their calculation.

Let us mention at this point some basic details about the structural units used for the construction of digital filters at the hardware level. As is well known from basic theory, the basic elements of any digital system are the *logic gates* AND, OR, NAND, NOR and NOT (the XOR gate, although it is a basic component of many circuits, *is not* a basic gate because it can be constructed with the help of other gates). The above gates, the set of which is depicted in Fig. 12.7 along with their symbol and their truth table, implement the homonymous operations of *Boolean algebra* and therefore, their function is unambiguously determined by the corresponding truth tables. Although generally speaking, by combining the appropriate gates in each case, we can construct *any* digital circuit, often it is possible to use the circuit with a single type of gate. Regarding to the type of digital circuits that can be constructed, in practice there are two possibilities, more specifically: (a) *the construction of circuits that result from simple combinations of elementary gates and whose output, which depends only on the current and not on the previous inputs, is calculated from them with simple application of the rules of Boolean algebra* and (b) *the sequential circuits that are characterized by the existence of memory, which is implemented using special circuits called flip-flops and which allows the correlation of the current output, not only with the current input, but also with past inputs that have been applied to the system at previous times*. Usually, the operation of these circuits is guided by the beats of a clock, which determine the times of reading and writing to and from the system's memories and allows the simultaneous change of the state of all the elements of the circuit.

A technique that is particularly popular as it leads to a significant increase in the performance of the underlying digital arrangements is the technique of *pipelining* (pipeline), which characterizes the way in which all modern processors operate, but is also used in simpler arrangements, such as the *adders* and the *multipliers* of digital filters. Consider a combinational circuit that implements a set of digital processes whose execution is quite time consuming. We can speed up the process by subdividing the circuit into a multitude of smaller circuits and implement the complex process as a *sequence of elementary processes*. In this case, each part of the circuit is characterized by less time delay than with the original complex circuit and therefore, this new modified structure, will operate under a regime of higher clock frequency. In subsequent sections we present adder and multiplier circuits that use this pipelining technique.

12.5 Representation of Numbers in Digital Systems and Filters

As described in detail in the relevant section, the quantization process *rounds* the value of each sample up or down, so that it is identified with the value of some quantization level, in order to finally be encoded in the binary system. In addition, the fact that all digital systems are characterized by processors that include registers of *finite length*, imposes the realization of all kinds of calculations, not with infinite but with *finite* accuracy, which depends on the *length of the registers* used for storing the numbers, the *type of representation* that has been adopted (for example, representation of *signed numbers* or *two's complement*), as well as the type of arithmetic used (for example, fixed- or floating-point arithmetic). For this reason, and despite the fact that the analytical presentation of the types of numerical representation in a computer can be found in any introductory computer science book, let us briefly present at this point, for completeness, the basic theory related to these two types of representation.

Name	Graphic symbol	Algebraic function	Truth table
AND	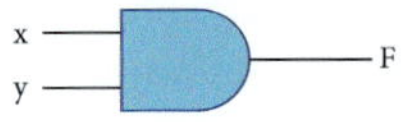	F = x - y	x y | F 0 0 | 0 0 1 | 0 1 0 | 0 1 1 | 1
OR	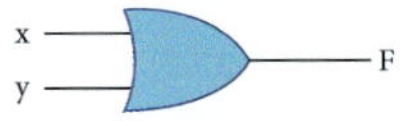	F = x + y	x y | F 0 0 | 0 0 1 | 1 1 0 | 1 1 1 | 1
Inverter	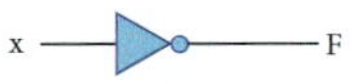	F = x′	x | F 0 | 1 1 | 1
Buffer	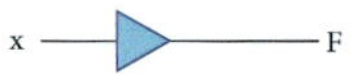	F = x	x | F 0 | 0 1 | 1
NAND	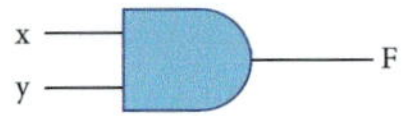	F = [xy]′	x y | F 0 0 | 1 0 1 | 1 1 0 | 1 1 1 | 0
NOR	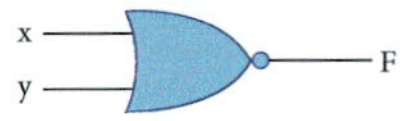	F = [x + y]′	x y | F 0 0 | 1 0 1 | 0 1 0 | 0 1 1 | 0
Exclusive-OR (XOR)	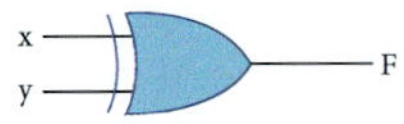	F = xy′ + x′y = x⊕y	x y | F 0 0 | 0 0 1 | 1 1 0 | 1 1 1 | 0
Exclusive-NOR or equivalence	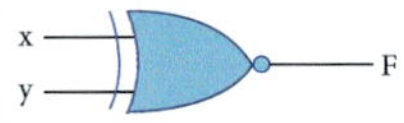	F = xy + x′y′ = [x⊕y]	x y | F 0 0 | 1 0 1 | 0 1 0 | 0 1 1 | 1

Fig. 12.7 The set of available logic gates along with their symbol and their truth table

12.5.1 Fixed-Point Representation

From basic mathematics, it is known that the digits that appear in the string that describes each number constitute the *coefficients of the linear combination of the powers of the base of the positional system in which this number is expressed* and the value of which is the value of the number. We also know that in computer science, apart from the *decimal* system that is used everywhere and always, the *binary* numbering system is primarily used, in addition to the *octal* and the *hexadecimal* systems also frequently being used. In the general case, the numbers used have both an *integer* and a *fractional* part, with the digits that are located before the decimal point symbol (which should be noted, *is not implemented in a natural way* on a computer) are associated with positive powers of 2 (in this description, our study is limited exclusively to the binary system, although what is mentioned here applies with the appropriate modifications in any numerical system) and the digits that are located after this symbol are associated with *negative* powers of 2 (which of course applies to all numerical systems and for the base that describes them). Considering for example the number 101010.110101 this, as a linear combination of powers of 2, is formulated as

$$\mathbf{1} \times 2^5 + \mathbf{0} \times 2^4 + \mathbf{1} \times 2^3 + \mathbf{0} \times 2^2 + \mathbf{1} \times 2^1 + \mathbf{0} \times 2^0 + \mathbf{1} \times 2^{-1} + \mathbf{1} \times 2^{-2} + \mathbf{0} \times 2^{-3} + \mathbf{1} \times 2^{-4} + \mathbf{0} \times 2^{-5} + \mathbf{1} \times 2^{-6},$$

and it is not difficult to find out that it corresponds to the decimal number 42.828125. Generalizing the above description, if a number L with a length of K bits expressed in a numerical system with base r consists of M digits for the integer part (which are located to the left of the decimal point) and from N digits for the fractional part (which are located to the right of the decimal point), it can be represented in the form

$$L = (d_{-M}, d_{-M+1}, \ldots, d_{-1}, d_0, d_1, d_2, \ldots, d_{N-1}, d_N) = \sum_{i=-M}^{N} d_i r^{-i}, \tag{12.10}$$

where b_i is any of the digits of the used numerical system of a system that takes values in the range $0 \leq d_i \leq (r-1)$, whereas r is the *base* of the system. In the case of the decimal system, it is $r = 10$, whereas the digits used are $\{0, 1, 2, 3, 4, 5, 6, 7, 8, 9\}$. In the case of the binary system it is $r = 2$ and the digits used are $\{0, 1\}$. In computer science terminology, each digit is referred to as a (*bit binary digit*), and each binary number is represented by a *finite-length string of KK bits*. The storage of each bit relies on a hardware element known as a *flip-flop*, which can exist in one of two stable states: one representing 0 and the other representing 1. It is not difficult to understand that the registers in which these values are stored consist of a group of such arrangements placed one after the other, *with the number of these arrangements dictating the length of the register*. From the set of bits that form a binary number, the first bit from the left has the *smallest* contribution to the value of the number, which is $2^0 = 1$, and for this reason it is known as the *least significant bit*, whereas the first bit from the right has *the largest* contribution to the value of the number, which is equal to 2^{K-1}, and is known as the *most significant bit*. In the simplest case, all K bits of the number are used for the representation of its value, whereas it does not support a sign and fractional part, so $K = M$ and $N = 0$. These numbers are known as *unsigned integers*. The number of possible combinations that can arise from K bits is 2^K and therefore, the range of values for the numbers encoded with K bits is $[0, 2^K - 1]$. For example, for the value $K = 8$ we have $2^8 = 256$ and therefore the unsigned integer numbers that are encoded with $K = 8$ bits, belong to the range of values [0, 255]. On the other hand, if no integer is used at all, but *only a fractional part*, that is, if it is $K = N$ and $M = 0$, the range of values to which the numbers encoded in this way belong is the $[0, 1 - 2^{-N}]$. Finally, in the case where the number has *both an integer and a fractional part*, the situation is similar, although in each case, we can represent the number using the representation of only the fractional part, dividing it by the coefficient 2^M. In cases where the value of the number is *greater* than the maximum value or *smaller* than the minimum value that can be encoded with K bits, a situation arises that is respectively characterized as *overflow* or *underflow* and that is a source of error.

On the other hand, the representation of negative numbers can be done in three different ways, which in general are as follows:

- The *sign-magnitude representation*, in which the rightmost bit, i.e., the most significant bit is used for the representation of the sign (with the value 0 indicating that the number is positive and the value 1 indicating that the number is negative) and consequently, the remaining $K - 1$ bits are used for the representation of the number's value. In other words, in this representation, a number x is expressed as $s_x x_1 x_2 x_3 \cdots x_n$ with the first bit s_x representing the sign of the number and the bits $x_1 x_2 x_3 \cdots x_n$ representing its value. For reasons discussed later, numbers are usually expressed as *fractional numbers*—that is, they only have a fractional part or equivalently, they have been scaled appropriately to take values in the interval $(-1, 1)$—and therefore, in this type of representation, positive numbers are represented as $0.xxxxxx$ whereas negative numbers are represented as $1.xxxxxx$, where $xxxxxx$ is a string of bits of the appropriate length. In other words, the binary number $s_x x_1 x_2 x_3 \cdots x_n$ has a value of $x = x_1 2^{-1} + x_2 2^{-2} + \cdots + x_n 2^{-n}$ if $s_x = 0$ and a value of $x = -(x_1 2^{-1} + x_2 2^{-2} + \cdots + x_n 2^{-n})$ if $s_x = 1$.
- The *one's complement representation*, in which negative numbers x are represented as $2 - 2^{-L} - |x|$, where L is the so-called word length defined as the number of bits located to the right of the decimal point. Taking into account that the binary representation of the number $2 - 2^{-L}$ is a binary number of length $L + 1$ bits, all of which have the value 1, as can easily be proven, the complement of a negative number can be calculated *by expressing the number as a number of the length $L + 1$ bits* (which means that if it is represented by fewer bits, we should add the appropriate number of zero values before the most significant bit), so as to make its length equal to $L + 1$ and then proceeding to *invert* its digits (which means that the bits with value 1 are set to the value 0, and vice versa). In an equivalent description, the representation of the complement of a negative number consists of *inverting the bits used to represent its absolute value* and using an additional bit for the sign in the way we did previously.
- The *representation of the complement to* 2 , in which negative numbers N are represented as $2 - |x|$ and can be easily calculated *by adding one to the least significant bit of the complement to* 1. In this representation, the sign bit s_x is used

as before. Observing that for negative numbers it is $|x| = -x$, we have $2 - |x| = 2 + x$ and therefore, the number x can be expressed in the form

$$x = -2+1+x_1 2^{-1}+x_2 2^{-2}+\cdots+x_n 2^{-n} = -1+x_1 2^{-1}+x_2 2^{-2}+\cdots+x_n 2^{-n} = -s_x+x_1 2^{-1}+x_2 2^{-2}+\cdots+x_n 2^{-n},$$

which is particularly useful for presenting the operation of *multiplication* between two numbers represented using the complement to 2. From the above relationship, it also follows that

$$\frac{x}{2} = -s_x 2^{-1} + x_1 2^{-2} + x_2 2^{-3} + \cdots + x_n 2^{-n-1},$$

and since $-s_x 2^{-1} = -s_x + s_x 2^{-1}$, the above relationship becomes

$$\frac{x}{2} = -s_x + s_x 2^{-1} + x_1 2^{-2} + x_2 2^{-3} + \cdots + x_n 2^{-n-1}. \tag{12.11}$$

From the last expression, it follows that if the representation of the complement to 2 of a number x is the $s_x x_1 x_2 x_3 \cdots x_n$, then the representation of the complement to 2 of the number $x/2$ is the $s_x s_x x_1 x_2 x_3 \cdots x_n$. In other words, the *division by* 2 of a number x that is expressed in the representation of the complement to 2 simply consists of *shifting all the bits of the number one position to the right and repeating the sign bit in the first position from the left*. This property significantly simplifies the operation of multiplication used in this type of representation. Generally speaking, the representations of the complement to 1 and to 2 are more effective for performing *additions*, whereas the sign magnitude representation is more effective for performing *multiplications*.[6] However, in our time, the type of representation that has been established is the complement representation with respect to 2.

As already mentioned, these three different types of representation, concern *only negative numbers*, whereas positive numbers are represented in these three forms of representation in the way we previously described. In practice, the representation of the complement with respect to 2 is mainly used, because it is proven to be more suitable for performing calculations. A characteristic of this representation is that *if during the addition a carry arises from the most significant bit, this is ignored in certain cases* (as happens in the subtraction $a - b$, which is performed as $a + (-b)$ and is $a \geq b$, but not in the subtraction $b - a$), something that *does not apply* in the representation of the complement with respect to 1, where the carry that arises is added to the least significant bit. On the other hand, in the sign magnitude representation things are more complicated, since among other things, the sign of the sum is required to be checked. Of course, the advantage of this type of representation, is that the implementation of multiplications is a simple process that consists *in multiplying the values of the two numbers and adding the sign*, whereas in the representations of the complement with respect to 1 and 2, things are more complex and there may be cases in which special handling is required.

12.5.2 Floating-Point Representation

What we mentioned in the previous section constitutes the so-called *fixed-point representation*, which is depicted in Fig. 12.8. The disadvantage of this representation, which is generally used when the digital filter is implemented in hardware and therefore uses specialized equipment such as *DSP processors*, is that the number of numbers that can be represented, *is not very large*. Specifically, the smallest number that can be represented using the complement representation to 2 is -1, whereas the largest is $1 - 2^{-L}$. Two other disadvantages of this representation are that *the percentage error resulting from the application of quantization processes based on truncation and rounding* (see the next section) *increases with the increase in the size of the number*, whereas the resolution it supports is constant and equal to $x_{\max} - x_{\min}/(2^{\beta} - 1)$, where $x_{\max}$ and $x_{\min}$ are the maximum and minimum values of the numbers we want to encode, whereas β is the number of bits used for this encoding. However, we know that a number can be written in many different ways, if we move the decimal point to the left or right as appropriate and correspondingly multiply or divide the value of the number by the appropriate power of 10. For example, the number 107.32 can also be expressed as 10.732×10, 1.0732×100, but also as $1073.2/100$ and $10732/100$. The situation in the binary system, but also in all numerical systems, is similar. Because these equivalent ways of

[6] See Hwang K., *Computer Arithmetic: Principles, Architecture and Design*. New York, Wiley, 1979.

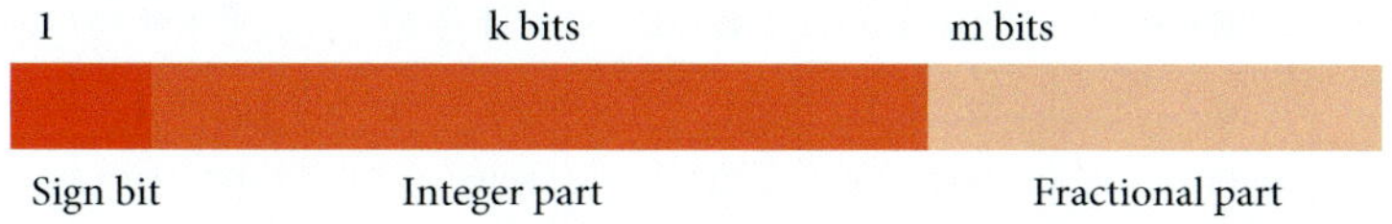

Fig. 12.8 Fixed-point representation

writing the same number, consist of *moving the decimal point to the left or to the right by the appropriate number of digits*, this representation is known as *floating-point representation*. This type of representation is used in all modern electronic computers and therefore, it is used in all digital filters that are implemented in software.

In a more detailed description, in the *floating-point representation* and limiting ourselves to the binary system that interests us, each number L is formulated in the form $L = M \times 2^E$, where M is a *coefficient* that takes values in the area $1/2 \leq M < 1$ and E is the *exponent* of the base of the binary system, which can be a *positive or a negative integer*. Of the two above numbers, the *coefficient* M is a fractional number, which can be considered to correspond to the fractional numbers of fixed-point representation, whereas the *exponent* is an integer number that represents the number of positions at which the decimal point appearing in the coefficient (to the left or to the right) should be moved, in order to produce the original number. In this representation, in which the decimal point is after the second most significant bit of the coefficient, since the first bit represents the sign, both these numbers require a bit for the representation of the sign and therefore, a more detailed definition of it, is the $L = (-1)^S M \times 2^E$ with the bit S representing the sign and having the value zero for the positive numbers and the value 1 for the negative numbers. Using this type of representation, the numbers 0.00110101 and 1001.11 are represented as 0.110101×2^{-2} and 0.100111×2^4 respectively. Based on the above type of description, it is not difficult to understand that the registers in which these numbers are stored *are divided into sections in which the sign bit and the values of the coefficient and the exponent are stored* with the length of these sections in bits being dictated by whether the representation is half, single or double precision according to what we mention in the next section.

One of the characteristics of floating-point representation is that *the density on the line of real numbers represented in this way is not constant*, owing to the existence of the term 2^E used in the defining equation of this representation. On the contrary, what is constant is the number of floating-point numbers that are between two consecutive powers of 2. For example, the number of numbers that are between the numbers $16 (= 2^4)$ and $32 (= 2^5)$ is *the same* as the number of numbers that are between the numbers 128 $(= 2^7)$ and 256 $(= 2^8)$. However, since the difference $256 - 128 = 128$ is greater than the difference $32 - 16 = 16$, the distance between two consecutive numbers in the second range of values is greater than the distance between two numbers that belong to the first range. Generalizing this observation, we can formulate the claim that *the larger the values of floating-point numbers, that is, the larger the value of the exponent E, the greater the distance between two such numbers*.

In order to understand this characteristic in practice, let us try to enumerate the natural numbers using floating-point representation with 4 bits for storing the coefficient (the number of bits for the exponent E does not concern us, because the example involves only a few numbers). Initially, the value of the exponent is $E = 0$ and therefore the coefficient M will be multiplied by $2^0 = 1$. Starting from 0 and starting to count, we reach the value 15, which is the maximum value that can be encoded with 4 bits. At this point, the *initial cycle* of the process is completed, which could be characterized as the *zero cycle*. In the next stage, we increase the value of the coefficient by 1, that is, we set $E = 1$, which means that from this point on, the value of the coefficient will be multiplied by $2^1 = 2$. As for the value of the coefficient, it is initialized at 1000. This value in the decimal system is the value 8, which, multiplied by 2, gives us the value 16, which is the first value created in this first cycle. Continuing the enumeration, the next values are 1001, 1010, 1011 (i.e., 9, 10, 11, and 12), which, multiplied by 2, will give us the numbers 18, 20, 22, and 24. The last number in this first cycle is 1111 (i.e., 15), which, if multiplied by 2, will give us the number 30. At this point, the *second cycle* is completed and it is not difficult to realize that because in this cycle the values of the coefficient are multiplied by 2 on the one hand it is not possible to represent all numbers (for example, odd numbers are not represented); on the other hand, the distance between successive numbers has doubled. We then increase the value of the coefficient E by one unit, that is, we assign it the value $E = 2$. Consequently, in this cycle, the possible values of the coefficient, which is reinitialized to the value 1000, and therefore, are the same as before, that is, 1000 (8), 1001 (9), 1010 (10), etc., up to the value 1111 (15), will be multiplied by 4 leading to the numbers 32, 36, 40, 44, 48, 52, 56, and 60. Continuing in the same way, the values that will result in the third cycle are the products of the above numbers by $2^3 = 8$ and are 64, 72, 80, 88, 96, 104, 112, and 120. The *fourth cycle* follows, in which the resulting values are 128, 144, 160, 176, 192, 208, 224, and 240 (here the multiplication is performed by $2^4 = 16$), etc. Therefore, in the p-th cycle and for a coefficient of 4 bits, the resulting values are the products of the eight above numbers by 2^p. If a different number of bits is used to encode

E = 0		E = 1		E = 2		E = 3		E = 4	
M	L	M	L	M	L	M	L	M	L
0000	0	1000	16	1000	32	1000	64	1000	128
0001	1	1001	18	1001	36	1001	72	1001	144
0010	2	1010	20	1010	40	1010	80	1010	160
0011	3	1011	22	1011	44	1011	88	1011	176
0100	4	1100	24	1100	48	1100	96	1100	192
0101	5	1101	26	1101	52	1101	104	1101	208
0110	6	1110	28	1110	56	1110	112	1110	224
0111	7	1111	30	1111	60	1111	120	1111	240
1000	8	1st cycle		2nd cycle		3rd cycle		4th cycle	
1001	9	x 2		x 4		x 8		x 16	
1010	10								
1011	11								
1100	12								
1101	13								
1110	14								
1111	15								

Fig. 12.9 The process of enumerating natural numbers expressed in floating-point representation

the value of the coefficient, the situation is adjusted accordingly. This process is depicted in Fig. 12.9 and as can be easily verified, *the number of numbers that are created in each cycle, that is, essentially, between two successive powers of* 2 *is constant* and in this case equal to 8. In addition, the fact that some numbers are not represented, this *should not* lead to the erroneous conclusion that there are numbers that cannot be represented with floating-point representation (although strictly speaking, this is something that is true, since the number of floating-point numbers that can be created is obviously *a subset* of the infinite number of real numbers included in the uncountable set $\mathbb{R}$). What we mean is that the numbers that are not represented in the above example can be represented if a larger number of bits is used. For example, if for the coefficient we use 5 bits, then in the zero cycle the numbers 0 to 31 will be created, including the odd numbers that are between 16 and 30 and that cannot be represented with 4 bits.

The comparison between the *floating-point representation* and the *fixed-point representation*, under the assumption that they use the same number of bits, leads to the conclusion that the floating-point representation allows the encoding of numbers that belong *to a much larger range of values*, adjusting the resolution appropriately for the range of values that interest us. Conversely, floating-point representation can handle a dynamic range of a specific size using fewer bits than fixed-point arithmetic. The above analysis is reduced with the increase in the size of two successive numbers and consequently, the distance between two real numbers that are encoded according to the floating-point representation increases as the numbers increase in size. In an alternative description, *if we want to cover the same dynamic range using both the fixed-point representation and the floating-point representation, the latter offers greater resolution for small numbers but less resolution for larger numbers, whereas fixed-point representation offers the same resolution for the entire range of values*. However, on the other hand, floating-point representation is *more complex*, since before performing additions and multiplications, the *quantization* of the coefficient value M must take place, whereas in fixed-point representation, the quantization process is only required after performing the multiplications. As for the behavior of these two representations with the criterion of rounding noise, the relevant research on this subject[7] has shown how floating-point representation is *superior* to fixed-point representation only for filters characterized by a high value of the *quality factor* Q and characterized by a high gain value, something that is particularly true for filters that result from the series connection of simpler stages. However, as we explain elsewhere, in these filters it is very common to scale the signals that are circulated between the various stages. For this reason, many of these digital arrangements use a fixed-point numerical constant.

[7] For the study of the behavior of floating-point representation regarding rounding noise, see Liu B. & Kaneko T., *Error Analysis of Digital Filters Realized with Floating-Point Arithmetic*, Proc. IEEE, vol. 57, pp. 1735–1747, Oct. 1969. For a similar study on fixed-point arithmetic, see Jackson L.B., *Roundoff-Noise Analysis for Fixed-Point Digital Filters Realized in Cascade or Parallel Form*, IEEE Tran. Audio Electroacoust vol. AU-18, pp. 107–122, June 1970 whereas for a comparison of the above results see Weinstein C. & Oppenheim A.V., *A Comparison of Roundoff Noise in Floating Point and Fixed Point Digital Filter*, Proc. IEEE, vol. 57, DD. 1181–1183. June 1969.

Considering two numbers that are encoded according to floating-point representation, their multiplication is done very easily, *by multiplying the coefficients and adding the exponents*, whereas the addition can only be done if we previously transform the numbers so that both are characterized by the same exponent. This is something that is done by modifying the value of the coefficient in the appropriate way, a process that leads to a *reduction in accuracy*, although it is possible to lead to the appearance of overflow and inadequacy. Because the type of floating-point representation is the one that is mainly used in scientific applications that are run on computers, the scientific union of IEEE has defined a standard and more specifically the standard *IEEE 754*, which dictates the form of the numbers that follow this type of representation. According to this standard, a number has the form $N = S \times 2^{E+B}$. In this notation, the coefficient S is a *signed mixed number* that has both an integer and a fractional part and takes a value in the range from 1 to $2 - 2^{-L}$, where L is the number of significant bits allowed in the fractional part of S, whereas the size B is known as *bias* and only takes integer values. It is interesting to note that if the number is modified in the appropriate way so that the value of S belongs to the range of values from 1 to $2 - 2^{-L}$, then its binary representation will always have the form $1.xxxxxx$ and since it is guaranteed that the integer part of it *is always equal to one*, this bit *does not need to be stored* since it always has the same known value. This capability is very important, as we can increase the accuracy by one bit, without any additional hardware cost. The numbers that are expressed in this form are characterized as *normalized*. On the other hand, the use of bias frees us from the need to deal with negative numbers, which *significantly simplifies* the arithmetic although it makes the implementation of multiplications and divisions more complex, since the presence of the number B in the exponent of the number implies that in the case of a multiplication the exponent of the result will contain the number $2B$ (since in multiplication the exponents are added) and therefore, we should subtract one B, whereas in the case of division, the polarization *is eliminated* (since in division the exponents are subtracted) and therefore we should add one B. Of course, in each case we can use the value $B = 0$, so as not to use this feature at all. According to the IEEE754 standard, for floating-point representation, *three different forms* can be used, with the first form using 16 bits (1 bit for the sign, 5 bits for the exponent, and 10 bits for the coefficient) and is known as *half-precision representation*; the second form uses 32 bits (1 bit for the sign, 8 bits for the exponent, and 23 bits for the coefficient) and is known as *single-precision representation*; and the third form uses 64 bits (1 bit for the sign, 11 bits for the exponent, and 52 bits for the coefficient) and is known as *double-precision representation*. The structure of the number that follows these three different types of representation is shown in Fig. 12.10, in which the groups of bits containing the values of the exponent and the coefficient for these three cases are shown. More details about this topic, such as the range of values supported, as well as related information, can be found in the literature. We should note that despite the fact that floating-point representation *supports a larger dynamic range and leads in increasing the accuracy of processing*, it does require *more expensive hardware and complex software*. It also leads *to a reduction in processing speed* as it requires processing of the values of both the coefficient and the exponent.

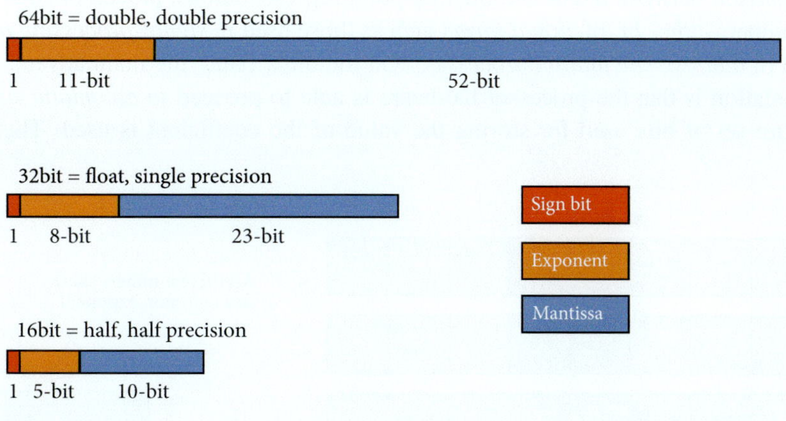

Fig. 12.10 Half-, single-, and double-precision floating-point representation according to the IEEE 754 standard

12.5.3 Choosing the Appropriate Type of Representation

The choice of the type of arithmetic to be used for the implementation of the digital filter's difference equation (i.e., whether fixed- or floating-point representation will be used), is one of the most important factors taken into account in the design of such a system. Usually, most analyses regarding the impact of noise rely on the use of fixed-point arithmetic because this type of arithmetic is *simpler* in its implementation. When the underlying digital system uses this type of representation, the number of significant digits allocated to internal signals determines the filter's *dynamic range*. Since signal values may vary significantly between different stages of the system, appropriate *signal scaling* must be applied at each stage. This ensures that the full available range of each stage's dynamic range is utilized, while also preventing overflow. On the other hand, if the filter uses *floating-point arithmetic*, the implementation is naturally more complex but the advantage is that this scaling, where required, is done in an *automatic* way. Specifically, because in this type of representation each number as we have seen is represented by a coefficient and an exponent, the scaling factors used for this purpose are represented by the exponents of the numbers. It turns out that the *SNRs* in filters using floating-point arithmetic are than in filters that use fixed-point arithmetic, whereas if the filter uses block floating-point representation an intermediate situation arises. The precision offered by each representation—defined as its ability to distinguish between different numerical values—is a key factor in selecting the appropriate representation format. In fixed-point representation, precision depends *on the word length*. In contrast, for floating-point representation, it is determined *by the number of bits used for the significant* and is defined as *the smallest difference between two numbers that share the same exponent*.

Let us mention for completeness and some things about this last type of representation. The basic characteristic and at the same time advantage of the floating-point representation of a group, is the achievement of the dynamic range (which is defined as the ratio between the smallest and the largest number that can be represented in this processor), supported by floating-point processors, but using fixed-point processors. Note that although this is something that can be achieved, using simulation software of floating-point calculations on fixed-point processors, this solution is characterized by high computational requirements and is only suitable in cases where only a small part of the total calculations require an extensive dynamic range. Specifically, in this type of representation proposed by Oppenheim, the numbers used are placed in a group and then their exponents are adjusted in such a way that all these numbers share the same common exponent, which is stored for a single time in a single word and is used for all these numbers (see Fig. 12.11). The value of this exponent is determined by the element of the group that is characterized by the largest size. This solution, which is used in cases where reliable data representation is not possible through fixed-point arithmetic it finds many applications in the field of digital signal processing, with the most characteristic of these being related to the implementation of the FFT (see for example David and Iovescu, 2003). Because this approach is based on software, the conversion from fixed-point representation to floating-point group representation is the programmer's task.

Returning to the comparison between fixed- and floating-point representations, processors that use fixed-point representation allow the use of numbers *either in fractional form* (such as those used in *signal-processing applications*), or in *integer form* (such as those used in *address-calculation processes*). On the other hand, the main advantage of processors that rely on floating-point representation is that the processor hardware is able to proceed *to automatic scaling* of the values of the numbers, so that the entire set of bits used for storing the value of the coefficient is used. This means that if a value is

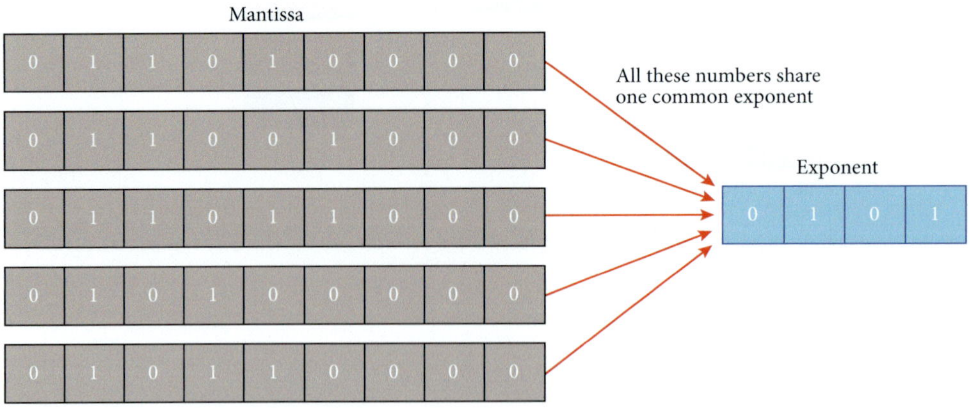

Fig. 12.11 The floating-point group representation of data

excessively large and cannot be represented by the number of bits used, the hardware is able to proceed to *downscaling*, performing the appropriate shift to the right, whereas in the opposite case, where the value of the coefficient does not use the total available number of bits, *upscaling* takes place through the appropriate shift to the left. In this case, the exponent, as mentioned, retains the value of the number of shifts of the fraction to the appropriate direction.

12.6 The Implementation of Additions and Multiplications

As already mentioned, the three fundamental operations used for the implementation of the difference equation of any digital filter are the *addition* (the *subtraction*, as is well known, is implemented through addition), the *multiplication*, and the *time delay*. Of these operations, *the simplest* in its implementation and for the case of serial arithmetic is the *time delay* (since it is implemented very simply with the help of a *shift register* with a simple input and a simple output), whereas addition and multiplication are implemented with the help of *simple circuits*, which are known as *adders* and *multipliers* and which together constitute the *arithmetic unit* of the digital filter. In this section, we present in detail these two operations, as well as the structures through which they are implemented. Specifically, as we comment later, these structures are implemented *both in serial and in parallel form*, with the serial form leading to *simple and economical solutions* and the parallel form leading *to an increase in speed* and consequently to the *performance* of the filter.

12.6.1 The Operations of Addition and Subtraction

The simplest but also most fundamental operation that can be performed between two numbers expressed in any number system is the *addition*. In the case of the binary system, which contains only two digits, things are very simple, since there are only four possibilities and more specifically: (a) $0 + 0 = 0$, (b) $0 + 1 = 1$, (c) $1 + 0 = 1$, and (d) $1 + 1 = 2$. Of course, in the last case, the number 2 should be expressed in the *binary system*, that is, written as 10. Therefore, in the last case we have $1 + 1 = 10$, or as we would say in the addition that we perform in the decimal system, $1 + 1 = 0$ and 1 is the carry. The above four possibilities refer to the case of adding two binary numbers of one bit in length, whereas in the general case where we wish to add numbers of L bits in length, we write the numbers one below the other, draw a horizontal line under the second number and then, we add the corresponding bits, in exactly the same way we perform addition in the decimal system. If the two numbers do not have the same length, then, before performing the addition, we usually *add to the left of the number with the smaller length, the appropriate number of zeros, so that the two numbers acquire the same number of bits*. In general, the operation of addition between the corresponding bits *will use a carry that has resulted from the addition of the bits located in the previous positions of the two numbers and will create a carry that will be used in the addition of the bits located in the next positions of the two numbers*, as we scan the numbers from right to left, adding the corresponding bits. If at least one of the most significant bits of the two numbers has a value equal to 1, then if the carry that has resulted from the previous addition is equal to 1, the sum of the two numbers will have a length of $L + 1$ bits, whereas the same will apply in the case where both these bits have a value equal to 1. In this case, the phenomenon of *overflow* may occur or not, depending on the length of the registers of the underlying digital system that implements the operation of addition.

The above description refers to the simple case of adding two *unsigned integers*, whereas in the general case, the operation of *addition* is performed in a different way, depending on the type of the representation used for encoding numbers in the binary system (remember, in all types of representation, positive numbers are represented in the same way, whereas negative numbers vary depending on the representation used). Specifically, in the *sign magnitude representation*, the situation is *quite complex*, as there are four possible cases for addition (which have to do with the fact that for each of the two values there are two versions regarding the sign) and another four cases for subtraction, although these eight different versions can be reduced, if we first proceed to *check the signs of the two numbers*. The process that takes place is depicted in the flow diagram of Fig. 12.12 and includes the use of the *logical XOR operation*, in order to determine whether the signs of the two numbers A and B are *the same or different* (note that if instead of addition we have subtraction this is calculated as the addition $A - B = A + (-B)$ and therefore, we should initially change the sign of the second number). In the first case (where the signs are the same), we *simply add the magnitudes of the two numbers* and use as the sign in the result, the *common sign* of the two numbers, simultaneously checking if there has been an *overflow*, whereas in the second case (where the signs are different), *we subtract the second number from the first* (note, instead of performing the subtraction bit by bit, we add to the first number the two's complement of the second number) and use as the sign in the result, *the sign of the number with the*

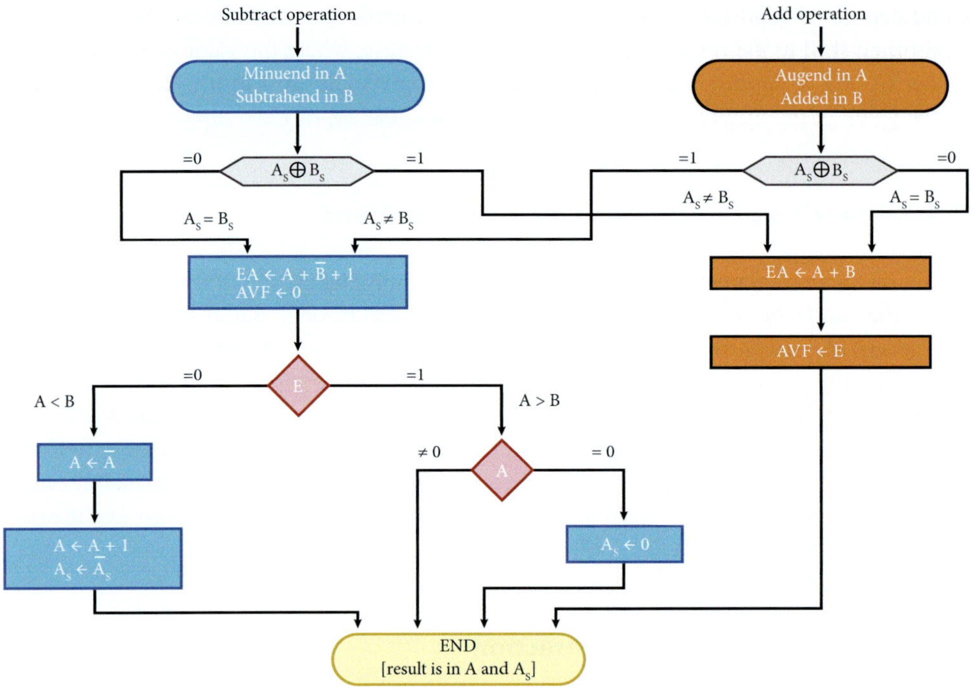

Fig. 12.12 The processes of addition and subtraction of integers that are expressed in the form of signed magnitude (Chauhan P, *Addition and Subtraction*, https://www.naukri.com/code360/library/addition-and-subtraction)

largest magnitude. Depending on the relationship between the magnitudes of the two numbers, the phenomenon of overflow may or may not occur, a situation that must be addressed.

On the other hand, *in the representations of complement with respect to* 1 *and with respect to* 2, positive numbers are represented as before, whereas for negative numbers the complement with respect to 1 and with respect to 2 of the corresponding positive number is used respectively. The addition of the two numbers is performed between the corresponding bits including the sign bit and since the addition constitutes a subtraction, since, as we aforementioned, the subtraction $A - B$ is reduced to the addition $A + (-B)$, we distinguish two cases depending on the relationship that exists in the measures of the numbers A and B. Specifically, in the case of the complement with respect to 1, if it is $A > B$, the subtraction is performed by *adding the complement of number B to number A* and if a carry occurs after the most significant bit, it *is added* to the result, whereas if $A < B$ the operation is performed as in the case of $A > B$, with the difference that since the result of the operation is expressed in the complement of 1, *we must take the complement of* 1 *of the resulting value, with a negative sign*. Conversely, in the case of adding two positive numbers that are expressed in the two's complement representation, if there is a carry after the sign bit it is ignored, whereas if the result is negative (as happens for example in the addition of two negative numbers), we must take the two's complement of the resulting value. Conversely, in the case of subtraction, we add to the first number the two's complement of the second number.

In order to understand in practice the above arithmetic processes, let us limit ourselves to the representations of the complement and consider the numbers $A = (57)_{10} = (111001)_2$ and $B = (43)_{10} = (101011)_2$. The operation of addition does not present any difficulty and the result is $A + B = (57)_{10} + (43)_{10} = (111001)_2 + (101011)_2 = (1100100)_2 = (100)_{10}$. For the subtraction $A - B = 57 - 43$, which will be performed as $A + (-B) = 57 + (-43)$ the number A is used as is, whereas for the number $-B = -43$ the one's complement of the number $B = (43)_{10} = (101011)_2$ will be used, which results from simply reversing the bits, and leads to the binary number (010100). Consequently, the subtraction $(111001)_2 - (101011)_2$ is reduced to the addition $(111001)_2 + (010100)_2 = (1001101)_2$. We observe that the result of the addition has *one extra bit* compared with the two numbers (the first bit from the left), which is removed from there and added to the most significant bit of the number remaining after this removal, which is $(001101)_2$. The result of this addition is the number $(001101)_2 + (000001)_2 = (001110)_2 = (14)_{10}$, which, of course, is correct, since in the decimal system it is $57 - 43 = 14$. Finally, for the subtraction $B - A = 43 - 55$, which will be performed as $B + (-A) = 43 + (-57)$, the number B is used as is, whereas for the number $-A = -57$ the complement to 1 of the number $A = (57)_{10} = (111001)_2$ will be used, which results from *simple inversion* of the bits and leads to the number $(000110)_2$. Consequently, the subtraction

$(101011)_2 - (111001)_2$ is reduced to the addition $(101011)_2 + (000110)_2 = (110001)_2$. We observe that in this result the first bit is 1 and therefore the result is *negative*. In order to determine which number this is, we must take the complement to 1 of the binary number containing the remaining bits, which is $(10001)_2$. This complement resulting from the inversion of the bits is the number $(01110)_2 = (14)_{10}$ and therefore the final result is -14, which is certainly correct.

How do the above differ, if instead, we use *the representation of the complement to* 2? In this case, the process will be applied as follows (with the results, of course, being the same). The positive numbers $A = (57)_{10} = (111001)_2$ and $B = (43)_{10} = (101011)_2$ will be used as they are, whereas for the negative numbers the complement to 2 of the opposite positive numbers will be used, which results from the complement to 1 by adding the unit to the least significant bit. Using the results we constructed previously, the number -43 is represented as $(010100)_2 + (000001)_2 = (010101)_2$, whereas the number -57 is represented as $(000110)_2 + (000001)_2 = (000111)_2$. As for the operations, the addition will be done exactly as before. The subtraction $A - B$ will be performed as $A + (-B)$ and therefore the subtraction $(111001)_2 - (101011)_2$ is reduced to the addition $(111001)_2 + (010101)_2 = (1001110)_2$. We observe that as before, the result of the addition has *one extra bit* in relation to the two numbers (the first bit from the left), which, however, *is not added* to the number consisting of the remaining bits, but is simply *ignored*. Therefore, the final result is $(001110)_2 = (14)_{10}$. Similarly, for the case of the subtraction $B - A = 43 - 57$, which will be performed as $B + (-A) = 43 + (-57)$, the number B is used as it is, whereas for the number $-A = -57$ will be used as the complement to 1 of the number A, which is $(000111)_2$. Therefore, the subtraction $(101011)_2 - (111001)_2$ is reduced to the addition $(101011)_2 + (000111)_2 = (110010)_2$. We observe that in this result the first bit is 1 and therefore the result is *negative*. In order to determine which number this is, we must take the complement with respect to 2 of the binary number that contains the remaining bits and is $(10010)_2$. This complement, which results from adding to the complement with respect to 1, which is (01101), 1 has the value $(01110)_2 = (14)_{10}$. Therefore, the result of the subtraction in the decimal system is the number -14, which is certainly the same as before.

The *representation of the complement with respect to* 2, is characterized by two very important properties regarding the implementation of addition, which make it an attractive solution, as well as the *dominant way of representation* of numbers in binary form in modern electronic computers. First, in the case of serial addition, which consists in adding the pairs of corresponding bits of the numbers being added, one pair at a time, *the operation of addition can be performed, starting from the addition of the least significant bits of the two numbers, without knowing either the signs of the two numbers or the correlation between them regarding their values*. Second, during the operation of adding more than two numbers, *if the value of the correct final sum is such that it is possible to represent it by the number of bits that characterizes the architecture of the underlying digital system, this sum is guaranteed to be calculated correctly, regardless of the order with which the addition of the numbers will take place* and indeed, even in the case in which during the calculation of some of the intermediate partial sums, the phenomenon of overflow (note that this second property also characterizes the representation of the complement to 1). This in other words means that the system has the ability *to recover from overflow errors* that have occurred in earlier times of the system's operation. The basic consequence of this property is that in those systems that use the representation of the complement to 1, or the representation of the complement to 2, *the only points at which we need to ensure that overflow will not occur are the inputs of the multipliers* and this is the reason why, as we see in the relevant section, the act of scaling the signal that ensures the prevention of overflow phenomena is applied exactly at these points.

12.6.2 Serial and Parallel Adders

At the hardware level, the operations of addition and subtraction are performed by a digital circuit, which is known as an *adder*, the various forms of which are depicted in Fig. 12.13. This adder is characterized as *full* in order to be distinguished from the *half-adder*, which accepts two bits at its input, adds them, and displays at its output both the result of the addition and the carry bit, which has a value equal to 0 if no carry arises and a value of 1 in the opposite case (remember that the carry appears only when both input bits have the value 1). In other words, the half-adder circuit constitutes *a system of two inputs and two outputs*, which simply implements the truth table of the binary addition operation. Conversely, the circuit of the *full adder* accepts an additional input, which contains the carry from the previous addition, if any. In other words, the full adder accepts three inputs and more specifically, the two bits that are to be added *together with the carry from the previous addition* (which will have the value 0 or the value 1), as well as two outputs, which are the same as before. In Fig. 12.13, the top left image is the schematic diagram of the full adder, the top right image is one of the many equivalent logical circuits of the full adder, which operates according to the truth table of the bottom right image, whereas the bottom left image, depicts the corresponding actual circuit diagram. For the structure of the circuits that implement the gates used in this, the reader is referred to the bibliography.

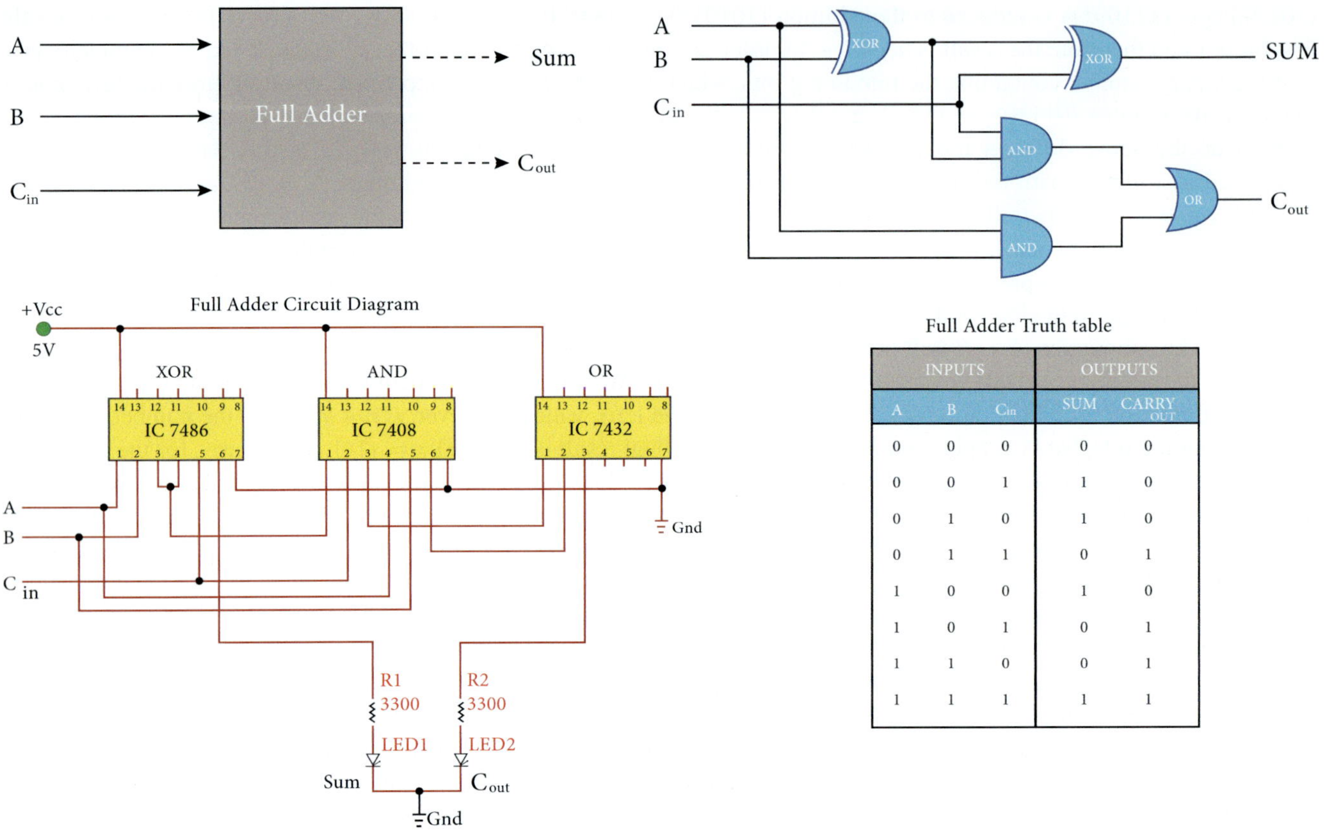

INPUTS			OUTPUTS	
A	B	C_{in}	SUM	CARRY OUT
0	0	0	0	0
0	0	1	1	0
0	1	0	1	0
0	1	1	0	1
1	0	0	1	0
1	0	1	0	1
1	1	0	0	1
1	1	1	1	1

Fig. 12.13 The structural diagram of the full adder along with the logic diagram and the corresponding logic circuit

The simplest case of an adder circuit is the *serial adder*, which sequentially adds the pairs of corresponding bits of two numbers, one after the other. This system is an excellent choice in terms of features, such as the *simplicity of the structure* of the hardware and the *power consumption*; however, it is characterized by a very low processing speed. This adder receives input from two *shift registers*, which contain the bits of the two binary numbers that are to be added, with the corresponding bits of the two numbers to *be progressively transmitted* to the inputs of the adder, starting from the *least significant bit*, in order to perform the addition operation between them. The operation of the system is guided with the help of a clock, the period of which dictates the *timing of sending the pairs of bits to the circuit of the adder*, one after the other. The carry-out bit that appears at one of the two outputs of the adder after the completion of the addition operation between the current pair of bits being added is stored in a D flip-flop, so that, before the addition of the next pair of bits, it is sent to the third input of the adder, in order to be used as the carry-in during the execution of the next addition. The carry bit is transmitted to an AND logic gate along with a control signal, which triggers the *initialization* of the flip-flop, essentially *zeroing* the carry bit, when the adder is about to start adding another pair of numbers. This circuit of the adder that performs addition of the complement to 2 is depicted in the left image of Fig. 12.14, whereas the right image depicts the circuit that performs *the subtraction* of two numbers. Considering that subtraction is performed using the complement to 2 for the negative number, this circuit contains a NOT logic gate located after the *shift register*, which contains the bits of the negative number, thus causing their inversion, whereas subsequently, to this inverted value that has emerged, the unit is added (as dictated by the way of calculating this complement), which comes from a NAND gate, replacing the AND gate in the adder's circuit. In this way, at the start of the process, the carry input bit is set to the value 1, which is added to the *inverted* number, so as to calculate the complement to 2. However, this process requires the use of a second NOT logic gate, which *inverts* the value of the carry output stored in the flip-flop. Indeed, since the NAND gate causes *inversion (or reversal) of value*, it also requires a second such reversal, in order to calculate and use the correct value. If the numbers are not represented according to the complement of 2, but according to the *sign-magnitude representation*, the adder is more complex, since the rules for addition and subtraction are different for this type of representation and for this reason, usually the above adder is used along with a system for converting the value of the number from one representation to the other.

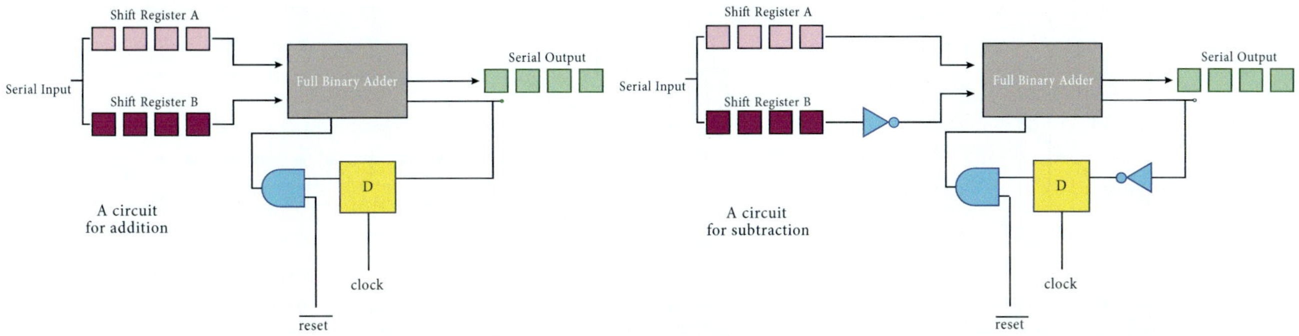

Fig. 12.14 The structural diagram of the serial adder (left) and the modifications that need to be made so that instead of addition it performs subtraction (right)

As we mentioned earlier, the main disadvantage of the serial adder is *its very low operating speed*, since in each clock cycle the addition between the corresponding bits of the two numbers is performed. For this reason, in practice *parallel adders* are used, which result *from the serial connection of the appropriate number of serial adders*. For example, by connecting in a series eight one-bit serial adders, we can construct an eight-bit parallel adder, which can calculate the sum of two eight-bit numbers each, with each adder calculating *the sum of the corresponding bits of the two inputs*. This process significantly accelerates the procedure, since for an adder of numbers of length N bits driven by a clock of frequency f, the parallel adder resulting from the serial connection of N serial adders is able to perform f additions per second, whereas the corresponding performance of the serial adder is only f/N additions per second, with the increase, however, of speed, obviously also requiring the *increase in hardware complexity*. However, the main disadvantage of the parallel adder has to do with the handling of carries. Remember, that in each of the additions that are made between the corresponding bits of the two numbers that are added, a carry-in is used, whereas a *carry-out* is created, with the values of these carries having to traverse *the entire sequence* of these adders in such a way that the *carry-out* of the serial adder that performs the addition between the pair under number $i-1$ of the corresponding bits, becomes the carry-in of the serial adder that performs the addition between the under number i pair of bits. This process of carry propagation from one stage to the next is, if nothing else, quite time consuming and for this reason efforts have been made to smooth out this problem and generally increase the performance of these calculations of sums, with the obvious cost of increasing the degree of complexity of the involved material. A first attempt refers to the so-called *carry-save adders*, which reduce the problem of adding three numbers of length k bits to the problem of adding two numbers. These arrangements constitute *a parallel collection of k full adders*, among which *there are no horizontal connections* (see the left image of Fig. 12.15) that accept three integer numbers A, B, and C at their input and produce two integer numbers C' and S at their output, such that $D+S=A+B+C$, with the ith bit of the number S and the $i+1$th bit of the number D being calculated as $S_i = A_i \otimes b_i \otimes C_i$ and $D_{i+1} = A_i B_i + A_i C_i + B_i C_i$, where $\otimes$ is the operation of exclusive logical disjunction (XOR). The main feature of these arrangements is that *the total propagation delay is equal to the sum of the propagation delays of three gates regardless of the number of bits of the numbers being added*, whereas the carry is not propagated from right to left at the same level as in the usual architecture of parallel adders but *from top to bottom*, until the last level of addition, keeping it at each level. In other words, each intermediate addition in the carry-save process combines three input numbers and computes a sum in which carries are generated from the same bit level but only based on previous levels. These carries are then propagated to the next level. A second attempt concerns the adders that use the *mechanism of pipelining* (see the right image of Fig. 12.15)[8] in which *there is no requirement for synchronization between the bits of a word*, whereas the carry of some of the stages of pipelining is not required for the operation of the next stage, but only during the next time slot. A third attempt is *the pre-calculation* of all the carries that characterize the realization of an addition (carry look-ahead), which are therefore all known from the beginning and there is no reason for them to be transmitted from one stage to the next. This method works in the following way. Let us symbolize with x_i and y_i the bits number i of the two numbers we want to add and suppose that in this addition, which will be performed by the serial adder number i, we will use the carry c_i that has arisen during the addition of the bits x_{i-1} and y_{i-1}, which has been performed by the stage number $i-1$. In this case, resorting to the truth table of Fig. 12.13 that describes the full adder, it is not difficult to verify that the above quantities are related to each other as

[8] For adders of this type see Dadda L. & Piuri V., *Pipelined Adders*, IEEE Transactions on Computers, Vol 45, No. 3, pp. 348–356, March 1996.

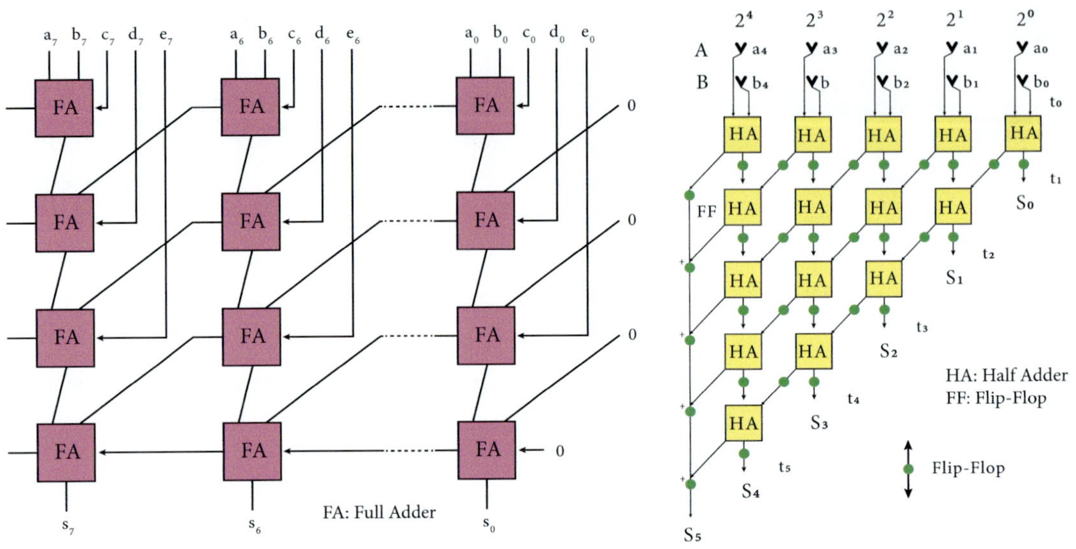

Fig. 12.15 Carry-save adder (left) and pipelined adder (right)

$$c_i = x_i \cdot y_i + (x_i \otimes y_i) \cdot c_{i-1}$$

with the symbols $\cdot$ and $+$ describing the operations of logical conjunction (AND) and logical disjunction (OR) and the symbol $\otimes$ describing the logical operation of exclusive disjunction (XOR). Constructing the above equation for each of the $N-1$ carryovers involved in the addition of two binary numbers of length N bits, we end up with a set of $N-1$ independent equations from which we can calculate the values of the carryovers. Considering, for example, a four-bit parallel adder, the iterative application of the above relationship gives us

$$\begin{aligned} c_1 &= x_1 \cdot y_1 \\ c_2 &= x_2 \cdot y_2 + (x_2 \otimes y_2) \cdot c_1 \\ c_3 &= x_3 \cdot y_3 + (x_3 \otimes y_3) \cdot c_2. \end{aligned}$$

Although the equations in the above form are not particularly useful, since the carryover c_3 cannot be calculated without c_2, which in turn requires the calculation of c_1, however, if we rearrange them and formulate them as follows:

$$\begin{aligned} c_1 &= x_1 \cdot y_1 \\ c_2 &= x_2 \cdot y_2 + (x_2 \otimes y_2) \cdot x_1 \cdot y_1 \\ c_3 &= x_3 \cdot y_3 + (x_3 \otimes y_3) \cdot x_2 \cdot y_2 + (x_3 \otimes y_3) \cdot (x_2 \otimes y_2) \cdot x_1 \cdot y_1 \end{aligned}$$

We can calculate *in advance* all the carryovers of the addition operation from the values of the bits x_i and y_i ($i = 1, 2, \ldots, N$), which therefore, *do not need to be transferred from one adder to the next*, as happens in the above implementation. However, the cost of this approach is *the increase in the degree of complexity of the circuit*. If the addition of longer words is required, then the delay due to the operation of the gates *does not increase*; however, the cost of the hardware increases sharply, a fact that makes this method practically inapplicable for long words. This situation is demonstrated in Fig. 12.16, with the left image depicting the combinatorial circuit for calculating the carryovers for the addition of binary numbers of length $N = 3$ bits and the right image depicting the corresponding circuit for the case of numbers of length $N = 4$ bits. Notice how the increase in the length of the number by a single bit led to a noticeable increase in the number of gates and generally, the complexity. For this reason, an intermediate solution is *the division of words into groups of shorter length words* and the transmission of carryovers only between these groups and not for each bit separately, whereas in the general case, as usual, we resort to a kind of *compromise* between the additional hardware and the time delays resulting from the use of the gates.

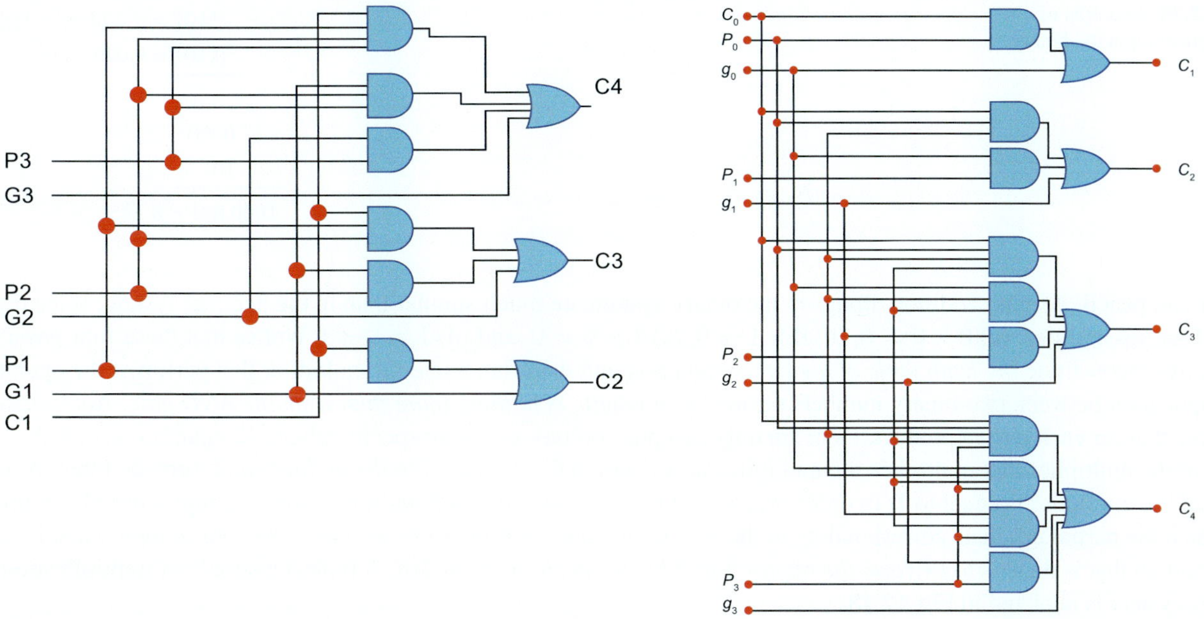

Fig. 12.16 The combinational circuit that allows the pre-calculation of the carries for the addition of numbers with length $N = 3$ bits (left) and $N = 4$ bits (right). The inputs p_i and g_i used in these circuits are defined as $p_i = x_i \cdot y_i$ [carry generate] and $g_i = x_i \otimes y_i$ $(i = 1, 2, \ldots, N)$ [carry propagate]

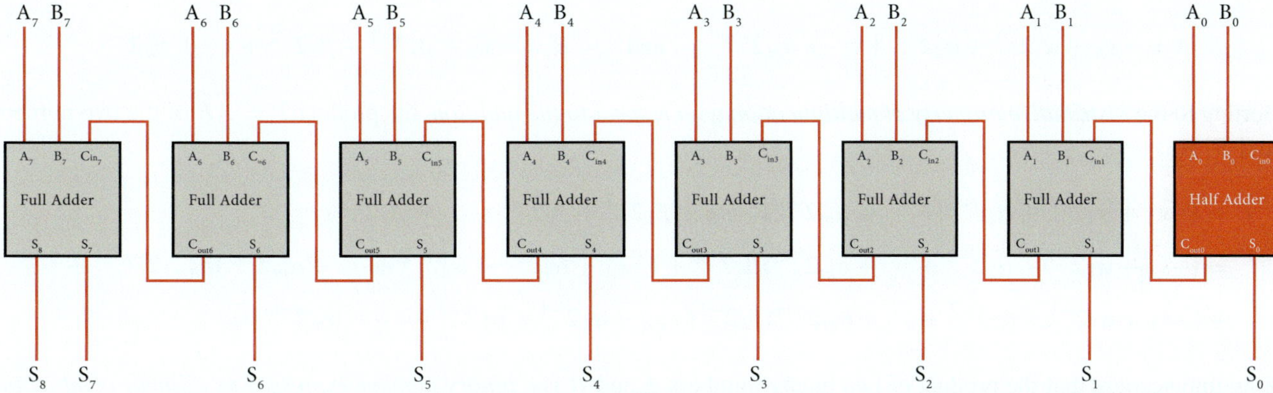

Fig. 12.17 Schematic diagram of a parallel adder that allows the addition of binary numbers of eight bits in length

A typical example of a parallel adder that allows the addition of numbers of eight bits each is shown in Fig. 12.17. Note that because in the addition of the least significant bits, no carry is used from a previous addition, since this is the first addition that is made, a *half-adder* can be used instead of a full adder, which does not include an input for the carry-in. However, usually this first adder is also a full adder, since in some architectures, such as MIPS, the carry-in line of the first adder that is not used, provides the value 1 that will be added to the one's complement of the subtrahend of a subtraction operation, so as to calculate the two's complement that will be used for the execution of this operation.

12.6.3 The Operation of the Multiplication

Using the operation of addition, we can perform the operation of *multiplication*, which is performed *exactly as in the decimal system*. In other words, *we multiply the multiplicand by each digit of the multiplier, moving from right to left and at the end of the process we add the partial products, which are written shifted by one position to the left*, as required by the fact that the numerical system we use is a positional system. This shift by one bit to the left is done with the help of a flip-flop, whereas the addition of the partial products will be performed one product at a time and not in a single step, as we do when we use

Fig. 12.18 Example of multiplication in the binary system

```
    1101 =    13 = data word
   0.101 = 0.625 = coefficient
   -----
    1101
   0000
  1101
--------
1000.001 = 8.125 = product.
```

paper and pencil. As with addition, things in the binary system are much simpler than in the decimal system, since there are only four possibilities: (a) $0 \times 0 = 0$, (b) $0 \times 1 = 0$, (c) $1 \times 0 = 0$, and (d) $1 \times 1 = 1$. Notice that these four possibilities collectively constitute *the truth table of logical conjunction* and therefore a one-bit multiplier that performs the operation of multiplication between two binary numbers of one bit in length, *is nothing more than a simple AND gate*. Additionally, we observe that for each partial product, there are only two possibilities and more specifically, *to be equal to zero*, if the current digit of the multiplier that multiplies the multiplicand is equal to 0, or *is equal to the multiplicand*, provided that the current digit of the multiplier is equal to 1. In order to speed up the process, advanced multiplication techniques have been invented, in which the partial products corresponding to the zero digit value multiplier *are not performed*, since their value is equal to zero and all that is done is *to perform the required shift by one position to the left*. A typical example of multiplication in the binary system is depicted in Fig. 12.18.

It is not difficult to understand that the way in which multiplications are performed depends on the type of representation used and let us present at this point the process for the case of the complement to 2 (see Diniz et. al, page 676). Let us consider for this purpose, two numbers A and B of n and m bits in length respectively, which in this type of representation are formulated as $s_A\alpha_1\alpha_2\cdots\alpha_m$ and $s_B\beta_1\beta_2\cdots\beta_n$ or equivalently, as we have commented, as

$$A = -s_A + \alpha_1 2^{-1} + \alpha_2 2^{-2} + \cdots + \alpha_m 2^{-m} \qquad \text{and} \qquad B = -s_B + \beta_1 2^{-1} + \beta_2 2^{-2} + \cdots + \beta_n 2^{-n}.$$

Resorting to the *distributive property of multiplication with respect to the addition*, the product $P = AB$ of the two numbers A and B, is

$$\begin{aligned} P &= (-s_A + \alpha_1 2^{-1} + \alpha_2 2^{-2} + \cdots + \alpha_m 2^{-m})(-s_B + \beta_1 2^{-1} + \beta_2 2^{-2} + \cdots + \beta_n 2^{-n}) \\ &= (-s_A + \alpha_1 2^{-1} + \alpha_2 2^{-2} + \cdots + \alpha_m 2^{-m})\beta_n 2^{-n} + (-s_A + \alpha_1 2^{-1} + \alpha_2 2^{-2} + \cdots + \alpha_m 2^{-m})\beta_{n-1} 2^{-n+1} + \cdots \\ &+ (-s_A + \alpha_1 2^{-1} + \alpha_2 2^{-2} + \cdots + \alpha_m 2^{-m})\beta_1 2^{-1}(-s_A + \alpha_1 2^{-1} + \alpha_2 2^{-2} + \cdots + \alpha_m 2^{-m})s_B. \end{aligned}$$

Taking into account that the product of two binary numbers A and B is *a binary number expressed as a linear combination of powers of* 2, we should perform the above operation in such a way that it is possible to express the result in the form of such a linear combination. This is something that can be done *step by step*, multiplying the number A by each term of the number B and performing the appropriate factorizations. Specifically, in the first step we multiply the number A by the last two terms of B, which are the terms with coefficients β_n and β_{n-1}. It will then be

$$\begin{aligned} C &= \beta_n 2^{-n} A + \beta_{n-1} 2^{-n+1} A \\ &= (-s_A + \alpha_1 2^{-1} + \alpha_2 2^{-2} + \cdots + \alpha_m 2^{-m})\beta_n 2^{-n} + (-s_A + \alpha_1 2^{-1} + \alpha_2 2^{-2} + \cdots + \alpha_m 2^{-m})\beta_{n-1} 2^{-n+1} \\ &= 2^{-n+1}\Big[\beta_n 2^{-1}(-s_A + \alpha_1 2^{-1} + \alpha_2 2^{-2} + \cdots + \alpha_m 2^{-m}) + \beta_{n-1}(-s_A + \alpha_1 2^{-1} + \alpha_2 2^{-2} + \cdots + \alpha_m 2^{-m})\Big]. \end{aligned}$$

We now observe that the first term within the above brackets is the

$$M = \beta_n 2^{-1}(-s_A + \alpha_1 2^{-1} + \alpha_2 2^{-2} + \cdots + \alpha_m 2^{-m}) = \frac{1}{2}\beta_n(-s_A + \alpha_1 2^{-1} + \alpha_2 2^{-2} + \cdots + \alpha_m 2^{-m}) = \frac{N}{2},$$

where $N = \beta_n(-s_A + \alpha_1 2^{-1} + \alpha_2 2^{-2} + \cdots + \alpha_m 2^{-m})$ and according to Eq. (12.11) (see page 1138) it will be represented as

$$M = \beta_n(-s_A + s_A 2^{-1} + \alpha_1 2^{-2} + \alpha_2 2^{-2} + \cdots + \alpha_m 2^{-m-1}).$$

Substituting in the above expression, we get

$$C = 2^{-n+1}\left[\beta_n(-s_A + s_A 2^{-1} + \alpha_1 2^{-2} + \alpha_2 2^{-2} + \cdots + \alpha_m 2^{-m-1}) + \beta_{n-1}(-s_A + \alpha_1 2^{-1} + \alpha_2 2^{-2} + \cdots + \alpha_m 2^{-m})\right].$$

Performing the operations and grouping the terms that correspond to the same power of 2, the above expression can be reformulated as

$$C = 2^{-n+2}(-s_C + \gamma_1 2^{-1} + \gamma_2 2^{-2} + \gamma_3 2^{-3} + \cdots + \gamma_{m+1} 2^{-m-1} + \gamma_{m+2} 2^{-m-2})$$

with the coefficients γ_i $(i = 1, 2, \ldots, m+2)$ depending both on the coefficients α_i $(i = 1, 2, \ldots, m)$ and on the coefficients β_{n-1} and β_n, whereas, because the sum of two positive numbers is always positive, and the sum of two negative numbers is always negative, it follows that $s_C = s_A$.

In the next step of the process we should add to the number C the product $\beta_{n-2}2^{-n+2}A$ and working in the same way we find

$$\begin{aligned} D &= (-s_A + \gamma_1 2^{-1} + \gamma_2 2^{-2} + \cdots + \gamma_{m+2} 2^{-m-2})2^{-n+2} + (s_A + \alpha_1 2^{-1} + \alpha_2 2^{-2} + \cdots + \alpha_m 2^{-m})\beta_{n-2}2^{-n+2} \\ &= 2^{-n+3}(-s_A + \delta_1 2^{-1} + \delta_2 2^{-2} + \cdots + \delta_{m+2} 2^{-m-2} + \delta_{m+3} 2^{-m-3}). \end{aligned}$$

This process continues *iteratively* until all terms appearing in the expansion of B as a linear combination of powers of 2, multiplied by A, are added. Special mention should be made of the last term, which must be added to complete the process, i.e. $(-s_A + \alpha_1 2^{-1} + \alpha_2 2^{-2} + \cdots + \alpha_m 2^{-m})s_B$. Specifically, if the number B is positive, the value of s_B is zero and the process has essentially been completed in the previous step, whereas conversely, if this number is negative, then the value of s_B is equal to 1 and the addition of the term $-s_A + \alpha_1 2^{-1} + \alpha_2 2^{-2} + \cdots + \alpha_m 2^{-m}$ must take place. It is worth noting that the product of two binary numbers of $(m+1)$ bits and $(n+1)$ bits in length (where we have also included the sign bit), is a number of $m+n+1$ bits in length and if we wish to represent the final result using fewer bits, we must proceed to a *quantization process*, which can be performed either by *truncation* or by *rounding*, as we describe in the relevant section (see Sect. 12.8 on page 1156). If, for example, we wish to represent the final result using $m+1$ bits, as many as are used for the representation of the number A, we must remove the bits located at positions $m+1, m+2, \ldots, m+n$. In the case of truncation, all we have to do is *ignore* these bits, whereas in the case of rounding, we should add to the result of the multiplication the value $\Delta = 0.000\cdots 1$, with the number of digits after the decimal point being equal to m, that is, essentially, *to add the unit to the bit of the product of the two numbers* that is in position $(m+1)$, that is, to that bit that is n positions away from the bit number $m+n$. In reality, it does not matter if this bit is added to the last or the first sum of the above series of sums and an equivalent process to perform this rounding is to add the number $1.000\cdots 0$, with the number of zeros being equal to $n-1$ to the first partial sum. Also note that since we will ultimately only keep the $m+1$ bits from the final product, we can keep the first $m+1$ bits from each partial sum.

12.6.4 Serial and Parallel Multipliers

The execution of the multiplication operation between two binary numbers is carried out with the help of suitable *multipliers*, the structure and operation of which in the general case can be characterized by significant variations. The simplest such arrangement is the so-called *serial multiplier*, in which *each partial sum is calculated with the help of a serial adder*, whereas the summation of the partial sums in the way imposed by the way of performing the multiplication operation is carried out with the help of *delay circuits*. A typical example of the structure of such a serial multiplier that performs multiplication between a coefficient and a data signal, a situation that constantly appears in practice, is depicted in Fig. 12.19. Note that although the handling of data values is done with *serial mode*, the handling of the values of the coefficients is done in parallel and for this reason, this multiplier, in reality, is characterized as *serial-parallel multiplier*. However, despite this parallelism, this arrangement is characterized by two inherent problems, more specifically: (a) *the serial connection of the adders requires the signals to pass through many gates during each bit period, which implies significant delays*, and (b) *although in the general case there is the requirement for the input and output signals to have the same length* (this requirement stems from the fact that the values of successive samples are adjacent in time and there are no more available time slots for storing a larger number of bits), that is, their values are encoded by the same number of bits; *however, this is not possible, since we know that the length of the product is equal to the sum of the lengths of the two numbers being multiplied.* The first problem

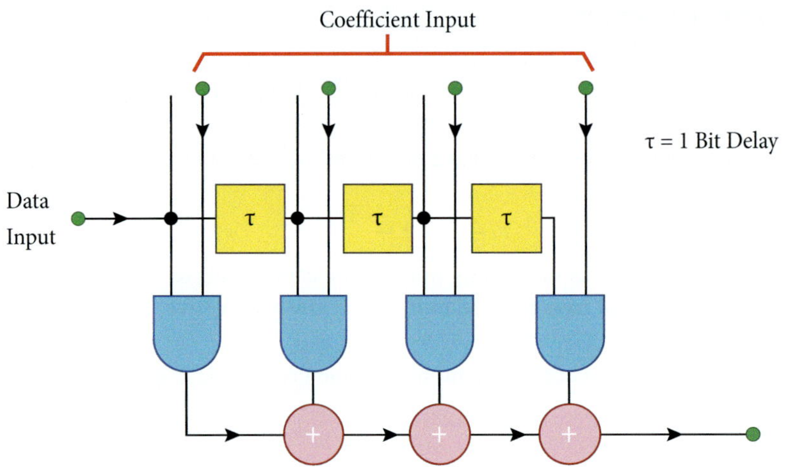

Fig. 12.19 Schematic diagram of a serial multiplier (Freeny, Figure 11)

is solved by resorting again to the mechanism of *pipelining*, which is based on the use of time-delay elements at each stage of the serial multiplier, which offer the ability to start calculating the value of a partial sum, as soon as the first bit of the value of the previous partial sum becomes available. This ability is something that, if nothing else, significantly speeds up the process of calculations. Indeed, although in the general case, the multiplication of two numbers $m + 1$ and $n + 1$ bits in length using a serial multiplier, it is proven that it requires at least $m + n + 2$ clock cycles even in the case where the final result has been quantized to $m + 1$ bits, the use of the pipelining mechanism makes it possible to calculate a quantized product of $m + 1$ bits in length, in just $m + 1$ clock cycles. On the other hand, the second problem is solved by proceeding to *rounding the value of the product* with the help of the appropriate signal. Note that although the above circuit operates using *specific values* for the filter coefficients, it can be easily modified, so as to support the use of a different coefficient for each input value, with the different values of the coefficients to be read from external memories and stored in registers that exist specifically for this purpose. Finally, although the above circuit operates for multiplication of numbers expressed in the *sign-magnitude* representation, it can, with only minor modifications, be used for the multiplication of numbers, the encoding of which is based on the representation of the *two's complement*.

Even greater acceleration can be achieved, using the so-called *parallel multipliers*, in which the involved data are circulated in the circuit in the *horizontal*, the *vertical*, and the *diagonal* direction, in an efficient way. Although these multipliers lead to *significant acceleration*, they are characterized by *relatively large size* and by *high power consumption* and for this reason they are used only in cases in which the speed and processing time significantly affect the performance of the underlying digital system. In this type of multiplier, initially the multiplication of the multiplicand with each bit of the multiplier is performed for the calculation of the value of the partial products, which are then added in parallel for the calculation of the final result. A typical example of a parallel multiplier is depicted in the left image of Fig. 12.20. In this figure, each of the processing elements is known as a *Hoffmann cell* and a typical structure of it is depicted in the right image of Fig. 12.20. In this array, the number of rows equals the number of bits of the coefficient value, with each of these bits controlling a line of elements. Specifically, if the value of this bit is equal to zero, then the elements of this line *simply forward* the sum of the partial products (moving in the vertical direction) without the slightest processing, whereas in the opposite case, the current input value (moving diagonally) *is added to the appropriate position of the sum of the partial products* from the line elements, which operate as a parallel adder. This structure can be modified to perform the appropriate quantization of the product value, but also to be used with both positive and negative numbers.

It is interesting to note that, apart from the above traditional approaches to implementing the multiplication operation, In the literature, other techniques for performing this operation can be found. An approach of particular interest,[9] involves the computation of the sums of the products between the filter coefficients and the internal signals of the digital system, *without the need for physical implementation of multiplier structures*. To demonstrate this idea, let us consider a set of numbers α_i $(i = 1, 2, \ldots, N)$, which are assumed to represent *filter coefficients*, as well as a second set of numbers x_i $(i = 1, 2, \ldots, N)$

[9] See for example Peled A. & Liu B., *A New Hardware Realization of Digital Filters*, IEEE Transactions on Acoustics, Speech, and Signal Processing, ASSP-22, 456–462, 1974 and Wanhammar L., *An Approach to LSI Implementation of Wave Digital Filters*, PhD thesis, Linköping University, Linköping, Sweden.

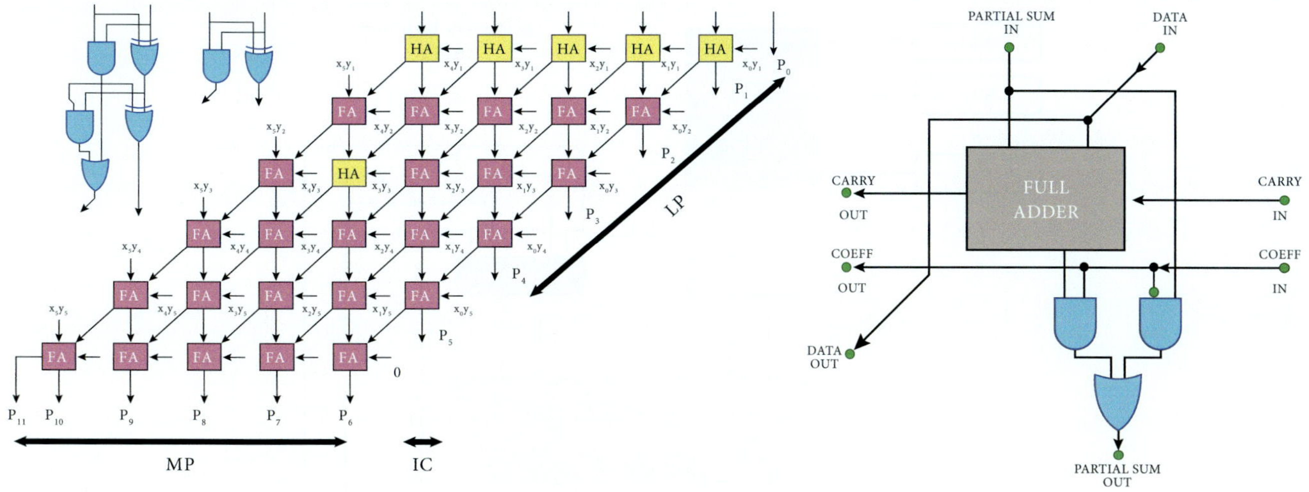

Fig. 12.20 Schematic diagram of a parallel multiplier (left) and Hoffmann cell (right) (Freeny, Figure 15)

that have been mapped to the interval $(-1, 1)$, which are assumed to represent the values of some internal signal. Let us also assume that these two sets of numbers are encoded in the binary system, using a total of $K + 1$ bits for their encoding, of which, as usual, the K bits represent the value of the number, whereas one bit represents the sign. In this case, the sum of the products $\alpha_i x_i$ $(i = 1, 2, \ldots, N)$ can be successively written as

$$w = \sum_{i=1}^{N} \alpha_i x_i = \sum_{i=1}^{N} \alpha_i \left(\sum_{j=1}^{K} x_{ij} 2^{-j} - x_{i0} \right) = \sum_{j=1}^{K} \left(\sum_{i=1}^{N} \alpha_i x_{ij} \right) 2^{-j} - \sum_{i=1}^{N} \alpha_i x_{i0}$$

$$= \sum_{j=1}^{K} g(x_{1j}, x_{2j}, \ldots, x_{Nj}) 2^{-j} - g(x_{10}, x_{20}, \ldots, x_{N0}) = \{[(g_K 2^{-1} + g_{K-1}) 2^{-1} + g_{K-2}] 2^{-1} + \cdots + g_1\} 2^{-1} - g_0.$$

In the above expression, where the *representation of the complement with respect to* 2 is used, the quantity x_{ij} $(i = 1, 2, \ldots, N$ and $j = 1, 2, \ldots, K)$ expresses the jth bit of the number x_i, the quantity x_{i0} expresses the first bit of this number, which, as is well known, represents the sign, whereas we have defined

$$g(z_1, z_2, \ldots, z_N) = \sum_{i=1}^{N} \alpha_i z_i .$$

Furthermore, for the sake of brevity in writing, we use the notation $g_j = g(x_{1j}, x_{2j}), \ldots, x_{Nj}$. According to the above equation, we can calculate the value of the sum of products w, *without resorting to the use of a multiplier system*, in the following way. Initially, we calculate the value of g_K and then divide the result by 2, a process that is carried out very easily and quickly, by performing a *right shift* of the value of g_K by one bit. Then, we add to the result that will emerge the value g_{k-1}, divide the result by 2, add to the value that will emerge the value of g_{k-2}, and continue in the same way. At the end of the process and in order to calculate the value of w, we should subtract the value of g_0 from the last result that will emerge. The advantage of this method has to do with the fact that the functions $g(z_1, z_2, \ldots, z_N)$ can take at most 2^N different values (since the quantities z_i $(i = 1, 2, \ldots, N)$ are binary numbers of one bit and therefore can only take the values 0 and 1), which of course also depend on the values of the coefficients α_i $(i = 1, 2, \ldots, N)$ and which can be calculated *in advance* and stored in a *ROM*, in order to be used in the calculations. The above process can be performed by a circuit like the one depicted in Fig. 12.21 (see Diniz, page 687) in which the values $x_1, x_2, \ldots, x_N$ are stored in the *shift registers* $R_1, R_2, \ldots, R_N$, which have a length of $k + 1$ bits and the N outputs of which the $x_{1,K-j}, x_{2,K-j}, \ldots, x_{N,K-j}$ collectively define the address of the ROM where the value of the function g_{K-j} is stored. After reading this value from the ROM, it is copied to register A in order to be added to the value that is already in register B and which is nothing else, from the *current cumulative value* of the final product that has been calculated up to this point. Then, the sum that is calculated, according to what we mentioned

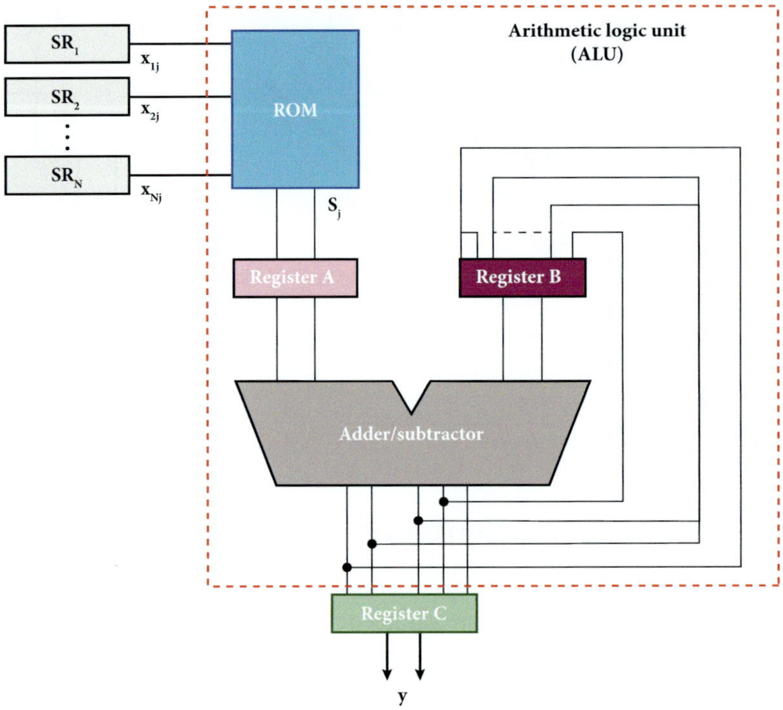

Fig. 12.21 Circuit of the distributed computation of product sums

previously, is divided by 2. At the start of the calculations, register A is initialized to the value g_K, whereas register B is initialized to the zero value. The final result, which is the value of w, is obtained by subtracting from the last partial result that has been calculated the value g_0 and is stored in register C. The above architecture can be improved in terms of its performance, by incorporating into it characteristics of *pipelining* or alternatively, reading the values of the quantities g_j not sequentially but *in parallel* and using *parallel adders* for the simultaneous calculation of the required partial sums. It is proven that the time interval required for the calculation of a sample of the output depends exclusively on the length K of each word and not on the number N of the partial products involved in the calculation of the value of w. Specifically, this time interval corresponds to $K + 1$ clock cycles, with the duration of each cycle defined as $\max(t_1, t_2)$, where t_1 is the time required for the access of the ROM memory and t_2 is the time required for the execution of an addition or subtraction. On the other hand, the size of the ROM obviously depends on the length K of the words, whereas the value of N *should not be too large*, in order not to have a long ROM access time. This type of distributed arithmetic computation is used for the implementation of many digital filter structures (see for example Diniz, pages 688–691).

12.7 The Appearance of the Overflow Phenomenon

The *overflow*, as a concept, is encountered in almost all applications that rely on the execution of calculations and is defined as *the appearance of a value that is greater than the maximum value that can be stored in a register, owing to its finite length* (note, however, that overflow can also occur in encoding processes when a value of the input signal is greater than the maximum value supported by the encoding system). For example, as we mentioned in a previous section, the maximum value of an unsigned integer that can be stored in a register of length L bits, is equal to $2^L - 1$ (for the value $L = 8$, this value is equal to 255). Therefore, if the addition of the numbers 200 and 125 takes place, the attempt to store the sum with a value of 325, in a register of length 8 bits, will lead to the appearance of *overflow*. The ability to detect this is therefore elevated to a factor of *capital importance*, in all the processes of designing digital filters. Overflow characterizes *all forms of number representation* used in digital systems and its detection can be carried out in the following way: (a) in the case of *addition*, overflow can occur *only when we add two numbers with the same sign* (i.e., two positive or two negative numbers), whereas it never appears when the two numbers have different signs. Taking into account that the sum of two positive numbers is always a positive number, whereas the sum of two negative numbers is always a negative number, it is not difficult to understand

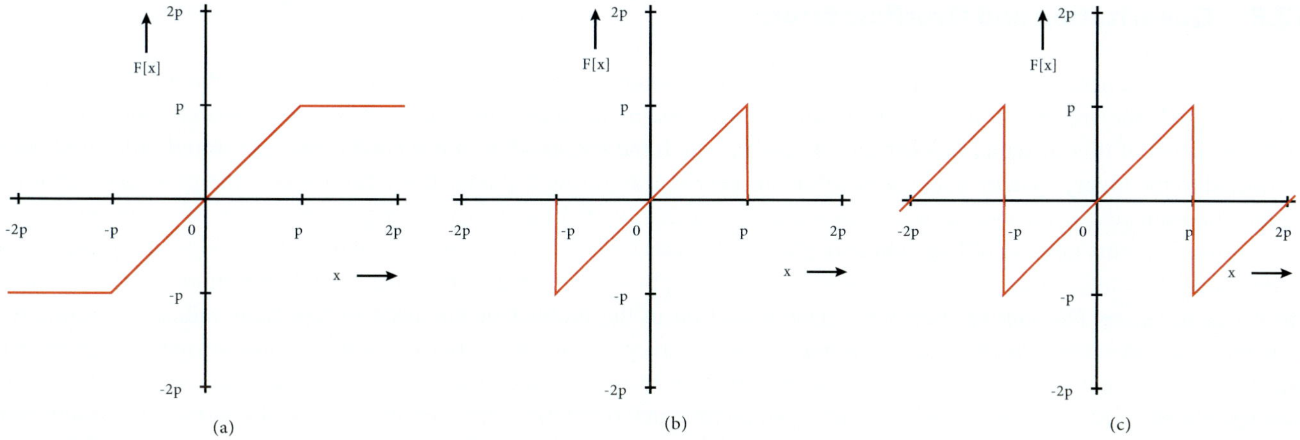

Fig. 12.22 Input–output characteristics for the three overflow recovery mechanisms, which are saturation (left), zeroing (middle) and the addition of a complement to 2 (right). (**a**) Saturation. (**b**) Zeroing. (**c**) Two's complement

that overflow occurs *when the sign of the sum is different from the sign of the numbers being added*. In other words, when the sum of two positive numbers is a negative number (i.e., when $(+A) + (+B) = -C$) or when the sum of two negative numbers is a positive number (i.e., when $(-A) + (-B) = +C$), this is a clear indication that overflow has occurred. (b) in the case of *subtraction*, if the two numbers used have different signs, overflow occurs *if and only if the remainder of the subtraction has the same sign as the subtrahend*. In other words, overflow occurs when $(+A) - (-B) = -C$ or alternatively, when $(-A) - (+B) = +C$.

The occurrence of overflow is a factor of *non-=linearity*, which affects the most significant bits of signal values and as we comment elsewhere, leads among other things to the appearance of *large-scale limit cycles*, which must be suppressed, as it can lead to quite serious problems. Speaking generally, there are three ways in which a system can recover from an overflow situation and which are described by characteristic overflow, similar to the characteristic quantization errors, and more specifically: (a) the *saturation*, which leads to the smallest deviation of the system from the conditions of normal operation, although during its manifestation, the filter ceases to be functional to a small or large extent; (b) the *nullification*, in which the output of the system takes the zero value if its input exceeds the threshold of overflow appearance and can be generalized so as to allow the reset of all states if even one of them leads to overflow; and (c) the *two's complement overflow*, which consists of *the periodic continuation of the straight line with a slope of* 45° and is characterized by the advantage that it allows *the automatic correction of errors* caused by overflow within the system. The common characteristic of all these overflow recovery techniques, is that the error induced owing to the manifestation of this phenomenon is reduced in such a way that *the output of the actual nonlinear digital system converges to the output of the ideal linear filter*. This type of behavior is known as the stability of the forced response and is studied in the relevant section (see Sect. 12.9.9 on page 1218).

The input–output characteristics for these three techniques for dealing with overflow are depicted in Fig. 12.22 and each of them is described by its own defining equation of the form $y = F(x)$ or $y = Q[x]$, where x and y are the input and output arguments, which have the same value, if there is no overflow and a different value, in the opposite case. The straight line with a slope of 45° that leaves the input value unaffected in the first case implies that these characteristics relate *exclusively to overflow*, since, as commented elsewhere, the processes of *overflow* and *quantization*, can be studied *separately and independently* from each other. Conversely, if we want to take into account these two processes, we should use the general characteristic curve that includes the contributions of both these processes. For example, the saturation characteristic is described by a behavior of the form $Q[x] = x$ if $|x| \leq M$ and $Q[x] = M$ if $|x| \geq M$. In other words, if the signal value is less than the maximum supported value M, then it remains as is, whereas in the opposite case, the signal value is cut off and becomes equal to M. The remaining two curves are described by similar equations. In the general case, these curves are approximated by the conditions $Q[x] = x$ if $|x| \leq M$ and $|Q[x]| \leq M$ if $|x| > M$. From the form of these curves, we find that the elements described by these input–output characteristics are *nonlinear* and therefore, a second reason for the appearance of nonlinear characteristics in a system is *the control of the unwanted characteristic of overflow*. However, in the case where it is $|x| \leq M$, then the nonlinear element operates in its linear region and behaves as an identity function—since it is $Q[x] = x$—and therefore, the nonlinear system is identified with the corresponding ideal linear system.

12.8 Quantization and Overflow Errors

As already mentioned, although in the description of digital filters we silently accept that the values of the filter's coefficients and the signals they receive at their input are known with *infinite accuracy*, practically this *is not possible* , since the length of the registers of digital signal processing systems in which the values of the filter coefficients are stored (after first being converted to the binary system, using some of the representation forms that have been described in the previous section), is *finite*. Consequently, if the system we use has registers of length K bits, then we must, by necessity, store the values that interest us using this number of bits, which in almost all cases, *is less than the number of bits of the original representation of the value*. It is interesting, how the knowledge of the length K of the registers of the digital system *uniquely determines the range of values that can be used by it*. This reduction in the number of bits used to represent values is essentially a process of *quantization*, which leads to a *reduction* in accuracy and the appearance of *quantization errors*. The process of quantization has been described in detail in the presentation sections of the process of converting an analog signal to digital (see specifically Sect. 9.8.3 on page 748) and as we commented there, the characteristic curve of the nonlinear arrangement of the quantizer, has the form depicted in Fig. 12.23. This quantizer is used for the quantization of values that follow the *fixed-point representation* and its operation is carried out as we have commented, in the following way: if the value received by the quantizer at its input is less than p (region A of the diagram), then its output is a value that is quite close to the input value and the differentiation that occurs between these values, lies in the fact that the input of the quantizer can be anything and is characterized by infinite accuracy, whereas its output is a value that can be represented by the underlying digital system. Conversely, if the input to the quantizer is greater than p (region B of the diagram), then its output cannot exceed the value p. From these two regions, region A corresponds *to the quantization of the values of the appropriate signal*, whereas region B is related to the correction that *is required*, when overflow phenomena occur. From the definition of these two processes, it is not difficult to understand that quantization affects *the least significant bits of numerical values*, whereas on the contrary, overflow affects *the most significant bits* and therefore, causes *much more severe distortion* to the signal compared with the process of quantization.

According to what we mentioned above, the representation of numerical values used in digital systems with finite accuracy requires the addition of *nonlinear arrangements* (which are known as *quantizers* and operate according to the characteristic of Fig. 12.23), at various points of the ideal linear and time-invariant digital system, which, therefore, acquires *nonlinear characteristics*. These characteristics make the study of the system *a relatively complex process* and for this reason, as we comment in detail later, in practice we replace the nonlinear arrangements of the quantizers, with *sources of linear noise*, a

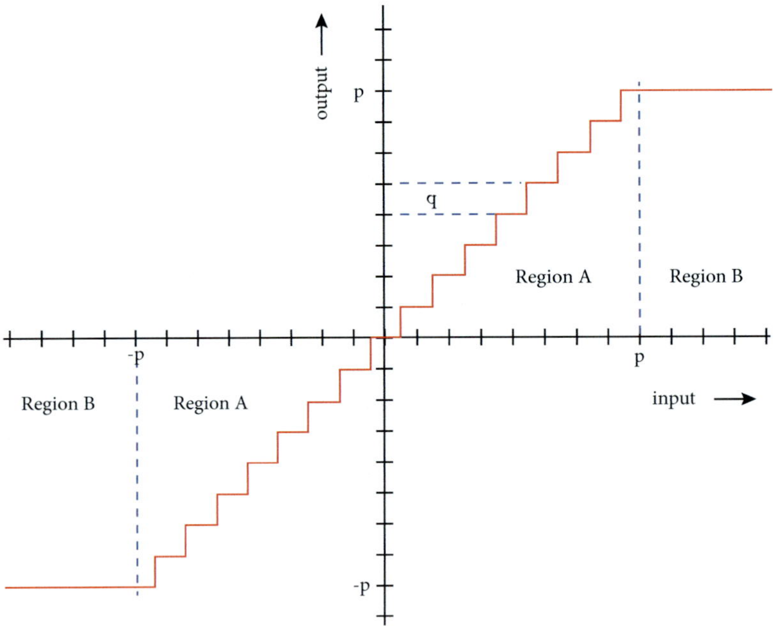

Fig. 12.23 Typical characteristic curve of a quantizer that allows the representation of the finite accuracy of numerical values so that their storage in finite-length registers is possible

process that significantly simplifies the analysis and makes it possible to calculate quantities, which under other conditions would be impossible. The question that arises quite reasonably at this point is *in which positions within the system should we add these quantizers?* The answer is easy: *at least one quantizer should be added to each feedback branch of the recursive digital system*. This is something that characterizes all the diagrams presented in the relevant sections. The other question that can also be asked has to do *with the way of studying the phenomena of quantization and overflow* that characterize the operation of a quantizer and are related to areas A and B of the Fig. 12.23. Should these phenomena be studied together, or separately and independently of each other? Although one would expect quantization and overflow, since they concern the same arrangement, to be studied together; in reality, as long as the values of the signals are represented by a sufficient number of binary digits, exactly the opposite happens. Indeed, it is interesting to note that if the values of the involved signals are very large, and more specifically, much larger than the quantization step, then in the study of overflow, *we can completely ignore the quantization process* and replace the stepped curve of region A of Fig. 12.23 with a straight line with a slope of 45° passing through the origin of the axes. Observing the characteristics of the three overflow recovery mechanisms depicted in Fig. 12.22, we find that things are indeed so. In other words, for signal values that are smaller than the overflow threshold, *the quantizer is considered a linear arrangement*, which essentially leaves the input signal unchanged, since, owing to the very large value of the signal and the very small quantization step, the difference between the initial value of the signal and the quantized value will be very small. Of course, for all the above to apply, as we commented, a *relatively large register length* must be used, which will lead to a correspondingly large value for the total number of quantization intervals. However, it should be noted that in some cases, this independent study of the two above phenomena *is not recommended*, since sometimes, the study of overflow phenomena requires the consideration and use of the stepped structure of region A of Fig. 12.23, as is, for example, the appearance of *limit cycles*, which are studied in the next section.

12.8.1 Quantization with Truncation and Rounding

Generally speaking, this quantization process can be carried out in two ways, specifically with *truncation* and with *rounding*. The truncation process, as its name suggests, *simply involves removing the extra bits*, so that only those that can be stored in the registers of the digital system are preserved. Considering, for example, a system with 8 bit registers and a number, which after the decimal point contains 12 bits, its storage is possible if we consider only the first 8 bits and ignore the remaining 4 bits (that is, in the general case, we remove the bits that are after the least significant bit of the representation, which in this example is the eighth bit after the decimal point). If, for example, this number is 0.1011010**1**1101 (for the reader's convenience, the eighth bit of interest is printed in bold) the result of this truncation process is the number 0.10110101. It is not difficult to understand that this process leads to *a reduction in accuracy and the appearance of quantization error*, since the decimal value of the number before truncation, is 0.710205, whereas the value that results after truncation is 0.707031, a difference that is significant. There are two types of truncation that depend on the way in which negative numbers are represented and more specifically, the *magnitude truncation* when the complement to 1 is used to represent negative numbers and the *value truncation*, when the complement to 2 is used to represent negative numbers, which is the one usually used in practice, although magnitude truncation is characterized by the advantage that it *does not allow the appearance of limit cycles*, as we comment in detail in the relevant section. These two truncation techniques are different from each other, since in magnitude truncation, the direction of quantization is *always towards zero* (therefore, for negative numbers it is upward, and for positive numbers it is downward-), whereas in value truncation, the direction of quantization is *always downward*. However, in *both* methods, the magnitude of the quantization error, i.e., the difference $E = Q(x) - x$ between the quantized value $Q(x)$ and the original unquantized value x is *always smaller* than the quantization step q.

On the other hand, the method of *rounding* consists of *reducing the size of a binary number*, but in such a way that the rounded value that results is *as close as possible to a value that can be represented in the digital system used*. This rounding process[10] is a combination of truncation and addition processes and is performed in the following way. Initially, the value of the number undergoes truncation, which is performed in the way we described, keeping only the appropriate number of bits, and then we examine the value of the bit, which in the initial value of the number (i.e., before the truncation) was *after the*

[10] It is interesting to note that the rounding process in some cases is performed under control conditions (controlled rounding, CR). In this case, the way in which the least significant bit is handled is not dictated by the value of the signal itself undergoing the quantization process, according to what we described above, but on the contrary is controlled by some other signal, which either is some external and in almost all cases, random signal, or one of the internal signals of the underlying digital system. In more complex cases, the decision about whether rounding is done upward or downward, is controlled by more than one control signal, one of which may be the signal itself undergoing quantization.

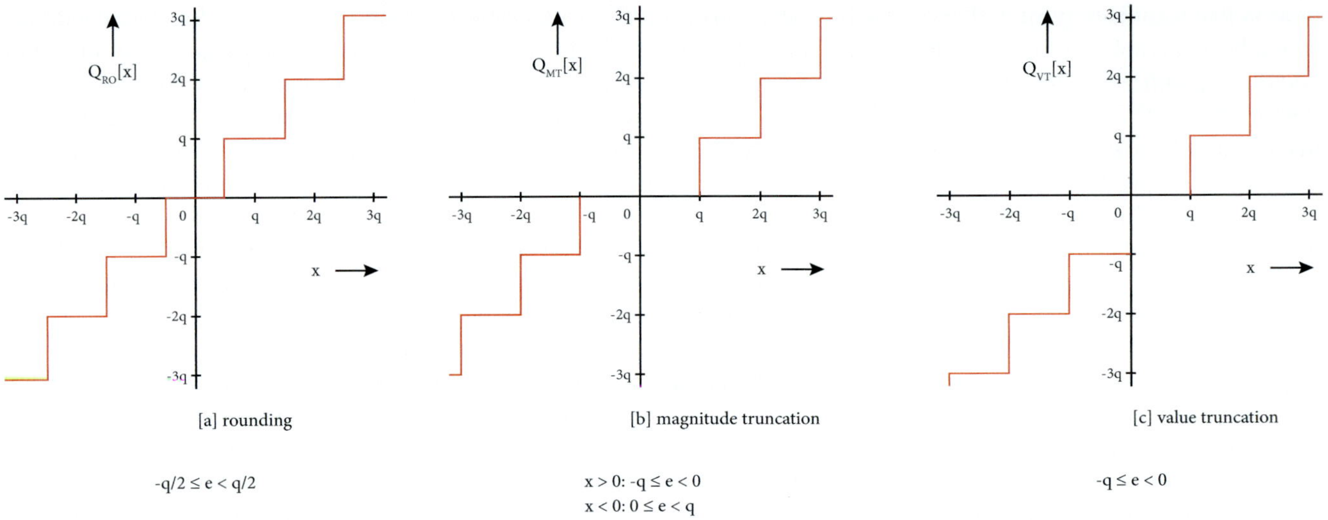

Fig. 12.24 The input–output characteristics for quantization with rounding (left) with value truncation (middle image) and with size truncation (right)

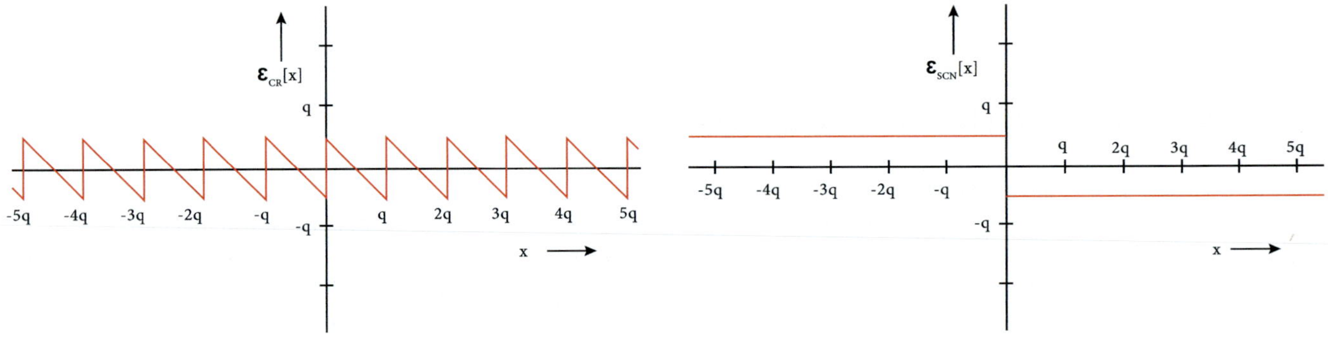

Fig. 12.25 The change in quantization error for rounding quantization (left) and for size truncation quantization (right) (Butterweck, 1979, Figure 2)

least significant bit of the number after the truncation. If this bit has a value 0 we do nothing (or equivalently, we add to the least significant bit the value 0), whereas if this bit is equal to 1, we add to the less significant bit the value 1. In the above example, the bit of the initial value that is after the eighth bit, that is, the ninth bit, has the value 1 and therefore, to perform the rounding, we must add to the eighth bit the value 1 (note that this eighth bit also has the same value). The result of this process is the binary number 0.10110110, which corresponds to the decimal number 0.7109375. Note that *although the resulting value is slightly larger than the original value before quantization, it is much closer to it.* Generally speaking, this rounding method leads to *better results than the truncation method,* since the measure of the quantization error is less than $q/2$ and not less than q, as happens in the two methods of truncation and for this reason, it is generally preferred in practice. It is interesting, however, to note that if we add to the original unquantized signal the value $q/2$, the methods of rounding and truncation of value *lead to the same result*. Figure 12.24 presents the input–output characteristics for these three quantization methods. On the other hand, Fig. 12.25 depicts the quantization error for rounding (left image) and for size truncation (right image).

Note that although in both cases, the situation that arises is *deterministic* in nature, in the sense that *each input signal is associated with a unique error signal*, generally speaking, we often tend to attribute to the rounding process a *degree of randomness* under the following sense. If we consider that the signal $x[n]$ represents a *stationary random process*(see Sect. 13.2.3 on page 1392) characterized by probability density $P(x)$ and autocorrelation function $r_{xx}[m] = E\{x[n]x[n-m]\}$, this process under the influence of the characteristic associated with the rounding error, transforms *into another random process* $y[n]$, which almost always has *white noise characteristics* with autocorrelation function $r_{yy}[n] = (\Delta^2/12)\delta[n]$ and uniform probability density in the range of values $-\Delta/2 \leq E \leq \Delta/2$, where Δ is the quantization step. This is particularly true for the case of *any non-periodic signal* the word length of which *is reduced* by a significant number of bits. In an

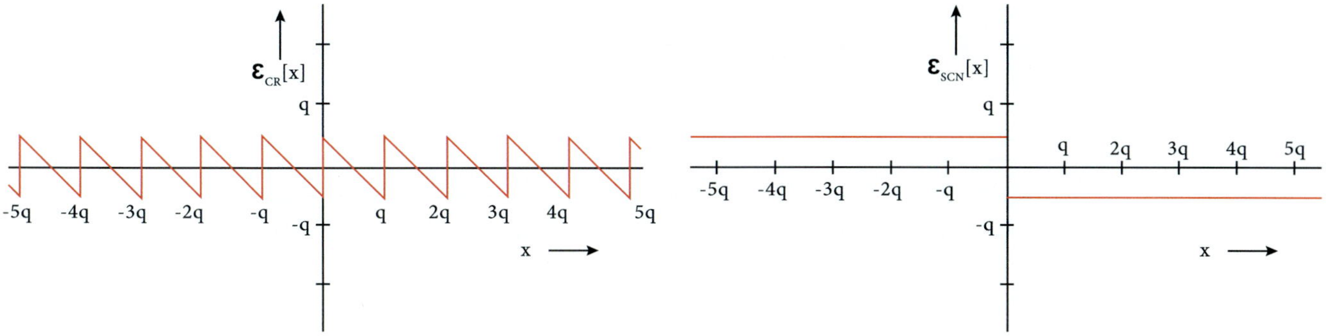

Fig. 12.26 The analysis of the characteristic error for the truncation of size into two parts, the first of which can be described by a white noise model whereas the second is non-linear and strongly related to the input signal (Butterweck, Ritzerfeld, and Werter, Figure 2.2)

equivalent description, the error signals generated by the quantizers can be modeled with quite high accuracy as *white noise*, which is *uncorrelated* both with all noise signals and with the signal being circulated in the filter, whereas the same applies to the noise appearing at the output of the filter. This property constitutes the basis on which *the statistical nature treatment of rounding error* is based, through a model *of white noise*, which is described in another section. On the other hand, when quantization is based on the technique of *size truncation*, then the individual noise sources correlate both with each other and with the circulating signal, whereas the same applies to the noise at the output of the filter. A solution to this problem, proposed by Claasen et al.[11] consists *in adding a linear distortion to the input signal* , which leads to a quantization noise signal with lower power than the original signal, which is orthogonal to the signal of the filter, but continues to be correlated with it. Butterweck went a step further, as by removing a linearly correlated signal from the quantized signal, he created a noise signal, which, *on the one hand, has even less power than the previous signal; on the other hand, it is uncorrelated with the filter signal.*[12]

On the other hand, the error characteristic for the case of size truncation, can be correlated with the rounding error characteristic, by analyzing it in two parts, both of which are depicted in Fig. 12.26 (note that the combination of the two curves of this figure, which consists in adding the values of the ordinates of the corresponding points of the two curves, leads to the error characteristic for size truncation depicted in the right image of Fig. 12.25). It is not difficult to observe, that the left image of Fig. 12.26, *looks very much like the characteristic error of rounding quantization* (since it is simply horizontally displaced by a distance $q/2$) and therefore, can be adopted for this, the *white noise model*. It is worth noting that the error of quantization through size truncation, *is strongly related* to the input signal, since these quantities have opposite signs, or in mathematical notation, $\text{sign}[e(x)] = -\text{sign}(x)$, although the addition of the term $(1/2)\text{sign}(x)$ to this error leads to the correlation of the characteristics we described previously. Therefore, rounding and size truncation*differ from each other essentially only in terms of the second part of the curve*, the sign-part, which is a nonlinear component that, as we commented, is highly related to the input signal. This component of the error is *strictly deterministic in nature*, in contrast to the other component, for which *it is not possible* to provide a statistical description, although in the case where the input signal is a purely random signal, this possibility exists for this error as well.

The white noise models for the errors related to the rounding and truncation of size are depicted in Fig. 12.27.

12.8.2 Value Ranges for Quantization Errors

Let us symbolize with E_t and E_r the quantization errors that result from the application of the truncation and rounding processes that are required for the representation of a binary number with value X of ℓ bits in length, using ℓ' bits with $\ell' < \ell$ and leading to a new value X'. Let us also consider a number of the form $0.xxxxxx$ encoded in the binary system, which has *only a fractional part*, since, as we mentioned, all numbers can be transformed into this form, when floating-point representation is used. In reality, the use of numbers expressed in this form is *mandatory*, since it is proven that *the use*

[11] See the reference Claasen T.A.C.M., Mecklenbräuker W.F.G., & Peek J.B.H., *Quantization Noise Analysis for Fixed Points Digital Filters Using Magnitude Truncation for Quantization*, IEEE Transactions on Circuits and Systems, vol. CAS-22, pp. 887–895, November 1975.

[12] See reference Butterweck H.J., *On the Quantization Noise Contributions in Digital Filters which are Uncorrelated with the Output Signal*, IEEE Transactions of Circuits and Systems, vol. CAS-26, No. 11, November 1979, pp. 901–910.

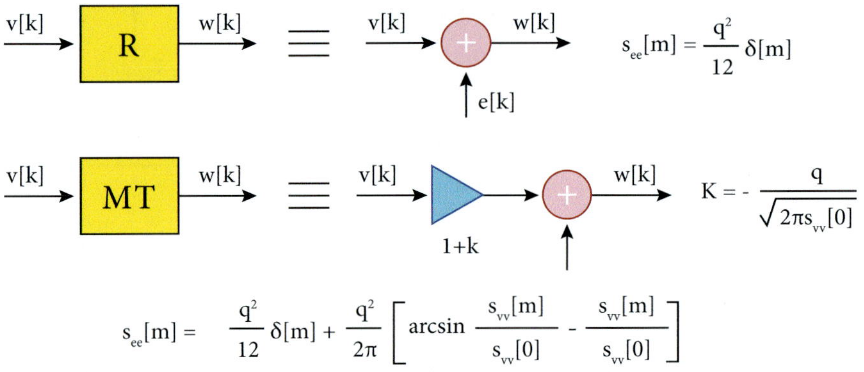

Fig. 12.27 The white-noise models for the quantization errors that are performed with rounding (top) and truncation of size (bottom)

of integers leads to a very large quantization error, whereas the use of *mixed* numbers (i.e., numbers that contain both an integer and a fractional part) *increases the degree of complexity regarding the multiplication operation*. It is not difficult to understand that the range of values for the above quantization errors that both are defined as $X' - X$ will be *different* for each of the types of representation used, and the value of this error will depend on the relationship between the values ℓ and ℓ' of the number of bits used before and after quantization. In the following description, we limit ourselves to representations of sign magnitude and two's complement.

Starting with the *sign-magnitude representation*, if the quantization is performed through *truncation* and the value of X is positive, then the new value will always be *less* than the original, whereas the error E_t will always have a *negative value* that falls within the area

$$-(2^{-\ell'} - 2^{-\ell}) \leq E_t \leq 0,$$

with the maximum value of this error corresponding to the case where the $\ell - \ell'$ number of bits that are removed from the number all have the value 1. On the other hand, if the value of X is *negative*, then the value of the error will be *positive*, since the truncation of fractional digits of a negative number leads *to an increase in its value*, with the range of values of the error E_t being the

$$0 \leq E_t \leq (2^{-\ell'} - 2^{-\ell}).$$

Combining all of the above, we come to the conclusion that for the case of positive and negative numbers for which the sign magnitude representation is used, the truncation error E_t takes values in a region that is *symmetric around the zero value*, since the combination of the two above conditions allows us to write that

$$-(2^{-\ell'} - 2^{-\ell}) \leq E_t \leq (2^{-\ell'} - 2^{-\ell}).$$

Conversely, in the case of *two's complement representation*, this symmetry *does not exist*. Specifically, for positive numbers the same as before applies (since as we commented, the differentiation regarding the type of representation is limited only to negative numbers), whereas for negative numbers the truncation process results in *an increase in the value of the negative number* (since the negative number arises by subtracting from 2 the corresponding positive number). In other words, the error E_t will always be *negative* and will take values in the area

$$-(2^{-\ell'} - 2^{-\ell}) \leq E_t \leq 0,$$

which also applies to the case of positive numbers, according to what we mentioned earlier.

In the case where the quantization method used is *not truncation but rounding*, the quantization error E_r is related *only to the magnitude of the number* and therefore, it is *independent* of the type of representation. In this case, the maximum error that will arise is equal to $\frac{1}{2}(2^{-\ell'} - 2^{\ell})$ and can be *both positive and negative*. Therefore, the range of error values in this case is

$$-\frac{1}{2}(2^{-\ell'} - 2^{\ell}) \leq E_t \leq \frac{1}{2}(2^{-\ell'} - 2^{\ell})$$

What we mentioned above refers to the *fixed-point representation*, whereas in the floating-point representation, things are different, since in this type of representation, *the resolution is proportional to the magnitude of the number* to which we apply the quantization process. Consequently, the quantization error will also be proportional to the value X, so that we can write that $E = X' - X = \lambda X$, with the parameter λ known as the *relative error*. It is not difficult to understand that in floating-point representation, the quantization process will be applied to the coefficient M and not to the exponent E and can be performed *both by truncation and by rounding*. Specifically, in the case of positive numbers and for the representation of the complement to 2, if the quantization of the coefficient is performed *by truncation*, it follows that $-2^E 2^{-\ell'} < \lambda x < 0$ and because it is $2^{E-1} \le X < 2^E$, it finally turns out that $-2^{-\ell'+1} < \lambda \le 0$. On the other hand, for the negative numbers we have $0 \le \lambda X < 2^E 2^{-\ell'}$ and therefore $0 \le \lambda < 2^{-\ell'+1}$. Conversely, if the quantization of the coefficient is performed *by rounding*, then the error will be *symmetrical* in relation to the zero value, with a maximum value equal to $\pm 2^{-\ell'}/2$. It will therefore be

$$-\frac{1}{2}2^{-E}2^{-\ell'} < \lambda X < \frac{1}{2}2^{-E}2^{-\ell'} \qquad \text{and therefore} \qquad -2^{-\ell'} < \lambda < 2^{-\ell'}$$

On the other hand, in contrast to quantization errors that *cannot be eliminated*, this *does not apply* to overflow, as it can be completely avoided if we use input signals with suitably selected small values. Indeed, in the literature, expressions can be found for the upper limit of the input signal values, which if not exceeded, then the phenomenon of overflow, guaranteed, *does not* appear. Some of these expressions, which are related to the characteristic of escalation, are presented in Sect. 12.9.8 (see page 1193). However, generally, in practice, it is possible to accept the possibility of overflow appearing in some special cases of input signals that occur very rarely, silently accepting that after the appearance of the overflow, the system will return, and indeed at high speed, to the usual operating conditions. This *recovery from the overflow state*, which quite reasonably can be considered as a type of *stability*, constitutes one of the most significant problems related to this topic and can be studied both under conditions of *zero input*, and under conditions of *nonzero input*. In the first case, stability is defined as *the complete absence of spontaneous and mainly periodic oscillations*, with the system behaving linearly and approaching exponentially a balance point, at which all state variables tend to take zero value. On the other hand, the second case is related to a specific category of input signals $x[n]$, which are defined with the help of the *ideal linear system* and are characterized by the property *that there is at least one set of initial conditions, for which the threshold related to the appearance of the overflow will never be approached*. In this case, the system has the ability to "*forget*" any overflows that have occurred took place in the past and to *fully recover*, continuing its operation.

12.9 The Impact of the Finite Length of Registers

As already mentioned, the *finite accuracy* that characterizes the storage and use of all kinds of numbers in numerical systems and that is due to the *finite length* of the registers used in them, leads to the appearance of *quantization errors and overflow*, which almost always lead to the appearance of erroneous results at the output of the filter. This quantization concerns *both the values of the filter coefficients and the samples of the input and output signals, as well as the intermediate results that arise as a result of the additions and multiplications required for the implementation of the difference equation that describes the behavior of the system*. Speaking generally, the impact of the quantization of the filter coefficients is studied separately from the impact of the quantization associated with the remaining numerical values and this is exactly the approach that we follow in the description that follows.

It is interesting to note that the reduction of the word length due to the finite length of the registers of the underlying physical system in reality affects *only the operation of the IIR filters*, whereas this process, strictly speaking, *is not required* in FIR-type filters. The only reason for which applies to the last category of filters, that is, to avoid the use of *very long words*, whereas the only consequence of this process is the appearance of an error signal that is associated with the input signal and can lead to the appearance of *quantization errors*. Conversely, in IIR-type filters, characterized by the existence of *feedback loops*, each of these loops (i.e., after each signal multiplication process with the appropriate coefficient), a *correction of the consequences* of quantization and overflow, which manifest in the form of *nonlinear phenomena*, should be made in order to avoid the unlimited increase in word length. Additionally, because, owing to the recursive nature of these filters, errors due to the existence of nonlinear elements are used in *feedback* processes, these filters are characterized by the existence of *limit cycles due to quantization and overflow*, which must be removed. The analytical presentation of these limit cycles is carried out in a subsequent section (see Sect. 12.9.9 on page 1198).

12.9.1 Quantization of Filter Coefficients

How is it possible to assess the quantization of the filter coefficients on the behavior of the transfer function? The simplest thing we can do is *to determine the quantized values of these coefficients, to replace them in the transfer function, using these in place of the theoretical values that have been calculated with infinite accuracy and then, to check whether the new transfer function that arises behaves in the desired way or if the effects of quantization are so severe that the system's behavior cannot be considered acceptable*. In reality, the application of this process, implies that the errors associated with the quantization of the filter coefficients are *deterministic* in nature and the study of the consequences of quantization is generally an easy process to implement. Let us consider for this purpose the general case of a digital IIR filter, which is described by the transfer function

$$\mathcal{H}(z) = \frac{N(z)}{D(z)} = \frac{\sum_{k=0}^{M} \beta_k z^{-k}}{1 + \sum_{k=1}^{N} \alpha_k z^{-k}} = H_0 \frac{\prod_{k=1}^{M} (1 - z_k z^{-1})}{\prod_{k=1}^{N} (1 - p_k z^{-1})}, \tag{12.12}$$

with the positions of the poles $\{p_k\}$ $(k = 1, 2, \ldots, N)$ and the zero values $\{z_k\}$ $(k = 1, 2, \ldots, M)$ depending on the values of the coefficients $\{\alpha_k\}$ $(k = 1, 2, \ldots, N)$ and $\{\beta_k\}$ $(k = 1, 2, \ldots, M)$ respectively. According to what we mentioned in a previous section, although during the analysis of the system, the values of the coefficients $\{\alpha_k\}$ and $\{\beta_k\}$ are assumed to be known with infinite accuracy, in practice, as we have explained, this accuracy *is limited by the length of the registers* that are used for storing these coefficients during the actual implementation of the filter. In other words, although the theoretical values of these coefficients are $\{\alpha_k\}$ and $\{\beta_k\}$, in practice *these* values are not used, but instead, the *quantized* values $\{\alpha_k^q\}$ and $\{\beta_k^q\}$, which are related to the above theoretical values through the relations $\alpha_k^q = \alpha_k + \Delta\alpha_k$ $(k = 1, 2, \ldots, N)$ and $\beta_k^q = \beta_k + \Delta\beta_k$ $(k = 1, 2, \ldots, M)$, where $\Delta\alpha_k$ and $\Delta\beta_k$ are the corresponding quantization errors, with the range of these errors depending on what we have mentioned, from the length of the registers and the way numbers are represented in the digital system used.

It is not difficult to understand that from the instant in the actual implementation of the filter, the theoretical values are not used but the quantized values characterized by finite accuracy, the positions of the poles and the zero values *will not be* the p_k $(k = 1, 2, \ldots, N)$ and z_k $(k = 1, 2, \ldots, M)$, which have been calculated so as to satisfy the specifications of the filter's operation, but some other values p_k^q and z_k^q, which are related to the actual values p_k and z_k through the relations $p_k^q = p_k + \Delta p_k$ and $z_k^q = z_k + \Delta z_k$, where Δp_k and Δz_k are the quantization errors associated with the positions of the poles and the zero values with this deviation naturally reflected in the *amplitude response*, which differs to some extent from the ideal. This situation is depicted in Fig. 12.28, with the top left image representing the ideal *pole-zero diagram* that was constructed under conditions of *infinite accuracy* and the remaining three images having arisen after quantization of the filter coefficients, using different numbers of bits. From the top right image, which corresponds to finite accuracy and which results from quantization with 4 bits, we find that the modification of the pole positions is *so violent*, that the system has become unstable, since some poles have moved outside the circumference of the unit circle. Conversely, in the bottom left and bottom right images, where the quantization is performed with 8 and 32 bits respectively, the situation is clearly better, with the image corresponding to the 32 bits being *almost identical* to that corresponding to the calculations of infinite accuracy. Consequently and in accordance with what we mentioned earlier, the transfer function that will be used during the actual implementation of the filter, either through hardware, or through use of the software, *will not be* the one we quoted above, but on the contrary

$$\mathcal{H}^q(z) = \frac{N^q(z)}{D^q(z)} = \frac{\sum_{k=0}^{M} \beta_k^q z^{-k}}{1 + \sum_{k=1}^{N} \alpha_k^q z^{-k}} = H_0^q \frac{\prod_{k=1}^{M} \left(1 - z_k^q z^{-1}\right)}{\prod_{k=1}^{N} \left(1 - p_k^q z^{-1}\right)}, \tag{12.13}$$

or equivalently, the

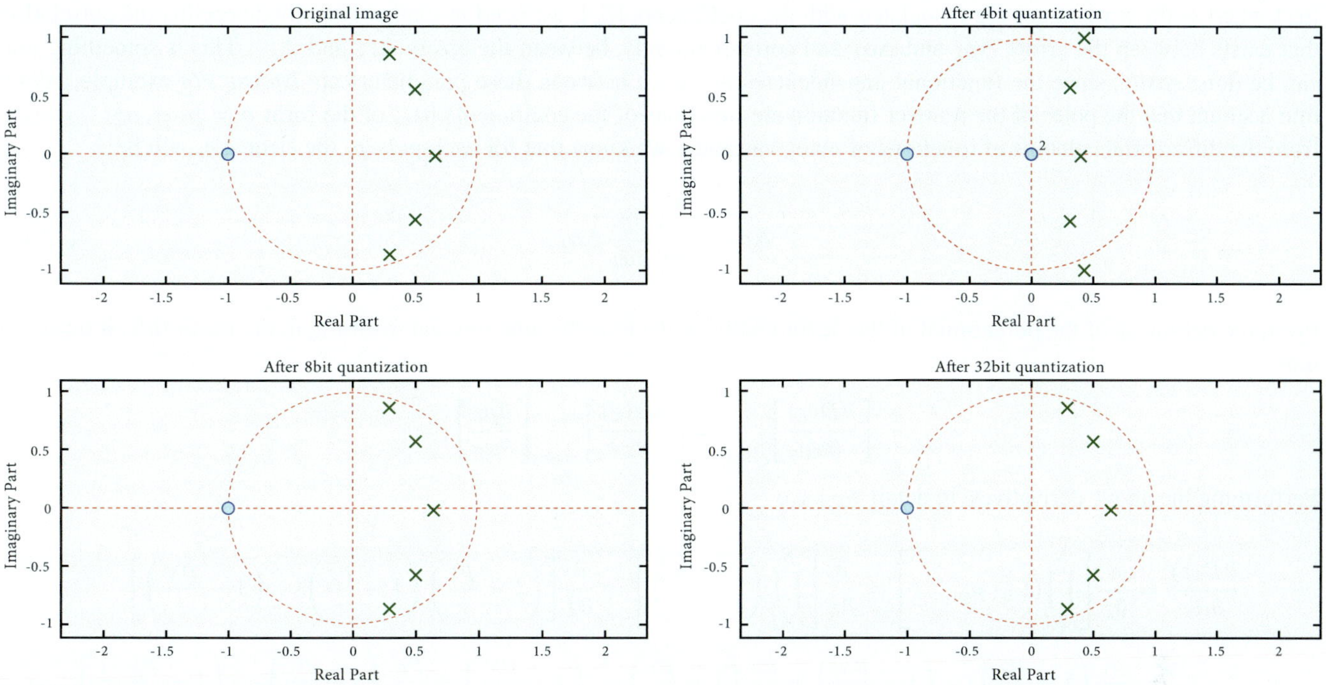

Fig. 12.28 Pole-zero plot for the case of infinite accuracy (top left), finite accuracy with 4 bits (top right), finite accuracy with 8 bits (bottom left), and finite accuracy with 32 bits (bottom right)

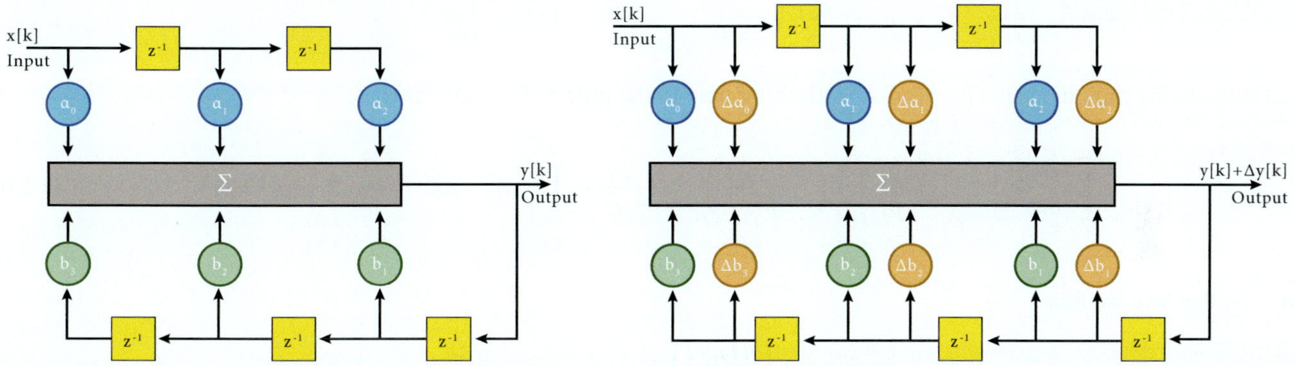

Fig. 12.29 The theoretical digital filter in which the values of the coefficients are known with infinite accuracy and the corresponding real filter in which these values have undergone quantization and therefore are characterized by finite accuracy for the values $M = 2$ and $N = 3$ (Source: Widrow and Kollár, page 568)

$$\mathcal{H}^q(z) = \frac{N^q(z)}{D^q(z)} = \frac{\sum_{k=0}^{M}(\beta_k + \Delta\beta_k)z^{-k}}{1 + \sum_{k=1}^{N}(\alpha_k + \Delta\alpha_k)z^{-k}} = H_0^q \frac{\prod_{k=1}^{M}[1 - (z_k + \Delta z_k)z^{-1}]}{\prod_{k=1}^{N}[1 - (p_k + \Delta p_k)z^{-1}]}, \tag{12.14}$$

with the corresponding theoretical diagram (of infinite accuracy) and real diagram (of finite accuracy) for the values $M = 2$ and $N = 3$ depicted in the left and right images of Fig. 12.29.

How can one quantitatively assess the sensitivity of the system to perturbations in the filter coefficients, represented by $\Delta\alpha_k$ and $\Delta\beta_k$, which are regarded as *disturbances*? Taking into account the fact that the poles are related to the coefficients

$\{\alpha_k\}$ whereas the zero values are associated with the coefficients $\{\beta_k\}$, we need to mathematically determine the correlation that exists between the errors Δp_k and $\Delta\alpha_k$ and correspondingly, between the errors Δz_k and $\Delta\beta_k$. This is something that can be done *easily*, since the functional dependencies that exist between these parameters are *known*. For example, taking into account that the poles of the transfer function are functions of the coefficients $\{\alpha_k\}$ of the form $p = p(\alpha_1, \alpha_2, \ldots, \alpha_N)$, from the differential calculus of functions of many variables, we know that for each pole p_i the error Δp_i will be[13]

$$\Delta p_i = \sum_{k=1}^{N} \frac{\partial p_i}{\partial \alpha_k}\Delta\alpha_k. \tag{12.15}$$

From the definition of the polynomial of the denominator of the transfer function and resorting to the *chain rule*, it turns out that

$$\left[\frac{\partial D(z)}{\partial \alpha_k}\right]_{z=p_i} = \left[\frac{\partial D(z)}{\partial z}\right]_{z=p_i}\left[\frac{\partial p_i}{\partial \alpha_k}\right].$$

Performing the above derivatives, in detail we have

$$\begin{aligned}\frac{\partial D(z)}{\partial \alpha_k} &= \frac{\partial}{\partial z}\left[\prod_{\ell=1}^{N}(1-p_\ell z^{-1})\right] = \frac{\partial}{\partial z}\left[\prod_{\ell=1}^{N}\left(1-\frac{p_\ell}{z}\right)\right] = \frac{\partial}{\partial z}\left[\left(1-\frac{p_1}{z}\right)\left(1-\frac{p_2}{z}\right)\cdots\left(1-\frac{p_N}{z}\right)\right]\\ &= \frac{p_1}{z^2}\left(1-\frac{p_2}{z}\right)\cdots\left(1-\frac{p_N}{z}\right)+\frac{p_2}{z^2}\left(1-\frac{p_1}{z}\right)\cdots\left(1-\frac{p_N}{z}\right)+\frac{p_N}{z^2}\left(1-\frac{p_1}{z}\right)\cdots\left(1-\frac{p_{N-1}}{z}\right)\\ &= \sum_{k=1}^{N}\frac{p_k}{z^2}\prod_{\substack{\ell=1\\ \ell\neq k}}^{N}(1-p_\ell z^{-1}) = \sum_{k=1}^{N}\frac{p_k}{z^2}\prod_{\substack{\ell=1\\ \ell\neq k}}^{N}\left(1-\frac{p_\ell}{z}\right) = \sum_{k=1}^{N}\frac{p_k}{z^2}\prod_{\substack{\ell=1\\ \ell\neq k}}^{N}\left(\frac{z-p_\ell}{z}\right)\end{aligned}$$

and considering only the value $k = i$, that is, limiting only to the pole $z = p_i$ we have

$$\left[\frac{\partial D(z)}{\partial \alpha_k}\right]_{z=p_i} = \frac{p_i}{p_i^2}\prod_{\substack{\ell=1\\ \ell\neq i}}^{N}\left(\frac{p_i-p_\ell}{p_i}\right) = \frac{p_i}{p_i^2 p_i^{N-1}}\prod_{\substack{\ell=1\\ \ell\neq i}}^{N}(p_i-p_\ell) = \frac{1}{p_i^N}\prod_{\substack{\ell=1\\ \ell\neq i}}^{N}(p_i-p_\ell). \tag{12.16}$$

In a similar way we find

$$\left[\frac{\partial D(z)}{\partial z}\right]_{z=p_i} = -p_i^{-k},$$

and therefore, we have

$$\frac{\partial p_i}{\partial \alpha_k} = \left\{\left[\frac{\partial D(z)}{\partial \alpha_k}\right]_{z=p_i}\right\}\bigg/\left\{\left[\frac{\partial D(z)}{\partial z}\right]_{z=p_i}\right\} = \frac{-p_i^{N-k}}{\prod_{\substack{\ell=1\\ \ell\neq i}}^{N}(p_i-p_\ell)}. \tag{12.17}$$

This partial derivative is known as the *sensitivity of the pole position* in the complex plane, in the quantization process of the coefficients $\{\alpha_k\}$. Replacing in our initial expression we get

[13] Note that this method of analysis can only be applied to linear systems, which of course are also those in which we are usually interested. Conversely, in the case of nonlinear systems, the implications of the quantization of the coefficients can only be studied through simulations that are performed on electronic computers.

$$\Delta p_i = \sum_{k=1}^{N} \frac{\partial p_i}{\partial \alpha_k} \Delta \alpha_k = - \sum_{k=1}^{N} \frac{-p_i^{N-k}}{\prod_{\substack{\ell=1 \\ \ell \neq i}}^{N} (p_i - p_\ell)} \Delta \alpha_k. \tag{12.18}$$

This result is particularly interesting, because it tells us that *the error induced in the position of any pole p_i on the complex plane owing to the quantized values of the coefficients $\{\alpha_k^q\}$ used in place of the real coefficients $\{\alpha_k\}$ is inversely proportional to the distance of all the other poles of the transfer function from this pole on the complex plane*, which is equal to the magnitude of the corresponding vector defined by these two poles. Similar results can also arise for the sensitivity of the positions of the zero values to the quantization of the coefficients $\{\beta_k\}$, since the functional dependence is *essentially the same*. Consequently, filters whose poles are in *neighboring positions* on the complex plane, are characterized by *greater sensitivity* to the quantization of the coefficients than those filters whose poles are further apart. This problem becomes even more acute when *increasing* the order of the filter, since this results in the appearance of *additional poles* in the interior of the unit circle (we are obviously referring to stable systems) and the further reduction of the distances between the poles. In order to maximize the lengths $p_i - p_\ell$ so as to reduce the sensitivity of the system, we can implement the filter *by combining first- and second-order steps in the appropriate way*, so as to avoid the appearance of complex values in the transfer function, the handling of which by the digital system is *more complex* than that of real values. In order to carry out this process, we proceed to the *factorization* of the polynomials of the numerator and the denominator, which can be carried out with great accuracy and which are combined, as we mentioned, in second-order steps, the coefficients of which subsequently undergo *quantization*, considering one step after the other. Note that the value of the derivative $\partial p_i / \partial \alpha_k$ depends on the order of the filter N and therefore, the second-order steps for which $N = 2$ are characterized by a *lower sensitivity value*. This is something that greatly improves the behavior of the system's frequency response and is a technique that is almost always used in practice.

However, it should be noted that the way of interconnecting the second-order steps (since, as we saw, the same transfer function can be implemented in many different ways), *also affects* the sensitivity of the system and the reason for this, is the existence of zero values, the sensitivity of which is analyzed in a similar way to that of the poles. Specifically, in their *series connection* of these stages (see Sect. 12.15.1 on page 1358) and in cases where the filter contains zero values on the circumference of the unit circle, each of the second-order stages will contain complex conjugate zero values, the positions of which are determined by the values of the coefficients $\{\beta_k\}$ and the system's sensitivity to the quantization of the values of these coefficients *is easy to check*, using for the representation of these values the appropriate number of bits. From the other side, in the *parallel connection* (see Sect. 12.15.2 on page 1361), this ability to directly control the system's sensitivity associated with zero values *is not possible*, since the coefficients that appear in the numerators of the fractions resulting from the system's analysis into partial fractions *do not allow the direct calculation of the positions of the zero values*, but *they do not affect them either*. In other words, in parallel implementations, the assessment of the consequences of the quantization of the coefficients, is *more difficult* than for the series implementations, which for this reason are used more often in practice.

In order to understand in practice all of the above, let us consider two typical implementations of second-order stages characterized by poles at positions $p_{1,2} = \pm r e^{\pm j\vartheta}$ and more specifically (a) the usual *straight form* and (b) the so-called *conjugate form*, the transfer functions of which are respectively the

$$\mathcal{H}_d(z) = \frac{1}{r^2 z^{-2} + (2r \cos\vartheta) z^{-1} + 1} \quad \text{and} \quad \mathcal{H}_c(z) = \frac{(r \sin\vartheta) z^{-1}}{r^2 z^{-2} + (2r \cos\vartheta) z^{-1} + 1}.$$

These implementations are depicted in the top left and top right images of Fig. 12.30 respectively. Although these two transfer functions have *the same poles*, the dependence between the magnitude and phase of the poles r and θ and the coefficients α_1 and α_2 is different, since in the case of straight implementation, it turns out that $\alpha_1 = 2r \cos\vartheta$ and $\alpha_2 = -r^2$, whereas in the second case, it is $\alpha_1 = r \cos\vartheta$ and $\alpha_2 = r \sin\vartheta$. It is not difficult to understand that *if the values of the coefficients α_1 and α_2 are known with infinite accuracy, then the corresponding values for the magnitude and phase of the two poles will be any, and therefore, there is no restriction regarding their choice in such a way as to satisfy the specifications of the filter that have been set in each case*. However, if the values of the filter coefficients undergo *quantization* and the (initially infinite) accuracy of the representation is limited to that supported by the underlying digital system, things *differ*, since now

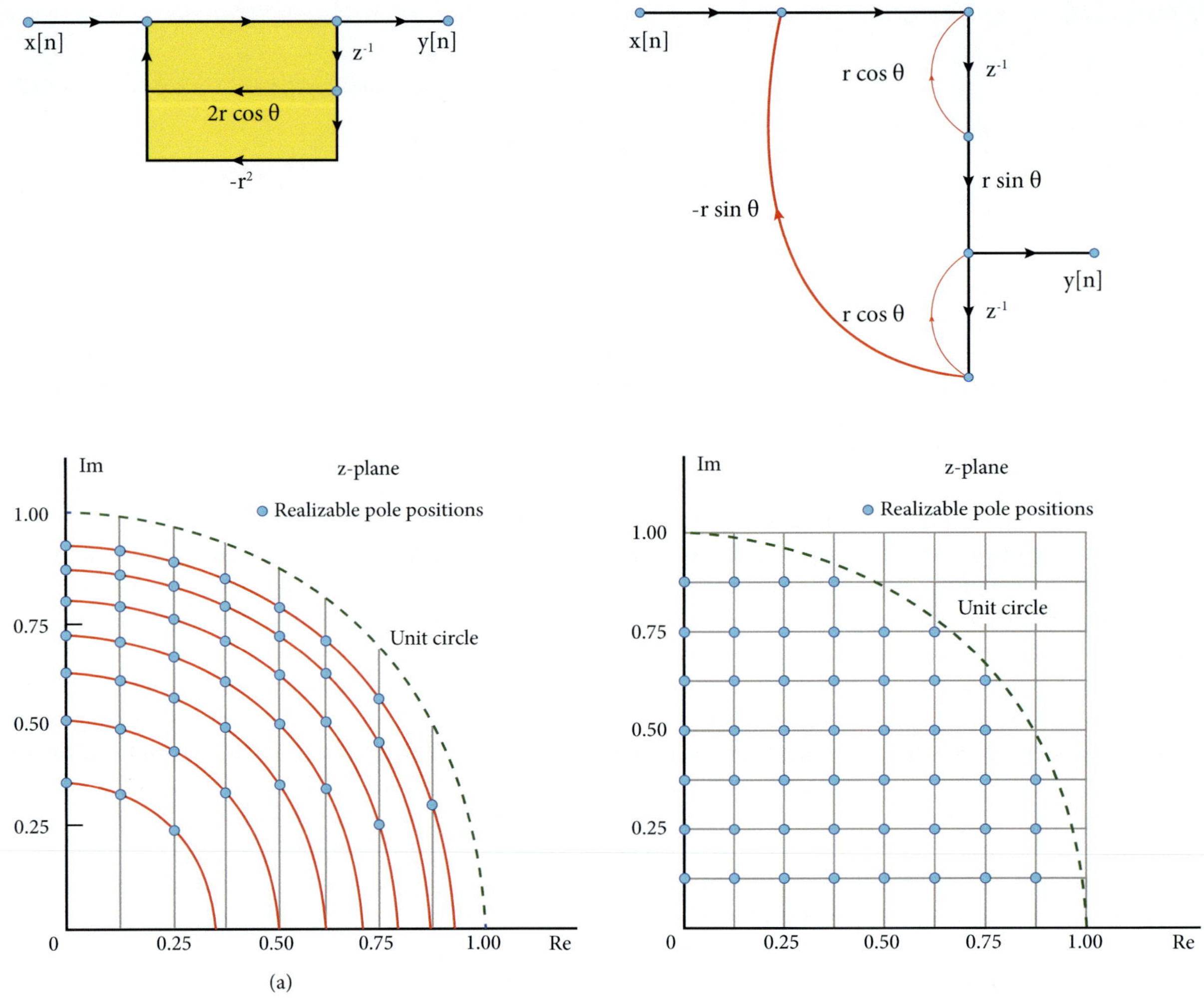

Fig. 12.30 The permissible positions of the poles of the direct form (left image) and the conjugate form (right image) in the first quadrant, for a second-order step with poles $p_{1,2} = \pm re^{\pm j\vartheta}$ after a quantization process with $\ell = 4$ bits

each of the coefficients α_1 and α_2 *cannot take any value* but only one from a set of $2^\ell - 1$ values, where ℓ is the length of the registers of the underlying digital system. If, for example, $\ell = 4$, then the coefficient α_k and the magnitude r of the poles of the transfer function can only take one from a set of 15 different values. In the general case and because each pole depends on both coefficients α_1 and α_2, the set of possible positions is equal to $(2^\ell - 1)^2$, whereas of course the cases $\alpha_1 = 0$ and $\alpha_2 = 0$ are not taken into account.

The permissible positions of the poles of the two above systems as they emerge *after a quantization process of the coefficients* α_1 and α_2 using $\ell = 4$ bits, are depicted in the lower left and lower right images of Fig. 12.30 for the direct and the conjugate form respectively and for the first (upper right) quadrant of the complex plane. The number of these pole positions is determined in the way we previously described, whereas their characteristics depend *on the functional dependence that exists between their magnitude and phase and the filter coefficients*. For example, in the case of the conjugate form, the coefficients α_1 and α_2 both constitute *linear functions* of r and therefore, their quantization leads to *discrete pole positions*, which *are equidistant* from each other, both horizontally and vertically, something that *does not apply* to the direct form, since in this, the above dependence *is not linear* (note that in this case, the process of quantization is not applied to the value of r, but to the value of r^2). If the number ℓ of bits takes a *higher value*, the positions that will emerge will be more, but they will be characterized *by the same type of distribution in the complex plane*, leading in this way to denser arrangements of positions, which, however, will follow the same pattern. It is not difficult to understand that in the case of the infinite value of the number ℓ of bits, which essentially corresponds *to the complete absence of quantization*, that is, to infinite accuracy, the above distribution from discrete will become *continuous*, since in this case the pole will be able to be placed in any of the infinite points of the first quadrant that are inside the unit circle.

This different way of distributing the positions of a pole in the complex plane naturally affects the behavior of the filter. For example, in the *direct form*, this specific type of distribution imposes on the pole positions that are close to the limits $\vartheta = 0$ and $\vartheta = \pi$, to be *more sparse* in relation to the other areas of the complex plane (which in the case of low-pass and high-pass filters is *particularly problematic*, since exactly these are the areas in which most of their poles are located), whereas this is not true in the case of the conjugate form, since there the distribution of the poles in the complex plane is, as we commented, uniform.

The abovementioned concerns IIR filters, whereas in the case of FIR-type filters that contain only zero values, the same things we mentioned for the zero values of IIR filters apply, since in both cases, *similar conclusions* can be drawn to those formulated for the case of poles. Of particular interest are the *FIR filters of generalized linear phase* (see Sect. 12.11 on page 1223), since their implementation according to their direct form *is not affected* by the quantization process of their coefficients, provided of course that this is done in such a way as to maintain the symmetry that characterizes the coefficients of these filters (remember, that the impulse response of this type of filters is *symmetric* or *antisymmetric*). In reality, their transfer function, which, as is well known, has the form $\mathcal{H}(z) = \pm z^{M-1}\mathcal{H}(z^{-1})$, where M is the order of the filter, *is the same*, whether the quantization of the filter takes place or not. In other words, *the quantization of the coefficients of the filter only affects the magnitude of the transfer function of the system and not its phase characteristics*. In the case where an increase in the order M of the filter takes place, all we have to do is *to increase the number ℓ of bits used for their storage*, so that the quantization error associated with the frequency response of the filter remains at the same levels.

It is not difficult to understand that *the modification of the characteristics of the transfer function regarding the positions of the poles and the zero values induced by the quantization of the filter coefficients will also characterize the frequency response function*, since this is nothing more than *the transfer function calculated on the periphery of the unit circle*. This means that although in the ideal case, the amplitude response of the filter is the $\mathcal{M}(\omega) = |\mathcal{H}(e^{j\omega})|$, in practice the amplitude response $\mathcal{M}^q(\omega) = |\mathcal{H}^q(z)|_{z=e^{j\omega}} = |\mathcal{H}^q(e^{j\omega})|$ will be used, which obviously *does not coincide* with the amplitude response $\mathcal{M}(\omega)$ corresponding to the infinite-accuracy representation. In other words, the quantization of the filter coefficients, leads to the appearance of an *error* defined as the difference between these two amplitude responses, that is,

$$\Delta\mathcal{M} = \mathcal{M}(\omega) - \mathcal{M}^q(\omega) = |\mathcal{H}(e^{j\omega})| - |\mathcal{H}^q(e^{j\omega})|.$$

If $\mathcal{M}_I(\omega) = |\mathcal{H}_I(e^{j\omega})|$ is the frequency response of the ideal filter, which is characterized by a constant width in the passband, zero width in the stopband and by an instantaneous transition from the passband to the stopband, which, as is well known, is not physically implementable, it is not difficult to understand, observing Fig. 12.31, that the maximum permissible value of the difference $|\Delta\mathcal{M}| = \Delta\mathcal{M}_{\max}(\omega)$ is defined as

$$\Delta\mathcal{M}_{\max}(\omega) = \begin{cases} \delta_p - |\mathcal{M}(\omega) - \mathcal{M}_I(\omega)| & \omega \le \omega_p \\ \delta_s - |\mathcal{M}(\omega) - \mathcal{M}_I(\omega)| & \omega \ge \omega_s, \end{cases} \tag{12.19}$$

where δ_p and δ_s are the widths of the fluctuations in the passband and stopband, whereas ω_p and ω_s are the limits of these two bands. In order for the quantization of the coefficients not to affect the behavior of the filter, the approximation error of the function $\mathcal{M}(\omega)$ by the function $\mathcal{M}^q(\omega)$, *should not be greater* than $\Delta\mathcal{M}_{\max}(\omega)$, that is, to be $|\Delta\mathcal{M}| \le \Delta\mathcal{M}_{\max}(\omega)$. As for the word length of the digital system that should be used, so that the above condition is satisfied, this can be determined *by applying the method of trial and error* and more specifically, by progressively changing the number of bits that dictate the length of the register, calculating the corresponding value of the quantity $|\Delta\mathcal{M}|$ and checking whether or not the condition in question is satisfied.

Statistical Calculation of the Optimal Length of the Registers

An alternative statistical approach to determining the optimal length of the registers leading to the calculation of the so-called *statistical word length* is the one proposed by Avenhaus (based on the concept of energy) and modified by Crochiere (who uses, on the contrary, the measure of frequency response—see also Antoniou, page 629). For the presentation of this method, which constitutes a *more realistic approach* than the one based on the worst-case scenario, according to which the change in the filter coefficients due to quantization will be *the worst possible*, both in terms of magnitude and direction, and which

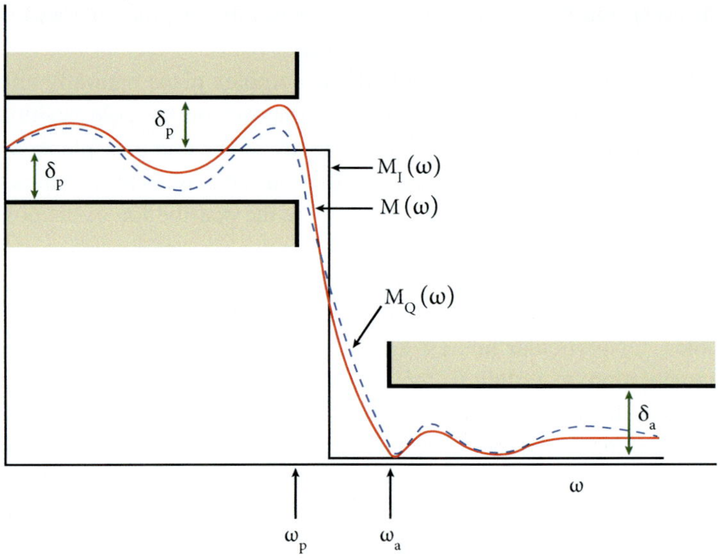

Fig. 12.31 The amplitude responses $\mathcal{M}(\omega) = |\mathcal{H}(e^{j\omega})|$ and $\mathcal{M}^q(\omega) = |\mathcal{H}^q(z)|_{z=e^{j\omega}} = |\mathcal{H}^q(e^{j\omega})|$ corresponding to the representations of the filter coefficients of infinite and finite accuracy. The same figure also depicts the amplitude response of the ideal filter, which is noncausal and therefore, not implementable

consists of and this in the calculation of a *sensitivity parameter*, which, as before, takes the form of a *partial derivative*, let us consider a digital system that uses *fixed-point arithmetic* and performs quantization through rounding. Considering an arbitrary number of coefficients c_k $(k = 1, 2, \ldots, m)$, let us assume as before, that the application of quantization leads to a change in the coefficient c_k $(k = 1, 2, \ldots, m)$ by $\Delta c_k = c_k^q - c_k$, where c_k and c_k^q are the values before and after the application of quantization, and let us suppose that the quantization error Δc_k is a random variable that takes any value in the range from $-Q/2$ to $Q/2$, where Q is the quantization step.[14] If we assume that the PDF $p(\Delta c_k)$ is *uniform* in this area the range of which is equal to Q, that is, that $p(\Delta c_k) = 1/Q$, when $-Q/2 \leq \Delta c_k \leq Q/2$ and $p(\Delta c_k) = 0$ anywhere else, then the average $\overline{\Delta c_k}$ and the variability $\sigma^2_{\Delta c_k}$ of Δc_k will be respectively equal to zero and $Q^2/12$ (see the relevant section). As for the change $\Delta\mathcal{M}$ of the frequency response measure $\mathcal{M}(e^{j\omega})$, this will also constitute a *random variable* and, working in the way we did in the transfer function, we can similarly write

$$\Delta\mathcal{M} = \sum_{k=1}^{m} \frac{\partial\mathcal{M}(e^{j\omega})}{\partial c_k} \Delta c_k,$$

with the partial derivative $S_{c_k}^{\mathcal{M}}(e^{j\omega}) = \partial\mathcal{M}(e^{j\omega})/\partial c_k$ constituting the *sensitivity of the amplitude response with respect to the change of the coefficient* c_k. Apart from the above sensitivity parameter, the *normalized sensitivity* is also defined in practice.

$$\overline{S}_{c_k}^{\mathcal{M}}(e^{j\omega}) = \lim_{\Delta c_k \to 0} \frac{\Delta\mathcal{M}(e^{j\omega})/\mathcal{M}(e^{j\omega})}{\Delta c_k / c_k} = \frac{c_k}{\mathcal{M}(e^{j\omega})} \frac{\partial\mathcal{M}(e^{j\omega})}{\partial c_k}, \tag{12.20}$$

which is used for the comparison of different structures, whereas another sensitivity measure is the partial derivative

[14] Although in the general case, the quantization step is symbolized by the letter Δ, in this description, the symbol Q is used for this size, in order not to cause confusion with the symbol Δ, which in this description is used to describe the changes in the values of the involved quantities, such as the change $\Delta\mathcal{M}$.

$$\widetilde{S}_{c_k}^{\mathcal{M}}(e^{j\omega}) = \frac{1}{\mathcal{M}(e^{j\omega})}\frac{\partial \mathcal{M}(e^{j\omega})}{\partial c_k}, \tag{12.21}$$

from which it follows that $\overline{S}_{c_k}^{\mathcal{M}} = c_k \widetilde{S}_{c_k}^{\mathcal{M}}$. In several textbooks, these expressions contain in the place of the measure of the frequency response function $\mathcal{M}(e^{j\omega})$ the frequency response function itself $\mathcal{H}(e^{j\omega})$ or the transfer function $\mathcal{H}(z)$. According to the literature (see Diniz, page 706) the sensitivity $S_{c_k}^{\mathcal{M}}$ is used when the values of the filter coefficients are encoded according to *fixed-point representation*; on the contrary, the sensitivity $\overline{S}_{c_k}^{\mathcal{M}}$ is used for the case of floating-point encoding and constitutes a measure of the relative change of the transfer function owing to the corresponding change of a filter coefficient due to quantization. The above expressions are related to the change of *only one filter coefficient* c_k, whereas in order to describe the contribution to the change of the measure of the transfer function $|\mathcal{H}(e^{j\omega})|$ of all the filter coefficients, we can define the size

$$\mathcal{S}(e^{j\omega}) = \sum_{k=1}^{m} |R_{c_k}^{\mathcal{M}}(e^{j\omega})|,$$

where $R_{c_k}^{\mathcal{M}}$ any of the sensitivities $S_{c_k}^{\mathcal{M}}$, $\overline{S}_{c_k}^{\mathcal{M}}$ and $\widetilde{S}_{c_k}^{\mathcal{M}}$ the measures of which, if these are defined for the frequency response function $\mathcal{H}(e^{j\omega}) = \mathcal{M}(e^{j\omega})e^{j\measuredangle\mathcal{H}(e^{j\omega})}$ are the

$$|S_{c_k}^{\mathcal{H}}(e^{j\omega})| = \sqrt{(S_{c_k}^{\mathcal{M}}(e^{j\omega}))^2 + |\mathcal{H}(e^{j\omega})|^2\left(\frac{\partial\measuredangle\mathcal{H}(e^{j\omega})}{\partial c_k}\right)^2} \tag{12.22}$$

$$|\widetilde{S}_{c_k}^{\mathcal{H}}(e^{j\omega})| = \sqrt{(\widetilde{S}_{c_k}^{\mathcal{M}}(e^{j\omega}))^2 + \left(\frac{\partial\measuredangle\mathcal{H}(e^{j\omega})}{\partial c_k}\right)^2} \tag{12.23}$$

$$|\overline{S}_{c_k}^{\mathcal{H}}(e^{j\omega})| = \sqrt{(S_{c_k}^{\mathcal{M}}(e^{j\omega}))^2 + |c_k|^2\left(\frac{\partial\measuredangle\mathcal{H}(e^{j\omega})}{\partial c_k}\right)^2}. \tag{12.24}$$

From the above expressions, it follows that $|R_{c_k}^{\mathcal{H}}(e^{j\omega})| \geq R_{c_k}^{\mathcal{M}}(e^{j\omega})|$ (where $R_{c_k}^{\mathcal{M}}$ is any of $S_{c_k}^{\mathcal{M}}$, $\overline{S}_{c_k}^{\mathcal{M}}$ and $\widetilde{S}_{c_k}^{\mathcal{M}}$) and consequently, the parameter $|R_{c_k}^{\mathcal{H}}(e^{j\omega})|$ can be used *as a conservative estimate* of the value of the parameter $R_{c_k}^{\mathcal{M}}(e^{j\omega})|$, in the sense that it *ensures that the variation of the transfer function will have a value less than a predetermined tolerance error*. Indeed, in the case where the filter has a pole that is *close to the circumference of the unit circle*, for frequencies that are *close to the frequency* of this pole, it can be shown that $|R_{c_k}^{\mathcal{H}}(e^{j\omega})| \approx R_{c_k}^{\mathcal{M}}(e^{j\omega})|$. In the general case, the calculation of sensitivity for the frequency-response function is *easier* than the calculation of sensitivity for the measure of this function.

It is not difficult to understand that the average change $\Delta\mathcal{M}$ will be equal to zero, that is, it follows that

$$\overline{\Delta\mathcal{M}} = \sum_{k=1}^{m}\frac{\partial\mathcal{M}(e^{j\omega})}{\partial c_k}\overline{\Delta c_k} = 0,$$

since $\overline{\Delta c_k} = 0$, whereas if we make the assumption that the quantization errors Δc_k are *uncorrelated*, that is, *statistically independent*, from the basic theory of statistical signal processing, we will have

$$\sigma_{\Delta M}^2 = \sum_{i=1}^{m}\sigma_{\Delta c_k}^2\left(S_{c_k}^{M}\right)^2 = \sum_{i=1}^{m}\frac{Q^2}{12}\left(S_{c_k}^{M}\right)^2 = \frac{Q^2}{12}\sum_{i=1}^{m}\left(S_{c_k}^{M}\right)^2 = \frac{Q^2}{12}\sum_{i=1}^{m}\left(\frac{\partial\mathcal{M}(e^{j\omega})}{\partial c_k}\right)^2 = \frac{Q^2}{12}S^2, \tag{12.25}$$

where

$$S^2 = S^2(\omega) = \sum_{k=1}^{m}\left(S_{c_k}^{\mathcal{M}}\right)^2 = \sum_{k=1}^{m}\left(\frac{\partial\mathcal{M}(e^{j\omega})}{\partial c_k}\right)^2.$$

The partial derivatives in the above expression can be calculated for *arbitrary filter structures*, resorting to the use of suitable methods. In cases where the quantization of some coefficients has been performed with greater accuracy than others, we can choose as quantization step Q the *smallest possible quantization step* and choose the change steps of the various coefficients c_k as *multiples* of this step.

In cases where the value of m is quite large, the random variable $\Delta\mathcal{M}$ can be considered to be described by a *Gaussian-type distribution*, according to the *central limit theorem* and consequently, the PDF $p(\Delta\mathcal{M})$ will have the form

$$p(\Delta\mathcal{M}) = \frac{1}{\sigma_{\Delta\mathcal{M}}\sqrt{2\pi}} \exp\left(-\frac{\Delta\mathcal{M}^2}{2\sigma^2_{\Delta\mathcal{M}}}\right), \quad -\infty \leq \Delta\mathcal{M} \leq \infty. \tag{12.26}$$

Therefore, according to the basic principles of statistics, the value of $\Delta\mathcal{M}$ will belong to the range of values $-\Delta\mathcal{M}_1 \leq \Delta\mathcal{M} \leq \Delta\mathcal{M}_1$ with the probability

$$y = P\left[\, |\Delta\mathcal{M}| \leq \Delta\mathcal{M}_1 \,\right] = \frac{2}{\sigma_{\Delta\mathcal{M}}\sqrt{2\pi}} \int_0^{\Delta\mathcal{M}_1} \exp\left(-\frac{\Delta\mathcal{M}^2}{2\sigma^2_{\Delta\mathcal{M}}}\right) d(\Delta\mathcal{M}). \tag{12.27}$$

Proceeding with the variable change $\Delta\mathcal{M} = x\sigma_{\Delta\mathcal{M}}$, $\Delta\mathcal{M}_1 = x_1\sigma_{\Delta\mathcal{M}}$ and the above function is transformed into the well-known *error function*

$$y = \frac{2}{\sqrt{2\pi}} \int_0^{x_1} e^{-x^2/2} dx,$$

which, as is well known, can be found in countless mathematics books expressed in table form. Therefore, if we select the appropriate *confidence factor* for each case, we can then use the data tables to calculate the corresponding value of x and, from there, determine the statistical limit. $\Delta\mathcal{M}_1 = x\sigma_{\Delta\mathcal{M}}$. Let it be noted that although this quantity is referred to as a limit, in reality, *it does not constitute a limit in the usual sense of the term*, but rather *a conservative estimate* regarding the possibility for the quantity $\Delta\mathcal{M}$ to take a value, which is greater than the value determined by this limit, something that happens with a probability equal to $1 - y$. Specifically, *if we choose a word length, such that the condition* $\Delta\mathcal{M}_1 \leq \Delta\mathcal{M}\text{max}(\omega)$ *is satisfied, then the filter's operating specifications will be met with a confidence factor of* y. In order to determine the quantization step that leads to results that *do not affect* the filter's operating specifications, let us combine the equations $\sigma^2_{\Delta\mathcal{M}} = (Q^2/12)S^2$ and $\Delta\mathcal{M}_1 = x\sigma_{\Delta\mathcal{M}}$ ending up in the expression $\Delta\mathcal{M}_1 = xQS/\sqrt{12}$. Replacing this expression in the inequality $\Delta\mathcal{M}_1 \leq \Delta\mathcal{M}\text{max}(\omega)$, it turns out that

$$x_1\frac{QS}{\sqrt{12}} \leq \Delta\mathcal{M}\text{max}(\omega) \qquad \text{and finally} \qquad q \leq \frac{\sqrt{12}}{x_1 S}\Delta\mathcal{M}\text{max}(\omega)$$

In other words, *if the quantization of the filter coefficient values is performed with a quantization step that satisfies the above inequality, then the filter's operating specifications are met with a confidence coefficient or factor equal to* y. Practice has shown that the value $x = 2$ leads to a confidence coefficient $y = 0.95$.

Finally, in order to calculate the *statistical word length*, which was also our initial goal, we must take into account that it should at least support the storage of the maximum coefficient c_k of the filter, with a value let us say c_k^{max}. If $Q[c_k^{\text{max}}]$ is the quantized value of this coefficient that will be used by the underlying digital system, its representation in the binary system will have as known the form

$$Q[c_k^{\text{max}}] = \sum_{i=-A}^{B} \beta_i 2^i,$$

with $\beta_B \leq 0$ and $\beta_{-A} \neq 0$. Taking into account that apart from the $A + B$ bits for the integer and the fractional part we also need a bit for the sign, the required length L in bits will be $L = A + B + 1$. However, $Q = 2^{-A}$ or equivalently $A = \log_2(1/Q)$ and consequently, the required *statistical register length* will be calculated as

$$L(\omega) = 1 + B + \log_2 \frac{x_1 S}{\sqrt{12\Delta\mathcal{M}_{\max}}},$$

where we have replaced the quantization step Q from the equation we calculated previously. We observe that the required length is a *function of the frequency* ω, as it depends on the quantity S, which is also a function of ω. In order to use a digital system that *does not violate* the operating specifications of the quantized system to an extent that *does not exceed* that dictated by the variable x_1 and the corresponding y, we must use a register length $L \geq L(\omega)$ with the length $L(\omega)$ being the *statistical register length*. Apart from the calculation of the statistical length, Crochiere also presents an optimization technique that consists in minimizing it and for which the reader is referred to the literature.[15]

Let us conclude this section by presenting the way of calculating the values of the partial derivatives $S_{c_k}^{\mathcal{M}} = \partial\mathcal{M}(e^{j\omega})/\partial c_k$. Considering a frequency response of the form $\mathcal{H}(e^{j\omega}) = \mathcal{M}(e^{j\omega})e^{j\Theta(\omega)}$, as well as a constant c, which expresses any multiplicative coefficient c_k $(k = 1, 2, \ldots, m)$, it easily follows that

$$\begin{aligned} S_c^{\mathcal{H}}(e^{j\omega}) = \frac{\partial\mathcal{H}(e^{j\omega})}{\partial c} &= \frac{\partial}{\partial c}\Big[\mathcal{M}(e^{j\omega})e^{j\Theta(\omega)}\Big] = \frac{\partial}{\partial c}\Big[\mathcal{M}(e^{j\omega})\cos[\Theta(\omega)] + j\mathcal{M}(e^{j\omega})\sin[\Theta(\omega)]\Big] \\ &= \frac{\partial}{\partial c}\Big[\mathcal{M}(e^{j\omega})\cos[\Theta(\omega)]\Big] + j\frac{\partial}{\partial c}\Big[\mathcal{M}(e^{j\omega})\sin[\Theta(\omega)]\Big]. \end{aligned}$$

Performing the derivatives, we arrive at the results

$$\begin{aligned} \Re\{S_c^{\mathcal{H}}(e^{j\omega})\} &= \frac{\partial}{\partial c}\Big[\mathcal{M}(e^{j\omega})\cos[\Theta(\omega)]\Big] = \frac{\partial\mathcal{M}(e^{j\omega})}{\partial c}\cos[\Theta(\omega)] + \mathcal{M}(e^{j\omega})\frac{\partial\cos[\Theta(\omega)]}{\partial c} \\ &= \cos[\Theta(\omega)]\frac{\partial\mathcal{M}(e^{j\omega})}{\partial c} - \mathcal{M}(e^{j\omega})\sin[\Theta(\omega)]\frac{\partial\Theta(\omega)}{\partial c} \qquad \text{and} \\ \Im\{S_c^{\mathcal{H}}(e^{j\omega})\} &= \frac{\partial}{\partial c}\Big[\mathcal{M}(e^{j\omega})\sin[\Theta(\omega)]\Big] = \frac{\partial\mathcal{M}(e^{j\omega})}{\partial c}\sin[\Theta(\omega)] + \mathcal{M}(e^{j\omega})\frac{\partial\sin[\Theta(\omega)]}{\partial c} \\ &= \sin[\Theta(\omega)]\frac{\partial\mathcal{M}(e^{j\omega})}{\partial c} + \mathcal{M}(e^{j\omega})\cos[\Theta(\omega)]\frac{\partial\Theta(\omega)}{\partial c}. \end{aligned}$$

The above two equations constitute a *system of two equations with unknowns the sensitivities of the amplitude response* $\mathcal{M}(e^{j\omega})$ *and the phase response* $\Phi(\omega)$ *to the change of the parameter* c, which are defined as

$$S_c^{\mathcal{M}} = \frac{\partial\mathcal{M}(e^{j\omega})}{\partial c} \quad \text{and} \quad S_c^{\Theta} = \frac{\partial\Theta(\omega)}{\partial c}.$$

The solution of the above system leads to the results

$$S_c^{\mathcal{M}} = \frac{\partial\mathcal{M}(e^{j\omega})}{\partial c} = \cos[\Theta(\omega)]\Re\{S_c^{\mathcal{H}}(e^{j\omega})\} - \sin[\Theta(\omega)]\Im\{S_c^{\mathcal{H}}(e^{j\omega})\} \tag{12.28}$$

$$S_c^{\Theta} = \frac{\partial\Theta(\omega)}{\partial c} = \frac{1}{\mathcal{M}(e^{j\omega})}\Big\{\cos[\Theta(\omega)]\Im\{S_c^{\mathcal{H}}(e^{j\omega})\} - \sin[\Theta(\omega)]\Re\{S_c^{\mathcal{H}}(e^{j\omega})\}\Big\}. \tag{12.29}$$

[15] See Crochiere R.E., *A New Statistical Approach to the Coefficient Word Length Problem for Digital Filters*, IEEE Transactions on Circuit and Systems, Vol CAS-22, No. 3, March 1975, page 191 and beyond.

We observe that the calculation of the quantities $S_c^{\mathcal{M}}$ and S_c^{Θ} requires knowledge of the functions $\Re\{S_c^{\mathcal{H}}(e^{j\omega})\}$ and $\Im\{S_c^{\mathcal{H}}(e^{j\omega})\}$, which in turn requires the knowledge of the partial derivative $S_c^{\mathcal{H}} = \partial\mathcal{H}(z)/\partial c$ that expresses the sensitivity of the system's transfer function to the change of the coefficient c. If we divide the digital filter into two parts, of which the first extends from the filter's input to the multiplier of the signal with the considered coefficient c and the second from the output of the multiplier to the output of the filter, it is proven that this sensitivity equals the product of the transfer functions of these two subsystems. After calculating the sensitivity $S_c^{\mathcal{H}}$ we can calculate the value of the coefficient $S_c^{\mathcal{M}}$ for each coefficient and from these values finally calculate the statistical word length.

12.9.2 Indicative Examples of Coefficient Quantization

In this section, two well-known examples from the literature are presented, representing the application of the process of quantizing the filter's coefficients and the inaccuracy that arises from the limited ability to represent the values of the coefficients owing to finite accuracy. Because of the complexity that characterizes both the calculations of parameter values and the process of designing the graphs involved, in practice these calculations and the process of creating diagrams are based on the use of specialized software such as the MATLAB application.

Example 12.9.1 (Diniz, da Silva, Netto, Example 11.6) Assuming that the polynomial of the denominator of the transfer function of a second-order step expressed in the direct form is the $D(z) = z^2 + \alpha_1 z + \alpha_2$, the representation of the coefficients α_1 and α_2 with 6 bits, which include the sign bit, is depicted in the left and right images of the inset figure for the direct and conjugate form respectively. Specifically, this figure depicts the possible positions of the poles of the transfer function in the first quadrant of the complex plane and inside the unit circle, whereas in the remaining quadrants the corresponding pole positions are characterized by symmetry and result from the reflection of the poles depicted here, in the appropriate way. Notice that the arrangement of the poles in the grid areas that are close to the positions $z = 0$ and $z = 1$ are characterized by very low density, leading, therefore, to a very sparse grid, a fact that is responsible for the limited accuracy of the filter design, especially in the case of applications that require high sampling rates.

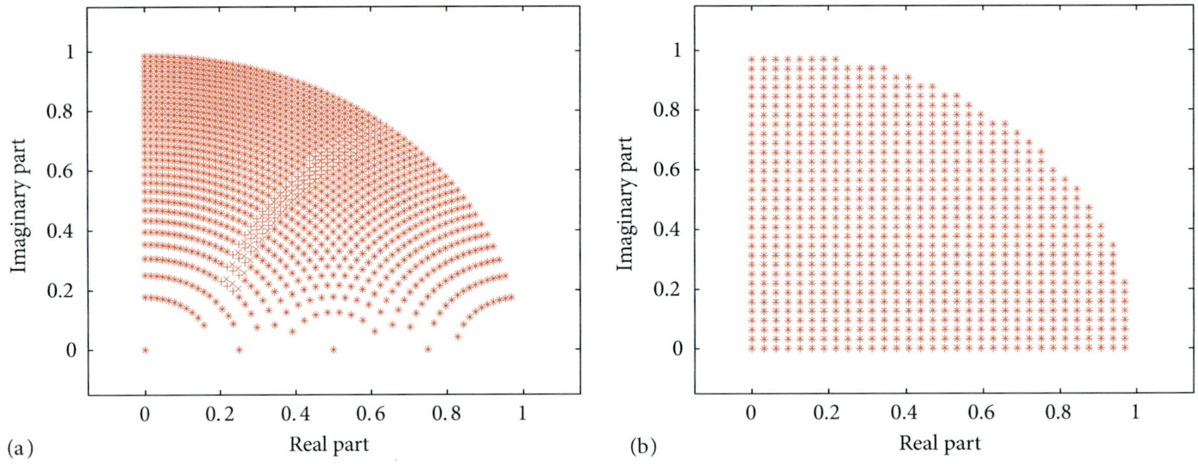

Example 12.9.2 (Proakis, Manolakis, Example 9.5.3) The attached figure shows the effects of the quantization process with 3 bits on the amplitude response of a linear phase band-pass FIR filter, with an order value of $M = 32$ that has been implemented according to the direct form. Observing the right image of the figure that presents the frequency response of the filter with the quantized coefficients in relation to the corresponding response of the nonquantized filter depicted in the left image, the increase in the sidelobes by several decibels is evident, something that does not apply to the case of coefficient quantization with 4 bits, for which the effects on the amplitude response are practically negligible. As Proakis & Manolakis characteristically comment, the reliable representation of the coefficients of such a filter requires the use of 12 to 14 bits with a minimum value for this number being 10 bits.

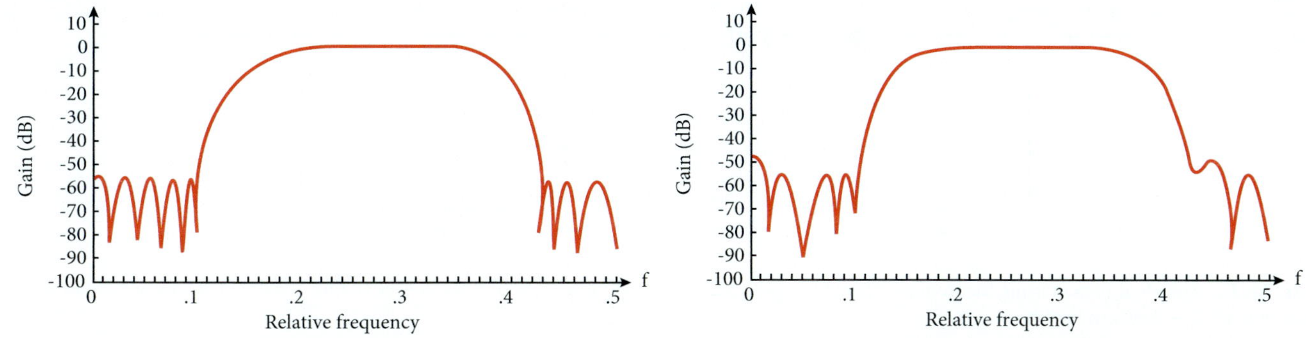

(a) No quantization

(b) Quantization to 3 digits

12.9.3 Rounding Errors and Quantization Noise

After presenting the consequences of the quantization of the filter coefficients on the behavior of the system and, more specifically, on the positions of the poles and the zero values of the transfer function, let us now study *the more general problem of the impact of finite accuracy on the operation of a real system.* In the following description, we limit ourselves to IIR systems and let us consider, for educational purposes, a simple pole system, which is described by a *simple first-order difference equation* of the form

$$y[n] = \alpha y[n-1] + x[n]$$

and uses *fixed-point arithmetic*. According to what we have mentioned, although theoretically the knowledge of the values of the parameters of the above equation, as well as the execution of the additions and multiplications, *is possible with infinite accuracy*, in practice this *cannot happen*, since the registers of real digital systems have a *finite length* of at least $K+1$ bits (and more specifically, K bits for the value of the number and one bit for the sign). This means that the value of the coefficient α and the previous output $y[n-1]$ will in fact be represented by binary numbers of $K+1$ bits in length, let us say α_q and $y^q[n-1]$, whereas for the same reason, the arithmetic operations, i.e., the additions and multiplications that will be performed by the binary system, will be applied to binary numbers characterized by this length in bits. For example, the multiplication of the parameter α^q with the value $y^q[n-1]$, both of which have a length of $K+1$ bits, will result in a product whose value will have a length of $2K+1$ bits. However, since the length of the system's registers is $K+1$ bits, we will have to proceed with the *truncation* of the K least significant bits of the product value, using for this purpose a *quantizer*. Consequently, the difference equation of the real system will be

$$y^q[n] = Q(\alpha^q y^q[n-1]) + x^q[n],$$

since the input and output signals $x[n]$ and $y[n]$ are also encoded, using *the same number of binary digits*.

Therefore, and in accordance with what we mentioned above, at the level of implementation, the real system *will not be represented by the ideal diagram of the left image of Fig. 12.32 but by the real diagram of the right image of this figure*, which includes quantizers that perform the quantization of numerical values used in the system. Generalizing the above description, in the general case, instead of the theoretical system described by the difference equation

$$y[n] = -\sum_{k=1}^{N} \alpha_k y[n-k] + \sum_{k=0}^{M} \beta_k x[n-k], \tag{12.30}$$

the real quantized system will be used, the difference equation of which is

$$y[n] = -\sum_{k=1}^{N} Q(\alpha_k^q y^q[n-k]) + \sum_{k=0}^{M} Q(\beta_k^q x^q[n-k]), \tag{12.31}$$

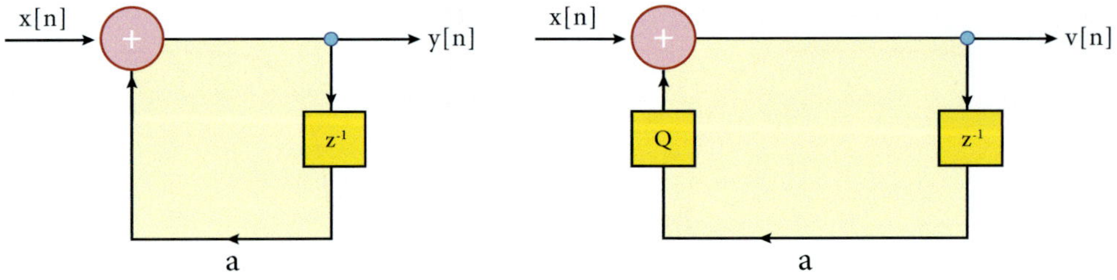

Fig. 12.32 The ideal system using theoretical values of infinite accuracy (left image) and the real system using quantized values, which are binary numbers of $K+1$ bits in length

and which arises, by adding to the direct form of the IIR system the appropriate number of quantizers, which are connected at the appropriate points, such as after the points of product calculation, like the one we commented on previously.

The main problem with this approach is that *whereas the ideal IIR system characterized by infinite accuracy representation is a linear and time-invariant discrete-time system, the system resulting from the addition of the quantizers does not fall into this category, as the quantizer is known to be a nonlinear arrangement*. If the system remains as is, it is *not easy* to model the nonlinear phenomena associated with the quantization of the product of two numbers and the truncation of the sum digits, with the aim of avoiding overflow, especially *for large systems, which include many multipliers and adders*. In order to address this problem, we replace each quantizer *with a noise source*, the value of which is considered to be *equal to the quantization error at the output of each quantizer*. This approach gives us the ability to study *in a quantitative way* the impact of quantization errors on the output of digital systems and the way it is implemented is depicted in Fig. 12.33. In this figure, the top image represents *the usual straight form* of the system characterized by *numerical values of infinite precision*, the middle image represents the non-linear quantized system, the numerical values of which are encoded with $K+1$ bits and therefore, are characterized by finite precision, whereas the lower image arises *by replacing each quantizer with a noise source*. As for the type of quantization noise analysis, this is carried out adopting two different approaches, the first of which is *deterministic* and the second *stochastic* in nature. The disadvantage of the deterministic approach that was studied in control systems of data originating from sampling (sampled-data system), is that the upper limit for the quantization error, *only applies to the case in which the quantization errors are combined with each other in the worst possible way*, thus leading to very pessimistic estimates that *do not always correspond* to reality. However, they are the only tool we have at our disposal for the study of systems, in which the prerequisites for the application of the second alternative approach, which as we mentioned is *stochastic in nature*, do not exist, as happens, for example, in systems characterized by the appearance of *limit cycles* and the *dead zone phenomenon* that are studied subsequently. In this second approach, each quantizer is replaced by a *source of additive noise*, with the signal generated by it being described by the expression

$$e[n] = Q[c_i x[n]] - c_i x[n]$$

and consequently, it is related to the quantization of the value of the product of two numbers. In other words, the output of a real multiplier characterized by finite word length will have the form

$$Q[c_i x[n]] = c_i x[n] + e[n],$$

with the system in question being represented by the schematic diagram of Fig. 12.34.

The basic prerequisite that must be met in order for the application of this second stochastic approach to be possible is that *the sequences of quantization noise originating from the various multipliers of the underlying physical system are statistically independent and uncorrelated signals, both among themselves, and with the input signal*. In this case, we can consider that *the sequence of quantization noise is described by a uniform probability density function*. Although the power of the above assumptions generally leads to quite accurate results with the reliability of this white noise model improving with the increase in the input signal values and the dispersion of the power spectrum that describes it, there are cases, such as *limit cycles* and more generally, *any periodic signal*, that are associated with a linear spectrum. In this case, the quantization error is *periodic* and therefore, this stochastic approach cannot be applied. In any case, as usually happens, the final judge for the characterization of accuracy and the degree of success of the application of the stochastic approach, is, as usual, the *experiment*.

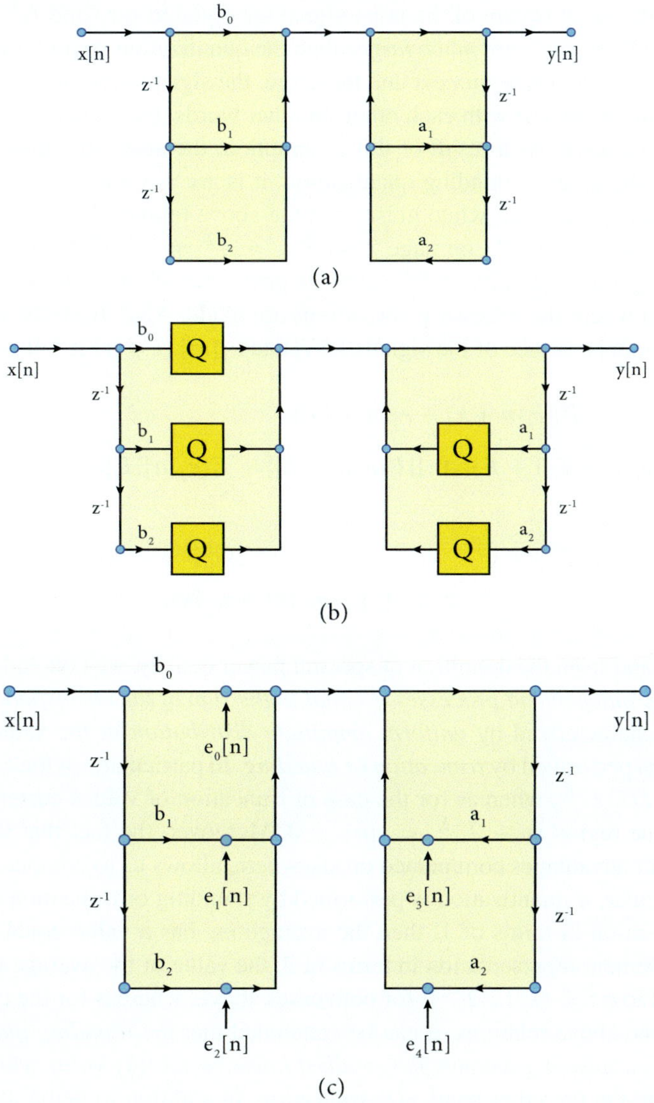

Fig. 12.33 The ideal system using theoretical values of infinite accuracy (top image), the real system using quantized values of finite accuracy (middle image), and the system resulting from the replacement of nonlinear quantizers by sources of linear noise (bottom image)

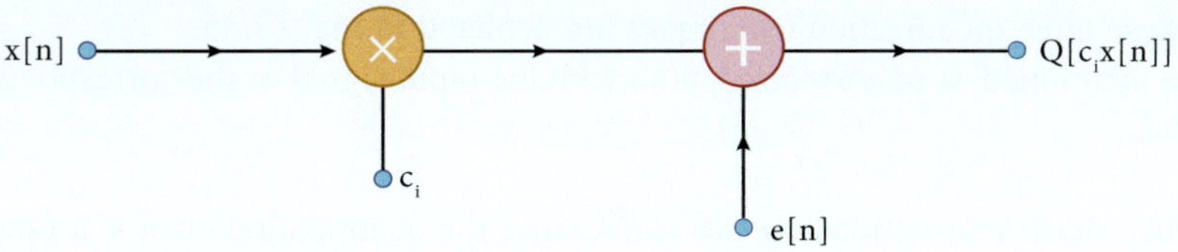

Fig. 12.34 Multiplier model with finite word length

Let us describe in detail this stochastic approach and its application prerequisites. Let us assume that in our system there are L quantizers in total, which are required for the representation of finite accuracy of all kinds of numerical values, which have been replaced by *equally numerous additive noise sources* $e_i[n]$ $(i =, 1, 2, \ldots, L)$. If we accept that the quantization process is carried out with *rounding*, these sources of additive noise are considered as *random processes with uniform probability density* the value of which is usually equal to $1/\Delta$ in the range of values $-\Delta/2 \leq e_i[n] \leq \Delta/2$, where Δ is the quantization step and equal to zero anywhere else. These additive noise sources are characterized by the following properties:

- The values of the mean and the mean square of the noise signal are equal to zero and $\Delta^2/12$ respectively.
- If the signal values across the filter range are *much larger* than the quantization step Q, then each noise source constitutes a *stationary in the broad sense white noise process* and therefore, the signals $e_i[n]$ and $e_i[m]$ for values $n \neq m$ as well as $e_i[n]$ and $e_j[m]$ $(i \neq j)$ are *uncorrelated* with each other. In other words, the signals are *uncorrelated*, both *from sample to sample* and *from source to source*. As a result of this assumption, the autocorrelation function for quantization noise is $r_{ee}[n] = \sigma^2\delta[n] + m_e^2$. In the case of rounding quantization, it is $m_e = 0$ (note that this condition applies even when the noise sequence is assumed to represent white noise) and the above relationship becomes $r_{ee}[n] = \sigma^2\delta[n]$ leading to spectral power density (see Exercise 3.3.23 on page 254) $\mathcal{R}_{ee} = \sigma^2 = \Delta^2/12$ for frequency values $\omega \leq \pi$. In other words, in the case of rounding, the variability and the average power *are identical to each other*, something that *does not hold* in the case of truncation where the necessary corrections are made. Also, from the definition of autocorrelation and the assumption of statistical independence of the signals $e_i[n]$ and $e_j[m]$ $(i \neq j)$ it follows that

$$\begin{aligned} r_{e_i+e_j}[k] &= E\{[e_i[n] + e_j[n]][e_i[n+k] + e_j[n+k]]\} \\ &= E\{[e_i[n]e_i[n+k]]\} + E\{e_i[n]\}E\{e_j[n+k]\} + E\{e_j[n]\}E\{e_i[n+k]\} + E\{[e_j[n]e_j[n+k]]\} \end{aligned}$$

  and finally

$$r_{e_i+e_j}[k] = r_{e_i}[k] + r_{e_j}[k].$$

  From the above relationship and from the definition of spectral power density, we conclude that *the spectral power density of the sum of two statistically independent processes is equal to the sum of the corresponding spectral power quantities.*
- Additive noise sources are characterized by *uniform amplitude distribution* in the value range dictated by the type of quantization, i.e., whether it is performed by *truncation* or *rounding*. In particular, for the case of rounding this value region is $-(1/2)2^{-K} < e[n] \leq (1/2)2^{-K}$, whereas for the case of truncation of values corresponding to the 2 representation of the complement, this value region is $-2^{-K} < e[n] \leq 0$. Moreover, the fact that the noise model used is a *linear model*, in addition to the other advantages commented on elsewhere, allows us to compute the values of quantities such as *mean* and *variance*. In particular, if quantization is performed by rounding or truncation of the magnitude corresponding to the complement representation in terms of 1, then the average m_e has a value equal to zero, whereas in the case of truncation and for the complement representation in terms of 2, the value of the average is equal to $-2^{-K}/2$. As for the value of *variance*, it is equal to $\sigma^2 = (2/12)2^{-2K}$ for both cases above, whereas for the case of *size truncation* it is equal to $\sigma^2 = (2/3)2^{-2K}$. From the above relations, it can be concluded that *the rounding quantization technique is superior to the other two techniques* because it generates *the smallest noise variability value while keeping the average value of the quantization error at zero*.On the other hand, size truncation, in addition to being associated with a larger volatility value, *leads to a higher degree of correlation between the signal and the quantization noise*,which, based on what we have mentioned, constitutes a *serious drawback*, but nevertheless, it is characterized by the very important advantage, *suppressing the occurrence of quantum limit cycles in digital filters*, as we discuss in another section. The probability density functions for these three quantization techniques are depicted in Fig. 12.35.
- The noise signal, as we mentioned, is *uncorrelated*, both with the input signal of the corresponding quantizer and with the system input signal $x[n]$.

Note that although in reality, these assumptions *are not valid*, since the quantization error is a function of the input signal of the quantizer, in general, practice has shown that, relying on these assumptions, *we have the ability to calculate accurately statistical quantities, such as the average value, variability, and correlation*, even for complex signals characterized by *wide bandwidth and rapid temporal changes*, such as *vocal signals*.

From the last image of Fig. 12.33, we find that for the general case of the IIR system, there are $M + 1$ quantizers for the quantization of the products $\beta_k x[n-k]$ $(k = 0, 1, 2, \ldots, M)$ and N quantizers for the calculation of the products $\alpha_k y[n-k]$ $(k = 1, 2, \ldots, N)$. If all these quantizers are replaced by an equal number of noise sources, they will appear in total in the circuit $(M + N + 1)$ noise signals $e_k[n]$ $(k = 0, 1, 2, \ldots, M + N)$. Under the regime of the power of the above assumptions, we can combine all the above noise signals into a simple noise signal of the form

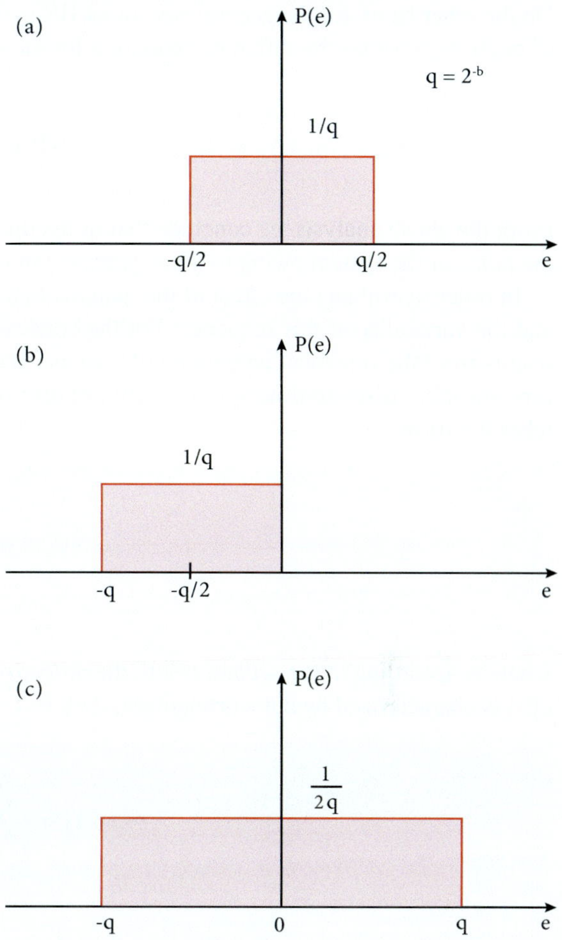

Fig. 12.35 The probability density functions for the quantization of the product value with truncation (**a**), value quantization (**b**), and magnitude quantization (**c**)

$$e[n] = \sum_{k=0}^{M+N} e_k[n] = e_0[n] + e_1[n] + \cdots + e_{M+N-1}[n] + e_{M+N}[n].$$

This noise signal comes from a source that is connected between the two subsystems, which are connected to each other to form the considered IIR system and with the simplified system that results, to be equivalent to the original (although we must note, that *the assumptions about the lack of correlation of noise signals, both from sample to sample and from source to source, do not apply in the case of size clipping, owing to the inherent correlation that exists between the signal and the quantization noise*, so the results are not particularly accurate). This noise signal, in which the contributions *from all sources of linear noise* have replaced an equal number of quantizers in the system, is characterized by a variability value that is equal to the sum of the corresponding variability values from the $M + N + 1$ noise sources, a property that is mathematically formulated as

$$\sigma_e^2 = (M + N + 1)\frac{2^{-2K}}{12}.$$

As for the output of the system, the fact that the noise signal $e[n]$ and the input sequence $x[n]$ are *uncorrelated between the signs* allows us to consider that it consists of *the sum of two responses* and more specifically, from the response of the system $y_x[n]$ to the input signal $x[n]$ and from the response of the system $y_e[n]$ to the noise signal $e[n]$. For the simple case of a single pole system, the difference equations for the responses $y_x[n]$ and $y_e[n]$ are

$$y_x[n] = \alpha y_x[n-1] + x[n]$$

$$y_e[n] = \alpha y_e[n-1] + e[n]$$

On the other hand, for the general case of an IIR system, the equation for the response $y_x[n]$ is the *usual difference equation* of order N, whereas the difference equation for the response $y_e[n]$ is the

$$y_e[n] = \sum_{k=1}^{N} \alpha_k y_e[n-k] + e[n].$$

From the above analysis we conclude that in the direct form I of the IIR systems, the properties of the noise depend only on the poles of the system owing to the appearance in the above equation of the coefficients α_k of the filter.

In order to evaluate the effect of the quantization noise at the output of the system, let us calculate the values of the mean and the variability of this sequence. For the construction of the equations for these two quantities, we resort to two known results from the statistical processing of signals. According to the first result, the *average value of the output* of a linear and time-invariant discrete-time system, with impulse response $h[n]$, when a random sequence $e[n]$ is transmitted to its input, takes the form

$$m_y = m_e \sum_{n=-\infty}^{\infty} h[n] = m_e \mathcal{H}(e^{j0}), \tag{12.32}$$

whereas according to the second result, the *autocorrelation* of the output sequence $y_e[n]$ of this filter, when the input signal $e[n]$ is characterized by autocorrelation $r_{ee}[n]$, is

$$r_{y_e y_e}[n] = \sum_{k=-\infty}^{\infty} \sum_{\ell=-\infty}^{\infty} h[k]h[\ell] r_{ee}[k-\ell+n], \tag{12.33}$$

where the autocorrelation function $r_{ee}[n]$ of the quantization noise, as we commented, takes the form $r_{ee}[n] = \sigma_e^2 \delta[n]$. It will therefore be,

$$r_{y_e y_e}[n] = \sum_{k=-\infty}^{\infty} \sum_{\ell=-\infty}^{\infty} h[k]h[\ell]\sigma_e^2 \delta[k-\ell+n] = \sigma_e^2 \sum_{k=-\infty}^{\infty} h[k]h[n+k]$$

since $\delta[n] = 1$ for $n = 0$ and therefore $\delta[k-\ell+n] = 1$ for $k-\ell+n = 0$ i.e., for $n = \ell - k$, from which it follows $\ell = k+n$. The *variability* of the output noise results from the function $r_{y_e y_e}[n]$ for the value $n = 0$—it will therefore be,

$$\sigma_{y_e}^2 = \sigma_e^2 \sum_{k=-\infty}^{\infty} |h[k]|^2 = \frac{\sigma_e^2}{2\pi} \int_{-\pi}^{\pi} |\mathcal{H}(e^{j\omega})|^2 d\omega,$$

where for the derivation of the last expression, we used *Parseval's theorem*. The ratio $\sigma_{y_e}^2/\sigma_e^2$ is known as *noise gain or relative noise variability* and constitutes a performance criterion for evaluating the impact of the roundoff noise, at the output of the digital filter. Using the expression for the variability σ_e^2 that we mentioned above we get[16]

[16] Results related to the analysis of roundoff noise for the case of floating-point arithmetic can be found in the references Smith L.M., Bomar B.W., Joseph R.D., & Yang G.C.-J., *Floating-Point Roundoff Noise Analysis of Second-Order State-Space Digital Filter Structures*, IEEE Transactions on Circuits and Systems II: Analog and Digital Signal Processing, 39, 90–98, 1992 and Bomar B.W., Smith L.M., & Joseph R.D., *Roundoff Noise Analysis of State-Space Digital Filters Implemented on Floating-Point Digital Signal Processors*, IEEE Transactions on Circuits and Systems II: Analog and Digital Signal Processing, 44, 952–955, 1997.

$$\sigma_{y_e}^2 = (M+N+1)\frac{2^{-2K}}{12}\sum_{k=-\infty}^{\infty}|h[k]|^2 = (M+N+1)\frac{2^{-2K}}{12}2\pi\int_{-\pi}^{\pi}|\mathcal{H}(e^{j\omega})|^2 d\omega. \tag{12.34}$$

The direct application of this expression for first- and second-order systems described respectively by the transfer functions

$$\mathcal{H}(z) = \frac{\beta}{1-\alpha z^{-1}}\ (|\alpha|<1) \quad \text{and} \quad \mathcal{H}(z) = \frac{\beta_0+\beta_1 z^2+\beta_2 z^{-2}}{(1-re^{j\vartheta}z^{-1})(1-re^{-j\vartheta}z^{-1})}$$

and contain respectively a simple pole $p=\alpha$ and two complex conjugate poles at the positions $p_{1,2}=re^{\pm j\vartheta}$ leads us to the results

$$\sigma_{y_e}^2 = 2\frac{2^{-2K}}{12}\left(\frac{1}{1-|\alpha|^2}\right) \quad \text{and} \quad \sigma_{y_e}^2 = 5\frac{2^{-2K}}{12}\left(\frac{1+r^2}{1-r^2}\right)\frac{1}{r^4+1-2r^2\cos 2\vartheta}$$

Indeed, in the first case it is $M=0$ and $N=1$ and therefore $M+N+1=2$, whereas in the second case, it is $M=2$ and $N=2$ and therefore $M+N+1=5$. The rest of the expression consists in the multiplicative factor $2^{-2K}/12$ and from an expression that arises from the calculation of the sum or the integral depending on which of the two expressions related to Parseval's theorem is used. The reader is invited as an exercise, to verify the correctness of these results, performing the required operations. Similar results arise if we use the direct form II using the appropriate equations related to it. Specifically, it is proven that in this case the expression applies

$$\sigma_{y_e}^2 = N\frac{2^{-2K}}{12}\sum_{k=-\infty}^{\infty}|h[k]|^2 + (M+1)\frac{2^{-2K}}{12} = N\frac{2^{-2K}}{12}\frac{1}{2\pi}\int_{-\pi}^{\pi}|\mathcal{H}(e^{j\omega})|^2 d\omega + (M+1)\frac{2^{-2K}}{12},$$

where once again we resorted to the use of *Parseval's theorem*. Based on all the above, we come to the conclusion that the direct forms I and II *are affected in different ways by the quantization of the values of the products during the implementation of the system's difference equations*, whereas the same applies to other implementations such as series and parallel implementations. In general, *it is difficult to determine* which of these implementations lead to the creation of less noise at the output. However, it is certain, as Kaiser showed[17] that an implementation of a filter that arises from the series or parallel connection of first- and second-order stages behaves *much better*, in terms of quantization noise, than direct implementations, and for this reason is preferred in practice.

12.9.4 Quantization of Product Number by Signal-Rounding Error

The simplest case of quantization used in practice is the *quantization of the product* $\alpha x[n]$, where α is a numerical real coefficient and $x[n]$ is a signal that is either read from the *input* of the filter or constitutes some *intermediate signal* that is circulated inside the filter. The way in which the quantization process is applied in this case is depicted in Fig. 12.36. Initially, the *quantization* of the signal value $x[n]$ takes place through its passage through a quantizer, which creates the quantized value $Q(x[n])$. Then, this value is passed to the multiplier, which calculates the product $\alpha x[n]$ (the coefficient α is assumed to be already expressed in the appropriate (binary) form and stored in the system, so it does not need quantization), whereas in the end the product $\alpha Q(x[n])$ *it is subject to new quantization*, with the final value created being $Q(\alpha Q(x[n]))$. On the other hand, if the product αx is calculated at some internal point of the filter, then the value of x is already quantized with $K+1$ bits (remember that this parameter is connected to the quantization step through the relation $\Delta = 2^{-K}$) and therefore it is described by a *discrete random variable* with a probability mass function $p_x(n\Delta)$ and characteristic function (see Sect. A.14.8 on page 1540)

[17] See Kaiser F., *Some Practical Considerations in the Realization of Linear Digital Filters*, in Proceedings of the 3rd Allerton Annual Conference on Circuit and System Theory, 1965.

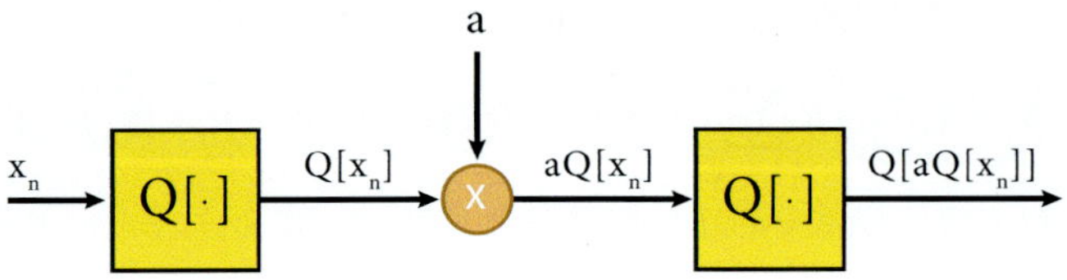

Fig. 12.36 The quantization process of the product $\alpha x[n]$

$$\Phi_x(\omega) = \sum_{n=-\infty}^{\infty} p_x(n\Delta)e^{-j\omega n\Delta}. \tag{12.35}$$

After calculating the value of the product αx, it undergoes *truncation* so that the first K bits are preserved and this is exactly the process that leads to the appearance of the rounding error.

The sequence of the quantization error that is created during the application of this process has the form

$$\begin{aligned} e(x[n], \alpha) \equiv e[n] &= Q(\alpha Q(x[n])) - \alpha x[n] = Q(\alpha Q(x[n])) - \alpha x[n] + \alpha Q(x[n]) - \alpha Q(x[n]) \\ &= [Q(\alpha Q(x[n])) - \alpha Q(x[n])] + \alpha(Q(x[n]) - x_n) = e(\alpha Q(x[n])) + \alpha e(x[n]), \end{aligned}$$

where $e(\alpha Q(x[n])) = Q(\alpha Q(x[n])) - \alpha Q(x[n])$ and $e(x[n]) = Q(x[n]) - x_n$ and its graphical representation is shown in Fig. 12.37. We are therefore led to the conclusion that *the error caused by the quantization of the product of a real coefficient times a signal consists of two terms, the first of which $e(\alpha Q(x[n]))$ is the rounding error associated with the quantization of the value of the product $\alpha Q(x[n])$, whereas the second term $\alpha e(x[n])$ is related to the usual quantization error, which appears during the conversion of the signal $x[n]$ from analog to digital*. Considering that the upper limit of the quantization error $e(x[n])$, when there is no overflow, is equal to $\Delta/2$, where Δ is the quantization step and considering that the same will apply to the rounding error $e(\alpha Q(x[n]))$, since this process essentially constitutes *quantization*, we conclude that the upper limit of the error $e[n]$ under the regime of absence of overflow will be equal to $\alpha\Delta/2 + \Delta/2 = (\alpha + 1)\Delta/2$. It is not difficult to understand that the error $\alpha\Delta/2$ that arises if we *first* proceed to the quantization of the signal $x[n]$ and *then* multiply the value $Q(x[n])$ by α depends on α and is greater than the error $\Delta/2$ that would arise if we *first* perform the multiplication $\alpha x[n]$ and *then* proceed to the quantization of the product, since regardless of the value that is quantized, the upper limit of the quantization error is always equal to $\Delta/2$. Therefore, $\alpha e(x[n]) > e(\alpha x[n])$. From the above expression, useful conclusions can also be drawn *regarding the contribution of each of the two above errors to the total quantization error*. We find, for example, that if the value of the multiplicative coefficient α is quite large, then the quantization error constitutes the *dominant error*, since its value is multiplied by α, whereas in the opposite case, the most significant of the two errors is the rounding error.

From the study of the graphical representation of Fig. 12.37 it is not difficult to ascertain that the rounding error $e(x, \alpha)$ is a *periodic function* of x of the form $e(x, \alpha) = -\alpha x$ for the range of values $-(1/2\alpha) \leq x < (1/2\alpha)$ and that it satisfies the property

$$e\left(x + \frac{1}{\alpha}, \alpha\right) = e(x, \alpha).$$

This error is also *periodic* with respect to the coefficient α with a period value equal to 1 and for this reason, it is sufficient to study the behavior of the error in the range of values [0, 1]. Note that despite the fact that the periodicity of the error allows us to express it as a Fourier series, the use of the corrective term we presented in the Footnote 14 (see page 1455) is required, since the Fourier series converges to zero at the points of discontinuity of the function $e(x, \alpha)$, whereas the required value at these points is equal to 0.5. The addition of this term significantly increases the complexity of the expressions that arise and this issue will not concern us further.

Regarding the *statistical properties* of the above errors, the values $E\{e(x_n)\}$, $E\{e^2(x_n)\}$ and $E\{e(x_n)e(x_{n+m})\}$ are given respectively by Eqs. (13.155), (13.156) and (13.158) (pages 1455 and 1456), whereas the corresponding expressions for the error $e(\alpha Q(x_n))$ are[18]

[18] In these expressions, we assume that the involved signal is characterized by a *continuous probability density* and therefore, *no correction term is required*, which we described in Footnote 14 (see page 1455). For the full version of these expressions in which this correction term is taken

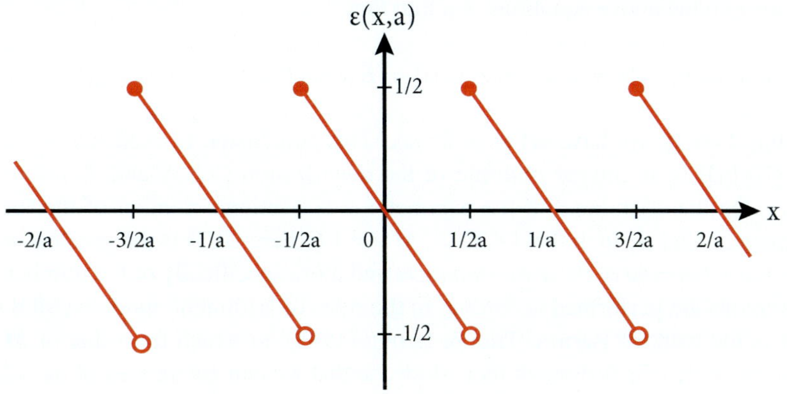

Fig. 12.37 The graphical representation of the rounding error function

$$E\{e(\alpha Q(x_n))\} = \frac{\Delta}{j2\pi} \sum_{\substack{k=-\infty \\ k\neq 0}}^{\infty} \sum_{\ell=-\infty}^{\infty} \frac{(-1)^k}{k} \Phi_x\left(\frac{2\pi(k\alpha+\ell)}{\Delta}\right) \operatorname{sinc}(k\alpha+\ell) \tag{12.36}$$

$$E\{e^2(\alpha Q(x_n))\} = \frac{\Delta^2}{12} + \frac{\Delta^2}{2\pi^2} \sum_{\substack{k=-\infty \\ k\neq 0}}^{\infty} \sum_{\ell=-\infty}^{\infty} \frac{(-1)^k}{k^2} \Phi_x\left(\frac{2\pi(k\alpha+\ell)}{\Delta}\right) \operatorname{sinc}(k\alpha+\ell) \tag{12.37}$$

$$E\{e(x_n), e(\alpha Q(x_n)\} = \frac{\Delta^2}{4\pi^2} \sum_{\substack{k=-\infty \\ k\neq 0}}^{\infty} \sum_{\substack{\ell=-\infty \\ \ell\neq 0}}^{\infty} \sum_{i=-\infty}^{\infty} \frac{(-1)^{k+\ell+1}}{k\ell} \Phi_{x_n}\left(\frac{2\pi(k\alpha+\ell+i)}{\Delta}\right) \operatorname{sinc}(k\alpha+i) \tag{12.38}$$

$$E\{e(x_n), e(\alpha Q(x_{n+m})\} = \frac{\Delta^2}{4\pi^2} \sum_{\substack{k=-\infty \\ k\neq 0}}^{\infty} \sum_{\substack{\ell=-\infty \\ \ell\neq 0}}^{\infty} \sum_{i=-\infty}^{\infty} \frac{(-1)^{k+\ell+1}}{k\ell} \Phi_{x_n,x_{n+m}}\left(\frac{2\pi k}{\Delta}, \frac{2\pi(\ell\alpha+i)}{\Delta}\right) \operatorname{sinc}(\ell\alpha+i) \tag{12.39}$$

and

$$E\{e(\alpha Q(x_n)), e(\alpha Q(x_{n+m})\} = \frac{\Delta^2}{4\pi^2} \sum_{\substack{k=-\infty \\ k\neq 0}}^{\infty} \sum_{\substack{\ell=-\infty \\ \ell\neq 0}}^{\infty} \sum_{r=-\infty}^{\infty} \sum_{s=-\infty}^{\infty} \frac{(-1)^{k+\ell+1}}{k\ell} \tag{12.40}$$

$$\Phi_{x_n,x_{n+m}}\left(\frac{2\pi(k\alpha+r)}{\Delta}, \frac{2\pi(\ell\alpha+s)}{\Delta}\right) \operatorname{sinc}(k\alpha+r)\operatorname{sinc}(\ell\alpha+s). \tag{12.41}$$

In the above expressions, the parameter α is a *real number*, whereas if it takes integer values we observe that

$$\Phi_{x_n}(\xi(k\alpha+\ell))\operatorname{sinc}(k\alpha+\ell) = \begin{cases} 1 & \text{if } k\alpha+r=\ell\alpha+s=0 \\ 0 & \text{in the opposite case,} \end{cases}$$

whereas similarly

$$\Phi_{x_n,x_{n+m}}(\xi(k\alpha+r), \xi(\ell\alpha+s))\operatorname{sinc}(k\alpha+r)\operatorname{sinc}(\ell\alpha+s) = \begin{cases} 1 & \text{if } k\alpha+\ell=0 \\ 0 & \text{in the opposite case.} \end{cases}$$

into account, as well as for their analytical proofs, see the work of Wong P.W., *Quantization Noise, Fixed-Point Multiplicative Roundoff Noise, and Dithering*, IEEE Transactions on Acoustics, Speech, and Signal Processing, Vol. 38, No. 2, pp 286–300, 1990.

If we substitute these values into the above equations, we find that

$$E\{e(\alpha Q(x_n))\} = E\{e^2(\alpha Q(x_n))\} = E\{e(\alpha Q(x_n)), e(\alpha Q(x_{n+m})\} = 0,$$

where for reasons of writing brevity, we have set $\xi = 2\pi/\Delta$. This conclusion is expected, since if the coefficient α is an integer, then the value $\alpha Q(x[n])$ is an integer multiple of the quantization step Δ and therefore, the error $E\{e(\alpha Q(x_n))\}$ is *identically* equal to zero. On the other hand, if the parameter α is a rational number of the form L/M, where L and M are prime to each other integers, that is, such that $\mathrm{GCD}(L, M) = 1$ (where GCD is the greatest common divisor) the above equations can be expressed *as a function of these two numbers* and more specifically of the number of the *denominator*, since the related arithmetic operations are performed according to the rules of arithmetic modulo (MOD) M. An interesting study of this case can be found in the work of Barnes, Tran & Lenug (1985), in which the value of M is a positive and integral power of 2. Let us present the basic conclusions of their study, so that we can get an idea of the situation and for the general case of a rational multiplier. Taking into account that in the case of rounding, the correction *always takes place in the positive direction*, it is not difficult to understand that the rounding error can take one of M different values of the form $e = \ell\Delta/M$, where $-(M/2) + 1 \leq \ell \leq (M/2)$. In order to express the rounding error as a function of the parameters x and α, let us assume that x has the form $x = n\Delta$ (which is of course *expected*, since after rounding, the new value of x will be an integral multiple of the quantization step Δ) and because it is $\alpha = L/M$, $\alpha x = (L/M)n\Delta$. If after quantization, the quantized value of the product αx is $Q(\alpha x) = k\Delta$ (for the reason we explained earlier), the rounding error will have the form

$$e = Q(\alpha x) - \alpha x = \ell\frac{\Delta}{M} = k\Delta - \frac{L}{M}n\Delta = kM\left(\frac{\Delta}{M}\right) - nL\left(\frac{\Delta}{M}\right) = (kM - nL)\frac{\Delta}{M}. \tag{12.42}$$

Therefore, $\ell = kM - nL$ and consequently $\ell + nL = kM$. Observing that the quantity $\ell + nL$ is an *integer multiple* of M, it is not difficult to understand that if numerical MOD M is used, then $\ell + nN = 0$ (MOD M) (since the remainder of the division kM/M is equal to zero).

The above equation is particularly interesting, as it determines the correlation that exists between the quantized value of x (through the coefficient n of the equation $x = n\Delta$) and the quantized value of the rounding error e (through the coefficient ℓ of the equation $e = \ell\Delta/M$). Specifically, for each value $n \in \mathbb{Z}$ there is *one and only one value* ℓ in the range of values $[-(M/2) + 1, M/2]$, for which the above equation is satisfied, with the uniqueness of this solution being ensured by the condition $\mathrm{GCD}(L, M) = 1$. If we symbolize this unique solution of the aforementioned equation with $\ell_0(n)$, then the value of the rounding error will be

$$e(x)\Big|_{x=n\Delta} = e(n\Delta) = \ell_0(n)\frac{\Delta}{M}.$$

Typical snapshots of this correlation are depicted in Fig. 12.38. From this figure, we observe that the rounding error $e(n\Delta)$ is *indeed periodic* with period M and with each possible value of the error appearing once and only once in each period. This periodic nature of the error allows us to formulate it as a Fourier series of the form

$$e(n\Delta) = \sum_{k=0}^{M-1} \varepsilon(k)\exp(j2\pi kn/M), \tag{12.43}$$

with Fourier coefficients

$$\varepsilon(k) = \frac{1}{M}\sum_{k=0}^{M-1} e(n\Delta)\exp(-j2\pi kn/M) = \frac{\Delta}{M^2}\sum_{n=0}^{M-1} \ell_0(n)\exp(-j2\pi kn/M), \tag{12.44}$$

which, after simple operations, can be formulated in the form

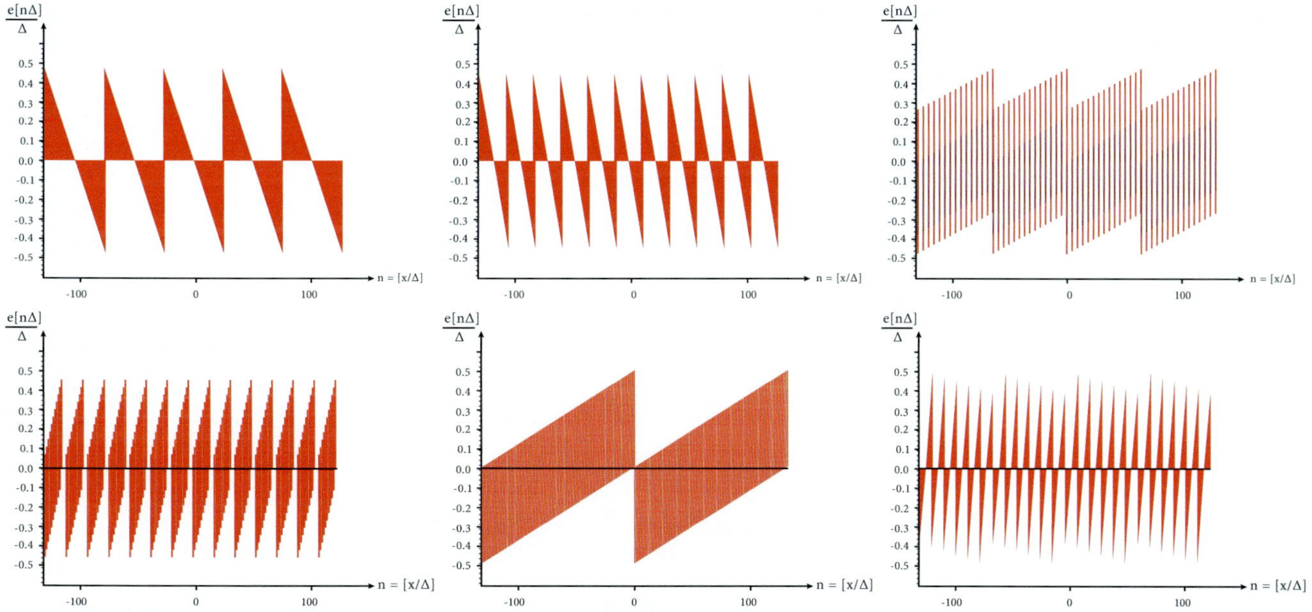

Fig. 12.38 The graphical representation of the rounding error $e(n\Delta)/\Delta$ as a function of $n = x\Delta$ from which its periodic nature is revealed. The value of α as we move from left to right and from top to bottom is respectively 5/256, 11/256, 63/256, 121/256, 127/256 and 235/256 (Source: Barnes C.W, Tran B.N & Lenug S.H, *On the Statistics of the Fixed Point Round-off Error*, IEEE Transactions on Acoustics, Speech and Signal Processing, Vol. ASSP-33, No. 3, June 1985, pp. 595–606, page 597, Figure 3)

$$\varepsilon(k) = \begin{cases} \dfrac{\Delta}{2M} & k = 0 \ (\text{MOD } m) \\ (-1)^k \dfrac{\Delta}{2M}\left[1 + j\cot\left(\dfrac{\pi n_0(k)}{M}\right)\right] & k \neq 0 \ (\text{MOD } m), \end{cases} \tag{12.45}$$

where $n_0(k)$ is the unique solution of the equation $n_0 L + k = 0$ (MOD m). Taking into account that $e = \ell\Delta/M$ for $-(M/2) < \ell \leq (M/2)$, $x = (n_0(\ell) + mM)\Delta$ for some $m \in \mathbb{Z}$ and consequently, the *probability mass function* of the error will have the form

$$p_e\left(\ell\frac{\Delta}{M}\right) = \sum_{m=-\infty}^{\infty} p_x((n_0(\ell) + mM)\Delta)$$

However, observing that

$$p_x(n\Delta) = \frac{\Delta}{2\pi}\int_0^{2\pi/\Delta} d\omega \Phi_x(\omega) e^{-j\omega n\Delta}$$

the above relationship is written

$$\begin{aligned} p_e\left(\ell\frac{\Delta}{M}\right) &= \frac{\Delta}{2\pi}\int_0^{2\pi/\Delta} d\omega \Phi_x(\omega) e^{-j\omega n_0(\ell)\Delta} \sum_{m=-\infty}^{\infty} e^{-j\omega\Delta m M} \\ &= \frac{1}{M}\int_0^{2\pi/\Delta} d\omega \Phi_x(\omega) e^{-j\omega n_0(\ell)\Delta} \sum_{m=-\infty}^{\infty} \delta\left(\omega - \frac{2\pi m}{M\Delta}\right) = \frac{1}{M}\sum_{m=0}^{M-1} \Phi_x\left(\frac{2\pi m}{M\Delta}\right) e^{-j2\pi n_0(\ell) m/M} \end{aligned} \tag{12.46}$$

for values $-M/2 < \ell \leq M/2$. In the above relationship, the term that corresponds to the value $m = 0$ represents a *uniform probability mass function*, whereas the higher-order terms represent *the degree of deviation* from this uniform function. If it is $\Phi_x(\omega) \approx 0$ for values $n \neq 0$, then $p_e(\ell\Delta/M) \approx (1/M)$. From this expression we conclude that *the rounding error will be characterized by a uniform distribution*, only if $\sigma_x \gtrsim M\Delta/2$, where σ_x is the standard deviation of the random variable x. This situation is completely different from that characterizing the quantization error during the quantization of a continuous random variable with a quantization step Δ, where the distribution is uniform if $\sigma_x \gtrsim \Delta/2$.

Similar conclusions and analogous equations can be found in the works of Barnes, Tran & Lenug (1985) and Tokaji & Barnes (1987). In fact, in the latter work, additional results can be found concerning *the characteristic function of the rounding error*, which is proven to be given by the equation

$$\Phi_e(\omega, \alpha) = E\{e^{j\omega e(x,\alpha)}\} = \sum_{m=-\infty}^{\infty} \operatorname{sinc}\left(m + \frac{\omega}{2\pi}\right) \Phi_x(2\pi m\alpha) + i \sin\left(\frac{\omega}{2}\right) \sum_{m=-\infty}^{\infty} p_x\left(\frac{2m+1}{2\alpha}\right),$$

where

$$\Phi_x(\omega) = E\{e^{j\omega x}\} = \sum_{n=-\infty}^{\infty} e^{j\omega n} p_x(n)$$

and $p_n(x)$ is the *mass density function* for the variable x, defined as $p_x(\xi) = P\{x = \xi\}$ when $\xi \in \mathbb{Z}$ and $p_x(\xi) = 0$ in the opposite case (i.e., when $\xi \notin \mathbb{Z}$). In the above expressions, the parameter α is assumed to be *integer*, whereas for the expressions that arise when α is a rational number of the form L/M, the reader is referred to the work of Tokaji and Barnes.

In full analogy with the conditions of Stripad & Snyder related to the *quantization error*, corresponding conditions can be formulated for the *rounding error*, in order for it to involve characteristics of *white and uniformly distributed noise*. Specifically, as Wong proved for the general case of a rational multiplier of the form $\alpha = L/M$, if the denominator M is an *odd number* whereas the property $\Phi_{x_n}(\xi k/M) = 0\,(\xi = 2\pi/\Delta)$ is satisfied for values $k \bmod M \leq 0$, then the *mean value* and the *variance* of the rounding error $e((L/M)Q(x_n))$ coincide with those concerning a random variable, *uniformly distributed* in the range of values $[-\Lambda, \Lambda]$, where $\Lambda = (\Delta/2)\sqrt{1-(1/M)^2}$. Additionally, if the condition $\Phi_{x_n,x_{n+m}}(\xi k/M, \xi\ell/M) = 0$ is satisfied for values $k \bmod M \leq 0$ and $\ell \bmod M \leq 0$, then the rounding error contains characteristics of *white noise*. On the other hand, if the value of M is *even*, then the above relations continue to hold, with the difference that the average value of the rounding error is *shifted* by the value $\Delta/2M$. Similar results can be found in the work of Barnes, Tran & Lenug (page 599), where it is mentioned that *these conditions are satisfied when the input signal is characterized by a relatively large dynamic range, so its (periodic) characteristic function is concentrated around the values* $\omega = 2\pi n/\Delta$ $(n \in \mathbb{Z})$ *and that knowledge of the range of this area is necessary in order for this approximation of white and uniformly distributed and uncorrelated noise to be valid and in order for us to be able to understand the statistical behavior of the rounding error in the opposite case.*

The comparison of the above conditions concerning the rounding error, with the conditions of Stripad & Snyder concerning the quantization error, leads to the conclusion that these two sets of conditions *are foreign to each other* and that *neither of them implies the other*. By observing how the arguments of the characteristic function in these two sets of conditions differ, it becomes evident that, in the case of rounding error, an *additional division* by the parameter MM, whose value is at least 2, is performed. It is not difficult to realize that the conditions that the rounding error must satisfy in order to be white and uniformly distributed within the appropriate range of values are *more stringent and restrictive* compared to the conditions set by Stripad & Snyder for quantization error. In practice, depending on the distribution of the input signal x, all four possible cases can occur, that is, the quantization error, or the rounding error, or both, or none of them, white and uniformly distributed in the appropriate range in each case.

12.9.5 Rounding Error and the Product of Complex Numbers

Having understood the quantization of the product αx, which leads to the quantized value $Q(\alpha x)$ and a rounding error with the characteristics we described, let us now see how these principles apply *to the more general problem of quantizing the product of two complex numbers*. Taking into account that the product of the complex numbers $x = \alpha + j\beta$ and $y = \gamma + j\delta$ has the form

$$xy = (\alpha + j\beta)(\gamma + j\delta) = (\alpha\gamma - \beta\delta) + j(\beta\gamma + \alpha\delta) = \epsilon + j\zeta,$$

it is not difficult to understand that this process can be carried out *by a MIMO system of four inputs and two outputs* (see Sect. 2.2.2 on page 69), which receives at its input the values of the parameters α, β, γ and δ and produces at its output, the real and imaginary parts of the product xy, which are $\epsilon = \alpha\gamma - \beta\delta$ and $\zeta = \beta\gamma + \alpha\delta$ respectively. Consequently, we need *four quantizers* for the quantization of the products $\alpha\gamma$, $\beta\delta$, $\beta\gamma$ and $\alpha\delta$ and the creation of the quantized values $Q(\alpha\gamma)$, $Q(\beta\delta)$, $Q(\beta\gamma)$ and $Q(\alpha\delta)$ and *two more quantizers* at the output terminals for quantization of the values $Q(\alpha\gamma) - Q(\beta\delta)$ and $Q(\beta\gamma) + Q(\alpha\delta)$, and the creation of the quantized versions of the parameters e and f, which are $e = Q(Q(\alpha\gamma) - Q(\beta\delta))$ and $f = Q(Q(\beta\gamma) + Q(\alpha\delta))$. According to what we have mentioned, if for the quantization of the parameters K bits are used, then for the storage of the values of the products involved in the process, $2K$ bits are required. After the quantization processes only the K most significant bits will remain, a truncation that also leads to the creation of the rounding error. The structure of the above system for the multiplication of complex numbers is depicted in Fig. 12.39.

According to what we mentioned in the relevant section, in order to transform the above nonlinear system (owing to the existence of the quantizers) into a linear system, *we should replace the quantizers that appear in this with pseudo quantization noise (PQN) sources* and let us symbolize with $\eta_{\alpha\gamma}$, $\eta_{\beta\delta}$, $\eta_{\beta\gamma}$ and $\eta_{\alpha\delta}$ the noise signals that correspond to the quantizers that calculate the values $Q(\alpha\gamma)$, $Q(\beta\delta)$, $Q(\beta\gamma)$, and $Q(\alpha\delta)$ respectively. In exactly the same way, the two quantizers at the output of the filter are also replaced by PQN sources that generate the noise signals η_ϵ and η_ζ. The above noise signals *are uncorrelated with each other* whereas the same applies to the real and the imaginary parts of the noise signal at the output. Note that the properties of the noise *depend on the characteristic of the quantizer*, which in turn *depends on the type of arithmetic used* (for example, if fixed-point arithmetic is used, we know that *there is no* rounding error after addition and therefore, the two quantizers at the system output are not necessary). After this replacement, the outputs of the linear system that arises in this way will be

$$\epsilon = (\alpha\gamma - \beta\delta) + \eta_{\alpha\gamma} - \eta_{\beta\delta} + \eta_\epsilon$$
$$\zeta = (\beta\gamma + \alpha\delta) + \eta_{\beta\gamma} - \eta_{\alpha\delta} + \eta_\zeta,$$

with the resulting state depicted in Fig. 12.40.

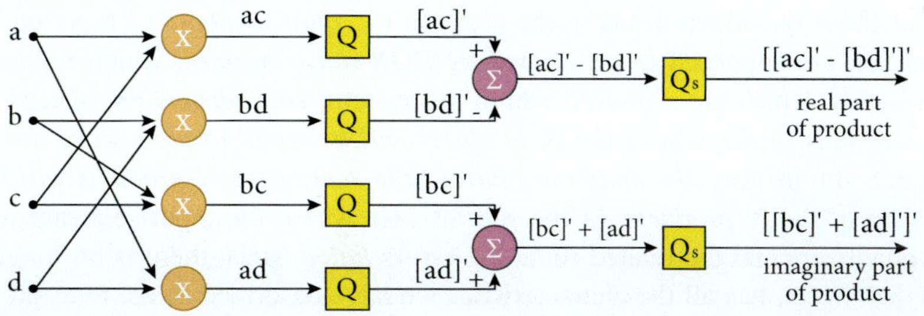

Fig. 12.39 The structure of the system for multiplying two complex numbers with the appropriate quantizers for truncating the values of the products to the appropriate number of bits

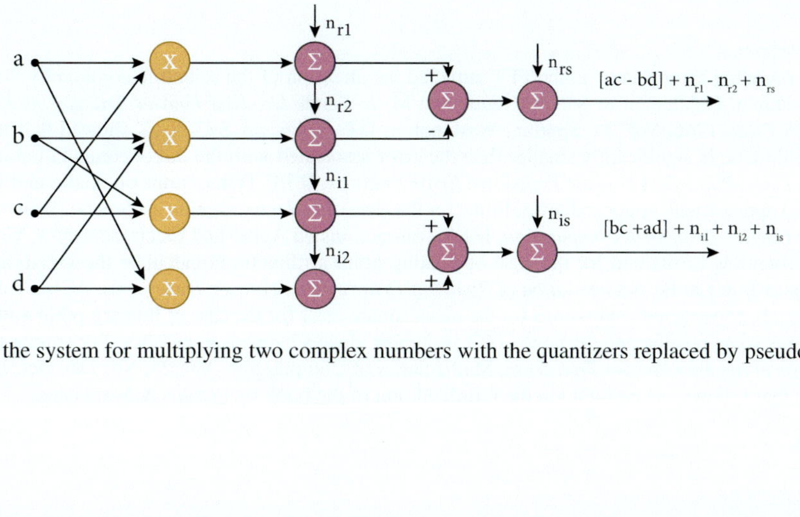

Fig. 12.40 The structure of the system for multiplying two complex numbers with the quantizers replaced by pseudo quantization noise sources

12.9.6 Rounding Error and Discrete Fourier Transform

The points mentioned in the previous section about the quantization of the product of complex numbers are particularly important because they find direct application in the process of calculating the direct and inverse discrete Fourier transform, which, as is well known, constitutes one of the most important tools for studying signals and systems in the frequency domain. Both the equations of the direct and the inverse transform that have the form

$$\mathcal{X}[k] = \sum_{n=0}^{N-1} x[n]W_N^{kn}, \quad k = 0, 1, 2, \ldots, N-1 \qquad \text{and} \qquad x[n] = \sum_{k=0}^{N-1} \mathcal{X}[k]W_N^{-kn}, \quad n = 0, 1, 2, \ldots, N-1$$

include complex quantities (since the parameter $W_N = \exp(-j2\pi/N)$ is a *complex number*, whereas the discrete sequence $x[n]$, in the general case, is of *complex nature*), with the products between complex numbers involved in the implementation of the above equations being carried out in the way we presented earlier.

Before we present the details related to this process,[19] we should mention that the multiplicative coefficient $1/N$ that appears in the equation of the inverse Fourier transform but not in that of the forward transform implies that the samples of the discrete Fourier transform $\mathcal{X}[k]$ have *significantly greater values* than the samples of the signal $x[n]$ and indeed, it is proven that the relation $\text{RMS}\{\mathcal{X}[k]\} = \sqrt{N}\text{RMS}\{x[n]\}$. Consequently, in the *constant underflow representation* and in order to avoid overflow appearance, we should proceed to the *downscaling* of the input signal values. Specifically, taking into account that the upper limit of $|\mathcal{X}[k]|$ is the sum $\sum_n |x[n]|$, with n taking values from 0 to $N-1$ and considering that the dynamic range used is the interval $(-1, 1)$, then the condition $|\mathcal{X}[k]| < 1$ requires the above sum to have a value $S < 1$, a condition that, however, leads *to a reduction in the SNR value*. For this reason, as an alternative solution, we can impose the condition $|x[n]| < 1$. For a discussion on these issues see Proakis & Manolakis, pages 550–551. It should also be commented that because the rounding errors that arise in relation to the values of the signal that occurs after the scaling *cannot be accepted, block floating-point arithmetic* is usually used. Specifically, if during the calculation of some value of the transform $\mathcal{X}[k]$, we find that we are led to overflow, the discrete sequence undergoes downscaling, usually by a factor with a value equal to 2 and with the value of the common exponent, increasing by one unit.

Taking into account that if a constant sub-sampling numerical constant is used, *no two quantizers are required* at the filter output, the total number of quantizers required for the quantization of the $4N$ products required for the calculation of the N products between complex numbers involved in the calculation of a discrete Fourier transform of N elements in length is equal to $4N$. The use of these quantizers leads to the creation of *equal numbers of rounding errors*, which, according to what we mentioned earlier, are supposed to be created by *PQN noise sources*, which replace these quantizers. If we assume that these noise sources, which are *scattered* within the system, are *independent of each other*, then, according to what we mentioned, *the real part of the sum of the N products will be equal to the sum of the noises of the real part of the N products*, whereas in a similar way, *the imaginary part of the sum of the N products will be equal to the sum of the noises of the imaginary part of the N products*. In the special case where the input sequence $x[n]$ is a sequence of real numbers, the number of quantizers and the related rounding errors *halve*, since there is no longer an imaginary part. The PQN used to replace the quantizers, has all the characteristics we have described in the relevant section, that is, described by *sets of random variables* uniformly distributed in the range $(-\Delta/2, \Delta/2)$ where $\Delta = 2^{-K}$ is the quantization step, with $4N$ the number of sources of such noise, creating noise signals that are uncorrelated both with each other and with the input signal $x[n]$. The variability of each error is equal to $\sigma_e^2 = \Delta^2/12 = 2^{-2K}/12$, whereas the variability of the rounding

[19] The issue of studying the rounding error related to the FFT attracted the attention of the scientific community very early on owing to its great practical importance. For example, Gentleman & Sande (Gentleman M. & Sande G., *Fast Fourier Transforms For Fun and Profit, Fall Joint Computer Conference AFIPS Proa*, 1966, Vol. 29, Spartan, Washington, D.C, 1966, pp. 563–578), showed that the cumulative rounding error when using floating-point arithmetic is significantly smaller than the error associated with the direct computation of the transform. On the other hand, Welch (Welch P.D., *A Fixed-Point Fast Fourier Transform Error Analysis*, IEEE Transactions on Audio and Electroacoustics, Vol. AU-17, 1969, pp. 151–157) calculated approximate upper and lower limits for the case of using fixed-point arithmetic. Weinstein (Weinstein C.J., *Roundoff Noise in Floating Point Fast Fourier Transform Computation*, IEEE Transactions on Audio and Electroacoustics, Vol. AU-17, 1969, pp. 209–215) used a statistical model for rounding errors and for the case of floating-point arithmetic to calculate the variability of the noise at the output. Finally, Kaneko & Liu (Kaneko T. & Liu B., *Accumulation of Roundoff Error in Fast Fourier Transforms*, ACM, Vol. 17, 1970, pp. 637–654) also resorting to a statistical approach, constructed expressions for the mean square error for the case of floating-point arithmetic. In all the above cases, the length N of the transform is some positive integer power of 2, as almost always happens in practice. For similar results see the work of Ramos G.U., *Roundoff Error Analysis of the Fast Fourier Transform*, Mathematics of Computation, Vol. 25, No. 116, Oct. 1971, pp. 757–768. The points mentioned in the description that follows can be found in the fourth edition of the book by Proakis & Manolakis.

error from the $4N$ multipliers is $\sigma_q^2 = 4N\sigma_e^2 = (N/3)2^{-2K}$ and as can be seen, it is proportional to the length N of the discrete Fourier transform. In the special case where the value of N is a power of 2 of the form $N = 2^\nu$, as is customary when for the calculation of the discrete Fourier transform the FFT algorithm is used, the above relationship takes the form $\sigma_q^2 = (2/3)2^{-2(K-\nu/2)}$. From this expression it follows that for every increase in the value of N by four times, an additional bit is required for the management of rounding errors.

On the other hand, if the calculation of the discrete Fourier transform is based on the FFT algorithm of Cooley & Tukey, the situation is completely different and despite the fact that this algorithm *requires fewer multiplications* compared with the classic way of calculation, the rounding errors are not smaller, as one would reasonably assume. In the description that follows we consider *the classic decimation in time algorithm*, which is implemented through a numerical constant, and its calculation for the value $N = 16$ is depicted in the butterfly diagram of Fig. 10.18 (see page 881).

In Fig. 10.18, the input to the algorithm consists of $N = 8$ samples $x[m]$ ($m = 0, 1, 2, \ldots, 7$) whereas its output is the transform $\mathcal{X}[k]$ ($k = 0, 1, 2, \ldots, 7$), the samples of which are defined *in the frequency domain* and are arranged according to the *bit-reversal order*. Each output sample is a complex number, with the distance between successive samples $\mathcal{X}[k]$ being equal to ω/N, where ω is the *sampling frequency*. As already mentioned, the basic computational unit in this algorithm is the so-called *butterfly structure* (see Fig. 10.17 on page 881), which implements a multiplication between two complex numbers, which, according to what we mentioned earlier, is reduced to *four multiplications between real numbers*, although some of these are *trivial*, in the sense that one of the two numbers being multiplied has a value of $+1$ or -1. Remember that the number of butterfly structures *halves* at each step of the calculation, with the first step containing $N/2$ such structures, the second step $N/4$ structures, the third step $N/8$ structures, and the last step one structure. For the value $N = 8$ that we consider here, there are *three steps* that contain 4, 2, and 1 butterfly structures respectively. Therefore, the number of these structures involved in the calculation of the value of each sample at the output is equal to

$$2^{\nu-1} + 2^{\nu-2} + \cdots + 2 + 1 = 2^{\nu-1}\left[1 + \left(\frac{1}{2}\right) + \cdots + \left(\frac{1}{2}\right)^{\nu-1}\right] = 2^\nu\left[1 - \left(\frac{1}{2}\right)^\nu\right] = N - 1.$$

According to what we have mentioned, each of the products $x[n]W_N^\ell$ where ℓ is the appropriate power in each case will undergo quantization, which will be carried out in the way we presented in a previous section. Remembering that the quantities W_N are defined as $W_N = \exp(-j2\pi/N)$ and therefore, the magnitude of the complex number W_N^ℓ, where ℓ ia any power is *equal to one*, it is not difficult to understand, how the multiplication $x[n]W_N^\ell$ will leave *unchanged* the magnitude of the sample $x[n]$ (since it will be multiplied by one), but it will lead to *a change in its phase*, a process that constitutes *rotation in the complex plane*. In order to prevent the increase in the number of bits required for the representation of numerical values, which is caused by these multiplications, each of the products that arise in the above way *must pass through a quantizer*, which will carry out the appropriate term truncation. On the contrary, the outputs of the butterfly structures constitute sums of complex numbers and as long as *fixed-point arithmetic* is used, subtraction, according to what we mentioned, *does not* require their quantization. As for the values of the signal $x[n]$ to which the FFT algorithm is applied, these are assumed to be already quantized.

Owing to the particular way in which the FFT algorithm operates, quantization errors that are created at each stage *are transmitted from the input to the output*. Specifically, the errors that are created at the first stage are transmitted through the remaining $\nu - 1$ stages, the errors that are created at the second stage are transmitted through the remaining $\nu - 2$ stages, and so on. As these errors are transmitted to the output through the various stages, they undergo, because of the multiplication with the coefficients W_N^{kn}, the appropriate phase shift. These phase shifts constitute *rotations on the complex plane*, which, however, *do not cause a change in the statistical properties of the rounding error and particularly the variability*, the value of which remains constant. If these rounding errors that are introduced in the various butterfly structures are considered *independent* of each other, then the value of each sample of the transform $\mathcal{X}[k]$ at the output will be affected by $4(N-1)$ quantization errors and according to what we mentioned, the *variability* of the total quantization error at the output will be

$$\sigma_q^2 = 4(N-1)\frac{\Delta^2}{12} \cong \frac{N\Delta^2}{3} = \frac{N}{3}2^{-2k},$$

(since it is $\Delta = 2^{-k}$), a value that is *exactly equal* to that calculated for the traditional discrete Fourier transform.

Note that, as in similar cases, the application of the appropriate *scaling* is required to avoid *overflow* phenomena. It is interesting that, owing to the particular way of implementing the FFT as a series of steps, we can, instead of the intense scaling we described previously, resort to *a more mild and gradual scaling process*, which is based on the observation that the values of the quantity $|\mathcal{X}_n[k]|$ associated with step number n satisfy the properties

$$\max[|\mathcal{X}_{n+1}[k]|, |\mathcal{X}_{n+1}[\ell]|] \geq \max[|\mathcal{X}_n[k]|, |\mathcal{X}_n[\ell]|]$$

$$\max[|\mathcal{X}_{n+1}[k]|, |\mathcal{X}_{n+1}[\ell]|] \leq 2\max[|\mathcal{X}_n[k]|, |\mathcal{X}_n[\ell]|].$$

The power of the above properties allows us to distribute the scaling to the various steps, in a way similar to that used in the series connection of FIR-type filters. Specifically, at each step, the sample value is multiplied by the coefficient $1/2$ so that, at the last step, which is step number ν, the total scaling is equal to $(1/2)^\nu = 1/2^\nu = 1/N$, exactly as required. This type of scaling does not change the signal level at the output but leads to a significant reduction of the rounding noise power, with the coefficient $1/2$ leading to a reduction of the noise variability to $1/4$ of its value. Therefore, the total variability of the noise at the output of the filter will be

$$\begin{aligned}\sigma_e^2 &= \frac{\Delta^2}{12}\left\{4\left(\frac{N}{2}\right)\left(\frac{1}{4}\right)^{\nu-1} + 4\left(\frac{N}{4}\right)\left(\frac{1}{4}\right)^{\nu-2} + 4\left(\frac{N}{8}\right)\left(\frac{1}{4}\right)^{\nu-3} + \cdots + \frac{1}{2} + 1\right\} \\ &= \frac{\Delta^2}{3}\left\{\left(\frac{1}{2}\right)^{\nu-1} + \left(\frac{1}{2}\right)^{\nu-2} + \cdots + \frac{1}{2} + 1\right\} = \frac{2\Delta^2}{3}\left[1 - \left(\frac{1}{2}\right)^{\nu}\right] \cong \frac{2}{3}2^{-2K},\end{aligned} \tag{12.47}$$

since the $4(N/2)$ rounding errors in the first step are reduced in terms of variability by the amount $(1/4)^{\nu-1}$, the $4(N/4)$ rounding errors in the second step are reduced in terms of variability by the amount $(1/4)^{\nu-2}$, etc. From the above relation, it follows that the variability σ_e^2 *is no longer proportional* to the length N of the signal and the discrete transform. Taking into account that the variability of the signal $x[n]$ is $\sigma_x^2 = 1/3N$, the value of the SNR is equal to

$$\text{SNR} = \frac{\sigma_x^2}{\sigma_e^2} \cong \frac{1}{3N}\frac{2}{3}2^{-2K} = \frac{1}{2N}2^{2K} = 2^{2K-\nu-1}.$$

12.9.7 Rounding Error in FIR and IIR Filters with Fixed-Point Arithmetic

As already mentioned, the quantization process *is not only applied during the conversion of an analog signal to digital*, so the input to the quantizer is a continuous-time signal, *but also after performing the appropriate in each case numerical operations*, which are applied to already quantized values, in order for these to be represented by the number of bits that are supported by the respective underlying architecture. In order to distinguish between these two cases, we characterize the difference between the output and the input of the quantizer, as *quantization error* in the first case and as *quantization noise* in the second case.

It is obvious that the process of quantization is applied *both to FIR-type filters and to IIR-type filters*, since both are digital arrangements and in both these structures, numerical operations take place, specifically, additions and multiplications. However, experience shows that *the effects of quantization are more complex in IIR-type filters*, which in contrast to FIR filters contain loops and operate with feedback. For this reason, the resulting rounding error will be studied separately for these two types of filters. In the following analysis, we assume that the values of the filter coefficients are known with infinite accuracy, something that of course is known to be untrue in reality and in another section we comment on the way in which the finite accuracy of the filter coefficient values affects its behavior and its overall operation.

Let us start our description with the *FIR-type filters*, which are known to be described by a transfer function that contains *only poles* and whose output is given by the convolution equation

$$y[n] = \sum_{k=0}^{N-1} h[k]x[n-k],$$

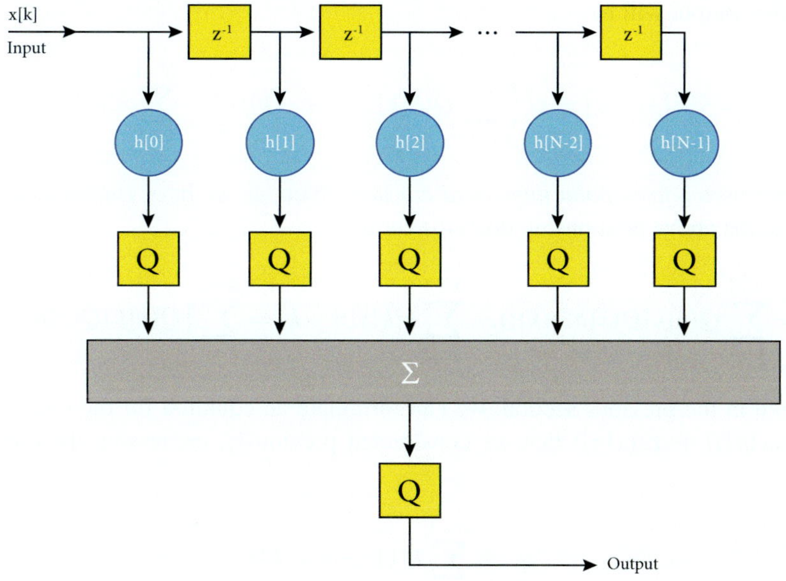

Fig. 12.41 Digital FIR-type filter, equipped with the appropriate quantizers for the quantization of the N products and their sum

where N is the number of filter coefficients, i.e., the samples of the impulse response, which is a *finite* discrete-time signal. If the filter is implemented and used with *fixed-point arithmetic*, then each of its coefficients will be a rational number of the form $h[n] = N_n\Delta = N_n/R = L_n/M_n$ ($n = 0, 1, 2, \ldots, N-1$), where R, M_n, N_n, and L_n are integers, $\Delta = 1/R$ is the *quantization step*, L_n, and M_n are *coprime* integers, whereas the condition $1 \leq M_n \leq R$ will still hold. We observe that the calculation of each sample of the filter output requires the calculation of the value of a sum of products and more specifically, a number of N products, which will then have to be added together. The way in which this calculation is performed, *depends on the underlying architecture*. For example, one way to do this is to *calculate the values of the N products, store them in memory, and then transfer them one after the other to the system's accumulator in order to add them*, whereas a more efficient way consists of *their direct transfer to the accumulator after their calculation without their intermediate storage in memory and with their sum value being progressively calculated*. Note that during the execution of these multiplications, it is very likely that the number of bits required to store the above products is *greater* than the length of the word in memory and consequently, each product, after its calculation, will pass through a quantizer in order for its value to be cut to the appropriate number of bits. In other words, we need a *(uniform) quantizer* for each of the N of these products, that is, a number of N quantizers. If the second approach is adopted, which *does not* include storage in the memory of these products, these quantizers are not necessary. However, in any case, a quantizer is definitely required at the output of the filter, since despite the fact that in the general case the length of the accumulator is greater than the length of the word in the memory, there is a possibility that the number of bits required for storage of the value of the sum of the N products is greater than the length of the accumulator. Figure 12.41 depicts the structure of an FIR-type filter, equipped with the $N+1$ above quantizers required for its proper operation, of which the N quantizers are used *only when memory is involved*, whereas the output quantizer is *always* used. Apart from the above quantizers, the digital system should also contain an additional quantizer that is outside the filter and is not depicted in the diagram, which ensures that the values applied to the input of the filter are *quantized* in the appropriate way.

Using mathematical notation, if we denote with $Q(x[n])$ the quantized input values of the filter, the N the number of products used in the convolution equation will be the $h[k]Q(x[n-k])$, whereas after their calculation they will also undergo quantization, which will lead to the values $Q(h[k]Q(x[n-k]))$ the application of which is the cause of the appearance of rounding errors. These quantized values will be added together in the accumulator of the FIR filter and then the final sum will be passed to the output quantizer in order to undergo the final quantization. Therefore, and according to the above, the real output of the filter is not the one given by the convolution equation, but the

$$\hat{y}[n] = Q\left(\sum_{k=0}^{N-1} Q(h[k]Q(x[n-k]))\right)$$

whereas the error at the filter output will be

$$e[n]=\hat{y}[n]-y[n]=Q\left(\sum_{k=0}^{N-1}Q(h[k]Q(x[n-k]))\right)-\sum_{k=0}^{N-1}h[k]x[n-k]$$

If the operation of the filter uses a *fixed-point numerical constant*, then, as we have commented, the quantizer at the filter output *is not necessary* and the above relationship takes the form

$$e[n]=\hat{y}[n]-y[n]=\sum_{k=0}^{N-1}Q(h[k]Q(x[n-k]))-\sum_{k=0}^{N-1}h[k]x[n-k]=\sum_{k=0}^{N-1}\big[Q(h[k]Q(x[n-k]))-h[k]x[n-k]\big].$$

Working in the way we did in the previous section, we can formulate an equation for the total error e_n at the filter output similar to $e[n] = e(\alpha Q(x[n])) + \alpha e(x[n])$ that we constructed previously, expressing the error in question as the sum $e_n = q_n + r_n$, where

$$q_n=\sum_{k=1}^{N-1}h[k]e(x[n-k]) \tag{12.48}$$

is the contribution of the *quantization error* and

$$r_n=\sum_{k=1}^{N-1}h[k]e(x[n-k]) \tag{12.49}$$

is the contribution of the *rounding error*. The description of the statistical properties of these two errors for FIR-type filters, results from the *direct generalization* of the results of the previous section, since in such a filter, there is a set of N multipliers, where N is the order of the filter. Provided that the conditions we previously stated are met, both quantization and rounding errors can be considered as *identically distributed and independent random processes* that are characterized by *uniform distribution* in the value range $[-\Delta/2, \Delta/2]$, where Δ is the *quantization interval*. This model of additive noise for both quantization and rounding errors leads to valid results if the appropriate conditions for the input signal are met (like those we mentioned earlier), whereas the satisfaction of additional conditions gives these errors *white noise characteristics*. Of particular interest is also the case where these characteristics appear if we add an additional signal to the input signal before its quantization. If this noise model for the two types of quantization error is *valid*, then the *power spectral density* at the output resulting as the *Fourier transform of the autocorrelation function of the noise*

$$E\{e(x_n),e(x_{n+m})\}=E\{e(\alpha Q(x_n)),e(\alpha Q(x_{n+m}))\}=\frac{\Delta^2}{12}\delta_{nm},$$

where δ_{nm} is the *Kronecker* δ, it is proven to have the form

$$S_e(f)=\frac{\Delta^2}{12}\left(|\mathcal{H}|^2+K\right),$$

where $\mathcal{H}(f)$ is the *frequency response function* of the filter. On the other hand, if the above conditions are *not met*, then this spectral power density must be calculated, with the resulting expressions being relatively complex.

As we commented in the previous sections, in order to simplify the study of the problem, we can replace the nonlinear arrangements of the quantizers with *additive PQN sources*, with the system resulting in this way, being depicted in Fig. 12.42. Of course, it is self-evident that in order for such a replacement to be permissible, the prerequisites of the appropriate theorem must be met, such as the *second quantization theorem* (see page 1451), so that the actual quantization noise (which in this case we characterize as *rounding noise*), is characterized by the same statistical properties as the PQN, that is, to be *uniformly distributed* in the appropriate range of values. This, for example, applies when numerical operations are performed with *fixed-point arithmetic*, the amplitude of the input signal is *very large* compared with the step of quantization and the number of quantization intervals is quite large. In this way, despite the fact that the FIR filter equipped with the quantizers undoubtedly

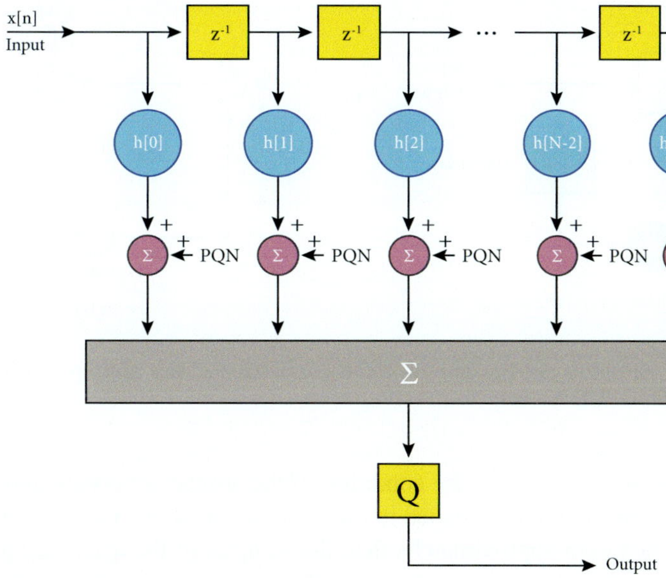

Fig. 12.42 Digital FIR-type filter, in which the nonlinear arrangements of the quantizer have been replaced by additive PQN sources

constitutes a *nonlinear* arrangement, after the above replacement, we can study it using the *usual linear theory*. However, it should be noted that although this approach generally leads to accurate results for parameters such as the SNR, this generally *does not apply* to power spectra related to the noise signal.

If the FIR filter is described by an impulse response of N samples in length, then the *convolution equation* used for calculating the filter output will contain the sum of N products of the form $h[k]x[n-k]$ $(0 \leq k < N)$, each of which must undergo quantization, passing through one of a multitude of an equal number quantizers, which as we commented, will be replaced by an equal number of noise sources. The total noise at the filter output, provided that these sources are not related to each other, will result from the *sum of the noises* that will arise from these N PQN noise sources, to which we must add the noise that is related to the quantization of the input signal, as well as the noise that is related to the quantizer that exists at the filter input and which propagates along the filter from the input to the output. Taking into account that the power of the noise resulting from the $N+1$ sources (i.e., the N sources related to the quantizers of the N products, as well as the source corresponding to the quantizer at the filter output) is equal to $\Delta^2/12 = 1/12$ for each source (with total power of all sources equal to $(N+1)/12$), whereas the noise that is related to the input signal and propagates along the filter from the input to the output is equal to $1/12$ times the sum of the squares of the impulse response samples. Therefore, the total power of the noise at the filter output will be equal to

$$P_e = \frac{1}{12}\left[\sum_{n=1}^{N-1} h^2[n] + (N+1)\right]. \tag{12.50}$$

We therefore find that the assumption of *noise independence from source to source* gives the analysis a significant degree of simplicity in the case where operations are performed using *fixed-point arithmetic*.

On the other hand, in IIR filters things are somewhat more complicated, since these structures have *feedback loops* and therefore, the noise originating from the PQN sources replacing the quantizers circulates within the system and combines with the input signal as well as with other noise signals. For example, in the case of the simple pole filter depicted in the left image of Fig. 12.43 the product $\alpha y[n-1]$ undergoes quantization by the quantizer located in the *feedback branch* of the system and then, the quantized signal that results combines with the input signal through the adder that exists at the input, in order to create the output signal. Owing to this particular operation of the filter characterized by the continuous repetitive passage of a signal through the multiplier by α, the number of bits required to store the products $\alpha y[n-1]$ for the various values of n, *continuously increases*, and therefore, the existence of the quantizer in the feedback loop is *necessary*.

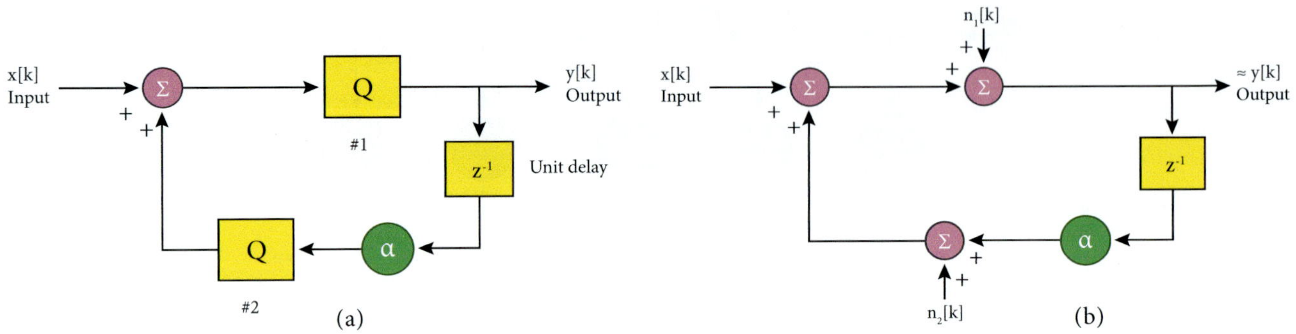

Fig. 12.43 (**a**) First-order IIR digital filter with the required quantizers (left image) and (**b**) the same structure, in which the nonlinear arrangements of the quantizer have been replaced by additive PQN sources

It is interesting to note that although at first glance the condition of the absence of correlation between the quantization noise and the input of the quantizer seems not to be satisfied in this case, owing to the recursive nature of the system, if the *second quantization theorem* is satisfied even approximately, then the samples of the quantization noise can be considered *independent*, a fact that allows us to replace the quantizer in the feedback loop with a *white noise source*, just as we did in similar cases (see the right image of Fig. 12.43). In reality, the correlation that exists between the quantization noise and the input of the quantizer is *exactly the same* as that which would arise if the arrangement of the quantizer were replaced by a source of independent additive PQN. We therefore find that the situation is *different* from that characterizing FIR-type filters, in which the absence of the feedback loop makes the quantization noise *completely uncorrelated* with the signal at the input of the quantizer. As for the second quantizer located in the upper branch of the circuit, it is only required in the case where the input signal has undergone quantization, with a *smaller quantization step* than that of the loop feedback quantizer. In this case, the value of the input signal will be represented with more bits than the value of the product $\alpha y[n-1]$, whereas the same will apply to the sum of the two signals, so a *truncation* of the number of bits to the value that is supported by the system is required. This truncation is performed by the quantizer of the upper branch, the step of which is certainly the same as the one that characterizes the quantizer of the feedback branch of the system. This additional quantizer can also be replaced by a PQN source, with the noise signals coming from these two sources being considered *uncorrelated* with each other, but without any guarantee that this will apply in all cases. These conclusions generalize for the *higher-order IIR systems*. For a discussion on the prerequisites for applying the additive noise model in the case of quantizers used in filters see Chapter 16 of the book by Widrow and Kollár, where the reader can also find the discussion we present here.

In cases where the conditions for considering the two sources of additive noise as uncorrelated with each other exist, *the power of the total noise at the filter output will be equal to the sum of the power values of the noise from these two sources*. Taking into account that the transfer function between the quantizer of the upper branch and the filter output is $\mathcal{H}(z) = 1/(1-\alpha z^{-1})$ leading to an impulse response $h[n] = \alpha^n u[n]$, the sum of the squares of the samples of the impulse response will be

$$\sum_{n=0}^{\infty} h^2[n] = \sum_{n=0}^{\infty} \alpha^{2n} = \frac{1}{1-\alpha^2}. \tag{12.51}$$

If the noise from the equivalent noise source is characterized by zero mean value and variability equal to $\Delta^2/12$, then as in the case of FIR filters, the power of the noise originating from the upper source will be equal to $(\Delta^2/12)/(1-\alpha^2)$. The exact same applies to the noise from the feedback branch source, the power of which is the same as the previous one and therefore, the power of the noise at the system output will be equal to the *sum* of these (equal to each other) power values and therefore will have a value of $(\Delta^2/6)/(1-\alpha^2)$. For more details and representative examples, see the book by Widrow & Kollár. Note that in the general case, this power value, apart from the filter coefficients, i.e., the samples of the impulse response, also depends on other factors, as does the *scaling criterion* used (see the relevant section), as well as the *filter structure*. More specifically, considering a specific filter structure, we can reduce the impact of the filter parameters using the technique of *error feedback*. Conversely, considering a specific linear transfer function, we can determine the optimal filter structure, that is, the structure that is characterized by the *minimum noise power at the output*. As for the scaling process, the more conservative the rule we use, the poorer the SNR, although on the other hand, if the scaling becomes too large, there is a risk of overflow.

12.9.8 Overflow and Scaling

In the previous sections, we defined overflow as the state in which the value of some internal signal of the circuit, that is, some sum or product calculated in the context of the implementation of its difference equation, *is greater than the maximum value supported by the number of bits that define the length of the registers*. The appearance of overflow, leads, as is well known, to *numerical errors* and therefore, care should be taken to avoid it or at least to reduce it, as much as possible. On the other hand, if the values of all the internal signals of the system are excessively small, the filter operation *does not perform correctly* either, whereas the value of the SNR *will not be satisfactory*. In order to avoid the appearance of these problematic situations, we must proceed to the appropriate *scaling* of the circuit signal values, in those systems where *fixed-point arithmetic* is used. This scaling can be performed by multiplying with the appropriate coefficient *either the input signal values or the filter values* and ensures that *the probability of overflow occurrence is the same at all internal nodes of the underlying digital system*. In such cases, the SNR takes its maximum value. According to what we mentioned in a previous section, owing to the property of complement representations with respect to 1 and with respect to 2, to recover from overflow errors related to the addition operation, the scaling operation needs to be applied *only at the input points of the multipliers* of the digital filter.

In order to avoid the phenomenon of overflow, the value of each number that is calculated at any point in the system *should not exceed one*, assuming that all the numbers used are *fractional numbers*, that is, they have a measure less than one. In order to formulate mathematically, considering the condition related to this situation, let us consider a node of the system, for example, the multiplier numbered k and let us symbolize with $y_k[n]$ the output of this node, when the input to the system is the signal $x[n]$. If the impulse response between the input signal and the considered node is $h_k[n]$, then the measure of the output $y_k[n]$ according to the convolution theorem will be

$$|y_k[n]| = \left| \sum_{m=-\infty}^{\infty} h_k[n]x[n-m] \right| \leq \sum_{m=-\infty}^{\infty} |h_k[m]||x[n-m]|$$

since as is well known, *the measure of the sum of some quantities is less than or equal to the sum of their measures*. If we assume that the input sequence $x[n]$ is *bounded* and more specifically, that $|x[n]| \leq x_{\max}$ for every n, the above relation becomes

$$|y_k[n]| \leq x_{\max} \sum_{m=-\infty}^{\infty} |h_k[m]|. \tag{12.52}$$

According to what we mentioned earlier, in order not to have overflow, it must be $|y_k[n]| < 1$. In order for this condition to be satisfied, the input $x[n]$ must be scaled in such a way that it is

$$x_{\max} < \frac{1}{\sum_{m=-\infty}^{\infty} |h_k[m]|}. \tag{12.53}$$

The above equation constitutes a *necessary and sufficient condition* for avoiding the appearance of overflow during the calculations, since *there is always* a signal that satisfies the Eq. (12.52) with respect to the equality symbol. Note that if Eq. (12.53) is not satisfied, we can enforce its validity by multiplying the input signal $x[n]$ with a constant β, before it enters the considered stage (see Fig. 12.44), so that Eq. (12.53) is satisfied by the value $\beta x_{\max}$, meaning that it is

$$\beta x_{\max} < \frac{1}{\sum_{m=-\infty}^{\infty} |h_k[m]|}. \tag{12.54}$$

In most practical implementations, the value of the constant β is a power of 2, such that the requirements that ensure the *prevention of overflow* are met, so this multiplication operation is easily performed through simple shift procedures.

The above applies to IIR filters, whereas for FIR filters, the sum in the denominator of the above relationship is defined from $m = 0$ to $M - 1$, where M is the number of samples of the impulse response. Note that although the above method

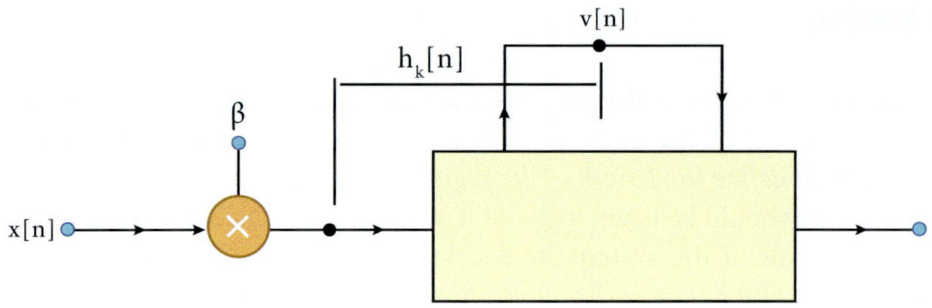

Fig. 12.44 Scaling of the input signal $x[n]$ through multiplication by a factor β, so as to satisfy the condition of the method. In this figure, the impulse response $h_k[n]$ is defined between the appropriate points of interest

guarantees that *there will be no occurrence of overflow* provided that the input signal is blocked, the problem is that its implementation *is not easy*, whereas the signal levels at the internal nodes of the filter may take small values and because quantization errors do not depend on the signal value, *small SNR values* may arise. In addition, practice has shown that if the input signal $x[n]$ is scaled to a very large extent in the above way, there may be a *reduction in accuracy* used for the representation of its samples and this is particularly true for sequences characterized by *narrow bandwidth*, such as the sine signal, on which the above scaling has a very large impact. In sequences of this type, the above scaling process *is not based* on the impulse response of the filter but on its frequency response.

More specifically, taking into account that the quantity $|\mathcal{H}(e^{j\omega})$ represents the *gain* of the system for a signal with frequency value ω_0 and considering such a signal of the form $x[n] = x_{\max}\cos(\omega_0 n)$, the output of each node of the system, as is well known from the basic theory of LTI systems, will have the form

$$y_k[n] = |\mathcal{H}(e^{j\omega_0})|x_{\max}\cos\left[\omega_0 n + \measuredangle\mathcal{H}(e^{j\omega_0})\right].$$

In order to avoid the phenomenon of overflow and taking into account that the measure of the function of the cosine is always less than or equal to one, it should be

$$\max_{k,|\omega|\le\pi} |\mathcal{H}(e^{j\omega_0})|x_{\max} < 1,$$

from which it follows that

$$\beta x_{\max} < \frac{1}{\max\limits_{k,|\omega|\le\pi} |\mathcal{H}(e^{j\omega_0})|}, \tag{12.55}$$

where we have also added the scaling factor β, if it is proven that its use is necessary.

The theory behind the construction of the above condition (which is nothing more than a special case of a more general result), is based on the use of the *norm function*, which is inextricably linked with the concept of the *vector space* (for information on these functions see page 1524). In the special case of a periodic function $\mathcal{A}(e^{j\omega})$ with period ω_s, the norm L_p for each $p \ge 1$ is defined as

$$\|\mathcal{A}\|_p = \sqrt[p]{\frac{1}{\omega_s}\int_0^{\omega_s} |\mathcal{A}(e^{j\omega})|^p d\omega},$$

and is defined when the integral found in the root of the above expression has a *finite value*. If the function $\mathcal{A}(e^{j\omega})$ is a continuous function, then as commented in the relevant section, the norm is also defined[20]

[20] For the proof of this relationship, see, for example, Margaris A.I., *Linear Algebra*, Tziola Publications, 2016, page 194.

$$\|\mathcal{A}\|_\infty = \lim_{p\to\infty} \|\mathcal{A}\|_p = \max_{0\le\omega\le\omega_p} |\mathcal{A}(e^{j\omega})|.$$

Adopting the symbolism we used earlier, let $x[n]$ be the input signal, which is a *deterministic and absolutely summable discrete sequence*, and $h_k[n]$ the impulse response between the input signal and an internal node of the system, the output of which is therefore

$$w_k[n] = \sum_{k=0}^{\infty} \beta h_k[n]x[n-k], \tag{12.56}$$

where in the above expression we have incorporated the *scaling coefficient* β. The $\mathcal{Z}$ transforms of the signals $h_k[n]$ and $x[n]$ are

$$\mathcal{H}_k(z) = \sum_{n=-\infty}^{\infty} h_k[n]z^{-n} \ \ (\alpha < |z| < \beta) \quad \text{and} \quad \mathcal{X}(z) = \sum_{n=-\infty}^{\infty} x[n]z^{-n} \ \ (\gamma < |z| < \beta)$$

respectively and according to the *convolution theorem*, $\mathcal{Z}$ transform of the system output $w_k[n]$ will have the form

$$\mathcal{W}_k(z) = \beta\mathcal{H}_k(z)\mathcal{X}(z) \ \ (\delta < |z| < \beta),$$

where $\delta = \max(\alpha, \gamma)$ with the inequalities accompanying the above relations defining the *regions of convergence* of the corresponding transforms, which, as known, constitute *annular regions (or union of such regions) centered at the origin in the complex plane*. In the above relations we accept that $\beta > 1$, whereas for the system to be stable it should be $\gamma < 1$. Taking the inverse $\mathcal{Z}$ transform of the function $\mathcal{W}_k(z)$, the output of the node numbered k of the system will be

$$w_k[n] = \frac{1}{2\pi j}\oint_\Gamma \beta\mathcal{H}_k(z)\mathcal{X}(z)z^{n-1}dz, \tag{12.57}$$

where Γ is a closed curve on the complex plane, which belongs to the region of convergence of the $\mathcal{Z}$ transform. Assuming that the system is *stable*, we can choose as curve Γ the unit circle $|z| = 1$ (this choice requires $\alpha < 1$), since this will belong to the region of convergence and taking into account that the points of the circumference of the unit circle are described by the expression $z = e^{j\omega}$, the above relation for frequency value $\omega_s = 2\pi$ takes the form

$$w_k[n] = \frac{1}{\omega_s}\int_0^{\omega_s} \beta\mathcal{H}_k(e^{j\omega})\mathcal{X}(e^{j\omega})e^{j\omega n}d\omega.$$

Taking the magnitude of the above expression we have

$$|w_k[n]| = \frac{1}{\omega_s}\int_0^{\omega_s} |\beta\mathcal{H}_k(e^{j\omega})\mathcal{X}(e^{j\omega})|d\omega,$$

since $|e^{j\omega n}| = 1$. In order to incorporate norm functions into our description, we observe that

$$|w_k[n]| \le \left[\max_{0\le\omega\le\omega_s} |\mathcal{X}(e^{j\omega})|\right]\frac{1}{\omega_s}\int_0^{\omega_s} |\beta\mathcal{H}_k(e^{j\omega})|d\omega$$

or alternatively, that

$$|w_k[n]| \le \left[\max_{0\le\omega\le\omega_s} |\beta\mathcal{H}_k(e^{j\omega})|\right]\frac{1}{\omega_s}\int_0^{\omega_s} |\mathcal{X}(e^{j\omega})|d\omega.$$

Referring to the definitions of the norm functions, we observe that

$$\|\mathcal{X}\|_\infty = \max_{0\le\omega\le\omega_s} |\mathcal{X}(e^{j\omega})|$$

$$\|\beta\mathcal{H}_k\|_1 = \frac{1}{\omega_s}\int_0^{\omega_s} |\beta\mathcal{H}_k(e^{j\omega})|d\omega$$

$$\|\beta\mathcal{H}_k\|_\infty = \max_{0\le\omega\le\omega_s} |\beta\mathcal{H}_k(e^{j\omega})|$$

$$\|\mathcal{X}\|_1 = \frac{1}{\omega_s}\int_0^{\omega_s} |\mathcal{X}(e^{j\omega})|d\omega$$

and consequently, the two above inequalities become

$$|w_k[n]| \le \|\mathcal{X}\|_\infty \|\beta\mathcal{H}_k\|_1 \quad \text{and} \quad |w_k[n]| \le \|\mathcal{X}\|_1 \|\beta\mathcal{H}_k\|_\infty.$$

On the other hand, the application of the *Schwarz inequality* to the initial expression for the measure $|w_k[n]|$ will give us

$$|w_k[n]|^2 \le \frac{1}{\omega_s^2}\int_0^{\omega_s} |\beta\mathcal{H}_k(e^{j\omega})|^2 d\omega \int_0^{\omega_s} |\mathcal{X}(e^{j\omega})|^2 d\omega. \tag{12.58}$$

It will therefore be

$$|w_k[n]| \le \sqrt{\frac{1}{\omega_s}\int_0^{\omega_s} |\beta\mathcal{H}_k(e^{j\omega})|^2 d\omega} \times \sqrt{\frac{1}{\omega_s}\int_0^{\omega_s} |\mathcal{X}(e^{j\omega})|^2 d\omega} \tag{12.59}$$

and using the defining equation of L_p, it finally turns out that

$$|w_k[n]| \le \|\mathcal{X}\|_2 \|\beta\mathcal{H}_k\|_2.$$

Taking into account that all the above inequalities are *special cases of the famous Hölder's inequality* $|w_k[n]| \le \|\mathcal{X}\|_q \|\beta\mathcal{H}_k\|_p$, with the numbers p and q satisfying the condition $(1/p) + (1/q) = 1$, there are many possibilities for scaling, depending on the values of p and q, as we comment later.

First of all, we should note that *Hölder's inequality* holds for any transfer function of the form $\beta\mathcal{H}_z(z)$, including $\beta\mathcal{H}_k(z) = 1$. In this case, the equation $\mathcal{W}_k(z) = \beta\mathcal{H}_k(z)\mathcal{X}(z)$ takes the simple form $\mathcal{W}_k(z) = \mathcal{X}(z)$, and from the inverse $\mathcal{Z}$ transform it follows that $w_k[n] = x[n]$ and $\|\beta\mathcal{H}_k\|_p = \|1\|_p = 1$ for every $p \ge 1$. Replacing the expressions $w_k[n] = x[n]$ and $\|\beta\mathcal{H}_k\|_p = 1$ in Hölder's inequality, this, for the special case $\beta\mathcal{H}_z(z) = 1$, takes the simple form $|x[n]| \le \|\mathcal{X}\|_q$ for every $q \ge 1$. If the input signal $x[n]$ is *bounded*, that is, if it is $|x[n]| \le x_{\max} = M$, then, from the last inequality, we have $|x[n]| \le \|\mathcal{X}\|_q \le x_{\max}$ and *Hölder's inequality* will give us $|w_k[n]| \le M\|\beta\mathcal{H}_k\|_p$. In order for $|w_k[n]| \le M$, which is also the desired property, the power of which ensures that *overflow will not occur*, the condition $\|\beta\mathcal{H}_k\|_p \le 1$ must be satisfied, from which it follows that $\beta \le (1/\|\mathcal{H}_k\|_p)$ provided that $|x[n]|_q \le M$, which is also the desired result. The above analysis concerns the case of a simple *multiplier*, whereas if there are L such nodes in the system with transfer functions $\mathcal{H}_k^1, \mathcal{H}_k^2, \ldots, \mathcal{H}_k^L$ the above condition takes the form

$$\beta \le \frac{1}{\max\{\|\mathcal{H}_k^1\|_p, \|\mathcal{H}_k^2\|_p, \ldots, \|\mathcal{H}_k^L\|_p\}}.$$

The above property holds for arbitrary values of the numbers p and q and therefore, by choosing specific values for these numbers, we can construct specific scaling conditions. Particularly interesting are the cases $p = 1$, $p = 2$ and $p = \infty$. Initially, we observe that because generally $\|\mathcal{H}_k\|_2 \le \|\mathcal{H}_k\|_\infty$, $(1/\|\mathcal{H}_k\|_2) \ge (1/\|\mathcal{H}_k\|_\infty)$ and therefore, the scaling that uses the L_2 norm usually leads to *larger scaling coefficients* than the scaling that uses the L_∞ norm. In addition, the case $p = q = 2$ is used with deterministic or random signals of finite energy. Taking into account that energy is related to the second power of the signal, since it is given by the expression

$$E = \sum_{n=-\infty}^{\infty} x^2[n] = \|\mathcal{X}\|_2^2 < \infty,$$

corresponds, in magnitude, to the value $p = 2$ and the L_2 norm, after the application of *Parseval's theorem*. On the contrary, for signals of infinite energy, the norm $\|\mathcal{X}\|_2$ does not exist and the L_2 scaling cannot be used. On the other hand, if the signal is *blocked* (something that means that the quantity $\|\mathcal{X}\|_\infty$ can be used exactly) and is characterized by *finite bandwidth*, the norm $\|\mathcal{X}\|_1$ and the condition corresponding to the values $p = \infty$ and $q = 1$ are used. If the signal is characterized by some *dominant* frequency component, as happens, for example, with a sine wave, it will have an impulse function in the frequency domain; but this means that neither the $\|\mathcal{X}\|_2$ nor the $\|\mathcal{X}\|_\infty$ is defined so only the L_1 norm can be used. Finally, taking into account that $\|\mathcal{H}_k\|_2 \leq \|\mathcal{H}_k\|_\infty \leq \|\mathcal{H}_k\|_1$, from the defining equations of these three norm functions it follows that

$$\sqrt{\sum_{n=-\infty}^{\infty} |h_k[n]|^2} \leq \max_{\omega} |\mathcal{H}_k(e^{j\omega})| \leq \sum_{n=-\infty}^{\infty} |h_k[n]|.$$

What we have mentioned in the above description, applies *only for deterministic and not for random signals*, since for the latter type of signals, the $\mathcal{Z}$ transform *is not* defined. In this case, we can construct the corresponding equations resorting to the use of quantities from the space of *statistical signal processing* and more specifically the autocorrelation of the signal $w[n]$, which is considered a wide-sense stationary process.[21]

In the special case of systems that have been implemented according to the *cascade structure*, that is, they result from the serial connection of smaller subsystems (see Sect. 12.15.1 on page 1358), the simplest thing we can do is to *directly scale* the input signal $x[n]$ that is applied to the filter input. This is something that reduces the signal amplitude from the beginning, thus requiring the stages from which the system is composed to operate *with a high gain value*, so as to jointly lead to a gain value for the entire system equal to unity. However, this is something that will lead to *large quantization noise*, since the gain that will be offered by each stage, in addition to the signal, will also be applied to the noise. Although in the ideal case, the total gain factor with a value less than 1 should be placed *at the end of the serial connections*, this may lead to overflow. For this reason, what is done in practice is *to distribute the value of the system's total gain to all stages* and ensure the avoidance of the overflow phenomenon at each stage separately. For example, if the filter we use is a sixth-order discrete-time system, which results from the serial connection of second-order stages with transfer functions $\mathcal{H}_1$, $\mathcal{H}_2$, and $\mathcal{H}_3$, whereas the total gain of the system is equal to g, then the transfer function of the composite system is not $\mathcal{H} = g\mathcal{H}_1\mathcal{H}_2\mathcal{H}_3$ but $\mathcal{H} = \mathcal{H}_1'\mathcal{H}_2'\mathcal{H}_3' = (g_1\mathcal{H}_1)(g_2\mathcal{H}_2)(g_3\mathcal{H}_3)$, where $g = g_1g_2g_3$. In other words, what we do is *to distribute the value of the system's gain to the individual stages*, whereas if we wish, we can incorporate the coefficients g_1, g_2 and g_3 into the coefficients of the numerator of each transfer function. In some cases and for some types of second-order stages, which are used in serial implementations, the required scaling factor associated with a stage *can be incorporated into the multipliers* located at the output of the previous stage and this is something that leads *to a reduction of quantization noise at the filter output*. Speaking generally, for second-order stages, it is possible to calculate the L_2 and L_∞ norms of the above internal transfer functions, in closed form.[22]

The mathematical analysis of the problem reveals that the power spectrum of the noise at the filter output depends *both on the way in which zeros are combined with poles for the construction of second-order stages, as well as on the order of the connection of these stages for the construction of the higher-order system*. Assuming that we have N stages, it is proven that there are $N!$ different ways of combining pole and zero values as well as $N!$ different ways of connecting the second-order stages, thus leading to a multitude of $(N!)^2$ different systems, which can be implemented using direct forms I and II. In the literature there are several practical rules, the application of which in practice has led to good results and systems with acceptable behavior. For example, according to a technique that is often used in practice, *we initially proceed to the classification of the poles in descending order based on the value of their magnitude and then combine the pair of poles that is as close as possible to the circumference of the unit circle, with the pair of zero values that is as close as possible to this pair*

[21] For a related discussion see, for example, Jackson L.B., *On the Interaction of Roundoff Noise and Dynamic Range in Digital Filters*, The Bell System Technical Journal, Vol. 49, pp-159–184, Feb 1970. In the same work, an application of the above concepts and techniques can also be found in the case of *direct form I*, which, although not recommended as it is characterized by the *high sensitivity of the filter parameters* to small changes in their values and by *greater rounding noise* at the output than with other implementations, it is nevertheless of theoretical interest.

[22] See for example Diniz P.S.R. & Antoniou A., *More Economical State-Space Digital-Filter Structures Which are Free of Constant-Input Limit Cycles*, IEEE Transactions on Acoustics, Speech, and Signal Processing, ASSP-34, 807–815, 1986 and Bomar B.W. & Joseph R.D., *Calculation of L_∞ Norms in Second-Order State-Space Digital Filter Sections*, IEEE Transactions on Circuits and Systems, CAS-34, 983–984, 1987.

of poles, so as to reduce the value of the gain associated with this stage (this technique is based on the observation that stages with high gain values are undesirable, since they may lead to overflow, while also causing and enhancing quantization noise). This process of pairing poles and zero values is performed *repeatedly* until all these points are exhausted. In the final step of the process, we sort the second-order steps that arise in this way either in ascending or descending order, with the criterion the value $\arg\max |\mathcal{H}(e^{j\omega})|$. The logic behind this approach is that if we do not want the spectrum of noise variability to be characterized by a large value around a pole that is close to the circumference of the unit circle, the frequency component corresponding to this pole *should not often appear in the power spectrum calculation of the noise at the filter output* and to do this, the poles with a large value Q should appear at the beginning of the cascade structure. Conversely, the poles that are close to the circumference of the unit circle should appear at the end of this structure, so as to avoid excessive reduction of the signal level as it passes through the first steps. More details on this subject can be found in the literature.

12.9.9 Limit Cycles and Their Handling

As already mentioned, the addition of quantizers to a digital system, which is imposed by the fact that such a system operates with *finite accuracy*, giving the system a certain degree of *nonlinearity*, which certainly affects the characteristics of the output signal. In the special case of IIR recursive systems, which operate under this finite accuracy regime and use *fixed-point arithmetic*, the nonlinearity induced by the addition of the quantizer, in order to perform the quantization applied to the last significant bit, may lead to *the appearance of periodic oscillations at the system output*, even in cases where the system input is either equal to zero or equal to a very small nonzero constant value. This is certainly something that *does not* apply to ideal recursive systems, which operate under the regime of infinite accuracy, in which, as is well known, if the system input takes a zero value for some time period, then the system output tends asymptotically to zero. These *parasitic oscillations* that appear in quantized recursive systems of finite accuracy and which are due, as we will comment immediately, to the fine texture of the quantized system, are known as *limit cycles*. Taking into account that these involved quantization and rounding errors affect *to different degrees* the behavior of different but equivalent implementations that are associated with the same transfer function, it is not difficult to understand that the creation of limit cycles and the characteristics that describe them are generally *different* for these different implementations.

Specifically, limit cycles appear in cases where *the signal levels take a very low value*, which may also be *constant* for short periods of time. In these cases, *the basic assumption on which the previous analysis was based does not apply* (according to which the signal values are much larger than the quantization step), a fact that allowed the consideration of the noise signals of quantization as *statistically independent* from each other both *from sample to sample* and *from source to source*. If this assumption does not hold, then all kinds of noise signals *are not* statistically independent but on the contrary, they are characterized by a *strong degree of correlation*, which can force the filter to enter an unstable operating state, through which the creation of a *steady-state oscillation* at the output will take place. This oscillation constitutes a *limit cycle*, whereas the phenomenon that appears in this case is called the *dead zone phenomenon*.

In order to understand the mechanism of the creation of limit cycles, we must take into account the fact that owing to the processes of *quantization* and *overflow*, each component of the state vector, that is, each *state variable*, can only take one of a multitude of discrete values, which are *integer multiples* of the quantization step q. Taking into account that the solutions of the system's state equation are *bounded* for $n \to \infty$ (since when the filter is stable, the state vector $\boldsymbol{q}[n]$ at the limit $n \to \infty$ tends toward a steady-state state vector $\boldsymbol{q}_{SS}[n]$), it is not difficult to understand that the number of different values that the state vector can take, or equivalently, the number of points in the state space through which this vector can pass, is *finite*. In other words, a quantized system, or equivalently a filter with quantized state variables, constitutes a *finite state machine*. Without the process of quantization, the state vector will move along a *spiral trajectory of decreasing radius* (as we commented during the description of oscillations due to overflow), whereas if this process is applied, then the quantized state vector will be modified to some extent and will correspond *to a point on the quantization grid* that will be *very close* to the point where it was before the quantization. However, since the number of different states of the system, as we commented, is *finite*, it is certain that at some point in time n_0, the system will be in a state $\boldsymbol{q}[n_0]$ from which it had passed at some previous point in time $n_0 - N$. It is not difficult to understand that under conditions of *zero input* and owing to the *deterministic nature* of the system, the system from this point onward will pass through the same states it had passed before and consequently, it will perform a *periodic oscillation* of period N, which is known as a *zero input limit cycle*. This means that the state vector $\boldsymbol{q}[n]$ will repeatedly pass through N points in the phase space. These cycles are characterized as *accessible*, if the system can enter such a cycle from a point outside the cycle and as *inaccessible* in the opposite case. Specifically, considering a limit cycle $\{y_i\}$ for values $0 \leq i \leq N-1$ with period N (where we now use for our description the system's output and the related difference equation for a second-order system, since the time series of the system's output will obviously also be a periodic sequence with period N), it is characterized as accessible, if for some index i there is a value $\overline{y}_i \neq y_i$ such that it is

$$y_{i+2} = Q[\alpha y_{i+1} + \beta y_i] = Q[\alpha y_{i+1} + \beta \overline{y}_i],$$

where, owing to periodicity, the indices are calculated using *numerical MOD N*. This means that the state (y_{i+2}, y_{i+1}) of the limit cycle can be approached either from the state (y_{i+1}, y_i) or from the state $(y_{i+1}, \overline{y}_i)$.[23] Note that in a digital filter there may be *more than one* such limit cycles, with the limit cycle that will eventually appear depending on the initial conditions used.[24]

The particular characteristic of these cycles is that they constitute *oscillations with constant amplitude and frequency*, the values of which are dictated by the nonlinear properties of the system. It is interesting that the amplitude of these oscillations is *independent of the initial conditions*, something that *does not* apply to the case of usual linear oscillations, the amplitude of which, as known, is *a function of the boundary conditions*. Of particular interest are the *zero-input limit cycles*, which can be considered as *spontaneous oscillations*, which appear when the system's input suddenly takes a zero value. There are three categories of limit cycles, the *stable* and the *unstable* limit cycles, which are characterized as such, if the trajectories in phase space over time converge to them or diverge from them respectively, as well as the *semi-stable* limit cycles, which characterize those systems, in which some trajectories converge to them, whereas some others diverge from them. Limit cycles may constitute a *particularly annoying problem*, as happens for example in voice signal processing applications, although in some cases they are useful, as happens in *digital oscillators*. However, as we comment in the description that follows, they can easily be eliminated, using a larger word length in digital systems and ignoring the less significant bits at the filter output. In cases where the cost of eliminating these cycles is very high, we can tolerate their existence, trying to minimize their negative effects.

Limit Cycles Due to Quantization in First-Order Systems

In order to understand in practice the appearance of marginal cycles, let us consider the *discrete single-pole system* described by the difference equation

$$y[n] = \alpha y[n-1] + x[n]$$

and that has a single pole at the position $z = \alpha$ with $|\alpha| < 1$, as required for a stable system. Although in the ideal case where the values of involved quantities (such as the value of the parameter α and the input and output signals) are known with *infinite accuracy*, whereas the system output $y[n] = \alpha^n u[n]$ tends to zero, at the limit $n \to \infty$, in practice, it *does not* happen. To understand the reason for this, let us assume that for storing the values of the signals and the filter coefficient, registers of $K + 1 = 4$ bits are used (of which the most significant bit is used for storing the sign whereas the remaining $K = 3$ bits are used for storing the value) and let us calculate the values of the output signal samples for the various values of n, in the way we did in Exercise 7.4.9 (see page 586). However, unlike that case, in which the values are characterized by *infinite* accuracy and therefore, the output tends to zero over time, in this case we proceed to the *quantization* of the values, since these are stored in registers with a length of five bits.

Initially, let us assume that the value of the coefficient α is $\alpha = 5/8$. Assume that the input signal is $x[n] = 0$ for every n. Since we are interested in the limit cycles of zero input, let us assume that $y[0] = 3/8$. If we perform the operations and calculate the output samples $y[n]$ in the way we did in Exercise 7.4.9 (see page 586), but this time, proceeding to quantize the values of the product $\alpha y[n-1]$ for each value of n, so as to represent it accurately with four bits, we find that if the quantization is performed with *rounding*, the system output is the sequence

$$y[n] = \left\{ \frac{3}{8}, \frac{1}{4}, \frac{1}{8}, \frac{1}{8}, \frac{1}{8}, \frac{1}{8}, \frac{1}{8}, \dots \right\},$$

[23] See Munson D.C., jr, *Accessibility of Zero-Input Limit Cycles* IEEE Trans. Acoust., Speech & Signal Process., vol. ASSP-29, pp. 1027–1032, 1981.

[24] See for example Jackson L.B., *An Analysis of Limit Cycles Due to Multiplication Rounding in Recursive Digital (Sub-) Filters*, Proc. 7th Annual Allerton Conf. Circuit Syst. Theory; Monticello, Illinois, USA, ed. Hadad A.H. et al. (Urbana, Illinois, USA: Dept. of Electrical Eng., Univ. of Illinois 1969), pp. 69–78, 1969; Parker S.R. & Hess S.F., *Limit Cycle Oscillations in Digital Filters*, IEEE Trans. Circuit Theory, vol. CT-18, pp. 687–697, 1971; Parker S.R., *The Phenomena of Quantization Error and Limit Cycles in Fixed Point Digital Filters*, Proc. 28th Nat. Electr. Conf, Chicago, USA, ed. Horton R.E. (Oak Brook, Illinois, USA: Nat. Electr. Conf. 1972), pp. 38–42, 1972; and Sandberg I.W., *A Theorem Concerning Limit Cycles in Digital Filters* Proc. 7th Annual Allerton Conf. Circuit Syst. Theory, Monticello, Illinois, USA, ed. Hadad A.H. et al. (Urbana, Illinois, Dept. of Electrical Eng, Univ. of Illinois 1969), pp. 63–68, 1969.

whereas, if the quantization is performed with *truncation*, the system output sequence is

$$y[n] = \left\{ -\frac{3}{8}, -\frac{1}{4}, -\frac{1}{8}, -\frac{1}{8}, -\frac{1}{8}, -\frac{1}{8}, -\frac{1}{8}, \cdots \right\}.$$

On the other hand, for the coefficient value $\alpha = -5/8$ and for $y[0] = 3/8$, the output sequence for quantization with *rounding* is

$$y[n] = \left\{ \frac{3}{8}, -\frac{1}{4}, \frac{1}{8}, -\frac{1}{8}, \frac{1}{8}, -\frac{1}{8}, \frac{1}{8}, \cdots \right\},$$

whereas the corresponding output sequence for quantization with *truncation* is

$$y[n] = \left\{ \frac{3}{8}, -\frac{1}{4}, \frac{1}{8}, 0, 0, 0, 0, \cdots \right\}.$$

We are therefore led to the conclusion that *the type of quantization applied leads to different sequences at the output*, whereas we furthermore observe that in the case of a *positive* value for the coefficient α and for the case of *rounding* (since generally speaking, truncation can lead to the elimination of such a cycle, as happens here), the output sequence (except for the first few samples) is *periodic* with a period of $N = 2$. This sequence constitutes a *limit cycle* and it is not difficult to understand that the existence of these limit cycles is an *undesirable consequence*, and more specifically, a *side effect* of the quantization of product values. Note that although what we mention here concerns *zero-input limit cycles*, which appear when the input $x[n]$ is equal to zero but the initial conditions set for the difference equation are nonzero, limit cycles also appear in cases where the *nonzero input signal of the system suddenly takes a zero value*, that is, the system stops receiving input. In this case, the system's output, after a small number of iterations, will enter a *limit cycle* from which it will exit when the system receives a nonzero input signal again. In the terminology of digital filters, the output amplitudes during a limit cycle take values in a region known as the dead zone of the filter.

For the case of the simple pole filter we used previously and for the rounding quantization process, the range of quantization error values is defined by the condition

$$|Q[\alpha^q y^q[n-1]] - \alpha^q y^q[n-1]| \leq \frac{1}{2}2^{-K}.$$

However, it is not difficult to realize that when the system's output is on the limit cycle, *the actual nonlinear system behaves like a linear system* with a pole at the position $z = 1$ when the pole is positive and at the position $z = -1$ when the pole is negative, that is, $Q[\alpha^q y^q[n-1]] = y^q[n-1]$ for $\alpha > 0$ and $Q[\alpha^q y^q[n-1]] = -y^q[n-1]$ for $\alpha < 0$. Indeed, observing the system's output when it is on a limit cycle we find that the sequence of values that arises has *the same absolute value but with alternating signs* so it will result from one of the above two conditions. Replacing the first of the two conditions in the above expression we get

$$|y^q[n-1]| - |\alpha^q y^q[n-1]| \leq \frac{1}{2}2^{-K}$$

or equivalently

$$|y^q[n-1]|(1 - |\alpha^q|) \leq \frac{1}{2}2^{-K}.$$

Therefore, the *dead zone* of the quantized simple pole filter is defined by the condition

$$|y^q[n-1]| \leq \frac{(1/2) \times 2^{-K}}{1 - |\alpha|}.$$

If $|\alpha| < 0.5$, then the filter *will not* exhibit unstable behavior, whereas if $|\alpha| \geq 0.5$ the response will tend to *zero* when the input signal $x[n]$ is removed and at some point the quantity $y^q[n-1]$ will take a value in the above area that defines the dead zone. In this case, the filter will become *unstable*. If the value of $y^q[n-1]$ exceeds the above limit value, the instability will be removed *temporarily*, but since there is no input signal, the output will weaken again and the system will enter *again* into a limit cycle, which is an oscillation with amplitude smaller or equal to the above value. The above applies to the case where quantization is performed with *rounding*. On the other hand, when quantization is performed with *size clipping*, the system *does not show* limit cycles of zero input, whereas when quantization is performed with *value clipping*, the only limit cycles of this type that can appear, have a period value of $N = 1$ (see Werter, 1986, pages 68–69).

Limit Cycles Due to Quantization in Second-Order Systems

Similar results can also appear in the case of second-order systems, such as systems that include two poles, although there the degree of complexity is greater and *multiple oscillations of different types can arise*. Specifically, considering a second-order system described by the linear difference equation

$$y[n] = \alpha_1 y[n-1] + \alpha_2 y[n-2] + x[n],$$

whereas the actual quantized system is described by the nonlinear difference equation

$$y^q[n] = Q[\alpha_1^q y^q[n-1]] + Q[\alpha_2^q y^q[n-2]] + x^q[n],$$

we know that if the condition $\alpha_1^2 < -4\alpha_2$ is satisfied, the system has *two complex conjugate poles* at the positions $p_{1,2} = re^{\pm j\vartheta}$ where $\alpha_2 = -r^2$ and $\alpha_1 = 2r\cos\vartheta$. In this case, when the system is in a limit cycle of zero input, as before, $Q[\alpha_2^q y^q[n-2]] = -y^q[n-2]$ and consequently, the system behaves as an oscillator that has *complex conjugate poles on the unit circle*. Working as before, the rounding of the product $\alpha_2^q y^q[n-2]$ leads to a dead zone that is defined by the condition

$$|y^q[n-1]| \leq \frac{(1/2) \times 2^{-K}}{1 - |\alpha_2|}$$

and that, interestingly, depends *only* on the coefficient α_2. On the other hand, the parameter $\alpha_1 = 2r\cos\vartheta$ defines the frequency of the oscillation. Let it also be noted that apart from the above limit cycle of zero input, it is proven that the system can also perform a *second type of oscillation*, which corresponds to an *equivalent* linear system with poles at positions $z = \pm 1$. In this case, the dead zone that also defines the amplitude of the oscillation is limited by the value $2^{-K}/(1 - |\alpha_1| - |\alpha_2|)$. In Fig. 12.45 the areas of the plane $\alpha_1 - \alpha_2$ that are associated with the creation of limit cycles due to quantization of the product are depicted. Notice that the values of the coefficients that correspond to the central area of the plane, *do not lead* to the creation of such cycles. Also notice that in the different areas, different oscillation amplitudes correspond, which have values $y_0 = 1$, $y_0 = 2$ and $y_0 = 4$.

In the literature, many other results can be found related to the limits of the oscillations that are associated with these limit cycles. Particular interest is found in the work of Parker & Hess (1971)[25], who investigated the problem for the case of *a recursive second-order digital filter containing two poles and no zeros*, with coefficients $\alpha_1 = \alpha$ and $\alpha_2 = \beta$. They constructed several different expressions for these limits, and let us describe the construction method for two of them. In the first case, the method is based on the *direct method of Lyapunov* and the use of a *Lyapunov function* of the form $V = \boldsymbol{y}^T \boldsymbol{Q} \boldsymbol{y}$, where $\boldsymbol{Q}$ *is a real symmetric and positive definite* matrix of dimensions 2×2 defined as

$$\boldsymbol{Q} = \begin{pmatrix} q_{11} & q_{12} \\ q_{21} & q_{22} \end{pmatrix} = \begin{pmatrix} 1 + \dfrac{2\beta^2(1+\beta)}{(1-\beta) + [(1+\beta)^2 - \alpha^2]} & \dfrac{2\alpha\beta}{(1-\beta) + [(1+\beta)^2 - \alpha^2]} \\ \dfrac{2\alpha\beta}{(1-\beta) + [(1+\beta)^2 - \alpha^2]} & \dfrac{2(1+\beta)}{(1-\beta) + [(1+\beta)^2 - \alpha^2]} \end{pmatrix} \tag{12.60}$$

---

[25] See among others Appel U., *Bounds on Second-Order Digital Filter Limit Cycles*, IEEE Trans. Circuits and Systems, Vol. CA-22. pp. 630–632. 1975 and Long J.L. & Trick T.N., *An Absolute Bound on Limit Cycles Due to Roundoff Errors in Digital Filters*, IEEE Trans. Audio and Electroacoust., vol. AU-21, pp. 27–30, 1973.

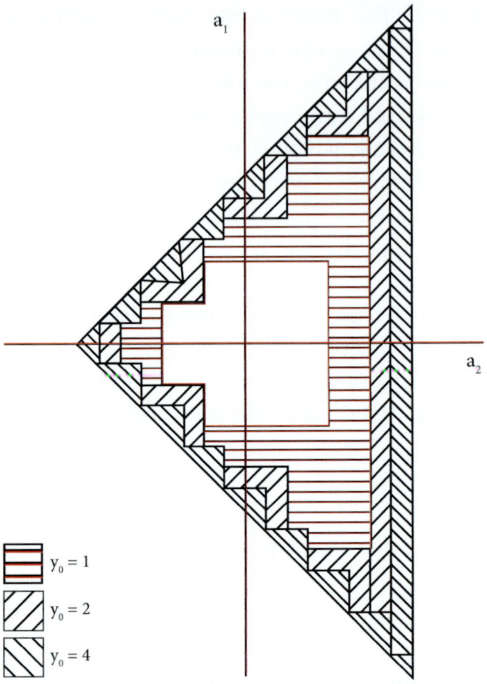

Fig. 12.45 The areas of the plane $\alpha_1 - \alpha_2$ that are associated with the creation of limit cycles due to quantization of the product (Antoniou, Figure 12.43)

and

$$\boldsymbol{y}[n+1] = \begin{pmatrix} 0 & 1 \\ -\beta & -\alpha \end{pmatrix} \boldsymbol{y}[n] + \begin{pmatrix} 0 \\ 1 \end{pmatrix} \boldsymbol{x}[n] \tag{12.61}$$

is the state equation of the system in the state space (see Sect. 9.7 on page 732), with state variables $y_1[n] = y[n-2]$ and $y_2[n] = y[n-1]$, state vector $y[n] = [\, y_1[n] \; y_2[n] \,]^T$ of dimensions 2×1, and input $x[n]$, which is assumed to be blocked for $n \geq 0$. Parker and Hess proved that an upper limit for the amplitude of oscillations of the zero-input limit cycles for this system is the

$$|y[n]| \leq \sqrt{\frac{\lambda_{\max} \boldsymbol{Q}}{\lambda_{\min} \boldsymbol{Q}}} \left[w + \sqrt{w^2 + q_{22}} \right],$$

where $\lambda_{\max}(\boldsymbol{Q})$ and $\lambda_{\min}(\boldsymbol{Q})$ are the *maximum* and the *minimum* eigenvalues of the matrix $\boldsymbol{Q}$, whereas the parameter w is defined as

$$w = \max(|\beta q_{22}|, |q_{12} - \alpha q_{22}|).$$

However, as they comment, because the matrix $\boldsymbol{Q}$ has been chosen *arbitrarily*, *there* is no guarantee that this constitutes the optimal limit that can be determined. As for the second method of calculating such a limit, this arises through the construction of a general expression, expressed in table form, for a limit cycle with period qT. Observing that according to the definition of the limit cycle and periodicity it will hold

$$y[n] = y[n-q]$$
$$y[n-1] = y[n-q-1],$$

etc., let us formulate the nonlinear difference equation of the quantized system after rounding, as

$$y[n] = -\alpha y[n-1] \pm [0.5 - \delta[n-1]] - \beta y[n-2] \pm [0.5 - \delta[n-2]],$$

where $\epsilon_\alpha = [0.5 - \delta[n-1]]$ and $\epsilon_\beta = [0.5 - \delta[n-2]]$ are the rounding noise sequences, which take values in the range ± 0.5. Reformulating the rounding noise sequences in the form

$$\varepsilon_i = \pm[0.5 - \delta[n-i]] \pm [0.5 - \delta[n-i-1]]$$

the iterative use of the difference equation and the condition of periodicity allow us to write the expression

$$\begin{pmatrix} 1 & \alpha & \beta & 0 & 0 & \cdots & 0 & 0 & 0 \\ 0 & 1 & \alpha & \beta & 0 & \cdots & 0 & 0 & 0 \\ 0 & 0 & 1 & \alpha & \beta & \cdots & 0 & 0 & 0 \\ \vdots & \vdots & \vdots & \vdots & \vdots & \ddots & \vdots & \vdots & \vdots \\ 0 & 0 & 0 & 0 & 0 & \cdots & 1 & \alpha & \beta \\ \beta & 0 & 0 & 0 & 0 & 0 & \cdots & 1 & \alpha \\ \alpha & \beta & 0 & 0 & 0 & 0 & 0 & \cdots & 1 \end{pmatrix} \begin{pmatrix} y[n] \\ y[n-1] \\ y[n-2] \\ \vdots \\ y[n-q+3] \\ y[n-q+2] \\ y[n-q+1] \end{pmatrix} = \begin{pmatrix} \varepsilon_1 \\ \varepsilon_2 \\ \varepsilon_3 \\ \vdots \\ \varepsilon_{q-2} \\ \varepsilon_{q-1} \\ \varepsilon_q \end{pmatrix},$$

or, more succinctly, in matrix form, $\boldsymbol{A}\boldsymbol{y} = \boldsymbol{\varepsilon}$, where $\boldsymbol{A}$ is the *coefficient matrix* of dimensions $q \times q$ and $\boldsymbol{x}$ and $\boldsymbol{\varepsilon}$ are vectors of dimensions $q \times 1$, which contain *the samples of the limit-cycle sequence* and *the rounding-noise samples* respectively. It is proven[26] that if a *symmetric* limit cycle is used, such that the samples corresponding to the first half of the period *have the same magnitude but opposite sign* compared with the samples located in the second half of the period, only $q/2$ equations are required. It is also proven that the matrix $\boldsymbol{A}$ is *nonsingular*, if and only if there is at least one error value $\varepsilon_i \neq 0$, whereas the matrix $\boldsymbol{A}$ is *singular* if and only if $\varepsilon_i = 0$ for every value of the index i. This last condition determines the values of the coefficients α and β, for which the system response is linear (without rounding error) and its samples, recorded for a time period of one oscillation period of the limit cycle, are all integers. Finally, it is proven that the satisfaction of this condition imposes in turn the requirement for $\beta = 1$ and consequently, the matrix cannot be idiosyncratic for values $\beta < 1.0$. In this case, by solving the above equation using some of the available methods, we can find a limit for the value $y[n-i]$ for quantization error due to rounding in the value range $0 \leq |\varepsilon_i| \leq 1.0$. For example, resorting to the well-known *Cramer's method*, we end up with the expression

$$y[n-i] = \frac{\varepsilon_1 \det(A_{1i}) + \varepsilon_2 \det(A_{2i}) + \cdots + \varepsilon_q \det(A_{qi})}{\det(A)}, \tag{12.62}$$

and therefore, we have

$$|y[n-i]| = \frac{|\varepsilon_1 \det(A_{1i}) + \varepsilon_2 \det(A_{2i}) + \cdots + \varepsilon_q \det(A_{qi})|}{|\det(A)|} \leq \frac{1}{|\det(A)|} \sum_{j=1}^{q} |\det(A_{ji})|. \tag{12.63}$$

It is interesting to note that owing to the cyclic nature of the equations involved, this limit has *the same numerical value* for $i = 1, 2, \ldots, q-1$, which, however, is *different* from the one determined by resorting to the *first Lyapunov method*, since in this case the period qT of the limit cycle should be determined. In addition, for large values of q, the above value is generally not easy to calculate.

Limit Cycles Due to Quantization in Higher-Order Systems with Parallel Implementations

Finally, in the general case of *higher-order* systems, which are characterized by *parallel* implementations, the outputs of the various second-order stages that they consist of are *independent of each other*, if the input to the system is equal to zero. In this case, some of the stages may lead to limit cycles, which will naturally appear in the final result, since in this case *the output of the system will result from the sum of the limit cycles of zero input owing to the individual stages*. Conversely, in serial implementations, the appearance of limit cycles *is more complex and more difficult to analyze*, since in this case, only the first stage can receive zero input (remember, in these implementations, the output of each stage becomes input to the next stage), with the stages that are connected sequentially one after the other either each displaying its own limit cycle or

[26] See the reference Hess S., *A Deterministic Analysis of Limit Cycle Oscillations in Recursive Digital Filters Due to Quantization*, Ph.D. dissertation, Naval Postgraduate School, Monterey, California, December 1970.

simply processing the limit cycle it has received as input from a previous stage. If the frequency of the limit cycle is *close to a resonance frequency* of one of the next stages, then there will be *amplification of the amplitude of the discrete sequence*, according to what is predicted by the phenomenon of *resonance*, a situation that should certainly be avoided. Finally, in cases where the higher-order systems have been implemented using more complex connections, the system's behavior regarding the appearance of limit cycles is certainly more complex.

Limit Cycles Due to Overflow

It is worth noting that, apart from the quantization process of the value of the products that arise as the result of multiplications, limit cycles can also appear as a result of *overflow*, which is related to the act of addition and arises when the calculated values are larger than the maximum value that can be represented by the word length used. The impact of this type of limit cycle, which seems to have been observed for the first time by Leland B. Jackson of Bell Laboratories, is *quite severe* and consists in the appearance of a *large error* at the filter output, resulting in the *oscillation* of the system output values between limits characterized by a large amplitude, which can be even equal to the maximum amplitude of the signal that can be used in the underlying digital system (note that this is something that *does not apply* to limit cycles due to quantization, which are due to the subtlety of the quantization process and therefore, are characterized by a small amplitude). These *large-scale limit cycles*are characterized as *oscillations due to overflow*. This type of limit cycle was studied by Ebert, Mazo, and Taylor and the most significant of the results that they concluded are presented later. The common feature of filters that exhibit such behavior is *the absence of a certain time instance, after which the overflow ceases to exist* (which is true for stable operation), whereas it should also be noted that the overflow will not necessarily appear immediately after the start of the system's operation for some initial conditions but after a certain period of time, whereby it will eventually end up in a state that will be *periodic*, *asymptotically periodic*, or even *chaotic*.

In order to understand the problem associated with these limit cycles, let us consider for simplicity's sake *a second-order digital filter with complex conjugate poles*, since practice has shown that systems of this type can exhibit a multitude of parasitic oscillations for large values of the Q factor. Limiting ourselves only to the *overflow* associated with the signals that are passed to the delay elements and ignoring all the other cases of overflow, we know that the state equation of the ideal linear system with system matrix $\boldsymbol{A} = \{\alpha_{ij}\}$ $(i, j = 1, 2)$ under zero input conditions $(x[n] = 0)$ is the $\boldsymbol{q}[n+1] = \boldsymbol{A}\boldsymbol{q}[n]$, whereas the corresponding equation for the nonlinear system in which nonlinear elements have been added to address the phenomenon of overflow is $\boldsymbol{q}[n+1] = Q(\boldsymbol{A}\boldsymbol{q}[n])$, where $y = Q(x)$ is the *input–output characteristic* of the nonlinear element. It is understood that this nonlinearity is applied component-wise to each coordinate of the state vector $\boldsymbol{q}[n]$. In the context of studying overflow situations, we accept that the values of the state variables $q_1[n]$ and $q_2[n]$ take values in the range $[-1, +1]$ and consequently, the conditions $-1 \le q_i[n] \le +1$ $(i = 1, 2)$ define the stability region $\mathcal{S}$ in the phase space, which is a *square with a side length* of 2 (since it extends from the value -1 to the value $+1$ on both axes related to the state variables q_1 and q_2) and the center is the origin of the coordinate system. This means that the system exhibits *stable* behavior when the point of the phase space corresponding to the state vector $\boldsymbol{q}[n]$ is *inside or on the outline of this region*, whereas points that are *outside* this region (which indicates that some of the variables $q_1[n]$ and $q_2[n]$ or both have values outside the interval $[-1, +1]$) correspond to the appearance of *overflow* and therefore, appropriate corrections should take place.

From basic theory we know that the solution of the state equation $\boldsymbol{q}[n+1] = Q(\boldsymbol{A}\boldsymbol{q}[n])$ of the nonlinear system has the form

$$\boldsymbol{q}[n] = \Re\{C(\xi_r + j\xi_i)\exp[(\gamma + j\omega)n + j\Phi]\} = Ce^{\gamma n}[\xi_r \cos(\omega n + \Phi) - \xi_i \sin(\omega n + \Phi)]. \tag{12.64}$$

In the above equation, the quantities $\exp(\gamma \pm j\omega)$ are the *complex eigenvalues* of the system matrix $\boldsymbol{A}$, the quantities $\xi_r \pm j\xi_i$ are the corresponding *eigenvectors*, which without loss of generality we assume have undergone the appropriate process of *normalization* and satisfy the *orthogonality condition* $\xi_r^t \xi_i = 0$, whereas C and Φ are integration constants, the values of which, as usual, depend on the initial conditions described by the vector $\boldsymbol{q}_0 = [\, q_1[0]\; q_2[0]\,]^T$. If we assume that $\gamma = 0$ and that the variable *n is not discrete but continuous*, then the trajectory traced by the state vector $\boldsymbol{q} = [\, q_1[n]\; q_2[n]\,]^T$ of the system in phase space is an *elliptical trajectory*, which is characterized by a major and minor semi-axis, the directions of which coincide with the directions of the eigenvectors ξ_r and ξ_i. Note, however, that in practice it is always $\gamma < 0$ (a condition that leads to poles of the system that are *inside the unit circle*) and consequently, the trajectory of the system is a *spiral trajectory* with a continuously decreasing radius, which approaches *progressively* the origin of the axes.[27] This spiral

[27] Note that this type of study in the space of phases can also be implemented for small-scale limit cycles related to the quantization process using the equation $\boldsymbol{q}[n+1] = Q(\boldsymbol{A}\boldsymbol{q}[n])$, with the difference that in this case, owing to quantization, the values of the state variables will be integer multiples of the quantization step q.

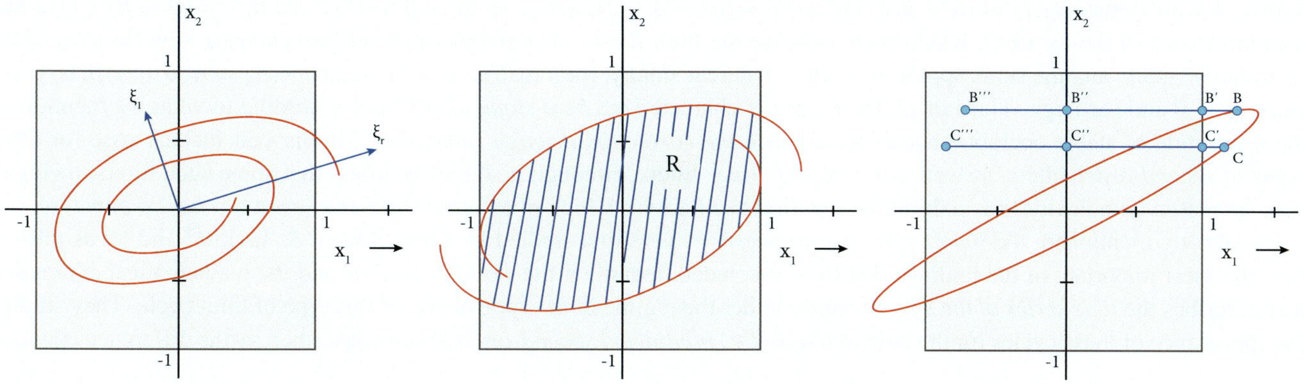

Fig. 12.46 The trajectory of the state vector in the state space and the square of stability (left), the area of initial conditions that do not lead to overflow (middle), and the overflow recovery corrections (right) (Butterweck, Ritzerfeld and Werter, Figures 3.1, 3.2 and 3.3)

trajectory, along with the stability region $\mathcal{S}$, is depicted in the left image of Fig. 12.46 whereas the initial conditions that are guaranteed *not to lead to overflow*, i.e., points outside the stability region, belong to the shaded area R of the middle image of Fig. 12.46 (see Butterweck et al., 1988).

On the other hand, if the vector $\boldsymbol{q}_0$ of the initial conditions is *outside* the area R, then the linearly determined state vector $\boldsymbol{q}[n]$ may exit the square of stability $\mathcal{S}$ (this means that one or both of the signals $q_1[n]$ and $q_2[n]$ have a value outside the range $[-1, +1]$); thus, an *overflow* arises and the appropriate *correction* is required. This correction is performed by *saturating*, by *nullifying*, or by using the *complement to* 2, processes that are performed by the appropriate nonlinear element, which is added to the system for this purpose and consists *in restoring the point determined by the current value of the state vector* $\boldsymbol{q}[n]$, *inside the square of stability*. This corrective process will lead *to either a reduction or an increase in the distance of the point from the origin of the coordinate system*, with the distance d not defined as the usual Euclidean distance, but through the expression

$$M^2 = \left(\frac{\boldsymbol{\xi}_r^T \boldsymbol{q}}{\boldsymbol{\xi}_r^T \boldsymbol{\xi}_r}\right)^2 + \left(\frac{\boldsymbol{\xi}_i^T \boldsymbol{q}}{\boldsymbol{\xi}_i^T \boldsymbol{\xi}_r}\right)^2 \tag{12.65}$$

(note that as can be proven, it is $M = Ce^{\gamma n}$) that constitutes a *quadratic form* of the state variables q_1 and q_2 and can be used as a *Lyapunov function* constituting one of the many available choices. The analysis of the problem reveals that the equation $M = \text{const}$ for the various values of M describes *a set of concentric ellipses with different lengths of major and minor axis, but with a common orientation in space*, which is determined by the directions of the eigenvectors ξ_r and ξ_i. Considering the point B of the right image of Fig. 12.46, which is located outside the square of stability and therefore, refers to *overflow* (from the position of this point, it is observed that the overflow is due to the value of $q_1[n]$, whereas the signal $q_2[n]$ has a value within the stability region), the application of the above corrective process transfers point B (a) to point B', which is *on the perimeter of the square of stability*, if it is performed with *saturation* (since we set $q_1[n] = 1$); (b) to point B'', which lies on the vertical axis, if it is performed with *nullification* (since we set $q_1[n] = 0$); and (c) to point B'', if it is performed with *complement to* 2. The problem is that for the random orientation of this example's ellipse the above corrections *increase the distance from the origin of the coordinate system* (which constitutes a *strongly undesirable characteristic* and can lead to *oscillations*), as is certified if we plot the ellipses that pass through these points and are concentric with respect to the ellipse of the scheme and take into account that the state vector moves on these ellipses. However, in the special case of ellipses whose axes coincide with the axes of the coordinate system, it is not difficult to realize that *all the above types of correction lead to a reduction of this distance and the preservation of the natural linear motion*, whereas for the case of an ellipse with a slope of 45°, this is achieved at least with the mechanism of *saturation*. It is interesting to note that the primary role in this *recovery process from overflow*, that is, the restoration of stable behavior, is played by the *eigenvectors*, as they determine the appearance of the elliptical orbit and the ratio of the lengths of the semi-axes of the ellipse, whereas the eigenvalues determine the *speed* at which the point determined by the state vector moves along the elliptical orbit, with the values of these parameters, in turn, depending *on the structure of the filter*, that is, essentially, *on the elements of the system matrix* $\boldsymbol{A}$. For example, for a matrix $\boldsymbol{A}$ with elements $\alpha_{11} = \alpha_{22} = \alpha/2$ and $\beta_{12} = -\alpha_{21} = \sqrt{-4\beta - \alpha^2}/2$, this ellipse *degenerates into a circle* and therefore, all the above correction techniques lead to stability. On the other hand, for a matrix $\boldsymbol{A}$ with elements $\alpha_{11} = (\alpha + \beta + 1)/2$, $\alpha_{12} = (\alpha + \beta - 1)/2$, $\alpha_{21} = (\alpha - \beta + 1)/2$ and $\alpha_{22} = (\alpha - \beta - 1)/2$ and for high values of the factor Q, which correspond to the limit $\beta \rightarrow -1$, *the axes of the ellipse coincide with the axes of the coordinate system* and the same applies as before, whereas finally, for a matrix $\boldsymbol{A}$ with elements $\alpha_{11} = \alpha$, $\alpha_{12} = \beta$, $\alpha_{21} = 1$ and $\alpha_{22} = 0$ and for large values of Q, the ellipse is characterized *by a slope of* 45° and according to what we mentioned, the restoration of stable operation is carried out *with the correction through saturation*. Similar conclusions arise for other types of *second-order* filters, as well as for *higher-order* filters, although it should be noted that some such filters, owing to their increased complexity, may exhibit large period oscillations, which in extreme cases may even be *chaotic* in nature.

As already mentioned, the *limit cycles due to overflow* were first studied by Ebert, Mazo, & Taylor[28] and let us present here the most important of the findings that they concluded, regarding the *stability regions* and the mathematical expression that describes the *time series of the system output* under the regime of the appearance of this type of limit cycle. They studied the appearance of these cycles for the case of a *real (i.e., nonlinear) second-order system* described by the difference equation

$$y[n+2] = f[\alpha y[n+1] + \beta y[n] + x[n+2]$$

and is characterized by *complex conjugate poles*. In the above expression, the function $y = f(x)$ describes the input/output characteristic of the adder that performs the addition operation, according to the complement to 2 (see the right image of Fig. 12.22), whereas the overflow management is considered with regard to how it is carried out *independently of the quantization process*. If we assume that the adder operates *in its linear region* and therefore, the input/output characteristic is described by the *identity* function $f(v) = v$, with $|v| < 1$ and $|f(v)| < 1$, then the above difference equation under conditions of *zero input* $x[n+2] = 0$ takes the simple form

$$y[n+2] = \alpha y[n+1] + \beta y[n]$$

and according to the relevant theory, it is described by the characteristic equation $\lambda^2 - \alpha\lambda - \beta = 0$, which has discriminant $\Delta = \alpha^2 + 4\alpha\beta$ and roots

$$\lambda_{1,2} = \frac{\alpha \pm \sqrt{\alpha^2 + 4\alpha\beta}}{2}.$$

In order to determine the *stability region in the parameter space* for this special regime of linear operation, we must impose the requirement $|\lambda_{1,2}| < 1$, a condition that, resorting to the above expression, is proven to be satisfied, if and only if the point (α, β) is *within the triangle* depicted in the left image of Fig. 12.47 and is defined by the points $A(1, 0)$, $B(-1, 2)$ and $C(-1, -2)$. In this case, each solution of the above difference equation *will tend to zero* after the passage of some time period. It should be noted, however, that for the above equation to constitute a valid simplification of the original equation, we must impose, in addition to the condition $|y[n+2]| < 1$, the additional condition $|\alpha y[n+1] + \beta y[n]| < 1$, since the expression $\alpha y[n+1] + \beta y[n]$ is used as an argument in the function f and according to what we mentioned above, in the area of linear operation, it is not enough to simply be $|f(v)| < 1$ but additionally it must be true that $|v| < 1$. Combining these two conditions, we end up with the condition $|\alpha| + |\beta| < 1$, which is satisfied by the points with coordinates (α, β) that are located *in the shaded area* of the left image of Fig. 12.47, which is a subset of the stability triangle. This last inequality is the *sufficient and necessary condition for the absence of oscillations*[29] and certainly *imposes restrictions* on the operating characteristics of digital filters, since now, the poles of the system *cannot be located at any point in the complex plane* but only in the shaded area of the right image of Fig. 12.47. Conversely, if the coefficients α and β of the filter have values that do not satisfy the above condition, but it holds that $|\alpha| + |\beta| > 1$, then the output of the filter is characterized by *periodic behavior*, that is, it takes the form of oscillations with some period value N.

---

[28] See the reference Ebert P.M., Mazo J.E., & Taylor M.G., *Overflow Oscillations in Digital Filters*, Bell System Technical Journal, 48(9), 2999–3020, 1969.

[29] For an alternative proof of this condition, with the help of the Lyapunov method, see Werter, M. J., *Suppression of Parasitic Oscillations Due to Overflow and Quantization in Recursive Digital Filters*, pages 49–50, whereas for proofs based on techniques applied in the frequency field, see Claasen T.A.C.M., Mecklenbräuker W.F.G., & Peek J.B.H., *Necessary and Sufficient Conditions for the Absence of Overflow Phenomena in a Second-Order Recursive Digital Filter*, IEEE Trans. Acoust. Speech, Signal Processing, vol. ASSP-23, pp. 509–515, 1975; Claasen T.A.C.M., Mecklenbräuker W.F.G., & Peek J.B.H., *Frequency Domain Criteria for the Absence of Zero-Input Limit Cycles in Nonlinear Discrete-Time Systems, with Applications to Digital Filters*, IEEE Trans, Circuits Syst, vol. CAS-22, pp. 232–239, 1975; and Claasen T.A.C.M., Mecklenbräuker W.F.G., & Peek J.B.H., *On the Stability of the Forced Response of Digital Filters with Overflow Nonlinearities*, IEEE Trans. Circuits Syst., vol. CAS-22, pp. 692–696, 1975.

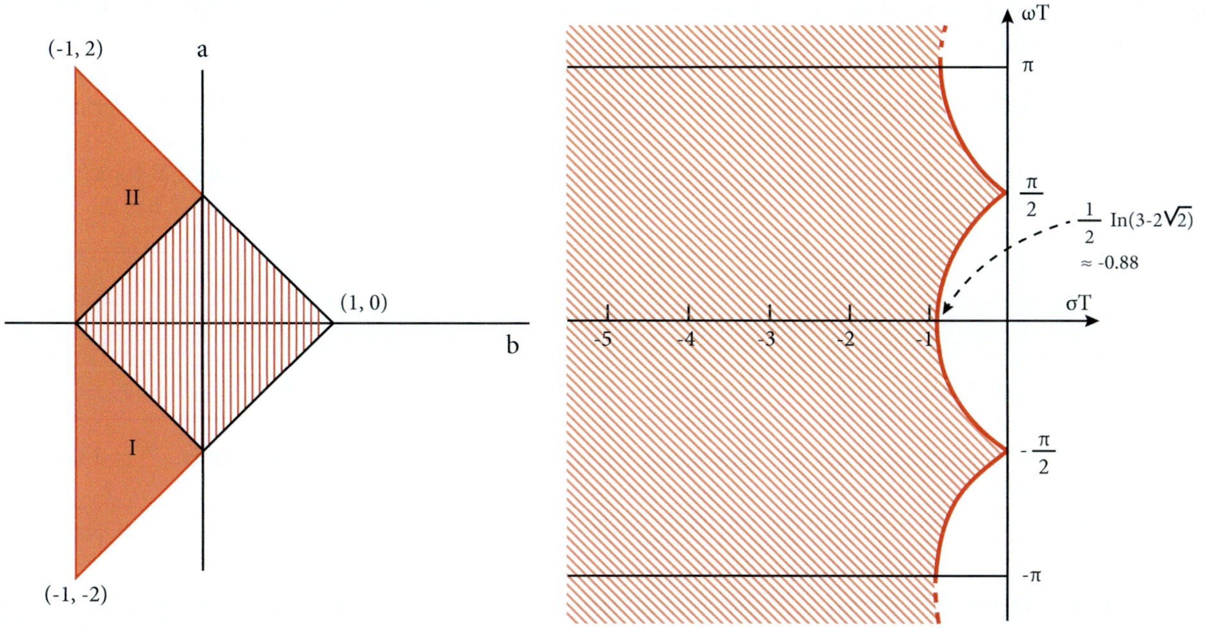

Fig. 12.47 (**a**) The stability triangle in the plane (α, β) and the area of the triangle in which the condition $|\alpha|+|\beta| < 1$ applies. (**b**) The permissible positions of the poles of the digital filter for which this condition is satisfied (right image) (Ebert, Mazo and Taylor, Figures 3 and 4)

The simplest case of such oscillations, the study of which is relatively easy and direct, are the oscillations with periods $N = 1$ and $N = 2$. In the first case, the output of the system has the constant value $y[n] = y$ for every value of n. Specifically, $y[n + 2] = y[n + 1] = y[n] = y$ and replacing in the difference equation, we get $y = f[(\alpha + \beta)y]$. Considering the case $y > 0$ (the other case is treated similarly) and studying the form of the characteristic of the adder, it is not difficult to find out that the above equation is satisfied if it is $(\alpha + \beta)y = y - 2$, from where it follows that $y = 2/(1 - \alpha - \beta)$. This value also constitutes *the output value of the system* for the period value $N = 1$ and since we are only interested in such values, so that $|y| < 1$, the condition $|1 - \alpha - \beta| > 2$ must apply, from which it follows that $1 - \alpha - \beta > 2$ or $-1 + \alpha + \beta > 2$. From the last two conditions, the second describes a region that is *outside the stability triangle* and does not interest us, whereas the first condition, along with the condition of linear stability, which leads to the stability triangle, defines region I of this triangle, which, therefore, constitutes the region of points (α, β) for which the output of the system is periodic with period $N = 1$. On the other hand, in the system output with period $N = 2$, the output samples again have a constant value y, but with *alternating signs*, that is, they have the form $\{y, -y, y, -y, y, -y, \dots\}$ and generally $y[n] = (-1)^n y$. So, if in the difference equation, we set $y[n + 2] = y[n] = y$ and $y[n + 1] = -y$, it takes the form $y = f[(\beta - \alpha)y]$. Referring to the characteristic of the adder one more time, we find that the above condition is required to be $y = 2/(1 + \alpha - \beta)$ for $\alpha > 0$, whereas the imposition on this solution, of the previous conditions, leads us to region II of the stability triangle, the points (α, β) of which therefore lead to *periodic outputs* of the system with period $N = 2$.

On the other hand, for the study of the problem in the general case, we formulate the initial nonlinear difference equation in the form

$$y[n + 2] - \alpha y[n + 1] - \beta y[n] = \sum_k \alpha_k u[n + 2 - k]. \tag{12.66}$$

In the above expression, the left side resembles the corresponding linear equation, whereas the right side contains a *linear combination of time-shifted square pulses* $u[t]$ *with unit height and duration, from the time point* $n = 0$ *to the time point* $n = 1$. The reason we formulate the equation in this way is to ensure the validity of the condition $|f(\upsilon)| < 1$, *regardless* of the value of the argument υ and this is something we achieve by choosing the appropriate values for the coefficients α_k. For example, from the form of the adder's characteristic, we find that if $|\upsilon| < 1$, then $|f(\upsilon)| < 1$ will also be. Observing that the required condition *is automatically satisfied*, the sum of the right-hand side is not required and therefore, $\alpha_k = 0$ will be. On the other hand, in the area $1 < \upsilon < 3$ we use the value $\alpha_k = -2$, whereas in the area $-3 < \upsilon < -1$, we use the value $\alpha_k = 2$. These are the only areas that interest us, since the condition $|y[n]| < 1$ along with the stability condition, requires $|\alpha| < 2$ and $|\beta| < 1$ to be. In other words, the coefficients α_k will take the values $\alpha_k = 0, \pm 2$, depending on whether the function $\upsilon[n] \equiv \alpha y[n+1] + \beta y[n]$ intersects or not the lines $\upsilon = \pm 1$. In order to calculate the output $y[n]$ of the system for each value of the period N, we consider a periodic sequence of coefficients $\alpha_0, \alpha_1, \alpha_2, \ldots, \alpha_{N-1}$ of period N, that is, such that $\alpha_{\ell N+k} = \alpha_k$ ($\ell = 0, 1, 2, \ldots,\ 0 \leq k < N-1$) and we solve the difference equation. However, we should not forget that we are only interested in the solutions for which $|y[n]| < 1$, a condition that in turn depends on the periodic sequence used α_k.

The calculation of the system's output for the general case is based on the well-known method of $\mathcal{Z}$ transform. Initially, as is well known, the $\mathcal{Z}$ transform of the periodic sequence $\{\alpha_k\}$ ($k = 0, 1, 2, \ldots, N-1$) of period N has the form

$$\mathcal{A}(z) = \sum_{k=-\infty}^{\infty} \alpha_k z^{-k} = \frac{1}{1-z^{-N}} \sum_{k=0}^{N-1} \alpha_k z^{-k} = \frac{1}{1-z^{-N}} \mathcal{A}_{N-1}\left(\frac{1}{z}\right). \tag{12.67}$$

The poles of this transform are the values of z that satisfy the equation $1 - z^{-N} = 0$ or equivalently $z^N = 1$ and therefore, constitute the nth roots of unity, whereas there are no other poles.

Now, taking the $\mathcal{Z}$ transform of Eq. (12.66) and using the above result, it turns out that

$$\mathcal{Z}\{y[n+2] - \alpha y[n+1] - \beta y[n]\} = \mathcal{Y}(z)z^2 - \alpha z\mathcal{Y}(z) - \beta\mathcal{Y}(z) = \mathcal{Y}(z)(z^2 - \alpha z - \beta) = \frac{1}{1-z^{-N}} \mathcal{A}_{N-1}\left(\frac{1}{z}\right)$$

and therefore, we have

$$\mathcal{Y}(z) = \frac{1}{(z^2 - \alpha z - \beta)(1 - z^{-N})} \mathcal{A}_{N-1}\left(\frac{1}{z}\right) = \frac{1}{D(z)(1 - z^{-N})} \mathcal{A}_{N-1}\left(\frac{1}{z}\right),$$

where we have defined $D(z) = z^2 - \alpha z - \beta$. In reality, $\mathcal{Z}$ transform we considered here is the unilateral $\mathcal{Z}$ transform, whereas *we did not* take into account the additional terms associated with the initial conditions, since these with the passage of time will eventually *be amortized* owing to the of existing linear stability. In order to determine the desired periodic output of the system, which constitutes a *steady-state response*, we must take into account that the transform $\mathcal{Y}(z)$ consists of the sum of two transforms $\mathcal{Y}_{\text{tr}}(z)$ and $\mathcal{Y}_{\text{SS}}(z)$ the inversion of which leads to the *transient response of the system* and the *steady-state response* respectively. Of these two terms, the first depends *on the poles of the system*, whereas the second depends *on the poles of the input signal* and consequently, in order to arrive at the desired result, we must construct the transform $\mathcal{Y}_{\text{SS}}(z)$ that uses the poles of the input signal, which, as we mentioned, are the nth roots of unity, and from there we must calculate the desired output $y_{\text{SS}}[n]$. Let us denote with r_i ($i = 1, 2, \ldots, N$) the nth roots of unity and let us define the auxiliary function

$$Q_i^{N-1}\left(\frac{1}{z}\right) = \sum_{k=0}^{N-1} \left(\frac{1}{r_i}\right)^{N-1-k} \left(\frac{1}{z}\right)^k = \left(\frac{1}{r_i}\right)^{N-1} \sum_{k=0}^{N-1} \left(\frac{r_i}{z}\right)^k = \left(\frac{1}{r_i}\right)^{N-1} \frac{1 - \left(\frac{r_i}{z}\right)^N}{1 - \frac{r_i}{z}} = \frac{1 - z^{-N}}{\frac{1}{r_i} - \frac{1}{z}}. \tag{12.68}$$

The last relation arises by observing that

$$Q_i^{N-1}\left(\frac{1}{r_i}\right) = Q_i^{N-1}\left(\frac{1}{z}\right)\Bigg|_{z=r_i} = \lim_{z\to r_i} \frac{1 - z^{-N}}{\frac{1}{r_i} - \frac{1}{z}} = \lim_{z\to r_i} \frac{d(1 - z^{-N})/dz}{d\left(\frac{1}{r_i} - \frac{1}{z}\right)/dz}$$

$$= \lim_{z\to r_i} \frac{Nz^{-N-1}}{1/z^2} = \lim_{z\to r_i} Nz^{-N+1} = \lim_{z\to r_i} \frac{Nz}{z^N} = \frac{Nr_i}{r_i^N} = Nr_i, \tag{12.69}$$

resorting to the use of the *L'Hospital's rule* to handle the indeterminate form $0/0$ and taking into account that $r_i^N = 1$. In this case, resorting to the defining equation of $\mathcal{Y}(z)$ it is not difficult to realize that

$$\mathcal{Y}_{SS}(z) = \sum_{i=1}^{N} \frac{1}{\left(\frac{1}{r_i} - \frac{1}{z}\right) \cdot (N r_i) \cdot D(r_i)} \mathcal{A}_{N-1}\left(\frac{1}{r_i}\right), \tag{12.70}$$

where we have calculated the contribution of each pole to the function $\mathcal{Y}_{SS}(z)$ and then added together all these contributions. From Eq. (12.68) it turns out that

$$\frac{1}{r_i} - \frac{1}{z} = \frac{1 - z^{-N}}{Q_i^{N-1}\left(\frac{1}{z}\right)}$$

and the above relationship becomes

$$\mathcal{Y}_{SS}(z) = \sum_{i=1}^{N} \frac{\mathcal{A}_{N-1}\left(\frac{1}{r_i}\right)}{(N r_i D(r_i))} \frac{Q_i^{N-1}\left(\frac{1}{z}\right)}{1 - z^{-N}} = \frac{1}{1 - z^{-N}} \frac{1}{N} \sum_{i=1}^{N} \frac{\mathcal{A}_{N-1}\left(\frac{1}{r_i}\right) Q_i^{N-1}\left(\frac{1}{z}\right)}{r_i D(r_i)}. \tag{12.71}$$

This relationship is our final result and expresses *the $\mathcal{Z}$ transform of the steady-state response* of the system $y_{SS}[n]$. However, we know that this signal constitutes a *periodic signal* with a period of N and comparing the above expression with the defining equation of such a periodic signal (see the previous discussion about the $\mathcal{Z}$ transform of the periodic sequence $\{\alpha_k\}$ $(k = 0, 1, 2, \ldots, N-1)$ of period N), it is not difficult to understand that the samples of the *steady-state response* $y_{SS}[n]$ will constitute the coefficients of z^{-k} $(k = 0, 1, 2, \ldots, N-1)$ in the expression

$$\frac{1}{N} \sum_{i=1}^{N} \frac{\mathcal{A}_{N-1}\left(\frac{1}{r_i}\right) Q_i^{N-1}\left(\frac{1}{z}\right)}{r_i D(r_i)}.$$

Resorting once again to Eq. (12.68) and taking into account that $r_i^N = 1$, we finally end up with the expression

$$y_{ss}[n] = \frac{1}{N} \sum_{i=1}^{N} \frac{\mathcal{A}_{N-1}\left(\frac{1}{r_i}\right)}{D(r_i)} r_i^k, \tag{12.72}$$

which is also the final result. This expression allows us to calculate the *steady-state response* of the system, for each sequence of coefficients $\{\alpha_k\}$ $(k = 0, 1, 2, \ldots, N-1)$. However, we should not forget that the solutions leading to self-sustaining oscillations of the digital filter are *only those* for which it holds that $|y[n]| < 1$. Using the above expression for the various values of N we can get the solutions we got previously for the values $N = 1$ and $N = 2$, as well as other solutions for larger values of N and in the work of Ebert, Mazo, & Taylor, results can be found for solutions with periods $N = 3$ and $N = 5$, as well as the areas of the parameter space in which these solutions are defined.

The Techniques of Eliminating Boundary Cycles

Ebert, Mazo, and Taylor also studied the issue of *elimination* of these boundary cycles due to overflow, which obviously relies on the ability to *detect overflow*. As they characteristically comment, although this problem can be addressed through the technique of zeroing, i.e., zeroing the signal value when it exceeds the overflow threshold, practice shows that this technique can still lead to the appearance of such oscillations. In reality, as they proved, the technique that *solves the problem and restores stable operation* is *saturation*, which consists in the replacement of the value that led to overflow, with one of the values $+1$ and -1, depending on the sign (and of course under the assumption that the filter operates in the range of values $-1, +1$). This proof is based on the observation that from the moment we know the difference equation of the linear system, as well as the values of the output $y[n_0]$ and $y[n_0+1]$ for some time $n = n_0$, *we can calculate a limit for the values of the subsequent samples* corresponding to times $n \geq n_0$. This limit for the case of the linear system tends to zero, whereas for the nonlinear case there is the possibility that it may be, for example $y[n+1] > +1$ for some n, a condition that will also signify *the activation of the nonlinear element to handle overflow*. In this case, according to the operation of saturation, the correction $y[n+1] = 1$ will be made, which can be considered as a kind of *compression*, since it sets the sample to a value smaller than the original. In order to prove what is desired, that is, how the technique of saturation *solves* the problem, we must show that the partial derivative of the above limit, with respect to the most recent sample $y[n+1]$, has *the same sign* as the value of the sample $y[n+1]$. In this case, according to the principles of mathematical analysis, *the reduction of a value that is very large in size will also reduce the value of this limit, with the speed approaching the zero value being the same as that which characterizes the linear case*. Ebert, Mazo, & Taylor proved this property (which confirms the claim that the technique of saturation leads to *stable behavior* every time it is predicted by linear theory), for all cases of second-order difference equations (i.e., with complex conjugates and real equal and unequal roots of the corresponding characteristic equation), with the mathematical details of the proof being found in their work.

It is interesting to note that, in contrast to the *quantization noise*, which is *impossible to remove and so is fully suppressed*, this is something that is possible for the *limit cycles of quantization*, but at the cost of *increasing* the degree of complexity of the noise analysis. For this reason, the problem of eliminating this type of limit cycle, which is based on quantization with size truncation, is studied *separately and independently* from the problem of noise optimization, which is based on rounding quantization. This ability to eliminate the limit cycles of quantization is particularly important in several cases. One such case is the *processing of voice signals*, in which the appearance of limit cycles every time the system does not receive an input signal may be interpreted as an *auditory signal*, thus leading to confusion. The characteristics of the structures that do not show limit cycles are presented below. It should also be noted that this characteristic of limit cycles characterizes exclusively the *recursive filters of type IIR* and not the FIR filters, since the latter *do not have feedback*. So in cases where it is not easy to remove these oscillations of limit cycles, we can simply resort to the use of FIR filters.

Starting our study from the elimination of the *small-scale limit cycles*, which are due to the phenomenon of quantization, these can be eliminated, either using *a sufficiently large word length*, which ensures that the width of the oscillation of the limit cycle is quite small such that to meet the specifications of the system that have been set by the appropriate applications in each case, or by designing the system in such a way that *the conditions for the appearance of limit cycles are not met*, which in this way *are suppressed from the beginning and forever*. Regarding the first method, several techniques for calculating limit values for this width can be found in the literature. For example, Sandberg and Kaiser (1972) determined an *upper limit* for the RMS value of the width of the limit cycles of period $K+1$ (where K is an arbitrary non-negative integer), the existence of which is due to *quantization owing to rounding*, which is applied to a second-order recursive filter implemented using fixed-point arithmetic. If α and β are the coefficients of the quantities $y[n-1]$ and $y[n-2]$ appearing in the system's difference equation, Sandberg and Kaiser proved that $\|P\| \leq \rho(\alpha, \beta)\|S\|$ where $\|P\|$ and $\|S\|$ are the RMS values of the system's response p_n when it is on the limit cycle and the rounding noise s_n at the output of the filter (both sequences are periodic with period $K+1$) defined as

$$\|P\| = \sqrt{\frac{1}{K+1}\sum_{n=0}^{K} p_n^2} \quad \text{and} \quad \|S\| = \sqrt{\frac{1}{K+1}\sum_{n=0}^{K} s_n^2},$$

whereas $\rho(\alpha, \beta)$ is a function of the filter coefficients α and β, which is defined as

$$\rho(\alpha,\beta) = \begin{cases} \dfrac{1}{(1-\beta)\sqrt{1-\dfrac{\alpha^2}{4\beta}}} & \text{for } \beta > 0 \text{ and } |\alpha| \leq \dfrac{4\beta}{1+\beta} \\ \dfrac{1}{1+\beta-|\alpha|} & \text{for } \beta > 0 \text{ and } \beta \leq 0 \text{ or } \beta > 0 \text{ and } |\alpha| \geq \frac{4\beta}{1+\beta}. \end{cases} \tag{12.73}$$

Long and Trick (1973), working with the above second-order digital filter, proved that *the absolute value of the rounding error*, is bounded by the quantity

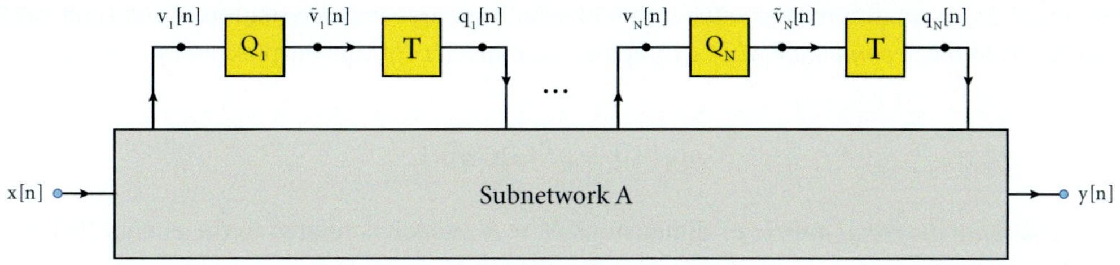

Fig. 12.48 Digital filter of order N that uses as state variables the outputs of the quantizers that are passed to the time-delay elements and allows the elimination of limit cycles of zero input quantization (Diniz and Antoniou, Figure 1)

$$|y[n]|\max = \begin{cases} \dfrac{1}{1-|\alpha|+\beta} & \text{for } \beta \le 0 \text{ or } \beta > 0 \text{ and } 2\sqrt{\beta} \le |\alpha| \\ \dfrac{1}{(1-\sqrt{\beta})^2} & \text{for } \beta > 0 \text{ and } 2\beta\sqrt{\dfrac{2}{\sqrt{\beta}}-1} \le |\alpha| \le 2\sqrt{\beta} \\ \dfrac{(1+\sqrt{\beta})}{(1-\beta)\sqrt{1-\dfrac{\alpha^2}{4\beta}}} & \text{for } \beta > 0 \text{ and } |\alpha| \le 2\beta\sqrt{\dfrac{2}{\sqrt{\beta}}-1}. \end{cases} \tag{12.74}$$

We should also mention the work of Green and Turner (1988), who determined *upper limits for the marginal cycles of zero and constant input* that appear in filters using *fixed-point arithmetic*, which are *independent* of the period of these cycles and which find application in systems with arbitrary order value, under the regime of certain preconditions.

On the other hand, the application of the second method leads to the *elimination of limit cycles due to quantization* by performing the quantization of signals in a specific way, as well as the elimination of limit cycles due to overflow, by incorporating into the arrangements for performing numerical operations the *ability to handle overflow* through saturation. An interesting method that we present at this point,[30] is based on the construction of a *Lyapunov function* related to stored power, which generally is as follows.

Let us consider a digital filter that uses *fixed-point arithmetic* and is characterized by finite precision, which means that in addition to the usual adders, multipliers, and delay elements, the filter also contains the appropriate number of quantizers (one quantizer in each internal feedback loop), which are *placed at the inputs of the delay elements and are responsible for the quantization of values and the handling of overflow*. For simplicity, let us completely ignore the adders and multipliers, along with their interconnections, and assume that they are included in the linear subsystem depicted at the bottom of Fig. 12.48. Let us also assume that the description of the order N system is carried out *in the state space* and let us use as N state variables the input signals of the N time-delay elements. In this case, the dynamic equations of the system will have the form

$$\boldsymbol{v}[n] = \boldsymbol{A}\boldsymbol{q}[n] + \boldsymbol{B}x[n] \qquad \text{and} \qquad y[n] = \boldsymbol{C}\boldsymbol{q}[n] + \boldsymbol{D}x[n].$$

Let us first examine the problem of eliminating the limit cycles of zero input, which appear in cases where the system input is equal to zero. If we set $x[n] = 0$ in the first of the above equations, we get

$$\boldsymbol{v}[n] = \boldsymbol{A}\boldsymbol{q}[n] \qquad \text{and} \qquad \boldsymbol{q}[n+1] = \overline{\boldsymbol{v}}[n] = Q[\boldsymbol{A}\boldsymbol{q}[n]],$$

where Q is a nonlinear and possibly time-varying function, which can be associated with quantization through *clipping* or *rounding*, or even with *overflow*. If the state vector $\boldsymbol{v}[n]$ has the form $\boldsymbol{v}[n] = \{v_1[n], v_2[n], \ldots, v_N[n]\}$, then the above nonlinear function will be applied *to each state variable separately*, or in mathematical notation, $\overline{v}_k[n] = Q_k[v_k[n]]$, with the function Q_k being associated with the kth quantizer, which is applied to the signal $v_k[n]$.

[30] See Meerkötter K., *Realization of Limit Cycle-Free Second-Order Digital Filters*, in Proc. IEEE Int. Symp. Circuits and Systems, pp. 295–298, 1976; Mills W.L., Mullis C.T., & Roberts R.A., *Digital Filter Realizations Without Overflow Oscillations*, IEEE Trans. Acoust., Speech, Signal Process., vol. 26, pp. 334–338, Aug. 1978; and Vaidyanathan P.P. & Liu V., *An Improved Sufficient Condition for Absence of Limit Cycles in Digital Filters*, IEEE Trans. Circuits Syst., vol. 34, pp. 319–322, Mar. 1987.

In order to determine those conditions, the satisfaction of which ensures the elimination of the limit cycles of zero-input quantization, let us resort to the *second method of Lyapunov* and consider for this purpose a *non-negative Lyapunov pseudo-energy function* of the form[31]

$$p[\boldsymbol{q}[n]] = \boldsymbol{q}^T[n]\boldsymbol{G}\boldsymbol{q}[n],$$

where $\boldsymbol{G}$ is a *positive definite diagonal matrix* of dimensions $N \times N$, which is related to the energy that is stored in the unit time-delay elements at each time instance n and such that the change of its content is in position to reveal information about the stability of the system under zero input conditions. Based on the above expression, the change in the energy of the system during a cycle of operation, that is, during a repetition, is

$$\Delta p[\boldsymbol{q}[n]] = p[\boldsymbol{q}[n+1]] - p[\boldsymbol{q}[n]] = \boldsymbol{q}^T[n+1]\boldsymbol{G}\boldsymbol{q}[n+1] - \boldsymbol{q}^T[n]\boldsymbol{G}\boldsymbol{q}[n] = \{Q[\boldsymbol{A}\boldsymbol{q}[n]]\}^T\boldsymbol{G}\{Q[\boldsymbol{A}\boldsymbol{q}[n]]\} - \boldsymbol{q}^T[n]\boldsymbol{G}\boldsymbol{q}[n].$$

If we add to the above expression the phrase

$$(\boldsymbol{A}\boldsymbol{q}[n])^T\boldsymbol{G}\boldsymbol{A}\boldsymbol{q}[n] - \boldsymbol{v}^T[n]\boldsymbol{G}\boldsymbol{v}[n],$$

the value of which is identically equal to zero, since the two terms of it are of the same sign (remember that we have defined $\boldsymbol{v}[n] = \boldsymbol{A}\boldsymbol{q}[n]$), it will take the form

$$\Delta p[\boldsymbol{q}[n]] = \{Q[\boldsymbol{A}\boldsymbol{q}[n]]\}^T\boldsymbol{G}\{Q[\boldsymbol{A}\boldsymbol{q}[n]]\} - \boldsymbol{q}^T[n]\boldsymbol{G}\boldsymbol{q}[n] - \boldsymbol{v}^T[n]\boldsymbol{G}\boldsymbol{v}[n] + (\boldsymbol{A}\boldsymbol{q}[n])^T\boldsymbol{G}\boldsymbol{A}\boldsymbol{q}[n]. \tag{12.75}$$

Resorting to our well-known from linear algebra *theory of the quadratic forms*, the first and third terms of the above expression are formulated as

$$\{Q[\boldsymbol{A}\boldsymbol{q}[n]]\}^T\boldsymbol{G}\{Q[\boldsymbol{A}\boldsymbol{q}[n]]\} = \overline{\boldsymbol{v}}[n]^T\boldsymbol{G}\overline{\boldsymbol{v}}[n] = \sum_{k=1}^{N}\overline{v}_k^2[n]g_{kk} \qquad \text{and} \qquad \boldsymbol{v}^T[n]\boldsymbol{G}\boldsymbol{v}[n] = \sum_{k=1}^{N}v_k^2[n]g_{kk}$$

respectively, where g_{kk} ($k = 1, 2, \ldots, N$) are the diagonal elements of the matrix $\boldsymbol{G} = \mathrm{diag}\{g_{11}, g_{22}, \ldots, g_{NN}\}$ (in reality, owing to the diagonal nature of this matrix, the two above quadratic forms are expressed in their *normal* or *diagonal* form), whereas taking into account the property of the *transposition of the product of two matrices* of suitable dimensions $\boldsymbol{P}$ and $\boldsymbol{Q}$, according to which $(\boldsymbol{P}\,\boldsymbol{Q})^T = \boldsymbol{Q}^T\boldsymbol{P}^T$, the fourth term of the above equation is written as

$$(\boldsymbol{A}\boldsymbol{q}[n])^T\boldsymbol{G}\boldsymbol{A}\boldsymbol{q}[n] = \boldsymbol{q}^T[n]\boldsymbol{A}^T\boldsymbol{G}\boldsymbol{A}\boldsymbol{g}[n].$$

Combining all of the above and rearranging the terms, our initial equation becomes

$$\begin{aligned}\Delta p[\boldsymbol{q}[n]] &= \boldsymbol{q}^T[n]\boldsymbol{A}^T\boldsymbol{G}\boldsymbol{A}\boldsymbol{g}[n] - \boldsymbol{q}^T[n]\boldsymbol{G}\boldsymbol{q}[n] + \sum_{k=1}^{N}\overline{v}_k^2[n]g_{kk} - \sum_{k=1}^{N}v_k^2[n]g_{kk} \\ &= -\boldsymbol{q}^T[n](\boldsymbol{G} - \boldsymbol{A}^T\boldsymbol{G}\boldsymbol{A})q[n] - \sum_{k=1}^{N}\left(v_k^2[n] - \overline{v}_k^2[n]\right)g_{kk}.\end{aligned} \tag{12.76}$$

It is not difficult to understand that the stability condition requires $\Delta p[\boldsymbol{q}[n]] \leq 0$, or equivalently $p[\boldsymbol{q}[n+1]] \leq p[\boldsymbol{q}[n]]$, so that the *energy per unit of time*, that is, the *power* stored in the elements of time delay, should *not* increase over time.

[31] See Willems J.L., *Stability Theory of Dynamical Systems*, London, Thomas Nelson and Sons, 1970. Note that the parameter p corresponds to the distance from the origin that we had used to describe the limit cycles of overflow in the state space, whereas the requirement $\Delta p[\boldsymbol{q}[n]] \leq 0$ or equivalently $p[\boldsymbol{q}[n+1]] \leq p[\boldsymbol{q}[n]]$ corresponds to the reduction of this distance. It is interesting to comment, that even in this case and considering a second-order system, the equation $p =$ constant describes an ellipse in the state space. If the matrix $\boldsymbol{G}$ appearing in the defining equation of the parameter p is such that the matrix $\boldsymbol{G} - \boldsymbol{A}^T\boldsymbol{G}\boldsymbol{A}$ is positive definite and at the same time the above condition holds, then it is a candidate to be characterized as an energy matrix and to be used for the definition of a suitable energy function. If for any of these matrices, the quantization process causes the reduction of the energy p, the system will be characterized by the guaranteed absence of marginal quantization cycles.

From the mathematical form of the expression for the parameter $\Delta p[\boldsymbol{q}[n]]$, it is found that for this to happen, it must be (a) $\boldsymbol{q}^T[n](\boldsymbol{G} - \boldsymbol{A}^T\boldsymbol{G}\boldsymbol{A})q[n] \geq 0$, a condition, which for a given system matrix $\boldsymbol{A}$, can be satisfied by choosing the matrix $\boldsymbol{G}$ in the appropriate way, and (b) $|\overline{v}_k[n]| \leq |v_k[n]|$ $(k = 1, 2, \ldots, N)$, a condition that is satisfied if the quantization of the numerical values used is performed using the technique of *magnitude truncation*. These two conditions, together constitute *a set of sufficient conditions for preventing the appearance of limit cycles of zero-input quantization*, since the signals $q[n]$ that constitute the state variables of the system, after a finite number of initial iterations, will either take a zero value or some of them will perform a temporary oscillation with *progressively decreasing amplitude* until they finally zero out as well. In the terminology of digital-signal processing, a digital-filter implementation that satisfies the property $\boldsymbol{q}^T(\boldsymbol{G} - \boldsymbol{A}^T\boldsymbol{G}\boldsymbol{A})q \geq 0$ is aforementioned to *support the elimination of limit cycles of zero-input quantization*. The above findings can be formulated in the form of a theorem as follows.

Theorem 12.9.1 *If a stable digital filter is characterized by a system matrix* $\boldsymbol{A}$ *of dimensions* $N \times N$ *and for every state vector* $\boldsymbol{q}$ *of dimensions* $N \times 1$ *there is a diagonal positive definite matrix* $\boldsymbol{G}$ *such that* $\boldsymbol{q}^T(\boldsymbol{G} - \boldsymbol{A}^T\boldsymbol{G}\boldsymbol{A})q \geq 0$, *then the limit cycles of zero input quantization can be eliminated, if the quantization of the values is performed by the method of size truncation.*

From the above analysis, the conclusion is drawn that the existence of a *Lyapunov function* for a quantized nonlinear digital filter ensures that this filter *is able to completely eliminate the unwanted limit cycles*. Indeed, assuming that the ideal linear system is a *stable system* and therefore, over time, the state of the system will approach the zero state for which the energy is equal to zero, it is not difficult to find out that *if the energy of the nonlinear system is continuously reduced approaching the zero value, the system will progressively approach the zero state and consequently, will be fully freed from the limit cycles*. It is also worth noting that the requirement for the positive definite matrix $\boldsymbol{G}$ is equivalent to the requirement for the matrix $\boldsymbol{F} = \boldsymbol{G} - \boldsymbol{A}^T\boldsymbol{G}\boldsymbol{A}$ to be a *positive semi-definite matrix*.[32] If the system matrix $\boldsymbol{A}$ describes a *stable* system, its eigenvalues will all be inside the unit circle and there will always be a positive definite matrix such that the matrix $\boldsymbol{F}$ is positive semi-definite. Let it also be noted that the requirement for the diagonal nature of the matrix $\boldsymbol{G}$ is of decisive importance because it leads *to a significant simplification of the system analysis*. Specifically, if the positive definite matrix $\boldsymbol{G}$ is diagonal, then, as happens in many similar cases, there is a decoupling between the system's quantizers, which operate *independently* of each other, whereas in the opposite case the operation of each quantizer affects (but is also affected by) the operation of the remaining quantizers, resulting in a *very large increase* in the degree of complexity of the system.[33]

In the literature, several references can be found with more specialized results, which concern simpler structures of digital filters. For example, for a matrix $\boldsymbol{A}$ with elements $\alpha_{11} = (\alpha + \beta + 1)/2$, $\alpha_{12} = (\alpha + \beta - 1)/2$, $\alpha_{21} = (\alpha - \beta + 1)/2$ and $\alpha_{22} = (\alpha - \beta - 1)/2$, the matrix $\boldsymbol{G}$ is a diagonal matrix, with elements $g_{11} = (1 + \alpha - \beta)/2$, $g_{22} = (1 - \alpha - \beta)/2$ and $g_{12} = g_{21} = 0$. It is proven that in this case, the ellipses described by the equation $p =$ constant have *axes parallel to the axes of the coordinate system*, whereas it holds that

$$p[\boldsymbol{q}[n+1]] - p[\boldsymbol{q}[n]] = -\frac{(1-\alpha-\beta)(1+\alpha-\beta)(1+\beta)}{4}(q_1[n] + q_2[n])^2 \leq 0.$$

We observe that for the case $q_1[n] + q_2[n] = 0$, the value of p remains constant, but from the next iteration, the condition $q_1[n] + q_2[n] \neq 0$ will apply and consequently, the energy starts to decrease again. The application of quantization with size truncation reduces the measures of both state variables and the parameter p, a fact that ensures the *elimination of the limit cycles of quantization*.[34] Even more interesting is the case of filters with system matrix $\boldsymbol{A}$, with elements $\alpha_{11} = \alpha_{22} = \alpha/2$

[32] In reality, this property of the positive of a bounded matrix applies *only for the case of equality* in the condition $\Delta p[\boldsymbol{q}[n]] \leq 0$, which *does not* lead to the progressive reduction of energy but to the stability of its value, which means that energy *is not reduced* as a result of the application of quantization. However, this case may lead to the appearance of *limit cycles*. In order to ensure the *complete absence* of limit cycles, the symbol of *inequality* should be used in the above condition, a case in which quantization leads to *energy reduction*.

[33] For more details on this topic, see the reference Meerkötter K., *Realization of Limit Cycle-Free Second-Order Digital Filters*, Proceedings of IEEE International Symposium on Circuits and Systems, pp. 295–298, 1976.

[34] See the references Fettweis A., *Pseudopassivity, Sensitivity and Stability of Wave Digital Filters*, IEEE Trans. Circuit Theory, vol. CT-19, pp. 668–673, 1972; Fettweis A. & Meerktötter K., *Suppression of Parasitic Oscillations in Wave Digital Filters*, Proc. IEEE Int. Symp. on Circuits and Systems, San Francisco, California (New York, USA: IEEE 1974), pp. 682–686, 1974; Fettweis A. & Meerktötter K., *Suppression of Parasitic Oscillations in Wave Digital Filters*, IEEE Trans. Circuits and Syst., vol. CAS-22, pp. 239–246, 1975; Meerkötter K. & Wegener W., *A New Second-Order Digital Filter Without Parasitic Oscillations* Arch. Elektron, Übertragungstech, vol. AEU-29, pp. 312–314, 1975; and Meerkötter

and $\beta_{12} = -\alpha_{21} = \sqrt{-4\beta - \alpha^2}/2$. In this case, the matrix $\boldsymbol{G}$ is the unit matrix $\boldsymbol{I}$ of dimensions 2×2, whereas the parameter p coincides with the known *Euclidean distance* from the origin of the coordinate system. In this case, it is proven that

$$p[\boldsymbol{q}[n+1]] - p[\boldsymbol{q}[n]] = -(1+\beta)(q_1[n] + q_2[n])^2 \leq 0,$$

whereas the equation $p =$ constant now describes *concentric circles*. The common characteristic of these two types of filters is that the matrix $\boldsymbol{G}$ is a *diagonal* matrix, whereas the equation $p =$ constant describes ellipses (remember, the circle is a special case of an ellipse, in which the major and minor semi-axes have the same length, which is the radius of the circle) the axes of which *coincide* with the axes of the coordinate system. As we commented previously, this diagonal nature of the matrix $\boldsymbol{G}$ significantly simplifies the calculations and therefore, quite reasonably the question arises which system matrices $\boldsymbol{A}$ can be associated with a diagonal matrix $\boldsymbol{G}$? This question has been answered by Mills et al.,[35] who proved that this property is satisfied by the matrices $\boldsymbol{A}$ that satisfy the condition $|\alpha_{11} - \alpha_{22}| + \det(\boldsymbol{A}) \leq 1$ such as the matrix with elements $\alpha_{11} = \alpha_{22} = \alpha$ and $\alpha_{12} = \alpha_{21} = -\beta$ with $\beta > 0$. In other words, all systems whose matrix $\boldsymbol{A}$ is characterized by this property *maintain the characteristic of stability under zero-input conditions* and under quantization processes of the values of the state variables with the technique of size clipping, but also under corrections for dealing with overflow. Finally, for filters with system matrix $\boldsymbol{A}$ with elements $\alpha_{11} = \alpha$, $\alpha_{12} = \beta$, $\alpha_{21} = 1$ and $\alpha_{22} = 0$, the matrix $\boldsymbol{G}$ is defined by the elements $g_{11} = g_{22} = 1 - \beta$ and $g_{12} = g_{21} = -\alpha$ and is *positive definite*, since it is easily found that $\text{Det}(\boldsymbol{G}) = (1 - \alpha - \beta)(1 + \alpha - \beta) > 0$ and $\text{Trace}(\boldsymbol{G}) = 2(1 - \beta) > 0$. In this case, it is proven that

$$p[\boldsymbol{q}[n+1]] - p[\boldsymbol{q}[n]] = -(1+\beta)[q_1[n+1] - q_2[n]]^2 + 2e[n][q_1[n+1] - q_2[n]] \leq 0$$

if now the size-clipping technique *is not able to operate without leading to an increase in energy*. An alternative solution is to resort to *controlled rounding*, since the above function constitutes a semi-energy Lyapunov function of the nonlinear system, if the condition $e[n][q_1[n+1] - q_2[n]] \leq 0$ is satisfied. Consequently, if it is $q_2[n] > q_1[n+1]$, then the quantization is performed *upward*, whereas if $q_2[n] < q_1[n+1]$, then quantization is performed *downward*. In other words, in this type of *conditional rounding*, the direction in which it is performed depends on the signal $q_2[n]$, which is considered a control signal. Although it is proven that in this case, *limit cycles of quantization* with a period of $N = 1$ or $N = 2$ may appear, something that happens when the condition $q_1[n+1] = q_2[n] - q_1[n-1]$ is satisfied, the advantage of this method (except for the problem we just mentioned), is that it *ensures not only zero-input stability but also steady-state input stability*, since the rule of quantization uses signal-value differences. Note that this problematic behavior (which can be corrected with additional hardware), characterizes the application of this method only in this special category of filters and not in the general case.

It is interesting to note that although the above theorem defines the conditions under which the digital filter *is not characterized* by the appearance of self-sustaining limit cycles of zero-input quantization, if the input is not equal to zero, but is characterized by a small constant value, then the probability of such cycles appearing *cannot* be ruled out. Diniz and Antoniou (1986) modified the structure of Fig. 12.48 in the way depicted in Fig. 12.49 and proved that this new modified filter (which, since it is an extension of the previous one, *already characterized by the ability to eliminate the marginal cycles of zero-input quantization*, it is able to eliminate the marginal cycles of constant input quantization as well. The elements of the vector $\boldsymbol{P} = [\, p_1 \; p_2 \; \cdots \; p_n \,]^T$ are defined in the way depicted in Fig. 12.49, whereas the vector itself is related to the matrices $\boldsymbol{A}$ and $\boldsymbol{B}$ of the system's representation in the state space through the relation $\boldsymbol{P} = (\boldsymbol{I} - \boldsymbol{A})^{-1}\boldsymbol{B}$, where $\boldsymbol{I}$ is the unit matrix of dimensions $N \times N$.

In order to present the way in which this modified digital filter is able to eliminate the marginal cycles of constant input quantization, we must observe that since it already has this property for the marginal cycles of zero-input quantization, the system $\boldsymbol{q}[n+1] = \boldsymbol{A}\boldsymbol{q}[n]$ will be characterized by the property $\lim \boldsymbol{q}[n] = [\,0\;0\;\cdots 0\,]^T$ in the limit $n \to \infty$. Considering now the constant input signal $x[n] = u_0$ and accepting that the vector $\boldsymbol{P}u_0$ can be represented by the word length of the underlying digital system, the *state equation of the modified system* can be written in the form

$$\boldsymbol{q}[n+1] = Q[\boldsymbol{A}\boldsymbol{q}[n] - \boldsymbol{P}u_0 + \boldsymbol{B}u_0] + \boldsymbol{P}u_0$$

K., *Realization of Limit Cycle-Free Second-Order Digital Filters*, Proc. IEEE Int. Symp. on Circuits and Systems, Munich, Germany (New York, USA: IEEE 1976), pp. 295–298, 1976.

[35] See the reference Mills W.L., Mullis C.T., & Roberts R.A., *Digital Filters Realizations Without Overflow Oscillations*, IEEE Transactions on Acoustics, Speech, and Signal Processing, ASSP-26, 334–338, 1978.

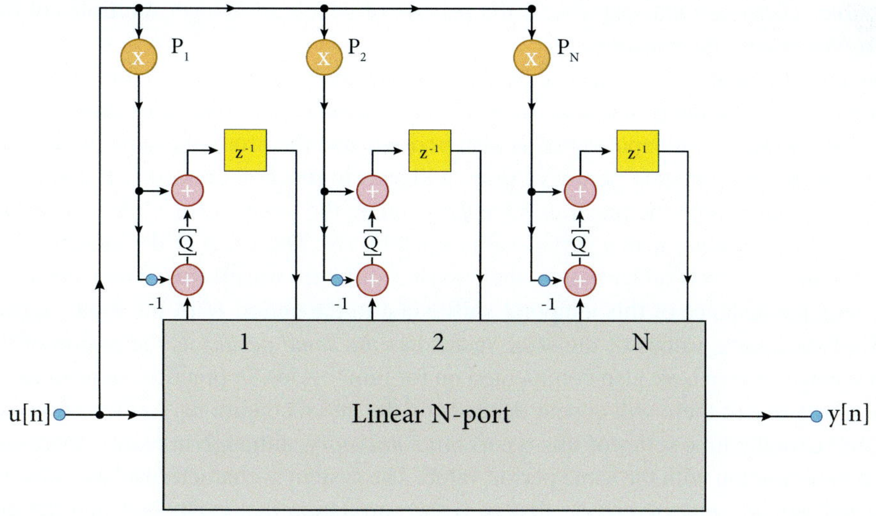

Fig. 12.49 Digital filter of order N, which uses as state variables the outputs of the quantizers that are passed to the time-delay elements and allows the elimination of marginal cycles of constant input quantization (Diniz and Antoniou, Figure 2)

$$= Q[\boldsymbol{A}\boldsymbol{q}[n] - \boldsymbol{I}(\boldsymbol{I}-\boldsymbol{A})^{-1}\boldsymbol{B}u_0 + (\boldsymbol{I}-\boldsymbol{A})(\boldsymbol{I}-\boldsymbol{A})^{-1}\boldsymbol{B}u_0] + \boldsymbol{P}u_0 = Q[\boldsymbol{A}(\boldsymbol{q}[n] - \boldsymbol{P}u_0)] + \boldsymbol{P}u_0. \quad (12.77)$$

The logic behind this process is *the conversion of a zero-input system*, which, as we saw, is characterized by the ability to completely eliminate the boundary quantization cycles *into a constant input system*, which will continue to have this ability. This can be done in the following way: for each quantizer Q_i, the addition of a signal v_i to the signal resulting from the output of the quantizer takes place, whereas the same signal is subtracted when the sum of these two signals passes through the unit time-delay element. This signal v_i is proportional to the input signal, that is, it has the form $v_i = p_i u_0$ and therefore, the modified system is described by the above equation. Now, proceeding to the change of variable $\boldsymbol{\xi}[n] = \boldsymbol{q}[n] - \boldsymbol{P}u_0$, the above equation takes the simple form $\boldsymbol{\xi}[n+1] = Q[\boldsymbol{A}\boldsymbol{\xi}[n]]$, which is the same as the one that describes the operation of the original system of Fig. 12.48, *except for the transformation of the state vector*. Under the condition, therefore, that the product $\boldsymbol{P}u_0$ can be represented by the underlying digital system, the above expression describes the operation of a system that *is capable of eliminating the limit cycles of constant input quantization*. In addition to the above structures, many other digital-filter structures can be found in the literature, which are capable of eliminating both the *limit cycles of zero input quantization*,[36] as well as the *limit cycles of constant input quantization*.[37]

Periodic Discrete Input and Subharmonics

A case of particular interest is the one in which the input signal passed to the quantized nonlinear system, is a *periodic discrete-time signal*, characterized by a certain period N. Unlike the ideal linear system, in which the output *tends asymptotically toward a steady state*, which is also periodic and characterized by the same period value, the quantized system that is characterized by the presence of *nonlinear elements* approaches a steady state, which depends on the initial conditions and whose period, in the general case, *does not coincide* with the period of the input signal. Specifically, after a finite number of periods N_0, the state vector $\boldsymbol{q}[n]$ will perform a periodic motion in the state space, repeatedly passing through a number of $M = SN$ points, where N is the period of the input signal and S is an integer, the value of which is

[36] See for example, Meerkötter K. & Wegener W., *A New Second-Order Digital Filter Without Parasitic Oscillations*, Archiv Elektrotechnik und Übertragungstechnik, 29, 312–314, 1975; Fettweis A. & Meerkötter K., *On Adaptors for Wave Digital Filters*, IEEE Transactions on Acoustics, Speech, and Signal Processing, ASSP-23, 516–525, 1975; and Diniz P.S.R. & Antoniou A., *Digital Filter Structures Based on the Concept of Voltage Generalized Immitance Converter*, Canadian Journal of Electrical and Computer Engineering, 13, 90–98, 1988.

[37] See for example Verkroost G. & Butterweck H.J., *Suppression of Parasitic Oscillations in Wave Digital Filters and Related Structures by Means of Controlled Rounding*, IEEE International Symposium on Circuits and Systems, Munich, West Germany, pp 628–629, 27–29 Apr. 1976; Liu K.S. & Turner L.E., *Stability, Dynamic-Range and Roundoff Noise in a New Second-Order Recursive Digital Filter*, IEEE Transactions on Circuits and Systems, CAS-30, 815–821, 1983; Diniz P.S.R., *Elimination of Constant-Input Limit Cycles in Passive Lattice Digital Filters*, IEEE Transactions on Circuits and Systems, 35, 1188–1190, 1988; and Sarcinelli F.M. & Diniz P.S.R., *Tridiagonal State-Space Digital Filter Structures*, IEEE Transactions on Circuits and Systems, CAS-36, 818–824, 1990.

taken as small as possible. These periodic sequences, the periods of which are integer multiples of the period of the input signal, are known as *subharmonics* (subharmonics).

The above special mode of operation is evidenced by the fact that since the system is a finite state machine, after the state vector has passed through all the available states, *it will necessarily start to pass through states from which it had passed in the past*. At the same time, however, the input signal is also *periodic* and therefore, the same will apply to the values of the samples of this signal, which after a certain period of time, will start to repeat themselves. But since according to the state equation, the state $\boldsymbol{q}[n+1]$ depends on the previous state $\boldsymbol{q}[n]$ and on the input signal $x[n]$, *it is certain, that at some point the system will enter a periodic motion*, with a period value equal to SN. The value of the integer coefficient S is known as the *order of the subharmonic*. For each subharmonic, there exist S different waveforms, which constitute *temporally shifted copies of each other*, with the distance of this temporal shift equal to the period N of the input signal. If $S = 1$, that is, if the periodic sequence of successive values of the state vector has *the same period* as the period of the input signal, these subharmonics *are not* created. As we have also commented on for limit cycles, in practice, sequences with a different period value may appear and which one of these will appear depends on the initial conditions. Let it be noted that the correlation of all initial conditions with periodic time series of this type*is not mandatory*, although in reality, there are *more than one* such conditions, which lead to oscillation with the same period value. The system is characterized as *stable* if all initial conditions lead *to the same one and only periodic sequence*, that is, if the correlation that exists between the input and the output is *unique*. More details on this subject can be found in the literature.[38] It is also worth noting that in cases where for some signal the system's state vector performs periodic motion, *the implementation of small changes in this vector may cause its transition from one type of oscillation to another*. This phenomenon is characterized as a *jump*.[39]

In order to come into contact with the issue of subharmonics, let us consider again the simple first-order recursive filter, in which the quantization process is performed with *rounding*. Let us assume that in this filter we transmit a periodic sequence input $x[n]$ with period N and let us define the difference $d[n] = q[n] - q[n+N]$, where $q[n]$ is the state vector, which in this case is a simple scalar. Resorting to the system's state equation

$$q[n+1] = \alpha q[n] + u[n] + e[n],$$

where the error $e[n]$ is defined as known as $e[n] = Q[\alpha q[n] + u[n]] - (\alpha q[n] + u[n])$, from the above equation we get

$$\begin{aligned} d[n+1] &= q[n+1] - q[n+N+1] = \alpha q[n] + u[n] + e[n] - \alpha q[n+N] - u[n+N] - e[n+N] \\ &= \alpha(q[n] - q[n+N]) + e[n] - e[n+N] = \alpha d[n] + e[n] - e[n+N], \end{aligned}$$

since the input signal is *periodic* and therefore, the terms $u[n]$ and $u[n+N]$ *cancel each other out*. Considering a normalized quantization step equal to the unit, which constitutes the *upper limit* of the error $e[n]$, we have (remember, how in the stable linear system it is $|\alpha| < 1$) $|d[n+1]| < |d[n]| + 1$ and since the values of $q[n]$ and $d[n]$ are *integer*, the above condition gives us $|d[n+1]| \leq |d[n]|$. Assuming that the sequence $q[n]$ is *periodic*, then, by definition, the same will apply to the sequence $d[n]$. Specifically, for $\alpha > 0$, $D[n] = D$ (for every n), whereas for $\alpha < 0$, $D[n] = (-1)^n D'$. Taking into account, as we commented, that the quantized digital filter is a *finite-state machine*, finally the change of the state vector $q[n]$ will lead to a periodic sequence of the form $q[n+SN] = q[n]$ for some satisfactorily large value of n and for some integer value S. Because such a periodic sequence satisfies the property

$$\sum_{i=0}^{S-1} d[n+iN] = \sum_{i=0}^{S-1} [q[n+iN] - q[n+(i+1)N]] = 0 \tag{12.78}$$

if it is $\alpha > 0$, combining this relationship with the $D[n] = D$, it turns out that $D = 0$, whereas if it is $\alpha < 0$, combining this relationship with the $D[n] = (-1)^n D'$ it turns out that $D' = 0$ for even values of N. Therefore, in this case, the subharmonics appear only when it is $\alpha < 0$ and N takes odd values. Observing that

[38] See for example Butterweck H.J., Lucassen F.H.R., & Verkroost G., *Subharmonics and Other Quantization Effects in Periodically Excited Recursive Digital Filters*, IEEE Trans. Circuits Syst., vol. CAS-33, pp. 958–964, 1986.

[39] See references Krishnan A. & Subramanian R., *Jump Phenomena in Digital Filters*, J. Inst. Electron. Telecomm. Eng. (India), vol. 25, pp. 373–376, 1979 and Kristiansson L.O.G., *Jump Phenomena in Digital Filters*, Electronics Letters, vol. 10, pp. 14–15, 1974.

$$q[n] - q[n + N] = d[n] = -d[n + N] = -q[n + N] + q[n + 2N]$$

that is, that $q[n] = q[n + 2N]$, it turns out that the order of the subharmonic is $S = 2$. Similar results arise if the quantization process is carried out using other methods.

Elimination of Boundary Cycles Through Rounding

In almost all the techniques we have presented up to this point, the quantization process relies on *size truncation*, which, as we mentioned, *is able to eliminate the marginal quantization cycles of both zero and constant input* (with appropriate modification of the system). On the other hand, if we resort to a quantization process through *rounding*, this, owing to the way it is used, generally leads to *signal enhancement* by a factor $c \le 2$, with the maximum value $c = 2$ corresponding to a signal value equal to half of the quantization step. In order to proceed with the elimination of marginal quantization cycles, when this is done with rounding, *we must offset the nonlinear increase in energy with an equal magnitude decrease in energy*, which is related to the operation of the linear filter and this is something that is done, imposing the condition $\|\boldsymbol{A}\| < 1/2$, where $\|\boldsymbol{A}\|$ is the norm function of the system matrix $\boldsymbol{A}$. In cases where this condition is not met by the operating specifications of the filter, we can proceed to the appropriate transformation and more specifically, calculate the matrix $\boldsymbol{A}^L$ for some value of L, so that it is $\|\boldsymbol{A}^L\| < 1/2$. In this case, we can apply to the process of quantization with *rounding*, without the risk of limit cycles appearing. However, for filters with a high value of coefficient Q, this method requires the use of complex hardware (Butterweck et al., 1988).

Elimination of Limit Cycles with Stochastic Methods

Let us complete the presentation of the techniques for eliminating this type of limit cycle with the observation that in the literature, techniques for implementing this process can be found, with *stochastic methods*, such as *random rounding*, in which the decision about how to handle the least significant bit is taken by an external random control signal.[40] A variation of this technique relies on handling the original unquantized signal through a random signal, which is characterized by *flat spectral distribution*, as well as by an amplitude distribution that is *uniform* in the area $[-q/2, q/2]$, where q is the *quantization step*. The disadvantage of these methods is the use of *additional noise sources* (more specifically, independent noise sources, one noise source for each quantizer), leading to the creation of *additional quantization noise*, the spectrum of which, however, is flat, in contrast to the narrow band-noise spectrum associated with magnitude-clipping quantization. Indeed, for filters with a high Q factor, the contribution of this noise to the filter output can be *smaller* than that associated with deterministic (nonstochastic) methods. More details about these stochastic methods can be found in the literature.

Elimination of Large-Scale Limit Cycles

Finally, as far as the so-called *large-scale limit cycles* are concerned, which are due to the characteristic of *overflow*, similar principles and techniques apply to those we presented for limit cycles due to quantization. For example, we can show that the quantized first-order IIR filter described by the state equation $q[n + 1] = Q[\alpha q[n] + x[n]]$ and the output equation $y[n] = q[n+1]$, where $y = Q[x]$ the input/output characteristic of the non-linear overflow correction element, the threshold of which has been set to the value $p = 1$, *does not exhibit zero-input oscillations* ($x[n] = 0$) due to *overflow*, whatever the input–output characteristic $y = Q[x]$ is (remember, that the corresponding linear system is characterized by stable behavior, as long as the condition $\alpha|| < 1$ is satisfied). To do this, let us define an energy function of the form

$$p[q[n]] = q^2[n] = (Q[\alpha q[n]])^2 = \alpha^2 q^2[n - 1] = \alpha^2 p[q[n - 1]].$$

Applying the above relation iteratively for the values $n = 0, 1, 2, \ldots$ we get

$$p[q[1]] = \alpha^2 p[q[0]]$$

$$p[q[2]] = \alpha^2 p[q[1]] = \alpha^4 p[q[0]]$$

$$p[q[3]] = \alpha^2 p[q[2]] = \alpha^6 p[q[0]]$$

[40] See the references Büttner M., *A Novel Approach to Eliminate Limit Cycles in Digital Filters with a Minimum Increase in the Quantization Noise*, Proc. IEEE Int. Symp. on Circuits and Systems. Munich, Germany (New York, USA: IEEE 1976), pp. 291–294, 1976 and Büttner M., *Elimination of Limit Cycles in Digital Filters with Very Low Increase in the Quantization Noise*, IEEE Trans. Circuits and Systems, vol. CAS-24, pp. 300–304, 1977.

and in the general case for each value of n, it holds that

$$p[q[n]] = \alpha^{2n} p[q[0]].$$

Taking into account the condition $|\alpha| < 1$, it is not difficult to realize that in the limit $n \to \infty$ we will have $\lim p[q[n] = 0$ and therefore, the function $p[q[n]]$ constitutes a *Lyapunov function* and according to what we mentioned previously, this system *is not characterized by zero-input overflow oscillations*, for all the input/output characteristics of the nonlinear element. Apart from the above conditions for eliminating the limit cycles of overflow, in practice other techniques are used such as the application, as we mentioned, of the appropriate type of *scaling*, in order to prevent the appearance of overflow. The problem with this approach is that since the levels of the signal circulating inside the filter are *relatively low*, the SNR will take a small value, which of course is *undesirable*. For this reason, a compromise solution that we can apply is to *allow the occasional display* of overflow (remember, representations of complement with respect to 1 and with respect to 2 can recover from overflow), but not the appearance of limit cycles.

Limit Cycles and Stability of Forced Response

A concept that is closely related to limit cycles due to overflow is the *stability of the forced response* of a non-linear system, that is, the response that is *independent* of the initial conditions and therefore, *is not affected* by possible overflow that has occurred in the past. This system, as we commented elsewhere, has arisen from the addition of *nonlinear elements* with input–output characteristic of the form $y = Q_k(x)$ $(k = 1, 2, \ldots, N)$ for the control of overflow in the corresponding ideal linear system. This stability is *inextricably linked* to the *recovery of the system from overflow*, in which, as commented elsewhere, the error that is induced owing to the manifestation of this phenomenon, is reduced in such a way that *the output of the real nonlinear digital system converges to the output of the ideal linear filter*. The analysis that is carried out in the context of the characterization of the system with respect to this stability is based on the use of signals such that their value in the ideal linear system *is not allowed to exceed the maximum value that leads to overflow*, provided that a certain time period has passed, which is defined by a value of discrete-time n_0. This requirement gives us the ability to check if the system has recovered from an overflow that has occurred at some earlier time, *without worrying about the possibility of new overflows during this investigation*. It is interesting to note that although the signals that have this feature undoubtedly constitute a special category of signals (which can be considered as a set U_0 that contains these signals),[41] which is rarely *encountered in practice* (indeed, for some filters, there may not even be such signals); however, it is proven (Claasen et al., 1975), that *if a system can recover from overflow owing to one of the signals of this particular category, then it can recover from overflow due to any signal, provided that the recovery period is shorter than the time interval between the occurrence of two successive overflows* (see also Diniz, page 729).

If we denote with $\boldsymbol{v}[n]$ and with $\boldsymbol{v}^q[n]$ the vectors of the state variables of the Fig. 12.48 before and after the addition of the quantizers, the *forced response* of the nonlinear system will be characterized as stable, if the condition is satisfied

$$\lim_{n\to\infty} \boldsymbol{e}[n] = \lim_{n\to\infty} [\boldsymbol{v}^q[n] - \boldsymbol{v}[n]] = [\,0\,0\;0\;\cdots\;0\,]^T.$$

In an equivalent formulation, the forced response of the nonlinear system is characterized as *stable*, in the case where if the ideal linear and the nonlinear systems receive *the same* input $x[n]$, the difference $y^q[n] - y[n]$ in the corresponding outputs tends to *zero*, in the limit $n \to \infty$; after resorting to the dynamic equations of the system, we can show that *the second condition implies the first*. Specifically, it is proven that *if the digital filter contains nonlinear elements, such as those used to handle overflow, and is stable under zero input conditions, then its forced response will be characterized by the same kind of stability, for the corresponding set of nonlinear elements*. According to Claasen et al., the above equivalence holds *regardless* of the type of characteristic of the nonlinear element and in reality, for all the characteristics that are included in the shaded area of the left image of Fig. 12.50, which for the special case of the parameter $m_k = 1$ is depicted in the right image of the figure. This area is described by the equations

[41] In a mathematical description, considering a signal $x[n]$, we say that $x[n] \in U_0$, if and only if there is an initial condition, corresponding to some time n_0, such that the linear filter *does not exceed the overflow threshold* for times $n \geq n_0$. In this case, a stable filter characterized by the ability to recover from overflow will always exhibit a finite number of overflows (which in some special cases, may be equal to zero), after the time $n = n_0$, the value of which depends on the time n_0. Assuming that $x[n] \in U_0$, there will be *at least one* initial condition (or in the best case, *a set* of initial conditions), for which no overflow will occur after the time instance $n = n_0$. If stability as defined here applies to every member of the signal set U_0, the filter is characterized by *forced response stability* with respect to the set U_0.

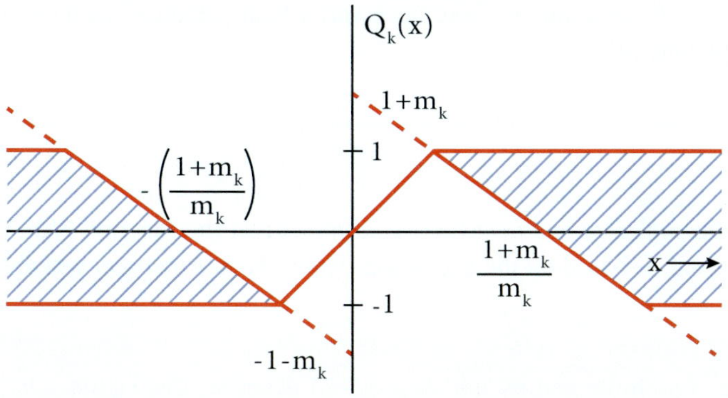

Fig. 12.50 Areas of the plane for the characteristics of the nonlinear elements of overflow handling for each value of m_k (left image) and for the value $m_k = 1$ (right image). Each nonlinear element, the input–output characteristic of which coincides with the identity function for the range of values $[-1, +1]$, whereas the rest of its segments belong to the shaded areas, is characterized by stable behavior (Claasen, Mecklenbräuker, and Peek, 1975, Figures 4 and 6

$$+1 + m_k - m_k x < Q_k(x) \leq +1 \quad x > +1$$
$$-1 - m_k - m_k x > Q_k(x) \geq -1 \quad x < -1$$

for the values $k = 1, 2, \ldots, N$. Studying these diagrams, we can draw useful conclusions about the characteristic of the nonlinear elements that ensure the validity of the above equivalence. For example, for the special case $m_k = 1$ depicted in the right image of Fig. 12.50, this study reveals that in the range of values $-1 \leq x \leq 1$ there is no overflow, in the range of values $1 \leq x \leq 3$ the equivalence is ensured if it is $2 - x \leq Q_k[x] \leq 1$, in the range of values $-3 \leq x \leq -1$ the equivalence is ensured if it is $-1 \leq Q_k[x] \leq -2 - x$, whereas in the ranges of values $x \geq 3$ and $x \leq -3$ the equivalence is ensured if it is $-1 \leq Q_k[x] \leq 1$. It should also be noted that this stability of the forced response *presupposes the elimination of limit cycles due to overflow* (see the previous description) and that the above analysis was based on the *unrealistic* assumption that the time interval between two successive overflows is of *infinite* duration. Taking into account that such a thing is not possible in practice, the above conditions *do not guarantee* the absence of limit cycles of overflow, if this appears during the recovery stage of the system from a previous overflow (see Antoniou, page 666).

As an example of the application of this method, let us return to the simple first-order IIR system we used before, let us show that *its forced response does not show overshoot oscillations if saturation quantization is used* (something that *does not* hold if complement quantization is used with respect to 2—note that in this case, the characteristic of the nonlinear element does not belong to the group of characteristics of Fig. 12.50, so this conclusion is expected). In other words, for the case of saturation quantization, the characteristic vector $q[n]$ tends *asymptotically* to the steady state $\overline{q}[n]$ of the ideal linear filter, regardless of the initial condition $q[0]$, as long as the input signal belongs to the class U_0 of signals for which $|\overline{q}[n]| \leq p$ for every n, where $p = 1$ is the threshold of overflow. To prove this result, let us define this time the Lyapunov function $p[q[n]] = e^2[n]$, where $e[n] = q[n] - \overline{q}[n]$. Defining $q_0[n] = \alpha q[n-1] + x[n-1]$ (i.e., as the value of the signal before the application of the overflow correction) and the function

$$p[q_0[n]] = (q_0[n] - \overline{q}[n])^2 = \alpha^2 p[q[n-1]] \tag{12.79}$$

if we find that correction is required ($q_0[n] > 1$), then, after the correction, $q[n] = 1$ and we will have

$$p[q[n] = (1 - \overline{q}[n])^2 (q_0[n] - \overline{q}[n])^2 = p[q_0[n]] = \alpha^2 p[q[n-1]] = \alpha^{2n} p[q[0]] \tag{12.80}$$

(with the last result arising in the way we showed on the previous page), whereas a similar conclusion arises for the case $q_0[n] < -1$. In this case, as before, $\lim p[q[n]] = 0$ at the limit $n \to \infty$ and therefore $\lim e[n] = 0$ at the limit $n \to \infty$ from which the desired result arises.

In cases where the filter is not stable but unstable, then for some input signal $x[n] \in U_0$, there is a set of initial conditions for which *overflow does not occur*. In the general case, there is a second set of initial conditions that leads to a *finite number of overflows*, whereas, owing to the instability of the system, there is also a third set of initial conditions that leads to an *infinite number of overflows* and this is exactly the situation in which unstable behavior manifests. If the input signal is periodic, then as we commented previously, the output will be *asymptotically periodic*, without necessarily being characterized by the same period as the input signal, whereas phenomena of jumps may arise.

12.10 Causality in Digital Filters

In full analogy with the analog filters, the determination of the *sufficient* and *necessary conditions* that the function $\mathcal{H}(e^{j\omega})$ must satisfy, so that the digital filter corresponding to it is causal, is based on the use of the *Paley-Wiener theorem* which *in its discrete form* is formulated in the following way.

Theorem 12.10.1 *If the impulse response of a filter* $h[n]$ *has* finite energy *and is a* causal discrete-time signal—*in other words, it satisfies the condition* $h[n] = 0$ *for* $n < 0$—*it is proven that*

$$\int_{-\pi}^{+\pi} |\ln|\mathcal{H}(e^{j\omega})||d\omega < \infty. \tag{12.81}$$

In this case, we can associate the function $|\mathcal{H}(e^{j\omega})|$ *with some* phase response $\measuredangle\mathcal{H}(e^{j\omega})$, *so that the filter described by the frequency-response function*

$$\mathcal{H}(e^{j\omega}) = \mathcal{H}_R(e^{j\omega}) + j\mathcal{H}_I(e^{j\omega}) = |\mathcal{H}(e^{j\omega})|e^{j\measuredangle\mathcal{H}(e^{j\omega})} \tag{12.82}$$

is a causal digital filter.

Considering a *causal digital filter* described by the function $\mathcal{H}(e^{j\omega})$, we can prove that the imposition of the causality condition allows the *complete determination* of the function $\mathcal{H}(e^{j\omega})$ knowing *only its real part*, $\mathcal{H}_R(e^{j\omega}) \equiv \Re\{\mathcal{H}(e^{j\omega})\}$, or equivalently, knowing *only its imaginary part*, $\mathcal{H}_I(e^{j\omega}) = \Im\{\mathcal{H}(e^{j\omega})\}$ *and the value of the sample* $h[0]$. Remember that a similar condition applies in the case of analog filters. In order to prove the truth of this claim, let us use again the known property of any discrete-time signal $x[n]$, according to which it can be written as *the sum of an even component* $x_e[n]$ *and an odd component* $x_o[n]$ (see Sect. 6.5 on page 518). Therefore, $h[n] = h_e[n] + h_o[n]$, where

$$h_e[n] = \frac{h[n] + h[-n]}{2} \quad \text{and} \quad h_o[n] = \frac{h[n] - h[-n]}{2}.$$

The above relations hold for the general case of any discrete impulse response $h[n]$. However, it is proven that if the discrete-time signal $h[n]$ is a causal signal—that is, it satisfies the condition $h[n] = 0$ for $n < 0$—then it can be recovered using *only its even component* $h_e[n]$, or alternatively using *only its odd component* $h_o[n]$ (along with the value $h[0]$), with the latter option only valid for time instances $n \neq 0$. This conclusion arises from the observation that since the discrete-time signal $h[n]$ is causal, $h[n] = 0$ for $n < 0$ and $h[-n] = 0$ for $n > 0$ and therefore the nonzero regions of the signals $h[n]$ and $h[-n]$ *do not overlap with each other*, except for the value $n = 0$ for which $h[n] = h[-n]$. Based on the above observations, the even component of the impulse response is calculated as

$$h_e[n] = \frac{h[n] + h[-n]}{2} = \begin{cases} \dfrac{h[n] + 0}{2} & \text{for } n > 0 \\ \dfrac{h[n] + h[n]}{2} & \text{for } n = 0 \end{cases} = \begin{cases} \dfrac{h[n]}{2} & \text{for } n > 0 \\ h[n] & \text{for } n = 0 \end{cases}$$

and therefore the impulse response will take the form

$$h[n] = \begin{cases} 2h_e[n] & \text{for } n > 0 \\ h_e[n] & \text{for } n = 0 \end{cases} = 2h_e[n]u[n] - h_e[0]\delta[n].$$

From the above relationship, we find that indeed we have the ability to recover all the samples of the signal $h[n]$ using *only its even component* (the process of recovering the signal from the samples of its odd component is commented on later).

In the next step of the process, let us calculate the *Fourier transform of discrete-time* of the two sides of the last equation. In this case, based on the *linear nature* of the transform, we are led to the relationship

$$\mathcal{F}\{h[n]\} = 2\mathcal{F}\{h_e[n]u[n]\} - h_e[0]\mathcal{F}\{\delta[n]\},$$

where obviously we used the symbolism $\mathcal{F}\{x[n]\} \equiv \mathcal{X}(e^{j\omega})$, to describe the Fourier transform of the discrete-time signal $x[n]$. However, we know that $\mathcal{F}\{h[n]\} = \mathcal{H}(e^{j\omega})$ and $\mathcal{F}\{\delta[n]\} = 1$ (see Exercise 10.3.7 on page 814), whereas from the property of the product of the Fourier transform, which is formulated as

$$\mathcal{F}\Big\{x[n]y[n]\Big\} = \frac{1}{2\pi}\int \mathcal{X}(e^{j\lambda})\mathcal{Y}(e^{j(\omega-\lambda)})d\lambda$$

where $\mathcal{X}(e^{j\omega}) = \mathcal{F}\{x[n]\}$ and $\mathcal{Y}(e^{j\omega}) = \mathcal{F}\{y[n]\}$—it holds that[42]

$$\mathcal{F}\Big\{h_e[n]u[n]\Big\} = \frac{1}{2\pi}\int_{-\pi}^{+\pi} \mathcal{H}_R(e^{j\lambda})\mathcal{U}(e^{j(\omega-\lambda)})d\lambda$$

with the function[43]

$$\mathcal{U}(e^{j\omega}) = \pi\delta(\omega) + \frac{1}{1-e^{-j\omega}} = \pi\delta(\omega) + \frac{1}{2} - j\frac{1}{2}\cot\left(\frac{\omega}{2}\right) \qquad -\pi \leq \omega \leq \pi$$

constitutes the *Fourier transform of the step function* $u[n]$ (this relation easily arises, if in the result of the Exercise 10.3.12 (see page 827) we set $k = 0$ and proceed to elementary trigonometric calculations). By substituting the above expressions into our initial equation, we get

$$\begin{aligned}\mathcal{H}(e^{j\omega}) &= 2\left\{\frac{1}{2\pi}\int_{-\pi}^{+\pi} \mathcal{H}_R(e^{j\lambda})\left[\pi\delta(\omega-\lambda) + \frac{1}{2} - \frac{j}{2}\cot\left(\frac{\omega-\lambda}{2}\right)\right]d\lambda\right\} - h_e[0] \\ &= \int_{-\pi}^{+\pi} \mathcal{H}_R(e^{j\lambda})\delta(\omega-\lambda)\,d\lambda + \frac{1}{2\pi}\int_{-\pi}^{+\pi} \mathcal{H}_R(e^{j\lambda})\,d\lambda - \frac{j}{2\pi}\int_{-\pi}^{+\pi} \mathcal{H}_R(e^{j\lambda})\cot\left(\frac{\omega-\lambda}{2}\right)d\lambda - h_e[0].\end{aligned}$$

It is not difficult to see that the first integral of the last equation is nothing but the *real part* $\mathcal{H}_R(e^{j\omega})$ of the frequency response of the filter $\mathcal{H}(e^{j\omega})$, since according to the *selection property* of the discrete impulse function δ (see Eq. 1.23)

$$\int_{-\pi}^{+\pi} \mathcal{H}_R(e^{j\lambda})\delta(\omega-\lambda)d\lambda = \mathcal{H}_R(e^{j\omega}). \tag{12.83}$$

On the other hand, using the *inverse transform* relation that connects the even component of the impulse response $h_e[n]$ with the real part $\mathcal{H}_R(e^{j\omega})$ of the frequency response of the filter $\mathcal{H}(e^{j\omega})$, which, as is well known, is the

---

[42] In the following description, we use the property of the even and real part of the discrete-time Fourier transform, according to which, if $\mathcal{F}\{x[n]\} = \mathcal{X}(e^{j\omega})$, then $\mathcal{F}\{\mathcal{E}\text{v}\{x[n]\}\} = \mathcal{X}_R(e^{j\omega})$ and $\mathcal{F}\{\mathcal{O}\text{d}\{x[n]\}\} = j\mathcal{X}_I(e^{j\omega})$, where $\mathcal{E}\text{v}\{x[n]\}$ and $\mathcal{O}\text{d}\{x[n]\}$ are the even and odd components of the signal $x[n]$ and $\mathcal{X}_R(e^{j\omega}) = \Re\{\mathcal{X}(e^{j\omega})\}$ and $\mathcal{X}_I(e^{j\omega}) = \Im\{\mathcal{X}(e^{j\omega})\}$ are the real and imaginary parts of the Fourier transform $\mathcal{X}(e^{j\omega})$. The application of this property in this case allows us to write that $\mathcal{F}\{h_e[n]\} = \mathcal{H}_R(e^{j\omega})$ and $\mathcal{F}\{h_o[n]\} = \mathcal{H}_I(e^{j\omega})$.

[43] Using Euler's formula $e^{-j\omega} = \cos\omega - j\sin\omega$ and the identity

$$\cot\left(\frac{\omega}{2}\right) = \frac{\sin\omega}{1-\cos\omega}$$

it is proven after simple algebraic operations that

$$\frac{1}{1-e^{-j\omega}} = \frac{1}{2} - j\frac{1}{2}\cot\left(\frac{\omega}{2}\right).$$

$$h_e[n] = \frac{1}{2\pi}\int_{-\pi}^{+\pi} \mathcal{H}_R(e^{j\omega})\, e^{j\omega n} d\omega$$

(see Footnote 42, page 1221) it is not difficult to see that the second integral of the last relation results from the above equation for the value $n = 0$; it will therefore be

$$\frac{1}{2\pi}\int_{-\pi}^{+\pi} \mathcal{H}_R(e^{j\lambda})\, d\lambda = h_e[0]$$

and therefore *the second and fourth terms of the frequency-response equation $\mathcal{H}(e^{j\omega})$ cancel each other out*. Using the above results in the aforementioned equation, it will finally take the form

$$\mathcal{H}(e^{j\omega}) = \mathcal{H}_R(e^{j\omega}) + j\left\{-\frac{1}{2\pi}\int_{-\pi}^{+\pi} \mathcal{H}_R(e^{j\lambda}) \cot\left(\frac{\omega-\lambda}{2}\right) d\lambda\right\}, \tag{12.84}$$

from which it immediately follows that the imaginary part $\mathcal{H}_I(\omega)$ of the frequency-response function $\mathcal{H}(e^{j\omega})$ can be calculated from its real part, $\mathcal{H}_R(\omega)$, through the equation

$$\mathcal{H}_I(e^{j\omega}) = -\frac{1}{2\pi}\int_{-\pi}^{+\pi} \mathcal{H}_R(e^{j\lambda}) \cot\left(\frac{\omega-\lambda}{2}\right) d\lambda. \tag{12.85}$$

The conclusion we have reached is very important, as it tells us that *the real and imaginary parts $\mathcal{H}_R(e^{j\omega})$ and $\mathcal{H}_I(e^{j\omega})$ of the transfer function of a causal digital filter are related to each other*. Therefore, the choice of one function affects the form of the other and consequently these functions cannot be chosen independently of each other. This property obviously also applies to the *magnitude* $|\mathcal{H}(e^{j\omega})|$ and the *phase* $\measuredangle\{\mathcal{H}(e^{j\omega})\}$, since these are calculated with the help of the functions $\mathcal{H}_R(e^{j\omega})$ and $\mathcal{H}_I(e^{j\omega})$. It is worth mentioning at this point that the integral of the above relation constitutes the defining equation of the *discrete Hilbert transform* and in this sense, *the function $\mathcal{H}_I(e^{j\omega})$ constitutes the discrete Hilbert transform of the function $\mathcal{H}_R(e^{j\omega})$*.

Repeating the process we used previously, we can calculate the function $\mathcal{H}_R(e^{j\omega})$ from the function $\mathcal{H}_I(e^{j\omega})$ through the equation

$$\mathcal{H}_R(e^{j\omega}) = h[0] + \frac{1}{2\pi}\int_{-\pi}^{+\pi} \mathcal{H}_I(e^{j\lambda}) \cot\left(\frac{\omega-\lambda}{2}\right) d\lambda. \tag{12.86}$$

To arrive at the above expression, we start from the relation[44]

$$h[n] = 2h_o[n]u[n] + h[0]\delta[n],$$

which allows the calculation of the discrete sequence $h[n]$ *only from its odd samples*, we calculate the Fourier transform of the two sides, as before, and after operations we end up with the equation written above. The implementation of this process is left as an exercise to the reader.[45] The equations

$$\mathcal{H}_I(e^{j\omega}) = -\frac{1}{2\pi}\int_{-\pi}^{+\pi} \mathcal{H}_R(e^{j\lambda}) \cot\left(\frac{\omega-\lambda}{2}\right) d\lambda \quad \text{and} \quad \mathcal{H}_R(e^{j\omega}) = +\frac{1}{2\pi}\int_{-\pi}^{+\pi} \mathcal{H}_I(e^{j\lambda}) \cot\left(\frac{\omega-\lambda}{2}\right) d\lambda + h[0]$$

---

[44] *Hint:* to derive this relation we should take into account the fact that since the signal $x_o[n]$ is characterized by odd symmetry, $x_o[0] = 0$ (see Sect. 6.5 on page 518). But this means that *it is not possible to calculate the signal only from the samples of its odd component*; on the contrary, to do this, we need the value of the signal $h[n]$ for the time instance $n = 0$. In reality, it is proven, as the equation $h[n] = 2h_o[n]u[n] + h[0]\delta[n]$ holds only for time instances $n \geq 1$.

[45] For the solution of the exercise use the property $\mathcal{F}\{\mathcal{O}d\{x[n]\}\} = j\mathcal{X}_I(e^{j\omega})$.

constitute a *pair of direct-inverse transform equations for the discrete Hilbert transform* and characterize the *real* and the *imaginary* parts of the Fourier transform of a *causal and stable discrete sequence of real numbers*. The calculation of the integrals that appear in these must be done in such a way that the final result is *finite* (observing the form of the above equations it is not hard to see that the difficulty of calculating the integrals that appear in these lies in the fact that at the position $\omega = \lambda$ the function to be integrated *becomes infinite*). To do this, we must treat the above integrals as *Cauchy principal values*. Therefore, we have

$$\mathcal{H}_I(e^{j\omega}) = -\frac{1}{2\pi}\mathcal{P}\int_{-\pi}^{+\pi} \mathcal{H}_R(e^{j\lambda})\cot\left(\frac{\omega-\lambda}{2}\right)d\lambda \tag{12.87}$$

$$\mathcal{H}_R(e^{j\omega}) = +\frac{1}{2\pi}\mathcal{P}\int_{-\pi}^{+\pi} \mathcal{H}_I(e^{j\lambda})\cot\left(\frac{\omega-\lambda}{2}\right)d\lambda + h[0], \tag{12.88}$$

from which we can calculate, for example, the function $\mathcal{H}_I(\omega)$ from the relation

$$\mathcal{H}_I(e^{j\omega}) = -\frac{1}{2\pi}\lim_{\varepsilon\to 0}\left[\int_{-\pi}^{\omega-\varepsilon} \mathcal{H}_R(e^{j\lambda})\cot\left(\frac{\omega-\lambda}{2}\right)d\lambda + \int_{\omega+\varepsilon}^{+\pi} \mathcal{H}_R(e^{j\lambda})\cot\left(\frac{\omega-\lambda}{2}\right)d\lambda\right]. \tag{12.89}$$

The above equations concern the Hilbert transform for the *real* and the *imaginary* parts of the frequency response function, whereas to determine the corresponding equations for the magnitude $|\mathcal{H}(e^{j\omega})|$ and the phase $\measuredangle\mathcal{H}(e^{j\omega})$, we must impose *the causality condition* on the discrete-time signal $\widetilde{h}[n]$ whose Fourier transform $\widetilde{\mathcal{H}}(e^{j\omega})$ is the *logarithm* of the Fourier transform of the signal $h[n]$. In other words, the signal $\widetilde{h}[n]$, which in the international literature is known as the *complex cepstrum* of the signal $h[n]$, is defined so that the correspondences hold

$$h[n] \rightleftarrows \mathcal{H}(e^{j\omega}) = |\mathcal{H}(e^{j\omega})|e^{j\measuredangle\mathcal{H}(e^{j\omega})}$$

$$\widetilde{h}[n] \rightleftarrows \widetilde{\mathcal{H}}(\omega) = \log[\mathcal{H}(e^{j\omega})] = \log(|\mathcal{H}(e^{j\omega})|) + j\measuredangle\mathcal{H}(e^{j\omega}).$$

Using the causality condition and working as before, we are led to the relations

$$\measuredangle\mathcal{H}(e^{j\omega}) = -\frac{1}{2\pi}\mathcal{P}\int_{-\pi}^{+\pi} \log(|\mathcal{H}(e^{j\lambda})|)\cot\left(\frac{\omega-\lambda}{2}\right)d\lambda \tag{12.90}$$

$$\log(|\mathcal{H}(e^{j\omega})|) = \frac{1}{2\pi}P\int_{-\pi}^{+\pi} \measuredangle\mathcal{H}(e^{j\lambda})\cot\left(\frac{\omega-\lambda}{2}\right)d\lambda + \widetilde{h}[0], \tag{12.91}$$

which are analogous to those that apply to analog filters (see Sect. 11.8 on page 979). The quantity $\widetilde{h}[0]$ that appears in the second of the above equations is calculated as

$$\widetilde{h}[0] = \frac{1}{2\pi}\int_{-\pi}^{+\pi} \log|\mathcal{H}(e^{j\omega})|d\omega.$$

It is proven that if all the zero values of the transfer function $\mathcal{H}(z)$ are inside the unit circle—in this case, the filter is known as a *minimum-phase filter*—the complex cepstrum $\widetilde{h}[n]$ is a causal signal.

12.11 Finite Impulse Response Filters

As already mentioned, in an FIR filter the number of samples of the discrete-time signal $h[n]$ is *finite* and equal to N (this size is known as the *order of the filter*) and therefore the output of the filter will be given by the equation

$$y[n] = h[n] * x[n] = \sum_{k=0}^{N-1} h[k]x[n-k] = \sum_{k=0}^{N-1} \beta_k x[n-k]. \tag{12.92}$$

From the above equation it follows that the coefficient number k of the filter will coincide with the corresponding sample of the distinctive impulse response, or in mathematical notation $\beta_k = h[k]$ (for the values $k = 0, 1, 2, \ldots, N-1$). From the two above sums, the first one resulted from the known *convolution equation*, whereas the second one from the *general difference equation of the system* for values $\alpha_k = 0$ ($k = 0, 1, 2, \ldots, N-1$), since the filters of the FIR type *are not recursive*.

Finite impulse response filters can be divided into two categories, *symmetrical* and *antisymmetrical* filters, whose impulse response satisfies the properties of symmetry

$$h[n] = +h[N-1-n] \qquad \text{and} \qquad h[n] = -h[N-1-n]$$

respectively, from which for the various values of n and for the two categories of filters it turns out that

for symmetrical filters	for antisymmetrical filters
$h[0] = +h[N-1]$	$h[0] = -h[N-1]$
$h[1] = +h[N-2]$	$h[1] = -h[N-2]$
$h[2] = +h[N-3]$	$h[2] = -h[N-3]$
....	....
$h[N-3] = +h[2]$	$h[N-3] = -h[2]$
$h[N-2] = +h[1]$	$h[N-2] = -h[1]$
$h[N-1] = +h[0]$	$h[N-1] = -h[0]$.

Therefore, the transfer function of the digital filter is calculated as

$$\mathcal{H}(z) = \sum_{n=0}^{N-1} h[n]z^{-n} = h[0] + h[1]z^{-1} + h[2]z^{-2} + \cdots + h[N-2]z^{-[N-2]} + h[N-1]z^{-[N-1]} =$$

$$= h[0] + h[1]z^{-1} + h[2]z^{-2} + \cdots \pm h[2]z^{-[N-3]} \pm h[1]z^{-(N-2)} \pm h[0]z^{-(N-1)}. \tag{12.93}$$

An interesting property that characterizes *symmetric* and *antisymmetric* filters *of FIR* is that the system's transfer function $\mathcal{H}(z)$ has roots that appear in *inverse pairs*. This means that if the quantity z_1 is a root of the function $\mathcal{H}(z)$, the same will happen with the quantity $1/z_1$. In order to prove the truth of this claim, let us consider the function $\mathcal{H}(1/z)$, which is obviously defined as

$$\mathcal{H}\left(\frac{1}{z}\right) = \sum_{n=0}^{N-1} h[n]\left(\frac{1}{z}\right)^{-n} = \sum_{n=0}^{N-1} h[n]z^{n}.$$

Multiplying by $z^{-(N-1)}$ and using the property

$$h[n] = \pm h[N-1-n]$$

we get

$$z^{-(N-1)}\mathcal{H}\left(\frac{1}{z}\right) = z^{-(N-1)}\sum_{n=0}^{N-1} h[n]z^{n} = \sum_{n=0}^{N-1} z^{-(N-1-n)}h[n]$$

$$= \pm \sum_{n=0}^{N-1} h[N-1-n]z^{-(N-1-n)}.$$

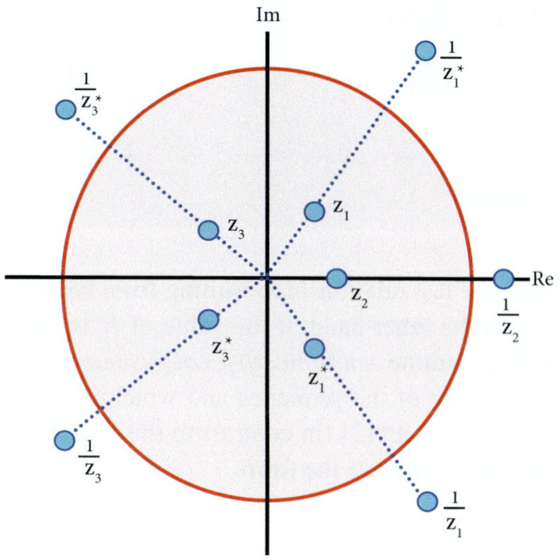

Fig. 12.51 Demonstration of the symmetry of the real and imaginary roots of the transfer function $\mathcal{H}(z)$ of an FIR filter

In the next step of the process, let us proceed to change the dummy index from n to $\ell = N-1-n$. For $n=0$, $\ell = N-1$, whereas for $n = N-1$, $\ell = 0$, and therefore the above relationship will take the form

$$z^{-(N-1)}\mathcal{H}\left(\frac{1}{z}\right) = \pm \sum_{\ell=N-1}^{0} h[\ell]z^{-\ell} = \mp \sum_{\ell=0}^{N-1} h[\ell]z^{-\ell} = \mp\mathcal{H}(z). \tag{12.94}$$

But the last equation confirms our initial claim, since the polynomials $\mathcal{H}(z)$ and $\mathcal{H}(1/z)$ have *common roots*. Therefore, *the roots of the transfer function of FIR filters appear in inverse pairs*, whereas additionally, if the impulse response of the digital filter is a *discrete-time signal of real values*, then the roots of the function $\mathcal{H}(z)$ appear in *pairs of complex conjugates*. This means that if the complex quantity z_1 is a root of the function $\mathcal{H}(z)$, the same will apply to the quantity z_1^*, whereas, owing to the property we proved above, the function $\mathcal{H}(z)$ will have as roots the quantities $1/z_1$ and $1/z_1^*$. This situation is presented diagrammatically in Fig. 12.51 for the real root z_2 and the complex roots z_1 and z_3.

The transfer function of *symmetric* and *antisymmetric* FIR-type filters is also characterized by the following two properties regarding the positions of its roots on the complex plane.

- At the positions $z_i = \pm 1$ there is an *arbitrary number of roots*, since in these positions the property $z_i^{-1} = \pm 1$ holds.
- On the circumference of the unit circle there is an *arbitrary number of pairs of conjugate complex roots*, since the complex numbers that belong to this curve satisfy the equation

$$(z - z_i)(z - z_i^*) = (z - e^{i\psi_i})(z - e^{-i\psi_i}) = \left(z - \frac{1}{z_i^*}\right)\left(z - \frac{1}{z_i}\right).$$

Returning now to the defining equation of the transfer function $\mathcal{H}(z)$ we can see that if the number N of samples of the impulse response is even, the quantity $h[n]$ $[n = 0, 1, 2, \ldots, (N/2) - 1]$ will appear in two terms; for example, the quantity $h[0]$ will appear in the first and last terms, the quantity $h[1]$ will appear in the second and penultimate terms, etc. Therefore, the above relationship is written

$$\begin{aligned}\mathcal{H}(z) &= h[0]\left(1 \pm z^{-(N-1)}\right) + h[1]\left(z^{-1} \pm z^{-(N-2)}\right) + \cdots + h\left[\frac{N}{2}\right]\left(z^{(\frac{N}{2}-1)} \pm z^{1-N+(\frac{N}{2}-1)}\right) \\ &= \sum_{n=0}^{\frac{N}{2}-1} h[n](z^{-n} \pm z^{1-N+k})\end{aligned} \tag{12.95}$$

or equivalently

$$\mathcal{H}(z)=z^{-(\frac{N-1}{2})}\sum_{n=0}^{\frac{N}{2}-1}h[n]\left(z^{\frac{N-1-2n}{2}}\pm z^{-\frac{N-1-2n}{2}}\right), \tag{12.96}$$

with the last relationship resulting from taking out the common factor of the quantity $z^{-[(N-1)/2]}$.

On the other hand, if the value of N is odd, all samples of the impulse response will appear, as before, *twice* (and thus will constitute *multiplicative coefficients* of two different powers of z), except for the sample $h[(N-1)/2]$, which is in the middle of the sequence and which will appear only once. For example, considering the value $N=5$, $h[0]=\pm h[4]$ and $h[1]=\pm h[3]$ (in contrast to the middle sample $h[2]$, which does not correspond to any other). Therefore, the transfer function will have the form

$$\mathcal{H}(z)=\sum_{n=0}^{4}h[n]z^{-n}=h[0]+h[1]z^{-1}+h[2]z^{-2}+h[3]z^{-3}+h[4]z^{-4}=h[0](1\pm z^{-4})+h[1](z^{-1}\pm z^{-3})+h[2]z^{-2},$$

whereas, in the general case and following similar reasoning, it turns out that

$$\mathcal{H}(z)=z^{-(\frac{N-1}{2})}\left\{h\left[\frac{N-1}{2}\right]+\sum_{n=0}^{\frac{N-3}{2}}h[n]\left(z^{\frac{N-1-2n}{2}}\pm z^{-\frac{N-1-2n}{2}}\right)\right\}. \tag{12.97}$$

Combining the above, we can write the transfer function of the filter for the *symmetric* and the *antisymmetric* filters and for the even and odd values of N in the form

$$\mathcal{H}(z)=\begin{cases} z^{-(\frac{N-1}{2})}\sum_{n=0}^{\frac{N}{2}-1}h[n]\left(z^{\frac{N-1-2n}{2}}\pm z^{-\frac{N-1-2n}{2}}\right) \\ z^{-(\frac{N-1}{2})}\left\{h\left[\frac{N-1}{2}\right]+\sum_{n=0}^{\frac{N-3}{2}}h[n]\left(z^{\frac{N-1-2n}{2}}\pm z^{-\frac{N-1-2n}{2}}\right)\right\}, \end{cases} \tag{12.98}$$

with the first equation corresponding to *even values* of N and the second equation corresponding to *odd values* of N. Having calculated the transfer function $\mathcal{H}(z)$, we can then determine the frequency-response function $\mathcal{H}(e^{j\omega})$ by calculating the transfer function on the circumference of the unit circle, which is known to be described by the equation $z=e^{j\omega}$. By substituting the above expression into the defining equation of the function $\mathcal{H}(z)$, we get the following results:

(a) Symmetric filters and even values of N: the frequency-response function is calculated as

$$\begin{aligned}\mathcal{H}(e^{j\omega})&=\exp\left(-j\omega\frac{N-1}{2}\right)\sum_{n=0}^{\frac{N}{2}-1}h[n]\left\{\exp\left(j\omega\frac{N-1-2n}{2}\right)+\exp\left(-j\omega\frac{N-1-2n}{2}\right)\right\}\\ &=2\left\{\sum_{n=0}^{\frac{N}{2}-1}h[n]\cos\left[\omega\left(\frac{N-1}{2}\right)-n\right]\right\}\exp\left(-j\omega\frac{N-1}{2}\right)=|\mathcal{H}(e^{j\omega})|e^{j\measuredangle\mathcal{H}(e^{j\omega})},\end{aligned} \tag{12.99}$$

where we replaced the cosine function from Euler's formula and defined the measure $|\mathcal{H}(e^{j\omega})|$ and the phase $\measuredangle\mathcal{H}(e^{j\omega})$ from the relations

$$|\mathcal{H}(e^{j\omega})| = 2\left\{\sum_{n=0}^{\frac{N}{2}-1} h[n]\cos\left[\omega\left(\frac{N-1}{2}\right)-n\right]\right\} \quad \text{and} \quad \measuredangle\mathcal{H}(e^{j\omega}) = -\omega\left(\frac{N-1}{2}\right).$$

The above relation for the phase response $\measuredangle\mathcal{H}(e^{j\omega})$ holds for the values $|\mathcal{H}(e^{j\omega})| > 0$. On the other hand, if $|\mathcal{H}(e^{j\omega})| < 0$, $-|\mathcal{H}(e^{j\omega})| > 0$ and the frequency-response function will be given by the relation

$$\mathcal{H}(e^{j\omega}) = -|\mathcal{H}(e^{j\omega})|e^{j\measuredangle\mathcal{H}(e^{j\omega})} = -|\mathcal{H}(e^{j\omega})|\cos[\measuredangle\mathcal{H}(e^{j\omega})] - j|\mathcal{H}(e^{j\omega})|\sin[\measuredangle\mathcal{H}(e^{j\omega})] =$$

$$= |\mathcal{H}(e^{j\omega})|\cos[\measuredangle\mathcal{H}(e^{j\omega})+\pi] + j|\mathcal{H}(e^{j\omega})|\sin[\measuredangle\mathcal{H}(e^{j\omega})+\pi] = |\mathcal{H}(e^{j\omega})|e^{j\mathcal{G}(e^{j\omega})},$$

where

$$\mathcal{G}(e^{j\omega}) = \measuredangle\mathcal{H}(e^{j\omega}) + \pi = -\omega\left(\frac{N-1}{2}\right) + \pi.$$

The *symmetric filters with an even number of samples N* are known as *type II FIR filters*. If we proceed to change the dummy index from n to $k = (N/2) - n$ and define the coefficients

$$\beta[k] = 2h\left[\frac{N}{2} - k\right] \quad k = 1, 2, \ldots, \frac{N}{2}$$

the above relation will take the form

$$|\mathcal{H}(e^{j\omega})| = \sum_{k=1}^{\frac{N}{2}} \beta[k]\cos\left[\omega\left(k - \frac{1}{2}\right)\right]. \tag{12.100}$$

In order to transform the last relation into a more compact form, we define new coefficients $\{\widetilde{\beta}[k]\}$ through the relations

$$\widetilde{\beta}[0] = \frac{1}{2}\beta[1], \qquad \widetilde{\beta}[k] = 2\beta[k] - \widetilde{\beta}[k-1], \qquad \widetilde{\beta}\left[\frac{N}{2} - 1\right] = 2\beta\left[\frac{N}{2}\right],$$

with the parameter k related to the second of the above relations taking values in the interval $[1, (N/2) - 2]$. In this case, the measure of the frequency-response function is formulated as

$$|\mathcal{H}(e^{j\omega})| = \cos\left(\frac{\omega}{2}\right)\sum_{k=0}^{\frac{N}{2}-1}\widetilde{\beta}(k)\cos(\omega k). \tag{12.101}$$

(b) Symmetric filters and odd values of N: following the methodology of the previous case, it is proven that the frequency-response function has the form $\mathcal{H}(e^{j\omega}) = |\mathcal{H}(e^{j\omega})|e^{j\measuredangle\mathcal{H}(e^{j\omega})}$. In this relation, the function $\measuredangle\mathcal{H}(e^{j\omega})$ is the same as the one we constructed previously, whereas the function $|\mathcal{H}(e^{j\omega})|$ is given by the relation

$$|\mathcal{H}(e^{j\omega})| = h\left[\frac{N-1}{2}\right] + 2\sum_{n=0}^{(N-3)/2} h[n]\cos\left[\omega\left(\frac{N-1}{2}\right) - n\right]. \tag{12.102}$$

The *symmetric filters with an odd number of samples* N are known as *type I FIR filters*. If we proceed to change the silent index from n to $k = [(N-1)/2] - n$ and define the coefficients

$$\alpha[k] = \begin{cases} h\left[\dfrac{N-1}{2}\right] & k = 0 \\ 2h\left[\dfrac{N-1}{2} - k\right] & k = 1, 2, \ldots, \dfrac{N-1}{2}, \end{cases} \tag{12.103}$$

the magnitude of the frequency-response function will take the form

$$|\mathcal{H}(e^{j\omega})| = \sum_{k=0}^{\frac{N-1}{2}} \alpha[k]\cos(\omega k). \tag{12.104}$$

Combining the above results we can formulate the conclusion that the frequency-response function of the *symmetric filters* FIR has the form

$$\mathcal{H}(e^{j\omega}) = |\mathcal{H}(e^{j\omega})|e^{j\measuredangle\mathcal{H}(e^{j\omega})},$$

where the function $|\mathcal{H}(e^{j\omega})|$ is given by the relation

$$|\mathcal{H}(e^{j\omega})| = \begin{cases} 2\displaystyle\sum_{n=0}^{\frac{N}{2}-1} h[n]\cos\left[\omega\left(\frac{N-1}{2}\right) - n\right] & \text{N even} \\ h\left[\dfrac{N-1}{2}\right] + 2\displaystyle\sum_{n=0}^{\frac{N-3}{2}} h[n]\cos\left[\omega\left(\frac{N-1}{2}\right) - n\right] & \text{N odd,} \end{cases} \tag{12.105}$$

whereas the function $\measuredangle\mathcal{H}(e^{j\omega})$ has *the same form* for even and odd filter orders and is calculated as

$$\measuredangle\mathcal{H}(e^{j\omega}) = \begin{cases} -\omega\left(\dfrac{N-1}{2}\right) & \text{for } |\mathcal{H}(e^{j\omega})| > 0 \\ -\omega\left(\dfrac{N-1}{2}\right) + \pi & \text{for } |\mathcal{H}(e^{j\omega})| < 0. \end{cases} \tag{12.106}$$

(c) Antisymmetric filters and odd values of N: following the process we applied previously, it is proven that the phase response of the filter $\measuredangle\mathcal{H}(e^{j\omega})$ is the same as the one we found for the even values of N, whereas the magnitude of the frequency response $|\mathcal{H}(e^{j\omega})|$ will be given by the relation

$$|\mathcal{H}(e^{j\omega})| = 2\sum_{n=0}^{\frac{N-3}{2}} h[n]\sin\left[\omega\left(\frac{N-1}{2}\right) - n\right]. \qquad (12.107)$$

The *antisymmetric filters with an odd number of samples N* are known as *type III FIR filters*. Let us proceed with changing the mute index from n to $k = [(N-1)/2] - n$ and define the coefficients

$$\gamma[k] = 2h\left[\frac{N-1}{2} - k\right] \quad k = 1, 2, \ldots, \frac{N-1}{2}. \qquad (12.108)$$

The magnitude of the filter's frequency response can be written in the form

$$|\mathcal{H}(e^{j\omega})| = \sum_{k=1}^{\frac{N-1}{2}} \gamma[k]\sin(\omega k). \qquad (12.109)$$

This relationship can be written with the help of the cosine function by defining the coefficients

$$\widetilde{\gamma}\left[\frac{N-3}{2}\right] = \gamma\left[\frac{N-1}{2}\right] \quad \widetilde{\gamma}\left[\frac{N-5}{2}\right] = 2\gamma\left[\frac{N-3}{2}\right]$$

$$\cdots\cdots\cdots\cdots \qquad \cdots\cdots\cdots$$

$$\widetilde{\gamma}(k-1) - \widetilde{\gamma}[k+1] = 2\gamma[k] \qquad \widetilde{\gamma}[0] + \frac{1}{2}\widetilde{\gamma}[2] = \gamma[1]$$

with the parameter k in the recursive from the above relations to take values in the interval $[2, (N-5)/2]$. In this case, it is proven that the magnitude of the filter's frequency response is formulated as

$$|\mathcal{H}(e^{j\omega})| = \sin(\omega)\sum_{k=0}^{\frac{N-3}{2}} \widetilde{\gamma}[k]\cos(\omega k). \qquad (12.110)$$

(d) Antisymmetric filters and even values of N: in this case, the filter's frequency response is calculated as

$$\begin{aligned}\mathcal{H}(e^{j\omega}) &= \exp\left(-j\omega\frac{N-1}{2}\right)\sum_{n=0}^{\frac{N}{2}-1} h[n]\left\{\exp\left(j\omega\frac{N-1-2n}{2}\right) - \exp\left(-j\omega\frac{N-1-2n}{2}\right)\right\} \\ &= 2j\left\{\sum_{n=0}^{\frac{N}{2}-1} h[n]\sin\left[\omega\left(\frac{N-1}{2}\right) - n\right]\right\}\exp\left(-j\omega\frac{N-1}{2}\right).\end{aligned}$$

However, it is

$$j = \cos\left(\frac{\pi}{2}\right) + j\sin\left(\frac{\pi}{2}\right) = e^{j\pi/2}$$

and the above relation is formulated as

$$\mathcal{H}(e^{j\omega}) = 2\left\{\sum_{n=0}^{\frac{N}{2}-1} h[n]\sin\left[\omega\left(\frac{N-1}{2}\right)-n\right]\right\}\exp\left[j\left(\frac{\pi}{2}-\omega\frac{N-1}{2}\right)\right]. \qquad (12.111)$$

If we write the function $\mathcal{H}(e^{j\omega})$ in the form $\mathcal{H}(e^{j\omega}) = |\mathcal{H}(e^{j\omega})|e^{j\measuredangle\mathcal{H}(e^{j\omega})}$, it is not difficult to understand that the measure of the function will be given by the relation

$$|\mathcal{H}(e^{j\omega})| = 2\sum_{n=0}^{\frac{N}{2}-1} h[n]\sin\left[\omega\left(\frac{N-1}{2}\right)-n\right], \qquad (12.112)$$

whereas its phase response will be calculated as

$$\measuredangle\mathcal{H}(e^{j\omega}) = \frac{\pi}{2}-\omega\left(\frac{N-1}{2}\right). \qquad (12.113)$$

The above relation holds for values $|\mathcal{H}(e^{j\omega})| > 0$, whereas in cases where it is $|\mathcal{H}(e^{j\omega})| < 0$ we can follow a similar methodology to the previous case and prove that

$$\measuredangle\mathcal{H}(e^{j\omega}) = \frac{3\pi}{2}-\omega\left(\frac{N-1}{2}\right). \qquad (12.114)$$

The *antisymmetric filters with an even number of samples N* are known as *type IV FIR filters*. Let us proceed to change the dummy index from n to $k = (N/2) - n$ and define the coefficients

$$\delta[k] = 2h\left[\frac{N}{2}-k\right] \quad k = 1, 2, \ldots, \frac{N}{2}. \qquad (12.115)$$

The magnitude of the frequency response function will take the form

$$|\mathcal{H}(e^{j\omega})| = \sum_{k=1}^{\frac{N}{2}} \delta[k]\sin\left[\omega\left(k-\frac{1}{2}\right)\right]. \qquad (12.116)$$

Following the methodology we used previously, we define the new coefficients

$$\widetilde{\delta}\left[\frac{N}{2}-1\right] = 2\delta\left[\frac{N}{2}\right]$$

$$\widetilde{\delta}[k-1] - \widetilde{\delta}[k] = 2\delta[k]$$

$$\widetilde{\delta}[0] - \frac{1}{2}\widetilde{\delta}[1] = \delta[1]$$

with the parameter k in the second of the above equations taking values in the interval $[2, (N/2) - 1]$. In this case, the magnitude of the frequency response of the filter is expressed as

$$|\mathcal{H}(e^{j\omega})| = \sin\left(\frac{\omega}{2}\right)\sum_{k=0}^{\frac{N}{2}-1}\widetilde{\delta}[k]\cos(\omega k).$$

Combining the above results, we can conclude that the amplitude response of the *antisymmetric filters* FIR will have the form

$$|\mathcal{H}(e^{j\omega})| = \begin{cases} 2\sum_{n=0}^{\frac{N}{2}-1} h[n]\sin\left[\omega\left(\frac{N-1}{2}\right)-n\right] & \text{N even} \\ 2\sum_{n=0}^{\frac{N-3}{2}} h[n]\sin\left[\omega\left(\frac{N-1}{2}\right)-n\right] & \text{N odd,} \end{cases} \tag{12.117}$$

whereas the phase response $\measuredangle\mathcal{H}(e^{j\omega})$ will have *the same form* for even and odd values of the length N of the impulse response and will be given by the relation

$$\measuredangle\mathcal{H}(e^{j\omega}) = \begin{cases} \frac{\pi}{2} - \omega\left(\frac{N-1}{2}\right) & \text{for } |\mathcal{H}(e^{j\omega})| > 0 \\ \frac{3\pi}{2} - \omega\left(\frac{N-1}{2}\right) & \text{for } |\mathcal{H}(e^{j\omega})| < 0. \end{cases} \tag{12.118}$$

From the above equations we find that *the phase response of symmetric and antisymmetric FIR-type filters is a linear function of frequency* and for this reason these filters are also called *linear-phase filters* or, referring to what we mentioned in Sect. 12.3.3 (see page 1123), generalized linear-phase filters. However, it should be commented that in contrast to the expression $\mathcal{H}(e^{j\omega}) = |\mathcal{H}(e^{j\omega})|e^{-j\alpha\omega}$, which is the polar representation of the frequency-response function $\mathcal{H}(e^{j\omega})$, the expression $\mathcal{H}(e^{j\omega}) = \mathcal{A}(e^{j\omega})e^{-j\alpha\omega+j\beta}$ used to describe the above filters is not a polar representation since the amplitude function $\mathcal{A}(e^{j\omega})$ can take both positive and negative values. Consequently, the function $f(\omega) = -\alpha\omega + \beta$ does not strictly constitute a phase-response function and in order to make this characteristic clear, we use the term generalized linear phase. Finally, by defining the appropriate coefficients we can describe the magnitude of the frequency response $\mathcal{H}_r(\omega)$ for the four categories of filters from the common equation

$$|\mathcal{H}(e^{j\omega})| = \mathcal{Q}(\omega)\mathcal{P}(\omega), \tag{12.119}$$

where the function $\mathcal{Q}(\omega)$ is defined as

$$\mathcal{Q}(\omega) = \begin{cases} \cos\left(\frac{\omega}{2}\right) & \text{for type I FIR filters} \\ 1 & \text{for type II FIR filters} \\ \sin\left(\frac{\omega}{2}\right) & \text{for type III FIR filters} \\ \sin(\omega) & \text{for type IV FIR filters} \end{cases} \tag{12.120}$$

whereas the function $\mathcal{P}(\omega)$ is given by the equation

$$\mathcal{P}(\omega) = \sum_{k=0}^{L} \xi[k]\cos(\omega k), \tag{12.121}$$

with the quantities $\xi[k]$ and L defined by the equations

$$\xi[k] = \begin{cases} \widetilde{\beta}(k) & \text{for type I FIR filters} \\ \alpha(k) & \text{for type II FIR filters} \\ \widetilde{\delta}(k) & \text{for type III FIR filters} \\ \widetilde{\gamma}(k) & \text{for type IV FIR filters} \end{cases}$$

and

$$L = \begin{cases} (N/2) - 1 & \text{for type I FIR filters} \\ (N-1)/2 & \text{for type II FIR filters} \\ (N/2) - 1 & \text{for type III FIR filters} \\ (N-3)/2 & \text{for type IV FIR filters.} \end{cases}$$

12.12 Design Finite Impulse Response Filters

According to what we mentioned in the previous chapters, the design of a *FIR filter* with the desired characteristics in each case is based on the determination of the frequency-response function, which in the general case has the form[46]

$$\mathcal{H}_d(e^{j\omega}) = \sum_{n=0}^{\infty} h_d[n]\, e^{-j\omega n}. \tag{12.122}$$

Having calculated this function, we can then determine the vector of the filter's impulse response from the equation of the *inverse Fourier transform*.[47]

[46] We remind the reader that in practice we are only interested in *causal filters*, the impulse response of which is defined only for time instances $n \geq 0$.

[47] For the case of the ideal low-pass filter with cutoff frequency $\omega = \omega_c$ and frequency response with *unit gain* and *linear phase* of the form

$$\mathcal{H}_d(e^{j\omega}) = \begin{cases} 1 \cdot e^{-j\alpha\omega} & |\omega| \leq \omega_c \\ 0 & \omega_c < |\omega| \leq \pi \end{cases},$$

the corresponding *ideal impulse response* of infinite length $h_d[n]$ is calculated as

$$h_d[n] = \frac{1}{2\pi}\int_{-\pi}^{+\pi} \mathcal{H}_d(e^{j\omega})e^{j\omega n}d\omega = \frac{1}{2\pi}\int_{-\pi}^{+\pi} 1 \cdot e^{-j\alpha\omega}e^{j\omega n}d\omega = \frac{1}{2\pi}\int_{-\omega_c}^{+\omega_c} e^{j\omega(n-\alpha)}d\omega = \frac{\sin[\omega_c(n-\alpha)]}{\pi(n-\alpha)}.$$

Taking into account that the spectrum $\mathcal{W}(e^{j\omega})$ of all *window functions* $w[n]$ of N samples in length is characterized by a *linear phase* $\angle\mathcal{W}(e^{j\omega}) = \omega(N-1)/2$, it is obvious that the construction of an FIR filter using a window function corresponds to the value $\alpha = (N-1)/2$, which is understood as a *delay in the time domain*.

$$h_d[n] = \frac{1}{2\pi} \int_{-\pi}^{+\pi} \mathcal{H}_d(e^{j\omega})\, e^{j\omega n} d\omega \qquad (12.123)$$

However, the discrete-time signal that will result in this way has an *infinite number of elements*. Therefore, in order to be able to use it for the representation of an FIR filter, we must transform it in some way, so that we only keep the first N samples. There are several methods of designing FIR filters, the most important of which are presented on the following pages.

12.12.1 Use of a Window Function

The easiest way to create a *finite length impulse response* $h[n]$ is the *multiplication* of the ideal infinite length impulse response $h_d[n]$ with a *window function* $w[n]$, of N samples in length,[48] so that it becomes a finite discrete-time signal. The basic theory of applying window functions for the general case of converting an infinite length signal $x[n]$ to its finite equivalent $y[n]$, as well as the study of the effect of the window spectrum $\mathcal{W}(\omega)$ on the spectrum $\mathcal{X}(j\omega)$ of the original signal $x[n]$, is presented in Sect. 10.5.6 (see page 886), which the reader is asked to study in detail before reading the following pages making the correspondences $x[n] \to h_d[n]$, $y[n] \to h[n]$, $\mathcal{X}(e^{j\omega}) \to \mathcal{H}_d(e^{j\omega})$, and $\mathcal{Y}(e^{j\omega}) \to \mathcal{H}(e^{j\omega})$. Applying what we mentioned in that section, in the more general problem of constructing an FIR type filter we can formulate the conclusion that the multiplication of the ideal impulse response $h_d[n]$ with the appropriate window function $w[n]$ in each case will lead to the FIR $h[n] = h_d[n]w[n]$ of length N elements and to the approximation of the ideal frequency response $\mathcal{H}_d(e^{j\omega})$ by the actual frequency response $\mathcal{H}(e^{j\omega})$, which is characterized by the following properties:

- The actual frequency response of the filter $\mathcal{H}(e^{j\omega}) = \mathcal{F}\{h[n]\}$ will result from the *convolution* defined between the *ideal frequency response* of the filter $\mathcal{H}_d(e^{j\omega}) = \mathcal{F}\{h_d[n]\}$ and the *spectrum of the window function* $\mathcal{W}(e^{j\omega}) = \mathcal{F}\{w[n]\}$, so that we can write that

$$\mathcal{H}(e^{j\omega}) = \frac{1}{2\pi} \int_{-\pi}^{+\pi} \mathcal{H}_d(e^{j\lambda})\mathcal{W}(e^{j(\omega-\lambda)})\, d\lambda \qquad (12.124)$$

  (see Eq. 10.146). For the ideal case of the *rectangular window*, which, as we have proven, leads *to the smallest mean square approximation error* ε^2 compared with all windows of the same length (see Eq. 10.3.2), the actual frequency response is calculated as

$$\mathcal{H}(e^{j\omega}) = \frac{1}{2\pi} \int_{-\pi}^{+\pi} \mathcal{H}_d(e^{j\lambda}) \frac{\sin[(\omega-\lambda)N/2]}{\sin[(\omega-\lambda)/2]} \exp\left[-j\,(\omega-\lambda)\frac{N-1}{2}\right] \qquad (12.125)$$

  and is diagrammatically presented in Fig. 12.52. From this figure, we observe that the ideal transfer function $\mathcal{H}_d(e^{j\omega})$ has undergone *smoothing*, resulting in the appearance of a *transition zone*, as well as *fluctuations of specific width*, which

[48] It is interesting to note that all window functions are *symmetrical* around the position $n = N/2$, a property that is mathematically formulated as $w[n] = w[N-n]$ $(0 \le n \le N-1)$ and allows the formulation of their spectrum as a function of the form $\mathcal{W}(e^{j\omega}) = \mathcal{W}_e(e^{j\omega})\exp[-j\omega(N-1)/2]$, where $\mathcal{W}_e(e^{j\omega})$ is an *even real function* of ω. In this case, if the desired impulse response $h_d[n]$ is a *symmetric signal* of the form $h_d[N-n] = h_d[n]$, the finite response $h[n] = h_d[n]w[n]$ will also be characterized by the same symmetry and the corresponding frequency response will be written as $\mathcal{H}(e^{j\omega}) = \mathcal{A}_e(e^{j\omega})\exp(-j\omega(N-1)/2)$, where $\mathcal{A}_e(e^{j\omega})$ is an *even real function* of the frequency ω. Conversely, if the desired impulse response is an *antisymmetric signal* of the form $h_d[M-n] = -h_d[n]$, the frequency response is formulated as $\mathcal{H}(e^{j\omega}) = j\mathcal{A}_o(e^{j\omega})\exp(-j\omega(N-1)/2)$, where $\mathcal{A}_o(e^{j\omega})$ is an *odd function* of the frequency ω and therefore is characterized by a *constant phase shift* equal to 90°. The proof of these properties is left as an exercise to the reader.

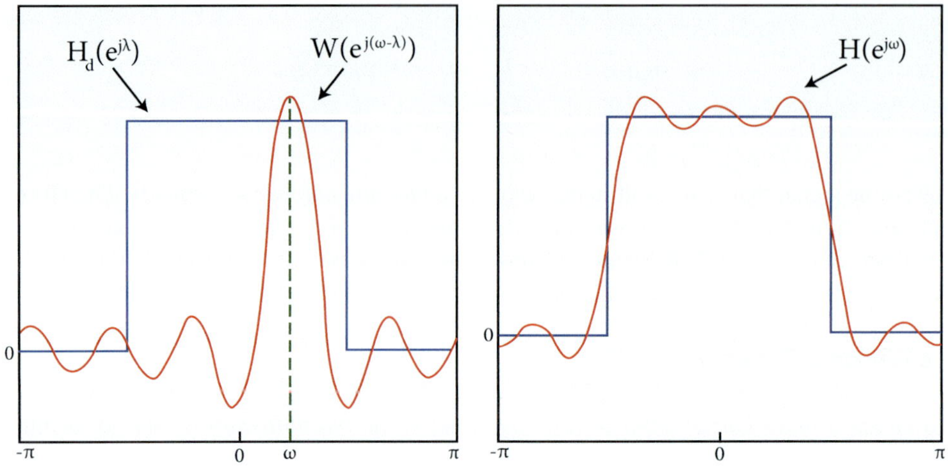

Fig. 12.52 Smoothing of the desired frequency response $\mathcal{H}_d(e^{j\omega})$, owing to its convolution with the spectrum $\mathcal{W}(e^{j\omega})$ of the rectangular window $w[n]$

are explained with the help of the *Gibbs phenomenon* (see Sect. 3.2.5 on page 200 and the discussion that follows). The existence of these characteristics characterizes *every transfer function* associated with a real filter.

- The calculation of the spectrum of a window function reveals the existence of a *central lobe*, the *width* of which is *inversely proportional to the length of the window*, as well as a series of *sidelobes*, the *height* of which varies *proportional to the number of samples of the discrete sequence* (this means that *the increase in length of the window leads to the reduction of the width of the mainlobe* and simultaneously *to the increase in the height of the sidelobes*, which changes in such a way that *the area of these lobes remains constant and independent of the order of the filter*). According to Eq. (10.147), the effect of the window's spectrum $\mathcal{W}(e^{j\omega})$ on the ideal frequency response of the filter $\mathcal{H}_d(e^{j\omega})$ consists *in the increase of the width of the ideal frequency response by the width of the spectrum of the window* (a phenomenon known as *spectral smearing*), while at the same time there is also *extension of the spectrum of the filter into frequency areas where initially its magnitude was equal to zero*, (a phenomenon known as *leakage*, which is due to the existence of the *sidelobes*). It is not difficult to realize (see also the discussion in Sect. 10.5.6 page 886, where these concepts are presented in greater detail) that we can reduce the phenomenon of spectral smearing *by increasing the length of the window* (see Eq. 10.148), a process that will, however, result in *the increase of the degree of the leakage phenomenon*. This in other words means that *we cannot change* simultaneously *and in the desired way the width of the mainlobe and the width of the sidelobes* and for this reason we are forced to resort to some kind of *compromise*, choosing the type of window characterized as the *most suitable* for each specific problem. The most well-known window functions along with their basic characteristics are presented in Table 10.6 (see page 897), whereas their graphical representation and their spectrum are presented in Fig. 10.31 (see page 898). In order for the following discussion to be understood, the reader is encouraged to study and understand all the mathematical details related to this material.

As already mentioned, although the rectangular window leads to the *minimum possible approximation error* and the *minimum possible spectral leakage*, it is not the best choice, since its spectrum contains *large sidelobes*. However, for educational purposes only and in order to describe the way in which the discontinuity it shows at the position $n = N - 1$ affects its spectral characteristics, let us consider an ideal low-pass FIR filter, the frequency response of which is related to this window, and let us study the change of the measure of the function $\mathcal{W}(e^{j\omega})$, which is presented in Fig. 12.53 for a filter order value of $N = 7$. In this figure both the *central lobe* of the filter (its bandwidth), which determines the degree of smoothing of the frequency response, and the $N - 1$ *sidelobes*, the height of which increases with the increase in the filter order and in the way we described in Sect. 10.5.6, are shown. The reduction of the degree of smoothing of the frequency response and the change in the width of the central lobe and the height of the sidelobes with the increase in the filter order value, is presented in Figs. 12.54 and 12.55.

The existence of the sidelobes we previously described leads to the appearance of *fluctuations* both in the *passband* of the filter and in the *stopband* (see Fig. 12.56 which refers to the case of a *low-pass discrete-time filter*, although the situation is completely analogous for the other three categories of frequency selection filters). The main characteristic of these fluctuations is the way in which their amplitude changes, which has quite a large value *at the cutoff frequency limit* and gradually decreases as we move away from this. It is shown that the width of these fluctuations depends *on the area*

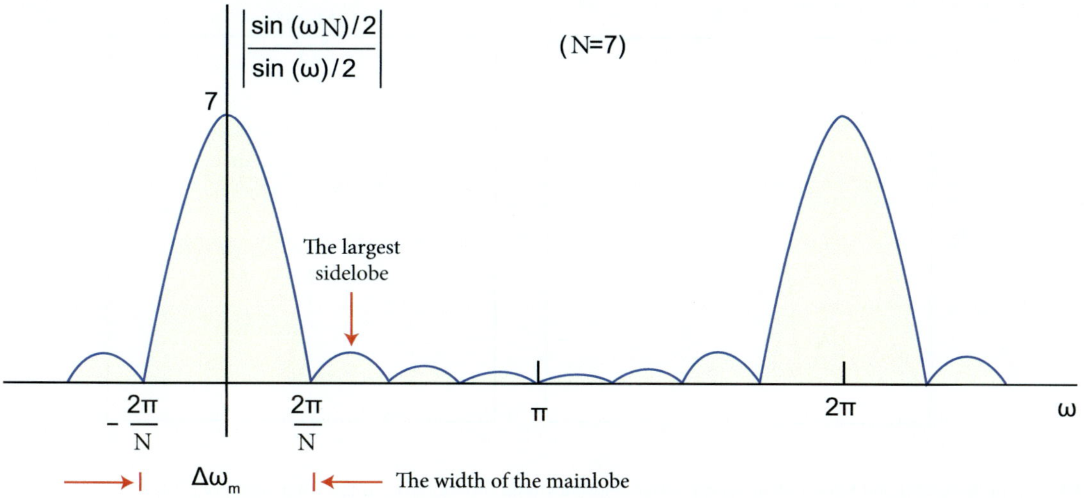

Fig. 12.53 The change of the spectrum measure $\mathcal{W}(e^{j\omega})$ of the rectangular window for $N = 7$

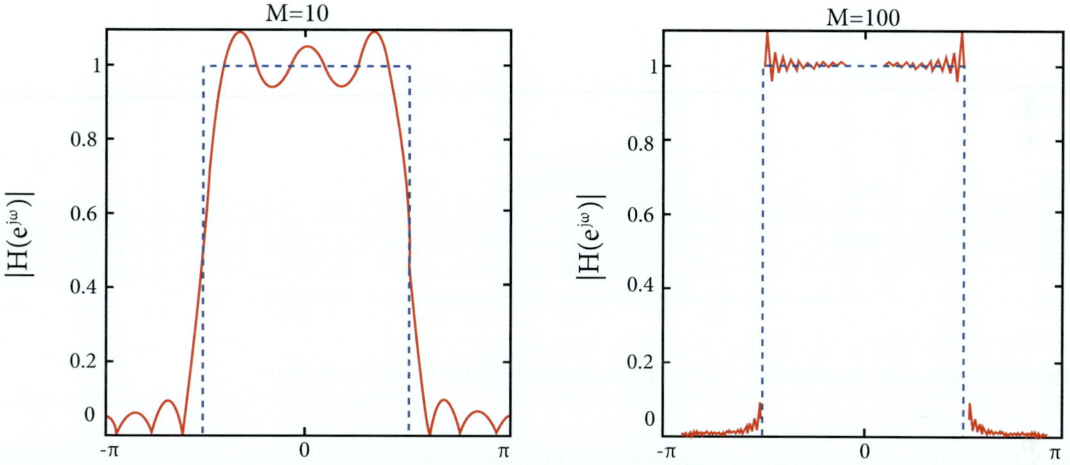

Fig. 12.54 Reduction of the degree of smoothing of the actual frequency response $\mathcal{H}(e^{j\omega})$ with the increase in the order of the digital filter N

of the surface defined by the lateral lobes and the horizontal axis and remains *constant* and *independent* of the value of N, since the area in question is characterized by the same type of dependence. If we increase the order of the filter, we will find that these oscillations occur *increasingly and with greater frequency*, but remain *unchanged in terms of their width*. This undesirable phenomenon is our well-known *Gibbs phenomenon* (see Sect. 3.2.5 on page 200), the appearance of which affects the techniques of digital filter design, as it cannot be completely eliminated (we remind that the existence of this phenomenon is due to the removal of the samples of the ideal impulse response $h_d[n]$ for values $n \geq N$ by multiplying it with the function of the *rectangular window* $w[n]$).

In order to understand this behavior of the filter's frequency response, we must take into account that the desired frequency-response function $\mathcal{H}_d(e^{j\omega})$ is the *Fourier transform* of the desired impulse response $h_d[n]$. From the basic theory of Fourier analysis, it is known that the Fourier series of a function $f(\omega)$ *converges to the value of the function* for all points at which the function is *continuous* and in this sense any function can be fully reconstructed by adding together all the infinite terms that appear in its Fourier series representation. If however the function presents *discontinuity* at some point $\omega = \omega_0$, as happens, for example, at the cutoff frequency ω_c, then it is treated as *piecewise continuous* and the Fourier series calculated at the point ω_0 does not converge to the value $f(\omega)$ but to the value $[f(\omega_0^+) + f(\omega_0^-)]/2$. The representation, however, of a function as piecewise continuous requires the removal of several terms from the Fourier series, which, however, *no longer constituting a complete set*, will be evaluated *erroneously* at the point of discontinuity, leading to a result greater than the actual one. Taking into account that the multiplication of the ideal impulse response $h_d[n]$ with the rectangular window function is equivalent to the removal of a very large number of samples of it and that this removal will cause in turn the

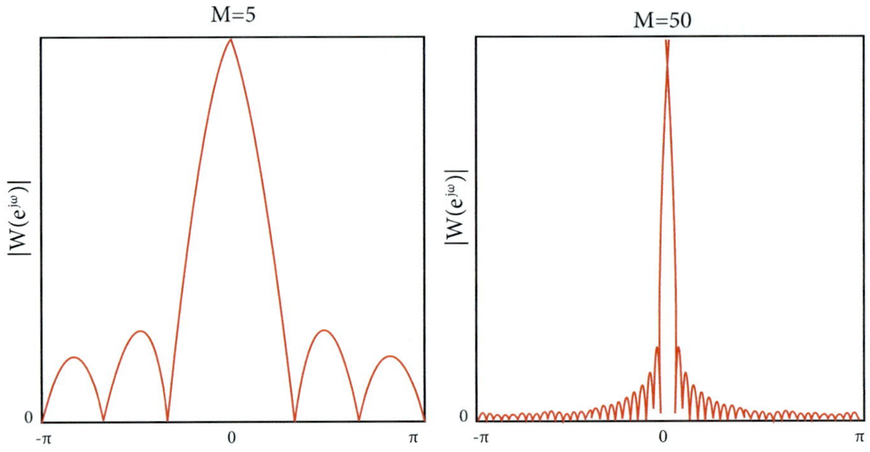

Fig. 12.55 Change in the width and height of the central and sidelobes with the increase in the filter order N

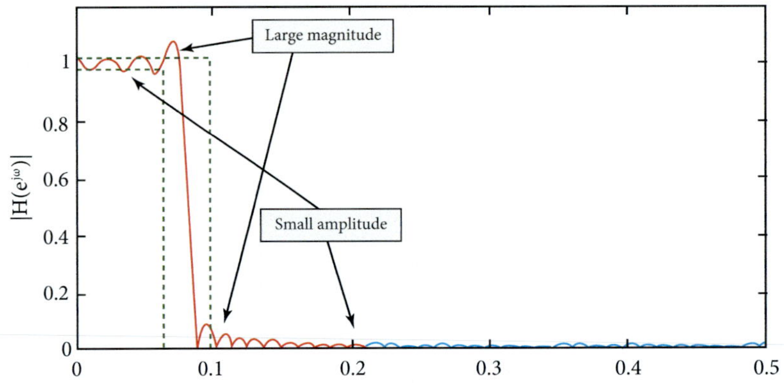

Fig. 12.56 The fluctuations of the function $|\mathcal{H}(e^{j\omega})|$ due to the appearance of the Gibbs phenomenon

cutoff of many terms from its representation according to Fourier series—which is none other than the frequency response $\mathcal{H}_d(e^{j\omega})$—the appearance of the Gibbs phenomenon in areas near the point of discontinuity $\omega = \omega_c$ is fully explained.

The appearance of all the above undesirable characteristics associated with the function of the rectangular window forces us to finally resort to the use of *window functions* that we presented in the previous chapter, with the choice of the appropriate window in each case depending on the requirements that exist in each case. The basic common feature of all these windows is that their representation in the frequency domain is characterized by *lower sidelobes* than the *rectangular window*, whereas at the same time their *central lobe is characterized by a larger width*—for the same natural value of the parameter N—so as to lead to a *greater degree of smoothing* of the desired frequency response function $\mathcal{H}_d(e^{j\omega})$. The last feature, of course, is not desirable, since we want this smoothing to be as small as possible, but nevertheless the problem is addressed with an appropriate choice of the parameter N. What is difficult to deal with in practice is the reduction of the *residual variance ratio*, which is defined as

$$r = 100 \times \left(\frac{\text{width of the maximum sidelobe}}{\text{width of the central lobe}} \right) \tag{12.126}$$

or alternatively as

$$r = 20 \log \left(\frac{\text{width of the maximum sidelobe}}{\text{width of the central lobe}} \right) dB \tag{12.127}$$

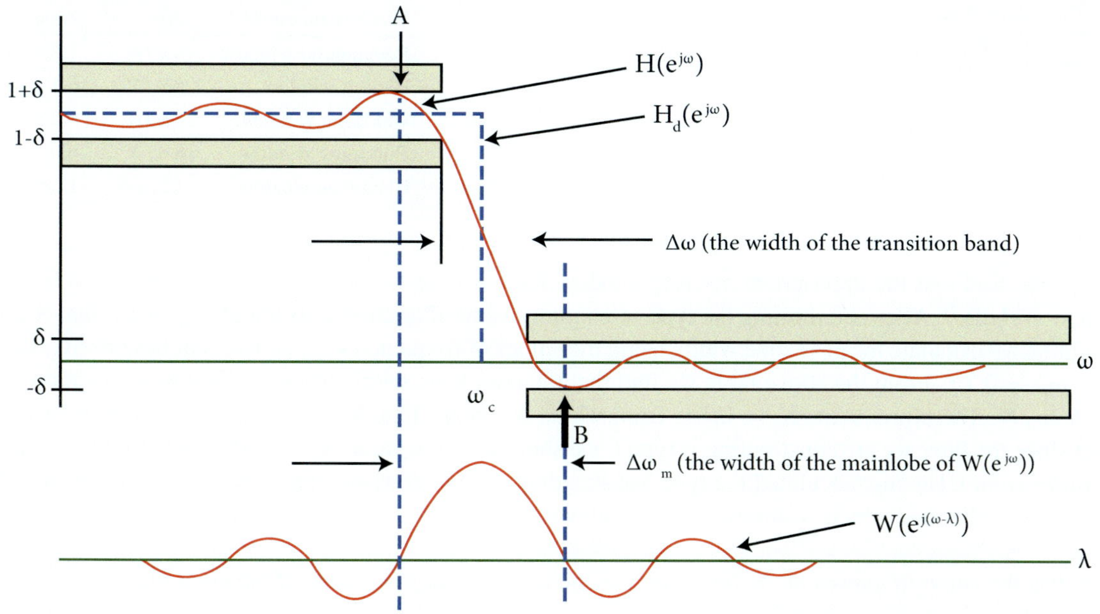

Fig. 12.57 The dependence of the characteristics of the desired frequency response from the spectrum of the applied window

and which for several types of windows is *independent of the order of the filter* N. The width of the central lobe and the attenuation level of the highest sidelobe (in units db/octave) for the most well-known window functions is presented in Table 10.6 (see page 897).

12.12.2 Selection of the Window Function and Construction of the Discrete-Time Filter

The way in which the characteristics of the spectrum of the applied window affect the amplitude response of the filter to be constructed is presented in Fig. 12.57. In this figure the shaded areas are dictated by the *filter design specifications*, which, in full analogy with analog filters, include *the limits of the pass and cutoff zones* ω_p and ω_s, *the width of the fluctuations in the pass zone* A_p, and *the minimum attenuation in the cutoff zone* A_s, with the last two quantities being calculated, as we have seen, from the values of the oscillations δ_p and δ_s in the two above zones. Studying the above figure, we come to the following conclusions:

(a) The ideal cutoff frequency ω_c bisects *exactly* the transition zone defined from the frequencies ω_p and ω_s and therefore can be calculated as the average $\omega_c = (\omega_p + \omega_s)/2$.
(b) The transition band width $\Delta\omega$ is *proportional to the central lobe width of the spectrum* $\mathcal{W}(e^{j\omega})$ of the applied window $\Delta\omega_m$, but always less than this. The values of the parameters $\Delta\omega$ and $\Delta\omega_m$ along with the minimum attenuation in the stopband $A_s^{\min}$ for the most known window functions are presented in Table 12.1.
(c) The *maximum passband ripple* δ (point A) is equal to the corresponding maximum stopband ripple (point B), with the distance between these points being equal to the width $\Delta\omega_m$. As already mentioned, the value of the ripple δ is *independent of the window length* and depends only on its shape (i.e., its type).

The choice of the appropriate window function in each case is dictated by the *attenuation characteristics* provided by it and by the specifications of the filter to be constructed in each case. Let us consider, for example, that we want to construct a type I FIR filter with frequency values $\omega_p = 0.3\pi$ and $\omega_s = 0.2\pi$ and ripple width in the passband and stopband $\delta_1 = \delta_2 = 0.01$. Observing that the ripple width δ_2 in the stopband[49] corresponds to a value $A_s^{\min} = -20\log_{10}\delta = -40\,\text{dB}$ and referring

[49] Since the filters that result from the window method are characterized from *equal amplitude fluctuations in the pass and cutoff zones*, we can use either of the two we want, ending up with the same results. However, it should be noted that this is not always a desirable feature as the specifications in the cutoff zone are *stricter* than those of the pass zone.

Table 12.1 The values of $\Delta\omega$, $\Delta\omega_m$ and $A_s^{\min}$ for the most well-known types of discrete windows

Window function	$\Delta\omega$	$\Delta\omega_m$	$A_s^{\min}$
Rectangular window	$4\pi/N$	$1.8\pi/N$	21 dB
Bartlett window	$8\pi/N$	$6.1\pi/N$	25 dB
von Hann window	$8\pi/N$	$6.2\pi/N$	44 dB
Hamming window	$8\pi/N$	$6.6\pi/N$	53 dB
Blackman window	$12\pi/N$	$11\pi/N$	74 dB

to Table 12.1, we find that the appropriate discrete window for our case is the *window of von Hann*, as it offers a value $A_s^{\min}$ equal to -44 dB.[50] After determining the type of window, we need to choose its length, a process that is carried out as follows. Initially, we calculate the pass zone width, using the values of frequencies ω_p and ω_s that have been given; therefore, $\Delta\omega = \omega_s - \omega_p = 0.1\pi$. From the above table we find that the pass-zone width corresponding to the window of von Hann is equal to $6.2\pi/N$. Therefore, $6.2\pi/N = 0.1\pi$, from which it follows that $N = 80$. Therefore, the order of the filter is $N = 79$, but since the filter we are constructing is type I, we should use the value $N = 80$ (the type I filters are characterized by *perfect order value*). Having calculated the type and length of the window, we can then construct the desired filter in the way we presented earlier and which is summarized as follows:

(a) We calculate the *cutoff frequency* $\omega_c = (\omega_p + \omega_s)/2$ and the *frequency response of the ideal filter*

$$\mathcal{H}_d(e^{j\omega}) = \begin{cases} 1 & |\omega| \leq \omega_c \\ 0 & \omega_c < |\omega| < \pi. \end{cases}$$

For the frequency values of the example it is $\omega_c = (0.2\pi + 0.3\pi)/2 = 0.25\pi$ and therefore

$$\mathcal{H}_d(e^{j\omega}) = \begin{cases} 1 & |\omega| \leq 0.25\pi \\ 0 & 0.25\pi < |\omega| < \pi. \end{cases}$$

(b) We calculate the *ideal (noncausal) impulse response of infinite length* from the relation

$$h_d[n] = \frac{1}{2\pi}\int_{-\pi}^{+\pi} \mathcal{H}_d(e^{j\omega})\, e^{j\omega n} d\omega = \frac{1}{2\pi}\int_{-\pi}^{+\pi} e^{j\omega n} d\omega = \frac{\omega_c}{\pi}\frac{\sin\omega_c n}{\omega_c n}.$$

In this case it is $\omega_c = 0.25\pi$ and therefore

$$h_d[n] = \frac{\omega_c}{\pi}\frac{\sin\omega_c n}{\omega_c n} = \frac{0.25\pi}{\pi}\frac{\sin(0.25\pi n)}{0.25\pi n} = \frac{\sin(0.25\pi n)}{\pi n}.$$

Although in this case the above calculation can be performed relatively easily, in practice and in cases where the desired frequency response $\mathcal{H}_d(e^{j\omega})$ is characterized by a high degree of complexity, the ideal impulse response is calculated with the help of software that implements the *inverse Fourier transform*.

(c) We make the ideal impulse response a *causal signal*, shifting it in time by $N/2$ samples. In this particular example, this process will lead to the discrete-time signal

[50] It should be noted that no one forbids us to use a window that leads to greater attenuation in the cutoff zone, such as the Hamming and Blackman windows; however, this would lead to an *undesirable increase in the transition zone width*.

$$h_d[n] = \frac{\sin[0.25\pi(n-40)]}{\pi(n-40)}.$$

(d) We multiply the *ideal impulse response* with the *window function* we have chosen, making it a *finite discrete-time signal*. In this case, the multiplication of the signal $h_d[n]$ with the discrete sequence of the von Hann window will lead to the result

$$h[n] = h_d[n]w[n] = \frac{1}{2}\frac{\sin[0.25\pi(n-40)]}{\pi(n-40)} \times \left[1 - \cos\frac{2\pi n}{79}\right] \quad 0 \le n < 80.$$

(e) We calculate the *actual frequency response* as the *discrete Fourier transform of the FIR* $h[n]$. Initially, we calculate the discrete-time Fourier transform

$$\mathcal{H}(e^{j\omega}) = \sum_{n=-\infty}^{+\infty} h[n]e^{-j\omega n} = \frac{1}{2\pi}\sum_{n=0}^{79} \frac{\sin[0.25\pi(n-40)]}{n-40}\left[1 - \cos\frac{2\pi n}{79}\right]e^{j\omega n}$$

and then the discrete Fourier transform $\mathcal{H}[k]$ as the samples of the function $\mathcal{H}(e^{j\omega})$ calculated at the positions of the frequencies $\omega = 2\pi k/N = \pi k/40$ ($k = 0, 1, \ldots, 79$). Because generally these calculations are difficult to compute analytically, in almost all cases we resort to the computation of the discrete Fourier transform and the graphical representation of the results with the help of software applications, such as, for example, the MATLAB application. The impulse response $h[n]$ along with the magnitude of the frequency response designed on a linear and logarithmic scale are presented in Fig. 12.58, from where it appears that the filter's operating specifications have been approximated to a satisfactory degree. If this is not the case, we can suitably adjust the length or even the type of the applied window and in some cases the values of the filter's operating frequencies and repeat the process as many times as needed until we achieve the desired approximation accuracy.

12.12.3 The Kaiser Window Method

As mentioned in the previous chapter, the *Kaiser window* offers an effective method of designing FIR-type filters, which is based on the fact that *the ratio of residual variance affects the width of the variances in the pass and cutoff regions*, whereas *the order N of the filter only affects the width of the transition region*. Therefore, the value of the parameter α that appears in the defining equation of this window can be selected with the help of an *empirical relationship*—which was formulated by Kaiser after extensive experimentation—so as to achieve *the desired width of variance in the pass and cutoff regions*, whereas subsequently with a second empirical equation we can choose the order N of the filter, so as to achieve the desired width of the transition region.

In order to present the *Kaiser window method* for the construction of digital filters with specific specifications, let us assume that we want to construct a *low-pass FIR filter*, the characteristics of which are presented in Fig. 12.59 and include *the value* δ *of the width of the variance* (which, as we have already mentioned, is the same for the pass and cutoff regions), *the values of parameters* $\tilde{A}_p$ *and* $\tilde{A}_s$ and *the values of the frequencies* ω_p *and* ω_s, which lead to a frequency value *cutoff* $\omega_c = (\omega_p + \omega_s)/2$ (this relationship arises from the symmetry that characterizes the approximation of the ideal impulse response at the point of discontinuity). In this case, the application of the Kaiser window method includes the implementation of the following steps.

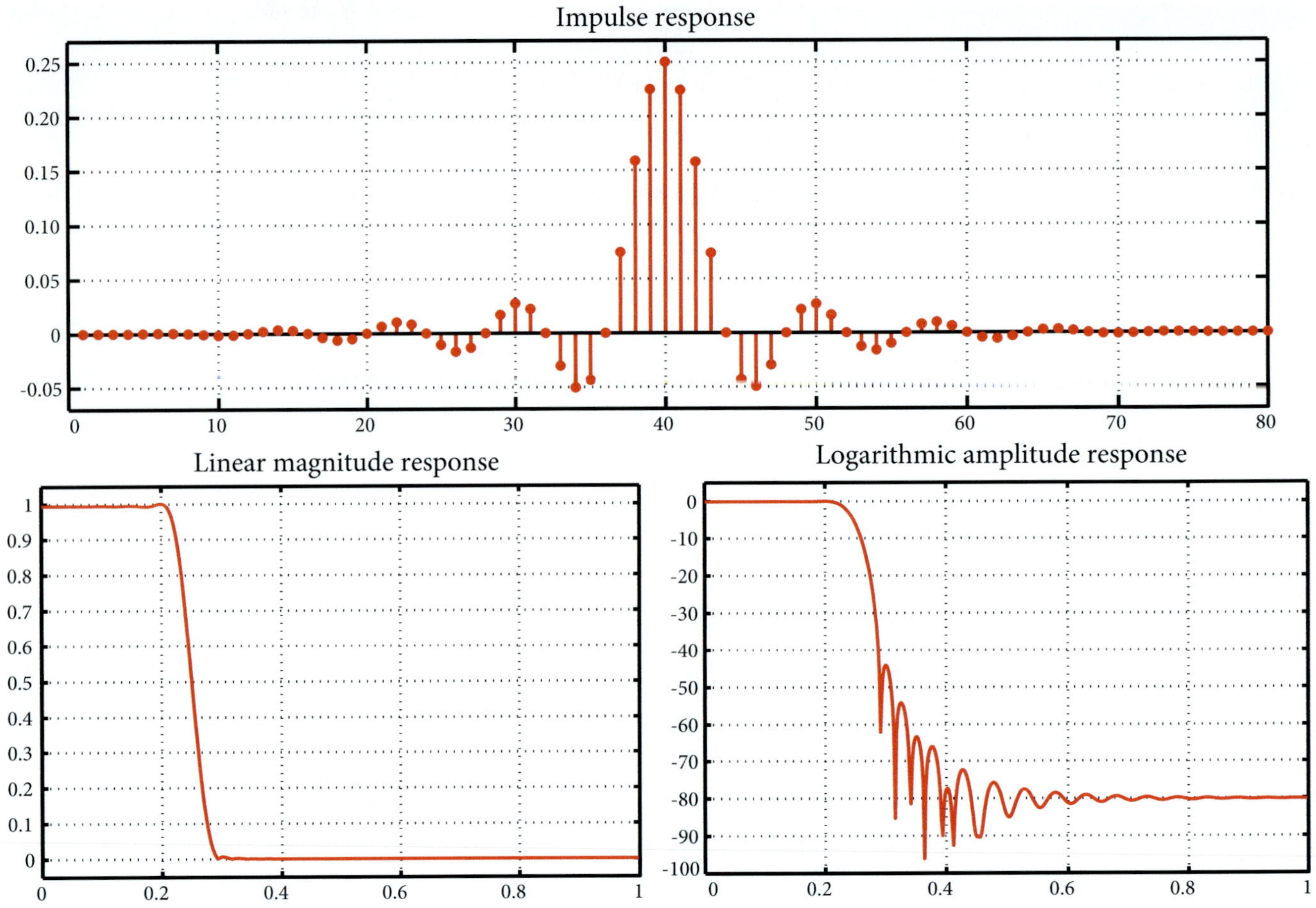

Fig. 12.58 The impulse response $h[n]$ and the magnitude of the frequency response designed on a linear and logarithmic scale of a low-pass filter with operating specifications $\omega_p = 0.3\pi$, $\omega_s = 0.2\pi$ and $\delta_1 = \delta_2 = 0.01$ that has been implemented with the help of the von Hann window

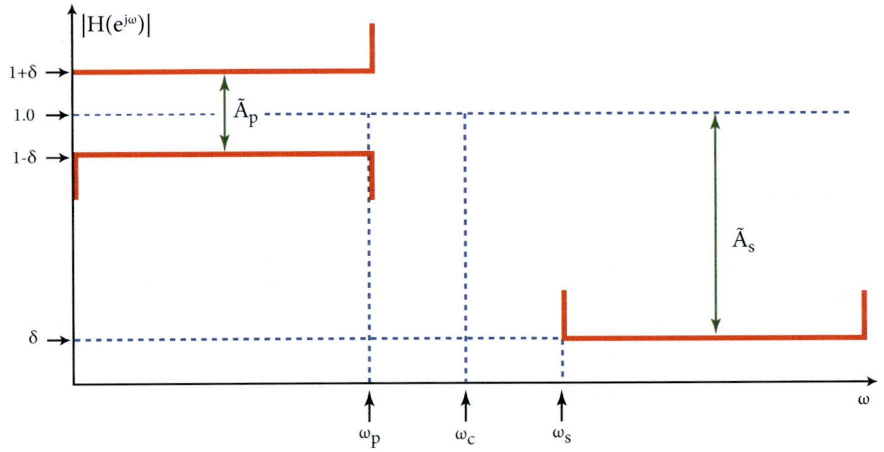

Fig. 12.59 Application of the Kaiser window method for the construction of a low-pass filter

1. We consider the *ideal low-pass filter*, the frequency response of which is given by the relationship

$$H_d(e^{j\omega}) = \begin{cases} 1 \text{ for } |\omega| \leq \omega_c \\ 0 \text{ for } \omega_c < |\omega| < \pi. \end{cases} \tag{12.128}$$

2. We calculate the *ideal impulse response* $h_d[n]$ from the *inverse Fourier transform of the ideal frequency response*, leading to the now known result

$$h_d[n] = \frac{\omega_c}{\pi}\frac{\sin(\omega_c n)}{\omega_c n} = \frac{\omega_c}{\pi}\mathrm{sinc}(\omega_c n). \tag{12.129}$$

3. We choose the value of the parameter δ so that *the actual bandwidth of the deviation in the passband A_p is smaller or equal to the desired bandwidth $\tilde{A}_p$ and at the same time the actual minimum attenuation A_s is greater than or equal to the desired minimum attenuation $\tilde{A}_s$*. A suitable value for the parameter δ is

$$\delta = \min(\tilde{\delta}_p, \tilde{\delta}_s), \tag{12.130}$$

where the quantities $\tilde{\delta}_p$ and $\tilde{\delta}_s$ are given by the empirical relations

$$\tilde{\delta}_p = \frac{10^{0.05\tilde{A}_p} - 1}{10^{0.05\tilde{A}_p} + 1} \quad \text{and} \quad \tilde{\delta}_s = 10^{-0.05\tilde{A}_s}. \tag{12.131}$$

4. Having determined the value of the parameter δ, we can calculate the value of the *actual attenuation A_s* from the relation

$$A_s = -20\log_{10}(\delta). \tag{12.132}$$

5. Based on the value of the parameter A_s that we calculated in the previous step, we calculate the parameter α of the Kaiser window from the empirical relation

$$\alpha = \begin{cases} 0 & \text{for } A_s \le 21 \\ 0.5842(A_s - 21)^{0.4} + 0.07886(A_s - 21) & \text{for } 21 \le A_s \le 50 \\ 0.1102(A_s - 8.7) & \text{for } A_s > 50. \end{cases} \tag{12.133}$$

6. We calculate the value of the *auxiliary parameter D* from the relation

$$D = \begin{cases} 0.9222 & \text{for } A_s \le 21 \\ \dfrac{A_s - 7.95}{14.36} & \text{for } A_s > 21. \end{cases} \tag{12.134}$$

7. We choose the *order of the filter N* as the smallest integer value that satisfies the inequality

$$N \ge 1 + \left[\frac{\Omega_s D}{\omega_s - \omega_p}\right], \tag{12.135}$$

where Ω_s is the *sampling frequency of the input signal* and $[x]$ is the integer part of the real quantity x. If we do not know information about the sampling frequency, we consider the value of the frequency Ω_s to be equal to the *maximum possible sampling frequency*, which, according to the *Nyquist theorem*, is equal to 2π.

8. Having calculated the values of the parameters α and N, we construct the discrete-time signal $w[n]$ from the defining equation of the *Kaiser window*.
9. We calculate the *transfer function of the filter* from the relation

$$\mathcal{H}(z) = z^{-(N-1)/2}\mathcal{Z}\{h_d[n]w[n]\}. \tag{12.136}$$

10. We calculate the frequency-response function of the filter $\mathcal{H}(e^{j\omega})$ as the transfer function $\mathcal{H}(z)$ calculated *on the circumference of the unit circle*, that is, for values $z = e^{j\omega}$.

In order to understand the above process, suppose we want to construct a *low-pass filter* with parameter values $\omega_p = 0.4\pi$, $\omega_s = 0.6\pi$ and $\delta_1 = \delta_2 = 0.001$. In this case, the application of the Kaiser window algorithm will be carried out in the following way:

- We calculate the value of the *cutoff frequency* from the relation

$$\omega_c = \frac{\omega_s + \omega_s}{2} = \frac{0.4\pi + 0.6\pi}{2} = 0.5\pi.$$

- We calculate the values of the parameters δ and $\Delta\omega$, with the results

$$\delta = \delta_1 = \delta_2 = 0.001 \qquad \text{and} \qquad \Delta\omega = \omega_s - \omega_p = 0.6\pi - 0.4\pi = 0.2\pi.$$

- We calculate the value of the *actual attenuation* from the relation

$$A_s = -20\log_{10}(\delta) = -20\log_{10}(10^{-3}) = (-20)(-3) = 60$$

  It is $A_s = 60 > 50$ and therefore the value of the parameter α is calculated as

$$\alpha = 0.1102(A_s - 8.7) = 0.1102(60 - 8.7) = 0.1102 \cdot 51.3 = 5.653260.$$

- We calculate the value of the parameter D from the relation

$$D = \frac{A_s - 7.95}{14.36} = \frac{60 - 7.95}{14.36} = \frac{52.05}{14.36} = 3.624651.$$

- Because *we do not know information about the sampling frequency*, we use the maximum value of $\Omega_s = 2\pi$. Therefore, the *order of the filter* is calculated as

$$N \geq 1 + \left[\frac{2\pi \cdot 3.624651}{0.2\pi}\right] \geq 1 + [36.24651] \geq 37$$

  and therefore the order of the filter is $N = 37$.

Having calculated the value of the parameter α and the order of the filter N, we can calculate the *discrete-time signal of the impulse response* by applying the usual methodology. We repeat from the basic theory that the application of the order N filter causes a delay equal to $N/2$ in the *ideal frequency response*, which therefore, will have the form

$$\mathcal{H}_{lp}(e^{j\omega}) = \begin{cases} e^{-j\omega N/2} & \text{for } |\omega| < \omega_c \\ 0 & \text{anywhere else} \end{cases}$$

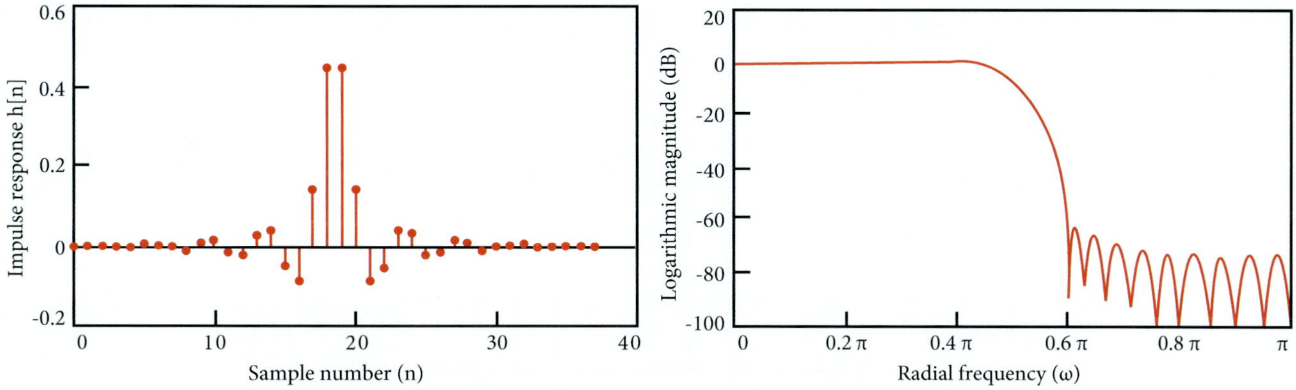

Fig. 12.60 The impulse response $h[n]$ and the amplitude response $|\mathcal{H}(e^{j\omega})|$ for the example application of the Kaiser window method

and therefore the ideal impulse response is calculated from the *inverse Fourier transform* of the above relationship, leading to the result

$$h_{lp}[n] = \frac{1}{2\pi}\int_{-\omega_c}^{+\omega_c} e^{-j\omega N/2} e^{j\omega n} d\omega = \frac{\sin[\omega_c(n-(N/2))]}{\pi[n-(N/2)]},$$

as can be verified. In the next step of the process, we calculate the actual impulse response from the relationship $h[n] = h_{lp}[n]w[n]$. Using the defining equation of the Kaiser window and the parameter values $\alpha = 5.653260$, we finally get

$$h[n] = \begin{cases} \dfrac{\sin[\omega_c(n-18.5)]}{\pi(n-18.5)} \times \dfrac{I_0\left[5.653260\sqrt{1-\left(\dfrac{n-18.5}{18.5}\right)^2}\right]}{I_0(5.653260)} & 0 \le n \le 37 \\ 0 & \text{elsewhere.} \end{cases}$$

In the last step of the process, we can calculate the frequency response of the actual filter by calculating the Fourier transform of the above relationship. The impulse response of the filter $h[n]$ and the magnitude of the frequency response function $\mathcal{H}(e^{j\omega})$ are shown in Fig. 12.60.

The above process can be very easily generalized, so as to allow the design of all other types of filters, namely *high-pass*, *band-pass*, and *band-stop* filters. Let us assume, for example, that we want to design a *band-pass filter* with parameter values $\tilde{A}_p$ for the remaining variance ratio in the passband and $\tilde{A}_s$ for the attenuation in the stopband, which allows the passage of frequencies $\omega_{p_1} \le \omega \le \omega_{p_2}$ and cuts off the frequencies $\omega_{s_1} \le \omega \le \omega_{s_2}$. The frequency response of the above digital filter is presented in Fig. 12.61. In this figure, the frequencies ω_{c_1} and ω_{c_2} define the passband of the ideal band-pass filter, whereas the parameter Ω_s represents the sampling frequency of the input signal.

The filter design methodology will be based on the frequency response function of the ideal band-pass filter, which for the filter of the above section has the form

$$\mathcal{H}(e^{j\omega}) = \begin{cases} 1 \text{ for } -\omega_{c_2} \le \omega \le -\omega_{c_1} \\ 1 \text{ for } +\omega_{c_1} \le \omega \le +\omega_{c_2} \\ 0 \text{ anywhere else.} \end{cases}$$

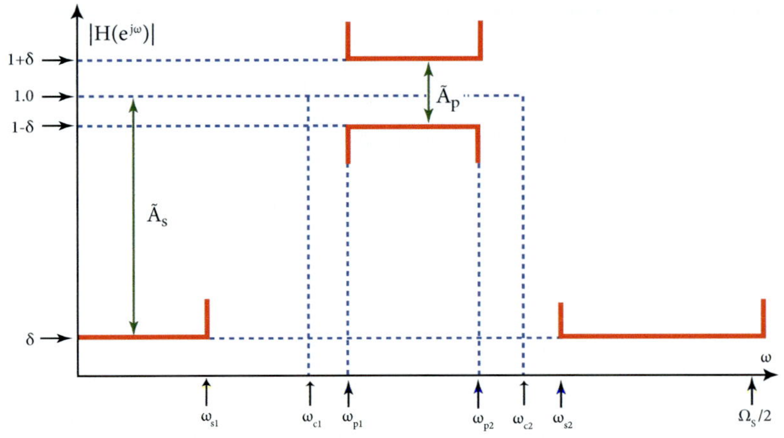

Fig. 12.61 Application of the Kaiser window method for the construction of a band-pass filter

In this case, the frequencies ω_{c_1} and ω_{c_2} are given by the relations $\omega_{c_1} = \omega_{p_1} - (B_t/2)$ and $\omega_{c_2} = \omega_{p_2} + (B_t/2)$ with the quantity B_t defined as $B_t = \min[(\omega_{p_1} - \omega_{s_1}), (\omega_{s_2} - \omega_{p_2})]$.

Having calculated the above parameters, the design methodology of the band-pass filter will be carried out as before. The above method is presented in the next exercise with the design of a digital filter that will meet specific operating specifications.

Problem 12.12.1 *Using the* Kaiser window *method, construct a* band-pass digital filter *characterized by a* generalized linear phase *and meeting the following specifications:*

$$|\mathcal{H}(e^{j\omega})| \leq 0.01 \; for \; 0 \leq |\omega| \leq 0.25\pi$$
$$0.95 \leq |\mathcal{H}(e^{j\omega})| \leq 1.05 \; for \; 0.35\pi \leq |\omega| \leq 0.6\pi$$
$$|\mathcal{H}(e^{j\omega})| \leq 0.01 \; for \; 0.65\pi \leq |\omega| \leq \pi.$$

(a) Determine the order of the filter N and the parameter α of the Kaiser window; (b) determine the value of the delay caused by the application of the filter; and (c) determine the ideal impulse response on which the window function w[n] will be applied, in order to calculate the impulse response of the real filter (Oppenheim, Schafer, and Buck, Exercise for resolution 7.5).

Answer: (a) From the filter specifications, it immediately follows that $\delta = \delta_p = \delta_s = 0.01$ and therefore the actual attenuation will be given by the relation

$$A_s = -20\log_{10}(\delta) = -20\log_{10}(10^{-2}) = (-20)(-2) = 40.$$

We observe that $A_s = 40 < 50$ and therefore the *filter parameter* is calculated as

$$\alpha = 0.5842(A_s - 21)^{0.4} + 0.07886(A_s - 21) = 0.5842(40 - 21)^{0.4} + 0.07886(40 - 21) = 3.395320,$$

as can be verified. On the other hand, the range of the two transition zones is determined by the frequencies $(\omega_{s_1}, \omega_{p_1}) = (0.25\pi, 0.35\pi)$ and $(\omega_{p_2}, \omega_{s_2}) = (0.60\pi, 0.65\pi)$, from which the values $\Delta_{\omega_1} = 0.35\pi - 0.25\pi = 0.10\pi$ and $\Delta\omega_2 = 0.65\pi - 0.60\pi = 0.05\pi$ are derived. Therefore, we have

$$\Delta\omega = \min(\Delta\omega_1, \Delta\omega_2) = \min(0.10\pi, 0.05\pi) = 0.05\pi,$$

meaning that the cutoff frequencies are calculated as

$$\omega_{c_1} = \omega_{p_1} - \frac{\Delta\omega}{2} = 0.350\pi - 0.025\pi = 0.325\pi \quad \text{and} \quad \omega_{c_2} = \omega_{p_2} + \frac{\Delta\omega}{2} = 0.600\pi + 0.025\pi = 0.625\pi.$$

The value of the auxiliary constant D for the value $A_s = 40$ is calculated as

$$D = \frac{A_s - 7.95}{14.36} = \frac{40 - 7.95}{14.36} = 2.231894$$

and therefore the order of the filter will result from the equation

$$N \geq 1 + \left[\frac{2\pi D}{\Delta\omega}\right] \geq 1 + \left[\frac{2\pi \cdot 2.231894}{0.05\pi}\right] \geq 1 + [89.27576] \geq 90,$$

from which it follows that $N = 90$. We observe that not having information about the sampling frequency Ω_s, we resorted to the use of the maximum permissible value for this frequency, which is the value 2π.

(b) Taking into account that the filter to be designed is a *linear-phase filter* with an order value of $M = 90$, the value of the time delay on the impulse response due to its application will be equal to $N/2 = 45$ samples.

(c) From the basic theory, it is known that the impulse response of the ideal band-pass filter with a frequency passband $(\omega_{c_1}, \omega_{c_2})$ is given by the equation

$$h[n] = \frac{\omega_2}{\pi}\frac{\sin(n\omega_2)}{n\omega_2} - \frac{\omega_1}{\pi}\frac{\sin(n\omega_1)}{n\omega_1}.$$

In this case, the cutoff frequencies ω_{c_1} and ω_{c_2} are $\omega_{c_1} = 0.325\pi$ and $\omega_{c_2} = 0.625\pi$. On the other hand, the use of the first $N = 90$ samples, as is well known, causes a delay in the ideal impulse response equal to $N/2 = 45$ samples. If we use the above values of the frequencies ω_{c_1} and ω_{c_2} in the last equation and simultaneously make the replacement $n \to n - 45$, we arrive at the desired result, which is

$$h[n] = \frac{\sin[0.625\pi(n-45)] - \sin[0.325\pi(n-45)]}{\pi(n-35)}.$$

Problem 12.12.2 *Using the* Kaiser window *method, construct a filter with an* FIR of real values and generalized linear phase *that meets the following specifications:*

$$0.90 < |\mathcal{H}(e^{j\omega})| < 1.10 \textit{ for } 0 \leq |\omega| \leq 0.2\pi$$
$$-0.06 < |\mathcal{H}(e^{j\omega})| < 0.06 \textit{ for } 0.3\pi \leq |\omega| \leq 0.475\pi$$
$$1.90 < |\mathcal{H}(e^{j\omega})| < 2.10 \textit{ for } 0.525\pi \leq |\omega| \leq \pi$$

and to approximate the behavior of an ideal filter with a real impulse response, *the desired frequency response of which will have the form*

$$\mathcal{H}_d(e^{j\omega}) = \begin{cases} 1 \textit{ for } 0 \leq |\omega| \leq 0.25\pi \\ 0 \textit{ for } 0.25\pi \leq |\omega| \leq 0.5\pi \\ 2 \textit{ for } 0.5\pi \leq |\omega| \leq \pi \end{cases}$$

(Oppenheim, Schafer, and Buck, Problem 7.6).

Answer: The behavior of the frequency response of the real filter in relation to the frequency response of the ideal filter is presented diagrammatically in Fig. 12.62. From this figure, we can see that the filter to be designed can be considered as being constructed *as the sum of two simpler filters*: a low-pass filter corresponding to a frequency-response measure $|\mathcal{H}(e^{j\omega})| = 1$ and a deviation value δ, and a high-pass filter characterized by a frequency-response measure $|\mathcal{H}(e^{j\omega})| = 2$ and a deviation value 2δ. The sum of these two errors in the frequency range $0.3\pi \leq \omega \leq 0.475\pi$ should not be less than 0.06, as required by the specifications of the real filter. Therefore, we have

$$\delta + 2\delta \leq 0.06 \Rightarrow 3\delta \leq 0.06 \Rightarrow \delta \leq 0.02.$$

Therefore, $\delta_{max} = 0.02$, and consequently the filter's attenuation is calculated as

$$A_s = -20\log_{10}(\delta_{max}) = -20\log_{10}(0.02) = (-20)(-1.698970) = 33.979400.$$

We observe that $A_s < 50$, and consequently the filter's parameter α is calculated as

$$\alpha = 0.5842(33.9794 - 21)^{0.4} + 0.07886(33.9794 - 21) = 2.652338.$$

From the operating specifications of the real filter it follows that

$$(\omega_{s_1}, \omega_{p_1}) = (0.2\pi, 0.3\pi) \quad \text{and} \quad (\omega_{p_2}, \omega_{s_2}) = (0.475\pi, 0.525\pi).$$

Therefore,

$$\Delta\omega_1 = \omega_{p_1} - \omega_{s_1} = 0.3\pi - 0.2\pi = 0.1\pi \quad \text{and} \quad \Delta\omega_2 = \omega_{s_2} - \omega_{p_2} = 0.525\pi - 0.475\pi = 0.05\pi$$

and consequently the value of the parameter Δ_ω is calculated as

$$D\omega = \min(\Delta\omega_1, \Delta\omega_2) = \min(0.1\pi, 0.05\pi) = 0.05\pi.$$

From the last parameter we can calculate the cutoff frequencies ω_{c_1} and ω_{c_2}, which have values

$$\omega_{c_1} = \omega_{p_1} - \frac{\Delta\omega}{2} = 0.300\pi - 0.025\pi = 0.275\pi \quad \text{and} \quad \omega_{c_2} = \omega_{p_2} + \frac{\Delta\omega}{2} = 0.475\pi + 0.025\pi = 0.500\pi.$$

The value of the auxiliary constant D for the value $A_s = 33.9794$ is calculated as

$$D = \frac{A_s - 7.95}{14.36} = \frac{33.9794 - 7.95}{14.36} = 1.812632$$

and therefore the order of the filter will result from the equation

$$N \geq 1 + \left\lceil \frac{2\pi D}{\Delta\omega} \right\rceil \geq 1 + \left\lceil \frac{2\pi \cdot 1.812632}{0.05\pi} \right\rceil \geq 1 + [72.50528] \geq 73$$

from which the value $N = 73$ results. For the calculation of the value of the last parameter, the maximum value of the sampling frequency was used, which is equal to 2π. After the calculation of the above parameters, the filter-design process continues and is completed in the way we saw in the previous exercise.

Problem 12.12.3 *Suppose we want to design a low-pass filter with an FIR with the following specifications:*

$$0.98 < |\mathcal{H}(e^{j\omega})| < 1.02 \; 0 \leq |\omega| \leq 0.63\pi$$
$$-0.15 < |\mathcal{H}(e^{j\omega})| < 0.15 \; 0.65\pi \leq |\omega| \leq \pi$$

by applying the function $w[n]$ of the Kaiser window on the impulse response $h_d[n]$ of the ideal low-pass filter, the cutoff frequency of which is equal to $\omega_c = 0.64\pi$. Calculate the values of the parameter α of the Kaiser window and the order of the filter N (Oppenheim, Schafer, and Buck, Exercise for solution 7.16).

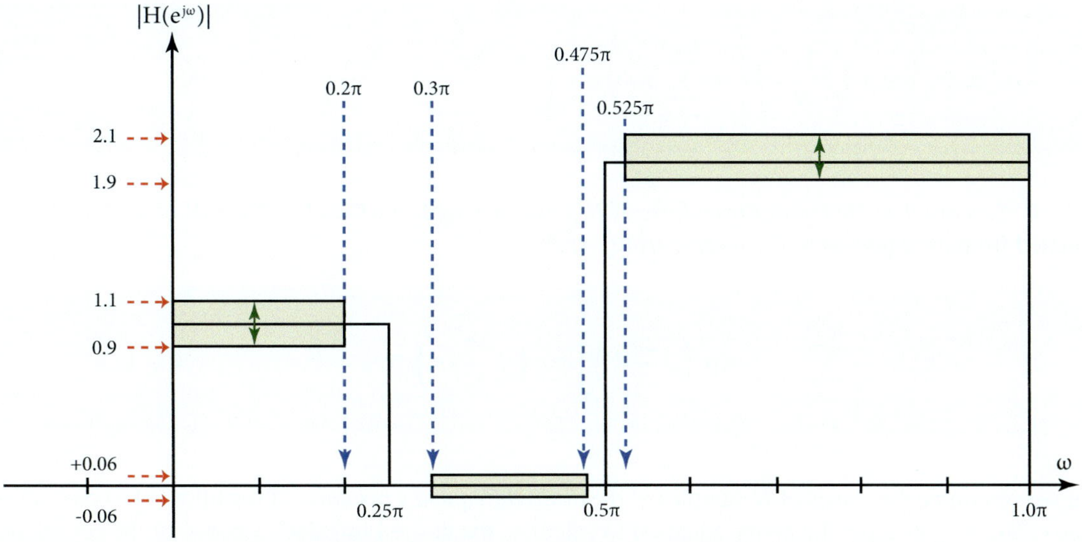

Fig. 12.62 The frequency response of the filter of Exercise 12.12.2 in relation to the ideal response

Answer: It is $\delta_1 = \delta_2 = \delta = 0.02$ and therefore the filter's attenuation is calculated as

$$A_\alpha = -20\log_{10}(\delta_{max}) = -20\log_{10}(0.02) = (-20)(-1.698970) = 33.979400.$$

We observe that $A_\alpha < 50$ and therefore the parameter α of the filter is calculated as

$$\alpha = 0.5842(33.9794 - 21)^{0.4} + 0.07886(33.9794 - 21) = 2.652338,$$

as can be proven. On the other hand, the value of the parameter $\Delta\omega$ is equal to

$$\Delta\omega = \omega_\alpha - \omega_p = 0.65\pi - 0.63\pi = 0.02\pi. \tag{12.137}$$

Therefore,

$$D = \frac{A_\alpha - 7.95}{14.36} = \frac{33.9794 - 7.95}{14.36} = 1.812632 \tag{12.138}$$

and therefore the order of the filter will result from the equation

$$M \geq 1 + (int)\left(\frac{2\pi D}{\Delta\omega}\right) \geq 1 + (int)\left(\frac{2\pi \cdot 1.812632}{0.02\pi}\right) \geq 1 + (int)(181.2632) \geq 182$$

and the value of the filter's order is $M = 182$, which is the desired result.

12.12.4 The Method of Frequency Sampling

The *frequency sampling method* is a simple and directly applicable method of designing an FIR-type filter and consists in determining the values of the *desired impulse response* $\mathcal{H}_d(e^{j\omega})$ at the k positions of frequencies of the form $\omega_k = 2\pi k/N$, where N is the order of the filter to be constructed, whereas the index k takes the values $k = 0, 1, 2, \ldots, L$ with the value of the parameter L being equal to $(N-1)/2$ for even order values and equal to $(N/2) - 1$ for odd order values. In this case, as is well known, we have

$$\mathcal{H}[k] \equiv \mathcal{H}_d\left[\exp\left(j\frac{2\pi}{N}k\right)\right] = \sum_{n=0}^{N-1} h[n]\exp\left(-j\frac{2\pi}{N}kn\right) \quad k = 0, 1, 2, \ldots, N-1, \tag{12.139}$$

an equation that refers to a *discrete Fourier transform* of N points in length and therefore the FIR of the filter to be constructed can be calculated from the equation of the *inverse transform*[51]

$$h[n] = \frac{1}{N}\sum_{k=0}^{N-1} \mathcal{H}_d\left[\exp\left(j\frac{2\pi}{N}k\right)\right]\exp\left(j\frac{2\pi}{N}kn\right), \quad k = 0, 1, 2, \ldots, N-1. \tag{12.140}$$

Therefore, if we determine the values of N samples of the *desired frequency response* of the filter to be constructed at *equally spaced frequencies*, we can apply the above equation to calculate the desired impulse response of the considered filter. The impulse response calculated in this way is related to the ideal impulse response $h_d[n]$, which is a noncausal signal of infinite length through the relation

$$h[n] = \sum_{k=-\infty}^{+\infty} h_d[n + kN], \quad 0 \leq n \leq N-1 \tag{12.141}$$

(see Eq. 10.75 and the argument we developed during the presentation of the *discrete Fourier transform*). After calculating the impulse response, we should apply the above transform to it, in order to calculate the desired frequency response $\mathcal{H}(e^{j\omega})$. This process is diagrammatically presented in Fig. 12.63.

The main advantage of the above method is that it allows the construction of an FIR filter with an *arbitrary frequency response*, whereas the above sampling process concerns both the *magnitude* and the *phase* of the desired function $\mathcal{H}_d(e^{j\omega})$ (although Fig. 12.63 only depicts the magnitude of the frequency response). Therefore, the values of the samples we determine are *complex in nature*, whereas if we choose to use *zero-phase values* for these samples (so as to work with real values and not with complex quantities that are related to the linear phase response), the impulse response that will result will be *symmetrical* with respect to the position $n = 0$ and for this reason it should be *shifted* by L samples, so as to become a *causal signal*. When applying this method, we should take special care to ensure that the amplitude response is characterized by *even symmetry*, whereas the phase response should be characterized by *odd symmetry*, so that the filter that will result is causal and is characterized by a *linear phase* and *constant coefficients*. When applying the discrete Fourier transform of the impulse response, we can check the accuracy of the results using *a sufficiently large value for the number of samples*, so as to increase the *discrimination ability*, whereas *limiting* the length of the signal to N samples, we can resort to the use of some *window function* in the way we described in the previous section.

---

[51] This relationship arises by multiplying both sides of the previous expression by the exponential quantity $e^{j2\pi km/N}$ $(m = 0, 1, 2, \ldots, N-1)$ so that we get

$$\mathcal{H}_d\left[\exp\left(j\frac{2\pi}{N}k\right)\right]\exp\left(j\frac{2\pi}{N}km\right) = \exp\left(j\frac{2\pi}{N}km\right)\sum_{n=0}^{N-1} h[n]\exp\left(-j\frac{2\pi}{N}kn\right) \quad k = 0, 1, 2, \ldots, N-1.$$

Adding up with respect to k and changing the order of summation in the sum of the second term, we get

$$\sum_{k=0}^{N-1} \mathcal{H}_d\left[\exp\left(j\frac{2\pi}{N}k\right)\right]\exp\left(j\frac{2\pi}{N}km\right) = \sum_{k=0}^{N-1}\exp\left(j\frac{2\pi}{N}km\right)\sum_{n=0}^{N-1} h[n]\exp\left(-j\frac{2\pi}{N}kn\right) = \sum_{n=0}^{N-1} h[n]\exp\left[j\frac{2\pi}{N}k(m-n)\right].$$

As we have proven in Exercise 6.7.2 (see page 526), the second sum of the second term has a value equal to 0 for $m \neq n$ and a value equal to N for $m = n$. Consequently, the second term of the above relation is equal to $Nh[m]$, so by proceeding to replace the dummy index from m to n and solving for $h[n]$, we arrive at the desired expression.

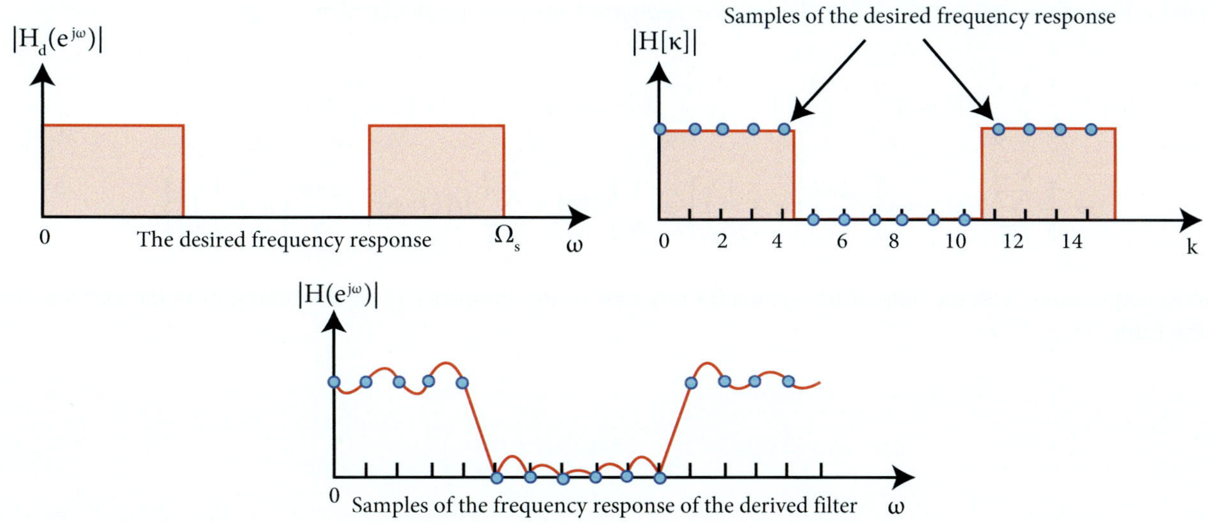

Fig. 12.63 Diagrammatic demonstration of the *frequency-sampling method*, in which the impulse response of the filter to be constructed is calculated from the samples of the desired frequency response

Taking into account that the impulse response $h[n]$ is *a signal of real values*, from the *property of symmetry* of discrete Fourier transform, we know that $\mathcal{H}[k] = \mathcal{H}^*[M-k]$, a property that in combination with the *symmetry conditions* that exist for the impulse response allows us to calculate the samples of the signal $h[n]$ using only $(N+1)/2$ samples for odd N and $N/2$ samples for even N. Combining the equations for the *symmetric* and the *antisymmetric* case as

$$\mathcal{H}_d(e^{j\omega}) = |\mathcal{H}_d(e^{j\omega})| \exp\left[j\left(\frac{\alpha\pi}{2} - \frac{\omega(N-1)}{2}\right)\right] \tag{12.142}$$

with the constant α being equal to 0 for the *symmetric signal* and equal to 1 for the *antisymmetric signal*, the values of the samples of the frequency response at the positions of the frequency values $\omega_k = 2\pi k/N$ are calculated as

$$\mathcal{H}[k] = \mathcal{H}_d\left[\exp\left(j\frac{2\pi}{N}k\right)\right] = \left|\mathcal{H}_d\left[\exp\left(j\frac{2\pi}{N}k\right)\right]\right| \exp\left[j\left(\frac{\alpha\pi}{2} - \frac{2\pi k}{N}\frac{(N-1)}{2}\right)\right]. \tag{12.143}$$

Defining the real samples $\mathcal{G}[k]$ from the relation $G[k] = (-1)^k|\mathcal{H}_d(e^{j2\pi k/N})|$ $(k = 0, 1, 2, \ldots, N-1)$ and eliminating the function $|\mathcal{H}_d(e^{j\,2\pi k/N})|$ from the last two relations, it follows that

$$\mathcal{H}[k] = \mathcal{G}[k]\exp(j\pi k)\exp\left[j\left(\frac{\alpha\pi}{2} - \frac{2\pi k}{N}\frac{(N-1)}{2}\right)\right] \tag{12.144}$$

and consequently the symmetry conditions for the samples $\mathcal{H}[k]$ are transformed into corresponding symmetry conditions for the samples $\mathcal{G}[k]$. For the symmetric case $(\alpha = 0)$, $\mathcal{G}[k] = -\mathcal{G}[N-k]$ and

$$\mathcal{H}[k] = \mathcal{G}[k]\exp(j\pi k)\exp\left[-j\pi k\left(\frac{N-1}{N}\right)\right] = \mathcal{G}[k]\exp\left(j\frac{\pi k}{N}\right), \tag{12.145}$$

whereas for the antisymmetric case $(\alpha = 1)$, $\mathcal{G}[k] = \mathcal{G}[N-k]$ and

$$\mathcal{H}[k] = \mathcal{G}[k]\exp(j\pi k)\exp\left[-j\frac{\pi}{2} - j\pi k\left(\frac{N-1}{N}\right)\right] = \mathcal{G}[k]\exp\left(j\frac{\pi}{2}\right)\exp\left(j\frac{\pi k}{N}\right) \tag{12.146}$$

Using the values of the real samples $\mathcal{G}[k]$, the *impulse response* $h[n]$ will be calculated as

$$h[n] = \frac{1}{N}\sum_{k=0}^{N-1}\mathcal{H}[k]\exp\left(j\frac{2\pi k}{N}n\right) = \frac{1}{N}\sum_{k=0}^{N-1}\mathcal{G}[k]\exp\left(j\frac{\pi k}{N}\right)\exp\left(j\frac{2\pi k}{N}n\right)$$

$$= \frac{1}{N}\sum_{k=0}^{N-1}\mathcal{H}[k]\exp\left[j\frac{2\pi k}{N}\left(n+\frac{1}{2}\right)\right] = \frac{1}{N}\left\{\mathcal{G}[0] + \sum_{k=1}^{N-1}\mathcal{H}[k]\exp\left[j\frac{2\pi k}{N}\left(n+\frac{1}{2}\right)\right]\right\}. \quad (12.147)$$

The above expression, with the help of the symmetry relations of the samples $\mathcal{G}[k]$ and working as in the previous section, takes the form

$$h[n] = \frac{1}{N}\left\{\mathcal{G}[0] + 2\sum_{k=1}^{L}\mathcal{G}[k]\cos\left[\left(\frac{2\pi k}{N}\right)\left(n+\frac{1}{2}\right)\right]\right\} \quad (12.148)$$

for the *symmetric* case, whereas for the *antisymmetric* case we have

$$h[n] = \begin{cases} -\dfrac{2}{N}\displaystyle\sum_{k=1}^{L}\mathcal{G}[k]\sin\left[\left(\frac{2\pi k}{N}\right)\left(n+\frac{1}{2}\right)\right] & N \text{ odd} \\ \dfrac{1}{N}\left\{(-1)^{n+1}\mathcal{G}\left(\dfrac{N}{2}\right) - 2\displaystyle\sum_{k=1}^{L}\mathcal{G}[k]\sin\left[\left(\frac{2\pi k}{N}\right)\left(n+\frac{1}{2}\right)\right]\right\} & N \text{ even.} \end{cases} \quad (12.149)$$

The above method can be generalized by sampling the ideal frequency response $\mathcal{H}_d(e^{j\omega})$ at the positions $\omega_k = 2\pi(k+m)/M$, with the parameter m taking one of the values $m = 0$ and $m = 1/2$. In this case, the defining equations of the functions $\mathcal{H}[k]$ and $h[n]$ take the form

$$\mathcal{H}[k] \equiv \mathcal{H}_d\left\{\exp\left[j\frac{2\pi}{N}(k+m)\right]\right\} = \sum_{n=0}^{N-1}h[n]\exp\left(-j\frac{2\pi}{N}(k+m)\,n\right) \quad k = 0, 1, 2, \ldots, N-1 \quad (12.150)$$

$$h[n] = \frac{1}{M}\sum_{k=0}^{M-1}\mathcal{H}_d\left\{\exp\left[j\frac{2\pi}{N}(k+m)\right]\right\}\exp\left(j\frac{2\pi}{N}(k+m)\,n\right) \quad k = 0, 1, 2, \ldots, N-1, \quad (12.151)$$

which for the value $m = 0$ simplify to those we presented earlier.

We conclude the presentation of the *frequency-sampling* method with the observation that if the samples of the desired frequency response define a very narrow—or even zero-width—transition band, then the desired frequency response that will result will be characterized by a damping value in the cutoff band, which *will not be satisfactory*. However, this behavior can be improved by *redefining* the samples of the function $\mathcal{H}_d(e^{j\omega})$ so as to lead to a smoother transition from the passband to the cutoff band. In other words, instead of defining a sharp transition of the form $\{\ldots, 1, 1, 1, 0, 0, 0, 0, 0, \ldots\}$, we can interpolate a transitional sample, so that, for example, we have $\{\ldots, 1, 1, 1, 0.5, 0, 0, 0, 0, \ldots\}$, or even two transitional samples, so that, for example, we have $\{\ldots, 1, 1, 1, 0.67, 0.33, 0, 0, 0, \ldots\}$. After this redefinition, we of course need to repeat the process from the beginning. It should be noted that although this method leads to better results that translate into a lower height of the maximum sidelobe in the spectrum of the filter to be constructed, it is not certain that these will emerge immediately. On the contrary, it is very likely that we will need to repeat this process several times, until we finally end up with an optimal result.[52]

Problem 12.12.4 *Find the frequency response of a linear-phase FIR filter with symmetrical impulse response of $N =$ 15 samples in length, when the samples of the magnitude of the desired frequency response $\mathcal{H}_d(e^{j\omega})$ are known with values*

$$\left|\mathcal{H}_d\left[\exp\left(j\frac{2\pi k}{15}\right)\right]\right| = \begin{cases} 1 & k = 0, 1, 2, 3 \\ 0.4 & k = 4 \\ 0 & k = 5, 6, 7 \end{cases}$$

(Proakis & Manolakis, Exercise for solution 8.7).

Solution Taking into account that the impulse response $h[n]$ is a symmetrical signal while the order N of the filter is an odd number, the amplitude response of the filter, according to Eq. (12.98), will have the form

$$|\mathcal{H}_d(e^{j\omega})| = h\left[\frac{N-1}{2}\right] + 2\sum_{n=0}^{(N-3)/2} h[n]\cos\left[\omega\left(\frac{N-1}{2} - n\right)\right] = h[7] + 2\sum_{n=0}^{6} h[n]\cos[\omega(7-n)]$$

for the value $N = 15$. On the other hand, since the impulse response is a symmetrical signal, it will satisfy the relation $h[n] = h[N-1-n]$, which for the value $N = 15$ takes the form $h[n] = h[14-n]$.

The sampling of the above frequency response at the positions of frequencies $\omega_k = 2\pi k/15$ (for the values $k =$ 0, 1, 2, 3, 4, 5, 6, 7) will lead to the expression

$$\left|\mathcal{H}_d\left[\exp\left(j\frac{2\pi k}{15}\right)\right]\right| = h[7] + 2\sum_{n=0}^{6} h[n]\cos\left[\frac{2\pi k}{15}(7-n)\right] \quad (k = 0, 1, 2, 3, 4, 5, 6, 7),$$

which, for the above values of k is formulated as

$\mathbf{k=0:}\ h[7] + 2\sum_{n=0}^{6} h[n] = 1$	$\mathbf{k=4:}\ h[7] + 2\sum_{n=0}^{6} h[n]\cos\left[\frac{8\pi}{15}(7-n)\right] = 0.4$
$\mathbf{k=1:}\ h[7] + 2\sum_{n=0}^{6} h[n]\cos\left[\frac{2\pi}{15}(7-n)\right] = 1$	$\mathbf{k=5:}\ h[7] + 2\sum_{n=0}^{6} h[n]\cos\left[\frac{10\pi}{15}(7-n)\right] = 0$
$\mathbf{k=2:}\ h[7] + 2\sum_{n=0}^{6} h[n]\cos\left[\frac{4\pi}{15}(7-n)\right] = 1$	$\mathbf{k=6:}\ h[7] + 2\sum_{n=0}^{6} h[n]\cos\left[\frac{12\pi}{15}(7-n)\right] = 0$
$\mathbf{k=3:}\ h[7] + 2\sum_{n=0}^{6} h[n]\cos\left[\frac{6\pi}{15}(7-n)\right] = 1$	$\mathbf{k=7:}\ h[7] + 2\sum_{n=0}^{6} h[n]\cos\left[\frac{14\pi}{15}(7-n)\right] = 0.$

Developing the above sums for the values $k = 0, 1, 2, 3, 4, 5, 6, 7$, we successively have

$$h[7] + 2\left\{h[0] + h[1] + h[2] + h[3] + h[4] + h[5] + h[6] + h[7]\right\} = 1$$

$$h[7] + 2\left\{h[0]\cos\left(\frac{14\pi}{15}\right) + h[1]\cos\left(\frac{12\pi}{15}\right) + h[2]\cos\left(\frac{10\pi}{15}\right) + h[3]\cos\left(\frac{8\pi}{15}\right) + h[4]\cos\left(\frac{6\pi}{15}\right)\right.$$

[52] For the description of a process of selecting the samples of the function $\mathcal{H}_d(e^{j\omega})$ in the transition zone, so as to obtain an optimal result regarding the attenuation in the cutoff zone, see the textbook Proakis J., Manolakis D., *Digital Signal Processing - Principles, Algorithms and Applications*, 3rd Edition, 1996, ISBN 0-13-373762-4, pp. 636-637.

$$+ h[5]\cos\left(\frac{4\pi}{15}\right) + h[6]\cos\left(\frac{2\pi}{15}\right) + h[7]\Bigg\} = 1$$

$$h[7] + 2\Bigg\{h[0]\cos\left(\frac{28\pi}{15}\right) + h[1]\cos\left(\frac{24\pi}{15}\right) + h[2]\cos\left(\frac{20\pi}{15}\right) + h[3]\cos\left(\frac{16\pi}{15}\right) + h[4]\cos\left(\frac{12\pi}{15}\right)$$

$$+ h[5]\cos\left(\frac{8\pi}{15}\right) + h[6]\cos\left(\frac{4\pi}{15}\right) + h[7]\Bigg\} = 1$$

$$h[7] + 2\Bigg\{h[0]\cos\left(\frac{42\pi}{15}\right) + h[1]\cos\left(\frac{36\pi}{15}\right) + h[2]\cos\left(\frac{30\pi}{15}\right) + h[3]\cos\left(\frac{24\pi}{15}\right) + h[4]\cos\left(\frac{18\pi}{15}\right)$$

$$+ h[5]\cos\left(\frac{12\pi}{15}\right) + h[6]\cos\left(\frac{6\pi}{15}\right) + h[7]\Bigg\} = 1$$

$$h[7] + 2\Bigg\{h[0]\cos\left(\frac{56\pi}{15}\right) + h[1]\cos\left(\frac{48\pi}{15}\right) + h[2]\cos\left(\frac{40\pi}{15}\right) + h[3]\cos\left(\frac{32\pi}{15}\right) + h[4]\cos\left(\frac{24\pi}{15}\right)$$

$$+ h[5]\cos\left(\frac{16\pi}{15}\right) + h[6]\cos\left(\frac{8\pi}{15}\right) + h[7]\Bigg\} = 0.4$$

$$h[7] + 2\Bigg\{h[0]\cos\left(\frac{70\pi}{15}\right) + h[1]\cos\left(\frac{60\pi}{15}\right) + h[2]\cos\left(\frac{50\pi}{15}\right) + h[3]\cos\left(\frac{40\pi}{15}\right) + h[4]\cos\left(\frac{30\pi}{15}\right)$$

$$+ h[5]\cos\left(\frac{20\pi}{15}\right) + h[6]\cos\left(\frac{10\pi}{15}\right) + h[7]\Bigg\} = 0$$

$$h[7] + 2\Bigg\{h[0]\cos\left(\frac{84\pi}{15}\right) + h[1]\cos\left(\frac{72\pi}{15}\right) + h[2]\cos\left(\frac{60\pi}{15}\right) + h[3]\cos\left(\frac{48\pi}{15}\right) + h[4]\cos\left(\frac{36\pi}{15}\right)$$

$$+ h[5]\cos\left(\frac{24\pi}{15}\right) + h[6]\cos\left(\frac{12\pi}{15}\right) + h[7]\Bigg\} = 0$$

$$h[7] + 2\Bigg\{h[0]\cos\left(\frac{98\pi}{15}\right) + h[1]\cos\left(\frac{84\pi}{15}\right) + h[2]\cos\left(\frac{70\pi}{15}\right) + h[3]\cos\left(\frac{56\pi}{15}\right) + h[4]\cos\left(\frac{42\pi}{15}\right)$$

$$+ h[5]\cos\left(\frac{28\pi}{15}\right) + h[6]\cos\left(\frac{14\pi}{15}\right) + h[7]\Bigg\} = 0.$$

By substituting the values of the cosines, performing the operations and rearranging the terms, we end up with a system of eight equations with eight unknowns of the form

$$2h[0] + 2h[1] + 2h[2] + 2h[3] + 2h[4] + 2h[5] + 2h[6] + h[7] = 1.0$$

$$-1.956295h[0] - 1.618033h[1] - h[2] - 0.209056h[3] + 0.618033h[4] + 1.338261h[5] + 1.827090h[6] + h[7] = 1.0$$

$$1.827090h[0] + 0.618033h[1] - h[2] - 1.956295h[3] - 1.618033h[4] - 0.209056h[5] + 1.338261h[6] + h[7] = 1.0$$

$$-1.618033h[0] + 0.618033h[1] + 2h[2] + 0.618033h[3] - 1.618033h[4] - 1.618033h[5] + 0.618033h[6] + h[7] = 1.0$$

$$1.338261h[0] - 1.618033h[1] - 0.999999h[2] + 1.827090h[3] + 0.618033h[4] - 1.956295h[5] - 0.209056h[6] + h[7] = 0.4$$

$$-0.999999h[0] + 2h[1] - h[2] - 0.999999h[3] + 2h[4] - h[5] - h[6] + h[7] = 0.0$$

$$0.618033h[0] - 1.618033h[1] + 2h[2] - 1.618033h[3] + 0.618033h[4] + 0.618033h[5] - 1.618033h[6] + h[7] = 0.0$$

$$-0.209056h[0] + 0.618033h[1] - 0.999999h[2] + 1.338261h[3] - 1.618033h[4] + 1.827090h[5] - 1.956295h[6] + h[7] = 0.0.$$

The solution to the above system can be found with the help of specialized software applications and the solution that emerges is

$$h[0] = -0.014129, \quad h[1] = -0.001945, \quad h[2] = 0.040000, \quad h[3] = 0.012234$$

$$h[4] = -0.091388, \quad h[5] = -0.018089, \quad h[6] = 0.313317, \quad h[7] = 0.519999.$$

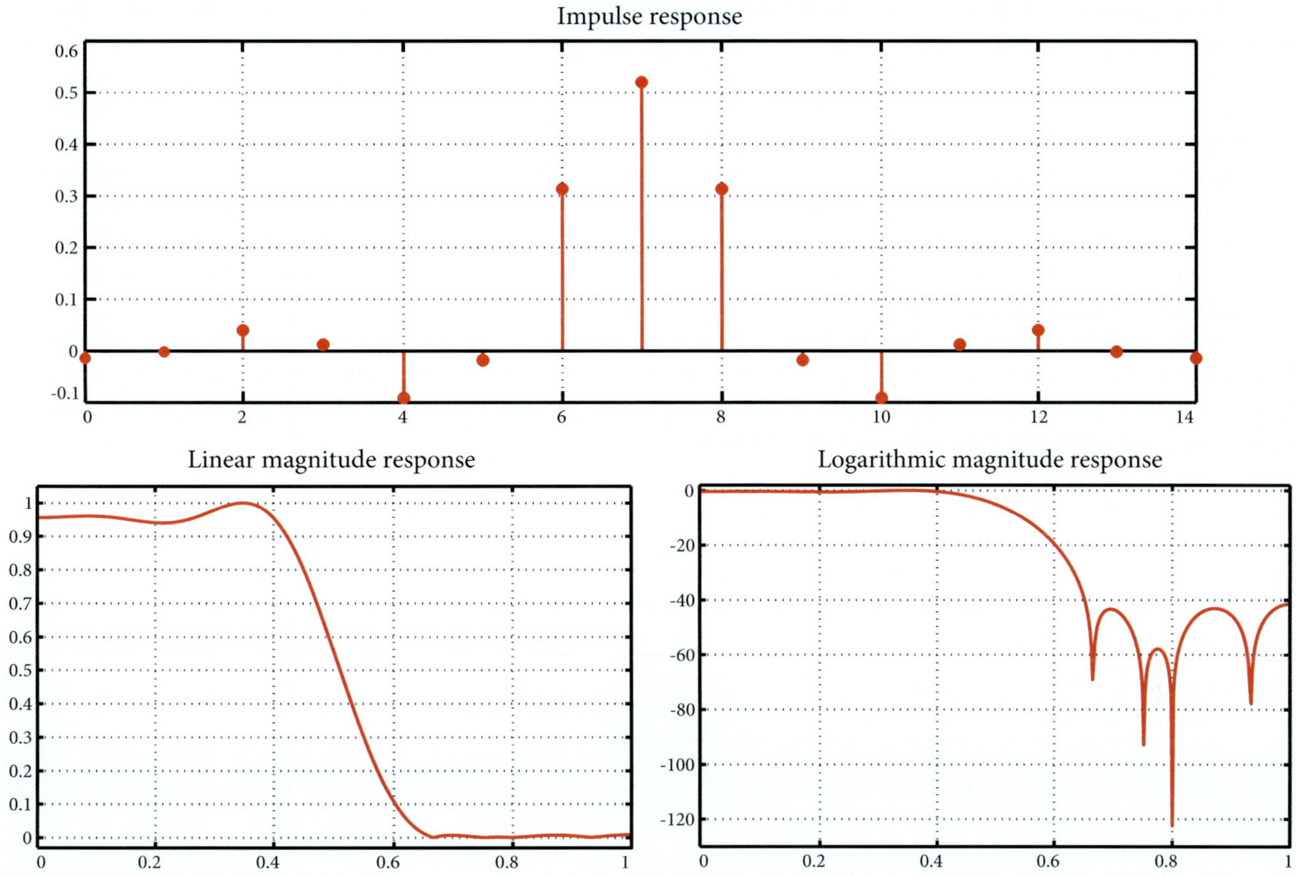

Fig. 12.64 The impulse response of Exercise 12.12.4, along with the measure of the calculated frequency response, designed on a linear and logarithmic scale

As for the remaining samples of the impulse response, these will be calculated from the above values, using the symmetry condition $h[n] = h[14 - n]$. The results are

$$h[08] = h[14 - 08] = h[6] = +0.313317 \qquad h[09] = h[14 - 09] = h[5] = -0.018089$$
$$h[10] = h[14 - 10] = h[4] = -0.091388 \qquad h[11] = h[14 - 11] = h[3] = +0.012234$$
$$h[12] = h[14 - 12] = h[2] = +0.040000 \qquad h[13] = h[14 - 13] = h[1] = -0.001945$$
$$h[14] = h[14 - 14] = h[0] = -0.014129.$$

After calculating the impulse response, we can calculate the desired response we seek through the calculation of the Fourier transform of the signal $h[n]$. If the result we get is not satisfactory, we should apply the process we presented above. The impulse response of this exercise, along with the measure of the calculated frequency response designed on a linear and logarithmic scale, are presented in the images (a), (b), and (c) of Fig. 12.64.

12.12.5 Optimal Equiripple Filters—The Remez Algorithm

The main disadvantage of the method of designing digital FIR filters with the help of the window function is *the lack of ability to precisely determine the pass and cutoff frequencies* ω_p and ω_s, since these depend on the *type of window* used and on the *order of the filter* N. In addition, the solution that is constructed *is not the best*, in the sense that the order of the resulting filter *is not the minimum possible*, which results in *an increase in the number of calculations* and *a reduction in efficiency and execution speed of the related algorithms*. An alternative method of designing FIR-type filters is based on the assumption that the approximation error of the ideal from the actual frequency response *is not related to specific frequencies, but is distributed across all frequencies that belong to both the passband and the stopband.* In addition, in contrast to the simple criterion of mean square error, in which all frequencies are considered *equivalent* and are associated with a statistical weight value *equal to one*, here, each frequency band is associated *with a generally different weight function* that also determines its contribution to the value of the final approximation error. This method is known as the *weighted Chebyshev method* and is based *on the description of the error function with the help of a linear combination of cosine functions*, which is then *minimized* with the help of an efficient multivariable optimization algorithm, which is known as the *Remez exchange algorithm*. Although this algorithm is characterized by high computational cost, nowadays this is not a problem, since modern computational systems are characterized by high computational power and low cost. It should be noted that this digital filter design algorithm of the FIR type is particularly efficient, because it allows us to define *exactly* the boundaries of the areas of interest and the relative sizes of the error function in each of them.

The result of applying this method to the design of FIR-type filters is the construction of *optimal* (in terms of their order) digital filters characterized by the appearance of fluctuations both in the passband and in the stopband. Limiting ourselves to the simple case of low-pass filters with cutoff and pass frequencies ω_s and ω_p respectively, the measure of the frequency-response function of the real filter will satisfy the relations

$$1-\delta_p \le |\mathcal{H}(e^{j\omega})| \le 1+\delta_p \quad \text{for } |\omega| \le \omega_P \qquad \text{and} \qquad -\delta_s \le |\mathcal{H}(e^{j\omega})| \le \delta_s \quad \text{for } |\omega| > \omega_s,$$

where δ_p and δ_s *are the amplitudes of the fluctuations of the amplitude response in the pass and stop regions*. This measure, as we have proven in Sect. 12.11 (see page 1223), is also given for the four categories of filters (types I, II, III, and IV) by the relation

$$|\mathcal{H}(e^{j\omega})| = \mathcal{Q}(\omega)\mathcal{P}(\omega) \tag{12.152}$$

with the form of the functions $\mathcal{Q}(\omega)$ and $\mathcal{P}(\omega)$ depending on the type of filter used in each case (see Eqs. 12.119 and 12.120). Taking into account that the function of the real filter $\mathcal{H}(e^{j\omega})$ is nothing more than an *approximation* of the function $\mathcal{H}_d(e^{j\omega})$ of the corresponding *ideal* filter, the design method we describe is based *on minimizing the approximation error* $\mathcal{E}(\omega)$ *of the ideal function* $\mathcal{H}_d(e^{j\omega})$ *from the actual function* $\mathcal{H}(e^{j\omega})$. Over time, several such error functions have been developed, such as the *mean absolute*, the *mean square* and the *maximum error*, which are given respectively by the equations

$$\mathcal{E}(\omega) = \frac{1}{2\pi}\int_{-\pi}^{+\pi} |\mathcal{H}_d(e^{j\omega}) - \mathcal{H}(e^{j\omega})|d\omega$$

$$\mathcal{E}(\omega) = \frac{1}{2\pi}\int_{-\pi}^{+\pi} |\mathcal{H}_d(e^{j\omega}) - \mathcal{H}(e^{j\omega})|^2\omega$$

$$\mathcal{E}(\omega) = \max_{\omega} |\mathcal{H}_d(e^{j\omega}) - \mathcal{H}(e^{j\omega})|$$

and which, as we mentioned, consider *equivalent* all the frequencies, as well as the *weighted mean square error* and the *weighted maximum error* that relate the various frequency areas with different *weight functions* $\mathcal{W}(\omega)$ and are defined as

$$\mathcal{E}(\omega) = \frac{1}{2\pi}\int_{-\pi}^{+\pi} \mathcal{W}(\omega)|\mathcal{H}_d(e^{j\omega}) - \mathcal{H}(e^{j\omega})|^2 d\omega \qquad \text{and} \qquad \mathcal{E}(\omega) = \max_{\omega} |\mathcal{W}(\omega)[\mathcal{H}_d(e^{j\omega}) - \mathcal{H}(e^{j\omega})]|$$

respectively. The method we present is based on the use of the *weighted maximum error*, which for a non-negative weight function $\mathcal{W}(\omega) \ge 0$ and using the properties of the frequency response and the relations we mentioned above can be formulated as follows:

$$\mathcal{E}(\omega) = \max_{\omega}\left\{\mathcal{W}(\omega)|\mathcal{H}_d(e^{j\omega}) - \mathcal{H}(e^{j\omega})|\right\} = \max_{\omega}\left\{\mathcal{W}(\omega)\left||\mathcal{H}_d(e^{j\omega})|e^{j\measuredangle\mathcal{H}_d(e^{j\omega})} - |\mathcal{H}(e^{j\omega})|e^{j\measuredangle\mathcal{H}(e^{j\omega})}\right|\right\}$$

$$= \max_{\omega}\left\{\mathcal{W}(\omega)\,\big|\,|\mathcal{H}_d(e^{j\omega})| - |\mathcal{H}(e^{j\omega})|\,\big|\right\} = \max_{\omega}\left\{\mathcal{W}(\omega)[|\mathcal{H}_d(e^{j\omega})| - \mathcal{Q}(\omega)\mathcal{P}(\omega)]\right\}$$

$$= \max_{\omega}\left\{\mathcal{W}(\omega)\mathcal{Q}(\omega)\left[\frac{|\mathcal{H}_d(e^{j\omega})|}{\mathcal{Q}(\omega)} - \mathcal{P}(\omega)\right]\right\}.$$

In the above equation, the function $|\mathcal{H}_d(e^{j\omega})|$, which expresses the measure of the *desired frequency response* has a value equal to the unit in the passband and zero value in the stopband and therefore is defined as

$$|\mathcal{H}_d(e^{j\omega})| = \begin{cases} 1\ 0 \le \omega \le \omega_p \\ 0\ \omega_s \le \omega \le \pi, \end{cases} \tag{12.153}$$

whereas the weight function $\mathcal{W}(\omega)$ is calculated as[53]

$$\mathcal{W}(\omega) = \begin{cases} 1 & 0 \le \omega \le \omega_p \\ \delta_p/\delta_s & \omega_s \le \omega \le \pi \end{cases} \tag{12.154}$$

Defining $\widetilde{\mathcal{W}}(\omega) = \mathcal{W}(\omega)\ \mathcal{Q}(\omega)$ and $|\widetilde{\mathcal{H}}_d(e^{j\omega})| = |\mathcal{H}_d(e^{j\omega})|/\mathcal{Q}(\omega)$ and using the equation

$$\mathcal{P}(\omega) = \sum_{k=0}^{L} \xi[k]\cos(\omega k) \tag{12.155}$$

the above relationship takes the form

---

[53] This relation arises from the definition of *the maximum value of the error function* in the passband and stopband

$$\mathcal{E}_{\max}(\omega) = \begin{cases} \delta_p \text{ for } 0 \le \omega \le \omega_p \\ \delta_s \text{ for } \omega_s \le \omega \le \pi \end{cases}$$

in combination with the relation $|\mathcal{E}(\omega)| \le |\delta_p|$. This relation holds for the case of *low-pass filters*, whereas for the three remaining categories of filters it is proven in the same way that

$$|\mathcal{H}_d(e^{j\omega})| = \begin{cases} 0 \text{ for } 0 \le \omega \le \omega_p \\ 1 \text{ for } \omega_s \le \omega \le \pi \end{cases} \quad \text{and} \quad \mathcal{W}(\omega) = \begin{cases} \delta_p/\delta_s & \text{for } 0 \le \omega \le \omega_p \\ 1 & \text{for } \omega_s \le \omega \le \pi \end{cases}$$

for the *high-pass* filters,

$$|\mathcal{H}_d(e^{j\omega})| = \begin{cases} 0 \text{ for } 0 \le \omega \le \omega_{s_1} \\ 1 \text{ for } \omega_{p_1} \le \omega \le \omega_{p_2} \\ 0 \text{ for } \omega_{s_2} \le \omega \le \pi \end{cases} \quad \text{and} \quad \mathcal{W}(\omega) = \begin{cases} \delta_p/\delta_s & \text{for } 0 \le \omega \le \omega_{s_1} \\ 1 & \text{for } \omega_{p_1} \le \omega \le \omega_{p_2} \\ \delta_p/\delta_s & \text{for } \omega_{s_2} \le \omega \le \pi \end{cases}$$

for the *band-pass* filters, and

$$|\mathcal{H}_d(e^{j\omega})| = \begin{cases} 1 \text{ for } 0 \le \omega \le \omega_{s_1} \\ 0 \text{ for } \omega_{p_1} \le \omega \le \omega_{p_2} \\ 1 \text{ for } \omega_{s_2} \le \omega \le \pi \end{cases} \quad \text{and} \quad \mathcal{W}(\omega) = \begin{cases} 1 & \text{for } 0 \le \omega \le \omega_{s_1} \\ \delta_p/\delta_s & \text{for } \omega_{p_1} \le \omega \le \omega_{p_2} \\ 1 & \text{for } \omega_{s_2} \le \omega \le \pi \end{cases}$$

for the *band-stop* filters.

$$\mathcal{E}(\omega) = \max_{\omega} \left\{ \widetilde{\mathcal{W}}(\omega) \left[|\widetilde{\mathcal{H}}_d(e^{j\omega})| - \sum_{k=0}^{L} \xi[k] \cos(\omega k) \right] \right\}.$$

Consequently, the process of designing the digital filter is reduced to determining the coefficients $\xi[k]$, which minimize the maximum absolute value of the measure $|\mathcal{E}(\omega)|$ calculated on the desired frequency range in each case. In mathematical writing, the above minimization problem is formulated as[54]

$$\min_{\boldsymbol{x}} \left[\max_{\omega} |E(\omega)| \right] = \min_{\boldsymbol{x}} \left\{ \max_{\omega} \left| \widetilde{\mathcal{W}}(\omega) \left(|\widetilde{\mathcal{H}}_d(e^{j\omega})| - \sum_{k=0}^{L} \xi[k] \cos(\omega k) \right) \right| \right\}, \tag{12.156}$$

where $\boldsymbol{x} = [\ \xi[0]\ \xi[1]\ \ldots\ \xi[L]\ ]^T$ and $\omega \in S$, with the set S resulting *from the union of all frequency regions for which the above minimization process*—in practice this set consists of the frequencies that belong to the *pass and cutoff regions* and therefore will be a genuine subset of the interval $[0, \pi]$, or in mathematical notation $S \subset [0, \pi]$. After calculating the coefficients $\xi[k]$ we can calculate the samples of the *impulse response* of the filter to be constructed using Eq. (12.143) and the definition relations of the coefficients that appear in it. Therefore, the problem we are dealing with is reduced to the process of calculating the samples of the impulse response that minimize the weighted maximum error $\max_{\omega} |\mathcal{W}(\omega)\ [\mathcal{H}_d(e^{j\omega}) - \mathcal{H}(e^{j\omega})]|$.

The above process for the case of approximating an *ideal* frequency response, characterized by the existence of *three frequency bands of interest*, is presented in Fig. 12.65. In image (a) of the figure the magnitude of the desired frequency response $|\mathcal{H}_d(e^{j\omega})|$ is depicted, which has a value of 1 in the first frequency band, a value of 0 in the second band, and a value

[54] The above problem is nothing more than a special case of the more general problem of *approximating a random function $f(x)$ by polynomials of the Chebyshev type*. We remember from the basic theory that, given a set of N linearly independent functions $\varphi_i(x)$ $(i = 1, 2, \ldots, N)$ defined on a closed interval $[\alpha, \beta]$, each of which has at most $N - 1$ zero values in the above interval (these functions constitute the so-called *Chebyshev system*) the *Chebyshev approximation* (or *minmax approximation*) of any function is defined as the function

$$P_N(x) = c_0\varphi_0(x) + c_1\varphi_1(x) + \cdots + c_N\varphi_N(x) = \sum_{i=0}^{N} c_i\varphi_i(x),$$

which *minimizes the maximum error*

$$\varepsilon = \max_{\alpha \le x \le b} |f(x) - P_N(x)|.$$

The most classic approximation method is the *polynomial approximation of degree N*, which is based on the use of functions $\varphi_i(x) = x^i$ $(i = 1, 2, \ldots, N)$, whereas if $\varphi_i(x) = T_i(x)$ $(i = 1, 2, \ldots, N)$ (where $T_i(x)$ is the *Chebyshev polynomial of the first kind of order i*) we have the *Chebyshev approximation*. It is proven that for smoothly behaving functions, the optimal polynomial leads to an error function that oscillates between two values $+\varepsilon$ and $-\varepsilon$, with the number of oscillations being equal to $N + 2$, where N is the degree of the polynomial $P_N(x)$. In the Chebyshev function approximation the polynomial $P_N(x)$ is determined as we have mentioned, developing the function *in a series of Chebyshev polynomials* and then keeping only those of the appropriate degree. The Chebyshev polynomials $T_N(x)$ are nothing but orthogonal polynomials (see also the homonymous section of the mathematical appendix) that are solutions of the differential Chebyshev equation

$$(1 - x^2)\frac{d^2y(x)}{dx^2} - x\frac{dy(x)}{dx} + \eta^2 y(x) = 0$$

and constitute a *complete set of eigenfunctions*. Therefore, every function $f(x)$ can be written as the *linear combination*

$$f(x) \sim \sum_{n=1}^{\infty} c_n T_n(x) \qquad \text{where} \qquad T_N(x) = \begin{cases} \cos(N\cos^{-1}(x)) & -1 < x < +1 \\ \cosh(N\cosh^{-1}(x)) & x \ge +1 \\ (-1)^N \cosh(N\cosh^{-1}(x)) & x \le -1. \end{cases}$$

After expanding the function in the above way, we can keep only the appropriate number of terms, so as to retain only the polynomials characterized by the desired degree N in each case. It is proven that *for functions whose series expansion converges very quickly, the total error that results from truncating the above expansion up to the polynomial $T_m(x)$ is almost equal to the term that is immediately after the last term that has been kept in this truncation process and therefore is proportional to $T_{m+1}(x)$*. This fact makes the Chebyshev polynomials *ideal* for approximating functions of any form, such as in this case the frequency response function of the ideal filter we wish to approximate.

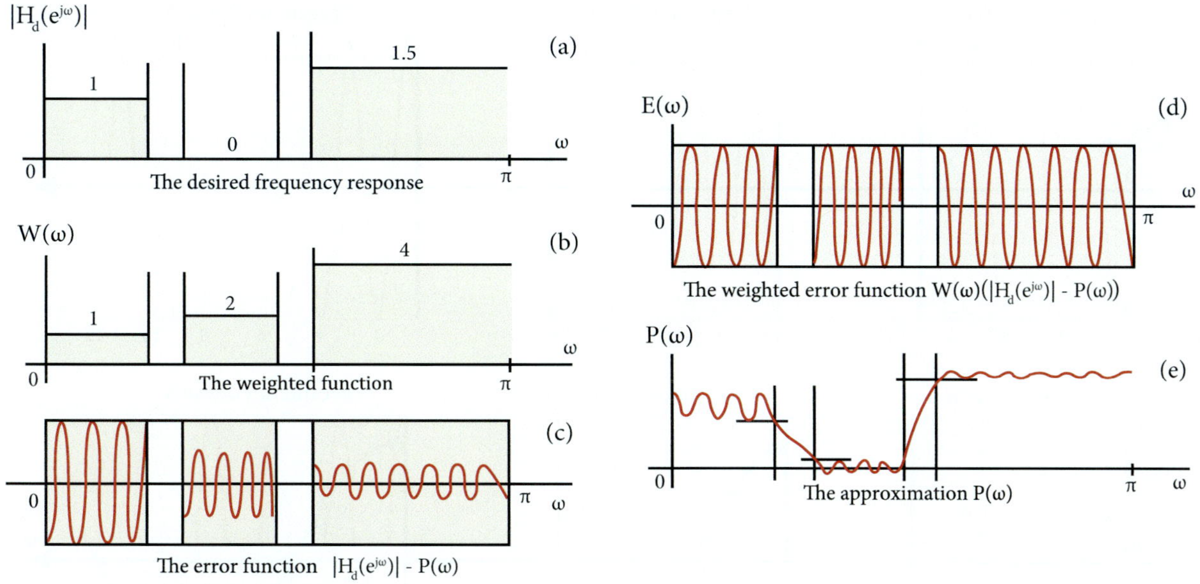

Fig. 12.65 Approach of ideal frequency response with weighted error function

of 1.5 in the third band, whereas in image (b) the corresponding *weight function* $\mathcal{W}(\omega)$ is depicted, which assigns different weight values to these three frequency bands. In images (c) and (d) the functions $|\mathcal{H}_d(e^{j\omega})| - \mathcal{P}(\omega)$ and $\mathcal{W}(\omega)[|\mathcal{H}_d(e^{j\omega})| - \mathcal{P}(\omega)]$ are presented, whereas in image (e) the approximation $\mathcal{P}(\omega)$ of the ideal function $|\mathcal{H}_d(e^{j\omega})|$ is shown. The conditions that must be met in order for the *minimization* of the maximum error $\mathcal{E}(\omega)$ to be possible are dictated by the *alternation theorem* presented in the next section.

The Alternation Theorem

The *alternation theorem* allows the determination of a set of conditions that must be met, so that the filter to be constructed is *the best possible* from the perspective of *minimizing the maximum error* and therefore is used for the calculation of the coefficients $\xi[k]$, which minimize the *maximum absolute error* $|\mathcal{E}(\omega)|$. This theorem is formulated in the following way.

Theorem 12.12.1 *Let S be a compact subset of the interval* $[0, \pi]$ *resulting from the union of a number of smaller closed intervals expressing the frequency bands of interest and within which the* desired frequency response $|\widetilde{\mathcal{H}}_d(e^{j\omega})|$ *and the* positive weight function $\mathcal{W}(\omega)$ *are defined. In this case, a sufficient and necessary condition for the polynomial*

$$\mathcal{P}(\omega) = \sum_{k=0}^{L} \xi[k]\cos(\omega k) \tag{12.157}$$

the unique optimal weighted approximation by Chebyshev *of the function* $|\widetilde{\mathcal{H}}_d(e^{j\omega})|$ *in the set S is for the error function* $\mathcal{E}(\omega) = \widetilde{\mathcal{W}}(\omega)[|\widetilde{\mathcal{H}}_d(e^{j\omega})| - \mathcal{P}(\omega)]$ *to have* at least $L+2$ extreme frequency values *in the set S. In other words, there should be at least* $L+2$ *frequencies* $\omega_i \in S$ *such that*

- $\omega_1 < \omega_2 < \cdots < \omega_L < \omega_{L+1} < \omega_{L+2}$
- $\mathcal{E}(\omega_i) = -\mathcal{E}(\omega_{i+1})$ $(i = 1, 2, \ldots, L+2)$*, that is* $\mathcal{E}(\omega_1) = -\mathcal{E}(\omega_2) = \mathcal{E}(\omega_3) = -\mathcal{E}(\omega_4) = \ldots$
- $|\mathcal{E}(\omega_i)| = \max_{\omega\in S} |\mathcal{E}(\omega)|, \quad i = 1, 2, \ldots, L+2.$

The name of this theorem is given because of the sign of the function $\mathcal{E}(\omega)$ alternates *between two successive extreme frequencies.*

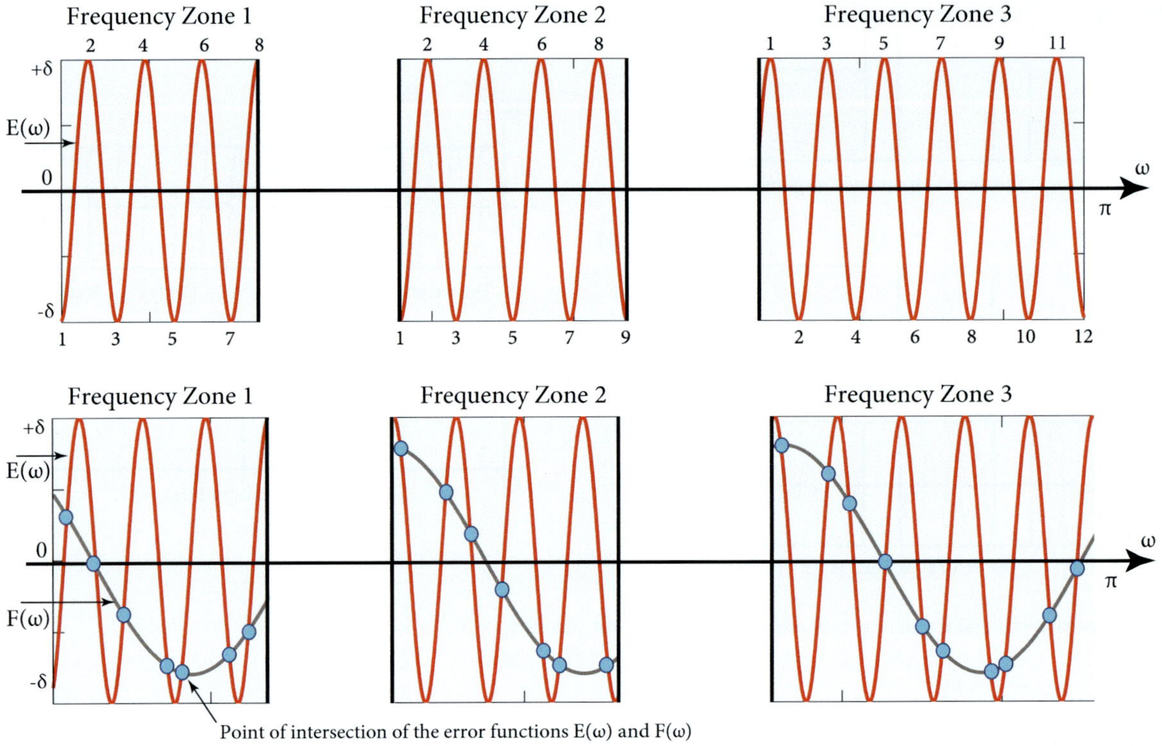

Fig. 12.66 Examples of error functions that fit exactly in the shaded areas corresponding to the frequency zones of interest

Proof For the proof of the theorem[55] we accept the *existence* of the polynomial $\mathcal{P}(\omega)$[56] and we will show that the conditions of the theorem lead to the characterization of $\mathcal{P}(\omega)$ as the *unique optimal solution*. Let us consider, then, that this polynomial constitutes the *weighted equiripple solution* with at least $L+2$ *extreme* frequency values. Consequently, the weighted error function $\mathcal{E}(\omega)$ is equiripple and fits exactly in the shaded areas of image (a) of Fig. 12.66, with each area expressing a *desired passband or stopband*. Assuming that there are K such zones of interest in the range of values $[0, \pi]$, the total number of extreme frequency values for the entire above area is equal to

$$N_1 + N_2 + \cdots + N_i + \cdots + N_K = \sum_{i=1}^{K} N_i,$$

where N_i is the number of extreme frequency values in zone number i. For the example of this specific figure, it is not difficult to ascertain that $N_1 = 8$, $N_2 = 9$ and $N_3 = 14$.

Assuming that the necessary condition dictated by the theorem is satisfied, let us say that the function $\mathcal{E}(\omega)$ has at least $L+2$ extreme frequency values, so

$$\sum_{i=1}^{K} N_i \geq L + 2.$$

Let us now consider a second approximation $\mathcal{R}(\omega)$ of the desired function "$|\widetilde{\mathcal{H}}_d(e^{j\omega})|$, which is given by the relation

$$\mathcal{R}(\omega) = \sum_{k=0}^{L} \zeta[k] \cos(\omega k)$$

[55] See the reference Vaidyanathan P. & Nguyen T., *A Simple Proof of the Alternation Theorem* in *Conference Record of the* 41th *Asilomar Conference on Signals, Systems and Computers*, IEEE, Piscataway, NJ, pages 1111–1115, ISBN 978-1-4244-2109-1, 2007.

[56] For the proof of the existence of the polynomial $\mathcal{P}(\omega)$ see the reference Rice J.R., *The Approximation of Functions*, Volume 1, Linear Theory, ISBN 978-0201064308, Addison-Wesley Publishing Company, Reading, MA, 1964.

and for which the maximum error is *at least as small* as that associated with the previous approximation. In this case, the corresponding error function will have the form

$$\mathcal{F}(\omega) = \mathcal{W}(\omega)\left[|\widetilde{\mathcal{H}}_d(e^{j\omega})| - \sum_{k=0}^{L} \zeta[k]\cos(\omega k)\right], \tag{12.158}$$

whereas, according to the previous assumption, $\max_{\omega \in S} |\mathcal{F}(\omega)| \leq \max_{\omega \in S} |\mathcal{E}(\omega)|$. But this means that the curve of the error function $\mathcal{F}(\omega)$ also exactly matches the shaded areas corresponding to the zones of interest, as shown in image (b) of Fig. 12.66.

From Fig. 12.66β we find that the curves of the functions $\mathcal{E}(\omega)$ and $\mathcal{F}(\omega)$ *intersect each other*. Observing that the curve $\mathcal{E}(\omega)$ *oscillates* between the values δ and $-\delta$, we can observe that within the zone of interest i the curve of $\mathcal{F}(\omega)$ intersects the curve of $\mathcal{E}(\omega)$ at $N_i - 1$ points and therefore the difference

$$\begin{aligned}\Delta\mathcal{E}(\omega) = \mathcal{E}(\omega) - \mathcal{F}(\omega) &= \mathcal{W}(\omega)\left[|\widetilde{\mathcal{H}}_d(e^{j\omega})| - \sum_{k=0}^{L} \xi[k]\cos(\omega k)\right] - \mathcal{W}(\omega)\left[|\widetilde{\mathcal{H}}_d(e^{j\omega})| - \sum_{k=0}^{L} \zeta[k]\cos(\omega k)\right] \\ &= W(\omega)\sum_{k=0}^{L}(\zeta[k] - \xi[k])\cos(\omega k)\end{aligned}$$

It will have at least $N_i - 1$ zero values in the i-numbered zone, whereas the number N_z of zero values of the function $\Delta\mathcal{E}(\omega)$ for the entire value range $[0, \pi]$ will satisfy the relation

$$N_z \geq (N_1 - 1) + (N_2 - 1) + \cdots + (N_i - 1) + \cdots + (N_K - 1) \equiv \sum_{i=1}^{K} N_i - K.$$

Using the known trigonometric expression

$$\cos(\omega k) = \sum_{n=0}^{k} \beta_{nk}(\cos\omega)^n$$

the function $\Delta\mathcal{E}(\omega)$ is formulated as

$$\begin{aligned}\Delta\mathcal{E}(\omega) &= \mathcal{W}(\omega)\sum_{k=0}^{L}(\zeta[k] - \xi[k])\cos(\omega k) = \mathcal{W}(\omega)\sum_{k=0}^{L}(\zeta[k] - \xi[k])\sum_{n=0}^{k}\beta_{nk}(\cos\omega)^n \\ &= \mathcal{W}(\omega)\sum_{k=0}^{L} p[k](\cos\omega)^k = \mathcal{W}(\omega)\mathcal{A}(\omega),\end{aligned}$$

where

$$p[k] = (\zeta[k] - \xi[k])\sum_{n=0}^{k}\beta_{nk} \quad \text{and} \quad \mathcal{A}(\omega) = \sum_{k=0}^{L} p[k](\cos\omega)^k.$$

Although the error function $\mathcal{W}(\omega)$ is not defined in the entire frequency range $[0, \pi]$ but only *in the union of the frequency zones of interest*, we can assume that in the areas where this function is not defined it has some arbitrary positive value, so that finally the product $\mathcal{W}(\omega)\mathcal{A}(\omega)$ is defined in the entire range $[0, \pi]$. Since now the weight function $\mathcal{W}(\omega)$ is positive $[\mathcal{W}(\omega) > 0)]$, we can formulate the claim that *the zero values of the function $\Delta\mathcal{E}(\omega)$ come entirely from the function $\mathcal{A}(\omega)$* and therefore the parameter N_z that we defined above expresses nothing but *the number of zero values of the function $\mathcal{A}(\omega)$*. From image (b) of Fig. 12.66 we also observe that the difference function $\Delta\mathcal{E}(\omega)$ is characterized by a *change in sign* as we move from the last extreme frequency of a zone to the first extreme frequency of the zone of interest immediately following and therefore, for the frequencies ω_{e_1} and ω_{e_2} depicted in the figure $\Delta\mathcal{E}(\omega_{e_1}) \geq 0$ and $\Delta\mathcal{E}(\omega_{e_2}) \leq 0$.

The power of the above property regarding the alternation in the sign of the function $\Delta\mathcal{E}(\omega)$ allows us to formulate the conclusion that this function *becomes zero* when the frequency increases from ω_{e_1} to ω_{e_2}. Consequently, the function $\Delta\mathcal{E}(\omega)$

becomes zero not only within the K zones of interest *but also within the frequency areas that are interposed between these zones* and are defined by the last extreme of one zone and the first extreme of the next zone. For the case of K zones of interest there are $K - 1$ such zeros in the intermediate areas and therefore the total number of zero values of the function $\mathcal{A}(\omega)$ in the range of values $[0, \pi]$ will satisfy the condition

$$N_z \geq \sum_{i=1}^{K} N_k - K + (K-1) = \sum_{i=1}^{K} N_k - 1 \geq L + 1 \tag{12.159}$$

with the last inequality resulting from the assumption that in the above range of values, there are $L+2$ extreme values. But on the other hand, we know that a polynomial of the form $\sum_{k=0}^{K} p[k]\, x^k$ has L zero values. Therefore, if we set $x = \cos\omega$ and observe that the frequency ω is characterized by a *monotonic reduction* in the interval $[0, \pi]$, we conclude that the function $\mathcal{A}(\omega)$ *cannot have more than L zero values in this interval*, a conclusion that contradicts what we proved above. Therefore, the above relationship cannot hold, unless we assume that the function $\mathcal{A}(\omega)$ is *identically equal to zero*. But for this to happen, it must be $\zeta[k] = \xi[k]$ and therefore $\mathcal{R}(\omega) = \mathcal{P}(\omega)$, that is, the function $\mathcal{P}(\omega)$ constitutes *the unique solution that minimizes the maximum weighted error* $\mathcal{E}(\omega)$.

The above proof concerns the validity of the necessary condition, whereas for the proof of the sufficient condition, according to which *if the function* $\mathcal{P}(\omega)$ *constitutes the optimal approximation of the desired frequency response* $|\widetilde{\mathcal{H}}_d(e^{j\omega})|$, *then the error function has* $L+2$ *extreme frequency values*, the reader is referred to the literature.[57]

According to the *alternation theorem*, the coefficients $\xi[k]$ of Eq. (12.153) will correspond to a filter that constitutes *the optimal approximation according to Chebyshev of an ideal low-pass filter* with a variance ratio $\delta_p\delta_s = K_1/K_2 = K$ and with passband and cutoff limits ω_p and ω_s, *if and only if the error function* $\mathcal{E}(\omega)$ *displays at least* $L+2$ *alternations within the closed interval S that results from the union of all zones of interest*, that is, if and only if the aforementioned function takes at its $L+2$ extremal frequencies its maximum value but with alternating signs. This situation for a parameter value $L = 7$ is shown in Fig. 12.67. From this figure, we see that in the set of $L+2$ extremal frequencies, the frequencies ω_p and ω_s are included, with the value of the function $|\widetilde{\mathcal{H}}_d(e^{j\omega})|$ at their positions being equal to $1 - \delta_p$ and δ_s respectively, a property that holds in all cases. In order to prove the truth of this assertion, let us assume that the frequency ω_p has a slightly lower value than the one shown in the figure (it is, for example, between the frequencies ω_2 and ω_3), so that the value of $|\widetilde{\mathcal{H}}_d(e^{j\omega})|$ at the position $\omega = \omega_p$ *does not constitute an extremum of the error function*. It is not difficult to see that in this case, the curve would display *two successive extrema* at the positions ω_2 and ω_s with the same sign and consequently the number of extrema associated with sign changes would be less than $L+2$, resulting in the violation of the alternation theorem. The same argument can be applied in the case where the frequency that deviates from the reference set is ω_s. The above also applies to high-pass filters but not to band-pass filters and filters with multiple zones of interest, in which the situation may be different.

In order to understand how this theorem is applied to the design of digital filters, let us consider the case of a *low-pass filter* with a passband $0 \leq \omega \leq \omega_p$ and a stopband $\omega_s \leq \omega \leq \pi$. By differentiating the defining equation of the approximation error $\mathcal{E}(\omega) = \mathcal{W}(\omega)[|\mathcal{H}_d(e^{j\omega})| - |\mathcal{H}(e^{j\omega})|]$, we get

$$\frac{d\mathcal{E}(\omega)}{d\omega} = \frac{d\mathcal{W}(\omega)}{d\omega} \times \left[|\mathcal{H}_d(e^{j\omega})| - |\mathcal{H}(e^{j\omega})|\right] + \mathcal{W}(\omega) \times \left[\frac{d|\mathcal{H}_d(e^{j\omega})|}{d\omega} - \frac{d|\mathcal{H}(e^{j\omega})|}{d\omega}\right]. \tag{12.160}$$

However, since the functions $\mathcal{W}(\omega)$ and $|\mathcal{H}_d(e^{j\omega})|$ are *piecewise constant* in their domain (see images (a) and (b) of Fig. 12.65)

$$d\mathcal{W}(\omega)/d\omega = d|\mathcal{H}_d(e^{j\omega})|/d\omega = 0.$$

Furthermore, since we are only interested in frequency values ω_i at which the error function $\mathcal{E}(\omega)$ exhibits *extremes* (maximum or minimum), we should take into account that at these frequency positions, $\dot{\mathcal{E}}(\omega_i) = 0$. Substituting the above relations into the last equation, it will take the form

[57] See Cheney E.W., *Introduction to Approximation Theory*, AMS Chelsea Publishing, ISBN 0-8218-1374-9, 1966, pages 75–161.

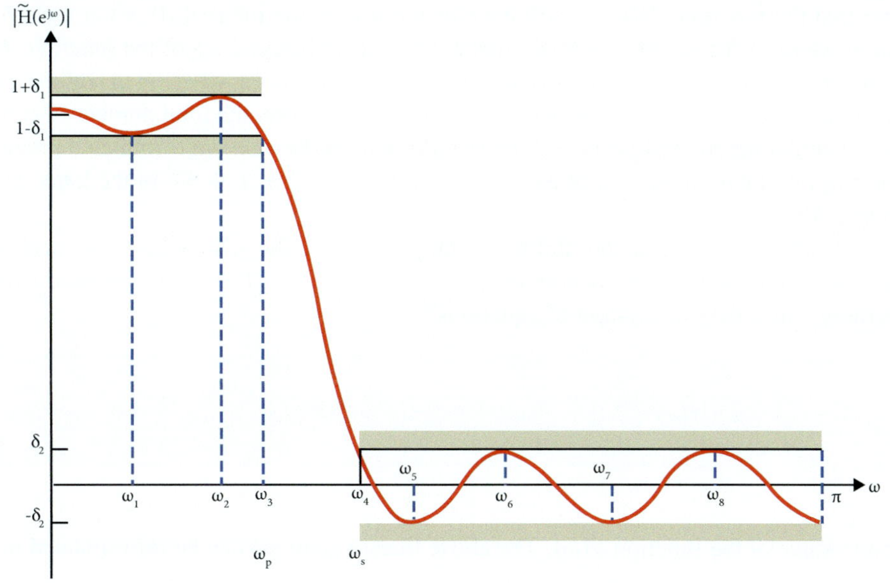

Fig. 12.67 Demonstration of the consequences of the *alternation theorem* for the optimal approximation according to Chebyshev of an ideal low-pass filter and for a parameter value of $L = 7$

$$dE(\omega)/d\omega = -d|\mathcal{H}(e^{j\omega})|/d\omega = 0,$$

from which we conclude that *the frequencies* ω_i *at which the error function* $\mathcal{E}(\omega)$ *exhibits maximum or minimum values correspond to extreme values at which the frequency response* $|\mathcal{H}(e^{j\omega})|$ *of the actual filter meets the tolerance in the error that has been set*. If we now assume that the considered low-pass filter is a digital FIR filter of type I, its frequency response will be, as known, a polynomial function of degree L of the form[58]

$$\begin{aligned}|\mathcal{H}(e^{j\omega})| &= \sum_{k=0}^{L}\alpha[k]\cos(\omega k) = \sum_{k=0}^{L}\alpha[k]\left[\sum_{n=0}^{k}\beta_{nk}(\cos\omega)^n\right]\\ &= \sum_{k=0}^{L}\alpha[k]\left(\beta_{0k}+\beta_{1k}\cos\omega+\beta_{2k}(\cos\omega)^2+\beta_{3k}(\cos\omega)^3+\cdots+\right) = \sum_{k=0}^{L}\alpha'[k](\cos\omega)^k, \end{aligned} \quad (12.161)$$

(where $\alpha'[k] = \alpha[k]\,\beta_{nk}$), with derivative at the positions of the extreme frequencies

$$\frac{d|\mathcal{H}(e^{j\omega})|}{d\omega} = \frac{d}{d\omega}\left[\sum_{k=0}^{L}\alpha'[k](\cos\omega)^k\right] = -\sin\omega\sum_{k=0}^{L}k\alpha'[k](\cos\omega)^{k-1} = -\sin\omega\sum_{k=0}^{L-1}(k+1)\,\alpha'_{k+1}(\cos\omega)^k = 0$$

(for the derivation of the last expression we adjusted the index of the sum in the way we did in previous cases—see the Footnote 51 on page 1248).

From the above relation it is not difficult to see that the zeroing of the expression requires either $\sin\omega = 0$ (from where it follows that the function $|\mathcal{H}(e^{j\omega})|$ has two extreme frequencies at the positions $\omega = 0$ and $\omega = \pi$) or

$$\sum_{k=0}^{L-1}(k+1)\,\alpha'_{k+1}(\cos\omega)^k = 0$$

[58] For the derivation of this relation we replaced the function $\cos(\omega k)$ from its equivalent formulation as a linear combination of powers of $\cos\omega$ (see the proof of the alternation theorem).

(from where it follows that the function $|\mathcal{H}(e^{j\omega})|$ will have in addition in the interval $(0, \pi)$ at most $L-1$ *local extremes*, which can be *maxima* or *minima*). Therefore, the total number of extreme frequencies of the function $|\mathcal{H}(e^{j\omega})|$ will be equal to $L+1$. As for the error function $\mathcal{E}(\omega)$, this will also have $L+1$ extreme values, whereas in addition, the positions $\omega = \omega_s$ and $\omega = \omega_p$ also constitute extreme values, as we proved previously. Therefore, the total number of extremes of the function $\mathcal{E}(\omega)$ is equal to $L+3$. Combining the last property with the *alternation theorem* we formulated previously,[59] we conclude that the error function $\mathcal{E}(\omega)$ will have a number of extremals equal to $L+2$ or $L+3$—in the latter case the filter is known as the *maximum variance filter*.

Using the *alternation theorem* we have the ability to uniquely solve the *Chebyshev optimization problem* that was formulated in the previous paragraphs. If we denote by $\{\omega_n\}$ $(n = 0, 1, \ldots, L+1)$ the desired extremal frequencies of the digital filter, then these will satisfy the system of equations[60]

$$\widetilde{\mathcal{W}}(\omega_n)[|\widetilde{\mathcal{H}}_d(e^{j\omega_n})| - \mathcal{P}(\omega_n)] = (-1)^n\delta \quad (n = 0, 1, \ldots, L+1),$$

where δ is the maximum value of the function $\mathcal{E}(\omega)$. The above linear equations can be reformulated in the form

$$|\widetilde{\mathcal{H}}_d(e^{j\omega_n})| = \mathcal{P}(\omega_n) + \frac{(-1)^n\delta}{\widetilde{\mathcal{W}}(\omega_n)} = \sum_{k=0}^{L} \alpha[k]\cos(\omega_n k) + \frac{(-1)^n\delta}{\widetilde{\mathcal{W}}(\omega_n)}$$

$(n = 0, 1, \ldots, L+1)$—for the derivation of the last relation the defining equation of the function $\mathcal{P}(\omega)$ was used—or in matrix form[61] $\boldsymbol{AX} = \boldsymbol{B}$, with the matrices $\boldsymbol{A}$, $\boldsymbol{X}$ and $\boldsymbol{B}$ defined as

$$\boldsymbol{A} = \begin{bmatrix} 1 & \cos(\omega_0) & \cos(2\omega_0) & \ldots \cos(L\omega_0) & \dfrac{1}{\widetilde{\mathcal{W}}(\omega_0)} \\ 1 & \cos(\omega_1) & \cos(2\omega_1) & \ldots \cos(L\omega_1) & \dfrac{1}{\widetilde{\mathcal{W}}(\omega_1)} \\ 1 & \cos(\omega_2) & \cos(2\omega_2) & \ldots \cos(L\omega_2) & \dfrac{1}{\widetilde{\mathcal{W}}(\omega_2)} \\ \ldots\ldots & \ldots & \ldots\ldots & \ldots\ldots\ldots \\ 1 & \cos(\omega_L) & \cos(2\omega_L) & \ldots \cos(L\omega_L) & \dfrac{1}{\widetilde{\mathcal{W}}(\omega_L)} \\ 1 & \cos(\omega_{L+1}) & \cos(2\omega_{L+1}) & \ldots \cos(L\omega_{L+1}) & \dfrac{1}{\widetilde{\mathcal{W}}(\omega_{L+1})} \end{bmatrix} \quad \boldsymbol{X} = \begin{bmatrix} \alpha(0) \\ \alpha(1) \\ \alpha(2) \\ \ldots \\ \alpha(L) \\ \delta \end{bmatrix} \quad \boldsymbol{B} = \begin{bmatrix} |\widetilde{\mathcal{H}}_d(e^{j\omega_0})| \\ |\widetilde{\mathcal{H}}_d(e^{j\omega_1})| \\ |\widetilde{\mathcal{H}}_d(e^{j\omega_2})| \\ \ldots \\ |\widetilde{\mathcal{H}}_d(e^{j\omega_L})| \\ |\widetilde{\mathcal{H}}_d(e^{j\omega_{L+1}})|. \end{bmatrix}$$

By solving the above system, we can determine the coefficients $\{\alpha_k\}$ $(k = 0, 1, \ldots, L)$ and finally design the digital filter. Indeed, after calculating the values of the coefficients $\alpha[k]$ from the relation $\boldsymbol{X} = \boldsymbol{A}^{-1}\boldsymbol{B}$—it is proven that in all cases the matrix $\boldsymbol{A}$ of dimensions $(L+2) \times (L+2)$ *is not singular* and therefore its inverse exists—we can calculate the values of the

[59] The alternation theorem ensures that the number of extremal frequency values is *at least equal* to $L+2$ but does not exclude the appearance of more extremals, as is the case here, where their number is equal to $L+3$.

[60] The existence of the term $(-1)^n\delta$ in the second member of this relationship justifies the characterization of the filters that result from this method as *equiripple*, since the error values at the positions of the optimal extremal frequencies are characterized by *the same magnitude, but alternating signs*.

[61] In reality, this matrix equation constitutes an interpolation problem, in which $L+1$ coefficients $\alpha[k]$ and $L+2$ interpolation points appear, something that does not create a problem, however, since the parameter δ also constitutes a variable. It is proven that this system has a unique solution under the condition that *the frequencies ω_k are distinct from each other* and in reality the problem of finding the filter coefficients that minimize the maximum error is reduced to the problem of determining the appropriate number of points that will be used in this interpolation problem.

impulse response $h[k]$ and finally the function of the frequency response $\mathcal{H}(e^{j\omega})$ from the Fourier transform of the discrete-time signal of the impulse response. It should be noted that because initially we know neither the extreme frequencies $\{\omega_n\}$ nor the filter parameters $\alpha[k]$ and the parameter δ, the above system is solved with the help of an *iterative method* known as the *Remez exchange algorithm*[62] and which is applied as follows: starting from an initial set of values for the extreme frequencies, we determine the function $\mathcal{P}(\omega)$ and the parameter δ, and from there the value of the error function $\mathcal{E}(\omega)$. In the next step of the process, using the value of this function, we calculate the new values of the $L+2$ extreme frequencies and apply the process iteratively *until finally the sequences of these frequencies converge to their optimal value*. Although this process can solve the problem, it is characterized by very poor performance, since the matrix inversion process is of order $O(L^3)$, whereas the Fourier transform that is used in the process of calculating the function $\mathcal{H}(e^{j\omega})$ is of order $N\log_2 N$, where N is the order of the digital filter.

The Remez Algorithm

The *Remez algorithm* is defined as the next *iterative process of many variables*:

Step 1 We start from an initial set of $L+1$ *extreme frequencies* ω_i $(i=1,2,\ldots,L+1)$ having ensured that *such a frequency corresponds in each zone of interest* . The simplest way to initialize these frequencies is to assume that they *are distributed uniformly* in the frequency bands of interest. Assuming that there are K such zones that are defined by a minimum frequency ω_i^ℓ and a maximum frequency ω_i^h $(i=1,2,\ldots,K)$, leading to a bandwidth value $B_i=\omega_i^h-\omega_i^\ell$ $(i=1,2,\ldots,K)$ for zone number i, and that the frequencies ω_i^ℓ and ω_i^h are located at the *left* and *right* end of each zone respectively, then the *total bandwidth of interest* $B=B_1+B_2+\cdots+B_K$ should be divided into $L+2-K$ intervals and therefore the average distance between two successive extreme frequencies will be equal to

$$\Delta\omega=\frac{B}{L+2-K}=\frac{B_1+B_2+\cdots+B_K}{L+2-K}=\frac{1}{L+2-K}\sum_{i=1}^{K}B_i=\frac{1}{L+2-K}\sum_{i=1}^{K}\left(\omega_i^h-\omega_i^\ell\right). \tag{12.162}$$

Therefore, zone number i with bandwidth B_i will be divided into

$$\frac{B_i}{\Delta\omega}=\frac{\Delta\omega_i}{\Delta\omega}=\frac{\omega_i^h-\omega_i^\ell}{\Delta\omega} \tag{12.163}$$

equal intervals. Note that because in the general case the value of the quantity $B_i/\Delta\omega$ is not an integer but a decimal number, *rounding to the nearest integer is required*, a process that leads to the calculation of the integer number of intervals for zone number i

$$m_i=\left[\left(\frac{B_i}{\Delta\omega}\right)+0.5\right]=\left[\left(\frac{\omega_i^h-\omega_i^\ell}{\Delta\omega}\right)+0.5\right]\quad(i=1,2,\ldots,K), \tag{12.164}$$

where the operator $[x]$ returns the *integer part* of the decimal quantity x. The above equation allows the calculation of the number of intervals for all the zones of interest *except the last one* (i.e., the zone numbered K), for which the value of the number of intervals m_K is calculated as

[62] In reality, the Remez algorithm constitutes *a family of iterative multivariable approximation algorithms* of any rational function (see Remez, E.Y. General Computational Methods of Chebyshev Approximation in Problems with Linear Real Parameters, AEC-tr-4491, Books 1 and 2, English translation by US AEC.) and the application of it in the optimization processes of digital filters is based *on the second optimization method of Remez*, which was implemented in a computer program by *Parks and McClellan*. A detailed description of the Remez algorithm can also be found in the work Ralston, A. Rational Chebyshev Approximation by Remez's Algorithms, Numer. Math. 7 (1965) pp. 322–330.

$$m_K = L + 1 - \sum_{i=1}^{K-1} (m_j + 1). \tag{12.165}$$

After calculating the parameters m_j $(j = 1, 2, \ldots, K)$, we can recalculate the distance of the initial extreme frequency values for each zone of interest from the relation

$$\delta\omega_i = \frac{B_i}{m_i} \quad (i = 1, 2, \ldots, K). \tag{12.166}$$

Step 2 Starting from the set of $L + 2$ frequencies $\{\omega_0, \omega_1, \ldots, \omega_L, \omega_{L+1}\}$, we determine analytically the value of the parameter δ from the relation

$$\delta = \frac{\gamma_0|\widetilde{\mathcal{H}}_d(e^{j\omega_0})| + \gamma_1|\widetilde{\mathcal{H}}_d(e^{j\omega_1})| + \cdots + \gamma_L|\widetilde{\mathcal{H}}_d(e^{j\omega_L})| + \gamma_{L+1}|\widetilde{\mathcal{H}}_d(e^{j\omega_{L+1}})|}{\dfrac{\gamma_0}{\widetilde{\mathcal{W}}(\omega_0)} - \dfrac{\gamma_1}{\widetilde{\mathcal{W}}(\omega_1)} + \cdots + \dfrac{(-1)^{L+1}\gamma_{L+1}}{\widetilde{\mathcal{W}}(\omega_{L+1})}} = \frac{\sum\limits_{k=0}^{L+1} \gamma_k|\widetilde{\mathcal{H}}_d(e^{j\omega_k})|}{\sum\limits_{k=0}^{L+1} \dfrac{(-1)^k\gamma_k}{\widetilde{\mathcal{W}}(\omega_k)}}$$

with the parameter γ_k $(k = 1, 2, \ldots, L + 1)$ defined as

$$\gamma_k = \prod_{n=0, n\neq k}^{L+1} \frac{1}{\cos(\omega_k) - \cos(\omega_n)}. \tag{12.167}$$

This relationship arises by solving the matrix equation $\boldsymbol{AX} = \boldsymbol{B}$ with respect to $\boldsymbol{X}$, thus

$$\boldsymbol{X} \equiv \left[\alpha[0]\ \alpha[1] \ldots \delta\right]^T = \boldsymbol{A}^{-1} \cdot \boldsymbol{B} \tag{12.168}$$

and considering only the expression for the last element of this vector, which is the parameter δ; in this case, it is proven that this expression has the form we quoted above.

In the next step of the process, taking into account that (a) the function

$$\mathcal{P}(\omega) = \sum_{k=0}^{L} \alpha[k]\,(\cos\omega)^k = \sum_{k=0}^{L} \alpha[k]\,x^k \quad (\text{where } x = cos\omega)$$

is a polynomial of degree L and (b) the values of the function $\mathcal{P}(\omega)$ at the positions $x_k = \cos(\omega_k)$ can be calculated (which is done using the value of δ that we calculated above) from the relationship

$$\mathcal{P}(\omega_k) = |\widetilde{\mathcal{H}}_d(e^{j\omega_k})| - \frac{(-1)^k\delta}{\widetilde{\mathcal{W}}(\omega_k)} \quad (k = 0, 1, 2, \ldots, L) \tag{12.169}$$

we have the ability to calculate the function $\mathcal{P}(\omega)$ at the positions of the $L + 1$ frequencies $\{\omega_0, \omega_1, \omega_2, \ldots, \omega_L\}$ using *the barycentric form of the interpolation method of Lagrange*

$$\mathcal{P}(\omega) = \begin{cases} \mathcal{P}(\omega_k) & \omega = \omega_k \quad (k = 0, 1, 2, \ldots, L) \\ \left\{ \dfrac{\displaystyle\sum_{k=0}^{L} \dfrac{w_k \mathcal{P}(\omega_k)}{\cos\omega - \cos\omega_k}}{\displaystyle\sum_{k=0}^{L} \dfrac{w_k}{\cos\omega - \cos\omega_k}} \right\} & \text{in the opposite case} \end{cases} \tag{12.170}$$

for weight values $w_k = \gamma_k$, where γ_k are the parameters that appear in the calculation equation of the parameter δ. Using, finally, the calculated value of $\mathcal{P}(\omega)$, we can calculate the error function $\mathcal{E}(\omega)$ from the relation $\mathcal{E}(\omega) = \widetilde{W}(\omega)[|\widetilde{H}_d(e^{j\omega})| - \mathcal{P}(\omega)]$. This process of calculating the function $\mathcal{E}(\omega)$ is necessary for the realization of the next step described below[63]

Note The above process of calculating the parameter δ and the functions $\mathcal{P}(\omega)$ and $\mathcal{E}(\omega)$ requires *the recording of the values of the indices of the first and last extreme frequencies for each zone of interest*, which gives the algorithm a high degree of complexity. An alternative way of performing these calculations is to assume that the initial $L + 2$ frequencies do not define a one-dimensional array of dimensions $L + 2$ but are characterized by a *two-dimensional organization* of the form

[63] Considering $N + 1$ discrete real points x_i and equal values f_i $(i = 0, 1, 2, \ldots, N)$ that may or may not be samples of a function $f(x)$, the finding of a polynomial $p \in \Pi_N$ (where Π_N is the vector space of all polynomials of degree $d \leq N$) such that $p(x_i) = f_i$ $(i = 0, 1, 2, \ldots, N)$, is based on the *method of Lagrange interpolation*, which is described by the equation

$$P(x) = \sum_{i=0}^{N} f_i \ell_i(x) = \sum_{i=0}^{N} f_i \left\{ \prod_{k=0,k\neq i}^{N} (x - x_k) \right\} \Big/ \left\{ \prod_{k=0,k\neq i}^{N} (x_i - x_k) \right\}$$

with the *Lagrange polynomials* $\ell_i(x)$ satisfying the property

$$\ell_i(x_j) = \begin{cases} 1 & i = j \\ 0 & i \neq j. \end{cases}$$

Since the complexity of this method is equal to $\mathcal{O}(N^2)$ in terms of the number of additions and multiplications required, in practice, the *barycentric form*, the complexity of which is of the order $\mathcal{O}(N)$, that is, it is equal to that of the interpolation method of Newton. To transform the method into the above form, we formulate the i-th Lagrange polynomial as

$$\ell_i(x) = \ell(x)\frac{w_i}{x - x_i} \quad \text{where} \quad \ell(x) = (x - x_0)(x - x_1)(x - x_2)\ldots(x - x_N) = \prod_{i=1}^{N}(x - x_i) \quad \text{and} \quad w_i = \left\{ \prod_{k=1,k\neq i}^{N} (x_i - x_k) \right\}^{-1},$$

so the polynomial $p(x)$ will take the form

$$p(x) = \ell(x) \sum_{i=0}^{N} \frac{w_i}{x - x_i} f_i.$$

Let us now consider the interpolation process for the constant function $p(x) = 1$, which is based on the use of the sample $f_i = 1$. In this case, the above equation is written

$$1 = \sum_{i=0}^{N} \ell_i(x) = \ell(x) \sum_{i=0}^{N} \frac{w_i}{x - x_i}.$$

Dividing the last two equations term by term and eliminating the common factor $\ell(x)$, we are led to the equation

$$p(x) = \left(\sum_{i=0}^{N} \frac{w_i}{x - x_i} f_i \right) \Big/ \left(\sum_{i=0}^{N} \frac{w_i}{x - x_i} \right),$$

which constitutes the *barycentric form* of the interpolation method of Lagrange. For more information on this topic, the reader is referred to the literature (see, for example, the reference Berrut J.P. & Trefethen L.N., *Barycentric Lagrange Interpolation*, SIAM Review, Vol. 46, No. 3, pp. 501–517, 2004.)

$$\boldsymbol{\Omega} = \begin{bmatrix} \hat{\omega}_{11} & \hat{\omega}_{12} & \dots & \hat{\omega}_{1i} & \dots & \hat{\omega}_{1K} \\ \hat{\omega}_{21} & \hat{\omega}_{22} & \dots & \hat{\omega}_{2i} & \dots & \hat{\omega}_{2K} \\ \vdots & \vdots & \vdots & \vdots & \vdots & \vdots \\ \hat{\omega}_{\mu_1 1} & \hat{\omega}_{\mu_2 2} & \dots & \hat{\omega}_{\mu_i i} & \dots & \hat{\omega}_{\mu_K K}, \end{bmatrix} \quad (12.171)$$

where μ_i $(i = 1, 2, \dots, N)$ is the number of extreme frequency values of zone number i. In this case, the above relations must of course be modified accordingly, so as to reflect this new two-dimensional formulation of the problem. It is not difficult to see that this new formulation now requires the use of two indices ℓ and m, with the index m describing *the zone of interest number* m (and therefore will take values in the area $1 \le m \le K$) and the index ℓ describing *the extreme frequency number* ℓ *within the zone number* m (and therefore will take values in the area $1 \le \ell \le \mu_m$, where μ_m is the number of extremes of the zone number m of interest). These two indices are used together to describe the extreme frequency number ℓ of the zone number m, and they take values in the set

$$\boldsymbol{K} = \{\{\ell, m\} : (1 \le \ell \le \mu_m) \text{ and } (1 \le m \le K)\}.$$

Denoting with ω_k the frequency number k of the one-dimensional representation $(0 \le k \le L+1)$, this, according to the two-dimensional formalism, will correspond to some frequency $\omega_{\ell m}$ and consequently the indices k and $\{\ell, m\}$ will be related to each other through a mapping of the form $k \to \{\ell, m\}$. For values of k such that $0 < k < L+1$, the parameter γ_k, which for the one-dimensional case is defined by the relation

$$\gamma_k = \prod_{n=0,\, n\neq k}^{L+1} \frac{1}{x_k - x_n} = \frac{1}{x_k - x_0} \cdot \frac{1}{x_k - x_1} \cdot \frac{1}{x_k - x_2} \cdots \frac{1}{x_k - x_{k-1}} \frac{1}{x_k - x_{k+1}} \cdots \frac{1}{x_k - x_L} \frac{1}{x_k - x_{L+1}}, \quad (12.172)$$

where we defined the variable $x_x = \cos\omega_k$, will correspond for the two-dimensional case to the parameter

$$\gamma_{\ell m} = \frac{1}{x_{\ell m} - x_{11}} \frac{1}{x_{\ell m} - x_{21}} \cdots \frac{1}{x_{\ell m} - x_{(\ell-1)m}} \frac{1}{x_{\ell m} - x_{(\ell+1)m}} \cdots \frac{1}{x_{\ell m} - x_{(\mu_K-1)K}} \frac{1}{x_{\ell m} - x_{\mu_K K}}$$

or equivalently

$$\gamma_{\ell m} = \frac{1}{x_{\ell m} - x_{\mu_K K}} \prod_{(i,j)\in \boldsymbol{I}'} \frac{1}{x_{km} - x_{ij}}, \quad (12.173)$$

where $\boldsymbol{I}'$ is the set of index pairs

$$\boldsymbol{I}' = \{\{i, j\} : (1 \le i \le h) \text{ and } (1 \le j \le K) \text{ and } (i \ne k \text{ or } j \ne m)\},$$

with the parameter h being equal to $\mu_K - 1$ for $m = K$ and μ_m for $m \ge K$. On the other hand, for the value $k = L+1$ (i.e., for the $L+2$ extreme frequency) by analogy we will have

$$\gamma_{\ell m} = \frac{1}{x_{\ell m} - x_{11}} \cdot \frac{1}{x_{\ell m} - x_{21}} \cdots \frac{1}{x_{\ell m} - x_{(\ell-1)m}} \cdot \frac{1}{x_{\ell m} - x_{(\ell+1)m}} \cdots \frac{1}{x_{\ell m} - x_{(\mu_K-1)K}} = \prod_{(i,j)\in \boldsymbol{I}'} \frac{1}{x_{km} - x_{ij}}. \quad (12.174)$$

Combining the above results, for the cases $0 < k < L+1$ and $k = L+1$ the parameters $\gamma_{\ell m}$ are defined as

$$\gamma_{\ell m} = \begin{cases} w_{\ell m} & \ell = \mu_K \text{ and } m = K \\ \dfrac{w_{\ell m}}{\cos\omega_{\ell m} - \cos\omega_{\mu_K K}} & \text{in the opposite case} \end{cases} \quad \text{where} \quad w_{km} = \prod_{\{i,j\}\in \boldsymbol{I}_{r-1}} \frac{1}{\cos\omega_{\ell m} - \cos\omega_{ij}}$$

and consequently the corresponding new expressions for the parameters δ and $\mathcal{P}(\omega)$ will have the form

$$\delta = \frac{\sum\limits_{\{\ell,m\}\in \boldsymbol{K}} \gamma_{\ell m} |\widetilde{\mathcal{H}}_d(e^{j\omega_{\ell m}})|}{\sum\limits_{\{\ell,m\}\in \boldsymbol{K}} \dfrac{(-1)^q \gamma_{\ell m}}{\widetilde{\mathcal{W}}(\omega_{\ell m})}} \quad \text{and} \quad \mathcal{P}(\omega) = \begin{cases} \mathcal{P}(\omega_{\ell m}) & \omega \in \boldsymbol{\Omega} \\ \dfrac{\sum\limits_{\{\ell,m\}\in \boldsymbol{K}'} \dfrac{w_{\ell m}\mathcal{P}(\omega_{\ell m})}{\cos\omega - \cos\omega_{\ell m}}}{\sum\limits_{\{\ell,m\}\in \boldsymbol{K}'} \dfrac{w_{\ell m}}{\cos\omega - \cos\omega_{\ell m}}} & \text{in the opposite case,} \end{cases}$$

where $\boldsymbol{K}'$ is a suitably defined set of indices of the form

$$\boldsymbol{K}' = \{\{k, m\} : (1 \leq \ell \leq \nu) \text{ and } (1 \leq m \leq K)\}$$

with the parameter ν being equal to $\mu_K - 1$ for $m = K$ and equal to μ_m for $m \neq K$, whereas the parameter q has a value equal to $\ell - 1$ for the first zone (i.e., for the value $m = 1$), whereas it is equal to

$$q = k - 1 + \sum_{i=1}^{m-1} \mu_i \tag{12.175}$$

for the remaining zones (i.e., for values $m \geq 2$). This result arises from the fact that the formation of the two-dimensional table from its one-dimensional equivalent is done by *placing in column number m of the table* $\boldsymbol{\Omega}$ *the extreme frequencies of zone number m* (in other words, this table *is not constructed of rows but of columns*). Consequently, the correspondence $\omega_{\ell m} \leftarrow \omega_\ell$ will give us $\omega_{11} \leftarrow \omega_0, \omega_{21} \leftarrow \omega_1, \ldots, \omega_{k1} \leftarrow \omega_{k-1}, \ldots \omega_{\mu_1 1} \leftarrow \omega_{\mu_1 - 1}$ for $m = 1$, $\omega_{12} \leftarrow \omega_{\mu_1}, \omega_{22} \leftarrow \omega_{\mu_1+1}, \ldots, \omega_{k2} \leftarrow \omega_{\mu_1+k-1}, \ldots \omega_{\mu_1 2} \leftarrow \omega_{\mu_1+\mu_2-1}$ for $m = 2$ and generally $\omega_{\ell m} = \omega_{\mu_1+\mu_2+\cdots+\mu_{m-1}+\ell-1}$ for each m, from which the desired result is derived.

Finally, in full analogy with the previous case, the samples $\mathcal{P}(\omega_{\ell m})$ will be calculated from the samples of the desired frequency response $|\widetilde{\mathcal{H}}_d(e^{j\omega_{\ell m}})|$ through the expression

$$\mathcal{P}(\omega_{\ell m}) = |\widetilde{\mathcal{H}}_d(e^{j\omega_{\ell m}})| - \frac{(-1)^q \delta}{\widetilde{\mathcal{W}}(\omega_{\ell m})}. \tag{12.176}$$

Step 3 This step is the heart of the recursive algorithm of Remez and consists *in determining the frequencies* $\tilde{\omega}_0, \tilde{\omega}_1, \ldots, \tilde{\omega}_\rho$ *(for value* $\rho > L + 1$*) at the positions of which the magnitude of the error function* $|\mathcal{E}(\omega)|$ *is maximized at the ends of the zones of interest*, satisfying the property $|\mathcal{E}(\tilde{\omega}_i)| \geq \delta$ (these frequencies are potential extremals for the next iteration). This determination process is based on the calculation of the function $|\mathcal{E}(\omega)|$ at the positions of a *relatively dense set of frequency values*, the number S of which should be satisfactory, so as to ensure relatively good accuracy. In practice, it has been proven that a satisfactory value is $S = 16$ frequency values per deviation and therefore the distance between two successive frequency values is equal to $\delta\omega_i = B_i/(m_i S) = B_i/(16 m_i)$. As for the process of calculating the above frequency values, this is done in the following way:

1. For the selected value of the parameter S and for the i-th zone of interest ($i = 1, 2, \ldots, K$) we calculate the distance between successive frequencies $\delta\omega_i = B_i/N_i$ where $N_i = m_i S$ and set $\xi = 0$ where ξ is an auxiliary parameter used as a *counter*, in the sense that at the end of the process it will contain the number of frequencies that have been initialized in the next step.
2. For each zone of interest i ($i = 1, 2, \ldots, K$) and for each of the $N_i + 1$ frequencies $\omega_i^1 = \omega_i^\ell, \omega_i^2 = \omega_i^\ell + \delta\omega_i, \ldots, \omega_i^j = \omega_i^\ell + (j-1)\delta\omega_i, \ldots, \omega_i^{N_i+1} = \omega_i^h$ we set $\omega_\xi = \omega_i^j$ and increase the value of the counter ξ by one unit (i.e., we set $\xi = \xi + 1$), if it is $|\mathcal{E}(\omega_i^j)| \geq |\delta|$ and one of the following conditions applies:
   (a) The function $|\mathcal{E}(\omega_i^j)|$ is maximized at the position $\omega_i^j = \omega_i^\ell$, i.e., $|\mathcal{E}(\omega_i^\ell)| > |\mathcal{E}(\omega_i^\ell + \varepsilon)|$ (this condition applies to the first frequency $\omega_i^j = \omega_i^\ell$).
   (b) The function $|\mathcal{E}(\omega_i^j)|$ is maximized at the position ω_i^j, i.e., $|E(\omega_i^j - \delta\omega_i)| < |\mathcal{E}(\omega_i^j)| > |\mathcal{E}(\omega_i^j + \delta\omega_i)|$ (this condition applies to the intermediate frequency values $\omega_i^\ell < \omega_i^j = \omega_i^h$).

(c) The function $|\mathcal{E}(\omega_i^j)|$ is maximized at the position $\omega_i^j = \omega_i^h$, i.e., $|\mathcal{E}(\omega_i^h)| > |\mathcal{E}(\omega_i^h - \varepsilon)|$ (this condition applies to the last frequency $\omega_i^j = \omega_i^h$).

In the above expressions, the parameter ε is a constant with a *very small value* of the order of $0.01\delta\omega_i$.

The above search process relies on the use of the *first* and *second derivative* of the function $|\mathcal{E}(\omega)|$, which, as derived from the equation $\mathcal{E}(\omega) = \widetilde{W}(\omega)[|\widetilde{\mathcal{H}}_d(e^{j\omega})| - \mathcal{P}(\omega)]$, is[64]

$$\frac{d|\mathcal{E}(\omega)|}{d\omega} = \operatorname{sgn}\,[|\widetilde{\mathcal{H}}_d(e^{j\omega})| - \mathcal{P}(\omega)] \times \left[\frac{d|\widetilde{\mathcal{H}}_d(e^{j\omega})|}{d\omega} - \frac{d\mathcal{P}(\omega)}{d\omega}\right] \quad \text{and}$$

$$\frac{d^2|E(\omega)|}{d\omega^2} = \operatorname{sgn}\,[|\widetilde{\mathcal{H}}_d(e^{j\omega})| - \mathcal{P}(\omega)] \times \left[\frac{d^2|\widetilde{\mathcal{H}}_d(e^{j\omega})|}{d\omega^2} - \frac{d^2\mathcal{P}(\omega)}{d\omega^2}\right],$$

where sgn(x) is the *sign function*. Using for reasons of greater efficiency the formulation of the function $\mathcal{P}(\omega)$ dictated by the *two-dimensional formalism* (see Eq. 12.170), we can calculate its derivative for the various cases and from there the *first* and *second* derivatives of the error function's magnitude. It is interesting to note that in some cases it is possible *to fail to find a maximum frequency* from the above algorithm, when, for example, the function $|\mathcal{E}(\omega)|$ zeros immediately before or immediately after the right and left ends of a zone of interest, respectively (this problem is addressed by using in the processes 2(a) and 2(b) of the current step (Step 3) the conditions $|\mathcal{E}(\omega_i^\ell - \varepsilon)| > |\mathcal{E}(\omega_i^\ell|$ and $|\mathcal{E}(\omega_i^h)| < |\mathcal{E}(\omega_i^h + \varepsilon)|$). But this failure to determine a maximum is very easy to detect, if after the end of the process we find that the value of the parameter ξ that contains the number of maxima that were detected is less than the minimum value $L + 2$ predicted by the *alternation theorem*. If we find that there are maxima that have not been found, we continuously double the value of S and repeat the process as many times as necessary, until we finally determine all the maxima.

The above *exhaustive step-by-step search process* of maximum frequencies is quite time consuming and is characterized by a large computational cost, since a filter order equal to N requires $S(N+1)/2$ evaluations of the function $\mathcal{P}(\omega)$ with each of these requiring the execution of $(N-1)$ additions, $(N+1)/2$ multiplications, and $(N+1)/2$ divisions per iteration, whereas to the above calculations we should add those required for the calculation of the values of the parameters δ, γ_k and w_k. Taking into account that the number of iterations of the Remez algorithm ranges between the values 4 and 12, depending on the type of filter to be constructed, it is not difficult to understand that the effort to speed up this process is an imperative need. There are several algorithms that have been proposed regarding this issue and in this section we present two from those that are used either separately or in combination with each other, significantly accelerating the determination of the frequencies of the maximum.[65]

The first of these algorithms is an improved version of simple exhaustive search and is known as *selective search*. This algorithm is based on the assumption that *in most common cases there is a one-to-one correspondence between the extreme frequencies and the maxima of the error function measure*, which can therefore be determined by calculating the slope (i.e., the first derivative) of the function $|\mathcal{E}(\omega)|$ at points of interest. Indeed, if the first derivative of the function is positive (or negative), this means that there is some maximum of the function $|\mathcal{E}(\omega)|$ on the right (or left) side of the current extreme frequency, which can be determined by *continuously increasing (decreasing) the frequency ω with a step change $\delta\omega_i$ and calculating the value of the error function measure, until it begins to decrease (or increase)*. The above process obviously concerns internal frequencies, i.e., frequencies $\omega_i^j \neq 0$ and $\omega_i^j \neq \pi$ respectively, whereas if it is $\omega_i^j = 0$ or $\omega_i^j = \pi$, this process is applied if the second derivative of the error function measure has a *positive value* (and therefore corresponds to a minimum). If no maximum is found in the frequency range $\omega_i^j \leq \omega \leq (\omega_{i+1}^j$ or $\omega_i^h)$ (or correspondingly in the frequency range $(\omega_i^\ell$ or $\omega_{i-1}^j) \leq \omega \leq \omega \leq \omega_i^j)$, the search is interrupted. This behavior of the function in combination with the *symmetry* that characterizes the fluctuations it displays allows us to limit the search *only to half of the total frequency range occupied by all the zones of interest* (since the distance between a peak and the neighboring maximum point will be less than half the period of the corresponding fluctuation), and therefore, to speed up the process. It should be noted that over

[64] For the mathematical expression of the derivative of the function $P(\omega)$ for the various cases of interest, see the book Antoniou A., *Digital Signal Processing - Signals, Systems and Filters*, McGraw Hill, 2005, pp. 695–696.

[65] See Antoniou A., *Digital Signal Processing - Signals, Systems and Filters*, McGraw Hill, 2005, pp. 683–689.

time and with the implementation of the algorithm iterations one after the other, the above distance progressively decreases and therefore the area of the frequencies in which the search is conducted progressively shrinks, resulting in the number of calculations becoming smaller by more than 70% compared with the simple exhaustive search method.

An alternative method of calculating the extreme frequency values is the *numerical method of cubic interpolation*, which, however, is not suitable for application in frequency areas near the limits of the zones of interest, since there the error function *does not behave smoothly*. Assuming that the function $|\mathcal{E}(\omega)|$ is described by a *cubic polynomial* of the form

$$|\mathcal{E}(\omega)| = \alpha_0 + \alpha_1\omega + \alpha_2\omega^2 + \alpha_3\omega^3 = \sum_{i=0}^{3}\alpha_i\omega^i. \tag{12.177}$$

we know from analysis that the extreme values will emerge as the roots of the *quadratic equation*

$$\frac{d|\mathcal{E}(\omega)|}{d\omega} = \alpha_1 + 2\alpha_2\omega + 3\alpha_3\omega^2 = 0, \tag{12.178}$$

which, as it turns out, have the form

$$\omega_{1,2} = \frac{-\beta \pm \sqrt{\Delta}}{2\alpha} = \frac{-2\alpha_2 \pm 2\sqrt{\alpha_2^2 - 3\alpha_1\alpha_3}}{6\alpha_3} = \frac{1}{3\alpha_3}\left(-\alpha_2 \pm \sqrt{\alpha_2^2 - 3\alpha_1\alpha_3}\right), \tag{12.179}$$

whereas the *maxima* of the function satisfy the condition

$$\frac{d^2|\mathcal{E}(\omega)|}{d\omega^2} = 2\alpha_2 + 6\alpha_3\omega < 0 \tag{12.180}$$

from which it follows that $\omega < -\alpha_2/3\alpha_3$. Considering, therefore, three different frequency values $\omega_1, \omega_2, \omega_3$, the value of the error function's magnitude at these frequency positions is calculated as

$$\begin{aligned}
|\mathcal{E}(\omega_1)| &= \alpha_0 + \alpha_1\omega_1 + \alpha_2\omega_1^2 + \alpha_3\omega_1^3 \\
|\mathcal{E}(\omega_2)| &= \alpha_0 + \alpha_1\omega_2 + \alpha_2\omega_2^2 + \alpha_3\omega_2^3 \\
|\mathcal{E}(\omega_3)| &= \alpha_0 + \alpha_1\omega_3 + \alpha_2\omega_3^2 + \alpha_3\omega_3^3,
\end{aligned}$$

whereas the *first derivative* at the frequency position $\omega = \omega_1$ will obviously be equal to

$$\frac{d|\mathcal{E}(\omega_1)|}{d\omega} = \alpha_1 + 2\alpha_2\omega_1 + 3\alpha_3\omega_1^2. \tag{12.181}$$

The solution of the system of the above four equations for specific frequency values allows the calculation of the values of the *coefficients* α_i $(i = 0, 1, 2, 3)$ and then the *roots* $\omega_{1,2}$ of the quadratic equation $\mathcal{E}'(\omega) = 0$. Of the two values that emerge, we keep only the one that satisfies the maximum condition $\mathcal{E}''(\omega) < 0$. This is an alternative way of calculating the maximum frequencies. The solution of the above system and the analytical calculation of the values of the coefficients α_i $(i = 0, 1, 2, 3)$ is a trivial process and is left as an exercise to the reader. Specifically, the parameters α_2, α_3 and α_4 required for the calculation of the roots $\omega_{1,2}$ are shown to have the form

$$\alpha_3 = \frac{\varphi_1 - \varphi_2}{\varphi_3 - \varphi_4}, \quad \alpha_2 = \varphi_1 - \varphi_3\alpha_3, \quad \alpha_1 = \mathcal{E}'(\omega_1) - 2\alpha_2\omega_1 - 3\alpha_3\omega_1^2$$

with the auxiliary quantities φ_i $(i = 1, 2, 3, 4)$ being defined as

$$\varphi_1 = \frac{[\mathcal{E}(\omega_2) - \mathcal{E}(\omega_1)] + \mathcal{E}'(\omega_1)(\omega_1 - \omega_2)}{(\omega_1 - \omega_2)^2}, \quad \varphi_2 = \frac{[\mathcal{E}(\omega_3) - \mathcal{E}(\omega_1)] + \mathcal{E}'(\omega_1)(\omega_1 - \omega_3)}{(\omega_1 - \omega_3)^2}$$

$$\varphi_3 = \frac{2\omega_1^2 - \omega_2(\omega_1 + \omega_2)}{\omega_1 - \omega_2}, \quad \varphi_4 = \frac{2\omega_1^2 - \omega_4(\omega_1 + \omega_4)}{\omega_1 - \omega_4}.$$

From the algorithmic side of the issue, the *search via cubic interpolation* is performed as follows.

For the current extreme frequency ω_i^j:

1. We initialize the three frequencies ω_1, ω_2 and ω_3 by setting

$$\omega_1 = \omega_i^j, \quad \omega_3 = \begin{Bmatrix} \omega_i^j + \frac{Q}{2}(\omega_{i+1}^j - \omega_i^j) \text{ if } \mathcal{E}'(\omega_1) > 0 \\ \omega_i^j - \frac{Q}{2}(\omega_i^j - \omega_{i-1}^j) \text{ if } \mathcal{E}'(\omega_1) < 0 \end{Bmatrix}, \quad \omega_2 = \frac{1}{2}(\omega_1 + \omega_3).$$

The choice $\omega_1 = \omega_i^j$ is justified by the fact that since ω_i^j is an *extreme frequency*, the value of the function $\mathcal{E}(\omega_i^j)$ at its position is known and equal to $|\delta|$ and does not need to be recalculated. On the other hand, the definition of the frequency ω_3 is justified by the fact that *the frequencies* ω_1 *and* ω_3 *should enclose the new maximum value* $\omega_{\max}$, with the parameter Q causing the *progressive approximation* of the frequencies ω_1 and ω_3 as the algorithm evolves and the frequency $\omega_{\max}$ increasingly approaches the frequency ω_1 (the convergence parameter Q defined in *Step 5* of the algorithm and calculated *in each iteration cycle* is characterized by *monotonic reduction* and under convergence conditions takes a value *very close to zero*). Finally, the frequency ω_2 is defined as the average of the frequencies ω_1 and ω_3.
2. We calculate the values of the auxiliary parameters φ_i and the coefficients α_i $(i = 1, 2, 3, 4)$. If it is $3\alpha_1\alpha_3 > \alpha_2^2$ (which means that *it does not exist at its maximum*, since the radicand of the equation defining the roots $\omega_{1,2}$ takes a negative value), the attempt is abandoned and repeated anew using the *exhaustive* or the *selective* search.
3. We calculate the roots $\omega_{1,2}$ and choose one of them, depending on which of the two satisfies the inequality $\omega < -\alpha_2/3\alpha_3$. If the frequency value calculated in this way *does not belong to the value range* $[\omega_1, \omega_3]$, the attempt is abandoned and repeated anew using the exhaustive or the selective search.
4. We set $\omega = \omega_i^j + [(\omega/\omega_i^j) + 0.5]$ with the operator [ ] returning the value of the *integer part*. Then, we set $\ell = \ell + 1$ and $\omega_\ell^j = \omega$, and terminate the process.

It is proven that the efficiency of cubic interpolation is *constant* during the iterations of the Remez algorithm, although over time it becomes *inferior* in relation to the efficiency of the selective search, since the number of calculations of the latter *decreases progressively as the optimal solution is approached*. For this reason, we can start with the *cubic interpolation* and resort to the selective search when certain performance criteria are met. In practice, the combined use of these two methods can lead to a total reduction in the number of calculations even by 75% compared with the simple exhaustive step-by-step search.

Step 4 Although the solution of the matrix equation $\boldsymbol{AX} = \boldsymbol{B}$ requires knowledge of *exactly* $L + 2$ *extreme frequency values*, something that applies to the case of filters with a single and unique zone of interest, in practice, the filters we construct usually have *more than one zone*[66] As a result, the number of extreme values increases *by 2 values per zone for each zone of interest*, since the two ends of each zone are considered as *potential maximum positions*. Consequently, for the case of a frequency response with K zones of interest, the number of maxima will be in the worst case equal to $T = L + 2K$, thus causing the appearance of $T - (L + 2) = 2K - 2$ redundant extremes, which must be removed.

[66] Band-pass filters are an example of this case, as they are characterized by the existence of *two cutoff zones and one transition zone* and therefore have $K = 3$ zones (which of course also applies to band-stop filters). In the general case, we can consider three different deviations δ_1, δ_2 and δ_3 for these zones (in Footnote 53 (page 1255) we consider the simple case in which the two cutoff zones are characterized *by the same deviation*) and consequently as a weight function we can choose one of the

$$\mathcal{W}(\omega) = \begin{cases} 1 & \text{for } 0 \le \omega \le \omega_{s_1} \\ \delta_1/\delta_2 & \text{for } \omega_{p_1} \le \omega \le \omega_{p2} \\ \delta_1/\delta_3 & \text{for } \omega_{s_2} \le \omega \le \pi \end{cases} \quad \mathcal{W}(\omega) = \begin{cases} \delta_2/\delta_1 & \text{for } 0 \le \omega \le \omega_{s_1} \\ 1 & \text{for } \omega_{p_1} \le \omega \le \omega_{p2} \\ \delta_2/\delta_3 & \text{for } \omega_{s_2} \le \omega \le \pi \end{cases} \quad \mathcal{W}(\omega) = \begin{cases} \delta_3/\delta_1 & \text{for } 0 \le \omega \le \omega_{s_1} \\ \delta_3/\delta_2 & \text{for } \omega_{p_1} \le \omega \le \omega_{p2} \\ 1 & \text{for } \omega_{s_2} \le \omega \le \pi. \end{cases}$$

Let it be noted that because the alternation theorem *does not impose restrictions regarding the number of zones of interest*, the minimum number of extreme values for the optimal solution will still be equal to $L + 2$. However, the maximum number of extremes will now be equal to $L + 5$, since there are $L - 1$ local extremes in the open interval $(0, \pi)$, whereas each zone is characterized by two frequencies that define the beginning and the end of it and all these points are likely to be characterized as *alternation points*.

Although the simplest solution is to identify and remove the redundant extreme frequency values that lead to the minimum value of the measure $|\mathcal{E}(\omega_i)|$ *recounting the remaining frequencies*,[67] the problem with this is that if a frequency value is characterized as *extreme* in *two consecutive repetitions*, then the value of the measure $|\mathcal{E}(\omega)|$ will remain stationary during these repetitions—taking the value δ during the second repetition, as shown by the equation $\widetilde{W}(\omega_n)[|\widetilde{\mathcal{H}}_d(e^{j\omega_n})| - \mathcal{P}(\omega_n)] = (-1)^n\delta$—and thus leading to an *increase* in the number of repetitions required. For this reason, in practice, an iterative algorithm for eliminating extreme frequencies is used, which initially *sorts in ascending order the frequency bands based on the average band error* and then *rejects one extreme frequency for each band*, traversing the bands in a cyclic manner and starting from the band with the smallest average error. This process in algorithmic form is formulated as follows:

1. We calculate the *average band error* for band number i ($i = 1, 2, \ldots, K$) using the relation

$$\mathcal{E}_i = \frac{1}{\upsilon_i} \sum_{\tilde{\omega}_i \in \Omega_i} |\mathcal{E}(\tilde{\omega}_i)| \quad (i = 1, 2, \ldots, K), \tag{12.182}$$

   where Ω_i is the set of *potential extreme frequencies* of band i defined as $\Omega_i = \{\tilde{\omega}_i : \omega_i^\ell \leq \omega_i \leq \omega_i^h\}$ and υ_i ($i = 1, 2, \ldots, K$) the number of these extreme values for band i.
2. We sort the K bands *in ascending order* according to the average zone error we calculated in the previous step, thus constructing the *sorted list* of zones $Z_1, Z_2, \ldots, Z_K$ with zones Z_1 and Z_K characterized by the *minimum* and *maximum zone error* respectively.
3. We traverse the sorted list in a *circular manner*, that is, in the order $Z_1, Z_2, Z_3, \ldots, Z_i, \ldots, Z_K, Z_1, Z_2, Z_3, \ldots$, rejecting a maximum frequency ω_i per zone, until all redundant maximum frequencies are rejected. The frequency that is rejected is the one that leads to the smallest value of the measure $|\mathcal{E}(\omega_i)|$, but it does not coincide with the two frequencies that constitute the limits of the zone.

The result of the above elimination process is the creation of a set of $L + 2$ extreme frequencies that constitute the new reference set for the next iteration.

Step 5 The Remez algorithm is designed in such a way that with the *increase* in the number of iterations *the value of δ increases monotonically, whereas the maximum error value continuously decreases* (however, it should be noted that this does not always happen, as in some cases it is possible to observe an *increase* in the value of $E_{\max}$ for some iterations, but this does not affect the convergence of the algorithm). The algorithm terminates *when the maximum error value coincides with the current absolute value of the parameter δ*, or equivalently, when the value of the *parameter convergence Q*, which is defined as[68] $Q = (E_{\max} - |\delta|)/E_{\max}$—for the current values of $E_{\max}$ and δ—and expresses the "distance" of the current from the optimal solution, takes a value less than ε, where ε is a convergence parameter with a very small value (less than, for example, 0.0001). Conversely, if it is found that $Q > \varepsilon$, we return to *Step 2* for the next iteration.

Step 6 In the last step of the Remez algorithm we replace the current values of the reference set (which are related to the *optimal solution*) and the corresponding values of the functions $\widetilde{W}(\omega)$ and $|\widetilde{\mathcal{H}}(e^{j\omega})|$ in the matrix equation and solve it with respect to the vector $\boldsymbol{X}$. In this case, we will find that the absolute value of the parameter δ that will be calculated in this way coincides with that which was calculated with the help of the iterative process, which, of course, is expected. As for the rest of the elements of the vector $\boldsymbol{X}$, they contain *the values of the coefficients* $\alpha[k]$ from which we calculate an equal number of samples of the impulse response (we remind the reader that in this description the ideal low-pass filter that we approximate is a type I filter). The remaining samples of the impulse response will emerge through the use of the *symmetry condition* that

[67] This process is based on the fact that *the fewer the maximum points within a zone of interest, the greater the approximation error for this zone and vice versa.*

[68] In some textbooks, the *convergence parameter* is defined as $Q = (\max|E(\omega_i)| - \min|E(\omega_i)|)/\max|E(\omega_i)|$ ($i = 0, 1, \ldots, T-1$), where T is the number of extreme frequencies that have been identified during the current iteration.

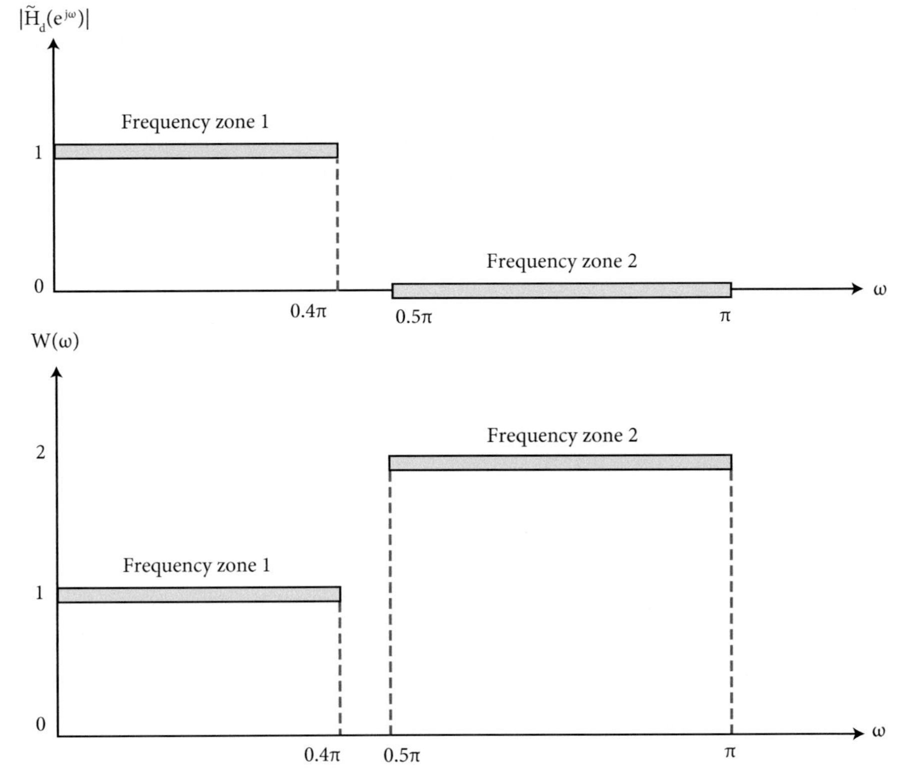

Fig. 12.68 The operating specifications of the low-pass filter type I of Exercise 12.12.5

is satisfied by the discrete-time signal in question. As for the corresponding frequency response, it will emerge, as is well known, *from the Fourier transform of the calculated impulse response*.

An alternative way of calculating the frequency response, which applies to any type of filter, is based on the calculation (for one last time) of the function $\mathcal{P}(\omega)$ from the samples $\mathcal{P}(\Omega_m)$ $(m = 1, 2, \ldots, L+2)$. This calculation will be carried out with the help of the *numerical interpolation method of Lagrange* with the samples $\mathcal{P}(\Omega_m)$ $(m = 1, 2, \ldots, L+2)$ calculated at the positions of the optimal extreme frequency values that led to the convergence of the iterative process. Then, from the relation $|\mathcal{H}(e^{j\omega})| = \mathcal{P}(\omega)\ \mathcal{Q}(\omega)$ we calculate *the magnitude of the actual frequency response*, with the function $\mathcal{Q}(\omega)$ depending on the type of filter to be designed (see Eq. 12.116), whereas, immediately after, we determine the values of the L samples $|\mathcal{H}(e^{j\omega_k})|$ (for the values $\omega_k = 2\pi k/N, k = 0, 1, 2, \ldots, L$). After this calculation process, the samples $h[n]$ of the impulse response will emerge through the application of the *inverse discrete Fourier transform* of the samples $|\mathcal{H}(e^{j\omega_k})|$ using the equations we constructed in the frequency sampling method. This process for the case of a type I FIR filter is presented in the next exercise. It should be noted that the discrete-time signal that emerges in this way is *noncausal* and therefore it must become causal, something that is achieved by applying to this the *time-shift transformation* $h[n] \longleftarrow h[n-L]$ $(n = 0, 1, 2, \ldots, N-1)$. Finally, the frequency response of the filter to be constructed will emerge as the *Fourier transform* of the discrete-time signal $h[n]$. We observe that in this method the calculation of the function $\mathcal{H}(e^{j\omega})$ does not require knowledge of the values of the parameters $\alpha[k]$; therefore, it is not mandatory to determine it.

The above iterative process of the Remez algorithm is demonstrated in the next exercise.

Problem 12.12.5 *Using the* iterative Remez algorithm *design a low-pass filter type I of order* $N = 13$ *with the operating specifications presented in Fig. 12.68.*

Solution From Fig. 12.68 we find that the filter to be constructed is characterized by the existence of *two zones of interest*: zone 1 defined by the frequencies $\omega_1^\ell = 0$ and $\omega_1^h = 0.4\pi$ and zone 2 defined by the frequencies $\omega_2^\ell = 0.5\pi$ and $\omega_2^h = \pi$. The range of frequencies of these two zones is $B_1 = \omega_1^h - \omega_1^\ell = 0.4\pi$ and $B_2 = \omega_2^h - \omega_2^\ell = 0.5\pi$. In these frequency areas the desired frequency response has values $|\widetilde{\mathcal{H}}_d(e^{j\omega})| = 1$ and $|\widetilde{\mathcal{H}}_d(e^{j\omega})| = 0$, whereas the corresponding values for the weight function are $\mathcal{W}(\omega) = 1$ and $\mathcal{W}(\omega) = 2$. Conversely, the area $0.4\pi \leq \omega \leq 0.5\pi$ is not involved in the filter operation and constitutes an *indifference zone*.

Taking into account that the filter to be constructed is a type I filter, the value of the parameter L of the *alternation theorem* is equal to $L = (N-1)/2 = 6$ where $N = 13$ is the order of the filter; therefore, the number of extreme frequency values that will be used at each step of the iterative process of Remez will be $R = L + 2 = 8$. For the initialization of the algorithm, we consider that the frequency area $[0, \pi]$ undergoes discretization by defining within it $G = 1000$ *discrete* frequency values that are at a distance $\Delta\omega = 0.00314159$. This set of 1000 frequency values is also used to define the functions $|\widetilde{\mathcal{H}}(e^{j\omega})|$ and $\mathcal{W}(\omega)$ in the above way.

Although the initialization of the $R = 8$ frequencies can be done by uniform distribution in the way we previously described, in practice it has been proven that *the algorithm converges very quickly, regardless of what the initial starting frequencies are*. For this reason, for simplicity's sake, let us choose some random frequencies that span the entire range of interest zones and which are $\Omega_1 \equiv \omega_{51} = 0.05\pi = 0.157079$, $\Omega_2 \equiv \omega_{101} = 0.1\pi = 0.314159$, $\Omega_3 \equiv \omega_{341} = 0.34\pi = 1.068141$, $\Omega_4 \equiv \omega_{361} = 0.36\pi = 1.130973$, $\Omega_5 \equiv \omega_{531} = 0.53\pi = 1.665044$, $\Omega_6 \equiv \omega_{671} = 0.67\pi = 2.104867$, $\Omega_7 \equiv \omega_{701} = 0.7\pi = 2.199114$ and $\Omega_8 \equiv \omega_{851} = 0.85\pi = 2.670353$. In this notation, the $R = 8$ extreme frequencies used in each iteration are the Ω_m $(m = 1, 2, \ldots, 8)$ and coincide with equal frequencies from the set of $G = 1000$ frequencies ω_k $(k = 1, 2, \ldots, 1000)$. It is not difficult to see that the desired frequency response at these eight frequency positions has values $|\widetilde{\mathcal{H}}_d(e^{j\Omega_1})| = |\widetilde{\mathcal{H}}_d(e^{j\Omega_2})| = |\widetilde{\mathcal{H}}_d(e^{j\Omega_3})| = |\widetilde{\mathcal{H}}_d(e^{j\Omega_4})| = 1$ and $|\widetilde{\mathcal{H}}_d(e^{j\Omega_5})| = |\widetilde{\mathcal{H}}_d(e^{j\Omega_6})| = |\widetilde{\mathcal{H}}_d(e^{j\Omega_7})| = |\widetilde{\mathcal{H}}_d(e^{j\Omega_8})| = 0$, whereas the corresponding values of the error function are $\mathcal{W}(\Omega_1) = \mathcal{W}(\Omega_2) = \mathcal{W}(\Omega_3) = W(\Omega_4) = 1$ and $\mathcal{W}(\Omega_5) = \mathcal{W}(\Omega_6) = \mathcal{W}(\Omega_7) = \mathcal{W}(\Omega_8) = 2$.

According to what we have mentioned in the previous section, for the current set of extreme frequencies Ω_m $(m = 1, 2, \ldots, m)$ we will calculate the parameters γ_m $(m = 1, 2, \ldots, 8)$ and from there the current value of the parameter δ. Executing the first iteration, the cosines of the current extreme frequencies are calculated as $cos(\Omega_1) = 0.987688$, $cos(\Omega_2) = 0.951056$, $cos(\Omega_3) = 0.0481753$, $cos(\Omega_4) = 0.425779$, $cos(\Omega_5) = -0.094108$, $cos(\Omega_6) = -0.509041$, $cos(\Omega_7) = -0.587785$, and $cos(\Omega_8) = -0.891006$, and therefore the parameters γ_m $(m = 1, 2, \ldots, 8)$ will have values

$$\gamma_1 = \frac{1}{x_1 - x_2} \times \frac{1}{x_1 - x_3} \times \frac{1}{x_1 - x_4} \times \frac{1}{x_1 - x_5} \times \frac{1}{x_1 - x_6} \times \frac{1}{x_1 - x_7} \times \frac{1}{x_1 - x_8} = +20.036625$$

$$\gamma_2 = \frac{1}{x_2 - x_1} \times \frac{1}{x_2 - x_3} \times \frac{1}{x_2 - x_4} \times \frac{1}{x_2 - x_5} \times \frac{1}{x_2 - x_6} \times \frac{1}{x_2 - x_7} \times \frac{1}{x_2 - x_8} = -25.599676$$

$$\gamma_3 = \frac{1}{x_3 - x_1} \times \frac{1}{x_3 - x_2} \times \frac{1}{x_3 - x_4} \times \frac{1}{x_3 - x_5} \times \frac{1}{x_3 - x_6} \times \frac{1}{x_3 - x_7} \times \frac{1}{x_3 - x_8} = +89.819272$$

$$\gamma_4 = \frac{1}{x_4 - x_1} \times \frac{1}{x_4 - x_2} \times \frac{1}{x_4 - x_3} \times \frac{1}{x_4 - x_5} \times \frac{1}{x_4 - x_6} \times \frac{1}{x_4 - x_7} \times \frac{1}{x_4 - x_8} = -93.315063$$

$$\gamma_5 = \frac{1}{x_5 - x_1} \times \frac{1}{x_5 - x_2} \times \frac{1}{x_5 - x_3} \times \frac{1}{x_5 - x_4} \times \frac{1}{x_5 - x_6} \times \frac{1}{x_5 - x_7} \times \frac{1}{x_5 - x_8} = +18.097472$$

$$\gamma_6 = \frac{1}{x_6 - x_1} \times \frac{1}{x_6 - x_2} \times \frac{1}{x_6 - x_3} \times \frac{1}{x_6 - x_4} \times \frac{1}{x_6 - x_5} \times \frac{1}{x_6 - x_7} \times \frac{1}{x_6 - x_8} = -39.586226$$

$$\gamma_7 = \frac{1}{x_7 - x_1} \times \frac{1}{x_7 - x_2} \times \frac{1}{x_7 - x_3} \times \frac{1}{x_7 - x_4} \times \frac{1}{x_7 - x_5} \times \frac{1}{x_7 - x_6} \times \frac{1}{x_7 - x_8} = +32.279576$$

$$\gamma_8 = \frac{1}{x_8 - x_1} \frac{1}{x_8 - x_2} \frac{1}{x_8 - x_3} \frac{1}{x_8 - x_4} \frac{1}{x_8 - x_6} \frac{1}{x_8 - x_6} \frac{1}{x_8 - x_7} = -01.731981,$$

where for reasons of simplifying the writing we resorted to the symbolism $x_m = \cos\Omega_m$ $(m = 1, 2, \ldots, 8)$. After calculating the quantities γ_m the current value of the parameter δ is found to be equal to

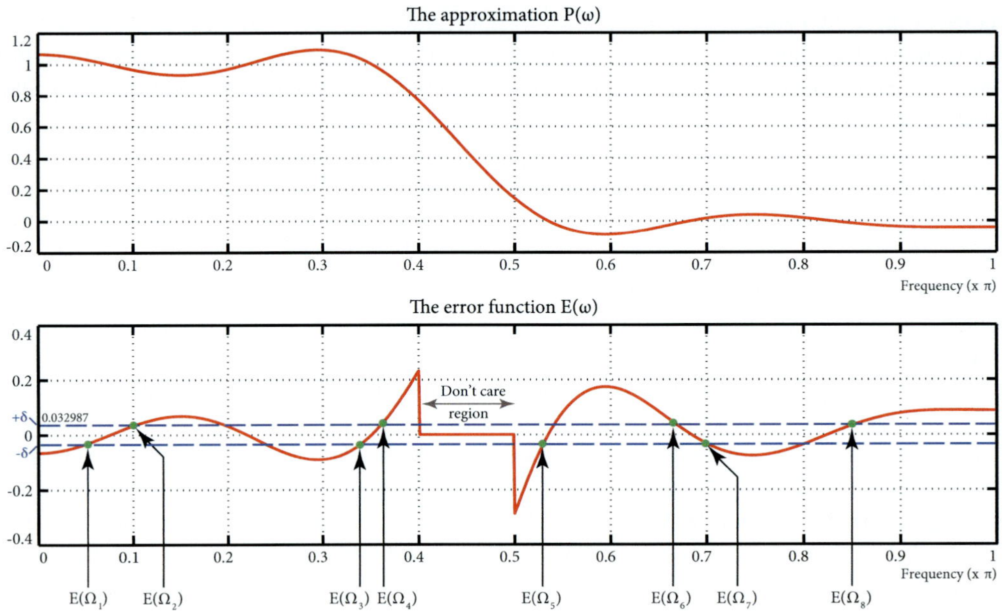

Fig. 12.69 The approximation $P(\omega)$ and the error function $E(\omega)$ after the end of the first iteration. From the figure we observe that the error function at the position of the current reference set takes the same value but with alternating signs, in full agreement with the alternation theorem

$$\delta = \frac{\gamma_1|\widetilde{\mathcal{H}}_d(e^{j\Omega_1})| + \gamma_2|\widetilde{\mathcal{H}}_d(e^{j\Omega_2})| + \gamma_3|\widetilde{\mathcal{H}}_d(e^{j\Omega_3})| + \gamma_4|\widetilde{\mathcal{H}}_d(e^{j\Omega_4})|}{\frac{\gamma_1}{\widetilde{W}(\Omega_1)} + \frac{\gamma_2}{\widetilde{\mathcal{W}}(\Omega_2)} + \frac{\gamma_3}{\widetilde{\mathcal{W}}(\Omega_3)} + \frac{\gamma_4}{\widetilde{\mathcal{W}}(\Omega_4)} + \frac{\gamma_5}{\widetilde{\mathcal{W}}(\Omega_5)} + \frac{\gamma_6}{\widetilde{\mathcal{W}}(\Omega_6)} + \frac{\gamma_7}{\widetilde{\mathcal{W}}(\Omega_7)} + \frac{\gamma_8}{\widetilde{\mathcal{W}}(\Omega_8)}}$$

$$+ \frac{\gamma_5|\widetilde{\mathcal{H}}_d(e^{j\Omega_5})| + \gamma_6|\widetilde{\mathcal{H}}_d(e^{j\Omega_6})| + \gamma_7|\widetilde{\mathcal{H}}_d(e^{j\Omega_7})| + \gamma_8|\widetilde{\mathcal{H}}_d(e^{j\Omega_8})|}{\frac{\gamma_1}{\widetilde{W}(\Omega_1)} + \frac{\gamma_2}{\widetilde{\mathcal{W}}(\Omega_2)} + \frac{\gamma_3}{\widetilde{\mathcal{W}}(\Omega_3)} + \frac{\gamma_4}{\widetilde{\mathcal{W}}(\Omega_4)} + \frac{\gamma_5}{\widetilde{\mathcal{W}}(\Omega_5)} + \frac{\gamma_6}{\widetilde{\mathcal{W}}(\Omega_6)} + \frac{\gamma_7}{\widetilde{\mathcal{W}}(\Omega_7)} + \frac{\gamma_8}{\widetilde{\mathcal{W}}(\Omega_8)}} =$$

$$= \frac{\gamma_1 + \gamma_2 + \gamma_3 + \gamma_4}{-\gamma_1 + \gamma_2 - \gamma_3 + \gamma_4 - \frac{1}{2}(\gamma_5 - \gamma_6 + \gamma_7 - \gamma_8)} = 0.032987.$$

Using this value of the parameter δ and the values of the functions $|\widetilde{\mathcal{H}}(e^{j\Omega_m})|$ and $\mathcal{W}(\Omega_m)$ $(m = 1, 2, \ldots, 8)$, we can, using Eq. (12.163) to calculate the corresponding values of the function $\mathcal{P}(\Omega_m)$ $(m = 1, 2, \ldots, 8)$, resulting in the results $\mathcal{P}(\Omega_1) = 1.032987$, $\mathcal{P}(\Omega_2) = 0.967012$, $\mathcal{P}(\Omega_3) = 1.032987$, $\mathcal{P}(\Omega_4) = 0.967012$, $\mathcal{P}(\Omega_5) = 0.016493$, $\mathcal{P}(\Omega_6) = -0.016493$, $\mathcal{P}(\Omega_7) = 0.016493$, and $\mathcal{P}(\Omega_8) = -0.016493$, and the corresponding values of the error function $\mathcal{E}(\Omega_m)$ $(m = 1, 2, \ldots)$, from the relation $\mathcal{E}(\Omega_m) = \widetilde{\mathcal{W}}(\Omega_m) \times [|\widetilde{\mathcal{H}}_d(e^{j\Omega_m})| - \mathcal{P}(\omega_m)]$. Performing the operations, it turns out that $\mathcal{E}(\Omega_1) = \mathcal{E}(\Omega_3) = \mathcal{E}(\Omega_5) = \mathcal{E}(\Omega_7) = -0.032987$ and $\mathcal{E}(\Omega_2) = \mathcal{E}(\Omega_4) = \mathcal{E}(\Omega_6) = \mathcal{E}(\Omega_8) = +0.032987$. We observe, therefore, that *at the positions of the extreme frequencies* the error function has a *constant value* equal to $\delta = 0.032987$, but with *alternating signs*, in full agreement with the alternation theorem. This feature is demonstrated in Fig. 12.69, which depicts the approximation $\mathcal{P}(\omega)$ and the error function $\mathcal{E}(\omega)$ after the end of the first iteration. The error function has emerged with the help of the values $\mathcal{P}(\Omega_m)$ $(m = 1, 2, \ldots, 8)$ using the *interpolation method of Lagrange* in the way we presented in the theoretical section.

The interpolation process relies on the use of *a denser set of frequencies*. Although according to the theory the number of this set is of the order of $16N$, where N is the order of the filter, the above process was used for reasons of greater accuracy and simplification of calculations, and the 1000 discrete frequency values. The result we get is equal values for the error function $\mathcal{E}(\omega)$, which are scanned one after the other, in order to identify the frequency values for which *the absolute value of the function* $\mathcal{E}(\omega)$ *is greater than the current absolute value of the parameter* δ. If these frequency values are more than $R = 8$, we should only keep the eight of them (removing the rest), which will constitute *the reference set for the next iteration*. The result of this process, which is based on the simple but computationally demanding and time-consuming

Table 12.2 The results of applying the *Remez algorithm* (Iterations 1,2 and 3)

Initialization and 1st iteration ($\delta = 0.032987$, $E_{\max} = 0.232995$, $Q = 0.858421$)					
$m(k)$	$\Omega_m = \omega_k$	γ_m	$P(\omega_m)$	$newK$	$E\ (newK)$
1(051)	$\Omega_1 = \omega_{051} = 0.050\,\pi = 0.157079$	$\gamma_1 = +20.036625$	$P(\omega_1) = +1.032987$	001	$E(001) = -0.162307$
2(101)	$\Omega_2 = \omega_{101} = 0.100\,\pi = 0.314159$	$\gamma_2 = -25.599676$	$P(\omega_2) = +0.967012$	151	$E(162) = +0.165878$
3(341)	$\Omega_3 = \omega_{341} = 0.340\,\pi = 1.068141$	$\gamma_3 = +89.819272$	$P(\omega_3) = +1.032987$	296	$E(315) = -0.176768$
4(361)	$\Omega_4 = \omega_{361} = 0.360\,\pi = 1.130973$	$\gamma_4 = -93.315063$	$P(\omega_4) = +0.967012$	401	$E(401) = +0.162305$
5(531)	$\Omega_5 = \omega_{531} = 0.530\,\pi = 1.665044$	$\gamma_5 = +18.097472$	$P(\omega_5) = +0.016493$	501	$E(501) = -0.162310$
6(671)	$\Omega_6 = \omega_{671} = 0.670\,\pi = 2.104867$	$\gamma_6 = -39.586226$	$P(\omega_6) = -0.016493$	595	$E(569) = +0.203751$
7(701)	$\Omega_7 = \omega_{701} = 0.700\,\pi = 2.199114$	$\gamma_7 = +32.279576$	$P(\omega_7) = +0.016493$	748	$E(712) = -0.212916$
8(851)	$\Omega_8 = \omega_{851} = 0.850\,\pi = 2.670353$	$\gamma_8 = -01.731981$	$P(\omega_8) = -0.016493$	952	$E(890) = +0.215631$
$E(\omega_1) = E(\omega_3) = E(\omega_5) = E(\omega_7) = -\delta = -0.032987$, $E(\omega_2) = E(\omega_4) = E(\omega_6) = E(\omega_8) = +\delta = +0.032987$					
2nd iteration ($\delta = 0.162307$, $E_{\max} = 0.215631$, $Q = 0.247291$)					
$m(k)$	$\Omega_m = \omega_k$	γ_m	$P(\omega_m)$	$newK$	$E\ (newK)$
1(001)	$\Omega_1 = \omega_{001} = 0.000\,\pi = 0.000000$	$\gamma_1 = +07.613378$	$P(\omega_1) = +1.162307$	001	$E(001) = -0.162307$
2(151)	$\Omega_2 = \omega_{151} = 0.150\,\pi = 0.471238$	$\gamma_2 = -17.224201$	$P(\omega_2) = +0.837692$	162	$E(162) = +0.165878$
3(296)	$\Omega_3 = \omega_{296} = 0.295\,\pi = 0.926769$	$\gamma_3 = +26.719955$	$P(\omega_3) = +1.162307$	315	$E(315) = -0.176768$
4(401)	$\Omega_4 = \omega_{401} = 0.400\,\pi = 1.256637$	$\gamma_4 = -35.145559$	$P(\omega_4) = +0.837692$	401	$E(401) = +0.162305$
5(501)	$\Omega_5 = \omega_{501} = 0.500\,\pi = 1.570796$	$\gamma_5 = +30.029444$	$P(\omega_5) = +0.081153$	501	$E(501) = -0.162310$
6(595)	$\Omega_6 = \omega_{595} = 0.594\,\pi = 1.866106$	$\gamma_6 = -14.748618$	$P(\omega_6) = -0.081153$	569	$E(569) = +0.203751$
7(748)	$\Omega_7 = \omega_{748} = 0.747\,\pi = 2.346769$	$\gamma_7 = +03.410799$	$P(\omega_7) = +0.081153$	712	$E(712) = -0.212916$
8(952)	$\Omega_8 = \omega_{952} = 0.951\,\pi = 2.987654$	$\gamma_8 = -00.653197$	$P(\omega_8) = -0.081153$	890	$E(890) = +0.215631$
$E(\omega_1) = E(\omega_3) = E(\omega_5) = E(\omega_7) = -\delta = -0.162307$, $E(\omega_2) = E(\omega_4) = E(\omega_6) = E(\omega_8) = +\delta = +0.162307$					
3rd iteration ($\delta = 0.170784$, $E_{\max} = 0.200581$, $Q = 0.148550$)					
$m(k)$	$\Omega_m = \omega_k$	γ_m	$P(\omega_m)$	$newK$	$E\ (newK)$
1(001)	$\Omega_1 = \omega_{001} = 0.000\,\pi = 0.000000$	$\gamma_1 = +06.787708$	$P(\omega_1) = +1.170784$	001	$E(001) = -0.170784$
2(162)	$\Omega_2 = \omega_{162} = 0.161\,\pi = 0.505796$	$\gamma_2 = -16.992145$	$P(\omega_2) = +0.829151$	162	$E(162) = +0.170784$
3(315)	$\Omega_3 = \omega_{315} = 0.314\,\pi = 0.986460$	$\gamma_3 = +38.796252$	$P(\omega_3) = +1.170784$	314	$E(315) = -0.170797$
4(401)	$\Omega_4 = \omega_{401} = 0.400\,\pi = 1.256637$	$\gamma_4 = -56.720906$	$P(\omega_4) = +0.829215$	401	$E(401) = +0.170782$
5(501)	$\Omega_5 = \omega_{501} = 0.500\,\pi = 1.570796$	$\gamma_5 = +54.690810$	$P(\omega_5) = +0.085392$	501	$E(501) = -0.170788$
6(569)	$\Omega_6 = \omega_{569} = 0.568\,\pi = 1.784424$	$\gamma_6 = -30.655893$	$P(\omega_6) = -0.085392$	569	$E(567) = +0.170721$
7(712)	$\Omega_7 = \omega_{712} = 0.711\,\pi = 2.233672$	$\gamma_7 = +04.781521$	$P(\omega_7) = +0.085392$	712	$E(701) = -0.175791$
8(890)	$\Omega_8 = \omega_{890} = 0.889\,\pi = 2.792875$	$\gamma_8 = -00.687346$	$P(\omega_8) = -0.085392$	890	$E(859) = +0.200581$
$E(\omega_1) = E(\omega_3) = E(\omega_5) = E(\omega_7) = -\delta = -0.170784$, $E(\omega_2) = E(\omega_4) = E(\omega_6) = E(\omega_8) = +\delta = +0.170784$					

exhaustive search, leads to the nine frequencies $\Omega_1 \equiv \omega_1 = 0\pi = 0$, $\Omega_2 \equiv \omega_{151} = 0.150\pi = 0.471238$, $\Omega_3 \equiv \omega_{296} = 0.295\pi = 0.926769$, $\Omega_4 \equiv \omega_{401} = 0.4\pi = 1.256637$, $\Omega_5 \equiv \omega_{501} = 0.5\pi = 1.570796$, $\Omega_6 \equiv \omega_{595} = 0.594\pi = 1.866106$, $\Omega_7 \equiv \omega_{748} = 0.747\pi = 2.346769$, $\Omega_8 \equiv \omega_{952} = 0.951\pi = 2.987654$, and $\Omega_9 = \pi = 3.14159$ with error values at these $\mathcal{E}_1 = -0.065204$, $\mathcal{E}_{151} = 0.067694$, $\mathcal{E}_{296} = -0.090620$, $\mathcal{E}_{401} = 0.232995$, $\mathcal{E}_{501} = -0.288182$, $\mathcal{E}_{595} = 0.173365$, $\mathcal{E}_{748} = -0.076018$, $\mathcal{E}_{952} = 0.087959$, and $\mathcal{E}_{1001} = 0.086691$. We observe that the algorithm has determined nine extreme values, of which we keep the first eight and reject the last one, since it *does not satisfy the alternation theorem* (because the error values for the last two (frequencies have the same sign). From the remaining error values, the maximum value is $E_{\max} = \mathcal{E}_{401} = 0.232995$, which corresponds to the frequency $\Omega_4 \equiv \omega_{401} = 0.4\pi = 1.256637$, as indeed is shown in the figure. We observe that $Q = (E_{\max} - |\delta|)/E_{\max} = 0.858421$ and we continue to the next iteration. The above process continues iteratively *until the value of the parameter Q, which also constitutes the convergence criterion, becomes extremely small, that is, practically until the maximum value of the error function $\mathcal{E}(\omega)$ equals the current absolute value of the parameter δ.*

The results of the calculations (along with those concerning the initial iterations) are included in Tables 12.2 and 12.3. For each iteration, the *positions and values of the eight reference frequencies, the values of the parameters γ_m, $\mathcal{P}(\omega_m)$, $\mathcal{E}(\omega_m)$ $(m = 1, 2, \ldots, 8)$, and δ, the indices of the eight frequencies newK that will constitute the reference set in the next iteration and the corresponding error values, as well as the value of the maximum error $E_{\max}$ and the coefficient Q* are recorded. In all cases, the algorithm gives nine extreme frequencies, the last of which is rejected for the reason we explained earlier.

Table 12.3 The results of applying the *Remez algorithm* (Repetitions 4 and 5)

4th iteration ($\delta = 0.170956$, $E_{\max} = 0.172286$, $Q = 0.0077223$)					
$m(k)$	$\Omega_m = \omega_k$	γ_m	$P(\omega_m)$	$newK$	$E(newK)$
1(001)	$\Omega_1 = \omega_{001} = 0.000\,\pi = 0.000000$	$\gamma_1 = +07.119945$	$P(\omega_1) = +1.170956$	001	$E(001) = -0.170956$
2(162)	$\Omega_2 = \omega_{162} = 0.161\,\pi = 0.505796$	$\gamma_2 = -17.925648$	$P(\omega_2) = +0.829043$	162	$E(162) = +0.170956$
3(314)	$\Omega_3 = \omega_{314} = 0.313\,\pi = 0.983318$	$\gamma_3 = +40.727571$	$P(\omega_3) = +1.170956$	314	$E(314) = -0.170956$
4(401)	$\Omega_4 = \omega_{401} = 0.400\,\pi = 1.256637$	$\gamma_4 = -60.353777$	$P(\omega_4) = +0.829043$	401	$E(401) = +0.170954$
5(501)	$\Omega_5 = \omega_{501} = 0.500\,\pi = 1.570796$	$\gamma_5 = +61.140712$	$P(\omega_5) = +0.085478$	501	$E(501) = -0.170959$
6(567)	$\Omega_6 = \omega_{567} = 0.566\,\pi = 1.778141$	$\gamma_6 = -35.817370$	$P(\omega_6) = -0.085478$	568	$E(568) = +0.170973$
7(701)	$\Omega_7 = \omega_{701} = 0.700\,\pi = 2.199114$	$\gamma_7 = +05.958046$	$P(\omega_7) = +0.085478$	701	$E(701) = -0.170956$
8(859)	$\Omega_8 = \omega_{859} = 0.858\,\pi = 2.695486$	$\gamma_8 = -00.849478$	$P(\omega_8) = -0.085478$	853	$E(853) = +0.172286$
$E(\omega_1) = E(\omega_3) = E(\omega_5) = E(\omega_7) = -\delta = -0.170956$, $E(\omega_2) = E(\omega_4) = E(\omega_6) = E(\omega_8) = +\delta = +0.170956$					
5th iteration ($\delta = 0.170961$, $E_{\max} = 0.170961$, $Q = 8.837230 \times 10^{-8}$					
$m(k)$	$\Omega_m = \omega_k$	γ_m	$P(\omega_m)$	$newK$	$E(newK)$
1(001)	$\Omega_1 = \omega_{001} = 0.000\,\pi = 0.000000$	$\gamma_1 = +07.132941$	$P(\omega_1) = +1.170961$	001	$E(001) = -0.170961$
2(162)	$\Omega_2 = \omega_{162} = 0.161\,\pi = 0.505796$	$\gamma_2 = -17.958622$	$P(\omega_2) = +0.829038$	162	$E(162) = +0.170961$
3(314)	$\Omega_3 = \omega_{314} = 0.313\,\pi = 0.983318$	$\gamma_3 = +40.795853$	$P(\omega_3) = +1.170961$	314	$E(314) = -0.170961$
4(401)	$\Omega_4 = \omega_{401} = 0.400\,\pi = 1.256637$	$\gamma_4 = -60.409278$	$P(\omega_4) = +0.829038$	401	$E(401) = +0.170961$
5(501)	$\Omega_5 = \omega_{501} = 0.500\,\pi = 1.570796$	$\gamma_5 = +60.800264$	$P(\omega_5) = +0.085480$	501	$E(501) = -0.170961$
6(568)	$\Omega_6 = \omega_{568} = 0.567\,\pi = 1.781283$	$\gamma_6 = -35.615504$	$P(\omega_6) = -0.085480$	568	$E(568) = +0.170961$
7(701)	$\Omega_7 = \omega_{701} = 0.700\,\pi = 2.199114$	$\gamma_7 = +06.169203$	$P(\omega_7) = +0.085480$	701	$E(701) = -0.170961$
8(853)	$\Omega_8 = \omega_{853} = 0.852\,\pi = 2.676636$	$\gamma_8 = -00.914774$	$P(\omega_8) = -0.085480$	853	$E(853) = +0.170961$
$E(\omega_1) = E(\omega_3) = E(\omega_5) = E(\omega_7) = -\delta = -0.170961$, $E(\omega_2) = E(\omega_4) = E(\omega_6) = E(\omega_8) = +\delta = +0.170961$					

From the above tables, we find that the value of the convergence parameter convergence Q takes a value very close to zero during Repetition 5, since, as we can see, the new reference set that emerges in the way we described coincides with the current set and therefore the optimal frequency values have been determined. The continuation of repetitions from there on *will not offer anything substantial; it may just slightly improve the accuracy of the calculations regarding the number of decimal digits we consider in each case*. It is not difficult to see that the optimal solution leads to an approximation error that satisfies *exactly* the alternation theorem (see the last column for Repetition 5). The curve of functions $\mathcal{P}(\omega)$ and $E(\omega)$ for repetitions 2, 3, 4, and 5 are presented in images (a), (b), (c), and (d) of Fig. 12.70 respectively.

In the last step of the process, we replace the eight optimal extreme frequencies

$$\Omega_m = \omega_k, \quad (m,k) = \{(1,101), (2,162), (3,314), (4,401), (5,501), (6,568), (7,701), (8,801)\}$$

in the matrix equation $\boldsymbol{AX} = \boldsymbol{B}$, along with the values of functions $\widetilde{W}$ and $|\widetilde{\mathcal{H}}(e^{j\omega})|$ at the positions of these frequencies, which are $\widetilde{W}_1 = \widetilde{W}_{162} = \widetilde{W}_{314} = \widetilde{W}_{401} = 1$, $\widetilde{W}_{501} = \widetilde{W}_{568} = \widetilde{W}_{701} = \widetilde{W}_{853} = 2$ and $|\widetilde{\mathcal{H}}(e^{j\omega})|^1 = |\widetilde{\mathcal{H}}(e^{j\omega})|^{162} = |\widetilde{\mathcal{H}}(e^{j\omega})|^{314} = |\widetilde{\mathcal{H}}(e^{j\omega})|^{401} = 1$, $|\widetilde{\mathcal{H}}(e^{j\omega})|^{501} = |\widetilde{\mathcal{H}}(e^{j\omega})|^{568} = |\widetilde{\mathcal{H}}(e^{j\omega})|^{701} = |\widetilde{\mathcal{H}}(e^{j\omega})|^{853} = 0$, resulting in it taking the form

$$\begin{bmatrix} 1 & 1.000000 & 1.000000 & 1.000000 & 1.000000 & 1.000000 & 1.000000 & 1.0 \\ 1 & 0.874788 & 0.530511 & 0.053381 & -0.437115 & -0.818149 & -0.994300 & -1.0 \\ 1 & 0.554263 & -0.385583 & -0.981693 & -0.702649 & 0.202787 & 0.927445 & 1.0 \\ 1 & 0.309016 & -0.809016 & -0.809016 & 0.309016 & 1.000000 & 0.309016 & -1.0 \\ 1 & 0.000000 & -1.000000 & -0.000000 & 1.000000 & 0.000000 & -1.000000 & 0.5 \\ 1 & -0.208935 & -0.912691 & 0.590323 & 0.666011 & -0.868631 & -0.303035 & -0.5 \\ 1 & -0.587785 & -0.309016 & 0.951056 & -0.809016 & -0.000000 & 0.809016 & 0.5 \\ 1 & -0.893841 & 0.597904 & -0.175023 & -0.285019 & 0.684547 & -0.938733 & -0.5 \end{bmatrix} \begin{bmatrix} \alpha[0] \\ \alpha[1] \\ \alpha[2] \\ \alpha[3] \\ \alpha[4] \\ \alpha[5] \\ \alpha[6] \\ \delta \end{bmatrix} = \begin{bmatrix} 1 \\ 1 \\ 1 \\ 1 \\ 0 \\ 0 \\ 0 \\ 0 \end{bmatrix}$$

The solution of the above equation leads to the values $\alpha[0] = 0.444585$, $\alpha[1] = 0.629278$, $\alpha[2] = 0.098964$, $\alpha[3] = -0.177803$, $\alpha[4] = -0.094090$, $\alpha[5] = 0.103976$, $\alpha[6] = 0.166049$ and $\delta = -0.170961$. We observe that the magnitude of the quantity δ calculated above *is exactly the same as the one we determined from the application of the iterative process* (see the value of δ after the completion of Iteration 5), which, of course, is *expected*. Using now the values of the coefficients

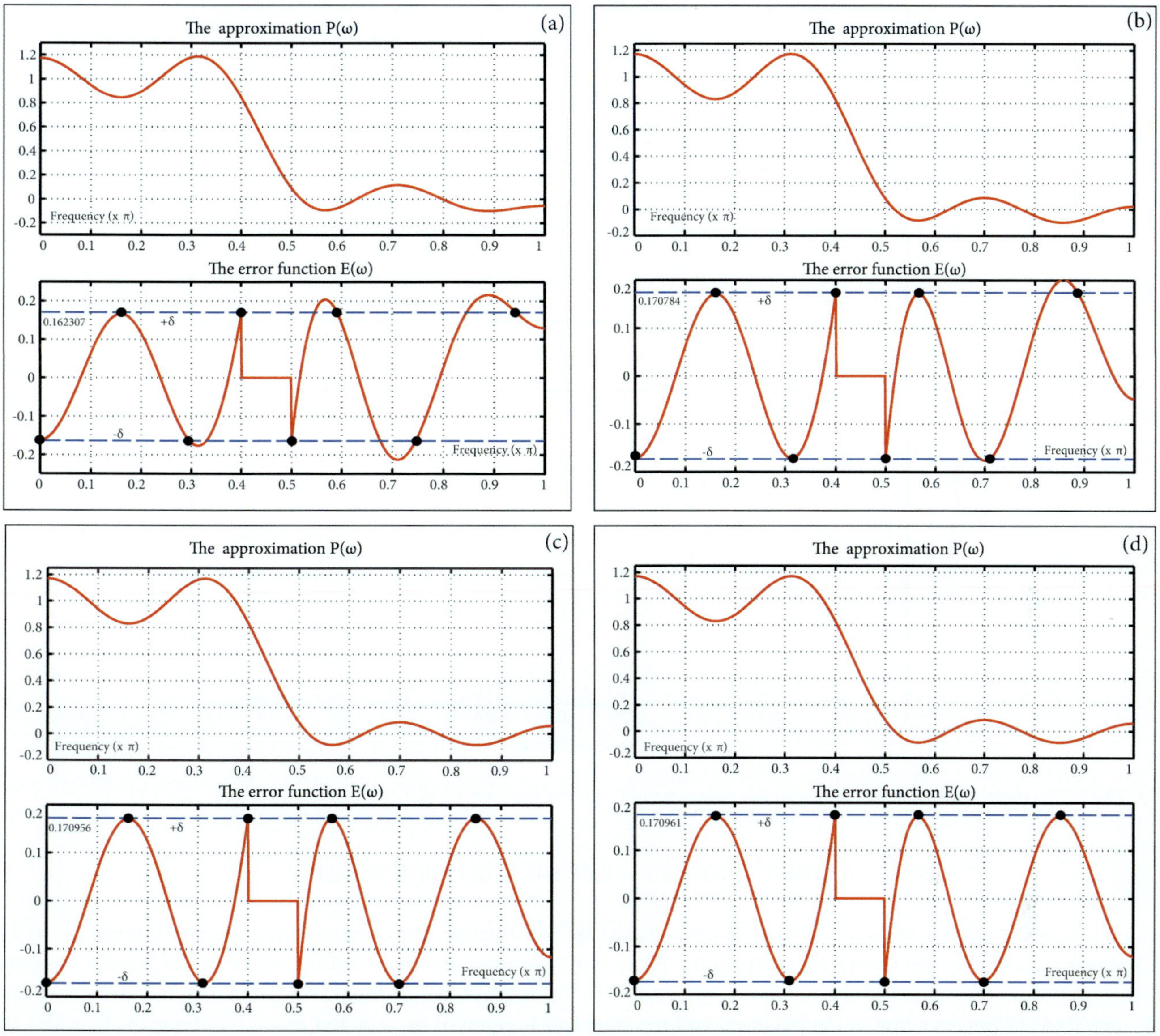

Fig. 12.70 The curve of the functions $\mathcal{P}(\omega)$ and $\mathcal{E}(\omega)$ for the iterations 2 [image (a)], 3 [image (b)], 4 [image (c)], and 5 [image (d)]

$\alpha[k]$ ($k = 0, 1, 2, \ldots, 6$), we can determine the samples of the impulse response *of the optimal approximation of the given ideal low-pass filter*, taking into account that the filter to be constructed is a type I filter and therefore the coefficients $\alpha[k]$ and the samples $h[n]$ will be related to each other through the expression

$$\alpha[n] = \begin{cases} h[(N-1)/2] & n = 0 \\ 2h[(M-1)/2 - n] & n = 1, 2, \ldots, (N-1)/2. \end{cases}$$

This relationship for a filter order $N = 13$ gives us $h[6] = \alpha[0]$ and $h[6-n] = \alpha[n]/2$ ($n = 1, 2, 3, 4, 5, 6$). From the above relationships we calculate the values $h[n]$ ($k = 0, 1, 2, \ldots, 6$), whereas the remaining samples corresponding to the values $n = 7, 8, 9, 10, 11, 12$ will be calculated from the *property of symmetry* $h[n] = h[N-1-n]$ (since the impulse response of type I FIR filters is a *symmetric signal*), which for the value $N = 13$ takes the form $h[n] = h[12-n]$. The implementation of the above calculations leads us to the discrete-time signal

$$h[n] = \{0.083024, 0.051988, -0.047045, -0.088901, 0.049482, 0.314639, 0.444585, 0.314639,$$

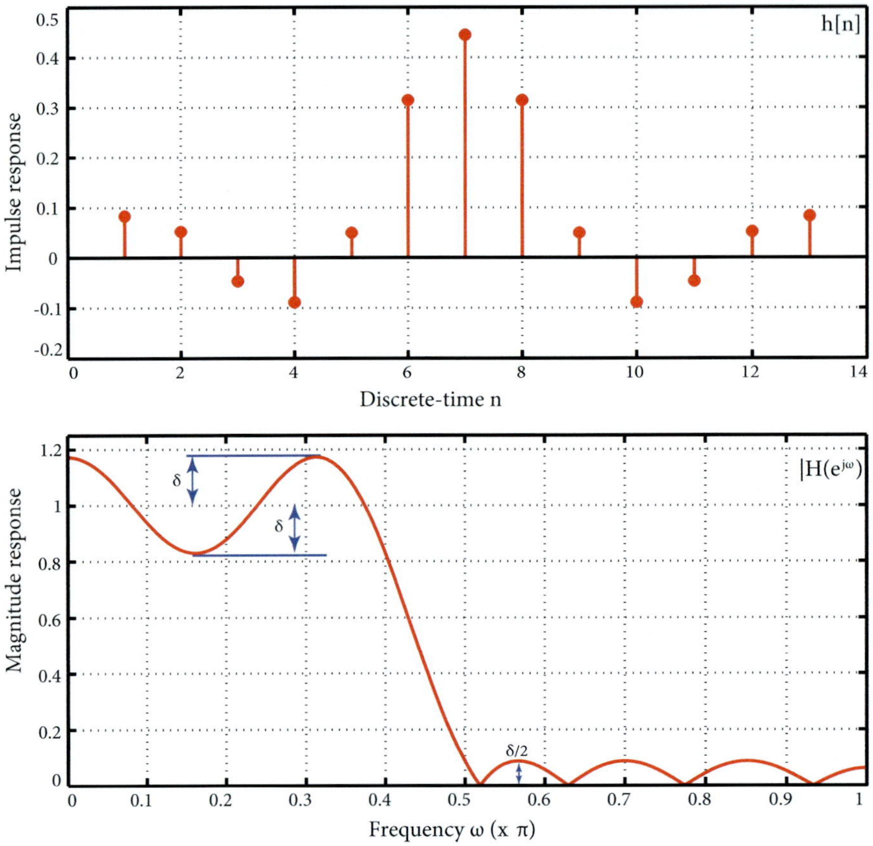

Fig. 12.71 The impulse response and the amplitude response of the filter of Exercise 12.12.5

$$0.049482, -0.088901, -0.047045, 0.051988, 0.083024\},$$

which is also the impulse response of the low-pass filter to be constructed. The corresponding frequency response will be calculated according to the known by applying the *Fourier transform* on the discrete-time signal of the impulse response. The graphical representation of the functions $h[n]$ and $|\mathcal{H}(e^{j\omega})|$ is presented in Fig. 12.71, from which it is apparent that the amplitude of the deviation in the passband and the stopband is equal to $\delta_1 = \delta$ and $\delta_2 = \delta/2$ respectively, a fact that is expected,[69] since the ratio of the error values that have been determined according to the filter specifications is equal to $K = K_s/K_p = 2$.

On the other hand, the alternative way of calculating the impulse response, *without the need to determine the values of the coefficients* $\alpha[k]$, requires the calculation of the function $\mathcal{P}(\omega)$ from its m samples $\mathcal{P}(\Omega_m)$ $(m = 1, 2, 3, 4, 5, 6, 7, 8)$, calculated at the positions of the eight best of extreme frequencies with the help of the *Lagrange interpolation method*. In the next step of the process, we calculate the value of the function in question at the positions $\omega_\ell = 2\pi\ell/N$ $(\ell = 1, 2, 3, 4, 5, 6)$ for $N = 13$. Since the accuracy of the calculations is limited, in the sense that the function $\mathcal{P}(\omega)$ has been calculated at the positions of 1000 uniformly distributed frequencies in the interval $[0, \pi]$, we choose the samples that correspond to frequency values that are as close as possible to the desired frequencies ω_ℓ. The results of this process are presented in Table 12.4, which contains from left to right the values of ℓ and the frequencies $\omega_\ell = 2\pi\ell/13$, the closest to these frequency values ω_k along with their position k within the table of 1000 uniformly distributed available frequencies as well as the value of the function $\mathcal{P}(\omega)$ at the positions of these frequencies. Taking into account that the measure of the desired frequency response $|\mathcal{H}(e^{j\omega})|$ is given for all filter categories by the relation $|\mathcal{H}(e^{j\omega})| = \mathcal{P}(\omega)\,\mathcal{Q}(\omega)$ and that for type I filters like the one we are designing is $\mathcal{Q}(\omega) = 1$, it is obvious that the samples of the function $\mathcal{P}(\omega)$ resulting from the above process are actually *samples of*

[69] As already mentioned, the value of the weight function $\mathcal{W}(\omega)$ is equal to 1 in the passband and equal to δ_1/δ_2 in the stopband, where $\delta_1 = \delta_s$ and $\delta_2 = \delta_p$ are the deviations in the passband and stopband respectively. In the general case where the weight function in these two bands has values K_p and K_s respectively, $\delta_p/\delta_s = K_s/K_p$.

Table 12.4 Calculation of the samples of the optimal approximation $\mathcal{P}(\omega)$ at the positions $\omega_\ell = 2\pi\ell/N (\ell = 1, 2, 3, 4, 5, 6)$ for filter order $N = 13$

ℓ	$\omega_\ell = 2\pi\ell/13$	ω_k	k	$P(\omega_k)$
0	0.000000	0.000000	001	+ 1.170961
1	0.483321	0.483805	155	+ 0.830808
2	0.966643	0.967610	309	+ 1.169736
3	1.449965	1.448274	462	+ 0.342986
4	1.933287	1.932074	616	− 0.028782
5	2.416609	2.415884	770	+ 0.009439
6	2.899931	2.899690	924	− 0.016118

the measure of the desired frequency response and therefore can be used directly for the calculation of the impulse response samples $h[n]$.

After calculating the samples of $P(\omega_k)$ we can calculate the samples of the impulse response $h[n]$ from the relation

$$h[n] = \frac{1}{N}\left[\mathcal{P}(0) + 2\sum_{k=1}^{L}\mathcal{P}(\omega_k)\cos(n\omega_k)\right] \quad -L \le n \le L$$

for the values $L = 6$ and $N = 13$. It is not difficult to realize that the signal that arises in this way *is not causal* (since it is defined for negative values of n) and for this reason it should undergo the operation of time shifting. So, by calculating the samples of the signal $h[n]$ from the above relation and then proceeding to the transformation $h[n] \longrightarrow h[n-L]$ $(n = 0, 1, 2, \ldots, N-1)$, for the values $L = 6$ and $N = 13$, we end up with the result

$$h[n] = \{0.0832788, 0.053559, -0.046537, -0.089750, 0.048549, 0.314692, 0.445161, 0.314642,$$
$$0.048549, -0.089750, -0.046537, 0.053559, 0.083788\},$$

which is practically the same as the one we determined with the previous method. The small deviation observed from the third decimal digit and beyond can be attributed to the fact that the samples of the function $\mathcal{P}(\omega)$ were not calculated *exactly* at the positions of the frequencies $\omega_\ell = 2\pi\ell/N$ $(\ell = 1, 2, 3, 4, 5, 6)$, but at *neighboring frequencies*, which were extremely close to the above values. If the requirements of the application require extremely high calculation accuracy, we can calculate exactly the values of the function $\mathcal{P}(\omega)$ at the frequencies $\omega_\ell = 2\pi\ell/N$ $(\ell = 1, 2, 3, 4, 5, 6)$ using the method of *polynomial interpolation*.

12.12.6 The Parks-McClellan Algorithm

The use of the *Remez algorithm* for the design of digital filters, which are characterized by the maximum degree of optimality for a given filter order and a given set of operating specifications, is a complex process, as shown by the description of the algorithm and Exercise 12.12.5 (see page 1272). For this reason, it soon became imperative to implement a *computer application* for the realization of this process. Such an application was implemented in 1972 by *Parks* and *McClellan*[70] for the design of *low-pass, high-pass*, and *band-pass* filters, as well as special-type filters, such as *digital differentiators* and *Hilbert transformers*. The original application of Parks and McClellan accepts as input the *order* N of the filter, its *type*, the *number* (with a minimum value of 2 for a *low-pass filter* and a maximum value of 10 for a *multi-band filter*—see Fig. 12.72) and the *limits* of the zones of interest (two frequencies for each zone), the *density of frequency values* used in numerical interpolation, as well as the values of the functions $\widetilde{W}(\omega)$ and $|\widetilde{\mathcal{H}}_d(e^{j\omega})|$ for each zone of interest for the filter to be constructed.

[70] See references (1) Parks T.W. & McClellan J.H., *Chebyshev Approximation for Nonrecursive Digital Filters with Linear Phase*, IEEE Transactions on Circuit Theory, Vol 9, pp. 189–194, March 1972 and (2) Parks T.W. & McClellan J.H., *A Program for the Design of Linear Phase Finite Impulse Response Digital Filters*, IEEE Transactions on Audio and Electroacoustics, Vol 20, pp. 195–199, August 1972.

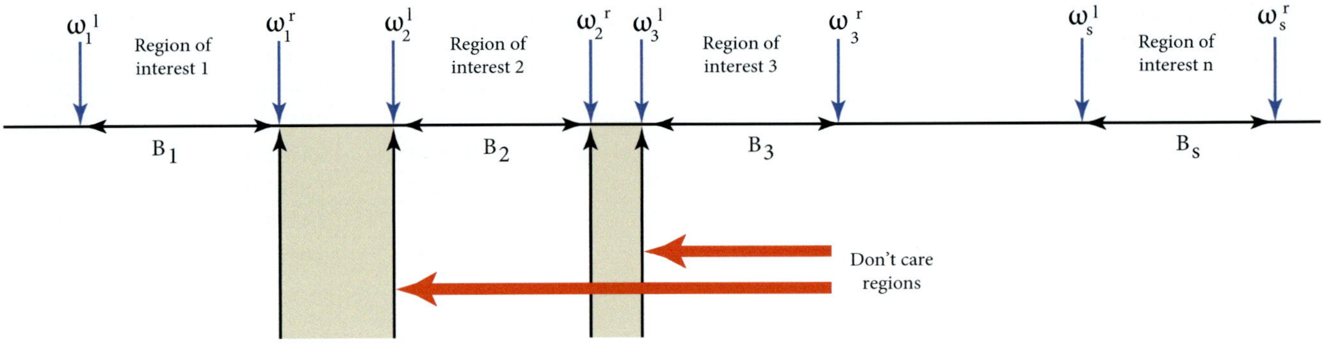

Fig. 12.72 The zones of interest and the zones of indifference of a digital filter of the FIR type

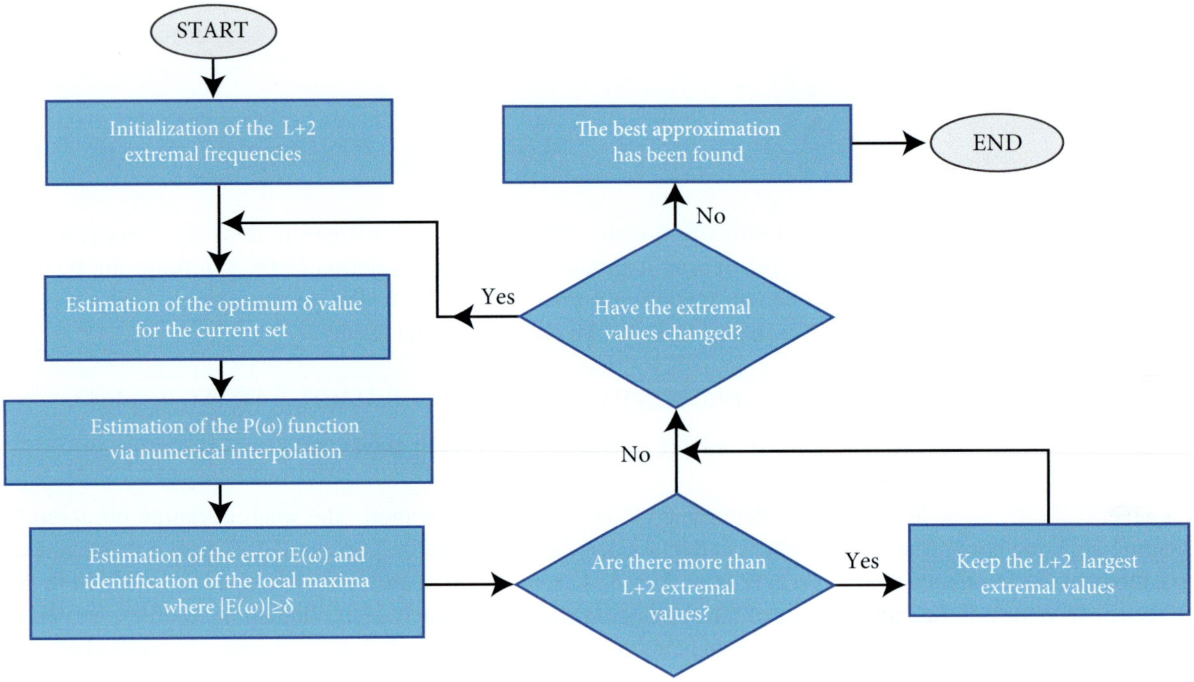

Fig. 12.73 The flow diagram of the Parks-McClellan method (Oppenheim, Schafer, Buck, Figure 7.41)

The *flow diagram* of the *Parks-McClellan method* is presented in Fig. 12.73. It is proven that the cost of calculating the value of δ is of the order of $O(L^2)$, whereas the cost of applying the interpolation method of Lagrange is of the order of $O(16L^2)$, a value that ultimately constitutes the cost of executing each iteration, which is obviously much smaller than the cost of the direct application of the Remez method—$O(L^3)$—especially for large values of the parameter L. Of course, the cost of calculating the samples $h[n]$ is of the order of $O(L^3)$ but this calculation is done once and only at the end of the process.

In practice, the operating specifications taken into account during the design of a digital filter include *the order N of the filter, the limits of the pass and cutoff zones ω_p and ω_s, as well as the values of the corresponding deviations δ_p and δ_s.* Generally speaking, these parameters are not independent of each other; on the contrary, knowing the values of some of them allows the calculation of the values of the rest, through an iterative process or some approximate equations. For example, Herrmann, Schüssler, & Hofstetter[71] developed algorithms in which the values of L, δ_p and δ_s remain constant, whereas the

[71] See references (1) Herrmann O., *On the Design of Nonrecursive Digital Filters with Linear Phase*, Electronic Letters, Vol. 6, pp. 328–329, 1970, (2) Herrmann O. & Schüssler W., *Design of Nonrecursive Digital Filters with Minimum Phase*, Electronic Letters, Vol. 6, No.1, pp. 329–330, 1970, and (3) Hofstetter E., Oppenheim A.V., & Siegel J., *On Optimum Nonrecursive Digital Filters*, Proceedings of 9th Allertan Conference on Circuit and System Theory, October 1971.

values of ω_p and ω_s vary; on the other hand, Parks, McClellan, & Rabiner[72] developed algorithms in which the values of L, ω_p, ω_s, and the ratio δ_p/δ_s remain constant, but the value of δ_p or δ_s changes. We mention for example, that if the values of δ_p, δ_s, ω_p and ω_s are known, then the order of the filter N can be approximately calculated from the relation

$$N \approx \frac{-20\log_{10}(\sqrt{\delta_p\delta_s}) - 13}{14.6\dfrac{\omega_s - \omega_p}{2\pi}} + 1, \tag{12.183}$$

which was constructed by Kaiser or from the more accurate relation

$$N = \frac{D_\infty(\delta_p, \delta_s) - f(\delta_p, \delta_s)(\Delta F)^2}{\Delta F} + 1, \tag{12.184}$$

where

$$\begin{aligned} D_\infty(\delta_p, \delta_s) &= [0.005309(\log_{10}\delta_p)^2 + 0.07114\log_{10}\delta_p - 0.4761]\log_{10}\delta_p \\ &\quad - [0.00266(\log_{10}\delta_p)^2 + 0.5941\log_{10}\delta_p + 0.4278] \\ f(\delta_p, \delta_s) &= 11.01217 + 0.51244(\log_{10}\delta_p - \log_{10}\delta_s) \end{aligned}$$

with the parameter ΔF defined as $\Delta F = (\omega_s - \omega_p)/2\pi$, which was constructed by Herrmann and his collaborators.[73] The above equation applies in the case of $\delta_s < \delta_p$, whereas in the opposite case we should *alternate* the role of these parameters. If the value of N that we calculate in this way leads to a δ value greater than δ_s, this *increases progressively* and the calculations are repeated until it finally approaches *the desired attenuation level.* In the case where the filter to be constructed has many zones of interest, we calculate from the above equation the value of N for *each* of the transition zones that are interposed between a passband and a stopband and then we choose for the order of the filter, *the maximum of the values* that will emerge in this way. This process, formulated in algorithmic form, includes the following steps:

1. We calculate the value of the filter order N from Eq. (12.177). If the N that will emerge is an even number, we make it odd by setting $N = N + 1$.
2. We design the order N filter using the *Remez algorithm* and calculate the maximum error value δ_{max}.
   (A) If $\delta_{max} > \delta_p$
      (a) We set $N = N + 2$, redesign the filter for the new value of the order N, and recalculate the maximum error value δ_{max}.
      (b) If for the new error value the relation $\delta_{max} \leq \delta_p$ holds, we proceed to Step 3 of the algorithm, otherwise we return to Step 2(A)α.
   (B) If $\delta_{max} < \delta_p$
      (a) We set $N = N - 2$, redesign the filter for the new value of the order N, and recalculate the maximum error value δ_{max}.
      (b) If for the new error value the relation $\delta_{max} > \delta_p$ holds, we proceed to Step 4 of the algorithm, otherwise, we return to Step 2(A)β.
3. We use the last set of extreme frequency values and the corresponding value for N to calculate the impulse response and the frequency response in the way we described in the previous section.
4. We perform the process of Step 3, using, however, not the last but the penultimate set of extreme frequency values.

[72] See the previous references by Parks-McClellan as well as references (1) McClellan J.J. & Parks T.W., *A Unified Approach to the Design of Optimum FIR Linear Phase Digital Filters*, IEEE Transactions of Circuit Theory, Vol CT-20, pp. 697–701, November 1973, (2) Rabiner L.R., *The Design of Finite Impulse Response Digital Filters Using Linear Programming Techniques*, Bell System Technical Journal, pp. 1177–1198, July-August 1972, and (3) Rabiner L.R., Linear Programming Design on Finite Impulse (FIR) Digital Filters, IEEE Transactions on Audio & Electroacoustics, Vol AU-20, No. 4, pp. 280–288, October 1972.

[73] See the reference Herrmann O., Rabiner L.R., & Chan D.S.K., *Practical Design Rules for Optimum Finite Impulse Response Low-pass Digital Filters*, The Bell Technical Journal, Vol. 52, pp. 769–799, 1973.

12.12.7 Digital Differentiators and Hilbert Transformers

The basic function of a *digital differentiator* is the calculation of the derivative of a signal, with the frequency response of the ideal filter given by the equation

$$\mathcal{H}(e^{j\omega}) = j\omega, \quad -\pi \leq \omega \leq \pi \tag{12.185}$$

leading to an impulse response of the form

$$h[n] = \frac{1}{2\pi}\int_{-\pi}^{+\pi} \mathcal{H}(e^{j\omega})\, e^{j\omega n} = \frac{1}{2\pi}\int_{-\pi}^{+\pi} j\omega e^{j\omega n} = \frac{\cos(\pi n)}{n} \quad -\infty < n < +\infty \quad (n \neq 0). \tag{12.186}$$

We observe that the ideal differentiator is characterized by an *antisymmetric impulse response* of the form $h[n] = -h[-n]$ and therefore $h[n] = 0$.

On the other hand, the *Hilbert transformer* is an *all-pass filter*, which causes a phase shift of 90° to the input signal, with the frequency response of the ideal filter having the form

$$\mathcal{H}(e^{j\omega}) = \begin{cases} -j & 0 < \omega \leq \pi \\ +j & -\pi < \omega < \pi, \end{cases} \tag{12.187}$$

leading to a noncausal response of infinite duration that is formulated as

$$h[n] = \frac{1}{2\pi}\int_{-\pi}^{+\pi} \mathcal{H}(e^{j\omega})\, e^{j\omega n} d\omega = \frac{1}{2\pi}\left(\int_{-\pi}^{0} je^{j\omega n} d\omega - \int_{0}^{\pi} je^{j\omega n} d\omega\right) = \begin{cases} \dfrac{2}{\pi}\dfrac{\sin^2(\pi n/2)}{n} & n \neq 0 \\ 0 & n = 0. \end{cases} \tag{12.188}$$

We observe that the impulse response is, as in the previous case, an *antisymmetric signal*.

It is not difficult to ascertain that both filters will be either *type III* or *type IV*, depending on whether the order N of the filter is *odd* or *even* respectively. Owing to the antisymmetric nature of the impulse response, as we have already mentioned, $h[0] = 0$, whereas additionally for the case of an odd order filter N, $|\mathcal{H}(e^{j\pi})| = 0$ will also be the case. But this means that *it is impossible to construct a differentiator that operates throughout the value range* $[0, \pi]$—something that is rarely needed in practice—as well as a *fully operational Hilbert transformer*. Conversely, in all practical applications, the Hilbert transformer operates *only in the frequency range defined by the input signal bandwidth*: if, therefore, f_1 and f_2 are the upper and lower limits of this bandwidth, then the desired frequency response of a *Hilbert transformer* will have a value of $|\mathcal{H}_d(e^{j\omega})| = 1$ and will be defined in the range $2\pi f_1 \leq \omega \leq 2\pi f_2$. Analogously, the desired frequency response of the digital differentiator needs to be a *linear function of frequency*, only within the value range $0 \leq \omega \leq 2\pi f_p$, where f_p is the so-called *bandwidth of the differentiator*.

In order to achieve a *constant bandwidth deviation in the differentiator's passband* and taking into account that the *desired frequency response is proportional to the frequency*, it is not difficult to ascertain that the *weight function* $\mathcal{W}(\omega)$ that will be used for the design of the differentiator through the use of the *approximate method by Chebyshev* will vary *inversely proportionally* to the frequency, so that we can write that $\mathcal{W}(\omega) = 1/\omega$ for the range of values $0 \leq \omega \leq 2\pi f_p$. Consequently, the absolute error between the *desired* and the *approximate* frequency response will indeed increase as the frequency ω varies between the values 0 and $2\pi f_p$, but nevertheless the weight function ensures that the value of the parameter

$$\delta = \max_{0\leq\omega\leq 2\pi f_p}\left\{\mathcal{W}(\omega)\left[\omega - |\mathcal{H}(e^{j\omega})|\right]\right\} = \max_{0\leq\omega\leq 2\pi f_p}\left[1 - \frac{|\mathcal{H}(e^{j\omega})|}{\omega}\right] \tag{12.189}$$

remains constant within the passband of the differentiator. The corresponding relationship for the case of a Hilbert transformer is

$$\delta = \max_{0 \leq \omega \leq 2\pi f_p} \left[|\mathcal{H}_d(e^{j\omega})| - |\mathcal{H}(e^{j\omega})| \right] = \max_{0 \leq \omega \leq 2\pi f_p} \left[1 - |\mathcal{H}(e^{j\omega})| \right], \tag{12.190}$$

with the filter coefficients being calculated so as to *minimize* the value of the quantity δ. In this case, the weight function obviously has the constant value $\mathcal{W}(\omega) = 1$.

More details about these two special types of digital filters can be found in the literature.[74]

12.13 Design of Infinite Impulse Response Filters

After the description of the design techniques for *FIR filters*, let us now move on to the corresponding description for *IIR filters*, which in the international literature are known as *IIR*. As we have already mentioned in the introductory section of this chapter, there are several methods of designing filters of this type, the most well-known of which consists of *converting the analog IIR filters we studied in the previous sections (i.e., the Butterworth, Chebyshev I & II, Bessel, and Cauer filters) into their digital equivalent*. This conversion process can take place in many ways, such as converting the *differential equation* of the analog filter into the *difference equation* of the corresponding digital filter, as well as constructing the digital filter so that *its impulse response is the sampled version of the impulse response of the corresponding analog filter*. Taking into account the fact that the convergence area of an analog filter is *the left half-plane of the complex plane*, whereas the convergence area of the corresponding digital filter is *the interior of the unit circle*, it is obvious that these conversion processes should represent *the imaginary axis of the s domain in the unit circle of the z domain* and transform one region into another, so that *a stable pole of the analog filter is transformed into a stable pole of the corresponding digital filter*. Although the above techniques allow such a representation, it is characterized by several restrictions and is only applicable to *low-pass filters*, as well as certain categories of band-pass filters. For this reason, in practice, this representation is performed by a special transformation known as *bilinear transformation*. It is obvious that in every case the transfer function of the digital filter must meet all the known properties, so that it can be characterized as *implementable*, such as, for example, its formulation in explicit form (i.e., as a ratio of polynomial functions of *real coefficients*) with the degree of the numerator polynomial being less than or equal to the degree of the denominator polynomial.

A second way to design recursive filters IIR is the use of *optimization methods*, similar to those used for the design of type filters FIR. In other words, these methods allow the construction of a frequency response function that constitutes the *optimal approximation of an ideal function* $\mathcal{H}(e^{j\omega})$, from the perspective of *minimizing* some *error function*. Typical methods of designing digital filters that belong to this category are the various variations of the *Newton's minimum finding method*, as well as the minmax algorithms that operate in a similar way to the *Remez algorithm*. The detailed description of these recursive digital filter design techniques is the subject of the following pages.

12.13.1 Approximation of Derivatives from Difference Equations

As is well known from basic theory, a linear and time-invariant continuous-time system is described by a differential equation of the form

$$\sum_{k=0}^{N} \alpha_k \frac{d^k y(t)}{dt^k} = \sum_{k=0}^{M} \beta_k \frac{d^k x(t)}{dt^k}, \tag{12.191}$$

[74] See references (1) Rabiner L.R. & Schafer R.W., *On the Behavior of Minimax Relative Error FIR Digital Differentiators*, Bell System Technical Journal, Vol. 53, pp. 333–362, February 1974 and (2) Rabiner L.R. & Schafer R.W., *On the Behavior of Minimax Digital Hilbert Transformers*, Bell System Technical Journal, Vol. 53, pp. 363–394, February 1974.

which contains derivatives up to the order of k (with the maximum value of k being equal to M for the input and equal to N for the output). This *differential equation* can be approximated by a *difference equation* that is solved *numerically* with the help of computers. Although the *method of finite differences* (as this approach is characteristically called), was proposed by Euler in the seventeenth century, it attracted the interest of scientists only after 1945, when electronic computers began to be used for solving numerical problems.

The basic idea behind this study method is *the approximation of the derivative of a function by the corresponding difference* (see also the homonymous section of the mathematical appendix). Taking into account the fact that the first derivative of a continuous function $x = x(t)$ is defined as the limit

$$x'(t) = \frac{dx(t)}{dt} = \lim_{\Delta t \to 0} \frac{x(t+\Delta t) - x(t)}{\Delta t}, \tag{12.192}$$

its approximation in the field of discrete-time is based on the use of the so-called *backward difference* and is given by the relation[75]

$$x'(t) \approx \frac{x(t+\Delta t) - x(t)}{\Delta t}, \tag{12.193}$$

where now the interval Δt is not *infinitesimal*, as in the continuous case, but *finite* and with a naturally small value (it is obvious that *the smaller the value of the interval Δt the more accurate is the approximation of the continuous derivative from its discrete equivalent, with the two quantities coinciding at the limit $\Delta t \to 0$*. In the case of continuous-time systems that we are interested in and which are known to be transformed into their *discrete equivalents* through *sampling processes*, this time interval Δt can be considered as the sampling period $T_s \equiv T$, whereas the quantities appearing in the above approximation can be considered as *samples of the function $x(t)$* corresponding to time instances $t_n = nT_s$. Based on all these assumptions, the above relationship can take the form

$$\left.\frac{x(t)}{dt}\right|_{t=nT} = \frac{x[nT] - x[nT-T]}{T} = \frac{x[nT] - x[(n-1)T]}{T_s} = \frac{x[n] - x[n-1]}{T},$$

where, for simplicity of writing, we adopted the notation $x[n] \equiv x[nT]$. Now, using this approximation for the first derivative, the second derivative of the function $x(t)$ will be approximated analogously from the equation

$$\begin{aligned}\left.\frac{d^2x(t)}{dt^2}\right|_{t=nT} &= \left[\frac{d}{dt}\left(\frac{dx(t)}{dt}\right)\right]_{t=nT} = \frac{x'[n] - x'[n-1]}{T} \\ &= \frac{\dfrac{x[n]-x[n-1]}{T} - \dfrac{x[n-1]-x[n-2]}{T}}{T} = \frac{x[n] - 2x[n-1] + x[n-2]}{T^2}\end{aligned}$$

with the expression $x[n] - 2x[n-1] + x[n-2]$ being known as the *second difference* of the function $x[n]$.

In order to determine the correlation that exists between the *continuous derivative* and its *discrete equivalent* from the perspective of the transfer function, let us consider a continuous-time system LTI that accepts as input a continuous-time signal $x(t)$ and produces at its output its derivative, that is, the signal $y(t) = x'(t)$. This system is known as a *differentiator* and its discrete version was studied in Sect. 12.12.7 (see page 1282). It is not difficult to realize that the differential equation of this system arises from the general expression for parameter values $N = 0$, $M = 1$, $\alpha_0 = \beta_1 = 1$ and $\beta_0 = 0$. Taking the *Laplace transform* of the two terms and using the property of *differentiation*, we will have $\mathcal{L}\{y(t)\} = \mathcal{Y}(s)$ and $\mathcal{L}\{x'(t)\} = s\mathcal{X}(s)$, and therefore the transfer function of the system will have the form $\mathcal{H}(s) = \mathcal{Y}(s)/\mathcal{X}(s) = s$. On the other hand, the equation for calculating the discrete approximation of the continuous first-order derivative can be considered as the output of a system that accepts at its input the discrete-time signal $x[n]$ and produces at its output the discrete-time signal

[75] In a more accurate formulation this relation is written as

$$x'(t) = \frac{x(t+\Delta t) - x(t)}{\Delta t} + \mathcal{O}(\Delta t)$$

with the quantity $\mathcal{O}(\Delta t)$, which is of the order of Δt expressing the *approximation error*. This relationship arises by expanding the function $x(t)$ *by Taylor* around the point $x(t + \Delta t)$.

$y[n] = (x[n] - x[n-1])/T$. But the latter relationship constitutes a special case of Eq. (7.1) for parameter values $N = 0$, $M = 1$, $\alpha_0 = 0$, and $\beta_0 = \beta_1 = 1/T$. $\mathcal{Z}$ transform of the two terms and using the known property $\mathcal{Z}\{x[n-k]\} = z^{-k}\mathcal{X}(z)$, the above relationship is written

$$\mathcal{Y}(z) = \frac{1}{T}\mathcal{X}(z) - \frac{1}{T}z^{-1}\mathcal{X}(z) = \frac{1-z^{-1}}{T}\mathcal{X}(z)$$

leading to a transfer function of the form

$$\mathcal{H}(z) = \frac{\mathcal{Y}(z)}{\mathcal{X}(z)} = \frac{1-z^{-1}}{T}.$$

From the above analysis, the conclusion emerges that *the discretization of the continuous-time system that calculates the continuous first derivative and is described by the transfer function* $\mathcal{H}(s) = s$ *can be achieved by proceeding to the substitution* $s = (1-z^{-1})/T$*, where* T *is a suitably selected sampling period with a relatively small value* (for example, less than or equal to 0.1). In this case, the equivalent discrete-time system that will emerge and which calculates the *discrete equivalent* of the continuous first derivative, that is, the expression $(y[n] - y[n-1])/T$, is described by the transfer function $\mathcal{H}(z)$ that we calculated above.

The application of the above process to the continuous-time system of calculation of the second derivative and its discrete equivalent, which, as we have proven, are described by the relations

$$y(t) = \frac{d^2x(t)}{dt^2} \quad \text{and} \quad y[n] = \frac{x[n] - 2x[n-1] + x[n-2]}{T^2}$$

respectively, will lead to the transfer functions

$$\mathcal{H}(s) = \frac{\mathcal{Y}(s)}{\mathcal{X}(s)} = s^2 \quad \text{and} \quad \mathcal{H}(z) = \frac{\mathcal{Y}(z)}{\mathcal{X}(z)} = \frac{1-2z^{-1}+z^{-2}}{T^2} = \left(\frac{1-z^{-1}}{T}\right)^2$$

and consequently the function $\mathcal{H}(z)$ of the discrete-time system can arise from the transfer function $\mathcal{H}(s)$ of the corresponding continuous-time system by making the substitution $s^2 = [(1-z^{-1})/T]^2$. Applying the same process, we can analogously prove that the continuous computation system of the continuous derivative of order k is transformed into the equivalent discrete-time system, by making the substitution in the transfer function $\mathcal{H}(s)$

$$s^k = \left(\frac{1-z^{-1}}{T}\right)^k. \tag{12.194}$$

Based on the above analysis, we can conclude that the transfer function $\mathcal{H}(z)$ of a digital IIR system is related to the transfer function $\mathcal{H}(s)$ of the equivalent (from the perspective of finite differences) continuous-time system, according to the relation

$$\mathcal{H}(z) = \mathcal{H}(s)\Big|_{s=\frac{1-z^{-1}}{T}} \tag{12.195}$$

for a suitably chosen (small) value of the sampling interval T.

In order to check whether the above transformation maps the imaginary axis $s = j\omega$ of the s domain onto the unit circle in the z domain, let us solve the transformation equation for z, which will take the form $z = 1/(1-sT)$. Proceeding now with the substitution $s = j\omega$, we get

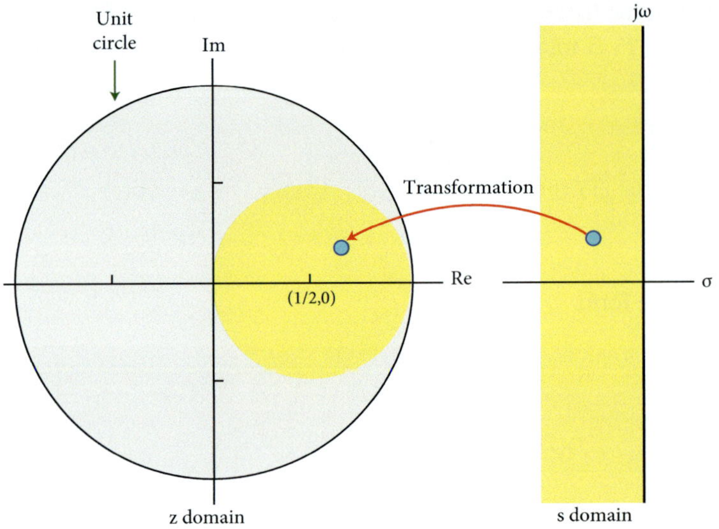

Fig. 12.74 Representation of the s domain in the z domain under the application of the transformation based on the *approximation of derivatives by finite differences* (Proakis and Manolakis, Figure 8.30)

$$z=\frac{1}{1-sT}=\frac{1}{1-j\omega T}=\frac{1}{2}\frac{(1-j\omega T)+(1+j\omega T)}{1-j\omega T}=\frac{1}{2}\left[1+\frac{1+j\omega T}{1-j\omega T}\right].$$

Taking into account that $1\pm j\omega T=(\sqrt{1+\omega^2T^2})\ \exp[\pm j\tan^{-1}(\omega T)]$, the image of the variable $s=j\omega$ in the z domain takes the form

$$z=\frac{1}{2}\left[1+\frac{1+j\omega T}{1-j\omega T}\right]=\frac{1}{2}\left[1+\frac{(\sqrt{1+\omega^2T^2})\ \exp[+j\tan^{-1}(\omega T)]}{(\sqrt{1+\omega^2T^2})\ \exp[-j\tan^{-1}(\omega T)]}\right]=\frac{1}{2}\left\{1+\exp\left[j\,2\tan^{-1}(\omega T)\right]\right\} \tag{12.196}$$

with real and imaginary parts

$$\Re\{z\}=\frac{1}{2}+\frac{1}{2}\cos\left[2\tan^{-1}(\omega T)\right]\quad\text{and}\quad\Im\{z\}=\frac{1}{2}\sin\left[2\tan^{-1}(\omega T)\right].$$

Solving the above equations for $\cos\vartheta$ and $\sin\vartheta$ (where $\vartheta=2\tan^{-1}(\omega T)$) and taking into account the relation $\cos^2\vartheta+\sin^2\vartheta=1$ it turns out that

$$\left(\Re\{z\}-\frac{1}{2}\right)^2+\left(\Im\{z\}\right)^2=\left(\frac{1}{2}\right)^2, \tag{12.197}$$

which expresses the circumference of a circle with its center at the point $(1/2,0)$ and its radius equal to $1/2$. We observe, therefore, that the above transformation *does not map the imaginary axis of the s domain to the circumference of the unit circle of the z domain* but to the circumference we defined above. Exception to this rule, are small frequency values ω for which we can make the approximation $\omega T\to 0$. In this case, $\tan^{-1}(\omega T)\to 0$ and consequently $\cos\left[2\tan^{-1}(\omega T)\right]\to 1$ and $\sin\left[2\tan^{-1}(\omega T)\right]\to 0$ from where $\Re\{z\}\to 1$ and $\Im\{z\}\to 0$ follow, coordinates that belong to the circumference of the unit circle; however, this is the *exception* and not the rule.

On the other hand, in order to determine whether the above transformation *preserves stability*, that is, if *the points of the left complex half-plane of the s domain are mapped inside the unit circle in the z domain*, let us consider a point of the left half-plane $sT=\alpha+j\beta$ ($\alpha<0,\alpha,\beta\in R$). In this case, this point is mapped to the point

$$z = \frac{1}{1-\alpha-j\beta} = \frac{1}{\sqrt{(1-\alpha)^2+\beta^2}} \exp\left[j \tan^{-1}\left(\frac{\beta}{1-\alpha}\right)\right]. \tag{12.198}$$

Observing that the modulus of the complex quantity $|z|$ is less than one, namely

$$|z| = \frac{1}{\sqrt{(1-\alpha)^2+\beta^2}} < 1,$$

we can state the claim that *the transformation of the analog to digital filter using the method of backward difference preserves stability*, since a stable point of the s domain is mapped to a stable point of the z domain. Specifically, it can be proven that the points of the left complex half-plane are mapped inside the circle with the center point $(1/2, 0)$ and radius $r = 1/2$ that we encountered previously, whereas the points of the right complex half-plane of the s domain are mapped outside the above circle. However, the aforementioned transformation *does not preserve the frequency characteristics of the analog system*, since the points of the imaginary axis $s = j\omega$ are not depicted on the circumference of the unit circle in the z domain. Taking into account that only for very small values of ω is its depiction in the z domain done correctly (since, as we saw, in this case, the image in the z domain belongs to the neighborhood of the point $z = 1$, which lies on the circumference of the unit circle), we can conclude that this method works correctly *only for certain low-pass filters, as well as for band-pass filters with very low-resonance frequency*, whereas it cannot be used for the construction, let us say, of high-pass filters, in which the more general *bilinear transformation* is used, which is presented in a subsequent section. The depiction of the s domain in the z domain through the application of the finite-difference method is presented diagrammatically in Fig. 12.74.

Example 12.13.1 In order to understand the application of this method, let us consider an analog filter with a transfer function

$$\mathcal{H}(s) = \frac{1}{s^2+s+4.25}.$$

In this case, the substitution $s = (1-z^{-1})/T$ leads to the transfer function

$$\mathcal{H}(z) = \frac{1}{\left(\dfrac{1-z^{-1}}{T}\right)^2 + \dfrac{1-z^{-1}}{T} + 4.25} = \frac{T^2}{(4.25T^2+T+1)-(T+2)z^{-1}+z^{-2}}$$

$$= \frac{T^2/(4.25T^2+T+1)}{1-\dfrac{T+2}{4.25T^2+T+1}z^{-1}+\dfrac{1}{4.25T^2+T+1}z^{-2}},$$

which has two poles that are roots of the quadratic equation

$$1-\frac{T+2}{4.25T^2+T+1}z^{-1}+\frac{1}{4.25T^2+T+1}z^{-2} = 0$$

with discriminant

$$\Delta = \left(-\frac{T+2}{4.25T^2+T+1}\right)^2 - 4\frac{1}{4.25^2+T+1} = \frac{-16T^2}{(4.25T^2+T+1)^2}$$

and roots

$$z_{(1,2)}^{-1} = \frac{(T+2)/(4.25T^2+T+1) \pm j\,(4T)/(4.25T^2+T+1)}{2/(4.25T^2+T+1)} = \frac{(T+2)\pm j\,4T}{2} = \left(T+\frac{1}{2}\right) \pm j\,2T.$$

The reader is asked to determine the transfer function $\mathcal{H}(z)$ and its poles for various values of the parameter T and then to construct the frequency response of the digital filter, in order to ascertain the way in which the value of T affects the frequency characteristics of the system.

Example 12.13.2 As a second example of the application of this method, let us consider the first-order resistor-conductor RC circuit of Exercise 2.4.1 (see page 81), which, as we have proven, is described by the differential equation

$$\frac{dy(t)}{dt} + \frac{1}{RC}y(t) = \frac{1}{RC}x(t),$$

whereas its transfer function and its impulse response have the form

$$\mathcal{H}(s) = \frac{1}{RCs+1} \qquad \text{and} \qquad h(t) = \frac{1}{RC}e^{-t/RC}u(t)$$

(see Exercise 4.8.6 on page 365 and Fig. 4.13 on page 366). Applying the method of derivative approximation, let us approximate the derivative $y'(t)$ from the first difference $y'(t) \approx (y[n] - y[n-1])/T$. In this case, using the simplified notation $f(nT) \equiv f[n]$, the evaluation of the differential equation for the time $t = nT$ will give us

$$y[n] = \frac{1}{(Rc/T)+1}x[n] + \frac{RC}{RC+T}y[n-1] = \frac{T}{RC+T}x[n] + \frac{RC}{RC+T}y[n-1].$$

Taking the $\mathcal{Z}$ transform of the two terms of the above difference equation we get

$$\mathcal{Y}(z) = \frac{T}{RC+T}\mathcal{X}(z) + \frac{RC}{RC+T}z^{-1}\mathcal{Y}(z) \qquad \text{or equivalently} \qquad \mathcal{Y}(z)\left(1 - \frac{RC}{RC+T}z^{-1}\right) = \frac{T}{RC+T}\mathcal{X}(z)$$

and consequently the transfer function of the equivalent discrete-time filter will have the form

$$\mathcal{H}(z) = \frac{\mathcal{Y}(z)}{\mathcal{X}(z)} = \left(\frac{T}{RC+T}\right)\Big/\left(1 - \frac{RC}{RC+T}z^{-1}\right).$$

Replacing in the above equation the value $z = e^{j\omega}$, the frequency response of the discrete-time filter expressed in the *amplitude-phase representation* is proven to have the form

$$\mathcal{H}(e^{j\omega}) = \frac{\dfrac{T}{RC+T}}{\sqrt{1 - 2\dfrac{RC}{RC+T}\cos\omega + \dfrac{R^2C^2}{(RC+T)^2}}} \exp\left[j\left(\omega - \tan^{-1}\frac{\sin\omega}{\cos\omega - [RC/(RC+T)]}\right)\right].$$

On the other hand, the amplitude response of the aforementioned analog filter is given by the equation

$$|\mathcal{H}(e^{j\omega})| = \frac{1}{\sqrt{1+\omega^2R^2C^2}}.$$

The reader is asked to estimate the values of the above expressions for indicative values of the circuit elements R, C and the sampling period T and to draw conclusions about the characteristics of the analog and the corresponding digital filter.

In order to remove the restrictions regarding the application of the above approach, we can resort to the use of the order difference L by proceeding to the replacement

$$\left.\frac{dy(t)}{dt}\right|_{t=nT} = \sum_{k=1}^{L}\alpha_k\frac{y[n+k]-y[n-k]}{T} \qquad (12.199)$$

with the coefficients α_k being chosen so that *they optimize the approximation of the derivative from the equivalent finite difference*. In this case, the transfer function of the continuous differentiator of the left side will still be $\mathcal{H}(s) = s$, whereas for the discrete-time system of the right side that is described by the difference equation

$$y[n] = \sum_{k=1}^{L} \alpha_k \frac{x[n+k] - x[n-k]}{T}$$

the application of the $\mathcal{Z}$ transform to the two sides in combination with the property of the *time shift* will give us

$$\mathcal{Y}(z) = \mathcal{Z}\{y[n]\} = \sum_{k=1}^{L} \alpha_k \frac{\mathcal{Z}\{x[n+k]\} - \mathcal{Z}\{x[n-k]\}}{T} = \sum_{k=1}^{L} \alpha_k \frac{\mathcal{X}(z)(z^k - z^{-k})}{T} = \frac{\mathcal{X}(z)}{T} \sum_{k=1}^{L} \alpha_k (z^k - z^{-k}). \quad (12.200)$$

Therefore, the transfer function of the discrete-time system is the

$$\mathcal{H}(z) = \frac{\mathcal{Y}(z)}{\mathcal{X}(z)} = \frac{1}{T} \sum_{k=0}^{L} (z^k - z^{-k}) \quad (12.201)$$

and is related to the function $\mathcal{H}(s)$ of the corresponding continuous-time system through the transformation

$$s = \frac{1}{T} \sum_{k=1}^{L} \alpha_k (z^k - z^{-k}). \quad (12.202)$$

In order to check whether this new but more complex transformation leads to better results, let us determine *the geometric locus of the s domain* in which the points of the circumference of the unit circle in the z domain are depicted. Substituting in the last relation the expression $z = e^{j\omega}$ (which, as known, constitutes the defining equation of the unit circle), we get

$$s = \frac{1}{T} \sum_{k=1}^{L} \alpha_k (e^{j\omega k} - e^{-j\omega k}) = j\frac{2}{T} \sum_{k=1}^{L} \alpha_k \sin(\omega_k). \quad (12.203)$$

We observe that the quantity of the right member is a purely imaginary number. Therefore, if we define the frequency Ω of the s domain from the relation

$$\Omega = \frac{2}{T} \sum_{k=1}^{L} \alpha_k \sin(\omega_k) \quad (12.204)$$

the above relation takes the form $s = j\Omega$ and obviously describes the imaginary axis of the s domain. We find, then, that this transformation *allows the depiction of the imaginary axis of the s domain on the unit circle of the z domain*, whereas additionally it is proven that *the left half-plane of the space of s is mapped inside the unit circle in the space of z* and therefore the stability of the filter *is preserved*. However, the problem of the appropriate choice of the values of the coefficients α_k

remains, which, let it be noted, is not trivial. For this reason, this method is not used very often, especially when there are more efficient and user-friendly mapping techniques, such as *bilinear transformation*, which is presented in a subsequent section.

12.13.2 The Method of Invariant Impulse Response

Let us consider an analog filter with an impulse response $h_\alpha(t)$, the operation of which meets some specifications. Starting from this analog filter, we can construct a digital filter in such a way that its impulse response $h[n]$ is related to the impulse response $h_\alpha(t)$ of the analog filter through a *sampling* process of period T_s and therefore is described by a relationship of the form

$$h[n] = h_\alpha(nT_s), \quad n = 0, 1, 2, \ldots. \tag{12.205}$$

This way of constructing the digital filter is known as the *method of invariant impulse response*.

As we have mentioned in Sect. 6.2 (see page 501), the spectra $\mathcal{X}(j\Omega) = \mathcal{F}\{x_s(t)\}$ and $\mathcal{X}(e^{j\omega}) = \mathcal{F}\{x[n]\}$ of the train of impulse functions $x_s(t)$ and the equivalent discrete-time signal $x[n]$ that arise from sampling are related to the spectrum $\mathcal{X}_\alpha(j\Omega) = \mathcal{F}\{x(t)\}$ of the signal being sampled as

$$\mathcal{X}_s(j\Omega) = \frac{1}{T_s}\sum_{k=-\infty}^{\infty} \mathcal{X}_\alpha\left[j\left(\Omega - \frac{2k\pi}{T_s}\right)\right] \quad \text{and} \quad \mathcal{X}(e^{j\omega}) = \frac{1}{T_s}\sum_{k=-\infty}^{\infty} \mathcal{X}_\alpha\left[j\left(\frac{\omega}{T_s} - \frac{2k\pi}{T_s}\right)\right], \tag{12.206}$$

(where T_s is the sampling period) and therefore constitute *periodic repetitions* of the scaled spectrum $\mathcal{X}_\alpha(e^{j\Omega})/T_s$, with the construction of the spectrum $\mathcal{X}(e^{j\omega})$ also including frequency scaling (in this and the following sections we use again the symbol Ω to describe the frequency in the s domain) with a period $F_s = 1/T_s$ (see Fig. 6.3 on page 510). Using the second of the above expressions for the spectra $\mathcal{H}_s(e^{j\omega}) = \mathcal{F}\{h[n]\}$ and $\mathcal{H}_\alpha(j\Omega) = \mathcal{F}\{h_\alpha(t)\}$, by analogy we have

$$\mathcal{H}_s(e^{j\omega}) = \frac{1}{T_s}\sum_{k=-\infty}^{\infty} \mathcal{H}_\alpha\left[j\left(\frac{\omega}{T_s} - \frac{2k\pi}{T_s}\right)\right], \tag{12.207}$$

with the correlation between them being similar to that of Fig. 6.3 and the frequencies ω and Ω being related to each other through the relation $\omega = \Omega T_S$. From this figure and based on the discussion we had in Sect. 6.2, it is obvious that the digital filter that arises with this method preserves the frequency characteristics of the equivalent analog filter, *provided that the sampling period T_s is small enough so that there is no overlap of the successive repetitions of the spectrum $\mathcal{H}_\alpha(\Omega)$* (which, as is well known, leads to the appearance of folded (or aliased) frequencies), or if such a thing is inevitable, at least to minimize the side effects of this overlap. But this means that this method *is not suitable for the construction of high-pass filters*, the operating frequencies of which do not satisfy the Nyquist theorem.

In this method of constructing digital filters, the mapping of the s domain to the z domain is done by an expression of the form $z = e^{sT_s}$, which satisfies the desired mapping properties. For example, considering that the complex variable s has the form $s = \sigma + j\Omega$ and expressing the complex variable z in its polar form $z = re^{i\omega}$, the above relationship is written as

$$z = re^{j\omega} = e^{(\sigma+j\Omega)T_s} = e^{\sigma T_s}e^{j\Omega T_s},$$

from which it follows that $r = e^{\sigma T_s}$ and obviously $\omega = \Omega T_s$. It is not difficult to see that for $\sigma = 0$ (i.e., for $s = j\Omega$), $r = 1$ and therefore *the imaginary axis of the s domain is mapped to the circumference of the unit circle of the z domain*. On the other hand, if $\sigma < 0$, $0 < r < 1$ and therefore *the points of the left of the complex half-plane of the space of s are depicted inside the unit circle of the space of* z. Finally, for the case $\sigma > 0$, $r > 1$ and therefore *the points of the right complex half-plane of the space of* s *are depicted outside the unit circle of the space of* z. However, let us note that the mapping of the imaginary axis $s = j\Omega$ on the circumference of the unit circle of the space of z *is not 1-to-1*, since in the range of values $-\pi \leq \omega \leq \pi$ of the discrete frequencies are depicted *many ranges of values of the continuous frequencies* of the form $(2k-1)\pi/T_s \leq \Omega \leq (2k+1)\pi/T_s$ for various values of the integer parameter k. This many-to-1 correspondence of continuous to discrete frequencies is due to the appearance of *alias frequencies*, which is also one of the disadvantages of the method and is related to the fact that in practice there are no analog filters that are characterized by a *strictly limited bandwidth spectrum*, so that there is no overlap of the repeated snapshots of the analog spectrum, the periodic repetition of which leads, as we commented previously, to the spectrum of the corresponding digital filter. However, if the spectrum of the analog filter approaches the zero value in the area of high frequencies, then the degree of this overlap is relatively small and the effect of the spurious frequencies is significantly reduced. One way to address this problem is to define for the analog filter that we use operating specifications that cover those we really want, especially for the cutoff zone (i.e., to define a cutoff frequency Ω_s greater than the one we really need).

Assuming that the transfer function of the standard analog filter $\mathcal{H}_\alpha(s)$ has N *discrete poles* p_k $(k = 1, 2, \ldots, N)$, it can be developed into a *partial fraction series* of the form

$$\mathcal{H}_\alpha(s) = \frac{C_1}{s-p_1} + \frac{C_2}{s-p_2} + \cdots + \frac{C_N}{s-p_N} = \sum_{k=1}^{N} \frac{C_k}{s-p_k}. \tag{12.208}$$

In this case (see Sect. 5.2 on page 376), the impulse response of the analog filter will emerge as the *inverse Laplace transform* of the function $\mathcal{H}_\alpha(s)$, which is formulated as

$$h_\alpha(t) = C_1 e^{p_1 t} u(t) + C_2 e^{p_2 t} u(t) + \cdots + C_N e^{p_N t} u(t) = \left(\sum_{k=1}^{N} C_k e^{p_k t} \right) u(t) \tag{12.209}$$

with the step function $u(t)$ being used to make the impulse response $h(t)$ a *causal signal*. Consequently, the impulse response $h[n]$ of the digital filter will take the form

$$h[n] = h_\alpha(nT_s) = \sum_{k=1}^{N} C_k e^{p_k n T_s} \tag{12.210}$$

whereas the transfer function $\mathcal{H}(z)$ will emerge as the *Z transform of the discrete impulse response* $h[n]$ (see Sect. 7.9 on page 608) and will be, therefore,

$$\mathcal{H}(z) = \sum_{n=0}^{\infty} h[n] z^{-n} = \sum_{n=0}^{\infty} \left(\sum_{k=1}^{N} C_k e^{p_k n T_s} \right) z^{-n} = \sum_{k=1}^{N} C_k \sum_{n=0}^{\infty} (e^{p_k T_s} z^{-1})^n = \sum_{k=1}^{N} \frac{C_k}{1 - e^{p_k T_s} z^{-1}}. \tag{12.211}$$

The last expression was derived using the appropriate expression from the mathematical typology (see Table 6.2 on page 500) and taking into account that since the analog filter $\mathcal{H}_\alpha(s)$ is stable, the poles p_k will have a *negative real part*, which ensures both the *convergence* of the sum in which they appear and the *stability of the poles* $z_k = e^{p_k T_s}$ of the corresponding digital filter, which are located *inside the unit circle* in the z domain, since they will be $|z_k| = |e^{p_k T_s}| = |e^{(\sigma_k + j\Omega_k)T_s}| = e^{\sigma_k T_s} < 1$. The details of the mapping of the s domain to the z domain are presented in Fig. 12.75, from which we observe that the transformation $z = e^{sT_s}$ maps *zones of the left complex half-plane of width* $2\pi/T_s$ *of the s domain into the interior of the unit*

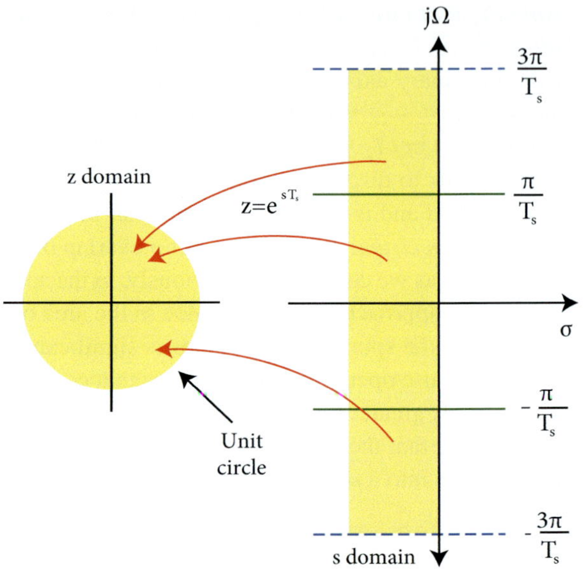

Fig. 12.75 Mapping of the s domain to the z domain under the application of the transformation $z = e^{sT}$ used in the *method of invariant impulse response* (Proakis and Manolakis, Figure 8.32)

circle in the z domain. It is interesting to note, however, that this correspondence between the poles of the analog and digital filters *does not apply to their zero values*, which, being a function of the poles p_k and the coefficients C_k, *are not related through the above representation*. The above analysis concerns the *discrete poles* of the transfer function, whereas for the case of higher-order poles we apply the same process, using, however, the equations we constructed in Sect. 5.3.1 (see page 389)—this process is left as an exercise to the reader.

The main advantage of this method is the *linear correlation* that exists between continuous and discrete frequencies through the relation $\omega = \Omega T_s$, which leads to complete identification by setting $T_s = 1$, something that *is permissible*, since it is proven that the digital filter that results is *independent* of the value of the parameter T_s. Its disadvantage is the negative impact of the appearance of folded (or aliased) frequencies. Our main motivation for constructing digital filters using this method is the assumption that *if the analog filter we use is characterized by a strictly limited bandwidth spectrum, then the frequency response of the digital filter that will result will be a good approximation of the frequency response of the corresponding analog filter*. Let us note that a similar technique can be applied using not the *impulse* but the *step* response of the analog filter, leading to the *method of the invariant step response*, which allows the construction of the step response of the digital filter from the *sampling* of the continuous step response of the corresponding analog filter.

Problem 12.13.1 *Use the method of the invariant impulse response to construct a digital filter IIR with the help of an analog filter Butterworth with operating specifications* $\omega_p = 0.15\pi$, $\omega_s = 0.35\pi$, $A_p = 3\,dB$ *and* $A_s = 20\,dB$.

Solution Adopting the symbolism of the theoretical description of the subject, let us symbolize with Ω_p and Ω_s the *pass* and *cutoff* frequencies of the analog filter Butterworth. In this case, they will be

$$\Omega_p = \frac{\omega_p}{T_s} = \frac{0.15\pi}{T_s} = \frac{0.471238}{T_s} \quad \text{and} \quad \Omega_s = \frac{\omega_s}{T_s} = \frac{0.35\pi}{T_s} = \frac{1.099557}{T_s}.$$

Observing that $\Omega_p/\Omega_s = (\omega_p/T_s)/(\omega_s/T_s) = \omega_p/\omega_s$, it is concluded that the value of the parameter T_s does not affect the order of the filter, N. This order value for the specific values of the parameters ω_p, ω_s, A_p and A_s is calculated as

$$N = \frac{1}{2}\frac{\log\left(\dfrac{10^{A_s/10}-1}{10^{A_p/10}-1}\right)}{\log\left(\dfrac{\Omega_s}{\Omega_p}\right)} = \frac{1}{2}\frac{\log\left(\dfrac{10^{20/10}-1}{10^{3/10}-1}\right)}{\log\left(\dfrac{0.35}{0.15}\right)}$$

$$= \frac{1}{2}\frac{\log\left(\dfrac{99}{0.995262}\right)}{\log\left(\dfrac{0.35}{0.15}\right)} = \frac{1}{2}\frac{log(99.471294)}{log(2.333333)} = \frac{1}{2}\frac{1.997697}{0.367976} = 2.714434.$$

This value should be rounded to the next higher integer and therefore the order of the analog Butterworth filter is $N = 3$.

On the other hand, the cutoff frequency will be calculated using one of the relations

$$\omega_c = \frac{\omega_p}{\sqrt[2N]{10^{A_p/10} - 1}} \quad \text{or} \quad \omega_c = \frac{\omega_s}{\sqrt[2N]{10^{A_s/10} - 1}}$$

depending on whether we want to exactly satisfy the restrictions of the passband or cutoff band respectively. Choosing the satisfaction of the restrictions related to the *passband*, the first of the above relations will give us[76]

$$\omega_c = \frac{\omega_p}{\sqrt[2N]{10^{A_p/10} - 1}} = \frac{0.471238}{\sqrt[6]{10^{3/10} - 1}} = \frac{0.471238}{0.999208} = 0.471611$$

and therefore the cutoff frequency of the analog filter will be $\Omega_c = 0.471611/T_s$. Using this value of the cutoff frequency, the three stable poles of the transfer function are

$$p_2 = \Omega_c\left[\cos\left(\frac{2\pi}{3}\right) + j\sin\left(\frac{2\pi}{3}\right)\right] = \frac{0.471611}{T_s}(-0.5 + 0.866025j) = -\frac{0.235805}{T_s} + j\frac{0.408416}{T_s}$$

$$p_3 = -\Omega_c = -\frac{0.471611}{T_s}$$

$$p_4 = \Omega_c\left[-\cos\left(\frac{2\pi}{3}\right) - j\sin\left(\frac{2\pi}{3}\right)\right] = \frac{0.471611}{T_s}(-0.5 - 0.866025j) = -\frac{0.235805}{T_s} - j\frac{0.408416}{T_s}$$

and therefore the transfer function of the analog filter is formulated as

$$\mathcal{H}(s) = \frac{\Omega_c^N}{(s - p_2)(s - p_3)(s - p_4)} = \frac{(0.471611/T_s)^3}{\left(s + \dfrac{0.471611}{T_s}\right)\left(s + \dfrac{0.235805}{T_s} - j\dfrac{0.408426}{T_s}\right)\left(s + \dfrac{0.235805}{T_s} + j\dfrac{0.408426}{T_s}\right)}$$

In the next step of the process, we develop the function $\mathcal{H}(s)$ into a *sum of partial fractions*. Following the usual methodology, the above relation is developed as

$$\mathcal{H}(s) = \frac{(0.471611/T_s)^3}{\left(s + \dfrac{0.471611}{T_s}\right)\left(s + \dfrac{0.235805}{T_s} - j\dfrac{0.408426}{T_s}\right)\left(s + \dfrac{0.235805}{T_s} + j\dfrac{0.408426}{T_s}\right)} =$$

$$= \frac{C_1}{s + \dfrac{0.471611}{T_s}} + \frac{C_2}{s + \dfrac{0.235805}{T_s} - j\dfrac{0.408426}{T_s}} + \frac{C_3}{s + \dfrac{0.235805}{T_s} + j\dfrac{0.408426}{T_s}}.$$

From the above equation, we observe that all the poles of the function $\mathcal{H}(s)$ are complex numbers divided by T_s, whereas as we have proven, in the equation defining the poles of the transfer function of the digital filter, $z_k = e^{p_k T_s}$ these poles are

[76] If we chose the satisfaction of the restrictions related to the *cutoff band*, then the cutoff frequency was

$$\omega_c = \frac{\omega_s}{\sqrt[2N]{10^{A_s/10} - 1}} = \frac{1.099557}{\sqrt[6]{10^{20/10} - 1}} = \frac{1.099557}{2.150828} = 0.511224$$

and by analogy we would have $\Omega_c = 0.511224/T_s$. In reality, any value could be used for frequency in the range of values $0.511224/T_s \leq \Omega_c \leq 0.471238/T_s$.

multiplied by T_s. But this means that the choice of the value of the parameter T_s does not affect the final result at all and for this reason, let us set for simplicity $T_s = 1$. Applying the method of partial fraction expansion (for the value $T_s = 1$) in the way we know, we end up with the system of three equations

$$\begin{aligned}
&C_1 + C_2 + C_3 = 0\\
&0.471610C_1 + (0.707416 + 0.408426j)C_2 + (0.707416 - 0.408426j)C_3 = 0\\
&0.222415C_1 + (0.111208 + 0.192618j)C_2 + (0.111208 - 0.192618j)C_3 = 0.104894
\end{aligned}$$

with solution $C_1 = 0.471611$, $C_2 = -0.235805 - 0.136143j$, and $C_3 = -0.235805 + 0.136143j$, and therefore

$$\mathcal{H}(s) = \frac{0.471611}{s + 0.471611} + \frac{-0.235805 - 0.136413j}{s + 0.235805 - 0.408426j} + \frac{-0.235805 + 0.136143j}{s + 0.235805 + 0.408426j} \approx \frac{0.105}{s^3 + 0.9432s^2 + 0.448s + 0.105}.$$

Taking into account that the above coefficients also appear in the partial fraction expansion of the transfer function $\mathcal{H}(z)$ of the digital filter, the only thing left to do is to determine its poles from the relation $z_k = e^{p_k}$ (for the value $T_s = 1$). Using the trigonometric identity $e^{x \pm iy} = e^x e^{\pm iy} = e^x(\cos y \pm \sin y)$, the poles of the function $\mathcal{H}(z)$ are calculated as

$$\begin{aligned}
z_2 &= e^{p_2} = e^{-0.235805+0.408426j} = e^{-0.235805}[\cos(0.408426) + \sin(0.408426)j] = 0.724960 + 0.313734j\\
z_3 &= e^{p_3} = e^{-0.471238} = 0.623996\\
z_4 &= e^{p_4} = e^{-0.235805-0.408426j} = e^{-0.235805}[\cos(0.408426) - \sin(0.408426)j] = 0.724960 - 0.313734j
\end{aligned}$$

and therefore the required digital function has the form

$$\begin{aligned}
\mathcal{H}(z) &= \frac{0.471611}{1 - 0.623996z^{-1}} + \frac{-0.235805 - 0.136413j}{1 - (0.724960 + 0.313734j)z^{-1}} + \frac{-0.235805 + 0.136413j}{1 - (0.724960 - 0.313734j)z^{-1}}\\
&= \frac{0.471611}{1 - 0.623996z^{-1}} + \frac{-0.471611 + 0.427494z^{-1}}{1 - 1.449992z^{-1} + 0.623996z^{-2}} = \frac{0.037944z^{-1} + 0.027529z^{-2}}{1 - 2.073988z^{-1} + 1.528785z^{-2} - 0.389371z^{-3}}.
\end{aligned}$$

The impulse response of the analog Butterworth filter will emerge known from the inverse Laplace transform of the transfer function $\mathcal{H}(s)$. Using the formulation of this function in partial-fraction expansion and taking into account that $-0.235805 \pm 0.136413j = 0.272420e^{\pm 2.617134}$, we have

$$\begin{aligned}
h(t) = \mathcal{L}^{-1}\{\mathcal{H}(s)\} &= 0.471611\mathcal{L}^{-1}\left\{\frac{1}{s + 0.471611}\right\} + 0.272420e^{-2.617134j}\mathcal{L}^{-1}\left\{\frac{1}{s + 0.235805 - 0.408426j}\right\}\\
&+ 0.272420e^{2.617134j}\mathcal{L}^{-1}\left\{\frac{1}{s + 0.235805 + 0.408426j}\right\} = 0.471611e^{-0.471611t}\\
&+ 0.272420e^{-2.617134j}e^{-(0.235805-0.408426j)t} + 0.272420e^{2.617134j}e^{-(0.235805+0.408426j)t}\\
&= 0.471611e^{-0.471611t} + 0.544840e^{-0.235805t}\cos(0.408426t + 2.617134)
\end{aligned}$$

as it emerges after operations. The corresponding impulse response of the digital filter Butterworth will result from the inverse $\mathcal{Z}$ transform of the digital transfer function $\mathcal{H}(z)$ and by performing the operations, we end up with the expression

$$\begin{aligned}
h[n] = \mathcal{Z}^{-1}\{\mathcal{H}(z)\} &= 0.471611\mathcal{Z}^{-1}\left\{\frac{1}{1 - 0.623996z^{-1}}\right\} + 0.272420e^{-2.617134j}\mathcal{Z}^{-1}\left\{\frac{1}{1 - (0.724960 + 0.313734j)z^{-1}}\right\}\\
&+ 0.272420e^{2.617134j}\mathcal{Z}^{-1}\left\{\frac{1}{1 - (0.724960 - 0.313734j)z^{-1}}\right\} = 0.471611(0.623996)^n\\
&+ 0.272420e^{-2.617134j}(0.724960 + 0.313734j)^n + 0.272420e^{2.617134j}(0.724960 - 0.313734j)^n\\
&= 0.471611(0.623996)^n + 0.544840(0.789934)^n\cos(0.408425n + 2.617134).
\end{aligned}$$

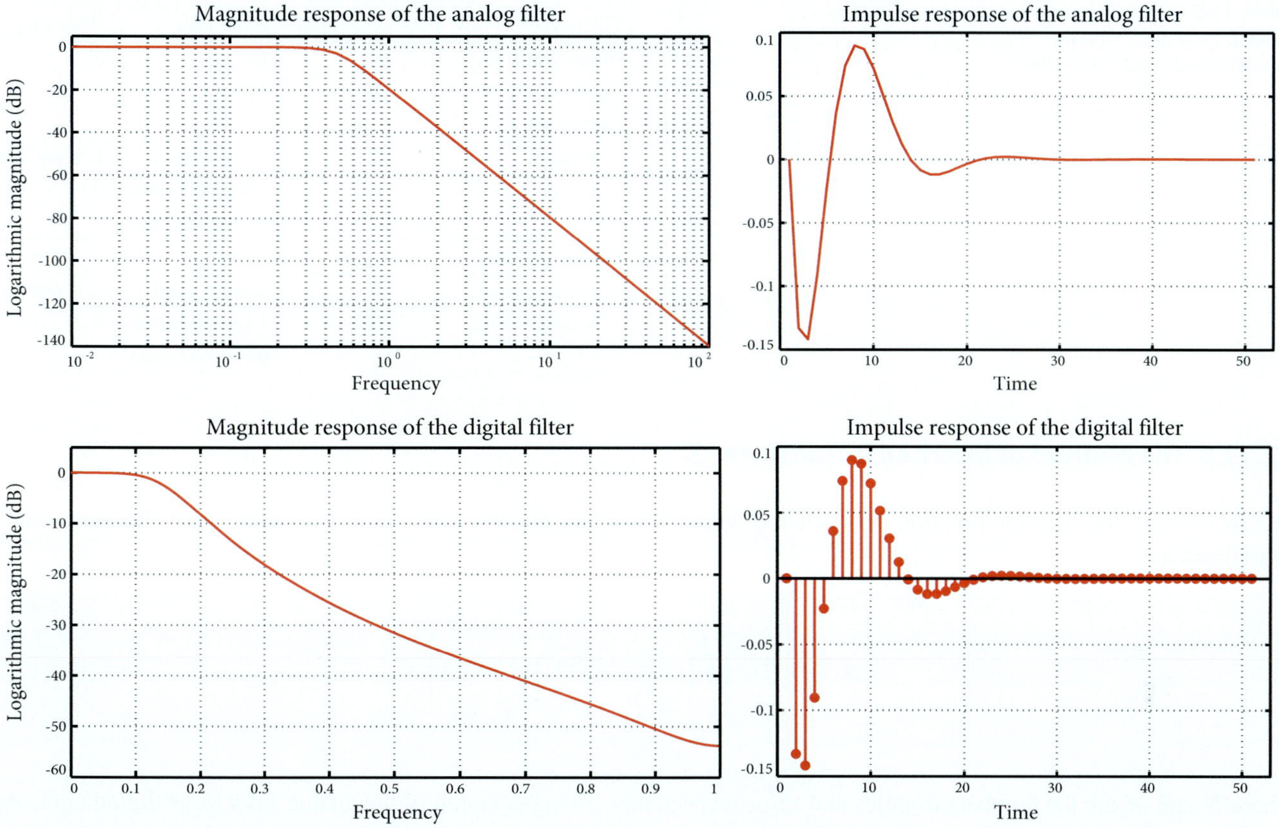

Fig. 12.76 The frequency and impulse response of the analog filter and the corresponding digital filter of Exercise 12.13.1

The frequency and impulse response of the analog filter and the corresponding digital filter (along with their respective amplitude responses) are depicted in Fig. 12.76, from where it is observed that the impulse responses $h(t)$ and $h[n]$ are related to each other through a *sampling process*.

In order to verify the correctness of the result, let us calculate the values of the magnitude of the frequency response function of the digital filter, at the positions of the frequencies ω_p and ω_s which, according to the statement, should be equal to -3 dB and -20 dB. We remind the reader that the frequency ω_c was chosen so that *the specifications for the passband are exactly met*; therefore, let us see if this is confirmed from the results we received. If in the digital transfer function we make the substitution $z = e^{j\omega}$, it will take the form

$$\mathcal{H}(e^{j\omega}) = \frac{0.037944e^{-j\omega} + 0.027529e^{-2j\omega}}{1 - 2.073988e^{-j\omega} + 1.528785e^{-2j\omega} - 0.389371e^{-3j\omega}}$$

and replacing in this the values $\omega_p = 0.15\pi = 0.471238$ and $\omega_s = 0.35\pi = 1.099557$ it turns out that

$$\mathcal{H}(e^{j\omega_p}) = -0.499756 - 0.502233j = 0.708515e^{-2.353722j}$$

$$\mathcal{H}(e^{j\omega_s}) = -0.061399 + 0.048940j = 0.078518e^{2.468627j}.$$

From the above relations we observe that

$$|\mathcal{H}(e^{j\omega_p})| = 0.708515 = -2.993009\,\text{dB} \qquad \text{and} \qquad |\mathcal{H}(e^{j\omega_s})| = 0.078518 = -22.100587\,\text{dB}$$

values that confirm our predictions. We observe that the attenuation in the cutoff zone is greater than what we had specified in the operating specifications of the filter and therefore the final result is acceptable.

Table 12.5 The value of the factor L associated with the method of *matched Z transform*

Filter type	Low-pass	High-pass	Band-pass	Band-stop
Butterworth	N	0	$N/2$	0
Chebyshev I	N	0	$N/2$	0
Chebyshev II				
N odd	1	0	It is not used	It is not used
N even	0	0	1 for (N/2) even 0 for (N/2) odd	0
Elliptic (Cauer)				
N odd	1	0	It is not used	It is not used
N even	0	0	1 for (N/2) even 0 for (N/2) odd	0

12.13.3 The Method of Matched Z Transform $\mathcal{Z}$

In cases where the transfer function $\mathcal{H}(s)$ of the analog filter has *besides poles and zero values* and therefore can be formulated as

$$\mathcal{H}(s) = H_0 \left\{ \prod_{i=1}^{M} (s - z_i) \Big/ \prod_{i=1}^{N} (s - p_i) \right\}, \tag{12.212}$$

where N and M are the numbers of poles and zeros respectively. The conversion of the analog filter to its digital equivalent can be carried out with the help of a transform known as *matched $\mathcal{Z}$ transform*. This transform can be considered as a *generalization of the invariant impulse response method*, in the sense that *the mapping of the zeros of the analog filter to the corresponding zeros of the digital filter is carried out in the same way as the mapping of the poles*. In other words, if the analog filter has N poles s_{p_i} $(i = 1, 2, \ldots, N)$ and M zeros s_{z_i} $(i = 1, 2, \ldots, M)$, then the corresponding poles and zeros of the digital filter will be calculated from the relations $z_{p_i} = e^{s_{p_i} T_s}$ and $z_{z_i} = e^{s_{z_i} T_s}$, where T_s is a suitably chosen *sampling period*, whereas the corresponding transfer function will have the form

$$\mathcal{H}(z) = (z+1)^L H_0 \times \frac{\prod_{i=1}^{M} (1 - e^{z_{z_i} T_s} z^{-1})}{\prod_{i=1}^{N} (1 - e^{z_{p_i} T_s} z^{-1})} = (z+1)^L H_0 \times \frac{\prod_{i=1}^{M} (z - e^{z_{z_i} T_s})}{\prod_{i=1}^{N} (z - e^{z_{p_i} T_s})} \tag{12.213}$$

with the corrective factor $(z+1)^L$ *improving* the behavior of the filter, so as to make possible the application of the transform for the construction of *low-pass* and *band-pass* filters. The value of the factor L appearing in the above relationship is equal to the number of zero values of the transfer function $\mathcal{H}(s)$ at the position $s = \infty$,[77] and for the four known types of analog filters is included in Table 12.5.

It is obvious that the poles that arise through the application of this method are the same as those that arise with the method of the invariant impulse response; however, the two methods *differ* in terms of *the positions of the zero values*. The

[77] In reality, the form of this factor is due to the fact that *all the zero values of the transfer function corresponding to the position $s = \infty$ are depicted at the position $z = -1 = e^{i\pi}$ which corresponds to the maximum frequency of the z domain*. Therefore, if there are L zero values of the function $\mathcal{H}(s)$ at the position $s = \infty$, these lead to the creation of L zero values of the transfer function $\mathcal{H}(z)$ (at the position $z = -1$), which contribute the term $(z+1)(z+1)\ldots(z+1) = (z+1)^L$.

main disadvantage of the matched $\mathcal{Z}$ transform (which is also *vulnerable* in terms of the consequences of the appearance of *folded (or aliased) frequencies*, which can, however, be limited by choosing a satisfactorily small value for the parameter T_s) is the introduction of *quite a large error* in terms of the *attenuation* in the passband, whereas generally it does not offer control of the value of the level of the amplitude response of the digital filter. This means that if the maximum gain of the analog filter is equal to the unit—as usually happens—the corresponding gain of the digital filter is generally different from the unit and for this reason it requires *redefinition of the normalization coefficient*.

12.13.4 Bilinear Transformation

Although the method of *invariant impulse response* is relatively simple, as it leads to *linear mapping* of continuous to corresponding discrete frequencies of the form $\omega = \Omega T_s$, it is greatly affected by the *undesirable appearance of aliasing frequencies* and for this reason its application is limited to the design of *low-pass*, as well as some types of *band-pass filters* (this limitation, as we have mentioned, also characterizes the method of *derivative approximation*). In order to lift these restrictions, in practice we resort to the use of *conformal mapping*, which maps the imaginary axis $s = j\Omega$ of the s domain to the periphery of the unit circle of the z domain *only once*, also maintaining the *stability* in the sense that we described in the previous sections. This representation is known as *bilinear transformation*.[78]

In order to construct the equation that maps the space of s to the space of z, we consider for simplicity a system of *first order* described by the transfer function

$$\mathcal{H}(s) = \frac{\mathcal{Y}(s)}{\mathcal{X}(s)} = \frac{\beta}{s+\alpha}. \tag{12.214}$$

Reformulating the above relation in the form $s\mathcal{Y}(s) + \alpha\mathcal{Y}(s) = \beta\mathcal{X}(s)$ and taking the *inverse Laplace transform*, we have

$$\mathcal{L}^{-1}\{s\mathcal{Y}(s)\} + \alpha\mathcal{L}^{-1}\{\mathcal{Y}(s)\} = \beta\mathcal{L}^{-1}\{\mathcal{X}(s)\} \qquad \text{or equivalently} \qquad \frac{dy(t)}{dt} + \alpha y(t) = \beta x(t).$$

The expression constitutes the differential equation of the system in the field of continuous-time. At this point, instead of replacing the derivative $y'(t)$ with the corresponding *first difference* as we did in the derivative approximation, let us proceed to replace the function $y(t)$ with *the integral of its first derivative* calculated in the value range $[t_0, t]$ with initial condition $y_0 \equiv y(t_0)$—therefore,

$$y(t) = \int_{t_0}^{t} \frac{dy(\tau)}{d\tau} d\tau + y(t_0) = \int_{t_0}^{t} y'(\tau)\, d\tau + y(t_0) \tag{12.215}$$

and the differential equation is written

$$y'(t) + \alpha\left(\int_{t_0}^{t} y'(\tau)\, d\tau + y(t_0)\right) = \beta x(t). \tag{12.216}$$

[78] Let it be noted that *bilinear transformation* is not exclusively related to the design of digital filters, but is an integral part of the theory of *complex analysis* and *conformal mappings*. A bilinear transformation is defined as *the quotient of two linear transformations* (from where its name derives) of the form $w = (\alpha z + \beta)/(\gamma z + \delta)$ ($\alpha\delta - \beta\gamma \neq 0$ and $z \neq -\delta/\gamma$), and can arise from the successive application of the elementary transforms of *translation*, *scaling*, and *inversion*. In the theory of conformal mappings, bilinear transformation is also known as *Moebius mapping*.

At this point, let us approximate the value of the above defined integral using the *trapezoid rule*[79] for the values $\alpha = nT_s - T_s = (n-1)T_s$ and $\beta = nT_s$, leading to the expression

$$y(nT_s) = \frac{1}{2}\Big[y'(nT_s) + y'[(n-1)T_s]\Big][nT_s - (n-1)T_s] + y[(n-1)T_s]$$

or equivalently

$$y[n] = \frac{T_s}{2}\Big(y'[n] + y'[n-1]\Big) + y[n-1],$$

where, for reasons of brevity in writing, we adopted the symbolism $f(nT) \equiv f[n]$. If we now solve the differential equation of the system with respect to $y'(t)$ and calculate it at the positions $t = nT_s$ and $t = (n-1)T_s$ it turns out that

$$y'[n] = -\alpha y[n] + \beta x[n] \qquad \text{and} \qquad y'[n-1] = -\alpha y[n-1] + \beta x[n-1].$$

Replacing the above expressions in the previous equation, it takes the form

$$\begin{aligned} y[n] &= \frac{T_s}{2}\Big(-\alpha y[n] + \beta x[n] - \alpha y[n-1] + \beta x[n-1]\Big) + y[n-1] \\ &= -\frac{\alpha T_s}{2} y[n] + \frac{\beta T_s}{2} x[n] - \frac{\alpha T_s}{2} y[n-1] + \frac{\beta T_s}{2} x[n-1] + y[n-1], \end{aligned}$$

which, after simple mathematical operations, is reformulated as

$$\left(1 + \frac{\alpha T_s}{2}\right) y[n] - \left(1 - \frac{\alpha T_s}{2}\right) y[n-1] = \frac{\beta T_s}{2}\Big(x[n] + x[n-1]\Big)$$

and constitutes the *difference equation of the equivalent discrete-time system*. Now using the known methodology (see, for example, Exercise 9.2.1 on page 701), we can determine the analytical form of the transfer function of $\mathcal{H}(z)$. Therefore, taking the *transform* $\mathcal{Z}$ of the two terms and using the property of *time shift* $\mathcal{Z}\{x[n-1]\} = z^{-1}\mathcal{X}(z)$ the above relationship is written

$$\mathcal{Z}\left\{\left(1 + \frac{\alpha T_s}{2}\right) y[n]\right\} - \mathcal{Z}\left\{\left(1 - \frac{\alpha T_s}{2}\right) y[n-1]\right\} = \frac{\beta T_s}{2}\mathcal{Z}\{x[n]\} + \frac{\beta T_s}{2}\mathcal{Z}\{x[n-1]\}$$

or equivalently

$$\left[\left(1 + \frac{\alpha T_s}{2}\right) - \left(1 - \frac{\alpha T_s}{2}\right) z^{-1}\right]\mathcal{Y}(z) = \frac{\beta T_s}{2}(1 + z^{-1})\,\mathcal{X}(z)$$

and therefore the transfer function $\mathcal{H}(z)$ of the discrete-time system will have the form

$$\mathcal{H}(z) = \frac{\mathcal{Y}(z)}{\mathcal{X}(z)} = \frac{(\beta T_s/2)(1 + z^{-1})}{\left(1 + \dfrac{\alpha T_s}{2}\right) - \left(1 - \dfrac{\alpha T_s}{2}\right) z^{-1}} = \frac{\beta}{\dfrac{2}{T_s}\left(\dfrac{1 - z^{-1}}{1 + z^{-1}}\right)}. \tag{12.217}$$

[79] From mathematical analysis it is known that the value of the definite integral of the function $f(x)$ in the value range $[\alpha, \beta]$ can be approximated by the area of the trapezium defined by the horizontal axis, the straight lines $x = \alpha$ and $x = \beta$ and the straight line that connects the points $[\alpha, f(\alpha)]$ and $[\beta, f(\beta)]$, or in mathematical writing

$$\int_\alpha^\beta f(x)\,dx \approx (\beta - \alpha) \times \frac{f(\alpha) + f(\beta)}{2}.$$

If we compare the transfer functions of the continuous and the discrete-time system

$$\mathcal{H}(s) = \frac{\beta}{s+\alpha} \qquad \text{and} \qquad \mathcal{H}(z) = \frac{\beta}{\frac{2}{T}\left(\frac{1-z^{-1}}{1+z^{-1}}\right)}$$

it immediately follows that the mapping *from the s domain to the z domain* takes place through the correspondence

$$s = \frac{2}{T}\left(\frac{1-z^{-1}}{1+z^{-1}}\right). \tag{12.218}$$

But the above equation is an example of *bilinear transformation*, from which this method gets its name. Although the proof we gave was based for simplicity on the use of a first-order differential equation, it is valid *for the general case of a differential equation of order N*.

After constructing the equation of the transformation of the s domain in the z domain, let us see if it has the desired mapping properties, as these were described in Sect. 12.13 (see page 1283) and in the related subsections. For this reason, let us solve the above equation for s, so we get

$$s = \frac{2}{T}\left(\frac{1-z^{-1}}{1+z^{-1}}\right) = \frac{2}{T}\left(\frac{z-1}{z+1}\right).$$

Expressing the complex variable z in its polar form $z = re^{j\omega}$ and substituting it into the above equation, it is successively transformed as

$$\begin{aligned} s = \sigma + j\Omega &= \frac{2}{T_s}\frac{re^{j\omega}-1}{re^{j\omega}+1} = \frac{2}{T_s}\frac{(r\cos\omega - 1) + jr\sin\omega}{(r\cos\omega + 1) + jr\sin\omega} = \frac{2}{T_s}\frac{[(r\cos\omega - 1) + jr\sin\omega][(r\cos\omega + 1) - jr\sin\omega]}{[(r\cos\omega + 1) + jr\sin\omega][(r\cos\omega + 1) - jr\sin\omega]} \\ &= \frac{2}{T_s}\frac{r^2 + j2r\sin\omega - 1}{r^2 + 2r\cos\omega + 1} = \frac{2}{T_s}\frac{r^2-1}{r^2 + 2r\cos\omega + 1} + j\frac{2r\sin\omega}{r^2 + 2r\cos\omega + 1} \end{aligned} \tag{12.219}$$

from which it follows that

$$\sigma = \frac{2}{T_s}\frac{r^2-1}{r^2 + 2r\cos\omega + 1} \qquad \text{and} \qquad \Omega = \frac{2r\sin\omega}{r^2 + 2r\cos\omega + 1}.$$

It is proven that for values $r < 1$ it is $\sigma < 0$, whereas for values $r > 1$ we have $\sigma > 0$. But this means that *the points of the left complex half-plane of the s domain are mapped inside the unit circle of the z domain*, whereas conversely *the points of the right half-plane are mapped outside the aforementioned circle*. Therefore, the bilinear transformation *preserves stability*, in the sense that *a stable system remains stable after the transformation*. The pole of the analog filter is depicted at a stable point in the z domain. As for the points on the unit circle of the z domain for which $r = 1$, these correspond to the values

$$\sigma = 0 \qquad \text{and} \qquad \Omega = \frac{2}{T_s}\left(\frac{2\sin\omega}{2+2\cos\omega}\right) = \frac{2}{T_s}\left(\frac{\sin\omega}{1+\cos\omega}\right) = \frac{2}{T_s}\tan\frac{\omega}{2},$$

where to derive the last equation we used the half-angle tangent formula. So, $s = j\Omega$ and therefore *the circumference of the unit circle of the z domain is depicted on the imaginary axis of the s domain*, which is certainly desirable, with the frequency Ω defined by the last of the above equations. Solving this equation for ω, it turns out that

$$\omega = 2\tan^{-1}\left(\frac{\Omega T_s}{2}\right) \tag{12.220}$$

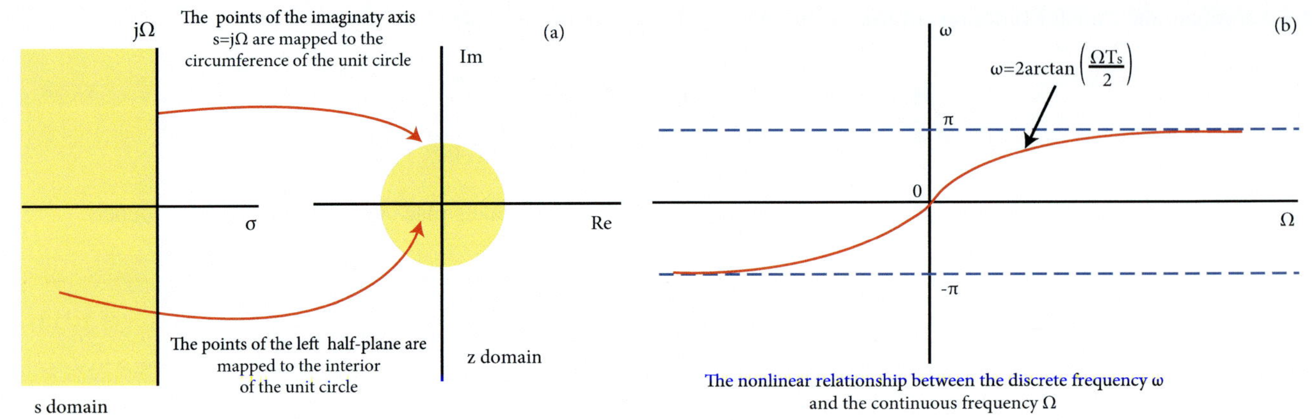

Fig. 12.77 The mapping of the s domain to the z domain with the help of the *bilinear transformation* and the *nonlinear correlation* that exists between the continuous frequency Ω and the discrete frequency ω (Oppenheim, Schafer, and Buck, Figures 7.6 and 7.7)

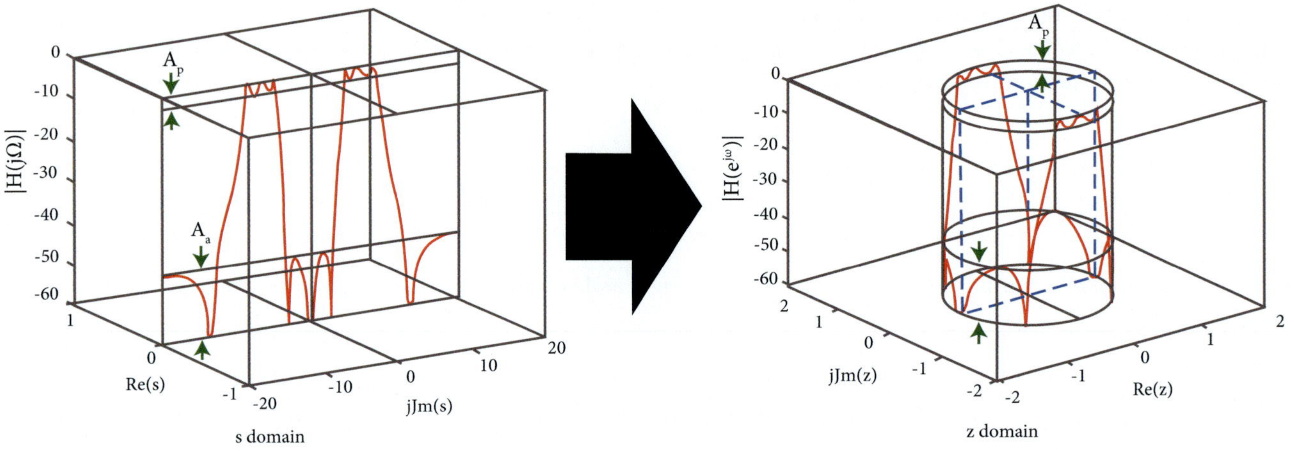

Fig. 12.78 The representation of the amplitude response of the analog filter in the z domain, under the influence of bilinear transformation (Antoniou, Figure 11.7)

an equation that is also the desired expression that correlates *the frequencies* Ω *and* ω *in the field of continuous and discrete-time*. We observe that in contrast to the method of invariant impulse response that led to a linear relationship of the form $\omega = \Omega T_s$, the method of bilinear transformation leads to a *nonlinear correlation* of the frequencies ω and Ω, which adds a degree of complexity to the design process of the digital filter. However, on the other hand, this transformation causes *a unique representation* of the frequency range $-\infty < \Omega < \infty$ in the area $-\pi \leq \omega \leq \pi$ (since, as can be proven, for $\Omega = 0$ it is $\omega = 0$, for $\Omega \to +\infty$ it is $\omega \to +\pi$, and for $\Omega \to -\infty$ it is $\omega \to -\pi$), in contrast to the method of invariant impulse response that depicted *multiple zones* of the left half-plane of the s domain inside the unit circle. For this reason, the design of digital filters using bilinear transformation is free from the problems that the appearance of *folded (or aliased) frequencies* causes. The mapping of the s domain to the z domain under the application of bilinear transformation is diagrammatically presented in Fig. 12.77, whereas the mapping of the amplitude response of the analog filter to the z domain is presented in Fig. 12.78. It is not difficult to realize that *the pass and cutoff zones of the analog filter will be depicted in the pass and cutoff zones of the digital filter* and that this mapping will preserve all the characteristics that make the *discrete transfer function implementable* (such as the *real nature of the filter coefficients* and the known correlation that exists between the degree of the polynomials of the numerator and the denominator of the rational transfer function).

As already mentioned, the main disadvantage of bilinear transformation is the *frequency distortion* ω due to the nonlinear nature of Eq. (12.214). This is particularly evident in Fig. 12.79, in which a set of frequency regions of the same width uniformly distributed on the frequency axis and their corresponding zones of frequency in the discrete space, under the application of bilinear transformation, are depicted. It is not difficult to see that *this frequency distortion manifests in the form of compression that makes the images of the frequency regions* Ω *increasingly smaller as we move toward the high-*

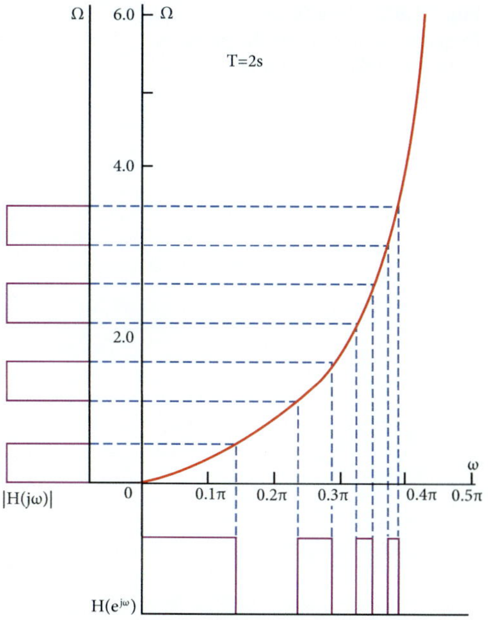

Fig. 12.79 Frequency distortion due to bilinear transformation (Antoniou, Figure 11.8)

frequency areas. This obviously does not happen in the method of the *invariant impulse response*, where the corresponding curve $\omega = f(\Omega) = \Omega T$ is a *straight line* and therefore *the frequency zones in the* ω *space are exact images of the frequency zones in the* Ω *space.* Let us note that in cases where we are only interested in the amplitude response of the digital filter, the effect of frequency distortion can be completely removed *warping* the frequency values associated with the operating specifications of the analog filter. In order for this process to be understood, let us consider a simple analog low-pass filter with pass-through and cutoff frequencies Ω_p and Ω_s, which is used to construct a digital filter, the operating frequencies of which will be associated with the frequencies Ω_p and Ω_s through the relations

$$\omega_p = \frac{2}{T_s}\tan^{-1}\left(\frac{\Omega_p T_s}{2}\right) \quad \text{and} \quad \omega_s = \frac{2}{T_s}\tan^{-1}\left(\frac{\Omega_s T_s}{2}\right).$$

In this case, the warping process consists in characterizing as operating frequencies of the analog filter, not the Ω_p and Ω_s but the

$$\tilde{\Omega}_p = \frac{2}{T_s}\tan\left(\frac{\omega_p T_s}{2}\right) \quad \text{and} \quad \tilde{\Omega}_s = \frac{2}{T_s}\tan\left(\frac{\omega_s T_s}{2}\right),$$

since, in this case, the application of the transformation will lead exactly to the frequencies ω_p and ω_s detailed in the filter specifications. Indeed, in this case

$$\tilde{\omega}_p = \frac{2}{T_s}\tan^{-1}\left(\frac{\tilde{\Omega}_p T_s}{2}\right) = \frac{2}{T_s}\tan^{-1}\left[\frac{T_s}{2}\frac{2}{T_s}\tan\left(\frac{\omega_p T_s}{2}\right)\right] = \frac{2}{T_s}\tan^{-1}\left[\tan\left(\frac{\omega_p T_s}{2}\right)\right] = \frac{2}{T_s}\frac{\omega_p T_s}{2} = \omega_p \tag{12.221}$$

$$\tilde{\omega}_s = \frac{2}{T_s}\tan^{-1}\left(\frac{\tilde{\Omega}_s T_s}{2}\right) = \frac{2}{T_s}\tan^{-1}\left[\frac{T_s}{2}\frac{2}{T_s}\tan\left(\frac{\omega_s T_s}{2}\right)\right] = \frac{2}{T_s}\tan^{-1}\left[\tan\left(\frac{\omega_s T_s}{2}\right)\right] = \frac{2}{T_s}\frac{\omega_s T_s}{2} = \omega_p \tag{12.222}$$

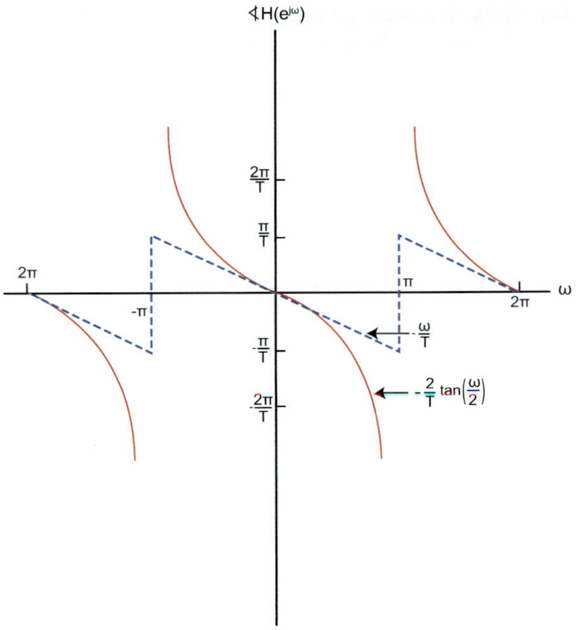

Fig. 12.80 The effect of frequency distortion on the phase response of the digital filter

and therefore the operating frequencies of the digital filter to be constructed *will be exactly those that have been determined during the specification stage.*

The effect of frequency distortion on the phase response of the digital filter is shown in Fig. 12.80. In this figure, the dashed lines represent the *linear phase* of the phase factor e^{-s} of the ideal filter, whereas the continuous lines represent the phase resulting from the application of *bilinear transformation*.

Problem 12.13.2 *Using the method of bilinear transformation, construct a digital filter IIR with the help of an analog filter Butterworth with specifications* $\omega_p = 0.15\pi,\ \omega_s = 0.35\pi,\ A_p = 3\,dB$ *and* $A_s = 20\,dB$.

Solution This exercise is identical to the previous one, with the difference that the digital filter Butterworth will be constructed using the *bilinear transformation* method. Adopting the usual symbolism, the *pass* and *cutoff* frequencies of the analog filter Ω_p and Ω_s are calculated as

$$\Omega_p = \frac{2}{T_s}\tan\left(\frac{\omega_p}{2}\right) = \frac{2}{T_s}\tan\left(\frac{0.471238}{2}\right) = \frac{0.480157}{T_s} \quad \text{and}$$

$$\Omega_s = \frac{2}{T_s}\tan\left(\frac{\omega_s}{2}\right) = \frac{2}{T_s}\tan\left(\frac{1.099557}{2}\right) = \frac{1.225601}{T_s}$$

and therefore the order of the filter will be calculated from the equation

$$N = \frac{1}{2}\frac{\log\left(\dfrac{10^{A_s/10}-1}{10^{A_p/10}-1}\right)}{\log\left(\dfrac{\Omega_s}{\Omega_p}\right)} = \frac{1}{2}\frac{\log\left(\dfrac{10^{20/10}-1}{10^{3/10}-1}\right)}{\log\left(\dfrac{1.225601}{0.480157}\right)}$$

$$= \frac{1}{2}\frac{\log\left(\dfrac{99}{0.995262}\right)}{\log\left(\dfrac{1.225601}{0.480157}\right)} = \frac{1}{2}\frac{\log(99.471294)}{\log(2.552498)} = \frac{1}{2}\frac{1.997697}{0.406965} = 2.454470.$$

We observe that in this case, the ratio of frequencies Ω_s/Ω_p is larger than the corresponding frequency ratio of the previous case, owing to the *frequency warping* caused by the *nonlinear nature of the transformation*. Let us note, of course, that this does not affect the value of the filter order, which, in full analogy with the previous case, is $N = 3$.

On the other hand, the cutoff frequency of the digital filter will be calculated in the way we explained previously. Choosing the frequency value that satisfies *exactly* the passband specifications (although this is not particularly important when using bilinear transformation), $\omega_c = 0.471611$ (see the previous exercise) and therefore the corresponding frequency of the analog filter will be

$$\Omega_c = \frac{2}{T_s}\tan\left(\frac{\omega_c}{2}\right) = \frac{2}{T_s}\tan\left(\frac{0.471611}{2}\right) = \frac{0.480551}{T_s}$$

Using this cutoff frequency value, the three stable poles of the transfer function are

$$p_2 = \Omega_c\left[\cos\left(\frac{2\pi}{3}\right) + j\sin\left(\frac{2\pi}{3}\right)\right] = \frac{0.480551}{T_s}(-0.5 + 0.866025j) = -\frac{0.240275}{T_s} + j\frac{0.416169}{T_s}$$

$$p_3 = -\Omega_c = -\frac{0.480551}{T_s}$$

$$p_4 = \Omega_c\left[-\cos\left(\frac{2\pi}{3}\right) - j\sin\left(\frac{2\pi}{3}\right)\right] = \frac{0.480551}{T_s}(-0.5 - 0.866025j) = -\frac{0.240275}{T_s} - j\frac{0.416169}{T_s}$$

and therefore the transfer function of the analog filter is formulated as

$$\mathcal{H}(s) = \frac{\Omega_c^N}{(s-p2)(s-p_3)(s-p_4)} = \frac{(0.480551/T_s)^3}{\left(s + \frac{0.480551}{T_s}\right)\left(s + \frac{0.240275}{T_s} - j\frac{0.416169}{T_s}\right)\left(s + \frac{0.240275}{T_s} + j\frac{0.416169}{T_s}\right)}$$

$$= \frac{(0.480551/T_s)^3}{s^3 + \frac{0.961102}{T_s}s^2 + \frac{0.461856}{T_s^2}s + \frac{0.110972}{T_s^3}}.$$

Having determined the transfer function of the analog filter $\mathcal{H}(s)$, we can determine the transfer function of the digital filter $\mathcal{H}(z)$ using the bilinear transformation equation—it will therefore be

$$\mathcal{H}(z) = \frac{(0.480551/T_s)^3}{\left(\frac{2}{T_s}\frac{1-z^{-1}}{1+z^{-1}}\right)^3 + \frac{0.961102}{T_s}\left(\frac{2}{T_s}\frac{1-z^{-1}}{1+z^{-1}}\right)^2 + \frac{0.461856}{T_s^2}\left(\frac{2}{T_s}\frac{1-z^{-1}}{1+z^{-1}}\right) + \frac{0.110972}{T_s^3}}.$$

It is not difficult to realize that the parameter T_s can be *simplified*, since it appears raised to the third power in the denominators of all the expressions of the numerator and the denominator and therefore its value does not affect the design process (in some cases the value $T = 2_s$ is used, in order to simplify the expression of bilinear transformation). By simplifying the parameter T_s (or alternatively, setting $T_s = 1$), the above relation after simple algebraic operations is written

$$\mathcal{H}(z) = \frac{0.110972}{8\left(\frac{1-z^{-1}}{1+z^{-1}}\right)^3 + 3.844408\left(\frac{1-z^{-1}}{1+z^{-1}}\right)^2 + 0.923712\left(\frac{1-z^{-1}}{1+z^{-1}}\right) + 0.110972}$$

$$= \frac{0.008616 + 0.025848z^{-1} + 0.025848z^{-2} + 0.008616z^{-3}}{1 - 2.064414z^{-1} + 1.519112z^{-2} - 0.385767z^{-3}},$$

which is the desired result. If we compare this relation with the result we reached through the use of the invariant impulse response method, we find that the denominators are more or less the same, whereas the numerator of the function that emerged with the help of bilinear transformation contains a term z^{-3} that is multiplied by a very small number, as well as a constant term with the same (very small value)—we expect, therefore, the two filters to exhibit more or less the same behavior. Proceeding to the substitution $z = e^{j\omega}$, the frequency response of the digital filter is calculated as

$$\mathcal{H}(e^{j\omega}) = \frac{0.008616 + 0.025848e^{-j\omega} + 0.025848e^{-2j\omega} + 0.008616e^{-3j\omega}}{1 - 2.064414e^{-j\omega} + 1.519112e^{-2j\omega} - 0.385767e^{-3j\omega}}$$

and replacing in this the values $\omega_p = 0.15\pi = 0.471238$ and $\omega_s = 0.35\pi = 1.099557$, it turns out that

$$\mathcal{H}(e^{j\omega_p}) = -0.499559 - 0.501599j = 0.707928e^{-2.354156j}$$

$$\mathcal{H}(e^{j\omega_s}) = -0.043476 + 0.041591j = 0.060166e^{2.378348j}.$$

From the above relations we observe that

$$|\mathcal{H}(e^{j\omega_p})| = 0.707928 = -3.000216 \text{ dB} \quad \text{and} |\mathcal{H}(e^{j\omega_s})| = 0.060166 = -24.412897 \text{ dB},$$

values that verify our predictions. We observe that the attenuation in the cutoff zone is greater than the one we had specified in the operating specifications of the filter and therefore the final result is acceptable.

12.13.5 The Method of Pole-Zero Placement

As is well known from basic theory, the contribution of a *pair of conjugate complex zeros* $z = re^{\pm j\vartheta}$ (where r is the magnitude and $\pm\vartheta$ the polar angle) to the transfer function $\mathcal{H}(z)$ of a discrete-time system is the appearance of the term $(z - re^{j\vartheta})(z - re^{-j\vartheta})$ in the numerator representation. In the special case where the zeros belong *on the circumference of the unit circle*, $r = 1$ and the above contribution takes the simplified form $(z - e^{j\vartheta})(z - e^{-j\vartheta})$. The corresponding contribution to the frequency-response function that arises by setting $z = e^{j\omega}$ will obviously be $(e^{j\omega} - e^{j\vartheta})(e^{j\omega} - e^{-j\vartheta})$, from which it immediately follows that the magnitude of the frequency response at the position $\omega = \vartheta$ will be equal to zero. On the other hand, a *pair of conjugate complex poles* at the positions $p = re^{\pm j\vartheta}$ contributes to the denominator of the transfer function the term $(z - re^{j\vartheta})(z - re^{-j\vartheta})$, whereas the corresponding contribution to the frequency-response function that arises by setting again $z = e^{j\omega}$ is $(e^{j\omega} - re^{j\vartheta})(e^{j\omega} - re^{-j\vartheta})$. Therefore, at the frequency position $\omega = \vartheta$ the magnitude of the frequency response is *inversely proportional* to the term $(e^{j\vartheta} - re^{j\vartheta})(e^{j\omega} - re^{-j\vartheta}) = (1 - r)e^{j\vartheta}(e^{j\omega} - re^{-j\vartheta})$ and consequently *it is larger the closer the pole is to the circumference of the unit circle and vice versa*. If the pole of the transfer function belongs to the circumference of the unit circle, the measure of the frequency response at the position of this pole takes, as known, an *infinite value*.

From the above analysis, we find that if we want to reduce the measure of the frequency response at the position of a certain frequency, *we must place a zero value in a neighboring position to it*, whereas to cause an increase in the measure, we must place *a pole* in this position. This situation has been presented in detail in Sect. 10.4.3 and on page 842 (see also Fig. 10.10 on page 844 and Exercise 10.4.8 on page 845). Consequently, we have the ability to determine the appropriate frequency response in each case, by placing in the appropriate way (i.e., in the appropriate positions) the required number of poles and zero values on the complex plane. This process for the case of the four basic types of filters is presented below.

Low-pass and High-pass Filters

Starting our description from the *low-pass* and *high-pass* filters and limiting ourselves for simplicity to *first-order filters* described by a simple transfer function of the form

$$\mathcal{H}(z) = H_0 \frac{z - z_1}{p - p_1}, \tag{12.223}$$

these are known to be characterized by a *cutoff frequency* $f_c = \omega_c/2\pi$, the value of which dictates and correspondingly defines *the positions of the poles and zeros on the complex plane*. As we have mentioned in a previous section, the

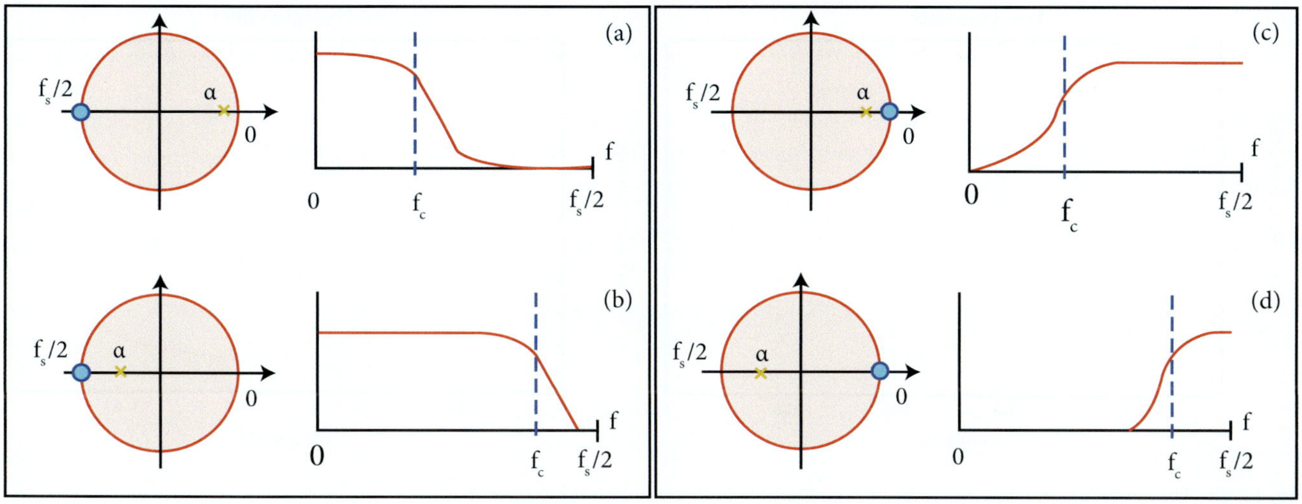

Fig. 12.81 The *pole-zero plot* and the amplitude response of low-pass and high-pass filters designed with the *pole-zero placement method* for the cases $f_c < f_s/4$ (images **a** and **c**) and $f_c > f_s/4$ (images **b** and **d**)

decisive role regarding the behavior of the filter is played by the *poles*, whereas the *zero values* allow the determination of characteristics such as the behavior of the frequency response at the positions of specific frequencies. It is not difficult to find out that in the case where the filter is implemented by a system *of first order*, the values of the pole and the zero value that appear in the transfer function are *real numbers*. As we see later, the determination of the positions of the pole and the zero value requires the knowledge of *the type of filter* (high-pass or low-pass), as well as the values of the *cutoff frequency* $f_c = \omega_c/2\pi$ and the *sampling frequency* F_s.

The pole-zero diagram and the amplitude response of a low-pass filter with cutoff frequency f_c is presented in images (a) and (b) of Fig. 12.81 for the cases $f_c > f_s/4$ and $f_c < f_s/4$. From the figure we observe that the zero value is *always* placed at the position $z = -1$ (which corresponds to the maximum frequency $\omega = \pi$ or equivalently to the folding frequency $F_s/2$), so as to ensure that *the frequency response of the filter at this position will be equal to zero*. On the other hand, the real pole $z = \alpha$ is placed *either on the positive or on the negative real semi-axis* (but always of course on the *inside* of the unit circle, so as to ensure the *stability*) depending on the relationship that exists between the frequencies f_c and f_s. More specifically, if $f_c < f_s/4$ the pole is placed on the *positive real semi-axis*, whereas if it is $f_c > f_s/4$ the pole is placed on the *negative real semi-axis*. As for the value of the position α of the pole, this for the two above cases is approximately calculated from the equations

$$\alpha \approx 1 - 2\pi\frac{f_c}{f_s} \qquad \text{and} \qquad \alpha \approx -\left(1 - \pi + 2\pi\frac{f_c}{f_s}\right)$$

respectively, when the pole is very close to the circumference of the unit circle (for example, one of the intervals [0.9, 1) and (−1, −0.9] respectively). Finally, as for the transfer function of the filter, this will have the form $\mathcal{H}(z) = H_0(z+1)/(z-\alpha)$. The value of the constant H_0 will result from the requirement to be $\mathcal{H}(z) = 1$ for $z = 1$ (that is, for frequency $\omega = 0$, since $z = e^{j\omega}$), from where it follows that $H_0 = (1-\alpha)/2$.

On the other hand, the design of *high-pass filters* is carried out in a similar way and the pole-zero diagram for the cases $f_c < f_s/4$ and $f_c > f_s/4$ is presented in images (c) and (d) of Fig. 12.81. We observe that the determination of the pole position is done as before, but in this case, the zero value is placed in the position $z = 1$. The equations for calculating the value of the pole position α are the same as the previous ones, but the transfer function of the high-pass filter will now have the form $\mathcal{H}(z) = H_0(z-1)/(z-\alpha)$ with the value of the constant H_0 this time being equal to $H_0 = (1+\alpha)/2$.

Example 12.13.3 In order to demonstrate the use of the pole-zero placement method, let us design a low-pass filter with a cutoff frequency of $f_c = 110$ Hz and a sampling rate of $f_s = 9000$ Hz. Starting from the zero value, this will be placed at the

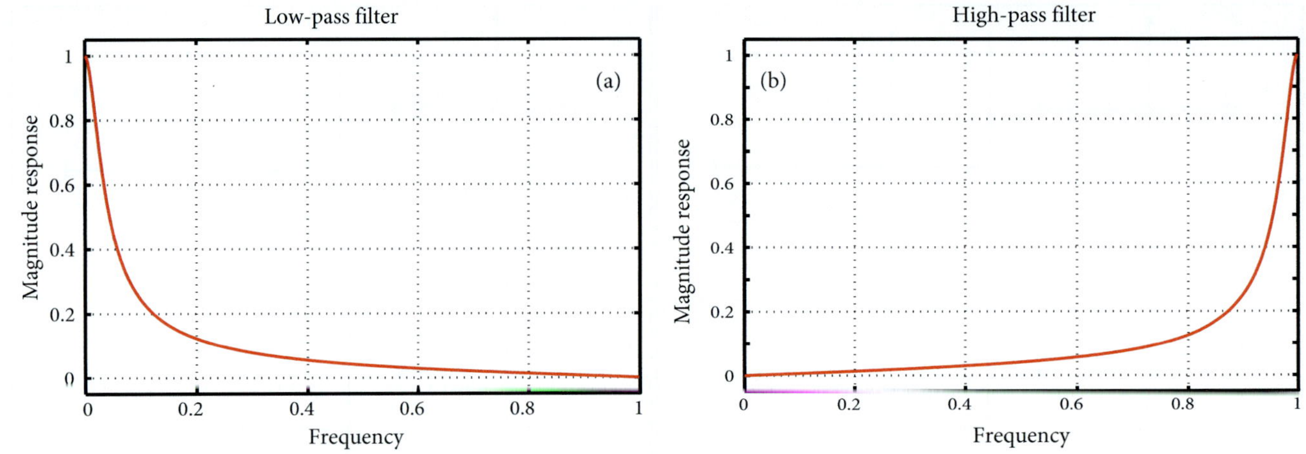

Fig. 12.82 The amplitude response of the low-pass and of the high-pass filter of Example 3

position $z = -1$. In order to determine the position where we should place the pole, we use the first of the two approximate equations—after we find that $f_c < f_c/4$—and therefore, we have

$$\alpha \approx 1 - 2\pi \frac{f_c}{f_s} = 1 - 2 \times 3.14159 \times \frac{110}{9000} = 0.923205.$$

We observe that this value is greater than 0.9 and therefore the above approximation is considered satisfactory. The proportionality constant H_0 corresponding to this value of α is calculated as $H_0 = (1-\alpha)/2 = (1-0.923205)/2 = 0.038397$ and consequently the transfer function of the low-pass filter will be

$$\mathcal{H}(z) = \frac{0.038397(z+1)}{z - 0.923295} = \frac{0.038397 + 0.038397z^{-1}}{1 - 0.923205z^{-1}}.$$

Let us also design, using this method, a high-pass filter with a cutoff frequency $f_c = 3900$ Hz and a sampling rate $f_s = 8000$ Hz. Starting from the zero value, this will now be placed at the position $z = 1$. On the other hand, to determine the pole placement position, we use the second of the two approximate equations—since in this case, $f_c > f_c/4$—and therefore we will have

$$\alpha \approx -\left(1 - \pi + 2\pi \frac{f_c}{f_s}\right) - \left(1 - 3.14159 + 2 \times 3.14159 \times \frac{3900}{8000}\right) = -0.921460.$$

We observe that this value is greater than 0.9 and therefore the above approximation is considered satisfactory. The proportionality constant H_0 corresponding to this value of α, in this case, is calculated as $H_0 = (1+\alpha)/2 = (1 - 0.921460)/2 = 0.039269$. Consequently, the transfer function of the high-pass filter will be

$$\mathcal{H}(z) = \frac{0.039269(z-1)}{z + 0.921460} = \frac{0.039269 - 0.039269z^{-1}}{1 + 0.921460z^{-1}}.$$

The amplitude response of the low-pass and high-pass filters for this example are presented in images (a) and (b) of Fig. 12.82.

Band-pass and Band-stop Filters After the description of the low-pass and high-pass filters, let us move on to the description of the *band-pass* and *band-stop* filters, which, being *second-order systems*, are characterized by the existence of *complex conjugate poles and zero values*. Starting with the case of *band-pass filters*, we should initially place *two zero values* at the positions $z = 1$ (which corresponds to the value of the *minimum frequency* $\omega = 0$) and $z = -1$ (which corresponds to the position of the *maximum frequency* $\omega = \pi$—that is, the folding frequency).[80] As for the poles of the function, these have

[80] The band-pass filters presented in this section are a special case of a category of discrete-time systems known as *digital resonators*. These systems are implemented in the form of band-pass filters characterized by the existence of *two complex conjugate poles* $p_{1,2} = re^{\pm j\vartheta}$, which are *close to the circumference of the unit circle*, and their name comes from the fact that the measure of their frequency response has a high value at

the form $p = re^{\pm j\vartheta}$ with the modulus r and the angle ϑ being a function of the filter's operating parameters (which are the *bandwidth* BW corresponding to attenuation 3 dB, the *center frequency* f_0 and the *sampling frequency* f_s) and are given by the approximate relations

$$r \approx 1 - \frac{\mathrm{BW}}{f_s}\pi \quad (0.9 \le r < 1.0) \qquad \text{and} \qquad \vartheta = \frac{f_0}{f_s} \times 360^\circ.$$

The transfer function of the *band-pass filter* that is constructed in this way has the form

$$\mathcal{H}(z) = H_0\frac{(z+1)(z-1)}{(z-re^{j\vartheta})(z-re^{-j\vartheta})} = H_0\frac{z^2-1}{(z-re^{j\vartheta})(z-re^{-j\vartheta})} = H_0\frac{1-z^{-2}}{(1-re^{j\vartheta}z^{-1})(1-re^{-j\vartheta}z^{-1})} \tag{12.224}$$

and consequently the frequency response of the filter that arises from the above relationship for $z = e^{j\omega}$ is formulated as

$$\mathcal{H}(e^{j\omega}) = \frac{H_0 \times (1-e^{-2j\omega})}{(1-re^{j\vartheta}e^{-j\omega})(1-re^{-j\vartheta}e^{-j\omega})} = \frac{H_0 \times (1-e^{-2j\omega})}{1-re^{-j\omega}(e^{j\vartheta}+e^{-j\vartheta})+r^2e^{-2j\omega}} = \frac{H_0 \times (1-e^{-2j\omega})}{1-2r\cos\vartheta e^{-j\omega}+r^2e^{-2j\omega}}. \tag{12.225}$$

The proportionality constant H_0 that appears in the above relationship is determined so as to correspond to *unit gain in the passband* and at the position $\omega = \vartheta$—in this case, the above relationship is reduced to the equation

$$\mathcal{H}(e^{j\vartheta}) = H_0\frac{1-e^{-2j\vartheta}}{1-re^{-j\vartheta}(e^{j\vartheta}+e^{-j\vartheta})+r^2e^{-2j\vartheta}} = H_0\frac{1-e^{-2j\vartheta}}{(1-r)(1-re^{-2j\vartheta})} \tag{12.226}$$

as can be proven after simple algebraic operations. We observe that the modulus of the numerator and the denominator of the last relationship are calculated as

$$|1-e^{-2i\vartheta}| = |([1-\cos(2\vartheta)] + j\sin(2\vartheta)| = \sqrt{[1-\cos(2\vartheta)]^2+\sin^2(2\vartheta)} = \sqrt{2-2\cos(2\vartheta)} = \sqrt{4\sin^2\vartheta} = 2|\sin\vartheta|$$

$$\begin{aligned}|(1-r)(1-re^{-2j\vartheta})| &= (1-r)|[1-r\cos(2\vartheta] + j[r\sin(2\vartheta)])| \\ &= (1-r)\sqrt{[1-r\cos(2\vartheta)]^2+r^2\sin^(2\vartheta)} = (1-r)\sqrt{1-2r\cos(2\vartheta)+r^2}.\end{aligned}$$

Combining the above and taking into account that for $\omega = \vartheta$ it is $|\mathcal{H}(e^{j\vartheta})| = 1$, we have

---

the pole position (and very close neighboring frequencies), thus referring to a *resonance phenomenon*. There are two variations of these types of filters, one of which is the one described in this section that is characterized by the existence of two zeros at the positions $z = +1$ $z = -1$ that cause the nullification of the amplitude response at the corresponding frequencies $\omega = 0$ and $\omega = \pi$, whereas the other has two zero values at the position $z = 0$ and is therefore characterized by a transfer function of the form

$$\mathcal{H}(z) = \frac{H_0}{(z-re^{j\vartheta})(z-re^{-j\vartheta})} = \frac{H_0}{(1-re^{j\vartheta}z^{-1})(1-re^{-j\vartheta}z^{-1})}.$$

Working, as in this section, we find that the amplitude response of this second category of digital differentiators has the form

$$\mathcal{H}(e^{j\omega}) = \frac{H_0}{1-re^{-j\omega}(e^{j\vartheta}+e^{-i\vartheta})+r^2e^{-2j\omega}} = \frac{H_0}{1-2r\cos\vartheta e^{-j\omega}+r^2e^{-2j\omega}} \qquad \text{where} \qquad H_0 = (1-r)\sqrt{1-2r\cos(2\vartheta)+r^2}.$$

$$|\mathcal{H}(e^{j\vartheta})| = \frac{2|\sin\vartheta|}{(1-r)\sqrt{1-2r\cos(2\vartheta)+r^2}} = 1 \quad \text{and therefore} \quad H_0 = \frac{(1-r)\sqrt{1-2r\cos(2\vartheta)+r^2}}{2|\sin\vartheta|}.$$

On the other hand, for the case of *band-stop filters*,[81] the poles are placed *in the same positions as those of band-pass filters*, whereas the zeros are placed *on the circumference of the unit circle under the same angles in relation to the poles*. The bandwidth corresponding to attenuation 3 dB and the central frequency f_0 are determined with the help of the measures and the angles of the complex conjugate poles. In the present case, the pole calculation equations are the

$$r \approx 1 - \frac{\text{BW}}{f_s}\pi \quad (0.9 \le r < 1.0) \qquad \text{and} \qquad \vartheta = \frac{f_0}{f_s} \times 360^\circ.$$

The *transfer function* of the band-pass filter that is constructed in this way is the

$$\mathcal{H}(z) = H_0 \frac{(z-e^{j\vartheta})(z-e^{-j\vartheta})}{(z-re^{j\vartheta})(z-re^{-j\vartheta})} = H_0 \frac{(1-e^{j\vartheta}z^{-1})(1-e^{-j\vartheta}z^{-1})}{(1-re^{j\vartheta}z^{-1})(1-re^{-j\vartheta}z^{-1})} = H_0 \frac{1-2\cos\vartheta z^{-1}+z^{-2}}{1-2r\cos\vartheta z^{-1}+r^2z^{-2}}. \quad (12.227)$$

The *frequency response* of the filter will result from the above relation for $z = e^{j\omega}$ and is given by the relation

$$\mathcal{H}(e^{j\omega}) = H_0 \frac{1-2\cos\vartheta e^{-j\omega}+e^{-2j\omega}}{1-2r\cos\vartheta e^{-j\omega}+r^2e^{-2j\omega}}. \quad (12.228)$$

The proportionality constant H_0 results from the requirement to be $|\mathcal{H}(e^{j\omega})| = 1$ for $\omega = 0$ and, as it turns out quite easily, it can be calculated from the equation

$$H_0 = \frac{1-2r\cos\vartheta + r^2}{2(1-\cos\vartheta)}. \quad (12.229)$$

The pole-zero diagram and the frequency response of a typical band-pass and band-stop filter of second order constructed with this method are presented in Fig. 12.83.

Example 12.13.4 In this example, we use the equations we constructed in the previous section for the construction of a typical band-pass and band-stop filter. Starting with the case of the band-pass filter, let us assume that it is characterized by a central frequency $f_0 = 1200$ Hz, bandwidth corresponding to attenuation 3 dB equal to 220 Hz and sampling frequency $f_s = 8500$ Hz. In this case, from the equations for calculating the values of r and ϑ we get

$$r \approx 1 - \frac{\text{BW}}{f_s}\pi = 1 - 3.14159 \times \frac{220}{8500} = 0.918688 \quad \text{and} \quad \vartheta = \frac{f_0}{f_s} \times 360^\circ = \frac{220}{8500} \times 360^\circ = 50.823529^\circ$$

We observe that the magnitude of the poles is greater than 0.9 and therefore the approximation is considered satisfactory. Consequently, the poles of the digital filter are located at

$$p_{1,2} = 0.918688 e^{\pm 50.823529^\circ j} = 0.918688 e^{\pm 50.823529\pi/180 j} = 0.918688 e^{\pm 0.887037 j} = 0.580345 \pm 0.712170 j$$

[81] The filters described in this section are also known as *notch filters*, owing to the shape of the amplitude response function, which nullifies at the positions of specific frequency values. These filters are particularly useful in cases where for some reason it is required to *remove certain components from the input signal*. In order to remove a specific frequency ω_0, we need to place a pair of complex conjugate zero values at the positions $z_{1,2} = e^{\pm j\omega_0}$. It should be noted that in the case of *FIR notch filters*, their bandwidth is quite extensive, resulting in *undesirable significant attenuation of adjacent frequency components*. In order to address this problem, we can place poles in the appropriate positions to enhance the signal amplitude. However, these poles may lead to the appearance of small *fluctuations* due to *resonance phenomena*, which can undergo suppression, using additional poles and working with the *method of trial and error* until we finally get an acceptable desired result.

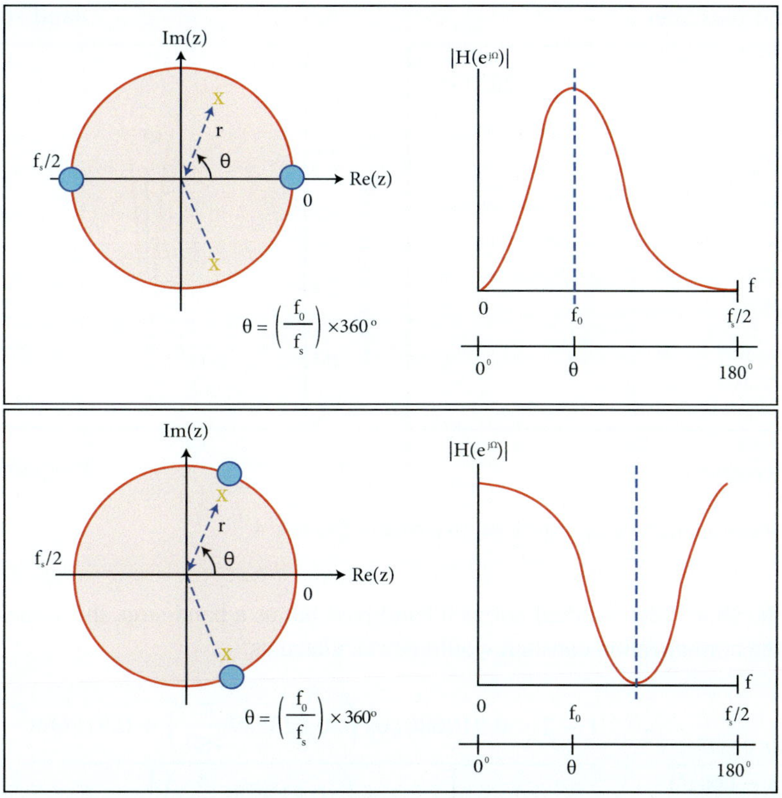

Fig. 12.83 The *pole-zero diagram* and the *frequency response* of a band-pass and band-stop filter of second order constructed with the *pole-zero placement method*

and they contribute to the denominator of the transfer function the term

$$(z - 0.580345 - 0.712170j)(z - 0.580345 + 0.712170j) = z^2 - 1.160905z + 0.843987.$$

Finally, the normalization constant is calculated as

$$H_0 = \frac{(1-r)\sqrt{1-2r\cos(2\vartheta)+r^2}}{2|\sin\vartheta|} = \frac{(1-0.918688)\sqrt{1-2\times 0.918688\cos\left(2\times 50.823529\dfrac{\pi}{180}\right)}}{2\left|\sin\left(2\times 50.823529\dfrac{\pi}{180}\right)\right|} = 0.078052.$$

Combining the above results, the transfer function of the filter is calculated as

$$\mathcal{H}(z) = \frac{0.078052(z^2-1)}{z^2 - 1.160690z^{-1} + 0.843987} = \frac{0.078052 - 0.078052z^{-2}}{1 - 1.160690z^{-1} + 0.843987z^{-2}},$$

whereas the corresponding frequency response function results from the substitution $z = e^{j\omega}$ and is

$$\mathcal{H}(e^{j\omega}) = \frac{0.078052 - 0.078052e^{-2j\omega}}{1 - 1.160690e^{-j\omega} + 0.843987e^{-2j\omega}}.$$

The amplitude response of this digital filter is presented in image (a) of Fig. 12.84, from where it can be seen that the filter approximates the operating specifications to a satisfactory degree. For example, the frequency of the digital filter corresponding to the central frequency f_0 for which the amplitude response is maximized is $\omega_0 = \pi f_0/(f_s/2) = 0.282352\pi$, which is depicted in the figure.

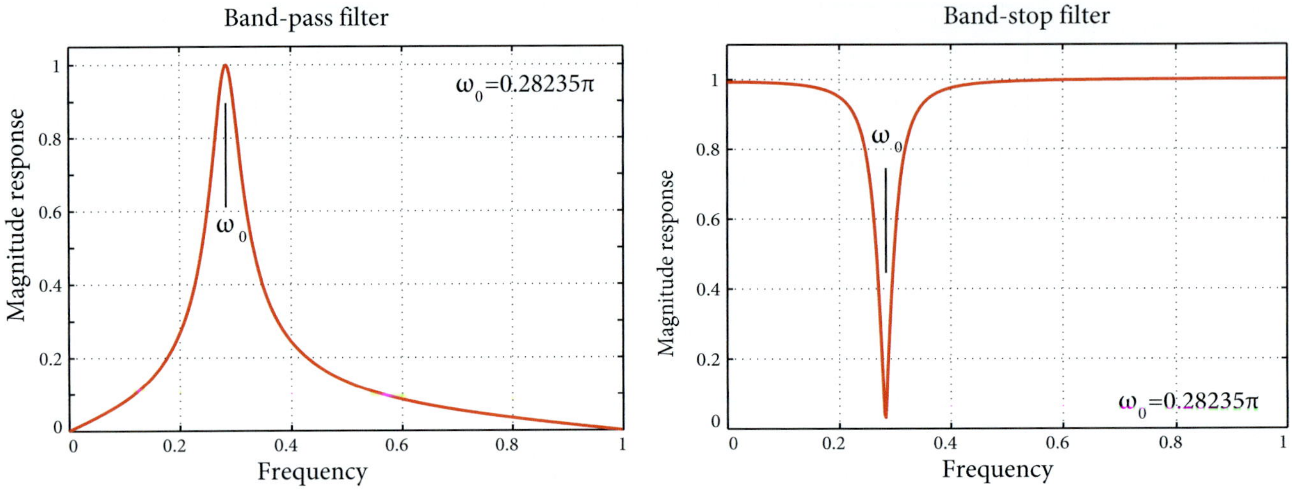

Fig. 12.84 The amplitude response of the band-pass and band-stop filter of Example 4

On the other hand, if the above filter worked not as a band-pass but as a band-stop, the values of r and ϑ would be the same as before, whereas the normalization constant would be calculated as

$$H_0 = \frac{1 - 2r\cos\vartheta + r^2}{2(1-\cos\vartheta)} = \frac{1 - 2\times 0.918688\cos\left(50.823529\dfrac{\pi}{180}\right) + 0.918688^2}{2\left[1-\cos\left(50.823529\dfrac{\pi}{180}\right)\right]} = 0.927664.$$

Consequently, the transfer function and the frequency response of the filter will have the form

$$\mathcal{H}(z) = H_0\frac{1-2\cos\vartheta z^{-1}+z^{-2}}{1-2r\cos\vartheta z^{-1}+r^2z^{-2}} = 0.927664\frac{1-1.263442z^{-1}+z^{-2}}{1-1.160690z^{-1}+0.843987z^{-2}}$$
$$= \frac{0.927664-1.172031z^{-1}+0.927664z^{-2}}{1-1.160690z^{-1}+0.843987z^{-2}}.$$

The amplitude response of this digital filter is presented in image (b) of Fig. 12.84. ■

We conclude the description of the filter design method with direct pole-zero placement, with the observation that although it is relatively easy to use, it is not suitable for the construction of filters for use in demanding applications. Although using the above relations we can predict the correct positions of the poles and zeros of the transfer function, in many cases the results we get may not be very satisfactory. In this case, we must resort to the *method of trial and error*, changing the positions of the poles and zero values, until we finally arrive at acceptable results, a process that is generally ineffective and quite time consuming, especially when there are more efficient methods of designing digital recursive filters.

12.13.6 Design with the Least-Squares Methods

In the previous sections, design techniques for *recursive digital filters* (such as the *derivative approximation method*, the *invariant impulse response method*, and *bilinear transformation*) were presented, which are based on the construction of the desired digital filter with the help of a prototype analog filter. However, in addition to these design techniques, there are other alternative methods that allow the design of the filter directly in the z domain (a characteristic example of such a method is the *pole-zero placement method* that was presented in Sect. 12.13.5—see page 1304). In the following pages, we describe a set of recursive digital filter design techniques that are based on *least-squares techniques*, that is, on the minimization of the suitable in every case of approximation error. Typical examples of methods that belong to this category are the *Padé method*, *Prony's method* and the *Shanks method*.

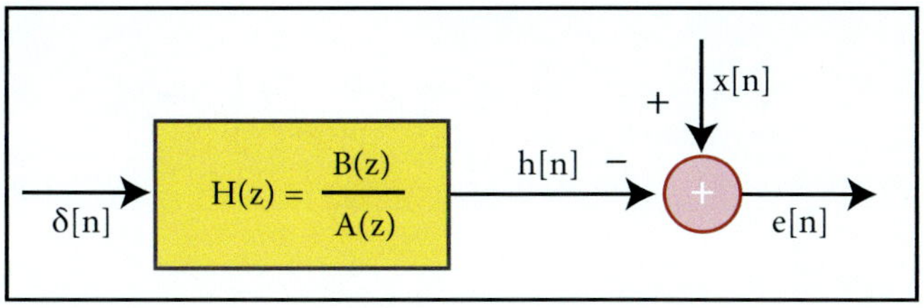

Fig. 12.85 Modeling of a discrete-time signal $x[n]$ as the response of a causal, linear and time-invariant discrete-time system with a genuine rational transfer function

The design methods of recursive digital filters with least-squares techniques are special cases of the more general problem of *modeling* a discrete sequence $x[n]$ with the help of a *causal, linear and time-invariant discrete-time system* that is described by a *proper rational transfer function* $\mathcal{H}(z)$ with N poles and M zeros of the form

$$\mathcal{H}(z)=\sum_{n=0}^{\infty}h[n]z^{-n}=\frac{\mathcal{B}(z)}{\mathcal{A}(z)}=\frac{\beta_0+\beta_1 z^{-1}+\beta_2 z^{-2}+\cdots+\beta_M z^{-M}}{\alpha_0+\alpha_1 z^{-1}+\alpha_2 z^{-2}+\cdots+\alpha_N z^{-N}}=\left\{\sum_{n=0}^{M}\beta_n z^{-n}\right\}\Big/\left\{1+\sum_{n=1}^{N}\alpha_n z^{-n}\right\} \tag{12.230}$$

or equivalently from the *difference equation*

$$y[n]+\alpha_1 y[n-1]+\alpha_2 y[n-2]+\cdots+\alpha_N y[n-N]=\beta_0 x[n]+\beta_1 x[n-1]+\beta_2 x[n-2]+\cdots+\beta_M x[n-M], \tag{12.231}$$

which is related to the transfer function $\mathcal{H}(z)$ in the way we described in Sect. 9.2 (see page 699). Speaking generally, *the modeling of a signal is defined as the representation of the signal with the help of the parameters of some model* and as a process it is extremely significant, as it is widely used in many signal-processing applications, such as *compression*, *prediction*, and *signal reconstruction*. In this case, the discrete-time signal $x[n]$ will be modeled as the *response of a system* like the one we presented above for the *impulse function* $\delta[n]$. It is obvious that in the case of *perfect* and *error-free signal representation* $x[n]$ in the above way, it will *coincide* with the impulse response of the system $h[n]$. In practice, of course, this does not happen and in all cases there is an *approximation error* between the signals $x[n]$ and $h[n]$. Taking into account that this approximation error is a function of the *parameters* of the filter β_k and α_k that appear in its transfer function, it is not difficult to realize that the modeling process consists of *calculating the values of the coefficients, so as to minimize the approximation error that exists between the desired (or ideal) output of the system $h[n]$ and the actual output $x[n]$*. This situation is presented diagrammatically in Fig. 12.85. As for the error $\mathcal{E}$ that is minimized in the way we described, this, according to the analysis techniques of the *least-squares method*, is defined as a *quadratic error* given by the relation

$$\mathcal{E}=\sum_{n=0}^{\infty}|e[n]|^2=\sum_{n=0}^{\infty}|x[n]-h[n]|^2. \tag{12.232}$$

The above analysis refers to *the modeling of the signal in the time domain*, whereas if we work *in the frequency domain*, the signals $x[n]$ and $h[n]$ are given by the *inverse Fourier transforms of discrete-time*

$$x[n]=\frac{1}{2\pi}\int_{-\pi}^{+\pi}\mathcal{X}(e^{j\omega})\,e^{j\omega n}d\omega \quad \text{and} \quad h[n]=\frac{1}{2\pi}\int_{-\pi}^{+\pi}\mathcal{H}(e^{j\omega})\,e^{j\omega n}d\omega=\frac{1}{2\pi}\int_{-\pi}^{+\pi}\frac{\mathcal{B}(e^{j\omega})}{\mathcal{A}(e^{j\omega})}e^{j\omega n}d\omega$$

and the above error is formulated as

$$\mathcal{E} = \sum_{n=0}^{\infty} \left| \frac{1}{2\pi} \int_{-\pi}^{+\pi} \left[\mathcal{X}(e^{j\omega}) - \frac{\mathcal{B}(e^{j\omega})}{\mathcal{A}(e^{j\omega})} \right] e^{j\omega n} d\omega \right|^2. \tag{12.233}$$

Example 12.13.5 In order to understand the problem of modeling a discrete-time signal $x[n]$ from a discrete LTI system, let us consider the simplest of cases in which the system is described by a simple transfer function *with only one pole* of the form $\mathcal{H}(z) = \beta_0/(1 + \alpha_1 z^{-1})$, leading to an impulse response

$$h[n] = \beta_0 \mathcal{Z}^{-1}\left\{\frac{1}{1 + \alpha_1 z^{-1}}\right\} = \beta_0(-\alpha_1)^n u[n].$$

From the above relationships, we observe that the determination of the transfer function requires the calculation of the parameters β_0 and α_1, which will be determined as *the solution of a system of two algebraic equations with two unknowns*. Taking into account that the output of the system in the ideal case will be the discrete-time signal $x[n] = h[n]$, we can construct these two equations by setting $x[0] = h[0]$ and $x[1] = h[1]$. Using the above equation for the impulse response $h[n]$, it turns out that $x[0] = \beta_0$ and $x[1] = -\beta_0\alpha_1$, from which it follows that $\beta_0 = x[0]$ and $\alpha_1 = -x[1]/x[0]$. Therefore, the transfer function of the system, the response of which to the impulse function is the discrete-time signal $x[n]$, is

$$\mathcal{H}(z) = \frac{x[0]}{1 - (x[1]/x[0])z^{-1}} = \frac{x^2[0]}{1 - x[1]z^{-1}}$$

Let us now see what results we end up with, if the degree of complexity of the system increases. Let us assume, then, that the system to be constructed, in addition to a pole, contains a zero value and therefore its transfer function has the form

$$\mathcal{H}(z) = \frac{\beta_0 + \beta_1 z^{-1}}{1 + \alpha_1 z^{-1}} \qquad \text{leading to impulse response} \qquad h[n] = \beta_0(-\alpha_1)^n u[n] + \beta_1(-\alpha_1)^{n-1} u[n-1].$$

We observe that, in this case, the determination of the system's transfer function requires the determination of the values *of three parameters*, β_0, β_1, and α_1, which will arise as the solution of an algebraic system *of three equations with three unknowns*. Taking into account that the output of the system in the ideal case will be the discrete-time signal $x[n] = h[n]$, we can construct these three equations, setting $x[0] = h[0]$, $x[1] = h[1]$, and $x[2] = h[2]$. Using the above equation for the impulse response $h[n]$, it turns out that $\beta_0 = x[0]$, $\beta_1 = x[1] - (x[0]x[2])/x[1]$, and $\alpha_1 = -x[2]/x[1]$. Therefore, the transfer function of the system, whose response to the impulse function is the discrete-time signal $x[n]$, is

$$\mathcal{H}(z) = \frac{[x[0] + (x[1] - x[2]/x[1])]z^{-1}}{1 - (x[2]/x[1])z^{-1}}.$$

We observe that the resulting system is obviously more complex, whereas the determination of the coefficients of the transfer function requires the solution of equations that this time are nonlinear. Although in this equation the solution of these nonlinear equations is possible and relatively easy, in practice this does not happen when the system is quite complex, that is, when it contains a large number of poles and zero values.

From the above example we conclude that in the general case of a discrete-time system LTI, the transfer function of which contains N poles and M zero values, the construction of the system's transfer function so as to create the desired discrete-time signal $x[n]$ at its output, requires the calculation of the $M + N + 1$ coefficients $\{\alpha_k\}$ $(k = 1, 2, \ldots, N)$ and $\{\beta_k\}$ $(k = 1, 2, \ldots, M)$ by solving an equal number of algebraic equations, which constitute a *nonlinear algebraic system*. It is not difficult to ascertain that if we use the signal $h_d[n]$ that expresses the *desired impulse response of a digital filter* with transfer function $\mathcal{H}_d(z) = \mathcal{Z}\{h_d[n]\}$ in place of any discrete-time signal $x[n]$, this technique can be used *to approximate the behavior of a digital f when the transfer function of $\mathcal{H}_d(z)$ or its impulse response $h_d[n]$ is known*. In the following description, we use the desired impulse response $h_d[n]$ in place of the signal $x[n]$, so that what we describe can be directly applied to the *design of digital filters*, knowing, however, that what will be aforementioned applies to any discrete-time signal

$x[n]$. Consequently, the square error $\mathcal{E}$, which we will minimize in the next section, according to the *method of least squares* and what we mentioned previously, is formulated as

$$\mathcal{E} = \sum_{n=0}^{M+N} \left[h_d[n] - h[n] \right]^2, \tag{12.234}$$

where we have set the upper limit of the sum to the value $M + N$, so that the identification of the signals $h_d[n]$ and $h[n]$ is limited to their first $M + N$ samples.

The Padé Method of Approximation

A technique that we can resort to in order to avoid solving nonlinear algebraic systems is to determine the coefficients $\{\alpha_k\}$ $(k = 1, 2, \ldots, N)$ and $\{\beta_k\}$ $(k = 1, 2, \ldots, M)$ using not the transfer function of the system but the *difference equation* that describes it, expressed in the form

$$y[n] = -\alpha_1 y[n-1] - \alpha_2 y[n-2] - \cdots - \alpha_N y[n-N] + \beta_0 x[n] + \beta_1 x[n-1] + \beta_2 x[n-2] + \cdots + \beta_M x[n-M].$$

If we input the *impulse function* $\delta[n]$ into the system, the output of the system will be the well-known *impulse response* $h[n]$ and therefore the above relationship is formulated as

$$\begin{aligned} h[n] &= -\alpha_1 h[n-1] - \alpha_2 h[n-2] - \cdots - \alpha_N h[n-N] + \beta_0 \delta[n] + \beta_1 \delta[n-1] + \beta_2 \delta[n-2] + \cdots + \beta_M \delta[n-M] \\ &= -\alpha_1 h[n-1] - \alpha_2 h[n-2] - \cdots - \alpha_N h[n-N] + \beta_n \end{aligned}$$

with the last relationship resulting from the fact that the function $\delta[n]$ has a value *equal to one* for the value $n = 0$ and *zero value* in any other case. The above relationship holds for the range of values $0 \le n \le M$, whereas for $M < n < N$ the coefficients β_n do not exist and the above relationship is written as

$$h[n] = -\alpha_1 h[n-1] - \alpha_2 h[n-2] - \cdots - \alpha_N h[n-N]. \tag{12.235}$$

The above relationships, which in condensed form are formulated as

$$y[n] + \sum_{k=0}^{N} \alpha_k h[n-k] = \begin{cases} \beta_n & n = 1, 2, \ldots, M \\ 0 & n = M+1, M+2, \ldots, M+N \end{cases} \tag{12.236}$$

constitute a *system of linear equations*, the solution of which will also give us the desired values of the coefficients α_n and β_n.

The calculation of the coefficients $\{\alpha_k\}$ $(k = 1, 2, \ldots, N)$ and $\{\beta_k\}$ $(k = 1, 2, \ldots, M)$ with the help of the above equations is done in two stages: *first, we calculate the coefficients* $\{\alpha_k\}$ $(k = 1, 2, \ldots, N)$ *and then the values that are calculated are used for the determination of the values of the coefficients* $\{\beta_k\}$. To calculate the coefficients $\{\alpha_k\}$, we work in the range of values $M+1 \le n \le M+N$ (using therefore the second of the above equations) proceeding to the substitution $h[n] = h_d[n]$. Taking into account that the coefficients β_n in this range of values are *equal to zero*, this process leads to the system of equations

$$\begin{bmatrix} h_d[M+1] & h_d[M] & h_d[M-1] & \ldots & h_d[M-N+1] \\ h_d[M+2] & h_d[M+1] & h_d[M] & \ldots & h_d[M-N+2] \\ h_d[M+3] & h_d[M+2] & h_d[M+1] & \ldots & h_d[M-N+3] \\ \vdots & \vdots & \vdots & \vdots & \vdots \\ h_d[M+N] & h_d[M+N-1] & h_d[M+N-2] & \ldots & h_d[M] \end{bmatrix} \begin{bmatrix} 1 \\ \alpha_1 \\ \alpha_2 \\ \vdots \\ \alpha_n \end{bmatrix} = \begin{bmatrix} 0 \\ 0 \\ 0 \\ \vdots \\ 0 \\ \cdot \end{bmatrix} \tag{12.237}$$

The above system can alternatively be formulated as

$$\underbrace{\begin{bmatrix} h_d[M] & h_d[M-1] & \dots h_d[M-N+1] \\ h_d[M+1] & h_d[M] & \dots h_d[M-N+2] \\ h_d[M+2] & h_d[M+1] & \dots h_d[M-N+3] \\ \vdots & \vdots & \ddots \quad \vdots \\ h_d[M+N-1] & h_d[M+N-2] \dots & h_d[M] \end{bmatrix}}_{X_M} \underbrace{\begin{bmatrix} \alpha_1 \\ \alpha_2 \\ \alpha_3 \\ \vdots \\ \alpha_N \end{bmatrix}}_{\boldsymbol{\alpha}} = - \underbrace{\begin{bmatrix} h_d[M+1] \\ h_d[M+2] \\ h_d[M+3] \\ \vdots \\ h_d[M+N] \end{bmatrix}}_{X_{M+1}} \tag{12.238}$$

or in *condensed matrix form* $\boldsymbol{X}_M \cdot \boldsymbol{\alpha} = \boldsymbol{X}_{M+1}$, with the matrices $\boldsymbol{X}_M$, $\boldsymbol{\alpha}$ and $\boldsymbol{X}_{M+1}$ being of dimensions $N \times N$, $N \times 1$ and $N \times 1$ respectively. For the solution of the above algebraic system, what we know from linear algebra applies and more specifically the following cases exist:

- The matrix $\boldsymbol{X}_M$ is *nonsingular* and the system *has a solution*. In this case, the *inverse matrix* $\boldsymbol{X}_M^{-1}$ exists and the solution of the system is unique and equal to $\boldsymbol{\alpha} = -\boldsymbol{X}_M^{-1}\boldsymbol{X}_{M+1}$.
- The matrix $\boldsymbol{X}_M$ is *singular* and the system *has a solution*. In reality, in this case, the system has *multiple solutions*, since if the vector $\boldsymbol{\alpha}$ constitutes a solution of the system, then the vector $\tilde{\boldsymbol{\alpha}} = \boldsymbol{a} + z$, where z are the nonzero solutions of the homogeneous equation $\boldsymbol{X}_m z = \boldsymbol{0}$ will also be a solution.
- The matrix $\boldsymbol{X}_M$ is *singular* and the system *does not have a solution*. In this case, we set $\alpha_0 = 0$ and solve the system $\boldsymbol{X}_M \cdot \boldsymbol{\alpha} = \boldsymbol{0}$, instead of the $\boldsymbol{X}_M \cdot \boldsymbol{\alpha} = \boldsymbol{X}_{M+1}$.

After calculating the coefficients α_n these are used for the calculation of the coefficients β_n using the equations we previously constructed for the range of values $0 \le n \le M$. These equations constitute an algebraic system, which in matrix form is formulated as

$$\underbrace{\begin{bmatrix} h_d[0] & 0 & 0 & \dots & 0 \\ h_d[1] & h_d[0] & 0 & \dots & 0 \\ h_d[2] & h_d[1] & h_d[0] & \dots & 0 \\ \vdots & \vdots & \vdots & \ddots & \vdots \\ h_d[M] & h_d[M-1] & h_d[M-2] & \dots & h_d[M-N] \end{bmatrix}}_{X_0} \underbrace{\begin{bmatrix} 1 \\ \alpha_1 \\ \alpha_2 \\ \vdots \\ \alpha_N \end{bmatrix}}_{\boldsymbol{\alpha}} = \underbrace{\begin{bmatrix} \beta_0 \\ \beta_1 \\ \beta_2 \\ \vdots \\ \beta_M \end{bmatrix}}_{\boldsymbol{\beta}} \tag{12.239}$$

or in matrix form $\boldsymbol{X}_0 \cdot \boldsymbol{a} = \boldsymbol{b}$ with the matrices $\boldsymbol{X}_0$, $\boldsymbol{a}$ and $\boldsymbol{b}$ being of dimensions $M \times N$, $N \times 1$ and $M \times 1$.

The approximate method we described in this section is known as the *Padé approximation*. Its main disadvantage is that *the accuracy of the approximation it offers increases with the increase in the complexity of the filter and therefore the filters for which the resulting outcome is considered acceptable are characterized by a high degree of complexity, that is, by a large number of poles and zero values*. For this reason, the *approximation of Padé* is not used very often in practice.

Note In the simple case where the transfer function of the filter has only poles, $M = 1$ and the coefficients α_k will be calculated from the equation

$$\underbrace{\begin{bmatrix} h_d[0] & 0 & 0 & \dots & 0 \\ h_d[1] & h_d[0] & 0 & \dots & 0 \\ h_d[2] & h_d[1] & h_d[0] & \dots & 0 \\ \vdots & \vdots & \vdots & \ddots & \vdots \\ h_d[N-1] & h_d[N-2] & h_d[N-3] & \dots & h_d[0] \end{bmatrix}}_{X_0} \underbrace{\begin{bmatrix} \alpha_1 \\ \alpha_2 \\ \alpha_3 \\ \vdots \\ \alpha_N \end{bmatrix}}_{\boldsymbol{a}} = - \underbrace{\begin{bmatrix} h_d[1] \\ h_d[2] \\ h_d[3] \\ \vdots \\ h_d[N] \end{bmatrix}}_{X_1} \tag{12.240}$$

or in matrix form $\boldsymbol{X}_0 \cdot \boldsymbol{a} = -\boldsymbol{X}_1$. It is not difficult to find that in this case, the matrix $\boldsymbol{X}_0$ is a *lower triangular Toeplitz matrix*[82] and therefore, the coefficients α_k can be calculated directly from the relation

$$\alpha_k = -\frac{1}{h_d[0]}\left[h_d[k] + \sum_{\ell=1}^{k-1} \alpha_\ell h_d[k-\ell]\right]. \tag{12.241}$$

Example 12.13.6 In order to see in practice the application of the Padé approximation, let us consider the construction of a filter, the desired impulse response of which is $h_d[n] = 4(1/3)^n u[n]$. For simplicity, let us consider the simple case of a transfer function with one pole and one zero value, from which it follows that $M = N = 1$. Applying the methodology of the previous section, the matrix equation of the Padé approximation is formulated as

$$\begin{bmatrix} h_d[0] & 0 \\ h_d[1] & h_d[0] \\ h_d[2] & h_d[1] \end{bmatrix} \begin{bmatrix} 1 \\ \alpha_1 \end{bmatrix} = \begin{bmatrix} \beta_0 \\ \beta_1 \\ 0 \end{bmatrix} \quad \text{or equivalently} \quad \begin{bmatrix} 4 & 0 \\ 4/3 & 4 \\ 4/9 & 4/3 \end{bmatrix} \begin{bmatrix} 1 \\ \alpha_1 \end{bmatrix} = \begin{bmatrix} \beta_0 \\ \beta_1 \\ 0 \\ , \end{bmatrix}$$

where we have replaced the values of the impulse response samples $h_d[0] = 4$, $h_d[1] = 4/3$ and $h_d[2] = 4/9$ from the defining equation. From the last line of the product matrix, it follows that $\alpha_1 = -1/3$. Replacing this value of α_1 in the above equation and using the first two lines, we are led to the matrix equation

$$\begin{bmatrix} h_d[0] & 0 \\ h_d[1] & h_d[0] \end{bmatrix} \begin{bmatrix} 1 \\ \alpha_1 \end{bmatrix} = \begin{bmatrix} \beta_0 \\ \beta_1 \end{bmatrix} \quad \acute{\eta}\ \iota\sigma o\delta\acute{\upsilon}\nu\alpha\mu\alpha \quad \begin{bmatrix} 4 & 0 \\ 4/3 & 4 \end{bmatrix} \begin{bmatrix} 1 \\ -1/3 \end{bmatrix} = \begin{bmatrix} \beta_0 \\ \beta_1 \\ , \end{bmatrix}$$

from which it follows that $\beta_0 = 4$ and $\beta_1 = 0$. Consequently, the desired transfer function will be

$$\mathcal{H}(z) = \frac{4}{1-(1/3)z^{-1}}.$$

[82] A *square matrix* of order N is called a *Toeplitz matrix* when its elements are constant along all of its *negatively sloped* diagonals (i.e., the diagonals that run *from left to right*). Given $2N-1$ numbers α_k $(k = -n+1, \ldots, -1, 0, 1, \ldots, n-1)$, a Toeplitz matrix is defined as

$$T = \begin{bmatrix} \alpha_0 & \alpha_{-1} & \alpha_{-2} & \ldots & \alpha_{-n+1} \\ \alpha_1 & \alpha_0 & \alpha_{-1} & \ddots & \vdots \\ \alpha_2 & \alpha_1 & \alpha_0 & \ddots & \alpha_{-2} \\ \vdots & \ddots & \ddots & \ddots & \alpha_{-1} \\ \alpha_{N-1} & \ldots & \alpha_2 & \alpha_1 & \alpha_0 \end{bmatrix}$$

and consequently, its elements satisfy the relation $\alpha_{ij} = \alpha_{j-i}$ for all values of i and j. If a *Toeplitz matrix* is also a *lower triangular matrix*, it will have the form

$$T = \begin{bmatrix} \alpha_0 & 0 & 0 & \ldots & 0 \\ \alpha_1 & \alpha_0 & 0 & \ddots & 0 \\ \alpha_2 & \alpha_1 & \alpha_0 & \ddots & 0 \\ \vdots & \ddots & \ddots & \ddots & 0 \\ \alpha_{N-1} & \alpha_{N-2} & \alpha_{N-3} & \ldots & \alpha_0. \end{bmatrix}$$

If such a matrix appears in an equation, like $\boldsymbol{X}_0 \cdot \boldsymbol{a} = -\boldsymbol{X}_1$, it is proven that the solution of the system arises through the application of the *backward substitution* method. More details on this topic can be found in the literature (see, for example, the book G.H. Golub, C.F van Loan, *Matrix Computations*, The John Hopkins University Press, ISBN 0-8018-4513X, 1996, p. 193.)

Note that this result coincides with that obtained by calculating the transfer function as the $\mathcal{Z}$ transform of the desired impulse response. In this case

$$\mathcal{H}(z) = \sum_{n=-\infty}^{+\infty} h_d[n]z^{-n} = \sum_{n=-\infty}^{+\infty} 4\left(\frac{1}{3}\right)^n u[n]z^{-n} = 4\sum_{n=-\infty}^{+\infty}\left(\frac{1}{3z}\right)^n = \frac{4}{1-(1/3)z^{-1}},$$

which is exactly the same as the one we obtained with the *Padé approximation*.

Although the Padé method gives excellent results when the filter to be constructed is described by a *rational transfer function with a known number of poles and zeros*, in practice the desired impulse response $h_d[n]$ is determined *indirectly* as the *inverse Fourier transform* of a desired amplitude response that meets certain specifications. In this case, the method does not always lead to acceptable results. It is common practice to design the filter with the trial and error method, applying the method for several combinations of the parameters M and N (i.e., the number of poles and the number of zero values) until we achieve an acceptable approximation error, a process that naturally makes the method *extremely complex and time consuming*. This characteristic of the method is demonstrated in the following example.

Example 12.13.7 Let us use the Padé approximation, to approximate the behavior of an *ideal low-pass filter* (zero phase), the amplitude response of which has the form

$$|\mathcal{H}(e^{j\omega})| = \begin{cases} e^{-5j\omega} & |\omega| < \pi/2 \\ 0 & \text{in the opposite case.} \end{cases}$$

Calculating the inverse Fourier transform of discrete-time of the above function, the desired impulse response of the system has the form

$$h_d[n] = \frac{1}{2\pi}\int_{-\pi/2}^{+\pi/2} \mathcal{H}(e^{j\omega})e^{i\omega n}d\omega = \frac{1}{2\pi}\int_{-\pi/2}^{+\pi/2} e^{j\,(n-5)\omega}d\omega$$

$$= \frac{\exp\left[j\,(n-5)\frac{\pi}{2}\right] - \exp\left[-j\,(n-5)\frac{\pi}{2}\right]}{\pi(n-5)\times(2j)} = \frac{\sin\left[(n-5)\frac{\pi}{2}\right]}{\pi(n-5)}.$$

Assuming that the discrete-time system we design is characterized by a transfer function with $M = 5$ zero values and $N = 5$ poles, the number of coefficients $\{\alpha_k\}$ and $\{\beta_k\}$ that we need to calculate is equal to $M + N + 1 = 11$ and therefore for the application of the Padé method, knowledge of the first 11 samples of the desired impulse response is required, the values of which, with the help of the above equation, are

$$h_d[n] = \{0.063661, 0, -0.106103, 0, 0.318309, 0.5, 0.318309, 0, -0.106103, 0, 0.063661\}.$$

Applying the equations of the *Padé approximation*, let us first calculate the coefficients $\{\alpha_k\}$ $(k = 1, 2, 3, 4, 5)$ from the equation

$$\begin{bmatrix} h_d[5] & h_d[4] & h_d[3] & h_d[2] & h_d[1] \\ h_d[6] & h_d[5] & h_d[4] & h_d[3] & h_d[2] \\ h_d[7] & h_d[6] & h_d[5] & h_d[4] & h_d[3] \\ h_d[8] & h_d[7] & h_d[6] & h_d[5] & h_d[4] \\ h_d[9] & h_d[8] & h_d[7] & h_d[6] & h_d[5] \end{bmatrix} \begin{bmatrix} \alpha_1 \\ \alpha_2 \\ \alpha_3 \\ \alpha_4 \\ \alpha_5 \end{bmatrix} = \begin{bmatrix} h_d[6] \\ h_d[7] \\ h_d[8] \\ h_d[9] \\ h_d[10], \end{bmatrix}$$

which, substituting the values of the desired impulse response samples, takes the form

$$\begin{bmatrix} 0.500000 & 0.318309 & 0.000000 & -0.106103 & 0.000000 \\ 0.318309 & 0.500000 & 0.318309 & 0.000000 & -0.106103 \\ 0.000000 & 0.318309 & 0.500000 & 0.318309 & 0.000000 \\ -0.106103 & 0.000000 & 0.318309 & 0.500000 & 0.318309 \\ 0.000000 & -0.106103 & 0.000000 & 0.318309 & 0.500000 \end{bmatrix} \begin{bmatrix} \alpha_1 \\ \alpha_2 \\ \alpha_3 \\ \alpha_4 \\ \alpha_5 \end{bmatrix} = \begin{bmatrix} 0.318309 \\ 0.000000 \\ -0.106103 \\ 0.000000 \\ 0.063661. \end{bmatrix}$$

The solution of the above system leads to the result

$$\boldsymbol{a} = \left[\alpha_0\ \alpha_1\ \alpha_2\ \alpha_3\ \alpha_4\ \alpha_5\right]^T = \\ = \left[1.000000\ -2.524810\ 3.675606\ -3.483043\ 2.128893\ -0.702628\right]^T.$$

In the second stage of the process, we use the coefficients $\{\alpha_k\}$ that we determined above, to calculate and the coefficients $\{\beta_k\}$ from the equation

$$\begin{bmatrix} h_d[0] & 0 & 0 & 0 & 0 & 0 \\ h_d[1] & h_d[0] & 0 & 0 & 0 & 0 \\ h_d[2] & h_d[1] & h_d[0] & 0 & 0 & 0 \\ h_d[3] & h_d[2] & h_d[1] & h_d[0] & 0 & 0 \\ h_d[4] & h_d[3] & h_d[2] & h_d[1] & h_d[0] & 0 \\ h_d[5] & h_d[4] & h_d[3] & h_d[2] & h_d[1] & h_d[0] \end{bmatrix} \begin{bmatrix} 1 \\ \alpha_1 \\ \alpha_2 \\ \alpha_3 \\ \alpha_4 \\ \alpha_5 \end{bmatrix} = \begin{bmatrix} \beta_0 \\ \beta_1 \\ \beta_2 \\ \beta_3 \\ \beta_4 \\ \beta_5, \end{bmatrix}$$

which, substituting the values of the signal samples and the coefficients α_k, is formulated as

$$\begin{bmatrix} 0.063661 & 0.000000 & 0.000000 & 0.000000 & 0.000000 & 0.000000 \\ 0.000000 & 0.063661 & 0.000000 & 0.000000 & 0.000000 & 0.000000 \\ -0.106103 & 0.000000 & 0.063661 & 0.000000 & 0.000000 & 0.000000 \\ 0.000000 & -0.106103 & 0.000000 & 0.063661 & 0.000000 & 0.000000 \\ 0.318309 & 0.000000 & -0.106103 & 0.000000 & 0.063661 & 0.000000 \\ 0.500000 & 0.318309 & 0.000000 & -0.106103 & 0.000000 & 0.063661 \end{bmatrix} \begin{bmatrix} 1.000000 \\ -2.524810 \\ 3.675606 \\ -3.483043 \\ 2.128893 \\ -0.702628 \end{bmatrix} = \begin{bmatrix} \beta_0 \\ \beta_1 \\ \beta_2 \\ \beta_3 \\ \beta_4 \\ \beta_5. \end{bmatrix}$$

The solution of the above system leads to the result

$$\boldsymbol{\beta} = \left[\beta_0\ \beta_1\ \beta_2\ \beta_3\ \beta_4\ \beta_5\right]^T = \\ = \left[0.063661\ -0.160731\ 0.127889\ 0.046155\ 0.063843\ 0.021161\right]^T.$$

Therefore, the transfer function of the system that approximates the ideal low-pass filter of the example is given by the equation

$$\mathcal{H}(z) = \frac{0.063661 - 0.160731z^{-1} + 0.127889z^{-2} + 0.046155z^{-3} + 0.063843z^{-4} + 0.021161z^{-5}}{1.000000 - 2.524810z^{-1} + 3.675606z^{-2} - 3.483043z^{-3} + 2.128893z^{-4} - 0.702628z^{-5}}$$

and is characterized by the amplitude response depicted in Fig. 12.86. It is not difficult to realize that the solution we constructed *is far from being characterized as even a satisfactory approximation*, a failure that is obviously due to the fact that the choice of values $M = 5$ and $N = 5$ was completely random. For this reason, as we mentioned previously, the method is applied *iteratively* for different combinations of the parameters M and N, until it finally leads to an acceptable result. This process that follows *the approach of trial and error* is left as an exercise to the reader, who is asked to repeat the above calculations for different combinations of M and N, until the amplitude response received approaches to a satisfactory degree that of the given ideal low-pass filter.

Design of Optimal Inverse Filters

The *Padé approximation*, which we presented in the previous section, is based on the *minimization of the error between the desired and the actual impulse response*, by identifying as many samples as possible. Another method that is based on the same logic also involves minimizing a type of error, only this time it does not concern the impulse response $h[n]$ but is defined *between the system output* $y[n]$ *and the impulse response* $\delta[n]$. In order to understand the application of this method, let us remember that *if we construct a composite system that results from the serial connection of a system with its inverse, the system that will result from this process will be the identity system, which will reproduce at its output its input*, a feature

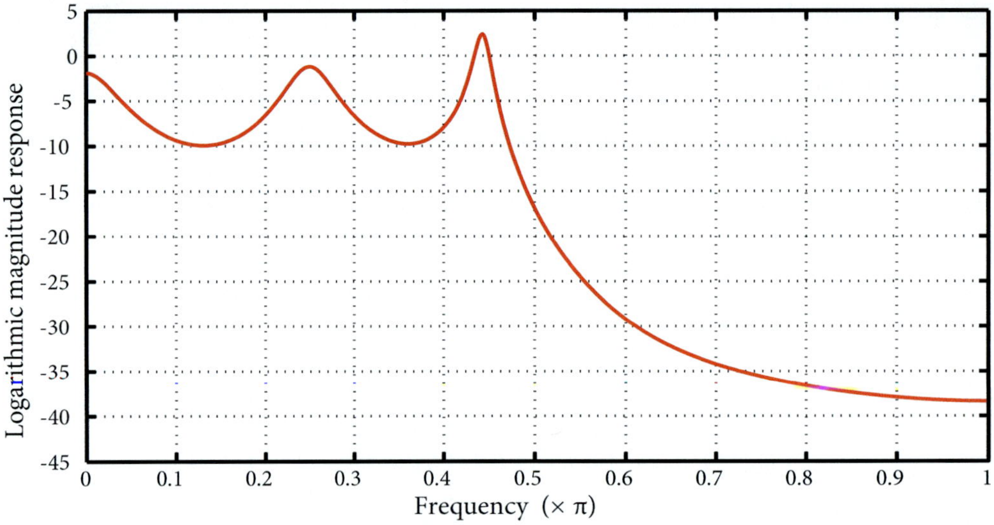

Fig. 12.86 The amplitude response of the approximate low-pass filter that was constructed by applying the *Padé method* for parameter values $M = 5$ and $N = 5$

that mathematically is expressed as $y[n] = x[n]$. It is clear that if the *impulse function* $\delta[n]$ is used as the input to the first system (which is the desired system to be constructed), its output, which will serve as the input to the second system, will be the *desired impulse response* $h_d[n]$. The output of the second system, being an identity system, will simply be the *impulse function* $\delta[n]$. For a detailed description of all these concepts, see Sect. 9.5 on page 714 and Fig. 9.5 on page 715.

Based on the above description, let us consider Fig. 12.87, in which the desired discrete and to be constructed system $\mathcal{H}_d(z)$ is connected *in series* with an inverse system, the transfer function of which has the form $1/\mathcal{H}(z)$. If this second system is the inverse of the system and therefore its transfer function is $1/\mathcal{H}_d(z)$, then the response of the composite system to the impulse function $\delta[n]$ will be $y[n] = \delta[n]$. Conversely, if the second system in series *is not exactly the inverse of the first* but approximates it to some degree, then the system output $y[n]$ *will not be exactly the impulse function* but only an approximation of it, with the approximation error mathematically defined as

$$\mathcal{E} = \sum_{n=1}^{\infty} y^2[n]. \tag{12.242}$$

Consequently, the design of the desired system using this method consists in the *minimization of the error* $\mathcal{E}$, based on the logic that the smaller this error, the greater will be the degree of similarity between the signals $y[n]$ and $\delta[n]$ and the more the function $1/\mathcal{H}(z)$ will calculate the inverse of the desired function $1/\mathcal{H}_d(z)$. If, therefore, we determine *the filter coefficients* $\{\alpha_k\}$ and $\{\beta_k\}$ that lead to the best possible approximation of the function $1/\mathcal{H}_d(z)$, we can calculate this function and then invert it, to get the desired function $\mathcal{H}_d(z)$.

Assuming for simplicity that the system to be constructed contains *only poles* and is therefore described by the transfer function

$$\mathcal{H}(z) = \beta_0 \Big/ \left\{ 1 + \sum_{k=1}^{N} \alpha_k z^{-k} \right\},$$

the transfer function of the inverse system will be

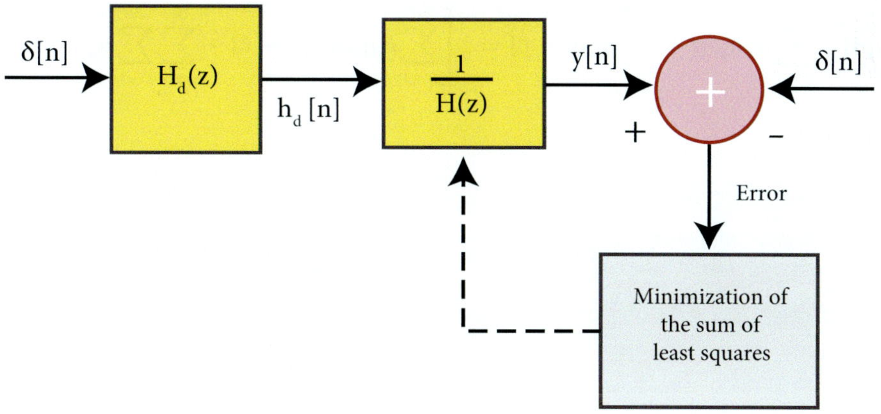

Fig. 12.87 Design of an *inverse filter* using the least-squares method

$$\frac{1}{\mathcal{H}(z)}=\left\{1+\sum_{k=1}^{N}\alpha_k z^{-k}\right\}\Big/\beta_0=\frac{\mathcal{X}(s)}{\mathcal{Y}(s)} \quad \text{from which it follows that} \quad \left(\sum_{k=1}^{N}\alpha_k z^{-k}\right)\mathcal{Y}(s)=\beta_0\mathcal{X}(s).$$

Taking the inverse $\mathcal{Z}$ transform of the above equation and using *the property of time shifting* in the way we did in the previous section, it follows that the output of the inverse system $y[n]$ (which is also the final output of the composite system) is

$$y[n]=\frac{1}{\beta_0}\Big[x[n]+\alpha_1 x[n-1]+\alpha_2 x[n-2]+\cdots+\alpha_N x[n-N]\Big]=\frac{1}{\beta_0}\left[x[n]+\sum_{k=1}^{N}\alpha_k x[n-k]\right]. \tag{12.243}$$

Therefore, if we input the impulse function $x[n]=\delta[n]$ into the system, the system output will be

$$y[n]=\frac{1}{\beta_0}\left[h_d[n]+\sum_{k=1}^{N}\alpha_k h_d[n-k]\right] \tag{12.244}$$

It is not difficult to see that for the values $n=0$ and $\beta_0=h_d[0]$, $y[n]=0$. On the contrary, for values $n>0$ the desired system output $\delta[n]$ will be *equal to zero* and therefore the discrete-time signal $y[n]$ will represent *the approximation error between the desired output* $\delta[n]=0$ $(n>0)$ *and the actual output* $y[n]$ (this is also the reason why the error defining equation $\mathcal{E}$ appears). Based on all the above, we can formulate the claim that the design of the filter consists in determining the values of the coefficients $\{\alpha_k\}$, which minimize the approximation error defined as

$$\mathcal{E}=\sum_{n=1}^{\infty}\left[h_d[n]+\sum_{k=1}^{N}\alpha_k h_d[n-k]\right]^2. \tag{12.245}$$

Following the usual way of calculating function extremes, *we differentiate the error function with respect to the coefficients* $\{\alpha_k\}$ *and set the resulting expression to zero*. In this case, we will get

$$\frac{d\mathcal{E}}{d\alpha_k} = 2\sum_{n=1}^{\infty}\left[h_d[n] + \sum_{k=1}^{N}\alpha_k h_d[n-k]\right]h_d[n-\ell] = 2\left[\sum_{n=1}^{\infty}h_d[n]h_d[n-\ell] + \sum_{n=1}^{\infty}\sum_{k=1}^{N}\alpha_k h_d[n-k]h_d[n-\ell]\right] = 0$$

from which it follows that

$$\sum_{k=1}^{N}\alpha_k\sum_{n=1}^{\infty}h_d[n-k]h_d[n-\ell] = -\sum_{n=1}^{\infty}h_d[n]h_d[n-\ell]. \tag{12.246}$$

Defining the function

$$\xi(k,\ell) = \sum_{n=1}^{\infty}h_d[n-k]h_d[n-\ell] = \sum_{n=1}^{\infty}h_d[n]h_d[n+k-\ell] = \xi(k-\ell) \tag{12.247}$$

(where for the construction of the last expression we proceeded to replace the dummy index in the function $h_d[n-k]$ from $n-k$ to n)

$$\xi(\ell,0) = \sum_{n=1}^{\infty}h_d[n-\ell]h_d[n] \quad \text{and the above relation writes} \quad \sum_{k=1}^{N}\alpha_k\xi(k,\ell) = -\xi(k,0), \quad \ell = 1,2,\ldots,N$$

or in expanded form

$$\begin{aligned}
&\alpha_1\xi(1,1) + \alpha_2\xi(2,1) + \alpha_3\xi(3,1) + \cdots + \alpha_N\xi(N,1) = -\xi(1,0)\\
&\alpha_1\xi(1,2) + \alpha_2\xi(2,2) + \alpha_3\xi(3,2) + \cdots + \alpha_N\xi(N,2) = -\xi(2,0)\\
&\alpha_1\xi(1,3) + \alpha_2\xi(2,3) + \alpha_3\xi(3,3) + \cdots + \alpha_N\xi(N,3) = -\xi(3,0)\\
&\cdots\cdots\cdots\cdots\cdots\cdots\cdots\cdots\cdots\cdots\cdots\cdots\cdots\cdots\\
&\alpha_1\xi(1,N) + \alpha_2\xi(2,N) + \alpha_3\xi(3,N) + \cdots + \alpha_N\xi(N,N) = -\xi(N,0).
\end{aligned}$$

The above expressions constitute *a system of N equations with N unknowns*, the solution of which are the unknown coefficients $\alpha_1, \alpha_2, \ldots, \alpha_N$. Having calculated the coefficients α_N, we can then calculate the transfer function of the inverse system $1/\mathcal{H}(z)$ and from there the desired transfer function $\mathcal{H}(z)$.

We conclude the presentation of this recursive digital filter design method with the observation that in the case where the number L of available samples of the desired impulse response $h_d[n]$ is much larger than the number of poles N, which happens very often in practice, the function $\xi(k-\ell)$—which is recognized as a *correlation function* is calculated as

$$\xi(k-\ell) = \sum_{n=0}^{L-|k-\ell|}h_d[n]h_d[n+k-\ell], \quad 0 \le k-\ell \le N. \tag{12.248}$$

Example 12.13.8 In order to understand the application of this method, let us approximate the ideal low-pass filter using the previous example and the first 11 samples of the *desired impulse response* $h_d[n]$, which define the *discrete-time signal*

$$h_d[n] = \{0.063661, 0, -0.106103, 0, 0.318309, 0.5, 0.318309, 0, -0.106103, 0, 0.063661\}.$$

Following the methodology of the theory presented above, let us construct an *approximate model* that contains $N = 5$ poles (without zero values). In this case, the constant β_0 will be set, as mentioned, to the value $\beta_0 = h_d[0] = 0.063661$, whereas the constants $\{\alpha_k\}$ $(k = 1, 2, 3, 4, 5)$ will result from the solution of the linear system of the previous section, which in *matrix form* (for $N = 5$) is formulated as

$$\begin{bmatrix} \xi(1,1) & \xi(2,1) & \xi(3,1) & \xi(4,1) & \xi(5,1) \\ \xi(1,2) & \xi(2,2) & \xi(3,2) & \xi(4,2) & \xi(5,2) \\ \xi(1,3) & \xi(2,3) & \xi(3,3) & \xi(4,3) & \xi(5,3) \\ \xi(1,4) & \xi(2,4) & \xi(3,4) & \xi(4,4) & \xi(5,4) \\ \xi(1,5) & \xi(2,5) & \xi(3,5) & \xi(4,5) & \xi(5,5) \end{bmatrix} \begin{bmatrix} \alpha_1 \\ \alpha_2 \\ \alpha_3 \\ \alpha_4 \\ \alpha_5 \end{bmatrix} = - \begin{bmatrix} \xi(1,0) \\ \xi(2,0) \\ \xi(3,0) \\ \xi(4,0) \\ \xi(5,0) \end{bmatrix}$$

with the quantities $\xi(k, \ell)$ $(k, \ell = 1, 2, 3, 4, 5)$ being calculated from Eq. (12.241). The implementation of the above calculations leads to the system

$$\begin{bmatrix} 0.483262 & 0.318309 & 0.027019 & -0.053052 & -0.013510 \\ 0.318309 & 0.483262 & 0.318309 & 0.027019 & -0.053052 \\ 0.027019 & 0.318309 & 0.483262 & 0.318309 & 0.027019 \\ -0.053052 & 0.027019 & 0.318309 & 0.483262 & 0.318309 \\ -0.013510 & -0.053052 & 0.027019 & 0.318309 & 0.483262 \end{bmatrix} \begin{bmatrix} \alpha_1 \\ \alpha_2 \\ \alpha_3 \\ \alpha_4 \\ \alpha_5 \end{bmatrix} = \begin{bmatrix} -0.318309 \\ -0.027019 \\ 0.053052 \\ 0.013510 \\ -0.031830, \end{bmatrix}$$

the solution of which leads to the coefficient vector

$$\boldsymbol{a} = \left[\alpha_0\ \alpha_1\ \alpha_2\ \alpha_3\ \alpha_4\ \alpha_5\right]^T =$$
$$= \left[1.000000\ -0.008577\ -1.596067\ 2.780640\ -2.568049\ 1.294707\right]^T.$$

Consequently, the desired transfer function takes the form

$$\mathcal{H}(z) = \frac{0.063661}{1.000000 - 0.008577z^{-1} - 1.596067z^{-2} - 2.780640z^{-3} - 2.568049z^{-4} + 1.294707z^{-5}}$$

leading to an amplitude response, which, as can be seen, cannot be characterized as a good approximation, since the function we used *does not have any zero values*, whereas the number of poles was chosen *completely randomly*, just to demonstrate the method of use. It is obvious that the above equations give good results, *only when the filter to be constructed does not contain zeros*, whereas for the general case of filters that contain both poles and zero values it will need *generalization of the above equations*. In full analogy with the *Padé approximation*, this method requires systematic experimentation and use of many different values for the parameter N with the *trial and error method*, until we finally arrive at an acceptable result. ■

Note The method we describe in this section can also be used for the construction of *FIR* filters, simply by truncating the calculated impulse response of the inverse filter, so that it becomes a *finite signal of N samples in length*. Alternatively, we can use the least-squares method for the calculation of the $N + 1$ coefficients β_k that appear in the transfer function of the inverse system, defining the error function

$$\mathcal{E} = \sum_{n=0}^{\infty} \left[\zeta[n] - \sum_{k=0}^{N} \beta_k h[n-k] \right]^2,$$

where $\zeta[n]$ is the desired output sequence. Following the analysis method we describe in this section, the coefficients β_k will be calculated as the solution of the linear algebraic system

$$\sum_{k=0}^{N} \beta_k \xi(k-\ell) = \chi(\ell), \quad \ell = 0, 1, \ldots, N,$$

with the functions $\xi(\ell)$ and $\chi(\ell)$ defined as

$$\xi(\ell) = \sum_{k=0}^{\infty} h[n]h[n-\ell] \qquad \text{and} \qquad \chi(\ell) = \sum_{k=0}^{\infty} \zeta[n]h[n-\ell]$$

(these functions are recognized as *autocorrelation* and *cross-correlation* functions—see Sect. 6.10.3 on page 553). The *optimal inverse filters* constructed in this way are known as *Wiener filters* after the mathematician *Norbert Wiener*, who invented them.[83]

If we input the impulse function $\zeta[n] = \delta[n]$ into the system, $\chi[\ell] = h[0]\delta[\ell]$ and the coefficients $\{\beta_k\}$ $(k = 0, 1, 2, \ldots, N)$ will emerge as the solution of the system

$$\begin{bmatrix} \xi(0) & \xi(1) & \xi(2) & \ldots\, \xi(N) \\ \xi(1) & \xi(0) & \xi(1) & \ldots\, \xi(N-1) \\ \xi(2) & \xi(1) & \xi(0) & \ldots\, \xi(N-2) \\ \vdots & \vdots & \vdots & \ddots\, \vdots \\ \xi(N) & \xi(N-1) & \zeta(N \;\; 2) & \ldots\, \xi(0) \end{bmatrix} \begin{bmatrix} \beta_0 \\ \beta_1 \\ \beta_2 \\ \vdots \\ \beta_N \end{bmatrix} = \begin{bmatrix} h[0] \\ 0 \\ 0 \\ \vdots \\ 0 \end{bmatrix}$$

Once again, we are led to a *Toeplitz matrix* (which is expected, since finite impulse response filters contain *only poles*), the inversion of which and ultimately the solution of the system and the calculation of the coefficients $\{\beta_k\}$ $(k = 0, 1, 2, \ldots, N)$ can be performed with the help of efficient algorithms, such as the *Levinson-Durbin algorithm*.[84]

Prony's Method

In contrast to the previous design method, which uses the concept of the *inverse filter*, the *Prony's method*[85] described in this section directly approaches the desired digital filter, following a different approach from the one used in the *Padé method*. More specifically, whereas the *Padé method* tries to exactly match the first $M + N + 1$ samples of the desired and actual impulse responses $h_d[n]$ and $h[n]$, ignoring what happens beyond that, the Prony's method *takes into account the remaining samples* and instead of proceeding to zero the error for the first $M + N + 1$ values, as the Padé method does, it tries to minimize an approximation error defined *over all samples of discrete-time signals* with the help of the *least-squares method*.

In order to describe the *Prony's*, let us assume that the transfer function of the considered filter is characterized by the existence of M zero values and N poles and therefore the difference equation that describes it will have the form

$$y[n] = -\sum_{k=1}^{N} \alpha_k y[n-k] + \sum_{k=0}^{M} \beta_k x[n-k]. \tag{12.249}$$

In this case, if we transmit the impulse function $x[n] = \delta[n]$ at the input of the filter, then its output will be the impulse response $h[n]$ and the above relation is formulated as

$$h[n] = -\sum_{k=1}^{N} \alpha_k h[n-k] + \sum_{k=0}^{M} \beta_k \delta[n-k]. \tag{12.250}$$

However, we know that from the M *time-shifted impulse functions* that appear in the above relation, the only one that has a nonzero value (and equal to the unit) is the one that corresponds to the index value $k = n$ and therefore will be

$$h[n] = \begin{cases} -\displaystyle\sum_{k=1}^{N} \alpha_k h[n-k] + \beta_n & 0 \le n \le M \\ -\displaystyle\sum_{k=1}^{N} \alpha_k h[n-k] & n > M \end{cases} \tag{12.251}$$

[83] See the book Wiener N., *Extrapolation, Interpolation & Smoothing of Stationary Time Series with engineering applications*, John Wiley & Sons, New York, 1949.

[84] See references (1) Levinson N., *The Wiener RMS Error Criterion in Filter Design and Prediction*, Journal of Mathematical Physics, Vol. 25, pp. 261–278, 1947 and (2) Durbin J., *Efficient Estimation of Parameters in Moving-Average Models*, Biometrika, Vol. 46, pp. 306–316, 1959.

[85] This method was invented in 1795 by the French mathematician and engineer *Gaspard Riche de Prony* (1755–1839) but only gained practical utility in the last century with the advent of electronic computers.

(where we silently assumed that $M < N$ and therefore the coefficients β_n for $n > M$ will obviously be equal to zero, since they are not defined for this range of values). The above expressions concern *the approximation of the ideal transfer function* $\mathcal{H}_d(z)$ *by the real function* $\mathcal{H}(z)$. Let us note that although it is not certain that the function $\mathcal{H}_d(z)$ will contain poles and zero values, like its approximation $\mathcal{H}(z)$, we can at least for the range of values $n > M$ approximate the equivalent desired impulse response (which is the inverse $\mathcal{Z}$ transform of the desired function $\mathcal{H}_d(z)$) by defining the *discrete-time signal*

$$\hat{h}_d[n] = -\sum_{k=1}^{N} \alpha_k h_d[n-k] \qquad \text{and the discrete error} \qquad e[n] = h_d[n] - \hat{h}_d[n] = h_d[n] + \sum_{k=1}^{N} \alpha_k h_d[n-k].$$

Consequently, the values of the coefficients α_k can be calculated as those values that *minimize the square approximation error* of the desired impulse response $h_d[n]$ from the real impulse response $h[n]$ and which for the range of values $n > M$ is formulated as

$$\mathcal{E} = \sum_{n=M+1}^{\infty} |e[n]|^2 = \sum_{n=M+1}^{\infty} \left| h_d[n] - \hat{h}_d[n] \right|^2 = \sum_{n=M+1}^{\infty} \left| h_d[n] + \sum_{k=1}^{N} \alpha_k h_d[n-k] \right|^2. \tag{12.252}$$

Working as before, that is, *differentiating the error function* with respect to α_k and equating the result to zero, we end up with the equation

$$\frac{\partial \mathcal{E}}{\partial \alpha_k} = \frac{\partial}{\partial \alpha_k} \left\{ \sum_{n=M+1}^{\infty} |e[n]|^2 \right\} = 2 \sum_{n=M+1}^{\infty} e[n] \frac{\partial e[n]}{\partial \alpha_k} = 0. \tag{12.253}$$

However,

$$\frac{\partial e[n]}{\partial \alpha_k} = h_d[n-k] \qquad \text{and since} \qquad e[n] = h_d[n] + \sum_{k=1}^{N} \alpha_k h_d[n-k]$$

the above relation is written as

$$\sum_{n=M+1}^{\infty} \left[h_d[n] + \sum_{k=1}^{N} \alpha_k h_d[n-k] \right] h_d[n-\ell] = 0$$

and therefore

$$\sum_{k=1}^{N} \alpha_k \left[\sum_{n=M+1}^{\infty} h_d[n-k] h_d[n-\ell] \right] = - \sum_{n=M+1}^{\infty} h_d[n] h_d[n-k].$$

Defining the function

$$\xi(k,\ell) = \sum_{n=M+1}^{\infty} h_d[n-k] h_d[n-\ell] \tag{12.254}$$

the above relation takes the simple form

$$\sum_{\ell=1}^{N} \alpha_\ell \xi(k,\ell) = -\xi(k,0), \quad k = 1, 2, \ldots, N \tag{12.255}$$

or equivalently, using *matrix notation*,

$$\underbrace{\begin{bmatrix} \xi(1,1) & \xi(1,2) & \xi(1,3) & \dots & \xi(1,N) \\ \xi(2,1) & \xi(2,2) & \xi(2,3) & \dots & \xi(2,N) \\ \xi(3,1) & \xi(3,2) & \xi(3,3) & \dots & \xi(3,N) \\ \vdots & \vdots & \vdots & \ddots & \vdots \\ \xi(N,1) & \xi(N,2) & \xi(N,3) & \dots & \xi(N,N) \end{bmatrix}}_{\boldsymbol{\xi}} \underbrace{\begin{bmatrix} \alpha_1 \\ \alpha_2 \\ \alpha_3 \\ \vdots \\ \alpha_5 \end{bmatrix}}_{\boldsymbol{\alpha}} = - \underbrace{\begin{bmatrix} \xi(1,0) \\ \xi(2,0) \\ \xi(3,0) \\ \vdots \\ \xi(N,0) \end{bmatrix}}_{\boldsymbol{\gamma}} \tag{12.256}$$

If the matrix $\boldsymbol{\xi}$ is *nonsingular*, its inverse exists and the system has *one and only one solution*, which is given by the relation $\boldsymbol{\alpha} = -\boldsymbol{\xi}^{-1} \cdot \boldsymbol{\gamma}$. In the opposite case, we apply procedures similar to those we described in the *Padé approximation*.

After calculating the coefficients α_k we can calculate the coefficients β_n in the way we did in the Padé approximation, that is, by *zeroing the error* $e[n]$ for the range of values $0 \leq n \leq M$. In this way we end up with the system of equations

$$b_n = h_d[n] + \sum_{k=1}^{N} \alpha_k h_d[n-k], \quad 0 \leq n \leq M, \tag{12.257}$$

which in *matrix form* is formulated as

$$\underbrace{\begin{bmatrix} h_d[0] & 0 & 0 & \dots & 0 \\ h_d[1] & h_d[0] & 0 & \dots & 0 \\ h_d[2] & h_d[1] & h_d[0] & \dots & 0 \\ \vdots & \vdots & \vdots & \ddots & \vdots \\ h_d[M] & h_d[M-1] & h_d[M-2] & \dots & h_d[M-N] \end{bmatrix}}_{\boldsymbol{X}_0} \underbrace{\begin{bmatrix} 1 \\ \alpha_1 \\ \alpha_2 \\ \vdots \\ \alpha_N \end{bmatrix}}_{\boldsymbol{\alpha}} = \underbrace{\begin{bmatrix} \beta_0 \\ \beta_1 \\ \beta_2 \\ \vdots \\ \beta_M \end{bmatrix}}_{\boldsymbol{\beta}} \tag{12.258}$$

or equivalently $\boldsymbol{X}_0 \cdot \boldsymbol{a} = \boldsymbol{b}$, with the matrices $\boldsymbol{X}_0$, $\boldsymbol{a}$, and $\boldsymbol{b}$ being of dimensions $M \times N$, $N \times 1$, and $M \times 1$ respectively.

Example 12.13.9 In order to understand the application of Prony's method, let us approach the ideal low-pass filter of Example 12.13.2 (see page 1288) using the first 11 samples of the desired impulse response $h_d[n]$, which define the discrete-time signal

$$h_d[n] = \{0.063661, 0, -0.106103, 0, 0.318309, 0.5, 0.318309, 0, -0.106103, 0, 0.063661\}.$$

Following the methodology of the theory presented above, let us construct an approximate model that contains $N = 5$ poles and $M = 5$ zero values. Using the equation for calculating the coefficients α_k with an upper limit value for the sum of $L = 11$, it turns out that

$$\begin{bmatrix} \xi(1,1) & \xi(1,2) & \xi(1,3) & \xi(1,4) & \xi(1,5) \\ \xi(2,1) & \xi(2,2) & \xi(2,3) & \xi(2,4) & \xi(2,5) \\ \xi(3,1) & \xi(3,2) & \xi(3,3) & \xi(3,4) & \xi(3,5) \\ \xi(4,1) & \xi(4,2) & \xi(4,3) & \xi(4,4) & \xi(4,5) \\ \xi(5,1) & \xi(5,2) & \xi(5,3) & \xi(5,4) & \xi(5,5) \end{bmatrix} \begin{bmatrix} \alpha_1 \\ \alpha_2 \\ \alpha_3 \\ \alpha_4 \\ \alpha_5 \end{bmatrix} = - \begin{bmatrix} \xi(1,0) \\ \xi(2,0) \\ \xi(3,0) \\ \xi(4,0) \\ \xi(5,0) \\ , \end{bmatrix}$$

with the quantities $\xi(k, \ell)$ $(k, \ell = 1, 2, 3, 4, 5)$ being calculated as the solution of the system

$$\begin{bmatrix} 0.362578 & 0.318309 & 0.067547 & -0.106103 & -0.067547 \\ 0.318309 & 0.463899 & 0.318309 & 0.033774 & -0.106103 \\ 0.067547 & 0.318309 & 0.452641 & 0.318309 & 0.067547 \\ -0.106103 & 0.033774 & 0.318309 & 0.463899 & 0.318309 \\ -0.067547 & -0.106103 & 0.067547 & 0.318309 & 0.362578 \end{bmatrix} \begin{bmatrix} \alpha_1 \\ \alpha_2 \\ \alpha_3 \\ \alpha_4 \\ \alpha_5 \end{bmatrix} = \begin{bmatrix} -0.159155 \\ -0.060792 \\ 0.053052 \\ 0.047283 \\ -0.031830 \end{bmatrix}$$

and to be equal to

$$\boldsymbol{a} = \left[\alpha_0 \; \alpha_1 \; \alpha_2 \; \alpha_3 \; \alpha_4 \; \alpha_5\right]^T =$$
$$= \left[1.000000 \; -2.517033 \; 3.657151 \; -3.459241 \; 2.110133 \; -0.694543\right]^T .$$

Following the technique of the *Padé approximation*, we use the coefficients $\{\alpha_k\}$ that we determined above, to calculate the coefficients $\{\beta_k\}$ from the equation

$$\begin{bmatrix} h_d[0] & 0 & 0 & 0 & 0 & 0 \\ h_d[1] & h_d[0] & 0 & 0 & 0 & 0 \\ h_d[2] & h_d[1] & h_d[0] & 0 & 0 & 0 \\ h_d[3] & h_d[2] & h_d[1] & h_d[0] & 0 & 0 \\ h_d[4] & h_d[3] & h_d[2] & h_d[1] & h_d[0] & 0 \\ h_d[5] & h_d[4] & h_d[3] & h_d[2] & h_d[1] & h_d[0] \end{bmatrix} \begin{bmatrix} 1 \\ \alpha_1 \\ \alpha_2 \\ \alpha_3 \\ \alpha_4 \\ \alpha_5 \end{bmatrix} = \begin{bmatrix} \beta_0 \\ \beta_1 \\ \beta_2 \\ \beta_3 \\ \beta_4 \\ \beta_5, \end{bmatrix}$$

which, substituting the values of the signal samples and the coefficients α_k, takes the form

$$\begin{bmatrix} 0.063661 & 0.000000 & 0.000000 & 0.000000 & 0.000000 & 0.000000 \\ 0.000000 & 0.063661 & 0.000000 & 0.000000 & 0.000000 & 0.000000 \\ -0.106103 & 0.000000 & 0.063661 & 0.000000 & 0.000000 & 0.000000 \\ 0.000000 & -0.106103 & 0.000000 & 0.063661 & 0.000000 & 0.000000 \\ 0.318309 & 0.000000 & -0.106103 & 0.000000 & 0.063661 & 0.000000 \\ 0.500000 & 0.318309 & 0.000000 & -0.106103 & 0.000000 & 0.063661 \end{bmatrix} \begin{bmatrix} 1.000000 \\ -2.517033 \\ 3.657151 \\ -3.459241 \\ 2.110133 \\ -0.694543 \end{bmatrix} = \begin{bmatrix} \beta_0 \\ \beta_1 \\ \beta_2 \\ \beta_3 \\ \beta_4 \\ \beta_5. \end{bmatrix}$$

The solution to the above system leads to the result

$$\boldsymbol{\beta} = \left[\beta_0 \; \beta_1 \; \beta_2 \; \beta_3 \; \beta_4 \; \beta_5\right]^T =$$
$$= \left[0.063661 \; -0.160236 \; 0.126714 \; 0.046846 \; 0.064607 \; 0.021626\right]^T$$

Therefore, the transfer function of the system that approximates the ideal low-pass filter of the example is given by the equation

$$\mathcal{H}(z) = \frac{0.063661 - 0.160236z^{-1} + 0.126714z^{-2} + 0.046846z^{-3} + 0.064607z^{-4} + 0.021626z^{-5}}{1.000000 - 2.517033z^{-1} + 3.657151z^{-2} - 3.459241z^{-3} + 2.110133z^{-4} - 0.694543z^{-5}}.$$

The result we arrived at is almost the same as the one we got in the Padé approximation and its graphical representation is similar to that of Example 12.13.2 (see page 1288).

Formulating the approximation error in the form

$$\mathcal{E} = \sum_{n=M+1}^{\infty} |e[n]|^2 = \sum_{n=M+1}^{\infty} e[n]\left[h_d[n] + \sum_{k=1}^{N} \alpha_k h_d[n-k]\right] \tag{12.259}$$

and using the orthogonality condition

$$\sum_{n=M+1}^{\infty} h[n]h_d[n-k] = 0, \tag{12.260}$$

which is related to the *optimal* (in the sense of *minimization*) solution, it is proven that the *minimum value of the approximation error* is given by the relation

$$\mathcal{E}_{\min} = \sum_{n=M+1}^{\infty} e[n]h_d[n] = \sum_{n=M+1}^{\infty} \left[h_d[n] + \sum_{k=1}^{N} \alpha_k h_d[n-k] \right] h_d[n] = \xi(0,0) + \sum_{k=1}^{N} \alpha_k \xi(0,k). \tag{12.261}$$

The application of this equation to the system of Example 9 leads to a minimum error value equal to 0.137372.

The Shanks Method

the The basic characteristic of the *Prony's method* that was presented in the previous section is that the coefficients β_k found in the numerator of the transfer function of the filter are calculated *by the Padé approximation*, that is, by zeroing the value of the error $e[n]$ between the signals $h[n]$ and $h_d[n]$ for the range of values $0 \le n \le M$. This is something that leads to a very accurate representation for this range of values of M, *without, however, taking into account what happens for larger values of* N. A more accurate approximation method is one in which the coefficients β_k of the transfer function of the filter to be designed are also calculated *by minimizing an error* that is calculated on *all* the available samples of the desired impulse response $h_d[n]$, using *least-squares techniques*. This more general approximation method is known as the *Shanks method*.

In order to describe the *Shanks method of approximation*, let us assume that the filter to be designed arises from the series connection of a filter $\mathcal{H}_1(z)$ that contains only poles and a filter $\mathcal{H}_2(z)$ that contains only zero values—therefore,

$$\mathcal{H}_1(z) = \frac{1}{1 + \sum_{k=1}^{N} \alpha_k z^{-k}} \quad \text{and} \quad \mathcal{H}_2(z) = \sum_{k=0}^{M} \beta_k z^{-k}.$$

In the first step of the process, we calculate the coefficients α_k using the *least-squares method* in the way we did in the previous section. Having determined the coefficients α_k, we can construct the transfer function $\mathcal{H}_1(z)$ and from there, the difference equation of the system, which, as known, has the form

$$y[n] = -\sum_{k=1}^{N} \alpha_k y[n-k] + x[n] \quad (n \ge 0). \tag{12.262}$$

Consequently, if in the first system we use as input the *impulse function* $\delta[n]$, the output $\upsilon[n]$ that is generated from this will be the

$$\upsilon[n] = -\sum_{k=1}^{N} \alpha_k \upsilon[n-k] + \delta[n] \quad (n \ge 0). \tag{12.263}$$

In the next step of the process, the discrete sequence $\upsilon[n]$ is used as input in the second system, the transfer function of which contains the coefficients to be calculated β_k. Taking into account that the difference equation constructed from the function $\mathcal{H}_2(z)$ is the

$$y[n] = \sum_{k=0}^{M} \beta_k x[n-k] \tag{12.264}$$

the system's response to the input $\upsilon[n]$ will be the signal

$$\hat{h}_d[n] = \sum_{k=0}^{M} \beta_k \upsilon[n-k], \tag{12.265}$$

thus allowing the definition of an *approximation error* $e[n]$ between the *actual* and the *desired impulse response*, which is defined as

$$e[n] = h_d[n] - \hat{h}_d[n] = h[n] - \sum_{k=0}^{M} \beta_k \upsilon[n-k]. \tag{12.266}$$

Consequently, the coefficients β_k can be determined with the help of *least-squares techniques*, as the coefficients that minimize the error

$$\mathcal{E} = \sum_{n=0}^{\infty} \left[h_d[n] - \sum_{k=0}^{M} \beta_k \upsilon[n-k] \right]^2. \tag{12.267}$$

Following the development method of the previous section, *we differentiate with respect to β_k and set the derivative to a zero value*. In this case, the result of the process, as can be proven after simple operations, takes the form

$$\sum_{k=0}^{M} \beta_k \left[\sum_{n=0}^{\infty} \upsilon[n-\ell]\upsilon[n-k] \right] = \sum_{n=0}^{\infty} h_d[n]\upsilon[n-k], \quad k = 0, 1, 2, \ldots, M \tag{12.268}$$

Therefore, defining the functions

$$\xi(k,\ell) = \sum_{n=0}^{\infty} \upsilon[n-\ell]\upsilon[n-k] = \xi(k-\ell) \quad \text{and} \quad \xi(k) = \sum_{n=0}^{\infty} h_d[n]\upsilon[n-k] \tag{12.269}$$

the above relation is written

$$\sum_{k=0}^{M} \beta_k \xi(k,\ell) = \xi(k), \quad k = 0, 1, 2, \ldots, M \tag{12.270}$$

or in matrix form

$$\underbrace{\begin{bmatrix} \xi(0,0) & \xi(0,1) & \xi(0,2) & \ldots & \xi(0,M) \\ \xi(1,0) & \xi(1,1) & \xi(1,2) & \ldots & \xi(1,M) \\ \xi(2,0) & \xi(2,1) & \xi(2,2) & \ldots & \xi(2,M) \\ \vdots & \vdots & \vdots & \ddots & \vdots \\ \xi(M,0) & \xi(M,1) & \xi(M,2) & \ldots & \xi(M,M) \end{bmatrix}}_{\boldsymbol{\xi}} \underbrace{\begin{bmatrix} \beta_0 \\ \beta_1 \\ \beta_2 \\ \vdots \\ \beta_M \end{bmatrix}}_{\boldsymbol{\beta}} = \underbrace{\begin{bmatrix} \xi(0) \\ \xi(1) \\ \xi(2) \\ \vdots \\ \xi(M) \end{bmatrix}}_{\boldsymbol{\gamma}} \tag{12.271}$$

the solution of which for the case of *non-diagonal coefficient matrix* is *unique* and is given by the relation $\boldsymbol{\beta} = \boldsymbol{\xi}^{-1} \cdot \boldsymbol{\gamma}$. Finally, working as in the Prony method, the minimum square error in the Shank method is found to be equal to

$$\mathcal{E}_{\min} = \sum_{n=0}^{\infty} x^2[n] - \sum_{k=0}^{M} \beta_k \left\{ \sum_{n=0}^{\infty} h_d[n]h_d[n-k] \right\} = \xi(0) - \sum_{k=0}^{M} \beta_k \xi(k). \tag{12.272}$$

Example 12.13.10 Let us consider once again the problem of approximating the ideal low-pass filter of Example 12.13.2 (see page 1288), the first 11 samples of which are

$$h_d[n] = \{0.063661, 0, -0.106103, 0, 0.318309, 0.5, 0.318309, 0, -0.106103, 0, 0.063661\}$$

from a digital filter with $M = 5$ zero values and $N = 5$ poles. For the application of the Shanks method, we must initially determine the values of the coefficients α_k by applying the least-squares method, something we did in the previous example, leading to the result

$$\begin{aligned} \boldsymbol{a} &= \begin{bmatrix} \alpha_0 \; \alpha_1 \; \alpha_2 \; \alpha_3 \; \alpha_4 \; \alpha_5 \end{bmatrix}^T = \\ &= \begin{bmatrix} 1.000000 \; -2.517033 \; 3.657151 \; -3.459241 \; 2.110133 \; -0.694543 \end{bmatrix}^T. \end{aligned}$$

Consequently, the transfer function $H_1(z)$ has the form

$$\mathcal{H}_1(z) = \frac{1}{1 - 2.517033z^{-1} + 3.657151z^{-2} - 3.459241z^{-3} + 2.110133z^{-4} - 0.694543z^{-5}}.$$

Developing the function $\mathcal{H}_1(z)$ in a series of partial fractions, this is formulated as

$$\begin{aligned} \mathcal{H}_1(z) = &\frac{0.918789}{1 - 0.888896z^{-1}} + \frac{-0.357887 - 0.201647j}{1 - (0.170284 + 0.955239j)z^{-1}} + \frac{-0.357887 + 0.201647j}{1 - (0.170284 - 0.955239j)} \\ &+ \frac{0.398492 - 0.716669j}{1 - (0.643784 + 0.644565j)z^{-1}} + \frac{0.398492 + 0.716669j}{1 - (0.643784 - 0.644565j)z^{-1}} \end{aligned}$$

Expressing the complex quantities of the above equation in their polar form and taking the inverse $\mathcal{Z}$ transform of the above transfer function, the impulse response of the first system is calculated as (for a similar calculation see Exercise 12.13.1 on page 1292)

$$\begin{aligned} h[n] &= 0.918789\,(0.888896)^n + 0.821578\,(0.970298)^n \cos\,(1.394385n - 2.628491) \\ &+ 0.164014\,(0.911000)^n \cos\,(0.786004n - 1.063332) \end{aligned}$$

and its contents for length $N = 11$ (to solve the exercise we use the same number of samples as the desired impulse response, which is $N = 11$) define the discrete signal

$$\begin{aligned} h[n] = \{&0.999994, 2.517026, 2.678297, 0.995439, -0.692505, -0.7353806, 0.221702, 0.611565, \\ &0.337321, 0.450156, 1.036401\}. \end{aligned}$$

If we transmit the impulse function $\delta[n]$ as input to the first of the two systems, its output will be the signal $h[n]$, which is then applied as input to the second system. So, setting $v[n] = h[n]$ and calculating the values of the method parameters, the equation for calculating the values of the coefficients β_k

$$\begin{bmatrix} \xi(0,0) & \xi(0,1) & \xi(0,2) & \xi(0,3) & \xi(0,4) & \xi(0,5) \\ \xi(1,0) & \xi(1,1) & \xi(1,2) & \xi(1,3) & \xi(1,4) & \xi(1,5) \\ \xi(2,0) & \xi(2,1) & \xi(2,2) & \xi(2,3) & \xi(2,4) & \xi(2,5) \\ \xi(3,0) & \xi(3,1) & \xi(3,2) & \xi(3,3) & \xi(3,4) & \xi(3,5) \\ \xi(4,0) & \xi(4,1) & \xi(4,2) & \xi(4,3) & \xi(4,4) & \xi(4,5) \\ \xi(5,0) & \xi(5,1) & \xi(5,2) & \xi(5,3) & \xi(5,4) & \xi(5,5) \end{bmatrix} \begin{bmatrix} \beta_0 \\ \beta_1 \\ \beta_2 \\ \beta_3 \\ \beta_4 \\ \beta_5 \end{bmatrix} \begin{bmatrix} \xi(0) \\ \xi(1) \\ \xi(2) \\ \xi(3) \\ \xi(4) \\ \xi(5) \end{bmatrix}$$

takes the form

$$\begin{bmatrix} 18.333646 & 12.541575 & 2.693488 & -2.434441 & -1.675774 & 0.722504 \\ 12.541575 & 17.259519 & 12.075034 & 2.343888 & -3.068268 & -1.905546 \\ 2.693488 & 12.075034 & 17.056878 & 11.923187 & 2.068588 & -3.168068 \\ -2.434441 & 2.343888 & 11.923187 & 16.943093 & 11.716893 & 1.993803 \\ -1.675774 & -3.068268 & 2.068588 & 11.716893 & 16.569081 & 11.581308 \\ 0.722504 & -1.905546 & -3.168068 & 1.993803 & 11.581308 & 16.519929 \end{bmatrix} \begin{bmatrix} \beta_0 \\ \beta_1 \\ \beta_2 \\ \beta_3 \\ \beta_4 \\ \beta_5 \end{bmatrix} \begin{bmatrix} -0.707878 \\ -0.566769 \\ 1.021664 \\ 2.574157 \\ 2.516937 \\ 1.148755. \end{bmatrix}$$

The solution of the above system is

$$\boldsymbol{\beta} = \left[\beta_0\ \beta_1\ \beta_2\ \beta_3\ \beta_4\ \beta_5\right]^T =$$
$$= \left[0.063661\ -0.160239\ 0.126720\ 0.046836\ 0.0646205\ 0.021616\right]^T,$$

leading practically to the same transfer function as the one we calculated using the previous methods.

12.13.7 Design with Optimization Methods

The construction of recursive digital filters using *optimization methods* lifts the restriction about the *piecewise continuous desired amplitude response* that characterizes the previous methods of formulating the transfer function in closed form with the help of the filter's coefficients or the zero values and the poles of it. Optimization methods allow the construction of transfer functions *with arbitrary amplitude and phase responses*, but nevertheless they are characterized by *high computational cost*, being *iterative* in their nature. In full analogy with the homonymous methods presented in the section of the filters FIR, the *optimization methods* used for the construction of filters IIR are based *on the approximation of the desired response amplitude and/ the phase from the corresponding real function and on the optimization of an objective function that is defined in terms of the difference in the two above responses (i.e., the desired and the real).* This optimization process is based on the calculation of the coefficients of the filter of the real transfer function, the values of which *minimize the above objective function*, a process similar to the one we described in the section on the design of FIR filters.

For the mathematical description of the problem, let us consider for reasons the simplicity that the filter under construction is characterized by an *optimal order value* N and therefore will be described by a transfer function of the form

$$\mathcal{H}(z) = H_0 \prod_{i=1}^{L} \frac{\alpha_{0i} + \alpha_{1i} z + z^2}{\beta_{0i} + \beta_{1i} z + z^2}, \tag{12.273}$$

where $L = N/2$, α_{ij} and β_{ij} ($i = 0, 1$ and $j = 1, 2, \ldots, L$) are the *real coefficients* of the elementary second-order filters from the series connection of which the filter of order N under construction and H_0 is the *multiplicative normalization constant* calculated in the way we saw in the previous sections and examples. It is not difficult to understand that the full specification of the above function requires the specification of the coefficients α_{ij} and β_{ij} ($i = 0, 1$ and $j = 1, 2, \ldots, L$) and the normalization constant H_0, which define a vector $\boldsymbol{\xi}$ in the space of $4L + 1$ dimensions of the form

$$\boldsymbol{\xi} = \left[\alpha_{01} \ldots \alpha_{0L}\ \alpha_{11} \ldots \alpha_{1L}\ \beta_{01} \ldots \beta_{0L}\ \beta_{11} \ldots \beta_{1L}\ H_0.\right] \tag{12.274}$$

Following the description of the corresponding optimization methods for FIR filters, let us use the notation $\mathcal{H}_{\mathrm{dr}}(\omega) = |\mathcal{H}_d(e^{j\omega})|$ to describe the *desired amplitude response* of the filter under construction, and the notation

$$\mathcal{H}_r(\boldsymbol{\xi}, \omega) = |\mathcal{H}(e^{j\omega})| \tag{12.275}$$

to describe the *actual amplitude response*, with the use of the unknown vector $\boldsymbol{\xi}$ as one of the arguments of the function implies that *this vector constitutes the desired solution to the optimization problem, which minimizes the difference between the functions* $\mathcal{H}_r(\boldsymbol{\xi}, \omega)$ *and* $\mathcal{H}_{\mathrm{dr}}(\omega)$. Defining the *approximation error* from the expression

$$e(\boldsymbol{\xi}, \omega) = H_r(\boldsymbol{\xi}, \omega) - \mathcal{H}_{\mathrm{dr}}(\omega) \tag{12.276}$$

and assuming for simplicity that the desired amplitude response $\mathcal{H}_{\mathrm{dr}}(\omega)$ is *piecewise continuous* in the frequency region where it is defined, the *sampling of the approximation error* $e(\boldsymbol{\xi}, \omega)$ at the positions of K discrete frequency values $\omega_1, \omega_2, \ldots, \omega_K$ leads to the creation of the discrete-time signal

$$\boldsymbol{\mathcal{E}}(\boldsymbol{\xi}) = \left[e_1(\xi)\ e_2(\xi)\ e_3(\xi)\ \ldots\ e_K(\xi)\right] = \left[e\ (\xi, \omega_1)\ e\ (\xi, \omega_2)\ e\ (\xi, \omega_3)\ \ldots\ e\ (\xi, \omega_K),\right] \tag{12.277}$$

where we used the notation $e_i(\boldsymbol{\xi}) = e\ (\boldsymbol{\xi}, \omega_i)\ (i = 1, 2, \ldots, K)$.

But what will be the form of the *objective function to be minimized* for this particular problem? Assuming that this problem has some solution and taking into account the fact that this function should be characterized by properties of *metric function* (for example, it should be a *positive graded and differentiable quantity*), we can use for it the most general measure function L_p, which for the case of the discrete-time signals is defined as

$$\Psi(\boldsymbol{\xi}) = L_p = \|\boldsymbol{\mathcal{E}}(\boldsymbol{\xi})\| = \sqrt[p]{\sum_{i=1}^{K} |e_i(\boldsymbol{\xi})|^p} = \sqrt[p]{\sum_{i=1}^{K} |e(\boldsymbol{\xi}, \omega_i)|^p}, \tag{12.278}$$

(where p is an integer quantity) and which for $p = 1$ and $p = 2$ degenerates into the measures

$$L_1 = \sum_{i=1}^{K} |e_i(\boldsymbol{\xi})| = \sum_{i=1}^{K} |e(\boldsymbol{\xi}, \omega_i)| \quad \text{and} \quad L_2 = \sqrt{\sum_{i=1}^{K} |e_i(\boldsymbol{\xi})|^2} = \sqrt{\sum_{i=1}^{K} |e(\boldsymbol{\xi}, \omega_i)|^2}.$$

It is not difficult to understand that the least-squares methods presented in the previous section *constitute a special case of this general optimization problem* for which it is $\Psi(\boldsymbol{\xi}) = \varphi(L_2^2)$, whereas in the general case this problem is formulated as

$$\underset{\boldsymbol{\xi}}{\text{minimize}}\ \Psi(\boldsymbol{\xi}) \tag{12.279}$$

and consists of *calculating the vector of unknowns* $\boldsymbol{\xi}$, *so as to minimize the objective function* $\Psi(\boldsymbol{\xi})$. Finally, if $p = \infty$, the corresponding metric function is defined as

$$L_\infty = \max_{1 \le i \le K} |e_i(\boldsymbol{\xi})| = \max_{1 \le i \le K} |e(\boldsymbol{\xi}, \omega_i)| \tag{12.280}$$

and the optimization problem is known as the *minimax problem*, since it consists *in minimizing the maximum element* of the function $\boldsymbol{\mathcal{E}}(\xi)$. The analytical presentation, as well as the demonstration of an example of the application of the IIR filter design methods that are based on the above optimization techniques, is the subject of the following pages.

The Method of Newton and Its Variations

The *method of Newton* is a well-known *general method of finding minimum values of convex functions* and is based on the use of the *Taylor expansion* of a general function N variables $y = f(\boldsymbol{x})$ (where $\boldsymbol{x} = [\ x_1\ x_2\ \ldots\ x_N\ ]^T$) with respect to the point $\tilde{\boldsymbol{x}} = \boldsymbol{x} + \boldsymbol{\delta x}$ with the vector $\boldsymbol{\delta x} = [\ \delta x_1\ \delta x_2\ \ldots\ \delta x_N\ ]^T$ to describe a change in the vector $\boldsymbol{x}$. If this function has *continuous second-order derivatives*, then its Taylor expansion will be given by the well-known relation

$$f(\boldsymbol{x}+\boldsymbol{\delta x}) = f(\boldsymbol{x}) + \sum_{i=1}^{N} \frac{\partial f(\boldsymbol{x})}{\partial x_i}\delta x_i + \frac{1}{2}\sum_{i=1}^{N}\sum_{j=1}^{N} \frac{\partial^2 f(\boldsymbol{x})}{\partial x_i \partial x_j}\delta x_i \delta x_j + \mathcal{O}(\|\boldsymbol{\delta x}\|_2^2). \tag{12.281}$$

Under the assumption that the higher-order terms $\mathcal{O}(\|\boldsymbol{\delta x}\|_2^2)$ are *negligible*, so that they can be ignored, the determination of a *stationary point* of the function $f(\boldsymbol{x})$ at the position of some point $\tilde{\boldsymbol{x}}$ it can be found by the known *nullification* of the derivatives of the function $f(\boldsymbol{x}+\boldsymbol{\delta x})$ with respect to the parameters δx_k $(k = 1, 2, \ldots, N)$. In this case, we end up with the system of equations

$$\frac{\partial f(\boldsymbol{x}+\boldsymbol{\delta x})}{\partial(\delta x_k)} = \frac{\partial f(\boldsymbol{x})}{\partial x_k} + \sum_{i=1}^{N} \frac{\partial}{\partial x_k}\left[\frac{\partial f(\boldsymbol{x})}{\partial x_i}\right]\delta x_i = \frac{\partial f(\boldsymbol{x})}{\partial x_k} + \sum_{i=1}^{N} \frac{\partial^2 f(\boldsymbol{x})}{\partial x_i \partial x_k}\delta x_i = 0$$

or equivalently

$$\frac{\partial f(\boldsymbol{x})}{\partial x_k} = -\sum_{i=1}^{N} \frac{\partial^2 f(\boldsymbol{x})}{\partial x_i \partial x_k}\delta x_i$$

(for the values $k = 1, 2, \ldots, N$), which in expanded form is expressed as

$$\frac{\partial f(\boldsymbol{x})}{\partial x_1} = -\sum_{j=1}^{N} \frac{\partial^2 f(\boldsymbol{x})}{\partial x_1 \partial x_j}\delta x_1 = -\left(\frac{\partial^2 f(\boldsymbol{x})}{\partial x_1^2}\delta x_1 + \frac{\partial^2 f(\boldsymbol{x})}{\partial x_1 \partial x_2}\delta x_1 + \cdots + \frac{\partial^2 f(\boldsymbol{x})}{\partial x_1 \partial x_N}\delta x_1\right)$$

$$\frac{\partial f(\boldsymbol{x})}{\partial x_2} = -\sum_{j=1}^{N} \frac{\partial^2 f(\boldsymbol{x})}{\partial x_2 \partial x_j}\delta x_2 = -\left(\frac{\partial^2 f(\boldsymbol{x})}{\partial x_2 \partial x_1}\delta x_2 + \frac{\partial^2 f(\boldsymbol{x})}{\partial x_2^2}\delta x_2 + \cdots + \frac{\partial^2 f(\boldsymbol{x})}{\partial x_2 \partial x_N}\delta x_2\right)$$

$$\cdots\cdots\cdots\cdots\cdots\cdots\cdots\cdots\cdots\cdots\cdots\cdots\cdots\cdots\cdots\cdots\cdots\cdots$$

$$\frac{\partial f(\boldsymbol{x})}{\partial x_N} = -\sum_{j=1}^{N} \frac{\partial^2 f(\boldsymbol{x})}{\partial x_N \partial x_j}\delta x_N = -\left(\frac{\partial^2 f(\boldsymbol{x})}{\partial x_N \partial x_1}\delta x_N + \frac{\partial^2 f(\boldsymbol{x})}{\partial x_N \partial x_2}\delta x_N + \cdots + \frac{\partial^2 f(\boldsymbol{x})}{\partial x_N^2}\delta x_N\right).$$

The above system of equations formulated *in matrix form* is written as

$$\begin{bmatrix} \frac{\partial f(\boldsymbol{x})}{\partial x_1} \\ \frac{\partial f(\boldsymbol{x})}{\partial x_2} \\ \vdots \\ \frac{\partial f(\boldsymbol{x})}{\partial x_N} \end{bmatrix} = \begin{bmatrix} \frac{\partial^2 f(\boldsymbol{x})}{\partial x_1^2} & \frac{\partial^2 f(\boldsymbol{x})}{\partial x_1 \partial x_2} & \cdots & \frac{\partial^2 f(\boldsymbol{x})}{\partial x_1 \partial x_N} \\ \frac{\partial^2 f(\boldsymbol{x})}{\partial x_2 \partial x_1} & \frac{\partial^2 f(\boldsymbol{x})}{\partial x_2^2} & \cdots & \frac{\partial^2 f(\boldsymbol{x})}{\partial x_2 \partial x_N} \\ \vdots & \vdots & \ddots & \vdots \\ \frac{\partial^2 f(\boldsymbol{x})}{\partial x_N \partial x_1} & \frac{\partial^2 f(\boldsymbol{x})}{\partial x_N \partial x_2} & \cdots & \frac{\partial^2 f(\boldsymbol{x})}{\partial x_N^2} \end{bmatrix} \begin{bmatrix} \delta x_1 \\ \delta x_2 \\ \vdots \\ \delta x_N \end{bmatrix} \tag{12.282}$$

or equivalently $\nabla f(\boldsymbol{x}) = -\boldsymbol{H}\cdot\boldsymbol{\delta x}$, where

$$\nabla f(\boldsymbol{x}) = \begin{bmatrix} \frac{\partial f(\boldsymbol{x})}{\partial x_1} & \frac{\partial f(\boldsymbol{x})}{\partial x_2} & \cdots & \frac{\partial f(\boldsymbol{x})}{\partial x_N} \end{bmatrix}^T \tag{12.283}$$

the *gradient of the function* $f(\boldsymbol{x})$, $\boldsymbol{\delta x} = [\ \delta x_1\ \delta x_2\ \ldots\ \delta x_N\ ]$ the *change of the vector* $\boldsymbol{x}$, and

$$\boldsymbol{H} = \begin{bmatrix} \frac{\partial^2 f(\boldsymbol{x})}{\partial x_1^2} & \frac{\partial^2 f(\boldsymbol{x})}{\partial x_1 \partial x_2} & \cdots & \frac{\partial^2 f(\boldsymbol{x})}{\partial x_1 \partial x_N} \\ \frac{\partial^2 f(\boldsymbol{x})}{\partial x_2 \partial x_1} & \frac{\partial^2 f(\boldsymbol{x})}{\partial x_2^2} & \cdots & \frac{\partial^2 f(\boldsymbol{x})}{\partial x_2 \partial x_N} \\ \vdots & \vdots & \ddots & \vdots \\ \frac{\partial^2 f(\boldsymbol{x})}{\partial x_N \partial x_1} & \frac{\partial^2 f(\boldsymbol{x})}{\partial x_N \partial x_2} & \cdots & \frac{\partial^2 f(\boldsymbol{x})}{\partial x_N^2} \end{bmatrix} \tag{12.284}$$

the *Hessian matrix*, which is defined as the *matrix of the second-order derivatives* of the function $f(\boldsymbol{x})$, which, according to what we mentioned above, *exist* and are considered *continuous functions*. If the aforementioned matrix is *invertible* and therefore its inverse $\boldsymbol{H}^{-1}$ exists, then, according to the theory of algebraic systems, the above system has *one and only one solution*, which is calculated as

$$\boldsymbol{\delta x} = -\boldsymbol{H}^{-1} \cdot \nabla f(\boldsymbol{x}). \tag{12.285}$$

Although the Newton method is general in nature and can be applied to *any convex function* $f(\boldsymbol{x})$, the way it is applied depends on the *analytical form of the function*. In the simplest of cases where the function $f(\boldsymbol{x})$ is a *quadratic function* of its arguments and therefore all second-order derivatives will have *constant values*, the Hessian matrix $\boldsymbol{H}$ will be a *constant symmetric matrix*, whereas all third- and higher-order derivatives will be equal to zero. In this case, the two prerequisites for applying the algorithm (i.e., ignoring higher-order terms and the existence of the inverse Hessian matrix) are satisfied, the system has the unique solution $\boldsymbol{\delta x} = -\boldsymbol{H}^{-1} \cdot \nabla f(\boldsymbol{x})$, and the stationary point is calculated as $\tilde{\boldsymbol{x}} = \boldsymbol{x} + \boldsymbol{\delta x}$. On the other hand, if the function $f(\boldsymbol{x})$ *is not quadratic*, we can consider that in the neighborhood of the stationary point with radius ϵ defined as $\|\boldsymbol{x} - \tilde{\boldsymbol{x}}\|_2 < \epsilon$ the higher-order terms are *negligible* and the elements of the Hessian matrix are *approximately constant*, so that we can work as before. Finally, in the most general case of *any convex function* $f(\boldsymbol{x})$ of an *arbitrarily selected point* $\boldsymbol{x}$ it may not satisfy one or both of the two conditions, so the system *either will have no solution at all or will have infinite solutions*. In this case, we resort to the use of an iterative process during which the value of the function *is progressively reduced* by applying *a series of corrections* to the value of the vector $\boldsymbol{x}$, until finally we end up at a point that is in the neighborhood of the solution. The above iterative process stops when a point $\boldsymbol{x}$ is found for which the higher-order terms of the Taylor expansion of the function $f(\boldsymbol{x})$ have *negligible* value, so then the solution is calculated from the relation $\boldsymbol{\delta x} = -\boldsymbol{H}^{-1} \cdot \nabla f(\boldsymbol{x})$. It should be noted, however, that this method *is not guaranteed to lead to the optimal solution*, since it relies on the assumption that the function $f(\boldsymbol{x})$ has *one and only one* minimum value, which generally does not hold in practice. For this reason, the problem is usually considered to have been solved *when the solution found satisfies some specifications regarding the desired values of the approximation error*.

The main disadvantage of the classical Newton method is that in each iteration it requires *the calculation of the first- and second-order derivatives of the function* $f(\boldsymbol{x})$ for the determination of the gradient vector $\nabla f(\boldsymbol{x})$ and the Hessian matrix $\boldsymbol{H}$, as well as *the calculation of the inverse matrix* $\boldsymbol{H}^{-1}$, provided that it exists, which is ensured when the matrix $\boldsymbol{H}$ is *positively defined* and therefore *nonsingular*. The check of the validity of the property of the nonsingular matrix constitutes an additional complexity of this method. For this reason, in practice a variation of this classic Newton method known as a *quasi Newtonalgorithm* is used and is presented in the continuation. As for the details of the above iterative process, this in general is based on a fundamental property of matrix algebra, according to which *if the matrix* $\boldsymbol{H}$ *is a positively defined matrix, then the same applies to its inverse*, as well as the fact that the direction defined by the vector $-\boldsymbol{H}^{-1} \cdot \nabla f(\boldsymbol{x})$ and which is known as *Newton direction* constitutes *the direction of reduction of the value of the function* $f(\boldsymbol{x})$. If, therefore, for some value of the vector $\boldsymbol{x}$ the matrix $\boldsymbol{H}$ is *positively defined*, we can reduce the value of the function $f(\boldsymbol{x})$ by applying on it of the vector $\boldsymbol{x}$ a correction of the form $\boldsymbol{\delta} = \alpha \boldsymbol{d}$, where α is a *correction factor* with a positive value and $\boldsymbol{d} = -\boldsymbol{H}^{-1} \cdot \nabla f(\boldsymbol{x})$. Conversely, if the matrix $\boldsymbol{H}$ is not positive definite, then some *transformation* is applied to it in order to become a positive definite matrix. In order to achieve the *maximum possible reduction* of the function's value $f(\boldsymbol{x})$ in the direction $\boldsymbol{d}$, we must choose the value of the variable α, which minimizes the value of the function $f(\boldsymbol{x} + \alpha \boldsymbol{d})$, something that can be done using one of the available minimization algorithms. This iterative process terminates when the magnitude of the change of $\boldsymbol{x}$ (i.e., the L_2 norm of the quantity $\alpha_k \boldsymbol{d}_k$ for the iteration numbered k) becomes smaller than a *permissible tolerance error* ϵ, the value of which is dictated by the application considered in each case. The above algorithm, which constitutes the classic Newton optimization algorithm in a concise formulation, is as follows:

Step 1 We initialize the *starting vector* $\boldsymbol{x}_0$ and the *tolerance error value* ϵ and set $k = 0$.

Step 2 We calculate the *gradient vector* $\nabla f(\boldsymbol{x}_k) \equiv \nabla f(\boldsymbol{x})_{\boldsymbol{x}=\boldsymbol{x}_k}$ and the *Hessian matrix* $\boldsymbol{H}_k$, which becomes *positive definite*, if it is not already expressed in this form.

Step 3 We calculate the *inverse Hessian matrix* $\boldsymbol{H}^{-1}$ and the vector $\boldsymbol{d}_k = -\boldsymbol{H}_k^{-1} \cdot \nabla f(\boldsymbol{x}_k)$.

Step 4 We determine the value of α, which *minimizes* the function $f(\boldsymbol{x}_k + \alpha \boldsymbol{d}_k)$

Step 5 We set $\boldsymbol{x}_{k+1} = \boldsymbol{x}_k + \boldsymbol{\delta}_k$ where $\boldsymbol{\delta}_k = \alpha_k \boldsymbol{d}_k$ and calculate the value $f_{k+1} \equiv f(\boldsymbol{x}_{k+1})$.

Step 6 If $\|\alpha_k \boldsymbol{d}_k\| < \epsilon$, we calculate and return the values $\tilde{\boldsymbol{x}} = \boldsymbol{x}_{k+1}$ and $f(\tilde{\boldsymbol{x}}) = f(\boldsymbol{x}_{k+1})$, and terminate the algorithm, otherwise we return to Step 2.

The above implementation of the Newton algorithm can be improved in terms of its complexity by using information about the gradient of the function $f(\boldsymbol{x})$ at various points and thus avoiding the calculation of the *second-order derivatives* and consequently the *Hessian matrix* $\boldsymbol{H}$. In order to prove the truth of this claim, let us consider two points $\boldsymbol{x}_k$ and $\boldsymbol{x}_{k+1} = \boldsymbol{x}_k + \boldsymbol{\delta}_k$ in the N dimensional space in which the multivariable function $f(\boldsymbol{x})$ is defined and let us assume that the gradient of this function at these points is described by the vectors $\nabla f(\boldsymbol{x}_{k+1}) \equiv \nabla f(\boldsymbol{x})_{\boldsymbol{x}=\boldsymbol{x}_{k+1}}$ and $\nabla f(\boldsymbol{x}_k) \equiv \nabla f(\boldsymbol{x})_{\boldsymbol{x}=\boldsymbol{x}_k}$ respectively. In this case, the components of the gradient vector $\nabla f(\boldsymbol{x}_{k+1})$ will be related to the components of the gradient vector $\nabla f(\boldsymbol{x}_k)$ through a *Taylor expansion* of the form

$$\nabla f(\boldsymbol{x}_{k+1})_m = \nabla f(\boldsymbol{x}_k)_m + \sum_{i=1}^{N} \frac{\partial \nabla f(\boldsymbol{x}_k)_m}{\partial x_{ki}} \delta_{ki} + \frac{1}{2} \sum_{i=1}^{N} \sum_{j=1}^{N} \frac{\partial^2 \nabla f(\boldsymbol{x}_k)_m}{\partial x_{ki}\, \partial x_{kj}} \delta_{ki}\, \delta_{kj} + \mathcal{O}(\|\boldsymbol{\delta}\|_2^2) \tag{12.286}$$

(for the values $m = 1, 2, \ldots, N$). Considering for simplicity the case of a *quadratic* function $f(\boldsymbol{x})$ with respect to its variables, the second- and higher-order derivatives of the function $f(\boldsymbol{x})$ will have *constant* and *zero* values respectively and consequently the second and higher order derivatives of the m-th component of the gradient $\nabla f(\boldsymbol{x}_k)_m$ will be equal to zero. Taking into account that according to the definition of the gradient vector $\nabla f(\boldsymbol{x}_k)_m = \partial f_k / \partial x_{km}$, the above relation is written

$$\nabla f(\boldsymbol{x}_{k+1})_m = \nabla f(\boldsymbol{x}_k)_m + \sum_{i=1}^{N} \frac{\partial \nabla f(\boldsymbol{x}_k)_m}{\partial x_{ki}} \delta_{ki} = \nabla f(\boldsymbol{x}_k)_m + \sum_{i=1}^{N} \frac{\partial}{\partial x_{ki}} \left(\frac{\partial f_k}{\partial x_{km}} \right) \delta_{ki}$$

and finally

$$\nabla f(\boldsymbol{x}_{k+1})_m = \nabla f(\boldsymbol{x}_k)_m + \sum_{i=1}^{N} \frac{\partial^2 f_k}{\partial x_{ki} \partial x_{km}} \delta_{ki} \tag{12.287}$$

(for the values $m = 1, 2, \ldots, N$), or in matrix form

$$\nabla f(\boldsymbol{x}_{k+1}) = \nabla f(\boldsymbol{x}_k) + \boldsymbol{H} \cdot \boldsymbol{\delta}_k = \nabla f(\boldsymbol{x}_k) + \boldsymbol{\gamma}_k, \tag{12.288}$$

where $\boldsymbol{H}$ is the *Hessian matrix* of the function $f(\boldsymbol{x})$ that we defined earlier, whereas the vector $\boldsymbol{\gamma}_k$ is defined as $\boldsymbol{\gamma}_k = \boldsymbol{H} \cdot \boldsymbol{\delta}_k$. We observe, therefore, that *without resorting to the calculation of the second-order derivatives of the function* $f(\boldsymbol{x})$, we have the ability to calculate the *Hessian matrix* $\boldsymbol{H}$ with the help of the vectors $\boldsymbol{\gamma}_k = \nabla f(\boldsymbol{x}_{k+1}) - \nabla f(\boldsymbol{x}_k)$ and $\boldsymbol{\delta}_k = \boldsymbol{x}_{k+1} - \boldsymbol{x}_k$ using *only coordinates of points and gradient data*. Taking into account that the equation of matrices $\boldsymbol{\gamma}_k = \boldsymbol{H} \cdot \boldsymbol{\delta}_k$ $(k = 1, 2, \ldots, N)$ leads to a system of N equations, whereas the real symmetric matrix $\boldsymbol{H}$ has $N(N+1)/2$ independent elements, it is obvious that the above equations *are not sufficient* for the determination of all the elements of the matrix and for this reason the above system of equations must be supplemented with other $N+1$ equations, which are

$$\boldsymbol{\delta}_0 = \boldsymbol{x}_1 - \boldsymbol{x}_0, \qquad \boldsymbol{\delta}_1 = \boldsymbol{x}_2 - \boldsymbol{x}_1, \qquad \boldsymbol{\delta}_2 = \boldsymbol{x}_3 - \boldsymbol{x}_2, \qquad \ldots \qquad \boldsymbol{\delta}_{N-1} = \boldsymbol{x}_N - \boldsymbol{x}_{N-1},$$

where $(\boldsymbol{\delta}_0, \boldsymbol{\delta}_1, \boldsymbol{\delta}_2, \ldots, \boldsymbol{\delta}_{N-1})$ is a set of N linearly independent vectors. In this case, the relation $\boldsymbol{\gamma}_k = \boldsymbol{H} \cdot \boldsymbol{\delta}_k$ can be formulated in the expanded form

$$\left[\boldsymbol{\gamma}_0\ \boldsymbol{\gamma}_1\ \boldsymbol{\gamma}_2 \ldots \boldsymbol{\gamma}_{N-1}\right] = \boldsymbol{H}\left[\boldsymbol{\delta}_0\ \boldsymbol{\delta}_1\ \boldsymbol{\delta}_2 \ldots \boldsymbol{\delta}_{N-1}\right] \tag{12.289}$$

from which the Hessian matrix can be uniquely determined as

$$\boldsymbol{H} = \left[\boldsymbol{\gamma}_0\ \boldsymbol{\gamma}_1\ \boldsymbol{\gamma}_2 \ldots \boldsymbol{\gamma}_{N-1}\right]\left[\boldsymbol{\delta}_0\ \boldsymbol{\delta}_1\ \boldsymbol{\delta}_2 \ldots \boldsymbol{\delta}_{N-1}\right]^{-1}. \tag{12.290}$$

Note, however, that although this modification of the classic Newton algorithm is free from the need to calculate second-order derivatives, it still requires the calculation and inversion of the Hessian matrix $\boldsymbol{H}$, whereas we also need to determine the set of N linearly independent vectors $(\boldsymbol{\delta}_0, \boldsymbol{\delta}_1, \boldsymbol{\delta}_2, \ldots, \boldsymbol{\delta}_{N-1})$. Consequently, the benefits we gain from using this modified version of the original algorithm are not so significant.

The Quasi-Newton Method From the above description, it is not difficult to understand that the substantial improvement of the original Newton algorithm will come from a modification that will allow the direct calculation of the inverse Hessian matrix $\boldsymbol{H}^{-1}$ from the available data, since ultimately this matrix is used to determine the solution of the problem. Assuming as before that the data we have are the position vectors $\boldsymbol{x}_k$ and slope $\boldsymbol{g}_k$, we construct the groups of vectors

$$\boldsymbol{\delta}_k = -\boldsymbol{S}_k \cdot \nabla f(\boldsymbol{x}_k), \qquad \boldsymbol{x}_{k+1} = \boldsymbol{x}_k + \boldsymbol{\delta}_k, \qquad \boldsymbol{\gamma}_k = \nabla f(\boldsymbol{x}_{k+1}) - \nabla f(\boldsymbol{x}_k),$$

where $\boldsymbol{S}_k$ is a matrix of suitable dimensions, which during the iterative application of the algorithm undergoes a series of corrections of the form $\boldsymbol{S}_{k+1} = \boldsymbol{S}_k + \boldsymbol{C}_k$ $(k = 0, 1 \ldots, N-1)$. If the correction matrix $\boldsymbol{C}_k$ is such that the matrix $\boldsymbol{S}_{k+1}$ satisfies the condition $\boldsymbol{S}_{k+1} \cdot \boldsymbol{\gamma}_i = \boldsymbol{\delta}_i$ $(0 \le i \le k)$, whereas the vectors $(\boldsymbol{\delta}_0, \boldsymbol{\delta}_1, \boldsymbol{\delta}_2, \ldots, \boldsymbol{\delta}_{N-1})$ and $(\boldsymbol{\gamma}_0, \boldsymbol{\gamma}_1, \boldsymbol{\gamma}_2, \ldots, \boldsymbol{\gamma}_{N-1})$ constitute sets of *linearly independent vectors*, then for the value $k = N-1$, $\boldsymbol{S}_N \cdot \boldsymbol{\gamma}_i = \boldsymbol{\delta}_i$ $(0 \le i \le k)$. The above relationship for the various values of i leads to the set of N equations

$$\boldsymbol{S}_N \cdot \boldsymbol{\gamma}_1 = \boldsymbol{\delta}_1 \qquad \boldsymbol{S}_N \cdot \boldsymbol{\gamma}_2 = \boldsymbol{\delta}_2 \qquad \ldots \qquad \boldsymbol{S}_N \cdot \boldsymbol{\gamma}_{N-1} = \boldsymbol{\delta}_{N-1},$$

which, expressed in matrix form, is formulated as

$$\boldsymbol{S}_N\left[\boldsymbol{\gamma}_0\ \boldsymbol{\gamma}_1\ \boldsymbol{\gamma}_2 \ldots \boldsymbol{\gamma}_{N-1}\right] = \left[\boldsymbol{\delta}_0\ \boldsymbol{\delta}_1\ \boldsymbol{\delta}_2 \ldots \boldsymbol{\delta}_{N-1}\right] \tag{12.291}$$

from which it immediately follows that

$$\boldsymbol{S}_N = \left[\boldsymbol{\delta}_0\ \boldsymbol{\delta}_1\ \boldsymbol{\delta}_2 \ldots \boldsymbol{\delta}_{N-1}\right]\left[\boldsymbol{\gamma}_0\ \boldsymbol{\gamma}_1\ \boldsymbol{\gamma}_2 \ldots \boldsymbol{\gamma}_{N-1}\right]^{-1} = \boldsymbol{H}^{-1}, \tag{12.292}$$

where for the last expression we used the equation for calculating the Hessian matrix that we constructed previously. Combining at this point the relationship $\boldsymbol{\delta}_k = -\boldsymbol{S}_k \cdot \nabla f(\boldsymbol{x}_k)$ for the value $k = N$ with the relation $\boldsymbol{S}_N = \boldsymbol{H}^{-1}$, which we constructed previously, it immediately follows that $\boldsymbol{\delta}_N = -\boldsymbol{H}^{-1}\nabla f(\boldsymbol{x}_N)$ from which the solution of the problem is finally calculated as

$$\tilde{\boldsymbol{x}} = \boldsymbol{x}_{N+1} = \boldsymbol{x}_N - \boldsymbol{H}^{-1}\nabla f(\boldsymbol{x}_N). \tag{12.293}$$

The main feature of this variation of the Newton algorithm, known as the *quasi-Newton algorithm*, is that *it not only does not require the calculation of the second order derivatives of the function* $f(\boldsymbol{x})$—something we had already achieved in the previous modification—*but also frees us from determining both the linear independent vectors* $(\boldsymbol{\delta}_0, \boldsymbol{\delta}_1, \boldsymbol{\delta}_2, \ldots, \boldsymbol{\delta}_{N-1})$ *and the inverse Hessian matrix* $\boldsymbol{H}^{-1}$, since this is calculated by applying successive corrections to the matrix $\boldsymbol{S}$. In order to apply this iterative process of applying successive corrections, we start from the value $\boldsymbol{S}_0 = \boldsymbol{I}_N$, where $\boldsymbol{I}_N$ is the *unit matrix of order N* and at the appropriate step of the process we update the contents of the matrix $\boldsymbol{S}$, choosing one of the many types of corrections that have been developed in recent years. For reasons of completeness, we present two such types, which have proven quite effective and useful in practice and are known as *the correction types of Davidon-Fletcher-Powell and Broyden-Fletcher-Goldfarb-Shanno*—these types are given correspondingly from the equations

$$S_{k+1} = S_k + \frac{\delta_k \delta_k^T}{\gamma_k^T \delta_k} - \frac{S_k \gamma_k \gamma_k^T S_k}{\gamma_k^T S_k \gamma_k} \quad \text{and} \quad S_{k+1} = S_k + \left(1 + \frac{\gamma_k^T S_k \gamma_k}{\gamma_k^T \delta_k}\right) \frac{\delta_k \delta_k^T}{\gamma_k^T \delta_k} - \frac{\delta_k \gamma_k^T S_k + S_k \gamma_k \delta_k^T}{\gamma_k \delta_k}.$$

Let it be noted, of course, that in each iteration of the algorithm it should be checked, *if the property of the positive definite for the matrix* **S** *holds*, and the transfer to this form should take place in case it is found that the matrix is not expressed in it.

The main feature of the classic Newton algorithm and all its variations is *the determination of the value of the parameter α that minimizes the value of the function $f(\boldsymbol{x}_k + \alpha \boldsymbol{d}_k)$* through the application of a *search process*, which, although characterized by high-accuracy computations, leads to significant computational cost. In recent years, many alternative forms of such a search have been developed, which is known as *line search or straight-line search* and consists of *the continuous change of the parameter α along the direction of the gradient vector until finally an optimal value is found according to the tolerance value that has been set*. Information about such search processes can be found in manuals dealing with issues of *optimization* and *nonlinear programming*.[86] Let it be noted, however, that because the algorithms quasi-Newton *are not affected too much* by the possible lack of accuracy of this search process (provided of course that *they continue to converge*), in practice, algorithms of limited accuracy search are used, which nevertheless lead to the desired correct result.

In order to briefly describe the way of applying this process, let us assume for simplicity that the function $f(\boldsymbol{x}_{k+1})$ is characterized by *a single minimum*, which corresponds to some positive parameter $\alpha = \tilde{\alpha}$. In this case, the *Taylor expansion* of the function $f(\boldsymbol{x}_{k+1})$ can be approximated by the equation

$$f(\boldsymbol{x}_{k+1}) = f(\boldsymbol{x}_k) + \alpha \nabla f(\boldsymbol{x}_k)^T \boldsymbol{d}_k = f(\boldsymbol{x}_k) + \alpha \left. \frac{df(\boldsymbol{x}_k + \alpha \boldsymbol{d}_k)}{d\alpha} \right|_{\alpha=0}, \tag{12.294}$$

(where $\boldsymbol{d}_k$ is the *direction of decrease* of the function $f(\boldsymbol{x})$ and $\alpha \nabla f(\boldsymbol{x}_k)^T \boldsymbol{d}_k$ is the *slope* of the function $f(\boldsymbol{x}_k + \alpha \boldsymbol{d}_k)$, whereas of course it is $\boldsymbol{x}_{k+1} = \boldsymbol{x}_k + \alpha \boldsymbol{d}_k$), which, being a *linear approximation of the Taylor expansion*, constitutes a straight-line equation. To describe the search method, we need two more such straight-line equations, one of which is $f(\boldsymbol{x}_{k+1}) = f(\boldsymbol{x}_k) + \rho \alpha \nabla f(\boldsymbol{x}_k)^T \boldsymbol{d}_k$, which for the range of values $0 \le \rho \le 0.5$ represents a straight line with a slope in the interval $[0, 0.5 \nabla f(\boldsymbol{x}_k)^T \boldsymbol{d}_k]$, whereas the other is $\nabla f(\boldsymbol{x}_{k+1})^T \boldsymbol{d}_k = \sigma \nabla f(\boldsymbol{x}_k)^T \boldsymbol{d}_k$ $(0 < \sigma < 1$ and $\sigma \ge \rho)$, which correlates the derivative of the function $f(\boldsymbol{x}_{k+1})$ at the position $\alpha = \alpha_1$ with the derivative of the aforementioned function at the position $\alpha = 0$. The three straight lines represented by the three above equations are the lines A, B, and C of Fig. 12.88 respectively. It is obvious that because $0 < \sigma < 1$, it will also be $0 < \alpha_1 < \tilde{\alpha}$, where $\tilde{\alpha}$ is the value of the parameter α for which the function $f(\boldsymbol{x})$ presents its (unique) minimum value.

From Fig. 12.88 we observe that the intersection points of the lines C and B with the curve of the function $f(\boldsymbol{x})$ have abscissas α_1 and α_2, which are *on either side of the global minimum* $\tilde{\alpha}$ and therefore the equations $f(\boldsymbol{x}_{k+1}) = f(\boldsymbol{x}_k) + \rho \alpha \nabla f(\boldsymbol{x}_k)^T \boldsymbol{d}_k$ and $\nabla f(\boldsymbol{x}_{k+1})^T \boldsymbol{d}_k = \sigma \nabla f(\boldsymbol{x}_k)^T \boldsymbol{d}_k$ can be used *as termination conditions of the iterative search process*, which will terminate when the value of α calculated in some iteration simultaneously satisfies the relations $\alpha_0 \le \alpha_2$ and $\alpha_0 \ge \alpha_1$—that is, it belongs to the closed interval $[\alpha_1, \alpha_2]$—where α_0 is the current estimate of the minimum position $\tilde{\alpha}$. It is not difficult to see that the first inequality is satisfied if $f(\boldsymbol{x}_{k+1}) \le f(\boldsymbol{x}_k) + \rho \alpha \nabla f(\boldsymbol{x}_k)^T \boldsymbol{d}_k$, whereas the second inequality is satisfied if $\nabla f(\boldsymbol{x}_{k+1})^T \boldsymbol{d}_k \ge \sigma \nabla f(\boldsymbol{x}_k)^T \boldsymbol{d}_k$, and therefore if both these relations are simultaneously satisfied for the current value of α_0, the new (reduced) value of the function $f(\boldsymbol{x}_{k+1})$ is considered *acceptable* and the process terminates. Conversely, if either of the two above conditions is not satisfied and therefore α_0 does not belong to the closed interval $[\alpha_1, \alpha_2]$, as required by the algorithm, then the iterative process continues with the calculation of a new value for α_0,

[86] See for example (1) Luenberger D.G., Ye Y., *Linear and Nonlinear Programming*, ISBN 978-3-319-37439-0, Springer, Switzerland, October 2016 and (2) Himmelblau D.M., *Applied Non Linear Programming*, New York, McGraw Hill, 1972.

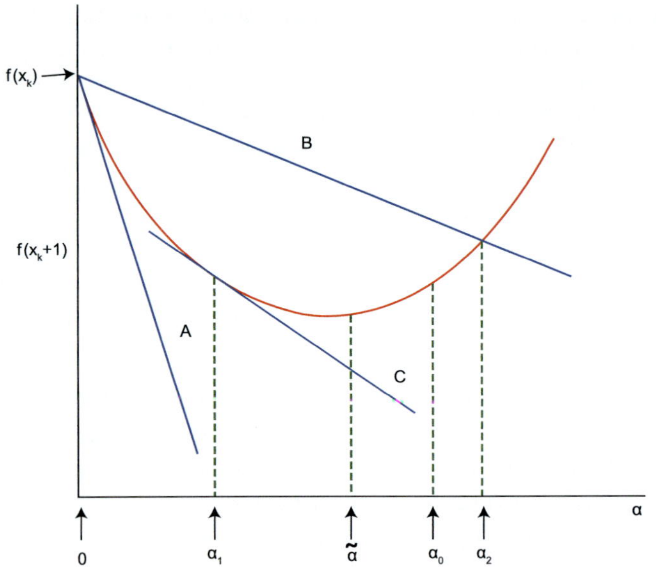

Fig. 12.88 The lines A, B and C used for the description of the *linear search* algorithm. In the example of the figure, the current value of the tangent of the minimum point α_0 is between the points α_1 and α_2 and for this reason the corresponding reduction of the function $f(\boldsymbol{x})$ is considered *acceptable* (Antoniou, Figure 16.2a)

which takes place through the application of the numerical methods of *interpolation* and *extrapolation*. More specifically, if the condition $f(\boldsymbol{x}_{k+1}) \leq f(\boldsymbol{x}_k) + \rho\alpha\nabla f(\boldsymbol{x}_k)^T \boldsymbol{d}_k$ is violated, then $\alpha_0 > \alpha_2$ (and therefore $0 < \tilde{\alpha} < \alpha_0$) and the new value of α_0 is calculated from the *interpolation equation*.

$$\tilde{\alpha}_0 = \alpha_L + \frac{(\alpha_0 - \alpha_L)^2 f'L}{2[f_L - f_0 + (\alpha_0 - \alpha_L) f'_L]}, \tag{12.295}$$

whereas conversely if the condition $\nabla f(\boldsymbol{x}_{k+1})^T \boldsymbol{d}_k \geq \sigma \nabla f(\boldsymbol{x}_k)^T \boldsymbol{d}_k$ is violated, then it is $0 < \alpha_0 < \alpha_1$ and the new value of α_0 is calculated from the *interpolation equation*

$$\tilde{\alpha}_0 = \alpha_0 + \frac{(\alpha_0 - \alpha_L) f'_0}{(f'_L - f'_0)} \tag{12.296}$$

The use of the above equations assumes that the function $f(\boldsymbol{x}_k + \alpha \boldsymbol{d}_k)$ is (for simplicity reasons) *quadratic with respect to its parameters*, whereas the values of the function and its first derivative with respect to α at the positions $\alpha = \alpha_L$ and $\alpha = \alpha_0$ are also known. The remaining expressions that appear in the above relations are defined as

$$f_L = f(\boldsymbol{x}_k + \alpha_L \boldsymbol{d}_k),\ f'_L = f'(\boldsymbol{x}_k + \alpha_L \boldsymbol{d}_k) = \nabla f(\boldsymbol{x}_k + \alpha_L \boldsymbol{d}_k)^T \boldsymbol{d}_k \tag{12.297}$$

$$f_0 = f(\boldsymbol{x}_k + \alpha_0 \boldsymbol{d}_k),\ f'_0 = f'(\boldsymbol{x}_k + \alpha_0 \boldsymbol{d}_k) = \nabla f(\boldsymbol{x}_k + \alpha_0 \boldsymbol{d}_k)^T \boldsymbol{d}_k. \tag{12.298}$$

After calculating the new value of α_0 the above conditions are checked again for their validity and the process is repeated until both are satisfied *both*. At this point the process terminates and the value of $\tilde{\alpha}$ that has been found is assumed to be the one that *minimizes* the desired multivariable function. This method is characterized by *limited accuracy*, in the sense that it may not determine exactly the right value, since the iterative algorithm terminates when the current value of α_0 belongs to the closed interval $[\alpha_1, \alpha_2]$. However, it is proven that *the range of this interval depends on the values of the parameters* ρ *and* σ and therefore by initializing these two parameters in the appropriate way, we can achieve the *desired* level of accuracy in each case.

Combining all of the above, a practical version of the *quasi-Newton algorithm* that can be used—among other things—for the design of recursive digital filters is as follows:

Step 1 (Algorithm Initialization)

1. We initialize the input vector $\boldsymbol{x}_0$ and the tolerance value ε_1.
2. We set $k = m = 0$ with the parameter k expressing *the vector numbered* k $(0 \le k < N)$ and the parameter m *the numbered* m *function value computation*.
3. We initialize the parameters of the algorithm to the appropriate values in each case, for example, $\rho = 0.1$, $\sigma = 0.7$, $\tau = 0.1$, $\chi = 0.75$, $M = 500$ and $\varepsilon_2 = 10^{-10}$. We remind the reader that the values of the parameters ρ and σ control *the range of the interval* $[\alpha_1, \alpha_2]$ and consequently *the accuracy of the linear search*. Small values of ρ and σ, such as the values $\rho = \sigma = 0.1$, lead to a *high-accuracy search*, whereas accuracy is limited if larger values are used, such as the values $\rho = 0.15$ and $\sigma = 0.9$. The parameters χ and τ are used by the linear search method to update the value of the parameter α_0 through the application of the numerical methods of *extrapolation* and *interpolation* respectively. The value of M expresses *the maximum number of function value calculations of the algorithm* (this means that the index m that we initialized above takes values in the range $0 \le m < M$), whereas the tolerance ε_2 is part of the condition that dictates whether the processes of *interpolation* and *extrapolation* will be carried out during the linear search.
4. We set $\boldsymbol{S}_0 = \boldsymbol{I}_N$ initializing the matrix $\boldsymbol{S}$, which, through successive corrections, eventually becomes equal to the inverse Hessian matrix $\boldsymbol{H}^{-1}$.
5. We calculate the values of f_0 and $\boldsymbol{g}_0$ and set $m = m + 2$, $f_{00} = f_0$, and $\Delta f_0 = f_0$.

Step 2 (Initialization of the Linear Search)

1. We set $\boldsymbol{d}_k = -\boldsymbol{S}_k \cdot \nabla f(\boldsymbol{x}_k)$.
2. We set $\alpha_L = 0$ and $\alpha_U = 10^{99}$ (a very large value).
3. We set $f_L = f_0$ and calculate the value of the parameter $f_L' = \nabla f(\boldsymbol{x}_k + \alpha_k \boldsymbol{d}_k)^T \boldsymbol{d}_k$.
4. We calculate the value of the parameter α_0 in the following way: if $|f_L'| > \varepsilon_2$ we set $\alpha_0 = -2\Delta f_0 / f_L'$, whereas in the opposite case we set $\alpha_0 = 1$. If the value of α_0 that will emerge is such that $\alpha_0 \le 0$ or $\alpha_0 > 1$, we set $\alpha_0 = 1$.

Step 3 We set $\boldsymbol{\delta}_k = \alpha_0 \boldsymbol{d}_k$ and calculate the value of the expression $f_0 = f(\boldsymbol{x}_k + \boldsymbol{\delta}_k)$. Also, we set $m = m + 1$.

Step 4 (Interpolation) If the conditions $f_0 > f_L + \rho(\alpha_0 - \alpha_L) f_L'$, $|(f_L - f_0)| > \varepsilon_2$ and $m < M$ are simultaneously satisfied, we perform the following procedures:

1. If $\alpha_0 < \alpha_U$, we set $\alpha_U = \alpha_0$.
2. We calculate the value of the parameter $\tilde{\alpha_0}$ using Eq. (12.289).
3. We calculate the value $\tilde{\alpha}_{0L} = \alpha_L + \tau(\alpha_U - \alpha_L)$. If $\tilde{\alpha}_0 < \tilde{\alpha}_{0L}$, we set $\tilde{\alpha}_0 = \tilde{\alpha}_{0L}$.
4. We calculate the value $\tilde{\alpha}_{0U} = \alpha_U - \tau(\alpha_U - \alpha_L)$. If $\tilde{\alpha}_0 > \tilde{\alpha}_{0U}$, we set $\tilde{\alpha}_0 = \tilde{\alpha}_{0U}$.
5. We set $\alpha_0 = \tilde{\alpha}_0$ and return to Step 3.

Step 5 We calculate the value of the quantity $f_0' = \nabla f(\boldsymbol{x}_k + \alpha_0 \boldsymbol{d}_k)^T \boldsymbol{d}_k$ and set $m = m + 1$.

Step 6 (Extrapolation) If the conditions $f_0' < \sigma f_L'$, $|(f_L - f_0)| > \varepsilon_2$ and $m < M$ are simultaneously satisfied, we perform the following procedures:

1. We calculate the quantity $\Delta\alpha_0 = (\alpha_0 - \alpha_L) f_0' / (f_L' - f_0')$ using Eq. (12.290).
2. If $\Delta\alpha_0 \le 0$, we set $\tilde{\alpha}_0 = 2\alpha_0$, whereas in the opposite case we set $\tilde{\alpha}_0 = \alpha_0 + \Delta\alpha_0$.

3. We calculate the value of the expression $\tilde{\alpha}_{0U} = \alpha_0 + \chi(\alpha_U - \alpha_0)$. If it is $\tilde{\alpha}_0 > \tilde{\alpha}_{0U}$, we set $\tilde{\alpha}_0 = \tilde{\alpha}_{0U}$.
4. We set $\alpha_L = \alpha_0$, $\alpha_0 = \tilde{\alpha}_0$, $f_L = f_0$, $f'_L = f'_0$ and return to Step 3.

Step 7 (Checking Termination Criteria and Returning Results)

1. We set $\boldsymbol{x}_{k+1} = \boldsymbol{x}_k + \boldsymbol{d}_k$.
2. We set $\Delta f_0 = f_{00} - f_0$.
3. If the conditions $\|\boldsymbol{\delta}\|_2 < \varepsilon_1$ and $|\Delta f_0| < \varepsilon_1$ are simultaneously satisfied or alternatively if $m \geq M$, we return the values $\tilde{\boldsymbol{x}} = \boldsymbol{x}_{k+1}$ and $f(\tilde{\boldsymbol{x}}) = f_{k+1}$ and terminate.
4. We set $f_{00} = f_0$.

Step 8 (Preparation for the Next Iteration)

1. We calculate the value of the gradient vector $\nabla f(\boldsymbol{x}_{k+1})$ and set $\boldsymbol{\gamma}_k = \nabla f(\boldsymbol{x}_{k+1}) - \nabla f(\boldsymbol{x}_k)$.
2. We calculate the value of the parameter $D = \boldsymbol{\delta}_k^T \boldsymbol{\gamma}_k$. If $D \leq 0$, we set $\boldsymbol{S}_{k+1} = \boldsymbol{I}_N$, whereas in the opposite case we calculate the matrix $\boldsymbol{S}_{k+1}$ using the *Davidon-Fletcher-Powell equation* or the *Broyden-Fletcher-Goldfarb-Shanno equation*.
3. We set $k = k + 1$ and return to Step 2.

It is interesting to note that the technique of *linear search* that has been incorporated into the above algorithm is slightly different than the one we presented earlier, without this affecting in any way the efficiency of the algorithm, which is efficient and reliable and can be used for the design of digital recursive filters in the way we will show later. For a brief commentary on some of the steps and calculations that are included in the above implementation of the quasi-Newton algorithm, the reader is referred to the manual Antoniou A., *Digital Signal Processing - Signals, Systems & Filters*, Section 16.4.4, pp. 734–736.

Minimax Optimization Algorithms

The *minimax optimization algorithms* used for the construction of *digital recursive filters* follow the same logic as the corresponding digital filter design algorithms FIR and allow the calculation of the filter coefficients as *those parameter values that minimize some objective function*. As mentioned in the introductory description of optimization algorithms, this function is the L_∞ measure function and the filters that result in this way are characterized as *equiripple filters* in the sense that *the approximation error is distributed in the various frequency zones of interest in a uniform way*.

A characteristic algorithm for designing digital filters IIR that belongs to this category of algorithms and is described by the name least-p is an iterative process that accepts as input an *initial vector* $\boldsymbol{x}_0$, an *error tolerance* ε_1, and a *set of K frequencies* $(\omega_1, \omega_2, \ldots, \omega_K)$ at the positions at which *the error function is sampled* and during the iteration under the number k tries to minimize the objective function

$$\Psi_k(\boldsymbol{x}) = \hat{\mathcal{E}}(\boldsymbol{x}) \sqrt[p]{\sum_{i=1}^{K} \left[\frac{|e_i(\boldsymbol{x})|}{\hat{\mathcal{E}}(\boldsymbol{x})}\right]^p}, \qquad \text{where} \qquad \hat{\mathcal{E}}(\boldsymbol{x}) = \max_{1 \leq i \leq K} |e_i(\boldsymbol{x})| \tag{12.299}$$

using some minimization algorithm (such as, the *quasi-Newton algorithm* that was presented in the previous section). To carry out this minimization process we need the gradient of the function $\Psi_k(\boldsymbol{x})$, which after the appropriate differentiation it turns out to be equal to

$$\nabla_k \Psi(\boldsymbol{x}) = \sqrt[\frac{1}{p}-1]{\sum_{i=1}^{K} \left[\frac{|e_i(\boldsymbol{x})|}{\hat{\mathcal{E}}(\boldsymbol{x})}\right]^p} \times \sum_{i=1}^{K} \left[\frac{|e_i(\boldsymbol{x})|}{\hat{\mathcal{E}}(\boldsymbol{x})}\right]^{p-1} \nabla |e_i(\boldsymbol{x})|. \tag{12.300}$$

In the above iterative process, the initial value of the vector $\tilde{\boldsymbol{x}}_{k+1}$ of the previous iteration is used, whereas the result is the new value of the vector $\tilde{\boldsymbol{x}}_k$, which is obviously used as a value in the next iteration. After calculating the vector $\boldsymbol{x}_k$ We define the error $\mathcal{E}_k = \mathcal{E}(\boldsymbol{x}_k)$ and check the value of the expression $|\mathcal{E}_{k-1} - \mathcal{E}_k|$. If this value is less than the tolerance error ε_1, the algorithm terminates and the vector $\boldsymbol{x}_k$ is returned as the desired solution, whereas in the opposite case the algorithm proceeds to the next iteration (which is done by setting $k = k + 1$) using as the new value for the parameter p the value $p' = \mu p$. Indicative initial values of the algorithm parameters, which in practice have been proven to lead to good results, are the values $\mu = p = 2$ and $\hat{\mathcal{E}}_0 = 10^{99}$ (or generally a very large value).

An alternative approach to which we can resort and which is characterized by greater efficiency and lower computational cost than the previous one is based on the use of the objective function

$$\Psi(\boldsymbol{x}, \boldsymbol{\lambda}, \xi) = \frac{1}{2}\left\{\sum_{i\in I_1} \lambda_i[\varphi_i(\boldsymbol{x}, \xi)]^2 + \sum_{i\in I_2}[\varphi_i(\boldsymbol{x}, \xi)]^2\right\}, \tag{12.301}$$

where $\varphi_i(\boldsymbol{x}, \xi) = |e_i(\boldsymbol{x})| - \xi$, ξ and λ_i constants ($i = 1, 2, \ldots, K$), whereas the sets I_1 and I_2 are defined as

$$I_1 = \{i : \varphi_i(\boldsymbol{x}, \xi) > 0 \text{ and } \lambda_i > 0\} \qquad \text{and} \qquad I_2 = \{i : \varphi_i(\boldsymbol{x}, \xi) > 0 \text{ and } \lambda_i = 0\}.$$

Without being able to give all the details of this algorithm, it is proven that as long as certain conditions regarding the constants ξ and λ_i are met, the vector $\tilde{\boldsymbol{x}}$ constitutes a *strong local minimum* of the function $\Psi(\boldsymbol{x}, \boldsymbol{\lambda}, \xi)$ and therefore the minimization of the error $\hat{\mathcal{E}}(\boldsymbol{x})$ consists in the minimization of the function $\Psi(\boldsymbol{x}, \boldsymbol{\lambda}, \xi)$ with respect to the vector $\boldsymbol{x}$. The gradient of the function $\Psi(\boldsymbol{x}, \boldsymbol{\lambda}, \xi)$ required for its minimization is calculated as

$$\nabla\Psi(\boldsymbol{x}, \boldsymbol{\lambda}, \xi) = \sum_{i\in I_1} \lambda_i \varphi(\boldsymbol{x}, \xi)\nabla|e_i(\boldsymbol{x})| + \sum_{i\in I_2} \varphi(\boldsymbol{x}, \xi)\nabla|e_i(\boldsymbol{x})|, \tag{12.302}$$

whereas the minimization is done as before.[87]

The basic common feature of the two above algorithms is the use of the error values $e_i(\boldsymbol{x})$ ($i = 1, 2, \ldots, K$), which arise from the samples of the error function $e_i(\boldsymbol{x})$ at the positions of the K frequencies $\omega_1, \omega_2, \ldots, \omega_K$. It is not difficult to understand that *the denser the samples of the error function* (i.e., *the more frequency values at which it is sampled*), *the greater the accuracy of the approximation*, as in the opposite case the error function may present undesirable fluctuations, such as *abrupt changes*. Although this problem can be addressed by using a sufficiently large value for the multitude K of sampling frequencies, in practice, a technique *of non-uniform sampling of frequency bands of interest* is used, which solves the problem without the need for a large value for the parameter K and which in general is carried out as follows:

Step 1 For a given value of the parameter K we proceed to *sampling the error function* $e(\boldsymbol{x})$ at the positions L frequency values $\tilde{\omega}_1, \tilde{\omega}_2, \ldots, \tilde{\omega}_L$ (for a sufficiently large value of L), which are uniformly distributed in the zones of interest, according to the relation $\tilde{\omega}_i = \tilde{\omega}_{i-1} + \Delta\omega$ ($i = 1, 2, \ldots, L$), where $\Delta\omega = (\tilde{\omega}_L - \tilde{\omega}_1)/(L - 1)$.

Step 2 We subdivide the zone of interest into K *equal frequency intervals* $\Omega_1, \Omega_2, \ldots, \Omega_K$ in such a way that the intervals Ω_1 and Ω_K have a range of $\Delta\omega/2$, the intervals Ω_2 and Ω_{K-1} have a range of $\ell\Delta\omega$, and the intervals Ω_i ($i = 2, 3, 4, \ldots, K - 2$) have a range of $2\ell\Delta\omega$ (where ℓ is an integer variable that is initialized in the appropriate way in each case). This partition is carried out in the following way:

$$\Omega_1 = \left\{\omega : \tilde{\omega}_1 \leq \omega < \tilde{\omega}_1 + \frac{1}{2}\Delta\omega\right\}$$

$$\Omega_2 = \left\{\omega : \tilde{\omega}_1 + \frac{1}{2}\Delta\omega \leq \omega < \tilde{\omega}_1 + \left(\ell + \frac{1}{2}\right)\Delta\omega\right\}$$

[87] For more information regarding the implementation details of this method, see the reference Charalambous C., *Acceleration of the Least p_{th} Algorithm for Minimax Optimization with Engineering Applications*, Mathematical Programming, Vol. 17, pp. 270–297, 1979.

....

$$\Omega_i = \left\{\omega : \tilde{\omega}_1 + \left[(2i-5)\ell + \frac{1}{2}\right]\Delta\omega \le \omega < \tilde{\omega}_1 + \left[(2i-3)\ell + \frac{1}{2}\right]\Delta\omega\right\} \quad (i = 3, 4, \ldots, K-2)$$

....

$$\Omega_\sigma = \left\{\omega : \tilde{\omega}_1 + \left[(2K-7)\ell + \frac{1}{2}\right]\Delta\omega \le \omega < \tilde{\omega}_1 + \left[(2K-6)\ell + \frac{1}{2}\right]\Delta\omega\right\}$$

$$\Omega_K = \left\{\omega : \tilde{\omega}_1 + \left[(2K-6)\ell + \frac{1}{2}\right]\Delta\omega \le \omega \le \tilde{\omega}_L\right\}, \quad \tilde{\omega}_L = \tilde{\omega}_1 + [(2K-6)\ell + 1]\Delta\omega,$$

(where $\sigma = K-1$) for specific values of the parameters K and ℓ. A necessary prerequisite for the use of this partitioning scheme of the frequency band of interest is the selection of the value for the parameter L, so that it satisfies the relation $L = (2K-6)\ell + 2$. Using the above partitioning scheme, the intervals Ω_1 and Ω_k contain only one sample, the intervals Ω_2 and Ω_{K-1} contain ℓ samples, whereas the intervals $\Omega_3, \Omega_4, \ldots, \Omega_{K-2}$ contain 2ℓ samples (this is also the reason why this sampling process is characterized as *non-uniform*).

Step 3 After determining the intervals $\Omega_1, \Omega_2, \ldots, \Omega_K$, we calculate for each of these the frequency value $\hat{\omega}$ that corresponds to the maximum error.

Step 4 The set of frequencies $\hat{\omega}_i$ $(i = 1, 2, \ldots, K)$ that results in Step 3 is used for the calculation of the *objective function*. It is not difficult to understand that using the values at positions where the error function exhibits *high values*, these *progressively decrease*, thus leading to the *suppression* of the creation of high error values.

We conclude the description of this process with the observation that the application of the algorithm we presented above requires the determination of the desired amplitude response $\mathcal{H}_{\mathrm{dr}}(\omega)$ at the positions L values of frequency. If S is the number of samples of the function $\mathcal{H}_{\mathrm{dr}}(\omega)$ we have $(S < L)$, then the L samples of the desired amplitude response required for the application of the algorithm can arise from the S available samples using the method of *cubic interpolation*, which has been presented in Sect. 12.12.5 (see page 1253). Applying this method for the values $j = 1, 2, \ldots, S-3$ and describing the amplitude response from the cubic polynomial

$$\mathcal{H}_{\mathrm{dr}}(\omega) = \alpha_{0j} + \alpha_{1j}\omega + \alpha_{2j}\omega^2 + \alpha_{3j}\omega^3$$

the coefficients $\alpha_{0j}, \alpha_{1j}, \alpha_{2j}$ and α_{3j} are determined by calculating the quantities $\tilde{\omega}_m, \tilde{\omega}_m^2$ and $\tilde{\omega}_m^3$ for the values $m = j, j+1, j+2$ and $j+3$. In this case, we end up with a system of equations, which, expressed in matrix form, is formulated as $\boldsymbol{\Omega}_j \boldsymbol{\alpha}_j = \tilde{\boldsymbol{M}}_{0j}$, where

$$\begin{aligned} \boldsymbol{\alpha}_j &= \left[\alpha_{0j}\ \alpha_{1j}\ \alpha_{2j}\ \alpha_{3j}\right] \\ \tilde{\boldsymbol{M}}_{0j} &= \left[M_0(\omega_j)\ M_0(\omega_{j+1})\ M_0(\omega_{j+2})\ M_0(\omega_{j+3})\right] \end{aligned} \quad \text{and } \boldsymbol{\Omega}_j = \begin{bmatrix} 1 & \tilde{\omega}_j & \tilde{\omega}_j^2 & \tilde{\omega}_j^3 \\ 1 & \tilde{\omega}_{j+1} & \tilde{\omega}_{j+1}^2 & \tilde{\omega}_{j+1}^3 \\ 1 & \tilde{\omega}_{j+2} & \tilde{\omega}_{j+2}^2 & \tilde{\omega}_{j+2}^3 \\ 1 & \tilde{\omega}_{j+3} & \tilde{\omega}_{j+3}^2 & \tilde{\omega}_{j+3}^3 . \end{bmatrix}$$

The solution of the above system is calculated from the equation $\boldsymbol{\alpha}_j = \tilde{\boldsymbol{\Omega}}_j^{-1} \tilde{\boldsymbol{M}}_{0j}$ $(j = 1, 2, \ldots, S-3)$ and then used for the calculation of the frequencies $\tilde{\omega}_1, \tilde{\omega}_2, \ldots, \tilde{\omega}_L$. It should be noted that in order to maximize the accuracy of the polynomial interpolation, each of the above polynomials should be used for frequency values that are *in the center of the frequency range of interest*.

The above process in algorithmic form is formulated as follows:

Step 1 We initialize the input vector $\boldsymbol{x}_0$, the tolerance value ε_1, the frequencies $\tilde{\omega}_1, \tilde{\omega}_2, \ldots, \tilde{\omega}_S$, and the values of the samples of the desired frequency response $M(\tilde{\omega}_1), M(\tilde{\omega}_2), \ldots, M(\tilde{\omega}_S)$. We set $p = \mu = 2$ (values that in practice have been proven to lead to good results), $\hat{\mathcal{E}}_0 = 10^{99}$ (or generally a very large value) and $k = 1$, whereas we also initialize the variable K.

Step 2 We calculate the values of the parameters L and $\Delta\omega$ from the equations

$$L = (2K - 6)\ell + 2 \qquad \text{and} \qquad \Delta\omega = (\tilde{\omega}_L - \tilde{\omega}_1)/(L - 1)$$

for specific values of the parameters ℓ, $\tilde{\omega}_1$ and $\tilde{\omega}_L$ as well as for specific values of the coefficients α_{ij} ($i = 0, 1, 2, 3$ and $j = 1, 2, \ldots, S - 3$) using the equations of cubic interpolation that we presented above.

Step 3 We calculate the desired amplitude response at the positions of the frequency values $\tilde{\omega}_1, \tilde{\omega}_2, \ldots, \tilde{\omega}_L$ from the *cubic polynomial equation* $M_0(\omega) = \alpha_{0j} + \alpha_{1j}\omega + \alpha_{2j}\omega^2 + \alpha_{3j}\omega^3$.

Step 4 We set $\omega_1 = \tilde{\omega}_1$, $\omega_2 = \tilde{\omega}_{1+\ell}$, $\omega_i = \tilde{\omega}_{2(i-2)\ell+1}$ ($i = 3, 4, \ldots, K - 2$), $\omega_{K-1} = \tilde{\omega}_{L-1}$ and $\omega_K = \tilde{\omega}_L$.

Step 5 Using the vector $\tilde{\boldsymbol{x}}_{k-1}$ as an initial value, we minimize the *objective function*

$$\Psi_k(\boldsymbol{x}) = \hat{\mathcal{E}}(\boldsymbol{x}) \sqrt[p]{\sum_{i=1}^{K} \left[\frac{|e_i(\boldsymbol{x})|}{\hat{\mathcal{E}}(\boldsymbol{x})}\right]^p} \qquad \text{where} \qquad \hat{\mathcal{E}}(\boldsymbol{x}) = \max_{1 \le i \le K} |e_i(\boldsymbol{x})|$$

with respect to $\boldsymbol{x}$, to calculate the new value of the vector $\tilde{\boldsymbol{x}}_k$. We set $\hat{\boldsymbol{\mathcal{E}}}_k = \hat{\boldsymbol{\mathcal{E}}}(\boldsymbol{x})$.

Step 6 We calculate the values $|e_i(\tilde{\boldsymbol{x}}_k)| \equiv |e(\tilde{\boldsymbol{x}}_k, \omega_i)|$ ($i = 1, 2, \ldots, L$) using their defining equation $e(\tilde{\boldsymbol{x}}_k, \omega_i) = \mathcal{H}_r(\tilde{\boldsymbol{x}}_k, \omega_i) - \mathcal{H}_{\text{dr}}(\omega)$.

Step 7 We calculate the frequencies $\hat{\omega}_i$ ($i = 1, 2, \ldots, K$) and the value $\hat{P}_k = \hat{P}(\tilde{\boldsymbol{x}}_k) = \max|e_i(\tilde{\boldsymbol{x}}_k|$ (for the values $i = 1, 2, \ldots, K$).

Step 8 We set $\omega_i = \hat{\omega}_i$ ($i = 1, 2, \ldots, K$).

Step 9 If it is $|\hat{\boldsymbol{E}}_{k-1} - \hat{\boldsymbol{E}}_k| < \varepsilon_1$ and $|\hat{\boldsymbol{P}}_k - \hat{\boldsymbol{E}}_k| < \varepsilon_1$, we return the current values of $\tilde{\boldsymbol{x}}_k$ and $\boldsymbol{E}_k$ and *terminate the algorithm*, whereas in the opposite case we return to Step 5.

Design of Digital Recursive Filters The design of digital recursive filters with the *method of Newton* is based on the direct use of the equations we constructed in the previous sections simply replacing the function $f(\boldsymbol{x})$ that we minimized in the way we described with the objective function

$$\Psi(\boldsymbol{\xi}) = \sqrt[p]{\sum_{i=1}^{K} |e(\boldsymbol{\xi}, \omega_i)|^p}, \tag{12.303}$$

which, of course, should be formulated as a function of the frequency ω. Considering for simplicity reasons a filter of *even order* N, the transfer function of which has the form

$$\mathcal{H}(z) = H_0 \prod_{i=1}^{L} \frac{\alpha_{0i} + \alpha_{1i}z + z^2}{\beta_{0i} + \beta_{1i}z + z^2}, \tag{12.304}$$

the amplitude response of it will arise as known from the measure of the above function calculated *on the perimeter of the unit circle* $z = e^{j\omega}$—therefore,

$$|\mathcal{H}(e^{j\omega})| = |\mathcal{H}(z)|_{z=e^{j\omega}} = H_0 \prod_{i=1}^{L} \frac{|\alpha_{0i} + \alpha_{1i}z + z^2|_{z=e^{j\omega}}}{|\beta_{0i} + \beta_{1i}z + z^2|_{z=e^{j\omega}}}, \qquad \text{where} \tag{12.305}$$

$$|\alpha_{0i} + \alpha_{1i}z + z^2|_{z=e^{j\omega}} = |\alpha_{0i} + \alpha_{1i}e^{j\omega} + e^{2j\omega}| = \sqrt{1 + \alpha_{0i}^2 + \alpha_{1i}^2 + 2\alpha_{0i}\cos(2\omega) + 2\alpha_{1i}[1 + \alpha_{0i}\cos(\omega)]} \quad \text{and}$$

$$|\beta_{0i} + \beta_{1i}z + z^2|_{z=e^{j\omega}} = |\beta_{0i} + \beta_{1i}e^{j\omega} + e^{2j\omega}| = \sqrt{1 + \beta_{0i}^2 + \beta_{1i}^2 + 2\beta_{0i}\cos(2\omega) + 2\beta_{1i}[1 + \beta_{0i}\cos(\omega)]}.$$

The above relations result from the combined use of *Euler's formula*, the *defining equation of the modulus of a complex number*, and the identity $\cos(\alpha - \beta) = \cos\alpha\cos\beta - \sin\alpha\sin\beta$. Therefore,

$$\mathcal{H}_r(\boldsymbol{x}, \omega) = H_0 \prod_{i=1}^{L} \sqrt{\frac{1 + \alpha_{0i}^2 + \alpha_{1i}^2 + 2\alpha_{0i}\cos(2\omega) + 2\alpha_{1i}[1 + \alpha_{0i}\cos(\omega)]}{1 + \beta_{0i}^2 + \beta_{1i}^2 + 2\beta_{0i}\cos(2\omega) + 2\beta_{1i}[1 + \beta_{0i}\cos(\omega)]}}. \tag{12.306}$$

If we know the values of the desired amplitude response $\mathcal{H}_{dr}(\omega_i)$ at the positions $\omega_1, \omega_2, \ldots, \omega_K$, the corresponding error values will be the

$$e_\ell(\boldsymbol{x}) = e(\boldsymbol{x}, \omega_\ell) = \mathcal{H}_r(\boldsymbol{x}, \omega) - H_{dr}(\omega_\ell) = H_0 \prod_{i=1}^{L} \sqrt{\frac{1 + \alpha_{0i}^2 + \alpha_{1i}^2 + 2\alpha_{0i}\cos(2\omega_\ell) + 2\alpha_{1i}(1 + \alpha_{0i})\cos(\omega_\ell)}{1 + \beta_{0i}^2 + \beta_{1i}^2 + 2\beta_{0i}\cos(2\omega_\ell) + 2\beta_{1i}(1 + \beta_{0i})\cos(\omega_\ell)}} - H_{dr}(\omega_\ell)$$

and the objective function for minimization takes the form

$$\Psi(\boldsymbol{\xi}) = \sqrt[p]{\sum_{\ell=1}^{K} \left| H_0 \prod_{i=1}^{L} \sqrt{\frac{1 + \alpha_{0i}^2 + \alpha_{1i}^2 + 2\alpha_{0i}\cos(2\omega_\ell) + 2\alpha_{1i}(1 + \alpha_{0i})\cos(\omega_\ell)}{1 + \beta_{0i}^2 + \beta_{1i}^2 + 2\beta_{0i}\cos(2\omega_\ell) + 2\beta_{1i}(1 + \beta_{0i})\cos(\omega_\ell)}} - H_{dr}(\omega_\ell) \right|^p}. \tag{12.307}$$

Having formulated the objective function for minimization as a function of frequency ω, we can construct the various quantities required by the algorithm to which we will resort for determining the solution of the problem. For example, for the calculation of the gradient vector $\nabla\Psi(\boldsymbol{\xi})$, we first determine the *gradient of the error function* $\nabla e_\ell(\boldsymbol{\xi}) = \nabla e(\boldsymbol{\xi}, \omega_\ell)$ by directly deriving the above relation. Therefore,

$$\frac{\partial e_\ell(\xi)}{\partial \alpha_{0i}} = \frac{\alpha_{0i} + \alpha_{1i}\cos(\omega_\ell) + \cos(2\omega_\ell)}{1 + \beta_{0i}^2 + \beta_{1i}^2 + 2\beta_{0i}\cos(2\omega_\ell) + 2\beta_{1i}(1 + \beta_{0i})\cos(\omega_\ell)} \times \mathcal{H}_r(\boldsymbol{x}, \omega_\ell) \tag{12.308}$$

$$\frac{\partial e_\ell(\xi)}{\partial \alpha_{1i}} = \frac{\alpha_{1i} + (1 + \alpha_{0i})\cos(\omega_\ell)}{1 + \beta_{0i}^2 + \beta_{1i}^2 + 2\beta_{0i}\cos(2\omega_\ell) + 2\beta_{1i}(1 + \beta_{0i})\cos(\omega_\ell)} \times \mathcal{H}_r(\boldsymbol{x}, \omega_\ell) \tag{12.309}$$

$$\frac{\partial e_\ell(\xi)}{\partial \beta_{0i}} = -\frac{\beta_{0i} + \beta_{1i}\cos(\omega_\ell) + \cos(2\omega_\ell)}{1 + \beta_{0i}^2 + \beta_{1i}^2 + 2\beta_{0i}\cos(2\omega_\ell) + 2\beta_{1i}(1 + \beta_{0i})\cos(\omega_\ell)} \times \mathcal{H}_r(\boldsymbol{x}, \omega_\ell) \tag{12.310}$$

$$\frac{\partial e_\ell(\xi)}{\partial \beta_{1i}} = -\frac{\beta_{1i} + (1 + \beta_{0i})\cos(\omega_\ell)}{1 + \beta_{0i}^2 + \beta_{1i}^2 + 2\beta_{0i}\cos(2\omega_\ell) + 2\beta_{1i}(1 + \beta_{0i})\cos(\omega_\ell)} \times \mathcal{H}_r(\boldsymbol{x}, \omega_\ell) \tag{12.311}$$

$$\frac{\partial e_\ell(\xi)}{\partial H_0} = \frac{1}{H_0} \times \mathcal{H}_r(\boldsymbol{x}, \omega). \tag{12.312}$$

After calculating the gradient of the function $e_{\boldsymbol{\xi},\omega_i}$ *the gradient of the absolute value of this error* is calculated as $\nabla|e(\boldsymbol{\xi}, \omega_\ell)| = \text{sgn}\{e(\boldsymbol{\xi}, \omega_\ell)\} \times \nabla e(\boldsymbol{\xi}, \omega_\ell)$, where sgn is our known sign function. If we choose to solve the problem using the *classic method of Newton*, something that, of course, *is not recommended* for the reasons we mentioned above, we need to calculate the *Hessian matrix* $\boldsymbol{H}$ by deriving the above relations and calculating all the second-order derivatives of the objective function to minimize that are included in this matrix. As for the gradient of the objective function $\Psi(\boldsymbol{\xi})$, which is also needed for the construction of the recursive filter, this can be calculated with the help of the gradient of the error function.

Example of the Application of Optimization Methods After the analytical presentation of the methods of constructing *digital recursive filters* using *optimization methods*, let us now move on to the presentation of an example of their application in practice. Owing to the inherent complexity of these algorithms, it is not easy to analytically present their *direct implementation* using the relations derived in the previous section. Hence, MATLAB functions are employed. The basic tool we use in this presentation is the function of MATLAB *Dig_Filter*, which was developed by *Iman Moazzen* of the

Department of Electrical Engineering of the University of Victoria under the supervision of Professors *Andreas Antoniou* and *Panagiotis Agathoklis* and which is available along with its accompanying functions from the address

http://www.mathworks.com/matlabcentral/fileexchange/30321-digfilter

This function allows the construction of digital *FIR* and *IIR*, which can be *low-pass, high-pass, band-pass*, or *band-stop* digital filters. The characteristics of the filter to be constructed are passed to the function as arguments, with the argument *Filter_Length* taking one of the values 'FIR' and 'IIR' depending on the *number of samples of the desired impulse response* (finite or infinite) and the parameters w_s, A_p, A_s and *Filter_Order*, allowing respectively the determination of the *sampling frequency*, the *maximum deviation in the passband*, the *minimum attenuation in the stopband*, and the *order of the filter*, which must be an *even number*. The *type* of the filter (low-pass, high-pass, band-pass, or band-stop) is determined by the argument *Type*, which takes one of the values 'Low-pass', 'High-pass', 'Band-pass', and 'Band-stop', whereas the *construction method* of the digital filter is dictated by its type. More specifically, for the case of a FIR-type filter this argument takes the value 'Kaiser', if the filter is to be constructed with the *Kaiser window method*, and the value 'Optimization' if the filter will be constructed using *optimization methods*. On the other hand, for the case of *digital recursive filters*, the Method argument can take one of the values 'Butterworth', 'Chebyshev', and 'Elliptic' if the construction of the digital filter uses the *corresponding prototype analog filter*, and the value 'Optimization' if one of the *optimization methods* presented in the previous section is used. In the latter case, the optimization method to be used is determined by the value of the argument Optimization_Method, which can either be 'fminunc', which minimizes the objective function $\Psi(\boldsymbol{x})$ by calling the *minimization function* of MATLAB,[88] or one of the values 'bfgs' and 'dfp', which use the *Broyden-Fletcher-Goldfarb-Shanno* and *'Davidon-Fletcher-Powell'* equations presented previously.

The determination of the *operating frequencies* of the filter to be constructed is determined by the contents of the argument edges, which is a frequency table containing the limits of the one and only transition zone that characterizes the *low-pass* and *high-pass filters* or the boundaries of the two transition zones that characterize the operation of the *band-pass* and *band-stop* filters (this means that this table contains *two* frequency values for the low-pass and the high-pass filters and *four* frequency values for the band-pass and band-stop filters). The order in which these frequency values are recorded is (w_p, w_s) for the low-pass filters, (w_s, w_p) for the high-pass filters, $(w_{s1}, w_{p1}, w_{p2}, w_{s2})$ for the band-pass filters, and $(w_{p1}, w_{s1}, w_{s2}, w_{p2})$ for the band-stop filters. The application requires the recording of the above frequency values *sorted in ascending order*, and care should be taken, so that *the maximum of these frequencies is not greater than half the sampling frequency*, in order not to violate the *Nyquist theorem*.

As already mentioned in previous sections, in practice, it is not at all certain that the filter that will result will satisfy *obligatorily* the operating specifications that have been set, something that happens when *the attenuation in the cutoff zone is greater or equal to the value of the argument* A_s and simultaneously *the maximum deviation in the passband is less than or equal to the value of the argument* A_p. However, in the case of iterative optimization methods, the filter's parameter values, on which the results depend, are influenced, among other factors, by the random values of the initial parameter vector used as input data in the iterative process. If the function fails to construct a satisfactory filter in terms of its performance, it repeats the filter construction attempt with different random values of the initial parameter vector. In each case, if the filter that was constructed meets our requirements, the function terminates and returns the filter coefficients, otherwise it repeats the attempt from the beginning. The number of attempts to construct the desired filter is determined by the argument *Number_Of_iterations*, the default value of which (if not set by the user) is equal to 5. The application stores in a table the operating parameters it calculates in each such attempt and, if after exhausting all attempts it has not managed to construct a satisfactory filter, it returns the coefficients of the best of the filters it constructed. Finally, the arguments eps1 and eps2 correspond to the parameters ε_1 and ε_2 of the optimization iterative algorithms that were presented in the previous sections and the default value of both is the value 0.01. The above filter construction processes use all the equations, methods, and techniques that were presented throughout the chapter.

After the detailed presentation of all the function's arguments *Dig_Filter* and the way it is defined and used, let us present its prototype, which is characterized by a syntax of the form

```
[b,a] = Dig_Filter (Filter_Length, Type, Method, Edges, ws, Ap, As, Optimization_Method,
                    Filter_Order, Number_Of_Iterations, eps1, eps2)
```

[88] This function relies on the use of the iterative algorithm *trust region Newton method*—see the reference Lin C.-J. et al., *Trust Region Newton Method for Logistic Regression*, Journal of Machine Learning Research, Vol. 9, pp. 627–650, June 2008.

with the vectors b and a containing *the coefficients of the numerator and the denominator* of the *rational transfer function* of the filter constructed by the function. The following description presents an example of using the *Dig_Filter* function to construct digital recursive filters using optimization methods.

Example 12.13.11 Suppose we want to construct a low-pass recursive filter using the optimization method with transition zone limits $w_p = 500\,\text{Hz}$, $w_s = 800\,\text{Hz}$, sampling frequency $\Omega_s = 2000\,\text{Hz}$, operating parameters $A_p = 0.25\,\text{dB}$, $A_s = 45\,\text{dB}$, and even order $N = 8$. Although the solution can also be obtained with the default tolerance error values $\varepsilon_1 = \varepsilon_2 = 0.01$, this will be found in only a few iterations (which of course in practice *is desirable*) and for this reason we will use the much smaller values $\varepsilon_1 = \varepsilon_2 = 0.0000001$, so that the algorithm is executed with more iterations in order to be understood (consequently, this is something that is done exclusively for *educational purposes* only).

The call of the *Dig_Filter* function for the creation of the filter to be constructed has the form

[b,a] = Dig_Filter ('IIR,' 'Low-pass,' 'Optimization,'[500 800],2000,0.25,45,'fminunc',8,3,0.0000001,0.0000001)

from which it appears that for the minimization of the objective function $\Psi(\boldsymbol{x})$ the minimization function fminunc of the MATLAB programming environment will be used, whereas for the determination of the appropriate filter, a total of three attempts will be made. The function operates in the way we described above and returns to the arguments b and a the values of the coefficients of the recursive digital filter that approximates to the greatest possible degree the transfer function of the desired filter.

Taking into account that the order of the filter N is $N = 8$, it is not difficult to realize that the vector of unknowns $\boldsymbol{\xi}$ will contain 17 parameters and will be the

$$\boldsymbol{\xi} = [\alpha_{01} \quad \alpha_{11} \quad \beta_{01} \quad \beta_{11} \quad \alpha_{02} \quad \alpha_{12} \quad \beta_{02} \quad \beta_{12} \quad \alpha_{03} \quad \alpha_{13} \quad \beta_{03} \quad \beta_{13} \quad \alpha_{04} \quad \alpha_{14} \quad \beta_{04} \quad \beta_{14} \quad H_0].$$

This vector, before the start of the iterative process, is *initialized to random values*, constituting the vector $\boldsymbol{x}_0$ that is used as input to the algorithm. The application initializes at this point the values of the L frequencies $\hat{\omega}_1, \hat{\omega}_2, \ldots, \hat{\omega}_L$, with the value of L being 20 times greater than the length of the vector $\boldsymbol{\xi}$, so that the algorithm is applied to a sufficiently dense set of frequencies. In the general case where the frequency vector of interest edge contains four frequencies e_i $(i = 1, 2, 3, 4)$ and taking into account that $0 < e_1 < e_2 < e_3 < e_4 < (w_s/2)$, the initialization of these L frequencies is based on (a) the *separation* of the frequency axis into the zones of interest $[0, e_1]$, $[e_2, e_3]$ and $[e_4, w_s/2]$, (b) the *determination* of the number of samples m_1, m_2 and m_3 for these three zones according to the relations

$$m_1 = \left[\frac{e_1}{\omega_0} + 0.5\right] \quad m_1 = \left[\frac{e_3 - e_2}{\omega_0} + 0.5\right], \quad m_3 = L - (m_1 + m_2), \quad \text{where} \quad \omega_0 = \frac{e_1 + (e_3 - e_2) + [(w_s/2) - e_4]}{L - 2},$$

and (c) the *calculation* of the $L = m_1 + m_2 + m_3$ frequency values with the values of two successive frequency values differing from each other by a distance of ω_0 (in the equations defining the parameters m_1 and m_2 the brackets [ ] are used for the calculation of the *integer part* of the quantity enclosed in them). After the above initialization processes, the algorithm enters its main operational loop, which includes successive minimizations of the objective function with the help of the minimization function fminunc.

After the calculation of the vector $\boldsymbol{x}_k$ during the iteration numbered k, the error values $e_i(\boldsymbol{x}_k) \equiv e(\boldsymbol{x_k}, \omega_i) = w_m[M(\boldsymbol{x}, \omega_i) - M_0(\omega_i)]$ are calculated for each of the L frequencies $\tilde{\omega}_i$ $(i = 1, 2, \ldots, L)$, with the value of the desired amplitude response $M_0(\omega_i)$ being *equal to one in the passband* and *equal to zero in the stopband*. We observe that for the calculation of the error, the function *Dig_Filter* does not use the simple difference equation $e_i(\boldsymbol{x}_k) = M(\boldsymbol{x}, \omega_i) - M_0(\omega_i)$ but instead multiplies the above difference with a *weight value* in order to make the contribution of each frequency band of interest different in the final result. The logic of using these weights is similar to the one we described during the presentation of optimization algorithms that allow the construction of FIR-type filters. For the special case of the low-pass filter that we are dealing with in this exercise, the weights w_1, w_2 and w_3 associated with the frequency bands $[0, e_1]$, $[e_2, e_3]$, and $[e_3, w_s/2]$ are calculated as $w_1 = 1$ and $w_2 = w_3 = \varepsilon_p/\varepsilon_s$, where

$$\varepsilon_p = \frac{10^{0.05A_p} - 1}{10^{0.05A_p} + 1} \quad \text{and} \quad \varepsilon_s = 10^{-0.05A_s}.$$

After calculating the above values, the algorithm calculates the *slope of the objective function* $\Psi(\boldsymbol{x}_k)$ at the position of the current vector $\boldsymbol{x}_k$ and the iterative algorithm terminates when the condition $|\hat{\mathcal{E}}_k - \hat{\mathcal{E}}_{k-1}| < \varepsilon_1$ is satisfied.

The results of the execution of this algorithm are presented in the following tables. For the given execution conditions, the iterative algorithm was executed a total of 13 times, thus leading to an equal number of successive updates of the parameter vector. $\boldsymbol{x}_k$ $(k = 1, 2, \ldots, 13)$, the contents of which are included in Table 12.6. It is not difficult to understand that during the evolution of the algorithm, the *correction* of the vector $\boldsymbol{x}$ becomes increasingly smaller, as shown in the second column of Table 12.7. For each of the 13 iterations of the algorithm, the third column of Table 12.7 contains the *number of iterations* performed by the function fminunc (it is important to understand at this point that this function contains *its own internal loop*, whereas of course it is called *repeatedly* by the optimization algorithm), and the fourth column contains the *current minimum value of the objective function*. In order to understand the way of minimizing this function, Table 12.8 contains the successive values of this function as it is minimized by fminunc for the last six iterations of the optimization algorithm. From this table we observe that on the one hand the value of the function *is progressively reduced* as the algorithm passes from one iteration to the next, and on the other hand that the successive changes of the function within the same iteration *become increasingly smaller* (note that in the last two iterations the function has essentially approached its minimum value and does not undergo the slightest change within the minimization function fminunc). The final solution of the problem that leads to the minimum value of the function equal to 0.001068 is the contents of the vector $\boldsymbol{x}_{13}$—thus

$$\boldsymbol{\xi} = [\ 0.997040\ 1.640905\ 1.872144\ 1.060528\ 0.981364\ 1.803155\ 1.904817\ 1.075352\ 3.058014\ 2.306290$$
$$1.918875\ 2.498418\ 0.223229\ 1.126887\ 0.079042\ 0.117681\ 0.363767\ ]$$

and consequently the coefficients of the transfer function of the filter and the normalization constant will have the values $\alpha_{01} = x[1] = 0.997040$, $\alpha_{11} = x[2] = 1.640905$, $\beta_{01} = x[3] = 1.872144$, $\beta_{11} = x[4] = 1.060528$, $\alpha_{02} = x[5] = 0.981364$, $\alpha_{12} = x[6] = 1.803155$, $\beta_{02} = x[7] = 1.904817$, $\beta_{12} = x[8] = 1.075352$, $\alpha_{03} = x[9] = 3.058014$, $\alpha_{13} = x[10] = 2.306290$, $\beta_{03} = x[11] = 1.918875$, $\beta_{13} = x[12] = 2.498418$, $\alpha_{04} = x[13] = 0.223229$, $\alpha_{14} = x[14] = 1.126887$, $\beta_{04} = x[15] = 0.079042$, $\beta_{14} = x[16] = 0.117681$, $H_0 = x[17] = 0.363767$.

The above results are graphically represented in Figs. 12.89 and 12.90. In Fig. 12.89, the *change of* $|\boldsymbol{\delta x}|$ *and the minimum value* $\Psi(\boldsymbol{x}_k)_{\min}$ *as a function of the number of iterations* is presented, whereas in Fig. 12.90, the *progressive approximation of the minimum value of the function during the iterative process* performed by the function fminunc *as a function of the number of iterations of the optimization algorithm* is graphically represented (the data of the last six iterations presented in Table 12.8 are not clearly visible in the graph, because, owing to the extremely small changes, one curve falls on top of the other). The image (a) of the figure also shows the change in the function during the first iteration, which is very fast and starts from a high value, since the initial vector $\boldsymbol{x}_0$ has been initialized randomly. However, as fminunc continues to execute, the function begins to decrease progressively until it finally approaches its minimum value. For better preview purposes and owing to the high value of the function during the first iteration, the vertical axis of the image (a) is designed in *logarithmic scale*. In image (b) of the figure, the curve related to the first iteration is not depicted and therefore both axes are designed in the usual linear scale.

Having calculated the coefficients of the filter with the above optimization process, the transfer functions $H_i(z)$ $(i = 1, 2, 3, 4)$ of the *four second-order filters*, the series connection of which will give us the filter to be designed (see Fig. 12.91), are formulated as[89]

$$\mathcal{H}_1(z) = \frac{\alpha_{01} + \alpha_{11}z + z^2}{\beta_{01} + \beta_{11}z + z^2} = \frac{0.997040 + 1.640905z + z^2}{1.872144 + 1.060528z + z^2}$$

$$\mathcal{H}_2(z) = \frac{\alpha_{02} + \alpha_{12}z + z^2}{\beta_{02} + \beta_{12}z + z^2} = \frac{0.981364 + 1.803155z + z^2}{1.904817 + 1.075352z + z^2}$$

$$\mathcal{H}_3(z) = \frac{\alpha_{03} + \alpha_{13}z + z^2}{\beta_{03} + \beta_{13}z + z^2} = \frac{3.058014 + 2.306290z + z^2}{1.918875 + 2.498418z + z^2}$$

[89] In order for these functions to be expressed as a function of the variable z^{-1}, as usually happens with digital filters, we should divide the representations of the numerator and the denominator by the quantity z^2.

Table 12.6 The contents of the vector xk during the execution of the optimization algorithm

x_1	x_2	x_3	x_4	x_5	x_6	x_7
0.994511	0.996523	0.9950130	0.996968	0.997227	0.997298	0.997182
1.653819	1.644478	1.6399628	1.641118	1.641005	1.640993	1.640913
1.902734	1.886956	1.8798829	1.874406	1.873722	1.873700	1.871863
1.148689	1.116231	1.0824146	1.068104	1.067057	1.067033	1.062260
0.979680	0.982455	0.9811280	0.980540	0.981049	0.981857	0.981221
1.824238	1.813584	1.8076312	1.804477	1.804188	1.804145	1.803377
1.932708	1.918119	1.9118069	1.906901	1.906259	1.906238	1.904525
1.160622	1.129603	1.0966316	1.082726	1.081720	1.081697	1.077048
3.054917	−3.057526	−3.0584384	−3.058829	−3.058639	−3.058625	−3.058122
2.348101	−2.329994	−2.3159172	−2.311719	−2.310627	−2.310593	−2.306874
1.979131	−1.957030	−1.9352164	−1.922712	−1.922126	−1.922132	−1.919417
−2.433153	−2.457420	−2.4807894	−2.490907	−2.492292	−2.492325	−2.497493
0.184599	0.192672	0.2110780	0.221038	0.221040	0.221003	0.222403
1.076457	1.098800	1.1151335	1.123457	1.124013	1.124017	1.126298
0.076819	0.077562	0.0780923	0.078865	0.078878	0.078873	0.079101
−0.086322	−0.100889	−0.1095784	−0.113520	−0.114285	−0.113989	−0.117447
0.405703	0.388693	0.3746885	0.367023	0.366486	0.366470	0.364243
x_8	x_9	x_{10}	x_{11}	x_{12}	x_{13}	
0.997029	0.997065	0.997047	0.997043	0.997040	0.997040	
1.640900	1.640895	1.640897	1.640902	1.640904	1.640905	
1.872142	1.872142	1.872143	1.872143	1.872144	1.872144	
1.060536	1.060529	1.060529	1.060528	1.060528	1.060528	
0.981137	0.981392	0.981374	0.981369	0.981365	0.981364	
1.803158	1.803144	1.803145	1.803150	1.803154	1.803155	
1.904815	1.904815	1.904816	1.904816	1.904816	1.904817	
1.075361	1.075354	1.075353	1.075353	1.075353	1.075352	
−3.058016	−3.058013	−3.058014	−3.058014	−3.058014	−3.058014	
−2.306299	−2.306291	−2.306291	−2.306290	−2.306290	−2.306290	
−1.918878	−1.918877	−1.918876	−1.918876	−1.918875	−1.918875	
−2.498408	−2.498417	−2.498418	−2.498418	−2.498418	−2.498418	
0.223229	0.223228	0.223229	0.223229	0.223229	0.223229	
1.126883	1.126885	1.126886	1.126886	1.126887	1.126887	
0.079036	0.079041	0.079041	0.079042	0.079042	0.079042	
−0.117726	−0117667	−0.117679	−0.117680	−0.117681	−0.117681	
0.363775	0.363771	0.363771	0.363769	0.363767	0.363767	

Table 12.7 The details of the optimization algorithm. For the iteration numbered k (where $k = 2, 3, \ldots, 13$) are displayed *the values* $\boldsymbol{\delta x}_k = \boldsymbol{x}_k - \boldsymbol{x}_{k-1}$, the *number of iterations* N_k of the algorithm *trust region Newton method* used by the function Fminunc, and the *minimum value of the objective function* $\Psi(\boldsymbol{x}_k)_{\min}$ for the current iteration

k	$\lvert\delta x_k\rvert$	N_k	$\Psi\ (x_k)_{\min}$
2	0.071892138	77	0.010932
3	0.066924025	13	0.003294
4	0.031425360	11	0.001801
5	0.002843203	09	0.001350
6	0.000868151	07	0.001196
7	0.011136135	02	0.001138
8	0.002975547	08	0.001094
9	0.000265183	07	0.001077
10	0.000002837	03	0.001072
11	0.000000994	03	0.001069
12	0.000000728	12	0.001069
13	0.000000224	09	0.001068

Table 12.8 The last six iterations of the optimization algorithm regarding the *successive values of the function* $\Psi(\boldsymbol{x_k})$. From the table, we observe that as the number of iterations increases, the value of the function *progressively decreases*, whereas the distance between its successive values becomes increasingly smaller. The last value of each iteration is the minimum value of the function for that iteration, which is shown in the third column of Table 12.7

$N_8 = 8$	$N_9 = 7$	$N_{10} = 3$	$N_{11} = 3$	$N_{12} = 12$	$N_{13} = 9$
0.001113	0.001083	0.001072	0.001070	0.001069	0.001068
0.001113	0.001083	0.001072	0.001070	0.001069	0.001068
0.001113	0.001083	0.001072	0.001069	0.001069	0.001068
0.001113	0.001083			0.001069	0.001068
0.001099	0.001083			0.001069	0.001068
0.001097	0.001078			0.001069	0.001068
0.001094	0.001077			0.001069	0.001068
0.001094				0.001069	0.001068
				0.001069	0.001068
				0.001069	
				0.001069	
$k = 8$	$k = 9$	$k = 10$	$k = 11$	$k = 12$	$k = 13$

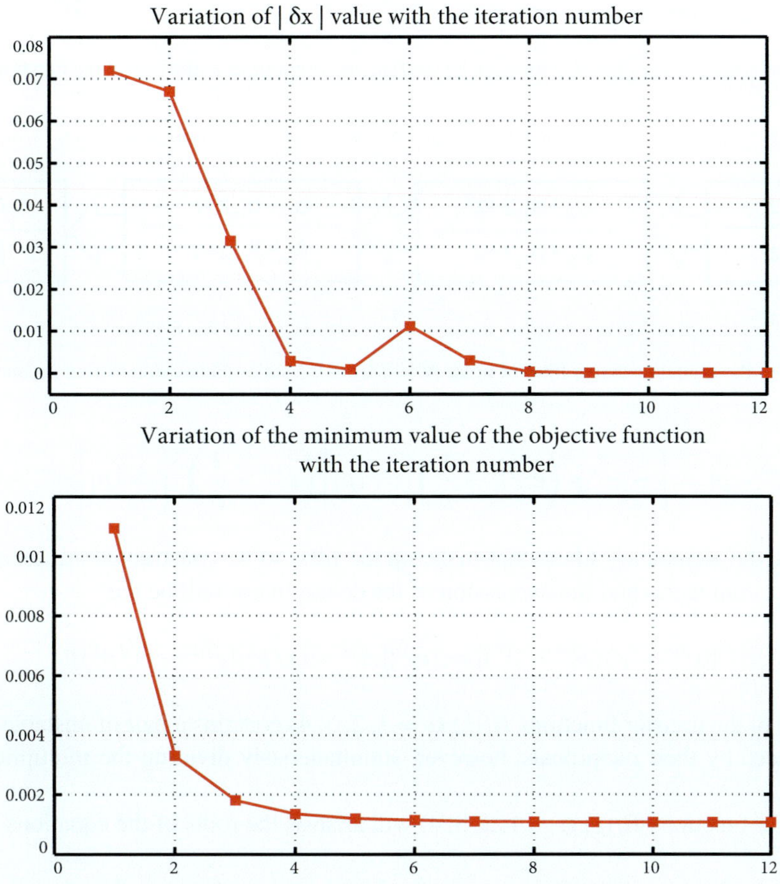

Fig. 12.89 The change in the measure of the difference $|\boldsymbol{\delta x}|$ of the successive vectors and the minimum value of the objective function $\Psi(\boldsymbol{x}_k)_{\text{min}}$ as a function of the iteration number k

$$\mathcal{H}_4(z) = \frac{\alpha_{04} + \alpha_{14}z + z^2}{\beta_{04} + \beta_{14}z + z^2} = \frac{0.223229 + 1.126887z + z^2}{0.079042 + 0.117681z + z^2}.$$

However, before we proceed with the construction of the final filter function, we need to check *if the poles of the four above transfer functions are stable*, that is, their measure is less than one and therefore they are *inside the unit circle*. If this is not the case, we can simply *replace the unstable pole p with its reciprocal* $1/p$, adjusting the value of the multiplicative constant H_0 accordingly and thus constructing a filter that has *the same characteristic loss as the original*. The proof of this property is relatively simple, since for some pole p of the transfer function it holds that

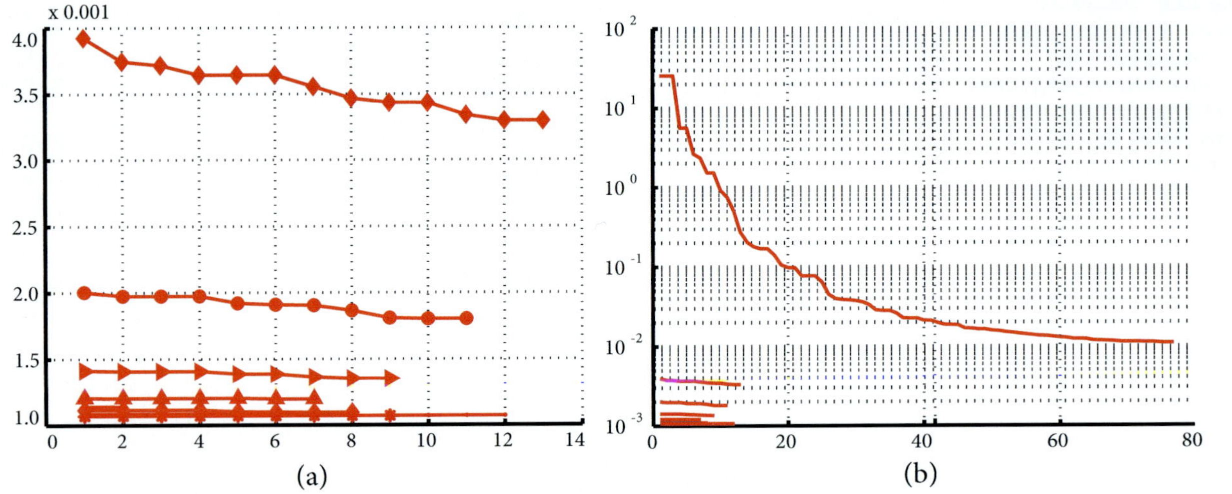

Fig. 12.90 Approach of the minimum of the objective function by the function fminunc as a function of the iteration number k of the optimization algorithm

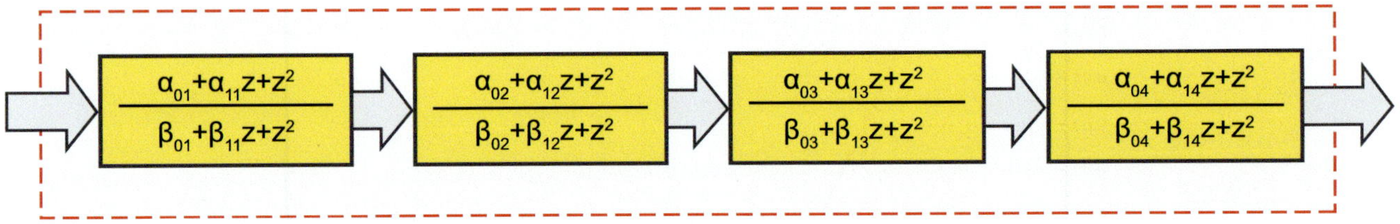

Fig. 12.91 The composition of the digital recursive filter of order $N = 8$ from the series connection of four second-order filters

$$|e^{j\omega} - p| = \left| -e^{j\omega} p \left(e^{-j\omega} - \frac{1}{p} \right) \right| = |p| \left| \left(e^{j\omega} - \frac{1}{p} \right)^* \right| = |p| \left| e^{j\omega} - \frac{1}{p} \right|$$

In the case at hand where the elementary filters that make up the filter to be constructed are second-order filters, the poles appear in pairs of complex conjugates and the expression of the denominator will be the

$$|(e^{j\omega} - p)(e^{j\omega} - p^*)| = |p|^2 |[e^{j\omega} - (1/p)][e^{j\omega} - (1/p^*)]|.$$

But this means that if any of the transfer functions $H_i(z)$ $(i = 1, 2, 3, 4)$ contains a pair of unstable complex conjugate poles, they can simply be replaced by their reciprocals, however, simultaneously dividing the multiplicative constant H_0 by the quantity $|p|^2$.

The poles of the transfer functions $H_i(z)$ $(i = 1, 2, 3, 4)$ will arise as the roots of the equations

$$1.872144 + 1.060528z + z^2 = 0 \quad \text{for the transfer function } \mathcal{H}_1(z),$$

$$1.904817 + 1.075352z + z^2 = 0 \quad \text{for the transfer function } \mathcal{H}_2(z),$$

$$1.918875 + 2.498418z + z^2 = 0 \quad \text{for the transfer function } \mathcal{H}_3(z), \text{ and}$$

$$0.079042 + 0.117681z + z^2 = 0 \quad \text{for the transfer function } \mathcal{H}_4(z)$$

and their values are calculated as

$$z_{1,2}^{\text{align}} = -0.530264 \pm 1.261334j = 1.368263e^{\pm 1.968763j} \quad \text{for the transfer function } \mathcal{H}_1(z),$$

$$z_{1,2}^{(2)} = -0.537676 \pm 1.271110j = 1.380151e^{\pm 1.970969j} \quad \text{for the transfer function } \mathcal{H}_2(z),$$

$$z_{1,2}^{(3)} = -1.249209 \pm 0.598624j = 1.385234e^{\pm 2.694720j} \quad \text{for the transfer function } \mathcal{H}_3(z), \text{ and}$$

$$z_{1,2}^{(4)} = -0.058840 \pm 0.274917j = 0.281114e^{\pm 1.781644j} \quad \text{for the transfer function } \mathcal{H}_4(z).$$

We observe, therefore, that three out of the four elementary filters have *pairs of unstable complex conjugate poles*, which, according to what we mentioned earlier, should be replaced by their inverses, dividing the constant ratio H_0 by the square of their magnitude. Therefore, the new poles of these three filters and the new value of the constant H_0 will have values

$$y_{1,2}^{(1)} = \frac{1}{z_{1,2}^{(1)}} = -0.283239 \mp 0.673737j = 0.730853e^{\mp 1.968763j} \quad \text{and} \quad H_0 \longrightarrow \frac{H_0}{|z_{1,2}^{(1)}|^2} = \frac{0.363767}{(1.368263)^2} = 0.194305$$

$$y_{1,2}^{(2)} = \frac{1}{z_{1,2}^{(2)}} = -0.282271 \mp 0.667313j = 0.724558e^{\mp 1.970969j} \quad \text{and} \quad H_0 \longrightarrow \frac{H_0}{|z_{1,2}^{(2)}|^2} = \frac{0.194305}{(1.380151)^2} = 0.102007$$

$$y_{1,2}^{(3)} = \frac{1}{z_{1,2}^{(3)}} = -0.651011 \mp 0.311966j = 0.721899e^{\mp 2.694720j} \quad \text{and} \quad H_0 \longrightarrow \frac{H_0}{|z_{1,2}^{(3)}|^2} = \frac{0.102007}{(1.385234)^2} = 0.053159.$$

On the other hand, the complex conjugate poles of the transfer function $\mathcal{H}_4(z)$ are stable and do not undergo the slightest change.

In the next step of the process, we need to recalculate the polynomials of the denominator $D_i(z)$ of the functions $H_i(z)$ $(i = 1, 2, 3)$ for the new poles. They will be, therefore,

$$D_1(z) = (z + 0.283239 + 0.673737j)(z + 0.283239 - 0.673737j) = 0.534145 + 0.566478z + z^2$$

$$D_2(z) = (z + 0.282271 + 0.667313j)(z + 0.282271 - 0.667313j) = 0.524983 + 0.564542z + z^2$$

$$D_3(z) = (z + 0.651011 + 0.311966j)(z + 0.651011 - 0.311966j) = 0.521138 + 1.302022z + z^2.$$

Therefore, the final form of the four elementary transfer functions will be

$$\mathcal{H}_1(z) = \frac{\alpha_{01} + \alpha_{11}z + z^2}{\beta_{01} + \beta_{11}z + z^2} = \frac{0.997040 + 1.640905z + z^2}{0.534145 + 0.566478z + z^2}$$

$$\mathcal{H}_2(z) = \frac{\alpha_{02} + \alpha_{12}z + z^2}{\beta_{02} + \beta_{12}z + z^2} = \frac{0.981364 + 1.803155z + z^2}{0.524983 + 0.564542z + z^2}$$

$$\mathcal{H}_3(z) = \frac{\alpha_{03} + \alpha_{13}z + z^2}{\beta_{03} + \beta_{13}z + z^2} = \frac{3.058014 + 2.306290z + z^2}{0.521138 + 1.302022z + z^2}$$

$$\mathcal{H}_4(z) = \frac{\alpha_{04} + \alpha_{14}z + z^2}{\beta_{04} + \beta_{14}z + z^2} = \frac{0.223229 + 1.126887z + z^2}{0.079042 + 0.117681z + z^2},$$

whereas the final vector-solution of the problem is

$$\xi = [\ 0.997040\ 1.640905\ 0.534146\ 0.566478\ 0.981364\ 1.803155\ 0.524984\ 0.564543\ 3.058014\ 2.306290$$
$$0.521138\ 1.302022\ 0.223229\ 1.126887\ 0.079042\ 0.117681\ 0.0.053159\ ]$$

from which it follows that $\alpha_{01} = x[1] = 0.997040$, $\alpha_{11} = x[2] = 1.640905$, $\beta_{01} = x[3] = 0.534146$, $\beta_{11} = x[4] = 0.566478$, $\alpha_{02} = x[5] = 0.981364$, $\alpha_{12} = x[6] = 1.803155$, $\beta_{02} = x[7] = 0.524984$, $\beta_{12} = x[8] = 0.564543$, $\alpha_{03} = x[9] = 3.058014$, $\alpha_{13} = x[10] = 2.306290$, $\beta_{03} = x[11] = 0.521138$, $\beta_{13} = x[12] = 1.302022$, $\alpha_{04} = x[13] = 0.223229$, $\alpha_{14} = x[14] = 1.126887$, $\beta_{04} = x[15] = 0.079042$, $\beta_{14} = x[16] = 0.117681$, $H_0 = x[17] = 0.053159$.

At this point, the solution has been completed and all that remains is the calculation of the transfer function of the composite filter from the defining equation. Taking into account that *the multiplication of two polynomials algebraically is equivalent to the operation of the convolution*,[90] the coefficients of the numerator and the denominator of the transfer function will be calculated from the equation

$$H = H_0 \frac{\left[\left((1, \alpha_{11}, \alpha_{01}) * (1, \alpha_{12}, \alpha_{02})\right) * (1, \alpha_{13}, \alpha_{03})\right] * (1, \alpha_{14}, \alpha_{04})}{\left[\left((1, \beta_{11}, \beta_{01}) * (1, \beta_{12}, \beta_{02})\right) * (1, \beta_{13}, \beta_{03})\right] * (1, \beta_{14}, \beta_{04})}.$$

Replacing the values of the filter coefficients and performing the operations, we finally get

$$H = \frac{[\ 0.053159\ 0.365587\ 1.203611\ 2.369421\ 2.978941\ 2.408685\ 1.202804\ 0.329768\ 0.035516\ ]}{[\ 1\ 2.550724\ 3.738048\ 3.572970\ 2.396580\ 1.224218\ 0.366439\ 0.070727\ 0.115509\ ]}.$$

The above result refers to the coefficients of the powers of the z^{-1}, which, if added, lead to the transfer function

$$H(z) = \frac{0.053 + 0.365z^{-1} + 1.203z^{-2} + 2.369z^{-4} + 2.978z^{-4} + 2.408z^{-5} + 1.202z^{-6} + 0.329z^{-7} + 0.035z^{-8}}{1 + 2.550z^{-1} + 3.738z^{-2} + 3.572z^{-3} + 2.396z^{-4} + 1.224z^{-5} + 0.366z^{-6} + 0.070z^{-7} + 0.115z^{-8}},$$

which is also the final result. The amplitude response of this filter is presented in Fig. 12.92, from which it can be seen that although it has many of the desired characteristics, it is not the best possible choice. If this happens, the function, as we have already mentioned, repeatedly tries to construct better filters using different (random) values for the vector $\boldsymbol{x}_0$, and if it fails to do so within the attempts set by the user, *it returns the parameters of the best filter it constructed*. The reader is invited as an exercise to calculate the values of the parameters A_p and A_a of the filter and to check if they meet the design requirements. Let it be noted that the function does not return all the information that has been presented in the previous description (printing all this information requires intervention in the source code and therefore good programming skills, as well as good knowledge of MATLAB) and the only thing it prints are the final values of the coefficients listed above. The calculation of the frequency response of the filter is based on the use of the function freqz. The way to use it can be found in MATLAB's documentation files.

12.14 Description of Digital Filters with Signal-Flow Graphs

The signal-flow graphs for the case of continuous-time systems are presented in detail in Sect. 5.10 (see page 455) and let us repeat at this point, for the sake of the autonomy of the text, the basic definitions, since what we mentioned there also applies to discrete-time systems. A signal flow graph is an *alternative representation of a difference equation*, which is *completely equivalent* to the structural diagrams we presented in a previous section. In reality, the only apparent difference between these two diagrammatic techniques is that in the signal-flow graphs, *a node represents both an adder and a branching point*, just as happens with the corresponding structures used in the field of continuous-time.

As already mentioned, a signal-flow graph is nothing other than a *network* consisting of *directed branches*, that is, branches in which the signal that is supposed to flow through this network *moves only in one direction*. Apart from branches, a signal-flow graph also contains *nodes*, which are interconnected via the branches, just as happens in an electrical circuit, with each branch w being associated with a weight value w_k or more precisely $w_k[n]$, in order to emphasize the fact that in digital filters the variables of the nodes of such a graph constitute *discrete sequences*. In order to show that a branch connects node

[90] Considering two polynomials

$$A(x) = \sum_{i=0}^{N} \alpha_i x^i \quad \text{and} \quad B(x) = \sum_{i=0}^{M} \beta_i x^i$$

their product is the polynomial

$$C(x) = \sum_{k=0}^{N+M} \gamma_k x^k, \quad \text{where} \quad \gamma_k = \sum_{i=1}^{k} \alpha_i \beta_{k-i} \quad 0 \le k \le M + N.$$

But the last relation is recognized as *the defining equation of convolution*, so we can state the claim that the vector $(\gamma_0, \gamma_1, \dots, \gamma_{M+N})$ is the convolution of the vectors $(\alpha_0, \alpha_1, \dots, \alpha_N)$ and $(\beta_0, \beta_1, \dots, \beta_M)$. Therefore, from the algebraic side of the subject, *the convolution of two vectors is equivalent to the multiplication of the polynomials, the coefficients of which are the elements of the vectors that are convolved.*

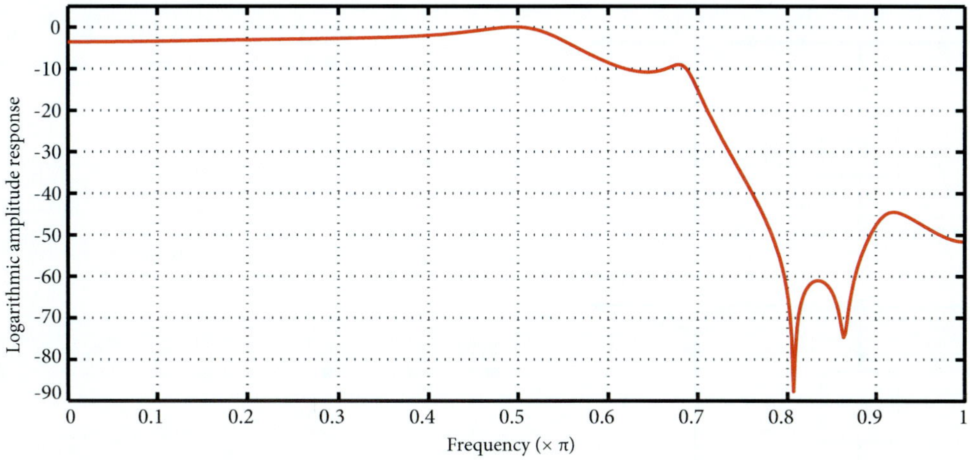

Fig. 12.92 The amplitude response of the digital recursive filter of Example 16

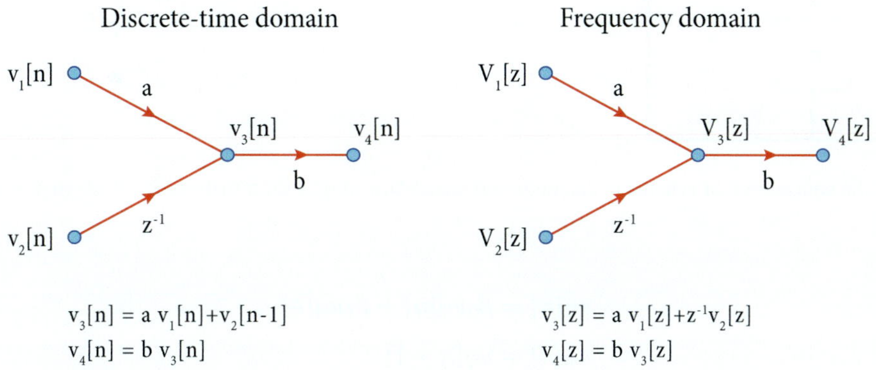

Fig. 12.93 Examples of signal flow graphs in the fields of time and frequency

x with node y, we use the notation (x, y). Taking into account that *a branch allows the passage of a signal through it, which in general can be modified within the branch*, it is fully described by an input signal and an output signal, with the direction of the signal being indicated by a small arrow. The most representative category of signal-flow graphs that we examine here are the *linear graphs*, in which the output of each branch results from applying to the input of the branch a *linear transformation*, with the simplest case of such a transformation obviously being the multiplication of the branch input by a constant, the value of which is written next to the arrow. If a branch is not accompanied by such a symbol, it is implied that the input signal *is not modified* during its passage through the branch (in other words, the branch output signal is identical to the input signal) and therefore, the branch applies to the input signal the so-called *identity transformation*, which leaves the input signal unchanged. If there are no branches leading to a node, but only branches that start from it, it is characterized as a *source node*; on the contrary, if for a node there are no branches that start from it but only branches that end up in it, then it is characterized as a *sink node*. The above concepts are diagrammatically demonstrated in Fig. 12.93, from which it is found that these graphs can be used in the fields of both discrete-time and frequency. In this figure, the three fundamental operations that allow the construction of any linear system described by a linear difference equation with constant coefficients, are the *addition of signals*, the *multiplication of a signal by a constant*, and the operation of *time delay*.

The equivalence of schematic diagrams and signal-flow graphs is demonstrated in Fig. 12.94, with the left image showing the schematic diagram, which refers to the *normal form* of the simple system described by a first-order difference equation (see Fig. 7.9 on page 611) and the right image showing the equivalent signal-flow graph. From the study of the signal-flow graph, it is found that the difference equations represented in it are the following:

$$w_1[n] = \alpha w_4[n] + x[n]$$

$$w_2[n] = w_1[n]$$

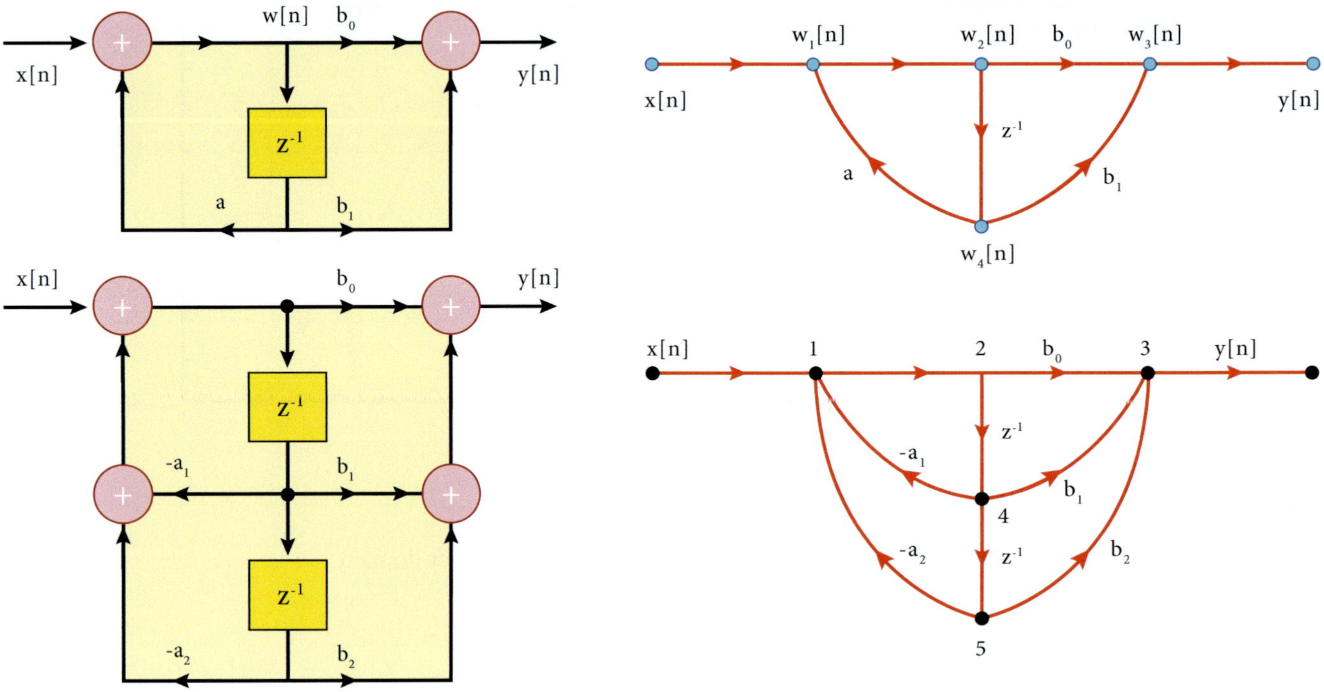

Fig. 12.94 Example of the equivalence of schematic diagrams and signal-flow graphs for a first-order system (top images) and a second-order system (bottom images)

$$w_3[n] = \beta_0 w_2[n] + \beta_1 w_4[n]$$
$$w_4[n] = w_2[n-1]$$
$$y[n] = w_3[n].$$

The construction of these equations is generally easy and arises from the *simple visual examination* of the graph. For example, the node $w_1[n]$ receives two incoming signals, the first of which is the signal $x[n]$, whereas the second is the signal $w_4[n]$, multiplied by the coefficient α. These two signals at the node $w_1[n]$ are added together (note how in a signal flow graph, a node represents both branching points and signal adders, whereas in schematic diagrams, an additional symbol is used for signal adders). Therefore, $w_1[n] = \alpha w_4[n] + x[n]$. Between the node $w_1[n]$ and the node $w_2[n]$, no kind of signal processing takes place (since the arrow of the branch defined by these two nodes is not accompanied by any label) and therefore, $w_2[n] = w_1[n]$. The same applies to the branch defined by the nodes $w_3[n]$ and $y[n]$ and therefore $y[n] = w_3[n]$. Between the nodes $w_2[n]$ and $w_4[n]$, a time-delay element is interposed and since the direction of the arrow is from top to bottom, $w_4[n] = w_2[n-1]$. Finally, the node $w_3[n]$ has as input signals the signal $w_2[n]$ multiplied by the coefficient β_0 and the signal $w_4[n]$ multiplied by the coefficient β_1. Thus, $w_3[n] = \beta_0 w_2[n] + \beta_1 w_4[n]$. In this way, we construct all the difference equations that are implemented by this graph. Notice that for each node of the graph, we construct a difference equation, in the same way as is done in an electrical circuit. Speaking generally, the above difference equations should be used in a specific order, since if the result of one equation is used in another equation, the first of the two equations must of course be evaluated first. Let us also mention that because in practice integrated circuits that perform addition of signals accept as input two signals and not more, the input signals to a node of a signal-flow graph are at most two.

In many cases, it is possible to *combine* the above equations with each other and therefore, to simplify the description of the system. For example, the first, second, and fourth equations can be combined into the equation

$$w_2[n] = \alpha w_2[n-1] + x[n],$$

whereas the third and fifth equations can be combined into the equation

$$y[n] = \beta_0 w_2[n] + \beta_1 w_2[n-1].$$

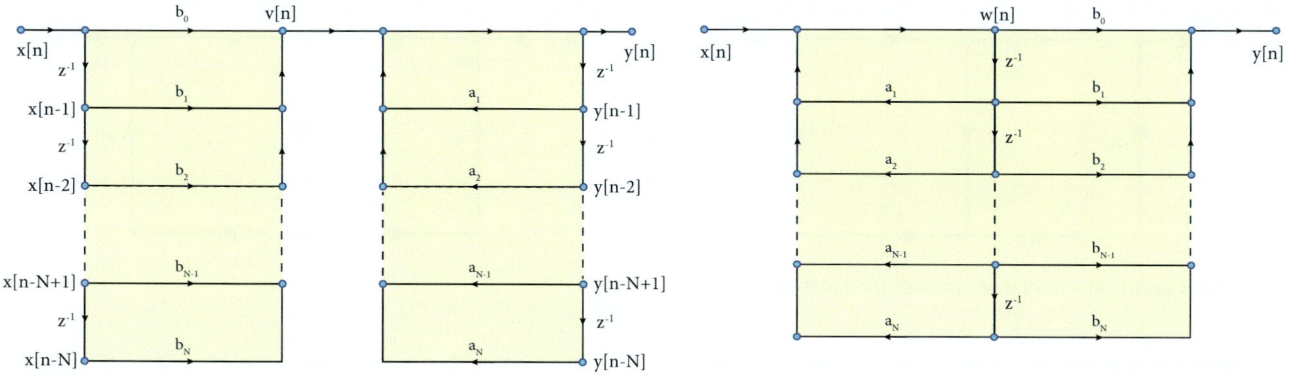

Fig. 12.95 The non-normal form (left image] and the normal form (right image] for the general case of a discrete recursive system LTI using signal-flow graphs for the case $M = N$

Observing that in the first of the above equations the input signal $x[n]$ appears, whereas in the second equation the output signal $y[n]$ appears, we can calculate the transfer function of the system by resorting to the $\mathcal{Z}$ transform method. Indeed, from the first equation we have

$$\mathcal{W}_2(z) = \alpha z^{-1}\mathcal{W}_2(z) + \mathcal{X}(z) \quad \Longrightarrow \quad (1 - \alpha z^{-1})\mathcal{W}_2(z) = \mathcal{X}(z) \quad \Longrightarrow \quad \mathcal{W}_2(z) = \frac{\mathcal{X}(z)}{1 - \alpha z^{-1}}.$$

Taking the $\mathcal{Z}$ transform of the second equation and using the expression for the function $\mathcal{W}_2(z)$ that we constructed previously, we have

$$\mathcal{Y}(z) = \beta_0\mathcal{W}_2(z) + \beta_1 z^{-1}\mathcal{W}_2(z) = (\beta_0 + \beta_1 z^{-1})\mathcal{W}_2(z) = \frac{\beta_0 + \beta_1 z^{-1}}{1 - \alpha z^{-1}}\mathcal{X}(z)$$

and consequently, the transfer function of the system will have the form

$$\mathcal{H}(z) = \frac{\mathcal{Y}(z)}{\mathcal{X}(z)} = \frac{\beta_0 + \beta_1 z^{-1}}{1 - \alpha z^{-1}},$$

a result that is the same as the one we have concluded, setting $\alpha = -\alpha_1$. Having calculated the transfer function, we can calculate the impulse response and the difference equation of the system in the way we saw in the previous chapters.

For every schematic diagram that is formulated in either of the two straight forms, the equivalent signal-flow graph can be constructed, which is simpler and more easily manageable, and for this reason all alternative forms of representing a system that will be presented subsequently do not use schematic diagrams but rather signal-flow graphs. Figure 12.95 depicts the signal-flow graphs corresponding to the straight forms I and II that we previously presented for the special case $M = N$.

Inversion of Signal-Flow Graphs

An interesting property of signal-flow graphs, which is characterized by widespread practical application, is that *they remain unchanged under the operation of inversion*, which consists of (a) *reversing the directions of all branches*, (b) *replacing branching nodes with sum nodes and vice versa*, and (c) *switching the input and output nodes* (which is simply due to the convention that the input is on the left and the output is on the right of the circuit). According to the famous *theorem of the transposition*,[91] the signal flow graph that arises after performing the above processes *is equivalent to the original*, in the sense that they describe the same transfer function, that is, *the same* system.

[91] For the proof of the transposition theorem, see Claasen T.A. & Mecklenbräuker W.F., *On the Transposition of Linear Time-Varying Discrete-Time Networks and Its Application to Multirate Digital Systems*; Philips J. Res., 23:78–102, 1978; or Crochiere R.E. & Rabiner L.R., *Multirate Digital Signal Processing*, Prentice Hall, Englewood Cliffs, New Jersey, 1983, whereas for a proof based on the concept of state space see Jackson L.B., *On the Interaction of Roundoff Noise and Dynamic Range in Digital Filters*, Bell System Technical Journal, 49:159–184, February 1970a and Jackson L.B., *Digital Filters and Signal Processing*, Kluwer Academic Publishers, Boston, 3rd edition, 1996.

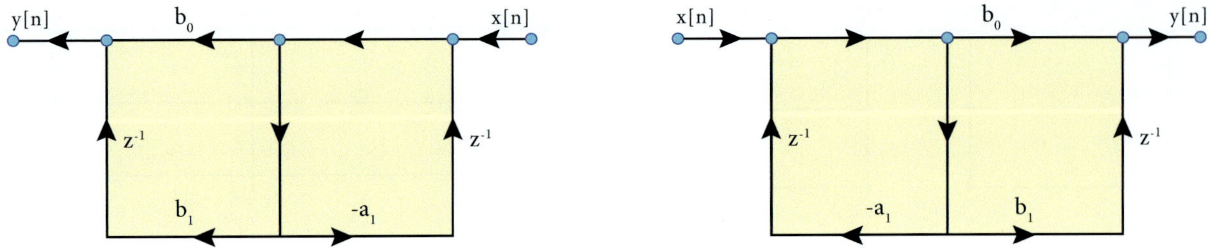

Fig. 12.96 Construction of inverse structure for a first-order system

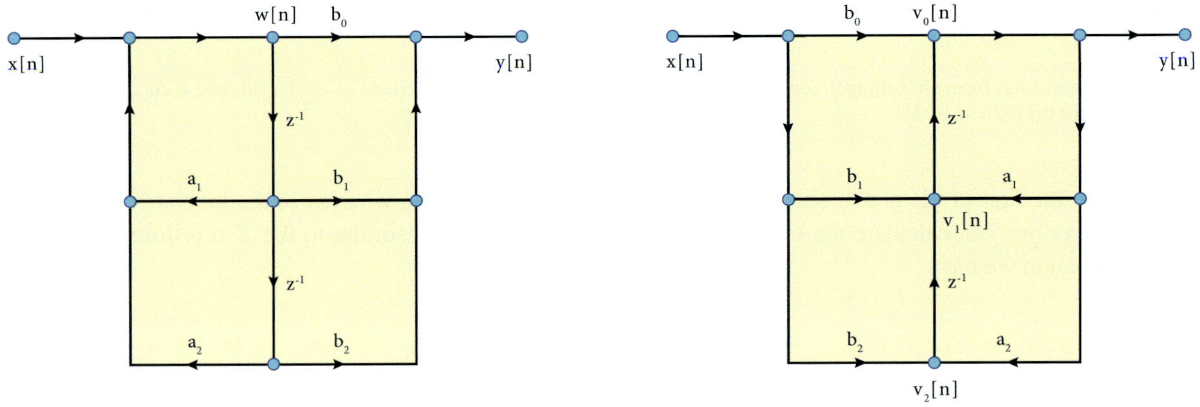

Fig. 12.97 Construction of inverse structure for a second-order system, which is also the one used for the construction of more complex systems in series and parallel structure—see the relevant sections

The process of reversing the signal flow graph for a first-order IIR system is depicted in Fig. 12.96, whereas the corresponding process for a second-order IIR system is depicted in Fig. 12.97, with the final result in each case being depicted in the right image of the figure. Note that the number of additions, multiplications, as well as memory positions required for the calculation of the system's output is the same as that characterizing the direct structure. In order to prove that the reverse diagram that has emerged in the above way is equivalent to the original, let us calculate the transfer function of the system it describes. Considering, for example, the case of the first-order system, the difference equations corresponding to the left and right sum nodes are

$$v[n] = x[n] - \alpha_1 v[n-1],$$

from which it follows that

$$\mathcal{V}(z) = \frac{1}{1+\alpha_1 z^{-1}} X(z)$$

and the

$$y[n] = \beta_0 v[n] + \beta_1 v[n-1],$$

from which it follows that

$$\mathcal{Y}(z) = (\beta_0 + \beta_1 z^{-1}) V(z).$$

Consequently, the transfer function of the system described by the *inverse* signal-flow diagram will be

$$\mathcal{H}(z) = \frac{\mathcal{Y}(z)}{\mathcal{X}(z)} = \frac{\mathcal{Y}(z)}{\mathcal{V}(z)} \frac{\mathcal{V}(z)}{\mathcal{X}(z)} = \frac{\beta_0 + \beta_1 z^{-1}}{\alpha_0 + \alpha_1 z^{-1}}.$$

But this is nothing more than the transfer function of the system described by the original signal-flow diagram and therefore, the two systems are indeed equivalent. In exactly the same way, for the case of Fig. 12.97, the difference equations of the original system are[92]

$$w[n] = \alpha_1 w[n-1] + \alpha_2 w[n-2] + x[n]$$
$$y[n] = \beta_0 w[n-1] + \beta_1 w[n-1] + \beta_2 w[n-2]$$

whereas the difference equations of the inverse system are

$$y[n] = b_0 x[n] + v_1[n-1]$$
$$v_1[n] = \alpha_1 y[n] + \beta_1 x[n] + v_2[n-1]$$
$$v_2[n] = \alpha_2 y[n] + \beta_2 x[n].$$

Working in the same way, we can show that these two sets of equations are *equivalent* in the sense that they lead to the same transfer function and therefore describe the same system. Indeed, calculating the transforms $\mathcal{Z}$ of the above expressions, we have

$$\mathcal{V}(z) = b_0 \mathcal{X}(z) + z^{-1}\mathcal{V}_1(z)$$
$$\mathcal{V}_1(z) = a_1 \mathcal{Y}(z) + b_1 \mathcal{X}(z) + z^{-1}\mathcal{V}_2(z)$$
$$\mathcal{V}_2(z) = a_2 \mathcal{Y}(z) + b_2 \mathcal{X}(z).$$

Replacing the third equation in the second, we are led to the expression

$$\mathcal{V}_1(z) = \alpha_1 \mathcal{Y}(z) + \beta_1 \mathcal{X}(z) + z^{-1}\alpha_2 \mathcal{Y}(z) + z^{-1}\beta_2 \mathcal{X}(z),$$

which if in turn replaced in the first equation gives us

$$\mathcal{Y}(z) = \beta_0 \mathcal{X}(z) + z^{-1}\alpha_1 \mathcal{Y}(z) + z^{-1}\beta_1 \mathcal{X}(z) + z^{-2}\alpha_2 \mathcal{Y}(z) + z^{-2}\beta_2 \mathcal{X}(z).$$

It will therefore be

$$\mathcal{Y}(z)(1 - \alpha_1 z^{-1} - \alpha_2 z^{-2}) = \mathcal{X}(z)(\beta_0 + \beta_1 z^{-1} + \beta_2 z^{-2})$$

and therefore

$$\mathcal{H}(z) = \frac{\mathcal{Y}(z)}{\mathcal{X}(z)} = \frac{\beta_0 + \beta_1 z^{-1} + \beta_2 z^{-2}}{1 - \alpha_1 z^{-1} - \alpha_2 z^{-2}},$$

which is nothing but the transfer function of the system described by the original signal-flow diagram. As for the equations that describe the inverse form II for the general case, they have the form

$$y[n] = v_1[n-1] + \beta_0 x[n]$$
$$v_k[n] = v_{k+1}[n-1] - \alpha_k y[n] + \beta_k x[n], \quad k = 1, 2, \ldots, N-1$$
$$v_N[n] = \beta_N x[n] - \alpha_N y[n],$$

[92] Notice that the diagrams of Figs. 12.96 and 12.97 use different symbolism regarding the signs and the use of variables, which is deliberately maintained, in order to demonstrate the various conventions regarding the construction of these diagrams, which are used in the literature. It is not difficult to understand that the issue in this case is not so much the convention followed regarding the construction of these structures but rather the correct conversion of these graphs into difference equations so that we finally get the correct result.

which, if combined, lead to the well-known general difference equation of the system and therefore to the same transfer function. If we set $\alpha_k = 0$ ($k = 1, 2, \ldots, N - 1$) in the above equations, they simplify to the corresponding equations of FIR-type filters, which are

$$\begin{aligned} y[n] &= v_1[n-1] + \beta_0 x[n] \\ v_k[n] &= v_{k+1}[n-1] + \beta_k x[n], \quad k = 1, 2, \ldots, N-1 \\ v_N[n] &= \beta_N x[n] \end{aligned}$$

The above structures are also used in the case of FIR-type systems whose difference equation and impulse response have the form

$$y[n] = \sum_{k=0}^{M} b_k x[n-k] \qquad \text{and} \qquad h[n] = \begin{cases} b_n & n = 0, 1, \ldots, M \\ 0 & \text{otherwise} \end{cases}$$

Taking into account that in FIR-type filters, the denominator of the transfer function is equal to the *unit*, it is not difficult to understand that *the direct form I and the direct form II for these systems lead to the same result*, with the structure that emerges being presented in Fig. 12.98. In this figure, the top and middle images depict the direct and inverse forms of such a filter respectively. From the figure we observe that this form is characterized by the existence of a multitude of unit time-delay elements, which form the so-called delay line, with the values that emerge being linearly combined among them to form the input, with the coefficients of this linear combination being the finite number of samples of the system's impulse response. The corresponding schematic diagram for the direct form I is depicted in the bottom image of Fig. 12.98, whereas the schematic diagram for the inverse form is constructed in the way we presented previously.

From the study of the above structures, we observe that these emerge directly from the convolution equation

$$y[n] = \sum_{k=0}^{M-1} h[k]x[n-k],$$

which correlates the input and output of the system, with the involved computation process requiring $M-1$ memory positions for storing the $M - 1$ previous samples and with the computation of the value of each sample $y[n]$ requiring the execution of M multiplications and $M - 1$ additions.

In the special case where the FIR system is characterized by a *generalized linear phase*, as we comment in the relevant section, the impulse response is characterized *either by the condition of symmetry, or by the condition of antisymmetry*, that is, it is such that $h[n] = \pm h[M - 1 - n]$ leading to the four types of FIR filters that are known as types I, II, III, and IV filters (see Sect. 12.11 on page 1223). In this case, the use of the existing symmetry allows us to construct simpler direct structures for these systems, which lead approximately to the *halving* of the number of multiplications required for the calculation of each value of the output signal. A typical example of a direct structure for an even and symmetric filter is depicted in Fig. 12.99, with the left image showing the schematic diagram and the right image the corresponding signal flow graph. Similar diagrams can be constructed for the other three filter categories.

Feedback Loops

The *feedback loops* are a characteristic feature of almost all signal-flow graphs. A feedback loop is defined *as a closed path, which starts from a node of the graph and returns back to the same node*, allowing the traversal of a branch for a *specific* time, which is indicated by an *arrow*. These feedback loops are *a necessary but not sufficient condition* for the creation of impulse responses of infinite length, which characterize IIR systems. Indeed, if a signal-flow graph does not have feedback loops, *the greatest possible delay that can appear in such a system equals the number of different delay elements and corresponds to a path that passes through all these delay elements*. In this case, the transfer function will contain *only* zero values (except for the poles corresponding to the position $z = 0$), with the number of these zero values *not being greater* than the number of time-delay elements. In order to achieve greater time delays and eventually lead to impulsive responses with infinite length, the signal-flow graph must necessarily have feedback loops.

Examples of feedback loops are shown in the left and right images of Fig. 12.100. The diagram of the left image represents the signal-flow diagram for a discrete-time system that is described by the difference equation

$$y[n] = ay[n-1] + x[n]$$

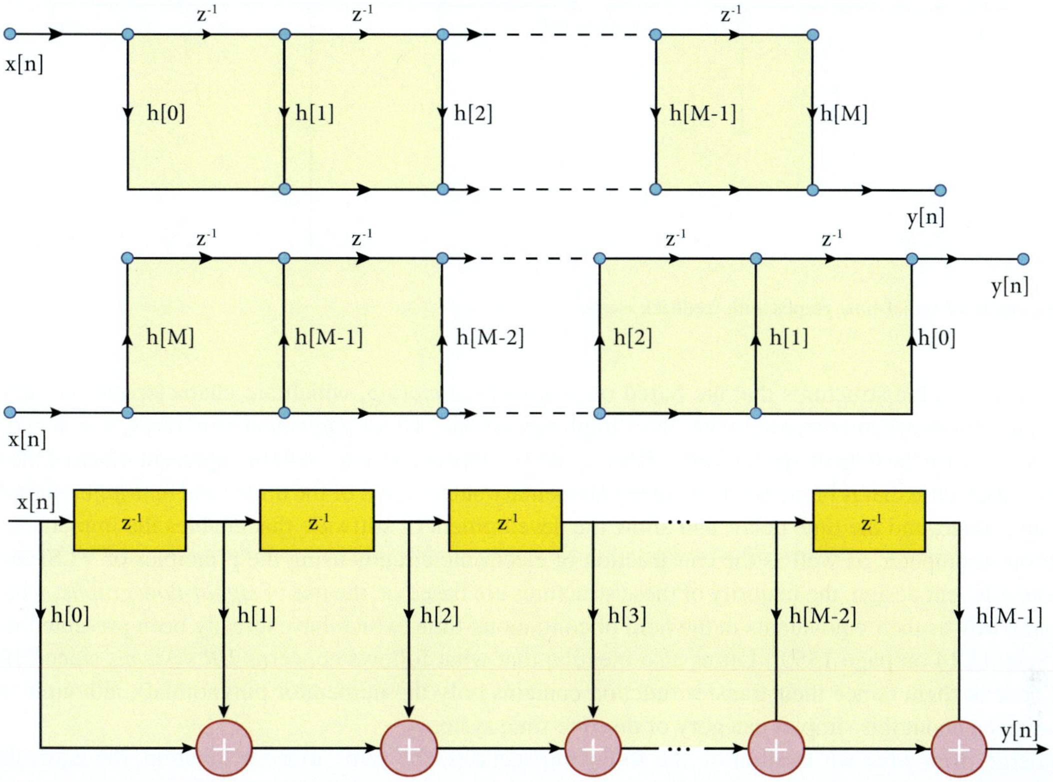

Fig. 12.98 Structures for FIR-type systems. Signal-flow graphs for the direct form (top image) and for the reverse form (middle image), as well as a schematic diagram for the direct form (bottom image)

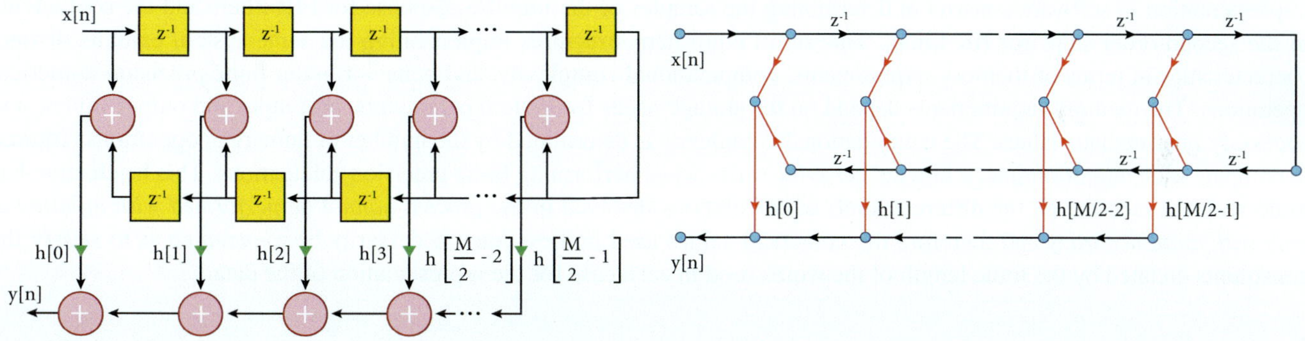

Fig. 12.99 Schematic diagram and signal-flow graph for the direct form of an even and symmetric digital filter with FIR

and its impulse response is $h[n] = \alpha^n u[n]$, whereas the diagram of the right image refers to the system with transfer function

$$H(z) = \frac{1 - a^2 z^{-2}}{1 - az^{-1}} = \frac{(1 - az^{-1})(1 + az^{-1})}{1 - az^{-1}} = 1 + az^{-1}.$$

12.15 Implementations of Digital Filters

As we commented in detail in Sect. 7.10 (see page 608), although the direct implementations are *the simplest and most direct* in their construction of discrete-time systems, however, they are *almost never* used, owing to the many limitations that characterize their use. For this reason, in practice, we resort to the use of *more efficient* but of course completely *equivalent* implementations that include among others the *serial structures* or *cascade structures*, the *parallel structures*, and the *grid structures*, which are characterized by *resilience* in terms of the finite length of the representation of the data used in the

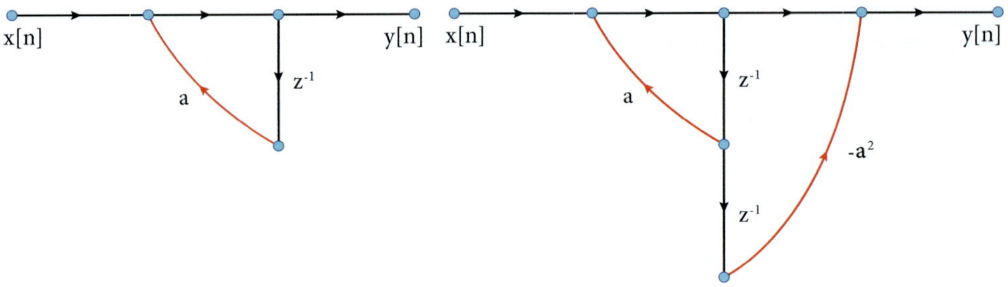

Fig. 12.100 Examples of signal-flow graphs with feedback loops

calculations, as well as the structures that are based on *frequency sampling*, which are characterized by high *efficiency* in terms of the calculations when compared with other implementations. These equivalent structures, which can be converted from one to another with the help of specialized software, can be considered as schematic representations of the system under consideration in each case that is being built, using the elementary subsystems of the *adder* and *multiplier* signals, the scaling of the signal amplitude, and the time delay, and allow the development of software that enables the implementation and use of the system on a computer, as well as the construction of electronic circuits using the principles of VLSI technology. For easier and more efficient design, the majority of these structures are based on the use of *signal-flow graphs*, which are used in exactly the same way as their equivalents in the field of continuous-time, which have already been presented in the previous section (see Sect. 12.14 on page 1350). Let us also mention that what follows concerns *IIR systems*, since FIR systems are but a special case of them (since their transfer function contains only the numerator polynomial), although certainly some details will be given about this simpler category of discrete-time systems.

As will emerge from what we mention in the following sections, the same transfer function, the equivalent, the same impulse response, can be described by a set of *difference equations*, which describe the same input/output correlation and offer an algorithm or alternatively a structure, which ultimately leads to the implementation of the system either in hardware or in software, with the implementation in hardware leading to a real circuit that implements the behavior of the filter and the implementation in software consists in determining the samples of the impulse response for FIR filters and the coefficients of the second-order steps for IIR filters. This set of equivalent structures implementing the same system exhibits distinct characteristics in terms of memory requirements, computational complexity, and behavior under finite precision numerical operations. The memory requirements depend on the storage needs for system parameters, past input and output values, and necessary intermediate values. The computational complexity is determined by the number of numerical operations required to compute each output sample, while the behavior varies when performing finite precision calculations. This last feature has to do with the fact that all the different kinds of calculations involved in the process *cannot be performed with infinite but only with finite accuracy* and therefore the numerical values used in these must necessarily be *rounded* so as to satisfy the constraints dictated by the finite length of the words used in each case for the representation of the data.

12.15.1 Serial Form or Cascade Form

The *waterfall form* arises by factoring the polynomials of the variable z^{-1} found in the numerator and in the denominator of the transfer function, that is, formulating the transfer function in the form

$$H(z) = A\frac{\prod_{k=1}^{M_1}(1-f_kz^{-1})\prod_{k=1}^{M_2}(1-g_kz^{-1})(1-g_k^*z^{-1})}{\prod_{k=1}^{N_1}(1-c_kz^{-1})\prod_{k=1}^{N_2}(1-d_kz^{-1})(1-d_k^*z^{-1})}, \tag{12.313}$$

where $M = M_1 + 2M_2$ and $N = N_1 + 2N_2$. If the value of N_1 (or correspondingly, if the order of M_1) is an odd number, we must include a pole (or correspondingly, a zero value at the position $z = 0$), since the analysis of the system into a multitude of second-order systems requires these parameters to have even values. In the above expression, the first-order factors contain *real zeros* located at the positions f_k and *real poles* located at the positions c_k, whereas the second-order factors contain *complex conjugate pairs of zero values* located at the positions g_k and g_k^* and *complex conjugate pairs of poles*, located at the positions d_k and d_k^*. The formulation of the transfer function in this form allows us to represent the structure of the system as a *cascade structure*, which consists of the series connection of first- and second-order systems. It should be noted that the method offers considerable freedom of choice in the way in which we carry out the above analysis, as well as in the way in which we connect the above first- and second-order systems.

If we combine between them the terms of the real and complex conjugate values into second-order terms with real coefficients, we can reformulate the above transfer function in the form

$$H(z) = G \prod_{k=1}^{K} \frac{b_{0k} + b_{1k} z^{-1} + b_{2k} z^{-2}}{1 - a_{1k} z^{-1} - a_{2k} z^{-2}} = G \prod_{k=1}^{K} \mathcal{H}_k(z), \tag{12.314}$$

where $K = \lfloor (N+1)/2 \rfloor$ (note, however, that this is only true for the case $M \leq N$, which is the one that interests us in practice, whereas if $M = N$, then $N = M = 2K$). In the above expression, the magnitude G expresses the total gain of the system, whereas the transfer functions $\mathcal{H}_k(z)$ $(k = 1, 2, \ldots, K)$ describe second-order systems with gain values A_k $(k = 1, 2, \ldots, K)$ such that $A_1 A_2 \cdots A_K = A$. If the value of N is *odd*, then one of the coefficients α_{2k} and α_{2k} $(k = 1, 2, \ldots, K)$ will have a value equal to zero, whereas if $N > M$, then the numerator coefficients in some of the transfer functions $\mathcal{H}_k(z)$ will also have a zero value. The obvious advantage of this form is that since the system is broken down into a set of identical second-order systems, with different, of course, coefficients, it can be implemented very easily and efficiently using *the same circuit* (for hardware implementations) or function (for software implementations), whereas additionally, for its implementation, it requires *significantly fewer multiplications* for the zeros that are on the periphery of the unit circle and therefore, it is suitable for a variety of band-pass and band-stop filters. As for the second-order systems that are connected in series to form the composite system, these can be used implemented both in direct form I and in direct form II, although the latter is usually preferred (and indeed the reverse form), as this leads, as we commented, to a minimum number of time-delay elements. Note that in order to form structures, the transfer functions of which are characterized by real coefficients, *the poles and zero values must be combined with their complex conjugate quantities* (unless we refer to zero values and poles that lie on the real axis, so that they can be combined in any way), although subsequently, a pair of complex conjugate poles and a pair of complex conjugate zero values can be combined in an arbitrary way to form a second-order system described by some of the functions $\mathcal{H}_k(z)$ (with the representations constructed from these pairs of poles and zero values appearing in the numerator and the denominator of this rational transfer function respectively). As for the way in which these elementary second-order subsystems are connected in series with each other, this can again be *arbitrary*, thus leading to many different variations of the composite system. It is interesting to note that although all these different variations lead to systems that are *theoretically equivalent*, this is true only when using numerical infinite accuracy. Conversely, in the case where the accuracy of the calculations is *finite*, as naturally happens in practice, the way of sequentially connecting second-order structures can significantly affect the behavior of the system.

Figure 12.101 presents the *cascade structure* for a sixth-order system. In the general case, the difference equations for the cascade structure of a system of order N that is analyzed into second-order system structures are

$$\begin{aligned}
y_0[n] &= x[n] \\
w_k[n] &= a_{1k} w_k[n-1] + a_{2k} w_k[n-2] + y_{k-1}[n], \quad k = 1, 2, \ldots, N_s \\
y_k[n] &= b_{0k} w_k[n-1] + b_{2k} w_k[n-2] + y_{k-1}[n], \quad k = 1, 2, \ldots, N_s \\
y[n] &= y_{N_s}[n].
\end{aligned}$$

In the special case of FIR systems, which as is well known, are characterized by a transfer function with a denominator equal to the unit, the situation is certainly simpler. In this case, the transfer function of the system is factored as

$$\mathcal{H}(z) = G \prod_{k=1}^{K} \mathcal{H}_k(z),$$

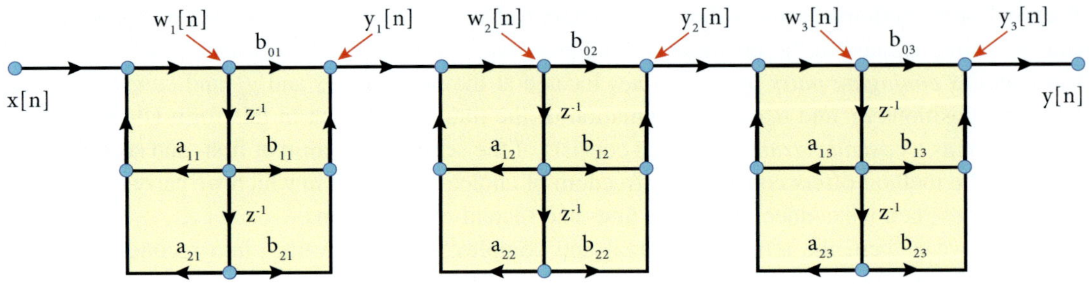

Fig. 12.101 Cascade structure for a sixth-order system characterized by the sequential connection of second-order system structures for each of which the normal form is used

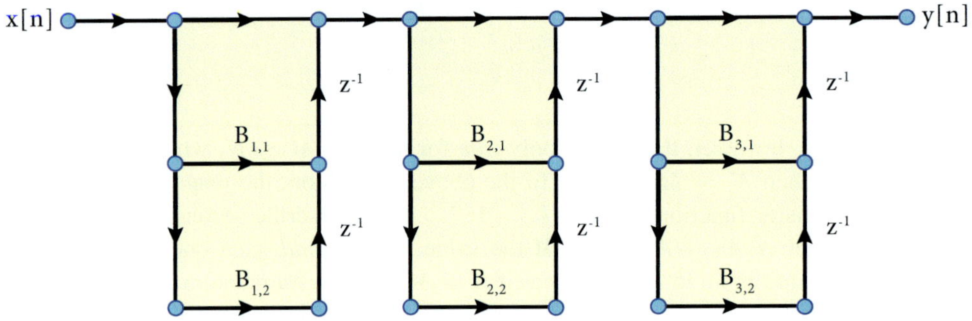

Fig. 12.102 Cascade structure for a sixth-order FIR system, characterized by the series connection of second-order system structures for each of which the inverse normal form is used

where

$$\mathcal{H}_k(z) = \beta_{k0} + \beta_{k1} z^{-1} + \beta_{k2} z^{-2} \ (k = 1, 2, 3, \ldots, K),$$

whereas the condition $M = 2K$ holds. If the value of M is an odd number, one of the coefficients β_{k2} will have a value equal to zero. The value of the coefficient G can be associated with one of the filters that will be connected in series, or distributed to all these filters, that is, to assign to each of them a different value, with the product of all these values being equal to the value of G. As in the general case, these filters will be connected in series, with the way of their connection being characterized by all that we aforementioned in the previous general description. In Fig. 12.102 the series connection of three second-order systems is depicted, which are implemented according to the *inverse normal form* and lead to an FIR filter with order value $M = 6$.

Finally, if the FIR systems we consider are characterized by a *generalized linear phase* and therefore, belong to one of the four types of filters, I, II, III, and IV, things are simplified even more. In this case, the factorization of the transfer function of the system $\mathcal{H}(z)$ and the combination of the individual factors, should be implemented in such a way as to result in structures that are characterized by the symmetry of the appropriate type. Remember, in these filters, the *symmetry* that characterizes the impulse response $h[n]$ implies that the corresponding transfer function $\mathcal{H}(z)$ will also have its own symmetry, mainly as far as the positions of the zero values are concerned, which as we have seen, are in mirror positions on the complex plane. Specifically, as we have shown, if g_k and g_k^* constitute a pair of complex conjugate zero values of the transfer function, then the same applies to $1/g_k$ and $1/g_k^*$ and consequently, in the general case, the zero values should be considered *in quadruples*, unless they are on the periphery of the unit circle or on the real axis, where they are considered in pairs. Based on the above symmetry, we can further simplify the problem, constructing subsystems for serial connection, not second-order, as we did previously, but *fourth-order*, with each of the functions $\mathcal{H}_k(z)$ we defined previously to have the form

$$\mathcal{H}_k(z) = c_{k0}(1 - g_k z^{-1})(1 - g_k^* z^{-1})(1 - z^{-1}/g_k)(1 - z^{-1}/g_k^*) = c_{k0} + c_{k1} z^{-1} + c_{k2} z^{-2} + c_{k1} z^{-3} + c_{k0} z^{-4}$$

and with the coefficients $\{c_{k1}\}$ and $\{c_{k2}\}$ being a function of the zeros g_k. All we have to do next is to combine the two pairs of poles to finally construct a fourth-order subsystem. In this way, the number of required multiplications is further reduced by

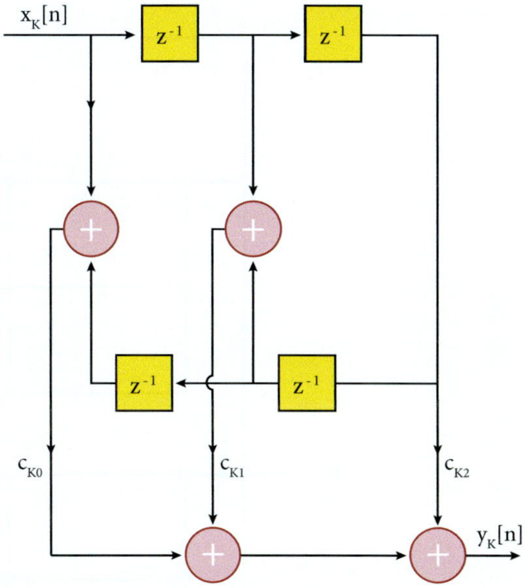

Fig. 12.103 The basic fourth-order structural unit used in the cascade structure of a generalized linear phase FIR system

a factor equal to two, that is, it is *halved*. The basic fourth-order structural unit used in the cascade structure of a generalized linear phase FIR system is depicted in Fig. 12.103.

12.15.2 Parallel Form

In the parallel form, instead of resorting to the factorization of the polynomials of the numerator and the denominator, as we do in the cascade form, we develop the transfer function into a *sum of partial fractions*, that is, we formulate it in the form

$$H(z) = \sum_{k=0}^{N_p} C_k z^{-k} + \sum_{k=1}^{N_1} \frac{A_k}{1 - c_k z^{-1}} + \sum_{k=1}^{N_2} \frac{B_k(1 - e_k z^{-1})}{(1 - d_k z^{-1})(1 - d_k^* z^{-1})}, \tag{12.315}$$

where $N = N_1 + 2N_2$. If $M \geq N$, then $N_p = M - N$, whereas in the opposite case, the first of the three sums of the above equation is not used. If the coefficients a_k and b_k in the equation defining the transfer function are real numbers, then the same applies to the quantities A_k, B_k, C_k, c_k, and e_k. It is not difficult to observe that, expressed in this form, the transfer function describes a system consisting of a set of first- and second-order systems, which are connected *in parallel* and which may include N_p elements of time delay, scaled by the appropriate coefficient. Alternatively, we can group the real poles into pairs of values and express the function in the form

$$H(z) = \sum_{k=0}^{N_p} C_k z^{-k} + \sum_{k=1}^{N_s} \frac{e_{0k} + e_{1k} z^{-1}}{1 - a_{1k} z^{-1} - a_{2k} z^{-2}}, \tag{12.316}$$

where the parameter N_s is defined as in the case of the cascade form, as $N_s = \lfloor (N+1)/2 \rfloor$. If $N_p = M - N < 0$, the first sum does not appear. Figure 12.104 presents such a parallel form for the case $M = N = 6$. In the general case, the difference equations for this parallel form using second-order subsystems and expressed in normal form are

$$w_k[n] = a_{1k} w_k[n-1] + a_{2k} w_k[n-2] + x[n] \qquad k = 1, 2, \ldots, N_s$$

$$y_k[n] = e_{0k} w_k[n] + e_{1k} w_k[n-1] \qquad k = 1, 2, \ldots, N_s$$

$$y[n] = \sum_{k=0}^{N_p} C_k x[n-k] + \sum_{k=1}^{N_s} y_k[n].$$

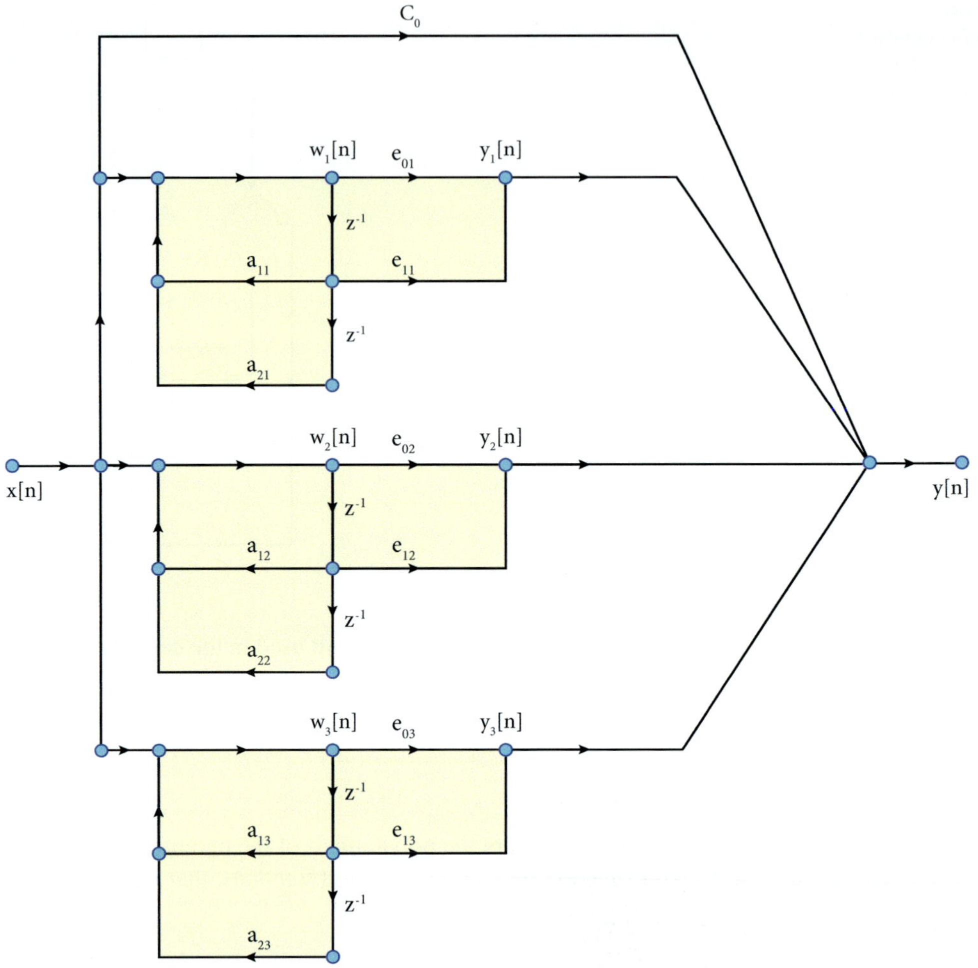

Fig. 12.104 Parallel form for the case $M = N = 6$ and with the real and complex poles grouped in pairs

As before, in the last equation the first sum is not included, if the condition $N_p < 0$ applies. Note that, as in the serial form, the second-order systems used for the implementation of the composite system, it has been implemented according to the direct form II or normal form, whereas as before, *the order in which this connection is made affects the behavior of the composite system*, since in practice, finite numerical accuracy is always used.

12.15.3 Frequency Sampling Form

The concept of *sampling in the frequency domain* has already been used for the construction of the discrete Fourier transform (see Sect. 10.5.1 on page 847) and the eponymous design technique *FIR filters* (see Sect. 12.12.4 on page 1247) and at this point we see how it can be used for the implementation of this type of filter. It is not difficult to realize that from the moment the process of implementing a filter is based on the transfer function $\mathcal{H}(z)$, whereas the *frequency sampling* consists in the calculation of the samples $\mathcal{H}[k]$ of the impulse response at the positions $\omega_k = 2\pi k/M$, the implementation of a filter based on this frequency-sampling technique requires the formulation of the transfer function $\mathcal{H}(z)$ as a function of frequency values $\omega_k = 2\pi k/M$. This process was presented in Exercise 10.5.2 (see page 852 for the general case of the transform $\mathcal{X}(z)$ of a signal $x[n]$. If we use the impulse response of the filter to be designed $h[n]$ as signal $x[n]$, so the transform $\mathcal{Z}\{h[n]\}$ coincides with the transfer function $\mathcal{H}(z)$, the application of the same process but applied for the frequency values $\omega_k = 2\pi(k+\alpha)/M$, with the value of α being either equal to zero or equal to $1/2$, leads us to the result

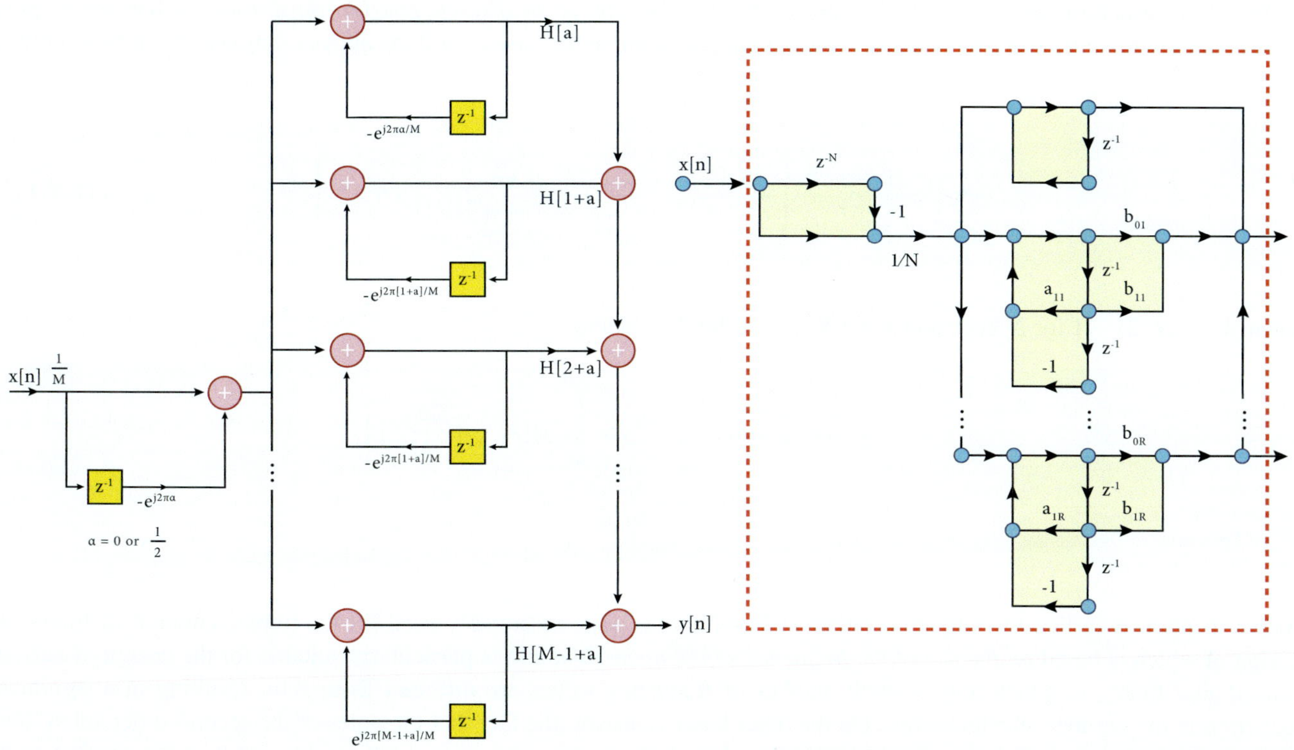

Fig. 12.105 Schematic diagram and signal-flow graph for an FIR filter implemented through the application of the frequency sampling method

$$\mathcal{H}(z) = \frac{1 - z^{-M} e^{j2\pi\alpha}}{M} \sum_{k=0}^{M-1} \frac{\mathcal{H}[k]}{1 - \frac{1}{z}\exp\left(j\frac{2\pi(k+\alpha)}{M}\right)} \tag{12.317}$$

(for the way of constructing this expression refer to the mathematical details that are demonstrated in detail in Exercise 10.5.2, since the process is exactly the same, whereas it is not difficult to realize that for the case where $\alpha = 0$, the problem is reduced to the construction of the correlation between the discrete Fourier transform and the $\mathcal{Z}$ transform). Observing the above expression, it is not difficult to find out that the transfer function $\mathcal{H}(z)$ is a product of the form $\mathcal{H}(z) = \mathcal{H}_1(z)\mathcal{H}_2(z)$, where

$$\mathcal{H}_1(z) = \frac{1 - z^{-M} e^{j2\pi\alpha}}{M} \quad \text{and} \quad \mathcal{H}_2(z) = \sum_{k=0}^{M-1} \frac{\mathcal{H}[k]}{1 - \frac{1}{z}\exp\left(j\frac{2\pi(k+\alpha)}{M}\right)}$$

and consequently, the considered filter can be constructed *from the series connection of two simpler systems* with transfer functions $\mathcal{H}_1(z)$ and $\mathcal{H}_2(z)$, of which the first contains only zero values that are located at equidistant positions on the circumference of the unit circle and are calculated as $z_k = \exp[j2\pi(k+\alpha)/M]$ $(k = 0, 1, 2, \ldots, M-1)$, whereas the second consists of a set of filters, each of which contains a simple pole located at the position $p_k = \exp[j2\pi(k+\alpha)/M]$ $(k = 0, 1, 2, \ldots, M-1)$ and which are connected in parallel, as derived from the sum symbol contained in the transfer function $\mathcal{H}_2(z)$. Note that the positions of the zeros z_k of the filter $\mathcal{H}_1(z)$ and the poles p_k of the filter $\mathcal{H}_2(z)$ *coincide with each other* and correspond to the frequency values $\omega_k = 2\pi(k+\alpha)/M$, at the positions of which the samples of the frequency response function are calculated. The structure of the filter that emerges through the application of the above technique is presented in Fig. 12.105.

The above structure can be simplified by exploiting *the symmetry of the discrete Fourier transform* as well as *the property of pole conjugation* of the above systems. Considering for example the case $\alpha = 0$ the above relationship can be simplified in the form

$$\mathcal{H}(z) = \frac{1 - z^{-M}}{M} \left\{ \frac{\mathcal{H}[0]}{1 - z^{-1}} + \frac{\mathcal{H}[M/2]}{1 - z^{-2}} + \sum_{k=0}^{K} 2|\mathcal{H}[k]|\mathcal{H}_k(z) \right\}, \tag{12.318}$$

where $k = (N/2) - 1$ for N even and $k = (N - 1)/2$ for N odd and

$$\mathcal{H}_k(z) = \frac{\cos(\measuredangle\mathcal{H}[k]) - z^{-1} \cos\left[\measuredangle\mathcal{H}[k] - \left(\frac{2\pi k}{M}\right)\right]}{1 - 2\cos\left(\frac{2\pi k}{M}\right) z^{-1} - z^{-2}}. \tag{12.319}$$

As we mentioned earlier, this implementation of FIR filters is quite reasonably used for the implementation of filters, the design of which is based *on the use of the frequency sampling method* and is particularly suitable for the design of narrow-band digital filters, in which only a small number of frequency values are different from zero, resulting in a significant reduction in the number of calculations. On the other hand, however, the fact that the poles of the second-order subsystems are located *on the circumference of the unit circle* makes these systems *only marginally stable* and therefore, owing to the finite accuracy that characterizes real implementations, some of these poles may be found *outside the circumference of the unit circle*, thus leading to an *unstable* system. In order to face this problem, we can proceed to a *radial displacement of the poles*, so that they are in a position that is inside the unit circle and at a very short distance from its circumference, so that they never end up in areas outside of it. Although certainly in this case the system that emerges constitutes, theoretically at least, an approximation of the real system, the samples of the output that are calculated are very close to their real values.

12.15.4 Lattice Form

The *lattice form* is an alternative method of implementing digital FIR and IIR filters particularly suitable for their use in *applications of digital voice signal processing*. As a structure, it is equivalent to the direct form of these filters, in the sense that it is designed to give exactly the same output as them.

Lattice Structures for FIR-Type Filters

In order to present the way of implementing this form let us start for simplicity reasons from the *FIR filters* and let us consider a sequence of M such filters with transfer functions of the form

$$\mathcal{H}_m(z) = 1 + \sum_{k=1}^{m} h_m[k]z^{-k} \qquad 1 \leq m \leq M - 1, \tag{12.320}$$

whereas for $m = 0$ it is $\mathcal{H}_0(z) = 1$. Also, for each value of m, we define $h_m[0] = 1$. Notice that the upper limit in the above sum is *variable* and equal to m and therefore, the impulse responses of these filters for each value of m have a length the value of which differs by one unit in relation to the length of the impulse response of the filter corresponding to the previous value of m. Specifically, it is $h_1 = \{h_1[1]\}$, $h_2 = \{h_2[1], h_2[2]\}$, $h_3 = \{h_3[1], h_3[2], h_3[3]\}$, etc. The index m in the transfer functions $\mathcal{H}_m(z)$, which are *polynomial*, since in FIR-type filters, the denominator of these functions is equal to the unit, indicates the degree of each of these polynomials, that is, essentially, the order of the filter.

If we transmit a discrete-time signal $x[n]$ to the filter number m from the above filters, then the output of the system according to the *convolution theorem* (the above filters are obviously *linear and time invariant*) will have the form

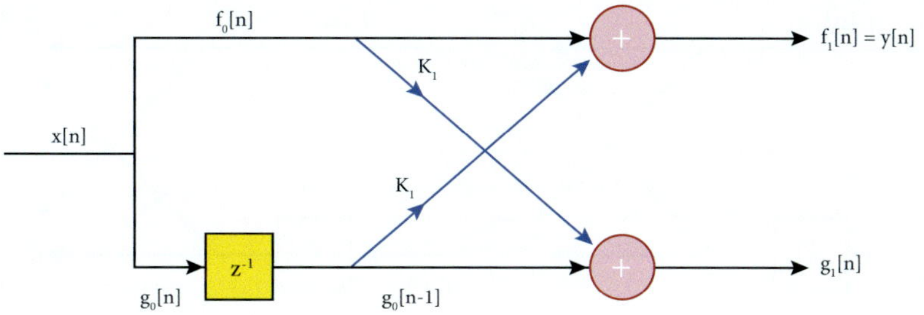

Fig. 12.106 The structure of a simple mesh stage

$$y[n] = h_m[n] * x[n] = x[n] + \sum_{k=1}^{m} h_m[k]x[n-k] \tag{12.321}$$

and therefore, the outputs of the filters corresponding to the values $m = 1$, $m = 2$, and $m = 3$, will be respectively

$$y[n] = x[n] + h_1[1]x[n-1]$$
$$y[n] = x[n] + h_2[1]x[n-1] + h_2[2]x[n-2]$$
$$y[n] = x[n] + h_3[1]x[n-1] + h_3[2]x[n-2] + h_3[3]x[n-3].$$

This output is calculated in the same way for the remaining values of m.

Is there the possibility of getting the same outputs using another system characterized by a *different* structure but operating in the *same* way? The answer is *affirmative*, with the systems used for this purpose known as *lattice structures*. A lattice structure is nothing more than *a set of identical elementary subsystems, known as stages, which are two-input and two-output systems*. These stages are connected *in series* in such a way that *the outputs of one stage become inputs to the stage immediately following it* in their series connection, whereas their number is equal to the order of the filter, that is, the value of m used in each case. In other words, (a) to get the output $y[n]$ that appears in the first of the above equations and corresponds to the value $m = 1$ we use a simple stage, (b) to get the output $y[n]$ that appears in the second of the above equations and corresponds to the value $m = 2$ we use two such stages connected in series, and (c) to get the output $y[n]$ that appears in the third of the above equations and corresponds to the value $m = 3$, we use three such stages connected in series, etc.

In order to understand the process, let us start with the use of a simple stage that generates the output for the value $m = 1$ and its structure is depicted in Fig. 12.106. From what we observe in the diagram, this stage, which resembles very much the *butterfly structure* used in the calculation process of the FFT (see Fig. 10.17 on page 881), is characterized by two inputs $f_0[n]$ and $g_0[n]$ and by two outputs $f_1[n]$ and $g_1[n]$, which are related to each other in the way depicted in the diagram. Notice that the input $f_0[n]$ is used *as is*, whereas the input $g_0[n]$ passes through an element *of unit time delay*, which produces at its output the sample $g_0[n-1]$. Then, the signals $f_0[n]$ and $g_0[n-1]$, *are both multiplied* by the constant K_1, with the signals resulting from these multiplications being directed toward the two adders that are located on the opposite branches, so that the outputs $f_1[n]$ and $g_1[n]$ both contain the inputs. As for the way of calculating these outputs, from the study of the diagram it is not difficult to find out that these will be calculated from the equations

$$f_1[n] = f_0[n] + K_1 g_0[n-1] \quad \text{and} \quad g_1[n] = K_1 f_0[n] + g_0[n-1].$$

Since this stage is the only one used, both inputs $f_0[n]$ and $g_0[n]$ receive the signal $x[n]$, which is supposed to be the input signal of the filter. Replacing in the above expressions the inputs $f_0[n]$ and $g_0[n]$ with the input signal $x[n]$, they take the form

$$f_1[n] = x[n] + K_1 x[n-1] \quad \text{and} \quad g_1[n] = K_1 x[n] + x[n-1]$$

Limiting ourselves to the output $f_1[n]$ of the upper branch and considering it as the desired output $y[n]$ (that is, we set $y[n] = f_1[n]$) and defining $K_1 = h_1[1]$, the first of the above equations takes the form

$$y[n] = x[n] + h_1[1]x[n-1],$$

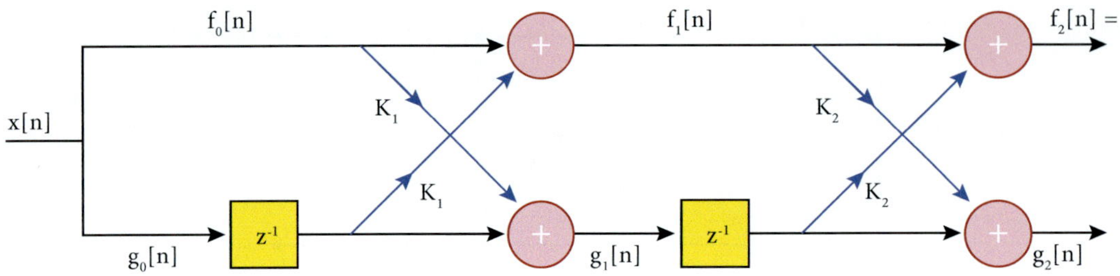

Fig. 12.107 The structure of a double-ladder mesh

which is nothing but the output of the FIR filter corresponding to the value $m = 1$. Therefore, *a first-order FIR filter is equivalent to a simple lattice structure*, in the sense that if they receive the same input, they calculate the same output.

On the other hand, in order to construct the output $y[n]$ for the second-order FIR filter corresponding to the value $m = 2$, *we need to connect two simple stages in series*, with the resulting structure being depicted in Fig. 12.107. The fact that these two stages are connected *in series* means that the outputs $f_1[n]$ and $g_1[n]$ of the first stage will be transmitted as inputs to the second stage, so that the two outputs $f_2[n]$ and $g_2[n]$ of this stage can be calculated. Note that because the new stage is considered as *the next of the first simple stage*, all indices have increased by one unit and consequently, the coefficient with which the inputs of the second ladder in the way we previously described is no longer symbolized with K_1 but with K_2. This is certainly something that characterizes all the ladders that are connected in series for the case of filters with a higher-order value, with the inputs of the m ladder being the $f_{m-1}[n]$ and $g_{m-1}[n]$, the corresponding outputs being the $f_m[n]$ and $g_m[n]$, and the multiplicative coefficient being K_m. In the terminology of digital filters, these multiplicative coefficients are known as *reflection coefficients*.

How does this two-ladder mesh filter work, which receives at its inputs $f_0[n]$ and $g_0[n]$, the discrete input signal $x[n]$, and generates its outputs $f_2[n]$ and $g_2[n]$? The inputs of the first ladder as we saw previously are calculated from the equations

$$f_1[n] = x[n] + K_1x[n-1] \quad \text{and} \quad g_1[n] = K_1x[n] + x[n-1].$$

These outputs of the first ladder become inputs to the second ladder, the outputs of which, working in the exact same way, are the

$$f_2[n] = f_1[n] + K_2g_1[n-1] \quad \text{and} \quad g_2[n] = K_2f_1[n] + g_1[n-1].$$

Replacing in the second equations the expressions of $f_1[n]$ and $g_1[n]$ from the first equations, it easily follows that

$$\begin{aligned}
f_2[n] &= x[n] + K_1x[n-1] + K_2(K_1x[n-1] + x[n-2]) = x[n] + K_1x[n-1] + K_2K_1x[n-1] + K_2x[n-2] \\
&= x[n] + K_1(1+K_2)x[n-1] + K_2x[n-2] \qquad \text{and} \\
g_2[n] &= K_2(x[n] + K_1x[n-1]) + K_1x[n-1] + x[n-2] = K_2x[n] + K_2K_1x[n-1] + K_1x[n-1] + x[n-2] \\
&= K_2x[n] + K_1(1+K_2)x[n-1] + x[n-2].
\end{aligned}$$

Therefore, defining $h_2[2] = K_2$ and $h_2[1] = K_1(1 + K_2)$, from where it follows that $K_1 = h_2[1]/(1 + h_2[2])$ and limiting ourselves as before, to the output of the upper branch $f_2[n]$, which is considered as the desired output $y[n]$, the first of the above equations becomes

$$y[n] = x[n] + h_2[1]x[n-1] + h_2[2]x[n-2],$$

which is nothing but the output of the second-order FIR filter, which corresponds to the value $m = 2$ and is characterized by the transfer function $\mathcal{H}_2(z)$.

The above process continues in exactly the same way. For example, if we connect three stages in series and work as before, the output of the upper branch of the three-stage lattice filter coincides with the output of the third-order FIR filter, which corresponds to the value $m = 3$ and is characterized by the transfer function $\mathcal{H}_2(z)$, etc. If we generalize this conclusion and resort to the method of *mathematical induction*, we can show that the direct structure of an FIR filter of order m described

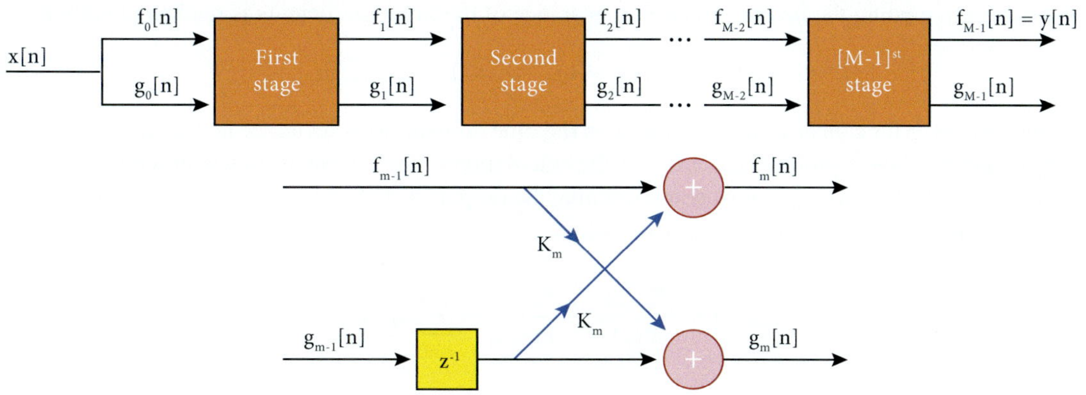

Fig. 12.108 The lattice structure of a filter of order $M-1$, along with the structure of the step that corresponds to some value of the parameter m

by the transfer function $\mathcal{H}_m(z)$ is equivalent to a lattice filter of m stages connected in series, or in a different formulation, with a filter of order m. As for the equations for calculating the outputs for the general case, these are

$$f_0[n] = g_0[n] = x[n]$$
$$f_m[n] = f_{m-1}[n] + K_m g_{m-1}[n-1], \quad m = 1, 2, \ldots, M-1$$
$$g_m[n] = K_m f_{m-1}[n] + g_{m-1}[n-1], \quad m = 1, 2, \ldots, M-1.$$

Figure 12.108 depicts the *lattice structure* of a filter of order $M-1$, along with the structure of the step that corresponds to some value of the parameter m.

From the above analysis, we observe that if we know the coefficients of the filter, that is, the values of the samples of the impulse response $h_k[n]$ $(m = 1, 2, \ldots, M-1)$ of the direct implementation (remember that $h_m[0] = 1$ for every value of m), we can calculate the values of the coefficients $K_1, K_2, \ldots, K_m$. In addition, from the expressions of the two outputs $f_m[n]$ and $g_m[n]$ $(m = 1, 2, \ldots, M-1)$ of the lattice structure, we find that the filter coefficients that are used in these appear *in reverse order*. Indeed, the outputs $f_1[n]$ and $g_1[n]$ are the

$$f_1[n] = x[n] + K_1 x[n-1] = x[n] + h_1[1]x[n-1]$$
$$g_1[n] = K_1 x[n] + x[n-1] = h_1[1]x[n] + x[n-1]$$

and consequently, the sets of filter coefficients for these two outputs, which are the $\alpha_1 = \{\alpha_1[0], \alpha_1[1]\} = \{1, h_1[1]\}$ and $\beta_1 = \{\beta_1[0], \beta_1[1]\} = \{h_1[1], 1\}$, are *ordered inversely to each other*. In a similar way, studying the form of the outputs $f_2[n]$ and $g_2[n]$, which are

$$f_2[n] = x[n] + K_1(1+K_2)x[n-1] + K_2 x[n-2] = x[n] + h_2[1]x[n-1] + h_2[2]x[n-2]$$
$$g_2[n] = K_2 x[n] + K_1(1+K_2)x[n-1] + x[n-2] = h_2[2]x[n] + h_2[1]x[n-1] + x[n-2],$$

we find that the sets of filter coefficients, which are $\alpha_2 = \{\alpha_2[0], \alpha_2[1], \upsilon_2[2]\} = \{1, h_2[1], h_2[2]\}$ and $\beta_2 = \{\beta_2[0], \beta_2[1], \beta_2[2]\} = \{h_2[2], h_2[1], 1\}$, are also characterized by the inverse order. Taking into account the equivalence that exists between the direct structure of a filter of order m and a mesh structure of m stages and resorting to the convolution equation, we can formulate the outputs $f_m[n]$ and $g_m[n]$ as

$$f_m[n] = \alpha_m[n] * x[n] = \sum_{k=0}^{m} \alpha_m[k]x[n-k] \;\; (\alpha_m[0] = 1), \qquad m = 1, 2, \ldots, M-1$$
$$g_m[n] = \beta_m[n] * x[n] = \sum_{k=0}^{m} \beta_m[k]x[n-k] \;\; (\beta_m[m] = 1), \qquad m = 1, 2, \ldots, M-1.$$

The above observation regarding the inverse order of appearance of the filter coefficients is mathematically formulated as

$$\beta_m(k) = \alpha_m(m-k), \quad k = 0, 1, 2, \ldots, m.$$

What we have mentioned so far concerns the description in the time domain. Now let us see now exactly how all the above are described in the complex plane, a process that involves the calculation of the $\mathcal{Z}$ transform for all the involved quantities and equations. Starting from the last two equations above defining the outputs $f_m[n]$ and $g_m[n]$, the application of the *convolution theorem* allows us to write $\mathcal{F}_m(z) = \mathcal{A}_m(z)\mathcal{X}(z)$, from which it follows that

$$\mathcal{A}_m(z) = \frac{\mathcal{F}_m(z)}{\mathcal{X}(z)} = \frac{\mathcal{F}_m(z)}{\mathcal{F}_0(z)} = \sum_{k=0}^{m} \alpha_m[k]z^{-k}, \tag{12.322}$$

since $f_0[n] = x[n]$ and therefore, $\mathcal{F}_0(z) = \mathcal{X}(z)$. In exactly the same way we have $\mathcal{G}_m(z) = \mathcal{B}_m(z)\mathcal{X}(z)$ and therefore, we have

$$\mathcal{B}_m(z) = \frac{\mathcal{G}_m(z)}{\mathcal{X}(z)} = \sum_{k=0}^{m} \beta_m[k]z^{-k}. \tag{12.323}$$

Observing that $\beta_m[k] = \alpha_m[m-k]$, the last relation becomes

$$\mathcal{B}_m(z) = \sum_{k=0}^{m} \beta_m[k]z^{-k} = \sum_{k=0}^{m} \alpha_m[m-])z^{-k}. \tag{12.324}$$

Proceeding to change the dummy index from k to $\ell = m - k$, $k = m - \ell$. For $k = 0$, $\ell = m$, whereas for $k = m$, $\ell = 0$ and the above relation becomes

$$\mathcal{B}_m(z) = \sum_{\ell=m}^{0} \alpha_m[\ell]z^{-(m-\ell)} = \sum_{\ell=0}^{m} \alpha_m[\ell]z^{\ell-m} = z^{-m}\sum_{\ell=0}^{m} \alpha_m[\ell](z^{-1})^{-\ell} = z^{-m}\mathcal{A}_m(z^{-1}). \tag{12.325}$$

But the relation $\mathcal{B}_m(z) = z^{-m}\mathcal{A}_m(z^{-1})$ leads us to the conclusion that the zero values of the FIR filter with transfer function $\mathcal{B}_m(z)$ $(m = 1, 2, \ldots, M-1)$ are *the inverses of the zero values of the transfer function* $\mathcal{A}_m(z)$ $(m = 1, 2, \ldots, M-1)$. In reality, the polynomial $\mathcal{B}_m(z)$ (remember that the transfer functions of the FIR filters do not have a denominator and therefore are not rational but polynomial functions), is the inverse polynomial of the polynomial $\mathcal{A}_m(z)$.

Finally, the application of the $\mathcal{Z}$ transform to the group of three equations that allow the calculation of the system's outputs will give us

$$\begin{aligned}
&\mathcal{F}_0(z) = \mathcal{G}_0(z) = \mathcal{X}(z)\\
&\mathcal{F}_m(z) = \mathcal{F}_{m-1}(z) + K_m z^{-1}\mathcal{G}_{m-1}(z), \quad m = 0, 1, 2, \ldots, M-1\\
&\mathcal{G}_m(z) = K_m\mathcal{F}_{m-1}(z) + z^{-1}\mathcal{G}_{m-1}(z), \quad m = 0, 1, 2, \ldots, M-1.
\end{aligned}$$

Dividing the above expressions by the function $\mathcal{X}(z)$ and taking into account that $\mathcal{F}_m(z) = \mathcal{A}_m(z)\mathcal{X}(z)$ and $\mathcal{G}_m(z) = \mathcal{B}_m(z)\mathcal{X}(z)$ we arrive at the result

$$\begin{aligned}
&\mathcal{A}_0(z) = \mathcal{A}_0(z) = 1\\
&\mathcal{A}_m(z) = \mathcal{A}_{m-1}(z) + K_m z^{-1}\mathcal{B}_{m-1}(z), \quad m = 0, 1, 2, \ldots, M-1\\
&\mathcal{B}_m(z) = K_m\mathcal{A}_{m-1}(z) + z^{-1}\mathcal{B}_{m-1}(z), \quad m = 0, 1, 2, \ldots, M-1.
\end{aligned}$$

The first of the above equations corresponds to the value $m = 0$, whereas the remaining two equations, which correspond to the values $m = 1, 2, \ldots, M-1$, can be combined into a matrix equation of the form

$$\begin{bmatrix} \mathcal{A}_m(z) \\ \mathcal{B}_m(z) \end{bmatrix} = \begin{bmatrix} 1 & K_m \\ K_m & 1 \end{bmatrix} \begin{bmatrix} \mathcal{A}_{m-1}(z) \\ z^{-1}\mathcal{B}_{m-1}(z). \end{bmatrix} \tag{12.326}$$

Let us complete the presentation of the mesh structures for the filters of FIR type, noting that the equivalence that exists between filters of order m and mesh structures of m stages makes it possible to calculate the coefficients of the filter $\alpha_m[k]$ and $\beta_m[k]$ from the reflection coefficients K_m, whereas of course the inverse also holds. This calculation process, which is *recursive in nature*, is performed in the following way. From the equation

$$\mathcal{A}_m(z) = \mathcal{A}_{m-1}(z) + K_m z^{-1}\mathcal{B}_{m-1}(z)$$

and using the property $\beta_m[k] = \alpha_m[m-k]$ it follows that

$$\sum_{k=0}^{m} \alpha_m[k]z^{-k} = \sum_{k=0}^{m} \alpha_{m-1}[k]z^{-k} + K_m \sum_{k=0}^{m} \alpha_{m-1}[m-1-k]z^{-(k+1)}. \tag{12.327}$$

Equating the equal powers of z^{-1} and taking into account that $\alpha_m[0] = 1$ $(m = 1, 2, \ldots, M-1)$, we end up with the equations $\alpha_m[0] = 1$, $\alpha_m[m] = K_m$ and

$$\alpha_m[k] = \alpha_{m-1}[k] + K_m\alpha_{m-1}[m-k] = \alpha_{m-1}[k] + \alpha_m[m]\alpha_{m-1}[m-k],$$

where $1 \le k \le m-1$ and $m = 1, 2, \ldots, M-1$. In the terminology of digital signal processing, these equations are known as *Levinson-Durbin recursive equations*.

On the other hand, for the reverse process of calculating the values of the coefficients K_m, when the filter coefficients are known, we work in the following way: starting from the mesh structure of the m stages, the knowledge of the polynomial $\mathcal{A}_m(z)$ allows us to calculate the coefficient K_m as $K_m = \alpha_m[m]$. In order to subsequently calculate the value of the coefficient K_{m-1}, we need the polynomial $\mathcal{A}_{m-1}(z)$ and by repeating the same process moving backward, we can finally calculate all the coefficients up to the value $m = 1$. From the mathematical side of the issue, combining the equations

$$\mathcal{A}_m(z) = \mathcal{A}_{m-1}(z) + K_m z^{-1}\mathcal{B}_{m-1}(z)$$

$$\mathcal{B}_m(z) = K_m\mathcal{A}_{m-1}(z) + z^{-1}\mathcal{B}_{m-1}(z)$$

we are led to the expression

$$\mathcal{A}_m(z) = \mathcal{A}_{m-1}(z) + K_m[\mathcal{B}_m(z) - K_m\mathcal{A}_{m-1}(z)],$$

from which it follows that

$$\mathcal{A}_{m-1}(z) = \frac{\mathcal{A}_m(z) - K_m\mathcal{B}_m(z)}{1 - K_m^2}, \quad m = M-1, M-2, \ldots, 2, 1. \tag{12.328}$$

In this way, starting from the system $\mathcal{A}_m(z)$, we calculate all the polynomials of lower degree of the form $\mathcal{A}_{m-1}(z)$, $\mathcal{A}_{m-2}(z)$ and finally $\mathcal{A}_1(z)$ and for each value of m we calculate the coefficient K_m as $K_m = \alpha_m[m]$. Of course, from the form of the above relation, the above process can only be applied when $|K_m| \neq 1$ for $m = 1, 2, \ldots, M-1$, as in the opposite case, the denominator tends to infinity.

Grid Structures for IIR Filters Containing Only Poles

The theory of the lattice structures presented so far, exclusively concern the case of FIR filters (which are characterized by *finite impulse response*, with a transfer function that contains *only zero values*, whereas they lack any form of recursion). Let us now look at the case of IIR filters, the transfer function of which contains both poles and zero values. For the presentation of the theory, we start from the simple case of IIR filters that contain *only poles* and then we see what additionally applies to the case of the more complex IIR filters, which contain *both poles and zero values*. There are two reasons that advocate for the development of the theory in this way. The first reason is that the lattice structures for IIR filters that contain *only poles* can be used as *the basic building units for the construction of lattice structures for IIR filters*, which in addition to poles, contain zero values. The second reason is that the development of the theory of IIR filters that contain only poles does not need to be done from scratch, since as we see, we can rely on the conclusions we formulated in the previous section for FIR filters, because the lattice structures of IIR filters that contain only poles use the same coefficients K as the FIR-type structures.

In order to prove the truth of this claim, let us consider an IIR filter that contains only poles and that therefore is described by a transfer function of the form

$$\mathcal{H}(z) = 1 \Bigg/ \left\{ 1 + \sum_{k=1}^{N} \alpha_N[k] z^{-k} \right\} = \frac{1}{\mathcal{A}_N(z)}$$

and that, as it easily emerges, working in the now familiar way, corresponds to the difference equation

$$y[n] = -\sum_{k=1}^{N} \alpha_N[k] y[n-k] + x[n].$$

The design of mesh structures for IIR pole-only filters, from the structures corresponding to FIR-type filters, is based on the observation that if we alternate the input and output signals in the above equation, that is, if we replace $x[n]$ with $y[n]$ and vice versa, the resulting equation is

$$x[n] = -\sum_{k=1}^{N} \alpha_N[k] x[n-k] + y[n]$$

or equivalently

$$y[n] = x[n] + \sum_{k=1}^{N} \alpha_N[k] x[n-k].$$

But the last equation is easily recognized as the equation that describes *the input/output correlation* of an FIR filter with transfer function $\mathcal{H}(z) = \mathcal{A}_N(z)$, whereas the initial equation describes an IIR system with transfer function $\mathcal{H}(z) = 1/\mathcal{A}_N(z)$. We are therefore led to the conclusion that an IIR system that contains only poles can be constructed by starting from an FIR system and simply alternating the input and output of the system.

This observation is particularly important because it gives us the ability to implement mesh structures for IIR pole-only filters, using the homonymous structures we built for FIR-type filters. Indeed, since to get the equation for IIR filters, we alternated the roles of input and output, it is not difficult to comprehend that the description of IIR filters is the same as the description of FIR filters, provided, however, that this is done *in reverse*, or equivalently, *in the opposite direction*. For example, the input of an IIR filter of only poles of order N, corresponds to the output of an FIR filter and therefore, will be calculated as $x[n] = f_N[n]$, whereas the output of the IIR filter that corresponds to the input of an FIR filter will be calculated as $y[n] = f_0[n]$. Note that this process of *reversal*, which in a sense consists in the alternation of the role of inputs and outputs, should also be applied to all the calculations involved. For example, the quantities $\{f_m[n]\}$ will be calculated in reverse order, compared with that characterizing the FIR filters, that is, not as $\{f_0[n], f_1[n], \ldots, f_{N-1}[n], f_N[n]\}$ but as $\{f_N[n], f_{N-1}[n], \ldots, f_1[n], f_0[n]\}$ and this is something that can be done by reformulating the recursive equation for the function $f_m[n]$, in the form

$$f_{m-1}[n] = f_m[n] - K_m g_{m-1}[n], \quad m = N, N-1, N-2, \ldots, 2, 1,$$

that is, expressing the function $f_{m-1}[n]$ as a function of $f_m[n]$. Conversely, the equation for the function $g_m[n]$ remains as is. Based on the above, the equations of the lattice structure for IIR filters that contain *only poles* are

$$f_N[n] = x[n]$$

$$f_{m-1}[n] = f_m[n] - K_m g_{m-1}[n], \quad m = N, N-1, N-2, \ldots, 2, 1$$

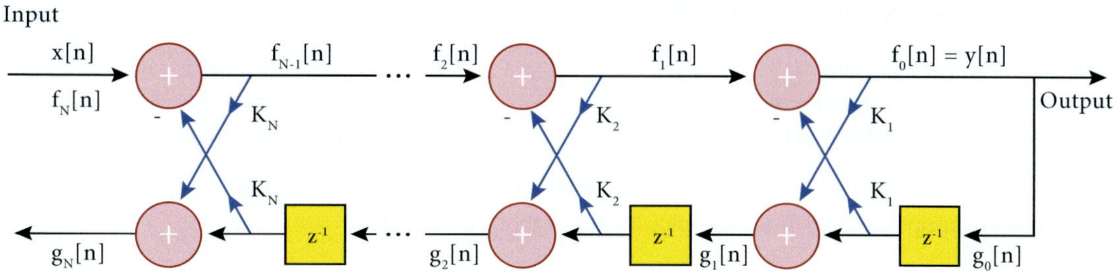

Fig. 12.109 The lattice structure of a pole-only IIR filter of order N (Proakis and Manolakis, Figure 7.24)

$$g_m[n] = K_m f_{m-1}[n] + g_{m-1}[n-1], \quad m = N, N-1, N-2, \ldots, 2, 1$$
$$y[n] = f_0[n] = g_0[n],$$

whereas the structure that implements the above behavior and corresponds to a lattice structure of N stages for a pole-only IIR filter, is depicted in Fig. 12.109. From the study of the diagram, it is observed that as in the case of lattice structures for FIR filters, the corresponding lattices of IIR filters emerge *from the series connection of identical copies of a simple stage*, which in the diagram is enclosed within a frame.

As already mentioned, the calculation of the functions $f_m[n]$ is performed *in reverse order*, that is, the function f_{m-1} is calculated based on the function $f_m[n]$, whereas the functions $g_m[n]$ are calculated as before, which means that the calculation of the function $g_m[n]$ is based on the function g_{m-1}. This characteristic is also reflected in Fig. 12.109, from which it appears that *the direction of the calculations in the upper branch is from left to right, whereas the direction of the calculations in the lower branch is from right to left*. In reality, this diagram describes a system with *feedback*, since after the calculation of the output $f_0[n] = y[n]$, this is transmitted to the lower branch as the signal $g_0[n]$ (remember that $y[n] = f_0[n] = g_0[n]$) to be used for the execution of calculations related to this branch. Of course, the existence of this feedback is expected, since it characterizes all IIR systems (both those that contain only poles and those that contain both poles and zero values) as in reality, this feedback is exactly *the cause of the existence of the poles*. For this to be understood, let us remember the way of transforming a simple difference equation, let us say

$$y[n] + \alpha_1 y[n-1] = x[n] + \beta_1 x[n-1],$$

into a transfer function. Taking the $\mathcal{Z}$ transform of the two terms and working in the known way, we successively have

$$\mathcal{Y}(z) + \alpha_1 z^{-1}\mathcal{Y}(z) = \mathcal{X}(z) + \beta_1 z^{-1}\mathcal{X}(z) \Longrightarrow \mathcal{Y}(z)(1+\alpha_1 z^{-1}) = \mathcal{X}(z)(1+\beta_1 z^{-1}) \Longrightarrow \mathcal{H}(z) = \frac{\mathcal{Y}(z)}{\mathcal{X}(z)} = \frac{1+\beta_1 z^{-1}}{1+\alpha_1 z^{-1}},$$

with the simple pole of the transfer function being calculated as a function of the parameter α_1. But this parameter is nothing but the coefficient of the term $y[n-1]$, which indicates the existence of *feedback*. If there was no feedback, this term would not appear and therefore $\alpha_1 = 0$ would. In this case, the denominator of the function $\mathcal{H}(z)$ would be equal to the unit and the system would not be of the IIR type but of the FIR type. In other words, *IIR systems are inextricably linked to the characteristic of feedback* and therefore it is completely reasonable for it to appear in all structures used for their description.

Working as in the case of FIR-type filters, we can construct the equations for a *simple mesh structure single-stage IIR filter-only poles* (i.e., for the value $N = 1$), which have the form

$$x[n] = f_1[n]$$
$$f_0[n] = f_1[n] - K_1 g_0[n-1]$$
$$g_1[n] = K_1 f_0[n] + g_0[n-1] = K_1 y[n] + y[n-1]$$
$$y[n] = f_0[n] = x[n] - K_1 y[n-1],$$

whereas correspondingly, for the value $N = 2$ we have

$$\begin{aligned} f_2[n] &= x[n] & f_0[n] &= f_1[n] - K_1 g_0[n] \\ f_1[n] &= f_2[n] - K_2 g_1[n-1] & g_1[n] &= K_1 f_0[n] + g_0[n-1] \\ g_2[n] &= K_2 f_1[n] + g_1[n-1] & y[n] &= f_0[n] = g_0[n]. \end{aligned}$$

In order to construct the transfer functions, we must determine the forms of the functions $y[n]$ and $g_2[n]$, combining the above equations in the appropriate way. For example, for the signal $y[n]$ we have

$$\begin{aligned} y[n] = f_0[n] = f_1[n] - K_1 g_0[n-1] &= f_2[n] - K_2 g_1[n-1] - K_1 g_0[n-1] \\ &= x[n] - K_2[K_1 f_0[n-1] + g_0[n-2]] - K_1 g_0[n-1] \\ &= x[n] - K_2 K_1 g_0[n-1] - K_2 g_0[n-2] - K_1 g_0[n-1] \\ &= -K_1(1+K_2) g_0[n-1] - K_2 g_0[n-2] + x[n], \end{aligned}$$

and finally

$$y[n] = \alpha_2[0] y[n] = -K_1(1+K_2)y[n-1] - K_2 y[n-2] + x[n] = -\alpha_2[1]y[n-1] - \alpha_2[2]y[n-2] + x[n],$$

or equivalently

$$y[n] + \alpha_2[1]y[n-1] + \alpha_2[2]y[n-2] = x[n],$$

whereas in a similar way we have

$$g_2[n] = K_2 y[n] + K_1(1+K_2)y[n-1] + y[n-2] = \beta_2[0]y[n] + \beta_2[1]y[n-1] + \beta_2[2]y[n-2],$$

where in the last expressions we have used the correlations existing between the *reflection coefficients* and the *filter coefficients*. These two expressions constitute the input/output relations for the upper and lower branches of the circuit, with the upper branch receiving as input the signal $f_2[n] = x[n]$ and producing at its output the signal $y[n]$ and the second branch receiving as input the signal $g_0[n] = y[n]$ and producing at its output the signal $g_2[n]$.

In order to calculate the transfer function for these two branches, we must, as known, take the $\mathcal{Z}$ transform of the above equations. Working in the now familiar way, it easily follows that

$$\mathcal{Y}(z)(\alpha_2[0] + \alpha_2[1]z^{-1} + \alpha_2[2]z^{-2}) = \mathcal{X}(z).$$

Therefore, for the upper branch we have

$$\mathcal{H}_{\text{top}}(z) = \frac{\mathcal{Y}(z)}{\mathcal{X}(z)} = \frac{\mathcal{F}_0(z)}{\mathcal{F}_2(z)} = \frac{1}{\mathcal{A}_2(z)} = \frac{1}{\sum_{k=0}^{2} \alpha_2[k]z^{-k}} \tag{12.329}$$

whereas similarly for the lower branch we can write

$$\mathcal{G}_2(z) = (\beta_2[0] + \beta_2[1]z^{-1} + \beta_2[2]z^{-2})\mathcal{Y}(z),$$

from which it follows that

$$\mathcal{H}_{\text{bottom}}(z) = \frac{\mathcal{G}_2(z)}{\mathcal{Y}(z)} = \sum_{k=0}^{2} \beta_2[k]z^{-k} = \mathcal{B}_2(z) = z^{-2}\mathcal{A}_2(z^{-1}). \tag{12.330}$$

The above relations hold for the case $m = 2$, that is *for a lattice structure with two stages*, whereas for the general case, the corresponding expressions are

$$\mathcal{H}_{\text{top}}(z) = \frac{\mathcal{Y}(z)}{\mathcal{X}(z)} = \frac{\mathcal{F}_0(z)}{\mathcal{F}_m(z)} = \frac{1}{\mathcal{A}_m(z)} = \frac{1}{\sum_{k=0}^{m} \alpha_m[k]z^{-k}} \tag{12.331}$$

and

$$\mathcal{H}_{\text{bottom}}(z) = \frac{\mathcal{G}_m(z)}{\mathcal{Y}(z)} = \sum_{k=0}^{2} \beta_m[k]z^{-k} = \mathcal{B}_m(z) = z^{-m}\mathcal{A}_m(z^{-1}). \tag{12.332}$$

From the mathematical form of the two above transfer functions, it follows that *the lattice structure of an IIR system-only poles, contains a recursion that is related to a structure that represents an FIR system, which contains only zero values*. In fact the lower branch of this structure, is isomorphic with the branch of the lattice structure of FIR filters, as can easily be found. This path, which is therefore common in these two types of lattice structure, is described by the transfer function $\mathcal{H}_{\text{bottom}}(z) = \mathcal{B}_m(z)$. Note also that the lattice structures of FIR filters and of IIR filters-only poles, are both described *by the same coefficients reflection* $K_1, K_2, K_3, \ldots$, whereas they only differ in the way in which they are connected.

Mesh Structures for IIR Filters That Contain Both Poles and Zero Values

Using the mesh structures of the IIR pole-only filters, we can construct the homonymous structures for the general case of IIR filters that contain both poles and zero values. This capability, at least, seems reasonable, since the function $\mathcal{A}_N(z)$ that appears in the transfer function of the IIR pole-only filters, is part of the transfer function

$$\mathcal{H}(z) = \frac{\mathcal{C}_M(z)}{\mathcal{A}_N(z)} = \frac{\sum_{k=0}^{M} \gamma_M[k]z^{-k}}{1 + \sum_{k=1}^{N} \alpha_N[k]z^{-k}}, \tag{12.333}$$

which applies in the general case (for reasons related to the stability of the system, we usually consider that $N \geq M$). For the development of the relevant theory, we use the direct form II, which, as we have commented on in detail (see Sect. 7.10 on page 608) is also the one usually used in practice, since it is characterized by *the minimum number of unit time delay elements*, in which, as we have also already commented on in detail, the considered system is considered to arise *from the series connection* of an IIR pole-only system with the input/output equation

$$w[n] = -\sum_{k=1}^{N} \alpha_N[k]w[n-k] + x[n]$$

and of an FIR system that contains only zero values and is described by the equation

$$y[n] = \sum_{k=0}^{M} \gamma_M[k]w[n-k].$$

Studying the first of the two above equations, we find that the output of the IIR pole/zero value system is *a linear combination of time-delayed values* $w[n-k]$ of the output $w[n]$ of the IIR pole-only system, with the coefficients of this linear combination being the quantities $\gamma_M[k]$. Note that these coefficients are related to the polynomial $\mathcal{C}_M(z)$, which is found *in the numerator*

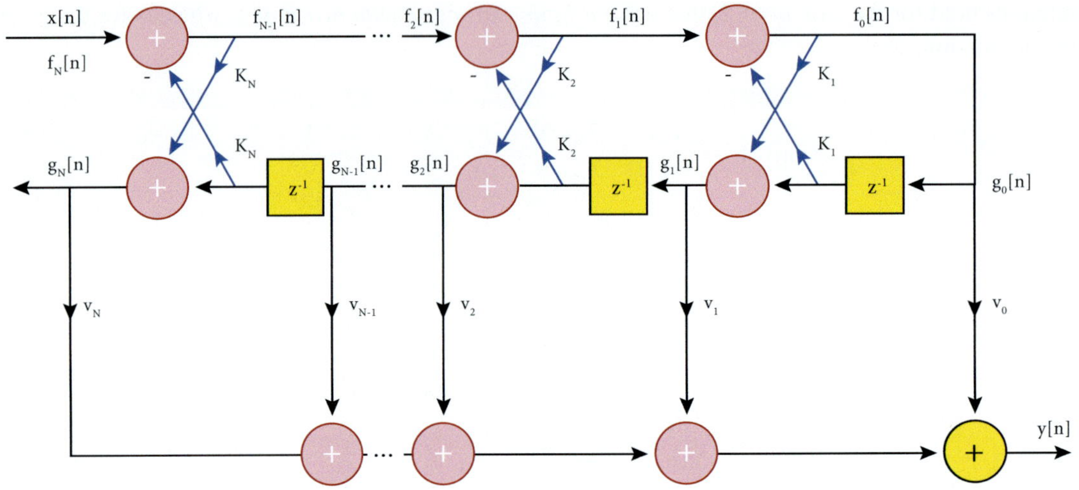

Fig. 12.110 The mesh structure of an IIR pole/zero value filter (Proakis and Manolakis, Figure 7.27)

of the transfer function $\mathcal{H}(z)$ of the IIR pole/zero value system and the roots of which constitute the zero values for this system. Based on this observation, we can construct the mesh structure for the IIR pole/zero value system, using the mesh structure of the IIR pole-only system and more specifically, extending it so as to incorporate the zero values. The result of this extension is depicted diagrammatically in Fig. 12.110, the upper part of which is nothing but the mesh structure of the IIR pole-only system, as can easily be verified.

What is the logic behind the construction of this shape? Initially, it is expected that some part of this shape *will constitute the mesh structure of the IIR pole-only system* and this is immediately observed by noticing that the upper part of the shape does indeed constitute such a structure. The extension of this shape so that in addition to the poles, the zero values are also supported, is based on the conclusion we previously formulated, according to which the zero values of the IIR pole system are related to linear combinations of the previous outputs of the IIR pole-only system. In order to exploit the existing structure and get the desired diagram, simply by extending it, we observe that the functions $g_m[n]$ that appear in it are indeed linear combinations of the current and previous outputs and therefore are the most suitable to be used in this extension. Consequently, the last relation regarding the output of the system $y[n]$ can be formulated in the form

$$y[n] = \sum_{m=0}^{M} v_m g_m[n].$$

In other words, the outputs $g_i[n]$ $(i = 1, 2, \ldots, M)$ of the mesh structure of the IIR pole-only system, are multiplied by the coefficients v_i and then all these products are added together to give the output $y[n]$. Observing the lower part of Fig. 12.110, we find that it performs exactly this calculation.

Taking the $\mathcal{Z}$ transform of the above equation, we end up with the expression

$$\mathcal{Y}(z) = \sum_{m=0}^{M} v_m \mathcal{G}_m(z)$$

and therefore, the transfer function of the system is

$$\mathcal{H}(z) = \frac{\mathcal{Y}(z)}{\mathcal{X}(z)} = \sum_{m=0}^{M} v_m \frac{\mathcal{G}_m(z)}{\mathcal{X}(z)} = \sum_{m=0}^{M} v_m \frac{\mathcal{G}_m(z)}{\mathcal{G}_0(z)} \frac{\mathcal{F}_0(z)}{\mathcal{F}_N(z)} = \sum_{m=0}^{M} v_m \frac{\mathcal{B}_m(z)}{\mathcal{A}_N(z)} = \frac{1}{\mathcal{A}_N(z)} \sum_{m=0}^{M} v_m \mathcal{B}_m(z), \tag{12.334}$$

where we took into account that $x[n] = f_N[n]$ (so it is $\mathcal{X}(z) = \mathcal{F}_N(z)$), $f_0[n] = g_0[n]$ (so it is $\mathcal{F}_0(z) = \mathcal{G}_0(z) = \mathcal{Y}(z)$), as well as the expressions $\mathcal{G}_m(z) = \mathcal{B}_m(z)\mathcal{Y}(z) = \mathcal{B}_m(z)\mathcal{G}_0(z)$ and $\mathcal{F}_N(z) = \mathcal{A}_N(z)\mathcal{F}_0(z)$. Comparing the last expression with the equation $\mathcal{H}(z) = \mathcal{C}_M(z)/\mathcal{A}_N(z)$ that we formulated previously, we get

$$\mathcal{C}_M(z) = \sum_{m=0}^{M} v_m \mathcal{B}_m(z). \tag{12.335}$$

In order to calculate the values of the coefficients v_m, we reformulate the above relation as

$$\mathcal{C}_m(z) = \sum_{k=0}^{m-1} v_k \mathcal{B}_k(z) + v_m \mathcal{B}_m(z), \tag{12.336}$$

which, expressed in recursive form, is formulated as $C_m(z) = C_{m-1}(z) + v_m \mathcal{B}_m(z)$. Therefore, using this recursive expression repeatedly, we can calculate the quantities $C_m(z)$ from the polynomials $\mathcal{B}_m(z)$ $(m = 1, 2, \ldots, M)$. To carry out these calculations, we should take into account that $v_m = \gamma_m(m)$ $(m = 0, 1, 2, \ldots, m)$. So, by reformulating the above expression as

$$C_{m-1}(z) = C_m(z) - v_m \mathcal{B}_m(z) \tag{12.337}$$

and repeatedly using this recursive expression for the values $m = M, M-1, M-2, \ldots, 2, 1$ we can calculate the coefficients $\gamma_m(m)$ and from there the coefficients v_m. As for the polynomials $\mathcal{B}_m(z)$ required in the above calculation, these are calculated along with the coefficients $\{K_m\}$ in the way we previously described, provided that the polynomials $C_M(z)$ and $A_N(z)$ are known.

12.16 Digital Filter Transformations

Digital filters, like their analog equivalents, allow *frequency transformations* that enable their conversion from one form to another. In full analogy with the corresponding analog filters, digital filters are constructed *in their low-pass form* in the way we presented in the previous sections and then the resulting digital low-pass filter is transformed into the appropriate filter in each case (*low-pass, high-pass, band-pass*, or *band-stop*. In order to describe these transformation processes, let us assume that the low-pass filter to be transformed is defined in the space of z, whereas the transformed digital filter is defined in the space of $\tilde{z}$. In this case, the transformation in question can be considered as a function $z = f(\tilde{z})$, which must satisfy the following properties: (a) to be a *rational and real function* of the complex variable $\tilde{z}$, (b) to *preserve stability* (i.e., *to map the interior, the periphery, and the exterior of the unit circle of the z domain to the corresponding regions in the $\tilde{z}$ space*), and (c) to *have an inverse function*, so that we can write that $\tilde{z} = f^{-1}(z)$.

Let us assume that the desired function has the form $f(\tilde{z}) = r(\omega)e^{j\varphi(\omega)}$ for the variable $\tilde{z} = e^{j\omega}$, where $r(\omega)$ and $\varphi(\omega)$ are the magnitude and the phase of it. It is not difficult to understand that stability is preserved (in the sense we described above), only if it is $r(\omega) = 1$. But this means that both the variable $\tilde{z}$ and its image z will belong *both* to the periphery of the unit circle and consequently the application of the transformation *simply causes the displacement of the variable $\tilde{z}$ on the periphery of the unit circle by an arc corresponding to some rotation angle ϑ*. Using the maximum modulus theorem and taking into account that the function $f(\tilde{z})$ should satisfy the above conditions, it is proven that the function $f(\tilde{z})$ should have the form[93]

$$z = f(\tilde{z}) = e^{j\vartheta} \prod_{i=1}^{M} \frac{\tilde{z} - \lambda_i}{1 - \lambda_i^* \tilde{z}}, \tag{12.338}$$

[93] See the reference Constantinides A.G., *Spectral Transformations for Digital Filters*, Proceedings of IEE, Vol. 117, No. 8, pp. 110–115, August 1970.

where $\lambda_i = \mu_i e^{j\nu_i}$ $(i = 1, 2, \ldots, M)$ *are complex coefficients* such that $\mu_i < 1$. Let us note that the function $f(\tilde{z})$ defined by the above relation *is not generally a real function*, a property that is ensured if (a) the coefficients λ_i appear *in pairs of complex conjugates*, and (b) the angle ϑ is *an integer multiple of* π of the form $\vartheta = \xi\pi$ $(\xi = 1, 2, 3, \ldots)$, so that $e^{j\vartheta} = \pm 1$. Considering that the angle ϑ constitutes the *rotation angle* of the variable $\tilde{z}$ on the circumference of the unit circle, it is not difficult to understand that *the only permissible rotations in these digital filter transformation processes are those corresponding to integer multiples of* π. We return to this issue later.

Combining all of the above, we can state the claim that *the general function of the spectral transformation of digital filters with real coefficients* will be described by the equation

$$z = f(\tilde{z}) = \pm \prod_{i=1}^{M} \left(\frac{\tilde{z} - \lambda_i}{1 - \lambda_i^* \tilde{z}} \times \frac{\tilde{z} - \lambda_i^*}{1 - \lambda_i \tilde{z}} \right), \tag{12.339}$$

where now the order of the filter has *doubled*, whereas the parameters λ_i *are no longer necessarily complex numbers*. As for the value of the parameter M, it dictates the nature of the mapping performed by the above function. Considering that the function $f(\tilde{z})$ maps the unit circle to itself, *the value of M corresponds to the number of z values that correspond to each value of $\tilde{z}$*. As we see later, the value of the parameter M is equal to 1 for the low-pass $\longrightarrow$low-pass and low-pass $\longrightarrow$high-pass transformations and equal to 2 for the low-pass $\longrightarrow$band-pass and low-pass $\longrightarrow$band-stop transformations.

The processes of mapping a digital low-pass filter to a digital low-pass, high-pass, and band-pass filter are presented in images (a), (b), and (c) of Fig. 12.111 respectively, whereas for the transformation *from low-pass to band-stop filter* this is similar to the transformation to a band-pass filter and is described by a diagram similar to that of image (c) but *with a different frequency mapping*. In this diagram, the positive hemisphere of the unit circle corresponds to the frequency range $[0, \pi]$, whereas the negative hemisphere corresponds to the range of frequencies $[0, -\pi]$. This means that the point A with value $z = 1$ corresponds to the frequency $\omega = 0$, whereas the point D with value $z = -1$ corresponds to the values $\pm\pi$. The segments of the circumference of the circle with a bold line correspond to *pass bands*, whereas the rest with a normal (thin) line correspond to *stop bands*.

Transformation of Low-pass to Low-pass Digital Filter

The transformation of a *digital low-pass filter* $\mathcal{H}(z)$ with cutoff frequency Ω_p to a digital low-pass filter $\mathcal{H}(\tilde{z})$ with cutoff frequency ω_p, *preserves the characteristics of the amplitude response* and constitutes a *1:1 mapping* between the unit circle and its image. This means that a full rotation of $\tilde{z}$ corresponds to a full rotation of z and therefore $M = 1$. Consequently, the low-pass $\longrightarrow$low-pass transformation involves characteristics of *bilinear transformation*. Observing image (a) of Fig. 12.111 we find that the points $z = 1$ and $z = -1$ (points A and D) should correspond to the points $\tilde{z} = 1$ and $\tilde{z} = -1$ (points A' and D')—something that means that points A and D remain unchanged—whereas on the contrary, the arcs (AB) and (AC) extend or shrink according to the relationship between the frequencies Ω_p and ω_p. Replacing the above values of z and $\tilde{z}$ in the transformation equation, it turns out that

$$e^{j\xi\pi} \frac{1 - \lambda^*}{1 - \lambda} = e^{j\xi\pi} \frac{1 + \lambda^*}{1 + \lambda} = 1.$$

and therefore $\xi = 0$ and $\lambda = \lambda^* = \alpha$, where α is a real constant. Consequently, the *transformation of low-pass to low-pass filter* will be given by the equation

$$z = \frac{\tilde{z} - \alpha}{1 - \alpha\tilde{z}}. \tag{12.340}$$

The constant α will arise from the requirement that *the cutoff frequency $\Omega = \Omega_p$ of the original filter should correspond to the frequency $\omega = \omega_p$ of the transformed filter*. Replacing in the above relation the corresponding values of z and $\tilde{z}$, which are $z = e^{j\Omega}$ and $\tilde{z} = e^{j\omega}$, we will get

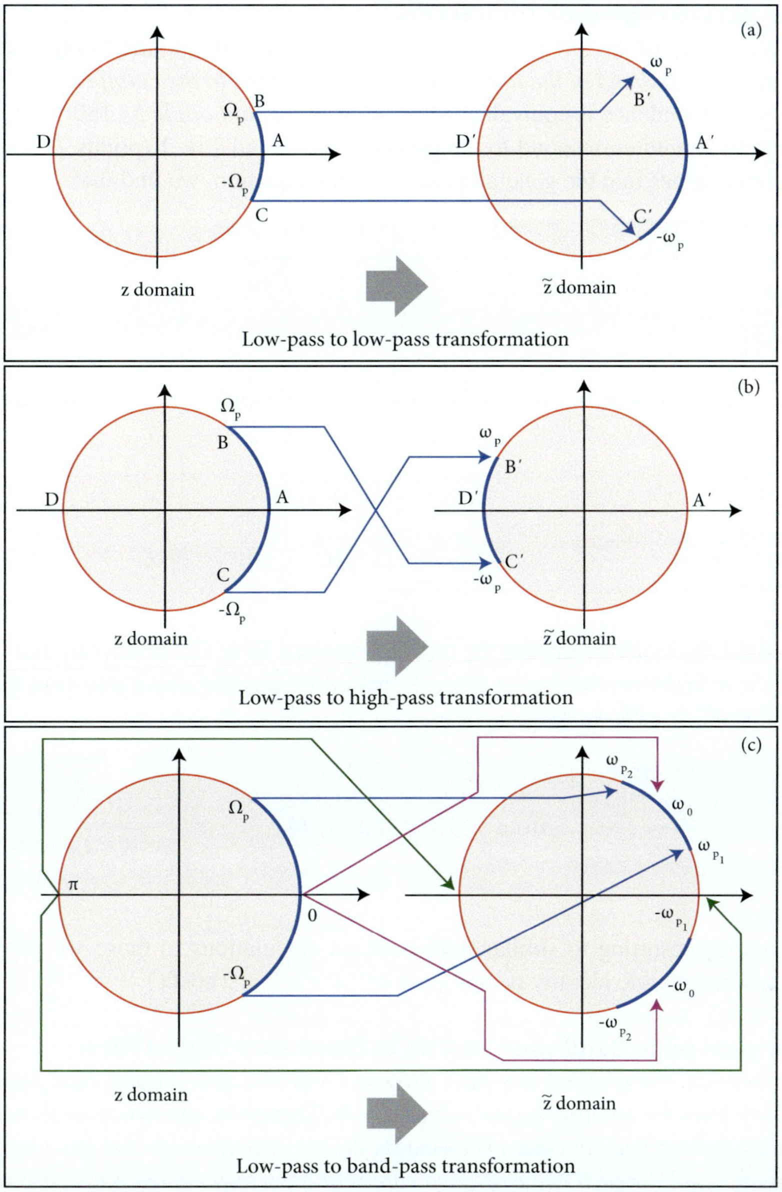

Fig. 12.111 Frequency transformations on the unit circle for the conversion of low-pass to low-pass (**a**), high-pass (**b**), and band-pass (**c**) filters respectively

$$e^{j\Omega_p} = \frac{e^{j\omega_p} - \alpha}{1 - \alpha e^{j\omega_p}} \quad \text{from which it follows that} \quad \alpha = \frac{e^{j\omega_p} - e^{j\Omega_p}}{1 - e^{j(\Omega_p + \omega_p)}}.$$

Using the known algebraic expression $e^{jx} - e^{-jx} = 2j\sin(x)$ and after simple algebraic operations, the value of α is successively formulated as

$$\alpha = \frac{e^{j\omega_p} - e^{j\Omega_p}}{1 - e^{j(\Omega_p+\omega_p)}} = \frac{e^{j\Omega_p} - e^{j\omega_p}}{e^{j(\Omega_p+\omega_p)} - 1} = \frac{e^{j\frac{\Omega_p+\omega_p}{2}}}{e^{j\frac{\Omega_p+\omega_p}{2}}} \left\{ \frac{e^{j\frac{\Omega_p-\omega_p}{2}} - e^{j\frac{\omega_p-\Omega_p}{2}}}{e^{j\frac{\Omega_p+\omega_p}{2}} - e^{-j\frac{\Omega_p+\omega_p}{2}}} \right\} = \frac{e^{j\frac{\Omega_p-\omega_p}{2}} - e^{-j\frac{\Omega_p-\omega_p}{2}}}{e^{j\frac{\Omega_p+\omega_p}{2}} - e^{-j\frac{\Omega_p+\omega_p}{2}}} = \frac{\sin[(\Omega_p - \omega_p)/2]}{\sin[(\Omega_p + \omega_p)/2]}. \tag{12.341}$$

Transformation of Low-pass to High-pass Digital Filter

In this case, the *cutoff frequency* Ω_p of the low-pass filter corresponds, as before, to the *cutoff frequency* ω_p of the *high-pass* filter, with the difference that *the passband of the low-pass filter corresponds to the cutoff band of the transformed high-pass filter and vice versa*. This correspondence is equivalent to *rotation of the unit circle by* 180° and therefore the points $z = 1$ and $z = -1$ (points A and D) should correspond to the points $\tilde{z} = -1$ and $\tilde{z} = 1$ (points D' and A'), whereas, as before, $M = 1$. Substituting the above values into the general transformation equation, we find that

$$e^{j\xi\pi}\frac{1+\lambda^*}{1+\lambda} = e^{j\xi\pi}\frac{1-\lambda^*}{1-\lambda} = -1$$

and therefore $\xi = 1$ and $\lambda = \lambda^* = \alpha$, where α is a *real constant*. Consequently, the *transformation of low-pass to high-pass filter* will have the form

$$z = -\frac{\tilde{z}-\alpha}{1-\alpha\tilde{z}}. \tag{12.342}$$

The constant α will arise from the requirement that the cutoff frequency $\Omega = \Omega_p$ of the original filter should correspond to the frequency $\omega = \omega_p$ of the *transformed high-pass filter*. Substituting into the above equation the corresponding values of z and $\tilde{z}$, which are $z = e^{j\Omega}$ and $\tilde{z} = e^{j\omega}$, we get

$$e^{j\Omega_p} = -\frac{e^{j\omega_p}-\alpha}{1-\alpha e^{j\omega_p}} \quad \text{from which it follows that} \quad \alpha = \frac{\cos[(\Omega_p-\omega_p)/2]}{\cos[(\Omega_p+\omega_p)/2]}.$$

The value of α is calculated by resorting to similar mathematical calculations to those we demonstrated in the previous section, where the well-known algebraic identity is now used $e^{jx} + e^{-jx} = 2\cos(x)$.

Transformation of Low-pass (a) to Band-pass and (b) to Band-stop Digital Filter

In these transformation processes, *the original low-pass filter is converted into a band-pass filter for rotation angle value* $\vartheta = \pi$ *and into a band-stop filter for rotation angle value* $\vartheta = 0$. Therefore, the two transformations are constructed in a similar way and the equation that implements them is based on the correspondences that exist between the spaces Ω and ω. Taking into account that in this case *for each value of* $\tilde{z}$ *two values of* z *correspond* (as is the case with analog filters), $M = 2$ and from the general transformation equation we have

$$z = f(\tilde{z}) = \pm\prod_{i=1}^{2}\frac{\tilde{z}-\lambda_i}{1-\lambda_i^*\tilde{z}} = \pm\frac{\tilde{z}-\lambda_1}{1-\lambda_1^*\tilde{z}}\frac{\tilde{z}-\lambda_2}{1-\lambda_2^*\tilde{z}} = \pm\frac{\tilde{z}^2-(\lambda_1+\lambda_2)\tilde{z}+\lambda_1\lambda_2}{1-(\lambda_1^*+\lambda_2^*)\tilde{z}+\lambda_1^*\lambda_2^*\tilde{z}^2} = \pm\frac{\tilde{z}^2+\alpha_1\tilde{z}+\alpha_2}{1+\alpha_1\tilde{z}+\alpha_2\tilde{z}^2}, \tag{12.343}$$

where we assumed that λ_1 and λ_2 are *real numbers*, so that $\lambda_1 = \lambda_1^*$ and $\lambda_2 = \lambda_2^*$, and we set $\alpha_1 = -(\lambda_1+\lambda_2) = -(\lambda_1^*+\lambda_2^*)$ and $\alpha_2 = \lambda_1\lambda_2 = \lambda_1^*\lambda_2^*$.

Starting from the low-pass $\longrightarrow$ band-pass transformation equation, this is characterized by the existence of the *negative sign*, as it is associated with a rotation angle $\vartheta = \pi$. Limiting ourselves to the semi-axis of positive frequencies of the ω space, this transformation maps the frequency $\Omega = 0$ to the *central frequency* of the band-pass filter $\omega = \omega_0$, the frequency $\Omega = -\Omega_p$ *to the lower limit of the passband* of the band-pass filter $\omega = \omega_{p1}$, the frequency $\Omega = \Omega_p$ *to the upper limit of the passband* of the band-pass filter $\omega = \omega_{p2}$, and the frequency $\Omega = \pi$ (or the frequency $\Omega = \Omega_s/2$, if the *sampling frequency* is known) to the value $\omega = \pi$ (or to the value $\Omega = \Omega_s/2$). This situation is presented in image (c) of Fig. 12.111. Now, using the above correspondences of frequencies of the Ω space to the ω space, the algebraic equation that implements the above transformation is constructed as described in the following.

Initially, we observe that the frequency $\Omega = 0$ corresponds to the frequency $\omega = \omega_0$ and therefore the corresponding variables of the z domain will be $z = 1$ and $\tilde{z} = e^{j\omega_0}$. Substituting the above expressions into the general equation of the transformation (without forgetting the negative sign), it takes the form

$$1 = -\frac{e^{2j\omega_0} + \alpha_1 e^{j\omega_0} + \alpha_2}{1 + \alpha_1 e^{j\omega_0} + \alpha_2 e^{2j\omega_0}} \quad \text{or equivalently} \quad \frac{e^{2j\omega_0} + \alpha_1 e^{j\omega_0} + \alpha_2}{1 + \alpha_1 e^{j\omega_0} + \alpha_2 e^{2j\omega_0}} + 1 = 0,$$

from which it follows that

$$(\alpha_2 + 1)\, e^{2j\omega_0} + 2\alpha_1 e^{j\omega_0} + (\alpha_2 + 1) = 0 \quad \text{or equivalently} \quad (\alpha_2 + 1)(1 + e^{2j\omega_0}) + 2\alpha_1 e^{j\omega_0} = 0.$$

The above equation, by factoring the exponential term $e^{j\omega_0}$, takes successively the form

$$e^{j\omega_0}\left[(\alpha_2 + 1)(e^{-j\omega_0} + e^{j\omega_0}) + 2\alpha_1\right] = 0 \Longrightarrow 2e^{j\omega_0}[(\alpha_2 + 1)\cos(\omega_0) + \alpha_1] = 0 \Longrightarrow (\alpha_2 + 1)\cos(\omega_0) + \alpha_1 = 0.$$

We observe, therefore, that the parameter α_1 can be expressed as a function of the parameter α_2 through the expression $\alpha_1 = -(\alpha_2 + 1)\cos(\omega_0)$ and consequently the transformation equation is formulated as

$$z = f(\tilde{z}) = -\frac{\tilde{z}^2 - (\alpha_2 + 1)\cos(\omega_0)\tilde{z} + \alpha_2}{1 - (\alpha_2 + 1)\cos(\omega_0)\tilde{z} + \alpha_2\tilde{z}^2} = -\frac{\tilde{z}\left(\dfrac{\tilde{z} - \alpha}{1 - \alpha\tilde{z}}\right) + \alpha_2}{1 + \alpha_2\tilde{z}\left(\dfrac{\tilde{z} - \alpha}{1 - \alpha\tilde{z}}\right)}, \tag{12.344}$$

where for simplicity we defined the parameter $\alpha = \cos(\omega_0)$. We observe that the function enclosed within the parentheses of the numerator and the denominator has the form of the elementary functions that appear in the product of Eq. (12.308) and that the magnitude of the parameter α is less than unity. Under the regime of the power of this assumption, functions of this form are known as *unitary functions*.

Defining the auxiliary functions[94]

$$t = t(\tilde{z}) = \tilde{z}\frac{\tilde{z} - \alpha}{1 - \alpha\tilde{z}} \quad \text{and} \quad s = s(\tilde{z}) = -\frac{\tilde{z} + \alpha_2}{1 + \alpha_2\tilde{z}},$$

the transformation equation is formulated as a composition of functions of the form

$$z = f(\tilde{z}) = s\left(\tilde{z}\frac{\tilde{z} - \alpha}{1 - \alpha\tilde{z}}\right) = s[t(\tilde{z})] = -\frac{t(\tilde{z}) + \alpha_2}{1 + \alpha_2 t(\tilde{z})}.$$

It will then be

$$1 - z = 1 - f(\tilde{z}) = 1 + \frac{t(\tilde{z}) + \alpha_2}{1 + \alpha_2 t(\tilde{z})} = \frac{1 + \alpha_2 t(\tilde{z}) + t(\tilde{z}) + \alpha_2}{1 + \alpha_2 t(\tilde{z})} = \frac{(1 + \alpha_2) + t(\tilde{z})(1 + \alpha_2)}{1 + \alpha_2 t(\tilde{z})} = \frac{(1 + \alpha_2)[1 + t(\tilde{z})]}{1 + \alpha_2 t(\tilde{z})}$$

$$1 + z = 1 - f(\tilde{z}) = 1 - \frac{t(\tilde{z}) + \alpha_2}{1 + \alpha_2 t(\tilde{z})} = \frac{1 + \alpha_2 t(\tilde{z}) - t(\tilde{z}) - \alpha_2}{1 + \alpha_2 t(\tilde{z})} = \frac{(1 - \alpha_2) - t(z)(1 - \alpha_2)}{1 + \alpha_2 t(\tilde{z})} = \frac{(1 - \alpha_2)[1 - t(\tilde{z})]}{1 + \alpha_2 t(\tilde{z})}$$

[94] Taking into account that the parameter α_2 is characterized by a magnitude *less than the unit*, since it is defined as the product of two stable zero values that are located inside the unit circle, we can formulate the claim that the function $s = s(\tilde{z})$ is also a *unit function*.

and dividing the equations term by term, we get

$$\frac{1-z}{1+z}=\frac{1+\alpha_2}{1-\alpha_2}\left[\frac{1+t(\tilde{z})}{1-t(\tilde{z})}\right]=\frac{1+\alpha_2}{1-\alpha_2}\left(\frac{\tilde{z}^2-2\alpha\tilde{z}+1}{1-\tilde{z}^2}\right), \tag{12.345}$$

since as it is proven

$$\frac{1+t(\tilde{z})}{1-t(\tilde{z})}=\frac{1+\tilde{z}\dfrac{\tilde{z}-\alpha}{1-\alpha\tilde{z}}}{1-\tilde{z}\dfrac{\tilde{z}-\alpha}{1-\alpha\tilde{z}}}=\frac{1-\alpha\tilde{z}+\tilde{z}^2+\alpha\tilde{z}}{1-\alpha\tilde{z}-\tilde{z}^2+\alpha\tilde{z}}=\frac{\tilde{z}^2-2\alpha\tilde{z}+1}{1-\tilde{z}^2}. \tag{12.346}$$

At this point we can use the correspondence of the space of Ω to the space of ω dictated by the above transformation and according to which for $\Omega=-\Omega_p$ is $\omega=\omega_{p1}$, whereas for $\Omega=\Omega_p$ it is $\omega=\omega_{p2}$. Starting from the first correspondence, in the space of z this is formulated as $z=e^{-j\Omega_p}\longrightarrow\tilde{z}=e^{j\omega_{p1}}$ and with substitution of the values of z and $\tilde{z}$ in the last relation, this takes the form

$$\frac{1-e^{-j\Omega_p}}{1+e^{-j\Omega_p}}=\frac{1+\alpha_2}{1-\alpha_2}\left(\frac{e^{2j\omega_{p1}}-2\alpha e^{j\omega_{p1}}+1}{1-e^{2j\omega_{p1}}}\right). \tag{12.347}$$

However, we observe that the exponential expressions of the left and right sides are formulated as

$$\frac{1-e^{-j\Omega_p}}{1+e^{-j\Omega_p}}=\frac{e^{-j\frac{\Omega_p}{2}}\left(e^{j\frac{\Omega_p}{2}}-e^{-j\frac{\Omega_p}{2}}\right)}{e^{-j\frac{\Omega_p}{2}}\left(e^{j\frac{\Omega_p}{2}}+e^{-j\frac{\Omega_p}{2}}\right)}=\frac{2j\sin(\Omega_p/2)}{2\cos(\Omega_p/2)}=j\tan\frac{\Omega_p}{2} \tag{12.348}$$

and

$$\frac{e^{2j\omega_{p1}}-2\alpha e^{j\omega_{p1}}+1}{1-e^{2j\omega_{p1}}}=\frac{e^{j\omega_{p1}}(e^{j\omega_{p1}}-2\alpha+e^{-j\omega_{p1}})}{e^{j\omega_{p1}}(e^{-j\omega_{p1}}-e^{j\omega_{p1}})}=\frac{2\cos(\omega_{p1})-2\alpha}{-2j\sin(\omega_{p1})}=j\frac{\cos(\omega_{p1})-\alpha}{\sin(\omega_{p1})}, \tag{12.349}$$

and with direct substitution it immediately follows that

$$-\tan\frac{\Omega_p}{2}=\frac{\alpha_2+1}{\alpha_2-1}\left[\frac{\cos(\omega_{p1})-\alpha}{\sin(\omega_{p1})}\right]. \tag{12.350}$$

On the other hand, the application of the second correspondence, which in the z domain is formulated as

$$z=e^{j\Omega_p}\longrightarrow\tilde{z}=e^{j\omega_{p2}}$$

leads us analogously to the relation

$$\frac{1-e^{j\Omega_p}}{1+e^{j\Omega_p}}=\frac{1+\alpha_2}{1-\alpha_2}\left(\frac{e^{2j\omega_{p2}}-2\alpha e^{j\omega_{p2}}+1}{1-e^{2j\omega_{p2}}}\right), \tag{12.351}$$

which, if we work as above, finally takes the form

$$\tan\frac{\Omega_p}{2}=\frac{\alpha_2+1}{\alpha_2-1}\left[\frac{\cos(\omega_{p2})-\alpha}{\sin(\omega_{p2})}\right]. \tag{12.352}$$

Combining the last two relations we get

$$\frac{\cos(\omega_{p2})-\alpha}{\sin(\omega_{p2})}=\frac{\alpha-\cos(\omega_{p1})}{\sin(\omega_{p1})}, \tag{12.353}$$

from which it follows that

$$\alpha = \frac{\cos(\omega_{p2})\sin(\omega_{p1}) + \sin(\omega_{p2})\cos(\omega_{p1})}{\sin(\omega_{p1}) + \sin(\omega_{p2})} = \frac{\sin(\omega_{p1} + \omega_{p2})}{\sin(\omega_{p1}) + \sin(\omega_{p2})}$$

$$= \frac{2\sin\left(\dfrac{\omega_{p1} + \omega_{p2}}{2}\right)\cos\left(\dfrac{\omega_{p1} + \omega_{p2}}{2}\right)}{2\sin\left(\dfrac{\omega_{p1} + \omega_{p2}}{2}\right)\cos\left(\dfrac{\omega_{p1} - \omega_{p2}}{2}\right)} = \frac{\cos\left(\dfrac{\omega_{p1} + \omega_{p2}}{2}\right)}{\cos\left(\dfrac{\omega_{p1} - \omega_{p2}}{2}\right)}.$$

The above relation expresses *the functional dependence of the parameter α on the frequencies ω_{p1} and ω_{p2} of the passband of the band-pass filter to be constructed*, whereas its dependence on the central frequency ω_0 is described by the equation $\alpha = \cos(\omega_0)$. Now, defining the auxiliary variable

$$\kappa = -\frac{\alpha_2 + 1}{\alpha_2 - 1} = \frac{\sin(\omega_{p2})\cos\left(\dfrac{\omega_{p1} - \omega_{p2}}{2}\right)}{\cos\left(\dfrac{\omega_{p1} + \omega_{p2}}{2}\right) - \cos(\omega_{p2})\cos\left(\dfrac{\omega_{p1} - \omega_{p2}}{2}\right)}\tan\frac{\Omega_p}{2}$$

$$= \frac{\cos\left(\dfrac{\omega_{p1} - \omega_{p2}}{2}\right)}{\sin\left(\dfrac{\omega_{p1} - \omega_{p2}}{2}\right)}\tan\frac{\Omega_p}{2} = \cot\left(\frac{\omega_{p1} - \omega_{p2}}{2}\right)\tan\frac{\Omega_p}{2}$$

(where we substituted the value of α from the defining equation and used the *trigonometric identity of the sum for cosine*), the parameter α_2 is formulated as $\alpha_2 = (k - 1)/(k + 1)$, whereas the parameter α_1 takes the form

$$\alpha_1 = -(1 + \alpha_2)\cos(\omega_0) = -\left(1 + \frac{\kappa - 1}{\kappa + 1}\right)\alpha = -\frac{2\alpha\kappa}{\kappa + 1}.$$

Substituting the above expressions into the general equation of transformation, we finally get

$$z = f(\tilde{z}) = -\left\{\tilde{z}^2 - \frac{2\alpha\kappa}{\kappa + 1}\tilde{z} + \frac{\kappa - 1}{\kappa + 1}\right\} \Big/ \left\{1 - \frac{2\alpha\kappa}{\kappa + 1}\tilde{z} + \frac{\kappa - 1}{\kappa + 1}\tilde{z}^2\right\}. \quad (12.354)$$

Finally, regarding the equation of the transformation *from low-pass to band-stop filter*, it is constructed in exactly the same way with the difference that: (a) in the starting equation a *positive sign* (+) is used (since the rotation angle ϑ, in this case, is equal to zero); and (b) the correspondences of the frequencies of the Ω space to the ω space are different and more specifically the frequency $\Omega = 0$ corresponds to the frequency $\omega = 0$, the frequency $\Omega = -\Omega_p$ corresponds to the frequency $\omega = \omega_{p_2}$, the frequency $\Omega = \Omega_p$ corresponds to the frequency $\omega = \omega_{p_1}$, whereas the frequency $\Omega = \pi$ (or $\Omega = \Omega_s/2$) corresponds to the central frequency $\omega = \omega_0$. Working, as before, it is proven in the same way that the defining equation of the low-pass $\longrightarrow$ band-stop transformation is given by the relation

$$z = f(\tilde{z}) = \left\{\tilde{z}^2 - \frac{2\alpha}{\kappa + 1}\tilde{z} + \frac{1 - \kappa}{1 + \kappa}\right\} \Big/ \left\{1 + \frac{2\alpha}{\kappa + 1}\tilde{z} + \frac{1 - \kappa}{1 + \kappa}\tilde{z}^2\right\}, \quad (12.355)$$

with the parameters α and κ calculated as

$$\alpha = \cos\omega_0 = \frac{\cos[(\omega_{p2} + \omega_{p1})/2]}{\cos[(\omega_{p2} - \omega_{p1})/2]} \quad \text{and} \quad \kappa = \tan\left(\frac{\omega_{p2} - \omega_{p1}}{2}\right)\tan\frac{\Omega_p}{2}.$$

The process of constructing the low-pass $\longrightarrow$ band-stop transformation equation is left as an exercise to the reader.

Problem 12.16.1 *The transfer function*

$$\mathcal{H}(z) = \frac{0.008616 + 0.025848z^{-1} + 0.025848z^{-2} + 0.008616z^{-3}}{1 - 2.064414z^{-1} + 1.519112z^{-2} - 0.385767z^{-3}}$$

describes a low-pass digital filter with operating frequencies $\Omega_p = 0.15\pi$ *and* $\Omega_s = 0.35\pi$ *and specific attenuation levels in the pass and cutoff zones. Using the digital filter transformations that were presented in the previous section to transform the above low-pass filter into: (a) a low-pass filter with cutoff frequency* $\omega_p = 0.65\pi$, *(b) a high-pass filter with cutoff frequency* $\omega_p = 0.65\pi$, *(c) a band-pass filter with passband limits* $\omega_{p1} = 0.45\pi$ *and* $\omega_{p2} = 0.45\pi$, *and (d) a band-stop filter with stopband limits* $\omega_{p1} = 0.45\pi$ *and* $\omega_{p2} = 0.45\pi$.

Solution (a) The constant α of the low-pass $\longrightarrow$low-pass transformation for the specific frequency values has value

$$\alpha = \frac{\sin[(\Omega_p - \omega_p)/2]}{\sin[(\Omega_p + \omega_p)/2]} = \frac{\sin[(0.15\pi - 0.65\pi)/2]}{\sin[(0.15\pi + 0.65\pi)/2]} = -\frac{\sin(0.25\pi)}{\sin(0.40\pi)} = -0.743496$$

and applying the defining equation of the transformation, it results after operations that

$$\mathcal{H}_{\text{low-pass}}(\tilde{z}) = \mathcal{H}(z)\Big|_{z=\frac{\tilde{z}-\alpha}{1-\alpha\tilde{z}}} = \frac{0.008616 + 0.025848z^{-1} + 0.025848z^{-2} + 0.008616z^{-3}}{1 - 2.064414z^{-1} + 1.519112z^{-2} - 0.385767z^{-3}}\Bigg|_{z=\frac{\tilde{z}+0.743496}{1+0.743496\tilde{z}}}$$
$$= \frac{0.045663 + 0.136990z^{-1} + 0.136990z^{-2} + 0.045663z^{-3}}{0.013501 + 0.078281z^{-1} + 0.127228z^{-2} + 0.146312z^{-3}}.$$

The amplitude response of the low-pass filter resulting in this way is depicted in image (a) of Fig. 12.112.

(b) The constant α of the low-pass $\longrightarrow$high-pass transformation for the specific frequency values is calculated as

$$\alpha = \frac{\cos[(\Omega_p - \omega_p)/2]}{\cos[(\Omega_p + \omega_p)/2]} = \frac{\cos[(0.15\pi - 0.65\pi)/2]}{\cos[(0.15\pi + 0.65\pi)/2]} = \frac{\cos(0.25\pi)}{\cos(0.40\pi)} = 2.288245$$

and applying the defining equation of the transformation, the result after operations is that

$$\mathcal{H}_{\text{high-pass}}(\tilde{z}) = \mathcal{H}(z)\Big|_{z=-\frac{\tilde{z}-\alpha}{1-\alpha\tilde{z}}} = \frac{0.008616 + 0.025848z^{-1} + 0.025848z^{-2} + 0.008616z^{-3}}{1 - 2.064414z^{-1} + 1.519112z^{-2} - 0.385767z^{-3}}\Bigg|_{z=-\frac{\tilde{z}-2.288245}{1-2.288245\tilde{z}}}$$
$$= \frac{-0.3063362 + 0.919008z^{-1} - 0.919008z^{-2} + 0.306336z^{-3}}{-4.262331 - 3.694887z^{-1} - 2.275097z^{-2} - 0.391745z^{-3}}.$$

The amplitude response of the high-pass filter that has this result is depicted in image (b) of Fig. 12.112.

(c) In the case of the low-pass $\longrightarrow$band-pass transformation there are two constants, the values of which for the specific values of the operating frequencies are calculated as

$$\alpha = \frac{\cos\left(\dfrac{\omega_{p1} + \omega_{p2}}{2}\right)}{\cos\left(\dfrac{\omega_{p1} - \omega_{p2}}{2}\right)} = \frac{\cos\left(\dfrac{0.45\pi + 0.75\pi}{2}\right)}{\cos\left(\dfrac{0.45\pi - 0.75\pi}{2}\right)} = \frac{\cos(0.60\pi)}{\cos(0.15\pi)} = -0.346817$$

and

$$\kappa = \cot\left(\frac{\omega_{p1} - \omega_{p2}}{2}\right)\tan\frac{\Omega_p}{2} = \cot\left(\frac{0.45\pi - 0.75\pi}{2}\right)\tan\frac{0.15\pi}{2} = \cot(-0.15\pi)\tan(0.075\pi) = -0.471181.$$

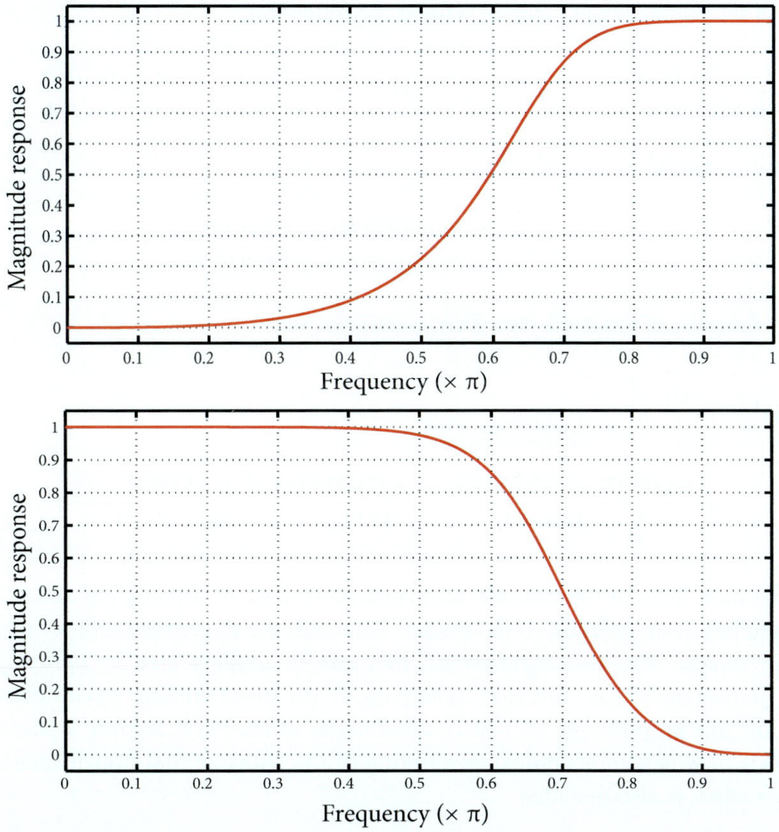

Fig. 12.112 The amplitude response of the low-pass and high-pass filters that have resulted from the application of the low-pass ⟶low-pass and low-pass ⟶high-pass transformations to the prototype digital low-pass filter

Therefore, the constants α_1 and α_2 of the low-pass ⟶band-pass transformation are calculated as

$$\alpha_1 = -\frac{2\alpha\kappa}{k+1} = \frac{-2 \times (0.346817) \times (-0.471181)}{-0.471181+1} = 0.618032$$

$$\alpha_2 = \frac{\kappa - 1}{\kappa + 1} = \frac{-0.471181 - 1}{-0.471181 + 1} = -2.782012,$$

whereas the equation of the transformation takes the form

$$z = -\frac{\tilde{z}^2 + 0.618032\tilde{z} - 2.782012}{1 + 0.618032\tilde{z} - 2.782012\tilde{z}^2}.$$

Therefore, the transfer function of the band-pass filter is calculated as

$$\mathcal{H}_{\mathrm{P}}(\tilde{z}) = \mathcal{H}(z)\Big|_{z=-\frac{\tilde{z}^2+0.618032\tilde{z}-2.782012}{1+0.618032\tilde{z}-2.782012\tilde{z}^2}}$$

$$= \frac{0.008616 + 0.025848z^{-1} + 0.025848z^{-2} + 0.008616z^{-3}}{1 - 2.064414z^{-1} + 1.519112z^{-2} - 0.385767z^{-3}}\Bigg|_{z=-\frac{\tilde{z}^2+0.618032\tilde{z}-2.782012}{1+0.618032\tilde{z}-2.782012\tilde{z}^2}}.$$

(d) Finally, in the case of the low-pass ⟶band-stop transformation the constant α is calculated as before, and has a value of $\alpha = -0.346817$, whereas the value of the constant κ is now calculated from the equation

$$\kappa = \tan\left(\frac{\omega_{p2} - \omega_{p1}}{2}\right)\tan\frac{\Omega_p}{2} = \tan\left(\frac{0.45\pi - 0.75\pi}{2}\right)\tan\frac{0.15\pi}{2} = \tan(-0.15\pi)\tan(0.075\pi) = -0.122326.$$

Therefore, the values of the parameters α_1 and α_2 will be the following:

$$\alpha_1 = -\frac{2\alpha}{k+1} = \frac{-2 \times (0.346817)}{-0.122326 + 1} = -0.790309 \quad \text{and} \quad \alpha_2 = \frac{1-\kappa}{1+\kappa} = \frac{1+0.122326}{1-0.122326} = \frac{1.122326}{0.877674} = 1.278750,$$

whereas the transformation equation takes the form

$$z = \frac{\tilde{z}^2 - 0.790309\tilde{z} + 1.278750}{1 - 0.790309\tilde{z} + 1.278750\tilde{z}^2}.$$

Therefore, the transfer function of the band-pass filter is calculated as

$$\begin{aligned}\mathcal{H}_{\text{band-pass}}(\tilde{z}) &= \mathcal{H}(z)\Big|_{z=\frac{\tilde{z}^2-0.790309\tilde{z}+1.278750}{1-0.790309\tilde{z}+1.278750\tilde{z}^2}} \\ &= \frac{0.008616 + 0.025848z^{-1} + 0.025848z^{-2} + 0.008616z^{-3}}{1 - 2.064414z^{-1} + 1.519112z^{-2} - 0.385767z^{-3}}\Bigg|_{z=\frac{\tilde{z}^2-0.790309\tilde{z}+1.278750}{1-0.790309\tilde{z}+1.278750\tilde{z}^2}}\end{aligned}$$

The construction of the final form of the last two transformations and the calculation and design of the corresponding amplitude responses are left as an exercise to the reader. Although the implementation of these operations is relatively complex, *because they involve algebraic expressions, they can be implemented by applying successive convolutions to the corresponding polynomials.* The function zmapping of the programming environment of MATLAB has been implemented exactly for this purpose and its source code exists in the book *Digital Signal Processing Using Matlab V.4* by Ingle and Proakis in which an analytical description of the above digital transformations, along with indicative examples of their application, is also included.

12.17 Problems

1. Convert the numbers 15, 246, 1358, 8192, 11384, and 35627 (a) using the appropriate number of bits and (b) using 16 bits.
2. Determine the range of numbers covered by a set of 40 bits (a) for the unsigned case and (b) for the signed case.
3. Perform the additions 457 + 1046, 12549 + 192345 and 675 + 763 + 12340 + 37621 in the binary system.
4. Convert the binary numbers 1101101110, 0110110101, 10011011101111101, and 1101000111101010101101 to the decimal system, assuming that they describe unsigned integers.
5. Convert the numbers 2543, 267, and 3468 to the binary system and then determine the complement to 1 and the complement to 2.
6. Calculate the values of the complement to 1 and the complement to 2 for the binary numbers of Problem 4.
7. Perform the subtraction $(1110)_2 - (1011)_2$ in the binary system using the representation (a) complement to 1 and (b) complement to 2.
8. Design a circuit that determines in a serial manner the complement to 2 of a binary number.
9. Describe the number $-0/00057342$ in the floating-point representation and with the coefficient of the number encoded according to the representation of the complement to 2.
10. Convert the binary numbers of Problem 4 to the decimal system, assuming the sign-magnitude representation.
11. Perform the multiplications $(1011)_2 \times (01101)_2$ and $(1111010)_\times(111011)_2$.
12. Convert the numbers 55, -85, and -3500 to the binary system using (a) 6 bits, (b) 8 bits, (c) 16 bits, and (d) $32 bits$ and report the cases in which overflow occurs.
13. Prove that the encoding of the number -53851 of the decimal system to the binary system according to the IEEE 754 single-precision standard leads to the binary number 11000111010100100101101100000000.
14. Prove that the encoding of the number -0.3 of the decimal system to the binary system according to the IEEE 754 single-precision standard leads to the binary number 10111110100110011001100110011001.
15. Prove that the encoding of the number 2.375 of the decimal system to the binary system according to the IEEE 754 single-precision standard leads to the binary number 01000000000110000000000000000000.

16. Prove that the encoding of the number 2.375 of the decimal system to the binary system according to the IEEE 754 double-precision standard leads to a binary number of length 64 bits, the value of which is

$$0100000000000011000000000000000000000000000000000000000000000000.$$

17. Determine the minimum number of bits of a multiplier that are required in order for it to maintain at its input an SNR ratio with a value above 80 dB, if the quantization process is carried out using the rounding technique.
18. Design the structure of a parallel multiplier when the representation of numbers uses the complement to 2.
19. Repeat the process of Sect. 12.12.2 (see page 1237) using the *rectangular window* function, as well as the *Bartlett*, *Hamming*, and *Blackman* windows and comment on the results you obtain regarding the characteristics of the amplitude response in relation to those we obtained using the von Hann window. For the calculation of the amplitude response of the filter to be constructed use the software application of your choice.
20. Repeat the process of Sect. 12.12.2 for the construction of a low-pass FIR filter with a frequency response function of the form $0.95 < \mathcal{H}(e^{j\omega}) < 1.05$ for the value range $0 \le |\omega| \le 0.25\pi$ and $-0.1 < \mathcal{H}(e^{j\omega}) < 0.1$ for the value range $0.35\pi \le |\omega| \le \pi$. Determine the type and length of the appropriate window and design the impulse and frequency response of the resulting filter.
21. Design an FIR filter with linear phase response and order value $N = 25$ that approximates the ideal frequency response

$$\mathcal{H}_d(e^{j\omega}) = \begin{cases} 1 \text{ for } |\omega| \le \pi/6 \\ 0 \text{ for } \pi/6 \le |\omega| \le \pi. \end{cases}$$

Determine the coefficients of the digital filter and design the amplitude response and the phase response using the rectangular window, as well as the Bartlett and Hamming windows.
22. Suppose we want to design a band-pass filter that meets the following specifications:

- Sampling frequency $F_s = 44.1$ KHz
- 1^{η} cutoff zone: $0 - 3$ KHz with attenuation 90 dB
- Passband: $4 - 6$ KHz with variation 0.1 dB
- 2^{η} cutoff zone: $8 - 10$ KHz with attenuation 90 dB

Determine the parameter α and the order M of the filter as well as the value of the time delay.
23. Repeat Exercise 12.12.5 (see page 1272) for a low-pass filter of type I with order $N = 31$, with parameter values $\omega_p = 0.35\pi$, $\omega_s = 0.75\pi$ and error values $K_s = 1$ and $K_p = 4$.
24. Repeat Exercise 12.13.1 (see page 1292) for the case of a digital Butterworth filter with operating specifications $0.89765 \le |\mathcal{H}(e^{j\omega})| \le 1$ for $0 \le |\omega| \le 0.3\pi$ and $|\mathcal{H}(e^{j\omega})| \le 0.175342$ for $0.4 \le |\omega| \le \pi$.
25. Repeat Exercise 12.13.1 for the case of a digital filter with operating specifications $\omega_p = 0.12\pi$, $\omega_s = 0.38\pi$, $A_p = 2.5$ dB, and $A_s = 25$ dB using as a prototype filter (a) a Chebyshev II filter and (b) an elliptic filter.
26. Assuming that the discrete transfer function is

$$\mathcal{H}(z) = \frac{5}{1 - e^{-0.3}z^{-1}} - \frac{1}{1 - e^{-0.5}z^{-1}}.$$

It has emerged through the application of the invariant impulse response method to determine the function $\mathcal{H}(s)$ of the analog filter from which it has emerged. Is this function unique? For the solution of the exercise, consider the value $T = 2$.
27. Transform the analog filter with transfer function

$$\mathcal{H}(s) = \frac{s+2}{s^2 + 8s + 15}$$

into a digital filter, using the invariant impulse response method. For the solution of the exercise, consider the value $T = 0.1$.

28. Repeat Exercise 12.13.2 (see page 1302) for the case of a Butterworth digital filter with operating specifications $0.89765 \leq |\mathcal{H}(e^{j\omega})| \leq 1$ for $0 \leq |\omega| \leq 0.3\pi$ and $|\mathcal{H}(e^{j\omega})| \leq 0.175342$ for $0.4 \leq |\omega| \leq \pi$.
29. An analog low-pass elliptic filter is characterized by specifications $\omega_p = 0.65$ rad/sec, $\omega_s = 1.55$ rad/sec, $A_p = 0.5$ dB, and $A_s = 36.5$ dB. Construct the corresponding digital filter with the help of bilinear transformation.
30. Using the pole-zero placement method, design a band-pass filter, the amplitude response of which presents a maximum at the position $f_0 = 100$ Hz. Determine the transfer function, its frequency response, and the difference equation that describes it, if its poles are placed at a distance $r = 0.995$ from the center of the unit circle. What is the bandwidth of the filter corresponding to a 3-dB attenuation? (*Hint:* For the calculation of the filter bandwidth corresponding to a 3-dB attenuation, use the approximate relation $\Delta\omega \approx 2(1-r)$, which holds for pole positions very close to the circumference of the unit circle).
31. Using the pole-zero placement method, design a band-pass filter with central frequency $\omega_0 = \pi/2$ and bandwidth corresponding to a 3-dB attenuation $\Delta\omega = \pi/8$. Determine the transfer function and its frequency response, as well as the difference equation that describes its operation. What is the impulse response of the filter to be constructed?
32. Using the pole-zero placement method, design a band-stop filter with central frequency $\omega_0 = \pi/10$ and bandwidth corresponding to a 3-dB attenuation $\Delta\omega = \pi/20$. Determine the transfer function and its frequency response, as well as the difference equation that describes its operation. What is the impulse response of the filter to be constructed?
33. Suppose we want to design a low-pass FIR filter with a generalized linear phase using the window method, starting from a low-pass filter with impulse response

$$h_{\text{low-pass filter}}[n] = \frac{1}{\pi n} \sin\left(\frac{\pi n}{4}\right)$$

The impulse response of such a filter, as is well known, is given by the relation

$$h[n] = \frac{1}{\pi\left(n - \frac{N-1}{2}\right)} \sin\left[\frac{\pi}{4}\left(n - \frac{N-1}{2}\right)\right]$$

for values $0 \leq n \leq N-1$, where $w[n]$ is the appropriate window function. Calculate and plot the quantities $h[n]$, $|\mathcal{H}_d(e^{j\omega})|$ and $\measuredangle\mathcal{H}_d(e^{j\omega})$ (a) for a rectangular window of $N = 21$ in length, (b) for a Hamming window of $N = 21$ in length, and (c) for a Hamming window of $N = 21$ in length and compare the results. To create the functions $|\mathcal{H}_d(e^{j\omega})|$ and $\measuredangle\mathcal{H}_d(e^{j\omega})$, which you will plot in the value range $[0, \pi]$, use a discrete Fourier transform of 4096 points in length.
34. Using the frequency sampling method, construct a low-pass FIR filter of a generalized linear phase of $N = 121$ in length with cutoff frequencies $\omega_c = \pi/4$ and then calculate and plot the frequency and impulse response of the filter.
35. Design a FIR filter of generalized linear phase of length $N = 25$, which will approximate the ideal frequency response

$$\mathcal{H}_d(e^{j\omega}) = \begin{cases} 1 & |\omega| < \pi/6 \\ 0 & \pi/6 \leq |\omega| \leq \pi \end{cases}$$

using the window method and the Hamming and von Hann windows as well as the rectangular window.
36. Resorting to the use of bilinear transformation, convert the analog filter with the transfer function

$$\mathcal{H}(s) = \frac{s+1}{s^2 - s + 2}$$

in a digital filter for parameter value $T_s = 1$ and then determine the transfer function of the digital filter.
37. Using the window method, design a low-pass type I filter with operating specifications $\omega_p = 0.2\pi$, $\omega_s = 0.3\pi$, $\delta_p = 0.01$, and $\delta_s = 0.01$.
38. Design a band-pass FIR filter of $N = 5$ in length with cutoff frequencies 1600 Hz and 1800 Hz and a sampling rate equal to 8 KHz using the Hamming window function.
39. Design a band-stop FIR filter of $N = 5$ in length with cutoff frequencies 1200 Hz and 1900 Hz and a sampling rate equal to 7 KHz using the Hamming window function.
40. Construct the transfer function and the difference equation and design the amplitude response for the filters of Problems 38 and 39 for frequency values $\omega = 0$, $\omega = \pi/4$, $\omega = \pi/2$, $\omega = 3\pi/4$, and $\omega = \pi$ (rad).

41. Suppose we want to design a low-pass FIR filter with operating specifications $f_p = 0 \sim 800\,\text{Hz}$, $f_s = 1200 \sim 4000\,\text{Hz}$, $\delta_p = 0.1\,\text{dB}$, and $A_s = 40\,\text{dB}$, and a sampling rate of 8 kHz. Determine the type and appropriate length of the window function to be used, as well as the cutoff frequency of the filter.
42. Repeat Problem 41 for operating specifications $f_p = 0 \sim 1500\,\text{Hz}$, $f_s = 2000 \sim 4000\,\text{Hz}$, $\delta_p = 0.02\,\text{dB}$, and $A_s = 60\,\text{dB}$, and a sampling rate of 7 kHz.
43. Using the Parks-McClellan method, design a digital low-pass filter with a passband in the frequency range $0 \sim 200\,\text{Hz}$, a stopband in the frequency range $300 - 500\,\text{dB}$, and with parameter values $\delta_p = 1\,\text{dB}$ and $A_s = 40\,\text{dB}$. The sampling rate for this filter is equal to 1 kHz.
44. Design a digital high-pass filter with a passband in the frequency range $6 \sim 8\,\text{kHz}$, a stopband in the frequency range $0 - 5.5\,\text{kHz}$, a passband ripple $-1\,\text{dB} \leq |\mathcal{H}(j\omega)| \leq 0\,\text{dB}$ for $\Omega_p \leq |\Omega| \leq 8\,\text{kHz}$, and stopband attenuation $|\mathcal{H}(j\omega)| \leq -60\,\text{dB}$ for $|\Omega| \leq \Omega_s$.
45. Assume that the amplitude response of an ideal filter has the form

$$\mathcal{H}_d(e^{j\omega}) = \begin{cases} 1 \; \dfrac{\omega_s + \omega_p}{2} \leq |\omega| \leq \pi \\ 0 \; 0 \leq |\omega| \leq \dfrac{\omega_s + \omega_p}{2}. \end{cases}$$

(a) What is the ideal impulse response $h_d[n]$? (b) What is the optimal window function that should be used? Determine the optimal type of window as well as its length. (c) Determine the impulse response $h[n]$ of the actual filter.
46. Repeat problem 45 using the Kaiser window function. What is the minimum length of the impulse response and the value of the Kaiser window parameter α so that the operating specifications of the filter are met? What is the phase delay for this filter?
47. Implement a digital low-pass filter using a Butterworth filter as a prototype and resorting to the use of bilinear transformation. The operating specifications of the filter to be designed are $\omega_p = 0.2\pi$, $\omega_s = 0.6\pi$, passband deviations $-3\,\text{dB} \leq |\mathcal{H}(j\omega)| \leq 0\,\text{dB}$ for $0 < |\omega| < \omega_p$, and attenuation in the stopband $|\mathcal{H}(j\omega)| \leq -25\,\text{dB}$ for $\omega_s \leq |\omega| \leq \pi$. (a) What are the frequencies Ω_c, Ω_p and Ω_s in the continuous-time domain? (b) What is the order N of the analog filter? (c) What are the transfer functions $\mathcal{H}(s)$ and $\mathcal{H}(z)$ for the continuous and discrete cases?
48. Provide two different types of implementation for digital filters with transfer functions

$$\mathcal{H}(z) = 0.0047 + 0.0117z^{-2} + 0.0025z^{-4} + 0.0157z^{-6} \qquad \text{and}$$

$$\mathcal{H}(z) = \left(\frac{z^2 - 1.359z + 1}{z^2 - 1.919z + 0.937}\right)\left(\frac{z^2 - 1.875z + 1}{z^2 - 1.937z + 0.953}\right).$$

49. Construct the parallel implementation of the transfer function

$$\mathcal{H}(z) = \frac{z^3 + 5z^2 + (12/7)z + (5/6)}{\left(z^2 + \frac{1}{3}z + \frac{1}{2}\right)\left(z + \frac{1}{5}\right)}.$$

50. Prove that a transfer function of the form

$$\mathcal{H}(z) = \frac{(r^{-1}z)^{-(M+1)} - 1}{re^{j2\pi/(M+1)}z^{-1} - 1}$$

corresponds to a digital filter with FIR.
51. Show that the L_2 norm of the transfer function

$$\mathcal{H}(z) = \frac{b_1 z + b_2}{z^2 + a_1 z + a_2}$$

has the form

$$\|\mathcal{H}(z)\|_2^2 = \frac{(b_1^2 + b_2^2)(1 + a_2) - 2b_1 b_2 a_1}{(1 - a_1^2 + a_2^2 + 2a_2)(1 - a_2)}.$$

52. Considering the transfer function

$$\mathcal{H}(z) = \frac{1 - (az^{-1})^{M+1}}{1 - az^{-1}}, \qquad |\alpha| < 1,$$

calculate the scaling coefficients using the norms L_2 and L_∞.

53. (Diniz, da Silva and Netto) Prove that the norms L_2 and L_∞ of the transfer function

$$\mathcal{H}(z) = \frac{\gamma_1 z + \gamma_2}{z^2 + m_1 z + m_2}$$

are given by the equations

$$\|\mathcal{H}(e^{j\omega})\|_2^2 = \frac{\gamma_1^2 + \gamma_2^2 - 2\gamma_1\gamma_2\left(\frac{m_1}{m_2+1}\right)}{(1 - m_2^2)\left[1 - \left(\frac{m_1}{m_2+1}\right)^2\right]} \quad \text{and}$$

$$\|\mathcal{H}(e^{j\omega})\|_\infty^2 = \max\left\{\left(\frac{\gamma_1 + \gamma_2}{1 + m_1 + m_2}\right)^2, \left(\frac{-\gamma_1 + \gamma_2}{1 - m_1 + m_2}\right)^2, \frac{\gamma_1^2 + \gamma_2^2 + 2\gamma_1\gamma_2\zeta}{4m^2[(\zeta - \eta)^2 + \upsilon]}\right\},$$

with the quantities appearing in the last expression defined as

$$\eta = \frac{-m_1(1 + m_2)}{4m_2}, \qquad \upsilon = \left(1 - \frac{m_1^2}{4m_2}\right)\frac{(1 - m_2)^2}{4m_2} \qquad \zeta = \begin{cases} \operatorname{sat}(\eta) & \gamma_1\gamma_2 = 0 \\ \operatorname{sat}\left[v\left(\sqrt{\left(1 + \frac{\eta}{v}\right)^2 + \frac{\upsilon}{v^2}} - 1\right)\right] & \gamma_1\gamma_2 \neq 0, \end{cases}$$

where

$$v = \frac{\gamma_1^2 + \gamma_2^2}{\gamma_1\gamma_2} \quad \text{and} \quad \operatorname{sat}(x) = \begin{cases} 1 & x > 1 \\ -1 & x < -1 \\ x & -1 \le x \le 1. \end{cases}$$

54. Considering the transfer function

$$\mathcal{H}(z) = \frac{0.5z^2 - z + 1}{(z^2 - z + 0.5)(z + 0.5)},$$

formulate it as a composite system that arises (a) from the series connection and (b) from the parallel connection of simple stages that constitute FIR type I filters expressed in their direct form.

55. Considering the transfer function

$$\mathcal{H}(z) = -z^{-2N} + \cdots - z^{-6} + z^{-4} - z^{-2} + 1,$$

construct an implementation that uses the minimum possible number of adders.

13 Elements of Statistical Signal Processing

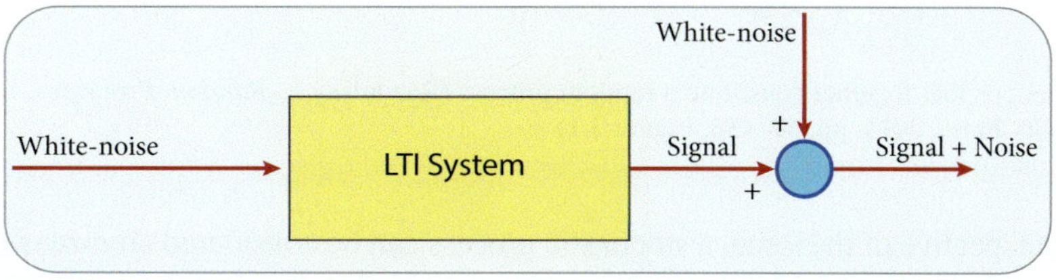

13.1 Statistical Signal Processing

Although almost all of the analysis techniques presented in this book are applied to signals characterized as *deterministic*, a characterization that is also transferred to these techniques themselves, in practice, we also encounter so-called *random signals*, which are signals that can be associated with an *infinite* number of waveforms and which appear in very important and well-known areas of digital signal processing, such as *digital telecommunications*. The most characteristic example of such a random signal is noise of all kinds, which appears in many well-known cases, such as the noise generated by an amplifier or noise *quantization* and *rounding*. It is not difficult to understand that, due to their particular nature, these random signals *cannot be studied analytically*, and for this reason, their analysis is carried out with the help of *statistical properties* whose values can be calculated from the content of the signal. The set of all these statistical techniques used to characterize this special category of random signals constitutes the field of *statistical signal processing*. Readers interested in delving deeper into these techniques and methods are referred to more specialized textbooks specifically on this subject.

13.2 Stochastic or Random Processes

A concept that occupies a central position in statistical signal processing, as well as other fields, is the stochastic or random process. This process is defined as the *mathematical model* of a phenomenon that, from an observer's point of view, seems to evolve over time in a *random* and *unpredictable* way. This is substantiated by the structure and characteristics of the time series recorded in the context of the study. The time series may comprise data of various kinds, including *voltage and temperature measurements, stock market index values, sequences of binary digits* generated by a digital filter or a telecommunication system, and so forth. The statistical processing of these signals allows us to describe these time series, as well as the involved phenomena, through the calculation of *statistical indicators* and the construction of mathematical models, which on the one hand constitute *reliable and reasonable approximations* of real phenomena and, on the other hand, allow for the effective management of those phenomena, which in many cases consist of an ability to make predictions. For this reason, the random process is one of the most useful study tools, especially at the moment that allows for the modeling of natural phenomena without even having to resort to *sampling processes*. The most characteristic cases, which are described with the help of random processes, are the *modeling* of noise that appears in various natural systems, as well as all kinds of random phenomena, which, on the one hand, do not constitute noise and, on the other hand, possess a nature and traits that are unknown, as happens, for example, with the *transmission of multimedia signals* (i.e., image, sound, and video) in telecommunications systems. The only way to describe these processes is to model them as random processes.

A. I. Margaris, *Continuous and Discrete-Time Signals and Systems*,
https://doi.org/10.1007/978-3-031-81209-5_13

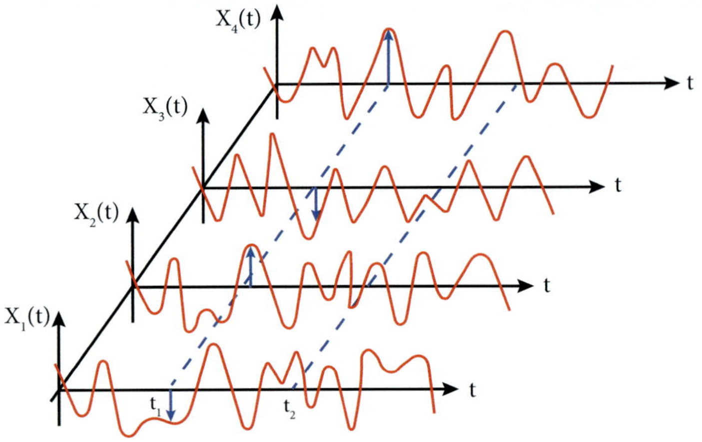

Fig. 13.1 A collection of functions that together constitute a random process (Razdolsky L, *Random Processes*, in *Probability-Based Structural Fire Load*, Cambridge University Press, 2014, pp. 89–136, Figure 3.1)

From a mathematical perspective of the issue, a stochastic process can be considered an *extension* or *generalization* of the concept of the *random discrete sequence* (which in turn constitutes a *generalization* of the concept of the *random variable*) and, more specifically, an ordered collection of random variables whose arrangement is based on the moment with which they are associated. This generalization essentially constitutes *a transition from the domain of discrete-time to the domain of continuous-time*, although in reality, as continuous variables, they can use other quantities besides time, such as, for example, an angle. Specifically, considering an experiment whose results constitute the elements $\{\xi_1, \xi_2, \ldots, \xi_n\}$ of a *sample space* Ω, the random variable is defined as a *rule* that assigns a number $\boldsymbol{x}(\xi)$ to each element $\xi_i \equiv \xi \in \Omega$. The random process constitutes an *extension* of this situation, in the sense that what corresponds to the element ξ *is not a simple number* $\boldsymbol{x}(\xi)$, *but a function* $\boldsymbol{x}(t, \xi)$ and, therefore, constitutes a *family of functions* of time t, which also depend on the parameter ξ, with the set from which the variable t takes values being the set $\mathbb{R}$ of real numbers. Apart from the set of these temporal functions, the definition of the random process also includes the related *probability distributions*, which usually constitute combined or marginal probability density functions (PDFs) for all random variables that jointly define the process at each point in time. It is interesting that, although from the mathematical side of the issue, the extension of the concept of a random variable to the concept of a random process is relatively simple; the real difficulty lies *in the definition of correlation* that exists between the mathematical representation of random variables and the physical properties of the process.

In the most general case, the arguments ξ and t of a random process take various values. On the other hand, if the variable ξ is constant and equal to ξ_0 and *only* the time t varies, then the function $\boldsymbol{x}(t, \xi_0)$ is characterized as a *sample* of the random process, with the set of these samples corresponding to an equal number of random variables (which collectively define a random vector) whose statistical characterization is based on the use of a joint PDF $f(x_1, x_2, \ldots, x_n; t_1, t_2, \ldots, t_n)$ for each value of n. If we know this *combined probability density function*, then we have the ability to statistically describe the random process, in the same way we did in the case of random sequences. In the field of signal processing, the samples of a random process are referred to as *random signals*. It is worth emphasizing that, in the general case, only a single realization of the process is available and is used for its analysis. This condition leads directly to the concept of *ergodicity*, which is a particularly important property for the practical exploitation of the statistical characteristics of the process. Conversely, if time t assumes a fixed value $t = t_0$, the random process $\boldsymbol{x}(t_0, \xi)$ degenerates into a *random variable*. Finally, if both arguments t and ξ have the constant values t_0 and ξ_0, the function $\boldsymbol{x}(t_0, \xi_0)$ is just a simple number.

An example of a random process is depicted in Fig. 13.1. In this figure, the collection of the depicted functions constitutes the random process and corresponds to various values of the variable ξ, while each of these processes constitutes a sample function.

The *stochastic processes* encountered in practice can be categorized in several different ways, with the criterion being the characteristic that is suitable in every case, and the most important of these categorizations are presented below.

13.2.1 Continuous and Discrete Processes

A random process $\boldsymbol{x}(t)$ is aforementioned to be *continuous* if its random variables can take any value from a predefined continuous range of values, which can be *finite or infinite*, that is, if both the random variable $\boldsymbol{x}$ and the time t take *continuous*

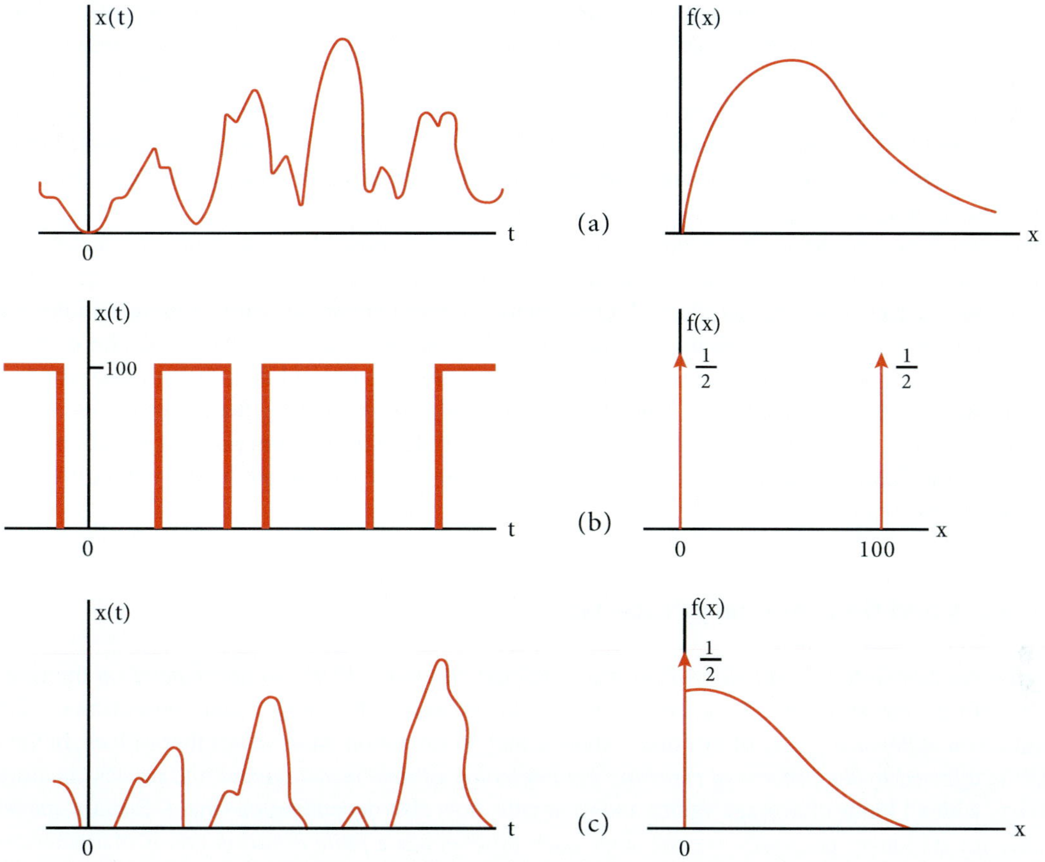

Fig. 13.2 Examples of (**a**) continuous, (**b**) discrete, and (**c**) mixed random processes. For each type of process, the left image depicts a typical function, a sample of the process, while the right image depicts the corresponding PDF

values. In a stricter description, a stochastic process is aforementioned to be *continuous* if it is described by a *continuous cumulative distribution probability*. Typical examples of a continuous process are (a) the *thermal noise* that appears in all kinds of electronic circuits (since its amplitude can take any positive or negative value) and (b) the *quantization error* that appears during the process of converting analog values to digital values (since this can take any value within a finite range of values, which is determined by the size of the quantization step). In both these cases, the PDF is a continuous function, which in the first case is defined over the entire line of real numbers, while in the second case it is uniformly distributed throughout the error value range.

On the other hand, a random process $\boldsymbol{x}[t]$ is aforementioned to be *discrete* if its random variables can take *only discrete* values, that is, specific values that belong to a set of values that can even be infinite in number. However, time t can take *any value* and so constitutes a *continuous* variable. A typical example of such a process is (a) *the number of active telephone conversations* at any given time (whose value, obviously, is an integer number) and (b) the result of the quantization process of a discrete sequence (since the values that will result from this process belong to a specific set of values). The basic characteristic of discrete random processes is that the *PDFs* that describe them contain *impulsive functions*, something that does not apply to continuous random processes. It is also worth noting that, apart from continuous and discrete random processes, there are also *mixed random processes* that contain both continuous and discrete components, such as, for example, the result of the rectification process of a function of time, a feature that naturally characterizes the corresponding PDF. Typical examples of continuous, discrete, and mixed random processes along with their corresponding PDFs are depicted in Fig. 13.2a–c.

13.2.2 Causal and Noncausal Processes

In the vast majority of cases, each sample function of a random process constitutes a *random function* of time, and therefore, an exact prediction of future values based on the values that have already been observed and recorded is *essentially impossible*. Such a process is aforementioned to be *noncausal*, and it is a fact that *almost all processes observed in nature are characterized as noncausal* since the underlying mechanism leading to their creation either *cannot be observed* or is *so complex* that it is impossible to draw any conclusions. A classic example of such a random process from the field of physics, in which statistics plays a dominant role, is the *motion of the molecules of a gas within a closed container*, which is one of the most important problems of thermodynamics. In this case, the function $\boldsymbol{x}(t, \xi)$ describes the *motion of an entire set of gas molecules*, while the function $\boldsymbol{x}(t, \xi_i)$ describes the motion of the ith molecule and, according to our earlier discussion, constitutes an example of a random process.

On the other hand, however, there are random processes in which it is possible to accurately predict the future values of each sample function, based on the knowledge of the values of the samples that have already been recorded. An example of such a process is the trend $\boldsymbol{x}(r) = \boldsymbol{r}\cos(\omega t + \boldsymbol{\varphi})$ of an alternating current generator, with amplitude values $\boldsymbol{r}$ and frequency ω being constant and the phase $\boldsymbol{\varphi}$ being a *random variable* described by some *PDF*. In other words, for each sample function $\boldsymbol{x}(t, \xi) = \boldsymbol{r}(\xi)\cos[\omega t + \boldsymbol{\varphi}(\xi)]$, the angle φ has the same value for all time instances t but different values for the other samples of the collection. Note that the aforementioned random process is *periodic*, something that, however, *is not* a prerequisite in order for a random process to be deterministic, as happens, for example, with the process $\boldsymbol{x}(t) = A\exp -\beta t$, in which the parameters A and β are random variables, which have a constant value for each sample function but different values for the different functions of this type.

13.2.3 Stationary and Nonstationary Processes

A process is aforementioned to be *stationary* if the marginal and combined PDFs *do not depend* on the time point chosen as the beginning of time measurement. As a consequence of this property, all average values and instances defined for each process depend on the *difference* $t_2 - t_1$ of two time values t_1 and t_2 and not on these values themselves. In the opposite case, where the PDF *is affected by the process of choosing the beginning of time measurement*, the process is aforementioned to be *nonstationary*, with all kinds of average values and moments, now also depending on time t. Strictly speaking, in nature, *it is not possible* for stationary processes to exist since each process has a *finite duration* and is characterized by a certain moment of *start* and *termination*, although in practice, if during the observation of a process its characteristics seem to *not change significantly* over time, then we can approximately characterize the process as *stationary*, so that we can study it, using the techniques and tools of the field, since the study of nonstationary processes is very difficult.

The characterization of a process as stationary greatly facilitates its study. The reason for this is that all kinds of statistical properties and characteristics of the process must generally be calculated *for each time instance t*, and consequently, we must calculate *as many sets of properties as there are time instances of interest*. However, if the process is *stationary*, that is, its statistical properties *do not depend* on each individual time instance but rather, as we will see, *on the time distance between two time instances*, then we can study the process, considering only a set of parameters that significantly facilitates the study.

Generally speaking, causal processes can be considered stationary *only under the regime of the validity of certain special conditions*, conditions whose existence is *usually* accepted but often simply assumed. On the other hand, in the case of noncausal processes, whether a process is stationary or not, depends *on the underlying mechanism of the process creation*, as well as *on the duration of the observation period*. The common practice is to accept here, too, that the process we observe is stationary, unless there are clear indications to the contrary. For example, the *thermal noise* generated by the movement of the behavior of electrons within a resistance can, with a fair degree of truth, be characterized as *stationary*, something that *does not hold*; however, when there is a sudden increase in the intensity of the electric current flowing through the resistance, as in this case, the behavior of the noise will change dramatically. Due to the great importance of stationary processes, these will be studied in detail in the relevant section, where we will define *two different versions of stationarity*: *strict-sense stationarity* (SSS), in which all time instances related to the process are involved, and *wide-sense stationarity* (WSS), in which the requirement of independence from the start of time measurement is limited only to the *mean*, the *mean square*, the *variance*, and the *correlation coefficient*.

13.2.4 Ergodic and Nonergodic Processes

A special category of stationary processes are the so-called *ergodic processes*. In these processes, *each sample function is characterized by the same statistical behavior as that characterizing the entire collection of samples*, and therefore, it is possible to determine the statistical behavior of this collection, which is essentially the process it defines, by examining *a single such function*. It is not difficult to understand that in the case of ergodic processes, tendencies of all kinds can be calculated *both as temporal averages of a simple sample* (note that, although in the simple case, the calculation requires a period of *infinite duration*, in practice a period of *finite* duration is always used, thereby leading to an *approximation* of the actual value) and as average values of an entire collection of samples. A necessary condition for a process to be characterized as ergodic is to be stationary, so *processes that are not stationary are not ergodic*, although it should be emphasized that there are also stationary processes that are not ergodic. In the general case, *it is difficult, if not impossible*, to prove that ergodicity is a characteristic of a physical process since, in practice, usually a single sample function is recorded, so it is *impossible* to show whether or not the property dictated by the definition of ergodicity applies. However, it is customary, as with stationarity, to accept that the process being studied is *ergodic*, unless there are indications that argue in favor of the opposite. Ergodic processes and the conditions of ergodicity are presented in detail in another section (Sect. 13.2.8, page 1404).

For more examples of known random processes that often appear in engineering sciences, such as *asynchronous binary signaling*, *random arrival processes with a Poisson distribution*, *random telegraphic signals*, *digital modulation with frequency shift keying*, and *Wiener and Markov processes*, see Stark and Woods, pages 406–430, as well as many other works given in the references.

13.2.5 Stochastic Processes and Their properties

Generally speaking, a stochastic process is a *noncountable collection of infinitely many random variables* for various values of t. For a specific value of t, the function $\boldsymbol{x}(t)$ is a *random variable* with cumulative distribution function

$$F(x, t) = P\{\boldsymbol{x}(t) \le x\}$$

which is defined as the *probability* of the event $\{\boldsymbol{x}(t) \le x\}$ consisting of all samples ξ, such that, at the considered time t, the samples $\boldsymbol{x}(t, \xi)$ of the process *do not exceed* the value x. The partial derivative of this function with respect to x, that is,

$$f(x, t) = \frac{\partial F(x, t)}{\partial x}$$

is the *first-order density* of $\boldsymbol{x}(t)$. On the other hand, the *second-order distribution* of $\boldsymbol{x}(t)$[1] is defined as the joint distribution

$$F(x_1, x_2; t_1, t_2) = P\{\boldsymbol{x}(t_1) \le x_1, \boldsymbol{x}(t_2) \le x_2\}$$

of the random variables $\boldsymbol{x}(t_1)$ and $\boldsymbol{x}(t_2)$, while the corresponding density is

$$f(x_1, x_2; t_1, t_2) = \frac{\partial^2 F(x_1, x_2; t_1, t_2)}{\partial x_1 \partial x_2}$$

The preceding equations can be generalized for the case of statistical order n, which is related to functions of n random variables with the *PDF* of order n defined as

$$f(x_1, x_2, \cdots, x_N; t_1, t_2, \ldots, t_N) = \frac{\partial^2 F(x_1, x_2, \ldots, x_N; t_1, t_2, \ldots, t_N)}{\partial x_1 \partial x_2 \cdots \partial x_N}$$

If this combined PDF (and, correspondingly, the cumulative distribution function) can be expressed as the product

$$f(x_1, x_2, \cdots, x_N; t_1, t_2, \ldots, t_N) = f(x_1; t_1) f(x_2; t_2) \cdots f(x_N; t_N)$$

where $f(x_i; t_i)$ $(i = 1, 2, \ldots, N)$ is the PDF of the ith random variable, then the random (or stochastic) process is aforementioned to be an *independent* process.

---

[1] In the following description, for simplicity of symbolism, we will omit the dependence of the function $\boldsymbol{x}(t, \xi)$ on the variable ξ and simply write $\boldsymbol{x}(t)$, in exactly the same way as a random variable is symbolized, not as $\boldsymbol{x}(\xi)$, but simply as $\boldsymbol{x}$.

The most common properties used for the *statistical characterization* of a random or stochastic process $\boldsymbol{x}(t)$ are those that characterize the random variables, with the difference that these quantities are now *functions of time* due to the way random processes are defined. Considering, for example, the magnitude of the *correlation* defined between two random variables, this, as mentioned earlier, constitutes the *average value of the product* of the two variables, and because in the general case these variables are not necessarily related to functions of time, the value of the correlation is a simple *number*. However, if the two variables are samples of a random process, which correspond to two different time points t_1 and t_2, then this expected value depends *on the rate of change of these time functions*. Specifically, if the time points t_1 and t_2 are very close to each other, it is logical to assume that the random variables *will be highly correlated* since there is not enough time for a significant change of the function to take place. Conversely, if the distance between these time points is large, it is reasonable to surmise that the correlation that exists between these random variables should be relatively small. Also, it is logical to assume that since the correlation depends *on how quickly these time functions change*, it will also be a function of time itself and, more specifically, a function of the time points t_1 and t_2 and not a simple number, as in the preceding simple case. In full analogy with the case of random variables, if the two functions constitute *samples of the same random process*, then the correlation is characterized as an *autocorrelation*, while if, in contrast, they constitute samples of different processes, the correlation is known as a *cross-correlation*. Similar conclusions can be drawn for the average, which will also be a function of time, since its value, if calculated using a different time interval in each instance, will naturally be different.

From the mathematical side of the subject, the *average* $\mu(t)$ of the random process $\boldsymbol{x}(t)$ is given by the equation

$$\mu_{\boldsymbol{x}}(t) = E\{\boldsymbol{x}(t)\} = \int_{-\infty}^{\infty} x f(x, t) dx, \qquad -\infty < t < \infty \tag{13.1}$$

while, in a similar way, the *mean square* $E\{\boldsymbol{x}^2(t)\}$ is defined as[2]

$$\mu_{\boldsymbol{x}}(t) = E\{\boldsymbol{x}^2(t)\} = \int_{-\infty}^{\infty} x f(x, t) dx, \qquad -\infty < t < \infty \tag{13.2}$$

On the other hand, the *autocorrelation function* $R_{XX}(t_1, t_2)$, which, as mentioned previously, constitutes the expected value of the product $\boldsymbol{x}(t_1)\boldsymbol{x}(t_2)$, is given by the equation

$$R_{\boldsymbol{xx}}(t_1, t_2) = E\{\boldsymbol{x}(t_1)\boldsymbol{x}(t_2)\} = \int_{-\infty}^{\infty} dx_1 \int_{-\infty}^{\infty} x_1 x_2 f(x_1, x_2; t_1, t_2) dx_2, \qquad -\infty < t_1, t_2 < \infty \tag{13.3}$$

while the *cross-correlation* defined between two random processes $\boldsymbol{x}(t)$ and $\boldsymbol{y}(t)$ takes the form

$$R_{\boldsymbol{xy}}(t_1, t_2) = E\{\boldsymbol{x}(t_1)\boldsymbol{y}(t_2)\} = \int_{-\infty}^{\infty} dx_1 \int_{-\infty}^{\infty} x_1 y_2 f(x_1, y_2; t_1, t_2) dy_2, \qquad -\infty < t_1, t_2 < \infty \tag{13.4}$$

where we have defined $x_1 = \boldsymbol{x}(t_1)$ and $y_2 = \boldsymbol{y}(t_2)$. Similarly, the *cross-correlation* function $R_{\boldsymbol{yx}}(t_1, t_2)$ is defined as

$$R_{\boldsymbol{yx}}(t_1, t_2) = E\{\boldsymbol{y}(t_1)\boldsymbol{x}(t_2)\} = \int_{-\infty}^{\infty} dy_1 \int_{-\infty}^{\infty} y_1 x_2 f(y_1, x_2; t_1, t_2) dx_2, \qquad -\infty < t_1, t_2 < \infty \tag{13.5}$$

[2] Using the appropriate exponent k in this equation, we can, as is well known, define the kth moment of the random variable.

A classic example of the application of the cross-correlation function is its use in systems that receive more than one random signal as inputs or the case in which we wish to compare between them the *random voltage and current values* that appear at different points within a circuit.

Two other interesting quantities used to describe random processes are the *average power*, defined as the value $R(t,t)$ for $t_1 = t_2 = t$, or, in mathematical notation, $R_x(t,t) = E\{\boldsymbol{x}^2(t)\}$, and the *autocovariance* $C_{\boldsymbol{xx}}(t_1,t_2)$, defined as the covariance of $\boldsymbol{x}(t_1)$ and $\boldsymbol{x}(t_2)$, that is,

$$C_{\boldsymbol{xx}}(t_1,t_2) = E\{(\boldsymbol{x}(t_1) - \mu_{\boldsymbol{x}}(t_1))(\boldsymbol{x}(t_2) - \mu_{\boldsymbol{x}}(t_2))^*\} = R_{\boldsymbol{xx}}(t_1,t_2) - \mu_{\boldsymbol{x}}(t_1)\mu_{\boldsymbol{x}}(t_2) \tag{13.6}$$

with the value $C_{\boldsymbol{xx}}(t_1,t_2)$ for $t_1 = t_2 = t$ representing the *variability* of the stochastic process $\boldsymbol{x}(t)$, or, in mathematical notation, $C_{\boldsymbol{xx}}(t,t) = E\{|\boldsymbol{x}(t)|^2\}$. In the case where the process $\boldsymbol{x}(t)$ is of a *complex nature*, the autocorrelation function is defined as $R(t_1,t_2) = E\{\boldsymbol{x}(t_1)\boldsymbol{x}^*(t_2)\}$, from which it follows that $R_{\boldsymbol{xx}}(t_2,t_1) = E\{\boldsymbol{x}(t_1)\boldsymbol{x}^*(t_1)\} = R^*_{\boldsymbol{xx}}(t_1,t_2)$, while the same will apply to the *autocovariance*, so that we can write that $C_{\boldsymbol{xx}}(t_2,t_1) = C^*_{\boldsymbol{xx}}(t_1,t_2)$. In other words, these two functions are characterized by *Hermitian symmetry*, while additionally they are *diagonally dominant*, satisfying respectively the conditions

$$|R_{\boldsymbol{xx}}(t,s)| \le \sqrt{R_{\boldsymbol{xx}}(t,t)R_{\boldsymbol{xx}}(s,s)} \quad \text{and} \quad |C_{\boldsymbol{xx}}(t,s)| \le \sqrt{C_{\boldsymbol{xx}}(t,t)C_{\boldsymbol{xx}}(s,s)}$$

for every value of the parameters t and s, while in addition they satisfy the property of *positive semidefiniteness*, that is, they satisfy, respectively, the conditions

$$\sum_{i=1}^{N} \alpha_i\alpha_j^* R_{\boldsymbol{xx}}(t_i,t_j) \ge 0 \quad \text{and} \quad \sum_{i=1}^{N} \alpha_i\alpha_j^* C_{\boldsymbol{xx}}(t_i,t_j) \ge 0$$

for every $N > 0$, for every $t_1 < t_2 < \cdots < t_N$, and for every set of complex constants $\alpha_1, \alpha_2, \ldots, \alpha_N$. Of the two aforementioned functions, the autocorrelation function is particularly important because it constitutes *a measure of the dependence that exists between the instantaneous values of a signal* at the time instances $t = t_1$ and $t = t_2$ and allows for the description of the process in the frequency domain.

If for every t_1 and t_2 it holds that $R_{xy}(t_1,t_2) = 0$, then the processes $\boldsymbol{x}(t)$ and $\boldsymbol{y}(t)$ are aforementioned to be *mutually orthogonal*, while if for every t_1 and t_2 it holds that $C_{xy}(t_1,t_2) = 0$, then the processes are aforementioned to be *uncorrelated*. In this case, we have $R_{\boldsymbol{xx}}(t_1,t_2) = \mu_{\boldsymbol{x}}(t_1)\mu_{\boldsymbol{x}}(t_2)$. Note that two processes are orthogonal *if they are uncorrelated and at least one of them has a zero mean value*—in reality, this concept of orthogonality is useful only when the given random processes have a zero mean value, so the concept of orthogonality becomes *equivalent* to the concept of being uncorrelated. If for every t_i and t_j with $t_i \ne t_j$ the values $\boldsymbol{x}(t_i)$ and $\boldsymbol{x}(t_j)$ are uncorrelated between them, that is, if $C(t_i,t_j) = 0$ $(t_i \ne t_j)$, then the process is known as a *white noise process* and is characterized by the property $C(t_1,t_2) = q(t_2)\delta(t_1 - t_2)$ $(q(t) \ge 0)$. On the other hand, if the random variables $\boldsymbol{x}(t_1)$ and $\boldsymbol{x}(t_2)$ are not only uncorrelated but also independent of each other, then the process is characterized as *strict white noise*. The mean value of a white noise process is equal to zero. This very important process is described in detail in the next section.

Example: Statistical Means of Derivative Processes

Considering that a random process is described by a function of time $x(t)$, it is not difficult to understand that if its derivative $\dot{x}(t)$ is defined, this will constitute a *random process* that will also be characterized by its own statistical means. These statistical means are easily calculated from the statistical means of the original process since the operator of the *expected value* as a *linear* operator can undergo interchange (in terms of the order of its application) with the also linear operator of the *differentiation*, an interchange that, as we have seen several times so far, can be performed, in almost all cases, between the operator of the expected value and the operator of integration. Specifically, symbolizing the mean value and the covariance of the process $x(t)$ by $\mu_x(t)$ and $C_{xx}(t)$, the corresponding functions $\mu_{\dot{x}}(t)$ and $C_{\dot{x}\dot{x}}(t)$ are calculated as

$$\mu_{\dot{x}(t)} = E\{\dot{x}(t)\} = E\left\{\frac{dx(t)}{dt}\right\} = \frac{d}{dt}[E\{x(t)\}] = \frac{d\mu_x(t)}{dt} = \dot{\mu}_x(t) \tag{13.7}$$

while the *covariability* of the derivative process $\dot{x}(t)$ is also calculated by a similar derivation.

13.2.6 Returning to Stationary Processes

A stochastic process is aforementioned to be *stationary* (stationary) *in the strict sense* (SSS) (if the statistical properties are preserved *unchanged under the action of time displacement*, that is, if the processes $\boldsymbol{x}(t)$ and $\boldsymbol{x}(t+T)$ are characterized by the same statistical order N for every value of the parameter T. In the simple case where this property concerns *only the mean value*, and therefore it holds that $f_x(x_1; t_1) = f_x(x_1; t_1 + T)$, the process is aforementioned to be a *first-order stationary process*, while if the same holds for the *combined density* $f_(x_1, x_2; t_1, t_2) = f_x(x_1, x_2; t_1 + T, t_2 + T)$, then the process is aforementioned to be a *second-order stationary process*. In the general case where this property concerns the corresponding function of order N, the corresponding condition is

$$f_x(x_1, x_2, \ldots, x_n; t_1, t_2, \ldots, t_n) = f_x(x_1, x_2, \ldots, x_n; t_1 + T, t_2 + T, \ldots, t_n + T)$$

for every value of the parameters T and n, while, of course, a similar expression can be used for the *cumulative distribution function*; so we will have

$$F_x(x_1, x_2, \ldots, x_n; t_1, t_2, \ldots, t_n) = F_x(x_1, x_2, \ldots, x_n; t_1 + T, t_2 + T, \ldots, t_n + T)$$

It is not difficult to understand that the mean value, that is, the first-order moment of a stationary process, is *constant*. Indeed, the condition $f_x(x; t) = f_x(x; t + T)$ for $T = -t$ gives us $f(x; t) = f(x; 0)$, and therefore $\mu_x(t) = \mu_x(0) = \text{constant}$. Similarly, for the second-order moment, we will have $f(x_1, x_2; t_1, t_2) = f(x_1, x_2; t_1 + T, t_2 + T)$ and for $T = -t_2$ (note that such an assignment is permissible since stationary processes do not depend on the origin of time measurement), it follows that $f(x_1, x_2; t_1, t_2) = f(x_1, x_2; t_1 - t_2, 0)$, from which it in turn follows that

$$E\{x(t_1)x^*(t_2)\} = R_{xx}(t_1, t_2) = R_{xx}(t_1 - t_2, 0)$$

Consequently, in the case of stationary processes, we can use a simplified notation, defining the variable $\tau = t_1 - t_2$ and formulating the autocorrelation as

$$R_{xx}(t_1 - t_2) = R_{xx}(\tau, 0) = E\{\boldsymbol{x}(t + \tau)\boldsymbol{x}(t)\}$$

Observing that the second argument in the autocorrelation function is always *zero*, we can, for the sake of brevity, omit it and simply write $R_{xx}(\tau, 0) = R_{xx}(\tau)$. We therefore find that, indeed, the autocorrelation function for the case of stationary processes *depends not on specific time points* t_1 and t_2 *but on their difference* $t_1 - t_2$. Conversely, if the process *is not stationary* and, therefore, *is not characterized* by the foregoing property, then we use the notation $R_{xx}(t_1, t_2)$. Therefore, the symbol by which a process is represented allows us to draw conclusions about its character, as far as its stationarity is concerned.

On the other hand, a process is aforementioned to be *stationary in the broad sense* (WSS) if the aforementioned requirement about the invariance of statistical properties under displacement *is limited only to first- and second-order densities* and not to higher-order densities. In other words, a WSS type of process is simply a *second-order stationary process*. Consequently, and according to the preceding discussion, a random process $\boldsymbol{x}(t)$ is stationary in the broad sense if its mean value is constant, while its autocorrelation depends *only on the difference* $\tau = t_1 - t_2$, that is, if it is $E\{\boldsymbol{x}(t)\} = \mu = \text{constant}$ and

$$E\{\boldsymbol{x}(t_1)\boldsymbol{x}(t_2)\} = R(t_1 - t_2) = R(\tau)$$

Taking into account that $t_1 = t_2 + T$, the function $R(T)$ can be formulated in its *symmetric* form

$$R(\tau) = E\left\{\boldsymbol{x}\left(t + \frac{\tau}{2}\right)\boldsymbol{x}^*\left(t - \frac{\tau}{2}\right)\right\} \tag{13.8}$$

while for $\tau = 0$, it turns out that $E\{[\boldsymbol{x}(t)]^2\} = R(0)$. In other words, the *average power* of a stationary process is *independent* of τ and equal to $R(0)$. On the other hand, for values $\tau \neq 0$, the autocorrelation function $R(\tau)$ can be treated *as a similarity measure* that exists between the waveforms $x(t)$ and $x(t + \tau)$. To understand this feature, let us consider a function $x(t)$ that is a sample of a stationary random process with zero mean value, and let us define the function

$$y(t) = x(t) - \rho x(t + \tau)$$

To define a measure of the *degree of similarity* that exists between the waveforms $x(t)$ and $x(t+\tau)$ and determines the part of the function $x(t)$ that is contained in the function $x(t+\tau)$, we need to determine the value of the coefficient ρ that minimizes the *mean square value* of the function $y(t)$. To do this, we need to determine the variance of the function $y(t)$, equate its derivative with respect to ρ with the zero value, and solve for ρ the equation that arises. The variance of the function $y(t)$ according to its definition is

$$E\{y^2(t)\} = E\{[x(t) - \rho x(t+\tau)]^2\} = E\{x^2(t) - 2\rho x(t) + \rho^2 x^2(t+\tau)\} = E\{x^2(t)\} - 2\rho E\{x(t)x(t+\tau)\} + \rho^2 E\{x^2(t+\tau)\}$$

or, equivalently,

$$\sigma_y^2 = \sigma_x^2 - 2\rho R_x(\tau) + \rho^2 \sigma_x^2$$

We will therefore have

$$\frac{d(\sigma_y^2)}{d\rho} = -2R_x(\tau) + 2\rho\sigma_x^2 = 0 \quad \text{from where it follows that} \quad \rho = \frac{R_x(\tau)}{\sigma_x^2}$$

The parameter ρ is simply the *correlation coefficient* that we defined in another unit and that can be considered *the fraction of the waveform $x(t)$ that remains after the passage of a time interval of τ seconds* and, therefore, is also contained in the waveform $x(t+\tau)$. As mentioned previously, the coefficient ρ takes values in the interval $[-1, 1]$, with the value $\rho = 1$ referring to the *complete identity* of the waveforms $x(t)$ and $x(t+\tau)$, the value $\rho = 0$ indicating that these two waveforms are characterized by *zero degree of identity* or, equivalently, that they are *completely uncorrelated* with each other and the value $\rho = -1$ to indicate that the two waveforms are *identical*, but with a reversal in sign, and therefore, the waveform $x(\tau + t)$ is the negative of the waveform $x(t)$. Finally, from the relation $R_x(\tau)$ it is not difficult to understand that *the autocorrelation function depends both on the degree ρ of correlation and on the variability σ_x^2 of the underlying process.* Therefore, to estimate the significance of a specific value of the function $R_x(\tau)$, we need to know the values of both of the preceding quantities.

As mentioned previously, in practice we are *mainly interested in stationary processes* because the study of nonstationary processes is quite complex. The function of *autocorrelation* of these types of processes is characterized by the following properties:

- If $\tau = 0$, then, as noted elsewhere, it holds that $R_x(0) = \overline{x^2}$. This conclusion arises immediately from the definition of *autocorrelation*, which for $\tau = 0$ gives us $R_x(0) = E\{x(t)x(t)\} = E\{x^2(t)\} = \overline{x^2}$. Consequently, the value $R_x(0)$ is equal to the value of the *mean square* of the process. Note that this is something that applies *whether the process is characterized by a nonzero mean value or not.* Specifically, in the second case, in which the process is characterized by a zero mean value, the mean square value of the process is equal to the value of the *variance*.
- The function of *autocorrelation* is characterized by *even symmetry* or, in mathematical notation, $R_x(\tau) = R_x(-\tau)$. For the proof of this property, we observe that if in the defining equation of autocorrelation $R_x(\tau) = E\{x(t)x(t+\tau)\}$ we replace τ with $-\tau$, it will take the form $R_x(-\tau) = E\{x(t)x(t-\tau)\}$. Now defining the variable $u = t - \tau$ results in $t = u + \tau$, and the preceding relationship is written

$$R_x(-\tau) = E\{x(u+\tau)x(u)\} = E\{x(u)x(u+\tau)\} = R_x(\tau)$$

which is the desired result. This property, which can intuitively be understood from the fact that the operator of the expected value applies *to the same product of functions*, regardless of the direction toward which the shift is made, is particularly useful because it allows us to calculate the value of autocorrelation *only for positive values* of τ that correspond to a shift in the function $x(t)$ in a specific direction. Next, taking advantage of this property of symmetry, we extend the preceding result to the case of negative values of τ.

- If the random stationary process $z(t)$ is defined as the sum $z(t) = x(t) + y(t)$, then we will have

$$R_{zz}(\tau) = R_{xx}(\tau) + R_{xy}(\tau) + R_{yx}(\tau) + R_{yy}(\tau) \tag{13.9}$$

Indeed, from the definition of the functions of *autocorrelation* and *cross-correlation* it immediately follows that

$$\begin{aligned} R_{zz}(\tau) &= E\{z(t)z(t+\tau)\} = E\{[x(t)+y(t)][x(t+\tau)+y(t+\tau)]\} \\ &= E\{x(t)x(t+\tau) + x(t)y(t+\tau) + y(t)x(t+\tau) + y(t)y(t+\tau)\} \\ &= E\{x(t)x(t+\tau)\} + E\{x(t)y(t+\tau)\} + E\{y(t)x(t+\tau)\} + E\{y(t)y(t+\tau)\} \\ &= R_{xx}(\tau) + R_{xy}(\tau) + R_{yx}(\tau) + R_{yy}(\tau) \end{aligned}$$

- The maximum value of the correlation sequence corresponds to the value $\tau = 0$, and therefore, $|R_x(\tau)| \leq R_x(0)$. Although this value may appear for other values of τ, as happens, for example, when the autocorrelation function is *periodic*, it will *never* have a value greater than this. This property is proven, noting that (recall that $R_x(0) = E\{x^2(t)\}$)

$$E\{(x_1 \pm x_2)^2\} = E\{x_1^2 + x_2^2 \pm 2x_1x_2\} \geq 0 \tag{13.10}$$

from which it follows that

$$E\{x_1^2 + x_2^2\} = 2R_x(0) \geq |E\{2x_1x_2\}| = |2R_x(\tau)|$$

and finally $R_x(0) \geq |R_x(\tau)|$.

- If the function $x(t)$ has a continuous component or average value, then the function $R_x(\tau)$ will be characterized by a *constant component*. Specifically, if $x(t) = \overline{x} + N(t)$, where $N(t)$ is a component with a zero average value, then we will have

$$R_x(\tau) = E\{[\overline{x} + N(t)][\overline{x} + N(t+\tau)]\} = E\{(\overline{x})^2 + \overline{x}N(t) + \overline{x}N(t+\tau) + N(t)N(t+\tau)\} = (\overline{x})^2 + R_N(\tau) \tag{13.11}$$

since, based on our assumption, we have $E\{N(t)\} = E\{N(t+\tau)\} = 0$.

- If the function $x(t)$ is *periodic*, then the autocorrelation $R_x(\tau)$ will have a *periodic component with the same period*. Indeed, if the period of the process is equal to T, then we will have $x(t) = x(t \pm T)$ and $x(t+\tau) = x(t+\tau \pm T)$, and from the definition of autocorrelation we will have

$$R_{xx}(\tau \pm T) = E\{x(t)x(t+\tau \pm T)\} = E\{x(t)x(t+\tau)\} = R_x x(\tau)$$

which is also the desired result. For the extraction of the final result, we took into account the fact that the process is of the WSS type, so we will have $x(t+\tau \pm T) = x(t+\tau)$.

Specifically, if $x(t) = A\cos(\omega t + \vartheta)$, then from the definition of the autocorrelation function we have

$$\begin{aligned} R_x(\tau) &= E\{A\cos(\omega t + \vartheta)A\cos(\omega t + \omega\tau + \vartheta)\}] = \frac{A^2}{2}E\left\{\cos(2\omega t + \omega\tau + 2\vartheta) + \cos(\omega\tau)\right\} \\ &= \frac{A^2}{2}\int_0^{2\pi}\frac{1}{2\pi}[\cos(2\omega t + \omega\tau + 2\theta) + \cos(\omega\tau)] = \frac{A^2}{2}\cos\omega\tau \end{aligned}$$

while in the general case of the process $x(t) = A\cos(\omega t\vartheta) + N(t)$ with the quantities ϑ and $N(t)$ being statistically independent for each t and working in the same way, we find

$$R_x(\tau) = \frac{A^2}{2}\cos\omega\tau + R_N(\tau) \tag{13.12}$$

The foregoing property can be directly extended to the case of *random variables* that include *any number* of periodic functions. If the random variables associated with the periodic components are *statistically independent*, then the autocorrelation of the sum of the periodic components *will be equal to the sum of the periodic autocorrelations* associated with each of these components. In the case where each sample of the random process is a *periodic function* and, therefore, can be represented by a Fourier series, then the same will apply to the autocorrelation function, although in the case where the involved random variables are not statistically independent, the Fourier expansion of the autocorrelation will contain *other terms*, apart from the autocorrelations of the individual terms.

- If the process $x(t)$ is *ergodic with a nonzero mean*, that is, if it holds that $E\{x(t)\} = \overline{x} \neq 0$, while *it does not have* a periodic component, then in the limit $|\tau| \to \infty$, $\lim R_x(\tau) = \overline{x}^2$. Indeed, from the definition of autocorrelation we will have $R_{xx}(\tau) = E\{x(t)x(t+\tau)\}$. Because, now, the process does not have a periodic component, in the limit $|\tau| \to \infty$, that is, for large values of τ (since the influence of previous values grows increasingly weaker over time, random variables tend to become *statistically independent*. Therefore, $\lim R_{xx}(\tau) = E\{x(t)\}E\{x(t+\tau)\} = \overline{x}^2$ at the limit $|\tau| \to \infty$, with the desired result emerging by observing that since the process is ergodic, it will hold that $E\{x(t)\} = E\{x(t+\tau)\} = \overline{x}$. As a consequence of this result, if $\overline{x} = 0$, then $\lim R_{xx}(\tau) = 0$ at the limit $|\tau| \to \infty$.

On the other hand, the two versions of the cross-correlation function, in the case where both processes involved in its definition are jointly WSS processes, have the form

$$R_{xy}(\tau) = E\{x_1 y_2\} = \int_{-\infty}^{\infty} dx_1 \int_{-\infty}^{\infty} x_1 y_2 f(x_1, y_2) dy_2 \qquad \text{and} \qquad R_{yx}(\tau) = E\{y_1 x_2\} = \int_{-\infty}^{\infty} dy_1 \int_{-\infty}^{\infty} y_1 x_2 f(y_1, x_2) dx_2$$

where in the first case $x_1 = x(t_1)$ and $y_2 = y(t_1+\tau)$, while in the second case $y_1 = y(t_1)$ and $x_2 = x(t_1+\tau)$. Note that in this case, *both of these cross-correlation functions depend on the time distance* τ because the involved processes are jointly of the WSS type. Note that if each of the processes is stationary in the wide sense, but not jointly stationary, then the preceding functions *do not depend* only on the time distance τ but, on the contrary, on the time t as well.
The cross-correlation function is characterized by the following interesting properties:

- From the definition of the preceding functions, it is easily observed that $R_{xy}(0) = R_{yx}(0)$, but these quantities have no physical significance, as is the case, for example, with autocorrelation, where the quantity $R_{xx}(0)$ represents the *mean square* of the process.
- The cross-correlation function generally *is not* an even function of τ, although it is proven that $R_{xy}(\tau) = R_{yx}(-\tau)$. Working as in the case of autocorrelation (see the property about even symmetry), we observe that $R_{yx}(\tau) = E\{y(t)x(t+\tau)\}$, and, replacing the parameter τ with the parameter $-\tau$, we get $R_{yx}(-\tau) = E\{y(t)x(t-\tau)\}$. Proceeding with the variable change $u = t - \tau$ we have $t = u + \tau$, thus yielding

$$R_{yx}(-\tau) = E\{y(t+\tau)x(t)\} = E\{x(t)y(t+\tau)\} = R_{xy}(\tau)$$

  This property can intuitively be understood observing that *the shift of the function $y(t)$ in one direction is equivalent to the shift of the function $x(t)$ in the opposite direction.*
- Although in the general case the maximum value of cross-correlation does not correspond to the position $\tau = 0$, as is the case with autocorrelation, it is proven that

$$|R_{xy}(\tau)| \leq \sqrt{R_{xx}(0)R_{yy}(0)} \tag{13.13}$$

  while a similar condition applies to the $|R_{xy}(\tau)|$. Indeed, starting from the inequality

$$E\left\{\left(\frac{x(t)}{\sqrt{R_{xx}(0)}} + \frac{y(t+\tau)}{\sqrt{R_{yy}(0)}}\right)^2\right\} \geq 0 \tag{13.14}$$

  and expressing it in the form

$$\frac{E\{x^2(t)\}}{R_{xx}(0)} + \frac{E\{y^2(t)\}}{R_{yy}(0)} \pm 2\frac{E\{x(t)y(t+\tau)\}}{\sqrt{R_{xx}(0)R_{yy}(0)}} \geq 0 \tag{13.15}$$

we observe that since $E\{x^2(t)\} = R_{xx}(0)$ and $E\{y^2(t)\} = R_{yy}(0)$, each of the first two terms of the left-hand side has a value equal to 1, and therefore the sum of these two terms is equal to 2. If we transfer the third term to the second member, simplify the resulting inequality, removing the coefficient 2 from both sides, and take into account that $R_{xy}(\tau) = Ex(t)y(t+\tau)$, we are led directly to the desired result.

- The *cross-correlation* function satisfies the property

$$|R_{xy}(\tau)| \leq \frac{1}{2}\left[R_{xx}(0) + R_{yy}(0)\right]$$

This property is proven directly, taking into account that the geometric mean $\sqrt{ab}$ of two numbers a and b *cannot* be greater than their arithmetic mean $(a+b)/2$. Using this property, together with the inequality that we proved previously, we will have

$$|R_{xy}(\tau)| \leq \sqrt{R_{xx}(0)R_{yy}(0)} \leq \frac{1}{2}[R_{xx}(0) + R_{yy}(0)] \tag{13.16}$$

from which the desired result follows.

- If two random stationary processes are *statistically independent*, it holds that

$$R_{xy}(\tau) = E\{x_1 y_2\} = E\{x_1\}E\{y_2\} = \overline{x}\overline{y} = R_{yx}(\tau) \tag{13.17}$$

Furthermore, if one of the two processes is characterized by a *zero mean value*, then the cross-correlation function *is equal to zero for any value of the argument* τ, without, however, the converse being true. In other words, if the cross-correlation has zero value and one of the processes has a mean value equal to zero, this *does not mean* that the two processes are statistically independent, except in the case of random variables with a combined Gaussian probability density.

- If $x(t)$ is a *stationary random process* with derivative $\dot{x}(t)$, it is proven that $R_{x\dot{x}}(\tau) = R_x'(\tau)$, as easily follows from the definition of the derivative. Indeed, we will have

$$R_{x\dot{x}}(\tau) = E\{x(t)\dot{x}(t+\tau)\} = E\left\{\lim_{e\to 0}\frac{x(t)x(t+\tau+e) - x(t)x(t+\tau)}{e}\right\} = \lim_{e\to 0}\frac{R_x(\tau+e) - R_x(\tau)}{e} = \frac{dR_x(\tau)}{d\tau}$$

Repeating the same process, we find that $R_{\dot{x}}(\tau) = -R_x''(\tau)$.

It is also proven that, in the case of a WSS process, the function *autocovariance* and the correlation coefficient have the form $C(T) = R(T) - |\mu|^2$ and $r(t) = C(T)/C(0)$, while from the preceding definitions it is not difficult to see that *a SSS process is simultaneously also a WSS process, while the opposite is not true.*

If the process is complex in nature, that is, it has the form $z(t) = \boldsymbol{x}(t) + j\boldsymbol{y}(t)$, then it will be characterized as stationary if the processes $\boldsymbol{x}(t)$ and $\boldsymbol{y}(t)$ are jointly stationary. Finally, if the processes $\boldsymbol{x}(t)$ and $\boldsymbol{y}(t)$ are WSS processes, they will be characterized as jointly WSS if they satisfy the properties $R_{xy}(T) = E\{\boldsymbol{x}(t+T)\boldsymbol{y}^*(t)\}$ and $C_{xy}(T) = R_{xy}(T) - \mu_x\mu_y^*$.

Finally, the correlation matrix for sample functions $\boldsymbol{x}(t)$ that undergo sampling at the time instances $t_1, t_2, \ldots, t_N$, leading to the values $x(t_1), x(t_2), \ldots, x(t_N)$ that jointly define the vector $\boldsymbol{x}(t) = [x(t_1), x(t_2), \ldots, x(t_N)]^T$, has the form

$$R_x = E\{\boldsymbol{x}\boldsymbol{x}^T\} = \begin{pmatrix} x(t_1)x(t_1) & x(t_1)x(t_2) & \cdots & x(t_1)x(t_N) \\ x(t_2)x(t_1) & x(t_2)x(t_2) & \cdots & x(t_2)x(t_N) \\ \vdots & \vdots & \ddots & \vdots \\ x(t_N)x(t_1) & x(t_N)x(t_2) & \cdots & x(t_N)x(t_N) \end{pmatrix} = \begin{pmatrix} R_x(t_1,t_1) & R_x(t_1,t_2) & \cdots & R_x(t_1,t_N) \\ R_x(t_2,t_1) & R_x(t_2,t_2) & \cdots & R_x(t_2,t_N) \\ \vdots & \vdots & \ddots & \vdots \\ R_x(t_N,t_1) & R_x(t_N,t_1) & \cdots & R_x(t_N,t_1) \end{pmatrix}$$

where for the construction of the last matrix we applied *the operator of the expected value* to each of the elements of the first matrix and took into account the definition of *autocorrelation*. In the special case in which the considered process is a random WSS-type process, the autocorrelation will depend, as is well known, *only on the time distance* between the time instances $t_1, t_2, \ldots, t_N$. If it is taken into account that these time instances are separated by a distance equal to Δt, that is, that $t_2 = t_1 + \Delta t$, $t_3 = t_1 + 2\Delta t$, and generally $t_i = t_1 + (i-1)\Delta t$ $(i = 2, 3, \ldots, N)$, then the preceding table takes the form

$$R_x = \begin{pmatrix} R_x(0) & R_x(\Delta t) & \cdots & R_x[(N-1)\Delta t] \\ R_x(\Delta t) & R_x(0) & \cdots & R_x[(N-2)\Delta t] \\ \vdots & \vdots & \ddots & \vdots \\ R_x[(N-1)\Delta t) & R_x[(N-2)\Delta t) & \cdots & R_x(0) \end{pmatrix} \tag{13.18}$$

It is not difficult to understand that this table is a *symmetric matrix* (something that, let us note, is true even in the case where the process *is not stationary*), while as a consequence of the concept of stationarity, both *the main diagonal* and all other diagonals of the matrix that are *parallel* to it have the same elements.

On the other hand, the *covariance matrix* will contain, as before, the values of *variability* and *covariability* of the random variables, while it is defined by the equation

$$\boldsymbol{C}_x = E\{(\boldsymbol{x} - \overline{\boldsymbol{x}})(\boldsymbol{x}^T - \overline{\boldsymbol{x}}^T)\} \tag{13.19}$$

with the elements of this matrix being calculated from the expression

$$E\{[x(t_i) - \overline{x(t_i)}][x(t_i)^T - \overline{x(t_i)}^T]\} = \sigma_i \sigma_j \rho ij$$

Using the foregoing expression, the *covariance matrix* takes the form

$$C_x = \sigma^2 \begin{pmatrix} 1 & \rho_1 & \rho_2 & \cdots & \rho_{N-2} & \rho_{N-1} \\ \rho_1 & 1 & \rho_1 & \cdots & \rho_{N-3} & \rho_{N-2} \\ \vdots & \vdots & \vdots & \ddots & \vdots & \\ \rho_{N-2} & \rho_{N-3} & \rho_{N-4} & \cdots & 1 & \rho_1 \\ \rho_{N-1} & \rho_{N-2} & \rho_{N-3} & \cdots & \rho_1 & 1 \end{pmatrix}$$

and from its structure it is not difficult to ascertain that it constitutes a *Toeplitz matrix*.

Apart from the case where the elements of the vector $\boldsymbol{X}$ are samples of the same process that arises through a sampling process and are recorded at discrete-times, another case in which the foregoing tables are used is that in which the elements of the vector $\boldsymbol{X}$ come from N *different random processes* x_i $(i = 1, 2, \ldots, N)$, and therefore, the vector $\boldsymbol{X}$ is defined as $\boldsymbol{X}(t) = [x_1(t), x_2(t), \ldots, x_N(t)]^T$. In this case, the *correlation matrix* is defined as

$$R_X(\tau) = E\{\boldsymbol{X}(t)\boldsymbol{X}^T(t+\tau)\} = \begin{pmatrix} R_1(\tau) & R_{12}(\tau) & \cdots & R_{1N}(\tau) \\ R_{21}(\tau) & R_2(\tau) & \cdots & R_{2N}(\tau) \\ \vdots & \vdots & \ddots & \vdots \\ R_{N1}(\tau) & R_{N2}(\tau) & \cdots & R_{NN}(\tau) \end{pmatrix} \tag{13.20}$$

where for brevity's sake we used the symbolism $R_{ii}(\tau) = R_i(\tau)$ $(i = 1, 2, \ldots, N)$ while defining $R_i(\tau) = E\{x_i(t)x_i(t+\tau)\}$ and $R_{ij}(\tau) = E\{x_i(t)x_j(t+\tau)\}$. Note that in this case, the elements of this matrix *are not simple numbers*, as previously, but, conversely, functions of the variable τ. Finally, in the most general of cases, where we consider two vectors of the forms $\boldsymbol{X}(t) = [x_1(t), x_2(t), \ldots, x_N(t)]^T$ and $\boldsymbol{Y}(t) = [y_1(t), y_2(t), \ldots, y_N(t)]^T$, the corresponding matrix that arises and is known as the *cross-correlation matrix* is defined as $R_{XY}(\tau) = E\{\boldsymbol{X}(t)\boldsymbol{Y}^T(t+\tau)\}$ and has the same form as before, with its elements being defined by the expressions $R_i(\tau) = E\{x_i(t)y_j(t+\tau)\}$ and $R_{ij}(\tau) = E\{x_i(t)y_j(t+\tau)\}$.

13.2.7 Periodic and Cyclostationary Processes

A random process $\boldsymbol{x}(t)$ is aforementioned to be *periodic in the broad sense* if $T > 0$ such that

$$\mu_x(t) = \mu_x(t+T), \ \forall t \qquad \text{and} \qquad C_{xx}(t_1, t_2) = C_{xx}(t_1+T, t_2) = C_{xx}(t_1, t_2+T), \ \forall t_1, t_2$$

The smallest number T that satisfies the foregoing properties is called a *period*. Note that the function $C_{xx}(t_1, t_2)$ is *periodic with period T*, towards *both* axes. If, in addition, the foregoing process is also *a WSS process*, then it is aforementioned to be *periodic stationary in the broad sense*. These processes are characterized by an *average periodic squareness* (see also the discussion that follows).

On the other hand, a random process $\boldsymbol{x}(t)$ is aforementioned to be *cyclostationary in the broad sense* if there $T > 0$ such that

$$\mu_x(t) = \mu_x(t+T), \ \forall t \qquad \text{and} \qquad C_{xx}(t_1, t_2) = C_{xx}(t_1+T, t_2+T), \ \forall t_1, t_2$$

It is worth noting that when a process is *cyclostationary*, all its statistical properties are *periodic*, but this does not apply to the process itself, as happens in periodic processes. For this reason, *special attention* is required, so that these two concepts are not confused, as often happens in practice.

Example 13.2.1 (Stark and Woods, page 405) Considering the random process $\boldsymbol{x}(t) = A\sin(\omega_0 t + \theta)$, where A and θ are independent random variables and with the angle θ being uniformly distributed in the range of values $[-\pi, \pi]$, the mean value $\mu_x(t)$ is equal to zero, since from the definition of this quantity we have

$$\mu_x(t) = E\{A\sin(\omega_0 t+\theta)\} = E\{A\}E\{\sin(\omega_0 t+\theta)\} = \mu_A \frac{1}{2\pi}\int_{-\pi}^{\pi} \sin(\omega_0 t+\theta) = \mu_A \cdot 0 = 0$$

while for the calculation of the autocorrelation function, we have

$$\begin{aligned} R_{xx}(t_1, t_2) &= E\{\boldsymbol{x}_1(t)\boldsymbol{x}_2^*(t)\} = E\{A^2\sin(\omega_0 t_1+\theta)\sin(\omega_0 t_2+\theta)\} = E\{A^2\}E\{\sin(\omega_0 t_1+\theta)\sin(\omega_0 t_2+\theta)\} \\ &= \frac{1}{2}E\{A^2\}\Big[E\{\cos[\omega_0(t_1-t_2)]\} - E\{\cos[\omega_0(t_1+t_2)] + 2\theta\}\Big] = \frac{1}{2}E\{A^2\}\cos[\omega_0(t_1-t_2)] \end{aligned}$$

where for the derivation of the final result we used the known trigonometric identity for the product of sines $\sin\alpha\sin\beta = (1/2)[\cos(\alpha-\beta)-\cos(\alpha+\beta)]$ and we took into account that since the random variable θ is uniformly distributed in the value range $[-\pi, \pi]$, the result related to the expected value of the second term on the right-hand side of the preceding equation, is equal to zero. Consequently, according to the preceding definitions, this process is a stationary process.

The preceding concept of *stationarity* can be extended to the case of other processes. For example, considering the two processes $\boldsymbol{x}(t)$ and $\boldsymbol{y}(t)$, these are aforementioned to be *jointly stationary* if the joint statistics of the processes $\boldsymbol{x}(t)$ and $\boldsymbol{y}(t)$ are the same as the joint statistics of the processes $\boldsymbol{x}(t+T)$ and $\boldsymbol{y}(t+T)$ for every value of T. For the special case of WSS-type processes, the processes $x(t)$ and $y(t)$ are considered *jointly WSS processes* if the conditions (a) $E\{x(t)\} = \mu_x =$ constant, (b) $E\{y(t)\} = \mu_y =$ constant, and (c) the cross-correlation function of the two processes *does not depend* on the absolute time but *only* on the difference $\tau = t_2 - t_1$, and therefore, it holds that $R_{xy}(t_1, t_2) = R_{xy}(\tau) = E\{x(t)y(t+\tau)\}$.

Considering two random processes $\boldsymbol{x}(t)$ and $\boldsymbol{y}(t)$ and the sets of random variables $\boldsymbol{x}(t_i)$ $(i = 1, 2, \ldots, n)$ and $\boldsymbol{y}(\tau_j)$ $(j = 1, 2, \ldots, m)$, with $t_1 > t_2 > \cdots > t_n$ and $\tau_1 > \tau_2 > \cdots > \tau_m$ related to them, these two sets are characterized by the *combined PDF* $f(x_1, x_2, \ldots, x_n; y_1, y_2, \ldots, y_m)$, where $x_i \equiv \boldsymbol{x}(t_i)$ $(i = 1, 2, \ldots, n)$ and $y_j = \boldsymbol{y}(\tau_j)$ $(j = 1, 2, \ldots, m)$ for every set of time instances $\{t_i\}$ and $\{\tau_j\}$ and for every positive integer value of the parameters m and n. Based on these symbols, the *correlation* (cross-correlation) R_{xy} of the random processes $\boldsymbol{x}(t)$ and $\boldsymbol{y}(t)$ has the form

$$R_{xy}(t_1, t_2) = E\{\boldsymbol{x}(t_1)\boldsymbol{y}(t_2)\} = \int_{-\infty}^{\infty}\int_{-\infty}^{\infty} x_1 y_2 f(x_1, y_2)dx_1 dy_2 \tag{13.21}$$

while the *covariance* is defined as

$$C_{xy}(t_1, t_2) = R_{xy}(t_1, t_2) - \mu_x(t_1)\mu_y(t_2)$$

If the processes $\boldsymbol{x}(t)$ and $\boldsymbol{y}(t)$ are stationary, both individually and jointly, then $R_{xy}(t_1, t_2) = R_{xy}(t_1 - t_2)$ and $C_{xy}(t_1, t_2) = C_{xy}(t_1 - t_2)$, while additionally we will have $R_{xy}(-T) = R_{xy}(T)$ $(T = t_1 - t_2)$, and consequently, the function R_{xy} is an even function. If the condition

$$f(x_1, x_2, \ldots, x_n; y_1, y_2, \ldots, y_m) = f(x_1, x_2, \ldots, x_n)f(y_1, y_2, \ldots, y_m)$$

applies, then the processes $\boldsymbol{x}(t)$ and $\boldsymbol{y}(t)$ are *statistically independent*, while if $R_{xy}(t_1, t_2) = \mu_x(t_1)\mu_y(t_2)$, then we will have $C_{xy}(t_1, t_2)$, and the two processes are aforementioned to be *uncorrelated*.

Considering a random process $\boldsymbol{x}(t)$ with mean value $\mu_x(t)$, the process defined by the difference $\tilde{\boldsymbol{x}}(t) = \boldsymbol{x}(t) - \mu_x(t)$ is aforementioned to be *process shifted with respect to the mean value* (centered process). It is not difficult to ascertain that for this process the properties

$$E\{\tilde{\boldsymbol{x}}(t)\} = E\{\boldsymbol{x}(t) - \mu_x(t)\} = E\{\boldsymbol{x}(t)\} - \mu_x(t) = 0 \tag{13.22}$$

will apply (since by definition $E\{\boldsymbol{x}(t)\} = \mu_x(t)$) and $R_x(t_1, t_2) = C_x(t_1, t_2)$. Specifically, in the case where the process $\boldsymbol{x}(t)$ is *stationary with respect to covariance*, that is, it holds that $C_x(t_1, t_2) = C_x(t_1 - t_2)$, then the process shifted with respect to the mean value $\tilde{\boldsymbol{x}}(t)$ is a WSS process.

In the science of statistics, processes characterized by a *zero mean value*, like the previous ones, are particularly useful, and for this reason it is worth trying to construct a representation of them in the frequency domain. To do this, let us start with the simple process $\boldsymbol{x}(t) = A\exp(j2\pi ft)$, where f is a real constant value and A is a real random variable with mean value $E\{A\} = 0$ and finite mean power $E\{A^2\}$. In this case, from the equation defining the first- and second-order moments, it is not difficult to ascertain that $E\{\boldsymbol{x}(t)\} = E\{A\exp(j2\pi ft)\} = E\{A\}\exp(j2\pi ft) = 0$ and

$$E\{\boldsymbol{x}(t+\tau)\boldsymbol{x}^*(t)\} = E\{A\exp[j2\pi f(t+\tau)]A\exp(j2\pi ft)\} = E\{A^2\}\exp(j2\pi f\tau) = R_{xx}(\tau) \tag{13.23}$$

Generalizing the preceding example, we can construct a representation of the *random processes with zero mean* in the frequency domain (see Stark and Woods, page 439). Indeed, defining the process

$$\boldsymbol{x}(t) = \sum_{k=1}^{M} A_k \exp j2\pi f_k t \tag{13.24}$$

with the complex variables A_k $(k = 1, 2, \ldots, M)$ being *uncorrelated with each other* and characterized by zero mean value and dispersion value σ_k^2 $(k = 1, 2, \ldots, M)$, it is not difficult to see that, based on the foregoing discussion, the process $\boldsymbol{x}(t)$ will be characterized by a zero mean value and by an autocorrelation function

$$R_{xx}(\tau) = \sum_{k=1}^{M} \sigma_k^2 \exp(j2\pi f_k \tau) \tag{13.25}$$

where the coefficients $\{A_k\}$ $(k = 1, 2, \ldots, M)$ can be treated as the representation of the process in the frequency domain.

Another interesting category of processes are those characterized by the so-called *mean square periodicity*, a property that is mathematically formulated as $E\{|\boldsymbol{x}(t+T) - \boldsymbol{x}(t)|^2\} = 0$ for every time value t, from which it follows that for every specific time instance t, it will hold that $\boldsymbol{x}(t+T) = \boldsymbol{x}(t)$ with probability $P = 1$. However, the preceding condition *does not imply* that the set of outcomes ξ for which $\boldsymbol{x}(t+T, \xi) = \boldsymbol{x}(t, \xi)$ holds for every t is characterized by a probability value equal to one. According to an interesting theorem, *a sufficient and necessary condition for a process $\boldsymbol{x}(t)$ to be characterized by a mean square periodicity is to have an autocorrelation function that is doubly periodic*, that is, periodic with respect to both arguments of a property that is mathematically formulated as $R_{xx}(t_1 + mT, t_2 + nT) = R_{xx}(t_1, t_2)$ (see Papoulis and Pillai, page 393).

13.2.8 Returning to Ergodic Processes

As already previously, one of the central problems in the field of stochastic processes is *the calculation of the value of all kinds of statistical parameters using recorded values of real data*, with most parameters being *average values* of some quantity related to the process. Consequently, the problem of calculating the average value

$$\hat{\mu}_x(t) = \frac{1}{n} \sum_{i=1}^{n} \boldsymbol{x}(t, \xi_i) \tag{13.26}$$

from a set of n samples $\boldsymbol{x}(t, \xi_i)$ of the process $\boldsymbol{x}(t)$ is elevated to a problem of great importance. The preceding value can be considered a *a reliable estimate of the average of the process* μ_x, but under the condition that we have *a large number of samples* $\boldsymbol{x}(t, \xi_i)$. However, in some cases, the number of samples will be very small, while in the extreme case we will have at our disposal *a single and unique sample*.

Referring to the theorem of ergodicity, according to which in an ergodic process, if all the statistical average values of it can be determined from a simple sample of it, with probability $P = 1$, in the sense that they coincide with the values that arise from the set of the entire collection, *we have the ability to calculate the average values for the entire collection of samples, using only a simple sample of the process*. These temporal averages include the *usual average value*

$$\overline{x} = \lim_{T \to \infty} \frac{1}{2T} \int_{-T}^{T} x(t)dt \tag{13.27}$$

the temporal average of the product $x(t)x(t + \tau)$ (*temporal autocorrelation*), which is defined as

$$\hat{R}_{xx}(\tau) = \lim_{T \to \infty} \frac{1}{2T} \int_{-T}^{T} x(t)x(t + \tau)d\tau \tag{13.28}$$

the temporal average of the product $x(t)y(t + \tau)$ (*temporal cross-correlation*), which is defined as

$$\hat{R}_{xy}(\tau) = \lim_{T \to \infty} \frac{1}{2T} \int_{-T}^{T} x(t)y(t + \tau)d\tau \tag{13.29}$$

and the *temporal average square*, which is defined as

$$\overline{x^2} = \lim_{T \to \infty} \frac{1}{2T} \int_{-T}^{T} x^2(t)dt \tag{13.30}$$

In practice, we are mainly interested in the first two moments, that is, the *average value* and the *autocorrelation*, and for this reason we will examine the ergodic processes only for these two simple cases.

Ergodic Processes with Respect to the Average

Starting from the continuous case, let us consider a process $\boldsymbol{x}(t)$ with average value $\mu_x = E\{\boldsymbol{x}(t)\}$, and let us define the size

$$\mu_T = \frac{1}{2T} \int_{-T}^{T} x(t)dt$$

which is a random variable with average value

$$E\{\mu_T\} = \frac{1}{2T} \int_{-T}^{T} E\{x(t)\}dt$$

in the sense that it constitutes an *estimate of the statistical mean* whose value varies with the different samples of the random process. If the variance $\text{Var}(\mu_T)$ tends to zero in the limit $T \to \infty$, then we will have $\mu_T \to \mu_x$ (in MS sense). In this case, the time average $\mu_T(\xi)$ calculated from a simple sample of the process $\boldsymbol{x}(t)$ will be very close to the value of μ_x, with probability P very close to unity. If this property holds, then we will say that *the process is ergodic with respect to the mean value*. In this case, the condition $\lim \mu_t = \mu_x$ holds in the limit $T \to \infty$.

How do we calculate the variance $\mathrm{Var}(\mu_T)$ required for the preceding characterization? Defining the *moving average* of the process $\boldsymbol{x}(t)$ as

$$\boldsymbol{w}(t) = \frac{1}{2T} \int_{t-T}^{t+T} \boldsymbol{x}(\xi) d\xi$$

The process $\boldsymbol{w}(t)$, which—as is well known—constitutes *the output of a linear system with input $\boldsymbol{x}(t)$ and an impulse response corresponding to a pulse centered at $t = 0$*, is a *stationary* random process whose autocovariance assumes the value

$$C_{ww}(t) = \frac{1}{2T} \int_{-2T}^{2T} C(\tau - \xi) \left(1 - \frac{|\xi|}{2T}\right) d\xi \tag{13.31}$$

where $C(\tau)$ is the autocovariance of the process $\boldsymbol{x}(t)$. Observing now that $\mu_T = \boldsymbol{w}(0)$, it will hold that $\mathrm{Var}(\mu_T) = \mathrm{Var}(\boldsymbol{w}(0)) = C_{ww}(0)$ and that $C(-\xi) = C(\xi)$, and we will have

$$\mathrm{Var}(\mu_T) = \frac{1}{2T} \int_{-2T}^{2T} C(\xi) \left(1 - \frac{|\xi|}{2T}\right) d\xi = \frac{1}{T} \int_{0}^{2T} C(\xi) \left(1 - \frac{\xi}{2T}\right) d\xi \tag{13.32}$$

If the preceding expression tends to zero in the limit $T \to \infty$, then the process is aforementioned to be *ergodic with respect to the mean value*.

The situation is similar for the case of random processes. Let us consider a *stationary random process* $x[n]$ with mean value $\mu_x = E\{x[n]\}$, and let us define the corresponding time average as

$$\mu_N = \frac{1}{2N+1} \sum_{n=-N}^{N} x[n] \tag{13.33}$$

As before, since the value of the statistical average varies for different samples of the random process, this quantity is a *random variable*. The average value for all possible samples of the process $x[n]$ is

$$E\{\mu_N\} = E\left\{\frac{1}{2N+1} \sum_{n=-N}^{N} x[n]\right\} = \frac{1}{2N+1} \sum_{n=-N}^{N} E\{x[n]\} = \frac{1}{2N+1} \sum_{n=-N}^{N} \mu_x = \mu_x \tag{13.34}$$

We observe that the average value of the estimate is equal to the *statistical average*, and this is something we describe, saying that the estimate μ_N is unbiased.

Working as before, let us calculate the *variance* in order to formulate the criterion of ergodicity. In this case we have

$$\begin{aligned} E\{|\mu_N|^2\} &= \frac{1}{(2N+1)^2} \sum_{n=-N}^{N} \sum_{k=-N}^{N} E\{x^*[n]x[k]\} \\ &= \frac{1}{(2N+1)^2} \sum_{n=-N}^{N} \sum_{k=-N}^{N} R_{xx}[k-n] = \frac{1}{2N+1} \sum_{m=-2N}^{2N} \left(1 - \frac{|m|}{2N+1}\right) R_{xx}[m] \end{aligned} \tag{13.35}$$

It will therefore be that

$$\mathrm{Var}(\mu_N) = E\{|\mu_N|^2\} - |\mu_x|^2 = \frac{1}{2N+1} \sum_{m=-2N}^{2N} \left(1 - \frac{|m|}{2N+1}\right) R_{xx}[m] - |\mu_x|^2 = \frac{1}{2N+1} \sum_{m=-2N}^{2N} \left(1 - \frac{|m|}{2N+1}\right) C_{xx}[m]$$

Consequently, the discrete process $x[n]$ *will be ergodic with respect to the mean if the limit of the foregoing expression when* $N \to \infty$ *is equal to zero*. In this case and at the limit $N \to \infty$, *the estimate of the mean is equal to the mean value*. For more conditions of ergodicity with respect to the mean and related theorems, see Papoulis and Pillai, pages 528 and 530.

Ergodic Processes with Respect to Autocorrelation

In the case where the property of ergodicity concerns *the magnitude of autocorrelation*, the situation is entirely analogous. Limiting ourselves to the *discrete case* and working in the way we did for the case of the mean, let us define the estimate

$$r_{xx}[m] = \frac{1}{2N+1} \sum_{n=-N}^{N} x[n]x[n+m] \tag{13.36}$$

which, as before, constitutes a *random variable* for each value of m since it depends on the sample used. The expected value, that is, the mean for all samples, is

$$E\{r_{xx}[m]\} = \frac{1}{2N+1} \sum_{n=-N}^{N} E\{x[n]x[n+m]\} = \frac{1}{2N+1} \sum_{n=-N}^{N} R_{xx}[m] = R_{xx}[m] \tag{13.37}$$

and therefore, it constitutes an *unbiased estimate of the autocorrelation* $R_{xx}[m]$. On the other hand, to construct the expression for the variance $\mathrm{var}\{r_{xx}[m]\}$, we find that

$$\begin{aligned} E\{|r_{xx}[m]|^2\} &= \frac{1}{(2N+1)^2} \sum_{n=-N}^{N} \sum_{k=-N}^{N} E\{x^*[n]x[n+m]x[k]x^*[k+m]\} \\ &= \frac{1}{(2N+1)^2} \sum_{n=-N}^{N} \sum_{k=-N}^{N} R^m_{ww}[n-k] = \frac{1}{2N+1} \sum_{n=-2N}^{2N} \left(1 - \frac{|n|}{2N+1}\right) R^m_{ww}[n] \end{aligned} \tag{13.38}$$

since the average value $E\{x^*[n]x[n+m]x[k]x^*[k+m]\}$ appearing in the second sum of the preceding equation is simply the *autocorrelation* of the random process $w_m[n] = x^*[n]x[n+m]$. Consequently, the *variance* of the random variable $r_{xx}[m]$ will have the form

$$\mathrm{var}\{r_{xx}[m]\} = E\{|r_{xx}[m]|^2\} - |R_{xx}[m]|^2 = \frac{1}{2N+1} \sum_{n=-2N}^{2N} \left(1 - \frac{|n|}{2N+1}\right) R^m_{ww}[n] - |R_{xx}[m]|^2 \tag{13.39}$$

and if in the limit $N \to \infty$ it is the case that $\mathrm{var}\{r_{xx}[m]\} \to 0$, then the estimate $r_{xx}[m]$ converges to *statistical autocorrelation* $R_{xx}[m]$ with probability $P = 1$, and the process is ergodic with respect to autocorrelation.

The aforementioned discussion pertains to *continuous-time random signals*. The characterization of discrete-time signals follows a similar rationale, either by assigning discrete values to the outcomes of the underlying experiments or, equivalently, by performing *uniform sampling* of continuous-time random processes. The result of this process is a *discrete-time random process*, with the collection of signals that constitute its samples now being *discrete-time signals*. In other words, a stochastic process of discrete-time is defined as a sequence $\boldsymbol{x}_n$ of random variables. The average and the mean square in this case are defined as

$$\mu_x = E\{\mathbf{x}[n]\} = \int_{-\infty}^{\infty} x f(x,n)dx \quad \text{and} \quad E\{\mathbf{x}^2[n]\} = \int_{-\infty}^{\infty} x^2 f(x,n)dx$$

while as before, using the variable x of the integral, the appropriate exponent k, we can calculate the corresponding k order moment of the discrete random process $\boldsymbol{x}[n]$. On the other hand, the *autocorrelation function* is defined as

$$R_x[n,k] = E\{\mathbf{x}[n]\mathbf{x}^*[k]\} = \int_{-\infty}^{\infty} x_n x_k f(x_n, x_k; n, k)dx_n dx_k \tag{13.40}$$

while the *autocovariance function* is defined as

$$C_x[n,k] = R_x[n,k] - \mu[n]\mu^*[k] \tag{13.41}$$

In the case where the discrete process is a *stationary process*, then, by analogy with the continuous case, its statistical properties will remain *invariant under the action of shifting the origin of time*. Defining as before the time difference $m = n - k$, we will have

$$R_x[n,k] = R_x[n-k] = R_x[m] \quad \text{and} \quad C_x[n,k] = C_x[n-k] = C_x[m] = R_x[n] - \mu_x^2$$

If the stationary process is of a complex nature, in the form $z[n] = x[n] + jy[n]$, then the autocorrelation function is defined as

$$R_z[m] = R_x[m] + R_y[m] + j[R_{yx}[m] - R_{xy}[m]]$$

while the correlation between the complex and jointly stationary processes $z[n] = x[n] + jy[n]$ and $w[n] = u[n] + jv[n]$ is given by the expression

$$R_{zw}[m] = R_{xu}[m] + R_{yv}[m] + j[R_{yu}[m] - R_{xv}[m]]$$

A process $\boldsymbol{x}[n]$ is a *strictly white noise process* if the random variables $x[n_i]$ are *independent*, while in the case where these variables are uncorrelated, the process is simply characterized as a *white noise process*. The autocorrelation of a white noise process with zero mean value is defined as $R[n_1, n_2] = q[n_1]\delta[n_1 - n_2]$, where $\delta[n]$ is the discrete impulse function and $q[n] = E\{\boldsymbol{x}^2[n]\}$. If additionally the process $\boldsymbol{x}[n]$ is *stationary*, then $R[m] = q\delta[m]$, and consequently, *a WSS-type white noise process is simply a sequence of mutually independent random variables characterized by the same probability distribution, as well as by a variability value equal to* q.

13.3 Representation of Random Signals in Frequency Domain

The representation of a signal in the frequency domain that was presented in great detail in the relevant chapters constitutes one of the most important tools in the field of engineering science, since by resorting to the use of the convolution theorem, we can transform the problem of calculating convolutions in the time domain to the much simpler problem of calculating products in the frequency domain. The function that holds central importance in this type of representation is the *Fourier transform*, with the related representation usually based on the use of quantities such as *the amplitude and phase of the signal, as well as the energy spectrum*. The question, then, of the possibility of using the aforementioned concepts and the tools related to them for the description of random signals is more than reasonable. Fortunately, the answer to this question is affirmative, provided that the appropriate modifications are made.

Although it would be reasonable to define the Fourier transform of random signals in the way we did for deterministic (nonrandom) signals, this is not possible—for two main reasons. First, the Fourier transform *will be characterized by a different value for each of the sample functions* that are members of the collection of functions of the random process an,d therefore, will be a random variable. This means that representation in the frequency domain *is not possible for the process itself but only for one of its samples*. Note that, although we could use some average value of the transform from the collection, this *is not feasible*, for the second reason, which is related to the fact that the *condition of absolute integrability*, which is a necessary prerequisite for the convergence of the Fourier transform, *is not satisfied by any nonzero function, which is a sample of a random WSS-type process*. Moreover, in most cases, the analytical form of the function describing the underlying process *is not known*, so the possibility of using this transform is not even raised. On the other hand, the use of the *Laplace transform*, which converges in more cases compared to the Fourier transform due to the presence of an exponential function with a negative exponent, proves to be *not* particularly effective, and for this reason the definition of the Fourier transform in the case of random processes requires *the appropriate modification of the sample functions*. Let us also mention that although in the case of deterministic signals the Fourier transform arises *from the superposition of exponential functions*, in the case of random signals, the concept of the spectrum admits *two* interpretations. According to the first interpretation, which is of a deterministic nature, *the spectrum is related to the transform of average values*, while according to the second interpretation, *the process under examination is modeled as a parallelism of exponential functions with random coefficients*.

13.3.1 The Power Spectral Density of Random Signals

Although the modification of the sample function done in such a way that the prerequisites of the Fourier transform for random signals are met can be achieved in several ways, the simplest of these consists *in defining a second function of finite duration, with the help of the original function*. Let us consider a sample function $x(t)$ and a second function $x_T(t)$ of finite duration equal to $2T$, which coincides with the signal $x(t)$ in the value range $[-T, T]$, while it has zero value anywhere else. Note that this new function satisfies the property of *absolute integrability* (although in reality, it satisfies the stricter condition of *absolute square integrability*), provided that the value of the parameter T is *finite* and that the stationary process of which this function is a sample is characterized by a *finite* value regarding the mean square. Consequently, this function meets the convergence prerequisites of the Fourier transform so that we can write that

$$\mathcal{X}_T(j\omega) = \mathcal{F}\{x_T(t)\} = \int_{-\infty}^{\infty} x_T(t)e^{-j\omega t}dt$$

under the assumption that $T < \infty$. However, as we will see later, the expected value of the quantity $|\mathcal{X}(j\omega)|^2$ in the limit $T \to \infty$ exists, despite the fact that this *does not* hold for the Fourier transform itself. To prove this claim, let us resort to the use of *Parseval's theorem*, which is known to have the form

$$\int_{-T}^{T} x_T^2(t)dt = \frac{1}{2\pi}\int_{-\infty}^{\infty} |\mathcal{X}(j\omega)|^2 d\omega \tag{13.42}$$

with the limits of the integral of the left-hand side being due to the fact that for $t > |T|$ it holds that $x_T(t) = 0$.

Since the quantity we want to calculate is *the average power as a function of frequency*, let us divide the two sides of the foregoing expression by the quantity $2T$, which is the time duration for which the function $x_T(t)$ is defined. Then

$$\frac{1}{2T}\int_{-T}^{T} x_T^2(t)dt = \frac{1}{4\pi T}\int_{-\infty}^{\infty} |\mathcal{X}(j\omega)|^2 d\omega \tag{13.43}$$

with the left-hand side of the preceding expression appearing to correspond to the *average power of the sample function* for the time interval $[-T, T]$. Note that although one might be tempted to take the limit of the preceding expression when $T \to \infty$, which leads to the value of the mean square of the considered process, this in reality *is not permissible*, since in the limit $T \to \infty$, the Fourier transform of the function $x(t)$ is not defined. But because, as noted earlier, the Fourier transform itself is a random variable (since it has a different value for the various functions—samples related to the process $x(t)$)—we can take the limit when $T \to \infty$ not of the preceding function but *of the expected value of this function*. Applying to the foregoing expression the operator of the *expected value*, it takes the form

$$E\left\{\frac{1}{2T}\int_{-T}^{T} x_T^2(t)dt\right\} = E\left\{\frac{1}{4\pi T}\int_{-\infty}^{\infty} |\mathcal{X}(j\omega)|^2 d\omega\right\} \tag{13.44}$$

Taking the limit of the foregoing expression when $T \to \infty$ and alternating the processes of applying the limit and the expected value, we are led to the expression

$$\lim_{T\to\infty} \frac{1}{2T}\int_{-T}^{T} E\{x_T^2(t)\}dt = \lim_{T\to\infty} \frac{1}{4\pi T}\int_{-\infty}^{\infty} E\{|\mathcal{X}(j\omega)|^2\}d\omega \tag{13.45}$$

and finally

$$\left\langle \overline{x^2} \right\rangle = \frac{1}{2\pi} \int_{\infty}^{\infty} \lim_{T \to \infty} \frac{E\{|\mathcal{X}(j\omega)|^2\}}{2T} d\omega \tag{13.46}$$

If the considered process is a stationary process, then $\left\langle \overline{x^2} \right\rangle = \overline{x^2}$, and the foregoing relation takes the form

$$\overline{x^2} = \frac{1}{2\pi} \int_{\infty}^{\infty} \lim_{T \to \infty} \frac{E\{|\mathcal{X}(j\omega)|^2\}}{2T} d\omega = \frac{1}{2\pi} \int_{\infty}^{\infty} S_x(j\omega) d\omega \tag{13.47}$$

with the magnitude

$$S_x(j\omega) = \lim_{T \to \infty} \frac{E\{|\mathcal{X}(j\omega)|^2\}}{2T}$$

known as the *spectral density of a random process*. This size is particularly important, as it plays for random processes the same role that the Fourier transform plays in common (nonrandom) processes.

Based on the foregoing discussion, we can formulate the assertion that the quantity $|\mathcal{X}_T(j\omega)|^2/2T$ expresses *the average power per unit bandwidth*, and for this reason, it is known as the *power spectral density* (PSD) of the signal $x_T(t)$. The physical meaning of the PSD can be easily understood, considering as an example a resistance of 1 Ohm, to which a voltage $x(t)$ is applied. Just as the function $\overline{x^2}$ expresses *the average power consumed in the resistance*, so does the function $S_x(j\omega)$ express the average power associated with a bandwidth equal to 1 Hz, which is centered at the position $\omega/2\pi$ Hz. Because of this correlation that exists between the PSD and the average power of a random process, the PSD is often characterized as a *power spectral density*.

13.3.2 The Wiener–Khinchin Theorem

A very important result that finds widespread application in all processes of analyzing random processes is that *the PSD constitutes the Fourier transform of the autocorrelation function*. To verify this claim, let us restate the defining equation of the function $S_x(j\omega)$ as

$$\begin{aligned} S_x(\omega) &= \lim_{T \to \infty} E\left\{\frac{|\mathcal{X}_T(j\omega)|^2}{2T}\right\} = \lim_{T \to \infty} E\left\{\frac{\mathcal{X}_T(j\omega)\mathcal{X}_T^*(j\omega)}{2T}\right\} = \lim_{T \to \infty} \frac{1}{2T} E\left\{\int_{-T}^{T} \boldsymbol{x}(t_2) e^{-j\omega t_2} dt_2 \int_{-T}^{T} \boldsymbol{x}(t_1) e^{j\omega t_1} dt_1\right\} \\ &= \lim_{T \to \infty} \frac{1}{2T} \int_{-T}^{T} \int_{-T}^{T} E\{\boldsymbol{x}(t_1)\boldsymbol{x}(t_2)\} e^{-j\omega(t_2 - t_1)} dt_1 dt_2 \end{aligned} \tag{13.48}$$

At this point, it is not difficult to see that the quantity $E\{\boldsymbol{x}(t_1)\boldsymbol{x}(t_2)\}$ appearing in the last expression constitutes the *autocorrelation function* $R_x(t_1, t_2)$ *of the random process* $\boldsymbol{x}(t)$, which, if the process is considered *WSS*, depends, as we mentioned, only on the difference $t_2 - t_1$. We will therefore have $E\{\boldsymbol{x}(t_1)\boldsymbol{x}(t_2)\} = R_x(t_2 - t_1)$, and the foregoing relation takes the form

$$S_x(\omega) = \lim_{T \to \infty} \frac{1}{2T} \int_{-T}^{T} \int_{-T}^{T} R_x(t_2 - t_1) e^{-j\omega(t_2 - t_1)} dt_1 dt_2 = \lim_{T \to \infty} \frac{1}{2T} \int_{-T}^{T} \int_{-T}^{T} f(t_2 - t_1) dt_1 dt_2 \tag{13.49}$$

with the auxiliary function $f(t_2 - t_1) = R_x(t_2 - t_1)e^{-j\omega(t_2 - t_1)}$ depending *only* on the difference $t_2 - t_1$ since the same applies to the quantities involved in its definition.

To calculate the value of this integral, we must observe that it constitutes *a double integral* with respect to the variables t_1 and t_2, with the region of the plane (t_1, t_2) involved in this integration process a *square* with the origin of the axes at its

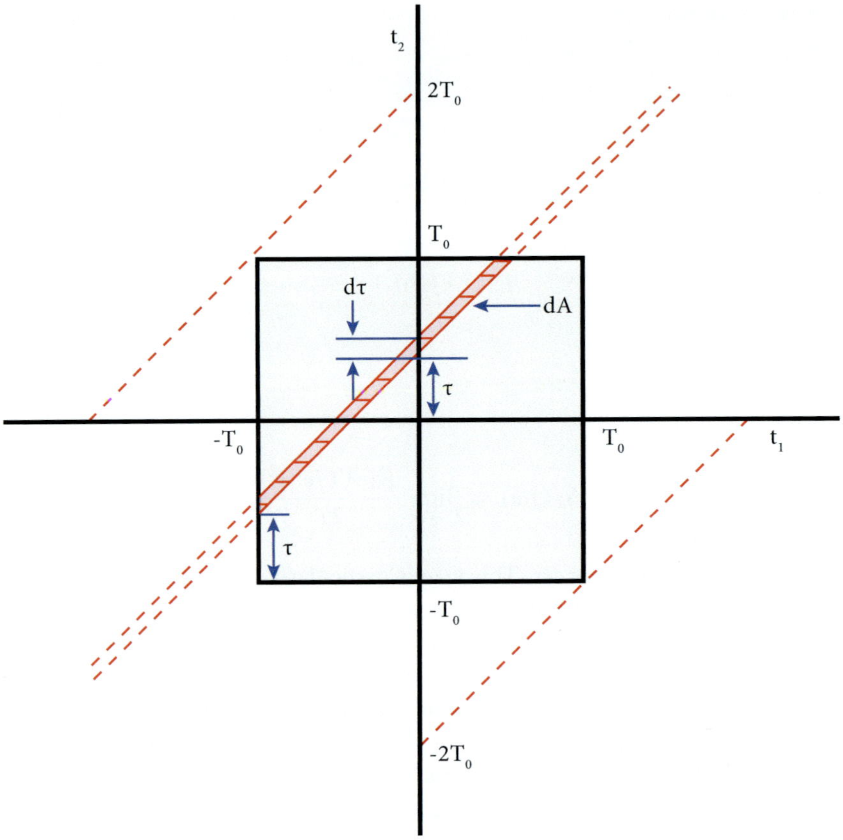

Fig. 13.3 Geometry of problem of calculating power spectral density (Antoniou, Figure 13.4)

center and side length $2T$, since integration with respect to t_1 and t_2 is from $-T$ to T (see Fig. 13.3—Antoniou, Digital Signal Processing, page 607). On the other hand, the function $t = f(t_2 - t_1)$ defines a *three-dimensional surface* and it is not difficult to see that the value of the double integral of the foregoing expression *is equal to the three-dimensional volume V defined by the three-dimensional surface* $y = f(t_2 - t_1)$ *and the square region of the plane* (t_1, t_2). How do we calculate this volume? Initially, we should note that since the function $f(t_2 - t_1)$ depends on the difference $t_2 - t_1$, it will have a constant value at all points $c = t_2 - t_1$ that define straight lines in the plane (t_1, y_2) with equations of the form $t_2 = t_1 + c$ (see the *diagonal dashed lines* in that plane for the various values of c). Let us now consider an element of length $d\tau$, as well as the lines $t_2 = t_1 + \tau$ and $t_2 = t_1 + \tau + d\tau$, which have the form $t_2 = t_1 + c$ for the values $c = \tau$ and $c = \tau + d\tau$ (recall that the value of f along them is constant), which together define a *surface element* dA on the plane (t_1, t_2) (see shaded area of the figure). This surface element dA cuts from the volume V we defined earlier a *vertical diagonal slice* whose volume can be considered approximately constant.

From the geometry of the figure we find that for values $\tau \geq 0$, the surface element dA is calculated as[3]

$$dA = \frac{1}{2}(2T - \tau)^2 - \frac{1}{2}[2T - (\tau + d\tau)]^2 = (2T - \tau)d\tau + \frac{1}{2}(d\tau)^2 \approx (2T - \tau)d\tau$$

while, in the same way, for $\tau < 0$ it turns out that $dA \approx (2T + \tau)d\tau$. Therefore, combining these results in the limit $d\tau \to 0$ leads to $dA \approx (2T + |\tau|)d\tau$. Consequently, the volume dV of the vertical slice described earlier will be $dV = f(\tau)dA = f(\tau)(2T - |\tau|)d\tau$. Now, to calculate the entire volume below the surface $y = f(t_2 - t_1)$ and above the square area with the origin at its center and side length $2T_0$, we must integrate from $-2T$ to $2T$ (see the geometry of the figure and note that the double integral became a simple one with the integration variable τ). So we will have

[3] Notice that the area of the element dA is simply the difference in the areas of the two overlapping equilateral rectangular triangles with a common right angle at the upper left corner of the square area, subtending the diagonal lines $t_2 = t_1 + \tau$ and $t_2 = t_1 + \tau + d\tau$ and side lengths $2T - \tau$ and $2T - (\tau + d\tau)$, respectively.

$$S_x(\omega) = \lim_{T\to\infty} \frac{1}{2T} \int_{-T}^{T} f(\tau)(2T - |\tau|)d\tau = \int_{-2T}^{2T} f(\tau) \lim_{T\to\infty} \left(1 - \frac{|\tau|}{2T}\right) d\tau = \int_{-\infty}^{\infty} f(\tau)d\tau \tag{13.50}$$

where for the construction of the last expression *we have alternated the operations of limit calculation and integration*. But this expression is easily recognized as the *Fourier transform* of the function $f(\tau)$, which is simply the *autocorrelation* $R_x(t_1, t_2) = R_x(t_2 - t_1) = R_x(\tau)$, as required for a WSS process. In other words, *the PSD of a random WSS process is equal to the Fourier transform of the autocorrelation function*, or, in mathematical notation, $S_x(e^{j\omega}) = \mathcal{F}\{R_x(\tau)\}$. In the literature in the field, this result is known as the *Wiener–Khinchin theorem* and is considered particularly important, as it correlates the time domain, in which the autocorrelation function is included, with the frequency domain, in which the PSD function is included. Because of the uniqueness of the Fourier transform, if the function $R_x(\tau)$ is *absolutely integrable*, it will in turn constitute *the inverse Fourier transform of the PSD*, so that we can write

$$R_x(\tau) = E\{\boldsymbol{x}(t)\boldsymbol{x}(t+\tau)\} = \frac{1}{2\pi} \int_{-\infty}^{\infty} S_x(\omega)e^{j\omega\tau} d\omega \tag{13.51}$$

Of course, for the foregoing expression to hold, which allows one to calculate the *autocorrelation function* from the PSD, the underlying process must be a *stationary* process, as only then does the relation $\langle \overline{x^2} \rangle = \overline{x^2}$ hold. For this reason, in the following analysis, we will only consider stationary processes of the WSS type.

Taking into account that the autocorrelation function is an *even* function, the same will apply, as is well known, to its Fourier transform, from which it follows that for the case of a *real random* process $\boldsymbol{x}(t)$, the function $S_x(j\omega)$ will be a *real and even* positive function (see the properties of the transform). Of particular interest, however, are those spectral density functions that are *rational* functions of frequency, i.e., they are defined as the *ratio of two polynomial functions*. Moreover, in this case, due to the *even* symmetry of the function in question, the polynomials of the numerator and the denominator contain only *even* powers of the frequency ω, and therefore, the function $S_x(j\omega)$ for this rational case has the form

$$S_x(j\omega) = S_0 \frac{\omega^{2n} + \alpha_{n-2}\omega^{2n-2} + \alpha_{n-4}\omega^{2n-4} + \cdots + \alpha_2\omega^2 + \alpha_0}{\omega^{2n} + \beta_{n-2}\omega^{2n-2} + \beta_{n-4}\omega^{2n-4} + \cdots + \beta_2\omega^2 + \beta_0} \tag{13.52}$$

with the degree of the polynomial of the denominator being *greater* than the degree of the polynomial of the numerator if the mean square value of the random process is *finite* (with the exception of the case of white noise, which is characterized by constant power value for all frequencies). In the foregoing expressions, the PSD is expressed as a function of frequency ω. However, as is often the case, it is often useful to express this function with the help of the *complex frequency s* by proceeding to the replacement $\omega \mapsto -js$, so that we can study it using the known function study tools and to determine *its poles and zero values*. In the case where the PSD is a *rational* function, we should replace the quantity ω^2 with the quantity $(-js)(-js) = -s^2$.

Taking into account that the PSD function represents *the average power per unit bandwidth*, it is not difficult to understand that a *continuous spectral power* cannot describe a process that contains a constant and/or periodic component. Indeed, a constant component of a random process represents a *finite power that belongs to zero bandwidth* and that corresponds to *infinite PSD*. Taking into account that the frequency of a constant component has a value of $\omega = 0$, we expect the PSD to have an infinite value at the position $\omega = 0$ and a zero value anywhere else. But this description refers to an *impulse function* $\delta(\omega)$ at the position $\omega = 0$ while the same applies to each of the discrete frequencies which characterize the periodic functions. This result, although it can be understood intuitively and in accordance with what we mentioned, is also mathematically proven.

The fact that the autocorrelation function is an *even* function allows us to simplify the foregoing expression, writing it in the form

$$S_x(\omega) = \int_{-\infty}^{\infty} R_x(\tau)\cos(\omega\tau)d\tau = 2\int_{0}^{\infty} R_x(\tau)\cos(\omega\tau)d\tau \tag{13.53}$$

from which, by inversion, it turns out that

$$R_x(\tau) = \frac{1}{2\pi} \int_{-\infty}^{\infty} S_x(\omega)\cos(\omega\tau)d\omega = \frac{1}{\pi} \int_{0}^{\infty} S_x(\omega)\cos(\omega\tau)d\omega \tag{13.54}$$

If we set $\omega = 0$, then the function $S_x(0)$ will be equal to the *integral of the autocorrelation function*, and therefore, the area defined by the curve of the autocorrelation function and the horizontal axis will be equal to the *PSD* at the position $\omega = 0$. On the other hand, if the process $\boldsymbol{x}(t)$ is *complex*, then we will have $R_x(\tau) = R_x^*(-\tau)$ and, resorting to the properties of the Fourier transform, we find that $S_x^*(\omega) = S_x(\omega)$. Therefore, in this case, the function $S_x(\omega)$ is an *even* function.

In several cases, as noted previously, the PSD is expressed using the *complex frequency s*, and therefore, the function in question is related to the autocorrelation function not using the Fourier transform but with the help of the Laplace transform, with the direct and inverse relationship having the form

$$S_x(s) = \int_{-\infty}^{\infty} R_x(\tau)e^{-s\tau}d\tau \qquad \text{and} \qquad R_x(\tau) = \frac{1}{2\pi j}\int_{-j\infty}^{j\infty} S_x(s)e^{s\tau}ds$$

Note that the Laplace transform used is the bilateral (two-sided) and not the unilateral (one-sided) *transform* since the autocorrelation function due to its even symmetry *cannot take zero values* for $\tau < 0$. As for the curve along which the preceding integral is calculated, this is the entire $j\omega$ axis, since the PSD is characterized by a *finite* mean square value and therefore *will not* have poles on the vertical axis.

Considering an *arbitrary integrable and nonnegative real function* $S(\omega)$, it is proven that a stationary random process $\boldsymbol{x}(t)$ can *always* be determined whose PSD is the function $S(\omega)$. To prove the truth of this claim, let us consider a process $\boldsymbol{x}(t) = \alpha e^{j(\omega t - \varphi)}$, where α is a real constant, ω a random variable with PDF $f_\omega(\omega)$, and φ is a second random variable that is independent of ω and uniformly distributed in the range of values $[0, 2\pi]$. This process is a WSS process characterized by zero mean value and by an autocorrelation function of the form

$$R_x(t_1, t_2) = E\{\boldsymbol{x}(t_1)\boldsymbol{x}^*(t_2)\} = E\{\alpha e^{j(\omega t - \varphi)}\alpha^* e^{j(\omega t - \varphi)}\} = \alpha^2 E\{e^{j\omega(t_1 - t_2)}\} = \alpha^2 E\{e^{j\omega\tau}\} = \alpha^2 \int_{-\infty}^{\infty} f_\omega(\omega)e^{j\omega\tau}d\omega = R_x(\tau)$$

Taking into account that the autocorrelation function $R_x(\tau)$ constitutes the *inverse Fourier transform of the PSD* $S_x(\omega)$, we will have

$$R_x(\tau) = \frac{1}{2\pi}\int_{-\infty}^{\infty} S_x(\omega)e^{j\omega t}d\omega \tag{13.55}$$

and from the comparison of the two foregoing relations, it follows that

$$\alpha^2 f_\omega(\omega) = \frac{S_x(\omega)}{2\pi} \qquad \text{or equivalently} \qquad S_x(\omega) = 2\pi\alpha^2 f_\omega(\omega)$$

If we therefore define the quantity

$$f_\omega(\omega) = \frac{S(\omega)}{2\pi\alpha^2} \qquad \text{where} \qquad \alpha^2 = \frac{1}{2\pi}\int_{\infty}^{\infty} = R_x(0)$$

for every given function $S(\omega)$, then the function $f_\omega(\omega)$ will constitute a *probability density*, while $S_x(\omega) = S(\omega)$. The only thing left to do to complete the determination of the function $\boldsymbol{x}(t)$ is to construct a *random variable* with PDF $S(\omega)/2\pi\alpha^2$ and incorporate it into the equation $\boldsymbol{x}(t) = \alpha e^{j(\omega t - \varphi)}$. In the special case where the function $S(\omega)$ is characterized by *even* symmetry, that is, it is such that $S(\omega) = S(-\omega)$, then the process $\boldsymbol{x}(t)$ just determined is a *real* process, while the inverse also holds. In conclusion, *the validation of a power spectral function is much easier than the corresponding validation of an autocorrelation function* since all we have to examine is whether its value is nonnegative, while in the case of autocorrelation, we must verify that it constitutes a positive semidefinite function. In reality, it is proven that *the property of a function being positive semidefinite is equivalent to the property of its Fourier transform taking nonnegative values*, and in this sense it constitutes a *sufficient and necessary condition* for an autocorrelation or autocovariance function to be a valid function.[4] Just as the power spectral density is the Fourier transform of the autocorrelation function, the characteristic function is the Fourier transform of the probability density function.

[4] For a proof of this equivalence see Stark and Woods, page 454.

13.3.3 Power Spectral Density and Jointly Stationary Processes

Note that the preceding definition of PSD can be extended to the case of *two jointly stationary processes*, $\boldsymbol{x}(t)$ and $\boldsymbol{y}(t)$. If $\mathcal{X}(j\omega)$ and $\mathcal{Y}(j\omega)$ are the *Fourier transforms* of the two functions, then, we can define

$$S_{xy}(\omega) = \lim_{T\to\infty} \frac{E\{\mathcal{X}(-j\omega)\mathcal{Y}(j\omega)\}}{2T} \quad \text{and} \quad S_{yx}(\omega) = \lim_{T\to\infty} \frac{E\{\mathcal{Y}(-j\omega)\mathcal{X}(j\omega)\}}{2T}$$

From the foregoing expressions we observe that these functions in the general case *are not real, positive, and even*. On the contrary, these functions, generally speaking, are complex in nature and, moreover, satisfy the property $S_{xy}(\omega) = S^*_{yx}(\omega)$, where $*$ is the operator of complex conjugation. However, it should be noted that the real and imaginary parts of these functions are symmetric functions, since the functions $\Re\{\mathcal{X}(j\omega)\}$ and $\Re\{\mathcal{Y}(j\omega)\}$ are characterized by even symmetry, while the functions $\Im\{\mathcal{X}(j\omega)\}$ and $\Im\{\mathcal{Y}(j\omega)\}$ are characterized by odd symmetry. If the cross-correlation functions $R_{xy}(\tau)$ and $R_{yx}(\tau)$ are known, then the functions $S_{xy}(\omega)$ and $S_{yx}(\omega)$ can, as before, be calculated as the Fourier transforms

$$S_{xy}(\omega) = \int_{-\infty}^{\infty} R_{xy}(\tau)e^{-j\omega\tau}d\omega \quad \text{and} \quad S_{yx}(\omega) = \int_{-\infty}^{\infty} R_{yx}(\tau)e^{-j\omega\tau}d\omega$$

with inverse relations

$$R_{xy}(\tau) = \frac{1}{2\pi}\int_{-\infty}^{\infty} S_{xy}(\omega)(\tau)e^{j\omega\tau}d\omega \quad \text{and} \quad R_{yx}(\tau) = \frac{1}{2\pi}\int_{-\infty}^{\infty} S_{yx}(\omega)(\tau)e^{j\omega\tau}d\omega$$

Finally, in the case where the complex frequency $s = j\omega$ is used instead of the simple frequency ω, the foregoing relations are correspondingly modified as

$$S_{xy}(s) = \int_{-\infty}^{\infty} R_{xy}(\tau)e^{-s\tau}d\omega$$

$$S_{yx}(s) = \int_{-\infty}^{\infty} R_{yx}(\tau)e^{-s\tau}d\omega$$

and

$$R_{xy}(\tau) = \frac{1}{2\pi j}\int_{-j\infty}^{j\infty} S_{xy}(s)(\tau)e^{s\tau}ds$$

$$R_{yx}(\tau) = \frac{1}{2\pi j}\int_{-j\infty}^{j\infty} S_{yx}(s)(\tau)e^{s\tau}ds$$

If the processes $\boldsymbol{x}(t)$ and $\boldsymbol{y}(t)$ are of a complex nature, we will have $S^*_{xy}(\omega) = S_{yx}(-\omega)$, while in the case of real stationary processes, it is proven that $S_{yx}(\omega) = S_{xy}(-\omega)$. The proof of these properties is carried out as in the chapter on continuous-time Fourier transforms and is left as an exercise to the reader.

The Case of Discrete-Time Signals

The extension of the foregoing concepts to the field of *discrete* time is direct. Initially, as in the continuous case, *a random discrete-time process is characterized by infinite energy but finite average power*, which is equal to the value of the autocorrelation $R_x(0)$. Referring to the *Wiener–Khinchin theorem*, the power spectral density is given by the *discrete-time Fourier transform* of the autocorrelation sequence, and therefore, it has the form

$$S_x(e^{j\omega}) = \sum_{m=-\infty}^{\infty} R_x[m]e^{-j\omega m}$$

On the other hand, reversing the foregoing relationship, it turns out that

$$R_x[m] = \frac{1}{2\pi}\int_{-\pi}^{\pi} S_x(e^{j\omega})e^{j\omega m}d\omega \geq 0 \tag{13.56}$$

The function $S_x(\omega)$, as in the continuous case, describes the *distribution of power* and, therefore, constitutes *the power spectrum of the random process* $\boldsymbol{x}[n]$. Consequently, in the domain of discrete-time, *the discrete-time Fourier transform, which is simply the* $\mathcal{Z}$ *transform calculated on the circumference of the unit circle*, is *numerically equal* to the PSD of the discrete process $\boldsymbol{x}[n]$. According to our earlier discussion in the chapter on the discrete-time Fourier transform, the function $S_x(e^{j\omega})$ is a periodic function with period $T = 2\pi$ and with Fourier expansion coefficients the quantities $R[m]$. As happens in all similar cases, usually the calculation of this function is limited only to the interval $|\omega| < \pi$. If the process $\boldsymbol{x}[n]$ constitutes a continuous process, then $R[m] = R[-m]$ and the foregoing relationship, as known, takes the simple form

$$S_x(e^{j\omega}) = R[0] + 2\sum_{m=0}^{\infty} R_x[m]\cos(\omega m) \tag{13.57}$$

13.3.4 Understanding the Power Spectral Density

To understand the *PSD*, let us consider a process $\boldsymbol{x}(t)$ of the WSS type, and let us define through it the process $\boldsymbol{x}_T(t) = \boldsymbol{x}(t)\boldsymbol{w}_T(t)$, where $\boldsymbol{w}_T(t) = 1$ for $-T < t < T$, and $\boldsymbol{w}_T(t) = 0$ anywhere else. In this case, the Fourier transform of the function $\boldsymbol{x}_T(t)$ has the form

$$\mathcal{X}_T(j\omega) = \int_{-T}^{T} \boldsymbol{x}(t)e^{-j\omega t}dt$$

from which it follows that

$$|\mathcal{X}_T(j\omega)|^2 = \mathcal{X}_T(j\omega)\mathcal{X}_T^*(j\omega) = \int_{-T}^{T}\int_{-T}^{T} \boldsymbol{x}(t_1)\boldsymbol{x}^*(t_2)e^{-j\omega(t_1-t_2)}dt_1dt_2$$

Dividing both sides of the foregoing expression by the quantity $2T$ and taking the expected value, it follows that

$$\frac{1}{2T}E\left\{|\mathcal{X}_T(j\omega)|^2\right\} = \frac{1}{2T}E\left\{\mathcal{X}_T(j\omega)\mathcal{X}_T^*(j\omega)\right\} = \frac{1}{2T}E\left\{\int_{-T}^{T}\int_{-T}^{T} \boldsymbol{x}(t_1)\boldsymbol{x}^*(t_2)e^{-j\omega(t_1-t_2)}dt_1dt_2\right\}$$

or, equivalently,

$$\frac{1}{2T}E\left\{|\mathcal{X}_T(j\omega)|^2\right\} = \frac{1}{2T}\int_{-T}^{T}\int_{-T}^{T} E\{\boldsymbol{x}(t_1)\boldsymbol{x}^*(t_2)\}e^{-j\omega(t_1-t_2)}dt_1dt_2 = \frac{1}{2T}\int_{-T}^{T}\int_{-T}^{T} R_x(t_1-t_2)e^{-j\omega(t_1-t_2)}dt_1dt_2$$

since if we move the operator inside the *double integral* in the second member, we will have $E\{\boldsymbol{x}(t_1)\boldsymbol{x}^*(t_1)\} = R_x(t_1, t_2) = R_x(t_1, t_2)$, where we used the definition of autocorrelation and took into account that the input process is a WSS process. Proceeding now to the change of variables $s = t_1 + t_2$ and $\tau = t_1 - t_2$, the square area of integration R, of length $2T$, of the left image of Fig. 13.4 is transformed into area S of the right image of Fig. 13.4, which has a side length of $T\sqrt{2}$ and has undergone a rotation of 45° in the opposite direction of the clock hands. In this case, the preceding integral is transformed into[5]

$$\frac{1}{2T}E\left\{|\mathcal{X}_T(j\omega)|^2\right\} = \frac{1}{4T}\iint_S R_{xx}(\tau)e^{-j\omega\tau}d\tau ds = \frac{1}{4T}\left\{\int_{-2T}^{0} R_{xx}(\tau)e^{-j\omega\tau}\left[\int_{-(2T+\tau)}^{2T+\tau} ds\right]d\tau\right\}$$

[5] From the calculus of several variables, we know that the double integral $\iint_R f(t_1, t_2)dt_1dt_2$, under the change of variables $s = s(t_1, t_2)$ and $\tau = \tau(t_1, t_2)$, is transformed into the integral $\iint_S f[s(t_1, t_2), \tau(t_1, t_2)]|J|dsd\tau$, where $|J|$ is the *Jacobian determinant* of the transformation.
The new limits of integration, which define the shape of the image of the original integration region under this transformation, must be determined accordingly. In this case, the *Jacobian determinant* equals $1/2$, which justifies the division by the quantity $4T$. The transformed integrals are then expressed with the new limits.

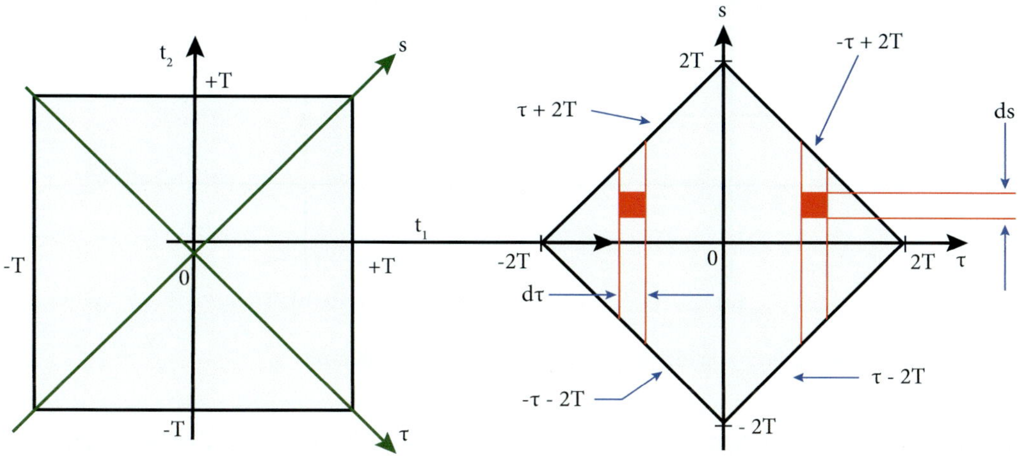

Fig. 13.4 The geometry of the problem of understanding the PSD (Stark and Woods, Figure 7.5–1)

$$+\frac{1}{4T}\left\{\int_{0}^{2T} R_{xx}(\tau)e^{-j\omega\tau}\left[\int_{-(2T-\tau)}^{2T-\tau} ds\right]d\tau\right\} = \int_{-2T}^{2T}\left(1-\frac{|\tau|}{2T}\right)R_{xx}(\tau)e^{-j\omega\tau}d\tau \qquad (13.58)$$

From the last relation we observe that in the limit $T \to \infty$, the contribution of the triangular term in parentheses in the last expression tends to zero, and what remains *is simply the Fourier transform of the autocorrelation function*, which is also the function of the *PSD*. Based on all the preceding considerations, we can write that

$$S_x(\omega) = \lim_{T\to\infty}\frac{1}{2T}E\left\{|\mathcal{X}_T(j\omega)|^2\right\} \qquad (13.59)$$

Consequently, the Power Spectral Density (PSD) is related to the average power at the frequency ω.

Examples of Calculating the Power Spectral Density (PSD)

Taking into account that the PSD is simply the Fourier transform of the autocorrelation function, it is not difficult to understand that the calculation of the first function, when the second is known, *simply constitutes a problem of calculating continuous-time Fourier transforms* like those we presented in the homonymous section. Reproducing indicative results that we had proven in that section, if the autocorrelation function is $S_x(\tau) = \exp(-\alpha|\tau|)$ $(-\infty < \tau < \infty)$, then the PSD is $S_x(\omega) = \mathcal{F}\{\exp(-\alpha|\tau|)\} = 2\alpha/(\alpha^2+\omega^2)$, while if the autocorrelation function is the triangular pulse $R_{xx}(\tau) = \max[1-(|\tau|/T), 0]$, then the PSD is $S_x(\omega) = \mathcal{F}\{\max[1-(|\tau|/T), 0]\} = T\,\text{sinc}^2(\omega T/2)$ (recall that *the triangular pulse is the convolution of two square pulses*, and therefore the Fourier transform of the triangular pulse, according to the convolution theorem, *will be equal to the product of the Fourier transforms of the two square pulses*).

White Noise Processes

Having defined the concept of *PSD*, we can at this point define the so-called *white noise process* $\boldsymbol{x}_{\text{WN}}(t)$ as a process characterized by a *constant PSD for all frequency values* ω or, in mathematical notation, $S_x(\omega) = \sigma^2 =$ constant for $-\infty < \omega < \infty$. Referring to the *Wiener–Khinchin theorem*, it is not difficult to see that the autocorrelation function for this process, which is the inverse transform of the PSD function, has a value

$$R_x(\tau) = \mathcal{F}^{-1}\{S_x(\omega)\} = \mathcal{F}^{-1}\{\sigma^2\} = \sigma^2\mathcal{F}^{-1}\{1\} = \sigma^2\delta(t)$$

and therefore, the autocorrelation function for a *white noise process* is simply a *unit impulse* at position $\tau = 0$. This condition implies that for any nonzero value of τ, the samples of white noise *statistically do not relate to each other*. The *white noise* (which took its name from white light, in which all color components contribute *the same power value*) does not exist in practice but is approximated through algorithms for generating pseudorandom numbers. Due to the exceptional simplicity

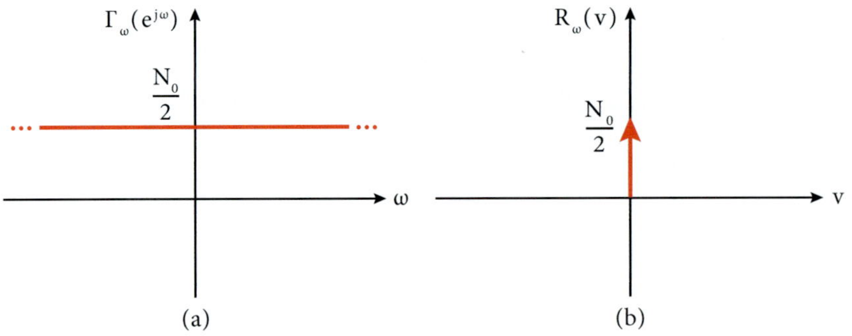

Fig. 13.5 Representation of white noise process in (**a**) time domain and (**b**) frequency domain

of the representation of white noise in both the time and frequency domains (Fig. 13.5), this constitutes a valuable tool for modeling stochastic processes for which there is no possibility of prediction.

13.4 Systems with Stochastic Inputs

As already mentioned, a continuous- or discrete-time system can be considered a transform $\mathcal{T}$ that accepts an input signal $x(t)$—let us start with the continuous case—and creates as its output a second signal $y(t) = \mathcal{T}[x(t)]$ (Sect. 2.1, page 67). Exactly the same applies in the case of *stochastic processes*, with the system accepting at its input a sample $\boldsymbol{x}(t, \xi_i)$ of a stochastic process $\boldsymbol{x}(t)$ and produces as its output a second signal $\boldsymbol{y}(t, \xi_i)$, which can be considered a sample of another process $\boldsymbol{y}(t)$. These two processes, $\boldsymbol{x}(t)$ and $\boldsymbol{y}(t)$, are characterized, quite reasonably, as the input and output processes of the considered system, while the foregoing mode of operation is mathematically formulated as $\boldsymbol{y}(t) = \mathcal{T}[\boldsymbol{x}(t)]$. In practice, systems that operate in the manner described above can be classified as either deterministic or stochastic, depending on whether or not they take the second argument ξ_i into account ξ_i that appears in the samples $x(t, \xi_i)$ and $y(t, \xi_i)$. If this argument is *ignored* or, rather, is treated as a *parameter* and the system only acts on the variable t representing time, then the system is aforementioned to be *deterministic*, while in the case where the system *acts* both on the time variable t and on the argument ξ_i, the system is aforementioned to be *stochastic*. This differentiation dramatically affects the operation of the system, since in the first case, if the input and output signals $\boldsymbol{x}(t, \xi_1)$ and $\boldsymbol{x}(t, \xi_2)$ are *identical* with respect to time t, then the same will apply to the corresponding output samples, while in the second case, the output samples do not coincide with each other. In the following description, we will study deterministic systems.

However, before we proceed with the description, it is important to mention that *regardless* of whether the input to the system is deterministic or random, the correlation that exists between the input and output of the system and that is determined using the known analysis techniques in the time and frequency domain is *a single and unique correlation*. Of course, in the case of random input, because we do not have an exact representation of the input signal, *it is not possible* to construct a corresponding expression for the output of the system, and for this reason, we are obliged to resort to the use of either a *probabilistic* description, which leads to the determination of certain probability functions, or a *statistical* description, which leads to the calculation of quantities, such as the *average*, *variance*, and *autocorrelation*. Of these two different types of description, the most useful description and the one that is widely used in practice is the *statistical description*, while the *probabilistic description* is feasible only for a very limited category of problems and generally will not occupy us further. In addition, for reasons of simplicity, we will limit our description to *naturally implementable linear and (usually) time-invariant (LTI) systems*, which are characterized by *BIBO stability* and accept as input signals that are sample functions of stationary random processes. Conversely, the case of nonlinear and/or time-varying systems whose inputs are related to processes that are not stationary is particularly complex and is not examined in this chapter.

13.4.1 Description of System Using Convolution

According to our earlier discussion in the chapter on continuous-time systems, continuous-time systems can be characterized in a variety of ways, based on the type of their operation, such as, for example, their characterization as *static or dynamic* (based on whether or not they have memory), as *linear or nonlinear* (based on whether or not they satisfy the principle of superposition), as *time-varying or time-invariant* (based on whether their mode of operation and their behavior changes over time), and so forth. Of particular interest are systems that are *simultaneously* linear and time-invariant, since in this case *the output of such systems results from the convolution between the input of the system and the impulse response*, defined as *the response of a system to a unit impulse function*. The same applies in the case where the inputs and outputs of a system *are not deterministic but random signals*, that is, they are samples of *stochastic* processes, with the difference that now, to characterize a system, we should, as mentioned previously, resort to the use of *statistical properties*, such as expected values. For example, in the case of a static system, in which the output of the system at each time point depends only on the input of the system at the same time and not from previous inputs, or, in mathematical notation, $\mathbf{y}(t) = g[\mathbf{x}(t)]$, we will have

$$E\{\mathbf{y}(t)\} = \int_{-\infty}^{\infty} g(x) f_x(x;t)dx \quad \text{and} \quad E\{\mathbf{y}(t_1)\mathbf{y}(t_2)\} = \int_{-\infty}^{\infty}\int_{-\infty}^{\infty} g(x_1)g(x_2)f(x_1,x_2;t_1t_2)dx_1dx_2$$

In the case of linear systems, the *principle of superposition* allows us to write

$$\mathcal{T}[\alpha_1\mathbf{x}_1(t) + \alpha_2\mathbf{x}_2(t)] = \alpha_1\mathcal{T}[\mathbf{x}_1(t)] + \alpha_2\mathcal{T}[\mathbf{x}_2(t)] \tag{13.60}$$

for every α_1, α_2 and for every $x_1(t), x_2(t)$, while if the system is time-invariant and $\mathbf{y}(t) = \mathcal{T}[\mathbf{x}(t)]$, then $\mathbf{y}(t+c) = \mathcal{T}[\mathbf{x}(t+c)]$ will also hold. Referring to the property of *linearity* of the operator of *mathematical expectation*, we can formulate a fundamental property that characterizes all LTI systems and according to which

$$\mu_y = E\{\mathbf{y}(t)\} = E\{\mathcal{T}[\mathbf{x}(t)]\} = \mathcal{T}[E\{\mathbf{x}(t)\}] = \mathcal{T}[\mu_x(t)] \tag{13.61}$$

This very important property tells us that *the average value of the output of the system* $\mathbf{y}(t)$ *is equal to the response of the system when the average value of the input* $\mathbf{x}(t)$ *is transmitted to its input* (Fig. 13.6). This property is proven by observing that[6]

$$\mu_y = E\{\mathbf{y}(t)\} = E\{\mathcal{T}[\mathbf{x}(t)]\} = E\left\{\int_{-\infty}^{\infty} \mathbf{x}(t)h(t,\lambda)d\lambda\right\} = \int_{-\infty}^{\infty} E\{x(t)\}h(t,\lambda)d\lambda = \mathcal{T}[\mu_x(t)] \tag{13.62}$$

since $\mu_x(t) = E\{\mathbf{x}(t)\}$. Notice that in order to arrive at the foregoing result, we proceeded *to the alternation of the order of application of the operation of integration and the operator of mathematical expectation*. In other words, while initially we applied the operator $E\{...\}$ to the integral of the system's output calculation, we then transferred the operator inside the integral in order for it to be applied only to the input signal $x(t)$, leading ultimately to the foregoing result. In the general case, considering a function $g(t)$, which is a sample function of a random process or even a function of it, $h[g(t)]$, as well as a second deterministic (i.e., nonrandom) function, $f(t)$, the *alternation* of the operations of applying integration and mathematical expectation, which allows us to write that

$$E\left\{\int_{t_1}^{t_2} g(t)f(t)dt\right\} = \int_{t_1}^{t_2} E\{g(t)\}f(t)dt$$

is permissible if (a) the condition $\int E\{|g(t)||f(t)|\} < \infty$ with the limits t_1 and t_2 of the integral can take the infinite values $-\infty$ and $+\infty$ and (b) the function $g(t)$ is *bounded* in the range of values from t_1 to t_2. Note that for the preceding exchange to be permissible, it is *not* mandatory for the function $g(t)$ to constitute a *sample of a stationary process*. It is proven that these conditions are satisfied in almost all cases of practical interest and for this reason, every time it is necessary to proceed with such an exchange, we will do it immediately, silently accepting that the preceding conditions are satisfied everywhere and always.

[6] Although usually in the integral of convolution, we use the variable τ, since in the case of random stationary processes the variable τ represents the time difference $t_2 - t_1$, in equations containing convolution, instead of the variable τ, we will use the variable λ. Additionally, in many textbooks, integrals are calculated not from $-\infty$ to ∞, as is done here, but from 0 to ∞, a condition we can impose, silently accepting that the impulse response $h(t)$ constitutes a causal signal.

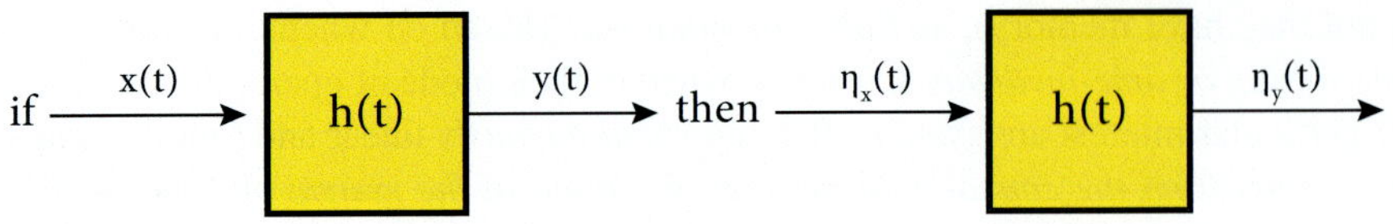

Fig. 13.6 An interesting property of LTI systems (Papoulis and Pillai, Figure 9.8α)

In the special case where the linear system is *time-invariant*, $h(t, \lambda) = h(t - \lambda)$, the foregoing expression constitutes a *convolution equation*, which allows us to write that

$$\mu_y = E\{\mathbf{y}(t)\} = E\left\{\int_{-\infty}^{\infty} x(t)h(t-\lambda)d\lambda\right\} = \int_{-\infty}^{\infty} E\{x(t)\}h(t-\lambda)d\lambda = E\{\mathbf{x}(t)\} * h(t) = \mu_x(t) * h(t) \quad (13.63)$$

As a consequence of the foregoing property, if the input process of a LTI system is *shifted in terms of the mean value*, that is, it is of the form $\tilde{\mathbf{x}}(t) = \mathbf{x}(t) - \mu_x(t)$, then the same applies to the output process, as it is easily found that

$$\mathcal{T}[\tilde{\mathbf{x}}(t)] = \mathcal{T}[\mathbf{x}(t) - \mu_x(t)] = \mathcal{T}[\mathbf{x}(t)] - \mathcal{T}[\mu_x(t)] = \mathbf{y}(t) - \mu_y(t) = \tilde{\mathbf{y}}(t)$$

If the input process $\mathbf{x}(t)$ is a *WSS process*, its average value will be constant and equal to $E\{x(t)\} = \mu_x(t) = \mu_x$, and consequently, it can be taken out of the convolution integral. In this case and proceeding with the appropriate handling of the integration variable, we get

$$\mu_y = \int_{-\infty}^{\infty} E\{x(t)\}h(t-\lambda)d\lambda = \mu_x(t)\int_{-\infty}^{\infty} h(\lambda)d\lambda = \mu_x\mathcal{H}(0) \quad (13.64)$$

where $\mathcal{H}(e^{j\omega})$ is the *frequency response* of the system. Therefore, in the case of a LTI system that receives a WSS process as input, the average value of the output is equal to the average value of the input, multiplied by the value $\mathcal{H}(0)$, which constitutes the dc gain of the system (i.e., the gain at zero frequency), since it corresponds to the value of the transfer function for frequency $\omega = 0$ and is equal to the area defined by the impulse response curve and the horizontal axis. From this relationship it is not difficult to understand that *if the random input process is characterized by a zero average value, then the same will apply to the random output process, while if the system does not allow the passage of a continuous current, so* $\mathcal{H}(0) = 0$, *then the system's output will be characterized by a zero average value.*

Let us now see how the function of *correlation* is transformed under the influence of the preceding system, which, apart from being linear, we consider it to be time-invariant as well, so the input and output are related through the operation of convolution. Taking into account that the input and output signals of the system are related through the convolution equation (in other words, we consider that the system is LTI) the *cross-correlation* $R_{xy}(t_1, t_2)$ is defined as

$$\begin{aligned} R_{xy}(t_1, t_2) &= E\{x(t_1)y(t_2)\} = E\left\{x(t_1)\int_{-\infty}^{\infty} x(t_2-\lambda)h(\lambda)d\lambda\right\} \\ &= \int_{-\infty}^{\infty} E\{x(t_1)x(t_2-\lambda)\}h(\lambda)d\lambda = \int_{-\infty}^{\infty} R_{xx}(t_1, t_2-\lambda)h(\lambda)d\lambda \end{aligned} \quad (13.65)$$

while for the *autocorrelation* $R_{yy}(t_1, t_2)$ similarly we will have

$$\begin{aligned} R_{yy}(t_1, t_2) &= E\{y(t_1)y(t_2)\} = E\left\{\left(\int_{-\infty}^{\infty} x(t_1-\lambda)h(\lambda)d\lambda\right)y(t_2)\right\} \\ &= \int_{-\infty}^{\infty} E\{x(t_1-\lambda)y(t_2)\}h(\lambda)d\lambda = \int_{-\infty}^{\infty} R_{xy}(t_1-\lambda, t_2)h(\lambda)d\lambda \end{aligned} \quad (13.66)$$

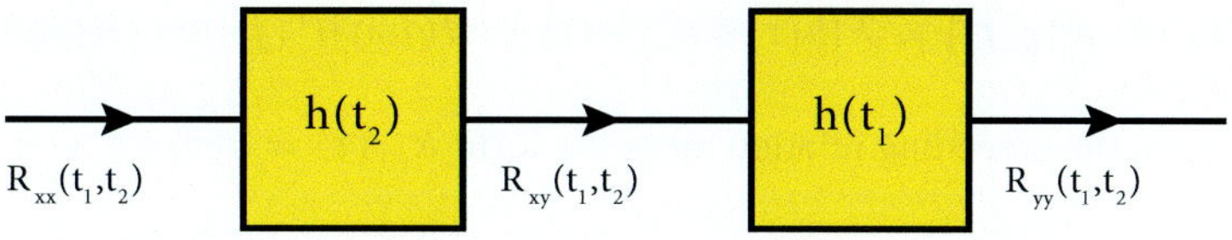

Fig. 13.7 Correlation between functions $R_{xx}(t_1, t_2)$ and $R_{yy}(t_1, t_2)$ (Papoulis and Pillai, Figure 9.8β)

According to the foregoing expressions, the functions $R_{xy}(t_1, t_2)$ and $R_{xx}(t_1, t_2)$ are related through a *convolution equation*, and therefore, the function $R_{xy}(t_1, t_2)$ can be considered the output of a LTI system, with impulse response $h(t_2)$ (note that in the integral of the convolution, the parameter λ in the function $R_{xx}(t_1, t_2 - \lambda)$ is subtracted from the variable t_2) and input the function $R_{xx}(t_1, t_2)$, while in a similar way, the functions $R_{yy}(t_1, t_2)$ and $R_{xy}(t_1, t_2)$ are related through a convolution equation, and therefore, the function $R_{yy}(t_1, t_2)$ can be considered the output of a LTI system, with impulse response $h(t_1)$ (note that this time, in the convolution integral, the parameter λ in the function $R_{xx}(t_1 - \lambda, t_2)$ is subtracted from the variable t_1) and input the function $R_{xy}(t_1, t_2)$. If we always take into account that *each convolution is applied to the appropriate variable* and we know which one it is, we can formulate the foregoing results in the form $R_{xy}(t_1, t_2) = R_{xx}(t_1, t_2) * h(t_2)$ and $R_{yy}(t_1, t_2) = R_{xy}(t_1, t_2) * h(t_1)$, while in the case where the processes $\boldsymbol{x}(t)$ and $\boldsymbol{y}(t)$, as well as the impulse response $h[n]$, are *of a complex nature*, the foregoing relations take the form $R_{xy}(t_1, t_2) = R_{xx}(t_1, t_2) * h^*(t_2)$ and $R_{yy}t_1, t_2 = R_{xy}(t_1, t_2) * h(t_1)$. If we connect *in series* these two systems, that is, if the output of the first system $R_{xy}(t_1, t_2)$ is used as input to the second system, then a composite system will emerge, which will accept as input the function $R_{xy}(t_1, t_2)$ and will produce as its output the function $R_{yy}(t_1, t_2)$, thereby implementing the desired correlation. This situation is depicted in Fig. 13.7. On the mathematical side of the issue, to construct the equation that implements this correlation, we need do nothing more than substitute the second of the preceding equations into the first equation, resulting in the expression,[7]

$$R_{yy}(t_1, t_2) = \int_{-\infty}^{\infty} \int_{-\infty}^{\infty} R_{xx}(t_1 - \lambda_1, t_2 - \lambda_2) h(\lambda_1) h(\lambda_2) d\lambda_1 d\lambda_2 \tag{13.67}$$

which is similar to the one that applies to the discrete case. In the special case where the random process is of the WSS type, which is usually of interest in practice, the two foregoing relations simplify as

$$R_{xy}(\tau) = \int_{-\infty}^{\infty} R_{xx}(\tau - \lambda) h(\lambda) d\lambda \quad \text{and} \quad R_{yy}(\tau) = \int_{-\infty}^{\infty} \int_{-\infty}^{\infty} R_{xx}(\lambda_2 - \lambda_1 - \tau) h(\lambda_1) h(\lambda_2) d\lambda_1 d\lambda_2$$

where $\tau = t_2 - t_1$, while the last relation for $\tau = 0$ gives us

$$R_{yy}(0) = \int_{-\infty}^{\infty} \int_{-\infty}^{\infty} R_{xx}(\lambda_2 - \lambda_1) h(\lambda_1) h(\lambda_2) d\lambda_1 d\lambda_2 \tag{13.68}$$

a value that is simply the *mean square of the process*, as is easily verified if we calculate the last value from the definition of the quantity $\overline{y^2} = E\{y^2(t)\}$. Working in the same way, we find that the second function, the cross-correlation $R_{yx}(\tau)$ for a WSS process is defined as

$$R_{yx}(\tau) = \int_{-\infty}^{\infty} R_{xx}(\tau + \lambda) h(\lambda) d\lambda \tag{13.69}$$

The foregoing relations expressed as *convolutions* take the simplest form

$$R_{xy}(\tau) = h^*(-\tau) * R_{xx}(\tau) \quad \text{and} \quad R_{yy}(\tau) = h(\tau) * h^*(-\tau) * R_{xx}(\tau) = g(\tau) * R_{xx}(\tau)$$

where $g(\tau) = h(\tau) * h^*(-\tau)$, with the function $g(\tau)$ being *positively semidefinite* with the Fourier transform

[7] In the general case, where the system is *linear but not time-invariant*, immutable the preceding relations take the most general form $R_{xy}(t_1, t_2) = \mathcal{T}_2[R_{xx}(t_1, t_2)]$ and $R_{yy}(t_1, t_2) = \mathcal{T}_1[R_{xy}(t_1, t_2)]$, with the operator $\mathcal{T}_1$ applied to the variable t_1 and the operator $\mathcal{T}_2$ applied to the variable t_2 (if the functions are of a *complex nature*, a more general expression that also contains the operator of complex conjugation will be used), while their combination leads to the expression $R_{yy}(t_1, t_2) = \mathcal{T}_1[\mathcal{T}_2(R_{xx}(t_1, t_2))]$.

$$G(\omega) = \mathcal{F}\{g(\tau)\} = \mathcal{F}\{h(\tau) * h^*(-\tau)\} = \mathcal{H}(j\omega)\mathcal{H}^*(j\omega) = |\mathcal{H}(j\omega)|^2$$

In the case where $R_{xx}(\tau) = \delta(\tau)$, the preceding relation takes the form $R_{yy}(\tau) = h(\tau) * h^*(-\tau)$ and is known as an *impulse response autocorrelation.*

Combining the foregoing results, it turns out that $S_{yy}(\omega) = |\mathcal{H}(j\omega)|^2 S_{xx}(\omega)$. In the special case where the input process $\boldsymbol{x}(t)$ is a white noise process with average power q, we will have $R_{xx}(\tau) = q\delta(\tau)$, and therefore, $S_{xx}(\omega) = q$ and $S_{yy}(\omega) = q|\mathcal{H}(j\omega)|^2$, from which it follows that $R_{yy}(\tau) = qh(\tau) * h^*(-\tau)$.

Finally, applying the preceding equations to the processes $\tilde{\boldsymbol{x}}(t) = \boldsymbol{x}(t) + \mu_x(t)$ and $\tilde{\boldsymbol{y}}(t) = \boldsymbol{y}(t) + \mu_y(t)$, we get

$$C_{xy}(t_1, t_2) = C_{xx}(t_1, t_2) * h(t_2) \qquad \text{and} \qquad C_{yy}(t_1, t_2) = C_{xy}(t_1, t_2) * h(t_1)$$

Example 13.4.1 Let us calculate the values of the mean square of the system's output, the autocorrelation $R_{yy}(\tau)$ and the cross-correlations $R_{xy}(\tau)$ and $R_{yx}(\tau)$ for the simple case in which the input to the system is a random stationary white noise process, with autocorrelation $R_{xx}(\tau) = (N_0/2)\delta(\tau)$, while the system is the simple low-pass RC filter with impulse response $h(t) = (1/RC)\exp(-t/RC)$. Starting from the quantity $\overline{y^2}$, this is calculated as

$$\overline{y^2} = \int_0^\infty \int_0^\infty R_{xx}(\lambda_2 - \lambda_1)h(\lambda_1)h(\lambda_2)d\lambda_1 d\lambda_2 = \frac{N_0}{2}\int_0^\infty \int_0^\infty \delta(\lambda_2 - \lambda_1)h(\lambda_1)h(\lambda_2)d\lambda_1 d\lambda_2 = \frac{N_0}{2}\int_0^\infty h^2(\lambda)d\lambda$$

while for the special case of the RC circuit, from the foregoing relation we get

$$\overline{y^2} = \frac{N_0}{2}\int_0^\infty h^2(\lambda)d\lambda = \frac{N_0}{2R^2C^2}\int_0^\infty \exp(-2\lambda/RC)d\lambda = -\frac{N_0}{2R^2C^2}\frac{RC}{2}\left[\exp\left(-\frac{2\lambda}{RC}\right)\right]_0^\infty = \frac{N_0}{4RC} \tag{13.70}$$

On the other hand, the autocorrelation of the output is calculated from the expression

$$R_{yy}(\tau) = \int_0^\infty \int_0^\infty R_{xx}(\lambda_2 - \lambda_1 - \tau)h(\lambda_1)h(\lambda_2)d\lambda_1 d\lambda_2 \tag{13.71}$$

$$= \frac{N_0}{2}\int_0^\infty \int_0^\infty \delta(\lambda_2 - \lambda_1 - \tau)h(\lambda_1)h(\lambda_2)d\lambda_1 d\lambda_2 = \frac{N_0}{2}\int_0^\infty h(\lambda_1)h(\lambda_1 + \tau)d\lambda_1 \tag{13.72}$$

and consequently, for the case of white noise, the output autocorrelation is equal to the temporal autocorrelation of the impulse response. For the case of the RC circuit, the foregoing relationship gives us

$$R_{yy}(\tau) = \frac{N_0}{2R^2C^2}\int_0^\infty \exp\left(-\frac{\lambda_1}{RC}\right)\exp\left(-\frac{\lambda_1 + \tau}{RC}\right)d\lambda_1 \tag{13.73}$$

$$= \frac{N_0}{2R^2C^2}\frac{RC}{2}\exp\left(-\frac{\tau}{RC}\right)\left[\exp\left(-\frac{2\lambda}{RC}\right)\right]_0^\infty = \frac{N_0}{4RC}\exp\left(-\frac{\tau}{RC}\right) \tag{13.74}$$

The preceding expression does not hold for $\tau \geq 0$ as well as it does for $\tau < 0$, since the impulse response is a causal signal, and therefore for $\tau < 0$ it holds that$h(\tau) = 0$. By appropriately changing the limits of the integral, it turns out that

$$R_{yy}(\tau) = \frac{N_0}{2R^2C^2}\int_{-\tau}^\infty \exp\left(-\frac{\lambda_1}{RC}\right)\exp\left(-\frac{\lambda_1 + \tau}{RC}\right)d\lambda_1$$

$$= \frac{N_0}{2R^2C^2}\frac{RC}{2}\exp\left(-\frac{\tau}{RC}\right)\left[\exp\left(-\frac{2\lambda}{RC}\right)\right]_{-\tau}^\infty = \frac{N_0}{4RC}\exp\left(-\frac{\tau}{RC}\right)\exp\left(\frac{2\tau}{RC}\right) = \frac{N_0}{4RC}\exp\left(\frac{\tau}{RC}\right)$$

Combining the two preceding results, we arrive at the expression

$$R_{yy}(\tau) = \frac{N_0}{4RC}\exp\left(\frac{|\tau|}{RC}\right), \quad -\infty < \tau < \infty \tag{13.75}$$

Finally, the cross-correlation functions $R_{xy}(\tau)$ and $R_{yx}(\tau)$ are calculated as

$$R_{xy}(\tau) = \int_0^\infty R_{xx}(\tau-\lambda)h(\lambda)d\lambda = \frac{N_0}{2}\int_0^\infty \delta(\tau-\lambda)h(\lambda)d\lambda = \frac{N_0}{2}h(\tau) \tag{13.76}$$

$$R_{yx}(\tau) = \int_0^\infty R_{xx}(\tau+\lambda)h(\lambda)d\lambda = \frac{N_0}{2}\int_0^\infty \delta(\tau+\lambda)h(\lambda)d\lambda = \frac{N_0}{2}h(-\tau) \tag{13.77}$$

Note that the preceding results only apply for values $\tau \geq 0$ for the function $R_{xy}(\tau)$ and $\tau \leq 0$ for the function $R_{yx}(\tau)$, while for the remaining values of τ, these functions have a value equal to zero. In the preceding integrals, as integration limits, 0 and ∞ were used, due to the causal nature of the involved signals.

The next quantity of interest is the *PSD* and, more specifically, the *correlation* that exists between the corresponding quantities of the input and output signals. Recall that the PSD of a process is a measure of the way in which the average power of the process is distributed as a function of frequency and is *the basic tool for studying stationary random processes in the frequency domain*, since as we noted, the Fourier transform of these processes *is not defined*. For this reason, as we noted, when describing this quantity, we conduct the appropriate analysis in each case, considering that the process is defined only for a finite time period of duration $2T$, and after we take the expected value, we examine what happens in the limit $T \to \infty$. In practice, however, we can work much more easily relying on a very important result that we previously used and according to which *the PSD is the Fourier transform of the autocorrelation function*.

Therefore, using the defining equation of autocorrelation $R_{yy}(\tau)$, its Fourier transform is calculated as

$$\begin{aligned} S_{yy}(\omega) = \mathcal{F}\{R_{yy}(\tau)\} &= \int_{-\infty}^{\infty}\left[\int_{-\infty}^{\infty} d\lambda_1 \int_{-\infty}^{\infty} R_{xx}(\lambda_2-\lambda_1-\tau)h(\lambda_1)h(\lambda_2)d\lambda_1 d\lambda_2\right]e^{-j\omega\tau}d\tau \\ &= \int_{-\infty}^{\infty} d\lambda_1 \int_{-\infty}^{\infty} h(\lambda_1)h(\lambda_2)d\lambda_2 \int_{-\infty}^{\infty} R_{xx}(\lambda_2-\lambda_1-\tau)e^{-j\omega\tau}d\tau \\ &= \int_{-\infty}^{\infty} d\lambda_1 \int_{-\infty}^{\infty} h(\lambda_1)h(\lambda_2)S_{xx}(j\omega)e^{-j\omega(\lambda_2-\lambda_1)}d\lambda_2 = S_{xx}(\omega)\int_{-\infty}^{\infty} h(\lambda_1)e^{j\omega\lambda_1}d\lambda_1 \int_{-\infty}^{\infty} h(\lambda_2)e^{-j\omega\lambda_2}d\lambda_2 \end{aligned}$$

and finally

$$S_{yy}(\omega) = S_{xx}(\omega)\mathcal{H}(-j\omega)\mathcal{H}(j\omega) = S_{xx}(\omega)|\mathcal{H}(j\omega)|^2 \tag{13.78}$$

where to derive the final result, we resorted to the *symmetry* property $R_{xx}(-\tau) = R_{xx}(\tau)$. The corresponding expression if the complex frequency ω is used as an argument is $S_y y(s) = \mathcal{H}(s)\mathcal{H}(-s)S_{xx}(s)$, and from the last relation we observe that *the quantity $\mathcal{H}(s)\mathcal{H}(-s)$ correlates the PSDs of the input and output, in exactly the same way as the transfer function $\mathcal{H}(s)$ correlates the Laplace transforms $\mathcal{X}(s)$ and $\mathcal{Y}(s)$ for the deterministic (nonrandom) case.*

Having calculated the function $S_{yy}(s)$, we can calculate the mean square of the output $y(t)$ as the *inverse Laplace transform*

$$\overline{y^2} = \frac{1}{2\pi}\int_{-j\infty}^{j\infty} \mathcal{H}(s)\mathcal{H}(-s)S_{xx}(s)ds \tag{13.79}$$

Finally, the spectral densities $S_{xy}(s)$ and $S_{yx}(s)$ constitute the Fourier transforms of the cross-correlation functions $R_{xy}(s)$ and $R_{yx}(s)$ and are related to the transfer function of the system $\mathcal{H}(s)$ through the relations $S_{xy}(s) = \mathcal{H}(s)S_{xx}(s)$ and $S_{yx}(s) = \mathcal{H}(-s)S_{xx}(s)$.

13.4.2 Description Using Differential Equations

The points mentioned in the previous sections are related to the description of the input/output of a continuous-time LTI system through the *convolution equation*. However, similar conclusions can be formulated, and in the case where the system is described using a *differential equation* of order N with constant coefficients of the form

$$\alpha_n \frac{d^n \boldsymbol{y}(t)}{dt^n} + \alpha_{n-1}\frac{d^{n-1}\boldsymbol{y}(t)}{dt^{n-1}} + \cdots + \alpha_2 \frac{d^2\boldsymbol{y}(t)}{dt^2} + \alpha_1 \frac{d\boldsymbol{y}(t)}{dt} + \alpha_0 \boldsymbol{y}(t) = \boldsymbol{x}(t)$$

with the functions $\boldsymbol{x}(t)$ and $\boldsymbol{y}(t)$ now describing *stochastic processes*. Consequently, the differential equations are now aforementioned to be *stochastic differential equations*, while it is assumed that the systems they describe are *BIBO stable*, so that if the input process is *stable in the strict or broad sense*, the same applies to the output process. In the following description, we will limit ourselves to the *zero-state response*, which is defined as *the solution of a differential equation under the regime of zero initial conditions* of the form $\boldsymbol{y}(0) = \boldsymbol{y}'(0) = \boldsymbol{y}''(0) = \cdots = \boldsymbol{y}^{n-1}(0) = 0$, and let us construct the differential equations that are satisfied by the two basic statistical quantities of the mean value and the correlation.

The construction of a differential equation for the quantities of the *mean value* arises easily, applying to both sides of the preceding equation the operator of the *mathematical expectation* and then limiting the application of the operator *only to the derivatives* of the differential equation, a process that is permissible since the mathematical expectation is a *linear* operator. We will therefore have

$$\alpha_n E\left\{\frac{d^n \boldsymbol{y}(t)}{dt^n}\right\} + \alpha_{n-1}E\left\{\frac{d^{n-1}\boldsymbol{y}(t)}{dt^{n-1}}\right\} + \cdots + \alpha_2 E\left\{\frac{d^2\boldsymbol{y}(t)}{dt^2}\right\} + \alpha_1 E\left\{\frac{d\boldsymbol{y}(t)}{dt}\right\} + \alpha_0 E\{\boldsymbol{y}(t)\} = E\{\boldsymbol{x}(t)\} \tag{13.80}$$

Taking into account now that $\mu_x(t) = E\{\boldsymbol{x}(t)\}$, $\mu_y(t) = E\{\boldsymbol{y}(t)\}$, and $E\{\boldsymbol{k}^{(k)}(t)\}]\mu_y^{(k)}(t)$, the foregoing relation becomes

$$\alpha_n \frac{d^n \mu_y(t)}{dt^n} + \alpha_{n-1}\frac{d^{n-1}\mu_y(t)}{dt^{n-1}} + \cdots + \alpha_2 \frac{d^2\mu_y(t)}{dt^2} + \alpha_1 \frac{d\mu_y(t)}{dt} + \alpha_0 \mu_y(t) = \mu_x(t) \tag{13.81}$$

and it is the differential equation that correlates the mean values $\mu_y(t)$ and $\mu_x(t)$ and must be solved with initial conditions $\mu_y(0) = \mu_y'(0) = \mu_y''(0) = \cdots = \mu_y^{n-1}(0) = 0$. We could have arrived at the same result more quickly based on the observation that if $\boldsymbol{y}(t) = \mathcal{T}\{\boldsymbol{x}(t)\}$, then we will also have $\mu_y(t) = \mathcal{T}\{\mu_x(t)\}$, and consequently, these two mean values are related through the differential equation of the system.

The situation is similar for the case of *autocorrelations*, since, as we proved, the pairs of functions $[R_{xx}(t_1,t_2), R_{xy}(t_1,t_2)]$ and $[R_{xy}(t_1,t_2), R_{yy}(t_1,t_2)]$ are related to each other by an input/output relationship. Consequently, they will also satisfy a similar differential equation, only in this case this differential equation will be a *partial differential equation*, since the functions of autocorrelation are functions of two variables, of t_1 and t_2. In reality, according to our earlier discussion about which variable is used in the convolution equation, the partial differential equation that correlates the functions $R_{xx}(t_1,t_2)$ and $R_{xy}(t_1,t_2)$ will contain the partial derivatives with respect to the variable t_2, while the partial differential equation that correlates the functions $R_{xy}(t_1,t_2)$ and $R_{yy}(t_1,t_2)$ will contain the partial derivatives with respect to the variable t_1. Consequently, the functions $R_{xx}(t_1,t_2)$ and $R_{xy}(t_1,t_2)$ will be related to each other as follows:

$$\alpha_n \frac{\partial^n R_{xy}(t_1,t_2)}{\partial t_2^n} + \alpha_{n-1}\frac{\partial^{n-1}R_{xy}(t_1,t_2)}{\partial t_2^{n-1}} + \cdots \alpha_2 \frac{\partial^2 R_{xy}(t_1,t_2)}{\partial t_2^2} + \alpha_1 \frac{\partial R_{xy}(t_1,t_2)}{\partial t_2} + \alpha_0 R_{xy}(t_1,t_2) = R_{xx}(t_1,t_2)$$

with initial conditions

$$R_{xy}(t_1,0)=\frac{\partial R_{xy}(t_1,0)}{\partial t_2}=\frac{\partial^2 R_{xy}(t_1,0)}{\partial t_2^2}=\cdots=\frac{\partial^{n-1}R_{xy}(t_1,0)}{\partial t_2^{n-1}}=0$$

Similarly, the functions $R_{xy}(t_1,t_2)$ and $R_{yy}(t_1,t_2)$ will be related to each other as follows:

$$\alpha_n\frac{\partial^n R_{yy}(t_1,t_2)}{\partial t_1^n}+\alpha_{n-1}\frac{\partial^{n-1}R_{yy}(t_1,t_2)}{\partial t_1^{n-1}}+\cdots\alpha_2\frac{\partial^2 R_{yy}(t_1,t_2)}{\partial t_1^2}+\alpha_1\frac{\partial R_{yy}(t_1,t_2)}{\partial t_1}+\alpha_0 R_{yy}(t_1,t_2)=R_{xy}(t_1,t_2)$$

with initial conditions

$$R_{yy}(0,t_2)=\frac{\partial R_{yy}(0,t_2)}{\partial t_2}=\frac{\partial^2 R_{yy}(0,t_2)}{\partial t_2^2}=\cdots=\frac{\partial^{n-1}R_{yy}(0,t_2)}{\partial t_2^{n-1}}=0$$

Similar conclusions can be drawn for higher-order moments.

13.5 Response of a Linear System to a Random Sequence

Conclusions similar to those we formulated for the continuous case can be formulated for the *discrete case*, in which the input of a linear system is a *random sequence*. In full analogy with the continuous case, if the system implements a transform $\mathcal{T}$, then the system's response to the *random input sequence* $Y[n,\xi]$, for every $\xi\in\Omega$, is the *random output sequence* $Y[n,\xi]=\mathcal{T}\{X[n,\xi]\}$. Because this condition holds *for every* $\xi\in\Omega$, we can omit the variable ξ and simply write $Y[n]=\mathcal{T}\{X[n]\}$, just as we did for the *usual discrete sequences*. Indeed, as mentioned previously, the random sequence $X[n,\xi]$ in reality describes a whole family of sequences that are indexed using the parameter ξ, and therefore, for a constant parameter value $\xi=\xi_0$, the random sequence degenerates into a *simple discrete sequence*.

Generally speaking, the calculation of the probability distribution of the system's output $Y[n,\xi]$, based on the probabilistic description of its input, is *very difficult* since the impulse response has *very large* or even *infinite* length, and therefore, if the output of the system depends on the k previous samples of its input, the calculation of the *first-order probability density* of the output requires knowledge of the k-order probability density of the input. However, when calculating the moments and under the assumption that the underlying system is *linear*, things are easier, and for this reason, let us focus on this calculation, determining the *mean value* and the *autocorrelation* of the output signal from the corresponding quantities of the input signal, just as we did for the continuous case. From the definition of the mean value and the correlation of the output with the input, which characterizes the linear systems of discrete-time, it is not difficult to determine that (using the simplified symbolism, i.e., omitting the argument ξ)

$$E\{Y[n]\}=E\{\mathcal{T}\{X[n]\}\}=E\left\{\sum_{k=-\infty}^{\infty}h[n,k]X[k]\right\}=\sum_{k=-\infty}^{\infty}h[n,k]E\{X[k]\}=\mathcal{T}\{E\{X[n]\}\}\tag{13.82}$$

Thus, $E\{Y[n]\}=\mathcal{T}\{E\{X[n]\}\}$, and taking into account that $\mu_Y=\{E\{Y[n]\}\}$ and $\mu_X=\{E\{X[n]\}\}$, we can write

$$\mu_y[n]=\sum_{k=-\infty}^{\infty}h[n,k]\mu_X[k]=\mathcal{T}\{\mu_X[n]\}\tag{13.83}$$

We are led to the conclusion that *the system's response to the mean value of the input is simply the mean value of the output*, a conclusion that, as we saw earlier, also applies to the continuous case. However, we must take into account that the interchange of the linear operator $\mathcal{T}$ and the operator of the expected value, which ultimately led us to the foregoing result, *is not always permissible*, only in cases where the impulse response of the considered system is *absolutely summable*.

In the special case where the input sequence is a wide-sense stationary (WSS) random process, its mean value—that is, the quantity $E\{x[n-k]\}$—will be constant and equal to $\mu_x[n]=E\{X[n]\}=M$ for all values of n. Moreover, if the underlying linear system is shift-invariant, its impulse response satisfies the relation $h[n, k] = h[n - k]$. Under these conditions, the previous expression is significantly simplified, since the constant mean value $\mu_x[n]$ can be taken outside the summation. We then obtain:

$$\mu_y[n] = M \sum_{k=-\infty}^{\infty} h[n-k] = M \sum_{k=-\infty}^{\infty} h[k] = M\mathcal{H}(z)|_{z=1} \tag{13.84}$$

as it is not difficult to ascertain that the expression of the sum, after the appropriate index change from $n - k$ to k (the limits, of course, as infinite, remain as they are), is simply the $\mathcal{Z}$ transform of the signal $X[n]$, calculated at the position $z = 1$. Of course, once again, the impulse response must be *absolutely summable*, which will be true if the considered system is *BIBO stable*. As in the continuous case, the value of this sum can be considered the *gain* of the system, and therefore, for a discrete LTI system, which receives as input a random WSS sequence, *the statistical average of the output signal is equal to the product of the statistical average of the input signal times the gain of the system, while its value is constant, as the product of two constant values.*

13.5.1 Calculation of Autocorrelation of System Output

On the other hand, the calculation of the autocorrelation of the output $R_{YY}[m, n]$, as a function of the autocorrelation of the input $R_{XX}[m, n]$, is based, as in the continuous case, on the use of the *cross-correlation* $R_{XY}[m, n]$, which for the general case of complex random sequences is defined as

$$R_{XY}[m, n] \triangleq E\{X[m]Y^*[n]\} = E\big\{X[m][\mathcal{T}\{X[n]\}]^*\big\}$$

Note that in the preceding definition of the function $R_{XY}[m, n]$, *two indices* are involved, m and n, and let us define, as before, two operators, the operator L_m that acts on the index m, considering that the index n is constant and is related to the impulse response $h[m, k]$, and the operator L_n^* that acts on the index n, considering that the index m is constant and is related to the impulse response $h^*[n, k]$ (note that this operator is related to the *conjugate operator*, which is known from linear algebra). In this case, it is proven that[8]

$$R_{XY}[m, n] = L_n^*\{R_{XX}[m, n]\} \qquad \text{and} \qquad R_{YY}[m, n] = L_m\{R_{XY}[m, n]\}$$

and combining these relations leads us to the expression

$$R_{YY}[m, n] = L_m\{R_{XY}[m, n]\} = L_m\{L_n^*\{R_{XX}[m, n]\}\} = \sum_{k=-\infty}^{\infty} h[m, k]\left(\sum_{\ell=-\infty}^{\infty} h^*[n, \ell]R_{XX}[k, \ell]\right)$$

where we took into account that

[8] Indeed, for the first expression, we observe that $X[m]Y^*[n] = X[m]L_n^*\{X^*[n]\} = L_n^*\{X[m]X^*[n]\}$, and therefore

$$R_{XY}[m, n] = E\{X[m]Y^*[n]\} = E\{L_n^*\{X[m]X^*[n]\}\} = L_n^*\{E\{X[m]X^*[n]\}\} = L_n^*\{R_{XX}[m, n]\}$$

similarly, for the second relation, we observe that $Y[m]Y^*[n] = L_m\{X[m]\}Y^*[n] = L_m\{X[m]Y^*[n]\}$, and therefore

$$R_{YY}[m, n] = E\{Y[m]Y^*[n]\} = E\{L_m\{X[m]Y^*[n]\}\} = L_m\{E\{X[m]Y^*[n]\}\} = L_m\{R_{XY}[m, n]\}.$$

$$R_{XY}[m,n] = L_n^*\{R_{XX}[m,n]\} = \sum_{\ell=-\infty}^{\infty} h^*[n,\ell]R_{XX}[m,\ell] \quad \text{and}$$

$$R_{YY}[m,n] = L_m\{R_{XY}[m,n]\} = \sum_{k=-\infty}^{\infty} h[m,k]R_{XX}[k,n]$$

In the special case where the input sequence is of the *WSS type*, the situation simplifies, as the autocorrelation of the system output now takes the form $R_{XX}[k+m,k] = E\{X[k+m]X^*[k]\} = R_{XX}[m]$. In this case, using the *Schwarz inequality*, it turns out that $|R_{XY}[m]| \leq \sqrt{R_{XX}[0]R_{YY}[0]}$, and therefore it will hold that $|R_{XX}[m]| \leq R_{XX}[0]$. Additionally, it is proven that $R_{XX}[m] = R_{XX}^*[-m]$[9] and that the autocorrelation function is *positively semidefinite*, or, in mathematical notation,

$$\sum_{n=1}^{N}\sum_{k=1}^{N} \alpha_n\alpha_k^* R_{XX}[n-k] \geq 0 \tag{13.85}$$

As for the correlation that exists between the functions $R_{XX}[m,n]$ and $R_{YY}[m,n]$ for the case of random WSS input sequences and a discrete LTI system, from the input/output relationship of the system given by the convolution equation, it turns out that

$$R_{XY}[m,n] = E\{X[m]Y^*[n]\} = \sum_{k=-\infty}^{\infty} h^*[n-k]E\{X[m]X^*[k]\} \tag{13.86}$$

$$= \sum_{k=-\infty}^{\infty} h^*[n-k]R_{XX}[m-k] = \sum_{k=-\infty}^{\infty} h^*[-\ell]R_{XX}[(m-n)-\ell] \tag{13.87}$$

where $\ell = k - m$. This relationship leads us to the conclusion that if the random input sequence is of the WSS type, then the cross-correlation R_{XY} is *invariant to displacement*. Therefore, for simplicity, we define $R_{XX}[m] = R_{XX}[m,0]$, that is, essentially, ignoring the index n, the foregoing expression takes the form

$$R_{XY}[m] = \sum_{k=-\infty}^{\infty} h^*[-\ell]R_{XX}[m-\ell] = h^*[-\ell] * R_{XX}[m]$$

The foregoing expression correlates the functions $R_{XX}[m,n]$ and $R_{XY}[m,n]$, while in a similar way we have

$$R_{YY}[n+m,n] = E\{Y[m+n]Y^*[m]\} = \sum_{k=-\infty}^{\infty} h[k]E\{X[n+m-k]Y^*[m]\} = \sum_{k=-\infty}^{\infty} h[k]R_{XY}[m-k] = h[m] * R_{XY}[m]$$

and since, once again, the resulting function is *independent of displacement*, for simplicity, we define $R_{YY}[m] = R_{YY}[m,0]$, the foregoing relationship is written

$$R_{YY}[m] = h[m] * R_{XY}[m]$$

Therefore, combining the two foregoing expressions, we are led to the expression

$$R_{YY}[m] = h[m] * h^*[-m] * R_{XX}[m] = g[m] * R_{XX}[m]$$

where $g[m] = h[m] * h^*[-m]$ is the so-called *autocorrelation impulse response* (since the foregoing relationship constitutes an input/output relationship of a discrete LTI system, with the function $g[m]$ playing the role of the *impulse response*. In the special case where the random sequence is not only WSS but also *independent* (i.e., if the random variables that constitute it are *independent of each other at any time* $n_1, n_2, \ldots, n_N$ for any $N > 0$), it is proven that the *autocorrelation function* of the

[9] Indeed, from the definition we have $R_{XX}[m] = E\{X[m+\ell]X^*[\ell]\} = E\{X[\ell]X^*[\ell-m]\} = E\{X[\ell-m]X_{XX}^*[-m]\}$.

input sequence has the form $R_{XX}[m] = A\delta[m]$, where A is a positive constant. Then $R_{YY}[m] = Ag[m]*\delta[m] = Ag[m]$, and for the value $A = 1$ it will hold that $R_{YY}[m] = g[m]$. But this means that the function $g[m]$ constitutes an autocorrelation function and satisfies all the properties that characterize such a function, while from the defining equation of $g[m] = h[m] * h^*[-m]$ it follows that it depends on the impulse response of the system $h[m]$. However, knowledge of the function $g[m]$ *is not sufficient* on its own to determine the impulse response of the system. Indeed, if we take the Fourier transform of the relation $g[m] = h[m] * h^*[-m]$ and refer to the convolution theorem, we will have $\mathcal{F}g[m] = \mathcal{F}h[m]\mathcal{F}h^*[-m]$ or, equivalently, $\mathcal{G}(e^{j\omega}) = \mathcal{H}(e^{j\omega})\mathcal{H}^*(e^{j\omega}) = |\mathcal{H}(e^{j\omega})|^2$, from which it follows that $|\mathcal{H}(e^{j\omega})| = \sqrt{\mathcal{G}(e^{j\omega})}$. We therefore find that from the foregoing equation we can calculate *only the magnitude* of the transfer function $|\mathcal{H}(e^{j\omega})|$ and *not its phase* $\measuredangle\mathcal{H}(e^{j\omega})$, the determination of which requires knowledge and other information.

13.5.2 Calculation of PSD of System Output

Having calculated the relationship between the autocorrelation functions $R_{XX}[m]$ and $R_{YY}[m]$ of the input and output sequences, we can now determine the correlation between the PSDs $S_x(e^{j\omega})$ and $S_y(e^{j\omega})$ of these two random sequences. The most direct way to determine this correlation is to start from the defining equation of the function $S_Y(e^{j\omega})$ as *the Fourier transform of the autocorrelation* $R_{YY}[m]$ and then express the autocorrelation $R_{YY}[m]$ as a function of the autocorrelation $R_{XX}[m]$ and proceed to the construction of such a correlation, taking into account that the input and output of the system are related to each other through the *convolution equation* (for this type of proof see Diniz, pages 125–126). Alternatively, to eliminate the somewhat complex index management that characterizes the foregoing direct calculation method, we can use the correlation we determined between the functions $R_{YY}[m]$ and $R_{XX}[m]$ (resorting again to the convolution equation) and then take the Fourier transform of the expression that will result. The advantage of this approach is that since the relationship between the functions $R_{YY}[m]$ and $R_{XX}[m]$, as we have proven, has the form of a convolution, we can exploit the *convolution theorem* and more easily arrive at the final result (for this type of proof see Antoniou, pages 610-611, and Stark & Woods, page 351). Indeed, using the convolution theorem with the equation $R_{YY}[m] = g[m] * R_{XX}[m]$ it easily follows that

$$\mathcal{F}\{R_{YY}[m]\} = \mathcal{F}\{g[m]\}\mathcal{F}\{R_{XX}[n]\} \qquad \text{or, equivalently,} \qquad S_Y(e^{j\omega}) = \mathcal{G}(e^{j\omega})S_X(e^{j\omega})$$

and using the expression $\mathcal{G}(e^{j\omega}) = \mathcal{H}(e^{j\omega})\mathcal{H}^*(e^{j\omega}) = |\mathcal{H}(e^{j\omega})|^2$, we finally find that

$$S_y(e^{j\omega}) = |\mathcal{H}(e^{j\omega})|^2 S_x(e^{j\omega}) \tag{13.88}$$

which is also our final result. According to this equation, *the PSD of the output* $y[n]$ *of the system is equal to the PSD function of the input* $x[n]$ *of the system, multiplied by the square of the system's amplitude response*. This expression constitutes the correlation between the input and the output of a discrete-time LTI system for the case of random signals and corresponds to the input/output relationship $\mathcal{Y}(e^{j\omega}) = \mathcal{H}(e^{j\omega})\mathcal{X}(e^{j\omega})$ used in the case of *deterministic* signals.

In full analogy with the continuous case, the PSD for the case of discrete sequences is characterized by the following properties: (a) The function $S_X(\omega)$ is a *real function*, because the autocorrelation function $R_{XX}(\omega)$ is conjugate-symmetric; (b) if the random sequence $X[n]$ is a *sequence of real values*, then the function $S_{XX}(\omega)$ is an *even function* of ω; (c) it always holds that $S_{XX}(\omega) \geq 0$ for every ω, regardless of whether the random sequence $X[n]$ is a sequence of real or complex values; and (d) if for some finite integer $N > 0$ it is the case that $R_{XX}[n] = 0$ for every $|n| > N$, then the function $S_X(\omega)$ is an *analytic function* of ω, and therefore, its *Taylor expansion* exists if its value and the values of its derivatives at some point $\omega = \omega_0$ are known.

What is the meaning behind the *PSD*? From the definition of this function, as *the Fourier transform of the autocorrelation function* of the random sequence $X[n]$, it is logical to conclude that it should express some kind of *average value* of the square $|\mathcal{X}(e^{j\omega})|^2$. The problem is that a random signal $X[n]$ of the WSS type has a constant average power $R_{XX}[0]$ for all values of n, and therefore, the Fourier transform *is not defined*. To bypass the problem, we use the *rectangular window* function, which has a nonzero value $w_N[n] = 1$ in the range $-N \leq n \leq N$ and a zero value anywhere else. We define the signal $x_N[n] = w_N[n]X[n]$ and take the Fourier transform of this signal, which is defined as

$$\mathcal{X}_N(e^{j\omega}) = \mathcal{F}\{x_N[n]\} = \mathcal{F}\{w[n]X[n]\}$$

In this case, working in the way we did when we defined the power of discrete-time signals, we construct the quantity

$$\frac{1}{2N+1}E\{|\mathcal{X}_N(e^{j\omega})|^2\} = \frac{1}{2N+1}E\{\mathcal{X}_N(e^{j\omega})\mathcal{X}_N^*(e^{j\omega})\} = \frac{1}{2N+1}E\left\{\sum_{k=-N}^{N}\sum_{\ell=-N}^{N} X[k]X^*[\ell]e^{-j\omega k}e^{j\omega\ell}\right\}$$

$$= \frac{1}{2N+1}\sum_{k=-N}^{N}\sum_{\ell=-N}^{N} E\{X[k]X^*[\ell]\}e^{-j\omega(k-\ell)} = \frac{1}{2N+1}\sum_{k=-N}^{N}\sum_{\ell=-N}^{N} R_{XX}[k-\ell]e^{-j\omega(k-\ell)}$$

$$= \sum_{m=-2N}^{2N} R_{XX}[m]\left(1-\frac{|m|}{2N+1}\right)e^{-j\omega m}$$

where for the derivation of the last expression we took into account that the input sequence is a *WSS sequence*, we defined the index $m = k - \ell$, and we relied on the fact that the function value was *constant* along the diagonal $k - \ell = m$ of the square grid of points of dimensions $(2N+1) \times (2N+1)$ defined on the plane (k, ℓ). We now observe that in the limit $N \to \infty$, the triangular function $1 - [|m|/(2N+1)]$ exerts *less and less influence* on the value of the foregoing expression if in the limit $|m| \to m$ it holds that $|R_{XX}[m]| \to 0$. It is not difficult to understand that *if the influence of this triangular term is nullified, the remaining expression is simply the Fourier transform of discrete-time of the autocorrelation function $R_{XX}[m]$, which by definition is the PSD $S_X(\omega)$* (of course, for this to happen, the value of the expression $|R_{XX}[m]|$ to decrease quite rapidly, so that the condition $\sum |m|R_{xx}[m] < \infty$ is satisfied with the parameter m taking values from $-\infty$ to ∞). Based on all the preceding considerations, we can define the PSD as the limit

$$S_X\omega = \lim_{N\to\infty}\frac{1}{2N+1}E\{|\mathcal{X}(e^{j\omega})|^2\} \tag{13.89}$$

that is, in a similar way to how we defined the power of discrete-time signals.

13.5.3 Construction of a Transfer Function from Autocorrelation or from a PSD Function

How is it possible to construct a transfer function of a system whose output will be characterized by a specific autocorrelation function $P_{YY}(\omega)$ or PSD $S_Y(\omega)$? Considering the general difference equation of a discrete LTI system of order N, we know that the frequency response function in the general case is a *rational function* of the form

$$\mathcal{H}(e^{j\omega}) = \frac{\mathcal{Y}(e^{j\omega})}{\mathcal{X}(e^{j\omega})} = \frac{\mathcal{B}(e^{j\omega})}{\mathcal{A}(e^{j\omega})} = \left\{\sum_{k=0}^{M}\alpha_k e^{j\omega k}\right\}\Big/\left\{\sum_{k=0}^{N}\beta_k e^{j\omega k}\right\} \tag{13.90}$$

If as the input of this system we transmit a *white noise sequence* $W[n]$, with power value $E\{|W[n]|^2\} = \sigma_W^2$, then, according to our earlier discussion, the autocorrelation of the system's output will be

$$S_{YY}(e^{j\omega}) = |\mathcal{H}(e^{j\omega})|^2\sigma_W^2 = \frac{|\mathcal{B}(e^{j\omega})|^2}{|\mathcal{A}(e^{j\omega})|^2}\sigma_W^2 = \frac{\mathcal{B}(e^{j\omega})\mathcal{B}^*(e^{j\omega})}{\mathcal{A}(e^{j\omega})\mathcal{A}^*(e^{j\omega})}\sigma_W^2 \tag{13.91}$$

with the corresponding expression for the complex variable z being

$$S_{YY}(z) = \frac{\mathcal{B}(z)\mathcal{B}(z^{-1})}{\mathcal{A}(z)\mathcal{A}(z^{-1})}\sigma_W^2 = \mathcal{H}(z)\mathcal{H}(z^{-1})\sigma_W^2 \tag{13.92}$$

In the foregoing expression, the variable z is limited to the *unit circle*, and therefore, the full definition of this function requires *its extension*, so that z can take values across the entire complex plane, but in such a way that the explicit function of z that will result from the process has characteristics that comply with the specifications that have been set, as regards the PSD $R_{YY}[n]$.

In the general case of constructing a system that can generate a discrete sequence $X[n]$, which is characterized by a *specific* PSD $R_{XX}(z)$ with a finite number of poles and zeros that are not located at infinity, it is proven that *it is always possible to realize the aforementioned spectral factorization* (spectral factorization), as it is typically called, correlating the poles and the zero values that are located inside the unit circle with the function $\mathcal{H}(z)$, while the poles and zero values that are located outside the unit circle are associated with the function $\mathcal{H}(z^{-1})$. Of course, the foregoing process must be carried out in such a way that *the properties that characterize the spectral density functions of power are not violated.* For example, if during the implementation of this process a zero appears that is on the circumference of the unit circle, it is proven that this must be *of even order*, since in the opposite case the function will take a negative value, something which, of course, is not permissible. This is something that gives us the ability to correlate half the number of zero values with the function $\mathcal{H}(z)$ and the remaining zero values with the function $\mathcal{H}(z^{-1})$. In addition, because the poles of the function $\mathcal{H}(z)$ are *inside the unit circle*, the system will be *BIBO stable*, while this stability will characterize also the *inverse* function, except in the case of the appearance of zeros on the circumference of the unit circle. As for the second function $\mathcal{H}(z^{-1})$, its poles are *outside the unit circle*, and therefore, this system is *stable* in the anticausal sense. Finally, if we symbolize by $\rho_{\max}$ the largest pole measure located inside the unit circle, the function $S_{XX}(z)$ is proven to be *analytic*, that is, it will not have abnormal points in the ring of the complex plane defined by the equation $\rho_{\max} < |z| < 1/\rho_{\max}$.

Applying the foregoing process, we can determine the function $\mathcal{H}(z)$ of a system, which, when it receives at its input a sequence of white noise $W[n]$, creates as its output a random sequence $X[n]$ with a specific PSD function $S_{XX}(\omega)$. This process can be easily simulated on a computer, with the sequence of white noise being constructed using the random number generator that all programming languages have.

Example 13.5.1 (Stark and Woods, pages 357–358) Let us consider the system described by the difference equation

$$y[n] = \alpha y[n-1] + \beta x[n]$$

with impulse response $h[n] = \beta\alpha^n u[n]$ and transfer function $\mathcal{H}(z) = \beta/(1-\alpha z^{-1})$, and suppose we want to determine the parameters α and β, so that when the sequence of white noise $w[n]$ with zero mean and unit power is used as input, the system output $y[n]$ is characterized by average power $R_{xx}[0] = \sigma^2$ and autocorrelation $R_{xx}[1] = \rho\sigma^2$. Since the mean value of the output signal is zero, we will naturally proceed to calculate the autocovariance, which, in full analogy with the autocorrelation, will have the form

$$C_{yy}[m] = h[m] * h[-m] * C_{yy}[m] = h[m] * h[-m] = \beta^2(\alpha^m u[m])(\alpha^{-m}u[-m]) = \beta^2 \sum_{k=-\infty}^{\infty} \alpha^k u[k]\alpha^{m+k}u[m+k]$$

$$= \beta^2\alpha^m \sum_{m=m_0}^{\infty} \alpha^{2k} = \frac{\beta^2}{1-\alpha^2}\alpha^{|m|}, \quad -\infty < m < \infty, \quad m_0 = \max 0, -m$$

For the values $m = 0$ and $m = 1$ and referring to the specifications of the statement, the foregoing relations become

$$C_{xx}[0] = \frac{\beta^2}{1-\alpha^2} = \sigma^2 \quad \text{and} \quad C_{xx}[1] = \frac{\alpha\beta^2}{1-\alpha^2} = \rho\sigma^2$$

from which it follows that $\alpha = \rho$ and $\beta = \sigma^2(1-\rho)^2$. Thus,

$$S_{yy}(\omega) = \frac{\beta}{|1-\alpha e^{-j\omega}|^2} = \frac{\sigma^2(1-\rho^2)}{1-2\rho\cos\omega+\rho^2}$$

Example 13.5.2 (Correlated Noise Process) The correlated noise process is the output $x[n]$ of a discrete LTI system described by the difference equation

$$x[n] = \alpha x[n-1] + w[n], \quad (|\alpha| < 1)$$

where $w[n]$ is a Bernoulli sequence. Developing the foregoing expression, we can formulate the output of the system $x[n]$ in the form

$$x[n] = \sum_{m=1}^{n} \alpha^{n-m} w[m], \quad (n \geq 1)$$

This sequence represents noise with mean value

$$E\{x[n]\} = E\left\{\sum_{m=1}^{n} \alpha^{n-m} w[m]\right\} = \sum_{m=1}^{n} \alpha^{n-m} E\{w[m]\} = p\sum_{m=1}^{n} \alpha^{n-m} = p\frac{1-\alpha^n}{1-\alpha} \tag{13.93}$$

However, this sequence does not constitute a sequence of independent random variables; indeed, considering, for example, the samples $x[1] = w[1]$ and $x[2] = \alpha w[1] + w[2]$, which result from the foregoing expression for the values $n = 1$ and $n = 2$, we find that

$$E\{x[2]x[1]\} = E\{(\alpha w[1] + w[2])w[1]\} = E\{\alpha w^2[1] + w[1]w[2]\} = \alpha E\{w^2[1]\} + E\{w[1]w[2]\} = \alpha p + p^2 \quad \text{and}$$

$$E\{x[2]\}E\{x[1]\} = E\{\alpha w[1] + w[2]\}E\{w[1]\} = (\alpha E\{w[1]\}) + E\{w[2]\}E\{w[1]\} = (\alpha p + p)p = \alpha p^2 + p^2 = (\alpha + 1)p^2$$

and therefore it will hold that $E\{x[2]x[1]\} \neq E\{x[2]\}E\{x[1]\}$. Of course, we can arrive at the same result from the difference equation since the sample $x[n]$ depends on the previous sample $x[n-1]$. Consequently, even though the input of this system is the uncorrelated Bernoulli sequence $w[n]$, the noise that arises at the output of the system is correlated. The variance of this output is easily calculated as

$$\mathrm{Var}\{x[n]\} = \sum_{m=1}^{n} \mathrm{Var}\{w[n]\} = \sum_{m=1}^{n} \alpha^{2(n-m)} \mathrm{Var}\{w[m]\} = \frac{1-\alpha^{2n}}{1-\alpha^2} \tag{13.94}$$

Example 13.5.3 (Random Walk Process) A second process that is based on the concept of the Bernoulli sequence is the random walk process. This process was invented as a model for the random motion of gas molecules (and for this reason it is also known as Brownian motion) and is described by the equation

$$x[n] = \sum_{k=1}^{n} w[k], \quad (x[0] = 0) \tag{13.95}$$

where $w[n]$ is the Bernoulli sequence, which is defined as $w[k] = +s$ for $\xi = K$ and $w[k] = -s$ for $\xi = \Gamma$. This sequence simulates the motion of a particle, which starts from the position $x[0] = 0$ corresponding to the time $n = 0$. To make the next step of the motion, a Bernoulli experiment is performed, for example, the toss of a coin and for each value of n (which progressively increases by one unit each time), the particle moves either to the right by s, if the outcome of the experiment is K, or to the left by s, if the outcome of the experiment is Γ. In this way and after performing n movements, the particle will be in some position rs, which obviously will be an integer multiple of s, since in each movement a distance s is covered to the left or to the right. If during the performance of the n coin tosses the result K appeared k times and the result Γ appeared $n-k$ times, we will have $rs = ks + (n-k)s = (2k-n)s$, from which it follows that $r = 2k - n$. We will have, therefore, $k = (r+n)/2$ for values of r, for which the quantity $2k - n$ is an integer. Taking into account the definition of the sequence $x[n]$ as $x[n] = w[1] + w[2] + \cdots + w[n]$ and the fact that the samples of the sequence $w[n]$ are statistically independent, it is proven that $E\{x[n]\} = 0$ and $E\{x^2[n]\} = ns^2$.

Example 13.5.4 (Wiener Process and White Noise) A process that is inextricably linked to the random walk process is the Wiener process, which is defined as

$$x_T(t) = \sum_{k=1}^{\infty} w[k]u(t - k\tau) \tag{13.96}$$

where $w[k]$ is a Bernoulli sequence. The foregoing equation defines a continuous process, which is piecewise continuous in time intervals of duration T. If we sample in the form $x(nT) = x[n]$, the resulting sequence is a random walk process. In reality, the Wiener process corresponds to the limiting case of this process, when the step change and the time interval T tend to zero. Referring to previous results, we can write that $E\{x(nT)\} = 0$ and $\text{Var}\{x_T(t)\} = E\{x_T^2(nT)\} = ns^2 = s^2(t/T)$. Defining $s^2 = \alpha t$, where $\alpha > 0$ in the limit $T \to 0$, and we will have $\text{Var}(x_T(t)) \to \alpha t$. It is interesting to note that because the Wiener process is the sum of an infinite number of identically distributed random steps, the PDF that characterizes it, according to the central limit theorem, will tend to the normal distribution. Specifically, we will have

$$f_x(x;t) = \frac{1}{\sqrt{2\pi\alpha t}} \exp\left(-\frac{x^2}{2\alpha t}\right), \quad t > 0 \tag{13.97}$$

It should also be noted that because the random variables defined as the differences $x[n_1]$, $x[n_2] - x[n_1]$, $x[n_3] - x[n_2]$, and finally $x[n_N] - x[n_{N-1}]$ (where $n_1 < n_2 < \cdots < n_N$ are integer parameters), due to their definition, are independent of each other, the Wiener process will also have this property, since it is a limiting case in the limit $T \to \infty$. Finally, it is shown that the covariance of the Wiener process is given by the expression

$$C_{xx}(t_1, t_2) = \alpha \min(t_1, t_2) \qquad (\alpha > 0)$$

This is something that is proven by observing that for $t_1 \geq t_2$, the difference $x(t_1) - x(t_2)$ is independent of the value $x(t_2)$, while both of these quantities are characterized by zero mean value. Therefore, we will have

$$E\{[x(t_1) - x(t_2)]x(t_2)\} = E\{x(t_1) - x(t_2)\}E\{x(t_2)\} = E\{x(t_1)x(t_2)\} - E\{x^2(t_2)\} = 0$$

and therefore, it will hold that

$$C_{xx}(t_1, t_2) = E\{x(t_1)x(t_2)\} = E\{x^2(t_2)\} = \alpha t_2$$

On the other hand, if $t_2 > t_1$ using the same argument, we find that $C_{xx}(t_1, t_2) = \alpha t_1$, and combining these two results, we arrive at the desired result. The foregoing expression is based on the silent assumption that the beginning of the time measurement is the time $t = 0$, while if this beginning is some time $t = t_0$, the covariance of the Wiener process takes the form

$$C_{xx}(t_1, t_2) = \alpha \min(t_1 - t_0, t_2 - t_0) \qquad (\alpha > 0)$$

A random process of particular interest is the derivative $\dot{x}(t)$ of a Wiener process $x(t)$. Using the results of a previous example, the mean value of the process $\dot{x}(t)$ is $\mu_{\dot{x}}(t) = d\mu_x(t)/dt = 0$ since for the Wiener process $\mu_x(t) = 0$, while the covariance of the process $\dot{x}(t)$ is calculated as

$$\begin{aligned} C_{\dot{x}\dot{x}}(t_1, t_2) &= \frac{\partial}{\partial t_1}\left(\frac{\partial}{\partial t_2} C_{xx}(t_1, t_2)\right) = \frac{\partial}{\partial t_1}\left(\frac{\partial}{\partial t_2}\alpha \min(t_1, t_2)\right) = \frac{\partial}{\partial t_1}\left(\frac{\partial}{\partial t_2}\begin{cases}\alpha t_2 & t_2 < t_1 \\ \alpha t_1 & t_2 \geq t_1\end{cases}\right) \\ &= \frac{\partial}{\partial t_1}\begin{cases}\alpha & t_2 < t_1 \\ 0 & t_2 > t_1\end{cases} = \frac{\partial}{\partial t_1}\begin{cases}0 & t_1 < t_2 \\ \alpha & t_1 > t_2\end{cases} = \frac{\partial}{\partial t_1}\alpha\begin{cases}0 & t_1 < t_2 \\ 1 & t_1 > t_2\end{cases} = \alpha\frac{\partial}{\partial t_1}u(t_1 - t_2) = \alpha\delta(t_1 - t_2) \end{aligned}$$

We are therefore led to the conclusion that the covariance function of the time derivative $\dot{x}(t)$ of a Wiener process is an impulse function. Taking into account that this process is characterized by zero mean value, we will have

$$R_{\dot{x}\dot{x}}(t_1, t_2) = C_{\dot{x}\dot{x}}(t_1, t_2) = \sigma^2\delta(t_1 - t_2) \tag{13.98}$$

where we have set $\alpha = \sigma^2$, so as to be in agreement with the established symbolism. Taking the Fourier transform of this autocorrelation function, we are led to a PSD of the form

$$S_{\dot{x}\dot{x}}(\omega) = \mathcal{F}\{R_{\dot{x}\dot{x}}(t_1, t_2)\} = \sigma^2 = S_0$$

the value of which is constant and equal to S_0, for all values of frequency ω. As we will comment later, a process characterized by this property is known as a white noise process, and therefore, the time derivative of a process Wiener is a process of white noise and, indeed, of the Gaussian type, since the Wiener process is a Gaussian type process. Let us dwell on this topic for a while. Although white noise has been defined in the preceding way as the time derivative of the Wiener process, the problem is that the Wiener function is not differentiable, and in reality, white noise constitutes the formal derivative of the Wiener process. Nevertheless, we have the ability, applying the appropriate arguments, to define white noise in the way we would if we took the limit involved in the process of defining the derivative of a random process and, more specifically, if we considered white noise as the limit that was approached by a sequence of second-order processes. To do this, we must take into account that considering a difference $\Delta = x(t) - x(\tau)$ for each $t > \tau$, we will have $E\{x(t) - x(\tau)\} = E\{\Delta\} = 0$ and $E\{[x(t) - x(\tau)]^2\} = \alpha(t - \tau)$ for each $t > \tau$ or, equivalently, $E\{[x(t + \Delta) - x(t)]^2\} = \alpha\Delta$. Now defining the function

$$x_\Delta(t) = \frac{x(t + \Delta) - x(t)}{\Delta}$$

(note that this function is used in the definition of the derivative), we will have

$$E\{x_\Delta^2(t)\} = \frac{E\{[x(t + \Delta) - x(t)]^2\}}{\Delta^2} = \frac{\alpha\Delta}{\Delta^2} = \frac{\alpha}{\Delta}$$

and

$$\begin{aligned} E\{x_\Delta(t_1)x_\Delta(t_2)\} &= \frac{E\{[x(t_1 + \Delta) - x(t_1)][x(t_2 + \Delta) - x(t_2)]\}}{\Delta^2} \\ &= \frac{E\{x(t_1 + \Delta)x(t_2 + \Delta)\} - E\{x(t_1 + \Delta)x(t_2)\} - E\{x(t_1)x(t_2 + \Delta)\} + E\{x(t_1)x(t_2)\}}{\Delta^2} \end{aligned}$$

Considering that the average value of the Wiener process is equal to zero, the correlation and covariance functions coincide, and therefore, we can calculate the preceding expected product values using the result $C_{xx}(t_1, t_2) = \alpha \min(t_1, t_2)$. We distinguish two cases:

- The time points t_1 and t_2 are at a distance greater than Δ, or, in mathematical notation, $t_1 > t_2 + \Delta$. In this case it will hold that

$$\begin{aligned} &E\{x(t_1 + \Delta)x(t_2 + \Delta)\} = \alpha \min(t_1 + \Delta t, t_2 + \Delta) = \alpha(t_2 + \Delta) \\ &E\{x(t_1 + \Delta)x(t_2)\} = \alpha \min(t_1 + \Delta, t_2) = \alpha t_2 \\ &E\{x(t_1)x(t_2 + \Delta)\} = \alpha \min(t_1, t_2 + \Delta) = \alpha(t_2 + \Delta) \\ &E\{x(t_1)x(t_2)\} = \alpha \min(t_1, t_2) = \alpha t_2 \end{aligned}$$

  We will therefore have

$$E\{x_\Delta(t_1)x_\Delta(t_2)\} = \frac{\alpha(t_2 + \Delta) - \alpha t_2 - \alpha(t_2 + \Delta) + \alpha t_2}{\Delta^2} = 0 \tag{13.99}$$

  and consequently, any two samples that are at a time distance greater than Δ are uncorrelated with each other.
- The time points t_1 and t_2 are at a distance less than Δ, or, in mathematical notation, $t_2 \leq t_1 \leq t_2 + \Delta$. In this case, we will have

$$\begin{aligned} &E\{x(t_1 + \Delta)x(t_2 + \Delta)\} = \alpha \min(t_1 + \Delta t, t_2 + \Delta) = \alpha(t_2 + \Delta) \\ &E\{x(t_1 + \Delta)x(t_2)\} = \alpha \min(t_1 + \Delta, t_2) = \alpha t_2 \\ &E\{x(t_1)x(t_2 + \Delta)\} = \alpha \min(t_1, t_2 + \Delta) = \alpha t_1 \\ &E\{x(t_1)x(t_2)\} = \alpha \min(t_1, t_2) = \alpha t_2 \end{aligned}$$

and therefore we will have

$$E\{x_\Delta(t_1)x_\Delta(t_2)\} = \frac{\alpha(t_2+\Delta) - \alpha t_2 - \alpha t_1 + \alpha t_2}{\Delta^2} = \frac{\alpha\Delta - \alpha(t_1 - t_2)}{\Delta^2} = \frac{\alpha}{\Delta}\left(1 - \frac{t_1 - t_2}{\Delta}\right)$$

Taking into account that (a) the expression $E\{x_\Delta(t_1)x_\Delta(t_2)\}$ is simply the autocorrelation $R_{\Delta\Delta}(t_1, t_2)$, (b) this expression depends on the time difference $\tau = t_1 - t_2$, so that we can write that

$$R_{\Delta\Delta}(\tau) = \max\left\{1 - \frac{|\tau|}{\Delta}, 0\right\} \tag{13.100}$$

and (c) the process is characterized by zero mean, it is not difficult to ascertain that the process $x_\Delta(t)$ constitutes a WSS-type process. The defining equation and the graphical representation of the autocorrelation function $R_{\Delta\Delta}(\tau)$ lead us to the conclusion that this function describes a triangular pulse that, as is well known, is characterized by a Fourier transform of the form

$$S_{\Delta\Delta}(\omega) = \alpha\left(\frac{\sin(\omega\Delta/2)}{\omega\Delta/2}\right)^2 \tag{13.101}$$

which also constitutes the PSD function of the process. We now observe that as the parameter Δ takes increasingly smaller values, (a) the dispersion $E\{x_\Delta^2(t)\} = \alpha/\Delta$ becomes increasingly large, (b) the distance between the uncorrelated points for which it holds that $t_1 > t_2 + \Delta$ becomes increasingly smaller, and (c) the PSD is increasingly approaching the value $\alpha = \sigma^2$. In the limit $\Delta \to 0$, the autocorrelation function $R_{\Delta\Delta}(\tau)$ tends to the Dirac function $\delta(\tau)$, with area σ^2, the dispersion (or, equivalently, the power) tends to infinity, the PSD tends to the constant value $S_0 = \sigma^2$, while the process is characterized by infinite bandwidth. This continuous process that corresponds to the limit $\Delta \to 0$ is simply the continuous white noise process that we have already encountered and is the formal derivative of an uncorrelated step process. The discrete version of the continuous white noise process is the white noise sequence characterized by an autocorrelation function

$$R_{xx}[i, j] = E\{x[i]x^*[j]\} = \sigma_x^2\delta[i - j]$$

and by a PSD $S_{xx}(e^{j\omega}) = \sigma_x^2 \ \omega \in [-\pi, \pi]$. Of course, in this case, the average power of the white noise sequence is not infinite, as in the continuous case, but finite.

13.5.4 Vector Processes

Earlier we referred to *simple SISO systems* that receive as input a stochastic process $\boldsymbol{x}(t)$ and produce as output another stochastic process $\boldsymbol{y}(t)$. However, they can be easily generalized to the case of *MIMO system dimensions* $M \times N$, which according to our earlier discussion in the corresponding section, accept as input a *vector process* of M elements in the form $\boldsymbol{X}(t) = (\boldsymbol{x}_1(t), \boldsymbol{x}_2(t), \ldots, \boldsymbol{x}_M(t))^T$ and produce as their output a *vector process* of N elements in the form $\boldsymbol{Y}(t) = (\boldsymbol{y}_1(t), \boldsymbol{y}_2(t), \ldots, \boldsymbol{y}_N(t))^T$, with elements

$$y_j(t) = \sum_{i=1}^{M} h_{ji}(t) * x_i(t), \quad j = 1, 2, \ldots, N \tag{13.102}$$

where h_{ji} ($j = 1, 2, \ldots, N$ and $i = 1, 2, \ldots, M$) is the impulse response between the ith input terminal and the jth output terminal. Remember that in this case, the system's impulse response is a matrix of dimensions $M \times N$, while the same applies to the autocorrelation functions $\boldsymbol{R}_{\boldsymbol{xx}}(\tau)$ and $\boldsymbol{R}_{\boldsymbol{yy}}(\tau)$, with the processes $\boldsymbol{x}_i(t)$ ($i = 1, 2, \ldots, M$) and $\boldsymbol{y}_i(t)$ ($i = 1, 2, \ldots, N$) being *jointly stationary processes*. Working as in the simple case, we can show that

$$\boldsymbol{R}_{\boldsymbol{yy}}(\tau) = \boldsymbol{h}(\tau) * \boldsymbol{R}_{\boldsymbol{xx}}(\tau) * \boldsymbol{h}^\dagger(-\tau) \quad \text{and} \quad \boldsymbol{S}_{\boldsymbol{yy}}(\omega) = \boldsymbol{H}(\omega)\boldsymbol{S}_{\boldsymbol{xx}}(\omega)\boldsymbol{H}^\dagger(\omega)$$

where $\boldsymbol{H}(\omega) = \mathcal{F}\{\boldsymbol{h}(\tau)\}$ and $\boldsymbol{S}(\omega) = \mathcal{F}\{\boldsymbol{R}(\tau)\}$, or, in component form,

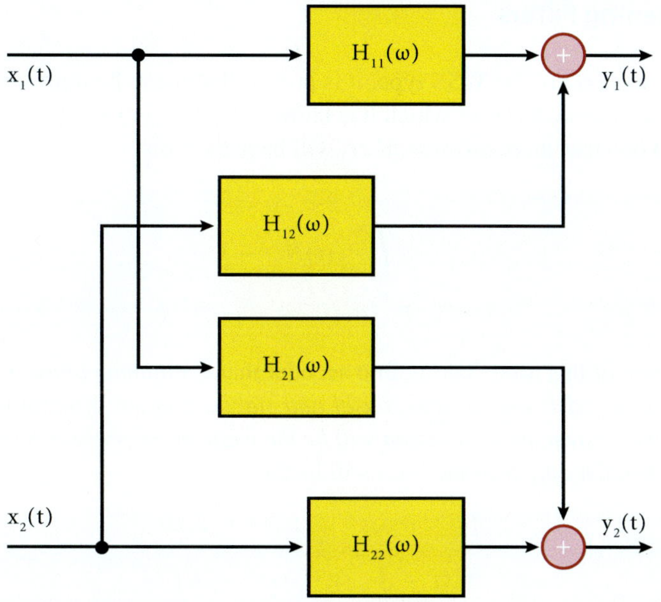

Fig. 13.8 Structure of a dual-channel system (Stark and Woods, Figure 7.7–2)

$$r_{y_k y_\ell}(\tau) = \sum_{i=1}^{M}\sum_{j=1}^{M} h_{ki}(\tau) * r_{x_i x_j}(\tau) * h_{\ell j}(-\tau) \qquad \text{and} \qquad s_{y_k y_\ell}(\omega) = \sum_{i=1}^{M}\sum_{j=1}^{M} \mathcal{H}_{ki}(\omega) * S_{x_i x_j}(\omega) * \mathcal{H}^*_{\ell j}(\omega)$$

For the special case of a *MIMO system of dimensions* 2×2, the foregoing quantities are defined respectively as

$$\boldsymbol{X}(t) = (\boldsymbol{x}_1(t), \boldsymbol{x}_2(t))^T \qquad \text{and} \qquad \boldsymbol{Y}(t) = (\boldsymbol{y}_1(t), \boldsymbol{y}_2(t))^T$$

$$\boldsymbol{h}(t) = \begin{bmatrix} h_{11}(t) & h_{12}(t) \\ h_{21}(t) & h_{22}(t) \end{bmatrix} \qquad \boldsymbol{R}_{\boldsymbol{xx}}(\tau) = \begin{bmatrix} R_{x_1x_1}(\tau) & R_{x_1x_2}(\tau) \\ R_{x_2x_1}(\tau) & R_{x_2x_2}(\tau) \end{bmatrix} \qquad \boldsymbol{R}_{\boldsymbol{yy}}(\tau) = \begin{bmatrix} R_{y_1y_1}(\tau) & R_{y_1y_2}(\tau) \\ R_{y_2y_1}(\tau) & R_{y_2y_2}(\tau) \end{bmatrix}$$

The structure of the system for the case of the 2×2 system depicted in Fig. 13.8 is known as a *dual-channel system* and finds many applications in the field of engineering science. In the special case where $\mathcal{H}_{12}(\omega) = \mathcal{H}_{21}(\omega) = 0$, the two channels of the system *do not interact with each other* and are independent of each other, in the sense that the output $y_1(t)$ is related *only* to the input $x_1(t)$ while the output $x_2(t)$ is related *only* to the input $y_2(t)$. In this case, the foregoing relations take the simple form

$$S_{y_1y_1}(\omega) = |\mathcal{H}_{11}(\omega)|^2 S_{x_1x_1}(\omega)$$
$$S_{y_2y_2}(\omega) = |\mathcal{H}_{22}(\omega)|^2 S_{x_2x_2}(\omega)$$
$$S_{y_1y_2}(\omega) = \mathcal{H}_{11}(\omega)\mathcal{H}^*_{22}(\omega) S_{x_1x_2}(\omega)$$

while it still holds that $|M_{y_1y_2}(\omega)|^2 = |M_{x_1x_2}(\omega)|^2$, where

$$M_{xy}(\omega) = \frac{S_{xy}(\omega)}{\sqrt{S_{xx}(\omega)S_{yy}(\omega)}} = M^*_{yx}(\omega) \qquad (13.103)$$

the *spectral correlation coefficient*.

13.5.5 Renewal and Whitening Filters

Considering a real random process $\boldsymbol{x}(t)$ of the WSS type, it is proven that it can be represented as *the output of a minimum phase system* when a *white noise process* $N(t)$ for which it is known that $R_{NN}(\tau) = \delta(\tau)$ is transmitted to its input. In other words, the output of this filter, known as an *innovation filter*, will have the form

$$\boldsymbol{x}(t) = \int_0^{\infty} i(t)N(t-\tau)d\tau \tag{13.104}$$

where $i(t)$ is the impulse response of this filter. Taking into account that *a minimum phase system is defined as a causal and stable system whose inverse system exists and is also causal and stable*, it is not difficult to understand that *if the process $\boldsymbol{x}(t)$ is used as input to the inverse system, then its output will be the white noise process $N(t)$*. In other words, if the impulse response of this filter, is $w(t)$, then the output of the filter will be the

$$N(t) = \int_0^{\infty} w(t)x(t-\tau)d\tau \tag{13.105}$$

If we connect these two systems in series so that *the output of the whitening filter becomes the input to the renewal filter*, the composite system resulting from the *series* connection of these two filters is the *identical* system that accepts as input the process $\boldsymbol{x}(t)$ and produces as output this same process. Consequently, if $\mathcal{I}(s) = \mathcal{L}\{i(t)\}$ and $\mathcal{W}(s) = \mathcal{L}\{w(t)\}$ are the transfer functions of these two filters, it will hold that $\mathcal{I}(s)\mathcal{W}(s) = 1$, and therefore, $\mathcal{I}(s) = 1/\mathcal{W}(s)$.

The processes represented in the preceding way are *linearly equivalent* to *white noise processes* characterized by a power spectrum of the form $\mathcal{S}(s) = \mathcal{I}(s)\mathcal{I}(-s)$, from which it follows that $\mathcal{S}(j\omega) = |\mathcal{I}(j\omega)|^2$, with the function $\mathcal{I}(s)$ being, according to our earlier discussion, a *minimum phase function*. Given a positive and even function $\mathcal{S}(\omega)$ with *finite area*, the determination of the renewal filter consists in finding a minimum phase function $\mathcal{I}(s)$ such that $|\mathcal{I}(j\omega)|^2 = \mathcal{S}(j\omega)$. It is proven that the problem of finding such a function has a solution if the function $S(\omega)$ satisfies the conditions dictated by the *Paley–Wiener theorem* (refer to corresponding section).

Things are completely analogous in the discrete case. Specifically, considering a real process $x[n]$ of the WSS type, it is aforementioned to be a *normal* if its spectrum $\mathcal{S}(z)$ can be represented as the product $\mathcal{S}(z) = \mathcal{I}(z)\mathcal{I}(1/z)$, from which it follows that $\mathcal{S}(e^{j\omega}) = |\mathcal{I}(e^{j\omega})|^2$. In this case, the outputs of the two preceding filters have the forms

$$N[n] = \sum_{k=0}^{\infty} w[k]x[n-k] \quad \text{and} \quad x[n] = \sum_{k=0}^{\infty} i[k]N[n-k]$$

where $w[k]$ and $i[k]$ are the impulse responses of the *whitening filter* and the *renewal filter*, respectively, while to have a solution to this problem, the function $S(e^{j\omega})$ must satisfy the conditions of the discrete version of the *Paley–Wiener* theorem.

To prove this property, let us consider a process $\boldsymbol{x}[n]$ of the WSS type, with the autocorrelation sequence $R_{xx}[m]$ and PSD $S_{xx}(e^{j\omega})$, which is a *real and continuous function* for every frequency value ω that belongs to the value range $-\pi \leq \omega \leq \pi$ and results from the $\mathcal{Z}$ transform of the autocorrelation function

$$S_{xx}(z) = \sum_{m=-\infty}^{\infty} R_{xx}[m]z^{-m} \tag{13.106}$$

calculated on the *periphery of the unit circle*, that is, at the positions $z = e^{j\omega}$. If we assume that the logarithm $\log S_{xx}(z)$ of the preceding function is an *analytic function* in a ring-shaped region of the complex plane that is defined by the equation $r_1 < |z|r_2$, with $r_1 < 1$ and $r_2 > 1$, which includes the unit circle, so that the function $S_{xx}(e^{j\omega})$ exists, then, according to the theory of complex variable functions, the *Laurent expansion* of the preceding function is defined and has the form

$$F(z) = \log S_{xx}(z) = \sum_{m=-\infty}^{\infty} v[m]z^{-m} \tag{13.107}$$

and is easily recognized as a $\mathcal{Z}$ transform. If we calculate the aforementioned function on the *periphery of the unit circle* and reverse the resulting Fourier transform, we can calculate the coefficients $v[m]$ from the known inverse transform equation

$$v[m] = \frac{1}{2\pi} \int_{-\pi}^{\pi} [\log S_{xx}(e^{j\omega})] e^{j\omega m} d\omega, \quad m = 0, \pm 1, \pm 2, \ldots \tag{13.108}$$

which are known as *cepstral coefficients*, while the sequence $\{v[m]\}$ they define is known as the *cepstrum* of the autocorrelation sequence $R_{xx}[m]$. Taking into account that the function $S_{xx}(e^{j\omega})$ is an *even real function*, the coefficients $v[m]$, according to the properties of the Fourier transform, will be characterized by *even symmetry*—they will therefore be $v[-m] = v[m]$. We now observe that the foregoing sum of the Laurent expansion is analyzed as follows:

$$\sum_{m=-\infty}^{\infty} v[m] z^{-m} = \sum_{m=-\infty}^{-1} v[m] z^{-m} + v[0] + \sum_{m=1}^{\infty} v[m] z^{-m}$$

If in the first of the preceding sums we proceed to change the index from m to $m' = -m$, then the two limits of the sum become $m' = 1$ and $m' = \infty$, while the coefficients of the powers of z become $v[-m]$, and finally $v[m]$, due to the aforementioned even symmetry. Consequently, the first term of the sum successively transforms as

$$\sum_{m=-\infty}^{-1} v[m] z^{-m} = \sum_{m'=1}^{\infty} v[-m'] z^{-(-m')} = \sum_{m=1}^{\infty} v[m] z^{m} = \sum_{m=1}^{\infty} v[m] (z^{-1})^{-m}$$

where we have reintroduced the index m, and therefore, the initial sum is written

$$\sum_{m=-\infty}^{\infty} v[m] z^{-m} = \sum_{m=1}^{\infty} v[m] (z^{-1})^{-m} + v[0] + \sum_{m=1}^{\infty} v[m] z^{-m}$$

Defining, then, the quantity $\sigma_w^2 = \exp(w[0])$ and the function

$$\mathcal{I}(z) = \exp\left(\sum_{m=1}^{\infty} v[m] z^{-m} \right)$$

it is not difficult to ascertain that

$$\exp\left(\sum_{m=1}^{\infty} v[m] (z^{-1})^{-m} \right) = \mathcal{I}(z^{-1})$$

and the function $S_{xx}(s)$ can take the form

$$\begin{aligned} S_{xx}(s) &= \exp[\log S_{xx}(z)] = \exp\left(\sum_{m=-\infty}^{\infty} v[m] z^{-m} \right) = \exp\left(\sum_{m=1}^{\infty} v[m] (z^{-1})^{-m} + v[0] + \sum_{m=1}^{\infty} v[m] z^{-m} \right) \\ &= \exp\left(\sum_{m=1}^{\infty} v[m] (z^{-1})^{-m} \right) \exp(w[0]) \exp\left(\sum_{m=1}^{\infty} v[m] z^{-m} \right) = \sigma_w^2 \mathcal{I}(z^{-1}) \mathcal{I}(s) \end{aligned} \tag{13.109}$$

Calculating, then, the preceding expression for the value $z = e^{j\omega}$, the *PSD* is calculated as

$$S_{xx}(e^{j\omega}) = \sigma_w^2 \mathcal{I}(e^{j\omega}) \mathcal{I}^*(e^{j\omega}) = \sigma_w^2 |\mathcal{I}(e^{j\omega})|^2 \tag{13.110}$$

The filter defined in the preceding manner is characterized by a transfer function $\mathcal{I}(z)$, which is *analytical* in the region of the complex plane of a ring type described by the equation $|z_1| > r_1$, where $r_1 < 1$. The output of this filter, when a sequence of white noise with PSD σ_w^2 is transmitted to its input, is a *stationary random process* $\boldsymbol{x}[n]$, with *PSD* $S_{xx}(\omega) = \sigma_w^2 |\mathcal{I}(e^{j\omega})|^2$. As noted at the beginning of this section, if we use the inverse of this filter, which is causal and stable (as a minimum phase system) and is characterized by a transfer function $\mathcal{W}(z) = 1/\mathcal{I}(z)$, with the input of a random process $\boldsymbol{w}[n]$ of the WSS type

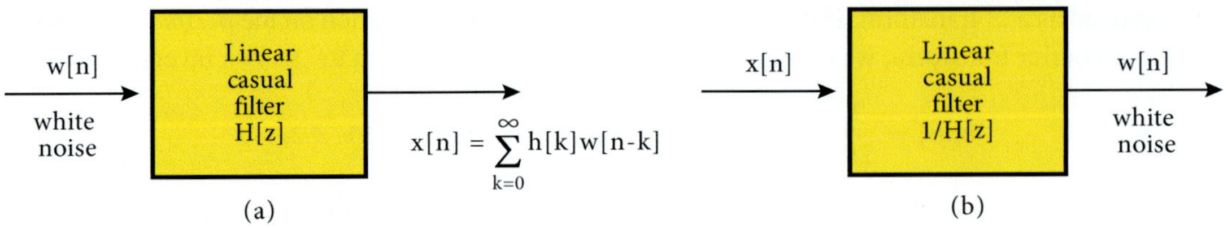

Fig. 13.9 Whitening and renewal filters

with PSD $S_{xx}(e^{j\omega})$, the output of the filter *will be a white noise process*. Consequently, *this filter causes the transformation of a process into a white noise process, and for this reason, it is known as a whitening filter*. The output of this filter, that is, the white noise sequence $w[n]$, is known as the *innovation* of the process $\boldsymbol{x}[n]$. This term was proposed by Wiener since this white noise process is supposed to contain *new information* about the process, $\boldsymbol{x}[n]$, and since then it has been established and is widely used. These two whitening and renewal filters are depicted in Fig. 13.9.

The main advantage of this approach is that since the white noise process is generally characterized by simple properties compared to usual processes, we can significantly simplify the complexity of the problem, *transforming the random stationary process into a linearly equivalent white noise process* using the whitening filter (with the resulting process being characterized as a *renewal* process), and after the appropriate processing in each case, we can reformulate the results that have arisen as a function of the original random process using the *renewal filter*. In the terminology of statistical signal processing, this way of representing a stochastic process $\boldsymbol{x}(t)$, as the output of a filter with a transfer function $\mathcal{I}(z)$ defined in the way we saw, is known as a *Wold representation*.

From the foregoing discussion it is not difficult to understand that we can characterize and generally describe a process using the *renewal filter* $\mathcal{I}(z)$. The reason for this is that, due to the nature of the underlying system, as a *minimum phase system*, the random process uniquely determines the statistical properties of the linearly equivalent white noise, and vice versa. Considering *the most general cases*, in which the PSD $S_{xx}(s)$ of the random process $\boldsymbol{x}[n]$ is a *rational* function of z, the same obviously applies to the renewal filter $\mathcal{I}(z)$, and resorting to the known tools for studying continuous- and discrete-time systems, *we can factorize the polynomials of the numerator and the denominator of the rational transfer function to determine the positions of the poles on the complex plane*, while we can use the *inverse transform* method to construct the *difference equation* of the system. This issue was discussed in detail in the relevant section, where, depending on the form of the transfer function, we characterized the filters as *autoregressive* (AR), *moving average* (MA), and *autoregressive moving average* (ARMA), terms now attributed to the processes whose renewal filter falls into the corresponding category from the preceding discussion. In other words, if the renewal filter of the process with *PSD* of the form

$$S_{xx}(z) = \sigma_w^2 \frac{\mathcal{B}(z)\mathcal{B}(z^{-1})}{\mathcal{A}(z)\mathcal{A}(z^{-1})} \tag{13.111}$$

is characterized by a transfer function of the form

$$\mathcal{I}(z) = \frac{\mathcal{B}(z)}{\mathcal{A}(z)} = \left\{\sum_{k=0}^{M} \beta_k z^{-k}\right\} / \left\{1 + \sum_{k=1}^{N} \alpha_k z^{-k}\right\} \tag{13.112}$$

which leads to the difference equation

$$x[n] + \sum_{k=1}^{N} \alpha_k x[n-k] = \sum_{k=0}^{M} \beta_k w[n-k] \tag{13.113}$$

the process will fall into one of the following categories:

- AR if $\beta_0 = 1$ and $\beta_k = 0$ $(k > 0)$, with transfer function

$$\mathcal{I}(z) = \frac{1}{\mathcal{A}(z)} = \frac{1}{1 + \alpha_1 z^{-1} + \alpha_2 z^{-2} + \cdots + \alpha_N z^{-N}}$$

 and difference equation

$$x[n] + \alpha_1 x[n-1] + \cdots + \alpha_N x[n-N] = \beta_0 w[n]$$

 Consequently, *the renewal filter* $\mathcal{I}(z)$ *is a filter of only poles*, while the inverse filter, that is, *the whitening filter* that generates the white noise process, is a filter *of only zeros*, that is, a filter *of finite impulse response*.
- MA if $\alpha_k = 0$ $(k = 0, 1, 2, \ldots, M)$. In this case, the transfer function of the renewal filter is a polynomial of the form

$$\mathcal{I}(z) = \mathcal{B}(z) = \beta_0 + \beta_1 z^{-1} + \cdots + \beta_M z^{-M}$$

 which leads to a difference equation

$$x[n] = \beta_0 w[n] + \beta_1 w[n-1] + \cdots + \beta_M w[n-M]$$

 In other words, the *renewal filter is a zero-value-only filter*, while the inverse filter, that is, the *whitening filter*, is a *pole-only filter*.
- ARMA if the transfer function has the general form $\mathcal{I}(z) = \mathcal{B}(z)/\mathcal{A}(z)$, which was mentioned previously. In this case, *the filter contains both poles and zero values* and is described by the general difference equation, while *the same applies* to the inverse filter, that is, to the *whitening filter*.

How are the coefficients $\{\alpha_k\}$ $(k = 1, 2, \ldots, N)$ and β_k $(k = 0, 1, 2, \ldots, N)$ of the transfer function $\mathcal{I}(z)$ of the renewal filter of the process $x[n]$ related to the autocorrelation sequence $R_{xx}[m]$? Working as in similar cases, we multiply both sides of the general difference equation by the quantity $x^*[n-k]$, apply to both sides the *expected value operator*, and apply the definition of autocorrelation and cross-correlation. It will then be the case that

$$E\{x[n]x^*[n-k]\} = -\sum_{k=1}^{N} \alpha_k E\{x[n-k]x^*[n-m]\} + \sum_{k=0}^{M} \beta_k E\{w[n-k]x^*[n-m]\}$$

or, equivalently,

$$R_{xx}[m] = -\sum_{k=1}^{N} \alpha_k R_{xx}[m-k] + \sum_{k=0}^{M} \beta_k R_{wx}[m-k]$$

We observe that the cross-correlation function $R_{wx}[m]$ that appears in the second term of the foregoing equation is defined using the input $w[n]$ and the output $x[n]$ of the renewal filter, with these two processes being related to the *convolution equation*, since the considered filter is a discrete LTI system. We will therefore have

$$x[n] = \sum_{k=0}^{\infty} h[k]w[n-k] \qquad \text{from which it follows that} \qquad x^*[n] = \sum_{k=0}^{\infty} h[k]w^*[n-k]$$

Consequently, the cross-correlation $R_{wx}[m]$ takes the form

$$R_{wx}[m] = E\{x^*[n]w[n+m]\} = E\left\{\sum_{k=0}^{\infty} h[k]w^*[n-k]e[n+m]\right\} \tag{13.114}$$

Taking into account that the process $w[n]$ is a *white noise process* with statistical properties $E\{|w[n]|^2\} = \sigma_w^2$ and $E\{w^*[n]w[k] = 0\}$ for $k \neq n$, it is not difficult to understand that the *nonzero terms* of the preceding sum are those for which the condition $n - k = n + m$, from which it follows that $k = -m$. Thus, we will have $h[k] = h[-m]$ and $E\{w^\lceil n - k]w[n + k]\} = E\{w^\lceil n + m]w[n + k]\} = \sigma_w^2$, from which it follows that $R_{wx} = \sigma_w^2 h[-m]$ for $m \leq 0$ (the impulse response is a *real causal signal*) and $R_{wx}[m] = 0$ for $m < 0$. Consequently, the second sum of our initial expression will take the form (after suitable handling of the indices)

$$\sum_{k=0}^{M} \beta_k R_{wx}[m-k] = \sigma_w^2 \sum_{k=0}^{M} \beta_k h[k-m] = \sigma_w^2 \sum_{k=0}^{M-m} \beta_{k+m} h[k] \tag{13.115}$$

Incorporating this expression into the autocorrelation equation $R_{xx}[m]$, we are led to the result

$$R_{xx}[m] = \begin{cases} -\sum_{k=1}^{N} \alpha_k R_{xx}[m-k] & m > M \\ -\sum_{k=1}^{N} \alpha_k R_{xx}[m-k] + \sigma_w^2 \sum_{k=0}^{M-m} \beta_{k+M} h[k] & m > M \\ R_{xx}^*[-m] & m < 0 \end{cases} \tag{13.116}$$

This result applies to the general case of ARMA processes, while for the cases of the two other types of processes, namely, the AR and MA processes, which are special cases of ARMA processes, the situation is as follows. In the case of AR processes, we know that $\beta_0 = 1$ and $b_k = 0$ $k > 0$, while also $M = 0$, and the preceding expression takes the simplest form:

$$R_{xx}[m] = \begin{cases} -\sum_{k=1}^{N} \alpha_k R_{xx}[m-k] & m > 0 \\ -\sum_{k=1}^{N} \alpha_k R_{xx}[m-k] + \sigma_w^2 & m = 0 \\ R_{xx}^*[-m] & m < 0 \end{cases} \tag{13.117}$$

We observe, therefore, that *in the case of AR processes, the relationship that exists between the coefficients of the renewal filter and the autocorrelation sequence is of a linear nature, something that does not apply in the general case of ARMA processes, where this relationship is nonlinear*. Developing the preceding expression, we are led to a linear system of equations of the form

$$\begin{aligned} &R_{xx}[0] + \alpha_1 R_{xx}[-1] + \alpha_2 R_{xx}[-2] + \cdots + \alpha_N R_{xx}[-N] = \sigma_w^2 \\ &R_{xx}[1] + \alpha_1 R_{xx}[0] + \alpha_2 R_{xx}[-1] + \cdots + \alpha_N R_{xx}[-N+1] = 0 \\ &R_{xx}[2] + \alpha_1 R_{xx}[1] + \alpha_2 R_{xx}[0] + \cdots + \alpha_N R_{xx}[-N+2] = 0 \\ &\cdots\cdots\cdots\cdots\cdots\cdots\cdots\cdots\cdots\cdots\cdots\cdots\cdots\cdots \\ &R_{xx}[N] + \alpha_1 R_{xx}[N-1] + \alpha_2 R_{xx}[-2] + \cdots + \alpha_N R_{xx}[0] = 0 \end{aligned}$$

which, expressed as an *array equation*, is written

$$\begin{bmatrix} R_{xx}[0] & R_{xx}[-1] & R_{xx}[-2] & \cdots & R_{xx}[-N] \\ R_{xx}[1] & R_{xx}[0] & R_{xx}[-1] & \cdots & R_{xx}[-N+1] \\ R_{xx}[2] & R_{xx}[1] & R_{xx}[0] & \cdots & R_{xx}[-N+2] \\ \vdots & \vdots & \vdots & \ddots & \vdots \\ R_{xx}[N] & R_{xx}[N-1] & R_{xx}[N-2] & \cdots & R_{xx}[0] \end{bmatrix} \begin{bmatrix} 1 \\ \alpha_1 \\ \alpha_2 \\ \vdots \\ \alpha_N \end{bmatrix} = \begin{bmatrix} 1 \\ \sigma_w^2 \\ 0 \\ \vdots \\ 0 \end{bmatrix}$$

The coefficient matrix of the preceding linear system is recognized as a *Toeplitz type matrix*,[10] the inversion of which can be performed *with fast and efficient algorithms*. The preceding equations allow us to express the parameters of the ARMA process as a function of the equal values of the autocorrelation sequence, while, conversely, if the transfer function $\mathcal{I}(z)$ is known, that is, the filter coefficients, then we can determine the values of autocorrelation from the preceding system of equations. Finally, the case of MA processes arises from the equation of the general ARMA model, setting $\alpha_k = 0$ $(k = 0, 1, 2, \ldots, M)$ and $h[k] = \beta_k$ $(k = 1, 2, \ldots, N)$, leading to the result

$$R_{xx}[m] = \begin{cases} \sigma_w^2 \sum_{k=0}^{M} \beta_k \beta_{k+M} & 0 \leq m \leq M \\ 0 & m > M \\ R_{xx}^*[-m] & m < 0 \end{cases} \tag{13.118}$$

13.6 White Noise and White Noise Processes

In the terminology of statistical signal processing, the term *white noise* is used to describe *any process with zero mean value and with PSD whose value is constant $S_{xx}(\omega) = S_0$ for all frequencies ω*. If this property does not hold, then the noise is aforementioned to be *colored*. The name of this noise comes from physics, from which it is known that *the spectral power of the spectrum of natural sunlight is considered to be equally distributed at all frequencies, although in reality this is not absolutely true*. As we will note later, due to the *constant value* that characterizes the PSD of white noise, the autocorrelation function for white noise is the *impulse function*, with area equal to the *spectral power* S_0. Specifically, this autocorrelation has the form $R_x(\tau) = S_0\delta(\tau) = (N_0/2)\delta(\tau)$, where N_0 is a *real positive constant*, with the use of the variable τ, indicating that the white noise process is considered to be a WSS-type process. Indeed, such an autocorrelation function leads to a constant PSD value, as it is easily observed that

$$S_{xx}(\omega) = \int_{-\infty}^{\infty} R_{xx}(\tau)e^{-j\omega t}d\tau = S_0 \int_{-\infty}^{\infty} \delta(\tau)e^{-j\omega t}d\tau = S_0 \tag{13.119}$$

Consequently, *the samples of a white noise process are all uncorrelated with each other*. It is not difficult to understand that such a white noise process *cannot exist in nature* since the surface defined by the PSD curve (which is a horizontal line of infinite length extending to all frequencies from $-\infty$ to $+\infty$) is *infinite*, while the same applies to the *mean square value*, which, as is well known, is equal to $R_{xx}(0) = S_0\delta(0)$, with the value $\delta(0)$ being *infinite*. Nevertheless, white noise processes are an important tool for signal and system analysis, as several significant noise sources, such as *thermal noise* to be discussed later, are characterized by essentially flat power spectrum, for a very large frequency range. Specifically, in cases where the random input signal is characterized by a bandwidth that is *much larger than the system's passband*, the assumption that the PSD has white noise characteristics significantly simplifies the calculation of the system's response and without the appearance of significant error.

Another concept that also finds wide application is that of *band-limited white noise*, in which the PSD is *constant* and equal to S_0, *but not* across the entire infinite frequency range $-\infty < f < \infty$ but in a *finite frequency range* with a bandwidth of W, which means that this noise has *finite* and not infinite power. Conversely, in any other frequency range, it will hold that $S_x x(f) = 0$. In other words, this density is defined as $S_{xx}(\omega) = S_0$ for $|f| \leq W$, and $S_{xx}(f) = 0$ anywhere else. The reason we are forced to make this assumption is that *the infinite bandwidth that characterizes white noise does not allow us to handle it in the way we do for the usual deterministic and random signals, which can undergo sampling at the appropriate frequency in each case, so that their reconstruction is possible*. Indeed, the adoption of such an assumption is facilitated by the fact that *if this finite bandwidth of noise is greater than the bandwidth of the underlying system, then the output of the system is essentially the same as what we would get if the noise were characterized by infinite bandwidth*.

[10] In honor of the German mathematician Otto Toeplitz (1881–1940).

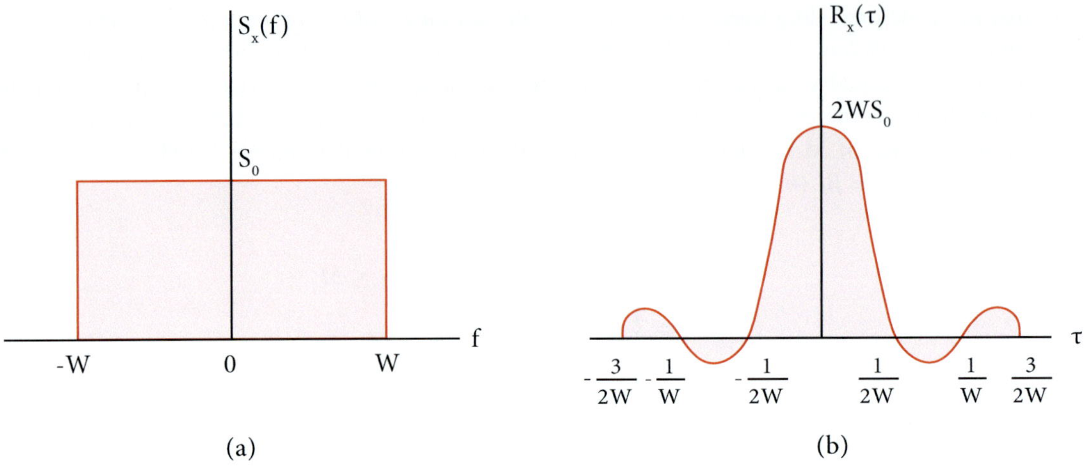

Fig. 13.10 (**a**) PSD for white noise of limited bandwidth and (**b**) corresponding autocorrelation function

The graphical representation of this PSD, along with the corresponding autocorrelation function, is depicted in Fig. 13.10. Notice that the function $S_{xx}(\omega)$ has the shape of a *rectangular pulse* of the rect type and consequently, the autocorrelation function that constitutes the inverse Fourier transform of $S_{xx}(\omega) = 0$ will be a function of the *sinc type*, in full analogy with our earlier discussion in the relevant chapter, since, indeed, it will hold that

$$R_{xx}(\tau) = \mathcal{F}^{-1}\{S_{xx}(f)\} = \mathcal{F}^{-1}\left\{S_0\text{rect}\left(\frac{f}{2W}\right)\right\} = 2WS_0\text{sinc}(2\pi W\tau) = 2WS_0\frac{\sin(2\pi W\tau)}{2\pi W\tau} \tag{13.120}$$

It is not difficult to understand that in the limit $\tau \to 0$, $R_{xx}(\tau) \to \delta(\tau)$, which is naturally expected. Note that, although in this case the mean square value *is not infinite but, on the contrary, is finite* and equal to $2WS_0$, then not even in the regime of this power of this more realistic assumption can such a white noise process exist in practice. Of course, it is proven that such a process can be approached with *arbitrary accuracy*, and for this reason it also constitutes an important tool for the analysis of many problems. Also notice from Fig. 13.10b that the autocorrelation function $R_{xx}(\tau)$ takes *zero value* at the positions $\tau_n = n/2W$ $(n = \pm 1, \pm 2, \ldots)$, and therefore, the random variables that constitute samples of a continuous white noise process, whose sampling is carried out with a sampling period $T = 1/2W$, that is, when the temporal distance between them has some of the values τ_n $(n = 0, \pm 1, \pm 2, \ldots)$, they appear *uncorrelated* with each other. Therefore, if a limited-bandwidth process is to be represented by samples characterized by the aforementioned property, it seems that these are *statistically uncorrelated among themselves*, a fact that significantly simplifies the analysis since the *covariance matrix* is a diagonal matrix, or, in mathematical notation, $\boldsymbol{C}_{xx} = \text{diag}(\sigma_1^2, \sigma_2^2, \ldots, \sigma_n^2)$. If in addition the random variables are *identically distributed*, then it will hold that $\sigma_1^2 = \sigma_2^2 = \cdots = \sigma_n^2 = \sigma^2$ and the covariance matrix will take an even simpler form, $\boldsymbol{C}_{xx} = \sigma^2\boldsymbol{I}_n$, where $\boldsymbol{I}_n$ is the *unit matrix* of dimensions $n \times n$. The discrete sequence that results in this way is aforementioned to be a *white noise sequence*, and the way in which it is related to the continuous white noise process from the sampling of which it arose is clear and not characterized by indeterminacies. However, if the noise bandwidth is *greater than twice the sampling frequency*, ambiguities arise as a result of the appearance of the phenomenon of *frequency folding*.

The foregoing definition of white noise, based on the concept of *correlation*, describes white noise *under the broad concept*, while a more restrictive definition of white noise, in the strict sense, describes *(strictly) white noise* as a *random signal* whose random variables are *statistically independent* and therefore satisfy the property

$$f_{\boldsymbol{x}}(\boldsymbol{x}; \boldsymbol{t}) = \prod_{i=1}^{n} f_{x_i}(x_i) \tag{13.121}$$

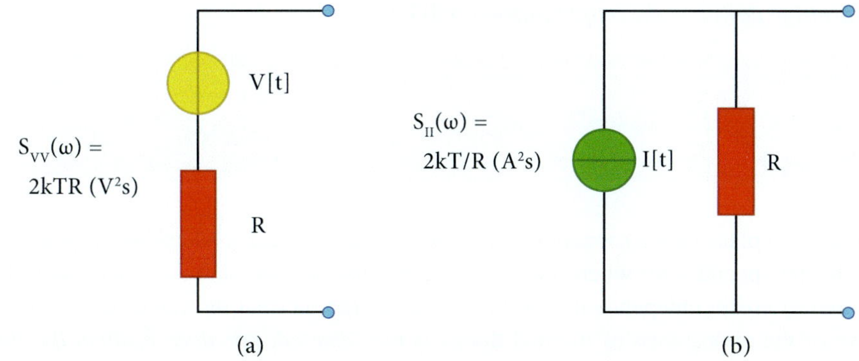

Fig. 13.11 Resistance model characterized by thermal noise. Thevenin equivalent circuit (**a**) and Norton equivalent circuit (**b**)

The reason this definition of white noise is considered *more strict* than the previous one is because *if two random variables are independent, then, as we have noted, they will be simultaneously uncorrelated, while conversely, two variables that are uncorrelated are not necessarily also independent, since they may be dependent* (an exception to this rule is the case of random variables, whose combined distribution is of a *Gaussian type*, since, as we noted in this case, *if variables are uncorrelated, then they are also independent of each other*).

A classic example of white noise is the *thermal noise* due to the *thermal motion* (or *Brownian motion*) of free electrons that characterize each electrical conductor. Specifically, considering an electrical resistance R that is at absolute temperature T, it is found that a voltage develops between its open terminals, which is due to noise and is characterized by a *Gaussian-type PDF* with a mean value equal to zero and a PSD defined as

$$S_{vv}(\omega) = \frac{Rh|\omega|}{\pi\left[\exp\left(\frac{h|\omega|}{2\pi kT}\right) - 1\right]} \ \mathrm{V^2 s} \tag{13.122}$$

where $k = 1.38 \times 10^{-23}$ J/K is Boltzmann's constant and $h = 6.63 \times 10^{-34}$ J $\cdot$ s is Planck's constant. It is proven that for a wide range of frequencies extending up to 10^{12} Hz, the value of the function $S_{vv}(\omega)$ has almost a constant value, which for higher frequency values progressively decreases until it reaches the zero value. Therefore, for this frequency range, which includes significant areas of the electromagnetic spectrum, with great practical value, such as radio waves and microwaves, this thermal noise is considered white noise. Developing the exponential function of the preceding expression in a power series, we end up with the much simpler equation $S_{vv}(\omega) = 2kTR$ $\mathrm{V^2s}$ for the PSD of thermal noise.

It is not difficult to understand that, since an electrical resistance is a source of noise, which is an *inherent property of the material* that cannot be erased, the system output will also have a *noise* component. To incorporate this noise into system analysis (since, as is well known, in the usual analysis of electrical circuits, circuit elements are considered to be *noise-free*, an assumption that, as we will see, *does not* hold in practice), we can simply connect in series an ideal resistance, free from noise, and a noise voltage source $V(t)$ whose PSD is $S_{vv}(\omega) = 2kTR$. In the terminology of electrical circuits, this circuit is known as a *Thevenin equivalent circuit*, while by resorting to the appropriate transformation, we can construct a *Norton equivalent circuit*, in which the ideal resistance is connected in series with a noise current source $I(t)$ with PSD $S_{ii}(\omega) = 2kT/R$ $\mathrm{A^2s}$. These two equivalent circuits are depicted in Fig. 13.11.

To understand the effect of noise on the system output, let us consider the RC circuit with frequency response function $\mathcal{H}(j\omega) = 1/(1 + j\omega RC)$ (see Eq. 3.66, page 289) and suppose that we replace the resistance R, which is considered *ideal and noise-free*, with the Thevenin equivalent circuit of Fig. 13.11, so as to take into account the effect of noise. In this case, as we mentioned previously, the *PSD of the system's output* will be

$$S_{yy}(\omega) = S_{xx}(\omega)|\mathcal{H}(j\omega)|^2 = \frac{2kTR}{1 + \omega^2 R^2 C^2} \ \mathrm{V^2 s} \tag{13.123}$$

and consequently, the *average power of the output process* will be

$$P_v = R_{yy}(0) = \frac{1}{2\pi}\int_{-\infty}^{\infty} S_{xx}(\omega)|\mathcal{H}(j\omega)|^2 d\omega = \frac{1}{2\pi}\int_{-\infty}^{\infty} \frac{2kTR}{1+\omega^2R^2C^2} d\omega = \frac{kT}{C}\ \mathrm{V}^2\mathrm{s} \tag{13.124}$$

Similar conclusions arise if, in place of the capacitor C, another circuit element is added, such as another *resistance* or some *operational amplifier*. In the special case where the system contains *more resistances*, then each of them will obviously contribute *its own amount of noise*, independently of the rest, since the thermal motion of the electrons in each resistance, which is also the cause of the appearance of thermal noise, *is not affected, nor does it affect the thermal motions of the electrons of the other resistances*. Consequently, at the system's output, we must sum all the different spectra that arise from the various resistances. More details on this topic can be found in the literature.[11]

Example 13.6.1 Let us consider a discrete LTI system described by the difference equation

$$x[n] = \alpha x[n-1] + w[n],\quad \alpha \in \mathbb{R} \tag{13.125}$$

with initial condition $x[-1] = 0$, whose output is known to have the form

$$x[n] = \sum_{k=0}^{n} \alpha_{n-k} w[k] \tag{13.126}$$

Developing the preceding expression for the various values of n, we will get

$$\begin{aligned}
x[0] &= w[0]\\
x[1] &= \alpha w[0] + w[1]\\
x[2] &= \alpha^2 w[0] + \alpha w[1] + w[2]\\
&\cdots\cdots\cdots\cdots\cdots\cdots\\
x[n] &= \alpha^n w[0] + \alpha_{n-1} w[n-1] + \alpha_{n-2} w[n-2] + \cdots + \alpha w[n-1] + w[0]
\end{aligned}$$

or, in matrix form,

$$\underbrace{\begin{bmatrix} x[0]\\ x[1]\\ x[2]\\ \vdots\\ x[n]\end{bmatrix}}_{\boldsymbol{x}} = \underbrace{\begin{bmatrix} 1 & 0 & 0 & \cdots & 0\\ \alpha & 1 & 0 & \cdots & 0\\ \alpha^2 & \alpha & 1 & \cdots & 0\\ \vdots & \vdots & \vdots & \ddots & \vdots\\ \alpha^n & \alpha_{n-1} & \alpha_{n-2} & \cdots & 1\end{bmatrix}}_{\boldsymbol{A}} \underbrace{\begin{bmatrix} w[0]\\ w[1]\\ w[2]\\ \vdots\\ w[n]\end{bmatrix}}_{\boldsymbol{w}} \tag{13.127}$$

or, more succinctly, $\boldsymbol{x} = \boldsymbol{A}\boldsymbol{w}$. If this system receives at its input a sequence of Gaussian white noise, that is, a sequence of independent variables, with zero mean value $\boldsymbol{\mu}_w = \boldsymbol{0}$, variance σ^2, and covariance matrix $\boldsymbol{C}_{ww} = \sigma^2\boldsymbol{I}$, which is described by a Gaussian-type PDF, then the mean value and the covariance matrix of the system output will be, respectively, $\boldsymbol{\mu}_x = \boldsymbol{A}\boldsymbol{\mu}_w = \boldsymbol{0}$ and $\boldsymbol{C}_{xx} = \boldsymbol{A}\boldsymbol{C}_{ww}\boldsymbol{A}^T$, with the latter matrix being calculated analytically as

[11] See, for example, van Etten W.C, *Introduction to Random Signals and Noise*, Chapter 6.

$$
\boldsymbol{C}_{xx} = \begin{bmatrix} 1 & 0 & 0 & \cdots & 0 \\ \alpha & 1 & 0 & \cdots & 0 \\ \alpha^2 & \alpha & 1 & \cdots & 0 \\ \vdots & \vdots & \vdots & \ddots & \vdots \\ \alpha^n & \alpha_{n-1} & \alpha_{n-2} & \cdots & 1 \end{bmatrix} \begin{bmatrix} \sigma^2 & 0 & 0 & \cdots & 0 \\ 0 & \sigma^2 & 0 & \cdots & 0 \\ 0 & 0 & \sigma^2 & \cdots & 0 \\ \vdots & \vdots & \vdots & \ddots & \vdots \\ 0 & 0 & 0 & \cdots & \sigma^2 \end{bmatrix} \begin{bmatrix} 1 & \alpha & \alpha^2 & \cdots & \alpha^n \\ 0 & 1 & \alpha & \cdots & \alpha^{n-1} \\ 0 & 0 & 1 & \cdots & \alpha^{n-2} \\ \vdots & \vdots & \vdots & \ddots & \vdots \\ 0 & 0 & 0 & \cdots & 1 \end{bmatrix}
$$

or, after operations,

$$
\frac{\sigma^2}{1-\alpha^2} \begin{bmatrix} 1 & \alpha & \alpha^2 & \cdots & \alpha^n \\ \alpha & 1 & \alpha & \cdots & \alpha^{n-1} \\ \alpha^2 & \alpha & 1 & \cdots & \alpha^{n-2} \\ \vdots & \vdots & \vdots & \ddots & \vdots \\ \alpha^n & \alpha^{n-1} & \alpha^{n-2} & \cdots & 1 \end{bmatrix} - \frac{\sigma^2\alpha^2}{1-\alpha^2} \begin{bmatrix} 1 & \alpha & \alpha^2 & \cdots & \alpha^n \\ \alpha & \alpha^2 & \alpha^3 & \cdots & \alpha^{n+1} \\ \alpha^2 & \alpha^3 & \alpha^4 & \cdots & \alpha^{n+2} \\ \vdots & \vdots & \vdots & \ddots & \vdots \\ \alpha^n & \alpha^{n+1} & \alpha^{n+2} & \cdots & \alpha^{2n} \end{bmatrix}
$$

In the preceding expression, the first of the matrices on the right-hand side is recognized as a Toeplitz matrix, while the second matrix is recognized as a Hankel matrix. Of these two matrices, the Toeplitz matrix describes the stationary component of the process, which appears as the output of the foregoing discrete LTI system, when the preceding sequence of white noise appears at its input, while the Hankel matrix corresponds to the transient response component, which for values $|\alpha| < 1$ weakens over time and eventually nullifies. To remove this transient component, we can define as the variance value for the noise sequence $w[n]$, not σ^2 but $\sigma'^2 = \sigma^2 + \sigma^2\alpha^2/(1-\alpha^2)$. In this case, the upper left element of the new covariance matrix $\boldsymbol{C}_{ww'}$ becomes equal to σ'^2, while the product $\boldsymbol{AC}'_{ww}\boldsymbol{A}^T$ contains an additional term that cancels the term of the transient response, as can be easily proven.

13.7 Statistical Analysis of Quantization Error

Although in the previous sections we presented the quantization process *without* resorting to the use of statistical methods, both in this and in similar cases, it is particularly helpful to base our study, *not on the signals involved themselves, but on the probability distributions that characterize them*, ending up, in this way, with results of a *statistical nature*. This is particularly true for the quantization process, in which the use of the quantizer, which, as is well known, constitutes a *nonlinear arrangement*, making it *difficult* to analyze a situation using the usual analytical methods. Conversely, the statistical approach leads to *a significant reduction in the complexity of the description*, since, as we will see, the distribution of the probability density of the output of the quantizer results from the application of a linear sampling process in the corresponding distribution of the input signal. The statistical analysis of the quantization error that we will present in this section was formulated by Bernard Widrow[12] in 1956 and subsequently extended by other researchers, as we briefly discuss at the end of this section.

Recall that a quantizer constitutes a *nonlinear arrangement* of the form $y = Q(x)$, where Q is a *nonlinear operator* described by the input/output characteristic of Fig. 13.12. In these quantization characteristics, the diagonal dashed line constitutes the characteristic of the quantizer for marginal cases in which the quantization step Δ tends toward the value of zero. Consequently, the nonlinearity of the quantizer is due precisely to the fact that the step of quantization *does not have a zero value*, no matter how small it is. The slope of this discontinuous line is considered the *average gain* of the quantizer and is not difficult to understand that for the usual characteristic like the one depicted in Fig. 13.12, this average gain has a value equal to *unity*. In practice, we can define, if necessary, many other similar characteristics, combining the arrangement of the quantizer with the appropriate *multipliers* and *adders*. Such an example of a quantizer is depicted in Fig. 13.13, with the characteristic that is depicted there, having an average slope, as we defined previously, with a value equal to 2. If Δ is the quantization step, then the shape of the characteristic curve of the quantizer allows for the definition of square regions with side length Δ, with each sample that is located in such a region being depicted under the application of the quantization process in the center of this region.

[12] The description that follows can be found in Widrow's research publications, which are cited in the bibliography and the full text of which is available on his personal website at MIT, although the reader is referred to the book by Widrow and Kollár that was published 52 years later and presents the theory in a more systematic and understandable way. The material presented is in this very book.

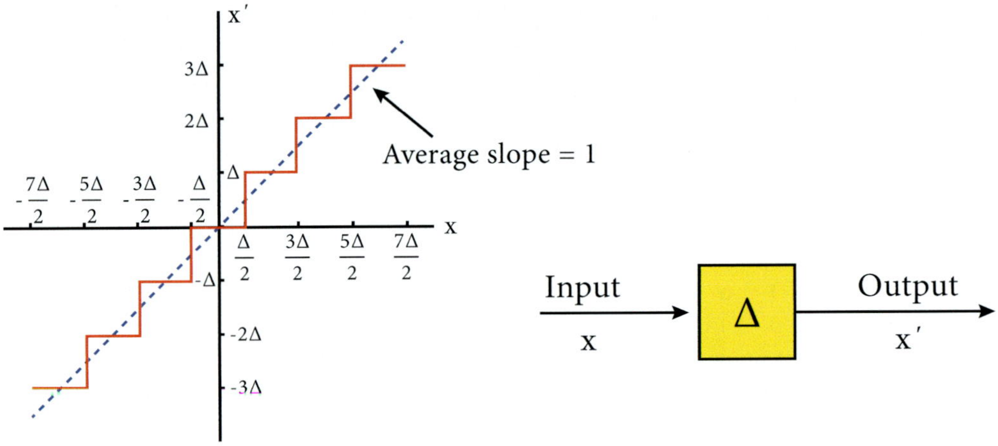

Fig. 13.12 Typical characteristic curve of a quantizer, with the operating area being approximated by a straight line $y = x$, with a slope equal to unity, at the limit $\Delta \to 0$

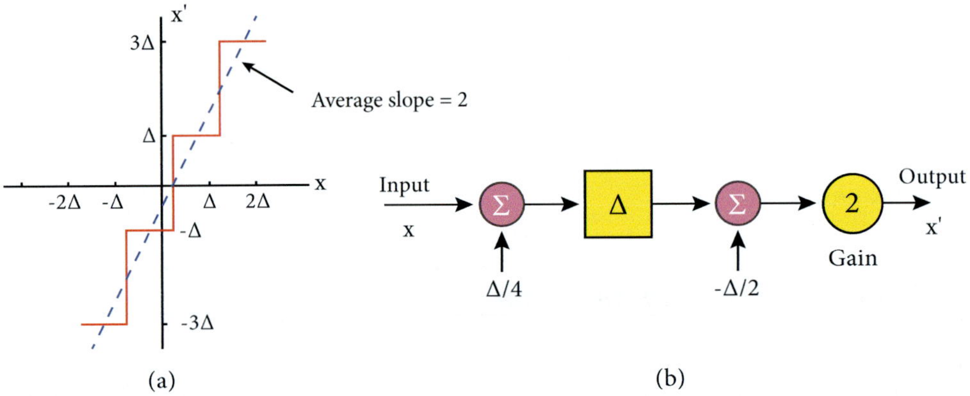

Fig. 13.13 Characteristic curve of quantizer with different features that arise by combining a quantizer with the appropriate elements (Widrow and Kollár, Figure 1.3)

The errors of *quantization* or *rounding* $e = y - x = Q(x) - x$ take values in the area $[-\Delta/2, +\Delta/2]$, where Δ is the step of quantization, and therefore, *the smaller this step, the smaller the value of the quantization error*. Of course, *the smaller* the quantization step, *the larger* the number of quantization intervals for the same range of values, which leads to an increase in both *the number of bits* required to represent the quantized values and *their storage requirements* in memory or on disk. Therefore, some compromise is required between the accuracy of representation and resource savings.

Note that the quantizer can be used *both with continuous-time signals*, leading to a *piecewise continuous output*, and with data that arise from *sampling*. In the latter case, the macroscopic situation is similar to the continuous case if we ignore the effects of sampling. Note that, due to the *inherent nonlinearity* that characterizes the arrangement of the quantizer, the addition of a quantizer to a linear system makes it nonlinear, while in the case where the system already has nonlinear elements, the degree of nonlinearity of the system is increasing.

The operation of the quantizer is characterized by noise to some extent, and in comparison with similar cases, *we can represent a real quantizer using an ideal quantizer without noise, as the input of which we have connected a source of noise*. This way of dealing with quantization error constitutes a *stochastic method*, in contrast to the analysis of the limit cycles of Parker and Hess (refer to page 1201), which is a deterministic method. This situation can be seen in Fig. 13.14. Note that this noise, in full analogy with the thermal noise that characterizes electrical circuits, is an *inherent characteristic* of the arrangement and *cannot* be eliminated.

In cases where the quantizer is described by the curve of Fig. 13.12, the value of its output when the input has values in the area $-\Delta/2 < x < \Delta/2$ is equal to zero. For this reason, the preceding area of values is known as a *dead zone*, in the sense that for these input values of the quantizer, its output has *zero value*. Due to this particular mode of operation, this

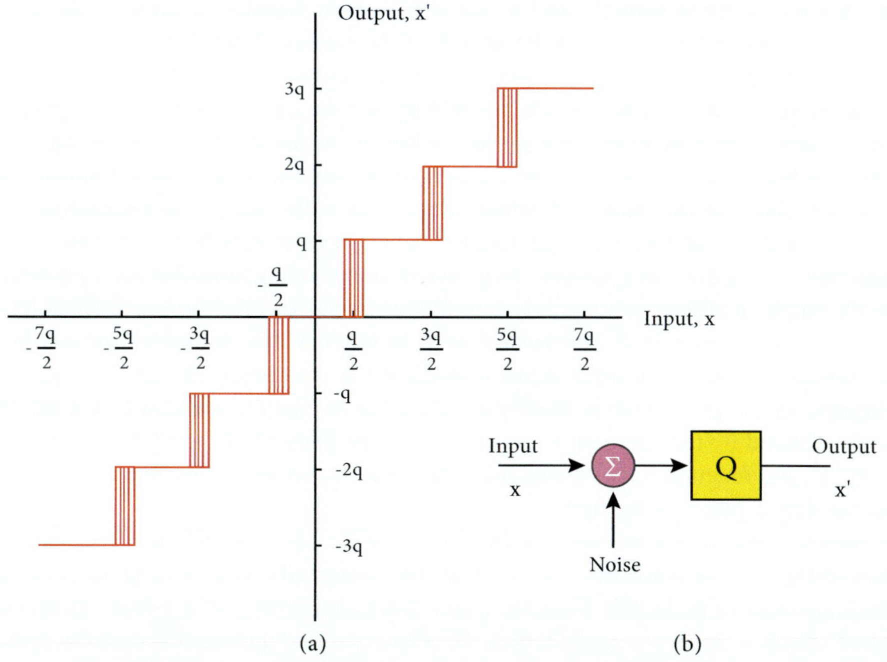

Fig. 13.14 Real quantizer characterized by the presence of noise (Widrow and Kollár, Figure 1.7)

type of quantizer (*midtread*) is used in all kinds of *measuring arrangements*. On the other hand, the quantizer described by a characteristic like that of Fig. 13.13, which *does not have* a quantization level with a zero value, is known as a *midriser* quantizer and is used in *control system applications* because its output *oscillates* when its input oscillates around the zero value, offering the possibility of creating an output with an average value equal to zero, with the help of a feedback-based controller, even in the case where the discrimination ability is limited.

A subtle distinction that should be made is that which exists between *quantization* and *numerical rounding*. This distinction is characterized as *subtle*, since in both cases, the input value is depicted in some neighboring value. In practice, we characterize as *quantization* a process in which the input to the quantizer comes from an *analog signal*, as happens in analog-to-digital conversion processes, and as *numerical rounding* a process in which the input to the quantizer is *an already quantized sample*. This situation occurs in almost every case in which numerical operations take place, mainly multiplication.

The processes of sampling and quantization can undergo *alternation* in terms of the order of their application, and consequently, the digital signal that will result from their combined use is the same, *whether sampling is first performed and then quantization or quantization first and then sampling*. Both processes lead to a *degradation* of the original signal, in the sense that they cause a *loss of information*, with the error accompanying the reconstructed signal being characterized by a *dynamic component* whose value is *proportional* to the amplitude and is due to *sampling*, as well as a static component whose value is *practically independent of the signal amplitude* and is due to *quantization*.

13.7.1 Calculation of PDFs

In the case where the samples of a continuous variable x are *random and statistically independent of each other*, the process is fully described by a first-order PDF $f_x(x)$, with the *characteristic function* $\mathcal{W}_x(u)$ defined as the Fourier transform

$$\mathcal{W}_x(u) = \int_{-\infty}^{\infty} f_x(x) e^{-jux} dx \tag{13.128}$$

Taking into account that the output variable of the quantizer takes values in terms of its amplitude from a *continuous* range of values, while the output is characterized by *discrete* amplitude values, it is not difficult to understand that the PDF $f_y(y)$ of the output $y = Q(x)$ of the quantizer will be *a train of impulse functions uniformly arranged on the horizontal axis*, which represents the *width* of the input variable. In this waveform, each impulse function is located in the center of each quantization interval, while its measure is equal to the area E of the surface defined by (a) the curve of the PDF $f_x(x)$ of the input x, (b) the horizontal axis of width, and (c) two vertical lines drawn perpendicular to the horizontal axis and at the positions of the limits of each quantization interval. The probability that the quantized value $y = Q(x)$ will coincide with any of the quantized values defined during quantization and that are known as *quantization levels* is equal to the total probability *that the quantizer input will belong to the corresponding quantization interval*. This train of impulse functions $f_y(y)$, which constitutes the PDF of the output of the quantizer, which, according to the foregoing discussion, results from the *linear sampling* of the PDF $f_x(x)$ of the input x of the quantizer, will have a Fourier transform, i.e., *characteristic function* $\Phi_y(u)$, which is a *periodic* function, as it applies to a *periodic* sequence of impulse functions that are equidistant from each other Δ.

The PDF $f_y(y)$ of the output y of the quantizer arises from the *sampling of the surface* defined *by the curve of the PDF $f_x(x)$ of the input x of the quantizer and the horizontal axis*. In other words, the quantizer can be considered a device *sampling* this surface, which receives as an input signal (considered as a sampler) the function $f_x(x)$. Consequently, each impulse function of the density $f_y(y)$ is characterized by a surface that is equal to the area E defined previously. Taking into account that the surface enclosed by the curve of $f_x(x)$ is equal to the integral of the PDF, which is simply the *cumulative distribution function* $F_x(x)$, the function $f_y(y)$ arises from the sampling of the difference $F_x(x + \Delta/2) - F_x(x - \Delta/2)$, which is performed in the way depicted in the figure.

Studying the issue from a statistical point a view, the PDF $f_x(x)$ of the input variable x of the quantizer is described by the *continuous distribution* of Fig. 13.15a, while the PDF $f_y(y)$ of the quantized output $y = Q(x)$ of the quantizer is described by the *train of impulse functions* of Fig. 13.15b. From the characteristic input/output of the quantizer, we find that if the input x belongs to the range of values $-\Delta/2 \leq x \leq \Delta/2$, then the output y of the quantizer is *equal to zero*, and according to the definition of the PDF, the probability of it being $y = 0$ is given by the integral

$$P(y = 0) = \int_{-\Delta/2}^{\Delta/2} f(x)dx \tag{13.129}$$

The PDF of the output y of the quantizer for this range of values is an *impulse function* at the position $y = 0$, whose area is equal to the value $P(y = 0)$, which is calculated from the preceding integral. Referring to the definition of an integral whose value is equal to the area of the surface defined by (a) the PDF of the input x, (b) the horizontal axis, and (c) the (vertical) lines $x = -\Delta/2$ and $x = \Delta/2$, it is not difficult to see that the value of the impulse function at the position $y = 0$ is equal to the area E_0 of the shaded strip. Similarly, the probability that the output y of the quantizer is equal to Δ, which is true, when the input belongs to the region $\Delta/2 \leq x \leq 3\Delta/2$ is equal to the area of the shaded region E_1 in Fig. 13.15a, while the corresponding PDF of the output y of the quantizer will contain for this range of values an impulse function at the position $y = \Delta$, with the corresponding surface E_1 of image (a) being defined as

$$P(y = \Delta) = \int_{\Delta/2}^{3\Delta/2} f(x)dx \tag{13.130}$$

Observe how the vertical strips with surfaces E_i $(i = 0, \pm1, \pm2, \dots)$ defined by the curve of $f_x(x)$ *compress into impulse functions*, which jointly define the PDF $f_y(y)$ of the output y. Taking into account that the events corresponding to finding the input x in one of the quantization intervals defined by the subdivisions of the horizontal axis are *foreign to each other*, that is, *mutually exclusive*, since obviously each value of x will belong to one and only one of these intervals, the PDF $f_y(y)$ will be defined as the sum

$$\begin{aligned} f_y(y) &= \cdots + \delta(y + \Delta) \int_{-3\Delta/2}^{\Delta/2} f_x(x)dx + \delta(y) \int_{-\Delta/2}^{\Delta/2} f_x(x)dx + \delta(y - \Delta) \int_{3\Delta/2}^{3\Delta/2} f_x(x)dx + \cdots \\ &= \sum_{m=-\infty}^{\infty} \delta(x - m\Delta) \int_{m\Delta-(\Delta/2)}^{m\Delta+(\Delta/2)} f_x(x)dx \end{aligned}$$

The construction of the function $f_y(y)$ is depicted in Fig. 13.16, which essentially results from the *superposition* of Fig. 13.15a, b placed on top of one another. This figure reveals that *each impulse function is associated with the area of the surface of the vertical strip to which it belongs*, a fact that justifies its multiplication by the corresponding integral (which

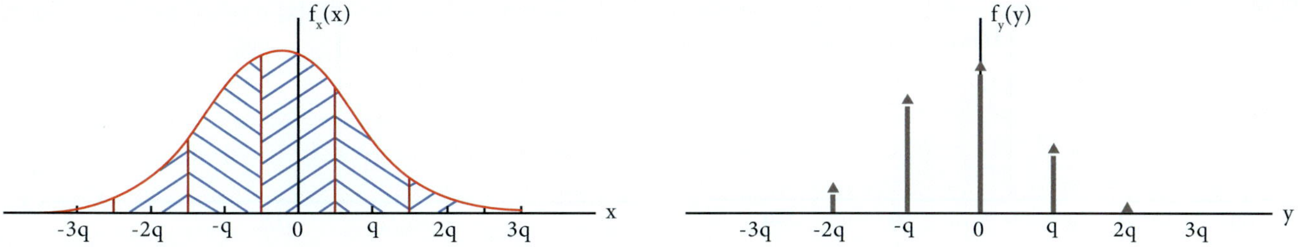

Fig. 13.15 The PDFs $f_x(x)$ [image (a)] and $f_y(y)$ [image (b)] (Widrow and Kollár, Figures 2.8γ and 2.8δ)

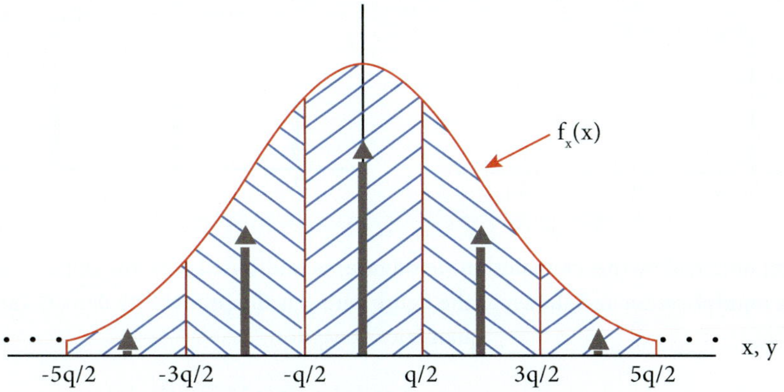

Fig. 13.16 Method of constructing the PDF $f_y(y)$ (Widrow and Kollár, Figure 4.1)

expresses this area) in the defining equation of $f_y(y)$, which results from the *sum* of these products. This process of surface sampling leading to the creation of the function $f_y(y)$ is depicted in Fig. 13.17.

Based on all the preceding considerations, we come to the conclusion that the construction of the function $f_y(y)$ from the function $f_x(x)$ does indeed resemble a *sampling process* since the continuous distribution $f_x(x)$ is depicted under the application of this function to the discrete distribution $f_y(y)$. However, it should be understood that this sampling *does not constitute* the usual sampling of values that we have encountered up to this point, since in this case the values of the function $f_y(y)$ *are not related* to the samples of $f_x(x)$, but, rather, *to the areas of the vertical strips* described previously. In other words, this sampling process indeed constitutes *surface sampling*, which, as we will now demonstrate, arises *from the convolution of the PDF* $f_x(x)$*, with a window function, followed by a regular sampling*. Note also that, although the quantization process, as noted earlier, is *nonlinear*, this surface sampling constitutes a *linear process*.

To prove the foregoing property, let us consider a *rectangular window function*, which is defined as $w(x) = 1/\Delta$ for values $-\Delta/2 < x < \Delta/2$ and zero value anywhere else and with the area defined by this function to be equal to unity, as can easily be verified. If we apply the operation of *convolution* between this window and the function $f_x(x)$, we get

$$w(x) * f_x(x) = \int_{-\infty}^{\infty} w(x-\alpha) f_x(\alpha) d\alpha = \int_{-\infty}^{\infty} \frac{1}{\Delta} f_x(\alpha) d\alpha \tag{13.131}$$

If we multiply the preceding convolution with a *train of impulse functions* defined as

$$c(x) = \sum_{n=-\infty}^{\infty} \Delta\delta(x - m\Delta)$$

the result is

$$[w(x) * f_x(x)] \cdot c(x) = \sum_{m=-\infty}^{\infty} \Delta\delta(x - m\Delta) \int_{x-(\Delta/2)}^{x+(\Delta/2)} \frac{1}{\Delta} f_x(\alpha) d\alpha = \sum_{m=-\infty}^{\infty} \delta(x - m\Delta) \int_{x-(\Delta/2)}^{x+(\Delta/2)} f_x(\alpha) d\alpha$$

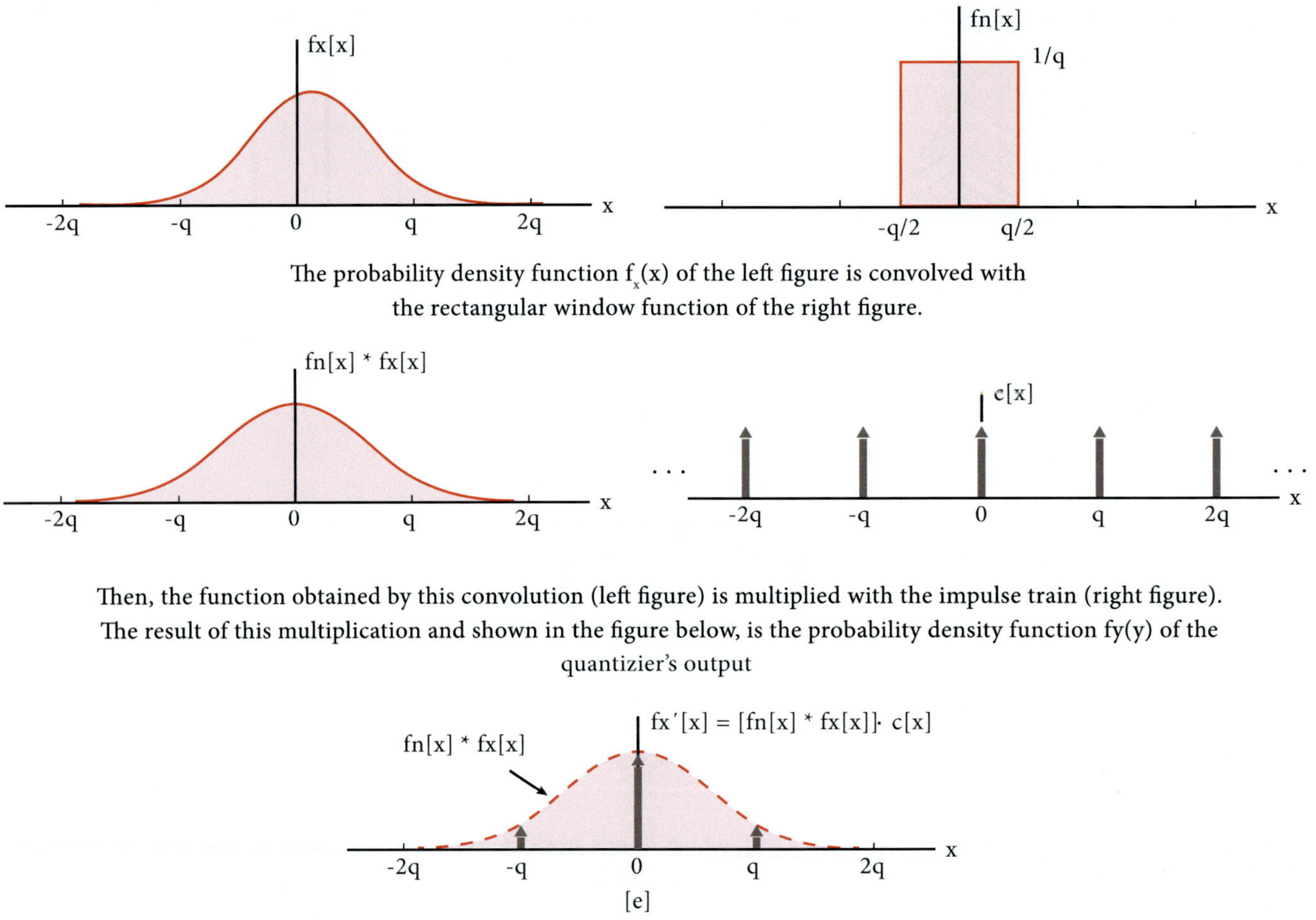

Fig. 13.17 The process of surface sampling (Widrow and Kollár, Figure 4.2)

Now referring to the property of the function $\delta(x)$ according to which

$$\delta(x - x_0) f(x) = \delta(x - x_0) f(x_0)$$

the foregoing product takes the form

$$[w(x) * f_x(x)] \cdot c(x) = \sum_{m=-\infty}^{\infty} \delta(x - m\Delta) \int_{m\Delta-(\Delta/2)}^{m\Delta+(\Delta/2)} f_x(\alpha) d\alpha$$

but the last expression is simply the function $f_y(y)$, which is the *PDF* of the quantizer's output y. Therefore, we will have

$$f_y(y) = [w(x) * f_x(x)] \cdot c(x) \tag{13.132}$$

and it is this very equation that correlates the PDFs $f_x(x)$ and $f_y(y)$. We are thus led to the conclusion that *surface sampling involves a convolution with a rectangular window function*, an operation that is followed by a *regular sampling* (i.e., *not* surface sampling).

13.7.2 Calculation of Characteristic Functions

Having determined the PDFs $f_x(x)$ and $f_y(y)$ of the input and output of the quantizer, as well as the correlation between them, let us now determine the corresponding correlation between their *characteristic* functions of these PDFs, which are

$$\Phi_x(u) = E\{e^{jux}\} = \int_{-\infty}^{\infty} f_x(x)e^{jux}dx \tag{13.133}$$

and

$$\Phi_y(u) = E\{e^{juy}\} = \cdots + e^{ju\Delta}\int_{-3\Delta/2}^{-\Delta/2} f_x(x)dx + \int_{-\Delta/2}^{\Delta/2} f_x(x)dx + e^{-ju\Delta}\int_{\Delta/2}^{3\Delta/2} f_x(x)dx + \cdots \tag{13.134}$$

which is analogous to the correlation that exists between the Fourier transform of a signal and the Fourier transform of its samples. To determine this correlation, we take the Fourier transform of Eq. (13.132). We will then have

$$\begin{aligned}\Phi_y(u) &= \mathcal{F}\{f_y(y)\} = \mathcal{F}\{[w(x) * f_x(x)]\cdot c(x)\} = \mathcal{F}\{[w(x) * f_x(x)]\} * \mathcal{F}\{c(x)\}\\ &= \left(\mathcal{F}\{w(x)\}\mathcal{F}\{f_x(x)\}\right) * \mathcal{F}\{c(x)\} = [\mathcal{W}(ju)\Phi_x(ju)] * \mathcal{C}(ju)\end{aligned}$$

where we referred to the *property of the product of the Fourier transform* and the *convolution theorem*. In the foregoing expression, the function $\Phi(ju) = \mathcal{F}\{f_x(x)\}$ is the Fourier transform of the PDF $f_x(x)$, the function $\mathcal{W}(ju)$ is the Fourier transform of the function $w(x)$ of the rectangular window, that is, a sinc function, which in this case has the form[13]

$$\begin{aligned}\mathcal{W}(ju) &= \int_{-\infty}^{\infty} w(x)e^{jux}dx = \frac{1}{\Delta}\int_{-\infty}^{\infty} e^{jux}dx = \frac{1}{ju\Delta}\Big[e^{jux}\Big]_{-\Delta/2}^{\Delta/2} = \frac{\exp(ju\Delta/2) - \exp(-ju\Delta/2)}{ju\Delta}\\ &= \frac{2}{u\Delta}\,\frac{\exp(ju\Delta/2) - \exp(-ju\Delta/2)}{2j} = \frac{\sin(u\Delta/2)}{u\Delta/2} = \operatorname{sinc}\left(\frac{u\Delta}{2}\right)\end{aligned} \tag{13.135}$$

while $\mathcal{C}(ju)$ is *the Fourier transform of the train of impulse functions*, which is calculated as (Exercise 3.3.21, page 252)

$$\mathcal{C}(ju) = \mathcal{F}\{c(x)\} = \mathcal{F}\left\{\sum_{m=-\infty}^{\infty}\Delta\delta(x - m\Delta)\right\} = \sum_{\ell=-\infty}^{\infty}\delta\left(u + \ell\frac{2\pi}{\Delta}\right) \tag{13.136}$$

Replacing the preceding expressions and referring once again to the properties of the *impulse function*, we are led to the equation

$$\Phi_y(u) = \sum_{\ell=-\infty}^{\infty}\Phi_x\left(u + \ell\frac{2\pi}{\Delta}\right)\operatorname{sinc}\left(\frac{\Delta[u + \ell(2\pi/\Delta)]}{2}\right) = \sum_{\ell=-\infty}^{\infty}\Phi_x(u + \ell\xi)\operatorname{sinc}\left(\frac{\Delta(u + \ell\xi)}{2}\right) \tag{13.137}$$

where $\xi = 2\pi/\Delta$, which is also our final result, i.e., the equation that correlates the characteristic functions $\Phi_x(u)$ and $\Phi_y(u)$ of the input and output of the quantizer. This relationship leads us to the conclusion that the characteristic function $\Phi_y(u)$ is defined *as a sum of infinite terms, each of which is a positive copy of a basic waveform that repeats periodically*, which is expected since the function $\Phi_y(u)$ is *the Fourier transform of a periodic train of impulse functions, uniformly arranged on the horizontal axis of width*. According to this expression, the variable $\xi = 2\pi/\Delta$ can be seen as a "*radial quantization frequency*" expressed in *radians per unit width* and correlated to the sampling frequency Ω, which is expressed in *radians per second*. Similarly, the quantization period Δ corresponds to the *sampling period* T. We therefore observe that between the processes of sampling and quantization there is a *complete correspondence*, a finding that *agrees* with (and actually is a consequence of) the fact that *quantization constitutes a surface sampling*.

[13] Note the difference in the sign of the exponent, which has been noted elsewhere.

13.7.3 PQN Noise Model

Let us now assume that the PDF $w(x)$ represents the *PDF* $w_n(n)$ *of a random noise signal* and, more specifically, of an *artificial noise*, which *does not* coincide with the quantization noise and for this reason is aforementioned to be *pseudo-quantization noise* (PQN). This artificial noise is *uniformly distributed* in the value range $[-\Delta/2, \Delta/2]$ (something that *does not* apply to the real quantization noise, which can have *any* distribution in the foregoing value range) is characterized by zero value and by a dispersion equal to $\Delta^2/12$ and constitutes the *ideal* case of noise, in the sense that *it has those statistical characteristics that would be desirable for real quantization noise to have*, for reasons we explain elsewhere. The conditions under which real quantization noise has such characteristics are presented later. In contrast to the real quantization noise $\nu = y - x = Q(x) - x$, which is a function of the input signal x and leads to an output $y = x + \nu$, which is characterized by a distinct PDF that takes the form of a *train of impulse functions*, the PQN is *independent of the input signal* x, with the signal $x + n$ characterized by a *continuous PDF*. What we essentially do at this point is *compare* the system of the real quantizer, that is, an ideal quantizer to which a source of real noise has been connected, with an adder, which adds to the input signal a signal of ideal noise (with uniform distribution and the aforementioned statistical properties), and then we *determine the conditions* under which the real quantization noise exhibits the characteristics of ideal noise. This comparison is particularly important because, as noted elsewhere, to simplify the study of real systems that, due to the existence of quantizers, exhibit nonlinear characteristics, we *replace the quantizers with PQN sources*. Therefore, if we determine the conditions under which the real quantization noise exhibits the statistical characteristics of PQN, then the error of the approximation is the *minimum possible*.

If the foregoing PQN signal is added to the input signal x to form the signal $x + n$, then the PDF of this signal will be

$$f_{x+n}(x) = w_n(x) * f_x(x) \tag{13.138}$$

Comparing this relationship with Eq. (13.132), which gives the probability density of the quantizer's output y, it turns out that

$$f_y(y) = f_{x+n}(x)c(x)$$

and since the multiplication of the function $f_{x+n}(x)$ with the periodic train of impulse functions $c(x)$ is simply *the sampling of those function* with a sampling period T equal to the period of the train $c(x)$, we come to the conclusion that the function $f_y(y)$ consists of samples of the density of the signal $x + n$.

The characteristic function of the adder's output $x + n$ is given by the equation

$$\Phi_{x+n}(u) = \Phi_x(u)\Phi_n(u) = \Phi_x(u)\text{sinc}\left(\frac{u\Delta}{2}\right) \tag{13.139}$$

and is obviously an *aperiodic function*, which coincides with the function $\Phi_y(u)$ for $n = 0$ and constitutes the basic waveform that is repeated to form its graph.

13.7.4 Formulating the First Theorem and the Second Quantization

The careful reader will have already perceived the close analogy that exists between this situation and that which characterizes the sampling process, something that is certainly expected, since, as we have shown, *quantization constitutes a special type of sampling*. Consequently, we can formulate, for the case of quantization, conclusions similar to those we formulated for the case of sampling. For example, we can formulate the so-called *quantization theorem*, which is *analogous* to the *Nyquist sampling theorem*, according to which, if the characteristic function of the input signal $\Phi_x(u)$ is characterized by *limited bandwidth*, that is, it is equal to zero for values $|u| > \pi/\Delta$, then repeated copies of it that appear in the characteristic function $\Phi_y(u)$ *do not overlap* with each other, a fact that allows for the *reconstruction* of the function $\Phi_x(u)$ from the function $\Phi_y(u)$, as well as the PDF $f_x(x)$ from the density $f_y(y)$. Note that in order for this quantization theorem to be

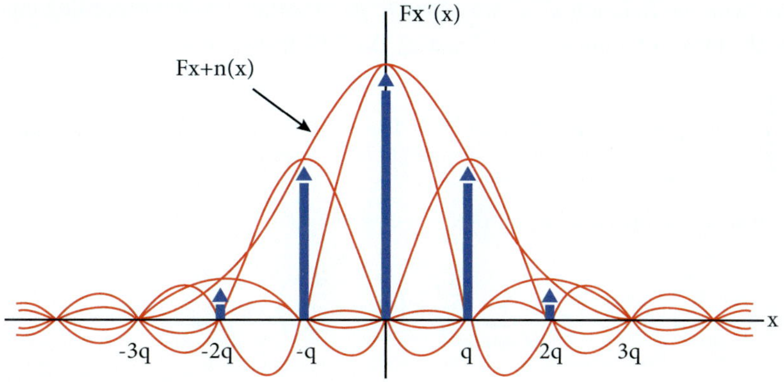

Fig. 13.18 Reconstruction of PDF $f_x(x)$ of input x of quantizer, from PDF $f_y(y)$ of output of quantizer $y = Q(x)$

satisfied, *the "frequency" of quantization must be greater than twice the largest "frequency component" of the function* $\Phi_x(u)$, just as it applies to the case of sampling, and in order for this to happen, the step Δ of quantization must be *quite small* since the smaller this value is, the greater is the "quantization frequency" and the smaller is the degree of overlap. It is interesting that the condition imposed by this first theorem is *sufficient but not necessary*, in the sense that there are input signals for which the distribution of the quantization error is *uniform*, without these signals being characterized by a limited bandwidth, such as, for example, an input with a PDF $f_x(x) = 1/\Delta$ for $-\Delta/2 \leq x \leq \Delta/2$ and with zero value anywhere else and with a characteristic function $\Phi_x(u) = \text{sinc}(\Delta u/2)$. Let it also be noted that if $\Phi_x(u) = 0$ for $|u| > 2\pi/\Delta$, then it is also possible to calculate the moments of the input x from the moments of the output y, a claim that can be considered a *second quantization theorem*, which accompanies the first. This second theorem characterizes *exclusively* the process of quantization, in the sense that *there is no analogous theorem* for the sampling process, as is the case with the first theorem.

In full analogy with the theory of reconstructing a signal from its samples, if the repeated copies included in the curve of the characteristic function $\Phi_y(u)$ *do not overlap* with each other, we can recover the PDF $f_x(x)$ of the quantizer input, that is, of the original nonquantized signal x, from the PDF $f_y(y)$ of the quantizer output, that is, of the quantized signal $y = Q(x)$, provided, of course, we have at our disposal the *statistical properties* of the original data. This reconstruction process is based on the known process of *interpolation*, through the use of the sinc function, depicted in Fig. 13.18, which is the equivalent of Fig. 6.5 on page 513 that demonstrates the reconstruction of an analog signal from its samples, for the case of conventional sampling. Taking into account that $f_{x+n}(x) = f_n(x) * f_x(x)$, it is not difficult to understand how the determination of the density $f_x(x)$ is carried out through the *deconvolution* of $f_{x+n}(x)$ and $f_n(x)$, which can be carried out *in both the time and frequency domains*. For a diagrammatic procedure of calculating the PDF $f_x(x)$ see Widrow and Kollár, pages 71–74.

13.7.5 Characteristic Function and Probability Density Function of Quantization Noise

The PDF of the actual quantization noise is determined by adding the probabilities of the value of the quantizer's input x belonging to each of the *mutually exclusive and infinitely many* quantization intervals of length Δ, into which the horizontal axis of the amplitude is divided. These events, treated as sets, are mutually exclusive, and therefore, the probabilities are added. It will therefore be the case that

$$f_\nu(x) = \sum_{m=-\infty}^{\infty} f_x(-x + m\Delta) \tag{13.140}$$

for the range of values $-\Delta/2 \leq x \leq \Delta/2$ and $f_\nu(x) = 0$ in any other area. On the other hand, the *characteristic function of real quantization noise* arises by taking the Fourier transform of the preceding expression. To do this, we note that if we replace ν with $-\nu = x - y$, the preceding relationship is written

$$f_{-\nu}(x) = w(x) \sum_{m=-\infty}^{\infty} f_x(x + m\Delta) \tag{13.141}$$

where multiplication by the window function $w(x)$ was done to incorporate into the preceding equation the information that this expression is valid for the range of values $-\Delta/2 \leq x \leq \Delta/2$. Consider now

$$\mathcal{F}\left\{\sum_{m=-\infty}^{\infty} f_x(x+m\Delta)\right\} = \Phi_x(u)(\cdots + e^{j2\Delta u} + e^{j\Delta u} + 1 + e^{-j\Delta u} + e^{-j2\Delta u} + \cdots) = \Phi_x(u)\sum_{m=-\infty}^{\infty} e^{-jm\Delta u} \quad (13.142)$$

Comparing this relationship with the previous equations, we observe that

$$\sum_{m=-\infty}^{\infty} e^{-jm\Delta u} = \sum_{\ell=-\infty}^{\infty} \frac{2\pi}{u}\delta\left(u - \ell\frac{2\pi}{\Delta}\right) = \sum_{\ell=-\infty}^{\infty} \xi\delta(u-\ell\xi) \quad (13.143)$$

Thus, we will have

$$\mathcal{F}\left\{\sum_{m=-\infty}^{\infty} f_x(x+m\Delta)\right\} = \Phi_x(u)\sum_{\ell=-\infty}^{\infty} \xi\delta(u-\ell\xi) \quad (13.144)$$

and taking into account that $\mathcal{F}\{f_x(-x)\} = \Phi_x(-u)$ we will have

$$\mathcal{F}\left\{\sum_{m=-\infty}^{\infty} f_x(-x+m\Delta)\right\} = \Phi_x(u)\sum_{\ell=-\infty}^{\infty} \xi\delta(-u-\ell\xi) \quad (13.145)$$

Calculating the convolution between the preceding function and the Fourier transform of the rectangular window

$$\mathcal{W}(u) = \mathcal{F}\{w(x)\} = \Delta\text{sinc}\left(\frac{\Delta u}{2}\right) \quad (13.146)$$

and dividing by 2π, as required by the properties of the Fourier transform, we will have

$$\begin{aligned}\mathcal{W}_\nu(u) &= \frac{1}{2\pi}\Big(\Phi_x(u)\sum_{m=-\infty}^{\infty}\xi\delta(-u-\ell\xi)\Big) * \Delta\text{sinc}\left(\frac{u\Delta}{2}\right) = \frac{1}{2\pi}\int_{-\infty}^{\infty}\Big(\Phi_x(-\lambda)\sum_{\ell=-\infty}^{\infty}\xi\delta(-\lambda-\ell\xi)\Big)\Delta\text{sinc}\left[\frac{\Delta(u-\lambda)}{2}\right]d\lambda \\ &= \sum_{\ell=-\infty}^{\infty}\Phi(\lambda\xi)\text{sinc}\left[\frac{\Delta(u+\lambda\xi)}{2}\right] = \sum_{\ell=-\infty}^{\infty}\Phi(-\lambda\xi)\text{sinc}\left[\frac{\Delta(u-\lambda\xi)}{2}\right]\end{aligned} \quad (13.147)$$

an expression that constitutes the *characteristic function of the quantization noise*. By reversing the preceding expression, we can construct an alternative equation for the probability density of quantization noise, which is

$$f_\nu(\nu) = f_n(\nu)\sum_{\ell=-\infty}^{\infty}\Phi_x(\ell\xi)e^{j\xi\nu} = f_n(\nu) + f_n(\nu)\sum_{\substack{\ell=-\infty \\ \ell\neq 0}}^{\infty}\Phi_x(\ell\xi)e^{j\xi\nu} \quad (13.148)$$

which constitutes the *Fourier series representation* of the probability density of quantization noise for the range of values $\Delta/2 \leq \nu \leq \Delta/2$.

13.7.6 Formulating the Third Theorem of Quantization

The foregoing relationship correlates the PDF $f_\nu(\nu)$ of the PQN with the PDF $f_n(\nu)$ and is particularly important for the following reason. According to our earlier discussion, the *PQN constitutes an ideal noise, in the sense that it has the highly desirable characteristics of zero mean value and uniform distribution in the range of values* $[-\Delta/2, \Delta/2]$, *in contrast to real quantization noise, which can have any distribution in the foregoing range of values*. The question that arises at this point is this: *Under what conditions is it possible for the real quantization noise to be described by a uniform distribution in the preceding range, a feature that would be particularly interesting and desirable*? For this to happen, it must be that $f_\nu(\nu) = f_n(\nu)$, and from the preceding relationship it follows that this must be

$$\sum_{\substack{\ell=-\infty \\ \ell\neq 0}}^{\infty} \Phi_x(\ell\xi)e^{j\xi\nu} = 0 \tag{13.149}$$

However, since the exponential functions that appear in the foregoing sum are *orthogonal to each other*, this can take a zero value if the condition $\Phi_x(\ell\xi) = 0$ is satisfied for the values $\ell = \pm 1, \pm 2, \ldots$. It is proven that this condition is a *sufficient and necessary condition*, which ensures that the actual quantization noise will be *uniformly distributed* in the value range $[-\Delta/2, \Delta/2]$ with zero mean value, dispersion value $\Delta^2/12$, and characteristic function

$$\Phi_\nu(u) = \text{sinc}\left(\frac{\Delta u}{2}\right) \tag{13.150}$$

which has *one and only one maximum* corresponding to the value $\ell = 0$, and probability density, with the value $f_\nu(x) = 1/\Delta$ for the value range $|x| \leq \Delta/2$, and $f_\nu(x) = 0$ in any other area. This very important conclusion was formulated by Stripad and Snyder in 1977 and can be considered a *third quantization theorem*, which is even *weaker* in relation to the two previous theorems. This means that if the conditions for satisfying the first quantization theorem are met, then it is proven that the second quantization theorem is automatically satisfied, while if the conditions for satisfying the second quantization theorem are met, then it is proven that the third quantization theorem is automatically satisfied. We are therefore led to the conclusion that *if the conditions of the preceding quantization theorems are met, then the actual quantization noise is a uniformly distributed noise in the value range* $[-\Delta/2, \Delta/2]$ (with the condition of the third quantization theorem being sufficient and necessary), so it will hold that $f_\nu(\nu) = f_n(\nu)$, while in the opposite case, in which these conditions are not absolutely met, the actual noise *will not be perfectly uniformly distributed*, with the deviation from the ideal behavior corresponding to the uniform distribution, which is described by the sum of the right-hand member of Eq. (13.148).

13.7.7 Statistical Properties of Quantization Noise

A second characteristic that is desirable for quantization noise to have, apart from its uniform distribution in the appropriate range of values, is its characterization as *white noise* (Sect. 13.6, page 1439). Since the definition of noise involves the concept of *correlation*, the formulation of the conditions under which quantization noise is *white noise* requires the use of second-order moments that arise from the direct generalization of first-order moments. Therefore, by extending the PDF of the quantization error to two dimensions, that is, expressing it as a function of two variables, it is proven that it is given by the equation

$$f_{\nu_1,\nu_2}(\nu_1,\nu_2) = \frac{1}{\Delta^2} + \frac{1}{\Delta^2}\sum_{\substack{\ell=-\infty \\ \ell\neq 0}}^{\infty}\sum_{\substack{k=-\infty \\ k\neq 0}}^{\infty} \Phi_{x_1,x_2}\left(\frac{2\pi\ell}{\Delta}, \frac{2\pi k}{\Delta}\right)\exp\left[-j\frac{2\pi}{\Delta}(\ell\nu_1 + k\nu_2)\right] \tag{13.151}$$

for the range of values $-\Delta/2 \leq E_i \leq \Delta/2$ $(i = 1, 2)$, and $f_{\nu_1,\nu_2}(\nu_1,\nu_2) = 0$ in any other case. The preceding relation is proven as was done for the one-dimensional case, and the full proof along with the whole theory related to the application of the aforementioned concepts in two dimensions can be found in the literature (e.g., Chapters 8, 9, and 10 in the book by Widrow and Kollár). Although the analytical description of this theory is impossible due to the limited volume of the textbook, let us at least formulate the three quantization theorems for the case of two variables, with the generalization to N variables being direct. According to these theorems, if the *combined characteristic function* $\Phi_{x_1,x_2}(x_1,x_2)$ is *of finite bandwidth*, that is, it satisfies the condition $\Phi_{x_1,x_2}(x_1,x_2) = 0$ for $|x_i| > \xi_i$ $(i = 1, 2)$ (for the first theorem) or $\Phi_{x_1,x_2}(x_1,x_2) = 0$ for $|x_i| > \xi_i$ $(i = 1, 2)$ (for the second theorem), then the characteristic function of the quantization noise has the form

$$\Phi_{\nu_1,\nu_2}(x_1, x_2) = \operatorname{sinc}\left(\frac{\Delta_1 x_1}{2}\right) \operatorname{sinc}\left(\frac{\Delta_2 x_2}{2}\right) \tag{13.152}$$

(where Δ_1 and Δ_2 are the *quantization steps* in two dimensions), and since the foregoing characteristic is expressed as the product of the characteristic functions of the variables ν_1 and ν_2, these variables are *independent of each other and uniformly distributed* in the range of values $[-\Delta_i/2, \Delta_i/2]$ $(i = 1, 2)$. On the other hand, the sufficient and necessary condition of Stripad and Snyder for the case of *two dimensions* tells us that the *combined* characteristic function of the random input variables x_1 and x_2 satisfies the condition

$$\Phi_{x_1,x_2}\left(\frac{2\pi\ell}{\Delta}, \frac{2\pi k}{\Delta}\right) = 0 \tag{13.153}$$

for every $\ell \neq 0$ and for every $k \neq 0$ if and only if the combined probability density of the noise has the form $f_{\nu_1,\nu_2} = 1/\Delta^2$ for the range of values $[-\Delta_i/2, \Delta_i/2]$ $(i = 1, 2)$, and a zero value in any other case. In this case, the random variables ν_1 and ν_2 are *statistically independent and uniformly distributed* in the foregoing regions, satisfying the property $f_{\nu_1,\nu_2}(\nu_1, \nu_2) = f_{\nu_1}(\nu_1) f_{\nu_2}(\nu_2)$ and therefore constitute *white noise*. Several distributions satisfy the preceding conditions, such as (a) any *uniform distribution* in the range of values $[-n\Delta/2, n\Delta/2]$ for any positive integer value n (recall that this distribution does not satisfy the first theorem of quantization, and therefore, the condition of Stripad and Snyder is indeed weaker than this theorem), (b) any number of *successive convolutions* between distributions of this type, (c) any *convex combination* of the distributions of the second category (i.e., a linear combination whose coefficients are nonnegative and have a sum equal to one), and (d) the *result of the convolution* between any probability density and in any of the densities of the three aforementioned types. It is worth noting that the distributions of the fourth category are particularly important for the following reason. Let us consider a discrete-time process $x[n]$, which can be formulated as a sum of the form $y[n] + d[n]$, with the processes $y[n]$ and $d[n]$ being *independent* of each other and the process $d[n]$ being characterized by *uniform distribution*, with samples uniformly distributed and with *marginal density* belonging to any of the types (a), (b), and (c) mentioned earlier, so in fact the process $x[n]$ is characterized by a distribution that belongs to type (d). In this case, we can consider the process $y[n]$ the signal on which we wish to apply the process of quantization and the process $d[n]$ as a "*dithering signal*," which is added to the signal $y[n]$ to impart to the quantization noise characteristics of *white noise*. In other words, if we add to the signal $y[n]$ the signal $d[n]$ and then proceed to the quantization of the sum $x[n] = y[n] + d[n]$, then the resulting quantization noise will be *white, uniformly distributed, and independent of the signal* $y[n]$. However, on the other hand, this approach *leads to a reduction in the range of signal values that fall within the operating range of the quantizer*, since now the value that will be used will be, not $y[n]$, but $x[n] = y[n] + d[n]$. In order for this increased value to be able to undergo the process of quantization, the maximum value of the signal $y[n]$ must necessarily be *smaller*. Moreover, this process leads to an increase in the power of the noise at the output and, consequently, to a reduction in the value of the signal-to-noise ratio (SNR), which can sometimes, but not always, be dealt with.

On the other hand, a very well-known and widely used category of signals that do not satisfy the conditions of the theorems of quantization are those characterized by a *Gaussian distribution*, since the characteristic function that describes them *is not characterized by a finite bandwidth*, as required by these theorems, and in reality, *none of the conditions* required by the theorems of quantization are satisfied. However, because the characteristic function *decreases at a rapid rate*, the theorems of quantization are satisfied even *approximately* for a wide range of quantization step values Δ. The significant parameter for the study of these signals is the ratio Δ/σ, where σ is the *standard deviation* of the input signal. For a statistical description of the quantized signals that follow a Gaussian distribution and the construction of the equations of the characteristic function, of the moments of quantization noise, and of the covariance between the input and the quantization noise for the one-dimensional and two-dimensional cases, refer to Chapter 11 of Widrow and Kollár's book (pages 225–252).

In cases where the conditions of the quantization theorems are satisfied, and therefore the quantization noise is characterized by a *uniform distribution* in the region $[-\Delta/2, \Delta/2]$, it automatically has all the characteristics associated with this distribution, while in the opposite case, the noise *is not white, but colored*. For example, the odd-order moments are all equal to zero, the second-order moment is equal to $\Delta^2/12$, the fourth-order moment is equal to $\Delta^4/80$, and so forth. Conversely, if the conditions of the foregoing theorems *are not satisfied*, then the noise *is not* characterized by a uniform distribution, and therefore, the foregoing moments must be calculated *through derivatives of the characteristic function of the quantization noise*, since it is known that

$$E\{x^k\} = \frac{1}{j^k}\frac{d^k\Phi_k(u)}{du^k}\bigg|_{u=0}$$

Using the preceding expression for the values $k = 1$ and $k = 2$, the first- and second-order moments $E\{\nu\}$ and $E\{\nu^2\}$ of the quantization noise are calculated as[14]

$$E\{\nu\} = \frac{1}{j}\sum_{\ell=-\infty}^{\infty}\Phi_x(\ell\xi)\left[\frac{d}{du}\mathrm{sinc}\left(\frac{\Delta(u+\ell\xi)}{2}\right)\right]_{u=0} = \sum_{\substack{\ell=-\infty\\ \ell\neq 0}}^{\infty}\frac{\Phi_x(\ell\xi)}{j}\frac{\Delta}{2}\frac{(-1)^\ell}{\ell\pi} \tag{13.155}$$

$$= \frac{\Delta}{2\pi}\sum_{\substack{\ell=-\infty\\ \ell\neq 0}}^{\infty}\frac{\Phi_x(\ell\xi)}{j}\frac{(-1)^\ell}{\ell} = 0 + \frac{\Delta}{\pi}\sum_{\ell=1}^{\infty}\Im\{\Phi_x(\ell\xi)\}\frac{(-1)^\ell}{\ell}$$

and

$$E\{\nu^2\} = \frac{1}{j^2}\sum_{\ell=-\infty}^{\infty}\Phi_x(\ell\xi)\frac{d^2}{du^2}\left[\mathrm{sinc}\left(\frac{\Delta u}{2}+\ell\pi\right)\right]_{u=0} = -\Phi_x(0)\left(\frac{\Delta}{2}\right)^2\left(-\frac{1}{3}\right) - \sum_{\substack{\ell=-\infty\\ \ell\neq 0}}^{\infty}\Phi_x(\ell\xi)\left(\frac{\Delta}{2}\right)^2\frac{2(-1)^{\ell+1}}{\ell^2\pi^2} \tag{13.156}$$

$$= \frac{\Delta^2}{12} + \frac{\Delta^2}{2\pi^2}\sum_{\substack{\ell=-\infty\\ \ell\neq 0}}^{\infty}\Phi_x(\ell\xi)\frac{(-1)^\ell}{\ell^2} = \frac{\Delta^2}{12} + \left(\frac{2^2}{\pi^2}\sum_{\ell=1}^{\infty}\Re\{\Phi_x(\ell\xi)\}\frac{(-1)^\ell}{\ell^2}\right)$$

The preceding relations can alternatively arise by applying the operator of the *expected value* (and then referring to the definition of the *characteristic function*) in the equations

$$\nu = \nu(x) = \frac{\Delta}{j2\pi}\sum_{\substack{\ell=-\infty\\ \ell\neq 0}}^{\infty}\frac{(-1)^\ell}{\ell}\exp\left(j\frac{2\pi\ell x}{\Delta}\right) \quad \text{and} \quad \nu^2 = \nu^2(x) = \frac{\Delta^2}{12} + \frac{\Delta^2}{2\pi^2}\sum_{\substack{\ell=-\infty\\ \ell\neq 0}}^{\infty}\frac{(-1)^\ell}{\ell^2}\exp\left(j\frac{2\pi\ell x}{\Delta}\right)$$

of which the first constitutes the *Fourier series expansion* of the periodic quantization error, while the second equation arises from the first (Wong 1990, page 288).

From the two foregoing moments, the *variability of the quantization noise* is calculated as

$$\mathrm{Var}\{\nu\} = E\{\nu^2\} - (E\{\nu\})^2 = \frac{\Delta^2}{2} + \frac{\Delta^2}{2}\sum_{\substack{\ell=-\infty\\ \ell\neq 0}}^{\infty}\Phi_x(\ell\xi)\frac{(-1)^\ell}{(\pi\ell)^2} - \left(\frac{\Delta}{2\pi}\sum_{\substack{\ell=-\infty\\ \ell\neq 0}}^{\infty}\frac{\Phi_x(\ell\xi)}{j}\frac{(-1)^\ell}{\ell}\right)^2$$

---

[14] Regarding Eq. (13.155), Wong comments that because the function $e(x)$ is discontinuous at the points $u = (2k+1)\Delta/2$ for every integer value k, we should add to this expression a *corrective term* of the form

$$\frac{\Delta}{2}\sum_{k=-\infty}^{\infty}\mathbf{1}_{\{u=(2k+1)\Delta/2\}} \tag{13.154}$$

where $\mathbf{1}_A$ is a function that takes a value of 1 when condition A is satisfied and a value of 0 in the opposite case. However, if the input signal x is a signal with *continuous* width, that is, it is characterized by a *continuous probability density*, then the probability that x is exactly equal to $(2k+1)\Delta/2$ is *exactly zero* for all values of k, so this correction term is not needed. Because in the theory developed here, the input signals are characterized by such distributions, the use of this term is not necessary.

$$= \frac{\Delta^2}{12} + \left(\frac{\Delta^2}{\pi^2} \sum_{\ell=1}^{\infty} \Re\{\Phi_x(\ell\xi)\} \frac{(-1)^\ell}{\ell^2} - \left(\frac{\Delta}{\pi} \sum_{\ell=1}^{\infty} \Im\{\Phi_x(\ell\xi)\} \frac{(-1)^\ell}{\ell^2} \right)^2 \right) \tag{13.157}$$

while the *autocorrelation function* has the form (defining $\nu_1 = \nu(x[n])$ and $\nu_2 = \nu(x[n+m])$)

$$E\{\nu_1, \nu_2\} = \frac{\Delta^2}{4\pi^2} \sum_{\substack{k=-\infty \\ k\neq 0}}^{\infty} \sum_{\substack{\ell=-\infty \\ \ell\neq 0}}^{\infty} \frac{(-1)^{k+\ell+1}}{k\ell} \Phi_{\nu_1,\nu_2}\left(\frac{2\pi k}{\Delta}, \frac{2\pi \ell}{\Delta} \right) \tag{13.158}$$

Higher-order moments are calculated in the same way, and the reader is invited to calculate the moments $E\{\nu^3\}$ and $E\{\nu^4\}$ as an exercise (see Problems 14 and 15 on page 1462). Note that these moments constitute *the sum of two terms*, of which the first refers to *the value of the moment of uniform noise* and the second to *the deviation from this value*, which appears when the noise *is not characterized* by a uniform distribution. Conversely, if the conditions of the quantization theorems are met and the noise is characterized by a uniform distribution, then the second term takes a value equal to zero and we end up with the known value of the uniform distribution.

Finally, the equations of *cross-correlation* and *covariance* that exist between the quantization noise ν, the input signal x, and the output $y = \nu + x$ of the quantizer are interesting. For reasons of economy, we only present the final results; detailed proofs can be found in Chapter 6 of the book by Widrow and Kollár with reference to Appendix B of the aforementioned textbook.

The *cross-correlation* between the signals x and ν has the form

$$E\{x\nu\} = \frac{\Delta}{\pi} \sum_{\ell=1}^{\infty} \Re\{\dot{\Phi}_x(\ell\xi)\} \frac{(-1)^{\ell+1}}{\ell} \tag{13.159}$$

If the conditions of the second quantization theorem are met, it will hold that $E\{x\nu\} = 0$.

The *covariance* between the signals x and ν is given by the expression

$$\text{cov}\{x, \nu\} = E\{x\nu\} - E\{x\}E\{\nu\} = \frac{\Delta}{\pi} \sum_{\ell=1}^{\infty} \Re\{\dot{\Phi}_x(\ell\xi)\} \frac{(-1)^{\ell+1}}{\ell} - E\{x\} \frac{\Delta}{\pi} \sum_{\ell=1}^{\infty} \Im\{\Phi_x(\ell\xi)\} \frac{(-1)^\ell}{\ell} \tag{13.160}$$

If the conditions of the second quantization theorem are satisfied, then $\text{cov}\{x, \nu\} = 0$.

The *cross-correlation* between the signals y and ν takes the form

$$E\{\nu y\} = \frac{\Delta^2}{12} + \frac{\Delta}{\pi} \sum_{\ell=1}^{\infty} \Re\{\dot{\Phi}_x(\ell\xi)\} \frac{(-1)^{\ell+1}}{\ell} + \frac{\Delta^2}{\pi^2} \sum_{\ell=1}^{\infty} \Re\{\Phi_x(\ell\xi)\} \frac{(-1)^\ell}{\ell} \tag{13.161}$$

If the conditions of the second quantization theorem are satisfied, then $E\{\nu y\} = \Delta^2/12$.

The *covariance* between the signals y and ν is given by the expression

$$\begin{aligned} \text{cov}\{\nu, y\} = E\{\nu y\} - E\{\nu\}E\{y\} &= \frac{\Delta^2}{12} + \frac{\Delta}{\pi} \sum_{\ell=1}^{\infty} \Re\{\dot{\Phi}_x(\ell\xi)\} \frac{(-1)^{\ell+1}}{\ell} \\ &+ \frac{\Delta^2}{\pi^2} \sum_{\ell=1}^{\infty} \Re\{\Phi_x(\ell\xi)\} \frac{(-1)^\ell}{\ell} - \left(\frac{\Delta}{\pi} \sum_{\ell=1}^{\infty} \Im\{\Phi_x(\ell\xi)\} \frac{(-1)^\ell}{\ell} \right) \left(E\{x\} + \frac{\Delta}{\pi} \sum_{\ell=1}^{\infty} \Im\{\Phi_x(\ell\xi)\} \frac{(-1)^\ell}{\ell} \right) \end{aligned} \tag{13.162}$$

If the conditions of the second quantization theorem are satisfied, then $\text{cov}\{\nu, y\} = \Delta^2/12$.

The *cross-correlation* between the signals x and y takes the form

$$E\{xy\} = E\{x(x+\nu)\} = E\{x^2\} + E\{x\nu\} = E\{x^2\} + \frac{\Delta}{\pi} \sum_{\ell=1}^{\infty} \Re\{\dot{\Phi}_x(\ell\xi)\} \frac{(-1)^{\ell+1}}{\ell} \tag{13.163}$$

If the conditions of the second quantization theorem are satisfied, it will hold that $E\{xy\} = E\{x^2\}$.

Finally, the *covariance* between signals x and y is given by the expression

$$\mathrm{cov}\{x, y\& = E\{xy\} - E\{x\}E\{y\} = \mathrm{var}\{x\} + \frac{\Delta}{\pi}\sum_{\ell=1}^{\infty}\Re\{\dot{\Phi}_x(\ell\xi)\}\frac{(-1)^{\ell+1}}{\ell} - E\{x\}\frac{\Delta}{\pi}\sum_{\ell=1}^{\infty}\Im\{\Phi_x(\ell\xi)\}\frac{(-1)^{\ell}}{\ell} \tag{13.164}$$

If the conditions of the second quantization theorem are satisfied, it will hold that $\mathrm{cov}\{x, y\} = \mathrm{var}\{x\}$.

Let us conclude this section on the statistical study of quantization error, with the observation that Widrow's theory, which models this error as an *additive random variable*, along with the theorems that accompany it, apply *only in the case of uniform quantizers*, in which both the quantization step Δ (i.e., the length of the subdivisions of the horizontal axis of the quantizer's input/output characteristic) and the distance Q between successive quantization levels (i.e., the length of the subdivisions of the vertical axis) have *constant values* (and, as we saw, *not necessarily equal to each other*, although this is usually the case). In addition, from the study of the equations that allow for the calculation of the statistical properties of quantization noise, we find that these include *sums of infinite terms*, with the number of terms that need to be calculated for finding the appropriate value in each case being *quite large*, while additionally, the use of a finite number of such terms leads to the appearance of *error*, which must be added to the error due to the *finite length* of the registers in which the sample values are stored. For this reason, after the initial theory of Widrow, efforts were made to describe the statistical properties in such a way that they apply not only to uniform quantizers but also to the general case of quantizers and allow the calculation of the values of the statistical properties using not infinite but finite sums, such as the results of Pacut & Hejn, who *did not use the characteristic function* but relied on *analytical* methods, while the theory they formulated finds application *not only in stochastic but also in deterministic signals*. More details on these issues can be found in the literature.[15]

13.8 Floating-Point Quantizers

As was already mentioned, the representation of numerical values according to the standard of *floating point* allows for the use of a *very large dynamic range* using a *relatively small* number of bits. Due to this wide use of this representation, it was expected that quantizers would emerge that operate in the way we described in previous sections, but with the quantized value that appears as their output *not being a fixed-point number, as previously, but rather a floating-point number*. It is not difficult to see how the differences that exist between these different forms of representation will lead to variations in the operation of the quantizers that use them. The two most important differences between these two types of quantizers are that *in contrast to fixed-point quantizers in which quantization is uniform, with the distance between any two successive quantization levels being constant and equal to* Δ, *while the quantization error takes values in the range* $[-\Delta/2, \Delta/2]$, *in floating-point quantizers, quantization is not uniform, while the error quantization is not limited to the foregoing range but, in reality, is proportional to the amplitude of the signal undergoing quantization.*

Both of these differences originate from the properties of floating-point representation, specifically from the fact that the number of representable floating-point numbers between two successive powers of two remains constant. As a result, the spacing between successive numbers increases with each increment in the exponent EE of the representation. This is something that was presented in the section on floating-point representation (Sect. 12.5, page 1135), where we saw that if we try to enumerate the natural numbers, there are numbers that *cannot be represented* and do not appear at all, with the final result depending on the number of bits used to represent the coefficient M. As we noted there, if p is the value of this number, then the distance between two successive cycles is equal to 2^p, while if the minimum step change is equal to Δ (and not 1 as in this example), this distance is $D = 2^p\Delta$. Taking into account that for the enumeration of natural numbers encoded with a floating point we used $p = 4$ bits to accompany the value of the coefficient M, we will have $D = 16\Delta$.

The preceding characteristic of the floating-point representation dramatically differentiates the behavior and characteristics of the operation of a floating-point quantizer in relation to the uniform quantizer of constant subtraction. Indeed, from the moment when the numbers that appear as the output of the floating-point quantizer *not only are not equal to each other*, but the distances between two such numbers *increase as their size increases*, uniform quantization *is not possible*, while the

[15] Hejn and Pacut, "Generalized model of the quantization error - a unified approach," *IEEE Transactions on Instrumentation and Measurement*, Vol. 45, No. 1, pp. 41–44, Feb. 1996; and Pacut and Hejn, "Analog-to-digital converters: Towards a generalization of Widrow's theorem," *Proceedings of the IEEE Instrumentation and Measurement Technology Conference*, 2:1190–1197, Vol. 2.

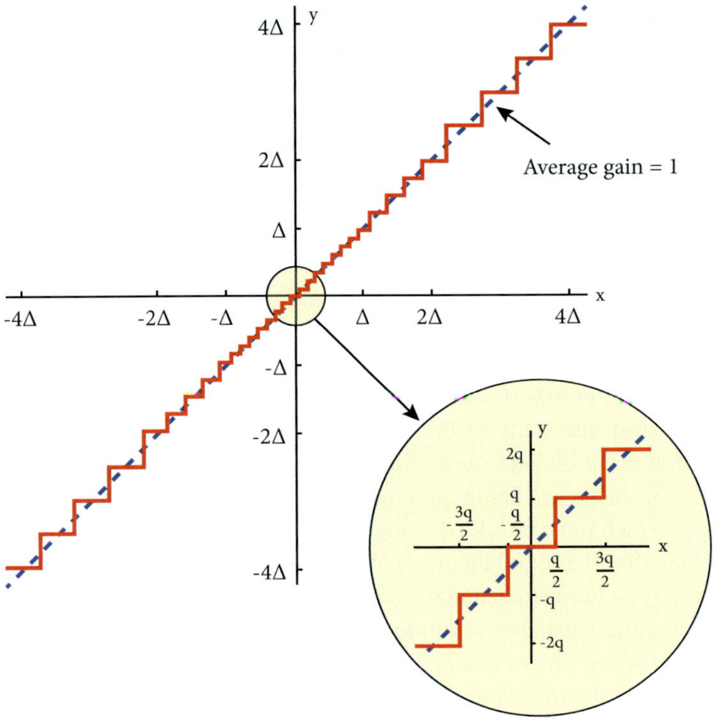

Fig. 13.19 Typical input/output characteristic of a moving sub-dial quantizer (Widrow and Kollár, Figure 12.3)

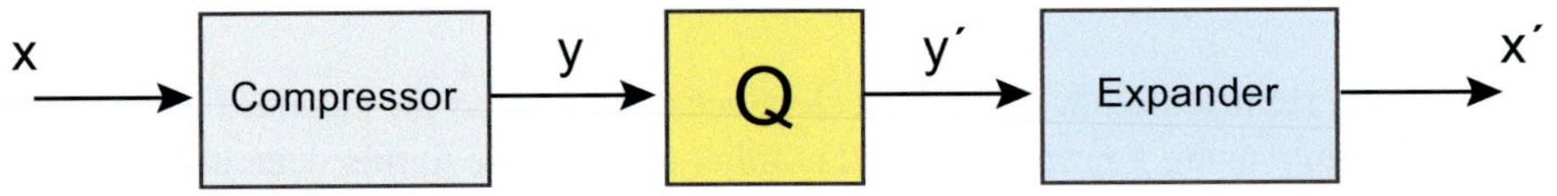

Fig. 13.20 Typical structure of a floating-point quantizer that emerges from the series connection of a compressor, a uniform quantizer, and an expander (Widrow and Kollár, Figure 12.6α)

characteristic input/output of the quantizer is similar to that of a uniform quantizer of constant sub-dial but has a form like that depicted in Fig. 13.19.

To describe a floating point quantizer that is described by the characteristics of Fig. 13.19, we can use a *uniform quantizer* like the one described in the previous section, which accepts input from a *compressor* and creates *quantized output*, which is then transmitted as input to an *expander*. These three subsystems, which of course are *nonlinear*, if connected in series create a (*nonlinear*) *floating point quantizer* (Fig. 13.20), the input x of which is the *input of the compressor*, while its output y is the *output of the expander*. As for the intermediate quantizer Q_u, it accepts as input the output y_c of the compressor and produces the output signal $y_u = Q_u(y_c)$, which is then transmitted as input to the expander. In the preceding notation, the internal (hidden) quantizer is denoted by Q_u to indicate the fact that the quantization it performs is *uniform*, while the output of the composite quantizer is denoted by $y = Q(x)$, with the value y being expressed using *floating-point representation*. It is important to note that the compressor and expander used in this model are each the inverse of the other, and therefore, if they are connected in series without an intermediate quantizer, they lead to a system with unit gain.

The *rounding noise* associated with the operation of a nonlinear quantizer is defined as the difference $v = y - x = Q(x) - x$. The *PDF* of this noise is calculated as in the case of a uniform quantizer, but because the input/output characteristic of a floating-point quantizer *is not characterized by a constant step change*, as in the case of a uniform quantizer, the resulting distribution is not characterized by the uniformity of the noise distribution for uniform quantization but is generally more complex.

The foregoing floating point model is particularly useful when the quantization noise $v_u = y_u - y_c$ of the internal uniform quantizer *is consistent with the PQN model*, which means that *the conditions imposed by the first two quantization theorems are satisfied*. However, the verification of the validity of these conditions is now much more complex due to the nonlinear nature of the arrangement of the preceding compressor. Experience has shown that *these conditions are almost never met* by the input to the internal quantizer, but nevertheless generally *are satisfied approximately*, and therefore, the quantization noise introduced by the internal quantizer can be characterized as *uniform* with a very good approximation and *uncorrelated* with the input of the quantizer.

After the preceding theoretical introduction, let us now move on to the statistical description of the noise $v = y - x$ of the moving subband quantizer, calculating the mean value, the mean square of the noise, and its correlation with the input x.

The internal quantizer Q_u creates noise v_u, which is added to its output y_u and which, through the expander, is also added to the output y of the moving subband quantizer. The noise v of the composite quantizer is $v = y - x$. If the noise v_u of the internal quantizer is small compared to the input y_c of the internal quantizer, it will lead *to only a slight increase* in the value of y_c, which in turn will lead *to a slight increase* in the noise v at the output of the mobile quantizer's subdecimal point.

It is proven that the derivative dy/dy_c calculated at the position of the value y_c coincides with the value of the derivative for the input x (this derivative is a function of both y_c and x) and that the noise v expressed as a function of the noise v_u of the internal quantizer and the foregoing derivative takes the form

$$v = v_u \left(\frac{dy}{dy_u}\right)_{y_c} = v_u 2^{Q_u\left(\left(\log_2 \frac{|x|}{\Delta}\right)\right)+0.5}$$

with the quantization of the quantity $log_2(\mid x \mid /\Delta)+0.5$ being performed with a unit step of quantization (Widrow & Kollár, page 275).

Taking into account that the noise v is uncorrelated both with the quantity y_c and with any function of y_c, such as the quantities dy/dy_w and x, and relying on the validity of the condition $E\{v_u\} = 0$ (since the noise v_u follows the noise model PQN), it turns out that

$$E\{vx\} = E\left\{v_u \left(\frac{dy}{dy_u}\right)_{y_c} x\right\} = E\{v_u\}E\left\{\left(\frac{dy}{dy_u}\right)_{y_c} x\right\} = 0 \tag{13.165}$$

and

$$E\{v\} = E\left\{v_u \left(\frac{dy}{dy_u}\right)_{y_c}\right\} = E\{v_u\}E\left\{\left(\frac{dy}{dy_u}\right)_{y_c}\right\} = 0 \tag{13.166}$$

Now, observing that

$$Q_u\left(\log\left(\frac{|\,x\,|}{\Delta}\right) + 0.5\right) = \left(\log\left(\frac{|\,x\,|}{\Delta}\right) + 0.5\right) + v_{exp} \tag{13.167}$$

where v_{exp} is the noise associated with the quantization of the preceding expression, which is the exponent to which the number 2 is raised (note that this noise is blocked, as it takes values in the interval $(-0.5, +0.5)$), the preceding relationship becomes

$$v = v_u\, 2^{\left(\log_2\left(\frac{|x|}{\Delta}\right)+0.5\right)} + v_{exp} = v_u \frac{|\,x\,|}{\Delta} 2^{0.5} 2^{v_{exp}} = \sqrt{2} v_u \frac{|\,x\,|}{\Delta} 2^{v_{exp}}$$

where $\Delta = 2^P q$. Therefore, it will hold that

$$E\left\{v^2\right\} = 2E\left\{v_u^2 \frac{x^2}{\Delta^2} 2^{2v_{exp}}\right\} = 2E\left\{v_u^2\right\} E\left\{\frac{x^2}{\Delta^2} 2^{2v_{exp}}\right\}$$

where the factorization is due to the *consideration of PQN*. Consequently, the noise v at the output of the floating-point quantizer is a function of the input x of the compressor, and since it holds that $0.5 \leq v_{exp} \leq 0.5$, from the preceding relationship we get

$$E\left\{v_u^2\right\} E\left\{\frac{x^2}{\Delta^2}\right\} \leq E\left\{v^2\right\} \leq 4E\left\{v^2\right\} E\left\{\frac{x^2}{\Delta^2}\right\} \tag{13.168}$$

If we consider that the PQN model can be applied to the exponent noise and, therefore, the noise v_{exp} is characterized by zero mean value and statistical properties $E\left\{v_{exp}^2\right\} = 1/12$ and $E\left\{v_{exp}^4\right\} = 1/80$ (recall that the noise v_{exp} resulted from the quantization that was performed with a unit step) and is uncorrelated with the function $\log_2(|x|/\Delta)$, while it is characterized by zero moments of odd order formulating the expression for the noise v in the form

$$v = \sqrt{2}v_u \frac{|\,x\,|}{\Delta} e^{v_{exp}\ln 2} \tag{13.169}$$

and by expanding the exponential function in a series, we are led to the expression (Widrow and Kollár, page 276)

$$E\{v^2\} = E\{v_u^2\}E\left\{\frac{x^2}{\Delta^2}\right\}\left(2+4\left(\frac{1}{12}\right)(\ln 2)^2 + \left(\frac{4}{3}\right)\left(\frac{1}{80}\right)(\ln 2)^4 + \cdots\right) = 2.16E\{V_u^2\}E\left\{\frac{x^2}{\Delta^2}\right\} \tag{13.170}$$

with the resulting value being in the middle of the distance between the upper and lower limit of $E\{V^2\}$. Taking into account that $E\{v_u^2\} = q^2/12$ and $q/\Delta = 2^{-p}$, we will have

$$\Delta = \frac{q}{2^{-p}} \Rightarrow \frac{1}{\Delta} = \frac{2^{-p}}{q} \Rightarrow E\left\{\frac{x^2}{\Delta^2}\right\} = \frac{2^{-2p}}{q^2}E\left\{x^2\right\} \tag{13.171}$$

and the preceding relationship becomes

$$E\{v^2\} = \frac{2.16}{12}q^2\frac{2^{-2p}}{q^2}E\{x^2\} = 0.180 \cdot 2^{-2p}E\{x^2\} \tag{13.172}$$

Therefore, the *SNR* for the floating-point quantizer is equal to

$$\text{SNR} = \frac{E\{x^2\}}{E\{v^2\}} = \frac{1}{0.180}2^{2p} = 5.55 \cdot 2^{2p} \tag{13.173}$$

where we assume that both noises v and v_{exp} follow the *PQN model*. The preceding ratio expressed in decibels takes the form

$$\text{SNR} = 10\log\left((5.55)2^{2p}\right)\ \text{dB} = 7.44 + 6.02p$$

On the other hand, if only the noise v follows the PQN model (and not v_{exp}), then the limits of the SNR are

$$12 \cdot 2^p \geq \text{SNR} \geq 3 \cdot 2^{2p} \qquad \text{or} \qquad 4.77 + 6.02p \leq \text{SNR} \leq 10.79 + 6.02p$$

in decibels, and consequently, the increase of bits used for the representation of the floating-point number coefficient value improves the situation regarding the noise characteristic.

Practice has shown that the PQN model for the internal quantizer leads to *very good results* when the floating-point representation coefficient is encoded with 8 or more bits and when the dynamic range of the input x is much larger than the value of Δ. Conversely, the corresponding conditions for the exponent quantizer that creates the PQN are much more restrictive. The issue is analyzed in detail in the book by Widrow and Kollár for several known distributions of the input signal (Gaussian, triangular, uniform, and sinusoidal) to which the interested reader is referred for more information.

13.9 Problems

1. Given a probability space, $((R), \mathcal{B}(\mathcal{R}), m)$ where m is a probability measure associated with the uniform PDF $f(r)$ in the value range [0, 1] (that is, $f(r) = 1$ for $r \in [0, 1]$ and $f(r) = 0$ otherwise), determine the PDF for the following random variables defined in this space: (a) $X(r) = r^2$, (b) $\mathcal{Y}(r) = \sqrt{|r|}$, (c) $Z(r) = \ln|r|$ and $V(r) = \alpha r + \beta$, where α and β are constants.

2. Repeat Problem 1 assuming that the function $f(r)$ follows an exponential distribution.
3. Repeat Problem 1 assuming that the function $f(r)$ follows a normal distribution.
4. (Stark and Woods, page 407) The process of asynchronous binary signaling is described by the equation

$$\boldsymbol{x}(t) = \sum_n X_n w\left[\frac{t - D - nT}{T}\right] \tag{13.174}$$

where $w(t) = 1$ for $|t| \leq 1/2$, and $w(t) = 0$ elsewhere, a rectangular window function. In the preceding relation, the parameter T expresses the duration of the pulse, while X_n is a random variable that represents the height of the pulse number n, which takes with equal probability one of the values α or β. On the other hand, the magnitude D expresses the time shift of the pulse that follows the uniform distribution $U(-T/2, T/2)$, while the heights of the different pulses are independent variables. Show that

$$R_{xx}(t) = \frac{1}{4}(\alpha - \beta)^2\left(1 - \frac{\tau}{T}\right) w\left(\frac{\tau}{2T}\right) + \left(\frac{\alpha + \beta}{2}\right)^2 \tag{13.175}$$

5. (Stark and Woods, page 416) In digital modulation methods, the frequency of a signal is shifted to a region of the electromagnetic spectrum, which is suitable for transmission. In the phase shift frequency keying (PSK) method, the binary data described by the Bernoulli-type binary sequence $B[n]$, in which the values 0 and 1 are characterized by the same probability, are depicted bit by bit in a sequence $\Theta[n]$ that contains phase values and that is then used for the modulation of the carrier signal that has the form $\cos(2\pi f_c t)$, where f_c is the carrier frequency. This second sequence is defined as $\Theta[n] = \pi/2$ if $B[n] = 1$ and $\Theta[n] = -\pi/2$ if $B[n] = 0$. Defining the function $\Theta_\alpha(t) = \Theta[k]\ kT \leq t < (k+1)T$, the resulting modulated signal has the form $X(t) = \cos(2\pi f_c t + \Theta_\alpha(t))$, where T is the transmission time of a bit, which is usually equal to $1/f_c$. Show that $\mu_x = 0$ and how, for $0 \leq t_1 \leq T$ and $0 \leq t_2 \leq T$, it holds that

$$R_{xx}(t_1, t_2) = \sin(2\pi f_c t_1)\sin(2\pi f_c t_2)$$

6. (Cooper and McGillem, pp 265–266) Show that the PSD of the WSS random process defined as

$$x(t) = A + B\cos(2\pi f_0 t + \vartheta)$$

where A, B, and f_0 are constants and ϑ is a random variable that is uniformly distributed in the range of values $[0, 2\pi]$, is given by the relation

$$S_x(f) = A^2\delta(t) + \frac{B^2}{4}[\delta(f + f_0) + \delta(f - f_0)]$$

7. (Cooper and McGillem, pp 269–270) Show that if the PSD of the WSS random process is $S_x(j\omega)$, then the corresponding function for the process $x'(t) = dx(t)/dt$ is $S_{x'}(j\omega) = \omega^2 S_x(\omega)$.
8. (Stark and Woods, page 462) Assume that the PSD of the PSK process, provided it is considered a WSS process, is given by the equation

$$S_{xx}(\omega) = \frac{T}{4}\left\{\text{sinc}^2\left[(\omega - 2\pi f_c)\frac{T}{2}\right] + \text{sinc}^2\left[(\omega + 2\pi f_c)\frac{T}{2}\right]\right\} \tag{13.176}$$

for values $f_c \gg 1$ (see also Problem 5 in this group of problems).
9. Show that the first- and second-order densities of the static system $\boldsymbol{y}(t) = \boldsymbol{x}^2(t)$ have the respective forms

$$f_y(y;t) = \frac{1}{2\sqrt{y}}[f_x(\sqrt{y};t) + f_x(-\sqrt{y};t)] \quad \text{and} \quad f_y(y_1, y_2; t_1, t_2) = \frac{1}{4\sqrt{y_1 y_2}}\sum f_x(\pm\sqrt{y_1}, \pm\sqrt{y_2}; t_1, t_2)$$

10. Considering the system $g(x) = 1$ for $x > 0$ and $g(x) = -1$ for $x < 0$ that receives at its input a normal process with zero average value, show that

$$R_y(\tau) = \frac{2}{\pi}\arcsin\frac{R_x(\tau)}{R_x(0)}$$

11. Show that if the input of a static system $\boldsymbol{y} = g(\boldsymbol{x})$ is a normal process with zero average value, then it holds that $R_{xy}(\tau) = KR_{xx}(\tau)$, where $K = E\{g'[\boldsymbol{x}(t)]\}$. This result is known as Baussgang's theorem.
12. (Cooper and McGillem, pp. 329, 333, 337) Considering a RC circuit with transfer function $\mathcal{H}(s) = b/(s+b)$ and impulse response $h(t) = be^{-bt}$, where $b = 1/RC$, show that (a) if the input signal comes from a white noise process with spectral power q and function of autocorrelation $R_{xx}(\tau) = q\delta(\tau)$, then it holds that $R_{yy}(\tau) = (bq/2)\exp(-b|\tau|)$ and $R_{xy}(\tau) = qh(\tau)u(\tau)$ and (b) if the autocorrelation function of the input signal is $R_{xx}(\tau) = (bS_0/2)\exp(-b|\tau|)$ $(-\infty < \tau < \infty)$, then it will hold that

$$R_{yy}(\tau) = \frac{b^2\beta S_0}{2(b^2-\beta^2)}\left(e^{-\beta|\tau|} - \frac{\beta}{b}e^{-b|\tau|}\right) \quad \text{and} \quad R_{xy}(\tau) = \begin{cases} \beta b S_0\left[\frac{\beta}{\beta^2-b^2}e^{-b\tau} - \frac{e^{-\beta\tau}}{2(\beta-b)}\right] & \tau > 0 \\ \frac{\beta b S_0}{2(\beta+b)}e^{\beta\tau} & \tau < 0 \end{cases}$$

(c) Referring to the symmetry property $R_{yx}(\tau) = R_{xy}(-\tau)$, calculate the cross-correlation function $R_{yx}(\tau)$ for the two preceding cases.
13. (Stark and Woods, pages 358–362) Show that (a) if $y[n] = x[2n]$, then

$$S_{yy}(\omega) = \frac{1}{2}\left[S_{xx}\left(\frac{\omega}{2}\right) + S_{xx}\left(\frac{\omega-2\pi}{2}\right)\right] \tag{13.177}$$

and (b) if $x_e[n] = x[n/2]$ (only for even values of n, while for odd values we have $x_e[n] = 0$) and $h[n] = \text{sinc}(\pi n/2)$ the impulse response of a LTI system, then the PSD of the output $y[n] = h[n] * x_e[n]$ is

$$S_{yy}(\omega) = 2S_{xx}(2\omega), \quad |\omega| \leq \frac{\pi}{2}$$

while for values of ω outside this range, we have $S_{yy}(\omega) = 0$.
14. Show that the third-order moment of quantization noise is given by the equation

$$E\{v^3\} = 0 + \frac{\Delta^3}{4}\sum_{\ell=1}^{\infty}\Im\{\Phi_x(\ell\xi)\}(-1)^\ell\left(\frac{1}{\pi\ell} - \frac{6}{(\pi\ell)^3}\right)$$

15. Show that the fourth-order moment of quantization noise is given by the equation

$$E\{v^v\} = \frac{\Delta^4}{80} + \frac{\Delta^4}{8}\sum_{\ell=1}^{\infty}\Re\{\Phi_x(\ell\xi)\}(-1)^\ell\left(\frac{1}{(\pi\ell)^2} - \frac{6}{(\pi\ell)^4}\right)$$

Fundamental Mathematical Concepts

A

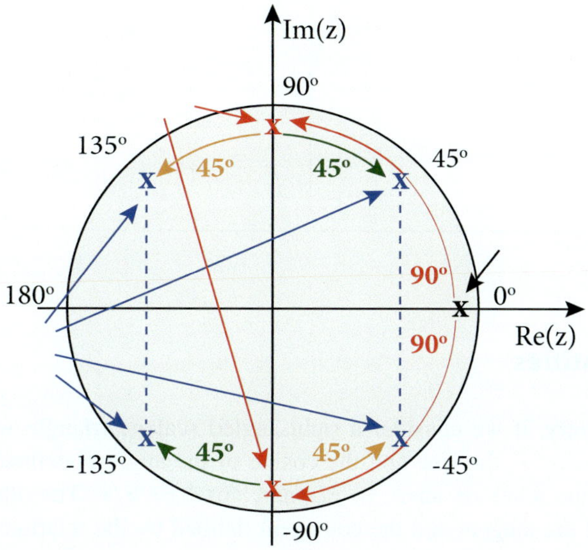

This appendix contains a concise presentation of the most important mathematical concepts used throughout the book.

This appendix presents basic concepts of mathematics used throughout the book whose knowledge on the part of the reader constitutes the minimum prerequisite for understanding the concepts presented in the book. These concepts include, among others, trigonometric functions, basic theory of complex numbers and functions of complex variables, polynomial division, expansion of rational functions into partial fractions, Chebyshev polynomials, elliptic integrals and elliptic functions, techniques for solving differential equations, basic theory of vector spaces, and elements of probability theory (random variables and their functions, continuous and discrete probability distributions, generating functions and characteristic functions, and random vectors and sequences).

A.1 Introduction

The aim of this appendix is the concise presentation of concepts from the field of mathematics, a very good knowledge of which is considered a necessary prerequisite for understanding the material of the book. These introductory concepts include among others the theory of *functions of complex variables*, the methods of *polynomial long division* and of *expansion of rational function into partial fractions*, the techniques for solving *difference equations*, and the basic theory of Chebyshev polynomials and elliptic functions. Also presented are the most important concepts of probability theory, which are used in the material that is related to the field of statistical signal processing. For more details regarding these issues, the reader is referred to the relevant literature.

A. I. Margaris, *Continuous and Discrete-Time Signals and Systems*,
https://doi.org/10.1007/978-3-031-81209-5

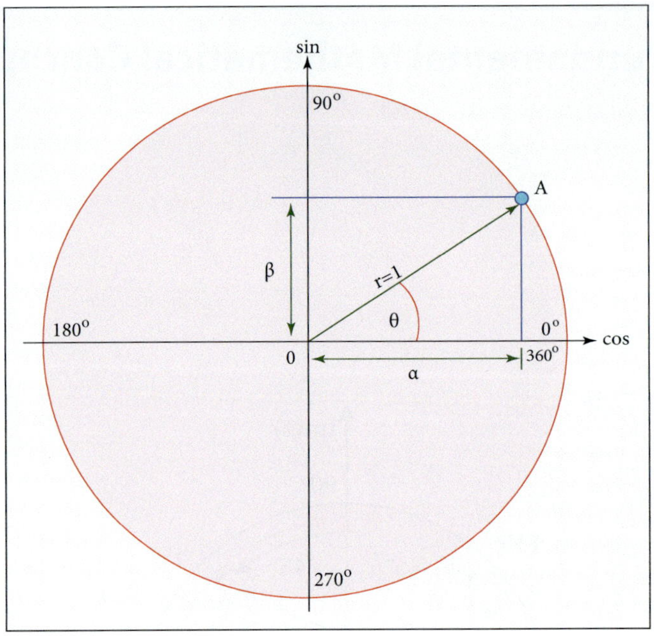

Fig. A.1 The trigonometric circle

A.2 Trigonometric Quantities

As is well known from trigonometry, if we consider a right-angled scalene triangle with lengths of perpendicular sides x and y and length of the hypotenuse r, the sine and the cosine of the angle ϑ defined by the longer perpendicular side and the hypotenuse are given by the relations $\sin\vartheta = y/r$ and $\cos\vartheta = x/r$. The other two basic functions defined for such an angle are the functions of the tangent and the cotangent defined by the relations $\tan\vartheta = \sin\vartheta/\cos\vartheta = y/x$ and $\cot\vartheta = \cos\vartheta/\sin\vartheta = x/y$.

The trigonometric functions of the sine and the cosine are usually represented with the help of the trigonometric circle, which has a radius equal to one and its center is the origin O of a coordinate system whose horizontal axis represents the cosine while the vertical axis represents the sine. This means that if we consider a point A on the circumference of the unit circle, the cosine of the angle ϑ is defined by the straight-line segment OA and the horizontal axis is equal to the projection α of this segment on the horizontal axis, whereas the sine of this angle is defined by the projection of the straight-line segment on the vertical axis. This situation is presented in a diagrammatic way in Fig. A.1. If we apply the Pythagorean theorem to the right triangle defined by the two projections and the straight-line segment OA, we will get $\alpha^2 + \beta^2 = r^2$ or equivalently $\cos^2\vartheta + \sin^2\vartheta = 1$ (since $r = 1$), which is one of the fundamental identities of trigonometry. From the above figure it is also easily proven that the sine function is an odd function (i.e., it satisfies the relation $sin(-x) = -\sin x$), whereas the cosine function is an even function (i.e., it satisfies the relation $\cos(-x) = \cos x$).

The fundamental functions of sine and cosine are periodic functions with a fundamental period of $T = 2\pi$, whereas between them there is a phase difference equal to $\pi/2$. These properties of periodicity are described by the mathematical relations

$$\cos(x) = \sin\left(x + \frac{\pi}{2}\right)$$

$$\cos(x) = \cos(x + 2\pi n), \quad \forall n \in N$$

$$\sin(x) = \sin(x + 2\pi n), \quad \forall n \in N$$

$$\sin(x) = \cos\left(x - \frac{\pi}{2}\right)$$

$$\cos(x) = \cos[x + (2n+1)\pi], \quad \forall n \in N$$

$$\sin(x) = -\sin[x + (2n+1)\pi], \quad \forall n \in N.$$

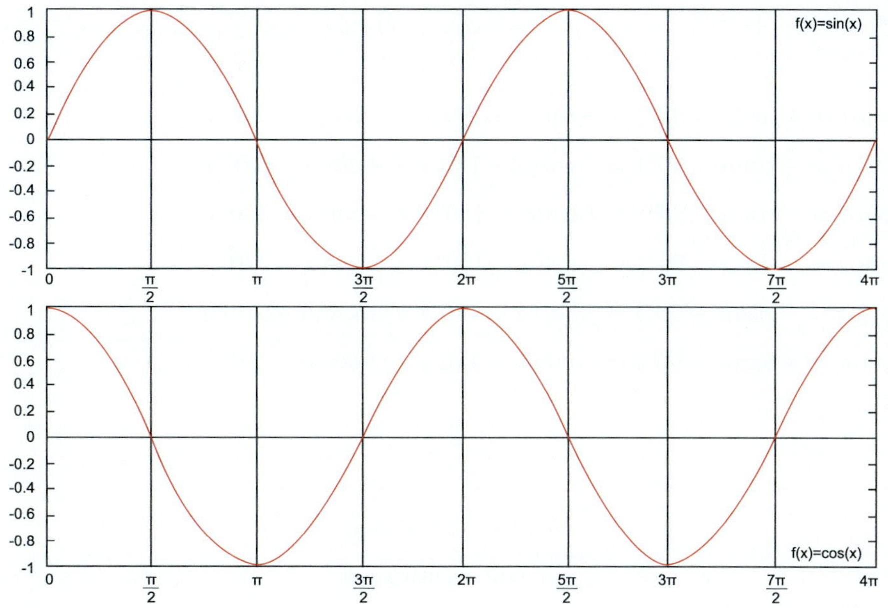

Fig. A.2 The graphical representations of the sine and cosine functions

In the above equations, the variable x is considered as the independent variable; however, usually the functions of sine and cosine are used to describe time-varying signals in which time t is considered as the independent variable. In these cases, we proceed to the replacement $x = \omega t$, where $\omega = 2\pi/T = 2\pi f$ is the angular frequency of the signal. For a detailed description of continuous-time signals, refer to Chap. 1.

The graphical representations of the sine and cosine functions are presented in Fig. A.2. From this figure, we observe that the functions are periodic with a period of 2π, and have a phase difference equal to $\pi/2$ and amplitude equal to the unit.

The following are categories of trigonometric identities, many of which are frequently used in various sections of the book.

Relations Between Trigonometric Functions

$$\sin x = 1\ \csc x \qquad \text{and vice versa} \qquad \csc x = 1/\sin x$$

$$\cos x = 1/\sec x \qquad \text{and vice versa} \qquad \sec x = 1/\cos x$$

$$\tan x = \frac{1}{\cos x} = \frac{\sin x}{\cos x} \qquad \text{and} \qquad \cot x = \frac{1}{\tan x} = \frac{\cos x}{\sin x}$$

$$\sin^2 x + \cos^2 x = 1 \qquad 1 + \tan^2 x = \sec^2 x \qquad 1 + \cot^2 x = \csc^2 x$$

$$\sin x = \pm\sqrt{1 - \cos^2 x} \qquad \cos x = \pm\sqrt{1 - \sin^2 x} \qquad \tan x = \pm\sqrt{\sec^2 x - 1}$$

$$\cot x = \pm\sqrt{\csc^2 x - 1} \qquad \sec x = \pm\sqrt{\tan^2 x + 1} \qquad \csc x = \pm\sqrt{\cot^2 x + 1}.$$

$$\sin x = \cos(90^\circ - x) = +\sin(180^\circ - x)$$

$$\cos x = \sin(90^\circ - x) = -\cos(180^\circ - x)$$

$$\tan x = \cot(90^\circ - x) = -\tan(180^\circ - x)$$

$$\cot x = \tan(90^\circ - x) = -\cot(180^\circ - x).$$

Reduction Formulas

$$\begin{aligned}
\sin\alpha &= +\cos(\alpha - 90^\circ) = -\sin(\alpha - 180^\circ) = -\cos(\alpha - 270^\circ)\\
\cos\alpha &= -\sin(\alpha - 90^\circ) = -\cos(\alpha - 180^\circ) = +\sin(\alpha - 270^\circ)\\
\tan\alpha &= -\cot(\alpha - 90^\circ) = +\tan(\alpha - 180^\circ) = -\cot(\alpha - 270^\circ)\\
\cot\alpha &= -\tan(\alpha - 90^\circ) = +\cot(\alpha - 180^\circ) = -\tan(\alpha - 270^\circ)\\
\sec\alpha &= -\csc(\alpha - 90^\circ) = -\sec(\alpha - 180^\circ) = +\csc(\alpha - 270^\circ)\\
\csc\alpha &= +\sec(\alpha - 90^\circ) = -\csc(\alpha - 180^\circ) = -\sec(\alpha - 270^\circ).
\end{aligned}$$

Product Relations

$$\begin{aligned}
\sin\alpha &= \tan\alpha\cos\alpha\\
\cos\alpha &= \cot\alpha\sin\alpha\\
\tan\alpha &= \sin\alpha\sec\alpha\\
\cot\alpha &= \cos\alpha\csc\alpha\\
\sec\alpha &= \csc\alpha\tan\alpha\\
\csc\alpha &= \sec\alpha\cot\alpha.
\end{aligned}$$

Sum and Difference of Angles Relations

$$\begin{aligned}
\sin(\alpha+\beta) &= \sin\alpha\cos\beta + \cos\alpha\sin\beta\\
\sin(\alpha-\beta) &= \sin\alpha\cos\beta - \cos\alpha\sin\beta\\
\cos(\alpha+\beta) &= \cos\alpha\cos\beta - \sin\alpha\sin\beta\\
\cos(\alpha-\beta) &= \cos\alpha\cos\beta + \sin\alpha\sin\beta\\
\tan(\alpha+\beta) &= \frac{\tan\alpha + \tan\beta}{1 - \tan\alpha\tan\beta}\\
\tan(\alpha-\beta) &= \frac{\tan\alpha - \tan\beta}{1 + \tan\alpha\tan\beta}\\
\cot(\alpha+\beta) &= \frac{\cot\beta\cot\alpha - 1}{\cot\beta + \cot\alpha}\\
\cot(\alpha-\beta) &= \frac{\cot\beta\cot\alpha + 1}{\cot\beta - \cot\alpha}.
\end{aligned}$$

Transformations of Products into Sums or Differences

$$\begin{aligned}
\sin\alpha\sin\beta &= \frac{1}{2}\cos(\alpha-\beta) - \frac{1}{2}\cos(\alpha+\beta)\\
\cos\alpha\cos\beta &= \frac{1}{2}\cos(\alpha-\beta) + \frac{1}{2}\cos(\alpha+\beta)
\end{aligned}$$

$$\sin\alpha\cos\beta = \frac{1}{2}\sin(\alpha+\beta) + \frac{1}{2}\sin(\alpha-\beta)$$

$$\cos\alpha\sin\beta = \frac{1}{2}\sin(\alpha+\beta) - \frac{1}{2}\sin(\alpha-\beta).$$

Sum and Difference Relations

$$\sin\alpha + \sin\beta = +2\sin\left[\frac{1}{2}(\alpha+\beta)\right]\cos\left[\frac{1}{2}(\alpha-\beta)\right]$$

$$\sin\alpha - \sin\beta = +2\cos\left[\frac{1}{2}(\alpha+\beta)\right]\sin\left[\frac{1}{2}(\alpha-\beta)\right]$$

$$\cos\alpha + \cos\beta = +2\cos\left[\frac{1}{2}(\alpha+\beta)\right]\cos\left[\frac{1}{2}(\alpha-\beta)\right]$$

$$\cos\alpha - \cos\beta = -2\sin\left[\frac{1}{2}(\alpha+\beta)\right]\sin\left[\frac{1}{2}(\alpha-\beta)\right]$$

$$\tan\alpha + \tan\beta = \frac{\sin(\alpha+\beta)}{\cos\alpha\cos\beta}$$

$$\tan\alpha - \tan\beta = \frac{\sin(\alpha-\beta)}{\cos\alpha\cos\beta}$$

$$\cot\alpha + \cot\beta = \frac{\sin(\alpha+\beta)}{\sin\alpha\sin\beta}$$

$$\cot\alpha - \cot\beta = \frac{\sin(\beta-\alpha)}{\sin\alpha\sin\beta}.$$

Half-Angle Formulas

$$\sin\left(\frac{\alpha}{2}\right) = \pm\sqrt{\frac{1-\cos\alpha}{2}}$$

$$\cos\left(\frac{\alpha}{2}\right) = \pm\sqrt{\frac{1+\cos\alpha}{2}}$$

$$\tan\left(\frac{\alpha}{2}\right) = \pm\sqrt{\frac{1-\cos\alpha}{1+\cos\alpha}} = \frac{1-\cos\alpha}{\sin\alpha} = \frac{\sin\alpha}{1+\cos\alpha}$$

$$\cot\left(\frac{\alpha}{2}\right) = \pm\sqrt{\frac{1+\cos\alpha}{1-\cos\alpha}} = \frac{1+\cos\alpha}{\sin\alpha} = \frac{\sin\alpha}{1-\cos\alpha}.$$

Double-Angle Formulas

$$\sin(2\alpha) = 2\sin\alpha\cos\alpha = \frac{2\tan\alpha}{1+\tan^2\alpha}$$

$$\tan(2\alpha) = \frac{2\tan\alpha}{1-\tan^2\alpha}$$

$$\cot(2\alpha) = \frac{\cot^2\alpha - 1}{2\cot\alpha}$$

$$\cos(2\alpha) = \cos^2\alpha - \sin^2\alpha = 2\cos^2\alpha - 1 = 1 - 2\sin^2\alpha = \frac{1-\tan^2\alpha}{1+\tan^2\alpha}.$$

A.3 The Basic Theory of Complex Numbers

From complex analysis it is known that a complex number z has the form $z = \alpha + i\beta$, where the parameters $\alpha = Re(z)$ and $\beta = Im(z)$ express the real and the imaginary parts of it, whereas the quantity i—called the complex unit—is defined by the relation $i = \sqrt{-1}$. An alternative notation for this quantity is $z = (\alpha, \beta)$, in which the representation (α, β) is considered an ordered pair (the characterization of this pair as ordered is justified by the fact that if we consider the pair (β, α), it represents the complex number $z = \beta + i\alpha$, which is obviously different from the original). A complex number can be represented as a vector $\vec{z} = (\alpha, \beta)$, can be depicted in a special type of plane called the complex plane, and is symbolized by the letter M. This plane is described as a coordinate system, in which the horizontal axis is called the real axis and is used to depict the real part of a complex number, whereas the vertical axis is called the imaginary axis and is used to depict its imaginary part. The representation of a complex number on the complex plane is shown in Fig. A.3.

Closely related to the concept of the complex plane are the following definitions, which are used very often in the theory of complex variable functions:

- *Region* $\varpi(z_0, \varepsilon)$ of a point $z_0 \in S$ is called the set of points located inside a circle with the center at point z_0 and the radius equal to ε. This definition in mathematical form is written as

$$\varpi(z_0, \varepsilon) = \{z \in M :| z - z_0 |< \varepsilon, \varepsilon > 0, \varepsilon \in R\}.$$

- *Isolated point* of the point set S is called a point $z_0 \in S$ if there is a region of $\varpi(z_0, \varepsilon)$ that does not contain any point belonging to S.
- *Boundary point* or *accumulation point* of the point set S is called a point z_0 if every region $\varpi(z_0, \varepsilon)$ of z_0 contains at least one point of S different from z_0.
- *Interior point* of the set S is called a point z_0 if there is a region $\varpi(z_0, \varepsilon)$ that contains only points of S. Every interior point of the point set S is also an accumulation point of it, but the converse does not hold.
- *Boundary point* of the point set S is called a point z_0 if every region $\varpi(z_0, \varepsilon)$ of z_0 contains both points that belong to S and points that do not belong to S.
- *Exterior point* of the point set S is called a point z_0 if there is a region $\varpi(z_0, \varepsilon)$ such that no point of it belongs to S.

Using the above definitions, we can define the following concepts, which are applied not to individual points of a set of points but to entire point sets:

- The *boundary* of a point set S is called the set of its boundary points.
- A set is called *open* if all its points are internal.
- A set is called *closed* when it contains its accumulation points.

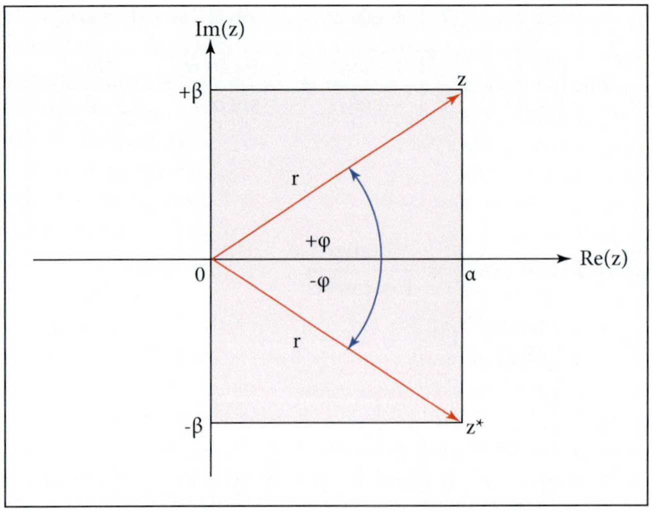

Fig. A.3 Representation of a complex number on the complex plane

- The *closure* or *envelope* of a set is called the union of a set with its external points.
- A set is called *bounded* if there is a constant number M such that $\mid z \mid < M$ for every $z \in S$.
- A set is called *compact* when it is closed and bounded.
- An *open place* is called a non-empty set S that is open and connected, that is, any two points of it can be connected with a continuous polygonal line, all points of which belong to S, or equivalently, that it is impossible to be represented as a union of two open non-empty and foreign sets.
- A place is called *closed* if it results from the closure of an open place.
- A place is called *simply connected* if all the points enclosed by a closed line of the place belong to it.
- A place is called *multiply connected* if there are closed lines of the place that enclose both points that belong to the place and points that do not belong to it.

Forms of Representation of Complex Numbers The writing of a complex number as $z = \alpha + i\beta$ constitutes the so-called *Cartesian* or *rectangular* form; however, there are also many other ways of representation. If we define the quantities $\alpha = r \cos \varphi$ and $\beta = r \sin \varphi$—these equations are nothing more than the conversion equations of Cartesian to polar coordinates—and replace them in the rectangular form of the complex number, it is written as

$$z = r \cos \varphi + ir \sin \varphi = r(\cos \varphi + i \sin \varphi),$$

which is known as *trigonometric* or *polar* form. In this form the parameters $r = \mid z \mid$ and φ refer to the *modulus* and the *phase* of the complex number and are given by the relations

$$r = \sqrt{\alpha^2 + \beta^2} \qquad \text{and} \qquad \varphi = \tan^{-1}(\beta/\alpha),$$

as can be very easily proven. If we use the relation of Euler $e^{i\varphi} = \cos \varphi + i \sin \varphi$ the polar form is reduced to the expression $z = re^{i\varphi}$, which constitutes the so-called *exponential* form. A final way of representing a complex number is the *representation of measure phase*, which takes the form $z = r\angle\varphi$ with the magnitudes r and φ defined in the way we described previously.

The type of representation of complex numbers that we use in each case depends on the type of processing that we apply to them. For example, the operations of addition and subtraction are very easily applied by expressing the complex numbers in their square form, whereas the operations of multiplication, division, and finding roots are significantly simplified by using the exponential representation of complex numbers.

Conjugate Complex Numbers Considering a complex number $z = \alpha + i\beta$, we can define its conjugate number from the relation $z^* = \alpha - i\beta$, from which it follows that the conjugate complex number arises from the original by changing the sign of its imaginary part. From the above expression, it is not difficult to see that the trigonometric form of the conjugate complex number is $z^* = r(\cos \varphi - i \sin \varphi)$, whereas its exponential form is $z^* = re^{-i\varphi}$. Therefore, the vector representation of these two numbers in the complex plane will be characterized by the existence of two vectors symmetric with respect to the real axis. This situation is presented in Fig. A.3.

The Basic Operations of Complex Numbers in comparison with the set R of real numbers, the set C of complex numbers supports the four fundamental operations defined for real numbers, which are addition, subtraction, multiplication, and division. Considering two complex numbers $z_1 = \alpha_1 + i\beta_1$ and $z_2 = \alpha_2 + i\beta_2$, these four operations are defined as follows:

- Addition: $z_1 + z_2 = \alpha_1 + i\beta_1 + \alpha_2 + i\beta_2 = (\alpha_1 + \alpha_2) + i(\beta_1 + \beta_2)$.
- Subtraction: $z_1 - z_2 = \alpha_1 + i\beta_1 - \alpha_2 - i\beta_2 = (\alpha_1 - \alpha_2) + i(\beta_1 - \beta_2)$.
- Multiplication: $z_1 * z_2 = (\alpha_1 + i\beta_1)(\alpha_2 + i\beta_2) = (\alpha_1\alpha_2 - \beta_1\beta_2) + i(\alpha_1\beta_2 + \alpha_2\beta_1)$.
- Division: $\dfrac{z_1}{z_2} = \dfrac{\alpha_1 + i\beta_1}{\alpha_2 + i\beta_2} = \dfrac{(\alpha_1 + i\beta_1)(\alpha_2 - i\beta_2)}{(\alpha_2 + i\beta_2)(\alpha_2 - i\beta_2)} = \dfrac{\alpha_1\alpha_2 + \beta_1\beta_2}{\alpha_2^2 + \beta_2^2} + i\dfrac{\alpha_2\beta_1 - \alpha_1\beta_2}{\alpha_2^2 + \beta_2^2}$.

We observe, therefore, that the result of the four fundamental operations is also a complex number. If we apply these operations between a complex number and its conjugate, we arrive at a number of useful conclusions. Thus, the operation of addition will lead to the result

$$z + z^* = (\alpha + i\beta) + (\alpha - i\beta) = 2\alpha,$$

from which it follows that $\alpha = (z + z^*)/2$. Therefore, the real part of a complex number arises from the semi-sum of this number and its conjugate. In a completely analogous way, from the operation of subtraction, the imaginary part of the complex number arises as $\beta = (z - z^*)/2i$. Finally, if we apply the operation of multiplication, we get

$$zz^* = (\alpha + i\beta)(\alpha - i\beta) = \alpha^2 + \beta^2 =| z |^2,$$

that is, the product of a complex number with its conjugate equals the square of its modulus. It is obvious that the measures of these two numbers are equal to each other, since

$$| z^* |= \sqrt{\alpha^2 + (-\beta)^2} = \sqrt{\alpha^2 + \beta^2} =| z | .$$

It is interesting to mention that the above property of conjugation can be extended to the sum, product, and quotient of two complex numbers, and therefore we can write the relations

$$(z_1 + z_2)^* = z_1^* + z_2^*, \quad (z_1 z_2)^* = z_1^* z_2^*, \quad \left(\frac{z_1}{z_2}\right)^* = \frac{z_1^*}{z_2^*}.$$

The implementation of basic arithmetic operations is significantly facilitated by transforming the complex numbers we use into their exponential form. Using, therefore, a symbolism of the form $z_1 = r_1 e^{i\varphi_1}$ and $z_2 = r_2 e^{i\varphi_2}$ it is not difficult to find that

$$z_1 z_2 = (r_1 e^{i\varphi_1})(r_2 e^{i\varphi_2}) = (r_1 r_2) e^{i(\varphi_1+\varphi_2)} \quad \text{and} \quad \frac{z_1}{z_2} = \frac{r_1 e^{i\varphi_1}}{r_2 e^{i\varphi_2}} = \frac{r_1}{r_2} e^{i(\varphi_1-\varphi_2)}.$$

Finally, regarding the inverse of a complex number $z = re^{i\varphi}$, this, in polar form, is given by the relation $(1/z) = (1/r)e^{-i\varphi}$, whereas in rectangular form it is expressed as

$$\frac{1}{z} = \frac{1}{\alpha + i\beta} = \frac{\alpha - i\beta}{(\alpha + i\beta)(\alpha - i\beta)} = \frac{\alpha}{\alpha^2 + \beta^2} - i\frac{\beta}{\alpha^2 + \beta^2}.$$

Power Operation in full analogy with real numbers, complex numbers can be raised to any power, an operation that is easily defined by transferring the complex number to its polar form. Considering, for example, the complex number $z = re^{i\varphi} = r(\cos\varphi + i\sin\varphi)$, raising it to the power of number k will lead to the result

$$z^k = (re^{i\varphi})^k = r^k e^{ik\varphi} = r^k[\cos(k\varphi) + i\sin(k\varphi)].$$

Using the trigonometric expression of the complex number $z = r(\cos\varphi + i\sin\varphi)$, raising it to the power of number k will lead to the result $z^k = r^k(\cos\varphi + i\sin\varphi)^k$. Comparing the last two expressions for the quantity z^k it is not difficult to ascertain that

$$(\cos\varphi + i\sin\varphi)^k = \cos(k\varphi) + i\sin(k\varphi),$$

an identity known as De Moivre's rule.

Roots of Complex Numbers in order to define the roots of a complex number it is important to mention that as far as this refers to its trigonometric or exponential representation, it is not unambiguous, but there are infinite angles that determine its position on the complex plane. This property is based on the fact that the functions of the sine and cosine are periodic with period $T = 2\pi = 360°$ and therefore satisfy the relations $\cos\varphi = \cos(\varphi + 2n\pi)$ and $\sin\varphi = \sin(\varphi + 2n\pi)$. Therefore, considering the complex number $z = re^{i\varphi} = r(\cos\varphi + i\sin\varphi)$ it can equivalently be represented in the form $z = re^{i(\varphi+2n\pi)} = [\cos(\varphi + 2n\pi) + i\sin(\varphi + 2n\pi)]$ $(n = 0, 1, 2, \ldots)$.

Based on the above property of complex numbers, we can define the kth root of a complex number z from the relation

$$z_k = \sqrt[k]{z} = \sqrt[k]{re^{i(\varphi+2n\pi)}} = \sqrt[k]{r}[e^{i(\varphi+2n\pi)}]^{1/k} = \sqrt[k]{r}\exp\left(i\frac{\varphi + 2n\pi}{k}\right).$$

$(k = 1, 2, 3 \ldots)$. Therefore, for each complex number $z = re^{i\varphi}$ there are k kth roots, which result from the above relation for the values $n = 0, 1, 2, \ldots, k-1$.

In order to understand the above process, let us calculate the cubic roots of the complex number $z = 125e^{i75^\circ}$. There are three such roots that result from the above relation for values of $n = 0, 1, 2$ and which are calculated as

$$z_1 = \sqrt[3]{125}\exp\left(i\frac{75^\circ}{3}\right) = 5e^{i25^\circ},\ z_2 = \sqrt[3]{125}\exp\left(i\frac{75^\circ+360^\circ}{3}\right) = 5e^{i145^\circ},\ z_3 = \sqrt[3]{125}\exp\left(i\frac{75^\circ+720^\circ}{3}\right) = 5e^{i265^\circ}.$$

Inner and Outer Product of Complex Numbers let us consider two complex numbers of the form $z_1 = \alpha_1 + i\beta_1$ and $z_2 = \alpha_2 + i\beta_2$ and let ϑ be the angle defined by the vectors that describe them in the complex plane. In this case, we can define the inner (or dot) and the outer (or cross) product of the two numbers from the relations

$$z_1 \circ z_2 = \mid z_1 \mid\mid z_2 \mid \cos\vartheta = \alpha_1\alpha_2 + \beta_1\beta_2 = Re(z_1^* z_2) = \frac{1}{2}(z_1^* z_2 + z_1 z_2^*)$$

$$z_1 \times z_2 = \mid z_1 \mid\mid z_2 \mid \sin\vartheta = \alpha_1\beta_2 - \alpha_2\beta_1 = Im(z_1^* z_2) = \frac{1}{2i}(z_1^* z_2 - z_1 z_2^*)$$

respectively. If we add the above relations term by term we get

$$z_1^* z_2 = (z_1 \circ z_2) + i(z_1 \times z_2) = \mid z_1 \mid\mid z_2 \mid e^{i\vartheta}.$$

Logarithms of Complex Numbers the natural logarithm of a complex number arises very easily by transferring it to its exponential form $z = re^{i\varphi}$. By taking the logarithm of the above relation we get

$$\ln z = \ln(re^{i\varphi}) = \ln r + \ln(e^{i\varphi}) = \ln r + i\varphi.$$

From the result we arrived at, it follows that the natural logarithm of a complex number is also a complex number.

A.3.1 Examples of Using Complex Numbers

After a brief description of the basic theory related to complex numbers, some exemplary solved exercises follow for the reader to familiarize themselves with their use.

Problem A.3.1 Convert the complex numbers $z_1 = 2 + 2\sqrt{3}i$, $z_2 = -5 + 5i$, $z_3 = -\sqrt{6} - \sqrt{2}i$ and $z_4 = -3i$ to their polar and exponential form and represent them in the complex plane.

Answer (a) From the basic theory it is known that the polar and exponential form of a complex number $z = \alpha + i\beta$ is $z = r(\cos\varphi + i\sin\varphi)$ and $z = re^{i\varphi}$ respectively, where $r = \sqrt{\alpha^2 + \beta^2}$ and $\varphi = \tan^{-1}(\beta/\alpha)$. In detail:

- For the complex number $z_1 = 2 + 2\sqrt{3}i$ we have $\alpha = 2$ and $\beta = 2\sqrt{3}$ and therefore

$$r = \sqrt{2^2 + (2\sqrt{3})^2} = \sqrt{4+12} = \sqrt{16} = 4 \quad \text{and} \quad \varphi = \tan^{-1}[2\sqrt{3}/2] = \tan^{-1}(\sqrt{3}) = \pi/3 = 60^\circ.$$

  So, we have $z_1 = 4e^{i60^\circ} = 4[\cos(60^\circ) + i\sin(60^\circ)]$.
- For the complex number $z_2 = -5 + 5i$: we have $\alpha = -5$ and $\beta = 5$ and therefore

$$r = \sqrt{(-5)^2 + 5^2} = \sqrt{25+25} = \sqrt{2 * 25} = 5\sqrt{2} \quad \text{and} \quad \varphi = \tan^{-1}[5/(-5)] = \tan^{-1}(-1) = 3\pi/4 = 135^\circ.$$

  So, we have $z_2 = 2\sqrt{5}e^{i135^\circ} = 2\sqrt{5}[\cos(135^\circ) + i\sin(135^\circ)]$.

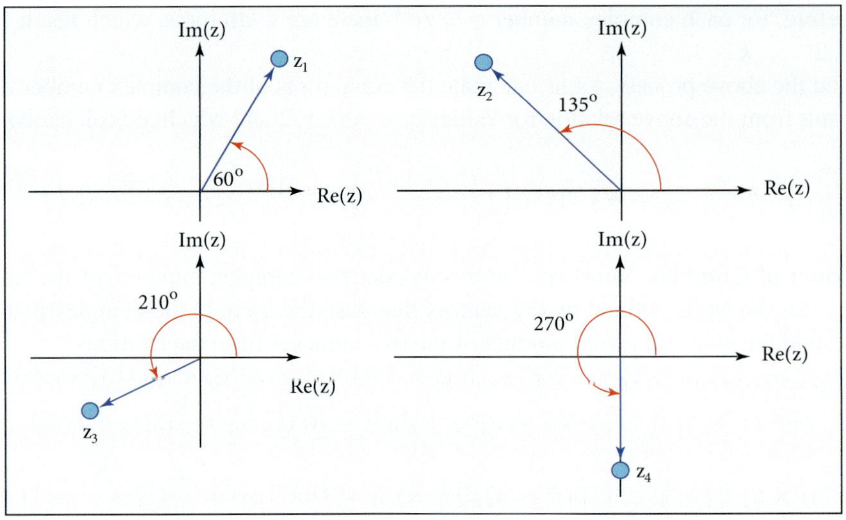

Fig. A.4 Representation of complex numbers on the complex plane

- For the complex number $z_3 = -\sqrt{6} - i\sqrt{2}$ we have $\alpha = -\sqrt{6}$ and $\beta = -\sqrt{2}$ and therefore

$$r = \sqrt{(-\sqrt{6})^2 + (-\sqrt{2})^2} = \sqrt{6+2} = \sqrt{8} = \sqrt{2*4} = 2\sqrt{2}, \quad \text{and}$$
$$\varphi = \tan^{-1}[(-\sqrt{2})/(-\sqrt{6})] = \tan^{-1}[\sqrt{2}/(\sqrt{2}\sqrt{3})] = \tan^{-1}(1/\sqrt{3}) = \tan^{-1}(\sqrt{3}/3) = 7\pi/6 = 210^\circ$$

  Therefore, $z_3 = 2\sqrt{2}e^{i210^\circ} = 2\sqrt{2}[\cos(210^\circ) + i\sin(210^\circ)]$.
- For the complex number $z_4 = -3i$ it is $\alpha = 0$ and $\beta = -3$ and therefore

$$r = \sqrt{(-3)^2} = 3 \quad \text{and} \quad \varphi = \tan^{-1}[(-3)/0] = \tan^{-1}(\infty) = 3\pi/2 = 270^\circ$$

  Therefore, $z_4 = 3e^{i270^\circ} = 3[\cos(270^\circ) + i\sin(270^\circ)]$.

The representation of the four complex numbers in the complex plane is shown in Fig. A.4. In this figure only the angle that the vector representing the complex number forms with the horizontal axis is depicted, whereas the length of the aforementioned vector expresses the modulus of the complex number that has been calculated for each of the four cases.

Problem A.3.2 Let the complex quantities $z_1 = r_1(\cos\varphi_1 + i\sin\varphi_1)$ and $z_2 = r_2(\cos\varphi_2 + i\sin\varphi_2)$. Show that $z_1z_2 = r_1r_2[\cos(\varphi_1+\varphi_2) + i\sin(\varphi_1+\varphi_2)]$ and $z_1/z_2 = (r_1/r_2)[\cos(\varphi_1-\varphi_2) + i\sin(\varphi_1-\varphi_2)]$.

Answer (a) These two identities can be proven very easily with direct multiplication and division of the complex numbers z_1 and z_2. Thus

$$\begin{aligned} z_1z_2 &= r_1r_2(\cos\varphi_1 + i\sin\varphi_1)(\cos\varphi_2 + i\sin\varphi_2) \\ &= r_1r_2(\cos\varphi_1\cos\varphi_2 + i\cos\varphi_1\sin\varphi_2 + i\sin\varphi_1\cos\varphi_2 - \sin\varphi_1\sin\varphi_2) \\ &= r_1r_2[(\cos\varphi_1\cos\varphi_2 - \sin\varphi_1\sin\varphi_2) + i(\cos\varphi_1\sin\varphi_2 + \sin\varphi_1\cos\varphi_2)] = r_1r_2[\cos(\varphi_1+\varphi_2) + i\sin(\varphi_1+\varphi_2)], \end{aligned}$$

whereas in a similar way we have

$$\frac{z_1}{z_2} = \frac{r_1}{r_2}\frac{\cos\varphi_1 + i\sin\varphi_1}{\cos\varphi_2 + i\sin\varphi_2} = \frac{r_1}{r_2}\frac{(\cos\varphi_1 + i\sin\varphi_1)(\cos\varphi_2 - i\sin\varphi_2)}{(\cos\varphi_2 + i\sin\varphi_2)(\cos\varphi_2 - i\sin\varphi_2)}$$

$$= \frac{r_1}{r_2}\frac{\cos\varphi_1\cos\varphi_2 - i\cos\varphi_1\sin\varphi_2 + i\sin\varphi_1\cos\varphi_2 + \sin\varphi_1\sin\varphi_2}{\cos^2\varphi_2 + \sin^2\varphi_2}$$

$$= \frac{r_1}{r_2}[(\cos\varphi_1\cos\varphi_2 + \sin\varphi_1\sin\varphi_2) + i(\sin\varphi_1\cos\varphi_2 - \cos\varphi_1\sin\varphi_2)] = \frac{r_1}{r_2}[\cos(\varphi_1 - \varphi_2) + i\sin(\varphi_1 - \varphi_2)].$$

The above identities can be proven much more easily if we express the two complex numbers in their exponential form, that is, we write them as $z_1 = r_1 e^{i\varphi_1}$ and $z_2 = r_2 e^{i\varphi_2}$. In this case, it immediately follows that

$$z_1 z_2 = r_1 r_2 e^{i(\varphi_1+\varphi_2)} = r_1 r_2[\cos(\varphi_1 + \varphi_2) + i\sin(\varphi_1 + \varphi_2)], \quad \text{and}$$

$$z_1/z_2 = (r_1/r_2)e^{i(\varphi_1-\varphi_2)} = (r_1/r_2)[\cos(\varphi_1 - \varphi_2) + i\sin(\varphi_1 - \varphi_2)].$$

Problem A.3.3 Prove that $\cos\varphi = (e^{i\varphi} + e^{-i\varphi})/2$ and $\sin\varphi = (e^{i\varphi} - e^{-i\varphi})/2i$.

Answer (a) From the basic theory it is known that $e^{i\varphi} = \cos\varphi + i\sin\varphi$ and $e^{-i\varphi} = \cos\varphi - i\sin\varphi$ (these equations are known as Euler's formulas). Adding the two equations term by term we get $e^{i\varphi} + e^{-i\varphi} = 2\cos\varphi$, from which it follows that $\cos\varphi = (e^{i\varphi} + e^{-i\varphi})/2$, whereas subtracting term by term we get $e^{i\varphi} - e^{-i\varphi} = 2i\sin\varphi$, from which it follows that $\sin\varphi = (e^{i\varphi} - e^{-i\varphi})/2i$.

Problem A.3.4 Find the cubic roots of the complex number $z_1 = -1 + i$, the square roots of the complex number $z_2 = -15 - 8i$, and the fourth-order roots of the complex number $z_3 = -2\sqrt{3} - 2i$.

Answer (a) The modulus and the phase of the complex number z_1 are calculated as

$$r = \sqrt{(-1)^2 + 1^2} = \sqrt{2} \qquad \text{and} \qquad \varphi = \tan^{-1}(-1) = 3\pi/4 = 135^\circ$$

and therefore this number in its polar form is written as $z_1 = \sqrt{2}e^{i(3\pi/4)} = \sqrt{2}e^{i\,135^\circ}$. According to the theory we presented in the previous pages, there are three cubic roots for the complex number z_1, which are defined as

$$z_n = (\sqrt{2})^{1/3}\exp\left[i\left(\frac{(3\pi/4) + 2\pi n}{3}\right)\right] = \sqrt[6]{2}\exp\left(i\frac{3\pi + 8\pi n}{12}\right) \qquad n = 0, 1, 2$$

and they have the values $z_0 = \sqrt[6]{2}e^{i(\pi/4)}$, $z_1 = \sqrt[6]{2}e^{i(11\pi/12)}$, $z_2 = \sqrt[6]{2}e^{i(19\pi/12)}$.
(b) The modulus of the complex number z_2 is easily calculated as $r = \sqrt{289} = 17$ and therefore this number can be written as $z_2 = 17(\cos\varphi + i\sin\varphi) = 17e^{i(\varphi+2\pi n)}$, where $\cos\varphi = -15/17$ and $\sin\varphi = -8/17$, as can easily be proved by representing the complex number z_2 in the complex plane. Therefore, the two square roots it has will be $z_n = \sqrt{17}\exp[i(\varphi + 2\pi n)/2]$ $(n = 0, 1)$ and will be calculated as

$$z_0 = \sqrt{17}\,e^{i(\varphi/2)} = \sqrt{17}\left(\cos\frac{\varphi}{2} + i\sin\frac{\varphi}{2}\right)$$

$$z_1 = \sqrt{17}\,e^{i[(\varphi/2)+\pi]} = -\sqrt{17}\,e^{i(\varphi/2)} = -\sqrt{17}\left(\cos\frac{\varphi}{2} + i\sin\frac{\varphi}{2}\right),$$

since from trigonometry it is known that

$$\exp\left(\frac{\varphi}{2}+\pi\right)=\cos\left(\frac{\varphi}{2}+\pi\right)+i\cos\left(\frac{\varphi}{2}+\pi\right)=-\left(\cos\frac{\varphi}{2}+i\sin\frac{\varphi}{2}\right)=-\exp\left(\frac{\varphi}{2}\right).$$

In order to calculate the values of the quantities $\cos(\varphi/2)$ and $\sin(\varphi/2)$ and taking into account that we know the values of $\cos\varphi$ and $\sin\varphi$, we can use the known trigonometric formulas of the half angle

$$\cos\frac{\varphi}{2}=\pm\sqrt{\frac{1+\cos\vartheta}{2}} \quad \text{and} \quad \sin\frac{\varphi}{2}=\pm\sqrt{\frac{1-\cos\vartheta}{2}}.$$

So, using the values $\cos\varphi=-15/17$ and $\sin\varphi=-8/17$ we find that $\cos(\varphi/2)=\pm(1/\sqrt{17})$ and $\sin(\varphi/2)=\pm(4/\sqrt{17})$. In order to choose the correct sign for these two quantities, we must take into account that since the angle φ belongs to the third quadrant, the angle $\varphi/2$ should be in the second quadrant, where the cosine takes negative values whereas the sine takes positive values. Therefore, we have $\cos(\varphi/2)=-(1/\sqrt{17})$ and $\sin(\varphi/2)=4/\sqrt{17}$ and substituting into the defining equations of the two square roots, we get

$$z_0=+\sqrt{17}\left(-\frac{1}{\sqrt{17}}+i\frac{4}{\sqrt{17}}\right)=-1+4i \quad \text{and} \quad z_1=-\sqrt{17}\left(-\frac{1}{\sqrt{17}}+i\frac{4}{\sqrt{17}}\right)=1-4i,$$

which is also the desired result.

(c) The modulus and phase of the complex number z_3 are easily calculated as $r=4$ and $\varphi=\tan^{-1}(\sqrt{3}/3)=7\pi/6=210°$ (as derived from the position of the complex number on the complex plane) and therefore this number in its polar form will be written as $z_3=4e^{i(7\pi/6)}=4e^{i(210°+2\pi n)}$. According to the theory we presented in the previous pages, there are four fourth-order roots for the complex number z_3, which are defined as

$$z_n=\sqrt[4]{4}\exp\left[i\left(\frac{(7\pi/6)+2\pi n}{4}\right)\right]=\sqrt{2}\exp\left(i\frac{7\pi+12\pi n}{24}\right) \quad (n=0,1,2,3)$$

and have the values $z_0=\sqrt{2}e^{i(7\pi/24)}$, $z_1=\sqrt{2}e^{i(19\pi/24)}$, $z_2=\sqrt{2}e^{i(31\pi/24)}$, and $z_3=\sqrt{2}e^{i(43\pi/24)}$.

Problem A.3.5 Given the complex numbers $z_1=3-4i$ and $z_2=-4+3i$, calculate the inner product $z_1\circ z_2$ and the outer product $z_1\times z_2$.

Answer We have $z_1^*=3+4i$ and therefore

$$z_1^*z_2=(3+4i)(-4+3i)=-12+9i-16i-12=-24-7i.$$

Therefore, according to the theory, we have $z_1\circ z_2=Re(z_1^*z_2)=-24$ and $z_1\times z_2=Im(z_1^*z_2)=-7$.

A.4 Complex Variable Functions

In full analogy with our known real variable functions, a complex variable function is defined as $w=f(z)$, with the numbers w and z belonging to the set C of complex numbers. If we express these numbers in their rectangular form, that is, we write $z=\alpha+j\beta$ and $w=u+jv$, the above relation will take the form $w=u+jv=f(z)=f(\alpha+j\beta)=f(\alpha,\beta)$, from which it follows that $u=u(\alpha,\beta)$ and $v=v(\alpha,\beta)$. According to the above equations, the real and imaginary parts of the returned value are a function of the real and imaginary parts of the input argument, which, of course, is expected.

There are many categories of complex variable functions that are generalizations of the corresponding real functions. We mention as examples the polynomial functions

$$w=\alpha_n z^n+\alpha_{n-1}z^{n-1}+\alpha_{n-2}z^{n-2}+\cdots+\alpha_1 z+\alpha_0=P(z),$$

the rational functions $w = P(z)/Q(z)$, where the functions $P(z)$ and $Q(z)$ are polynomial functions with $Q(z) \neq 0$, and the basic trigonometric functions $\cos z = (e^{jz} + e^{-jz})/2$ and $\sin z = (e^{jz} - e^{-jz})/2j$. A special category of complex functions are the sequences of complex numbers denoted as $f(n)$ or u_n and are a function of the integer variable n.

It is obvious that since complex functions are a generalization of the corresponding real functions, they will be characterized by the same properties as them. In order to demonstrate this claim, let us prove that $e^{z_1} e^{z_2} = e^{z_1+z_2}$, a property that is known to hold for the case of real numbers. Thus

$$\begin{aligned} e^{z_1} e^{z_2} &= e^{\alpha_1 + j\beta_1} e^{\alpha_2 + j\beta_2} = e^{\alpha_1} (\cos\beta_1 + j\sin\beta_1)\, e^{\alpha_2} (\cos\beta_2 + j\sin\beta_2) \\ &= e^{\alpha_1+\alpha_2} (\cos\beta_1\cos\beta_2 + j\cos\beta_1\sin\beta_2 + j\sin\beta_1\cos\beta_2 - \sin\beta_1\sin\beta_2) \\ &= e^{\alpha_1+\alpha_2}[\cos(\beta_1+\beta_2) + j\sin(\beta_1+\beta_2)] = e^{\alpha_1+\alpha_2} e^{j(\beta_1+\beta_2)} = e^{[(\alpha_1+j\beta_1)(\alpha_2+j\beta_2)]} = e^{(z_1+z_2)}. \end{aligned}$$

In a similar way, it is proven that $\cos^2 z + \sin^2 z = 1$, just as it holds for the case of real numbers. Indeed, we have

$$\begin{aligned} \cos^2 z + \sin^2 z &= \left(\frac{e^{jz} + e^{-jz}}{2}\right)^2 + \left(\frac{e^{jz} - e^{-jz}}{2j}\right)^2 = \frac{e^{2z} + e^{-2z} + 2e^z e^{-z}}{4} + \frac{e^{2z} + e^{-2z} - 2e^z e^{-z}}{4j^2} \\ &= \frac{e^{2z} + e^{-2z} + 2e^z e^{-z} - e^{2z} - e^{-2z} + 2e^z e^{-z}}{4} = \frac{2+2}{4} = \frac{4}{4} = 1. \end{aligned}$$

A.4.1 Differentiation of Complex Functions and Analytic Functions

In this section, a set of useful definitions and theorems related to the differentiation of complex functions are presented in a summary manner. The theorems are presented without proof; for more details, the reader is referred to any book on complex analysis.

Limit of the single-valued complex function $f(z)$ $(z \in S)$ at the point $z = z_0 \in M$ (where the function $f(z)$ may not be defined) is called a number $W_0 \in M$ such that

$$\forall \varepsilon > 0 \quad \exists\, \delta(z_0, \varepsilon) > 0 : \forall z \in S \text{ with } |\, z - z_0 \,| < \delta \text{ it is } |\, f(z) - W_0 \,| < \varepsilon.$$

In order to show that the limit of a complex function $f(z)$ at the point $z = z_0$ is the point W_0 we use the notation $\lim\limits_{z \to z_0} f(z) = W_0$.

A complex function $f(z)$ is called continuous at the point $z = z_0$, if and only if the following three conditions are met: (a) the function $f(z)$ is defined at the point $z = z_0$, (b) the limit W_0 of the function $f(z)$ exists at the point $z = z_0$, and (c) the relation $W_0 = f(z_0)$ holds. A function is called continuous in its domain of definition S if it is continuous at every point $z \in S$.

Theorem A.4.1 *If the function $f(z) = u(x, y) + jv(x, y)$ is continuous at the point $z_0 = x_0 + jy_0$, then the real and the imaginary parts of it will be continuous functions at the point $z_0 = x_0 + jy_0$ and vice versa.*

Definition A.4.1 (Derivative of a Complex Function) The derivative $f'(z_0)$ of the function $f(z)$ with respect to the complex variable z at the point $z = z_0 \in D \subseteq M$ is called the limit

$$f'(z_0) = \lim_{\Delta z \to 0} \frac{\Delta W}{\Delta z} = \lim_{\Delta z \to 0} \frac{f(z_0 + \Delta z) - f(z_0)}{\Delta z} \tag{A.1}$$

if this limit exists and has a finite value. In this case, the function $f(z)$ is aforementioned to be differentiable at the point $z = z_0$.

The rules for differentiating complex functions are similar to those characterizing the differentiation of real functions. They will, therefore, be

$$(a)\quad [f(z) \pm g(z)]' = f'(z) \pm g'(z) \qquad (b)\quad [cf(z)]' = cf'(z)$$

$$(c)\quad [f(z)g(z)]' = f(z)g'(z) + f'(z)g(z)$$

$$(d)\quad [f(z)/g(z)]' = [g(z)f'(z) - f(z)g'(z)]/[g(z)]^2 \quad (g(z) \neq 0).$$

A complex function $f(z)$ is called *analytic* or *holomorphic* at the point z_0, if it has a derivative at z_0 as well as at every point of a neighborhood of z_0. If the function $f(z)$ is analytic at every point of a place, then it is characterized as analytic in that place.

Regular point of a complex function $f(z)$ is called a point at which the function is analytic.

Irregular point of a complex function $f(z)$ is called a point at which the function $f(z)$ is not analytic, whereas it can be analytic at every other point of a region of the considered point.

Theorem A.4.2 *The sufficient and necessary conditions for the complex function $f(z) = u(x, y) + jv(x, y)$ to be analytic at the point $z_0 = x_0 + jy_0$ are the*

$$\frac{\partial u}{\partial x} = \frac{\partial v}{\partial y} \quad \textit{and} \quad \frac{\partial u}{\partial y} = -\frac{\partial v}{\partial x},$$

which are known as Cauchy-Riemann conditions.

It is proven that the derivative $f'(z)$ of an analytic function $f(z)$ with respect to the complex variable z is given by the relation

$$f'(z) = \frac{\partial u}{\partial x} + j\frac{\partial v}{\partial x} = \frac{\partial v}{\partial y} - j\frac{\partial u}{\partial x}.$$

Theorem A.4.3 *The real and the imaginary parts of an analytic complex function of the form $f(z) = u(x, y) + jv(x, y)$ satisfy the Laplace's equation in two dimensions, that is*

$$\nabla^2 u(x, y) = \frac{\partial^2 u}{\partial x^2} + \frac{\partial^2 u}{\partial y^2} = 0 \quad \textit{and} \quad \nabla^2 v(x, y) = \frac{\partial^2 v}{\partial x^2} + \frac{\partial^2 v}{\partial y^2} = 0.$$

A real function $u(x, y)$ with continuous partial derivatives of second order that satisfies Laplace's equation is known as a harmonic function.

Problem A.4.1 Calculate the derivative of the complex function defined by the equation $f(z) = (1 + z)/(1 - z)$.

Answer (a) Applying the rules of differentiation of complex functions we get

$$\frac{df(z)}{dz} = \frac{d}{dz}\left(\frac{1+z}{1-z}\right) = \frac{(1-z)\dfrac{d(1+z)}{dz} - (1+z)\dfrac{d(1-z)}{dz}}{(1-z^2)} = \frac{1-z+1+z}{(1-z)^2} = \frac{2}{(1-z)^2}$$

Problem A.4.2 Calculate the derivative of the function $f(z) = e^z$.

Answer In order to solve the exercise we write the function $f(z)$ in the form

$$f(z) = e^z = e^{(x+iy)} = e^x e^{iy} = e^x(\cos y + i \sin y) = u(x, y) + iv(x, y),$$

where $u(x, y) = e^x \cos(y)$ and $v(x, y) = e^x \sin(y)$. Thus

$$\frac{df(z)}{dz} = \frac{\partial u(x,y)}{\partial x} + i\frac{\partial v(x,y)}{\partial x} = \frac{\partial}{\partial x}[e^x \cos(y)] + i\frac{\partial}{\partial y}[e^x \sin(y)]$$
$$= e^x \cos(y) + ie^x \sin(y) = e^x [\cos(y) + i \sin(y)] = e^x e^{iy} = e^{x+iy} = e^z.$$

A.4.2 Integration of Complex Functions

The basic characteristic of integrals of complex functions is that they are *always calculated along curves on the complex plane* and in this sense they are *curve integrals*. In order to present the basic theory, let us consider a complex number $z = x + jy$ and let us assume that the coordinates x and y that refer to the real and imaginary parts of it are continuous functions of an independent variable t defined in some interval $[t_a, t_b]$, that is, they have the form $x = x(t)$ and $y = y(t)$. In this case, the equations $x = x(t)$ and $y = y(t)$ are called *parametric equations of the curve* $z(t) = x(t) + jy(t)$, which is known as a continuous curve. If, in addition, this curve is characterized by the property for $t_1 \neq t_2$ to be and $[x(t_1), y(t_1)] \neq [x(t_2), y(t_2)]$ or equivalently $z(t_1) \neq z(t_2)$, then it is known as the *Jordan arc or curve*. Closely connected with the theory of curves of the complex plane are also the following definitions (let us note that these definitions can be extended, so as to describe arbitrary curves defined in the space of n dimensions R^n).

Definition A.4.2 (Simply Closed Curve) A simply closed curve is called a continuous curve, if for two values $t_a \neq t_b$ it is $x(t_a) = x(t_b)$ and $y(t_a) = y(t_b)$ and there are no other two values of t that correspond to the same point (x, y). A typical example of a curve of this type is the circumference of a circle of the radius $r = \alpha$, the parametric equations of which have the form $x(t) = \alpha \cos(t)$ and $y(t) = \alpha \sin(t)$ $(0 \leq t \leq 2\pi)$.

Definition A.4.3 (Smooth or Regular Curve) A smooth or regular curve is a continuous curve, if the functions $x(t)$ and $y(t)$ have continuous first-order derivatives that do not simultaneously become zero for any value of t.

Definition A.4.4 (Path) A continuous curve is called a path, if it can be divided into a finite number of smooth curves. If a path is also a simple closed curve, then it is called a closed path. It is proven that the length of a path is given by the relation

$$L = \int_a^b \sqrt{[x'(t)]^2 + [y'(t)]^2 dt}. \tag{A.2}$$

Examples of curves of the complex plane are presented in Fig. A.5 with parametric equations $(x(t) = t^2, y(t) = \sqrt{t})$ for the left curve and $(x(t) = t^2, y(t) = \sin t)$ for the right curve.

Let us now consider a complex function $f(z)$, which is defined at every point $z_1, z_2, \ldots, z_n$ of the path C of Fig. A.6α—which in the general case is described by equations of the form $x = x(t)$ and $y = y(t)$ $(t_a \leq t \leq t_b)$. In this case, the integral of the function $f(z)$ along the path C (for the strict definition of the integral refer to a book on complex analysis) from point A to point B is written

$$S = \int_C f(z)\,dz = \int_A^B f(z)\,dz = \int_C \Big\{u(x, y)\,dx - v(x, y)\,dy\Big\} + j\int_C \Big\{v(x, y)\,dx + u(x, y)\,dy\Big\}, \tag{A.3}$$

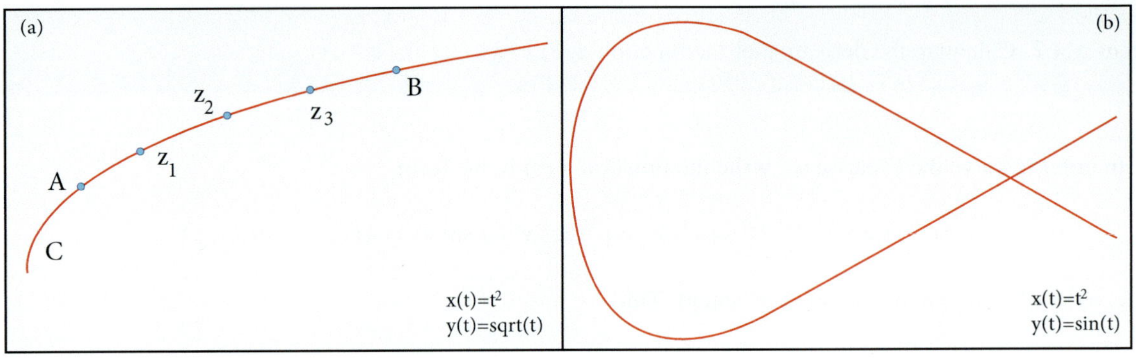

Fig. A.5 Typical examples of parametric curves defined on the complex plane

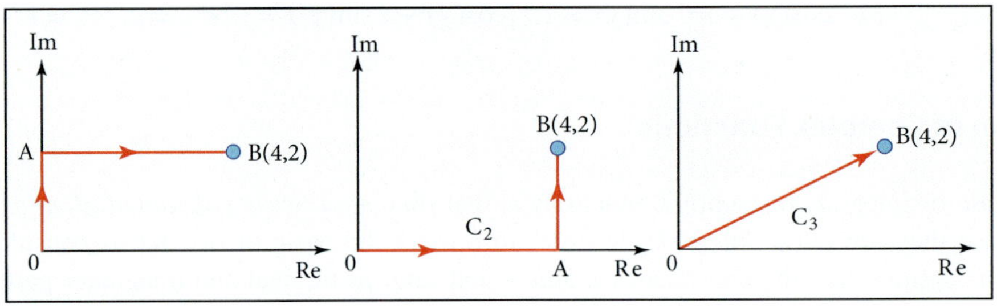

Fig. A.6 Examples of paths for the calculation of complex integrals

with the last relation having arisen by replacing the function $f(z)$ with its equivalent notation $f(z) = u(x, y) + jv(x, y)$. If we use the parametric equations of the path C along which the integral is calculated, it is obvious that

$$\frac{dz}{dt} = \frac{d}{dt}\Big\{x(t) + jy(t)\Big\} = \frac{dx(t)}{dt} + j\frac{dy(t)}{dt}. \tag{A.4}$$

Therefore, the above relation is written

$$\begin{aligned} S &= \int_C f(z)\,dz = \int_{t_a}^{t_b}\left(u(x,y)\frac{dx}{dt} - v(x,y)\frac{dy}{dt}\right)dt + j\int_{t_a}^{t_b}\left(v(x,y)\frac{dx}{dt} + u(x,y)\frac{dy}{dt}\right)dt \\ &= \int_{t_a}^{t_b}[u(x,y) + jv(x,y)]\frac{dz}{dx}dt = \int_{t_a}^{t_b} f[z(t)]\,z'(t)dt, \end{aligned} \tag{A.5}$$

that is, the integral S is reduced from a curve to a simple integral.

It is obvious that in the general case the value of the integral S of a complex function $f(z)$ along a path C on the complex plane and from a point A to a point B depends on the path that connects the two points. However, if the function $f(z)$ can be expressed as the derivative of some other function $F(z)$ (that is, $f(z) = F'(z)$) (in this case, as is well known, the function $F(z)$ is called *derivative function* of $f(z)$) the above property does not exist and the integral becomes path-independent. Indeed, in this case

$$f[z(t)]\,z'(t) = \frac{dF(z)}{dz}\frac{dz}{dt} = \frac{dF(z(t))}{dt}$$

and the last relation becomes

$$S = \int_{t_a}^{t_b} f[z(t)]\,z'(t)dt = \int_{t_a}^{t_b}\frac{dF(z(t))}{dt}dt = F(B) - F(A),$$

from which it follows that the value of the integral equals the difference of the values of the function $F(z)$, calculated at the initial and at the final point, and does not depend on the path C, which connects the two points.

Problem A.4.3 Calculate the complex integral $S = \int_C z^* dz$ along the paths C_1, C_2 and C_3, depicted in Fig. A.6.

Answer (a) The path C_1 consists of the straight-line segments $\overline{OA}$ and $\overline{OB}$ and therefore the integral to be calculated will have the form

$$S = \int_{C_1} f(z)\, dz = \int_O^B z^* dz = \int_O^A z^* dz + \int_A^B z^* dz = S_{OA} + S_{AB}.$$

To calculate the first of the two integrals, we need to find the parametric equations $x(t)$ and $y(t)$ of the straight-line segment $\overline{OA}$. To do this, we assume that these equations will have the form $x = x(t) = \alpha t + \beta$ and $y = y(t) = \gamma t + \delta$ $(0 \le t \le 1)$, with the values of the parameters α, β, γ and δ calculated from the requirement for $t = 0$ to yield the point O with coordinates $(x, y) = (0, 0)$, whereas for $t = 1$ to yield the point A with coordinates $(x, y) = (0, 2)$. Therefore, (a) $\alpha * 0 + \beta = 0$ and $\gamma * 0 + \delta = 0$, from which it follows that $\beta = \delta = 0$, and (b) $\alpha * 1 + 0 = 0$ and $\gamma * 1 + 0 = 2$, from which it follows that $\alpha = 2$ and $\gamma = 2$. Therefore, the parametric equations of the path OA will be $x(t) = 0$ and $y(t) = 2t$, meaning that the function $f[z(t)]$ will have the form $f[z(t)] = z^*(t) = x(t) - jy(t) = -2jt$. It will also be $z'(t) = x'(t) + jy'(t) = 2i$ and therefore the line integral along the path OA will be calculated as

$$S_{OA} = \int_{t_a}^{t_b} f[z(t)]\, z'(t) dt = \int_0^1 (-2jt)\,(2j)\, dt = -4j^2 \int_0^1 t dt = 4 \left[\frac{t^2}{2}\right]_0^1 = 2.$$

In a similar way, the calculation of the integral along the straight-line segment $\overline{AB}$ will take place. Considering the parametric equations $x = x(t) = \alpha t + \beta$ and $y = y(t) = \gamma t + \delta$ $(0 \le t \le 1)$ and imposing the requirement for $t = 0$ and $t = 1$ to yield the points $A(0, 2)$ and $B(4, 2)$, we end up with the values $\beta = \gamma = 0$, $\alpha = 4$, and $\delta = 2$. Therefore, the parametric equations of the path AB will be $x(t) = 4t$ and $y(t) = 2$, meaning that the function $f[z(t)]$ will have the form $f[z(t)] = z^*(t) = x(t) - jy(t) = 4t - 2j$. It will also be $z'(t) = x'(t) + jy'(t) = 4$ and therefore the line integral along the path AB is calculated as

$$S_{AB} = \int_{t_a}^{t_b} f[z(t)]\, z'(t) dt = \int_0^1 (4t - 2j)\,(4)\, dt = 16 \int_0^1 t dt - 8j \int_0^1 dt = 16 \left[\frac{t^2}{2}\right]_0^1 - 8i,$$

and finally $S_{AB} = 8 - 8i$. Therefore, by adding the values of the integrals along the paths OA and AB we end up with the value $S = 2 + 8 - 8j = 10 - 8j$, which is the desired result.

(b) The methodology for solving the problem is the same as the previous one and for this reason the description of the solution will be brief. In this case, the path C_2 consists of the straight line segments $\overline{OA}$ and $\overline{AB}$ defined by the points $O(0, 0)$, $A(4, 0)$ and $B(0, 2)$. The parametric equations of these two paths are $[x(t), y(t)] = (4t, 0)$ for the path $\overline{OA}$ and $[x(t), y(t)] = (4, 2t)$ for the path $\overline{AB}$ and therefore the function $f[z(t)] = z^*(t)$ for these two cases will be $f[z(t)] = 4t$ and $f[z(t)] = 4 - 2jt$ respectively. The derivatives $z'(t)$ will be given by the relations $z'(t) = 4$ for the path $\overline{OA}$ and $z'(t) = 2j$ for the path $\overline{AB}$ and using the equation for calculating the line integral that we used previously, we find $S_{OA} = 8$ and $S_{AB} = 8j + 2$. So, $S = S_{OA} + S_{AB}$ and by substituting the values we find $S = 8 + 8j + 2 = 10j + 2$, which is the desired result.

(c) The path C_3 consists only of the straight-line segment OB, the parametric equations of which are $x(t) = 4t$ and $y(t) = 2t$. So, $f[z(t)] = z^*(t) = 4t - 2jt$, $z'(t) = x'(t) + jy'(t) = 4 + 2j$ and from the equation for calculating the line integral we get

$$S = \int_{t_a}^{t_b} f[z(t)]\, z'(t) dt = \int_0^1 (4t - 2jt)\,(4 + 2j)\, dt = \int_0^1 20t dt = 10.$$

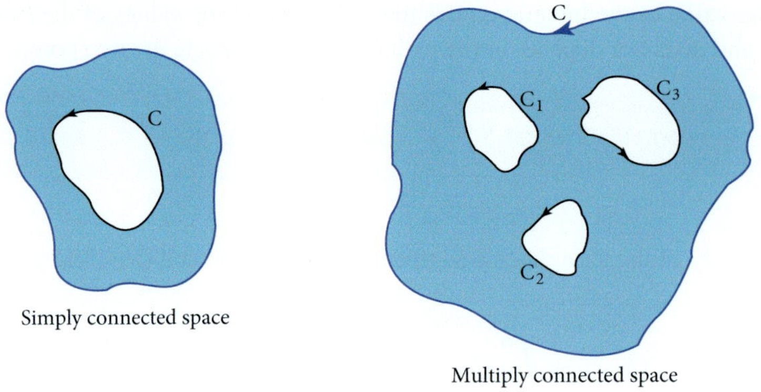

Fig. A.7 Simply and multiply connected space

We observe that the results we obtained for these three cases of calculating the contour integral are different than each other. This happens because the function $f(z) = z^*$ does not have a derivative function and therefore the value of the contour integral depends on the path C along which the calculation took place.

Theorem A.4.4 *If a function $f(z)$ is analytic on a closed path C, as well as at every point of a simply connected region that is enclosed by this path, it is proven that*

$$\oint_C f(z)dz = 0. \tag{A.6}$$

The above theorem is known as Gauss's theorem for simply connected regions and its main consequence is that *the integral of a function $f(z)$ between two points A and B is independent of the path that connects the two points, provided that it is analytic in the region that is enclosed by two different paths that connect the two points, as well as on these paths.* If the region in which the function is defined is not simply but *multiply connected*, Gauss's theorem is formulated as follows.

Theorem A.4.5 *If $f(z)$ is an analytic function in a closed multiply connected region that is externally bounded by the path C and internally by the paths $C_1, C_2, \ldots, C_n$, then the integral of the function $f(z)$ along the outer path C is equal to the sum of the integrals of $f(z)$ along the inner paths $C_1, C_2, \ldots, C_n$, with the integration along all paths being done at the same time. This property in mathematical writing is rendered by the equation*

$$\oint_C f(z)dz = \sum_{i=1}^{n} \oint_{C_j} f(z)dz. \tag{A.7}$$

The way of calculating the integral of the function $f(z)$ for simply and multiply connected places is graphically presented in Fig. A.7.

Theorem A.4.6 *If $f(z)$ is an analytic function on a closed path C and at every point of the place enclosed by this path and z is an inner point of the path C, then it is proven that the function in question satisfies the Gauss type.*

$$f(z) = \frac{1}{2\pi j} \oint_C \frac{f(z')}{(z' - z)} dz'. \tag{A.8}$$

A.4.3 Power Series of Complex Functions and Residue Calculation

It is proven that an analytic complex function $f(z)$ can be developed into a *Taylor series*, as well as into a *Laurent* series. The equations defining these power series are described by the following theorems.

Theorem A.4.7 *A function $f(z)$ that is analytic inside a circle with the center at point z_0 and the radius r_0 can be represented at every internal point z of the circle as the Taylor expansion*

$$f(z) = \sum_{n=0}^{\infty} \frac{1}{n!} f^{(n)}(z_0)(z - z_0)^n. \tag{A.9}$$

For $z_0 = 0$ the above relation is reduced to the simplified form

$$f(z) = \sum_{n=0}^{\infty} \frac{f^{(n)}(0)}{n!} z^n, \tag{A.10}$$

which is known as the Maclaurin expansion of the function $f(z)$.

Problem A.4.4 Find the Taylor expansion of the function $f(z) = \sin(z)$ around the point $z = \pi/4$.

Answer From the defining equation of the Taylor expansion it follows that

$$f(z) = \sum_{n=0}^{\infty} \frac{1}{n!} f^{(n)}(z_0)(z - z_0)^n =$$

$$= f\Big(\frac{\pi}{4}\Big) + f'\Big(\frac{\pi}{4}\Big)\Big(z - \frac{\pi}{4}\Big) + \frac{1}{2!} f''\Big(\frac{\pi}{4}\Big)\Big(z - \frac{\pi}{4}\Big)^2 + \frac{1}{3!} f'''\Big(\frac{\pi}{4}\Big)\Big(z - \frac{\pi}{4}\Big)^3 + \frac{1}{4!} f^{iv}\Big(\frac{\pi}{4}\Big)\Big(z - \frac{\pi}{4}\Big)^4 + \ldots,$$

where $f(z) = \sin(z), f'(z) = \cos(z), f''(z) = -\sin(z), f'''(z) = -\cos(z), f^{iv}(z) = \sin(z), \ldots$ and for $z = \pi/4$, $f(\pi/4) = \sqrt{2}/2$, $f'(\pi/4) = \sqrt{2}/2, f''(\pi/4) = -\sqrt{2}/2$, $f'''(\pi/4) = -\sqrt{2}/2$, $f^{iv}(\pi/4) = \sqrt{2}/2, \ldots$. Therefore, substituting in the above equation, we get

$$\begin{aligned} f(z) &= \frac{\sqrt{2}}{2} + \frac{\sqrt{2}}{2}\Big(z - \frac{\pi}{4}\Big) - \frac{1}{2!}\frac{\sqrt{2}}{2}\Big(z - \frac{\pi}{4}\Big)^2 - \frac{1}{3!}\frac{\sqrt{2}}{2}\Big(z - \frac{\pi}{4}\Big)^3 + \frac{1}{4!}\frac{\sqrt{2}}{2}\Big(z - \frac{\pi}{4}\Big)^4 + \ldots \\ &= \frac{\sqrt{2}}{2}\left(1 + \Big(z - \frac{\pi}{4}\Big) - \frac{1}{2!}\Big(z - \frac{\pi}{4}\Big)^2 - \frac{1}{3!}\Big(z - \frac{\pi}{4}\Big)^3 + \frac{1}{4!}\Big(z - \frac{\pi}{4}\Big)^4 + \ldots\right). \end{aligned}$$

Theorem A.4.8 *If a function $f(z)$ is analytic inside and at the boundaries of a circular ring with radii R and r (with $R > r$), then it can be represented at every internal point of the ring as a Laurent series of the form*

$$f(z) = \sum_{n=0}^{\infty} \alpha_n (z - z_0)^n + \sum_{n=1}^{\infty} \frac{\beta_n}{(z - z_o)^n}, \tag{A.11}$$

where $\alpha_n = \dfrac{1}{2\pi j} \oint_{C_R} \dfrac{f(z')dz'}{(z' - z_0)^{n+1}}$ $(n = 0, 1, 2, \ldots)$ *and* $\beta_n = \dfrac{1}{2\pi j} \oint_{C_r} \dfrac{f(z')dz'}{(z' - z_0)^{-n+1}}$ $(n = 1, 2, \ldots)$,

and C_R, C_r are the circular paths defined by the peripheries $\mid z \mid = R$ and $\mid z \mid = r$. The integration along the paths C_r and C_R is done in the positive direction.

Problem A.4.5 Find the Laurent series of the complex function given by the equation $f(z) = e^{2z}/(z-1)^3$ around the point $z = 1$.

Answer Setting $z - 1 = u$ will be $z = u + 1$ and the above function will take the form

$$f(u) = \frac{e^{2(u+1)}}{u^3} = \frac{e^{2u+2}}{u^3} = \frac{e^2}{u^3}e^{2u}.$$

From the basic theory, however, it is known that the exponential function can be developed into an infinite series of terms of the form

$$e^x = 1 + x + \frac{x^2}{2!} + \frac{x^3}{3!} + \frac{x^4}{4!} + \dots.$$

From the above relation for $x = 2u$ we get

$$e^{2u} = 1 + 2u + \frac{(2u)^2}{2!} + \frac{(2u)^3}{3!} + \frac{(2u)^4}{4!} + \dots$$

and substituting into the defining equation of the function $f(u)$, this will take the form

$$f(u) = \frac{e^2}{u^3}\left(1 + 2u + \frac{(2u)^2}{2!} + \frac{(2u)^3}{3!} + \frac{(2u)^4}{4!} + \dots\right).$$

In the last step we replace the variable u with the expression $z - 1$ and we are led to the equation

$$f(z) = \frac{e^2}{(z-1)^3} + \frac{2e^2}{(z-1)^2} + \frac{2e^2}{z-1} + \frac{4e^2}{3} + \frac{2e^2}{3}(z-1) + \dots,$$

which is the desired result.

The nature and characteristics of the Laurent expansion of a complex function $f(z)$ allows us to classify its isolated singular points into three categories, which in general are defined as follows:

- A *removable singular point* is called an isolated singular point, if in the Laurent expansion of the complex function $f(z)$ in the interval $0 < |z| < r$ does not include terms with negative powers of n, that is, $\beta_n = 0$ for every $n \in N$.
- *Pole of order m* of the complex function $f(z)$ is called an isolated singular point $z = z_0$ if in the Laurent expansion of $f(z)$ in the interval $0 < |z| < r$ there is an integer m such that $\beta_n = 0$ for $\forall n > m$, but $\beta_m \neq 0$. This property is mathematically formulated as

$$f(z) = \sum_{n=0}^{\infty} \alpha_n (z - z_0)^n + \frac{\beta_1}{z - z_0} + \frac{\beta_2}{(z - z_0)^2} + \dots + \frac{\beta_m}{(z - z_0)^m} \quad (\beta_m \neq 0). \tag{A.12}$$

  If the relation $m = 1$ holds, then the isolated singular point that satisfies the above property is called a *first-order pole*.
- *Isolated essential singular point* is called a point $z = z_0$, if in the Laurent expansion of the function $f(z)$ in a ring of the form $0 < |z - z_0| < r$ there are infinite nonzero coefficients β_n.

Definition A.4.5 (Residue) *Residue* or simply *residue* of the complex function $f(z)$ at the isolated singular point $z = z_0$ (or even at a regular point) different than $z = \infty$ is called the coefficient β_1 of the Laurent expansion of the function $f(z)$ in the ring $0 < |z - z_0| < r$, where the function is analytic and is denoted by Res $[f(z_0)]$. Thus

$$\text{Res}\,[f(z_0)] = \beta_1 = \frac{1}{2\pi j}\oint_C f(z)dz, \tag{A.13}$$

where C is any closed path in the positive direction located in the place where the function is analytic and the only anomalous point it contains is the point $z = z_0$.

Theorem A.4.9 *If the point $z = z_0$ is a pole of order m for the function $f(z)$, then the residue of $f(z)$ at the point z_0 is given by the formula*

$$Res\,[f(z_0)] = \begin{cases} \lim\limits_{z\to z_0} [(z - z_0) f(z)] & \text{if } m = 1 \\ \dfrac{1}{(m-1)!} \lim\limits_{z\to z_0} \dfrac{d^{m-1}}{dz^{m-1}} \Big[(z - z_0)^m f(z) \Big] & \text{if } m > 1. \end{cases} \tag{A.14}$$

Theorem A.4.10 *If the function $f(z)$ is a fractional function of the form $f(z) = g(z)/h(z)$ and the functions $g(z)$ and $h(z)$ are analytic in the vicinity of z_0, whereas simultaneously the relations $g(z_0) \neq 0$, $h(z_0) = 0$ and $h'(z_0) \neq 0$ hold (i.e., z_0 is a simple root of the function $h(z)$), the integral residue of $f(z)$ at the simple pole z_0 will be given by the relation $Res\,[f(z_0)] = g(z_0)/h'(z_0)$.*

Problem A.4.6 Find the integral residues of the function

$$f(z) = \frac{z^2 - 2z}{(z+1)^2(z^2+4)}.$$

Answer From the analytic form of the function we find that it has a second-order pole at the point $z = -1$ and two simple poles at the points $z = \pm 2j$. Therefore, by direct application of the defining equation of the residue, we get

$$\mathrm{Res}\,[f(-1)] = \frac{1}{(2-1)!} \lim_{z\to -1} \frac{d^{2-1}}{dz^{2-1}} \left[(z+1)^2 \frac{z^2 - 2z}{(z+1)^2(z^2+4)} \right] = \lim_{z\to -1} \frac{d}{dz} \left(\frac{z^2 - 2z}{z^2+4} \right) = \lim_{z\to -1} \frac{2z^2 + 8z - 8}{(z^2+4)^2} = -\frac{14}{25}$$

$$\mathrm{Res}\,[f(+2j)] = \lim_{z\to +2j} \left[(z - 2j) \frac{z^2 - 2z}{(z+1)^2(z+2j)(z-2j)} \right] = \lim_{z\to +2j} \frac{z^2 - 2z}{(z+1)^2(z+2j)} = \frac{-4-4j}{(2j+1)^2(+4j)} = \frac{7+j}{25}$$

$$\mathrm{Res}\,[f(-2j)] = \lim_{z\to -2j} \left[(z + 2j) \frac{z^2 - 2z}{(z+1)^2(z+2j)(z-2j)} \right] = \lim_{z\to -2j} \frac{z^2 - 2z}{(z+1)^2(z-2j)} = \frac{-4+4j}{(-2j+1)^2(-4j)} = \frac{7-j}{25}.$$

Problem A.4.7 Find the residues of the following function at all its poles on the complex plane.

$$f(z) = \frac{1+z}{z(2-z)}.$$

Answer The function has two simple poles at the positions $z = 0$ and $z = 2$. Therefore, we have

$$\mathrm{Res}\,[f(0)] = \lim_{z\to 0} \left[z \frac{1+z}{z(2-z)} \right] = \lim_{z\to 0} \left[\frac{1+z}{(2-z)} \right] = \frac{1}{2} \quad \text{and}$$

$$\mathrm{Res}\,[f(2)] = \lim_{z\to 2} \left[(z-2) \frac{1+z}{z(2-z)} \right] = -\lim_{z\to 2} \left[\frac{1+z}{z} \right] = -\frac{3}{2}.$$

Problem A.4.8 Find the residues of the following function at all its poles on the complex plane.

$$f(z) = \frac{e^z}{(z-1)(z+3)^2}.$$

Answer From the analytic form of the function we find that it has a simple pole at $z = 1$ and a double pole at $z = -3$. Therefore,

$$\text{Res}\,[f(+1)] = \lim_{z\to 1}\left[(z-1)\frac{e^z}{(z-1)(z+3)^2}\right] = \lim_{z\to 1}\left[\frac{e^z}{(z+3)^2}\right] = \frac{e}{16} \quad \text{and}$$

$$\text{Res}\,[f(-3)] = \frac{1}{(2-1)!}\lim_{z\to -3}\frac{d}{dz}\left[(z+3)^2\frac{e^z}{(z-1)(z+3)^2}\right] = \lim_{z\to -3}\frac{d}{dz}\left[\frac{e^z}{z-1}\right] = \lim_{z\to -3}\left[\frac{e^z(z-2)}{(z-1)^2}\right] = -\frac{5}{16e^3}.$$

Theorem A.4.11 *If C is a closed path and $f(z)$ is a function that is analytic on the path C and inside it except for a finite number of isolated singular points $z_1, z_2, \ldots, z_n$ inside the place C, then it is proven that*

$$\oint_C f(z)dz = 2\pi j \sum_{k=1}^{n} Res\,[f(z_k)]. \tag{A.15}$$

The above theorem is known as the *residue theorem*; it is one of the most important theorems of complex analysis and is used very often for the calculation of contour integrals of complex functions on the complex plane.

Problem A.4.9 Calculate the contour integral

$$I = \oint_C \frac{e^z}{(z-1)(z+3)^2}\,dz,$$

where C is the circumference of a circle with equation $|\, z \,| = 10$.

Answer As we have proven in the previous exercise, the function $f(z) = e^z/[(z-1)(z+3)^2]$ has a simple pole at position $z = 1$ and a double pole at position $z = -3$, with residue values at these points $\text{Res}\,[f(1)] = e/16$ and $\text{Res}\,[f(-3)] = -5/(16e^3)$. Since both these points are located within the closed path C, the application of the residue theorem will lead to the equation

$$I = \oint_C \frac{e^z}{(z-1)(z+3)^2}dz = 2\pi j\left\{\text{Res}\,[f(1)] + \text{Res}\,[f(-3)]\right\} = \frac{\pi j(e^4-5)}{8e^3}.$$

Problem A.4.10 Calculate the contour integral

$$I = \frac{1}{2\pi j}\oint_C \frac{e^{zt}}{z^2(z^2+2z+2)}\,dz,$$

where C is the circumference of a circle with equation $|\, z \,| = 3$.

Answer The equation $z^2 + 2z + 2 = 0$ is a quadratic algebraic equation with respect to the variable z with discriminant $\Delta = \beta^2 - 4\alpha\gamma = 4 - 8 = -4$ and roots $z_1 = -1 + j$ and $z_2 = -1 - j$. These two values constitute the two simple poles of the function to be integrated, which also has an additional double pole at position $z = 0$. It is not difficult to ascertain that the set of poles of the function to be integrated are within the closed path $\mid z \mid = 3$ and therefore the calculation of the value of the integral requires the calculation of the residues of the function for all these poles. In detail, then, we have

$$\begin{aligned}
\text{Res}\,[f(0)] &= \frac{1}{(2-1)!}\lim_{z\to 0}\frac{d^{2-1}}{dz^{2-1}}\left[z^2\frac{e^{zt}}{z^2\,(z^2+2z+2)}\right] = \lim_{z\to 0}\frac{d}{dz}\left[\frac{e^{zt}}{z^2+2z+2}\right]\\
&= \lim_{z\to 0}\left\{\frac{e^{zt}\,[tz^2+(2t-2)z+(2t-2)]}{(z^2+2z+2)^2}\right\} = \frac{t-1}{2}\\
\text{Res}\,[f(-1+j)] &= \lim_{z\to -1+j}\left[(z+1-j)\frac{e^{zt}}{z^2\,(z+1-j)\,(z+1+j)}\right] = \lim_{z\to -1+j}\left[\frac{e^{zt}}{z^2\,(z+1+j)}\right] = \frac{1}{4}e^{(-1+j)t}\\
\text{Res}\,[f(-1-j)] &= \lim_{z\to -1-j}\left[(z+1+j)\frac{e^{zt}}{z^2\,(z+1-j)\,(z+1+j)}\right] = \lim_{z\to -1-j}\left[\frac{e^{zt}}{z^2\,(z+1-j)}\right] = \frac{1}{4}e^{(-1-j)t}.
\end{aligned}$$

Applying the residue theorem, we get

$$\begin{aligned}
\oint_C \frac{e^{zt}}{z^2\,(z^2+2z+2)}\,dz &= 2\pi j\left\{\text{Res}\,[f(0)] + \text{Res}\,[f(-1+j)] + \text{Res}\,[f(-1-j)]\right\}\\
&= 2\pi j\left[\frac{t-1}{2} + \frac{1}{4}e^{-t}(e^{jt}+e^{-jt})\right] = 2\pi j\left[\frac{t-1}{2} + \frac{1}{2}e^{-t}\cos(t)\right]
\end{aligned}$$

and therefore

$$I = \frac{1}{2\pi j}2\pi j\left(\frac{t-1}{2} + \frac{1}{2}e^{-t}\cos(t)\right) = \frac{t-1}{2} + \frac{1}{2}e^{-t}\cos(t).$$

Problem A.4.11 Calculate the next contour integral along the path C, which is the circumference of a circle with equation $z = e^{j\vartheta}$.

$$I = \oint_C \frac{1+z}{z(2-z)}dz.$$

Answer In a previous exercise, we had shown that the complex function $f(z) = (1+z)/[z(2-z)]$ has two simple poles at positions $z = 0$ and $z = 2$, with residue values $\text{Res}\,[f(0)] = 1/2$ and $\text{Res}\,[f(2)] = -3/2$ respectively. However, of these poles, only the pole $z = 0$ is enclosed by the closed path $z = e^{j\vartheta}$ for which the relation $\mid z \mid = 1$ holds. Therefore, by applying the residue theorem, we get

$$I = \oint_C \frac{1+z}{z(2-z)}dz = 2\pi j\;\text{Res}\,[f(z)] = 2\pi j \times \frac{1}{2} = \pi j.$$

A.5 The Method of Polynomial Long Division

The method of *polynomial long division* is quite useful in processes of calculating inverse transforms and allows the division of polynomials in a way similar to that used for the division of integer numbers. Considering two polynomial functions $f(x)$ and $d(x)$ of degrees p and q respectively, with $q \leq p$, it is proven that there are unique polynomials $q(x)$ and $r(x)$ such that

$$\frac{f(x)}{d(x)} = q(x) + \frac{r(x)}{d(x)}. \tag{A.16}$$

If k is the degree of the polynomial function $r(x)$, it is obvious that the relation $k \leq q$ must hold. In full analogy with the division of integers, the polynomials $d(x)$, $q(x)$, and $r(x)$ are known as the divisor, the quotient and the remainder of the division. In the special case where the quotient $r(x)$ is equal to zero, we say that the polynomial $d(x)$ divides the polynomial $f(x)$ exactly.

In order to understand the operation of polynomial division, let us use an example and assume that we want to calculate the value of the expression

$$\frac{3x^3 - 2x^2 + 4x - 3}{x^2 + 3x + 3}.$$

In the first step of the process, we write the two polynomials side by side in the form

$$\begin{array}{r|llll} x^2 + 3x + 3 \lceil & 3x^3 & -2x^2 & +4x & -3, \end{array}$$

that is, we first place the polynomial—divisor and then the polynomial—dividend. Then, we divide the first term of the dividend by the first term of the divisor (i.e., we perform the operation $3x^3/x^2$) and place the quotient of the division (i.e., the expression $3x$) directly above the first term of the dividend, as shown in the next expression:

$$\begin{array}{r|llll} & 3x & & & \\ \hline x^2 + 3x + 3 \lceil & 3x^3 & -2x^2 & +4x & -3. \end{array}$$

Then, (a) we multiply the quotient of the division by the polynomial-divisor (i.e., we perform the multiplication $3x(x^2 + 3x + 3)$), (b) we write the result of the operation (which is the polynomial $3x^3 + 9x^2 + 9x$) exactly below the dividend, as shown below:

$$\begin{array}{r|llll} & 3x & & & \\ \hline x^2 + 3x + 3 \lceil & 3x^3 & -2x^2 & +4x & -3 \\ & 3x^3 & +9x^2 & +9x, & \end{array}$$

and (c) we subtract the second line from the first, so as to eliminate the term of the dividend that is raised to the highest power

$$\begin{array}{r|llll} & 3x & & & \\ \hline x^2 + 3x + 3 \lceil & 3x^3 & -2x^2 & +4x & -3 \\ & 3x^3 & +9x^2 & +9x & \\ \hline & & -11x^2 & -5x & -3. \end{array}$$

At this point, we have reduced the order of the dividend by one unit, ending up with the quadratic polynomial $-11x^2 - 5x - 3$. In order to proceed, we just have to repeat the same process, but considering as the dividend not our original polynomial but the second-degree polynomial with which we have ended up. In other words, we perform the division $-11x^2/x^2$ and place the remainder of it (which is the value -11) at the top of the table and next to the remainder of the previous division, which was equal to $3x$. Then, we multiply this value of the remainder with the polynomial divisor (i.e., we perform the operation $-11(x^2 + 3x + 3)$) and write down its result (which is the polynomial $-11x^2 - 33x - 33$) under the last line, ending up with the result that is presented below:

$$\begin{array}{r|llll} & 3x & -11 & & \\ \hline x^2 + 3x + 3 \lceil & 3x^3 & -2x^2 & +4x & -3 \\ & 3x^3 & +9x^2 & +9x & \\ \hline & & -11x^2 & -5x & -3 \\ & & -11x^2 & -33x & -33. \end{array}$$

In the last step of the process, we subtract, as before, the last line from the previous one and we arrive at the result

$$\begin{array}{rrrrr} & 3x & -11 & & \\ \cline{2-5} x^2+3x+3 \,\lceil & 3x^3 & -2x^2 & +4x & -3 \\ & 3x^3 & +9x^2 & +9x & \\ \cline{2-5} & & -11x^2 & -5x & -3 \\ & & -11x^2 & -33x & -33 \\ \cline{3-5} & & & +28x & +30. \end{array}$$

If we attempt to repeat the process, we find that the quotient of the division $28x/x^2$ does not lead to a polynomial expression, since the value of the power of x that results is negative. This means that the process has been completed. Therefore, the result of the division of the polynomial $3x^3 - 2x^2 + 4x - 3$ by the polynomial $x^2 + 3x + 3$ led to a quotient value equal to $3x - 11$, whereas the remainder of the division is equal to $28x + 30$. Using the notation we presented during the theoretical introduction of the method of long polynomial division, the above result is written in the form

$$\frac{3x^3 - 2x^2 + 4x - 3}{x^2 + 3x + 3} = (3x - 11) + \frac{28x + 30}{x^2 + 3x + 3}.$$

Problem A.5.1 Calculate the result of the division $\dfrac{x^3 - 3x^2 + 5x - 3}{x - 1}$.

Answer Applying the methodology of the previous paragraph, we get

$$\begin{array}{rrrrr} & x^2 & -2x & +3 & \\ \cline{2-5} x-1 \,\lceil & x^3 & -3x^2 & +5x & -3 \\ & x^3 & -x^2 & & \\ \cline{2-5} & & -2x^2 & +5x & -3 \\ & & -2x^2 & +2x & \\ \cline{3-5} & & & +3x & -3 \\ \cline{3-5} & & & +3x & -3 \\ \cline{4-5} & & & & 0\;. \end{array}$$

From the process, it turns out that the remainder of the division is equal to zero and therefore the polynomial $d(x) = x - 1$ divides the polynomial $f(x) = x^3 - 3x^2 + 5x - 3$ exactly, so we can write that $x^3 - 3x^2 + 5x - 3 = (x - 1)(x^2 - 2x + 3)$.

Problem A.5.2 Calculate the result of the division $\dfrac{x^4 + 4}{x^2 - 5}$.

Answer Applying the methodology of the previous paragraph, we get

$$\begin{array}{rrrr} & x^2 & +5 & \\ \cline{2-4} x^2-5 \,\lceil & x^4 & & +4 \\ & x^4 & -5x^2 & \\ \cline{2-4} & & +5x^2 & +4 \\ & & +5x^2 & -25 \\ \cline{3-4} & & & +29. \end{array}$$

Therefore, we have

$$\frac{x^4 + 4}{x^2 - 5} = (x^2 + 5) + \frac{29}{x^2 - 5}$$

A.6 Expansion of a Rational Function into a Sum of Partial Fractions

The technique of expanding a rational function $g(x) = f(x)/d(x)$ as a sum of partial fractions is very useful in processes of analytical calculation of integrals of rational functions. However, it also constitutes the basis of one of the techniques for calculating inverse transforms and for this reason it is briefly described in this section. Let us note that when the degree of the polynomial function $f(x)$ is greater than or equal to the degree of the function $d(x)$, the application of this method presupposes the prior use of the operation of polynomial division that was presented in the previous section, so that a polynomial of a lower degree in relation to the original one emerges. A typical example of the application of the method of expansion into partial fractions is the writing of the rational function

$$g(x) = (x^3 + 4x^2 - 1)/[x^3(x-1)^2]$$

as a sum of fractions of the form

$$g(x) = \frac{f(x)}{d(x)} = \frac{x^3 + 4x^2 - 1}{x^3(x-1)^2} = \frac{1}{x} - \frac{2}{x^2} - \frac{1}{x^3} - \frac{1}{x-1} + \frac{4}{(x-1)^2}.$$

In a more detailed description, the application of the method of partial fraction decomposition is carried out in the following way:

(a) Provided that the degree of the function $f(x)$ is greater or equal to the degree of the function $d(x)$, the method of polynomial division is applied and two polynomials $q(x)$ and $r(x)$ are determined, such that

$$\frac{f(x)}{d(x)} = q(x) + \frac{r(x)}{d(x)}. \tag{A.17}$$

If, however, the above condition does not hold, this stage is ignored.

(b) The polynomial function $d(x)$ that is in the denominator of the rational function $g(x)$ undergoes the process of factorization, that is, it is written as a product of two or more polynomials of lower degree that can be both linear (i.e., first degree with respect to x) and non-linear (i.e., of degree greater than or equal to two). This process in mathematical form is expressed as

$$d(x) = \prod_{i=1}^{n} p_i(x, q_i), \tag{A.18}$$

where n is the number of polynomials that result from the factorization and q_i $(i = 1, 2, \ldots, n)$ is the degree of each of these.

(c) The rational function $g(x)$ is written as a sum of fractions of the form

$$g(x) = \frac{f(x)}{g(x)} = \frac{s_1(x, q_1 - 1)}{p_1(x, q_1)} + \cdots + \frac{s_n(x, q_n - 1)}{p_n(x, q_n)} = \sum_{i=1}^{n} \frac{s_i(x, q_i - 1)}{p_1(x, q_i)}. \tag{A.19}$$

In other words, in the denominator of each fraction are the polynomials that resulted from the factorization—one polynomial in each fraction—whereas the degree of the polynomials that appear in the numerator is one unit less than the degree of the corresponding polynomial of the denominator. Thus, if in the denominator there is a linear term (i.e., a first-degree polynomial), in the numerator there will be a zero-degree polynomial (i.e., essentially a constant); if in the denominator there is a second-degree polynomial of the form $\alpha x^2 + \beta x + \gamma$, in the numerator there will be a first-degree polynomial of the form $\lambda x + \mu$, etc. In order to determine the coefficients of the polynomials that appear in each case, we combine the partial fractions into one simple fraction, the denominator of which is identical to the function $d(x)$ and we impose the requirement for the polynomial of the numerator to be the same as the function $f(x)$. There are four basic cases that usually appear in practice, the detailed description of which, along with indicative examples that will help the reader to understand the method of use, are presented in the following pages.

Case 1 the function $d(x)$ can be expressed as a product of n linear polynomials

$$d(x) = (x - x_1)(x - x_2)\ldots(x - x_n), \tag{A.20}$$

where obviously $x_1, x_2, \ldots, x_n$ are the roots of the polynomial $d(x)$. In this case, the rational function $g(x)$ is written as a sum of fractions of the form

$$g(x) = \frac{f(x)}{d(x)} = \frac{\alpha_1}{x - x_1} + \frac{\alpha_2}{x - x_2} + \cdots + \frac{\alpha_n}{x - x_n} \tag{A.21}$$

with the coefficients $\alpha_1, \alpha_2, \ldots \alpha_n$ emerging in the way we previously described. In order to understand the way of applying the method for this case, let us consider the following example.

Example suppose we want to develop into a sum of partial fractions the function

$$g(x) = \frac{3x+5}{x^2+3x+2}.$$

We observe that the quadratic equation $x^2+3x+2=0$ has discriminant $\Delta = 1$ and roots $x_1 = -1$ and $x_2 = -2$, that is, the function $d(x)$ is written as a product of the form $d(x) = (x+1)(x+2)$. In this case, the rational function $g(x)$ is developed into a sum of partial fractions as follows:

$$\begin{aligned}\frac{3x+5}{x^2+3x+2} &= \frac{3x+5}{(x+1)(x+2)} = \frac{A}{x+1} + \frac{B}{x+2} = \frac{A(x+2)+B(x+1)}{(x+1)(x+2)}\\ &= \frac{Ax+2A+Bx+B}{(x+1)(x+2)} = \frac{(A+B)x+(2A+B)}{x^2+3x+2}.\end{aligned}$$

If we compare the first with the last fraction, we find that in order for these to be equal to each other, the relation must hold

$$(A+B)x + (2A+B) = 3x+5.$$

By equating the coefficients of the corresponding powers of x the system of equations $A+B=3$ and $2A+B=5$ arises, from which it follows that $A=2$ and $B=1$. Therefore, the rational function $g(x)$ is written as a sum of partial fractions of the form

$$g(x) = \frac{f(x)}{d(x)} = \frac{3x+5}{x^2+3x+2} = \frac{2}{x+1} + \frac{1}{x+2}.$$

Case 2 the function $d(x)$ can be expressed as a product of n linear polynomials, some of which (or possibly all) may be repeated, i.e., raised to some power. This representation in mathematical form is formulated as

$$d(x) = (x-x_1)^{q_1}(x-x_2)^{q_2}\ldots(x-x_n)^{q_n} = \prod_{i=1}^{n}(x-x_i)^{q_i}. \tag{A.22}$$

In this case, the function $g(x)$ is developed into a sum of partial fractions of the form

$$\begin{aligned}g(x) = &\frac{\alpha_1^1}{x-x_1} + \frac{\alpha_2^1}{(x-x_1)^2} + \cdots + \frac{\alpha_{q_1}^1}{(x-x_1)^{q_1}} + \frac{\alpha_1^2}{x-x_2} + \frac{\alpha_2^2}{(x-x_2)^2} + \ldots\\ &\frac{\alpha_{q_2}^2}{(x-x_2)^{q_2}} + \cdots + \frac{\alpha_1^n}{x-x_n} + \frac{\alpha_2^n}{(x-x_n)^2} + \cdots + \frac{\alpha_{q_n}^n}{(x-x_n)^{q_n}}.\end{aligned} \tag{A.23}$$

Example suppose we want to develop into a sum of partial fractions the function

$$g(x) = \frac{x^3+4x^2-1}{x^5-2x^4+x^3}.$$

In this case, we observe that the function of the denominator can be developed into the product

$$d(x) = x^5 - 2x^4 + x^3 = x^3(x^2-2x+1) = x^3(x-1)^2$$

and is a special case of the general expression

$$d(x) = (x-x_1)^{q_1}(x-x_2)^{q_2}\ldots(x-x_n)^{q_n}$$

with parameter values $n = 2$, $x_1 = 0$, $x_2 = 1$, $q_1 = 3$, and $q_2 = 2$. Therefore, according to what we mentioned earlier, the function $g(x)$ can be written as the sum of partial fractions

$$g(x) = \frac{x^3 + 4x^2 - 1}{x^5 - 2x^4 + x^3} = \frac{x^3 + 4x^2 - 1}{x^3(x-1)^2} = \frac{A}{x} + \frac{B}{x^2} + \frac{C}{x^3} + \frac{D}{x-1} + \frac{E}{(x-1)^2}.$$

If we combine the partial fractions of the above equation into a simple fraction and perform the operations, we will be led to the result

$$g(x) = \frac{A}{x} + \frac{B}{x^2} + \frac{C}{x^3} + \frac{D}{x-1} + \frac{E}{(x-1)^2} = \frac{Ax^2(x-1)^2 + Bx(x-1)^2 + C(x-1)^2 + Dx^3(x-1) + Ex^3}{x^3(x-1)^2}$$

$$= \frac{(A+D)x^4 + (B - 2A - D + E)x^3 + (A - 2B + C)x^2 + (B - 2C)x + C}{x^3(x-1)^2}.$$

Combining the last two relations, we find that it must be

$$x^3 + 4x^2 - 1 = (A+D)x^4 + (B - 2A - D + E)x^3 + (A - 2B + C)x^2 + (B - 2C)x + C$$

and by equating the coefficients of the corresponding powers of x, we get the system of equations

$$A + D = 0, \quad B - 2A - D + E = 1, \quad A - 2B + C = 4, \quad B - 2C = 0, \quad -C = 1$$

with the solution $A = 1$, $B = -2$, $C = -1$, $D = -1$ and $E = 4$. Therefore, the rational function $g(x)$ is written as a sum of partial fractions of the form

$$g(x) = \frac{f(x)}{d(x)} = \frac{x^3 + 4x^2 - 1}{x^3(x-1)^2} = \frac{1}{x} - \frac{2}{x^2} - \frac{1}{x^3} - \frac{1}{x-1} + \frac{4}{(x-1)^2}.$$

Case 3 the function $d(x)$ can be expressed as a product of n polynomials, of which m are linear and the remaining k are nonlinear (i.e., of at least degree two). In the simplest of cases where the nonlinear polynomials are of second degree with respect to the independent variable x, the above property is mathematically described by the equation

$$d(x) = (x - x_1) \cdots (x - x_m) \cdot (\alpha_1 x^2 + \beta_1 x + \gamma_1) \cdots (\alpha_k x^2 + \beta_k x + \gamma_k) = \prod_{i=1}^{m} (x - x_i) \prod_{j=1}^{k} (\alpha_j x^2 + \beta_j x + \gamma_j). \quad \text{(A.24)}$$

In this case, in the sum of partial fractions, the linear polynomials will correspond to fractions, the numerator of which will contain the value of a constant, whereas the second-degree polynomials will correspond to fractions, the numerator of which will contain a first-degree polynomial (this happens because, according to what we have described in previous pages, the degree of the polynomial in the numerator is one unit less than the degree of the polynomial in the denominator). Therefore, the function $g(x)$ will be represented by a sum of partial fractions of the form

$$g(x) = \frac{\lambda_1}{x - x_1} + \cdots + \frac{\lambda_m}{x - x_m} + \frac{q_1 x + r_1}{\alpha_1 x^2 + \beta_1 x + \gamma_1} + \cdots + \frac{q_k x + r_k}{\alpha_k x^2 + \beta_k x + \gamma_k}. \quad \text{(A.25)}$$

Example suppose we wish to develop in a sum of partial fractions the function

$$g(x) = \frac{x^2 - 9x - 9}{x^3 - 2x^2 + x - 2}.$$

In this case, we observe that the function of the denominator can be factored in the following way:

$$d(x) = x^3 - 2x^2 + x - 2 = x(x^2 + 1) - 2(x^2 + 1) = (x - 2)(x^2 + 1).$$

In the above equation, the polynomial $h(x) = x^2 + 1$ is of the second degree with respect to x and cannot be factored further. Therefore, according to the theory, the function $g(x)$ can be written as a sum of partial fractions of the form

$$g(x) = \frac{x^2 - 9x + 9}{x^3 - 2x^2 + x - 2} = \frac{x^2 - 9x + 9}{(x-2)(x^2+1)} = \frac{Ax + B}{x^2 + 1} + \frac{C}{x - 2}$$
$$= \frac{(Ax + B)(x - 2) + C(x^2 + 1)}{(x^2 + 1)(x - 2)} = \frac{(A + C)x^2 + (B - 2A)x + (C - 2B)}{x^3 - 2x^2 + x - 2}.$$

Comparing the first with the last fraction we find that these are equal to each other, if

$$x^2 - 9x + 9 = (A + C)x^2 + (B - 2A)x + (C - 2B).$$

Equating the coefficients of the corresponding powers of x, we end up with the system of equations $A + C = 1$, $2A - B = 9$ and $C - 2B = 9$ with solution $A = 2$, $B = -5$, and $C = -1$. Therefore, the function $g(x)$ can be expressed as the sum of partial fractions

$$g(x) = \frac{f(x)}{d(x)} = \frac{x^2 - 9x + 9}{x^3 - 2x^2 + x - 2} = \frac{2x - 5}{x^2 + 1} - \frac{1}{x - 2}.$$

Case 4 the function $d(x)$ can be expressed as a product of n polynomials, both linear and nonlinear some of which may be repeated, that is, they may be raised to some power. In full analogy with the previous case, we consider that these nonlinear factors cannot be simplified further, that is, they cannot be written as a product of polynomials of a lower degree.

Assuming for simplicity that the nonlinear factors are second-degree polynomials of the independent variable x and ignoring the existence of the linear factors that may exist, the above representation of the function $d(x)$ is mathematically rendered by the equation

$$d(x) = (\alpha_1 x^2 + \beta_1 x + \gamma_1)^{q_1} \cdot (\alpha_2 x^2 + \beta_2 x + \gamma_2)^{q_2} \cdots (\alpha_n x^2 + \beta_n x + \gamma_n)^{q_n} = \prod_{i=1}^{n} (\alpha_i x^2 + \beta_i x + \gamma_i)^{q_i}, \tag{A.26}$$

where q_i is the degree of the second-degree term $\alpha_i x^2 + \beta_i x + \gamma_i$ $(i = 1, 2, \ldots, n)$. According to the theory of partial-fraction expansion, the function $g(x)$ in this case will be expanded as a sum of fractional terms of the form

$$\begin{aligned} g(x) = {} & \frac{\lambda_1^1 x + \mu_1^1}{\alpha_1 x^2 + \beta_1 x + \gamma_1} + \frac{\lambda_2^1 x + \mu_2^1}{(\alpha_1 x^2 + \beta_1 x + \gamma_1)^2} + \cdots + \frac{\lambda_{q_1}^1 x + \mu_{q_1}^1}{(\alpha_1 x^2 + \beta_1 x + \gamma_1)^{q_1}} \\ & + \frac{\lambda_1^2 x + \mu_1^2}{\alpha_2 x^2 + \beta_2 x + \gamma_2} + \frac{\lambda_2^2 x + \mu_2^2}{(\alpha_2 x^2 + \beta_2 x + \gamma_2)^2} + \cdots + \frac{\lambda_{q_1}^2 x + \mu_{q_1}^2}{(\alpha_2 x^2 + \beta_2 x + \gamma_2)^{q_2}} \\ & + \frac{\lambda_1^n x + \mu_1^n}{\alpha_n x^2 + \beta_n x + \gamma_n} + \frac{\lambda_2^n x + \mu_2^n}{(\alpha_n x^2 + \beta_n x + \gamma_n)^2} + \cdots + \frac{\lambda_{q_n}^n x + \mu_{q_n}^n}{(\alpha_n x^2 + \beta_n x + \gamma_n)^{q_n}}. \end{aligned} \tag{A.27}$$

Example suppose we want to expand the function into a sum of partial fractions

$$g(x) = \frac{x^4 + x^3 + 2x^2 - x + 4}{x(x^2 + 2)(x^2 + 1)^3}.$$

In this case, we observe that the function of the denominator (which is assumed to have already been factored in some way) contains both linear and nonlinear terms, some of which are raised to some power. Therefore, according to the theory, the function $g(x)$ can be written as a sum of partial fractions of the form

$$g(x) = \frac{x^4 + x^3 + 2x^2 - x + 4}{x(x^2 + 2)(x^2 + 1)^3} = \frac{A}{x} + \frac{Bx + C}{x^2 + 2} + \frac{Dx + E}{x^2 + 1} + \frac{Fx + G}{(x^2 + 1)^2} + \frac{Hx + I}{(x^2 + 1)^3}.$$

If we perform the operations and apply the methodology that we used in the previous examples, we find the values $A = 2$, $B = 2$, $C = 3$, $D = -4$, $E = -3$, $F = 0$, $G = 3$, $H = -3$ and $I = -2$. Therefore, the function $g(x)$ can be written as the sum of partial fractions

$$g(x) = \frac{x^4 + x^3 + 2x^2 - x + 4}{x(x^2+2)(x^2+1)^3} = \frac{2}{x} + \frac{2x+3}{x^2+2} - \frac{4x+3}{x^2+1} + \frac{3}{(x^2+1)^2} - \frac{3x+2}{(x^2+1)^3}.$$

Problem A.6.1 Given the rational function

$$g(x) = \frac{x^4 + 5x^3 + 16x^2 + 26x + 22}{x^3 + 3x^2 + 7x + 5},$$

develop this function into a sum of partial fractions.

Answer From the analytic form of the function $g(x)$ we find that the degree of the polynomial of the numerator is greater than the degree of the polynomial of the denominator. Therefore, before the expansion of the function into partial fractions, we should perform the operation of polynomial division so as to get a polynomial of a lower degree. Therefore, we have

$$\begin{array}{r|rrrrr}
 & x & +2 & & & \\
\hline
x^3 + 3x^2 + 7x + 5 & x^4 & +5x^3 & +16x^2 & +26x & +22 \\
 & x^4 & +3x^3 & +7x^2 & +05x & \\
\hline
 & & +2x^3 & +9x^2 & +21x & +22 \\
 & & +2x^3 & +6x^2 & +14x & +10 \\
\hline
 & & & +3x^2 & +7x & +12
\end{array}$$

and the rational function $g(x)$ is therefore formulated as

$$g(x) = \frac{f(x)}{d(x)} = (x+2) + \frac{3x^2 + 7x + 12}{x^3 + 3x^2 + 7x + 5}.$$

Consequently, the process of expansion into partial fractions will not be applied to our initial function, but to the function

$$\frac{r(x)}{d(x)} = \frac{3x^2 + 7x + 12}{x^3 + 3x^2 + 7x + 5}.$$

In order to factorize the function of the denominator $d(x) = x^3 + 3x^2 + 7x + 5$, we find that the value $x = -1$ is a root of the function—that is, the relation $d(-1) = 0$ holds—and therefore the polynomial $d(x)$ is exactly divided by the polynomial $(x + 1)$. Therefore, applying the operation of polynomial division between these two polynomials—the intermediate operations are omitted for brevity—we find that $x^3 + 3x^2 + 7x + 5 = (x + 1)(x^2 + 2x + 5)$. Thus

$$\begin{aligned}
\frac{r(x)}{d(x)} &= \frac{3x^2 + 7x + 12}{x^3 + 3x^2 + 7x + 5} = \frac{3x^2 + 7x + 12}{(x+1)(x^2 + 2x + 5)} = \frac{A}{x+1} + \frac{Bx + C}{x^2 + 2x + 5} \\
&= \frac{(A+B)x^2 + (2A + B + C)x + (5A + C)}{x^3 + 3x^2 + 7x + 5},
\end{aligned}$$

from which it follows that $A = 2$, $B = 1$ and $C = 2$. Therefore, our initial function will be expanded as a sum of fractions of the form

$$g(x) = \frac{x^4 + 5x^3 + 16x^2 + 26x + 22}{x^3 + 3x^2 + 7x + 5} = (x+2) + \frac{2}{x+1} + \frac{x+2}{x^2 + 2x + 5}.$$

A.7 Difference Equations and Their Solution Techniques

Let us consider the bilateral sequence of real or complex numbers

$$\Big\{ \dots, y[-2],\ y[-1],\ y[0],\ y[1],\ y[2],\ \dots,\ y[n-1], y[n],\ \dots \Big\} \tag{A.28}$$

and suppose there is a relationship between the terms of the form

$$f\,\{y[n+r],\ y[n+r-1],\ \dots,\ y[n-1], y[n]\} = 0. \tag{A.29}$$

Such an equation is called a *difference equation* and any sequence similar to the one we defined earlier is called a solution to this equation. Difference equations are fundamental mathematical tools for studying recursive discrete-time systems and for this reason, the most important properties that describe them are presented on the following pages.

In a difference equation, the quantity $\Delta y_n = y_{n+1} - y_n$—for brevity in writing we use the notation $y[n] = y_n$—is known as the *first difference*, with the operator Δ being referred to as the *difference operator*. Another important operator that is often used in equations of this type is the *forward difference operator* defined as $Ey_n = y_{n+1}$. Combining the last two relations, we get $(1+\Delta)y_n = y_{n+1} = Ey_n$, from which it follows that the two aforementioned operators are related through the equation $E = 1 + \Delta$. Having defined the first difference, we can define the *second difference* from the equation

$$\Delta^2 y_n = \Delta(\Delta y_n) = \Delta(y_{n+1} - y_n) = (y_{n+2} - y_{n+1}) - (y_{n+1} - y_n) = y_{n+2} - 2y_{n+1} + y_n \tag{A.30}$$

and inductively the *difference of m-th order* as $\Delta^m y_n = \Delta(\Delta^{m-1} y_n)$.

Finally, in a completely analogous way we can define the difference $E^2 y_n = E(Ey_n) = Ey_{n+1} = y_{n+2}$ and in the general case write that $E^m y_n = E(E^{m-1} y_n) = y_{n+m}$.

A.7.1 First-Order Linear Difference Equations

A difference equation of the form

$$\alpha_r(n)y[n+r] + \alpha_{r-1}(n)y[n+r-1] + \cdots + \alpha_1(n)y[n+1] + \alpha_0(n)y[n] = f(n) \tag{A.31}$$

is called a *linear difference equation* of order r. If in the above general equation we set $r = 1$, we get the *first-order linear difference equation*

$$\alpha_1(n)y[n+1] = -\alpha_0(n)y[n] + f(n) \qquad \text{or equivalently} \qquad y[n+1] = -\frac{\alpha_0(n)}{\alpha_1(n)}y[n] + \frac{f(n)}{\alpha_1(n)} = p(n)y[n] + q(n),$$

where $p(n) = -[\alpha_0(n)/\alpha_1(n)]$ and $q(n) = [f(n)/\alpha_1(n)]$, the solution of which is given by the relation

$$y[n] = \left(\sum_{j=1}^{n-1} p(j) \right) y[1] + \sum_{k=1}^{n-1} \left(\prod_{j=k+1}^{n-1} p(j) \right) q(k). \tag{A.32}$$

If $q(k) = 0$ $(k = 0, 1, 2, \dots)$, the above equation is called *homogeneous* and its solution is

$$y[n] = \left(\sum_{j=1}^{n-1} p(j) \right) y[1]. \tag{A.33}$$

Problem A.7.1 Solve the difference equation $y[n+1] = \dfrac{n}{n+1} y[n] + n \quad (n \geq 1)$.

Answer The linear difference equation to be solved has the general form $y[n+1] = p(n)y[n]+q(n)$ with $p(n) = n/(n+1)$ and $q(n) = n$. Therefore, we have

$$\prod_{j=1}^{n-1} p(j) = \prod_{j=1}^{n-1} \frac{j}{j+1} = \frac{1}{2}\cdot\frac{2}{3}\cdot\frac{3}{4}\cdots\frac{n-2}{n-1}\cdot\frac{n-1}{n} = \frac{1}{n}$$

$$\prod_{j=k+1}^{n-1} p(j) = \prod_{k-1}^{n-1} \frac{j}{j+1} = \frac{k+1}{k+2}\cdot\frac{k+2}{k+3}\cdot\frac{k+3}{k+4}\cdots\frac{n-1}{n} = \frac{k+1}{n}$$

and since $q(k) = k$ $(k = 0, 1, 2, \ldots)$, the second term of the solution of the difference equation is written as

$$\sum_{k=1}^{n-1}\left[\prod_{j=k+1}^{n-1} p(j)\right] q(k) = \sum_{k=1}^{n-1} \frac{k+1}{n} k = \frac{1}{n}\sum_{k=1}^{n-1} k(k+1) = \frac{1}{n}\left(\sum_{k=1}^{n\ 1} k^2 + \sum_{k=1}^{n-1} k\right)$$

$$= \frac{1}{n}\left[\frac{n(n-1)(n-2)}{6} + \frac{n(n-1)}{2}\right] = \frac{1}{3}(n^2-1).$$

Therefore, the solution of the difference equation will be given by the relation $y[n] = \frac{1}{n}y[1] + \frac{1}{3}(n^2-1) \quad (n \geq 1)$.

A.7.2 Linear Difference Equations of Higher Order

In order to determine the solution of *linear difference equations of higher order*, we must first define the concept of *linear independence among sequences*. Considering two sequences $x[n]$ and $y[n]$, they are considered *linearly dependent*, if there are constants C_1 and C_2 different than zero such that $C_1x[n] + C_2y[n] = 0$ $(n = 0, 1, 2, \ldots)$. Otherwise, the two sequences are called *linearly independent*.

An important quantity that is used very often for the characterization of discrete sequences is the Casorati determinant , which is defined as

$$C[x(n), y(n)] = \begin{vmatrix} x[n] & y[n] \\ x[n+1] & y[n+1] \end{vmatrix} \tag{A.34}$$

and is used in the same way as the Wronski determinant is used for the characterization of solutions $y_1(x)$ and $y_2(x)$ of a second-order differential equation as linearly independent. If we consider a complete second-order difference equation of the form

$$y[n+2] + p(n)y[n+1] + q(n)y[n] = f(n) \quad (q(n) \neq 0) \tag{A.35}$$

and two solutions of it $y_1[n]$ and $y_2[n]$, then these two solutions will be linearly independent, when the Casorati determinant is different from zero, a property that in mathematical formulation is written as $C[y_1[n], y_2[n]] \neq 0$. In this case, the general solution of the homogeneous difference equation (which arises from the complete equation by setting $f(n) = 0$) is any linear combination of the form $y[n] = C_1y_1[n] + C_2y_2[n]$, with the constants C_1 and C_2 being determined with the help of the initial conditions. Having determined the general solution of the homogeneous equation, all we have to do is to determine any partial solution $y_p[n]$ of the corresponding complete equation. In this case, *the general solution of the complete equation will result from the sum of the general solution of the homogeneous equation and the partial solution of the complete equation*, or in mathematical writing

$$y[n] = C_1y_1[n] + C_2y_2[n] + y_p[n]. \tag{A.36}$$

After a brief description of the linear independence of sequences, let us now see how the solution of higher-order difference equations is possible. In the following description, we deal with *second-order difference equations* (the generalization of the methodology for higher-order equations is obvious), whereas for simplicity reasons we assume that the coefficients of the terms of the difference equation are constant (the case of variable coefficients will not concern us in this section). According to the theory associated with the methods of solving difference equations, a *homogeneous linear equation of second-order differences* of the form

$$y[n+2] + \alpha_1 y[n+1] + \alpha_0 y[n] = 0 \quad (\alpha_1, \alpha_2 \in R, \alpha_2 \neq 0)$$

admits the exponential solutions $y[n] = \lambda^n$, where λ is a real or complex number different than zero. Thus

$$y[n+1] = \lambda^{n+1} \qquad \text{and} \qquad y[n+2] = \lambda^{n+2}$$

and substituting in the difference equation, it will take the form

$$\lambda^{n+2} + \alpha_1 \lambda^{n+1} + \alpha_2 \lambda^n = 0 \Rightarrow \qquad \text{(dividing by } \lambda^n\text{)} \qquad \Rightarrow \lambda^2 + \alpha_1 \lambda + \alpha_2 = 0.$$

The last equation is a quadratic equation with discriminant $\Delta = \beta^2 - 4\alpha\gamma = \alpha_1^2 - 4\alpha_2$ and roots

$$\lambda_1 = \frac{-\beta + \sqrt{\Delta}}{2\alpha} = \frac{-\alpha_1 + \sqrt{\alpha_1^2 - 4\alpha_2}}{2} \qquad \text{and} \qquad \lambda_2 = \frac{-\beta - \sqrt{\Delta}}{2\alpha} = \frac{-\alpha_1 - \sqrt{\alpha_1^2 - 4\alpha_2}}{2},$$

the nature and characteristics of which determine the form of the general solution of the homogeneous difference equation. We distinguish the following cases:

- The two roots are different from each other ($\lambda_1 \neq \lambda_2$). In this case, the solutions that arise from these are $y_1[n] = \lambda_1^n$ and $y_2[n] = \lambda_2^n$ and the general solution of the homogeneous equation has the form

$$y[n] = C_1 \lambda_1^n + C_2 \lambda_2^n. \tag{A.37}$$

- The two roots are equal to each other ($\lambda_1 = \lambda_2$). In this case, the solutions that arise from the double root λ_1 are $y_1[n] = \lambda_1^n$ and $y_2[n] = n\lambda_2^n$ and the general solution of the homogeneous equation is

$$y[n] = C_1 \lambda_1^n + nC_2 \lambda_1^n = (C_1 + nC_2)\lambda_1^n. \tag{A.38}$$

- The two roots are conjugate complex numbers of the form $\lambda_1 = \alpha + j\beta = re^{j\vartheta}$ and $\lambda_2 = \alpha - j\beta = re^{-j\vartheta}$, where $r = \sqrt{\alpha^2 + \beta^2}$, $\cos\vartheta = \alpha/r$, and $\sin\vartheta = \beta/r$ $(-\pi \leq \vartheta \leq \pi)$. In this case, the general solution of the homogeneous equation will have the form

$$\begin{aligned} y[n] &= C_1 \lambda_1^n + C_2 \lambda_2^n = C_1(\alpha + j\beta)^n + C_2(\alpha - j\beta)^n = C_1 r^n e^{jn\vartheta} + C_2 r^n e^{-jn\vartheta} \\ &= r^n[C_1(\cos(n\vartheta) + j\sin(n\vartheta)) + C_2(\cos(n\vartheta) + j\sin(n\vartheta))] \\ &= r^n[(C_1 + C_2)\cos(n\vartheta) + j(C_1 - C_2)\sin(n\vartheta)] = r^n[A\cos(n\vartheta) + B\sin(n\vartheta)], \end{aligned} \tag{A.39}$$

  where $A = C_1 + C_2$ and $B = j(C_1 - C_2)$. In the above expression, the constants C_1 and C_2 are real numbers, although they can generally be complex quantities; in the latter case, the general solution of the homogeneous equation is given by some other expressions that arise in a similar way.

Problem A.7.2 Solve the difference equation $y[n+2] - y[n+1] - 12y[n] = 0$.

Answer Considering an exponential solution of the form $y[n] = \lambda^n$, the difference equation is written as $\lambda^2 - \lambda - 12 = 0$, which is a quadratic equation with discriminant $\Delta = \beta^2 - 4\alpha\gamma = 1 - 4(-12) = 49$ and roots

$$\lambda_1 = \frac{-\beta + \sqrt{\Delta}}{2\alpha} = \frac{1+7}{2} = 4 \qquad \text{and} \qquad \lambda_2 = \frac{-\beta - \sqrt{\Delta}}{2\alpha} = \frac{1-7}{2} = -3.$$

Therefore, the general solution of the difference equation is $y[n] = C_1(-3)^n + C_2(4)^n$.

Problem A.7.3 Solve the difference equation

$$y[n+2] - 6y[n+1] + 9y[n] = 0.$$

Answer Considering an exponential solution $y[n] = \lambda^n$ the equation is written as $\lambda^2 - 6\lambda + 9 = 0$. The above equation has a discriminant $\Delta = \beta^2 - 4\alpha\gamma = (-6)^2 - 4 * 9 = 0$ and has a double root $\lambda = -\beta/2\alpha = 3$. Therefore, the general solution of the equation is

$$y[n] = C_1(3)^n + nC_2(3)^n = (C_1 + nC_2)(3)^n.$$

Problem A.7.4 Solve the difference equation $y[n+2] + 2y[n+1] + 2y[n] = 0$.

Answer Considering an exponential solution of the form $y[n] = \lambda^n$ the equation is written as $\lambda^2 + 2\lambda + 2 = 0$. The above equation has a discriminant $\Delta = \beta^2 - 4\alpha\gamma = 4 - 8 = -4$ and has two conjugate complex roots, which are

$$\lambda_1 = \frac{-\beta + \sqrt{\Delta}}{2\alpha} = \frac{-2 + 2j}{2} = -1 + j \quad \text{and} \quad \lambda_2 = \frac{-\beta - \sqrt{\Delta}}{2\alpha} = \frac{-2 - 2j}{2} = -1 - j.$$

We observe that the modulus of the conjugates is $r = \sqrt{2}$, whereas it is also $\cos\vartheta = -1/\sqrt{2}$ and $\sin\vartheta = -1/\sqrt{2}$, from which it follows that $\vartheta = 3\pi/4$. Therefore, according to the theory we presented earlier, the solution of the difference equation will have the form

$$y[n] = (\sqrt{2})^n \left[C_1 \cos\left(\frac{3n\pi}{4}\right) + C_2 \sin\left(\frac{3n\pi}{4}\right) \right],$$

which is the desired result. ■

After determining the general solution of the homogeneous equation, we should determine a partial solution of the full equation, so that we end up with the general solution as the sum of the two above solutions. The methodologies and techniques with which we can determine this general solution are presented in the seventh chapter of the book, where difference equations are presented together with the *recursive discrete-time systems* of which they constitute basic study tools. Let it be noted here that in the general case of the full linear second-order difference equation

$$y[n+2] + \alpha_1 y[n+1] + \alpha_2 y[n] = f(n) \tag{A.40}$$

with the function $f(n)$ having the form

$$f(n) = \sum_i P_i(n)(\beta_i)^n [A_i \cos(\gamma_i n) + B_i \sin(\gamma_i n)], \tag{A.41}$$

where $P_i(n)$ is a polynomial of n and $A_i, B_i, \beta_i, \gamma_i$ are constant quantities, the partial solution of the full equation is the

$$\gamma(n) = \sum_i (n)^{\rho_i} (\beta_i)^n [A_i(n) \cos(\gamma_i n) + B_i(n) \sin(\gamma_i n)], \tag{A.42}$$

where $A_i(n)$ and $B_i(n)$ are polynomials of the variable n of the same degree with $P_i(n)$, whereas the value of the parameter ρ_i is equal to zero or equal to one depending on whether the quantity $\beta_i(\cos\gamma_i + j\sin\gamma_i)$ is or is not a root of the characteristic equation $\lambda^2 + \alpha_1\lambda + \alpha_2 = 0$ respectively. However, in practice, the form of the function $f(n)$ is generally simpler. As mentioned in the seventh chapter, if the form of the function $f(n)$ belongs to some specific categories (e.g., exponential, polynomial), then the partial solution has a similar form and therefore it is relatively easy to determine (see Table 7.1 on page 593).

A.8 Chebyshev Rational Function Approximation

Let us consider a continuous function $f(x)$ defined in the interval $[\alpha, \beta]$ and let us assume that we are looking for a rational function of the form

$$R_{mk}(x) = \frac{P_m(x)}{Q_k(x)} = \frac{\sum\limits_{n=0}^{m} c_n x^n}{\sum\limits_{n=0}^{k} d_n x^n} \tag{A.43}$$

such that the quantity

$$r_{mk} = \max_{[\alpha,\beta]} \| w(x)[f(x) - R_{mk}(x)] \| \tag{A.44}$$

—where the weighting function $w(x)$ is positive in the interval (α, β)—is minimized for all functions $R_{m'k'}(x) = P_{m'}(x)/Q_{k'}(x)$ with degrees $m' \leq m$ and $k' \leq k$ for the polynomials $P_{m'}(x)$ and $Q_{k'}(x)$ respectively. The determination of this function is based on the application of the following theorem, which has been proven by Chebyshev:

Theorem A.8.1 *The quantity r_{mk} is minimized by a unique function $R^*_{mk}(x)$ of the form $R_{mk}(x)$. If we transform this function into the form*

$$R^*_{mk}(x) = \frac{P^*_m(x)}{Q^*_k(x)} = \frac{\sum\limits_{n=0}^{m-\nu} c^*_{n+\nu} x^n}{\sum\limits_{n=0}^{k-\mu} d^*_{n+\mu} x^n} \tag{A.45}$$

[$+0 \leq \mu \leq k$, $0 \leq \nu \leq m$, $c^*_m, d^*_k \neq 0$] and assume that the expression $P^*_m(x)/Q^*_k(x)$ is expressed in a form that cannot be simplified further, it is proven that if the relation $r^*_{mk} = \max \| w(x)[f(x) - R^*_{mk}(x)] \| \neq 0$ holds, then the number of consecutive points in the interval $[\alpha, \beta]$, at which the function $E(x) = w(x)[f(x) - R^*_{mk}(x)]$ takes its maximum value $E_{max}(x) = r^*_{mk}$ with an alternating sign, cannot be less than $L = N + 2 - d$, where $N = m + k$ and $d = \min(\mu, \nu)$.

In the following description, for simplicity, we consider the case $d = 0$—if the value of d is positive, the function $R_{mk}(x)$ is called degenerate—whereas without loss of generality, we assume that the interval $[\alpha, \beta]$ includes the value $x = 0$. Closely related to the following description is the next theorem.

Theorem A.8.2 *Let us assume that the function*

$$S_{mk}(x) = \frac{\sum\limits_{n=0}^{m-\nu} c^*_{n+\nu} x^n}{\sum\limits_{n=0}^{k-\mu} d^*_{n+\mu} x^n} \tag{A.46}$$

is expressed in its simplest form and that the difference $w(x)[f(x) - S_{mk}(x)]$ is finite in the interval $[\alpha, \beta]$. Let us also consider $L = N + 2 - d$ points of the interval $[\alpha, \beta]$ such that $x_1 < x_2 <, \ldots, x_L$ and let us define the quantities $w(x_i)[f(x_i) - S_{mk}(x_i)] = (-1)^i \lambda_i$ $(i = 1, 2, \ldots, L)$, where $\lambda_i > 0$ $(\forall i)$. In this case, it is proven that

(continued)

$$\upsilon_{mk} = \max_{[\alpha,\beta]} \|w(x)[f(x) - V_{m'k'}(x)]\| \geq \min_i \lambda_i, \tag{A.47}$$

where $V_{m'k'}(x)$ any rational function with polynomial degrees numerator and denominator $m' \leq m$ and $k' \leq k$.

A.9 The Chebyshev Polynomials

The Chebyshev polynomials[1] are the solution of the differential equation of Chebyshev

$$(1-x^2)\frac{d^2y(x)}{dx^2} - x\frac{dy(x)}{dx} + N^2y(x) = 0, \quad N = 1, 2, 3, \ldots, \tag{A.48}$$

which, by making the substitution $x = \cos t$, takes the form

$$\frac{d^2y(t)}{dt^2} + N^2y(t) = 0 \tag{A.49}$$

with a general solution

$$y(t) = A\cos(Nt) + B\sin(Nt) = A\cos(N\cos^{-1}x) + B\sin(N\cos^{-1}x) = AT_N(x) + BU_N(x), \quad |x| < 1, \tag{A.50}$$

where $T_N(x) = \cos(N\cos^{-1}x)$ and $U_N(x) = \sin(N\sin^{-1}x)$ are the Chebyshev polynomials of degree N of the first and second kind respectively. The above expressions concern the range of values $|x| < 1$, whereas for the range of values $|x| > 1$ we use the substitution $x = \cosh t$. In this case, the differential equation of Chebyshev is formulated as

$$\frac{d^2y(t)}{dt^2} - N^2y(t) = 0 \tag{A.51}$$

with a general solution

$$y(t) = A\cosh(Nt) + B\sinh(Nt) = A\cosh(N\cosh^{-1}x) + B\sinh(N\cosh^{-1}x) = AT_N(x) + BU_N(x), \quad |x| > 1, \tag{A.52}$$

where $T_N(x) = \cosh(N\cosh^{-1}x)$ and $U_N(x) = \sinh(N\sinh^{-1}x)$ are the Chebyshev polynomials of degree N of the first and second kinds respectively. Combining the above results for these two types of polynomials, we have

$$T_N(x) = \begin{cases} \cos(N\cos^{-1}x) & |x| < 1 \\ \cosh(N\cosh^{-1}x) & |x| > 1 \end{cases} \quad \text{and} \quad U_N(x) = \begin{cases} \sin(N\cos^{-1}x) & |x| < 1 \\ \sinh(N\cosh^{-1}x) & |x| > 1. \end{cases}$$

The Chebyshev polynomials of the first kind can arise either from Rodrigues' formula

$$T_N(x) = \frac{(-2)^N N!}{(2N!)}\sqrt{1-x^2}\frac{d^N}{dx^N}\left\{(1-x^2)^{N-\frac{1}{2}}\right\} \tag{A.53}$$

or from the generating function

$$G(x) = \frac{1-tx}{1-2tx+t^2} = \sum_{n=0}^{\infty} T_N(x)t^N, \tag{A.54}$$

[1] P. L. Chebyshev, *Théorie des mécanismes connus sous le nom de parallelogrammes*, Oeuvres, Vol I, St. Petersburg, 1899.

Table A.1 The Chebyshev polynomials of the first kind $T_N(x)$ for order values $N = 1, 2, \ldots, 10$

$T_0(x) = 1$
$T_1(x) = x$
$T_2(x) = 2x^2 - 1$
$T_3(x) = 4x^3 - 3x$
$T_4(x) = 8x^4 - 8x^2 + 1$
$T_5(x) = 16x^5 - 20x^3 + 5x$
$T_6(x) = 32x^6 - 48x^4 + 18x^2 - 1$
$T_7(x) = 64x^7 - 112x^5 + 56x^3 - 7x$
$T_8(x) = 128x^8 - 256x^6 + 160x^4 - 32x^2 + 1$
$T_9(x) = 256x^9 - 576x^7 + 432x^5 - 120x^3 + 9x$
$T_{10}(x) = 512x^{10} - 1280x^8 + 1120x^6 - 400x^4 + 50x^2 - 11x$

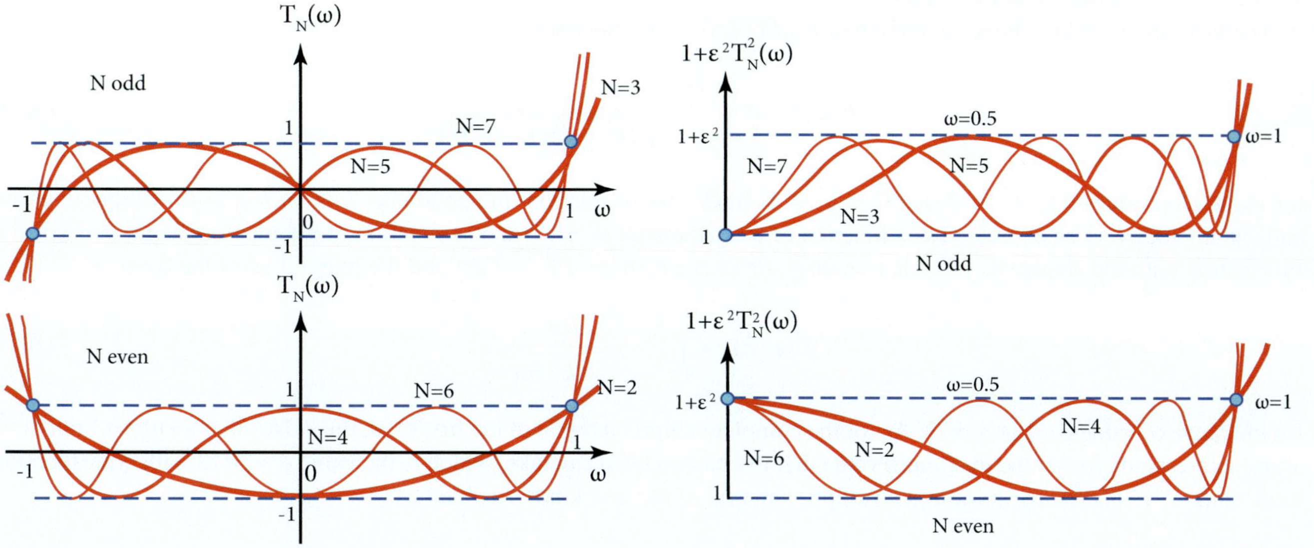

Fig. A.8 The Chebyshev polynomials of the first kind of degree N for various even- and odd-order values and the corresponding function $1 + T_N^2(\omega)$

although they are usually calculated much more easily from the recursive relation

$$T_{N+1}(x) = 2xT_N(x) - T_{N-1}(x) \tag{A.55}$$

starting from the expressions $T_0(x) = 1$ and $T_1(x) = x$. The Chebyshev polynomials for order values $N = 1, 2, \ldots, 10$ are included in Table A.1, whereas many of them for various even- and odd-order values are graphically depicted in Fig. A.8. It is not difficult to see that in the value range $-1 < x < 1$ the polynomials $T_N(x)$ oscillate between the values -1 and 1, whereas outside this range they tend to $+\infty$ or $-\infty$. We also observe that for the value $x = 0$ the odd-order polynomials pass through the origin, which of course is expected, whereas those corresponding to an even-order value have a value equal to -1 or +1. Conversely, for the value $x = 1$ it will always be $T_N(1) = 1$, regardless of their order value. The same figure also depicts the function $1 + \varepsilon^2 T_N^2(\omega)$, the value of which at the position $x = 0$ depends on the corresponding value of the Chebyshev polynomials and is equal to 1 for an odd-order value and equal to $1 + \varepsilon^2$ for an even-order value. This characteristic finds application in the study and design of analog filters of the Chebyshev I and II type (see Sect. 11.9.6 on page 1011 and Sect. 11.9.7 on page 1028).

The analytical description of the properties and applications of the Chebyshev polynomials is beyond the scope of this book and the interested reader is referred to mathematical physics books, such as those by Arfken and Riley, Robson and Bence. Some of the properties of these polynomials are presented in the following in the form of exercises to be solved for anyone interested in delving into their properties and how to use them.

A.10 Elliptic Integrals and Elliptic Functions

The *elliptic integrals* first appeared in the works of Wallis and Newton and were systematically used by Legendre for the study of mechanical problems, such as the motion of the simple pendulum. They are considered generalizations of inverse trigonometric functions and have the general form

$$f(x) = \int \frac{\alpha(x) + \beta(x)}{\gamma(x) + \delta(x)\zeta(x)}$$

with the functions of the right-hand side being polynomials with respect to x with the degree of the polynomial $\zeta(x)$ being equal to 3 or 4. In practice, these functions depend on two parameters, one of which is related to the upper limit of integration. There are three types of elliptic integrals, the integrals of *first*, *second*, and *third* kind, with each kind being able to be characterized as *complete* or *incomplete*.

The *incomplete elliptic integral of the first kind* is defined by the equation

$$\boldsymbol{F}(\varphi, k) = \int_0^{\sin\varphi} \frac{dt}{\sqrt{(1-t^2)(1-k^2t^2)}} \tag{A.56}$$

for parameter values $0 \leq k^2 \leq 1$ and $0 \leq \varphi \leq 1$. In the above relation, the parameter k is known as the elliptic measure and can alternatively be expressed as a function of the parameter $m = k^2$ or as a function of the measure angle $\alpha = \sin^{-1} k$. Proceeding with the change of variable $t = \sin\vartheta$, $dt = \cos\vartheta d\vartheta = \sqrt{1-t^2}\, d\vartheta$ and the integral takes the form

$$\boldsymbol{F}(\varphi, k) = \int_0^{\varphi} \frac{d\vartheta}{\sqrt{1-k^2\sin^2\vartheta}} \tag{A.57}$$

for the range of values $0 \leq \varphi \leq \pi/2$. As for the complete elliptic integral of the first kind, this arises if as an upper limit in the above expressions we set the maximum value that the parameter φ can take, which is the value $\varphi = \pi/2$ with corresponding limit $\sin\varphi = 1$. Therefore, the complete elliptic integral of the first kind is defined as

$$\boldsymbol{F}(k) = \int_0^1 \frac{dt}{\sqrt{(1-t^2)(1-k^2t^2)}} = \int_0^{\pi/2} \frac{d\vartheta}{\sqrt{1-k^2\sin^2\vartheta}}. \tag{A.58}$$

On the other hand, the *incomplete* and *complete elliptic integrals of the second kind* are defined as

$$\boldsymbol{E}(\varphi, k) = \int_0^{\sin\varphi} \frac{\sqrt{1-k^2t^2}}{\sqrt{1-t^2}dt} = \int_0^{\varphi} \sqrt{1-k^2\sin^2\vartheta} \quad \text{and} \quad \boldsymbol{E}(k) = \int_0^1 \frac{\sqrt{1-k^2t^2}}{\sqrt{1-t^2}dt} = \int_0^{\pi/2} \sqrt{1-k^2\sin^2\vartheta},$$

whereas the corresponding expressions for the *incomplete* and the *complete elliptic integrals of the third kind* are

$$\boldsymbol{\Pi}(\varphi, \eta, k) = \int_0^{\sin\varphi} \frac{dt}{(1+\eta t^2)\sqrt{(1-t^2)(1-k^2t^2)}} = \int_0^{\varphi} \frac{d\vartheta}{(1+\eta\sin^2\vartheta)(1-k^2\sin^2\vartheta)} \quad \text{and}$$

$$\boldsymbol{\Pi}(\eta, k) = \int_0^1 \frac{dt}{(1+\eta t^2)\sqrt{(1-t^2)(1-k^2t^2)}} = \int_0^{\pi/2} \frac{d\vartheta}{(1+\eta\sin^2\vartheta)(1-k^2\sin^2\vartheta)}$$

respectively. In the above expressions, the parameter η is known as the *characteristic of the integral* and can take any values, regardless of the values of the other parameters.

In the international literature, the complete elliptic integrals of the first and second kind are symbolized by $\boldsymbol{K}$ and $\boldsymbol{E}$ respectively—they will, therefore, be

$$\boldsymbol{F}\left(\varphi = \frac{\pi}{2}, k\right) = \boldsymbol{F}(\sin\varphi = 1, k) = \boldsymbol{K}(K) = \boldsymbol{K}$$

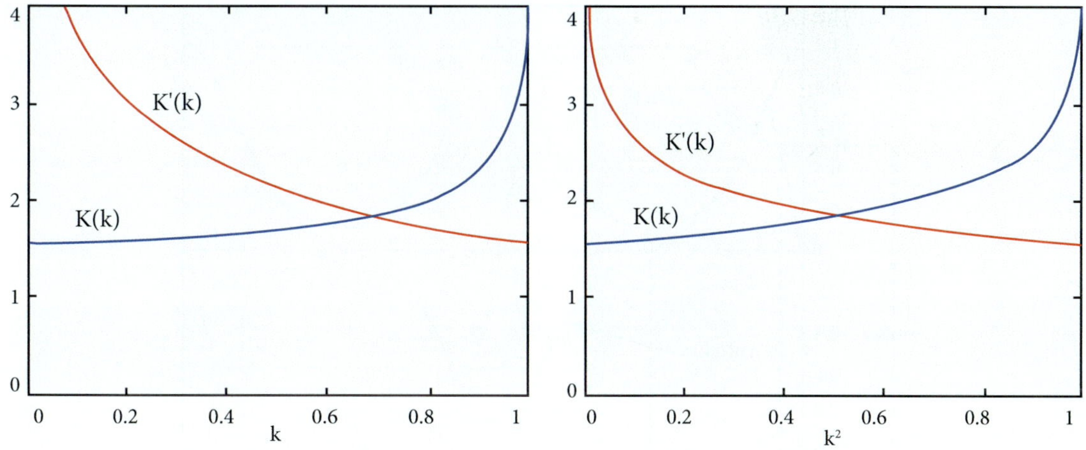

Fig. A.9 Graphical representations of the integrals $\boldsymbol{K}$ and $\boldsymbol{K}'$ as a function of the parameters k and k^2 for the value range $0 \leq k \leq 1$

and

$$\boldsymbol{E}\left(\varphi = \frac{\pi}{2}, k\right) = \boldsymbol{E}(\sin\varphi = 1, k) = \boldsymbol{E}(K) = \boldsymbol{E}.$$

Defining the new parameter $k' = \sqrt{1-k^2}$ and making the substitution $\upsilon = \tan\vartheta$, the above integrals are formulated in their so-called *complementary* form, which for the complete integrals leads to the corresponding expressions

$$\boldsymbol{F}\left(\varphi = \frac{\pi}{2}, k'\right) = \boldsymbol{F}(\sin\varphi = 1, k') = \boldsymbol{K}(K') = \boldsymbol{K}'$$

and

$$\boldsymbol{E}\left(\varphi = \frac{\pi}{2}, k'\right) = \boldsymbol{E}(\sin\varphi = 1, k') = \boldsymbol{E}(K') = \boldsymbol{E}'.$$

The complete integrals $\boldsymbol{K}$ and $\boldsymbol{K}'$ are known as a quarter of the period, and their graphical representations as a function of the elliptic measure k, as well as its square k^2, are presented in Fig. A.9. From this figure, we observe that at the position $k = 0$ it is $\boldsymbol{K} = \pi/2$ and $\boldsymbol{K}' = \infty$, whereas at the position $k = 1$ it is $\boldsymbol{K} = \infty$ and $\boldsymbol{K}' = \pi/2$. Based on the defining equations of the integrals $\boldsymbol{K}$ and $\boldsymbol{K}'$, we can prove a set of useful properties and correlations, which, of course, cannot be presented here. Some of these properties are listed below in the form of unsolved exercises for the interested reader, in several of which the auxiliary integral is used

$$\boldsymbol{D}(\varphi, k) = \frac{\boldsymbol{F} - \boldsymbol{E}}{k^2} = \int_0^{\varphi} \frac{\sin^2\vartheta}{\sqrt{1 - k^2\sin^2\vartheta}}\, d\vartheta \tag{A.59}$$

Elliptic Functions Closely related to elliptic integrals are the *elliptic functions*, which are associated with the inversion of elliptic integrals. There are many types of such functions, of which here we will briefly present only the Jacobi elliptic functions, which arise from the inversion of the incomplete elliptic integral of the first kind. Expressing the aforementioned integral in the form

$$u = F(\varphi, k) = \int_0^{\varphi} \frac{d\vartheta}{\sqrt{1 - k^2\sin^2\vartheta}} \tag{A.60}$$

the upper limit of integration is known as the Jacobi amplitude and is denoted by amp(u, k)—therefore, $\varphi(u, k) = F^{-1}(u, k) = \text{amp}(u, k)$. In this case, we can define the functions of the elliptic sine, the elliptic cosine, and the elliptic delta width from the relations

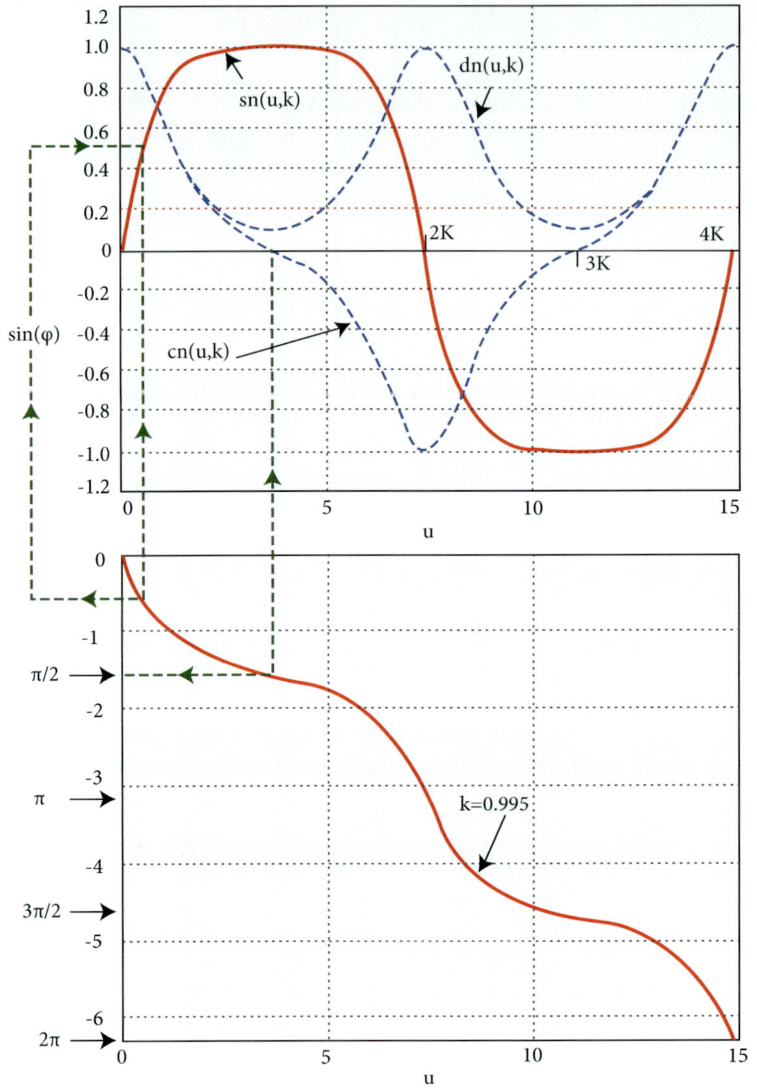

Fig. A.10 Graphical representations of the elliptic functions $\text{sn}(u, k)$, $\text{cn}(u, k)$, and $\text{dn}(u, k)$

$$
\begin{aligned}
\text{sn}(u, k) &= \sin \varphi(u, k) = \sin[\text{amp}(u, k)] \\
\text{cn}(u, k) &= \cos \varphi(u, k) = \cos[\text{amp}(u, k)] \\
\text{dn}(u, k) &= \sqrt{1 - k^2 \sin^2 \varphi(u, k)} \\
&= \sqrt{1 - k^2 \sin^2[\text{amp}(u, k)]}.
\end{aligned}
$$

These functions, the graphs of which are presented in Fig. A.10, are considered generalizations of the known trigonometric and hyperbolic functions for the values $k = 0$ and $k = 1$, since, as it emerges from the defining equations, they will be $\text{sn}(u, 0) = \sin u$, $\text{cn}(u, 0) = \cos u$, $\text{dn}(u, 0) = 1$, $\text{cd}(u, 0) = \cos u$, and $\text{sn}(u, 1) = \tanh u$, $\text{cn}(u, 1) = \text{sech} u$, $\text{dn}(u, 1) = \text{sech} u$, and $\text{cd}(u, 1) = 1$. Using the parameter $k' = \sqrt{1 - k^2}$, it is proven that the above functions satisfy the equations

$$
\begin{aligned}
&\text{sn}^2(u, k) + \text{cn}^2(u, k) = 1, \quad k^2\text{sn}^2(u, k) + \text{dn}^2(u, k) = 1, \\
&\text{dn}^2(u, k) - k^2\text{cn}^2(u, k) = k'^2, \ \text{sn}^2(u, k) = \frac{1 - \text{cd}^2(u, k)}{1 - k^2\text{cd}^2(u, k)},
\end{aligned}
$$

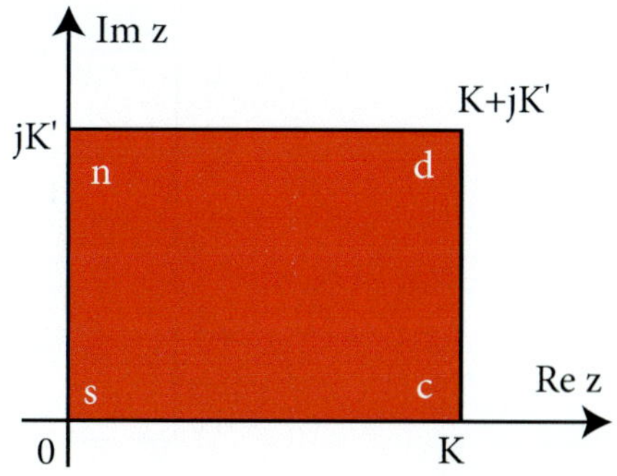

Fig. A.11 The fundamental rectangle of elliptic functions

and $k'^2\text{sn}^2(u,k) + \text{cn}^2(u,k) = \text{dn}^2(u,k)$.

These properties are easily proven using the defining equations of the above functions and this is left as an exercise to the reader.

The unified description of the elliptic functions in the complex plane is based on the use of the so-called fundamental rectangle that is drawn on the complex plane and is defined by the points $s(0,0)$, $c(\boldsymbol{K},0)$, $n(0,j\boldsymbol{K}')$, and $d(\boldsymbol{K},j\boldsymbol{K}')$ (see Fig. A.11). These four points are used in the naming of the elliptic functions, the name of which is in the form $\text{pq}(u,k)$ with the letter p being one of the s,c,n,d and the letter q one of the remaining three letters (the use of two similar letters is prohibited). So, there are 12 different combinations of such letters and therefore 12 different elliptic functions, three of which have already been mentioned in the previous paragraphs, whereas the remaining nine are

$$\text{cd}(u,k) = \frac{\text{cn}(u,k)}{\text{dn}(u,k)}, \qquad \text{dc}(u,k) = \frac{\text{dn}(u,k)}{\text{cn}(u,k)}$$

$$\text{ns}(u,k) = \frac{1}{\text{sn}(u,k)}, \qquad \text{sd}(u,k) = \frac{\text{sn}(u,k)}{\text{dn}(u,k)}$$

$$\text{nc}(u,k) = \frac{1}{\text{cn}(u,k)}, \qquad \text{ds}(u,k) = \frac{\text{dn}(u,k)}{\text{sn}(u,k)}$$

$$\text{nd}(u,k) = \frac{1}{\text{dn}(u,k)}, \qquad \text{sc}(u,k) = \frac{\text{sn}(u,k)}{\text{cn}(u,k)}$$

$$\text{cs}(u,k) = \frac{\text{cn}(u,k)}{\text{sn}(u,k)}$$

(note that the name of each function consists of two letters, the first of which is the first letter of the function of the numerator and the second is the first letter of the function of the denominator, whereas the name of the functions in the numerator of which there is a unit is the reversed name of the function of the denominator). Each of these functions is characterized by a zero value at the point p and by a pole at the point q. For example, the function $\text{sn}(u,k)$ has zero value at the point $s(0,0)$ and a pole at the point $n(0,j\boldsymbol{K}')$, the function $\text{cn}(u,k)$ has a zero value at the point $c(\boldsymbol{K},0)$ and a pole at the point $n(0,j\boldsymbol{K}')$, etc.

Of all these elliptic functions, in the design of elliptic filters we only need $\text{sn}(u,k)$ and $\text{cd}(u,k)$, of which the first is characterized by even symmetry and the second is characterized by odd symmetry, and they are related to each other through the relations $\text{cd}(u,k) = \text{sn}(u+\boldsymbol{K},k) = \text{sn}(\boldsymbol{K}-u,k)$. Other interesting properties of $\text{cd}(u,k)$, which generally find application in elliptic filter design processes, are the following:

$$\text{cd}\,[u+(2m-1)\boldsymbol{K},k] = (-1)^m\text{sn}(u,k) \qquad (\forall m \in N)$$

$$\text{cd}\,[u+2m\boldsymbol{K},k] = (-1)^m\text{cd}(u,k) \qquad (\forall m \in N)$$

$$\text{cd}\,[u+j\boldsymbol{K}',k] = 1/[k\text{cd}(u,k)] \qquad (\forall u \in N)$$

$$\text{cd}\,[ju,k] = 1/\text{dn}(u,k') \qquad (\forall u \in N).$$

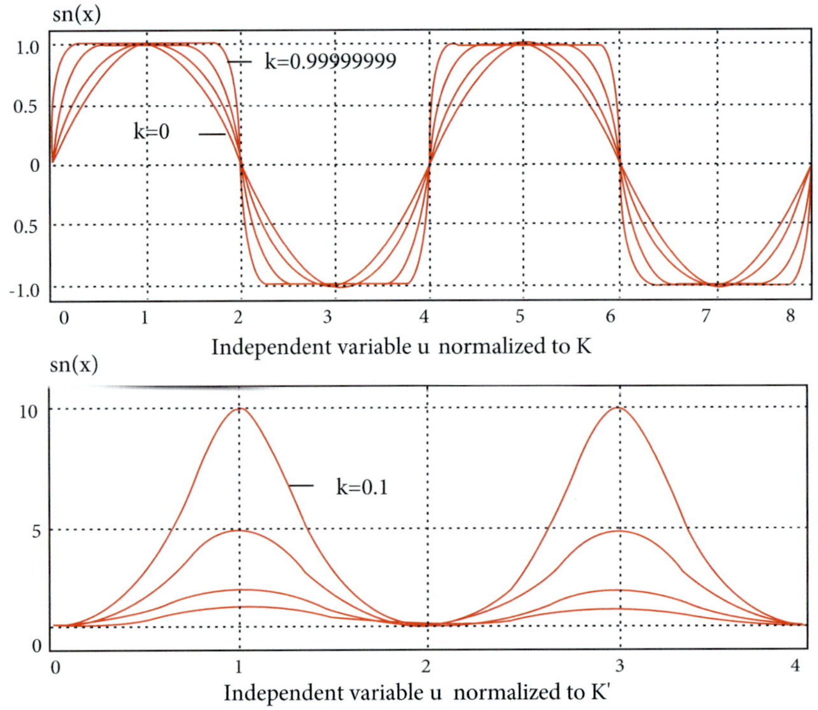

Fig. A.12 The period of the elliptic sine function with respect to the real and imaginary axis for various values of the elliptic modulus k

The elliptic functions of Jacobi are doubly periodic, that is, they are periodic both along the real axis and along the imaginary axis. It is proven that the period of these functions along the real axis is equal to $4\boldsymbol{K}$, whereas the period along the imaginary axis is equal to $2j\boldsymbol{K}'$—for example, for the functions $\mathrm{sn}(u,k)$ and $\mathrm{cn}(u,k)$ and for any value of the argument u

$$\mathrm{sn}(u+4m\boldsymbol{K},k)=\mathrm{sn}(u,k) \qquad \text{and} \qquad \mathrm{cn}(u+2jm\boldsymbol{K}',k)=\mathrm{cn}(u,k),$$

where m is an integer. The graph of the elliptic sine function with respect to the real variable u is shown in image (a) of Fig. A.12 for a range of values spanning two periods and for the values $k=0, 0.9, 0.999$ and 0.99999999. From the figure, we observe that for $k=0$ the function simplifies to our familiar trigonometric sine—since it is $\mathrm{sn}(u,0)=\sin u$—and that as the value of k increases, the shape of the function's curve becomes increasingly squared. In the above figure, the values of u are normalized, i.e., they have all been divided by the corresponding value of the integral $\boldsymbol{K}$, which for the four above values of k are equal to 1.5708, 2.2805, 4.4956 and 10.2501 respectively. We observe, therefore, that the value of the elliptic measure k affects the shape of the curve (since the period of the function increases with k), whereas the normalized value $u=1$ corresponds in all cases to 1/4 of the period, which is expected. The corresponding graph of the function with respect to the imaginary argument ju is shown in image (b) of Fig. A.12, which depicts the elliptic sine of Jacobi for the values $k=0.1, 0.2, 0.4$ and 0.6 respectively. In this case, the values of the complementary integral $\boldsymbol{K}'$ for these values of k are equal to 3.6956, 3.0161, 2.3593, and 1.9953 respectively. Taking into account that the period of the function that is depicted—which in reality is $\mathrm{sn}(\boldsymbol{K}+ju,k)$—is equal to $2\boldsymbol{K}'$, it is expected that the normalized value $u=1$ will correspond to half of this period.

Using the definitions of the integrals $\boldsymbol{K}$ and $\boldsymbol{K}'$, we can prove that

$$\mathrm{sn}[(2m+1)\boldsymbol{K}+j(2n+1)\boldsymbol{K}',k]=\pm\frac{1}{k} \qquad \text{and} \qquad \mathrm{sn}[2m\boldsymbol{K}+j(2n+1)\boldsymbol{K}',k]=\pm\infty, \tag{A.61}$$

where m and n are integers. Therefore, by constructing the graph of the function $\mathrm{sn}(u+j\boldsymbol{K}',k)$ as a function of the real parameter k, we find that it has the form depicted in Fig. A.13. From this figure, the conclusion follows that at the positions $u=2m\boldsymbol{K}$ (which correspond to even multiples of $\boldsymbol{K}$) the above function shows poles, whereas at the positions $u=(2m+1)\boldsymbol{K}$ (which correspond to odd multiples of $\boldsymbol{K}$) the function shows minima that depend on the value of k and more specifically

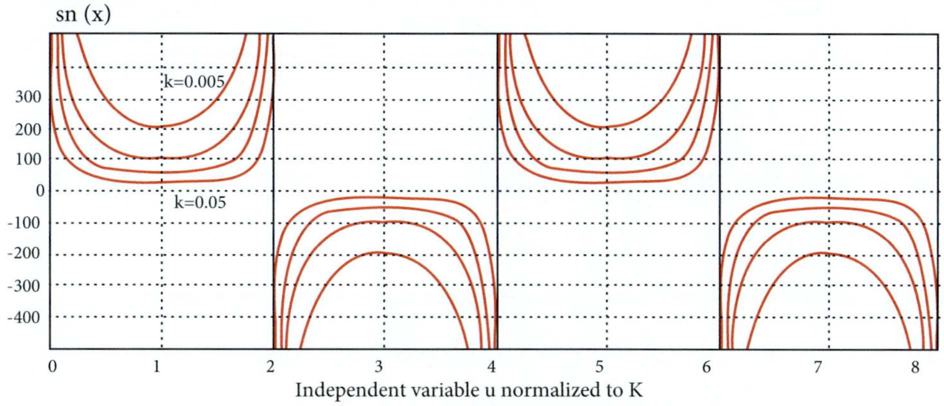

Fig. A.13 The variation of the function $\mathrm{sn}(u + j\boldsymbol{K}', k)$ with respect to the real variable u for various values of the elliptic measure k

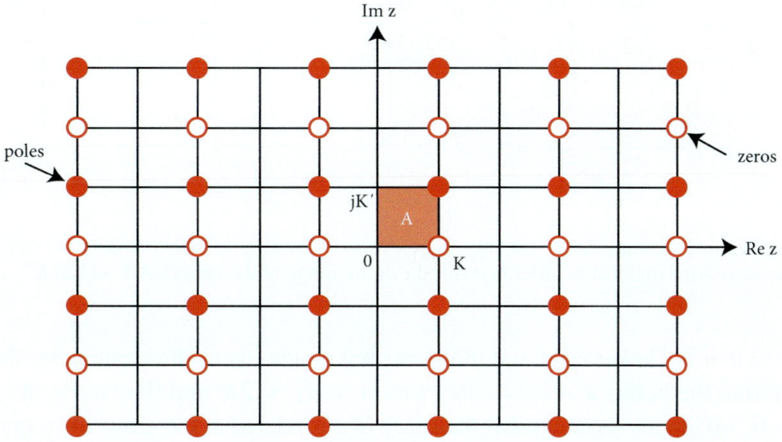

Fig. A.14 The pole-zero diagram of the function $\mathrm{cd}(u, k)$ with area A representing the fundamental rectangle on the complex plane

are equal to $1/k$. In Fig. A.13, the elliptic modulus k takes the values $k = 0.005, 0.01, 0.02$, and 0.05 for which the values of the integrals $\boldsymbol{K}$ and $\boldsymbol{K}'$ are 1.5708, 1.5708, 1.5710, 1.5718 and 6.6846, 5.9916, 5.2987, 4.3841 respectively.

In a more general description and using the concept of the fundamental rectangle that we presented above, we can prove that the zero values and the poles of the functions pq lie on grid nodes on the complex plane, with their distances dictated by the dimensions of the fundamental rectangle. For example, the poles and zero values of the function $\mathrm{cd}(u, k)$ are defined by the relations

$$z_{mn} = (2m + 1)\boldsymbol{K} + 2nj\boldsymbol{K}' \qquad \text{and} \qquad p_{mn} = (2m + 1)\boldsymbol{K} + (2n + 1)j\boldsymbol{K}',$$

where m and n are integer parameters (see Fig. A.14).

The elliptic functions $\omega = \mathrm{sn}(u, k)$ and $\omega = \mathrm{cd}(u, k)$ implement a conformal mapping[2] of the plane u to the plane ω, with the smallest area of the plane u being depicted in the entire plane ω known as the fundamental region. For each elliptic function $\mathrm{pq}(u, k)$, this region consists of four fundamental rectangles surrounding some zero value p, with each of these rectangles being depicted in one of the four quadrants of the plane ω. Figure A.15 presents the fundamental regions of the elliptic functions $\mathrm{cd}(u, k)$ and $\mathrm{sn}(u, k)$, which have as center the points c and s respectively and are defined as $0 \leq \Re\{u\} \leq 2\boldsymbol{K},\ -\boldsymbol{K}' \leq \Im\{u\} \leq \boldsymbol{K}'$ for the function $\mathrm{cd}(u, k)$ and $-\boldsymbol{K} \leq \Re\{u\} \leq \boldsymbol{K},\ -\boldsymbol{K}' \leq \Im\{u\} \leq \boldsymbol{K}'$ for the function $\mathrm{cd}(u, k)$. In this figure, the quadrants of the plane ω in which the quadrants of the plane u are depicted are symbolized by 1, 2, 3, and 4.

[2] Considering two planes Oxy and Ouv, the equations $u = u(x, y)$ and $v = v(x, y)$ define a known transformation or mapping between the points of these two planes. If two smooth curves C_1 and C_2, which intersect at the point $z_0 = x_0 + jy_0$ of the plane Oxy, are mapped with the above transformation to the curves C_1' and C_2' of the plane Ouv, so that the angles ϑ_1 and ϑ_2 under which the pairs of curves (C_1, C_2) and (C_1', C_2') intersect are equal to each other in magnitude and direction of deletion, then this representation is called conformal mapping at the point z_0.

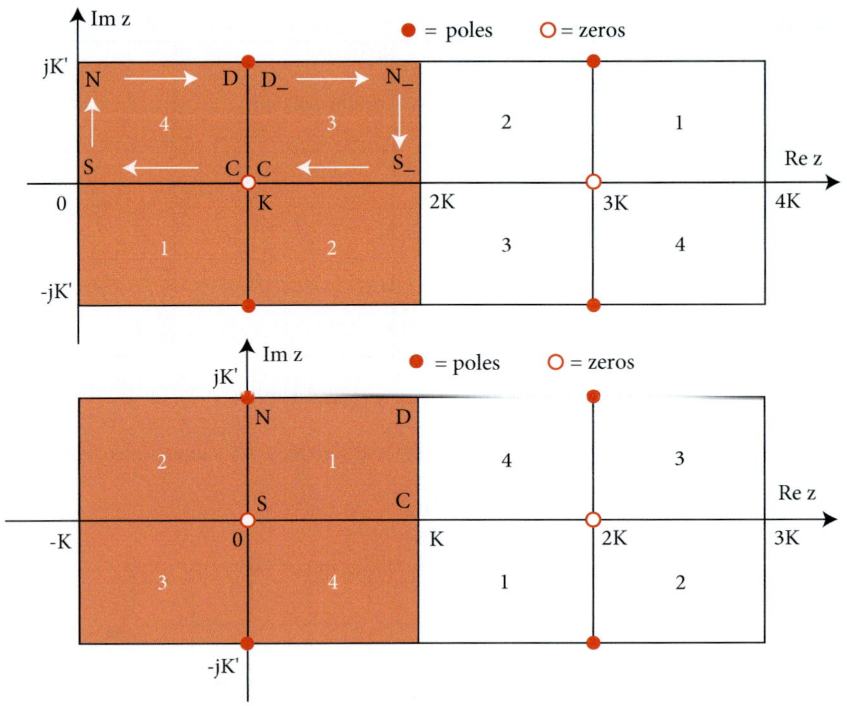

Fig. A.15 The fundamental region and the conformal mappings in quadrants of the elliptic functions cd(u, k) (top) and sn(u, k) (bottom). In this figure, the fundamental regions have been doubled in order to cover the entire range of the periods $4\boldsymbol{K}$ and $2j\boldsymbol{K}'$

From the figure we observe that the lower right and the lower left quadrants of the plane u are depicted in quadrants 1 and 2 of the plane ω and therefore the points $u = u_1 - ju_2$ with $0 \le u_1 \le 2K$ and $0 < u_2 < \boldsymbol{K}'$ are depicted in the points $\omega = \omega_1 + j\omega_2$ with $\omega_2 > 0$. Taking into account that the space of s is related to the space of frequencies through the relation $s = j\omega$, we can formulate the conclusion that the first and second quadrants of the ω space are depicted in the left half-plane of the s domain—since $s = j\omega = j(\omega_1 + j\omega_2) = -\omega_2 + j\omega_1$—a property that is used in the construction of the poles of elliptic filters that belong to the left complex half-plane.

A final property of elliptic functions related to the design of elliptic filters concerns the path c → s → n → d that we must follow to move from the zero value c to the pole d, which is depicted in Fig. A.15. It is proven that this path under the application of the above transformation is depicted on the positive semi-axis of ω in the following way:

$$\begin{aligned}
&\text{path c} \to \text{s:} \quad 0 \le u \le 1, \ \ z = \boldsymbol{K}' - ku \Longrightarrow 0 \le \omega \le 1\\
&\text{path s} \to \text{n:} \quad 0 \le u \le 1, \ \ z = j\boldsymbol{K}'u \Longrightarrow 1 \le \omega \le 1/k\\
&\text{path n} \to \text{d:} \quad 0 \le u \le 1, \ \ z = \boldsymbol{K}u + j\boldsymbol{K}' \Longrightarrow 1/k \le \omega \le \infty.
\end{aligned}$$

As noted in the presentation section of elliptical low-pass filters, these three value ranges of ω (for the normalized frequency ω/ω_p) correspond to the pass, transition, and cut-off zones of these filters. In a completely analogous way, the continuation of the path d → n → s → c that brings us back to the zero value c is depicted on the negative semi-axis of frequencies.

The Functions Nome and ϑ Using the complete integrals $\boldsymbol{K}$ and $\boldsymbol{K}'$, we can define the nome of the Jacobi function, as well as its complement, from the relations

$$q = q(k) = \exp\left(-\pi\frac{\boldsymbol{K}'}{\boldsymbol{K}}\right) = \exp\left(-\pi\frac{\boldsymbol{K}(\sqrt{1-k^2})}{\boldsymbol{K}(k)}\right) \qquad \text{and}$$

$$q_c = q_c(k) = \exp\left(-\pi\frac{\boldsymbol{K}}{\boldsymbol{K}'}\right) = \exp\left(-\pi\frac{\boldsymbol{K}(k)}{\boldsymbol{K}(\sqrt{1-k^2})}\right),$$

which, as can easily be verified, satisfy the equation $\ln(q) \times \ln(q_c) = \pi^2$. The calculation of the function $q(k)$ is performed by defining the variable $k = \sin\alpha$, as well as the auxiliary parameter

$$\varepsilon = \varepsilon(k) = \frac{1}{2}\left(\frac{1-\sqrt{\cos\alpha}}{1+\sqrt{\cos\alpha}}\right) = \frac{1}{2}\left(\frac{1-\sqrt[4]{1-k^2}}{1+\sqrt[4]{1-k^2}}\right). \tag{A.62}$$

In this case, it is proven that the function $q(k)$ can be developed into a series of the form

$$q(k) = \varepsilon(k) + 2\varepsilon^5(k) + 15\varepsilon^9(k) + 150\varepsilon^{13}(k) + 1707\varepsilon^{17}(k), \tag{A.63}$$

which is considered a very good approximation of the function $q(k)$ and converges for the range of values $0 \le \alpha \le \pi/4$ or $0 \le k \le 1/\sqrt{2}$. On the other hand, the complementary function $q_c(k)$ can be calculated as the value of the series

$$q_c(k) = \varepsilon_c(k) + 2\varepsilon_c^5(k) + 15\varepsilon_c^9(k) + 150\varepsilon_c^{13}(k) + 1707_c\varepsilon^{17}(k), \tag{A.64}$$

which converges in the range of values $\pi/4 \le \alpha \le \pi/2$ or $1/\sqrt{2} \le k \le 1$, with the function $\varepsilon_c(k)$ being defined as

$$\varepsilon_c = \varepsilon_c(k) = \frac{1}{2}\left(\frac{1-\sqrt{\sin\alpha}}{1+\sqrt{\sin\alpha}}\right) = \frac{1}{2}\left(\frac{1-\sqrt{(k)}}{1+\sqrt{k}}\right). \tag{A.65}$$

The calculation of the nome of the Jacobi function allows the calculation of the functions $\vartheta_i(u, q)$ (for the values $i = 1, 2, 3, 4$), which are defined as

$$\vartheta_1(u, q) = 2\sqrt[4]{q}\sum_{n=0}^{\infty}(-1)^n q^{n(n+1)} \sin[(2n+1)\pi u]$$

$$\vartheta_2(u, q) = 2\sqrt[4]{q}\sum_{n=0}^{\infty} q^{n(n+1)} \cos[(2n+1)\pi u]$$

$$\vartheta_3(u, q) = 1 + 2\sum_{n=1}^{\infty} q^{n^2} \cos(2n\pi u)$$

$$\vartheta_4(u, q) = 1 + 2\sum_{n=1}^{\infty}(-1)^n q^{n^2} \cos(2n\pi u).$$

The ϑ functions are quite useful, as it is proven that all the elliptic Jacobi functions can be expressed as the ratio of two ϑ functions. The mathematical expressions that describe these types of correlations contain the complete first kind integral expressed with the help of the *arithmetic-geometric mean* $M(x, y)$, which is defined as the common limit of convergence of the sequences $\alpha_n = (\alpha_{n-1} + \beta_{n-1})/2$ and $\beta_n = \sqrt{\alpha_{n-1}\beta_{n-1}}$ with $\alpha_0 = x$ and $\beta_0 = y$. In this case, it is proven that the complete elliptic integral of the first kind can be calculated from the equation

$$\boldsymbol{k}(k) = \int_0^{\pi/2} \frac{d\vartheta}{\sqrt{1-k^2\sin^2\vartheta}} = \frac{\pi}{2M(1, \sqrt{1-k^2})} = \frac{\pi}{M(1-k, 1+k)} \tag{A.66}$$

and then the function of the elliptic sine as the ratio of functions θ

$$\mathrm{sn}(u, k) = \frac{1}{\sqrt{k}} \frac{\vartheta_1\left[\dfrac{u}{2\boldsymbol{K}(k)}, q(k)\right]}{\vartheta_4\left[\dfrac{u}{2\boldsymbol{K}(k)}, q(k)\right]} = \frac{1}{\sqrt{k}} \frac{\vartheta_1\left[u\dfrac{M(1, \sqrt{1-k^2})}{\pi}, \exp\left(-\pi\dfrac{M(1, \sqrt{1-k^2})}{M(1, k)}\right)\right]}{\vartheta_4\left[u\dfrac{M(1, \sqrt{1-k^2})}{\pi}, \exp\left(-\pi\dfrac{M(1, \sqrt{1-k^2})}{M(1, k)}\right)\right]}. \tag{A.67}$$

Similar expressions apply to the remaining elliptic functions.

A.11 Fourier Transform of Many Dimensions

Although the theory presented in the previous sections refers to the expansion and the Fourier transform of functions of one variable, these concepts can be extended to describe functions of n independent variables. Considering for simplicity a function of two variables $f(x, y)$, we say that this is periodic with periods (T_1, T_2) if for every pair (x, y) the relations hold

$$f(x, y) = f(x + T_1, y) \qquad \text{and} \qquad f(x, y) = f(x, y + T_2).$$

If the above relations hold, then it is proven that

$$f(x, y) = f(x + T_1, y + T_2) \qquad \text{or more generally} \qquad f(x, y) = f(x + n_1 T_1, y + n_2 T_2)$$

for every pair of parameters (n_1, n_2). In this case, it is proven that the function $f(x, y)$ can be developed into an exponential Fourier series of the form

$$f(x, y) = \sum_{n=-\infty}^{+\infty} \sum_{m=-\infty}^{+\infty} A_{mn} \exp\left[i2\pi\left(\frac{mx}{T_1} + \frac{ny}{T_2}\right)\right] \tag{A.68}$$

with the Fourier coefficients A_{mn} calculated as

$$A_{mn} = \frac{1}{T_1 T_2} \int_0^{T_1} \int_0^{T_2} f(x, y) \exp\left[-i2\pi\left(\frac{mx}{T_1} + \frac{ny}{T_2}\right)\right] dxdy. \tag{A.69}$$

The above mathematical expressions for the general case of a periodic function of n independent variables with periods $(T_1, T_2, \ldots, T_n)$ have the form

$$\begin{aligned} f(\boldsymbol{x}) &= \sum_{k_1=-\infty}^{+\infty} \sum_{k_2=-\infty}^{+\infty} \cdots \sum_{k_n=-\infty}^{+\infty} A_{\boldsymbol{k}} \exp\left[i2\pi\left(\frac{k_1 x_1}{T_1} + \frac{k_2 x_2}{T_2} + \cdots + \frac{k_n x_n}{T_n}\right)\right] \\ &= \sum_{k_1=-\infty}^{+\infty} \sum_{k_2=-\infty}^{+\infty} \cdots \sum_{k_n=-\infty}^{+\infty} A_{\boldsymbol{k}} \exp\left(i2\pi \sum_{i=1}^{n} \frac{k_i x_i}{T_i}\right) \end{aligned} \tag{A.70}$$

and

$$\begin{aligned} A_{\boldsymbol{k}} &= \frac{1}{T_1 T_2 \ldots T_n} \int_0^{T_1} \int_0^{T_2} \cdots \int_0^{T_n} f(\boldsymbol{x}) \exp\left[-i2\pi\left(\frac{k_1 x_1}{T_1} + \frac{k_2 x_2}{T_2} + \cdots + \frac{k_n x_n}{T_n}\right)\right] dx_1 dx_2 \ldots dx_n \\ &= \frac{1}{T_1 T_2 \ldots T_n} \int_0^{T_1} \int_0^{T_2} \cdots \int_0^{T_n} f(\boldsymbol{x}) \exp\left(-i2\pi \sum_{i=1}^{n} \frac{k_i x_i}{T_i}\right) dx_1 dx_2 \ldots dx_n, \end{aligned} \tag{A.71}$$

where $\boldsymbol{x} = \{x_1, x_2, \ldots, x_n\}$ is the vector of the independent variables of the function $f(\vec{x})$ and we have also resorted to the use of the symbolism $A_{\boldsymbol{k}} = A_{k_1 k_2 \ldots k_n}$.

Similar generalizations to those we presented for the Fourier series apply to the homonymous transform. Thus, the Fourier transform for two and three dimensions is defined by the equations

$$F(\omega, \xi) = \int_{-\infty}^{+\infty} \int_{-\infty}^{+\infty} f(x, y)\, e^{-i(\omega x + \xi y)} dxdy \tag{A.72}$$

$$F(\omega, \xi, \zeta) = \int_{-\infty}^{+\infty} \int_{-\infty}^{+\infty} \int_{-\infty}^{+\infty} f(x, y, z)\, e^{-i(\omega x + \xi y + \zeta z)} dxdydz \tag{A.73}$$

respectively, whereas for the general case of n dimensions, the above relations take the form

$$F(\boldsymbol{\omega}) = \int_{-\infty}^{+\infty} \int_{-\infty}^{+\infty} \cdots \int_{-\infty}^{+\infty} f(\boldsymbol{t})\, e^{-i\boldsymbol{\omega}\cdot\boldsymbol{t}} dt_1 dt_2 \ldots dt_n \tag{A.74}$$

with the inverse transform being calculated as

$$F(\boldsymbol{t}) = \frac{1}{(2\pi)^n} \int_{-\infty}^{+\infty} \int_{-\infty}^{+\infty} \cdots \int_{-\infty}^{+\infty} F(\boldsymbol{\omega})\, e^{+i\boldsymbol{\omega}\cdot\boldsymbol{t}} dt_1 dt_2 \ldots dt_n, \tag{A.75}$$

where $\boldsymbol{\omega} = \{\omega_1, \omega_2, \ldots, \omega_n\}$ and $\boldsymbol{t} = \{t_1, t_2, \ldots, t_n\}$. Similar generalizations characterize the properties of the Fourier transform that we proved in the homonymous section. If we use the symbolism $\mathcal{F}\{f(\boldsymbol{t})\} = F(\boldsymbol{\omega})$ to show that the Fourier transform of the function $f(\boldsymbol{t})$ is the function $F(\boldsymbol{\omega})$ the validity of the following properties is demonstrated in a similar way:

$$\mathcal{F}\{f(\alpha\boldsymbol{t})\} = \frac{1}{|\alpha|} F\left(\frac{\boldsymbol{\omega}}{\alpha}\right) \qquad \mathcal{F}\{f(\boldsymbol{t}\mathbf{A})\} = \frac{1}{|\mathbf{A}|} F(\boldsymbol{\omega}\mathbf{A}^{-T})$$

$$\mathcal{F}\{f(\boldsymbol{t} - \boldsymbol{t}_0)\} = e^{-i\boldsymbol{\omega}\cdot\boldsymbol{t}_0} F(\boldsymbol{\omega}) \qquad \mathcal{F}\{e^{i\boldsymbol{\omega}_0\cdot\boldsymbol{t}} f(\boldsymbol{t})\} = F(\boldsymbol{\omega} - \boldsymbol{\omega}_0)$$

$$\mathcal{F}\left\{\frac{\partial f}{\partial t_k}\right\} = i\omega_k \frac{\partial F}{\partial \omega_k} \qquad \mathcal{F}\{t_k f(\boldsymbol{t})\} = i\frac{\partial F}{\partial \omega_k}$$

$$\mathcal{F}\{f(\boldsymbol{t}) * G(\boldsymbol{t})\} = F(\omega)G(\omega) \qquad \mathcal{F}\{f^*(\boldsymbol{t})g(\boldsymbol{t})\} = F^*(\boldsymbol{\omega})G(\omega),$$

whereas the generalization of the convolution theorem in n dimensions takes the form

$$\mathcal{F}\{f(\boldsymbol{t})g(\boldsymbol{t})\} = \frac{1}{(2\pi)^n}\Big(F(\omega) * G(\omega)\Big).$$

A particular category of functions of practical interest are the functions of many variables in which the independent variables can be separated from each other, thus allowing the function to be expressed as a product of simple functions of one variable. In this case, it is proven that this separation of variables extends to the Fourier transform, which is characterized by the same property. Considering, for example, a separable function of two variables of the form $f(x, y) = f_1(x) f_2(y)$, the transform of it according to Fourier, as calculated from the defining equation, will have the form

$$\begin{aligned} F(\omega, \xi) &= \int_{-\infty}^{+\infty} \int_{-\infty}^{+\infty} f(x, y) e^{-i(\omega x + \xi y)} dx dy = \int_{-\infty}^{+\infty} \int_{-\infty}^{+\infty} f_1(x) f_2(y) e^{-i\omega x} e^{-i\xi y} dx dy \\ &= \left(\int_{-\infty}^{+\infty} f_1(x) e^{-i\omega x} dx\right)\left(\int_{-\infty}^{+\infty} f_2(y) e^{-i\xi y} dy\right) = F_1(\omega) F_2(\xi), \end{aligned} \tag{A.76}$$

that is, the property of variable separation extends to the Fourier transform of the function $f(x, y)$, which is also written as the product of two single-variable functions. Extending the above property to n dimensions, it is proven that the Fourier transform of a separable function

$$f(\boldsymbol{x}) = f(x_1, x_2, \ldots, x_n) = f_1(x_1) f_2(x_2) \ldots f_n(x_n) = \prod_{i=1}^{n} f_i(x_i) \tag{A.77}$$

is formulated as

$$F(\boldsymbol{\omega}) = F(\omega_1, \omega_2, \ldots, \omega_n) = F_1(\omega_1) F_2(\omega_2) \ldots F_n(\omega_n) = \prod_{i=1}^{n} F_i(\omega_i). \tag{A.78}$$

A.12 Solving Differential Equations

The expansion and the Fourier transform are valuable tools for solving differential equations with partial derivatives, which are parts of more general boundary value problems. Writing the solution of such an equation in a Fourier series of sines and cosines is inextricably linked to the method of separation of variables, applied to problems defined in finite domains. On the other hand, for problems defined in infinite domains, the sine and cosine Fourier transforms are used as solution tools for problems with Dirichlet-type boundary conditions or the Laplace transform for problems with boundary conditions of the Cauchy type.[3] The method of separation of variables is a classic method for solving partial differential equations and is based on the assumption that the solution of such an equation, which is generally a function of n variables $\Psi(x_1, x_2, \ldots, x_n)$, can be written as a product of n functions of the form

$$\Psi(x_1, x_2, \ldots, x_n) = f_1(x_1) f_2(x_2) \ldots f_n(x_n) = \prod_{i=1}^{n} f_i(x_i), \tag{A.79}$$

each of which is a function of only one variable. In this case, the partial differential equation is reduced to a set of n ordinary differential equations for the functions $f_1(x_1), f_2(x_2), \ldots, f_n(x_n)$.

In order to demonstrate the application of the method in practice, let us consider for simplicity a function of two variables $u(x, t)$, which obeys the partial differential equation $(L_x + L_t)u(x, t) = 0$ with the operators L_x and L_t acting on the variables x and t respectively. Assuming that the function $u(x, t)$ can be written as a product of functions of the form $u(x, t) = X(x)T(t)$, the above differential equation can be transformed in the following way:

$$(L_x + L_t)u(x, t)0 \Rightarrow (L_x + L_t)X(t)T(t) = 0 \Rightarrow T(t)L_x X(x) + X(x)L_t T(t) = 0 \Rightarrow \frac{T_t L_x X(x)}{X(x)T(t)} + \frac{X(x)L_t T(t)}{X(x)T(t)} = 0,$$

and finally

$$\frac{L_x X(x)}{X(x)} + \frac{L_t T(t)}{T(t)} = 0 \quad \text{or equivalently} \quad \frac{L_x X(x)}{X(x)} = -\frac{L_t T(t)}{T(t)} = \lambda,$$

where the parameter λ is known as the separation constant. From the last equation, it is not difficult to see that the partial differential equation $(L_x + L_t)u(x, t) = 0$ has been transformed into the ordinary differential equations

$$L_x X(x) = \lambda X(x) \quad \text{and} \quad L_t T(t) = -\lambda T(t),$$

which are recognized as eigenvalue equations since they have the general form $Ly(x) = \lambda y(x)$. Therefore, the application of the method of separation of variables for the case of the differential equation $(L_x + L_t)u(x, t) = 0$ led to the creation of two eigenvalue equations for the operators L_x and L_y, the eigenvalues of which are equal and opposite to each other. The generalization of this method for the case of three variables and more generally, of n variables, is obvious.

In practice, the most difficult point of application of the method of separation of variables concerns the spatial and not the temporal variables. For this reason, usually, the separation of variables is done in two stages: in the first we separate the spatial from the temporal variable, writing the solution of the differential equation in the form $u(x, y, t) = U(x, y)T(t)$, whereas in the second we separate the spatial variables between them, writing the spatial function in the form $U(x, y) = X(x)Y(y)$. Initially, then, our differential equation will take the form

$$(L_x + L_y + L_t)u(x, y, t) = 0 \Rightarrow (L_{xy} + L_t)U(x, y)T(t) = 0 \Rightarrow T(t)L_{xy}U(x, y) + U(x, y)L_t T(t) = 0,$$

and finally

$$\frac{T(t)L_{xy}U(x, y)}{U(x, y)T(t)} + \frac{U(x, y)L_t T(t)}{U(x, y)T(t)} = 0 \quad \text{or equivalently} \quad \frac{L_{xy}U(x, y)}{U(x, y)} = -\frac{L_t T(t)}{T(t)} = \lambda,$$

[3] From the theory of partial differential equations, it is known that the boundary conditions of a problem with a solution being a function of many variables Ψ can be grouped into three different categories:

- Dirichlet conditions, which consist in specifying the value of the function Ψ at each point on the boundary
- Neumann conditions, which consist in specifying the value of the function Ψ as well as the normal component of the gradient of the function, $(\vec{\nabla}\Psi)_n$, at each point on the boundary.
- Cauchy conditions, which consist in specifying the quantities Ψ and $(\vec{\nabla}\Psi)_n$ at each point on the boundary.

from which the eigenvalue equation of the temporal function arises

$$L_t T(t) + \lambda T(t) = 0,$$

whereas subsequently setting $U(x, y) = X(x)Y(y)$ we get

$$L_{xy}U(x, y) = \lambda U(x, y) \Rightarrow (L_x + L_y)X(x)Y(y) = \lambda X(x)Y(y) \Rightarrow$$

$$\Rightarrow Y(y)L_x X(x) + X(x)L_y Y(y) = \lambda X(x)Y(y) \Rightarrow \frac{L_x(x)}{X(x)} + \frac{L_y Y(y)}{Y(y)} = \lambda,$$

from which it follows that

$$\frac{L_x(x)}{X(x)} = \mu \qquad \text{and} \qquad \frac{L_y Y(y)}{Y(y)} = \lambda - \mu.$$

Therefore, ultimately, the application of the method of separation of variables led to the three eigenvalue equations

$$L_x X(x) - \mu X(x) = 0 \qquad L_y Y(y) - (\lambda - \mu)Y(y) = 0 \qquad L_t T(t) + \lambda T(t) = 0,$$

the solution of which will ultimately lead to the solution $u(x, y, t)$ of our differential equation. It is proven that in order for the method of separation of variables to be applicable, the boundary conditions of the problem must be linear and homogeneous so that they can be associated only with the spatial and not with the time function. The more general eigenvalue problem is known as the Sturm-Liouville theory, the full description of which is beyond the goals of this book.

A.12.1 The One-Dimensional Heat-Diffusion Equation

As an example of the application of Fourier analysis techniques in combination with the method of separation of variables, let us solve the one-dimensional heat-diffusion equation in both finite and infinite space. Starting from our first case, which is also the simplest, let us consider a metallic plate with finite thickness L and practically infinite surface, which is immersed and heated in a cooling liquid of zero temperature. In this case, as is well known, there will be a transfer of heat from the metallic plate to the liquid; however, in the description that follows we assume that the volume of the liquid is enormous, so that this heat transfer does not lead to a change in its temperature. It is obvious that, owing to the practically infinite surface of the plate, the temperature inside it will depend only on the dimension of its thickness and not on the other two dimensions. Therefore, defining a coordinate system such that the dimension of the thickness of the plate coincides with the x axis, whereas its infinite surface extends along the y and z axes, it is obvious that the temperature inside the plate will be described by the function $u = u(x, t)$ and will satisfy the one-dimensional heat-transfer equation $\partial u/\partial t = \sigma \partial^2 u/\partial x^2$, where σ is the coefficient of thermal conductivity. The boundary conditions of the problem will be of the form $u(0, t) = u(L, t) = 0$, whereas the initial conditions describing the distribution of temperature inside the cooling liquid at the moment of immersion will be described by the equation $u(x, 0) = f(x)$.

To solve the differential heat-transfer equation, we use the method of separation of variables. Let us consider a solution of the form $u(x, t) = X(t)T(t)$. Substituting this solution into our differential equation we get

$$\frac{\partial}{\partial t}\left[X(x)T(t)\right] = \sigma \frac{\partial^2}{\partial x^2}\left[X(x)T(t)\right] \Rightarrow X(x)\frac{d^2T(t)}{dt^2} = \sigma T(t)\frac{d^2X(x)}{dx^2} \Rightarrow \frac{X(x)}{X(x)T(t)}\frac{dT(t)}{dt} = \frac{\sigma T(t)}{X(x)T(t)}\frac{d^2X(x)}{dx^2}$$

and finally

$$\frac{1}{T(t)}\frac{dT(t)}{dt} = \frac{\sigma}{X(x)}\frac{d^2X(x)}{dx^2} = -\lambda,$$

from which two ordinary differential equations of the form arise

$$\frac{dT(t)}{dt} + \lambda T(t) = 0 \quad \text{and} \quad \frac{d^2X(x)}{dx^2} + \frac{\lambda}{\sigma} X(x) = 0.$$

From the above equations, the first one concerning the time function $T(t)$ is easily solved as it is a separable variables equation. Therefore, we have

$$\frac{dT(t)}{dt} = -\lambda T(t) \Rightarrow \frac{dT(t)}{T(t)} = -\lambda dt \Rightarrow \int \frac{dT(t)}{dt} = -\lambda \int dt$$

and finally

$$\ln T(t) = -\lambda t + C \Rightarrow e^{\ln T(t)} = e^{-\lambda t + C} \Rightarrow T(t) = Ae^{-\lambda t},$$

where $A = e^C$. In the above equation, the eigenvalue λ does not take continuous but discrete values, which will arise from the solution of the second differential equation, which can be written in the form $X''(x) + k^2 X(x) = 0$, where $k^2 = \lambda/\sigma$. In order to determine the form of the function $X(x)$ we must use the boundary conditions. Therefore, we have

$$u(0,t) = X(0)T(t) = 0 \Rightarrow X(0) = 0 \quad \text{and} \quad u(0,L) = X(L)T(t) = 0 \Rightarrow X(L) = 0.$$

In order for the function $X(x)$ to intersect the horizontal axis at points $x = 0$ and $x = L$ it must be periodic, and therefore it must have the form $X(x) = A\cos(kx) + B\sin(kx)$ with the coefficients A and B being determined by the boundary conditions. Indeed, for $x = 0$, $X(0) = A\cos(0) + B\sin(0) = 0$ and therefore $A = 0$. Substituting into the solution of the function, it will take the form $X(x) = B\sin(kx)$. Using the second boundary condition we get $X(L) = B\sin(kL) = 0 \Rightarrow \sin(kL) = 0 \Rightarrow kL = n\pi$ $(n = 1, 2, \ldots)$. From the last relation we find that the permitted values of k are $k = k_n = n\pi/L$, that is, the function $X(x)$ will take the form

$$X(x) \equiv X_n(x) = B_n \sin\left(\frac{n\pi x}{L}\right) \quad n = 1, 2, \ldots. \tag{A.80}$$

Having determined the permitted values of the parameter k, we can also determine the permitted values of the parameter λ using the equation $\lambda = \lambda_n = \sigma k_n^2 = \sigma n^2\pi^2/L^2$. Substituting this value of λ into the defining equation of the function $T(t)$ we get the equation

$$T(t) \equiv T_n(t) = A_n e^{-\lambda_n t} = A_n e^{-\sigma \frac{n^2\pi^2}{L^2} t}. \tag{A.81}$$

Therefore, the solution of the differential equation for a given value of the parameter n will be of the form

$$u_n(x,t) = X_n(x)T_n(t) = B_n \sin\left(\frac{n\pi x}{L}\right) A_n e^{-\sigma \frac{n^2\pi^2}{L^2} t} = C_n \sin\left(\frac{n\pi x}{L}\right) e^{-\sigma \frac{n^2\pi^2}{L^2} t},$$

where $C_n = A_n B_n$, whereas its general solution for all values of the parameter n will be written as

$$u(x,t) = \sum_{n=1}^{\infty} u_n(x,t) = \sum_{n=1}^{\infty} C_n \sin\left(\frac{n\pi x}{L}\right) e^{-\sigma \frac{n^2\pi^2}{L^2} t}. \tag{A.82}$$

The coefficients A_n appearing in the above relation can be determined with the help of the initial conditions $u(x,0) = f(x)$ and using the orthogonality conditions of the sine functions. If we set $t = 0$ in the above equation, it will take the form

$$u(x,0) = \sum_{n=1}^{\infty} C_n \sin\left(\frac{n\pi x}{L}\right) = f(x). \tag{A.83}$$

Multiplying the above relation by $\sin(m\pi x/L)$ and integrating over the interval $[0, L]$ we get

$$\sum_{n=1}^{\infty} C_n \sin\left(\frac{n\pi x}{L}\right) \sin\left(\frac{m\pi x}{L}\right) = f(x) \sin\left(\frac{m\pi x}{L}\right)$$

or equivalently

$$\sum_{n=1}^{\infty} C_n \int_0^L \sin\left(\frac{n\pi x}{L}\right) \sin\left(\frac{m\pi x}{L}\right) dx = \int_0^L f(x) \sin\left(\frac{m\pi x}{L}\right) dx.$$

However, as is well known, the sine functions are orthogonal and it is obvious that from all the terms of the sum of the left-hand side, the only nonzero term is the one corresponding to the value $n = m$. Therefore, we have

$$C_m \int_0^L \sin^2\left(\frac{m\pi x}{L}\right) dx = C_m \frac{L}{2} = \int_0^L f(x) \sin\left(\frac{m\pi x}{L}\right) dx,$$

from which we finally get that

$$C_n = \frac{2}{L} \int_0^L f(x) \sin\left(\frac{n\pi x}{L}\right) dx, \tag{A.84}$$

where we silently changed the dummy index from m to n—to derive the last relation we used the definite integral

$$\int_0^L \sin^2\left(\frac{m\pi x}{L}\right) dx = \left[\frac{x}{2} - \frac{L}{4m\pi} \sin\left(\frac{2m\pi x}{L}\right)\right]_0^L = \frac{L}{2}.$$

Substituting the above value of the coefficient C_n into the general solution $u(x, t)$, it will take the form

$$u(x,t) = \sum_{n=1}^{\infty} u_n(x,t) = \sum_{n=1}^{\infty} \left[\frac{2}{L} \int_0^L f(x) \sin\left(\frac{n\pi x}{L}\right) dx\right] \sin\left(\frac{n\pi x}{L}\right) e^{-\sigma \frac{n^2\pi^2}{L^2} t}, \tag{A.85}$$

which is also the final result.

After solving the heat-diffusion problem for the case of a finite domain of length L, let us now study the way in which the solution process differs, by studying the problem in an infinite domain, that is, considering the limiting case $L \to \infty$. It is obvious that the differential equation to be solved will be the one we solved previously, whereas the boundary and initial conditions of the problem will now take the form $u(0, t) = 0$ and $u(x, 0) = f(x)$ respectively. Applying the method of separation of variables and following the process we applied in the previous case, it is not difficult to find that the solution of the problem will have the form $u(x, t) = A \sin(\lambda x) e^{-\sigma\lambda^2 t}$, where A is a suitable constant and λ is the eigenvalue of the problem.

The main difference that characterizes the nature of the problem in infinity in relation to the past region is that the spectrum of permissible eigenvalues is no longer discrete but continuous. The power of this assertion is not difficult to verify if we take into account that the discrete eigenvalues for the past region have the form $\lambda_n = n\pi/L$ and therefore in the limit $L \to \infty$ the distance between them tends to zero, thus making them a continuous quantity. But this means that if we use the property of superposition to determine the general solution of the differential equation for each value of the parameter λ, we must, instead of the symbol of the sum that we used for the discrete spectrum of eigenvalues, use the symbol of the integral and write the general solution for the infinite region in the form

$$u(x,t) = \int_0^{\infty} D(\lambda) \sin(\lambda x)\, e^{-\sigma\lambda^2 t} d\lambda. \tag{A.86}$$

From the above relation it is obvious that the constant of proportionality of the discrete case has now become a function of the continuous eigenvalue λ and its value can be determined using the initial condition $u(x, 0) = f(x)$ in the following way: setting in the above equation $t = 0$ and relying on the aforementioned condition, we are led to the equation

$$u(x,0) = \int_0^{\infty} D(\lambda) \sin(\lambda x)\, d\lambda = f(x). \tag{A.87}$$

But the above relation is nothing but the sinusoidal Fourier transform of the function $f(x)$. Therefore, the required function $D(\lambda)$ is nothing but the Fourier coefficients. $D(\lambda)$ of the aforementioned function, which, as is well known, will have the form

$$D(\lambda) = \frac{2}{\pi}\int_0^\infty f(x)\sin(\lambda x)\,dx. \tag{A.88}$$

Substituting the above relation into the general solution of our differential equation, it is written as

$$u(x,t) = \frac{2}{\pi}\int_0^\infty \int_0^\infty f(\xi)\sin(\lambda\xi)\sin(\lambda x)\,e^{-\sigma\lambda^2 t}d\xi d\lambda. \tag{A.89}$$

In the last step of the process, using the known trigonometric identity

$$\sin(\lambda\xi)\sin(\lambda x) = \frac{1}{2}\Big\{\cos[\lambda(\xi - x)] - \cos[\lambda(\xi + x)]\Big\}$$

we formulate the last equation in the form

$$\begin{aligned} u(x,t) &= \frac{1}{\pi}\int_0^\infty \int_0^\infty f(\xi)\Big\{\cos\,[\lambda(\xi - x)] - \cos\,[\lambda(\xi + x)]\Big\}e^{-\sigma\lambda^2 t}d\xi\,d\lambda \\ &= \frac{1}{\pi}\int_0^\infty \Big\{\int_0^\infty e^{-\sigma\lambda^2 t}\cos\,[\lambda(\xi - x)]d\lambda - \int_0^\infty e^{-\sigma\lambda^2 t}\cos\,[\lambda(\xi + x)]\,d\lambda\Big\}\,d\xi \end{aligned}$$

and using the definite integral

$$\int_0^\infty e^{-\alpha^2}\cos(\beta\lambda)d\lambda = \frac{1}{2}\sqrt{\frac{\pi}{\alpha}}\exp\left(-\frac{\beta^2}{4\alpha}\right)$$

we finally arrive at the equation

$$u(x,t) = \frac{1}{2\sqrt{\pi\sigma t}}\Big\{\int_0^\infty f(\xi)\exp\Big[-\frac{(\xi - x)^2}{4\sigma t}\Big]d\xi - \int_0^\infty f(\xi)\exp\Big[-\frac{(\xi + x)^2}{4\sigma t}\Big]d\xi\Big\}, \tag{A.90}$$

which is the desired result.

A.12.2 The Two-Dimensional Wave Equation

As a second example of the application of Fourier analysis for solving partial differential equations, let us consider the two-dimensional wave equation

$$\frac{\partial^2 u}{\partial x^2} + \frac{\partial^2 u}{\partial y^2} = \frac{1}{c^2}\frac{\partial^2 u}{\partial t^2}, \tag{A.91}$$

which is defined for the finite region $(0 < x < L, 0 < y < L)$ and has a solution of the form $u = u(x, y, t)$. Without intending to present all the steps in detail one by one as we did previously—this process is left as an exercise to the reader—we consider a separable solution of the form $u(x, y, t) = U(x, y)T(t)$—that is, initially we separate the spatial from the time variable—and we are led to the system of equations

$$\frac{\partial^2 U}{\partial x^2} + \frac{\partial^2 U}{\partial y^2} = -k^2 U \qquad \text{and} \qquad \frac{d^2 T}{dt^2} + (kc)^2 T = 0.$$

The last equation has solutions of the form $T(t) = \alpha\cos(kct) + \beta\sin(kct)$ with the permissible values of the parameters α and β resulting from the solution of the spatial differential equation, which, expressed in the form $(\nabla^2 + k^2)U(x, y) = 0$, is known as the Helmholtz equation. To solve it, we again consider separable solutions of the form $U(x, y) = X(x)Y(t)$ and by applying the method of separation of variables we separate it into the ordinary differential equations

$$\frac{d^2X}{dx^2} + k_x^2 X = 0 \quad \text{and} \quad \frac{d^2Y}{dy^2} + k_y^2 Y = 0$$

with solutions

$$X_n(x) = B_n \sin\left(\frac{n\pi x}{L}\right) \quad \text{and} \quad Y_m(y) = C_m \sin\left(\frac{m\pi y}{L}\right)$$

and permitted eigenvalues $k_x = (n\pi/L)$ and $k_y = (m\pi/L)$ $(m, n = 1, 2, 3, \ldots)$. It is proven that the parameter k appearing in the spatial equation has the form

$$k = \sqrt{k_x^2 + k_y^2} = \frac{\pi}{L}\sqrt{n^2 + m^2}$$

and defining the parameter $\omega_{nm} = ck = (c\pi/L)\sqrt{n^2 + m^2}$ the time function is written as

$$T(t) = \alpha_{nm} \cos(\omega_{nm} t) + \beta_{nm} \sin(\omega_{nm} t).$$

Therefore, the partial solution of the differential equation for a given pair of index values (m, n) will have the form $u_{mn}(x, y, t) = U_{mn}(x, y)T_{mn}(t) = X_n(x)Y_m(y)T_{mn}(t)$, whereas the general solution for every (m, n) will have the form

$$u(x, y, t) = \sum_{n=1}^{\infty}\sum_{m=1}^{\infty} \sin\left(\frac{n\pi x}{L}\right) \sin\left(\frac{m\pi y}{L}\right)\Big[\alpha_{nm} \cos(\omega_{nm} t) + \beta_{nm} \sin(\omega_{nm} t)\Big]. \tag{A.92}$$

Differentiating the above relation with respect to time and using the initial conditions $u(x, t, 0) = f(x, y)$ and $u_t(x, y, 0) = g(x, y)$ and the orthogonality conditions of the trigonometric functions, it is proven that the coefficients α_{nm} and β_{nm} appearing in the above equation are given by the relations

$$\alpha_{nm} = \frac{4}{L^2}\int_0^L \int_0^L f(x, y) \sin\left(\frac{n\pi x}{L}\right) \sin\left(\frac{m\pi y}{L}\right) dx\, dy \tag{A.93}$$

$$\beta_{nm} = \frac{4}{L^2\omega_{nm}}\int_0^L \int_0^L g(x, y) \sin\left(\frac{n\pi x}{L}\right) \sin\left(\frac{m\pi y}{L}\right) dx\, dy \tag{A.94}$$

A.12.3 The Method of the Fourier Transform

The basic method of solving differential equations with the help of the Fourier transform is very simple and is based on the property of differentiation of this transformation, which mathematically is formulated as $\mathcal{F}\{f^{(n)}(t)\} \leftrightarrows (i\omega)^n F(\omega)$. If we take the Fourier transform of all the terms of our differential equation and solve the equation that will arise for the function $F(\omega)$, we can then determine the solution of the differential equation $f(t)$ by simply calculating the inverse transform of the function $F(\omega)$. The above technique can be applied to both ordinary and partial differential equations, whereas a similar method of solving these equations is based on the use of the Laplace transform.

Problem A.12.1 Using the method of Fourier transform to solve the differential equation

$$\frac{dy(t)}{dt} + y(t) = \frac{1}{2}e^{-|t|}$$

Solution Taking the transform of the two sides of the above equation and using the property of differentiation we get

$$\mathcal{F}\left\{\frac{dy(t)}{dt}\right\} + \mathcal{F}\{y(t)\} = i\omega Y(\omega) + Y(\omega) = \mathcal{F}\left\{\frac{1}{2}e^{-|t|}\right\},$$

where

$$\mathcal{F}\left\{\frac{1}{2}e^{-|t|}\right\} = \frac{1}{2}\int_{-\infty}^{+\infty} e^{-|t|}e^{-i\omega t}dt = \frac{1}{2}\int_{-\infty}^{0} e^{(1-i\omega)t}dt + \frac{1}{2}\int_{0}^{\infty} e^{-(1+i\omega)t}dt$$

$$= \frac{1}{2}\left[\frac{e^{(1-i\omega)t}}{1-i\omega}\right]_{-\infty}^{0} + \frac{1}{2}\left[\frac{e^{-(1+i\omega)t}}{(1+i\omega)}\right]_{\infty}^{0} = \frac{1}{2}\left[\frac{1}{1-i\omega} + \frac{1}{1+i\omega}\right] = \frac{1}{\omega^2+1}$$

By substituting the last expression into our final equation and solving $Y(\omega)$ we get

$$(1+i\omega)Y(\omega) = \frac{1}{\omega^2+1} \quad \text{and finally} \quad Y(\omega) = \frac{1}{(\omega^2+1)(i+i\omega)} = \frac{1}{(1+i\omega)^2(1-i\omega)},$$

hence the desired solution $y(t)$ will result from the determination of the inverse transform

$$y(t) = \frac{1}{2\pi}\int_{-\infty}^{+\infty} Y(\omega)e^{i\omega t}d\omega = \frac{1}{2\pi}\int_{-\infty}^{+\infty} \frac{e^{i\omega t}}{(1+i\omega)^2(1-i\omega)}.$$

To solve the exercise we use the method of complex integration and the basic theorem of residue calculus

$$\oint_C f(z)\,dz = 2\pi i \sum_{k=1}^{N} \text{Res}\left[f(z)\right]_{z=z_k},$$

where the complex function $f(z)$ is assumed to have N poles at the points $z = z_k$ ($k = 1, 2, 3, \ldots, N$), which are located within the closed path C—in this case the function $f(z)$ has the form

$$f(z) = \frac{e^{izt}}{(1+iz)^2(1-iz)}$$

and has a double pole at the position $z = i$ as well as a simple pole at the position $z = -i$, as can be observed. Starting from the case $t > 0$, we define the closed path C as the curve that is defined by the horizontal axis as well as by a semicircle of infinite radius C_R with the center at the point $z = 0$ extending into the positive half-plane. According to Jordan's lemma, the contour integral along the semicircle C_R and for the limit $R \to \infty$ is equal to zero. Furthermore, of the two poles of the function $f(z)$, the only one that is within the closed path C is the double pole at the position $z = i$. The residue of the function $f(z)$ at this pole is calculated as

$$\text{Res}\left[f(z)\right]_{z=i} = \lim_{z\to i}\left\{\frac{d}{dz}\left[(z-i)^2\frac{e^{itz}}{i(z-i)^2(z+i)}\right]\right\} = \frac{te^{-t}}{2i} + \frac{e^{-t}}{4i},$$

from which it finally follows that

$$y(t) = \frac{1}{2\pi}(2\pi i)\left[\frac{te^{-t}}{2i} + \frac{e^{-t}}{4i}\right] = \frac{e^{-t}}{4}(2t+1).$$

On the other hand, for the case $t < 0$ the closed path C will be defined so as to consist of the horizontal axis and the semicircle of infinite radius with the center at the point $z = 0$ extending into the negative half-plane. According to Jordan's lemma, the contour integral of the function $f(z)$ will again be equal to zero, whereas the residue of the aforementioned function at the position of the simple pole $z = -i$—which is also the only pole that belongs to the closed path C—is calculated as

$$\mathrm{Res}\left[f(z)\right]_{z=-i} = \lim_{z\to -i}\left[(z+i)\frac{e^{itz}}{i(z+i)(z-i)^2}\right] = \frac{e^t}{4i},$$

as can be verified, that is, the function $y(t)$ for this case will be given by the relation

$$y(t) = \frac{1}{2\pi}(-2\pi i)\left(-\frac{e^t}{4i}\right) = \frac{e^t}{4},$$

which is also the desired result. The negative sign in the above relation is due to the fact that the contour integral along the real axis was calculated by traversing the axis in the negative direction, that is, from the positive to the negative semi-axis. ■

The Fourier transform can be used in a similar way to solve partial differential equations, the solution of which is characterized by specific initial and boundary conditions. Considering for simplicity a function of two variables $u(x,t)$, its Fourier transform with respect to the spatial variable x is given by the relation

$$\mathcal{F}_x\{u(x,t)\} \equiv U_x(\omega,t) = \int_{-\infty}^{+\infty} u(x,t)e^{-i\omega x}dx, \tag{A.95}$$

which is similar to the classic equation of the definition,[4] considering the variable t as constant. Using the above equation, the Fourier transform with respect to x of the partial derivative $u_t(x,t) = \partial u(x,t)/\partial t$ is calculated as

$$\mathcal{F}_x\left\{\frac{\partial u(x,t)}{\partial t}\right\} = \int_{-\infty}^{+\infty}\frac{\partial u(x,t)}{\partial t}e^{-i\omega x}dx = \frac{\partial}{\partial t}\left\{\int_{-\infty}^{+\infty}u(x,t)\,e^{-i\omega x}dx\right\} = \frac{\partial U_x(\omega,t)}{\partial t},$$

whereas in a similar way we get

$$\mathcal{F}_x\left\{\frac{\partial u(x,t)}{\partial x}\right\} = \int_{-\infty}^{+\infty}\frac{\partial u(x,t)}{\partial x}e^{-i\omega x}dx.$$

Using the method of integration by parts, we immediately find that

$$\frac{\partial u(x,t)}{\partial x}e^{-i\omega x} = \frac{\partial}{\partial x}\left[u(x,t)e^{-i\omega x}\right] + (i\omega)\,u(x,t)e^{-i\omega x}$$

and considering that the function $u(x,t)$ tends to zero at the limit $x \to \pm\infty$

$$\mathcal{F}_x\left\{\frac{\partial u(x,t)}{\partial x}\right\} = \frac{\partial}{\partial x}\int_{-\infty}^{+\infty}u(x,t)\,e^{-i\omega x}dx + i\omega\int_{-\infty}^{+\infty}u(x,t)\,e^{-i\omega x}dx = i\omega U_x(\omega,t).$$

The above equations concern the Fourier transform of the first-order partial derivatives of the function $u(x,t)$, whereas for the general case of derivatives of order n it turns out that

$$\mathcal{F}_x\left\{\frac{\partial^n u(x,t)}{\partial t^n}\right\} = \frac{\partial^n U_x(\omega,t)}{\partial t^n} \quad \text{and} \quad \mathcal{F}_x\left\{\frac{\partial^n u(x,t)}{\partial x^n}\right\} = (i\omega)^n U_x(\omega,t).$$

On the other hand, if we consider the spatial variable x as a constant, we can define the Fourier transform with respect to the time variable t as the function

$$\mathcal{F}_t\{u(x,t)\} \equiv U_t(x,\omega) = \int_{-\infty}^{+\infty} u(x,t)e^{-i\omega t}dt. \tag{A.96}$$

---

[4] The discussion we had about the value of the multiplicative factor placed in front of the integral of the transform (i.e., whether it will be equal to one, equal to $(1/2\pi)$, or equal to $(1/\sqrt{2\pi})$) applies here as well and in the solved exercises that follow we use the convention that is followed in the books in which these have been presented.

Following the methodology of the previous paragraph, we can prove that

$$\mathcal{F}_t\left\{\frac{\partial u(x,t)}{\partial t}\right\} = i\omega U_t(x,\omega) \quad \text{and} \quad \mathcal{F}_t\left\{\frac{\partial u(x,t)}{\partial x}\right\} = \frac{\partial U_t(x,\omega)}{\partial x},$$

whereas, in the general case, the relations

$$\mathcal{F}_t\left\{\frac{\partial^n u(x,t)}{\partial t^n}\right\} = (i\omega)^n U_t(x,\omega) \quad \text{and} \quad \mathcal{F}_t\left\{\frac{\partial^n u(x,t)}{\partial x^n}\right\} = \frac{\partial^n U_t(x,\omega)}{\partial x^n} \text{ hold.}$$

The use of the Fourier transform in boundary value problem-solving processes is presented in detail through the exemplary solved exercises that follow.

Problem A.12.2 Using the Fourier transform method, solve the differential equation of sound wave transmission, which is emitted from a spherical source of radius α, the surface of which expands uniformly at an expansion speed equal to $\upsilon_0\delta(t)$.

Solution The differential equation to be solved is the three-dimensional wave equation, which, based on the property of spherical symmetry that characterizes the problem, can be expressed in the simple form

$$\frac{1}{r^2}\frac{\partial}{\partial r}\left(r^2\frac{\partial u}{\partial r}\right) = \frac{1}{c^2}\frac{\partial^2 u}{\partial t^2}$$

with the boundary condition

$$\left.\frac{\partial u}{\partial r}\right|_{r=\alpha} = -\rho\upsilon_0\delta(t),$$

where c is the wave propagation speed and ρ is the speed of the fluid in which the wave is transmitted. Assuming that the Fourier transform of the problem's solution $u(x,t)$ with respect to the time variable t is the function $U_t(r,\omega)$, the inverse transform equation will have the known form

$$u(r,t) = \frac{1}{2\pi}\int_{-\infty}^{+\infty} U_t(r,\omega)\, e^{i\omega t}\, d\omega.$$

Calculating the partial derivatives $\partial u(x,t)/\partial r$ and $\partial^2 u(x,t)/\partial t^2$ it turns out that

$$\frac{\partial u(x,t)}{\partial r} = \frac{1}{2\pi}\int_{-\infty}^{+\infty} \frac{dU_t(r,\omega)}{dr} e^{i\omega t} d\omega \quad \text{and} \quad \frac{\partial^2 u(x,t)}{\partial t^2} = -\frac{\omega^2}{2\pi}\int_{-\infty}^{+\infty} U_t(r,\omega)\, e^{i\omega t}\, d\omega,$$

and substituting into the differential equation to be solved, it will take the form

$$\frac{1}{2\pi}\int_{-\infty}^{+\infty}\left\{\frac{1}{r^2}\frac{d}{dr}\left[r^2\frac{dU_t(r,\omega)}{dr}\right] + \frac{\omega^2}{c^2}U_t(r,\omega)\right\} e^{i\omega t} d\omega = 0,$$

whereas with the exact same process, the boundary condition is formulated as

$$\frac{1}{2\pi}\int_{-\infty}^{+\infty}\left[\left.\frac{dU_t(r,\omega)}{dr}\right|_{r=\alpha} + p\upsilon_0\right] e^{i\omega t}\, d\omega = 0.$$

Since the above equations must hold for every time instance t it must be

$$\frac{1}{r^2}\frac{d}{dr}\left[r^2\frac{dU_t(r,\omega)}{dr}\right]+\frac{\omega^2}{c^2}U_t(r,\omega)=0 \quad \text{and} \quad \left.\frac{dU_t(r,\omega)}{dr}\right|_{r=\alpha}+\rho\upsilon_0=0.$$

It is not difficult to see that the first of the above equations can be written as

$$\frac{d^2U_t(r,\omega)}{dr^2}+\frac{2}{r}\frac{dU_t(r,\omega)}{dr}+\frac{\omega^2}{c^2}U_t(r,\omega)=0,$$

and according to the theory of differential equations, its solution requires a substitution of the form $U_t(r,\omega)=\xi(r)/r$ (the parameter ω is omitted for simplicity as it is considered constant anyway). Differentiating the last relation twice with respect to r we get

$$\frac{dU_t(r,\omega)}{dr}=\frac{r\xi'(r)-\xi(r)}{r^2} \quad \text{and} \quad \frac{d^2U_t(r,\omega)}{dr^2}=\frac{r^2\xi''(r)-2r\xi'(r)+2\xi(r)}{r^3}$$

and substituting the above expressions into our differential equation, it transforms as

$$\frac{d^2\xi(r)}{dr^2}+\frac{\omega^2}{c^2}\xi(r)=0,$$

with characteristic polynomial $k^2+(\omega^2/c^2)=0$ and roots $k=\pm i(\omega/c)$. Therefore, the general solution will be

$$\xi(r)=Ae^{i(\omega/c)r}+Be^{-i(\omega/c)r},$$

and returning to our original function, it is written as

$$U_t(r,\omega)=A(\omega)\frac{e^{i\frac{\omega}{c}r}}{r}+B(\omega)\frac{e^{-i\frac{\omega}{c}r}}{r},$$

where now the coefficients A and B have become functions of the cyclic frequency ω. The physical interpretation of the above relationship is that the solution to the problem arises from the superposition of two waves, of which the one corresponding to the first term approaches the spherical surface of radius α, whereas the one corresponding to the second term moves away from it. It is obvious that the existence of the first wave requires an energy source at infinite distance, which, however, does not exist—the statement of the exercise does not imply such a thing—and therefore the first term of the general solution $U_t(r,\omega)$ should be removed, which is done by setting $A(\omega)=0$. Therefore, finally the function $U_t(r,\omega)$ will have the form

$$U_t(r,\omega)=B(\omega)\frac{e^{-i\frac{\omega}{c}r}}{r}$$

Differentiating the above equation with respect to r we find that

$$\frac{dU_t(r,\omega)}{dr}=-\frac{B(\omega)}{r^2}e^{-i\frac{\omega}{c}r}\left(\frac{i\omega r+c}{c}\right)$$

and substituting the above equation into the boundary condition for the value $r=\alpha$ we get

$$\frac{B(\omega)}{\alpha^2}e^{-i\frac{\omega}{c}\alpha}\left(\frac{i\omega\alpha+c}{c}\right)=\rho\upsilon_0 \quad \text{from which it follows that} \quad B(\omega)=\frac{\rho\upsilon_0\alpha^2c}{i\omega\alpha+c}e^{i\frac{\omega}{c}\alpha}.$$

Therefore, the Fourier transform of the solution to the problem $u(r,t)$ will have the form

$$U_t(r,\omega)=\frac{\rho\upsilon_0\alpha^2c}{i\omega\alpha+c}e^{i\frac{\omega}{c}\alpha}\frac{e^{-i\frac{\omega}{c}r}}{r}=\frac{\rho\upsilon_0\alpha^2c}{i\omega\alpha+c}e^{-i\omega\left(\frac{r-\alpha}{c}\right)},$$

whereas the function itself will emerge from the equation of the inverse transform

$$u(r,t)=\frac{1}{2\pi}\int_{-\infty}^{+\infty}U_t(r,\omega)\,e^{i\omega t}d\omega=\frac{\rho\upsilon_0\alpha^2c}{2\pi r}\int_{-\infty}^{+\infty}\frac{e^{i\omega(t-\frac{r-\alpha}{c})}}{i\omega\alpha+c}\,d\omega.$$

The calculation of the above integral will be carried out using the method of residue calculus and the theorem of Cauchy. Considering the complex function

$$f(z)=\frac{e^{iz(t-\frac{r-\alpha}{c})}}{iz\alpha+c}$$

of the complex variable $z=\omega+i\xi$ it has a simple pole at the position $z=ic/\alpha$, which is located on the positive imaginary semi-axis. Starting from the range of values $t>(r-\alpha)/c$ for which the value of the exponent in the above function is positive, the closed path C required by the residue theorem consists of the real semi-axis and from a semicircle of infinite radius that has the center at the point $z=0$. According to Jordan's lemma, the value of the integral along this semicircle is zero. On the other hand, within the closed path C, there is the simple pole $z=ic/\alpha$ with the residue value

$$\text{Res}\Big[f(z)\Big]_{z=ic/\alpha}=\lim_{z\to ic/\alpha}\left[\left(z-\frac{ic}{\alpha}\right)\frac{e^{iz(t-\frac{t-\alpha}{c})}}{i\alpha[z-(ic/\alpha)]}\right]=\frac{1}{i\alpha}\lim_{z\to(ic/\alpha)}e^{iz(t-\frac{t-\alpha}{c})}=\frac{1}{i\alpha}e^{\frac{1-ct-r}{\alpha}},$$

as can be verified. Therefore, we have

$$\oint_C f(z)\,dz=\int_{-\infty}^{+\infty}\frac{e^{i\omega(t-\frac{r-\alpha}{c})}}{i\omega\alpha+c}\,d\omega=(2\pi i)\frac{1}{i\alpha}e^{\frac{1-ct-r}{\alpha}}=\frac{2\pi}{\alpha}e^{\frac{1-ct-r}{\alpha}}$$

and the desired solution will have the form

$$u(r,t)=\frac{\rho\upsilon_0\alpha^2c}{2\pi r}\frac{2\pi}{\alpha}e^{\frac{1-ct-r}{\alpha}}=\frac{\rho\upsilon_0\alpha c}{r}e^{\frac{1-ct-r}{\alpha}}.$$

The above relation holds for the area $t>(r-\alpha)/c$ whereas for $t<(r-\alpha)/c$ the closed path C will close with a semicircle of infinite radius and center at $z=0$ extending to the negative half-plane. However, within this closed path there are no poles of the function $f(z)$ and according to Cauchy's theorem, the value of the above integral as well as the function $u(r,t)$ will be equal to zero.

Problem A.12.3 Using the Fourier transform method to solve the Laplace differential equation

$$\nabla^2\varphi(x,y)=\frac{\partial^2\varphi(x,y)}{\partial x^2}+\frac{\partial^2\varphi(x,y)}{\partial y^2}=0$$

for the range of values $(y>0)$ with the solution the function $\varphi=\varphi(x,y)$ and boundary conditions of the form $\varphi(x,0)=f(x)$ and $|\varphi(x,\infty)|<M$—the last relation means that the function $\varphi(x,y)$ is bounded at the limit $y\to\infty$.

Solution The Fourier transform of the function $\varphi(x,y)$ with respect to the variable x and considering the parameter y constant, is given as known from the relation

$$\mathcal{F}_x\{\varphi(x,y)\}\equiv\Phi_x(\omega,y)=\int_{-\infty}^{+\infty}\varphi(x,y)\,e^{-i\omega x}dx.$$

Taking the Fourier transform of all the terms of the above equation, it is written as

$$\mathcal{F}_x\left\{\frac{\partial^2\varphi(x,y)}{\partial x^2}\right\}+\mathcal{F}_x\left\{\frac{\partial^2\varphi(x,y)}{\partial y^2}\right\}=0,$$

where

$$\mathcal{F}_x\left\{\frac{\partial^2\varphi(x,y)}{\partial x^2}\right\}=-\omega^2\Phi(\omega,y) \quad \text{and} \quad \mathcal{F}_x\left\{\frac{\partial^2\varphi(x,y)}{\partial y^2}\right\}=\frac{\partial^2\Phi_x(\omega,y)}{\partial y^2}$$

according to the properties of the Fourier transform that we presented in a previous section. Replacing the above relations in the differential equation of Laplace it will take the form

$$\frac{d^2\Phi_x(\omega,y)}{dy^2}-\omega^2\Phi_x(\omega,y)=0,$$

where we used the notation of total derivative since the cyclic frequency ω is constant.

The above equation is a second-order linear differential equation with constant coefficients and its solution has the form

$$\Phi_x(\omega,y)=A(\omega)e^{|\omega|y}+B(\omega)e^{-|\omega|y}=B(\omega)e^{-|\omega|y},$$

where we used the boundary condition $|\varphi(x,\infty)|<M$ from which it follows that $A(\omega)=0$. On the other hand, if we take the Fourier transform of the other boundary condition $\varphi(x,0)=f(x)$, we get $\Phi_x(\omega,0)=\mathcal{F}\{f(x)\}$. However, from the above solution of the differential equation, the assignment of the value $y=0$ will lead to the result $\Phi_x(\omega,y)=B(\omega)$. Comparing the last two relations it immediately follows that $B(\omega)=\mathcal{F}\{f(x)\}$, that is, the function $\Phi_x(\omega,y)$ will finally take the form

$$\Phi_x(\omega,y)=\mathcal{F}\{f(x)\}e^{-i|\omega|y}.$$

The determination of the desired solution of the problem $\varphi(x,y)$ can now be carried out using the convolution theorem. Assuming that the expression $e^{-i|\omega|y}$ constitutes the Fourier transform $\mathcal{F}\{g(x,y)\}$ of some unknown function $g(x,y)$, then, according to the convolution theorem, the desired solution would have the form

$$\varphi(x,y)=\mathcal{F}^{-1}\{\mathcal{F}\{f(x)\}\mathcal{F}\{g(x,y)\}\}=f(x)*g(x,y),$$

that is, it would result from the convolution between the function $f(x,y)$ appearing in the boundary condition and the inverse Fourier transform of the expression $e^{-i|\omega|y}$. Using the properties of the Fourier transform

$$\mathcal{F}\{e^{-|x|}\}=\frac{2}{1+\omega^2} \quad \text{and} \quad \mathcal{F}\{f(\alpha x)\}=\frac{1}{\alpha}F\left(\frac{\omega}{\alpha}\right),$$

which we proved in a previous section, it immediately follows that

$$g(x,y)=\mathcal{F}^{-1}\left\{e^{-|k|y}\right\}=\frac{y}{\pi(y^2+x^2)}.$$

Therefore, we have

$$\varphi(x,y)=f(x)*g(x,y)=\int_{-\infty}^{+\infty}f(\xi)g(x-\xi)d\xi=\int_{-\infty}^{+\infty}\frac{yf(\xi)}{\pi[y^2+(x-\xi)^2]}d\xi=\frac{y}{\pi}\int_{-\infty}^{+\infty}\frac{f(\xi)}{y^2+(x-\xi)^2}d\xi$$

and therefore the desired solution depends on the form of the function $f(x)$. For example, in the special case where the function $f(x)$ has the form

$$f(x)=\begin{cases}1 \text{ for } -1<x<1\\ 0 \text{ elsewhere}\end{cases}$$

from the above equation it follows that

$$\varphi(x, y) = \frac{y}{\pi}\int_{-1}^{+1}\frac{d\xi}{y^2+(x-\xi)^2} = \frac{y}{\pi}\left[\frac{1}{y}\tan^{-1}\left(\frac{\xi-x}{y}\right)\right]_{-1}^{+1} = \frac{1}{\pi}\left[\tan^{-1}\left(\frac{1-x}{y}\right)+\tan^{-1}\left(\frac{1+x}{y}\right)\right],$$

which is the desired result.

A.13 The Basic Theory of Vector Spaces

Vector spaces are inextricably linked to the space of continuous-time signals, as they form the basis on which the Fourier development of a signal is based, which in turn leads to the Fourier transform used for the transition of the signal representation from the field of continuous-time to the field of frequency. For this reason, in this section we present briefly the basic theory of vector spaces; for more details the reader is referred to any linear algebra textbook.

From linear algebra it is known that a linear vector space $V[\mathbb{F}]$ is defined as a quartet of the form $(V, \mathbb{F}, +, *)$ where V is a *non-empty* set of vectors equipped with the operation $+$ of vector addition, $(\mathbb{F}, \star, \bullet)$ an algebraic body equipped with the operations $\star$ and $\bullet$ and $* : \mathbb{F}\times V \to V$ an external operation characterized by the properties $\alpha * (\boldsymbol{u}+\boldsymbol{v}) = \alpha * \boldsymbol{u} + \alpha * \boldsymbol{v}$, $(\alpha \star \beta) * \boldsymbol{u} = \alpha * \boldsymbol{u} + \beta * \boldsymbol{u}$, $(\alpha \bullet \beta) * \boldsymbol{u} = \alpha * (\beta * \boldsymbol{u})$, and $1 * \boldsymbol{u} = \boldsymbol{u}$ $(1 \in \mathbb{F})$ $(\forall \boldsymbol{u} \in V$ and $\forall \alpha, \beta \in \mathbb{F})$. Using the operations $\star$ and $\bullet$ as the usual addition (symbolized by $+$) and the usual multiplication (symbolized by $\cdot$) and referring to the definition of the algebraic body, the above vector space will have the following properties:

1. The vectors $\boldsymbol{u}+\boldsymbol{v}$ and $\alpha\cdot\boldsymbol{u}$ will belong to the vector space V for every $\boldsymbol{u}, \boldsymbol{v} \in V$ and for every $\alpha \in \mathbb{F}$.
2. $\forall \boldsymbol{u}, \boldsymbol{v}, \boldsymbol{w} \in V : (\boldsymbol{u}+\boldsymbol{v})+\boldsymbol{w} = \boldsymbol{u}+(\boldsymbol{v}+\boldsymbol{w})$ (associative property of addition).
3. $\forall \boldsymbol{u}, \boldsymbol{v} \in V : \boldsymbol{u}+\boldsymbol{v} = \boldsymbol{v}+\boldsymbol{u}$ (commutative property of addition).
4. $\exists \boldsymbol{0} \in V : \forall \boldsymbol{u} \in V : \boldsymbol{u}+\boldsymbol{0} = \boldsymbol{0}+\boldsymbol{u} = \boldsymbol{u}$ (neutral element for the operation of addition).
5. $\forall \boldsymbol{u} \in V : \exists(-\boldsymbol{u}) \in V : \boldsymbol{u}+(-\boldsymbol{u}) = (-\boldsymbol{u})+\boldsymbol{u} = \boldsymbol{0}$ (symmetric (opposite) element).
6. $\forall \boldsymbol{u}, \boldsymbol{v}$ and $\forall \alpha \in \mathbb{F} : \alpha * (\boldsymbol{u}+\boldsymbol{v}) = \alpha * \boldsymbol{u} + \alpha * \boldsymbol{v}$.
7. $\forall \boldsymbol{u} \in V$ and $\forall \alpha, \beta \in \mathbb{F} : (\alpha+\beta) * \boldsymbol{u} = \alpha * \boldsymbol{u} + \beta * \boldsymbol{u}$.
8. $\forall \boldsymbol{u} \in V$ and $\forall \alpha, \beta \in \mathbb{F} : (\alpha\cdot\beta) * \boldsymbol{u} = \alpha * (\beta * \boldsymbol{u})$.
9. $\forall \boldsymbol{u} \in V : 1 * \boldsymbol{u} = \boldsymbol{u}$ where $1 \in \mathbb{F}$.

The elements defined within it not only include the usual vectors but also other entities such as functions—in this case, the vector space is known as a functional space—as well as matrices. If the field $\mathbb{F}$ is the field $\mathbb{R}$ of real numbers, the space V is characterized as a *real vector space*, whereas if the field $\mathbb{F}$ is the field $\mathbb{C}$ of complex numbers, the space V is known as a *complex vector space*. Classic examples of vector spaces include (a) the *Euclidean vector space* of n dimensions $\mathbb{R}^n$ defined with the help of the Cartesian product of sets as $\mathbb{R}^n = \{(x_1, x_2, x_3, \ldots, x_n) \mid x_1, x_2, x_3, \ldots, x_n \in \mathbb{R}\} = \mathbb{R}\times\mathbb{R}\times\mathbb{R}\times\cdots\times\mathbb{R}$ (n terms) with the internal operation the vector addition and the external operation $* : (\mathbb{R}\times\mathbb{R}^n) \to \mathbb{R}^n$ we define the *product* of the number $\lambda \in \mathbb{R}$ times the vector $\boldsymbol{x}_i = \{x_{i1}, x_{i2}, x_{i3}, \ldots, x_{in}\}$; (b) the set of matrices $\mathbb{F}^{m\times n}$ of dimensions $m \times n$ $(m, n \in \mathbb{F},\ m, n \geq 1)$ with elements from the field $\mathbb{F}$ and with composition laws the *addition* and the *multiplication* of matrices; (c) the *functional spaceF*(x), which is defined as the set of functions $V = \{\ f(x) \mid f : (\alpha, \beta) \to \mathbb{R}\ \}$ with the domain the interval (α, β), with the internal operation the *sum* of functions, and the *external operation* the *multiplication of a number by a function*; (d) the set L of functions $f : \mathbb{R}^2 \to \mathbb{R}$, which are *absolutely integrable*, that is, such that

$$\int_{-\infty}^{+\infty}\int_{-\infty}^{+\infty}|f(x, y)|\,dx\,dy < \infty,$$

equipped with the usual operations of *addition* and *multiplication*; (e) the set of solutions of the *homogeneous linear system* of equations $\boldsymbol{AX} = \boldsymbol{0}$ and (st) the set of *signals of limited spectrum*, which is presented in detail in the relevant section. It is interesting to note that the *functional vector spaces* have not only an *analytical* but also an *algebraic* structure, which makes them a valuable mathematical tool in many branches of physics, such as quantum mechanics.

Linear Dependence and Independence Considering a *vector space* V over a field $\mathbb{F}$ and a set of n vectors of the space $\boldsymbol{u}_1, \boldsymbol{u}_2, \boldsymbol{u}_3, \ldots, \boldsymbol{u}_n$ different from each other, if there are numbers $\alpha_1, \alpha_2, \alpha_3, \ldots, \alpha_n \in \mathbb{F}$ that are not all zero, the vectors $\boldsymbol{u}_1, \boldsymbol{u}_2, \boldsymbol{u}_3, \ldots, \boldsymbol{u}_n$ are characterized as *linearly dependent*, if they satisfy the equation

$$\alpha_1 \boldsymbol{u}_1 + \alpha_2 \boldsymbol{u}_2 + \alpha_3 \boldsymbol{u}_3 + \cdots + \alpha_n \boldsymbol{u}_n = \boldsymbol{0},$$

where $\boldsymbol{0}$ is the *zero vector* of the vector space. Conversely, if the above relation implies that $\alpha_1 = \alpha_2 = \alpha_3 = \cdots = \alpha_n = 0$, the vectors are characterized as *linearly independent*. The basic consequence of the linear independence of a group of vectors is that *none of them can be expressed as a linear combination of the rest*, where the linear combination refers to the ability to formulate a vector $\boldsymbol{v} \in V$ in the form

$$\boldsymbol{v} = \beta_1 \boldsymbol{u}_1 + \beta_2 \boldsymbol{u}_2 + \beta_3 \boldsymbol{u}_3 + \cdots + \beta_m \boldsymbol{u}_m,$$

where $\boldsymbol{u}_1, \boldsymbol{u}_2, \boldsymbol{u}_3, \ldots, \boldsymbol{u}_m \in V$ and $\beta_1, \beta_2, \beta_3, \ldots, \beta_m \in \mathbb{F}$, nonzero numbers. Conversely, if a vector $\boldsymbol{v}$ can be expressed in the above way, we say that it constitutes a *linear combination* of the vectors $\boldsymbol{u}_1, \boldsymbol{u}_2, \boldsymbol{u}_3, \ldots, \boldsymbol{u}_m \in V$, with the vectors $\boldsymbol{v}, \boldsymbol{u}_1, \boldsymbol{u}_2, \boldsymbol{u}_3, \ldots, \boldsymbol{u}_m \in V$ being linearly dependent.

Using the concepts of *linear dependence and independence*, we can characterize a set $S = \{\boldsymbol{u}_1, \boldsymbol{u}_2\, \boldsymbol{u}_3, \ldots, \boldsymbol{u}_m\}$ of nonzero vectors of a vector space V as *linearly dependent* or *independent* depending on the nature of the vectors that belong to this. Two other characteristics that can be attributed to a set of such vectors are the characteristic *orthogonal*, if these vectors are pairwise orthogonal (i.e., perpendicular) to each other and the characteristic *orthonormal*, if the vectors are pairwise orthogonal and in addition their measure is equal to the unit. As is well known to all of us, the unit vectors $(\hat{\boldsymbol{i}}, \hat{\boldsymbol{j}})$ of the axes in two dimensions and the vectors $(\hat{\boldsymbol{i}}, \hat{\boldsymbol{j}}, \hat{\boldsymbol{k}})$ of the axes in three dimensions of the Euclidean space constitute an *orthonormal linearly independent* set of vectors (and according to which we mention later that they form the basis of) and therefore, it is impossible to express any of these as a function of the remaining ones. Note that the concepts of orthogonality and orthonormality can, in addition to the usual vectors, be attributed to sets of other mathematical objects, such as matrices and functions, depending on the nature of the vector space to which they belong.

Linear Extension, Base, and Dimension Considering a vector space V defined over a body $\mathbb{F}$, we characterize as *linear extension* span$\{\boldsymbol{u}_1, \boldsymbol{u}_2, \boldsymbol{u}_3, \ldots, \boldsymbol{u}_m\}$ of the vectors of space V in the body $\mathbb{F}$ the set *of all possible linear combinations* of the vectors of space V, or in mathematical notation

$$\begin{aligned}\operatorname{span}\{\boldsymbol{u}_1, \boldsymbol{u}_2, \boldsymbol{u}_3, \ldots, \boldsymbol{u}_m\} &= \left\{\sum_{i=1}^{m} \alpha_i \boldsymbol{u}_i \mid \alpha_i \in \mathbb{F}, \boldsymbol{u}_i \in V\right\} \\ &= \{\alpha_1 \boldsymbol{u}_1 + \alpha_2 \boldsymbol{u}_2 + \alpha_3 \boldsymbol{u}_3 + \cdots + \alpha_m \boldsymbol{u}_m\}\ \forall \alpha_i \in \mathbb{F}, \boldsymbol{u}_1, \boldsymbol{u}_2, \ldots, \boldsymbol{u}_m \in V\end{aligned}$$

$(i = 1, 2, \ldots, m)$. If every vector of the vector space V can be expressed as a linear combination of the vectors $\{\boldsymbol{u}_1, \boldsymbol{u}_2, \boldsymbol{u}_3, \ldots, \boldsymbol{u}_m\}$, that is, it can be written as

$$\boldsymbol{v} = \alpha_1 \boldsymbol{u}_1 + \alpha_2 \boldsymbol{u}_2 + \alpha_3 \boldsymbol{u}_3 + \cdots + \alpha_m \boldsymbol{u}_m\ (\alpha_1, \alpha_2, \ldots, \alpha_m \in \mathbb{F}),$$

then we say that the linear extension span$\{\boldsymbol{u}_1, \boldsymbol{u}_2, \boldsymbol{u}_3, \ldots, \boldsymbol{u}_m\}$ *generates* the space V, with the vectors $\{\boldsymbol{u}_1, \boldsymbol{u}_2, \boldsymbol{u}_3, \ldots, \boldsymbol{u}_m\}$ being characterized as *generators* of the aforementioned vector space. In the special case where the set of generators $S = \{\boldsymbol{u}_1, \boldsymbol{u}_2, \boldsymbol{u}_3, \ldots, \boldsymbol{u}_m\}$ constitutes a linearly independent set of vectors, we say that these vectors form a basis of the vector space , with the number $d = m$ of the vectors of the basis of the space V being known as the *dimension* of this space and is denoted by dim(V). The basis of a vector space can be considered as a *generalization* of the concept of a *system coordinate*, which is known from geometry. If the number of vectors in the base is *finite*, we say that the space V has *finite dimension*, whereas, if the base of the space has *infinite* vectors we say that the space is *infinite-dimensional*, or equivalently that it has *infinite* dimension. A vector space is *a space of finite dimension*, if and only if it has *a finite base*, whereas it is proven that such a space will contain *at least* one base. Considering a vector space of n dimensions, *any set of n linearly independent vectors constitutes a base of this space*.

A particularly interesting method that is related to the bases of vector spaces and has been widely applied is the *Gram-Schmidt orthogonalization method*, which allows us to construct an orthogonal base of vectors $\{\boldsymbol{w}_1, \boldsymbol{w}_2, \boldsymbol{w}_3, \ldots, \boldsymbol{w}_n\}$ for a vector space $V[\mathbb{F}]$, starting from a random base of $\{\boldsymbol{u}_1, \boldsymbol{u}_2, \boldsymbol{u}_3, \ldots, \boldsymbol{u}_n\}$.

The base and the dimension of a vector space are characterized by a number of useful properties. For example, it is proven that the expansion of a vector with respect to a base of the vector space V to which it belongs is unique and that all bases of a vector space V have the same dimension, that is, they contain the same number of linearly independent vectors.

Considering a vector $\boldsymbol{v}$ that belongs to a vector space U and a base $\boldsymbol{B} = \{\boldsymbol{v}_1, \boldsymbol{v}_2, \ldots, \boldsymbol{v}_n\}$ of this space, the coefficients of the linear combination $\boldsymbol{v} = \alpha_1\boldsymbol{v}_1 + \alpha_2\boldsymbol{v}_2 + \cdots + \alpha_n\boldsymbol{v}_n$ are called coordinates of the vector $\boldsymbol{v}$ with respect to the base $\boldsymbol{B}$ and are denoted as $[\boldsymbol{v}]_{\boldsymbol{B}} = (\alpha_1, \alpha_2, \ldots, \alpha_n)^T$. Taking into account that in a vector space *infinite bases* are defined, it is obvious that the coordinates of the vector $\boldsymbol{v}$ with respect to each of these bases will be different and therefore the same vector of a vector space can be represented in many different ways. In the terminology of linear algebra, the process of determining the coordinates of a vector $\boldsymbol{v}$ with respect to a base $\boldsymbol{B}'$, when the coordinates with respect to another base $\boldsymbol{B}$ are known, is known as *base change*.

Vector Norms The *norm* of a vector of n elements $\boldsymbol{v} \in V[\mathbb{F}]$, is a function $f(\boldsymbol{v}) : V \rightarrow \mathbb{F}$, which assigns to each vector a *length* or equivalently a *measure* and is usually denoted as $\|\boldsymbol{v}\|$ in order to be distinguished from the function of *absolute value*. Considering a function $f(\boldsymbol{v})$, this constitutes a *valid vector norm function*, if it satisfies the following properties:

1. For every vector $\boldsymbol{v} \neq \mathbf{0}$ it is $f(\boldsymbol{v}) = \|\boldsymbol{v}\| \geq 0$, whereas it is $f(\boldsymbol{v}) = \|\boldsymbol{v}\| = 0$, if and only if it is $\boldsymbol{v} = \mathbf{0}$.
2. For every constant $\lambda \in \mathbb{F}$ and for every vector $\boldsymbol{v} \in V$ it is $f(\lambda\boldsymbol{v}) = |\lambda| f(\boldsymbol{v})$, or equivalently $\|\lambda\boldsymbol{v}\| = |\lambda|\|\boldsymbol{v}\|$.
3. For every pair of vectors $\boldsymbol{v}_1, \boldsymbol{v}_2 \in V$ it is $f(\boldsymbol{v}_1 + \boldsymbol{v}_2) \leq f(\boldsymbol{v}_1) + f(\boldsymbol{v}_2)$, or equivalently $\|\boldsymbol{v}_1 + \boldsymbol{v}_2\| \leq \|\boldsymbol{v}_1\| + \|\boldsymbol{v}_2\|$.

The most well-known vector norm function is the one known to us from analytical geometry, the *Euclidean norm*, which for the most general case of a vector of n complex values $\boldsymbol{v} \in \mathbb{C}^{n\times 1}$ is defined as

$$\|\boldsymbol{v}\| = \sqrt{\sum_{i=1}^{n} |v_i|^2} = \sqrt{\boldsymbol{v}^*\boldsymbol{v}}.$$

A generalization of the Euclidean norm is the p *norm*, which for values $p \geq 1$ and for every vector $\boldsymbol{v} \in \mathbb{C}^{n\times 1}$ is defined as

$$\|\boldsymbol{v}\|_p = \sqrt[p]{\sum_{i=1}^{n} |v_i|^p} = \left(\sqrt{\sum_{i=1}^{n} |v_i|^p}\right)^{1/p}.$$

It is not difficult to see that the Euclidean norm arises from the p norm for $p = 2$ and for this reason it is often also symbolized as $\|\boldsymbol{v}\|_2$. The most important p norms are those that correspond to the values $p = 1$, $p = 2$ and $p = \infty$, and which are therefore defined as

$$\|\boldsymbol{v}\|_1 = |v_1| + |v_2| + \cdots + |v_n| \qquad \|\boldsymbol{v}\|_2 = \sqrt{|v_1|^2 + |v_1|^2 + \cdots + |v_n|} = \sqrt{\boldsymbol{v}^T\boldsymbol{v}}$$

$$\|\boldsymbol{v}\|_\infty = \lim_{p\to\infty} \|\boldsymbol{v}\|_p = \lim_{p\to\infty} \left\{\left(\sqrt{\sum_{i=1}^{n} |v_i|^p}\right)^{1/p}\right\} = \max_{1\leq i\leq n} |v_i|.$$

The function of a vector's norm is characterized by certain interesting properties, which are usually formulated in the form of inequalities. Two interesting and widely used inequalities are the Cauchy-Bunyakovsky-Schwarz inequality $|\boldsymbol{v}^*\boldsymbol{w}| \leq \|\boldsymbol{w}\|\|\boldsymbol{v}\|$, where $\boldsymbol{v}, \boldsymbol{w} \in \mathbb{C}^{n\times 1}$ are two non-zero vectors, the well-known triangular inequality $\|\boldsymbol{v} + \boldsymbol{w}\| \leq \|\boldsymbol{v}\| + \|\boldsymbol{w}\|$, which applies to every pair of vectors $\boldsymbol{v}, \boldsymbol{w} \in \mathbb{C}^{n\times 1}$ and the Holder inequality, which is formulated as $\sum_{i=1}^{n} |v_i w_i| \leq \|\boldsymbol{v}\|_p\|\boldsymbol{w}\|_q$, where p and q are numbers such that $(1/p) + (1/q) = 1$. Finally, for the case of vectors of n elements, it is proven that $\|\boldsymbol{v}\|_2 \leq \|\boldsymbol{v}\|_1 \leq \sqrt{n}\|\boldsymbol{v}\|_2$, $\|\boldsymbol{v}\|_\infty \leq \|\boldsymbol{v}\|_2 \leq \sqrt{n}\|\boldsymbol{v}\|_\infty$ and $\|\boldsymbol{v}\|_\infty \leq \|\boldsymbol{v}\|_1 \leq n\|\boldsymbol{v}\|_\infty$. More information about norm functions can be found in the literature.

Inner Product and Distance Function

(1) Considering a *vector space* $V[\mathbb{F}]$, the *inner product* constitutes a *single-valued mapping* $\langle\,,\rangle : V \times V \rightarrow \mathbb{F}$ of the Cartesian product $V \times V$ to the field $\mathbb{F}$ such that

1. $\langle \boldsymbol{u}, \boldsymbol{v}\rangle = \langle \boldsymbol{v}, \boldsymbol{u}\rangle^*$, $\forall \boldsymbol{u}, \boldsymbol{v} \in V$ (Hermitian symmetry).
2. $\langle \alpha\boldsymbol{v} + \beta\boldsymbol{u}, \boldsymbol{w}\rangle = \alpha^*\langle \boldsymbol{v}, \boldsymbol{w}\rangle + \beta^*\langle \boldsymbol{u}, \boldsymbol{w}\rangle$ and $\langle \boldsymbol{u}, \alpha\boldsymbol{v} + \beta\boldsymbol{w}\rangle = \alpha^*\langle \boldsymbol{u}, \boldsymbol{v}\rangle + \beta^*\langle \boldsymbol{u} + \boldsymbol{w}\rangle$, $\forall \boldsymbol{u}, \boldsymbol{v} \in V$

3. $\langle \boldsymbol{v}, \boldsymbol{v} \rangle \geq 0$ and $\langle \boldsymbol{v}, \boldsymbol{v} \rangle = 0 \Leftrightarrow \boldsymbol{v} = \mathbf{0}, \forall \boldsymbol{v} \in V$.
It is obvious that the form of the inner product depends on the vector space with which it is associated. For example, in the case of the Euclidean vector space $\mathbb{R}^n$, a mapping of the form $\langle\, , \rangle : \mathbb{R}^n \times \mathbb{R}^n \to \mathbb{R}$ constitutes an inner product function, if and only if there is a symmetric positive definite matrix $\boldsymbol{M}$ such that $\langle \boldsymbol{u}, \boldsymbol{v} \rangle = \boldsymbol{u}^T \boldsymbol{M} \boldsymbol{v}$ for every $\boldsymbol{u}, \boldsymbol{v} \in \mathbb{R}^n$. If the matrix $\boldsymbol{M}$ is the identity matrix, then the inner product simplifies to the usual product of vectors $\langle \boldsymbol{u}, \boldsymbol{v} \rangle = \boldsymbol{u}^T \boldsymbol{v} = \sum_{i=1}^{n} u_i v_i = u_1 v_1 + u_2 v_2 + \cdots + u_n v_n$, which allows us to formulate mathematically the relationship of orthogonality as

$$\langle \boldsymbol{u}_i, \boldsymbol{u}_j \rangle = \delta_{ij} = \begin{cases} 1 \text{ when } i = j \\ 0 \text{ when } i \neq j \end{cases}$$

with the symbol δ_{ij} known as the *Kronecker delta*. In the case where the vector space is defined in the field of complex numbers $\mathbb{C}^n$, the above function is defined as $\langle \boldsymbol{u}, \boldsymbol{v} \rangle = \boldsymbol{v}^\dagger \boldsymbol{M} \boldsymbol{u}$ for every $\boldsymbol{u}, \boldsymbol{v} \in \mathbb{C}^n$, with the matrix $\boldsymbol{M}$ this time being a Hermitian positive definite matrix and the vector $\boldsymbol{v}^\dagger$ is the conjugate transpose of the vector $\boldsymbol{v}$. On the other hand, as we see in the next section, the product function used in the Hilbert functional spaces of infinite dimensions, is defined as

$$\langle f, g \rangle = \int_{\alpha}^{\beta} \overline{f(x)} g(x) dx$$

with the function $\overline{f(x)}$ being the complex conjugate of the function $f(x)$. In this case, a group of functions $\varphi_n(t)$ $(n = 1, 2, 3, \ldots)$ is aforementioned to constitute an *orthogonal set* of functions over a range (α, β) if the inner product of any two functions of the group $\varphi_m(t)$ and $\varphi_k(t)$ satisfies the equation

$$< \varphi_m(t), \varphi_k(t) >= \int_{\alpha}^{\beta} \varphi_m(t)\, \overline{\varphi_k(t)}\, w(t)\, dt = \begin{cases} 0 \text{ for } m \neq k \\ \alpha \text{ for } m = k \end{cases} = \alpha \delta_{mk},$$

where δ_{mk} is the δ *of Kronecker* and $w(t)$ is a suitably chosen weight function. The constant α that appears in the above relation is calculated as the integral

$$\alpha = \int_{\alpha}^{\beta} w(t)\, [\varphi_m^2(t)]\, dt$$

and if its value is equal to one, the set of functions $\varphi_n(t)$ $(n = 1, 2, 3, \ldots)$ is *orthonormal*.

(2) Considering two vectors $\boldsymbol{u} = \{u_1, u_2, u_3, \ldots, u_n\}$ and $\boldsymbol{v} = \{v_1, v_2, v_3, \ldots, v_n\}$ of the Euclidean space, the distance of the points defined by them is nothing but the measure of the difference of the two vectors, or in mathematical notation

$$d(\boldsymbol{u}, \boldsymbol{v}) = \|\boldsymbol{u} - \boldsymbol{v}\| = \sqrt{\sum_{i=1}^{n} (u_i - v_i)} = \sqrt{(u_1 - v_1)^2 + (u_2 - v_2)^2 + (u_3 - v_3)^2 + \cdots + (u_n - v_n)^2}$$

The distance between the vectors $\boldsymbol{u}$ and $\boldsymbol{v}$ is *greater than or equal to zero*, unless it is $\boldsymbol{u} = \boldsymbol{v}$. Therefore, in general, $d(\boldsymbol{u}, \boldsymbol{v}) \geq 0$. In addition, the distance d will satisfy the relations $d(\boldsymbol{u}, \boldsymbol{v}) = d(\boldsymbol{v}, \boldsymbol{u})$ and $d(\boldsymbol{u}, \boldsymbol{v}) \leq d(\boldsymbol{u}, \boldsymbol{w}) + d(\boldsymbol{w}, \boldsymbol{v})$ (for any triplet of vectors $\boldsymbol{u}, \boldsymbol{v}, \boldsymbol{w} \in \mathbb{F}^n$), of which the first is a direct consequence of its definition, whereas the second is a consequence of the known triangular inequality $\|\boldsymbol{u} + \boldsymbol{v}\| \leq \|\boldsymbol{u}\| + \|\boldsymbol{v}\|$, expressed in an n-dimensional space.

(3) Considering an inner product $\langle \boldsymbol{v}, \boldsymbol{v} \rangle$ defined on a vector space $V[\mathbb{F}]$, it is proven that the function

$$\|\boldsymbol{v}\| = \sqrt{\langle \boldsymbol{v}, \boldsymbol{v} \rangle}$$

constitutes a *vector norm function* in the space V, since it fulfills *all the properties* of a norm function.

Normed and Metric Vector Spaces A vector space $V[\mathbb{F}]$ that is equipped with a *vector norm function* $\| \star \| : V \to R$ is characterized as *normed* and is denoted by $(V, \| \star \|)$. *Any suitably defined vector norm function* can be used as such a function, although in practice the three known vector norm functions $\|\boldsymbol{v}\|_1$, $\|\boldsymbol{v}\|_2$, and $\|\boldsymbol{v}\|_\infty$ are used. If the norm function that is used is associated with a *metric* or *distance function* d, which is defined as a function $d : V \times V \to \mathbb{R}$ such that

$d(\boldsymbol{v}, \boldsymbol{w}) = \|\boldsymbol{v} - \boldsymbol{w}\|$, then the vector space is called a *metric* and is denoted as (V, d). This distance function is characterized by the three properties that we previously mentioned and which arise from the corresponding vector norm functions. Note that for a space to be normed, it *must have the structure and properties of a vector space*, a requirement that *is not necessary* for a metric space, which is not necessarily associated with any *algebraic structure*. Conversely, this can be defined as a simple set of points described by a position vector with respect to some origin of a coordinate system and that is equipped with a *distance function*, with the help of which we can define fundamental concepts, such as the concepts of *continuity*, *convergence*, and *compactness*.

If the vector space is equipped with the operation of the *inner product*, we can define a vector norm function as $\|\boldsymbol{v}\| = \sqrt{\langle \boldsymbol{v}, \boldsymbol{v} \rangle}$, and a distance function as

$$d(\boldsymbol{v}, \boldsymbol{w}) = \|\boldsymbol{v} - \boldsymbol{w}\| = \sqrt{\langle \boldsymbol{v} - \boldsymbol{w}, \boldsymbol{v} - \boldsymbol{w} \rangle}.$$

Consequently, *an inner product vector space is both normed and metric*, whereas the opposite is not always true. In the terminology of linear algebra, a space of the inner product V is often characterized as a pre-Hilbert space and is symbolized as $[V, \langle \star, \star \rangle]$. A *pre-Hilbert space* is *real* or *complex* depending on the nature of the *underlying vector space*.

Sequences of Vectors, Convergence of Sequences, and Completeness of Vector Space Considering a *vector space* V and a set of vectors $\{\boldsymbol{v}_1, \boldsymbol{v}_2, \boldsymbol{v}_3, \ldots, \boldsymbol{v}_n\}$ of this space, these can be considered a sequence $\{\boldsymbol{v}_n\} = \{\boldsymbol{v}_n : n \in \mathbb{N}\}$ with indices from the set $\mathbb{N}$ of natural numbers. If the vector space V is *normed*, that is, equipped with a function *vector norm* $\| \star \|$, we say that the sequence $\{\boldsymbol{v}_n\}$ *converges* to the vector $\boldsymbol{v}$ and we write $\lim \boldsymbol{v}_n = \boldsymbol{v}$ in the limit $n \to \infty$ or more simply $\boldsymbol{v}_n \to \boldsymbol{v}$ if and only if the real sequence $\|\boldsymbol{v}_n - \boldsymbol{v}\|$ tends to zero. It is proven that the limit to which such a sequence of vectors converges is unique.

A sequence $\{\boldsymbol{v}_n\}$ defined in a normed vector space is called a Cauchy sequence , if and only if the property $\|\boldsymbol{v}_n - \boldsymbol{v}_m\| \to 0$ holds in the limit $n, m \to \infty$ or in more strict mathematical writing, $\forall \varepsilon > 0,\ \exists M > 0 : n, m > M \Rightarrow \|\boldsymbol{v}_n - \boldsymbol{v}_m\| < \varepsilon$. In a *normed* vector space *every converging sequence of vectors is a Cauchy sequence*, although the converse of this proposition generally does not hold.

The concepts of the sequence of vectors and the convergence of sequences can be reformulated for a metric space, with the corresponding definitions based on the concept of the *distance function*. In a metric vector space V, a sequence of vectors $(\boldsymbol{v}_n)$ converges to some vector $\boldsymbol{v} \in V$, if for every $\varepsilon > 0$ there is $\ell \in \mathbb{N}$ such that $d(\boldsymbol{v}_n, \boldsymbol{v}) < \varepsilon$ for every $n \geq \ell$. If for every $\varepsilon > 0$ there is $\ell \in \mathbb{N}$ such that for every $m, n \geq \ell$ it is $d(\boldsymbol{v}_m, \boldsymbol{v}_n) < \varepsilon$, the sequence of vectors is a Cauchy sequence.

If every *Cauchy sequence* of the metric space (V, d) converges to some vector of the space V, then the space V is characterized as a *complete vector space*, whereas in the opposite case it is characterized as incomplete or unfinished. It is proven that every finite-dimensional vector space is a complete vector space and that *every vector space V that is not complete can always be extended to a larger complete metric vector space*, which is known as the *completion* of the vector space V.

Banach Space In complete analogy with metric vector spaces, a normed vector space $(V, \| \star \|)$ is called *complete*, if and only if *every Cauchy sequence that is defined in space V converges on V* and therefore its limit constitutes some vector $\boldsymbol{x} \in V$. In the terminology of mathematical analysis, a *complete normed vector space* is known as a *Banach space*, in honor of the Polish mathematician Stefan Banach (1892–1945). Note that although every finite-dimensional normed vector space is a Banach space, this concept of space is usually used to describe vector *spaces of infinite dimensions*.

Examples of *Banach spaces* include (a) the Euclidean space $\mathbb{R}^n$ equipped with the function of the p norm

$$\|\boldsymbol{x}\|_p = \sqrt[p]{\sum_{i=1}^{n} |x_i|^p} = \sqrt[p]{|x_1|^p + |x_2|^p + |x_3|^p + \cdots + |x_n|^p}$$

and (b) the *weighted vector space* $C([\alpha, \beta])$ of *continuous real or complex functions* defined in the interval $[\alpha, \beta]$ with the norm function $\|f\|_\infty = \sup |f(x)|$. However, the vector space $L^p([\alpha, \beta])$ $(1 \leq p < \infty$ and $[\alpha, \beta] \subset \mathbb{R})$ of Lebesgue measurable functions $f : [\alpha, \beta] \to \mathbb{R}$ (or $\mathbb{C}$), which are described by the norm function

$$\|f\|_p = \sqrt[p]{\int_\alpha^\beta |f(x)|^p dx}.$$

The integral that appears in the root of the above equation is known as the Lebesgue integral and cannot take an infinite value. For the value $p = \infty$ the space $L^{\infty}([\alpha, \beta])$ contains the *Lebesgue measurable functions* $f : [\alpha, \beta] \to \mathbb{R}$ (or $\mathbb{C}$) that are *essentially bounded* in the closed interval $[\alpha, \beta]$, that is, they are bounded in a subset of $[\alpha, \beta]$, the complement of which has zero measure. In the general case, let Ω be a *measurable subset* of the space $\mathbb{R}^n$ and $L^p(\Omega)$ the set of *Lebesgue measurable functions* $f : \Omega \to \mathbb{R}$ (or $\mathbb{C}$). In this case it is proven that for values $1 \le p \le \infty$ the space $L^p(\Omega)$ is a *Banach space of infinite dimensions*.

Hilbert Space In the terminology of functional analysis, the Hilbert space, which took its name from the great German mathematician David Hilbert (1862–1943), is described as a *complete metric vector space* V, the norm function of which *is induced by an inner product* and therefore is defined as $\|\boldsymbol{v}\| = \sqrt{\langle \boldsymbol{v}, \boldsymbol{v} \rangle}$ for every vector $\boldsymbol{v} \in V$. In other words, *the Hilbert space is a Banach space equipped with an inner product that induces its norm function*. It is proven that *a Hilbert space is a complete pre-Hilbert space*.

Considering the vector space $C[(\alpha, \beta)]$ of *Riemann integrable complex functions* $f : [\alpha, \beta] \to \mathbb{C}$ defined by the equation

$$C[(\alpha, \beta)] = \left\{ f : [\alpha, \beta] \to \mathbb{C} \;\middle|\; \int_{\alpha}^{\beta} \overline{f(x)} g(x) dx < \infty \right\}$$

equipped with the inner product function

$$\langle f, g \rangle = \int_{\alpha}^{\beta} \overline{f(x)} g(x) dx$$

it is proven that this space *is not complete* and therefore *does not constitute a Hilbert space*. However, the completion of this space with respect to the norm function

$$\|f\| = \sqrt{\langle f, g \rangle} = \sqrt{\int_{\alpha}^{\beta} \overline{f(x)} g(x) dx}$$

leads to the Hilbert vector space $L^2([\alpha, \beta])$, which arises from the space $L^p([\alpha, \beta])$ we presented previously, for the value $p = 2$. This space is particularly interesting because it is used in the construction of the trigonometric Fourier development, which plays a fundamental role in the theory of continuous-time signals.

Orthogonal Bases of Hilbert Space and Fourier Development Considering a *orthonormal set of vectors* $U = \{\boldsymbol{x}_n \,|\, n \in \mathbb{I}\}$ of a *Hilbert vector space*, this set is characterized as complete if it satisfies any of the following properties that are proven to be equivalent to each other: (1) The power of the equation $\langle \boldsymbol{x}_n, \boldsymbol{x} \rangle = 0$ for each index $n \in \mathbb{I}$ implies that $\boldsymbol{x} = 0$. (2) For each $\boldsymbol{x} \in H$ it is proven that $\boldsymbol{x} = \sum \langle \boldsymbol{x}_n, \boldsymbol{x} \rangle \boldsymbol{x}_n$ $(n \in \mathbb{I})$. (3) For each $\boldsymbol{x} \in H$ it is proven that $\|\boldsymbol{x}\|^2 = \sum |\langle \boldsymbol{x}_n, \boldsymbol{x} \rangle|^2$ $(n \in \mathbb{I})$. This relationship is known as *Parseval's identity*. (4) The set of vectors U generates the space H, since it is proven that the *closed linear extension*

$$\overline{\text{span}}\, \{U\} = \left\{ \sum_{\boldsymbol{x} \in U} c_n \boldsymbol{x} \mid c_u \in C : \text{ the sum } \sum_{\boldsymbol{x} \in U} c_n \boldsymbol{x} \text{ converges} \right\}$$

is the *Hilbert vector space*. (5) The set U contains *the maximum number of orthonormal vectors* and therefore *there is no genuine superset of U that also constitutes an orthonormal set*. Taking into account now that the set U constitutes a *linearly independent set of vectors* and that *every vector of space H can be expressed as a linear combination of these vectors*, it is obvious that the set of words constitutes an *orthonormal basis* of the Hilbert space and therefore, each vector $\boldsymbol{x} \in H$ can be written as a linear combination of the form

$$\boldsymbol{x} = \sum_{n=1}^{d} \langle \boldsymbol{x}_n, \boldsymbol{x} \rangle \boldsymbol{x}_n = \langle \boldsymbol{x}_1, \boldsymbol{x} \rangle \boldsymbol{x}_1 + \langle \boldsymbol{x}_2, \boldsymbol{x} \rangle \boldsymbol{x}_2 + \langle \boldsymbol{x}_3, \boldsymbol{x} \rangle \boldsymbol{x}_3 + \cdots + \langle \boldsymbol{x}_d, \boldsymbol{x} \rangle \boldsymbol{x}_d.$$

Similar arguments can be formulated for the case of a Hilbert vector space of *infinite dimensions*. Using the terminology we introduced in previous sections, we can state the claim that the *coordinates* of the arbitrary vector $\boldsymbol{x} \in H$ with respect to the orthonormal basis U are the $\xi_n = \langle \boldsymbol{x}_n, \boldsymbol{x} \rangle$ $(n = 1, 2, \ldots, d)$. These coefficients are known as *Fourier coefficients* of the vector $\boldsymbol{x}$, with the above relation known as the Fourier expansion or the Fourier series of the vector $\boldsymbol{x}$ with respect to the orthonormal basis U. We therefore find that in Hilbert spaces of finite dimension the Fourier expansion *decomposes* the arbitrary vector $\boldsymbol{x} \in H$ into n mutually orthogonal vectors $\langle \boldsymbol{x}_n, \boldsymbol{x} \rangle \boldsymbol{x}_n$ $(n = 1, 2, \ldots, d)$.

Parseval's Identity Considering a vector space V of finite dimension n and two vectors $\boldsymbol{v}$ and $\boldsymbol{w}$ with coordinates v_i and w_i $(i = 1, 2, \ldots, n)$, it is proven that

$$\langle \boldsymbol{w}, \boldsymbol{v} \rangle = \sum_{i=1}^{n} \langle \boldsymbol{e}_i, \boldsymbol{w} \rangle^* \langle \boldsymbol{e}_i, \boldsymbol{v} \rangle = \sum_{i=1}^{n} \langle \boldsymbol{w}, \boldsymbol{e}_i \rangle \langle \boldsymbol{e}_i, \boldsymbol{v} \rangle.$$

This relationship is known as *Parseval's identity*. In the special case where $\boldsymbol{v} = \boldsymbol{w}$ this relationship takes its simplest form

$$\langle \boldsymbol{v}, \boldsymbol{v} \rangle = \sum_{i=1}^{n} \langle \boldsymbol{v}, \boldsymbol{e}_i \rangle \langle \boldsymbol{e}_i, \boldsymbol{v} \rangle = \sum_{i=1}^{n} \langle \boldsymbol{e}_i, \boldsymbol{v} \rangle^* \langle \boldsymbol{e}_i, \boldsymbol{v} \rangle = \sum_{i=1}^{n} |\langle \boldsymbol{e}_i, \boldsymbol{v} \rangle|^2.$$

The above relationship involves *the set of vectors of the orthonormal basis*. Conversely, if in the above sum the full set of n unit vectors does not appear but only k of them $(k < n)$ the above equality transforms into the inequality

$$\langle \boldsymbol{v}, \boldsymbol{v} \rangle \geq \sum_{i=1}^{n} |\langle \boldsymbol{e}_i, \boldsymbol{v} \rangle|^2,$$

which is known as *Bessel's inequality*.

A.14 Elements of Probability Theory

A significant concept related to the field of *statistical signal processing*, while also holding a central position in other mathematical sciences and in the science of engineering, is the concept of *probability*, which is defined for both *continuous* and *discrete* cases. From the mathematical side of the subject, probability, as a measure, constitutes a *measure* and therefore, the theory of probability is a *measure theory*, that is, a theory that deals with *the study of functions that assign numerical values to sets of objects*, which are defined in the appropriate way in each case *abstract mathematical space*. However, in contrast to the known measures from the field of physics and mathematics, such as *area* and *volume* in two and three dimensions, as well as *mass* and *weight*, which can take any value, the maximum probability value is equal to one, a value that corresponds to the *certainty of appearance* of a possible event from the set of all possible events that can occur in practice. Let us also mention that the process of assigning probabilities, *does not* assign probabilities to individual points on the real number line $\mathbb{R}$, but to *whole intervals*, that is, to *sets of such points*, since the assignment of probabilities to simple points, for the continuous case, has no practical application. These sets of points of the set $\mathbb{R}$, in which the assignment of probabilities takes place, are known as *events* or *occurrences*, whereas the contents of these sets, that is, the points of the line, are known as *samples*.

A.14.1 Defining Probability

How is it possible to define the concept of *probability*, which, let it be noted, has appeared in the field of mathematics for several centuries? In reality, this can be defined in *three different* ways and more specifically, (a) a *practical* or *empirical* way, which is based on the implementation of *experiments* and therefore, gives the size of probability, a *physical substance*, (b) a *theoretical* way that *does not* rely on the experiment, and (c) an *axiomatic* way, which defines the probability through a *set of axioms*. Of these three definitions, the *practical/empirical* definition formulated by Von Mises in 1957 describes the probability of an outcome *as a relative frequency the value of which is proportional to the ratio of the number of appearances of this result to the total number of experiments conducted*. From the mathematical side of the subject, if we consider an infinite sequence of the form $x = (x_0, x_1, x_2, \ldots, x_n)$ containing the results of an experiment, the *relative frequency of appearance* of some value α as the result of this experiment, that is, to be $x_n = \alpha$, where x_n any term of the sequence is defined as

$$r_a(x) = \lim_{n\to\infty} \frac{\text{number of indices } k \in (0, 1, 2, \ldots, n-1) \text{ for which } x_k = \alpha \text{ holds}}{n}.$$

The above definition is justified by the fact that since probabilities are used to describe *relative frequencies of appearance*, it is reasonable to define them as *the limits of such frequencies*. Specifically, in the limit $n \to \infty$, this practical definition of probability as a *relative frequency of occurrence* leads to *the same* result as the classic theoretical definition. This classic definition *is not based* on the experiment, but on the contrary, the probability $P(A)$ of the occurrence of an event A is calculated *theoretically in advance* and without any experiment being conducted, as the ratio N_A/N, where N is the number of all possible outcomes of the experiment and N_A is the number of outcomes that favor the occurrence of event A. For example, in the dice experiment, the number of possible outcomes is equal to $N = 6$ (since a die has six faces), whereas the number of outcomes that favor the appearance of an even number is $N_A = 3$ (since three out of the six numbers written on the faces of the die are even). Consequently, without even throwing the die, we claim that the probability of an even number appearing is $P(\text{even}) = 3/6$. On the contrary, in the approach of relative frequency, we *repeatedly conduct the experiment*, that is, we *repeatedly throw* the die on a table, record the number of even number appearances and divide this number by the total number n of throws that were made. In this case, the theoretical probability $P(\text{even}) = 3/6$ will coincide with the value of the relative frequency in the limit $n \to \infty$, that is, after an infinite number of repetitions.

Finally, according to the *axiomatic definition of probability*, which was formulated by Kolmogorov in 1933 and is based on concepts such as the *sample space*, the *random experiment*, and the *event*, probability is defined through a set of four axioms, which describe properties that every function that aspires to constitute a valid probability function, or more accurately, a valid probability measure, should satisfy, and more specifically:

- It must hold that $P(A) \geq 0$ for every $A \in \mathcal{F}$. According to this property, *the values of probability can only be positive or zero*. There are no negative values of probability, something that also applies to other properties that express measures, such as area, volume, weight, and mass.
- It must hold that $P(\Omega) = 1$. According to this property, *the probability of an event with elements from the sample space* Ω (see the description that follows) *is equal to* 1, which, unlike the values of other measures, constitutes the maximum value that probability can take and corresponds to certainty. This conclusion is a direct consequence of the definition of the sample space Ω as the set of all possible outcomes of an experiment.
- *The probability of the union of two mutually exclusive sets or events is equal to the sum of the probabilities of the individual events*, just as happens with the other measures, since the same applies to the sum of weights, masses, volumes, etc. In other words, if $A \cap B \neq \emptyset$, then we will have $P(A \cup B) = P(A + B) = P(A) + P(B)$. This property is known as the additive property of probability.
- The *additive property* must hold for an infinite set of events, according to which

$$P(\bigcup_{i=1}^{\infty} F_i) = \sum_{i=1}^{\infty} P(F_i),$$

where F_i $(i = 1, 2, \ldots)$ is a *countable infinite set of mutually exclusive events*. This last property is necessary because in cases related to infinity, such as the throwing of a coin or a die, infinite times, the handling of the problem, and the formulation of the appropriate theory require the use of limits applied to sequences of sets, which in turn requires the use of relations like the last one.

A second concept that holds a central position in the theory of probabilities is the *probability space*, which is defined as a triad of the form $(\Omega, \mathcal{F}, \mathcal{P})$ that includes the following elements:

- An abstract mathematical space Ω known as the *sample space* and its contents (which are called *samples*, or *elementary events* or *outcomes*) are *all possible outcomes* that may arise during the execution of an experiment.
- An *event space* or possible $\mathcal{F}$, which is defined as *a non-empty collection of subsets* of the space Ω that includes *all the possible outcomes* related to the execution of an experiment and therefore we can assign a probability value to them.
- A *probability measure* $\mathcal{P}$ with a value in the range of values $[0, 1]$, which can be assigned to each event, that is to each subset/member A of the event space $\mathcal{F}$.

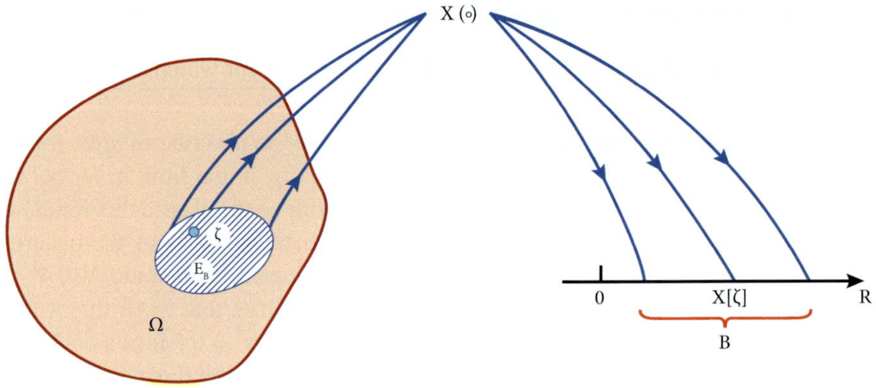

Fig. A.16 A random variable constitutes a mapping of the sample space Ω to the line $\mathbb{R}$ of real numbers (Stark and Woods, Figure 2.2–1)

An abstract space Ω for which there is the possibility of defining a probability measure for each set $\mathcal{F}$ of subsets of Ω is characterized as a *measurable space* and is symbolized as $(\Omega, \mathcal{F})$. In the literature, many examples can be found of spaces that are used as sample spaces Ω, that is, as sets of all possible outcomes of an experiment. Indicative examples of such spaces include (a) *finite and infinite length sequences*, (b) *the real number line* $\mathbb{R}$ *as well as any interval defined on it*, and (c) *the space of functions or waveforms of the form* $x(t)$ *with values from any of the above sets of numbers and with the continuous-time variable* t *taking values in the area* $(-\infty, +\infty)$. More details about all the above spaces can be found in the literature.

A.14.2 Random Variables and Probability Density and Mass Functions

In the terminology of probability theory, we characterize as *random variable* $\boldsymbol{x}(\xi)$, a *mapping* of the sample space Ω (which in general is not numerical), to a *numerical space* (which in almost all cases, is the set $\mathbb{R}$ of real numbers), which assigns to each outcome $\xi \in \Omega$ of an experiment a simple number and therefore has the form $\boldsymbol{x} : \Omega \mapsto \mathbb{R}$. This way of defining a random variable is depicted diagrammatically in Fig. A.16.

In probability theory, the concept of the *random variable* is inextricably linked with the fundamentally important concept of the *probability density* (probability density function, PDF), which makes possible the description of the correlation that exists between the results of experiments (or observations), which are points of the sample space, and the probability of their occurrence. Specifically, *the probability density, is defined as a function, the value of which for each point of the sample space (i.e., for each sample), determines the relative probability (likelihood) of finding the value that has emerged from the conduct of an experiment, close to the value of this point*. In other words, the probability density (which for the case of the one dimension we are dealing with can be considered as the probability per unit length), *allows the calculation of the probability that some experimental measurement belongs to a range of values* $x_1 \leq x \leq x_2$. Essentially, this very density of probability is what makes possible the formulation and use of all principles and techniques of the science of statistics, since the absolute probability of the appearance of any of the infinite values that belong to the considered continuous range of values is *equal to zero*. Knowing the PDF $f(x)$, we can calculate the probability $P\{x_1 \leq \boldsymbol{x} \leq x_2\}$ of the appearance of a result in the range of values $x_1 \leq x \leq x_2$ *by taking the integral of this function and integrating the variable* x *which is supposed to record the resulting value from* x_1 *to* x_2. It will therefore be

$$P\{x_1 \leq \boldsymbol{x} \leq x_2\} = \int_{x_1}^{x_2} f(x)dx. \tag{A.97}$$

A second function that is inextricably linked to the PDF is the *cumulative distribution function* (CDF). This function is defined as

$$F_x(x_i) = P\{\boldsymbol{x} \leq x_i\} = \int_{-\infty}^{x_i} f(x)dx, \tag{A.98}$$

whereas using this new function the previous equation is written as

$$P\{x_1 \leq \boldsymbol{x} \leq x_2\} = F_x(x_2) - F_x(x_1) = \int_{-\infty}^{x_2} f(x)dx - \int_{-\infty}^{x_1} f(x)dx = \int_{x_1}^{x_2} f(x)dx. \quad \text{(A.99)}$$

From the above definition it immediately follows that $F_x(-\infty) = 0$ and $F_x(+\infty) = 1$, whereas for other interesting properties of this function, the reader is referred to the literature. Having defined all of the above, it is not difficult to understand that *the PDF at each point x_i is defined as the derivative of the CDF at the position $x = x_i$* or in mathematical writing

$$f_x(x_i) = \frac{dF_x(x)}{dx}|_{x=x_i} \geq 0,$$

under the condition, of course, that the function $F_x(x)$ is a *continuous* and *differentiable* function, whereas it will also be

$$P\{x_1 \leq \boldsymbol{x} \leq x_2\} = F(x_2) - F(x_1) = \int_{x_1}^{x_2} f_x(x)dx, \quad \text{(A.100)}$$

as can easily be seen. Based on all of the above, we can characterize a random variable $\boldsymbol{x}$ as *a continuous type variable* if the CDF $F_x(x)$ is *continuous* for every x and its derivative is defined at every point except for a countable set of points. Specifically, at those points where the first derivative of the cumulative distribution is defined, the probability density of the random variable x is calculated as $f_x(x) = F_x'(x)$, whereas conversely, at those points where the function $F_x(x)$ is continuous, but the first derivative $F_x'(x)$ is discontinuous, we can assign to the probability density *any positive value*. On the other hand, if the function $F_x(x)$ has a *stepwise form*, or equivalently, is *piecewise continuous*, that is, if it has a constant value everywhere, except for a number of jump discontinuities, then the variable $\boldsymbol{x}$ is characterized as a *discrete type variable*. In an equivalent formulation, a random variable is discrete, *if its set of values constitutes a countable, i.e., discrete set*, although this definition does not reflect reality, as often for both types of random variables, we consider the line $\mathbb{R}$ of real numbers as the range, assigning zero probability values to those points that do not constitute images of the arguments of the random variable. It should also be noted that there are some random variables that *cannot* be classified either as continuous or as discrete.

In the case where the considered random variable is *discrete*, then in place of the PDF, the *probability mass function* $P_x(\boldsymbol{x}) = P\{\boldsymbol{x} \leq x\} - P\{\boldsymbol{x} < x\}$ is used. Consequently, $P_x(x) = 0$ at every point where $F_x(x)$ is continuous, whereas it will be characterized by a *nonzero but finite value*, only at those points where the PDF shows *discontinuities*. It is interesting that although for a discrete random variable, the function $F_x(x)$ *is not* continuous and therefore, strictly speaking, the PDF *is not defined*, resorting to the use of an *impulse function* $\delta(t)$, we can assign a CDF to the random variables, as

$$F_x(x) \triangleq P\{\boldsymbol{x} \leq x\} = \sum_{x_i \leq x} P_x(x_i) = \sum_{i=-\infty}^{\infty} P_x(x_i)u(x - x_i), \quad \text{(A.101)}$$

where $P_x(x_i) = F_x(x_i) - R_x(x_i^-)$ and with the parameters x_i in the above expression representing *the points of discontinuity* of the CDF. In the above expression, the unit step function ensures that the sum is limited only to the points x_i for which $x_i \leq x$. Differentiating this expression we get

$$f_x(x) = \frac{dF_x(x)}{dx} = \sum_{i=-\infty}^{\infty} P_x(x_i)\delta(x - x_i). \quad \text{(A.102)}$$

This *discrete version* of the PDF is known as the *mass density function* and takes its name from the analogy that exists between this function and the known concept of mass in physics, since, as is well known, the calculation of the mass of a body results from the sum of the values of the point masses, just as happens here. It becomes apparent that, owing to the discrete nature of the sample space, the concept of probability *is no longer defined for sets of points*, such as the interval [0, 1], but for *individual points*, since these are distinct and distinguishable from each other. In order to define the mass density function, let us consider a sample space Ω and let $\mathcal{F}$ be the power set of this space, which as is well known is defined as *the set of all subsets* of Ω and contains 2^N elements, where N is the number of elements in the space Ω. Let us also consider a function $g(\omega)$, which assigns a real number to each element $\omega \in \Omega$ and is such that $\rho(\omega) \geq 0$ for each $\omega \in \Omega$ and $\sum_{\omega\in\Omega} \rho(\omega) = 1$. Based on all the above, let us define the function

$$P(F) = \sum_{\omega\in F} \rho(\omega)$$

for each set $F \in \mathcal{F}$. It is proven that the function $P(F)$ satisfies *all the necessary conditions* that a function that constitutes a probability measure should satisfy and therefore constitutes such a measure. The function $\rho(\omega)$ is known as the *probability mass function*. Using the *CDF*, we can define the mass PDF as

$$F_x(x_i) = \sum_{\omega\leq\omega_i} \rho(\omega).$$

A.14.3 Known Continuous and Discrete Distributions

In the literature of statistical signal processing, there are many well-known PDFs or, more simply, distributions that are used in practice. The most well-known of these are the following:

- *Uniform distribution*. This is defined by the equation $f(x) = 1/(\beta - \alpha)$ with $\beta > \alpha$ and $x \in [\alpha, \beta]$ and corresponds to a CDF $F_x(x) = 1$ for $x \geq \beta$, $F_x(x) = (x - \alpha)/(\beta - \alpha)$ for $\alpha \leq x < \beta$ and $F_x(x) = 0$ for $x < \alpha$. This distribution appears in problems related to queue theory.
- *Exponential distribution*. It is defined by the equation $f(x) = \lambda e^{-\lambda x} u(x)$, where x is the unit step function. It is used in problems involving waiting times, lifetimes of arrangements, and in problems related to the changes in the intensity of nonsynchronous light.
- *Double exponential distribution* or *Laplace distribution*. This is defined by the equation $f(x) = (\lambda/2)e^{-\lambda|x|}$ $(\lambda > 0,\ x \in \mathbb{R})$ and finds wide application in problems of modeling voice signals and grayscale tones in image-processing applications. In several cases, this distribution is formulated in the form

$$f(x) = \frac{1}{\sqrt{2}\sigma} \exp\left(-\frac{\sqrt{2}|x|}{\sigma}\right), \tag{A.103}$$

  where $\sigma > 0$ is the standard deviation of the random variable that follows this distribution.
- *Normal distribution* or *Gaussian distribution*. This is defined by the equation

$$f(x) = \frac{1}{\sqrt{2\pi\sigma^2}} \exp\left(-\frac{(x - m)^2}{2\sigma^2}\right). \tag{A.104}$$

  In the above expression, the parameters m and σ^2 represent the *mean value* and the *variance*, quantities that are known from statistics. The *CDF* that corresponds to this *PDF* is calculated as

$$F_x(x) = \int_{-\infty}^{x} f(u)du = \frac{1}{\sigma\sqrt{(2\pi)}} \int_{-\infty}^{x} exp\left(-\frac{(u - \mu)^2}{2\sigma^2}\right)du', \tag{A.105}$$

and finally

$$F_x(x) = \frac{1}{2}\left[1 + erf\left(\frac{x-\mu}{\sigma\sqrt{2}}\right)\right],$$

where the function $erf(x)$ is known as the *error function* (error function), given by the relation

$$\text{erf}(x) = \frac{1}{\sqrt{2\pi}} \int_0^x \exp(-x^2/2)dx \tag{A.106}$$

(although other definitions for this function appear in the literature) and can be found very easily in many books, in the form of data organized in tables. From the above relation it follows that for the interval $(x_1, x_2]$ we have

$$P\{x_1 < \mathbf{x} \le x_2\} = \frac{1}{\sqrt{2\pi\sigma^2}} \int_{x_1}^{x_2} \exp\left(-\frac{(x-m)^2}{2\sigma^2}\right) = erf\left(\frac{x_2-\mu}{\sigma\sqrt{2}}\right) - erf\left(\frac{x_1-\mu}{\sigma\sqrt{2}}\right).$$

This normal distribution is one of the most important distributions of the science of statistics and characterizes the way in which many natural phenomena operate, such as *the distribution of the velocities of the molecules of a gas*. It is proven that as long as some very general conditions are satisfied, the marginal distribution of the average of any number of independent and ideally distributed random variables constitutes a *normal distribution*.

- *Rayleigh distribution*. The Rayleigh distribution with the parameter σ^2 is defined by the equation

$$f_x(x) = \frac{x}{\sigma^2} e^{-x^2/2\sigma^2} u(x), \tag{A.107}$$

where $u(x)$ is the unit step function (this means that the distribution in question is defined only for values $x \ge 0$). This distribution is used in several cases such as in communication systems to describe the amplitude of signals received randomly.

- *Beta distribution*. This is defined by the equation (for the range of values $0 < x < \beta$)

$$f_x(x) = \frac{1}{B(\alpha, \beta)} x^{\alpha-1}(1-x)^{\beta-1}, \tag{A.108}$$

where α and β are non-negative parameters, whereas the function $B(\alpha, \beta)$ is defined as

$$B(\alpha, \beta) = \int_0^1 x^{\alpha-1}(1-x)^{\beta-1}dx = 2\int_0^{2\pi} (\sin\vartheta)^{2\alpha-1}(\cos\vartheta)^{2\beta-1}d\vartheta = \frac{\Gamma(\alpha)\Gamma(\beta)}{\Gamma(\alpha+\beta)},$$

where $\Gamma(x)$ is the Γ function presented below.

- *Gamma distribution*. This is defined by the equation

$$f_x(x) = \frac{x^{\alpha-1}}{\beta^\alpha \Gamma(\alpha)} e^{-x/\beta} \tag{A.109}$$

($x \geq 0, \alpha > 0, \beta > 0$), where

$$\Gamma(\alpha) = \int_0^{\infty} e^{-x} x^{\alpha-1} dx.$$

If the parameter α has the integer variable n, then according to a well-known property of this function, it will hold $\Gamma(n) = (n-1)\Gamma(n-1) = (n-1)!$ and the above relationship takes the form (for $x \geq 0$ and $\beta = 1/\lambda$)

$$f_x(x) = \frac{\lambda^n x^{n-1}}{(n-1)!} e^{-\lambda x}.$$

Completing the above equation, we end up with a CDF of the form

$$F_x(t) = \int_0^t f_x(x) dx = 1 - \sum_{k=0}^{n-1} \frac{(\lambda t)^k}{k!} e^{-\lambda t}. \quad \text{(A.110)}$$

- *Cauchy distribution*. This is defined by the equation (for $|x| < \infty$)

$$f_x(x) = \frac{\alpha/\pi}{(x-\mu)^2 + \alpha^2}. \quad \text{(A.111)}$$

- *Logistic distribution*. This is defined by the equation

$$f(x) = \frac{e^{x/\lambda}}{\lambda(1 + e^{x/\lambda})^2} \quad \text{(A.112)}$$

  with ($x \in \mathbb{R}$) and $\lambda > 0$.
- *Maxwell distribution*. This is defined by the equation (for $x \geq 0$)

$$f_x(x) = \frac{4}{\alpha^3 \sqrt{\pi}} x^2 e^{-x^2/\alpha^2}. \quad \text{(A.113)}$$

- *Weibull distribution*. This is defined by the equation ($x > 0, \alpha > 0, \beta > 0$)

$$f(x) = \frac{\beta x^{\beta-1}}{\alpha^{\beta}} e^{-(x/\alpha)^{\beta}}. \quad \text{(A.114)}$$

  If $\beta = 2$, this distribution transforms into the Rayleigh distribution.

On the other hand, known examples of distributions for the discrete case are the following:

- *Uniform distribution*. This is defined for the sample space $\Omega = \{0, 1, 2, \ldots, n-1\}$ and is given by the expression $\rho(k) = 1/n$ ($k \in \Omega$). It is the appropriate PDF for describing the experiment of throwing a fair dice (where all possibilities are equally likely).
- *Binary distribution* or *Bernoulli distribution*. This is defined for the sample space $\Omega = \{0, 1\}$ and is characterized by the properties $\rho(0) = 1 - p$ and $\rho(1) = p$, where p is a parameter that takes values in the interval $(0, 1)$. It is the appropriate distribution for describing the coin-toss experiment. In a sequence of n independent Bernoulli experiments (which is similar to the *coin-toss* experiment), if the random variable $\boldsymbol{r}$ represents the total number of favorable outcomes, it is characterized as a *binomial random variable* with the corresponding distribution described below.
- *Poisson distribution*. This distribution is largely related to *binomial distribution* and in reality, it is its limit when $n \gg 1$ and $p \ll 1$ but with the product np being constant and equal to μ. It is described by the equation

$$\rho(k) = P\{x = k\} = \frac{\mu^k}{k!} e^{-\mu} \tag{A.115}$$

with the parameter μ taking values in the open interval $(0, \infty)$.

- *Geometric distribution.* This is defined for the sample space $\Omega = 1, 2, 3, \ldots$ and is given by the expression $\rho(k) = P\{\boldsymbol{x} = k\} = q^{k-1} p$ $(k = 1, 2, \ldots)$ where p is a parameter and $q = 1 - p$. This distribution is related to *the number of repetitions* required for the first successful result to appear in a series of *Bernoulli-type* experiments.
- *Negative binomial distribution* or *Pascal distribution.* The random variable $\boldsymbol{y}$ is aforementioned to follow the negative binomial distribution, if the condition

$$P\{\boldsymbol{y} = k\} = \binom{k-1}{r-1} p^r q^{k-r}$$

holds for the values $k = r, r+1, r+2, \ldots$. If n or fewer attempts are required for r successful outcomes, then the number of successes in n attempts must be at least equal to r. It will therefore be $P\{\boldsymbol{y} \leq n\} = P\{\boldsymbol{x} \geq r\}$, where $\boldsymbol{x}$ is a random variable that follows *binomial distribution.* Because the *negative binomial distribution* represents *the waiting time for the appearance of the rth successful result*, this distribution is often also characterized as *waiting-time distribution.*

For the graphical representations of the above distributions, as well as for other additional properties, the reader is referred to the very rich bibliography and countless books on statistics.

A.14.4 Moments and Expected Values

Let us consider a discrete random variable $\boldsymbol{x} = \boldsymbol{x}(\xi)$ and suppose that during the repetitive conduct of an experiment, the outcomes ξ_i arise, which are depicted by the random variable $\boldsymbol{x}$ in the values $\boldsymbol{x}(\xi_i) = x_i$. If we assume that the variable $\boldsymbol{x}$ can take values in the set $\{x_1, x_2, \ldots, x_M\}$ and that the experiment is repeated N times and in such a way so that the value x_1 appears n_1 times, the value x_2 appears n_2 times, etc. (where $n_1 + n_2 + \cdots + n_M = N$), resorting to the above definition of the average we can define the *mean or expected value* $E\{\boldsymbol{x}\}$ from the expression

$$E\{\boldsymbol{x}\} = \frac{1}{N}\sum_{i=1}^{N} x^{(i)} = \frac{1}{N}\sum_{i=1}^{N} n_i x_i = \sum_{i=1}^{M} x_i \left(\frac{n_i}{M}\right) = \sum_{i=1}^{M} x_i P[\boldsymbol{x} = x_i] = \sum_{i=1}^{M} x_i P_x(x_i), \tag{A.116}$$

where here we used the probability in the sense of *relative frequency* and with the function $P_x(x_i) = P[\boldsymbol{x} = x_i]$ being the *mass density function.* The corresponding expression for the continuous case is the

$$E\{\boldsymbol{x}\} = \int_{-\infty}^{\infty} x f_x(x) dx$$

with the *mass density function* being replaced by the *PDF* $f_x(x)$. By using a function $g(x)$ in place of the simple variable x, the above integral takes the form

$$E\{g\} = \int_{-\infty}^{\infty} g(x) f(x) dx$$

and consequently, the expected or average value results from the above expression, if we choose the identity function $g(x) = x$ as function $g(x)$. Note that from this relationship the probability P itself can arise, using the constant function $g(x) = 1$. In the terminology of the science of statistics, this expression constitutes the defining equation of the *mathematical expectation* or *expectation* of the random variable $\boldsymbol{y} = g(\boldsymbol{x})$ $(g : \Omega \mapsto \mathbb{R})$ with respect to the PDF $f_x(x)$. By using other more complex functions in place of this random variable, we can define very important and useful quantities that constitute basic tools of probability theory and that are known as *moments*. Specifically, there are *three basic categories* of such moments that can be defined, in which the function $g(x)$ constitutes a quantity raised to a power k, which is called the order of the moment. These three categories of moments include the *simple moments*

$$m_k = E\{x^k\} = \int_{-\infty}^{\infty} x^k f(x)dx, \tag{A.117}$$

the *central moments*, which are defined as

$$\mu_k = E\{(x-a)^k\} = \int_{-\infty}^{\infty} (x-a)^k f(x)dx, \tag{A.118}$$

and the *absolute moments*, which are defined as

$$m_k^\alpha = E\{|x|^k\} = \int_{-\infty}^{\infty} |x|^k f(x)dx. \tag{A.119}$$

From the above equations, it follows that the *mean values* for the continuous and the discrete cases are defined respectively by the equations

$$E\{x\} = \int_{-\infty}^{+\infty} xf(x)dx \qquad \text{and} \qquad E\{x\} = \sum_i p_i x_i, \quad (p_i = P\{\mathbf{x} = x_i\}),$$

whereas the *variance*, the square root of which $\sigma_x = \sqrt{E\{(x-\mu)^2\}}$ is known as the *standard deviation* of the variable x, is defined by the expression

$$\text{Var}\{x\} = \sigma_x^2 = E\{(x-\mu)^2\} = \int_{-\infty}^{+\infty} (x-\mu)^2 f(x)dx.$$

As is well known from statistics, the standard deviation represents *the mean square value of the random variable x around the mean value* μ and the reason for its use is that *the mean value alone is not sufficient to describe the PDF of a random variable*, since it is very likely that these two functions are characterized *by the same mean value but by a different degree of dispersion around the mean value.* From the defining equation of the variance, we find that it constitutes the mean value of the random variable $(x-\mu)^2$, that is, $E\{(x-\mu)^2\}$ and alternatively, as proven, it can also be calculated through the expression

$$\text{Var}\{x\} = \sigma_x^2 = E\{(x-\mu)^2\} = E\{x^2 - 2x\mu + \mu^2\} = E\{x^2\} - 2\mu E\{x\} + \mu^2 = E\{x^2\} - \mu^2 = E\{x^2\} - (E\{x\})^2. \tag{A.120}$$

The mean and standard deviation of known continuous and discrete distributions are included in Table A.2 and their verification is left as an exercise to the interested reader.

A.14.5 Combined Cumulative Distribution and Probability Density

Considering the random variables $\boldsymbol{x}(\xi)$ and $\boldsymbol{y}(\xi)$, which jointly define a point $(\boldsymbol{x}, \boldsymbol{y})$ of the plane $\mathbb{R}^2$, let us define the event $\{\boldsymbol{x} \le x, \boldsymbol{y} \le y\} \triangleq \{\boldsymbol{x} \le x\} \cap \{\boldsymbol{y} \le y\} = \{(x, y) \in D\}$ as the set of all elements ξ of the sample space Ω for which $\boldsymbol{x}(\xi) \le x$ and $\boldsymbol{y}(\xi) \le y$. In other words, this event occurs when the point (x, y) belongs to a region D of the plane $\mathbb{R}^2$. Having defined

Table A.2 The mean and standard deviation of known continuous and discrete distributions

Random variable	Probability density function	Mean value $E\{X\}$	Variance Var $\{X\}$
Gauss	$\frac{1}{\sqrt{2\pi\sigma^2}}\exp\left(-\frac{(x-\mu)^2}{2\sigma^2}\right)$	μ	σ^2
Exponential	$\lambda e^{-\lambda x},\ x\geq 0,\ \lambda>0$	$\frac{1}{\lambda}$	$\frac{1}{\lambda^2}$
Gamma	$\frac{x^{\alpha-1}}{\Gamma(\alpha)\beta^{\alpha}}e^{-x/\beta},\ x\geq 0,\ \alpha>0,\ \beta>0$	$\alpha\beta$	$\alpha\beta^2$
Erlang-k	$\frac{(k\lambda)^k}{(k-1)!}x^{k-1}e^{-k\lambda x}$	$\frac{1}{\lambda}$	$\frac{k}{\lambda^2}$
Chi-squared	$\frac{x^{n/2-1}}{2^{n/2}\Gamma(n/2)}e^{-x/2},\ x\geq 0$	n	$2n$
Rayleigh	$\frac{x}{\sigma^2}\exp\left(-\frac{x^2}{2\sigma^2}\right),\ x\geq 0$	$\frac{\pi}{2}\sigma$	$(2-\pi/2)\sigma^2$
Uniform (continuous)	$\frac{1}{\beta-\alpha},\ \alpha<x<\beta$	$\frac{\alpha+\beta}{2}$	$\frac{(\beta-\alpha)^2}{12}$
Beta	$\frac{\Gamma(\alpha+\beta)}{\Gamma(\alpha)\Gamma(\beta)}x^{\alpha-1}(1-x)^{\beta-1},\ 0<x<1,\ \alpha>0,\ \beta>0$	$\frac{\alpha}{\alpha+\beta}$	$\frac{\alpha\beta}{(\alpha+\beta)^2(\alpha+\beta+1)}$
Bernoulli	$P\{x=1\}=p, P\{x=0\}=1-p=q$	p	$p(1-p)$
Binomial	$\binom{n}{k}p^k q^{n-k},\ k=0,1,2,,\ldots,n,\ p+q=1$	np	npq
Poisson	$e^{-\lambda}\frac{\lambda^k}{k!},\ k=0,1,2,\ldots,\infty$	λ	λ
Pascal	$\binom{k-1}{r-1}p^r q^{k-r},\ k=r,\ r+1,\ \ldots,\infty,\ p+q=1$	$\frac{rq}{p}$	$\frac{rq}{p^2}$
Uniform (discrete)	$\frac{1}{n}, k=1,2,\ldots,n$	$\frac{n+1}{2}$	$\frac{n^2-1}{12}$
Geometric	$pq^k, k=0,1,2,\ldots,\infty$	$\frac{q}{p}$	$\frac{q}{p^2}$

this set, we can extend the concept of the CDF for a simple variable $\boldsymbol{x}$ to the *combined* or *joint* CDF $F_{xy}(x, y)$ of the random variables $\boldsymbol{x}(\xi)$ and $\boldsymbol{y}(\xi)$, which is defined as

$$F_{xy}(x, y) = P\{\mathbf{x} \leq x, \mathbf{y} \leq y\}$$

and which, in turn, allows us to define the function of the *joint PDF* $f_{xy}(x, y)$ as the *partial derivative*

$$f(x, y) = \frac{\partial^2 F(x, y)}{\partial x \partial y} \geq 0$$

from which it follows that

$$F(x, y) = \int_{-\infty}^{x}\int_{-\infty}^{y} f(\alpha, \beta)d\alpha d\beta, \tag{A.121}$$

whereas it will also be

$$f_{xy}(x, y) = P\{x \leq \mathbf{x} \leq x + dx, y \leq \mathbf{y} \leq y + dy\}.$$

This function of the *joint probability density* satisfies the *normalization condition*

$$\int_{-\infty}^{\infty}\int_{-\infty}^{\infty} f(x, y)dxdy = 1$$

and is related to the homonymous functions $f_x(x)$ and $f_y(y)$ of the random variables x and y through the expressions

$$f_x(x) = \frac{dF_{xy}(x, y)}{dx} = \int_{-\infty}^{\infty} f(x, y)dy \quad \text{and} \quad f_y(y) = \frac{dF_{xy}(x, y)}{dy} = \int_{-\infty}^{\infty} f(x, y)dx,$$

whereas conversely we have

$$F_x(x) = F_{xy}(x, \infty) = \int_{-\infty}^{x} d\alpha \int_{-\infty}^{\infty} d\beta f(\alpha, y) \quad \text{and} \quad F_y(y) = F_{xy}(\infty, y) = \int_{-\infty}^{y} d\beta \int_{-\infty}^{\infty} d\alpha f(\alpha, y).$$

Other useful properties that are proven through the above definitions are the

$$\begin{aligned} &P\{x_1 \le \mathbf{x} \le x_2, \mathbf{y} \le y\} = F(x_2, y) - F(x_1, y) \\ &P\{\mathbf{x} \le x, y_1 < \mathbf{y} \le y_2\} = F(x, y_2) - F(x, y_1) \qquad \text{and} \\ &P\{x_1 < \mathbf{x} \le x_2, y_1 < \mathbf{y} \le y_2\} = F(x_2, y_2) - F(x_1, y_2) - F(x_2, y_1) + F(x_1, y_1). \end{aligned}$$

One of the most well-known combined PDFs is the *combined Gaussian probability density*, which is defined as

$$f_{xy}(x, y) = \frac{1}{2\pi\sigma_x\sigma_y\sqrt{1-\rho^2}} \exp\left\{-\frac{1}{2(1-\rho^2)}\left[\left(\frac{x-\mu_x}{\sigma_x}\right)^2 - 2\rho\frac{(x-\mu_x)(y-\mu_y)}{\sigma_x\sigma_y} + \left(\frac{y-\mu_y}{\sigma_y}\right)^2\right]\right\}, \tag{A.122}$$

where $\rho = E\{(x-\mu_x)(y-\mu_y)\}/\sigma_x\sigma_y$ is the *correlation coefficient*, which takes values in the range $[-1, 1]$. If $\rho = 0$, then the above equation takes the form

$$\begin{aligned} f_{xy}(x, y) &= \frac{1}{2\pi\sigma_x\sigma_y} \exp\left[-\frac{(x-\mu_x)^2}{2\sigma_x^2} - \frac{(y-\mu_y)^2}{2\sigma_y^2}\right] \\ &= \left\{\frac{1}{\sigma_x\sqrt{2\pi}} \exp\left[-\frac{1}{2}\left(\frac{x-\mu_x}{\sigma_x}\right)^2\right]\right\}\left\{\frac{1}{\sigma_y\sqrt{2\pi}} \exp\left[-\frac{1}{2}\left(\frac{y-\mu_y}{\sigma_y}\right)^2\right]\right\} = f_x(x)f_y(y) \end{aligned}$$

with the probability densities $f_x(x)$ and $f_y(y)$ being defined as

$$f_x(x) = \frac{1}{\sigma_x\sqrt{2\pi}} \exp\left[-\frac{1}{2}\left(\frac{x-\mu_x}{\sigma_x}\right)^2\right] \quad \text{and} \quad f_y(y) = \frac{1}{\sigma_y\sqrt{2\pi}} \exp\left[-\frac{1}{2}\left(\frac{y-\mu_y}{\sigma_y}\right)^2\right].$$

We are thus led to the very important conclusion that, *if the two variables associated with the joint Gaussian probability density function are uncorrelated* (that is, $\rho = 0$), *then they are also independent*—since we showed that $f_{xy}(x, y) = f_x(x)f_y(y)$, something, which let it be noted, does not hold in the general case.

A.14.6 Independence of Random Variables

Considering two random variables $\mathbf{x}$ and $\mathbf{y}$ these are characterized as *statistically independent* if the events $A = \{\mathbf{x} \le x\}$ and $B = \{\mathbf{y} \le y\}$ are *independent* for every combination of x and y meaning that they are such that $P(AB) = P(A)P(B)$. Taking into account that the *CDF* is defined as $F_x(x) = P\{\mathbf{x} \le x\} = P(A)$ and $F_y(y) = P\{\mathbf{y} \le y\} = P(B)$, $F_{xy}(x, y) =$

$F_x(x)F_y(y)$ for every x, y if and only if the random variables $\boldsymbol{x}$ and $\boldsymbol{y}$ are *independent of each other*. Another condition for the independence of $\boldsymbol{x}$ and $\boldsymbol{y}$ related to the CPD is given by the equation

$$f_{xy}(x, y) = \frac{\partial^2 F_{xy}(x, y)}{\partial x \partial y} = \frac{\partial F_x(x)}{\partial x}\frac{\partial F_y(y)}{\partial y} = f_x(x)f_y(y).$$

From the definition of conditional probability and the definition of independence, it is not difficult to understand that

$$F_x(x|\boldsymbol{y} \le y) = \frac{F_{xy}(x, y)}{F_y(y)} = F_x(x) \quad \text{and} \quad F_y(y|\boldsymbol{x} \le x) = \frac{F_{xy}(x, y)}{F_x(x)} = F_y(y).$$

But these conditions mean that the probability of the occurrence of each event *does not depend on whether or not the other event has occurred*, something that is expected since the two events are *statistically independent*.

A.14.7 Combined (Joint) Moments

The *combined* moment ξ_{ij} of the random variables $\boldsymbol{x}$ and $\boldsymbol{y}$ for the continuous and the discrete case is defined respectively as

$$\xi_{ij} \triangleq E\{\boldsymbol{x}^i \boldsymbol{y}^j\} = \int_{-\infty}^{\infty}\int_{-\infty}^{\infty} x^i y^j f_{xy}(x, y)dxdy \quad \text{and} \quad \xi_{ij} \triangleq \sigma_\ell \sigma_m x_\ell^i y_m^j P_{xy}(x_\ell, y_m).$$

The above expressions refer to the *simple* or usual moment, whereas the corresponding *central* moment is defined as

$$m_{ij} \triangleq E\{(\boldsymbol{x} - \mu_x)^i (\boldsymbol{y} - \mu_y)^j\}, \tag{A.123}$$

where $\mu_x = E\{\boldsymbol{x}\}$ and $\mu_y = E\{\boldsymbol{y}\}$. Both of the above categories of moments are characterized as *moments of order* $i+j$. The most important of these moments are the *second-order moments*, which are the *simple* moments $\xi_{02} = E\{\boldsymbol{y}^2\}$, $\xi_{20} = E\{\boldsymbol{x}^2\}$ and $\xi_{11} = E\{\boldsymbol{xy}\}$, and the *central* moments $m_{02} = E\{(\boldsymbol{y} - \mu_y)^2\}$, $m_{20} = E\{(\boldsymbol{x} - \mu_x)^2\}$ and $m_{11} = E\{(\boldsymbol{y} - \mu_y)(\boldsymbol{y} - \mu_y)\} = E\{\boldsymbol{xy}\} - \mu_x\mu_y$. In the terminology of statistics, the simple moment ξ_{11}, that is, the expected value of the product of the two variables is characterized as the *correlation*, whereas the corresponding *central* moment

$$C_{xy} = \mu_{11} = E\{(\boldsymbol{x} - \mu_x)(\boldsymbol{y} - \mu_y)\} = \int_{-\infty}^{\infty}\int_{-\infty}^{\infty} (\boldsymbol{x} - \mu_x)(\boldsymbol{y} - \mu_y)f(x, y)dxdy. \tag{A.124}$$

It is characterized as *covariance*, by analogy with the variance that refers to one variable. From the above definition, it is easily observed that

$$\text{Cov}^2(\boldsymbol{x}, \boldsymbol{y}) \le E\{(\boldsymbol{x} - \mu_x)^2\}E\{(\boldsymbol{y} - \mu_y)^2\}.$$

Closely connected with the above concepts is the *correlation coefficient* ρ_{xy}, which is defined as

$$\rho_{xy} = \frac{m_{11}}{\sqrt{m_{20}m_{02}}} = \frac{C_{xy}}{\sigma_x\sigma_y} = E\left\{\left[\frac{x - \mu_x}{\sigma_x}\right]\left[\frac{y - \mu_y}{\sigma_y}\right]\right\} = \int_{-\infty}^{\infty}\int_{-\infty}^{\infty} \frac{x - \mu_x}{\sigma_x} \cdot \frac{y - \mu_y}{\sigma_y} f(x, y)dxdy$$
$$= \int_{-\infty}^{\infty}\int_{-\infty}^{\infty} \frac{xy - \mu_x y - \mu_y x + \mu_x\mu_y}{\sigma_x\sigma_y} = \frac{E\{\boldsymbol{xy}\} - \mu_x\mu_y}{\sigma_x\sigma_y}.$$

It is proven that *when the value of autocorrelation is equal to one, the variable* $\boldsymbol{y}$ *is a linear function of the variable* $\boldsymbol{x}$ *and these variables are fully correlated*. On the other hand, if $\rho_{xy} = 0$ then the two variables are characterized as *uncorrelated*. In this case, $C_{cy} = 0$ and therefore $E\{\boldsymbol{x}\boldsymbol{y}\} = E\{\boldsymbol{x}\}E\{\boldsymbol{y}\}$. If the variables are *independent*, then they will also be *uncorrelated*. Finally, if $E\{\boldsymbol{x}\boldsymbol{y}\} = 0$, the variables are characterized as *orthogonal*, a property symbolized as $\boldsymbol{x} \perp \boldsymbol{y}$.

A.14.8 Moment Generating and Characteristic Functions

The *moment-generating function* of a random variable $\boldsymbol{x}$ (if it exists) is defined as

$$\theta(s) \triangleq E\{e^{sx}\} = \int_{-\infty}^{\infty} e^{sx} f_x(x)dx \tag{A.125}$$

for the *continuous* case and as

$$\theta(s) = \sum_i e^{sx_i} P_x(x_i) \tag{A.126}$$

for the *discrete* case, where s is a complex variable. The above expression (for the continuous case) except for the change in the sign of the exponential function, is recognized as *the Laplace transform of the PDF $f_x(x)$*. Knowing the PDF we can determine the *moment-generating function* and vice versa. This function is particularly important, because it allows us to calculate the various moments of the random variable $\boldsymbol{x}$ (from which its name originates) but also to deal with very important concepts such as the *central limit theorem*.[5] Indeed, by replacing the exponential function in the above expression with its Taylor expansion and using the linearity property of mathematical expectation, it follows that

$$E\{e^{sx}\} = E\left\{1 + sx + \frac{(sx)^2}{2!} + \cdots + \frac{(sx)^n}{n!} + \cdots\right\} = 1+sE\{x\}+\frac{s^2}{2!}+\cdots+\frac{s^n}{n!}E\{x^n\}+\cdots = 1+s\mu_x+\frac{s^2}{2!}m_2+\cdots+\frac{s^n}{n!}m_n,$$

where $m_k = E\{x^k\}$ is the kth moment. Consequently, in a case where the function $\theta(t)$ exists, the moments of the random variable $\boldsymbol{x}$ are calculated through differentiation, as it is easily observed that

$$m_k = \theta^{(k)}(0) = \left.\frac{d^k\theta(s)}{ds^k}\right|_{s=0}.$$

The above concepts extend directly for a larger number of variables. For example, for the case of the two random variables $\boldsymbol{x}$ and $\boldsymbol{y}$ we have respectively

$$\theta_{xy}(s_1, s_2) \triangleq E\{e^{s_1x+s_2y}\} = \int_{-\infty}^{\infty}\int_{-\infty}^{\infty} e^{(s_1x+s_2y)} f_{xy}(x, y)dxdy. \tag{A.127}$$

Expanding the above expression *by Taylor* it takes the form

---

[5] According to the **central limit theorem**, first proved by De Moivre in 1933 for the special case of *Bernoulli random variables*, whereas for the general case it was proved in 1922 by J.W. Lindeberg, the normalized sum of a large number of *mutually independent random variables* $\boldsymbol{x}_1, \boldsymbol{x}_2, \ldots, \boldsymbol{x}_n$ with *zero mean values* and *finite variance values* $\sigma_1^2.\sigma_2^2, \ldots, \sigma_n^2$, tends to the *standard normal PDF*, under the assumption that the individual dispersion values σ_i^2 $(i = 1, 2, \ldots, n)$ are small compared with the value of the sum $\sum_i \sigma_i^2$ $(i = 1, 2, \ldots, n)$ (this restriction is known as the Lindeberg condition). In mathematical formulation, if $\boldsymbol{x}_1, \boldsymbol{x}_2, \ldots, \boldsymbol{x}_n$ are a set of *mutually exclusive random variables* with CDFs $F_{x_i}(x_i)$ $(i = 1, 2, \ldots, n)$ respectively, with mean value $E\{x_i = 0\}$ $(i = 1, 2, \ldots, n)$ and dispersion $\mathrm{Var}\{\boldsymbol{x}_i\} = \sigma_i^2$ $(i = 1, 2, \ldots, n)$, if for a given $\varepsilon > 0$ and a sufficiently large number of variables n the parameters σ_i $(i = 1, 2, \ldots, n)$ satisfy the condition $\sigma_i < \varepsilon s_n$ $(i = 1, 2, \ldots, n)$, where $s_n = \sigma_1^2 + \sigma_2^2 + \cdots + \sigma_n^2$, then the normalized sum $Z_n \triangleq (x_1 + x_2 + \cdots + x_n)/s_n$ converges to the *standard normal distribution*, or in mathematical writing $\lim F_{z_n}(z) = (1/2) + \mathrm{erf}(z)$ in the limit $n \to \infty$. The great importance of this theorem lies in the observation that *in cases where we encounter natural situations that arise from the combined action of many events, such as the thermal motion of electrons in a conductor or atmospheric noise, we can expect that the recorded observations will always follow the normal distribution regardless of the PDFs of the individual phenomena, which in the general case may not even be known*; it is astonishing that in almost all cases, this is something that *is verified* in practice.

$$\theta_{xy}(s_1, s_2) = \sum_{i=0}^{\infty}\sum_{j=0}^{\infty}\frac{s_1^i s_2^j}{i!j!}\xi_{ij} \tag{A.128}$$

and consequently, we have

$$\xi_{in}\theta_{xy}^{(i,n)}(0,0) = \left.\frac{\partial^{i+n}\theta_{xy}(s_1, s_2)}{\partial s_1^i s_2^n}\right|_{s_1=s_2=0}, \tag{A.129}$$

with the first- and second-order moments, which are also the ones usually used in practice, being $\theta_{xy}^{1,0}(0,0) = E\{x\} = \mu_x$, $\theta_{xy}^{0,1}(0,0) = E\{y\} = \mu_y$, $\theta_{xy}^{2,0}(0,0) = E\{x^2\}$, $\theta_{xy}^{0,2}(0,0) = E\{y^2\}$, and $\theta_{xy}^{1,1}(0,0) = E\{xy\} = \mathrm{Cov}(x,y) + \mu_x\mu_y$. The generalization for a larger number of variables is done in the same way.

On the other hand, the *characteristic function* $\Phi_x(\omega)$ of a random variable $\boldsymbol{x}$, is defined as the *mathematical expectation* of the quantity $e^{j\omega}x$ or equivalently as the *Fourier transform* (with the exception, as before, of the sign of the exponent) *of the PDF*. It will therefore be

$$\Phi_x(\omega) = E\{e^{i\omega x}\} = \int_{-\infty}^{\infty} f(x)e^{j\omega x}d\omega dx. \tag{A.130}$$

Taking into account that $f(x) \geq 0$, the function $\Phi_x(\omega)$ presents a maximum at the position $\omega = 0$, or in mathematical writing, $|\Phi_x(\omega)| \leq \Phi_x(0) = 1$, since for the value $x = 0$ the above integral has a value equal to the unit as imposed by the definition of probability density.

Developing the function $\Phi_x(\omega)$ by Taylor will take the form

$$\Phi_x(\omega) = \int_{-\infty}^{+\infty}\left(\sum_{k=0}^{\infty}\frac{x^k(j\omega)^k}{k!}\right)\rho_x(x)dx = \sum_{k=0}^{\infty}E\{x^k\}\frac{(j\omega)^k}{k!} \tag{A.131}$$

from which it follows that the coefficients of the above transform are the moments $E\{x^k\}$ of the variable x (assuming that these exist). For this reason, the function $\Phi_x(\omega)$ is also known as the *moment-generating function* of the random variable $\boldsymbol{x}$,[6] whereas its natural logarithm $\Psi(\omega)\ln\Phi_x(\omega)$ is known as the *second characteristic function* of the random variable $\boldsymbol{x}$. The above applies to continuous random variables, whereas for discrete variables the above function is defined as

$$\Phi_x(\omega) = \sum_i p_i e^{j\omega x_i}.$$

If we know the *characteristic function* $\Phi_x(\omega)$, we can calculate the *PDF* $f(x)$ from the equation

$$f(x) = \frac{1}{2\pi}\int_{-\infty}^{\infty}\Phi_x(\omega)e^{-j\omega x}dx$$

and it is not difficult to understand how these two functions are related by the *direct-inverse Fourier transform*.

The characteristic functions for the distributions of Table A.2 are included in Table A.3 and the reader is invited as an exercise to proceed with their calculation in the way we did in the above examples.

[6] In reality, this is nothing more than a *special case* that arises from the *moment-generating function* we presented previously, in the same way that the Fourier transform arises from the Laplace transform, that is, by setting $s = j\omega$. Consequently, *the techniques for calculating the characteristic function used in the following examples can be applied to calculate the moment-generating function, which can then be transformed into the characteristic function*, by making the above substitution. Owing to the above correlation, the characteristic function of a random variable within the context of our discussion is characterized by the same properties that describe the moment-generating function.

Table A.3 The characteristic function of known continuous and discrete distributions

Random variable	Probability density function	Characteristic function
Gauss	$\sqrt{\frac{1}{2\pi\sigma^2}}\exp\left(-\frac{(x-\mu)^2}{2\sigma^2}\right)$	$\exp\,(j\mu\omega-\sigma^2\omega^2/2)$
Exponential	$\lambda e^{-\lambda x},\ x\geq 0,\ \lambda>0$	$(1-j\omega/\lambda)^{-1}$
Gamma	$\frac{x^{\alpha-1}}{\Gamma(\alpha)\,\beta^{\alpha}}e^{-x/\beta},\ x\geq 0,\ \alpha>0,\ \beta>0$	$(1-j\omega\beta)^{-\alpha}$
Erlang-k	$\frac{(k\lambda)^k}{(k-1)!}x^{k-1}e^{-k\lambda x}$	$(1-j\omega/k\lambda)^{-k}$
Chi-squared	$\frac{x^{n/2-1}}{2^{n/2}\Gamma(n/2)}e^{-x/2},\ x\geq 0$	$(1-j\,2\omega)^{-n/2}$
Rayleigh	$\frac{x}{\sigma^2}\exp\left(-\frac{x^2}{2\sigma^2}\right),\ x\geq 0$	$\left(1+j\sqrt{\frac{\pi}{2}}\sigma\omega\right)e^{-\sigma^2\omega^2/2}$
Uniform (continuous)	$\frac{1}{\beta-\alpha},\ \alpha<x<\beta$	$\frac{e^{j\beta\omega}-e^{j\alpha\omega}}{\beta-\alpha}$
Beta	$\frac{\Gamma(\alpha+\beta)}{\Gamma(\alpha)\Gamma(\beta)}x^{\alpha-1}(1-x)^{\beta-1},\ 0<x<1,\ \alpha>0,\ \beta>0$	---
Bernoulli	$P\{x=1\}=p,\ P\{x=0\}=1-p=q$	$pe^{j\omega}+q$
Binomial	$\binom{n}{k}p^kq^{n-k},\ k=0,1,2,,\ldots,n,\ p+q=1$	$(pe^{j\omega}+q)^n$
Poisson	$e^{-\lambda}\frac{\lambda^k}{k!},\ k=0,\ 1,\ 2,\ \ldots,\ \infty$	$e^{\lambda(1-e^{j\omega})}$
Pascal	$\binom{k-1}{r-1}p^rq^{k-r},\ k=r,\ r+1,\ \ldots,\ \infty,\ p+q=1$	$\left(\frac{p}{1-qe^{-j\omega}}\right)^r$
Uniform (discrete)	$\frac{1}{n},\ k=1,2,\ldots,n$	$\exp\left(\frac{j\,(N+1)\,\omega}{2}\frac{\sin(N\omega/2)}{\sin(\omega/2)}\right)$
Geometric	$pq^k,\ k=0,1,2,\ldots,\infty$	$\frac{p}{1-qe^{j\omega}}$

A.14.9 Functions of Random Variables

Having defined the concept of a *random variable*, we can define the *function of a random variable* as a new random variable of the form $\boldsymbol{y}=g(\boldsymbol{x})$, which is defined in the following way: for some element $\xi\in\Omega$ we know that the quantity $\boldsymbol{x}(\xi)$ is a number. In exactly the same way, the quantity $g[\boldsymbol{x}(\xi)]$ is also a number, which is defined with the help of the functions $\boldsymbol{x}(\xi)$ and $g(x)$ and has the value $\boldsymbol{y}(\xi)=g(\boldsymbol{x}(\xi))$. Although there are other different but equivalent definitions for a function of a random variable (see for example Stark and Woods, page 119), the definition used here is *particularly important*, because it models the function g as a system that accepts as input a random process $\boldsymbol{x}$ and produces as its output another random process $\boldsymbol{y}$. Indeed, as we comment in another section, when such a system receives a random variable at its input, then its output will also be a random variable, the statistical characterization of which can be carried out with the help of the statistical characteristics of its input and the function $g(x)$.

In a more technical description, the random variable $\boldsymbol{y}$ that results from the application of the function g to the random variable $\boldsymbol{x}$ will also be characterized by its own *CDF* $F_y(y)$, which is defined as the probability of the event $\{\boldsymbol{y}\leq y\}$ consisting of the elements $\xi\in\Omega$ such that $\boldsymbol{y}(\xi)=g[\boldsymbol{x}(\xi)]\leq y$, or in mathematical writing,

$$F_y(y)=P\{\boldsymbol{y}\leq y\}=P\{g(\boldsymbol{x})\leq y\},$$

as well as by its own *PDF* $f_y(y)$, which is the *derivative of the CDF*. The above characterization refers to the continuous case, whereas in the case of discrete random variables, the known ones are used *mass density functions*. It is not difficult to understand that the determination of the functions $F_y(y)$ and $f_y(y)$ is based on the knowledge of the functions $F_x(x)$ and $f_x(x)$ and the function $g(x)$, and can be very easy but also very difficult depending on the case. In the second case, where the degree of difficulty in determining these functions is *very high*, we are forced to resort to descriptors that contain less information in relation to the CDFs and probability density and which are nothing more than our well-known *means* or *expected* values.

There are *three basic categories of problems* that are encountered in practice and consist in the calculation of the functions $F_y(y)$ and $f_y(y)$ from the functions $F_x(x)$ and $f_x(x)$, and more specifically: the problems of type $\boldsymbol{y} = g(\boldsymbol{x})$, with the function g accepting only one argument, the problems of type $z = g(\boldsymbol{x}, \boldsymbol{y})$ in which the function g includes two input arguments, as well as the more complex problems that involve the use of two functions of the two variables and therefore, have the form $z = g(\boldsymbol{x}, \boldsymbol{y})$ and $\boldsymbol{w} = h(\boldsymbol{x}, \boldsymbol{y})$ (this last case is generalized directly, so as to include the use of more functions with more variables). For each of these three types of problems, the corresponding theory has been developed that allows the calculation of the above quantities, both for specific very well-known and widely used functional correlations, as well as for the general case of arbitrary correlations.

Indicative examples for the first category of problems $\boldsymbol{y} = g(\boldsymbol{x})$ include *linear dependence* $\boldsymbol{y} = \alpha\boldsymbol{x} + \beta$ for which it is proven that for $\alpha > 0$ it is

$$F_y(y) = F_x\left(\frac{y-\beta}{\alpha}\right) \quad \text{and} \quad f_y(y) = \frac{1}{\alpha} f_x\left(\frac{y-\beta}{\alpha}\right),$$

whereas for $\alpha < 0$ it is

$$F_y(y) = 1 - F_x\left(\frac{y-\beta}{\alpha}\right) \quad \text{and} \quad f_y(y) = \frac{1}{|\alpha|} f_x\left(\frac{y-\beta}{\alpha}\right), \quad \alpha \neq 0$$

and the *quadratic dependence* $\boldsymbol{y} = \boldsymbol{x}^2$ for which the functions $F_y(y)$ and $f_y(y)$ are defined only for $y > 0$, whereas they have zero value for $y < 0$ and are given by the expressions

$$F_y(y) = F_x(\sqrt{y}) - F_x(-\sqrt{y}) \quad \text{and} \quad f_y(y) = \frac{1}{\sqrt{y}}[f_x(\sqrt{y}) + F_x(-\sqrt{y})].$$

For more results on this topic, refer to the very rich bibliography. In the general case, it is proven that the PDF $f_y(y)$ is expressed in terms of the PDF $f_x(x)$ according to the expression (see Papoulis and Pillai pages 130–131)

$$f_y(y) = \frac{f_x(x_1)}{|g'(x_1)|} + \frac{f_x(x_2)}{|g'(x_2)|} + \cdots + \frac{f_x(x_n)}{|g'(x_n)|} = \sum_{i=1}^{n} \frac{f_x(x_i)}{g'(x_i)}, \tag{A.132}$$

where $x_1, x_2, \ldots, x_n$ are the real roots of the function $y = g(x)$ and $g'(x)$ is the derivative of the function $g(x)$. Similar results apply to the discrete case. Note that in order for the function $\boldsymbol{g}(x)$ to be considered a *valid random variable*, (a) its domain *must include* the range of the random variable $\boldsymbol{x}$ as required by the fact that the function $\boldsymbol{y}(\xi) = g(\boldsymbol{x}(\xi))$ constitutes a *composite function*, (b) it must be a *Borel function*, which means that the set R_y that contains the $\boldsymbol{x}$ for which $g(\boldsymbol{x}) \leq \boldsymbol{y}$ must be a *Borel set* (Borel sets jointly define the so-called *Borel body*, which contains all subsets of the real line that can be constructed from countable sequences of operations on intervals), and (c) the event $\{g(\boldsymbol{x}) = \pm\infty\}$ must be characterized by a probability value equal to *zero*.

On the other hand, for the case of problems of the type $z = g(\boldsymbol{x}, \boldsymbol{y})$, among the most well-known cases are $z = \boldsymbol{x} + \boldsymbol{y}$ and $z = \boldsymbol{x}\boldsymbol{y}$, with the most classic example of the first case being *a signal* $\boldsymbol{x}$ *to which a noise signal* $\boldsymbol{y}$ *has been added*, whereas the second case appears in the operation of a *signal multiplier* that multiplies two signals $\boldsymbol{x}$ and $\boldsymbol{y}$. Speaking generally, these types of problems are similar to the problems $\boldsymbol{y} = g(\boldsymbol{x})$, with the CDF $F_z(z)$ being defined as

$$F_z(z) = \iint_{(x,y)\in D} f_{xy}(x, y)dxdy. \tag{A.133}$$

Specifically, this problem consists in determining the region D of the plane (x, y), for which the event $\{\xi : \mathbf{z}(\xi) \leq z\}$ coincides with the event $\{\xi : [\mathbf{x}(\xi), \mathbf{y}(\xi)] \in D\}$. Limiting ourselves to the case of the sum $\mathbf{z} = \mathbf{x} + \mathbf{y}$, which is also the most classic, the CDF is given by the relation

$$F_z(z) = P\{\mathbf{x} + \mathbf{y} \leq z\} = \iint_{x+y\leq z} f_{xy}(x, y)dxdy = \int_{-\infty}^{\infty} \left(\int_{-\infty}^{z-y} f_{xy}(x, y)dx \right) dy,$$

whereas the PDF is calculated as

$$f_z(z) = \frac{dF_z(z)}{dz} = \int_{-\infty}^{\infty} \left(\frac{\partial}{\partial z} \int_{-\infty}^{z-y} f_{xy}(x, y)dx \right) dy = \int_{-\infty}^{\infty} f_{xy}(z - y, y)dy.$$

In the case where the variables $\mathbf{x}$ and $\mathbf{y}$ are *independent*, which often occurs in practice, $f_{xy}(x, y) = f_x(x) f_y(y)$ and the above relation is written

$$f_z(z) = \int_{-\infty}^{\infty} f_{xy}(z - y, y)dy = \int_{-\infty}^{\infty} f_x(z - y) f_y(y)dy = \int_{-\infty}^{\infty} f_x(x) f_y(z - x)dx.$$

A category of problems that presents particular interest is that in which the random variables $\mathbf{x}$ and $\mathbf{y}$ are *statistically independent Gaussian variables*, with PDFs

$$f_x(x) = \frac{1}{\sigma_x \sqrt{2\pi}} \exp\left[-\frac{(x - \mu_x)^2}{2\sigma_x^2} \right] \quad \text{and} \quad f_y(y) = \frac{1}{\sigma_y \sqrt{2\pi}} \exp\left[-\frac{(y - \mu_y)^2}{2\sigma_y^2} \right].$$

In this case, the *probability density of the sum* $\mathbf{z} = \mathbf{x} + \mathbf{y}$ will have the form

$$f_z(z) = \frac{1}{2\pi\sigma_x\sigma_y} \int_{-\infty}^{\infty} \exp\left[-\frac{(x - \mu_x)^2}{2\sigma_x^2} \right] \exp\left[-\frac{(z - x - \mu_y)^2}{2\sigma_y^2} \right] dx = \frac{1}{\sqrt{2\pi(\sigma_x^2 + \sigma_y^2)}} \exp\left[-\frac{[z - (\mu_x + \mu_y)]^2}{2(\sigma_x^2 + \sigma_y^2)} \right],$$

as it emerges after performing the integration, which is left as an exercise to the reader. This equation tells us that *the sum of two independent Gauss-type variables is also a Gauss-type variable with a mean equal to the sum of the means of the two variables and variance equal to the sum of the variances of the two variables*. This result can be generalized for the sum of *any number of independent Gauss-type variables*, whereas similar conclusions can be formulated for the difference of such functions. PDFs characterized by this property are referred to as *reproducible*, with Gauss-type functions being just one example of such functions.

Finally, for the third category of problems that have the form $\mathbf{z} = g(\mathbf{x}, \mathbf{y})$ and $\mathbf{w} = h(\mathbf{x}, \mathbf{y})$, where x and y are two random variables with a combined PDF $f_{xy}(x, y)$, the CDF $F_{zw}(z, w)$ is given by the equation

$$F_{zw}(z, w) = P\{\mathbf{z}(\xi) \leq z, \mathbf{w}(\xi) \leq w\} = P\{g(x, y) \leq z, h(x, y) \leq w\} = P\{(x, y) \in D_{zw}\} = \iint_{(x,y)\in D_{zw}} f_{xy}(x, y)dxdy,$$

where $D_{zw} = \{(x, y) : g(x, y) \leq z, h(x, y) \leq w\}$ is the region of the xy plane where the inequalities $g(x, y) \leq z$ and $h(x, y) \leq w$ are simultaneously satisfied. For example, considering the functions $\mathbf{z} = \mathbf{x} + \mathbf{y}$ and $\mathbf{w} = \mathbf{x} - \mathbf{y}$, the set D_{zw} is defined as the region of the plane where the inequalities $x + y \leq z$ and $x - y \leq w$ are simultaneously satisfied. Resorting to the above equation, we can show that

$$F_{zw}(z, w) = \int_{-\infty}^{(z+w)/2} \left(\int_{x-w}^{z-x} f_{xy}(x, y)dy \right) dx,$$

an expression that leads to a *PDF* of the form

$$f_{zw}(z, w) = \frac{\partial^2 F_{zw}(z, w)}{\partial x \partial y} = \frac{1}{2} f_{xy} \left(\frac{z + w}{2} - \frac{z - w}{2} \right).$$

For an alternative and interesting method of directly calculating the function $f_{zw}(z, w)$ from the function $f_{xy}(x, y)$, which is based on the use of inverse transforms and Jacobian determinants, as well as related examples, see Papoulis and Pillai (pages 199–201) or Stark and Woods (pages 154–156).

Considering the function of the random variable $\boldsymbol{y} = g(\boldsymbol{x})$ (with the variable $\boldsymbol{x}$ being both real and complex), its average value is calculated as[7]

$$E\{\boldsymbol{y}\} = E\{g(\boldsymbol{x})\} = \int_{-\infty}^{\infty} y f_y(y) dy = \int_{-\infty}^{\infty} g(\boldsymbol{x}) f_x(x) dx$$

and is characterized by the property of linearity

$$E\{\alpha_1 g_1(\boldsymbol{x})\} + E\{\alpha_2 g_2(\boldsymbol{x})\} + \cdots + E\{\alpha_n g_n(\boldsymbol{x})\} = \alpha_1 E\{g_1(\boldsymbol{x})\} + \alpha_2 E\{g_2(\boldsymbol{x})\} + \cdots + \alpha_n E\{g_n(\boldsymbol{x})\}.$$

In full analogy with the functions of a random variable, the functions $z = g(\boldsymbol{x}, \boldsymbol{y})$ of two random variables $\boldsymbol{x}$ and $\boldsymbol{y}$ are characterized by a *CDF* of the form

$$F_z(z) = P\{z(\xi) \le z\} = P\{g(\boldsymbol{x}, \boldsymbol{y}) \le z\} = P\{(\boldsymbol{x}, \boldsymbol{y}) \in D_z\} = \int\int_{x,y \in D_z} f_{xy}(x, y) dx dy,$$

where D_z is the region of the plane where the condition $g(\boldsymbol{x}, \boldsymbol{y}) \le z$ is satisfied and by a *PDF* $f_z(z) = dF_z(z)/dz$. On the other hand, the *expected value* $E\{z\}$ is proven to have the form[8]

$$E\{\boldsymbol{z}\} = \int_{-\infty}^{\infty} z f_z(z) dz = \int_{-\infty}^{\infty} g(\boldsymbol{x}, \boldsymbol{y}) f_{xy}(x, y) dx dy.$$

Resorting to this equation, it is easily found that

$$E\{\boldsymbol{x}\} = \int_{-\infty}^{\infty}\int_{-\infty}^{\infty} x f_{xy}(x, y) dx dy = \int_{-\infty}^{\infty}\left(\int_{-\infty}^{\infty} f_{xy}(x, y) dy\right) x dx = \int_{-\infty}^{\infty} x f_x(x) dx \tag{A.134}$$

$$E\{\boldsymbol{y}\} = \int_{-\infty}^{\infty}\int_{-\infty}^{\infty} y f_{xy}(x, y) dx dy = \int_{-\infty}^{\infty}\left(\int_{-\infty}^{\infty} f_{xy}(x, y) dx\right) y dy = \int_{-\infty}^{\infty} y f_y(y) dy \tag{A.135}$$

and using the above expressions we are led to the expression

$$E\{\boldsymbol{x}+\boldsymbol{y}\} = \int_{-\infty}^{\infty}\int_{-\infty}^{\infty} (x+y) f_{xy}(x, y) dx dy = \int_{-\infty}^{\infty}\left(\int_{-\infty}^{\infty} f_{xy}(x, y) dy\right) x dx + \int_{-\infty}^{\infty}\left(\int_{-\infty}^{\infty} f_{xy}(x, y) dx\right) y dy = E\{\boldsymbol{x}\} + E\{\boldsymbol{y}\},$$

which generalizes as $E\{\alpha_1 \boldsymbol{x}_1 + \alpha_2 \boldsymbol{x}_2 + \cdots + \alpha_n \boldsymbol{x}_n\} = \alpha_1 E\{\boldsymbol{x}_1\} + \alpha_2 E\{\boldsymbol{x}_2\} + \cdots + \alpha_n E\{\boldsymbol{x}_n\}$.

The above concepts can be extended to describe the case of *multiple dimensions*. Specifically, considering a *graded* or *vector* function $g(\boldsymbol{x})$ of the vector variable $\boldsymbol{x}$, we can define the *mathematical expectation* of the function $g(\boldsymbol{x})$, from the relation

$$E\{g(\boldsymbol{x})\} = \langle g(\boldsymbol{x}) \rangle = \int_{-\infty}^{+\infty} g(\boldsymbol{x}) \rho_{\boldsymbol{x}}(\boldsymbol{x}) d\boldsymbol{x}, \tag{A.136}$$

with the above integral being calculated *for all the components* of the vector $\boldsymbol{x}$. The above relation refers to a *continuous random variable* x, whereas the corresponding equation for a *discrete variable* is

[7] For a rigorous proof of this theorem, which is one of the most important theorems of statistical science, see Feller W., *An Introduction to Probability Theory and Its Applications*, Vol. 2, 2nd edition, New York: John Wiley, 1971 (page 5) or Davenport W.B. Jr, *Probability and Random Processes*, New York: McGraw-Hill (page 99), whereas for an informal proof see Stark and Woods, pages 172–173.

[8] See Stark & Woods, page 176, and for an alternative proof, Helstrom C.W., Probability and Stochastic Processes for Engineers, 2nd Edition, New York, MacMillan, 1991.

$$E\{g(\boldsymbol{x})\} = \langle g(\boldsymbol{x})\rangle = \sum_x g(\boldsymbol{x})\rho_{\boldsymbol{x}}(\boldsymbol{x}). \quad \text{(A.137)}$$

In cases of functions of many variables $f(x_1, x_2, \ldots, x_n)$, the value of the above quantity is given by the relation

$$E\{f(x_1, x_2, \ldots, x_n)\} = \langle f(x_1, x_2, \ldots, x_n)\rangle = \sum_{x_1, x_2, \ldots, x_n} f(x_1, x_2, \ldots, x_n)\rho_{x_1, x_2, \ldots, x_n}(x_1, x_2, \ldots, x_n)$$

for the discrete case and

$$E\{f(x_1, x_2, \ldots, x_n)\} = \langle f(x_1, x_2, \ldots, x_n)\rangle = \int f(x_1, x_2, \ldots, x_n)\rho_{x_1, x_2, \ldots, x_n}(x_1, x_2, \ldots, x_n)dx_1dx_2\ldots dx_n$$

for the continuous case. If the relation $g(\boldsymbol{x}) = \boldsymbol{x}$ holds, then the above equation reduces to the defining equation of the expectation of the vector $\boldsymbol{x}$, known as the *mean vector* $\boldsymbol{m_x}$ of $\boldsymbol{x}$ or *first-order moment* and is defined by the relation

$$\boldsymbol{m_x} = E\{\boldsymbol{x}\} = \int_{-\infty}^{+\infty} \boldsymbol{x}\rho_{\boldsymbol{x}}(\boldsymbol{x})d\boldsymbol{x} \ . \quad \text{(A.138)}$$

If the vector $\boldsymbol{x}$ is defined in the space of n dimensions $\mathbb{R}^n$, the vector $\boldsymbol{m_x}$ will also have n components, with the component number i given by the relation

$$m_{x_i} = E\{x_i\} = \int_{-\infty}^{+\infty} x_i\rho_{\boldsymbol{x}}(\boldsymbol{x})d\boldsymbol{x} = \int_{-\infty}^{+\infty} x_i\rho_{x_i}(x_i)dx_i.$$

Statistical Independence Two random variables $\boldsymbol{x}$ and $\boldsymbol{y}$ are characterized as *statistically independent*, if the events $\{\boldsymbol{x} \in A\}$ and $\{\boldsymbol{y} \in B\}$ are *independent* of each other, which mathematically is expressed as

$$P\{\boldsymbol{x} \in A\ \boldsymbol{y} \in B\} = P\{\boldsymbol{x} \in A\}P\{\boldsymbol{y} \in B\},$$

where A and B are *two arbitrary intervals* of the x and y axes respectively. In this case, $F(x, y) = F_x(x)F_y(y)$ and therefore $f(x, y) = f_x(x)f_y(y)$. It is proven that if the random variables $\boldsymbol{x}$ and $\boldsymbol{y}$ are *independent* of each other, then the same will apply to the variables $\boldsymbol{z} = g(\boldsymbol{x})$ and $\boldsymbol{w} = h(\boldsymbol{y})$. The above applies to the *continuous* case, whereas for the *discrete* case, if $P\{\boldsymbol{x} = x_i\} = p_i$ and $P\{\boldsymbol{y} = y_k\} = q_k$, then the joint or combined probability is defined as $P = \{\boldsymbol{x} = x_i, \boldsymbol{y} = y_k\} = p_{ik}$ with $\sum_{i,k} p_{ik} = 1$, whereas $p_i = \sum_k p_{ik}$ and $q_k = \sum_i p_{ik}$.

Characteristic Function Another quantity that extends directly for the case of two random variables is the *characteristic function*, with its *combined* or *joint* version (joint characteristic function) given by the equation

$$\Phi(\omega_1, \omega_2) = E\{e^{j(\omega_1 x + \omega_2 y)}\} = \int_{-\infty}^{\infty}\int_{-\infty}^{\infty} f(x, y)e^{j(\omega_1 x + \omega_2 y)}dxdy, \quad \text{(A.139)}$$

from which it follows that

$$f(x, y) = \frac{1}{4\pi^2}\int_{-\infty}^{\infty}\int_{-\infty}^{\infty} \Phi(\omega_1, \omega_2)e^{-j(\omega_1 x + \omega_2 y)}d\omega_1 d\omega_2 \quad \text{(A.140)}$$

and with the characteristic functions of the variables $\boldsymbol{x}$ and $\boldsymbol{y}$, related to the *joint characteristic function* as $\Phi_x(\omega) = \Phi(\omega, 0)$ and $\Phi_y(\omega) = \Phi(0, \omega)$.

Considering the variable $\boldsymbol{z} = \alpha\boldsymbol{x} + \beta\boldsymbol{y}$ we have

$$\Phi_z(\omega) = E\{e^{j(\alpha x+\beta y)\omega}\} = \Phi(\alpha\omega, \beta\omega).$$

If the random variables $\boldsymbol{x}$ and $\boldsymbol{y}$ are *independent* of each other, then

$$E\{e^{j(\omega_1 x+\omega_2 y)}\} = E\{e^{j\omega_1 x}\}E\{e^{j\omega_2 y}\},$$

and therefore, $\Phi(\omega_1, \omega_2) = \Phi_x(\omega_1)\Phi_y(\omega_2)$, whereas of course the opposite is also true. In addition, if $\boldsymbol{z} = \boldsymbol{x} + \boldsymbol{y}$ then $E\{e^{j\omega z}\} = E\{e^{j\omega(\boldsymbol{x}+\boldsymbol{y})}\} = E\{e^{j\omega x}\}E\{e^{j\omega y}\}$ so $\Phi_z(\omega) = \Phi_x(\omega)\Phi_y(\omega)$. In this case, it is proven that the PDF $f_z(z)$ equals the *convolution* of the functions $f_x(x)$ and $f_y(y)$, or in mathematical writing,

$$f_z(z) = \int_{x=-\infty}^{\infty} f_x(z-y)f_y(y)dy = \int_{x=-\infty}^{\infty} f_x(x)f_y(z-x)dx.$$

We are therefore led to the conclusion that *the characteristic function of the convolution of two PDF functions equals the product of the characteristic functions of the two PDFs, a conclusion that is expected, since essentially the characteristic function is nothing but a Fourier transform.* In the general case, if $\boldsymbol{z} = \boldsymbol{x}_1 + \boldsymbol{x}_2 + \cdots + \boldsymbol{x}_n$, then it will also be $f_z(z) = f_{x_1}(z) + f_{x_2}(z) + \cdots + f_{x_n}(z)$, as well as $\Phi_z(\omega) = \Phi_{x_1}(\omega)\Phi_{x_2}(\omega)\cdots\Phi_{x_n}(\omega)$. The proof of these properties is based on the method of *mathematical induction*, with the basic step of the method concerning the case of two variables being what we proved previously.

Generalizing the above for the case of n variables, the characteristic function is defined as

$$\Phi_{x_1x_2\cdots x_n}(\omega, \omega_2, \ldots, \omega_n) = E\left\{\exp\left(j\sum_{i=1}^{n}\omega_i x_i\right)\right\}. \tag{A.141}$$

Taking into account that the *characteristic function is the Fourier transform* (except for the reversal in the sign of the exponent) of the *joint PDF* $f_{x_1x_2\cdots x_n}(x_1, x_2, \ldots, x_n)$, this can be calculated from the equation of the *inverse Fourier transform* (with the exception of the reversal in the sign of the exponent), that is, as

$$f_{x_1x_2\cdots x_n}(x_1, x_2, \ldots, x_n) = \frac{1}{(2\pi)^n}\int_{-\infty}^{\infty}\int_{-\infty}^{\infty}\cdots\int_{-\infty}^{\infty}\Phi_{x_1x_2\cdots x_n}(\omega, \omega_2, \ldots, \omega_n)\exp\left(-j\sum_{i=1}^{n}\omega_i x_i\right)d\omega_1 d\omega_2\cdots d\omega_n, \tag{A.142}$$

whereas similar results can be formulated for the discrete case, replacing the integrals with sums.

A.14.10 The Concept of Estimation

In cases where the CDF of a random variable is of the form $F_x(x, \lambda)$, that is, it depends on a parameter λ and its functional form is known, the question arises of *estimating the value* of the parameter λ, which can be both a *scalar* and a *vector* quantity. The classic way to proceed with this estimation is to perform *many repetitions of the related experiment* and to record the sequence of experimental outcomes $x_1, x_2, \ldots, x_n$ that result from this process, which as commented elsewhere, they jointly define a vector $X = (x_1, x_2, \ldots, x_n)^T$, which is understood as a *measurement* or *observation* vector. Based on all the above, we can define two types of estimation of the value of the parameter λ, the *point estimate* and the *interval estimate*, which are defined in the following way.

The *point estimate* is defined as a function of the vector X of the form $\tilde{\lambda} = h(X)$, with the corresponding random variable $\tilde{\boldsymbol{\lambda}} = h(\boldsymbol{X})$ constituting the *point estimator*. If the condition $E\{\tilde{\boldsymbol{\lambda}}\} = \lambda$ holds, where λ is the parameter the value of which we wish to estimate, then the estimator is characterized as *unbiased* (unbiased estimator), whereas in the opposite case where $E\{\tilde{\boldsymbol{\lambda}}\} \neq \lambda$, the estimator is characterized as *biased*, with bias value $b = E\{\tilde{\boldsymbol{\lambda}}\} - \lambda$. In addition, if the estimation error $\tilde{\boldsymbol{\lambda}} - \lambda$ tends to zero in the limit $n \to \infty$ where n is the number of repetitions of the experiment, the estimator $\tilde{\boldsymbol{\lambda}} = h(\boldsymbol{X})$ is characterized as *consistent*. As for the goal of the estimation process, it consists of *determining the function* $h(\boldsymbol{X})$, *which minimizes the approximation error* $h(\boldsymbol{X}) - \lambda$, whereas in the special case where the function is chosen so as to minimize the mean square error

$$e = E\{[h(\boldsymbol{X}) - \lambda]^2\} = \int_R [h(\boldsymbol{X}) - \lambda]^2 f(X, \lambda) d\lambda \tag{A.143}$$

then the estimator $\tilde{\boldsymbol{\lambda}} = h(\boldsymbol{X})$ is characterized as an *optimal* estimator.

As an example of estimation, let us consider the estimation $\tilde{\boldsymbol{x}}$ of the mean value $\overline{\boldsymbol{x}}$ of a random *ergodic* process. Defining this estimation as

$$\tilde{\boldsymbol{x}} = \frac{1}{T} \int_0^T \boldsymbol{x} t dt, \tag{A.144}$$

where $\boldsymbol{x}(t)$ is any sample of the process, it is not difficult to understand that although the value of $\tilde{x}$ for each sample is a simple number, this value will be *different* if a different time interval or a different sample of the considered process is selected. In other words, this parameter is a *random variable*, whereas it is obvious that since this quantity is only an *estimate of the real value* of the parameter $\overline{\boldsymbol{x}}$, it does not coincide with it, with the estimation process having *practical value* when the values of the parameters $\tilde{\boldsymbol{x}}$ and $\overline{\boldsymbol{x}}$ are *very close* to each other. Calculating the *mean value* of the random variable $\tilde{\boldsymbol{x}}$, we easily find that

$$E\{\tilde{\boldsymbol{x}}\} = E\left\{\frac{1}{T} \int_0^T \boldsymbol{x} t dt\right\} = \frac{1}{T} \int_0^T E\{\boldsymbol{x} t\} dt = \frac{1}{T} \int_0^T \overline{\boldsymbol{x}} dt = \overline{\boldsymbol{x}}.$$

In other words, *the value of the parameter* $\tilde{\boldsymbol{x}}$ *coincides with the real mean value* $\overline{\boldsymbol{x}}$ *and consequently, the corresponding estimator is characterized as unbiased*. Of course, in practice, the mathematical form of the function $\boldsymbol{x}$ is generally *not known* and therefore the above integration *is rarely performed analytically*. Instead, for the above calculation we resort to *numerical integration*, using samples of the function $\boldsymbol{x}$ that are equidistant in time. In this case, the estimate $\tilde{\boldsymbol{x}}$ is defined as

$$\tilde{\boldsymbol{x}} = \frac{1}{N} \sum_{i=1}^{N} x_i,$$

and, working as before, we find

$$E\{\tilde{\boldsymbol{x}}\} = E\left\{\frac{1}{N} \sum_{i=1}^{N} x_i\right\} = \frac{1}{N} \sum_{i=1}^{N} E\{x_i\} = \frac{1}{N} \sum_{i=1}^{N} \overline{\boldsymbol{x}} = \frac{N\overline{\boldsymbol{x}}}{N} = \overline{\boldsymbol{x}}, \tag{A.145}$$

which is the same result as before. On the other hand, the *expected value* for the mean square is calculated as

$$E\left\{(\tilde{x})^2\right\} = E\left\{\left(\frac{1}{N} \sum_{i=1}^{N} x_i\right)^2\right\} = E\left\{\left(\frac{1}{N} \sum_{i=1}^{N} x_i\right)\left(\frac{1}{N} \sum_{i=1}^{N} x_i\right)\right\} = \frac{1}{N^2} \sum_{i=1}^{N} \sum_{j=1}^{N} E\{\boldsymbol{x}_i(t) \boldsymbol{x}_j(t)\}, \tag{A.146}$$

where $\boldsymbol{x}_i(t) = \boldsymbol{x}(t_i)$ and $\boldsymbol{x}_j(t) = \boldsymbol{x}(t_j)$. Taking into account that the samples of the process are *statistically independent*, we have $E\{\boldsymbol{x}_i(t)\boldsymbol{x}_j(t)\} = E\{\boldsymbol{x}_i(t)\}E\{\boldsymbol{x}_j(t)\} = (\overline{x})^2$ for $i \neq j$ and $E\{\boldsymbol{x}_i(t)\boldsymbol{x}_j(t)\} = E\{\boldsymbol{x}(t)^2\} = \overline{\boldsymbol{x}^2}$ for $i = j$ and the above expression becomes

$$E\left\{(\tilde{\boldsymbol{x}})^2\right\} = \frac{1}{N^2}\left(\sum_{i=1}^{N} E\{\boldsymbol{x}_i^2\} + \sum_{i=1}^{N}\sum_{j=1}^{N} E\{\boldsymbol{x}_i(t)\boldsymbol{x}_j(t)\}\right) = \frac{1}{N^2}\left(\sum_{i=1}^{N}\overline{\boldsymbol{x}^2} + \sum_{i=1}^{N}\sum_{j=1}^{N}(\overline{\boldsymbol{x}})^2\right)$$

$$= \frac{1}{N^2}\left(N\overline{\boldsymbol{x}^2} + N(N-1)(\overline{\boldsymbol{x}})^2\right) = \frac{1}{N}\overline{\boldsymbol{x}^2} + \left(1 - \frac{1}{N}\right)(\overline{\boldsymbol{x}})^2 = \frac{1}{N}\sigma_x^2 + (\overline{\boldsymbol{x}})^2.$$

Therefore, we have

$$\mathrm{Var}(\tilde{\boldsymbol{x}}) = E\left\{(\tilde{\boldsymbol{x}})^2\right\} - [E\{(\tilde{\boldsymbol{x}})\}]^2 = \frac{1}{N}\sigma_x^2 + (\overline{\boldsymbol{x}})^2 - (\overline{\boldsymbol{x}})^2 = \frac{1}{N}\sigma_x^2.$$

We are led, therefore, to the conclusion that *the variability of the estimate of the mean value of the process is equal to the variability of the process multiplied by the value* $1/N$. Therefore, in order to improve this estimate, we can increase the value of N, recording a *larger number of samples* and calculate the average using this new larger value of N.

In a similar way, it is possible to estimate the value of the *autocorrelation of stationary processes* from experimental data, a process that has particular value and practical significance, since the autocorrelation function is one of the most important tools for characterizing random processes. Working as before, we can, at least theoretically, construct an estimator of the magnitude of the autocorrelation as

$$\hat{R}_x(\tau) = \frac{1}{T-\tau}\int_0^{T-\tau} x(t)x(t+\tau)dt, \quad \tau \ll T, \tag{A.147}$$

although the above integral generally *cannot be calculated analytically*, since the form of the function $x(t)$ is not known. Consequently, we must resort again to *numerical methods*, proceeding to *uniform sampling* of the function $x(t)$ with a sampling period Δt and approximate the above integral from the sum

$$\hat{R}_x(\tau) = \frac{1}{N-n+1}\sum_{k=0}^{N-n} x_k x_{k+n} \tag{A.148}$$

for values $n = 0, 1, 2, \ldots, M$, where $M \ll N$. This estimator is a *random variable* for which the calculation of the *mean value* and the *variance* can take place so that we can proceed to draw conclusions about its nature and characteristics. For more details on this topic see Cooper and McGillem, page 222.

Having determined the estimator of the autocorrelation function, let us now move on to the problem of estimating the *power spectral density*, which, as we commented, constitutes *the Fourier transform of the autocorrelation function*. This problem is *particularly important*, since the knowledge of the power spectral density of an unwanted signal can contribute to the determination of its origin, a fact that *potentially allows its elimination*, or if such a thing is not possible, *the construction of a filter whose application will lead to the reduction of its impact*. To demonstrate the process, let us consider a continuous-time signal $x(t)$ of duration T, which is a sample of a *random ergodic process*. According to what we mentioned previously, the estimator of the autocorrelation function is given by Eq. (A.148), with the maximum allowable value of τ, so that $\tau \ll T$, being $\tau_{\max}$. Consequently, the estimator of the autocorrelation, will lead to the value $\hat{R}_x(\tau)$ if $|\tau| \leq \tau_{\max}$, whereas it will have zero value for the values $|\tau| > \tau_{\max}$. In order to modify the function $\hat{R}_x(\tau)$ so as to incorporate the above restriction into its equation, we should, as known, multiply it by a *window function* $w(\tau)$ of duration $2\tau_{\max}$, which should be in the form $w(\tau) = 0$ for $|\tau| > \tau_{\max}$ and nonzero value for $|\tau| \leq \tau_{\max}$, which depends on the function of the window used. Consequently, the estimator of the autocorrelation function should be formulated in the form

$$\hat{R}_x^w(\tau) = \hat{R}_x(\tau)w(\tau) = \frac{w(\tau)}{T-\tau}\int_0^{T-\tau} x(t)x(t+\tau)dt, \quad \tau \ll T. \tag{A.149}$$

If we take the *Fourier transform* of the above expression and use the property of the *product* of the aforementioned transform, we get

$$\hat{S}_x^w(\tau) = \mathcal{F}\{\hat{R}'_x(\tau)w(\tau)\} = \mathcal{F}\{\hat{R}'_x(\tau)\} * \mathcal{F}\{w(\tau)\} = \hat{S}'_x(f) * \mathcal{W}(f), \tag{A.150}$$

where $\mathcal{W}(f) = \mathcal{F}\{w(\tau)\}$ is the *Fourier transform* of the window function $w(t)$, whereas the emphasized notation in the function $\hat{R}'_x(\tau)$ simply indicates that this function is now defined *for every value* of τ, as is now allowed by the above modification and the use of the window function. As for the *selection* of the appropriate window function, all that we presented in detail in the relevant section applies. Specifically, as we commented there, *the rectangular window, despite the fact that it leads to the minimum approximation error in relation to all windows of the same length, it is not the most suitable, owing to the large size of the sidelobes, and for this reason, in practice, other window functions are used, such as the Hamming window*. For the defining equation, the properties, and the spectral characteristics of these windows, refer to the relevant section (see Sect. 10.5.6 on page 900).

In practice, as we have commented, the calculation of the autocorrelation function is based *on the recording of the samples $R_x(n\Delta t)$ of this function through a uniform sampling process*, with a sampling period Δt, and for this reason, in the estimation problem *the continuous-time Fourier transform will not be used but the discrete Fourier transform*, for which there are very efficient implementations, such as the *fast Fourier transform (FFT)*, which allows for the fastest calculation in the electronic systems of computers (see the relevant section). Consequently, the estimation of the spectral power density takes the form $\hat{S}_x(k\Delta t) = \Delta t \mathcal{F}\{\tilde{R}_x(n\Delta t)\}$ $(k = 0, 1, 2, \ldots, M)$ with the calculation of the Fourier transform, based on the use of the FFT algorithm. In practice, this process can alternatively be implemented with the help of the *numerical integration* technique, through the equation

$$\hat{S}_x(k\Delta\omega) = \Delta t \left[\hat{R}_x(0) + 2\sum_{n=1}^{M-1} \tilde{R}_x(n\Delta t)\cos\left(\frac{\pi nk}{M}\right) + \tilde{R}_x(M\Delta t)\cos(\pi k) \right], \tag{A.151}$$

with the existence of the cosine in the above expression, owing to the fact that the autocorrelation function is characterized by *even* symmetry.

The autocorrelation function R_{xy}, defined between the input $x(t)$ and the output $y(t)$ of a linear system, when these constitute samples of random stationary processes, can be used as a means of estimating the *impulse response $h(t)$ of the system*, when the bandwidth of the signal is large in relation to the bandwidth of the system. The process consists in the application of *uniform sampling* of the period Δt (the value of which depends on the value of the number M of points at which we want to calculate the samples of the impulse response as well as the length of the time interval in which the signal in question has significant value) to the input and output functions, leading to the sequences $x_k = x(k\Delta t)$ $k = 1, 2, \ldots, N$ and $y_k = y(k\Delta t)$ $k = 1, 2, \ldots, N$. In this case, it is possible to estimate the sample number n of the *autocorrelation function* according to the expression

$$\hat{R}_x(n\Delta t) = \frac{1}{N-n+1}\sum_{k=n}^{N} x_{k-n}y_k, \quad n = 0, 1, 2, \ldots, \quad M \ll N. \tag{A.152}$$

How can the above expression be used to estimate the samples of the *impulse response*? To do this, we must correlate the variance of the process samples $x(t)$ with the power density of this process. If we assume that the bandwidth of the signal is *sufficiently large* and that the samples of the process are *statistically independent*, then these samples can be considered to arise from a *white-noise process of limited bandwidth*, with a value of $1/2\Delta t$ (see Cooper and McGillem, page 343). Since the variance of such a process is equal to $2S_0W$, $S_0 = \sigma_x^2\Delta t$ and from Eq. (A.152) we have

$$\hat{h}(n\Delta t) = \frac{1}{\sigma_x^2\Delta t}\tilde{R}_{xy}(n\Delta t) = \frac{1}{\sigma_x^2\Delta t(N-n+1)}\sum_{k=n}^{N} x_{k-n}y_k, \quad n = 0, 1, 2, \ldots, \quad M \ll N \tag{A.153}$$

with a dispersion value for this estimator

$$\mathrm{Var}[\hat{h}(n\Delta t)] \leq \frac{2}{N}\sum_{k=0}^{M} h^2(k\Delta t). \tag{A.154}$$

On the other hand, the *interval estimation* is related to the *uncertainty* or the *error* that is inextricably linked with every measurement process. If, for example, the parameter λ expresses some *angle* or some *length*, the value of which is calculated by recording a sequence of measurements $x_1, x_2, \ldots, x_n$, the values of this sequence will have the form $x_i = \lambda + e_i$ $(i = 1, 2, \ldots, n)$, where e_i $(i = 1, 2, \ldots, n)$ is the error associated with the ith measurement, which led to the recording of the value x_i. These values can be considered as *samples* of a random variable $\boldsymbol{x} = \lambda + \boldsymbol{e}$. In this case, although it is *impossible* to assert with certainty that the real value of the parameter λ is equal to the point estimate $\hat{\lambda}$, we can assert with a high degree of certainty that this is something that holds, but *within certain specific tolerance limits* λ_1 and λ_2, which define the interval (λ_1, λ_2). This range of values, the limits of which are functions of the form $\lambda_1 = h_1(X)$ and $\lambda_2 = h_2(X)$, constitutes the interval estimate of the parameter λ, with the corresponding random vector $(\boldsymbol{\lambda}_1, \boldsymbol{\lambda}_2)$ known as the *interval estimator* of the parameter λ. Furthermore, we will say that the interval (λ_1, λ_2) is a *confidence interval* γ of the parameter λ, if the condition $P\{\boldsymbol{\lambda}_1 < \lambda < \boldsymbol{\lambda}_2\} = \gamma$ holds, with the constant γ constituting the *confidence coefficient* of the estimate and the difference $\delta = 1 - \gamma$ constituting the confidence level. Based on the above definitions, it is not difficult to understand that the confidence coefficient is a (purely *subjective*) measure of our confidence that the (unknown) value of the parameter λ belongs to the range of values (λ_1, λ_2). As for the goal of the estimation process for this interval estimation, it consists in determining the functions $h_1(X)$ and $h_2(X)$, which *minimize* the range of the interval (λ_1, λ_2).

In the above definition of estimation, which can be considered as the classic definition, the parameter λ of the function $F_x(x, \lambda)$ is considered an *unknown constant*, whereas the estimation is based *exclusively* on the recording of the experimental values $x_1, x_2, \ldots, x_n$. However, there are cases where the parameter λ is *not completely unknown* and therefore, in the problem of estimation, the *a priori probability* associated with the parameter λ is also involved. In this type of estimation, known as *Bayesian estimation*, the unknown parameter λ is considered another *random variable* $\boldsymbol{\lambda}$, whereas the distribution function of the random variable $\boldsymbol{x}$ is interpreted as the *conditional cumulative distribution* $F_x(x|\lambda)$ of the variable $\boldsymbol{x}$, assuming that $\boldsymbol{\lambda} = \lambda$. In other words, the random variable $\boldsymbol{\lambda}$ is associated with some *PDF* $f_\lambda(\lambda)$ and the problem is reduced to calculating the value λ of the random variable $\boldsymbol{\lambda}$, with the help of the sequence of recorded values $x_1, x_2, \ldots, x_n$ of the random variable $\boldsymbol{x}$ and the density $f_\lambda(\lambda)$. More details about this type of estimation can be found in the bibliography.

A.14.11 Random Vectors and Sequences

The fundamental concepts of statistics that have been presented so far apply to both the *continuous* and the *discrete* case and can be extended to collections of random variables, which are defined for both these cases. Specifically, in the discrete case, these collections of variables can have both *finite* and *infinite* length. In the first case, these collections are characterized as *random vectors*, whereas in the second case they are characterized as *random sequences*. In the following pages, we generalize the concepts and tools we defined in the previous sections, so that they describe the random vectors and the random sequences, whereas in another chapter of the book, we deal with the description of the above concepts for the continuous case, with the objects involved in this description known as stochastic or random processes.

Random Vectors

Starting with the random vectors, these constitute *finite collections* of n random variables of the form $\boldsymbol{X} = (\boldsymbol{x}_1, \boldsymbol{x}_2, \ldots, \boldsymbol{x}_n)^T$, which can be considered as defined in a space of n dimensions. The *CDF* of the vector $\boldsymbol{X}$ constitutes an *immediate extension* of the homonymous functions for the case of one and two variables and is defined as

$$F_X(\boldsymbol{x}) \triangleq P\{\boldsymbol{X} \leq \boldsymbol{x}\} \equiv P\{\boldsymbol{x}_1 \leq x_1, \boldsymbol{x}_2 \leq x_2, \cdots, \boldsymbol{x}_n \leq x_n\},$$

where the event $\{\boldsymbol{X} \leq \boldsymbol{x}\}$ constitutes a more concise notation of $\{\boldsymbol{x}_1 \leq x_1, \boldsymbol{x}_2 \leq x_2, \cdots, \boldsymbol{x}_n \leq x_n\}$. The boundary conditions for this multidimensional function, are $F_X(\infty) = 1$ (certain event) and $F_X(\mathbf{0}) = 0$ (impossible event), whereas the *PDF*, is defined as

$$f_X(\boldsymbol{x}) \triangleq \frac{\partial^n F_X(\boldsymbol{x})}{\partial x_1 \partial x_2 \cdots \partial x_n}, \tag{A.155}$$

from which the function $F_X(\boldsymbol{x})$ can be defined as the integral

$$F_X(\boldsymbol{x}) = \int_{-\infty}^{x_1}\int_{-\infty}^{x_2}\cdots\int_{-\infty}^{x_n} f_X(\boldsymbol{x}')dx_1'dx_2'\cdots dx_n' = \int_{-\infty}^{\boldsymbol{x}} f_X(\boldsymbol{x}')d\boldsymbol{x}'. \tag{A.156}$$

The *CDF* under the condition of the vector $\boldsymbol{X}$, knowing that some event M has already occurred, is defined as

$$F_{X|M}(\boldsymbol{x}|M) \triangleq P\{\boldsymbol{X} \leq \boldsymbol{x}|M\} = \frac{P\{\boldsymbol{X} \leq \boldsymbol{x}, M\}}{P(M)}, \tag{A.157}$$

with a corresponding *PDF*

$$f_{X|M}(\boldsymbol{x}|M) = \frac{\partial^n F_{X|M}(\boldsymbol{x}|M)}{\partial x_1 \partial x_2 \cdots \partial x_n}. \tag{A.158}$$

Considering a set of n events M_i, which jointly define a *partition* of the sample space Ω, we can, in full analogy with the one-dimensional case, write

$$F_X(\boldsymbol{x}) = F_{X|M_1}(\boldsymbol{x}|M_1)P(M_1) + F_{X|M_2}(\boldsymbol{x}|M_2)P(M_2) + \cdots + F_{X|M_n}(\boldsymbol{x}|M_n)P(M_n) = \sum_{i=1}^{n} F_{X|M_i}(\boldsymbol{x}|M_i)P(M_i),$$

from which it immediately follows that

$$f_X(\boldsymbol{x}) = f_{X|M_1}(\boldsymbol{x}|M_1)P(M_1) + f_{X|M_2}(\boldsymbol{x}|M_2)P(M_2) + \cdots + f_{X|M_n}(\boldsymbol{x}|M_n)P(M_n) = \sum_{i=1}^{n} f_{X|M_i}(\boldsymbol{x}|M_i)P(M_i).$$

Considering two random vectors $\vec{X}$ and $\vec{Y}$ defined in the spaces $\mathbb{R}^n$ and $\mathbb{R}^m$, we can define the *joint CDF* of the vectors $\vec{X}$ and $\vec{Y}$ from the relation

$$F_{XY}(\boldsymbol{x}, \boldsymbol{y}) = P(\boldsymbol{X} \leq \boldsymbol{x}, \boldsymbol{Y} \leq \boldsymbol{y}), \tag{A.159}$$

where $\boldsymbol{x}$ and $\boldsymbol{y}$ are constant vectors that have the same dimensions as the vectors $\boldsymbol{X}$ and $\boldsymbol{Y}$ respectively. In a completely analogous way, the *joint density function* is defined as the *derivative of the joint distribution function*,

$$f_{XY}(\boldsymbol{x}, \boldsymbol{y}) = \frac{\partial^{(n+m)} F_{XY}(\boldsymbol{x}, \boldsymbol{y})}{\partial x_1 \partial x_2 \cdots \partial x_n \partial y_1 \partial y_2 \cdots \partial y_m} \tag{A.160}$$

so that the latter can be defined as the integral

$$F_{XY}(\boldsymbol{x}, \boldsymbol{y}) = \int_{-\infty}^{\boldsymbol{x}}\int_{-\infty}^{\boldsymbol{y}} f_{XY}(\boldsymbol{\xi}, \boldsymbol{n})d\boldsymbol{n}d\boldsymbol{\xi}, \tag{A.161}$$

whereas in full analogy with the previous simpler cases we have

$$f_X(\boldsymbol{x}) = \int_{-\infty}^{+\infty} P_{XY}(\boldsymbol{x}, \boldsymbol{n})d\boldsymbol{n} \text{ and } f_Y(\boldsymbol{y}) = \int_{-\infty}^{+\infty} P_{XY}(\boldsymbol{\xi}, \boldsymbol{y})d\boldsymbol{\xi}.$$

The average or expected value of the vector $\boldsymbol{X} = (\boldsymbol{x}_1, \boldsymbol{x}_2, \ldots, \boldsymbol{x}_n)^T$ is a vector $\boldsymbol{\mu} = E\{\boldsymbol{X}\}$ the elements of which μ_i $(i = 1, 2, \ldots, n)$ are defined as

$$\mu_i = \int_{-\infty}^{\infty}\int_{-\infty}^{\infty} \cdots \int_{-\infty}^{\infty} x_i f_X(x_1, x_2, \ldots, x_n)dx_1dx_2 \cdots dx_n = \int_{-\infty}^{\infty} x_i f_{x_i}(x_i)dx_i, \tag{A.162}$$

where

$$f_{x_i}(x_i) = \int_{-\infty}^{\infty}\int_{-\infty}^{\infty} \cdots \int_{-\infty}^{\infty} f_X(\boldsymbol{x})dx_1dx_2 \cdots dx_{i-1}dx_{i+1} \cdots dx_n.$$

Another important quantity that needs to be defined is the *correlation* or *second-order moment* between pairs of components of a random vector $\boldsymbol{X}$. Considering two components $\boldsymbol{x}_i$ and $\boldsymbol{x}_j$ of such a vector, the correlation r_{ij} between them is given by the relation

$$r_{ij} = E\{\boldsymbol{x}_i\boldsymbol{x}_j\} = \int_{-\infty}^{+\infty} \boldsymbol{x}_i\boldsymbol{x}_j f_x(\boldsymbol{X})d\boldsymbol{X} = \int_{-\infty}^{+\infty}\int_{-\infty}^{+\infty} \boldsymbol{x}_i\boldsymbol{x}_j f_{x_i,x_j}(\boldsymbol{x}_i, \boldsymbol{x}_j)d\boldsymbol{x}_jd\boldsymbol{x}_i. \tag{A.163}$$

If we calculate the correlation between all the different components of the vector $\boldsymbol{X}$, we form the so-called *autocorrelation matrix*, which has dimensions $n \times n$ and is defined by the relation $R_X = E\{\boldsymbol{X}\boldsymbol{X}^T\}$. Considering a random vector $\boldsymbol{X} = [\,\boldsymbol{x}_1\, \boldsymbol{x}_2\, \cdots\, \boldsymbol{x}_n\,]^T$, we define as the *covariance matrix* or *dispersion* the matrix

$$C_X = E\{(\boldsymbol{X} - \boldsymbol{m}_X)(\boldsymbol{X} - \boldsymbol{m}_X)^T\}, \tag{A.164}$$

whereas the elements of this matrix, which have the form

$$C_{Xij} = E\{(x_i - m_i)(x_j - m_j)\} \tag{A.165}$$

are called *covariances* or *co-variabilities* and are defined as the *central moments* that correspond to the correlations r_{ij} we previously described. The *covariance* and *autocorrelation* matrices R_X and C_X are related through the relation $R_X = C_X + \boldsymbol{m}_X\boldsymbol{m}_X^T$ and coincide with each other when the relation $\boldsymbol{m}_X = \boldsymbol{0}$ holds. Note that

$$C_{Xij} = E\{x_i - \mu_i)(x_j - \mu_j)\} = E\{x_j - \mu_j)(x_i - \mu_i)\} = C_{Xij}$$

$(i, j = 1, 2, \ldots, n)$, as well as that $C_{Xii} = \sigma_i^2$. Also note that the diagonal elements of the table are nothing more than the *values of the dispersion of the components of the random vector*. The condition $C_{Xij} = C_{Xji}$ implies that the covariance matrix is *symmetric* and since its elements are *real numbers*, which happens when the elements of the vector $\boldsymbol{X}$ are real numbers, the matrix C_X belongs to the broader category of *Hermitian matrices* and is characterized by all the properties that characterize these matrices, such as the property of being *positive (semi) definite*. This matrix is nothing more than a $n \times n$ dimension matrix, of the form

$$C_X = \begin{pmatrix} \sigma_1^2 & C_{X12} & \cdots & C_{X1n} \\ C_{X21} & \sigma_2^2 & \cdots & C_{X2n} \\ \ddots & \ddots & \vdots & \vdots \\ C_{Xn1} & C_{Xn2} & \cdots & \sigma_n^2 \end{pmatrix}.$$

Considering a pair of vectors $(\boldsymbol{X}, \boldsymbol{Y})$, the *autocorrelation matrix* R_X generalizes to the *cross-correlation matrix* $R_{XY} = E\{\boldsymbol{X}\boldsymbol{Y}^T\}$, whereas the *covariance matrix* C_X generalizes to the *cross-covariance matrix*, which is defined as

$$C_{XY} = E\{(X - m_x)(Y - m_y)\}. \tag{A.166}$$

Because in the general case, the dimensions of the vectors X and Y *are not necessarily the same*, the matrices R_{XY} and C_{XY} *are not always square*. However, the power of the relations $R_{XY} = R_{YX}^T$ and $C_{XY} = C_{YX}^T$ is easily proven, whereas the two matrices coincide, that is, $R_{XY} = C_{XY}$ holds, when the relation $m_x = m_y = 0$ holds. Finally, it is proven that for the case of the sum $X + Y$, the matrix C_{X+Y} is given by the relation

$$C_{X+Y} = C_X + C_{XY} + C_{YX} + C_Y.$$

Considering two random vectors X and Y, these are characterized as *uncorrelated* (uncorrelated) when the *cross covariance* matrix is equal to zero, or in mathematical writing,

$$C_{XY} = E\{(X - m_x)(Y - m_y)\} = 0,$$

whereas an equivalent condition is

$$R_{XY} = E\{XY^T\} = E\{X\}E\{Y^T\} = m_x m_y^T.$$

Note that in the case where the variables x and y are components of *the same* vector X, a relation of the form $C = 0$ is not possible, as each component *correlates with itself*. For this reason, all we can do is assume that the components of the vector x are *mutually uncorrelated* (mutually uncorrelated), leading to the definition of the *covariance matrix* C_X.

A special category of random vectors is the *white random vectors*, which are characterized by the properties $m_X = 0$ and $R_X = C_X = I$. In the above notation, the symbol I expresses the unit matrix of dimensions $n \times n$.

Extending the concept of independence to the space of n dimensions, the random variables $\{x_1, x_2, \ldots, x_n\}$ are characterized as *(mutually) independent*, if the events $\{x_1 \le x_1\}$, $\{x_2 \le x_2\}$, $\cdots$, $\{x_n \le x_n\}$ are independent. In this case, generalizing the conclusions we formulated for the case of two variables we will have

$$F(x_1, x_2, \ldots, x_n) = F(x_1)F(x_2)\cdots F(x_n) \quad \text{and} \quad f(x_1, x_2, \ldots, x_n) = f(x_1)f(x_2)\cdots f(x_n).$$

The above property of statistical independence in the space of n dimensions allows us to describe the *combined probability densities* as the product of the simple densities, according to the last of the above expressions, and the most characteristic and important of these functions is the *combined normal distribution or Gauss distribution*. Specifically, considering the random vector $X = (x_1, x_2, \ldots, x_n)^T$ of independent random variables $x_1, x_2, \ldots, x_n$, the combined probability density for the normal distribution in n dimensions is defined as

$$f_X(x) = f_X(x_1, x_2, \ldots, x_n) = \prod_{i=1}^{n} f_{x_i}(x_i) = \frac{1}{(2\pi)^{n/2}\sigma_1\sigma_2\cdots\sigma_n} \exp\left[-\frac{1}{2}\sum_{i=1}^{n}\left(\frac{x_i - \mu_i}{\sigma_i}\right)^2\right] \tag{A.167}$$

$$= \frac{1}{(2\pi)^{n/2}\sqrt{\det(C_{XX})}} \exp\left[-\frac{1}{2}(X - \mu_X)^T C_{XX}^{-1}(X - \mu_X)\right], \tag{A.168}$$

where $f_{x_i}(x_i)$ is the *univariate normal distribution*, whereas μ_i and σ_i^2 $(i = 1, 2, \ldots, n)$ are the *mean value* and the *variance* of the random variables x_i $(i = 1, 2, \ldots, n)$ respectively. On the other hand, the matrix

$$C_{XX} = \mathrm{Cov}(X) = E\{(X - \mu_X)(Y - \mu_Y)\} = \mathrm{diag}\{\sigma_1^2, \sigma_2^2, \cdots, \sigma_n^2\}$$

is the *covariance matrix*, which, owing to the independence characterizing the random variables $x_1, x_2, \ldots, x_n$, is a *diagonal matrix* with determinant equal to the product of the diagonal elements—it will therefore be $\det(C_X) = \sigma_1^2\sigma_2^2\cdots\sigma_n^2$—whereas $\mu_X = E\{X\} = [\mu_{x_1}, \mu_{x_2}, \ldots, \mu_{x_n}]$ is the vector defined by the mean values of random variables x_i $(i = 1, 2, \ldots, n)$. The fact that the covariance matrix is *diagonal*, as imposed by the condition of independence of the components of the vector, *simplifies to a great extent* the calculation of the inverse matrix, which according to linear algebra is the diagonal matrix $C_X^{-1} = \text{diag}\{\sigma_1^{-2}, \sigma_2^{-2}, \cdots, \sigma_n^{-2}\}$. It is proven that the function $f_{XX}(x)$ is a *valid PDF* (see, for example, Stark and Woods, page 271).

The above refers to the case of real variables, whereas for the case of complex variables the reader is referred to the literature (see, for example, the book by Schreier and Scharf).

Example: Conditional Probability Density for Gauss-Type Variables

As an example of *conditional probability density*, let us consider two sets of random variables $\{x_1, x_2, \dots, x_n\}$ and $\{y_1, y_2, \dots, y_m\}$ *of the Gauss type*, which define respectively the vectors

$$\boldsymbol{X} = [x_1, x_2, \dots, x_n]^T \in \mathbb{R}^n \qquad \text{and} \qquad \boldsymbol{Y} = [y_1, y_2, \dots, y_m]^T \in \mathbb{R}^m$$

and the vector $\boldsymbol{Z} = [\boldsymbol{X}\ \boldsymbol{Y}]^T \in \mathbb{R}^{N+M}$ that results from their union and which is also characterized by combined Gauss-type probability density with mean vector and covariance matrix defined as

$$\boldsymbol{\mu}_Z = E\{\boldsymbol{Z}\} = \begin{bmatrix} \boldsymbol{\mu}_X \\ \boldsymbol{\mu}_Y \end{bmatrix} \qquad \text{and} \qquad \boldsymbol{C}_{ZZ} = E\{(\boldsymbol{Z}-\boldsymbol{\mu}_Z)(\boldsymbol{Z}-\boldsymbol{\mu}_Z)^T\} = \begin{pmatrix} \boldsymbol{C}_{XX} & \boldsymbol{C}_{XY} \\ \boldsymbol{C}_{YX} & \boldsymbol{C}_{YY} \end{pmatrix}.$$

Note that the above *covariance matrix* is a *composite* matrix, that is, a matrix whose elements are smaller matrices. Referring to any book of linear algebra, we find that the inverse of this matrix is the matrix

$$\boldsymbol{C}_{ZZ}^{-1} = \begin{pmatrix} \boldsymbol{T}_{XX} & \boldsymbol{T}_{XY} \\ \boldsymbol{T}_{YX} & \boldsymbol{T}_{YY} \end{pmatrix} = \begin{pmatrix} \boldsymbol{C}_{XX}^{-1} + \boldsymbol{C}_{XX}^{-1}\boldsymbol{C}_{XY}\boldsymbol{T}_{YY}\boldsymbol{C}_{YX}\boldsymbol{C}_{XX}^{-1} & -\boldsymbol{C}_{XX}^{-1}\boldsymbol{C}_{YX}\boldsymbol{C}_{XX}^{-1} \\ -\boldsymbol{T}_{YY}\boldsymbol{C}_{YX}\boldsymbol{C}_{XX}^{-1} & \boldsymbol{C}_{YY}^{-1} + \boldsymbol{C}_{YY}^{-1}\boldsymbol{C}_{YX}\boldsymbol{T}_{XX}\boldsymbol{C}_{XY}\boldsymbol{C}_{YY}^{-1} \end{pmatrix}.$$

Resorting to *Bayes' rule* it turns out that

$$f_{X|Y}(\boldsymbol{X}|\boldsymbol{Y}) = \frac{f_{XY}(\boldsymbol{x}, \boldsymbol{y})}{f_Y(\boldsymbol{y})} = \frac{|2\pi \boldsymbol{C}_{ZZ}|^{-1/2} \exp\left(-\frac{1}{2}(\boldsymbol{Z}-\boldsymbol{\mu}_Z)^T \boldsymbol{C}_{ZZ}^{-1}(\boldsymbol{Z}-\boldsymbol{\mu}_Z)\right)}{|2\pi \boldsymbol{C}_{YY}|^{-1/2} \exp\left(-\frac{1}{2}(\boldsymbol{Y}-\boldsymbol{\mu}_Y)^T \boldsymbol{C}_{YY}^{-1}(\boldsymbol{Y}-\boldsymbol{\mu}_Y)\right)}$$

$$= \sqrt{\frac{|2\pi \boldsymbol{C}_{YY}|}{|2\pi \boldsymbol{C}_{ZZ}|}} \exp\left[-\frac{1}{2}\left((\boldsymbol{Z}-\boldsymbol{\mu}_Z)^T \boldsymbol{C}_{ZZ}^{-1}(\boldsymbol{Z}-\boldsymbol{\mu}_Z) + (\boldsymbol{Y}-\boldsymbol{\mu}_Y)^T \boldsymbol{C}_{YY}^{-1}(\boldsymbol{Y}-\boldsymbol{\mu}_Y)\right)\right].$$

It is proven that this combined probability density is of the Gauss type with mean value $\boldsymbol{\mu}_X + \boldsymbol{C}_{XY}\boldsymbol{C}_{YY}^{-1}(\boldsymbol{Y}-\boldsymbol{\mu}_Y)$ and variability $\boldsymbol{C}_{XX} - \boldsymbol{C}_{XY}\boldsymbol{C}_{YY}^{-1}\boldsymbol{C}_{YX}$. The proof can be found in the literature.

Characteristic Function of Random Vectors

The next concept, the definition of which in the space of n dimensions is a simple extension of the definition of one dimension, is the *characteristic function*, which for a random vector $\boldsymbol{X} = (\boldsymbol{x}_1, \boldsymbol{x}_2, \dots, \boldsymbol{x}_n)^T$ is defined as

$$\Phi_X(\boldsymbol{\omega}) = E\{e^{j\boldsymbol{\omega}^T \boldsymbol{X}}\}, \tag{A.169}$$

where $\boldsymbol{\omega} = [\omega_1, \omega_2, \dots, \omega_n]^T$ is a vector of parameters with real values. The above relation for the continuous case is formulated as

$$\Phi_X(\boldsymbol{\omega}) = \int_{-\infty}^{\infty} f_X(\boldsymbol{x}) e^{j\boldsymbol{\omega}^T \boldsymbol{x}} d\boldsymbol{x}, \tag{A.170}$$

whereas for the discrete case it takes the form

$$\Phi_X(\boldsymbol{\omega}) = \sum_{i_1}\sum_{i_2}\cdots\sum_{i_n} \exp(j(\omega_1 x_{i1} + \omega_2 x_{i2} + \cdots + \omega_n x_{in}))P\{\boldsymbol{x}_1 = x_1, \boldsymbol{x}_2 = x_2, \cdots, \boldsymbol{x}_n = x_n\}$$

./As an example of a characteristic function for the case of n dimensions, let us consider this function for the *combined Gaussian probability density* we defined previously and which, according to what we mentioned above, will have the form

$$\Phi_X(\boldsymbol{\omega}) = \int_{-\infty}^{\infty} f_X(\boldsymbol{x})e^{j\boldsymbol{\omega}^T\boldsymbol{x}}d\boldsymbol{x} = \int_{-\infty}^{\infty} \frac{d\boldsymbol{x}}{(2\pi)^{n/2}\sqrt{\det(C_{XX})}} \exp\left(j\boldsymbol{\omega}^T\boldsymbol{X} - \frac{1}{2}(\boldsymbol{X} - \boldsymbol{\mu}_X)^T C_{XX}^{-1}(\boldsymbol{X} - \boldsymbol{\mu}_X)\right),$$

or after the appropriate mathematical operations,

$$\Phi_X(\boldsymbol{\omega}) = \exp\left(j\boldsymbol{\omega}^T\boldsymbol{\mu}_X - \frac{1}{2}\boldsymbol{\omega}^T\boldsymbol{C}_{XX}\boldsymbol{\omega}\right). \quad \text{(A.171)}$$

The above equation allows, as in simpler cases, the calculation of the moments of the random vector $\boldsymbol{X}$ through the expression

$$E\{\boldsymbol{x}_1k_1\boldsymbol{x}_2k_2\cdots\boldsymbol{x}_nk_n\} = j^{k_1+k_2+\cdots+k_n} \left.\frac{\partial^{k_1+k_2+\cdots+k_n}\Phi_X(\omega_1,\omega_2,\ldots,\omega_n)}{\partial\omega_1^{k_1}\partial\omega_2^{k_2}\cdots\partial\omega_n^{k_n}}\right|_{\omega_1=\omega_2=\cdots=\omega_n=0}, \quad \text{(A.172)}$$

with the function $\Phi_X(\boldsymbol{\omega})$ being formulated in the form

$$\Phi_X(\boldsymbol{\omega}) = \sum_{k_1=0}^{\infty}\sum_{k_2=0}^{\infty}\cdots\sum_{k_n=0}^{\infty} E\{\boldsymbol{x}_1k_1\boldsymbol{x}_2k_2\cdots\boldsymbol{x}_nk_n\}\frac{(j\omega_1)^{k_1}}{k_1!}\frac{(j\omega_2)^{k_2}}{k_2!}\cdots\frac{(j\omega_n)^{k_n}}{k_n!}, \quad \text{(A.173)}$$

where once again we expanded the function *by Taylor* and used the *linearity* of the mathematical expectation operator. In the case where the variables are independent of each other, then we can, once again, *formulate the combined PDF as the product of the homonymous functions of the individual variables*, thus resulting in simpler expressions.

Considering the random variables $\boldsymbol{y}_i = g_i(\boldsymbol{X})$ $(i = 1, 2, \ldots, n)$, which collectively define the random vector $\boldsymbol{Y}$, where $g_i(\boldsymbol{X})$ $(i = 1, 2, \ldots, n)$ are suitably defined functions, the PDF $f_y(\boldsymbol{Y}) = f_y(y_1, y_2, \ldots, y_n)$ and $f_x(\boldsymbol{X}) = f_x(x_1, x_2, \ldots, x_n)$ are related to each other through the expression

$$f_y(y_1, y_2, \ldots, y_n) = \frac{f_x(x_1, x_2, \ldots, x_n)}{|J(x_1, x_2, \ldots, x_n)|}, \quad \text{(A.174)}$$

where $J(x_1, x_2, \ldots, x_n)$ is the *Jacobian* of the transformation $\boldsymbol{y}_i = g_i(\boldsymbol{X})$ $(i = 1, 2, \ldots, n)$.

Finally, the property of *statistical independence* can easily be generalized for the case of *vector variables*. Considering a set of such variables $\boldsymbol{X}, \boldsymbol{Y}, \boldsymbol{Z}, \ldots$, which are generally defined in spaces of different dimensions, we can describe the property of *statistical independence* with the help of equations

$$f_{X,Y,Z,\ldots}(\boldsymbol{X}, \boldsymbol{Y}, \boldsymbol{Z}, \ldots) = f_X(\boldsymbol{X})f_Y(\boldsymbol{Y})f_Z(\boldsymbol{Z})\ldots$$

and

$$E\{g_X(\boldsymbol{X}), g_Y(\boldsymbol{Y}), g_Z(\boldsymbol{Z})\ldots\} = E\{g_X(\boldsymbol{X})\}E\{g_Y(\boldsymbol{Y})\}E\{g_Z(\boldsymbol{Z})\}\ldots,$$

where $g_X(\boldsymbol{X}), g_Y(\boldsymbol{Y}), g_Z(\boldsymbol{Z})\ldots$ are arbitrary functions of the vector variables $\boldsymbol{X}, \boldsymbol{Y}, \boldsymbol{Z}, \ldots$, which satisfy the above relations.

Random Sequences

A *random sequence* can be considered a *vector of random variables with infinite dimensions*. Specifically, considering a probability space $(\Omega, \mathcal{F}, P)$, an element $\xi \in \Omega$ and a mapping $X[n, \xi]$ of the space Ω to the space of complex discrete sequences, the elements of which are indicated with the help of a set of indices $\mathbb{I}$, which is usually the set $(-\infty, \infty)$, then if for every constant value $n \in \mathbb{I}$, the quantity $X[n, \xi]$ constitutes a random variable, this quantity is characterized as a *random sequence*.

For every constant value of ξ, the quantity $X[n, \xi]$ constitutes a *usual discrete sequence of numbers*, that is, a function of the index n. If the value of n is constant, whereas the value of ζ varies, then the result is a *random variable*. If the random variables that constitute a random sequence are *independent* of each other at every time $n_1, n_2, \ldots, n_N$ for every $N > 0$, then the random sequence is characterized as *independent*. The treatment of random sequences requires the use of the *fourth axiom of probability theory*, which is nothing but the *additive property of probability* for the case of the union of an *infinite set of mutually exclusive events*.

Considering a random sequence $X[n]$, it is statistically determined by the CDF

$$F_X(x_n, x_{n+1}, x_{n+2}, \ldots, x_{n+N-1}; n, n{+}1, n{+}2, \ldots, n{+}N{-}1) = P\{X[n] \le x_n, X[n{+}1] \le x_{n+1}, \cdots, X[n{+}N{-}1] \le x_{n+N-1}\}$$

for all integers $N \geq 1$ and for all times $n, n+1, n+2, \ldots, n+N-1$, which indicate the positions of the N random variables. For each value of N, we should know the joint cumulative distribution function for all infinite the set of values $-\infty < n < \infty$ and consequently, there are *infinitely many* such functions. Also, we can define the above function $F(X_{n_1}, X_{n_2}, \ldots, X_{n_N})$ for a *finite number N* of nonconsecutive time instances $n_1, n_2, \ldots, n_N$, where for simplicity we have omitted the arguments of the function after the question mark symbol. Using the above CDF, we can define the *PDF* as

$$\begin{aligned} &f_X(x_n, x_{n+1}, x_{n+2}, \ldots, x_{n+N-1}; n, n+1, n+2, \ldots, n+N-1) \\ &\quad = \frac{F_X(x_n, x_{n+1}, x_{n+2}, \ldots, x_{n+N-1}; n, n{+}1, n{+}2, \ldots, n{+}N{-}1)}{\partial x_n \partial x_{n+1} \partial x_{n+2} \cdots \partial x_{n+N-1}}. \end{aligned}$$

The *mean value* of a random sequence $X[n]$ for the continuous and discrete case respectively is defined as

$$\mu_X[n] = E\{X[n]\} = \int_{-\infty}^{\infty} x f_X(x; n) dx = \int_{-\infty}^{\infty} x_n f_X(x_n) dx_n \quad \text{and} \quad \mu_X[n] = E\{X[n]\} = \sum_{k=-\infty}^{\infty} x_k P[x[k] = x_k],$$

whereas the *autocorrelation and auto-covariance functions* are given by the equations

$$R_{XX}[k, \ell] = E\{X[k]X^*[\ell]\} = \int_{-\infty}^{\infty} x_k x_\ell^* f_X(x_k, x_\ell; k, \ell) dx_k dx_\ell \quad \text{and}$$

$$C_{XX}[k, \ell] = E\{(X[k] - \mu_X[k])(X[\ell \mu_X[\ell])^*\} = R_{XX}[k, \ell] - \mu_X[k]\mu_X^*[\ell].$$

If the above combined PDF is *of Gaussian type*, then the random sequence is characterized as a *Gaussian random sequence* and is one of the most widely used sequences of this type in the field of engineering science, and more specifically, in the field of control and communications.

In many cases it is particularly useful and desirable to characterize a random sequence $X[n]$ based solely on the value of its mean and covariance. Specifically, considering a random sequence $X[n]$ $(-\infty < n < \infty)$, this is called *wide sense stationary (WSS)*, if (a) the mean value of $x[n]$ is constant and equal to $\mu_X[n] = \mu_X[0]$ for all integers n $(-\infty < n < \infty)$ and (b) for each pair of integers k and ℓ $(-\infty < k, \ell < \infty)$ and for each integer n $(-\infty < n < \infty)$, the covariance function is independent of the shift n, or in mathematical notation, $C_{XX}[k, \ell] = C_{XX}[k + n, \ell + n]$. It turns out that all stationary random sequences are WSS sequences without the converse being true.

More information about random sequences that find wide application in the field of engineering science can be found in the very rich related literature.

Bernoulli Random Sequences One of the most well-known examples of random sequences with wide application in many branches of statistical science, are the *infinite-length Bernoulli sequences*, which are constructed by repeatedly performing an experiment for which there are *only two possibilities* and recording the result that arises for each of these experiments. For simplicity, let us use as an example the very well-known to all of us coin-toss experiment, which is carried out by tossing a coin into the air and recording the outcome that emerged, which can be either heads (H) or tails (T). Taking into account that the sample space for a single toss is $\Omega = \{H, T\}$ (heads or tails), the sample space for two tosses is

$\Omega^2 = \Omega \times \Omega = HH, HT, TH, TT$, where the notation AB is a more concise writing of the ordered pair (A, B), whereas the sample space for an experiment that includes infinite tosses would be analogously

$$\Omega_\infty = \Omega \times \Omega \times \Omega \times \cdots \text{ infinite terms } = \times_{n=1}^{\infty} \Omega_n = \{(\alpha_i)_{i=1}^{\infty} \mid \alpha_i \in \{H, T\}\}$$

(where Ω_n is the *sample space* corresponding to the nth coin toss with $\Omega_1 = \Omega_2 = \cdots = \Omega$), which contains *infinite-length strings* containing the letters H and T. As for the probability measure for a single toss, it is defined as $P(H) = p$ and $P(T) = q = 1 - p$. To describe the situation, let us define the random variable $\boldsymbol{w}$ as $\boldsymbol{H} = 1$ and $\boldsymbol{w}(T) = 0$. In this case, the Bernoulli sequence is defined *as a sequence of independent and identically distributed Bernoulli experiments* $W[n, \xi] = W(\xi_n) = W[n]$ $(n \geq 1)$, that is, a sequence of values that are equal to 0 or 1 and such that for each value of n, the probability of $\boldsymbol{w}[n] = 1$ being the same is given. Of course, in order to fully define the problem, we must also define the other two elements of the triplet $(\Omega_\infty, F_\infty, P_\infty)$ and more specifically, the *event space* F_∞ and the *probability measure* P_∞. In the event space F_∞ we must add the appropriate *subsets* of the space Ω_∞ along with the appropriate generalized intersections and unions, whereas owing to the statistical independence that characterizes the problem, the probability measure P_∞ must be defined as a product of the form $P_\infty(A_\infty) = \prod_{n=1}^{\infty} P(A_n)$, where $A_\infty = \bigcap_{i=n}^{\infty} A_n$ (with A_n being an event K or Γ that appears in the nth attempt). Having defined the probability measure $P_\infty(A_\infty)$ we can extend it to the entire event space, using the *fourth axiom of probability*. The Bernoulli sequence defined in the above way is one of the most well-known discrete sequences in the field of statistics.

A.15 Problems

1. If $z_1 = 1 - i$, $z_2 = -2 + 4i$ and $z_3 = \sqrt{3} - 2i$ calculate the values of the following expressions: (a) $z_1^2 + 2z_1 - 3$, (b) $\mid 2z_2 - 3z_1 \mid^2$, (c) $(z_3 - z_3^*)^5$, (d) $\mid z_1 z_2^* + z_2 z_1^* \mid$, (e) $\mid (z_1 + z_2 + 1)/(z_1 - z_2 + i) \mid$, (f) $(1/2)[(z_3/z_3^*) + (z_3^*/z_3)]$, (g) $\mid z_1^2 + (z_2^*)^2 \mid^2 + \mid (z_3^*)^2 - z_2^2 \mid$, (h) $Re(2z_1^3 + 3z_2^2 - 5z_3^2)$, and (i) $Im(z_1 z_2/z_3)$.
2. Perform the following operations analytically and graphically (i.e., representing the complex numbers on the complex plane): (a) $(2+3i)+(4-5i)$, (b) $3(1+2i) - 2(2-3i)$, (c) $3(1+i) + 2(4-3i) - (2+5i)$, (d) $(7+i) - (4-2i)$, and (e) $(1/2)(4-3i) + (3/2)(5+2i)$.
3. Express the following complex numbers in trigonometric and polar form: (a) $2 - 2i$, (b) $2\sqrt{2} + 2\sqrt{2}i$, (c) $-i$, (d) -4, (e) $-2\sqrt{3} - 2i$, (f) $\sqrt{2}i$, and (g) $(\sqrt{(3)}/2) - (3/2)i$.
4. Show that $r_1 e^{i\varphi_1} + r_2 e^{i\varphi_2} = r_3 e^{i\varphi_3}$, where

$$r_3 = \sqrt{r_1^2 + r_2^2 + 2r_1 r_2 \cos(\varphi_1 - \varphi_2)}$$

$$\varphi_3 = \tan^{-1}\left(\frac{r_1 \sin\varphi_1 + r_2 \sin\varphi_2}{r_1 \cos\varphi_1 + r_2 \cos\varphi_2}\right).$$

5. Calculate analytically the following roots and determine graphically the points they represent: (a) $(2\sqrt{3} - 2i)^{1/2}$, (b) $(-4+4i)^{1/5}$, and (c) $(2 + 2\sqrt{3}i)^{1/3}$.
6. Calculate analytically the following roots and determine graphically the points that they represent: (a) $(-16i)^{1/4}$ and (b) $(64)^{1/6}$.
7. If $z_1 = 2 + 5i$ and $z_2 = 3 - i$ calculate the values of the expressions: (a) $z_1 \circ z_2$, (b) $z_1 \times z_2$, (c) $z_1 \circ z_1$, (d) $z_2 \times z_1$, (e) $\mid z_1 \circ z_2 \mid$, (f) $\mid z_1 \times z_2 \mid$, (g) $\mid z_1 \circ z_1 \mid$, and (h) $\mid z_2 \times z_1 \mid$
8. If $z_1 = r_1 e^{i\varphi_1}$ and $z_2 = r_2 e^{i\varphi_2}$ prove that $z_1 \circ z_2 = r_1 r_2 \cos(\varphi_2 - \varphi_1)$ and $z_1 \times z_2 = r_1 r_2 \sin(\varphi_2 - \varphi_1)$.
9. Prove that on the circumference $z = Re^{i\vartheta}$ it is $\mid e^{iz} \mid = e^{-R\sin\vartheta}$.
10. Find the square roots of the complex numbers $z_1 = 5 - 12i$ and $z_2 = 8 + 4\sqrt{5}i$.
11. Find the cubic roots of the complex number $z_1 = -11 - 2i$.
12. Show that: (a) $\mid e^z \mid = e^\alpha$ and $\mid e^{iz} \mid = e^{-\beta}$, where $z = \alpha + i\beta$, (b) $e^{(z_1/z_2)} = e^{z_1 - z_2}$, (c) $\sin(-z) = -\sin z$, $\cos(-z) = \cos z$, $\tan(-z) = -\tan z$, (d) $\sin(2z) = 2\sin z \cos z$, (e) $\cos(2z) = \cos^2 z - \sin^2 z$, and (f) $\sin(z_1 + z_2) = \cos z_1 \sin z_2 + \sin z_1 \cos z_2$.
13. Calculate the derivative of the functions

$$f_1(z) = 3z^2 + 4iz - 5 + i \qquad f_2(z) = \frac{2z - i}{z + 2i} \qquad f_3(z) = \frac{3}{z^2}.$$

14. Prove that

$$\frac{d}{dz}\left\{\sqrt[3]{(1+z^2)^2}\right\} = 3z\sqrt{1+z^2} \qquad \text{and} \qquad \frac{d}{dz}\left\{\sqrt[3]{z+2\sqrt{z}}\right\} = \frac{\sqrt{z}+1}{3\sqrt{z}\sqrt[3]{(z+2\sqrt{z})^2}}.$$

15. Calculate the complex integral $S = \int_C z^2 dz$ along the paths C_1, C_2, and C_3 depicted in Fig. A.6 (refer to page 1478). In this case, the result will be the same for all three cases: indeed, the function $f(z) = z^2$ arises from the derivative of the antiderivative function $F(z) = z^3/3$, that is, the value of the integral equals the difference of the values of the function $F(z)$ calculated at points O and B.
16. Calculate the integral

$$I = \int_{(0,1)}^{(2,5)} (3x+y)dx + (2y-x)dy$$

along: (a) the curve $y = x^2 + 1$, (b) the straight-line segment that connects the points $(0, 1)$ and $(2, 5)$, (c) of the line segments from $(0, 1)$ to $(0, 5)$, and then from $(0, 5)$ to $(2, 5)$, and (d) of the line segments from $(0, 5)$ to $(2, 1)$, and then from $(2, 1)$ to $(2, 5)$.
17. Find the Maclaurin expansion of the functions e^z, $\sin(z)$, $\cos(z)$ and $\tan(z)$.
18. Find the Laurent expansion of the following function for the region $|\, z-1 \,| < 1$.

$$f(z) = \frac{1}{[z-1)(z-2)}.$$

19. Determine the type of singular points of the functions

$$\frac{1}{z^2+\alpha^2}, \quad \frac{1}{(z^2+\alpha^2)^2}, \quad \frac{z^2}{(z^2+\alpha)^2}, \quad \frac{ze^{iz}}{z^2+\alpha^2}, \quad \frac{ze^{iz}}{z^2-\alpha^2}, \quad \frac{e^{iz}}{z^2-\alpha^2}$$

($\alpha > 0$) and calculate the residues of the functions at these points.
20. Calculate the contour integral

$$I = \oint_C \frac{e^z}{\cosh(z)} dz,$$

where C is the circumference of a circle with equation $|\, z \,| = 5$.
21. Calculate the contour integral

$$I = \oint_C \exp\left(-\frac{1}{z}\right) \sin\left(\frac{1}{z}\right) dz,$$

where C is the circumference of a circle with equation $|\, z \,| = 1$.
22. Calculate the contour integral

$$I = \oint_C \frac{2z^2+5}{(z+2)^3(z^2+4)z^2} dz,$$

where C is (α) the circumference with equation $|\, z-2i \,| = 6$ and (β) the perimeter of a square with vertices at the points $1+i, 2+i, 2+2i, 1+2i$.
23. Calculate the contour integral

$$I = \oint_C \frac{2+3\sin(\pi z)}{z(z-1)^2} dz,$$

where C is the contour of a square with vertices at the points $3+3i$, $3-3i$, $-3+3i$ and $-3-3i$.
24. Calculate the contour integral

$$I = \frac{1}{2\pi i} \oint_C \frac{e^{zt}}{z(z^2+1)} dz \quad (t > 0),$$

where C is the contour of a square with vertices at the points $1+i,\ -1+i,\ -1-i$, and $1-i$.

25. Prove that the Chebyshev polynomials satisfy the recursive equation

$$T_{N+1}(x) = 2xT_N(x) - T_{N-1}(x).$$

26. Prove that the derivative of the Chebyshev polynomials satisfies the equation

$$\frac{dT_N(x)}{dx} = \frac{NT_{N-1}(x) - NxT_N(x)}{1-x^2}.$$

27. Prove that $T_N(-x) = (-1)^N T_N(x)$, $T_{2N}(x) = (-1)^N$, and $T_{2N+1}(0) = 0$.
28. Using the orthogonality condition of trigonometric functions, prove that

$$\int_{-1}^{+1} \frac{T_M(x)T_N(x)}{\sqrt{1-x^2}} = \begin{cases} 0 & M \neq N \\ \pi/2 & M = N \neq 0 \\ \pi & M = M = 0 \end{cases}$$

29. Proceeding with the change of variable $x = \cos\vartheta$, prove that

$$T_N(x) = \frac{1}{2}\left[(x + j\sqrt{1-x^2})^N + (x - j\sqrt{1-x^2})^N\right].$$

30. By differentiating the defining equations of the polynomials $T_{N+1}(x)$ and $T_{N-1}(x)$, prove that

$$\int T_N(x)dx = \frac{1}{2}\left[\frac{T_{N+1}(x)}{N+1} - \frac{T_{N-1}(x)}{N-1}\right] + C.$$

31. Using the Rodrigues formula, determine the expressions of the polynomials $T_0(x)$, $T_1(x)$, and $T_2(x)$.
32. Using the recursive expression, construct the polynomials $T_2(x)$, $T_3(x)$, and $T_4(x)$, starting from the expressions $T_0(x) = 1$ and $T_1(x) = x$.
33. Prove that the complete elliptic integrals $\boldsymbol{K}$, $\boldsymbol{K}'$, $\boldsymbol{E}$, and $\boldsymbol{E}'$ satisfy the relation

$$\boldsymbol{K}\boldsymbol{E}' + \boldsymbol{K}'\boldsymbol{E} - \boldsymbol{K}\boldsymbol{K}' = \frac{\pi}{2}.$$

34. Prove that the derivatives of the integrals $\boldsymbol{E}$ and $\boldsymbol{K}$ with respect to the variable k are

$$\frac{d\boldsymbol{E}}{dk} = \frac{\boldsymbol{E} - \boldsymbol{K}}{k} \quad \text{and} \quad \frac{d\boldsymbol{K}}{dk} = \frac{\boldsymbol{E} - (k')^2\boldsymbol{K}}{k(k')^2}.$$

35. Prove that

$$\frac{d\boldsymbol{F}(\varphi,k)}{d\varphi} = \frac{d\boldsymbol{E}(\varphi,k)}{d\varphi} = \frac{1}{\sqrt{1-k^2\sin^2\varphi}}.$$

36. Prove that $\boldsymbol{E}(0,k) = \boldsymbol{F}(0,k) = \boldsymbol{\Pi}(0,-n,k) = 0$ and $\boldsymbol{K}(0) = \boldsymbol{K}'(1) = \boldsymbol{E}(0) = \boldsymbol{E}'(1) = 0$.
37. Prove that

$$\int_0^{\varphi} \frac{\cos^2\vartheta}{\sqrt{1-k^2\sin^2\vartheta}}\,d\vartheta = \boldsymbol{F} - \boldsymbol{D}, \quad \int_0^{\varphi} \frac{\tan^2\vartheta}{\sqrt{1-k^2\sin^2\vartheta}}\,d\vartheta = \frac{\tan\varphi\sqrt{1-k^2\sin^2\vartheta} - \boldsymbol{E}}{(k')^2}$$

$$\int_0^{\varphi} \frac{d\vartheta}{\cos^2\vartheta\sqrt{1-k^2\sin^2\vartheta}} = \frac{\tan\varphi\sqrt{1-k^2\sin^2\vartheta} + k^2(\boldsymbol{D} - \boldsymbol{F})}{(k')^2}.$$

38. Prove that

$$\int_0^{\varphi} \frac{\sin^2\vartheta}{1-k^2\sin^2\vartheta}d\vartheta = \frac{\boldsymbol{F}-\boldsymbol{D}}{(k')^2} - \frac{\sin\varphi\cos\varphi}{(k')^2\sqrt{1-k^2\sin^2\vartheta}}, \quad \int_0^{\varphi} \frac{\cos^2\vartheta}{1-k^2\sin^2\vartheta}d\vartheta = \boldsymbol{D} + \frac{\sin\varphi\cos\varphi}{\sqrt{1-k^2\sin^2\vartheta}}$$

$$\int_0^{\varphi} \tan^2\vartheta\sqrt{1-k^2\sin^2}d\vartheta = \tan\varphi\sqrt{1-k^2\sin^2\vartheta} + \boldsymbol{F} - \boldsymbol{E}.$$

39. Prove that $\boldsymbol{E}(\varphi, 1) = \sin\varphi$ and $\boldsymbol{F}(\varphi, 1) = \ln(\tan\varphi + \sec\varphi)$.

40. Prove that

$$\boldsymbol{\Pi}(\varphi, \alpha^2, 1) = \begin{cases} \dfrac{\ln(\tan\varphi + \sec\varphi) - \alpha\ln\sqrt{\dfrac{1+\alpha\sin\varphi}{1-\alpha\sin\varphi}}}{1-\alpha^2} & \text{for} \quad \alpha^2 > 0, \alpha^2 \neq 1 \\ \dfrac{\ln(\tan\varphi + \sec\varphi) + \mid\alpha\mid\tan^{-1}(\mid\alpha\mid\sin\varphi)}{1-\alpha^2} & \text{for} \quad \alpha^2 < 0. \end{cases}$$

41. Prove that

$$\frac{\partial \boldsymbol{F}(\varphi,k)}{\partial k} = \frac{k}{(k')^2}\left[\boldsymbol{F}(\varphi,k) - \boldsymbol{D}(\varphi,k) - \frac{\sin\varphi\cos\varphi}{\sqrt{1-k^2\sin^2\vartheta}}\right], \quad \frac{\partial \boldsymbol{E}(\varphi,k)}{\partial k} = -k\boldsymbol{D}(\varphi,k)$$

$$\frac{\partial \boldsymbol{D}(\varphi,k)}{\partial k} = \frac{1}{k(k')^2}\left[\boldsymbol{F}(\varphi,k) - \boldsymbol{D}(\varphi,k) - \frac{\sin\varphi\cos\varphi}{\sqrt{1-k^2\sin^2\vartheta}} - \frac{\boldsymbol{D}(\varphi,k)}{k}\right].$$

42. Prove that

$$\int \boldsymbol{F}(\varphi,k)\boldsymbol{K}(k)dk = \boldsymbol{E}(\varphi,k) - (k')\boldsymbol{F}(\varphi,k) - (1-\sqrt{1-k^2\sin^{\vartheta}})\cot\varphi, \quad \int \boldsymbol{D}(\varphi,k)\boldsymbol{K}(k)dk = -\boldsymbol{E}(\varphi,k)$$

$$\int \boldsymbol{E}(\varphi,k)\boldsymbol{K}(k)dk = \frac{(1+k^2)\boldsymbol{E}(\varphi,k)}{3} - (k')^2\boldsymbol{F}(\varphi,k) - (1-\sqrt{1-k^2\sin^{\vartheta}})\cot\varphi.$$

43. Prove that

$$\frac{d\mathrm{sn}(u,k)}{du} = \mathrm{cn}(u,k)\mathrm{dn}(u,k)$$

$$\frac{d\mathrm{cn}(u,k)}{du} = -\mathrm{sn}(u,k)\mathrm{dn}(u,k)$$

$$\frac{d\mathrm{dn}(u,k)}{du} = -k^2\mathrm{sn}(u,k)\mathrm{cn}(u,k).$$

44. Prove that

$$\mathrm{sn}(u+\upsilon,k) = \frac{\mathrm{sn}(u,k)\mathrm{cn}(u,k)\mathrm{dn}(u,k) + \mathrm{sn}(\upsilon,k)\mathrm{cn}(\upsilon,k)\mathrm{dn}(\upsilon,k)}{1-k^2\mathrm{sn}^2(u,k)\mathrm{sn}^2(\upsilon,k)}$$

$$\mathrm{cn}(u+\upsilon,k) = \frac{\mathrm{cn}(u,k)\mathrm{cn}(u,k) + \mathrm{sn}(\upsilon,k)\mathrm{sn}(\upsilon,k)\mathrm{dn}(\upsilon,k)\mathrm{dn}(u,k)}{1-k^2\mathrm{sn}^2(u,k)\mathrm{sn}^2(\upsilon,k)}$$

$$\mathrm{dn}(u+\upsilon,k) = \frac{\mathrm{dn}(u,k)\mathrm{dn}(u,k) - k^2\mathrm{sn}(\upsilon,k)\mathrm{sn}(\upsilon,k)\mathrm{dn}(\upsilon,k)\mathrm{cn}(u,k)}{1-k^2\mathrm{sn}^2(u,k)\mathrm{sn}^2(\upsilon,k)}.$$

45. Using the methodology of solving the one-dimensional heat propagation equation in an infinite domain, solve the two-dimensional wave equation in the infinite two-dimensional domain $-\infty < x, y < \infty$.

46. Using the defining equation of the Laplace operator in spherical coordinates

$$\nabla^2 = \frac{1}{\rho}\frac{\partial^2}{\partial \rho^2}\rho + \frac{1}{\rho^2 \sin\vartheta}\frac{\partial}{\partial\vartheta}\sin\vartheta\frac{\partial}{\partial\vartheta} + \frac{1}{\rho^2\sin^2\vartheta}\frac{\partial^2}{\partial\varphi^2}$$

and applying the method of separation of variables, prove that the Laplace equation $\nabla^2 V = 0$ with a solution of the form $V = V(\rho, \vartheta, \varphi) = R(\rho)\,\Theta(\vartheta)\,\Phi(\varphi)$ is transformed into the system of differential equations

$$\rho^2\frac{d^2R(\rho)}{d\rho^2} + 2\rho\frac{dR(\rho)}{d\rho} - \lambda R(\rho) = 0$$

$$\frac{d^2\Phi(\varphi)}{d\varphi^2} + \mu\Phi(\varphi) = 0$$

$$\frac{d^2\Theta(\vartheta)}{d\vartheta^2} + \frac{\cos\vartheta}{\sin\vartheta}\frac{d\Theta(\vartheta)}{d\vartheta} + \left(\lambda - \frac{\mu}{\sin^2\vartheta}\right)\Theta(\vartheta) = 0$$

47. Use the method of Fourier transform to solve the differential equation

$$\frac{\partial^2 u(x,t)}{\partial x\partial t} = \frac{\partial^2 u(x,t)}{\partial x^2}$$

with the initial condition $u(x,0) = \sqrt{\pi/2}e^{-|x|}$.

48. Use the method of Fourier transform to solve the differential equation

$$\frac{\partial^2 u(x,t)}{\partial t^2} = -c^2\frac{\partial^4 u(x,t)}{\partial x^4}$$

with the initial condition $u(x,0) = f(x)$.

49. Use the method of Fourier transform to solve the differential equation

$$t\frac{\partial u(x,t)}{\partial x} + \frac{\partial u(x,t)}{\partial t} = 0$$

with the initial condition $u(x,0) = f(x)$.

50. Use the method of Fourier transform to prove that the solution of the differential equation

$$\frac{\partial u(x,t)}{\partial t} = k\frac{\partial^2 u(x,t)}{\partial x^2} + c\frac{\partial u(x,t)}{\partial x}, \qquad -\infty < x < +\infty$$

with the initial condition $u(x,0) = f(x)$ has the general form

$$u(x,t) = \int_{-\infty}^{+\infty}\frac{f(s)}{\sqrt{4\pi kt}}\exp\left(-\frac{(x+ct-s)^2}{4kt}\right)ds,$$

which, for the initial condition $f(x) = \delta(x)$, takes the simplest form

$$u(x,t) = \frac{1}{\sqrt{4\pi kt}}\exp\left(-\frac{(x+ct)^2}{4kt}\right).$$

Bibliography

1. Aasnaes H.B & Kailath T, *An Innovations Approach to Least Squares Estimation, Part VII: Some Applications of Vector Autoregressive - Moving Average Models*, IEEE Transactions on Automatic Control, Vol AC-18, No. 6, December 1973, pp. 601–607.
2. Abramowitz M & Stegun I.A, *Handbook of Mathematical Functions*, Dover, ISBN 486-61272-4, 1972.
3. Aiello G.L & Angelo P.M, *Transitional Legendre - Thomson Filters*, IEEE Transactions on Circuits and Systems, Vol. CAS-21, No.1, pp. 159–162, 1974.
4. Alexander C & Sadiku M, *Fundamentals of Electric Circuits*, McGraw Hill, ISBN 978-0072493504, 2003.
5. Antoniou A, *Digital Signal Processing - Signals, Systems and Filters*, ISBN 978-0-07-145425-4, 2005.
6. Appel U, *Bounds on Second-Order Digital Filter Limit Cycles*, IEEE Transactions on Circuits and Systems, Vol. CA-22. pp. 630–632, 1975.
7. Arfken G, *Mathematical Methods for Physicists*, Academic Press, Inc, ISBN 0-12-059820-5, 1985.
8. Avenhaus, E, *On the Design of Digital Filters with Coefficients of Limited Word Length*, IEEE Transactions on Audio and Electroacoustics, Vol Au-20, No.3 August 1972.
9. Baraniuk R (Ed), *Signals and System*, Online Document available at http://cnx.org/content/col10064/1.12/, Connexions, Rice University, Houston, Texas, 2010.
10. Barnes C.W, Tran B.N & Lenug S.H, *On the Statistics of the Fixed Point Round-off Error*, IEEE Transactions on Acoustics, Speech and Signal Processing, Vol. ASSP-33, No.3, June 1985, pp. 595–606.
11. Bennett W.R, *Spectra of Quantized Signals*, Bell System Technical Journal, New York, Vol 27, pp. 446–472, July 1948.
12. Bode H,W, *Network Analysis and Feedback Amplifier Design*, D. Van Nostrand Company, Inc, Princeton, New Jersey, 1945.
13. Boas M.L, *Mathematical Methods in the Physical Sciences*, John Wiley & Sons, Inc, ISBN 0-471-19826-9, 2006.
14. Boulet B, *Fundamentals of Signals and Systems*, Charles River Media, ISBN-13 978-1-58450-381-1, 2006.
15. Butterweck H.J, *On the Quantization Noise Contributions in Digital Filters which are Uncorrelated with the Output Signal*, IEEE Transactions of Circuits and Systems, vol. CAS-26 , No. 11, pp. 901–910, November 1979.
16. Butterweck H.J, Lucassen F.H.R & Verkroost G, *Subharmonics and Other Quantization Effects in Periodically Excited Recursive Digital Filters*, IEEE Transactions of Circuits and Systems, vol. CAS-33, pp. 958–964 , 1986.
17. Butterweck, H. J., Ritzerfeld, J. H. F. & Werter, M. J, *Finite Wordlength Effects in Digital Filters : A Review*, EUT report. E, Fac. of Electrical Engineering; Vol. 88-E-205, Eindhoven University of Technology, 1988.
18. Büttner M, *A Novel Approach to Eliminate Limit Cycles in Digital Filters with a Minimum Increase in the Quantization Noise*, Proceedings of the IEEE International Symposium on Circuits and Systems, Munich, Germany (New York, USA: IEEE 1976), pp. 291–294, 1976.
19. Büttner M, *Elimination of LKimit Cycles in Digital Filters with Very Low Increase in the Quantization Noise*, IEEE Transactions on Circuits and Systems, Vol. CAS-24, pp. 300–304, 1977.
20. Chaparro L, *Signals and Systems Using MATLAB*, Elsevier Inc, ISBN 978-0-12-374716, 2011.
21. Charalambous C, *Acceleration of the Least p_{th} Algorithm for Minimax Optimization with Engineering Applications*, Mathematical Programming, Vol. 17, pp. 270–297, 1979.
22. Chen C-T, *Linear System Theory and Design*, Holt, Rinehart and Winston Publishing Company Inc., ISBN 0-03-060289-0, 1984.
23. Chen W-K (Ed), *The Circuits and Filters Handbook*, 2nd edition, ISBN 978-0849309120, 2002, CRC-Press, Chapter 82, pp. 2495–2515.
24. Chen W-k (Ed), *Passive, Active, and Digital Filters*, CRC Press, ISBN 978-1-4200-5885-7, 2009.
25. Cheney E.W, *Introduction to Approximation Theory*, AMS Chelsea Publishing, ISBN 0-8218-1374-9, 1966.
26. Chow T.L, *Mathematical Methods for Physicists - A Concise Introduction*, Cambridge University Press, 1st Edition, ISBN 978-0521-655446, 2000.
27. Claasen T.A.C.M, Mecklenbrauker W.F.G & Peek J.B.H, *Necessary and Sufficient Conditions for the Absence of Overflow Phenomena in a Second-Order Recursive Digital Filter*, IEEE Transactions on Acoustics, Speech, and Signal Processing, Vol. ASSP-23, pp. 509–515, 1975.
28. Claasen T.A.C.M, Mecklenbrauker W.F.G & Peek J.B.H, *Frequency Domain Criteria for the Absence of Zero-Input Limit Cycles in Nonlinear Discrete-Time Systems, with Applications to Digital Filters*, IEEE Transactions on Circuits and Systems, Vol. CAS-22, pp. 232–239, 1975.
29. Claasen T.A.C.M, Mecklenbräuker W.F.G& Peek J.B.H, *On the Stability of the Forced Response of Digital Filters with Overflow Nonlinearities*, IEEE Transactions on Circuits and Systems, vol. 22, pp. 692–696, August 1975.
30. Claasen T.A.C.M, Mecklenbräuker W.F.G& Peek J.B.H, *Quantization Noise Analysis for Fixed Points Digital Filters Using Magnitude Truncation for Quantization*, IEEE Transactions on Circuits and Systems, Vol. CAS-22, pp. 887–895, November 1975.
31. Claasen T.A & Mecklenbräuker W.F, *On the Transposition of Linear Time-Varying Discrete-Time Networks and Its Application to Multirate Digital Systems*, Philips J. Res., 23:78–102, 1978.
32. Constantinides A.G, *Spectral Transformations for Digital Filters*, Proceedings of IEE, Vol. 117, No. 8, pp. 110–115, August 1970.
33. Cooper G & McGillem C.D, *Probabilistic Methods of Signal and System Analysis*, 3rd Edition, ISBN 978-0195123548, 1998.

A. I. Margaris, *Continuous and Discrete-Time Signals and Systems*,
https://doi.org/10.1007/978-3-031-81209-5

34. Creese T.M, and Haralick R.M, *Differential Equations for Engineers*, McGraw Hill Book Company, ISBN 0-07-013510-X, 1978.
35. Crochiere R.E, *A New Statistical Approach to the Coefficient Word Length Problem for Digital Filters*, IEEE Transactions on Circuit and Systems, Vol CAS-22, No.3, March 1975.
36. Crochiere R.E & Rabiner L.R, *Multirate Digital Signal Processing*, Prentice Hall, Englewood Cliffs, New Jersey, 1983
37. Daniels R.W, *Approximation Methods for Electronic Filter Design*, ISBN 978-0070153080, McGraw-Hill, New York, 1974.
38. Davenport W.B Jr, *Probability and Random Processes*, New York, McGraw-Hill, ISBN 978-0070154407, 1970.
39. Davenport W. B. Jr & Root W.L, *An Introduction to the Theory of Random Signals and Noise*, John Wiley and Sons, ISBN 0-87942-235-1, 1987.
40. David E, Iovescu C, *A Block Floating Point Implementation for an N-Point FFT on the TMS320C55x DSP*, Application report, TMS320C5000 Software Applications, Texas Instruments, SPRA948, September 2003.
41. de Queiroz, A.C.M, *Chebyshev Filters*, Wiley Encyclopedia of Electrical and Electronics Engineering, 1999.
42. Deliyannis T, *High-Q Factor Circuit with Reduced Sensitivity*, Electronics Letters, Vol.4, pp. 567, 1968.
43. Deller J.R. & Hansen J.H.L, *Discrete Time Processing of Speech Signals*, MacMillan Publishing Company, ISBN 0-02-328301-7, 1993.
44. Dimopoulos H.G, *Analog Electronic Filters: Theory, Design and Synthesis*, 1st Edition, Springer, ISBN 978-9400721890, 2012.
45. Diniz P.S.R & Antoniou A, *More Economical State-Space Digital-Filter Structures Which are Free of Constant-Input Limit Cycles*, IEEE Transactions on Acoustics, Speech, and Signal Processing, ASSP-34, 807–815, 1986.
46. Diniz P.S.R, *Elimination of Constant-Input Limit Cycles in Passive Lattice Digital Filters*, IEEE Transactions on Circuits and Systems, Vol. 35, pp. 1188–1190, 1988.
47. Diniz P.S.R, da Silva E.A.B & Netto S.L, *Digital Signal Processing - System Analysis and Design*, Cambridge University Press, ISBN 978-0-521-88775-5, 2010.
48. Dishal M, *Gaussian-Response Filter Design*, Electrical Communication, Vol 36, No 1, pp. 3–26, 1959.
49. Dorf R.C & Bishop R.H, *Modern Control Systems*, 14th Edition, Prentice Hall, ISBN 978-1-292-42237-4, 2007.
50. Duffy D.G, *Green's Functions with Applications*, Chapman and Hall/CRC, ISBN 1-58488-110-0, 2001.
51. Durbin J, *Efficient Estimation of Parameters in Moving-Average Models*, Biometrika, Vol. 46, pp. 306–316, 1959.
52. Ebert P.M, Mazo J. E & Taylor M.G, *Overflow Oscillations in Digital Filters*, Bell System Technical Journal, 48(9), 2999–3020, 1969.
53. Feller W, *An Introduction to Probability Theory and its Applications*, Vol. 1, 3rd edition, ISBN 978-0-471-25708-0, Wiley, 1968.
54. Feller W, *An Introduction to Probability Theory and Its Applications*, Vol. 2, 2nd edition, ISBN 978-0-471-25709-7, Wiley, 1971.
55. Fettweis A, *Pseudopassivity, Sensitivity and Stability of Wave Digital Filters*, IEEE Transactions on Circuit Theory, vol. CT-19, pp. 668–673, 1972.
56. Fettweis A & Meerkötter K, *Suppression of Parasitic Oscillations in Wave Digital Filters*, Proceedings of the IEEE International Symposium on Circuits and Systems, San Francisco, California (New York, USA: IEEE 1974), pp. 682–686, 1974.
57. Fettweis A & Meerkötter K, *Suppression of Parasitic Oscillations in Wave Digital Filters*, IEEE Transactions on Circuits and Systems, Vol. CAS-22, pp. 239–246, 1975.
58. Freeny, S.L, *Special-Purpose Hardware for Digital Filtering*, Proceedings of the IEEE, 63, 633–48, 1975.
59. Frost P.A & Kailath T, *An Innovations Approach to Least Squares Estimation, Part III: Non Linear Estimation in White Gaussian Noise*, IEEE Transactions on Automatic Control, Vol AC-16, No. 3, June 1971, pp. 217–226.
60. Gevers M & Kailath T, *An Innovations Approach to Least Squares Estimation, Part VI: Discrete Time Innovations Representations & Recursive Estimation*, IEEE Transactions on Automatic Control, Vol AC-18, No. 6, December 1973, pp. 588–600.
61. Goertzel G, *An Algorithm for Evaluation of Finite Trigonometric Series*, The American Mathematical Monthly, Vol. 65, pp. 34–35, 1958.
62. Gollub G.H & van Loan C.F, *Matrix Computations*, The Johns Hopkins University Press, 3rd Edition, ISBN 0-8018-5413-X, 1996.
63. Grebennikov A, *RF and Microwave Transmitter Design*, Wiley, ISBN 978-0-470-92930-8, 2011.
64. Gray R.M & Davisson L.D, *An Introduction to Statistical Signal Processing*, 1st Edition, Cambridge University Press, 2004.
65. Green B.D & Turner L.E, *New Limit Cycle Bounds for Digital Filters*, IEEE Trans. Circuits Syst., vol. 35, pp. 365–374, April 1988.
66. Grypaios M & Masen S, *Functions of Complex Variable and Special Functions*, Aristotle University Press, Thessaloniki, 1984.
67. Halpern P.H, *Optimum Monotonic Low-Pass Filters*, IEEE Transactions on Circuit Theory, Vol. CT-16, No.2, pp.240–242, 1969.
68. Harris F, *On the Use of Windows for Harmonic Analysis with the Discrete Fourier Transform*, Proceedings of IEEE, Volume 66, pp.51–83, January 1978.
69. Haugen F, *Dynamic Systems: Modeling, Analysis and Design*, Tapir Academic Press, 1st Edition, ISBN-13 978-8251919265, 2004.
70. Hayes M.H, *Digital Signal Processing*, Schaum's Outline Series, McGraw Hill, ISBN 0-07-027389-8, 1999.
71. Haykin S, *Communication Systems*, 4th Ed, John Wiley and Sons Ltd, ISBN 0-471-17869-1, 2000.
72. Hayt W.H & Kemmerly J.E, *Engineering Circuit Analysis*, McGraw Hill, ISBN-13 978-0-07-352957-8, 2012.
73. Heinzel G, Rüdiger A & Schilling R, *Spectrum and Spectral Density Estimation by the Discrete Fourier Transform (DFT)*, Max-Planck-Institut für Gravitationsphysik, Document Id, 395068.0, 2002.
74. Hejn K & Pacut A, *Generalized Model of the Quantization Error - A Unified Approach*, IEEE Transactions on Instrumentation and Measurement, Vol. 45, No. 1, pp. 41–44, February 1996.
75. Helstrom C.W, *Probability and Stochastic Processes for Engineers*, 2nd Edition, ISBN 978-0023535710, MacMillan, 1991.
76. Herrmann O, Rabiner L.R & Chan D.S.K, *Practical Design Rules for Optimum Finite Impulse Response Lowpass Digital Filters*, The Bell Technical Journal, Vol. 52, pp. 769–799, 1973.
77. Hess, S, *A Deterministic Analysis of Limit Cycle Oscillations in Recursive Digital Filters Due to Quantization*, Ph.D dissertation, Naval Postgraduate School, Monterey, California, December 1970.
78. Himmelblau D.M, *Applied Nonlinear Programming*, ISBN 978-0070289215, New York, McGraw Hill, 1972.
79. Horn R.A & Johnson C.R, *Topics in Matrix Analysis*, Cambridge University Press, ISBN 978-0-511-84037-1, 1991.
80. Hsu H.P, *Signals and Systems*, Schaum's Outline Series, McGraw Hill, ISBN 0-07-030641-9, 1995.
81. Hwang, K, *Computer Arithmetic: Principles, Architecture and Design*, ISBN 978-0471034964, New York, Wiley, 1979.
82. Ifpachov E.C & Jerris B.W, *Digital Signal Processing - A Practical Approach*, 2nd Edition, Pearson Education, ISBN 0-201-59619-9, 2002.
83. Jackson L.B, Kaiser J.F & McDonald H.R, *An Approach to the Implementation of Digital Filters*, IEEE Trans. Audio Electroacoust., Vol. AU-16, No.3, pp. 413–421, September 1968.

84. Jackson L.B, *An Analysis of Roundoff Noise in Digital Filters*, Sc.D. dissertation, Stevens Inst. Tech., Hoboken, N. J., 1969.
85. Jackson L.B, *An Analysis of Limit Cycles Due to Multiplication Rounding in Recursive Digital (Sub-)filters*, Proceedings of the 7th Annual Allerton Conf. Circuit Syst. Theory, Monticello, Illinois, USA, ed. A.H. Hadad et al. (Urbana, Illinois, USA : Dept. of Electrical Eng., Univ. of Illinois 1969), pp. 69–78, 1969.
86. Jackson L.B, *On the Interaction of Roundoff Noise and Dynamic Range in Digital Filters*, The Bell System Technical Journal, Vol.49, pp-159–184, February 1970.
87. Jackson L.B, *Roundoff-Noise Analysis for Fixed-Point Digital Filters Realized in Cascade or Parallel Form*, IEEE Trans. Audio Electroacoust., vol. AU-18, pp. 107–121, June 1970.
88. Jackson L.B, *Digital Filters and Signal Processing*, Kluwer Academic Publishers, Boston, 3rd edition, 1996.
89. Johnson D.E & Johnson J.R, *Low Pass Filters Using Ultraspherical Polynomials*, IEEE Transactions on Circuit Theory, Vol CT-13, No.4, pp. 364–369, 1966.
90. Jury E.I, *Theory and Applications of the Z Transform Method*, New York, John Wiley and Sons, Inc, 1964.
91. Kailath T, *An Innovations Approach to Least Squares Estimation, Part I: Linear Filtering in Additive White Noise*, IEEE Transactions on Automatic Control, Vol AC-13, No. 6, pp. 646–655, December 1968.
92. Kailath T & Frost P.A, *An Innovations Approach to Least Squares Estimation, Part II: Linear Smoothing in Additive White Noise*, IEEE Transactions on Automatic Control, Vol AC-13, No. 6, pp. 655–660, December 1968.
93. Kailath T & Geesey R.A, *An Innovations Approach to Least Squares Estimation, Part IV: Recursive Estimation Given Lumped Covariance Functions*, IEEE Transactions on Automatic Control, Vol AC-16, No. 6, pp. 720–727, December 1971.
94. Kailath T & Geesey R.A, *An Innovations Approach to Least Squares Estimation, Part V: Innovations Representations and Recursive Estimation in Colored Noise*, IEEE Transactions on Automatic Control, Vol AC-18, No. 5, pp. 435–453, October 1973.
95. Kaneko T & Liu B, *Roundoff Error of Floating Point Digital Filters*, Proceedings of Sixth Annual Allerton Conference on Circuit and System Theory, Monticello, Illinois, pp. 219–27, October 1968.
96. Karayiannis J & Tzitzirachou K, *Introduction to Signals and Systems*, Papasotiriou Publications, ISBN 960-7530-32-2, 2003.
97. Karris S.T, *Signals and Systems with MATLAB Applications*, Orchard Publications, ISBN 0-9709511-8-3, 2003.
98. Kay S.M, *Fundamentals of Statistical Signal Processing, Volume I, Estimation Theory*, Prentice Hall, ISBN 978-013-345-7117, 1993.
99. Kay S.M, *Fundamentals of Statistical Signal Processing, Volume II, Detection Theory*, Prentice Hall, ISBN 978-013-504-1352, 1998.
100. Kay S.M, *Fundamentals of Statistical Signal Processing, Volume III, Practical Algorithm Development*, Prentice Hall, ISBN 978-013-487-8409, 2017.
101. Kerwin W.J, Huelsman L.P & Newcomb R.W, *State-Variable Synthesis for Insensitive Integrated Circuit Transfer Functions*, IEEE Journal of Solid-State Circuits, Vol. 2, No. 3, pp. 87–92, September 1967.
102. King A.C, Billingham J & Otto S.R, *Differential Equations*, Cambridge, ISBN 0-521-81658, 2003.
103. Krall H.L & Frink O, *A New Class of Orthogonal Polynomials: The Bessel Polynomials*, Transactions of American Mathematical Society, Vol.65, pp.100–115, 1949.
104. Krishnan A & Subramanian R, *Jump Phenomena in Digital Filters*, Journal of Electronics and Telecommunication Engineering, Vol. 25, pp. 373–376, 1979.
105. Kristiansson L.O.G, *Jump Phenomena in Digital Filters*, Electronics Letters, Vol. 10, pp. 14–15, 1974.
106. Ku Y.H & Drubin M, *Network Synthesis using Legendre and Hermite Polynomials*, Journal of Franklin Institute, Vol. 273, No.2, pp. 138–157, 1962.
107. Kugelstadt, T, *Active Filter Design Techniques* from Carter B. & Mancini R. (Eds), *Op Amps for Everyone* (Chapter 16), 3rd Ed., ISBN 978-1856175050, 2009, Texas Instruments.
108. Kyo F.F, *Network Analysis and Synthesis*, 2nd Edition, John Wiley and Sons Inc, Library of Congress Catalog Number 66-16127, 1966.
109. Lacanette K, *A Basic Introduction to Filters - Active, Passive & Switched Capacitor*, National Semiconductor Corporation, Application Note 779, 21 April 2010.
110. Lathi B.P, *Linear Systems and Signals*, Oxford University Press, ISBN-13 978-0-19-515833-5, 2005.
111. Levine W.S, *The Control Handbook*, CRC-Press, 1st Edition, ISBN-13 978-0849385704, 1996.
112. Levinson N, *The Wiener RMS Error Criterion in Filter Design and Prediction*, Journal of Mathematical Physics, Vol. 25, pp. 261–278, 1947.
113. Levitin A, *Introduction to the Design and Analysis of Algorithms*, 3rd Edition, Pearson, ISBN 978-0132316811, 2011.
114. Lewis A.D, *A Mathematical Approach to Classical Control*, freely available eBook (https://mast.queensu.ca/~andrew/teaching/pdf/332-notes.pdf)
115. Lin C-J et al, *Trust Region Newton Method for Logistic Regression*, Journal of Machine Learning Research, Vol. 9, pp. 627–650, June 2008.
116. Lindner D.K, *Introduction to Signals and Systems*, McGraw-Hill, 1st Ed, ISBN 0-256-25259-9, 1999.
117. Lindquist G.C, *Active Network Design with Signal Filtering Applications*, Steward & Sons, Long Beach, ISBN 0-917144-01-5, 1977.
118. Liu B & Kaneko T, *Error Analysis of Digital filters Realized with Floating-Point Arithmetic*, Proceedings of the IEEE, vol. 57, pp. 1735–1747, October 1969.
119. Liu K.S & Turner L.E, *Stability, Dynamic-Range and Roundoff Noise in a New Second-Order Recursive Digital Filter*, IEEE Transactions on Circuits and Systems, CAS-30, pp. 815–821, 1983.
120. Long J.L & Trick T.N, *An Absolute Bound on Limit Cycles Due to Roundoff Errors in Digital Filters*, IEEE Transactions on Audio and Electroacoust, Vol. AU-2l, pp. 27–30, 1973.
121. Luenberger D.G, Ye y, *Linear and Nonlinear Programming*, ISBN 978-3-319-37439-0, Springer, Switzerland, October 2016
122. Ly U-L, *Linear Systems Theory*, Online Book (PDF Format), University of Washington, 2005.
123. Lynn P.A & Fuerst W, *Introductory Digital Signal Processing with Computer Applications*, 2nd Edition, John Wiley and Sons Ltd., ISBN 0-471-94374, 1994.
124. Lyons R.D, *Understanding Digital Signal Processing*, Prentice Hall, ISBN 0-201-63467-8, 2001.
125. Lyons R.D, *Reducing FFT Scalloping Loss Without Multiplication*, IEEE Signal Processing Magazine, pp. 112–116, March 2012.
126. Madisetti V.K & Williams D.B, *Digital Signal Processing Handbook*, CRC Press LLC, 1999.

127. Mandal M. & Asif M, *Continuous and Discrete Time Signals and Systems*, Cambridge University Press, 1st Edition, ISBN 978-0521854559, 2007.
128. Manolakis D & Ingle V, *Applied Digital Signal Processing - Theory and Practice*, Cambridge University Press, ISBN 978-0-521-11002-0, 2011.
129. Margaris N, *Analysis of Electric Circuits*, Tziolas Publications, ISBN 978-960-418-202-2, 2004.
130. Masen S, *Mathematical Methods of Physics - Fourier Analysis*, Aristotle University Press, Thessaloniki, 2001.
131. McClellan J.J & Parks T.W, *A Unified Approach to the Design of Optimum FIR Linear Phase Digital Filters*, IEEE Transactions of Circuit Theory, Vol CT-20, pp. 697–701, November 1973
132. McMahon D, *Signals and Systems Demystified*, McGraw Hill, ISBN 0-07-147578-8, 2007.
133. Meerkötter K & Wegener W, *A New Second-Order Digital Filter Without Parasitic Oscillations*, Arch. Elektron Übertragungstech, Vol. AEU-29, pp. 312–314, 1975.
134. Meerkötter K, *Realization of Limit Cycle-Free Second-Order Digital Filters*, in Proceedings of IEEE International Symposium on Circuits and Systems, pp. 295–298, 1976.
135. Mills W.L, Mullis C.T & Roberts R.A, *Digital Filter Realizations Without Overflow Oscillations*, IEEE Transactions on Acoustics, Speech, and Signal Processing, vol. 26, pp. 334–338, August 1978.
136. Moler C & Van Loan C, *Nineteen Dubious Ways to Compute the Exponential of a Matrix*, SIAM Review, Vol. 20, No.4, pp. 801–836, 1978.
137. Moustakidis G, *Fundamental Techniques of Digital Signal Processing*, Tziolas, ISBN 960-418-016-9, 2004.
138. Nandintsetseg G & Sun X-E, *The Finite Precision Numerical Effects Analysis of Infinite Impulse Response*, Open Access Library Journal, Vol. 6, 2019, e4719.
139. Oldham K.B et. al, *An Atlas of Functions*, 2nd Ed, Springer, ISBN 978-0-387-48807-3_62, 2009.
140. Oppenheim A. V, *Block Floating Point Realization of Digital Filters*, MIT Lincoln Laboratory, Technical Note 1969–19 (20 March 1969).
141. Oppenheim A.V, Schafer R.W & Buck J.R, *Discrete Time Signal Processing*, 2nd Edition, Prentice Hall, ISBN 0-13-754920-2, 1999.
142. Oppenheim A.V, Willsky A.S & Nawab S.H, *Signals and Systems*, 2nd International Edition, Prentice Hall Signal Processing Series, ISBN 0-13-651175-9, 1997.
143. Paarmann L, *Design and Analysis of Analog Filters - A Signal Processing Perspective*, Springer, 1st Edition, ISBN 978-0792373735, 2001.
144. Pacut A & Hejn K, *Analog-to-Digital Converters: Towards a Generalization of Widrow's Theorem*, IEEE Instrumentation and Measurement Technology Conference, Vol 2, pp.1190–1197, June 1998.
145. Paley R. & Wiener N, *Fourier Transforms in the Complex Domain*, American Mathematical Society, Colloquium Publication, Volume XIX, 1934.
146. Papoulis A, *Optimum Filters with Monotonic Response*, Proceedings of IRE, Vol. 46, No.3, pp. 606–609, 1958
147. Papoulis A, *On Monotonic Response Filters*, Proceedings of IRE, Vol. 47, pp 332–333, 1959.
148. Papoulis A. & Pillai S.U, *Probability, Random Variables and Stochastic Processes*, McGraw-Hill, ISBN 0-07-112256-7, 2002.
149. Parker S.R & Hess S.F *Limit-Cycle Oscillations in Digital Filters*, IEEE Transactions on Circuit Theory, Vol. CT-18, No. 6, November 1971.
150. Parker S.R, *The Phenomena of Quantization Error and Limit Cycles in Fixed Point Digital Filters*, Proceedings of the 28th Nat. Electr. Conf, Chicago, USA, ed. R.E . Horton (Oak Brook, Illinois , USA: Nat. Electr. Conf. 1972), pp. 38–42, 1972.
151. Parks T.W & McClellan J.H, *Chebyshev Approximation for Nonrecursive Digital Filters with Linear Phase*, IEEE Transactions on Circuit Theory, Vol 9, pp. 189–194, March 1972.
152. Parks T.W & McClellan J.H, *A Program for the Design of Linear Phase Finite Impulse Response Digital Filters*, IEEE Transactions on Audio and Electroacoustics, Vol. 20, pp. 195–199, August 1972.
153. Peles Y & Murakami T, *Analysis and Synthesis of Transitional Butterworth-Thomson Filters and Bandpass Amplifiers*, RCA Review, Vol. 18, No 1, pp. 60–94, 1957.
154. Polderman J.W & Willens J.C, *Introduction to Mathematical Systems Theory: A Behavioral Approach*, Text in Applied Mathematics, Volume 26, Springer Verlag, ISBN 978-0-387-35763-8, 2009.
155. Poularikas A, *The Transforms and Applications Handbook*, 2nd Edition, CRC Press, ISBN 0-8493-8595-4, 1999.
156. Proakis J.P & Manolakis D.K, *Digital Signal Processing*, Prentice Hall, 4th Edition, ISBN-10 8120330307, 2006.
157. Rabiner L.R, *The Design of Finite Impulse Response Digital Filters Using Linear Programming Techniques*, Bell System Technical Journal, pp. 1177–1198, July-August 1972.
158. Rabiner L.R, *Linear Programming Design on Finite Impulse (FIR) Digital Filters*, IEEE Transactions on Audio & Electroacoustics, Vol AU-20, No. 4, pp. 280–288, October 1972.
159. Rabiner L.R & Schafer R.W, *On the Behavior of Miniman Relative Error FIR Digital Differentiators*, Bell System Technical Journal, Vol. 53, pp. 333–362, February 1974.
160. Rabiner L.R & Schafer R.W, *On the Behavior of Miniman Digital Hilbert Transformers*, Bell System Technical Journal, Vol. 53, pp. 363–394, February 1974.
161. Rabiner L.R & Juang B-H, *Fundamentals of Speech Recognition*, Prentice Hall Signal Processing Series, ISBN 0-13-015157-2, 1993.
162. Ralston A, *Rational Chebyshev Approximation by Remez's Algorithms*, Numerical Mathematics, Vol. 7, pp. 322–330, 1965.
163. Raut K, and Swamy M.N, *Modern Analog Filters Analysis and Design*, Wiley, ISBN 978-3-527-40766-8, 2010.
164. Remez E.Y, *General Computational Methods of Chebyshev Approximation*, in Problems with Linear Real Parameters, AEC-tr-4491, Books 1 and 2, English translation by US AEC.
165. Retzeperis P, *Introduction to Fourier Analysis with Physics Application*, Giachoudi-Giapouli Publications, Thessaloniki, 1988.
166. Rice J.R, *The Approximation of Functions*, Volume 1, Linear Theory, ISBN 978-0201064308, Addison-Wesley Publishing Company, Reading, MA, 1964.
167. Riley K.F, Hobson M.P & Bence S.J, *Mathematical Methods for Physics and Engineering*, 2nd Edition, Cambridge University Press, ISBN 0-521-81372-7, 2002.
168. Rorabauch C, *Notes on Digital Signal Processing - Practical Recipes for Design, Analysis and Implementation*, Prentice Hall, ISBN 0-13-158334-4, 2011.
169. Sallen R.P & Key E.L, *A Practical Method of Designing RC Active Filters*, IRE Transactions on Circuit Theory, Vol. CT-2, pp. 74–85, 1955.

170. Sandberg I.W, *A Theorem Concerning Limit Cycles in Digital Filters*, Proceedings of the 7th Annual Allerton Conf. Circuit Syst. Theory, Monticello, Illinois, USA, ed. A.H. Hadad et al (Urbana, Illinois, Dept. of Electrical Eng, Univ. of Illinois 1969), pp. 63–68, 1969.
171. Sandberg I.W & Kaiser J.F, *A Bound on Limit Cycles in Fixed-Point Implementations of Digital Filters*, IEEE Transactions on Audio and Electroacoustics. Vol. 20, pp. 110–114, June 1972.
172. Sarcinelli F.M & Diniz P.S.R, *Tridiagonal State-Space Digital Filter Structures*, IEEE Transactions on Circuits and Systems, CAS-36, pp. 818–824, 1990.
173. Schlichtärle D, *Digital Filters - Basics and Design*, 2nd Edition, Springer Verlag, ISBN 978-3-642-14324-3, 2011.
174. Schreier P.J & Scharf L.L, *Statistical Signal Processing of Complex Valued Data*, Cambridge University Press, ISBN 978-0-511-67772-4, 2010.
175. Schoinas J, *Difference Equations and Special Functions*, Ziti Publications, Thessaloniki, 1990.
176. Shmaliy Y, *Continuous Time Signals*, Springer Verlag, ISBN 978-1-4020-4817-3, 2006.
177. Shmaliy Y, *Continuous Time Systems*, Springer Verlag, ISBN 978-1-4020-6271-1, 2007.
178. Smith S.W, *The Scientist and Engineer's Guide to Digital Signal Processing*, 2nd Ed, California Tech. Publishing, ISBN 0-9660176-6-8, 1999.
179. Smith S.W, *Digital Signal Processing: A Practical Guide for Engineers and Scientists*, Elsevier, 2003.
180. Spagnolini U, *Statistical Signal Processing in Engineering*, ISBN 978-111-929-3972, John Wiley and Sons, 2018.
181. Spiegel M, *Theory and Problems of Complex Variables*, 1st Ed, Schaum's Outline Series, McGraw Hill, ISBN 07-084382-1, 1974.
182. Sripad A.B & Snyder D.L, *A Necessary and Sufficient Condition for Quantization Error to be Uniform and White*, IEEE Transactions of Acoustics, Speech and Signal Processing, Vol ASSP-25, No. 5, October 1977, pp. 442–448.
183. Stark H & Woods J.W, *Probability and Random Processes with Applications to Signal Processing*, 3nd Edition, Pearson, ISBN 0-13-178457-9,2002.
184. Stein J.Y, *Digital Signal Processing - A Computer Science Perspective*, Wiley, ISBN 0-471-29546-9, 2000.
185. Su k, *Analog Filters*, 2nd Ed, Springer Verlag, ISBN 1402070330, 1995.
186. Temes G.C & LaPatra J.W, *Circuit Synthesis and Design, McGraw-Hill*, ISBN 978-0070634893, Kogakusha, Tokyo, 1977
187. Thede L, *Practical Analog and Digital Filter Design*, Artech House Inc, 2004.
188. Theodorides S, Berberides K, Kofidis E, *Introduction to the Theory of Signals and Systems*, Typothitw Publications, ISBN 960-402-114-1, 2005.
189. Therrien C.W, *Discrete Random Signals and Statistical Signal Processing*, Prentice Hall, ISBN 0-13-852112, 1992.
190. Thomas L, *The Biquad: Part I – Some Practical Design Considerations*, IEEE Transactions on Circuit Theory, Vol. CT-18, pp. 350—357, 1971.
191. Tokaji & Barnes C, *Roundoff error statistics for a continuous range of multiplier coefficients*, IEEE Transactions on Circuits and Systems, Vol. 34, No. 1, pp. 52–59, 1987.
192. Tolstov G, *Fourier Series*, Dover Publications, Inc, ISBN 978-0-486-63317-6, 1962.
193. Tow J, *A Step-by-Step Active-Filter Design*, IEEE Spectrum, Vol. 6, No. 12, pp. 64–68, December 1969.
194. Trahanas S, *Partial Differential Equations*, University of Crete, ISBN 960-524-090-4, 2004.
195. Tukey, J., *An Introduction to the Calculations of Numerical Spectrum analysis*, in Advanced Seminar on Spectral Analysis of Time Series, edited by B. Harris, pp. 25–46, John Wiley, New York, 1967.
196. van Etten Wim C, *Introduction to Random Signals and Noise*, John Wiley and Sons Ltd, ISBN 978-0-470-02411-9, 2000.
197. Vaidyanathan P.P. & Liu V, *An Improved Sufficient Condition for Absence of Limit Cycles in Digital Filters*, IEEE Transactions on Circuits and Systems, Vol. 34, pp. 319–322, March 1987.
198. Vaidyanathan P.P & Nguyen T, *A Simple Proof of the Alternation Theorem*, in Conference Record of the 41th Asilomar Conference on Signals, Systems and Computers, IEEE, Piscataway, NJ, pages 1111–1115, ISBN 978-1-4244-2109-1, 2007.
199. Vaseghi S.V, *Advanced Digital Signal Processing and Noise Reduction*, 2nd Edition, John Wiley and Sons Ltd, ISBN 0-471-62692-9, 2000.
200. Verkroost G & Butterweck H.J, *Suppression of Parasitic Oscillations in Wave Digital Filters and Related Structures by Means of Controlled Rounding*, IEEE International Symposium on Circuits and Systems, Munich, West Germany, pp 628–629, 27–29 April 1976.
201. Wanhammar L, *Analog Filters Using MATLAB*, Springer Verlag, ISBN 978-0-387-92766-4, 2009.
202. Watson N.G, *A Treatise on the Theory of Bessel Functions*, Cambridge University Press, ISBN 978-0-521-48391-9, 1995, p. 43.
203. Weinstein C & Oppenheim A.V, *A Comparison of Roundoff Noise in Floating Point and Fixed Point Digital Filter Realizations*, Proceedings of IEEE, Vol 57, No. 6, pp. 1181–1183, June 1969.
204. Welch P, *A Fixed-Point Fast Fourier Transform Error Analysis*, IEEE Transactions on Audio and Electroacoustics, Vol 17, No 2, pp. 151–157, 1969.
205. Werter M.J, *Suppression of Parasitic Oscillations Due to Overflow and Quantization in Recursive Digital Filters*, Ph.D Thesis, Technische Universiteit Eindhoven, ISBN 90-9002583-9, pp. 49–50, 1989.
206. Widrow B, *A Study of Rough Amplitude Quantization by Means of Nyquist Sampling Theory*, Ph.D Thesis, MIT, June 1956.
207. Widrow B, *A Study of Rough Amplitude Quantization by means of Nyquist Sampling Theory*, IRE Transactions on Circuit Theory, pp. 266–276, Vol. 3. Issue 4, December 1956.
208. Widrow B, *Statistical Analysis of Amplitude-Quantized Sampled-Data Systems*, Trans. Amer. Inst. Elec. Eng, Pt II, Applications & Industry, Vol. 79, pp. 555–568, January 1961.
209. Widrow B, Kollár I & Liu M.C, *Statistical Theory of Quantization*, IEEE Transactions on Instrumentation and Measurement, Vol. 45, No. 2, pp. 353–361, April 1996.
210. Widrow B & Kollár I, *Quantization Noise: Roundoff Error in Digital Computation, Signal Processing, Control, and Communications*, Cambridge University Press, ISBN 978-0-511-40990-5, 2008.
211. Wiener N, *Extrapolation, Interpolation & Smoothing of Stationary Time Series with Engineering Applications*, John Wiley & Sons, New York, 1949.
212. Wilamovski B.M, Irwin J.D, *Fundamentals of Industrial Electronics*, CRC Press, ISBN 978-1-4398-0279-3, 2011.
213. Wilkinson J.H, *Rounding Effects in Algebraic Processes*, Dover Publications, ISBN 978-0486679990, Reprint edition (7 June 1994).

214. Willems J.L, *Stability Theory of Dynamical Systems*, London, Thomas Nelson and Sons, ISBN 978-0177810060, 1970.
215. Winder S, *Analog and Digital Filter Design*, 2nd Edition, Newnes, ISBN 0-7506-7547-0, 1997.
216. Wong P. W, *Quantization Noise, Fixed-Point Multiplicative Roundoff Noise, and Dithering*, IEEE Transactions on Acoustics, Speech, and Signal Processing, Vol. 38, No. 2, pp 286–300, 1990.
217. Wong P.W, *Quantization and Roundoff Noises in Fixed Point FIR Digital Filters*, IEEE Transactions on Signal Processing, Vol 39, No 7, pp. 1552–1563, July 1991.
218. Wrede R. & Spiegel M.R, *Theory and Problems of Advanced Calculus*, 2nd Ed, Schaum's Outline Series, McGraw Hill, ISBN 0-07-139834-1, 2002.
219. Yarlagadda R.K, *Analog and Digital Signals and Systems*, Springer Verlag, ISBN 978-1-4419-0034-0, 2010.
220. Zumbahlen H, *Linear Circuit Designer Handbook*, Analog Devices Inc., ISBN 978-0750687034, 2008.

Index

A. I. Margaris, *Continuous and Discrete-Time Signals and Systems*,
https://doi.org/10.1007/978-3-031-81209-5

D

E

G

MIX
Papier aus verantwortungsvollen Quellen
Paper from responsible sources
FSC® C105338

If you have any concerns about our products,
you can contact us on
ProductSafety@springernature.com

In case Publisher is established outside the EU,
the EU authorized representative is:
Springer Nature Customer Service Center GmbH
Europaplatz 3, 69115 Heidelberg, Germany

Printed by Libri Plureos GmbH
in Hamburg, Germany